CRC HANDBOOK OF

Organic Photochemistry

Photobiology

AND

D1479380

CRC HANDBOOK OF
Organic Photochemistry
AND Photobiology

Editor
WILLIAM M. HORSPOOL
University of Dundee
Dundee, Scotland

Associate Editor
PILL-SOON SONG
University of Nebraska
Lincoln, Nebraska

CRC Press
Boca Raton   New York   London   Tokyo

Library of Congress Cataloging-in-Publication Data

CRC handbook of organic photochemistry and photobiology / editor,
    William M. Horspool ; associate editor, Pill-Soon Song.
        p.   cm.
    Includes bibliographical references and index.
    ISBN 0-8493-8634-9
    1. Photochemistry--Handbooks, manuals, etc.   2. Photobiology-
-Handbooks, manuals, etc.   I. Horspool, William M.   II. Song, Pill
-Soon.
QD719.C73   1994
547.1'35--dc20   94-12056
                                                                            CIP

No claim to original U.S. Government works
International Standard Book Number 0-8493-8634-9
Library of Congress Card Number 94-12056
Printed in the United States of America 1 2 3 4 5 6 7 8 9 0
Printed on acid-free paper

FOREWORD

Organic Photochemistry, of all the many branches of organic chemistry, has developed rapidly in the last 30 years and reached the full maturity of a scientific discipline.

The steady development of the subject has led to a wealth of novel reactions that can be of great use in synthetic chemistry. Within the *Handbook* the material is divided according to the chemical structure of the substrate undergoing irradiation. Thus there are major sections dealing with hydrocarbons, oxygen-containing compounds, sulphur-containing compounds, nitrogen-containing compounds, and halogenated compounds. Within these major groupings there are subsections corresponding to the more familiar functional groups. The various articles, written by international authorities at the forefront of the areas being described, are extensively referenced and should be of value to the beginner or to the specialist. All in all, the purpose of the organic section is to provide a compendium of readily accessible data on the many reactions of synthetic utility and interest available to the chemist within the blossoming area of organic photochemistry.

Such a *Handbook* could not have been compiled without the good will and cooperation of all the contributors, and I am truly indebted to everyone who has given of their valuable time to writing for this text. In addition, I express my sincere thanks to Professor Pill-Soon Song for his unstinting help in identifying and securing manuscripts from the authors for the Photobiology section.

Finally, my work has been made considerably easier by the invaluable help given to me by Dr. Linda Horspool, my daughter, who acted as my assistant and spent many hours editing various documents.

William M. Horspool
Dundee, Scotland

Photobiology is an interdisciplinary science in that its scientific development depends on the combined concepts and methods of biology, chemistry, and physics. Because the realm of photobiology is broad, both students and practitioners of the science often have found it difficult to keep up with the fast-paced progress in this field. For this reason, we believe this *Handbook* meets the need to readily find pertinent information and data that previously could be found only in a diversity of periodical literature and review monographs.

Major topics of photobiology usually include (1) spectroscopy, photophysics, and photochemistry of biological molecules, (2) phototechnology applicable to research in photobiology, (3) photo-signal transduction processes such as chronobiology, photomorphogenesis, photomovement, photoreception and vision, (4) photo-energy transduction processes such as photosynthesis, which converts solar radiation to chemical potential, and bioluminescence, which converts chemical energy (ATP) to light energy, (5) environmental photobiology, photocarcinogenesis, photosensitization and photomedicine, and (6) ultraviolet radiation-induced damage in biological systems including DNA damage.

In the Photobiology section, the editors have attempted to adequately represent these areas of photobiology. We believe we have succeeded in doing so, primarily because each of the topics presented has been written by the most active researchers in the field of photobiology today. As editors, we thank them for their contributions to the *Handbook*.

Finally, I would like to express gratitude to Ms. Twyla Strukoff, who assisted me expertly with the task of communicating with the contributors and gathering and sending their manuscripts to the publishers.

Pill-Soon Song
Lincoln, Nebraska

THE EDITORS

William M. Horspool is Reader in Organic Chemistry at the University of Dundee in Scotland where he has taught since 1965.

He obtained his Ph.D. from the University of Glasgow in 1964 and his D.Sc. from Dundee University in 1976. His undergraduate studies were carried out at the University of Strathclyde in Glasgow.

Dr. Horspool has carried out research in organic photochemistry for more than 30 years and has published extensively in this area. He is a regular contributor to *Photochemistry*, a publication of the Royal Society of Chemistry, has written two textbooks on the subject, and has edited another.

He is a Fellow of the Royal Society of Chemistry and the Royal Society of Edinburgh.

Pill-Soon Song, Dow Chemical Company Professor, has been Chairman of the Department of Chemistry at the University of Nebraska-Lincoln since 1987.

He received his Ph.D. from the University of California at Davis in 1964. After postdoctoral work at Iowa State University, he joined the faculty at Texas Tech University in 1965, where he became the Horn Professor of Chemistry in 1975.

From 1975 to 1994 he served as the Editor-in-Chief of *Photochemistry and Photobiology* of the American Society for Photobiology. He presently serves on a number of international publications and science advisory boards.

Contributors

Aboel-Magd A. Abdel-Wahab
Department of Chemistry
Assiut University
Assiut, Egypt

Waldemar Adam
Institute of Organic Chemistry
University of Wurzburg
Wurzburg, Germany

Angelo Albini
Department of Organic Chemistry
University of Pavia
Pavia, Italy

Atef M. Amer
Department of Chemistry
University of Pavia
Pavia, Italy

Wataru Ando
Department of Chemistry
University of Tsukuba
Tsukuba, Ibaraki
Japan

Diego Armesto
Department Quimica Organica
Universidad Complutense
Madrid, Spain

Alfons L. Baumstark
Department of Chemistry
Georgia State University
Atlanta, Georgia

René Beugelmans
Institute de Chimie des Substances
  Naturelles
C.N.R.S.
Gif-sur-Yvette, France

Robert R. Birge
Department of Chemistry
Syracuse University
Center for Science and Technology
Syracuse, New York

Cornelia Bohne
Department of Chemistry
University of Victoria
Victoria, British Columbia
Canada

David Budac
Department of Chemistry
University of Victoria
Victoria, British Columbia
Canada

Nigel J. Bunce
Department of Chemistry
University of Guelph
Guelph, Ontario
Canada

Drury Caine
Department of Chemistry
The University of Alabama
Tuscaloosa, Alabama

Howard A. J. Carless
Department of Chemistry
Birkbeck College
London, United Kingdom

Zhongping Chen
Department of Chemistry
Syracuse University
Center for Science and Technology
Syracuse, New York

Thomas P. Coohill
Ultraviolet Consultants
Bowling Green, Kentucky

Jan Cornelisse
Leiden University
Leiden, The Netherlands

Wesley D. Corson
Department of Ophthalmology
  and Pathology
Medical University of South
  Carolina
Charleston, South Carolina

David Creed
Department of Chemistry and
  Biochemistry
University of Southern Mississippi
Hattiesburg, Mississippi

Rosalie K. Crouch
Department of Ophthalmology and
  Pathology
Medical University of South
  Carolina
Charleston, South Carolina

Francesco Dall'Acqua
Department of Pharmaceutical
  Sciences
University of Padova
Padova, Italy

Gerard L. Descotes
Department of Organic Chemistry
University of Lyon
Villeurbanne, France

Dietrich Döpp
Organische Chemie
Universität Duisburg
Duisburg, Germany

Timothy M. Dore
Department of Chemistry
Stanford University
Stanford, California

Thomas J. Dougherty
Roswell Park Cancer Institute
Buffalo, New York

Heinz Dürr
Universität des Saarlandes
Organische Chemie
Saarbrucken, Germany

Thomas G. Ebrey
Department of Biophysics
University of Illinois
Urbana, Illinois

Elisa Fasani
Department of Chemistry
University of Pavia

Mauro Freccero
Department of Chemistry
University of Pavia
Pavia, Italy

Masaki Furuya
Advanced Research Laboratory
Hitachi Ltd.
Saitama, Japan

Francis P. Gasparro
Department of Dermatology
Yale University
New Haven, Connecticut

Andrew Gilbert
Department of Chemistry
University of Reading
Reading, United Kingdom

Rolf Gleiter
Organisch-Chemisches Institut der
   Universität Heidelberg
Heidelberg, Germany

John H. Golbeck
Department of Biochemistry
University of Nebraska
Lincoln, Nebraska

Deshan Govender
Department of Chemistry
Syracuse University
Center for Science and Technology
Syracuse, New York

Axel G. Griesbeck
Department of Organic Chemistry
University of Cologne
Cologne, Germany

Richard B. Gross
Department of Chemistry
Center for Science and Technology
Syracuse University
Syracuse, New York

Anna D. Gudmundsdottir
University of British Columbia
Vancouver, British Columbia
Canada

Donat-P. Häder
Institut für Botanik und
   Pharmzeutische Biologie der
   Universität Erlangen Nürnberg
Erlangen, Germany

Anthony Harriman
Center for Fast Kinetics Research
University of Texas
Austin, Texas

Harry G. Heller
Department of Applied Chemistry
Cardiff, Wales, United Kingdom

Hans-Georg Henning
Chemistry Section
Humboldt University
Berlin, Germany

Susan B. Horm
Department of Chemistry
Syracuse University
Center for Science and Technology
Syracuse, New York

Felix T. Hong
Department of Physiology
Wayne State University School of
   Medicine
Detroit, Michigan

William M. Horspool
Department of Chemistry
The University of Dundee
Dundee, Scotland, United Kingdom

Yorinao Inoue
Solar Energy Research Group
Institute for Physical and Chemical
   Research (RIKEN)
Wako-Shi, Saitama
Japan

K. Can Izgi
Department of Chemistry
Syracuse University
Center for Science and Technology
Syracuse, New York

H. J. C. Jacobs
Gorlaeus Laboratories
Leiden University
Leiden, The Netherlands

Carl Johnson
Department of Biology
Vanderbilt University
Nashville, Tennessee

Giulio Jori
Universita di Padova
Dipartimento di Biologia
Padova, Italy

Gerd Kaupp
Department of Chemistry
University of Oldenburg
Oldenburg, Germany

Richard E. Kendrick
Department of Plant Physiology
Wageningen Agricultural University
Wageningen, The Netherlands

Satinder V. Kessar
Department of Chemistry
Panjab University
Chandigarh, India

Tsugio Kitamura
Department of Applied Chemistry
Kyushu University
Fukuoka, Japan

Dong-Hoon Ko
Department of Chemisty
Florida State University
Tallahassee, Florida

Radomir Konjević
University of Belgrade
Belgrade, Yugoslavia

Maarten Koornneef
Department of Plant Physiology
Wageningen Agricultural University
Wageningen, The Netherlands

Paul J. Kropp
Department of Chemistry
University of North Carolina
Chapel Hill, North Carolina

Yasuo Kubo
Department of Chemistry
Shimane University
Shimane, Japan

Osamu Kuwata
Department of Applied Physics and
   Chemistry
University of Electro-
   Communications
Tokyo, Japan

W. H. Laarhoven
Department of Organic Chemistry
University of Nijmegen
Nijmegen, The Netherlands

Alain Lablache-Combier
Universite des Sciences et
   Techniques de Lille
Villeneuve d'Ascq, France

J. Clark Lagarias
Department of Biochemistry and
   Biophysics
University of California
Davis, California

Klaus Langer
Organisch-Chemisches Institut der
   Universitat Munster
Munster, Germany

William J. Leigh
Department of Chemistry
McMaster University
Hamilton, Ontario
Canada

D. G. E. Lemaire
University of Sherbrooke
Sherbrooke, Quebec
Canada

Francesco Lenci
C.N.R. Istituto Biofisica
Pisa, Italy

Jie Liang
Center for Biophysics and
 Department of Cell and Structural
   Biology
University of Illinois
Urbana-Champaign, Illinois

Chen-Chen Liao
Department of Chemistry
National Tsing Hua University
Hsinchu, Taiwan
Republic of China

Edward D. Lipson
Department of Physics
Syracuse University
Syracuse, New York

Robert S. H. Liu
Department of Chemistry
University of Hawaii
Honolulu, Hawaii

Anil K. Singh Mankotia
Department of Chemistry
Panjab University
Chandigarh, India

Patrick S. Mariano
Department of Chemistry
University of Maryland
College Park, Maryland

Kazuhiro Maruyama
Kyoto Institute of Technology
Kyoto, Japan

Paul Mathis
Section de Bioenergetique
CEA-SACLAY
Gif-sur-Yvette, France

Jochen Mattay
Organisch-Chemisches Institut der
   Universität Münster
Munster, Germany

Harald Mauder
Department of Materials and
   Interface
The Weizmann Institute of Science
Rehovot, Israel

Miguel A. Miranda
Departamento de Quimica
Universidad Politecnica de Valencia
Valencia, Spain

David L. Mitchell
University of Texas
M. D. Anderson Cancer Center
Science Park Research Division
Smithville, Texas

Kazuhiko Mizuno
Department of Applied Chemistry
University of Osaka Prefecture
Osaka, Japan

Felix Muller
Th. Goldschmidt AG
Essen, Germany

Richard Needleman
Department of Biochemistry
Wayne State University School of
   Medicine
Detroit, Michigan

Takehiko Nishio
Department of Chemistry
University of Tsukuba
Ibaraki, Japan

Chyongjin Pac
Kawamura Institute of Chemical
   Research
Sukura, Chiba
Japan

Kyung-Mi Park
Department of Chemistry
Florida State University
Tallahassee, Florida

William R. Parker
Department of Chemistry
University of Nebraska
Lincoln, Nebraska

Harun Parlar
Department of Analytical Chemistry
University of Kassel
Kassel, Germany

James W. Pavlik
Department of Chemistry
Worcester Polytechnic Institute
Worcester, Massachusetts

Jennifer G. Peak
Biological and Medical Research
   Division
Argonne National Laboratory
Argonne, Illinois

Meyrick J. Peak
Biological and Medical Research
   Division
Argonne National Laboratory
Argonne, Illinois

Jean-Pierre Pete
Université de Reims
Champagne-Ardenne, France

V. N. Rajasekharan Pillai
School of Chemical Sciences
Mahatma Ghandi Univesity and
   Jawaharlal Nehru
Centre for Advanced Scientific
   Research
Kottayam, Kerala
India

J. A. Pincock
Department of Chemistry
Dalhousie University
Halifax, Nova Scotia
Canada

Kenneth L. Poff
Plant Research Laboratory
Michigan State University
East Lansing, Michigan

V. Ramamurthy
Department of Chemistry
Tulane University
New Orleans, Louisiana

B. Nageshwer Rao
IBM Corporation
Technology Product Division
Endicott, New York

V. Pushkara Rao
Enichem America, Inc.
Research and Development Center
Monmouth Junction, New Jersey

Jean Rigaudy
Université Pierre et Marie Curie
Paris, France

Carlos Rivas
Centro de Quimica
Instituto Venezolano de
   Investigaciones Cientificas
Caracas, Venezuela

S. M. Roberts
Department of Chemistry
Exeter University
Exeter, United Kingdom

Augusto Rodriguez
Department of Chemistry
Clark Atlanta University
Atlanta, Georgia

Seth D. Rose
Department of Chemistry
Arizona State University
Tempe, Arizona

Kenneth J. Rothschild
Department of Physics
Boston University
Boston, Massachusetts

Mordecai B. Rubin
Department of Chemistry
Technion-Israel Institute
   of Technology
Haifa, Israel

Bela P. Ruzsicska
Department of Nuclear Medicine
University of Sherbrooke
Sherbrooke, Quebec
Canada

Coskun Sahin
Institute of Organic Chemistry
University of Wurzburg
Wurzburg, Germany

Jack Saltiel
Department of Chemistry
Florida State University
Tallahassee, Florida

Hugo Scheer
Botanisches Institut
Universität München
Munchen, Germany

John R. Scheffer
University of British Columbia
Vancouver, British Columbia
Canada

Arthur G. Schultz
Department of Chemistry
Rensselaer Polytechnic Institute
Troy, New York

David I. Schuster
Department of Chemistry
New York University
New York, New York

Donald F. Sears, Jr.
Department of Chemistry
Florida State University
Tallahassee, Florida

Robert Sheridan
Department of Chenistry
University of Nevada
Reno, Nevada

Sang Chul Shim
Korea Advanced Institute of
   Science and Technology
Taejon, Korea

Toshio Shimizu
Department of Chemistry
University of Tsukuba
Tsukuba, Ibaraki
Japan

Harry Smith
Department of Botany
University of Leicester
Leicester, United Kingdom

Sanjay Sonar
Department of Physics
Boston University
Boston, Massachusetts

Pill-Soon Song
Department of Chemistry
University of Nebraska
Lincoln, Nebraska

Jeffrey A. Stuart
Department of Chemistry
Syracuse University
Center for Science and Technology
Syracuse, New York

Hiroshi Suginome
Department of Chemical Process
   Engineering
Hokkaido University
Sapporo, Japan

Jack R. Tallent
Department of Chemistry
Syracuse University
Center for Science and Technology
Syracuse, New York

Björn Treptow
Organisch-Chemisches Institute der
   Universität Heidelberg
Heidelberg, Germany

James Trotter
University of British Columbia
Vancouver, British Columbia
Canada

Takashi Tsuchiya
Faculty of Pharmaceutical
   Science
Hokuriku University
Kanazawa, Japan

Shiao-Chun Tu
Department of Biochemistry
   and Biophysics
University of Houston
Houston, Texas

Franklin Vargas
Instituto Venezolano de
    Investigaciones Cientificas
Centro de Quimica
Caracas, Venezuela

Daniela Vedaldi
Department of Pharmaceutical
    Sciences
University of Padova
Padova, Italy

Bryan W. Vought
Department of Chemistry
Syracuse University
Center for Science and
    Technology
Syracuse, New York

Peter J. Wagner
Department of Chemistry
Michigan State University
East Lansing, Michigan

Peter Wan
Department of Chemistry
University of Victoria
Victoria, British Columbia
Canada

Masakatsu Watanabe
Research Support Facility
National Institute for Basic Biology
Okazaki, Japan

Alan C. Weedon
Department of Chemistry
University of Western Ontario
London, Ontario
Canada

Richard G. Weiss
Department of Chemistry
Georgetown University
Washington, D. C.

Paul A. Wender
Department of Chemistry
Stanford University
Stanford, California

Jie Yang
University of British
    Columbia
Vancouver, British Columbia
Canada

Paw-Hwa Yang
Department of Chemistry
National Tsing Hua University
Hsinchu, Taiwan
Republic of China

Tôru Yoshizawa
Department of Applied Physics
    and Chemistry
University of Electro-
    Communications
Tokyo, Japan

Howard E. Zimmerman
Department of Chemistry
University of Wisconsin
Madison, Wisconsin

Contents

SECTION II Photobiology

SECTION I
Organic Photochemistry

1

Cis-Trans Isomerization of Alkenes

Jack Saltiel
Florida State University

Donald F. Sears, Jr.
Florida State University

Dong-Hoon Ko
Florida State University

Kyung-Mi Park
Florida State University

1.1 Definition of the Reaction

Except as modified by steric interactions, cis-trans isomerization of alkenes involves 180° rotation about a double bond.

$$\underset{B}{\overset{A}{\diagdown}} C = C \underset{E}{\overset{D}{\diagup}} \;\rightleftharpoons\; \underset{B}{\overset{A}{\diagdown}} C = C \underset{D}{\overset{E}{\diagup}} \tag{1.1}$$

The reaction can be induced thermally, catalytically, or photochemically. From a synthetic perspective, the first two methods can be grouped together because the cis/trans equilibrium mixtures that characterize them are governed by the relative ground-state thermodynamic stabilities of each pair of isomers. Photochemical pathways, on the other hand, often have the advantage of giving equilibrium cis/trans mixtures (photostationary states) that are rich in thermodynamically unstable isomers. Compositions of photostationary states reflect primarily the relative rates of attainment of lowest electronically excited states from each isomer and the partitioning characteristics of these states, and are therefore not controlled by the energetics of ground-state potential energy surfaces.

The most notable early photoisomerization studies are those of Olson and coworkers.[1] Olson was first to discuss the reaction in terms of potential energy curves. He proposed that the barrier to rotation is much lower in the excited state and envisioned the isomerization as an adiabatic process in the lowest excited state. The study of Lewis, Magel, and Lipkin was the first to explore the complementarity of fluorescence and photoisomerization.[2] The synthetic utility of the reaction was exploited early by Zechmeister and co-workers in extensive investigations of the geometric isomers of polyenes.[3]

0-8493-8634-9/95/$0.00+$.50
© 1995 by CRC Press, Inc.

1.2 Mechanism

The quantum mechanical basis for expecting a perpendicular geometry for the alkene's vibrationally relaxed lowest excited singlet state is due to Mulliken's expansion[4] of Hückel's description[5] of the double bond in ethylene. Mulliken's potential energy curves for rotation about the CC bond of the lowest electronic states of ethylene guided early discussions of cis-trans photoisomerization. They were adopted for the lowest triplet state by Förster, who was first to suggest that photoisomerization was a consequence of intersystem crossing from the lowest excited states of the two isomers to a common triplet excited state.[6] During the resurgence of photochemical research in the late 1950s and early 1960s, Förster's triplet mechanism for cis-trans photoisomerization was generally assumed to be correct.[7–9] The pioneering studies of Fischer and co-workers are especially noteworthy in this connection.[8] Fischer's work quantitatively established, over a wide temperature range, the complementary relationship between fluorescence and photoisomerization in *trans*-stilbene[8,10] and between photoisomerization and dihydrophenanthrene formation in *cis*-stilbene and related molecules.[11] The introduction of triplet excitation transfer in photochemistry in Hammond's laboratory during the same period allowed the demonstration that efficient cis-trans photoisomerization of alkenes can be achieved via triplet state alkene intermediates.[12] However, soon after the feasibility of Förster's mechanism for the reaction was established by Hammond's studies, Saltiel and co-workers showed that lowest excited singlet states of alkenes seldom decay to triplet states.[13] Saltiel's conclusion that photoisomerization usually involves torsional relaxation within the lowest excited singlet state surface as a first step was based on comparative studies of singlet and triplet excited alkenes formed under direct and under triplet (sensitization) excitation conditions respectively.[14,15] Thus, 1,3-diene triplets were shown to undergo two-bond photoisomerization[16,17] while singlets isomerize at one bond only;[18] for example,

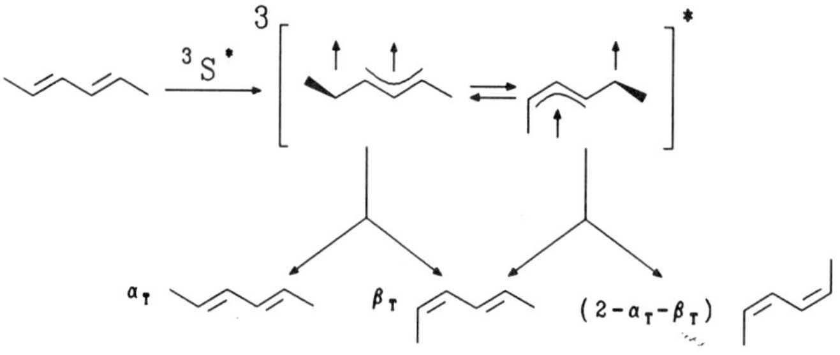

$$(1.2)$$

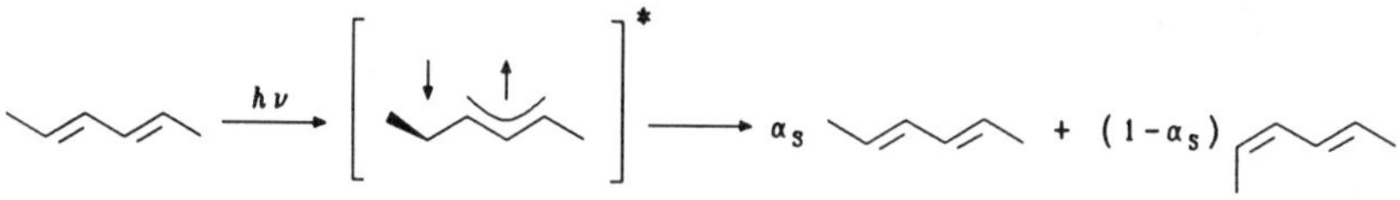

$$(1.3)$$

For the stilbenes, for which the decay fractions of the triplet and singlet excited states are nearly identical ($\alpha_T = \alpha_S$), comparative quenching studies and the use of rigid analogues were used to distinguish between proposed mechanisms.[19,20] Transient spectroscopic observations on stilbene triplets by Schulte-Frohlinde and co-workers in the late 1970s complemented Saltiel's steady-state observations and confirmed his conclusions.[21,22] Since then, a wealth of mechanistic detail has been uncovered, especially concerning the direct photoisomerization of the stilbenes through the application of laser flash spectroscopy employing excitation pulses in the ps and fs time scales.[23] The photoisomerization of the stilbenes is recognized as one of the best understood and most thoroughly studied photochemical reactions and is generally used as the model for cis-trans photoisomerization. Photoisomerization in either the lowest excited singlet or triplet states of stilbene occurs by twisting to common perpendicular intermediates, $^1p^*$ or $^3p^*$, on shallow torsional potential

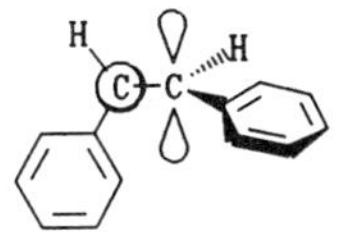

p

energy curves.[12,24–28] The decay fractions α_S and α_T characterize the partitioning of $^1p^*$ and $^3p^*$ to trans and cis ground-state isomers. Following direct excitation of the stilbenes, photoisomerization is confined entirely to the singlet excited state and competes with fluorescence on the trans side and photocyclization to dihydrophenanthrene, DHP, on the

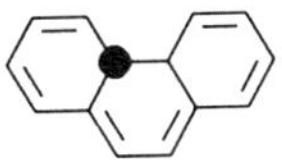

DHP

cis side. From a preparative standpoint, DHP should not interfere with the interconversion of the geometric isomers since it reverts to *cis*-stilbene both thermally and photochemically. However, in the presence of O_2 or other adventitious catalysts, it is converted to phenanthrene. In degassed pure dilute solutions, excitation with monochromatic light leads to photostationary trans/cis ratios, $([t]/[c])_s$, that depend on the ratio of the quantum yields for trans → cis and cis → trans isomerization, ϕ_{tc} and ϕ_{ct}, respectively, and the ratio of molar absorptivity coefficients of the two isomers at λ_{exc},

$$\left(\frac{[t]}{[c]}\right)_s = \frac{\epsilon_c}{\epsilon_t}\frac{\phi_{ct}}{\phi_{tc}} \tag{1.4}$$

In agreement with the absorption spectra of the two isomers, Figure 1.1 and the quantum yields, the cis content of the photostationary state of 0.01 *M* solutions in *n*-pentane increases from 48 to 91.7% as λ_{exc} is changed from 254 to 313 nm.[15] Preparative experiments employing direct excitation must be confined to low stilbene concentrations because the photodimerization of the trans isomer becomes a significant competing process as the stilbene concentration is increased.

Access to the triplet state photoisomerization pathway is gained when alkenes are excited by triplet excitation transfer from suitable energy donors (sensitizers). In many instances (the stilbenes included), photoisomerization is observed in the absence of side reactions and high alkene concentrations can be employed. Photostationary ratios are then given by

$$\left(\frac{[t]}{[c]}\right)_s = \frac{k_c}{k_t}\left(\frac{\alpha_T}{1-\alpha_T}\right) \tag{1.5}$$

where k_c/k_t, the ratio of excitation transfer rate constants to the two isomers, depends on the triplet energy of the donor.[12,14,29]

Recent studies have shown that as substituents on the CC double bond are changed or the number or the degree of conjugation is increased either in the substituent or in the number of double bonds in the olefinic portion of the molecule, new mechanisms of photoisomerization can come into play. For instance, heavy atom or nitro substituents in stilbene itself lead to participation of triplet states in the absence of triplet energy donors.[15,21,22] Alternatively, the excited-state surfaces may favor one of

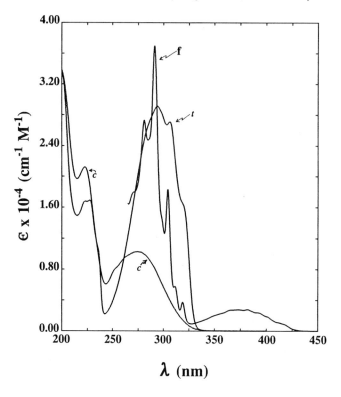

FIGURE 1   UV spectra of the stilbenes and fluorenone, f, in *n*-hexane.

the planar isomers over the other and over twisted intermediates and thus funnel decay through that isomer. In such instances, adiabatic one-way photoisomerizations may result. Notable examples of such triplet[30] and singlet[31] processes are shown in Equations 1.6 and 1.7, respectively.

$$(1.6)$$

$$(1.7)$$

One-way photoisomerizations were first reported by Liu and co-workers in their extensive studies of sensitized polyene photoisomerization.[32] Interestingly, in contrast to the above examples that are based on the decay characteristics of the participating lowest excited state, the early cases were based on selective formation of the excited state from one of the isomers. The quantitative conversion of *trans*-β-ionol using α-acetonaphthone, α-AN, as the

$$(1.8)$$

triplet excitation donor is a typical example.[33] It should be noted that Equation 1.8 gives the thermodynamically less stable isomer. Such reactions have been exploited by Liu in the synthesis of polyenes in the vitamin A series.[34]

Decay of p-type electronically excited intermediates usually occurs with nearly equal probabilities to cis and trans ground-state isomers. Thus, photochemical alkene interconversion lends itself to the formation of strained *trans*-cycloalkenes. For instance, the photoisomerization of *cis*-cyclooctene is a useful actinometer for far-UV studies.[34] More

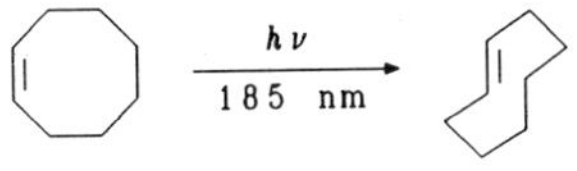

$$\text{(structure)} \xrightarrow[185 \text{ nm}]{h\nu} \text{(structure)} \qquad (1.9)$$

highly strained cycloalkenes formed either by direct or triplet sensitized excitation can serve as reactive intermediates to other products. An early example is *cis,trans*-1,3-cyclooctadiene, which is stable at low temperature but readily gives the *cis*-bicyclooctene at 80 °C.[35] Similarly, the triplet-sensitized conversion of *cis,cis*-1,1'-bicyclohexenyl cleanly to

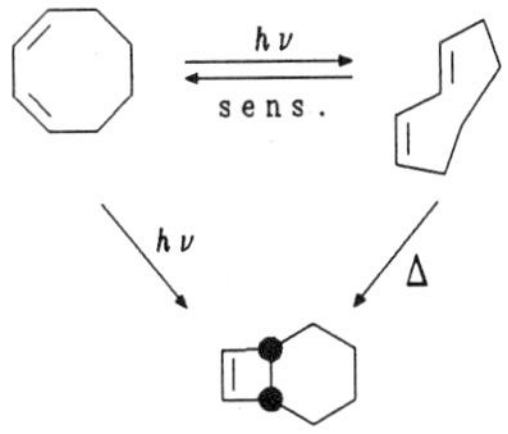

$$\text{(structure)} \xrightleftharpoons[sens.]{h\nu} \text{(structure)} \qquad (1.10)$$

the *cis*-cyclobutene tricyclo[6.4.0.0^{2,7}]-*cis*-7,8-dodec-1-ene[36] proceeds via the ground state of the short-lived transient *cis,trans*-1,1'-bicyclohexenyl.[37]

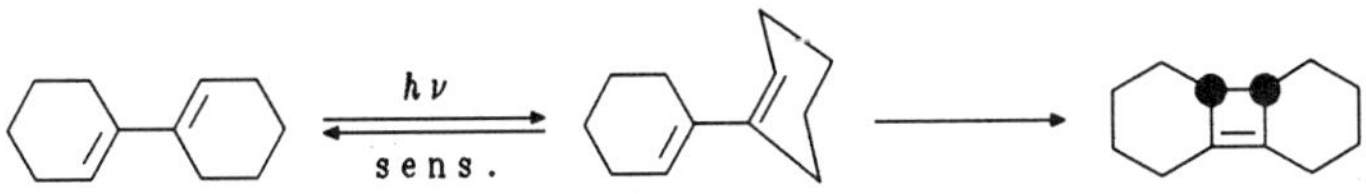

$$\text{(structure)} \xrightleftharpoons[sens.]{h\nu} \text{(structure)} \longrightarrow \text{(structure)} \qquad (1.11)$$

1.3 Synthetic Applications

cis-Stilbene

The photoisomerization of stilbene was first reported by Stoermer,[38] and chromatography was first applied to the separation of the isomers by Zechmeister and McNeely.[39] Ultrapure *cis*-stilbene samples, free from bibenzyl and *trans*-stilbene contamination, have recently been required in our laboratory for use in the determination of the enthalpy difference between cis and trans isomers[38] and especially for the determination of the fluorescence properties of *cis*-stilbene solutions.[28] Photostationary-state stilbene compositions for the benzophenone-sensitized reaction in benzene give 58.5%

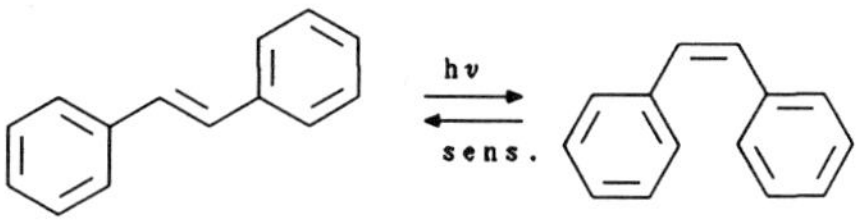

$$\text{(structure)} \xrightleftharpoons[sens.]{h\nu} \text{(structure)} \qquad (1.12)$$

cis, independent of stilbene concentration in the range 0.001 to 0.600 M.[13b] Benzil, the ketone sensitizer with the optimum k_c/k_t excitation ratio,[12,29] gives substantially higher cis content, 93%, at the photostationary state; however, when solutions are purged with N_2 that contains a small trace of O_2, it also gives biphenyl as a contaminant.[41] Fluorenone-sensitization, on the other hand, gives no side product and leads to a reasonably cis-rich 86%, photostationary state. The *trans*-stilbene (Aldrich 96%) starting material was contaminated by *cis*-stilbene and bibenzyl, both of which can be removed from small samples by careful chromatography on alumina.[28,40] Three recrystallizations from denatured ethanol starting with 50 g *trans*-stilbene containing 0.51% bibenzyl, GC, gave 27,0 g *trans*-stilbene with ≤0.01% bibenzyl. The 27.0 g *trans*-stilbene and 5.85 g fluorenone (Aldrich, purified by chromatography on alumina and recrystallized twice from methanol, 99.98% purity, GC) were dissolved in ~ 400 ml benzene (Fisher Certified ACS Spectranalyzed) to give 0.375 M *trans*-stilbene and 0.081 M fluorenone. This solution was irradiated in a cylindrical Pyrex Hanovia reactor equipped with a water-cooled Pyrex lamp probe, a 550-W Hanovia medium-pressure Hg lamp, and a fritted glass bubbler. A constant stream of N_2 gas was bubbled through the solution for 1 h prior to turning on the lamp and throughout the irradiation, and the progress of the reaction was monitored by GC. The photostationary state was nearly attained, 84.6% cis, after ~3 h irradiation, and the photoreaction was terminated after an additional 0.5 h of irradiation. The solution was concentrated to ~50 ml and chromatographed on 700 g neutral alumina using *n*-pentane (Fisher reagent, distilled) as eluent. The alumina (Fisher reagent) was first flushed with 1 l 6 N nitric acid, followed by 2 l deionized water and finally 500 ml methanol (Fisher reagent, distilled) and dried in an oven for 4 d at 130 °C. The first five 200-ml fractions contained mostly *cis*-stilbene. They were combined, concentrated on a rotary evaporator, and rechromatographed on 500 g alumina. *cis*-Stilbene, 15 g, ≥ 99.99% purity, GC, was obtained as a viscous colorless oil. Bibenzyl appeared as a trace contaminant in later *cis*-stilbene fractions. A slightly different procedure for the preparation of smaller samples of *cis*-stilbene[28c] is similar to that described below for *c*-NPE.

cis-1-(2-Naphthyl)-1-phenylethene (*c*-NPE)

trans-NPE, prepared using the Wittig reaction from 2-naphthaldehyde, was purified by column chromatography on alumina and recrystallization.[42] A benzene solution (350 ml) of a mixture of *t*-NPE (0.80 g, 0.010 M) and fluorenone (1.8 g, 0.03 M) was irradiated in a cylindrical Pyrex Hanovia reactor as described above except that the output of a 200-W Hanovia medium-pressure lamp filtered through a filter solution sleeve (~1 cm path length) was employed. The filter solution, prepared from $NaNO_2$ (30 g), $CuSO_4 \cdot 5H_2O$ (27 g) and concentrated NH_4OH (50 ml) diluted to 1 l with deionized water, transmits a band of light between 395 and 490 nm. As before, the benzene solution was purged with a stream of N_2 for 1 h prior and throughout the irradiation. A photostationary state composition, ~70% *c*-NPE,[43] was attained upon

$$ (1.13) $$

3-h irradiation. Following solvent evaporation, the mixture was chromatographed on activity I alumina deactivated with 2% water (w/w).[28c,44] Dichloromethane (2%, v/v) in *n*-hexane (Fisher HPLC) was used to elute the *c*-NPE from the column. The solvent was removed using a rotary evaporator and *c*-NPE was isolated as a colorless viscous oil, 0.50 g (63%, isolated yield, 99.0% purity by GC). ^{1}H-NMR (300 MHz, acetone-d_6): δ 7.18–7.82 (m, 12 H, aromatic), δ 6.74 (dd, 2 H, vinyl, J_{HH} = 2 Hz). See Figure 1.2 for UV spectra of *c*- and *t*-NPE.

1,6-Diphenyl-1,3,5-hexatrienes (DPH)

The trans → cis photoisomerization of *all-trans*-DPH, *ttt*, was first reported by Lunde and Zechmeister who isolated four isomers containing cis double bonds and tentatively assigned their structures based on the UV spectra and the kinetics of the iodine-catalyzed thermal reversion to *ttt*.[45] A qualitative comparative study of UV absorption changes under direct and triplet sensitized conditions led Görner to conclude that photoisomerization following direct excitation follows a singlet pathway.[46] This is consistent with transient spectroscopic measurements that show intersystem crossing to be inefficient in polyenes, $\phi_{is} \le 0.02$.[46–48] As part of a spectroscopic[49] and photochemical[50] investigation of the mechanism of DPH photoisomerization, we report here the isolation and characterization of the two major photoproducts: 1-*cis*-3-*trans*-5-*trans*-DPH, *ctt*, and 1-*trans*-3-*cis*-5-*trans*-DPH, *tct*. Quasiphotostationary states are attained by either direct or triplet-sensitized excitation in a variety of solvents.[50] The direct excitation of DPH in acetonitrile was selected because acetonitrile enhances both the efficiency of the reaction and the *ctt* content of the quasiphotostationary state.[50]

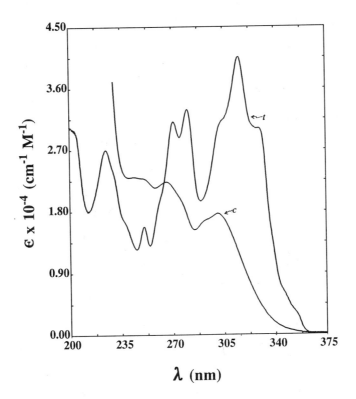

$$(1.14)$$

ttt-DPH (Aldrich, 98%) was recrystallized twice from *n*-hexane to 99.99% purity, HPLC. An acetonitrile (Fisher HPLC grade) solution (350 ml) of *ttt*-DPH (0.086 g, 1.06×10^{-3} *M*) was irradiated under the conditions employed for *c*-NPE so that excitation was limited to the onset of DPH absorption, Figure 1.3. A photostationary state mixture of 46.2% *ctt*, 11.3% *tct*, and 36.1% *ttt*

FIGURE 2   UV spectra of the NPEs in *n*-hexane.

was attained within 12 min of irradiation and remained stable for an additional 10 min of irradiation with the exception of a slow increase (to approximately 7.2%) of the secondary product *cct*-DPH. The solvent was removed under reduced pressure and the residue taken up in *n*-hexane (5 ml). The resulting *ctt* and *tct*-rich solution was chromatographed on activity I alumina (Wöelm) deactivated with 2% water (w/w)[28c,44] using 30% n-pentane in benzene as eluent. Collected fractions were concentrated with a rotary evaporator and further purified by HPLC. Although no destruction of the DPH isomers was observed during the chromatographies, yields (12.7% for *ctt* and 9.7% for *tct*) were relatively low because only isolated photoproducts of 99+% purity were considered. The UV spectra of *ctt* and *tct*, Figure 1.3, correspond closely to those previously reported.[45] The two photoproducts were characterized further by ¹H-NMR, *tct* (300 MHz, acetone-d_6): δ 6.26–6.30 (m, 2H$_c$), 6.64–6.72 (m, 2 H$_a$), 7.64–7.74 (m, Hb), 7.34 (m, 4 H), 7.58 (m, 4 H), 7.26 (m, 2 H); *ctt* (300 MHz, CDCl$_3$): δ 7.20–7.42 (m, 10 H), 6.82–6.96 (m, 2 H), 6.44–6.63 (m, 3 H), 6.34 (m, 1 H).

GC Analyses

GC analyses were performed using a Varian 3300 gas chromatograph equipped with a J & W Scientific DX-4 capillary column. For the stilbenes, instrumental conditions were as follows: detector and injector 250 °C; carrier gas ~20 psi; initial column temperature at 120 °C was held for 5.0 min, increased to 140 °C at 4.0 °C min⁻¹, held for 5.0 min, then increased to 160 °C at 4.0 °C

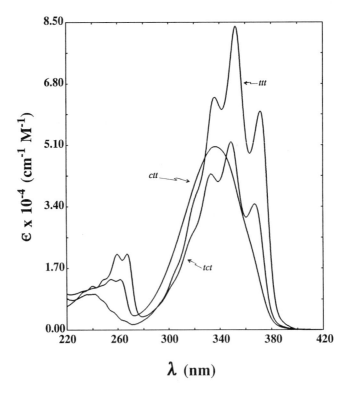

FIGURE 3   UV spectra of *ttt*-, *tct*-, and *ctt*-DPH in *n*-hexane.

min^{-1} and held for 5.0 min. Typical retention times: bibenzyl, 8.8 min; *cis*-stilbene, 9.3 min; *trans*-stilbene, 19.5 min; and fluorenone, 27.5 min. The same conditions were employed for the NPEs except that carrier gas pressure was ~16 psi, and the initial column temperature at 180 °C was held for 5 min, increased to 220 °C at 2 °C min^{-1}, and held there for 15 min. Typical retention times for *c*- and *t*-NPE were 16 and 29 min, respectively.

HPLC Analyses

HPLC analyses were performed using a Beckman 114M Solvent Delivery Module with Programmable Detector Module 166. Analyses of samples were performed on a normal-phase SI column, Beckman Ultrasphere, 4.6×250 mm with *n*-hexane as eluent. The flow rate was 0.8 ml min^{-1} and typical retention times were 20, 23, and 27 min for *ctt, tct,* and *ttt,* respectively. Anthracene was employed as internal standard. Absorbance was monitored at 350 nm. The photoproducts were purified on a normal semipreparative SI column, Beckman Ultrasphere, 10×250 nm, 5 μm with *n*-hexane as eluent. The flow rate was 1.5 ml min^{-1}.

To avoid inadvertent photoisomerization of photoproducts, all purification steps in the above procedures were carried out under red light in a darkened room.

Absorption Spectra

UV absorption spectra were measured using a Perkin-Elmer Lambda-5 Spectrophotometer. Spectra were measured at ambient temperature (~24 °C) using matched Hellma 1.00-cm Supracil quartz cells. Absorbance was recorded on a Dell Corporation 12 MHz 80286/87 microcomputer.

1.4 Industrial Applicability

The role of cis-trans photoisomerization in the photochemical production of vitamin D and related compounds and its potential use in the synthesis of vitamin A, *all-trans*-retinol, have been reviewed recently.[51] In both cases, the potential use of selective triplet sensitization to maximize the yield of desired products has been demonstrated.[51] Generally, previtamin Ds, formed by photochemical opening of the B rings of ergosterol and 7-dehydrocholesterol, give vitamins D by thermal [1,7]-sigmatropic rearrangement on moderate heating. The problem is that the previtamin Ds absorb light

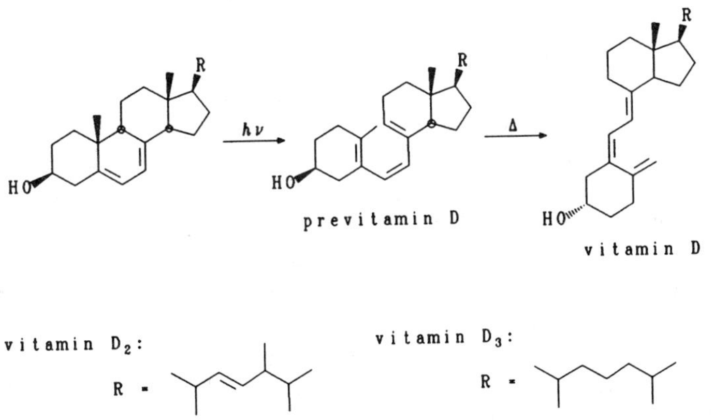

in the same spectral region as their precursors and undergo several competing photoreactions, including reversion to starting material and cis-trans photoisomerization to tachysterol. Since the photoreactions are λ_{exc} dependent, the yield of previtamin Ds can

previtamin D

tachysterol   (1.16)

be increased by judicious choice of excitation wavelength.[52–56] Especially promising is the possibility of using triplet energy donors with high previtamin D/tachysterol photostationary ratios in order to recycle tachysterol to previtamin D.[56–58] For instance, the photostationary state compositions for Equation 1.15 with fluorenone and 7,12-dimethylbenzanthracene as triplet energy donors are 81 and 94% previtamin D, respectively.[56–58] In alkenes such as NPE, DPH, or the vitamin precursors in Equation 1.15 for which a thermally equilibrated mixture of rotamers exists in the ground state, Havinga's NEER (nonequilibration of excited rotamers) principle[52] applies and Equations 1.4 and 1.5 must be expanded to include the contribution of each singlet or triplet[17] excited conformer.

References

1. Olson, A. R., The study of chemical reactions from potential energy diagrams, *Trans. Faraday Soc.,* 27, 69, 1931. Olson, A. R. and Hudson, F. L., The photostationary states of some geometrically isomeric acids, *J. Am. Chem. Soc.,* 55, 1410, 1933. Olson, A. R. and Maroney, W., The thermal and photochemical equilibria of the cis-trans isomers of dichloro- and dibromoethylene, *J. Am. Chem. Soc.,* 56, 1320, 1934.

2. Lewis, G. N., Magel, T. T., and Lipkin, D., The absorption and re-emission of light by *cis-* and *trans-*stilbene and the efficiency of their photochemical isomerization, *J. Am. Chem. Soc.,* 62, 2973, 1940.

3. Zechmeister, L., *Cis-trans Isomeric Carotenoids, Vitamin A, and Arylpolyenes,* Academic, New York, 1962.

4. Mulliken, R. S., Electronic structures of polyatomic molecules and valence. II. Quantum theory of the double bond, *Phys. Rev.,* 41, 751, 1932.

5. Hückel, E., Zur quantentheorie der doppelbindung, *Z. Physik,* 60, 423, 1930.

6. Förster, Th., Photochemische primärprozesse bei mehroatomigen molekülen. *Z. Elekrochem.,* 56, 716, 1952.

7. Dyck, R. H. and McClure, D. S., Ultraviolet spectra of stilbene, *p*-monohalogen stilbenes, and azobenzene and the trans to cis photoisomerization process, *J. Chem. Phys.,* 36, 2326, 1962.

8. (a) Malkin, S. and Fischer, E., Temperature dependence of photoisomerization. Part II. Quantum yields of cis $\rightleftharpoons$ trans isomerization in azo-compounds, *J. Phys. Chem.,* 66, 2482, 1962. (b) Malkin, S. and Fischer, E., Temperature dependence of photoisomerization. III. Direct and sensitized photoisomerization of stilbenes, *J. Phys. Chem.,* 68, 1153, 1964. (c) Gegiou, D., Muszkat, K. A., and Fischer, E., Temperature dependence of photoisomerization. VI. The viscosity effect, *J. Am. Chem. Soc.,* 90, 12, 1968. (d) Gegiou, D., Muszkat, K. A., and Fischer, E., The effects of substituents on the photoisomerization of stilbenes and azostilbenes, *J. Am. Chem. Soc.,* 90, 3907, 1968.

9. Stegemeyer, H., On the mechanism of photochemical cis $\rightleftharpoons$ trans isomerization, *J. Phys. Chem.,* 66, 2555, 1962, and references cited.

10. Sharafi, S. and Muszkat, K. A., Viscosity dependence of fluorescence quantum yields, *J. Am. Chem. Soc.,* 93, 4119, 1971.

11. (a) Muszkat, K. A. and Fischer, E., Structure, spectra, photochemistry, and thermal reactions of the 4a,4b-dihydrophenanthrenes, *J. Chem. Soc. B,* 662–678, 1967. (b) Wismontski-Knittel, T., Fischer, G., and Fischer, E., Temperature dependence of photoisomerization. Part VIII. Excited-state behavior

of 1-naphthyl-2-phenyl- and 1,2-dinaphthylethylenes and their photocyclization products and properties of the latter, *J. Chem. Soc., Perkin Trans.*, 2, 70, 1930, 1974. (c) Fischer, G., Seger, G., Muszkat, K. A., and Fischer, E., Emissions of sterically hindered stilbene derivatives and related compounds. Part IV. Large conformational differences between ground and excited states of sterically hindered stilbene: Implications regarding Stokes shifts and viscosity or temperature dependence of fluorescence yields, *J. Chem. Soc., Faraday Trans.*, 2, 71, 1569, 1975.

12. Hammond, G. S., Saltiel, J., Lamola, A. A., Turro, N. J., Bradshaw, J. S., Cowan, D. O., Counsell, R. C., Vogt, V., and Dalton, C., Mechanisms of photochemical reactions in solutions. XXII. Photochemical cis-trans isomerization, *J. Am. Chem. Soc.*, 86, 3197, 1964, and references cited.

13. (a) Saltiel, J., Perdeuteriostilbene. The role of phantom states in the cis-trans photoisomerization of stilbenes, *J. Am. Chem. Soc.*, 89, 1036, 1967. (b) Saltiel, J., Perdeuteriostilbene. The triplet and singlet paths for stilbene photoisomerization, *J. Am. Chem. Soc.*, 90, 6394, 1968.

14. For a review see Saltiel, J., D'Agostino, J., Megarity, E. D., Metts, L., Neuberger, K. R., Wrighton, M., and Zafiriou, O. C., The cis-trans photoisomerization of olefins, *Org. Photochem.*, 3, 1, 1973.

15. Saltiel, J., Marinari, A., Chang, D. W.-L., Mitchener, J. C., and Megarity, E. D., Trans-cis photoisomerization of the stilbenes and a reexamination of the positional dependence of the heavy-atom effect, *J. Am. Chem. Soc.*, 101, 2982, 1979.

16. Saltiel, J., Metts, L., and Wrighton, M., Benzophenone-sensitized photoisomerization of the 2,4-hexadienes in solution, *J. Am. Chem. Soc.*, 91, 5684, 1969; Saltiel, J., Rousseau, A. D., and Sykes, A., Temperature and viscosity effects on the decay characteristics of s-trans-1,3-diene triplets, *J. Am. Chem. Soc.*, 94, 5903, 1972.

17. Saltiel, J., Metts, L., Sykes, A., and Wrighton, M., The role of s-cis-1,3-diene triplets in sensitized cis-trans photoisomerization, *J. Am. Chem. Soc.*, 93, 5302, 1971.

18. Saltiel, J., Metts, L., and Wrighton, M., The direct *cis-trans* photoisomerization of 1,3-dienes in solutions, *J. Am. Chem. Soc.*, 92, 3227, 1970. Squillacote, M. E. and Semple, T. C., Polarization in the excited state of 1,3-pentadiene: Experimental evidence for an allyl cation-methylene anion species, *J. Am. Chem. Soc.*, 109, 892, 1987.

19. Saltiel, J. and Megarity, E. D., The mechanism of direct cis-trans photoisomerization of the stilbenes: The nature of the azulene effect, *J. Am. Chem. Soc.*, 91, 1265, 1969. Saltiel, J. and Megarity, E. D., Mechanism of direct cis-trans photoisomerization of the stilbenes. Solvent viscosity and the azulene effect, *J. Am. Chem. Soc.*, 94, 2742, 1972.

20. Saltiel, J., Zafiriou, O. C., Megarity, E. D., and Lamola, A. A., Tests of the singlet mechanism for *cis-trans* photoisomerization of the stilbenes, *J. Am. Chem. Soc.*, 90, 4759, 1968. DeBoer, C. D. and Schlessinger, R. H., The multiplicity of the photochemically reactive state of 1,2-diphenylcyclobutene, *J. Am. Chem. Soc.*, 90, 803, 1968.

21. Bent, D. V. and Schulte-Frohlinde, D., Laser flash photolysis of substituted stilbenes in solution, *J. Phys. Chem.*, 78, 446, 1974. Bent, D. V. and Schulte-Frohlinde, D., Evidence for the triplet route in the photochemical trans → cis isomerization of nitrostilbenes, *J. Phys. Chem.*, 78, 451, 1974. Görner, H. and Schulte-Frohlinde, D., Upper excited triplet state mechanism in the trans-cis isomerization of 4-bromostilbene, *J. Am. Chem. Soc.*, 101, 4388, 1979. Görner, H. and Schulte-Frohlinde, D., Observation of the triplet state of stilbene in fluid solution. Determination of the equilibrium constant ($^3t^* \rightleftharpoons ^3p^*$) and of the rate constant for intersystem crossing ($^3p^* \rightarrow ^1p^*$), *J. Phys. Chem.*, 85, 1835, 1981.

22. For a review see Saltiel, J. and Charlton, J. L., Cis-trans isomerization of olefins, *Rearrangements in Ground and Excited States*, Vol. 3, de Mayo, P., Ed., Academic, New York, 1980, 25.

23. For reviews, see (a) Hochstrasser, R. M., Picosecond processes in isomerism of stilbenes, *Pure Appl. Chem.*, 52, 2683, 1980. (b) Saltiel, J. and Sun, Y.-P., Cis-trans isomerization of C=C double bonds, in *Photochromism, Molecules and Systems*, Dürr, H. and Bouas-Laurent, H., Eds., Elsevier, Amsterdam, 1990, 64. (c) Waldeck, D. H., Photoisomerization dynamics of stilbenes, *Chem. Rev.*, 91, 415, 1991.

24. Saltiel, J., Rousseau, A. D., and Thomas, B., The energetics of twisting in the lowest stilbene triplet state, *J. Am. Chem. Soc.*, 105, 7631, 1983.

25. Saltiel, J., Marchand, G. R., Kirkor-Kaminska, E., Smother, W. S., Mueller, W. B., and Charlton, J. L., Nonvertical triplet excitation transfer to *cis*- and *trans*-stilbene, *J. Am. Chem. Soc.*, 106, 3144, 1984.

26. Ni, T., Caldwell, R. A., and Melton, L. A., The relaxed and spectroscopic energies of olefin triplets, *J. Am. Chem. Soc.*, 111, 457, 1989.

27. Caldwell, R. A., Riley, S. J., Gorman, A. A., McNeeney, S. P., and Unett, D. J., Relaxed triplet energies of phenylnorbornenes. The role of phenyl-vinyl torsions. On the origin of nonvertical triplet excitation transfer, *J. Am. Chem. Soc.*, 114, 4424, 1992.

28. (a) Saltiel, J., Waller, A., Sun, Y.-P., and Sears, D. F., Jr., *cis*-Stilbene fluorescence in solution. Adiabatic $^1c^* \rightarrow {}^1t^*$ conversion, *J. Am. Chem. Soc.*, 112, 4580, 1990. (b) Saltiel, J., Waller, A. S., and Sears, D. F., Jr., Dynamics of *cis*-stilbene photoisomerization. The adiabatic pathway to excited *trans*-stilbene, *J. Photochem. Photobiol. A: Chem.*, 65, 29, 1992. (c) Saltiel, J., Waller, A. S., and Sears, D. F., Jr., The temperature and medium dependencies of *cis*-stilbene fluorescence. The energetics for twisting in the lowest excited singlet state, *J. Am. Chem. Soc.*, 115, 1993.

29. Herkstroeter, W. G. and Hammond, G. S., Mechanisms of photochemical reactions in solutions. XXXIX. Study of energy transfer by kinetic spectrophotometry, *J. Am. Chem. Soc.*, 88, 4769, 1966.

30. Tokumaru, K. and Arai, T., Potential energy surfaces of one-way and two-way photoisomerization of olefins, *J. Photochem. Photobiol. A: Chem.*, 65, 1, 1992, and references cited.

31. Sandros, K., Sundahl, M., Wennerström, O., and Norinder, U., Cis-trans isomerization of a *p*-styrylstilbene, a one- and twofold adiabatic process, *J. Am. Chem. Soc.*, 112, 3082, 1990. Sundahl, M., Wennerström, O., Sandros, K., Arai, T., and Tokumaru, K., Triplet state Z/E isomerization of a *p*-styrylstilbene, a partly adiabatic process, *J. Phys. Chem.*, 94, 6731, 1990.

32. Liu, R. S. H. and Asato, A. E., Photochemistry and synthesis of stereoisomers of vitamin A, *Tetrahedron*, 40, 1931, 1984.

33. Ramamurthy, V., Butt, Y., Yang, C., Yang, P., and Liu, R. S. H., Preparation of 7-*cis*-ionly and -ionylidene derivatives and other sterically hindered olefins by one-way sensitized geometric isomerization, *J. Org. Chem.*, 38, 1247, 1973. Ramamurthy, V., Tustin, G., Yau, C. C., and Liu, R. S. H., Preparation of sterically hindered geometric isomers of 7-*cis*-β-ionyl and β-ionylidene derivatives in the vitamin A series, *Tetrahedron*, 31, 193, 1975. Ramamurthy, V. and Liu, R. S. H., The chemistry of polyenes. IX. Excitation, relaxation, and deactivation of dienes, trienes, and higher polyenes in the Vitamin A series in the sensitized isomerization reaction, *J. Am. Chem. Soc.*, 98, 2935, 1976.

34. Schuchmann, H.-P., von Sonntag, C., and Srinivasan, R., Quantum yields in the photolysis of *cis*-cyclooctene at 185 nm, *J. Photochem.*, 15, 159, 1981.

35. Liu, R. S. H., Photosensitized isomerization of 1,3-cyclooctadienes and conversion to bicyclo[4.2.0]oct-7-ene, *J. Am. Chem. Soc.*, 89, 112, 1967.

36. Dauben, W. G., Cargill, R. L., Coates, R. M., and Saltiel, J., The direct and sensitized irradiation of acyclic dienes, *J. Am. Chem. Soc.*, 88, 2742, 1966.

37. Saltiel, J., Marchand, G. R., and Bonneau, R., *cis-trans*-1,1'-Bicyclohexenyl, a strained ground state intermediate in the photocyclization of 1,1'-bicyclohexenyl to its isomeric *cis*-cyclobutene, *J. Photochem.*, 28, 367, 1985. Saltiel, J. and Marchand, G. R., *cis-trans*-1,1'-Bicyclohexenyl: Conformer dependent chemistry in benzene and in methanol, *J. Am. Chem. Soc.*, 113, 2702, 1991.

38. Stoermer, R., Rearrangement by means of ultraviolet light of stable ethylene compounds into labile stereoisomers, *Ber. Dtsch. Chem. Ges.*, 42, 4865, 1909.

39. Zechmeister, L. and McNeely, W. H., Separation of *cis* and *trans*-stilbenes by application of the chromatographic brush method, *J. Am. Chem. Soc.*, 64, 1919, 1942.

40. Saltiel, J., Ganapathy, S., and Werking, C., The ΔH for thermal *trans*-stilbene/*cis*-stilbene isomerization. Do S_0 and T_1 potential energy curves cross?, *J. Phys. Chem.*, 91, 2755, 1987.

41. Saltiel, J. and Curtis, H. C., The photooxidation of benzil, *Mol. Photochem.*, 1, 239, 1969.

42. Hammond, G. S., Shim, S. C., and Van, S. P., Mechanisms of photochemical reactions in solutions. LIX. Photoisomerization of β-styrylnaphthalene, *Mol. Photochem.*, 1, 89, 1969.

43. Saltiel, J. and Eaker, D. W., Lifetime and geometry of 1-phenyl-2-(2-naphthyl)ethene triplets. Evidence against the triplet mechanism for direct photoisomerization, *Chem. Phys. Lett.*, 75, 209, 1980. Görner, H., Eaker, D. W., and Saltiel, J., Analysis of the decay of 1-phenyl-2-(2-naphthyl)ethene triplets. A nanosecond pulsed laser study, *J. Am. Chem. Soc.*, 103, 7164, 1981.

44. Perrin, D. D. and Armarego, W. L. F., *Purification of Laboratory Compounds*, Vol. 3, Pergamon, New York, 1988.

45. Lunde, K. and Zechmeister, L., *cis-trans* Isomeric 1,6-diphenylhexatrienes, *J. Am. Chem. Soc.*, 76, 2308, 1954.

46. Görner, H., Singlet mechanism for *trans* → *cis* photoisomerization of α,ω-diphenylpolyenes in solution. Laser flash study of the triplet states of diphenylbutadiene, diphenylhexatriene and diphenyloctatetraene, *J. Photochem.*, 19, 343, 1982.

47. Chattopadhyay, S. K. and Das, P. K., Singlet-singlet absorption spectra of diphenylpolyenes, *Chem. Phys. Lett.*, 87, 145, 1982.

48. Goldbeck, R. A., Twarowski, A. J., Russell, E. L., Rice, J. K., Birge, R. R., Switkes, E., and Kliger, D. S., Excited state absorption spectroscopy and state ordering in polyenes. II. α,ω-diphenylpolyenes, *J. Chem. Phys.*, 77, 3319, 1982.

49. Saltiel, J., Sears, D. F., Jr., Sun, Y.-P., and Choi, J.-O., Evidence for ground state *s-cis*-conformers in the fluorescence spectra of *all-trans*-1,6-diphenyl-1,3,5-hexatriene, *J. Am. Chem. Soc.*, 114, 3607, 1992.

50. Saltiel, J. and Ko, D.-H., unpublished results.

51. Braun, A. M., Maurette, M.-T., and Oliveros, E., *Photochemical Technology*, Wiley, Chichester, 1991, 500–523.

52. Jacobs, H. J. C. and Havinga, E., Photochemistry of vitamin D and its isomers and of simple trienes, *Adv. Photochem.*, 11, 305, 1979.

53. Jacobs, H. J. C., Gielen, J. W. H., and Havinga, E., Importance of dual resonance susceptibilities for pi-donor and pi-acceptor substituents regarding a quantitative description of substituent effects. The case of basicity of pyridine-*N*-oxides, *Tetrahedron Lett.*, 21, 4013–4016, 1980.

54. Malatesta, V., Willis, C., and Hackett, P. A., Laser photochemical production of vitamin D, *J. Am. Chem. Soc.*, 103, 6781–6783, 1981.

55. (a) Dauben, W. G. and Phillips, R. B., Effects of wavelength on the photochemistry of provitamin D, *J. Am. Chem. Soc.*, 104, 5780–5781, 1982. (b) Dauben, W. G., Share, P. E., and Ollmann, R. R., Jr., Triene photophysics and photochemistry: Previtamin D_3, *J. Am. Chem. Soc.*, 110, 2548–2554, 1988. (c) Dauben, W. G., Disanayaka, B., Funhoff, D. J. H., Kohler, B. E., Schilke, D. E., and Zhou, B., Polyene 1^1A_g and 1^1B_u states and the photochemistry of previtamin D_3, *J. Am. Chem. Soc.*, 113, 8367–8374, 1991.

56. Snoeren, A. E., Daha, M. R., Lugtenburg, J., and Havinga, E., Vitamin D and related compounds. 21. Photosensitized reactions, *Rec. Trav. Chem., Pays-Bas*, 89, 261–264, 1970.

57. Eyley, S. C. and Williams, D. H., Photolytic production of vitamin D. The preparative value of a photosensitizer, *J. Chem. Soc., Chem. Commun.*, 858, 1975.

58. Denny, M. and Liu, R. S. H., Photochemistry of polyenes. 13. Photosensitized isomerization of previtamin D_2 and tachysterol and the related 1,2-*bis*-(1-cyclopentyl)ethylene: Torsional potential curves of triene triplets, *New. J. Chem.*, 2, 637–641, 1978.

2

Photorearrangement and Fragmentation of Alkenes

Paul J. Kropp
*University of North Carolina
at Chapel Hill*

2.1 Photochemical Behavior

On direct irradiation in nonhydroxylic solvents, alkenes typically undergo several competing photoreactions. In addition to $E \mathbf{1} Z$ isomerization, which is also exhibited on triplet sensitization, there are several processes that are usually observed only on direct irradiation such as rearrangement via carbene intermediates, double bond migration, alkyne formation, and hydrogen atom abstraction. This chapter is concerned with these latter processes. Emphasis is given to developments since the topic was reviewed extensively in 1979.[1] Portions of this material are covered in reviews that have appeared since then.[2,3] $E \mathbf{1} Z$ isomerization is discussed in Chapter 1 and the photobehavior of alkenes in hydroxylic media in Chapter 7.

Spectroscopic Properties

The UV absorption spectra of simple alkenes in solution typically consist of single, featureless bands with maxima at 185–195 nm ($\varepsilon = 5$–10×10^3) and tails extending to 210–255 nm.[4] Absorption extends to longer wavelengths with increasing degrees of substitution. The spectra are deceptively simple, as they consist of two or more overlapping bands arising from close-lying transitions, two of which are generally accepted as being $\pi \rightarrow \pi^*$ and $\pi \rightarrow R(3s)$.[5,6] Weak absorptions attributed to $\pi \rightarrow \sigma^*$ transitions have also been reported for alkenes in the liquid state,[5c] although their presence has not been independently confirmed. Gas-phase absorption spectra are generally similar to those in solution except for the presence of some structure in the bands and somewhat longer absorption tails. Due to the diffuseness of Rydberg orbitals, bands corresponding to Rydberg transitions are severely reduced in intensity and blue-shifted in the condensed phase compared to the gas phase at low pressure.

Alkenes exhibit only weak fluorescence, which is thought to involve emission from the $\pi, R(3s)$ state.[7] On excitation at 185 nm, the emission maximum moves from 231 to 263 nm with increasing alkyl substitution, and the fluorescence quantum yield increases from 1×10^{-6} to 1.5×10^{-4}. Ethene

0-8493-8634-9/95/$0.00+$.50
© 1995 by CRC Press, Inc.

itself does not fluoresce. The excitation maximum increases with increasing wavelength, moving from 195 nm for disubstituted to 229 nm for tetrasubstituted alkenes. The fluorescence lifetime is estimated at 15 ps.

Owing to the large energy gap between the singlet and triplet manifolds, alkenes do not readily undergo intersystem crossing. Thus, their behavior on direct irradiation arises from singlet excited states. It is useful to know which singlet excited state is lowest lying for a given alkene. The vertical excitation energy of the π,R(3s) singlet is generally below that of the π,π^* singlet. Having a somewhat electron-deficient core, the π,π^* excited state moves to lower energy with increasing alkyl substitution. Due to the large size of the 3s Rydberg orbital, the π,R(3s) state has an even more electron-deficient core and responds more strongly to alkyl substitution. Thus, the separation between the π,π^* and π,R(3s) vertical levels increases with increasing substitution. However, the π,π^* state undergoes substantial relaxation through pyramidalization and/or rotation about the central C—C bond. Thus, in alkenes that are not highly substituted, the relaxed π,π^* level is close to, if not lower than, that of the π,R(3s) state. With increasing substitution, there is increasing likelihood that the π,R(3s) state is lower lying.

Principal Pathways

Irradiations of alkenes are generally conducted with a low-pressure mercury lamp (185 nm), an ArF excimer laser (193 nm), or a zinc resonance lamp (214 nm). Highly substituted alkenes can also be irradiated with a medium-pressure lamp and quartz optics (>200 nm). Because of the presence of two or more close-lying singlet excited states, irradiation of alkenes usually results in several competing photoprocesses. Thus, for example, irradiation of 2,3-dimethyl-2-butene (1) in hydro-carbon or ether solvent results in rearrangement to the carbene-derived products 4 and 5 and the double-bond isomer 9.[8] E 1 Z isomerization also occurs but it is not apparent in this case because of the symmetry of alkene 1. Similar behavior is exhibited in the gas phase.[9] In alcoholic or aqueous media, the nucleophilic trapping products 7 and 8 are formed in competition with the carbene-derived products 4 and 5.[10]

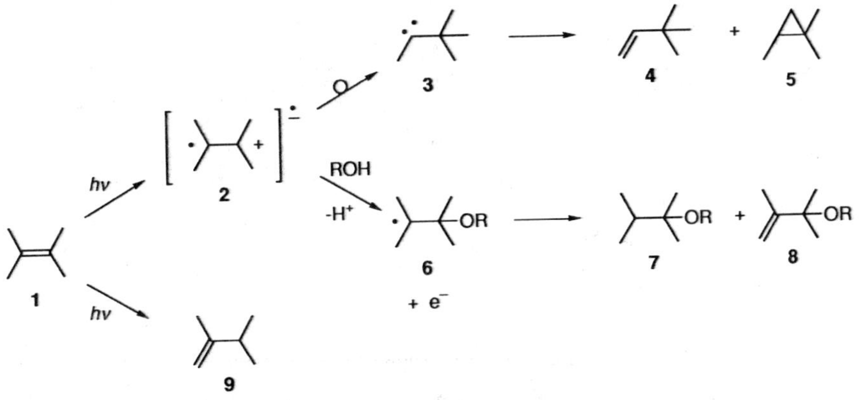

The monosubstituted alkene 1-octene (10), on the other hand, affords the double-bond migra-tion products (E)- and (Z)-2-octene (11) but no products obviously derived from a carbene intermediate.[11] However, a new type of product, alkyne 12, is formed, along with a small amount of the reduction product 13. In addition to these general types of photobehavior, some alkenes exhibit behavior that is specific to particular systems, as outlined below.

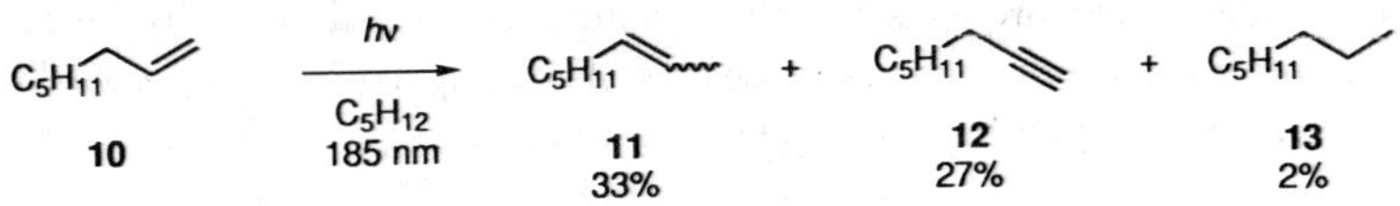

2.2   Mechanisms

Rearrangement via Carbene Intermediates

There is general agreement with the original proposal[8] that products such as the rearranged alkene 4 and the cyclopropane 5 arise via competing 1,2 and 1,3 insertion of a carbene intermediate (3) formed by rearrangement of the π,R(3s) Rydberg excited state (2). This state arises from promotion of one of the π-electrons to a large molecular orbital having approximately the size and shape of a 3s helium orbital, leaving an electron-deficient core with radical-cation character. The involvement of carbene intermediates is supported by formation of the same products in similar ratios on independent generation of the corresponding carbenes from a wide range of alkenes. Involvement of the π,R(3s) state is in keeping with its radical-cation character. Moreover, formation of the carbene-derived products 4 and 5 relative to the double-bond migration product 9 is wavelength dependent, being favored at higher wavelengths as expected for a low-lying π,R(3s) state.[9,12]

Further support for involvement of the π,R(3s) state in carbene formation has come from the photobehavior of 2-trifluoromethyl-2-norbornene (14), which afforded principally the 1,3-alkyl migration product 15 along with the related minor products 16 and 17.[13] This behavior is quite different from that of the parent, 2-norbornene (18), which affords the carbene-derived products 21 and 22 along with small amounts of the radical-derived products 24 to 26.[14,15] Spectroscopic studies of 14 showed that the strongly electron-withdrawing CF$_3$ substituent raises the energy of the π,R(3s) state, rendering the π,π^* state lower lying in solution.[4b] The absence of carbene-derived products from 14 thus implies that they arise from the π,R(3s) state.

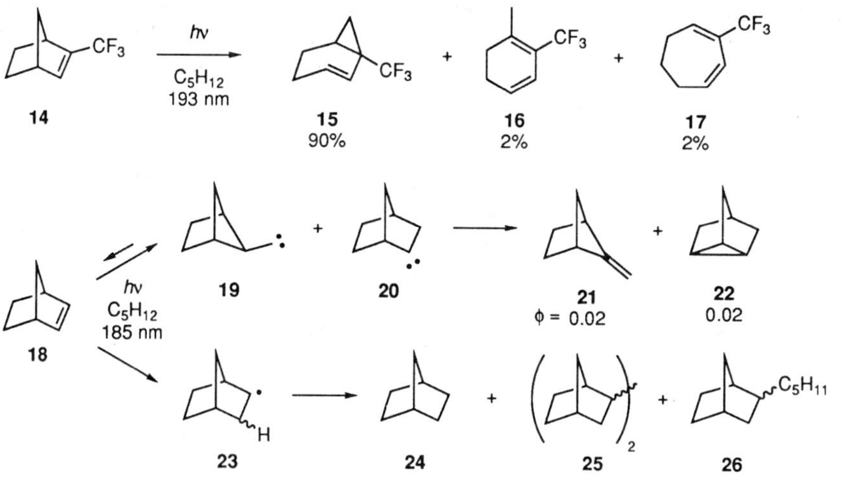

The formation of nortricyclene (22) from 2-norbornene (18) requires norbornylidene (20) as an intermediate. Thus, in alkenes that are not fully substituted hydrogen migration can compete with alkyl migration in carbene formation. In the case of cyclohexene (27), deuterium labeling studies showed that products 30 and 31 arise from carbenes 28 and 29.[14] It was concluded that these intermediates are formed in a ratio of 14:1, but this does take into account any differences in the rate of reformation of starting material. It was concluded that hydrogen migration to form cyclopentylidene is at most a minor process in the photobehavior of cyclopentene (32), which affords the carbene-derived products 36 and 37 along with the radical-derived products 34 and 35.[15,16] On the other hand, hydrogen migration obviously competes favorably with alkyl migration in 2-norbornene (18), perhaps because of the high strain associated with ring contraction.

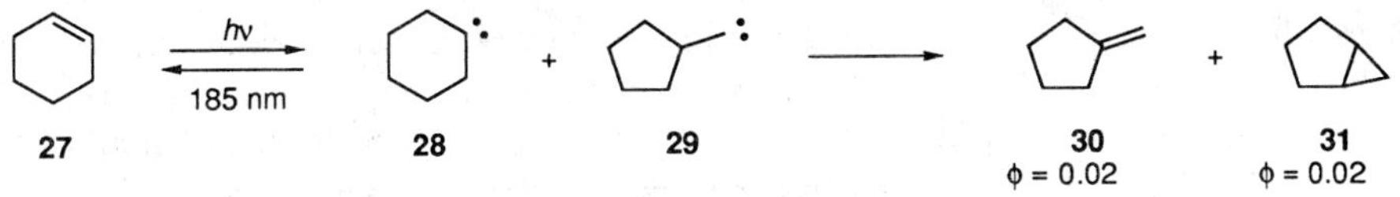

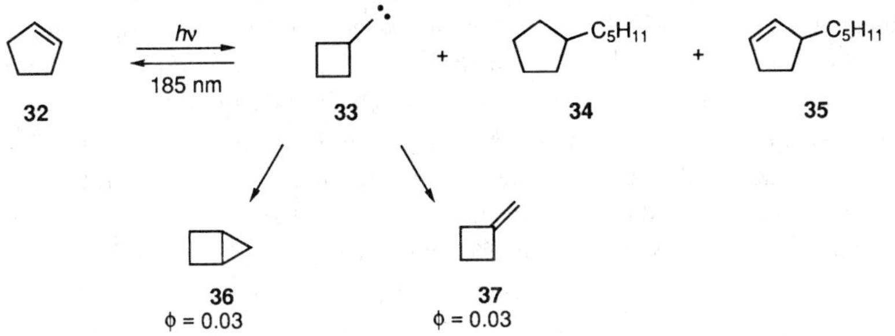

27   **28**   **29**   **30** $\phi = 0.02$ **31** $\phi = 0.02$

32   **33**   **34**   **35**

36 $\phi = 0.03$ **37** $\phi = 0.03$

Labeling studies on the related homobrexene (**38**), which afforded the carbene-derived products **40** and **42**, revealed that the migrating hydrogen atom had a 2.3:1 preference for assuming the axial position.[17] Interestingly, deuterium had an even greater preference (10.2:1) for the axial pathway. Some scrambling of labeled **38** occurred, presumably because of competing return of one or both of the carbene intermediates **39** and **41** to **38**. On independent generation, carbene **39** afforded an approximately 3:1 mixture of the insertion product **40** and homobrexene (**38**).

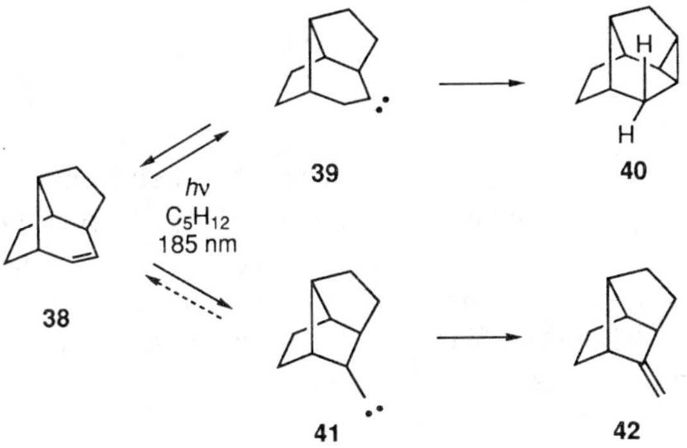

38   **39**   **40**   **41**   **42**

Sigmatropic Rearrangements

Double-bond migration generally competes with $E\,1\,Z$ isomerization and the formation of carbene-derived products, unless it is structurally inhibited as in 2-norbornene (**18**) and homobrexene (**38**). It presumably occurs in cyclohexene (**27**) and cyclopentene (**32**) but is not observable. In an alkene such as **43**, in which the double bond is exocyclic to a five-membered ring, migration is the dominant photoprocess and occurs preferentially into, rather than away from, the ring.[18] Labeling studies with **43**-d_4 showed that double-bond migration involves an intramolecular 1,3-sigmatropic migration of an allylic hydrogen atom and exhibits a deuterium isotope effect of 2.7.

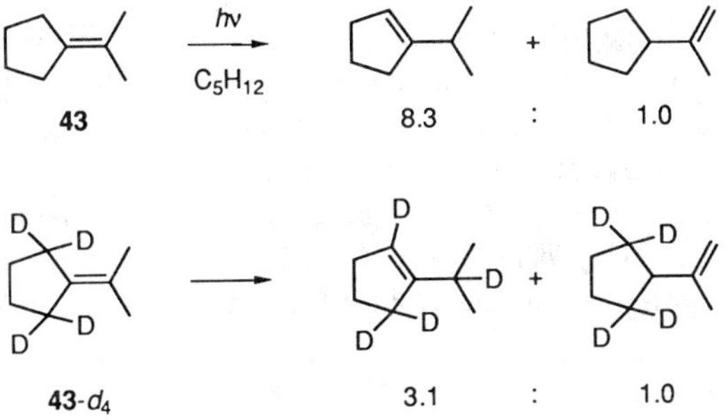

43   8.3   :   1.0

43-d_4 3.1   :   1.0

There is no current agreement as to which excited state is responsible for 1,3-hydrogen migration. As noted above, the ratio of double-bond migration to formation of carbene-derived products from alkene 1 is wavelength dependent, being greater at shorter wavelengths. Thus, the $\pi,R(3s)$ state is clearly not involved in 1,3-hydrogen migration. Similarly, irradiation of *(Z)*-2-octene [*(Z)*-11] at 185 nm afforded competing *E* 1 *Z* isomerization and rearrangement to a mixture of the double-bond isomers 10 and *(E)*- and *(Z)*-44, whereas at wavelengths greater than 200 nm only *E* 1 *Z* isomerization was observed.[12] Thus, it was concluded that 1,3-hydrogen migration does not involve the π,π^* state either, and the $\sigma_\pi(CH_2),\pi^*$ or $\pi(C{=}C),\sigma_\pi^*(CH_2)$ charge-transfer state was suggested as an alternative. However, it has been noted that the latter wavelength effect merely indicates that double-bond migration does not involve the π,π^* state in the case of a disubstituted alkene such as 11.[13] It is conceivable that the double-bond isomers 10 and 44 arise via the carbene intermediates 45 and thus are derived from the $\pi,R(3s)$ excited state, which is higher lying than the π,π^* state in a disubstituted alkene.

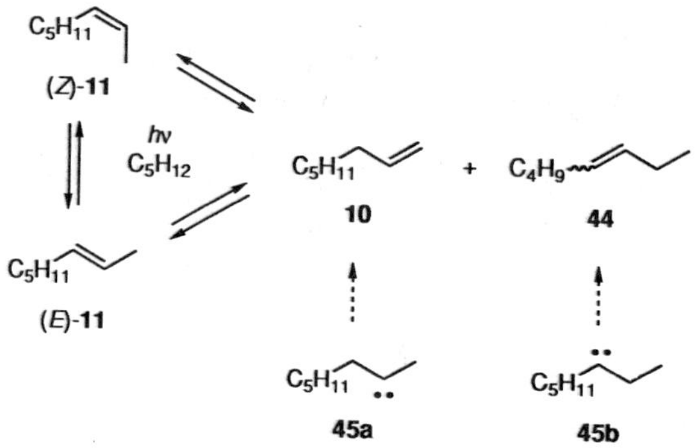

This leaves possible the involvement of the π,π^* state in the 1,3-hydrogen migrations in alkenes such as 1 and 43. Some support for this assignment is provided by the formation of the analogous 1,3-alkyl migration product 15 by the CF$_3$-substituted norbornene 14, which has a low-lying π,π^* state.[4b,13] Similar behavior was exhibited by the nitrile 46, which afforded a 20:1 mixture of the 1,3- and 1,2-alkyl shift products 47 and 48.[19] Since the lowest-lying excited singlet state of the conjugated alkene 46 is π,π^*, it has been proposed that 1,3-alkyl migration in norbornene 14 also involves the π,π^* state.[13]

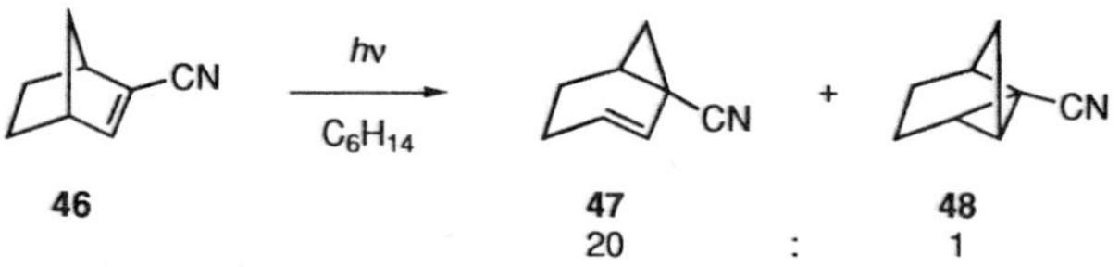

Alkyne Formation

As noted above, irradiation of 1-octene (10) afforded a substantial amount of 1-octyne (12).[11] Similarly, the principal photobehavior of ethene is formation of ethyne and H$_2$.[20] Since the 1,1-d_2 derivative 10-d_2 afforded alkyne 12 with an isotope effect ϕ_H/ϕ_D of 1.9 and almost complete loss of deuterium, the principal route to 12 apparently involves the vinylidene 49 as an intermediate. Alkyne formation is also a major pathway for the keto analog 55.[21] Interestingly, the γ,δ-unsaturated ketones 55, 59, and 62 exhibit photobehavior typical of alkenes at 185 nm. Similar to alkene 10-d_2, the terminally labeled derivative of enone 55 afforded alkyne 57 with 90% loss of deuterium. The 2,2-disubstituted alkene 50 afforded a mixture of the alkyne 52 and 1-methylcyclopentene (53), expected products from the vinylidene intermediate 51, along with the double-bond migration product 54, and similarly the keto analog 59 afforded alkyne 61 and the double-bond migration product **60**. Alkyne formation has also been observed in the 1,2-disubstituted analog **62**, which afforded a mixture of alkynes **64** and **65**, along with the double-bond migration product **63** and the 1,3-acyl migration product **66**. It has been proposed that alkyne formation involves the π,π^* excited state since it has not been observed for highly substituted alkenes, in which the $\pi,R(3s)$ state is lowest lying.[11,21]

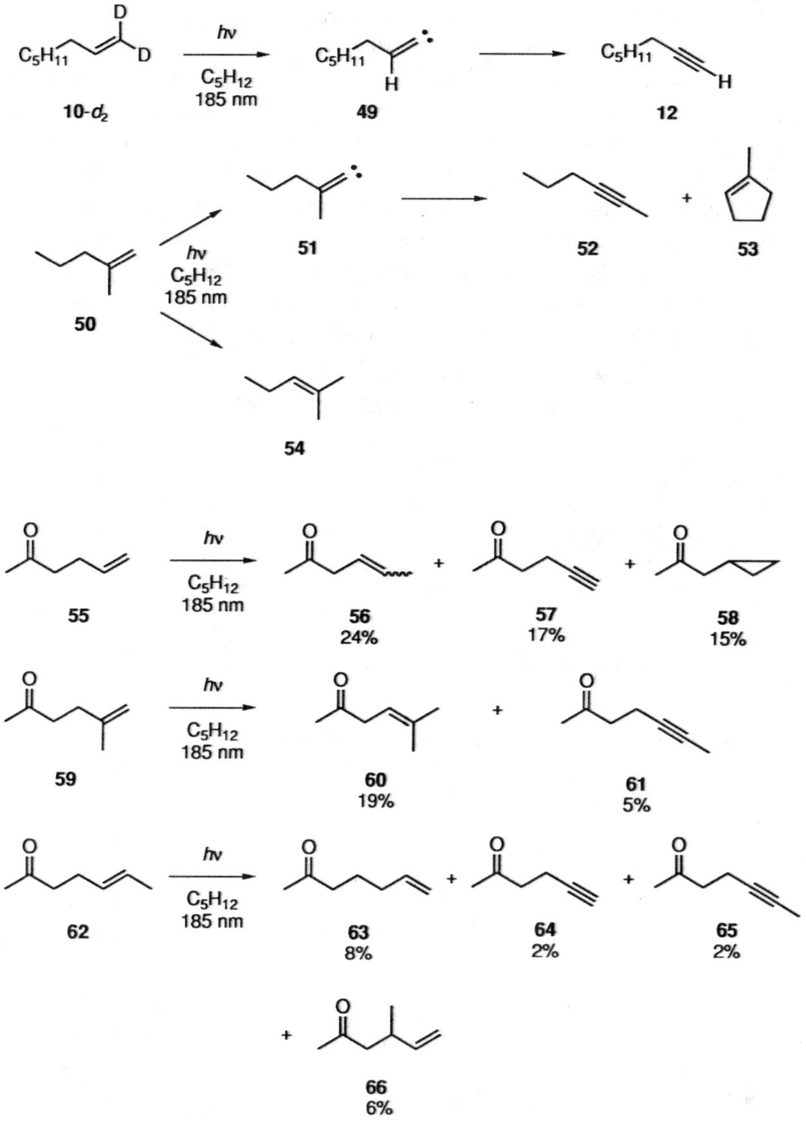

Hydrogen Atom Abstraction

As noted above, irradiation of 2-norbornene (**18**) in pentane afforded, in addition to the carbene-derived products **21** and **22**, a mixture of alkanes **24** to **26**, which appear to be derived from the radical **23** formed by hydrogen atom abstraction from the solvent.[15] Analogous behavior was observed for 1-octene (**10**), which afforded the reduction product octane (**13**), and cyclopentene (**32**), which afforded the radical-derived products **34** and **35**.[11,15] In each case, a mixture of branched decanes arising from coupling of solvent-derived pentyl radicals was also obtained. Hydrogen atom abstraction has also been attributed to the π,π^* excited state.[15]

2.3 Further Examples

Medium and Large Cycloalkenes

Medium- and large-ring systems are noted for undergoing transannular reactions. This is seen in the cycloalkenes **67**, which afforded, in addition to $E\,1\,Z$ isomerization, the *cis*-bicyclo[*n*.3.0]alkanes **70** resulting from 1,5-transannular insertion of the corresponding cycloalkylidene intermediates **68**.[22] Only the smallest member of the series, **67a**, afforded material (**72a**) arising apparently from a ring-contracted cycloalkylmethylene intermediate (**69a**). The largest members of the series, **67d** and **67e**, gave small amounts of the fragmentation products **74** and **75**. Double-bond migration presumably also occurs, but is an identity process in these cases. When generated independently, cycloalkylidenes **68** undergo substantial 1,2-insertion to form cycloalkenes **67**. This is particularly true for the two largest members of the series, **68d** and **68e**, which give only a small amount of transannular insertion when generated independently. Transannular insertion is magnified when the cycloalkylidene intermediates are generated by irradiation of the corresponding cycloalkenes since any 1,2-insertion simply regenerates the cycloalkene, which is recycled, whereas the bicyclic products **70** are photostable and accumulate in the irradiation mixture.

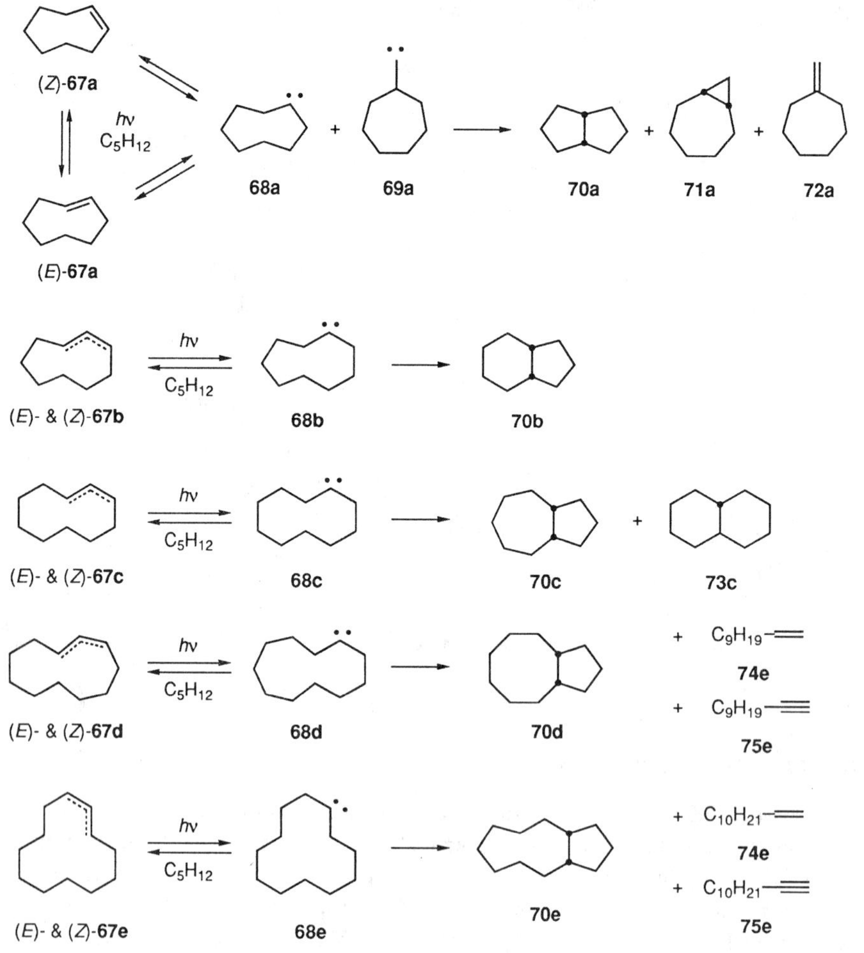

Cyclobutenes

Cyclobutenes have been extensively studied, particularly by Leigh and co-workers.[23] Irradiation of cyclobutene (**76**) itself afforded principally the ring opening product 1,3-butadiene (**77**), accompanied by smaller amounts of methylenecyclopropane (**78**) and the fragmentation products ethene and ethyne.[24] Under the irradiation conditions 1,3-butadiene (**77**) underwent a small amount of reversion to cyclobutene (**76**), and methylenecyclopropane (**78**) underwent secondary conversion to 1,3-butadiene (**77**) and the fragmentation products ethene and ethyne. It was proposed that 1,3-butadiene (**77**) arises via electrocyclic ring opening of the π,π^* excited singlet state and that the minor products are derived from the carbene intermediates **79** and **80** formed by rearrangement of the $\pi,R(3s)$ state.

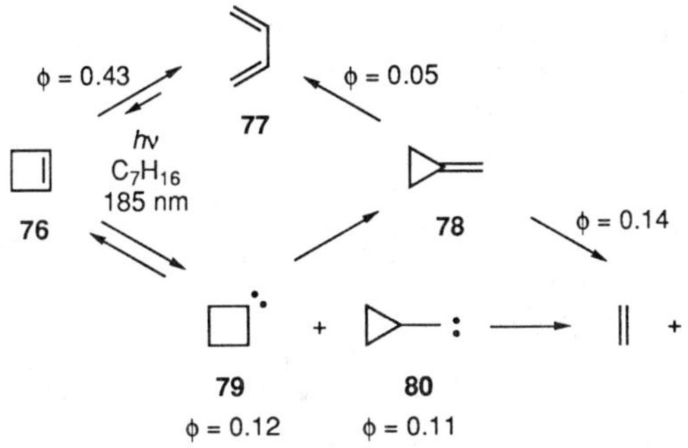

Similar behavior was exhibited by the bicyclo[*n*.2.1]alkenes **81**, except that the highly strained methylenecyclopropane **83a** was not formed.[25] Surprisingly, **81b** and **81c** afforded an *E,Z* mixture of the 1,3-dienes **82b** and **82c** under conditions in which the dienes do not undergo interconversion.[4c,25b,25c] Thus, ring opening of cyclobutenes is not stereospecific. Similar lack of specificity has been observed for the bicyclo[6.2.1]octenes **81d** and **85**,[25c] as well as the *cis*- and *trans*-isomers of cyclobutenes **86** and the pair of tetrasubstituted cyclobutenes **87**.[4a,4c,25b]

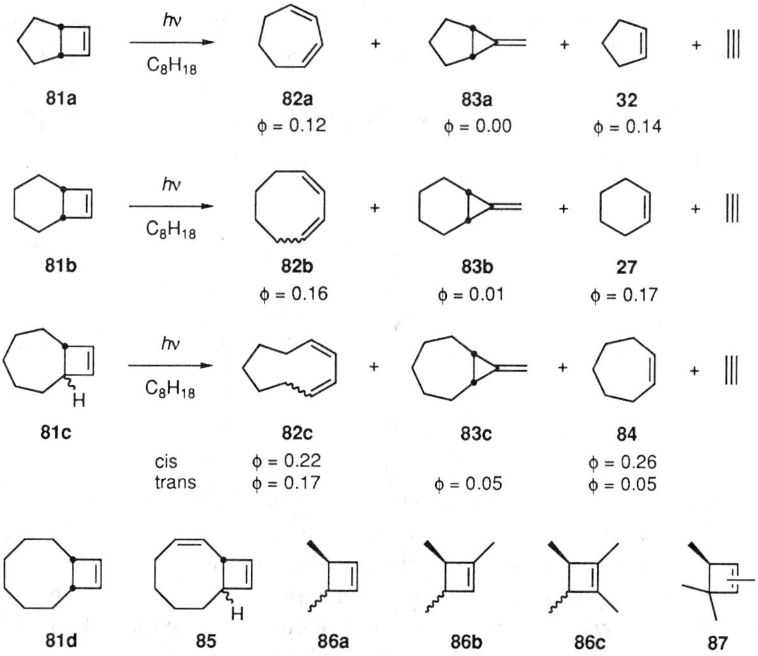

The bicyclo[4.2.0]oct-1(6)-ene system **88** comes closest to affording the disrotatory stereospecificity expected by orbital symmetry, with the *cis*-isomer giving mainly diene *(E,E)*-**89** and the *trans*-isomer giving, even more predominantly, the diene *(E,Z)*-**89**.[26] Similarly, cyclobutene *cis*-**90** undergoes ring opening to diene *(E,E)*-**91** with a substantially higher quantum yield than *trans*-**90**, which would have to afford the highly strained *E,Z* isomer of diene **91** if mid-opening had occurred in a disrotatory fashion.[27] Indeed, the *E,Z* isomer may be formed initially and undergo rapid isomerization to *(E,E)*-**91**. These observations suggest that orbital symmetry plays a role in the photochemical ring opening of cyclobutenes, but apparently only in the initial stages of reaction on the excited state potential energy surface.[23]

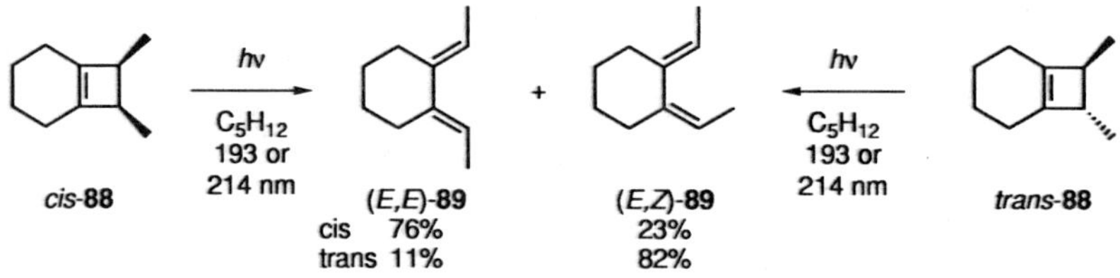

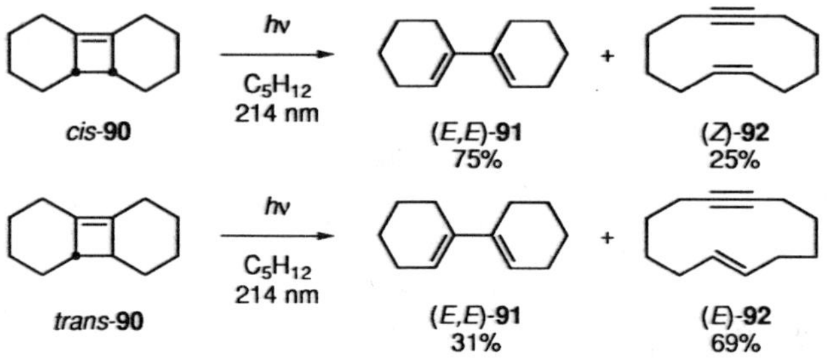

The assignment of the π,π* and π,R(3*s*) excited states for ring opening and fragmentation, respectively, of cyclobutenes has been supported by a study of the bicyclo[4.2.0]octenes **81b**.[4c] Spectroscopic studies indicate that the π,R(3*s*) state is the lowest lying singlet excited state in the parent **81b** (R = H) and the methyl derivative **81b** ($R = CH_3$), but is raised to a higher energy than the π,π* state in the trifluoromethyl derivative **81b** ($R = CF_3$). Accordingly, fragmentation predominates over ring opening for the methyl and unsubstituted derivatives, whereas ring opening is more efficient for the trifluoromethyl-substituted derivative. Moreover, fragmentation is favored at the longer wavelength for the first two compounds, but at shorter wavelength for the trifluoromethyl derivative.

		81b	**82b**		**27**
R = CH_3	193		1	:	1.49
	214		1	:	2.84
H	193		1	:	0.80
	214		1	:	1.10
CF_3	193		1	:	0.10
	214		1	:	0.06

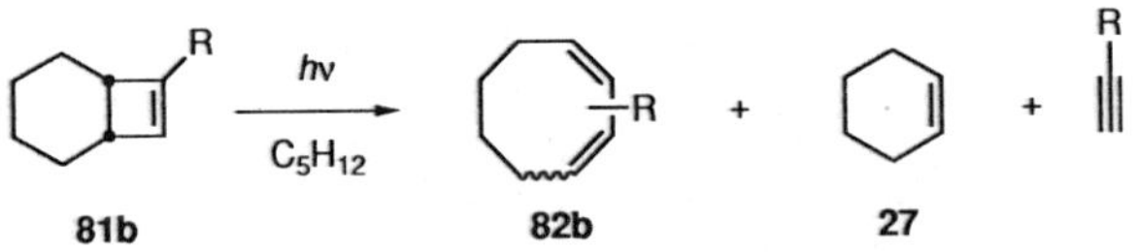

In contrast with ring opening, fragmentation is stereospecific. For example, the *cis*- and *trans*-epimers of cyclobutene **90** afford exclusively the *Z* and *E* isomers of enyne **92**, respectively.[27] Similarly, the *cis*-isomers of cyclobutenes **86** afford *(Z)*-2-butene and the *trans*-isomers afford *(E)*-2-butene with greater than 90% stereospecificity.[4c] Moreover, cyclobutenes **88**, which would have to afford the highly strained cyclohexyne, give only ring-opening products.[26] Although fragmentation is formally a retro [2 + 2]-cycloaddition, it apparently arises from rearrangement of the π,R(3*s*) excited state, as noted above. Some evidence that the required cyclopropylmethylene intermediates are indeed formed has come from the tetramethylcyclobutenes **87a** and **87b**, which undergo some interconversion on irradiation.[4a] Presumably, the cyclopropylmethylenes **93** and/or **94** are involved in the interconversion as well as fragmentation.

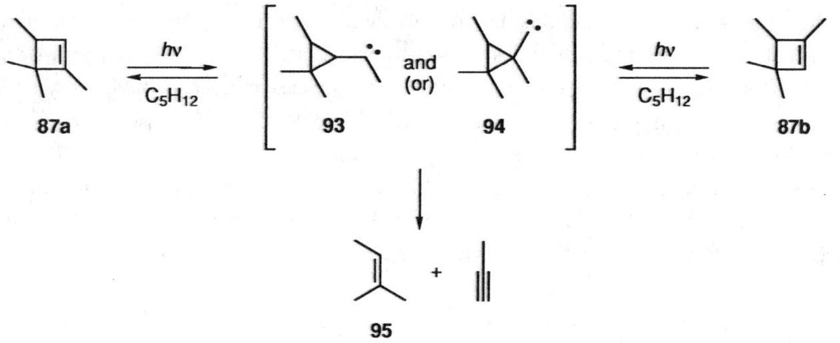

3-Oxacycloalkenes

The oxacycloalkenes **96** afforded aldehydes **98** and **99**, which apparently arise from the biradical intermediate **97** formed via cleavage of the allylic O–C bond.[28] The dihydropyran analog **96b** also afforded the carbene-derived product **100**. It was suggested that bond cleavage involves a π,σ^* excited state.

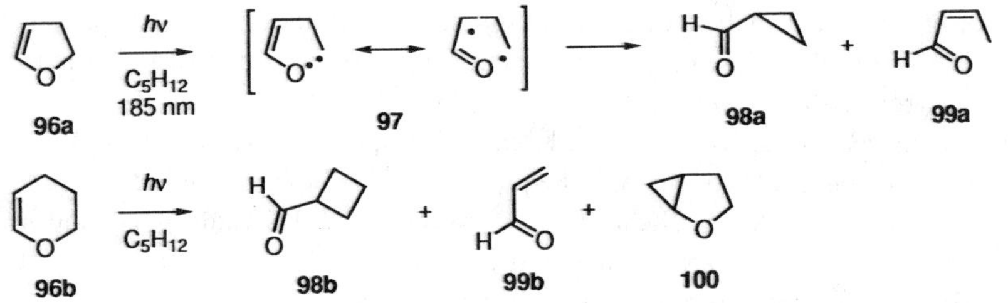

Homoallylic Alcohols

The homoallylic alcohols **101** afforded a mixture of aldehydes **103** and **104**.[29] Deuterium labeling showed that rearrangement occurs intramolecularly, presumably via the biradical **102**. Since in each case the *endo*-epimer afforded an excess of the Δ^3 isomer **103**, it was proposed that the *endo*-epimers undergo a competing concerted rearrangement as shown for *endo*-**101a**.

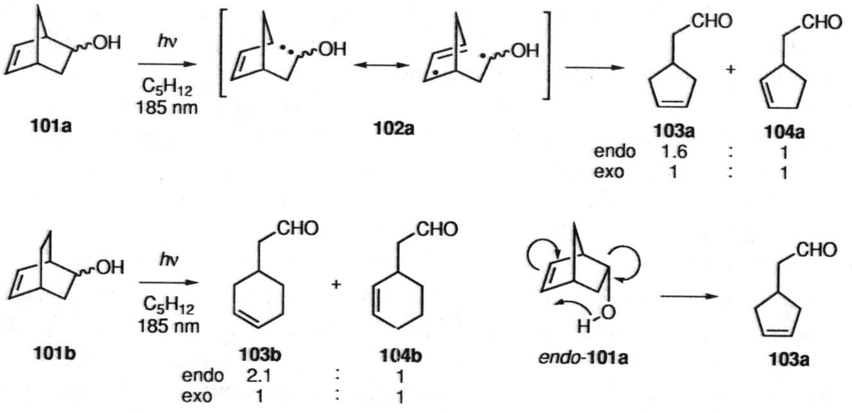

2.4 Summary

The photochemistry of alkenes is exceptionally rich, arising from two or more close-lying excited states that appear not to communicate well between themselves. The exclusive behavior attributed to the $\pi,R(3s)$ singlet excited state, aside from nucleophilic trapping in hydroxylic media,[10,30] is rearrangement to carbene intermediates. These intermediates, in turn, undergo intramolecular insertion to afford isomers of the starting alkene or, in the case of cyclopropylmethylenes derived from cyclobutenes, fragmentation to an alkene and alkyne. By contrast, a variety of behaviors has been ascribed to the π,π^* singlet excited state. In addition to its widely recognized role in $E\,1\,Z$ isomerization, the π,π^* state has been implicated in 1,3-alkyl migration,[13] alkyne formation,[11,21] hydrogen atom abstraction,[15] and ring opening of cyclobutenes.[23] By analogy to 1,3-alkyl migration, 1,3-hydrogen migration may also involve this state. However, much remains to be done to place these assignments on a firmer basis.

References

1. Kropp, P. J., Photochemistry of alkenes in solution, *Org. Photochem.*, 4, 1, 1979.
2. For recent reviews of the photobehavior of alkenes in solution, see (a) Adam, W. and Oppenländer, T., 185-nm Photochemistry of olefins, strained hydrocarbons, and azoalkanes in solution, *Angew. Chem., Int. Ed.*, 25, 661, 1986. (b) Steinmetz, M. G., Photochemistry with short UV light, *Org. Photochem.*, 8, 67, 1987.
3. For recent reviews of the photobehavior of alkenes in the gas phase, see (a) Collin, G. J., Ring contraction of cyclic olefins: Chemical processes specific to electronically excited states?, *J. Photochem.*, 38, 205, 1987. (b) Collin, G. J., Photochemistry of simple olefins: Chemistry of electronic excited states or hot ground state?, *Adv. Photochem.*, 14, 135, 1988.
4. For some recent examples, see (a) Clark, K. B. and Leigh, W. J., Cyclobutene photochemistry. Involvement of carbene intermediates in the photochemistry of alkylcyclobutenes, *Can. J. Chem.*, 66, 1571, 1988. (b) Wen, A. T., Hitchcock, A. P., Werstiuk, N. H., Nguyen, N., and Leigh, W. J., Studies of electronic excited states of substituted norbornenes by UV absorption, electron energy loss, and HeI photoelectron spectroscopy, *Can. J. Chem.*, 68, 1967, 1990. (c) Leigh, W. J., Zheng, K., and Clark, K. B., Cyclobutene photochemistry. Substituent and wavelength effects on the photochemical ring opening of monocyclic alkylcyclobutenes, *Can. J. Chem.*, 68, 1988, 1990. (d) Leigh, W. J., Zheng, K., and Clark, K. B., Cyclobutene photochemistry. The photochemistry of *cis*- and *trans*-bicyclo[5.2.0]non-8-ene, *J. Org. Chem.*, 56, 1574, 1991. (e) Leigh, W. J., Zheng, K., Nguyen, N. Werstiuk, N. H., and Ma, J., Cyclobutene photochemistry. Identification of the excited states responsible for the ring-opening and cycloreversion reactions of alkylcyclobutenes, *J. Am. Chem. Soc.*, 113, 4993, 1991. (f) Wiseman, J. R. and Kipp, J. E., *(E)*-Bicyclo[3.3.1]non-ene, *J. Am. Chem. Soc.*, 104, 4688, 1982.
5. (a) Merer, A. J. and Mulliken, R. S., Ultraviolet spectra and excited states of ethylene and its alkyl derivatives, *Chem. Rev.*, 69, 639, 1969. (b) Watson, F. H., Jr., Armstrong, A. T., and McGlynn, S. P., Electronic transitions in mono-olefinic hydrocarbons. I. Computational results, *Theoret. Chim. Acta*, 16, 75, 1970. (c) Watson, F. H., Jr. and McGlynn, S. P., Electronic transitions in mono-olefinic hydrocarbons. II. Experimental results, *Theoret. Chim. Acta*, 21, 309, 1971. (d) Robin, M. B., *Higher Excited States of Polyatomic Molecules*, Vol. II, Academic Press, New York, 1975, 22.
6. For a recent analysis, see Wiberg, K. B., Hadad, C. M., Foresman, J. B., and Chupka, W. A., Electronically excited states of ethylene, *J. Phys. Chem.*, 96, 10756, 1992.
7. (a) Hirayama, F. and Lipsky, S., Fluorescence of mono-olefinic hydrocarbons, *J. Chem. Phys.*, 62, 576, 1975. (b) Wickramaaratchi, M. A., Preses, J. M., and Weston, R. E., Jr., Lifetime and quenching rate constants for a fluorescent excited state of tetramethylethylene, *Chem. Phys. Lett.*, 120, 491, 1985. (c) Inoue, Y., Daino, Y., Tai, A., Hakushi, T., and Okada, T., Synchrotron-radiation study of weak fluorescence from neat liquids of simple alkenes: Anomalous excitation spectra as evidence for wavelength-dependent photochemistry, *J. Am. Chem. Soc.*, 111, 5584, 1989.

8. Fields, T. R. and Kropp, P. J., *Photochemistry of alkenes. 3. Formation of carbene intermediates, J. Am. Chem. Soc.*, 96, 7559, 1974.

9. (a) Collin, G. J., Deslauriers, H., and Wieckowski, A., Photolysis of gaseous tetramethylethylene between 185 and 230 nm, *J. Phys. Chem.*, 85, 944, 1981. (b) Collin, G. J. and Deslauriers, H., Photoisomerization of gaseous tetramethylethylene in the far ultraviolet, *Can. J. Chem.*, 61, 1510, 1983.

10. Kropp, P. J., Reardon, E. J., Jr., Gaibel, Z. L. F., Willard, K. F., and Hattaway, J. H., Jr., Photochemistry of alkenes. 2. Direct irradiation in hydroxylic media, *J. Am. Chem. Soc.*, 95, 7058, 1973.

11. Inoue, Y., Mukai, T., and Hakushi, T., Direct photolysis at 185 nm of 1-alkenes in solution. Molecular elimination of terminal hydrogens, *Chem. Lett.*, 1725, 1984.

12. Inoue, Y., Mukai, T., and Hakushi, T., Wavelength-dependent photochemistry of 2,3-dimethyl-2-butene and 2-octene in solution, *Chem. Lett.*, 1665, 1983.

13. Nguyen, N., Harris, B. E., Clark, K. B., and Leigh, W. J., The solution-phase photochemistry of 2-trifluoromethylnorbornene, *Can. J. Chem.*, 68, 1961, 1990.

14. (a) Srinivasan, R. and Brown, K. H., Organic photochemistry with high level (6.7 eV) photons: Duality of carbene intermediates from cyclic olefins, *J. Am. Chem. Soc.*, 100, 4602, 1978. (b) Quantum yields quoted from this report have been adjusted to reflect the subsequently revised value for the *E 1 Z* isomerization of cyclooctene. Schuchmann, H.-P., von Sonntag, C. and Srinivasan, R., Quantum yields in the photolysis of *cis*-cyclooctene at 185 nm, *J. Photochem.*, 15, 159, 1981.

15. Inoue, Y., Mukai, T., and Hakushi, T., Direct photolysis at 185 nm of cyclopentene and 2-norbornene. A novel reaction channel for π,π* excited singlet alkene, *Chem. Lett.*, 1045, 1982.

16. Adam, W., Oppenländer, T., and Zang, G., 185-nm photochemistry of bicyclo[2.1.0]pentane and cyclopentene, *J. Am. Chem. Soc.*, 107, 3924, 1985.

17. Nickon, A., Ilao, M. C., Stern, A. G., and Summers, M. F., Hydrogen trajectories in alkene to carbene rearrangements. Unequal deuterium isotope effects for the axial and equatorial paths, *J. Am. Chem. Soc.*, 114, 9230, 1992.

18. (a) Kropp, P. J., Fravel, H. G., Jr., and Fields, T. R., Photochemistry of alkenes. 4. Vicinally unsymmetrical olefins in hydroxylic media, *J. Am. Chem. Soc.*, 98, 840, 1976. (b) Kropp, P. J., Fields, T. R., Fravel, H. G., Jr., Tubergen, M. W., and Crotts, D. D., Photochemistry of Alkenes. 11. Rearrangements in nonnucleophilic media. Manuscript submitted.

19. Akhtar, I. A., McCullough, J. J., Vaitekunas, S., Faggiani, R., and Lock, C. J. L., Photorearrangement of 2-cyanobicyclo[2.2.1]hept-2-ene. Observation of 1,2- and 1,3-sigmatropic shifts, *Can. J. Chem.*, 60, 1657, 1982.

20. For a list of references and a theoretical treatment, see Evleth, E. M. and Sevin, A., A theoretical study of the role of valence and Rydberg states in the photochemistry of ethylene, *J. Am. Chem. Soc.*, 103, 7414, 1981.

21. Leigh, W. J. and Srinivasan, R., Organic photochemistry with 6.7-eV photons: γ,δ-unsaturated ketones, *J. Am. Chem. Soc.*, 104, 4424, 1982.

22. (a) Kropp, P. J., Mason, J. D., and Smith, G. F. H., Photochemistry of alkenes. 9. Medium-sized cycloalkenes, *Can. J. Chem.*, 63, 1845, 1985. (b) Haufe, G., Tubergen, M. W., and Kropp, P. J., Photochemistry of alkenes. 10. Photocyclization of cyclononene and cycloundecene, *J. Org. Chem.*, 56, 4292, 1991.

23. For a recent review see Leigh, W. J., Orbital symmetry and the photochemistry of cyclobutene, *Can. J. Chem.*, 71, 147, 1993.

24. Adam, W., Oppenländer, T., and Zang, G., 185-nm photochemistry of cyclobutene and bicyclo[1.1.0]butane, *J. Am. Chem. Soc.*, 107, 3921, 1985.

25. (a) Inoue, Y., Sakae, M., and Hakushi, T., Direct photolysis at 185 nm of simple cyclobutenes. Molecular elimination of acetylene, *Chem. Lett.*, 1495, 1983. (b) Clark, K. B. and Leigh, W. J., Cyclobutene photochemistry. Nonstereospecific photochemical ring opening of simple cyclobutenes, *J. Am. Chem. Soc.*, 109, 6086, 1987. (c) Dauben, W. G. and Haubrich, J. E., The 193-nm photochemistry of some fused-ring cyclobutenes: Absence of orbital symmetry control, *J. Org. Chem.*, 53, 600, 1988.

26. Leigh, W. J. and Zheng, K., Cyclobutene photochemistry. Partial orbital symmetry control in the photochemical ring opening of a constrained cyclobutene, *J. Am. Chem. Soc.*, 113, 4019, 1992 and 114, 796, 1992.

27. (a) Saltiel, J. and Lim, L.-S. N., Stereospecific photochemical fragmentation of cyclobutenes in solution, *J. Am. Chem. Soc.*, 91, 5404, 1969. (b) Leigh, W. J. and Zheng, K., Cyclobutene photochemistry. Reinvestigation of the photochemistry of *cis*- and *trans*-tricyclo[6.4.0.0^{2,7}]dodec-1-ene, *J. Am. Chem. Soc.*, 113, 2163, 1991.

28. Inoue, Y., Matsumoto, N., Hakushi, T., and Srinivasan, R., Photochemistry of 3-oxacycloalkenes, *J. Org. Chem.*, 46, 2267, 1981.

29. Studebaker, J., Srinivasan, R., Ors, J. A., and Baum, T., Organic photochemistry with 6.7-eV photons: Rigid homoallylic alcohols. An inverse Norrish type II rearrangement, *J. Am. Chem. Soc.*, 102, 6872, 1980.

30. Fravel, H. G., Jr. and Kropp, P. J., Photochemistry of alkenes. 4. Vicinally unsymmetrical olefins in hydroxylic media, *J. Org. Chem.*, 40, 2434, 1975.

3

[2+2]-Cyclobutane Synthesis (Liquid Phase)

Gerd Kaupp
University of Oldenburg

3.1 Introduction and Mechanistic Considerations

Photodimerizations of alkenes (virtually unlimited substitution) were exhaustively reviewed in 1975.[1] Major advances have been achieved since then in terms of synthetic applications, experimental engineering, and mechanistic investigation. In this chapter, the intermolecular cyclobutane syntheses from alkenes will be discussed. Solid-state photolyses of alkenes and enones are treated in Chapter 4. Other [2+2]-cycloadditions are dealt with in other sections.

Numerous mechanistic possibilities are currently under discussion and scrutiny. Adiabatic processes involving electronically excited cyclobutanes are energetically not feasible, unless two electronically excited alkenes combine in two-photon laser processes. The detection of kinetically active intermediates,[1] which behave like short-lived ground-state 1,4-biradicals,[2] permit the dismissal of concerted orbital-symmetry-allowed processes even in the situation where a singlet excited state is involved. The widespread concept of pericyclic minima does not take into account the fact that kinetically active intermediates are ground-state biradicals and their fluorescent states have to be reached by UV/VIS absorption.[2] Another common belief is that excimers or exciplexes involving the singlet or triplet state could be photointermediates on the path to cyclobutanes. This suggestion comes from the fact that, on occasion, excimer/exciplex fluorescence can be detected. It has been argued that such emissions do not necessarily prove the chemical action of such species.[1,3,4] Sometimes, biradicals are involved and are observed by loss of stereochemistry by internal rotations even in singlet state reactions.[1,4-6]

Currently, it has become popular to assume that single electron transfer (SET) processes are involved in sensitized photodimerizations. It should be pointed out that such intermediates shown as 3/4 or 8/9 in Scheme 3.1 can be more easily produced by electrolysis rather than photolysis. The disadvantage in such SET photolyses is the inefficiency due to the need for charge separation and the great complexity of the dimerization processes, even though there might be some chain

0-8493-8634-9/95/$0.00+$.50

character in the reaction once the radical ions (3 and 8) have diffused apart. This is not very efficient, and most arguments are usually equally supportive of biradical formation following energy transfer.[1]

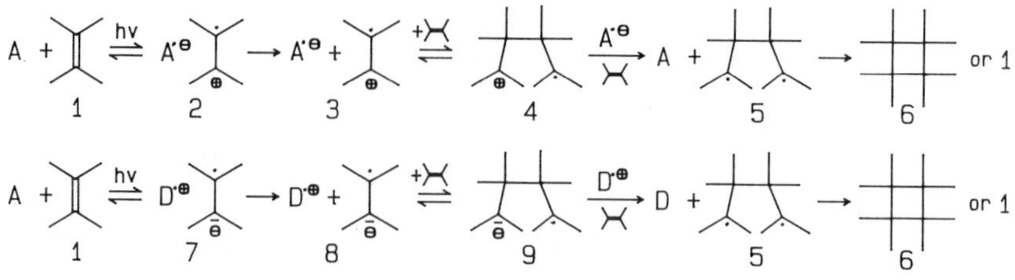

Interestingly, spectral evidence for the radical cation of stilbene has been obtained by pulsed laser-induced electron transfer and by secondary electron transfer.[7] Earlier experiments have shown that biphotonic excitation at 15 to 77K is required for the formation of radical anions of stilbenes and diaryl acetylenes.[8] In both cases, these processes do not lead to photodimerization but rather bring about cis/trans-isomerization[7] and photocycloreversion.[8]

Direct or sensitized singlet or triplet dimerizations are by far the most efficient. Both processes give rise to biradicals that will tend toward zwitterions in a situation where there is asymmetric polar substitution. Steric and polar effects permit the rationalization of both the orientations and the unproductive decay of the biradicals into alkenes.

3.2 Stilbenes

The interest in stilbene [2+2]-cycloadditions has been unabated since 1975. *trans*-Stilbene[1] and its derivatives form photodimers. Thus, **10** yields four stereoisomers, **11** through **14**. The effect of different solvents has been examined, and a hydrophobic association effect in water permits stereoselective photodimerization at very low concentration.[9]

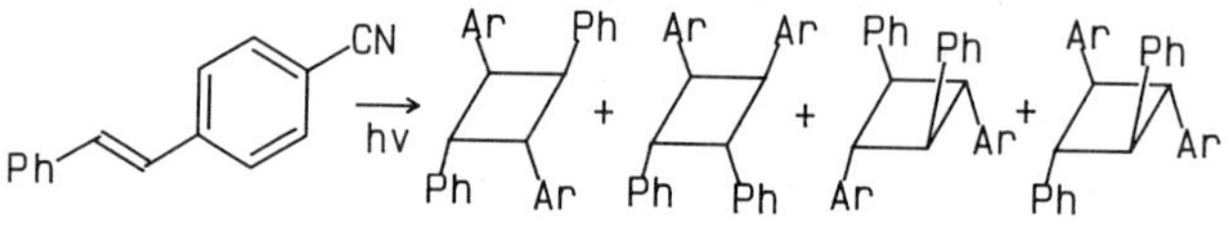

	10	11	12	13	14
(H₂O)		26	21	12	11 %
(MeOH)		22	12	4	3 %
(C₆H₆)		9	4	2	2 %

Styrylnaphthalenes also yield [2+2]-dimers **16** and **17** in yields of 7%.[10] The bislactone **18** yields the interesting tetraspirocyclobutane **19** (39%),[11] and the stereoisomeric propellanes **21** and **22** (1:2 ratio) are formed on irradiation of **20**.[12] The polystyrylarenes such as **23** give rise to interesting cyclophanes, such as **24**.[13] The effect of solvent polarity on the dimerization of **25** is shown in Table 3.1.[14]

Table 3.1   [2+2]-Photodimers of 25 in Various Media at 25 °C (%)

Solvent	*syn*-ht	*syn*-hh	*anti*-ht	*anti*-hh	*syn/anti*	ht/hh[a]
Benzene	33.0	31.8	17.0	18.2	1.84	1.00
Acetonitrile	38.3	36.5	19.2	9.0	2.55	1.20
0.2 *M* CTAB[b]	23.8	21.7	4.8	49.7	0.83	0.83

[a] hh = head to head, ht = head to tail orientation.
[b] Cetyltrimethylammonium bromide in water.

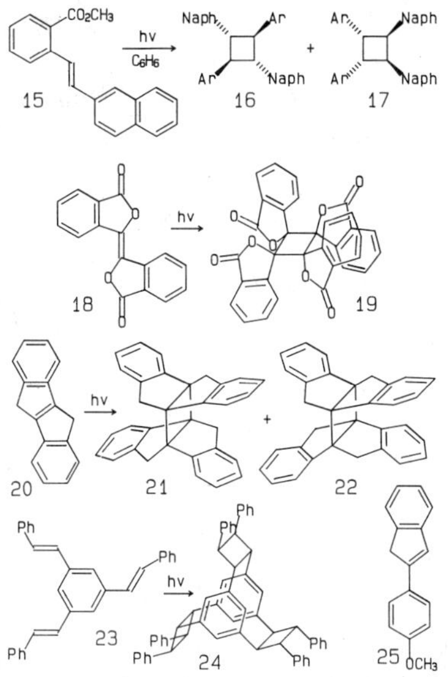

It has been suggested that reversible excimers are involved in the formation of these dimers. A similar explanation has been put forward to account for the observed temperature effects in the dimerization of 2-phenylindene.[14] However, it should be pointed out that it is more likely that biradicals are involved.[1-6]

Numerous *cis*-fused stilbenes photodimerize to give *syn*- and *anti*-cyclobutanes.[1] The mechanism of photodimerization of acenaphthylenes, which are formally related to *cis*-stilbenes, has been studied under various conditions. It is believed that a singlet mechanism gives the *syn*-dimers **27**, whereas a triplet mechanism gives the *anti*-dimers **28**.

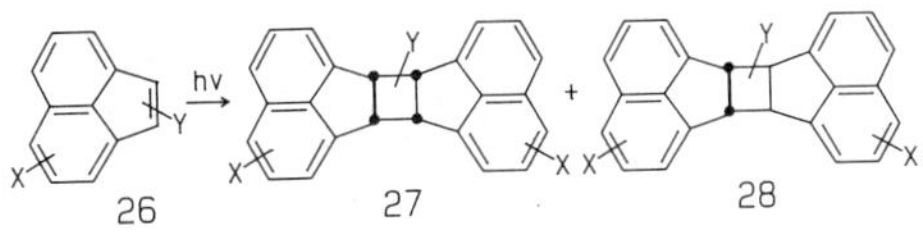

Magnetic fields (up to 13 kGauss) or micelles and solvent polarities brought about moderate changes in the ratio of dimers obtained. Dimerization following inclusion in γ-cyclodextrin (solution) or reaction in liquid crystalline (cholesteric) phase (smectic and nematic phases had no effect) lead to more than 99% of the *syn*-dimer **27** (X = Y = H).[15]

The results of cycloaddition to stilbene **29** are varied. Earlier claims of the formation of highly strained *trans*-fused and (or) *cis*-diphenylcyclobutanes in the reaction of **29*** with 3,4-dihydro-2H-pyran **30**[16] were dubious.[4] None of the presumed structures was easily accounted for by the singlet biradical mechanism, which in this instance requires *cis*-fusion and with the phenyl groups *trans*. The structures of the products were carefully reestablished[4] and as predicted the products were identified as **31** and **32** in 45 and 24% yield, respectively, supporting the biradical mechanism.[4]

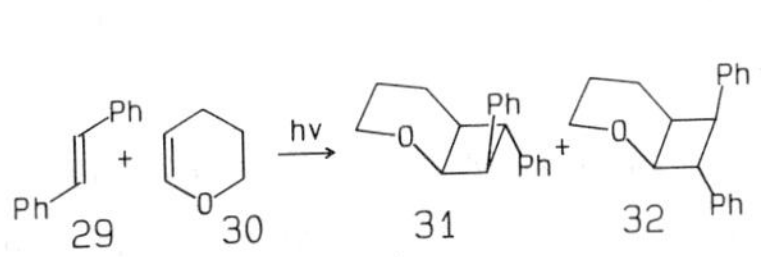

Diphenylacetylene also undergoes cycloaddition with electron-rich alkenes to form adducts (e.g., **33**).[17] The direct syntheses of highly hindered propellanes such as **34**,[18] **35**,[18] **37**,[11] **38**,[11] **40**, and **41**[19] can also be obtained by addition of stilbenes (to lane) to a variety of substrates. Also, *bis*-spirocyclobutanes **39** have been obtained. Again, a biradical mechanism was proposed. The preference for the formation of **40** rather than **41** has been rationalized by the semiempirical AM1 calculations on the caffeines **42** and of the intermediate biradicals in their reactive conformations.[19] Another propellane **44** can be formed by the addition of **20** with **43**,[20] and indolizine **45** provides, regio- and stereospecifically, the two cycloadducts **46** and **47**. This is again well rationalized by AM1 calculations.[21]

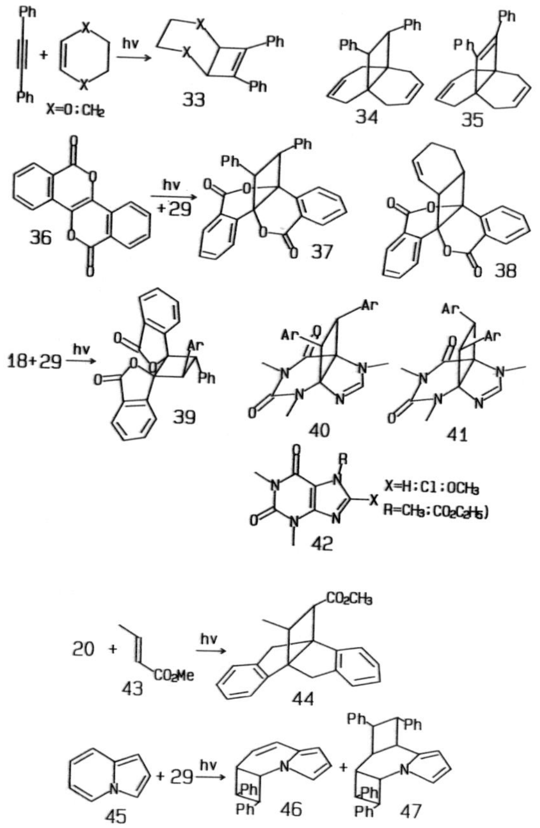

Stilbenes undergo photocycloaddition to open-chain 1,3-dienes[22] and to electron-poor olefins.[23] In all of these cases, an exciplex mechanism has been assumed.

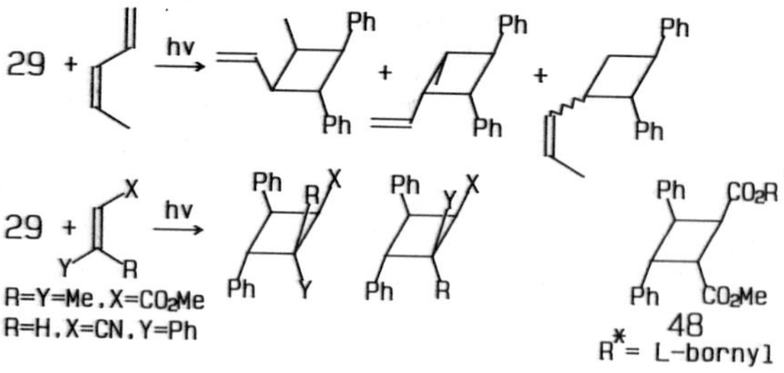

The reaction of stilbene **29** with fumaronitrile has been studied by picosecond laser flash photolysis,[24] and stereoinduction has been achieved using the L-bornyl ester of fumaric acid as the addend in the formation of **48**.[25]

The number of cyclobutanes from cyclic *cis*-stilbenes[1] is considerable and has been augmented in recent times. Thus, **49**,[26] **50**,[27] **51**,[28] **52**,[29] **53**,[30] and **54**[31] have been synthesized by irradiation of the corresponding *cis*-stilbene in the presence of the respective trapping olefin. Results of some importance have been obtained from the irradiation of various heterostilbenes[1] where mixtures of stereoisomers are generally obtained. This problem is reduced in solid-state photodimerizations (see Chapter 4); however, not all heterostilbenes are reactive in the solid state. The photodimerization of 55 has been studied by nonselective irradiation through quartz in benzene. The yields of the dimers obtained was strongly dependent on concentration and irradiation time (56, 57, 58, 59, 60: 4 h 25%, 1.5%, 25%, 0.8%, 0.0%; 18 h 50%, 6.5%, 7%, trace, 6%). Apparently, secondary photolysis of the dimers is important (see Chapter ?).[32] The stilbene analogue 61 has been transformed into 63 using TCNE,[33] while the dipyridyl ethene 64 undergoes dimerization to yield 65.[34] Dimerization of the stilbazolium cation 66 can be brought about by photolysis in water, yielding 67 in 60%. Dimerization in reversed micelles is also effective.[35] In this instance, 68 is formed in a yield of 72% in addition to 67 and 69.[35] The surfactant 70 forms monolayers on water and, upon photolysis, yields the mixture of dimers 71 and 72.[36]

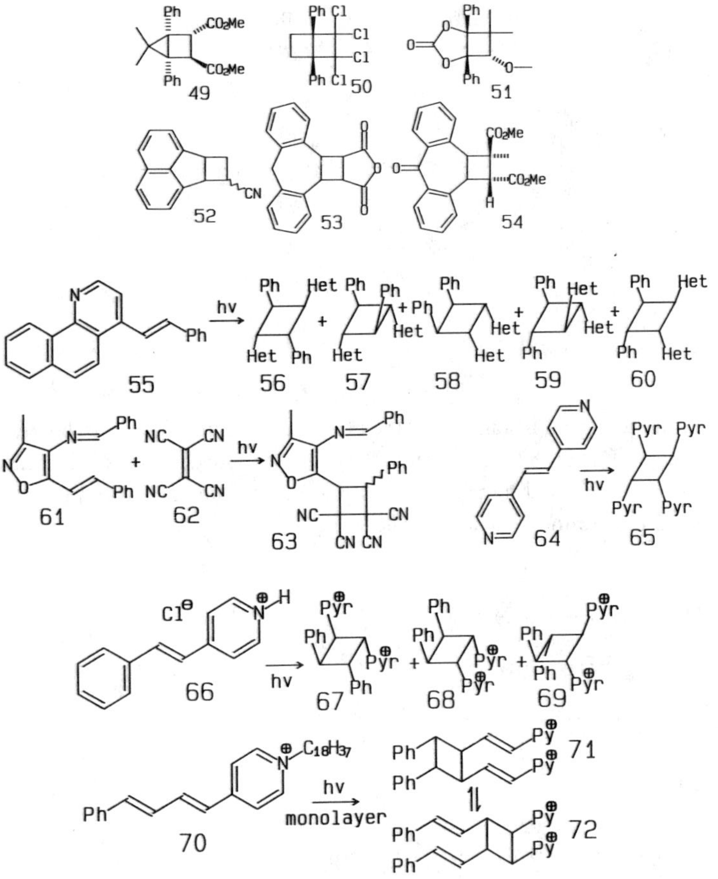

3.3 Styrenes

Single electron transfer (SET) mechanisms have been tried to bring about styrene photodimerizations.[1] Cyanoarenes or metal complexes were used as the electron-accepting sensitizers. However, simple examples such as 1,1-diphenylethylene or 3-methoxystyrene do not form cyclobutane dimers under SET conditions and, instead, Diels-Alder reactions take place.[37] Indenes do yield cyclobutanes under SET conditions and indene 73, using metal complexes[38] or N-methylacridinium salts[39] as sensitizer, brings about the formation of 74. Indene 75 gives 76 and 77 on cycloaddition to enol ethers with 1-cyanonaphthalene as the electron acceptor.[40] Some styrenes yield cyclobutanes particularly easily on direct irradiation and in the absence of sensitizers. Thus, numerous macrocycles and the macrocyclic crown ethers have been synthesized, such as the conversions of 78 into 79 (n = 2; 3; >75%).[41] Several vinylic heterocycles have been photodimerized via benzophenone sensitization, and 4-vinylpyridine 80 yields only 81.[42] A head-to-head cycloaddition is also reported with 82 and styrene 83 which, on direct irradiation, gives 84 as the *syn*-adduct. Transient spectra have been recorded for this process.[43]

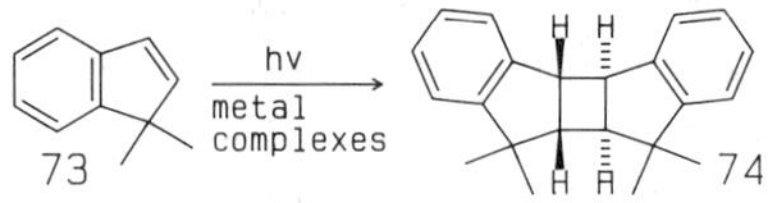

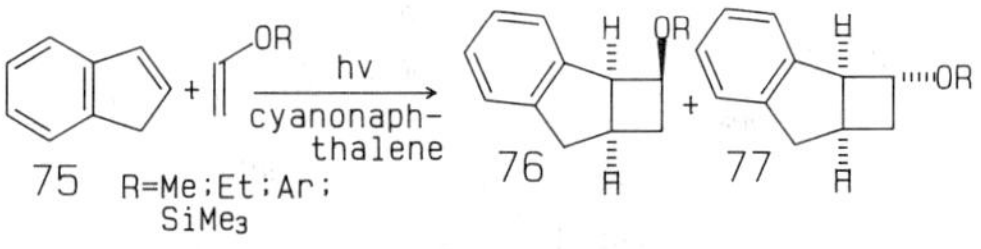

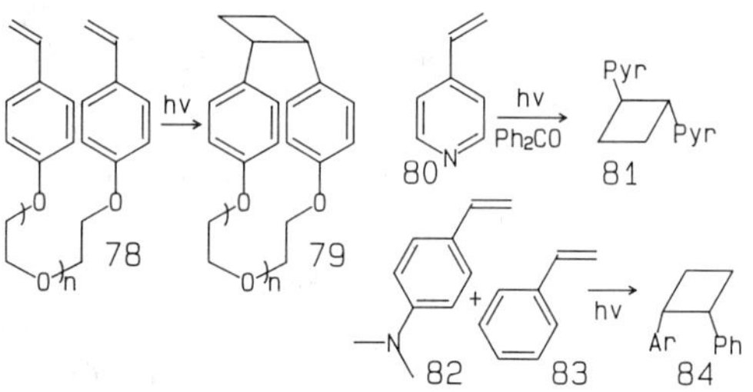

Three dimers are found on irradiation of the photoequilibrium of the anisyl derivative 85. These were identified as 86 (14%), 87 (13%), and 88 (73%). The mechanism of the reaction has been investigated and two long-lived transients, the *trans*-(8.5 ns) and the *cis*-singlets (6.1 ns), undergo cycloaddition with the ground-state isomers 85.[44]

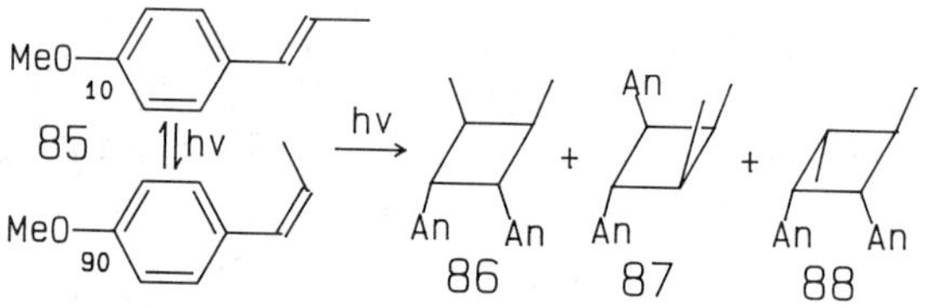

Numerous cyclic styrene-type molecules have been shown to photodimerize or undergo cycloaddition.[1] Thus, 89 gives 90 (55%) and 91 (40%).[45] Indenes and benzothiophene-1,1-dioxides have also been used for singlet or triplet [2+2]-photodimerization and cycloadditions to various alkenes.[1,46] All possible regio- and stereoisomers, such as 92, are formed from indene 75 and the chlorofluoroethylenes upon benzophenone sensitization.[47] Also, highly substituted cyclobutanes such as bislactone 94 (35%)[48] and 96 (80%, 2 isomers 1:2)[49] have been obtained by direct and Michler's ketone-sensitized irradiation.

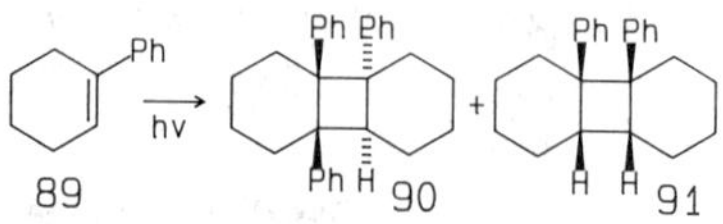

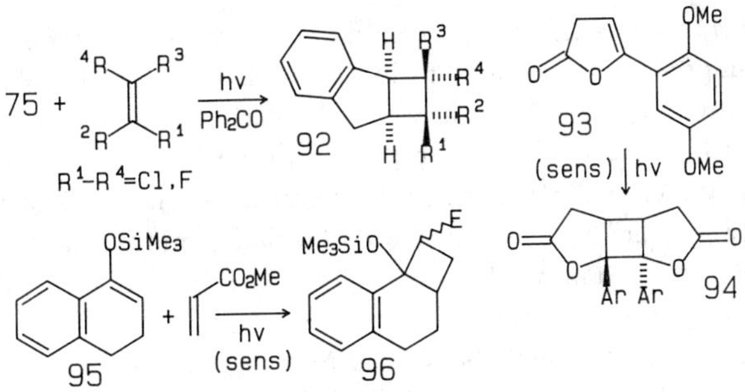

3.4 Cinnamates

In the field of cinnamic esters, the last of the six truxinic esters (i. e., ω-methyltruxinate) has been prepared and compared to the known methyl truxinates.[17] Tethered compounds such as **97**, also undergo [2+2]-cycloaddition under xanthene sensitization and with added LiClO$_4$ to form the macrocycles **98** (48%) and **99** (11%).[50]

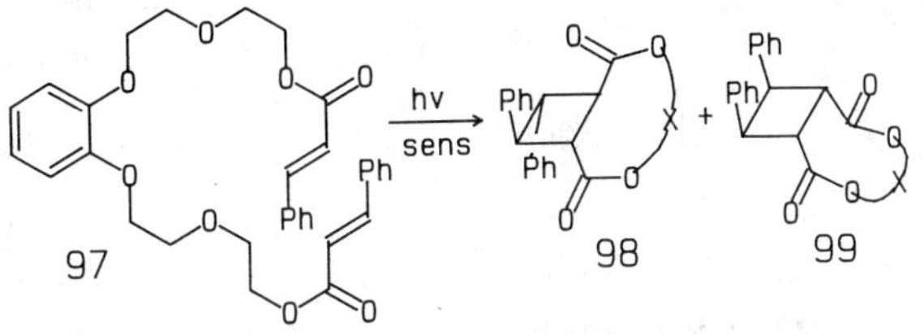

Considerable interest has been shown in the techniques of molecular self-organization in cinnamate derivatives. Thus, the salt **100** forms reversed micelles in CCl$_4$ and these give **101** (40%) and **102** (9%) upon photolysis.[51] Reversed micelles are also formed by lauric acid in cyclohexane and these help to accelerate the photodimerization of **103** and yield the dimer **104** stereoselectively. Other products are formed in addition to **104** when irradiation is carried out in isotropic solution.[52]

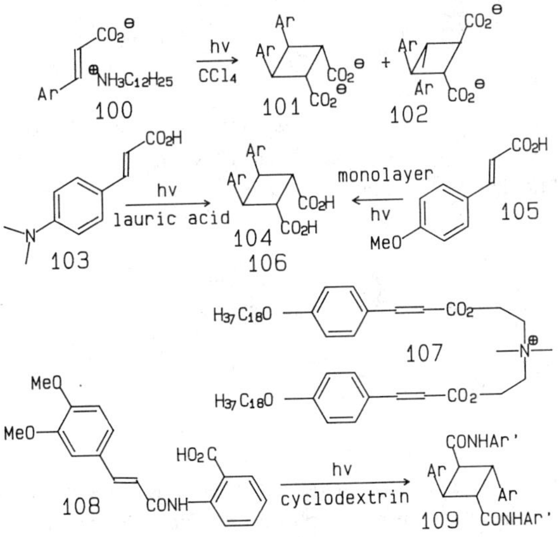

Two-dimensional crystalline clusters in monolayers have been detected and analyzed with synchroton X-rays set to grazing incidence. From these results, it is possible to explain the correspondence between monolayer and crystalline phase irradiations. Even *p*-methoxy cinnamic acid

105 forms monolayers at the water/air interface and 4,4′di-methoxy-β-truxinic acid 106 is the result of its photolysis. This product is also found on irradiation in the crystalline phase.[53] Monolayers of cinnamic acid derivatives, such as 107, at the water/air interface or their bilayer vesicles in aqueous solution give rise to the same products as in the irradiation in the crystalline phase. Intra- and intermolecular [2+2]-dimers of the truxinic type are formed.[54] A completely different stereoselection is obtained, if tranilast 108, at high concentration, is included into γ-cyclodextrin. This inclusion compound accommodates two molecules of 108 in its cage. Only the α-truxillic derivative 109 is formed under these conditions.[55]

The juxtaposition of remote double bonds can be achieved using hydrogen bonding via the anchor function of the 2-pyridone unit in 110. This process aligns the double bonds of the cinnamate esters in 110 and photodimerization in dilute benzene solution gives 111 and 112 in a combined yield of 35%. The free cinnamate groups in these dimers also undergo photochemical E/Z equilibration.[56] Cinnamic esters have been photodimerized neat in the absence of solvent, in micelles, in clathrates, catalyzed by BF$_3$ and by sensitization.[56]

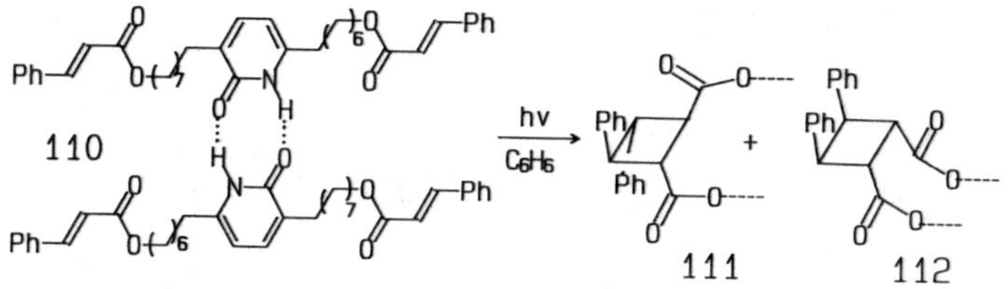

Heteroaryl cinnamates 113 undergo photodimerization in solution upon benzophenone sensitization, affording a 2:1-mixture of 114 and 115 (90%) with the furan derivative and a 1:1 mixture (25%) with the thiophene derivative.[57]

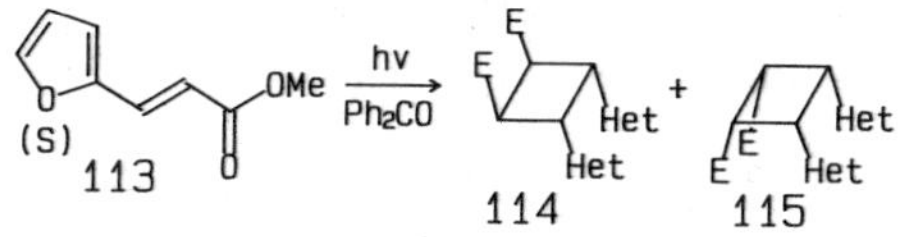

Cyclic cinnamates such as 116 are also photoreactive and yield photodimers on sensitized or direct photolyses. The exclusive [2+2]-photoadduct 118 is formed on thioxanthone sensitization of a mixture of 116 and 117. The adduct obtained is present as a 1:1 mixture with 119 formed by ring flipping upon heating.[58]

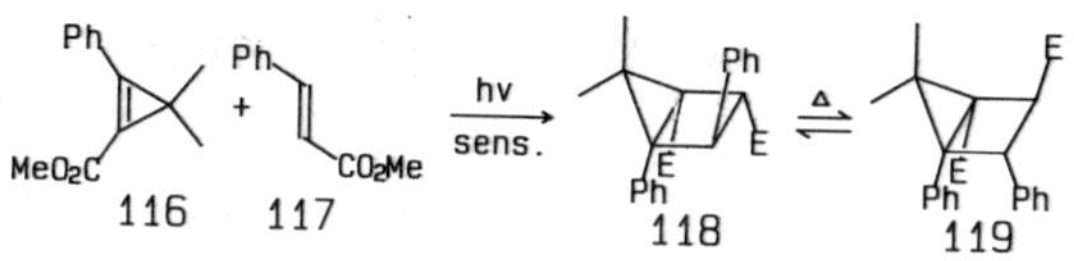

Quinolones-(2), coumarins (including furocoumarins), and thiocoumarins are internally ring-closed cinnamates, which have been used for photodimerizations and photocycloadditions. Some recent examples are the stereospecific synthesis of racemic 120 (64%),[59] 121 (70%),[59] 122,[60] 123,[61] and 124 (80%).[62] All of these are formed by direct irradiation of the mixture of enone and alkene. The influence of Lewis acids on the photodimerization and cross-cycloaddition of coumarin[1] has been studied.[63]

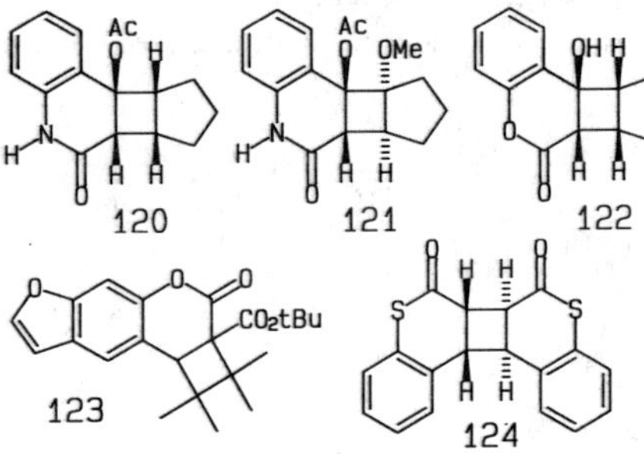

3.5 Acrylates

All kinds of α,β-unsaturated carboxylic acid derivatives have found widespread application in photochemical cyclobutane formation. Low-pressure mercury lamps (253.7 nm) or sensitization with aceto- and benzophenone are usually applied.[1] Photodimerization of methyl acrylate or acrylonitrile giving dimers in yields of 31 and 52%, respectively, have been induced by *bis*-triphenylphosphine-nickela-cyclopentane and other nickel complexes. The head-to-head *anti*-products **125** are formed specifically.[64] Cyclic acrylates[1] such as **126** have been used more frequently for these reactions. The dimers obtained have the *cis-anti-cis* configuration shown by **127** and the yields are moderate at 42%.[65] The formation of **129** from **128** is a further example of this type of reaction.[66] The acrylate may be part of a heterocyclic ring as in lactones and lactams[1] and, again, cycloaddition occurs; some examples include the synthesis of **131** (90%),[67] chiral **134** (major product, 25%) which is used in a total synthesis of (–)-β-bourbonene,[68] and **136** (7%) + **137** (16%).[69]

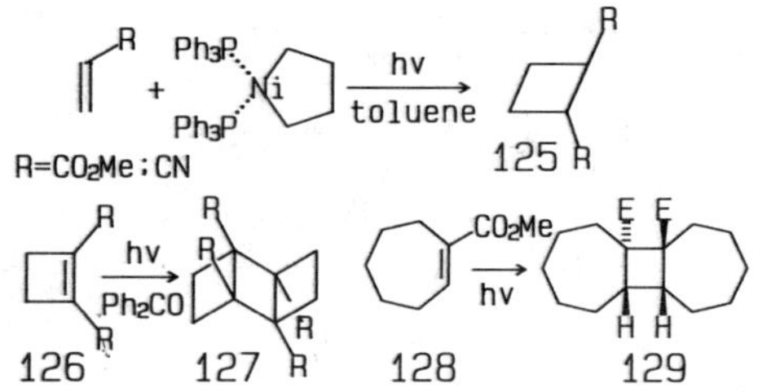

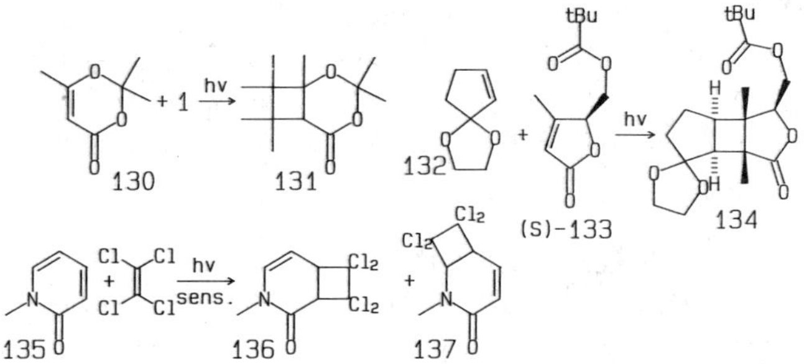

Cycloadditions to maleic anhydrides are extremely well studied, perhaps because their [2+2]-cycloadducts may be used as monomers for condensation polymers with diamines. The yield of the *anti*-photodimer of maleic anhydride[1] has been improved, with the reaction occurring quantitatively in CCl_4.[70] Other cycloadducts are also important such as **139**,[71] **140**,[72] **141**,[72] and numerous related derivatives that can be synthesized from **138** and the appropriate alkene via benzophenone sensitization (see also 210). 1,4,5,8-Tetrahydronaphthalene-2,3-anhydride has been used in sensitized triple [2+2]-photodimerization for asterane synthesis (see Chapter ?).[73]

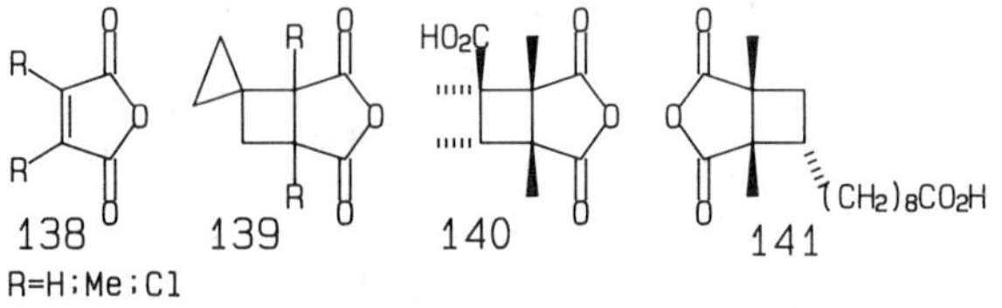

138   139   140   141

R=H;Me;Cl

3.6 Pyrimidines

Pyrimidine bases of nucleic acids could be considered as heterocyclic acrylic acid derivatives. Their importance in photodamage of biological material has long been recognized, and the culprit is the presence of cyclobutylpyrimidine dimers in double-stranded DNA.[1] Such dimerizations have been studied by two-photon absorption techniques using pulsed 532-nm laser radiation; under these conditions, strand breaks occur.[74] Frequently, thymidyl residues are the site of cyclobutane formation. Thus, the photochemically active antitumor antibiotic toromycin 142 forms intercalates with double-stranded calf thymus DNA in DMSO in the dark. Strand nicking and covalent modification occur only upon irradiation. After hydrolysis (0.1 M HCl, 100 °C, 2 h) and chromatography, 143 is isolated. During hydrolysis, the α-fucofuranose moiety is transformed into the β-fucopyranose moiety. Thus, the thymidyl residues can be extracted from the DNA and provide evidence that the thymidyl residues are the sites of photocycloaddition.[75]

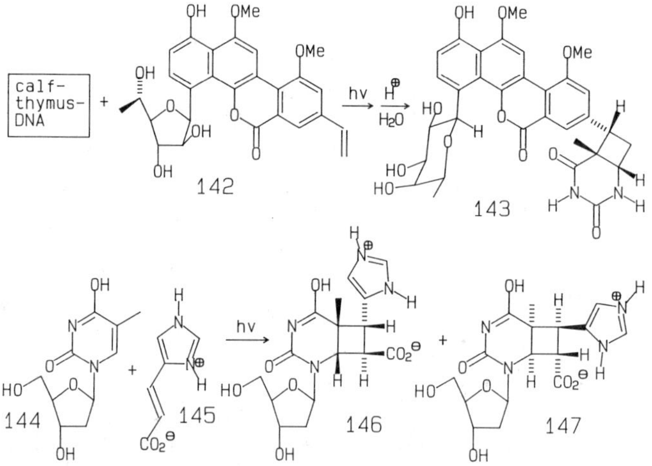

142   143

144   145   146   147

The photochemical reaction of thymidine 144 with urocanic acid 145 is important with regard to photocarcinogenesis. This reaction gives rise to the diastereoisomers 146 and 147. Other pyrimidine derivatives also undergo cycloaddition reactions. Thus 5-fluorouracil 148 adds to vinyl ethers to yield the cycloadducts 149 and 150 (77%, 1.8:1).[76] Quite interestingly, 151 does not behave like a heterostilbene, but cycloadds at the ring bond to give 152 upon selective irradiation.[77] The hydrolysis product of the pesticide pirimicarb 153, which resembles the "pyrimidine bases", has been photolyzed in water to give the dimers 154 and 155, probably from aggregates.[78] Two more dimers are obtained on irradiation in acetonitrile. Several trimethylene-bridged dinucleotides form cyclobutanes intramolecularly upon direct irradiation.[79]

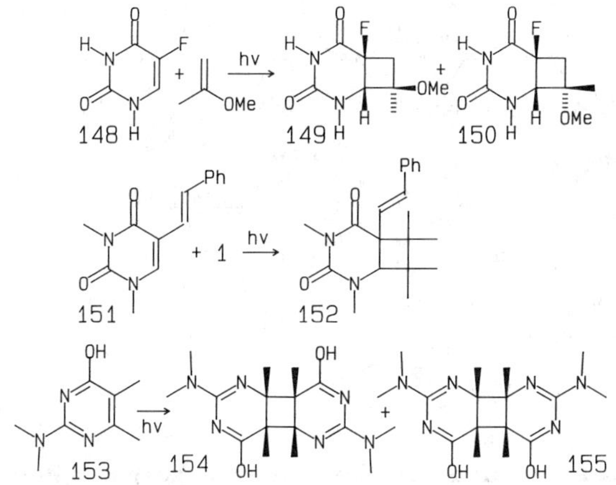

3.7 (Thio)Enolic Ethers(Esters), Enamides, but not Indoles

Single electron transfer (SET) mechanisms via electron transfer sensitization with acceptors are operative with enol ethers (also indoles, see next chapter, and 1,3-dienes, see also Reference 8, introduction). Thus, the dimerization of **156** may be sensitized with 1,4-naphthalenedinitrile[80] or with irradiated semiconductors (ZnO, CdS, TiO$_2$, Ta$_2$O$_5$, SnO$_2$, or Bi$_2$O$_3$) in various solvents.[81]

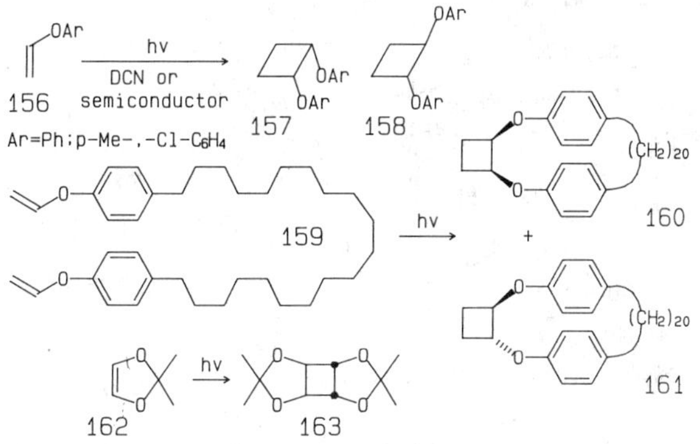

The cis:trans ratio of products from **156** (Ar = Ph) is 3:2 in solution and 1:2 in the semiconductor slurries (up to 10% on ZnO). Polyethyleneterephthalate has been applied as an insoluble electron transfer photosensitizer in this type of reaction.[82] Macrocyclic rings can be formed using dicyanoanthracene sensitization of tethered **159** when the two adducts **160** and **161** are formed in a ratio of 42:58 and a total yield of 66%.[83] Electron-rich alkenes can also form cyclobutanes without involvement of radical cations on direct or sensitized irradiation. 1,2-Bis-vinylic ethers and 1,3-dioxoles or furans have been used as addends for electronically excited acrylates.[1] However, **162** has been photodimerized to give the *cis-anti-cis* dimer **163**.[84] Dichlorovinylenecarbonate and vinylene carbonate undergo photodimerization and cross-photocycloaddition reactions to numerous alkenes.[1] Typically, products such as **164**,[85] **165**, and **166**[86] can be obtained. These are starting materials for hydrolytic syntheses of the mycotoxin moniliformin (semiquadratic acid). Both isomers of **167** (40% *anti-*, 32% *syn-*) have also been synthesized.[86] An interesting total synthesis of (±)biotin uses the photocycloaddition of the 1,2-*bis*-enamide **168** to the enol ether **169** to give **170** as the key step.[87] Ethylene can also be added photochemically at low temperatures; by this route, the adducts **171** and **172**[88] have been obtained.

164 165 166 167

168 169 170

171 172

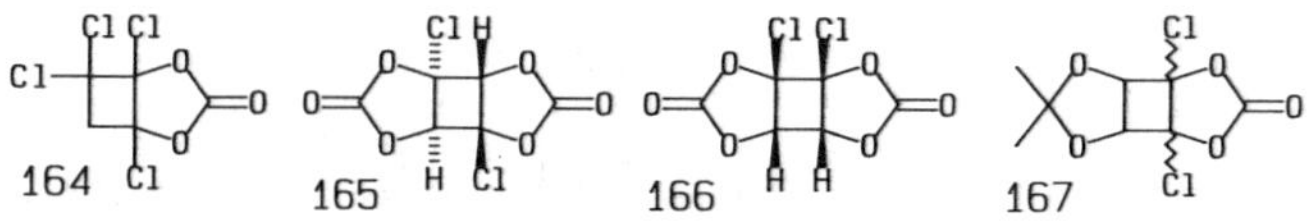

Benzofurans and benzothiophene(dioxide)s undergo photodimerization and cross-photocycloaddition.[1] Some typical examples are shown by the formation of **173**, **174**,[89] **175**,[90] **176**,[91] **177**, and **178**,[92] which have been obtained from the appropriate components. The formation of **179** from an isoquinolinone[93] is also of interest. The cyclic sulfone **180** gives **181** (18%), **182** (13%), and **183** (1%) upon photolysis in acetone.[94]

173 174 175 176 177 178 179

180 181 182 183

184 185 186 187

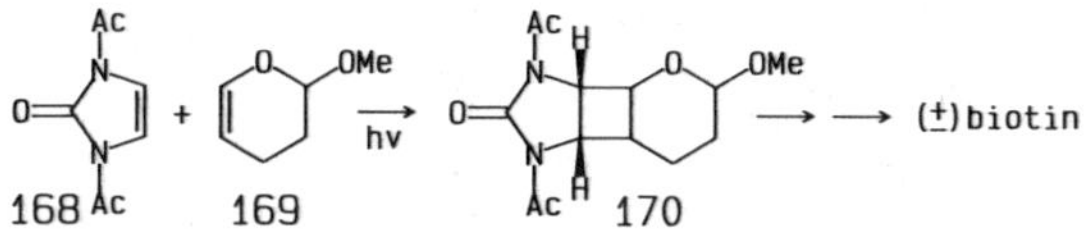

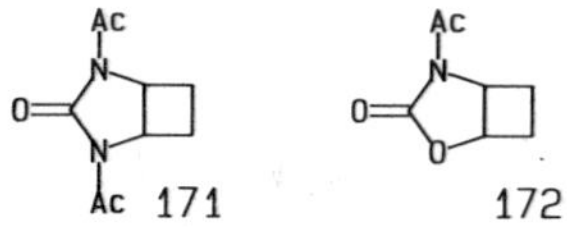

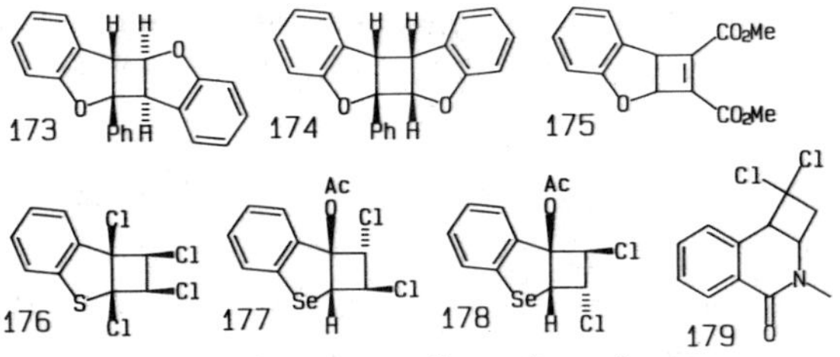

The photodimerizations of the triazoles **184** and **186** are remarkably selective. The formation of **185** (54%) from **184** involves only the enamide double bond of **184**, while **186** reacts both with its enamide and with its heteroacrylate double bond in the six-membered ring in the formation of **187** (70%).[95]

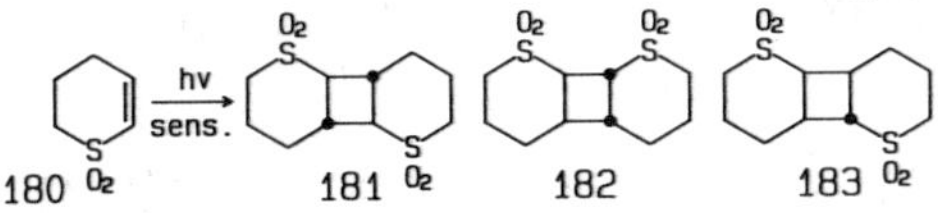

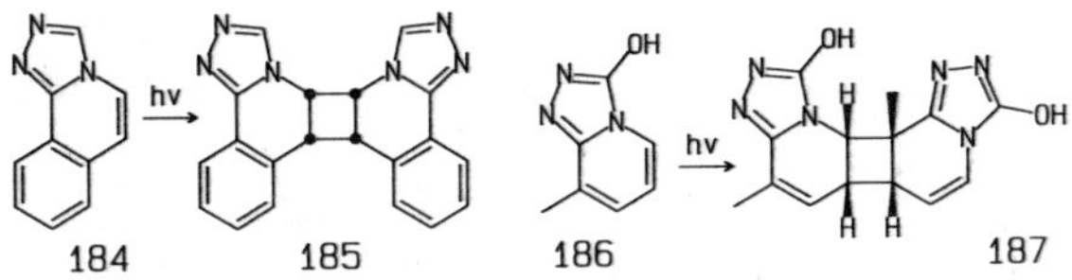

3.8 Simple Alkenes, Polyenes, and Cumulenes

Cyclohexene **188** has been photodimerized in *p*-xylene to give the dimers **189** (31%), **190** (51%), and **191** (18%). It has been argued that *trans*-cyclohexene might be an intermediate.[96] Cyclohexadiene **192** undergoes photoaddition to ethylene (10 to 50 bar) and acetylene (15 bar) using 2-acetonaphthone sensitization to give **193** and **194**, respectively.[97] The photodimerization of **192** has been reviewed[1] from a mechanistic standpoint. A more detailed study of the sensitization of addition reactions by arene nitriles has suggested that exciplexes and triplets might be intermediates in the formation of the *syn* and *anti* [2+2]-dimers **193** and **194**; whereas radical ion pairs, but not the free radical ions, are proposed as intermediates for the [4+2]-adducts that are also formed.[98] Only cyclovinylogous photoadditions have been described for cycloheptatriene trapping.[99] However, 195 gives only the [2+2]-dimer 196 (50%) upon Michler's ketone-sensitized irradiation, and 197 is transformed into 198.[100]

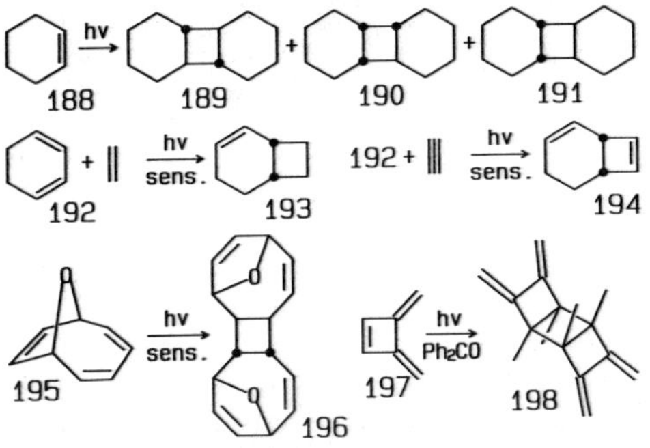

Enynes such as 199 react at the double bond in sensitized (benzaldehyde) dimerizations to give 200 and 201 (61%, 54:34).[101] 1,1-Dimethylallene 202 has been reported to yield the dimers 203 (31%) and 204 (16%).[102] Only the terminal double bonds are involved in the dimerization of butatriene 205 to give 206 on irradiation in stretched polyethylene or in nematic phase. Interestingly, 205 also gives 206 in solid-state photolysis, but not in isotropic solution.[103] 1-Phenyl-1,3,5-heptatriyne photoadds alkenes at each of its triple bonds.[103a]

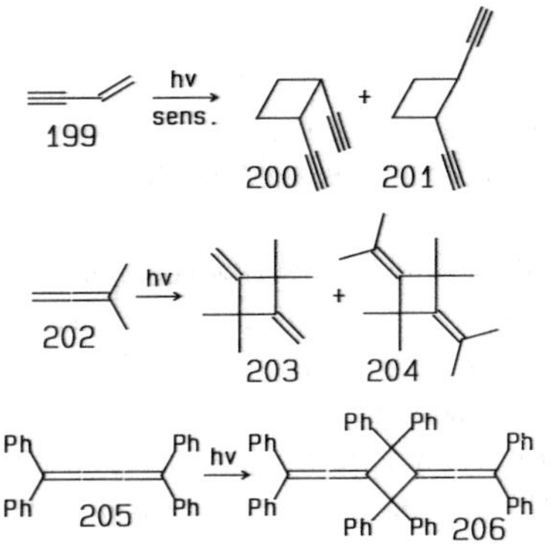

3.9 Industrial Applicability

Photochemical cyclobutane formations are versatile enough to allow for the introduction of substituents that could be useful intermediates for drug syntheses. Thus, compound 122 is a starting point for a furocoumarin by conversion of 207 into 208.[60] Furocoumarins are important drugs for phototherapy of psoriasis. Cyclobutane formations could also introduce the problem of phototoxicity in drugs such as 4,4'-diamidinostilbene,[1] pesticides, and certain plant materials, since many of these compounds can undergo [2+2]-cycloadditions in sunlight.

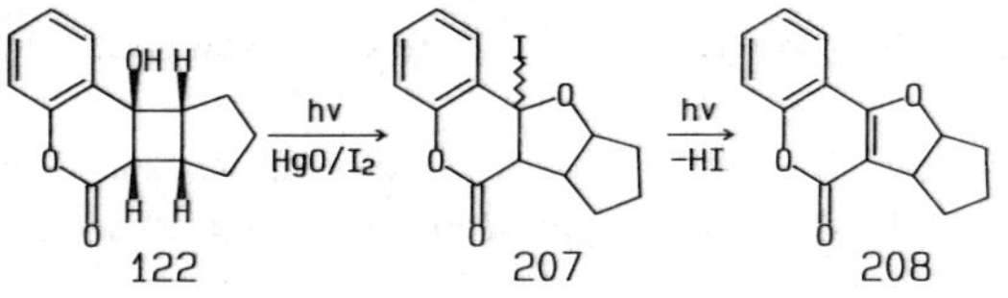

Furthermore, [2+2]-photocycloadducts of dimethylmaleic anhydride **138** have found synthetic applications. These have been used for the preparation of fungicides, herbicides, adhesives, coatings, photoresists, and photoimaging formulations.[104] Compound **210**, synthesized from **138** and **209**, can be used as a monomer for polyimide engineering resins.[105]

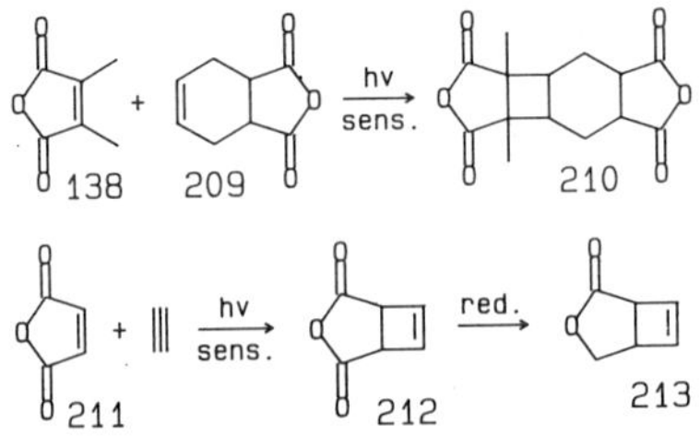

It has been pointed out that cyclobutene lactones **213**, which may be obtained via LiAlH$_4$ reduction from the cycloadduct **212** of maleic anhydride **211** and acetylene, are important intermediates for organic syntheses.[106] Much industrial work has been done in polymer cycloadditions. This has been reviewed exhaustively by Dilling,[107] and many patents have been lodged dealing with such systems. Generally speaking, bifunctional alkenes are photolyzed in a similar way as has been described for the monofunctional ones.

Quite different aspects are offered by large ring crown ethers like **79, 98**, and **99**. These may find use in large-scale processes for selective metal ion binding in ecotechnics or in marine mining, provided these molecules can be tightly immobilized in some way.

References

1. Kaupp, G., Dimerizations of alkenes; Cycloadditions between unequal alkenes; *Houben-Weyl Methoden der Organischen Chemie Photochemie*, Vol. IV/5a, b, Thieme, Stuttgart, 1975, 278 and 360.

2. First spectroscopic detection (UV/VIS and fluorescence) of photochemically generated 1,4-biradicals (ground state): Kaupp, G., Teufel, E., Hopf, H., First spectroscopic detection of diradicals in photocycloreversions, *Angew. Chem. Int. Ed.* 18, 216, 1979; this was not acknowledged in the recent review on the subsequent work of several groups on that subject, though with by far less well-resolved UV/VIS-spectra: L. J. Johnston, Spectroscopy of biradicals, Handbook of Photochemistry, Vol. 2 (Ed. J. C. Scaiano), CRC Press, Boca Raton, FL, USA, 1989, Chpt. 2.

3. Kaupp, G. and Teufel, E., Diene-catalyzed photodimerization via excimers (exciplexes) or diene addition to anthracene and 9-phenylanthracene?, *Chem. Ber.*, 113, 3669, 1980; Kaupp, G., Photo-Diels-Alder reactions with anthracene, *Liebigs Ann. Chem.*, 254, 1977; Saltiel, J., Dabestani, R., Schanze, K. S., Trojan, D., Townsend, D. E., Goedken, V. L., Photocycloaddition of anthracene to *trans, trans*-2,4-hexadiene, *J. Am. Chem. Soc.*, 108, 2674, 1986.

4. Kaupp, G., Stark, M., and Fritz, H., Configuration and *cis*-effect in diphenyl-(oxa)bicyclo[4.2.0]octanes, *Chem. Ber.*, 111, 3624, 1978.

5. Kaupp, G. and Stark, M., The *cis*-effect in photochemical cleavages of cyclobutanes, *Chem. Ber.*, 110, 3084, 1977.

6. Kaupp, G., Quantum yields in elucidation of photomechanisms, *Liebigs Ann. Chem.*, 844, 1973.

7. Lewis, F. D., Bedell, A. M., Dykstra, R. E., Elbert, J. E., Gould, I. R., and Farid, S., Photochemical generation, isomerization, and oxygenation of stilbene cation radicals, *J. Am. Chem. Soc.*, 112, 8055, 1990.

8. Kaupp, G. and Jostkleigrewe, E., Laser photolysis, a new path for charge separation, *Angew. Chem. Suppl.*, 1089, 1982; 1,4,5,8-Tetrahydronaphthalene donates the electron and photochemically deprotonates afterwards. *Angew. Chem. Int. Ed.*, 21, 436, 1982.

9. Ito, Y., Kajito, T., Kunimoto, K., and Matsuura, T., Accelerated photodimerization of stilbenes in methanol and water, *J. Org. Chem.*, 54, 587, 1989; further examples and references.

10. Jones, A. J. and Teitei, T., The photochemistry of alkyl 2-(naphthalenyl)ethenylbenzoates: Photodehydrocyclization and photodimerization products and their structural elucidation by ^{1}H and ^{13}C NMR, *Aust. J. Chem.*, 37, 561, 1984.

11. Kaupp, G. and Schmitt, D., Dispirocyclobutanes and propellanes from 1,2-bisenollactones, *Chem. Ber.*, 114, 1983, 1981.

12. Shim, S. C. and Chae, J. S., Photodimerization of indeno[2,1-α]indene, *Bull. Chem. Soc. Jpn.*, 55, 1310, 1982.

13. Winter, W., Langjahr, U., Meier, H., Merkuschew, J., and Juriew, J., Photochemistry of 1,3,5-tristyrylbenzene, *Chem. Ber.*, 117, 2452, 1984; further examples; see also Meier, H., Praβ, E., Zertani, R., and Eckes, H.-L., Naphthalenophanes by twofold photocyclodimerization reactions of distyrylnaphthalenes, *Chem. Ber.*, 122, 2139, 1989.

14. Wolff, T., Schmidt, F., and Volz, P., Regioselectivity and stereoselectivity in the photodimerization of rigid and semirigid stilbenes, *J. Org. Chem.*, 57, 4255, 1992.

15. Ichimura, K., Watanabe, S., and Sen'i Kobunshi, Effect of external magnetic fields on chemical reactions. I. Photodimerization of acenaphthylene, *Zairyo Kenkyusho Kenkyu Hokoku*, 108, 29, 1975; *Chem. Abstr.*, 84, 179277, 1976; Nerbonne, J. M., Weiss, R. C., Liquid crystalline solvents as mechanistic probes. 3. The influence of ordered media on the efficiency of the photodimerization of acenaphthylene, *J. Am. Chem. Soc.*, 101, 402, 1979; Castellan, A., Dumartin, G., and Bouas-Laurent, H., Photocycloaddition of polynuclear aromatic hydrocarbons in solution. IV. Study of the fluorescence and photoreactivity of acenaphthylene and 1-cyanoacenaphthylene, *Tetrahedron*, 36, 97, 1980; Mayer, H. and Sauer, J., Photodimerization of acenaphthylene derivatives in solutions and micelles, *Tetrahedron Lett.*, 24, 4091, 1983; Tong, Z. and Zhen, Z., Microenvironmental effects of cyclodextrins. II. Effects of cyclodextrins on photodimerization of acenaphthylene, *Youji Huaxue*, 44, 1986; *Chem. Abstr.*, 105, 152330, 1986.

16. The claims varied considerably: Rosenberg, H. M. Rondeau, R., and Servé, P., The photolysis of stilbene in the presence of 2,3-dihydropyran, *J. Org. Chem.*, 34, 471, 1969; Servé, P., The elucidation of the coupling constants for the cyclobutane ring of the *trans*-fused 2-oxabicyclo [4.2.0]octane system, *Can. J. Chem.*, 50, 3744, 1972; Servé, P., Rondeau, R. E., and Rosenberg, H. M., The application of *tris*-(1,1,1,2,2,3,3-heptafluoro-7,7-dimethyl-4,6-octanedionato)europium(III) in elucidation of structures of bicyclic ethers, *J. Heterocycl. Chem.*, 9, 721, 1972.

17. Kaupp, G. and Stark, M., Selectivities in the photolysis of diphenylcyclobutenes, *Chem. Ber.*, 111, 3608, 1978; 6,7-Diphenyl-2,3-dihydro-1,4-dioxin, *Angew. Chem. Int. Ed.*, 16, 552, 1977; further reactions with 33: Ref.1; Yamada, S. Mino, N. Nakayama, N. and Ohashi, M., Photochemical reactions of phenylacetylenes with ethylene trithiocarbonate. Synthesis of phenyl-substituted 2-thioxo-1,3-dithioles, *J. Chem. Soc. Perkin Trans. I*, 2497, 1984; ω-methyltruxinate via 33 and maleic anhydride: Kaupp, G., and Stark, M., The *cis*-effect in photochemical cleavages of cyclobutanes, *Chem. Ber.*, 110, 3084, 1977; similar reactions with diphenylbutadiyne: Shim, S. C. and Lee, T. S., Photocycloaddition reaction of some conjugated hexatriynes with 2,3-dimethyl-2-butene, *J. Org. Chem.*, 53, 2410, 1988; Shim, S. C., Lu, J. J., and Kwon, J. H., Photochemistry of conjugated polyacetylenes, *Proc. XVth Int. Conf. on Photochem.*, Paris, July 28 - August 2, 1992, III-34.

18. Kaupp, G. and Stark, M., A novel photochemical synthesis of propellanes, *Angew. Chem. Int. Ed.*, 17, 758, 1978; 1,2-Photoadditions of stilbenes and diarylacetylenes to bicyclic 1,4-cyclohexadienes: Propellanes and substitutive 1,2-adducts, *Chem. Ber.*, 114, 2217, 1981.

19. Kaupp, G. and Grüter, H. W., Photoreactions of stilbene with caffeine: Known, and new types of reaction, *Angew. Chem. Int. Ed.*, 19, 714, 1980; Known and new types of reactions in the photoreaction of stilbenes with cyclic imines, *Chem. Ber.*, 114, 2844, 1981; Kaupp, G. and Ringer, E., Multifunctional photoadditions of stilbene to derivatives of caffeine and benzothiazole, *Chem. Ber.*, 119, 1525, 1986; Selectivities in photoadditions with multifunctional caffeine derivatives, *Chem. Ber.*, 124, 339, 1991.

20. Shim, S. C., Chae, J. S., and Choi, J. H., Photochemical synthesis of some propellanes through [2+2]cycloaddition of indeno[2,1-*a*]indene with several olefins, *J. Org. Chem.*, 48, 417, 1983; numerous further derivatives there.

21. Kaupp, G. and Ringer, E., The first photocycloadditions of indolizine, *Tetrahedron Lett.*, 28, 6155, 1987.

22. Lewis, F. D., Hoyle, C. E., and Johnson, D. E., Abnormal regioselectivity in the photochemical cycloaddition of singlet *trans*-stilbene with conjugated dienes, *J. Am. Chem. Soc.*, 97, 3267, 1975; see Lewis, F. D., DeVoe, R. J., and MacBlane, D. B., Exciplex and radical ion intermediates in the photochemical reaction of α-phenylcinnamonitrile with 2,5-dimethyl-2,4-hexadiene, *J. Org. Chem.*, 47, 1392, 1982.

23. Lewis, F. D. and DeVoe, R. J., Photochemical cycloaddition of singlet *trans*-stilbene with α,β-unsaturated esters, *J. Org. Chem.*, 45, 948, 1980; Lewis F. D., Formation and reactions of stilbene exciplexes, *Acc. Chem. Res.*, 12, 152, 1979; Kitamura, T., Toki, S., and Sakurai, H., [2+2]-Cycloaddition and its cycloreversion of aromatic olefins, *Kokugaku Toronkai Koen Yoshishu*, 190, 1979; *Chem. Abstr.*, 93, 70806, 1980.

24. SET, $k_{et} = 1.9 \times 10^{12}$ M^{-1} s^{-1}, no cycloaddition: Peters, K. S., Angel, S. A., and O'Driscoll, E., Stilbene photocycloaddition reactions: Ion pair and electron transfer dynamics, *Pure Appl. Chem.*, 61, 629, 1989.

25. Tolbert, L. M. and Ali, M. B., High optical yields in a photochemical cycloaddition. Lack of cooperativity as a clue to mechanism, *J. Am. Chem. Soc.*, 104, 1742, 1982.

26. Arnold, D. R. and Morchat, R. M., Electronic excited states of small ring compounds. IV. Bicyclo[2.1.0]pentanes by the photocycloaddition of cyclopropenes to olefins, *Can. J. Chem.*, 55, 393, 1977; see also Wong, P. C. and Arnold, D. R., Electronic excited states of small ring compounds. VII. Bicyclo[2.1.0]pentanes by the photocycloaddition of 1,2,3-triphenylcyclopropene to fumaro- and maleonitrile, *Can. J. Chem.*, 57, 1037, 1979.

27. Penn, J. H., Gau, L. X., Chan, E. Y., Loesel, P. D., and Hohlneicher, G., Steric inhibition of photochemical reactions: The [2+2]-cycloaddition reaction, *J. Org. Chem.*, 54, 601, 1989.

28. Lewis, F. D., Hirsch, R. H., Roach, P. M., and Johnson, D. E., Photochemical cycloaddition of singlet and triplet diphenylvinylene carbonate with vinyl ethers, *J. Am. Chem. Soc.*, 98, 8438, 1976.

29. Nakamura, Y., Imakura, Y., and Morita, Y., Reactions using micellar system: Photocycloadditions of acenaphthylene with acrylonitrile and methyl acrylate, *Chem. Lett.*, 965, 1978; 969, 1978.

30. Willner, I. and Rabinovitz, M., Isomerizations of benzannelated C$_9$H$_{10}$ bicyclic systems into cyclononatetraenes. Insight into the behaviour of medium-sized conjugated rings, *Tetrahedron*, 35, 2359, 1979.

31. Fujiwara, Y., Sumino, M., Nozaki, A., and Okamoto, M., Studies on synthesis of 10,11-dihydro-5*H*-dibenzo[*a,d*]cycloheptane derivatives. IV. Photoreactions of 5-substituted-5*H*-dibenzo[*a,d*]cycloheptanes with 1,2-substituted olefins and the stereostructures of the cycloaddition products, *Chem. Pharm. Bull.*, 37, 1452, 1989; further derivatives there.

32. Andreani, F., Andrisano, R., Salvadori, G., and Tramontini, M., Photoreactivity of quinoline derivatives. Part 2. Photodimers of *trans*-4-styrylbenzo[*h*]quinoline, *J. Chem. Soc., Perkin Trans. I*, 1737, 1977.

33. Sailaja, S., Rajamarendar, E., and Murthy, M., Photocycloaddition of tetracyanoethylene to 4-benzalamino-5-styrylisoxazoles, *Ind. J. Chem.*, 25B, 191, 1986.

34. Horner, M. and Hünig, S., Reversible transition between a [4]radialene and a cyclobutadiene through electron transfer, *Angew. Chem. Int. Ed.*, 16, 410, 1977.

35. Takagi, K., Suddaby, B. R., Vadas, S. L., Backer, C. A., and Whitten, D. G., Topological control of reactivity by interfacial orientation: Excimer fluorescence and photodimerization of 4-stilbazolium cations in aerosol OT reversed micelles, *J. Am. Chem. Soc.*, 108, 7865, 1986.

36. Quina, F. H. and Whitten, D. G., Photochemical reactions in organized monolayer assemblies. 4. Photodimerization, photoisomerization, and excimer formation with surfactant olefins and dienes in monolayer assemblies, crystals, and micelles, *J. Am. Chem. Soc.*, 99, 877, 1977.

37. Neunteufel, R. A. and Arnold, D. R., Radical ions in photochemistry. I. The 1,1-diphenylethylene cation radical, *J. Am. Chem. Soc.*, 95, 4080, 1973; Yamamoto, M., Yoshikawa, H., Gotoh, T., and Nishijima, Y., Photodimerization mechanism of *m*-methoxystyrene *via* its cation radical. Open type dimer cation radical as its intermediate, *Bull. Chem. Soc. Jpn.*, 56, 2531, 1983; recently unspecified quantities of 1,1,2,2-tetraphenylcyclobutane[1] have been isolated under SET conditions as well: Mattes, S. L. and Farid, S., Photochemical electron-transfer reactions of 1,1-diarylethylenes, *J. Am. Chem. Soc.*, 108, 7356, 1986; review: Mattay, J. and Vondenhof, M., Contact and solvent-separated radical ion pairs in organic photochemistry, *Topics Curr. Chem.*, 159, 219, 1991; [4+2]- and [2+2]-cycloadducts with styrene or indene and 1,3-cyclohexadiene: Martiny, M., Steckhan, E., and Esch, T., Cycloaddition reactions initiated by photochemically excited pyrylium salts, *Chem. Ber.*, 126, 1671, 1993.

38. Mizuno, K., Ogawa, J., Kamura, M., and Otsuji, V., Photo-induced cyclodimerization of electron-rich aromatic olefins in the presence of transition metal complexes, *Chem. Lett.*, 731, 1979.

39. Todd, W. P., Dinnocenzo, J. P., Farid, S., Goodman, J. L., and Gould, I. R., Efficient photoinduced generation of radical cations in solvents of medium and low polarity, *J. Am. Chem. Soc.*, 113, 3601, 1991.

40. Mizuno, K., Kaji, R., and Otsuji, Y., Photocycloaddition between indene and electron-rich olefins through indene cation radical, *Chem. Lett.*, 1027, 1977.

41. Inokuma, S., Yamamoto, T., and Nishimura, J., Efficient intramolecular [2+2]photocycloaddition of styrene derivatives toward new crown ethers, *Tetrahedron Lett.*, 31, 97, 1990; references therein; more recent example: Fleming, S. A. and Ward, S. C., Stereocontrolled photochemical (2+2) cycloaddition, *Tetrahedron Lett.*, 33, 1013, 1992.

42. Nakano, T., Martin, A., Rivas, C., and Perez, C., Photosensitized cyclodimerizations of 4-vinylpyridine, 2-vinylpyridine, and 2-vinylquinoline, *J. Het. Chem.*, 14, 921, 1977.

43. Tsuchida, A., Yamamoto, M., and Nishijima, Y., Photocycloaddition of 4-dimethylaminostyrene with 1-vinylpyrene or styrene, *J. Chem. Soc., Perkin Trans. II*, 239, 1986; related examples: Tsubakiyama, K., Miyagawa, K., Kaizaki, K., Yamamoto, M., and Nishijima, Y., Photochemical reactions of 4-acylstyrenes, *Bull. Chem. Soc. Jpn.*, 65, 837, 1992.

44. Lewis, F. D. and Kojima, M., Photodimerization of singlet *trans-* and *cis*-anethole. Concerted or stepwise?, *J. Am. Chem. Soc.*, 110, 8660, 1988.

45. Dauben, W. G., van Riel, H. C. H. A., Robbins, J. D., and Wagner, G. J., Photochemistry of *cis*-1-phenylcyclohexene. Proof of involvement of trans isomer in reaction processes, *J. Am. Chem. Soc.*, 101, 6383, 1979.

46. Dekker, J., Martins, F. J. C., and Dekker, T. G., Photodimerization. Part IV. The photodimerization of 1-methylindene, *S. Africa J. Chem.*, 30, 21, 1977; *Chem. Abstr.*, 87, 52439, 1977; Wamhoff, H. and Hupe, H.-J., Photochemistry of heterocycles, 7. Photocycloadditions and photoinduced alkylations of dihalomaleimids with aromatic hydrocarbons and subsequent reactions, *Chem. Ber.*, 111, 2677, 1978; Majima, T., Pac, C., and Sakurai, H., Redox-photosensitized reactions. 5. Redox-photosensitized ring cleavage of 1,1a,2,2a-tetrahydro-7*H*-cyclobut[*a*]indene derivatives: Mechanism and structure-reactivity relationship, *J. Am. Chem. Soc.*, 102, 5265, 1980; Davies, W., Eunis, B. C., Mahavern, C., and Porter, Q. N., Thermal and photodimerization of benzo[*b*]thiophen-3-carboxylic acid 1,1-dioxide, *Aust. J. Chem.*, 30, 173, 1977; Hopkinson, M. J., Schlaman, W. W., Plummer, B. F., Wenkert, E. and Raju, M. Multipathway photodimerization of 2-methylthianaphthene,1,1-dioxide. An example of a heavy-atom effect on intersystem crossing in a 1,4 diradical, *J. Am. Chem. Soc.*, 101, 2157, 1979; Kirby, N. V. and Reid, S. T., Photocycloaddition of 3-acetoxybenzo[*b*]thiophen1,1-dioxide to cycloalkenes; a novel synthesis of substituted benzo[*b*]thiepinones., *J. Chem. Soc., Chem. Commun.*, 150, 1980; benzothiophene 1-oxide: El Faghi M., Al Amoudi, Geneste, P., and Olivé, J.-L., Photoreactivity of 2- and 3-substituted benzo[*b*]thiophene 1-oxides in solution, *J. Org. Chem.*, 46, 4258, 1981.

47. Kimoto, H., Takahashi, K., and Muramatsu, H., Photo-cycloaddition of indene to chlorofluoroethylenes and some reactions concerning the adducts, *Bull. Chem. Soc. Jpn.*, 53, 764, 1980; see also Sket, B. and Zupan, M., Fluorine control of stereospecificity and regioselectivity in

photocycloaddition reactions. Cycloaddition reactions of 1,1-diphenyl-2-fluoroethylene, *Tetrahedron Lett.*, 2607, 1978.

48. Martinez-Utrilla, R. and Miranda, M. A., Photochemistry of 5-aryl-2(3*H*)-furanones. A new route to the synthesis of chromones, *Tetrahedron*, 37, 2111, 1981.

49. Mizuno, K., Okamoto, H., Pac, C., Sakurai, M., Murai, S., and Sonoda, N., Photoreactions of silyl enol ethers of α-tetralone and α-indanone with electron deficient olefins; the photo-Michael reaction, *Chem. Lett.*, 237, 1975.

50. Kimura, K., Shimoyama, M., and Morosawa, S., The formation of aryltetralin derivatives in the photolysis of two *trans*-cinnamoyl moieties at both ends of a polyethylene glycol chain in the presence of lithium perchlorate, *J. Chem. Soc., Chem. Commun.*, 375, 1991; similar examples: Akabori, S., Kumagai, T., Habata, Y., and Sato, S., Preparation of photoresponsive cyclobutane-1,2-dicarbonyl-capped 2.*n* diazacrown ethers by intramolecular [2+2]photocycloaddition, and their highly selective complexation with lithium cation, *J. Chem. Soc., Chem. Commun.*, 661, 1988; The preparation of photoresponsive cyclobutanocrown ethers by means of intramolecular [2+2]photocycloaddition, *Bull. Chem. Soc. Jpn.*, 61, 2459, 1988.

51. Takagi, K., Fukaya, H., Miyake, N., and Sawaki, Y., Organized photodimerization of cinnamic acid in cationic reversed micelle, *Chem. Lett.*, 1053, 1988.

52. Jian, W. E., Xiong, L. Y., Zhen, L. M., and Hai, S. H., Photochemical behavior of N,N-dimethylamine substituted cinnamate in molecular aggregate, *Proc. 13th IUPAC Symp. Photochemistry*, Warwick, July 22–28, 1990, P276.

53. Weissbuch, I., Leiserowitz, L., and Lahav, M., Structured self-aggregates of 4-methoxy-*(E)*-cinnamic acid at the air/solution interface as detected by [2+2]-photodimerization and their role in the control of crystal polymorphism, *J. Am. Chem. Soc.*, 113, 8941, 1991.

54. Koch, H., Laschewsky, A., Ringsdorf, H., and Teng, K., Photodimerization and photopolymerization of amphiphilic cinnamic acid derivatives in oriented monolayers, vescicles, and solution, *Makromol. Chem.*, 187, 1843, 1986.

55. Hirayama, F., Utsuki, T., and Uekama K., Stoichiometry-dependent photodimerization of tranilast in a γ-cyclodextrin inclusion complex, *J. Chem. Soc., Chem. Commun.*, 887, 1991.

56. Beak, P. and Zeigler, J. M., Molecular organization by hydrogen bonding: Juxtaposition of remote double bonds for photocyclization in a 2-pyridone dimer, *J. Org. Chem.*, 46, 619, 1981; concerning the alternative techniques for cinnamic esters, *Tetrahedron*, 48, 2523, 1992.

57. D'Auria, M., Piancatelli, G., and Vantaggi, A., Photochemical dimerization of methyl-, 2-furyl-, and 2-thienylacrylate and related compounds in solution, *J. Chem. Soc., Perkin Trans. I*, 2999, 1990; D'Auria, M., D'Annibale, A., and Ferri, T., Photochemical behavior of furylidene carbonyl compounds, *Tetrahedron*, 48, 9323, 1992.

58. Padwa, A., Kennedy, G. D., Newkome, G. R., and Fronczek, F. R., Dimerization and cycloaddition reactions of a carbomethoxy-substituted cyclopropene, *J. Am. Chem. Soc.*, 105, 137, 1983; further examples there.

59. Suginome, H., Kobayashi, K., Itoh, M., Seko, S., and Furusaki, A., Photoinduced molecular transformations. 110. Formation of furoquinolinones via β-scission of cyclobutanoxyl radicals generated from [2+2]photoadducts of 4-hydroxy-2-quinolone and acyclic and cyclic alkenes, *J. Org. Chem.*, 55, 4933, 1980.

60. Suginome, H., Liu, C. F., Seko, S., Kobayashi, K., and Furusaki, A., Photoinduced molecular transformations. 100. Formation of furocoumarins and furochromones via a β-scission of cyclobutanoxyl radicals generated from [2+2]photoadducts from 4-hydroxycoumarin and acyclic and cyclic alkenes, *J. Org. Chem.*, 53, 5952, 1988; see also Kirpichenok, M. A., Mel'nikova, L. M., Denisov, L. K., and Grandberg, I. I., Photochemical reactions of 7-aminocoumarins. 1. [2+2]-Cycloadducts with vinyl butyl ether and acrylonitrile, *Khim. Geterotsikl. Soedin*, 1169, 1988; *Chem. Abstr.*, 111, 77774, 1989.

61. Ratiner, B. D., Elia, L. P., and Otsuki, T., Photochemical reactivity of 3-*t*-butoxycarbonyl furocoumarin, *Chem. Express*, 5, 225, 1990; *Chem. Abstr.*, 113, 78199, 1990; three [2+2]-dimers of

4,5′,8-trimethylpsoralen also including the furan moiety: Shim, S. C., Lee, S. S., and Choi, S. J., The C$_4$-photocyclodimers of 4,5′,8-trimethylpsoralen (TMP), *Photochem. Photobiol.*, 51, 1, 1990.

62. Karbe, C. and Margaretha, P., Photocycloadditions to 1-thiocoumarin, *J. Photochem. Photobiol. A Chem.*, 57, 231, 1991.

63. Lewis, F. D. and Barancyk, S. V., Lewis acid catalysis of photochemical reactions. 8. Photodimerization and cross-cycloaddition of coumarin, *J. Am. Chem. Soc.*, 111, 8653, 1989.

64. Miyashita, A., Ikezu, S., and Nohira, H., Stereospecific photodimerization of unsaturated compounds induced by nickel complexes, *Chem. Lett.*, 1235, 1985; rose bengal sensitization in conjugated acrylamides: Kumar, A., Krupadanam, G. L. D., and Srimannarayana, G., Photochemistry of *E,E*-piperylpiperidine, *Indian J. Chem. Sect. B*, 31, 784, 1992.

65. Cobb, R. L. and Mahan, J. E., Chemistry of cyclobutene-1,2-dicarbonitrile. 2. Cycloadducts, *J. Org. Chem.*, 42, 2597, 1977.

66. Gibson, T. W., Mujeti, S., and Barnett, B. L., Photochemistry of esters II. Effects of ring size on the photochemical behavior of α,β-unsaturated esters, *Tetrahedron Lett.*, 4801, 1976.

67. Baldwin, S. W. and Wilkinson, J. M., Four-carbon photochemical annelation of alkenes with 2,2,6-trimethyl-1,3-dioxolenone, *J. Am. Chem. Soc.*, 102, 3634, 1980; similar reactions with ethylene: Iwaoka, T., Katagiri, N., Sato, M., and Kaneko, C., Use of 1,3-dioxin-4-ones having a fluorine or trifluoromethyl group at the 5-position as π2 components in (2+2) photocycloaddition and Diels-Alder reactions, *Chem. Pharm. Bull.*, 40, 2319, 1992.

68. Tomiaka, K., Tanaka, M., and Koga, K., Stereoselective reactions. XVI. Total synthesis of (−)-β-bourbonene by employing asymmetric (2+2)photocycloaddition reaction of chiral butenolide, *Chem. Pharm. Bull.*, 37, 1201, 1989; cf. Ohga, K. and Matsuo, T., A study on the photochemistry of α,β-unsaturated lactones. III. Photodimerization of 2-penten-4-olides, *Bull. Chem. Soc. Jpn.*, 49, 1590, 1976; Hoffmann, N., Scharf, H. -D., and Runsink, R., Synthesis of chiral cyclobutane derivatives from (+)-5-menthyloxy-2-[5*H*]-furanone and ethylene, *Tetrahedron Lett.*, 30, 2637, 1989; for similar photoadditions: Hatsui, T., Kitashima, T., and Takeshita, H., The photoaddition of 4-methyl-2-oxo-γ-valerolactone to cycloalkenes, *Bull. Chem. Soc. Jpn.*, 67, 293, 1994; 1*H*-pyrrol-2(5*H*)-ones and tetramethylethylene: Ihlefeld, A. and Margaretha, P., Synthesis and photochemistry of 5,5-dimethyl-1H-pyrrol-2(5H)-one and of some N-substituted derivatives, *Helv. Chim. Acta*, 75, 1333, 1992.

69. Somekawa, K., Imai, R., Furukido, R., and Kumamoto, S., Photoadducts of 2-pyridones with chloroethylenes and their derivatives, *Bull. Chem. Soc. Jpn.*, 54, 1112, 1981; 2-pyridone and acetylene dicarboxylate: Somekawa, K., Okumura, Y., Uchida, K., and Shimo, T., Preparations of 2-azabicyclo[2.2.2]octa-5,7-dien-3-ones and 7-azabicyclo[4.2.0]octa-2,4-dien-8-ones from addition reactions of 2-pyridones, *J. Het. Chem.*, 25, 731, 1988.

70. Boule, P. and Lemaire, J., Photodimerization of maleic anhydride in carbon tetrachloride, *Tetrahedron Lett.*, 865, 1976.

71. Hartmann, W., Heine, H. -G., Hinz, J., and Wendisch, D., Spirohexane derivatives by photosensitized cycloadditions to methylenecyclopropane, *Chem. Ber.*, 110, 2986, 1977.

72. Winzenburg, M. L., Fields, E. K., and Sinclair, D. P., Photocycloadducts of dimethylmaleic anhydride with unsaturated acids and esters, *J. Org. Chem.*, 53, 2624, 1988.

73. Hoffmann, V. T. and Musso, H., Synthesis of a double tetraasterane: Nonacyclo-[10.8.0.0^{2,11}.0^{4,9}.0^{4,19}.0^{6,17}.0^{7,16}.0^{9,14}.0^{14,19}]eicosane, *Chem. Ber.*, 124, 103, 1991, and references therein.

74. Hefetz, H., Dunn, D. A., Deutsch, T. F., Buckly, L., Hillenkamp, F., and Kochevar, I. E., Laser photochemistry of DNA: Two-photon absorption and optical breakdown using high-intensity, 532-nm radiation, *J. Am. Chem. Soc.*, 112, 8528, 1990.

75. McGee, L. R. and Misra, R., Gilvocarcin photobiology. Isolation and characterization of the DNA photoadduct, *J. Am. Chem. Soc.*, 112, 2386, 1990; cf. **59/60** in chapter on "Stilbenes".

76. Swenton, J. S., Jurcak, J. G., and Wójtowicz, H., An unusual substituent effect on elimination vs. fragmentation reactions of the dianions of 5-fluorouracil-alkene photoadducts. Preparation of cyclobutane-annelated uracils, *J. Org. Chem.*, 53, 1530, 1988; cf. similar addition of **148** to

2-methylpropene: Wexler, A. and Swenton, J. S., Fluorine control of regioselectivity in photocycloaddition reactions. The direct functionalization of uracil via a novel 1,4-fragmentation, *J. Am. Chem. Soc.*, 98, 1602, 1976.

77. Shim, S. C. and Shin, E. J., Photocycloaddition reaction of 5(E)-styryl-1,3-dimethyluracil with some olefins, *Chem. Lett.*, 45, 1987; different isomers of type 152 with asymmetric alkenes.

78. Sen, D. and Wells, C. H. J., Photo-dimers from the photolysis of 2-dimethylamino-5,6-dimethylpyrimidin-4-ol in aqueous solution, *Pestic. Sci.*, 12, 339, 1981; *Chem. Abstr.*, 95, 187187t, 1981.

79. Wenska, G., Skalski, B., and Paszyc, S., Photophysical and photochemical properties of 4-(1,2,4-triazol-1-yl)-pyrimidin-2(1H)-ones, *J. Photochem. Photobiol. A Chem.*, 57, 279, 1991; Golankiewicz, K. and Wójtowicz, H., Influence of fluorine substitution on photochemical and spectroscopic properties of dinucleotide analogs, *Pol. J. Chem.*, 60, 943, 1986; see also Ahn, C. I., Choi, H. Y., and Hahn, B. S., Intramolecular [2+2]photocycloaddition of N^I(alkenyl)-pyrimidines: formation of diazatricyclodiones, *Heterocycles*, 31, 1737, 1990.

80. Mizuno, K., Ogawa, Z., Kamura, M., and Otsuji, Y., Photo-induced cyclodimerization of electron-rich aromatic olefins in the presence of transition metal complexes, *Chem. Lett.*, 731, 1979.

81. Draper, A. M., Ilyas, M., deMayo, P., and Ramamurthy, V., Surface photochemistry: Semiconductor photoinduced dimerization of phenyl vinyl ether, *J. Am. Chem. Soc.*, 106, 6222, 1984.

82. Albini, A. and Spreti, S., Poly(ethylene terephthalate) as an insoluble electron transfer photosensitizer, *J. Chem. Soc., Chem. Commun.*, 1426, 1986.

83. Mizano, K., Kagano, H., and Otsuji, Y., Regio- and stereoselective intramolecular photocycloaddition: Synthesis of macrocyclic 2,ω-dioxabicyclo[n.2.0] ring system, *Tetrahedron Lett.*, 24, 3849, 1983; shorter tethers work as well.

84. Scharf, H. -D. and Mattay, J., 2,2-Dimethyl-1,3-dioxol and Z-bis-trimethylsiloxy-ethylene as cyclophiles with nucleophilic character in photochemical (2+2) and thermal (4+2)-cycloadditions, *Tetrahedron Lett.*, 3509, 1976; there are numerous additions of 162 to acrylates.

85. Bellus, O., Fischer, H., Greuter, H., and Martin, P., Syntheses of moniliformin, a mycotoxine with a cyclobutenedione structure, *Helv. Chim. Acta*, 61, 1784, 1978.

86. Scharf, H. -D., Frauenrath, H., and Pinske, W., Synthesis and properties of semisquaric acid and its alkaline salts (moniliformin), *Chem. Ber.*, 111, 168, 1978.

87. Whitney, R. A., A total synthesis of (±)-biotin, *Can. J. Chem.*, 61, 1158, 1983.

88. Scholz, K. -H., Hinz, J., Heine, H. -G., and Hartmann, W., cis-1,2-Cyclobutanediamines by photosensitized cycloaddition of 1,3-diacetyl-4-imidazolin-2-ones to olefins, *Liebigs Ann. Chem.*, 248, 1981; Photosensitized [2+2]-cycloadditions with 4-oxazolin-2-ones, *Tetrahedron Lett.*, 1467, 1978.

89. Takamatsu, K., Ryang, H. -S., and Sakurai, H., Photochemical cyclodimerization of benzofurans, *J. Org. Chem.*, 41, 541, 1976.

90. Tinnemans, A. H. A. and Neckers, D. C., Photocycloaddition of dimethyl acetylenedicarboxylate and methyl propiolate to benzo[b]furans, *J. Org. Chem.*, 42, 2374, 1977.

91. Neckers, D. C. and Wagenaar, F. L., Synthesis of 3,4-benzo-2-thiabicyclo[3.2.0]hepta-1,3-diene, *J. Org. Chem.*, 46, 3939, 1981.

92. Minhy, T. Q., Christiaens, L., Grandelandon, P., and Lablache-Combier, A., Photoaddition reactions of benzo[b]selenophenes with dimethyl acetylenedicarboxylate and 1,2-dichloroethylene, *Tetrahedron*, 33, 2225, 1977.

93. Chiba, T., Takada, Y., Naito, T., and Kanoko, C., Novel methods for introducing a two-carbon unit at either the 3- or 4-position of the isoquinolone ring by means of photo[2+2]cycloaddition reaction, *Chem. Pharm. Bull.*, 38, 2335, 1990; similar additions of acrylonitrile: Suginome, H., Kajizuka, Y., Suzuki, M., Senboku, H., and Kobayashi, K., [2+2] Photoaddition of protected 4-hydroxy-1(2H)-isoquinoline with an electron deficient alkene, *Heterocycles*, 37, 283, 1994.

94. Kuhn, H. J., Defoin, R., Gollnick, K., Krüger, C., Tsay, Y. -H., Liu, L. -K., and Betz, P., Sensitized photocyclodimerization of α,β-unsaturated cyclic sulfones. Crystal structure analyses of the photodimers of 2-sulfolene and thia-2-cyclohexene-1,1-dioxide, *Tetrahedron*, 45, 1667, 1989.

95. Potts, K. T., Dunlap, W. C., and Apple, F. S., 1,2,4-Triazoles XXXIV/XXXI, *Tetrahedron,* 33, 1263, 1977; 1247, 1977.

96. Kropp, P. J., Snyder, J. J., Rawlings, P. C., and Fravel, H. G., Photochemistry of cycloalkenes. 9. Photodimerization of cyclohexene, *J. Org. Chem.,* 45, 4471, 1980.

97. Mirbach, M. F., Mirbach, M. J., and Saus, A., Photochemical 2+2-cycloadditions of ethylene and acetylene to 1,3-cyclohexadiene under high pressure, *Z. Naturforsch. B,* 32, 47, 1977; Photochemical telomerization of ethylene with malonic acid diethylester, *Z. Naturforsch. B,* 32, 58, 1977.

98. Mella, M., Fasani, E., and Albini, A., The photosensitized dimerization of 1,3-cyclohexadiene, *Tetrahedron,* 47, 3137, 1991.

99. Kaupp, G., Grüter, H. -W., and Teufel, E., Photoadditions of anthracenes to 2,4-hexadiene and 1,3,5-cycloheptatriene, *Chem. Ber.,* 116, 630, 1983; Cyclovinylogous additions of 1,3-diphenylbenzo[*c*]furan to 1,3,5-cycloheptatriene, *Chem. Ber.,* 116, 618, 1983.

100. Schmidt, H., Schweig, A., Anastassiou, A. G., and Wetzel, J. C., The dominant role of hyperconjugation in the 9-oxabicyclo[4.2.1]nona-2,4,7-triene series, *Tetrahedron,* 32, 2239, 1976; Borden, W. T., Gold, A., and Young, S. D., A serendipitous synthesis of 1,2,5,6-tetramethyl-3,4,7,8-tetramethylenetricyclo[3.3.0.0^{2,6}]octane, *J. Org. Chem.,* 43, 486, 1978.

101. Siegel, H., Eisenhuth, L., and Hopf, H., Photoaddition of vinylacetylene to other unsaturated hydrocarbons, *Chem. Ber.,* 118, 597, 1985.

102. Slobodin, Y. M., Photodimerization of 3-methyl-1,2-butadiene, *Zh. Org. Khim.,* 24, 1556, 1988; *Chem. Abstr.,* 110, 172721s, 1989.

103. Aviv, G., Sagiv, J., and Yogev, A., Photochemistry of oriented molecules. The photodimerization of tetraphenylbutatriene incorporated in polyethylene and nematic matrices, *Mol. Cryst. Liq. Cryst.,* 36, 349, 1976.

103a. Chung, C. B., Kwon, J. H., and Shim, S. C., Photoreaction of 1-phenyl-1,3,5-heptatriyne with some olefins, *Tetrahedron Lett.,* 34, 2143, 1993.

104. Berger, J., Light-curable polymers with thioether imidyl groups in the side position, Ger. Offen, DE 3,314,951, 3. Nov. 1983; *Chem. Abstr.,* 100, 53366, 1984; Roth, M., Water-soluble photocrosslinkable polymers and their use, Eur. Pat. Appl. EP 94,913, 23. Nov. 1983; *Chem. Abstr.,* 101, 31121, 1984; B. Mueller, Imidyl compounds, polymers thereof and their use in photoimaging, Eur. Pat. Appl. EP 72,780, 23. Feb. 1983; *Chem. Abstr.,* 99, 96843, 1983.

105. Fields, E. K. and Anderson, R. L., Certain substituted cyclobutane dicarboxylic acid anhydrides, U. S. US 4,388,470, 14. Jun. 1983; *Chem. Abstr.,* 99, 178841, 1983; Niy, S. T. and Fields, E. K., Polyimides-polyamides from tricyclo[4.2.1.0^{2,5}]nonane-3,4-dimethyl-3,4,7,8-tetracarboxylic acid dianhydride and dicarboxylic acids, U. S. US 4,391,967, 5. Jul. 1983; *Chem. Abstr.,* 99, 88771, 1983; Polyimides from bicyclo[4.2.0]octane-7,8-dimethyl-3,4,7,8-tetracarboxylic acid dianhydride and bicyclo[4.2.0]octane-2,5-diphenyl-7,8-dimethyl-3,4,7,8-tetracarboxylic dianhydride, U. S. US 4,362,859, 7. Dec. 1982; *Chem. Abstr.,* 98, 90128, 1983; Nonaromatic dianhydride and polyimides from tricyclo[4.2.1.0^{2,5}]nonane-3,4-dimethyl-3,4,7,8-tetracarboxylic acid dianhydride, U. S. US 4,358,582, 9. Nov. 1982; *Chem. Abstr.,* 98, 54702, 1983.

106. Lukas, L. K., Procedure for the preparation of cyclobutene lactones, Ger. Offen. DE 3613312; *Chem. Abstr.,* 109, 6399, 1988.

107. Dilling, W. L., Polymerization of unsaturated compounds by photocycloaddition reactions, *Chem. Rev.,* 83, 1, 1983.

4

Cyclobutane Synthesis
in the Solid Phase

Gerd Kaupp
University of Oldenburg

4.1 Introduction and Mechanistic Considerations

Solid-state photodimerizations of alkenes and enones have a long history. There are exhaustive reviews available.[1,2] These cover both synthetic and mechanistic aspects and show up the differences to solution photodimerizations. Therefore, the more recent developments and those that derive from exceptions and scrutiny of the topochemistry principle and advancements in the use of more varied solid phases will be discussed herein.

The early topochemical postulate assumed that the double bonds must be parallel and less distant than 4.2 Å in the crystal in order to allow photoreaction.[3] Numerous examples fulfilled these criteria and yielded specifically the expected single products, even though it had long been recognized that almost all of them did not proceed topotactically. Thus, phase rebuilding[3a,b] upon product formation was not included. Some help was provided by the reaction cavity principle of Cohen.[4] Thus, it is now widely assumed that dimerizations proceeding under lattice control do so with minimal change or distortion of the inner surface of the reaction cavity, while the molecules undergoing dimerization choose to react with minimal molecular and atomic motions. These assumptions are well accepted; however, there are numerous examples that deviate significantly from the claims of topochemistry.

Potentially reactive crystals exist which fulfill the above requirements but do not react at all, or not in the presumed sense. Some examples are (several) *cis*-cinnamic acids 1 (d < 4 Å), 2 (d = 4.14 Å), 3 (d = 3.78 Å), 4 (d = 3.79 Å), and others that do not photodimerize (further examples in References 2 and 5). Also in crystals of 6, there are centrosymmetric pairs (d = 4.12 Å) which do not react.[2]

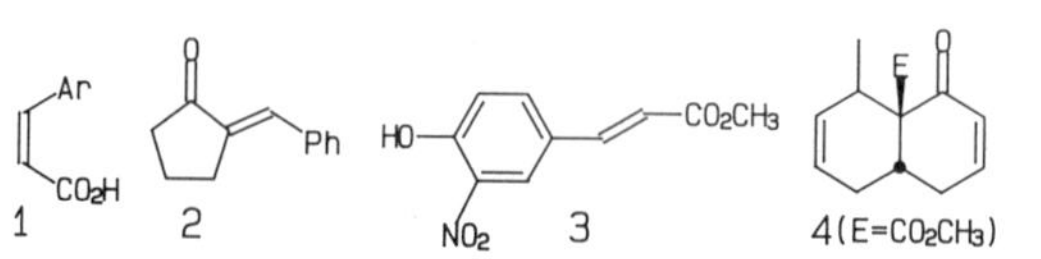

0-8493-8634-9/95/$0.00+$.50

On the other hand, there are examples of reacting crystals with larger distances than 4.2 Å in compounds such as **5** (d = 4.825 Å), **6** (d = 4.45 Å, translationally related); or **71** in **70** as a co-crystal (see below, d = 9.780, 8.929, and 7.780 Å); or with skew angles as in **7** (d = 3.83 Å, 65° skew; cf. also **9**: 49° and 131° skew); or with interplanar angles and lateral displacements as in **8** (d = 3.83, Θ = 35.45°, displacement 1.329 Å) that, nevertheless, are photoreactive (further examples in References 2, 5, 6, and 10). Thus, there might be considerable movement in order to get the molecules together or to increase the overlap between the double-bond electrons, or to internally rotate the 1,4-biradicals formed initially.[3b,8]

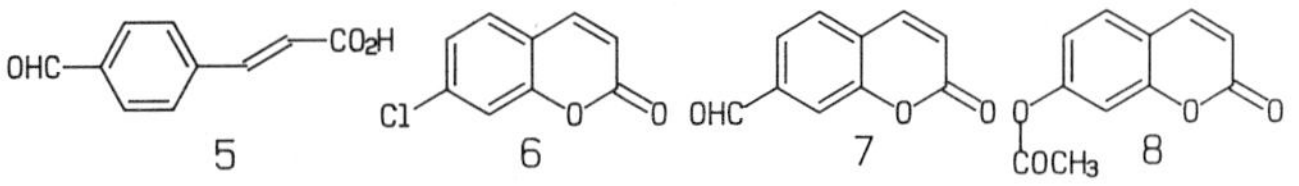

Further difficulties arise if more than one product is obtained in the photolysis of crystals. Thus, it has been shown that **9** yields three different products (**10**, **11**, and **12**) in comparable yields. These results are listed in Table 4.1 for various conditions, as there was some confusion from later claims of a British research group.[9] The double bonds in **9** (distances for **11**: d = 3.720 and 3.721 Å)[8] are 49° skew. Thus, nonconcerted photodimerization occurs as usual.[3b,8] The molecular arrangement for trimer **10** is excellent.[6,8] Finally, based on X-ray analysis, the conclusion is reached that the formation of **12** is also geometrically allowed even though the corresponding double bonds in **9** are 131° skew[3a,8,8a] (distances for **12** are d = 3.594 and 4.147 Å).[3a,8,8a] The formation of the initial biradical requires rotation, and it will have to further rotate internally for the formation of the cyclobutane ring to give **12**. The skew angle of 131° is by far the largest angle still permitting efficient [2 + 2]-photodimerization in crystalline material.

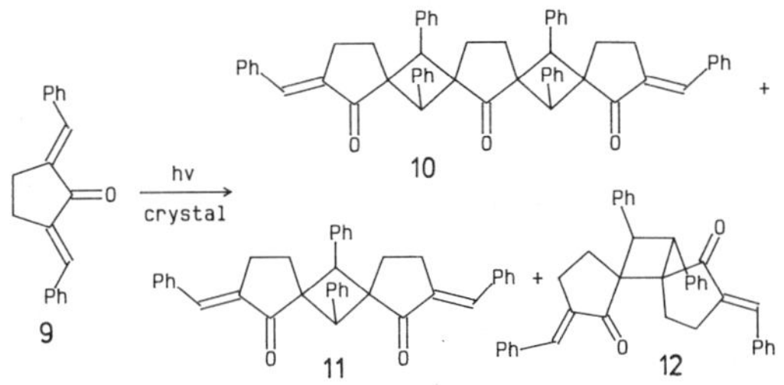

Table 4.1 Relative Yields in 10, 11, and 12 upon Photolysis of 9 in Crystalline Covers[7,8]

λ (nm)	Conversion (%)	Product Ratio		
		10	11	12
>380	>80	26	38	8
>300	>90	26	38	10
253.7	ca. 20	13	38	14
253.7	ca. 80[a]	25	38	5
>380[b]	20–80	26	38	18
>300[c]	92	4	38	19

[a] 48 h, low yield in dimers under these conditions.
[b] Recrystallized from methanol or from CH_2Cl_2
[c] Crystals from CH_3OH in CH_3OH/H_2O (10/90) suspension as in Reference 9.

The products are formed obviously in an amorphous state and no powder-X-ray-diffraction pattern could be obtained.[8,9] Further examples for the formation of several photoproducts in crystals are known (see also References 2 and 6). Thus, **13** yields **14** (42%), **16** (20%) (both with the unusual δ-truxinate configuration), and **15** (2%; β-truxinate).[10] X-ray analysis confirms that the strong electrostatic interactions in **13** enforce a head-to-head *anti*-orientation in the crystal.[10] More information could be obtained if comparison between a set of topotactic single crystal to single crystal [2+2]-photodimerizations was available. However, these could be achieved only in very favorable cases (**17** to **18**) where the cell dimensions of monomer and dimer crystals are nearly identical,[11] or when an unusual modification of the product was obtained by retaining the initial lattice.[11a]

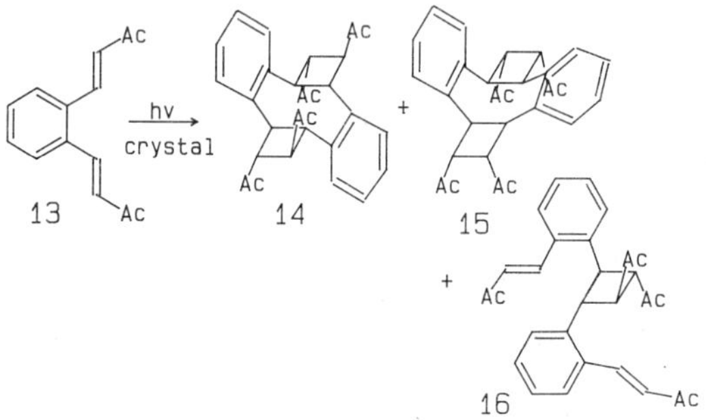

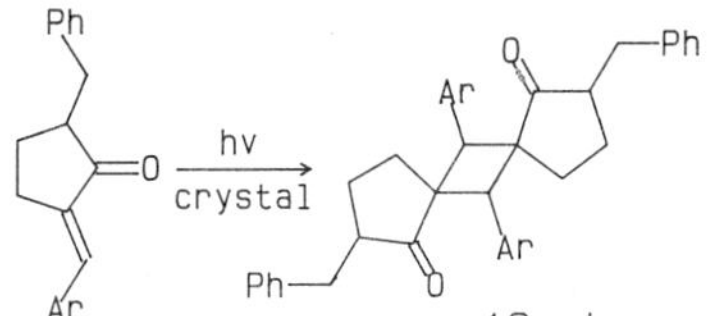

17 a: Ar=C₆H₅
   b: Ar=p-BrC₆H₄

Recently, the problem of phase transformation in the vast majority of cases of [2+2]-crystal photodimerizations has been studied by atomic force microscopy (AFM) on single-crystal surfaces.[3a,3b,8a,12–14] This technique has shown, even for the well-known standard reactions in topochemistry the photodimerization of α-*trans*- and β-*trans*-cinnamic acid **19** which give α-truxillic **20** and β-truxinic acid **21** highly selectively, that the reactions start from the surface region that initially absorbs all the incident light. This result is quite different from earlier claims. In fact, phase rebuilding involving distinct material transport phenomena brings about reaction of the crystal and the formation of new phases. The AFM study shows that epitactical layers are formed on the (010) face of α-**19**. These are divided by parallel furrows at 39° to c. Finally, perpendicularly layered plates are formed.[12–14] With β-**19**, the (100) face undergoes formation of craters (200 nm deep and 500 nm wide), and volcanoes (400 nm high) arise on the (010) face.[12–14] This has been interpreted in terms of monomer orientation, shrinking upon dimerization and partial rotation of the dimers in order to fit within the original lattice (mixed crystal) on the basis of semiempirical (PM3) van der Waal's surfaces. Upon continuation of irradiation, a thermal phase transition flattens the surfaces, still in a distinct way.[12–14]

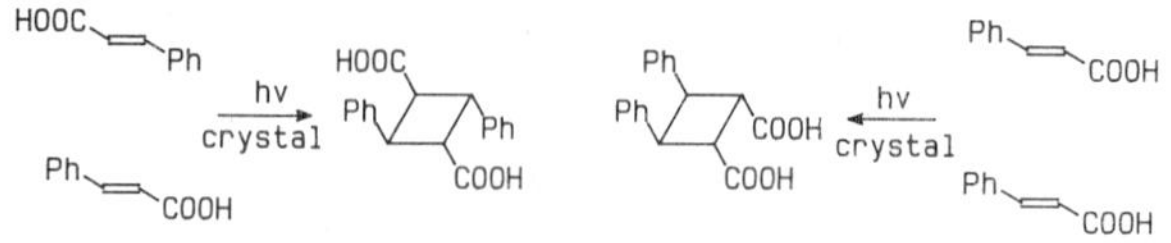

19 (alpha)   20   21   19 (beta)

Interestingly, the AFM of the photolysis of **9** (amorphous products) on (001) shows both volcanoes and steep valleys/ridges[3a,8a] and, in the case of **13**, the morphological prominent face (010) forms both volcanoes and craters, giving an egg-box structure. Further irradiation causes these to grow together into smooth surfaces around the craters, which are more than 50 nm deep.[14] Thus, it is hoped that a deeper insight into the salient features of [2+2]-solid-state photodimerizations will be gained from mechanistic considerations of the phase rebuilding processes (cf. also Section 4.5).

The importance of claiming crystal imperfections in apparently nontopochemical conversions is decreasing.[2] These claims were not helpful for predictions of photoreactivities.

4.2 Crystal Photolyses

It would be advantageous for the efficient formation of dimers if the monomers are favorably packed in terms of distance and electronic overlap. The packing arrangements may be engineered by deliberately applying steric and electrostatic effects to the monomers. Such effects will work on an empirical basis. The operation of these effects can be seen in the conversion of **22** into **23** as the only photodimer (in 76% yield) on photolysis of the crystal.[8] Apparently, steric hindrance of the bulky *t*-butyl group enforces head-to-tail *anti*-arrangement of **22** in the crystal lattice. However, if very strong polar effects are also applied to the system as in **24**, the centrosymmetric arrangement is enforced, affording **25** in 72% yield.[8]

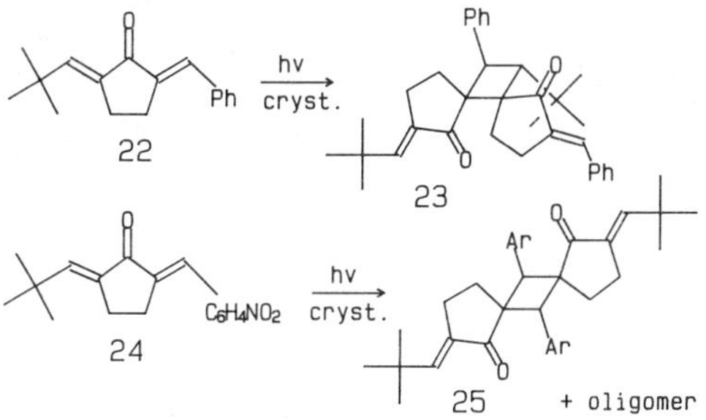

Some oligomers are also formed. Symmetrically disubstituted 2,5-*bis*arylidene cyclopentanones are photostable.[8]

Electrostatic effects play their role in the photodimerization of **26**, giving a 100% yield of **27** at a rate of 1 g h^{-1}.[8a,15] Thus, **26** is photoreactive, unlike the enone **2**. The electrostatic effects have been investigated in great detail and it is thought that C-H····O hydrogen bonding is involved.[16] AFM studies clearly show unidirectional behavior and the (010) face stays relatively smooth while some cracks form perpendicular to it; whereas, the (101) plane shows steep clefting even after short irradiation, making the sample unsuitable for AFM measurement.[8a,13] Interestingly, **27** is not formed in solution; instead, the rotational isomer **28** is produced inefficiently (12% yield with accompanying *E/Z* isomerization).[15]

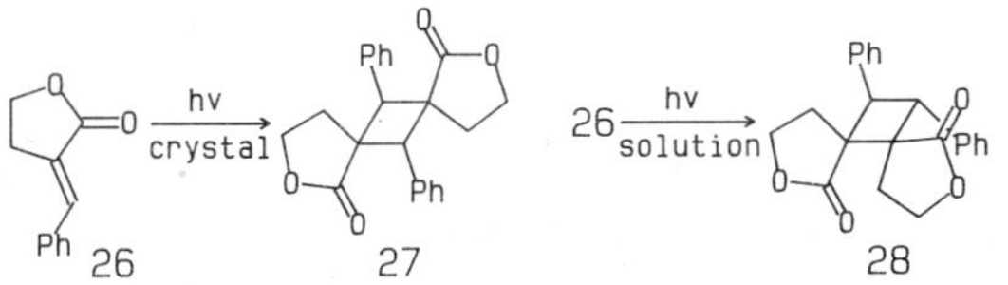

Symmetrical heterostilbenes give head-to-head *anti*-dimers in polymorphs with d < 4.2 Å.[17] Thus, compounds such as **29** yield **30**. In the case of the diquaternary salt **29b**, the counterion is of importance. Thus, the sulfate, but not the diiodide, yields the dimer **30**. In the asymmetric case **31** the head-to-tail *syn*-dimer **32** is formed.[17] This is in accord with well-studied electrostatic influences.[10]

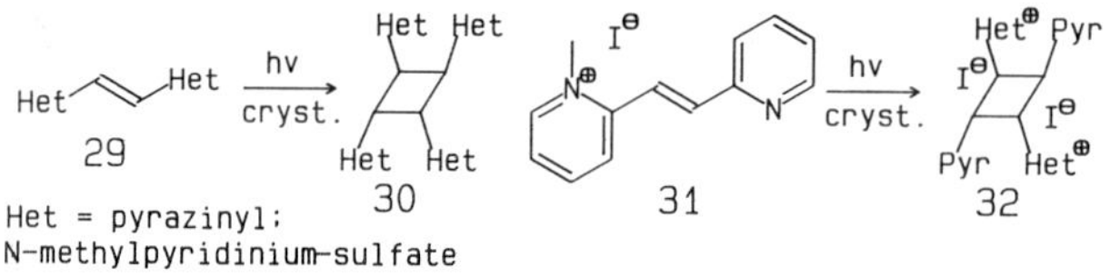

Het = pyrazinyl;
N-methylpyridinium-sulfate

In the case of monoheterostilbenes, the polar effects apparently dictate the crystal lattice arrangement and the photolysis outcome can be easily understood on that basis.[10] Thus, since the heterocycles are strongly electron withdrawing, irradiation of **33a** to **33f** gives **34** specifically as the head-to-tail *syn*-isomer. Such specificity is lost if the overall dipole moment of the monomer is decreased by less electron-withdrawing heterocyclic substituents (36% **34g**, no **34h**). In cases of intermediate polarity (**33i**, **35c**, or **33k**), there is no clear orientational preference in the crystal lattice, and both products **36i, c, k**, and **37i, c**[10] as well as the *all-trans*-dimer **37′k** are formed. The yields in **36k** and **37′k** are 85 and 15%, respectively. A *p*-cyanophenyl group (heterocycle **c** or **h**) enforces products **37e** and **37j** (CN instead of NO$_2$) exclusively. Thus, efficient crystal engineering is possible by application of dipolar effects.[10] Crystal engineering by chloro substitution has been studied also,[2] and photodimerization of bifunctional heterostilbenes has been reviewed.[17a]

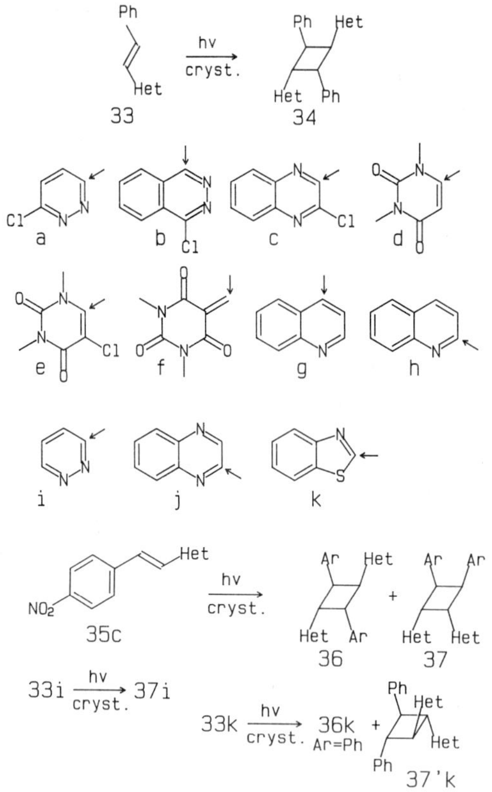

4.3 Asymmetric [2+2]-Photoreactions

The ethyl ester **38** crystallizes in its *cisoid* conformation, the two crystallographically independent molecules form chiral crystals (space group P2$_1$2$_1$2$_1$). In any batch of crystallization, one or the other of the enantiomeric crystals will be in excess (autoseeding, spontaneous resolution). By the use of seeding, large quantities of crystals **38** were obtained with the same chirality. Solid-state photolysis yields the optically active dimer **39** with high ee (>90%) as is enforced by the crystal lattice (β-packing; distances of reacting centers: 3.829 and 4.123 or 3.802 and 4.387 Å). This result constitutes a true absolute asymmetric synthesis.[18] Extensive literature references are available.[2,3a,5,18,19]

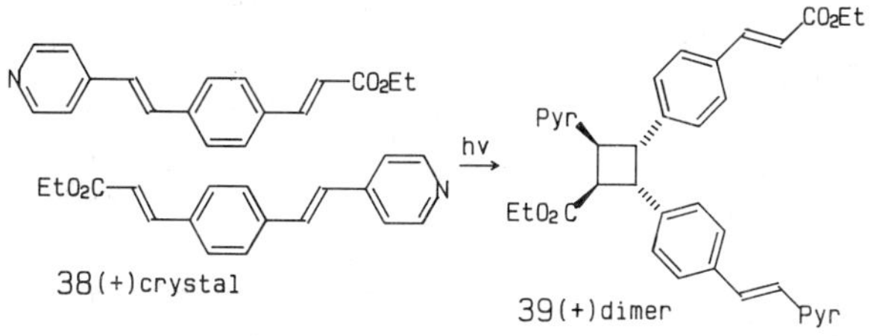

38(+)crystal   39(+)dimer

4.4 Miscellaneous [2+2]-Dimers from Crystals

The continuing importance of solid-state photochemistry is easily seen from a selection of more recent examples.[1,2,5,6] A photostable dimer 40 is formed topochemically from the appropriate monomer.[20] After crystallization of the dimer from propanol, it becomes photochemically reactive and it undergoes photopolycyclodimerization.[20] The dimers 41[21] and 42[22] are formed quantitatively on irradiation of the crystalline starting material. Quite different behavior is observed on irradiation in solution. Interesting heterocyclic systems are available in 43,[23] 44,[24] 45,[25] and 46/47.[26] Compound 49 is one of the photoproducts obtained from the sesquiterpene nootkatone.[26a] This dimer involves cycloaddition of two different double bonds. A cationic compound has also been photodimerized to yield 50.[27] (cf. 32). Most interestingly, surfactant monomers lead to *syn*-head-to-head dimers even if those are charged, and give 51,[28] while octadecylcinnamic ester yields 48 upon solid-state photolysis.[29]

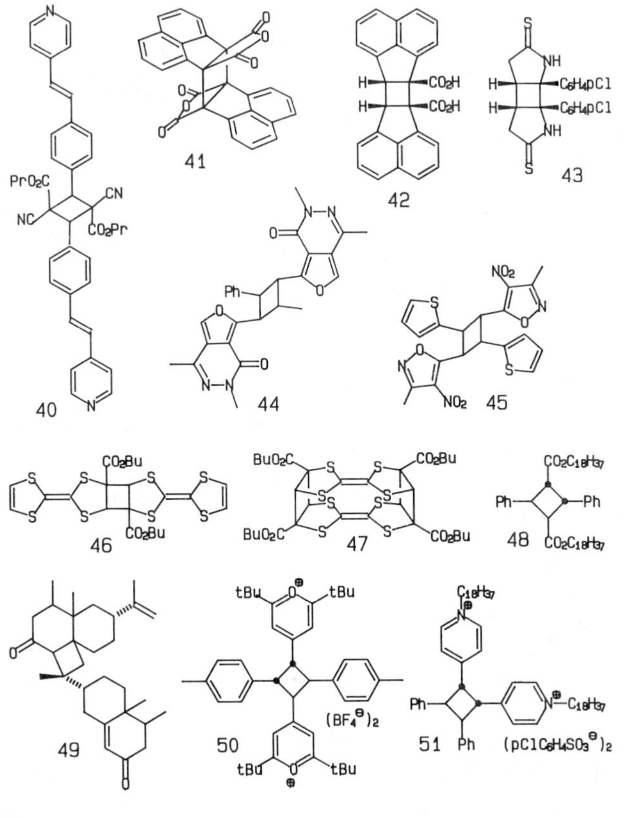

Table 4.2   Product Distribution (%) in the Photolysis
of Ethyl Cinnamate 52

Truxinate/	Liquid		Glass		
Truxillate	25 °C	−20 °C	−45 °C	−80 °C	−196 °C
53 (β)	24.0	35.3	36.0	34.6	36.0
54 (δ)	55.2	31.8	27.5	24.0	22.1
55 (α)	6.3	15.2	22.7	32.0	36.6
56 (neo)[a]	7.1	5.5	5.0	5.6	3.0
57 (ξ)[a]	4.4	6.2	4.2	2.6	

[a] This may only be formed after previous *E/Z* isomerization of the cinnamate.

4.5 Amorphous Solids (Frozen Aqueous Solutions)

trans-Ethyl cinnamate 52 does not photocyclodimerize in solution or in the crystalline state. However, photodimerization does occur efficiently on irradiation of the ester as a neat liquid or as a supercooled glassy solid at −20 to −196 °C.[30] There is not so much of a variation in product ratios (Table 4.2) even though there is a glass transition at −103 °C. Also, there is at least a similarity to the product ratio in the neat liquid at 25 °C. Up to 90% of the monomer undergoes the cyclodimerization, the head-to-head orientations being preferred. This highly photoreactive amorphous system represents a useful model of a photopolymer matrix. In both cases, there are decisive influences of local geometries.

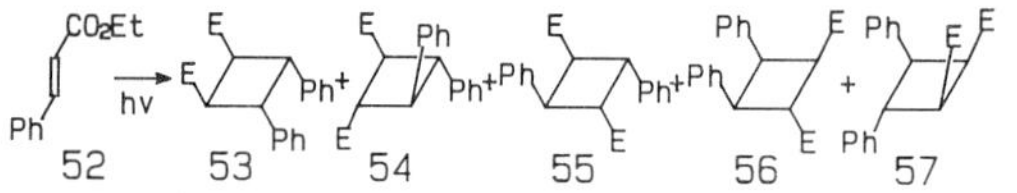

Similarly, although crystalline 58 does not photodimerize (even though there are favorable distances and overlaps between the double bonds, which are 4.06 Å apart), it does so as an amorphous solid in the form of 0.1 μ hydrosols.[31]

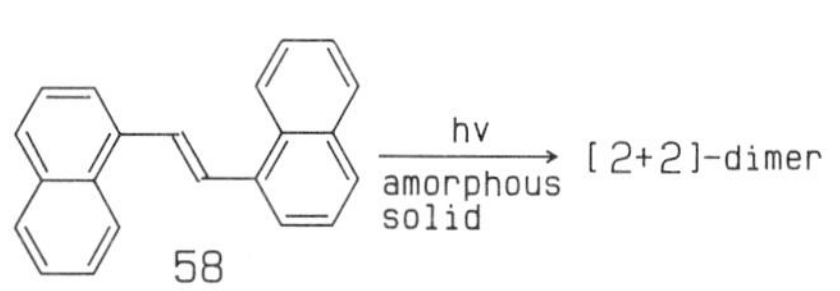

Solid aggregates of ill-defined structures are formed in frozen aqueous solutions (−20 °C through −80 °C) of pyrimidines (thymine, uracil, uridine, cytosine, bromouracil, 5-methylorotic acid, and mixtures therefrom). Irradiation of these yields photodimers, mixed dimers, or trimers efficiently. Some examples of these are 59, 60, and 61. It has been suggested that in some of these cases hydrogen bridges between the pyrimidine units and water align the molecules (e.g., thymine) in a favorable orientation for dimerization.[32] Solid films are far less efficient. Thymine and 8-methoxypsoralen mixtures form a mixed *syn*-dimer upon photolysis in frozen water/methanol.[33]

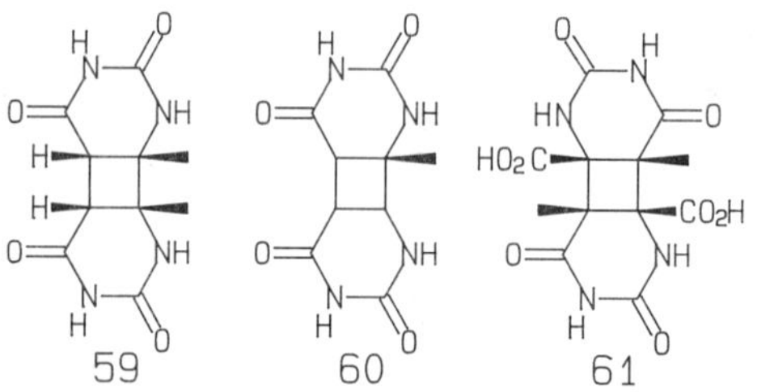

[2+2]-Photocrosslinking in polymers like **62** is less effective in the amorphous state than in the glassy (frozen) nematic films of this liquid crystalline material (nearly parallel stacks).[34] Reactant preordering is a general phenomenon in the [2+2]-photocuring of polymeric photoresists [e.g., poly(vinylcinnamates)] and mixtures of it with other material in the amorphous solid state.[34]

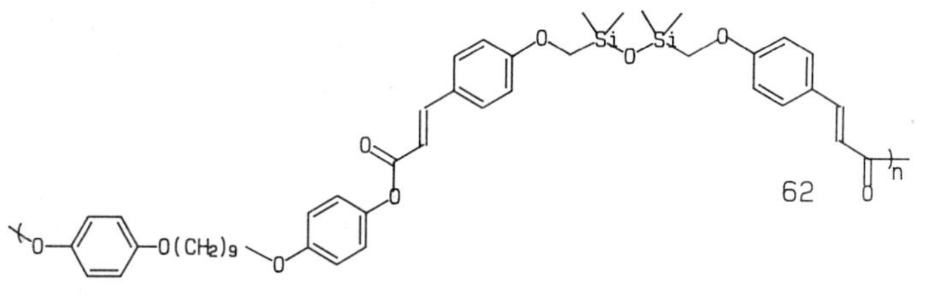

4.6 Mixed Crystals/Mixed Cycloadducts

If more or less stoichiometrically mixed crystals of different compounds can be made (usually from melts or solutions), it will be possible in some cases to obtain "mixed dimers" on photolysis. This extends the synthetic use of solid-state photochemistry; and again, topochemical principles have been applied on the basis of X-ray structures. Numerous examples from the fields of cinnamic acids, cinnamides, and stilbenes have been reviewed.[1,2]

A dry solid mixture of 3-carbethoxypsoralen and 2′-deoxycytidine (film from evaporated solution) yields two stable diastereomeric dimers **63** and **64** upon near-UV photolysis.[35] A 1:2 mixture of 6-cyano-1,3-dimethyluracil and acenaphthylene yields mixed crystals from their melt. Solid-state photolysis (Pyrex) gives **65** as the sole product in 50% yield.[36] Irradiation of co-crystals of the ketones **66** (photoactive modification) and **67** (crystallizes only in a photostable form) yield a mixed dimer.[37] Thus, the strategy of mixed crystal formation may be used to incorporate molecules into new and photoactive modifications.[38]

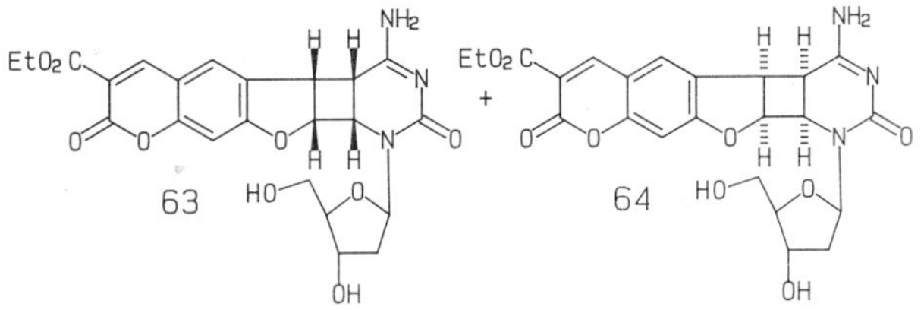

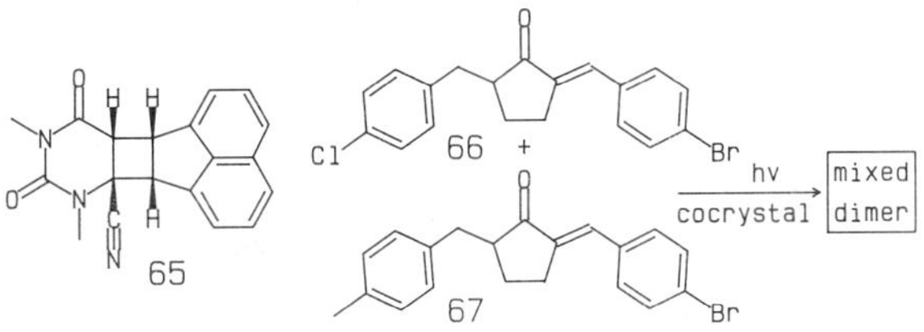

Very minor structural differences of the components may give rise to mixed crystals having a molecular arrangement quite different from each of the pure crystals. Thus, mixed crystals of **68a** and **68b** upon photolysis ($\lambda > 360$ nm) afforded the [2.2]paracyclophane **69** quantitatively. At $\lambda > 410$ nm it is possible to isolate the precursor with only one of the cyclobutane rings formed.[39]

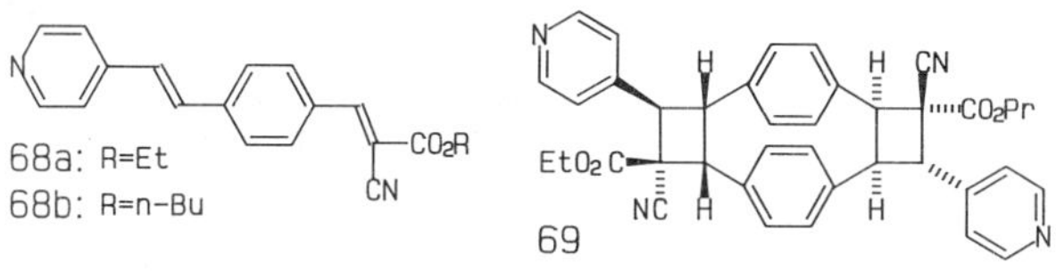

In the case of coumarin, which forms three [2+2]-photodimers in the crystalline state, though nontopochemically,[5] its homodimers (*syn + anti*) may also be obtained from mixed crystals with antipyrine, phenanthrene, and halogeno-1,3-dimethyluracils. It has been claimed that those substances control the crystal lattice and thus engineer the arrangement of the coumarin molecules.[40]

4.7 Solid (Inclusion) Complexes

The crystal packing of alkenes may be modified by complexation with Lewis acids ($HgCl_2$, BF_3, $SnCl_4$, or UO_2Cl_2). Thus, **52** forms a crystalline 2:1 complex with $SnCl_4$ (smallest double bond distance 4.023 and 4.125 Å, head-to-tail stacking), which upon photolysis yields α-truxillate **55** exclusively.[41] Inclusion complexes have a more general application. For example, the chiral host compound (R,R)-(–)**70** forms an inclusion complex with only one of the ring invertomers (enantiomers) of cycloocta-2,4-diene-1-one **71**. In the 3:2 host/guest crystal, the sites undergoing photodimerization are separated by 9.780 and 8.929 (twofold rotation) and 7.780 Å (unit translation toward the b axis). This is far outside the reach of topochemical postulates (<4.2 Å). Nevertheless, upon photolysis of the solid complex, the optically active dimer **72** is formed enantioselectively with 55% yield and 78% ee. Thus, there is a lot of controlled movement within the lattice upon photolysis.[42]

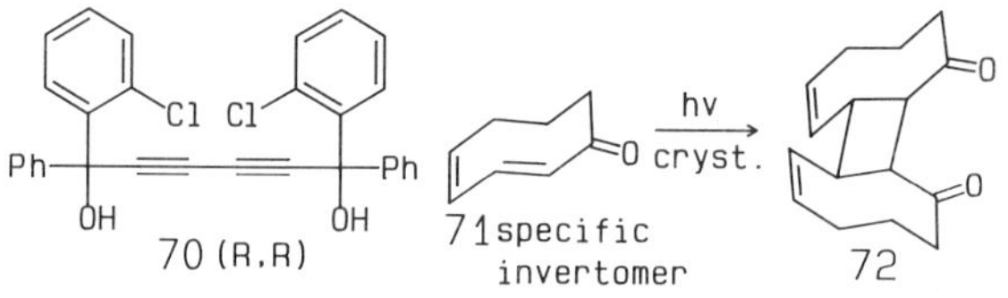

The solid host guest complex **73/74** upon photolysis yields 70% of the head-to-tail *syn* product **75**, whereas solid **74** is photostable and solutions of **74** give only the head-to-head *syn* dimer.[43] Coumarins have been photodimerized in solid cyclodextrin inclusion complexes.[5]

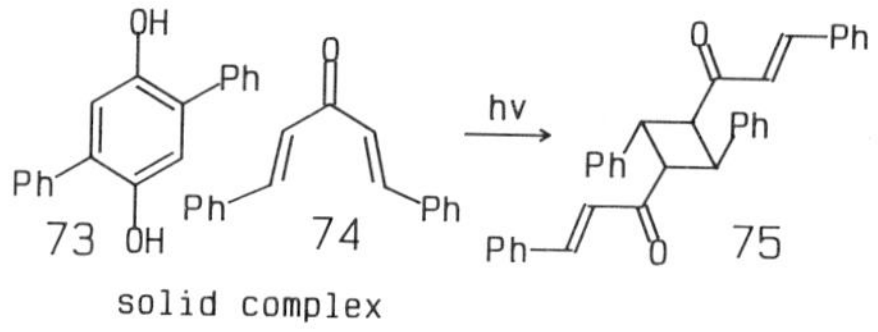

4.8 Adsorbed Media, Zeolites, Clays

If acenaphthylene is adsorbed on silica gel and photolyzed, the *syn/anti* dimer ratio varies with the coverage. Clearly, intra- and intergranular movement of monomers occurs and there are sites of preferred adsorption.[44] It has been shown that steroidal enones (e.g., **76**) adsorb from their less hindered α-side. Thus, upon photoaddition of ethylene, its attack to the more hindered β-side is favored (4β5β-/4α5α – **77** = 58/42; in CH_3OH [also –78 °C], this ratio is 12/82 and 6% are 4α5β – **77**).[45] Usually, the selectivities with adsorbed heterostilbenes are somewhat less pronounced than in crystal photolyses[46] of the same material and dry adsorbates are more selective than suspensions.[46] This is explained by either restriction of translational movements or by directionality in the molecular motions imposed by the SiO_2 surface.

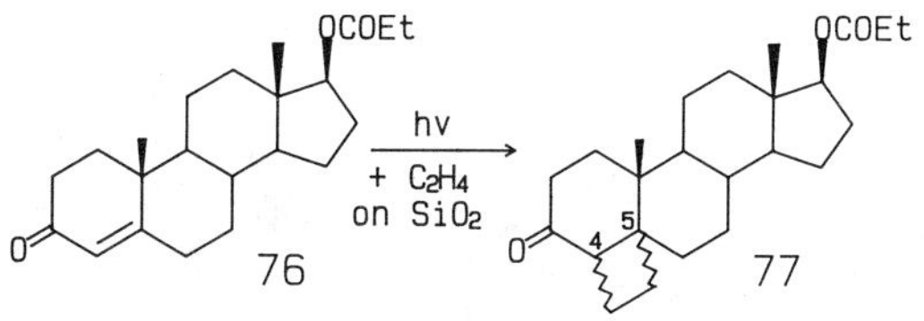

In zeolites (dry or hexane slurries) with large cages (faujasite), acenaphthylene has been photodimerized.[47] The *syn/anti* ratio of the dimers is perturbed by heavy cations and restricted mobility, probably of doubly occupied sites.

More studies on [2+2]-photodimerizations have been performed with clay intercalates.[48] Thus, co-intercalation of 78 and 79 into saponite yields a statistical distribution of crossed dimers 80, 81, and homo dimers upon irradiation of slurries. Interparticle migration of the alkenes is slow between clay particles in 10% aqueous methanol.[48] Also, the synthesis of photoactive cinnamate intercalates of double hydroxides has been described.[49]

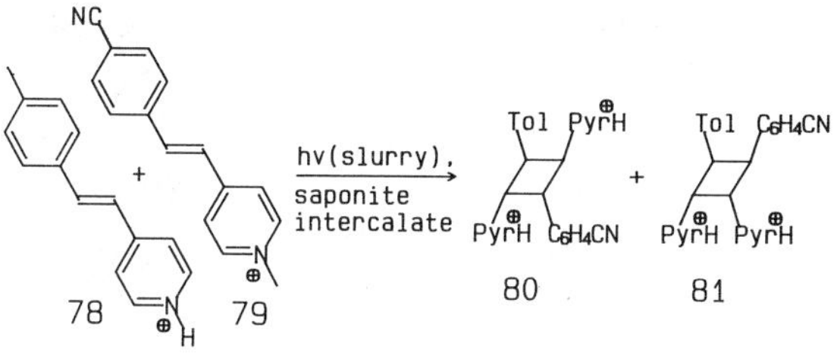

4.9 Industrial Applicability

If the monomers undergoing dimerization are colored, their photochromism may be used technically in various ways. Solar energy storage has been realized in several cases with [2+2]-cycloaddition; however, there seem to be no large-scale applications yet. Most industrial applications are in the field of photocuring of polymer layers via crosslinking (photoresists)[34] or by photopolymerization of doubly unsaturated crystalline alkenes. Numerous reviews have appeared that cover the field in great detail.[34,39,50] Any image forming is, of course, information storage. The different types of polymerization are schematically depicted in Scheme 4.1. In dissymmetric cases, there is the possibility of head-to-head (translation) and of head-to-tail (centrosymmetric) orientations.

Examples are known for all possible cases. Frequently, chain polymers of high regularity are obtained upon crystal photolyses under lattice control. The reactive monomer moieties include cinnamic acids, cinnamate esters, cinnamides, cinnamonitriles, styrenes, stilbenes, vinyl(hetero)arenes, maleimides, thymines, coumarins, and others.

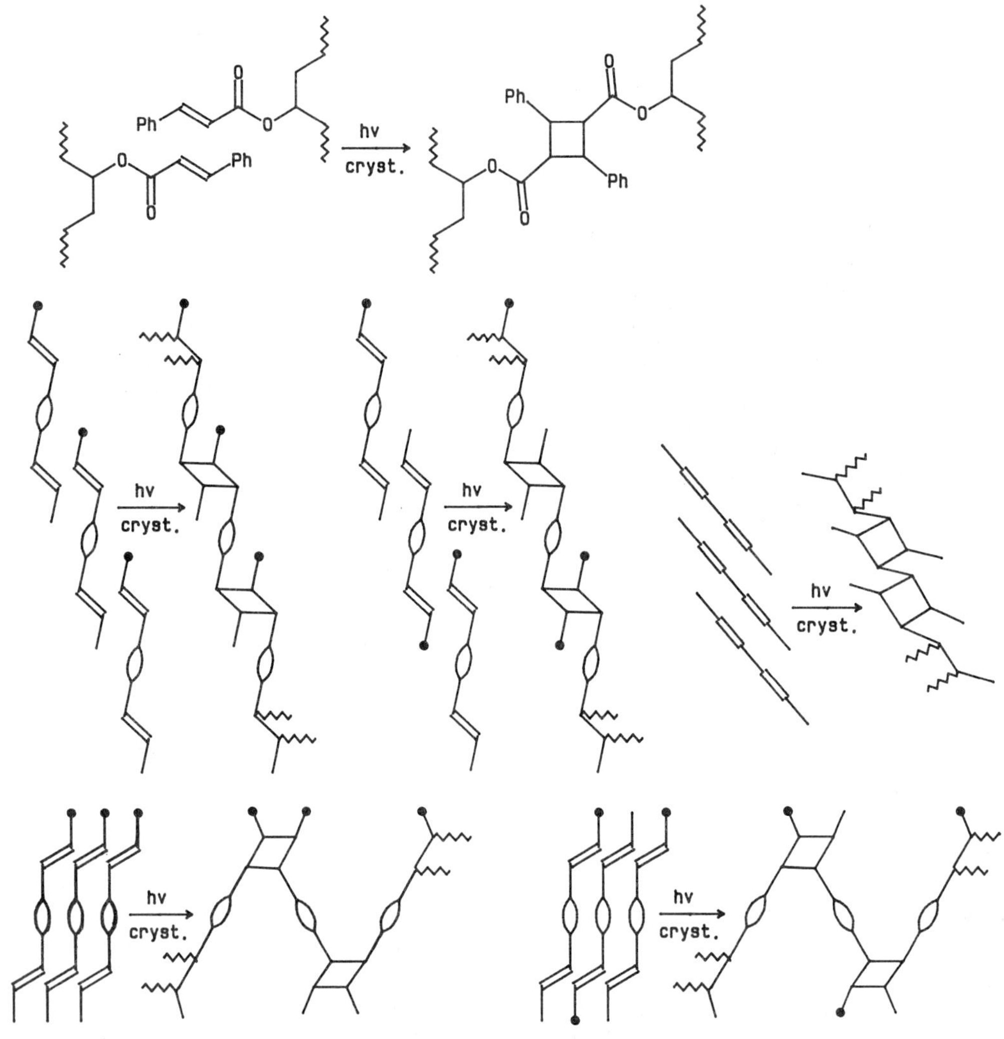

References

1. Kaupp, G., Dimerizations of alkenes; cycloadditions between unequal alkenes, *Houben Weyl Methoden der Organischen Chemie*, Thieme, Stuttgart, 1975, Vol. IV/5a, p. 278; 360.

2. Ramamurthy, V. and Venkatesan, K., Photochemical reactions of organic crystals, *Chem. Rev.*, 87, 433, 1987.

3. Schmidt, G. M. J., Photodimerization in the solid state, *Pure Appl. Chem.*, 27, 647, 1971.

3a. Kaupp, G., AFM and STM in photochemistry including photon tunneling, *Adv. Photochem.*, 19, 119, 1994.

3b. Kaupp, G. and Plagman, M., Atomic force microscopy and solid state photolyses: phase rebuilding, *J. Photochem. Photobiol. A: Chem.*, 80, 399, 1994.

4. Cohen, M. D., Excimers in crystals, *Mol. Cryst. Liq. Cryst.*, 50, 1, 1979; further concepts in Reference 2.

5. Ramamurthy, V., Organic photochemistry in organized media, *Tetrahedron*, 42, 5753, 1986; Moorthy, J. N., Venkatesan, K., and Weiss, R. G., Photodimerization of coumarins in solid cyclodextrin inclusion complexes, *J. Org. Chem.*, 57, 3292, 1992; Moorthy, J. N. and Venkatesan, K., *J. Mater. Chem.*, 2, 675, 1992.

6. Detailed discussion in Kearsley, S. K., The prediction of chemical reactivity within organic crystals using geometric criteria, *Studies in Organic Chemistry*, Vol. 32, Organic Solid State Chemistry, Desiraju, G. R., Ed., Elsevier, Amsterdam, 1987, 69.

7. Kaupp, G. and Zimmermann, I., First detection of a π-coupled 1,5-diradical *via* cycloaddition, *Angew. Chem. Int. Ed.*, 20, 1018, 1981; trimer 10 replaces the previously formulated spirooxepane.

8. Frey, H., Behmann, G., and Kaupp, G., Selectivity studies in solid-state photolyses of 2,5-*bis*(methylene)-cyclopentanones, *Chem. Ber.*, 120, 382, 1987.

8a. Kaupp, G., Atomic force microscopy in the photochemistry of chalcones, *J. Microsc.*, 17, 15, 1994.

9. Theocharis, C. R., Jones, W., Thomas, J. M., Motevalli, M., and Hursthouse, M. B., The solid-state photodimerization of 2,5-dibenzylidenecyclopentanone (DBCP): A topochemical reaction that yields an amorphous product, *J. Chem. Soc., Perkin Trans. II* 71, 1984; these authors used suspensions of crystals in CH_3OH/H_2O (10/90) and irradiated nonselectively.

10. Kaupp, G., Frey, H., and Behmann, G., Photoreactions in crystals with ethenes: Selectivities and crystal engineering, *Chem. Ber.*, 121, 2135, 1988; Kaupp, G., Gründken, E., and Matthies, D., Rearrangements and complex eliminations with 1,5-benzothiazepin-4-ones, *Chem. Ber.*, 119, 3109, 1986; the exact stereochemistry of 34k and 37′k has been established now.

11. Nakanishi, H., Jones, W., Thomas, J. M., Hursthouse, M. B., and Motevalli, M., Monitoring the crystallographic course of a single-crystal → single crystal photodimerization by X-ray diffractometry, *J. Chem. Soc. Chem. Commun.*, 611, 1980.

11a. Enkelmann, V. and Wegner, G., Single-crystal-to-single-crystal photodimerization of cinnamic acid, *J. Am. Chem. Soc.*, 115, 10390, 1993; *Mol. Cryst. Liq. Cryst.*, 240, 121, 1994, report 20 with P2$_1$/n space group; however, the samplings reported are conflicting in the two papers and the stable modification of photo-20 assumes Cc space group[12-14] (see Eanes, E. D. and Donnay, G., *Z. Kristauogr.*, 111, 368, 1959), a fact that has not been acknowledged.

12. Kaupp, G., Photodimerization of cinnamic acid in the solid state: New insights on application of atomic force microscopy, *Angew. Chem. Int. Ed.*, 31, 592, 1992.

13. Kaupp, G., Atomic force microscopy for the elucidation of photomechanisms in the organic solid state, *Proc. 14th IUPAC Symp. on Photochemistry*, Leuven, July 19–25, 1992, 258. We thank Prof. N. Karl and Dr. M. Möbus, University of Stuttgart, for the X-ray measurements.

14. Kaupp, G., Atomic force microscopy — Results with photodimerizations, *GIT Fachz. Lab.*, 93(284), 581, 1993.

15. Kaupp, G., Jostkleigrewe, E., and Hermann, H. J., Quantitative photodimerization of crystalline α-benzylidene-γ-butyrolactone, *Angew, Chem., Int. Ed.*, 21, 435, 1982.

16. Kearsley, S. K. and Desiraju, G. R., Determination of an organic crystal structure with the aid of topochemical and related considerations: Correlation of the molecular and crystal structure of α-benzylidene-γ-butyrolactone and 2-benzylidenecyclopentanone with their solid state photoreactivity, *Proc. Roy. Soc. London* A, 397, 157, 1985.

17. Vausant, J., Toppet, S., Smets, G., Declerq, J. P., Germain, G., and Van Meerssche, M., Azastilbenes. 2. Photodimerization, *J. Org. Chem.*, 45, 1565, 1980.

17a. Saigo, K. and Hasegawa, M., Determining factors of molecular arrangement and reaction course in the crystalline-state photoreaction of unsymmetrically substituted diolefins, in *Reactivity in Molecular Crystals*, Ohashi, Y., Ed., Kodansha, Tokyo/VCH, Weinheim, 1994, 203.

18. Hasegawa, M. and Hashimoto, Y., Kaleidoscopic photochemical behavior of unsymmetric diolefin crystals, *Mol. Cryst. Liq. Cryst.*, 219, 1, 1992; Kaupp, G. and Haak, M., Absolute asymmetric synthesis by irradiation of chiral crystals, *Angew Chem. Int. Ed.*, 32, 694, 1993.

19. Addadi, L., Mil, J. V., and Lahav, M., Engineering of chiral crystals for asymmetric [2+2]photopolymerization. Execution of an "absolute" asymmetric synthesis with quantitative enantiomeric yield, *J. Am. Chem. Soc.*, 104, 3422, 1982.

20. Hasegawa, M., Kato, S., Saigo, K., Wilson, S. R., Stern, C. L., and Paul, I. C., Topochemical photoreaction of unsymmetrically substituted diolefins. I. Topochemical [2+2]photopolymerization of *n*-propyl-α-cyano-4-[2-(4-pyridyl)ethenyl]cinnamate, *J. Photochem. Photobiol. A,* 41, 385, 1988.

21. Rieke, R. D., Page, G. O., Hudnale, P. M., Arhart, R. W., and Bouldin, T. W., A potential photochemically driven information storage system, *J. Chem. Soc. Chem. Commun.,* 38, 1990.

22. Desvergne, J. P. and Courseille, C., Topochemistry in crystalline imperfections, *Mol. Cryst. Liq. Cryst.,* 50, 59, 1979.

23. Nalini, V. and Desiraju, G. R., Crystal engineering through nonbonded contacts to sulphur. Structure and solid state photoreactivity of 4-(4′-chloro)-phenyl-Δ-4-thiazolene-2-thione, *Tetrahedron,* 43, 1313, 1987.

24. Chimichi, S., Ciciani, G., DalPiaz, V., DeSio, F., Sarti-Fantoni, P., and Torroba, T., Solid state photodimerization reaction of some 3-styrylisoxazolo[3,4-d]pyridazin-7(6*H*)-ones, *Heterocycles,* 24, 3467, 1986.

25. Baracchi, A., Chimichi, S., DeSio, F., Polo, C., Sarti-Fantoni, P., and Torroba, T., Solid state photodimerization of *(E)*-1-(3-methyl-4-nitroisoxazol-5-yl)-2-(2-thienyl)ethene, *Heterocycles,* 29, 2023, 1989.

26. Kreichberga, J., Liepius, E., Mazcika, I., and Neilands, O., Photodimerization of crystalline esters of tetrathiafulvalene-carboxylic acids, *Zh. Org. Khim.,* 22, 416, 1986; *Chem. Abstr.,* 106, 102123, 1987.

26a. Tateba, H., Morita, K., and Tada, M., Photochemical reaction of nootkatone under various conditions, *J. Chem. Res. Synop.,* 140, 1992; Reisch, J., Ekiz-Guecer, N., and Takacs, M., *Z. Naturforsch. B,* 44, 1102, 1989.

27. Hesse, K. and Hünig, S., Multistep reversible redox systems, LXII. [2+2]Photocycloadditions of α-styrylpyrylium salts to 4,4′-(1,3-cyclobutanediyl)*bis*(pyrylium)salts and their thermal and base-induced cycloreversions, *Liebigs Ann. Chem.,* 715, 1985.

28. Quina, F. H. and Whitten, D. G., Photochemical reactions in organized monolayer assemblies. 4. Photodimerization, photoisomerization, and excimer formation with surfactant olefines and dienes in monolayer assemblies, crystals, and micelles, *J. Am. Chem. Soc.,* 99, 877, 1977.

29. Bolt, J., Quina, F. H., and Whitten, D. G., Solid state photodimerization of surfactant esters of cinnamic acid, *Tetrahedron Lett.,* 30, 2595, 1976.

30. Egerton, P. L. Hyde, E. M., Trigg, J., Payne, A., Beynon, P. Mijovic, M. V., and Reiser, A., Photocycloaddition in liquid ethyl cinnamate and in ethyl cinnamate glasses. The photoreaction as a probe into the micromorphology of the solid, *J. Am. Chem. Soc.,* 103, 3859, 1981.

31. Rachinsky, A. G. and Razumov, V. F., Photochemical reactions in amorphous 1,2-di-(1-naphthyl)ethylene, *Proc. XIIIth IUPAC Symp. on Photochemistry,* Warwick, July 22–28, 1990, 164; Aldoskin, S. M., Alfimov, M. V., Atovmyan, L. O., Kaminsky, V. F., Ruzumov, V. F., and Rachinsky, A. G., Concomitant photochemical and phase rearrangements 2. Luminescent and X-ray studies on photochemistry of *cis-* and *trans-*1,2-di-(1-naphthyl)ethylenes in the crystalline state, *Mol. Cryst. Liq. Cryst.,* 108, 1, 1984.

32. Wang, S. Y., The mechanism for frozen aqueous solution irradiation of pyrimidines, *Photochem. Photobiol.,* 3, 395, 1964; Huber, C. P., Birnbaum, G. I., Post, M. L., Kulikowska, E., Gajewska, L., and Shugar, D., Structure and properties of a *trans-anti* photodimer of 5-methylorotate, *Can. J. Chem.,* 56, 824, 1978; Cheng, P. T., Hornby, V., Wong-Ng, W., Nyburg, S. C., and Weinblum, D., A photodimer of 1,2,3,6-tetrahydro-5-methyl-2,6-dioxo-4-pyrimidinecarboxylic acid (5-methylorotic acid); crystal structure of the trihydrated barium salt, *Acta Cryst.,* B32, 2251, 1976.

33. Land, E. J., Rushton, F. A. P., Beddoes, R. L., Bruce, J. M., Cernick, R. J., Dawson, S. C., and Mills, O. S., A [2+2]photo-adduct of 8-methoxypsoralen and thymine: X-ray crystal structure; a model for the reaction of psoralens with DNA in the phototherapy of psoriasis, *J. Chem. Soc. Chem. Commun.,* 22, 1982.

34. Creed, D., Griffin, A. C., Hoyle, C. E., and Venkataram, K., Chromophore aggregation and concomitant wavelength-dependent photochemistry of a main-chain liquid crystalline poly(aryl

cinnamate), *J. Am. Chem. Soc.*, 112, 4049, 1990; Lin, A. A., Chu, C. F., Huang, W. Y., and Reiser, A., Reactant preordering in solid photopolymers, *Proc. 14th IUPAC Symp. on Photochemistry,* Leuven, July 19–25, 1992, 23.

35. Vioturiez, L., Ulrich, J., Gaboriau, F., Viari, A., Vigny, P., and Cadet, J., Identification of the two *cis-syn* [2+2]cycloadducts resulting from the photoreaction of 3-carbethoxypsoralen with 2′-deoxycytidine and 2′-deoxyuridine, *Int. J. Radiat. Biol.*, 57, 903, 1990.

36. Meng, J., Zhu, Z., Wang, R., Yao, X., Ito, Y., Ihara, H., and Matsuura, T., Selectivity in the solid-state photoreaction of 6-cyanouracils with aromatic hydrocarbons, *Chem. Lett.*, 1247, 1990.

37. Theocharis, C. R., Desiraju, G. R., and Jones, W., The use of mixed crystals for engineering organic solid-state reactions: Application to benzylbenzylidenecyclopentanones, *J. Am. Chem. Soc.*, 106, 3606, 1984.

38. Green, B. S. and Heller, L., Solution and solid-state photodimerization of some styrylthiophenes, *J. Org. Chem.*, 39, 196, 1974.

39. Hasegawa, M. and Hashimoto, Y., Kaleidoscopic photochemical behavior of unsymmetric diolefin crystals, *Mol. Cryst. Liq. Cryst.*, 219, 1, 1992.

40. Meng, J., Fu, D., Yao, X.-K., Wang, R.-J., and Matsuura, T., Solid-state photodimerization of coumarin in the presence of a crystal lattice-controlling substance, *Tetrahedron*, 45, 6979, 1989.

41. Lewis, F. D. and Oxman, J. D., Photodimerization of Lewis acid complexes of cinnamate esters in solution and the solid state, *J. Am. Chem. Soc.*, 106, 466, 1984, and references cited therein.

42. Fujiwara, T., Nanba, N., Hamada, K., Toda, F., and Tanaka, K., Enantioselective photoreactions of cycloocta-2,4,6-trien-1-one and cycloocta-2,4-dien-1-one in their inclusion complexes with *(R,R)*-(−)-1,6-*bis*(*o*-chlorophenyl)-1,6-diphenylhexa-2,4-diyne-1,6-diol: Mechanistic study based on X-ray crystal structure analyses, *J. Org. Chem.*, 55, 4532, 1990; for further dimerizations in clathrates see Fujiwara, T., Stereoselectivity in reactions of clathrate crystals, in *Reactivity in Molecular Crystals*, Ohashi, Y., Ed., Kodansha, Tokyo/VCH, Weinheim, 1994, 236; and Tanaka, K., Toda, F., Ueda, Y., and Fukiwara, T., Regio- and enantio-selective photoreactions of cyclohex-2-enone, coumarin and acrylanilide as inclusion complexes with optically active host compounds, *Mol. Cryst. Liq. Cryst.*, 248, 43, 1994.

43. Kaftory, M., Tanaka, K., and Toda, F., Reactions in the solid state. 2. The crystal structures of the inclusion complexes of 1,1,6,6-tetraphenylhexa-2,4-diyne-1,6-diol with benzylideneacetophenone and 2,5-diphenylhydroquinone with dibenzylideneacetone, *J. Org. Chem.*, 50, 2154, 1985.

44. deMayo, P., Superficial photochemistry, *Pure Appl. Chem.*, 54, 1623, 1982.

45. Farwaha, R., deMayo, P., Schauble, J. H., and Toong, Y. C., Surface photochemistry: Enone photocycloaddition by adsorbed molecules on silica gel and alumina, *J. Org. Chem.*, 50, 245, 1985; Dave, V., Farwaha, R., deMayo, P., and Stothers, J. B., Surface photochemistry: Silica gel as a medium for photochemical cycloaddition of enones to allene, *Can. J. Chem.*, 63, 2401, 1985.

46. Donati, D., Fiorenza, M., and Sarti-Fantoni, P. J., Photochemical reactions of 3-methyl-4-nitro-5-styrylisoxazole in solution, in the solid state and adsorbed on silica gel, *J. Heterocycl. Chem.*, 16, 253, 1979.

47. Ramamurthy, V., Corbin, D. R., Turro, N., Zhang, Z., and Garcia-Garibay, M. A., A comparison between zeolite-solvent slurry and dry solid photolyses, *J. Org. Chem.*, 56, 255, 1991.

48. Takagi, K., Usami, H., Fukaya, H., and Sawaki, Y., Spatially controlled photocycloaddition of a clay-intercalated stilbazolium cation, *J. Chem. Soc. Chem. Commun.*, 1174, 1989; Takagi, K., Usami, H., Shiichi, T., and Sawaki, Y., Photodimerization of olefins on clay minerals, *Mol. Cryst. Liq. Cryst.*, 218, 109, 1992.

49. Valim J., Kariuki, B. M., King, J., and Jones, W., Photoreactivity of cinnamate intercalates of layered double hydroxides, *Mol. Cryst. Liq. Cryst.*, 211, 271, 1992.

50. Dilling, W. L., Polymerization of unsaturated compounds by photocycloaddition reactions, *Chem. Rev.*, 83, 1, 1983; Pappas, S. P., Photocrosslinking, *Comprehensive Polymer Science*, Vol. 6, 1989, chap. 5, 135; Turner, S. R. and Doley, R. C., Photochemical and radiation sensitive resists, *Comprehensive Polymer Science*, Vol. 6, 1989, chap. 7, 193; and references cited therein.

Photochemical Synthesis of Cage Compounds: Propellaprismanes and Their Precursors

Rolf Gleiter
Organisch-Chemisches Institut der Universität Heidelberg

Björn Treptow
Organisch-Chemisches Institut der Universität Heidelberg

5.1 Introduction

Cage compounds have fascinated chemists for many years.[1] Interest in these molecules ranges from purely esthetic reasons to their structural and chemical peculiarities. They represent ideal models for studying intramolecular interactions and intramolecular reactions because the geometry and the structural parameters have been established.[1] The term "cage compound" defies an exact definition and includes simple molecules such as bicyclo[m.n.o]alkanes as well as more complex structures like C_{60}. For obvious reasons which will be explained in the next section; this chapter will deal exclusively with prismane derivatives and their respective precursors. Scheme 5.1 illustrates the syntheses of a number of homologous prismanes in which photochemical key steps are involved.

The successful synthesis of [3]prismane 4(prismane) by Katz,[2] its hexamethyl derivative 7 almost synchronously by a German[3a] and an American group,[3b] [4]prismane 11 (cubane) and [5]prismane 17, both by Eaton,[4,5a] the latter also by Dauben,[5b] are milestones in the chemistry of strained cage compounds. The next higher congener, hexaprismane, is not known yet; only secohexaprismane[6] and *bis*homohexaprismane[7] have been obtained to date.

Our interest in prismanes was aroused by the synthesis of compound 18 (Scheme 5.2),[8] which consists of a cubane core enclosed in a cage of four trimethylene chains. For practical reasons (IUPAC: nonacyclo[10.8.0.0^{1,17}.0^{2,6}.0^{2,11}.0^{6,17}.0^{7,11}.0^{7,16}.0^{12,16}]eicosane) this compound was named propella[3$_4$]prismane because it not only possesses the structural features of a propellane (the bridges), but also those of a prismane (the cubane). According to cyclophane nomenclature, the

0-8493-8634-9/95/$0.00+$.50

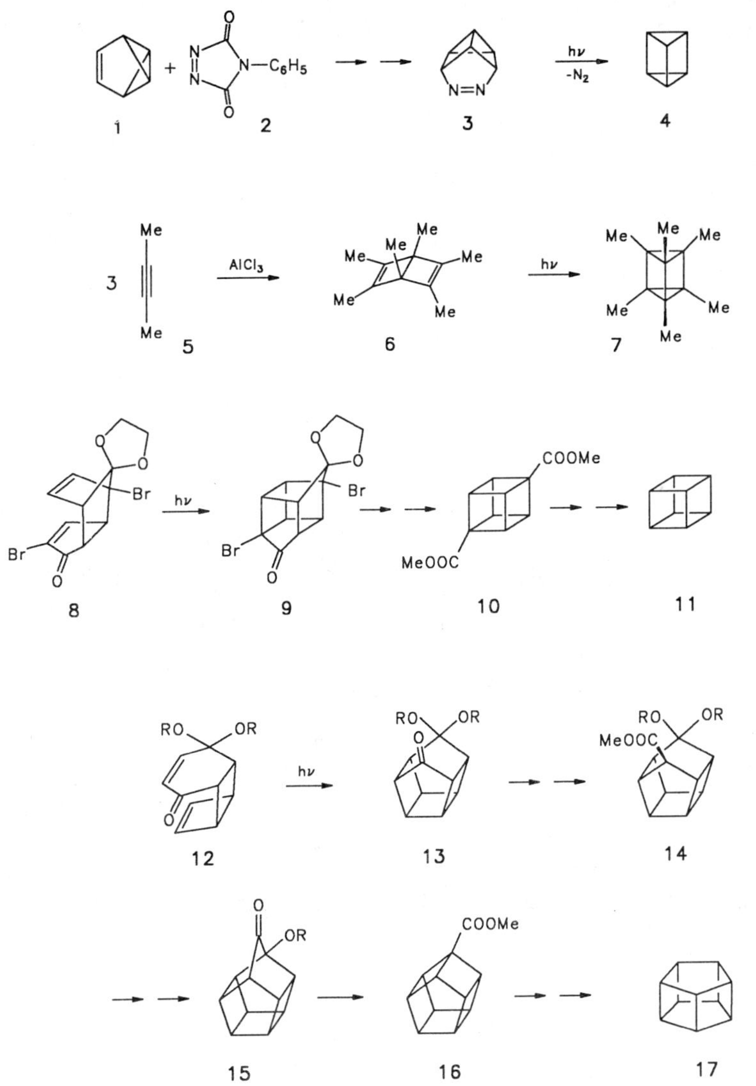

SCHEME 5.1

capital number 3 indicates the length of the bridges, while subscript 4 designates the number of the bridges (being identical with the order of the principle axis of the prismane moiety).

The exchange of a propellane's one-dimensional C-C-axis for the three-dimensional body of a prismane has structural implications in so far as the arrangement of the bridges leads to several conceivable isomers. If only adjacent centers of an [m]prismane are connected by m equivalent chains with chainlength n, the isomeric propella[n_m]prismanes 19 to 25 are obtained for the series of [3]- to [5]prismanes (they can be distinguished by their symmetry with the exception of 24 and 25). As 18 represented the first and only member of the family of propella[n_m]prismanes,[8] our interest was focused on the search for an access to further congeners of this series and particularly at the homologization of the central prismane body. Because the chemistry of [3]prismanes (from now on, simply denoted as prismanes) is more extensively studied than that of [5]prismanes, experiments were aimed at the synthesis of 19.

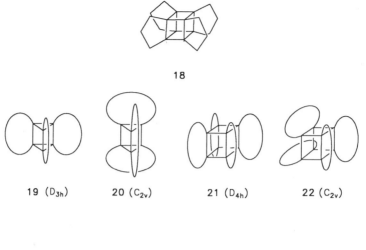

SCHEME 5.2

Since the photochemical generation of prismanes is closely related to the light-induced valence isomerization of benzene, a brief review of some facts of the photochemistry of benzene is given.[9] Irradiation of alkylbenzenes in solution at wavelengths around 254 nm usually results in a positional isomerization of the alkyl groups.[10] To rationalize this phenomenon, it is assumed that valence

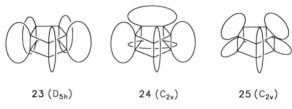

SCHEME 5.3

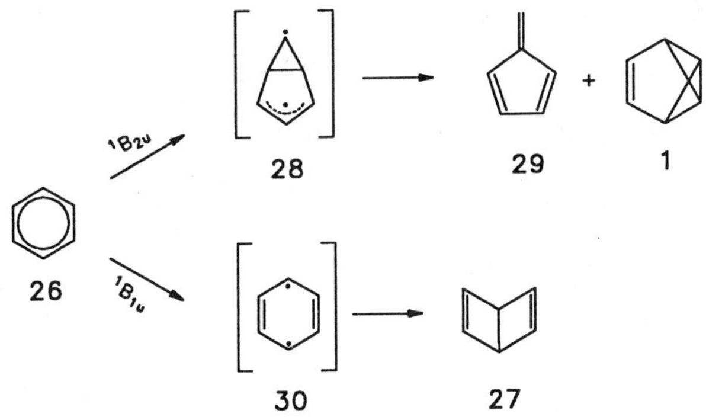

SCHEME 5.4

isomers of benzene, such as benzvalene **1**, Dewar benzene **27**, and prismane **4** are short-lived intermediates. The involvement of such isomers has been inferred from isotopic labeling experiments.[11] Possible ring transposition processes that give rise to apparent 1,2- and 1,3-shifts are shown schematically in Scheme 5.3.[12] It is generally assumed that excitation of the $^1B_{2u}$ state (S_1 state in benzene, 254-nm band) may produce the biradical **28** denoted "prefulvene" as a short-lived intermediate, which may pass over to benzvalene **1** and fulvene **29** (see Scheme 5.4).

Excitation of the $^1B_{1u}$ state (S_2 state in benzene, 204-nm band) is thought to generate the biradical **30** which again may pass over to Dewar benzene **27**.[12] The assumed mechanism is further supported by the observation that bulky substituents such as tertiary butyl groups allow the isolation of the postulated valence isomers such as **32**, **33**, **34**, and **36** as depicted in Scheme 5.5.[13]

To reach the synthetic goal, two independent routes were persued; the first involving a bridged Dewar benzene (Route 1 in Scheme 5.6) and the second with bridged bicyclopropenyl (Route 2 in Scheme 5.6) as key intermediates. Here, the *Dewar* benzene route is reported in detail because this route is further developed[14] than the alternative.[15]

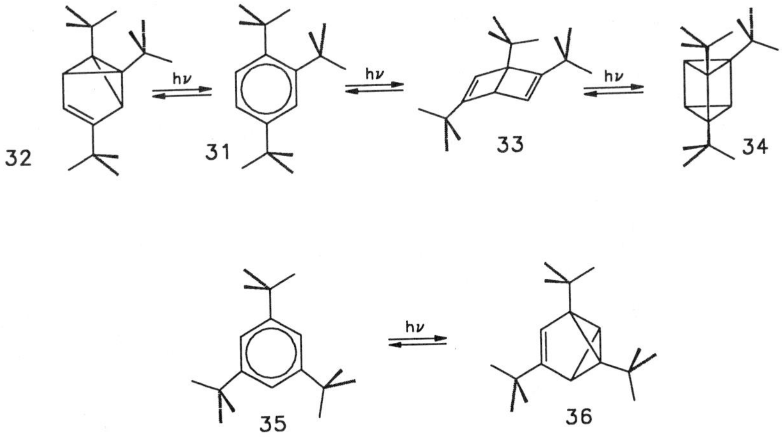

SCHEME 5.5

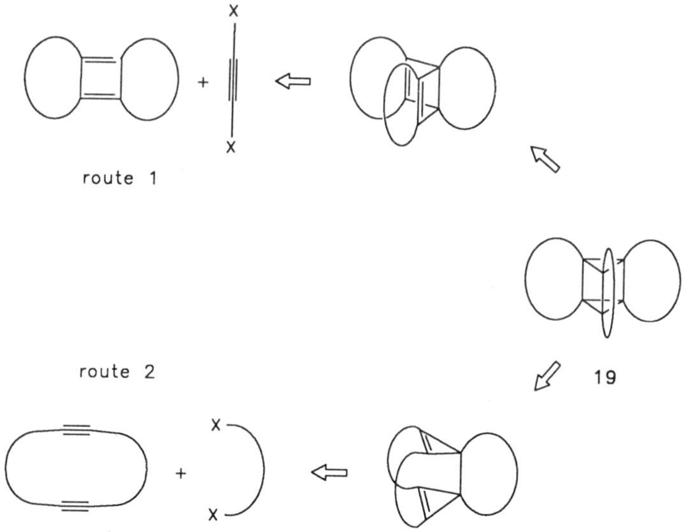

SCHEME 5.6

5.2 Prismanes Bridged by One and Two Chains

Prismanes with a Single Chain

Prismanes with one chain have been synthesized by Tobe et al.[16] and by Tochtermann et al.[17] The approaches of both groups adopted two different multistep syntheses of the corresponding [6]paracyclophanes 40 and 47, as shown in Schemes 5.7 and 5.8. They play a central role in the following photochemical sequences to be discussed in more detail.

Irradiation of Tobe's [6]paracylophane 40 through a Pyrex filter led to rapid formation of Dewar benzene 39 followed by a slow conversion of this to prismane 41, which in turn is converted to the rearranged aromatic 42 as the final product.[16]

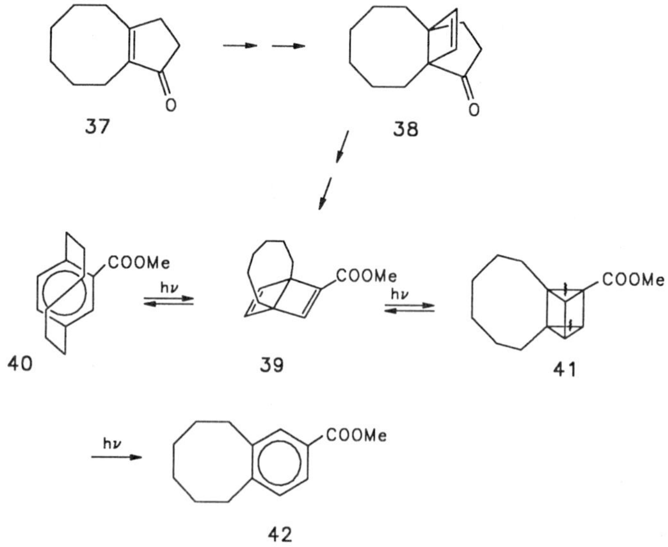

SCHEME 5.7

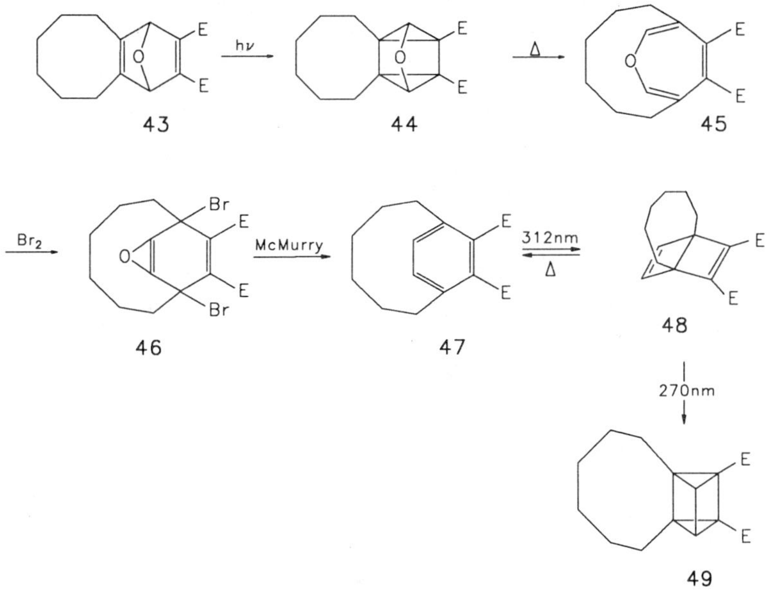

SCHEME 5.8

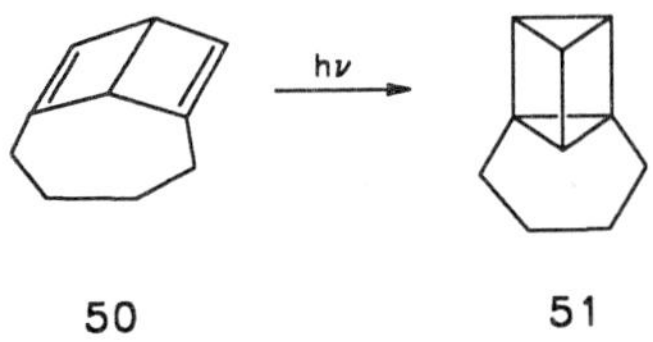

SCHEME 5.9

In the case of Tochtermann's compound, careful examination of the photochemical behavior of 47 revealed that product formation can be controlled by variation of the excitation wavelength. Irradiation in the long-wavelength band with $\lambda > 312$ nm yielded the 1,4-bridged Dewar benzene 48, quantitatively[18] while irradiation in the short wavelength band with $\lambda < 270$ nm afforded prismane 49 in 56% yield after work-up and only traces of 48.[17] In both prismanes 41 and 49, the chain spans a cyclobutane edge colinear to the principal axis of the prismane body. In 1987, Bickelhaupt et al. synthesized an isomeric prismane 51 (Scheme 5.9) with a tetramethylene chain bridging the edge of a cyclopropane unit.[19]

Prismanes with Two Chains

For the synthesis of twofold bridged Dewar benzenes, a route discovered in 1971[20] and extended widely since then was adopted.[21,22] 1,8-Cyclotetradecadiyne 52[23] and cyclooctyne 56[24] serve as starting materials. Reaction with $AlCl_3$ in CH_2Cl_2 yields the cyclobutadiene $AlCl_3$ σ-complexes 53b and 53c, respectively. Addition of dimethyl acetylenedicarboxylate (DMAD) and dimethyl sulfoxide (DMSO) results in the formation of the isomeric twofold bridged Dewar benzenes 54b, 55b and 54c, 55c (Scheme 5.10).[14] This strategy possesses the advantage that the Dewar benzenes are obtained in only one step from easily accessible educts. The only restriction of this route is that it cannot be extended to Dewar benzenes of type 54 with bridges containing less than five methylene units.

The Dewar benzenes with bridges spanning the 1,4- and 2,3-positions can be considered as [m.n.2]propellanes (54b: m = 7, n = 5; 54c: m = 8, n = 6). They are suitable model compounds for

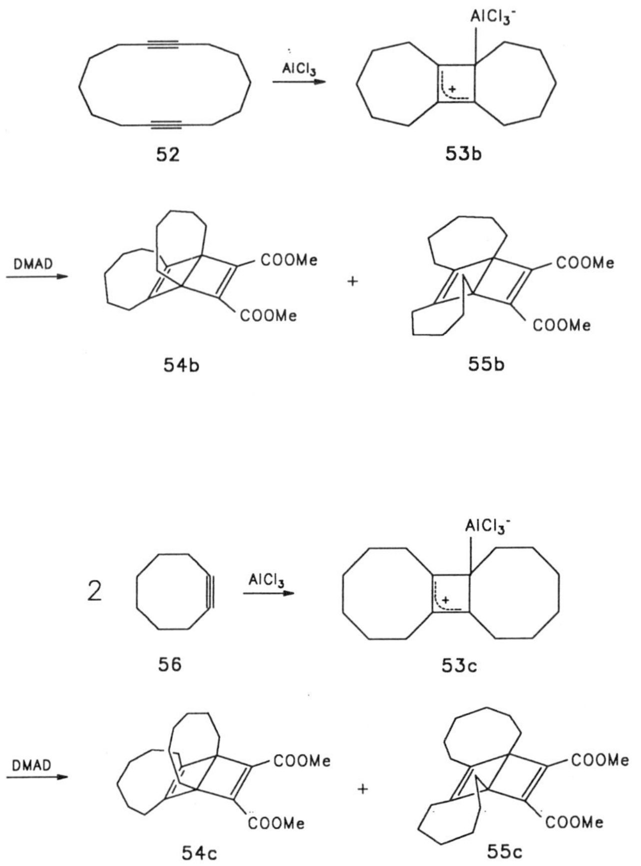

SCHEME 5.10

studying the effects that influence the generation of prismanes. The photochemical experiments with Dewar benzene **54c** reveal a photochemical equilibrium with [6]paracyclophane **57c**, which is independent of the wavelength used in the range 254 to 320 nm. The ratio of products obtained is 20:80 in favor of **57c**. Furthermore, there is a second photoequilibrium between **54c** and prismane **58c**. This one depends strongly on the applied wavelength. At λ = 280 nm, the prismane portion in the mixture is 2%; at shorter wavelengths no prismane **58c** is observed; while at λ > 320 nm, its concentration increases almost linearly with irradiation time, affording 30% **58c** after 12 h. These facts are summarized in Scheme 5.11.

To account for this behavior, it is supposed that the process of prismane formation is less efficient than the process of its decomposition. At wavelengths shorter than 280 nm, any generated prismane undergoes a rapid [2+2]-cycloreversion to Dewar benzene **54c**. Thus, the prismane equilibrium concentration is immeasurably low. Above 320 nm, prismane **58c** shows no noticeable UV absorption and thus does not participate in the photochemical equilibrium; hence, the increase of the compound in the mixture. The irradiation experiment at λ = 280 nm yields two further

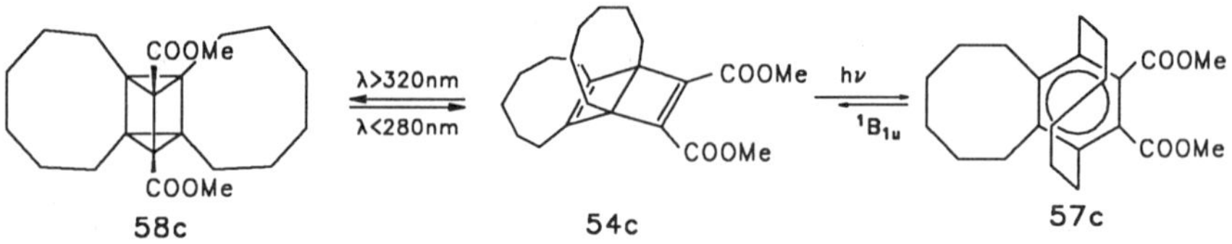

SCHEME 5.11

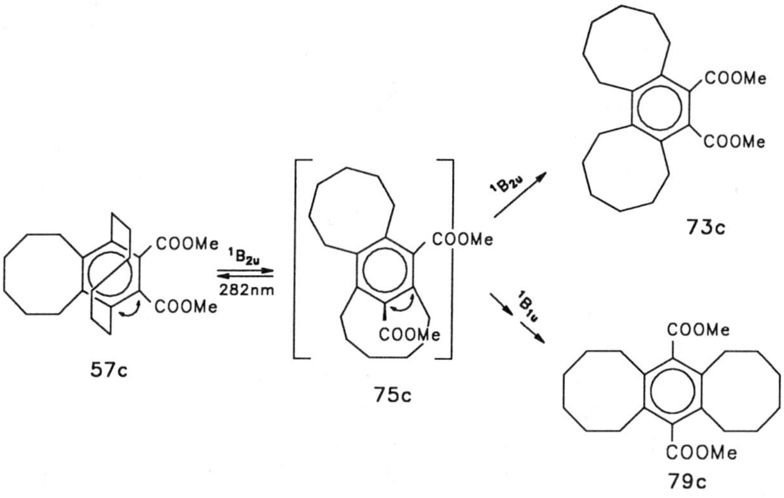

SCHEME 5.12

photoproducts: the aromatic species **73c** and **79c** in 6 to 12% yield after 12 h. These compounds are derived from [6]paracyclophane **57c** (Scheme 5.12). In this case, ring transposition via a [6]metacyclophane intermediate **75c** must have taken place. A possible mechanism for this rearrangement will be discussed later in connection with the photochemistry of Dewar benzenes **55b** and **55c**.

Irradiation of **54b**, the Dewar benzene with pentamethylene bridges, in diethyl ether with $\lambda > 320$ nm affords the expected prismane **58b** (Scheme 5.13). After 48 h, yields obtained are 15% **58b**, 30% **54b**, and 25% of an unexpected product with the structure **59**. The relatively low yield of **58b** — by comparison to its higher homologue **58c** — is probably due to the fact that **54b** is also transformed into the unstable [5]paracyclophane **57b**, which is removed from the equilibrium partly by reaction with the solvent (ether) involving cleavage of the *para*-bridge to give **59** and partly by decomposition. A species related to **54b** is Dewar benzene **60**. Bickelhaupt et al. have shown that upon photolysis at low temperature, there is indeed an equilibrium between Dewar benzene **60** and [5]paracyclophane **61**[22,25] (Scheme 5.14). Formation of prismane **62** was not reported, but this was not unexpected due to the short wavelengths used.

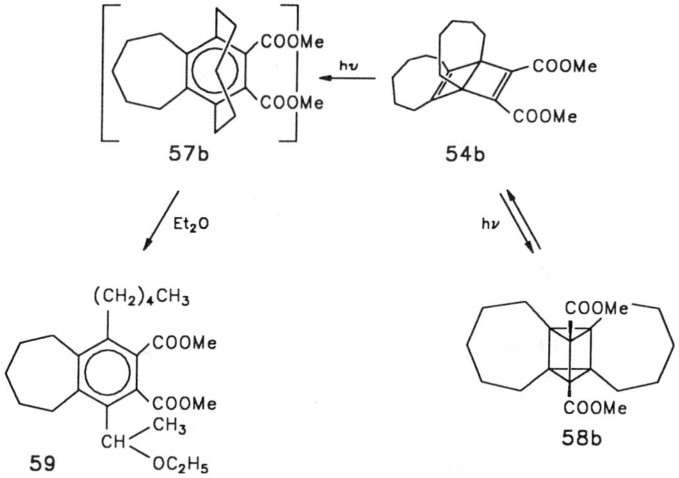

SCHEME 5.13

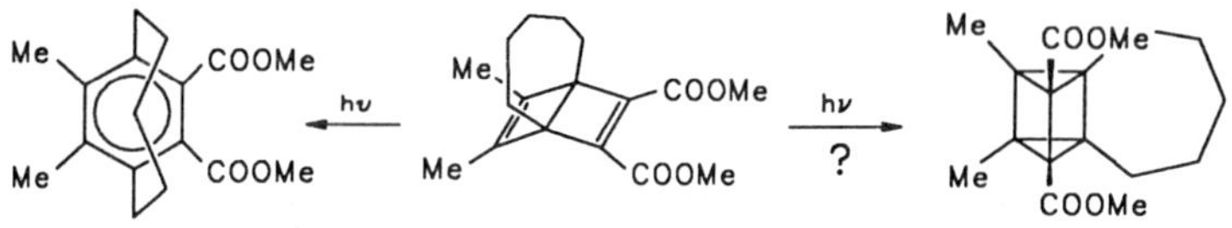

SCHEME 5.14

The prismanes 58b and 58c represent precursors to propella[n$_3$]prismanes 19 (chainlength n = 5,6) with D$_{3h}$ symmetry, where two of three bridges are already preformed and the third bridge may be constructed from the two adjacent functionalities.

There are two possible explanations to account for the formation of these prismanes. The first invokes a steric factor, caused by the short 1,4-bridges in Dewar benzenes 54b and 54c, which facilitates formation of the prismane for conformational reasons. The second can be denoted as an electronic factor. The ester carbonyl groups are in conjugation with the intervening double bond formally forming a triene system. The excitation of this triene unit (low energy n → π* transition and/or π → π* transition) may be more effective than that of an isolated double bond (high-energy π → π* transition). To judge the relative importance of the steric and the electronic effects, one of them has to be removed. This is achieved easily by conversion of the ester functions into hydroxymethyl groups by the sequence shown in Scheme 5.15.

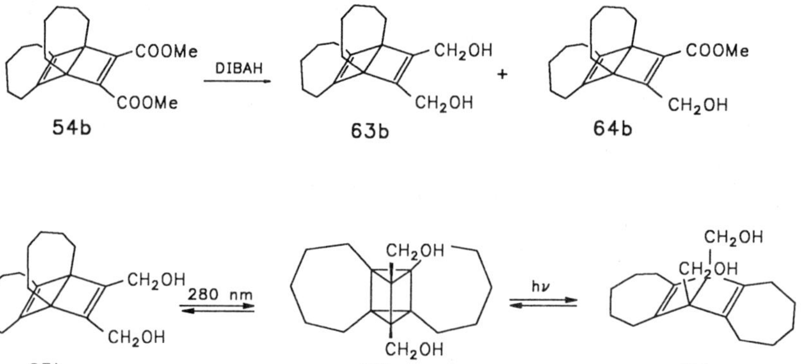

SCHEME 5.15

Reduction of diester 54b by the "ate"-complex, composed of butyllithium and diisobutyl aluminiumhydride, yields a mixture of the mono- and dialcohols 63b and 64b[49]. Photolysis (for 3h) of bis(hydroxymethylene) Dewar benzene 63b at λ = 280 nm affords the corresponding prismane 65b in an appreciable yield of 30%. If the irradiation time is extended, the yield of prismane diminishes in favor of the rearranged Dewar benzene 66b, where both bridges span the double bonds. This result leads to the conclusion that the steric factor is the main cause for prismane formation and electronic effects are small.

Photochemistry of Dewar-Phthalic Esters: An Unexpected Phthalic/Terephthalic Ester Rearrangement

The sequence depicted in Scheme 5.10 affords Dewar benzenes 55b and 55c, with the bridges spanning the 1,2- and 3,4-position as side products. They are potential precursors to propella[n$_3$]prismanes 20 (n = 5,6) with C$_{2v}$ symmetry (Scheme 5.2). However, as has been seen in the preceding section, a steric effect seems to be necessary for prismane formation. This contribution

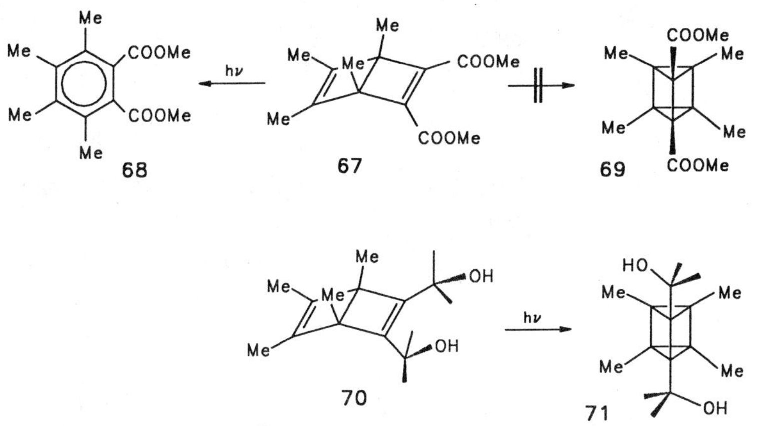

SCHEME 5.16

is obviously not present in **55**. Furthermore, a related compound, tetramethyl Dewar-phthalic ester **67**, could only be transformed into **68** and prismane **69** was not found.[26] After imposing a steric element as in **70**, the conversion to prismane **71** was successful (Scheme 5.16).[27]

Thus the expectations of observing interesting photochemistry with **55** were not very high. Indeed, irradiation of **55b** and **55c** as well as the lower homologue **55a** above 295 nm affords rapid and exclusive conversion to the phthalic esters **73a**, **73b**, and **73c**, respectively. No trace of the prismanes **72** can be detected, nor is there any indication of a photoequilibrium between **55** and **73**. Photolysis by unfiltered light from a high-pressure mercury lamp (254 nm), on the other hand, leads to an unexpected result: the phthalic esters **73b** and **73c** are transformed into their terephthalic isomers **79b** and **79c** (Scheme 5.17).[28] In the case of **73a**, no such transformation is observed. This unexpected outcome was an impetus for closer investigation. Irradiation of **73b** at 282 nm for 48 h yields, as well as **79b**, two further compounds (Scheme 5.18) prismane **77b** and Dewar benzene **78b**. These are consecutive precursors of **79b**. Selective [2+2]-cycloreversion of prismane **77b**, as indicated in Scheme 5.18, leads to Dewar benzene **78b**, which in turn isomerizes to **79b** by electrocyclic cleavage of the central 1,4-bond. This can be supported by irradiation of each compound at 282 nm.

It is notable that there is no direct path between prismane **77b** and phthalic ester **73b**. By comparison to the valence isomeric prismane **72b** (Scheme 5.17), one of the pentamethylene bridges in **77b** is displaced by one cyclopropane edge, and the symmetry is reduced to C_2. For that

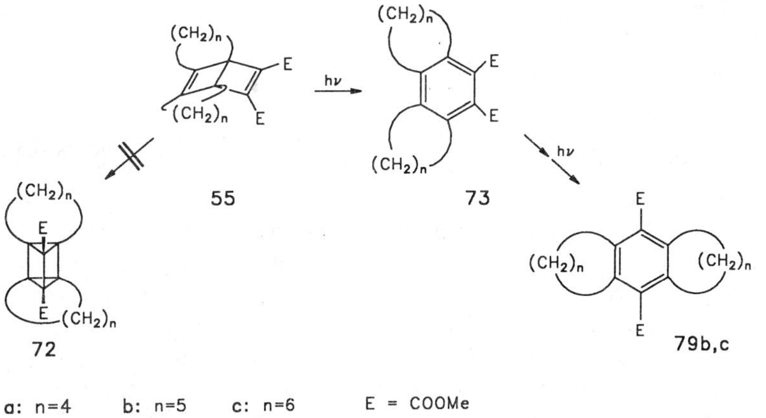

a: n=4   b: n=5   c: n=6   E = COOMe

SCHEME 5.17

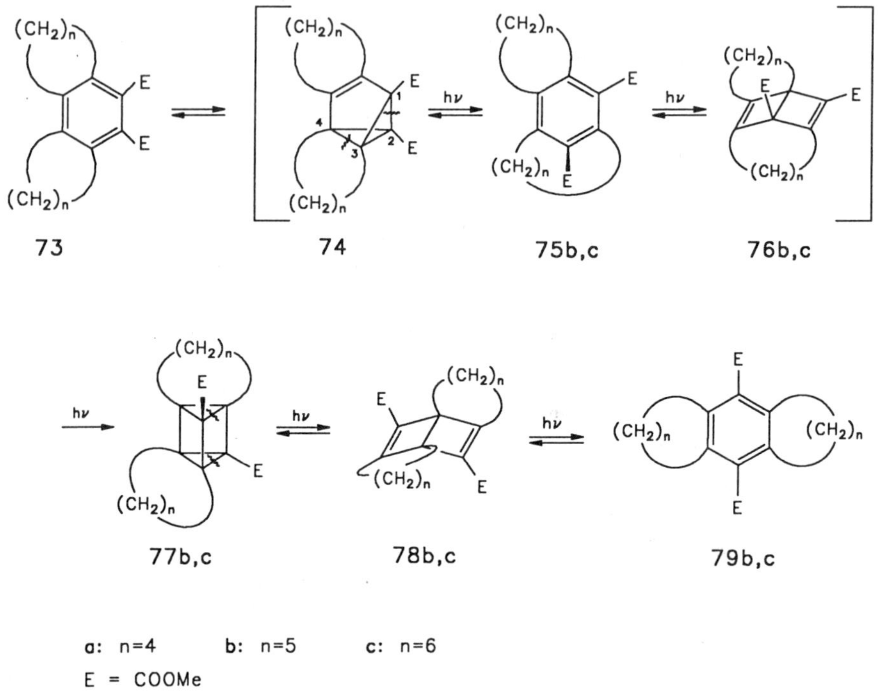

73   74   75b,c   76b,c

77b,c   78b,c   79b,c

a: n=4   b: n=5   c: n=6

E = COOMe

SCHEME 5.18

reason, the three further compounds (**74b, 75b,** and **76b**) are postulated as links. So far, none of them has been characterized due to both their low equilibrium concentration and their increased reactivity. The logic of this sequence can be rationalized as follows: the photochemical generation of benzvalenes is a very common reaction; they arise from the $^1B_{2u}$ state of benzenes (Scheme 5.4). Benzvalene **74b** possesses two possibilities for rearomatization. The trivial one is the back reaction; the alternative consists of a cleavage of the bonds between C1-C2 and C3-C4, resulting in [5]metacyclophane **75b**. Formally, an exchange between the substituents of the bridgehead of the bicyclo[1.1.0]butane moiety has taken place (see also Scheme 5.3, path a), resulting in a 1,2-shift of an ester group and the terminus of a pentamethylene bridge. The reverse sequence is usually invoked to explain the transformation of [5]metacyclophanes to 1,2-bridged arenes, although benzvalene intermediates have not been observed directly.[29] Prismane **77b** and metacyclophane **75b** are direct valence isomers connected via Dewar benzene **76b**. A schematic representation of the rearrangement sequence **75b** → **79b** was already introduced in Scheme 5.3 path b.

It is noteworthy that **74b** is the only reasonable benzvalene arising from **73b** that may yield sensible rearranged products. In Scheme 5.19, a compilation of all other possible benzvalene derivatives is given. It is obvious that **80b** and **81b** always lead back to **73b**. Bond cleavage as indicated in **80b** results in an interchange of the two ester groups and in **81b** in an interchange of two termini of a pentamethylene bridge, so **80b** and **81b** are not observable. Only **82b** might afford a highly strained system **83b** whose existence under the given conditions is very questionable.

In the sequence shown in Scheme 5.18, two interesting features will be regarded more thoroughly. First, in Section 5.2 (Prismanes with Two Chains), the necessity of a steric factor for prismane formation was postulated. Such a factor is present in metacyclophane **75b**, which is extremely strained as a result of the short 1,3-bridge. This strain is removed in the photochemically allowed sequence **75b** → **76b** → **77b**. Note that the geometry of both Dewar benzene **76b** and prismane **77b** is already inherent in the assumed unsymmetrical boat conformation of **75b**.[30] Two experiments support this line of argument: as already mentioned, the lower homologue phthalic

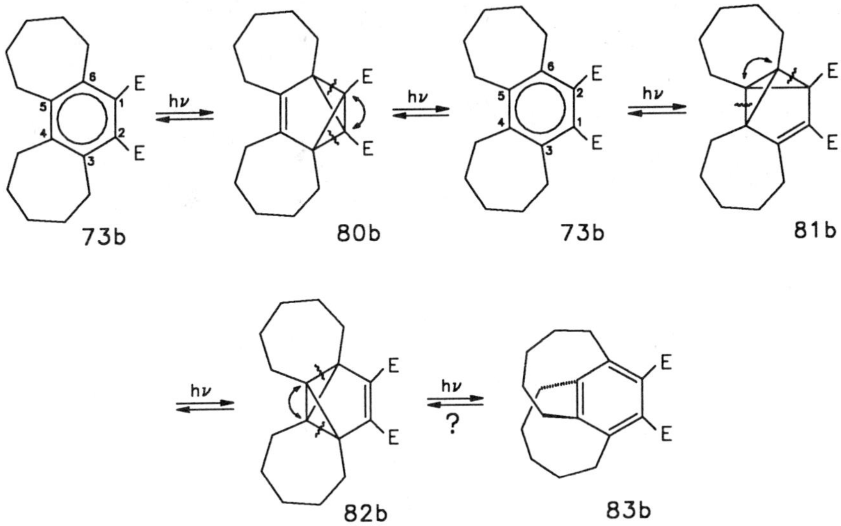

SCHEME 5.19

ester 73a, with tetramethylene bridges (n = 4 in Scheme 5.18), does not undergo this rearrangement. Even an extended irradiation and the use of shorter wavelengths (254 nm) does not afford 79a; instead, only slow polymerization is observed. This is probably because the potential [4]metacyclophane 75a does not exist under the given conditions due to the enormous strain energy imposed by the shortened *meta*-bridge. Consequently, the whole sequence is interrupted at the second step (74a → 75a). This result is in line with experiments by other groups that failed to produce a [4]metacyclophane.[19,31]

A second experiment proves that the higher homologue phthalic ester 73c, with hexamethylene bridges (n = 6 in Scheme 5.18), undergoes this rearrangement; but, under identical conditions, 79c is generated at a reduced rate and equilibrium concentrations of the intermediates 77c and 78c are lower. The explanation is simple. In [6]metacyclophane 75c the *meta*-bridge is extended by one additional methylene unit (in comparison to 75b); hence, it is less strained and the strain-releasing steps leading via 77c to 79c are less important. Scheme 5.12 shows that this phthalic/terephthalic ester rearrangement has already been encountered in connection with the irradiation of [6]paracyclophane 57c at exactly the same wavelength. The initial step is a 1,2-shift (via a benzvalene intermediate) converting the highly strained [6]paracyclophane 57c into the less strained [6]metacyclophane 75c.[32] With this compound as the connecting link, an entry to the rearrangement sequence of Scheme 5.18 is achieved and 75c can react in either direction, affording 73c and 79c, respectively.

The second noteworthy feature in this phthalic/terephthalic ester rearrangement is the irreversibility of prismane formation, step 76 → 77. Interaction of the carbomethoxy groups with the cyclopropane moieties of the prismane leads to stabilization of the distal bonds and destabilization of the vicinal bonds.[33] Photochemical [2+2]-cycloreversion takes place in a fashion by which only the destabilized bonds are cleaved. Prismane 77b possesses only one cyclobutane face with two destabilized cyclopropane edges (as indicated in Scheme 5.18). Ring opening occurs exclusively at this face, resulting in 78c; back reaction to 76c is not possible. Thus, the equilibrium is moved to the product side; in the case of 79b, the conversion amounts to over 90%. This reasoning can be corroborated by additional experiments using compounds 84a and 84b.

With hydroxy methylene or methyl groups as substituents, no conjugation is possible with the cyclopropane moieties of the hypothetical prismanes 86a and 86b; i.e., prismane formation is no longer irreversible as indicated in Scheme 5.20 by the equilibrium arrows in step 85 → 86. Hence, it is expected to obtain 88a and 88b in considerably lower yield. However, the variation of the electronic properties led to an even more dramatic change in reactivity and no products were found

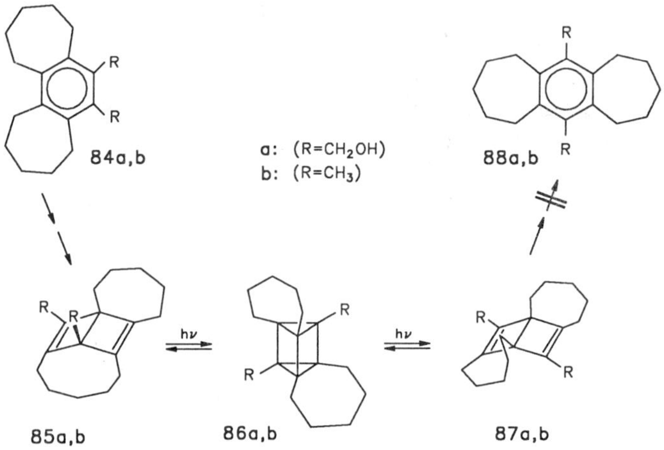

a: (R=CH$_2$OH)
b: (R=CH$_3$)

84a,b   88a,b   R

85a,b   86a,b   87a,b

SCHEME 5.20

at 282 or at 254 nm. In conclusion, it seems likely that a subtle interplay between steric and electronic factors is responsible for an unexpected photochemical rearrangement between twofold bridged phthalic esters into their isomeric terephthalic esters involving twofold bridged Dewar benzenes and prismanes as intermediates.

The Molecular Structure and Photochemistry of the Twofold Bridged Prismanes

To rationalize the photorearrangement shown in Scheme 5.18, it was assumed that the distal bonds in prismane 77b are shorter than the vicinal bonds.[33] To check this assumption, single crystals of 58b, 77b, and 65b were grown and their structure was investigated using X-ray crystallography. Figures 5.1–5.3 show ORTEP representations of all three molecules and list the most relevant C-C distances.[34] In the diester prismanes 58b and 77b, the vicinal bonds in each cyclopropane unit are significantly longer (1.55 to 1.56 Å) than the distal bonds (1.50 Å). Replacement of the ester groups

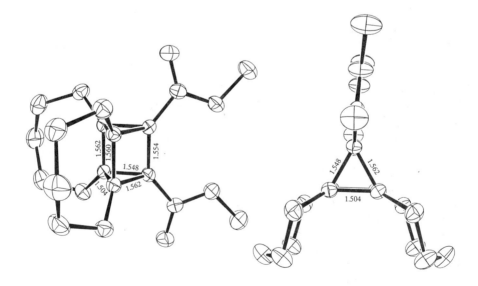

FIGURE 5.1   ORTEP plot of the molecular structure of 58b.

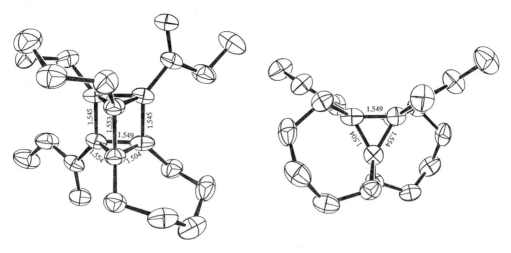

FIGURE 5.2   ORTEP plot of the molecular structure of **77b**.

by hydroxymethylene groups gives compounds with about equal bond lengths (1.54 Å) for all six C-C bonds in the cyclopropane units of **65b**, Figure 5.3. The bond alternation in the three-membered ring units of prismane can be rationalized in analogy to a cyclopropane ring[33,35] with an attached acceptor. In Figure 5.4, the two highest occupied MOs of prismane are depicted. Using first-order perturbation theory, it is concluded that electron density is transferred from the e_A-MO to the π^*-MO of the acceptor group, presuming a bisected conformation of the acceptor groups. This leads to a weakening of bonds 1–2, 1'-2', and 1–3, 1'-3', respectively, and to a strengthening of the distal bonds 2–3 and 2'-3'. By removing the acceptor property (e.g., by reduction), no such effect is anticipated.

Irradiation of **58b** at 295 nm yields 10% **54b** and 35% decomposition products (Scheme 5.21). The high ratio of decomposition products is probably a result of formation and decomposition of an unstable [5]paracyclophane **57b**. A striking fact is that neither **89b** nor **79b** could be detected (see Scheme 5.3, path b), which means that in fact only the vicinal bonds are cleaved. On the other hand, both anticipated Dewar benzenes **63b** and **66b** are generated from **65b** at 280 nm in the statistical ratio of 2:1 (because there are two vicinal faces and one distal face) as depicted in Scheme 5.15. This is in line with the finding that in **65b**, all bond lengths in the three-membered rings are equal within margin of error.

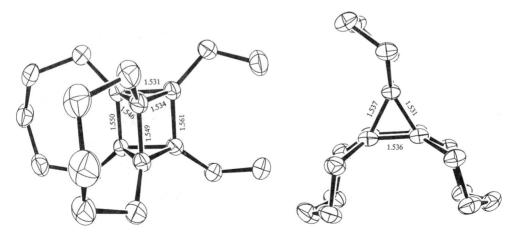

FIGURE 5.3   ORTEP plot of the molecular structure of **65b**.

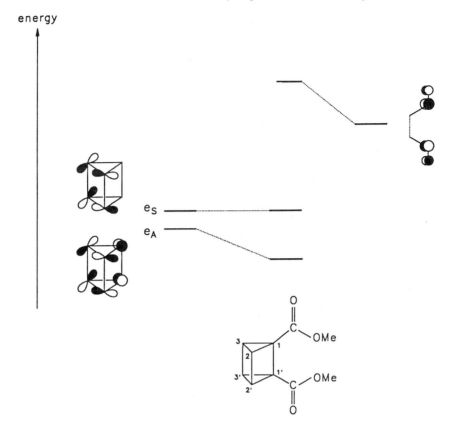

FIGURE 5.4   Interaction diagram between the HOMOs of prismane and the π*-MOs of two ester groups.

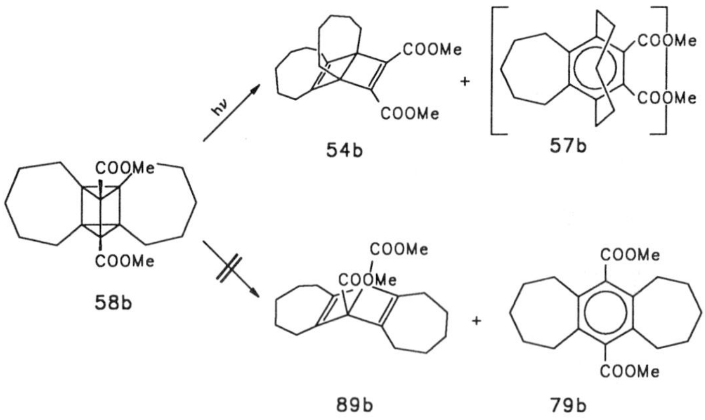

SCHEME 5.21

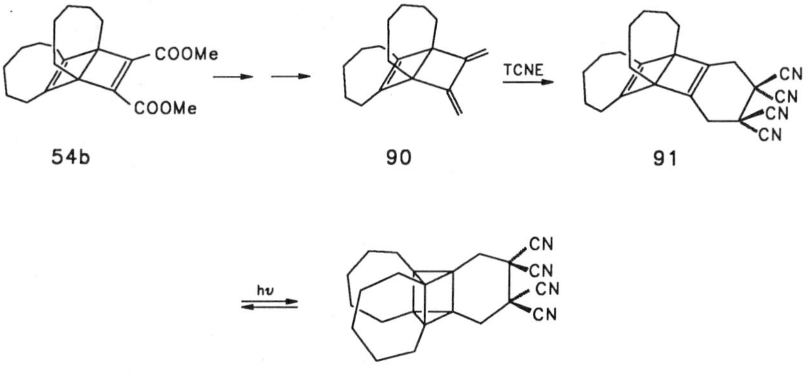

SCHEME 5.22

Conclusion and Outlook

In the preceding sections, it has been shown how properties and reactivity of Dewar benzenes and prismanes can be steered by the tethering with methylene chains of neighboring centers. This "rope trick", on the one hand, stabilizes otherwise sensitive intermediates and on the other, leads to unexpected reactions. A third effect is that the twofold bridged prismanes obtained show an unprecedented stability against X-rays so that precise molecular data could be obtained. Meanwhile, the sequence depicted in Scheme 5.22 has afforded threefold bridged Dewar benzene **91** and prismane **92**, the latter representing a propella[5.5.4₃]prismane, the first propellaprismane with a [3]prismane body.[36]

5.3 Propella[n₄]prismanes

A Cubane with a Single Chain

The synthesis of cubane **11** was already described in Scheme 5.1. Recent developments in cubane chemistry[1,37] have produced a cubane **94**, bridged by a urea group as shown in Scheme 5.23.[38] It can be prepared from the 1,2-diisocyanate of cubane **93**. Reaction with water forms a urea bridge. X-ray analysis shows a bond length of 1.526(3) Å for the C-C bond[37] which is spanned by the urea ring. This is considerably shorter than the bond length in cubane (1.551(3) Å.)[39]

Propella[3₄]prismane

The preparation of propella[3₄]prismane **18** is a short story if one leaves out the discussion of the mechanism of its formation. The synthesis of **18** is shown in Scheme 5.24. The key reaction is the

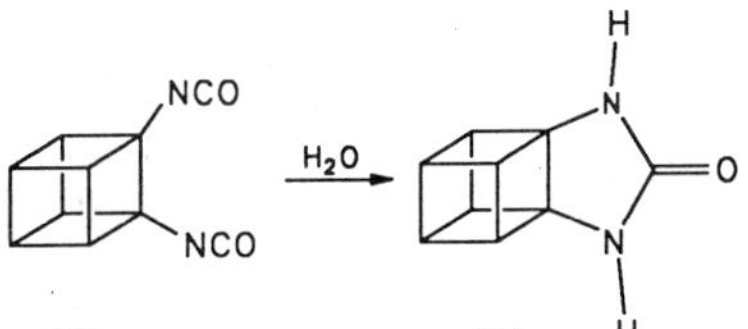

SCHEME 5.23

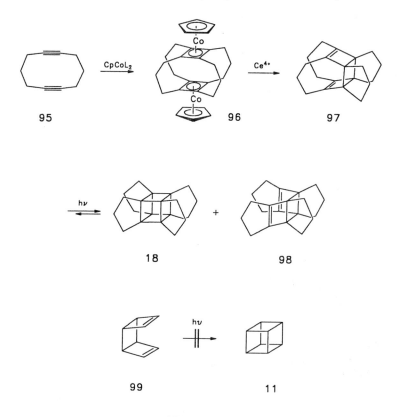

SCHEME 5.24

intermolecular reaction of 1,6-cyclodecadiyne **95** with CpCo(CO)$_2$ to afford superphane **96**.[40] Oxidation of **96** yields the fourfold bridged tricyclo[4.2.0.0^{2,5}]octa-3,7-diene **97**. Irradiation of the latter leads to propella[3$_4$]prismane **18** and the isomeric compound **98**.[8] Calculations on **18** using different levels of sophistication[41,42] predict a pinwheel conformation C$_{4h}$ for the bridges, as shown in Figure 5.5. The energy difference between the C$_{4h}$ and D$_{4h}$ conformation of **18** is predicted to be in the order of 10 to 12 kcal mol^{-1}.

The four propano chains alter considerably the properties of the tricyclooctadiene moiety in **97**.[43,44] While the parent compound **99** does not afford cubane **11** when irradiated,[45] **97** yields the cubane derivative **18**. We attribute this deviating behavior to the bridges by invoking steric and electronic effects.[8,43,44] The propano chains act as clamps diminishing the distance between the double bonds. In **97**, their distance amounts to 2.65 Å[8,42] in comparison to **99** with 3.05 Å.[42] As a result of this proximity, the frontier orbital arrangement in **97** is different from that of **99**, the sequence **97** → **18** becomes photochemically allowed.

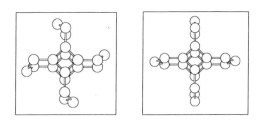

FIGURE 5.5   Calculated conformations of **18**.

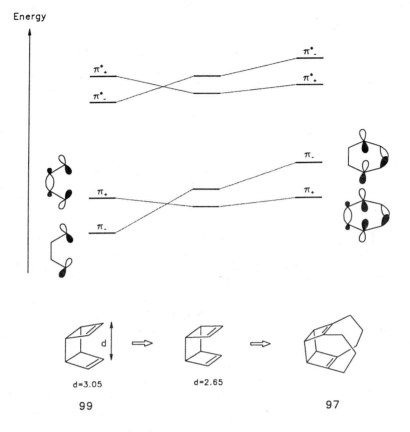

FIGURE 5.6   Correlation between the occupied and unoccupied π-MOs of free and squeezed tricyclo[4.2.0.0^{2,5}]octa-3,7-dienes (**99** → **97**).

A detailed explanation is given in Figure 5.6. In **99**, the sequence π_+ is above π_- due to strong through-bond interaction between the σ-bonds of the cyclobutane moiety and the two π-MOs. By squeezing the π-bonds together, through-space interaction dominates. This is amplified by the effect of the propano bridges as shown on the right side of Figure 5.6. Irradiation of **99** yields a π_+ → π^*_- transition that causes a considerable activation energy for the [2+2]-cycloaddition to **11**. In the case of **97**, however, where the first excited state can be described as a π_- → π^*_+ transition, the [2+2]-cycloaddition can occur smoothly.

Conclusion and Outlook

Propella[3$_4$]prismane **18** can be obtained very easily, although many questions dealing with the properties and the reactivity of **18** and its precursors **97** and **96** still remain to be answered. For example, the concept depicted in Figure 5.6 needs further elaboration by changing the nature and length of the chains. Therefore, the higher homologous superphanes with penta-[46] and heptamethylene chains,[47] respectively have been synthesized. In a second approach, the central methylene unit of the propano bridges was substituted by other functionalities like cyclopropyl- or isopropylidene groups.[48] The photochemical behavior of these superphanes and the resulting tricyclo[4.2.0.0^{2,5}]octadiene derivatives is currently under investigation. The interesting properties of the superphanes are being elucidated by means of photoelectron spectroscopy and cyclovoltammetry.

Acknowledgments

We are grateful to our colleagues and co-workers whose names appear in the references. Our work was supported by the Deutsche Forschungsgemeinschaft, the Volkswagenstiftung, the BASF Aktiengesellschaft, Ludwigshafen, DEGUSSA and Metallgesellschaft in Frankfurt. We thank K. T. for typing this manuscript.

References

1. Osawa, E.; Yonemitsu, O., Ed., *Carbocyclic Cage Compounds,* VCH Publishers, New York, 1992; Olah, G. A., Ed., *Cage Hydrocarbons,* John Wiley & Sons, New York, 1990.
2. Katz, T. J. and Acton, N., *J. Am. Chem. Soc.,* 95, 2738, 1973.
3. (a) Schäfer, W., Criegee, R., Askani, R., and Grüner, H., *Angew. Chem.,* 79, 54, 1967; *Angew. Chem. Int. Ed. Engl.,* 6, 78; 1967 (b) Lemal D. M.. and Lokensgard, J. P., *J. Am. Chem. Soc.,* 88, 5934, 1966.
4. Eaton, P. E. and Cole, T. W., Jr., *J. Am. Chem. Soc.,* 86, 3157, 1964.
5. Eaton, P. E., Or, Y. S., and Branca, S. J., *J. Am. Chem. Soc.,* 103, 2134, 1981; (b) Dauben, W. G. and Cunningham, A. F., Jr., *J. Org. Chem.,* 48, 2842, 1983.
6. Mehta, G. and Padma, S., *J. Am. Chem. Soc.,* 109, 2212, 1987.
7. Mehta, G. and Padma, S., *J. Am. Chem. Soc.,* 109, 7230, 1987.
8. Gleiter, R. and Karcher, M., *Angew. Chem.,* 100, 851, 1988; *Angew. Chem. Int. Ed. Engl.,* 27, 840, 1988.
9. Bryce-Smith, D. and Gilbert, A., in *Rearrangements in Ground- and Excited States,* Vol. 3, de Mayo, P., Ed., Academic Press, New York, 1980, 366; Wender, P. A. and van Geldern, T. W., in *Photochemistry in Organic Synthesis,* Coyle, J. D., Ed., The Royal Society of Chemistry, London, 1986, 226.
10. Wilzbach, K. E. and Kaplan, L., *J. Am. Chem. Soc.,* 86, 2307, 1964; Wilzbach, K. E., Harkness, A. L., and Kaplan, L., *J. Am. Chem. Soc.,* 90, 1116, 1968; Burgstrahler, A. W. and Chien P.-L., *J. Am. Chem. Soc.,* 86, 2940, 1964.
11. Kaplan, L., Wilzbach, K. E., Brown, W. G., and Yang, S. S., *J. Am. Chem. Soc.,* 87, 675, 1965.
12. Bryce-Smith, D. and Gilbert, A., *Tetrahedron,* 32, 1309, 1976.
13. Wilzbach, K. E. and Kaplan, L., *J. Am. Chem. Soc.,* 87, 4004, 1965.
14. Gleiter, R. and Treptow, B., *Angew. Chem.,* 102, 1452, 1990; *Angew. Chem. Int. Ed. Engl.,* 29, 1427, 1990.
15. Gleiter, R. and Merger, M., *Tetrahedron Lett.,* 33, 3473, 1992.
16. Tobe, Y., Kakiuchi, K., Odaira, Y., Hosaki, T., Kai, Y., and Kasai, N., *J. Am. Chem. Soc.,* 105, 1376, 1983.
17. Liebe, J., Wolff, C., and Tochtermann, W., *Tetrahedron Lett.,* 23, 2439, 1982; Liebe, J., Wolff, C., Krieger, C., Weiss, J., and Tochtermann, W., *Chem. Ber.,* 118, 4144, 1985.
18. Dreeskamp, H., Kapahnke, P., and Tochtermann, W., *Radiat. Phys. Chem.,* 32, 537, 1988.
19. Kostermans, G. B. M., Hogenbirk, M., Turkenburg, L. A. M., de Wolf, W. H., and Bickelhaupt, F., *J. Am. Chem. Soc.,* 109, 2855, 1987.
20. Kosters, J. B., Timmermans, G. J., and van Bekkum, H., *Synthesis,* 139, 1971; Schäfer, W. and Hellmann, H., *Angew. Chem.,* 79, 566, 1967; *Angew. Chem. Int. Ed. Engl.,* 6, 518, 1967.
21. Hogeveen, H. and Kok, D. M., in *The Chemistry of Triple-Bonded Functional Groups,* Suppl. C, Part 2, Patai, S. and Rappoport, Z., Eds., John Wiley & Sons, New York, 1983, chap. 23 and references therein.
22. Kostermans, G. B. M., de Wolf, W. H., and Bickelhaupt, F., *Tetrahedron,* 43, 2955, 1987.
23. Dale, J., Hubert, A. J., and King, G. S. D., *J. Chem. Soc.,* 73, 1963.
24. Bühl, H., Gugel, H., Kolshorn, H., and Meier, H., *Synthesis,* 536, 1978.
25. Kostermans, G. B. M., de Wolf, W. H., and Bickelhaupt, F., *Tetrahedron Lett.,* 27, 1095, 1986.
26. Criegee, R. and Zanker, F., *Chem. Ber.,* 98, 3838, 1965.

27. Criegee, R. and Askani, R., *Angew. Chem.*, 78, 494, 1966; *Angew. Chem. Int. Ed. Engl.*, 5, 519, 1966; Criegee, R., Askani, R., and Grüner, H., *Chem. Ber.*, 100, 3916, 1967.

28. Gleiter, R. and Treptow, B., *Angew. Chem.*, 104, 879, 1992; *Angew. Chem. Int. Ed. Engl.*, 31, 862, 1992.

29. Jenneskens, L. W., de Boer, H. J. R., de Wolf, W. H., and Bickelhaupt, F., *J. Am. Chem. Soc.*, 112, 8941, 1990; van Straten, J. W., de Wolf, W. H., and Bickelhaupt, F., *Tetrahedron Lett.*, 4667, 1977.

30. For an X-ray structure of a [5]metacyclophane, see: Jenneskens, L. W., Klamer, J. C., de Boer, H. J. R., de Wolf, W. H., Bickelhaupt, F., and Stam, C. H., *Angew. Chem.*, 96, 236, 1984; *Angew. Chem. Int. Ed. Engl.*, 23, 238, 1984.

31. Kostermans, G. B. M., van Dansik, P., de Wolf, W. H., and Bickelhaupt, F., *J. Am. Chem. Soc.*, 109, 7887, 1987; *J. Org. Chem.*, 53, 4531, 1988.

32. Hirano, S., Hara, H., Hiyama, T., Fujita, S., and Nozaki, H., *Tetrahedron*, 31, 2219, 1975.

33. Irngartinger, H., Kallfass, D., Litterst, E., and Gleiter, R., *Acta Crystallogr.*, C43, 266, 1987.

34. Gleiter, R., Treptow, B., Irngartinger, H., and Oeser, T., *J. Org. Chem.*, 59, 2787, 1994.

35. Hoffmann, R., *Tetrahedron Lett.*, 2907, 1970; Günther, H., *Tetrahedron Lett.*, 5173, 1970.

36. Gleiter, R. and Treptow B., unpublished results.

37. Eaton, P. E., *Angew. Chem.*, 104, 1447, 1992, *Angew-Chem. Int. Ed. Engl.*, 31, 1421, 1992.

38. Eaton, P. E., Pramond, K., and Gilardi, R., *J. Org. Chem.*, 55, 5746, 1990.

39. Fleischer, E. B., *J. Am. Chem. Soc.*, 86, 3889, 1964.

40. Gleiter, R., Karcher, M., Ziegler, M. L., and Nuber, B., *Tetrahedron Lett.*, 28, 195, 1987.

41. Osawa, E., Rudzinski, J. M., and Xun, Y.-M., *Struct. Chem.*, 1, 333, 1990.

42. Gleiter, R., Pfeifer, K.-H., and Koch, W., *J. Comput. Chem.*, in press.

43. Gleiter, R., Karcher, M., Kratz, D., Rittinger, S., and Schehlmann, V., in *Organometallics in Organic Synthesis*, Werner, H. and Erker, G., Eds., Springer, Berlin, 1989, 109.

44. Gleiter, R., *Angew. Chem.*, 104, 29, 1992; *Angew. Chem. Int. Ed. Engl.*, 31, 27, 1992.

45. Criegee, R., *Angew. Chem.*, 74, 703, 1962; *Angew. Chem. Int. Ed. Engl.*, 1, 519, 1962; Iwamura, H., Morio, K., and Kihara, H., *Chem. Lett.*, 457, 1973; but see Gleiter, R. and Brand, S., *Tetrahedron Lett.*, 35, 4969, 1994.

46. Gleiter, R., Treptow, B., Kratz, D., and Nuber, B., *Tetrahedron Lett.*, 33, 1733, 1992.

47. Gleiter, R. and Pflästerer, G., *Chem. Commun.*, in press.

48. Gleiter R., Merger, R., and Nuber, B., *J. Am. Chem. Soc.*, 114, 8921, 1992.

49. Gleiter, R. and Treptow, B., *J. Org. Chem.*, 58, 7740, 1993.

6

Copper(I)-Catalyzed Intra- and Intermolecular Photocycloaddition Reactions of Alkenes

Klaus Langer
Organisch-Chemisches Institut
der Universität Münster

Jochen Mattay
Organisch-Chemisches Institut
der Universität Münster

6.1 Introduction

Transition metals are known to participate as catalysts in many reactions. Their catalytic activity is based on several effects. First, different functional substituents are brought to each other by coordination to the metal. Moreover, the transition metal enhances the reactivity of the substrate by charge-transfer interaction and thus chemical transformations are facilitated. The interaction of a transition metal and a substrate generates a new "electronic system" which shows new properties, e.g., UV absorption bands. If such metal complexes are activated by irradiation into these bands, reactions can occur that are not observed under thermal conditions nor in absence of the transition metal, e.g., isomerizations, rearrangements, fragmentations, or cycloadditions.[1-4]

$$M + S \; \rightleftharpoons \; M\text{-}S \; \xrightarrow{h\nu} \; M\text{-}P \; \longrightarrow \; M + P$$

Due to their powerful synthetic potential, cycloadditions are of special interest. Copper is a transition metal found to catalyze transformations, especially of olefinic double bonds. Besides isomerizations and rearrangements, copper(I) salts catalyze selectively inter- and intramolecular cycloadditions, provided that the alkenes exhibit special features. Due to this selectivity, copper(I)-catalyzed photocycloadditions are powerful synthetic tools for the construction of complex organic structures and have already been applied in several syntheses. Below, mechanistic aspects as well as stereochemical behavior of inter- and intramolecular copper(I)-catalyzed photocycloadditions will be discussed. In addition, some well-known syntheses as well as conceivable applications are given.

0-8493-8634-9/95/$0.00+$.50
© 1995 by CRC Press, Inc.

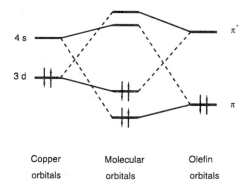

FIGURE 6-2.1   Schematic energy level diagram for copper(I)-olefin coordination.

6.2 The Mechanism of Catalysis

In the last 3 decades, many copper(I)-olefin complexes have been isolated.[5–9] In these complexes, the bonding between the olefin and the copper(I) ion is described qualitatively by the Dewar-Chatt-Duncanson model.[10,11] The copper(I)-olefin bonding interaction consists of two synergistic components. The first is the ligand-to-metal σ-donation (LMCT), which involves overlapping of an occupied π-orbital of the olefin with a vacant σ-type orbital of the copper. The second is the metal-to-ligand π-back donation where an occupied metal d-type orbital overlaps with a vacant antibonding π^*-orbital of the olefin. In the case of copper(I)-olefin complexes, the metal orbitals involved in the bonding interactions are the copper $3d_\pi$ orbital as a π-donor and the copper $4s$ orbital as a σ-acceptor (See Figure 6-2.1).[12,13]

The relative contribution of the two interactions depends on the olefin involved, as confirmed by ^{1}H-NMR investigations.[8,14] Copper(I) complexes of cyclic polyolefins show downfield shifts of the olefinic protons, indicating a dominance of the σ-donation of the olefin. Monoolefin complexes, however, exhibit the reverse behavior, which is presumed to be caused by predominance of the π-donation of the metal.[8,14] ^{13}C-NMR investigations generally exhibit an upfield shift for the sp^2-carbons, which is small for polyolefin ligands and higher for monoolefins. The various changes in chemical shifts of ^{1}H- and ^{13}C-NMR spectra are caused by different effects that are responsible for the shifting of the nuclei. However, the results of the NMR investigations are understandable in a qualitative fashion by the Dewar-Chatt-Duncanson model of the metal-olefin bonding.[13]

Solutions of copper(I)-olefin complexes exhibit UV absorption bands in the range of $\lambda = 250$ nm. Excitation of the complexes by irradiation into these bands can lead to photocycloadditions. Methanol solutions containing norbornene (NB) and copper(I) trifluoromethanesulfonate (CuOTf) show UV absorption bands at $\lambda = 236$ nm ($\varepsilon = 3400$) and $\lambda = 272$ nm ($\varepsilon = 2000$) (See Figure 6-2.2).[8,15]

NMR investigations demonstrate that both 1:1 and 1:2 copper(I)-olefin complexes exist that show the same UV absorption spectra.[15] Only the 1:2 complex leads to dimer production, as confirmed by quantum yield experiments, since the deactivation reaction of the 1:1 complex is much faster than the bimolecular association with a ground-state NB. These results presume the need for a ground-state coordination of both olefinic double bonds to the same copper(I) ion.[3] The mechanism of the copper(I)-catalyzed [2+2]-photocycloaddition of norbornene (NB) is given in Scheme 6-2.1.

$$\text{Cu-NB} + \text{NB} \; \overset{K}{\rightleftharpoons} \; \text{Cu-(NB)}_2$$

$$\text{Cu-NB} + h\nu \; \overset{\Phi_1}{\longrightarrow} \; \text{Cu-NB}^\bullet$$

$$\text{Cu-(NB)}_2 + h\nu \; \overset{\Phi_2}{\longrightarrow} \; \text{Cu-(NB)}_2^\bullet$$

$$\text{Cu-(NB)}_2^\bullet \; \longrightarrow \; \text{dimer}$$

SCHEME 6-2.1

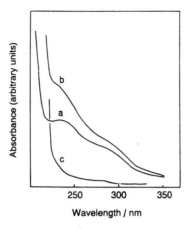

FIGURE 6-2.2   a) The absorption spectrum of CuOTf-norbornene complex in methanol. b) The spectrum of a) in the presence of a tenfold excess of norbornene with norbornene in the reference cell for balancing. c) The end absorption spectrum of norbornene. The vertical scale is arbitrary in a–c. *Source:* Salomon, R. G. and Kochi, J. K., *J. Am. Chem. Soc.*, 96, 1137, 1974. With permission.

The shorter wavelength absorption is presumed to be an MLCT absorption, while the longer one is probably an LMCT absorption.[3,16] Due to the lower electrophilicity of the copper in copper(I) halides, the second absorption is not observed in solutions containing halide anions. It has not been ascertained which excitation leads to photocycloaddition nor has there been an investigation of whether the reaction is concerted or sequential. Usually, a sequential pathway is assumed, occurring via radical (MLCT) or cationic (LMCT) copper alkyl intermediates (See Scheme 6-2.2).[3] Flash photochemical investigations of copper(I)-ethylene complexes[17] as well as Hartree-Fock-Slater calculations[12] support an MLCT mechanism, leading to the products via σ-bonded radical copper alkyl species; however, the alternative pathway via cationic intermediates has also been discussed.[15,18,19] However, investigations of intermolecular cycloadditions suggest a concerted pathway.[20]

SCHEME 6-2.2

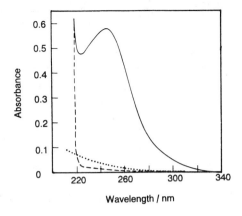

FIGURE 6-2.3   electronic absorption spectra: (…) ether saturated with CuCl (- -) 0.0163 mol l⁻¹ *cc*COD in ether and (————) 0.016 mol l⁻¹ *cc*COD in ether postsaturated with CuCl. Cell path length was 1 cm. *Source:* Grobbelaar, E., Kutal, C. and Orchard, W., *Inorg. Chem.*, 21, 414, 1982. With permission.

The first copper(I)-catalyzed photocycloaddition reaction of a diene was reported by Srinivasan in 1963, who irradiated *cis, cis*-1,5-cyclooctadiene (*cc*COD) 1 in the presence of CuCl and isolated tricyclo[3.3.0.0²,⁶]octane (TCO) 2.[21] See Scheme 6-2.3.

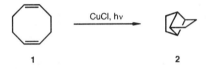

SCHEME 6-2.3

This reaction has been studied extensively concerning mechanistic aspects. Early investigations suggested that in a solution containing CuCl and *cc*COD, the uncomplexed *cc*COD absorbs most of the light, and it was considered that excited *cc*COD reacts with CuCl to form TCO.[22,23] This seems not to be very likely because the lifetime of an olefinic excited state of COD is probably too short to permit bimolecular processes. Spectral and quantum yield investigations of Kutal et al.,[24] who examined ether solutions containing CuCl and *cc*COD, indicated that the primary light-absorbing species is a complex containing two molecules of *cc*COD (Figure 6-2.3).

In addition, due to the fact that (CuCl-*cc*COD)₂ is a well-characterized complex that has also been isolated,[5] this complex is suggested to be the major light-absorbing species in solution. It was shown that during photolysis of *cc*COD in the presence of copper(I) halides, *ct*COD 3 and *tt*COD 4, which form very strong complexes with copper(I), are also formed.[23,24] While TCO is generated after a short induction period with a time-independent quantum yield, the concentration of *ct*COD rises rapidly in the beginning and reaches a steady state after several minutes (Figure 6-2.4). This leads to the conclusion that *ct*COD is a primary photoproduct, while TCO is built in a secondary process via *ct*COD. It is not clear whether TCO is formed directly from *ct*COD or via *tt*COD.[23,24] Irradiation of *tt*COD in the absence of copper(I) also leads to TCO, while *ct*COD is reconverted to *cc*COD. The proposed mechanism is outlined in Scheme 6-2.4[24]

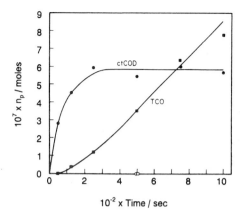

FIGURE 6-2.4   Time development of the amount of product (n_p) formed during irradiation of an ether solution containing 1.44×10^{-4} mol l^{-1} CuCl and 0.161 mol l^{-1} *cc*COD. *ct*COD, filled circles; TCO, filled squares. Open circle and square denote amounts of *ct*COD and TCO formed upon irradiating 0.161 mol l^{-1} *cc*COD in the absence of CuCl. *Source:* Grobbelaar, E., Kutal, C., and Orchard, W., *Inorg. Chem.*, 21, 414, 1982. With permission.

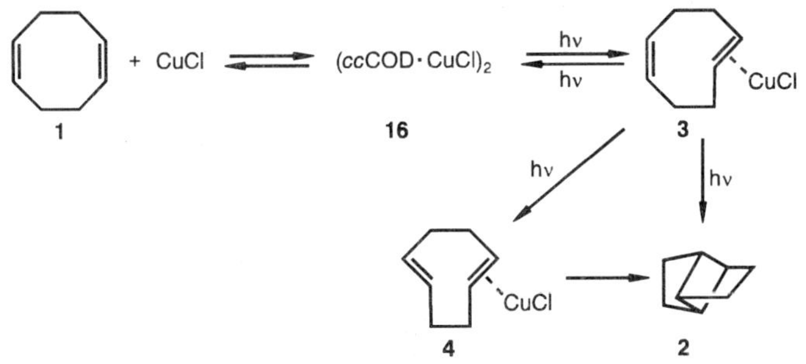

SCHEME 6-2.4

   Investigations with d$_1$-*cc*COD predict free radical intermediates during cyclization because loss of deuterium in the product is observed.[25] In contrast, the reaction of d$_{12}$-*cc*COD shows no loss of deuterium at all. This rules out free radical intermediates and supports an intramolecular process.[26] These contrary results were attributed to different radiation sources.

   Similar to this reaction, copper(I)-olefin complexes of highly strained cyclic monoolefins are supposed to play important roles as precursors for cycloaddition products in intermolecular reactions.[20,27,28] Cyclopentene, cyclohexene, or cycloheptene — but not cyclooctene or acyclic olefins — were dimerized. In the case of cyclohexene and cycloheptene, the thermodynamically less stable *trans,trans*-fused cycloadducts 7 and 10 are formed as the main products. While cyclopentene and cyclohexene yield dimeric products, the main product of the cycloheptene reaction, which was originally also presumed to be a dimer, is the *all trans*-fused trimer 10.[29] Since a 1:3 CuOTf-*trans*-cyclooctene complex has been isolated,[8] an analogous complex of *trans*-cycloheptene that could act as a precursor for the trimeric product seems to be reasonable. See Scheme 6-2.5.

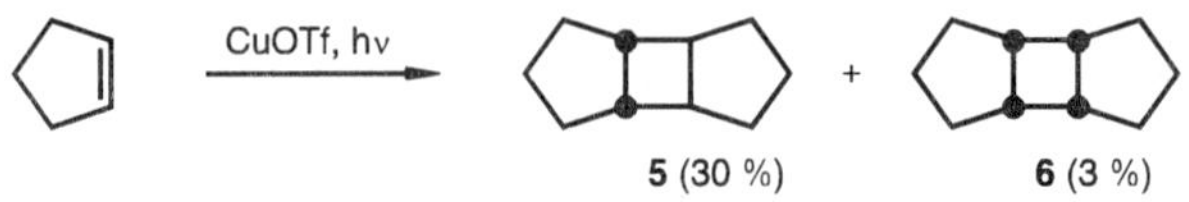

SCHEME 6-2.5

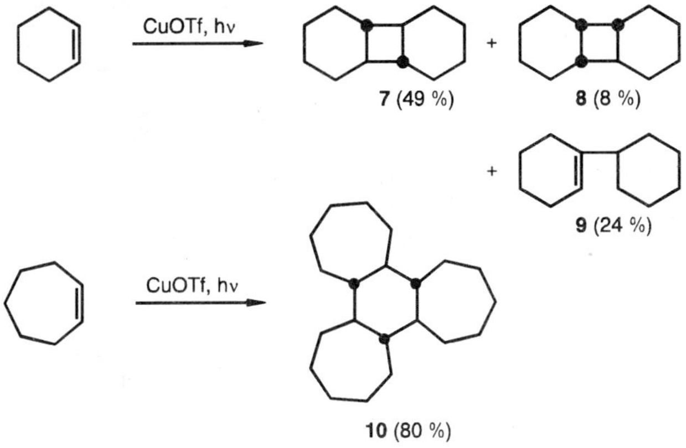

7 (49 %)   **8 (8 %)**

9 (24 %)

10 (80 %)

SCHEME 6-2.5 (continued)

Two different mechanisms are proposed to explain this behavior. The first mechanism involves photoinduced *cis-trans*-isomerization of the copper(I)-olefin complex **11**, followed by a concerted thermal $[2\pi_a+2\pi_s]$-cycloaddition of the highly strained and very reactive *trans*-cycloolefin **12** to another olefin; see Scheme 6-2.6.[20] The very reactive *trans*-olefins are probably stabilized by the copper(I) ion,[22] since a *trans*-cycloheptene-CuOTf complex has already been isolated.[9] In contrast to the reactive cycloolefins, cyclooctene or acyclic olefins do not form any highly strained intermediates, and consequently no dimerization reaction occurs. The second mechanism proposed involves an LMCT and a stepwise formation of the dimers via copper carbocation intermediates **13** and **14**;[3] see Scheme 6-2.7. While the first mechanism explains the stereochemistry of the cyclobutane principal products, the second mechanism explains the minor product **9**.

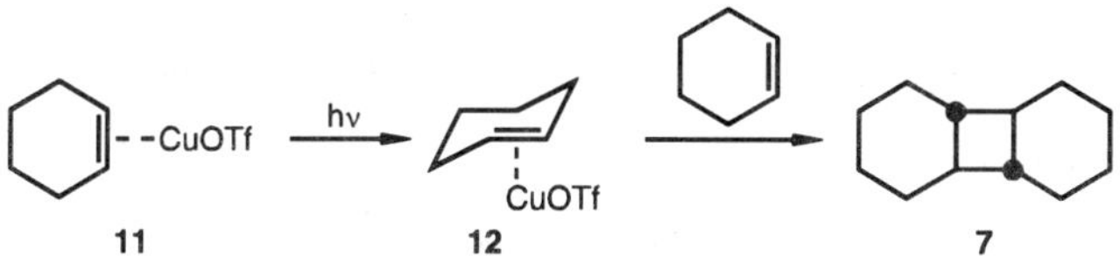

11   **12**   **7**

SCHEME 6-2.6

13   **7/8**

1,3 H-shift

14   **9**

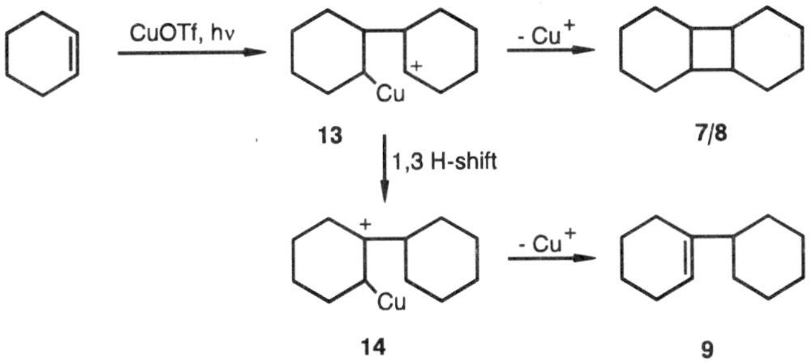

SCHEME 6-2.7

The earlier copper(I)-catalyzed photocycloadditions were carried out using copper(I) halides as catalysts. However, recent investigations use the copper(I) trifluoromethanesulfonate-benzene complex (CuOTf), which was first synthesized in 1972 by Salomon and Kochi.[7] Compared to copper(I) halides, this complex exhibits several advantages. It shows high solubility in many organic solvents and it is stable to photochemical reaction conditions. In the case of copper(I) halides, disproportion reactions occur and the reaction vessel must be cleaned repeatedly to remove the opaque copper coating formed during photolysis and new halide has to be added. In addition, quantum yields as well as isolated yields are raised by using CuOTf instead of copper(I) halides. The dimerization of norbornene **59** provides a 38% yield of dimer using CuBr in contrast to 88% yield using CuOTf as catalyst. Dicyclopentadiene **62** can be dimerized in the presence of CuOTf; while in the presence of CuBr, no reaction occurs.[30,31] This behavior is explained by the very weak coordination properties of the triflate anion. In contrast to the strongly associating halides, the olefins compete effectively with the triflate anion for the coordination sites of the copper(I) ion.[15] Hence, CuOTf shows a strong tendency to form 1:2 copper(I)-olefin complexes, which are required for cycloadditions as observed in the isolated copper(I) complexes of *cc*COD. Copper(I) chloride forms a halide-bridged 1:1 complex **15**, while CuOTf yields the 2:1 complex **16** (Scheme 6-2.8).[5,7] It is noteworthy that many copper(I)-catalyzed [2+2]-photocycloadditions occur if Cu(OTf)$_2$ is used instead of CuOTf.[32,33] This behavior is attributed to a photochemically initiated reduction of the copper(II) species, since copper(II) salts and *cc*COD form Cu(I)X-COD$_2$ complexes under photochemical reaction conditions.[34]

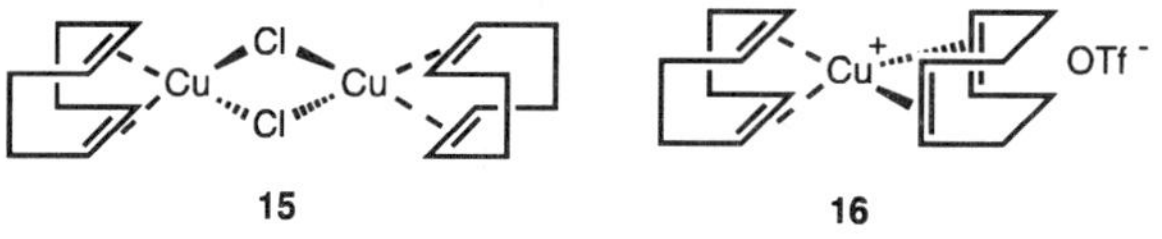

15   **16**

SCHEME 6-2.8

6.3 Intramolecular Photocycloadditions

In 1978, Evers and Mackor were the first to observe copper(I)-catalyzed intramolecular photocycloadditions between two acyclic double bonds. Irradiation of diallylether **17a** or 1,6-heptadien-4-ol **18** in the presence of CuOTf leads to bicyclo[3.2.0]heptane systems **19a** and **20**, respectively (Scheme 6-3.1).[34] The generality and the synthetic utility of this reaction was demonstrated by Salomon et al. who investigated manifold substituted diallyl ethers, homoallyl vinyl ethers, and 1,6-heptadienol derivatives. Diallyl ethers **17** (Scheme 6-3.2 and Table 6-3.1) are cyclized in fair yields to 3-oxabicyclo[3.2.0]heptanes **19**.[35–37]

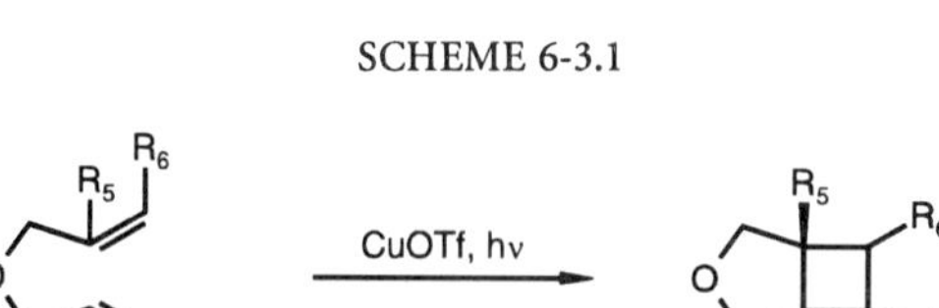

SCHEME 6-3.1

SCHEME 6-3.2

Table 6-3.1   Copper(I)-Catalyzed [2+2]-Photocycloaddition of 17 to 19 and Oxidation of 19 to 23 by RuO$_4$/NaIO$_4$

Entry	R$_1$	R$_2$	R$_3$	R$_4$	R$_5$	R$_6$	Yield of 19/%	Yield of 23/%
a	H	H	H	H	H	H	52	91
b	CH$_3$	CH$_3$	H	H	H	H	56	94
c	CH$_3$	CH$_3$	H	H	CH$_3$	H	54	44
d	CH$_3$	H	H	H	H	H	54	56
e	n-C$_4$H$_9$	H	H	H	CH$_3$	H	83	83
f	H	H	CH$_3$	CH$_3$	H	H	39	28/19[a]
g	H	H	H	H	CH$_3$	H	41	46/20[a]
h	H	H	H	H	H	CH$_2$OH	41[a]	
i	H	H	H	H	H	CH$_2$OAc	35[a]	
j	n-C$_6$H$_{11}$	H	H	H	H	CH$_2$OAc	21[a]	
k	H	H	-CH=CHCH$_3$	H	H	H	71[a]	
l	CH$_3$	H	-CH=CHCH$_3$	H	H	H	80[a]	
m	n-C$_4$H$_9$	H	-CH=CHCH$_3$	H	H	H	87[a]	
n	-CH=CH$_2$	H	H	H	H	H	70	
o	H	-(CH$_2$)$_2$-		H	H	H	47	73
p	H	-(CH$_2$)$_3$-		H	H	H	36	4/20[b]
q	H	-(CH$_2$)$_5$-		H	H	H	56	56
r	H	H	H	H	-(CH$_2$)$_4$-		35	44/34[a]
s	CH$_3$	-(CH$_2$)$_2$-		H	H	H	28	87
t	CH$_3$	-(CH$_2$)$_3$-		H	H	H	35	82

[a] Mixture of diastereoisomers.
[b] Main product (20%):1-hydroxy-2-oxatricyclo[4.3.1.0^{4,10}]decane.

The cyclizations of diallylethers bearing an α-alkyl substituent exhibit high stereoselectivity. Thus, the cyclization of **17d** yields exclusively the *exo*-product **19d** as confirmed by NMR data (Scheme 6-3.3).[36]

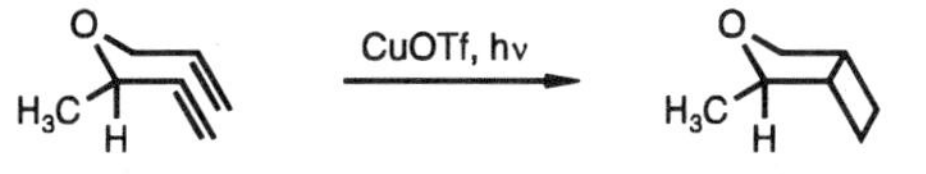

17d   CuOTf, hv   **19d**

SCHEME 6-3.3

As a coordination of both reacting olefinic bonds to the copper(I) catalyst is necessary, the observed stereoselectivity is assumed to arise from a preferential formation of the sterically less-crowded copper(I)-diene complex, leading to the *exo*-product.[36] However, the reaction does not occur stereoselectively with respect to the substituents on the alkenes. The configuration at the double bonds are not retained in the cyclobutane ring. Thus, the cyclization of pure *cis*-**17i** yields a mixture of the diastereoisomers **19i** (Scheme 6-3.4).[36] In a similar manner, homoallylvinyl ethers **21** are cyclized in moderate yields of 2-oxabicyclo[3.2.0]heptanes **22** (Scheme 6-3.5 and Table 6-3.2).[36,37]

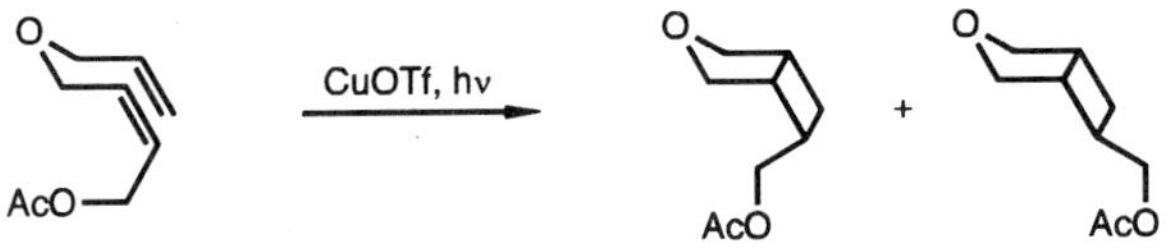

17i   CuOTf, hv   **19i**

SCHEME 6-3.4

Table 6-3.2 Copper (I)-Catalyzed [2+2]-Photocycloaddition of 21 to 22 and Oxidation of 22 to 24 by $RuO_4/NaIO_4$

Entry	R_1	R_2	R_3	R_4	R_5	Yield of 22/%	Yield of 24/%
a	H	H	H	H	H	60	
b	-CH=CH$_2$	H	H	H	H	58	
c	H	H	-(CH$_2$)$_2$-		H	92	78
d	CH$_3$	H	-(CH$_2$)$_2$-		H	50	71
e	H	H	H	-(CH$_2$)$_2$-		40	70
f	H	H	-(CH$_2$)$_3$-		H	48	85

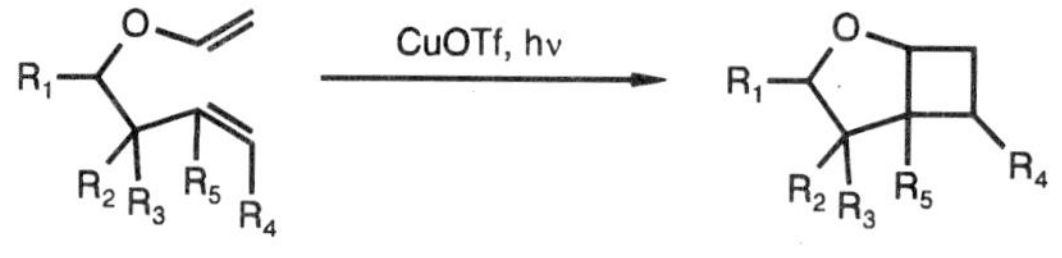

SCHEME 6-3.5

The synthesis of multicyclic carbon networks can be achieved by the cyclization of monocyclic diallyl- or homoallylvinyl ethers. Thus, a new unique synthetic route to a varity of polycyclic ethers is achieved.[36,37] The oxidation of the ether products by NaIO$_4$, using RuO$_4$ as catalyst, provides a new synthetic pathway to manifold substituted cyclobutanated γ-lactones 23 and 24 respectively,[36,37] as outlined in Tables 6-3.1 and 6-3.2 and Scheme 6-3.6. If the α-positions of the ether moiety are not equivalent and if they both can be attacked, mixtures of diastereoisomers are obtained as observed for 23f, g, p, and r. Besides homoallylvinyl ethers and diallyl ethers, 1,6-heptadienols have also been investigated thoroughly. 1,6-Heptadien-3-ols 25 are cyclized to bicyclo[3.2.0]heptan-2-ols 26 (Scheme 6-3.7 and Table 6-3.3).[38–41]

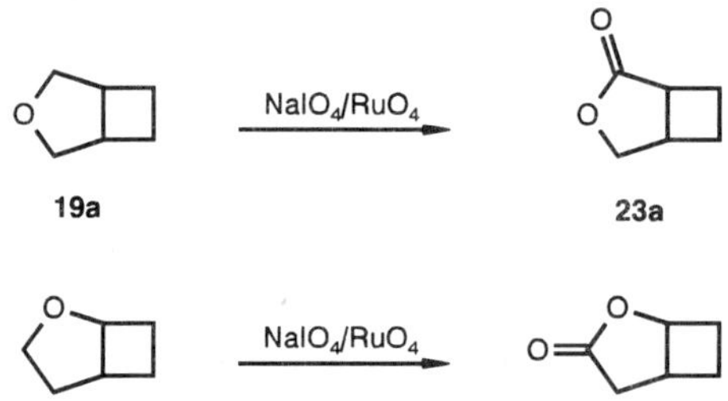

SCHEME 6-3.6

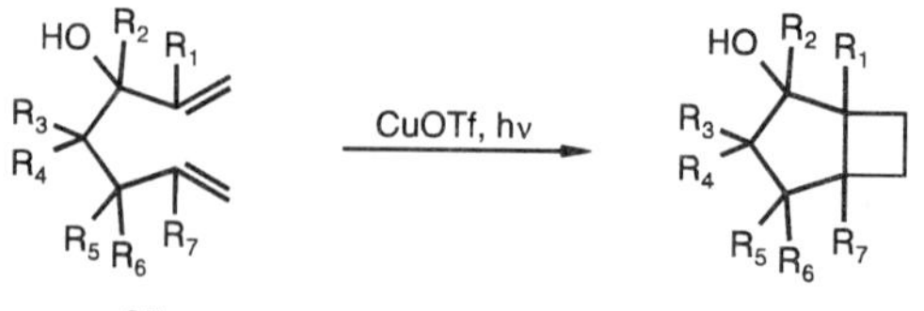

SCHEME 6-3.7

Table 6-3.3   Copper (I)-Catalyzed [2+2]-Photocycloaddition of 25 to 26, Oxidation of 26 to 31 by CrO_3/H_2SO_4, and Pyrolysis of 31 to the Corresponding Cyclopentenone 32

Entry	R_1	R_2	R_3	R_4	R_5	R_6	R_7	Yield of 26/%	*endo/exo* Ratio 26	Yield of 31/%	Yield of 32/%
a	H	H	H	H	H	H	H	86	9:1	78	70
b	CH$_3$	H	H	H	H	H	H	81	6:1	67	72
c	H	H	CH$_3$	CH$_3$	H	H	H	91	9:1	92	78
d	H	H	H	H	CH$_3$	CH$_3$	H	84	3:4	92	87
e	CH$_3$	H	CH$_3$	CH$_3$	H	H	H	83	3:2	93	54
f	CH$_3$	H	H	H	H	H	CH$_3$	a	>20:1[b]		
g	H	H	H	H	H	H	CH$_3$	78	5:1		
h	H	CH$_3$	H	H	H	H	H	89	>20:1[b]		

[a] 40% conversion after 24 h.
[b] No other epimer detected by ^{1}H-NMR measurements.

In clear contrast to the cyclizations of the ether derivatives, the cyclizations of the 1,6-heptadien-3-ols show high stereoselectivity to the thermodynamically less stable *endo*-isomer. This selectivity can be explained if the dienol is presumed to act preferably as a tridentate ligand 27 leading to the endo-isomer 26a n rather than as bidentate ligand 28. The *endo/exo*-ratio of the cyclization of 25g shows a remarkable dependence on the coordination capability of the solvent used (*endo/exo* = 6.7 in THF and 2.1 in *n*-hexane), which supports the tridentate coordination (Scheme 6-3.8).[33] The only exception to this is the reaction of 25d. In this case, the steric interaction of the methyl group at C5 favors the bidentate complex compared to the tridentate coordination. The cyclization of 1,6-heptadien-4-ol 18 is not stereoselective (Scheme 6-3.9).[39] In this case, it is unlikely that the dienol acts as a tridentate ligand 29, a fact confirmed by NMR investigations.[34] To achieve a tridentate coordination, a boat-like conformation must be presumed. Compared to the chair-like arrangement 30, this conformation is obviously not preferred.[39] However, the cyclization of 18 in various solvents shows only a small dependence of the *endo/exo*-ratio on the coordination capability of the solvent used (*endo/exo* = 0.87 in THF and 1.73 in *n*-hexane), and the opposite tendency, compared to 25g, is observed.[42] The bicyclo[3.2.0]heptan-2-ols 26 obtained (Scheme 6-3.10) can be oxidized to the corresponding ketones 31 in high yield using CrO_3/H_2SO_4. Pyrolysis of these ketones at 580 °C provides a new synthetic route to a series of substituted cyclopent-2-en-1-ones 32 which are important synthons for sesquiterpene synthesis.[39] Baeyer-Villiger oxidation of the ketones yields α,β-cyclobutanated δ-lactones 33, which can be pyrolyzed to the corresponding α,β-unsaturated δ-lactones 34.[38] By using monocyclic 1,6-heptadien-3-ols 35, tricyclic carbon networks can be synthesized. Treatment of the tricyclic compound 36 with trifluoroacetic acid brings about rearrangements via cationic intermediates. Saponification of the resulting trifluoroacetates yields polycyclic 7-hydroxynorbornenes 37 with very high stereoselectivity (Scheme 6-3.11).[43]

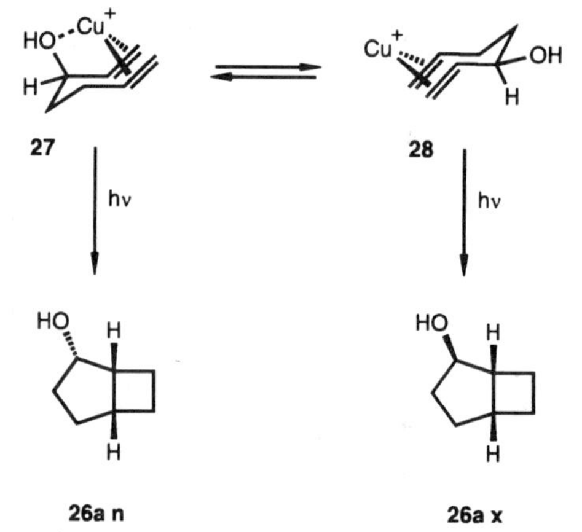

SCHEME 6-3.8

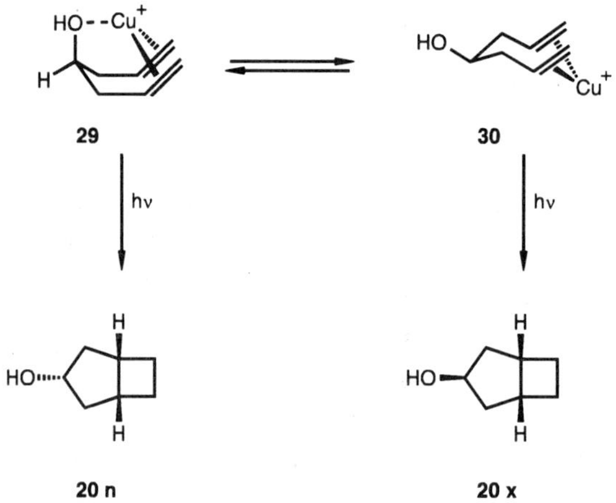

SCHEME 6-3.9

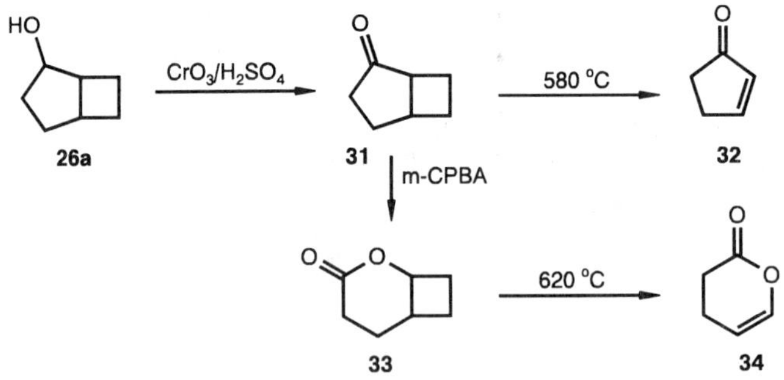

SCHEME 6-3.10

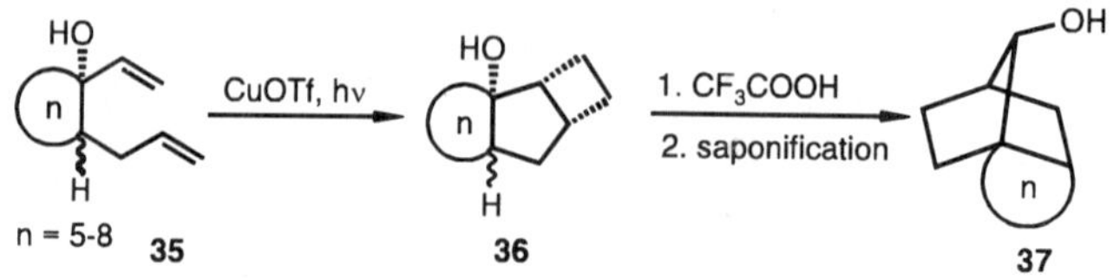

SCHEME 6-3.11

The intramolecular copper(I)-catalyzed photocycloaddition exhibits selectivity to 1,6-diene derivatives. 1,5-Dienes or 1,7-dienes, as well as 1,6-dien-3-ones or 1,6-enallenes, are unreactive.[34,39] If several double bonds are present in one compound, the 1,6-diene reacts exclusively as observed in 38 to 40.[32,35] In the case of 40, the products undergo further copper(I)-catalyzed reactions to yield 44–46. (Scheme 6-3.12).[32]

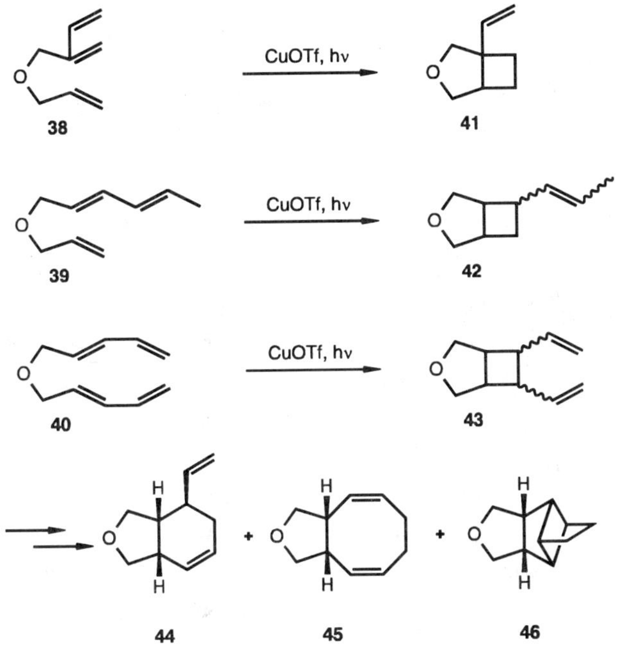

SCHEME 6-3.12

Besides the high selectivity of the reaction to the 1,6-diene derivatives, the synthetic value is increased due to the fact that many substituents are tolerated, which facilitate further transformations during synthesis; e.g., hydroxy, acetyl, phenyl, carbamate, ether, ester, vinyl, allyl, acetal, or lactone.[32,34–44] However, in the presence of an amine or an amide, no cyclization occurs.[34,44,45] The high chemo-, regio-, and stereoselectivity of this reaction as well as the tolerance to the presence of many functional substituents has led to several applications to the synthesis of natural products. The first synthesis that took advantage of this reaction (Scheme 6-3.13) was the synthesis of α- and β-panasinsene 49 and 50, respectively, two sesquiterpenes extracted from the roots of *Panax Ginseng*.[46]

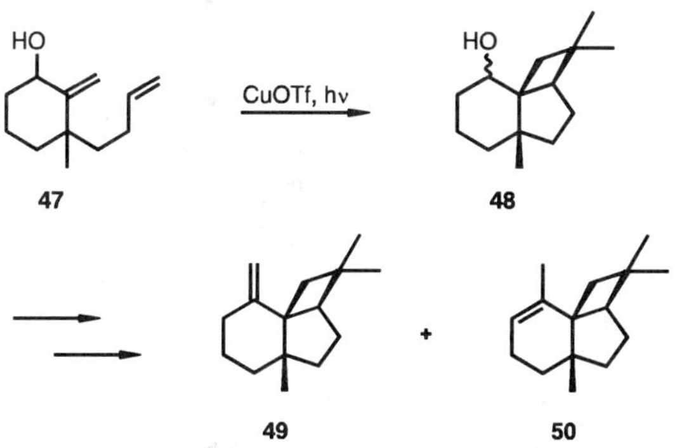

SCHEME 6-3.13

It is noteworthy that the copper(I)-catalyzed photocycloaddition of 47 generates the desired *all-cis* stereochemistry in the tricyclic product 48. The attempt to form the tricyclic compound by intermolecular photocycloaddition of *iso*-butene to the corresponding enone was not successful. Grandisol 51 (see Scheme 6-3.14), one of the four components of grandlure, the pheromone of the male boll weevil (*Anthonomus grandis*), can be synthesized in a similar manner.[33,47,48]

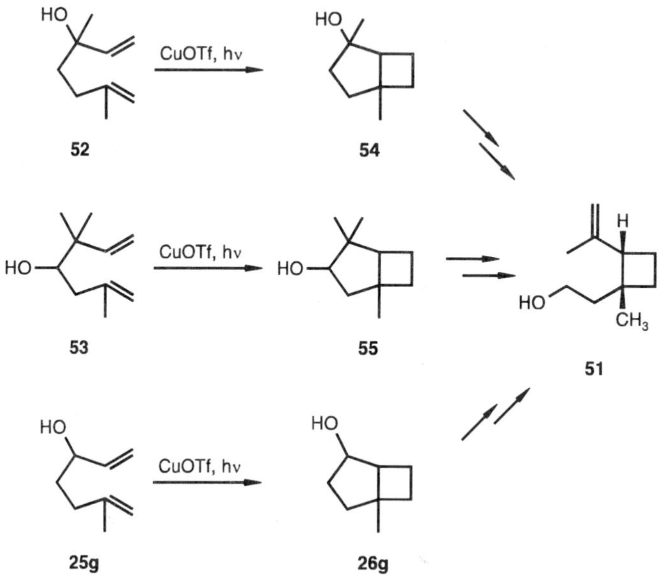

SCHEME 6-3.14

The copper(I)-catalyzed photocycloaddition of 25g, 52, or 53 generates two important features of grandisol. First, the cyclobutane structure is formed. Then, the *cis*-configuration of the two large substituents is fixed, due to the fact that the bicyclo[3.2.0]heptane structure formed is *cis*-fused. In addition, the proposed structures for robustadials A and B 56, isolated from leaves of *Eucalyptus robusta* and supposed to be an antimalaria agent, are synthesized using the copper(I)-catalyzed photocycloaddition as a key step in the conversion of 57 to 58 (Scheme 6-3.15).[49,50] A comparison of the spectra of the synthesized products with authentic robustadials shows clearly that the proposed structures are not correct. Further investigations suggest a bicyclo[3.3.1]heptane structure rather than a bicyclo[3.2.0]heptane structure.[51,52]

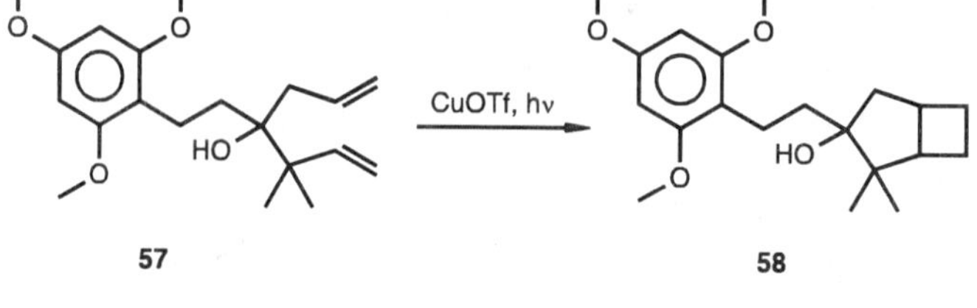

SCHEME 6-3.15

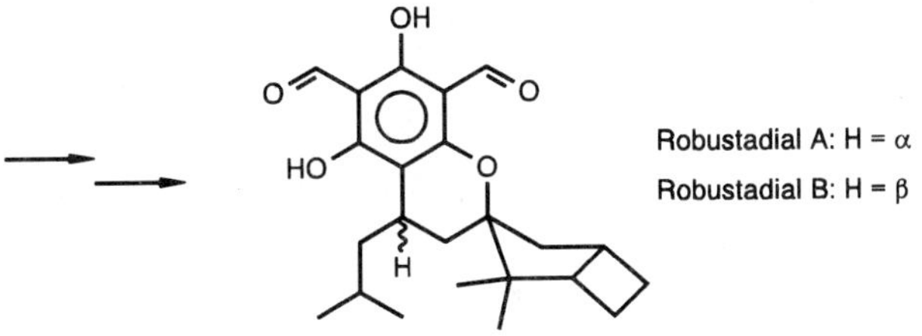
OH
O
O
HO
O
H
Robustadial A: H = α
Robustadial B: H = β

56

SCHEME 6-3.15 (continued)

6.4 Intermolecular Photocycloadditions

The dimerization of norbornene **59** was the first example of an intermolecular copper(I)-catalyzed [2+2]-photocycloaddition (Scheme 6-4.1).[53] Irradiation of norbornene in the presence of CuBr yields the *exo-trans-exo*-dimer **60** as the main product. Further investigations demonstrate that intermolecular cycloadditions between two acyclic double bonds do not occur.[20] Obviously, a strained, highly reactive double bond is necessary for photocycloaddition. Cyclopentene, cyclohexene, and cycloheptene are photochemically dimerized in the presence of CuOTf to form the cyclobutane derivatives **5** to **8** and **10** (see earlier).[20,31] In the case of *endo*-dicyclopentadiene **62**, the diene coordinates preferentially to the more strained 8,9-double bond as *exo*-monodentate ligand, leading to the dimer **63** (Scheme 6-4.2).[8,15] The second olefin for cycloaddition does not need to be highly reactive, since a mixed photocycloaddition occurs with norbornene and cyclooctene to yield **64**, while cyclooctene does not dimerize in the presence of CuOTf (Scheme 6-4.3).[15] It is remarkable that the mixed photolysis of cyclohexene with cycloheptene leads to the codimer **65** (yield: 37%), while in the presence of cyclopentene or cyclooctene only traces of co-dimers are detected (Scheme 6-4.4).[54]

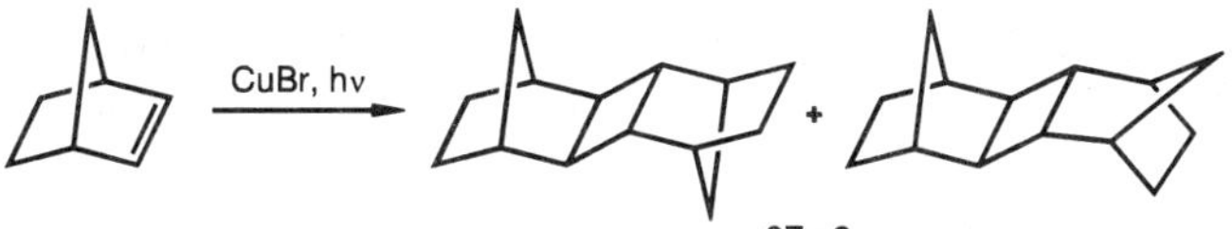
CuBr, hν
+

97 : 3

59 **60** **61**

SCHEME 6-4.1

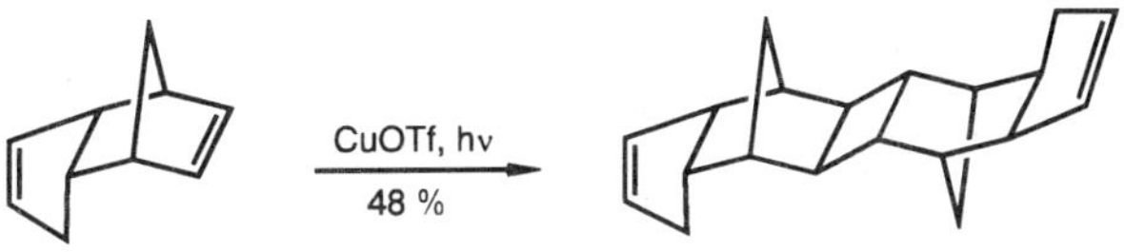
CuOTf, hν
48 %

62 **63**

SCHEME 6-4.2

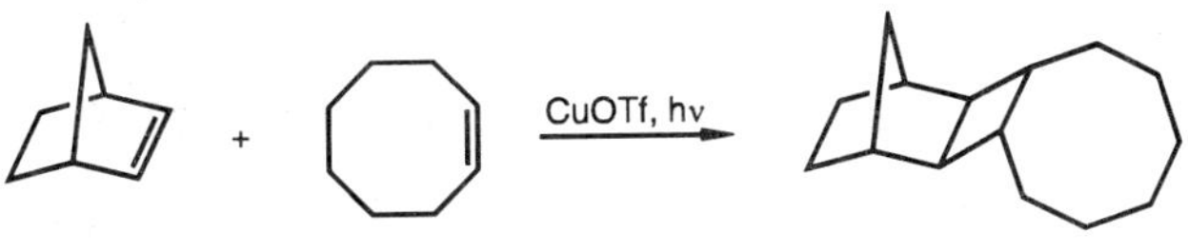
+
CuOTf, hν

59 **64**

SCHEME 6-4.3

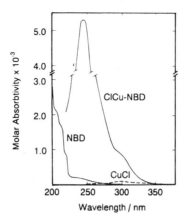

FIGURE 6-5.1 UV absorption spectra of a) 2×10^{-3} mol l^{-1} CuCl and 0.1 mol l^{-1} norbornadiene in ethanol, b) norbornadiene, and c) CuCl. *Source:* Schwendiman, D. P. and Kutal, C., *Inorg. Chem.,* 16, 719, 1977. With permission.

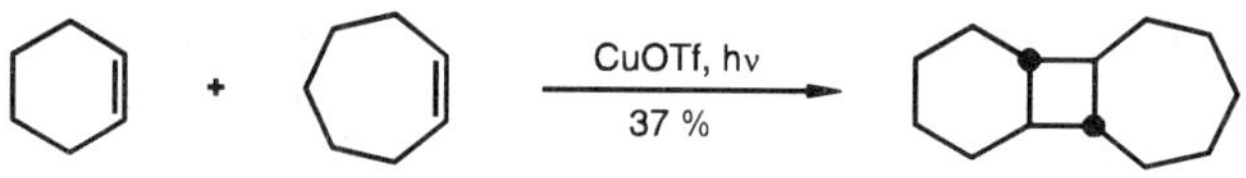

SCHEME 6-44

Cycloadditions of functionalized olefins are of special interest since further transformations are facilitated. CuOTf- catalyzed photolysis of norbornene in the presence of allyl alcohol produces the cyclobutane derivatives **66** and **67** (Scheme 6-4.5).[55] However, in the presence of cyclohexene, no dimerization occurs. Obviously, the very weak coordinating cyclohexene ligand cannot compete successfully with the allyl alcohol for the coordination sites on the copper(I) and the required intermediate complex is not formed.[31,55] Although no cyclization occurs in the presence of allyl alcohol, cyclohexene reacts with butadiene to the *trans*-fused cycloadducts **68, 69,** and **70** (Scheme 6-4.6).[27,28]

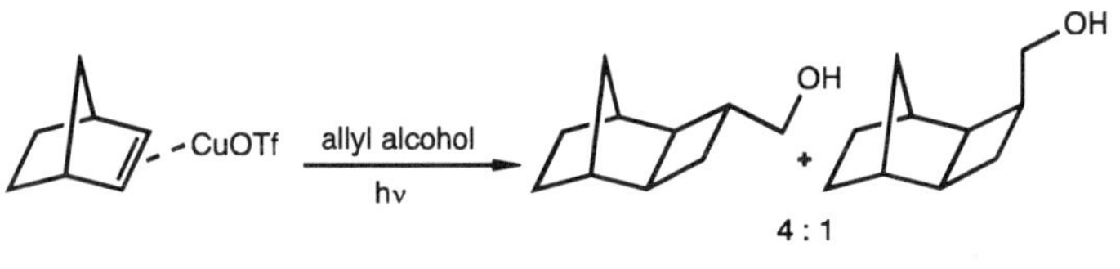

SCHEME 6-4.5

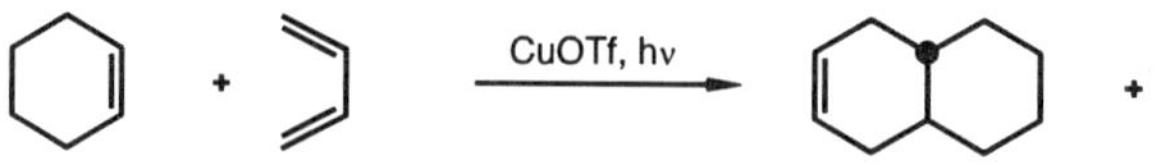

SCHEME 6-4.6

The reactivity of the dienes investigated increases with the number of electron-releasing substituents: butadiene < isoprene < 2,3-dimethyl-1,3-butadiene. Analogous to the dimerization path, a mechanism via *cis-trans*-isomerization of the alkene was proposed. The photochemically generated highly reactive CuOTf-*trans*-cyclohexene complex reacts in a thermal [4+2]-cycloaddition to the bicyclic adduct 68, which is converted to the tricyclic products 69 or 70 by repetitive addition of butadiene. The higher activity toward electron-rich dienes can be explained by an LMCT donation of the olefin to vacant Cu(I) *4s* orbitals. Cycloheptene shows the same reactions, but reacts much more slowly. Cyclopentene as well as cyclooctene and norbornene do not react with dienes under these conditions.

6.5 The Norbornadiene-Quadricyclane Isomerization

The photoassisted conversion of norbornadiene (NBD) 71 to quadricyclane (Q) 72 is considered for use as an energy storage system. The reaction is catalyzed effectively ($\Phi = 0.2$ to 0.4) by copper(I) chloride (Scheme 6-5.1).[56,57] Norbornadiene and copper(I) chloride form an *exo*-monodentate 1:1 complex with a formation constant $K > 10^2$, as confirmed by UV-absorption spectra and quantum yield experiments.[56–58] In ethanol the complex exhibits a UV absorption band at $\lambda = 248$ nm ($\varepsilon = 6300$) with a shoulder at about $\lambda = 300$ nm (Figure 6.5.1). Irradiation generates a charge-transfer excited state, but it is not evident if LMCT or MLCT occurs. Both directions of charge transfer will lead to a weakening of the bonds between C2-C3 as well as C5-C6, while the bonding between C2-C6 and C3-C5 is enhanced (Scheme 6-5.2).[56,57]

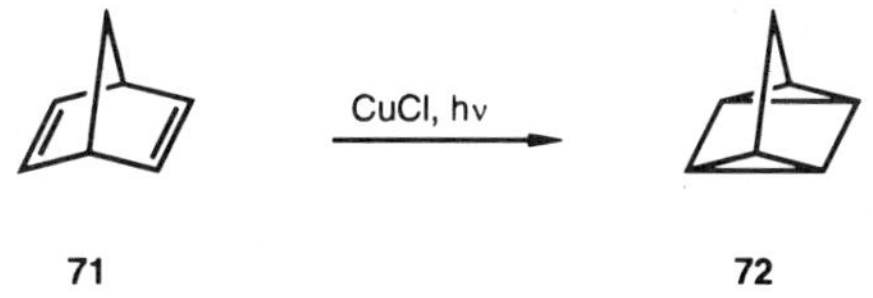

SCHEME 6-5.1

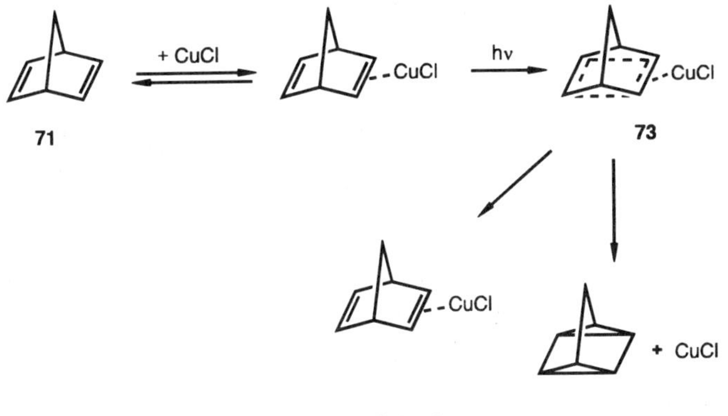

SCHEME 6-5.2

The subsequent relaxation of the excited state 73 is thought to occur via different pathways, leading to quadricyclane or back to NBD. The conversion may occur directly from the initially generated charge-transfer excited state or after intersystem crossing from the corresponding triplet state. Another pathway is by relaxation to a $\pi\pi^*$ triplet state of norbornadiene followed by conversion to quadricyclane. A completely alternative mechanism is the transformation of the excited state to a carbocation intermediate, which is converted to quadricyclane. This pathway seems to be unlikely, since no solvent effects on the quantum yield of the reaction are observed.[57]

In addition into copper(I) chloride, several copper(I) complexes of norbornadiene were investigated, and the reaction was shown to be remarkably insensitive to the presence and properties of the ligands on the copper. However, in the presence of strong coordinating ligands, the mechanism of the reaction changes.[59] If a ligand of the complex used is replaced very easily by norbornadiene, a copper(I)-norbornadiene ground-state complex is formed, and the reaction pathway is very similar to that in the case of copper(I) chloride. This behavior is observed in the case of **74** and **75**.[59,60] If very strong coordinating ligands are located at the copper(I) ion (**76** to **83**), a norbornadiene complex is not generated, as confirmed by UV, NMR, and IR spectroscopy and by quantum yield investigations.[59,61–64,66] Hence, the mechanism is fundamentally different from that observed for simple copper(I) salts. The reaction occurs via the bimolecular interaction of a photoexcited copper(I) complex with a ground-state norbornadiene molecule, presumably via electronic energy transfer or exciplex formation (Scheme 6-5.3).

$$Cu(I)L_4 \xrightarrow{\ h\nu\ } [Cu(I)L_4]^* \xrightarrow{\ NBD\ } [Cu(I)L_4{}^*NBD] \longrightarrow Q + Cu(I)L_4$$

SCHEME 6-5.3

Stern-Volmer type relationships suggest an electronic energy transfer process.[62,63] Since the copper(I) phosphine complexes **79** and **80** (Scheme 6-5.4) are known to undergo intersystem crossing with high quantum yield efficiency, ($\Phi_{isc} > 0.64$ resp. 0.78), a triplet energy transfer is likely to occur.[64] Thus, a $\pi\pi^*$ triplet excited state is generated, which undergoes an efficient conversion to quadricyclane.[65]

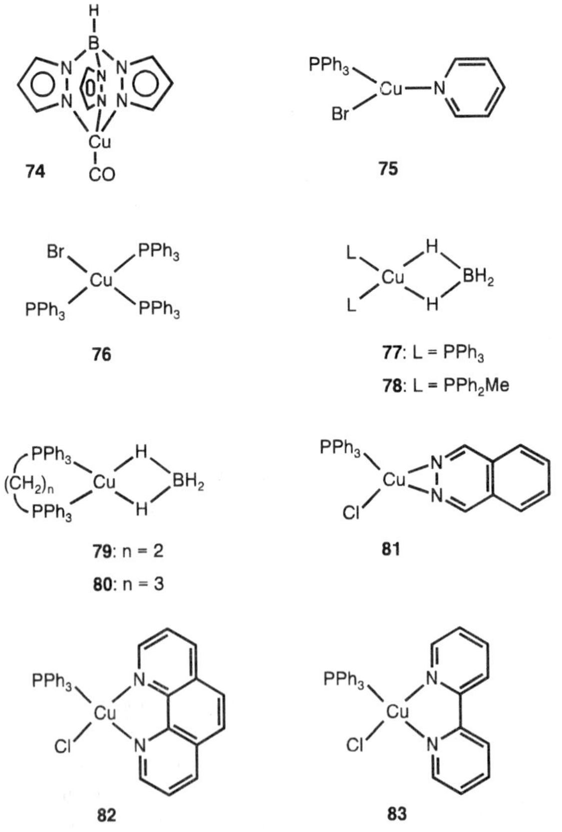

SCHEME 6-5.4

Besides copper(I) phosphine complexes, copper(I)-nitrogen ligand complexes **75** and **81** to **83** were investigated.[66] Isomerization is achieved by irradiation with $\lambda > 320$ nm, but the quantum yields ($\Phi < 0.2$) as well as the turnover numbers (<17) are quite low.

6.6 Conclusion

Copper(I)-catalyzed photocycloaddition reactions are valuable tools for special synthetic strategies. They exhibit high selectivity concerning the substrate, require small amounts of nontoxic reagents, tolerate many different functionalities at the substrate, and provide products in moderate to high yields. Although synthetic applications are well investigated, several mechanistic aspects remain uncertain, and the reasons for selectivities concerning the substrate have not been explained satisfactorily.

References

1. Koerner von Gustorf, E. A., Leenders, L. H. G., Fischler, I., and Perutz, R. N., Aspects of organo-transition-metal photochemistry and their biological implications, *Adv. Inorg. Chem. Radiochem.,* 65, 1976.
2. Hennig, H., Thomas, P., Wagener, R., Rehorek, D., and Jurdeczka, K., Koordinationsverbindungen als potentielle Photokatalysatoren, *Z. Chem.,* 17, 241, 1977.
3. Salomon, R. G., Homogeneous metal-catalysis in organic photochemistry, *Tetrahedron,* 39, 485, 1983, and references cited therein.
4. Kutal, C., Spectroscopic and photochemical properties of d^{10} metal complexes, *Coord. Chem. Rev.,* 99, 213, 1990, and references cited therein.
5. van den Hende, J. H. and Baird, W. C., The structure of the cuprous chloride-cyclooctadiene-1,5 complex, *J. Am. Chem. Soc.,* 85, 1009, 1963.
6. Ishino, Y., Ogura, T., Noda, K., Hirashima, T., and Manabe, O., Copper(I) perchlorate complexes of allyl, 2-methylallyl and *cis-* and *trans-*2-butenyl alcohols, *Bull. Chem. Soc. Jpn.,* 45, 150, 1972.
7. Salomon, R. G. and Kochi, J. K., Cationic benzene and olefin complexes of copper(I) trifluoromethanesulphonate, *J. Chem. Soc., Chem. Commun.,* 559, 1972.
8. Salomon, R. G. and Kochi, J. K., Cationic olefin complexes of copper(I). Structure and bonding in group Ib metal-olefin complexes, *J. Am. Chem. Soc.,* 95, 1889, 1973.
9. Evers, J. Th. M. and Mackor, A., Photocatalysis. IV. Preparation and characterization of a stable copper(I) triflate *trans-*cycloheptene complex, *J. R. Neth. Chem. Soc.,* 98, 423, 1979.
10. Dewar, M. J. S., A review of the π-complex theory, *Bull. Soc. Chim. Fr.,* C71, 1951.
11. Chatt, J. and Duncanson, L. A., Olefin coordination compounds. Part III. Infrared spectra and structure: Attempted preparation of acetylene complexes, *J. Chem. Soc.,* 2939, 1953.
12. Budzelaar, P. H. M., Timmermans, P. J. J. A., Mackor, A., and Baerends, E. J., Bonding in the ground state and excited states of copper alkene-complexes, *J. Organomet. Chem.,* 331, 397, 1987.
13. Salomon, R. G. and Kochi, J. K., Carbon-13 NMR spectra of olefin-copper(I) complexes, *J. Organomet. Chem.,* 64, 135, 1974.
14. Salomon, R. G. and Kochi, J. K., Structure and bonding in cationic olefin complexes of copper(I) by NMR, *J. Organomet. Chem.,* 43, C7, 1972.
15. Salomon, R. G. and Kochi, J. K., Copper(I) catalysis in photocycloadditions. I. Norbornene, *J. Am. Chem. Soc.,* 96, 1137, 1974.
16. Hurst, J. K. and Lane, R. H., Binuclear ions of copper(I) and certain transition metal complexes and kinetics of electron transfer between metal centers, *J. Am. Chem. Soc.,* 95, 1703, 1973.
17. Geiger, D. and Ferraudi, G., Photochemistry of Cu-olefin complexes: a flash photochemical investigation of the reactivity of Cu(ethylene)$^+$ and Cu(*cis,cis-*1,5-cyclooctadiene)$_2^+$, *Inorg. Chim. Acta,* 101, 197, 1985.

18. Salomon, R. G. and Salomon, M. F., Copper(I) catalysis of olefin photoreactions. Photorearrangement and photofragmentation of 7-methylenocarane, *J. Am. Chem. Soc.*, 98, 7454, 1976.

19. Salomon., R. G., Sinha, A., and Salomon, M. F., Copper(I) catalysis of olefin photoreactions. Photorearrangement and photofragmentation of methylenecyclopropanes, *J. Am. Chem. Soc.*, 100, 520, 1978.

20. Salomon, R. G., Folting, K., Streib, W. E., and Kochi, J. K., Copper(I) catalysis in photocycloadditions. II. Cyclopentene, cyclohexene, and cycloheptene, *J. Am. Chem. Soc.*, 96, 1145, 1974.

21. Srinivasan, R., Use of an olefin as a photochemical catalyst, *J. Am. Chem. Soc.*, 85, 3048, 1963.

22. Srinivasan, R., Photochemical transformations of 1,5-cyclooctadiene, *J. Am. Chem. Soc.*, 86, 3318, 1964.

23. Whitesides, G. M., Goe, G. L., and Cope, A. C., Irradiation of *cis,cis*-1,5-cyclooctadiene in the presence of copper(I) chloride, *J. Am. Chem. Soc.*, 91, 2608, 1969.

24. Grobbelaar, E., Kutal, C., and Orchard, W., Importance of ground state complex formation in the cuprous chloride sensitized photoisomerization of *cis,cis*-1,5-cyclooctadiene to tricyclo[3.3.0.0^{2,6}]-octane, *Inorg. Chem.*, 21, 414, 1982.

25. Baldwin, J. E. and Greeley, R. H., Cycloadditions. IV. Mechanism of the photoisomerization of *cis,cis*-1,5-cyclooctadiene to tricyclo[3.3.0.0.^{2,6}]octane, *J. Am. Chem. Soc.*, 87, 4515, 1965.

26. Haller, I. and Srinivasan, R., Mechanism of the photoisomerization of *cis,cis*-cyclooctadiene to tricyclo[3.3.0.0^{2,6}]octane, *J. Am. Chem. Soc.*, 88, 5084, 1966.

27. Evers, J. Th. M. and Mackor, A., Photocatalysis. II. Photochemical cycloadditions of cyclohexenes and cycloheptene with conjugated dienes catalyzed by copper(I) trifluoromethane sulphonate, *Tetrahedron Lett.*, 26, 2317, 1978.

28. Evers, J. Th. M. and Mackor, A., Photocatalysis. III. Photochemical isomerization of cyclohexenes and cycloheptene in the presence of copper(I) trifluoromethanesulphonate. Identification and acid-catalyzed isomerization of the products, *Tetrahedron Lett.*, 2321, 1978.

29. Evers, J. Th. M. and Mackor, A., Photocatalysis. V. Cyclotrimerization of cycloheptene, *Tetrahedron Lett.*, 415, 1980.

30. Trecker, D. J., Foote, R. S., Henry, J. P., and McKeon, J. E., Photochemical reactions of metal-complexed olefins. II. Dimerization of norbornene and derivatives, *J. Am. Chem. Soc.*, 88, 3021, 1966.

31. Salomon, R. G. and Kochi, J. K., Copper(I) triflate: A superior catalyst for olefin photodimerization, *Tetrahedron Lett.*, 2529, 1973.

32. Hertel, R., Mattay, J., and Runsink, J., Copper(I) catalyzed intramolecular diene-diene cycloaddition reactions and rearrangements, *J. Am. Chem. Soc.*, 113, 657, 1991.

33. Langer, K., Mattay, J., Heidbreder, A., and Möller, M., A new stereoselective synthesis of grandisol, *Liebigs Ann. Chem.*, 257, 1992.

34. Evers, J. Th. M. and Mackor, A., Photocatalysis. I. Copper(I) trifluoromethane sulphonate catalyzed photochemical reactions of unsaturated ethers and alcohols, *Tetrahedron Lett.*, 821, 1978.

35. Avasthi, K., Raychaudhuri, S. R., and Salomon, R. G., Synthesis of vinylcyclobutanes via copper(I) catalyzed intramolecular 2π + 2π photocycloadditions of conjugated dienes to alkenes, *J. Org. Chem.*, 49, 4322, 1984.

36. Ghosh, S., Raychaudhuri, S. R., and Salomon, R. G., Synthesis of cyclobutanated butyrolactones via copper(I) catalyzed intramolecular photocycloadditions of homoallyl vinyl or diallyl ethers, *J. Org. Chem.*, 52, 83, 1987.

37. Raychaudhuri, S. R., Ghosh, S., and Salomon, R. G., Copper(I) catalysis of olefin photoreactions. 11. Synthesis of multicyclic furans and butyrolactones via photobicyclization of homoallyl vinyl and diallyl ethers, *J. Am. Chem. Soc.*, 104, 6841, 1982.

38. Salomon, R. G., Coughlin, D. J., and Easler, E. M., Copper(I) catalysis of olefin photoreactions. 8. A stepwise olefin metathesis synthesis of cyclopent-2-en-1-ones via photobicyclization of 3-hydroxyhepta-1,6-dienes, *J. Am. Chem. Soc.*, 101, 3961, 1979.

39. Salomon, R. G., Coughlin, D. J., Ghosh, S., and Zagorski, M. G., Copper(I) catalysis of olefin photoreactions. 9. Photobicyclization of α, β, and γ-alkenylallyl alcohols, *J. Am. Chem. Soc.*, 104, 998, 1982.

40. Salomon, R. G., Ghosh, S., Zagorski, M. G., and Reitz, M., Copper(I) catalysis of olefin photoreactions. 10. Synthesis of multicyclic carbon networks by photobicyclization, *J. Org. Chem.*, 47, 829, 1982.

41. Salomon, R. G. and Ghosh, S., Copper(I) catalyzed photocycloaddition: 3,3-dimethyl-*cis*-bicyclo[3.2.0]heptan-2-one, *Org. Synth.*, 62, 125, 1984.

42. Hertel, R. and Mattay, J., unpublished results (Hertel, R., Diploma thesis, Aachen, 1987).

43. Salomon, R. G. and Avasthi, K., A copper(I) catalyzed photobicyclization route to *exo*-1,2-polymethylene- and 7-hydroxynorbornanes. Nonclassical 2-bicyclo[3.2.0]heptyl and 7-norbornyl carbenium intermediates, *J. Org. Chem.*, 51, 2556, 1986.

44. Langer, K. and Mattay, J., unpublished results.

45. Salomon, R. G., Ghosh, S., Raychaudhuri, S. R., and Miranti, T. S., Synthesis of multicyclic pyrrolidines via copper(I) catalyzed photobicyclization of ethyl N,N-diallyl carbamates, *Tetrahedron Lett.*, 25, 3167, 1984.

46. McMurry, J. E. and Choy, W., Total synthesis of α- and β-panasinsene, *Tetrahedron Lett.*, 21, 2477, 1980.

47. Rosini, G., Marotta, E., Petrini, M., and Ballini, R., Stereoselective total synthesis of racemic grandisol. An improved convenient procedure, *Tetrahedron*, 41, 4633, 1985.

48. Rosini, G., Geier, M., Marotta, E., Petrini, M., and Ballini, R., Stereoselective total synthesis of racemic grandisol via 3-oximo-1,4,4-trimethylbicyclo[3.2.0]heptane. An improved practical procedure, *Tetrahedron*, 42, 6027, 1986.

49. Lal, K., Zarate, E. A., Youngs, W. J., and Salomon, R. G., Total synthesis necessitates revision of the structure of robustadials, *J. Am. Chem. Soc.*, 108, 1311, 1986.

50. Lal, K., Zarate, E. A., Youngs, W. J., and Salomon, R. G., Robustadials. 2. Total synthesis of the bicyclo[3.2.0]heptane structure proposed for robustadials A and B, *J. Org. Chem.*, 53, 3673, 1988.

51. Mazza, S. M., Lal, K., and Salomon, R. G., Robustadials. 3. Total synthesis of camphane analogues, *J. Org. Chem.*, 53, 3681, 1988.

52. Jirousek, R. J., Mazza, S. M., and Salomon, R. G., Robustadials. 4. Molecular mechanism and nuclear magnetic resonance studies of conformational and configurational equilibria: 3,4-Dihydro[2H-1-benzopyran-2,2'-bicyclo[2.2.1]heptanes], *J. Org. Chem.*, 53, 3688, 1988.

53. Trecker, D. J., Henry, J. P., and McKeon, J. E., Photodimerization of metal-complexed olefins, *J. Am. Chem. Soc.*, 87, 3261, 1965.

54. Timmermans, P. J. J. A., de Ruiter, G. M. J., Tinnemans, A. H. A., and Mackor, A., Photocatalysis. VIII. (Stereo)selective photochemical codimerization of cyclohexene and cycloheptene, catalyzed by copper(I) triflate, *Tetrahedron Lett.*, 24, 1419, 1983.

55. Salomon, R. G. and Sinha, A., Copper(I) catalyzed 2π+2π photocycloadditions of allyl alcohol, *Tetrahedron Lett.*, 1367, 1978.

56. Schwendiman, D. P. and Kutal, C., Transition metal photoassisted valence isomerization of norbornadiene. An attractive energy storage reaction, *Inorg. Chem.*, 16, 719, 1977.

57. Schwendiman, D. P. and Kutal, C., Catalytic role of copper(I) in the photoassisted valence isomerization of norbornadiene, *J. Am. Chem. Soc.*, 99, 5677, 1977.

58. Baenziger, N. C., Haight, H. L., and Doyle, J. R., Metal olefin compounds. VII. The crystal and molecular structure of cyclo-tetra-μ-chloro-tetrakis[bicyclo[2.2.1]hepta-2π,5-dienecopper(I)], *Inorg. Chem.*, 3, 1535, 1964.

59. Borsub, N., Chang, S., and Kutal, C., Ligand control of the mechanism of photosensitation by copper(I) compounds, *Inorg. Chem.*, 21, 538, 1982.

60. Sterling, R. F. and Kutal, C., Photoconversion of norbornadiene to quadricyclene in the presence of a copper(I) carbonyl compound, *Inorg. Chem.*, 19, 1502, 1980.

61. Grutsch, P. A. and Kutal, C., Use of copper(I) phosphine compounds to photosensitize the valence isomerization of norbornadiene, *J. Am. Chem. Soc.*, 99, 6460, 1977.
62. Grutsch, P. A. and Kutal, C., Photobehavior of copper(I) compounds. Role of copper(I) phosphine compounds in the photosensitized valence isomerization of norbornadiene, *J. Am. Chem. Soc.*, 101, 4228, 1979.
63. Orchard, S. W. and Kutal, C., Photosensitation of the norbornadiene to quadricyclene rearrangement by an electronically excited copper(I) compound, *Inorg. Chim. Acta*, 64, L95, 1982.
64. Liaw, B., Orchard, S. W., and Kutal, C., Photobehavior of copper(I) compounds. 4. Role of the triplet state of (arylphosphine)copper(I) complexes in the photosensitized isomerization of dienes, *Inorg. Chem.*, 27, 1311, 1988.
65. Murov, S. and Hammond, G. S., Mechanisms of photochemical reactions in solutions. LVI. A singlet sensitized reaction, *J. Phys. Chem.*, 72, 3797, 1968.
66. Maruyama, K., Terada, K., Naruta, Y., and Yamamoto, Y., Photoisomerization of norbornadiene to quadricyclane in the presence of copper(I)-nitrogen ligand catalysts, *Chem. Lett.*, 1259, 1980.

7

Photoreactions of Alkenes in Protic Media

ul J. Kropp
iversity of North Carolina at
ipel Hill

7.1 Photochemical Behavior

Principal Pathways

On sensitized irradiation, alkenes undergo principally $E \, 1 \, Z$ isomerization, unless it is structurally inhibited. In weakly protic media, the highly strained E-isomers arising from cyclohexenes, -heptenes, and -octenes undergo protonation. Similar behavior is exhibited on direct irradiation. In addition, on direct irradiation, styrenes undergo protonation whether they are cyclic or acyclic, and alkenes undergo nucleophilic trapping in hydroxylic solvents to afford alcohols or ethers. This chapter is concerned principally with this latter process, as well as photoprotonation of cycloalkenes. Emphasis is given to developments since the topic was reviewed extensively in 1979.[1] $E \, 1 \, Z$ Isomerization is discussed in Chapter 1 and the photobehavior of alkenes on direct irradiation in nonhydroxylic media in Chapter 2.

Spectroscopic Properties

The UV absorption spectra of simple alkenes in solution consist typically of single, featureless bands with maxima at 185 to 195 nm ($\varepsilon = 5$–10×10^3) and tails extending to 210 to 255 nm.[2] Absorption extends to longer wavelengths with increasing degrees of substitution. Hydroxylic solvents, which are frequently used as proton sources, have cutoffs at 205 to 210 nm. Highly substituted alkenes can be irradiated in such media with a medium-pressure mercury lamp and quartz optics, but less highly substituted alkenes do not absorb light under these conditions. Since alkenes do not undergo intersystem crossing, sensitized irradiation is required to induce triplet photobehavior. Best results are obtained using aromatic sensitizers such as p-xylene or phenol, which have maximum absorption in the region 250 to 270 nm. For this, either a low- or medium-pressure arc and quartz or Vycor optics can be used. A preparative procedure has been described.[3] Phenol can serve as both the sensitizer and a proton source.[4] It has the added advantage of being easily removed from the irradiation mixture by extraction with base.

In the triplet manifold of alkenes, one excited state, π,π^*, is clearly low lying, separated from the $\pi,R(3s)$ Rydberg triplet by approximately 2 eV.[5] By contrast, as is detailed in Chapter 2, two or more excited states are low lying and close in energy in the singlet manifold — including the π,π^* and $\pi,R(3s)$ Rydberg states, which do not communicate effectively. Hence, the sensitized photobehavior of alkenes is usually much simpler than that resulting from direct irradiation.

7.2 Examples and Mechanisms

Sensitized Irradiation

On sensitized irradiation, acyclic and large-ring cyclic alkenes having a nine-membered ring or larger undergo only $E\ 1\ Z$ isomerization, whether in protic or aprotic media.[6] By contrast, sensitized irradiation of cyclohexenes, -heptenes, and -octenes in protic media results in photoprotonation.[7] Thus, for example, *p*-xylene-sensitized irradiation of 1-methylcyclohexene [(*Z*)-1b] in methanol affords a mixture of the exocyclic isomer 3b and the ether 4b.[8] The cycloheptyl and -octyl analogs (*Z*)-1c and -1d afford principally the corresponding ethers 4c-d, accompanied by small amounts of the exocyclic isomers 3c-d. Similar behavior is exhibited by the unsubstituted analogs (*Z*)-5b-d, which afford the ethers 6b-d.

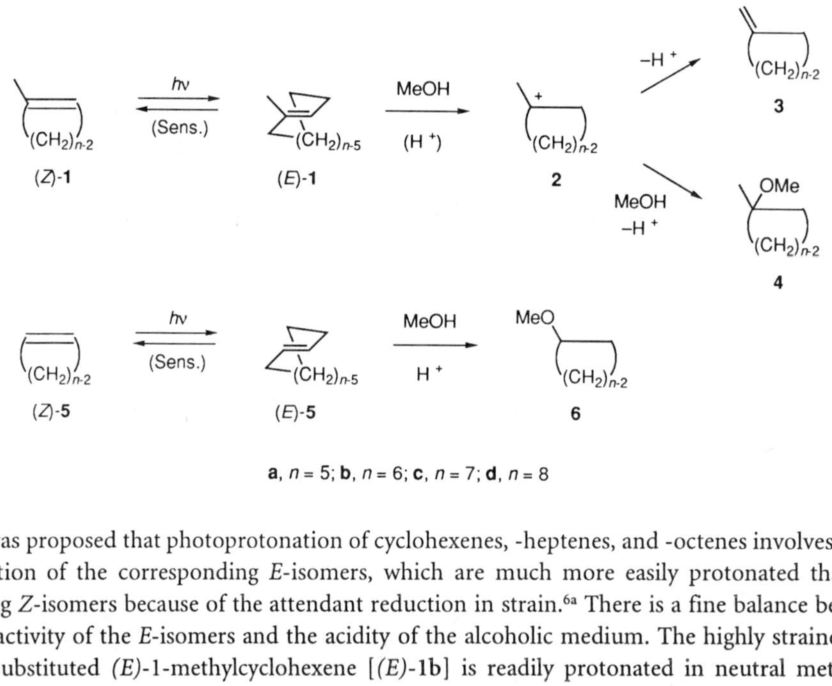

a, *n* = 5; b, *n* = 6; c, *n* = 7; d, *n* = 8

It was proposed that photoprotonation of cyclohexenes, -heptenes, and -octenes involves initial formation of the corresponding *E*-isomers, which are much more easily protonated than the starting *Z*-isomers because of the attendant reduction in strain.[6a] There is a fine balance between the reactivity of the *E*-isomers and the acidity of the alcoholic medium. The highly strained and alkyl-substituted (*E*)-1-methylcyclohexene [(*E*)-1b] is readily protonated in neutral methanol. However, the higher homologs (*E*)-1c-d, as well as the unsubstituted analogs (*E*)-5b-d, require the presence of small amounts of mineral acid (e.g., 0.3 to 0.6% H_2SO_4), conditions under which the *E*-isomers are protonated but the *Z*-isomers are stable.

Cyclooctenes

Substantial support for the proposed formation of *E*-cycloalkene intermediates on sensitized irradiation and their involvement in photoprotonation has been obtained. Irradiation of *Z*-cyclooctene [*Z*-5d] with the triplet sensitizers benzene, toluene, or xylene afforded a mixture of the *E*- and *Z*-isomers with a photostationary *E:Z* ratio of 0.05.[9] The use of methyl benzoate as the sensitizer, which involves the formation of a singlet exciplex followed by decay to the twisted π,π^*

singlet excited state, gave a higher *E:Z* ratio of 0.36:1.[10] The decay ratio to the *E-* vs. *Z-*isomer is higher for the singlet than for the triplet excited state, providing a simple method for the preparation of moderate amounts of the *E-*isomer. The highly strained *(E)*-5d was found to undergo acid-catalyzed addition of methanol 3000 times more rapidly than the *Z-*isomer.[11]

Cycloheptenes

Methyl benzoate-sensitized irradiation of *(Z)*-cycloheptene [*(Z)*-5c] in neutral methanol at −78 °C afforded similarly a mixture of the E- and Z-isomers with an estimated *E:Z* photostationary ratio of 0.25:1.[12] The *E-*isomer is stable in methanol at −78 °C but, on being warmed to room temperature in the absence of dioxygen, underwent unimolecular decay to the *Z-*isomer quantitatively with E_a = 17.4 ± 0.7 kcal mol⁻¹, $\Delta G\ddagger_{266}$ = 19.4 ± 1.4 kcal mol⁻¹, $\Delta H\ddagger_{266}$ = 17.0 ± 0.7 kcal mol⁻¹, and $\Delta S\ddagger_{266}$ = −9 ± 8 kcal K⁻¹ mol⁻¹. The lifetime in methanol at −10 °C was 38.3 min. In the presence of air, isomerization occurs by an alternative mechanism involving a termolecular complex between triplet dioxygen and two molecules of *(E)*-5c.[13]

Irradiation of *(Z)*-cycloheptene [*(Z)*-5c] in acidic methanol at −78 °C afforded ether 6c. When an irradiation was conducted in neutral methanol and the resulting irradiation mixture allowed to stand in the dark at −78 °C for 24 h before the addition of acid, ether 6c was obtained in a similar yield. This clearly rules out the protonation of an excited state in the formation of ether 6c. A kinetic study showed that *(E)*-cycloheptene [*(E)*-5c] undergoes protonation 7.7×10^8 times faster than its *Z-*isomer and 3.3×10^5 times faster than its higher homolog *(E)*-cyclooctene [*(E)*-5d].

Generation of *(E)*-cycloheptene [*(E)*-5c] by methyl benzoate irradiation has also permitted its ¹H-NMR and UV spectra to be obtained.[14] The UV maximum is shifted 40 to 50 nm to the red compared with the *Z-*isomer, suggesting substantial twist about the double bond; but the vinyl proton coupling shows that these atoms have very close to a 180° dihedral angle. Apparently, the vinyl carbon atoms mainly pyramidalize, rather than twist, to accommodate the strained ring.

Evidence has also been obtained for the intermediacy of the *(E)*-cycloheptene *(E)*-7 in the formation of methyl ethers 8 and 9 on direct, *p*-xylene-sensitized, and methyl benzoate-sensitized irradiation of the 3,7-disilacycloheptene *(Z)*-7 in acidified methanol.[15] Sensitized irradiation of *(Z)*-7 in methylcyclohexane solution at −75 °C followed by siphoning of the irradiation mixture into acidified methanol similarly afforded ethers 8 and 9, showing that a long-lived intermediate is involved. Addition of cyclopentadiene to the cold irradiation mixture in the dark afforded a Diels-Alder adduct shown to have a *trans* ring juncture.

Cyclohexenes

Although having much shorter lifetimes because of their higher strain, a number of *(E)*-cyclohexene derivatives have been observed on direct or photosensitized irradiation of the Z-isomers using flash spectroscopic techniques. The first observation of an *(E)*-cyclohexene derivative involved 1-phenylcyclohexene (*cis*-10), which afforded a transient having a lifetime of 9 μs in methanol at 25 °C.[16] On either direct or sensitized irradiation in methanol, *cis*-10 affords the ether 12.[17] On direct irradiation, a diastereisomeric mixture of the cyclobutane dimers 11 is also formed.[18] Kinetic studies showed that ether 12 is formed on direct and sensitized irradiation via a common intermediate having the same lifetime as the spectroscopic transient.[19] The proposed *trans*-stereochemistry for the transient was supported by the finding that direct or sensitized irradiation of *cis*-10 in neutral methanol at −75 °C afforded dimer 13, which was shown to have a *trans*-juncture by X-ray crystallographic analysis.[20] Time-resolved photoacoustic calorimetry has shown that the enthalpy of isomerization to the highly strained *trans* isomer is 45 kcal mol⁻¹.[21]

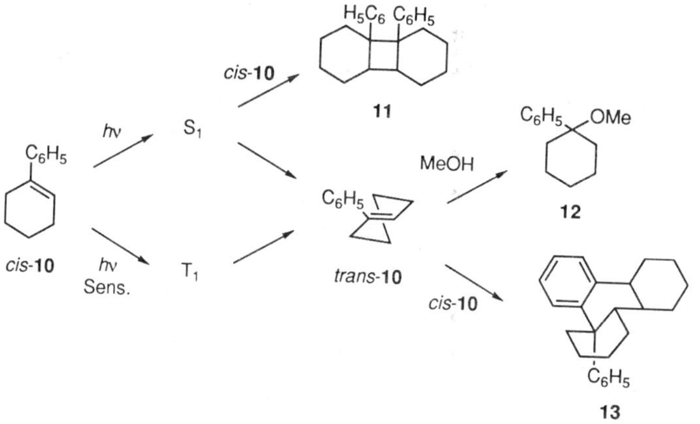

Either direct or triplet-sensitized irradiation of diene (E,E)-14 in methanol affords the cis-cyclobutene 15 and the methyl ethers 16 and 17.[8b,22] Laser flash spectroscopy revealed the formation of a ground-state intermediate having λ_{max} = 360 nm and a lifetime of 0.8 µs at 23 °C in methanol.[22] Based on the change of the quantum yields for the formation of ethers 16 and 17 with E_T for a variety of sensitizers, it was concluded that cyclobutene 15 and ether 16 arise from the s-Z conformation of isomer (E,Z)-14 and ether 17 from the s-E conformation.[23] In view of large deuterium isotope effects, k_H/k_D, of 8 ± 1 and 10 ± 2 that were observed for the formation of ethers 16 and 17, respectively, in CH₃OD, concerted formation of the C-H and C-O bonds in the transition states for ether formation was proposed.

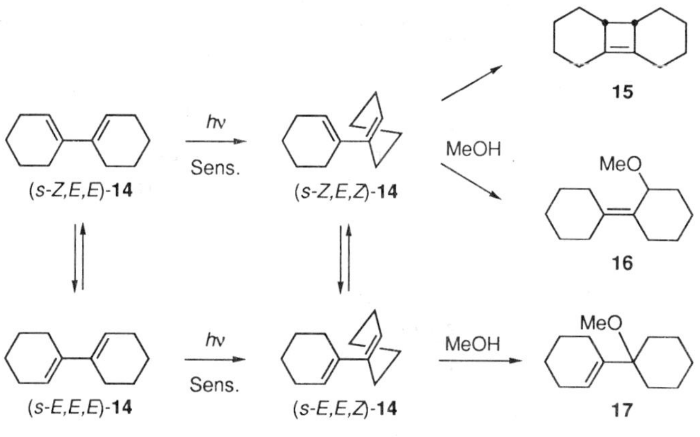

The spectroscopic and kinetic properties of five additional (E)-cyclohexene derivatives have recently been determined.[24] The properties of all of the (E)-cyclohexene derivatives that have been reported are remarkably constant: the π,π* maxima of the (E)-isomers are shifted generally to the red by about 15,000 cm⁻¹ relative to the Z-forms; the activation energy for the thermal E → Z isomerization is about 10 kcal mol⁻¹ and the corresponding preexponential factor is in the range 10¹² to 10¹³ s⁻¹; and they undergo extremely efficient protonation by acid. Ab initio calculations have confirmed that the parent (E)-cyclohexene [(E)-5b] corresponds to a local minimum, although the activation energy for its isomerization to the Z-isomer is small.[25]

Support for the involvement of (E)-cyclohexenes in photoprotonation has also come from the stereochemistry of protonation.[26] Direct or sensitized irradiation of (Z)-bicyclo[3.3.1]non-1-ene [(Z)-18] in neutral methanol afforded ether 19.[2a] On irradiation in CH₃OD, the labeled ether

19-*endo-2d* was obtained, whereas reaction of (Z)-18 with CH$_3$OD catalyzed by DCl gave ether 19-*exo-2d*. These results are consistent with preferential protonation of alkenes (Z)- and (E)-18 from their less-hindered faces. Protonation of (E)-18 was found to have a large kinetic isotope effect, k_H/k_D, of 8, indicating that proton transfer is about half completed in the transition state.

Cyclopentenes

In contrast with their higher homologs, cyclopentenes display no ionic behavior on sensitized irradiation in protic media. For example, sensitized irradiation of 1-methylcyclopentene [(Z)-1a] in either neutral or acidified methanol affords a mixture of the exocyclic isomer 3a, the saturated product 21, and ethane-1,2-diol — as expected from initial abstraction of a hydrogen atom from the solvent by the π,π^* excited state followed by disproportionation of the resulting radical 20.[8] In CH$_3$OD, the photoproducts 3a and 21 are formed with no detectable incorporation of deuterium. Thus, cyclohexene is apparently the smallest-sized cycloalkene that can accommodate an E double bond. Being unable to relax to an orthogonal conformation, the π,π^* triplet of cyclopentene intersystem crosses to the ground state more slowly and has a sufficiently long lifetime to undergo intermolecular reaction, in which it displays radical behavior.

Direct Irradiation

Alkenes

Alkenes undergo $E \rightleftharpoons Z$ isomerization on direct, as well as sensitized, irradiation. However, since there are two or more low-lying excited states in the singlet manifold, the singlet photobehavior of alkenes is complex. On direct irradiation in hydroxylic solvents, tetrasubstituted alkenes afford, in addition to $E \rightleftharpoons Z$ isomerization, mixtures of saturated and unsaturated alcohols or ethers.[8b,27] Thus, for example, 2,3-dimethyl-2-butene (22) affords a mixture of the alcohols or ethers 28 and 29, which apparently arise via nucleophilic trapping of the $\pi,R(3s)$ Rydberg excited state (23) followed by disproportionation of the resulting radical 25.[8b] Small amounts of the hydrocarbons 26 and 27 are also formed, apparently via trapping of hydrogen atoms formed by attack of free electrons on the solvent. Since formation of the $\pi,R(3s)$ state involves promotion of one of the π electrons to an orbital having approximately the size and shape of a 3s helium orbital, it has an electron-deficient core with radical cation character. Undergoing nucleophilic trapping in hydroxylic media was the first chemical property of this state to be established. In nonnucleophilic media, it rearranges to carbene intermediates, as detailed in Chapter 2.

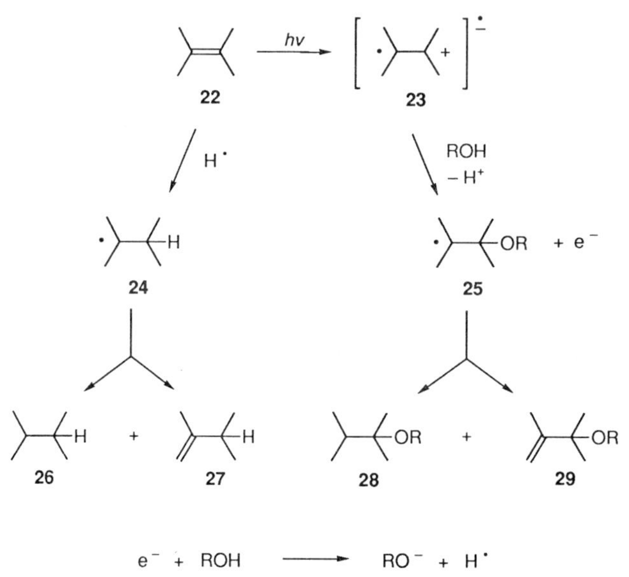

The cyclic analog (Z)-30, which on *p*-xylene-sensitized irradiation in methanol gives the epimeric ethers 31 and the exocyclic isomer 32, affords in addition the unsaturated ethers 34 and 35 on direct irradiation.[8b] There is apparently competing protonation of the *E*-isomer [-(E)-30] and nucleophilic trapping of the π,R(*3s*) excited state (33). This is consistent with the π,π* and π,R(*3s*) states being close lying and both readily populated in tetrasubstituted alkenes.

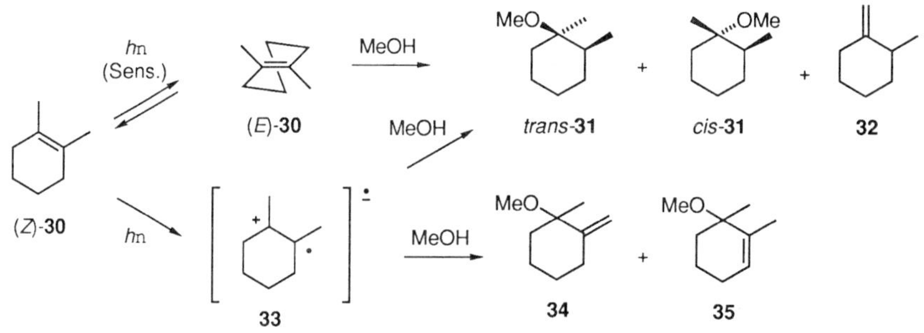

The trisubstituted alkene 36 affords similarly a mixture of the saturated and unsaturated ethers 37 and 38 and hydrocarbons 39 to 41 in methanol, but only on extended irradiation.[8b] By contrast, the cyclic analog 1-methylcyclohexene [(Z)-1b] affords only the exocyclic isomer 3b and the saturated ether 4b on direct irradiation in methanol.[8b] Similarly, the higher homologs (Z)-1c-d afford ethers 4b-c, accompanied by small amounts of the exocyclic isomers 3c-d, in acidified methanol. Apparently the π,R(*3s*) excited state, which is significantly more sensitive to alkyl substitution, lies higher than the π,π* state in trisubstituted alkenes. Disubstituted alkenes do not have sufficient absorption above the cutoffs of hydroxylic solvents to give photoproducts in these media.

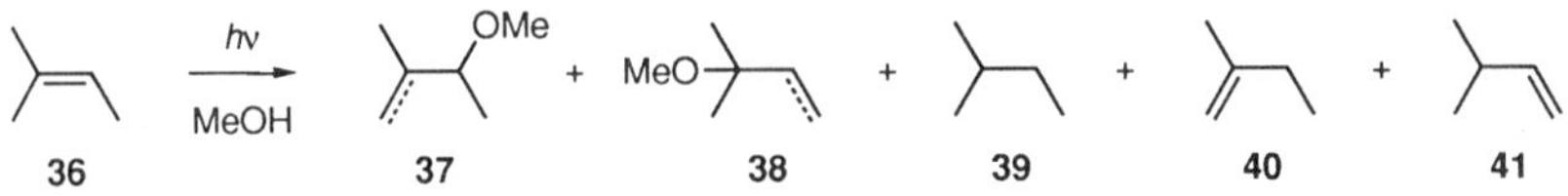

Styrenes

Acyclic styrenes (42) undergo Markovnikov addition on direct, but not sensitized, irradiation in water or methanol.[28] The reaction is acid-catalyzed but employs considerably lower acid concentrations or weaker acids than required for ground-state addition. The proposed mechanism is protonation of the first excited singlet state, which is estimated to occur 10^{11} to 10^{14} times more rapidly than protonation of the ground state. Similar behavior is exhibited by the cyclic analog 43, which on direct irradiation in aqueous or alcoholic media affords principally the alcohol or ether 44, accompanied by smaller amounts of an epimeric mixture of the reduction products 45.[29] Labeling studies showed that both types of products are formed via initial 3-*exo* protonation. Again, no reaction occurs on photosensitized irradiation. However, as noted above, the larger-ring analog *cis*-1-phenylcyclohexene (*cis*-10) undergoes addition of methanol on either direct or sensitized irradiation, with protonation in both cases involving the *trans*-isomer.[19] Thus, in contrast with acyclic and small-ring cyclic styrenes, in which the first singlet excited state undergoes protonation, the highly strained *trans*-isomers of medium-ring cyclic styrenes are apparently protonated more rapidly.

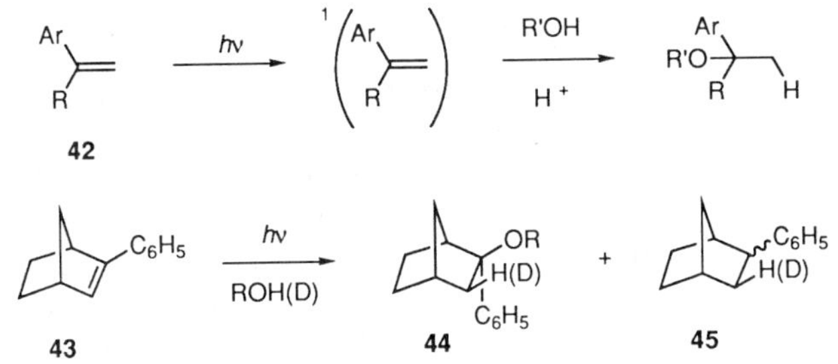

7.3 Synthetic Applications of Photoprotonation

There are several synthetically useful applications of the photoprotonation of cycloalkenes:

1. *Irradiation of 1-alkylcyclohexenes, -heptenes, and -octenes in non-nucleophilic, protic media is a useful method for effecting contrathermodynamic isomerization to the exocyclic isomer.* For example, sensitized irradiation of 1,2-dimethylcyclohexene [(Z)-30] in ether containing a trace of sulfuric acid afforded the exocyclic isomer 32 in excellent yield.[30] In the absence of competing nucleophilic trapping, the cycloalkyl cation 46 resulting from photoprotonation simply undergoes deprotonation to afford a mixture of exocyclic and endocyclic alkenes. Any endocyclic alkene formed undergoes photoprotonation again and is recycled. The exocyclic isomer, on the other hand, is photoinert under these conditions except for E 1 Z isomerization, which simply involves interchanging the positions of the vinyl hydrogen atoms and does not produce a strained E intermediate. Hence, there is a net isomerization of the double bond to the thermodynamically less stable exocyclic position.

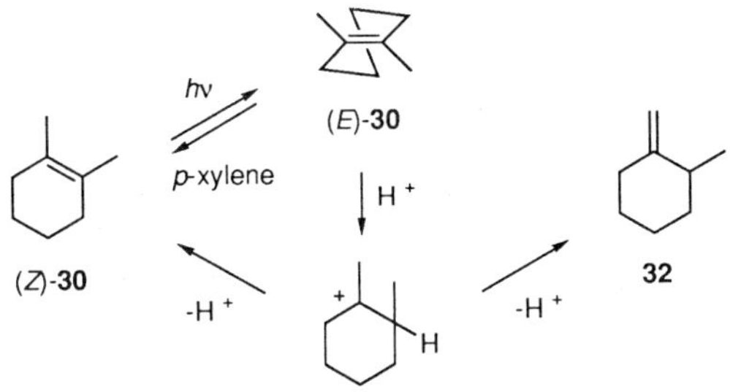

2. *Photoprotonation is a convenient method for effecting complete addition of water or an alcohol to a cyclohexene, -heptene, or -octene.* Acid-catalyzed addition is reversible and does not go to completion, whereas the photochemical method proceeds until all of the alkene has been consumed. Moreover, since photoprotonation can be effected at low acid concentrations, even less highly substituted cycloalkenes, which are not readily protonated in the ground state, undergo photoprotonation readily. In many cases, photoprotonation affords addition when acid-catalyzed addition fails. For example, sensitized irradiation of cyclohexene [(*Z*)-5b] in methanol containing 0.6% H₂SO₄ affords ether **6b** in excellent yield; in a preparative procedure, it was isolated in 70% yield.[31] By contrast, treatment of cyclohexene in the dark with methanol containing 10 times the concentration of acid afforded no detectable ether formation.

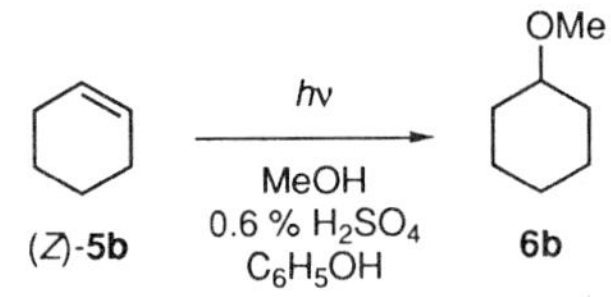

3. *The photochemical method permits the selective protonation of a cyclohexene, -heptene, or -octene in the presence of a second double bond located in an acyclic, exocyclic, or larger ring cyclic environment.* For example, light-induced addition of methanol to limonene (**47**) afforded specifically a mixture of the epimeric ethers **48**.[3] Acid-catalyzed, nonphotochemical additions to limonene (**47**) afford generally a mixture of products resulting from competing protonation of both double bonds. The selective photoreactivity stems, of course, from the fact that *E* 1 *Z* isomerization of the cyclohexene moiety affords a strained alkene, whereas isomerization of the second double bond does not.

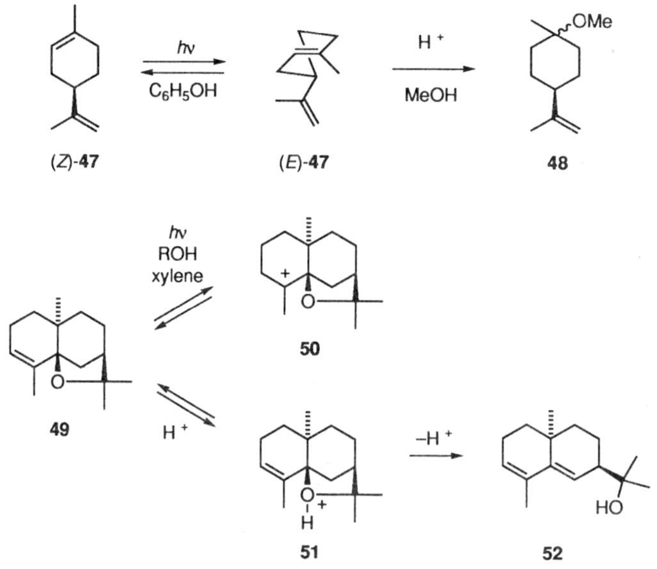

A similar type of selectivity is seen in the xylene-sensitized irradiation of the sesquiterpene α-agarofuran (**49**) in alcoholic media, which affords products resulting from selective protonation of the double bond to afford the cationic intermediate **50**.[32] By contrast, treatment with standard methods of protonation in the dark resulted in attack at oxygen (**51**) and opening of the oxide ring to afford alcohol **52** without rearrangement.[33]

These are just a few of the ways that a creative chemist can make use of the photoprotonation of cycloalkenes to induce selective reactivity in synthesis.

eferences

1. Kropp, P. J., Photochemistry of alkenes in solution, *Org. Photochem.*, 4, 1, 1979.
2. For some recent examples, see: (a) Wiseman, J. R. and Kipp, J. E., *(E)*-Bicyclo[3.3.1]non-ene, *J. Am. Chem. Soc.*, 104, 4688, 1982; (b) Clark, K. B. and Leigh, W. J., Cyclobutene photochemistry. Involvement of carbene intermediates in the photochemistry of alkylcyclobutenes, *Can. J. Chem.*, 66, 1571, 1988; (c) Wen, A. T., Hitchcock, A. P., Werstiuk, N. H., Nguyen, N., and Leigh, W. J., Studies of electronic excited states of substituted norbornenes by UV absorption, electron energy loss, and HeI photoelectron spectroscopy, *Can. J. Chem.*, 68, 1967, 1990; (d) Leigh, W. J., Zheng, K., and Clark, K. B., Cyclobutene photochemistry. Substituent and wavelength effects on the photochemical ring opening of monocyclic alkylcyclobutenes, *Can. J. Chem.*, 68, 1988, 1990; (e) Leigh, W. J., Zheng, K., and Clark, K. B., Cyclobutene photochemistry. The photochemistry of *cis*- and *trans*-bicyclo[5.2.0]non-8-ene, *J. Org. Chem.*, 56, 1574, 1991; (f) Leigh, W. J., Zheng, K., Nguyen, N., Werstiuk, N. H., and Ma, J., Cyclobutene photochemistry. Identification of the excited states responsible for the ring-opening and cycloreversion reactions of alkylcyclobutenes, *J. Am. Chem. Soc.*, 113, 4993, 1991.
3. Tise, F. P. and Kropp, P. J., Photoprotonation of cycloalkenes: Limonene to *p*-menth-8-en-1-yl methyl ether, *Org. Synth.*, 61, 112, 1983.
4. Guénard, D. and Beugelmans, R., Rôle du phénol dans les réactions photochimiques de type ionique subies par les alcools allyliques et homoallyliques en série stéroïde, *C. R. Seances Acad., Ser. C*, 280, 1033, 1975.
5. For a detailed treatment of the excited states of alkenes, see: Merer, A. J. and Mulliken, R. S., Ultraviolet spectra and excited states of ethylene and its alkyl derivatives, *Chem. Rev.*, 69, 639, 1969.
6. See, for example: Snyder, J. J., Tise, F. P., Davis, R. D., and Kropp, P. J., Photochemistry of alkenes. 7. $E\ 1\ Z$ isomerization of alkenes sensitized with benzene and derivatives, *J. Org. Chem.*, 46, 3609, 1981.
7. (a) Kropp, P. J., Photochemical behavior of cycloalkenes, *J. Am. Chem. Soc.*, 88, 4091, 1966; (b) Marshall, J. A. and Carroll, R. D., The photochemically initiated addition of alcohols to 1-menthene. A new type of photochemical addition to olefins, *J. Am. Chem. Soc.*, 88, 4092, 1966.
8. (a) Kropp, P. J. and Krauss, H. J., Photochemistry of cycloalkenes. 3. Ionic behavior in protic media and isomerization in aromatic hydrocarbon media, *J. Am. Chem. Soc.*, 89, 5199, 1967; (b) Kropp, P. J., Reardon, E. J., Jr., Gaibel, Z. L. F., Willard, K. F., and Hattaway, J. H., Jr., Photochemistry of alkenes. 2. Direct irradiation in hydroxylic media, *J. Am. Chem. Soc.*, 95, 7058, 1973.
9. Inoue, Y., Takamuku, S., and Sakurai, H., Direct and sensitized *cis-trans* photoisomerization of cyclooctene. Effects of spin multiplicity and vibrational activation of excited states on the photostationary *trans/cis* ratio, *J. Phys. Chem.*, 71, 3104, 1967.
10. Inoue, Y., Takamuku, S., Kunitomi, Y., and Sakurai, H., Singlet photosensitization of simple alkenes. Part 1. *cis-trans* Photoisomerization of cyclooctene sensitized by aromatic esters, *J. Chem. Soc., Perkin Trans. 2*, 1672, 1980.
11. Inoue, Y., Ueoka, T., and Hakushi, T., Relative rate of acid-catalyzed addition of methanol to *cis*- and *trans*-cyclooctenes, *J. Chem. Soc., Chem. Commun.*, 1076, 1982.
12. Inoue, Y., Ueoka, T., Kuroda, T., and Hakushi, T., Singlet photosensitization of simple alkenes. Part 4. *cis-trans* photoisomerization of cycloheptene sensitized by aromatic esters. Some aspects of the chemistry of *trans*-cycloheptene, *J. Chem. Soc., Perkin Trans. 2*, 983, 1983.
13. Inoue, Y., Ueoka, T., and Hakushi, T., A novel oxygen-catalyzed *trans-cis* thermal isomerization of *trans*-cycloheptene, *J. Chem. Soc., Perkin Trans. 2*, 2053, 1984.
14. Squillacote, M., Bergman, A., and De Felippis, J., *trans*-Cycloheptene: Spectral characterization and dynamic behavior, *Tetrahedron Lett.*, 30, 6805, 1989.
15. Steinmetz, M. G., Seguin, K. J., Udayakumar, B. S., and Behnke, J. S., Evidence for a metastable *trans*-cycloalkene intermediate in the photochemistry of 1,1,4,4-tetramethyl-1,4-disilacyclohept-2-ene, *J. Am. Chem. Soc.*, 112, 6601, 1990.

16. Bonneau, R., Joussot-Dubien, J., Salem, L., and Yarwood, A. J., A *trans*-cyclohexene, *J. Am. Chem. Soc.*, 98, 4329, 1976.

17. Kropp, P. J., Photochemistry of cycloalkenes. 5. Effects of ring size and substitution, *J. Am. Chem. Soc.*, 91, 5783, 1969.

18. Rosenberg, H. M. and Servé, M. P., The photolysis of 1-phenylcyclohexene in methanol, *J. Org. Chem.*, 37, 141, 1972.

19. Dauben, W. G., van Riel, H. C. H. A., Robbins, J. D., and Wagner, G. J., Photochemistry of *cis*-1-phenylcyclohexene. Proof of involvement of *trans*-isomer in reaction processes, *J. Am. Chem. Soc.*, 101, 6383, 1979.

20. Dauben, W. G., van Riel, H. C. H. A., Hauw, C., Jeroy, F., Joussot-Dubien, J., and Bonneau, R., Photochemical formation of *trans*-1-phenylcyclohexene. Chemical proof of structure, *J. Am. Chem. Soc.*, 101, 1901, 1979. See also: Cozens, F. L., McClelland, R. A., and Steenken, S., Observation of cationic intermediates in the photolysis of 1-phenylcyclohexene, *J. Am. Chem. Soc.*, 115, 5050, 1993.

21. Goodman, J. L., Peters, K. S., Misawa, H., and Caldwell, R. A., Use of time-resolved photoacoustic calorimetry to determine the strain energy of *trans*-1-phenylcyclohexene and the energy of the relaxed 1-phenylcyclohexene triplet, *J. Am. Chem. Soc.*, 108, 6803, 1986.

22. Saltiel, J., Marchand, G. R., and Bonneau, R., *cis,trans*-1,1'-Bicyclohexenyl: A strained ground state intermediate in the photocyclization of 1,1'-bicyclohexenyl to its isomeric *cis*-cyclobutene, *J. Photochem.*, 28, 367, 1985.

23. Saltiel, J. and Marchand, G. R., *cis,trans*-1,1'-Bicyclohexenyl: Conformer-dependent chemistry in benzene and in methanol, *J. Am. Chem. Soc.*, 113, 2702, 1991.

24. Bonneau, R., Some new examples of *"trans"* cyclohexenes: Properties characteristic of these species, *J. Photochem.*, 36, 311, 1987.

25. Verbeek, J., van Lenthe, J. H., Timmermans, P. J. J. A., Mackor, A., and Budzelaar, P. H. M., On the existence of *trans*-cyclohexene, *J. Org. Chem.*, 52, 2955, 1987.

26. For a review of some earlier studies on the stereochemistry of photoprotonation, see: Marshall, J. A., Photosensitized ionic additions to cyclohexenes, *Acc. Chem. Res.*, 2, 33, 1969.

27. Fravel, H. G., Jr. and Kropp, P. J., Photochemistry of alkenes. 4. Vicinally unsymmetrical olefins in hydroxylic media, *J. Org. Chem.*, 40, 2434, 1975.

28. For a review, see: Wan, P. and Yates, K., Photogenerated carbonium ions and vinyl cations in aqueous solution, *Rev. Chem. Intermed.*, 5, 157, 1984. See also: McEwen, J. and Yates, K., Photohydration of styrenes and phenylacetylenes. General acid catalysis and Brønsted relationships, *J. Am. Chem. Soc.*, 109, 5800, 1987. Wan, P., Davis, M. J., and Teo, M.-A., 3-Nitrostyrenes undergo anti-Markovnikov addition: Photoaddition of water and alcohols to 3-nitrostyrenes. Structure-reactivity and solvent effects, *J. Org. Chem.*, 54, 1354, 1989.

29. Kropp, P. J., Photochemistry of Cycloalkenes. 8. 2-Phenyl-2-norbornene and 2-phenyl-2-bornene, *J. Am. Chem. Soc.*, 95, 4611, 1973.

30. Kropp, P. J., Fields, T. R., Fravel, H. G., Jr., Tubergen, M. W., and Crotts, D. D., Photochemistry of Alkenes. 11. Rearrangements in nonnucleophilic media, manuscript submitted.

31. Tise, F. P. and Kropp, P. J., unpublished results.

32. (a) Marshall, J. A. and Pike, M. T., A stereoselective synthesis of α- and β-agarofuran, *J. Org. Chem.*, 33, 435, 1968; (b) Thomas, A. F. and Ozainne, M., A photoinitiated Wagner-Meerwein rearrangement, *Helv. Chim. Acta*, 59, 1243, 1976.

33. Asselin, A., Mongrain, M., and Deslongchamps, P., Syntheses of α-agarofuran and isodihydroagarofuran, *Can. J. Chem.*, 46, 2817, 1968.

8

The π-Cyclopropene Rearrangements

Howard E. Zimmerman
University of Wisconsin

8.1 Definition of the Reaction and Background

The π-Cyclopropene Rearrangements consist of a series of photochemical rearrangements of molecules having a π-moiety bonded to C-3 of a cyclopropene. These reactions have two mechanisms, depending on the multiplicity of the reacting excited state.

The discovery of the singlet π-Cyclopropene Rearrangement of vinylcyclopropenes came in 1976 simultaneously by Zimmerman[1] and Padwa.[2] The triplet rearrangement was discovered by Zimmerman[3] who showed it to lead to different products and to have a different reaction mechanism than the unsensitized counterpart. The basic rearrangement is depicted in Equation 8.1.

Still later, the acyl- and acylimino relatives were studied and these were found to afford furans and pyrroles, respectively. However, in this case, the singlet reaction was found to afford a multiplicity of photoproducts,[4] while the sensitized process led selectively to the five-ring heterocycles.[5]

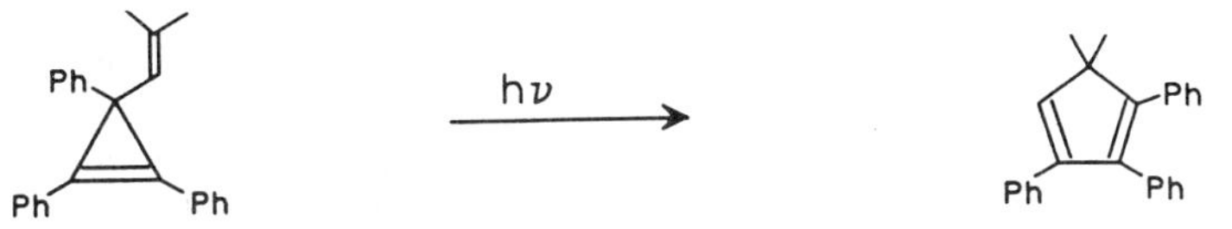

$$ (8.1) $$

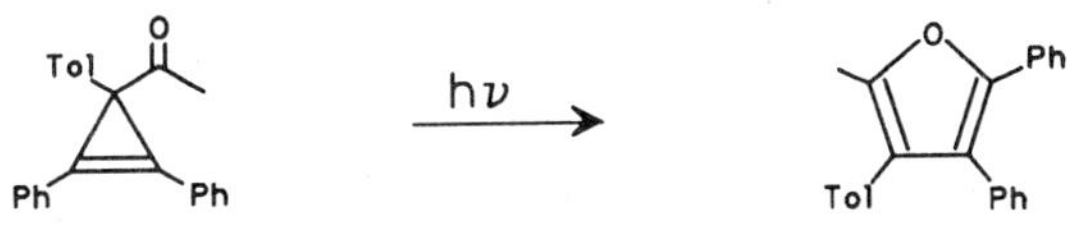

$$ (8.2) $$

One example is shown in Equation 8.2.

Thus, it was seen that there are two variants of the π-cyclopropene rearrangement in the cases of vinyl as well as acyl and acylimino substitution. These correspond to the singlet reaction resulting from direct irradiation and the triplet counterpart occurring on sensitization.

0-8493-8634-9/95/$0.00+$.50
© 1995 by CRC Press, Inc.

8.2 The Basic Reaction Mechanisms: The Vinylcyclopropene Singlet Rearrangement

In the case of direct irradiation of vinylcyclopropenes, two reaction mechanisms fit the structural course of the rearrangement. We term these mechanisms A and B. We will find that these are relevant to the related acyl and acylimino rearrangements as well. These two mechanisms are depicted in Schemes 8.1 and 8.2. Mechanism A involves vinyl-vinyl bridging in the singlet excited state to form bond "a" and a housane biradical; the biradical undergoes a 1,4-(2,3) fragmentation of bond "b" with formation of the cyclopentadiene photoproduct. Mechanism B begins by fission of a three-ring bond (i.e., "b") to yield a carbene, which then undergoes an electrocyclic ring closure forming bond *a* and affording cyclopentadiene product. Inspection of these two mechanisms reveals an interesting facet, namely that they differ only in chronology. Thus, the two steps — bond breaking (b) and bond forming (a) — are simply reversed in timing.

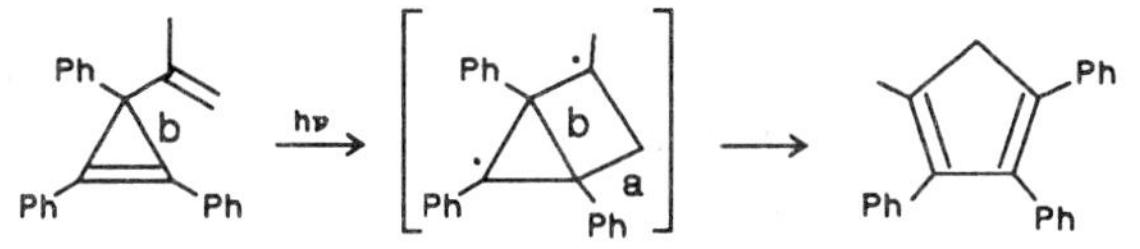

SCHEME 8.1   Reaction Mechanism A in the vinylcyclopropene rearrangement.

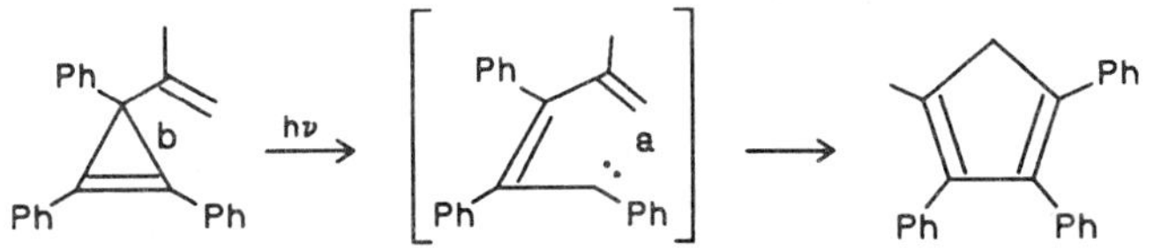

SCHEME 8.2   Reaction Mechanism B in the vinylcyclopropene rearrangement.

From the observation that the direct irradiations afford by-products characteristic of carbene reactions, one can conclude either that Mechanism B is operating in the S_1, direct irradiation or that the carbene is available concomitantly with the housane biradical.

However, the reaction regiochemistry is the reverse from that which one might anticipate naively from Mechanism B and is in accord with that expected from Mechanism A *(vide infra)*. It has been suggested by Padwa[6–8] that the regioselectivity may still fit the carbene mechanism. Thus, in the excited singlet, one expects aryl groups substituted on the three-ring double bond to be coplanar with that π-bond. This means that the aryl conformation is not suited to stabilize odd-electron density derived from the breaking σ-bond. The net result in this rationale is an inductive destabilization by phenyl substitution.

The interesting regiochemistry holds for a rather large number of examples, with the terminal carbon of the vinyl group invariably preferring to bond to the cyclopropene π-bond at the end not bearing a stabilizing group. Note Equation 8.3.

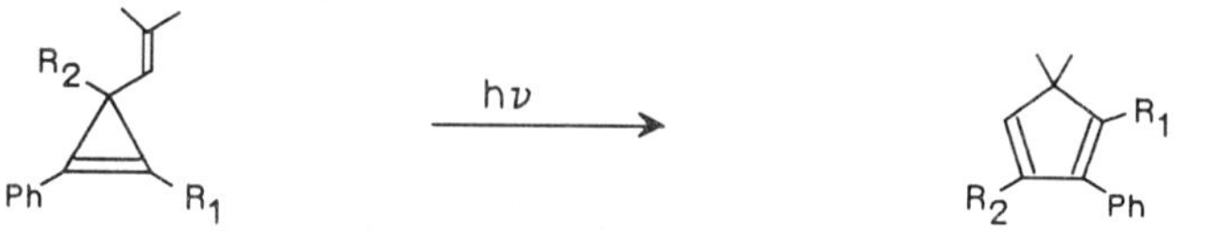

R_1	R_2	Reference
Me	Ph	8
t-Bu	Ph	1a,b

PREFERRED
REGIOISOMERIC
PRODUCT

(8.3)

One interesting mechanistic possibility results from the fact that the two alternative mechanisms do differ only in chronology of bond formation and bond fission. Thus, one can conceive of all mechanistic gradations between the two extremes, and it is also conceivable that the singlet housane biradical of Mechanism A and the singlet carbene of Mechanism B may interconvert. A further relevant aspect is that as bond "b" stretches in Mechanism B, the three-ring opened species is basically a singlet vinyl diradical (i.e., an S_1 vinyl carbene) rather than the usual ground state S_0 vinyl carbene. If Mechanism B is to be correct, radiationless decay of the S_1 excited carbene to the ordinary ground-state carbene should occur prior to cyclization for this to be a characteristic electrocyclic (five-ring) closure.

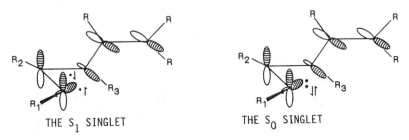

THE S_1 SINGLET   THE S_0 SINGLET

SCHEME 8.3 S_1 (excited sinlget) and S_0 (ground state) vinylcarbenes.

In addition, consideration of Scheme 8.3 provides a simple rationale as to why triplet cyclopropenes do not utilize the same mechanism. There is no triplet counterpart of the S_0 vinylcarbene to undergo an electrocyclic closure. A final point is that 3-arylcyclopropenes undergo the same basic reaction except that in this case, the photoproducts are indenes rather than cyclopentadienes.[1a,1b,2,9] The mechanisms available to the molecule are basically the same except for the aryl group replacing the vinyl moiety. However, when C-3 is substituted with both a vinyl and a phenyl group, some selectivity is seen, with a preference for interaction of the vinyl rather than the phenyl group. One might anticipate selectivity arising from avoidance of disruption of aromaticity. In some cases, complete selectivity is seen while in others, it is less pronounced. With only a phenyl group and no vinyl at C-3, indene formation is understandably the reaction seen.

8.3 The Basic Reaction Mechanisms: The Vinylcyclopropene Triplet Rearrangements

The triplet rearrangement of vinylcyclopropenes, with sensitization, is a marked contrast to the singlet process. The triplet reaction is exceptionally selective, shows few if any side reactions, and often gives totally different cyclopentadienes from those obtained from the singlet photochemistry.[3] Furthermore, often the photoproducts cannot be reached using Mechanisms A and B. While in organic photochemistry, one often encounters differences in singlet and triplet behavior, the case of vinylcyclopropenes is particularly intriguing.

First, we begin with the assumption, quickly verified, that carbenes are not intermediates in the triplet chemistry. Elegant studies by Pincock[10] have shown that, in contrast to the excited singlet (i.e., the S_1) state of cyclopropenes, the triplet counterpart has a large energy barrier to three-ring opening to the triplet carbene. Similarly, our own studies[11,12] have also demonstrated the reluctance of triplet cyclopropenes to ring open, and the same conclusion has been noted by Padwa.[13]

The formation of cyclopentadiene photoproducts that cannot be rationalized with Mechanisms A and B led to consideration of alternatives.[3] One mechanism uniquely accounted for those reaction products encountered and only those. This is presented in Scheme 8.4 and is termed "Mechanism C". It is seen that the first step of this mechanism is that of the di-π-methane rearrangement;[14] but after this step, the remainder of the mechanism is somewhat unusual. While the initial basis for postulating this mechanism was its ability to account for sensitized photoproducts not developed by the alternative mechanisms, more convincing evidence was obtained. This evidence derived from the triplet behavior of 1-phenyl-2-methyl-3-*p*-tolyl-3-isobutenylcyclopropene and its regioisomer 1-phenyl-2-*p*-tolyl-3-methyl-3-isobutenylcyclopropene. The 3-tolyl isomer afforded two cyclopentadienes as outlined in Scheme 8.5. Similarly, the 3-phenyl isomer led to two cyclopentadiene photoproducts; note Scheme 8.6. However, the 3-methyl isomer led to four cyclopentadienes. In this case, for simplicity, only two are depicted in Scheme 8.7. These are the two resulting from vinyl-vinyl bridging to give the slightly more stable tolyl-stabilized bicyclobutanyl biradical. In fact, however, relatively little selectivity was seen in this initial bridging step. The first item to note is that these mechanisms account for all of the observed reaction products and predict none which are not observed.

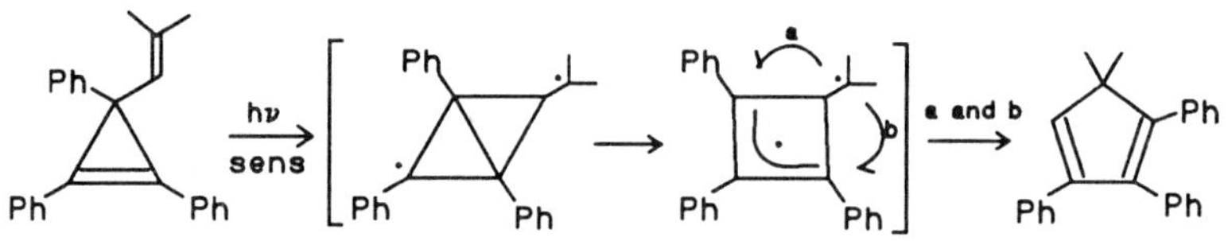

SCHEME 8.4   Mechanism C.

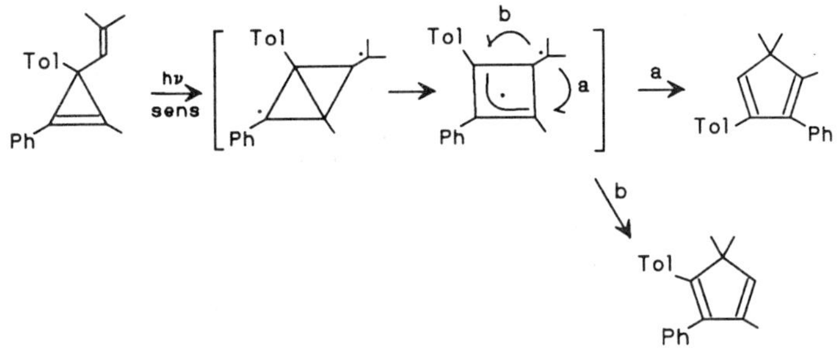

SCHEME 8.5   The triplet rearrangement of the 3-tolyl regioisomer.

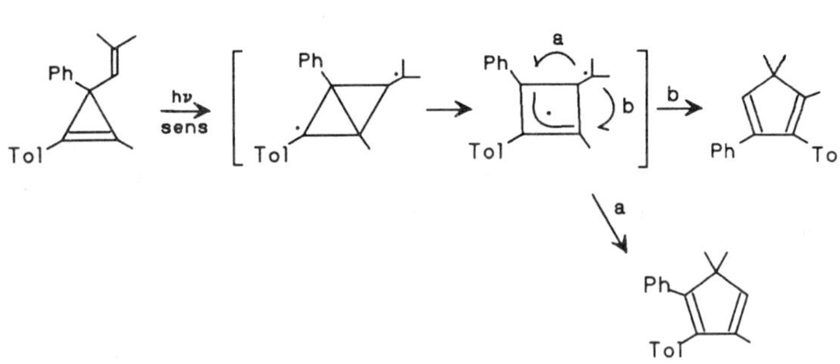

SCHEME 8.6   The triplet rearrangement of the 3-phenyl regioisomer.

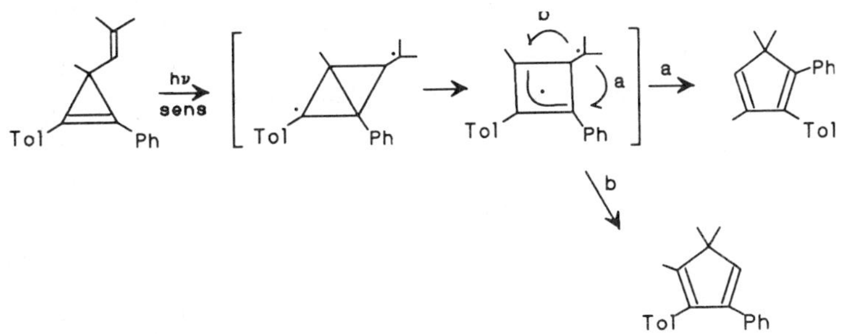

SCHEME 8.7 The triplet rearrangement of the 3-methyl regioisomer.

Still more strikingly, we note from Schemes 8.6 and 8.7 that *the same* cyclobutenylcarbinyl biradical results in the photoreactions of the two different vinylcyclopropenes, namely the 3-tolyl and 3-methyl isomers. This cyclobutenylcarbinyl biradical has two alternative ways of closing (labeled "a" and "b") to a housane and then onward to a cyclopentadiene. A severe test of Mechanism C is the identity of the photoproduct ratios (a:b) starting with the 3-tolyl and 3-methyl reactants. Indeed, it was observed that this test was satisfied.[3]

Final confirmation for this mechanism for the triplet rearrangement of vinylcyclopropenes came from the sensitized reaction of the azo precursor to the cyclobutenylcarbinyl biradical as shown in Equation 8.4. It is seen that this azo heterocycle, on sensitization, provides an independent route to the postulated cyclobutenylcarbinyl biradical. Experimentally, it was found that the cyclopentadienes resulted in the same a:b ratio, thus confirming that the same triplet cyclobutenylcarbinyl diradical is an intermediate.[3]

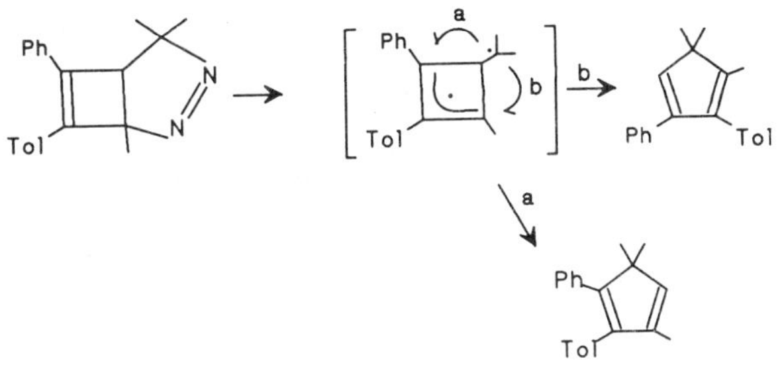

(8.4)

8.4 The Acylcyclopropene and Acylimino Singlet and Triplet Rearrangements

It was known that on direct irradiation, 3-acetyltriphenylcyclopropene afforded 2-methyl-3,4,5-triphenylfuran in addition to two by-products (an indene and a cyclopentenone), both resulting from carbene insertion involving the 3-phenyl group and the acetyl methyl, respectively.[4]

With the photochemical behavior of vinylcyclopropenes in mind, as discussed above, and knowledge of occurrence of the singlet acetylcyclopropene, it seemed of interest to pursue the triplet photochemistry of acylcyclopropenes. As in the case of the vinyl analogs, the triplet counterpart was less complex than the singlet photochemistry. For example, sensitization of 3-acetyl-1,2,3-triphenylcyclopropene led to a single photoproduct, namely 2-methyl-3,4,5-triphenylfuran.

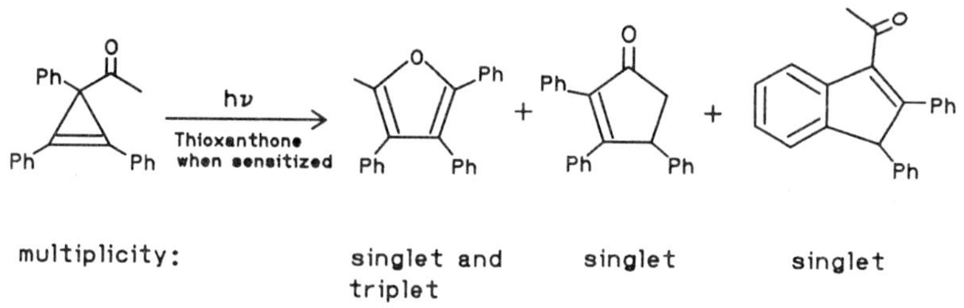

multiplicity: singlet and   singlet   singlet
                                 triplet

An initial guess might be that, as in the case of the vinylcyclopropenes, Mechanism C is being followed. Not only would use of Mechanism C be parallel to the vinylcyclopropene triplet case, but also the first step of bridging of the carbonyl carbon to the β,γ π-bond is characteristically a known oxa-di-π-methane process from a triplet.

A study[5] of the 3-acetyl-diphenyltolylcyclopropenes showed that the 3-tolyl reactant afforded one photoproduct, a furan, and that the 3-phenyl isomer led to two different furan products. Consideration of Scheme 8.8 that uses Mechanism C reveals that this mechanism is capable of leading each of the two photoreactants on to observed product(s). The course of the mechanism, starting with each acetylcyclopropene, is depicted in Scheme 8.8 as either a solid arrow or a dotted arrow, depending on which starting material is proceeding along this route.

It is readily seen that there is a cyclobutenyloxy biradical that derives from both reactants, and yet must react differently (dotted arrow vs. solid arrow) depending on which reactant formed it. Since the cyclobutenyloxy biradical is an essentially planar species, there is little reason to expect such a "memory effect". Hence, Mechanism C is not in accord with observation.

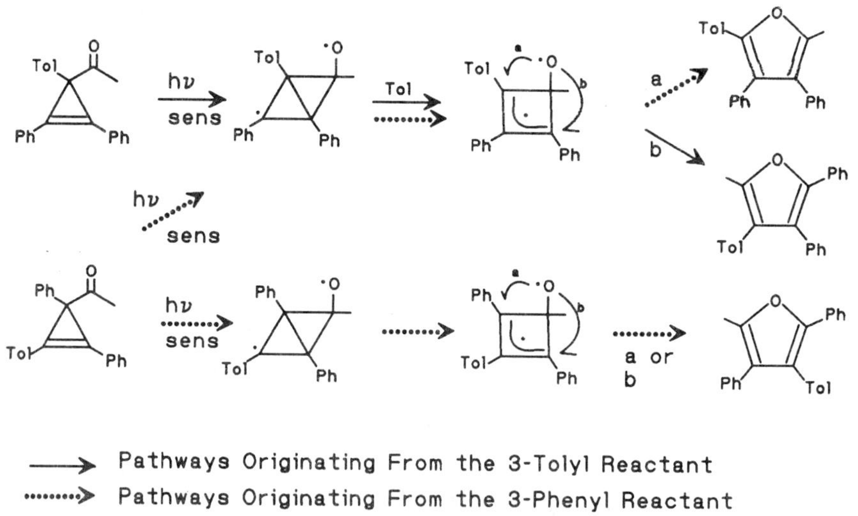

⟶   Pathways Originating From the 3-Tolyl Reactant
┈┈┈▷ Pathways Originating From the 3-Phenyl Reactant

SCHEME 8.8   Mechanism C applied to the acetyl diphenyl tolyl cyclopropenes.

Mechanism A, applied to these compounds in Scheme 8.9, does not suffer this deficiency and accounts for the reaction course. One of the most interesting facets of this chemistry is the differing mechanism observed for the acylcyclopropenes compared with the vinyl relatives.

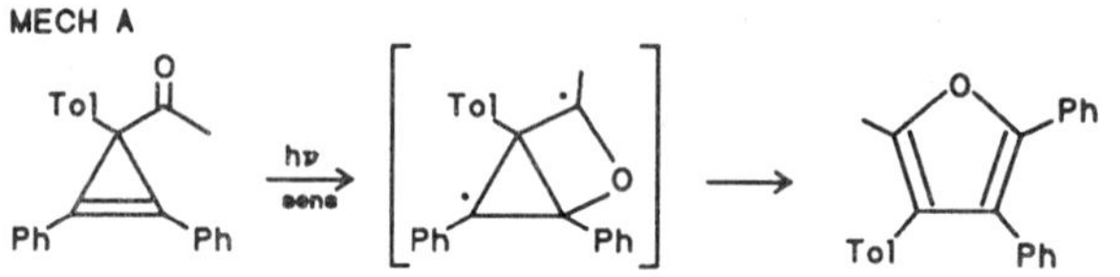

SCHEME 8.9   Mechanism A applied to the acetyltriphenylcyclopropene.

Finally, it is known that acyliminocyclopropenes also rearrange on sensitization to afford pyrroles; note Equation 8.5. The reaction pathway, whether Mechanism A or Mechanism C, has not been yet established.

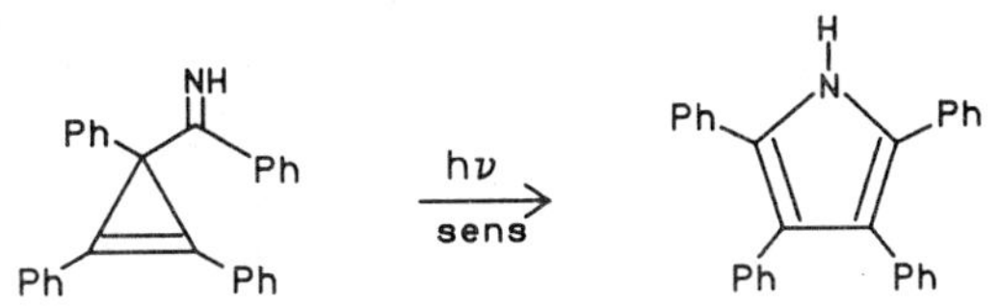

$$\xrightarrow[\text{sens}]{h\nu}$$

(8.5)

8.5 Concluding Comments — Some Philosophy

It is a curious phenomenon that the development of instrumental methodology and the study of the photophysics of molecules have advanced more rapidly than our understanding of the molecular details of excited-state transformations affording photochemical reactions of any real complexity. Reactions of appreciable complexity, where there are several molecular steps or where there are more than one or two bonds broken and formed, present exciting and challenging questions. These are the reactions where the excited state is not quickly presented with a mode of decay to ground-state product.

While recognizing the role of ground to excited-state degeneracies,[15] the present author long ago presented the view that in these cases, excited-state bond orders, electron densities, activation barriers, and energy changes on various molecular deformations provide insight and often permit understanding of photochemical reaction courses. While decay to photoproduct must occur somewhere, it most often is not the controlling feature for the more interesting and complex rearrangements.

This chapter has not delved into quantum mechanical features of the reactions as so common in the author's publications, but rather has focused on the course of these reactions and chemical ways of gaining insight into their mechanisms. So often, one can devise strategies permitting one to determine an excited-state reaction mechanism and obtain information not accessible by photophysical means.

Acknowledgment

The research studies upon which this chapter is based were supported by the National Science Foundation, and this support is gratefully acknowledged.

References

1. (a) Zimmerman, H. E. and Aasen, S. M., The photochemistry of vinylcyclopropenes: A new and general cyclopentadiene synthesis. Exploratory and mechanistic organic photochemistry, *J. Amer. Chem. Soc.*, 99, 2342, 1977. (b) Zimmerman, H. E. and Aasen, S. M., Vinylcyclopropene photochemistry: Photochemistry applied to organic synthesis. Exploratory and mechanistic organic photochemistry, *J. Org. Chem.*, 43, 1493, 1978.
2. Padwa, A., Blacklock, T., Getman, D., and Hatanaka, N., Regioselectivity of bond cleavage in the photochemical rearrangement of 3-vinylcyclopropenes, *J. Am. Chem. Soc.*, 99, 2344, 1978.
3. (a) Zimmerman, H. E. and Fleming, S. A., Vinylcyclopropene triplet rearrangement mechanisms: Mechanistic and exploratory organic photochemistry, *J. Am. Chem. Soc.*, 105, 622, 1983; (b) Zimmerman, H. E. and Fleming, S., Diradical pathways in vinylcyclopropene triplet rearrangements: Mechanistic and exploratory organic photochemistry, *J. Org. Chem.*, 50, 2539, 1985.
4. Padwa, A., Akiba, M., Chou, C. S., and Cohen, L., Photochemistry of cyclopropene derivatives. Synthesis and photorearrangement of a 3-acyl-substituted cyclopropene, *J. Org. Chem.*, 47, 183, 1982.

5. (a) Zimmerman, H. E. and Wright, C. W., A new photochemical reaction and its mechanism: Rearrangement of acyl and imino cyclopropenes, *J. Am. Chem. Soc.,* 114, 363, 1992; (b) Zimmerman, H. E. and Wright, C. W., Novel rearrangements of acyl and imino cyclopropenes: Multiplicity dependence and mechanism, *J. Am. Chem. Soc.,* 114, 6603, 1992.

6. Padwa, A., Blacklock, T. J., Getman, D., Hatanaka, N., and Loza, R., On the problem of regioselectivity in the photochemical ring-opening reaction of 3-phenyl and 3-vinyl substituted cyclopropenes to indenes and 1,3-cyclopentadienes, *J. Org. Chem.,* 43, 1481, 1978.

7. Photochemical Transformations of Cyclopropene Derivatives, Padwa, A., *Org. Photochem.,* Vol. 4, M. Dekker, New York, 1979, 261.

8. Excited state chemistry of cyclopropene derivatives, Padwa, A., *Accts. Chem. Res.,* 1979, 12, 310–317.

9. (a) Halton, B., Kulig, M., Battiste, M. A., Perreten, J., Gibson, D. M., and Griffin, G. W., Photocyclization of aryl-substituted acetylenes; Application of di-π-methane-like rearrangements to arylcyclopropene syntheses, *J. Am. Chem. Soc.,* 93, 2327, 1971; (b) Schrader, I., Zur Photolyse 3-Arylsubstituierter 3H-Pyrazole, *Chem. Ber.,* 104, 941, 1971; (c) Hartman, A., Welter, W., and Regitz, M., Intramolekulare Reaktionen von Vinyl- und Allyl-Phosphryl-Carbenen, *Tetrahedron Lett.,* 1825, 1974; (d) Kristinsson, H., Photochemische Bildung von Phenyl-methylcarben, *Tetrahedron Lett.,* 2343, 2345, 1966; (e) Kristinsson, H. and Griffin, G. W., Photochemistry of phenyloxiranes. II. New precursor for phenylcarbenes, *J. Am. Chem. Soc.,* 88, 1579, 1966.

10. (a) Pincock, J. A. and Boyd, R. J., Thermal and photochemical reactivity of cyclopropene derivatives: A semi-empirical molecular orbital study, *Can. J. Chem.,* 55, 2482, 1977; (b) Pincock, J. A. and Moutsokapas, A. A., An optically active cyclopropene as a mechanistic probe in cyclopropene photochemistry, *Can. J. Chem.,* 55, 979, 1977.

11. Zimmerman, H. E. and Bunce, R. A., Cyclopropene Photochemistry. Mechanistic and Exploratory Organic Photochemistry, *J. Org. Chem.,* 47, 3377, 1982.

12. Zimmerman, H. E. and Hovey, M. C., Mechanistic and Exploratory Organic Photochemistry. Cyclopropene Studies, *J. Org. Chem.,* 44, 2331, 1979.

13. Padwa, A., *Org. Photochem.,* 4, 261, 1979.

14. (a) Zimmerman, H. E. and Grunewald, G. L., The chemistry of barrelene. III. A unique photoisomerization to semibullvalene, *J. Am. Chem. Soc.,* 88, 183, 1966; (b) Zimmerman, H. E., Binkley, R. W., Givens, R. S., and Sherwin, M. A., Mechanistic organic photochemistry. XXIV. The mechanism of the conversion of barrelene to semibullvalene. A general photochemical process, *J. Am. Chem. Soc.,* 89, 3932, 1967; (c) Zimmerman, H. E., Binkley, R. W., Givens, R. S., Grunewald, G. L., and Sherwin, M. A., The barrelene to semibullvalene transformation. Correlation of excited state potential energy surfaces with reactivity. Mechanistic and exploratory organic photochemistry. XLIV, *J. Am. Chem. Soc.,* 91, 3316, 1969.

15. (a) Zimmerman, H. E., On molecular orbital correlation diagrams, Möbius systems, and factors controlling ground and excited state reactions. II. *J. Am. Chem. Soc.,* 88, 1566, 1966; (b) Michl, J., Photochemical reactions of large molecules. I. A simple physical model of photochemical reactivity, *Mol. Photochem.,* 4, 243, 1972; (c) Part II, Application of the model to organic photochemistry, *idem,* ibid, 257., (d) Part III, "Use of Correlation Diagrams for Prediction of Energy Barriers", *idem,* ibid, 287–314; (e) Teller, E., The Crossing of Potential Surfaces, *J. Phys. Chem.,* 41, 109, 1937; (f) Mechanisms of electron demotion. Direct measurement of internal conversion and intersystem crossing rates. Mechanistic organic photochemistry, Zimmerman, H. E., Kamm, K. S., and Werthemann, D. P., *J. Am. Chem. Soc.,* 97, 3718, 1975; (g) Zimmerman, H. E. and Factor, R. E., The bicycle rearrangement: Relationship to the di-π-methane rearrangement and control by bifunnel distortion. Mechanistic and exploratory organic photochemistry, *J. Am. Chem. Soc.,* 102, 3538, 1980.

9

Diene/Cyclobutene Photochemistry

William J. Leigh
McMaster University

9.1 Introduction

The photochemical interconversion of 1,3-butadiene and cyclobutene is one of the cornerstone pericyclic reactions in organic photochemistry.[1] In spite of this, the scope of the ring-opening process has only been delineated recently, and the mechanisms of both processes are still not fully understood.[2,3] In this chapter, the excited singlet state photochemistry of conjugated dienes and cyclobutenes will be discussed. Triplet state reactivity will be ignored because it generally is not involved in these interconversions;[4] to our knowledge, the only known exception to this generalization is the photochemical ring opening of Dewar aromatics.[5] While the central theme of the review is the photochemical interconversion of cyclobutenes and their isomeric dienes, other aspects of the excited singlet state chemistry of the two systems will also be discussed in the interest of completeness. A discussion of the mechanisms of these reactions is deferred to the final section of this chapter.

9.2 The Photochemistry of Conjugated Dienes

The photochemistry of 1,3-dienes has been studied in detail.[3,6] Their excited singlet-state behavior depends in large part on two factors: the ability of one or both of the double bonds to exhibit *cis,trans*-isomerism or undergo torsional relaxation, and the conformational properties of the diene system with respect to the flexibility of and the dihedral angle defined by the C2-C3 bond (1,3-butadiene numbering). In general, the main excited singlet-state deactivation pathways that are important in conjugated dienes are:

1. *cis,trans*-isomerization
2. *s-cis/s-trans*-conformer interconversion

0-8493-8634-9/95/$0.00+$.50
© 1995 by CRC Press, Inc.

3. Cyclobutene formation (from *s-cis*-dienes)
4. Bicyclo[1.1.0]butane formation (from *s-trans*-dienes)
5. [1,5]-H migration (in appropriately substituted *s-cis*-dienes)

Bicyclo[1.1.0]butane formation is, in general, extremely inefficient, but it can occur in useful chemical yields in certain cases.[7-8] Fluorescence and intersystem crossing are insignificant compared to reactive decay pathways in alkyl-substituted conjugated dienes.[3,9] Aryl-substituted dienes generally undergo *cis,trans*-photoisomerization efficiently and may exhibit weak fluorescence, but they are relatively unreactive toward cyclization to cyclobutenes or bicyclobutanes.[10]

Cis,trans-Photoisomerization, *s-cis/s-trans*-Interconversion, and Cyclobutene Formation

Direct irradiation of 1,3-butadiene 1 in hydrocarbon solution yields cyclobutene and bicyclo[1.1.0]butane in a ratio of about 10:1.[7] The quantum yield for cyclobutene formation has been reported to be ca. 0.04 with 254-nm excitation.[11] Irradiation of *s-cis*- or *s-trans*-1 in an argon matrix at 20K results in facile interconversion of the two conformers;[12] this process is substantially more efficient than cyclobutene formation for the *s-cis*-conformer.

In acyclic dienes, cyclobutene formation occurs only when there is a significant population of *s-cis*-conformer present in solution.[13-15] Thus, irradiation (254 nm) of *E,E*-2,4-hexadiene (*E,E*-2) in hydrocarbon solution yields *cis*-3,4-dimethylcyclobutene (*cis*-3; $\phi = 0.024$) in addition to *E,Z*-2 ($\phi = 0.37$), while irradiation of *E,Z*- and *Z,Z*-2 under similar conditions leads only to *cis,trans*-photoisomerization ($\phi_{EZ \rightarrow EE} = 0.17$, $\phi_{EZ \rightarrow ZZ} = 0.29$, $\phi_{ZZ \rightarrow EZ} = 0.41$).[15] It should be noted that ring closure proceeds stereospecifically, in disrotatory fashion, and that *cis,trans*-isomerization involves only one double bond per photon ($\phi_{EE \rightarrow ZZ} = \phi_{ZZ \rightarrow EE} < 0.01$).

The photochemistry of acyclic dienes often exhibits pronounced wavelength dependence,[16a] which has been attributed to the selective excitation of specific diene conformers.[3,17] *s-trans*-Dienes absorb typically at shorter wavelengths and with higher extinction coefficients than the analogous *s-cis*-dienes.[18] This is illustrated in Figure 9.1, which shows the UV absorption spectra of *s-trans*- and *s-cis*-1,3-butadiene (1) isolated in argon matrices at 20 K.[12] The relationship between the efficiencies of the various excited singlet-state decay pathways and conformation is best illustrated by the wavelength-dependent photochemistry of *E*- and *Z*-1,3-pentadiene (4) and *Z*-1-deuterio-*E*-1,3-pentadiene (4-*d*).[16,17] Irradiation of *E*- and *Z*-4 in solution with 254-nm light (where the *s-cis*-conformers are the primary absorbers) results in *E,Z*-isomerization and the formation of 3-methylcyclobutene 5 and 1,3-dimethylcyclopropene 6 in the quantum yields shown in Equations 9.1 and 9.2.[16] Irradiation with 229-nm light under similar conditions (where the *s-trans*-conformers are the primary absorbers) leads only to

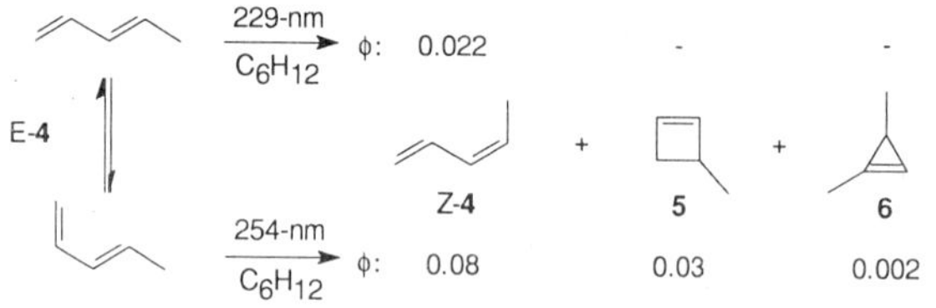

(9.1)

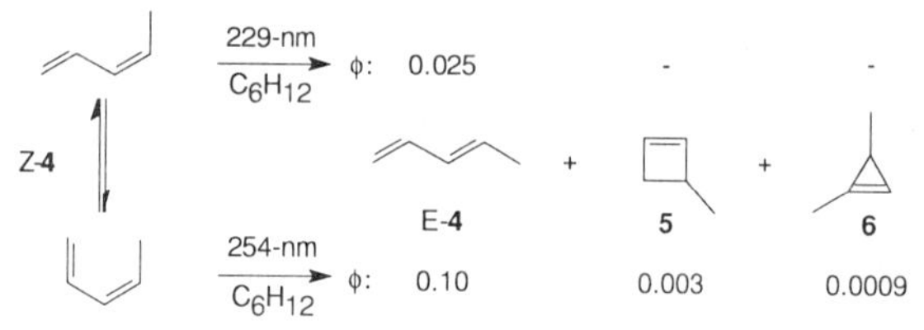

(9.2)

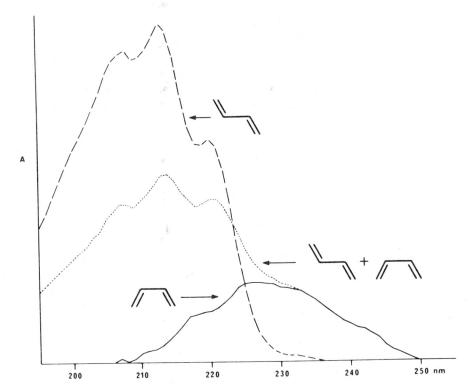

FIGURE 9.1 Ultraviolet absorption spectra of 1,3-butadiene conformers in argon matrices at 20K. (*Source:* Chapman, O. L., Pasto, D. J., Borden, G. W., and Griswold, A. A., *J. Am. Chem. Soc.*, 84, 1220, 1962. With permission.)

E,Z-isomerization; the cyclization products 5 and 6 are reportedly undetectable.[16a] The main decay process in *s-trans-E*-4 is torsional relaxation about the unsubstituted C=C bond, which occurs with ca. 25 times greater efficiency than *E,Z*-isomerization.[17]

Squillacote and co-workers have investigated the photochemistry of the individual *s-cis*- and *s-trans* conformers of 2-methyl-1,3-butadiene (7),[19,20] 2,4-hexadiene (2),[19] and 2,3-dimethyl-1,3-butadiene (8)[19,20] in argon and 3-methylpentane matrices at 10 to 20 K. The photochemistry of *s-cis*-8 has been reported in detail. The behavior of this compound is unusual in that *s-cis* → *s-trans* conversion is inefficient compared to cyclobutene formation, but this facet of its behavior has allowed a more definitive study of the photochemistry of conformationally labile *s-cis*-dienes, including the importance of torsional relaxation about the double bonds (employing *E,E*-1,4-d_2-8). The relative efficiencies of photochemical *E,Z*-isomerization, conformer interconversion, and cyclobutene formation in *s-cis*- and *s-trans-E,E*-1,4-d_2-8 are given in Equation 9.3. The quantum yield for cyclization of *s-cis*-8 in solution at room temperature is $\phi = 0.12$.[13]

$$(9.3)$$

Consideration of the photochemistry of cyclic, *s-cis*-dienes such as 1,3-cyclohexadiene derivatives (e.g., 9), 1,3-cycloheptadiene (10a), 1,3-cyclooctadiene (11a), and 1,3-cyclononadiene (12) is instructive since it provides an indication of the conformational requirements for electrocyclic ring closure and illustrates a second mechanism for the ring closure process that may occur in cyclic systems. The photochemistry of **9** is wavelength dependent (Equation 9.4); irradiation at 254 nm yields the *cis,cis,trans*-triene **13** (via photochemically allowed conrotatory ring opening) as the only detectable product, while irradiation at 300 nm slowly yields the tricyclic cyclobutene derivative **14** to the exclusion of **13**.[21] It was established that the **9/13** photostationary state at 300 nm is 99% **9**

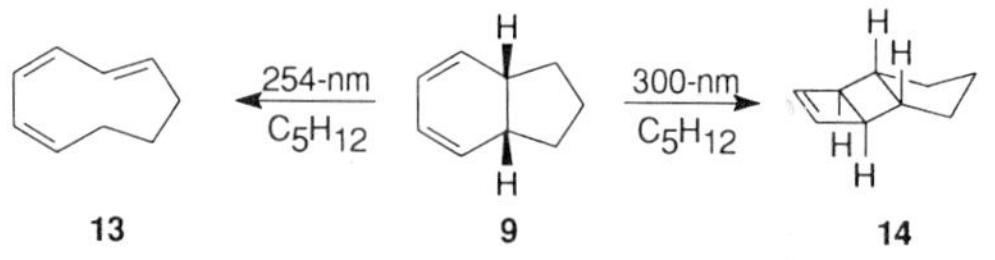

$$\text{13} \quad\xleftarrow[\text{C}_5\text{H}_{12}]{\text{254-nm}}\quad \text{9} \quad\xrightarrow[\text{C}_5\text{H}_{12}]{\text{300-nm}}\quad \text{14} \qquad (9.4)$$

and 1% **13**, so the possibility that the latter is in fact formed upon photolysis at this wavelength could not be ruled out. However, the fact that **14** is formed only upon excitation at long wavelengths was attributed to specific excitation of planar diene conformers; the ring opening to yield **13** was proposed to result from excitation of the most stable, twisted *s-cis*-conformer of the diene. The observation that the relative yield of **13** increases with decreasing temperature is consistent with this proposal. Dauben and co-workers have also reported several other examples of conformational control of photoreactivity of 1,3-cyclohexadiene systems.[22]

Irradiation of 1,3-cycloheptadienes (**10**) in solution results in efficient ring closure to the isomeric bicyclo[3.2.0]hept-6-enes (**15**; Equation 9.5).[23] The quantum yield for formation of **15** has been reported to be 0.35 for the parent compound (**10a**; $R^1 = R^2 = R^3 = H$).[24] While **15a** is formally the product of (symmetry-allowed)

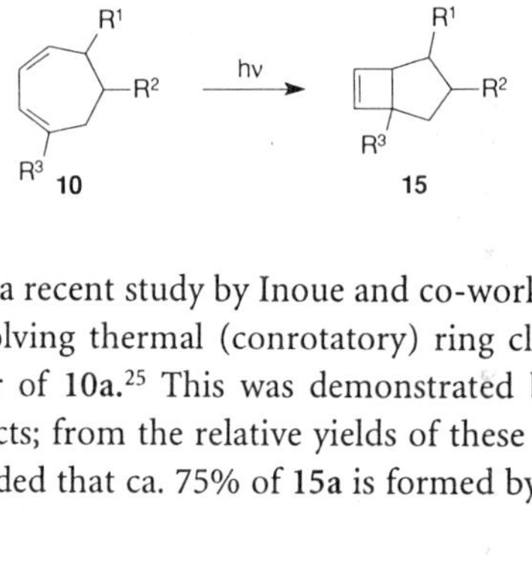

$$\text{10} \quad\xrightarrow{h\nu}\quad \text{15} \qquad (9.5)$$

disrotatory ring closure of **10a**, a recent study by Inoue and co-workers indicates that, in fact, it is formed largely by a route involving thermal (conrotatory) ring closure of the initially formed, highly strained *cis,trans*-isomer of **10a**.[25] This was demonstrated by acid-catalyzed trapping of *cis,trans*-**10a** as methanol adducts; from the relative yields of these and **15a** as a function of acid concentration, it can be concluded that ca. 75% of **15a** is formed by this pathway (Equation 9.6).

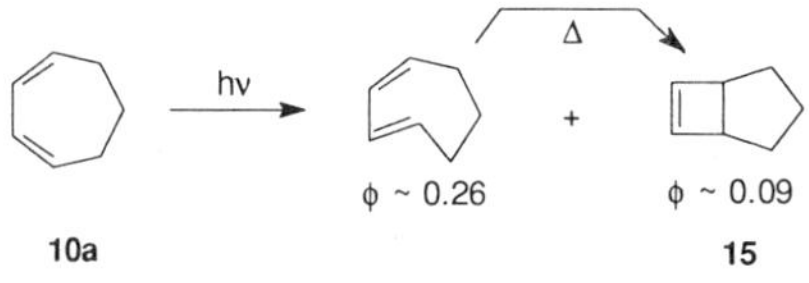

$$\text{10a} \quad\xrightarrow{h\nu}\quad \underset{\phi \sim 0.26}{\text{}} \quad+\quad \overset{\Delta}{\underset{\phi \sim 0.09}{\text{15}}} \qquad (9.6)$$

The efficiency of photochemical ring closure of larger-ring cyclic dienes decreases as ring size increases (and as the dihedral angle about the C2-C3 bond increases).[26,27] Quantum yields for *cis,trans*-photoisomerization and cyclobutene formation are shown in Equations 9.7–9.9 for *cis,cis*-1,3-cyclooctadiene (**11a**)[26] and the 1,3-cyclononadienes (**12**).[27b] Bicyclo[4.2.0]oct-7-ene (**16a**) is a true primary product of irradiation of **11a** in solution at room temperature; *cis,trans*-**11a** undergoes thermal cyclization to **16a** only at higher temperatures,[28] and photoisomerizes to the *cis,cis*-isomer with a quantum yield of 0.8.[26] Photochemical ring closure of *cis,cis*- and *cis,trans*-**12** proceeds with a high degree of disrotatory stereospecificity.[27]

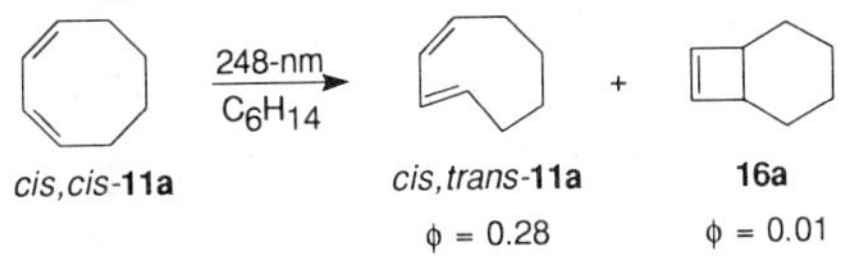

$$\text{cis,cis-11a} \xrightarrow[\text{C}_6\text{H}_{14}]{\text{248-nm}} \text{cis,trans-11a} + \text{16a} \tag{9.7}$$

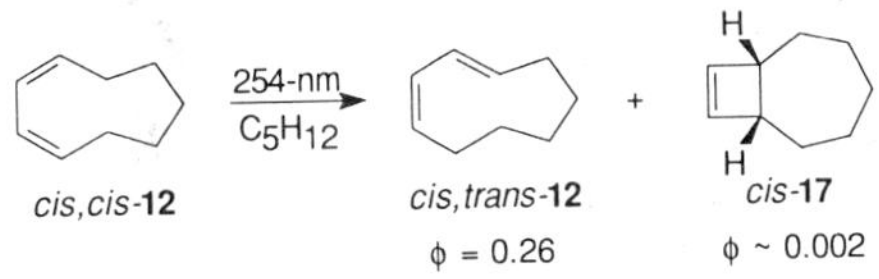

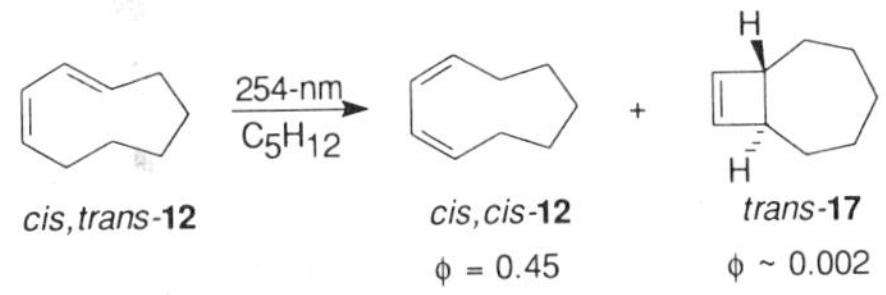

cis,cis-**11a**   cis,trans-**11a**   **16a**
$\phi = 0.28$ $\phi = 0.01$

$$\text{cis,cis-12} \xrightarrow[\text{C}_5\text{H}_{12}]{\text{254-nm}} \text{cis,trans-12} + \text{cis-17} \tag{9.8}$$

cis,cis-**12**   cis,trans-**12**   cis-**17**
$\phi = 0.26$ $\phi \sim 0.002$

$$\text{cis,trans-12} \xrightarrow[\text{C}_5\text{H}_{12}]{\text{254-nm}} \text{cis,cis-12} + \text{trans-17} \tag{9.9}$$

cis,trans-**12**   cis,cis-**12**   trans-**17**
$\phi = 0.45$ $\phi \sim 0.002$

The photochemistry of the 1,2-*bis*(methylidene)cycloalkanes (**18**; Equation 9.10)[29a] and the analogous 1,2-*bis*(ethylidene)cycloalkanes (**19**; Equations 9.11 and 9.12)[29b-31] provides additional details of the structural requirements for cyclobutene formation and *cis,trans*-photoisomerization in *s-cis*-dienes. Table 9.1 lists quantum yields for cyclobutene formation from both sets of compounds and quantum yields for *cis,trans*-photoisomerization of **19**.

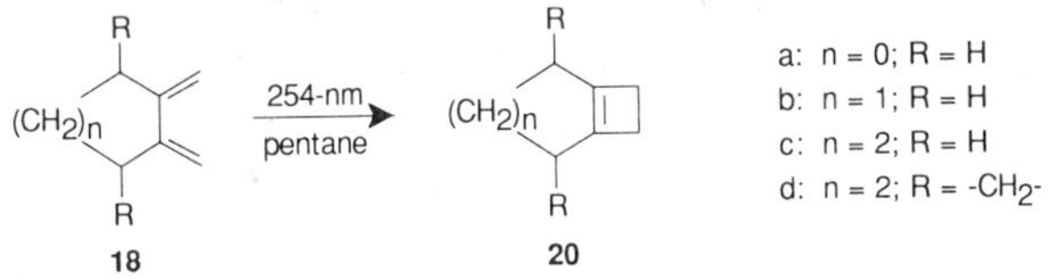

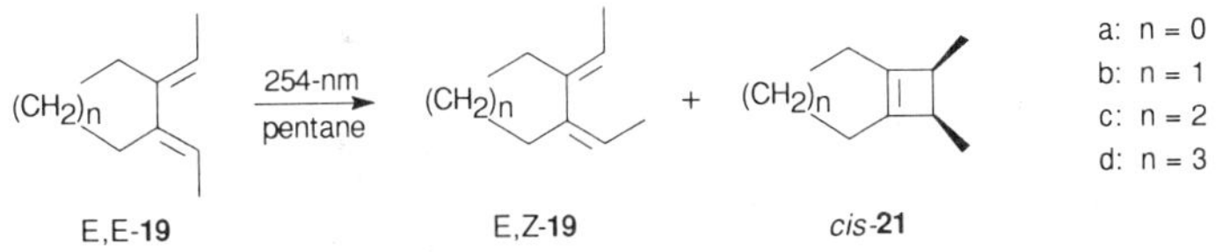

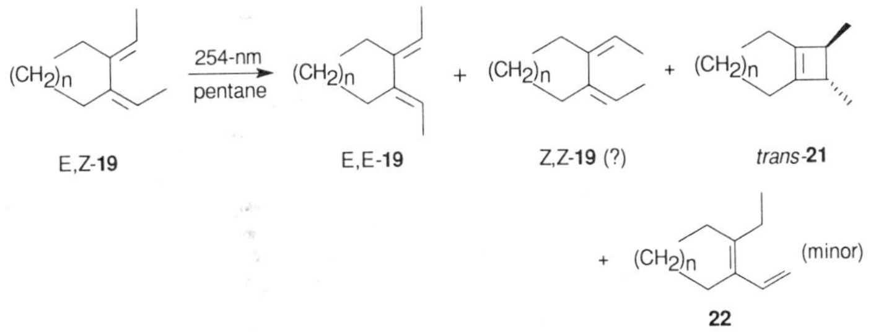

$$\mathbf{18} \xrightarrow[\text{pentane}]{\text{254-nm}} \mathbf{20} \tag{9.10}$$

a: $n = 0$; $R = H$
b: $n = 1$; $R = H$
c: $n = 2$; $R = H$
d: $n = 2$; $R = -CH_2-$

18   **20**

$$\text{E,E-}\mathbf{19} \xrightarrow[\text{pentane}]{\text{254-nm}} \text{E,Z-}\mathbf{19} + \text{cis-}\mathbf{21} \tag{9.11}$$

a: $n = 0$
b: $n = 1$
c: $n = 2$
d: $n = 3$

E,E-**19**   E,Z-**19**   cis-**21**

$$\text{E,Z-}\mathbf{19} \xrightarrow[\text{pentane}]{\text{254-nm}} \text{E,E-}\mathbf{19} + \text{Z,Z-}\mathbf{19}\,(?) + \text{trans-}\mathbf{21}$$

E,Z-**19**   E,E-**19**   Z,Z-**19** (?)   trans-**21**

$$+ \;\mathbf{22}\; (\text{minor}) \tag{9.12}$$

22

Aue and Reynolds attributed the trend in the efficiencies of cyclobutene formation from **18** (b ~ c > d > a) to product ring-strain effects since the activation energies for thermal ring opening of **20** decrease throughout the series in a similar fashion as the quantum yields for ring closure. This may indeed be the controlling factor, but recent theoretical results (*vide infra*) indicate that variations in the structures of the dienes may also play a role in governing the efficiencies of

Table 9.1   Quantum Yields for Cyclobutene Formation and *cis,*
trans-Photoisomerization from 254-nm Irradiation of Deoxygenated
0.02 *M* Pentane Solutions of 1,2-*bis*-Alkylidenecycloalkanes 18[29] and
E,E- and *E,Z*-19[30,31]

Compound	ϕ_{CB}	$\phi_{EZ\text{-}19}$	$\phi_{EE\text{-}19}$	$\phi_{ZZ\text{-}19}$
18a	<0.01[a]	—	—	—
18b	0.11	—	—	—
18c	0.11	—	—	—
18d	0.025	—	—	—
E,E-19a	<0.005	0.12 ± 0.02	—	—
E,Z-19a	<0.005	—	0.11 ± 0.03	0.01 ± 0.01
E,E-19b	0.10 ± 0.02	0.24 ± 0.02	—	—
E,Z-19b	0.13 ± 0.01	—	0.39 ± 0.04	—
E,E-19c	0.07 ± 0.01	0.20 ± 0.03	—	—
E,Z-19c	0.13 ± 0.01	—	0.29 ± 0.03	0.009 ± 0.002
E,E-19d	0.07 ± 0.02	0.21 ± 0.03	—	—
E,Z-19d	0.03 ± 0.01	—	0.14 ± 0.06	0.03 ± 0.01

[a] Product not detected.

cyclobutene formation (as well as *cis,trans*-photoisomerization).[32] The pertinent parameters (using 1,3-butadiene numbering and the results of AM1 theoretical calculations[29b]) are the C2-C3 dihedral angle (18c > 18b > 18d ~ 18a), the C1-C4 atomic distance (18c < 18b < 18d < 18a), and the degree of flexibility about the C2-C3 bond (18c > 18b > 18d > 18a). The quantum yields for ring closure of 19 follow the same trend as those of 18. Of particular significance, however, are the anomalously low quantum yields for *cis,trans*-photoisomerization obtained for the *bis*(ethylidene)cyclobutanes 19a. These may provide the first experimental indication that C2-C3 bond rotations accompany isomerization about the C1-C2 or C3-C4 bonds, as has been predicted by theory[32b] and suggested previously by other workers.[19,33]

Bicyclo[1.1.0]butane Formation

The formation of bicyclo[1.1.0]butane derivatives from irradiation of *s-trans*-dienes, though generally inefficient, has been well studied.[8,9,34,35] Bicyclobutane formation can be made to be the major product-forming process in inflexible *s-trans*-dienes that are incapable of undergoing *cis,trans*-isomerization, as exemplified in the well-known case of 3,5-cholestadiene (23; Equation 9.13).[8a,35] Irradiation of 23 in the presence of water yields the two alcohols 25a and 25b in chemical yields of 60 and 15%, respectively. These were shown to result from rapid hydrolysis of the bicyclobutane 24 formed initially.

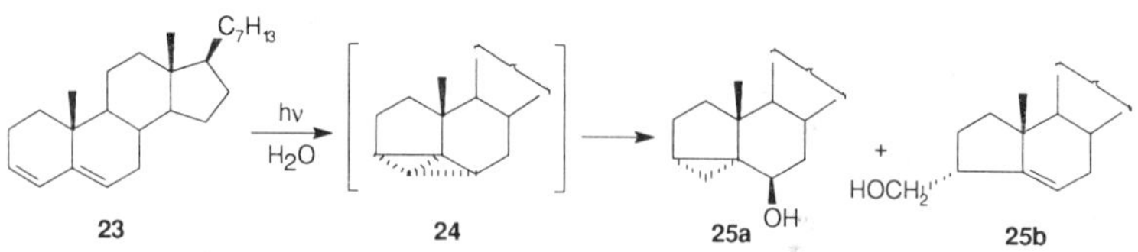

$$(9.13)$$

Dauben and Ritscher have studied the stereochemistry of bicyclobutane formation in an investigation of the photochemistry of the isomeric 3-ethylidenecyclooctenes 26.[8c] Irradiation of *E*- and *Z*-26 (Equations 9.14 and 9.15, respectively) in cyclohexane solution yielded the bicyclobutanes 27–29 and isomeric dienes from *cis,trans*-

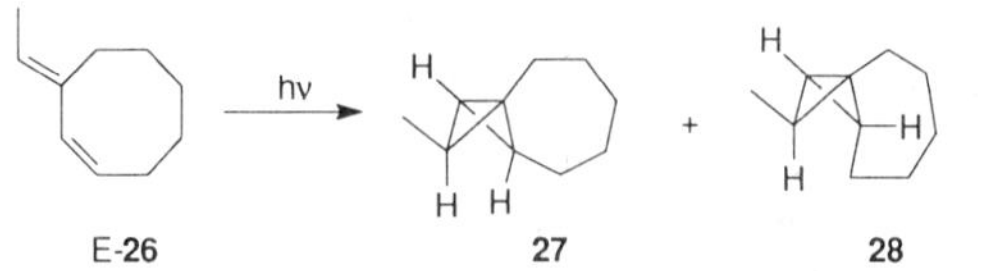

$$(9.14)$$

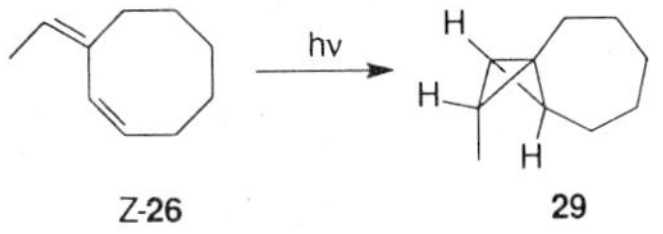

$$Z\text{-}26 \qquad 29 \tag{9.15}$$

photoisomerization. The relative yields of bicyclobutanes and C=C bond isomers were not reported, but it was noted that photoisomerization of the endocyclic C=C bond in 26 occurs 3 to 5 times faster than exocyclic C=C bond isomerization. The authors[8c] suggested that the results are best accommodated by a mechanism involving concerted conrotatory ring closure from the relaxed (allylmethylene) excited singlet state (in which the endocyclic C=C bond is twisted and the exocyclic C=C bond is incorporated in the allyl system).

[1,5]-Hydrogen Migration

Finally, direct irradiation of *s-cis*-1,3-pentadienyl systems can also lead to [1,5]-hydrogen migration in appropriately substituted cases. Since this process presumably occurs by a concerted (antarafacial) sigmatropic migration mechanism, the structural requirements for efficient [1,5]-H migration are fairly precise. Examples of this reaction include the photointerconversion of 4-methyl-1,3-pentadiene (30) and *Z*-2-methyl-1,3-pentadiene (31) (Equation 9.16)[14] and the photoisomerization of dienols such as 32 (Equation 9.17).[36] Irradiation of the *E,Z*-1,2-*bis*(ethylidene)cycloalkanes 19 yields minor amounts of the corresponding [1,5]-H migration products (22) in addition to double-bond isomers and cyclobutene derivatives (Equation 9.12).[31]

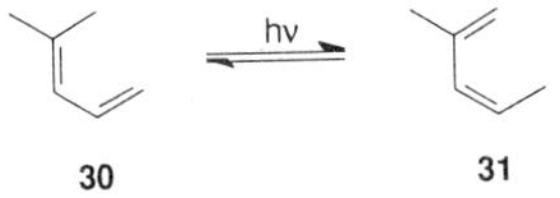

$$30 \qquad 31 \tag{9.16}$$

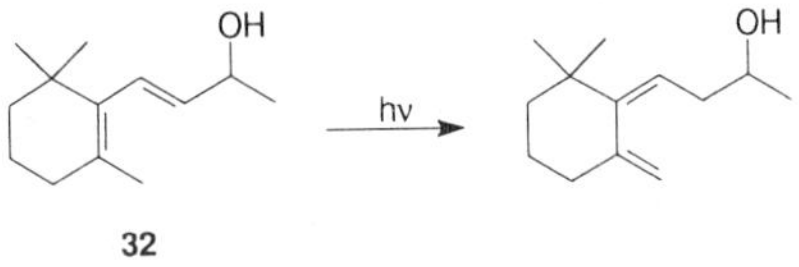

$$32 \tag{9.17}$$

9.3 The Photochemistry of Cyclobutenes

Aryl-Substituted Cyclobutenes

Simple aryl-substituted cyclobutenes fluoresce efficiently and undergo cycloreversion (Equation 9.18) upon direct photolysis in solution, to the exclusion of ring opening.[37-41] Certain aspects of their photochemistry is characteristic of the inflexible arylalkene chromophore.[3a,41] For example, in alcohol solution, both fluorescence and cycloreversion are quenched by the photoaddition of ROH.[39-41]

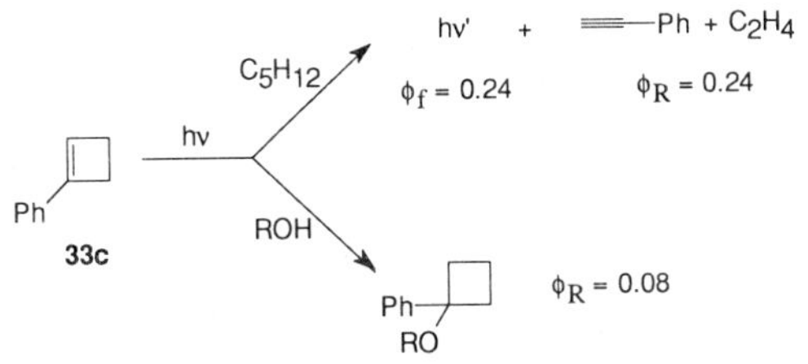

$$33c \tag{9.18}$$

The quantum yield of fluorescence (ϕ_F) depends on the degree of aryl substitution, but is unaffected by substituents on the aryl ring(s). For example, the fluorescence quantum yields of 1,2-diphenylcyclobutene (**34**) and 1-phenylcyclobutene (**33c**) in hydrocarbon solution are 0.9[38] and 0.24,[40,41] respectively. On the other hand, there is only slight variation in ϕ_F throughout the series of substituted 1-phenylcyclobutenes **33a-f**.[41]

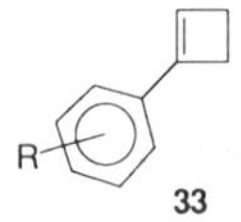

a. R = 4-OCH₃ d. R = 4-OCF₃
b. R = 4-CH₃   e. R = 3-CF₃
c. R = H   f. R = 4-CF₃

In contrast to the emission properties of **33a-f**, there are substantial substituent effects on the quantum yields of cycloreversion and photoaddition of methanol.[41] For example, **33a** exhibits quantum yields of cycloreversion and methanol addition of 0.20 and 0.13, respectively, while the corresponding values obtained for **33e** are 0.0045 and 0.0006. Thus, both processes are enhanced by electron-donating substituents on the aryl ring. Nanosecond laser flash photolysis studies have demonstrated that the first step in the photoaddition of alcohols to **33a** is protonation to form the corresponding 1-arylcyclobutyl carbocation; such behavior has ample precedent in the photochemistry of substituted styrenes and arylacetylenes.[42] In none of the cases studied does ring opening occur in detectable yield.

Photochemical ring opening in arylcyclobutenes can be promoted by the introduction of ring-strain, as in the case of benzocyclobutene derivatives.[43] Low-temperature studies have provided evidence that suggests that the ring opening *cis-* and *trans-*diphenylbenzocyclobutene (**35**) occurs nonstereospecifically (Equation 9.19).[43c]

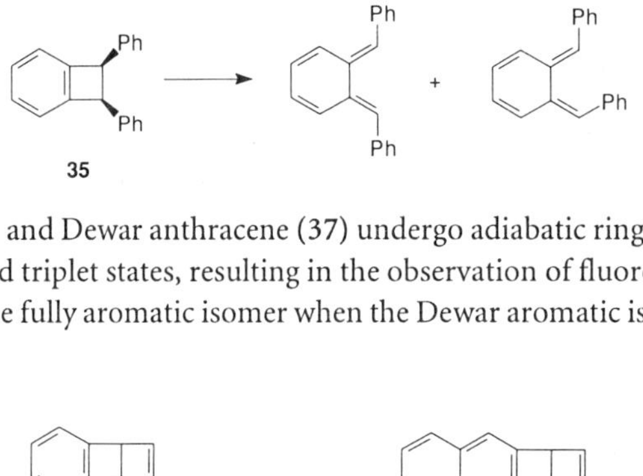

$$(9.19)$$

Dewar naphthalene (**36**) and Dewar anthracene (**37**) undergo adiabatic ring opening from both the lowest excited singlet and triplet states, resulting in the observation of fluorescence or phosphorescence emission due to the fully aromatic isomer when the Dewar aromatic is photolyzed in solution or in a glass at 77K.[5]

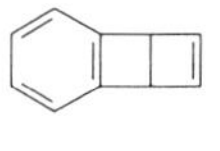

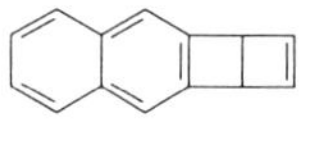

Alkyl-Substituted Cyclobutenes

The photochemistry of alkyl-substituted cyclobutene derivatives has been studied in detail over the last few years.[2] Direct irradiation of cyclobutene itself in hydrocarbon solution with a low-pressure mercury lamp (185 and 254 nm) leads to the product mixture shown in Equation 9.20.[44]

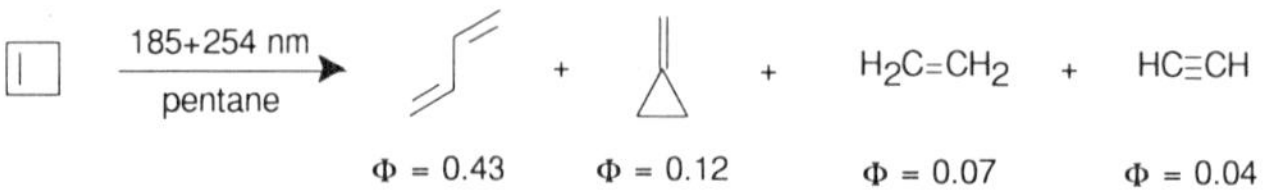

$$(9.20)$$

It has long been accepted that the photochemical ring/opening of cyclobutene derivatives obeys orbital symmetry selection rules, just as the thermal ring-opening does. This belief is evidently based on the 1969 report that irradiation of *cis*-tricyclo[6.2.0.0^{4,7}]dodec-1-ene (*cis*-38) yielded 1,1′-bicyclohexenyl (39) and *cis*-cyclododecen-7-yne (*cis*-40) in comparable yields, while irradiation of the *trans*-isomer yielded only the cycloreversion product *trans*-40.[45] The failure of *trans*-38 to yield 39 was rationalized as resulting from the fact that disrotatory ring opening is blocked since it would have to yield the unstable *cis,trans*-isomer of 39. A reinvestigation of the photochemistry of *cis*- and *trans*-38 has established that both isomers afford 39 upon direct irradiation in solution (Equation 9.21), although the quantum yield for diene formation is approximately 5 times lower from *trans*-38 than from the *cis*-isomer.[46]

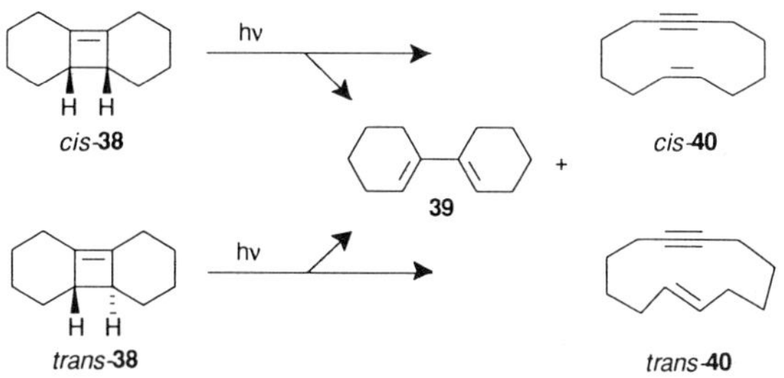

$$(9.21)$$

In general, ring opening of simple mono- and bicyclic alkyl-substituted systems proceeds nonstereospecifically with moderate efficiency ($\phi \sim 0.1$ to 0.3), in competition with formal cycloreversion (Equation 9.45).[2,27b,30,31,46–50] The latter occurs with a high degree of stereospecificity.[45,48,49] It can be concluded (from this and from the behavior of the aryl- and benzocyclobutenes discussed above) that ring opening requires a much higher excited singlet-state energy than cycloreversion does.

UV Spectra of Alkylcyclobutenes

The excited singlet/state manifold in alkylcyclobutenes is complicated by the presence of two low-lying excited states of similar energies: the $\pi,R(3s)$ Rydberg state and the π,π^* valence state.[51] The Rydberg transitions are heavily mixed with valence transitions (most likely, π,σ^*). As is the case with simple olefins, Rydberg absorptions are easily observable in gas-phase spectra, but are broadened and reduced in intensity in solution-phase spectra.[51]

A simple model olefinic Rydbergs depicts the species as being "semiionized" and having radical cation character at the C=C bond.[52] Accordingly, the energies of olefinic Rydbergs are substantially more sensitive to inductive substituent effects than are π,π^* state energies.[51] This is illustrated in Figure 9.2, which shows the gas-phase UV absorption spectra of bicyclo[4.2.0]oct-7-ene and the 7-methyl- and 7-trifluoromethyl-derivatives (16a-c).[50] The long-wavelength shoulders in the spectra of 16 and 41a have been assigned to Rydberg absorptions, while the more intense absorption bands are assigned to the valence (π,π^*) transitions. Compared to the parent compound (16), methyl substitution on the C=C bond lowers the energy of the Rydberg state, while trifluoromethyl substitution has the opposite effect.[50,53] Since the electronic effects of both these substituents are largely inductive, they have little effect on the energy of the π,π^* state.

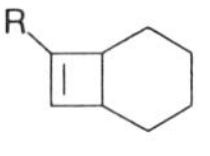

a: R = H
b: R = CH$_3$
c: R = CF$_3$

16

A second general feature of the UV absorption spectra of alkylcyclobutene derivatives is the variation in the relative intensities of π,π^* and $\pi,R(3s)$ absorptions as a function of stereochemistry at the C3 and C4 carbons. The gas- and solution-phase UV absorption spectra of *cis*- and *trans*-1,3,4-trimethylcyclobutene (41) (Figure 9.3)[49] illustrate these variations. The solution-phase spectrum

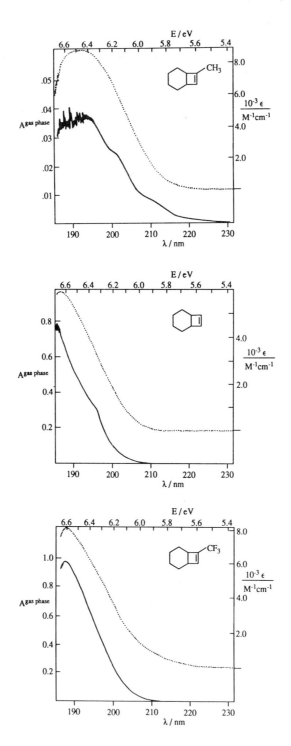

FIGURE 9.2   Gas-(·····) and solution-(———) phase UV absorption spectra of substituted bicyclo[4.2.0]oct-7-enes 16a-c. (*Source:* Leigh, W. J., Zheng, K., Nguyen, N., Werstiuk, N. H., and Ma, J., *J. Am. Chem. Soc.,* 113, 4993, 1991. With permission.)

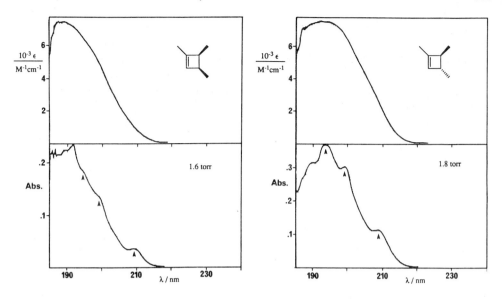

FIGURE 9.3 Solution (upper) and gas (lower) phase ultraviolet absorption spectra of *cis-* and *trans-*1,3,4-dimethylcyclobutene. (*Source:* Leigh, W. J., Zheng, K., and Clark, K. B., *Can. J. Chem.*, 68, 1988, 1990. With permission.)

of the *trans*-isomer is broader and red-shifted compared to that of the *cis*-isomer. The gas-phase spectra suggest that this may be due to a remote stereochemical effect on the relative magnitudes of the extinction coefficients of the Rydberg and valence transitions; *trans*-3,4-dialkyl substitution results in a relative increase in the intensities of the Rydberg transitions. Although the Rydberg transitions are broadened and reduced in intensity in the condensed phase, the end result is a broader, red-shifted absorption spectrum in solution compared to that of the corresponding *cis*-3,4-dimethyl isomer. This effect appears to be general,[27b,31,49] although its origins are not well understood.

The Photochemistry of Alkylcyclobutenes

A study of the photochemistry of the bicyclo[4.2.0]oct-7-enes **16a-c** has allowed identification of the specific excited singlet states that lead to ring-opening and cycloreversion processes[50] (it has been known for some time that the triplet states of alkylcyclobutenes are unreactive with respect to these unimolecular processes[4]). Direct irradiation of the three compounds in pentane solution with 193-nm light leads to ring opening and cycloreversion in each case (Equation 9.22). The quantum yields for formation of ring-opening products varies in the order **16c > 16a > 16b**, while those of the cycloreversion products vary in exactly opposite fashion. This suggests strongly that ring opening occurs from the π,π^* state, while cycloreversion occurs largely (if not entirely) from the

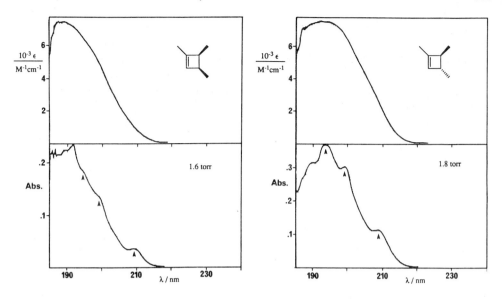

$$\text{(9.22)}$$

$\pi,R(3s)$ state. The relative yields of ring-opening and cycloreversion products also vary with excitation wavelength in a manner consistent with this conclusion.

By analogy with alkene photochemistry,[52] the mechanism of Rydberg-derived cycloreversion is most likely to involve initial ring contraction (with retention of stereochemistry at C3 and C4) to yield a cyclopropyl carbene intermediate, which undergoes stereospecific fragmentation to yield the alkyne and alkene.[44,47,54] Evidence for the involvement of cyclopropyl carbene intermediates in cyclobutene photochemistry was obtained from a study of the photochemistry of the isomeric cyclobutene derivatives **44** and **45**.[55] It was observed that in addition to yielding dienes and cycloreversion products, direct irradiation in solution leads to interconversion of the two isomers. The most likely mechanism for this interconversion is via a common cyclopropyl carbene intermediate as shown in Equation 9.23.

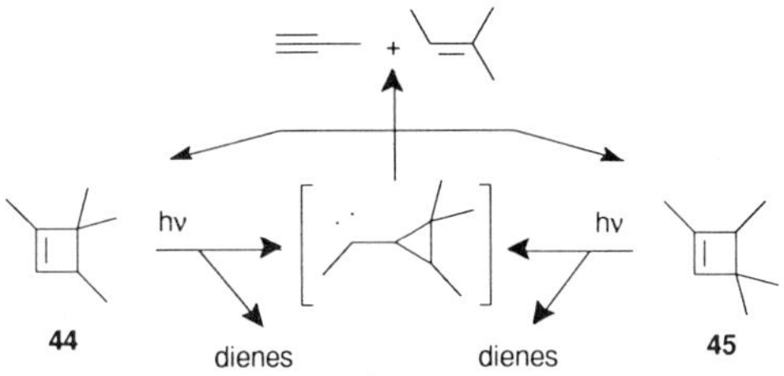

$$(9.23)$$

Direct irradiation of the series of bicyclo[n.2.0]alk-(n+3)-enes (**15–17**) in deoxygenated pentane solution with a filtered low-pressure mercury lamp (185 nm) yields the product mixtures summarized in Equation 9.24.[27b,47] Quantum yields for product formation, determined by cyclooctene actinometry, are also shown. These data reveal several interesting general features of the photochemistry of alkylcyclobutenes. First, cycloreversion and ring opening proceed with similar quantum yields when there is not a large difference in the energies of the π,π^* and $\pi,R(3s)$ states, and when the products of stereospecific cycloreversion are both relatively stable. Presumably, the formation of *cis*-cycloheptene from photolysis of *trans*-**17** is due to the initial formation of *trans*-cycloheptene,

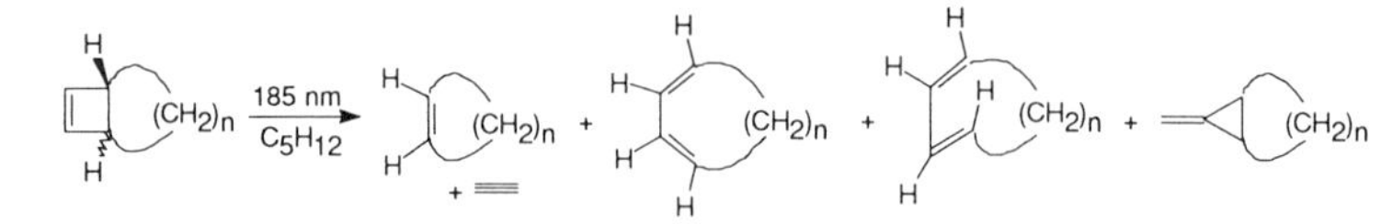

15: n=3 (*cis*)	0.12	0.14		
16a: n=4 (*cis*)	0.17	0.12	0.04	0.01
cis-**17**: n=5	0.26	0.06	0.16	~0.02
trans-**17**: n=5	0.05	0.04	0.13	0.05

$$(9.24)$$

which is relatively unstable; hence, cycloreversion is suppressed.[27b] While this has not been established conclusively, other examples exist that corroborate the idea that cycloreversion is suppressed when it leads to thermodynamically unstable products.[30,31] Secondly, the total quantum yield for ring opening does not appear to depend on the minor variations in the amount of cyclobutene ring-strain induced by the second ring in the bicyclic structure. This conclusion is tentative since it has not yet been established whether *cis,trans*-1,3-cycloheptadiene is a primary product of photolysis of **15**. This unstable diene is known to undergo rapid thermal ring closure to **15**,[25] so that under steady-state conditions at room temperature, its formation represents an energy-wasting step in the photochemistry of **15**.

The formation of *cis,cis*- and *cis,trans*-**12** in nearly common distributions from photolysis of *cis*- and *trans*-**17** is a particularly interesting feature of the photochemistry of these compounds. Similar behavior has been reported by Dauben and Haubrich for the photochemistry of *cis*- and *trans*-bicyclo[6.2.0]cyclodeca-2,9-diene (**46**).[48] The mechanistic implications of these results will be discussed later.

The formation of methylenecyclopropane derivatives (see Equation 9.24) is not a general feature of the photochemistry of alkylcyclobutenes, although it has been reported that the parent compound is formed in relatively high yield photolysis of cyclobutene itself in solution.[44] None of the other mono- or bicyclic cyclobutene derivatives studied thus far yield products of this type in significant yield.

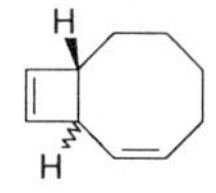

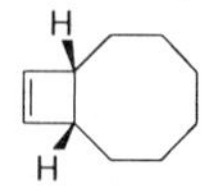

A final generalization that can be made is that the efficiency of the ring-opening reaction depends on the bond strength of the cyclobutene C3-C4 bond. This generalization is illustrated by the relative yields of ring-opening and cycloreversion products obtained from photolysis of *cis*-**46** (4.5) and *cis*-**47** (0.8).[48] According to molecular models, the C1-C8 bond in the latter compound should be weakened substantially owing to nearly perfect overlap with the cyclooctenyl C=C bond.[27b] A second example can be found in the photochemistry of *cis*- and *trans*-2,3-dimethylbicyclo[2.2.0]hex-1[4]-ene (**21a**), both of which yield isomeric 1,2-*bis*(ethylidene)cyclobutanes (**19a**) to the essential exclusion of the alternative, 1,2-dimethyl-3,4-*bis*(methylidene)cyclobutanes (**48**; Equation 9.25).[31]

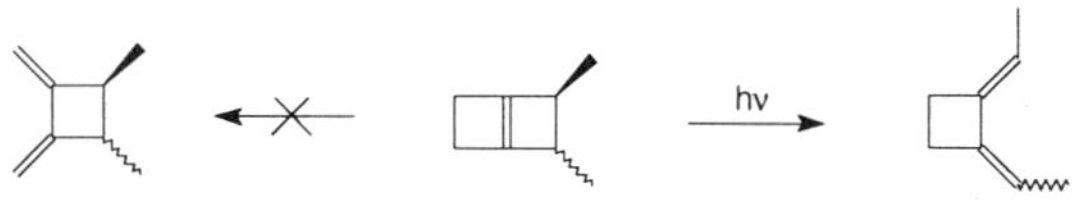

$$(9.25)$$

Out of the large number of mono-, bi-, and tricyclic alkylcyclobutenes whose solution-phase photochemistry has been studied, only one class of compounds undergoes photochemical ring opening with high degrees of disrotatory stereospecificity. These are the dimethyl-substituted bicyclo[n.2.0]alk-1-enes (**21**) shown in Equation 9.26.[30,31] Irradiation of the *cis*-dimethyl isomers in pentane solution with 193-nm light yields the corresponding *E,E*-1,2-*bis*(ethylidene)cycloalkanes (*E,E*-**19**) as the major products (70 to 90%), while the *trans*-dimethyl isomers yield the isomeric *E,Z*-1,2-*bis*(ethylidene)cycloalkanes (*E,Z*-**19**) predominantly (80 to 90%). None

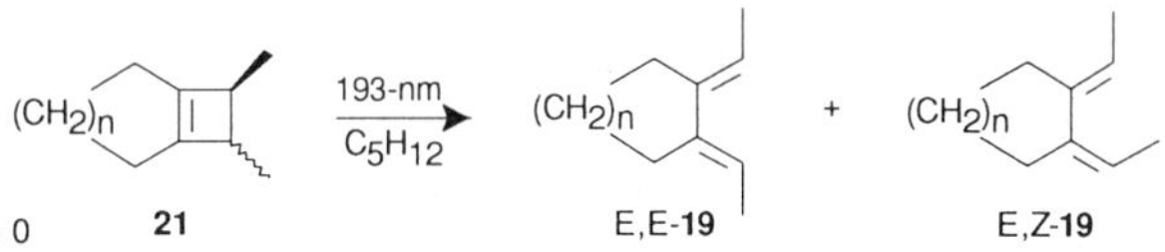

$$(9.26)$$

of these compounds undergoes cycloreversion, presumably due to the fact that the products of this reaction would be the relatively unstable C4-C7 cycloalkynes. Interestingly, the total quantum yields for ring opening of these compounds are anomalously high — a factor of 2 to 3 times higher than those of any other alkylcyclobutene studied to date.[30,31]

9.4 Theoretical Studies of the Photochemistry of 1,3-Butadiene and Cyclobutene

Avoided Surface Crossing Models

The commonly accepted mechanisms for the *cis,trans*-photoisomerization and ring closure reactions of conjugated dienes contain the basic assumption that, in the lowest excited singlet state, the C2-C3 bond possesses increased double-bond character compared to that in the ground state.[1,3a]

Thus, these processes are viewed as proceeding by torsional relaxation about the termini, with little torsional motion about the C2-C3 bond. Also common to these mechanisms is the idea that decay of the relaxed excited-state species to the ground-state reaction surface occurs at an avoided crossing between the excited- and ground-state surfaces. According to this model, *cis,trans*-photoisomerization and ring closure differ fundamentally from *s-cis/s-trans*-interconversion, which must involve torsion about the central (C2-C3) bond. These mechanisms are so well known[1] that they will receive only cursory discussion below.

Thus, *cis,trans*-isomerization is viewed as proceeding by torsional relaxation about one of the double bonds to the relaxed allylmethylene biradical (or zwitterion) geometry 49 (Equation 9.27), from which decay to the ground-state surface occurs.[3a,56,57] The geometry of the excited allylmethylene species corresponds to that of the transition state for thermal *cis,trans*-isomerization on the ground-state surface. Once the ground state is entered, the system undergoes further relaxation to the stable ground-state geometry of either the isomer or the original

$$(9.27)$$

diene. Obviously, two such pathways exist for asymmetrically substituted dienes. This model is consistent with the well-established fact that direct *cis,trans*-photoisomerization of conjugated dienes involves isomerization about only one of the two bonds per photon, and requires that there be no loss in efficiency when the diene is constrained to prevent rotation about the C2-C3 bond. As pointed out earlier, however, there is experimental evidence to suggest that such motions *are* important in the *cis,trans*-photoisomerization process.

Cyclobutene formation results from synchronous[58,59] (or asynchronous[60]) rotation about the C1-C2 and C3-C4 bonds as the system relaxes to energy minima corresponding to the geometries of the transition states for the thermal interconversions. Decay to the ground-state surface produces cyclobutene or regenerates diene. According to the theoretical calculations of Oosterhoff and Devaquet and their coworkers,[58,59b] both con- and disrotatory rotations are possible on the excited-state surface. However, the disrotatory pathway is preferred overall because the ground-state surface for this pathway is higher in energy than that of the conrotatory pathway; hence, internal conversion to the ground state is faster for the disrotatory process.[59] This mechanism is compatible with the high degree of disrotatory stereospecificity exhibited in the photochemical ring closure of numerous diene systems *(vide supra)*. However, it is incompatible with the well-established fact that, in general, the reverse reaction proceeds nonstereospecifically (unless ring opening proceeds partially via a hot ground-state pathway).

Conical Intersection Model

On the basis of *ab initio* (MCSCF) theoretical calculations, Bernardi and Robb and their co-workers have proposed recently a rather different mechanistic model to explain the photochemistry of 1,3-butadiene[32] and cyclobutene.[61] The basic feature of this model is that excited-to-ground state internal radiationless decay occurs at a *conical intersection* (CI) where the ground- and first excited-state potential energy surfaces are degenerate, rather than at an avoided surface crossing.[62] This theoretical concept is not new,[63] but it has only recently received detailed attention in explaining photochemical pericyclic reactions.[32,61,62,64,65] This model has several important ramifications regarding the mechanisms of photochemical reactions. First, radiationless decay from the excited state to the ground state is expected to be *fully* efficient at a CI. This is in contrast to that at an avoided crossing, where the decay rate is controlled by the magnitude of the energy gap between

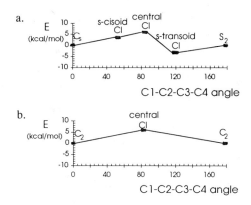

FIGURE 9.4 Reaction coordinate diagram for twisting about the 1,3-butadiene C2-C3 bond along the disrotatory (a) and conrotatory (b) *s-cis/s-trans*-excited state isomerization pathways. (*Source:* Olivucci, M., Ragazos, N., Bernardi, F., and Robb, M. A., *J. Am. Chem. Soc.*, in press. Adapted with permission.)

the two surfaces. Second, decay to the ground state at a single CI can result in the formation of a variety of different types of photoproducts, rather than a single product as predicted in the avoided crossing model for excited-state decay. One of the fundamental differences between the classical (i.e., avoided crossing) and conical intersection models is in which potential energy surface the selection between different types of possible photoproducts occurs. In the classical model, this selection is assumed to originate primarily on the excited-state surface. The conical intersection model predicts that this selection occurs largely on the ground-state surface after it has been entered from the CI, but is dependent upon from which direction the CI region is entered on the excited-state surface.

The calculations for 1,3-butadiene start with the assumption that the reactive state is the (doubly excited 2^1Ag state, which is thought to be accessed rapidly by internal conversion from the initially populated 1^1B_u (π,π^*) state,[9,66] Two equilibrium conformations are associated with each butadiene conformer on the 2^1A_g potential energy surface, one corresponding to ca. 40° conrotatory displacement of the termini (C_2 symmetry for both *s-cis*- and *s-trans*-conformers) and one corresponding to ca. 45° disrotatory displacement of the termini (S_2 symmetry for *s-trans* and C_s symmetry for *s-cis*). In all four structures, the carbon framework is planar and the C-C bond lengths are approximately equal. The two *s-cis*-conformers are separated by barriers of less than 1 kcal mol^{-1}, as are the two *s-trans*-conformers. The four equilibrium conformers are linked to three CI regions on the ^2Ag potential energy surface by asynchronous torsional motions about *all three* C-C bonds in the carbon framework of the molecule. Consequently, the structures of the molecule in the CI regions are twisted about all three C-C bonds such that they can be considered qualitatively as tetraradicaloid species. Their structures differ primarily in the twist angle about the C2-C3 bond: these angles are 52° ("*s-cisoid*"), 84° ("central"), and 119° ("*s-transoid*"). The location and energies of the excited-state minima and the three CIs with respect to the C2-C3 dihedral angle are illustrated in Figure 9.4. The *s-transoid* CI is the lowest energy point on the ^2Ag surface, lying 3.2 kcal mol^{-1} below the S_2 (*s-trans*) minimum, to which it is linked by a nearly barrierless pathway. The C_s (*s-cis*) minimum is linked to the *s-cisoid* CI which lies 3.7 kcal mol^{-1} higher in energy. The *s-cisoid* and *s-transoid* CIs are linked by the central CI, which lies 6 kcal mol^{-1} above the C_s and S_2 minima. Only the central CI is accessible from the C_2 minima of *s-cis*- and *s-trans*-1,3-butadiene. Additional calculations provide details of the nature of the ground-state reaction paths after the molecule has passed through one of the conical intersections, starting from the four excited-state equilibrium structures. These are summarized in Equations 9.28 to 9.30.

$$S_2 \text{ (s-trans)} \xrightarrow{-3.2 \text{ kcal/mol}} \text{s-transoid CI} \rightarrow \text{s-cis} \Longleftrightarrow \text{s-trans, cis} \Longleftrightarrow \text{trans isom, bicyclobutane} \tag{9.28}$$

$$C_S \text{ (s-cis)} \xrightarrow{\quad} \text{s-cisoid CI} \xrightarrow{} \text{s-cis<=>s-trans, cis<=>trans isom, cyclobutene (dis)} \qquad (9.29)$$
$$\phantom{C_S \text{ (s-cis)} \xrightarrow{\quad}} {\scriptstyle +3.7 \text{ kcal/mol}}$$

$$C_2 \text{ (s-trans)} \searrow$$
$$\qquad\qquad\qquad \text{central CI} \xrightarrow{} \text{s-cis<=>s-trans, cis<=>trans isom, cyclobutene (con), bicyclobutane}$$
$$C_2 \text{ (s-cis)} \nearrow \quad {\scriptstyle +6 \text{ kcal/mol}} \qquad\qquad\qquad\qquad\qquad\qquad\qquad\qquad (9.30)$$

The theory explains successfully most of the features of 1,3-butadiene photochemistry summarized in Section 9.1. For example, *s-cis,s-trans*-interconversion and *cis,trans*-photoisomerization (about one bond) are the main excited singlet state deactivation pathways because approach to the CI regions from either *s-cis*- or *s-trans*-dienes is dominated by C2-C3 bond rotations and *asynchronous* (disrotatory) rotations of the termini. Similarly, disrotatory cyclization of *s-cis*-dienes is preferred over the conrotatory pathway because the excited-state pathway to a conical intersection is energetically more favorable starting from the C_s minimum than from the C_2 minimum. Interestingly, the theory also predicts that *cis, trans*-photoisomerization and cyclobutene formation are intimately coupled with torsional motions about the central (C2-C3) bond. Experimental examples were presented earlier that suggest that the quantum yields of both these processes might, in fact, be significantly reduced in systems which are constrained to inhibit rotations about the central bond of the diene moiety. Clearly, this facet of diene photochemistry requires further study.

Analogous calculations have been reported recently on the photochemical ring opening of the alkylcyclobutene derivative **16b** in an effort to explain the nonstereospecificity associated with this reaction.[61] As was proposed for 1,3-butadiene, three tetraradicaloid CIs are accessible from the excited singlet state of **16b**, each twisted substantially about the initial cyclobutene double bond (the C2-C3 bond in 1,3-butadiene). The three CIs lead to different isomeric dienes once the ground-state surface is entered, which explains the formation of all three isomers of **11** in the photolysis of **16b**. It is believed that the photochemistry of the constrained cyclobutenes **21** verifies the gross features of these preliminary calculations: inhibiting rotation about the incipient central diene bond once ring opening has initiated leads to a high degree of stereospecificity in the reaction, presumably by inducing selectivity in the approach of the excited molecule to the CI responsible for the disrotatory component of the reaction.

9.5 Conclusions

It has long been established that orbital symmetry selection rules are rather precise in their predictions of the stereochemistry of ground-state pericyclic reactions, and equally so for most photochemical ones. They work adequately for the ring closure of 1,3-butadiene, but are borne out in cyclobutene photochemistry only in rather special cases. Still, there is evidence to suggest that ring opening *initiates* with disrotatory motions of the incipient termini of the product diene.[30,67] In the case of cyclobutene, it seems clear that the mechanism of the excited-state ring opening process is far more complex than the simple theory can handle, and more advanced approaches are necessary in order to obtain a complete understanding of the photochemistry of this simple molecule.

References

1. (a) Woodward, R. B. and Hoffman, R., *The Conservation of Orbital Symmetry*, Verlag Chemie, Weinheim, 1970; (b) Turro, N. J., *Modern Molecular Photochemistry*, 2nd ed., Benjamin-Cummings, New York, 1985, Chap. 7; (c) Gilbert, A. and Baggott, J., *Essentials of Molecular Photochemistry*, CRC Press, Boca Raton, FL, 1991.
2. Leigh, W. J., Orbital symmetry and the photochemistry of cyclobutene, *Can. J. Chem.*, 1993, in press.

3. Leigh, W. J., Techniques and applications of far-UV photochemistry in solution, *Chem. Rev.*, 1993, in press.

4. (a) Saltiel, J. and Charlton, J. L., *Cis,trans* isomerization of olefins, *Rearrangements in Ground and Excited States*, Vol. III, De Mayo, P., Ed., Academic, New York, 1980, 25, and references cited therein; (b) Saltiel, J., D'Agostino, J., Megarity, E. D., Metts, L., Neuberger, K. R., Wrighton, M., and Zafiriou, O. U., The *cis, trans* photoisomerization of olefins, *Org. Photochem.*, 3, 1, 1973.

5. (a) Yang, N. C., Carr, R. V., Li, E., McVey, J. K., and Rice, S. A., 2,3-Naphtho-2,5-bicyclo[2.2.0]hexadiene, *J. Am. Chem. Soc.*, 1974, 96, 2297; (b) Carr, R. V., Kim, B., McVey, J. K., Yang, N. C., Gerhartz, W., and Michl, J., Photochemistry and photophysics of 1,4-dewarnaphthalene, *Chem. Phys. Lett.*, 39, 57, 1976.

6. Srinivasan, R., Photochemistry of conjugated dienes and trienes, *Adv. Photochem.*, 4, 113, 1966.

7. (a) Srinivasan, R., A simple synthesis of bicyclo[1.1.0]butane and its relation to the internal conversion of electronic energy in 1,3-butadiene, *J. Am. Chem. Soc.*, 85, 4045, 1963; (b) Srinivasan, R. and Sonntag, F. I., Energy level of the first excited singlet state of 1,3-butadiene, *J. Am. Chem. Soc.*, 87, 3778, 1965.

8. (a) Dauben, W. G. and Wipke, W. T., Photochemistry of dienes, *Pure Appl. Chem.*, 9, 539, 1964; (b) Barltrop, J. A. and Browning, H. E., The photolysis of simple acyclic 1,3-dienes in methanol, *J. Chem. Soc., Chem. Commun.*, 1481, 1968. (c) Dauben, W. G. and Ritscher, J. S., Photochemistry of ethylidenecyclooctenes. Mechanism of bicyclobutane formation, *J. Am. Chem. Soc.*, 92, 2925, 1970.

9. Trulson, M. O. and Mathies, R. A., Excited-state structure and dynamics of isoprene from absolute resonance Raman intensities, *J. Phys. Chem.*, 94, 5741, 1990.

10. (a) Baldry, P. J., Formation of methyl ethers from photoaddition of methanol to phenyl-substituted butadienes, *J. Chem. Soc. Perkin I*, 1913, 1975; (b) Velsko, S. P. and Fleming, G. R., Photochemical isomerization in solution. Photophysics of diphenylbutadiene, *J. Chem. Phys.*, 76, 3553, 1982; (c) Yee, W. A., Hug, S. J., and Kliger, D. S., Direct and sensitized photoisomerization of 1,4-diphenylbutadienes, *J. Am. Chem. Soc.*, 110, 2164, 1988; (d) White, E. H. and Anhalt, J. P., Photochemical reactions of 2,3-diphenylbutadiene and 1,2-diphenylcyclobutene, *Tetrahedron Lett.*, 3937, 1965.

11. Boue, S., Rondelez, D., and Vanderlinden, P., *Excited States in Organic Chemistry and Biochemistry*, Pullman, B. and Goldblum, N., Eds., D. Reidel, Dordrecht, 1977, 199.

12. Squillacote, M. E., Sheridan, R. S., Chapman, O. L., and Anet, F. A. L., Planar *s-cis*-1,3-butadiene, *J. Am. Chem. Soc.*, 101, 3657, 1979.

13. Srinivasan, R., Kinetics of the photochemical dimerization of olefins to cyclobutane derivatives. I. Intramolecular addition, *J. Am. Chem. Soc.*, 84, 4141, 1962.

14. Crowley, K. J., Photoisomerizations. VI. Cyclobutene formation and diene migration in simple 1,3-dienes, *Tetrahedron*, 21, 1001, 1965, and references cited therein.

15. (a) Srinivasan, R., Mechanism of the photochemical valence tautomerization of 1,3-butadienes, *J. Am. Chem. Soc.*, 91, 4498, 1969; (b) Saltiel, J., Metts, L., and Wrighton, M., The direct *cis-trans* photoisomerization of 1,3-dienes in solution, *J. Am. Chem. Soc.*, 92, 3227, 1970.

16. (a) Vanderlinden, P. and Boué, S., Direct photolysis of the penta-1,3-dienes: Recognition of wavelength-dependent behaviour in solution, *J. Chem. Soc., Chem. Commun.*, 1975, 932; (b) Boué, S. and Srinivasan, R., Differences in reactivity between excited states of *cis*- and *trans*-1,3-pentadiene, *J. Am. Chem. Soc.*, 92, 3226, 1970.

17. Squillacote, M. E. and Semple, T. C., Polarization in the excited state of 1,3-pentadiene: Experimental evidence for an allyl cation-methylene anion species, *J. Am. Chem. Soc.*, 109, 892, 1987.

18. (a) Woodward, R. B., Structure and absorption spectra. III. Normal conjugated dienes, *J. Am. Chem. Soc.*, 1942, 64, 72; (b) Allinger, N. L. and Miller, M. A., Organic quantum chemistry. VII. Calculation of the near-ultraviolet spectra of polyolefins, *J. Am. Chem. Soc.*, 86, 2811, 1964.

19. Squillacote, M. and Semple, T. C., Photochemistry of *s-cis* acyclic 1,3-dienes, *J. Am. Chem. Soc.*, 112, 5546, 1990.

20. Squillacote, M. E., Semple, T. C., and Mui, P. W., The geometries of the *s-cis* conformers of some acyclic 1,3-dienes: Planar or twisted?, *J. Am. Chem. Soc.,* 107, 6842, 1985.

21. Dauben, W. G. and Kellogg, M. S., Photochemistry of *cis*-fused bicyclo[4.n.0]-2,4-dienes. Ground state conformational control, *J. Am. Chem. Soc.,* 102, 4456, 1980.

22. (a) Dauben, W. G., Rabinowitz, J., Vietmeyer, N. D., and Wendschuh, P. H., Photoequilibria between 1,3-cyclohexadienes and 1,3,5-hexatrienes. Photochemistry of 3-alkyl-6,6,9,9-tetramethyl-$\Delta^{3,5(10)}$-hexalins, *J. Am. Chem. Soc.,* 94, 4285, 1972; (b) Dauben, W. G., Kellogg, M. S., Seeman, J. I., Vietmeyer, N. D., and Wendschuh, P. H., Steric aspects of the photochemistry of conjugated dienes and trienes, *Pure Appl. Chem.,* 33, 197, 1973.

23. Chapman, O. L., Pasto, D. J., Borden, G. W., and Griswold, A. A., Photochemical transformations of conjugated cycloheptadienes, *J. Am. Chem. Soc.,* 84, 1220, 1962.

24. Dauben, W. G. and Cargill, R. L., Photochemical transformations. VI. Isomerization of cycloheptadiene and cycloheptatriene, *Tetrahedron,* 12, 186, 1961.

25. Inoue, Y., Hagiwara, S., Daino, Y., and Hakushi, T., *cis,trans*-Cyclohepta-1,3-diene as a transient intermediate in the photocyclization of *cis,cis*-cyclohepta-1,3-diene, *J. Chem. Soc., Chem. Commun.,* 1307, 1985.

26. Nebe, W. J. and Fonken, G. J., Photolysis of *cis,cis*-1,3-cyclooctadiene, *J. Am. Chem. Soc.,* 91, 1249, 1969.

27. (a) Shumate, K. M. and Fonken, G. J., Photochemical reactions of medium ring 1,3-dienes, *J. Am. Chem. Soc.,* 88, 1073, 1966; (b) Leigh, W. J., Zheng, K., and Clark, K. B., Cyclobutene photochemistry. The photochemistry of *cis*- and *trans*-bicyclo[5.2.0]non-8-ene, *J. Org. Chem.,* 56, 1574, 1991.

28. Liu, R. S. H., Photosensitized isomerization of 1,3-cyclooctadienes and conversion to bicyclo[4.2.0]oct-7-ene, *J. Am. Chem. Soc.,* 89, 112, 1967.

29. (a) Aue, D. H. and Reynolds, R. N., Photochemical synthesis and reactivity of strained polycyclic cyclobutenes. $\Delta^{2(5)}$-Tricyclo[4.2.1.0^{2,5}]nonene, *J. Am. Chem. Soc.,* 95, 2027, 1973 (b) Leigh, W. J. and Postigo, J. A., The role of central bond torsional motions in the direct *cis, trans*-photoisomerization of conjugated dienes, *J. Chem. Soc., Chem. Commun.,* 1836, 1993.

30. Leigh, W. J. and Zheng, K., Cyclobutene photochemistry. Partial orbital symmetry control in the photochemical ring opening of a constrained cyclobutene, *J. Am. Chem. Soc.,* 113, 4019, 1991; *Errata: J. Am. Chem. Soc.,* 114, 796, 1992.

31. Leigh, W. J., Postigo, A., and Zheng, K., Cyclobutene photochemistry. 8. Highly stereospecific photochemical ring-opening of *cis*- and *trans*-dimethylbicyclo[n.2.0]alk-1-enes, to be published.

32. (a) Bernardi, F., De, S., Olivucci, M., and Robb, M. A., Mechanism of ground-state-forbidden photochemical pericyclic reactions: Evidence for real conical intersections, *J. Am. Chem. Soc.,* 112, 1737, 1990; (b) Olivucci, M., Ragazos, N., Bernardi, F., and Robb, M. A., A conical intersection mechanism for the photochemistry of butadiene. A MC-SCF study, *J. Am. Chem. Soc.,* 115, 3710, 1993.

33. Liu, R. S. H. and Browne, D. T., A bioorganic view of the chemistry of vision: H.T.-n and B.P.-m,n mechanisms for reactions of confined, anchored polyenes, *Acc. Chem. Res.,* 19, 42, 1986, and references cited therein.

34. (a) Dauben, W. G. and Poulter, C. D., The addition of methanol to some bicyclobutanes, *Tetrahedron Lett.,* 3021, 1967; (b) Dauben, W. G. and Spitzer, W. A., The photochemistry of 4,4-dimethyl-1-methylene-2-cyclohexene, a methylene analog of a cyclohexenone, *J. Am. Chem. Soc.,* 90, 802, 1968.

35. (a) Dauben, W. G. and Ross, J. A., Photochemical transformations. V. The reaction of 3,5-cholestadiene, *J. Am. Chem. Soc.,* 81, 6521, 1959; (b) Dauben, W. G. and Willey, F. G., Photochemical transformations. XIII. The mechanism of the reaction of $\Delta^{3,5}$-cholestadiene, *Tetrahedron Lett.,* 893, 1962.

36. Mousseron, M., Mousseron-Canet, M., and Legendre, P., No. 10 - Isomérisation photochimique dans les séries de la β-ionone et de la déhydro β-ionone, *Bull. Soc. Chim. France,* 50, 1964.

37. DeBoer, C. D. and Schlessinger, R. H., The multiplicity of the photochemically reactive state of 1,2-diphenylcyclobutene, *J. Am. Chem. Soc.,* 90, 803, 1968.

38. Sakuragi, M. and Hasegawa, M., Photochemistry of 1,2-diphenylcyclobutene in protic solvents, *Chem. Lett.*, 29, 1974.

39. Kaupp, G. and Stark, M., Selektivitäten bei der photolyse von diphenylcyclobutenen, *Chem. Ber.*, 111, 3608, 1978.

40. Zimmerman, H. E., Kamm, K. S., and Werthemann, D. P., Mechanisms of electron demotion. Direct measurement of internal conversion and intersystem crossing rates. Mechanistic organic photochemistry, *J. Am. Chem. Soc.*, 97, 3718, 1975.

41. Leigh, W. J. and Postigo, A., Substituent effects on the photochemistry of 1-phenylcyclobutene, *Can. J. Chem.*, in press.

42. Wan, P. and Yates, K., Photogenerated carbonium ions and vinyl cations in aqueous solution, *Rev. Chem. Intermed.*, 5, 157, 1984.

43. (a) Quinkert, G., Opitz, K., Wiersdorff, W.-W., and Finke, M., Stereospezifische adduktbildungen bei benzocyclobutenen, *J. Liebigs Ann. Chem.*, 693, 45, 1966; (b) Quinkert, G., Wiersdorff, W.-W., Finke, M., Opitz, K., and von der Haar, F.-G., Darstellung und elektrocyclische isomerisierungen des tetraphenyl-*o*-chinodimethans, *Chem. Ber.*, 101, 2303, 1968; (c) Quinkert, G., Finke, M., Palmowski, J., and Wiersdorff, W.-W., Light induced reactions. VI. Low temperature study of the photochemical ring opening of symmetric benzocyclobutene derivatives, *Mol. Photochem.*, 1, 433, 1969; (d) Kolc, J. and Michl, J., Photochemical synthesis of matrix-isolated pleiadene, *J. Am. Chem. Soc.*, 92, 4147, 1970.

44. Adam, W., Oppenlander, T., and Zang, G., 185-nm Photochemistry of cyclobutene and bicyclo[1.1.0]butane, *J. Am. Chem. Soc.*, 107, 3921, 1985.

45. Saltiel, J. and Ng Lim, L.-S., Stereospecific photochemical fragmentation of cyclobutenes in solution, *J. Am. Chem. Soc.*, 91, 5404, 1969.

46. Leigh, W. J. and Zheng, K., Cyclobutene photochemistry. Reinvestigation of the photochemistry of *cis*- and *trans*-tricyclo[6.4.0.0^{2,7}]dodec-1-ene, *J. Am. Chem. Soc.*, 113, 2163, 1991.

47. Clark, K. B. and Leigh, W. J., Cyclobutene photochemistry. Nonstereospecific photochemical ring opening of simple cyclobutenes, *J. Am. Chem. Soc.*, 109, 6086, 1987.

48. Dauben, W. G. and Haubrich, J., The 193-nm photochemistry of some fused-ring cyclobutenes: Absence of orbital symmetry control, *J. Org. Chem.*, 53, 600, 1988.

49. Leigh, W. J., Zheng, K., and Clark, K. B., Cyclobutene photochemistry. Substituent and wavelength effects on the photochemical ring-opening of monocyclic alkylcyclobutenes, *Can. J. Chem.*, 68, 1988, 1990.

50. Leigh, W. J., Zheng, K., Nguyen, N., Werstiuk, N. H., and Ma, J., Cyclobutene photochemistry. Identification of the excited states responsible for the ring-opening and cycloreversion reactions of alkylcyclobutenes, *J. Am. Chem. Soc.*, 113, 4993, 1991.

51. (a) Loeffler, B. B., Eberlin, E., and Pickett, L. W., Far ultraviolet absorption spectra of small ring hydrocarbons, *J. Chem. Phys.*, 28, 345, 1958; (b) Robin, M. B., *Higher Excited States of Polyatomic Molecules*, Vol. II, Academic, New York, 1975, chap. IV.

52. Kropp, P. J., Photochemistry of alkenes in solution, *Org. Photochem.*, 4, 1, 1979, and references cited therein.

53. Wen, A. T., Hitchcock, A. P., Werstiuk, N. H., Nguyen, N., and Leigh, W. J., Studies of electronic excited states of substituted norbornenes by UV absorption, electron energy loss, and HeI photoelectron spectroscopy, *Can. J. Chem.*, 68, 1967, 1990.

54. Inoue, Y., Sakae, M., and Hakushi, T., Direct photolysis at 185 nm of simple cyclobutenes. Molecular elimination of acetylene, *Chem. Lett.*, 1495, 1983.

55. Clark, K. B. and Leigh, W. J., Cyclobutene photochemistry. Involvement of carbene intermediates in the photochemistry of alkylcyclobutenes, *Can. J. Chem.*, 66, 1571, 1988.

56. (a) Baraldi, I., Bruni, M. C., Momicchioli, F., Langlet, J., and Malrieu, J. P., Photochemical *cis-trans* isomerization of *s-trans*-1,3-pentadiene. A theoretical study, *Chem. Phys. Lett.*, 51, 493, 1977, and references cited therein; (b) Aoyagi, M. and Osmura, Y., A theoretical study of the potential energy surface of butadiene in the excited states, *J. Am. Chem. Soc.*, 111, 470, 1989, and references cited therein.

57. Bruckmann, P. and Salem, L., Coexistence of two oppositely polarized zwitterionic forms on the lowest excited singlet surface of terminally twisted butadiene. Two-funnel photochemistry with dual stereochemistry, *J. Am. Chem. Soc.*, 98, 5037, 1976.

58. Van der Lugt, W. Th. A. M. and Oosterhoff, L. J., Symmetry control and photoinduced reactions, *J. Am. Chem. Soc.*, 91, 6042, 1969.

59. (a) Kikuchi, O., Reaction paths in the photochemical isomerization between *cis*-butadiene and cyclobutene, *Bull. Chem. Soc. Jpn.*, 47, 1551, 1974; (b) Grimbert, D., Segal, G., and Devaquet, A., *Ab initio* SCF study of the photochemical disrotatory closure of butadiene to cyclobutene, *J. Am. Chem. Soc.*, 97, 6629, 1975; (c) Morihashi, K., Kikuchi, O., and Suzuki, K., Non-adiabatic coupling constants for the disrotatory and conrotatory isomerization paths between butadiene and cyclobutene, *Chem. Phys. Lett.*, 90, 346, 1982; (d) Morihashi, K. and Kikuchi, O., Reduced classical trajectory equations for electronically non-adiabatic transition on photochemical pericyclic reactions, *Theor. Chim. Acta*, 67, 293, 1985.

60. Pichko, V. A., Simkin, B. Ya., and Minkin, V. I., Asymmetrical reaction path of photocyclization of 1,3-butadiene into cyclobutene, *J. Mol. Struct. (Theochem.)*, 235, 107, 1991.

61. Bernardi, F., Olivucci, M., Ragazos, I. N., and Robb, M. A., Origin of the non-stereospecificity in the ring opening of alkyl-substituted cyclobutenes, *J. Am. Chem. Soc.*, 114, 2752, 1992.

62. Bernardi, F., Olivucci, M., and Robb, M. A., Predicting forbidden and allowed cycloaddition reactions: Potential surface topology and its rationalization, *Acc. Chem. Res.*, 23, 405, 1990.

63. (a) Zimmerman, H. E., On molecular orbital correlation diagrams, mobius systems and factors controlling ground and excited state reactions, *J. Am. Chem. Soc.*, 88, 1566, 1966; (b) Gerhartz, W., Poshusta, R. D., and Michl, J., Excited potential energy hypersurfaces for H_4. 2. "Triply right" (C_{2v}) tetrahedral geometries. A possible relation to photochemical "cross-bonding" processes, *J. Am. Chem. Soc.*, 99, 4263, 1977.

64. Bernardi, F., Olivucci, M., and Robb, M. A., Simulation of MC-SCF results on covalent multi-bond reactions: Molecular mechanics with valence bond (MM-VB), *J. Am. Chem. Soc.*, 114, 1606, 1992.

65. Bernardi, F., Olivucci, M., Robb, M. A., and Tonachini, G., Can a photochemical reaction be concerted? A theoretical study of the photochemical sigmatropic rearrangement of but-1-ene, *J. Am. Chem. Soc.*, 114, 5805, 1992.

66. (a) McDiarmid, R. and Sheybani, A.-H., Reinterpretation of the main absorption band of 1,3-butadiene, *J. Chem. Phys.*, 89, 1255, 1988; (b) Chadwick, R. R., Zgierski, M. Z., and Hudson, B. S., Resonance Raman scattering of butadiene: vibronic activity of a b_u mode demonstrates the presence of a ^1A_g symmetry excited electronic state at low energy, *J. Chem. Phys.*, 95, 7204, 1991; (c) Orlandi, G., Zerbetto, F., and Zgierski, M. Z., Theoretical analysis of spectra of short polyenes, *Chem. Rev.*, 91, 867, 1991; (d) Phillips, D. L., Zgierski, M. Z., and Myers, A. B., Resonance Raman excitation profiles of 1,3-butadiene in vapor and solution phases, *J. Phys. Chem.*, 97, 1800, 1993.

67. (a) Lawless, M. K., Wickham, S. D., and Mathies, R. A., Direct investigation of the photochemical ring-opening dynamics of cyclobutene with resonance Raman intensities, *J. Am. Chem. Soc.*, 116, 1593, 1994; (b) Leigh, W. J. and Postigo, J. A., Cyclobutene photochemistry. Steric effects on the photochemical ring-opening of alkylcyclobutenes, *J. Am. Chem. Soc.*, in press.

10

Photochemistry of Acyclic 1,3,5-Trienes and Related Compounds

W. H. Laarhoven
University of Nijmegen

H. J. C. Jacobs
Leiden University

10.1 Definition and History

The history of the research on the photochemistry of acyclic trienes is closely related to that of vitamin D (*vide infra*). Owing to the volatility of hexa-1,3,5-triene, the photochemical behavior of the triene in the gaseous state was examined early on.[1-3] In Scheme 10.1, the photoproducts under these conditions are given. Besides *E-Z*-isomerization and 1,3-hydrogen shift to yield hexa-1,2,4-triene, processes which also occur in solution, cyclization and degradation products are formed. The quantum yield of benzene formation at 253.7 nm and 25 °C increases with decreasing gas pressure. The study of the photochemistry of trienes in solution has led to a wealth of insight into, and understanding of, organic photochemistry in general and of the excited state of trienes in particular. Several kinds of reaction types have been identified: *E-Z*-isomerization, sigmatropic shifts, photocyclization via concerted or radical pathways, and photoadditions via direct or polar routes. Reviews on this topic have been published by Jacobs and Havinga,[4] Dauben and Michno,[5] and Laarhoven.[6]

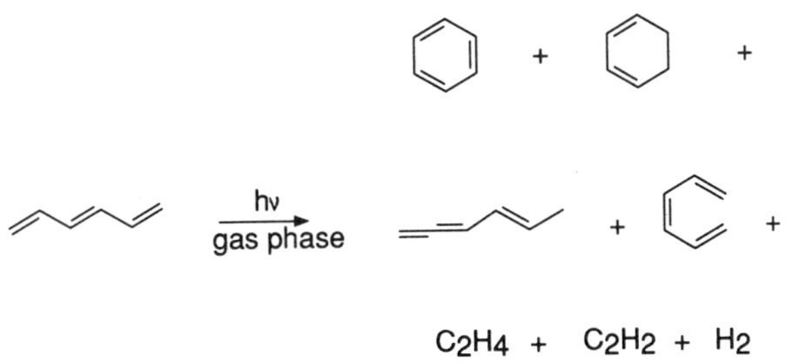

SCHEME 10.1

10.2 Mechanism

Irradiation of *Z*-hexa-1,3,5-triene (**1**) in solution with a low- or high-pressure mercury lamp mainly gives rise to the *E*-isomer (**2**) (Scheme 10.2). The isomerization occurs with a quantum yield of 0.034 (at 265 nm).[7] Other photoproducts formed are 3-vinylcyclobutene (**3**), bicyclo[3.1.0]hex-2-ene (**4**), and a small amount of hexa-1,2,4-triene (**5**). No, or only very little, cyclohexa-1,3-diene (**6**) is formed. On irradiation of **6**, both in the gas phase and in solution, the triene **1** is formed by conrotatory ring opening. Irradiation of **1** at wavelengths that do not allow its direct excitation leads to dimers.[8] Triplet-sensitized irradiation of **1** leads to dimerization and *Z-E*-isomerization.[7]

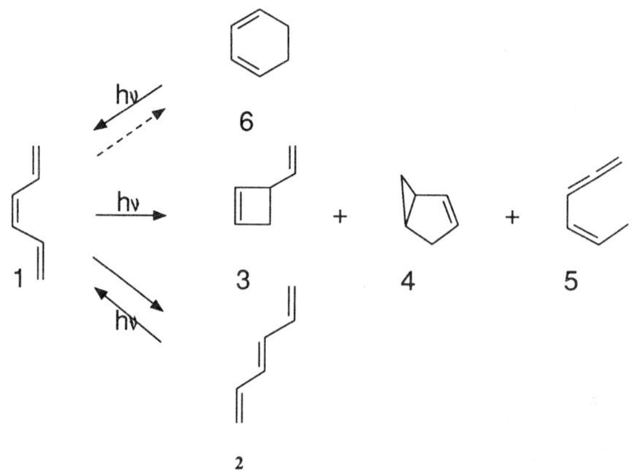

SCHEME 10.2

    Substituted *Z*-hexa-1,3,5-trienes behave similarly, but the ratio of the photoproducts can be quite different. Dauben et al.[9] studied the influence of the bulkyness of the substituent in 3-alkyl-6,6,9,9-tetramethyl-$\Delta^{3,5(10)}$-hexalins (**7**) on the product ratio (see Scheme 10.3). The ratio of the resulting vinylcyclobutenes (**9**) and bicyclo[3.1.0]hex-2-enes (**10**) changes with the size of the substituent R. The rate of disappearance of **7** is also dependent on the substituent R. The conclusion is that the substituent determines, to a high degree, the conformational equilibrium between *cZc*-**8** and *cZt*-**8**, and that the photochemistry depends on this ratio. Although **9** may arise from *cZc*-**8** as well as from *cZt*-**8**, the data suggest that *cZt*-**8** yields mainly **10**. Havinga et al.[10] established a similar relationship between ground-state conformational equilibrium and photoproduct composition in the case of simple mono- and dialkyl-substituted hexatrienes.[10]

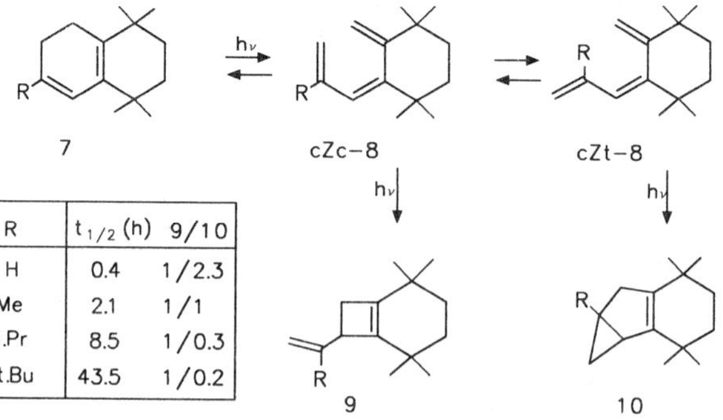

R	$t_{1/2}$ (h)	9/10
H	0.4	1/2.3
Me	2.1	1/1
i.Pr	8.5	1/0.3
t.Bu	43.5	1/0.2

SCHEME 10.3

Although these results suggest ground-state conformational control of photoproduct formation, it cannot be excluded *a priori* that equilibration between excited conformers results in a mixture of these species that is similar to that of the ground state. By time- and wavelength-dependent irradiation studies of the 2,5-dimethylhexatrienes, Z-11 and E-11 (Table 10.1), and restricting the conversion to about 10% in order to prevent the formation of secondary products, Havinga and Jacobs were able to ascertain that equilibration of conformers in the excited state does not occur.[11] The various planar conformations of Z- and E-hexatrienes are pictured in Scheme 10.4. In Table 10.1, the photoproduct composition in percentages at two wavelengths is given for the Z- and E-isomers of 2,5-dimethylhexa-1,3,5-triene, Z-11 and E-11.[11,12] It appears that at 313 nm, the photocyclization of Z-11 into 1,4-dimethylcyclohexa-1,3-diene (13) is the preferred reaction that dominates even over Z-E-isomerization. Compared with the absence of cyclohexa-1,3-diene as a photoproduct from 1, the preferred formation of 13 shows the influence of the methyl groups on the proportion of the *cZc*-conformation of Z-11. At 254 nm, the ratio of 13 to E-11 is reversed. This shows the influence of the absorption coefficient of the conformers at different wavelengths. The photoproduct composition of E-11 is quite different. Although the rate of the photoconversion is 10 to 100 times slower than that of the Z-isomer, the formation of 1-methyl-3-*iso*-propenyl-1-cyclobutene (12) is a primary photoprocess. 1-Methyl-2-isobutenylcyclopropene (16) is a product present only in the product mixture obtained from E-11. It is formed exclusively upon irradiation at 254 nm and may originate from a higher excited state, possibly via a carbene intermediate.

Table 10.1 Composition of the Photoproduct Mixture (%) of teh Z- and E-Isomers of 2,5-Dimethylhexa-1,3,5-triene after 10% Conversion at Two Wavelengths

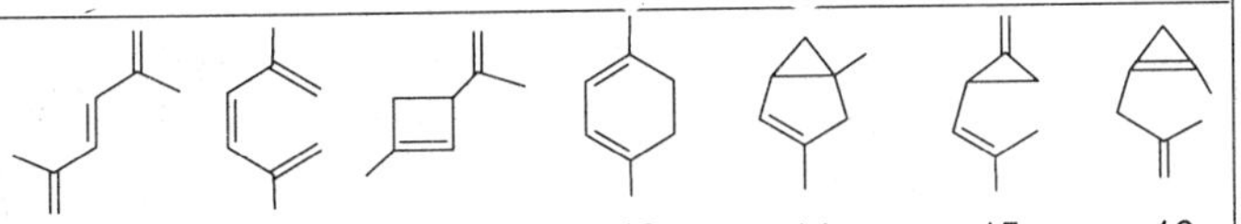

	λ(nm)	products						
		E-11	Z-11	12	13	14	15	16
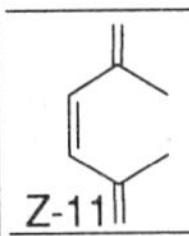 Z-11	313	1.13	~90	0.01	8.85	0.05	0.01	-
	254	8.05	~90	0.93	0.75	0.05	0.27	-
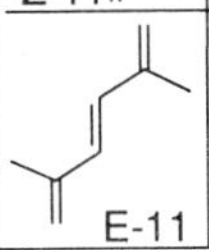 E-11	313	~90	6.23	0.59	0.34	-	-	-
	254	~90	4.55	3.50	0.15		0.54	0.20

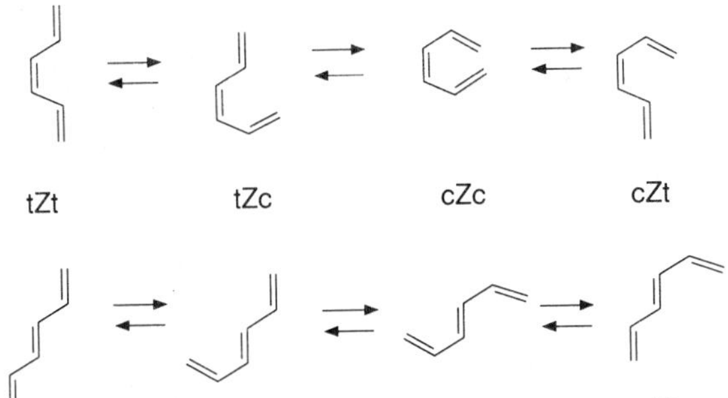

SCHEME 10.4

The correlation observed between the conformational equilibrium of the starting compound in its ground state and the composition of the photoproduct mixture is also found in several other photoreactions. Jacobs and Havinga[4] called this relationship the NEER principle (nonequilibration of excited rotamers). It is based on the increased π-bond order of the ground-state C-C single bonds and the short lifetime of the molecules in the first excited state. The NEER principle states that each conformer (A_n) of a polyene yields, in principle, its own specific photoproduct (P_nI), though, of course, different rotamers may lead to the same photoproduct and one excited rotamer may lead to several photoproducts (see Scheme 10.5). Fischer was able to demonstrate the validity of the NEER principle by showing the existence of different fluorescent species of *trans*-diarylethylenes by excitation at various wavelengths.[13] (Limitations of the principle have been discussed by Mazzucato.[14])

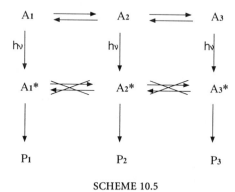

SCHEME 10.5

Another general principle that is important, especially for the photoproduct composition of trienes formed by irradiation of cyclohexa-1,3-dienes, is the *principle of least motion*. As a consequence of the balance between minimization of bond angle strain, torsional strain, and nonbonded repulsions, a cyclohexa-1,3-diene with a substituent at C5 can occur in two conformations, one in which the substituent is in a pseudoaxial (pa) position and one in which it is in a pseudoequatorial position (pe). *A priori*, in both conformers ring opening is possible via two conrotatory modes.[15,16] According to the principle of least motion, however, the favorable reactions are those in which the pa substituent moves into the internal position of the hexa-1,3,5-triene leading to a Z-configuration and the pe substituent moves into the external position, thus forming an E-configuration of the terminal double bond (see Scheme 10.6). A special case of this principle is the *accordancy principle*, which states that cyclic dienes with left- or right-handed twist follow a conrotatory mode of ring opening that is in accordance with the chirality of the diene. This mode requires the least motion of the orbitals involved. The validity of the principle was shown by Baldwin[17] from the linear relationship between the optical rotatory power of α-phellandrene in solution at various temperatures and the ratio of the two isomeric substituted hexatrienes formed by photochemical ring opening.

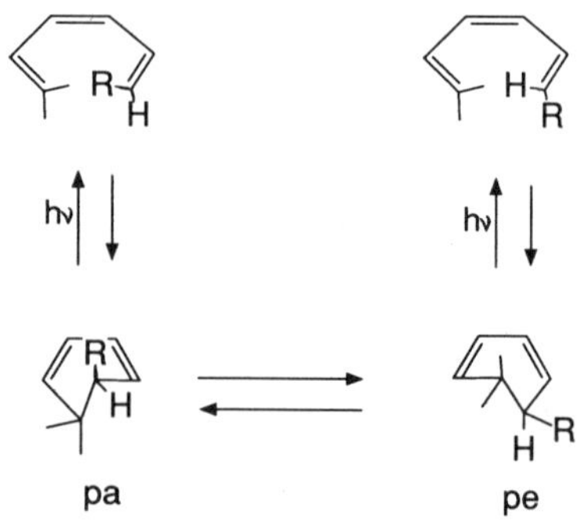

SCHEME 10.6

A change in the wavelength of irradiation can modify the product composition, as observed with compound 11 above. This can be brought about in several ways. Different conformers of the substrate may have different absorption wavelengths and/or extinction coefficients, and give different photoproducts according to the NEER principle. Using light of higher energy may give rise to other photoproducts than those corresponding to the ground-state conformer. This situation will occur when the potential energy surface of the excited state exhibits at least two minima corresponding to different conformers of the substrate. When the excess excitation energy exceeds the barrier between these minima, some excited molecules will be able to cross the barrier to a different conformer and eventually end up as a different product. Moreover, a wavelength effect on the photoproduct composition will occur when a primary photoproduct absorbs the incident light and gives rise to a secondary product. The composition of the equilibria of the ground-state conformers can be influenced by changing the solvent or the temperature. This can affect the photoproduct composition, irrespective of other effects of these factors on the photochemistry.

Another indication of the importance of the NEER principle is the observation that the irradiation of 2-methyl-5-*iso*-propylhexatriene (17) (Scheme 10.7) affords two derivatives of 2-vinyl-1-methylene cyclopropane, *viz.* 18 and 19, but not 20. This means that the carbon atom from which a hydrogen migrates does not take part in the cyclopropane ring formation. The product formation from *cZt*-17 dominates by a factor of about six over that from *tZc*-17, as revealed from the observed product ratio of the isomeric bicyclo[3.1.0]hexene derivatives and methylenecyclopropane derivatives.

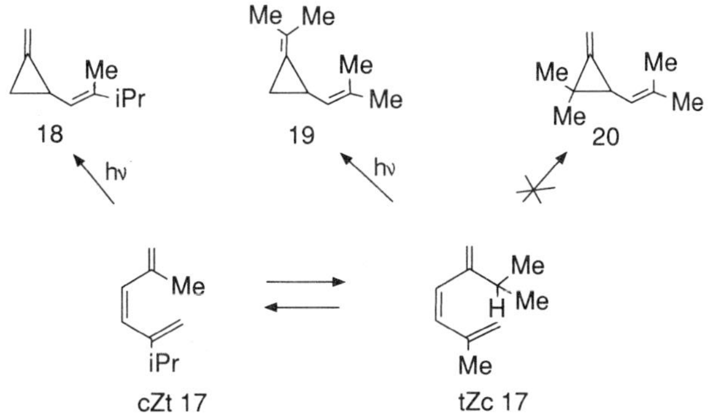

SCHEME 10.7

A strong argument in favor of extension of the NEER principle to the excited triplet state is the observation that the lifetimes of the triplet state of the *E*- and the *Z*-isomer of 11 are different.[18] Moreover, the resonance Raman spectra of the triplets of these isomers are different, whereas those of the triplets of *Z*- and *E*-1 are identical.[18]

In addition to the principles mentioned above, the mechanisms of product formation have also gained attention. In general, the Woodward-Hoffmann rules for electrocyclization and for sigmatropic shifts have been found to be of great value for the explanation and prediction of photoproducts. The formation of products of intramolecular cycloaddition has also been explained by using the concept of *sudden polarization* involving a route via polar intermediates. Calculations[19] indicate that the *bis*-allyl species resulting from excited *cZt*-hexatrienes may become polarized when the mutual orientation of the allylic moieties is nearly orthogonal. Two different zwitterionic forms may then be distinguished, ZT^- and ZT^+ (Scheme 10.8). The zwitterionic state, a linear combination of ZT^- and ZT^+, has been suggested as intermediate in the formation of photo-Diels-Alder products.[20]

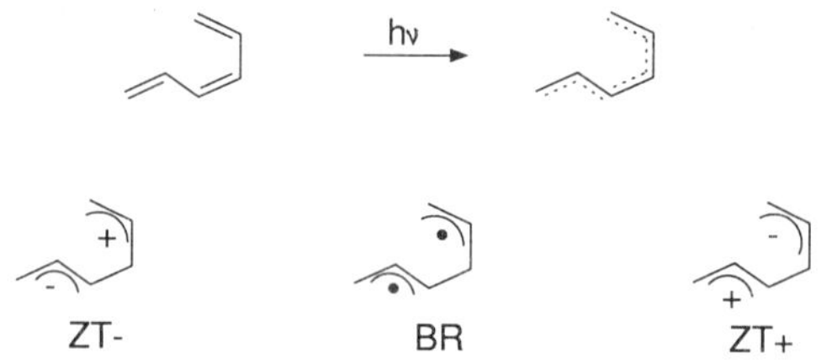

SCHEME 10.8

Irradiation of *Z,Z,Z*-1,2,6-triphenylhexa-1,3,5-triene (**21**) leads, via a fast isomerization to the *E,Z,Z*-isomer **22**, to 3,4-*exo*,6-*endo*-triphenylbicyclo[3.1.0]hex-2-ene (**23**). Analogously, the photoproduct from the *E,Z,E*-isomer **25** is 3,4-*exo*,6-*exo*-triphenylbicyclo[3.1.0]hex-2-ene (**26**)[21] (Scheme 10.9). In accordance with the Woodward-Hoffmann rules, the photoproduct formation corresponds with [π4s + π2a] photocycloadditions of **22** and **25**. However, the photoproduct formation from 1,5-diphenylhepta-1,3,5-triene (**28**) follows the [π4a + π2a] path to **29**, with the methyl group in the 4-*endo* position.[22,23] Careful kinetic analysis established that the 4-*exo*-tolyl-3,6-*exo*-biphenylbicyclo[3.1.0]hex-2-ene (**32**) is formed on irradiation of **31** in agreement with a [π4s + π2a] mechanism.[24] The contradictory results with the phenyl- and methyl-substituted derivatives are explained by postulating that, after the primary formation of the cyclopropane ring, orbital and steric factors may be nearly as effective for ring closure of the C1-C5 bond of the hexatriene.[24] Tanaka and Fukui attempted to reconcile the contradictory results by the concept of a [π4a + π2a] cross-bicyclization in excited trienes of low symmetry through interactions between nonneighbouring π-orbitals.[25]

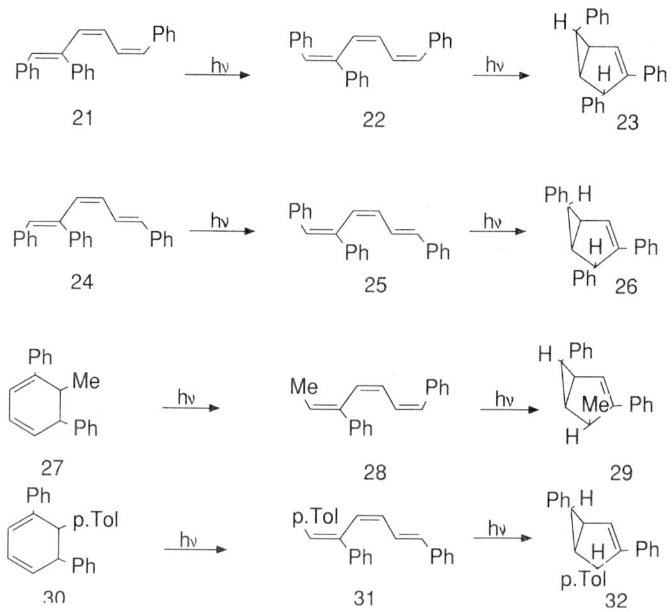

SCHEME 10.9

The ring-opened product **34**, readily formed on irradiation of 1,2-dihydronaphthalene (**33**) (Scheme 10.10), can be considered as a hexa-1,3,5-triene having a single bond in a fixed *s-cis* configuration. This feature is also present in vitamin D (*vide infra*) but, in the present case, the tendency to restore the double bond in the ring and thus regain aromaticity will be much larger. Therefore, the lifetime of **34** will be short, as ring closure can take place both thermally and

Table 10.2 Composition of the Photoproduct Mixture (in %) of *cis*- and *trans*-1,2-Dimethyl-1,2-dihydronaphthalene after Irradiation at Various Temperatures

			Products								
		C+t 36	37	38	39	40	41	42	43	44	45
cis-36	−100 °C	9	7	14	42	3	13	—	—	—	—
cis-36	+25 °C	45	0.2	—	—	—	—	19	7	19	9
trans-36	−100 °C	42	37	12	—	—	—	—	—	—	—
trans-36	+25 °C	71	—	—	—	—	—	7	1	11	5

photochemically. A broad spectrum light source had to be used for irradiation of 33 to realize a photoreaction of the hexatriene analogue. The main photoproduct of 34, under these conditions, is benzobicyclo[3.1.0]hex-2-ene (35).[26–28] When 33 is substituted with alkyl groups at C1 or C2, as in 36, the initial photochemical ring opening can be followed by thermal hydrogen shifts, thereby restoring the aromaticity.[29–32] As substituents at C1 or C2 can take two different positions, *viz.* pseudoaxial (pa) or pseudoequatorial (pe), the concerted ring opening can occur in two different conrotatory modes, governed by the principle of least motion (see above). In Scheme 10.11, the photoproducts of *cis*-1,2-dimethyl-1,2-dihydronaphthalene (36)[32] are given; and in Table 10.2, the yields of the products of both *cis*- and *trans*-36 at two different temperatures. From these results it was concluded that the photochemical Diels-Alder reaction yielding the benzobicyclo[3.1.0]hexenes does not occur stereospecifically. A disrotatory thermal ring closure of the triene occurs only if thermal 1,7-hydrogen shifts are sterically unfavorable. Photochemical 1,7-hydrogen shifts have never been observed with derivatives of 33, not even at low temperatures where thermal 1,7-H shifts are completely suppressed.[32] The high rate of thermal ring closure of the methyl derivative (50) of 34 is illustrated by the fact that irradiation of 46 at room temperature yields only 2-methylenetetralin (47). At lower temperatures, increasing amounts of reaction products from 50 are observed[33] (Scheme 10.12).

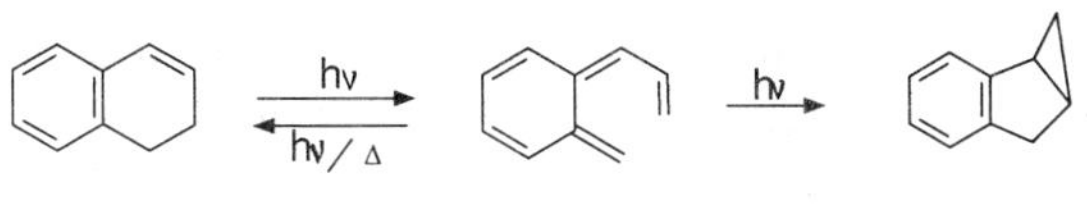

SCHEME 10.10

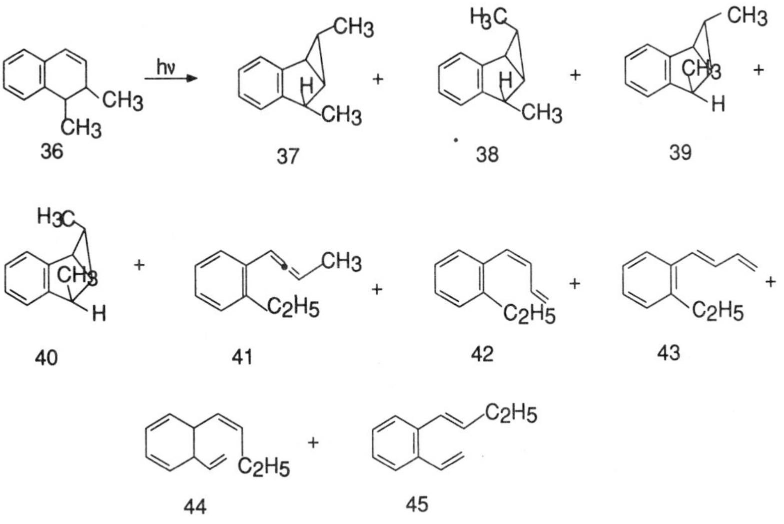

SCHEME 10.11

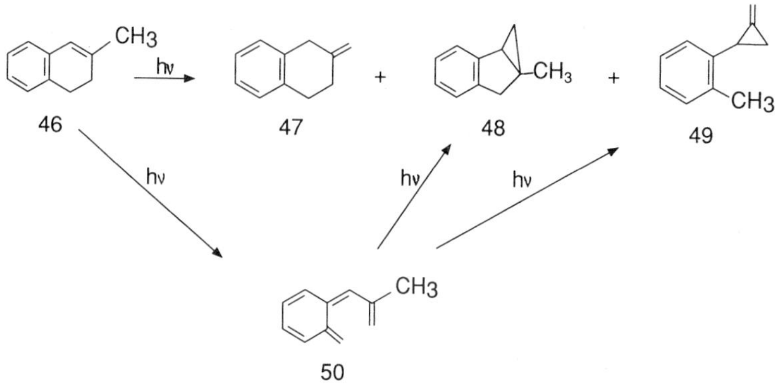

SCHEME 10.12

To prevent the hydrogen shifts in derivatives of **34**, phenyl substituents were introduced in **33**.[34] Although the ratio of the pe- and pa-conformers of 1-phenyl-1,2-dihydronaphthalene (**51**) in apolar solvents is 1:3, only one Diels-Alder photoproduct is formed, *viz.* **53** (Scheme 10.13). Assuming that interconversion between the hexatrienes **52** and **55** does not occur (at 254 nm, only **57** is formed[35]), a two-step mechanism for the formation of **53** from both isomers — starting with the formation of the cyclopropane ring and followed by closure of the five-membered ring into the thermodynamically most stable compound — is most probable. In an attempt to trap possible intermediates of a two-step process and to study the intermediacy of a sudden-polarized, zwitter-ionic state, the irradiation was performed in acidified methanol.[36] The initial rate of the formation of **53** turned out to be independent of the polarity of the solvent. The yield of **57** is increased in agreement with the increase of the ratio **51**pe:**51**pa to about 1.0. This result does not support the occurrence of sudden polarization. The main product from 2-phenyl-1,2-dihydronaphthalene (**58**) in apolar solvents is 6-phenylbenzobicyclo[3.1.0]hex-2-ene (**60**) (Scheme 10.14). Interestingly, this compound is formed via two competing reaction pathways: a photo-Diels-Alder reaction from the ring-opened primary photoproduct **59**, and a di-π-methane rearrangement of **58**, as was shown by a study of the reaction products from partially deuterated **58**.[37] Irradiation of **58** in methanol gives rise to the methanol addition product (**65**), in addition to the products obtained in apolar solvents. Although the latter product may arise from a zwitterionic state of **59** or **61**, it could be shown that it is formed by a di-π-methane rearrangement of the primary addition products **63** and **64**.[38] Irradiation of 3-phenyl-1,2-dihydronaphthalene does not lead to a ring-opened product.[34,39]

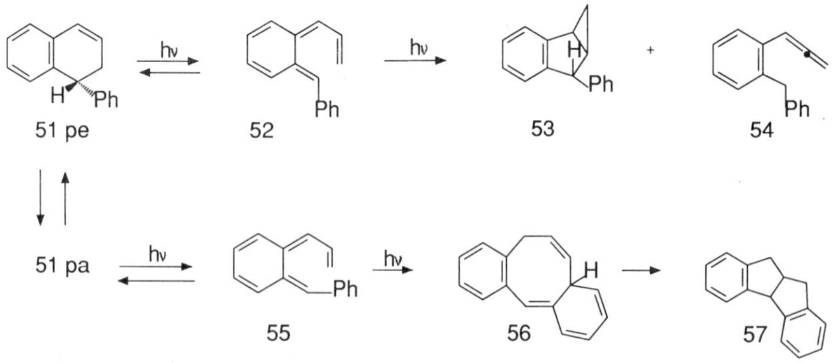

SCHEME 10.13

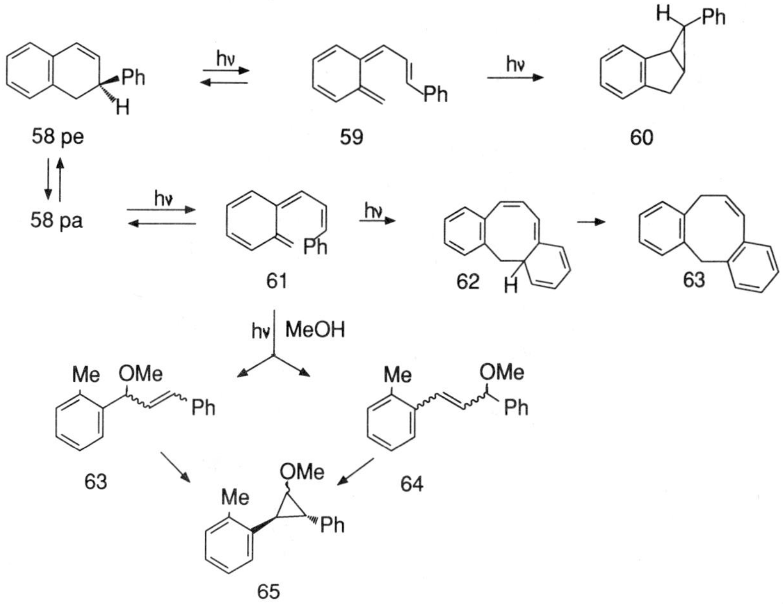

SCHEME 10.14

10.3 Synthetic Applications

The importance of the concepts and reactions treated in this chapter is not restricted to (substituted) hexatrienes, but is, *mutatis mutandis,* equally valid to systems in which one of the double bonds is part of an aromatic system.[40]

10.4 Industrial Applicability

The most important industrial application is found in the synthesis of vitamin D and its derivatives (*vide infra*).

Experimental UV Absorption Data of Hexa-1,3,5-triene and Some of Its Derivatives in Hydrocarbon Solvents

Compound	E-isomer		Compound	Z-isomer	
	λ(nm)	ε		λ(nm)	ε
	246	35600		247	30900
	256	51700		254	41000
	266	42800		266	31500
	248	31000		249	18800
	258	43000		259	22400
	268	35000		269	16600
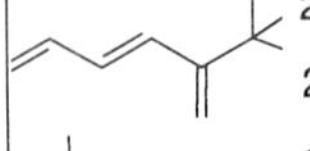	258	18000	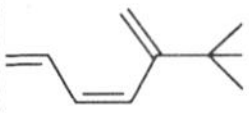	202	13000
				238	12700
	250	31600		207	9800
	260	43000		237	12500
	270	33800			
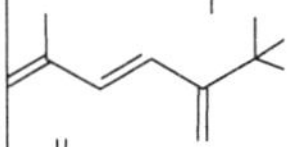	259	16900	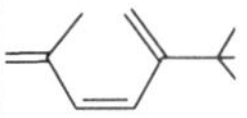	240	11900
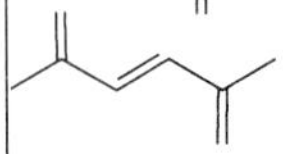	253	9600	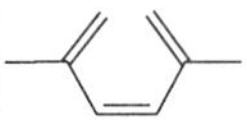	235	4000
	253	32500		252	27900
	263	45300		261	35900
	273	36900		271	27200
			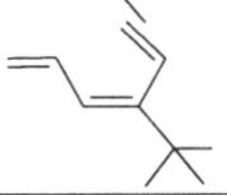	241	14600

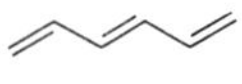

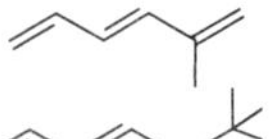

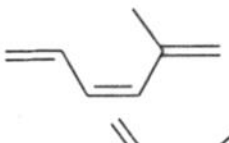

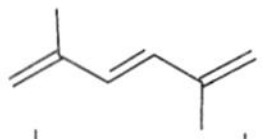

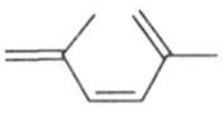

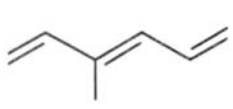

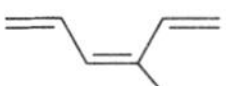

From A. M. Brouwer, Ph.D. thesis, Leiden.

Experimental UV Absorption Data of Some Derivatives of 1,2-Dihydronaphthalene (DHN)

Compound	Solvent	λ$_{max}$, nm (log ε)
1-Phenyl-DHN	Methanol	259 (3.72); 220 (4.08)
2-Phenyl-DHN	Methanol	270 (3.86)sh; 264 (3.89)sh; 257 (3.93)
3-Phenyl-DHN	Methanol	236 (3.67)sh; 217 (4.26); 302 (4.28); 230 (4.19); 206 (4.15)
4-Phenyl-DHN	Methanol	261 (3.79); 224 (4.21)sh; 218 (4.32)sh; 210 (4.23)

References

1. Srinivasan, R., The photochemistry of 1,3-butadiene and 1,3-cyclohexadiene, *J. Am. Chem. Soc.*, 82, 5063, 1960.
2. Srinivasan, R., The photochemistry of *cis, trans*-1,3,5-hexatriene, *J. Am. Chem. Soc.*, 83, 2806, 1961.
3. Srinivasan, R., Internal conversion in the photochemical system 1,3-cyclohexadiene:1,3,5-hexatriene, *J. Chem. Phys.*, 38, 1039, 1963.
4. Jacobs, H. J. C. and Havinga, E., Photochemistry of vitamin D and its isomers and of simple trienes, *Adv. Photochem.*, 11, 305, 1979.
5. Dauben, W. G., McInnis, E. L., and Michno, E. L., Photochemical rearrangements in trienes, in *Rearrangements in Ground and Excited States*, Vol. 3, de Mayo, P., Ed., Academic Press, New York, 1980, 91.
6. Laarhoven, W. H., Photocyclizations and intramolecular cycloadditions of conjugated olefins, in *Organic Photochemistry*, Vol. 9, Padwa, A., Ed., 1987, 129.
7. Minnaard, N. G. and Havinga, E., Some aspects of the solution photochemistry of 1,3-cyclohexadiene, (Z)- and (E)-1,3,5-hexatriene, *Recl. Trav. Chim., Pays Bas*, 92, 1315, 1973.

8. Bahurel, Y. L., MacGregor, D. J., Penner, T. L., and Hammond, G. S., Mechanisms of photochemical reactions in solution. LXXI. Photochemistry of 1,3-cyclohexadiene at long wavelength, *J. Am. Chem. Soc.*, 94, 637, 1972.

9. Dauben, W. G., Rabinowitz, J., Vietmeyer, N. D., and Wendschuh, P. H., Photoequilibria between 1,3-cyclohexadienes and 1,3,5-hexatrienes. Photochemistry of 3-alkyl-6,6,9,9-tetramethyl-$\Delta^{3,5(10)}$-hexalins, *J. Am. Chem. Soc.*, 94, 4285, 1972.

10. Vroegop, P. J., Lugtenburg, J., and Havinga, E., Conformational equilibrium and photochemistry of hexa-1,3,5-trienes, *Tetrahedron*, 29, 1393, 1973.

11. Gielen, J. W. J., Jacobs, H. J. C., and Havinga, E., Influence of wavelength on the photochemistry of *E*- and *Z*-hexa-1,3,5-triene, *Tetrahedron Lett.*, 3751, 1976.

12. (a) Brouwer, A. M. and Jacobs, H. J. C., Photoreactions of *(E)*- and *(Z)*-2,5-dimethylhexatriene. Formation of a cyclopropene derivative: A novel reaction in triene photochemistry, *Tetrahedron Lett.*, 27, 1395, 1986; (b) Brouwer, A. M., Bezemer, L., Cornelisse, J., and Jacobs, H. J. C., Photochemistry of 2,5-dialkyl-1,3,5-hexatrienes. The influence of the ground state conformation, controlled through steric substituent effects, *Recl. Trav. Chim., Pays Bas*, 106, 613, 1987.

13. Wismontski-Knittel, T., Das, P. K., and Fischer, E., Rotamerism in (2-anthryl)ethylenes. Evidence from fluorescence lifetimes and quenching studies, *J. Phys. Chem.*, 88, 1163, 1984.

14. Mazzucato, U. and Momicchioli, F., Ground-state rotamers of 1,2-diarylethylenes detected by techniques involving electronic excitation: Limits to the validity of the NEER principle, *EPA Newslett.*, 44, 31, 1992.

15. Butcher, S. S., Microwave spectrum of 1,3-cyclohexadiene, *J. Chem. Phys.*, 42, 1830, 1965.

16. Courtot, P., Rumin, R., and Salaün, J. Y., Photochemistry of polyenes. Control by orbital symmetry and ground state conformations?, *Pure Appl. Chem.*, 49, 317, 1977.

17. Baldwin, J. E. and Krueger, S. M., Stereoselective photochemical electrocyclic valence isomerizations of α-phellandrene conformational isomers, *J. Am. Chem. Soc.*, 91, 6444, 1969.

18. Langkilde, F. W., Jensen, N.-H., Wilbrandt, R., Brouwer, A. M., and Jacobs, H. J. C., UV absorption and Raman spectra of the ground states and time-resolved resonance Raman spectra of the lowest excited triplet states of *E* and *Z* isomers of 2,5-dimethyl-1,3,5-hexatriene. Indication of nonequilibration of excited rotamers in the lowest triplet states, *J. Phys. Chem.*, 91, 1029, 1987.

19. (a) Salem, L., The sudden polarization effect and its possible role in vision, *Acc. Chem. Res.*, 12, 87, 1979; (b) Malrieu, J. P., Nebot-Gil, I., and Sanchez-Marin, J., Neutral versus ionic excited states of conjugated systems; their role in photoisomerizations, *Pure Appl. Chem.*, 56, 1241, 1984.

20. Dauben, W. G., Kellogg, M. S., Seeman, J. I., Vietmeyer, N. D., and Wendschuh, P. H., Steric aspects of the photochemistry of conjugated dienes and trienes, *Pure Appl. Chem.*, 33, 197, 1973.

21. Padwa, A., Brodsky, L., and Clough, S., Photochemical transformations of small ring compounds. XLI. Orbital symmetry and steric control in the photorearrangement of 1,3,5-hexatrienes to bicyclo[3.1.0]hex-2-enes, *J. Am. Chem. Soc.*, 94, 6767, 1972.

22. Courtot, P., Rumin, R., and Salaün, J. Y., Photoisomérisation des quelques cyclohexadienes et heptatrienes. Photocycloaddition des hexatrienes, *Recl. Trav. Chim., Pays Bas*, 98, 192, 1979.

23. Courtot, P. and Salaün, J. Y., Effet de longueur d'onde et controle par la geometrie de l'etat fondamental de la photoisomerisation de derives de l'hexatriene-1,3,5., *Tetrahedron Lett.*, 1851, 1979.

24. Courtot, P. and Auffret, J., Observation of a [π4s + π2a] internal photocycloaddition of 2,6-diphenyl-1-*p*-tolylhexa-1,3,5-triene, *J. Chem. Res. (S)*, 304, 1981.

25. Tanaka, K. and Fukui, K., A stereoselective rule for the cross-bicyclization in linear conjugated polyenes, *Bull. Chem. Soc. Jpn.*, 51, 2209, 1978.

26. Cookson, R. C., de B. Costa, S. M., and Hudec, J., The photoisomerization of 1,2-dihydronaphthalenes, *J. Chem. Soc., Chem. Commun.*, 1272, 1969.

27. Salisbury, K., The photochemistry of 1,2-dihydronaphthalene, *Tetrahedron Lett.*, 737, 1971.

28. Seeley, D. A., Stereochemistry of the photochemical Diels-Alder reaction, *J. Am. Chem. Soc.*, 94, 4378, 1972.

29. Kleinhuis, H., Wijting, R. L. C., and Havinga, E., Photochemistry of some 1,2-dihydronaphthalenes, *Tetrahedron Lett.*, 255, 1971.

30. Heimgartner, H., Ulrich, L., Hansen, H. J., and Schmid, H., Photochemisches Verhalten von 1- und 2-alkylierten 1,2-Dihydronaphthalinen, *Helv. Chim. Acta*, 54, 2313, 1971.

31. Sieber, W., Heimgartner, H., Hansen, H. J., and Schmid, H., Photochemisches Verhalten von 1- und 2-alkylierten 1,2-Dihydronaphthalinen bei tiefen Temperaturen, *Helv. Chim. Acta*, 55, 3005, 1972.

32. Widmer, U., Heimgartner, H., and Schmid, H., Photochemie von 1,1-Dimethyl-4-phenyl- und 1-Methyl-1-phenyl-1,2-dihydronaphthalin. Nachweis einer photochemischen sigmatropischen [1,7] H-Verschiebung, *Helv. Chim. Acta*, 58, 2210, 1975.

33. Laarhoven, W. H. and Berendsen, N., Photochemistry of 3-methyl-1,2-dihydronaphthalene, *Recl. Trav. Chim., Pays Bas*, 105, 367, 1986.

34. Lamberts, J. J. M. and Laarhoven, W. H., The photochemistry of 1-, 3- and 4-phenyl-substituted 1,2-dihydronaphthalenes, *Recl. Trav. Chim., Pays Bas*, 103, 131, 1984.

35. Laarhoven, W. H., Lijten, F. A. T., and Smits, J. M. M., The photochemistry of 1-phenyl-1,2-dihydronaphthalene. A simple preparation of *cis*-dibenzobicyclo[3.3.0]octa-2,7-diene, *J. Org. Chem.*, 50, 3208, 1985.

36. Woning, J., Lijten, F. A. T., and Laarhoven, W. H., Photochemistry of 1-phenyl-1,2-dihydronaphthalene in methanol, *J. Org. Chem.*, 56, 2427, 1991.

37. Lamberts, J. J. M. and Laarhoven, W. H., Photochemistry of 2-phenyl-1,2-dihydronaphthalene. A competition between a singlet state di-π-methane rearrangement and a ring opening reaction, *J. Am. Chem. Soc.*, 106, 1736, 1984.

38. Woning, J., The Photochemistry of Stilbenes and 1,2-Dihydronaphthalenes in Methanol, Ph.D. Thesis, Nijmegen, 1988.

39. Keijzer, F., Stolte, S., Woning, J., and Laarhoven, W. H., Photoacoustic determination of the photostability of 3-phenyl-1,2-dihydronaphthalene, *J. Photochem. Photobiol. A*, 50, 401, 1990.

40. Laarhoven, W. H., Potocyclizations and intramolecular cycloadditions of conjugated arylolefines and related compounds, in *Organic Photochemistry*, Vol. 10, Padwa, A., Ed., 1989, 163.

11

Photochemistry of Vitamin D and Related Compounds

M. J. C. Jacobs
Leiden University

W. H. Laarhoven
University of Nijmegen

11.1 Definition and History

The first scientific medical description of the bone disease "the rickets", rachitis, appeared in a thesis from Leiden University in 1645.[1] A few years later, a report about the disease from an investigation commission of the Royal College of Physicians in England was published.[2] However, it took about 250 years before a correlation of rachitis and lack of sunlight was shown by Palm in 1890, and before Funk in 1914 suggested that the disease could be caused by deficiencies in the food. In the late 1920s, Reerink and colleagues at the Philips laboratories (Eindhoven, The Netherlands) were the first to isolate crystals of the antirachitic vitamin from irradiation mixtures of ergosterol.[3,4] Moreover, they inferred from their observations that vitamin D is not the primary photoproduct of ergosterol, and a thermally unstable precursor was postulated. For the production of vitamin D on a large scale, their observation that the optimum formation from ergosterol depends on the wavelength of irradiation was of significant importance. Between 1930 and 1940, Windaus and co-workers in Göttingen, Germany, elucidated the structures of vitamin D (D)[5a] and of the "precursors": ergosterol (E),[5b] lumisterol (L),[5c] and tachysterol (T).[5d] (Later, it was proven that an incorrect configuration had been assigned to L and T.) For the formation of D, Windaus proposed the reaction sequence: E → L → T → D, in which all steps are photochemical. Doubt about the correctness of this scheme arose on the basis of two facts: first, isolation of the thermal precursor, previtamin D, (P), by Velluz[6] in 1948 and its structure elucidation simultaneously accomplished by Velluz[7] and Havinga[8] in 1955; and second, the high quantum yield of conversion from E to D (0.3), which is not very probable with three consecutive photochemical steps. The reaction scheme as proposed by Velluz in France and Havinga in the Netherlands is given in Scheme 11.1. Much progress in the investigation of the chemistry of vitamin D and its isomers was made by these two groups during the 1950s and 1960s. Several reviews on the subject have appeared.[9]

0-8493-8634-9/95/$0.00+$.50
© 1995 by CRC Press, Inc.

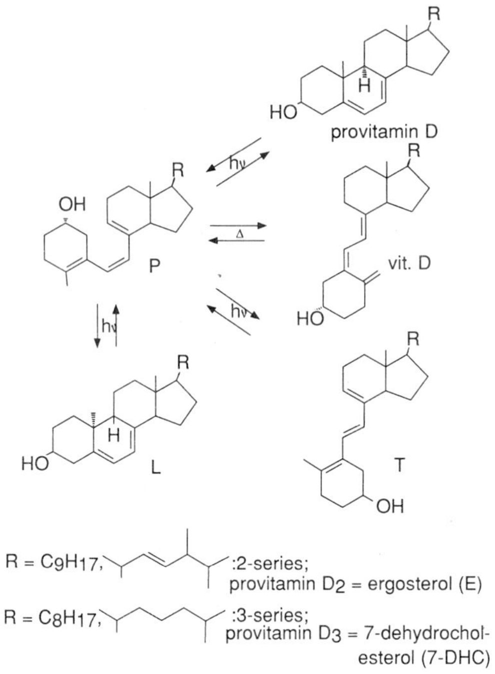

SCHEME 11.1

11.2 Mechanism

The starting material in the production of vitamin D is provitamin D: ergosterol (E) in the D_2-series, 7-dehydrocholesterol (7-DHC) in the D_3-series (there is no influence of the side chain R on the photochemical behavior). Photochemical conrotatory ring opening of the provitamin (Scheme 11.1) gives previtamin D (P). previtamin D, in turn, is reversibly photoisomerized to provitamin D and to lumisterol (L) by a conrotatory ring closure, and to tachysterol (T) by cis,trans-isomerization, resulting in a quasistationary state of four isomers. In Table 11.1, the proportion of the products in the photostationary state at 254 and 300 nm is given. Dauben and Phillips[10] observed a sudden change in the photostationary state composition between 295 and 305 nm, notably in the narrow wavelength range of 302.5 to 305 nm. At 302.5 nm, the proportion of ring-closed products (7-DHC and L) amounts to 21% and at 305 nm to 40%. According to these authors, this effect cannot be explained by a change in the molar extinction coefficient of the $c(-)Zc$ and $c(+)Zc$-conformers of P, leading to 7-DHC and L, respectively; a doubling of the extinction coefficient of both conformers within a 2.5-nm range would be required. The authors concluded that excited-state properties are involved in the wavelength effects observed in the reaction of P to 7-DHC and L. In a recent more detailed study, Dauben et al.[11] attributed the increase in photocyclization yield with decreasing photon energy to the involvement of both the 1B and 2A excited states of P.

At temperatures below 100 °C, previtamin D (P) is converted to vitamin D. The anticipated antarafacial stereochemistry of the [1,7]-sigmatropic hydrogen shift was demonstrated recently by Okamura[12] using a deuterium-labeled analogue of vitamin D. In equilibrium at 20 °C, the ratio vitamin D:previtamin D is 93:7.[13] Although usually a cyclohexane derivative with an exocyclic double bond is energetically less favored, the steric strain caused by the rings C and D in P shifts the equilibrium toward vitamin D formation.[14] At higher temperatures, P thermally isomerizes to pyro- and isopyrocalciferol by a disrotatory ring closure. These two isomers are converted on irradiation to photopyro- and photoisopyrocalciferol, respectively (Scheme 11.2).

Table 11.1 Photostationary State Composition
(in %) at 254 and 300 nm upon Irradiation of
Previtamin D and Its Isomers

Product	P	7-DHC	L	T	Ref.
254 nm	20	1.5	2.5	76	14
300 nm	68	5	8	19	10

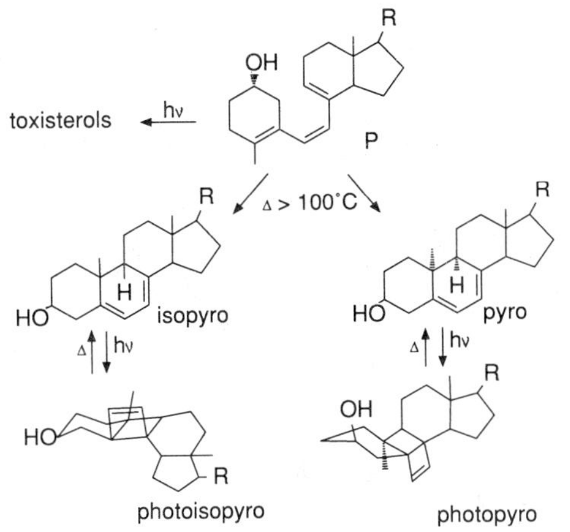

SCHEME 11.2

Prolonged irradiation of **P**, but also of **L, T, E,** and **7-DHC**, below 10 °C in aprotic solvents leads to products of overirradiation, the so-called "toxisterols". These are illustrated in Scheme 11.3. In addition to these products, some other toxisterols are formed when the irradiation is performed in an alcoholic solvent; see Scheme 11.4. These are alcohol adducts (toxisterols B), reduction products (toxisterols R), and trienes (toxisterols, D).[15] The toxisterols D are probably formed by loss of R'OH from the toxisterols B.

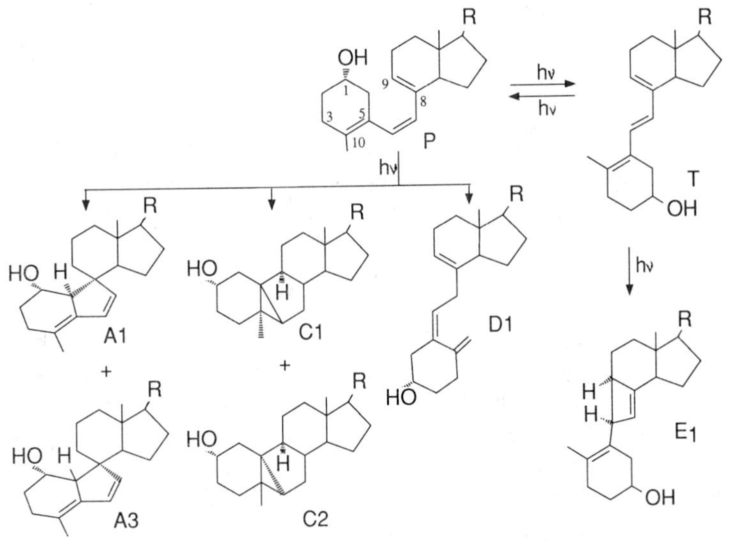

SCHEME 11.3

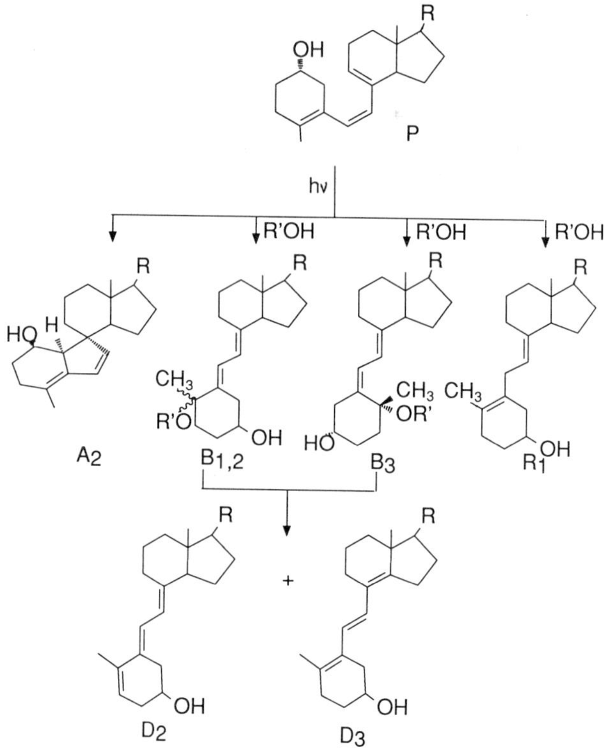

SCHEME 11.4

Similarly, long-term irradiation of vitamin D gives rise to a large number of photoproducts. These are called "suprasterols" and are illustrated in Scheme 11.7.

Comparing the type of photoproducts from **P** and **D** with those from hexa-1,3,5-triene derivatives, described in Chapter 10 (on hexatrienes), a good correspondence can be observed. It was, and is, a challenge to investigate if the compounds of the vitamin D series, which are larger and have more pathways of dissipating their vibrational energy in the excited state,[16] obey the same principles as those outlined for hexa-1,3,5-trienes.

Toxisterols

From irradiation experiments of 7-DHC in deuterated solvents (EtOD or MeOD), it could be concluded that the formation of the toxisterols A_1 and A_3 (Scheme 11.3) proceeds via an intramolecular mechanism; whereas, the formation of A_2 (Scheme 11.4) is solvent mediated, i.e., initiated by the addition of a proton from the solvent to the 9α position of **P**. The intramolecular hydrogen shift in the rearrangements to the toxisterols A_1 and A_3 occurs stereoselectively, *not* stereospecifically, as was confirmed by experiments using 4α-deutero-7-DHC. Therefore, a biradical intermediate is presumed in the formation of A_1 and A_3; whereas, in the formation of A_2, proton-addition at C9 of **P** and proton-abstraction from C4 is presumed[17] to give a Zwitterion that undergoes a disrotatory ring closure. The effect of temperature and solvent polarity on the photoproducts from **P** has also been investigated. In Table 11.2, these results are summarized.[17]

The data in Table 11.2 show that the ratio A_1:A_3 depends strongly on the polarity of the solvent. More polar solvents promote the formation of toxisterol A_1, the formation of which is enhanced by lowering the temperature. It was shown that viscosity effects are not important. An increase in the ratio $(A_1 + A_3)/C_1$ is effected mainly by increasing the temperature; solvent polarity has only a small effect. When the irradiation is performed with acetylated 7-DHC, the formation of the

Table 11.2 Ratios of the Toxisterols A_1, A_2, A_3, and C_1 at +20 and –20 °C in Solvents of Various Polarities

Solvent	A_1/A_3 20 °C	A_1/A_3 –20 °C	A_2/A_1 20 °C	A_2/A_1 –20 °C	$(A_1+A_3)/C_1$ 20 °C	$(A_1+A_3)/C_1$ –20 °C
Hexane	0.9	1.1	0	0	0.7	0.3
THF	1.7	2.1	0	0	0.8^5	0.3^5
DMF	2.4	3.1	0	0	0.9	0.4
MeOH	3.0	3.5	1.1	1.2	1.0	0.5

acetylated toxisterol A_3 is strongly reduced. This effect has been ascribed to a shift in the ground-state conformational equilibrium of **P** to (pseudo)equatorial conformers.[17]

To investigate the relationship between ground-state conformers and photoproduct composition, the preferred conformations of **P** and some derivatives were studied by UTAH and INDO calculations, and by solvent- and temperature-dependent UV absorption, NMR-LIS, and CD spectroscopy. The results indicated a considerable influence of the polarity of the solvent on the conformational equilibrium. It appears that, in polar solvents (and at low temperatures), a shift from *tZc*- to *cZc*-conformations occurs along with a shift in the ratio of conformers with pseudoaxial and pseudoequatorial hydroxyl groups. The correlation suggested between conformation and photoproducts is given in Scheme 11.5.[17] It is of interest to note that no indications for the occurrence of Zwitterionic structures were found.

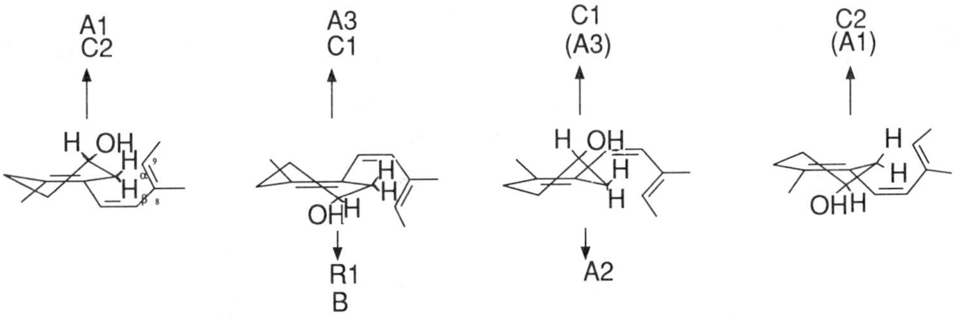

SCHEME 11.5

More recently, a theoretical evaluation of the conformations of **P** using MMP2 calculations has been offered.[18] This predicts that a *tZc* form with a pseudoequatorial hydroxyl group is energetically more favorable.

To obtain information about the photochemistry from molecules with some deviation of the conformations present in **P**, 10-demethyl analogues of pro- and previtamin D have been investigated.[19] In addition to the conformational changes, it was expected that, if polarized forms play a role in the relaxation process of excited trienes (as postulated in the "sudden polarization" concept), the absence of the electron-donating methyl group would destabilize the (sudden) polarized ZT^+ form and, consequently, products from this form would be diminished relative to **P**. Upon short-term irradiation of 10-demethyl-7-DHC (Scheme 11.6), the analogues of **P**, **L**, and **T** are formed. In addition, a bicyclo[2.2.0]hexene appeared to be present, obviously originating from a 4π-electrocyclization. Its formation is ascribed to the considerably diminished degree of puckering in the cyclohexadiene moiety, relative to 7-DHC.[20] The number of overirradiation products in both diethyl ether and in methanol surpasses that found in the case of 7-DHC. In methanol, alcohol addition products are formed. However, apart from those that are analogous to $B_{1,2,3}$, products of alcohol addition to the demethyl-7-DHC diene system are formed. No evidence could be gathered to support the occurrence of polarized forms in the excited state.

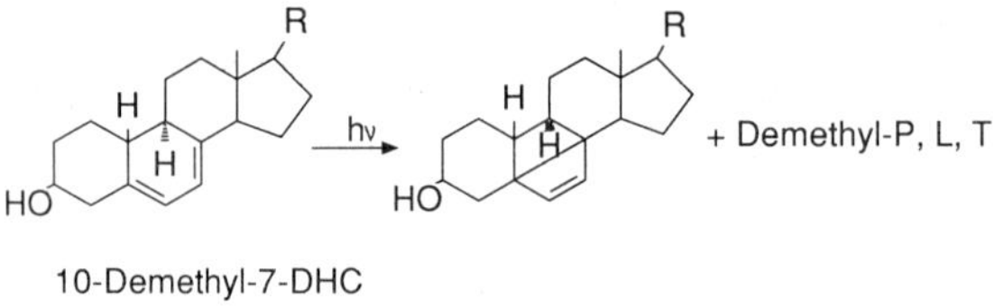

10-Demethyl-7-DHC

SCHEME 11.6

Suprasterols

Since vitamin D predominates over previtamin D (**P**) in the thermal equilibrium (see above) and has a larger extinction coefficient, its presence dominates the photochemistry between 10 °C and 100 °C. In addition to *trans*-vitamin D, it yields six overirradiation products — the "suprasterols". They are illustrated in Scheme 11.7. *trans*-vitamin D is only present on short-term irradiations. The photochemical formation of the suprasterols showed solvent, temperature, and wavelength dependence.[21] After complete conversion of vitamin D, at room temperature in ethanol at $\lambda > 250$ nm, the irradiation mixture was found to consist of 19% SI, 62% SII, 6% SIII, 2% SIV, and 11% of SV + SVI.[22] The type of photoproducts and their dependency on reaction parameters suggests again that ground-state properties play an important role in the photochemistry. From extensive spectroscopic measurements,[21,23,24] it is known that vitamin D occurs almost exclusively in the *cZt*-conformation. The A-ring can adopt two different chair conformations, resulting in two different vitamin D conformers, A(+) and A(−). The (+) or (−) indicates the sign of the torsion angle $\phi(4-5-10-1)$, which is identical to the sign of the torsion angle of the cisoid part $\phi(6-5-10-19)$. The equilibrium between these two conformers depends on the polarity of the solvent and on the temperature; its position is determined by a combination of electrostatic dipole-dipole, 1,3-diaxial steric, and orbital interactions.[25] The UV absorption spectrum does not change in polar or apolar solvents, or at different temperatures, indicating that the conformers of vitamin D have the same absorption characteristics.[21,25]

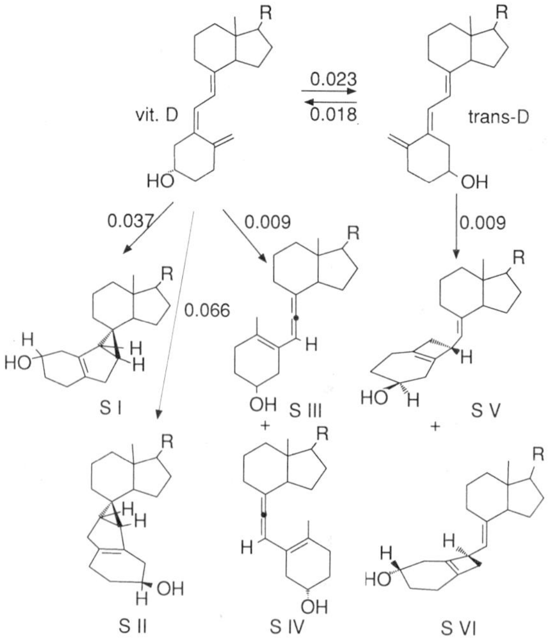

SCHEME 11.7

From measurement of the quantum yields of product formation and time-dependent irradiation of both vitamin D and *trans*-vitamin D, it became evident that the suprasterols SI, SII, SIII, and SIV are only formed from vitamin D; whereas, SV and SVI arise exclusively from the *trans*-isomer. The quantum yields of product formation at 313 nm in diethyl ether are included in Scheme 11.7.

The suprasterols SI and SII are good model compounds with which to study the photoformation of bicyclo[3.1.0]hex-2-ene derivatives. According to Dauben, who elucidated the structures of suprasterols SI and SII,[26] the reaction is not a concerted $[\pi4 + \pi2]$ process, but a two-step reaction.[27] By determination of the quantum yield of formation of SII from two labeled compounds using 313-nm irradiation — vitamin D_3-6-d and vitamin D_3-19,19-d_2 — the kinetic deuterium isotope effects were calculated.[25] From the results (isotope effect of D_3-6-d = 1.19; of D_3-19,19-d_2 = 1.02), it was concluded that the bicyclo[3.1.0]hex-2-ene moiety is formed by a multistep mechanism. By this mechanism, the cyclopropane ring is formed first, followed by the ring closure to the cyclopentene ring.

It is attractive to suppose that each of the two conformers of vitamin D gives rise to one of the suprasterols SI and SII. To test this hypothesis vitamin D has been irradiated in isooctane, diethyl ether and ethanol at various temperatures and wavelengths to elucidate them.[25] In Table 11.3, some results are given together with the ratios of the vitamin D conformers A(+) and A(−). Inspection of the data reveals that the ratio SI:SII is influenced by the wavelength of irradiation, the temperature, and the polarity of the solvent. The ratio decreases upon changing to more polar solvent, irrespective of wavelength and temperature, as is expected from the shift of the ground-state equilibrium to A(−). Irradiations at 313 nm, with low excess energy, revealed that in an apolar solvent, more SII is formed than corresponds with the proportion of A(−) in the ground-state equilibrium. In a polar solvent, on the other hand, more SI is formed. The temperature dependence of the ratio SI:SII at $\lambda > 302$ nm indicates that the barrier to formation of SI is higher than for formation of SII, both in apolar and polar solvents. However, in an apolar solvent, an increase of the excitation energy favors the formation of SI; whereas, in polar solvents, SII is favored. In polar solvents, the thermal energy barrier at shorter wavelengths is higher for SI formation than for SII formation; but in apolar solvents, the reverse relation holds true. It is proposed that irradiation with excess energy opens new decay channels.[25]

11.3 Synthetic Applications

The importance of the photoproducts and photoprocesses discussed in this chapter is obvious and is not easily overestimated.

11.4 Industrial Applicability

The synthesis of vitamin D is one of the most important industrial photochemical processes, the more so when only products of non-radical photoreactions are considered. The industrial production, using ergosterol and 7-dehydrocholesterol as starting materials, started in 1930, and is nowadays a continuous process. Usually, mercury lamps are used as light sources. Thanks to the vitamin D prepared synthetically, rachitis does not exist anymore in developed countries. The larger part of the vitamin D produced is used for livestock. Other industrial products are metabolites of vitamin D such as 25-hydroxycholecalciferol, 1α,25-dihydroxycholecalciferol, and 24R,25-dihydroxycholecalciferol.

Table 11.3a   Ratio of the Vitamin D Conformers A(+):A(−)

T (K)	Solvent		
	C_6D_{12} A(+):A(−)	THF-d_8 A(+):A(−)	C_2D_5OD A(+):A(−)
298	47:53	28:72	13:87
273	47:53	26:74	11:89
243	46:54	23:77	9:91
213	46:54	20:80	6:94

Table 11.3b   Ratio of the Suprasterols SI:SII upon Complete Conversion of Vitamin D

λ(nm)	T(K)	Solvent		
		Isooctane SI:SII	Diethyl ether SI:SII	Ethanol SI:SII
313	298	47:53	36:64	30:70
	273	46:54	33:67	28:72
	243	41:59	29:71	24:76
	213	30:70	25:75	20:80
253	298	51:49	24:76	22:78
	273	52:48	23:77	20:80
	243	53:47	19:81	17:83
	213	54:46	17:83	16:84

Experimental UV Absorption Data of Vitamin D_2 and Related Compounds in Diethylether (Et$_2$O) and Ethanol (EtOH)

Compound	Solvent	λ_{max}, nm (ε)	Ref.
Vitamin D_2	Et$_2$O	264.5 (18.800)	a
	EtOH	264.5 (18.800)	b
Ergosterol	Et$_2$O	293.5(6,900), 281.5(12,400), 271.5(11,600), 264(7,760)(i)	a
	EtOH	293(6,850), 282(12,000), 271.5(11,300), 262.5(8,000)(i)	b
Lumisterol$_2$	Et$_2$O	279(9,080)(sh), 272(9,680)	a
	EtOH	278.5(8,960)(sh), 271.5(9,420)	b
Previtamin D_2	Et$_2$O	261(9,120)	a
	EtOH	261(8,650)	b
Tachysterol$_2$	Et$_2$O	290(23,000)(i), 280(27,700), 272(23,400)(i)	a
	EtOH	290(22,800)(i), 280(28,000), 272.5(24,500)(i)	b

i = inflexion, sh = shoulder.

[a] *Source:* Havinga, E., Koevoet, A. L., and Verloop, A., Studies on vitamin D and related compounds. 4. The pattern of the photochemical conversion of the provitamins D. *Recl. Trav. Chim., Pays-Bas,* 74, 1230, 1955. With permission.

[b] *Source:* Rappoldt, M. P. and Havinga, E., Studies on vitamin D and related compounds. 11. The photoisomerization of ergosterol, *Recl. Trav. Chim. Pays-Bas,* 79, 369, 1960. With permission.

References

1. Whistler, D., De Morbo P. A., Quem Patrio Idiomate Indigenae Vocant "The Rickets", Thesis, Leiden, 1645.
2. Glisson, F., "De Rachitide sive Morbo Puerili qui vulgo "The Rickets" dicitur", London, 1650.
3. Reerink, E. H. and Van Wijk, A., The vitamin D problem. I. The photochemical reactions of ergosterol, *Biochem. J.,* 23, 1294, 1929.
4. Reerink, E. H., Van Wijk, A., and Van Niekerk, J., Physische methoden bij het vitamine onderzoek, *Chem. Weekbl.,* 29, 645, 1932.

5. (a) Windaus, A. and Grundmann, W., Ueber die konstitution des vitamin D_2. II., *Ann.*, 524, 295, 1936; (b) Windaus, A., Inhoffen, H. H., and Von Reichel, S., Ueber die konstitution des ergosterins, *Ann.*, 510, 248, 1934; (c) Dimroth, K., Ueber das lumisterin, *Ber.*, 69, 1123, 1936; (d) Grundmann, W., Beitrag zur konstitutionsermittlung des tachysterins, *Z. Physiol. Chem.*, 252, 151, 1938.

6. Velluz, L., Amiard, G., and Petit, A., Le précalciférol. Ses relations d'equilibre avec le calciférol, *Bull. Soc. Chim. Fr.*, 501, 1949.

7. Koevoet, A. L., Verloop, A., and Havinga, E., Studies on vitamin D and related compounds. 2. Preliminary communication on the interconversion of previtamin D and tachysterol, *Recl. Trav. Chim., Pays-Bas*, 74, 788, 1955.

8. Velluz, L., Amiard, G., and Goffinet, B., Le précalciférol. Structure et photochimie. Son rôle dans la génèse du calciférol et des photoisomères de l'ergostérol, *Bull. Soc. Chim. Fr.*, 1341, 1955.

9. (a) Sanders, G. M., Pot, J., and Havinga, E., Some recent results in the chemistry and stereochemistry of Vitamin D and its isomers, *Fortschr. Chem. Org. Naturst.*, 27, 131, 1969; (b) See Reference 149, (c) Norman, A. W., *Vitamin D: The Calcium Homeostatic Steroid Hormone*, Academic Press, New York, 1979; (d) Jacobs, H. J. C. and Havinga, E., Photochemistry of vitamin D and its isomers and of simple trienes, *Adv. Photochem.*, 11, 305, 1979; (e) Dauben, W. G., McInnis, E. L., and Michno, D. M., Photochemical rearrangements in trienes, in *Rearrangements in Ground and Excited States*, De Mayo, P., Ed., Vol. 3, Academic Press, New York, 1980, 91; (f) Havinga, E., Enjoying organic chemistry, 1927–1987, in *Profiles, Pathways and Dreams: Autobiographies of Eminent Chemists*, Seeman, J. I., Ed., American Chemical Society, Washington, D.C., 1991, 27.

10. Dauben, W. G. and Phillips, R. B., Effects of wavelength on the photochemistry of provitamin D_3, *J. Am. Chem. Soc.*, 104, 5780, 1982.

11. Dauben, W. G., Disanayaka, B., Funhoff, D. J. H., Kohler, B. E., Schilke, D. E., and Zgou, B., Polyene 2^1A_g and 1^1B_u states and the photochemistry of previtamin D_3, *J. Am. Chem. Soc.*, 113, 8367, 1991.

12. (a) Hoeger, C. A. and Okamura, W. H., On the antarafacial stereochemistry of the thermal [1.7]-sigmatropic hydrogen shift, *J. Am. Chem. Soc.*, 107, 268, 1985; (b) Hoeger, C. A., Johnston, A. D., and Okamura, W. H., Thermal [1.7]-sigmatropic hydrogen shifts: Stereochemistry, kinetics, isotope effects and π-facial selectivity, *J. Am. Chem. Soc.*, 109, 4690, 1987.

13. Hanewald, K. H., Rappoldt, M. P., and Roborgh, J. R., The antirachitic activity of previtamin D_3, *Recl. Trav. Chim., Pays-Bas*, 80, 1003, 1961.

14. Havinga, E., Vitamin D, example and challenge, *Experientia*, 29, 1181, 1973.

15. Boomsma, F., Jacobs, H. J. C., Havinga, E., and Van der Gen, A., Studies on vitamin D and related compounds. 24. New irradiation products of previtamin D_3. Toxisterols, *Tetrahedron Lett.*, 427, 1975; Boomsma, F., Jacobs, H. J. C., Havinga, E., and Van der Gen, A., Studies on vitamin D and related compounds. 26. The 'overirradiation products' of previtamin D and tachysterol: Toxisterols, *Recl. Trav. Chim., Pays-Bas*, 96, 104, 1977.

16. Jacobs, H. J. C., Gielen, J. W. J., and Havinga, E., Effects of wavelength and conformation on the photochemistry of Vitamin D and related compounds, *Tetrahedron Lett.*, 22, 4013, 1981.

17. Maessen, P. A., The Formation of Toxisterols from Previtamin D. Mechanistic Studies, Ph.D. Thesis, Leiden, 1983.

18. Dauben, W. G. and Funhoff, D. J. H., Theoretical evaluation of the conformations of previtamin D_3, *J. Org. Chem.*, 53, 5070, 1988.

19. Koolstra, R. B., Photochemistry and Spectroscopy of 10-Demethyl Analogues of Pro- and Previtamin D, Ph.D. Thesis, Leiden, 1988.

20. Koolstra, R. B., Cornelisse, J., and Jacobs, H. J. C., Photochemistry of estra-5,7-diene-3β,17β-diol 17-acetate and its 9,10-seco isomer, 10-desmethyl analogues of pro- and previtamin D. 1. "Reversible" isomerization reactions, *Recl. Trav. Chim., Pays-Bas*, 106, 526, 1987.

21. Gielen, J. W. J., The Photochemistry of Vitamin D and Related Compounds: Conformation and Wavelength Effects, Ph.D. Thesis, Leiden, 1981.

22. Bakker, S. A., Lugtenburg, J., and Havinga, E., Studies on vitamin D and related compounds. 22. New reactions and products in vitamin D_3 photochemistry, *Recl. Trav. Chim., Pays-Bas*, 91, 1459, 1972.

23. Helmer, B., Schnoes, H. K., and DeLuca, H. F., Proton nuclear magnetic resonance studies of the conformations of vitamin D compounds in various solvents, *Arch. Biochem. Biophys.*, 241, 608, 1985.

24. Berman, E., Friedman, N., Mazur, Y., Sheves, M., and Zaretskii, Z. V. I., in *Vitamin D, Basic Research and Its Clinical Application*, Norman, A. W., et al., Eds., Walter de Gruyter & Co, Berlin, 1979, 65.

25. Vroom, E. M., Conformational Analysis and Photochemistry of Vitamin D. A Mechanistic Study of Photochemical Bicyclo[3.1.0]hexene Formation, Ph.D. Thesis, Leiden, 1993.

26. Dauben, W. G. and Baumann, P., Photochemical transformations. IX. Total structure of suprasterol II, *Tetrahedron Lett.*, 565, 1965.

27. Dauben, W. G. and Kellogg, M. S., Mechanism of the photochemical transformation of a 1,3,5-hexatriene to a bicyclo[3.1.0]hexene, *J. Am. Chem. Soc.*, 94, 8951, 1972.

12

Photochemistry of Polyenes Related to Vitamin A

Robert S. H. Liu

University of Hawaii

Vitamin A, 1a, was isolated and characterized early this century.[1] The ensuing discoveries of its involvement in vision, i.e., formation of the visual pigment rhodopsin from 11-*cis*-retinal and the apoprotein opsin and identification of polyene *cis,trans*-isomerization as the primary photochemical process[2] provided an impetus for solution photochemical studies of retinal (1b).[3]

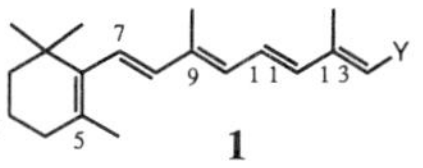

a: Y = CH$_2$OH; f: Y = CH=NH$^+$Bu
b: Y = CHO; g: Y = CO$_2$H
c: Y = CO$_2$Me; h: Y CH=NOH
d: Y = CN; i: 3-dehydro Y = CHO
e: Y = CH=NBu

It is not surprising that much of the early effort in this area centered around retinal (1b) photoisomerization. However, with the knowledge of the nature of the retinyl chromophore (a protonated Schiff base) in many photoactive retinal binding proteins and various end groups in other vitamin A-containing systems, recent studies have been extended to other derivatives. In the process, new types of photochemical reactions (e.g., concerted processes, bimolecular cycloaddition reactions) were identified. Below are summarized highlights in this field. Readers are, however, reminded of the many photochemical[4] and photophysical[5] reviews available, in which more detailed information can be found.

12.1 Photoisomerization

Scope and Mechanism

It is now recognized that the most thoroughly studied retinal system may not be the best model for isomerization of the visual or many other protein-bound retinyl chromophores. Retinal with a close lying n,π* state to the lowest π,π* state exhibits a high intersystem crossing efficiency (especially in a nonpolar solvent), while the protonated Schiff base (a common chromophore in retinal binding proteins) or retinol or many other retinyl end groups without low-lying n,π* states yield triplets in negligible amounts. Thus, the retinal photochemistry is complicated by competing triplet processes, while the visual chromophore and many other vitamin A derivatives exhibit exclusive

0-8493-8634-9/95/$0.00+$.50
© 1995 by CRC Press, Inc.

singlet-state processes. However, retinal isomers are easily separated on many commercially available *HPLC* columns,[4c] rendering accurate determination of product composition relatively easy. The accumulated photochemical information, however, has contributed to a better understanding of the excited properties of polyenes in general.

Of the 16 possible stereoisomers (structures shown below), the most readily available *all-trans-*,*13-cis-*, and *9-cis*-isomers and the visually important *11-cis*-isomer have traditionally been examined in most detail.[7] Recently, the synthesis of all 16 isomers of retinal has been completed.[8] Their characterization data and photochemical properties have been reported.[9] Some of them are summarized in Table 12.1. The UV/VIS absorption spectra of the more recently prepared isomers of retinal are shown in Figure 12.1. Also included in Table 12.1 are absorption data of the isomeric oximes, 1h,[6] useful for analysis of configurational purity of isomeric rhodopsins.

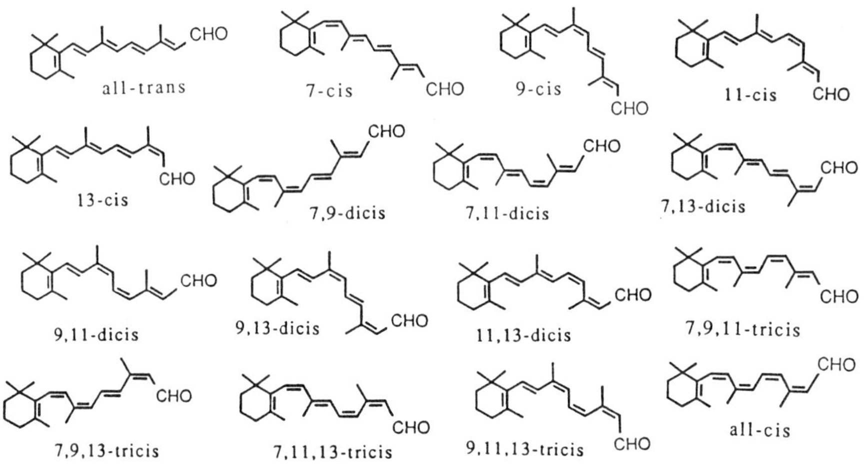

Direct irradiation of retinal isomers in a nonpolar solvent such as hexane results in reaction from both the triplet and the singlet states. The triplet contribution (E_T of *all-trans*-retinal = 149.0 or 35.6 kcal mol^{-1}),[10] as demonstrated via triplet sensitization (best results with porphyrin dyes),[7d] is thermodynamically controlled where the initially formed triplets (picosecond lifetimes for hindered isomers)[11] cascade to the more relaxed (unhindered) isomeric triplets within their lifetimes. Time-resolved resonance Raman studies showed, however, that the latter may not be fully equilibrated within their lifetimes (μsec),[10] thus giving non-identical mixtures when starting with different isomers.[12] Vertical decay generates a mixture of unhindered isomers (primarily *all-trans, 13-cis,* and *9-cis*), resembling those of the thermally equilibrated mixtures.[13] The net results correspond to those from one photon-multiple bond isomerization and quantum chain processes.[9b,c] On the other hand, the singlet-state processes appear to be kinetically controlled, favoring one bond isomerization particularly at the C13-C14 double bond. For severely hindered isomers, two-bond isomerization was also observed, possibly involving a concerted process rather than the adiabatic, stepwise mechanism as in the triplets.[9c]

It is well known that irradiation of retinal in a nonpolar solvent leads to selective formation of the unhindered *13-cis-* and *9-cis*-isomers, while irradiation in a polar solvent leads to nonselective formation of all isomers with a slight preference for isomerization at the central double bonds.[7a] The latter observation led to the accepted procedure, especially among vision researchers for conversion of the more readily accessible *all-trans*-isomer to small amounts of the visually important *11-cis*-isomer.[4a] The early workers primarily used hydroxylic solvents.[7a] Subsequently, dipolar aprotic solvents have also been employed.[14] However, in the latter cases, small amounts of water appear to be essential for formation of the hindered isomers.[15] These and related photochemical observations, supported further by absorption and emission characteristics of retinal in the presence of a proton

Table 12.1   Absorbance Characteristics and Major Primary Photoproducts from Direct Irradiation of All 16 Possible Stereoisomers of Retinal (1b) in Hexane and Absorption of Oxime (1h) Isomers

Isomer	λ_{max}, nm (ε)	ε_{360}	Primary products[h]	Oxime(ε_{360}) $E;Z$[i]
all-trans	368 (48,000)[a]	45,400	13c; 9c	54,900; 51,600[j]
7-cis	359 (45,100)[b]	45,100	trans; 7,13c	47,000; 46,200[k]
9-cis	363 (37,660)[a]	36,600	trans; 9,13c	39,300; 30,600[k]
11-cis	365 (26,360)[a]	24,300	trans; 11,13c	35,000; 29,600[j]
13-cis	363 (38,770)[a]	37,500	trans	49,000; 52,100[l]
7,9-cis	351 (42,500)[c]	39,800	trans; 7,9,13c; 9c; 13c	—
7,11-cis	355 (18,800)[c]	18,360	trans; 7,11,13c; 7c; 11c	—
7,13-cis	357 (36,000)[d]	36,000	7c; 13c; trans	32,800; 30,200[k]
9,11-cis	352 (30,600)[e]	29,800	trans; 9,11,13c; 9c; 9,13c	33,600; 32,300[k]
9,13-cis	359 (34,170)[a]	17,800	9c; 13c; trans	38,300; 30,900[k]
11,13-cis	302 (17,800)[f]	7,800	11c; 13c; trans	—
7,9,11-cis	345 (22,000)[g]	18,900	trans; all-c; 7,9c; 9c; 7,9,13c	—
7,9,13-cis	346 (36,600)[b]	30,600	7,9c; 13c; trans 9,13c; 9c	—
7,11,13-cis	289 (17,600)[h]	9,970	7,11c; 7,13c; 13c; trans; 7c	—
9,11,13-cis	302 (15,500)[f]	6,090	9,11c; 9,13c; 9c; trans; 13c	—
all-cis	287 (16,800)[g]	5,070	7,9,11c; 7,9,13c; 13c; trans	—

[a] Reference 4a.

[b] DeGrip, W. J., Liu, R. S. H., Ramamurthy, V., and Asato, A. E., *Nature*, 262, 416, 1976.

[c] Kini, A., Matsumoto, H., and Liu, R. S. H., *J. Am. Chem. Soc.*, 101, 5078, 1979.

[d] Reference 8.

[e] Kini, A., Matsumoto, H., and Liu, R. S. H., *Bioorg, Chem.*, 9, 406, 1980.

[f] Knudsen, C. G., Carey, S. C., and Okamura, W. H., *J. Am. Chem. Soc.*, 105, 6355, 1980.

[g] Reference 9a.

[h] Initial product distribution from direct irradiation in aerated solution: Reference 9c.

[i] Or, *syn/anti*, at the C15-C16 bond.

[j] Groenendijk, C. W. T., de Grip, W. J., and Daemen, F. J. M., *Anal. Biochem.*, 99, 304, 1979.

[k] Reference 6.

[l] Hamanaka, T., Hiraki, K., and Kito, Y., *Photochem. Photobiol.*, 44, 75, 1986.

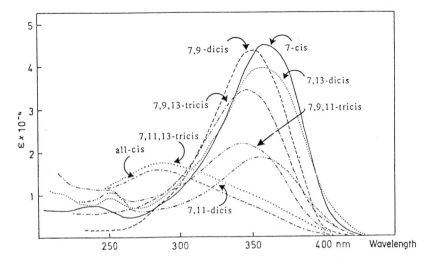

FIGURE 12.1   UV/VIS absorption spectra of selected hindered isomers of retinal (1b), taken in hexane. Complete data are listed in Table 12.1. For spectra of other more commonly available isomers, see References 4a, c.

Table 12.2   Absorption, Fluorescence, Triplet Yield (Φ_T)[a] and Initial Photoproducts from Direct Irradiation of Hexane or Alcohol Solutions of Derivatives of *all-trans*-Vitamin A at Room Temperature

Compound	Solvent	Temp. (K)	Abs$_{max}$ (nm)	Fl$_{max}$ (nm)	Φ_F	Φ_T	Major products
Retinol, 1a	3MePent	298	324	494	0.018		13-*cis*, 9-*cis*[b]
Retinal, 1b	Wet3MeP	77	383			0.2, 0.43	13-*cis*, 9-*cis*[c]
	Ethanol	295	382		0.005	0.1	13-*c*, 11-*c*, 9-*c*[c]
Me retinoate, 1c	3MePent	77	368	468	0.61		9-*cis*, 13-*cis*[d]
	DMSO						9-*c*, 13-*c*, 11-*c*[d]
Retinonitrile, 1d	Hexane						13-*cis*, 9-*cis*[e]
	Methanol						11-*c*, 13-*c*, 9-*c*[e]
Retinyl-SB, 1e	3MePent	77	365	500	0.05		13-*cis*[f]
(butyl)	Methanol	298	364		<0.0005	0.01	11-*cis*, 13-*cis*[f]
Retinyl-PSB, 1f	3MePent	77	490	672	0.14		11-*cis*[g]
(butyl)	Methanol	298	445		<0.0005	<0.01	11-*c*, 9-*c*, 13-*c*[f]
Retinoic acid, 1g	3MePent	77	381	470	0.44		13*c*, 11, 13*c*, 11*c*[h]
3-DehydroRet, 1i	Hexane						13-*cis*, 9-*cis*[5b,i]
	Ethanol						9*c*, 11*c*, 13*c*[5b,i]

    [a] Reference 5b.
    [b] Reference 33.
    [c] Reference 4c.
    [d] Reference 20. Pss data.
    [e] Reference 14.
    [f] Reference 23b.
    [g] Reference 23a.
    [h] Curley, Jr., R. W., Fowble, J. W., *Photochem. Photobiol.*, 47, 831, 1988. Pss data in ethanol.
    [i] Reference 22. Pss data.

donor,[16] led to the suggestion that the protonated retinal is responsible for the indiscriminate isomerization behavior around all four double bonds in polar solvents. It is clear that by invoking the polar character of the ^1B_u state of retinal[17] alone (postulated to play a role in regioselective twisting of the double bonds in other isomerization processes),[18] one cannot fully account for the unique solvent-dependent behavior. Nor is the "sudden polarization" concept[19] likely to be applicable here because any stabilization through solvent-solute interaction at this stage is too late to alter the course of the isomerization reaction, which is determined at the stage of selective twisting of the double bonds of the Franck-Condon species.

Prolonged irradiation led to secondary photoisomerization for *all-trans*-isomer, forming poly-*cis*-isomers. They, however, remain as minor products even in photostationary states.[14b]

Whereas other polyenes in the vitamin A series (Table 12.2), methyl retinoate (1c),[20] retinonitrile (1d),[16] retinyl Schiff bases (SB, 1e),[21] and 3-dehydroretinal (vitamin A₂)[22] exhibit the same solvent-dependent photochemical behavior as retinal, the protonated Schiff bases (PSB, 1f) simply demonstrate regioselective isomerization at the C11-C12 bond.[23] The low triplet yield of these derivatives (also for retinal in polar solvents) implies that the solvent-dependent characteristic is a reflection of the excited singlet-state properties. That such solvent-dependent photochemical behaviors are shown only in polyenes with potential proton accepting end groups is in agreement with the above suggestion that the possible involvement of a protonated substrate is essential.

Following Becker's assignment of state ordering of SB and PSB,[5b,21] one could make the general statement that this difference in photochemical properties of the free and hydrogen bonded vitamin A derivatives is due to the different extent of mixing of the upper ionic $^1B_u^+$ (allowed) state with the lower covalent $^1A_g^-$(forbidden) state. For PSB, the $^1B_u^+$ state simply dominates its excited-state properties.

Retinol methyl ether shows a different type of solvent dependence. While no 11-*cis*-isomer was formed during irradiation in a polar solvent, a large increase (3.5 times) of the 9-*cis*-isomer was observed when ethanol was used instead of hexane.[24] The preference for formation of the 9-*cis*-isomer must be a reflection of the effect of a polar solvent on relative ease of twisting double bonds in the excited Franck-Condon polyene, favoring the more centrally located double bond.

Synthetic Applications

In addition to preparation of 11-*cis*-retinal, photoisomerization has been incorporated as a key step in the synthesis of the hindered 7-*cis*-isomers. Thus, selective triplet sensitization (benzanthrone or Rose Bengal)[4c] is a useful method for introducing the severely crowded 7-*cis*-geometry of dienes and trienes in a quantitative manner.[25] (Failure to form such isomers in longer polyenes was attributed to the altered shape of the torsional potential curves of the polyene triplets.)[26] This procedure has been incorporated into schemes for preparation of all 16 isomers of vitamin A.[8] Furthermore, a better understanding of factors controlling torsional relaxation processes of the Franck-Condon polyene triplets led to additional stereo-control at the C9-C10 bond, giving selectively the 7-*cis*- or the 7,9-*dicis*-isomer.[27] Thus, with the 9-acetoxymethyl triene 2, the method was a key step in stereoselective synthesis of 7,9-*dicis*-C$_{15}$-aldehyde (3) which, upon chain extension, led to the 7,9-*dicis*- and 7,9,13-*tricis*-isomers of vitamin A.[28] In a separate scheme, selective triplet sensitization of triene derivatives with bulky substituents at the 10-position, such as sulfone 4, led to selective formation of the 7-*cis*-isomer. Subsequent base-catalyzed elimination yielded 7-*cis*-vitamin A derivatives. Similarly, starting with the 13-*cis*-isomer of 4, the 7,13-*dicis*-isomer of vitamin A was prepared.[29]

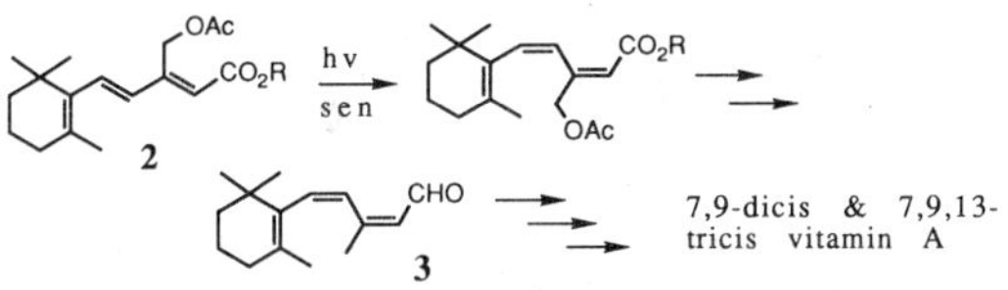

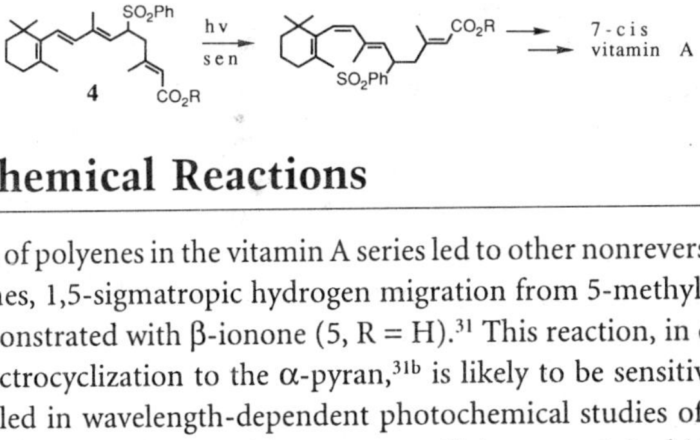

12.2 Other Photochemical Reactions

Prolonged irradiation of polyenes in the vitamin A series led to other nonreversible reactions.[4b,c] For lower member polyenes, 1,5-sigmatropic hydrogen migration from 5-methyl to C-8 is a common process,[30] as first demonstrated with β-ionone (5, R = H).[31] This reaction, in competition with the photochemical 6e-electrocyclization to the α-pyran,[31b] is likely to be sensitive to conformational perturbation as revealed in wavelength-dependent photochemical studies of 7-methyl-β-ionone. Prolonged irradiation of the latter with <347 nm light gave 1,5 shift along with other [2+2]-cycloaddition products while irradiation with >347 nm light gave instead a 2$_a$ + 4$_s$ internal cycloaddition product.[32]

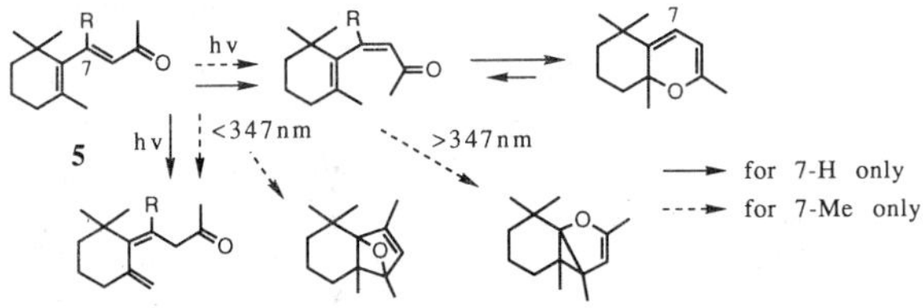

The sigmatropic process is also accompanied by 6e-electrocyclization involving the C5 to C10 unit (6) of a vitamin A polyene, as demonstrated for trienes, tetraenes,[29] and pentaenes (both for retinal, 6a, and retinoic acid).[33] So far, there are no examples in the literature of photocyclization involving other portions of the vitamin A polyene chromophore, although cases of thermal cyclization involving terminal portions of the polyene chain are well documented.[8,34] Expectedly, all these concerted processes proceed only in the singlet state.

Both under sensitized or direct irradiation, vitamin A and its acetate were reported to give C_4 (across the C13-C14 bonds) and C_6 (involving C11-C14 of one and the C13-C14 bond of another) dimers,[4b,35] the latter being structurally identical to the well-known natural product kitol. That this bimolecular process is limited to polyenes without an electron-withdrawing end group seems to be consistent with the notion that the longer-lived covalent $^1A_g^-$ is involved in the reaction as identified spectroscopically.[5b] When irradiated in the presence of oxygen and a triplet photosensitizer, vitamin A polyenes undergo oxygenation forming six-membered ring peroxides (involving the C5 to C8 portion of the molecule).[4b] With retinal, the polyene serves as a sensitizer as well as a substrate for singlet oxygen.[36]

Acknowledgments

The polyene photochemical research program at the University of Hawaii received generous support from the National Science Foundation (CHE-16500) and the U.S. Public Health Services (DK-17806).

References

1. Karrer, P., Morf, R., and Schopp, K., Zur Kenntuis des vitamins-A aus Fischtrauen II, *Helv. Chim. Acta.* 14, 1431, 1931.
2. (a) Wald, G., Carotenoids and the visual cycle, *J. Gen. Physiol.,* 19, 351, 1935; (b) Wald, G., The molecular basis of visual excitation, *Science,* 162, 230, 1968.
3. Hubbard, R. and Wald, G., *cis-trans* Isomers of vitamin A and retinene in the rhodopsin system, *J. Gen. Physiol.,* 36, 269, 1952–3.
4. (a) Hubbard, R., Brown, P. K., and Bownds, D., Methods of vitamin A and visual pigments, *Methods Enzymol.,* 18, 615, 1971; (b) Mousseron, M., Photochemical transformations of polyenic compounds, *Adv. Photochem.,* 4, 203, 1966; (c) Liu, R. S. H. and Asato, A. E., Photochemistry and synthesis of stereoisomers of vitamin A, *Tetrahedron,* 40, 1931, 1984.
5. (a) Ottolenghi, M., The photochemistry of rhodopsins, *Adv. Photochem.,* 12, 97, 1980; (b) Becker, R., The visual process: Photophysics and photoisomerization of model visual pigments and the primary reaction, *Photochem. Photobiol.,* 48, 369, 1988.
6. Trehan, A., Liu, R. S. H., Shichida, Y., Imamoto, Y., Nakamura, K., and Yoshizawa, T., On retention of chromophore configuration of rhodopsin isomers derived from three dicis retinal isomers, *Bioorg. Chem.,* 18, 30, 1990.
7. (a) Kropf, A. and Hubbard, R., The photochemistry of retinal, *Photochem. Photobiol.,* 12, 249, 1970; (b) Rosenfeld, T., Alchalel, A., and Ottolenghi, M., Donor energy effects on the triplet sensitized isomerization of 11-*cis* retinal, *J. Phys. Chem.,* 81, 1496, 1977; (c) Waddell, H. and Chihara, K., Activation barriers for *trans,cis* photoisomerization of *all-trans* retinal, *J. Am. Chem. Soc.,* 103, 7389, 1981; (d) Jensen, N., Wilbrandt, R., and Bensasson, R. V., Sensitized photoisomerization of *all-trans-* and 11-*cis*-retinal, *J. Am. Chem. Soc.,* 111, 7877, 1989.
8. Trehan, A., Mirzadegan, T., and Liu, R. S. H., The doubly hindered 7,11-*dicis,* 7,9,11-*tricis,* 7,11,13-*tricis* and *all-cis* isomers of retinonitrile and retinal, *Tetrahedron,* 46, 3769, 1990.
9. (a) Zhu, Y., Ganapathy, S., Trehan, A., Asato, A. E., and Liu, R. S. H., FT-IR spectra of all sixteen isomers of retinal, their isolation, and other spectroscopic properties, *Tetrahedron,* 48, 10061, 1992; (b) Ganapathy, S. and Liu, R. S. H., Quantum chain processes in photoisomerization of the

all-trans, 7-cis and 11-*cis* isomers of retinal, *J. Am. Chem. Soc.*, 114, 3459, 1992; (c) Ganapathy, S. and Liu, R. S. H., Photoisomerization of sixteen isomers of retinal. Initial product distribution in direct and sensitized irradiation, *Photochem. Photobiol.*, 56, 883, 1992.

10. Dawson, W. and Abrahamson, E. W., Population and decay of the lowest triplet states of polyenes with conjugated heteroatoms: Retinene, *J. Phys. Chem.*, 66, 2542, 1962.

11. Mukai, Y., Koyama, Y., Hirata, Y., and Magata, N., Configurational changes of retinal in the triplet state: psec time-resolved absorption spectroscopy on the 7-*cis*, 11-*cis* and 13-*cis* isomers and hplc analysis of photoisomerization, *J. Phys. Chem.*, 92, 4649, 1988.

12. (a) Hamaguchi, H., Okamoto, H., Tasumi, M., Mukai, Y., and Koyama, Y., Transient Raman spectra of *all-trans*, 7-, 9-, 11-, 13-mono-*cis* isomers of retinal and the mechanism of the *cis-trans* isomerization in the lowest excited triplet state, *Chem. Phys. Lett.*, 107, 355, 1984; (b) Wilbrandt, R., Jensen, N.-H., and Houee-Levin, C., Resonance Raman and absorption spectra of isomeric retinals in their lowest excited triplet-states, *Photochem. Photobiol.*, 41, 175, 1985.

13. (a) Rando, R. R. and Chang, A., Studies on the catalyzed interconversions of vitamin A derivatives, *J. Am. Chem. Soc.*, 105, 2879, 1983; (b) Sperling, W., Carl, P., Rafferty, C. N., and Dencher, N. A., Photochemistry and dark equilibrium of retinal isomers and bacteriorhodopsin isomers, *Biophys., Struct. Mech.*, 3, 79, 1977.

14. (a) Denny, M. and Liu, R. S. H., Sterically hindered isomers of retinal from direct irradiation of the *all-trans* isomer, *J. Am. Chem. Soc.*, 99, 4865, 1977; (b) Denny, M., Chun, M., and Liu, R. S. H., 9,11-*dicis*-Retinal from direct irradiation of *all-trans* retinal, *Photochem. Photobiol.*, 33, 267, 1981.

15. Zhang, B.-W. and Liu, R. S. H., Effect of hydrogen bonding on direct photoisomerization of retinoid: The origin of the solvent polarity effect, *Chinese Chem. Lett.*, 2, 9, 1991.

16. Takemura, T., Das, P. K., Hug, G., and Becker, R. S., Hydrogen bonding effects on fluorescence properties of retinal, *J. Am. Chem. Soc.*, 100, 2626, 1978.

17. Mathies, R. and Stryer, L., Retinal has a highly polar vertically excited singlet state. Implications for vision, *Proc. Natl. Acad. Sci., U.S.A.*, 73, 2169, 1976.

18. (a) Squillacote, M. E. and Semple, T. C., Polarization in the excited state of 1,3-pentadiene: Experimental evidence for an allyl cation-methylene anion species, *J. Am. Chem. Soc.*, 109, 892, 1987; (b) Muthuramu, K. and Liu, R. S. H., Regioselective photoisomerization of fluoro aryl dienes. Role of dipolar intermediates of excited dienes, *J. Am. Chem. Soc.*, 109, 6510, 1987; (c) Colmenares, L. and Liu, R. S. H., 11-Methyl-9-demethylretinal and 11-methyl-9,13-didemethylretinal. Effect of altered methyl substitution on polyene conformation, photoisomerization and formation of visual pigment analogs, *Tetrahedron Lett.*, 47, 3711, 1991.

19. Wulfman, C. E. and Kumei, S., Highly polarizable singlet excited states of alkenes, *Science*, 172, 1061, 1971.

20. Halley, B. A. and Nelson, E. C., Solvent effect on the time dependent photoisomerization of methyl retinoate, *Int. J. Vitam. Nutr. Res.*, 49, 347, 1979.

21. Freeman, K. A. and Becker, R. S., Comparative investigation of the photoisomerization of the protonated and unprotonated *n*-butylamine Schiff bases of 9-*cis*-, 11-*cis*-, 13-*cis*- and *all-trans*-retinals, *J. Am. Chem. Soc.*, 108, 1245, 1986.

22. Liu, R. S. H., Asato, A. E., and Denny, M., 7-*cis*-3-Dehydroretinal and 7-*cis*-3-dehydro-C_{18}-ketone from direct irradiation of the *trans* isomers in polar solvents, *J. Am. Chem. Soc.*, 99, 8095, 1977.

23. (a) Childs, R. F. and Shaw, G. S., A quantitative examination of the photoisomerization of retinal iminium salt by high field ^{1}H-NMR spectroscopy, *J. Am. Chem. Soc.*, 110, 3013, 1988; (b) Koyama, Y., Kubo, K., Komori, M., Yasuda, H., and Mukai, Y., Effect of protonation of the isomerization properties of *n*-butylamine Schiff base of isomeric retinal as revealed by direct hplc analyses: Selection of isomerization pathway by retinal proteins, *Photochem. Photobiol.*, 54, 433, 1991.

24. Unpublished results of X.-Y. Li. and R. S. H. Liu. But in a brief report, retinol was claimed not to exhibit solvent-dependent photochemical behavior: Tsukida, K., Kodama, A., Ito, M., Kawamoto, M., and Takahashi, K., The analysis of *cis-trans* isomeric retinals by high speed liquid chromatography, *J. Nutr. Sci. Vitaminol.*, 23, 263, 1977.

25. Ramamurthy, V., Tustin, G., Yau, C. C., and Liu, R. S. H., Preparation of sterically hindered geometric isomers of 7-*cis*-β-ionyl and β-ionylidene derivatives in the vitamin A series, *Tetrahedron*, 31, 193, 1975.

26. Ramamurthy, V. and Liu, R. S. H., Excitation, relaxation, and deactivation of dienes, trienes and higher polyenes in the vitamin A series in the sensitized isomerization reaction, *J. Am. Chem. Soc.*, 98, 2935, 1976.

27. Liu, R. S. H., Asato, A. E., and Denny, M., Medium- and substituent-directed stereoselective photoisomerization of polyenes in the vitamin A series. Application of the Dellinger-Kasha model, *J. Am. Chem. Soc.*, 105, 4829, 1983.

28. Asato, A. E. and Liu, R. S. H., 7-*cis*,9-*cis*- and 7-*cis*,9-*cis*,13-*cis*-Retinal. A stereoselective synthesis of 7-*cis*,9-*cis*-β-ionylideneacetaldehyde, *J. Am. Chem. Soc.*, 97, 4128, 1975.

29. Miller, D., Trammell, M., Kini, A., and Liu, R. S. H., Sulfone routes to sterically hindered 7-*cis* isomers of vitamin A., *Tetrahedron Lett.*, 22, 409, 1981.

30. Ramamurthy, V. and Liu, R. S. H., Sigmatropic hydrogen migration and electrocyclization processes in compounds in the vitamin A series, *J. Org. Chem.*, 41, 1862, 1976.

31. (a) de Mayo, P., Stothers, J. B., and Yip, R. W., The irradiation of β-ionone, *Can. J. Chem.*, 39, 2135, 1961; (b) Cerfontain, J. A. and Genevasen, J., The low temperature photochemistry of (Z)-β-ionone and its photo-isomers, *Tetrahedron*, 37, 1571, 1981.

32. Mathies, P., Nishio, T., Frei, B., and Jeger, O., Wavelength dependence of the photochemistry of 7-methyl-β-ionone, *Helv. Chim. Acta*, 72, 933, 1989.

33. (a) Halley, B. A. and Nelson, E. C., High performance liquid chromatography and proton nuclear magnetic resonance of eleven isomers of methyl retinoate, *J. Chromatogr.*, 175, 113, 1979; (b) Tsukida, K., Ito, M., and Kodama, A., Electrocyclized retinal, *J. Nutr. Sci. Vitaminol.*, 23, 375, 1977.

34. de Lera, A. R., Reischl, W., and Okamura, W. H., On the thermal behavior of Schiff bases of retinal and its analogues: 1,2-Dihydropyridine formation via six-π-electron electrocyclization of 13-*cis* isomers, *J. Am. Chem. Soc.*, 111, 4051, 1989.

35. Giannotti, C., Sur la photodimerisation de l'acetate de la vitamine A, *Can. J. Chem.*, 46, 3025, 1968.

36. (a) Tsujimoto, K., Hozoji, H., Ohashi, M., Watanabe, M., and Hottori, H., Wavelength dependent peroxide formation upon irradiation of *all-trans* retinal in aerated solution, *Chem. Lett., Chem. Soc., Japan*, 1673, 1984; (b) Baron, M. H., Coulange, M. J., Coupry, C., Baron, D., Favrot, J., and Abo Aly, M. M., *All-trans* retinal photoisomerization and photooxidation from UV laser radiation. Vibrational assignment of *all-trans* 5,8-peroxyretinal, *Photochem. Photobiol.*, 49, 739, 1989.

13

Fulgides and Related Systems

urry G. Heller
iversity of Wales

13.1 Photochromic Aromatic Fulgides

Over 30 years ago, the author chanced upon a review article by Hans Stobbe[1] on photochromic aryl-substituted *bis*-methylenesuccinic anhydrides. Stobbe called these compounds "fulgides" (from the Latin *fulgere,* to glisten and shine) because he obtained many of the compounds as beautiful glistening crystals. He noted that triphenylfulgide 1 showed a color change from orange to brown on exposure to sunlight, which reversed in the dark. He attributed the photochromic properties of fulgides to a crystal effect,[2] but this was untenable because fulgide 2 retained its photochromic properties in solution.[4] His second suggestion was that photochromism was due to different forms of the fulgide,[3] but this suggestion was invalid also because diphenylmethylene(isopropylidene)-succinic anhydride 2, which cannot exist as geometrical isomers, shows a yellow-to-red reversible color change.[4]

Stobbe noted that the photochromic properties were lost on prolonged irradiation of fulgides and that near colorless photoanhydrides were formed. Becker and Santiago[5] proposed that the photochromism of dibenzylidenesuccinic anhydride 3 was due to photocyclization to the 1,8a-DHN 4 to account for the formation of 1-phenylnaphthalic anhydride 6 on the photooxidation of fulgide 3. The reinvestigation[6] of the reactions of yellow *E*- and *Z*-benzylidene (diphenylmethylene)-succinic anhydrides 6 and 7 showed that they undergo reversible photochemical conrotatory ring closure to form red *cis*- and *trans*-1,8a-dihydronaphthalene intermediates (1,8a-DHNs) 8 and 9, respectively, which ring open by a disrotatory mode to yield *Z*- and *E*-fulgides 7 and 6 and rearrange irreversibly to the colorless *cis*- and *trans*-1,2-DHNs 10 and 11 in two competing thermal processes. Related studies on other fulgides have been reported.[7] On exposure to visible light, 1,8a-DHNs undergo photochemical conrotatory ring opening to the corresponding *E*-fulgides.

0-8493-8634-9/95/$0.00+$.50
© 1995 by CRC Press, Inc.

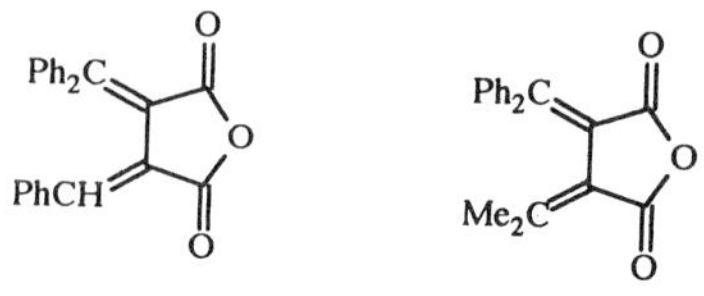

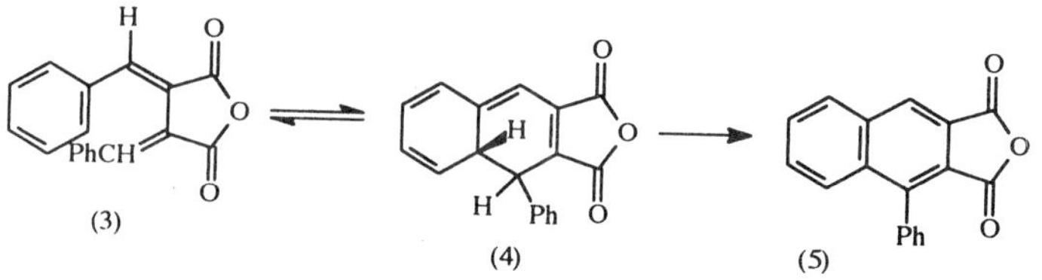

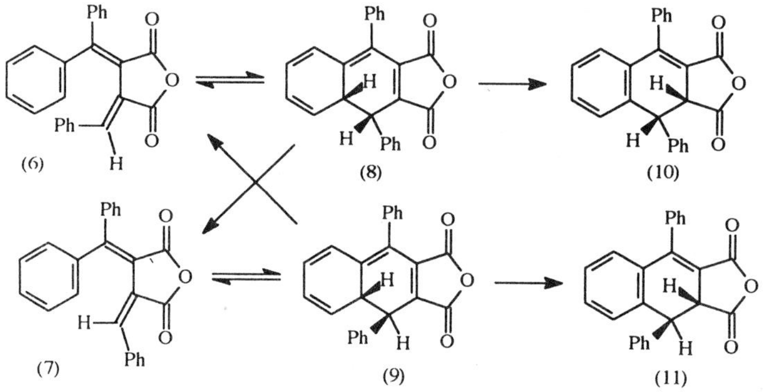

The weakly photochromic pale yellow E-fulgide **12** (R = H) photoisomerizes reversibly to the Z-fulgide **13** (R = H) and photocyclizes to the red 1,8a-DHN **14** (R = H), which undergoes a 1,5-H shift to the colorless 1,2-DHN **15** (R = H). Introduction of methoxy substituents in the 3- and 5-positions of the phenyl group results in a strongly photochromic fulgide **12** (R = OMe), which photocyclizes to the deep blue 1,8a-DHN **14** (R = OMe) which, in turn, undergoes a photochemical 1,7-H-shift to the colorless 1,4-DHN **16** (R = OMe) on prolonged irradiation (366 nm) in toluene, as well as undergoing the thermal 1,5-H shift to the 1,2-DHN **15** (R = OMe).[8]

3,4,5-Trimethoxyphenylfulgide **17** (R = H) is not photochromic, whereas fulgide **17** (R = OMe) is strongly photochromic, illustrating that photochromism is lost when the substituent R is hydrogen and there is a 2- or 4-methoxy group on the phenyl group.[9] Photopolychromic fulgides (i.e., which undergo more than one major reversible color change) can be prepared by the introduction of alkoxy groups in the 2- and 5-positions of the phenyl substituent. For example, the yellow fulgide **19** photocyclizes to the red and blue 1,8a-DHNs **20** and **21**, respectively. The latter fades thermally at ambient temperatures in the dark or can be bleached with light of wavelength longer than 590 nm, leaving the thermally stable red 1,8a-DHN **20** that does not absorb above 590 nm. When a trace of acid is added to the purple solution in toluene of a mixture of the two 1,8a-DHNs **19** and **20**, the solution turns blue due to rapid acid-catalyzed reversal of the red 1,8a-DHN **20**.[10]

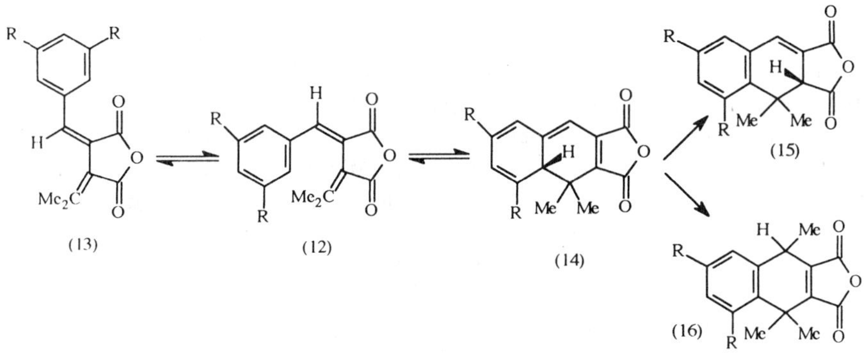

(13)   (12)   (14)   (15)   (16)

All the above photochromic fulgides have high intrinsic fatigue, namely photodehydrogenation to naphthalene derivatives or hydrogen-shift reactions to 1,2- or 1,4-dihydronaphthalene derivatives via their DHNs.

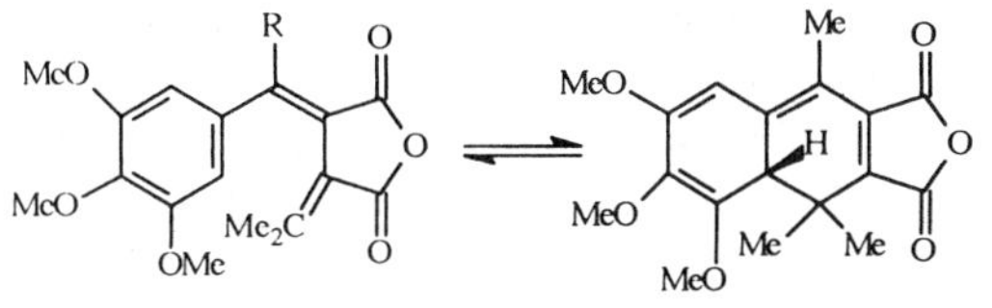

(17)   (18)

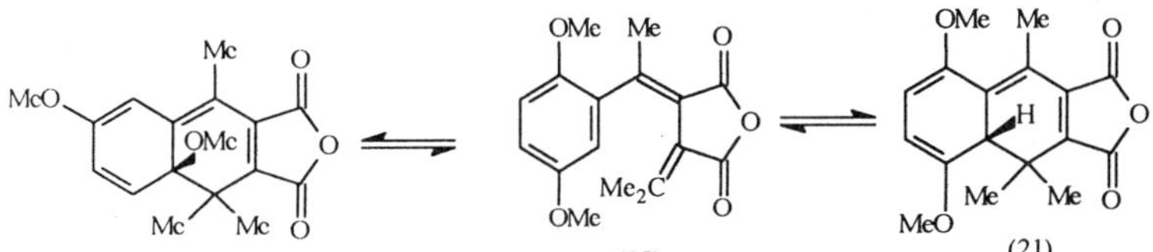

(20)   (19)   (21)

The challenge was to design thermally stable fatigue-resistant photochromic fulgides that potentially would be suitable for commercial applications, including optical recording and security printing. The compounds should have high quantum efficiencies for coloring and bleaching and high conversions into the colored forms.

The introduction of thermal stability and elimination of fatigue associated with the hydrogen-shift reactions was achieved simply by introducing methyl substituents into the 2- and the 6-positions of the phenyl substituent.[11] Fulgide 22 (R = H or OMe) photochemically ring-closed and ring-opened by conrotatory modes without steric interactions, while the thermal disrotatory ring opening of the 1,8a-DHN 23 (R = H or OMe) is prevented by the collision of methyl groups in the 1- and 8a-positions. However, fulgide 22 (R = H) has a low quantum efficiency for coloring and a low conversion to the red 1,8a-DHN 23 (R = H), whereas fulgide 21 (R = OMe) synthesized by Gonzenbach[12] has a high quantum yield for coloring and a high conversion to the purple 1,8a-DHN 23 (R = OMe) but involves a 14-stage synthesis.

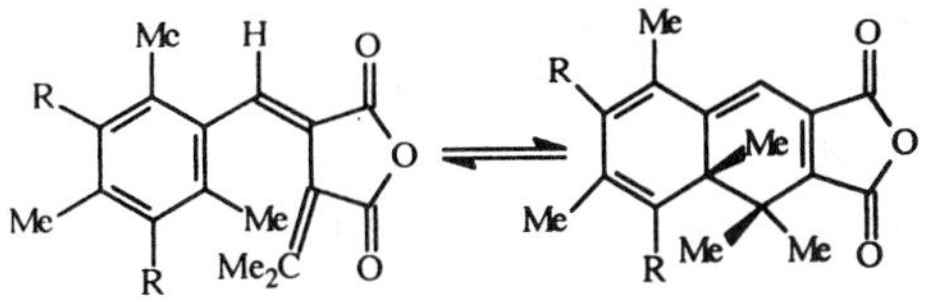

(22)   (23)

13.2 Photochromic Heterocyclic Fulgides

The synthesis of fulgide 22 (R = OMe) demonstrated that fatigue-resistant, thermally stable, strongly photochromic fulgides were an attainable goal. Replacement of the 2,4,6-trimethyl-3,5-dimethoxyphenyl substituent by the 2,5-dimethyl-3-furyl group was a logical development. Fulgide 24 (X = O) showed all the properties expected. In addition, because the colored form has a minimum absorption at 366 nm, the conversion into the colored form on irradiation at this wavelength for 1×10^{-4} mol^{-1} dm^{-3} solutions in most organic solvents was quantitative, allowing the spectrum of the colored·form 25 (X = O) to be measured and the quantum efficiencies for coloring and bleaching to be determined. Even in high concentrations, the conversion of pale yellow fulgide 24 (X = O) into the red 7,7a-dihydrobenzofuran derivative 25 (X = O) is nearly quantitative because the colored form does not act as an internal filter for the activating radiation. The quantum efficiency for coloring (20%) in the region 310 to 390 nm is wavelength and temperature independent for solutions in toluene and since the color change is reversible, solutions of the fulgide provide a convenient chemical actinometer for the UV region 310 to 390 nm. All that is required is that a known volume of a solution of the fulgide 24 (X = O) (5×10^{-3} mol^{-1} dm^{-3} in toluene) is irradiated for a known period of time and the absorbance change at 496 nm [the maximum of the long-wavelength band of the colored form (25) measured].[13] The actinometer can be used also in the visible region (436 to 536 nm) by measuring the decrease in absorbance of the colored form after irradiation. The quantum efficiency for bleaching is wavelength, temperature, and solvent dependent.[15]

The photochromic properties can be modified in several ways. Changing the heteroatom in fulgide 24 from oxygen to sulphur to nitrogen causes the color of the ring-closed form 25 to change from red to purple to blue, respectively1.[14,15]

(24)   (25)

When R^1 is hydrogen, the photochromic properties are lost or very poor, and the main photoreaction is *cis,trans*-isomerization.[16] The quantum efficiency for coloring increases with the increasing size of this substituent (e.g., 20% when R^1 is methyl and 62% when R^1 is isopropyl).[17] When R^5 is hydrogen, the photochromic system is more susceptible to photodegradation. A powerful electron-releasing substituent in this position causes a major bathochromic shift in the absorption band of the colored form and a large hyperchromic effect.[18] If R^2 is an aryl group, the photochromic properties are poor; and if R^2 is hydrogen, then a hydrogen shift occurs in the colored form and the thermal stability and fatigue resistance are lost. In fulgide 28, the corresponding 1,3- or 1,5-deuterium shifts do not occur even at 140 °C, but an unexpected reversible intramolecular hydrogen-deuterium exchange reaction takes place at room temperature.[19]

Replacement of the methyl groups at R^3 by cyclopropyl groups causes the fulgides to undergo a bathochromic shift of their long-wavelength absorption band.[20] Replacing the isopropylidene group by the bulky inflexible adamantylidene group causes a five- to ninefold increase in the quantum efficiency for bleaching, presumably due to the weakened 7,7a-sigma bond in the colored form 30 (R = methyl or isopropyl) by the spiroadamantane moiety. The quantum yield for bleaching is wavelength, temperature, and solvent dependent also.[15]

The bond-weakening effect of the spiroadamantane moiety has been used to synthesize a new class of *heliochromic* compounds that color rapidly in unfiltered sunlight and thermally fade at ambient temperatures in the absence of the activating radiation.

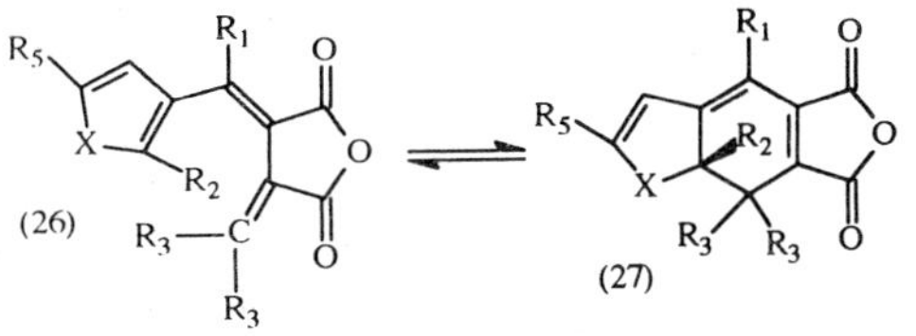

Fulgides 31 and 35 (X = S; R = Me) undergoes thermal ring closure followed by a 1,5-hydrogen shift to give the colorless dihydrobenzothiophene derivatives (DHBTs) 33 and 37 (X = S; R = Me), which on exposure to UV light, ring opens to the blue and purple forms 34 and 38 (X = S; R = Me), respectively, which have very low quantum efficiencies for bleaching with white light.

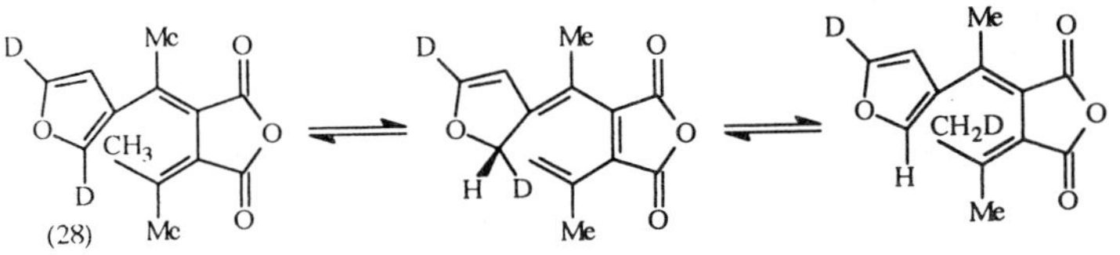

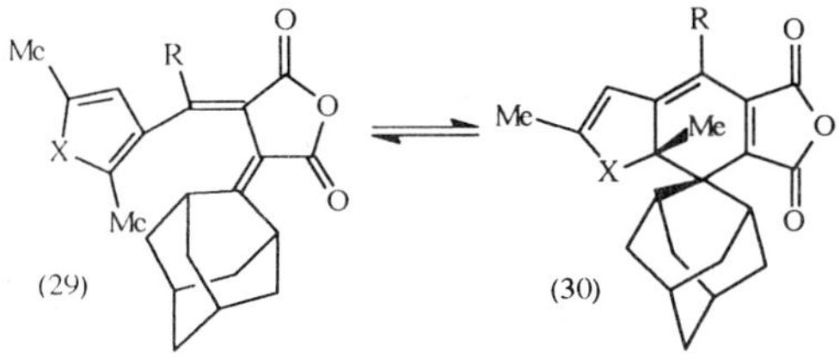

The color and the fade characteristics of the heliochromic compounds 33 and 39 can be modified by benzannellation, replacement of sulphur by oxygen, and by introduction of a substituent in the heteroaromatic group.[21,22] Molecular tailoring can be used to modify the photochromic change of the near-colorless fulgide to colors ranging from red to blue, and increase or decrease quantum yields for coloring and bleaching but not to produce a reversible color change from colorless to bright yellow or to produce a photochromic system that is IR active. These changes have been achieved by modification of the anhydride ring.

13.3 Systems Related to Fulgides

Fulgimides

The corresponding fulgimides[23] (39) are less colored than their corresponding fulgides (40), but show similar photochromic properties. Their corresponding colored forms 26 show broader absorption bands with a shift of the absorption maxima to longer wavelengths compared to the colored forms 27, resulting in a color shift toward blue. Fulgimides 40 are more resistant to acid- or base-catalyzed hydrolysis than the corresponding fulgides 26, but their resistance to photodegradation is not markedly improved. The importance of fulgimides is that the N-substituent can be used as a linking group to prepare photochromic Langmuir-Blodgett films,[24] photochromic liquid crystals,[25] photochromic diagnostic devices,[26] and photochromic copolymers.[27]

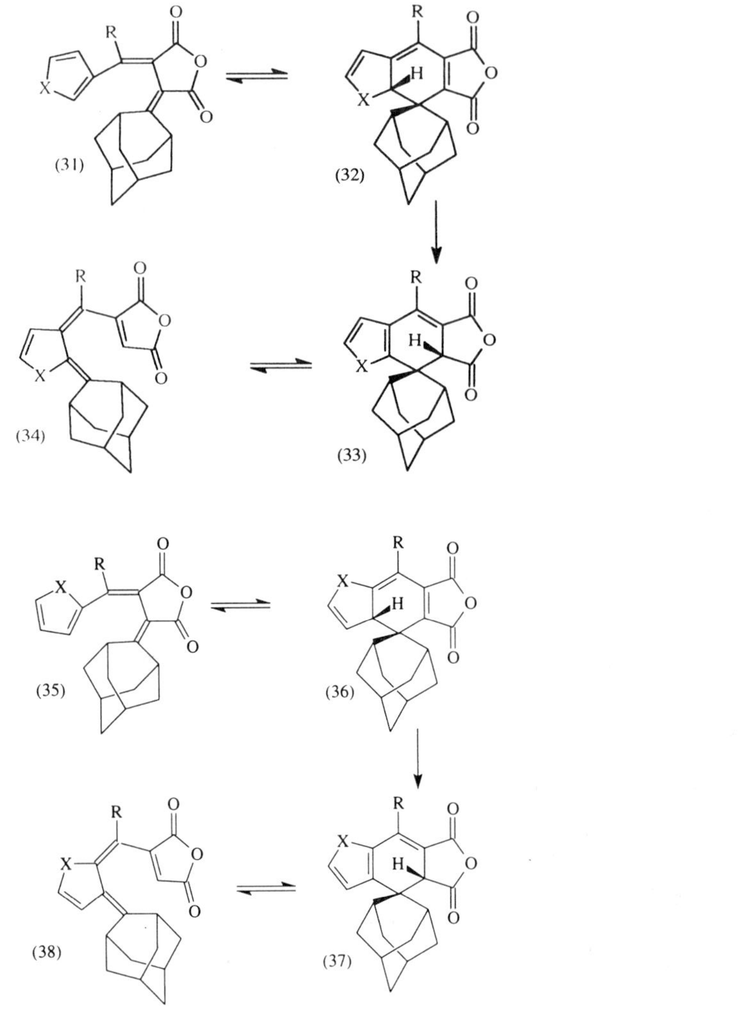

Isofulgimides

The term "isofulgimides" has been introduced to describe isomers of fulgimides in which a carbonyl group oxygen is replaced by nitrogen.[27] α-Isofulgimides have the remaining carbonyl group as part of the conjugated system in the corresponding cyclized form, whereas β-isofulgimides do not. Fulgide (Y = O) and the corresponding β-isofulgimide **43** [Y = N.N(Me)Ph] show reversible color changes from pale yellow to red. The cyclized form [Y = N.N(Me)Ph] has a molar extinction coefficient nearly 2.5 times great than the 7,7aDHBF (**44**; Y = O).[28]

The cyclized form **46** [Y = N.N(Me)Ph] of the α-isofulgimide **45** [Y = N.N(Me)Ph] shows a marked hypsochromic shift of its long-wavelength absorption band (λ_{max} = 438 nm) compared to that of the 7,7a-DHBF **46** (Y = O) (λ_{max} = 500 nm) and the cyclized form **44** [Y = N.N(Me)Ph] (λ_{max} = 527 nm).[28]

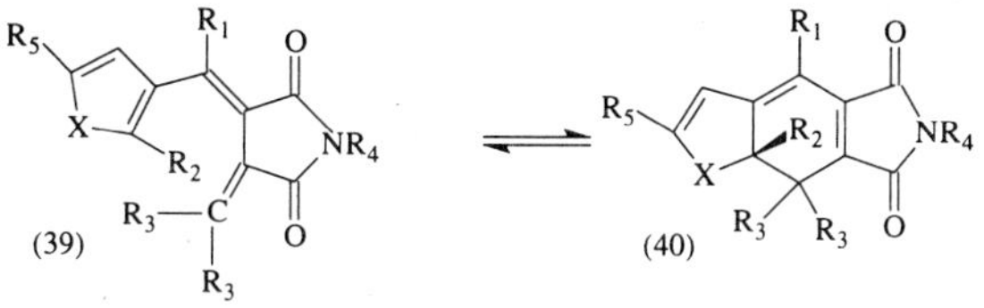

13.4 Photochromic Derivatives of Substituted 3,4-Bismethylene-5-dicyanomethylene tetrahydrofuran-2-ones

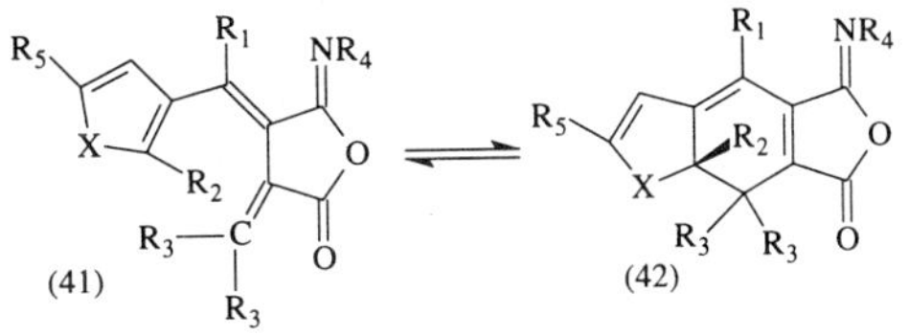

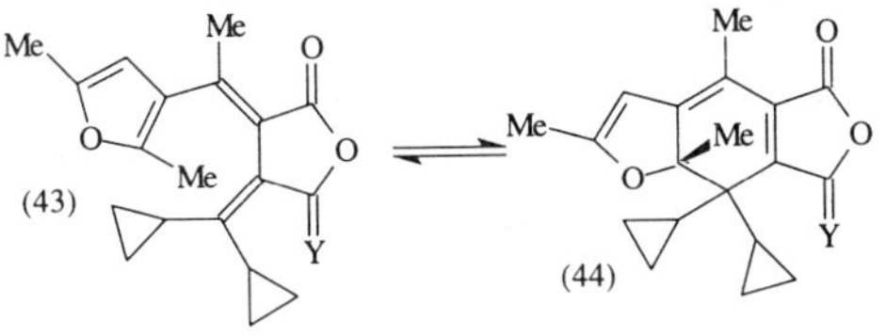

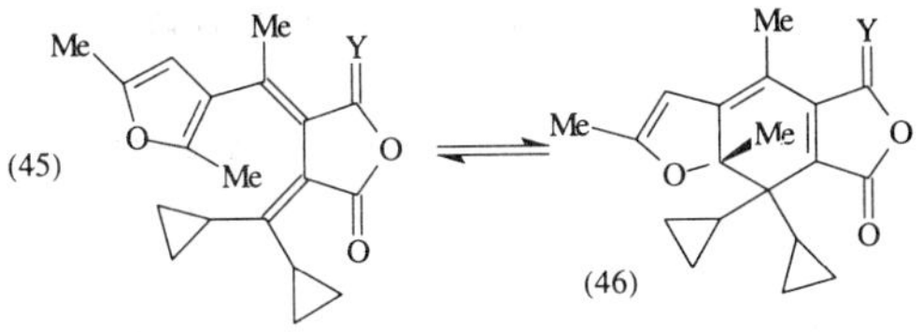

Replacement of the carbonyl oxygen in fulgide 43 (Y = O) by the dicyanomethylene group causes a bathochromic shift in the maximum of the long-wavelength absorption band of the colored form 44 [Y = C(CN)$_2$] (λ_{max} = 636 nm), resulting in a color change to blue.[28] Taking advantage of the other possibilities of molecular tailoring, e.g., by introducing a phenyl substituent into 5-position of the fulgide and replacing the furan ring by thiophene, it has been possible to prepare a thermally stable fatigue-resistant, photochromic, IR-active photochromic system. The cyclized form 48 [Y = C(CN)$_2$] of 5-dicyanomethylene-4-isopropylidene-3-(2-methyl-5-phenyl-3-thienylethylidene) tetrahydrofuran-2-one 47 [Y = C(CN)$_2$] is infrared active (λ_{max} = 699 nm, ε = 16,200 dm^3 mol^{-1} cm^{-1} [*cf.* (47; Y = O) (λ_{max} = 546 nm, ε = 12 000 dm^3 mol^{-1} cm^{-1})].[29]

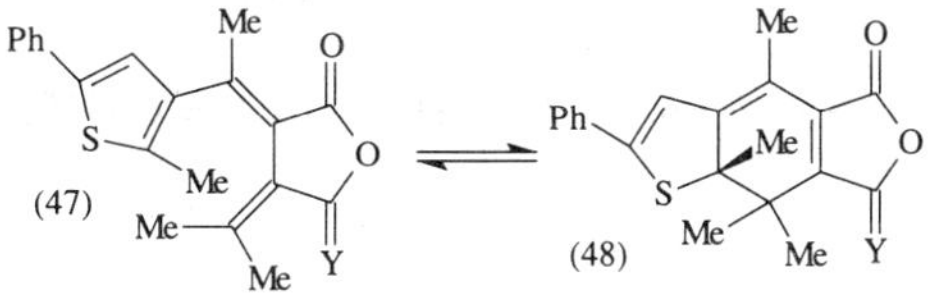

The optically active, IR-active, thermally stable, fatigue-resistant photochromic system has been designed by preparing the fulgide derivative 49 [X = C(CN)$_2$] in which the helical structure is determined by asymmetric induction by the R-3-methylcyclohexanylidene group in which the methyl group adopts only an equatorial conformation. The helix formed is the one with minimum steric interactions.[29]

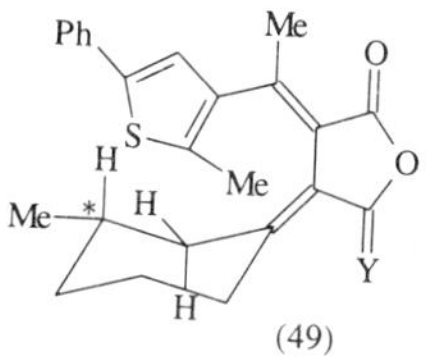

(49)

13.5 Syntheses of Fulgides

The Stobbe reaction was discovered accidentally in 1893,[30] when attempted Claisen condensation of acetone with diethyl succinate yielded ethyl isopropylidenesuccinate (50) instead of ethyl heptan-4,6-dionate (51), the expected product. Re-esterification of the half-ester 50, followed by a second condensation with an aromatic aldehyde or ketone, gave rise to substituted methylenesuccinic acids which cyclized, on treatment with acetyl chloride, to the fulgides. Condensations of aldehydes or ketones with succinic esters were carried out initially using sodium dissolved in ethanol. Improved yields are reported for reactions using potassium *t*-butoxide in toluene or sodium hydride in toluene, but the most convenient method is probably the condensation in the present of potassium *t*-butoxide suspended in toluene. The reaction mixture is poured into water and the required half-ester is obtained by acidification of the aqueous layer after separation of the toluene layer that contains reactants and neutral biproducts. The mechanism of the Stobbe condensation has been elucidated by Johnson and co-workers,[31] who demonstrated that the reaction proceeds via an intermediate lactone (52). The reaction has been reviewed by Daub and Johnson.[32] Most aromatic aldehydes condense to give half-esters with a *trans*-arrangement of aryl and carbonyl groups, attributed to "orbital overlap control".[33] It is only when half-esters are hydrolyzed to the diacids and cyclized that the five-membered anhydride ring causes the severe steric interactions characteristic of *E,E*-fulgides.

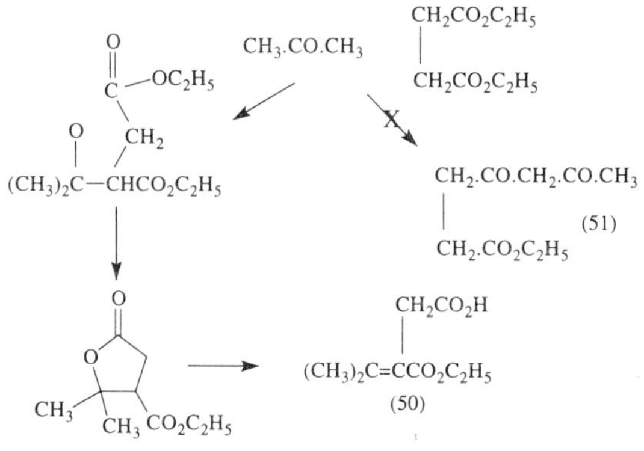

Many workers,[34–39] unable to accept the overcrowding of the *E,E*-isomers, assigned configurations that were incorrect even though *bis*-diphenylmethylenesuccinic anhydride (53)[1] and *bis*-fluorenylidenesuccinic anhydride (54)[40] were known to be stable compounds.

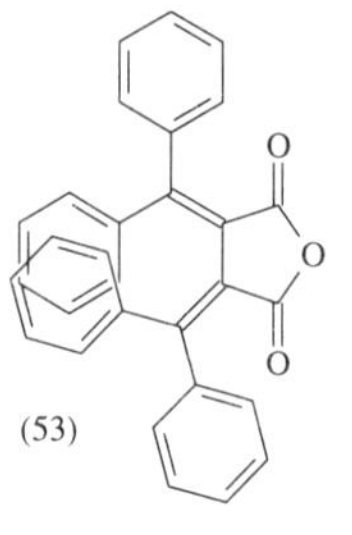

(53) 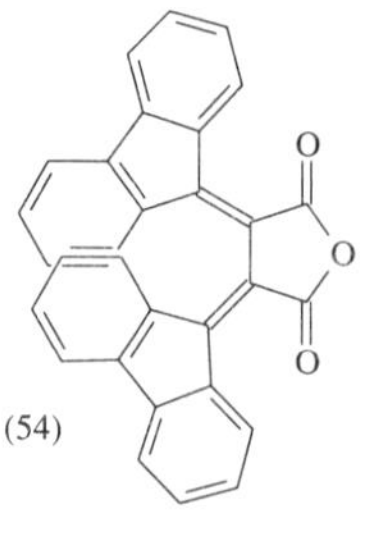   (54)

Proton NMR spectroscopy allows the stereochemistry to be established readily for many fulgides based on the deshielding effect of the carbonyl group on an adjacent substituent and the shielding effect of an aromatic ring on an adjacent group.[41,42] The assignments have been confirmed by X-ray crystallographic analyses of a number of fulgides.[43–46]

3.6 Applications

The studies on photochromic fulgides over the past 2 decades has helped to reduce the widely held view that organic photochromic compounds are unsuitable for commercial applications because of photodegradation, a myth that arose presumably from the failure to develop commercially the photochromic spironaphthopyrans, which followed Hirshberg's suggestion[47] that organic photochromic compounds could be used as an erasable recording medium. The potential of fulgides and related compounds include chemical actinometry, optical information storage and processing, security printing, and marking applications. Fulgides, dissolved in rigid plastic matrix, retain their photochromic properties so that optical recording is possible theoretically at the molecular level, the response time can be measured in picoseconds, the internal filter effect is low, photodegradation products do not markedly and adversely affect the photochromism of the system, and the potential for an erasable optical recording medium is high. Studies by Kurita[48] indicate that a writing speed of 10 GHz is possible. Kirkby and Bennion[49] have reported a recording potential of 100 million bits square centimeter (1000 line pairs per millimeter). Discs have been produced by Tsunoda and Susuki[50] that undergo several hundred color and reverse cycles without significant degradation. Nondestructive readout may be overcome by using the refractive index change between the two states, particulary for IR-active systems related to fulgides or using optically active fulgide and the difference in optical rotation between the two states.

References

1. Stobbe, H., Die Fulgide, **Annalen**, 380, 1, 1911.
2. Stobbe, H., Phototropiererscheinungen bei Fulgiden und anderen Stoffen, *Annalen*, 359, 1, 1908.
3. Stobbe, H., *Z. Electrochem.*, 14, 473, 1908.
4. Stobbe, H., Die Farbe der Fulgensauren und Fulgides, *Chem. Ber.*, 38, 3673, 1905.
5. Santiago, A. and Becker, R. S., Photochromic fulgides. Spectroscopy and mechanism of photoreactions, *J. Am. Chem. Soc.*, 90, 3654, 1968.
6. Heller, H. G. and Hart, R. J., Thermal and photochemical reactions of photochromic *(E)-* and *(Z)-*benzylidene(diphenylmethylene)succinic anhydrides and imides, *J. Chem. Soc., Perkin 1*, 1321, 1972.
7. Heller, H. G. and Szewczyk M., Photoreactions of photochromic a-phenylethylidene(substituted methylene)succinic anhydrides, *J. Chem. Soc., Perkin 1*, 1487, 1974.
8. Heller, H. G., Darcy, P. J., and Hart, R. J., Photochromic systems involving *(Z)*-1-methylpropylidene(diphenylmethylene)succinic and *(E)*-3,5-dimethoxybenzylidene(alkyl-substituted methylene)succinic anhydrides, *J. Chem. Soc., Perkin 1*, 571, 1978.
9. Heller, H. G. Darcy, P. J., Patharakorn, S., Piggott, R. D., and Whittall, J., Photochemical studies on *(E)*-2-isopropylidene-3-[1-(3,4,5-trimethoxyphenyl)-ethylidene]succinic anhydride and related compounds, *J. Chem. Soc., Perkin 1*, 315, 1986.
10. Heller, H. G., Crescente, O., and Patharakorn, S., Studies on the photoreactions and the photopolychromic properties of *(E)*-(2,5-dimethoxyphenyl-substituted)methylene isopropylidenesuccinic anhydrides, *J. Chem. Soc., Perkin 1*, 1599, 1986.
11. Heller, H. G. and Megit, R. J., Fatigue-free photochromic systems involving *(E)*-2-isopropylidene-3-(mesitylmethylene)succinic anhydride and N-phenylimide, *J. Chem. Soc., Perkin 1*, 923, 1974.
12. Gonzenbach, H. and Heller, H. G., unpublished work.

13. Heller, H. G. and Langan, J. R., The use of *(E)-α-*(2,5-dimethyl-3 furylethylidene)isopropylidene)succinic anhydride as a simple convenient chemical actinometer, *J. Chem. Soc., Perkin 2,* 341, 1981.

14. Heller, H. G., Harris, S. A., and Oliver, S. N., Rearrangement reactions of *(E)-α-*1,2,5-trimethyl-3-pyrrylethylidene(isopropylidene)succinic anhydride and related compounds, *J. Chem. Soc., Perkin 1,* 3259, 1991.

15. Heller, H. G., Glaze, A. P., and Whittall, J., *(E)-*Adamantylidene-α-[(2,5-dimethyl-3-furyl)ethylidene]succinic anhydride and derivatives: Model photochromic compounds for optical recording media, *J. Chem. Soc., Perkin 2,* 591, 1992.

16. Heller, H. G., Glaze, A. P., Harris, S. A., Johncock, W., Oliver, S. N., Strydom, P. J., and Whittall, J., *J. Chem. Soc., Perkin 1,* 957, 1985.

17. Kurita, Y., Yokoyama, Y., Goto, T., Inoue, T., and Yokoyama, M., Fulgides as efficient photochromic compounds. Role of the substituent on furylalkylylidene moiety of furylfulgides in the photoreaction, *Chem. Lett.,* 1049, 1988.

18. Wood, D., Studies on Fatigue-Resistant Photochromic Systems. Ph.D. Thesis, University of Wales, Cardiff, 1991.

19. Heller, H. G. and Oliver, S. N., Rearrangement reactions of *(E)-α-*3-furylethylidene(isopropylidene)succinic anhydride, *J. Chem. Soc., Perkin 1,* 197, 1981.

20. Heller, H. G., Oliver, S. N., Whittall, J., Johncock, W., Darcy, P. J., and Trundle, C., Photochromic Fused-ring Organic Compounds and their Use in Photoreactive Lenses, G.B. 214327A, 1985.

21. Heller, H. G., *Photochromics for the Future, 479–480, Electronic Materials,* Miller, L. S. and Mullin, J. B., Eds., Plenum Press, New York, 1991.

22. Heller, H. G., Darcy, P. J., Johncock, W., Oliver, S. N., Trundle, C., and Whittall, J., Photochromic Compounds and Their Uses in Photoreactive Lenses, EPA 0 140540, Environmental Protection Agency, Washington, D.C., 1984.

23. Heller, H. G., Hart, R. J., and Salisbury, K., The photochemical rearrangements of some photochromic fulgimides, *J. Chem. Soc., Chem. Commun.,* 1627, 1968.

24. Cabrera, I., Achim. D., and Ringsdorf, H., D.E 4007636A1.

25. Cabrera, I., Dittrich, A., and Ringsdorf, H., Thermally irreversible photochromic liquid crystal polymers, *Angew. Chem. Int. Ed.,* 30, 76, 1991.

26. Willmer, I., Rubin, S., Wonner, J., Effenberger, F., and Bauerle, P., Photoregulated binding of α-D-mannopyranose to concanavalin A modified by a thiophene dye, *J. Am. Chem. Soc.,* 114, 3150, 1992.

27. Elliot, C. C., Studies on a Photochromic Heterocyclic Isofulgimide, M.Sc, Thesis, University of Wales, Cardiff, 1990.

28. Koh, K., Studies on Isoimides, Ph.D. Thesis, University of Wales, Cardiff, 1993.

29. Al-Shihry, S., Optically Active Photochromic Fulgides, Ph.D. Thesis, University of Wales, Cardiff, 1993.

30. Stobbe, H., Eine neue Synthese der Teraconsaure, *Chem. Ber.,* 26, 2312, 1893.

31. Johnson, W. S., Dunningham, D. A., and McCloskey, A. L., The mechanism of the Stobbe condensation, *J. Am. Chem. Soc.,* 72, 514, 1950.

32. Johnson, W. S. and Daub, G. H., *Organic Reactions,* Vol. 6, chap. 1, 1951.

33. Heller H. G. and Szewczyk, M., Synthesis, stereochemistry, and isomerization of α-phenylethylidenesuccinic esters and related compounds, *J. Chem. Soc., Perkin 1,* 1483, 1974.

34. Chakaborty, D. P., Sleight, T., Stevenson, R., Swoboda, G. A., and Weinstein, B., Preparation and geometrical isomerism of dipiperonylidenesuccinic anhydride, *J. Org. Chem.,* 31, 3342, 1966.

35. Brunow, G. and Tylli, H., The photochemical *cis-trans* isomerization of substituted dibenzylidenesuccinic anhydride, *Acta Chem. Scand.,* 22, 590, 1968.

36. Swoboda, G. A., Wang, K. T., and Weinstein, B., The synthesis of Taiwanin A, *J. Chem. Soc. (B),* 161, 1967.

37. Abdel-Wahhab, S. M. and El-Assal S. L., The cyclization of 3-methoxycarbonyl-*cis*-4-(2-furyl)but-3-enoic acid and α,β-difurfurylidenesuccinic anhydride to the corresponding benzofuran derivatives, *J. Chem. Soc. (C)*, 867, 1968.

38. Harper, S. H., Kemp, A. D., and Tannock, J., Methoxynaphthaldehydes as constituents of the heartwood of *Dyospyros quiloensis* and their synthesis by the Stobbe condensations, *J. Chem. Soc. (C)*, 626, 1970.

39. Abdel-Wahhab, S. M. and El-Rayes, N. R., The cyclisation of *trans*-3-methoxycarbonyl-4-(2(thienyl)but-3-enoic acid and α,β-dithenylidenesuccinic anhydride to the corresponding benzothiophen derivatives, *J. Chem. Soc. (C)*, 3171, 1971.

40. Goldschmidt, S., Riedle, R., and Reichart, A., Uber die Bisdiphenylenfulgide und di Spaltung der Bisdiphenylenfugensaure in Optisch Aktive Komponenten, *Annalen*, 604, 121, 1957.

41. Heller, H. G., Darcy, P. J., Strydom, P. J., and Whittall, J., Electrocyclic reactions of (E)-α-2,5-dimethyl-3-furylethylidene(alkyl-substituted methylene)succinic anhydrides, *J. Chem. Soc., Perkin 1*, 202, 1981.

42. Ilge, H. D. and Schutz, J., Photochemistry of fulgides. 19. Proton nuclear magnetic resonance spectra of phenylfulgides, *J. Prakt Chem.*, 326, 863, 1984.

43. Cohen, M. D., Kaufmann, H. W., Sinnreich, D., and Schmidt, G. M. J., Photoreactions of di-*p*-anisylidenesuccinic anhydride, *J. Chem. Soc. (B)*, 1035, 1970.

44. Boeyens, J. C. A., Denner, L., and Perold, G. W., Stereochemistry and conformational analysis of *bis*-(3,4-dimethoxybenzylidene)succinic anhydride by X-ray crystallography and molecular mechanics, *J. Chem. Soc., Perkin 2*, 1749, 1988.

45. Kaftory, M., *Acta Crystallogr., Sect. C: Cryst. Struct. Commun.*, Photochromic and thermochromic compounds. I. Structures of *(E)* and *(Z)* isomers of 2-isopropylidene-3-[1-(2-methyl-5-phenyl-3-thiezyl)ethylidene]succinic anhydride, $C_{20}H_{18}O_3S$, and the photoproduct 7,7a-dihydro-4,7,7,7a-tetramethyl-2-phenyl benzo [b] thiophene-5,6-dicarboxylic anhydride (P), $C_{20}H_8O_3S$, C40, 1015, 1984.

46. Irie, M., Yoshioka, Y., Takanori, T., and Sawada, M., Molecular and crystal structures of *E*- and *Z*-isomers of 2,5-dimethyl-3-furylethylidene-(isopropylidene)succinic anhydride, *Chem. Lett. Jpn.*, 19, 1989.

47. Hirshberg, Y., Reversible formation and eradication of colours by irradiation at low temperatures. A photochemical memory model, *J. Am. Chem. Soc.*, 68, 2304, 1956.

48. Kurita, Y., Yokogama, Y., Hayata, H., and Ito, H., Photochromism of a furylfulgide, 2-[1-(2,5-dimethyl-3-furyl)ethylidene-3-isopropylidenesuccinic anhydride in solvents and polymer films, *Bull. Soc. Chem. Jpn.*, 63, 1607, 1990.

49. Kirkby, C. J. G. and Bennion, I., Organic ? photochromic for spatial light modulation Part J: Optoelectron, IEEE Proc. J., 133, 98, 1986.

50. Tsunoda, M. and Suzuki, Y., Jpn. Kokai Tokkyo Koho JP Rewritable optical recording materials, J.P. 01 08,092 1987.

14

The Di-π-Methane Rearrangement

Howard E. Zimmerman
University of Wisconsin

14.1 Definition of the Reaction and Background

The Di-π-Methane Rearrangement is a remarkable photochemical process that occurs from excited states of molecules having two π-moieties bonded to a single atom — most generally, carbon. The two π-groups may be aryl, vinyl, ethynyl, acyl, aroyl, etc. What is so remarkable is the very broad generality of both the reaction and the reaction mechanism. This mechanism may be simply shown as outlined in Equations 14.1 and 14.2. The same mechanism nicely predicts the products formed in an exceptionally wide spectrum of reactions.

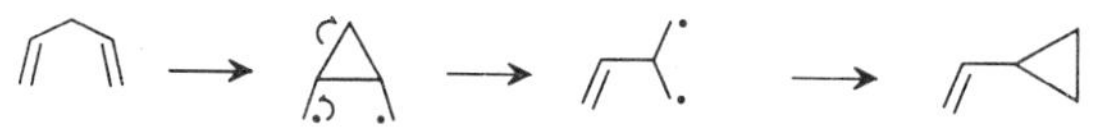

$$(14.1)$$

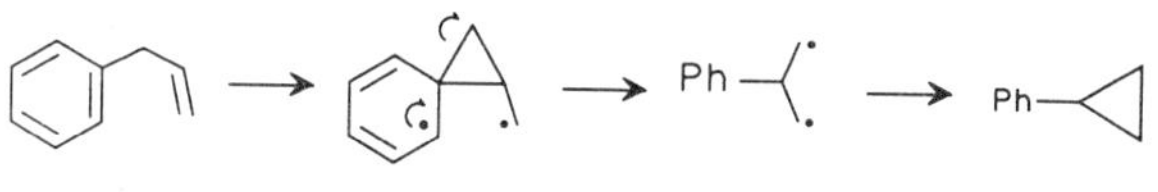

$$(14.2)$$

The discovery of the Di-π-Methane Rearrangement was serendipitous. Thus, in 1966, the rearrangement of barrelene to semibullvalene was observed (note Equation 14.3).[1a] Only in early 1967 was the mechanism established[1b] (note mechanistic discussion below). Once the reaction mechanism of this rearrangement was certain, it then became clear that this formulation could be applied to a vast variety of reactants. In fact, the literature contained some rearrangements that went with neither mechanism nor rationale, and our elucidation of the Di-π-Methane Rearrangement mechanism thus tied these few examples together.

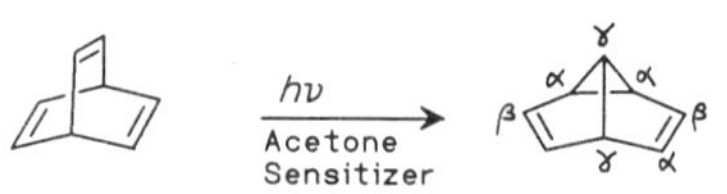

$$(14.3)$$

0-8493-8634-9/95/$0.00+$.50
© 1995 by CRC Press, Inc.

Additionally, it is true that in our earlier efforts we had uncovered general rearrangements of 2,5-cyclohexadienones and 4-aryl-substituted cyclohexenones that superficially appear to be Di-π-Methane Rearrangements (note Equations 14.4 and 14.5). However, these reactions arise from the n-π* triplets of the ketones and are mechanistically distinct from the Di-π-Methane Rearrangement while still being structurally similar. Nevertheless, they deserve mention here.[2,3] In the case of the cyclohexadienone rearrangement, our contribution again was in elucidation of the reaction mechanism as well as the generality of the photochemistry.[4]

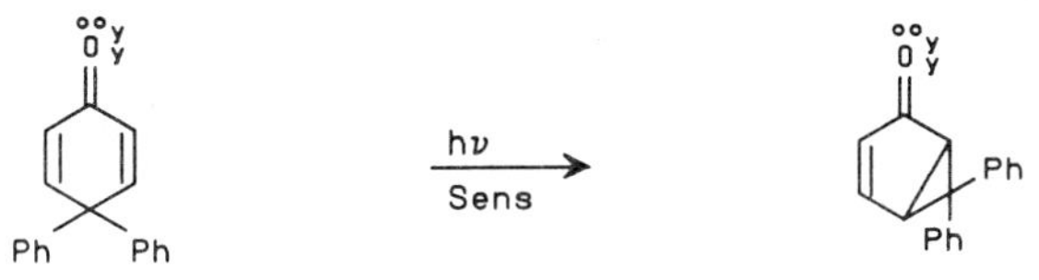

(14.4)

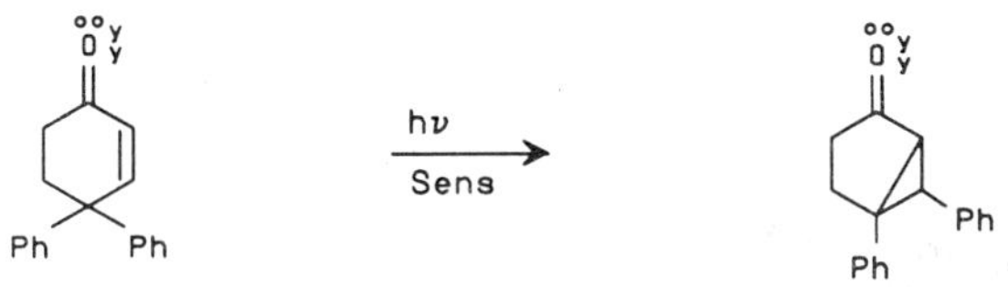

(14.5)

14.2 Initial Evidence for the Basic Reaction Mechanism

The initial evidence for the reaction mechanism outlined above in Equations 14.1 and 14.2 came, as noted, from the barrelene to semibullvalene rearrangement. In this study, hexadeuteriobarrelene, on rearrangement, afforded hexadeuterated semibullvalene with the protium label distributed at the α:β:γ-positions in the ratio of 1.5:0:0.5. If one applies the reaction mechanism of Equation 14.1 to the rearrangement of labeled barrelene to semibullvalene, the protium distribution resulting is precisely that cited above. (Note also Scheme 14.5 below).

With the Di-π-Methane Rearrangement encompassing such a broad spectrum of examples, mechanistic aspects became important in understanding and predicting the reaction multiplicity, stereochemistry, regiochemistry, and scope.

14.3 Multiplicity

The practical importance of multiplicity is in determining whether one should plan on a sensitized or unsensitized set of reaction conditions. However, multiplicity dependence is of intrinsic mechanistic interest since about equal numbers of Di-π-Methane Rearrangements are known from singlet and triplet excited states. Most often, one state is responsible for the reaction while the other is unreactive.

An approximate but simple generalization is that acyclic di-π-methane reactants with conformational flexibility tend to react from the S_1, while cyclic di-π-methanes tend to react via the triplet excited state. There are exceptions, and a discussion of the source of the multiplicity dependence reveals why.

In the case of acyclic di-π-methane reactants, most often the triplet undergoes radiationless decay via a "free-rotor" effect[5] in a process that is more rapid than the Di-π-Methane Rearrangement. Briefly, the free rotor mode of energy dissipation involves double-bond twisting with production of a degeneracy of T_1 with S_0 and thus rapid decay to ground state. One such example is illustrated in Equation 14.6.[6] In this case, photolysis with a sensitizer present is ineffective. However, there are cases where, with bulky substitution,[5e] the free-rotor effect may be inhibited; or with particularly effective stabilizing groups, the triplet will react faster than energy is dissipated (e.g., see Reference 7).

With cyclic and bicyclic reactants, one can understand the inhibition of π-bond rotation in the triplet and thus the enhanced reactivity of the triplet. What is less obvious is why the singlet excited states of many of these compounds do not undergo the Di-π-Methane Rearrangement. In a number

$$\text{(14.6)}$$

of instances, it is merely that often there are more rapid and competitive reactions available, such as electrocyclic cycloadditions. Thus, benzobarrelene affords benzocyclooctatetraene by such a competing reaction.[8]

Sometimes, however, when both the singlet and triplet excited states are reactive (e.g., see Reference 10), multiplicity may control the di-π-methane photoproduct regiochemistry (note Section 14.4).

14.4 Reaction Regioselectivity

It is seen in the basic mechanism given in Equations 14.1 and 14.2 that the first-formed species is a cyclopropyl dicarbinyl biradical. In examples of the type given in Equation 14.2, opening of the three-membered ring of this biradical has reason for occurring such that there is restoration of aromaticity. However, for examples of the type in Equation 14.1, if the cyclopropyl dicarbinyl biradical is unsymmetrically substituted, two alternative modes of three-ring opening exist. One example[9a] is given in Scheme 14.1. Here, two alternative three-ring opening processes are *a priori* possibilities corresponding to ring opening either by process *a* or *b*. Ring opening *b* utilizes odd-electron density, which is heavily localized at an isopropyl center, while opening *a* uses density at a benzhydryl center where the density is only partially available due to delocalization throughout the benzhydryl π-system. The more available electron density is utilized, with process *b* thus being favored. An alternative way of expressing the same basic idea is to note that ring-opening *b* leads to a 1,3-biradical that still retains the benzhydryl delocalization, while ring-opening *a* affords a much less stabilized 1,3-biradical in which the benzhydryl stabilization has been lost. Such simple control of regioselectivity is often seen.

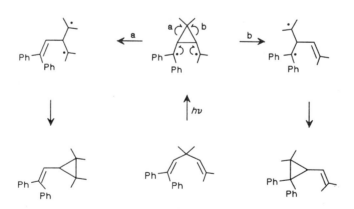

SCHEME 14.1   Regioselectivity control by delocalization maximization.

However, control is sometimes observed in cases where the source is less obvious. Where one has di-π-methane systems of the divinylmethane type with electron-donating or -withdrawing groups on one of the vinyl groups, a generalization is that electron donors appear on the residual double bond of photoproduct, while electron withdrawing groups are found on the three-ring of product. Two examples illustrating this control are given in Equations 14.7 and 14.8.[9b,c]

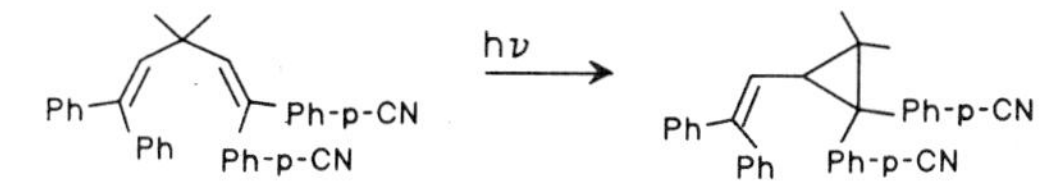

$$(14.7)$$

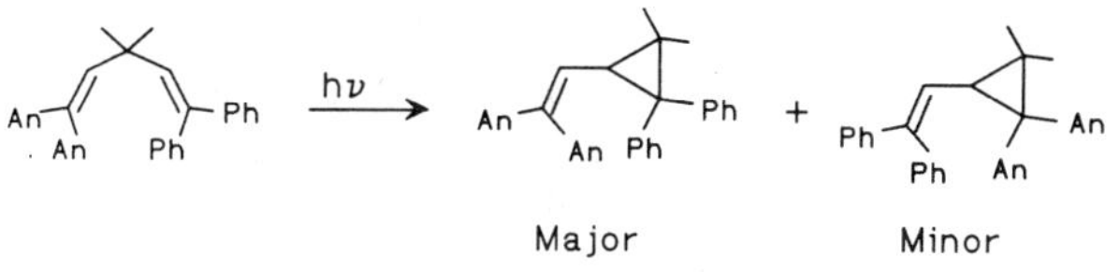

Major   Minor

$$(14.8)$$

This regioselectivity is quite general and one rationale is that the carbinyl carbons of the cyclopropyldicarbinyl biradical species (i.e., Diradical I) are electron-rich and thus dissipation of electron density at the carbinyl carbons with the better electron donors is favored.

Another factor encountered[10] is multiplicity. As illustrated in Scheme 14.2, an example was found where the regioselectivity of "unzipping" the three-membered ring of the cyclopropyldicarbinyl biradical depends on the multiplicity of that biradical. The observation (see Scheme 14.2) is that ring-opening regioselectivity (process "s") to afford odd-electron density at a carbomethoxy-substituted center is favored by the singlet, while the triplet prefers the process "t" in which the odd-electron density avoids the carbomethoxy-substituted center and affords a biradical with odd-electron density at the benzhydryl center. For reasons discussed below, the "s" ring opening is termed a "Small K reaction" and the "t" process is termed a "Large K process". K is a common symbol for the quantum mechanical "exchange integral" and 2K is the energetic separation between S_1 and T_1. The energetics of the two types of processes are depicted in Scheme 14.3. It is clear that a singlet will prefer the process on the left ("Small K"), and a triplet will prefer the situation on the right (a "Large K" reaction). This situation[10] has been noted to be more general than just the example in Scheme 14.2. We are left mainly with the matter of categorizing which processes will be of the "Small K" and "Large K" variety. This discussion belongs elsewhere,[10] but for reacting biradicals with strongly polar substituents, one has a "Small K" and for homopolar biradicaloids we have a "Large K" situation. In summary, singlets prefer "Small K" processes and triplets prefer "Large K" reactions.

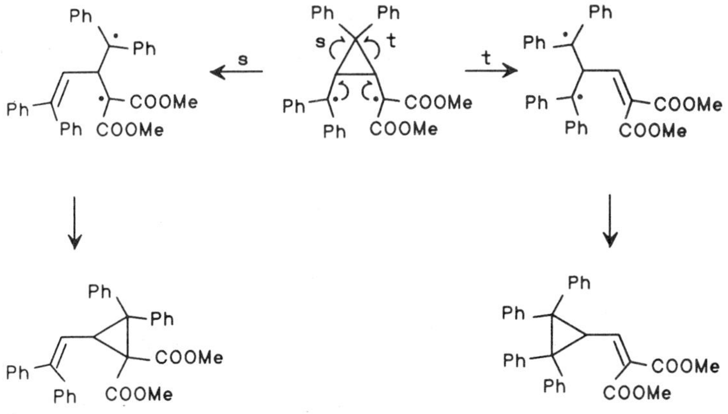

SCHEME 14.2   Multiplicity control of regioselectivity.

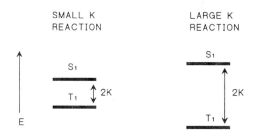

SCHEME 14.3　Large-K vs. small-K control of excited-state reactivity.

Another factor in regioselectivity is retention of aromaticity. With a choice of vinyl-vinyl vs. benzo-vinyl bridging generally, as found in the case of benzobarrelene (Equation 14.9), loss of aromaticity is avoided. In the case of a benzonorbornadiene, where disruption of aromaticity cannot be avoided, it occurs.

$$\xrightarrow[\text{Sens}]{h\nu}$$

(14.9)

Where there is a ring substituent, as in the case of the cyano and methoxy benzonorbornadienes,[12] there is a further question about regioselectivity. Note Equations 14.10 and 14.11. One interpretation of the regioselectivity is that the preferred bridging leads to that cyclopropyldicarbinyl biradical with the odd electron and electron density avoiding methoxy-bearing carbons and seeking out cyano-bearing carbons.

$$\xrightarrow[\text{Sens.}]{h\nu}$$

(14.10)

$$\xrightarrow[\text{Sens.}]{h\nu}$$

(14.11)

A particularly interesting example is that of the cyanodibenzobarrelene labeled at the vinyl group with deuterium atoms.[13] Here, there were four *a priori* reaction mechanisms, of which two are depicted in Scheme 14.4. On the front face of the molecule, benzo-vinyl bridging is possible to either benzo ring and, on the rear face, there are the same two options. Only the last two reaction mechanisms are depicted in Scheme 14.4 since the first two do not permit odd-electron stabilization by the cyano group at any stage of the reaction.

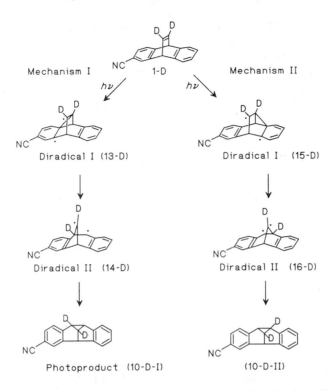

SCHEME 14.4 Two of the four possible mechanisms for the Di-π-Methane Rearrangement of cyanodibenzobarrelene.

In contrast, Mechanisms I and II do permit odd-electron interaction with the cyano group, but at different stages of the reaction. Mechanism I enjoys stabilization by cyano at the Biradical I stage, while Mechanism II has delocalization by cyano only when Biradical II is reached. Experimentally, it was found[13] that reaction Mechanism I is utilized, thus signifying that the reaction regioselectivity is controlled in formation of the lowest-energy Biradical I. Further, this confirms the importance of the benzo π-electron interaction with the vinyl group in forming Biradical I.

In support of this view, *ab initio* theoretical calculations were carried out, and these are summarized in Figure 14.1. Here, energy in Hartrees (627.5 kcal/mol per Hartree) is plotted vs. a reaction coordinate where point 0 on the abscissa (i.e., reaction coordinate) corresponds to barrelene, 1 has

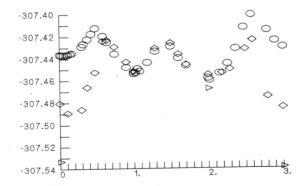

FIGURE 14.1 The energy hypersurface for the conversion of triplet barrelene to semibullvalene. The abscissa is 0 for barrelene, 1 for Biradical I, 2 for semi-bullvalene. The ellipses define the triplet surface, the diamonds refer to S_0 energies with triplet geometries, and the four triangles are optimized S_0 points.

Biradical I geometry, 2 has Biradical II geometry, and 3 corresponds to semibullvalene. Thus, there are four triplet minima. Hence, both Biradicals I and II are intermediates, and a 15-kcal mol^{-1} barrier separates Biradical I from Biradical II. Also included are energies of the S_0 configuration constrained to have the same geometry as corresponding geometry-optimized triplet species. The energies of several relaxed ground-state singlets are included as well. The corresponding mechanism in qualitative Lewis structure terms is given in Scheme 14.5.[1c]

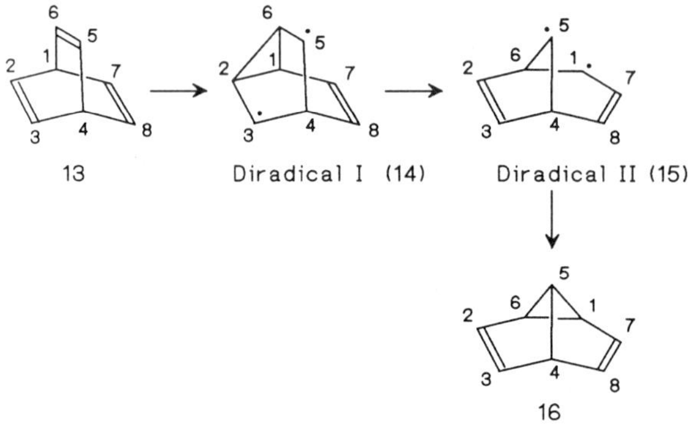

SCHEME 14.5   The barrelene to semibullvalene rearrangement mechanism.[1c]

It is seen that the triplet surface comes close to the geometry-unrelaxed S_0 surface near Biradicals I and II, and that the relaxed ground-state points are also close. Remarkably, these *ab initio* computations are qualitatively similar to much more approximate extended Hückel calculations carried out decades earlier.[1c,1d] It was noted there that a near approach of the excited triplet surface to S_0 led to an enhanced probability of intersystem crossing from T_1 to S_0. This general idea of HOMO-LUMO crossings controlling conversion of excited to ground states was made more generally by the present author in 1966.[14] Points where excited- and ground-state surfaces come together have been termed "funnels" by Michl[15] and "bifunnels" by the present author. The idea of funnel-shaped surfaces goes back to early theoretical studies by Teller.[16] Two further relevant points are that such bifunnels lead not only to reaction but also to radiationless decay to reactant, as has been noted in studies correlating molecular twisting of π-bonds to excited singlet lifetimes.[17] Second, such bifunnels are sometimes canted with the result of favoring a decay either toward reactant or toward product.[18]

14.5 Reaction Variations

It is not within the scope of this review to cover reactions that differ from the basic process, and these will just be mentioned. As noted above, the cyclohexadienone and cyclohexenone rearrangements differ in originating in an n-π* excited state. Beyond this, the dienone rearrangements proceed by relatively long-lived oxy-allyl Zwitterionic intermediates [2, references in the review in 19c].

In addition, a "Bora-Di-π-Methane", in which the central carbon has been replaced by boron, has been reported by Schuster.[20] However, the largest amount of work on Di-π-Methane Rearrangement variations has been on the "Oxa-Di-π-Methane Rearrangement" and its "Aza-Di-π-Methane" relative. Equations 14.12 and 14.13 (References 22a and 22b, respectively), provide typical examples.

The Oxa-Di-π-Methane Rearrangement was originally uncovered by Swenton.[23] The reaction mechanism is due to Givens,[24] who recognized the reaction as being a di-π-methane process. Finally, the name was given to the reaction by Dauben.[25] It is known that the reaction proceeds via the triplet excited state, and otherwise shows structural parallel with the ordinary Di-π-Methane Rearrangement. The reaction has been reviewed by Schuster[26a] and by Houk.[26b]

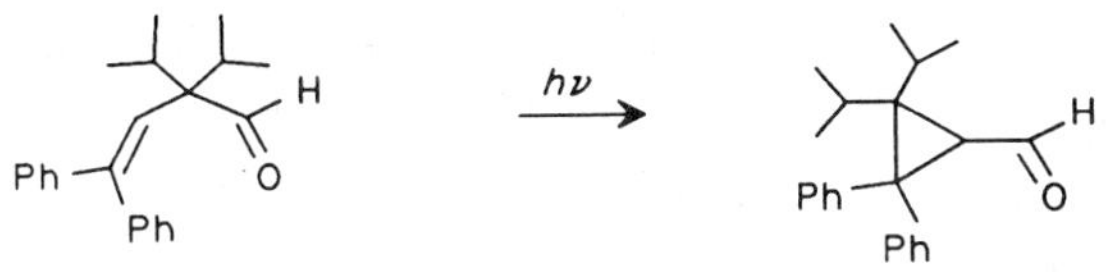

$$(14.12)$$

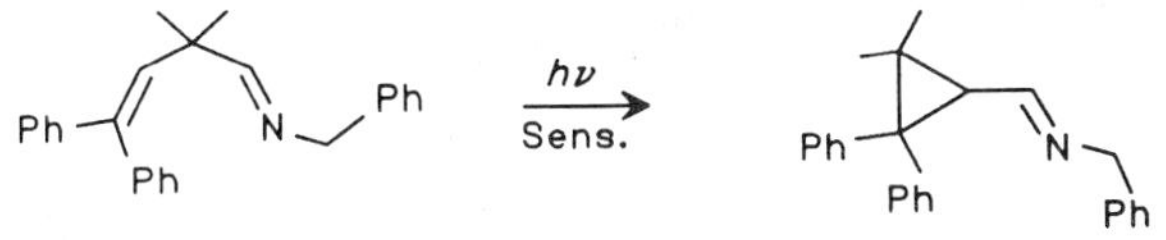

$$(14.13)$$

The Aza-Di-π-Methane Rearrangement is due to the work of Armesto and Horspool.[22,27] One of the main limitations of the reaction has been shown by these researchers to be the requirement of a modestly high ionization potential of the lone electron pair on nitrogen in order to suppress competing electron-transfer processes.

14.6 Dependence of Reaction Rate on Structure

In extensive studies, the rates of the singlet variation of the Di-π-Methane Rearrangement has been studied using single-photon counting.[28] In general, a tendency was found for the reaction rates to increase with increasing electron-withdrawing ability of substituents at the cyclopropyldicarbinyl biradical centers of Biradical I. Conversely, with diminished stabilization at these centers, the reaction rates decreased. Bridging to aryl groups in the aryl-vinyl variation led to slower rate constants, thus indicating that aromatic stabilization energy is still a factor in excited-state species.

For triplet rearrangements, less is known. However, relative rates can be deduced from product rates. For example, bridging between a vinyl and the α-position of a naphthyl group is preferred over bridging to a β-site.[29] The tendency of a vinyl moiety to bridge to an π-site increases in the order of (1) benzo and β-naphthyl to vinyl, (2) vinyl-vinyl, (3) α-naphthyl to vinyl.[30]

14.7 Concluding Comments

While the Di-π-Methane Rearrangement is one of the more exhaustively investigated photochemical rearrangements, it is only one of many intriguing photochemical reactions of interest. The number of known photochemical reactions is very small compared with their ground-state counterparts. This need not be the case. Electronically excited-state species follow reaction mechanisms surprisingly reminiscent of ground-state processes. For the more complex, and thus most interesting, photochemical reactions, much of the molecular transformation takes place while the molecule is still electronically excited; and radiationless decay to the ground-state surface, and on to photoproduct, occurs only rather late after considerable molecular reorganization has taken place. In recent years, much emphasis has been placed on the role of radiationless decay via bifunnels. While such decay is, by definition, a necessity, it generally is not the controlling factor in most cases as seen for the Di-π-Methane Rearrangement. This means that reactivity of electronically excited molecules can often be assessed in terms of electron densities, bond orders, and energy dependence on molecular motion. With modern quantum mechanical programming available and practical, one can most often assess these factors. In addition, one can often write explicit electronically excited-state structures, and thus a reaction mechanism using Lewis structure representation. These views are the same expressed by the author many years ago.[2b,31]

Acknowledgment

The research studies upon which this chapter is based were supported by the National Science Foundation, and this support is gratefully acknowledged.

References

1. (a) Zimmerman, H. E. and Grunewald, G. L., *J. Am. Chem. Soc.,* 88, 183, 1966; (b) Zimmerman, H. E., Binkley, R. W., Givens, R. S., and Sherwin, M. A., *J. Am. Chem. Soc.,* 89, 3932, 1967; (c) Zimmerman, H. E., Binkley, R. W., Givens, R. S., Grunewald, G. L., and Sherwin, M. A., *J. Am. Chem. Soc.,* 91, 3316, 1969.

2. (a) Zimmerman, H. E. and Schuster, D. I., *J. Am. Chem. Soc.,* 83, 4486, 1961; (b) Zimmerman, H. E. and Schuster, D. I., *J. Am. Chem. Soc.,* 84, 4527, 1962.

3. (a) Zimmerman, H. E. and Wilson, J. W., *J. Am. Chem. Soc.,* 86, 4036, 1964; (b) Zimmerman, H. E. and Hancock, K. G., *J. Am. Chem. Soc.,* 90, 3749, 1968.

4. Thus, the natural product Santonin had been known to undergo photochemical rearrangement in early Italian studies. However, it was the elegant work of Barton (Barton, D. H. R., DeMayo, P., and Shafiq, M., *Proc. Chem. Soc., London,* 205, 1958) which first clarified the structures of reactant and product.

5. (a) Zimmerman, H. E. and Pratt, A. C., *J. Am. Chem. Soc.,* 92, 1409, 1970; (b) Zimmerman, H. E. and Pratt, A. C., *J. Am. Chem. Soc.,* 92, 6267, 1970; (c) Zimmerman, H. E., Kamm, K. S., and Werthemann, D. P., *J. Am. Chem. Soc.,* 97, 3718, 1975; (d) Zimmerman, H. E., in *Rearrangements in Ground and Excited States,* Vol. 3, DeMayo, P., Ed., Academic Press, New York, 1980; (e) Zimmerman, H. E. and Schissel, D. N., *J. Org. Chem.,* 51, 196, 1986.

6. Zimmerman, H. E. and Mariano, P. S., *J. Am. Chem. Soc.,* 91, 1718, 1969.

7. Zimmerman, H. E., Armesto, D., Amezua, M. G., Gannett, T. P., and Johnson, R. P., *J. Am. Chem. Soc.,* 101, 6367, 1979.

8. Zimmerman, H. E. Givens R. S. and Pagni, R. *J. Am. Chem. Soc.,* 90, 6096, 1968.

9. (a) Zimmerman, H. E. and Pratt, A. C., *J. Am. Chem. Soc.,* 92, 6259, 1970; (b) Zimmerman, H. E. and Gruenbaum, W. T., *J. Org. Chem.,* 43, 1997, 1978; (c) Zimmerman, H. E. and Welter, T. R., *J. Am. Chem. Soc.,* 100, 4131, 1978.

10. Zimmerman, H. E. and Factor, R. E., *Tetrahedron,* 37, (Suppl.) 1, 1981.

11. Zimmerman, H. E., Givens, R. S., and Pagni, R. M., *J. Am. Chem. Soc.,* 90, 4191, 1968.

12. (a) Paquette, L. A., Cottrell, D. M., Snow, R. A., Gifkins, K. B., and Clardy, J., *J. Am. Chem. Soc.,* 97, 3275, 1975; (b) Paquette, L. A., Cottrell, D. M., and Snow, R. A., *J. Am. Chem. Soc.,* 99, 3723, 1977.

13. Zimmerman, H. E., Sulzbach, H. M., and Tollefson, M. B., *J. Am. Chem. Soc.,* 115, 6548, 1993.

14. Zimmerman, H. E., *J. Am. Chem. Soc.,* 88, 1566, 1966.

15. Michl, J., *Mol. Photochem.,* 243, 1972; 257, 1973; 287, 1973.

16. Teller, E., *J. Phys. Chem.,* 41, 109, 1937.

17. Zimmerman, H. E., Kamm, K. S., and Werthemann, D. P., *J. Am. Chem. Soc.,* 97, 3718, 1975.

18. Zimmerman, H. E. and Factor, R. E., *J. Am. Chem. Soc.,* 102, 3538, 1980.

19. (a) Zimmerman, H. E. and Pasteris, R. J., *J. Org. Chem.,* 45, 4864, 1980; (b) Zimmerman, H. E. and Lynch, D. C., *J. Am. Chem. Soc.,* 107, 7745, 1985; (c) Schaffner, K., in *Rearrangements in Ground and Excited States,* Vol. 3, DeMayo, P., Ed., Academic Press, New York, 1980.

20. (a) Schuster, G. B., *J. Am. Chem. Soc.,* 107, 7745, 1985; (b) Wilkey, J. D. and Schuster, G. B., *J. Am. Chem. Soc.,* 113, 2149, 1991.

21. Zimmerman, H. E. and Cassel, J. M., *J. Org. Chem.,* 54, 3800, 1989.

22. For leading references see: (a) Armesto, D., Horspool, W. M., Apoita, M., Gallego, M. G., and Ramos, A., *J. Chem. Soc., Perkin Trans. I,* 2035, 1989; (b) Armesto, D., Gallego, M. G., and Horspool, W. M., *Tetrahedron Lett.,* 2475, 1990.

23. Swenton, J. S., *J. Chem. Ed.,* 46, 217, 1969.

24. Givens, R. S. and Oettle, W. F., *J. Chem. Soc., Chem. Commun.,* 1164, 1969.

25. Dauben, W. G., Kellogg, M. S., Seeman, J. I., and Spitzer, W. A., *J. Am. Chem. Soc.,* 92 1786, 1970.

26. Note (a) Schuster, D. I., in *Rearrangements in Ground and Excited States,* Vol. 3, DeMayo, P., Ed., Academic Press, New York, 1980; (b) Houk, K. N., *Chem. Rev.,* 76, 1, 1976.

27. Armesto, D., Horspool, W. M., Mancheno, M. J., and Ortiz, M. J., *J. Chem. Soc., Perkin Trans. I*, 2325, 1992.

28. (a) Zimmerman, H. E., Werthemann, D. P., and Kamm, K. S., *J. Am. Chem. Soc.*, 95, 5094, 1973; (b) Zimmerman, H. E. and Welter, T. R., *J. Am. Chem. Soc.*, 100, 4131, 1978; (c) Zimmerman, H. E., Steinmetz, M. G., and Kreil, C. L., *J. Am. Chem. Soc.*, 100, 4146, 1978; (d) Zimmerman, H. E. and Swafford, R. L., *J. Org. Chem.*, 48, 3069, 1984.

29. Zimmerman, H. E. and Bender, C. O., *J. Am. Chem. Soc.*, 92, 4366, 1970.

30. Zimmerman, H. E. and Viriot-Villaume, M.-L., *J. Am. Chem. Soc.*, 95, 1274, 1973.

31. (a) Zimmerman, H. E., *Abstr. 17th Natl. Org. Symp. Am. Chem. Soc.*, Bloomington, IN, 1961, 31–41; (b) Zimmerman, H. E., *Tetrahedron*, 19 (Suppl. 2), 393, 1963; (c) Zimmerman, H. E., in *Advances in Photochemistry*, Vol. 1, Noyes, A., Jr., Hammond, G. S., and Pitts, J. N., Jr., Interscience, New York, 1963, 183.

15

Photorearrangements of Benzobarrelenes and Related Analogues

Chen-Chen Liao
National Tsing Hua University

Paw-Hwa Yang
*Development Center
for Biotechnology*

15.1 Introduction

Bicyclo[2.2.2]octa-2,5,7-triene **1**, named barrelene by Zimmerman's group because of the barrel-shaped molecular-orbital array of the π-system,[1] shows interesting photochemical behavior: direct irradiation in methylcyclohexane yields cyclooctatetraene **2**, whereas sensitized reaction in isopentane in the presence of acetone affords semibullvalene **3**.[2] The mechanisms of these transformations were proposed as depicted in Scheme 15.1.[2c] The first excited singlet state of barrelene undergoes an initial [2+2]-cycloaddition to give a quadricyclane-like intermediate **4**, followed by isomerization to **2**. In contrast, the triplet state of barrelene, generated by sensitization, yields a biradical-like intermediate **5**, followed by rearrangement to **6** and then ring closure to give **3**; this rearrangement is commonly called the Di-π-Methane Rearrangement.[3]

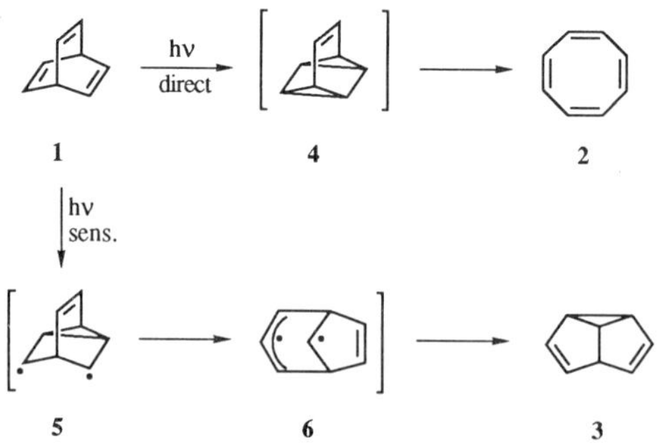

SCHEME 15.1

The effects of substituents on the photochemical behavior of barrelenes have attracted the attention of many chemists. In this chapter, we summarize the photochemistry of benzobarrelenes and related analogues in which one double bond of barrelene is incorporated into an aromatic ring.

15.2 Reaction Mechanism

Photorearrangements to Cyclooctatetraenes

Direct irradiation of parent benzobarrelene **7** leads to exclusive formation of benzocyclooctatetraene **8**.[4] This transformation occurs via an excited singlet state of **7** because triplet-sensitized reaction yields a different product, benzosemibullvalene **26** *(vide infra)*.[4] There are *a priori* two possible modes of the initial [2+2]-cycloadditions—benzo-vinyl and vinyl-vinyl cycloadditions. Zimmerman et al. elegantly used deuterated benzobarrelene **7a** to distinguish these possibilities; the results (Scheme 15.2) indicate that an initial benzo-vinyl cycloaddition prevails.[4c]

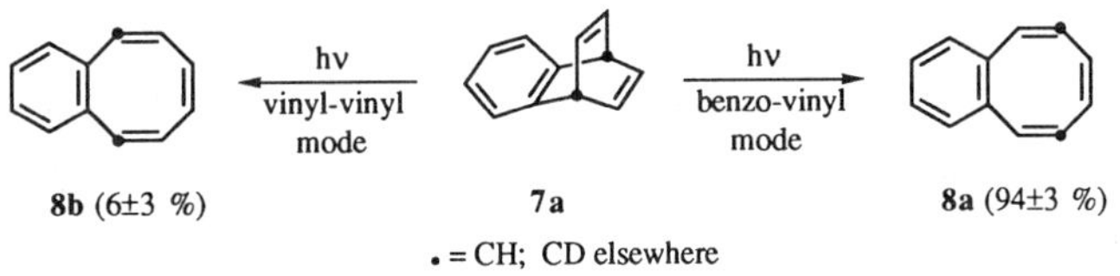

8b (6±3 %)   **7a**   **8a** (94±3 %)

• = CH; CD elsewhere

SCHEME 15.2

Introduction of substituents on the olefinic moiety and the benzene unit or changing the benzene ring into other aromatic systems may affect the efficiency of intersystem crossing, the reactivity, and modes of initial cycloadditions of an excited singlet state of ring-fused aromatic barrelenes. There are *a priori* three possible modes of initial cycloadditions (Scheme 15.3) for a barrelene having varied substituents on the three double bonds. Deuterated compounds are generally employed to distinguish these options. Whereas 5-cyanobenzobarrelene **9**[5] undergoes both benzo-cyanovinyl (route I) and vinyl-cyanovinyl (route IIIa) cycloadditions to give **10** (17%) and **11** (54%), respectively (Equation 15.1), 5,6-dimethoxycarbonylbenzobarrelene **13**[6] proceeds exclusively via route IIIa to yield **14** (Equation 15.2). Only the aromatic-vinyldiester cycloaddition (route I) occurs to produce **16** from **15**[7] in which, due to a methoxy substituent on the benzene moiety, a major difference occurs compared to **13**. The methoxy group also enhances the efficiency of intersystem crossing as the products (semibullvalenes) of the triplet-mediated Di-π-Methane Rearrangements also form upon direct irradiation of **15**.[7] In the case of **17**, with two methoxy groups on the benzene ring, [2+2]-cycloaddition is unable to compete with intersystem crossing because the direct excitation of **17** yields no cyclooctatetraene derivative.[7] Tetrafluorobenzobarrelene **18**[8] (Equation 15.3) and 5-cyanobenzobarrelene **9**[5] (Equation 15.1) also afford semibullvalenes and cyclooctatetraenes (17%:56% and 29%:81%, respectively) upon direct irradiation.

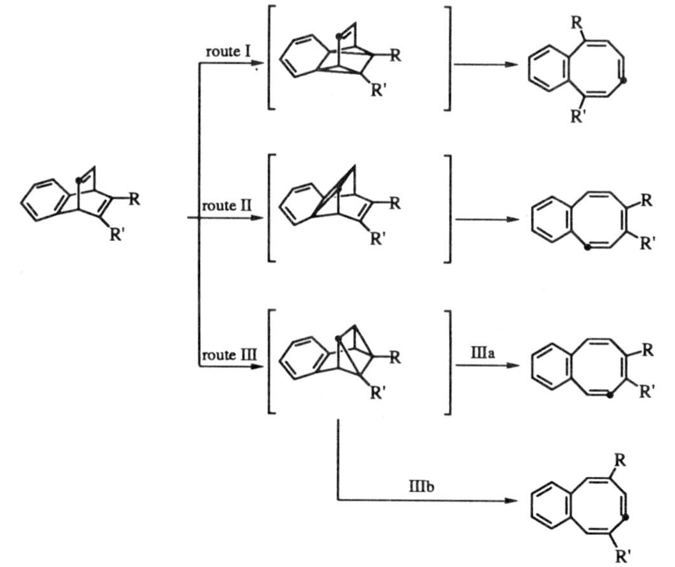

SCHEME 15.3

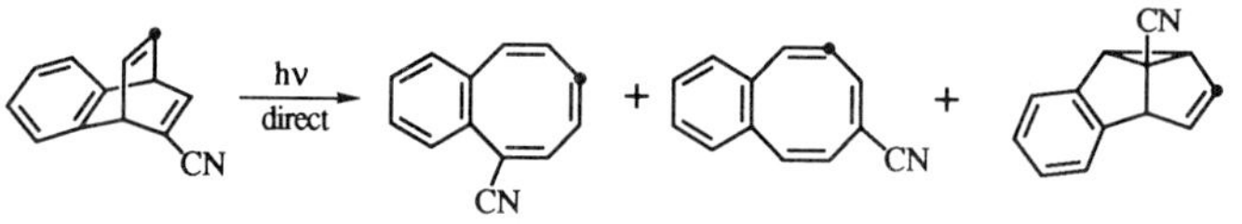

9 10 11 12 (15.1)

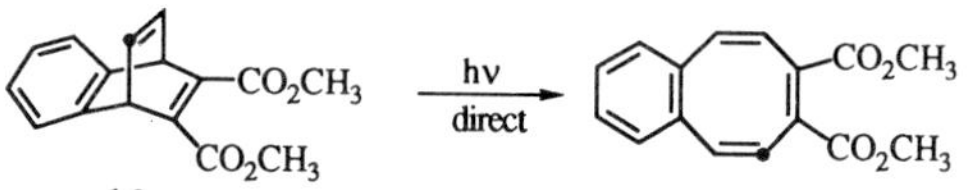

13 14

a • = CH b • = CD

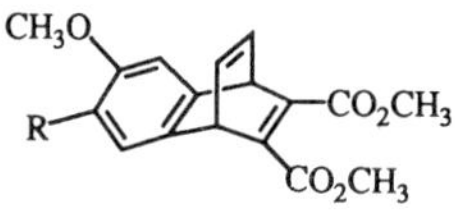

15 R = H
17 R = OCH_3

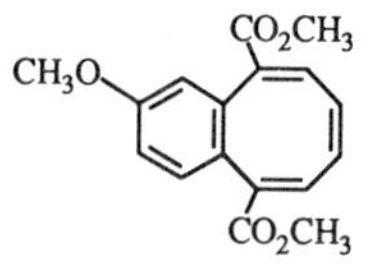

16 (15.2)

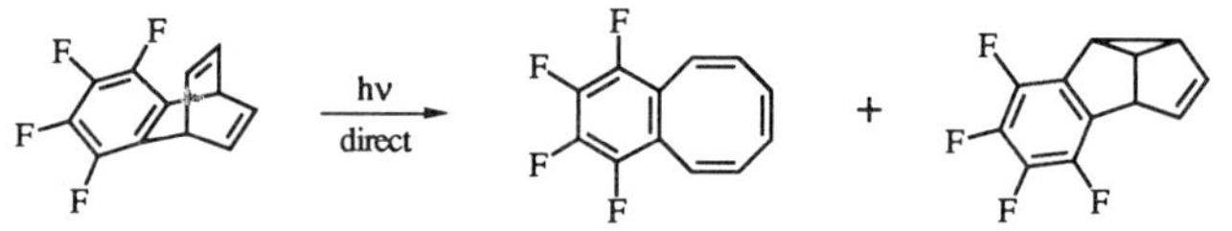

18 19 (56%) 20 (17%) (15.3)

Fused-ring polycyclic-aromatic and heteroaromatic barrelenes generally have greater yields of intersystem crossing than the parent benzobarrelene 7; hence, the singlet-mediated [2+2]-cycloadditions leading to cyclooctatetraenes become unfavorable processes, and semibullvalenes are obtained as exclusive products of direct irradiation *(vide infra)*. In contrast to 2′,3′-naphthobarrelene 21[9] which produces no cyclooctatetraene derivative upon direct excitation, 1′,2′-naphthobarrelene 22 undergoes an initial [2+2]-cycloaddition of naphtho-vinyl bonding (similar to route I in Scheme 15.3) to give naphthocyclooctatetraene 23. 5-Cyano-2′,3′-naphthobarrelene 24, although still showing efficient intersystem crossing, proceeds through initial vinyl-cyanovinyl bonding (route IIIa in Scheme 15.3) to afford 25 (4%);[10] the mechanisms were elucidated by the results of deuterium-labeling experiments Equations 15.4 and 15.5. The reactions appear to occur via pathways of minimum energy and not at the sites at which excitation energy is localized.

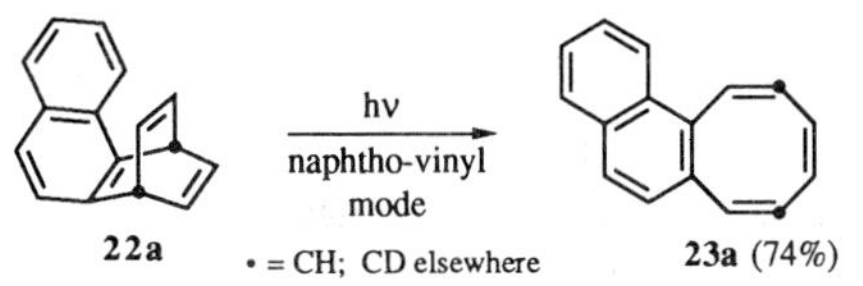

$$(15.4)$$

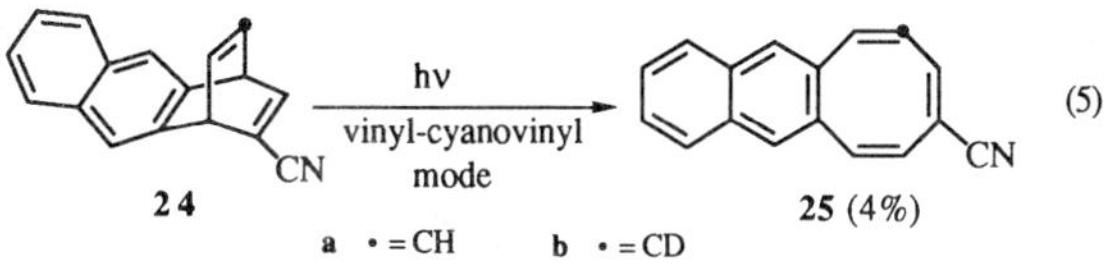

$$(15.5)$$

Di-π-Methane Photorearrangements

In contrast to direct irradiation, triplet-sensitized reaction of benzobarrelene **7** proceeds via a Di-π-Methane Rearrangement to benzosemibullvalene **26**[5] which, in principle, could be obtained by two reaction pathways — benzo-vinyl and vinyl-vinyl bridging (Scheme 15.4). Zimmerman and co-workers[6] used the deuterium-labeled compound **7a** to make the distinction; irradiation of **7a** in the presence of acetophenone produced benzosemibullvalenes **26a-c** in the ratio >90:<5:<5, and then showing that vinyl-vinyl bridging was the major process.

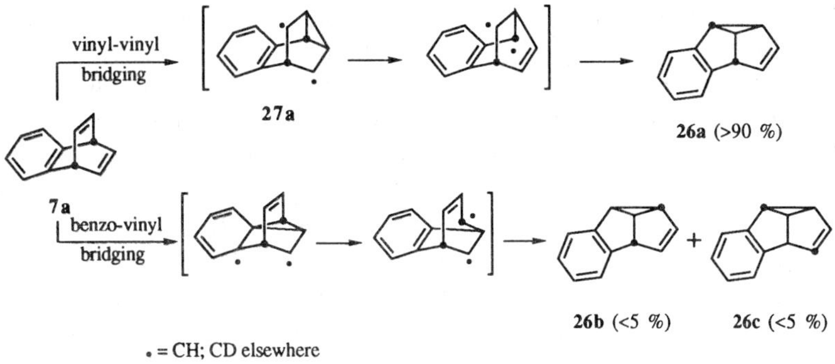

SCHEME 15.4

The biradical mechanism in Scheme 15.4 has been substantiated by independent generation of cyclopropyldicarbinyl biradical **27** from the azo compound **28** (Equation 15.6). The excited triplet state of **28** leads predominantly to benzosemibullvalene **26** whereas an excited singlet state furnishes benzobarrelene preferentially.[11]

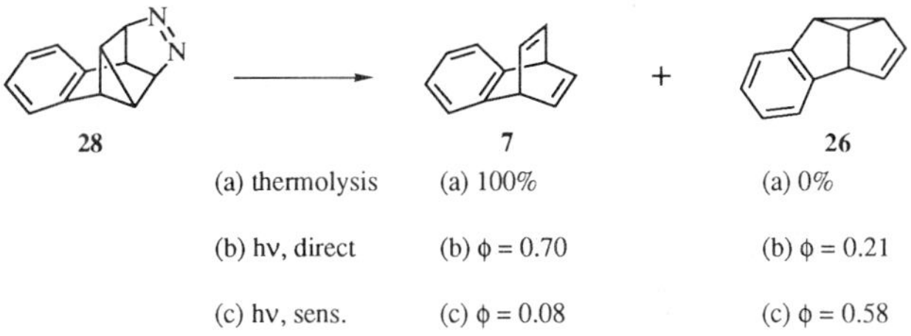

$$(15.6)$$

The possible reaction pathways of Di-π-Methane Rearrangement *a priori* increase with decreased symmetry of substituted benzobarrelenes; nevertheless, the reaction shows generally great site- and regio-selectivities, especially for substituents that are polar. There are three options — vinyl-vinyldiester, benzo-vinyl, and benzo-vinyldiester bridgings for 5,6-dimethoxycarbonylbenzobarrelene **13a**; however, the benzophenone-sensitized reaction affords **29** as the only product via vinyl-vinyldiester bridging.[6c] Similarly, **17** gives **32**, and **15** furnishes **30** and **31**. The Di-π-Methane Rearrangement of 5-cyanobenzobarrelene (**9c, d**), initiated either by sensitization with acetophenone or by direct irradiation, furnishes only benzosemibullvalene (**12c, d**) (Equation 15.7);[5] deuterium-labeling experiments reveal that the reaction proceeds via vinyl-cyanovinyl bridging between the less substituted carbon atoms[5b] because of the favorable stabilization by the cyano group of a biradical intermediate species.

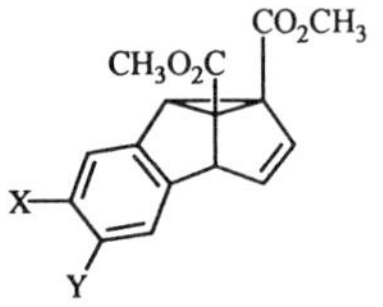

29 X = Y = H
30 X = H, Y = OCH$_3$
31 X = OCH$_3$, Y = H
32 X = Y = OCH$_3$

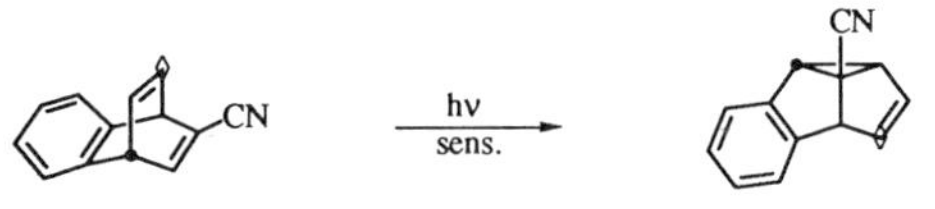

9c   • = H   ◊ = D
9d   • = D   ◊ = H

12c   • = H   ◊ = D
12d   • = D   ◊ = H

(15.7)

When the benzene ring is replaced by other polycyclic aromatics, the lowest triplet energies of the ring-fused aromatic barrelenes are expected to localize on the aromatic moieties; thus, the reaction modes are of interest from a mechanistic point of view. The results indicate that the modes of rearrangements are not dictated by the site at which the lowest triplet energy is localized, but follow the pathway of minimum energy. Both direct irradiation and benzophenone-sensitized reactions of 2′,3′-naphthobarrelene **21** afford 2′,3′-naphthosemibullvalene **33** as the sole product with equal quantum efficiency (Equation 15.8); the Di-π-Methane Rearrangement proceeds via its lowest triplet state with initial vinyl-vinyl bridging.[9] 5-Cyano-2′,3′-naphthobarrelene **24** undergoes the Di-π-Methane Rearrangement in the same manner as 5-cyanobenzobarrelene[10] (see Equation 15.9). In contrast to **21**, direct irradiation of 1′,2′-naphthobarrelene **22** gives 1′,2′-naphthocyclooctatetraene **23** and naphthosemibullvalenes (**34a** and **34b**) and (**35a** and **35b**), whereas benzophenone-sensitized reaction yields only the latter products **34a** and **35a**.[9] Deuterium-labeling experiments reveal that the sensitized Di-π-Methane Rearrangement proceeds exclusively with initial α-napho-vinyl bridging, whereas those with direct excitation occur through naphtho-vinyl (58%) and vinyl-vinyl (42%) bridgings (Scheme 15.5). The naphtho-vinyl bridging occurs via its lowest triplet state and high regioselectivity of exclusive α-bonding, whereas the vinyl-vinyl bridging is effected presumably from the S$_1$ state; the T$_2$ state cannot be excluded, and the α- and β-bonding paths are of equal facility.[9]

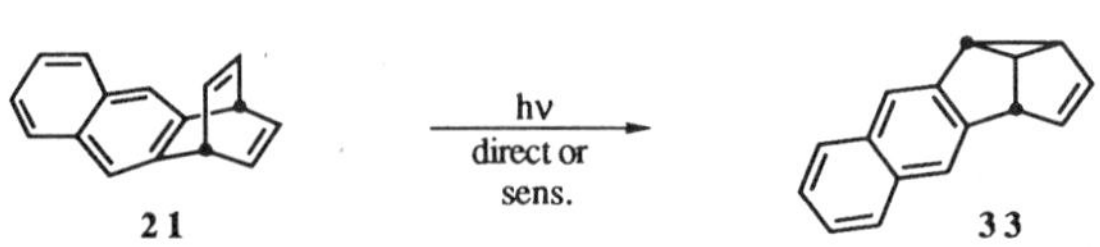

21

33

(15.8)

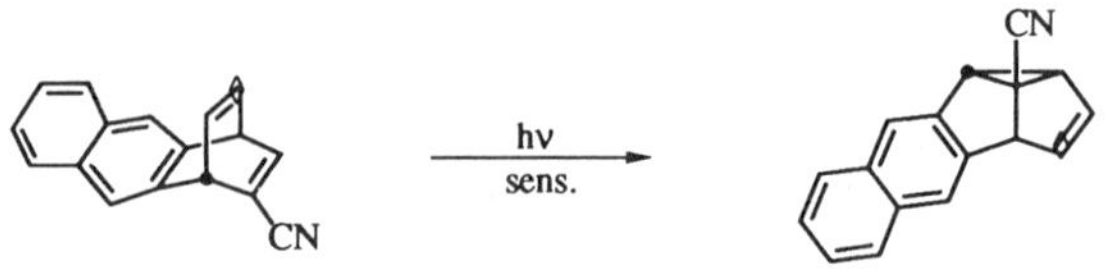

24c   • = H   ◊ = D
24d   • = D   ◊ = H

36c   • = H   ◊ = D
36d   • = D   ◊ = H

(15.9)

Scheme 5

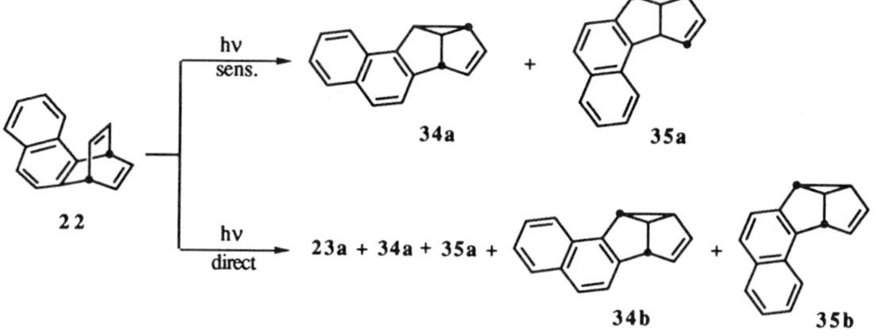

SCHEME 15.5

Barrelene **37** produces semibullvalene **38** (Equation 15.10) both by selectively exciting the naphtho moiety and by sensitization with benzophenone, although **37** emits naphthalene-like fluorescence and phosphorescence.[12] No reaction takes place when a mixture of benzobarrelene **7**, naphthalene and benzophenone is irradiated.[12] Anthrabarrelene **39**, which gives anthracene-like emissions, furnishes anthrasemibullvalene **40** with a quantum yield 0.25 upon direct irradiation (Equation 15.11); the rearrangement proceeds with initial vinyl-vinyl bridging, indicated by deuterium labeling.[13] The sensitizers fluorenone, acetophenone, and biacetyl do not bring about the Di-π-Methane Rearrangement. Thus, the Di-π-Methane Rearrangement of **39** occurs presumably via its T_2 or S_1 state.

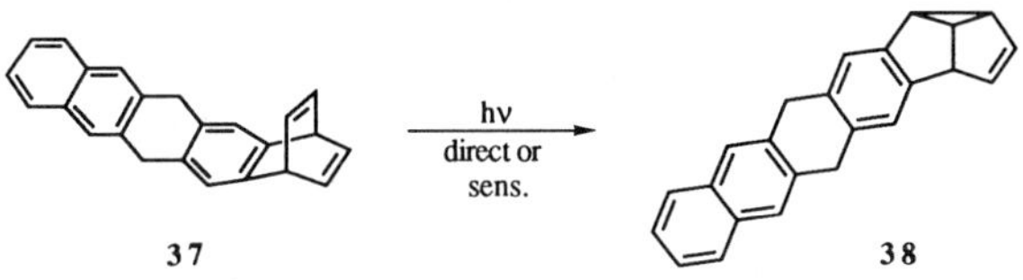

$$(15.10)$$

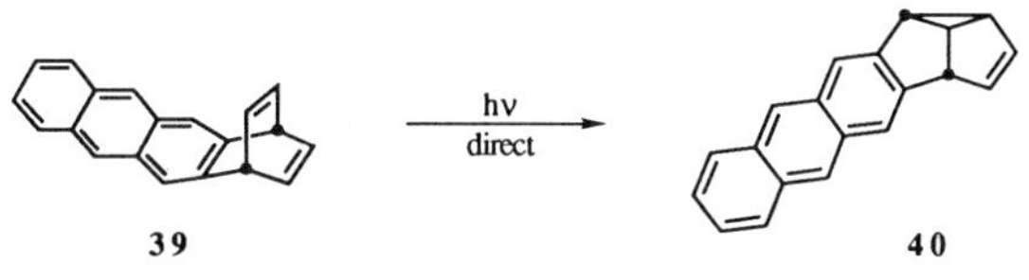

$$(15.11)$$

The photochemistry of fused-ring heteroaromatic barrelenes has also been investigated. The reaction mechanisms are expected to be even more complicated than those of the corresponding hydrocarbons because of the additional n,π^* states. Pyrazinobarrelene **41** affords products **42–44** by Di-π-Methane Rearrangement upon direct irradiation in benzene (Equation 15.12); no cyclooctatetraene derivative is found. The major product **42** is formed by the vinyl-vinyl bridging, whereas minor products **43** and **44** are derived from pyrazino-vinyl bridging.[14] In the case of dicyanopyrazinobarrelenes **45** and **48**, only Di-π-Methane Rearrangements by pyrazino-vinyl bridging take place. For **48** (Equation 15.13) that the a-a′ bridging is preferred to b-b′ bridging is presumably due to steric hindrance in the latter case during the initial bonding between the pyrazine and alkene moieties.[14]

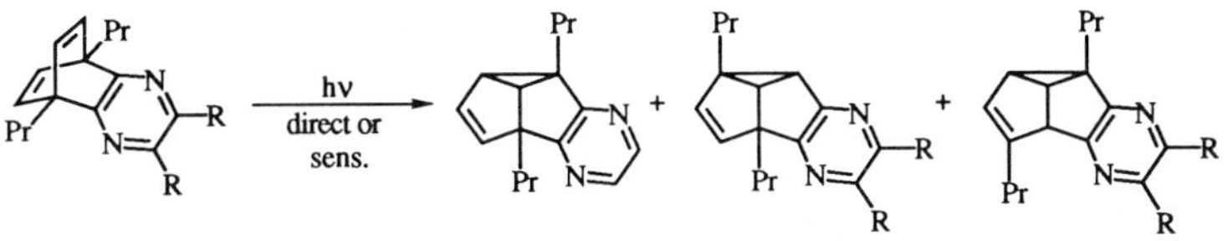

41 R = H
45 R = CN

42 R = H (53%)

43 R = H (25%)
46 R = CN (57%)

44 R = H (22%)
47 R = CN (43%)

$$(15.12)$$

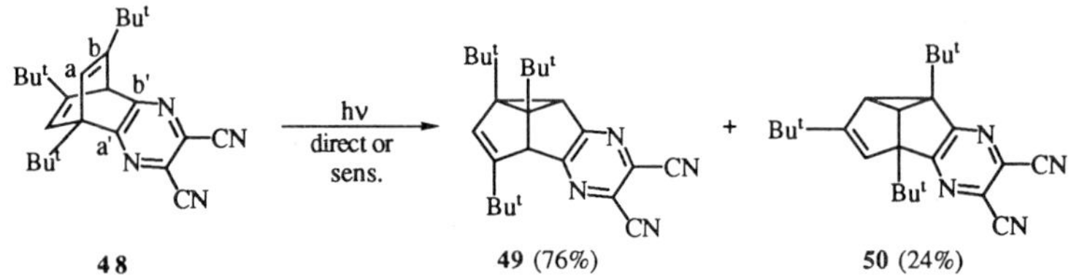

$$(15.13)$$

In contrast to 2',3'-naphthobarrelene 21, quinoxalinobarrelenes 51 and 55 undergo Di-π-Methane Rearrangements via both vinyl-vinyl and quinoxalino-vinyl bridgings, although the former is still predominant (Equations 15.14 and 15.15). The preferential a-a and a-a' bondings in 55 are due to steric effects. The triplet states of these fused-ring heteroaromatic barrelenes are involved in these Di-π-Methane Rearrangements because the same reactions can be sensitized with acetophenone and benzophenone; more rigorous evidence can be obtained from quantum yield determinations and quenching experiments. The nature of the triplet states, either n,π* or π,π*, responsible for these rearrangements is unclear. The facile pyrazino-vinyl bridging in 45 and 48 and quinoxalino-vinyl bridging in 51 and 55, compared with benzobarrelene 7 and naphthobarrelene 21, is explained in terms of the ability of nitrogen atoms to stabilize the biradical-like species along the reaction coordinates.[14,15]

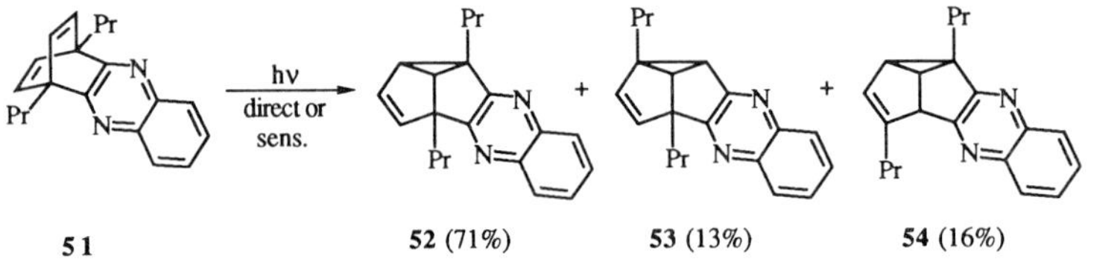

$$(15.14)$$

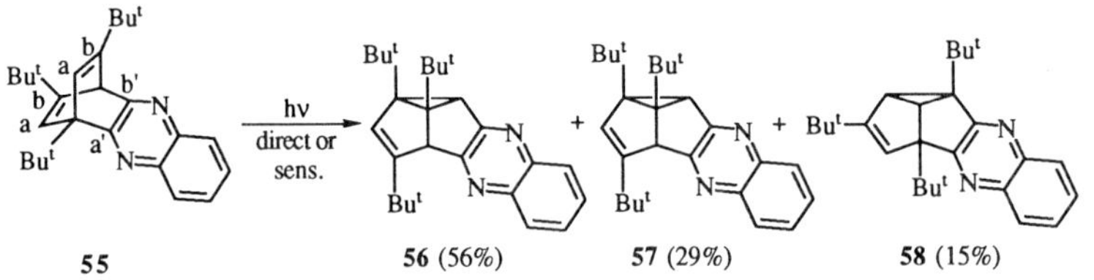

$$(15.15)$$

In contrast to anthrabarrelene 39, (benzo[g]quinoxalino)barrelene 59 and 60 do not undergo Di-π-Methane Rearrangement upon prolonged irradiation; starting material is recovered almost quantitatively, although the irradiated mixtures darken.[15] The reasons that these reactions are ineffective may be that the lowest triplet states of 59 and 60 have insufficient energy to initiate Di-π-Methane Rearrangement and that the excited singlet or upper triplet states are too transitory for chemical transformations because nitrogen atoms enhance the rate of radiationless processes. (See Equations 15.16 and 15.17.)

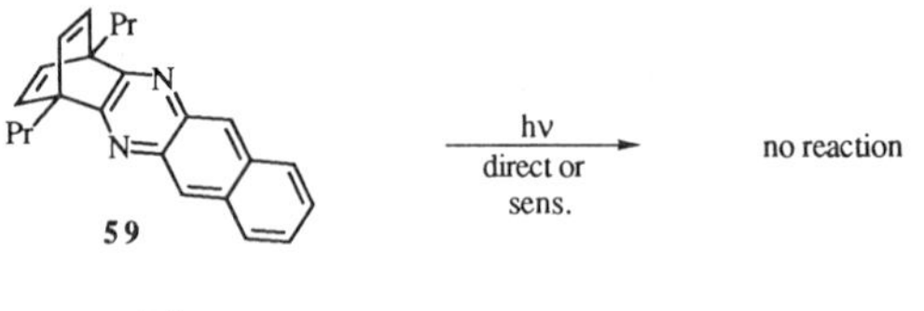

$$(15.16)$$

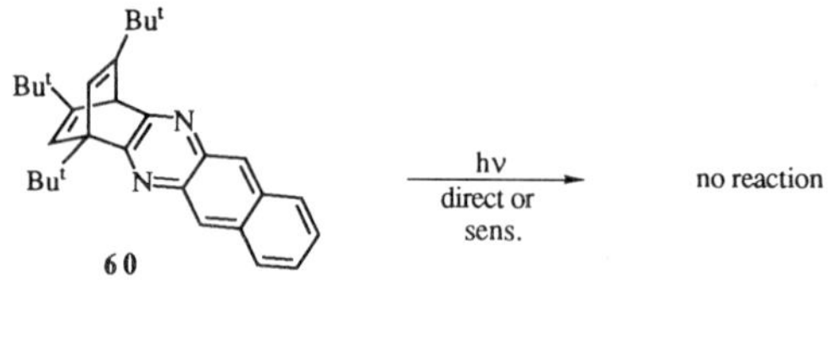

$$(15.17)$$

.5.3 Synthetic Applications

The above discussion indicates that the direct irradiation of benzobarrelenes and analogues may result in the formation of the corresponding cyclooctatetraenes if the intersystem crossing of the relevant barrelenes is less efficient than the [2+2]-cycloaddition leading to cyclooctatetraenes. Although these transformations are of limited synthetic application, special cyclooctatetraenes of physical or mechanistic interest may be prepared in this manner.

Triplet-sensitized reaction and, in some cases, direct irradiation of fused-ring aromatic barrelenes may yield the corresponding semibullvalenes. The reaction mixtures are generally simple for the symmetric and unsubstituted barrelenes such as **7**, **18**, and **21**, although the reaction mechanisms may be complicated. Despite the many possible *a priori* products for the substituted fused-ring aromatic barrelenes, the reactions are typically regioselective and may be clean; **13a**,[6] **15**,[7] **17**,[7] **61** (Equation 15.18),[16] **64** (Equation 15.19),[16] **66** (Equation 15.20),[17] and **68** (Equation 15.21)[18] are instances. Thus, the triplet-mediated Di-π-Methane Rearrangements of fused-ring aromatic barrelenes are good and unique methods to prepare benzosemibullvalenes and related compounds.

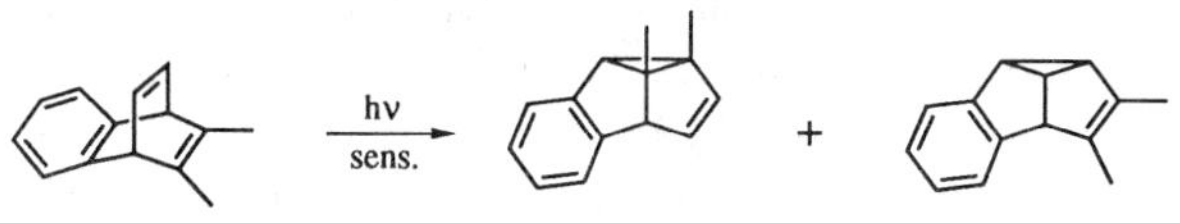

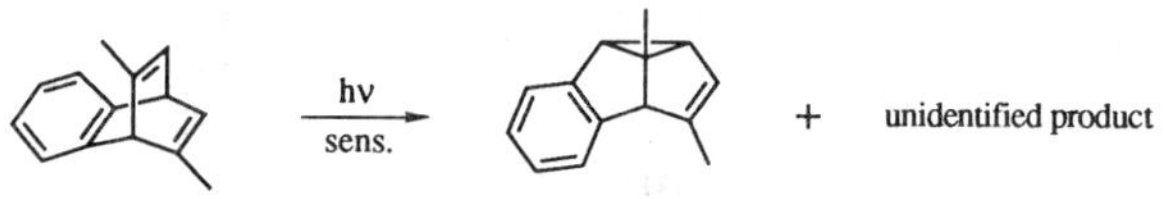

$$
\text{(15.18)}
$$

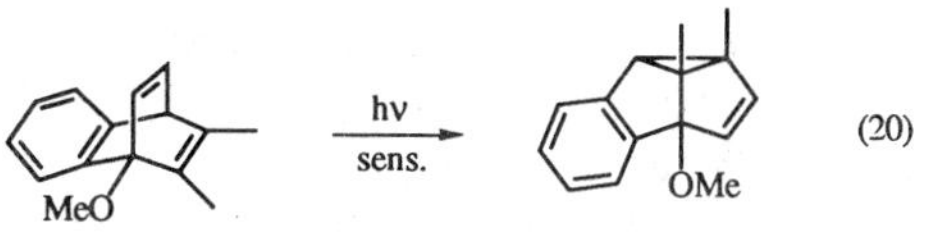

$$
\text{(15.19)}
$$

$$
\text{(15.20)}
$$

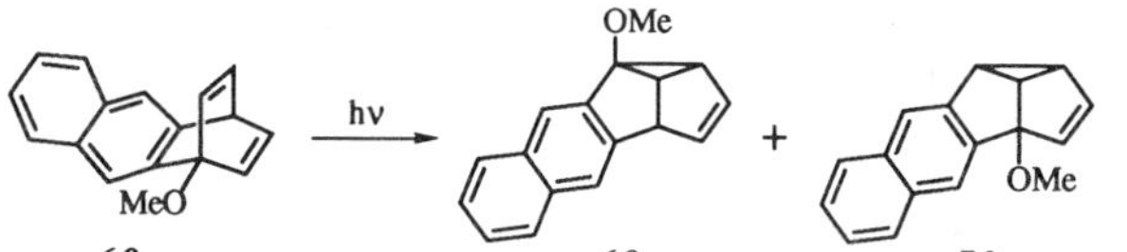

$$
\text{(15.21)}
$$

15.4 Spectroscopic Data

Spectroscopic data for **benzosemibullvalene 26:**

UV_{max} (95% ethanol): 281 nm (ε 930), 274 (1,100), 268 (sh, 930).

IR (CS_2, μ): strong bands 3.26 (sh), 3.28, 3.30 (sh), 3.38; weak bands 7.41 to 7.97, 8.21, 8.62 (br), 9.82. 10.03, 10.12, 10.46, 10.65, 10.79, 11.10, 11.46, 11.65, 12.10, 12.52; intense bands 12.85, 13.35, 13.58, 13.75, 14.72, 15.11.

^{1}H NMR (CCl_4): δ 7.33~6.80 (m, 4H, aromatic H), 5.57~5.38 (ABXq, 1H, vinyl H-6 next to benzylic H-5, $J_{6,7} = 5.0$ Hz, $J_{5,6} = 2.0$ Hz), 5.17~5.00 (ABXq, 1H, vinyl H-7 next to cyclopropyl H-8, $J_{6,7} = 5.0$ Hz, $J_{7,8} = 2.0$ Hz), 3.90~3.72 (ABXq, 1H, benzylic and allylic H-5, $J_{5,6} = 2.0$ Hz,

$J_{1,5}$ = 6.0 Hz), 3.38~2.78 (sex, 1H, interior cyclopropyl H-1; 1H, cyclopropyl and benzylic H-2), 2.75~2.45 (ABCX, dt, 1H, cyclopropyl and allylic H-8, $J_{7,8}$ = 2.0 Hz, $J_{1,8}$ = 6.5 Hz).
MS (70 eV): m/z (%) 154 (M$^+$), 153 (100)

Spectroscopic data for **benzocyclooctatetraene 16**:

UV$_{max}$ (cyclohexane): 224 nm (sh, ε 14,500), 271 (sh, 3,900).
IR (CCL$_4$, cm^{-1}): 1725 (br.).
^{1}H NMR (CDCl$_3$): δ 7.31~7.24 (m, 2H, C-6 and C-9), 6.99 (d, 1H, C-4, $J_{3,4}$ = 8.5 Hz), 6.85 (dd, 1H, C-3, $J_{3,4}$ = 8.5 Hz, $J_{1,3}$ = 2.5 Hz), 6.56 (d, 1H, C-1, $J_{1,3}$ = 2.5 Hz), 6.11~6.06 (m, 2H, C-7 and C-8), 3.77 (s, 3H, –OCH$_3$), 3.73 (s, 6H, –CO$_2$CH$_3$).
MS: m/z (%) 300 (M$^+$, 3), 58 (100).

Spectroscopic data for **pyrazinosemibullvalene 49**:

UV$_{max}$ (hexane): 289 nm (ε 9,500), 207 (15,000).
IR (CHCl$_3$, cm^{-1}): 2960, 2900, 2860, 2230, 1630, 1540, 1470, 1400, 1370, 1345, 1240, 869.
^{1}H NMR (CDCl$_3$): δ 5.49 (s, 1H, vinyl H), 4.28 (s, 1H, benzylic and allylic H), 3.32 (s, 1H, cyclopropyl H), 1.22 (s, 9H, *t*-Bu), 1.09 (s, 9H, *t*-Bu), 0.98 (s, 9H, *t*-Bu).
^{13}C NMR (CDCl$_3$): δ 165.0 (s), 160.7 (s), 154.4 (s), 131.3 (s), 127.7 (s), 126.1 (d), 113.9 (s), 113.7 (s), 82.7 (s), 67.6 (s), 60.1 (d), 40.3 (d), 34.2 (s), 34.0 (s), 33.9 (s), 30.8 (q), 30.5 (q), 29.2 (q).
MS (75 eV): m/z (%) 374 (M$^+$), 318 (18), 317 (13), 262 (15), 261 (19), 57 (100), 41 (30), 29 (20).

Spectroscopic data for **pyrazinosemibullvalene 50**:

UV$_{max}$ (hexane): 291 nm (ε 14,000), 207 (19,000).
IR (CHCl$_3$, cm^{-1}): 2960, 2900, 2860, 2240, 1475, 1460, 1390, 1365, 1320, 1240, 1085, 1045, 860.
^{1}H NMR (CDCl$_3$): δ 5.12 (s, 1H, vinyl H), 3.20 and 3.18 (ABq, J 6.4 Hz, 2H, cyclopropyl H), 1.18 (s, 9H, *t*-Bu), 1.12 (s, 9H, *t*-Bu), 0.87 (s, 9H, *t*-Bu).
^{13}C NMR (CDCl$_3$): δ 164.7 (s), 157.9 (s), 152.6 (s), 129.7 (s), 129.2 (s), 124.4 (d), 114.2 (s), 114.1 (s), 70.0 (s), 52.0 (s), 51.9 (d), 44.9 (d), 33.4 (s), 33.2 (s), 32.3 (s), 28.6 (q), 27.9 (q), 26.9 (q).
MS (75 eV): m/z (%) 374 (M$^+$), 319 (6), 318 (21), 304 (9), 303 (40), 261 (9), 247 (8), 57 (100), 41 (25), 29 (15), 18 (8).

References to the Spectroscopic Data

1. Zimmerman, H. E., Givens, R. S., and Pagni, R. M., *J. Am. Chem. Soc.,* 90, 6096, 1968.
2. Bender, C. O. and Wilson, J., *Helv. Chim. Acta,* 59, 1469, 1976.
3. Liao, C.-C., Hsieh, H.-P., and Lin, S.-Y., *J. Chem. Soc., Chem. Commun.,* 545, 1990.
4. Hsieh, H.-P., M.S. Thesis, National Tsing Hua University, June, 1987.

References

1. (a) Zimmerman, H. E. and Paufler, R. M., *J. Am. Chem. Soc.,* 82, 1514, 1960; (b) Zimmerman, H. E., Grunewald, G. L., Paufler, R. M., and Sherwin, M. A., *J. Am. Chem. Soc.,* 91, 2330, 1969.
2. (a) Zimmerman, H. E. and Grunewald, G. L., *J. Am. Chem. Soc.,* 88, 183, 1966; (b) Zimmerman, H. E., Binkley, R. W., Givens, R. S., and Sherwin, M. A., *J. Am. Chem. Soc.,* 89, 3932, 1967; (c) Zimmerman, H. E., Binkley, R. W., Givens, R. S., Grunewald, G. L., and Sherwin, M. A., *J. Am. Chem. Soc.,* 91, 3316, 1969.

3. (a) Hixson, S. S., Mariano, P. S., and Zimmerman, H. E., *Chem. Rev.,* 73, 531, 1973; (b) Zimmerman, H. E., in *Rearrangements in Ground and Excited States,* Vol. 3, de Mayo, P., Ed., Academic Press, New York, 1980, 131–166; (c) Adams, W., in *Comprehensive Organic Synthesis,* Vol. 5, Trost, B. M., Ed., Pergamon Press, Oxford, 1991, 193–214.

4. (a) Zimmerman, H. E., Givens, R. S., and Pagni, R. M., *J. Am. Chem. Soc.,* 90, 4191, 1968; (b) Rabideau, P. W., Hamilton, J. B., and Friedman, L., *J. Am. Chem. Soc.,* 90, 4465, 1968; (c) Zimmerman, H. E., Givens, R. S., and Pagni, R. M., *J. Am. Chem. Soc.,* 90, 6096, 1968.

5. (a) Bender, C. O. and Shugarman, S. S., *J. Chem. Soc., Chem. Commun.,* 934, 1974; (b) Bender, C. O. and King-Brown, E. H., *J. Chem. Soc., Chem. Commun.,* 878, 1976; (c) Bender, C. O., Brooks, D. W., Cheng, W., Dolman, D., O'Shea, S. F., and Shugarman, S. S., *Can. J. Chem.,* 56, 3027, 1978.

6. (a) Grovenstein, E., Jr., Campbell, T. C., and Shibata, T., *J. Org. Chem.,* 34, 2418, 1969; (b) Bender, C. O. and Brooks, D. W., *Can. J. Chem.,* 53, 1684, 1975; (c) Scheffer, J. and Yap, M., *J. Org. Chem.,* 54, 2561, 1989.

7. Bender, C. O. and Wilson, J., *Helv. Chim. Acta,* 59, 1469, 1976.

8. Brewer, J. P. N. and Heaney, H., *J. Chem. Soc., Chem. Commun.,* 811, 1967.

9. (a) Zimmerman, H. E. and Bender, C. O., *J. Am. Chem. Soc.,* 91, 7516, 1969; (b) Zimmerman, H. E. and Bender, C. O., *J. Am. Chem. Soc.,* 92, 4366, 1970.

10. Bender, C. O. and Burgess, H. D., *Can. J. Chem.,* 51, 3486, 1973.

11. (a) Zimmerman, H. E., Boettcher, R. J., Buehler, N. E., and Keck, G. E., *J. Am. Chem. Soc.,* 97, 5635, 1975; (b) Zimmerman, H. E., Boettcher, R. J., Buehler, N. E., Keck, G. E., and Steinmetz, M., *J. Am. Chem. Soc.,* 98, 7680, 1976.

12. Zimmerman, H. E., Amick, D. A., and Hemetsberger, H., *J. Am. Chem. Soc.,* 95, 4606, 1973.

13. Zimmerman, H. E. and Amick, D. R., *J. Am. Chem. Soc.,* 95, 3977, 1973.

14. Liao, C.-C., Hsieh, H.-P., and Lin, S.-Y., *J. Chem. Soc., Chem. Commun.,* 545, 1990.

15. Liao, C.-C., Lin, S.-Y., Hsieh, H.-P., and Yang, P.-H., *J. Chin. Chem. Soc. (Taipei),* 39, 275, 1992.

16. Bender, C. O., Bengtson, D. L., Dolman, D., Herle, C. E. L., and O'Shea, S. F., *Can. J. Chem.,* 60, 1942, 1982.

17. Bender, C. O., Cassis, I. M., Dolman, D., Heerze, L. D., and Schultz, F. L., *Can. J. Chem.,* 62, 2769, 1984.

18. Bender, C. O., Elder, J. L. M., and Miller, L. E., *Can. J. Chem.,* 50, 395, 1972.

16

The Photochemistry of Dibenzobarrelene (9,10-Ethenoanthracene) and Its Derivatives

John R. Scheffer
University of British Columbia

Jie Yang
University of British Columbia

Barrelene (bicyclo[2.2.2]octa-2,5,7-triene) and its mono-, di-, and tribenzo-derivatives have proved to be a treasure trove of interesting compounds for photochemical study. Almost without exception, they undergo novel and unusual photorearrangements that have led to a greatly enhanced understanding of the factors that govern excited-state chemical reactivity. In this chapter, the photochemical behavior of the *dibenzo* system is discussed; other chapters in this book discuss the lower and higher benzannulated members of the series.

16.1 The Preparation of Dibenzobarrelene and Derivatives

Also known as 9,10-ethenoanthracene, dibenzobarrelene and its derivatives are invariably prepared by Diels-Alder additions of acetylenic (and in some cases, ethylenic) dienophiles across the 9,10-positions of anthracene and substituted anthracenes. The first example of this reaction was provided

0-8493-8634-9/95/$0.00+$.50
© 1995 by CRC Press, Inc.

by Diels and Alder in 1931 with the successful addition of dimethyl acetylenedicarboxylate to anthracene to form dimethyl 9,10-dihydro-9,10-ethenoanthracene-11,12-dicarboxylate (compound 1, Scheme 16.1).[1] When ethylenic dienophiles are used, a subsequent elimination reaction must be employed to introduce the C11-C12 double bond. This was the strategy used in the first synthesis of dibenzobarrelene itself (compound 2, Scheme 16.1).[2]

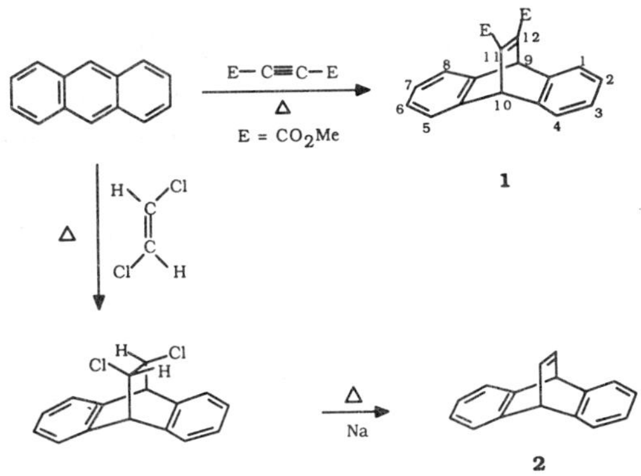

SCHEME 16.1

16.2 The Photochemistry of Dibenzobarrelene

Inasmuch as there are no substituent effects to complicate matters, this review starts by discussing the photochemistry of the parent molecule, dibenzobarrelene (2). The results obtained in this case illustrate many of the subtleties of the subject that will be encountered in later examples. Ciganek was the first to study the photochemistry of dibenzobarrelene.[3] He reported that irradiation of an acetone solution of compound 2 led to an 85% yield of 4b,8b,8c,8d-tetrahydrodibenzo-[a,f]cyclopropa[cd]pentalene (3) (Scheme 16.2). A few months earlier, Zimmerman and Grunewald had reported an analogous photorearrangement for barrelene itself, and had given the photoproduct the trivial name semibullvalene.[4] On this basis, photoproduct 3 and others like it have come to be known as dibenzosemibullvalenes.

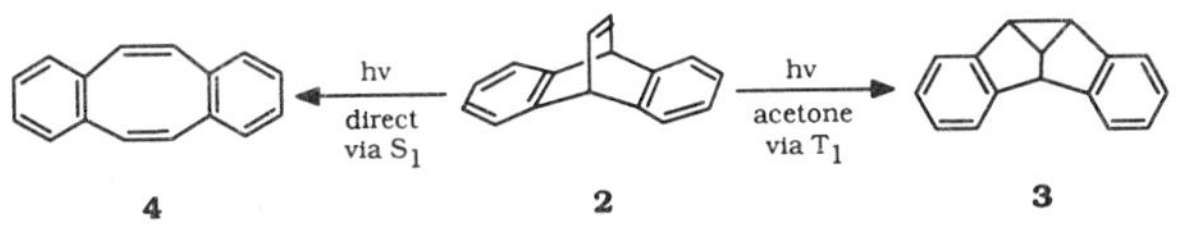

SCHEME 16.2

Ciganek also photolyzed dibenzobarrelene in benzene solution and reported only the slow formation of dibenzosemibullvalene.[3] However, in 1968, Rabideau et al.[5] showed that direct irradiation of dibenzobarrelene in cyclohexane or THF leads to a 75% yield of dibenzocyclooctatetraene (4) (Scheme 16.2) along with approximately 10% of dibenzosemibullvalene. These results parallel those of Brewer and Heany, who had shown earlier that direct irradiation of a *mono*benzobarrelene derivative leads to the formation of a mixture of benzosemibullvalene and benzocyclooctatetraene photoproducts.[6] Brewer and Heany suggested that the semibullvalene derivative is produced through the triplet excited state and the COT (cyclooctatetraene) derivative via the singlet, a multiplicity correlation that has stood the test of time for barrelene derivatives.

16.3 The Di-π-Methane Rearrangement

The barrelene to semibullvalene transformation, along with the related transformations of the benzannulated derivatives, was soon recognized by Zimmerman as an example of a general photochemical process termed the Di-π-Methane Rearrangement, whereby compounds containing two π-moieties attached to a common saturated or "methane" carbon atom are converted into vinylcyclopropane derivatives.[7] The π-moieties may be either aliphatic or aromatic in nature, as exemplified by dibenzobarrelene, which contains both types. The mechanism of the Di-π-Methane Photorearrangement has been studied in great detail by Zimmerman and co-workers,[8] and their mechanism, as applied to the dibenzobarrelene to dibenzosemibullvalene transformation, is shown in Scheme 16.3.

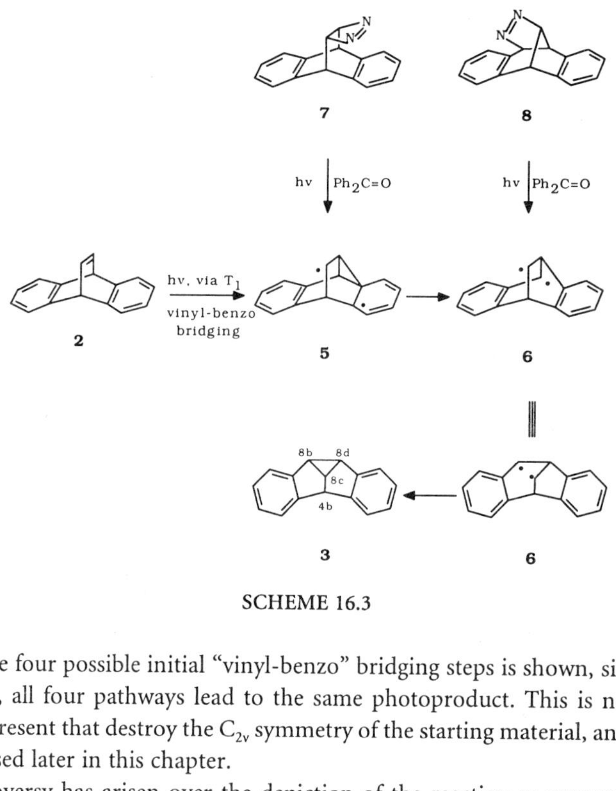

SCHEME 16.3

Only one of the four possible initial "vinyl-benzo" bridging steps is shown, since in the case of dibenzobarrelene, all four pathways lead to the same photoproduct. This is not the case when substituents are present that destroy the C_{2v} symmetry of the starting material, and many examples of this are discussed later in this chapter.

A slight controversy has arisen over the depiction of the reaction as proceeding through the cyclopropyldicarbinyl diradical species 5. It is conceivable that the triplet excited state of dibenzobarrelene could bypass this species and proceed directly to 1,3-biradical 6 via a 1,2-aryl shift. This possibility was raised by Paquette and Bay on the basis of some deuterium isotope effect studies.[9] To quote Paquette, however, "The structurally varied nature of the monodeuterated substrates examined and the relatively narrow range of their excited-state isotope effects must be construed to be an indication that this probe may not be as capable of distinguishing different mechanistic pathways as polar substituents." Theoretical calculations seem to support the intermediacy of cyclopropyldicarbinyl biradicals in the Di-π-Methane Rearrangement;[10] but to quote Zimmerman, "To the extent that the discussion is whether or not such diradicals are transition states or are intermediates, one has a valid question. To the extent that one is questioning the occurrence of the cyclopropyldicarbinyl biradical species on the hypersurface of the reaction, the point is not valid."[8b]

Another approach to assessing the possible intermediacy of the various biradical intermediates postulated for the Di-π-Methane Rearrangement has been to generate them independently by nitrogen extrusion from the appropriate azoalkane. Thus, in a study based on earlier work by Zimmerman and co-workers in the barrelene, benzobarrelene, and naphthobarrelene series,[11] Adam et al.[12] synthesized and photolyzed the radical precursors 7 and 8 in the dibenzo system (Scheme 16.3).[12] Triplet-sensitized nitrogen extrusion from these compounds was found to lead exclusively to dibenzosemibullvalene, a result that, in addition to providing support for the existence of the biradicals 5 and 6, suggests that the cyclopropyldicarbinyl biradical 5 does not revert to dibenzobarrelene in the triplet manifold. A similar conclusion concerning the essential irreversibility of benzo-vinyl bridging from the triplet excited state in the other systems was reached by Zimmerman et al.[8b]

6.4 Cyclooctatetraene Formation

The formation of dibenzocyclooctatetraene (4) in the direct (unsensitized) irradiation of dibenzobarrelene is an interesting reaction that has been assumed, on the basis of detailed studies of the photochemistry of the analogous benzo- and naphthobarrelene compounds,[13] to involve initial intramolecular [2+2]-cycloaddition through the singlet excited state followed by a thermal reorganization of the resulting cage compound 9 (Scheme 16.4). What seems to have escaped everyone's attention until now is that the commonly written mechanism involving *direct* thermal conversion of the quadricyclene-like intermediate 9 to dibenzocyclooctatetraene (4) is, like the conversion of quadricyclene itself into cycloheptatriene, a symmetry-forbidden process.[14] It seems more likely, therefore, that the initially formed [2+2]-photoproduct 9 undergoes a thermally allowed [4+2]-retrocycloaddition to afford intermediate 10, and that it is this species that rearranges to dibenzocyclooctatetraene *via* a symmetry-allowed electrocyclic process.

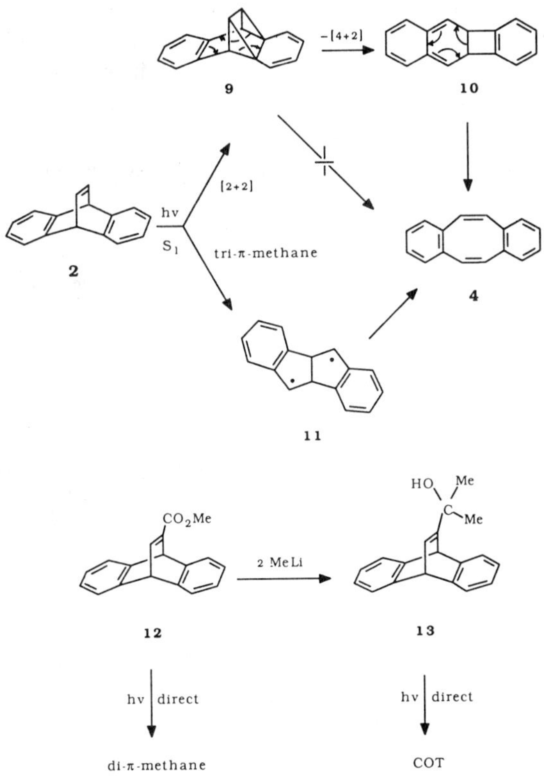

SCHEME 16.4

Recent research from this group has shown that there is yet another mechanism that must be considered for the dibenzobarrelene to dibenzocyclooctatetraene conversion.[15] In this mechanism, the key intermediate is biradical 11, most likely formed through sequential benzo-vinyl bridging involving *both* aromatic rings. A type of "tri-π-methane" process, this transformation is *antarafacial* with respect to the vinyl group and differs from the *suprafacial* [2+2]-mechanism discussed above. Once formed, 1,4-biradical 11 can undergo central bond cleavage to form COT derivative 4. The tri-π-methane and [2+2]-mechanisms can be differentiated in principle by deuterium labeling studies analogous to those first used by Zimmerman et al.[7] for barrelene; but such studies have not yet been carried out for dibenzobarrelene. As we shall see, examples of the operation of each mechanism have been discovered in dibenzobarrelenes containing electron-withdrawing substituents on the vinyl group.

Most dibenzobarrelene derivatives do not afford COTs upon direct irradiation, presumably because intersystem crossing to the triplet (and the Di-π-Methane Rearrangement) is faster than COT formation from the singlet. This is particularly true for dibenzobarrelenes in which there is a carbonyl-containing substituent attached to the vinyl group. Such compounds, like other α,β-unsaturated carbonyl compounds, undergo rapid intersystem crossing and, as a result, little or no COT formation is observed. An example of such behavior is found in the case of methyl dibenzobarrelene-11-carboxylate (12, Scheme 16.4), which reacts exclusively via the Di-π-Methane Rearrangement upon direct and triplet-sensitized photolysis.[3] However, when ester 12 is converted into the corresponding carbinol 13 by treatment with methyllithium and then photolyzed, rearrangement to a COT product is the only process observed.[5]

16.5 The Photochemistry of Substituted Dibenzobarrelenes: C$_{2v}$ Systems

When substituents are introduced into the dibenzobarrelene nucleus in such a way that its C$_{2v}$ molecular symmetry is maintained (compound 14, Scheme 16.5), only one di-π-methane regioisomer, 15, is possible. Ciganek was the first to study the photochemistry of such compounds.[3] He found that either direct or acetone-sensitized irradiation of dibenzobarrelene derivatives 14a-c led exclusively to the corresponding dibenzosemibullvalenes 15a-c; no COT derivatives were reported. This work was later extended by Scheffer and Garcia-Garibay to the diesters 14d-e[16] and by George et al.[17] to the dibenzoyl derivative 14f.

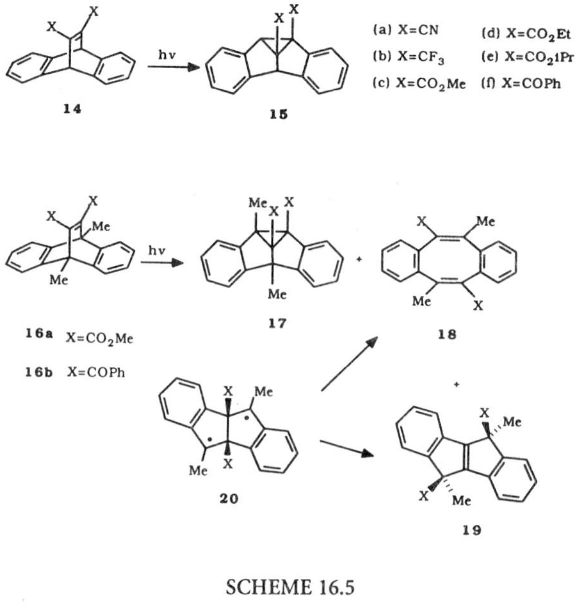

SCHEME 16.5

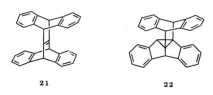

21   22

SCHEME 16.5 (continued)

The main goal of the work by Scheffer and co-workers was to investigate the photoreactions of such compounds in the *solid state;* a large body of evidence has been accumulated by this group during the past 10 years which indicates that, with few exceptions, dibenzobarrelene derivatives photorearrange in the crystalline phase as well as in solution.[18] The unique aspect of such research lies in the fact that solid-state reactions are governed by anisotropic crystal lattice forces that lead to increased regio-, stereo-, and enantioselectivity compared to their solution-phase counterparts. In the case of diester 14e, for example, the molecule crystallizes in a chiral space group, and photolysis of single, enantiomorphously pure crystals of this compound leads to product 15e in nearly quantitative enantiomeric excess.[16] By measuring the absolute configurations of the reactant and its photoproduct by X-ray crystallography, Scheffer et al.[19] were then able to chart the absolute steric course of the rearrangement and pinpoint the topochemical factors responsible for the high enantioselectivity. In contrast to the solid-state results, solution photolysis of compound 14e leads to racemic material.

The solution-phase photochemistry of bridgehead-substituted dibenzobarrelene derivative 16a proved to be particularly interesting. As outlined in Scheme 16.5, irradiation of this compound was found to lead to the di-π-methane product 17, along with COT 18, and the novel rearrangement product 19.[15] Compound 17 was the exclusive photoproduct in acetone, whereas 17 and 18 were formed in approximately equal amounts in benzene; the pentalene derivative 19 was formed only in the solid state. Photoproducts 18 and 19, whose structures were established through X-ray crystallography, were suggested to arise via a singlet-mediated tri-π-methane process leading to biradical 20, which can cleave to form 18 and undergo a double-ester migration to afford 19. It should be noted that the substituent pattern of COT 18 is different from that expected on the basis of the [2+2]-mechanism (see Scheme 16.4); it was this result that first led to consideration of the alternative tri-π-methane mechanism.[15] Following publication of these results, the photochemistry of the dibenzobarrelene derivative 16b was reinvestigated by George et al.[17] In work published in 1984, these authors had assigned different structures to photoproducts 18b and 19b; but upon reinvestigation in 1991, X-ray crystallography showed them to have structures exactly analogous to those of photoproducts 18a and 19a.[20] The unusual photobehavior of compounds 16a-b is evidently related to the presence of the bridgehead methyl groups, and it was suggested by Scheffer et al.[15] that the extra stability afforded biradical 20 by these substituents is a major factor in biasing the system in the direction of the tri-π-methane process. It was also suggested that the methyl groups sterically impede the competing process of [2+2]-benzo-vinyl photocycloaddition.

This section closes by mentioning the photochemistry of the interesting hydrocarbon dehydrojanusene (21, Scheme 16.5). This compound was prepared in 1971 by Cristol et al.[21] and its photochemistry reported in 1979 by Bartlett and co-workers.[22] Direct irradiation of dehydrojanusene in methylene dichloride solution was found to afford good yields of the di-π-methane product 22. In acetone, however, which is commonly used as both solvent and triplet energy sensitizer to promote di-π-methane photorearrangements, the yield of compound 22 was considerably lower than in methylene dichloride. Control experiments showed this to be due to a secondary photoreaction of dibenzosemibullvalene derivative 22 with acetone.[22]

16.6 The Photochemistry of Dibenzobarrelene Derivatives with C_s Symmetry: Compounds that Differ in Substitution at the Vinyl Positions

As outlined in Scheme 16.6, dibenzobarrelene derivatives that have different substituents X and Y at the vinyl positions can in principle lead to two regioisomeric di-π-methane photoproducts, 24 and 25. Only two of the four possible Zimmerman mechanism pathways are depicted, since involvement of the left-hand benzo ring leads to the enantiomers of the photoproducts shown.

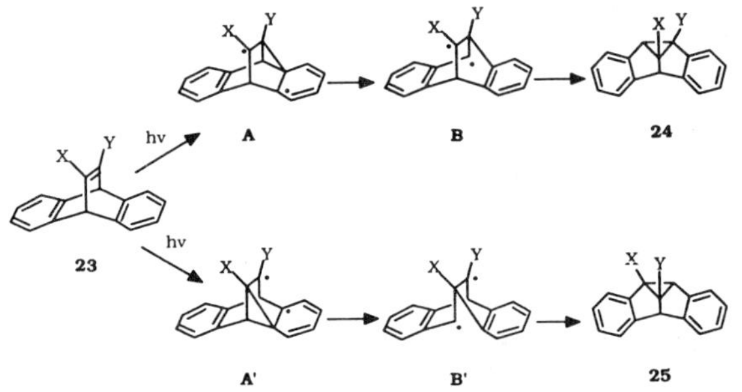

SCHEME 16.6

The first report of a photochemical study of this type, due to Ciganek,[3] dates back to 1966 and concerns the compound 23 for which substituent X = H and Y = COOMe. Since that time, several research groups around the world, including Scheffer's, have carried out similar work with dibenzobarrelenes containing a wide variety of vinyl substituents, and a literature search has uncovered a total of over 20 such compounds whose photochemistry has been investigated. The results of these studies, which are compiled in Table 16.1, refer to irradiations carried out in solution, usually acetone or benzene.

Traditionally, the regioselectivities summarized in Table 16.1 have been rationalized by suggesting that initial benzo-vinyl bridging to afford the more stable cyclopropyldicarbinyl diradical A or A′ (Scheme 16.6) is product-determining. Thus, in the case of entries 1–5, for example, photoproduct 25 predominates because biradical A′, in which one of the radical centers is conjugated to an acyl group, is preferred over biradical A (radical next to H). This picture is not quite so compelling, however, when one considers entries 6 and 16. Radical stabilization by CH₃ should clearly win out over H, yet the 24:25 ratio in this case is only 42:58 (entry 6).[26] In the case of entry 16, the relative radical-stabilizing abilities of methyl and ester are about equal,[32] yet the ester substituent dominates completely. To explain these results, use is made of a suggestion by Zimmerman,[8b] based on MO calculations, that the radical termini of the cyclopropyldicarbinyl biradicals become electron-rich during the Di-π-Methane Rearrangement. If this is the case, then both the polar nature as well as the radical-stabilizing ability of the substituents should be important in determining regioselectivity. A methyl group, therefore, despite its radical-stabilizing properties, is not very good at directing the Di-π-Methane Rearrangement because it is also a somewhat destabilizing electron-donating substituent. As a result, CH₃ and H have nearly equal directing power (entry 6), and the electron-withdrawing COOMe group dominates CH₃ completely (entry 16). Radical stability effects can evidently win out in some cases, however, as indicated by entry 15 (Ph > COOMe).

The relative di-π-methane directive influence of different acyl substituents is particularly interesting. The data in Table 16.1 reveal that ester substituents dominate both amides and carboxylate anions (entries 19–22). This is in accord with organic intuition, which tells us that carboxylate anions and amides, in which the carbonyl group is strongly involved in resonance with an adjacent heteroatom, should be poorer electron-withdrawing substituents than esters, and probably poorer radical-stabilizing groups as well; a simple PMO argument supports this view.[30] This effect may also be operating in the case of COOR vs. COOH (entries 17 and 18), where the former substituent

Table 16.1   Compounds 23 Studied and 24:25 Photoproduct Ratios

Entry	X	Y	24:25	Comments	Ref.
1	H	COOMe	0:100		3
2	H	COOEt	0:100		23
3	H	COPh	0:100		24
4	H	CONH$_2$	0:100		25
5	H	CONMe$_2$	0:100		25
6	H	Me	42:58	COT formed on direct hv	26
7	COOEt	COOMe	47:53		27
8	COO*i*Pr	COOMe	45:55		27
9	COO*s*Bu	COOMe	40:60	*sec*-Butyl group (±)	27
10	COO*t*Bu	COOMe	40:60		27
11	COOMenth	COOMe	44:56	Menth = (–)-Menthyl	28
12	COO*i*Pr	COOsBu	45:55	*sec*-Butyl group (±) or (+)	29
13	COOMe	CSOMe	0:100		25
14	COOMe	COSMe	0:100		25
15	COOMe	Ph	0:100		25
16	Me	COOMe	0:100		25
17	COOH	COOiPr	17:83	Ratio concentration dependent	30
18	COOH	COOsBu	≈10:90		29
19	CONHEt	COOMe	0:100		25
20	CONMe$_2$	COOMe	3:97		25
21	COO⁻Na⁺	COOiPr	10:90		30
22	COO⁻Prol⁺	COOEt	5:95	Prol = Proline	31

dominates (partial ionization of COOH), and in entries 7–11, where methyl esters slightly win out over esters bearing better electron-donating alkyl groups (increased importance of Zwitterionic resonance structures); steric factors may play a role here too. Entries 13 and 14 also deserve some comment. The observed regioselectivities indicate that there is a marked preference for the thioester substituents to occupy the nucleophilic radical terminus during photorearrangement. Such selectivity finds good analogy in the work of Wollowitz and Halpern, who showed that the COSEt group undergoes a 1,2-radical and a 1,2-anionic shift in preference to the COOEt substituent.[33]

A final point concerns entries 11, 12, and 22, where resolved chiral auxiliaries were introduced in an attempt to effect asymmetric induction. The results of such studies in solution, which gave diastereomeric excesses ranging from zero to ca. 40%, were not very encouraging. In the *solid state,* however, the diastereomeric excesses were generally much higher and, in some cases, near-quantitative optical purities were obtained. The "ionic chiral handle" approach exemplified by entry 22 is particularly promising, since the chiral handle is easily introduced and removed by simple acid-base chemistry and the resulting salt is a high-melting crystalline material suitable for solid-state photochemistry. In the case of entry 22, the use of (R)-(+)-proline led (after removal of the chiral handle and diazomethane workup) to levorotatory photoproduct 25 (80% ee), and (S)-(–)-proline gave (+)-25 in 76% ee; several other optically active amines were investigated as well.[31] For all the salts, irradiation in solution gave only racemic material. The greater optical purities obtained in the solid state were attributed to the fact that, in this medium, the chiral handle exerts as an asymmetric influence that is not present in isotropic liquid solvents; namely, it ensures the presence of a chiral environment for reaction through crystallization in a chiral space group.[31] Additional examples of the use of the ionic chiral handle approach to asymmetric induction in the Di-π-Methane Rearrangement will be presented in the sections that follow.

16.7 The Photochemistry of Dibenzobarrelene Derivatives that Differ in Substitution at the Bridgehead Positions

Dibenzobarrelenes of general structure 26 (Scheme 16.7) that have different substituents at the bridgehead positions can in principle form regioisomeric di-π-methane photoproducts 27 and 28,

Table 16.2   Compounds 26 Studied and 27:28 Photoproduct Ratios

Entry	X	Y	27:28	Comments	Ref.
1	H	D	53:47		9
2	H	CH$_3$	43:57		26
3	H	COOMe	67:33		3
4	COOMe	CH$_3$	71:29		34
			65:35		35
			76:24	COT observed	15a
5	COOMe	IPr	77:23		34
6	COOMe	*t*-Bu	100:0		34
7	COOMe	CH$_2$Cl	80:20	Novel rearrangement obsvd	36
8	COOMe	CH$_2$OAc	90:10	$\Phi_{total} = 0.12$	37
9	COOMe	CH$_2$OAc*	90:10	Ac* = opt act acyl group	28
10	COOMe	OAc	100:0		38
11	COOMe	OCOPh	100:0		34
12	COOMe	OMe	100:0		34
			85:15		35
13	COOMe	OEt	85:15		35
14	COOMe	OH	0:100		38
15	COOMe	NH$_2$	0:100		35
			30:70		39
16	COOMe	NHAc	71:29	COT obsvd	35
17	COOMe	NH$_3^+$A$^-$	100:0	A$^-$ is opt act	39
18	COOMe	Br	100:0		34,35
19	COOMe	Ac	71:29		34
20	COOMe	CHO	88:12		34
21	COOMe	NO$_2$	100:0		34
22	COOMe	Ph	0:100	COT obsvd	34,15
23	COOMe	COOMe	91:9		40
24	COOMe	D	54:46		9
25	COPh	CH$_3$	100:0		17

once again opening the way for interesting substituent effect studies. As summarized in Table 16.2, over 20 such investigations have been reported to date, the first of which was due to Ciganek, who photolyzed methyl dibenzobarrelene-9-carboxylate (entry 3).[3]

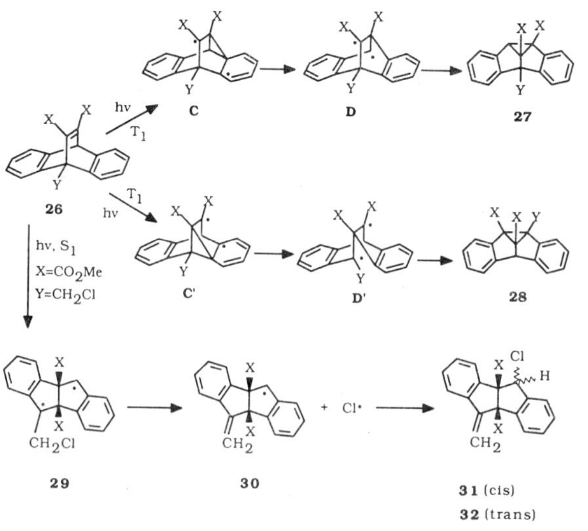

SCHEME 16.7

The interpretation of the product ratios presented in Table 16.2 is not straightforward, and no single theory seems capable of accommodating all the results. One thing that is clear, however, is that the nature of the second biradical intermediate in the Zimmerman mechanism is not the controlling factor in most cases, as this predicts, contrary to the experimental results, that intermediate D′ and photoproduct 28 would be favored on radical stability grounds. This point was recognized and discussed at some length by Paquette and Bay.[9] In this same paper, Paquette and Bay showed that second biradical stability *is* in accord with the regioselectivities observed for the di-π-methane photorearrangement of several bridgehead-substituted benzonorbornadienes.[9] Thus, different di-π-methane systems respond differently to bridgehead substitution.

The majority of the data presented in Table 16.2 comes from the work of Paddick et al.[35] and Iwamura and co-workers.[34] These authors interpreted their results in terms of the effect of the substituent Y on the relative stability of biradicals C and C′ (Scheme 16.7). Three effects were considered: (1) π-acceptors were suggested to *favor* biradical C′ by strengthening the newly formed benzo-vinyl bond in the same way that π-acceptors favor the norcaradiene form in the cycloheptatriene-norcaradiene equilibrium;[41] π-donors were suggested to have the opposite effect; (2) electronegative substituents were suggested to *disfavor* biradical C′, since there is evidence from the work of Zimmerman that the so-called methane carbon atom becomes electron-deficient during the di-π-methane process;[42] (3) hydrogen bonding effects (discussed later). Factors (1) and (2) work against one another, since most π-acceptors are also strongly electronegative. As a result, Iwamura[34] was forced to conclude that the electronegativity effect wins out in such cases (e.g., entries 19–21). Another problem with this interpretation is that alkyl substituents, which are neither π-donors nor electronegative (Taft G_I values of ca. −0.05[43]), nevertheless have the same effect on the photorearrangement as substituents that are.

Here, it is proposed that nonbonded repulsive and torsional effects play a major role in determining the regioselectivity of the Di-π-Methane Rearrangement of bridgehead-substituted dibenzobarrelenes. Both Iwamura[34] and Paquette[9] noted that steric effects seem to be important in the case of bridgehead alkyl substituents (entries 4–6), but there is no reason to think that such effects should be limited to these groups alone. Biradical C is considerably less strained than C′, the major difference being that eclipsing and nonbonded repulsions are relieved as the substituents X and Y in reactant 26 move apart from one another to form C, whereas eclipsing and steric crowding are maintained, if not increased, in C′. These ideas are supported by MM2 calculations recently carried out in by Scheffer.[44] If such steric effects are important, then reducing the size of the substituent X should decrease the regioselectivity. That this is so can be seen by comparing entries 2 and 4 as well as entries 3 and 23. In both instances, decreasing the size of X from COOMe to H leads to a considerable reduction in regioselectivity. Additional examples of apparent steric control of regioselectivity are discussed in later sections of this chapter.

Next, a discussion of the bridgehead substituents that do not conform to the picture presented above, namely Y = Ph, OH, and NH_2, is presented. In the case of Y = Ph, it may be that second biradical formation is product determining, owing to the extraordinary radical-stabilizing ability of the phenyl substituent. The molecule would therefore resemble the benzonorbornadiene derivatives studied by Paquette et al.,[9] which were suggested to photorearrange via a direct 1,2-aryl shift mechanism. Intramolecular hydrogen bonding to the X = COOMe substituent was advanced as the reason for the unusual behavior of the Y = OH and NH_2 derivatives.[35,38] Hydrogen bonding was suggested to decrease the radical-stabilizing ability of the ester group, thus favoring radical formation next to the non-hydrogen-bonded COOMe group (C′). Thus, it is thought that disruption of hydrogen bonding in forming C but not C′ may play a role here as well.

The asymmetric inductive effect of situating covalent (entry 9) as well as ionic (entry 17) chiral handles at the bridgehead position was studied by Scheffer and co-workers.[28,39] As was the case for the vinyl-substituted systems discussed earlier,[31] best results were obtained in the solid state, particularly for the ionic systems. Thus, for A⁻ = optically active 10-camphorsulfonate, di-π-methane photoproduct 27 was obtained (after removal of the chiral handle) in ca. 65% enantiomeric excess;[39] as before, solution-phase photolysis gave racemic material.

Three of the entries in Table 16.2 (4, 16, and 22) report COT formation as well as Di-π-Methane Rearrangement. In two of the cases (entries 4 and 22), the structures of the COTs were shown by X-ray crystallography to be those formed through a tri-π-methane process rather than [2+2]-cycloaddition;[15a] it seems likely that the structure of the COT formed in entry 16, which was assumed to be [2+2]-derived,[35] will need to be revised. The novel rearrangement mentioned in entry 7 represents a "diverted" tri-π-methane COT formation. Direct photolysis of compound 26 (X = COOMe, Y = CH$_2$Cl) gave, in addition to the major products 27 and 28, small amounts of the *exo*-methylene derivatives 31 and 32 (Scheme 16.7). These latter products were suggested to arise through the intermediacy of the tri-π-methane-derived biradical 29 which, rather than cleaving to COT, loses a chlorine atom to form monoradical 30, which then recombines with the liberated chlorine atom to form products 31 and 32.[36] To date, tri-π-methane photochemistry has been observed only for 9-substituted and 9,10-disubstituted dibenzobarrelene derivatives, a correlation that reinforces the idea that steric effects play an important role in this process as well as in the Di-π-Methane Rearrangement of these compounds.

16.8 The Photochemistry of Dibenzobarrelene Derivatives that Differ in Substitution at Both the Vinyl and Bridgehead Positions

The photochemistry of over 15 such compounds has been investigated. The results of these investigations are summarized in Table 16.3.

Entries 1 through 7 in Table 16.3 clearly indicate that vinyl substituents exert a stronger influence on the regioselectivity of the Di-π-Methane Rearrangement than do substituents at the bridgehead position. Consider entry 6 for example: based on the data presented earlier, the bridgehead ester group favors "distal" vinyl-benzo bridging and the vinyl ester substituent favors "proximal" bridging (the terms *proximal* and *distal* refer to bridging next to and remote from the bridgehead substituent, respectively). The fact that proximal bridging is observed experimentally indicates that the vinyl ester group is the dominant substituent, presumably acting through its direct polar and radical-stabilizing effect on the initial cyclopropyldicarbinyl biradical. Entry 1 indicates that even a hydrogen-bonded ester substituent that, according to Richards et al.,[38] possesses reduced radical stabilizing ability, dominates when situated in the vinyl position.

Entries 4 and 5 represent attempts at asymmetric induction through the influence of an optically active menthyl ester substituent. In solution, diastereomeric excesses (de) were 0 to 20%; but in the solid state, a 60% de was observed in the case of entry 4. The greater asymmetric influence of the chiral handle in this system was ascribed to its being closer to the reaction center than in the case of entry 5,[28] where the de was zero even in the solid state. The COTs of entries 4 and 5 were assigned [2+2]-derived structures on the basis of their spectra. A COT was also observed upon direct irradiation of the diester of entry 6. In this case, it was established beyond question by X-ray crystallography that the COT was [2+2]-derived.[15a] From this, it is tentatively concluded that steric deceleration of the [2+2]-process in such 9,11-disubstituted dibenzobarrelenes is insufficient to allow the tri-π-methane reaction to be competitive.

Entries 8–17 are concerned with dibenzobarrelene derivatives in which a vinyl carbon atom is connected to a bridgehead position by means of a ring. As illustrated in Scheme 16.8, such compounds can be prepared by intramolecular Diels-Alder methodology.[37,46] The di-π-methane regioselectivity observed in these cases (entries 8 to 12) seems once again to be heavily influenced by steric factors, although steric effects alone are not capable of rationalizing all the data. X-ray crystallography and molecular mechanics calculations indicate that compounds such as that shown in entry 8 experience severe steric crowding between the aromatic "peri" hydrogens and the methylene hydrogens of the lactone ring (Scheme 16.8). Scheffer et al.[37] have suggested that relief of this crowding is the factor that governs regioselectivity in entries 8 and 9. As indicated by molecular models and MM2 calculations, benzo-vinyl bridging on the side of the lactone ring

Table 16.3 Compounds Studied and Regioselectivity of Benzo-Vinyl Bridging (Dotted Lines)

Entry	Compound	Selectivity	Comments	Ref.
1	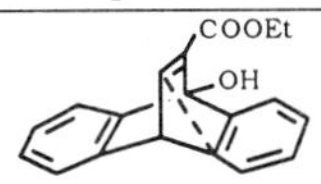 COOEt, OH	100%		35
2	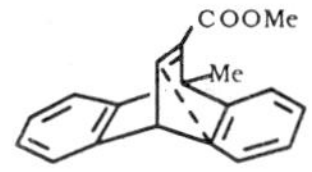 COOMe, Me	100%		35
3	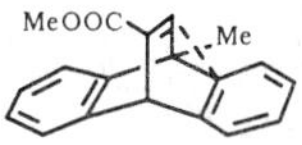 MeOOC, Me	100%		35
4	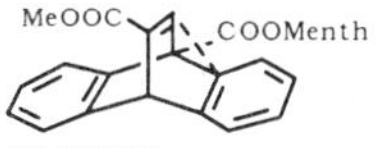 MeOOC, COOMenth	100%	20% de in solution 60% de in solid state COT observed	28
5	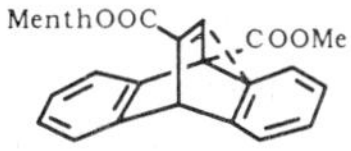 MenthOOC, COOMe	100%	0% de in solution and solid state COT observed	28
6	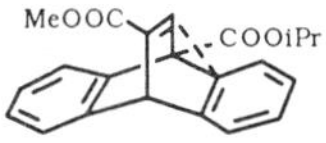 MeOOC, COOiPr	100%	Small amounts of [2+2]-derived COT formed on direct hv	15a
7	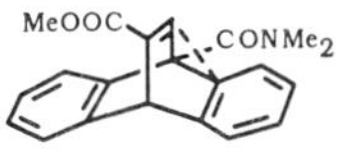 MeOOC, CONMe$_2$	100%		25
8	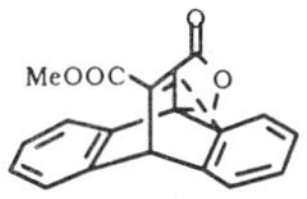 MeOOC, O	100%	$\Phi = 0.48$	37
9	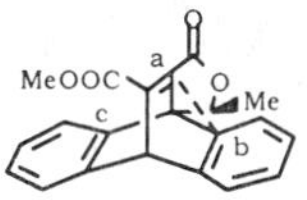 MeOOC, O, Me (a, b, c)	77% a-b 23% a-c	$\Phi_{total} = 0.78$	37
10	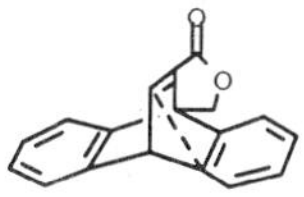 O, O	100%	$\Phi = 0.02$	37
11	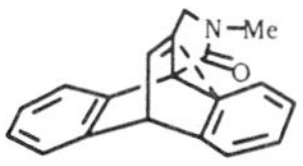 N→Me, O	62%		45
12	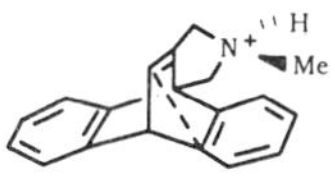 N$^+$, H, Me	80%		45
13	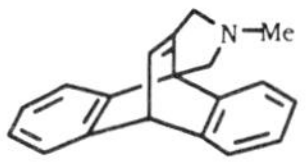 N→Me		Direct hv gives only COT Sensitized hv not reported	45
14	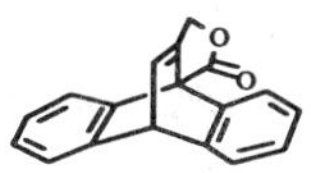 O, O		Direct hv gives only COT Sensitized hv not reported	45
15	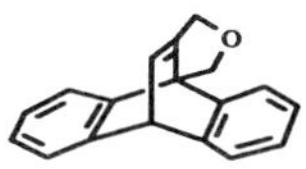 O		Direct hv gives only COT Sensitized hv not reported	45
16	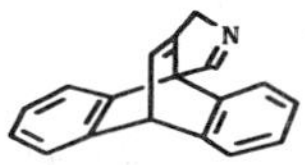 N		Direct hv gives only COT Sensitized hv not reported	45

Table 16.3 (continued)   Compounds Studied and Regioselectivity of Benzo-Vinyl Bridging (Dotted Lines)

Entry	Compound	Selectivity	Comments	Ref.
17	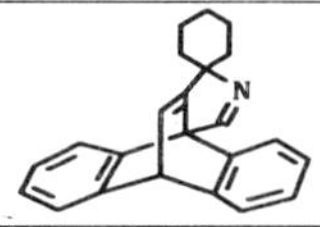		Direct hv gives only COT Sensitized hv not reported	45

relieves the unfavorable interactions to a greater extent than bridging on the other side, an effect that is particularly pronounced in the methylated lactone of entry 9. Here, the very severe CH_3/H interaction is best relieved by *a-b* bridging, and it was shown experimentally that this is the predominant pathway followed.[37] The high quantum yields in these cases, 0.48 for **8** and 0.78 for **9**, are proportional to the steric crowding and support the mechanistic interpretation. Also consistent with this picture is the fact that when the bridgehead substituent is noncyclic [e.g., CH_2OAc (entry 8, Table 16.2)], the pseudo-peri interaction in the reactant can be relieved by rotation, and the quantum yield goes down to 0.12. The normal torsional effects of the CH_2OAc group remain, however, with the result that the regioselectivity is reversed relative to the cyclic cases.

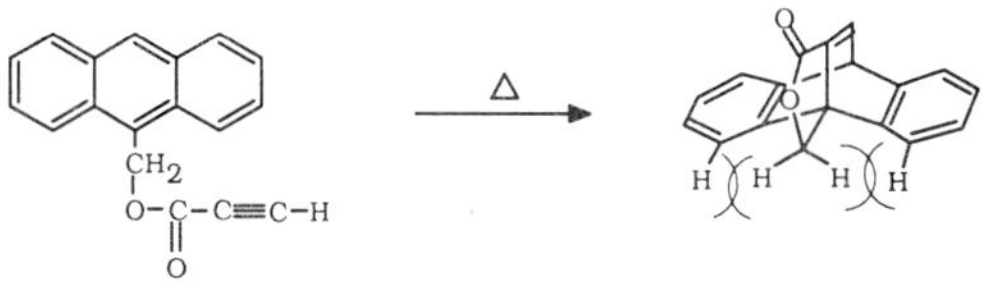

SCHEME 16.8

Consistent with entries 1 to 7, Table 16.3, radical stability effects clearly control regioselectivity in the case of entry 10. Entries 11 and 12, however, are more difficult to interpret. Ciganek suggested that distal bridging in 12 was favored by radical stability.[45] However, it has been observed that a methyl group exerts only a very modest directive influence (entry 6, Table 16.1), and it may be that, in this case, the CH_2NHMeR^+ substituent is much better than CH_3 at stabilizing an adjacent electron-rich radical center. Photolysis of the free amine would provide some evidence on this point, but this was not done.[45] Entry 11 is even more puzzling. Despite the fact that the pseudo-peri steric effects are reduced, proximal benzo-vinyl bridging is still preferred. There is nevertheless some solace in noting that the low regioselectivity in this case indicates that the opposing stereoelectronic forces must be very closely balanced.

For entries 13 to 17, only direct irradiation results leading to COT derivatives were reported.[45] Apparently, no di-π-methane-type photoproducts were formed under these conditions. Unfortunately, these results are not very informative, since both the [2+2] and the tri-π-methane mechanisms lead to the same COT in each case.

16.9 The Photochemistry of Dibenzobarrelene Derivatives that Lack Symmetry: Aryl-Substituted Systems

Placement of a substituent at one of the aryl positions in dibenzobarrelene destroys the symmetry of the system and leads to a reactant that can, once again, give two regioisomeric Di-π-Methane Rearrangement products. Rabideau et al. were the first to attempt such a study.[5] They showed that direct photolysis of 1-chlorodibenzobarrelene leads mainly to COT material; traces (0 to 5%) of dibenzosemibullvalene photoproduct(s) could be detected, but no structural assignments were made.

Table 16.4 Compounds 35 Studied and
36:37 Photoproduct Ratios (E = CO_2Me)

Entry	Z	36:37
1	CN	80:20
2	COOMe	35:65
3	Cl	38:62
4	OMe	86:14

Irradiation of dibenzobarrelenes formed by Diels-Alder addition of dimethyl acetylenedicarboxylate to 1-substituted or 2-substituted anthracene derivatives can in principle give four structurally distinct di-π-methane photoproducts corresponding to the four possible benzo-vinyl bridging pathways. While experiments of this type are attractive from the point of view of the large amount of information they provide, they should not be undertaken lightly, owing to the difficulties that can be anticipated in separating and identifying four very similar photoproducts. For this reason, the first foray into aryl substituent effects dealt with 1,5-disubstituted systems.[25] As outlined in Scheme 16.9, the product diversity (excluding enantiomers) in such systems is reduced to two, compounds 36 and 37.

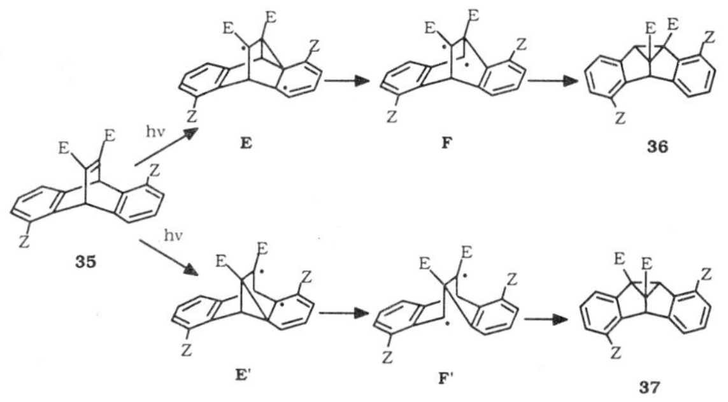

SCHEME 16.9

Table 16.4 summarizes the results of these studies; direct and acetone-sensitized photolysis gave identical product ratios within experimental error, and no COTs were observed. The product ratios in Table 16.4 are reminiscent of the results obtained by Paquette et al.[47] with *ortho*-substituted benzonorbornadienes. There, too, initial benzo-vinyl bridging occurred preferentially *ortho* to the cyano and methoxy substituents, and low regioselectivity (50:50) was observed in the case of halogen (fluoro). Paquette's results[47] were rationalized by Houk et al.[68] on the basis of perturbation molecular orbital theory.[48] The basic idea was that the substituents polarize the frontier molecular orbitals of the aromatic ring in such a way that the atomic orbital coefficients are largest at the *ortho*-position, thus leading to preferential benzo-vinyl bridging at this site. Strong donor substituents were postulated to act mainly on the HOMO, and strong acceptors perturb primarily the LUMO; weak donors such as halogen were expected to show low regioselectivity, since the HOMO is only weakly polarized in this case.

The main problem with interpreting the data in Table 16.4 lies in the behavior of the COOMe group, an electron acceptor that should, like the CN group, promote formation of photoproduct 36 through benzo-vinyl bridging at the *ortho*-position. It is suggested that steric effects reverse the product ratio in this case. MM2 calculations clearly indicate that formation of biradical E should be more difficult than formation of E' owing to an unfavorable nonbonded interaction developed in the former between the vinyl ester substituent and the Z group during vinyl-benzo bridging. For large substituents such as Z = COOMe (A value, 1.1 to 1.2), this effect could become dominant; whereas, for small substituents such as Z = CN (A value, 0.15 to 0.25), orbital polarization still prevails. Reduced orbital polarization in the case of Z = COOMe, a slightly poorer electron acceptor than CN, could also play a role here.

16.10 Summary

Four major new aspects of the photochemistry of dibenzobarrelene derivatives have been documented in this chapter. First of all, there are at least two distinct mechanisms by which COTs can be formed: the original mechanism involving initial [2+2]-benzo-vinyl photocycloaddition and a newly discovered tri-π-methane process. Secondly, the commonly assumed direct thermal coversion of the intramolecular [2+2]-cycloadduct to COT is symmetry forbidden and, although a nonconcerted process is possible, the transformation probably occurs in two symmetry-allowed concerted steps, the first being a retro Diels-Alder reaction and the second consisting of an electrocyclic ring-opening process. Scheffer and co-workers are currently attempting to trap the retro Diels-Alder intermediate (structure **10**, Scheme 16.4). Third, owing to "pseudo-peri" interactions in the reactants as well as the congested nature of the intermediates involved, steric effects are much more important in the photochemistry of dibenzobarrelenes than originally thought. This applies to both the di-π-methane reaction and COT formation. Fourth, for dibenzobarrelenes bearing substituents at the vinyl position(s), radical stability alone is insufficient to predict di-π-methane regioselectivity; the polar nature of the substituents must also be taken into account. Thus, for two substituents with comparable radical-stabilizing ability, the better electron-withdrawing group will control regioselectivity owing to the nucleophilic character of the radical termini in the cyclopropyldicarbinyl biradical intermediate.

Note Added in Proof

Pertinent to the discussion in Section 16.3, recent experimental and theoretical work by Zimmerman et al.[49,50] has demonstrated conclusively that cyclopropyldicarbinyl diradicals such as 5 (Scheme 3) are indeed true intermediates in the Di-π-Methane Rearrangement.

References

1. Diels, O. and Alder, K., Synthesen in der hydroaromatischen Reihe. IX. Dien-Synthesen des Anthracens, *Justus Liebigs Ann. Chem.*, 486, 191, 1931.
2. Cristol, S. J. and Hause, N. L., Mechanisms of elimination reactions. V. Preparation and elimination reactions of *cis*- and *trans*-11,12-dichloro-9,10-dihydro-9,10-ethanoanthracene, *J. Am. Chem. Soc.*, 74, 2193, 1952.
3. Ciganek, E., The photoisomerization of dibenzobicyclo[2.2.2]octatrienes, *J. Am. Chem. Soc.*, 88, 2882, 1966.
4. Zimmerman, H. E. and Grunewald, G. L., The chemistry of barrelene. III. A unique photoisomerization to semibullvalene, *J. Am. Chem. Soc.*, 88, 183, 1966.
5. Rabideau, P. W., Hamilton, J. B., and Friedman, L., Photoisomerization of mono- and dibenzobarrelenes, *J. Am. Chem. Soc.*, 90, 4465, 1968.
6. Brewer, J. and Heaney, H., Photoisomerization reactions of 5,6,7,8-tetrafluoro-1,4-dihydro-1,4-ethenonaphthalene: Tetrafluorobenzyne in mass spectrometry, *J. Chem. Soc., Chem. Commun.*, 811, 1967.
7. Zimmerman, H. E., Binkley, R. W., Givens, R. S., and Sherwin, M. A., Mechanistic organic photochemistry. XXIV. The mechanism of the conversion of barrelene to semibullvalene. A general photochemical process, *J. Am. Chem. Soc.*, 89, 3932, 1967.
8. (a) Zimmerman, H. E., The Di-π-Methane (Zimmerman) Rearrangement, in *Rearrangements in Ground and Excited States,* de Mayo, P., Ed., Academic Press, New York, 1980, chap. 16; (b) Zimmerman, H. E., The Di-π-Methane Rearrangement, in *Organic Photochemistry*, Vol. 11, Padwa,

A., Ed., Marcel Dekker, New York, 1991, chap. 1; (c) De Lucchi, O. and Adam, W., Di-π-methane photoisomerizations, in *Comprehensive Organic Synthesis*, Vol. 5, Trost, B. M., Fleming, I., and Paquette, L. A., Eds., Pergamon Press, Oxford, 1991, chap. 2.5.

9. Paquette, L. A. and Bay, E., Effective control of regioselectivity by a bridgehead substituent in the Di-π-Methane Rearrangement of dibenzobarrelenes and benzonorbornadienes, *J. Am. Chem. Soc.,* 106, 6693, 1984.

10. Quenemoen, K., Borden, W. T., Davidson, E. R., and Feller, D., Some aspects of the triplet Di-π-Methane Rearrangement: Comparison of the ring opening of cyclopropyldicarbinyl and cyclopropylcarbinyl, *J. Am. Chem. Soc.,* 107, 5054, 1985.

11. Zimmerman, H. E., Boettcher, R. J., Buehler, N. E., Keck, G. E., and Steinmetz, M. G., Independent generation of cyclopropyldicarbinyl diradical species of the Di-π-Methane Rearrangement. Excited singlet, triplet and ground-state hypersurfaces of barrelene photochemistry. Mechanistic and exploratory photochemistry, *J. Am. Chem. Soc.,* 98, 7680, 1976.

12. Adam, W., De Lucchi, O., Peters, K., Peters, E.-M., and von Schnering, H. G., Photochemical and thermal denitrogenations of azoalkanes as mechanistic probes for the diradical intermediates involved in the Di-π-Methane Rearrangement of dibenzobarrelene, *J. Am. Chem. Soc.,* 104, 5747, 1982.

13. (a) Zimmerman, H. E., Givens, R. S., and Pagni, R. M., The photochemistry of benzobarrelene. Mechanistic and exploratory organic photochemistry. XXXV., *J. Am. Chem. Soc.,* 90, 6096, 1968; (b) Zimmerman, H. E. and Bender, C. O., The Di-π-Methane Rearrangement of the naphthobarrelenes. Mechanistic and exploratory photochemistry. L., *J. Am. Chem. Soc.,* 92, 4366, 1970; (c) Bender, C. O. and Shugrman, S. S., Polar substituents in pericyclic reactions: Photochemistry of 2-cyanobenzobarrelene, *J. Chem. Soc., Chem. Commun.,* 934, 1974; (d) Bender, C. O. and Brooks, D. W., Polar substituents in pericyclic reactions: Mechanistic course of the 2π+2π photocycloaddition of dimethyl 1,4-dihydro-1,4-ethenonaphthalene-2,3-dicarboxylate, *Can. J. Chem.,* 53, 1684, 1975.

14. Woodward, R. B. and Hoffmann, R., The conservation of orbital symmetry, *Angew. Chem., Int. Ed.,* 8, 781, 1969.

15. (a) Pokkuluri, P. R., Scheffer, J. R., and Trotter, J., Novel photorearrangements of bridgehead-substituted dibenzobarrelene derivatives in solution and the solid state, *J. Am. Chem. Soc.,* 112, 3676, 1990; (b) Pokkuluri, P. R., Scheffer, J. R., and Trotter, J., Crystal structure correlations in the photochemistry of dimethyl 9,10-dimethyl-9,10-dihydro-9,10-ethenoanthracene-11,12-dicarboxylate, *Acta Cryst.,* B49, 107, 1993.

16. (a) Evans, S. V., Garcia-Garibay, M., Omkaram, N., Scheffer, J. R., Trotter, J., and Wireko, F., Use of chiral single crystals to convert achiral reactants to chiral products in high optical yield: Application to the di-π-methane and Norrish type II photorearrangements, *J. Am. Chem. Soc.,* 108, 5648, 1986; (b) Caswell, L., Garcia-Garibay, M. A., Scheffer, J. R., and Trotter, J., Optical activity can be created from "nothing", *J. Chem. Ed.,* 70, 785, 1993.

17. Kumar, C. V., Murty, B. A. R. C., Lahiri, S., Chackachery, E., Scaiano, J. C., and George, M. V., Photochemical transformations and laser flash photolysis studies of dibenzobarrelenes containing 1,2-dibenzoylalkane moieties, *J. Org. Chem.,* 49, 4923, 1984.

18. Chen, J., Scheffer, J. R., and Trotter, J., Differences in photochemical reactivity of 9,10-ethenoanthracene derivatives in liquid and crystalline media, *Tetrahedron,* 48, 3251, 1992.

19. Garcia-Garibay, M., Scheffer, J. R., Trotter, J., and Wireko, F., Determination of the absolute steric course of a solid-state photorearrangement by anomalous dispersion X-ray crystallography, *J. Am. Chem. Soc.,* 111, 4985, 1989.

20. Asokan, C. V., Kumar, S. A., Das, S., Rath, N. P., and George, M. V., Novel phototransformations of bridgehead-dimethyl-substituted dibenzobarrelene. Structure of the photoproducts, *J. Org. Chem.,* 56, 5890, 1991.

21. Cristol, S. J. and Imhoff, M. A., Bridged polycyclic compounds. LXVI. Electrophilic additions to dehydrojanusene and related reactions, *J. Org. Chem.,* 36, 1849, 1971.

22. Bartlett, P. D., Kimura, M., Nakayama, J., and Watson, W. H., Photochemical oxidation and rearrangement of dibenzobarrelene and dehydrojanusene, *J. Am. Chem. Soc.,* 101, 6332, 1979.

23. Cristol, S. J. and Hager, J. W., Photochemical transformations. 34. Some studies on the Di-π-Methane Rearrangement of 7-(ethoxycarbonyl)dibenzobarrelene, *J. Org. Chem.*, 48, 2005, 1983.

24. (a) Saxena, N. K., Venkataramani, M., and Venkataramani, P. S., Synthesis and photochemical transformation of 7-benzoyldibenzobarrelene, *Ind. J. Chem.*, 13, 1075, 1975; (b) Demuth, M., Amrein, W., Bender, C. O., Braslavsky, S. E., Burger, U., George, M. V., Lemmer, D., and Schaffner, K., Photochemical and thermal rearrangements of a benzoylnaphthobarrelene-like system, *Tetrahedron*, 37, 3245, 1981.

25. Rattray, G., Yang, J., Gudmundsdottir, A. D., and Scheffer, J. R., Substituent effects on the di-π-methane photorearrangement of 9,10-ethenoanthracene derivatives, *Tetrahedron Lett.*, 34, 35, 1993.

26. Cristol, S. J., Kaufman, R. L., Opitz, S. M., Szalecki, W., and Bindel, T. H., Photochemical transformations. 33. Some studies of the photorearrangements of dibenzobarrelenes. A novel excitation-transfer relay mechanism, *J. Am. Chem. Soc.*, 105, 3226, 1983.

27. Garcia-Garibay, M., Scheffer, J. R., Trotter, J., and Wireko, F., Intermolecular steric effects on unimolecular rearrangements in crystalline media, *Tetrahedron Lett.*, 29, 2041, 1988.

28. Chen, J., Garcia-Garibay, M., and Scheffer, J. R., Chiral handle-induced diastereoselectivity in an organic photorearrangement: Solution versus solid state results, *Tetrahedron Lett.*, 30, 6125, 1989.

29. Garcia-Garibay, M., Scheffer, J. R., Trotter, J., and Wireko, F., Generation of optical activity through solid state reaction of a racemic mixture that crystallizes in a chiral space group, *Tetrahedron Lett.*, 28, 4789, 1987.

30. Garcia-Garibay, M. A., Scheffer, J. R., and Watson, D. G., Prototropic control of organic photochemistry. Hydrogen bonding effects on the di-π-methane photorearrangement, *J. Org. Chem.*, 57, 241, 1992.

31. Gudmundsdottir, A. D. and Scheffer, J. R. Asymmetric induction in the solid state photochemistry of salts of carboxylic acids with optically active amines, *Tetrahedron Lett.*, 31, 6807, 1990.

32. Bordwell, F. G. and Lynch, T.-Y., Radical stabilization energies and synergistic (captodative) effects, *J. Am. Chem. Soc.*, 111, 7558, 1989.

33. Wollowitz, S. and Halpern, J., Free radical rearrangement involving the 1,2-migration of a thioester group. Model for the Coenzyme B_{12} dependent methylmalonyl-CoA mutase reaction, *J. Am. Chem. Soc.*, 106, 8319, 1984.

34. Iwamura, N., Tukada, H., and Iwamura, H., Contrasting photochemical bridging regioselectivity in bridgehead-substituted 9,10-etheno- versus 9,10-(*o*-benzeno)-9,10-dihydroanthracenes, *Tetrahedron Lett.*, 21, 4865, 1980.

35. Paddick, R. G., Richards, K. E., and Wright, G. J., Regioselective and regiospecific Di-π-Methane Rearrangements. Photoisomerization of 9-substituted 9,10-dihydro-9,10-ethenoanthracenes, *Aust. J. Chem.*, 29, 1005, 1976.

36. Chen, J., Pokkuluri, P. R., Scheffer, J. R., and Trotter, J., The novel solid state and solution phase photochemistry of dimethyl 9-chloromethyl-9,10-dihydro-9,10-ethenoanthracene-11,12-dicarboxylate, *J. Photochem. Photobiol. A: Chem.*, 57, 21, 1991.

37. Chen, J., Pokkuluri, P. R., Scheffer, J. R., and Trotter, J., Control of regioselectivity through relief of steric crowding in the di-π-methane photorearrangement of 9,10-ethenoanthracene derivatives, *Tetrahedron Lett.*, 33, 1535, 1992.

38. Richards, K. E., Tillman, R. W., and Wright, G. J., Di-π-methane photoisomerization of dimethyl 9-hydroxy- and 9-acetoxy-9,10-dihydro-9,10-ethenoanthracene-11,12-dicarboxylate, *Aust. J. Chem.*, 28, 1289, 1975.

39. Gudmundsdottir, A. D. and Scheffer, J. R., Asymmetric induction in the solid state photochemistry of ammonium salts, *Photochem. Photobiol.*, 54, 535, 1991.

40. Gudmundsdottir, A. D. and Scheffer, J. R., unpublished results.

41. (a) Wehner, R. and Güther, H., Direct observation of "Buchner's Acid" using ^{13}C and ^{1}H nuclear magnetic resonance spectroscopy, *J. Am. Chem. Soc.*, 97, 923, 1975; (b) Hoffmann, R., The norcaradiene-cycloheptatriene equilibrium, *Tetrahedron Lett.*, 2907, 1970.

42. (a) Zimmerman, H. E. and Welter, T. R., Control of regioselectivity and excited singlet reaction rates by substitution in the Di-π-Methane Rearrangement. Mechanistic and exploratory photochemistry, *J. Am. Chem. Soc.*, 100, 4131, 1978; (b) Zimmerman, H. E., Armesto, D., Amezua, M. G., Gannett, T. P., and Johnson, R. P., Unusual organic photochemistry effected by cyano and methoxy substitution. Exploratory and mechanistic organic photochemistry, *J. Am. Chem. Soc.*, 101, 6367, 1979.

43. Hine, J., *Structural Effects on Equilibria in Organic Chemistry*, John Wiley & Sons, New York, 1975, 98.

44. Scheffer, J. R., unpublished results.

45. Ciganek, E., Intramolecular Diels-Alder additions. 2. Photochemical and Wagner-Meerwein rearrangements of 9,12-bridged ethenoanthracenes, *J. Org. Chem.*, 45, 1505, 1980.

46. Ciganek, E., Intramolecular Diels-Alder additions. 1. Additions to anthracene and acridine, *J. Org. Chem.*, 45, 1497, 1980.

47. Snow, R. A., Cottrell, D. M., and Paquette, L. A., Demonstration and analysis of bridging regioselectivity operative during di-π-methane photorearrangement of *ortho*-substituted benzonorbornadienes and anti-7,8-benzotricyclo[4.2.2.0^{2,5}]deca-3,7,9-trienes, *J. Am. Chem. Soc.*, 99, 3734, 1977.

48. Santiago, C., Houk, K. N., Snow, R. A., and Paquette, L. A., Bridging regioselectivity in triplet-sensitized di-π-methane photorearrangements of *ortho*-substituted benzonorbornadienes. A case for the importance of benzene HOMO and LUMO polarization, *J. Am. Chem. Soc.*, 98, 7443, 1976.

49. Zimmerman, H. E., Sulzbach, H. M., Tollefson, M. B., Experimental and Theoretical Exploration of the Detailed Mechanism of the Rearrangement of Barrelenes to Semibullvalenes: Diradical Intermediates and Transition States, *J. Am. Chem. Soc.*, 115, 6548, 1993.

50. Zimmerman, H. E. Kutateladze, A. G., Maekawa, Y., Mangette, J. E., Excited State Reactivity as a Function of Diradical Structure. Evidence for Two Triplet Cyclopropyldicarbinyl Diradical Intermediates with Differing Reactivity, *J. Am. Chem. Soc.*, 116, 9795 (1994).

17

Valence Isomerization Between Norbornadiene and Quadricyclane Derivatives — A Solar Energy Storage Process

Kazuhiro Maruyama
Kyoto Institute of Technology

Yasuo Kubo
Shimane University

17.1 Definition of the Reaction

Valence photoisomerization of norbornadiene (1a) has been the prototype of reversible [2+2] cycloaddition reaction (Equation 17.1). The first example of this photoisomerization, involving the diacid derivative 1b, was reported by Cristol and Snell in 1954.[1] They named the product, which is a derivative of "quadricyclene" due to its olefin-like behavior. Dauben and Cargill examined direct photoisomerization of the parent compound 1a and reported that the product, "quadricyclane" (2a), was thermally labile and reverted readily to 1a.[2] The isomerization norbornadiene-quadricyclane, has considerable generality, readily including examples with a variety of substituent groups that control the efficiency of the isomerization and the degree of the thermal stability of the quadricyclane products.

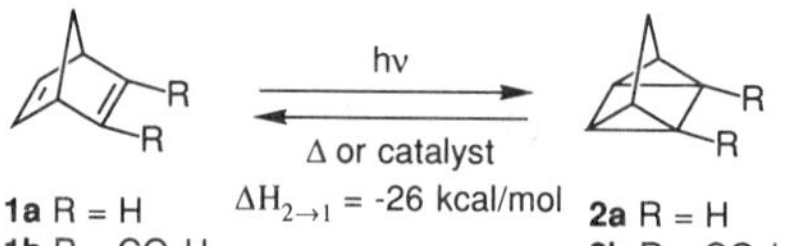

1a R = H
1b R = CO$_2$H

$\Delta H_{2\to1} = -26$ kcal/mol

2a R = H
2b R = CO$_2$H

$$(17.1)$$

In the 1970s, attention was refocused on the valence photoisomerization of norbornadiene derivatives due to the potential of the system for photochemical (solar) energy storage.[3-6] Among many systems examined, valence isomerization of norbornadiene derivatives has been one of the most promising systems for solar energy storage because of (1) the relative large enthalpy change of the isomerization, (2) quantitative chemical yields, (3) stability of the quadricyclane derivatives, and (4) easy and clean cycloreversion of quadricyclane derivatives by various methods.

0-8493-8634-9/95/$0.00+$.50

17.2 Mechanism

For norbornadiene derivatives with relatively simple substituents, the triplet-sensitized valence isomerization to quadricyclane derivatives occurred with much higher efficiency than direct (singlet) photoisomerization.[7] This observation was explained by the unsymmetrical potential energy surfaces that connected 1a and 2a.[8] The minimum on the S_1 (the first excited singlet state) energy surface corresponds to the biradicaloid 3, which can be stabilized better by Zwitterionic contributions than the biradicaloid 4. Interconversion from biradical 3 to S_0 (ground singlet state) favors formation of 1a. On the other hand, the minimum on the T_1 (the lowest triplet state) energy surface corresponds to biradical 4; intersystem crossing from 4 to S_0 favors formation of 2a. *ab initio* Molecular Orbital calculations supported the explanation.[9]

Other mechanistic features on valence isomerization between norbornadiene and quadricyclane derivatives, especially those involving photochemical electron transfer or excited complexes, have been reviewed recently by Jones.[10]

$\qquad$ **3** $\qquad\qquad\qquad$ **4**

17.3 Synthetic Applications

The radiation spectra of the sun reaching the earth has a maximum in the visible region; hence, the visible light is far stronger than the UV light. Consequently, one of the most important problems for the energy storage system is the efficient utilization of the longer wavelength region of sunlight. Since 1a itself does not absorb sunlight (> 300 nm), the research into solar energy storage has been directed toward the efficient photoisomerization of norbornadiene derivatives. Hitherto known approaches can be divided into two categories: (1) the use of organic sensitizers or transition metal catalysts, which assist the photoisomerization of 1a to 2a under sunlight; and (2) photoisomerization of norbornadiene derivatives substituted with an appropriate chromophore.

Use of Transition Metal Sensitizers

In the former, Kutal et al.[11,12] proposed transition metal complexes as sensitizers rather than more conventional organic sensitizers, such as acetophenone or benzophenone, which easily react with 1a and 2a to give undesired adducts.[7] The most typical sensitizer proposed by Kutal et al. was CuCl. However, the CuCl-1a complex does not absorb light at wavelengths longer than 350 nm. Since it was highly desirable to develop a new catalyst that works at longer wavelength from both practical and theoretical standpoints, Maruyama et al. proposed the use of stable $Ph_3PCuClL$ complexes 5,6, and 7 (L = 2,2′-bipyridine, *o*-phenanthroline, and phthalazine, respectively) as sensitizers.[13] These complexes can be obtained easily by the reaction of copper(I) with the corresponding nitrogen compounds, and they normally exhibit strong absorptions in near-UV with a tail into the visible region. These complexes were found to promote the valence isomerization of 1a to 2a without the formation of by-products. Among such complexes, 7 was the most effective. The catalyst 7 was, however, not as efficient as CuCl, as is shown by the following quantum yields values for deformation of quadricyclane ($\Phi = 0.13$ at 313 nm and $\Phi = 0.011$ at 366 nm in EtOH) compared with the quantum yields of 0.23 to 0.42 at 313 nm for CuCl.[11]

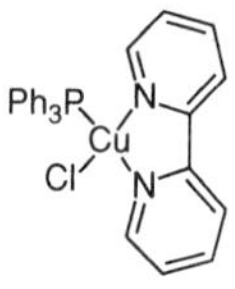

5 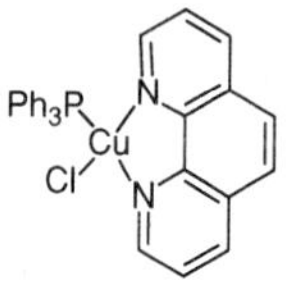   **6** 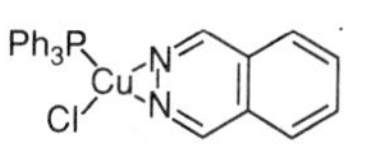   **7**

Norbornadiene Derivatives

A number of attempts to introduce several chromophores to one of the two double bonds of **1a** have been examined. Although norbornadiene derivatives with relatively simple substituents are known to isomerize photochemically to the corresponding quadricyclanes, almost all of them do not absorb light at wavelengths > 350 nm. For example, Kaupp and Prinzbach reported an example of valence isomerization of **8** that showed significant absorption in the near-UV (λ_{max} = 300 nm and a weak absorption to 350 nm).[14] Maruyama et al. proposed the introduction of arylcarbamoyl or arylimide groups to the norbornadiene skeleton (Scheme 17.1).[13] Compounds **10–13** had strong absorptions and a long tail in the visible region. Irradiation of acetonitrile solutions of **10–13** gave the corresponding quadricyclane derivatives in high yields. These derivatives were easily synthesized, but the quantum yields of the valence photoisomerization were relatively low.

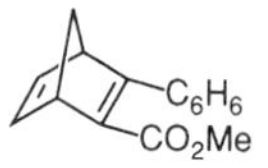

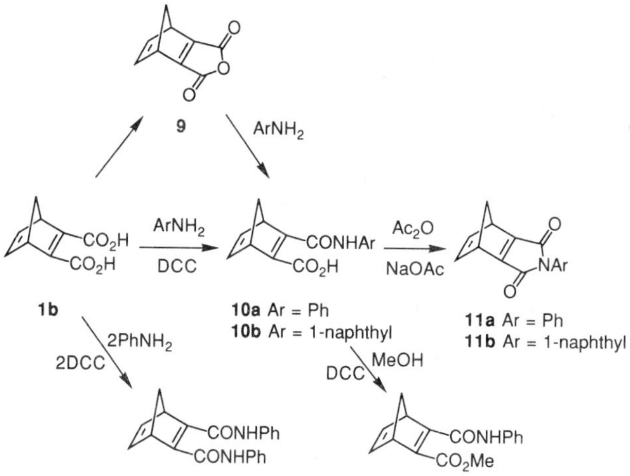

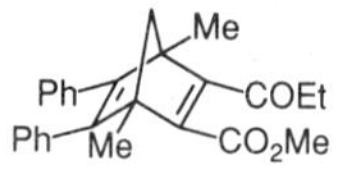

On the other hand, Yoshida,[15] Yonemitsu et al.,[16] and Kutal et al.[17] proposed the introduction of donor-acceptor chromophores to both of the double bonds of **1a**. With this pattern of substitution, norbornadienes such as **14** or **15** showed absorption in the visible region (λ_{max} = 340 to 400 nm and weak absorption at 420 to 500 nm). In these cases, the quantum yields of the valence isomerization were high (Φ = 0.75 for **14** in benzene;[16] 0.68 for **15** in acetonitrile),[17] but the quadricyclane derivatives were thermally less stable and the synthesis of these norbornadiene derivatives is

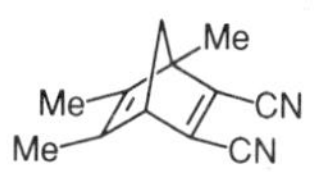

Table 17.1   Valence Photoisomerization of Water-Soluble
Norbornadiene Derivatives 10a,c-h

R^1	R^2	10	Yield (%)	$\Phi_{313\,nm}$	$\lambda_{\varepsilon=1}$ (nm)
Ph	H	10a	100	0.09	430
Ph	Me	10c	70	0.31	360
H	H	10d	100	0.59	365
Me	H	10e	100	0.52	365
CH_2Ph	H	10f	100	0.54	370
Me	Me	10g	100	0.45	335
CH_2CO_2H	H	10h	100	0.57	365

difficult. The finding that direct isomerization of 15 occurred with much higher efficiency than the triplet-sensitized process suggested that the lowest excited singlet state of 15 was significantly more reactive than the triplet state, even though the opposite reactivity pattern was observed in the case of 1a and in norbornadiene derivatives with relatively simple substituents. The influence of substituents and their effect on the cyclization was discussed in terms of the excited-state potential energy surfaces that interconnected the norbornadiene and quadricyclane structures.[17]

In all of the above systems, an organic solvent was used as the reacting medium. This was disadvantageous in a system for the practical storage of solar energy. Thus, Maruyama et al. first used water, a safe and inexpensive solvent, as the reaction medium.[18,19] Although water is usually reactive toward the cyclopropane and cyclobutane rings of 2a,[1] addition of base to keep the pH of the solution high was an effective method to stabilize quadricyclanes in water. Water-soluble norbornadiene derivatives 10 were prepared by easy high-yield procedures as shown in Scheme 17.1. This permitted the synthesis of 10 on a large scale. In an aqueous alkaline solution, 10a-h absorb at wavelengths > 300 nm and are sensitive to sunlight. Table 17.1 lists the absorption edges, that is, the wavelength where the extinction coefficient is unity. Upon irradiation with sunlight, 10 isomerized to 16 in high yields (Equation 17.2 and Table 17.1), even in the presence of air. The synthesized 17, with two norbornadiene chromophores in the same molecule gave increased solubility and an increase in the amount of energy being stored. The analysis of the spectral data and the photoreactivity of 17 showed no evidence of interference between the neighboring chromophores (Equation 17.3). Furthermore, Maruyama et al. synthesized a quadricyclane derivative 21, which was highly stable in water even without the addition of base.[20] Valence isomerization between 20 and 21 (Equation 17.4) was very clean and 21 was quite stable even in the hot water and could be stored for long periods.

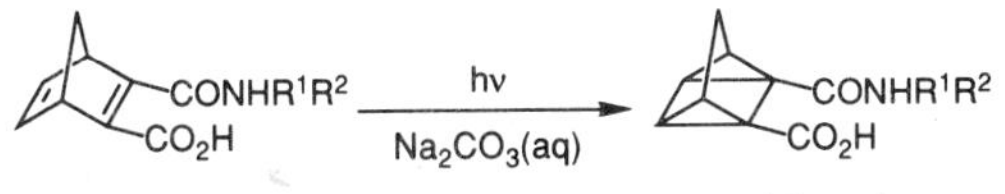

$$(17.2)$$

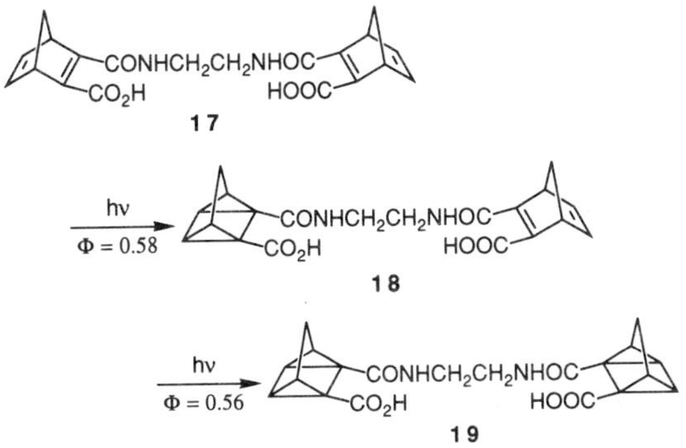

$$(17.3)$$

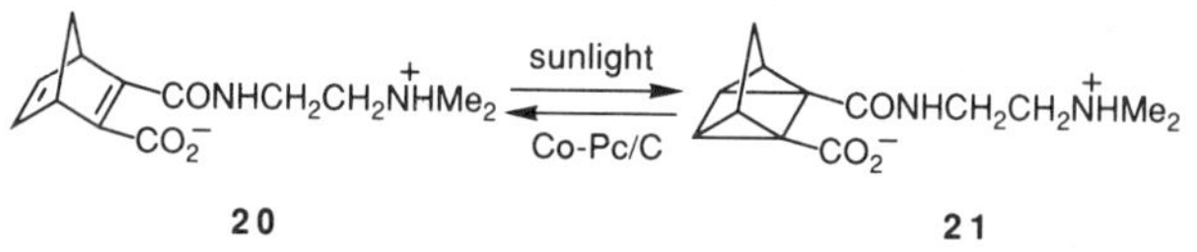

$$(17.4)$$

The cycloreversion of quadricyclanes to norbornadiene derivatives is also an important step to establish an efficient solar energy storage system. As **2a** was kinetically stable with respect to thermally induced cycloreversion, many workers in this field have examined many catalysts and have found that some induce rapid quantitative cycloreversion of quadricyclane to norbornadiene derivatives.[4,21] Efficient catalysts include metal compounds (Fe,[4] Co,[4,19] Ni,[4,13] Rh,[4,13] Pd,[13] Ag,[22] and Pt[13]) and oxidizing reagents.[23] Most of them are soluble in the reaction medium, and for such soluble catalysts a great many practical difficulties have to be overcome: for example, (1) control of the reaction, and (2) recovery of the catalyst for recycling. Insoluble catalysts have been studied therefore to overcome the above problems.[12]

In all of the catalytic cycloreversions investigated so far, an organic solvent was used and almost all of the catalysts were not only unstable in water but also induced addition of water to the quadricyclane moiety. Thus, Maruyama et al. used successfully the combination of stable water-soluble quadricyclane derivatives **16** and water-soluble cobalt-porphyrin complexes **22a-c** in aqueous alkaline solution (Equation 17.5).[24,25] Maruyama and Tamiaki also found that the catalytic action of $AgClO_4$ induced rapid and clean cycloreversion of the water-soluble quadricyclane to the norbornadiene derivative in aqueous ammonia even at room temperature.[26] The direction of attack of the cobalt-porphyrin complex **22c** and Ag(I) toward quadricyclanes (Chart 1) was examined by measuring the isomerization rate constants of the various methyl-substituted quadricyclane derivatives. Furthermore, Maruyama et al. have tried to prepare insoluble catalysts by immobilizing the catalysts in order to increase activities and stabilities of catalysts as well as their availabilities.[25] Among several immobilizing methods with some varieties of several supports, adsorption of cobalt-phthalocyanine on activated carbon (Co-Pc/C) was found to be most effective. The catalyst was stable in aqueous alkaline solution and no detachment of the complex from activated carbon during the reaction was observed.

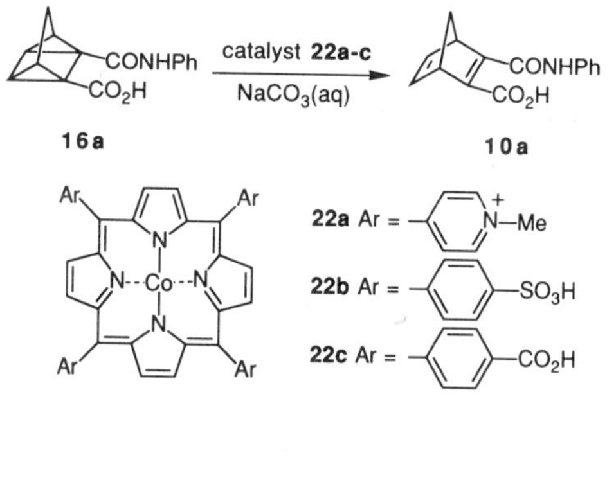

$$(17.5)$$

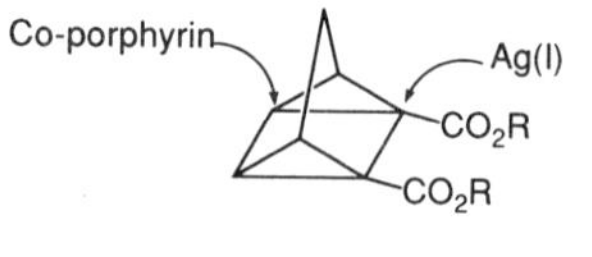

eferences

1. Cristol, S. J. and Snell, R. L., Synthesis of a nortetacyclene (tetracycloheptane) derivative, *J. Am. Chem. Soc.*, 76, 5000, 1954.
2. Dauben, W. G. and Cargill, R. L., Photochemical transformations. VIII. Isomerization of bicyclo[2.2.1]hepta-2,5-diene to quadricyclo[2.2.1.0^{2,6}.0^{3,5}]heptane (quadricyclene), *Tetrahedron*, 15, 197, 1961.
3. Sasse, W. H. F., *Solar Power and Fuels,* Academic Press, New York, 1977, chap. 8.
4. Hautala, R. R., King, R. B., and Kutal, C., *Solar Energy: Chemical Conversion and Storage,* Humana Press, Clifton, NJ, 1979.
5. Jones, G., II, Chiang, S., and Xuan, P. T., Energy storage in organic photoisomer, *J. Photochem.*, 10, 1, 1979.
6. Scharf, H.-D., Fleischhauer, J., Leismann, H., Ressler, I., Schlenker, W., and Weitz, R., Criteria for the efficiency, stability, and capacity of abiotic photochemical solar energy storage systems, *Angew. Chem., Int. Ed.*, 18, 652, 1979.
7. Hammond, G. S., Wyatt, P., DeBoer, C. D., and Turro, N. J., Photosensitized isomerization involving saturated centers, *J. Am. Chem. Soc.*, 86, 2532, 1964.
8. Turro, N. J., *Modern Molecular Photochemistry,* Benjamin/Cummings, Menlo Park, CA, 1978, chap. 11.
9. Raghavachari, K., Haddon, R. C., and Roth, H. D., Theoretical studies in the norbornadiene-quadricyclane system, *J. Am. Chem. Soc.*, 105, 3110, 1983.
10. Jones, G., II, Cycloaddition reactions involving 4n electrons: (2 + 2) cycloaddition; photochemical energy storage systems based on reversible valence photoisomerization, *Stud. Org. Chem. (Amsterdam)*, 40, 514, 1990.
11. Kutal, C., Use of transition metal compounds to sensitize an energy storage reaction, *Adv. Chem. Ser.*, 168, 158, 1978.
12. Grutsch, P. A. and Kutal, C., A silica-supported inorganic photosensitizer, *J. Chem. Soc., Chem. Commun.*, 893, 1982.
13. Maruyama, K., Terada, K., and Yamamoto, Y., Exploitation of solar energy storage systems. Valence isomerization between norbornadiene and quadricyclane derivatives, *J. Org. Chem.*, 46, 5294, 1981.
14. Kaupp, G. and Prinzbach, H., Photochemical conversions. XXX. Kinetics of the norbornadiene-quadricyclane photocycloaddition, *Helv. Chim. Acta*, 52, 956, 1969.
15. Yoshida, Z., New molecular energy storage systems, *J. Photochem.*, 29, 27, 1985.
16. Hirao, K., Ando, A., Hamada, T., and Yonemitsu, O., Valence isomerization between coloured acylnorbornadienes and quadricyclanes as a promising model for visible (solar) light energy conversion, *J. Chem. Soc., Chem. Commun.*, 1984, 300.
17. Ikezawa, H., Kutal, C., Yasufuku, K., and Yamazaki, H., Direct and sensitized valence photoisomerization of a substituted norbornadiene. Examination of the disparity between singlet- and triplet-state reactivities, *J. Am. Chem. Soc.*, 108, 1589, 1986.
18. Maruyama, K., Tamiaki, H., and Kawabata, S., Development of a solar energy storage process. Photoisomerization of a norbornadiene derivative to a quadricyclane derivative in an aqueous alkaline solution, *J. Org. Chem.*, 50, 4742, 1985.
19. Maruyama, K., Tamiaki, H., and Yanai, T., Valence isomerization between water-soluble norbornadiene and quadricyclane derivative, *Bull. Chem. Soc. Jpn.*, 58, 781, 1985.
20. Tamiaki, H. and Maruyama, K., A water-stable quadricyclane derivative, *Chem. Lett.*, 1875, 1988.
21. Bishop, K. C., III, Transition metal catalyzed rearrangements of small ring organic molecules, *Chem. Rev.*, 76, 461, 1976.
22. Maruyama, K. and Tamiaki, H., Isomerization of quadricyclanes to norbornadienes induced by silver(I) catalyst, *Chem. Lett.*, 683, 1987.
23. Yasufuku, K., Takahashi, K., and Kutal, C., Electrochemical catalysis of the valence isomerization of quadricyclene, *Tetrahedron Lett.*, 25, 4893, 1984.

24. Maruyama, K. and Tamiaki, H., Catalytic isomerization of water-soluble quadricyclane to norbornadiene derivatives induced by cobalt-porphyrin complexes, *J. Org. Chem.*, 51, 602, 1986.

25. Maruyama, K., Tamiaki, H., and Kawabata, S., Exothermic isomerization of water-soluble quadricyclanes to norbornadienes by soluble and insoluble catalysts, *J. Chem. Soc., Perkin Trans.*, 2, 543, 1986.

26. Maruyama, K. and Tamiaki, H., Silver(I)-catalyzed isomerization of water-soluble quadricyclanes, *J. Org. Chem.*, 52, 3967, 1987.

18

Ring Isomerization of Benzene and Naphthalene Derivatives

ndrew Gilbert
niversity of Reading

18.1 Definition of the Reaction and Historical Background

The benzene ring undergoes low quantum efficiency photoinduced isomerizations to give benzvalene 1, fulvene 2, and bicyclohexa-2,5-diene 3 (Dewar benzene).[1] Similar photoisomerization processes have also been described for naphthalenes and anthracenes as well as for heteroarenes: such reactions of the last class are considered in Chapter 23 of this text. Subsequent rearomatization of the valence bond isomers produced from di- and polysubstituted benzenes can, in principle, lead to positional isomers of the starting arene as shown in Scheme 18.1 for the conversion of *o-* to *m*-xylene.[2]

The report of the photoisomerization of benzene to fulvene in 1957 was an important event in the history of aromatic photochemistry for it was essentially this reaction[3] that discredited the widely held belief that the benzene ring was photostable and could even be used as a solvent for photochemical reactions. That report, and the account describing that attempts to trap the fulvene with maleic anhydride, led to the formation of the 2:1 photoadduct 4 of the dienophile and benzene[4] initiated considerable interest in aromatic photochemistry so that today the subject is a major area of study.[5]

0-8493-8634-9/95/$0.00+$.50
© 1995 by CRC Press, Inc.

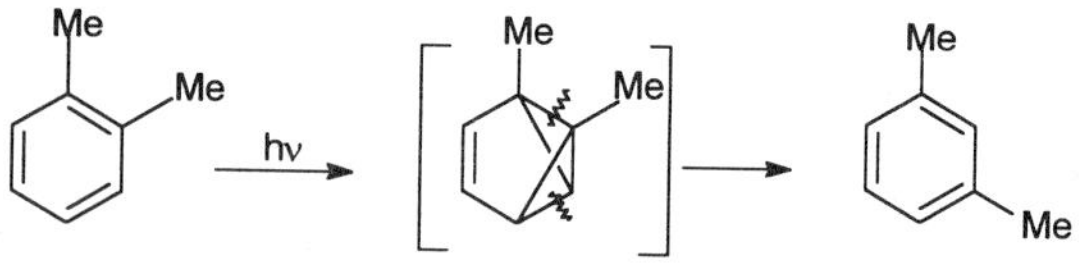

SCHEME 18.1

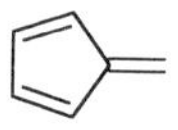

18.2 Mechanism

Isomerization of the Benzene Ring[1]

In its UV absorption spectrum, benzene has three maxima centered at 254, 203, and 180 nm, corresponding to electronic transitions from the S_0 to the S_1, S_2, and S_3 states, respectively. Irradiation in the $S_0 \rightarrow S_1$ band of benzene in the vapor or condensed phase gives a mixture of benzvalene (1) and fulvene (2); but, none of the Dewar isomer is detected.[6] The T_1 state is readily accessible by intersystem crossing ($\Phi = 0.23$ at 25 °C from a 0.112 M solution in C_6H_{12}), but it has been shown experimentally that this state is not involved and that the photoisomerization to 1 and 2 with 254-nm radiation arises solely from the S_1 state.

The S_1 state of benzene has $^1B_{2u}$ state symmetry (5) and is considered to be potentially bonding between the 1,3-positions.[1] This state is, however, a regular hexagon; and in order to induce the necessary deformation of the C_6 ring for the *meta*-bonding to occur, thermal activation or excitation to a high vibrational level of the S_1 state is required. In principle, the B_{2u} state may then undergo adiabatic transformation into the ground state of the nonplanar bicyclo[3.1.0]hex-2-enyl biradical (6) since this species, like 5, has an electronic configuration that is antisymmetric about the molecular plane. The biradical 6 has been termed "prefulvene," and its conversion into both fulvene and benzvalene can be readily envisaged. Definitive evidence for the involvement of 6 in the photoisomerization of benzene has not been obtained, but the requirement for thermal activation in the proposed mechanism is consistent with the reported increase in the isomerization efficiency in the vapor phase on decreasing the wavelength of excitation within the 253 to 237-nm band,[7] and in the solution phase by increasing the temperature over the range 9 to 55 °C.[8] The benzo analogue (7) of prefulvene has been generated both thermally and photochemically from the azo-compound 8.[9] In both series of experiments, the naphthalene and naphthvalene products were considered to arise from 7, but the precursor to benzfulvene formed on irradiation was deduced to be the diazo compound 9.

Quantum yields for the formation of benzvalene and fulvene from benzene are of the order of 0.01 to 0.03 and, while dilution with alkanes does increase the photoisomerization efficiency, limiting concentrations of each isomer of *ca.* 300 to 500 mg l^{-1} can be obtained from neat benzene.[10] Benzene only undergoes photoisomerization to give the Dewar isomer 3 when irradiated with wavelengths in the 200-nm region. Using an argon-iodine lamp, the S_2 (and not the S_3) excited state of the arene is selectively populated and this gives a 5:2:1 ratio of the isomers 1, 2, and 3, respectively.[11] From the results using 254-nm radiation and the absence of any influence of a triplet trap, it has been deduced that the precursor of the Dewar benzene is the $S_2(B_{1u})$ state and not the $T_1(B_{1u})$ state. The formation of 3 is noteworthy as it represents the first example of an upper excited singlet state leading to a nondissociative chemical reaction. That the photoisomerization to Dewar benzene arises from the S_2, but not the S_1, state is in accord with theoretical predictions from orbital symmetry analysis that show that one component of the B_{1u} state correlates with a low excited state of this isomer or with the ground state of the cyclohexa-2,4-dienyl biradical (10).[1] The optimum structure of the biradicaloid S_2 state of benzene has been calculated, and the detailed mechanism of its conversion into the Dewar isomer has been proposed.[12] However, 253.7-nm irradiation of benzene in an argon matrix does yield Dewar benzene.[13] This observation is rationalized by the proposal that the environment in the matrix induces mixing of the S_1 and S_2 states.

The formation of the photoisomers 1 and 2 using short-wavelength radiation probably arises from the S_1 state formed on decay from the upper level. There is, however, also the possibility that the S_2 state of benzene can undergo an adiabatic transformation into the S_1 state of benzvalene by way of an intermediate species that formally resembles Mobius benzene (11) (i.e., *cis, cis, trans*-cyclohexatriene); in this route, no thermal activation is apparently required.[1]

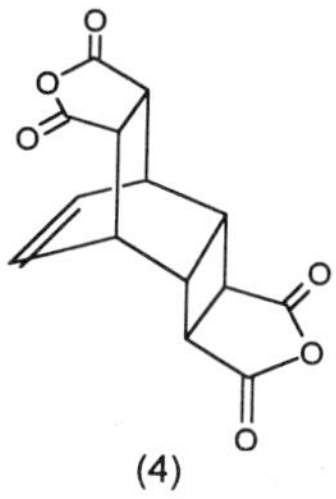

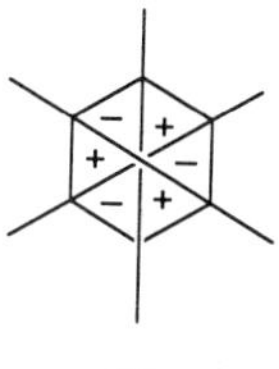

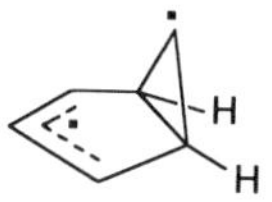

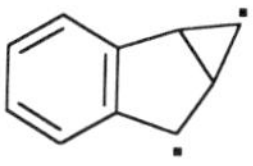

Phototranspositions

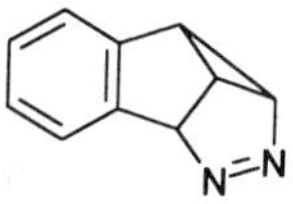

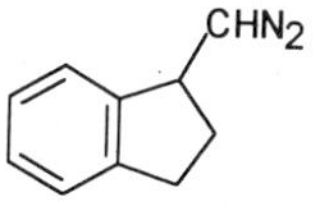

The mechanism of phototranspositions of carbon atoms in the benzene ring is widely acknowledged to involve the intermediacy and rearomatization of benzvalene and Dewar benzene isomers. However, it should be emphasized that only in a very few cases have these valence bond isomers been detected and their role in the transpositions been proven unambiguously. The involvement of such intermediates in preference to methyl group migration is inferred from labeling studies. Thus, in the photorearrangement of mesitylene 1,3,5-[14]C_3 to the 1,2,4-isomer, it is evident that throughout the reaction, the methyl substituent continues to reside on the ^{14}C-labeled ring atom.[14]

As shown for the general case in Scheme 18.2, phototranspositions *via* benzvalenes lead to apparent 1,2-shifts of the ring carbon atoms. Rearrangements that involve the Dewar isomers require either a secondary photoreaction to give the prismane structure 12 or an interconversion between Dewar isomers. In the former pathway, the prismane may cleave in two ways, thereby yielding different Dewar benzenes and hence different arenes. This mechanism actually involves *two* 1,2-shifts of the ring carbon atoms (i.e., C1-C2 and C3-C4 in route [i], and C1-C6 and C4-C5 in route [ii]). These double shifts result in either 1,2- or 1,3-transpositions, depending on the substitution pattern in the starting benzenoid compound (i.e., C1-C4 and C2-C3 in route [i], and C1-C4 and C5-C6 in route [ii]). Examples of phototranspositions involving interconverting Dewar isomers have not been reported for benzenoid compounds, but have been described in the conversion of perfluoroalkylpyridazines to pyrazines.[15]

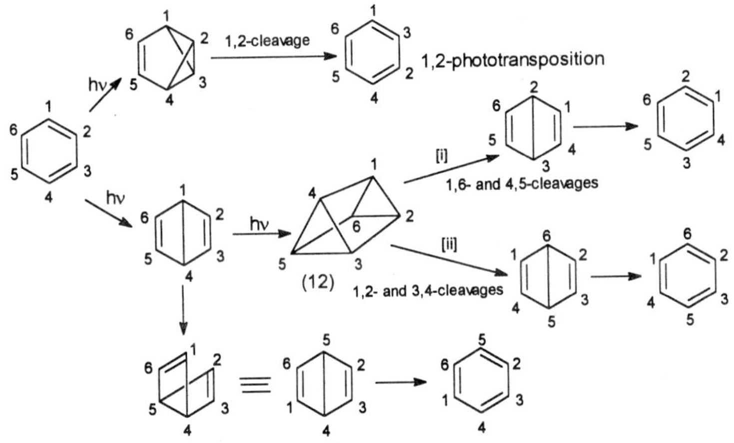

SCHEME 18.2

Ring transposition reactions have also been analyzed in terms of 12 possible ring permutation patterns:[16] this offers an alternative approach to that of intermediate valence bond isomers and has been particularly useful in interpreting and understanding the photoinduced rearrangements of five-membered ring heteroarenes.

18.3 Scope of Arene Photoisomerizations

In the 1960s and 1970s, photoisomerization of aromatic rings was a very active area of research and attracted considerable academic interest.[5] In recent years, the level of activity has declined markedly but research has continued into the factors that promote the formation and/or stabilize the photoisomers. However, despite the potential of this reaction to access highly strained and energy-rich molecules, this area of aromatic photochemistry has not been seriously exploited or applied.

The following account is intended to provide a brief overview of this area and give an outline of the types of compound that have been investigated and the molecular systems accessible by this route.

The isolation of nonaromatic isomers of the benzene ring has been achieved principally for three main classes of aromatic compound. These are: (1) arenes having bulky substituents; (2) cyclophanes; and (3) fluorobenzenes and perfluoroalkylbenzenes.

Arenes with Bulky Substituents

The stabilizing effect in such compounds arises both from the steric interaction between the substituents in the arene and their hindering influence on the rearomatization of the photoisomers. The compounds that have been most extensively studied are the tri-*t*-butylbenzenes and, as may be expected, the 1,3,5- and 1,2,4-substituted isomers display interesting photochemical variations as a result of the different degrees of steric interaction in these arenes and their photoisomers. Thus, 254-nm irradiation of the 1,3,5-isomer yields the 1,2,4-tri-*t*-butylbenzvalene 13 as the sole initial product with a quantum yield of 0.12.[17] On prolonged irradiation, rearomatization of 13 occurs and the fulvene 14, the Dewar benzene 15, the benzvalene 16, and the prismane 17 that are initially very minor photoproducts, all become significant. Indeed, since the prismane has the lowest extinction coefficient at 254 nm, it becomes the major product (65%) at the photostationary state. The same prismane also results from irradiation of the 1,2,4-tri-*t*-butylbenzene as well as the primary products 15 and 16. The interrelationship of these isomeric species is given in Scheme 18.3. To account for the experimental observations, the benzvalenes 13 and 16 must interconvert. This process apparently occurs directly without a rearomatization step and may arise by an ethenylcyclopropane-cyclopentene rearrangement. The formation of the Dewar-benzenes 15 and 18 from excitation within the S_0-S_1 absorption band is unexpected. Since the S_2 state is not accessible, it is tempting to suggest that the T_1 ($^3B_{1u}$) state is involved, but information on this aspect has not been reported. Not surprisingly, trimethylsilyl-substituted arenes undergo similar photoinduced rearrangements to give fulvenes, Dewar benzenes, and benzvalenes,[18] 1,2-phototranspositions have also been noted in this series[19] as well as for a variety of other substituted benzenes including di-*t*-butylphenols.[20] The irradiation of benzocyclobutene, however, provides an interesting variation on the reactions outlined above.[21] Again, a mechanism involving initial *meta*-bonding to give a prefulvene-type intermediate 19 is proposed. The formation of the pentalenes 20 and 21 is then accounted for by ring opening of 19 to the carbene and hydrogen shifts.

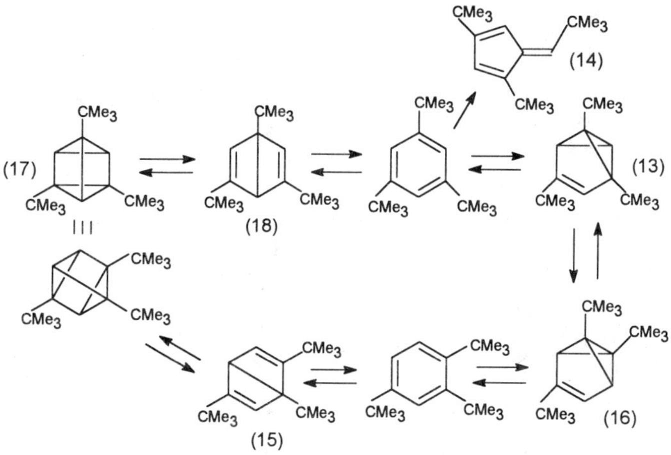

SCHEME 18.3

Condensed polyarenes are, in general, far less prone to photoisomerization than benzene, but substitution of naphthalene and anthracene with bulky groups does promote the formation of their stable photoisomers. In both series, the type of isomer formed appears to be markedly dependent on the degree of substitution. Thus, 1,2,4-tri-*t*-butylnaphthalene is reported to give the naphthvalene 22, whereas the dimethyl compound 23 undergoes 1,4-bonding to yield the Dewar-isomer 24.[22] The photoconversion of 9-*t*-butylanthracene to the Dewar-isomer 25 is so selective that the potential of the reaction as a solar energy storage mechanism has been investigated.[23] Introduction of a cyano group at the 10-position, however, has a marked influence, and the photorearrangement at −20 °C yields the dibenzobicyclo[2.2.1]heptane 26.[24] 9,10-Dialkoxy-octmethylanthracenes do undergo *para*-bonding between the 9,10-positions, but the Dewar-isomer formed at −78 °C undergoes rapid rearomatization on warming.[25] The influence of the degree of substitution on the photoreactivity of these systems is further illustrated by the 9,10-anthraquinones 27.[26] The 1,2-di-*t*-butyl compound displays characteristic anthraquinone photobehavior and yields no ring isomerization products; however, the presence of either a further *t*-butyl or a trimethylsilyl group at the 3-position not only induces a bathochromic shift and an increase in the extinction coefficient of the longest-wavelength absorption of the quinone, but also promotes 1,4-bonding in the excited state to give the photoisomer 28.

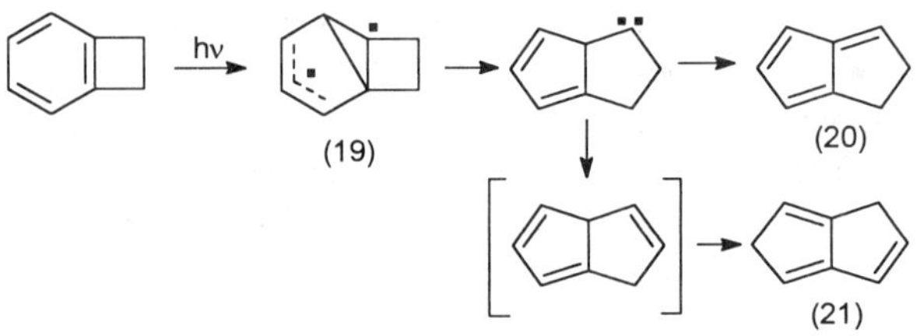

(19)   (20)   (21)

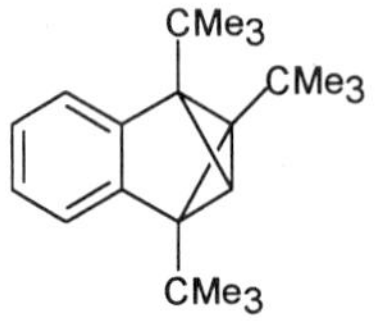

(22) 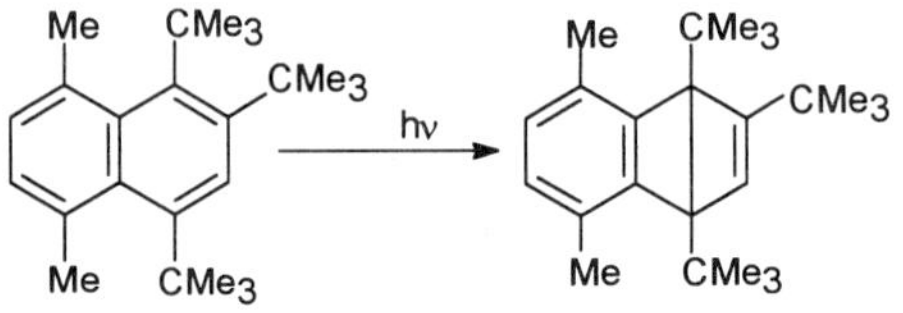   (23)   (24)

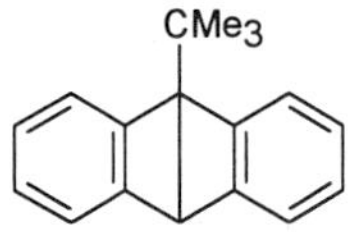

(25) 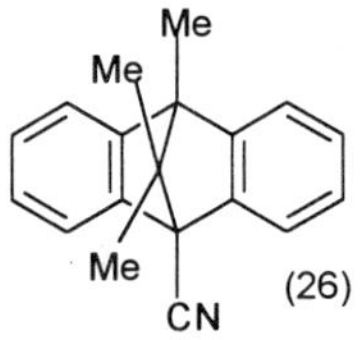   (26)

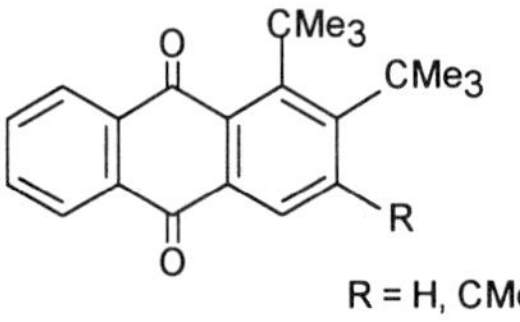

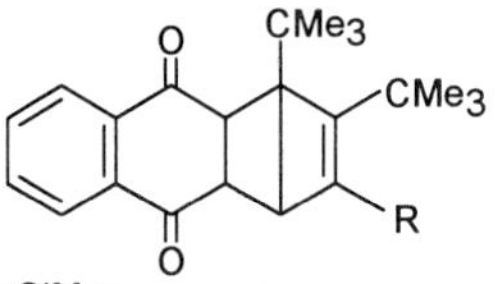

R = H, CMe3, or SiMe3

(27)   (28)   (29)

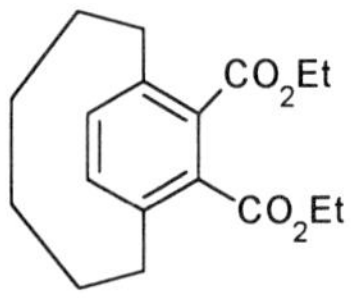

Cyclophanes

[6]-Paracyclophanes such as 29 show a marked propensity to undergo photoinduced 1,4-bonding to give Dewar-isomers 30.[27] The reaction can be very efficient and is reversible both thermally and photochemically. Conversion to the prismane may also occur on irradiation. The use of this photoisomerization as a method for the synthesis of bridges Dewar benzenes has been reviewed.[28]

Fluorobenzenes and Perfluoroalkylbenzenes

Hexafluorobenzene undergoes photoisomerization exclusively to the Dewar-isomer in both the vapor and liquid phases.[29] This tendency for 1,4- rather than 1,3-bonding is a characteristic of most fluorinated benzenes and alkylbenzenes. For example, 1,2,4-trifluorobenzene yields two of the possible three Dewar-isomers.[30] The isomerization is believed to be a singlet-state process, and it is interesting to note that no photoisomers have been reported from fluorobenzene, which has a triplet yield of 0.67 to 0.93 dependent on the wavelength of excitation.

The photoreactions of a variety of perfluoroalkylbenzenes have been investigated and, in some cases, the nonaromatic isomers have remarkable stabilities. For example, perfluorohexamethyl-benzene yields the benzvalene and Dewar-isomers as the primary photoproducts; the latter compound is so stable that it can also be formed *thermally* from the arene.[31]

8.4 Photoisomerization of Protonated Arenes

The tendency of the benzene ring to undergo 1,3-bonding on irradiation is not affected apparently by protonation, and methylated benzenonium ions such as **31** yield the corresponding bicyclo[3.1.0]hexenyl species **32**.[32] At one time, it was thought that irradiation of the parent, benzene, under acid conditions behaved similarly, but it was later shown that the bicyclo[3.1.0]hex-2-enes from this system arose from acidolysis of the benzvalene.[33] Triflic acid protonates phenols at the 4-position and irradiation of the resulting protonated cyclohexadienone, not surprisingly, yields bicyclo[3.1.0]hex-3-en-2-ones.[34] The intermediate is considered to be **33**, which is also formed on irradiation of the bicyclic enone under acid conditions: overall, by this route, positional isomers of the starting phenol may be formed. In a manner similar to protonated benzenes, tropylium fluoroborate isomerizes photochemically to the bicyclo[3.2.0]heptene ion **34**.[35]

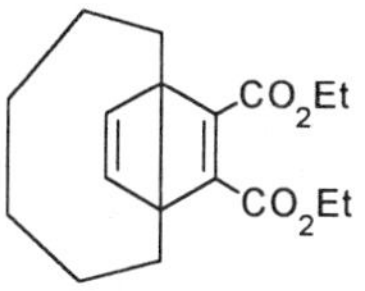

(30) 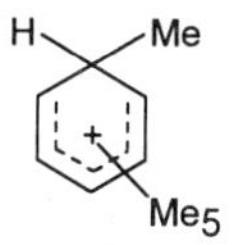   (31) 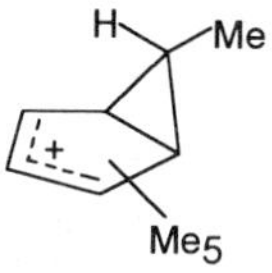   (32) 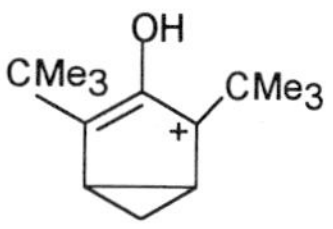   (33)

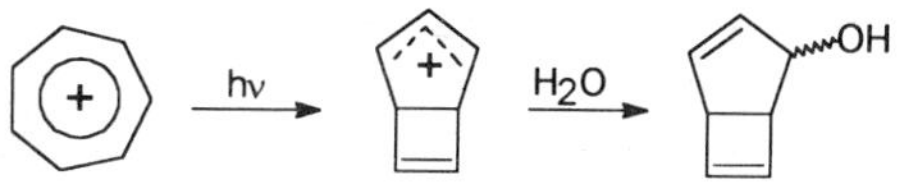

(34)

eferences

1. Bryce-Smith, D. and Gilbert, A., The organic photochemistry of benzene-1, *Tetrahedron, 32,* 1309, 1976.
2. Wilzbach, K. E. and Kaplan, L., Photoisomerization of dialkylbenzenes, *J. Am. Chem. Soc.,* 86, 2307, 1964.
3. Blair, J. M. and Bryce-Smith, D., The photoisomerization of benzene to fulvene, *Proc. Chem. Soc.,* 287, 1957.
4. Angus, H. J. F. and Bryce-Smith, D., Addition of maleic anhydride to benzene, *Proc. Chem. Soc.,* 326, 1959.
5. Reviews of aromatic compounds appear in annual volumes of *Photochemistry,* Specialist Periodical Reports, Vols. 1–23, The Royal Society of Chemistry, London, 1970–1993.
6. Kaplan, L., Rausch, D. J., and Wilzbach, K. E., Photosolvation of benzene, *J. Am. Chem. Soc.,* 94, 8638, 1972.
7. Wilzbach, K. E. and Kaplan, L., Photolysis of benzene vapour, *J. Am. Chem. Soc.,* 90, 3291, 1968.
8. See footnote on page 1314 of Reference 1.
9. Kjell, D. P. and Sheridan, R. S., Diazabenzosemibullvalene. A precursor to the benzoprefulvene biradical and indenylmethylene, *J. Am. Chem. Soc.,* 108, 4111, 1986.
10. Wilzbach, K. E., Ritscher, J. S., and Kaplan, L., Benzvalene, the tricyclic valence isomer of benzene, *J. Am. Chem. Soc.,* 89, 1031, 1967.
11. Bryce-Smith, D., Gilbert, A., and Robinson, D. A., Direct transformation of the second excited singlet state of benzene into Dewar benzene, *Angew. Chem. Int. Ed.,* 10, 745, 1971.
12. Meisl, M. and Janoschek, R., The dynamics of the low-lying excited states of benzene: The biradicaloid structure of the S_2 state, *J. Chem. Soc., Chem. Commun.,* 1066, 1986.
13. Johnstone, D. E. and Sodeau, J. R., Matrix-controlled photochemistry of benzene and pyridine, *J. Phys. Chem.,* 95, 165, 1991.

14. Kaplan, L., Wilzbach, K. E., Brown, W. G., and Yang, S. S., Phototransposition of carbon atoms in the benzene ring, *J. Am. Chem. Soc.,* 87, 675, 1965.

15. Chambers, R. D., Middleton, R., and Corbally, R. P., Transpositions in aromatic rings, *J. Chem. Soc., Perkin I,* 731, 1975.

16. Barltrop, J. A., Carder, R., Day, A. C., Harding, J. R., and Samuels, C., Ring permutations in the photochemistry of hydroxypyrylium cations, *J. Chem. Soc., Chem. Commun.,* 729, 1975.

17. Den Besten, I. E., Kaplan, L., and Wilzbach, K. E., Photoisomerization of tri-*t*-butylbenzenes. Photochemical interconversion of benzvalenes, *J. Am. Chem. Soc.,* 90, 5868, 1968.

18. West, R., Furue, M., and Mallikarjuna Rao, V. N., Photolysis of 1,2,4,5-tetrakis(trimethylsilyl)benzene, *Tetrahedron Lett.,* 911, 1973.

19. Ishikawa, M., Sakamoto, H., Kametani, F., and Minato, A., Synthesis and photochemical behaviour of *bis*(disilanyl)benzenes, *Organometallics,* 8, 2767, 1989.

20. Matsuura, T., Hiromoto, Y., Okada, A., and Ogura, K., The photochemical transposition of 2,6-di-*t*-butylphenol through their ketone tautomers, *Tetrahedron,* 29, 2981, 1973.

21. Turro, N. J., Zhang, Z., Trahanovsky, W. S., and Chou, C. H., Photochemistry of benzocyclobutene, *Tetrahedron Lett.,* 29, 2543, 1988.

22. Yoshida, Z., Miki, S., Kawamoto, F., Hijiya, T., and Ogoshi, H., Photovalence bond isomerization of naphthalenes, *Chem Abstr.,* 92, 163, 323; and, Valence bond isomerization of condensed polycyclic aromatic compounds, *Chem. Abstr.,* 197, 660, 1980.

23. Abdul-Ghani, A. J., Bashi, N. O. T., and Maree, S. N., Solar energy storage through catalysed reversible photovalence isomerization of 9-*t*-butylanthracene, *J. Solar Energy Res.,* 5, 53, 1987.

24. Meng, J., Yao, X., Wing, H., Matsuura, T., and Ito, Y., Novel photoisomerization of 9-*t*-butyl-10-cyanoanthracene involving 1,2-transfer of methyl group, *Tetrahedron,* 44, 355, 1988.

25. Meador, M. A. and Hart, H., Substituent effects on the photoisomerization of anthracenes to their 9,10-Dewar isomers, *J. Org. Chem.,* 54, 2336, 1989.

26. Miki, S., Matsuo, K., Yoshida, M., and Yoshida, Z., Novel anthraquinone derivatives undergoing photochemical valence isomerization, *Tetrahedron,* 29, 2211, 1988.

27. Liebe, J., Wolff, C., and Tochtermann, W., Reaktionen des [6]paracyclophan-8,9-dicarbonsaurediethylesters: Brom-addition und prisman-bildung, *Tetrahedron Lett.,* 2439, 1982; Dreeskamp, H., Kapahnke, P., and Tochtermann, W., Photovalence isomerization of sterically strained aromatic hydrocarbons: 8,9-dicarboethoxy[6]paracyclophane, *Radiat. Phys. Chem.,* 32, 537, 1988.

28. Bickelhaupt, F. and de Wolf, W. H., Bridged valence isomers of benzene and cyclophanes, *Rec. Trav. Chim., Pays-Bas,* 107, 459, 1988.

29. Camaggi, G., Gozzo, F., and Cevidalli, G., *para*-Bonded isomers of fluoroaromatic compounds, *J. Chem. Soc., Chem. Commun.,* 313, 1966; Haller, I., Photoisomerization of hexafluorobenzene, *J. Am. Chem. Soc.,* 88, 2070, 1966.

30. Semeluk, G. P. and Stevens, R. D. S., Photoisomerization of 1,2,4-trifluorobenzene, *J. Chem. Soc., Chem. Commun.,* 1720, 1970.

31. Clifton, E. D., Flowers, W. T., and Haszeldine, R. N., Thermal valence-bond isomerization of an aromatic compound to its *para*-bonded form: Preparation of hexakis(pentafluoroethyl)-bicyclo[2.2.0]hexa-2,5-diene, *J. Chem. Soc., Chem. Commun.,* 1216, 1969.

32. Childs, R. F., Sakai, M., and Winstein, S., The observation and behaviour of the pentamethylcyclopentadienylmethyl cation, *J. Am. Chem. Soc.,* 90, 7144, 1968.

33. Kaplan, L., Rausch, D. J., and Wilzbach, K. E., Photosolvation of benzene. Mechanism of formation of bicyclo[3.1.0]hex-3-en-2-yl and of bicyclo[3.1.0]hex-2-en-6-yl derivatives, *J. Am. Chem. Soc.,* 94, 8638, 1972.

34. Childs, R. F. and George, B. E., A quantitative examination of the photoisomerization of some protonated phenols, *Can. J. Chem.,* 66, 1343, 1988.

35. van Tamelen, E. E., Greeley, R. H., and Schumacher, H., Photolysis of carbocationic species. Nonbenzenoid aromatics, *J. Am. Chem. Soc.,* 93, 6151, 1971.

19

Phototransposition and Photo-Ring Contraction Reactions of 4-Pyrones and 4-Hydroxypyrylium Cations

James W. Pavlik
Worcester Polytechnic Institute

19.1 Definition and Historical Background

4-Pyrones and 4-hydroxypyrylium cations undergo mechanistically similar phototransposition and photo-ring contraction reactions to yield a variety of synthetically useful products.

The history of 4-pyrone photochemistry dates back to the early work of Paterno who reported that 2,6-dimethyl-4-pyrone 1a (Scheme 19.1) undergoes photodimerization.[1] The structure of this dimer was not correctly assigned,[2] however, until 1958 when Yates and Jorgenson[3,4] elucidated the structure of the product as the head-to-tail dimer 2. Later, Yates and Still[5] also observed that irradiation of 1a in dilute aqueous solution resulted in the formation of a low yield of 4,5-dimethyl-2-furaldehyde 4a. This monomeric product was suggested to arise via the intermediacy of epoxycyclopentenone 3a. More recently, 4-pyrones have been found to phototranspose to 2-pyrones, and again, epoxycyclopentenones were postulated as intermediates.[6-8] In support of these suggestions, 3,4-diphenylepoxycyclopentenone 3b was independently synthesized and shown to undergo efficient photorearrangement to 2-pyrone 5b.[9] These suggestions were substantiated when 2,5-dimethylepoxycyclopentenone 3c was found to be an isolable intermediate in the phototransposition of 3,5-dimethyl-4-pyrone 1c to 3,6-dimethyl-2-pyrone 5c.[10] Finally, a variety of studies have shown that 4-pyrones undergo photo-ring contraction with nucleophilic addition of solvent to afford 4-hydroxycyclopentenones such as 6 and 7 in a reaction that is competitive with epoxycyclopentenone and 2-pyrone formation.[11,12]

0-8493-8634-9/95/$0.00+$.50
© 1995 by CRC Press, Inc.

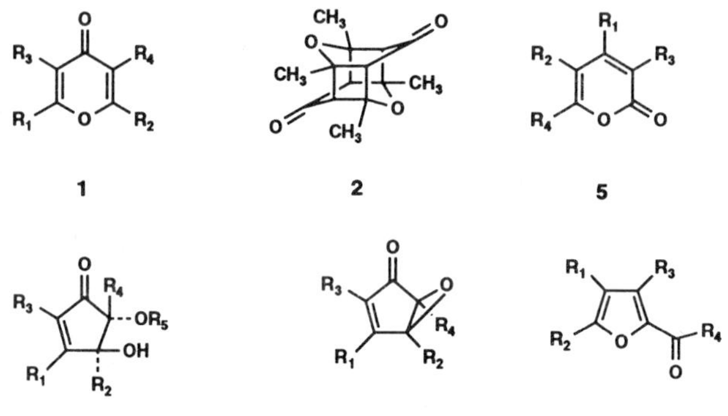

	R_1	R_2	R_3	R_4*
a:	CH_3	CH_3	H	H
b:	Ph	Ph	H	H
c:	H	H	CH_3	CH_3
d:	H	CH_3	H	CH_3
e:	CH_3	H	CH_3	H
f:	CH_3	CH_3	H	CH_3
g:	CH_3	CH_3	CH_3	CH_3
h:	H	CH_2CH_3	H	CH_3
i:	CH_2CH_3	H	CH_3	H
j:	tBu	tBu	H	H
k:	CH_3	H	H	CH_3
l:	H	CH_3	CH_3	H

*Unless otherwise specified, the substituent captions
apply throughout this manuscript.

SCHEME 19.1

The phototransposition of the 2,6-dimethyl-4-hydroxypyrylium cation $1aH^+$ (Scheme 19.2) to yield 4,6-dimethyl-2-hydroxypyrylium cation $7aH^+$ and the 4,5-dimethyl-2-hydroxypyrylium cation $5aH^+$ as primary photoproducts was the first reported example of a 4-hydroxypyrylium cation photorearrangement.[13] In this case, the initially formed $5aH^+$ was also observed to be in photoequilibrium with the 5,6-dimethyl-2-hydroxypyrylium cation $5dH^+$.[13,14] In addition to phototransposition to yield the 5,6-dimethyl-2-hydroxypyrylium cation $5dH^+$ and 3,4-dimethyl-2-hydroxypyrylium cation $5eH^+$, 2,3-dimethyl-4-hydroxypyrylium cation $1dH^+$ was observed to undergo photo-ring contraction upon irradiation in concentrated H_2SO_4 to yield furyl cation $4dH^+$.[15] The product distribution in this reaction was found to be remarkably dependent on the concentration of the H_2SO_4 solvent. Thus, whereas phototransposition products $5dH^+$ and $5eH^+$

constitute over 97% of the product mixture after irradiation of **1dH⁺** in 100% H_2SO_4, the photo-ring contraction product 5-methyl-2-acetylfuran **4d** is essentially the sole product obtained when the irradiation is carried out in 50% aqueous acid.[16]

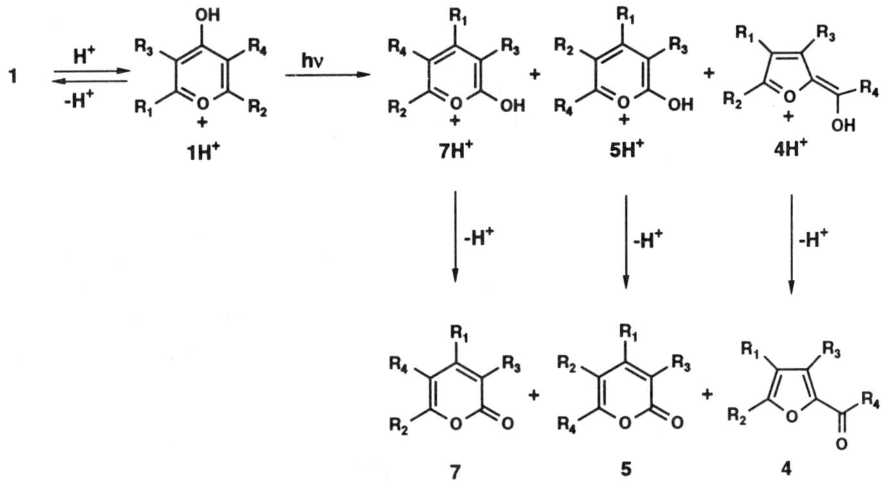

SCHEME 19.2

The effect of changing the acid concentration was found to be general since a variety of 4-pyrones that undergo phototransposition upon irradiation in concentrated H_2SO_4 were also observed to undergo photo-ring contraction to provide synthetically useful yields of furan derivatives upon photolysis in 50% aqueous H_2SO_4. Thus, whereas the 2,3,6-trimethyl-4-hydroxypyrylium cation **1fH⁺** undergoes regiospecific phototransposition to the 4,5,6-trimethyl-2-hydroxypyrylium cation **5fH⁺** in concentrated H_2SO_4, the ring contraction product, 4,5-dimethyl-2-acetylfuran **4f**, accounts for 94% of the product mixture when the irradiation is carried out in 50% H_2SO_4 at room temperature.[16]

Variable-temperature studies revealed that these furan derivatives are not primary products in these photoreactions.[16] Thus, irradiation of **1fH⁺** in 50% aqueous H_2SO_4 at 0 °C led to the formation of 4,5-dihydroxy-3,4,5-trimethylcyclopent-2-ene-1-one **6f** in greater than 60% yield. In 50% H_2SO_4 at room temperature, dihydroxycyclopentenone **6f** isomerizes to 4,5-dimethyl-2-acetylfuran **4f**, the product obtained when the irradiation is carried out at room temperature. In some cases, however, acid-catalyzed isomerization of the initially formed dihydroxycyclopentenone adduct is too rapid, even at 0 °C, to allow isolation.[16]

19.2 Mechanistic Studies

4-Pyrone Photoreactions

Mechanistic studies are consistent with the general mechanism for 4-pyrone photochemistry outlined in Scheme 19.3. Alkyl-substituted 4-pyrones exhibit intense absorption near 250 nm (Table 19.1) characteristic of $\pi \rightarrow \pi^*$ transitions. Thus, direct photochemical excitation at this wavelength results in the population of the $S_1(\pi,\pi^*)$ state, **1***, of the 4-pyrone,[17] which is suggested[6–8,10] to undergo symmetry-allowed electrocyclic ring closure to form oxyallyl bicyclic Zwitterion **8**. In one case (Scheme 19.4), the Zwitterion **8c**, formed by irradiation of 4-pyrone **1c**, has been trapped

Table 19.1   UV Absorption Maxima for 4-Pyrones and 4-Hydroxypyrylium Cations

ULTRA-VIOLET ABSORPTION MAXIMA
FOR 4-PYRONES AND 4-HYDROXYPYRYLIUM CATIONS

Compound	λ_{max} (nm)	ε	Solvent
2,6-dimethyl-4-pyrone	249 ; 209	14,800 ; 6,600	H_2O
	247 ; 244	16,100 ; 12,300	TFE
	257 ; 235	12,100 ; 11,400	96 % H_2SO_4
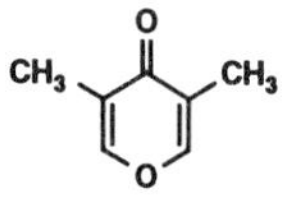 3,5-dimethyl-4-pyrone	253		96 % H_2SO_4
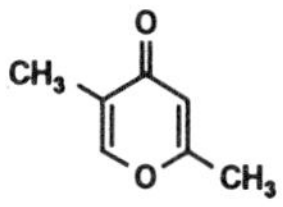 methyl/methyl pyrone	252 ; 211	13,400 ; 10,020	EtOH
dimethyl pyrone	254 ; 207	12,675 ; 9,150	H_2O
	262	11,000	96 % H_2SO_4
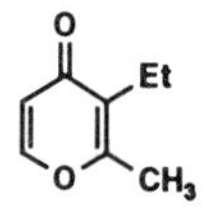 Et, CH_3 pyrone	263	12,000	96 % H_2SO_4
CH_3, Et pyrone	260 ; 215	10,650 ; 7,300	H_2O
	255 ; 210	13,600 ; 9,310	EtOH
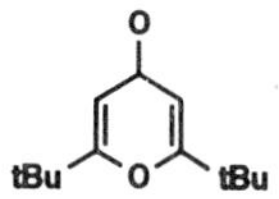 tBu, tBu pyrone	265 ; 240	7860 ; 7,400	96 % H_2SO_4
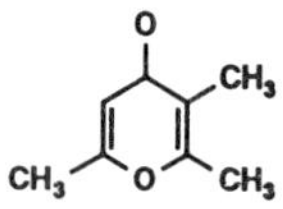 CH_3, CH_3, CH_3 pyrone	254 ; 211	13,100 ; 7,900	H_2O
Ph, CH_3, CH_3 pyrone	257	16,600	CH_3CN
	257 ; 227	13,800 ; 13,300	96 % H_2SO_4
CH_3, CH_3, CH_3, CH_3 pyrone	259 ; 212	12,700 ; 7900	H_2O
	268 ; 244	12,800 ; 8,000	96 % H_2SO_4
Ph, Ph, CH_3, CH_3 pyrone	255	15,500	CH_3CN
	290 ; 225	5,800 ; 18,600	96 % H_2SO_4
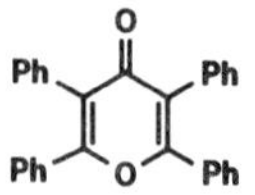 Ph, Ph, Ph, Ph pyrone	275	34,600	CH_3CN

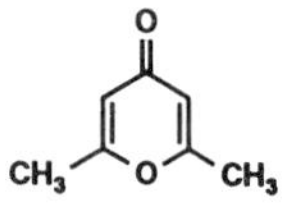

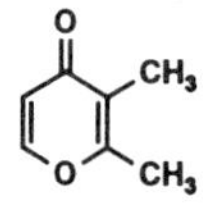

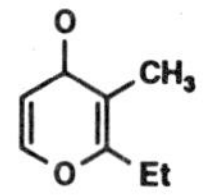

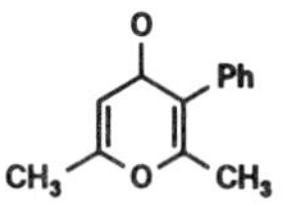

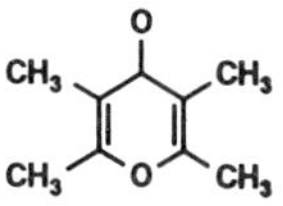

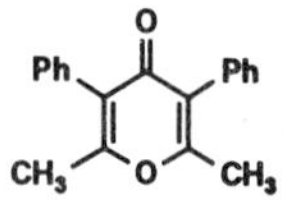

by cycloaddition of the oxyallyl system to furan to yield the [4+2]-adduct 9.[18] Interestingly, at higher furan concentrations, the yield of 9 decreases and is replaced by the formation of [2+2]-adduct 10.[18] This suggests the existence of a shallow potential energy minimum on the S_1 surface at approximately the S_0 equilibrium geometry, which preserves the excited state long enough for it to react with furan.[19] Alkynes have also been employed to trap the excited state of 2,6-dimethyl-4-pyrone 1a resulting in the formation of [2+2]-cyclobutene-4-pyrone adducts.[20] In the absence of such trapping agents and in the presence of weakly nucleophilic solvents, the major pathway (Scheme 19.3) for the bicyclic Zwitterion 8 involves intramolecular nucleophilic attack on the oxyallyl system by the lone pair on the epoxide oxygen leading to the formation of the epoxycyclopentenone 3 (and/or 11) possibly via an oxaniabenzvalene intermediate. Epoxycyclopentenone 3c, formed photochemically from 1c, is an isolable intermediate and has been shown to undergo efficient photoisomerization to 2-pyrone 5c. According to this interpretation, 2-pyrones are secondary products in these reactions and are formed by photochemical and/or thermal isomerization of the epoxycyclopentenones. Although, in most cases, the intermediate expoxycyclopentenones are photochemically and/or thermally too reactive to be isolated or detected, after short-duration irradiation of tetramethyl-4-pyrone 1g in 2,2,2-trifluoroethanol (TFE) solvent, tetramethyl-2-pyrone 5g is formed thermally from a photochemically generated species with a half-life of ~50 min.[11] This thermally labile precursor of 5g was postulated to be epoxycyclopentenone 3g. Addition of small amounts of methanol to the TFE solution of 1g before, but not immediately after, the short-duration irradiation results in a decrease in the final quantity of 5g formed with concomitant formation of tetramethyl-4-hydroxy-5-methoxycyclopentenone 7g. These trapping experiments show that methanol has trapped a species occurring earlier on the reaction pathway that has a lifetime far shorter than that observed for 3g, the immediate precursor of 5g. This short-lived species was suggested to be the Zwitterion 8g (Scheme 19.3). Methanol adduct 7g can be derived from nucleophilic attack on the oxyallyl Zwitterion 8g followed by opening of the epoxide ring in the enolate or the corresponding enol. In neat nucleophilic solvents such as water or methanol, the oxyallyl Zwitterions can be effectively trapped to provide 4-hydroxycyclopentenone adducts such as 6 and 13 or 7 and 14 (Scheme 19.3) as the major or only product observed.[11]

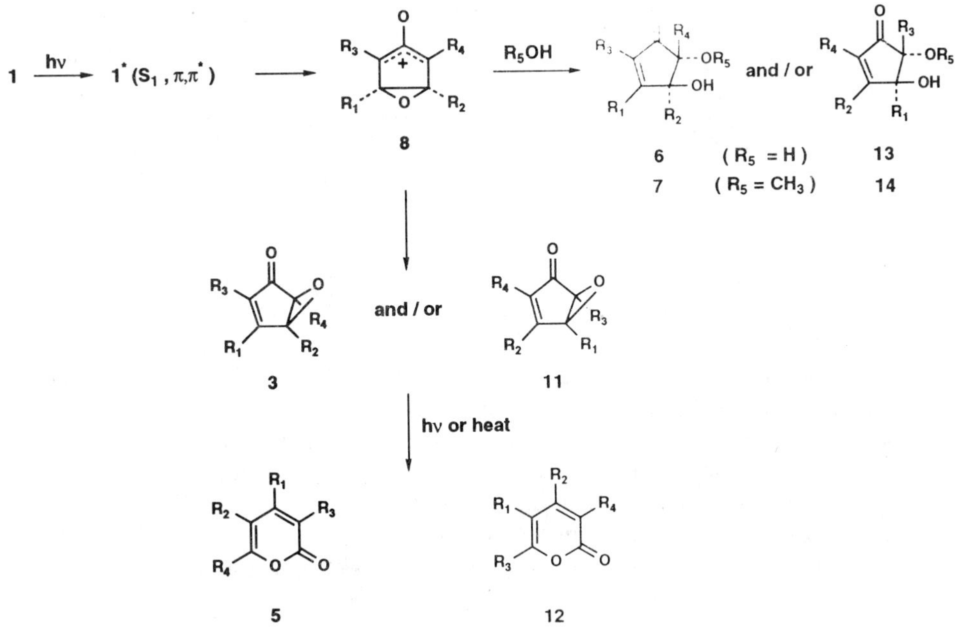

SCHEME 19.3

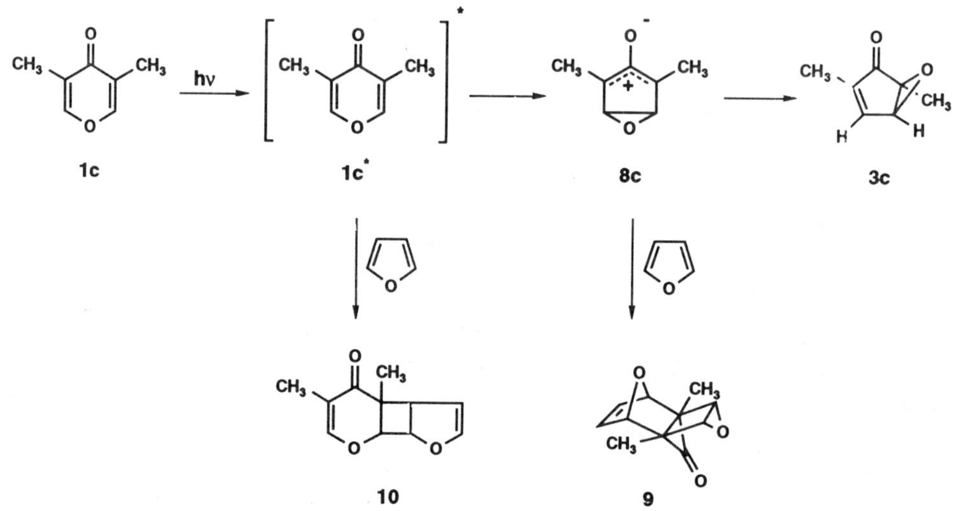

SCHEME 19.4

Regio- and Stereochemistry

The regiochemistry of the phototransposition of unsymmetrically substituted 4-pyrones appears to be controlled by the distribution of the positive charge in the oxyallyl system of the bicyclic Zwitterion intermediate. Thus, when R_1 = H and R_2 = CH_3 or Ph (Scheme 19.5), the phototransposition was observed to occur regiospecifically via Path A resulting in the 4,5,6-trisubstituted-2-pyrone.[8,11] These products are expected if the epoxide oxygen attacks the more highly substituted end of the oxyallyl system bearing the greater concentration of positive charge. When R_1 = Ph and R_2 = *para*-substituted phenyl, the distribution of isomeric 2-pyrones obtained is similarly sensitive to the electronic effects of the *para*-substituent in the phenyl ring. Thus, Path A predominates when the *para*-substituent is electron donating while Path B is the major route when the *para*-substituent is electron withdrawing.[21]

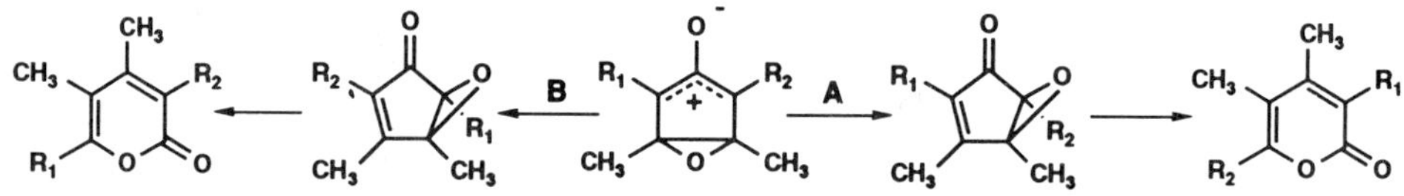

SCHEME 19.5

Intermolecular nucleophilic attack leading to photo-ring contraction yielding 4-hydroxycyclopentenone solvent adducts has also been reported to occur specifically at the more highly substituted side of the oxyallyl system to yield only the regioisomer bearing the hydroxyl group on the more highly substituted side of the molecule.[11,16,22] In one case, however, thorough analysis of the product mixture resulting from the irradiation of an aqueous solution of 2-ethyl-3-methyl-4-pyrone 1h led to the isolation of both regioisomers 6h and 13h in a ratio of 1.6 to 1.[23] These results show that intermolecular nucleophilic attack by water does occur at either end of the oxyallyl system, but mainly at the more highly substituted side.

In either case, it would be expected that the nucleophile would attack the bicyclic Zwitterion from the less-hindered side — *anti* to the epoxide ring — leading to the *trans*-stereochemistry observed in these cyclopentenone adducts.[11,12,23]

4-Hydroxypyrylium Cation Photoreactions

Experiments using deuterium and alkyl groups as positional labels indicate that the photochemistry of 4-hydroxypyrylium cations involves initial 2,6- and 2,5-bridging in the first excited state of the cation (Scheme 19.6).[24,25]

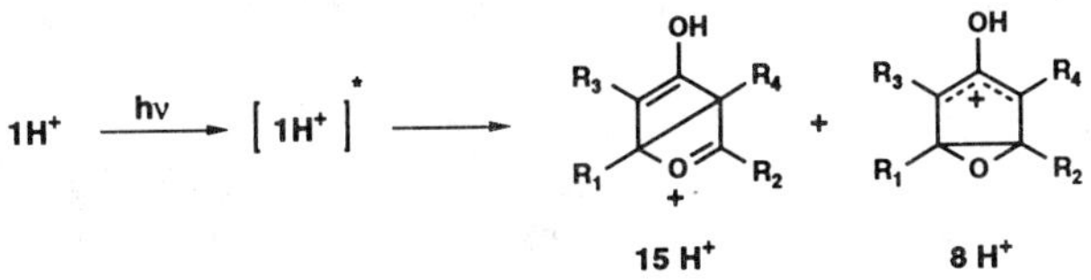

SCHEME 19.6

The 2,5-bridged 4-hydroxy-Dewar pyrylium cation $15H^+$ was then suggested[26] to undergo ground-state isomerization leading to the 2-hydroxy-Dewar pyrylium cation $17H^+$. As shown in Scheme 19.7, this rearrangement requires only rotation about the C2-C3 bond in $15H^+$ and could involve the intermediacy of the homoaromatic cyclobutenyl cation $16H^+$. Ring opening of the cation $17H^+$ leads directly to the 2-hydroxypyrylium cation $7H^+$ in which C2 and C6 of the reactant have transposed to ring positions 4 and 6 in the product.

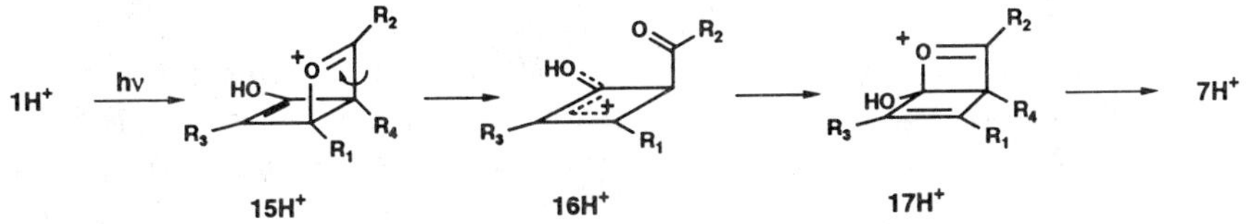

SCHEME 19.7

Both phototransposition and photo-ring contraction can be explained in terms of the 2,6-bridged 3-hydroxyoxabicyclohexenyl cation $8H^+$.[25,27] This species would be expected to undergo rapid inter- or intramolecular nucleophilic attack. At high H_2SO_4 concentrations (i.e., in the absence of high concentrations of a suitable external nucleophile), intramolecular attack (Path A in Scheme 19.8) of the epoxide ring on the oxyallyl system resulting in migration of the epoxide ring and formation of the more stable 2-hydroxyoxabicyclohexenyl cation $3H^+$ would take place. This latter cation is the conjugate acid of the epoxycyclopentenone 3, and such compounds are known to undergo acid-catalyzed rearrangement to 2-pyrones via a second epoxide ring migration.[28]

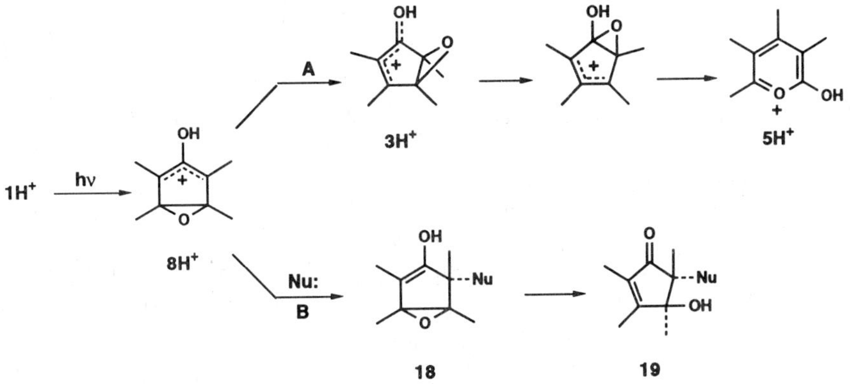

SCHEME 19.8

Alternatively, in 50% H_2SO_4, the rate of intermolecular nucleophilic attack on the oxyallyl system by H_2O or HSO_3O^- would be greatly increased leading eventually either to diol **19** (Nu = OH) or to cyclopentenone sulfate ester **19** (Nu = HSO_3O^-) that would be converted to the diol by subsequent hydrolysis. One such sulfate ester has been shown to be an isolable intermediate in the phototransposition of the 3,5-dimethyl-4-hydroxypyrylium cation **1cH+** in concentrated H_2SO_4[29] but does not appear to be formed upon photolysis of **1cH+** in 50% aqueous H_2SO_4.[27]

2,6-Bridging vs. 2,5-Bridging

Although the products arising via the 2,6-bridging pathway generally predominate over the products formed by 2,5-bridging by at least 20:1,[25] the former pathway is also more sensitive to the size of the substituents at C2 and C6. Thus, since these groups move closer together during 2,6-bridging than during 2,5-bridging, the former pathway is inhibited by bulky groups at these positions. This inhibition is dramatically illustrated in the case of the 2,6-di-*t*-butyl-4-hydroxypyrylium cation **1jH+** which phototransposes to yield the 4,6-di-*t*-butyl-2-hydroxypyrylium cation **7jH+**, the product of 2,5-bridging.[26]

Regiochemistry of Phototransposition

The regiochemistry of 4-hydroxypyrylium cation photochemistry is determined by the direction of epoxide ring migration in the 2,6-bridged oxabicyclohexenyl species, formed initially which in turn appears to be controlled both by the charge distribution in the oxyallyl system and by the relative stabilities of the possible 2-hydroxyoxabicyclohexenyl cations formed. Thus, in the oxabicyclohexenyl cation **8kH+**, photochemically generated from **1kH+**, both factors favor epoxide migration toward the more highly substituted side of the oxyallyl system. As expected, the 4,6-dimethyl-2-hydroxypyrylium cation **5kH+** was observed to predominate over the 3,5-dimethyl cation **5lH+** by a factor of almost 16.[24] In other cases, the position of the alkyl substituents may not favor migration in a single direction, and the phototransposition will exhibit less regioselectivity.[25]

19.3 Synthetic Applications

Phototransposition of 4-Hydroxypyrylium Cations

Phototransposition of 4-hydroxypyrylium cations to 2-hydroxypyrylium cations followed by neutralization of the acid solution allows isolation of 2-pyrones that are often tedious to prepare by multistep ground-state synthetic procedures. Although the use of concentrated sulfuric acid as solvent makes large-scale reactions impractical, 1-g size reactions requiring 90 to 100 ml concentrated H_2SO_4 as solvent can be carried out conveniently. As shown in Table 19.2, the synthetic usefulness of these reactions is somewhat limited by the fact that these phototranspositions generally lead to mixtures of 2-hydroxypyrylium cations and thus to mixtures of 2-pyrones that require chromatographic separation.

Phototransposition of 4-Pyrones

In highly polar alcohol solvents of low nucleophilicity, such as TFE, Zwitterion intermediates such as **8** are efficiently formed but not effectively trapped. As shown in Table 19.3, under such conditions phototransposition to the corresponding 2-pyrone predominates. 4-Hydroxy-5-(2,2,2-trifluoroethoxy)cyclopentenones are sometimes observed as minor products in these reactions.

Photo-Ring Contraction of 4-Hydroxypyrylium Cations

The photo-ring contraction of 4-hydroxypyrylium cations in 50% aqueous H_2SO_4 at room temperature can provide synthetically useful yields of 2-acetylfuran derivatives. At this time, however,

Table 19.2 Phototransposition Reactions in Concentrated H_2SO_4

Reactant	Primary Product (s) (after neutralization): major	minor	trace	Reference
				25
				24
				15 , 16
				15 , 25
				24
				24
				25
				25

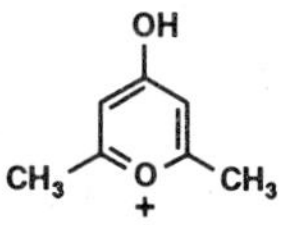

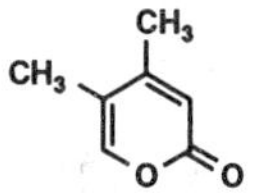

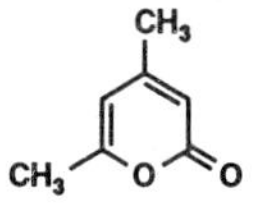

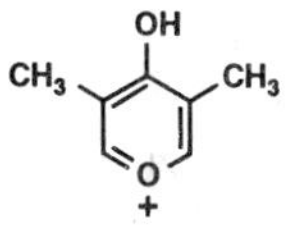

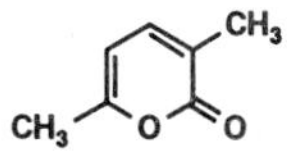

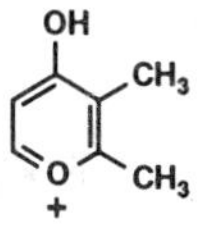

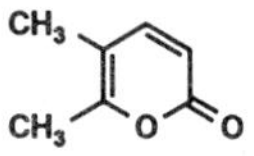

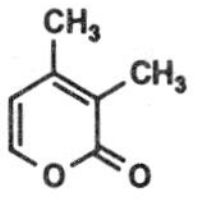

the reaction has been limited to the three examples shown in Table 19.4. These reactions are restricted to hydroxypyrylium cations with alkyl substituents at C3. In the absence of such a substituent, the rearrangement leads to 2-furaldehyde derivatives that have low stability in 50% H_2SO_4.

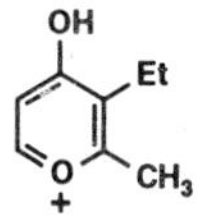

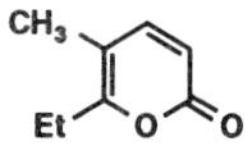

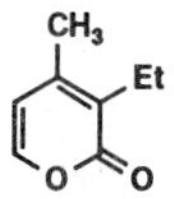

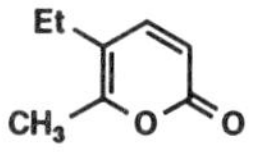

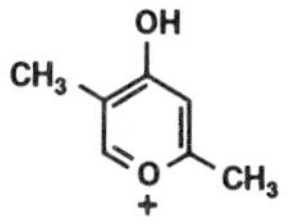

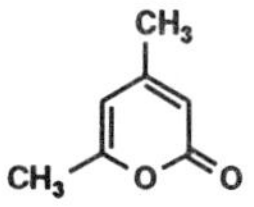

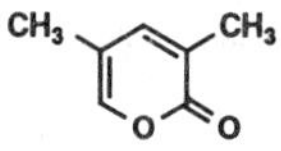

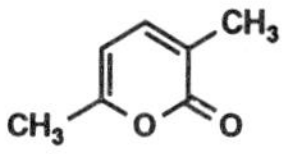

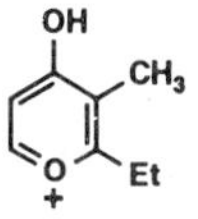

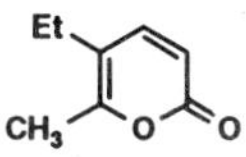

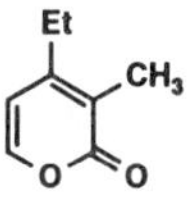

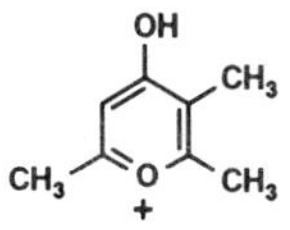

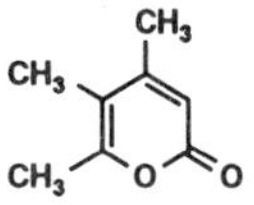

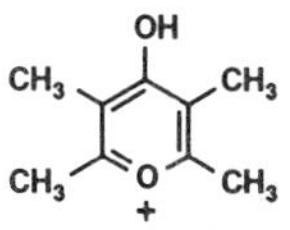

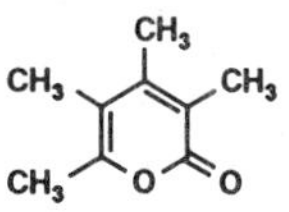

Photo-Ring Contraction of 4-Pyrones

The photo-ring contraction of 4-pyrones allows synthesis of a variety of highly functionalized 4-hydroxycyclopentenones of known stereochemistry. Barton has made use of this reaction in a synthesis of the mold metabolite, terrein, from kojic acid.[30]

Table 19.3   Phototransposition of Neutral 4-Pyrones

Reactant	Product	Yield %	Reference
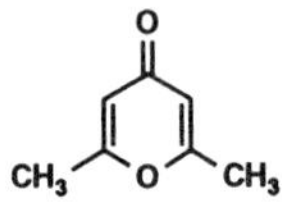	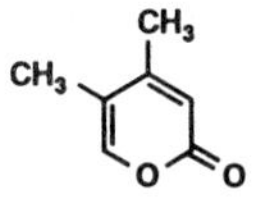	7	12
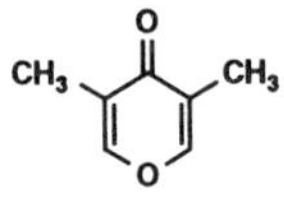	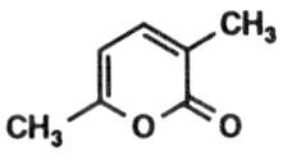	46	19
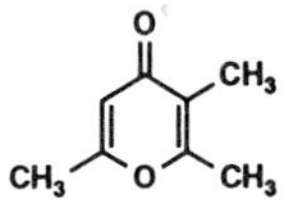	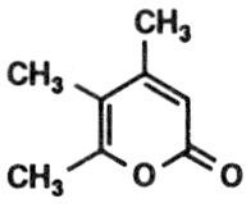	-----	11
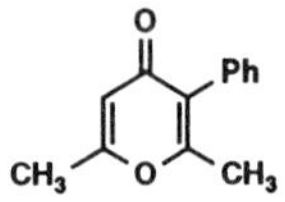	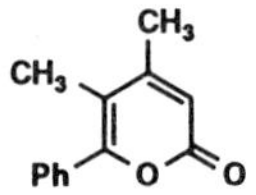	-----	8
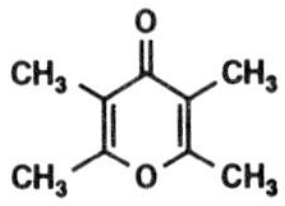	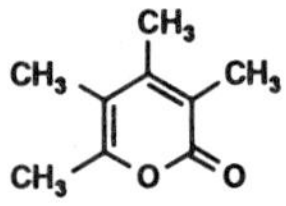	----	8 , 11
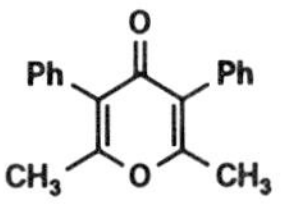	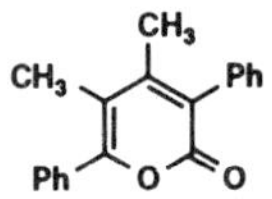	67	7
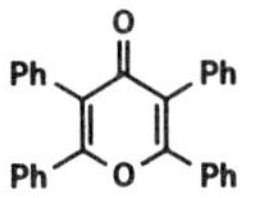	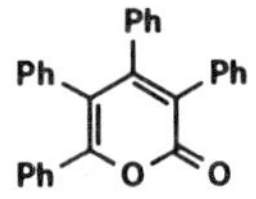	21	7

Table 19.4   Photo-Ring Contraction of 4-hydroxypyrylium Cations in 50% H_2SO_4 at Room Temperature

Reactant	Product (after neutralization)		Reference
	major	trace	
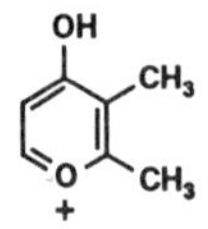	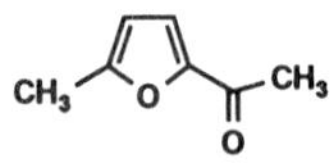	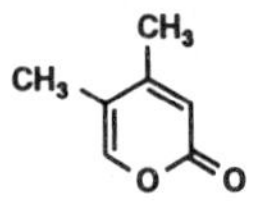	16 , 27
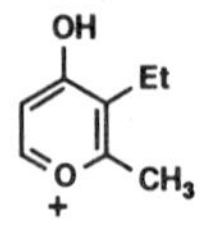	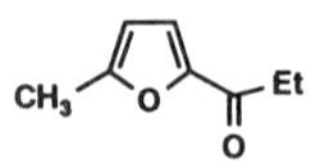		27
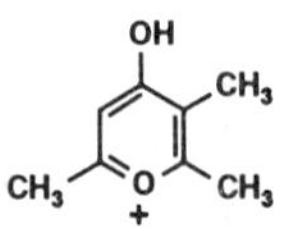	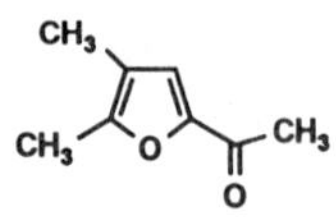	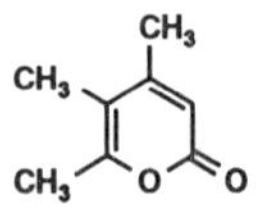	16 , 27

Table 19.5   Photo-Ring Contraction of 4-Pyrones in Water

Reactant	Product	Yield %	Reference
		51	19 , 22
		60	19
		30	16
		12	23
		7.6	23
		62	22
		71	22

Table 19.6   Intramolecular Photo-Ring Contraction of 4-Pyrones

Reactant	Product

	R_1	R_2	R_3	n	Yield%	Diastereomer Ratio	Reference
a:	CH_3	CH_3	H	1	75	—	31
b:	CH_3	CH_3	CH_3	1	84	1.9:1	31
c:	CH_3	CH_3	Ph	1	92	1:1	31
d:	H	CH_3	CH_3	1	75	1.6:1	31
e:	CH_3	CH_3	H	2	99	—	31
f:	CH_3	CH_3	iPr	2	61	1.5:1	31
g:	H	H	H	2	43	—	31

Intermolecular Reactions

Photo-ring contraction to provide 4,5-dihydroxy- or 4-hydroxy-5-alkoxycyclopentenones **6** or **7** are best carried out by direct irradiation of neutral 4-pyrones in aqueous or alcoholic solvent. Table 19.5 shows the dihydroxycyclopentenones formed by irradiation of various 4-pyrones in aqueous solution. Although 2-pyrones and 2-acetylfurans or 2-furaldehydes are often minor products in these reactions, these minor products can be easily removed by extracting the irradiated aqueous solution before evaporation of the water to obtain the crystalline diols. Many of the 4-pyrones shown in Table 19.5 also undergo photo-ring contraction upon irradiation in methanol solvent, resulting in the formation of the corresponding 4-hydroxy-5-methoxycyclopentenone adducts.[11]

Intramolecular Reactions

In addition to intermolecular trapping of the bicyclic Zwitterion by solvent, it has recently been shown that 4-pyrones bearing hydroxy-alkyl side chains undergo efficient 2,6-bridging to oxyallyl Zwitterions and subsequent intramolecular nucleophilic trapping to give the bicyclic cyclopentenone ethers shown in Table 19.6 in good to excellent yields.[31] These reactions thus allow formation of functionalized cyclopentenones of defined stereochemistry from planar heterocyclic precursors and have potential synthetic utility.

References

1. Paterno, E., Synthesis in organic chemistry by means of light. VIII. Various experiments, *Gazz. Chim. Ital.*, 44, I, 151, 1914.
2. Giua, M. and Civera, M., The dimer of 2,6-dimethyl-4-pyrone, *Gazz. Chim. Ital.*, 81, 875, 1951.
3. Yates, P. and Jorgenson, M. J., The photodimer of 2,6-dimethyl-4-pyrone, *J. Am. Chem. Soc.*, 80, 6150, 1958.
4. Yates, P. and Jorgenson, M. J., Photodimeric cage compounds. I. The structure of the photodimer of 2,6-dimethyl-4-pyrone, *J. Am. Chem. Soc.*, 85, 2956, 1963.
5. Yates, P. and Still, I. W. J., Photorearrangement of a 4-pyrone to a furan derivative, *J. Am. Chem. Soc.*, 85, 1208, 1963.
6. Ishibe, N., Odani, M., and Sunami, M., Photochemical rearrangement of 4H-pyran-4-ones to 2H-pyran-2-ones, *J. Chem. Soc., Chem. Commun.*, 1034, 1971.
7. Ishibe, N., Odani, M., and Sunami, M., Photoisomerization of 4H-pyran-4-ones to 2H-pyran-2-ones, *J. Am. Chem. Soc.*, 95, 463, 1973.
8. Pavlik, J. W. and Kwong, J., Photochemical rearrangements of neutral and protonated 4-pyrones, *J. Am. Chem. Soc.*, 95, 7914, 1973.
9. Padwa, A. and Hartman, R., Photochemical transformations of small ring carbonyl compounds. VIII. Photorearrangements in the cyclopentenone oxide series, *J. Am. Chem. Soc.*, 88, 1518, 1966.
10. Barltrop, J. A., Day, A. C., and Samuel, C. J., 4-Pyrone photochemistry. The intermediacy of a cyclopentadienone epoxide, *J. Chem. Soc., Chem. Commun.*, 598, 1977.
11. Pavlik, J. W. and Pauliukonis, L. T., Photoisomerization of 4-pyrones. Nucleophilic trapping of reactive intermediates, *Tetrahedron Lett.*, 1939, 1976.
12. Keil, E. B. and Pavlik, J. W., Photochemistry of 2,6-dimethyl-4-pyrone in trifluoroethanol, *J. Heterocyclic Chem.*, 13, 1149, 1976.
13. Pavlik, J. W. and Clennan, E. L., Photochemical rearrangements of pyrylium cations, *J. Am. Chem. Soc.*, 95, 1697, 1973.
14. Barltrop, J., Carder, J., Day, A. C., Harding, J. R., and Samuel, C., Ring permutations in the photochemistry of hydroxypyrylium cations, *J. Chem. Soc., Chem. Commun.*, 729, 1975.
15. Pavlik, J. W., Bolin, D. R., Bradford, K. C., and Anderson, W. G., Photoisomerization of 4-hydroxypyrylium cations. Furyl cation formation, *J. Am. Chem. Soc.*, 99, 2816, 1977.

16. Pavlik, J. W. and Spada, A., Photo-ring contraction reactions of 4-hydroxypyrylium cations, *Tetrahedron Lett.*, 4441, 1979.
17. Ishibi, N. and Yutaka, S., Heavy-atom effects in photoisomerization of 4-pyrones and 4-pyridones, *J. Org. Chem.*, 43, 2138, 1978.
18. Barltrop, J. A., Day, A. C., and Samuel, C. J., Evidence for a Zwitterionic 2,6-bonded intermediate in 4-pyrone photochemistry. Following the time evolution of an excited state, *J. Chem. Soc., Chem. Commun.*, 822, 1976.
19. Barltrop, J. A., Day, A. C., and Samuel, C. J., Heterocyclic photochemistry. 4-Pyrones. A mechanistic study, *J. Am. Chem. Soc.*, 101, 7521, 1979.
20. Hanifin, J. W. and Cohen, E., Preparation and pyrolysis of some 2,6-dimethyl-4-pyrone-alkyne photoadducts. Bicyclic Claisen rearrangement, *J. Org. Chem.*, 36, 910, 1971.
21. Ishibi, N., Yutaka, S., Masui, J., and Ihda, N., Substituent effects on selectivity in photoisomerization of 4-pyrones and 4-pyrones, *J. Org. Chem.*, 43, 2144, 1978.
22. Pavlik, J. W., Snead, T. E., and Tata, J. R., Photochemistry of 4-pyrones in water. Formation of dihydroxycyclopentenones and furan derivatives, *J. Heterocyclic Chem.*, 18, 1481, 1981.
23. Pavlik, J. W., Kirincich, S. J., and Pires, R. M., Regio- and stereochemistry of the photo-ring contraction of 4-pyrones, *J. Heterocyclic Chem.*, 28, 537, 1991.
24. Barltrop, J. A., Barrett, J. C., Carder, R. W., Day, A. C., Harding, J. R., Long, W. E., and Samuel, C. J., Heterocyclic photochemistry. I. Phototranspositions in hydroxypyrylium cations. Permutation pattern analysis and mechanistic studies, *J. Am. Chem. Soc.*, 101, 7510, 1979.
25. Pavlik, J. W., Patten, A. D., Bolin, D. R., Bradford, K. C., and Clennan, E. L., Photoisomerization of 4-hydroxypyrylium cations in concentrated sulfuric acid, *J. Org. Chem.*, 49, 4523, 1984.
26. Pavlik, J. W. and Dunn, R. M., Photoisomerization of 2,6-di-*t*-butyl-4-hydroxypyrylium cation. Evidence for a Dewar-type intermediate, *Tetrahedron Lett.*, 5071, 1978.
27. Pavlik, J. W., Spada, A. P., and Snead, T. E., Photochemistry of 4-hydroxypyrylium cations in aqueous sulfuric acid, *J. Org. Chem.*, 50, 3046, 1985.
28. Ullman, E. F., Photochemical valence tautomerization of 2,4,6-triphenylpyrylium 3-oxide, *J. Am. Chem. Soc.*, 85, 3529, 1963.
29. Barltrop, J. A., Day, A. C., and Samuel, C. J., Hydroxypyrylium photochemistry, *J. Chem. Soc., Chem. Commun.*, 823, 1976.
30. Barton, D. H. and Hulshof R., Photochemical transformation. Part 35. A simple synthesis of raceimic terrein, *J. Chem. Soc., Perkin Trans., 1*, 1103, 1977.
31. West, F. G., Fisher, P. V., and Willoughby, C. A., The photochemistry of pyran-4-ones: Intramolecular trapping of the Zwitterionic intermediate with pendant hydroxyl groups, *J. Org. Chem.*, 55, 5936, 1990.

20

Photochemical Aromatic Substitution

Jan Cornelisse
Leiden University

20.1 Introduction

This chapter deals mainly with photosubstitution reactions of homocyclic aromatic compounds. It will concentrate on reactions that start with excitation of the aromatic molecule and ignore (not because they are unimportant) reactions in which photogenerated radicals attack aromatic substrates. This brief review gives information on the most important mechanistic types of aromatic photosubstitution, and treats some representative examples of each type and a few synthetic applications. For a more comprehensive treatment, the reader is referred to the yearly *Specialist Periodical Reports, Photochemistry*.[1] In many of these reports, the section on aromatic substitution reactions begins with the reporter's remark that photosubstitution of aromatic rings remains a complicated area because of the diversity of mechanisms by which the reaction can occur,[2] which makes classification of the processes somewhat unrealistic.[3]

Aromatic substitution in the ground electronic state is one of the best known and most extensively studied organic reactions, having wide scope and enormous synthetic potential. The aromatic nucleus is electron-rich and it is not surprising, therefore, that electrophilic aromatic substitution is much more common than nucleophilic aromatic substitution. The majority of electrophilic aromatic substitutions proceed by one mechanism, called the arenium ion mechanism or S_EAr mechanism, in which an arenium ion or σ-complex occurs as the intermediate. Electrophilic aromatic substitutions proceeding via an S_E1 mechanism, in which the leaving group departs before the electrophile arrives, are rare. In most electrophilic aromatic substitutions, the leaving group is hydrogen; the reversal of Friedel-Crafts alkylation, the desulfonation, and the demetallation are among the exceptions and in those cases the substituent is replaced by hydrogen.

By far, the most important mechanism for aromatic nucleophilic substitution in the ground state is the S_NAr mechanism consisting of two steps, the first of which leads to a Meisenheimer complex or σ-complex. The reaction is accelerated by electron-withdrawing groups. The variety of leaving groups is much larger than in aromatic electrophilic substitution. A unimolecular S_N1 mechanism for aromatic nucleophilic substitution is very rare; the only important leaving group is N_2. A type of nucleophilic aromatic substitution proceeding via the $S_{RN}1$ mechanism can also be initiated photochemically, and it will be treated below.

0-8493-8634-9/95/$0.00+$.50

The most conspicuous difference between thermal and photochemical aromatic substitution is the fact that in the ground state, electrophilic substitution is the most frequently encountered process; whereas, in the excited state, nucleophilic substitution is the most common type. Indeed, examples of the electrophilic process are few and far between. A review has been published that deals with the use of deuterium and tritium in photochemical electrophilic aromatic substitution.[4] Anthracene undergoes a photochemical Friedel-Crafts reaction with benzoyl chlorides[5,6] via the excited singlet state from which an exciplex is formed with the aroyl chloride. Naphthalene undergoes proton-deuteron exchange when it is irradiated in the presence of trifluoroacetic acid,[7] and the same phenomenon has been observed and studied with 1-methoxynaphthalene in water(D_2O)-acetonitrile mixtures[8] and with tryptophan in D_2O.[9]

The abundance of nucleophilic aromatic photosubstitutions is of course related to the nature of the excited state of aromatic molecules. One electron has been promoted from an HOMO to a LUMO. On the one hand, this makes the molecule somewhat more electrophilic: it can accept an electron in its half-filled HOMO from a good electron donor, whereupon the radical anion undergoes further reactions leading to substitution, or it can react with a nucleophile and give a σ-complex. On the other hand, the electron in the half-filled LUMO can be taken up by a good electron acceptor, and the resulting radical cation can combine with a nucleophile. Thus, the reactive species in nucleophilic aromatic photosubstitution is a molecule in an electronically excited state or a radical anion or a radical cation (or, in the case of the $S_{RN}1$ reaction, a neutral radical). The relative reactivity of the various positions in such species is often different from that in the ground electronic state and, as a consequence, the orientation rules in photochemical nucleophilic substitution may differ from the familiar rules in the thermal reaction. The nitro group, which is *ortho/para*-directing in the ground state, is sometimes seen to be *meta*-directing in photochemical substitution reactions with nucleophiles, while the methoxy group is found to be activating and *ortho/para*-directing. Such features give rise to interesting possibilities for synthetic applications, particularly since the chemical yields from photonucleophilic aromatic substitutions using CN^-, OH^-, and amines are frequently good. Nevertheless, the remark made in 1986 by Gilbert[10] in the chapter "Aromatic Compounds: Substitution and Cyclisation" in J. D. Coyle's *Photochemistry in Organic Synthesis:* "... despite the potential of this route for the preparation of simple substituted arenes which may be difficult to obtain by conventional thermal means the process still lacks synthetic exploitation." is still applicable to today's situation. A similar statement was made in Gilbert and Baggott's textbook on molecular photochemistry published in 1991.[11]

20.2 Historical Background

In 1956, Havinga, De Jongh, and Dorst[12] reported that *meta*-nitrophenyl phosphate in alkaline aqueous solution undergoes photohydrolysis with a much higher quantum efficiency than the *para*- and the *ortho*-isomer. The quantum yield at 313 nm for the formation of free phosphate from a 2×10^{-4} M solution in 0.01 N NaOH is 0.055,[13] while that of the *para*-isomer under the same experimental conditions is 0.002. It was established that the reaction is a true aromatic substitution, proceeding via rupture of the C-O bond, by performing it in ^{18}O-enriched water (Equation 20.1).

$$(20.1)$$

meta-Nitrophenyl phosphate was also found to react with methylamine (1 M) in water, albeit with a low rate. The product of the reaction is *N*-methyl-3-nitroaniline, and the substitution has evidently taken place at the ring carbon atom[14] (Equation 20.2).

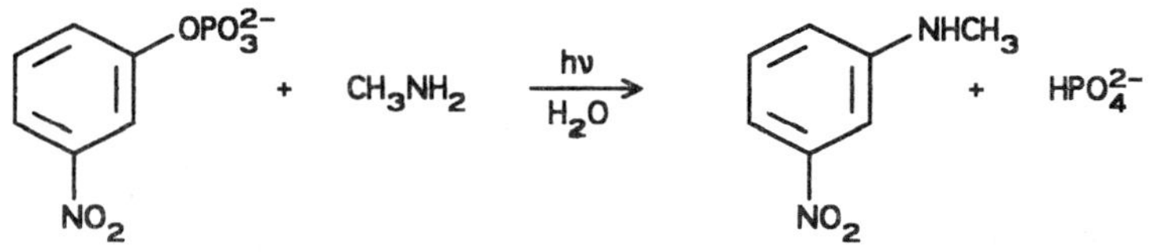

$$\text{(3-nitrophenyl-OPO}_3^{2-}) + CH_3NH_2 \xrightarrow[\text{H}_2\text{O}]{h\nu} \text{(3-nitrophenyl-NHCH}_3) + HPO_4^{2-}$$

(20.2)

The *meta*-directing effect of the nitro group became clear especially from studies on *meta*-nitroanisole and *para*-nitroanisole. Photohydrolysis of *meta*-nitroanisole in alkaline aqueous solution is efficient,[15] with the quantum efficiency of formation of *meta*-nitrophenolate reaching 0.23 (313 nm) at a hydroxide ion concentration of 7×10^{-3} M (Equation 20.3).

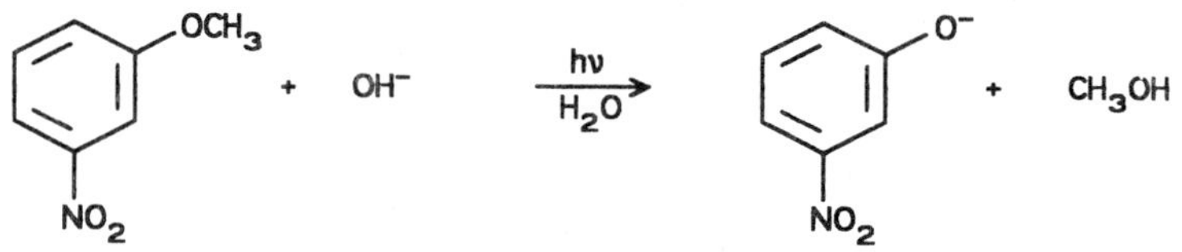

$$\text{(3-nitrophenyl-OCH}_3) + OH^- \xrightarrow[\text{H}_2\text{O}]{h\nu} \text{(3-nitrophenyl-O}^-) + CH_3OH$$

(20.3)

The main products from the irradiation of a solution of *para*-nitroanisole in water containing NaOH are *para*-nitrophenol and *para*-methoxyphenol[16,17] in a ratio of 1:4 (Equation 20.4).

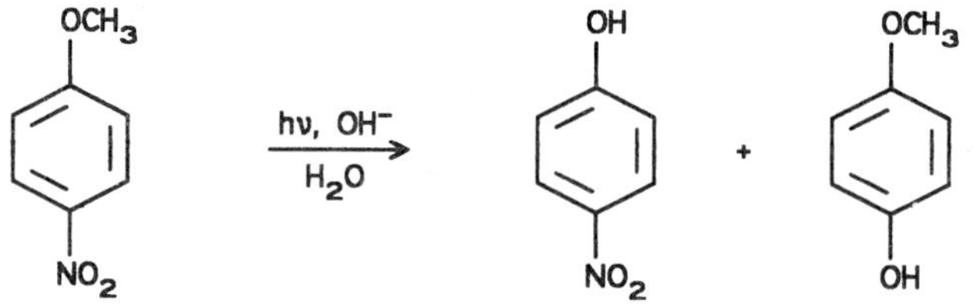

$$\text{(4-nitrophenyl-OCH}_3) \xrightarrow[\text{H}_2\text{O}]{h\nu,\ OH^-} \text{(4-nitrophenol)} + \text{(4-methoxyphenol)}$$

(20.4)

As in the case of *meta*-nitroanisole, the quantum yield of disappearance of *para*-nitroanisole depends on the hydroxide ion concentration. The value, 6×10^{-1} mol l^{-1}, at which the quantum yield becomes constant ($\varphi = 0.085$) is much higher than the corresponding value for *meta*-nitroanisole (7×10^{-3} mol l^{-1}), and the maximum quantum yield is about three times as low as that of the *meta*-compound. It was realized that the higher quantum yield of the photohydrolysis of the *meta*-isomer might not be caused by higher reactivity of the molecule in the excited state, but simply by a longer lifetime. Therefore, the thermal and the photochemical hydrolysis were studied with 4-nitroveratrole, a molecule that has one methoxy group *para* and another one *meta* with respect to the nitro group. Upon thermal hydrolysis, the *para*-methoxy group was replaced exclusively, but the photochemical reaction led only to substitution of the *meta*-methoxy group[18,19] (Equation 20.5). Irradiation of 4-nitroveratrole in the presence of ammonia, methylamine or cyanide ions likewise leads to replacement of the meta substituent.[18,20,21] Thus, the *meta*-directing effect of the nitro group in nucleophilic aromatic photosubstitution appeared to be established firmly.

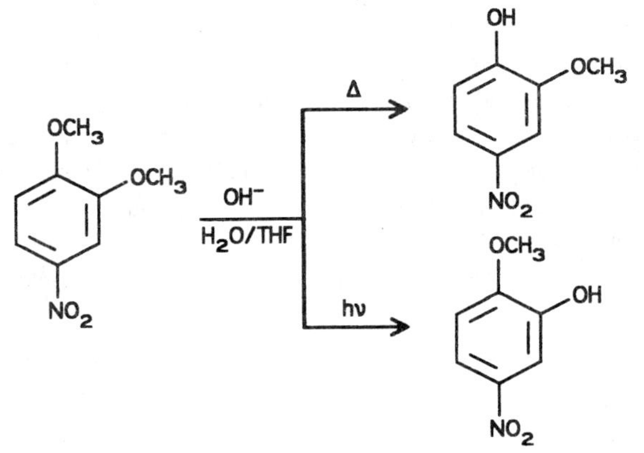

$$(20.5)$$

Many derivatives of benzene, naphthalene, azulene, and biphenyl containing nitro and methoxy groups were investigated subsequently and found to undergo nucleophilic photosubstitution reactions such as photohydrolysis, photocyanation, and photoamination.[18]

In 1970, El'tsov et al.[22] discovered that when solutions of *para*-chloro-, *para*-bromo-, and *para*-iodoaniline and *para*-chloro-*N,N*-dimethylaniline in methanol/water containing sodium nitrite are irradiated with UV light, the corresponding nitro compounds are formed by replacement of halogen by a nitro group in yields of 25, 18, 10, and 33%, respectively (Equation 20.6).

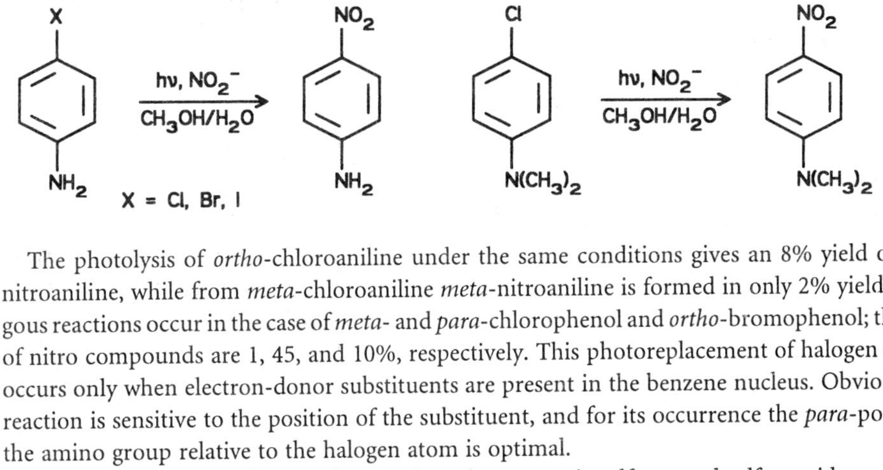

$$(20.6)$$

The photolysis of *ortho*-chloroaniline under the same conditions gives an 8% yield of *ortho*-nitroaniline, while from *meta*-chloroaniline *meta*-nitroaniline is formed in only 2% yield. Analogous reactions occur in the case of *meta*- and *para*-chlorophenol and *ortho*-bromophenol; the yields of nitro compounds are 1, 45, and 10%, respectively. This photoreplacement of halogen by nitro occurs only when electron-donor substituents are present in the benzene nucleus. Obviously, the reaction is sensitive to the position of the substituent, and for its occurrence the *para*-position of the amino group relative to the halogen atom is optimal.

It was found later by El'tsov and co-workers that aromatic sulfones and sulfonamides containing an activating *para*-amino group undergo photochemical replacement of the sulfur-containing group by various nucleophiles,[23] and that a chlorine atom *para* to an electron-donating substituent can be photosubstituted by a sulfonate group[24] (Equation 20.7).

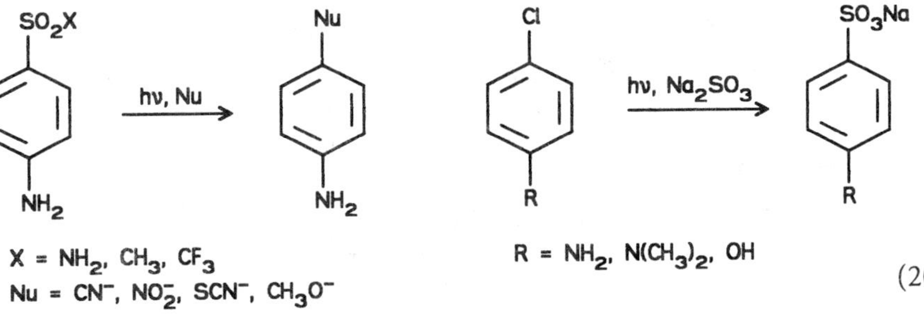

$$(20.7)$$

Nucleophilic aromatic photosubstitution in donor-substituted aromatic molecules was also found by Nilsson[25] and investigated extensively by Den Heijer et al.[26] and by Lok and Havinga.[20] Nilsson[25] irradiated anisole in the presence of cyanide ions in methanol and obtained *ortho*- and *para*-cyanoanisole in a ratio of 1:1 (Equation 20.8). This result illustrates the *ortho/para*-directing effect of the methoxy group in nucleophilic photosubstitution, an effect that is also evident particularly in the case of the dimethoxybenzenes. In the photocyanation of *ortho*- and *para*-dimethoxybenzene one of the methoxy groups is replaced; but in *meta*-dimethoxybenzene, a hydrogen atom (*ortho* with respect to one methoxygroup and *para* to the other) is substituted in preference to methoxy (Equation 20.9).

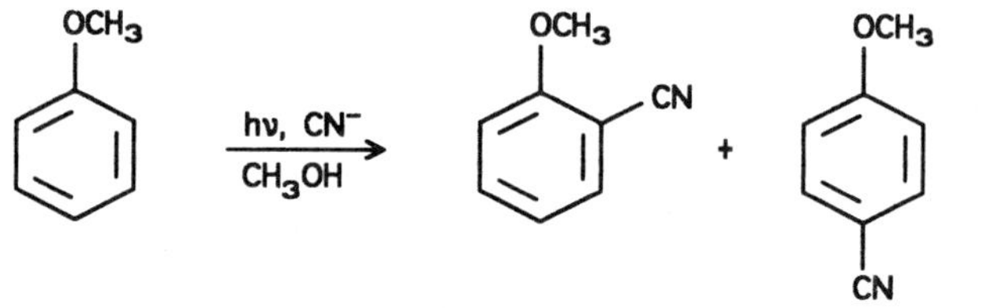

(20.8)

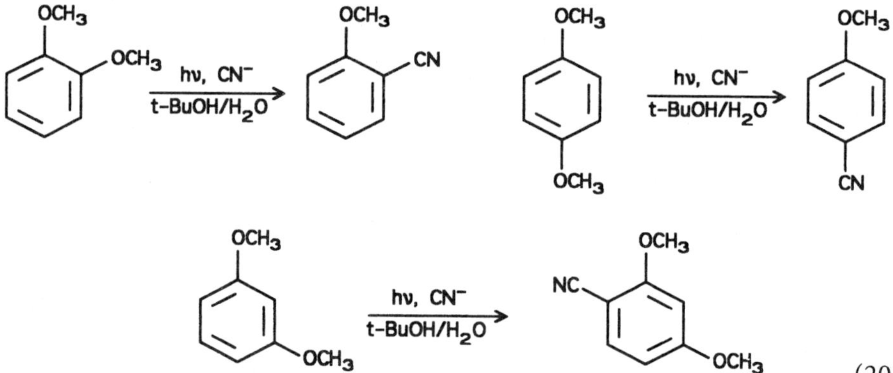

(20.9)

Nucleophilic photosubstitution does also occur in molecules that contain only nitro groups. Letsinger and co-workers[27,28] found that the nitro group in 1-nitronaphthalene can easily be replaced by the cyano group upon irradiation in CH_3CN/H_2O (1:9) containing KCN. 4-Nitrobiphenyl undergoes a similar reaction[29] (Equation 20.10).

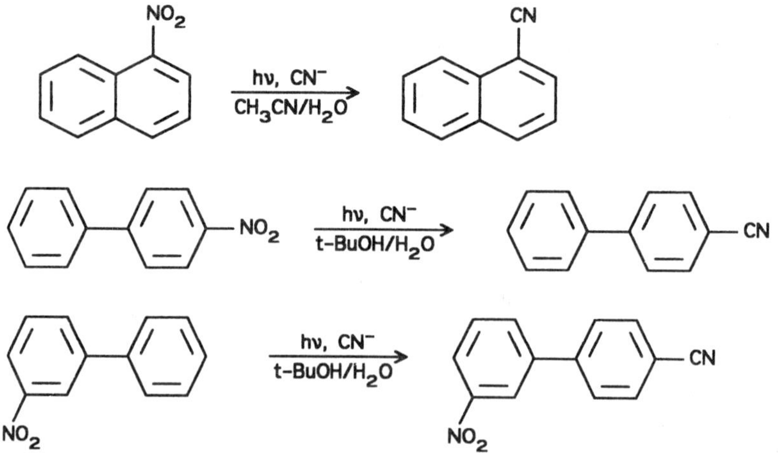

(20.10)

With 1-nitronaphthalene, substitution of the nitro group can also be effected with hydride ion (NaBH$_4$)[28] and methoxide ion.[30] If 3-nitrobiphenyl is irradiated in the presence of cyanide ions, the nitro group is not substituted but a hydrogen atom in the other ring is replaced.[29]

The nitro group of 1-nitroazulene can be photochemically replaced by methoxy or cyano.[30,31] Nitrobenzene, however, is not very reactive as far as photoinduced substitution is concerned.

In 1972, it was found that nucleophilic photosubstitution, especially cyanation, can also take place in unsubstituted aromatic hydrocarbons such as naphthalene, biphenyl, and azulene[32] (Equation 20.11).

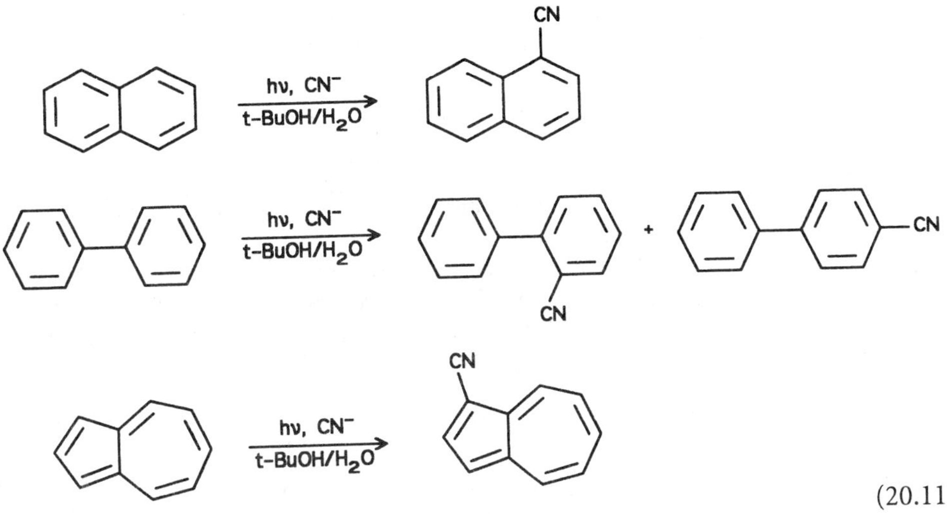

$$(20.11)$$

20.3 Mechanisms

In photosubstitutions proceeding via the S$_N$2Ar* mechanism (Equation 20.12), the nucleophile adds to the photoexcited molecule (usually in its triplet state), yielding a σ-complex that can either revert to the starting material or give rise to the substitution product by losing the leaving group. This mechanism was first proposed and discussed extensively by Havinga, Cornelisse, and co-workers.[33–35] Detailed investigations using time-resolved electronic absorption spectroscopy were performed by Varma et al.[36–38] on the systems 3,5-dinitroanisole and 3-nitroanisole with hydroxide ions. The rate-determining step in these reactions is addition of OH$^-$ to the methoxy-bearing carbon atom. This produces the σ-complex from which the substitution product is formed, next to other σ-complexes. The same mechanism was found for the photoreaction of 4-nitroveratrole with hydroxide ion.[39]

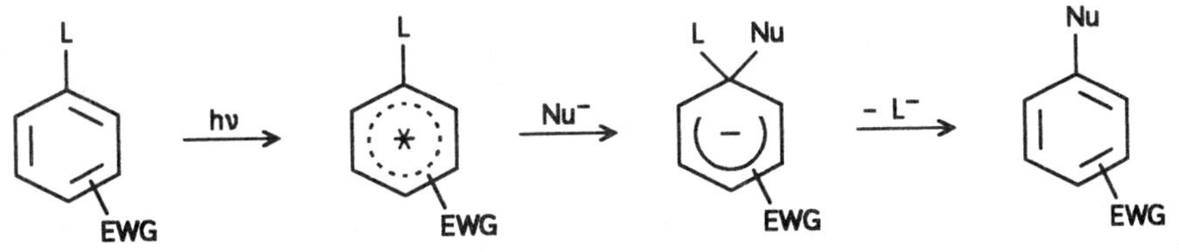

L: leaving group; EWG: electron–withdrawing group; Nu: nucleophile $$(20.12)$$

According to Mutai and co-workers,[40] the nucleophilic photosubstitutions that involve one-step formation of a σ-complex via direct interaction between an excited aromatic substrate and a nucleophile are HOMO controlled. The dominant orbital interaction is between the HOMO of the substrate and the HOMO of the nucleophile. This rule is in complete agreement with the familiar *meta*-directing effect of the nitro group in nucleophilic aromatic photosubstitution. It also agrees with the finding that if *meta*-substitution is disfavored (e.g., because the incoming nucleophile is a better leaving group that the substituent at the *meta*-position), *ipso* substitution at the nitro group occurs instead.

Van Riel, Lodder, and Havinga[41] have proposed that the regioselectivity is controlled by the size of the energy gap between the excited-state nucleophile encounter complex and the ground-state σ-complexes. The σ-complex resulting from attack *meta* to nitro lies 63–75 kJ mol^{-1} higher in energy than that from *para*- (or *ortho*-) attack, while the triplet of the *meta* σ-complex has a lower energy than that of the *para*-complex. The small energy gap in the *meta*-situation favors the transition from excited-state complex to ground-state complex. The "energy gap model" also rationalizes nicely the observation that the nitro group is much more successful in promoting nucleophilic aromatic substitution than other electron-withdrawing groups such as cyano and carbonyl. The triplet-state energies of nitroaromatic compounds are much lower than those of cyano- and carbonyl-substituted compounds.

Examples of reactions proceeding via the S_N2Ar^* mechanism are: the reactions of *meta*-nitroanisole with cyanide ion,[42] hydroxide ion,[15] and amines;[18,34,43] 4-nitroveratrole with cyanide ion,[33] hydroxide ion,[41] and primary amines;[44] 1-methoxy-4-nitronaphthalene with cyanide ion,[27] methoxide ion,[18] borohydride ion,[28] and primary amines;[45] and *para*-nitroanisole with cyanide ion,[46] primary amines,[47] hydroxide ion,[16] and pyridine.[16]

L: leaving group; EWG: electron–withdrawing group; Nu: nucleophile

$$(20.13)$$

In the $S_N(ET)Ar^*$ mechanism (Equation 20.13), the primary event following electronic excitation is transfer of an electron from the nucleophile to the photoexcited molecule, followed by coupling of the radical anion with the cationic or neutral radical formed from the nucleophile. A mechanism of this type was already suggested in 1969 by Van Vliet, Cornelisse, and Havinga.[48] Just as in the S_N2Ar^* mechanism, a σ-complex is formed; but the position of attack on the aromatic molecule is different because the dominant orbital interaction is now between the LUMO of the aromatic substrate (singly occupied in the radical anion) and the singly occupied HOMO of the nucleophile.[40] In many cases, this interaction leads to replacement of substituents *para* to the nitro group. Evidence for the occurrence of this mechanism was presented by Mutai and co-workers[49–51] in a study of the photo-Smiles rearrangement, an intramolecular nucleophilic aromatic photosubstitution in molecules of the type Ar-O-CH$_2$CH$_2$-NHR, in which Ar = *o*-, *m*-, or *p*-nitrophenyl and R = H, CH$_3$, or Ph. Mechanistic investigations of this rearrangement have also been performed by Wubbels and co-workers.[43,52,53]

Bunce et al.[45] have obtained evidence for the operation of the electron-transfer mechanism in the photoreaction of 1-methoxy-4-nitronaphthalene with secondary amines. Other examples are the reaction of *para*-nitroanisole with aliphatic amines[47] and anilines[49] and of 4-nitroveratrole derivatives with secondary amines and anilines.[51,54]

Marquet and co-workers[55] have found that 4-nitroveratrole is photosubstituted with *n*-hexylamine at the position *para* to the nitro group via the $S_N(ET)Ar^*$ mechanism and at the *meta*-position via the S_N2Ar^* mechanism. A dual mechanistic pathway was also found in the photoreaction of 4-nitroanisole with primary amines.[56,57] The reaction with *n*-hexylamine leads to regioselective replacement of methoxy via the electron-transfer mechanism, whereas ethyl glycinate replaces the nitro group exclusively via the S_N2Ar^* mechanism. The ionization potential of the amine appears to be a major factor in determining the course of the reaction.

L: leaving group; EDG: electron-donating group; Nu: nucleophile; ArL: ground-state substrate

(20.14)

Nucleophilic aromatic photosubstitution via the $S_{R+N}1Ar^*$ mechanism (Equation 20.14) starts with photoionization of the excited aromatic molecule. The radical cation combines with the nucleophile (e.g., CN^- or NO_2^-) and a neutral σ-complex is formed. If the leaving group is hydrogen, this atom may be removed by oxygen. If the leaving group is halogen or methoxy, the neutral σ-complex will dissociate into a halide or methoxide anion and the radical cation of the substitution product. This radical cation may then abstract an electron from the starting material, giving rise to a molecule of the product and a new radical cation of the starting material. The latter is formed in a nonphotochemical step and a chain reaction becomes possible. Photoionization was proposed by Nilsson[25] as a primary step in the photocyanation of anisole and by Den Heijer et al.[26] in the photoreactions of di- and trimethoxybenzenes and halogenoanisoles. It may also play a role in the photonitration of halogenoanilines.[58,59] Den Heijer et al.[26] have detected the formation of the solvated electron and the radical cation upon irradiation of 1,4-dimethoxybenzene in *t*-butyl alcohol/water (1:5) with a laser pulse (15 ns, 265 nm). Photoionization is supposed to occur from the singlet excited state. The formation of radical cations by photoinduced electron ejection from methoxylated benzenes in aqueous solution has been observed by Grabner et al.[60] Soumillion and De Wolf[61] have argued that in the photocyanation of *para*-chloroanisole, the radical cation is formed by dissociation of a triplet excimer into a radical pair. Photosubstitution via the $S_{R+N}1Ar^*$ mechanism is of course strongly promoted by the presence of electron-donating substituents such as methoxy or amino. These substituents have an *ortho/para*-directing effect on the attack by nucleophiles, in agreement with the charge distribution in the radical cation.

The photocyanation of unsubstituted aromatic compounds such as naphthalene,[32] biphenyl,[32] phenanthrene,[18] and fluorene[18] takes place preferentially at the positions that are also the most reactive sites in ground-state aromatic substitution, i.e., 1 in naphthalene, 2 and 4 in biphenyl, 9 in phenanthrene, and 2 in fluorene. These reactions also proceed via the $S_{R+N}1Ar^*$ mechanism. Bunce et al.[62] have established that photocyanation of naphthalene and biphenyl in the presence of oxidants such as *para*-dicyanobenzene and persulphate ion proceed by oxidation of the singlet excited hydrocarbon; whereas in the absence of oxidizing agents, a singlet excimer is involved, which dissociates into radical ions before the attack of cyanide ion.

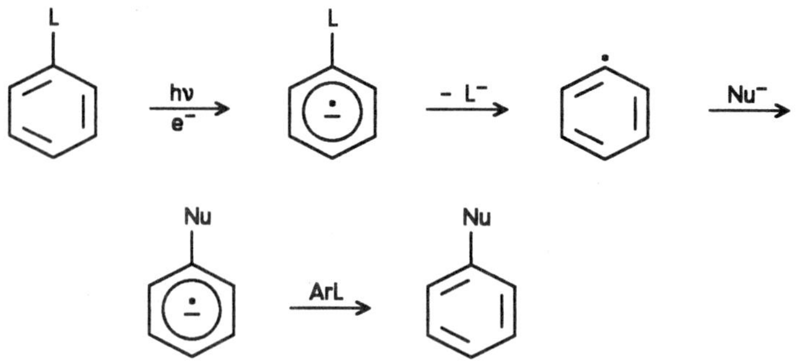

L: leaving group; Nu: nucleophile; ArL: ground–state substrate

$$(20.15)$$

In the $S_{R-N}1Ar$ mechanism (Equation 20.15), the first step consists of transfer of an electron from an electron donor to the aromatic molecule, in many cases an aryl halide. The donor may be the nucleophile or another strong chemical reductant. The reaction can be performed in the ground state, but in many cases it is initiated photochemically and then designated as $S_{R-N}1Ar^*$. It is, however, not always the aromatic molecule that is excited. The donor molecule or a ground-state complex between donor and aromatic substrate may also be the light-absorbing species, and in that case, it is perhaps better to classify the reaction as a photoinduced aromatic substitution instead of as an aromatic photosubstitution. Subsequently the radical anion loses the leaving group, usually a halide anion, and an aryl radical is created. This feature, loss of the leaving group before the nucleophile attacks, distinguishes this mechanism from the previous three. The radical then combines with the nucleophile, giving rise to the radical anion of the substitution product. This species can transfer its extra electron to a molecule of the starting material and thereby initiate a chain process.

The $S_{R-N}1Ar$ reaction, which has many synthetic applications, has been reviewed by Bunnett,[63] a pionier in this field, and by Rossi and de Rossi.[64]

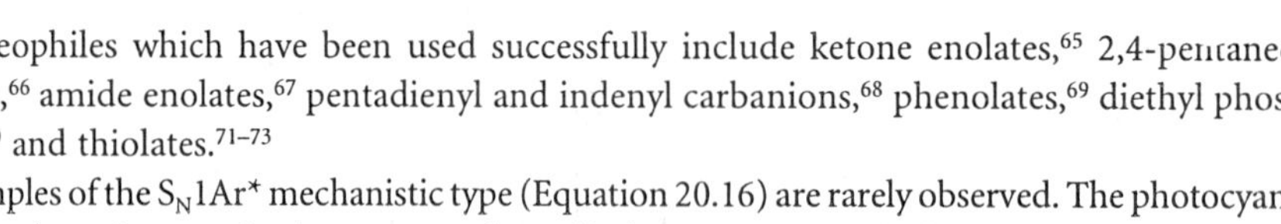

L: leaving group; Nu: nucleophile

$$(20.16)$$

Nucleophiles which have been used successfully include ketone enolates,[65] 2,4-pentanedione dianion,[66] amide enolates,[67] pentadienyl and indenyl carbanions,[68] phenolates,[69] diethyl phosphite anion,[70] and thiolates.[71–73]

Examples of the S_N1Ar^* mechanistic type (Equation 20.16) are rarely observed. The photocyanation of 2-nitrofuran in water leads to 2-cyanofuran.[74] The quantum yield of disappearance of the starting material (0.51 at 313 nm) is independent of the cyanide ion concentration. The quantum yield of formation of 2-cyanofuran, however, increases with increasing concentration of CN⁻, approaching the value of 0.51 at sufficiently high concentration. At lower concentrations, the disappearance of the starting material is supposed to be caused by reaction of a photochemically generated intermediate with water and, indeed, the presence of the tautomer of 2-hydroxyfuran has been demonstrated.

The reactive species is supposed to be the cation formed by dissociation of the photoexcited triplet-state substrate (Equation 20.17).

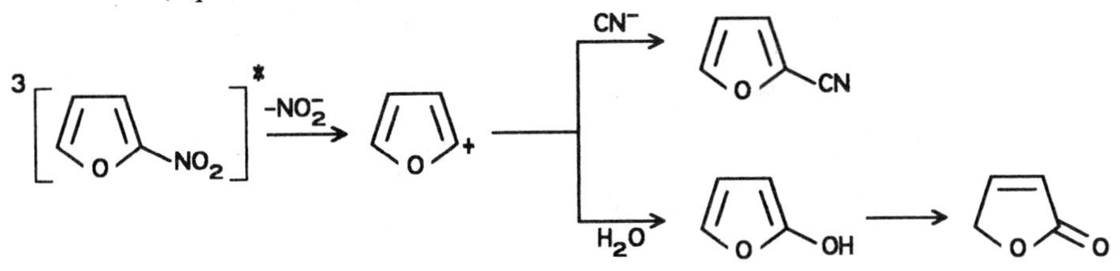

$$(20.17)$$

Mechanisms of aromatic photosubstitution reactions have been reviewed by Cornelisse, Lodder, and Havinga,[35] by Párkányi,[75] and recently, for compounds containing nitro groups, by Terrier.[76] The five mechanisms discussed above seem to be well established and, in each case, it is possible to mention representative examples of reactions that have been carefully studied. The number of aromatic photosubstitution reactions is, however, enormous and many different factors act together in determining the course of the reaction, e.g., the type of aromatic nucleus, the number and nature of the substituents, the multiplicity and the character ($n\pi^*$ or $\pi\pi^*$) of the excited state, the character of the leaving group, and the nature of the solvent. The stage has not been reached at which it is possible to make predictions for any given combination of aromatic substrate and nucleophile regarding the efficiency of the photosubstitution reaction and the nature of its products.

20.4 Further Examples and Possible Synthetic Applications

In many nucleophilic aromatic photosubstitutions, small anions such as CN^-, OH^-, and NO_2^- are used, and these reagents are best dissolved in aqueous media. For exploratory photochemistry and for the study of reaction mechanisms, low concentrations of the aromatic substrate are usually sufficient or even desirable, and mixtures of alcohols (methanol or *t*-butyl alcohol) and water have often been used as solvents. For synthetic purposes, one needs higher concentrations and two methods have been employed to make this possible. One is the use of crown ethers to dissolve the inorganic reaction partner in an anhydrous solvent and the other is the use of a phase-transfer agent in a two-component system of an organic solvent and water. Both methods have been used by Beugelmans et al. for the photochemical cyanation of aromatic hydrocarbons. The photocyanation of naphthalene, phenanthrene and biphenyl has been performed with KCN and 18-crown-6 in anhydrous acetonitrile,[77] with tetrabutylammonium cyanide in anhydrous CH_3CN or CH_2Cl_2 and with $Bu_4N^+CN^-/KCN$ in CH_2Cl_2/H_2O.[78,79] 1-Cyanonaphthalene was obtained from naphthalene and 4-cyanobiphenyl from biphenyl, while phenanthrene gave a mixture of two monocyano derivatives in which 9-cyanophenanthrene is predominant (Equation 20.18).

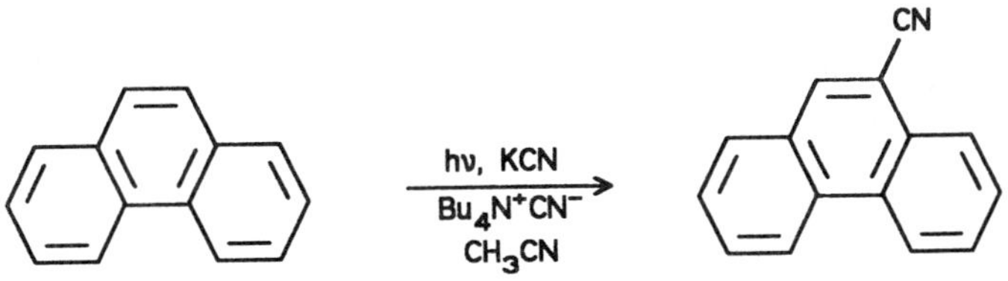

$$(20.18)$$

The direct introduction of cyano groups by photochemical means is attractive because this reaction is not easily performed in the ground electronic state. Photocyanation of anisole and the dimethoxybenzenes has been mentioned above. The reaction can also be performed with methoxynaphthalenes and the *para/ortho*-directing effect of the methoxy group is also manifest in these reactions. In 1-methoxynaphthalene, a cyano group is introduced at position-4 upon irradiation in the presence of KCN.[20] The yield of 1-cyano-4-methoxynaphthalene is 40%. 2-Methoxynaphthalene undergoes a clean photoinduced reaction with cyanide ion.[80] The sole product of the reaction is 1-cyano-2-methoxynaphthalene in 82% yield (Equation 20.19).

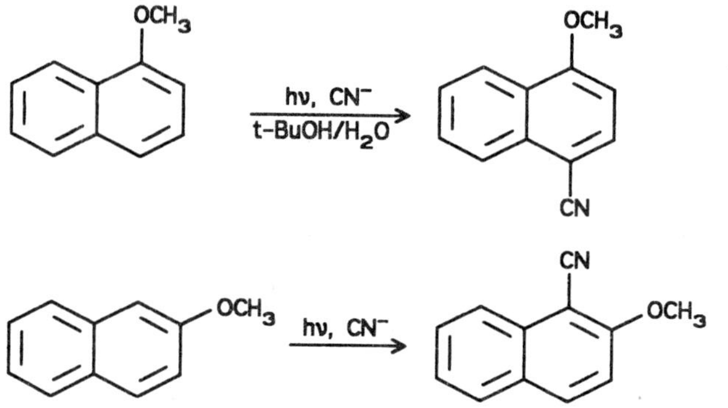

$$(20.19)$$

In the ground electronic state, nitration is a very common reaction; thermal nitration followed by photochemical substitution of the nitro group can sometimes be used to introduce substituents that are not easily attached to aromatic rings via nonphotochemical methods. Of the many examples,[18] only the preparation of 1-methoxyazulene will be mentioned here. 1-Nitroazulene can be prepared by nitration of azulene with cupric nitrate in acetic acid anhydride. Upon irradiation in the presence of sodium methoxide in methanol, the compound is easily converted into 1-methoxyazulene,[31] which is difficult to synthesize by classical methods (Equation 20.20).

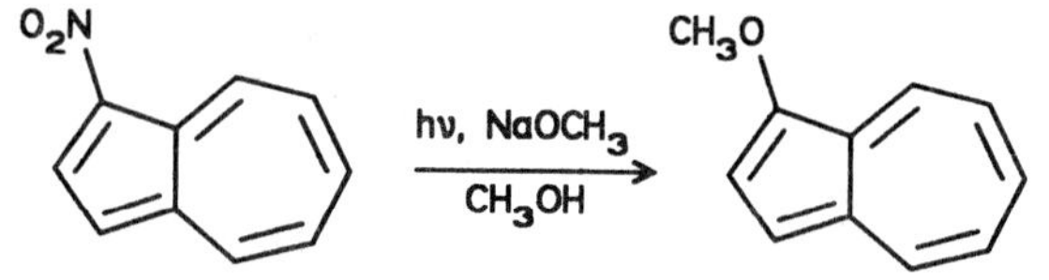

$$(20.20)$$

Most of the many reactions that have been discovered in the realm of aromatic photosubstitution have been performed only at small scale in studies of regioselectivities, quantum yields, intermediates, and other mechanistic aspects. Not much attention has been given to chemical yields and to the optimization of reactions conditions for synthetic purposes. An example of a reaction that may have synthetic interest, but for which thus far only low yields (4%) have been reported, is that of *para*-nitroanisole with cyanate ion. Substitution takes place at the position *meta* to nitro and a hydrogen atom is replaced.[81] Most probably, the primary product is the isocyanate, which in a secondary thermal reaction with water hydrolyzes to form the amino derivative (Equation 20.21). When the reaction is performed in methanol, the final product is the methyl ester of 2-methoxy-5-nitrophenylcarbamic acid.

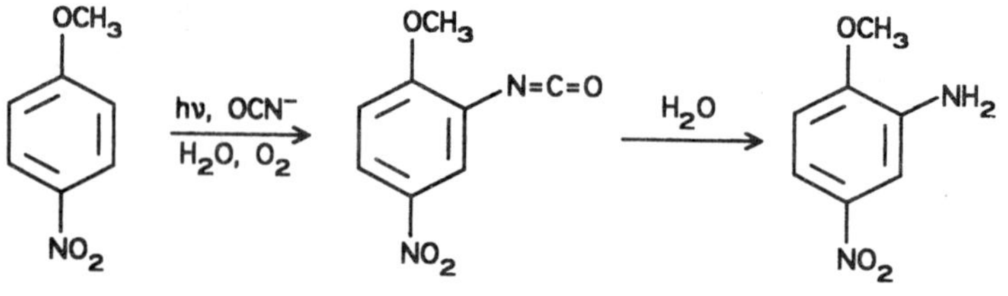

$$(20.21)$$

The photochemical hydrolyses of two dimethoxynitronaphthalenes in which the methoxy groups are in one ring and the nitro substituent in the other were investigated by Beijersbergen van Henegouwen and Havinga.[82] Irradiation of 2,3-dimethoxy-5-nitronaphthalene in acetonitrile/water in the presence of NaOH affords 3-methoxy-5-nitro-2-naphthol in high yield (77%). This result demonstrates that in the excited state, the nitro group extends its directing influence into the other ring and photosubstitution occurs at the carbon atom which is not the most reactive one in the ground state. The effect is of the same nature as the *meta*-directing influence of the nitro group in benzene derivatives. In 2,3-dimethoxy-6-nitronaphthalene, it is the methoxy group at position-3 which, according to the *meta*-activation principle, is expected to be the most reactive one in the excited state. Indeed, 3-methoxy-7-nitro-2-naphthol is the major product (66%) in the photohydrolysis (Equation 20.22).

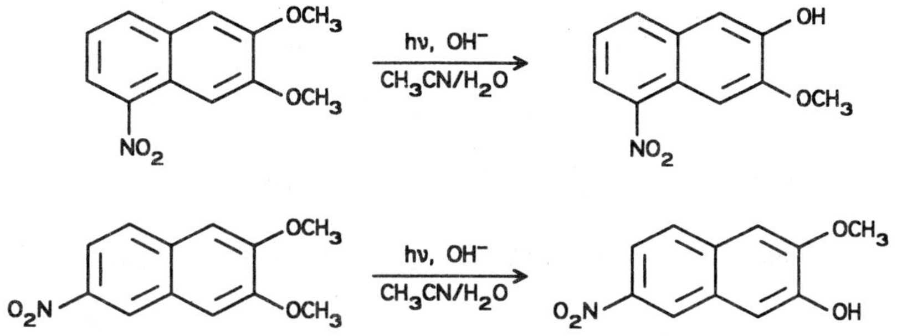

$$(20.22)$$

eferences

1. *Specialist Periodical Reports, Photochemistry*, Vol. 1–25, Bryce-Smith, D. and Gilbert, A., Eds., The Royal Society of Chemistry, Cambridge, U.K.
2. Weedon, A. C., Photochemistry of Aromatic Compounds in *Specialist Periodical Reports, Photochemistry*, Vol. 21, Bryce-Smith, D. and Gilbert, A., Eds., The Royal Society of Chemistry, Cambridge, U.K., 1990.
3. Gilbert, A., Photochemistry of Aromatic Compounds, in *Specialist Periodical Reports, Photochemistry*, Vol. 17, Bryce-Smith, D., Ed., The Royal Society of Chemistry, Cambridge, U.K., 1986.
4. Spillane, W. J., in *Isotopes in Organic Chemistry*, Vol. 4, Buncel, E. and Lee, C. C., Eds., Elsevier, Amsterdam, 51, 1978, 51.
5. Tamaki, T., Photo-induced Friedel-Crafts acylation of anthracene with aroyl chlorides, *Bull. Chem. Soc. Jpn.*, 51, 1145, 1978.
6. Tamaki, T., Photo-induced Friedel-Crafts acylation of anthracene with aroyl chlorides, *Kenkyu Hokoku-Sen'i Kobunshi Zairyo Kenkyusho*, 11, 1982; *Chem. Abstr.*, 101, 170372u, 1984.
7. Bunce, N. J., Kumar, Y., and Ravanal, L., Interaction of photoexcited naphthalenes with trifluoro-acetic acid, *J. Org. Chem.*, 44, 2612, 1979.
8. Shizuka, H. and Tobita, S., Proton-induced quenching and H-D isotope-exchange reactions of methoxynaphthalenes, *J. Am. Chem. Soc.*, 104, 6919, 1982.
9. Saito, I., Sugiyama, H., Yamamoto, A., Muramatsu, S., and Matsuura, T., Photochemical hydrogen-deuterium exchange reaction of tryptophan. The role in nonradiative decay of singlet tryptophan, *J. Am. Chem. Soc.*, 106, 4286, 1984.
10. Gilbert, A., Aromatic compounds: Substitution and cyclisation, in *Photochemistry in Organic Synthesis*, Coyle, J. D., Ed., The Royal Society of Chemistry, London, 1986, chap. 14.
11. Gilbert, A. and Baggott, J., *Essentials of Molecular Photochemistry*, Blackwell Scientific, Oxford, 1991, 386.
12. Havinga, E., De Jongh, R. O., and Dorst, W., Photochemical acceleration of the hydrolysis of nitrophenyl phosphates and nitrophenyl sulphates, *Recl. Trav. Chim. Pays-Bas*, 75, 378, 1956.
13. De Jongh, R. O. and Havinga, E., Photoreactions of aromatic compounds. XIV. The mechanism of the photosolvolysis of *m*-nitrophenyl phosphate, *Recl. Trav. Chim. Pays-Bas*, 87, 1318, 1968.

14. Havinga, E., Specificity of reactions of molecules in the excited state, *Kon. Ned. Akad. Wetensch., Afd. Natuurkunde,* 70, 52, 1961.

15. De Jongh, R. O. and Havinga, E., Photoreactions of aromatic compounds. VI. The mechanism of the photohydrolysis of *m*-nitroanisole, *Recl. Trav. Chim. Pays-Bas,* 85, 275, 1966.

16. Letsinger, R. L., Ramsay, O. B., and McCain, J. H., Photoinduced substitution. II. Substituent effects in nucleophilic displacement on substituted nitrobenzenes, *J. Am. Chem. Soc.,* 87, 2945, 1965.

17. De Vries, S. and Havinga, E., Photoreactions of aromatic compounds. V. Products isolated from the irradiation mixtures of *para*-nitroanisole and allyl *para*-nitrophenyl ether in 0.1 *N* NaOH, *Recl. Trav. Chim. Pays-Bas,* 84, 601, 1965.

18. Cornelisse, J. and Havinga, E., Photosubstitution reactions of aromatic compounds, *Chem. Rev.,* 75, 353, 1975.

19. Cornelisse, J., Photosubstitution reactions of aromatic compounds, *Pure Appl. Chem.,* 41, 433, 1975.

20. Lok, C. M. and Havinga, E., Photoreactions of aromatic compounds. XXXIV. Some orientation rules in nucleophilic aromatic substitution, *Proc. Kon. Ned. Akad. Wetensch. B,* 77, 15, 1974.

21. Fráter, Gy. and Cornelisse, J., in *Organic Photochemical Syntheses,* Vol. 1, Srinivasan, R., Ed., Wiley, New York, 1971, 71.

22. El'tsov, A. V., Frolov, A. N., and Kul'bitskaya, O. V., Photoreplacement of halogen by a nitro group in the aromatic series, *Zhur. Org. Khim.,* 6, 1943, 1970.

23. Frolov, A. N., Kul'bitskaya, O. V., and El'tsov, A. V., Photosubstitution of SO_2NH_2, SO_2CH_3, and SO_2CF_3 groups in the benzene nucleus, *Zhur. Org. Khim.,* 8, 432, 1972.

24. Frolov, A. N., Smirnov, E. V., and El'tsov, A. V., Photosubstitution of halogen and of SO_2X groups by the sulfite anion. Photosynthesis of aromatic sulfonic acids, *Zhur. Org. Khim.,* 10, 1686, 1974.

25. Nilsson, S., Direct cyanation of aromatic compounds. II. A comparison of isomer distributions from different cyanation methods, *Acta Chem. Scand.,* 27, 329, 1973.

26. Den Heijer, J., Shadid, O. B., Cornelisse, J., and Havinga, E., Photoreactions of aromatic compounds. XXXV. Nucleophilic photosubstitution of methoxy substituted aromatic compounds. Monophotonic ionization of the triplet, *Tetrahedron,* 33, 779, 1977.

27. Letsinger, R. L. and Hautala, R. R., Solvent effects in the photoinduced reactions of nitroaromatics with cyanide ion, *Tetrahedron Lett.,* 4205, 1969.

28. Petersen, W. C. and Letsinger, R. L., Photoinduced reactions of aromatic nitro compounds with borohydride and cyanide, *Tetrahedron Lett.,* 2197, 1971.

29. Vink, J. A. J., Verheijdt, P. L., Cornelisse, J., and Havinga, E., Photoreactions of aromatic compounds. XXVI. Photoinduced reactions of biphenyl and biphenyl derivatives with cyanide ion, *Tetrahedron,* 28, 5081, 1972.

30. Lok, C. M., Lugtenburg, J., Cornelisse, J., and Havinga, E., Photoreactions of aromatic compounds. XXII. Photo-induced reactions of 1-nitroazulene with nucleophiles, *Tetrahedron Lett.,* 4701, 1970.

31. Lok, C. M., Den Boer, M. E., Cornelisse, J., and Havinga, E., Photoreactions of aromatic compounds. XXVIII. Photo-induced reactions of 1-nitroazulene and derivatives with nucleophiles, *Tetrahedron,* 29, 867, 1973.

32. Vink, J. A. J., Lok, C. M., Cornelisse, J., and Havinga, E., Photoinduced reactions of some aromatic hydrocarbons with cyanide ion, *J. Chem. Soc., Chem. Commun.,* 710, 1972.

33. Havinga, E. and Cornelisse, J., Aromatic photosubstitution reactions, *Pure Appl. Chem.,* 47, 1, 1976.

34. Cornelisse, J., De Gunst, G. P., and Havinga, E., Nucleophilic aromatic photosubstitution, *Adv. Phys. Org. Chem.,* 11, 225, 1975.

35. Cornelisse, J., Lodder, G., and Havinga, E., Pathways and intermediates of nucleophilic aromatic photosubstitution reactions, *Rev. Chem. Intermed.,* 2, 231, 1979.

36. Varma, C. A. G. O., Plantenga, F. L., Van den Ende, C. A. M., Van Zeyl, P. H. M., Tamminga, J. J., and Cornelisse, J., Lifetime enhancement of the first electronic triplet state of 3,5-dinitroanisole by hydrogen bonding to clusters in mixed aqueous solvent systems, *J. Chem. Phys.,* 22, 475, 1977.

37. Varma, C. A. G. O., Tamminga, J. J., and Cornelisse, J., Mechanistic and kinetic aspects of the photoinduced OCH₃ substitution in 3,5-dinitroanisole. A probe for solvent effects in thermal reactions, *J. Chem. Soc., Faraday Trans.*, 78, 265, 1982.

38. Van Zeijl, P. H. M., Van Eijk, L. M. J., and Varma, C. A. G. O., Spectroscopic and kinetic study of the photoinduced methoxy substitution of 3-nitroanisole and 3,5-dinitroanisole, *J. Photochem.*, 29, 415, 1985.

39. Van Eijk, A. M. J., Huizer, A. H., Varma, C. A. G. O., and Marquet, J., Dynamic behavior of photoexcited solutions of 4-nitroveratrole containing OH⁻ or amines, *J. Am. Chem. Soc.*, 111, 88, 1989.

40. Mutai, K., Nakagaki, R., and Tukada, H., A rationalization of orientation in nucleophilic aromatic photosubstitution, *Bull. Chem. Soc. Jpn.*, 58, 2066, 1985.

41. Van Riel, H. C. H. A., Lodder, G., and Havinga, E., Photochemical methoxide exchange in some nitromethoxybenzenes. The role of the nitro group in S_N2Ar^* reactions, *J. Am. Chem. Soc.*, 103, 7257, 1981.

42. Letsinger, R. L. and McCain, J. H., Photoinduced substitution. III. Replacement of aromatic hydrogen by cyanide, *J. Am. Chem. Soc.*, 88, 2884, 1966.

43. Wubbels, G. G., Halverson, A. M., Oxman, J. D., and DeBruyn, V. H., Regioselectivity of photochemical and thermal Smiles rearrangements and related reactions of β-(nitrophenoxy)ethylamines, *J. Org. Chem.*, 50, 4499, 1985.

44. Kronenberg, M. E., Van der Heyden, A., and Havinga, E., Photoreactions of aromatic compounds. XIII. Photosubstitution of 4-nitroveratrole with methylamine; A convenient synthesis of *N*-methyl-4-nitro-*o*-anisidine, *Recl. Trav. Chim. Pays-Bas*, 86, 254, 1967.

45. Bunce, N. J., Cater, S. R., Scaiano, J. C., and Johnston, L. J., Photosubstitution of 1-methoxy-4-nitronaphthalene with amine nucleophiles: Dual pathways, *J. Org. Chem.*, 52, 4214, 1987.

46. Letsinger, R. L. and McCain, J. H., Photoreactions of nitroanisoles with cyanide ion. Studies of products and reaction sequence, *J. Am. Chem. Soc.*, 91, 6425, 1969.

47. Cervelló, J., Figueredo, M., Marquet, J., Moreno-Mañas, M., Bertrán, J., and Lluch, J. M., Nitrophenyl ethers as possible photoaffinity labels. The nucleophilic aromatic photosubstitution revisited, *Tetrahedron Lett.*, 25, 4147, 1984.

48. Van Vliet, A., Cornelisse, J., and Havinga, E., Photoreactions of aromatic compounds. XVII. Mechanistic aspects of the photoamination of nitrobenzene and derivatives, *Recl. Trav. Chim. Pays-Bas*, 88, 1339, 1969.

49. Yokoyama, K., Nakagaki, R., Nakamura, J., Mutai, K., and Nagakura, S., Spectroscopic and kinetic study of an intramolecular aromatic nucleophilic photosubstitution. Reaction mechanism of a photo-Smiles rearrangement, *Bull. Chem. Soc. Jpn.*, 53, 2472, 1980.

50. Mutai, K. and Kobayashi, K., Photoinduced intramolecular aromatic nucleophilic substitution (the photo-Smiles rearrangement) in amino ethers, *Bull. Chem. Soc. Jpn.*, 54, 462, 1981.

51. Mutai, K., Yokoyama, K., Kanno, S., and Kobayashi, K., Photoinduced intramolecular substitution. II. Absence of *meta*-favoring effect in nucleophilic photosubstitution of nitroveratrole derivatives, *Bull. Chem. Soc. Jpn.*, 55, 1112, 1982.

52. Wubbels, G. G., Halverson, A. M., and Oxman, J. D., Thermal and photochemical Smiles rearrangements of β-(nitrophenoxy)ethylamines, *J. Am. Chem. Soc.*, 102, 4848, 1980.

53. Wubbels, G. G., Sevetson, B. R., and Sanders, H., Competetive catalysis and quenching by amines of photo-Smiles rearrangement as evidence for a zwitterionic triplet as the proton-donating intermediate, *J. Am. Chem. Soc.*, 111, 1018, 1989.

54. Yokoyama, K., Nakamura, J., Mutai, K., and Nagakura, S., Substituent effect on a photo-Smiles rearrangement by laser photolysis, *Bull. Chem. Soc. Jpn.*, 55, 317, 1982.

55. Cantos, A., Marquet, J., and Moreno-Mañas, M., On the regioselectivity of the nucleophilic aromatic photosubstitution. The photoreaction of 4-nitroveratrole with *n*-hexylamine, *Tetrahedron Lett.*, 28, 4191, 1987.

56. Cantos, A., Marquet, J., and Moreno-Mañas, M., On the regioselectivity of the nucleophilic aromatic photosubstitution of 4-nitroanisole. A dual mechanistic pathway, *Tetrahedron Lett.*, 30, 2423, 1989.

57. Cantos, A., Marquet, J., Moreno-Mañas, M., González-Lafont, A., Lluch, J. M., and Bertrán, J., On the regioselectivity of 4-nitroanisole photosubstitution with primary amines. A mechanistic and theoretical study, *J. Org. Chem.*, 55, 3303, 1990.

58. Kul'bitskaya, O. V., Frolov, A. N., and El'tsov, A. V., UV-initiated substitution reactions in isomeric *N,N*-dimethyl-*p*-chloroanilines in the presence of oxidizing agents, *Zh. Org. Khim.*, 15, 440, 1979.

59. Frolov, A. N., Kul'bitskaya, O. V., and El'tsov, A. V., "Straight" and sensitized photosubstitution in the series of halogenoanilines. The role of radical cations in these processes, *Zh. Org. Khim.*, 15, 2118, 1979.

60. Grabner, G., Rauscher, W., Zechner, J., and Getoff, N., Photogeneration of radical cations from aqueous methoxylated benzenes, *J. Chem. Soc., Chem. Commun.*, 222, 1980.

61. Soumillion, J. Ph. and De Wolf, B., A link between photoreduction and photosubstitution of chloroaromatic compounds, *J. Chem. Soc., Chem. Commun.*, 436, 1981.

62. Bunce, N. J., Bergsma, J. P., and Schmidt, J. L., Mechanisms for aromatic photocyanation: naphthalene and biphenyl, *J. Chem. Soc., Perkin Trans. 2*, 713, 1981.

63. Bunnett, J. F., Aromatic substitution by the $S_{RN}1$ mechanism, *Acc. Chem. Res.*, 11, 413, 1978.

64. Rossi, R. A. and De Rossi, R. H., *Aromatic Substitution by the $S_{RN}1$ Mechanism*, American Chemical Society Monograph No. 178, A.C.S., Washington, D.C., 1983.

65. Semmelhack, M. F. and Bargar, T., Photostimulated nucleophilic aromatic substitution for halides with carbon nucleophiles. Preparative and mechanistic aspects, *J. Am. Chem. Soc.*, 102, 7765, 1980.

66. Bunnett, J. F. and Sundberg, J. E., Photostimulated arylation of ketone enolate ions by the $S_{RN}1$ mechanism, *J. Org. Chem.*, 41, 1702, 1976.

67. Rossi, R. A. and Alonso, R. A., Photostimulated reactions of *N,N*-disubstituted amide enolate anions with haloarenes by the $S_{RN}1$ mechanism in liquid ammonia, *J. Org. Chem.*, 45, 1239, 1980.

68. Rossi, R. A. and Bunnett, J. F., Arylation of several carbanions by the $S_{RN}1$ mechanism, *J. Org. Chem.*, 38, 3020, 1973.

69. Pierini, A. B., Baumgartner, M. T., and Rossi, R. A., Photostimulated reaction of aryl iodides with 2-naphthoxide ions by the $S_{RN}1$ mechanism, *Tetrahedron Lett.*, 29, 3429, 1988.

70. Bunnett, J. F. and Creary, X., Photostimulated condensation of aryl iodides with potassium dialkyl phosphites to form dialkyl arylphosphonates, *J. Org. Chem.*, 39, 3612, 1974.

71. Bunnett, J. F. and Creary, X., Arylation of arenethiolate ions by the $S_{RN}1$ mechanism. A convenient synthesis of diaryl sulfides, *J. Org. Chem.*, 39, 3173, 1974.

72. Bunnett, J. F. and Creary, X., "Nucleophilic" replacement of two halogens in dihalobenzenes without the intermediacy of monosubstitution products, *J. Org. Chem.*, 39, 3611, 1974.

73. Bunnett, J. F. and Creary, X., On fragmentation of aryl sulfide radical anions during aromatic $S_{RN}1$ reactions, *J. Org. Chem.*, 40, 3740, 1975.

74. Groen, M. B. and Havinga, E., Photoreactions of aromatic compounds. XXXIII. Nucleophilic photosubstitution of some heteroaromatic nitro compounds, *Mol. Photochem.*, 6, 9, 1974.

75. Párkányi, C., Aromatic photosubstitutions, *Pure Appl. Chem.*, 55, 331, 1983.

76. Terrier, F., *Nucleophilic Aromatic Displacement: The Influence of the Nitro Group*, VCH Publishers, Weinheim, 1991, chap. 6.

77. Beugelmans, R., Le Goff, M.-T., Pusset, J., and Roussi, G., Use of crown ethers for photochemical cyanation of aromatic hydrocarbons, *J. Chem. Soc., Chem. Commun.*, 377, 1976.

78. Beugelmans, R., Ginsburg, H., Lecas, A., Le Goff, M.-T., Pusset, J., and Roussi, G., Use of tetrabutylammonium cyanide for photocyanation of aromatic compounds: Phase transfer photochemistry, *J. Chem. Soc., Chem. Commun.*, 885, 1977.

79. Beugelmans, R., Ginsburg, H., Lecas, A., Le Goff, M.-T., and Roussi, G., Use of phase transfer agents for photocyanation of aromatic hydrocarbons, *Tetrahedron Lett.*, 3271, 1978.

80. Letsinger, R. L., quoted in *Eléments de Photochimie Avancée,* Courtot, P., Ed., Hermann, Paris, 1972, 343.
81. Hartsuiker, J., De Vries, S., Cornelisse, J., and Havinga, E., Photoreactions of aromatic compounds. XXI. Photoreaction of 4-nitroanisole with cyanate ion, *Recl. Trav. Chim. Pays-Bas,* 90, 611, 1971.
82. Beijersbergen van Henegouwen, G. M. J. and Havinga, E., Photoreactions of aromatic compounds. XX. The photosubstitution of nitronaphthalenes in alkaline media, *Recl. Trav. Chim. Pays-Bas,* 89, 907, 1970.

21

Photochemical Reactions of Arenes with Amines

Nigel J. Bunce
University of Guelph

21.1 Introduction

Many photochemical reactions involving amines and aromatic compounds have been described over the past 30 years, several of which have been recognized subsequently to be electron transfer processes. Amines have relatively low oxidation potentials, and it is therefore possible for an electron to be transferred from the amine as donor to the photoexcited aromatic compound as acceptor. Electron transfer is frequently possible even for unsubstituted arenes, and becomes more facile if the aromatic partner is substituted with electron-withdrawing groups. Since the oxidation potentials of amines cover a moderate range, it happens frequently that a given aromatic substrate may react by either electron transfer or some other mechanism depending on the identity of the amine.

Single electron transfer processes are generally described in terms of the Weller equation (Equation 21.1).

$$\Delta G(\text{electron transfer}) = {}^1E^* - \left(E_D^{ox} - E_A^{red}\right) - \left(e_o^2/\varepsilon r\right) \qquad (21.1)$$

In Equation 21.1, ${}^1E^*$ is the singlet-state excitation energy, E_D^{ox} and E_A^{red} are the donor oxidation potential and the acceptor reduction potential, respectively, and the term $(e_o^2/\varepsilon r)$ is the free energy gained when the separated ions are brought to the encounter distance r in a solvent of dielectric constant ε. Processes for which ΔG is negative usually occur at rates that approach the rate constant for diffusion in the solvent of interest. Full electron transfer to afford separated radical ions may occur in polar media, whereas in nonpolar media, partial electron transfer results in the formation of an amine-arene exciplex, which can be formed even when ΔG(electron transfer) is somewhat

0-8493-8634-9/95/$0.00+$.50
© 1995 by CRC Press, Inc.

positive. In this review, the term "electron transfer" will be used to describe both processes. For example,

$$ArH^* + R_3N \rightarrow R_3N^{\delta+} \cdot ArH^{\delta-} \text{ or } R_3N^{+\cdot} + ArH^{-}$$

In this review reactions are discussed in which the aromatic nucleus is reduced to a dihydro derivative, photosubstitutions at the aromatic nucleus, and reactions involving functional groups such as alkenes, carbonyl, nitro attached to the aromatic nucleus. In the latter category, amine-assisted dehalogenations of aryl halides will be specifically excluded, since they are described in Chapter 86. The general format will be to describe the early discoveries of each type of reaction, followed by key recent developments. Space limitations preclude a comprehensive coverage.

There have been several previous reviews of electron transfer photoreactions of aromatic compounds, in which the electron transfer reactions of amines feature prominently. These include Davidson (1984),[1] Mattay (1987,[2] 1989[3]), chapters in three recent volumes (vol. 156, 158, and 159) of *Topics in Current Chemistry*, and two pertinent chapters[4,5] in Part C of the book *Photoinduced Electron Transfer* (1988).

1.2 Additions and Reductions Involving the Aromatic Nucleus

Photoreactions of Arenes with Amines in the Absence of Other Reactants

To summarize the results of a large body of research, these reactions involve the amine quenching of the singlet excited state of the arene to form an amine-arene exciplex ^{1}(Amine$^{\delta+} \cdot$ ArH$^{\delta-}$). This may frequently be detected spectroscopically in nonpolar solvents, whereas in polar solvents it typically dissociates into radical ions. The final products of the reaction include the dihydroarene, the tetrahydrobiaryl coupling product, and amine-dihydroarene adducts, though not all products are seen in every case. Poorly oxidizable amines are inefficient both at quenching the arene fluorescence and at forming products. The dihydroarene-amine adducts from secondary amines usually involve N-H addition, while α-C-H adducts are formed from tertiary amines; aromatic amines frequently yield other adducts in which a new C-C bond is formed between the arene and a ring carbon of the aromatic amine. The issue of N-H vs. α-C-H addition in primary and secondary amines relates to whether the hydrogen is transferred "atom-like" or "proton-like". In a nonpolar exciplex having ^{1}(Amine $\cdot$ ArH) as the predominant structure, the hydrogen would be transferred effectively as if it were H$^\cdot$. However, most amine-arene exciplexes are highly polar ^{1}(Amine$^{\delta+} \cdot$ ArH$^{\delta-}$), in which case hydrogen is transferred as H$^+$. Lewis[6] has discussed the relative acidities of N-H vs. α-C-H protons in amine radical cations and suggests that the N-H may be kinetically more acidic. Finally, the exciplex may dissociate into radical ions; under these conditions, ArH$^{-\cdot}$ may be protonated either by the amine radical cation or by an external solvent such as water or an alcohol.

The photoreactions of anthracene with amines have been extensively studied and involve initial excitation of anthracene. In the original report,[7] irradiation of anthracene in the presence of triethylamine or *N,N*-diethylaniline gave 9,10-dihydroanthracene 1, tetrahydrobianthryl 2, and "aminated anthracenes". The reaction was more efficient in polar solvents such as acetonitrile. The dihydro-adduct from *N,N*-dimethylaniline (DMA) was shown[8] to be 9-(*p*-dimethylamino)-9,10-dihydroanthracene 3; subsequently,[9] the demethylated product 9-(*N*-methyl-*N*-phenylamino)-9,10-dihydroanthracene 4 was also found. The secondary amine *N*-methylaniline afforded a similar range of products:[10] 1, 2, 4, and 9-(*p-N*-methylamino)-9,10-dihydroanthracene 5; diethylamine gave 1, 2, and 9-diethylamino-9,10-dihydroanthracene 6. The primary aliphatic amine butylamine failed to react, and also failed to quench the fluorescence of anthracene.[7] However, the more easily oxidized primary aromatic amine aniline gave the photoadduct 9-phenylamino-9,10-dihydroanthracene 7.[11] All these reactions compete with the formation of the anthracene photodimer.

Fluorescence quenching of anthracene by amines in nonpolar solvents is accompanied by fluorescence from the amine-arene exciplex. With compounds such as 8 through 10, intramolecular exciplex formation strongly suppresses photodimerization of the anthracene moiety.[12,13] Several intramolecular dihydroadducts are formed in systems of this type.[14] Photolysis of the related 9-(ω-anilinoalkyl)-10-bromophenanthrenes leads to several products of intramolecular reaction, all of which are debrominated. Intramolecular electron transfer is followed by loss of HBr to afford an N,10-phenanthryl biradical, which can couple either at the 9- or 10-positions.[15]

In the intermolecular photoreaction between anthracene and DMA, both an exciplex and a higher aggregate such as a triplex[9,16] appear to be involved. Although maximal exciplex fluorescence quantum yield occurs at [DMA] ~ 0.2 M, products such as 2 and 4 are formed at higher [DMA] and thus originate with the triplex. The fate of the methyl group lost from 4 is formaldehyde, the collapse of the triplex to products being initiated by internal proton transfer.

$$^1\left(C_{14}H_{10} \cdot 2DMA\right) \leftrightarrow \left(C_{14}H_{10}{}^{\cdot -} \cdot DMA^{\cdot +} \cdot DMA\right) \rightarrow C_{14}H_{11}{}^{\cdot} \cdot PhN(CH_3)CH_2{}^{\cdot} + DMA$$

The mechanism of "in complex" proton transfer in arene-amine exciplexes is exemplified by the reaction between anthracene and diethylamine-d_1. Products 1, 2, and 6 all incorporated deuterium at the meso-position. In benzene, Et$_2$ND reacted slightly slower than Et$_2$NH, but there was no deuterium isotope effect in acetonitrile, consistent with complete rate-limiting electron transfer in that solvent.[10] In the corresponding reaction of tertiary amines, a proton is transferred from the α-C-H bond. In the photoreduction of anthracene by diethylhydroxylamine, it is suggested[17] that the O-H proton is transferred, on the grounds that O-ethylation (triethylhydroxylamine) inhibits both reduction and the quenching of anthracene fluorescence.

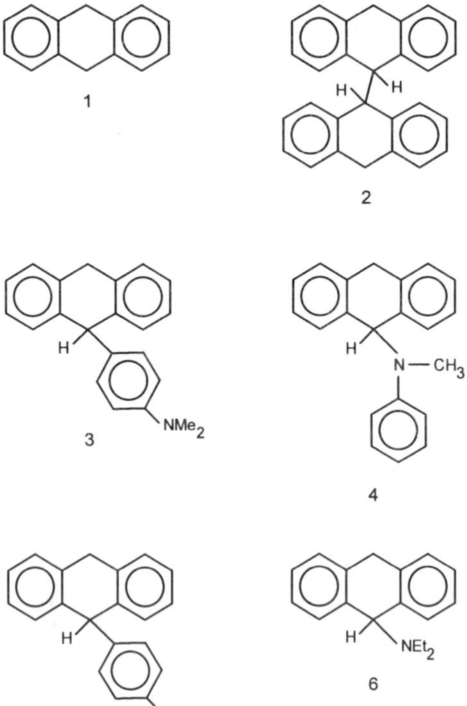

Str 1-10 (part 1)

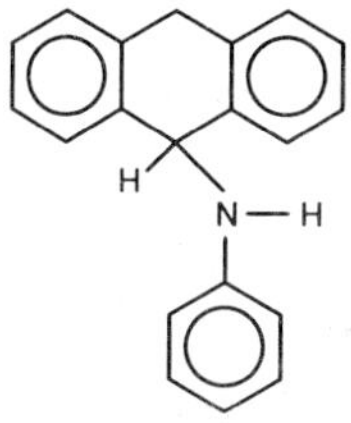

Str 1-10 (Part 2)

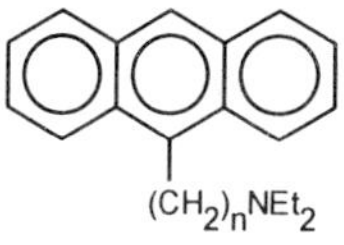

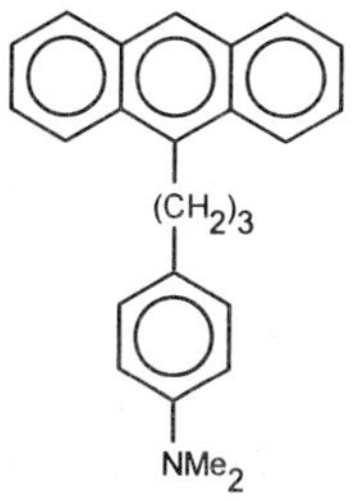

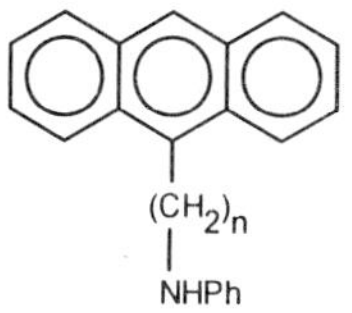

The principles outlined for the photoreactions of anthracene with amines also apply to phenanthrenes, naphthalenes, and benzenes. Barltrop and Owers[18] photoreduced naphthalene with triethylamine in aqueous acetonitrile, obtaining 1,4-dihydronaphthalene, the corresponding tetrahydrobinaphthyl, and 1-(1'-(N,N-diethylamino)ethyl)-1,4-dihydronaphthalene. These products were rationalized as arising through a series of electron transfers and protonations, supported by the observation that in acetonitrile-D_2O the dihydronaphthalene contained substantial deuterium at the 1,4-positions (Scheme 21.1).[19] In nonpolar media, a fluorescent exciplex is formed from naphthalene and triethylamine, rather than separated radical ions. The S_1 state of naphthalene was identified as the reactive excited state: benzophenone failed to sensitize the reduction, and there was concordance between the Stern-Volmer constant for quenching naphthalene fluorescence and the parameter intercept/slope from the plot of Φ^{-1} vs. $[Amine]^{-1}$. The reaction was extended to methyl-, methoxy-, and fluoronaphthalenes.

$$C_{10}H_8 + h\nu \rightarrow {}^1C_{10}H_8$$
$${}^1C_{10}H_8 + Et_3N \rightarrow C_{10}H_8{}^{\cdot-} + Et_3N^{\cdot+} \rightarrow C_{10}H_9{}^{\cdot} + Et_2NCH^{\cdot}CH_3$$
$$C_{10}H_8{}^{\cdot-} + H^+ \rightarrow C_{10}H_9{}^{\cdot}$$
$$C_{10}H_9{}^{\cdot} + C_{10}H_8{}^{\cdot-} \rightarrow C^{10}H_8 + C_{10}H_9{}^{\cdot-}$$
$$C_{10}H_9{}^{\cdot-} + H^+ \rightarrow C_{10}H_{10}$$

SCHEME 21.1

Flash photolysis of naphthalene/Et_3N/CH_3CN solutions showed absorption from an intermediate ($C_{10}H_8{}^{\cdot-}$ or $C_{10}H_9{}^{\cdot}$). This absorption was distinct from that of ${}^3C_{10}H_8$, and gradually replaced the absorption intensity of ${}^3C_{10}H_8$ as $[Et_3N]$ increased;[20] at $[Et_3N] > 0.1\ M$, no naphthalene triplet could be observed.[21]

In the presence of oxygen, an arene-sensitized oxidation of the aliphatic amine supplants the photoreduction of naphthalene.[22] Oxygen converts the α-aminoalkyl radical of Scheme 21.1 to $Et_2NCH(CH_3)OO^{\cdot}$, which oxidizes the reduced naphthalene species $C_{10}H_8{}^{\cdot-}$ or $C_{10}H_9{}^{\cdot}$ back to naphthalene. This makes the overall reaction catalytic in naphthalene.

$$Et_2NCH^{\cdot}CH_3 + O_2 \rightarrow Et_2NCH(CH_3)OO^{\cdot} \rightarrow CH_3CH{=}O + other\ products\ (ue3)$$

Unlike anthracene and naphthalene, which fail to photoreduce with primary amines, simple benzenes are reactive with all types of alkylamines. The photoreduction of benzene with triethylamine is entirely parallel to that of naphthalene[23] in terms of product types and the need for proton sources for efficient photoreduction. The mixed methylethylamines Me_2NEt and Et_2NMe both afford 1,4-dihydrobenzene adducts $C_6H_7CH_2NR_2$ in preference to $C_6H_7CH(CH_3)NR_2$, which is consistent with proton transfer from the amine to the aromatic; ω-dimethylaminophenylalkanes photocyclize through the methyl carbon to give similar cyclic, dihydroarene-amine adducts.[24] Fluorescence quenching, exciplex emission in cyclohexane,[25] and the depression of the quantum yield in the presence of xenon all support a singlet reactive state. Quoting from a review of benzene photochemistry:[26] "The mechanism resembles that of the thermal Birch reduction in that it involves electron transfer to benzene. Thus, sodium in liquid ammonia provides an electron donor sufficiently strong to transfer an electron to S_0 benzene, whereas the stronger electron acceptor S_1 benzene is required for electron transfer from amines." N-H dihydro-adducts form between benzene and 1,1-dimethylhydrazine,[27] morpholine, or piperidine.[28] The products of the photoreaction of fluorobenzene and triethylamine were similar to those from benzene; in addition, 1,2-dihydroarene adducts and *N,N*-diethylaniline (the latter representing the formal loss of ethyl fluoride from an initial adduct) were found.[29]

In cases where the amine is too poorly oxidizable to donate an electron to the singlet excited state of the parent arene, electron transfer may occur if the arene is substituted with electron-withdrawing groups such as cyano. For example, Ohashi and co-workers studied the photoreaction between aliphatic amines and the three dicyanobenzenes (DCB) in acetonitrile.[30] With Et_3N, the *o*- and *p*-DCB isomers were much more reactive than the *m*-isomer, and afforded significant yields of two substitution products, $NC \cdot C_6H_4 \cdot CH(CH_3)NEt_2$ and $NC \cdot C_6H_4 \cdot Et$, along with traces of benzonitrile. The first and last of these are rationalized as arising through HCN elimination from the initially formed α-C-H dihydro-adduct and dihydoarene, respectively. Ethylbenzonitrile is a secondary photoproduct formed by cleavage of the benzylic NEt_2 group of the α-C-H substituted product. DCB also was different from benzene in giving α-C-H adducts rather than N-H adducts even with primary and secondary alkylamines. Any initially formed N-adduct dissociates to reactants, and represents an energy-wasting process.

9,10-Dicyanoanthracene (DCA) has been photolyzed with primary, secondary, and tertiary aliphatic amines.[31] In benzene or methanol, 9-cyanoanthacene is the chief product, probably by loss of HCN from dihydro-DCA. In aqueous acetonitrile, 9-amino-10-cyanoanthracene is formed in high yield as the sole product. ^{15}N-labeling studies show that acetonitrile is the source of the 9-amino group, which is formed by hydrolysis of an intermediate imine $CH_3CH=N-DCA-H$. However, the photoaddition of CH_3NEt_2 and $Me_3SiCH_2NEt_2$ to DCA in acetonitrile led to dihydroarene and α-C-H adducts (which rearomatize by loss of HCN) as well as to reaction with acetonitrile.[32]

The course of reaction of dicyanoarenes with amines (α-C-H substitution vs. incorporation of CH_3CN) depends upon the reactivity of the cyanoarene radical anion. The radical anion from DCB is very reactive and attacks the amine radical cation before the radical ion pair can separate. The less reactive radical anion from DCA separates and then protonates in aqueous CH_3CN; it then accepts another electron from a second DCA·⁻, affording H-DCA⁻, which is the species that attacks acetonitrile by nucleophilic addition. 1,2,4,5-Tetracyanobenzene, which is intermediate in reduction potential between DCB and DCA, exhibits both types of reaction with hindered amines such as *N*-methylpyrrolidine or nicotine.[33] Finally, the addition of inert salts to DCA/Et_3N photolysis mixtures or further cyano substitution (2,6,9,10-tetracyanoanthracene/Et_3N) stabilizes the arene anion radicals so much that they become completely unreactive, even in the presence of water and acetonitrile.[34]

In a different application, photoinduced electron transfer to DCA was used to stimulate the fragmentation of an α-aminoketone. Recombination of the fragments gave a 9-acyl-9,10-dihydro DCA.[35]

The monocyanoarenes 9-cyanophenanthrene and 9-cyanoanthracene behave like the parent arenes upon irradiation with aliphatic amines in benzene. Primary and secondary amines afford dihydroarene and N-H dihydroarene-amine adducts (although the latter spontaneously rearomatize by loss of HCN).[36] Triethylamine afforded the dihydroarene alone, but with low quantum efficiency. It was argued that proton-like N-H transfer occurred within the exciplex for the primary and secondary amines, and that "α-C-H transfer is more efficient for the pure charge-transfer exciplex (radical ion pair) formed in polar solvent than for the heteroexcimer formed in nonpolar solvent".[37] Triplexes are formed upon irradiation of cyanoarene-diamine systems.[38]

Photoreactions of Arenes with Amines in the Presence of Another Reagent

The photoreactions between arenes and amines can be facilitated by the presence of a strong electron acceptor, such as a dicyanobenzene. In polar solvents such as aqueous acetonitrile, the initial excitation of the arene is followed by electron transfer to the electron acceptor. Thus in this reaction, the arene loses electron density, in contrast with the reactions of the previous section where the arene gained electron density from the amine.

$$^1ArH + DCB \rightarrow ArH^{\cdot+} + DCB^{\cdot-}$$

The electrophilicity of the arene radical cation promotes nucleophilic attack by the amine, and results in the overall addition of H-NR$_2$ to the arene to give a dihydroarene-amine adduct.[39] These reactions proceed in high yield between ammonia or primary aliphatic amines with a variety of arenes,[40] and hence complement the direct amine-arene photoreactions that are most successful for secondary, tertiary, and aromatic amines. In the DCB-assisted reactions, the latter amines undergo oxidation by the arene radical cation in preference to nucleophilic attack, and hence yield products of addition between the amine and DCB, rather than the amine-dihydroarene adduct.[41]

$$ArH^{\cdot+} + amine \rightarrow ArH + amine^{\cdot+}$$

The *m*-DCB-assisted photoaddition of primary amines to phenanthrene and anthracene occurs with highly selective *trans* stereochemistry.[42] Analogously, β-arylethylamines undergo intramolecular reaction in the presence of *p*-DCB, and cyclize to dihydroindoles in preparatively useful yields. When the aryl group is polymethoxyphenyl, the reaction provides a potentially useful entry into alkaloid synthesis.[43]

Several Japanese groups have examined the photocarboxylation of arenes using CO_2 in the presence of amine donors. The reactions are attractive as a means of functionalizing simple arenes, but so far have not been realized in high efficiency. The best yields to date have been achieved with phenanthrene/amine/CO_2 in dipolar aprotic solvents[44,45] where the products are 9,10-dihydrophenanthrene-9-carboxylic acid and amine-arene adducts. Carboxylation of the arene radical anion produced by single electron transfer is seen as a potential alternative to carboxylation of carbanion-like species such as Grignard reagents or electrochemically generated radical anions.

$$Amine + {}^1ArH \rightarrow ArH^{\cdot-} \xrightarrow{CO_2} \left(ArHCO_2\right)^{\cdot-} \xrightarrow{H^+, H^{\cdot}} ArH_2CO_2H$$

Photocarboxylation has been extended to 4-*H*-cyclopenta(*def*)phenanthrene[46] and naphthalene.[47] In the latter case, a variety of amine donors was used, most of which afforded 1-naphthoic acid as the chief isolable product (although the yields were very low), along with smaller amounts of 2-naphthoic acid and 1,4-dihydro-1-naphthoic acid.

21.3 Reactions Involving Functional Groups on the Aromatic Nucleus

Photoreactions of Amines with Arylalkenes

The photoreactions of amines with conjugated arylalkenes are singlet-state processes similar to those of the same amines with arenes such as anthracene, and are characterized by fluorescence quenching, reduction, and addition of the amine to the double bond. Cookson et al. reported the first examples of the photoreactions of styrene with primary and tertiary amines. Reduction, N-H addition, and α-C-H addition were all observed.[48] These reactions involve electron transfer from the amine to the singlet excited-state styrene derivative.[49] Chemical yields in these reactions are generally low, a result attributed to diffusion apart of radicals such as PhCH·CH$_3$ and R$_2$N· before they can recombine.[50] The analogous intramolecular reaction of Ph-CH = CH-(CH$_2$)$_n$-NH-R gives cyclization products in synthetically useful yield because the intermediate radical pair is held in close proximity and couples instead of diffusing apart.[51] In the case where n = 3 and R = CH$_3$, pyrrolidines and piperidines are formed; but with R larger than methyl, only the piperidine is formed.[52] A side reaction is the reduction of the double bond, with the formation of the N-alkylimine (inferred from the production of 5-phenylpentanal upon aqueous work-up). Photophysical studies have shown that the stability of the exciplex and the regiospecificity of cyclization depend on both the length of the polymethylene chain and the size of the R group (methyl vs. isopropyl).[50] Intramolecular α-C-H addition is not seen when the -NHR group is replaced with -NR$_2$, and this is also attributed to unfavorable stereochemistry in the exciplex.

1,1-Diphenyl-1-alkenes afford reduction and N-H addition products upon irradiation with primary alkenes in the presence of *p*-dicyanobenzene.[53] Since primary amines quench the fluorescence of arylalkenes inefficiently, the role of DCB is to generate the diphenylalkene radical cation. Other products include a formal acetonitrile adduct and the deconjugated arylalkene. In this reaction, the alkene becomes cationic and is then attacked by the amine acting as nucleophile, in contrast with the amine being the cationic partner as in the reactions discussed so far.

$$Ph_2C=CHR + DCB \rightarrow Ph_2C=CHR^{·+} + DCB^{·-}$$

Arylethylenes such as 1,1-diphenylethylene also undergo amine-assisted photocarboxylation, analogously with arenes, but the yields are very low.[54] The products include the reductive adduct of H-CO$_2$H across the double bond and reductive arylethylene/amine/CO$_2$ adducts.

Irradiation of diethylamine and the *cis*- and *trans*-stilbenes afford 1,2-diphenylethane and the H-NR$_2$ adduct 1-(*N,N*-diethylamino)-1,2-diphenylethane,[55] while triethylamine gives 1,2-diphenylethane; 1,2,3,4-tetraphenylbutane; and the α-C-H adduct 1-(1'-(*N,N*-diethylamino)-ethyl)-1,2-diphenylethane in polar solvents.[56] Fluorescent exciplexes can be seen with simple trialkylamines in nonpolar media; these exciplexes are highly polar and dissociate in polar solvents. Steric as well as redox effects govern exciplex stability. Back electron transfer within the exciplex suppresses product formation, leading to low quantum yields of product formation, even when the primary process (singlet quenching by the amine) is highly efficient. Lewis[6] has discussed the mechanism of proton transfer within stilbene-amine exciplexes: proton transfer is efficient in the initially formed contact radical ion pair, but does not occur in solvent-separated or free radical ions, which decay by back electron transfer. For this reason, rate constants for quenching stilbene fluorescence by amines are dependent upon the amine oxidation potential, but not upon the nature of the C-H bond which is ultimately broken (since bond breaking occurs after the rate-determining step).

The addition of secondary amines to 1,2-diphenylcyclobutene occurs entirely analogously to the reaction with *trans*-stilbene[57] (except, of course, that *cis,trans* isomerization cannot compete in the case of the cycloalkene). With the primary amine propylamine, *trans*-stilbene gave a variety of reductive cyclization products,[58] but fluorescence quenching was very inefficient with this amine. The various photochemical processes of the stilbenes, including their reactions with amines, have been reviewed.[59]

Many reactions of arylcyclopropanes resemble those of arylalkenes. 1,1-Diphenylcyclopropanes that are substituted in the 2- or 2,3-positions with cyano groups undergo reduction and reductive amination (α-C-H addition) upon irradiation with Et_3N in CH_3CN solvent. Fluorescence quenching and the similarity with amine-stilbene reactions suggested an electron transfer/proton transfer mechanism. With 1,1-diphenyl-2,2,3,3-tetracyanocyclopropane, fragmentation to give 1,1-diphenyl-2,2-dicyanoethylene occurred. The authors speculate that the latter product may arise through elimination from an unstable α-C-H adduct.[60]

Photoreduction of Carbonyl Compounds with Amines

In hydrogen donor solvents exemplified by isopropyl alcohol, aromatic ketones undergo photoreduction by way of the $^3n,\pi^*$ state, which has been likened to an alkoxy radical in terms of its hydrogen-abstracting capability (see Chapter ??). Photoreduction by this mechanism is inefficient for ketones whose lowest states are $^3\pi,\pi^*$, since the oxygen has little radical character. However, such substrates usually photoreduce smoothly in the presence of aliphatic amines, following electron transfer from the amine to the triplet excited ketone to give the ketyl anion radical. Protonation of the radical anion may involve the amine radical cation or an external proton source such as water or an alcohol.

Hydrogen abstraction pathway:

$$^3Ar \cdot CO \cdot R + R'H \rightarrow Ar \cdot \overset{\cdot}{C}(OH) \cdot R + R' \rightarrow \rightarrow \text{products}$$

Electron transfer pathway:

$$^3Ar \cdot CO \cdot R + \text{amine} \rightarrow \left(\text{amine}^{\delta+} \cdot Ar \cdot CO^{\delta-} \cdot R\right) \rightarrow \text{amine}^{\cdot+} + Ar \cdot CO^{\cdot-} \cdot R$$

$$Ar \cdot CO^{\cdot-} \cdot R + \left(H^+\right) \rightarrow Ar \cdot \overset{\cdot}{C}(OH) \cdot R \rightarrow \rightarrow \text{products}$$

Considerable interest has been shown in whether the amine radical cation donates H^+ to the ketone radical anion from the N-H group or the α-C-H position. To summarize the situation, the radical ion pair ($=CH-NH^{\cdot+}Ar_2CO^{\cdot-}$) can undergo three reactions: reverse electron transfer, N-H proton transfer, and α-C-H proton transfer. Only the last of these processes leads to products. Although N-H proton transfer occurs efficiently, the radical pair $R_2N \cdot Ar_2\overset{\cdot}{C}OH$ couples and then dissociates to reactants.[61] Tertiary alkylamines therefore photoreduce aromatic ketones most efficiently.[62] Occasionally, these quantum yields exceed unity, due to ground-state reduction of the substrate by the α-aminoalkyl radical;[63] for example,

$$Et_2NCH^{\cdot}CH_3 + Ar_2CO \rightarrow Et_2N-CH=CH_2 + Ar_2\overset{\cdot}{C}OH$$

Turro[64] summarized the experimental criteria for photoreduction of aryl ketones to proceed through electron transfer rather than hydrogen abstraction as follows.

"1. The rate constants for photoreduction by charge transfer are higher than those expected for radical-like hydrogen atom abstraction, based on an alkoxy radical model.

2. The quantum yields are solvent polarity dependent.

3. π,π^* states are more efficiently photoreduced by a hydrogen donor capable of reactivity via a charge transfer mechanism than by hydrogen donors capable only of hydrogen atom abstraction.

4. In favorable cases, direct spectroscopic evidence for radical cations, produced by complete electron abstraction, is available."

Photoreduction of Nitro Compounds with Amines

Like aromatic ketones, aromatic nitro compounds undergo photoreduction involving hydrogen abstraction by the $^3n,\pi^*$ state in hydrogen donor solvents. Substrates having lowest $^3\pi,\pi^*$ states (such as 1-nitronaphthalene), are unreactive by this mechanism, but can undergo electron transfer-assisted photoreduction in the presence of electron donors such as amines.[65] In some cases, the reaction takes a different course, depending on whether or not amines are present. For example, photoreduction of 4-nitro-4-diethylamino-azobenzene occurred at the azo linkage (hydrogen abstraction) in isopropyl alcohol, but at the nitro group (electron transfer) in the presence of butylamine.[66] Similarly, highly hindered nitrobenzenes such as 2,4,6-tri-*t*-butyl-nitrobenzene undergo intramolecular hydrogen abstraction by the nitro group from one of the *ortho*-*t*-butyl groups upon irradiation in neutral media; whereas, in the presence of tertiary aliphatic amines, the nitro group is converted to a nitrone. However, secondary aliphatic amines photoreduce these substrates directly to the aniline, but less efficiently.[67] The formation of the nitrone is proposed to involve electron transfer followed by proton exchange, with the resultant radical pair coupling to give, in the case of Et$_3$N, ArN(OH)·OCH(CH$_3$)·NEt$_2$, which fragments into ArNO, CH$_3$CHO, and Et$_2$NH. Further reduction of ArNO to ArNHOH is followed by coupling with CH$_3$CHO to yield the nitrone.

Aromatic nitro compounds are also photoreduced in the presence of aromatic amines such as DMA. For example, *m*-chloronitrobenzene underwent reduction to *m*-chloroaniline, while *N*-methylaniline and *N*-formyl-*N*-methylaniline[68] were formed from DMA. Note that, lacking the nitro group, chlorobenzene undergoes electron transfer-assisted dechlorination under similar conditions (Chapter 106). Methoxy substitution of the DMA enhances the quantum yield of electron transfer-assisted reduction of *m*-chloronitrobenzene, while cyano substitution has the opposite effect.[69] Isotope effects have been used to study the events following electron transfer from DMA to 3-chloronitrobenzene. k^H/k^D was unity for a mixture of DMA and DMA-d$_{11}$, ruling out hydrogen abstraction for which $k^H/k^D \sim 4.5$. Hence, hydrogen abstraction cannot be the rate-limiting step following photoexcitation. However, the product composition from PhN(CH$_3$)CD$_3$ indicated an isotope effect $k^H/k^D \sim 3$. This shows that while the quantum yield of photoreduction is determined by the rate of electron transfer within the DMA$^{\delta+}$· ArNO$_2$$^{\delta-}$ exciplex ($k^{DMA}/k^{DMA-d11} = 1$), the product distribution is determined by proton transfer from the methyl groups of DMA, for which $k^H/k^D = 3$.

Photoreduction of Benzylic Acetates with Amines

The photoreduction of benzylic acetates ArCH$_2$OAc to the corresponding toluene derivatives has been effected by various amines (in the order $\Phi_{tert} > \Phi_{sec} > \Phi_{prim}$).[70] In CH$_3$OD as solvent, deuterium is incorporated at the 9-position, but not completely. The intermediacy of free radicals is shown by the formation of the ArCH$_2$CH$_2$Ar as a by-product. 9-Fluorenyl acetate and 9-fluorenol are photoreduced to fluorene under similar conditions.

$$^1ArCH_2OAc + amine \rightarrow {}^1\left(ArCH_2OAc^{\delta-} \cdot amine^{\delta+}\right) \rightarrow ArCH_2OAc^{\cdot-} + amine^{\cdot+}$$

$$ArCH_2OAc^{\cdot-} \rightarrow ArCH_2^{\cdot} + OAc^- \text{ or } ArCH_2^- + OAc^{\cdot}$$

.4 Reactions Proceeding via Nucleophilic Substitution

Photochemical nucleophilic substitutions are covered elsewhere in detail in this volume. Van Riel et al. discussed three possible mechanisms for substitution:[71] (1) direct nucleophilic substitution (S$_N$2Ar*) to give an intermediate σ-complex; (2) electron transfer from the nucleophile to the aromatic compound, followed by ground-state reactions of the resulting radical ions; or (3) electron transfer from the aromatic partner to an electron acceptor, followed by attack of the nucleophile upon the aromatic radical cation. Reactions of the latter two types have already been encountered in Section 21.2. The present discussion focuses on amines as nucleophiles, with particular reference to whether the reaction involves the S$_N$2Ar* mechanism or electron transfer from the amine to the aromatic compound.

Early work on photochemical nucleophilic displacements was documented by Cornelisse and Havinga.[72] The most detailed studies have been made of methoxynitroarenes, among which it is well established that S$_N$2Ar* nucleophilic displacements occur most readily *meta* to NO$_2$. Like other nitroaromatic compounds (Section 21.3), methoxynitroarenes usually react from their triplet states; but the assignment of mechanism is complicated by the closeness in energy of the ^{3}n,π* and ^{3}π,π* states of methoxynitroarenes. Two concepts have been advanced to explain which of the two mechanisms (S$_N$2Ar* or electron transfer) will be followed in a given case: the ease of oxidation of the "nucleophile" or the nature of the lowest triplet state.

The case of 4-nitroveratrole (4-NV) illustrates the argument that the course of reaction is determined by the ease of oxidation of the nucleophile. Poorly oxidizable nucleophiles such as CN$^-$, OH$^-$, and CH$_3$NH$_2$ react by the S$_N$2Ar* mechanism, displacing the methoxy group *meta* to NO$_2$. Secondary aliphatic amines and aromatic amines undergo electron transfer, and the radical ions thus formed react together with eventual substitution *para* to NO$_2$. The S$_N$2Ar* reaction has been considered the default mechanism, which occurs only when electron transfer is energetically infeasible.[73] This is true for 4-NV and also for 1-nitro-4-methoxynaphthalene, although in the latter case, the S$_N$2Ar* reaction occurs at the *ipso*-position, since there is no substituent *meta* to NO$_2$. In nonpolar media, amine-assisted photoreduction (Section 21.3) competes with substitution by way of radical ions. Recent work shows that aromatic nucleophilic substitutions frequently afford σ complexes additional to those which can be inferred from the structures of the final products.[74]

The case of 2-fluoro-4-nitroanisole illustrates the argument that the course of the reaction is governed by the nature of the lowest triplet state.[75] Experiments with triplet quenchers and with radical scavengers indicated that the two substitution products obtained with 1-hexylamine derive from different states: replacement of fluoride (*meta* to NO$_2$) is an S$_N$2Ar* reaction of the ^{3}π,π* state, while replacement of methoxy (*para* to NO$_2$) is an electron transfer reaction of the ^{3}n,π* state.

The case of *m*-nitroanisole shows that the properties of both the nucleophile and the methoxynitroarene must be considered in order to explain the outcome of the reaction. *m*-Nitroanisole undergoes the S$_N$2Ar* reaction with nucleophiles such as OH$^-$, CN$^-$, and CH$_3$NH$_2$, but fails to form products with the more easily oxidized Et$_2$NH in polar solvents (no *para*-substituent to displace following electron transfer). However, it photoreduces smoothly in nonpolar media (those which favour a lowest ^{3}n,π* state).[76]

Mutai et al.[77] have discussed the difference between the S$_N$2Ar* and electron transfer mechanisms in terms of the frontier orbitals of the aromatic partner. In their paper, the terms "HOMO" and "LUMO" refer to the highest occupied and lowest unoccupied MOs of the *unexcited* aromatic, respectively. Poorly oxidizable nucleophiles (OH$^-$, CN$^-$, and primary aliphatic amines) form bonds at the carbon atom with the highest electron density in the (now singly occupied) HOMO. In the case of 4-NV, this is the position *meta* to the nitro group, and leads to the replacement of the *m*-OCH$_3$ by the incoming nucleophile. Conversely, when the nucleophile is easily oxidized (e.g., secondary or aromatic amine), the first step in the reaction is electron transfer from the amine to the aromatic, affording (ground-state) radical ions. Bond formation now takes place between the amine radical cation and the former LUMO of the aromatic, which in the case of 4-NV has the greatest electron density *para* to NO_2. Cantos et al.[78] have recently challenged the part of Mutai's explanation dealing with electron transfer substitutions, arguing in the case of 4-nitroanisole that the relative stabilities of the reaction intermediates are more important than the electron distributions in the frontier orbitals.

The photo-Smiles rearrangement of β-aryloxyethylamines to β-arylaminoethanols is the intramolecular variant of the photosubstitution of a nitroaryl ether with an amine. The *m*-nitroanisole analog having the -$OCH_2CH_2NH_2$ group *meta* to NO_2 undergoes intramolecular substitution at C-3, *meta* to NO_2 (S_N2Ar^*). The *para*-analog also undergoes attack *meta* to NO_2, with the initial formation of a dihydrobenzene, since H^- is a poor leaving group.[79] Replacement of the -$OCH_2CH_2NH_2$ group by -OCH_2CH_2NHPh changes the course of the reaction from S_N2Ar^* to electron transfer. Now, it is the *para*-isomer that undergoes the photo-Smiles rearrangement, and *meta*-isomers undergo cyclization at the *para*-position.

The photo-Smiles rearrangement of p-$NO_2 \cdot C_6H_4 \cdot OCH_2CH_2NHPh$ is unusual in that it is simultaneously accelerated and quenched by external amine bases. The initially formed electron transfer product p-$^{--}NO_2 \cdot C_6H_4 \cdot OCH_2CH_2NH^{\cdot +}Ph$ undergoes two independent reactions with the external amine, namely, deprotonation, leading to rearrangement, and quenching back to starting material.[80]

References

1. Davidson, R. S., The chemistry of excited complexes: A survey of reactions, *Adv. Phys. Org. Chem.*, 19, 1, 1983.
2. Mattay, J., Charge transfer and radical ions in photochemistry, *Angew. Chem. (Engl. Ed.)*, 26, 825, 1987.
3. Mattay, J., Photoinduced electron transfer in organic synthesis, *Synthesis*, 233, 1989.
4. Lewis, F. D., Carbon-carbon multiple bonds, *Photoinduced Electron Transfer, Part C*, Fox M. A. and Chanon, M., Eds., Elsevier, Amsterdam, 1988, 1.
5. Albini, A. and Sulpizio, A., Aromatics, *Photoinduced Electron Transfer, Part C*, Fox, M. A. and Chanon, M., Eds., Elsevier, Amsterdam, 1988, 88.
6. Lewis, F. D., Proton transfer reactions of photogenerated radical ion pairs, *Acc. Chem. Res.*, 19, 401, 1986.
7. Davidson, R. S., The photoreactions of aromatic hydrocarbons in the presence of amines. *J. Chem. Soc., Chem. Commun.*, 1450, 1969.
8. Pac, C. and Sakurai, H., The photochemical addition of aromatic amines to anthracene. *Tetrahedron Lett.*, 3829, 1969.
9. Yang, N. C., Shold, D. M., and Kim, B. Photochemistry of anthracene in the presence and absence of dimethylaniline, *J. Am. Chem. Soc.*, 98, 6587, 1976.
10. Yang, N. C. and Libman, J., Chemistry of exciplexes, photochemical addition of secondary amines to anthracene, *J. Am. Chem. Soc.*, 95, 5783, 1973.
11. Vaidyanathan, S. and Ramakrishnan, V., Exciplex state photochemistry. Solvent effect on the photoreaction between anthracene and aniline, *Ind. J. Chem.*, 13, 257, 1975.
12. Brimage, D. R. G. and Davidson, R. S., Effect of intramolecular exciplex formation upon the photoreactivity of naphthylalkylamines and anthrylalkylamines, *J. Chem. Soc., Chem. Commun.*, 1385, 1971.
13. Eisenthal, K. B., Crawford, M. K., Dupuy, C., Hetherington, W., Korenowski, G., McAuliffe, M. J., and Yang, Y., Photodissociation, short-lived intermediates, and charge transfer phenomena in liquids, *Springer Ser. Chem. Phys.*, 14, 220, 1980.
14. Sugimoto, S., Yamano, J., Suyama, K., and Yoneda, S., Photochemical behavior of 9-(2-anilinoethyl)- and 9-(3-anilinopropyl)-anthracenes in various solvents, *J. Chem. Soc., Perkin Trans. I*, 483, 1989.
15. Sugimoto, A., Hiraoka, R., Fukada, N., Kosaka, H., and Inoue, H., Photochemical reactions of 9-(ω-anilinoalkyl)-10-bromophenanthrenes, *J. Chem. Soc., Perkin Trans. I*, 2871, 1992.

16. Salteil, J., Townsend, D. E., Watson, B. D., Shannon, P., and Finson, S. L., Concerning the participation of the anthracene/*N,N*-dimethylaniline exciplex in anthracene photodimerization, *J. Am. Chem. Soc.*, 99, 884, 1977.

17. Lissi, E. A., Rubio, M. A., and Fuentealba, M., Photoreduction of anthracene by diethylhydroxylamine, *J. Photochem.*, 37, 205, 1987.

18. Barltrop, J. A. and Owers, R. J., A photochemical reduction of naphthalene and some of its derivatives, *J. Chem. Soc., Chem. Commun.*, 1462, 1970.

19. Barltrop, J. A., The photoreduction of aromatic systems, *Pure Appl. Chem.*, 33, 179, 1973.

20. Burrows, H. D., Flash photolytic study of the photoreduction of naphthalene by triethylamine, *Photochem. Photobiol.*, 19, 241, 1974.

21. Watkins, A. R., Electron-transfer fluorescence quenching by aliphatic amines, *Aust. J. Chem.*, 33, 177, 1980.

22. Bartholemew, R. F., Brimage, D. R. G., and Davidson, R. S. The photosensitized oxidation of amines. IV. The use of aromatic hydrocarbons as sensitizers, *J. Chem. Soc. (C)*, 3482, 1971.

23. Bellas, M., Bryce-Smith, D., Clarke, M. T., Gilbert, A., Klunklin, G., Krestonosich, S., Manning, C., and Wilson, S., The photoaddition of aliphatic amines to benzene, *J. Chem. Soc., Perkin Trans. I*, 2571, 1977.

24. Bryce-Smith, D., Gilbert, A., and Klunklin, G., *meta*-Photoaddition of phenylalkylamines, *J. Chem. Soc., Chem. Commun.*, 330, 1973.

25. Beecroft, R. A. and Davidson, R. S., Examples of competition between energy transfer and excited complex formation in the reactions of benzenoid compounds with tertiary amines, *Chem. Phys. Lett.*, 77, 77, 1981.

26. Bryce-Smith, D. and Gilbert, A., The organic photochemistry of benzene, II, *Tetrahedron*, 33, 2459, 1977.

27. Gilbert, A., Krestonosich, S., and Rivas, C., Photoreactions of aromatic compounds with 1,1-dimethylhydrazine, *Acta Cient. Venez.*, 39, 99, 1988, through *Chem. Abstr.*, 110, 148799u, 1989.

28. Gilbert, A., Krestonosich, S., Martinez, C., and Rivas, C., Photochemical reactions of heteroparaffins with benzene, *Acta Cient. Venez.*, 40, 189, 1989, through *Chem. Abstr.*, 112, 153805f, 1990.

29. Gilbert, A. and Krestonosich, S., Excited state substitution and addition reactions of aryl fluorides with aliphatic amines, *J. Chem. Soc., Perkin Trans. I*, 1393, 1980.

30. Ohashi, M., Miyake, K., and Tsujimoto, K., Photochemical reactions of dicyanobenzenes with aliphatic amines, *Bull. Chem. Soc. Jpn.*, 53, 1683, 1980.

31. Ohashi, M., Kudo, H., and Yamada, S., Photochemical reaction of dicyanoanthracene with acetonitrile in the presence of an aliphatic amine. A novel photochemical amination, *J. Am. Chem. Soc.*, 101, 2201, 1979.

32. Hasegawa, E., Brumfield, M. A., and Mariano, P. S., Photoadditions of ethers, thioethers, and amines to 9,10-dicyanoanthracene by electron transfer pathways, *J. Org. Chem.*, 53, 5435, 1988.

33. Yamada, S., Nakagawa, Y., Watbiki, O., Suzuki, S., and Ohashi, M., Photochemical amination of tetracyanobenzene with acetonitrile in the presence of aliphatic amine, *Chem. Lett.*, 361, 1986.

34. Kellett, M. A., Whitten, D. G., Gould, I. R., and Bergmark, W. R., Surprising differences in the reactivity of cyanoaromatic radical anions generated by photoinduced electron transfer, *J. Am. Chem. Soc.*, 113, 358, 1991.

35. Bergmark, W. R. and Whitten, D. G., Cooperative reactivity in photogenerated radical ion pairs: Photofragmentation of amino ketones, *J. Am. Chem. Soc.*, 112, 4042, 1990.

36. Lewis, F. D., Zebrowski, B. E., and Correa, P. E., Photochemical reactions of arenecarbonitriles with aliphatic amines. I. Effect of the arene structure on aminyl vs. α-aminoalkyl radical formation, *J. Am. Chem. Soc.*, 106, 187, 1984.

37. Lewis, F. D. and Correa, P. E., Photochemical reactions of arenecarbonitriles with aliphatic amines. II. Effect of amine structure on aminyl vs. α-aminoalkyl radical formation, *J. Am. Chem. Soc.*, 106, 194, 1984.

38. Schneider, S., Geiselhart, P., Seel, G., Lewis, F. D., Dykstra, R. E., and Nepras, M. J., Formation and decay of exciplexes between 9-cyanophenanthrene and mono- and diaminalkanes, *J. Phys. Chem.*, 93, 3112, 1989.

39. Yasuda, M., Yamashita, T., Matsumoto, T., Shima, K., and Pac, C., Synthetic applications of photochemical electron transfer to direct amination of arenes by ammonia and primary amines, *J. Org. Chem.*, 50, 3667, 1985.

40. Yasuda, M., Matsuzaki, Y., Shima, K., and Pac, C., Direct photoamination of arenes with ammonia and amines in the presence of electron acceptors, *J. Org. Chem.*, 52, 753, 1987.

41. Yasuda, M., Matsuzaki, Y., Shima, K., and Pac, C., Mechanisms for direct photoamination of arenes with ammonia and amines in the presence of *m*-dicyanobenzene, *J. Chem. Soc., Perkin Trans. II*, 745, 1988.

42. Yasuda, M., Shiomori, K., Hamasuna, S., Shima, K., and Yamashita, T., Stereochemical studies on the amination of arenes with ammonia and alkylamines via photochemical electron transfer, *J. Chem. Soc., Perkin Trans II*, 305, 1992.

43. Pandey, G., Sridhar, M., and Bhaleroa, U. T., Regiospecific dihydroindoles directly from β-arylethylamines by photoinduced SET reaction, *Tetrahedron Lett.*, 31, 5373, 1990.

44. Tazuke, S. and Ozawa, H., Photofixation of carbon dioxide: Formation of 9,10-dihydrophenanthrene-9-carboxylic acid from phenanthrene-amine-carbon dioxide systems, *J. Chem. Soc., Chem. Commun.*, 237, 1975.

45. Tazuke, S., Kazama, S., and Kitamoru, N., Reductive photocarboxylation of aromatic hydrocarbons, *J. Org. Chem.*, 51, 4548, 1986.

46. Minabe, M., Isozumi, K., Kawai, K., and Yoshide, M., An observation on carboxylation of 4-H-cyclopenta(*def*)phenanthrene, *Bull. Chem. Soc. Jpn.*, 61, 2063, 1988.

47. Tagaya, H., Onuki, M., Tomioka, Y., and Wada, Y., Photocarboxylation of naphthalene in the presence of carbon dioxide and an electron donor, *Bull. Chem. Soc. Jpn.*, 63, 3233, 1990.

48. Cookson, R. C., Costa, S. M. de B., and Hudec, J., Photochemical addition of amines to styrenes, *J. Chem. Soc., Chem. Commun.*, 753, 1969.

49. Brentnall, R. A., Crosby, P. M., and Salisbury, K., Excited state complex formation between styrene and tertiary amines, *J. Chem. Soc., Perkin Trans. II*, 2002, 1977.

50. Lewis, F. D., Reddy, G. D., Schneider, S., and Gahr, M., Photophysical and photochemical behavior of intramolecular styrene-amine exciplexes, *J. Am. Chem. Soc.*, 113, 3498, 1991.

51. Lewis, F. D., Reddy, G. D., Schneider, S., and Gahr, M., Intramolecular photochemical addition reactions of ω-styrylaminoalkanes, *J. Am. Chem. Soc.*, 111, 6465, 1989.

52. Lewis, F. D. and Reddy, G. D., Intramolecular photochemical reactions of *N*-alkyl-5-phenyl-4-penten-1-amines, *Tetrahedron Lett.*, 31, 5293, 1990.

53. Yamashita, T., Shiomori, K., Yasuda, M., and Shima, K., Photoinduced nucleophilic addition of ammonia and alkylamines to aryl-substituted alkenes in the presence of *p*-dicyanobenzene, *Bull. Chem. Soc. Jpn.*, 64, 366, 1991.

54. Ito, Y., Uozo, Y., and Matsuura, T., Photocarboxylation in the presence of aromatic amines and carbon dioxide, *J. Chem. Soc., Chem. Commun.*, 562, 1988.

55. Kawanisi, M. and Matsunaga, K., Novel photochemical reactions of diphenylacetylene and stilbenes in amines, *J. Chem. Soc., Chem. Commun.*, 313, 1972.

56. Lewis, F. D., Formation and reactions of stilbene exciplexes, *Acc. Chem. Res.*, 12, 152, 1979.

57. Sakuragi, M. and Sakuragi, H., Photochemical behavior of 1,2-diphenylcyclobutene in protic solvents. Addition of secondary amines, *Bull. Chem. Soc. Jpn.*, 50, 1802, 1977.

58. Buquet, A., Couture, A., and Lablache-Combier, A., Photocyclization reactions in primary amines. Convenient synthesis of 1,4-dihydrophenanthrene, *J. Org. Chem.*, 44, 2300, 1979.

59. Lewis, F. D., Bimolecular photochemical reactions of the stilbenes, *Adv. Photochem.*, 13, 166, 1986.

60. Tamioka, H. and Kanda, M., Photochemical reactions of 1,1-diphenylpolycyanocyclopropanes in the presence of amines. Formal extrusion of dicyanomethylene, *Chem. Lett.*, 2223, 1990.

61. Inbar, S., Linschitz, H., and Cohen, S. G., Primary quantum yields of ketyl radicals in photoreduction by amines. Abstraction of H from N, *J. Am. Chem. Soc.*, 102, 1419, 1980.

62. Parola, A. H. and Cohen, S. G. Primary quantum yields of ketyl radicals in photoreduction by amines. Abstraction of H from N, *J. Photochem.*, 12, 41, 1980.

63. Scaiano, J. C., Photochemical and free-radical processes in benzil-amine systems. Electron donor properties of α-aminoalkyl radicals, *J. Phys. Chem.*, 85, 2851, 1981.

64. Turro, N. J., *Modern Molecular Photochemistry*, Benjamin/Cummings, Menlo Park, CA., 1978, 383.

65. Döpp, D., Reactions of aromatic nitro compounds via excited triplet states, *Top. Curr. Chem.*, 55, 49, 1975.

66. Pacifici, J. G., Irick, G., and Anderson, C. G., Photochemistry of azo compounds. III. Evidence for an electron transfer process in amine solvents, *J. Am. Chem. Soc.*, 91, 5654, 1969.

67. Döpp, D. and Müller, D., Photochemistry of aromatic nitro compounds. XIII. Photoreduction of sterically hindered nitrobenzenes in aliphatic amines. *Rec. Trav. Chim.*, 98, 297, 1979.

68. Takami, M., Matsuura, T., and Saito, I., Photochemical reaction of aromatic nitro compounds with amines, *Tetrahedron Lett.*, 661, 1974.

69. Döpp, D. and Heufer, J., Photochemistry of aromatic nitro compounds. XVI. *N*-Demethylation of *N,N*-dimethylanilines by photoexcited 3-chloronitrobenzene, *Tetrahedron Lett.*, 23, 1553, 1982.

70. Ohashi, M., Furukawa, Y., and Tsujimoto, K., Photoinduced reduction of fluoren-9-ol and its acetate by aliphatic amines, *J. Chem. Soc., Perkin Trans. I*, 2613, 1980.

71. Van Riel, H. C. H. A., Lodder, G., and Havinga, E., Photochemical methoxide exchange in some nitromethoxybenzenes. The role of the nitro group in S_N2Ar^* reactions, *J. Am. Chem. Soc.*, 103, 7257, 1981.

72. Cornelisse, J. and Havinga, E., Photosubstitution reactions of aromatic compounds, *Chem. Rev.*, 75, 353, 1975.

73. Bunce, N. J., Cater, S R., Scaiano, J. C., and Johnston, L. J., Photosubstitution of 1-nitro-4-methoxynaphthalene with amine nucleophiles: Dual pathways, *J. Org. Chem.*, 52, 4214, 1987.

74. van Eijk, A. M. J., Huizer, A. H., Varma, C. A. G. O., and Marquet, J., Dynamic behavior of photoexcited systems of 4-nitroveratrole containing OH⁻ or amines, *J. Am. Chem. Soc.*, 111, 88, 1989.

75. Pleixats, R. and Marquet, J., The photosubstitution of 2-fluoro-4-nitroanisole with *n*-hexylamine. Evidence of two different triplet states in a dual mechanistic pathway, *Tetrahedron*, 46, 1343, 1990.

76. Cater, S. R., Ph.D. Thesis, University of Guelph, Guelph, Ontario, Canada, 1986.

77. Mutai, K., Nakagaki, R., and Tukada, H., A rationalization of orientation in nucleophilic aromatic photosubstitution, *Bull. Chem. Soc. Jpn.*, 58, 2066, 1985.

78. Cantos, A., Marquet, J., Moreño-Manas, Gonzalez-Lafont, A., Lluch, J. M., and Bertran, J., On the regioselectivity of 4-nitroanisole photosubstitution with primary amines. A mechanistic and theoretical study, *J. Org. Chem.*, 55, 3303, 1990.

79. Wubbels, G. G., Winitz, S., and Whitaker, C., NMR and ultraviolet spectral characterization of dihydrobenzene intermediates in the displacement of hydrogen by intramolecular nucleophilic aromatic substitution, *J. Org. Chem.*, 55, 631, 1990.

80. Wubbels, G. G., Sevetson, B. R., and Sanders, H., Competitive catalysis and quenching by amines of photo-Smiles rearrangement as evidence for a Zwitterionic triplet as the proton-donating intermediate, *J. Am. Chem. Soc.*, 111, 1018, 1989.

22

Intra- and Intermolecular Cycloadditions of Benzene Derivatives

Paul A. Wender
Stanford University

Timothy M. Dore
Stanford University

Arene-alkene photocycloadditions are a theoretically intriguing and synthetically unique class of reactions with enormous potential in complex-molecule synthesis. Identified during a fascinating decade of discovery from 1957 to 1967,[1-4] these transformations can proceed in three cycloaddition modes (Figure 22.1). In the *ortho* mode, an alkene attaches to the *ortho*- (or 1,2-) related carbons of the starting arene, while in the *meta*- and *para*-cycloadditions, attachment of the alkene occurs at the *meta*- (or 1,3-) and *para*- (or 1,4-) related atoms of the arene, respectively. Not unlike the Diels-Alder cycloaddition, these strategy-level reactions allow for the conversion of relatively simple, readily available starting materials into complex products in one operation, a feature that significantly shortens the length of a synthesis. Importantly, the cycloadducts or their immediate derivatives have ring systems that are commonly found in a variety of synthetically significant targets. Cycloadditions by the *ortho* and *para* route produce bicyclo[4.2.0]-[5] and bicyclo[2.2.2]-octadienes,[6] respectively, with up to four new stereogenic centers. The *meta*-cycloaddition provides tricyclic products with as many as six new stereogenic centers, a unique increase in complexity among cycloaddition processes. The *meta*-cycloadducts can also be readily converted to bicyclo[3.3.0]-[7-13] or [3.2.1]octane[14] derivatives, cycloheptanes,[15] cyclopentanes,[16] or fenestranes.[17,18] The synthetic benefit of harnessing the power of these processes is evident in numerous applications, typified by a three-step total synthesis of silphinene[19] (Figure 22.2). In this chapter, the salient features of arene-alkene cycloadditions are described, including the factors that influence the selectivity of these reactions and how they can be used advantageously in complex-molecule synthesis.

The utility of strategy-level reactions like the arene-alkene cycloaddition is directly related to the degree to which their selectivity can be controlled. For arene-alkene cycloadditions, there are four selectivities — mode (*ortho* vs. *meta* vs. *para*), regio-, *exo/endo*, and stereoselectivity. Without control over these selectivities, even a simple version of this reaction would be of little synthetic value since as many as 144 cycloadducts could be produced.

Several mechanisms have been proposed for the arene-alkene cycloaddition that provide a basis for analyzing and often predicting its selectivity. The present treatment represents an amalgamated mechanism, which is based on analyses by Cornelisse[20-24] and by Mattay,[25-29] and which incorporates and extends experimental and theoretical findings uncovered in their studies and those of other investigators including the groups of Bryce-Smith,[30] Gilbert,[31] Wilzbach,[32] Morrison,[33] Srinivasan,[20] Houk,[34] Sheridan,[35,36] and Wender.[37-40] According to this mechanism (Figure 22.3), the interaction of a ground-state alkene and an excited-state arene proceeds with initial polarization

0-8493-8634-9/95/\$0.00+\$.50
© 1995 by CRC Press, Inc.

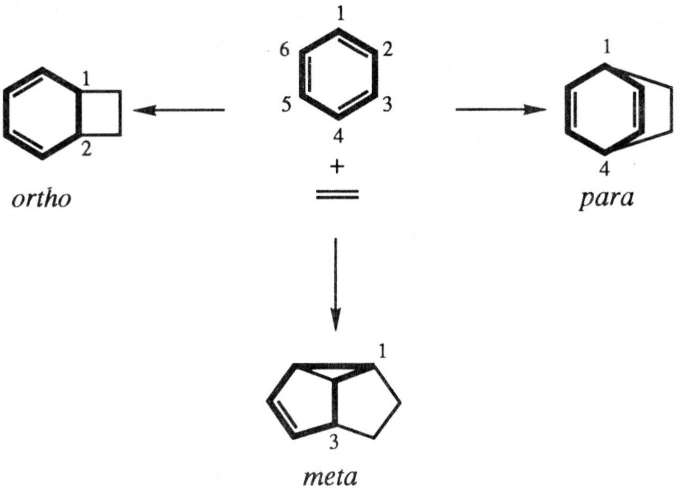

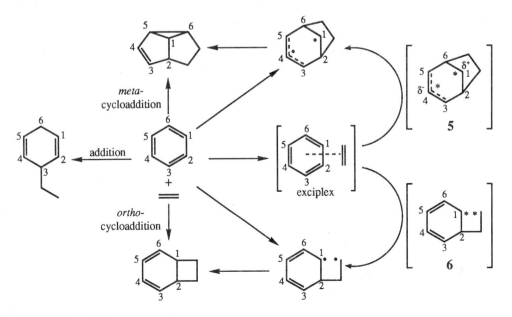

FIGURE 22.1   Possible modes of addition in the arene-alkene photocycloaddition reaction.

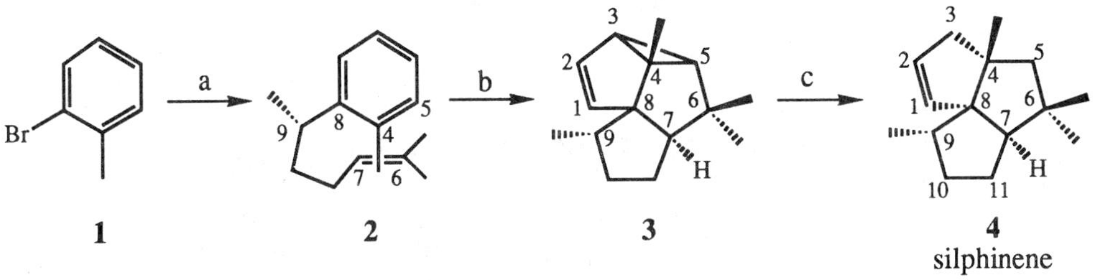

a. Li, Et$_2$O, rt;   , Δ; NH$_3$, -78°—33°C; NH$_4$Cl b. hv (medium pressure mercury arc lamp, vycor filter), pentane, rt c. Li, CH$_3$NH$_2$, -78°C

FIGURE 22.2   The three-step total synthesis of silphinene.

FIGURE 22.3   A mechanistic proposal for the arene-alkene photocycloaddition reaction.

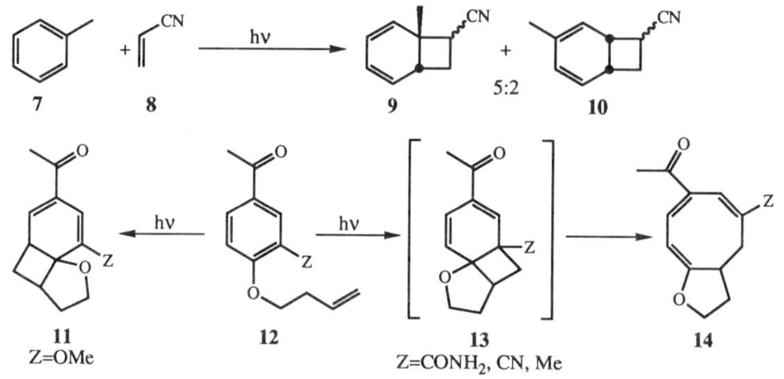

FIGURE 22.4   Regioselectivity of *ortho*-cycloadditions.

(5 and 6), possibly through an exciplex intermediate. In the *meta*-cycloaddition (and possibly the *ortho* as well), this polarization dissipates as the reaction proceeds, resulting in a biradical that collapses with bond formation to the cycloadduct. When the interaction between the arene and alkene involves extensive charge transfer[41–49] (as indicated by a favorable energy of electron transfer, $\Delta G_{ET} < 0$[50]) simple addition is expected; whereas, when electron transfer is disfavored, cycloaddition is preferred, with *ortho*-cycloaddition favored for ΔG_{ET} values from 0 to ~1.6 eV and *meta*-cycloaddition for more positive values. *Para*-cycloadditions are generally disfavored by poor orbital overlap and are primarily observed only with allenes,[4] butadienes,[51,52] and extended arenes.[53,54]

The regioselectivity attending arene-alkene cycloadditions is readily rationalized by the above mechanistic model on the basis of the stabilizing effects of substituents on polarized species 5 or 6 (Figure 22.3) or the analogous exciplexes formed in the initial stages of arene-alkene interactions. Thus, for the *ortho*-mode of cycloaddition, the alkene adds preferentially to C-2 in order to maximize partial charge stabilization by a C-1 substituent as indicated by 6.[34] In accord with this analysis, *ortho*-additions generally proceed with bond formation to positions C-1 and C-2, as illustrated by the selectivity observed in the reaction of toluene and acrylonitrile[55] (9 vs. 10, Figure 22.4). The contrasting regioselectivity observed for donor and acceptor groups in the intramolecular case with ketone 12 is also consistent with polarization effects arising in the initial stage of the reaction.[56] For *meta*-cycloadditions, partial positive charge is proposed to develop at position C-1 (Figure 22.3, 5), while negative charge builds up at C-3 and C-5. As such, *meta*-cycloadditions involving substituted arenes usually proceed with an electron-donating substituent at C-1 or with an electron-withdrawing group at C-3 or C-5 or both.[57–62] For arenes bearing two substituents, the stronger donor or acceptor generally directs the regioselectivity of the cycloaddition.[63–69]

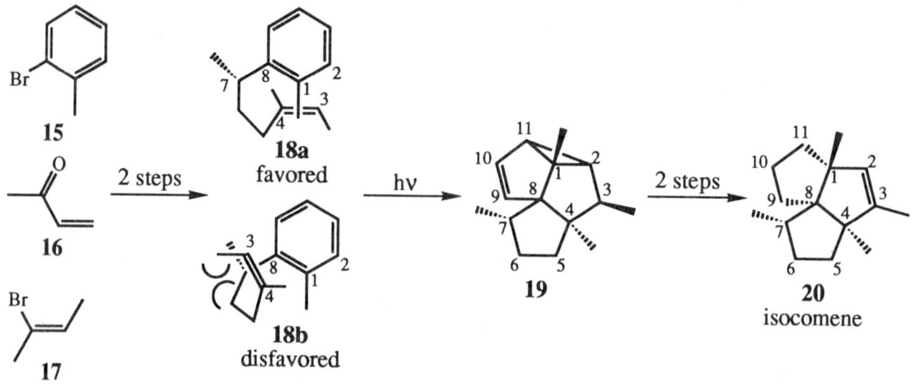

FIGURE 22.5   The total synthesis of isocomene.

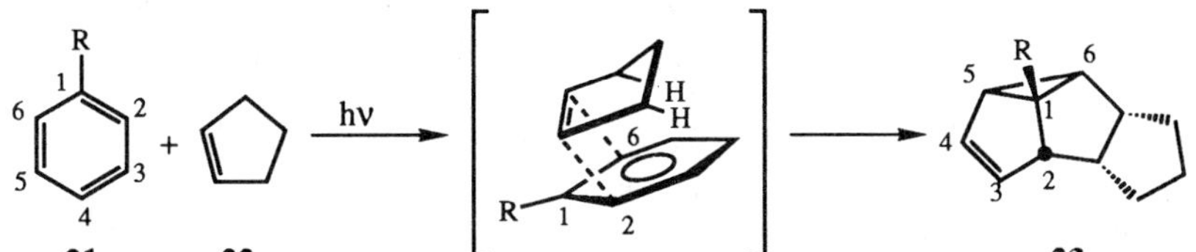

21 22 endo exciplex 23 endo product

FIGURE 22.6 *Exo/endo* selectivity in the intermolecular arene-alkene photocycloaddition reaction.

The influence of steric effects on regioselectivity can also be rationalized from the above mechanistic model.[70–74] For example, in the synthesis of isocomene[8] (Figure 22.5), which involves an arene with two approximately equivalent electron-donating groups, *meta*-cycloaddition is directed by the methyl donor at C-1 (18a) rather than the alkenyl chain at C-8 (18b). Computer models indicate that *meta*-cycloaddition directed by the alkenyl chain would lead to an unfavorable nonbonded interaction between the terminal methyl group and the homobenzylic carbon in 18b.[33] Of further note in this example is the retention of alkenyl stereochemistry during the cycloaddition process, a feature common to the vast majority of *meta*-cycloadditions.

While further work is needed to understand fully the range of factors contributing to *exo/endo* preferences, this selectivity has in many cases been rationalized by secondary orbital interactions and steric and electrostatic effects.[75–84] For example, *inter*molecular *meta*-cycloadditions of simple alkenes generally exhibit *endo* selectivity (Figure 22.6), a result attributed to stabilizing secondary orbital interactions between the allylic hydrogens of the alkene and the orbitals of the excited arene.[70,73,74] In contrast, the *intra*molecular *meta*-cycloaddition proceeds with *exo* selectivity (Figure 22.7).[14,33] Examination of models in this case reveals that better alignment of orbitals is achieved in the *exo* exciplex (or its equivalent) than in the *endo* exciplex (25 vs. 24). Thus, orbital alignment in the intramolecular process appears to override any benefit derived from secondary orbital interactions, in accord with the low stabilization energy attributed typically to the latter.

The stereoselectivity of arene-alkene cycloadditions has been successfully treated through the use of a parallel plane exciplex model.[14,21,34] In this model, the planes of the alkene and arene for each diastereomerically related cycloaddition are oriented in an approximately parallel fashion at a distance of ~2.5 Å; the lowest energy structure, as determined by inspection or computer modeling, generally corresponds to the experimentally favored product. As originally tested with substrate 25 (Figure 22.7) in the synthesis of cedrene,[14] this analysis indicates that the *meta*-cycloaddition should proceed in a fashion that minimizes interaction between the larger benzylic substituent (methyl > hydrogen) and the larger arene group *ortho* to the alkene chain (methoxy > hydrogen), thus favoring 25b over 25a.[3,13,19] Computer analysis based on the MM2 force field places the energy

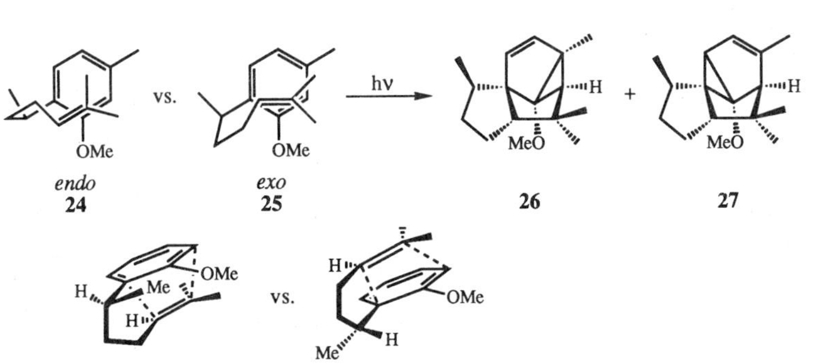

endo 24 *exo* 25 26 27

25a disfavored 25b favored

FIGURE 22.7 *Exo/endo* and stereoselectivity in the intramolecular arene-alkene photocycloaddition reaction.

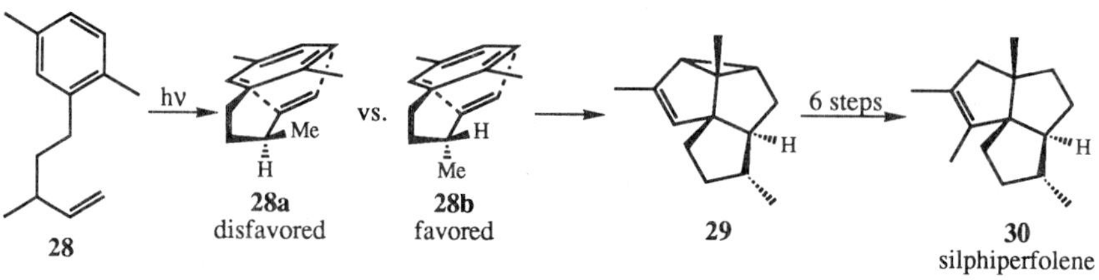

FIGURE 22.8   The synthesis of silphiperfolene: an illustration of allylic stereocontrol.

difference between these two possible arrangements at >1 kcal mol^{-1} in accord with the finding that the reaction proceeds with greater than 90% diastereoselectivity.

It follows from the above model that allylic substituents also would influence strongly the stereochemical course of the cycloaddition. Indeed, this is observed, as demonstrated in the synthesis of silphiperfolene[85] (Figure 22.8). Of the two possible diastereomeric pathways for cycloaddition, reaction occurs preferentially through the pathway (28b) that minimizes nonbonded interactions between the arene substituents and the allylic group (methyl vs. hydrogen). Asymmetry at the homobenzylic center also can influence predictably the diastereoselectivity of the cycloaddition, as indicated in studies related to the synthesis of grayanotoxin (Figure 22.9).[86] Exciplex 32 is favored over 31 due to the destabilizing nonbonded interaction between the alkoxy groups expected for the latter. Of further note in this example is the formation of only one cyclopropane isomer in the cycloadduct. While this selectivity has not been analyzed extensively,[87–89] it is often observed that the cyclopropane isomer exhibiting fewer nonbonded interactions is formed preferentially in these reactions. As noted by us and others, caution must be exercised in analyzing these and other selectivities because it is well known that cycloadducts can interconvert under some photochemical conditions and that some photoproducts suffer selective decomposition.

To illustrate how the above selectivities can be used in synthetic planning, this chapter closes with an analysis of the synthesis design for laurenene (Figure 22.10).[90] Laurenene possesses a commonly encountered triquinane ring system that, as noted previously, is derivable from an arene-alkene *meta*-cycloaddition. Screening a number of alkene-arene combinations and limiting consideration

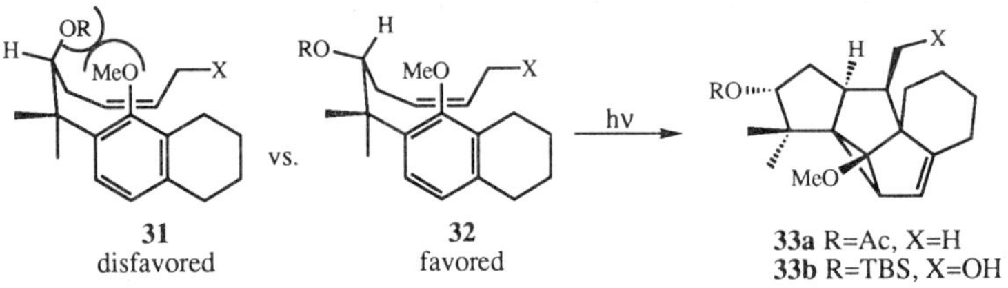

FIGURE 22.9   Synthetic studies on grayanotoxin: an illustration of homobenzylic stereocontrol.

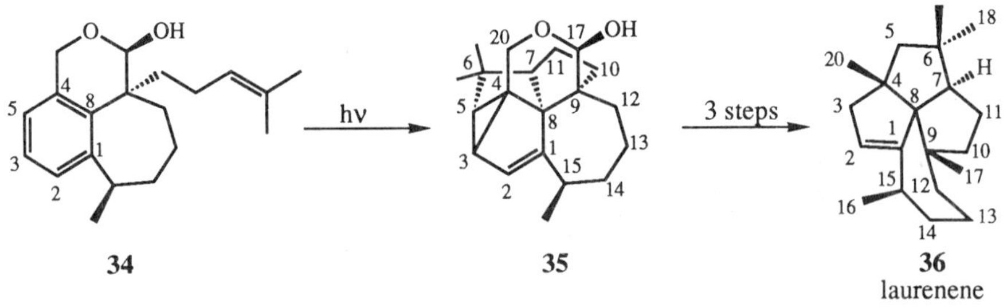

FIGURE 22.10   The total synthesis of laurenene.

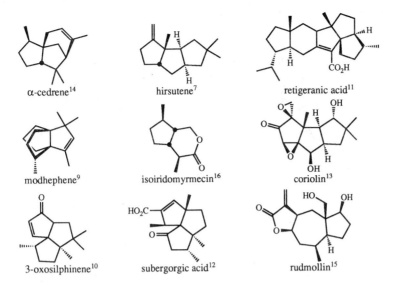

α-cedrene[14]　　　hirsutene[7]　　　retigeranic acid[11]

modhephene[9]　　　isoiridomyrmecin[16]　　　coriolin[13]

3-oxosilphinene[10]　　　subergorgic acid[12]　　　rudmollin[15]

FIGURE 22.11　Some compounds synthesized using the arene-alkene photocycloaddition reaction.

to simple alkyl substitution so as to favor the *meta*-mode of reaction — where ΔG_{ET} is positive — leads to the selection of arene-alkene **34**. Due to the limited reach of the tethered alkene in **34**, *meta*-additions across C-5, C-3, and C-2 are unlikely. Cycloaddition directed by C-8 is disfavored by steric interactions between the methyl on the alkene and the tether that would arise in the parallel plane exciplex. Of the remaining options (addition across C-4 or C-1), *meta*-addition across the sterically less demanding heterocyclic ring is expected. As noted above, for the intramolecular cycloaddition, an *exo* selective process is preferred, as it allows better alignment of the interacting orbitals relative to the *endo* process. Finally, closure of the biradical intermediate would favor cyclopropane isomer **35** since the alternative closure would proceed with an increase in nonbonded interactions between substituents at C-13, C-14, C-17, and C-20. In accord with this analysis, photolysis of **34** gave only one observed cycloadduct, **35** (51% yield), which was transformed into laurenene in three steps.[91] Figure 22.11 illustrates some of the range of other structural types that are accessible through arene-alkene cycloadditions.

　　In summary, arene-alkene cycloadditions, while relatively new, represent powerful processes for the synthesis of complex molecules and provide a fertile arena for the development of new theories and synthetic strategies in chemistry. The development of these and other strategy-level reactions that allow for a great increase in complexity are crucial to the advancement of synthesis from its current status in which complex molecules are rarely made in a practical fashion to a point in the future when such preparations would become routine.

UV DATA

Compound 18

UV: (hexane) 263.5 nm (logε=2.42), 271 nm (logε=2.36)

Compound 19

UV: (hexane) 220 nm (logε=3.72), 202 nm (logε=3.75)

Compound 32b

UV: (hexane) 266 nm (ε=4300)

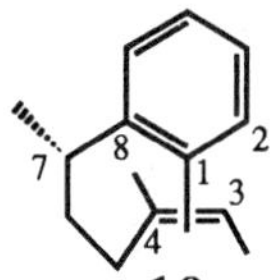

18 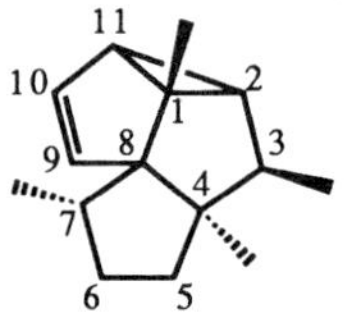   **19** 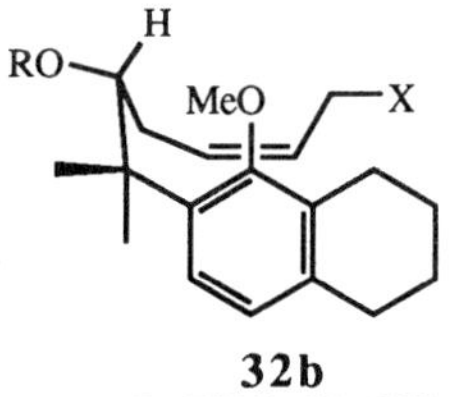   **32b**
R=TBS, X=OH

References

1. Bryce-Smith, D., Gilbert, A., and Orger, B. H., Photochemical 1,3-cycloaddition of olefins to aromatic compounds, *J. Chem. Soc., Chem. Commun.*, 512, 1966.
2. Wilzbach, K. S. and Kaplan, L., A photochemical 1,3-cycloaddition of olefins to benzene, *J. Am. Chem. Soc.*, 88, 2066, 1966.
3. Angus, H. J. F. and Bryce-Smith, D., Addition of maleic anhydride to benzene, *Proc. Chem. Soc.*, 326, 1959.
4. Kraft, K. and Klotzenburg, R., Photochemische Reaktionen von Aromaten mit Konjugierten Dienen. II. Neue Addukte des Benzols an Butadien-(1,3), *Tetrahedron Lett.*, 4357, 1967.
5. Gilbert, A., Taylor, G. N., and bin Samsudin, M. W., The photocycloaddition of enol ethers to benzene, *J. Chem. Soc., Perkin Trans. I*, 869, 1980.
6. Ipaktschi, J. and Iqbal, M. N., Photoaddition of isoprene to benzene: Synthesis of 3-methylenebicyclo[4.2.2]deca-7,9-diene, *Synthesis*, 633, 1977.
7. Wender, P. A. and Howbert, J. J., Synthetic studies on arene-olefin cycloadditions. III. Total Synthesis of (1)-hirsutene, *Tetrahedron Lett.*, 23, 3983, 1982.
8. Wender, P. A. and Dreyer, G., Synthetic studies on arene-olefin cycloadditions. II. Total synthesis of (1)-isocomene, *Tetrahedron*, 37, 4445, 1981.
9. Wender, P. A. and Dreyer, G., Synthetic studies on arene-olefin cycloadditions. 4. Total synthesis of (1)-modhephene, *J. Am. Chem. Soc.*, 104, 5805, 1982.
10. Stork, G., Presented at the National Organic Chemistry Symposium, Minneapolis, MN, June 1991.
11. Wender, P. A. and Singh, S., Synthetic studies on arene-olefin cycloadditions. 11. Total synthesis of (–)-retigeranic acid, *Tetrahedron Lett.*, 31, 2517, 1990.
12. Wender, P. A. and deLong, M. A., Synthetic studies of arene-olefin cycloadditions. XII. Total synthesis of (1)-subergorgic acid, *Tetrahedron Lett.*, 31, 5429, 1990.
13. Wender, P. A. and Howbert, J. J., Synthetic studies on arene-olefin cycloadditions. VI. Two syntheses of (1)-coriolin, *Tetrahedron Lett.*, 24, 5325, 1983.
14. Wender, P. A. and Howbert, J. J., Synthetic studies on arene-olefin cycloadditions. VI. Total synthesis of (1)-*a*-cedrene, *J. Am. Chem. Soc.*, 103, 688, 1981.
15. Wender, P. A. and Fisher, K., Seven membered ring synthesis based on arene-olefin cycloadditions: The total synthesis of (1)-rudmollin, *Tetrahedron Lett.*, 1857, 1986.
16. Wender, P. A. and Dreyer, G., Synthetic studies on arene-olefin cycloadditions, V. Total synthesis of (1)-isoiridomyrmecin, *Tetrahedron Lett.*, 4543, 1983.
17. Mani, J. and Keese, R., Synthesis of 2,3-dimethyltetracyclo(5.5.1.0.0)trideca-2,5-dien-9-one, a new derivative of all-*cis*-(5.5.5.5)fenestrane, *Tetrahedron*, 41, 5697, 1985.
18. Mani, J., Schttel, S., Zhang, C., Bigler, P., Mller, C., and Keese, R., Direct formation of a substituted [5.5.5.5]fenestrane by intramolecular arene-olefin photocycloaddition, *Helv. Chim. Acta*, 72, 487, 1989.

19. Wender, P. A. and Ternansky, R. J., Synthetic studies on arene-olefin cycloaddition. VII. A three step total synthesis of (1)-silphinene, *Tetrahedron Lett.*, 26, 2625, 1985.

20. Merritt, V. Y., Cornelisse, J., and Srinivasan, R., Photochemical addition of benzene to cyclopentene, *J. Am. Chem. Soc.*, 95, 8250, 1973.

21. van der Hart, J. A., Mulder, J. J. C., and Cornelisse, J., The *meta*-photocycloaddition of benzene to ethylene: Semiempirical calculations, *Theochem.*, 1987, 151, 1, 1987.

22. de Vaal, P., Lodder, G., and Cornelisse, J., Intramolecular deuterium isotope effect in the *meta* photocycloaddition of aromatic compounds to alkenes, *Tetrahedron Lett.*, 26, 4395, 1985.

23. de Vaal, P., Lodder, G., and Cornelisse, J., Intramolecular deuterium isoptope effects in the *meta* photocycloaddition of aromatic compounds to alkenes, *Tetrahedron*, 42, 4585, 1986.

24. de Vaal, P., Lodder, G., and Cornelisse, J., Concentration-dependent deuterium isotope effects on the quantum yield of arene-alkene *meta* photocycloaddition: The role of excimers, *J. Phys. Org. Chem.*, 3, 273, 1990.

25. Mattay, J., Selectivities in photocycloadditions of arenes to olefins, *J. Photochem.*, 37, 167, 1987.

26. Leismann, H. and Mattay, J., Zur Exciplexbildung von Benzol in Gegenwart von Endiolethern und Triethylamin, *Tetrahedron Lett.*, 4265, 1978.

27. Leismann, H., Mattay, J., and Scharf, H.-D., Intermolecular exciplex formation between benzene and some olefinic compounds, *J. Photochem.*, 9, 338, 1978.

28. Mattay, J., Leismann, H., and Scharf, H.-D., Photochemical cycloaddition of benzene to 1,3-dioxoles and 1,4-dioxene, *Mol. Photochem.*, 9, 119, 1979.

29. Leisman, H., Mattay, J., and Scharf, H.-D., Photochemical cycloaddition of olefins to aromatic compounds. 5. Formation and deactivation of exciplexes for singlet benzene or toluene and 1,3-dioxole, *J. Am. Chem. Soc.*, 106, 3985, 1984.

30. Bryce-Smith, D. and Longuet-Higgins, H. C., Photochemical transformation of the benzene ring, *J. Chem. Soc., Chem. Commun.*, 593, 1966.

31. Gilbert, A., The inter- and intramolecular photocycloaddition of ethylenes to aromatic compounds, *Pure Appl. Chem.*, 52, 2669, 1980.

32. Wilzbach, K. E. and Kaplan, L., Photoaddition of benzene to olefins. II. Stereospecific 1,2- and 1,4-cycloadditions, *J. Am. Chem. Soc.*, 93, 2073, 1971.

33. Ferree, W. Jr., Grutzner, B., and Morrison, H., Photochemistry of bichromophoric molecules. Intramolecular cycloaddition and *cis-trans*-isomerization of 6-phenyl-2-hexene in solution, *J. Am. Chem. Soc.*, 93, 5502, 1971.

34. Houk, K. N., Theory of cycloadditions of excited aromatics to alkenes, *Pure Appl. Chem.*, 54, 1673, 1982.

35. Sheridan, R. S., Azo precursors to a putative biradical in arene *meta*-photoadditions: Evidence for a novel concerted nitrogen expulsion, *J. Am. Chem. Soc.*, 105, 5140, 1983.

36. Reedich, D. E. and Sheridan, R. S., Independent generation of arene *meta* photoaddition biradicals, *J. Am. Chem. Soc.*, 107, 3360, 1985.

37. Wender, P. A. and von Geldern, T. W., Aromatic compounds: Isomerisation and cycloaddition, in *Photochemistry in Organic Synthesis*, Coyle, J. D., Ed., The Royal Society of Chemistry, London, 1986, 226.

38. Wender, P. A., Siggel, L., and Nuss, J. M., Arene-alkene photocycloaddition reactions, in *Organic Photochemistry*, Vol. 10, Padwa, A., Ed., Marcel Dekker: New York, 1989, 357.

39. Wender, P. A., Ternansky, R., deLong, M., Singh, S., Olivero, A., and Rice, K., Arene-alkene cycloadditions and organic synthesis, *Pure Appl. Chem.*, 62, 1597, 1990.

40. Wender, P. A., Siggel, L., and Nuss, J. M., [3 + 2] and [5 + 2] arene-alkene photocycloadditions, in *Comprehensive Organic Synthesis*, Vol. 5, Trost, B. M., Fleming, I., and Paquette L. A., Eds., Pergamon, Elmsford, NY, 1991, 645.

41. Bryce-Smith, D., Gilbert, A., Orger, B., and Tyrrell, H., Polar and stereochemical aspects of 1,2-photoaddition of ethylene to benzene, *J. Chem. Soc., Chem. Commun.*, 334, 1974.

42. Bryce-Smith, D., Foulger, B., Forrester, J., Gilbert, A., Orger, B., and Tyrrell, H. M., The photoaddition of ethylenic hydrocarbons to benzene, *J. Chem. Soc., Perkin Trans. I*, 55, 1980.

43. Mirbach, M. F., Mirbach, M. J., and Saus, A., Photochemical cycloadditions of ethene and propene to benzene, *Tetrahedron Lett.*, 959, 1977.

44. Bryce-Smith, D., Gilbert, A., and Mattay, J., Factors influencing the relative tendencies for *ortho* and *meta* cycloadditions of ethylenic compounds to the benzene ring, *Tetrahedron*, 42, 6011, 1986.

45. Mattay, J., Selectivity and charge transfer in photoreactions of arenes with olefins. 1. Substitution vs. cycloaddition, *Tetrahedron*, 41, 2393, 1985.

46. Mattay, J., Selectivity and charge transfer in photoreactions of arenes with olefins. 2. Mode of cycloaddition, *Tetrahedron*, 41, 2405, 1985.

47. Weber, G., Runsink, J., and Mattay, J., Selectivity and charge transfer in photoreactions of donor-acceptor systems. 10. *meta*-Photocycloadditions of *m-bis*(trifluoromethyl)benzene and trifluoromethylbenzonitrile to olefins, *J. Chem. Soc., Perkin Trans. I*, 2333, 1987.

48. Reedich, D. G. and Sheridan, R. S., Direct evidence for both thermal and photochemical stepwise cleavage in a pair of isomeric azo compounds, *J. Am. Chem. Soc.*, 110, 3697, 1988.

49. Mattay, J., Runsink, J., Rumbach, T., Ly, C., and Gersdorf, J., Selectivity and charge transfer in photoreactions of *a,a,a*-trifluorotoluene with olefins, *J. Am. Chem. Soc.*, 107, 2557, 1985.

50. Rehm, D. and Weller, A., Kinetics of fluorescence quenching by electron and H-atom transfer, *Isr. J. Chem.*, 8, 259, 1970.

51. Bryce-Smith, D., Foulger, B. E., and Gilbert, A., Photoaddition of allenes to benzene, *J. Chem. Soc., Chem. Commun.*, 664, 1972.

52. Berridge, J. C., Forrester, J., Foulger, B. E., and Gilbert, A., The photochemical reactions of benzene with 1,2-, 1,3-, and 1,4-dienes, *J. Chem. Soc., Perkin Trans. I*, 2425, 1980.

53. Arnold, D. R., Gillis, L. B., and Whipple, E. B., Photocycloaddition of 2-acetylnaphthalene to methylcinnamate. A [2 + 4] electron photocycloaddition, *J. Chem. Soc., Chem. Commun.*, 918, 1969.

54. Yang, N. C. and Libman, J., Chemistry of 9,10-benzotricyclo[4.2.2.2]dodeca-3,7,9-triene, a novel polycyclic compound, *J. Am. Chem. Soc.*, 94, 9228, 1972.

55. Gilbert, A. and Yianni, P., Regio- and stereo-selectivities of the *ortho* and *meta* photocycloaddition reactions of ethylenes to benzene and its simple derivatives, *Tetrahedron*, 37, 3275, 1981.

56. Wagner, P. J., Sakamoto, M., and Madkour, A. E., Regioselectivity in intramolecular cycloaddition of double bonds to triplet benzenes, *J. Am. Chem. Soc.*, 114, 7298, 1992.

57. Cornelisse, J., Merrit, V. Y., and Srinivasan, R., Photochemical 1,3 addition of benzene to olefins. Orientational specificity induced by methyl substituents on aromatic and olefinic rings, *J. Am. Chem. Soc.*, 95, 6197, 1973.

58. Ors, J. A. and Srinivasan, R. J., Photochemical 1,3-addition of anisole to olefins. Synthetic aspects, *J. Org. Chem.*, 42, 1321, 1977.

59. Osselton, E. M. and Cornelisse, J., The *meta* photocycloaddition of benzonitrile and *a,a,a*-trifluorotoluene to cyclopentene, *Tetrahedron Lett.*, 26, 527, 1985.

60. Neijenesch, H. A., Ridderikhoff, E. J., Ramsteijn, C. A., and Cornelisse, J., Intramolecular *meta* photocyloaddition of 6-arylhex-2-enes, *J. Photochem. Photobiol.*, 48, 336, 1989.

61. Gilbert, A. and Taylor, G. N., Intramolecular photocycloaddition reactions of phenyl-vinyl bichromophoric systems, *J. Chem. Soc., Perkin Trans. I*, 1761, 1980.

62. Blakemore, D. C. and Gilbert, A., Intramolecular photocycloaddition of 3-benzyloxyprop-1-enes, *J. Chem. Soc., Perkin Trans. I*, 2265, 1992.

63. Jans, A. W. H. and Cornelisse, J., The *meta*-photocycloaddition of cyclopentene to *ortho*-methylanisole, *Recl. Trav. Chim. Pays-Bas*, 100, 213, 1981.

64. Hoye, T. R., Steric impedance in the 1,3-photocycloaddition reaction between cyclopentene and anisole derivatives, *Tetrahedron Lett.*, 22, 2523, 1981.

65. Jans, A. W. H., van Dijk-Knepper J. J., and Cornellise, J., Photocycloaddition of ethylvinylether and cyclopentene to 3,5-dimethylanisole, *Tetrahedron Lett.*, 1111, 1982.

66. Ellis-Davies, G. C. R. and Cornelisse, J., Intermolecular and intramolecular photocycloaddition reactions of fluoroarenes, *Tetrahedron Lett.*, 1893, 1985.

67. Osselton, E. M., Lempers, E. L. M., Eyken, C. P., and Cornelisse, J., Substituent effects on the *meta* photocycloaddition of arenes to alkenes, *Recl. Trav. Chim. Pays-Bas*, 105, 171, 1986.

68. Osselton, E. M., Krijnen, E. S., Lempers, E. L. M., and Cornelisse, J., The *meta*-photocycloaddition of fluoroarenes to cyclopentene, *Recl. Trav. Chim. Pays-Bas*, 105, 375, 1986.

69. Osselton, E. M., Lempers, E. L. M., and Cornelisse, J., The *meta*-photocycloaddition of 3- and 4-cyanoanisole to cyclopentene, *Recl. Trav. Chim. Pays-Bas*, 104, 124, 1985.

70. Gilbert, A. and Heath, P., Steric effects in the *meta*-photocycloaddition of ethylene to alkylbenzenes, *Tetrahedron Lett.*, 2831, 1979.

71. Dadson, M., Gilbert, A., and Heath, P., On the mechanism of the *meta* photocycloaddition of ethylenes to benzenoid compounds, *J. Chem. Soc., Perkin Trans. I*, 1314, 1980.

72. Sheridan, R. S., A reinvestigation of the *meta*-photoadditions of cyclooctene to anisole and toluene, *Tetrahedron Lett.*, 23, 267, 1982.

73. Bryce-Smith, D., Fenton, G. A., and Gilbert, A., Mechanism of the photochemical *meta*-cycloaddition of alkenes to the benzene ring, *Tetrahedron Lett.*, 2697, 1982.

74. Bryce-Smith, D., Drew, M. G. B., Fenton, G. A., and Gilbert, A., Mechanism of photochemical *meta*-cycloaddition of alkenes to the benzene ring: A novel type of molecular interaction, *J. Chem. Soc., Chem. Commun.*, 607, 1985.

75. Mattay, J., Runsink, J., Leisman, H., and Scharf, H.-D., Photochemical cycloaddition of olefins to aromatic compounds. IV. On the mechanism of the photochemical *meta*-cycloaddition, *Tetrahedron Lett.*, 23, 4919, 1982.

76. Cornelisse, J., Merritt, V. Y., and Srinivasan, R., Photochemical 1,3 addition of benzene to olefins. Orientational specificity induced by methyl substituents on aromatic and olefinic rings, *J. Am. Chem. Soc.*, 95, 6197, 1973.

77. Jans, A. W. H., van Arkel, B., van Dijk-Knepper, J. J., and Cornelisse, J., The *meta*-photocycloaddition of ethylvinylether to 3-methylanisole and 3-fluoroanisole, *Tetrahedron Lett.*, 23, 3827, 1982.

78. Osselton, E. M., Eyken, C. P., Jans, A. W. H., and Cornelisse, J., The *meta* photocycloaddition of anisole and benzonitrile to 1,3-dioxol-2-one, *Tetrahedron Lett.*, 26, 1577, 1985.

79. Mattay, J., Runsink, J., Gersdorf, J., Rumbach, T., and Ly, C., Selectivity and charge transfer in photoreactions of *a,a,a*-trifluorotoluene with olefins, *Helv. Chim. Acta*, 69, 442, 1986.

80. Osselton, E. M. and Cornelisse, J., The *meta* photocycloaddition of anisole and benzene to 2-methyl-1,3-dioxole, *J. Photochem.*, 31, 381, 1985.

81. Mattay, J., Rumbach, T., and Runsink, J., Molecular recognition phenomena in the excited state: The *meta* photocycloaddition of *cis*- and *trans*-2,5-dihydro-2,5-dimethoxyfuran and *cis*-1,4-dimethoxycyclopent-2-ene to anisole, *J. Org. Chem.*, 55, 5691, 1990.

82. Bryce-Smith, D., Drew, M. G. B., Fenton, G., Gilbert, A., and Proctor, A. D., Stereochemical control in *meta* photocycloaddition of alkenes to the benzene ring: The anomalous behavior of cyclohexene, *J. Chem. Soc., Perkin Trans. I*, 779, 1991.

83. Mattay, J., Runsink, J., and Piccirilli, J. A., Photochemical cycloadditions of 1,3-dioxoles to anisole, *J. Chem. Soc., Perkin Trans. I*, 15, 1987.

84. Mattay, J., Runsink, J., Hertel, R., Kalbe, J., and Schewe, I., Selectivity and charge transfer in photoreactions of donor-acceptor systems. XI. Stereoselectivities in *meta* photocycloadditions of phenol ethers to olefins, *J. Photochem.*, 37, 335, 1987.

85. Wender, P. A. and Singh, S., Synthetic studies on arene-olefin cycloadditions. VII. Total synthesis of (1)-silphiperfol-6-ene, (1)-7aH-silphiperfol-5-ene, and (1)-7bH-silphiperfol-5-ene, *Tetrahedron Lett.*, 26, 5987, 1985.

86. Olivero, A., Ph.D. Thesis, Stanford University, 1988.

87. Cornelisse, J., Gilbert, A., and Rodwell, P. W., *meta* Photocycloaddition of *trans*-1,2-dichloroethylene to benzenoid compounds: Mechanistic considerations and photo and thermal labilities of the adducts, *Tetrahedron Lett.*, 27, 5003, 1986.

88. Drew, M. G. B., Gilbert, A., and Rodwell, P. W., Photoaddition reactions of *trans*-dichloroethylene with benzenoid compounds: Facile formation of tetracycyclo[3.3.0.0.0]octanes and semibullvalenes, *Tetrahedron Lett.*, 26, 949, 1985.

89. Cosstick, K. and Gilbert, A., Selective formation of linear and angular triquinane carbon skeleta by directed intramolecular photocycloaddition of 5-phenylpent-1-ene, *J. Chem. Soc., Perkin Trans. I*, 1541, 1989.

90. Levine, B. H., Ph.D. Thesis, Stanford University, 1990.

91. Wender, P. A., von Geldern, T. W., and Levine, B. H., Synthetic studies on arene-olefin cycloadditions. 10. A concise, stereocontrolled total synthesis of (1)-laurenene, *J. Am. Chem. Soc.*, 110, 4858, 1988.

23

Cyclization of Stilbene and its Derivatives

ndrew Gilbert

niversity of Reading

23.1 Definition of the Reaction

cis-Stilbene undergoes photoinduced 6π-electrocyclization of the type outlined in Chapter 12 to give the 12π-system 4a,4b-dihydrophenanthrene 1: this species either ring-opens back to the starting material or, under oxidizing conditions, is converted into phenanthrene. The formation of a compound having λ_{max} at 247nm from the irradiation of stilbene was first reported in 1934,[5] but it was not until the 1950s that the product was identified as phenanthrene.[6] Within 10 years, the generality of the photocyclization of 1,2-diarylethenes had been established and now its usefulness as a convenient access to polynuclear aromatic compounds and its potential application as a photochromic system have both been widely recognized.[7]

23.2. Mechanism

Formation of the 4a,4b-Dihydrophenanthrene

The 6π-photocyclization arises from the *cis*-isomer of the 1,2-diarylethene; but since *trans, cis*-photointerconversion is a facile process, many investigations and applications have used the more readily accessible *trans*-isomer. The formation of the cyclized product is not initiated by triplet sensitization nor quenched by triplet quenchers;[8] furthermore, the reaction is inhibited by substituents such as nitro and acetyl, which promote intersystem crossing.[9] The photocyclization is, therefore, considered to arise from the singlet π,π^* state of the stilbene. The photochemical symmetry-allowed conrotatory ring closure leads to the *trans*-4a,4b-dihydro compound 1; for *cis*-stilbene, this occurs with a quantum yield of 0.1. Evidence for the 4a,4b-dihydro species was initially provided by the appearance of a transient absorption with λ_{max} at 450nm from irradiated stilbene solutions (*ca.* 0.01 M to minimize [$2\pi+2\pi$] photodimerization).[10] For phenolic derivatives such as stilbestrol 2, the cyclized product undergoes ketonization and hence is trapped as the stable diketotetraene 3. Indeed, confirmation of the *trans*-configuration in the dihydrophenanthrene was provided by the ozonolysis of 3, which gave only the racemic mixture of 4 and not the *meso* form that would have resulted from the cis-4a,4b-dihydrophenanthrene.[11] The half-life of the trans-4a,4b-dihydro species is, as may be expected, very dependent upon structural features and varies from less than one second to a reported 36.8 days for 5 produced from 1,2-di(2-naphthyl)ethene.[12] Furthermore, the dihydroarene 6 from irradiation of the aryl substituted imidazole 7 is a "stable" crystalline solid (m.p. 227–228 °C!).[13]

Ring opening of 4a,4b-dihydrophenanthrenes to stilbenes occurs both photochemically and thermally. In the former case, this is symmetry allowed as a conrotatory process and, although formally forbidden thermally, is considered to arise because of the high energy content of the dihydrophenanthrene relative to the stilbene derivative.[14] Oxidative dehydrogenation of the dihydrophenanthrene is deduced to occur by initial formation of the benzyl radical 8 induced by oxidants (O_2, I_2) and π-electron acceptors such as tetracyanoethene.[12]

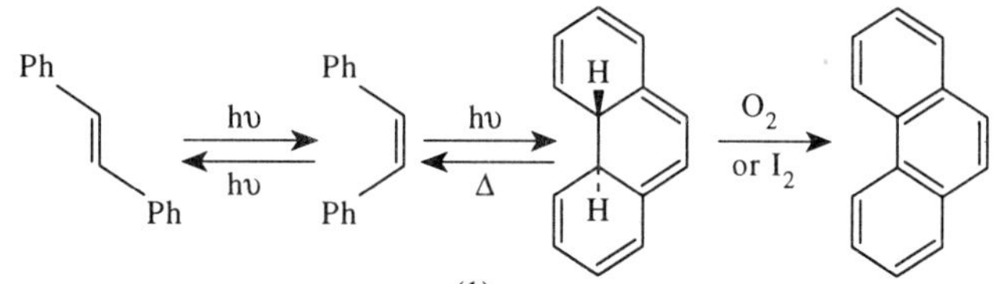

(1)

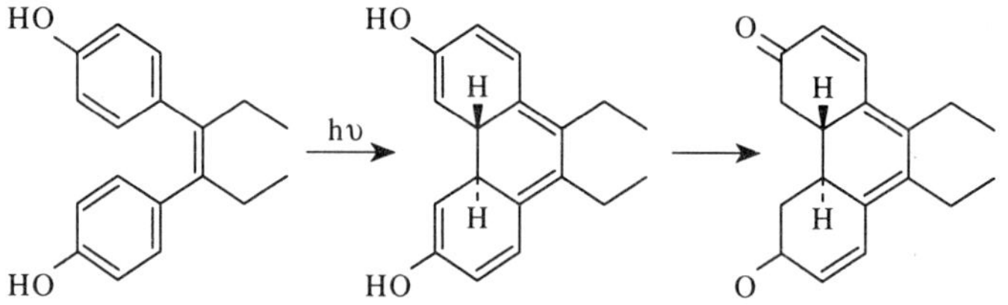

(2)

(3)

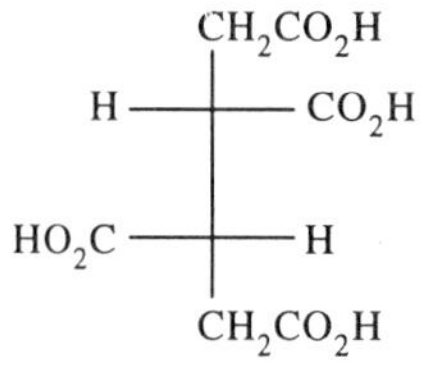

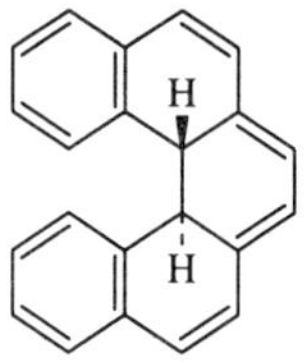

(4)

(5)

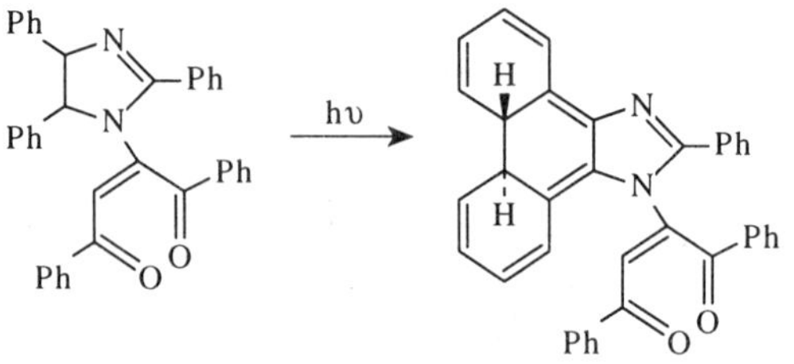

(7)

(6)

For synthetic purposes, it is important in 1,2-diarylethenes to be able to predict the arene positions between which the photocyclization will occur. One of the influences that controls this feature is aromatic resonance stabilization; thus, where alternative modes of cyclization exist, the preferred route is the one in which the dihydro species has the greater degree of aromaticity. For example, the dihydro intermediates 9 and 10 from the two possible directions of cyclization in the styryl benzophenanthrene 11 maintain aromaticity in two and three rings, respectively, and this is reflected in the formation of the helicene 12 as the sole product in 80% yield in the presence of iodine.[1]

Information concerning which dihydrophenanthrene is to be expected on photocyclization can also be obtained from the sum of the free valence indices (εF^*) of the arene carbon atoms between which the cyclization may occur, and from the localization energies L^*.[15] The latter is determined from the difference between the π-electron energies of the S_1 state of the 1,2-diarylethene and the S_0 state of the dihydrophenanthrene. Cyclization occurs between positions with εF^* greater than unity and those with L^* less than 3.45 (β). Thus, for example, the absence of cyclization products on irradiation of 4-distyrylbenzene is supported by an εF^* value of less than unity.[16] Other features that prevent the 6π-cyclization process are molecular constraints that cause the distance between the carbon atoms involved to be too great, as in the case of 9-benzylidenefluorene,[17] and the presence of low-lying $n\pi^*$ excited states. This latter feature is responsible for the lack of photocyclization of azobenzenes and benzylidene anilines: this photostability can, however, be readily overcome by complexation of the nitrogen compounds with Lewis acids,[18] or by carrying out the reaction in concentrated acids, since under both conditions the lowest excited state of the azobenzene and imine now has π,π^* character. The protonation can be intramolecular and thus the 2,2'-dicarboxylic acid of azobenzene photocyclizes in nonprotic solvents whereas under these conditions the 3,3'- and 4,4'-isomers are photostable.[19]

Substituent Effects

This 6π-photocyclization occurs for a very wide variety of substituted stilbenes and, in general, the presence of groups in the 3- or 4-positions do not inhibit the process:[20] under oxidizing conditions, phenanthrenes substituted in the 2- and/or 4-positions and in the 3-(or 6-)positions are formed respectively in yields of 60 to 90%. If the 3-substituent is an electron-withdrawing group (e.g., cyano, carbomethoxy), then the oxidative cyclization has an exceptionally high regioselectivity of *ca.* 4:1 favoring the 2-isomer: this has been explained by Molecular Orbital calculations.[21] For 3,3'-disubstituted stilbenes also, a statistical distribution of the disubstituted phenanthrenes is not observed and, in particular, the 4,5-disubstituted arene may be a very minor product. The features that influence isomer formation in such cases are steric rather than electronic and are illustrated by the cyclization of 3,3'-dimethylstilbene.[22] The dihydrophenanthrene 13 leading to the 4,5-di-methyl-substituted arene is highly strained compared to the two other possible cyclization product isomers, and this leads to an enhanced rate of ring opening of this particular dihydrophenanthrene to the *cis*-stilbene, thereby reducing the amount of the corresponding phenanthrene. Indeed, with oxygen as the oxidant, none of the 4,5-dimethylphenanthrene is observed. The proportions of this arene isomer in the product mixture do, however, increase with increasing iodine concentration, but even with sufficient oxidant present to ensure complete reaction of all the dihydrophenanthrene intermediates, the 18% proportion of the 4,5-dimethylphenanthrene is still significantly less than the statistical value. This feature is considered to result from steric crowding in the stilbene conformer 14, thus reducing the proportion of the intermediate dihydrophenanthrene 13.

The presence of substituents at the 2-positions of the phenyl rings of the stilbene can markedly influence the direction of cyclization and the stability of the resulting dihydrophenanthrene. The favored cyclization pathway is that which avoids the sites of substitution, and 1-substituted phenan-threnes then result from oxidation of the 4a,4b-dihydro intermediate. However, for 2,2-di- (and 2,2,2',2'-tetra-) substituted stilbenes, the 4a-position (and 4a,4b-positions) in the dihydrophenanthrene are inevitably substituted. This feature can impart appreciable stability to the intermediate (see Section 23.4), or elimination of the "blocking" group may occur with formation of the phenanthrene.[1-4] For example, dihydrophenanthrene intermediates **15** from stilbenes, having a potential leaving group such as a methoxy or bromo substituent in the 2-position, readily eliminate methanol and hydrogen bromide, respectively, from the dihydrophenanthrene. The favored direction of the cyclization would still be to avoid the substituted site, but as illustrated in Scheme 23.1, the reaction can be directed along different pathways, depending on the presence or absence of an oxidant.

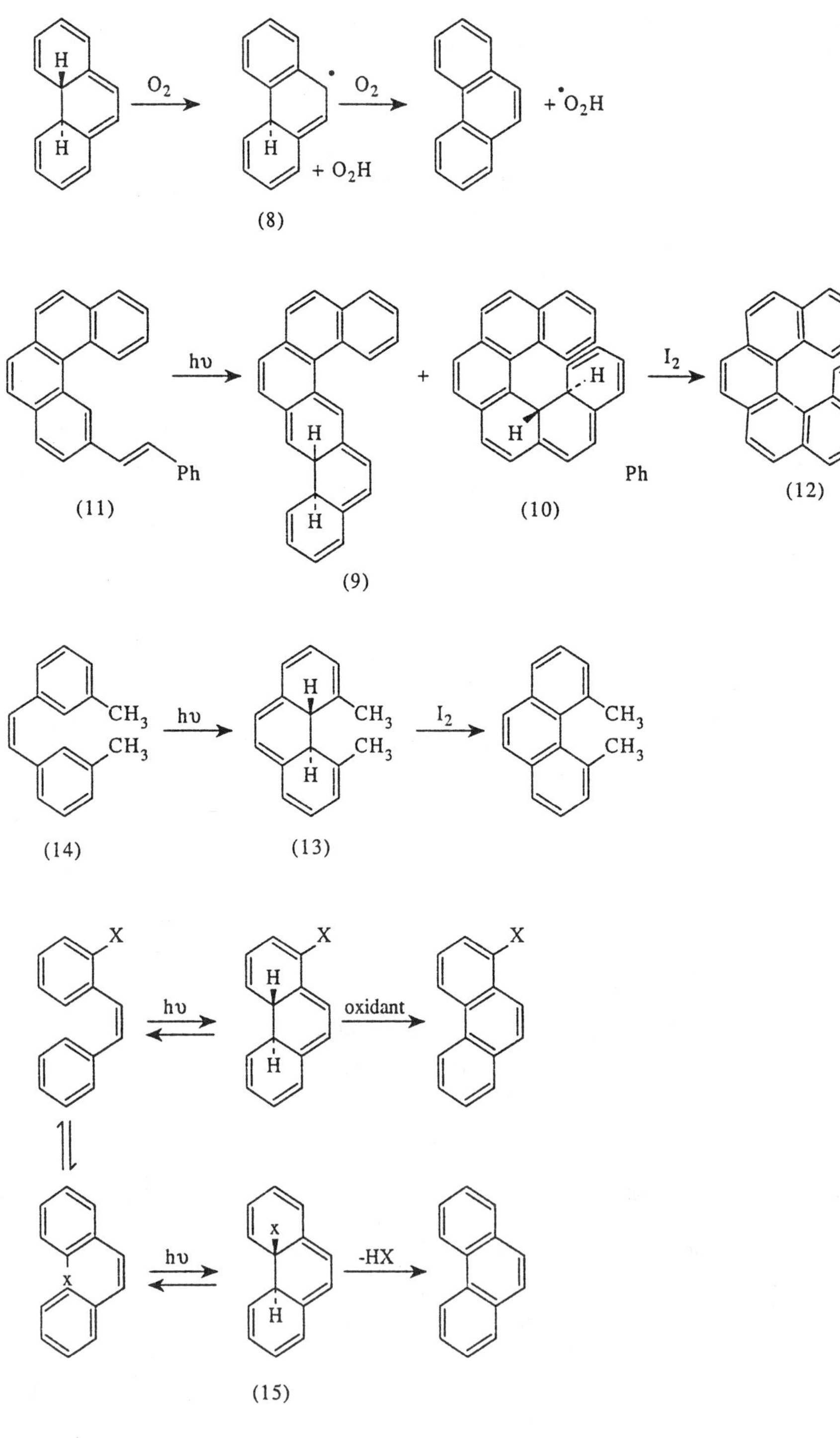

SCHEME 23.1

Formation of 9,10-Dihydrophenanthrenes

The 6π-photocyclization of stilbenes having α-substituents yields 9-(or 9,10-) substituted phenanthrenes under oxidizing conditions. However, for stilbenes having electron-withdrawing groups (e.g., cyano or carbomethoxy) on the α-position(s), then, in the absence of an oxidant, 9,10-dihydrophenanthrenes result. The ionic mechanism deduced for this deviation is outlined in Scheme 23.2 and is operable for enolizable α-substituted stilbenes in the presence of a proton source.[23] A similar type of dihydrophenanthrene is also formed from rearrangement under nonoxidizing conditions of the 4a,4b-intermediate 16 for hexahelicene. The resulting isomers 17 and 18 are stable and isolable, and the rearrangement is base catalyzed, with the ratios of 17 and 18 being base and solvent dependent.[24] 2-Vinylbiphenyls also yield 9,10-dihydrophenanthrenes by a 6π-photocyclization process.[25]

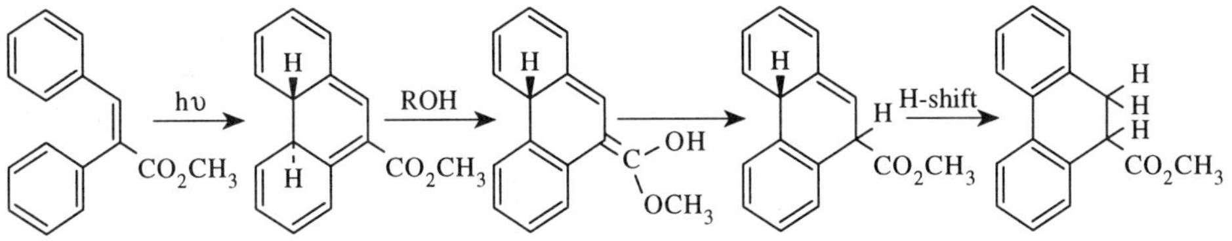

SCHEME 23.2

23.3 Scope and Synthetic Applications

The photochemical stilbene-to-phenanthrene conversion has enormous scope and application in the synthesis of polynuclear aromatic compounds. In general, the reaction is not adversely affected by polysubstitution in the stilbene or change of the aryl groups for five- and six-membered heteroarenes. The following small selection of examples has been chosen in an attempt to illustrate the breadth and versatility of the photocyclization and the types of molecular structures accessible by this route. Numerous substituted stilbenes have been reported to undergo the 6π-cyclization.[1-4] Examples include a novel approach to the synthesis of condensed (2.2)paracyclophanes by the cyclization of 19 in the presence of iodine and air to give 20,[26] and the formation of the tetraoxygenated methylphenanthrene skeleton 21 from 22 by rearomatization of the dihydro intermediate on elimination of methanol.[27]

Over the years, the reaction has been exploited as a convenient access to a range of helicenes. The double helicene, diphenanthro[4,3-a; 3′4′o]picene 23, can be obtained by this route,[28] and [7]helicene 24 is formed in equal amounts with 25 from irradiation of 26.[29] Direction of the cyclization is, however, achieved by the presence of a bromine substituent in the benzene ring of 26: this directs the cyclization away from the occupied position so that the [7]helicene is formed in 75% yield with 90% selectivity.[30]

The 6π-photocyclization is generally aided, and overall yields improved, for systems in which the ethene moiety of the stilbene is constricted into a *cis*-geometry. In several examples, the geometry is held by the ethene being part of a heterocyclic ring as in 27[31] and 28,[32] both of which photocyclize and yield the corresponding phenanthrenes. Analogous 6π-photooxidative cyclization occurs for a variety of 1,2-diarylarenes to give condensed systems: for example, 29[33] and 30[34] give 31 and 32, respectively. The required conformational rigidity can also be obtained by linking the aryl groups, as in the cyclophane 33. The bridged phenanthrene formed from the photocyclization is then reductively cleaved to provide an overall convenient access to cannithrene II 34.[35]

Altering one or both of the arene groups in the 1,2-diarylethenes for a heteroarene unit does not seemingly interfere with the smooth 6π-photocyclization and the subsequent formation of the condensed arene. Numerous examples of reaction with this type of compound have been reported for five-membered ring heteroarenes, their benzo-fused analogues, and six-membered ring units having a positively charged nitrogen atom. Cyclizations of these types of compounds are illustrated by the photoconversions of 35 to 36,[36] 37 to 38,[37] and 39 to 40.[38] The first of these transformations has been applied to the synthesis of antitumor antibiotics: a key aspect in this application is the incorporation of palladium to catalyze the aromatization of the dihydro intermediate and thereby avoid oxidation of the heteroarene.

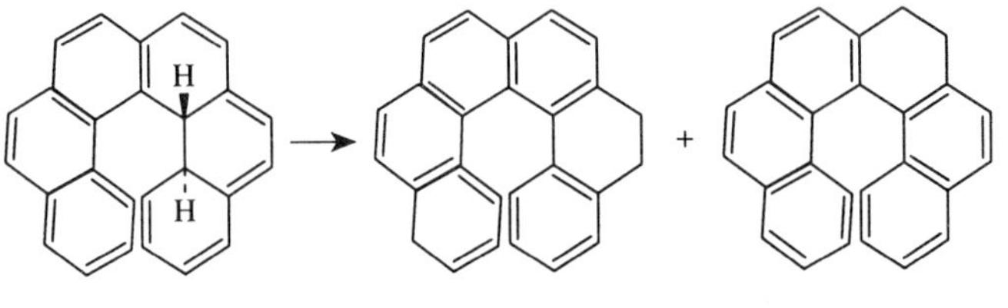

(16)   (17)   (18)

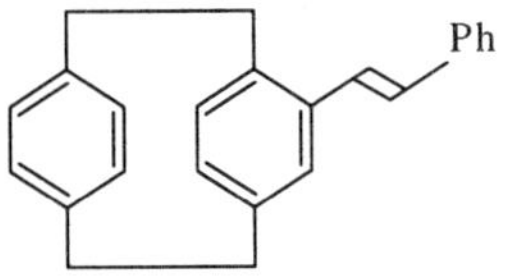

(19)

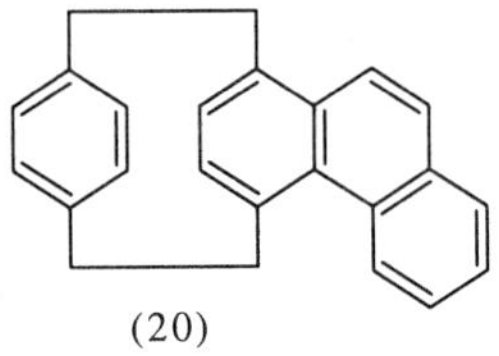

(20)

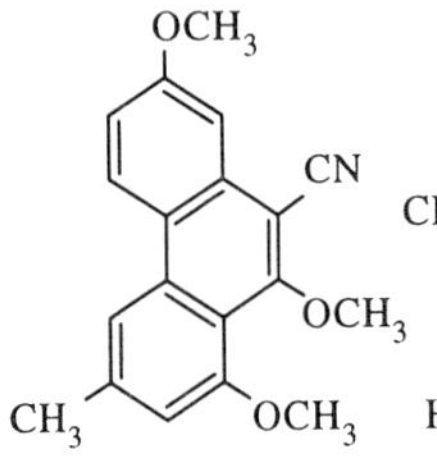

(21) 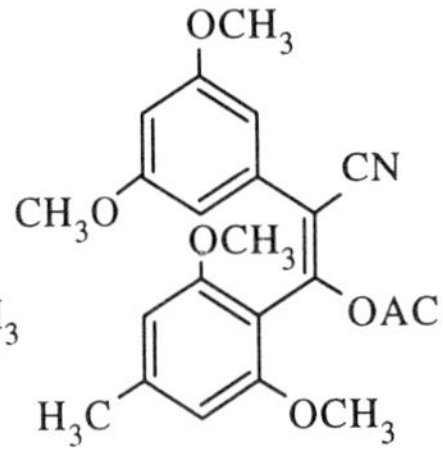   (22)

(24)

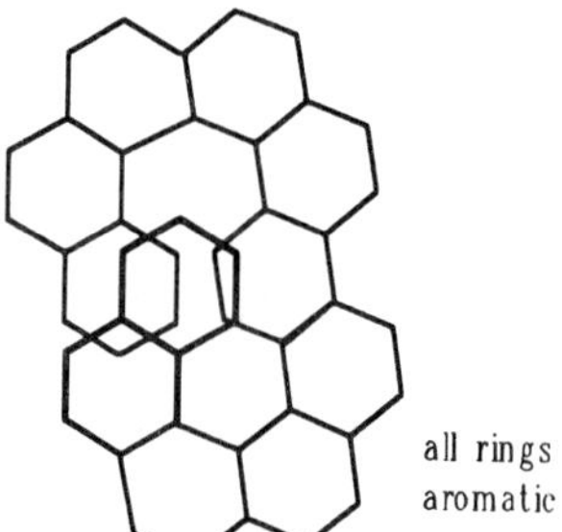

all rings aromatic

(23)

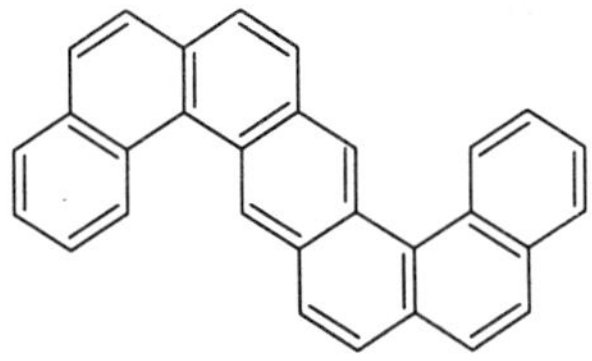

all rings aromatic

(25)

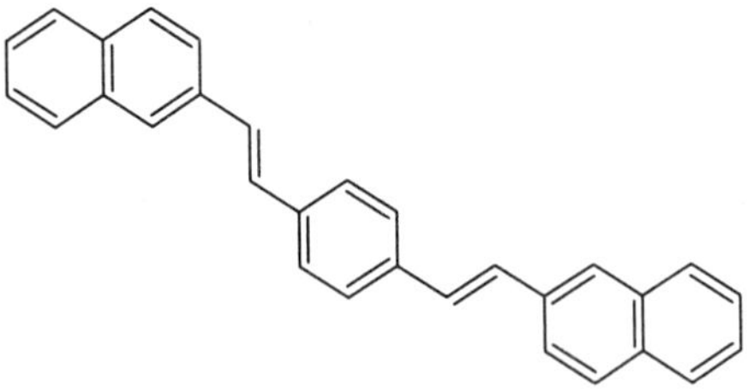

(26)

3.4 Industrial Applications

It has long been recognized that reversible photocyclization of 1,2-diaryl arenes to the 4a,4b-dihydrophenanthrene derivative has considerable potential as a photochromic system that may be capable of development into a data storage device. Of the range of compounds studied, the 1,2-bis(benzo[b]thiophen-3-yl)ethenes (41) would seem to offer the greatest promise for this application. The dihydro cyclized product 42 has λ_{max} at 544 nm, is thermally stable (no change after 21 days at 80 °C) and, furthermore, the photochromic properties appear to be fatigue free since the reversible conversion of the diarylethene to 42 has been cycled 10^4 times without apparent loss of starting material.[39]

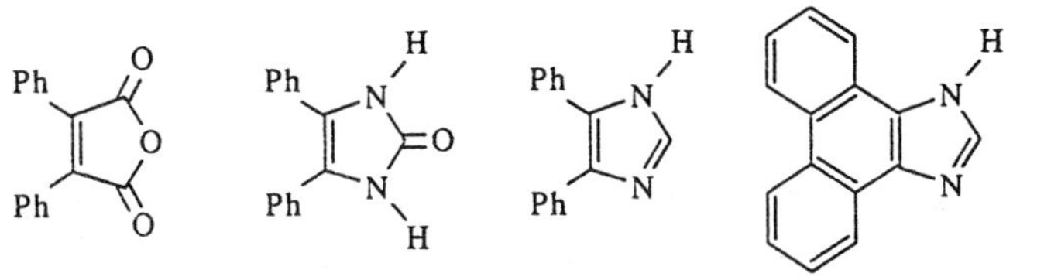

(27)   (28)   (29)   (31)

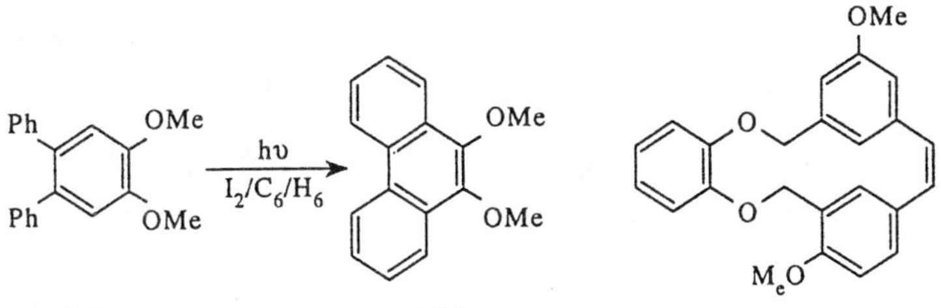

(30)   (32)   (33)

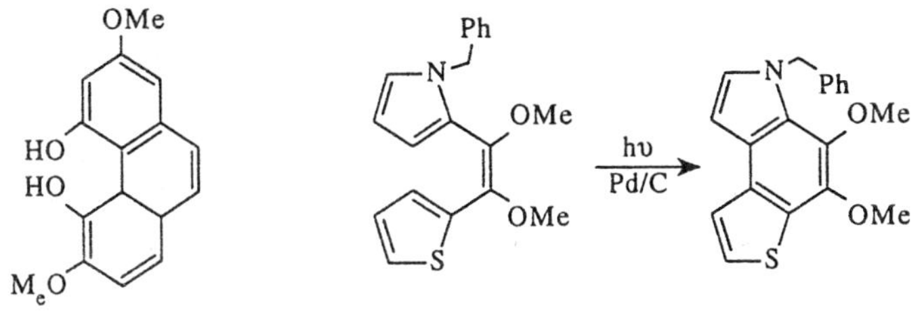

(34)   (35)   (36)

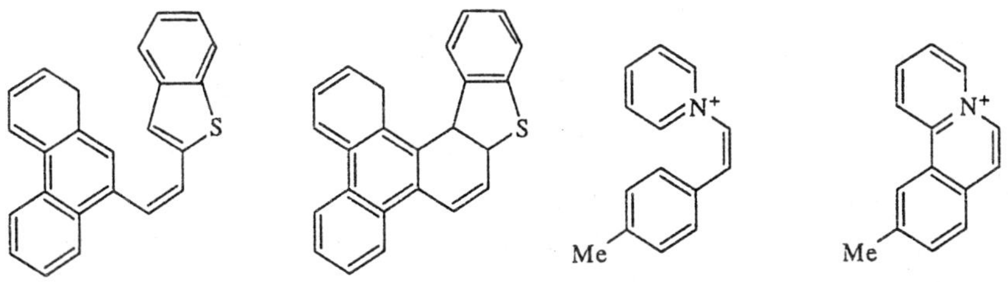

(37)   (38)   (39)   (40)

Str 27-42

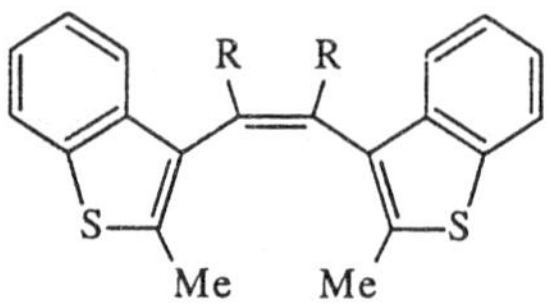

(41) R's = CN or R - R = -CO - O - CO- 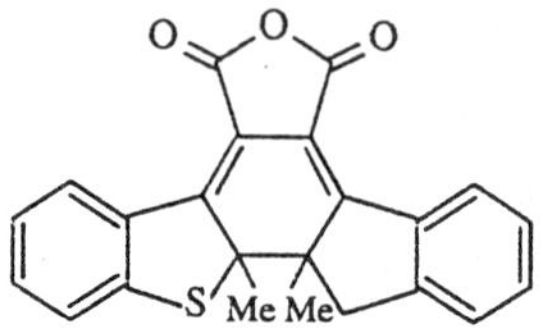   (42)

References

1. Laarhoven, W. H., Photochemical cyclizations and intramolecular cycloadditions of conjugated arylolefins. 1. Photocyclization with dehydrogenation, *Rec. Trav. Chim. Pays-Bas,* 102, 185, 1983.
2. Laarhoven, W. H., 2. Photocyclizations without dehydrogenation and photocycloadditions, *Rec. Trav. Chim. Pays-Bas,* 102, 241, 1983.
3. Mallory, F. B. and Mallory, C. W., Photocyclization of stilbene and related molecules, *Org. Reactions,* 30, 1, 1984.
4. Laarhoven, W. H., Photocyclizations and intramolecular photocycloadditions of conjugated arylolefins and related compounds, *Org. Photochem.,* Padwa, A., Ed., Marcel Dekker, New York, 1989.
5. Smakula, A., Photochemical transformation of *trans*-stilbene, *Z. Phys. Chem.,* B25, 90, 1934.
6. Buckles, R. E., Illumination of *cis*- and *trans*-stilbenes in dilute solution, *J. Am. Chem. Soc.,* 77, 1040, 1955.
7. Reviews of the photochemistry of aromatic compounds appear in the annual volumes of *Photochemistry,* Specialist Periodical Reports, Vol. 1–25, The Royal Society of Chemistry, London, 1970–1994.
8. Mallory, F. B., Wood, C. S., and Gordon, J. T., Photochemistry of stilbenes. III. Some aspects of the mechanism of photocyclization to phenanthrenes, *J. Am. Chem. Soc.,* 86, 3094, 1964; Hammond, G. S., Saltiel, J., Lamola, A. A., Turro, N. J., Bradshaw, J. S., Cowan, D. O., Corsell, R. C., Voet, V., and Dalton, C., Mechanisms of photochemical reactions in solution. XXII. Photochemical *cis-trans* isomerization, *J. Am. Chem. Soc.,* 86, 3197, 1964.
9. Wood, C. S. and Mallory, F. B., Photochemistry of stilbenes. IV. The preparation of substituted phenanthrenes, *J. Org. Chem.,* 29, 3373, 1964.
10. Moore, W. M., Morgan, D. D., and Stermitz, F. R., The photochemical conversion of stilbene to phenanthrene. The nature of the intermediate, *J. Am. Chem. Soc.,* 85, 829, 1963.
11. Doyle, T. D., Filipescu, N., Benson, W. R., and Banes, D., Photocyclization of α-diethyl-4,4′-stilbenediol. Isolation of a stable tautomer of the elusive dihydrophenanthrene, *J. Am. Chem. Soc.,* 92, 6371, 1970; Cuppen, T. J. H. M. and Laarhoven, W. H., Photodehydrocyclizations of stilbene-like compounds. VI. Evidence for an excited state mechanism, *J. Am. Chem. Soc.,* 94, 5914, 1972.
12. Muszkat, K. A., The 4a,4b-dihydrophenanthrenes, *Top. Curr. Chem.,* 88, 89, 1980.
13. Barik, R., Bhattacharyya, K., Das, P. K., and George, M. V., Photochemical transformations of 1-imidazolyl-1,2-dibenzoylalkenes, *J. Org. Chem.,* 51, 3420, 1986.
14. Muszkat, K. A. and Fischer, E., Structure, spectra, photochemistry, and thermal reactions of 4a,4b-dihydrophenanthrenes, *J. Chem. Soc.,* 602, 1967.
15. Scholz, M., Muhlstadt, M., and Dietz, F., Die richtung der photocyclisierung naphthalinsubstituierter athylene, *Tetrahedron Lett.,* 665, 1967.
16. Laarhoven, W. H., Cuppen, T. J. H. M., and Nivard, R. J. F., Photodehydrocyclizations in stilbene-like compounds. II. Photochemistry of distyrylbenzenes, *Tetrahedron,* 26, 1069, 1970.

17. Scholz, M., Dietz, F., and Muhlstadt, M., Photocycliserung von benzalanthronen, *Tetrahedron Lett.,* 2835, 1970.
18. Thompson, C. M. and Docter, S., Lewis acid promoted photocyclization of arylimines. Studies directed towards the synthesis of pentacyclic natural products, *Tetrahedron Lett.,* 29, 5213, 1988.
19. Joshua, C. P. and Pillai, V. N. R., Photochemical cyclodehydrogenation of Lewis acid conjugates of azobenzenes, *Tetrahedron,* 30, 3333, 1974.
20. Somers, J. B. M. and Laarhoven, W. H., The influence of substituents on the photodehydrocyclization of 1,2-diphenylcyclopentenes. I. *para* substituents, *J. Photochem. Photobiol.,* 40, 125, 1987; Somers, J. B. M. and Laarhoven, W. H., II. *meta* Substituents, *J. Photochem. Photobiol.,* 48, 353, 1989.
21. Muszkat, K. A., Kessel, H., and Sharafi-Ozeri, S., Electronic overlap population as a reactivity measure. VII. Substituent effects on the photocyclization of stilbenes, *Isr. J. Chem.,* 16, 291, 1977.
22. Mallory, F. B. and Mallory, C. W., Steric effects on the photocyclizations of some *meta* substituted stilbenes, *J. Am. Chem. Soc.,* 94, 6041, 1972.
23. op het Veld, P. H. G. and Laarhoven, W. H., The mechanism of the photoconversion of α-phenylcinnamic esters into 9,10-dihydrophenanthrenes, *J. Am. Chem. Soc.,* 99, 7221, 1977.
24. Prinsen, W. J. C. and Laarhoven, W. H., Mechanistic details of the rearrangement of the primary photocyclization product of 2-styrylbenzo[c]phenanthrene in the presence of base, *J. Org. Chem.,* 54, 3689, 1989.
25. Lapouyade, R., Manigand, C., and Nourmamode, A., Photocyclization of 2-vinylbiphenyls: stereochemistry of the triplet state cyclization, *Can. J. Chem.,* 63, 2192, 1985; Padwa, A., Doubleday, C., and Mazzu, A., Photocyclization reactions of substituted 2,2'-divinylbiphenyl derivatives, *J. Org. Chem.,* 42, 3271, 1977.
26. Hopf, H., Mlynek, C., El-Tamany, S., and Ernst, L., [2.2](1,4-)Phenanthrenoparacyclophane: synthesis and two-dimensional proton and carbon-13 NMR study, *J. Am. Chem. Soc.,* 107, 6620, 1985.
27. Finnie, A. A. and Hall, R. A., The synthesis of 1,5,7,10-tetraoxygenated 3-methylphenanthrenes, *J. Chem. Res. Synop.,* 78, 1987.
28. Martin, R. H., Eyndels, C., and Defay, N., Double helicenes, *Tetrahedron,* 30, 3339, 1974; Laarhoven, W. H., Cuppen, T. J. H. M., and Nivard, R. J. F., Photodehydrocyclizations of stilbene-like compounds, *Tetrahedron,* 30, 3343, 1974.
29. Martin, R. H., Marchant, M.-J., and Baes, M., Rapid synthesis of hexa- and heptahelicene, *Helv. Chim. Acta,* 54, 358, 1971.
30. Sudhakar, A. and Katz, T. J., Directive effect of bromine on stilbene photocycizations. An improved synthesis of [7] helicene, *Tetrahedron Lett.,* 27, 2231, 1986.
31. Fields, E. K., Behrend, S. J., Meyerson, S., Winzenburg, M. L., Ortega, B. R., and Hall, H. K., Diaryl substituted maleic anhydrides, *J. Org. Chem.,* 55, 5165, 1990.
32. Purushothaman, E. and Pillai, V. N. R., Regiospecific photochemical transformations of 4,5-diaryl-Δ⁴-imidazolin-2-ones: ketal formation and cyclodehydrogenation, *Proc. Ind. Acad. Sci. Chem. Sci.,* 101, 391, 1989.
33. Purushothaman, E. and Pillai, V. N. R., Photoreactions of 4,5-diarylimidazoles: singlet oxygenation and cyclodehydrogenation, *Ind. J. Chem.,* 28B, 290, 1989.
34. Bushby, R. J. and Hardy, C., Regiospecific synthesis of 2,3,6,7,10,11-hexasubstituted triphenylenes by oxidative cyclization of 3,3″, 4,4′, 4″, 5′-hexasubstituted 1,1′:2′1″-terphenyls, *J. Chem. Soc., Perkin Trans. I,* 721, 1986.
35. Ben, I., Castedo, L., Saa, J. M., Seijas, J. A., Suau, R., and Tojo, J., 4,5-O-substituted phenanthrenes from cyclophanes. The total synthesis of cannithrene II, *J. Org. Chem.,* 50, 2236, 1985.
36. Jones, R. J. and Cava, M. P., Photocyclization strategy for the synthesis of antitumour agent CC-1065: synthesis of the thiophene analogues of PDE-I and PDE-II, *J. Chem. Soc., Chem. Commun.,* 826, 1986; Rawal, V. H., Jones, R. J., and Cava, M. P., Synthesis of thiophene and furan analogues of dideoxy PDE-I and PDE-II, *J. Org. Chem.,* 52, 19, 1987.

37. Kudo, H., Tedjamulia, M. L., Castle, R. N., and Lee, M. L., Angular polycyclic thiophenes containing two thiophene rings, *J. Heterocyclic Chem.*, 21, 1833, 1984.

38. Arai, S., Yamazaki, M., and Hida, M., Synthesis and reaction of methylbenzo[a]quinolizinium salts, *J. Heterocyclic Chem.*, 27, 1073, 1990.

39. Uchida, K., Nakayama, Y., and Irie, M., Thermally irreversible photochromic systems. Reversible photocyclization of 1,2-*bis*(benzo[b]thiophen-3-yl)ethene derivatives, *Bull. Chem. Soc. Jpn.*, 63, 1311, 1990.

24

Ene Reactions with Singlet Oxygen

el G. Griesbeck
tut für Organische Chemie,
versity of Cologne

4.1 Introduction and Definition of the Reaction

The active species in Type II photooxygenation reactions is molecular oxygen in its first excited singlet state as postulated by Kautsky.[1] This is a clear result from experiments by Foote et al.,[2] where several reactions of photochemically generated singlet molecular oxygen were compared with corresponding reactions of the active oxygen formed in the hypochlorite/hydrogen peroxide reaction. The results from allylic oxidations with singlet molecular oxygen (referred to as $^1O_2$ from now on) are clearly different to those of autoxidative (triplet oxygen) pathways.[3] Several reaction modes of $^1O_2$ with unsaturated organic molecules are known (e.g., [4+2]- and [2+2]-cycloaddition, ene and silyl-ene reactions, and physical quenching) and have been summarized in excellent books on the subject.[4] For allylic activation of C-C double bonds, the ene-reaction is the most prominent. In the course of this reaction, $^1O_2$ attacks one center of a C-C double bond with abstraction of an allylic hydrogen atom (bound to carbon or oxygen) or an allylic silyl group (bound to oxygen) and simultaneous allylic shift of the double bond. As result of this reaction, allylic hydroperoxides or O-silylated allylic hydroperoxides are formed.

4.2 Mechanism of the Singlet Oxygen Ene Reaction

Theoretical Models

In the last 20 years, several mechanisms have been postulated for this reaction involving concerted or "concerted two-stage" mechanisms[5] as well as 1,4-biradicals,[6] 1,4-Zwitterions,[7] or perepoxides[8] as intermediates. A series of semiempirical and *ab initio* calculations has been published, but the exact mechanism of this reaction type is still in doubt.[9]

Experimental Conditions

Singlet molecular oxygen ($^1O_2$) can be generated by several methods.[10] From an experimental point of view, the only convenient method, however, is the photochemical route. This path is advantageous because of its flexibility concerning reaction temperature, solvent, sensitizer, and light source. In solution, $^1O_2$ is efficiently deactivated not only by chemical reaction with appropriate acceptor molecules, but also *physically* by the solvent and the sensitizer and in many cases even by the acceptor itself. Therefore, $^1O_2$ has to be used in high excess, and only the photosensitization can fulfill these conditions. Nevertheless, serious problems can arise from the wrong choice of parameters such as solvent and sensitizer. Competing reaction *modes* ("physical" quenching, cycloaddition, ene reaction) are influenced differently by the solvent polarity.[11] The lifetime of $^1O_2$ is also very sensitive to changes of the solvent.[12] Most serious is the possibility of changes in reaction *type*, e.g., "type II" and "type III". When powerful electron acceptors such as 9,10-dicyanoanthracene (DCA) or Methylene Blue (MB) are used as sensitizers, photostimulated electron transfer processes can be initiated, which can lead to different reactivity and product pattern.[13] Possibly, the superoxide anion radical, which also can be produced from singlet oxygen, is important in many of these reactions. Of course, the solvent polarity again plays an extremely important role in these "polar reactions".[14] Therefore, for "clean" $^1O_2$ reactions, some experimental rules should be borne in mind (these considerations are obviously valid for singlet oxygen [4+2]- and [2+2]-cycloaddition reactions as well):

1. Nonpolar solvents (halogenated or fluorinated hydrocarbons) suppress electron transfer reactions and increase the lifetime of singlet oxygen. In polar solvents, [2+2]-cycloadditions are considerably faster compared to reaction in nonpolar solvents.

2. Weak electron acceptors [tetraphenylporphine (TPP), metaloporphyrins (e.g., ZnTPP), or haematoporphyrin (HP)] with low triplet energies (MB: 33.5; TPP: 34.0; HP: 37.2; RB: 39.2–42.2, in kcal mol⁻)[15] should be used as sensitizers. The use of Rose Bengal (RB) is possible sometimes (in more polar solvents); but in some cases, even RB can undergo electron transfer photooxygenations.[16] Methylene Blue (MB) should be avoided.

3. Low-energy light should be used ($\lambda > 400$ nm). If high-pressure mercury lamps are used, filters should be used to exclude the high-energy UV light.[17] Sodium lamps are the most convenient and inexpensive.

4. In most cases, little activation energy is necessary for the transformations described (1–5 kcal mol⁻¹)[18] and the reactions can be carried through at low temperatures (which also prevents secondary reactions of the peroxidic products).

5. (+)-Limonene **1** is popular as a singlet oxygen acceptor giving a very characteristic pattern of regio- and stereoisomeric allylic hydroperoxides **2 to 5** as products. Under autoxidative conditions, this pattern is changed completely, and additional products without shift of the C-C double bond are formed.[19]

6. Regio- and stereoselectivity for a certain transformation should be determined directly at the peroxide stage. A number of examples are known, where further transformations (reduction, rearrangement, cleavage) clearly change the regio- as well as the stereochemistry of the products.

7. Addition of radical scavenging reagents (e.g., 2,6-di-*t*-butylphenol) inhibits radical autoxidation reactions, but does not alter the singlet oxygen reaction. As a test for the involvement of singlet oxygen, an efficient singlet oxygen quencher such as DABCO (1,4-diazabicyclo[2.2.2]octane) can be added.[20]

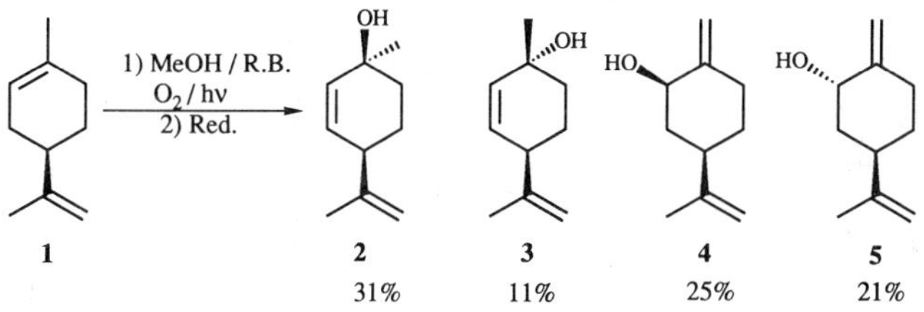

	2	3	4	5
	31%	11%	25%	21%

4.3 Selectivity

Chemoselectivity

Electron-rich alkenes could, in principle, interact with singlet oxygen in three ways: physical quenching, [2+2]-cycloaddition, and ene reaction. What reaction mode dominates depends strongly on the ionization potential of the double bond, the appropriate alignment of the allylic hydrogens, and the polarity of the solvent. Methyl-substituted acyclic enol ethers, for example, give ene products in several solvents,[22] whereas the corresponding cyclic enol ethers (2,3-dihydrofurans) give mixtures of allylic hydroperoxides (ene products) and 1,2-dioxetanes ([2+2]-cycloaddition products) with notable solvent dependency.[23]

Locoselectivity

In polyolefins, the locoselectivity of singlet oxygen ene reactions can be predicted by considering the ionization potentials of corresponding model monoolefins.[24] Additionally, several effects that moderate the reactivity can operate. However, as a rule of thumb, the addition of an additional alkyl group increases the ene reactivity by a factor of about 10 to 20.[25]

Regioselectivity

Several general effects controlling the regioselectivity of allylic oxidations of C-C double bonds by $^1O_2$ have been discovered in the last 20 years (Figure 24.1). These could be used for designing appropriate compounds for the regioselective introduction of hydroperoxy groups: the *cis-* or *syn-*effect, the "geminal effect", and the "large group effect". The *cis-*(or *syn-*) effect leads to highly selective hydrogen abstraction from the *side* of a trisubstituted C-C double bond with higher substitution, when both substitutents bear allylic hydrogens[6] or one substituent is alkoxy.[27] The geminal effect leads to highly selective abstraction of an allylic hydrogen from a substituent at the α-*position* of an α,β-unsaturated carbonyl compound.[28] The large group effect leads to selective (moderate) abstraction of an allylic hydrogen from the substituent with a large group at the same carbon atom of the C-C double bond.[29] In addition to these effects, a metal substitutent (R_3Si) at the C-C double bond directs singlet oxygen toward the nonsubstituted olefinic center.[30]

Stereoselectivity

Only a few systematic investigations concerning the induced stereoselectivity of singlet oxygen ene reactions (especially with acyclic alkenes) have been published. A large number of natural products, mostly terpenes and terpenoids, have been investigated and the inherent stereoselectivity is some-times very high. The most important examples are summarized in several reviews and are not mentioned here.[25]

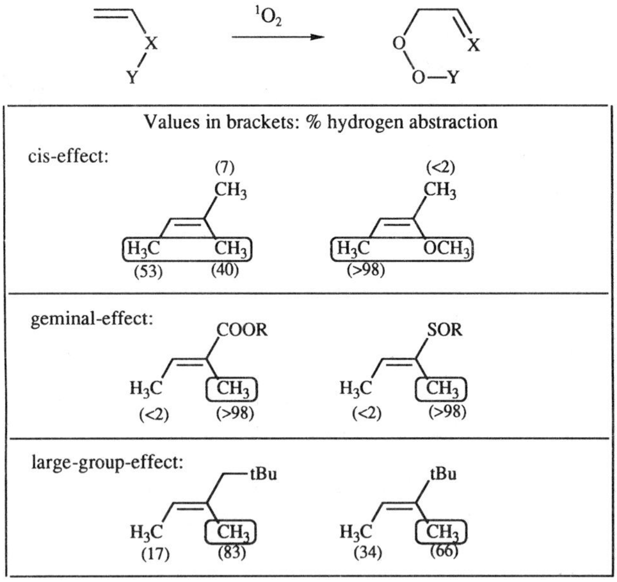

Figure 24-1   General effects controlling the regioselectivity of allylic oxidations of C-C double bonds by $^{\cdot}O_2$.

24.4   Substrates for Ene Reaction

Acyclic Substrates

1. Alkenes and Alkadienes

Stephenson[31] has shown, using the optically active monodeuterated olefin 6, that ene reactions are highly stereoselective suprafacial processes, in which the correct arrangement of the allylic hydrogens controls the stereochemical result. In this reaction, no isotopic discrimination is found and the allylic hydroperoxide with (R)-configuration and migrated deuterium 7 are formed in equal amounts with the corresponding (S)-hydroperoxide and migrated hydrogen 8. Similar results were obtained in the photooxygenation of silyl cyanohydrins of α,β-unsaturated aldehydes.[32] The trimethylsilyl and t-butyldimethylsilyl cyanohydrins of senecialdehyde 9 react with low regioselectivity (because of the *cis*-effect), but with high diastereoselectivity. In all cases, the two regioisomeric allylic hydroperoxides were formed with identical diastereoisomeric ratios, demonstrating clearly the necessity for a common symmetrical intermediate, presumably of the perepoxide type. The relative configuration of the photooxygenation products was not determined. Shimizu et al.[31] have shown that certain allylsilanes react with singlet oxygen in a Z-selective manner. The reaction of allyltrimethylsilane 10 (d.r. = 22.78) was compared with that of the corresponding hydrocarbon, 3-t-butyl-1-propene 11, where a diastereoisomeric ratio of >99:1 in favor of the *trans*-diastereoisomer was found. Chiral trisubstituted monoalkenes have been investigated by Kropf and Reichwaldt.[34] Diastereoisomeric ratios were reported as 1:4.4 to 1:5.3 *(threo:erythro)*; in one case, two regioisomeric products are formed with identical d.r., again showing the validity of the assumption of a stereoselective, suprafacial reaction mechanism.

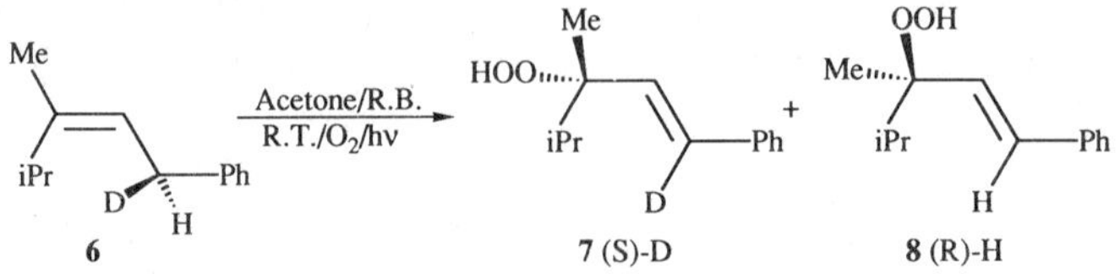

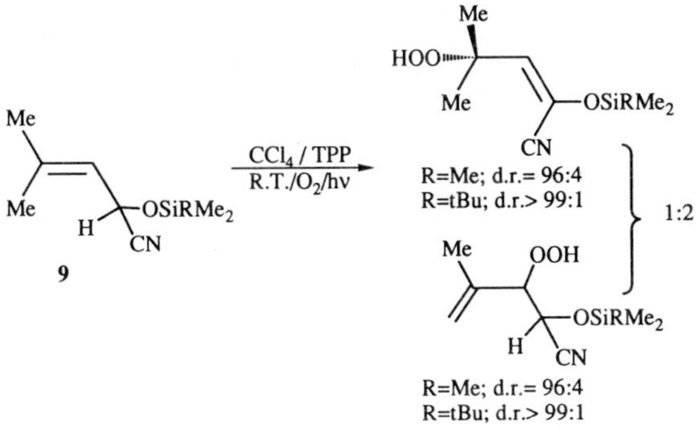

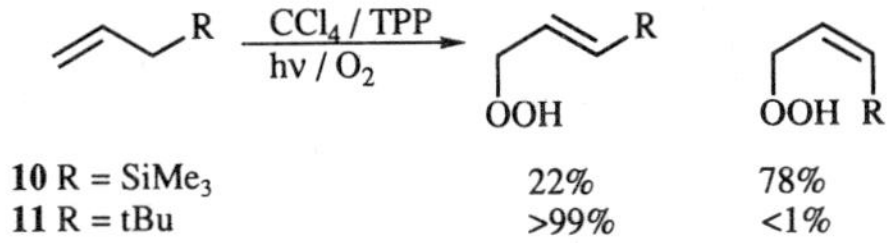

α,β-Unsaturated Carbonyl Compounds and Derivatives

The photooxygenation of α,β-unsaturated esters **12** with a stereogenic center at the γ-position leads to allylic hydroperoxides with >99% regiocontrol ("geminal effect" in operation) and preferred formation of the *ul* (unlike) products.[35] The diastereoselectivity is modest (d.r. = 65:35) when the substrate has Z-geometry and substantial (d.r. = 80:20) when the substrate has E-geometry, again demonstrating the importance of secondary orbital effects as postulated in the perepoxide mechanism. Adam and Nestler[36] have shown that for certain allylic alcohols, a remarkably high acyclic diastereoselectivity is obtained. This "hydroxy-directing" effect could be due to precomplexation of the incoming singlet oxygen with the allylic hydroxy group.

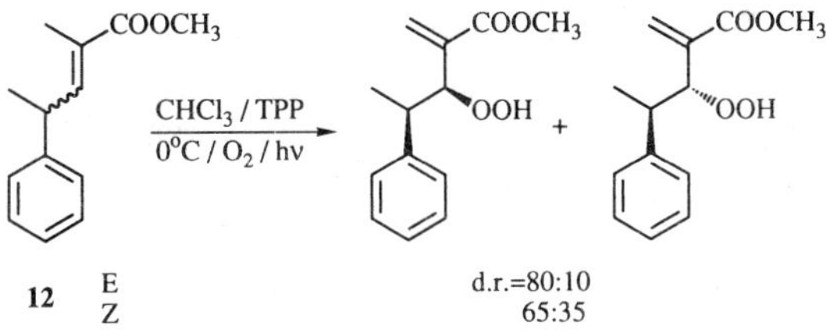

Cycloalkenes

With Exocyclic C-C Double Bond

The diastereofacial differentiation in *exo*-alkylidene norbornenes was investigated for 2-ethylidene bicyclo[2.2.1]hept-5-ene (**13**).[37] The *exo*- and *endo*-2-hydroxy-2-vinylnorborn-5-enes **14** and **15** were formed as principal products in a ratio of 35:12, with the regioisomeric 2-(1-hydroxyethyl)-norbornadiene **16** (23%) as the third component. Two ketones could be detected as secondary products. An important application in the total synthesis of the antimalaria drug *qinghaosu* was described by Schmid and Hofheinz.[38] The key step in the synthesis based on (–)-isopulegol is the ene reaction of singlet oxygen with an enol ether system. This reaction leads to an allylic hydroperoxide **17** when the reaction is performed in a nonpolar solvent such as CH_2Cl_2. The use of a more polar, protic solvent such as methanol inverts the regioselectivity and (presumably via trapping of an intermediate 1,4-Zwitterion) an α-hydroperoxy acetal **18** is formed, which could be cyclized to the target molecule **19** by acid catalysis.

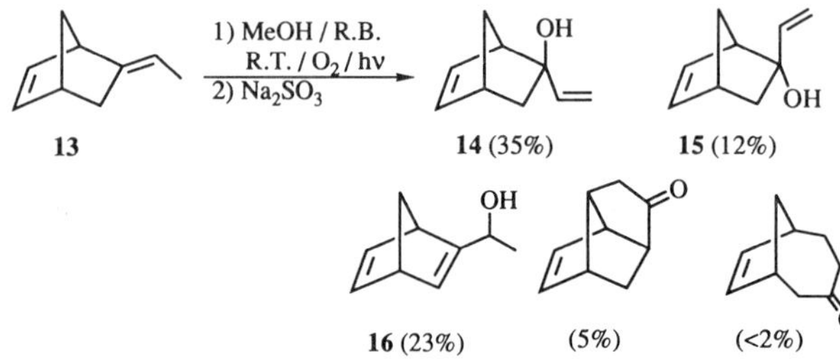

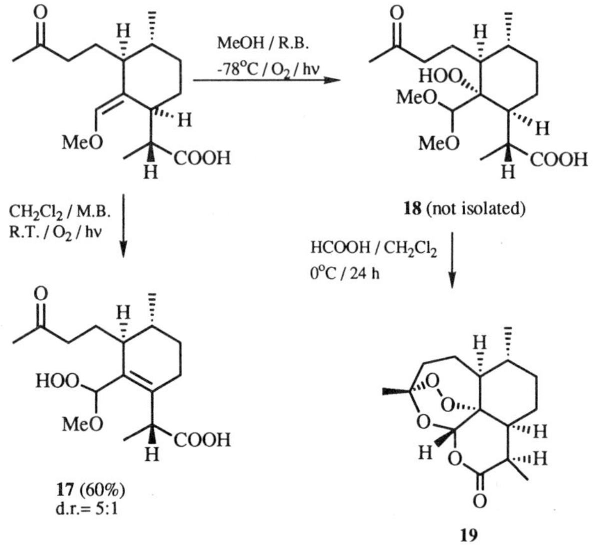

With Endocyclic C-C Double Bond

When 2-substituted (methyl or trimethylsiloxy) norborn-2-enes and 1,7,7-trimethylnorborn-2-enes are reacted with $^1O_2$, ene products are formed efficiently.[39] In the case of the trimethylsilyloxy substituted compounds 20, migration of the trimethylsilyl group occurs, and O-silylated α-hydroperoxyketones 21 are isolated and can be easily reduced to O-silylated α-hydroxyketones 22. High *exo*-selectivity is observed for the 7,7-unsubstituted substrates, and *endo*-selectivity for the 7,7-dimethyl compounds. The steric demand seems to be higher for the silyl-ene reaction, where *endo/exo*-ratios ranging from 0:100 to 94:6 were observed. 2-Methylnorborn-2-ene shows a 98.5:1.5 *exo*-diastereoselectivity.[40] Anellation of cyclopropane rings to some cycloalkenes alters the ground-state geometry of the ring system and can lead to considerable diastereotopic differentiation when singlet oxygen attacks the C-C double bond in an ene reaction. Several model systems were tested by Paquette and co-workers.[41] The photooxygenation of cholesterol (23) and cholest-5-en-3-one (24) has been investigated in detail by Schenck and Davies.[42,43] Whereas the reaction of $^1O_2$ with 23 results in the formation of the 5α-hydroperoxy compound 25 as the sole product, 24 gives a complex product mixture with a hemiperketal 26 arising from an 5α-attack as the principal product. *exo*-Methylene functionalities can easily be generated when the corresponding compounds with an *endo*-C-C double bond react with singlet oxygen.[44] This is an important application because the stereochemistry of the newly formed chiral center is perfectly controlled and the resulting products are synthetically valuable. Because bridgehead hydrogens cannot be abstracted, the reactions are also highly regioselective. Ene reactions with bicyclic systems are not limited to monoolefins; also, 1,3-dienes with suitable stereogenic centers can react in an ene fashion. This has been shown by Sasson and Labovitz[45] in their work on hexahydronaphthalene photooxygenation. None of the [4+2]-cycloadducts could be found in the reaction of 27 with $^1O_2$. However, the allylic hydroperoxide resulting from bridgehead-hydrogen abstraction is isolated as the major product.

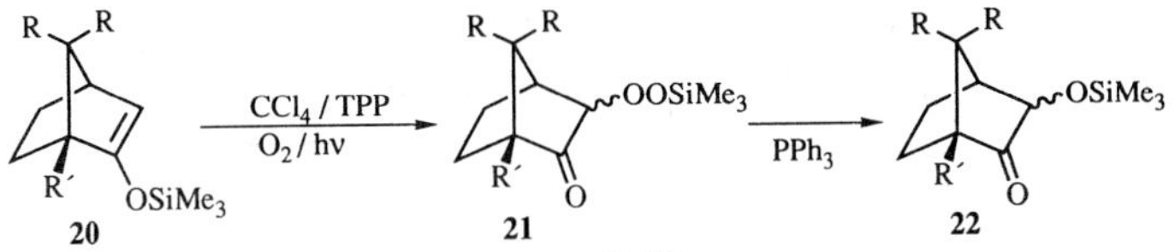

R=R′= H	exo/endo: 100 : 0
R=Me, R′=H	6 : 94
R=R′= Me	6 : 94

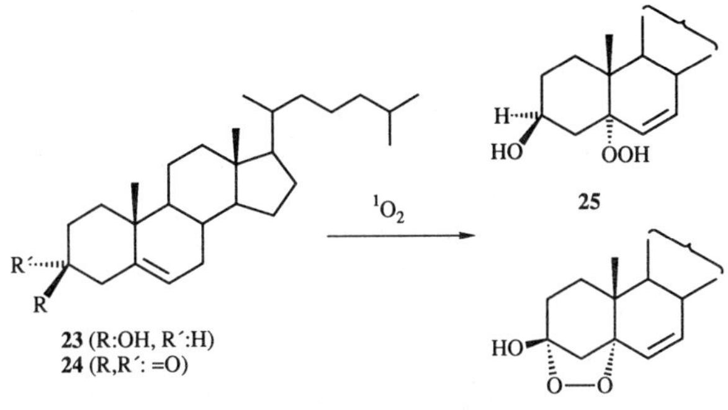

23 (R:OH, R′:H)
24 (R,R′: =O)

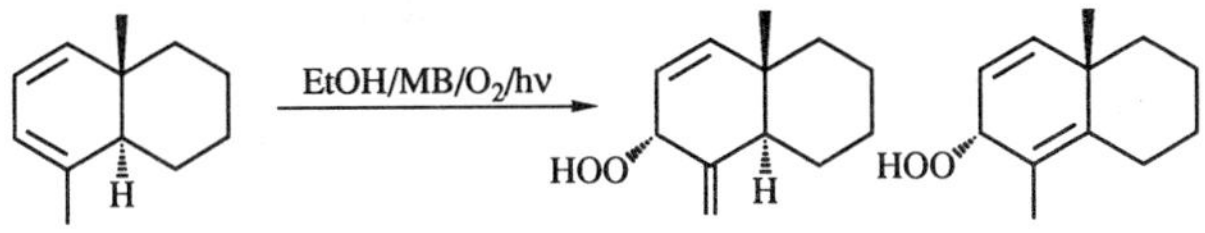

27   10-15%   85-90%

24.5 Further Transformations of Ene Products

Allylic hydroperoxides are easily reduced by a variety of reducing agents. Sodium sulfite or sodium iodide are most convenient for substrates that are not sensitive to water and further oxidation. For more sensitive hydroperoxides, dialkylsulfides or arylalkylsulfides in organic solvents can be used. Triphenylphosphine and triethyl- or triphenylphosphite have also been applied successfully. These reductions all preserve the given stereochemistry at the hydroperoxy-substituted carbon atom. The allylic alcohols produced in this way could be transformed to highly interesting compounds by a multitude of reactions. An efficient synthetic route to epoxy alcohols is the "one-pot" synthesis of epoxyalcohols via photooxygenation of olefins 28 in the presence of transition metal complexes derived from Ti, V, and Mo. This method has been described for a number of olefins.[46] The products are formed in good chemical yields and with high diastereoselectivity. In all cases, where acyclic substrates are used, *ul*-products 29 were formed preferentially.

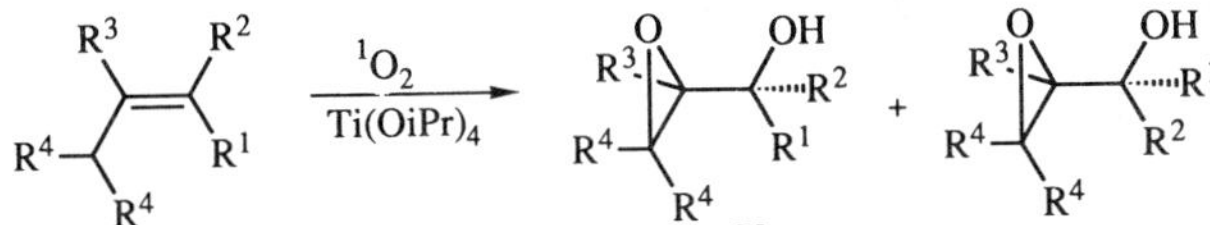

References

1. Kautsky, H., de Bruijn, H., Neuwirth, R., and Baumeister, W., Photo-sensibilisierte Oxydation als Wirkung eines aktiven, metastabilen Zustandes des Sauerstoff-Moleküls, *Chem. Ber.,* 66, 1588, 1933.

2. (a) Foote, C. S., Wexler, S., and Ando, W., Chemistry of singlet oxygen. III. Product selectivity, *Tetrahedron Lett.,* 4111, 1965; (b) Foote, C. S., Photosensitized oxygenations and the role of singlet oxygen, *Acc. Chem. Res.,* 1, 104, 1968.

3. Schenck, G. O., Gollnick, K., Buchwald, G., Schroeter, S., and Ohloff, G., Zur chemischen und sterischen Selektivität der photosensibilisierten O_2-Übertragung auf (+)-Limonen und (+)-Carvomenthen, *Justus Liebigs Ann. Chem.,* 674, 93, 1964.

4. (a) Wasserman, H. H. and Murray, R. W., Eds., *Singlet Oxygen,* Academic Press, New York, 1979; (b) *Oxygen and Oxy-Radicals in Chemistry and Biology,* Rodgers, M. A. J. and Powers, E. L., Eds., Academic Press, New York, 1981; (c) Frimer, A. A., Ed., *Singlet O_2,* CRC Press, Boca Raton, FL 1985, Vol. II and IV.

5. (a) Yamaguchi, K., Fueno, T., Saito, I., and Matsuura, T., On the mechanism of the ene reaction of electron-rich olefins with singlet oxygen. *Ab-initio* MO calculations, *Tetrahedron Lett.,* 21, 4087, 1980; (b) Yamaguchi, K., Fueno, T., Saito, I., Matsuura, T., and Houk, K. N., On the concerted mechanism of the ene reaction of singlet molecular oxygen with olefins. An *ab-initio* MO study, *Tetrahedron Lett.,* 22, 749, 1981.

6. Harding, L. B. and Goddard, W. A., III, The mechanism of the ene reaction of singlet oxygen with olefins, *J. Am. Chem. Soc.,* 102, 439, 1980.

7. Jefford, C. W., Kohmoto, S., Boukouvalas, J., and Burger, U., Reaction of singlet oxygen with enol ethers in the presence of acetaldehyde. Formation of 1,2,4-trioxanes, *J. Am. Chem. Soc.,* 105, 6498, 1983.

8. (a) Dewar, M. J. S. and Thiel, W., MINDO/3 study of reactions of singlet oxygen with carbon-carbon double bonds, *J. Am. Chem. Soc.,* 97, 3978, 1975; (b) Harding, L. B. and Goddard, W. A., III, The mechanism of the ene reaction of singlet oxygen with olefins, *J. Am. Chem. Soc.,* 102, 439, 1980; (c) Hotokka, B., Roos, P., and Siegbahn, CASSCF study of reaction of singlet molecular oxygen with ethylene. Reaction paths with C_{2v} und C_s symmetries, *J. Am. Chem. Soc.,* 105, 5263, 1983.

9. See, for example: Yamaguchi, K., Takada, K., Otsuji, Y., and Mizuno, K., Theoretical and general aspects of organic peroxides, in *Peroxides,* Ando, W., Ed., Wiley, Chichester, NY, 1992, 65.

10. Rosenthal, I., Chemical and physical sources of singlet oxygen, in *Singlet O_2,* Vol. I, Frimer, A. A., Ed., CRC Press, Boca Raton, FL, 1985, 13.

11. (a) Gollnick, K. and Griesbeck, A., Solvent dependence of singlet oxygen/substrate interactions in ene-reactions, [4+2]- and [2+2]-cycloaddition reactions, *Tetrahedron Lett.,* 25, 725, 1984; (b) Gollnick, K. and Griesbeck, A., Interactions of singlet oxygen with 2,5-dimethyl-2,4-hexadiene in polar and non-polar solvents. Evidence for a vinylog ene-reaction, *Tetrahedron,* 40, 3235, 1984.

12. Monroe, B. M., Singlet oxygen in solution: lifetimes and reaction rate constants, in *Singlet O_2,* Vol. I, Frimer, A. A., Ed., CRC Press, Boca Raton, FL, 1985, 177.

13. (a) Manring, L. E., Eriksen, J., and Foote, C. S., Electron-transfer photooxygenation. 4. Photooxygenation of *trans*-stilbene sensitized by methylene blue, *J. Am. Chem. Soc.,* 102, 4275, 1980; (b) Casarotto, M. G. and Smith, G. J., Methylene-blue-sensitized photooxidation of terpenes, *Photochem. Photobiol.,* A40, 87, 1987.

14. Rehm, D. and Weller, A., Kinetics of fluorescence quenching by electron and H-atom transfer, *Isr. J. Chem.,* 8, 259, 1970.

15. Pförtner, K. H., *Photochemistry in Organic Synthesis,* Coyle, J. D., Ed., Special Publication, The Royal Society of Chemistry, 57, 189, 1986.

16. Akasaka, T. and Ando, W., Stereospecific oxygenation of 3-adamantylidenetricyclo-[3.2.1.0^{2,4}]octane: Singlet oxygen vs. electron-transfer oxygenations, *J. Am. Chem. Soc.,* 109, 1260, 1987.

17. Calvert, J. G. and Pitts, J. N., *Photochemistry*, Wiley, New York, 1966.

18. Hurst, J. R., Wilson, S. L., and Schuster, G. B., The ene reaction of singlet oxygen: Kinetic and product evidence in support of a perepoxide intermediate, *Tetrahedron*, 41, 2191, 1985.

19. Schenck, G. O., Neumüller, O.-A., Ohloff, G., and Schroeter, S., Zur Autoxydation des (+)-Limones, *Justus Liebigs Ann. Chem.*, 687, 26, 1965.

20. Ouannes, C. and Wilson, T., Quenching of singlet oxygen by tertiary amines. Effect of DABCO, *J. Am. Chem. Soc.*, 90, 6527, 1968.

21. Paquette, L. A., Liotta, D. C., and Baker, A. D., Frontier molecular orbital basis for the structurally dependent regiospecific reactions of singlet oxygen with polyolefins, *Tetrahedron Lett.*, 2681, 1976.

22. Rousseau, G., LePerchec, P., and Conia, J. M., A novel synthesis of α,β-unsaturated aldehydes and esters by dye-photooxygenation of methyl enol ethers, *Synthesis*, 67, 1978.

23. Adam, W., Griesbeck, A. G., Gollnick, K., and Knutzen-Mies, K., 1,2-Dioxetanes derived from 4,5-dimethyl-2,3-dihydrofuran: Synthesis via photooxygenation, activation parameters, and excitation properties, *J. Org. Chem.*, 53, 1492, 1988.

24. Paquette, L. A., Liotta, D. C., and Baker, A. D., Frontier molecular orbital basis for the structurally dependent regiospecific reactions of singlet oxygen with polyolefins, *Tetrahedron Lett.*, 2681, 1976.

25. (a) Denny, R. W. and Nickon, A., Sensitized photooxygenation of olefins, *Org. Reactions*, 20, 133, 1973; (b) Gollnick, K., Ene-reaction with singlet oxygen, in *Singlet Oxygen*, Wasserman, H. H. and Murray, R. W., Eds., Academic Press, New York, chap. 8, 1979.

26. (a) Schulte-Elte, K. H., Muller, B. L., and Rautenstrauch, V., Preference for *syn* ene additions of $^1O_2$ to trisubstituted, acyclic olefins, *Helv. Chim. Acta*, 61, 2777, 1978; (b) Orfanopoulos, M., Grdina, M. B., and Stephenson, L. M., Site specificity in the singlet oxygen-trisubstituted olefin reaction, *J. Am. Chem. Soc.*, 101, 275, 1979; (c) Frimer, A. A. and Roth, D., Reaction of $^1O_2$ with strained olefins. 3. Photooxidation of vinylcyclopropanes, *J. Org. Chem.*, 44, 3882, 1979; (d) Jefford, C. W. and Rimbault, C. G., *syn* Regioselectivity of the hydroperoxidation of cyclo-alkenes with singlet oxygen, *Tetrahedron Lett.*, 22, 91, 1981; (e) Rautenstrauch, V., Thommen, W., and Schulte-Elte, K. H., Singlet oxygen ene reactions of (E)-4-propyl[1,1,1-$^2H_3$]oct-4-ene, *Helv. Chim. Acta*, 69, 1638, 1986.

27. (a) Rousseau, G., Le Perchec, P., and Conia, J. M., Stereochemical course in the addition of singlet oxygen to vinylcyclopropane derivatives, *Tetrahedron Lett.*, 2517, 1977; (b) Jefford, C. W., The role of Zwitterionic peroxides in controlling hydroperoxidation, *Tetrahedron Lett.*, 985, 1979.

28. (a) Ensley, H. E., Balakrishnan, P., and Ugarkar, B., Reaction of singlet oxygen with β-alkoxyenones, *Tetrahedron Lett.*, 24, 5189, 1983; (b) Orfanopoulos, M. and Foote, C. S., Regioselective reaction of singlet oxygen with α,β-unsaturated esters, *Tetrahedron Lett.*, 26, 5991, 1985; (c) Adam, W. and Griesbeck, A., Regioselective synthesis of 2-hydroperoxy-2-methylene-butanoic acid derivatives via photooxygenation of tiglic acid derivatives, *Synthesis*, 1050, 1986; (d) Akasaka, T., Misawa, Y., Goto, M., and Ando, W., Singlet oxygen and triazolinedione additions to α,β-unsaturated sulfoxides, *Tetrahedron*, 45, 6657, 1989; (e) Kwon, B. M., Kanner, R. C., and Foote, C. S., Reaction of singlet oxygen with 2-cyclopenten-1-ones, *Tetrahedron Lett.*, 30, 903, 1989.

29. (a) Clennan, E. L. and Chen, X., Reactions of an allylic sulfide, sulfoxide, and sulfone with singlet oxygen. The observation of a remarkable diastereoselective oxidation, *J. Am. Chem. Soc.*, 111, 5787, 1989; (b) Orfanopoulos, M., Stratakis, M., and Elemes, Y., Geminal selectivity of singlet oxygen ene reactions. The nonbonding large group effect, *J. Am. Chem. Soc.*, 112, 6417, 1990.

30. Fristad, W. E., Bailey, T. R., Paquette, L. A., Gleiter, R., and Böhm, M. C., Regiospecific photosensitized oxygenation of vinylsilanes. A method for converting saturated ketones to 1,2-transposed allylic alcohols. Possible role of silicon in directing the regioselectivity of epoxysilane cleavage reactions, *J. Am. Chem. Soc.*, 101, 4420, 1979.

31. (a) Stephenson, L. M., McClure, D. E., and Sysak, P. K., Stereochemistry of the singlet oxygen ene reaction with olefins, *J. Am. Chem. Soc.*, 95, 7888, 1973; (b) Orfanopoulos, M. and Stephenson, L. M., Stereochemistry of the singlet oxygen olefin-ene reaction, *J. Am. Chem. Soc.*, 102, 1417, 1980.

32. Adam, W., Catalani, L. H., and Griesbeck, A., Diastereoselective ene reaction in the photooxygenation of the silyl cyanohydrins of α,β-unsaturated aldehydes: Necessity for a common symmetrical intermediate of the perepoxide type, *J. Org. Chem.*, 51, 5494, 1986.

33. Shimizu, N., Shibata, F., Imazu, S., and Tsumo, Y., The ene reaction of allylsilanes with singlet oxygen. Unusual product stereoselectivity, *Chem. Lett.,* 1071, 1987.

34. Kropf, H. and Reichwaldt, R., Photo-oxygenation of phenyl-substituted propenes, but-2-enes, and pent-2-enes: Reactivity, regioselectivity, and stereoselectivity, *J. Chem. Res. (S),* 412, 1987.

35. Adam, W. and Nestler, B., Regio- and diastereoselective ene reactions of singlet oxygen with dialkyl-substituted acrylic esters, *J. Liebigs Ann. Chem.,* 1051, 1990

36. Adam, W. and Nestler, B., Photooxygenation of chiral allylic alcohols: Hydroxy-directed regio- and diastereoselective ene reaction of singlet oxygen, *J. Am. Chem. Soc.,* 114, 6549, 1992.

37. Adams, W. R. and Trecker, D. J., The dye-sensitized photooxygenation of 2-ethylidene bicyclo[2.2.1]hept-5-ene, *Tetrahedron,* 28, 2361, 1972.

38. Schmid, G. and Hofheinz, W., Total synthesis of qinghaosu, *J. Am. Chem. Soc.,* 105, 624, 1983.

39. (a) Jefford, C. W. and Rimbault, C. G., Reaction of singlet oxygen with norbornenyl ethers. Characterization of dioxetanes and evidence for Zwitterionic precursors, *J. Am. Chem. Soc.,* 100, 6437, 1978; (b) Jefford, C. W. and Boschung, A. F., Reaction of singlet oxygen with 2-methylnorborn-2-ene, 2-methylidenenorbornane, and their 7,7-dimethyl derivatives. The transition state geometry for hydroperoxidation, *Helv. Chim. Acta,* 57, 2242, 1974.

40. Jefford, C. W., Laffer, M. H., and Boschung, A. F., *exo-endo* Steric impediment in norbornene. Specification of the transition state for the reaction of singlet oxygen with 2-methylnorborn-2-ene and 2-methylenenorbornane, *J. Am. Chem. Soc.,* 94, 8904, 1972.

41. (a) Paquette, L. A. and Kretschmer, G., Stereoreversed electrophilic additions to 3-norcarenes. Insight into the relative steric demands of singlet oxygen in the ene reaction, *J. Am. Chem. Soc.,* 101, 4655, 1979; (b) Paquette, L. A. and Detty, M. R., Synthesis and protonation studies of *syn-* and *anti-*2,4-bishomotropone. Comparison with the behavior of epimeric 2,4-*bis*homocycloheptatrienols under long- and short-lived ionization conditions, *J. Am. Chem. Soc.,* 100, 5856, 1978.

42. Schenck, G. O., Neumüller, O.-A., and Eisfeld, W., Δ^5-Steroid-7α-hydroperoxide und -7-ketone durch Allylumlagerung von Δ^6-Steroid-5α-hydroperoxiden, *Liebigs Ann. Chem.,* 618, 202, 1958.

43. Dang, H.-S., Davies, A. G., and Schiesser, C. H., Allylic hydroperoxides formed by singlet oxygenation of cholest-5-en-3-one, *J. Chem. Soc., Perkin Trans. I,* 789, 1990.

44. (a) Piozzi, F., Nenturella, P., Bellino, A., and Marino, M. L., Partial synthesis of ent-Kaur-16-ene-15β,18-diol and ent-Kaur-16-ene-7α,15β,18-triol, *J. Chem. Soc., Perkin Trans. I,* 1164. 1973; (b) Büchi, G., Hauser, A., and Limacher, J., The synthesis of khusimone, *J. Org. Chem.,* 42, 3323, 1977.

45. Sasson, I. and Labovitz, J., Synthesis, photooxygenation, and Diels-Alder reactions of 1-methyl-4a,5,6,7,8,8a-*trans*-hexahydronaphthalene and 1,4a-dimethyl-4a,5,6,7,8,8a-*trans*-hexahydronaphthalene, *J. Org. Chem.,* 40, 3670, 1975.

46. (a) Adam, W., Griesbeck, A., and Staab, E., Ein einfacher Zugang zu 2-Epoxyalkoholen: Titan(IV)-katalysierter Sauerstofftransfer von Allylhydroperoxiden, *Angew. Chem.,* 98, 279, 1986; (b) Adam, W., Griesbeck, A., and Staab, E., A convenient "one-pot" synthesis of epoxy alcohols via photooxygenation of olefins in the presence of titanium(IV) catalyst, *Tetrahedron Lett.,* 27, 2839, 1986; (c) Adam, W., Braun, M., Griesbeck, A. G., Lucchini, V., Staab, E., and Will, B., Photooxygenation of olefins in the presence of titanium(IV) catalyst: A convenient "one-pot" synthesis of epoxy alcohols, *J. Am. Chem. Soc.,* 111, 203, 1989.

25

Photooxygenation of 1,3-Dienes

ldemar Adam
versity of Wurzburg

el G. Griesbeck
iversity of Cologne

5.1 Introduction and Definition of the Reaction

Since the early period of steroid chemistry, in which the first photooxygenation of a 1,3-diene was reported by Windaus and Brunken[1] using an ergosteryl derivative as substrate, hundreds of examples of this reaction type have been found. The reaction is quite similar to the (thermal) reaction of 1,3-dienes with electron-deficient dienophiles and Diels-Alder reaction mode ([4+2]-cycloaddition) is only one of several possibilities when these substrates are subjected to photooxygenation conditions (light, oxygen, a dyestuff, and the substrate). In contrast to ground-state reactions, the "ene" reaction, [2+2]-cycloaddition, and physical quenching of the electronically excited species (the dyestuff or singlet oxygen) can compete. The electronically excited dyestuff can interact with the diene, producing an electronically excited diene (energy transfer), a diene radical cation (electron transfer), or a pentadienyl radical (hydrogen abstraction) that is subsequently trapped by triplet oxygen. Another possibility is energy transfer from the excited dyestuff (the sensitizer) to triplet oxygen with formation of singlet oxygen. The subsequent reaction of singlet oxygen with an acceptor molecule is called a Type II process, whereas oxygenation reactions via an activated acceptor molecule are called Type I processes. In addition to this historically valuable nomenclature,[2] electron transfer processes involving alkene radical cations or superoxide radical anions have been designated as Type III photooxygenation reactions.[3] Most reactions discussed in this chapter are Type II reactions of singlet oxygen in its first electronically excited singlet state($^1\Delta_g$). There are numerous alternatives for performing photooxygenation reactions in respect to solvent, sensitizer, and reaction temperature. Some of the most useful sensitizers for singlet oxygen generation are Rose Bengal (RB), polymer-bound RB, Methylene Blue (MB), and eosine for aqueous or alcoholic solvents, and haematoporphyrin (HP) or tetraphenylporphin (TPP) for nonpolar organic solvents. The triplet energy of these sensitizers is low enough to guarantee selective energy transfer to oxygen; however, in some cases, the use of MB can be critical because of electron transfer side-reactions.

0-8493-8634-9/95/$0.00+$.50
© 1995 by CRC Press, Inc.

Sodium lamps are most convenient as the light source, but "traditional" light sources such as high-pressure mercury lamps could be applied if appropriate filters are used to cut off the short-wavelength region (λ < ca. 400 nm). Radical scavengers such as 2,5-di-*t*-butylphenol can be added to suppress radical reactions. On the other hand, typical singlet oxygen quenchers such as DABCO (diazobicyclo[2.2.2]octane) are useful as indicators for the nature of the reactive electronically excited species. We will concentrate here on the Diels-Alder-type additions of 1,3-dienes because the ene reaction and [2+2]-cycloadditions are being covered in other chapters of this book. Important reviews that cover recent developments in a greater detail are helpful for the reader as background material.[4]

25.2 Mechanism of the Singlet Oxygen Diene Photooxygenation

Theoretical Models

Several possibilities exist for the primary interaction of singlet oxygen with 1,3-dienes. As with the alkenes, the reversible formation of an exciplex may precede product formation. The exciplex can collapse to a perepoxide, a 1,4-biradical, or a 1,4-Zwitterion. Alternatively, the formation of the [4+2]-cycloadduct can take place in a concerted fashion without involvement of an intermediate. Should intermediates such as those mentioned above participate, other reaction pathways such as hydrogen transfer to give allylic hydroperoxides (ene reaction) or cycloaddition to give 1,2-dioxetanes ([2+2]-cycloaddition) may take place. MINDO/3 calculations suggested that a stable perepoxide intermediate for the butadiene/singlet oxygen reaction is feasible.[5] Orbital correlations characterize the [4+2]-cycloaddition reaction of singlet oxygen with 1,3-dienes as a symmetry-allowed, nonradical, concerted reaction.[6] The role of charge-transfer (CT) interactions in these reactions and a detailed discussion of various aspects of theoretical investigations is given in a review by Yamaguchi.[7]

Experimental Results

The stereospecificity of the additional singlet oxygen to a 1,3-diene was investigated originally by Rigaudy[8] by using the *s-cis*-1,3-diene 1. In agreement with the high degree of stereoselectivity observed in this case, Gollnick and Griesbeck[9] found that *trans,trans*-2,4-hexadiene 4 reacts with singlet oxygen chemo- and stereoselectivity to the *cis*-disubstituted endoperoxide 5. The stereoisomeric *cis,trans*-2,4-hexadiene 6, however, gave a mixture of the expected *trans*-endoperoxide 7 and the *cis*-isomer 5. This result was confirmed by NMR studies that clearly showed that singlet oxygen induces a *cis,trans*-isomerization of the starting 1,3-diene.[10] Such isomerization was also observed for reaction with the *cis,cis*-2,4-hexadiene isomer. A set of two *s-cis*-fixed 1,3-dienes has been investigated by Clennan.[11] These compounds, the *trans,trans*- and *cis,trans*-dioxolanes 8 and 9 react with a high degree of stereoselectivity to 10 and 11, respectively, besides giving appreciable amounts of dioxetane *(vide supra)*. Therefore, it can be deduced that the interaction of singlet oxygen with 1,3-dienes with conformationally flexible 2,3 single bonds leads to the intermediacy of 1,4-biradicals or 1,4-Zwitterions that can undergo retrocleavage to triplet oxygen (physical quenching) with concomitant *cis,trans* isomerization, whereas *s-cis*-fixed 1,3-dienes give endoperoxides in a highly stereoselective fashion. A detailed investigation of the kinetics of a series of Diels-Alder reactions between singlet oxygen and symmetrically and unsymmetrically substituted furans permitted the conclusion that these reactions proceed concertedly but asynchronously (at least for unsymmetrically substituted substrates) with the involvement of an exciplex as the primary reaction intermediate.[12] This assumption of the formation of an exciplex is in agreement with the near-zero activation energy for the singlet oxygen [4+2]-cycloadditions[13] and with the solvent insensitivity of reaction rates.[14]

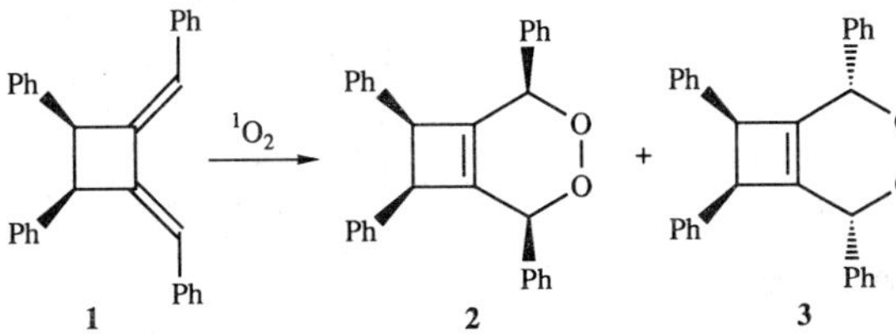

25.3 Substrates for Diene-Photooxygenation

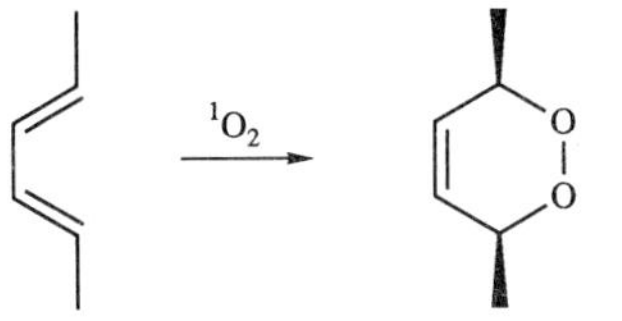

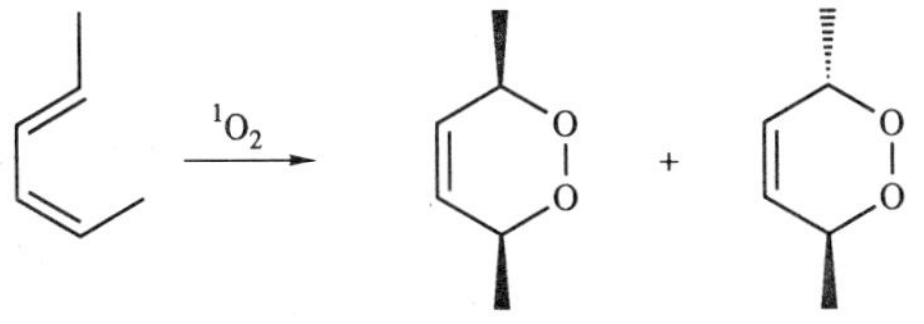

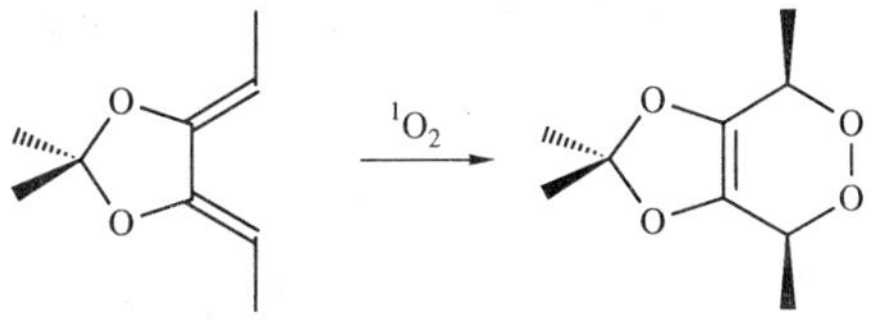

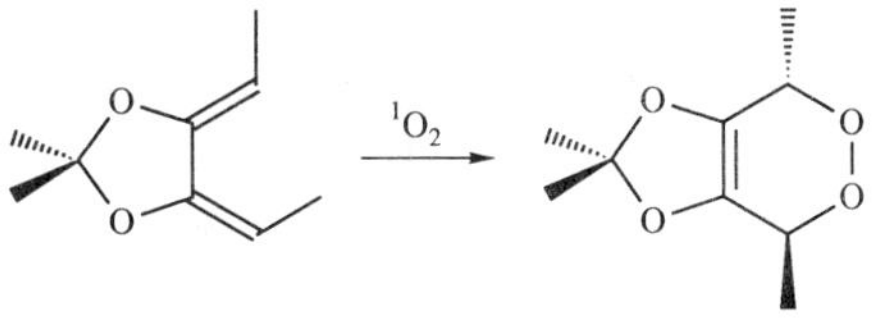

Carbocyclic Substrates

1. Unsubstituted Compounds

One of the most reactive substrates for endoperoxide formation is 1,3-cyclopentadiene. Rate constants of 1×10^8 (in chloroform)[15] and 3.9×10^7 (in toluene)[13] M^{-1} s^{-1} have been determined. The kinetics and the solvent dependence of the singlet oxygen [4+2]-cycloaddition with 1,3-cyclohexadiene have been investigated in detail.[14,15] There is about a tenfold decrease in the rate constant when the ring size is increased by one methylene unit (Table 25.1). Except for the endoperoxide from 1,3-cyclopentadiene,[16] the higher homologues are quite stable solid compounds that have been isolated in moderate to good yields under various photooxygenation conditions.[17–19] Cyclic 1,3,5-trienes, which can undergo valence tautomerization, give both types of cycloaddition

Table 25.1 Photooxygenation of Unsubstituted 1,3-Cyclodienes

n	$k_r \times 10^6$ (M^{-1}s^{-1})	Yield (%)
1	100,[a] 32[b]	86[c]
2	7.1[a]	21[d]
3	1.1[a]	29[e]
4	0.065[a]	53[f]

[a] In CHCl$_3$.[15]
[b] In toluene.[13]
[c] MeOH, RB, −100 °C.[16]
[d] *i*-PrOH, MB.[17]
[e] EtOH, eosin.[18]
[f] CH$_2$Cl$_2$, MB.[19]

Table 25.2 Photooxygenation of Unsubstituted 1,3,5-Cyclotrienes

X		Yields (%)	
CH$_2$	40[a]		3
(CH$_2$)$_2$	20[b]		5
CH=CH	26[c]		—

[a] CCl$_4$, TPP.[20]
[b] CH$_2$Cl$_2$, TPP.[21]
[c] Acetone, TPP.[22]

products (Table 25.2). For 1,3,5-cycloheptatriene[20] and 1,3,5-cyclooctatriene,[21] the respective bicyclic and tricyclic endoperoxides were obtained in ratios of 93:7 and 80:20, respectively. Cyclooctatetraene did not give photoadducts with its bicyclic valence tautomer, but 26% of the bicyclic [4+2]-adduct was isolated.[22] The valence tautomer of cyclooctatetraene, however, could be prepared independently and afforded the expected tetracyclic endoperoxide with singlet oxygen.[23]

Substituted Compounds

Electron-donating substituents activate the 1,3-cyclodiene toward the electrophilic singlet oxygen, which clearly demonstrates that the [4+2]-cycloaddition is controlled by HOMO (diene)-LUMO (singlet oxygen) interactions. For 1,3-cyclopentadiene derivatives, alkylation or arylation at positions 1 to 4 does not change the chemoselectivity, i.e., only bicyclic endoperoxides are formed.[24] This is not the case for the higher homologues: even α-terpinene 12, a classical singlet oxygen acceptor[25] and similarly reactive as 1,3-cyclopentadiene,[15] leads to 90% of the [4+2]-cycloadduct ascaridole 13 and about 5% of ene products 14–16 (besides 5% *p*-cumene).[26]

The result of the photooxygenation of the optically active terpene R-(−)-α-phellandrene 17 is most impressive.[27] In this case, a 3:2 face-selectivity (*cis* predominating) was observed for all products that are composed of 65% of the [4+2]-cycloadduct 18 and 30% of ene products 19–21.[28] The methyl group at position-2 is only marginally involved (*cf.* ene product 20) in the reaction, but does activate the trisubstituted double bond toward ene reaction (*cf.* ene product 19). Alkoxy substituents have been reported to exercise the same control.[29] An extreme example is the photooxygenation of a hexahydronaphthalene 22, where only the ene products 23 and 24 were isolated.[30] The latter case constitutes a combination of steric shielding and substituent activation that leads to high loco- and stereoselectivity.

Numerous examples are known for terpene and steroid substrates with 1,3-cyclohexadiene subunits where endoperoxides were isolated in good to excellent yields, e.g., dehydro-β-ionone 25 gives the endoperoxide 26 in 50% by using Rose Bengal in methanol.[31] Substituted 1,3,5-cycloheptatrienes and 1,3,5-cyclooctatrienes gave mixtures of bicyclic and tricyclic endoperoxides in ratios that are highly dependent on the electronic nature of the substituent.[21,22,32] Indene[33] and 1,2-dihydronaphthalene[34] derivatives were shown to accept two equivalents of singlet oxygen to give rise to *bis*-endoperoxides or their corresponding rearrangement products.[35] The photooxygenation of 6-methyl-5-phenylindene 27 in Freon 11/TPP gave the *bis*-endoperoxide 28 in 82% yield.[33]

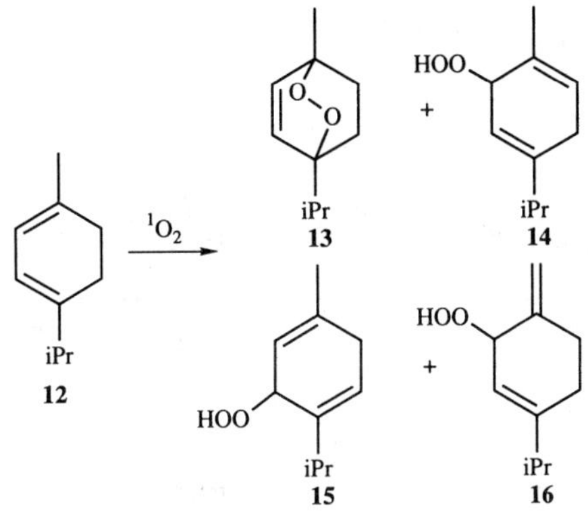

13 / 14 / 15 / 16 = 90 / 2.2 / 2.2 / 0.5

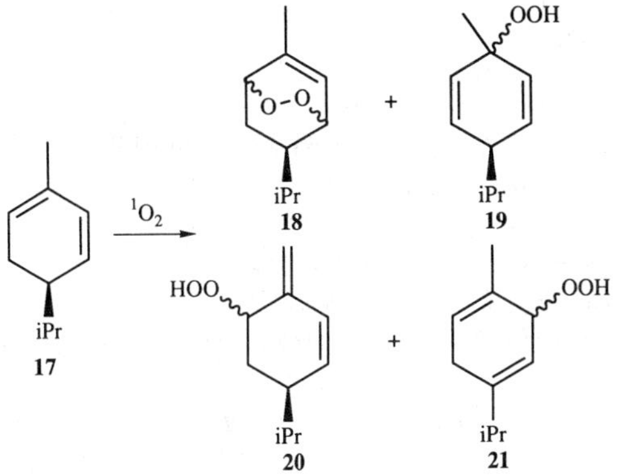

18 / 19 / 20 / 21 = 65 / 23 / 3.0 / 5.8

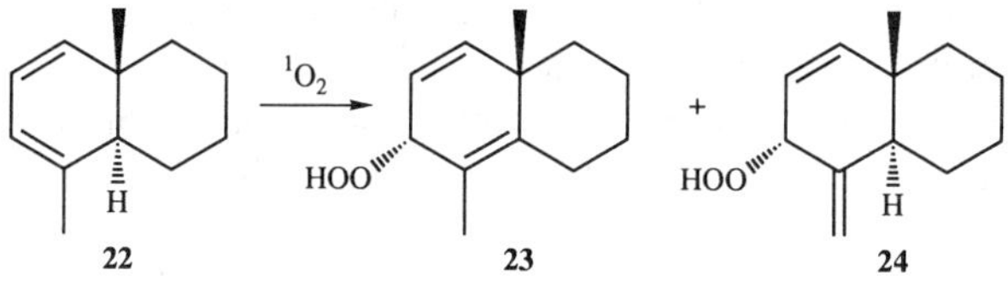

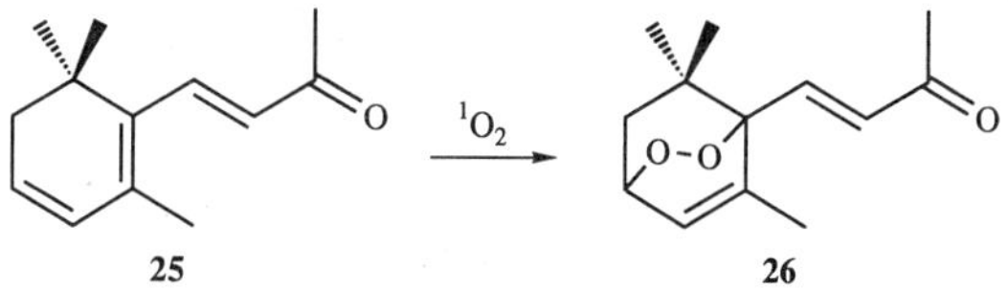

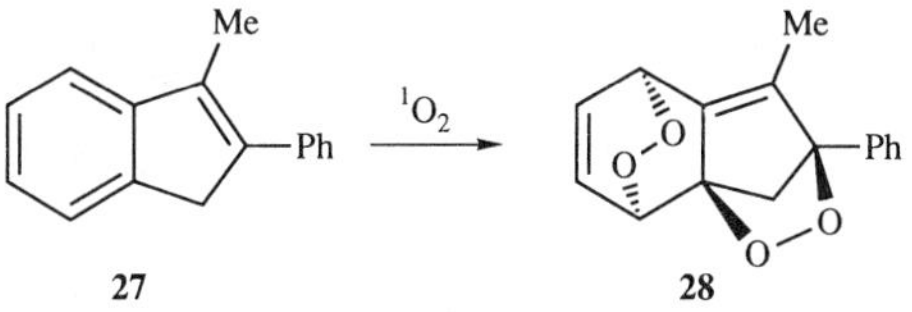

Table 25.3   Photooxygenation of Furans

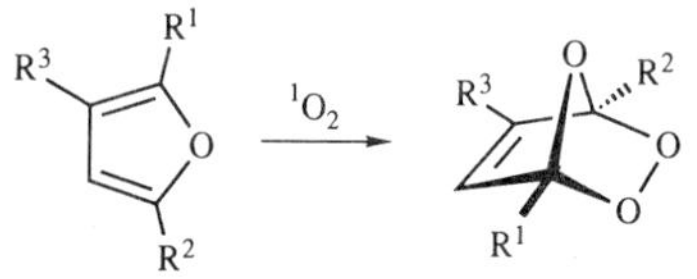

R^1	R^2	R^3	Decomp. Temp. (°C)
H	H	H	−15[a]
Me	Me	H	+20[b]
Me	Me	COOEt	+78[c]

[a] From Reference 39.
[b] From Reference 40.
[c] From Reference 41.

Heterocyclic Substrates

The photooxygenation of furans was investigated initially by Schenck in the mid-1940s.[36] A crystalline and *highly explosive* compound, 29, was isolated from furan itself by Schenck and Koch.[37] The rate constants of singlet oxygen reactions have been determined for a series of substituted furans by Koch,[38] Clennan,[12] and Gollnick.[39] Only a limited solvent dependence was observed, which indicates a concerted Diels-Alder process. More stable endoperoxides could be obtained from alkyl-substituted furans (e.g., 2,5-dimethylfuran)[40] or from furans with electron-withdrawing substituents.[41] Table 25.3 gives some representative examples of furan endoperoxides, together with their thermal stability. These furan endoperoxides constitute exceedingly valuable substrates for the synthesis of several derivatives (e.g., *bis*-epoxides, epoxy lactones, ene diones, *cis*-diacyloxiranes, enol esters, butenolides, and many more) as exemplified for the parent endoperoxide 29. Furthermore, the primary [4+2]-cycloadducts can rearrange into 1,2-dioxetanes[42] or 1,2-dioxolenes.[43] Unlike furan, thiophene was found to be unreactive toward photooxygenation. The alkylated derivative, 2,5-dimethylthiophene 30, however, is reactive[44] and the structure of the endoperoxide 31 could be proven unambiguously by low-temperature NMR studies[45] as well as by diimide reduction to the saturated endoperoxide 32.[46] Many more heterocyclic 1,3-dienes such as pyrroles,[47] thiazoles,[48] oxazoles,[49] etc. give endoperoxides on photooxygenation, most of them were not isolated or detected, but postulated as plausible reaction intermediates.

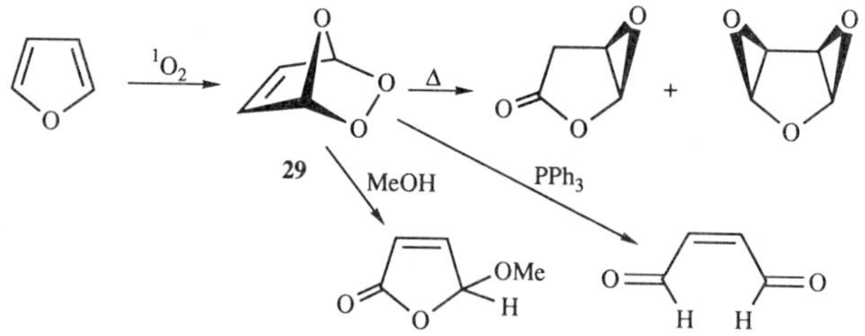

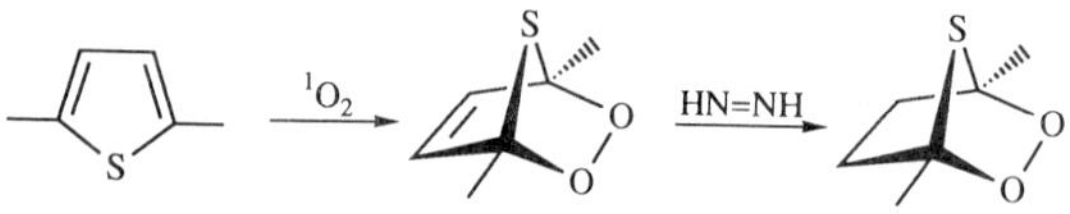

Table 25.4   Photooxygenation of 1-Vinylcycloalkenes

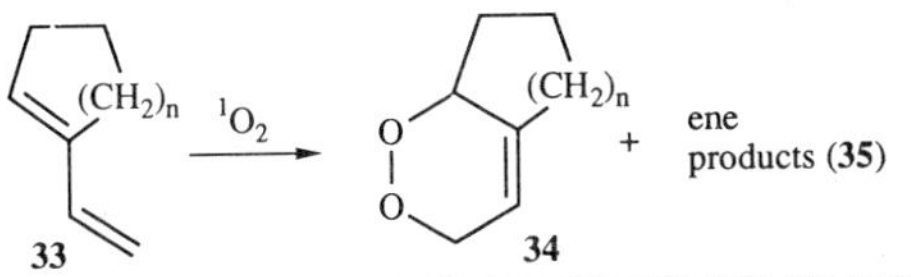

n	34:35 ([4+2] vs. ene reaction)
1	16:84
2	77:23
3	22:78
4	50:50
5	47:53
6	67:33

In 5% MeOH/CH$_2$Cl$_2$ mixture, RB as sensitizer. From Reference 51.

Acyclic/Cyclic Substrates

Chemoselectivity

As already mentioned in the Introduction, two main pathways, ene reactions and [2+2]-cycloadditions, can compete with the singlet oxygen [4+2]-cycloaddition. The formation of carbonyl products during photooxygenation of dienes is due probably to dioxetane decomposition in many cases. Another possibility is Hock-type cleavage of allylic hydroperoxides formed by the ene reaction.[50] Predominantly three factors control the chemoselectivity of the photooxygenation reactions with substrates having one (or more) endocyclic C-C double bond(s) and one conjugated exocyclic C-C double bond and also with acyclic 1,3-dienes: (1) the amount of *s-cis*-conformer in the equilibrium necessary for a concerted [4+2]-addition; (2) the relative reactivity difference of the C-C double bonds, i.e., an alkoxy substituent activates an exocyclic double bond strongly for [2+2]-addition; and (3) the appropriate alignment of allylic hydrogens for the ene reaction. Even the presence of a little of the *s-cis* conformer, however, can give rise to an appreciable amount of endoperoxide due to the very low activation energies of this reaction mode.

Substituted Compounds

As model compounds, 1-vinylcycloalkenes 33 with ring sizes from 3 to 8 have been investigated.[51] Obviously, factor (3) controls the ratio of [4+2]- vs. ene product (34:35) in these cases (Table 25.4). 1-Vinylnaphthalenes undergo preferentially [4+2]-cycloaddition when alkyl-substituted at the alkene terminus;[52] e.g., the dimethyl substrate 36 gives 37 in 78% yield.[53] In contrast to these results, the substrate 38 is prone to ene reaction because of steric (*peri*) interactions that depopulate the *s-cis*-conformer. The reaction of this compound is controlled by factor (1) and the allylic hydroperoxide 39 is formed in 90% yield.[52] An example in which the relative reactivity of the C-C double bond (factor 2) depends on the *s-cis* conformer population (factor 1) is given in the photooxygenation of the *E*- and *Z*-2′-methoxy-1 vinylnaphthalenes 40 and 41. In contrast to the latter compound, in which the [4+2]- and [2+2]-cycloaddition products 42 and 43 were formed in nearly equal amounts, the *E*-diastereoisomer gives nearly quantitative yields of endoperoxide 44.[54] There are many more examples of similar substrates: for example, substituted styrenes,[55] stilbenes,[56] and 1-vinylthiophenes.[57] In some cases, *bis*-endoperoxides were isolated similar to the indene and dihydronaphthalene series (*vide infra*). Terpene derivatives are a treasure house of polyfunctional substrates for which all types of reaction modes can be observed.[58]

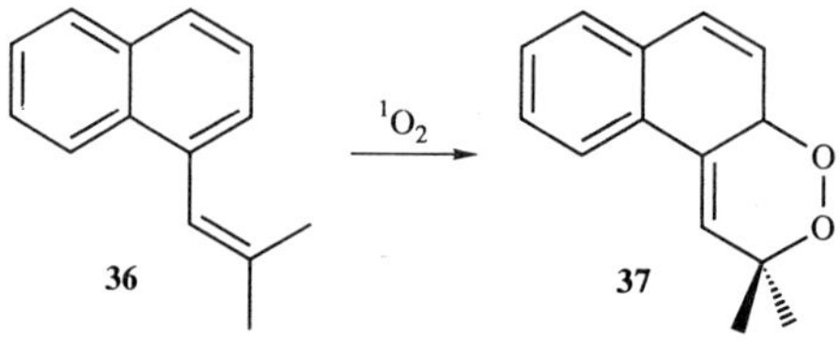

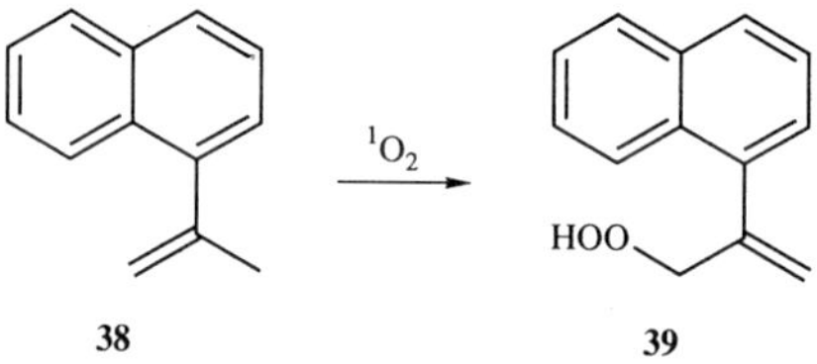

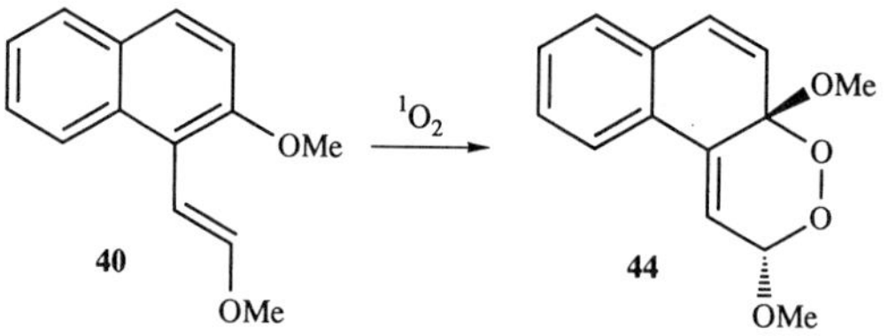

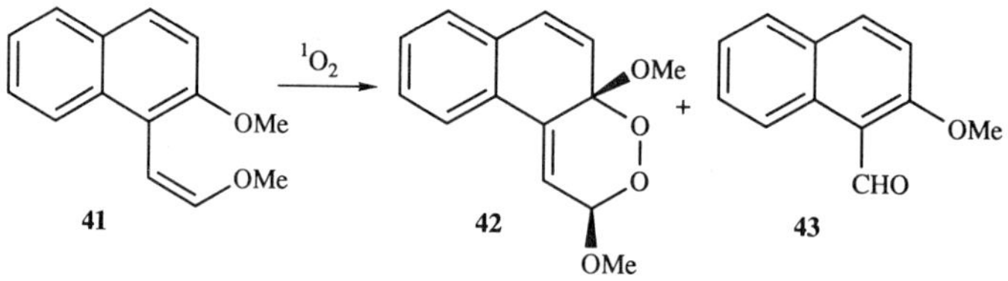

Acyclic Substrates

Chemo- and Regioselectivity

According to the rules in the section above on chemo- and regioselectivity, the chemo- and regioselectivity of singlet oxygen reactions with acyclic 1,3-dienes can be controlled. The parent compound, 1,3-butadiene, gives the endoperoxide in only 20% yield.[59] Electron-donating substituents activate 1,3-butadienes in their reaction with electrophilic singlet oxygen. Again, alkyl substituents with allylic hydrogens make ene reactions possible and alkoxy substituents activate the diene toward [2+2]-cycloadditions. In most cases, which is sometimes ignored, the main reaction pathway is physical quenching of singlet oxygen by the diene.[60]

Substituted Compounds

As a model substrate, 2,5-dimethyl-2,4-hexadiene 45 has been investigated intensively by several groups.[60-64] Depending on the solvent, five peroxidic products can be isolated in varying relative yields: the endoperoxide 46 ([4+2]-cycloaddition), the allylic hydroperoxide 47 (ene reaction), the 1,2-dioxetane 48 ([2+2]-cycloaddition), the diene hydroperoxide 49 (vinylogous ene reaction or radical-induced rearrangement of 47[63]), and the methoxy-substituted hydroperoxide 50 (methanol trapping product). Up to 23% of the [4+2]-cycloaddition product 46 could be detected in tetrachloromethane, whereas the dioxetane 48 dominates in polar solvents such as acetonitrile. A series of methyl- and phenyl-substituted 1,3-butadienes have been investigated by Matsumoto et al.[64] The proportion of [4+2]-cycloaddition to ene products is again determined by the factors mentioned above (Table 25.5). The terpenes α- and β-myrcene are classical examples for substrates

Table 25.5   Photooxygenation of Acyclic 1,3-Dienes

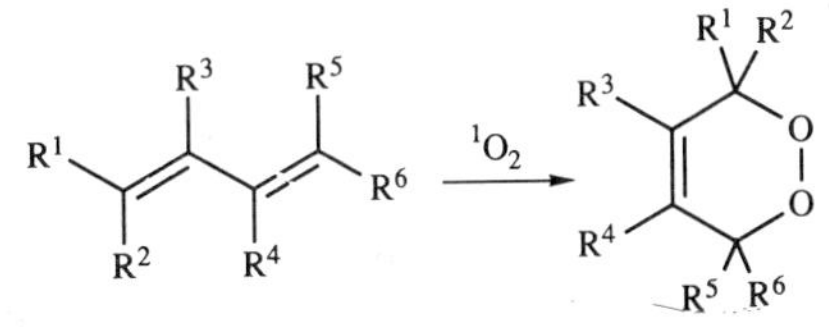

	Yield (%)
R^1–R^6 = H	20
R^1 = Me	31
R^3 = Me	50
$R^{1,5}$ = Me	56
$R^{1,4}$ = Me	58
$R^{3,4}$ = Me	10
$R^{1,2,4}$ = Me	41
$R^{1,5}$ = Ph	92

In CCl$_4$, TPP as sensitizer.

controlled by factor (2). In these polyenes, the reactivity of the nonconjugated C-C double bond controls the previous photochemical event. β-Myrcene **51**, which has a trisubstituted, nonconjugated double bond, undergoes ene reaction preferentially with subsequent [4+2]-cycloaddition,[65] whereas α-myrcene **52** reacts solely in a Diels-Alder fashion.[66] Alkoxy-substituted 1,3-dienes, which cannot undergo the ene reaction, have been shown to be excellent substrates for [2+2]-cycloaddition.[67,68] The 1,4-di-*t*-butoxy-1,3-butadienes **53** were the first substrates that showed nonstereoselective [2+2]-cycloaddition reactions. Only for the EZ-isomer **53** was formation of the endoperoxide observed, which indicates the pronounced activation effect of alkoxy groups for dioxetane formation.

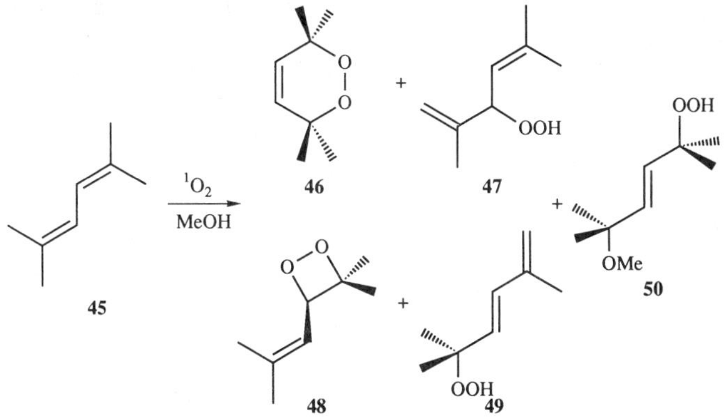

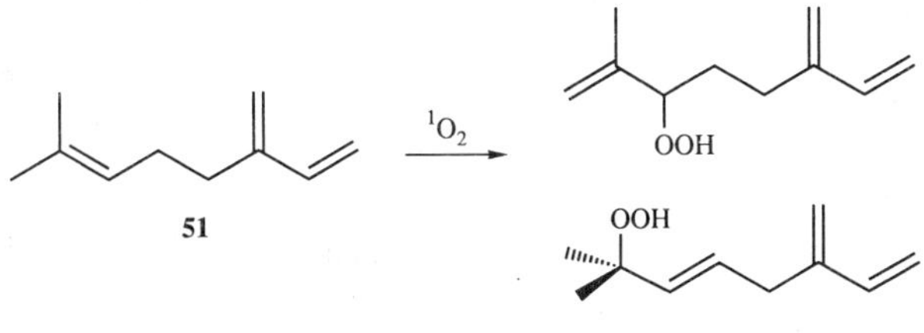

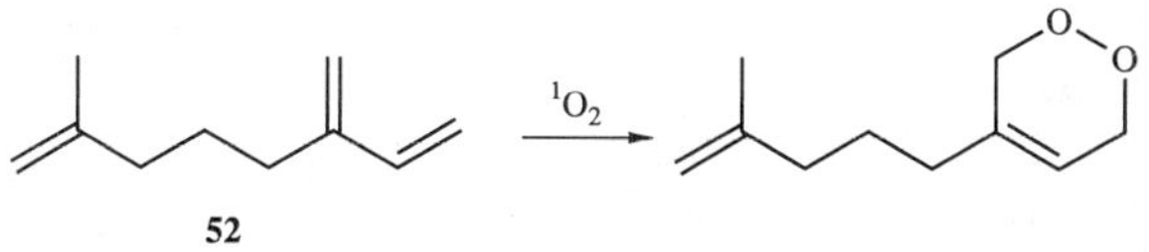

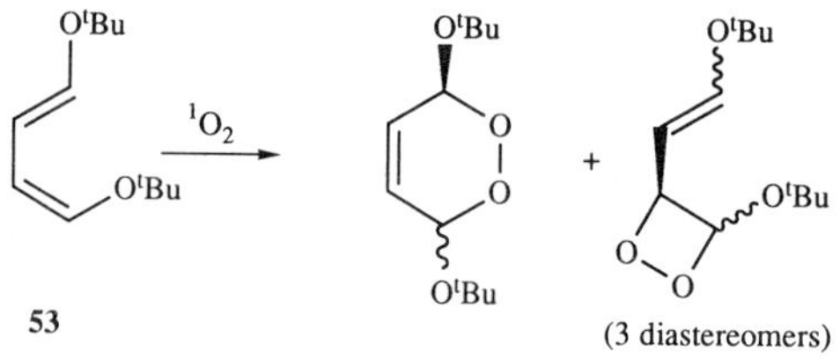

53

(3 diastereomers)

25.4 Applications of the Singlet Oxygen Diene Photooxygenation

Primary Products

Some peroxide products obtained in the photocycloaddition of 1,3-dienes with singlet oxygen are of interest either as pharmaceuticals (e.g., the legendary ascaridole **13**) or as model compounds for biochemical studies (e.g., prostaglandin biosynthesis).[69] These model compounds have been prepared directly from fatty acid precursors[70] or from 1,3-cyclopentadienes followed by reduction of the remaining C-C double bond.[71] For most preparative purposes, however, the photooxygenation reactions have been designed in such a way that either *in situ* or subsequent treatment of the peroxides formed initially leads to synthetically valuable building blocks for organic synthesis.

Secondary Products

Due to their special properties, endoperoxides (and even more so, 1,2-dioxetanes) are versatile starting materials for further transformations. The weak O-O single bond makes peroxidic substances sensitive to homolytic cleavage, but this feature makes most of them *hazardous!* Nucleophilic ring opening is another possibility for selective transformations that include reductions (to diols), oxidations (to dicarbonyl products), and rearrangements (to hydroxy carbonyl compounds, epoxy carbonyl compounds, *bis*-epoxides, ene diones, etc.). An impressive number of examples that are recommended for application in synthesis have been collected in several reviews.[4b,47,72–75]

References

1. Windaus, A. and Brunken, J., Über die photochemische oxydation des ergosterins, *Liebigs Ann. Chem.*, 460, 225, 1928.
2. Gollnick, K., Type II photooxygenation reactions in solution, *Adv. Photochem.*, 6, 1, 1968.
3. See, however: Foote, C. S., Definition of Type I and Type II photosensitized oxidation, *Photochem. Photobiol.*, 54, 659, 1991.
4. (a) Clennan, E. L., Synthetic and mechanistic aspects of 1,3-diene photooxidation, *Tetrahedron*, 47, 1343, 1991; (b) Bloodworth, A. J. and Eggelte, H. J., Endoperoxides, in *Singlet O₂*, Frimer, A. A., Ed., CRC Press, Boca Raton, FL, Vol. II, 1985, 93; (c) Gollnick, K. and Schenck, G. O., Oxygen as a dienophile, in *1,4-Cycloaddition Reactions,* Hamer, J., Ed., Academic Press, Orlando, FL, 1967, 255; (d) Frimer, A. A., The reaction of singlet oxygen with olefins: The question of mechanism, *Chem. Rev.*, 79, 359, 1979; (e) Kearns, D. R., Physical and chemical properties of singlet molecular oxygen, *Chem. Rev.*, 71, 395, 1971.
5. Dewar, M. J. S. and Thiel, W., MINDO/3 study on the addition of singlet oxygen ($^1\Delta_g$) to 1,3-butadiene, *J. Am. Chem. Soc.*, 99, 2338, 1977.
6. Yamaguchi, K., Yabushita, S., Fueno, T., and Houk, K. N., On the mechanism of photooxygenation reactions. Computational evidence against the diradical mechanism of singlet oxygen ene reactions, *J. Am. Chem. Soc.*, 103, 5043, 1981.
7. Yamaguchi, K., Theoretical calculations of singlet oxygen reactions, in *Singlet O₂*, Frimer, A. A., Ed., CRC Press, Boca Raton, FL, Vol. III, 1985, 119.

8. Rigaudy, J., Capdevielle, P., Cambrisson, S., and Maumy, M., Ouverture concertee des adduits du *bis*-benzylidene-1,2 *(E,E)*-diphenyl-3,4 cyclobutane *cis*. Rectification des structures des produits derives, *Tetrahedron Lett.*, 2757, 1974.

9. Gollnick, K. and Griesbeck, A., [4+2]-Cycloaddition of singlet oxygen to conjugated acyclic hexadienes: Evidence of singlet oxygen induced *cis-trans*-isomerization, *Tetrahedron Lett.*, 24, 3303, 1983.

10. O'Shea, K. E. and Foote, C. S., Chemistry of singlet oxygen. 51. Zwitterionic intermediates from 2,4-hexadienes, *J. Am. Chem. Soc.*, 110, 7167, 1988.

11. Clennan, E. L. and Nagraba, K., Reactions of singlet oxygen with alkoxy-substituted dienes. Formation of dioxetanes in the singlet oxygenations of *s-cis* fixed dienes *(Z,Z)*- and *(E,Z)*-4,5-diethylidene-2,2-dimethyl-1,3-dioxolanes, *J. Org. Chem.*, 52, 294, 1987.

12. (a) Clennan, E. L. and Mehrsheikh-Mohammadi, M. E., Addition of singlet oxygen to conjugated dienes. The mechanism of endoperoxide formation, *J. Am. Chem. Soc.*, 105, 5932, 1983; (b) Clennan, E. L. and Mehrsheikh-Mohammadi, M. E., Mechanism of endoperoxide formation. 2. Possibility of exciplexes on the reaction coordinates, *J. Org. Chem.*, 49, 1321, 1984; (c) Clennan, E. L. and Mehrsheikh-Mohammadi, M. E., Mechanism of endoperoxide formation. 3. Utilization of the Young and Carlsson kinetic techniques, *J. Am. Chem. Soc.*, 106, 7112, 1984.

13. Gorman, A. A., Lovering, G., and Rodgers, M. A. J., The entropy-controlled reactivity of singlet oxygen ($^1\Delta_g$) towards furans and indols in toluene. A variable-temperature study by pulse radiolysis, *J. Am. Chem. Soc.*, 101, 3050, 1979.

14. Gollnick, K. and Griesbeck, A., Solvent dependence of singlet oxygen/substrate interactions in ene reactions, [4+2]- and [2+2]-cycloaddition reactions, *Tetrahedron Lett.*, 25, 725, 1984.

15. Monroe, B. M., Rate constants for the reaction of singlet oxygen with conjugated dienes, *J. Am. Chem. Soc.*, 103, 7253, 1981.

16. (a) Schenck, G. O. and Dunlap, D. E., Photosynthese von cyclopentadien-endoperoxyd bei –100 °C und hydrierung von endoperoxyden mit thioharnstoff-verwendung von Na-dampflampen in der präparativen photochemie, *Angew. Chem.*, 68, 248, 1956; (b) Schulte-Elte, K. H., Willhalm, B., and Ohloff, G., Eine neue peroxidumlagerung: *cis*-4,5-epoxy-2-pentenal aus 1,4-epidioxy-2-cyclopenten, *Angew. Chem.*, 81, 1045, 1969.

17. Schenck, G. O., Probleme präparativer photochemie, *Angew. Chem.*, 64, 12, 1952.

18. Cope, A. C., Liss, T. A., and Wood, G. A., Proximity effects. X. *cis*-1,4-Cycloheptanediol from solvolysis of cycloheptene oxide, *J. Am. Chem. Soc.*, 79, 6287, 1957.

19. Horinaka, A., Nakashima, R., Yoshikawa, M., and Matsuura, T., Photosensitized oxygenation of unconjugated cyclic dienes, *Bull. Chem. Soc. Jpn.*, 48, 2095, 1975.

20. Adam, W. and Balci, M., Photooxygenation of 1,3,5-Cycloheptatriene: Isolation and characterization of endoperoxides, *J. Am. Chem. Soc.*, 101, 7537, 1979.

21. (a) Adam, W. and Erden, I., Cyclic peroxides. 85. Singlet oxygenation of cycloocta-1,3,5-triene: Formation of [4.2.2] and [2.2.2] cycloadducts, *Tetrahedron Lett.*, 2781, 1979; (b) Adam, W., Gretzke, N., Hasemann, L., Klug, G., Peters, E.-M., Peters, K., von Schnering, H. G., and Will, B., Cycloaddition of singlet oxygen and 4-phenyl-4H-1,2,4-trizole-3,5-dione to 7-substituted 1,3,5-cyclooctatrienes, *Chem. Ber.*, 118, 3357, 1985.

22. Adam, W., Klug, G., Peters, E.-M., Peters K., and von Schnering, H. G., Synthesis of endoperoxides derived from cyclooctatetraenes via singlet oxygenation, *Tetrahedron*, 41, 2045, 1985.

23. Adam, W., Cueto, O., DeLucchi, O., Peters, K., Peters, E.-M., and von Schnering, H. G., Synthesis of the endoperoxide *anti*-7,7-dioxatricyclo[4.2.2.0^{2,5}]deca-3,9-diene via singlet oxygenation of the bicyclic valence tautomer of cyclooctatetraene and its transformations, *J. Am. Chem. Soc.*, 103, 5822, 1981.

24. Rio, G. and Charifi, M., Le diphényl-1,2 cyclopentadiène et ses dérivés substitués en 4 par un méthyle ou un carboxyle. Photoxydation, *Bull. Soc. Chim. Fr.*, 3585, 1970.

25. Schenck, G. O. and Ziegler, K., Die synthese des ascaridols, *Naturwisschaften*, 32, 157, 1944.

26. Matusch, R. and Schmidt, G., Konkurrenz von endoperoxid- und hydroperoxidbildung bei der reaktion von singulettsauerstoff mit cyclischen, konjugierten dienen, *Angew. Chem.*, 100, 729, 1988.

27. Schenck, G. O., Kinkel, K. G., and Mertens, H.-J., Photochemische reaktionen I. Über die photosynthese des askaridols und verwandter endoperoxyde, *Justus Liebigs Ann. Chem.*, 584, 125, 1953.

28. Matusch, R. and Schmidt, G., Konkurrenz von endoperoxid- und hydroperoxid-bildung bei der umsetzung von singulett-sauerstoff mit cyclischen, konjugierten dienen, *Helv. Chim. Acta*, 72, 51, 1989.

29. Clennan, E. L. and L'Esperance, R. P., The addition of singlet oxygen to alkoxy and trimethylsilyloxy butadienes. The synthesis of novel new peroxides, *Tetrahedron Lett.*, 24, 4291, 1983.

30. Sasson, I. and Labovitz, J., Synthesis, photooxygenation and Diels-Alder reactions of 1-methyl-4a,5,6,7,8,8a-*trans*-hexahydronaphthalene and 1,4a-dimethyl-4a,5,6,7,8,8a-*trans*-hexahydro-naphthalene, *J. Org. Chem.*, 40, 3670, 1975.

31. Mousseron-Canet, M., Mani, J. C., Dalle, J. P., and Olivé, J. L., Photoxydation sensibilisée de quelques composés apparentés à la déhydro β-ionone, synthèse de l'ester méthylique de la (±) abscisine, *Bull. Soc. Chim. Fr.*, 3874, 1966.

32. Adam, W., Adamsky, F., Klärner, F.-G., Peters, E.-M., Peters, K., Rebello, H., Rüngeler, W., and von Schnering, H. G., Stereochemistry and product distribution in the singlet oxygen cycloaddition with 7,7-disubstituted 1,3,5-cycloheptatrienes, *Chem. Ber.*, 116, 1848, 1983.

33. (a) Jefford, C. W., Hatsui, T., Deheza, M. F., and Bernardinelli, G., Photo-oxygenation of indene and 1,2-dihydronaphthalene: formation of 1,2-dioxetanes and 1,2,4-trioxanes, *Chimia*, 46, 114, 1992; (b) Boyd, J. D. and Foote, C. S., Chemistry of singlet oxygen. 32. Unusual products from low-temperature photooxygenation of indenes and *trans*-stilbene, *J. Am. Chem. Soc.*, 101, 6758, 1979.

34. Burns, P. A. and Foote, C. S., Chemistry of singlet oxygen, XXIV. Low temperature photooxygenation of 1,2-dihydronaphthalenes, *J. Org. Chem.*, 41, 908, 1976.

35. (a) Burns, P. A., Foote, C. S., and Mazur, S., Chemistry of singlet oxygen. XXIII. Low temperature photooxygenation of indenes in aprotic solvent, *J. Org. Chem.*, 41, 899, 1976; (b) Jiancheng, Z. and Foote, C. S., Photooxygenation of substituted indenes at low temperature, *Tetrahedron Lett.*, 27, 6153, 1986.

36. Schenck, G. O., Autoxydation von furan und anderen dienen (Die synthese des ascaridols), *Angew. Chem.*, 56, 101, 1944.

37. Koch, E. and Schenck, G. O., Zur photosensibilisierten O_2-Übertragung auf furan: Isolierung und eigenschaften des ozonidartigen furanperoxids bei −100 °C, *Chem. Ber.*, 99, 1984, 1966.

38. Koch, E., Zur photosensibilisierten sauerstoffübertragung. Untersuchung der terminationschritte durch belichtungen bei tiefen temperaturen, *Tetrahedron*, 24, 6295, 1968.

39. Gollnick, K. and Griesbeck, A., Singlet oxygen photooxygenation of furans: Isolation and reactions of [4+2]-cycloaddition products (unsaturated sec.-ozonides), *Tetrahedron*, 41, 2057, 1985.

40. Gollnick, K. and Griesbeck, A., [4+2]-Cycloaddition von singulett-sauerstoff an 2,5-dimethylfuran: Isolierung und reaktionen des monomeren und dimeren endoperoxids, *Angew. Chem.*, 95, 751, 1983.

41. Graziano, M. L., Iesce, M. R., and Scarpati, R., Photosensitized oxidations of furans. 4. Influence of the substituents on the behaviour of the endoperoxides of furans, *J. Chem. Soc., Perkin Trans. I*, 2007, 1982.

42. Adam, W., Ahrweiler, M., and Sauter, M., Photosensitized oxygenation of a keto furan: Isolation of the first furan dioxetane by rearrangement of its endoperoxide and selected chemical transfor-mations, *Angew. Chem.*, 105, 104, 1993.

43. Graziano, M. L., Iesce, M. R., Cimminiello, G., and Scarpati, R., Photosensitized oxidation of furans. 14. Nature of intermediates in thermal rearrangement of some *endo*-peroxides of 2-alkoxyfurans: New rearrangement pathway of furan *endo*-peroxides, *J. Chem. Soc., Perkin Trans. I*, 241, 1989.

44. Skold, C. N. and Schlessinger, R. H., The reaction of singlet oxygen with a simple thiophene, *Tetrahedron Lett.*, 791, 1970.

45. Gollnick, K. and Griesbeck, A., Thiaozonide formation by singlet oxygen cycloaddition to 2,5-dimethylthiophene, *Tetrahedron Lett.*, 25, 4921, 1984.

46. Adam, W. and Eggelte, H. J., 2,3-Dioxa-7-thiabicyclo[2.2.1]heptan: Ein neues heterobicyclisches system mit thiaozonid-struktur, *Angew. Chem.*, 90. 811, 1978.

47. Matsumoto, M., Synthesis with singlet oxygen, in *Singlet O_2*, Vol. II, Frimer, A. A., Ed., CRC Press, Boca Raton, FL, 1985, 250.

48. Ando, W. and Takata, T., Photooxidation of sulfur compounds, in *Singlet O_2*, Vol. III, Frimer, A. A., Ed., CRC Press, Boca Raton, FL, 1985, 94.

49. (a) Wasserman, H. H., McCarthy, K. E., and Prowse, K. S., Oxazoles in carboxylate protection and activation, *Chem. Rev.*, 86, 845, 1986; (b) Gollnick, K., and Koegler, S., [4+2]-Cycloaddition of singlet oxygen to oxazoles — formation of oxazole endoperoxide, *Tetrahedron Lett.*, 29, 1003, 1988.

50. Sheldon, R. A., Synthesis and uses of alkyl hydroperoxides and dialkyl peroxides, in *The Chemistry of Peroxides*, Patai, S., Ed., Wiley, Chichester, 1983, 162.

51. Herz, W. and Juo, R.-R., Photooxygenation of 1-vinylcycloalkenes. The competition between "ene" reaction and cycloaddition of singlet oxygen, *J. Org. Chem.*, 50, 618, 1985.

52. Matsumoto, M. and Kondo, K., The 1,4-cycloaddition of singlet oxygen to 1-vinylnaphthalenes, *Tetrahedron Lett.*, 3935, 1975.

53. Matsumoto, M. and Kuroda, K., Solvent effect in reaction of singlet oxygen with conjugated dienes, *Synth. Commun.*, 11, 987, 1981.

54. Matsumoto, M., Kuroda, K., and Suzuki, Y., The 1,4-addition of singlet oxygen to 2,6-dimethoxy-1-(2-methoxyethenyl)benzene and 2-methoxy-1-(2-methoxyethenyl)naphthalene. The 1,4-*endo*-peroxides as equivalents of 6-oxo-2,4-cyclohexadienylidenacetates, *Tetrahedron Lett.*, 22, 3253, 1981.

55. Matsumoto, M., Dobashi, S., and Kondo, K., Sensitized photooxygenation of β,β-dimethylstyrenes; Synthesis of (±)-crotepoxide, *Tetrahedron Lett.*, 3361, 1977.

56. Matsumoto, M., Dobashi, S., and Kondo, K., The 1,4-cycloaddition of singlet oxygen to stilbenes and β-methylstyrenes, *Tetrahedron Lett.*, 2329, 1977.

57. Matsumoto, M., Dobashi, S., and Kondo, K., The 1,4-cycloaddition of singlet oxygen to 2-vinylthiophenes, *Tetrahedron Lett.*, 4471, 1975.

58. Gollnick, K. and Kuhn, H., Ene reactions with singlet oxygen, in *Singlet Oxygen*, Wasserman, H. H. and Murray, R. W., Eds., Academic Press, New York, 1979.

59. Kondo, T., Matsumoto, M., and Tanimoto, M., Microwave and NMR studies of the structure and the conformational isomerization of 3,6-dihydro-1,2-dioxin, *Tetrahedron Lett.*, 3819, 1978.

60. Gollnick, K. and Griesbeck, A., Interactions of singlet oxygen with 2,5-dimethyl-2,4-hexadiene in polar and non-polar solvents. Evidence for a vinylog ene-reaction, *Tetrahedron*, 40, 3235, 1984.

61. Hasty, N. M. and Kearns, D. R., Mechanisms of singlet oxygen reactions. Intermediates in the reaction of singlet oxygen with dienes, *J. Am. Chem. Soc.*, 95, 3380, 1973.

62. (a) Manring, L. E., Kanner, R. C., and Foote, C. S., Chemistry of singlet oxygen. 43. Quenching by conjugated olefins, *J. Am. Chem. Soc.*, 105, 4707, 1983; (b) Manring, L. E. and Foote, C. S., Chemistry of singlet oxygen. 44. Mechanism of photooxidation of 2,5-dimethylhexa-2,4-diene and 2-methyl-2-pentene, *J. Am. Chem. Soc.*, 105, 4710, 1983.

63. Adam, W. and Staab, E., Regio-controlled functionalization of 2,5-dimethyl-2,4-hexadiene into epoxy alcohols by photooxygenation in the presence of titanium(IV) or vanadium(V), *Tetrahedron Lett.*, 531, 1988.

64. Matsumoto, M., Dobashi, S., Kuroda, K., and Kondo, K., Sensitized photo-oxygenation of acyclic conjugated dienes, *Tetrahedron*, 41, 2147, 1985.

65. Kenney, R. L. and Fisher, G. S., Photosensitized oxidation of myrcene, *J. Am. Chem. Soc.*, 81, 4288, 1959.

66. Matsumoto, M. and Kondo, K., Sensitized photooxygenation of linear monoterpenes bearing conjugated double bonds, *J. Org. Chem.*, 40, 2259, 1975.

67. Clennan, E. L. and L'Esperance, R. P., The unusual reactions of singlet oxygen with isomeric 1,4-di-*tert*-butoxy-1,3-butadienes. A 2_s-2_a cycloaddition, *J. Am. Chem. Soc.*, 107, 5178, 1985.

68. Clennan, E. L. and Nagraba, K., Additions of singlet oxygen to alkoxy-substituted dienes. The mechanism of the singlet oxygen 1,2-cycloaddition reaction, *J. Am. Chem. Soc.*, 110, 4312, 1988.

69. Hamberg, M. and Samuelson, B., On the mechanism of the biosynthesis of prostaglandins E_1 and $F_1\alpha$, *J. Biol. Chem.*, 242, 5336, 1967.

70. Mihelich, E. D., Structure and stereochemistry of novel endoperoxides isolated from the sensitized photooxidation of methyl linoleate. Implications for prostaglandin biosynthesis, *J. Am. Chem. Soc.*, 102, 7141, 1980.

71. Adam, W. and Eggelte, H. J., Prostanoid endoperoxide model compounds: 2,3-Dioxabicyclo[2.2.1] heptane via selective diimide reduction, *J. Org. Chem.*, 42, 3987, 1977.

72. Wasserman, H. H. and Ives, J. L., Singlet oxygen in organic synthesis, *Tetrahedron*, 37, 1825, 1981.

73. Balci, M., Bicyclic endoperoxides and synthetic applications, *Chem. Rev.*, 81, 91, 1981.

74. Saito, I. and Nittala, S. S., in *The Chemistry of Peroxides*, Patai, S., Ed., Wiley, New York, 1983, 311.

75. Kropf, H., Ed., *Organische Peroxoverbindungen*, Houben-Weyl E13/2, Thieme Verlag, Stuttgart, 1988, 991.

26

Photorearrangements of Endoperoxides

an Rigaudy
iversité Pierre et Marie Curie

26.1 Introduction

Photorearrangements of endoperoxides are generally induced by excitation of the broad $\pi_{oo}{}^*\sigma_{oo}{}^*$ transition of low extinction coefficient of the peroxide chromophore. The spectral location and intensity of this transition appear to be very dependent on ring strain. As a result, the location of the transition ranges from 300 to about 350 nm (with $\varepsilon_{max} = 10\ M^{-1}\ cm^{-1}$) for unstrained peroxide groups;[1] but the transition can extend to about 400 nm (with $\varepsilon_{max} = 15\ M^{-1}\ cm^{-1}$) when the peroxide bridge is slightly strained in a five- or six-membered ring and even up to 450 nm (with $\varepsilon_{max} = 25\ M^{-1}\ cm^{-1}$) in the case of the considerably strained four-membered dioxetane ring.[2] For alicyclic endoperoxides, irradiations are usually carried out by medium- or high-pressure mercury arc lamps through a Pyrex filter.

From their behavior under irradiation, two classes of bicyclic endoperoxides may be distinguished according to whether or not they have a double bond attached to the bridgeheads of the system. Consequently, they will be treated separately. In a simplified way, it can be said that for both classes the photochemical behavior is, to a certain extent, rather similar to the thermal behavior. (For the various aspects of the chemistry of endoperoxides, see preceding reviews.[3,4])

26.2 Unsaturated Bicyclic Endoperoxides

Unsaturated bicyclic endoperoxides are readily attainable by photooxygenation of the corresponding 1,3-dienes or analogues. These endoperoxides are the most studied and the two main modes of decomposition are common to thermal and photochemical processes (Scheme 26.1). According to structures and conditions, one can observe the following.

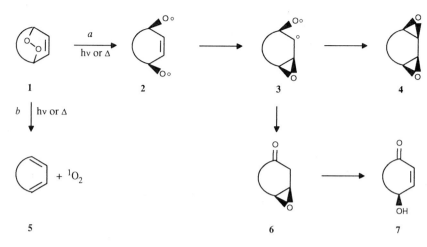

SCHEME 26.1

1. Cleavage of the O-O bond (path *a*) leading to *syn*-diepoxides **4**, frequently accompanied by the β-γ-epoxyketones **6** that are easily converted into the corresponding hydroxyenones **7**. The general outcome is then a **photoisomerization**.
2. Loss of molecular oxygen (path *b*) in a way which appears as the reverse of the formation process. This process is a **photodissociation**, and is often also called a *cycloreversion*.

From theoretical[5] and experimental[6,7] studies, it has been shown that path *a* originates from the dissociative $\pi_{oo}^*\sigma_{oo}^*$ transition of the peroxide bond already mentioned, occurring at long wavelengths; whereas path *b* results from excitation of a $\pi\pi^*$ transition of the unsaturated moiety, occurring at more or less shorter wavelengths. This explains why photoisomerization is general in the alicyclic series and also for endoperoxides of many polycyclic aromatic hydrocarbons, whereas photodissociation is restricted practically to those of the latter kind.

Photoisomerization

Alicyclic Endoperoxides

Photoisomerization of unsaturated endoperoxides was first shown to occur by Dufraisse et al.[8] with 2,3,4,5-tetraphenylfulvene endoperoxide **8a**, and later extended to a series of substituted 2,3,4,5-tetraphenylcyclopenta-2,4-diene endoperoxides **8b**.[9] (See Scheme 26.2.)

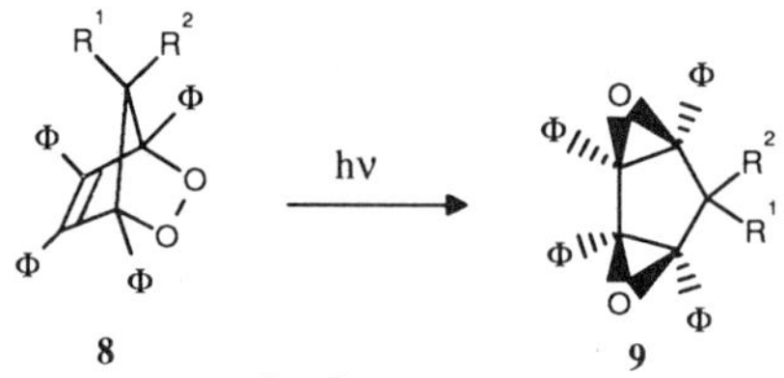

$$a:\ R^1, R^2 = CH_2=$$

$$b:\ \begin{cases} R^1 = H, R^2 = H, C_2H_5\ or\ C_6H_5 \\ R^1 = CH_3, R^2 = C_6H_5 \end{cases}$$

SCHEME 26.2

These endoperoxides differ from those with larger rings in giving exclusively and very efficiently the corresponding *syn*-dioxides 9a,b. It should be observed that thermal isomerization often leads to other rearranged products.[9]

In the cyclohexadiene series, the pioneering work was that of Maheshwari et al.[10] who established that irradiation through Pyrex or at 366 nm isomerizes ascaridole 10a into isoascaridole 11a, while other cyclohexadiene endoperoxides 10b afforded mixtures of diepoxides 11b and of β,γ-epoxyketones 12b arising from a 1,2-H shift (Scheme 26.3).

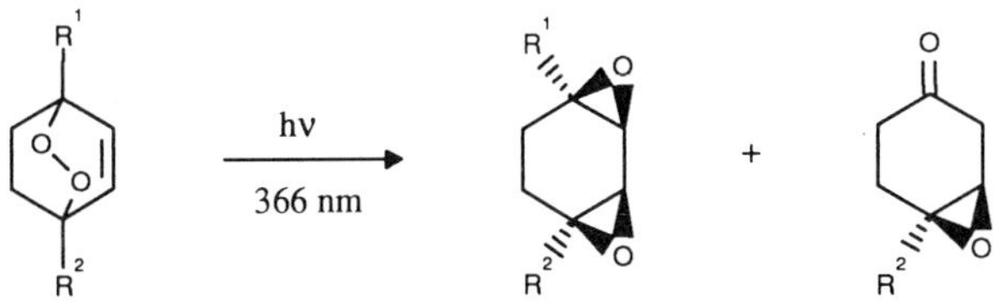

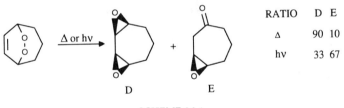

a: $R^1 = CH_3$, $R^2 = -CH(CH_3)_2$ b: $R^1 = H$, $R^2 =$ alkyl

SCHEME 26.3

More recently, a systematic study by Carless et al.[11] on various cyclohexadiene, cycloheptadiene, and cyclooctadiene endoperoxides has shown this behavior to be general, with the formation of epoxyketones taking precedence frequently over diepoxides in the irradiated mixtures. Furthermore, from a comparison of the results with those of the thermal isomerization that affords the same products, the authors concluded that "photolysis generally gave a greater proportion of epoxyketone than thermolysis." This effect was most remarkable in the [3.2.2] system as shown in Scheme 26.4.

RATIO	D	E
Δ	90	10
hv	33	67

D   E

SCHEME 26.4

Although conformational effects are assumed to play an important role in the partition between diepoxide and epoxyketone, to date the nature of the intermediates involved in these transformations is still unclear and needs further investigation. If a concerted $\pi_{2s} + \pi_{2s}$-cycloaddition, allowed in the singlet state, has been considered as a possible pathway to diepoxides,[9,10] a likely intermediate, at least in the triplet state, is a dioxyl radical such as 2 (Scheme 26.1).

Indeed, it has been found in several examples[10] that the reaction can be induced by triplet-triplet energy transfer and goes faster under these conditions (e.g., Michler's ketone or phenanthrene, λ_{exc} = 366 nm) than by direct irradiation. If, as suggested, the reaction proceeds in two steps via a triplet dioxyl radical, the spin inversion required for the second step may delay the closure to the diepoxide (*a*) sufficiently to allow the hydrogen migration leading to epoxyketone (*b*) to compete (Scheme 26.5).

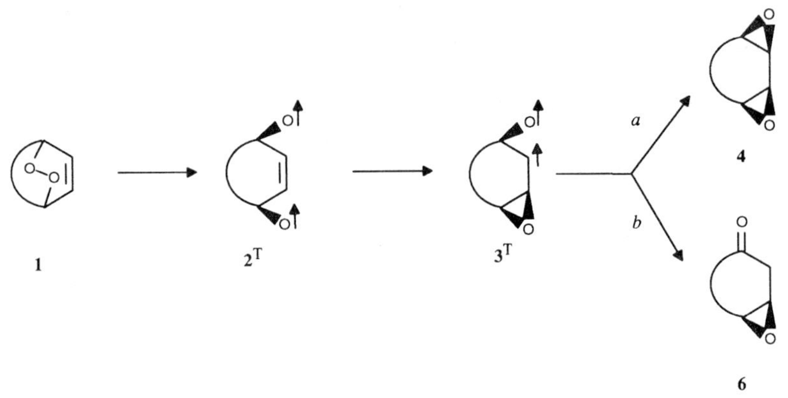

SCHEME 26.5

On the other hand, it has been shown[12] that these rearrangements may be selective in the case of bis(endoperoxides) such as 13, where it involves the more strained bicyclic peroxide bridge (Scheme 26.6).

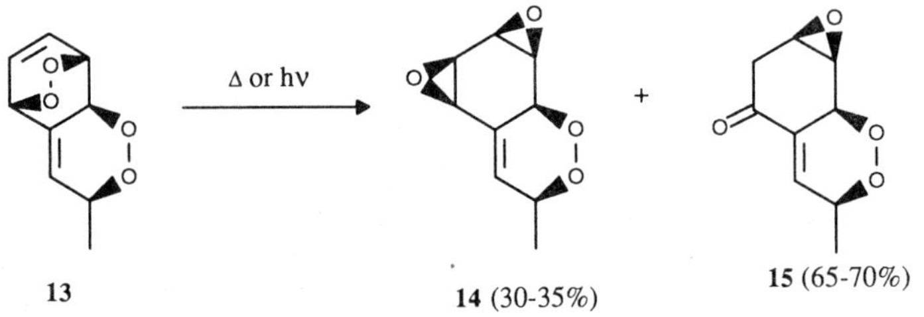

SCHEME 26.6

Turning now to multistep syntheses of more elaborate products than the preceding ones which include an endoperoxide-diepoxide rearrangement, one has to recognize that thermolysis has generally been preferred over photolysis to carry out the latter. Various examples may be found in a review on cyclic polyepoxides[13] to which we can add the synthesis of stemolide,[14] a natural *syn*-diepoxide with antileukemic activity. Also worthy of mention is the photolytic isomerization of 16, an endoperoxide derived from benzene, into a mixture of diepoxide 17 and β,γ-epoxyketones 18 and 19, valuable as inositol intermediates through opening to 20 and 21 with fixed configurations[15] (Scheme 26.7).

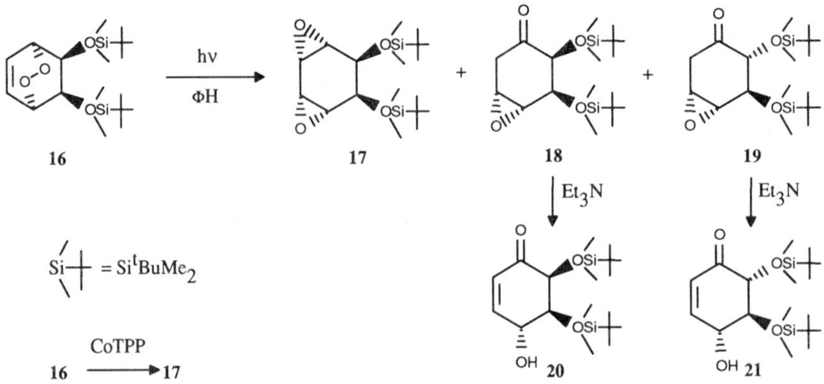

SCHEME 26.7

Finally, it must be said that the endoperoxide-diepoxide isomerization can be also carried out by chemical catalysis and that, in this area, the use of cobalt(II) tetraphenylporphyrin,[16a,b] more efficient than salts such as $FeSO_4$ or Cu_2Cl_2, is a serious competitor to photolytic or thermolytic procedures (see, for example, Scheme 26.7).

Aromatic Endoperoxides

With naphthalenic[17a,b] and anthracenic[18a,b] endoperoxides of the 1,4-type (*benzo*-endoperoxides) 22a, irradiation at long wavelengths (>366 nm) give exclusive formation of acid-sensitive *syn*-diepoxides 23a. In such cases, photolysis is the only way to arrive at diepoxides, as heating leads to complete dissociation. Competitive formation of the corresponding epoxyketones has been detected, up to now, only in the case of the tetramethyl derivative 22b and, again, it was found that when sensitized (by benzile or camphoroquinone, $\lambda_{exc} = 455$ nm), the reaction goes much faster than by direct irradiation and affords also higher proportions of epoxyketone 24b[19] (Scheme 26.8).

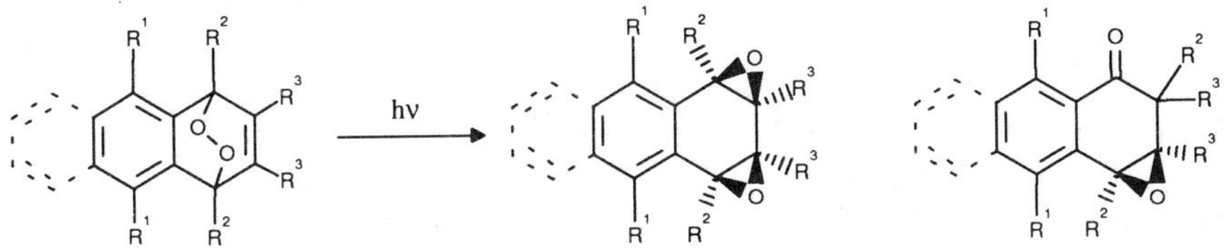

a: R^1= H or C_6H_5, R^2= CH_3 or OCH_3, R^3= H   b: R^1=H, R^2= R^3= CH_3

SCHEME 26.8

A useful synthetic application offered by the photoisomerization to diepoxides 23a is their acid-catalyzed isomerization leading readily to dihydroxy derivatives: 2,3-diphenols 25a[17b,18b] and 2,3-dihydroxynaphtho- or anthraquinones 26a[17a,18a] (Scheme 26.9).

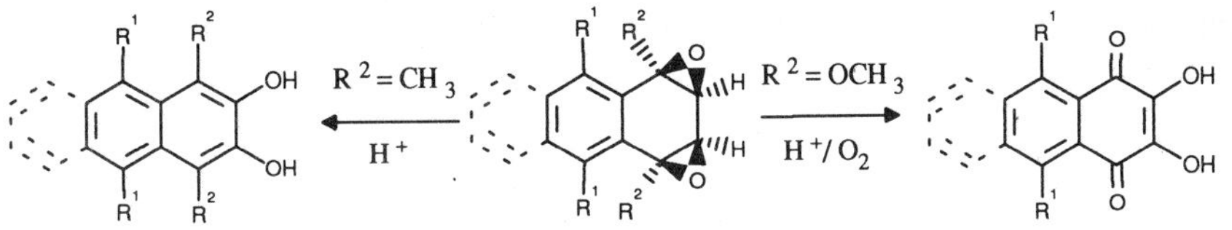

a: R^1= H or C_6H_5, R^2= CH_3 or OCH_3

SCHEME 26.9

Unstable *syn*-diepoxides 28 have been obtained from *meso*-anthracenic endoperoxides 27 by prolonged irradiation at the highest possible wavelengths ($\lambda > 400$ nm) and at low temperature.[20] They easily undergo thermal (or photochemical) isomerizations, affording among other products the benzocyclobutenic diethers 29 (Scheme 26.10). In limited cases (R = H or alkyl), diepoxides 28 can also result from thermolysis, but of course they cannot be isolated as such under these conditions.

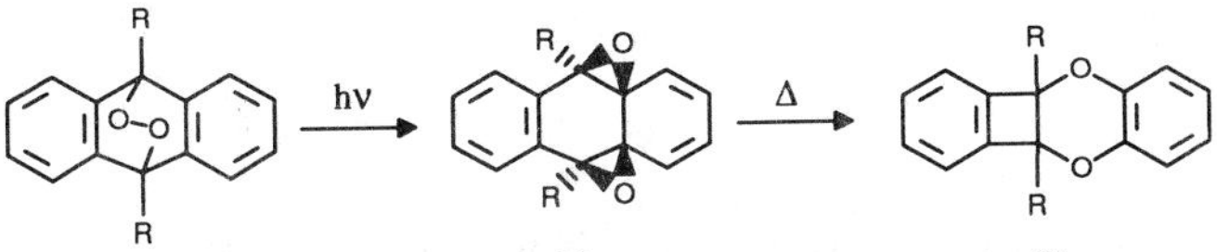

R=H or R=C_6H_5

SCHEME 26.10

Photodissociation

As mentioned earlier, photodissociation occurs with aromatic endoperoxides on irradiation at short wavelengths. More precisely, it has been shown[7] that irradiation of the endoperoxide of 9,10-diphenylanthracene 30a into the S_2 band ($\lambda < 300$ nm) leads to an adiabatic photocleavage into 9,10-diphenylanthracene and singlet molecular oxygen, whereas irradiation in the S_1 band ($\lambda > 300$ nm) brings about photoisomerization (Scheme 26.11).

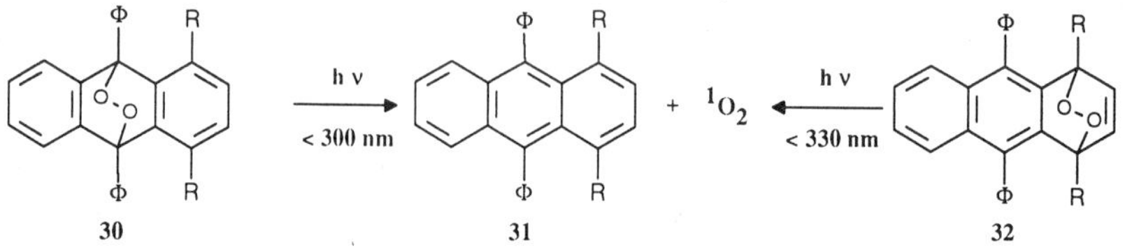

SCHEME 26.11

The same behavior has been demonstrated for several other *meso*-endoperoxides such as those of anthracene and 9,10-dimethylanthracene.[21] Moreover, it was established in the case of 32b, a *benzo*-endoperoxide, that cycloreversion may originate not only from the S_2 state, but also from upper excited singlet states S_3 and S_4,[22] a finding that contrasts with the conventional paradigm of very rapid deactivation of upper excited electronic states.

When the *meso*-phenyl substituents are linked to *peri*-positions by some sort of bridge, as in the endoperoxide of heterocoerdianthrone 33, the quantum yield of the irreversible reaction leading to decomposition products may become very small and the reversible photodissociation opens the way to a photochromic system.[23] In this area, the best compound is the blue dimethyl-homeocoerdianthrone 34, which does not undergo self-sensitized photooxygenation but affords, by external sensitization, a colorless endoperoxide 35 reverting back quantitatively to the colored starting material on UV irradiation at 313 nm[24] (Scheme 26.12). The system has been commercially proposed as an actinometer.

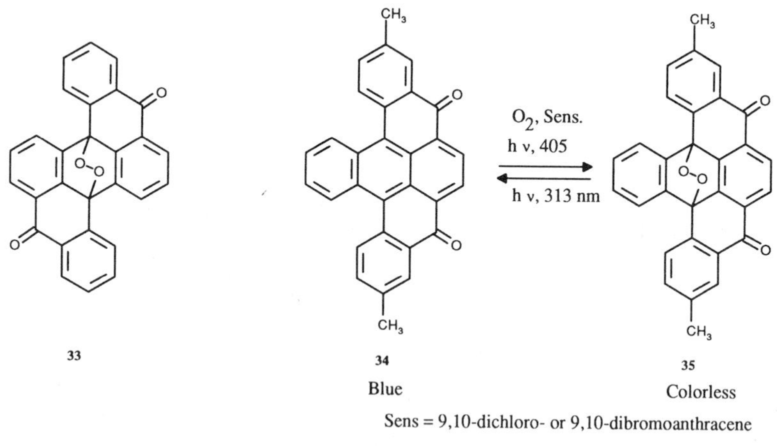

SCHEME 26.12

Kearns and Khan[5] originally proposed that C-O bonds are broken simultaneously in a concerted fashion. However, later work deduced from the relatively "slow" evolution of hydrocarbon ($\tau_2 = 1.6$ to 95 ps) compared to the very short lifetime of excited states ($\tau_1 \sim 3$ ps) that the photocycloreversion occurs in two steps. The first step is a very fast C-O bond rupture occurring in the excited state, followed by the "slow" decay of the resulting biradical to the ground state of the aromatic parent and singlet oxygen[25] (Scheme 26.13).

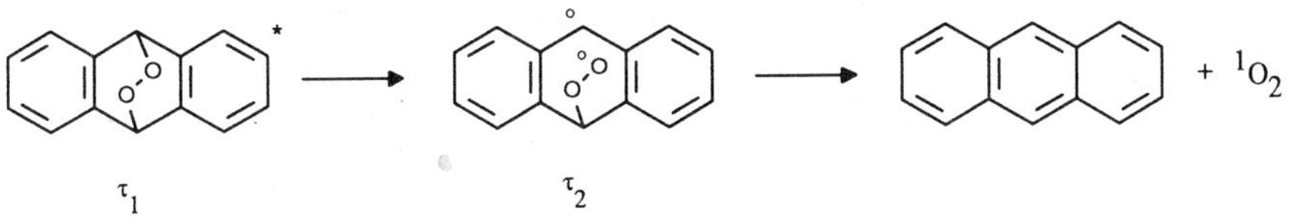

SCHEME 26.13

Laser-induced fluorescence[26] and transient absorption studies[27] have been used to measure the lifetimes of the intermediates, with the shortest ones being found for 35 ($\tau_1 < 0.35$ ps; $\tau_2 = 1.6 \pm 0.5$ ps).

26.3 Saturated Bicyclic Endoperoxides

Studies on the photolysis of saturated bicyclic endoperoxides are relatively recent compared to those on their thermolysis, which have provided a fair body of information on the behavior of thermally generated dioxyl radicals resulting from the homolysis of the O-O bond. The main contribution is that of Bloodworth et al.[28] in 1984. The outcome of both processes diverge in several cases, showing that the photolytically generated cycloalkanedioxyls may behave differently to the thermally derived species owing to different spin multiplicities and/or electronic configurations.

The major photoproduct from 2,3-dioxa [2.2.1]heptane was identified as 4,5-epoxypentanal, the same as that reported previously for thermolysis in the gas phase or in solvents of low polarity (Scheme 26.14).

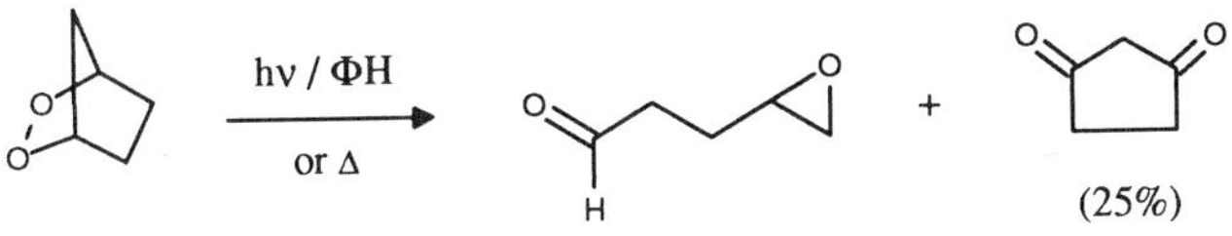

SCHEME 26.14

On the contrary, under the same photochemical conditions, the [n.2.1]peroxides (n = 3,4,5) afforded mainly the corresponding epoxyaldehydes, whereas thermolysis in the gas phase led to ketoaldehydes in over 90% yields (Scheme 26.15).

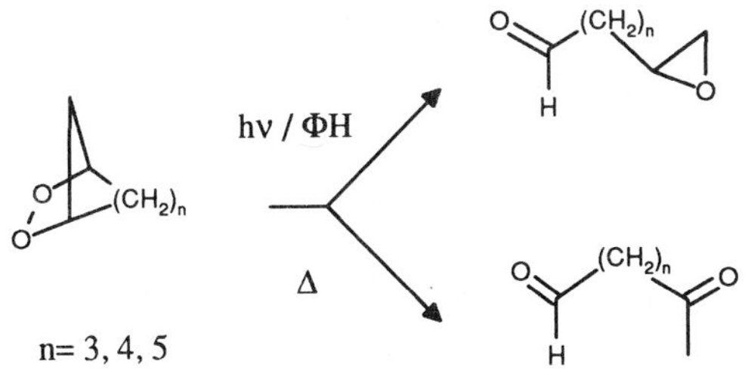

SCHEME 26.15

The most interesting case is that of the [3.2.1]peroxide **36a**, which differs from the others in giving, on benzophenone-sensitized photolysis, 6,8-dioxabicyclo[3.2.1]octane **38a** as the major product (50 to 60%). It is assumed that cleavage of the one-carbon bridge, after or in concert with O-O homolysis, affords a 1,3-biradical. In the direct photolysis, this biradical is a singlet species that can collapse to the epoxyaldehyde; whereas in the sensitized photolysis, it is a triplet species that has time for bond rotation and intramolecular addition to the aldehyde group (Scheme 26.16). The same effect (i.e., different products from different excited states) had been found earlier for the dimethyl analogue, 1,5-dimethyl-6,7-dioxabicyclo[3.2.1]octane 36b, whose benzophenone-sensitized irradiation provides an elegant photochemical synthesis of the pine beetle pheromone Frontalin 38b (Scheme 26.16).[29]

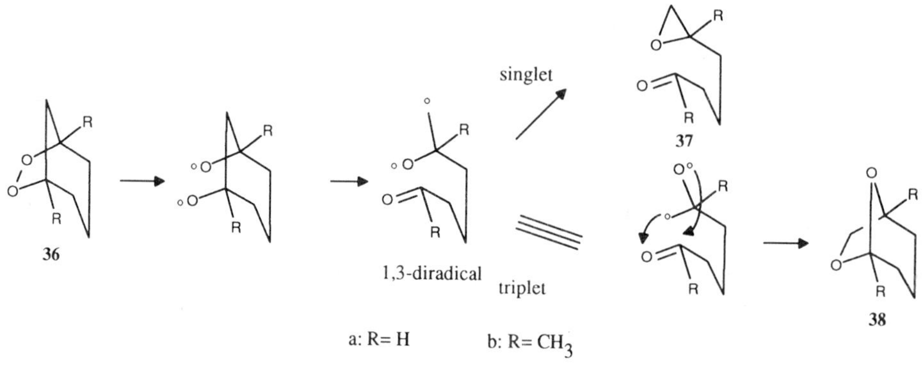

SCHEME 26.16

With [n.2.2]endoperoxides, the major pathway for photodecomposition is the same as that for thermolysis, affording 1,4-diketones (Scheme 26.17).

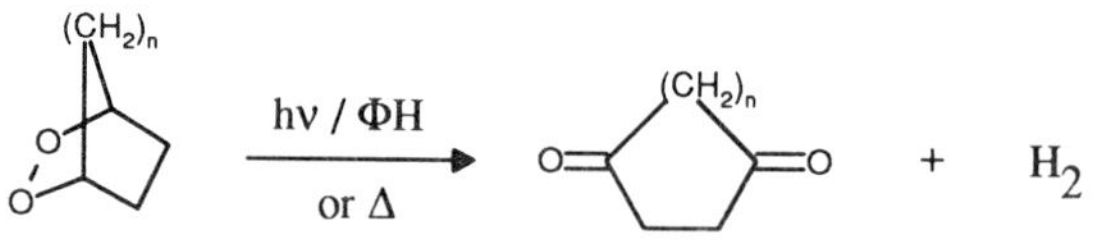

SCHEME 26.17

UV Characteristics of Substrates and Sensitizers

Substrates
   Alicyclic endoperoxides: λ = 300 to 400 nm (ε = 10 to 15 M^{-1} cm^{-1}). See Section 26.1.
   Aromatic endoperoxides:
      All present a flat tail of absorption extending to about 435 nm with very small extinction coefficients ($\varepsilon < 1$).
   Anthracene
      9,10-Endoperoxide (THF):
      λ_{max} nm (log ε): 278 (2.77), 270 (2.83), 254 (3.46).
      See also Reference 21.
   9,10-Diphenylanthracene
      9,10-Endoperoxide 30a (Et$_2$O):
      λ_{max} nm (log ε): 278 (2.60), 262 (3.08), 257 (3.16).
      See also Reference 7.

1,4-Dimethyl-9,10-diphenylanthracene

1,4-Endoperoxide 32b (Et$_2$O):

λ_{max} nm (log ε): 329 (2.87), 315 (2.91), 292 (3.97), 283 (4.10), 272 (4.04), 243 (4.67).

9,10-Endoperoxide 30b (Et$_2$O):

λ_{max} nm (log ε): 291 (3.02), 267 (3.20), 261 (3.22).

See also Reference 22.

Dimethylhomeocoerdianthrone 34 (toluene):

λ_{max} nm (ε): 647 (14100 M^{-1} cm^{-1}).

Endoperoxide 35 (toluene),

λ_{max} nm (ε): 400 (97 M^{-1} cm^{-1}), 345 (2800 M^{-1} cm^{-1}).

See Reference 24.

Sensitizers

Benzile (EtOH), λ_{max} nm: 370.

Camphoroquinone (EtOH), λ_{max} nm: 466.

eferences

1. Takezaki, Y., Miyazaki, T., and Nakahara, N., Photolysis of dimethylperoxide, *J. Chem. Phys.*, 25, 536, 1956.

2. Lechtken, P. and Steinmetzer, H.-C., Photofragmentierung von tetramethyl-1,2-dioxetan. Ein effizientes beispiel einer adiabatischen photoreaktion, *Chem. Ber.*, 108, 3159, 1975.

3. Balci, M., Bicyclic endoperoxides and synthetic applications, *Chem. Rev.*, 81, 91, 1981.

4. Saito, I. and Matsuura, T., in *Singlet Oxygen*, Wasserman, H. H. and Murray, R. W., Eds., Academic Press, New York, 1979, chap. 10.

5. Kearns, D. R. and Khan, A. U., Sensitized photooxygenation reactions and the role of singlet oxygen, *Photochem. Photobiol.*, 10, 193, 1969.

6. Srinivasan, R., Brown, K. H., Ors, J. A., White, L. S., and Adam, W., Organic photochemistry with 6–7 eV photons: Ascaridole, *J. Am. Chem. Soc.*, 101, 7424, 1979.

7. Drews, W., Schmidt, R., and Brauer, H.-D., The photolysis of the endoperoxide of 9,10-diphenylanthracene, *Chem. Phys. Lett.*, 70, 84, 1980.

8. Dufraisse, C., Rio, G., and Basselier, J.-J., Isophotooxydes cyclopentadiéniques, *C. R. Acad. Sci. Paris*, 246, 1640, 1958.

9. Basselier, J.-J. and Leroux, J. P., Réarrangements thermiques et photochimiques des photooxydes de tétraphényl-2,3,4,5 cyclopentadiènes-2,4. Influence des substituants méthéniques, *Bull. Soc. Chim. Fr.*, 4448, 1971.

10. Maheshwari, K. K., de Mayo, P., and Wiegand, D., Photochemical rearrangement of diene endoperoxides, *Can. J. Chem.*, 48, 3265, 1970.

11. Carless, H. A. J., Atkins, R., and Fekarurhobo, G. K., Thermal and photochemical reactions of unsaturated bicyclic endoperoxides, *Tetrahedron Lett.*, 26, 803, 1985.

12. Carless, H. A. J., Atkins, R., and Fekarurhobo, G. K., Polyoxygenated cyclohexenes from aromatic compounds: Selective reactions of *bis*(endoperoxides), *J. Chem. Soc., Chem. Commun.*, 139, 1985.

13. Adam, W. and Balci, M., Cyclic polyepoxides. Synthetic, structural and biological aspects, *Tetrahedron*, 36, 833, 1980.

14. van Tamelen, E. E. and Taylor, E. G., Total synthesis of stemolide, *J. Am. Chem. Soc.*, 102, 1202, 1980.

15. Carless, H. A. J., Billinge, J. R., and Oak, O. Z., Photochemical routes from arenes to inositol intermediates: The photo-oxidation of substituted *cis*-cyclohexa-3,5-diene-1,2,-diols, *Tetrahedron Lett.*, 30, 3113, 1989.

16. (a) Boyd, J. D., Foote, C. S., and Imagawa, D. K., Synthesis of 1,2; 3,4-diepoxides by catalyzed rearrangement of 1,4-endoperoxides, *J. Am. Chem. Soc.*, 102, 3641, 1980; (b) Balci, M. and Sutbeyaz, Y., CoTPP catalysed rearrangement of 1,4-endoperoxides, *Tetrahedron Lett.*, 24, 311, 1983.

17. (a) Rigaudy, J., Deletang, C., and Basselier, J.-J., Autoxydation photosensibilisée du diméthoxy-1,4 naphtalène: le photooxyde et ses produits de transformation, *C. R. Acad. Sci. Paris,* 268, Série C, 344, 1969; (b) Rigaudy, J., Maurette, D., and Nguyen Kim Cuong, Autoxydation photosensibilisée du diméthyl-1,4 naphtalène: le photooxyde et sa photo-isomérisation, *C. R. Acad. Sci. Paris,* 273, Série C, 1553, 1971.

18. (a) Rigaudy, J., Cohen, N. C., and Nguyen Kim Cuong, Photooxydes "benzo" des anthracènes dialcoxylés en 1–4. Leur photo-isomérisation en *bis*-époxydes, *C. R. Acad. Sci. Paris,* 264, Série C, 1851, 1967; (b) Rigaudy, J., Caspar, A., Lachgar, M., Maurette, D., and Chassagnard, C., Photo-isomérisations d' 1,4-endoperoxydes et de 1,2;3,4-diépoxydes dérivés de 1,4-diméthylanthracènes, *Bull. Soc. Chim. Fr.,* 129, 16, 1992.

19. Rigaudy, J., Lachgar, M., and Saad, M., Photoisomerization of 1,4-endoperoxides derived from 1,2,3,4-tetramethylanthracenes and 1,2,3,4-tetramethylnaphthalene, *Bull. Soc. Chim. Fr.,* 131, 177, 1994.

20. Defoin, A., Baranne-Lafont, J., and Rigaudy, J., Transformations photochimiques d'endoperoxydes dérivés d'hydrocarbures aromatiques polycycliques. II. Cas de l'endoperoxyde d'anthracène: Di- et tétraépoxydes dérivés, *Bull. Soc. Chim. Fr.,* 145, 1984 (II).

21. Schmidt, R., Schaffner, K., Trost, W., and Brauer, H.-D., Wavelength-dependent and dual photo-chemistry of the endoperoxides of anthracene and 9,10-dimethylanthracene, *J. Phys. Chem.,* 88, 956, 1984.

22. Schmidt, R., Brauer, H.-D., and Rigaudy, J., Reactions originating from different upper excited singlet states: The photocycloreversion of the endoperoxides of 1,4-dimethyl-9,10-diphenylanthracene, *J. Photochem.,* 34, 197, 1986.

23. Brauer, H.-D., Drews, W., and Schmidt, R., Ein neues photochromes system von ungewöhnlich hoher thermischer stabilität, *J. Photochem.,* 12, 293, 1980.

24. Schmidt, R., Drews, W., and Brauer, H.-D., Wavelength-dependent photostable or photoreversible photochromic system, *J. Phys. Chem.,* 86, 4909, 1982.

25. Jesse, K., Markert, R., Comes, F. J., Schmidt, R., and Brauer, H.-D., Picosecond photochemistry: The mechanism of photocycloreversion of aromatic endoperoxides, *Chem. Phys. Lett.,* 166, 95, 1990.

26. Jesse, K. and Comes, F. J., Rate parameters for the two-step photofragmentation of aromatic endoperoxides in solution, *J. Phys. Chem.,* 95, 1311, 1991.

27. Ernsting, N. P., Schmidt, R., and Brauer, H.-D., Subpicosecond transient absorption studies of the photocycloreversion of an aromatic endoperoxide, *J. Phys. Chem.,* 94, 5252, 1990.

28. Bloodworth, A. J. and Eggelte, H. J., Photolysis of saturated bicyclic peroxides, *Tetrahedron Lett.,* 25, 1525, 1984.

29. Wilson, R. M. and Rekers, J. W., Decomposition of bicyclic endoperoxides: An isomorphous synthesis of Frontalin via 1,5-dimethyl-6,7-dioxabicyclo[3.2.1]octane, *J. Am. Chem. Soc.,* 103, 206, 1981.

27

Photochemical Methods for the Synthesis of 1,2-Dioxetanes

Tons L. Baumstark
Georgia State University

Augusto Rodriguez
Clark Atlanta University

7.1 Introduction and Scope

1,2-Dioxetanes, four-membered cyclic peroxides, have been shown[1] to undergo a characteristic chemiluminescent thermal decomposition to two carbonyl fragments (Equation 27.1).

$$ \text{1,2 - Dioxetane} \xrightarrow[\Delta]{k_1} \quad + \quad + \quad CL $$

1,2 - Dioxetane   CL = chemiluminescence   (27.1)

Historically, 1,2-dioxetanes and α-peroxylactones (dioxetanones) had been proposed[2] as intermediates in chemiluminescent and bioluminescent processes.[3] The pioneering synthetic work of Kopecky[4a] on dioxetanes and Adam[4b] on α-peroxylactones opened many exciting, new areas of investigation. Interest in dioxetanes has expanded beyond purely theoretical aspects to include commercial applications.[5] Often, synthesis and isolation are the major difficulties encountered in these areas. Photochemical routes to dioxetanes can be of greater convenience and utility than thermal methods for the generation of a variety of compounds. The present review is focused on the direct synthesis of dioxetanes by photochemical methods. Multistep synthetic routes[1] that include a photochemical process (allylic hydroperoxide/endoperoxide intermediates) are not included. This chapter is not intended to be an exhaustive review of the literature. The treatment is

0-8493-8634-9/95/$0.00+$.50
© 1995 by CRC Press, Inc.

limited to illustrative examples of photochemical routes, including synthetic applications for this class of compounds.

27.2 Historical Background

Shortly after the discovery[4] of a synthetic (thermal) route for the preparation of 1,2-dioxetanes, the major photochemical route was discovered[6] independently by Bartlett's and Foote's research groups. The syntheses of cis-3,4-diethoxy-1,2-dioxetane[6a] (1) and tetramethoxy-1,2-dioxetane[6b] (2) by cycloaddition of singlet oxygen to the corresponding electron-rich alkenes (Equation 27.2) were reported in back-to-back communications.[6]

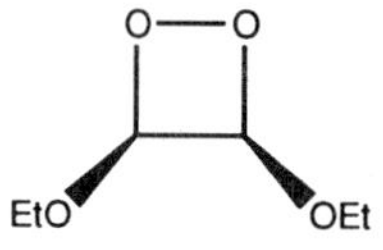

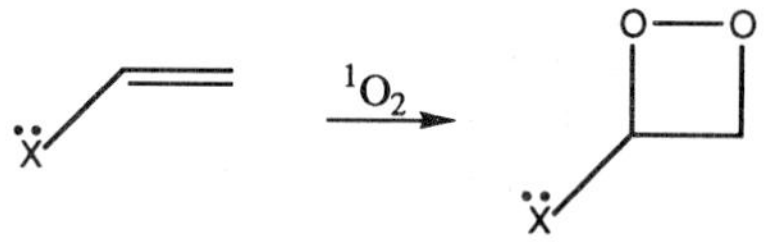

$$\text{(27.2)}$$

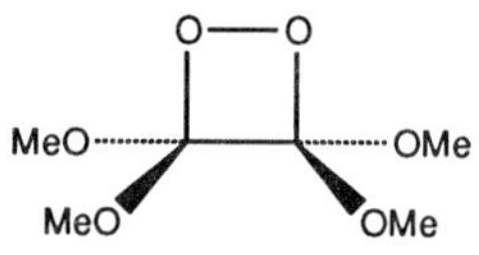

The reaction of alkoxy-substituted alkenes, in particular the work of Schaap[7] on 1,4-dioxenes, with $^1O_2$ has been shown to be of significant synthetic utility producing good-to-excellent yields of dioxetanes. Conjugated alkenes, including some aryl-substituted olefins, present problems in that [4+2]-addition (endoperoxide formation) may compete with the [2+2]-addition (dioxetane formation, Scheme 27.1).[8]

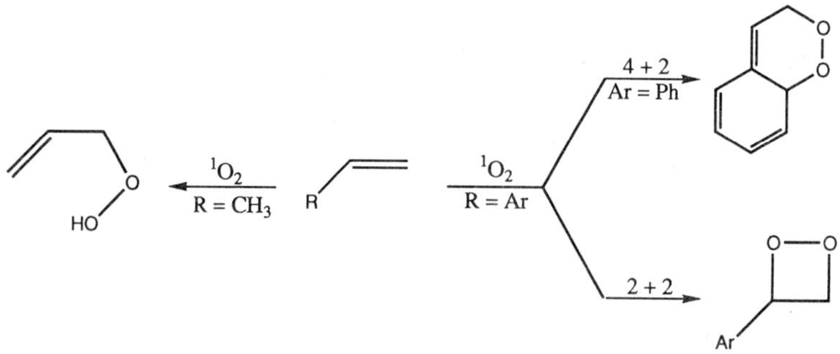

SCHEME 27.1

Alkenes with allylic (abstractable) hydrogen atoms, on the other hand, preferentially undergo the "ene" reaction to yield allylic hydroperoxides (Scheme 27.1) rather than the [2+2]-cycloaddition product.[9] Sterically hindered tetrasubstituted alkenes[10] like biadamatylidene[10a] do not undergo the ene reaction, but give good yields of dioxetane (Equation 27.3).

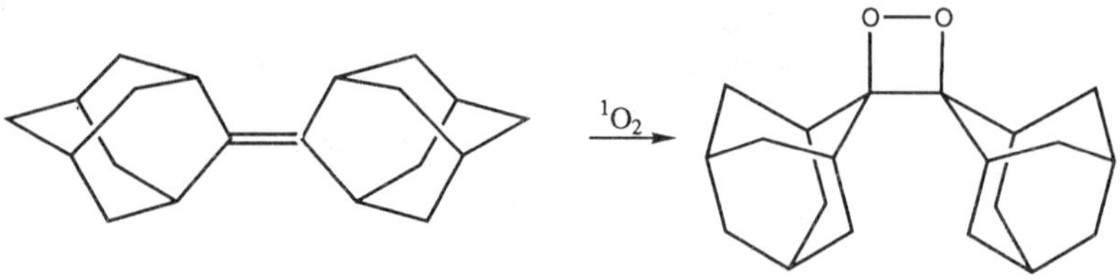

$$\text{(27.3)}$$

.3 Mechanistic Considerations

Although studied extensively,[1] the mechanism for the photochemical formation of dioxetanes has not been resolved.[11] At least five independent mechanisms for the reaction of singlet oxygen with alkenes to produce dioxetanes have been proposed: (1) $[2_s + 2_a]$ or $[2_s + 2_s]$ cycloaddition;[12] (2) perepoxide intermediate;[12b] (3) zwitterion intermediate;[13] (4) biradical intermediate;[14] and (5) charge-transfer complex.[13] It seems likely that there may be a variation in mechanism with substrate and/or reaction conditions.[15] The mechanism of the 9,10-dicyanoanthracene-sensitized process for the photochemical formation of dioxetanes is considered[16] to involve electron transfer steps rather than singlet oxygen chemistry.

7.4 Experimental Conditions for the Photochemical Synthesis of Dioxetanes

Most dioxetanes are pale yellow in color ($\lambda_{max} \approx 280$ nm and have a weak tail end absorption up to 450 nm).[1] Spectroscopic data are useful[1] in establishing the structure of dioxetanes. However, unequivocal proof of the dioxetane ring should include the observation of light emission during thermolysis in conjunction with product analysis. The spectral and physical properties of most dioxetanes require that samples and/or solvents remain metal ion-free (to avoid catalyzed decomposition to normal cleavage products via a "dark" pathway).[17] Solutions should be saturated with O_2 during synthesis to minimize autocatalytic induced decomposition as well as light-source and/or dye-related induced photolyses.[1] Pure samples of dioxetanes can be explosive under vacuum or in an inert atmosphere.[1] Subambient temperatures should be maintained whenever possible.

Sensitized Photooxygenation

Dioxetanes may be prepared[1] via the photooxygenation of a wide variety of electron-rich olefins such as vinyl ethers, ketene acetals, thioalkylsubstituted olefins, and enamines. For certain cases, Clennan has shown[18] that alkoxy dienes can be converted to dioxetanes via a singlet oxygen route (Equation 27.4).

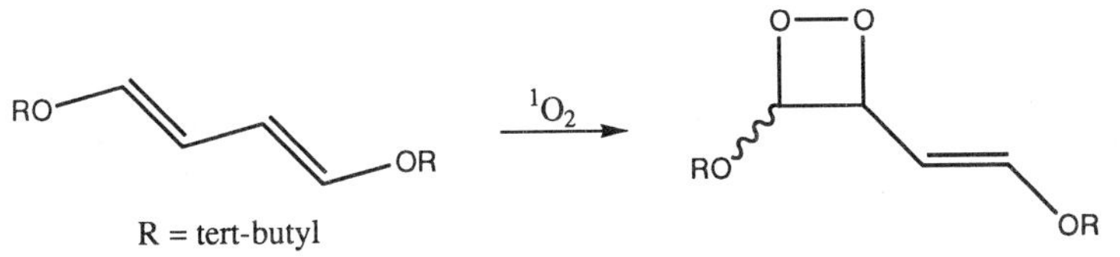

$$R = \text{tert-butyl} \qquad (27.4)$$

Singlet oxygen can be conveniently generated[19] using a combination of sensitizers/solvents: (1) tetraphenylporphine(TPP)/methylene chloride (for low temperatures) or carbon tetrachloride (for applications at −10 to 20°C); (2) Rose Bengal/acetone; (3) polymer-bound Rose Bengal/acetone; or (4) methylene blue/methanol.

In principle, the procedure for dioxetane generation is straightforward and only requires access to a light source (150-W sodium lamp, commercially available for under $200). The following procedure[20] can be adapted to produce dioxetanes from a host of electron-rich substrates. A solution containing 2 mmol olefin with a catalytic amount of sensitizer (1 to 2 mg) in 10 to 20 ml solvent is placed in a Pyrex reaction vessel equipped with an oxygen inlet tube. The vessel is immersed in an ethanol bath, cooled by means of a cryostat (temperature may vary from −10 to −78°C). A steady stream of oxygen is passed through the reaction mixture (O_2 saturation) while irradiating with a 150-W sodium lamp ($\lambda = 543$ nm). The progress of the reaction can be monitored by ^{1}H NMR spectroscopy. Upon complete consumption of the olefin, the solvent is evaporated and the residue can be purified by low-temperature column chromatography[1f] (silica gel) or, in cases where the dioxetane is extremely stable, by low-temperature flash chromatography.

Electron Transfer Routes

Dioxetanes can also be prepared via photochemical electron transfer reactions by using 9,10-dicyanoanthracene as a sensitizer.[1,16] This approach works well for the synthesis of stable dioxetanes. Typical experimental conditions[21] for this transformation include irradiation at 0°C (using a 1000-W mercury lamp and a $CuSO_4$ filter) of oxygenated solutions containing 2×10^{-4} M 9,10-dicyanoanthracene and 1×10^{-2} M of the appropriate alkene in anhydrous acetonitrile. For adamantyl enol ethers, this 9,10-dicyanoanthracene-catalyzed route led[21] to excellent yields of dioxetanes (Equation 27.5).

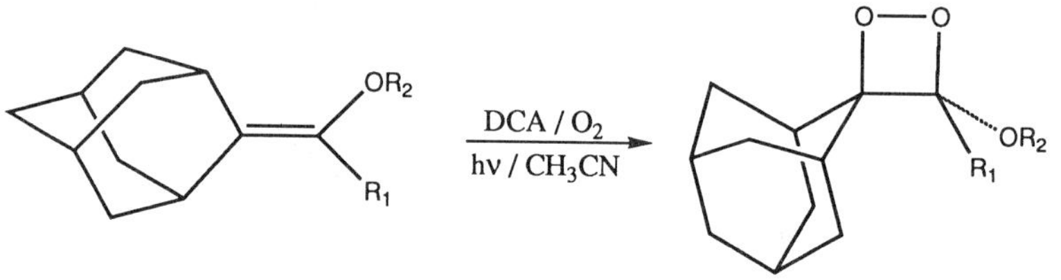

$$(27.5)$$

Direct Formation

In special cases, particularly for cumulenes, dioxetanes can be generated photochemically without the use of a sensitizer.[22] For example, irradiation of cummulene 3 in an oxygen/argon matrix (at 10 to 40 K) using a short ($\lambda > 300$ nm) or long ($\lambda > 500$ nm) wavelength was postulated[22] to give *tris*-dioxetane 4 as a transient intermediate which decomposed to acetone and carbon dioxide (Equation 27.6). Although chemiluminescence was not detected, additional evidence[22] for the formation of this dioxetane intermediate came from IR spectroscopic data and the isotopic distribution observed in the carbon dioxide when the overall process was carried out with labeled O_2.

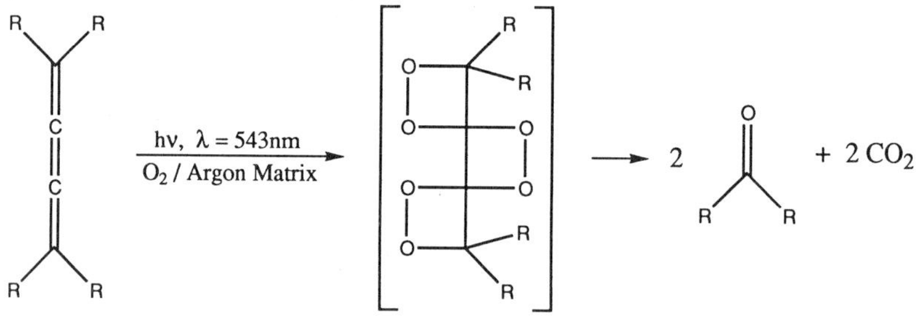

$$(27.6)$$

27.5 Synthetic Applications

Synthesis of Stable Dioxetanes for Biological Studies

An important property of most dioxetanes is the ability to generate blue light ($\lambda_{max} \sim 400$–430 nm) upon thermal decomposition to carbonyl fragments.[1] This chemical behavior has found many applications in studies of chemiluminescent phenomena,[23] bioluminescence,[24] and in studies aimed at elucidation of the mechanisms of excited-state formation via nonphotochemical means.[25] Recent work with 1,2-dioxetanes suggests that these species also have an important role in biological processes.[26] For example, a urethane-functionalized 1,2-dioxetane, 5, synthesized from the 3-hydroxymethyl-3,4,4-trimethyl-1,2-dioxetane has been found[27] to induce photochemical damage in bacterial and mammalian DNA.

The design and synthesis of thermally stable 1,2-dioxetanes that upon activation[28] produce highly efficient chemiluminescence has led to the development of dioxetanes as DNA probes.[29] The attractive features of this approach are that it provides a faster, safer, and more sensitive method of target DNA detection. This method of detection offers the advantage of being 2 to 3 orders of magnitude more sensitive than scintillation counting. One of the most common strategies is to immobilize the target DNA on a solid support, attach a DNA probe containing an enzyme (alkaline phosphatase) to the target DNA, followed by exposure to an enzyme-(phosphatase) triggerable stable dioxetane. The light emitted from the decomposition of the newly generated unstable dioxetane is then measured on a luminometer. At present, the most effective enzyme triggerable dioxetanes, 6, contain an adamantyl group (to provide kinetic stability via steric hinderance) and a fluorgenic protected phenolic substituent. In the protected form, often as a type of ester, these dioxetanes are thermally stable with half-lives of up to ~19 years at ambient temperatures. However, when the protecting group is removed (for phosphate esters by hydrolysis by alkaline phosphatase), the free phenoxide anion containing dioxetane is extremely unstable and rapidly undergoes decomposition via a chemically initiated electron exchange luminescence (CIEEL) mechanism resulting in efficient chemiluminescence[5] (quantum yields of approx. 10^{-5}).

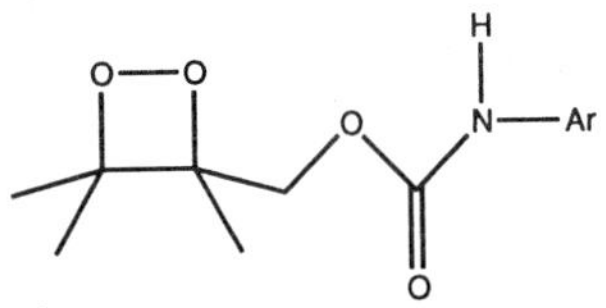

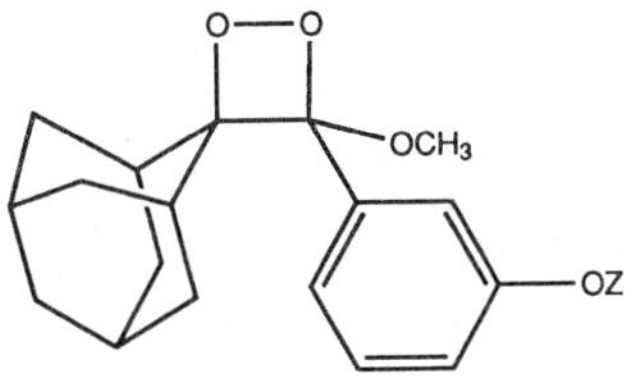

Rearrangement Reactions of Dioxetanes

Certain alkoxy-substituted dioxetanes are prone to rearrangement reactions that yield α-peroxy ketones and esters.[30] For example, siloxy dioxetane 7, generated by singlet oxygen addition to the corresponding alkene, was found[30] to rearrange to an α-silylperoxyester upon warming. The reaction is thought to proceed via formation of a Zwitterionic intermediate via heterolytic C-O bond cleavage (Scheme 27.2). Evidence for the formation of the intermediate was obtained by trapping with acetaldehyde to produce a trioxane. The trioxane was the sole reaction product when the photooxygenation was carried out in the presence of acetaldehyde.

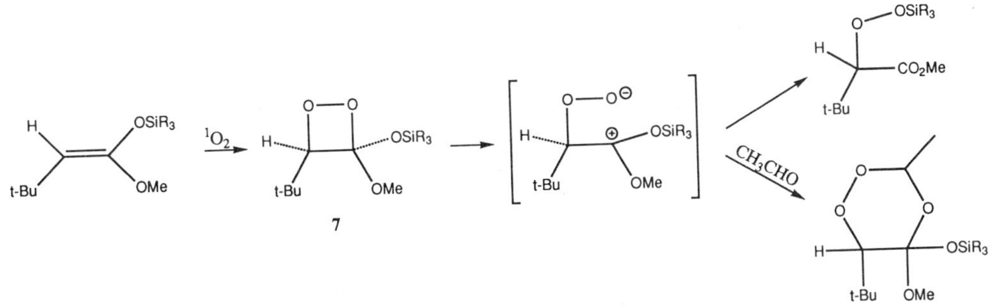

SCHEME 27.2

Perhaps, the most notable of these studies involves the total synthesis of the powerful antimalarial agent (+)Arteminisin and related analogs.[31] For these routes, the intermediate dioxetanes, **8**, can be generated by singlet oxygenation of the appropriate alkenes or via the ozonolysis of vinyl silanes. (The latter reaction appears to be general in scope, although the mechanism for the formation of dioxetanes from vinyl silanes is not understood.) The dioxetane intermediate is postulated to undergo acid-catalyzed opening to give Arteminisin directly (Equation 27.7). Presumably, hydrolysis of the siloxy substituted dioxetane **8** yields an α-peroxy aldehyde intermediate that undergoes a cascade of intramolecular cyclizations (peroxy hemiacetal and acetal formation) to produce the bioactive compound.

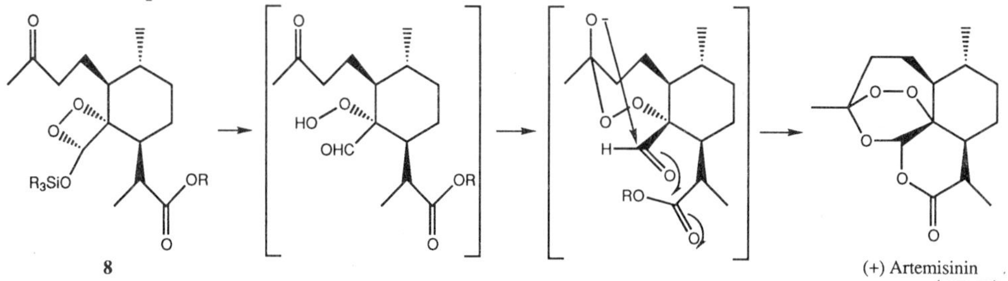

8   (+) Artemisinin

$$(27.7)$$

Dioxetanes Generated for the Purpose of Oxidative Cleavage

The [2+2]-addition of singlet oxygen followed by thermal or catalyzed decomposition of the dioxetane ring to carbonyl fragments represents an alternative to ozone for the cleavage of electron-rich olefins. The synthetic utility of this oxidative cleavage sequence can be exploited for the construction of large heterocyclic rings containing carbonyl groups.[32] For example, photooxygenation[32b] of an indole provides a convenient synthesis of a nine-membered ring system via a dioxetane intermediate (Equation 27.8).

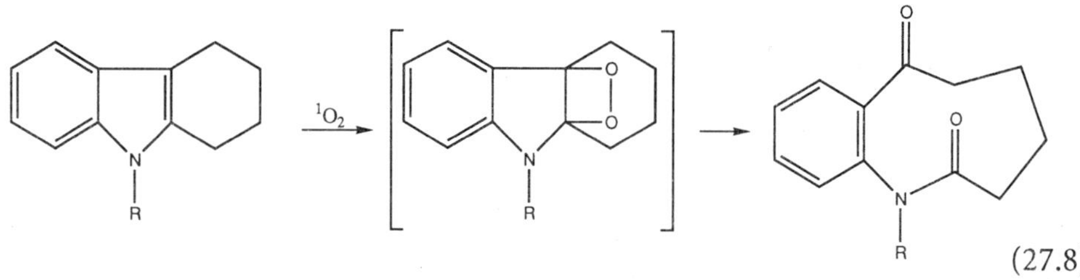

$$(27.8)$$

The oxidative cleavage of electron-rich double bonds using singlet oxygen has also been exploited[32a,33] in the synthesis of progesterone. A key step in the conversion of stigmasterol to progesterone involved the sensitized photooxygenation of a 22-aldehydo compound (Equation 27.9) in basic methanol. The reaction is thought to proceed via a [2+2]-addition of singlet oxygen to the enol (or enolate) to yield an intermediate dioxetane. Subsequent thermal decomposition of the dioxetanes renders formic acid and a 20-ketone compound.

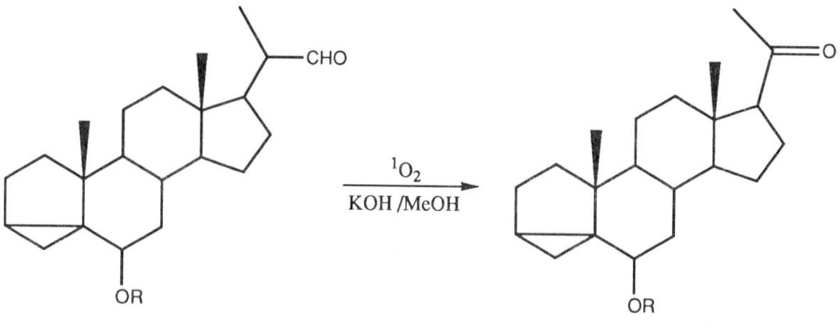

$$(27.9)$$

Other electron-rich olefins can be oxidatively cleaved[32a] with singlet oxygen and are thought to proceed via the formation of dioxetane intermediates (in some cases, these intermediates can be detected spectroscopically or chemiluminescence is observed). For example, enamines are converted[34] to ketones and amides, keteimines[35] to isocyanates, thioketene acetals[36] to dithioketones, and nitronates[37] to ketones by reaction with singlet oxygen.

A synthetically useful variation on this oxidative cleavage approach is based on the results[38] of Adam in the singlet oxygenation of a ketene thioacetal derivative. After treatment of a tetrathioethylene with singlet oxygen, a cyclic disulfide was isolated in low yield. The mechanism for this transformation was postulated to involve intermediate dioxetane formation via [2+2]-addition of singlet oxygen. The dioxetane appears to undergo homolytic cleavage of the O–O bond followed by β-scissions of the C–S bonds (rather than the usual C–C bond rupture) to give thio radicals, which, upon recombination, leads to cyclic disulfide (Scheme 27.3). A related reaction has been observed[39] for a dioxetane **9** generated in the oxidation of the corresponding ammonium salt with superoxide ion. The formation of the cyclic disulfide product (Equation 27.10) can be rationalized by a sequence analogous to that shown in Scheme 27.3.

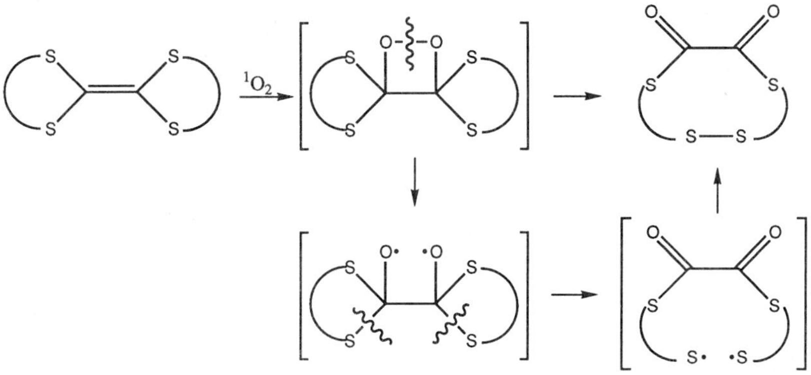

SCHEME 27.3

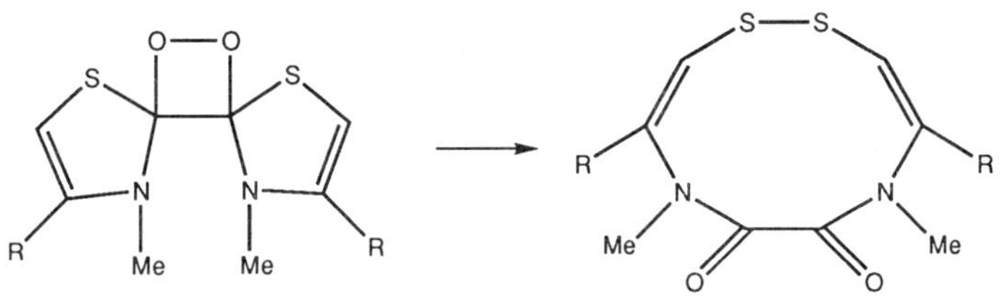

9

$$\tag{27.10}$$

Nucleophilic Transformations of Dioxetanes

The nucleophilic additions to the dioxetane ring can lead to carbonyl cleavage products, reduction to diols, and deoxygenation to epoxides. The labile peroxide bond in dioxetanes is prone to attack by reactive nucleophiles. For example, the reaction[40] of sulfides or phosphines with dioxetanes results in deoxygenation of the dioxetanes to epoxides. This process has been shown to involve phosphoranes[40a,b] for the trivalent phosphorus reactions (Equation 27.11). 1,2-Diols are obtained[41] when dioxetanes are treated with lithium aluminum hydride, mercaptans, or other reducing agents.[1] However, a recent report demonstrates the synthetic utility[42] of nucleophilic addition of heteroatom-substituted nucleophiles to a bromomethyl-substituted dioxetane for the synthesis of epoxy alcohols, cyclic carbonates, cyclic sulfites, and β-hydroxy hydroxylamines. These reactions are thought to proceed via an S_N2 attack at the less-hindered side of the peroxide bond in the dioxetane ring. Depending on the nature of the nucleophile, this adduct can undergo an intramolecular halide displacement to afford epoxy products or undergo 5-exo-trig cyclization to afford the cyclic carbonates and sulfites (Scheme 27.4).

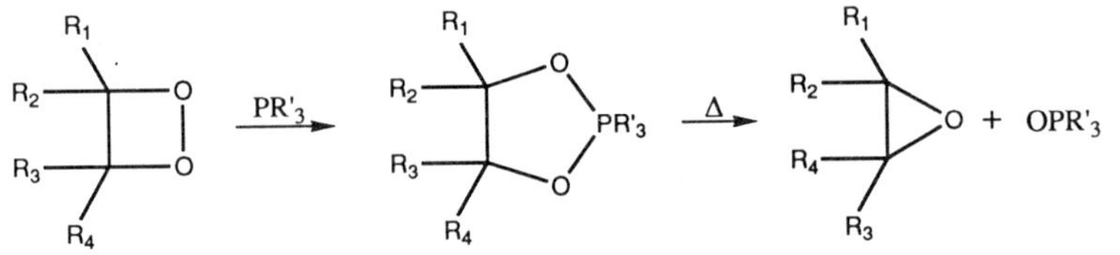

$$(27.11)$$

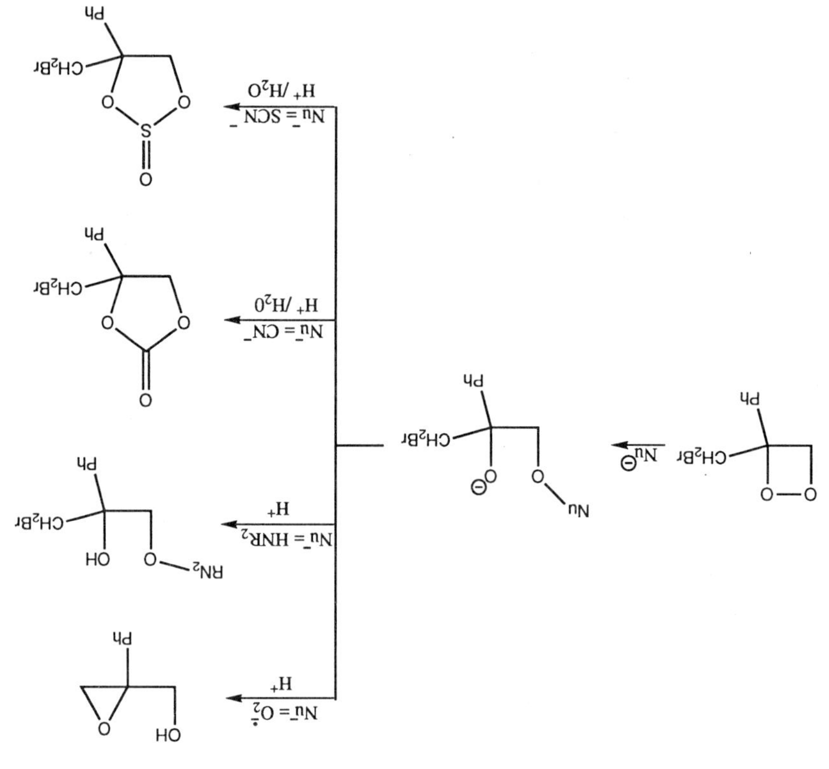

SCHEME 27.4

Acknowledgments

A. R. gratefully acknowledges support from the National Science Foundation (CHE-9014435), U.S. Army, CRDEC (#DAAA15–90-C-1076), and the RCMI program of Clark Atlanta University (NIH #RR03062). A.L.B. wishes to acknowledge the donors of the Petroleum Research Fund, administered by the American Chemical Society, for partial support of this work; the National Science Foundation (CHE-9017230); and the GSU Research Fund. We wish to express our deepest appreciation to our research students and postdoctoral fellows for their diligence and dedication. The authors thank Dr. Pedro C. Vasquez and Dan Adams for their invaluable assistance in the preparation of this manuscript.

References

1. For reviews see: (a) Wilson, T., Mechanisms of peroxide chemiluminescence, in *Singlet Oxygen,* Vol. II, Frimer, A., Ed., CRC Press, Uniscience, Boca Raton, FL, chap. 2, 1985; (b) Adam, W., Four-membered ring peroxides: 1,2-dioxetanes and α-peroxylactones, in *The Chemistry of Peroxides,* Patai, S., Ed., Wiley, New York, 1982, chap. 24, 829; (c) Schuster, G. B. and Schmidt, S. P., Chemiluminescence of organic compounds, *Adv. Phys. Org. Chem.*, 18, 187, 1982; (d) Bartlett, P. D. and Landis, M. E., The 1,2-dioxetanes, in *Singlet Oxygen,* Wasserman, H. H. and Murray, R. W., Eds., Academic Press, New York, 1979, 243; (e) Kopecky, K. R., Synthesis of 1,2-dioxetanes, in *Chemical and Biological Generation of Excited States,* Adam, W. and Cilento, G., Eds., Academic Press, New York, 1982, 85; (f) Baumstark, A. L., The 1,2-dioxetane ring system: Preparation, thermolysis and insertion reactions, in *Singlet Oxygen,* Vol. II, Frimer, A., Ed., CRC Press, Uniscience Series, Boca Raton, FL, chap. 1, 1985; (g) Baumstark, A. L., Thermolysis of 'alkyl'-1,2-dioxetanes, in *Advances in Oxygenated Processes,* Vol. I, Baumstark, A. L., Ed., JAI Press, Greenwich, CT, chap. 2, 1988.

2. See for example: (a) White, E. H. and Harding, M. J. C., Chemiluminescence in liquid solution. Chemiluminescence of lophine and its derivatives, *Photochem. Photobiol.*, 4, 1129, 1965; (b) McCapra, F., Richardson, D. G., and Chang, Y. C., Chemiluminescence involving peroxide decompositions, *Photochem. Photobiol.*, 4, 1111, 1965.

3. For historical perspective see: (a) McCapra, F., The chemistry of bioluminescence, *Endeavor*, 32, 139, 1973; (b) Rauhut, M. M., The chemiluminescence of concerted peroxide decomposition reactions, *Accts. Chem. Res.*, 2, 80, 1969; (c) White, E. H., The chemiluminescence of organic hydrazides, *Accts. Chem. Res.*, 3, 54, 1970; (d) McCapra, F., An application of the theory of electrolitic reactions to bioluminescence, *J. Chem. Soc., Chem. Commun.*, 155, 1968.

4. (a) Kopecky, K. R. and Mumford, C. K., Luminescence in the thermal decomposition of 3,3,4-trimethyl-1,2-dioxetane, *Can. J. Chem.*, 47, 709, 1969; (b) Adam, W. and Lui, J. C., An α-peroxylactone. Synthesis and chemiluminescence, *J. Am. Chem. Soc.*, 96, 2894, 1972.

5. Hummelen, J. C., Luider, T. M., Oudman, D., Koek, J. N., and Wynberg, H., 1,2-Dioxetanes: Luminescent and nonluminescent decomposition, chemistry, and potential applications, in *Luminescence Techniques in Chemical and Biochemical Analysis,* Vol. 12, Baeyens, W. R. G., Keukeleire, D. D., and Korkidis, K., Eds., Dekker, New York, 1991, chap. 18, 567.

6. (a) Bartlett, P. D. and Schaap, A. P., Stereospecific formation of 1,2-dioxetanes from *cis-* and *trans-*diethoxyethylenes by singlet oxygen, *J. Am. Chem. Soc.*, 92, 3223, 1970; (b) Mazur, S. and Foote, C. S., Chemistry of singlet oxygen. IX. A stable dioxetane from photooxygenation of tetramethoxyethylene, *J. Am. Chem. Soc.*, 92, 3225, 1970.

7. (a) Zalika, K. A., Kissel, T., Thayer, A. L., Burns, P. A., and Schaap, A. P., Mechanism of 1,2-dioxetane decomposition: The role of electron transfer, *Photochem. Photobiol.*, 30, 35 1979; (b) Schaap, A. P., Gugnon, S. D., and Zalika, K. A., Substituent effects on the decomposition of 1,2-dioxetanes: A Hammett correlation for substituted 1,6-diaryl-2,5,7,8-tetra oxabicyclo-[4.2.0]octanes, *Tetrahedron Lett.*, 2943 1982.

8. For a review of endoperoxide formation, see: Bloodworth, T. and Eggelte, H. J., Endoperoxides, in *Singlet Oxygen,* Vol. II, Frimer, A., Ed., CRC Press, Uniscience, Boca Raton, FL, 1985, chap. 4, 93.

9. For a review of the ene reaction, see: Frimer, A. A. and Stephenson, L. M., The singlet oxygen 'ene' reaction, in *Singlet Oxygen,* Vol. II, Frimer, A., Ed., CRC Press, Uniscience, Boca Raton, FL, 1985, chap. 3, 67.

10. (a) Wieringa, J. H., Strating, J., Wynberg, H., and Adam, W., Adamantylidene-adamantane peroxide, a stable 1,2-dioxetane, *Tetrahedron Lett.*, 169, 1972; (b) Bartlett, P. D. and Ho, M. S., *Bis*(1,4-cyclohexadiyl)-1,2-dioxetane and *bis*(1,4-cyclohexadiyl)oxirane from photooxidation of 7,7'-binorbornylidene, *J. Am. Chem. Soc.*, 96, 627, 1974; (c) Keul, J., Uber Konstitution und Entstehung der Ozonide von *Bis*-adamantyliden und von *Bis*-bicyclo[3.3.1]non-9-yliden, *Chem. Ber.*, 108, 1207, 1975.

11. For an overview, see: Frimer, A. A., Singlet oxygen in peroxide chemistry, in *The Chemistry of Peroxides*, Patai, S., Ed., Wiley, New York, 1982, chap. 7, 201.

12. Bartlett, P. D., Borderline cases of cycloaddition, *Pure Appl. Chem.*, 27, 597, 1971.

13. Foote, C. S., Mechanisms of addition of singlet oxygen to olefins and other substrates, *Pure Appl. Chem.*, 27, 639, 1971.

14. Harding, L. B. and Goddard, W. A., III, Intermediates in the chemiluminescent reaction of singlet oxygen with ethylene, *J. Am. Chem. Soc.*, 102, 439, 1980.

15. Frimer, A. A., The reaction of singlet oxygen with olefin: The question of mechanism, *Chem. Rev.*, 79, 359, 1979.

16. Schaap, A. P., Zaklika, K. A., Kaskar, B., and Fung, L. W.-M., Mechanisms of photooxygenation. 2. Formation of 1,2-dioxetanes via 9,10-dicyanoanthracene-sensitized electron-transfer process, *J. Am. Chem. Soc.*, 102, 389, 1981.

17. Wilson, T., Landis, M. E., Baumstark, A. L., and Bartlett, P. D., Solvent effects on the chemiluminescent decomposition of tetramethyl-1,2-dioxetane. Competitive dark pathways, *J. Am. Chem. Soc.*, 95, 4765, 1973.

18. (a) Clennan, E. L. and Lewis, K. K., Nucleophilic trapping of intermediates in the singlet oxygenations of isomeric 1,4-di-*tert*-butoxy-1,3-butadienes, *J. Org. Chem.*, 51, 3721, 1986; (b) Clennan, E. L., Singlet oxygenations of 1,3-butadienes, in *Advances in Oxygenated Processes*, Vol. I, Baumstark, A. L., Ed., JAI Press, Greenwich, CT, 1988, chap. 3.

19. (a) Rosenthal, I., Chemical and physical sources of singlet oxygen, in *Singlet Oxygen*, Vol. I, Frimer, A., Ed., CRC Press, Uniscience, Boca Raton, FL, 1985, chap. 2; (b) Frimer, A., Singlet oxygen in peroxide chemistry, in *The Chemistry of Peroxides*, Patai, S., Ed., Wiley, New York, 1982, chap. 7, 204–5.

20. (a) Burns, P. A. and Foote, C. S., Chemistry of singlet oxygen. XIX. Dioxetanes from indene derivatives, *J. Am. Chem. Soc.*, 96, 4339, 1974; (b) Zaklika, K. A., Thayer, A. L., and Schaap, A. P., Substituent effects on the decomposition of 1,2-dioxetanes, *J. Am. Chem. Soc.*, 100, 4916, 1978; (c) Jefford, C. W. and Pimbault, C. G., Characterization of a dioxetane deriving from norbornene and evidence for its Zwitterionic peroxide precursor, *J. Am. Chem. Soc.*, 100, 295, 1978.

21. Lopez, L., Troisi, L., Rashid, S. M. K., and Schaap, A. P., Synthesis of 1,2-dioxetanes via 9,10-dicyanoanthracene-sensitized chain electron-transfer photooxygenations, *Tetrahedron Lett.*, 485, 1989.

22. Sander, W. and Patyk, A., Photooxidation of 2,5-dimethyl-2,3,4-hexatriene. Matrix isolation of a tridioxetane, *Angew. Chem. Int. Ed.*, 26, 475, 1987.

23. Gundermann, K.-D. and McCapra, F., *Chemiluminescence in Organic Chemistry*, Springer-Verlag, Berlin, 1987.

24. Adam, W. and Yany, F., 1,2-Dioxetanes and α-peroxylactones, in *Heterocyclic Compounds*, Vol. 42, Part 3: Small Ring Heterocycles, Hassner, A., Ed., John Wiley, New York, 1985, 351.

25. Turro, N. J., Lechtken, P., Schore, N. E., Schuster, G., Steinmetzer, H.-C., and Yekta, A., Chemiexcitation and chemiluminescence of tetramethyl-1,2-dioxetane. A melting pot for ideas and playground for experiments in photochemistry, spectroscopy, chemical dynamics and reaction mechanisms, *Acc. Chem. Res.*, 7, 97, 1974.

26. Cilento, G., Electronic excitation in dark biological processes, in *Chemical and Biological Generation of Excited States*, Adam, W. and Cilento, G., Eds., Academic Press, New York, 1982, 278.

27. Cilento, G. and Adam, W., Photochemistry and photobiology without light, *Photochem. Photobiol.*, 48, 361, 1988.

28. (a) Schaap, A. P., Handley, R. S., and Giri, B. P., Chemical and enzymatic triggering of 1,2-dioxetanes. 1. Aryl esterase-catalyzed chemiluminescence from a naphthyl acetate-substituted dioxetane, *Tetrahedron Lett.*, 935, 1987; (b) Schaap, A. P., Chen, T.-S., Handley, R. S., DeSilva, R., and Giri, B. P., Chemical and enzymatic triggering of 1,2-dioxetanes. 2. Fluoride-induced chemiluminescence from *tert*-butyldimethylsilyloxy-substituted dioxetanes, *Tetrahedron Lett.*, 1155, 1987; (c) Schaap, A. P., Sandison, M. D., and Handley, R. S., Chemical and enzymatic triggering of

1,2-dioxetanes. 3. Alkaline phosphate-catalyzed chemiluminescence from an aryl phosphate-substituted dioxetane, *Tetrahedron Lett.*, 1159, 1987.

29. Beck, S. and Köster, H., Applications of dioxetane chemiluminescent probes to molecular biology, *Anal. Chem.*, 62, 2258, 1990.

30. (a) Jefford, C. W., Boukouvalas, J., Kohmoto, S., and Bernardinelli, G., New chemistry of Zwitterionic peroxides arising by photooxygenation of enol ethers, *Tetrahedron*, 41, 2081, 1985; (b) Adam, W. and Wang, X., Photooxygenation of silyl ketene acetals: Dioxetanes as precursors to α-silylperoxy esters in the silatropic ene reaction, *J. Org. Chem.*, 56, 4737, 1991.

31. (a) Avery, M. A., White-Jennings, C., and Chong, W. K. M., The total synthesis of (+)-artemisinin and (+)-9-desmethylartemisinin, *Tetrahedron Lett.*, 4629, 1987; (b) Avery, M. A., Chong, W. K. M., and Detre, G., Synthesis of (+)-8a,9-secoartemisinin and related analogs, *Tetrahedron Lett.*, 31, 1799, 1990; (c) Avery, M. A., Chong, W. K. M., and Jenning-White, C., Stereoselective total synthesis of (+)-Artemisinin, the antimalarial constituent of *Artemisia anna L.*, *J. Am. Chem. Soc.*, 114, 974, 1992; (d) Posner, G. and Oh, C. H., A regiospecifically oxygen-18 labeled 1,2,4-trioxane: A simple chemical model system to probe the mechanisms for the antimalarial activity of Artemisinin (qinghaosu), *J. Am. Chem. Soc.*, 114, 8328, 1992.

32. (a) Matsumoto, M., Synthesis with singlet oxygen, in *Singlet Oxygen*, Vol. II, Frimer, A., Ed., CRC Press, Uniscience, Boca Raton, FL, 1985, chap. 5, 206; (b) Otsuji, Y., Ohmura, N., Nakanishi, S., and Mizuno, K., Oxidation reactions of bicyclic pyrrole derivatives, *Chem. Lett.*, 1197, 1972 and see: Ninomiya, I. and Naito, T., Dioxetane intermediates for the synthesis of medium-sized heterocyclic rings, in *Photochemical Synthesis*, Academic Press, New York 1989, 196.

33. Sundararaman, P. and Djerassi, C., A convenient synthesis of progesterone from stigmasterol, *J. Org. Chem.*, 42, 3633, 1977.

34. (a) Wasserman, H. H. and Terao, S., Enamine-singlet oxygen reactions. α-diketones from intermediate amino dioxetanes, *Tetrahedron Lett.*, 1735, 1975; (b) Wasserman, H. H. and Ives, J. L., *J. Am. Chem. Soc.*, 98, 7868, 1976.

35. Inoue, Y. and Turro, N. J., Reaction of singlet oxygen with *trans*-cyclooctene. Evidence for a 1,2-dioxetane intermediate, *Tetrahedron Lett.*, 4327, 1980.

36. Geller, C. G., Foote, C. S., and Pechman, D. B., Chemistry of singlet oxygen. 41. Direct observation of dioxetane from the singlet oxygen photooxygenation of a thioketene acetal, *Tetrahedron Lett.*, 673, 1983.

37. Williams, J. R., Unger, L. R., and Moore, R. H., *J. Org. Chem.*, 43, 1271, 1978.

38. Adam, W. and Liu, J.-C., Photooxygenation (singlet oxygen) of tetrathioethylenes, *J. Am. Chem. Soc.*, 94, 1206, 1972.

39. Itoh, T., Naguta, K., Okada, M., Yamaguchi, K., and Ohsawa, A., The reaction of 3,3'-dimethyl-2,2'-bithiazolinium salts with superoxide, *Tetrahedron Lett.*, 33, 6983, 1992.

40. (a) Bartlett, P. D., Baumstark, A. L., and Landis, M. E., An insertion of triphenylphosphine with tetramethyl-1,2-dioxetane: Deoxygenation of a dioxetane to an epoxide, *J. Am. Chem. Soc.*, 95, 6486, 1973; (b) Baumstark, A. L. and Vasquez, P. C., Reaction of tetramethyl-1,2-dioxetane with phosphines: deuterium isotope effects, *J. Org. Chem.*, 49, 793, 1984; (c) Wasserman, H. H. and Saito, I., Trapping of intermediates in singlet oxygen reactions. Cleavage of dioxetanes by diphenyl sulfide, *J. Am. Chem. Soc.*, 97, 905, 1975; (d) Campbell, B. S., Denney, D. B., Denney D. Z., and Shih, L. S., Reaction of dioxetanes with sulfoxylates and sulfides. Preparation of novel tetraalkoxysulfuranes, *J. Am. Chem. Soc.*, 97, 3850, 1975.

41. (a) Kopecky, K. R., Filby, J. E., Mumford, C., Lockwood, P. A., and Ding, J.-Y., Preparation and thermolysis of some 1,2-dioxetanes, *Can. J. Chem.*, 53, 1103, 1975; (b) Adam, W., Epe, W., Schiffmann, D., Vargas, F., and Wild, D., Facile reduction of 1,2-dioxetanes by thiols as potential protective measures against photochemical damage of cellular DNA, *Angew. Chem. Int. Ed.*, 27, 249, 1988.

42. Adam, W. and Heil, M., Reaction of 1,2-dioxetanes with heteroatom nucleophiles: Adduct formation by nucleophilic attack at the peroxide bond, *J. Am. Chem. Soc.*, 114, 5591, 1992.

Oxidation of Aromatics

Angelo Albini
*University of Torino; currently
at the University of Pavia*

Mauro Freccero
University of Pavia

28.1 Introduction

The photoinduced oxidation of aromatics does not take place through a single mechanism. Indeed, aromatic derivatives are reactive substrates according to all of the three common mechanisms for photosensitized oxidation. These oxidation processes are distinguished from each other by the different role of the photoexcited sensitizer, as outlined in the following.

Type I: Activation of the substrate by energy or, more commonly, by hydrogen transfer. In this way, a radical — typically a benzylic radical — is formed and reacts with oxygen.[1,2]

$$Sens^* + Ar-X-H \rightarrow SensH\cdot + Ar-X\cdot \xrightarrow{O_2} Products \qquad (28.1)$$

Type II: Energy transfer from the sensitizer to oxygen and reaction of singlet excited-state oxygen with the substrate. Several aromatics are efficient sensitizers in this respect, and their irradiation in the presence of oxygen often causes a self-sensitized photooxidation.[2-5]

$$Sens^* + O_2 \rightarrow Sens + {}^1O_2 \qquad (28.2)$$

$$ {}^1O_2 + Ar-H \rightarrow Products \qquad (28.3)$$

Type III: Electron transfer from the aromatic substrate to the photoexcited sensitizer, and reaction of the radical cation with ground-state oxygen or the superoxide anion.[6,7]

$$Sens^* + ArH \rightarrow Sens^-\cdot + ArH^+\cdot \qquad (28.4)$$

$$Sens^-\cdot + O_2 \rightarrow Sens + O_2^-\cdot \qquad (28.5)$$

$$ArH^+\cdot + O_2 \text{ or } O_2^-\cdot \rightarrow Products \qquad (28.6)$$

Aromatic compounds played a special role in the development of our knowledge of singlet oxygen reactions. Thus, the bleaching of solutions of anthracene and its benzo homologues in the

0-8493-8634-9/95/$0.00+$.50
© 1995 by CRC Press, Inc.

presence of light and oxygen was observed in several laboratories. Early reports were somewhat blurred by the confusion between photodimerization and photooxidation, as well as by the fact that an erroneous structure had been attributed to the rubrenes, the photoreaction of which is particularly notable. However, in the mid-1930s, it had been established clearly that anthracene, naphthacene, and related hydrocarbons photochemically add oxygen to give endoperoxides as well-characterized crystalline solids, and that such products decompose upon heating (in some cases explosively) to give back the starting aromatic and molecular oxygen.[8,9] Later recognition that singlet oxygen was the active species led to the correct formulation of the process as a self-sensitized oxygenation (Equations 28.2 and 28.3, Sens = ArH).[10]

The mechanism is complicated by the fact that oxygen interacts with the ground state of aromatics (modifications of the UV spectrum due either to the formation of CT complexes or to enhancement of the S_0-T_1 transition are often observed),[11] as well as with both the singlet and the triplet excited states of these compounds. When both the S_1-T_1 and the T_1-S_0 energy gaps are larger than the energy of the $^1\Delta_g$ state of oxygen (22.4 kcal mol^{-1}), sensitization occurs according to Equations 28.7 and 28.8 with a limiting quantum yield of 2. This is not very important in practice, due to the short lifetime of the singlet state.[12-16]

$$ArH^{1*} + O_2 \rightarrow ArH^{3*} + {}^1O_2 \tag{28.7}$$

$$ArH^{3*} + O_2 \rightarrow ArH + {}^1O_2 \tag{28.8}$$

It should be further taken into account that the formation of exciplexes or electron transfer quenching of the singlet may result in an increase of the triplet (and thus of the singlet oxygen) yield.[16] Furthermore, the thermal retrocycloaddition of endoperoxides actually yields singlet oxygen rather than the ground-state molecule (see below, Equation 28.11),[17] and these compounds have become popular reagents for carrying out singlet-state oxygenations by way of a method alternative to photosensitization, either for preparative or, more usually, for mechanistic purposes. A large number of such endoperoxides have been prepared and characterized, and include compounds soluble in either organic solvents[17-19] or in water[20] as well as polymers,[21] with a large range of decomposition temperatures.

Aromatic compounds are common sensitizers in Type III photooxidations (Equation 28.4), and the same molecules may react either via a Type II or a Type III mechanism according to the reaction conditions.[15,22] Direct irradiation of a CT complex can also lead to a photochemical reaction.[11,23] From these facts, it is clear that a variety of processes is possible.

In the following sections, the observed reactions are discussed according to the structure of the substrate, while mentioning the mechanism involved in every case.

8.2 Oxidation of the Ring

Singlet oxygen is a typical dienophile, of small dimensions and minimal steric hindrance. It undergoes a facile [$4\pi+2\pi$]-cycloaddition with electron-rich aromatics. In the case of 1,3-dienes, it has been proposed that the cycloaddition process involves polar intermediates, such as a peroxide, rather than occurring concertedly through a six-membered transition state. The observed rates of reaction and regiochemistry of attack are rationalized reasonably with the Frontier Molecular Orbital approach.[24-27] The reaction with anthracene, naphthacene, and pentacene is very fast (1.5 $\times$ 10^5, 1.2 $\times$ 10^7, and 4.2 $\times$ 10^9 mol^{-1} s^{-1}, respectively, in benzene at 25 °C), while with phenanthrene, naphthalene, and, *a fortiori*, benzene derivatives, the rate is slower (e.g., 1.2 $\times$ 10^4 mol^{-1} s^{-1} for 1,4-dimethylnaphthalene under the same conditions).[27] Thus, the cycloaddition occurs with a preparatively useful quantum yield only with activated substrates.

Indeed, in the benzene series, the dye-sensitized reaction with singlet oxygen is significant only when electron-donating substituents are present and/or steric crowding hinders coplanarity and thus aromaticity. Thus, penta- and hexamethylbenzene undergo [4+2]-cycloaddition followed by ene addition on the allylic chromophore thus formed (Equation 28.9).[28] An endoperoxide is formed from [2.2.2.2](1,2,4,5)cyclophane (Equation 28.10).[29] Cleavage of a benzene ring following the cycloaddition has been observed with a benzo[3,4]cyclobuta[1,2-*b*]biphenylene.[30]

1-Methyl- (but not the 2-isomer) and all dimethyl-[31] and polyalkylnaphthalenes[19,32] undergo dye-sensitized oxidation to give 1,4-endoperoxides. With asymmetric substrates, the preferred attack is at the substituted ring, particularly if there is an α-substituent. Likewise, 1,4-dimethylphenanthrene adds oxygen at the substituted positions.[19]

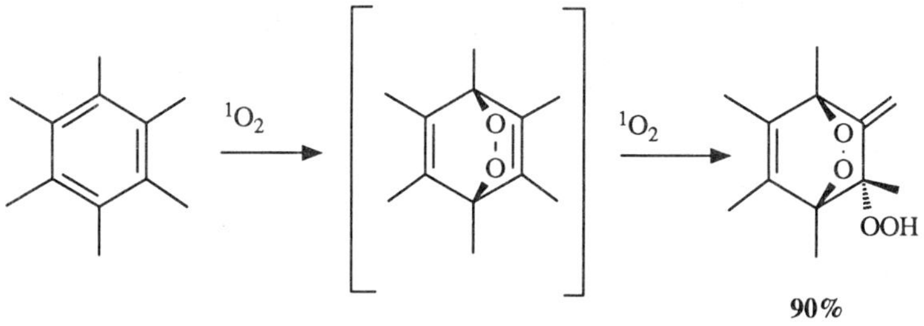

90%   (28.9)

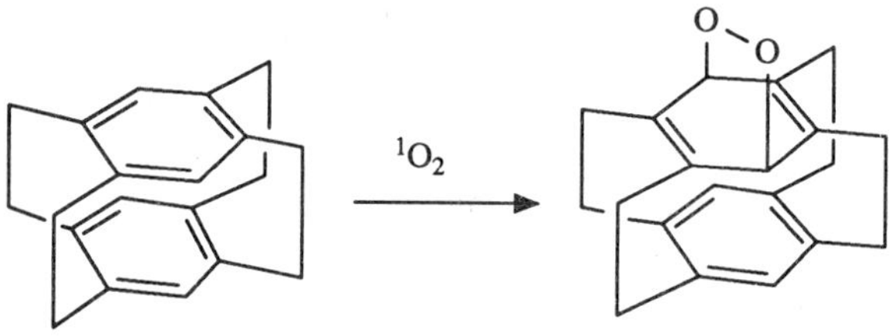

(28.10)

Anthracenes yield the 9,10-endoperoxides efficiently (Equation 28.11);[3,33] 1,4-dimethylanthracenes give a substantial yield of the 1,4-endoperoxides,[34,35] and reversion of the regiochemistry is complete when methoxy or dimethylamino groups are present at one of those positions.[3,9,26,36,37] No such effect is observed with substituents in the 2- or the 3-positions.[35,38] Naphthacenes[8,39–41] and pentacene[27] undergo fast self-sensitized oxidation. Although some other polycyclic aromatics have been reported to be unreactive with singlet oxygen,[2] several of them are photooxidized even if the endoperoxide has not always been characterized.[42,43] Heterocoerdianthrone yields an endoperoxide that undergoes adiabatic cycloreversion via the S_2 state when irradiated at $\lambda \leq 313$ nm. The analogous oxidation of *meso*-diphenylhelianthrene has been proposed as a reversible actinometer.[44]

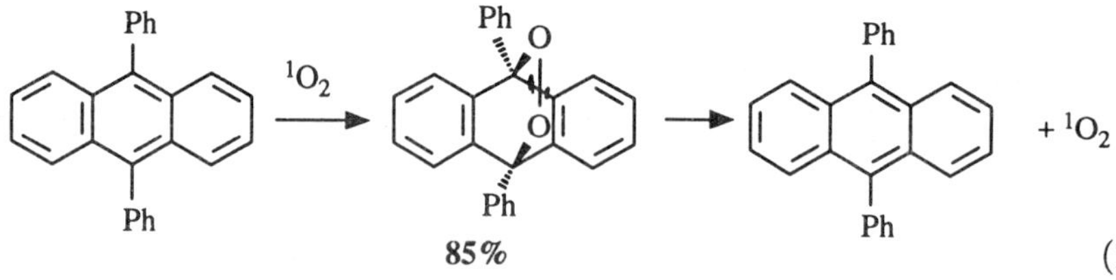

85%   (28.11)

Apart from the wavelength-dependent photocycloreversion,[45] the endoperoxides are useful synthetic intermediates in view of the variety of thermal or photochemical rearrangements they undergo.[25,33,37,46]

Oxygenations of arenes not involving singlet oxygen are less important, except for the side-chain functionalization discussed in Section 28.5. However, it should be mentioned that biphenyls are oxidized to benzoic acids under SET conditions with 9,10-dicyanoanthracene (DCA) as sensitizer in the presence of $Mg(ClO_4)_2$. Omission of the salt lead to the conversion of the sensitizer to anthraquinone.[47,48] Cyclopentadienecarboxyaldehyde is formed in the direct irradiation of oxygen-equilibrated benzene. This is presumed to involve photoisomerization of benzene to benzvalene followed by oxidation of the latter.[49]

8.3 Oxidation of Vinylarenes

Vinylarenes show a double reactivity, according to whether the singlet oxygen or the electron transfer paths is followed. As an example of the first path, dye-sensitized photooxygenation of *trans*-stilbene yields a dihydrobenzo-1,2-dioxin in the primary step. This is followed by rapid oxygenation of the cyclohexadiene chromophore to yield a diendoperoxide in 80% yield, along with minor amounts of benzaldehyde and styrene epoxide arising from alternative mechanisms (Equation 28.12).[50] Oxidation of *cis*- and *trans*-1-(β-methoxyvinyl)naphthalenes also yield mono endoperoxides in which the stereochemistry of substituent is retained.[51] β-Methoxystyrenes,[51-54] as well as the enol ethers (or silyl ethers) of phenylpyruvic acid,[55] phenylacetone, and related compounds,[56] also give endoperoxides that can be used for specific *ortho*-functionalization[56] or as equivalents of *o*-benzoquinone methides (Equation 28.13).[53,54]

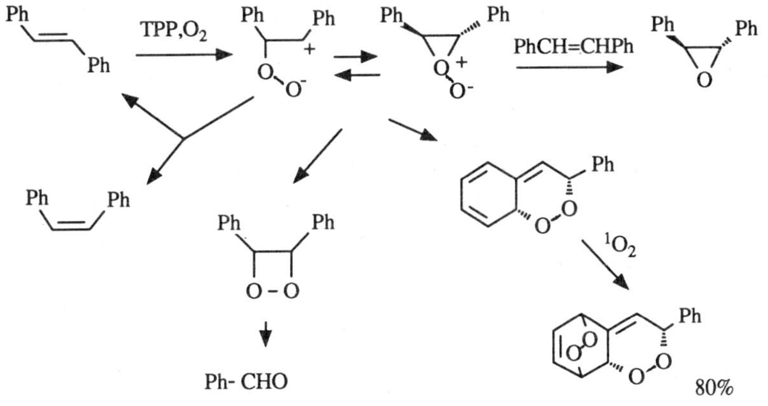

(28.12)

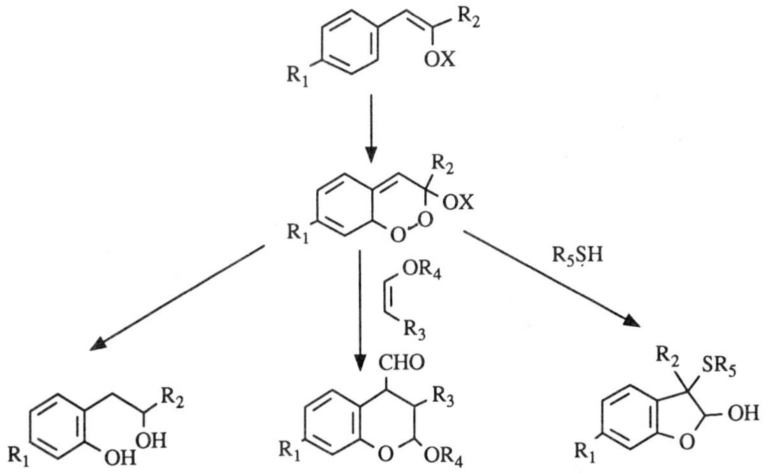

(28.13)

In other cases, reactions typical of the olefinic chromophore compete with the [4+2]-cycloaddition. Thus, with indene, [2+2]- and [4+2]-cycloaddition occur to different extents depending on the solvent used;[57] and with 1,2-dihydronaphthalenes, the competition between an ene reaction giving a hydroperoxide and a [4+2]-cycloaddition process followed by rearrangement and a second addition to yield a diepoxyendoperoxide depends on the substituents present.[58] The ene reaction occurs exclusively with α,α'-dimethylstilbene[59] and 1-phenylcyclopentene, whereas it competes with [4+2]-cycloaddition in the case of 1-phenylcyclohexene.[60] In the case of 1-phenylcyclobutene[61] the ene pathway, oxidative cleavage of the double bond (occurring also with phenylnorbornene),[62] and ring contraction are all observed.

The cleavage of steroidal ring C, observed in some estratetraene derivatives,[63,64] has been exploited for the synthesis of 11-oxaestrogens (Equation 28.14).[65]

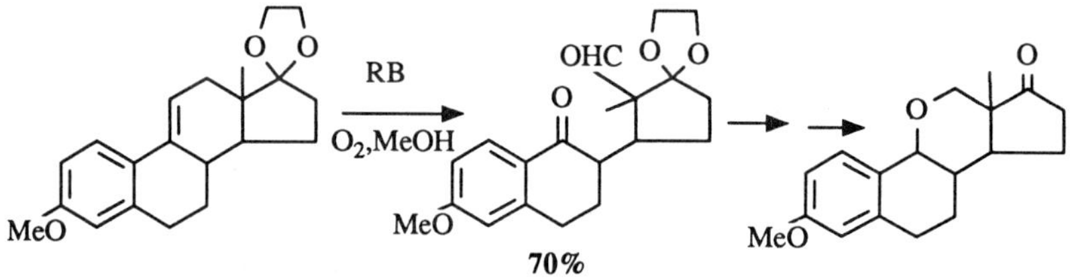

70%

$$(28.14)$$

As for the second path, electron transfer sensitized oxygenation of arylalkenes in the presence of DCA[66] or of heterogeneous sensitizers such as TiO_2[67] generally causes the oxidative cleavage of the double bond along with epoxidation as a minor path (diketones and benzoic acids are obtained from arylalkynes).[68] The mechanism involves reaction of the arylalkene radical cation either with the superoxide anion, formed by secondary electron transfer from the radical anion of the sensitizer, or with oxygen. This latter reaction mode is more likely due to kinetic factors.[69] This mechanism holds also for methoxyvinylarenes,[70,71] although in the case of alkoxyaryl-methyleneadamantanes, a stable dioxetane is obtained via a chain electron transfer mechanism.[72] A chain mechanism is also involved in the formation of 1,2-dioxanes (≥90% with electron-rich substrates), ketones, and cyclobutanes from 1,1-diarylethylenes (Equation 28.15).[73,74] The photo-oxidation of 3,3-dimethylindene in methanol gives methoxyhydroperoxides with the opposite regiochemistry when it is carried out under SET or $^1O_2$ conditions, due to the fact that oxygen and the nucleophile add in the opposite order in the two cases.[75]

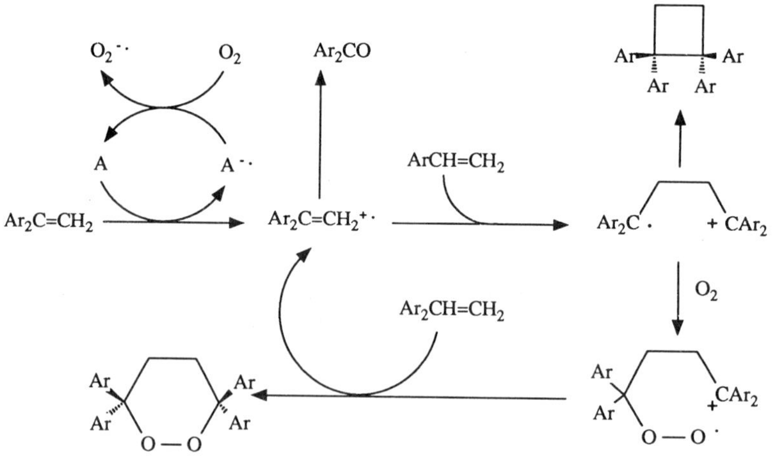

$$(28.15)$$

28.4 Oxidation of Phenols, Aromatic Amines, and Ethers

The dye-sensitized photooxidation of phenols proceeds via either a Type I or Type II mechanism. In fact, both the sensitizer triplet and $^1O_2$ can abstract a hydrogen from the O-H bond (in the latter case, probably via sequential electron and proton transfer). The resulting phenoxy radicals undergo coupling to yield a biphenyl or addition to ground-state oxygen to yield a 4-hydroperoxy-cyclohexadienone. On the other hand, the latter product can arise also through 1,4-addition onto the ring followed by nucleophilic addition or an ene reaction.[25,76-81]

Apart from the mechanistic problems, photosensitized oxidation of phenols followed by reduction of the hydroperoxide group affords *p*-quinols in fair yields. In other cases, or under different conditions, ring degradation results.[80,81] These reactions have been studied particularly as models for the degradation of biologically important phenols, e.g., *p*-hydroxyphenylacetic and pyruvic acids,[82] α-tocopherol,[83] and estrogenic steroids.[84] Naphthols[85] undergo related processes. Preparatively valuable applications have been found in the stereospecific oxidation of a tetracycline derivative (Equation 28.16)[86] as well as in the synthesis of 14-hydroxymorphinans.[87]

$$(28.16)$$

Polymethoxybenzenes yield diepoxides, epoxyenones, or epoxyquinones, all arising from the rearrangement of 1,4-endoperoxides formed initially.[88,89] Dye-sensitized photooxygenation of 9,10-dimethoxyphenanthrene yields, according to the conditions, phenanthrenequinone or a cleaved diester, probably via a dioxetane.[90] The SET-sensitized oxidation of 1,2-dimethoxybenzene also results in ring cleavage via a dioxetane.[91] Tertiary amines are efficient physical quenchers of singlet oxygen. However, chemical reaction (both Type I and Type II) also takes place with some electron-donor substituted dimethylanilines.[92] 1-Naphthylamine undergoes ring cleavage upon photooxygenation.[93]

28.5 Side-Chain Oxidation

Electron transfer photoinduced oxygenation of the side-chain is often efficient. Thus, aromatic radical cations bearing an electrofugal group at the α-position undergo fragmentation of the σ-bond, and the benzyl radicals thus formed add oxygen or the superoxide anion to yield hydroperoxides. Aldehydes or ketones are formed by this process and further oxidation can transform them into carboxylic acids. The cleaved cation is generally a proton, but a carbocation or a silyl cation work equally well. Thus, alkylaromatics,[94-99] benzyl alcohols or ethers,[100,101] bibenzyls,[94] arylpinacols,[94] and benzylsilanes[102] are photooxygenated in the presence of sensitizers such as DCA, quinones, or metallic compounds. The mechanism of the process with different sensitizers need not be uniform. With relatively good singlet oxygen acceptors, such as 1,4-dimethylnaphthalene, formation of the endoperoxide competes with side-chain oxidation to an extent dependent on the solvent polarity.[99] With benzyl ethers, the reaction may serve as a mild deprotecting procedure.[101]

The SET-induced oxygenation of arylated three-membered rings is a subgroup of the foregoing. Thus, cyclopropanes can be converted to dioxolanes,[103,104] oxirans to ozonides,[103,105] and aziridines to 1,2,4-dioxazolidines.[103] In each case, fragmentation of a strained C-C bond in the radical cation is the dominant process.

The reactions discussed thus far are based on the use of molecular sensitizers, usually DCA. However, side-chain oxidation, again through a SET path, can be carried out with heterogeneous sensitizers. Thus, oxidation of toluene[106] or 1-methylnaphthalene[107] (in the latter case, accompanied by ring-cleavage) or of benzyl ethers (to yield benzoates)[108] are typical examples. On the other hand, unsensitized reactions have also been observed, such as direct irradiation of toluene and polymethylbenzenes to yield alcohols, aldehydes or acids, as well as bibenzyls[109] via benzyl radicals or products from the trapping of benzyl cations (e.g., methyl ethers with MeOH).[23,110] Such reactions probably involve excitation of an EDA complex with oxygen followed by electron and proton transfer.[23,110]

28.6 Oxidations Not Involving Molecular Oxygen

Finally, one should take into account the several reported photooxidations of aromatics not involving molecular oxygen. As an example, ring hydroxylation occurs by reaction with photogenerated atomic ^3P oxygen in the gas phase,[111] or in solution by addition of hydroxy radicals formed by heterogeneous photocatalysis in water[112,113] or hydrogen peroxide splitting[114] or hydroperoxy radicals,[115] or again via interaction with excited heterocyclic *N*-oxides or products arising from their photolysis (see Chapter 70).[116–119] Excited nitroaromatics function likewise as oxygen donors and bring about oxidative ring cleavage of methoxynaphthalenes.[120]

Such reactions have some significance inasmuch as some photohydroxylations may mimic enzimatic paths.[117,119] In addition, they are important pathways for the photodegradation of pollutants such as haloaromatics,[121] in water. Thus, chlorobenzenes, chlorophenols, and dioxins are completely degraded (photomineralized) when irradiated in the presence of colloidal TiO_2.[122,123]

As has been shown in previous sections, photooxidation of aromatics offers several valuable synthetic paths, some of which lead to molecules of industrial significance. On the other hand, one should not underrate the application of this reaction to the degradation of pollutants in industrial or agricultural effluents.

References

1. Livingston, R., Photochemical autoxidation, in *Autoxidation and Antioxidants*, Vol. I, Lundberg, W. O., Ed., Interscience, New York, 1961, 249.
2. Gollnick, K., Photooxygenation reactions in solution, *Adv. Photochem.*, 6, 1, 1968; Gollnick, K. and Schenck, G. O., Oxygen as a dienophile, in *1,4-Cycloaddition Reactions*, Hamer, J., Ed., Academic Press, New York, 1967, 255.
3. Rigaudy, J., Photooxidation of aromatic derivatives, *Pure Appl. Chem.*, 16, 169, 1968.
4. Frimer, A. A., Ed., *Singlet Oxygen*, CRC Press, Boca Raton, FL, 1985.
5. Wasserman, H. H. and Murray, R. W., Eds., *Singlet Oxygen*, Academic Press, New York, 1979.
6. Fox, M. A., Activation of oxygen by photoinduced electron transfer, in *Photoinduced Electron Transfer*, Vol. D, Fox, M. A. and Chanon, M., Eds., Elsevier, Amsterdam, 1988, 1.
7. Lopez, L., Photoinduced electron transfer oxygenations, *Top. Curr. Chem.*, 156, 119, 1990.
8. Dufraisse, C. and Horclois, R., Naphthacenes, synthesis and photochemical peculiarities, *Bull. Soc. Chim. Fr. [5]*, 3, 1873, 1936. Synthesis of phenylnaphthacenes with the characteristics of rubrenes, *Bull. Soc. Chim. Fr. [5]*, 3, 1894, 1936.
9. Dufraisse, C. and Le Bras, J., Photooxides of *meso*-diphenylanthracenes: Formation, dissociation and properties, *Bull. Soc. Chim. Fr. [5]*, 4, 349, 1937.
10. Corey, E. J. and Taylor, W. C., Peroxidation of organic compounds by externally generated singlet oxygen molecules, *J. Am. Chem. Soc.*, 86, 3881, 1964.
11. Tsubomura, T. and Mulliken, R. S., UV absorption caused by the interaction of oxygen with organic molecules, *J. Am. Chem. Soc.*, 82, 5966, 1960.
12. Wu, K. C. and Trozzolo, A. M., Production of singlet molecular oxygen from the O_2 quenching of the lowest excited singlet state of rubrene, *J. Phys. Chem.*, 83, 2823, 1979.
13. Stevens, B., Marsch, K. L., and Barltrop, J. A., Sensitizer yields of O_2 $^1\Delta_g$, *J. Phys. Chem.*, 85, 3079, 1981.
14. Marsch, K. L. and Stevens, B., Dependence of the pyrene-sensitized O_2 $^1\Delta_g$ yield on pyrene concentration, *J. Phys. Chem.*, 87, 1765, 1983.
15. Albini, A. and Spreti, S., The photooxygenation of simple alkenes sensitized by cyanoanthracenes, *Gazz. Chim. Ital.*, 115, 227, 1985.
16. Davidson, R. S. and Pratt, J. E., Eximers and exciplex as sensitizers for photooxidation reactions, *Tetrahedron*, 40, 999, 1984.

17. Turro, N. J., Chow, M. F., and Rigaudy, D. L., Thermolysis of anthracene endoperoxides, *J. Am. Chem. Soc.*, 101, 1300, 1979.

18. Wasserman, H. H., Scheffer, J. R., and Cooper, J. L., Singlet oxygen reaction with 9,10-diphenylanthracene peroxide, *J. Am. Chem. Soc.*, 94, 4991, 1972.

19. Wasserman, H. H. and Larsen, D. L., Formation of 1,4-endoperoxides form the dye-sensitized photo-oxygenation of alkyl-naphthalenes, *J. Chem. Soc., Chem. Commun.*, 253, 1972.

20. Inoue, K., Matsuura, T., and Saito, I., Oxidation of electron-rich thioanisoles, *Tetrahedron*, 41, 2177, 1985.

21. Saito, I., Nagato, T., and Matsuura, T., Methyl-substituted poly(vinylnaphthalenes) as a reversible singlet oxygen carrier, *J. Am. Chem. Soc.*, 107, 6329, 1985.

22. Foote, C. S., Cyanoanthracenes sensitized photo-oxygenation of olefins, *Tetrahedron*, 41, 2221, 1985.

23. Onodera, K., Furusawa, G., Kojma, M., Tsuchiya, M., Aihara, S., Akaba, R., Sakuragi, H., and Tokumaru, K., Mechanistic considerations on the photochemical reactions of organic compounds via excitation of contact charge transfer complexes with oxygen, *Tetrahedron*, 41, 2215, 1985.

24. van den Heuvel, C. J. M., Verhoeven, J. W., and de Boer, T. J., A frontier orbital description of the reaction of singlet oxygen with simple aromatic systems, *Recl. Trav. Chim. Pays-Bas*, 99, 280, 1980.

25. Saito, I. and Matsuura, T., The oxidation of electron-rich aromatic compounds, in *Singlet Oxygen*, Wasserman, H. H. and Murray, R. W., Eds., Academic Press, New York, 1979, 511.

26. Chalvet, O., Daudel, R., Schmid, G. H., and Rigaudy, J., Theoretical treatment of the transition state. Two photochemical reactions, *Tetrahedron*, 26, 365, 1970.

27. Stevens, B., Perez, S. R., and Ors, J. A., O_2 $^1\Delta_g$ acceptor properties and reactivities, *J. Am. Chem. Soc.*, 96, 6846, 1974.

28. van den Heuvel, C. J. M., Hofland, A., Steinberg, H., and de Boer, T. J., The photo-oxidation of hexamethyl- and pentamethylbenzene by singlet oxygen, *Recl. Trav. Chim. Pays-Bas*, 99, 275, 1980.

29. Gray, R. and Boekelheide, V., Synthesis and properties of [2.2.2.2](1,2,4,5)cyclophane, *J. Am. Chem. Soc.*, 101, 2128, 1979.

30. Mestdagh, H. and Vollhart, K. P. C., Photo-oxidation of 2,3,7,8-tetrakis(trimethylsilyl)benzo[3,4]cyclobuta[1,2-*b*]biphenylene in the presence of oxygen: Unusual cleavage of the benzene ring to generate an alkyne unit, *J. Chem. Soc., Chem. Commun.*, 281, 1986.

31. van den Heuvel, C. J. M., Steinberg, H., and de Boer, T. J., The photo-oxidation of mono- and dimethylnaphthalenes by singlet oxygen, *Recl. Trav. Chim. Pays-Bas*, 99, 109, 1980.

32. Hart, H. and Oku, A., Octamethylnaphthalene 1,4-endoperoxide, *J. Chem. Soc., Chem. Commun.*, 254, 1972.

33. Rigaudy, J., Baranne-Lafont, J., Defoin, A., and Cuong, N. K., Chemical transformation of anthracene photo-oxides, *Tetrahedron*, 34, 73, 1978.

34. Rigaudy, J., Guillaume, J., and Maurette, D., Formation of isomeric 1,4 and 9,10 photo-oxides from 1,4-dimethylanthracenes, *Bull. Soc. Chim. Fr.*, 144, 1971.

35. Mellier, M. T., Effect of some substituents on the photo-oxidation of *meso*-diphenylanthracenes, *Ann. Chim. (Paris) [12]*, 10, 666, 1955.

36. Rigaudy, J., Dupont, R., and Cuong, N. K., Photo-oxidation of 1,4-*bis*(benzyloxy)anthracenes, *C. R. Hebd. Seances Acad. Sci., Ser. C*, 269, 416, 1969.

37. Rigaudy, J., Defoin, A., and Cuong., N. K., The photo-oxide of 1-dimethylamino-9,10-diphenylanthracene and its transformations, *C. R. Hebd. Seances Acad. Sci., Ser. C*, 271, 1258, 1970.

38. Panico, R., Thioethers of *meso*-diphenylanthracenes, *Ann. Chim. (Paris) [12]*, 10, 695, 1955.

39. Rigaudy, J. and Sparfel, D., Regioselectivity in the photo-oxidation and diene addition with 5,12-diphenylnaphthacene, *Bull. Soc. Chim. Fr.*, 742, 1977.

40. Sy, A. and Hart, H., Permethylnaphthacene, *J. Org. Chem.*, 44, 7, 1979.

41. (a) Aubry, J. M., Rigaudy, J., and Cuong, N. K., Kinetic studies of self-sensitized photo-oxygenation of a water soluble rubrene derivative, *Photochem. Photobiol.*, 33, 149, 1981; (b) A water soluble rubrene derivative: Synthesis, properties and $^1O_2$ trapping, *Photochem. Photobiol.*, 33, 155, 1981.

42. Brokmann, H. and Dicke, F., Studies of dibenzo[*a,o*]perylene photooxidation, *Chem. Ber.,* 103, 7, 1970.

43. Kajiwara, T., Fujisawa, S., Ohno, K., and Harada, Y., Photooxygenation product of dibenzo[*a,j*]perylene, *Bull. Chem. Soc. Jpn.,* 52, 2771, 1979.

44. (a) Schmidt, R., Drews, W., and Brauer, H. D., Photolysis of the endoperoxide of heterocoerdianthrone: A concerted, adiabatic cycloreversion originating from an upper excited singlet state, *J. Am. Chem. Soc.,* 102, 2791, 1980; (b) Schmidt, R. and Brauer, H. D., Self-sensitized photo-oxidation of aromatic compounds and photocycloreversion of endoperoxides: Application in chemical actinometry, *J. Photochem.,* 25, 489, 1984.

45. Gabriel, R., Schmidt, R., and Brauer, H. D., Wavelength-dependent and adiabatic photochemistry of the 1,4-endoperoxide of 1,4-dimethoxy-9,10-diphenylanthracene, *Z. Phys. Chem. (Munich),* 141, 41, 1984.

46. Balci, M., Bicyclic endoperoxides and synthetic applications, *Chem. Rev.,* 81, 91, 1981.

47. Mizuno, K., Ichinose, N., Tamai, T., and Otsuji, Y., Electron transfer mediated photooxygenation of biphenyl and its derivatives in the presence of $Mg(ClO_4)_2$, *Tetrahedron Lett.,* 5823, 1985.

48. Mizuno, K., Tamai, T., Nakanishi, I., Ichinose, N., and Otsuji, Y., Photooxygenation of cyanoanthracenes via their radical anions, *Chem. Lett.,* 2065, 1988.

49. (a) Kaplan, L., Wendling, L. A., and Wilzbach, K. E., Photooxidation of aqueous benzene to 1,3-cyclopentadien-1-carboxyaldehyde, *J. Am. Chem. Soc.,* 93, 3819, 1971; (b) Role of benzvalene in the formation of cyclopentadiencarboxyaldehyde, *J. Am. Chem. Soc.,* 93, 3821, 1971.

50. Kwon, B. M., Foote, C. S., and Khan, S. I., Reaction of singlet oxygen with *trans*-stilbene, *J. Org. Chem.,* 54, 3378, 1989.

51. Matsumoto, M. and Kuroda, K., Sensitized photooxygenation of β-methoxystyrene and 1-(β-methoxyvinyl)naphthalene, *Tetrahedron Lett.,* 1607, 1979.

52. Lerdal, D. and Foote, C. S., Directing effect of the methoxy group in additions to methoxystyrenes, *Tetrahedron Lett.,* 3227, 1978.

53. Matsumoto, M. and Kuroda, K., *o*-Benzoquinone monoformylmethides by sensitized photooxygenation of *cis*-β-methoxystyrene, *Angew. Chem. Int. Ed.,* 21, 382, 1982.

54. Matsumoto, M., Kurota, K., and Suzuki, Y., The 1,4-addition of singlet oxygen to 2,6-dimethoxy-1-(2-methoxyethenyl)benzene and 2-methoxy-1-(2-methoxyethenyl)naphthalene. The 1,4-endoperoxide as equivalent of 6-oxo-2,4-cyclohexadienyl acetates, *Tetrahedron Lett.,* 3253, 1981.

55. Kotsuki, H., Saito, I., and Matsuura, T., Photosensitized oxygenation of phenylpyruvic acid derivatives as a model for *p*-hydroxyphenylpyruvate dioxygenase, *Tetrahedron Lett.,* 469, 1981.

56. Saito, I., Nagata, R., Kotsuki, H., and Matsuura, T., Regiospecificity on the functionalization of substituted benzenes with singlet oxygen, *Tetrahedron Lett.,* 1717, 1982.

57. Zhang, J. and Foote, C. S., Photooxidation of substituted indenes at low temperature, *Tetrahedron Lett.,* 6153, 1986.

58. Burns, P. A. and Foote, C. S., Low-temperature photooxygenation of 1,2-dihydronaphthalenes, *J. Org. Chem.,* 41, 908, 1976.

59. Futamura, S., Ohta, H., and Kamiya, Y., Sensitized photooxidation of *cis*-α,α'-dimethylstilbene, *Chem. Lett.,* 697, 1983.

60. Jefford, C. W. and Rimbault, C. G., The reaction of singlet oxygen with 2-phenylcycloalkenes possessing small and common ring, *Tetrahedron Lett.,* 2479, 1976.

61. Sakuragi, M. and Sakuragi, H., Photosensitized oxidation of 1-phenylcyclobutene. The role of reactive active species other than singlet oxygen, *Chem. Lett.,* 1017, 1980.

62. Jefford, C. W., Boschung, A. F., and Rimbault, C. G., Reaction of singlet and triplet oxygen with 2-phenylnorbornene, *Helv. Chim. Acta,* 59, 2542, 1976.

63. Nowicki, A. W. and Turner, A. B., Photo-oxidation of an activated styrene, 17β-hydroxy-5-methoxy-de-*A*-oestra-5,7,9,14-tetraene, *J. Chem. Res. (S),* 110, 1981.

64. Planas, A., Lupon, P., Cascallo, M., and Bonet, J. J., Product characterization and kinetics of dye-sensitized photo-oxygenation of 9,11-didehydroestrone derivatives, *Helv. Chim. Acta,* 72, 715, 1989.

65. (a) Planas, A., Sala, N., and Bonet, J. J., Synthesis of 11-oxaestrogens via dye-sensitized photo-oxygenation of a 9,11-didehydroestrone derivative, *Helv. Chim. Acta,* 72, 725, 1989.

66. Eriksen, J. and Foote, C. S., Oxidation of phenyl-substituted alkenes sensitized by cyanoanthracenes, *J. Am. Chem. Soc.,* 102, 6083, 1980; (b) Griffin, G. W., Kirschenheuter, G. P., Vaz, C., Umrigar, P. P., Lankin, D. C., and Christensen, S., The sensitized photo-oxygenation of methyl substituted 1,2-diphenylcyclobutenes, *Tetrahedron,* 41, 2207, 2069.

67. (a) Fox, M. A. and Chen, C. C., Mechanistic features of the semiconductor photocatalyzed olefin-to-carbonyl oxidative cleavage, *J. Am. Chem. Soc.,* 103, 6757, 1981; (b) Sackett, D. A. and Fox, M. A., Effect of cosolvent additives on relative rates of photooxidation on semiconductor surfaces, *J. Phys. Org. Chem.,* 1, 103, 1988.

68. Mattes, S. L. and Farid, S., Photo-oxygenation via electron-transfer and its susceptibility to catalysis, *J. Chem. Soc., Chem. Commun.,* 457, 1980.

69. Tsuchiya, M., Ebbesen, T. W., Nishimura, Y., Sakuragi, H., and Tokumaru, K., Kinetic studies on electron transfer photooxygenation of aromatic olefins. Quenching rates of olefin radical cations by oxygen and superoxide anion, *Chem. Lett.,* 2121, 1987.

70. Steichen, D. S. and Foote, C. S., Indirect sensitized photooxygenation of aryl olefins, *J. Am. Chem. Soc.,* 103, 1855, 1981.

71. Lopez, L., Synthesis of Z-stilbenediol dibenzoate by sensitized photooxygenation of 2,3,5,6-tetraphenyl-*p*-dioxin, *Tetrahedron Lett.,* 4383, 1985.

72. Lopez, L., Troisi, L., Rashid, S. M. K., and Schaap, A. P., Synthesis of 1,2-dioxetanes via 9,10-dicyanoanthracene sensitized chain electron-transfer photooxygenations, *Tetrahedron Lett.,* 485, 1989.

73. (a) Gollnick, K. and Schnatterer, A., Formation of a 1,2-dioxane by electron-transfer photooxygenation of 1,1-di-(*p*-anisyl)ethylene, *Tetrahedron Lett.,* 185, 1984; (b) Formation of 1,2-dioxanes by electron-transfer photooxygenation of 1,1-disubstituted ethylenes, *Tetrahedron Lett.,* 2735, 1984.

74. Mattes, S. L. and Farid, S., Photochemical electron transfer reactions of 1,1-diarylethylenes, *J. Am. Chem. Soc.,* 108, 7356, 1986.

75. Mattes, S. L. and Farid, S., Photooxygenations via electron transfer. 1,1-Dimethylindene, *J. Am. Chem. Soc.,* 104, 1454, 1982.

76. Samsonova, L. V., Taimr, L., and Pospisil, J., Oxidation and photooxidation of alkyl 3-(3,5-di-*tert*-butyl-4-hydroxyphenyl)propionates, *Angew. Makromol. Chem.,* 65, 197, 1977.

77. Pfoetner, K. and Boese, D., The photosensitized oxidation of monohydric phenols to quinones, *Helv. Chim. Acta,* 53, 1553, 1970.

78. Matsuura, T., Yoshimura, N., Nishinaga, A., and Saito, I., Participation of singlet oxygen in the hydrogen abstraction from a phenol in the photosensitized oxygenation, *Tetrahedron,* 28, 4933, 1972.

79. Thomas, M. J. and Foote, C. S., Photooxygenation of phenols, *Photochem. Photobiol.,* 27, 683, 1978.

80. Saito, I., Yoshimura, N., Arai, T., Omura, K., Nishinaga, A., and Matsuura, T., Addition of singlet oxygen to 4,6-di-*t*-butylresorcinol and its derivatives, *Tetrahedron,* 28, 513, 1982.

81. Matsuura, T., Matsushima, H., Kato, S., and Saito, I., Photosensitized oxygenation of cathecol and hydroquinone derivatives: Non-enzymic models for the enzymatic cleavage of phenolic rings, *Tetrahedron,* 28, 5119, 1972.

82. Saito, I., Chujio, Y., Shimazu, H., Yamane, M., Matsuura, T., and Cahnmann, H. J., Non-enzymic oxidation of *p*-hydroxyphenylpyruvic acid with singlet oxygen to homogentisic acid. A model for the action of *p*-hydroxyphenylpyruvic hydroxylase, *J. Am. Chem. Soc.,* 97, 5272, 1975.

83. Clough, R. L., Yee, B. C., and Foote, C. S., The unstable primary product of tocopherol photooxidation, *J. Am. Chem. Soc.,* 101, 683, 1979.

84. Lupon, P., Gomez, J., and Bonet, J. J., The photooxygenation of estrogens: A new synthesis of 19-norsteroids, *Angew. Chem. Int. Ed.,* 22, 711, 1973.

85. (a) Griffiths, J., Chu, K. Y., and Hawkins, C., Photosensitized oxidation of 1-naphthols, *J. Chem. Soc., Chem. Commun.*, 676, 1976; (b) Bortolus, P., Monti, S., Albini, A., Fasani, E., and Pietra, S., Physical quenching and chemical reaction of singlet molecular oxygen with azo dyes, *J. Org. Chem.*, 54, 534, 1989.

86. von Wittenau, M. S., Preparation of tetracyclines by photooxidation of anhydrotetracyclines, *J. Org. Chem.*, 29, 2746, 1964.

87. Schwartz, M. A. and Wallace, R. A., Efficient synthesis of 14-hydroxymorphinans from codeine, *J. Med. Chem.*, 24, 1525, 1981.

88. (a) Saito, I., Imuta, M., and Matsuura, T., Reactivity of singlet oxygen toward methoxybenzenes, *Tetrahedron*, 28, 5307, 1972; (b) Arene peroxide intermediates in the photosensitized oxygenation of methoxybenzenes, *Tetrahedron*, 28, 5313, 1972.

89. Dureja, P., Devakumar, C., Walia, S., and Mukerjee, S. K., Evidence of similarity between biooxygenation and photooxygenation: Formation of quinone epoxides, *Tetrahedron*, 43, 1129, 1987.

90. Rio, G. and Berthelot, J., Sensitized photooxidation of 9,10-dimethoxyphenanthrene. A dioxetane rapidly dissociating at low temperatures, *Bull. Soc. Chim. Fr.*, 822, 1972.

91. Liang, J. J. and Foote, C. S., Dicyanoanthracene-sensitized photooxidation of *o*-dimethoxybenzene, *Tetrahedron Lett.*, 3039, 1982.

92. Saito, I., Abe, S., Takahashi, Y., and Matsuura, T., Dye-sensitized photooxygenation of dimethylamino-substituted benzenes. Cycloaddition of singlet oxygen in competition with Type I reaction, *Tetrahedron Lett.*, 4001, 1974.

93. Dubey, R. D., Gandhi, P. B., Ameta, S. C., and Bokadia, M. M., Dye-sensitized photo-oxygenation of β-naphthylamines by singlet oxygen, *Ind. J. Chem.*, 24B, 1186, 1985.

94. Albini, A. and Spreti, S., Photoinduced oxygenation of methylbenzenes, bibenzyls and pinacols in the presence of 1,4-dicyanonaphthalene, *J. Chem. Soc., Perkin Trans. II*, 1175, 1987.

95. Santamaria, J. and Ouchabane, R., 9,10-Dicyanoanthracene-sensitized photooxygenations. Formation of O_2^{-} and $^1O_2$, *Tetrahedron*, 42, 5559, 1986.

96. Juillard, M. and Chanon, M., Activation of aromatics towards oxygen by oxidative or reductive photosensitization, *Bull. Soc. Chim. Fr.*, 242, 1992.

97. Barbier, M., Selective photoinduced oxidation of benzylic methylene groups through UV irradiation in the presence of ferric chloride, *Helv. Chim. Acta*, 67, 866, 1984.

98. Graetzel, C. K., Kira, A., Jirousek, M., and Graetzel, M., Dimer cation formation in microemulsion media. Duroquinone-sensitized photooxidation of 2,6-dimethylnaphthalene and tetrathiofulvalene, *J. Phys. Chem.*, 87, 3983, 1983.

99. Bokobza, L. and Santamaria, J., Exciplex and radical ion intermediates in electron transfer reactions. Solvent effect on the photo-oxygenation of 1,4-dimethylnaphthalene sensitized by 9,10-dicyanoanthracene, *J. Chem. Soc., Perkin Trans. II*, 269, 1985.

100. Fukuzumi, S., Tanii, K., and Tanaka, T., Protonated pteridine and flavin analogues acting as efficient and substrate-selective photocatalysts in the oxidation of benzylic alcohol derivatives by oxygen, *J. Chem. Soc., Chem. Commun.*, 816, 1989.

101. Pandey, G. and Krishna, A., Photoinduced SET initiated oxidative cleavage of benzylic ethers protecting groups, *Synth. Commun.*, 18, 2309, 1988.

102. Tamai, T., Mizuno, K., Hashida, I., and Otsuji, Y., Photooxygenation of arylmethylsilanes via photoinduced electron transfer, *Chem. Lett.*, 781, 1992.

103. Schaap, A. P., Siddiqui, S., Prasad, G., Palomino, E., and Lopez, L., Cosensitized electron transfer photo-oxygenations. The photochemical preparation of 1,2,4-trioxolanes, 1,2-dioxolanes and 1,2,4-dioxazolidines, *J. Photochem.*, 25, 167, 1984.

104. (a) Ichinose, N., Mizuno, K., Tamai, T., and Otsuji, Y., Photooxygenation of 1-alkyl-2,3-diarylcyclopropanes via PET: Stereoselective formation of 4-alkyl-3,5-diaryl-1,2-dioxolanes and their conversion to 1,3-diols, *J. Org. Chem.*, 55, 4079, 1990; (b) Mizuno, K., Kamiyama, N., Ichinose, N., and Otsuji, Y., Photo-oxygenation of 1,2-diarylcyclopropanes via electron transfer, *Tetrahedron*, 41, 2214, 1985.

105. Schaap, A. P., Siddiqui, S., Prasad, G., Rahman, A. F. M. M., and Oliver, J. P., Stereospecific formation of *cis*-ozonides by electron transfer photooxygenation of naphthyl-substituted epoxides, *J. Am. Chem. Soc.*, 106, 6087, 1984.

106. Fujihira, M., Satoh, Y., and Osa, T., Photoelectrochemistry at semiconductor titanium dioxide/insulating hydrocarbon liquid interface, *J. Electroanal. Chem.*, 126, 277, 1981.

107. Fox, M. A., Chen, C. C., and Younathan, J. N. N., Oxidative cleavage of substituted naphthalenes induced by irradiated semiconductor powders, *J. Org. Chem.*, 49, 1969, 1984.

108. Pincock, J. A., Pincock, A. L., and Fox, M. A., Controlled oxidation of benzyl ethers on irradiated semiconductor powders, *Tetrahedron*, 41, 4107, 1985.

109. Sydnes, L. K., Burkow, I. C., and Hanson, S. H., Photooxidation of toluene and xylenes. Concurrent formation of products from photooxygenation and photodimerization, *Acta Chem. Scand.*, 39B, 829, 1985.

110. Onodera, K., Sakuragi, H., and Tokumaru, K., Effect of light wavelength on photooxygenation of hexamethylbenzene, *Tetrahedron Lett.*, 2831, 1980.

111. Grovenstein, E. and Mosher, A. J., Reaction of atomic oxygen with aromatic hydrocarbons, *J. Am. Chem. Soc.*, 92, 3812, 1970.

112. Fujihira, M., Satoh, Y., and Osa, T., Heterogeneous photocatalytic oxidation of aromatic compounds on titanium dioxide, *Nature*, 293, 206, 1981.

113. Shimamura, Y., Misawa, H., Oguchi, T., Kanno, T., Sakuragi, H., and Tokumaru, K., Titanium dioxide photocatalyzed oxidation of aromatic hydrocarbons. The role of water and oxygen in ring hydroxylation, *Chem. Lett.*, 1691, 1983.

114. Ogata, Y., Tomizawa, K., and Yamashita, Y., Photoinduced oxidation of benzoic acid with aqueous hydrogen peroxide, *J. Chem. Soc., Perkin Trans. II*, 616, 1980.

115. Skuratova, S. I., Mordvintsev, P. I., and Fomin, G. V., Photosensitized hydroxylation of aromatic compounds, *Zh. Fiz. Khim.*, 56, 2093, 1982.

116. Strub, H., Strehler, C., and Streith, J., Photoinduced nitrene, carbene, and atomic oxygen transfer reactions from the corresponding pyridinium *N*-, *C*-, and *O*-ylides, *Chem. Ber.*, 120, 355, 1987.

117. Akhatar, N. M., Boyd, D. R., Neill, J. D., and Jerina, D. M., Stereochemical and mechanistic aspects of sulfoxide, epoxide, arene oxide and phenol formation by photochemical oxygen atom transfer from aza-aromatics *N*-oxides, *J. Chem. Soc., Perkin Trans. I*, 1693, 1980.

118. Sako, M., Shimada, K., Hirata, K., and Maki, Y., Photochemical hydroxylation of benzene derivatives by pyrimido[5,4-*g*]pteridine *N*-oxide, *Tetrahedron Lett.*, 6493, 1985.

119. Maki, Y., Sako, M., Shimada, K., Murase, T., and Hiroto, K., Pyrimido[5,4-*g*]pteridine *N*-oxides as a simple chemical model of hepatic monooxygenase, *Stud. Org. Chem.*, 33, 465, 1988.

120. Saito, I., Takami, M., and Matsuura, T., Oxidation of aromatic methoxy compounds via photoexcited aromatic nitro compounds, *Bull. Chem. Soc. Jpn.*, 48, 2865, 1975.

121. Ollis, D. F., Pelizzetti, E., and Serpone, N., Heterogeneous photocatalysis in the environment, in *Photocatalysis*, Serpone, N. and Pelizzetti, E., Eds., Wiley, New York, 1989, 603.

122. Barbeni, M., Morello, M., Pramauro, E., Pelizzetti, E., Borgarello, E., and Serpone, N., Photodegradation of pentachlorophenol by semiconductor powder, *Chemosphere*, 14, 195, 1985.

123. Barbeni, M., Pramauro, E., Pelizzetti, E., Borgarello, E., Serpone, N., and Jamieson, M. A., Photodegradation of chlorinated dioxins, biphenyls, phenols, and benzenes, *Chemosphere*, 15, 1913, 1986.

Cyclobutane Photochemistry

Kazuhiko Mizuno
University of Osaka Prefecture

Chyongjin Pac
Kawamura Institute
of Chemical Research

29.1 Introduction

The cyclobutane ring has small, but significant π-bonding contributions in the σ-framework[1] and relatively high conformational rigidity,[2] thus showing high chemical reactivities of wide applicability.[3–5] The parent cyclobutane and alkyl derivatives are inert toward photochemical activation at $\lambda > 250$ nm using conventional light sources because of the absence of significant absorptions at $\lambda > 200$ nm, extremely high triplet energies, and very low electron-donating or electron-accepting capabilities. In most cases, therefore, cyclobutane photochemistry has been investigated with cyclobutanes bearing π- or n-chromophoric substituents that are susceptible to photoactivation by light absorption, triplet energy transfer, exciplex formation, and excited-state electron transfer. Exceptions to these generalizations are strained compounds of low oxidation potential in exciplex or electron transfer photochemistry (vide infra) and a few simple cyclobutanes employing 185-nm photochemistry.[6]

The photochemistry of chromophore-substituted cyclobutanes should be initiated by primary photochemical activation of the chromophore(s), followed by transfer of the activation to the cyclobutane ring through particular mechanisms. Orbital interactions between the π- or n-substituent(s) and the ring σ-bonds may open up channels for the energy transfer to control stereochemical courses and reactivities depending on the conformations of the substituents.[7] Figure 29.1 shows Salem's σ and σ* orbitals[8] of the cyclobutane ring as well as possible π-σ orbital interaction[9] in different conformations I and II of the two vicinal π-substituents. In case I, the symmetric π-combination (π_s) may interact with σ_s to give the highest occupied molecular orbital (HOMO), π_s-$\lambda_I\sigma_s$, and antisymmetric π_a and σ_a may give the HOMO, π_a-$\lambda_{II}\sigma_s$, in conformation II. Similarly, the lowest unoccupied MO may be given by $\pi_a^* + \lambda_I^*\sigma_a^*$ in case I or by $\pi_s^* + \lambda_{II}^*\sigma_s^*$ in case II. Such orbital interactions may open up routes for energy transfer from the initially activated π-substituents to the C1-C2 bond in case I or to the C1-C4 (C2-C3) bond in case II. It should be noted however that the two π-electron systems can couple more efficiently through the C1-C2 bond than across the separated C1-C4 and C2-C3 bonds, i.e., $\lambda_I > \lambda_{II}$ and $\lambda_I^* > \lambda_{II}^*$. In an aryl-substituted cage compound as an extreme case, strong through-bond coupling between two vicinal π-substituents

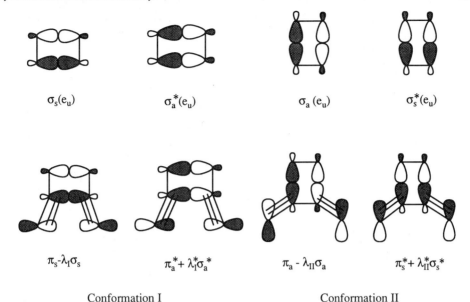

$$\sigma_s(e_u) \qquad \sigma_a^*(e_u) \qquad \sigma_a(e_u) \qquad \sigma_s^*(e_u)$$

$$\pi_s - \lambda_I \sigma_s \qquad \pi_a^* + \lambda_I^* \sigma_a^* \qquad \pi_a - \lambda_{II} \sigma_a \qquad \pi_s^* + \lambda_{II}^* \sigma_s^*$$

Conformation I   Conformation II

FIGURE 29.1 Salem's σ and σ^* orbitals of cyclobutane (top) and π-σ orbital interactions in different conformations of vicinal π-substituents.

occurs and results in unusual elongation of the relevant C-C bond due to significant mixing of σ_a^* in a lower bonding MO,[10] though usual aryl-substituted cyclobutanes have normal bond distances.[11] At any rate, it can be predicted that the C1-C2 bond should be inherently more reactive upon photoactivation of a π-substituent than the C1-C4 (C2-C3) bond, unless conformation II is "frozen" by steric constraint.

Typical cyclobutanes in this chapter are shown in Figure 29.2.

9.2 Cycloreversion

Direct Photolysis

Stereochemistry

Theory predicts that the concerted [$\sigma^2s + \sigma^2s$] cycloreversion is allowed to occur in the $^1\sigma,\sigma^*$ state of the cyclobutane ring.[12] However, little systematic investigation on the stereochemistry of cyclobutane cycloreversion has been carried out because of the difficulty in the selective excitation of the cyclobutane to its excited state, in contrast to extensive studies on thermal reactions.[3,13,14] It was reported that photolysis of some π-substituted cyclobutanes gave the corresponding olefins with high, but not complete, stereoretention[15–17] with one exception[18] (Figure 29.3). The reaction paths were interpreted in terms of the initial photoexcitation of the π-substituent, followed by crossing to the σ, σ^* reaction hypersurface through a state stabilized by orbital interactions.[15] The partial loss of stereointegrity would arise during the transition from the pericyclic minimum to the ground state[15,16] or from partial participation of a triplet mechanism.[18]

Regiochemistry

Photolytic cycloreversion of π-chromophore substituted cyclobutanes is specific or selective with respect to a ring C-C bond to which a π-chromophore (mostly the phenyl group) and a substituent are vicinally attached with *cis*-configuration (Equations 29.1 and 29.2), and "cis compounds" are more reactive than the "*trans*-analogues". This is called the "*cis* effect".[19] Other examples[15,20] are also in line with this formulation. Table 29.1 shows typical examples.

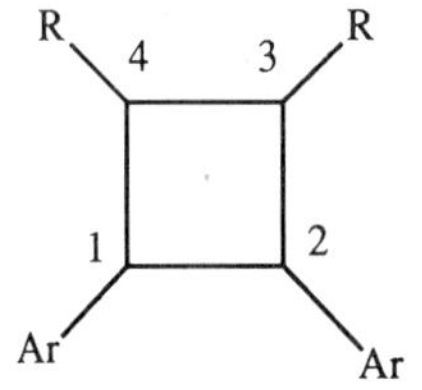
R
4
3
R
1
2
Ar
Ar

Ar	R	Configuration
Ph	H	*cis & trans*
Ph	Me	π-1,t-2,t-3,c-4
p-$MeOC_6H_4$	Me	π-1,c-2,t-3,t-4
p-$MeOC_6H_4$	H	*cis & trans*
2-Naphthyl	CO_2Me	π-1,c-2,t-3,t-4
9-Carbazolyl	H	*trans*
PhO	H	*cis & trans*

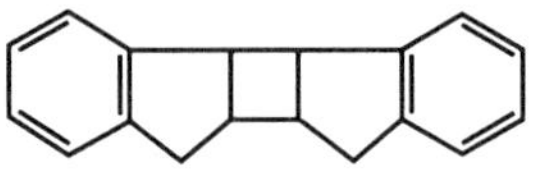

(*syn & anti*)

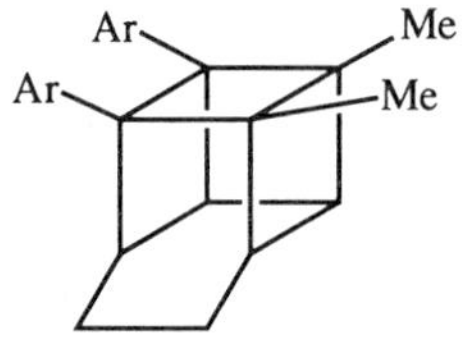
Ar
Me
Ar
Me

(Ar = p-X-C_6H_4 ; X = H,Cl,Me,MeO)

FIGURE 29.2 Typical cyclobutanes in cyclobutane photochemistry.

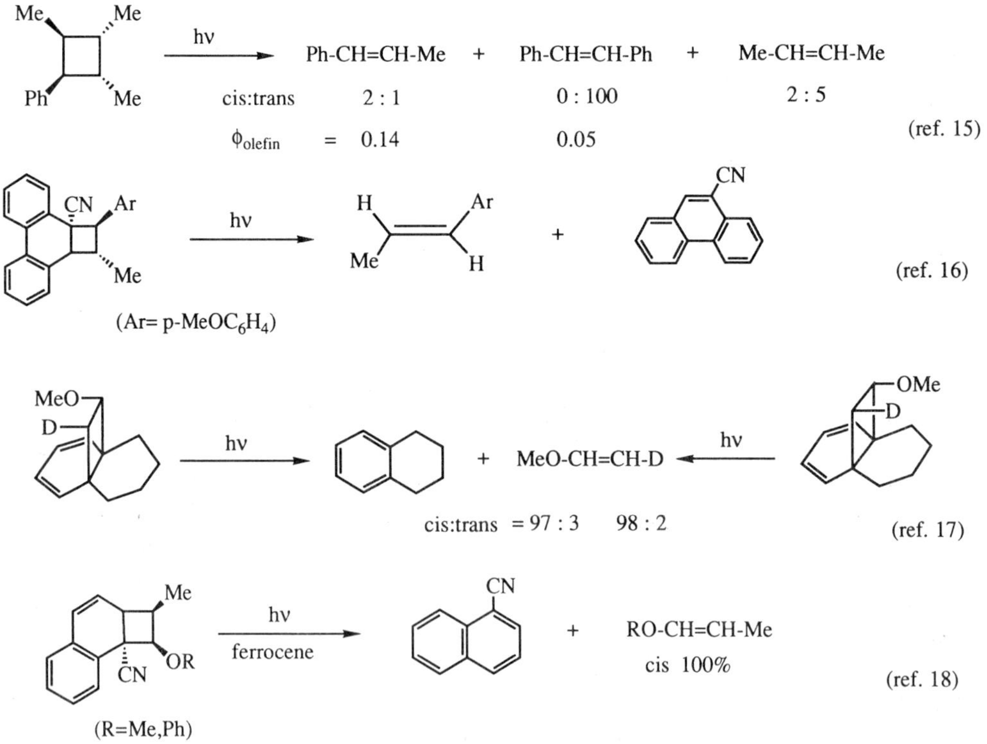
Me
Me
Ph
Me
hν
Ph-CH=CH-Me
+
Ph-CH=CH-Ph
+
Me-CH=CH-Me
cis:trans
2 : 1
0 : 100
2 : 5
ϕolefin = 0.14
0.05
(ref. 15)
CN
Ar
Me
hν
H
Ar
Me
H
+
CN
(Ar= p-MeOC6H4)
(ref. 16)
MeO
D
hν
+
MeO-CH=CH-D
hν
OMe
D
cis:trans = 97 : 3
98 : 2
(ref. 17)
Me
OR
CN
hν
ferrocene
CN
+
RO-CH=CH-Me
cis 100%
(R=Me,Ph)
(ref. 18)

FIGURE 29.3 Stereochemistry in photolytic ring splitting of some cyclobutanes.

Table 29.1 Regioselectivity in Photolytic Ring Splitting of
1,2-Diphenylcyclobutane and Its Derivatives ($\pi^1 = \pi^2 = $ Ph)

$R^3 = R^4$	Configuration	ϕ_a	ϕ_b	a/b	Ref.
H	cis-π^1,π^2	0.17	0.000	100/0	15, 19
H	$trans$-π^1,π^2	0.14	0.002	98/2	15, 19
Me	t-π^1-t-π^2-c-R^3-t-R^4	0.025	0.070	26/74	15
(Me)$_2$	$trans$-π^1,π^2	0.000	0.080	0/100	15
CO_2Me	t-π^1-c-π^2-t-R^3-c-R^4			97/3	19
CO_2Me	t-π^1-t-π^2-c-R^3-t-R^4			4/96	19

$$\pi^1\text{-CH=CH-R}^4 \ + \ \pi^2\text{-CH=CH-R}^3 \qquad (29.1)$$

$$\pi^1\text{-CH=CH-}\pi^2 \ + \ R^3\text{-CH=CH-R}^4 \qquad (29.2)$$

A reasonable interpretation of the the regiochemistry was made in terms of the conformation-controlled orbital interactions shown in Figure 29.1. Crystallographic X-ray analysis of several 1,2-diarylcyclobutanes[11] indicated that the conformations of the aryl groups are fixed by steric repulsion with the adjacent substituent in the *cis*-relationship or by polycyclic structures. For cyclobutanes with fixed conformation I, photoexcitation of the aryl groups results in the specific cycloreversion in direction **a** (Equation 29.1)[15] due to a mixing of the σ_a^* character by the π_a^*-σ_a^* interaction. On the other hand, the cleavage in direction **b** (Equation 29.2) can occur selectively in the fixed conformation II where the π_s^*-σ_s^* interaction is allowed. If the π-substituents can take either conformation I or II without steric constraint, however, the cycloreversion occurs dominantly in direction **a**[15,19] because the σ_a^* character can be more mixed inherently by coupling of the two π-electron systems through the C1-C2 bond rather than mixing with the σ_s^* character. For instance, *trans*-1,2-diphenylcyclobutane is photolyzed in direction **a** in 98 to 99% selectivity.[15,19] This is demonstrated by the net activation barriers ($\Delta H^{\ddagger}$) and preexponential factors (A) for the cycloreversion of aryl-substituted cyclobutanes; $\Delta H^{\ddagger} < 3$ kcal mol^{-1} and A $> 10^{11}$ s^{-1} for direction **a** and $\Delta H^{\ddagger} > 3$ kcal mol^{-1} and A $\leq 10^9$ s^{-1} in direction **b**.[15,21] These activation parameters were attributed to those for crossing from the local $^1\pi^*$ state to the relevant σ, σ^* hypersurface.[15]

Triplet Photosensitization

In most cases, triplet-sensitized cycloreversions proceed with loss of stereointegrity by the involvement of 1,4-biradicals. Typical examples are the photoreaction of *trans*-7,8-diacetoxybicyclo[4.2.0]-octa-2,4-diene in the presence of a triplet sensitizer, which gives *cis*- and *trans*-1,2-diacetoxyethenes,[22] and the benzophenone-photosensitized cycloreversion of 1,2, 2a,8b-tetrahydrocyclobuta[a]-naphthalene-2a-carbonitrile, which occurs with only partial stereoretention in contrast with the direct photolysis (Figure 29.3).[18] Reactions of *cis*-1,2-diphenylcyclobutane are photosensitized by acetone to give styrene and the *trans*-isomer of the cyclobutane, perhaps via 1,4-biradicals.[23] Photolytic splitting of *cis*- and *trans*-1,2-*bis*(4-acetylphenyl)cyclobutanes to 4-acetylstyrene in benzene is accompanied by the *cis, trans*-isomerization of the cyclobutanes.[24] A unique example is the photocleavage of naphthalene *peri*-fused cyclobutanes in rigid media that occurs in a stereospecific manner and with regioselectivities dependent on the cyclobutane structures and reaction media.[25] The involvement of an upper triplet state was implicated.

1,4-Diarylbiradicals were directly detected by laser flash photolysis of cyclobutanes and as *cis*- and *trans*-1,2-*bis*(4-acetylphenyl)cyclobutanes (biradical lifetimes = 208 to 235 ns),[24] the *anti*-dimer of acenaphthylene (330 ns),[26] *r*-1,c-2-dinaphthyl-*t*-3,*t*-4-dimethoxycarbonylcyclobutane in the presence of benzophenone (220 ns),[27] and 1,1,2,2-tetraphenylcyclobutane/acetone (~500 ns).[28] It was assumed that the *syn*-dimer of acenaphthylene in the triplet state gives triplet acenaphthylene adiabatically since a biradical intermediate was not detected.[26] In the triplet state, the dinaphthylcyclobutane gives both the triplet localized on one naphthalene chromophore and the excimer triplet formed between the vicinal chromophores. A 1,4-biradical is formed from the former triplet state.[27]

Exciplex Photosensitization

Pac et al. reported that the photosensitized ring-splitting reactions of diarylcyclobutanes in the presence of such electron acceptors as cyanoaromatics (singlet photosensitizer) and chloranil (triplet photosensitizer) in nonpolar solvents proceed via singlet and triplet exciplexes.[7,29–31] Reactivities were discussed in terms of combined contributions of charge transfer and excitation energy of exciplexes. Although the development of a positive charge on cyclobutanes by exciplex formation distorts the cyclobutane ring favorably for bond cleavage, large stabilization by strong charge-transfer interactions is energetically unfavorable for crossing from exciplexes to the reaction hypersurface on which ring splitting occur.

With 1,4-dicyanonaphthalene (DCN) used as photosensitizer in aromatic hydrocarbon solvents (ArH), DCN-ArH exciplexes are initially formed and they then interact with cyclobutanes to cause the ring-splitting reactions via hypothetical termolecular complexes, i.e., DCN-ArH-cyclobutane triplexes.[7] In the case of the *cis,syn*-indene dimer, reactivities depend on oxidation potentials of ArH, being attributed to population densities of positive charge on the cyclobutane. An adiabatic exciplex mechanism was shown for the ring-splitting reaction of diaryl-substituted cage compounds photosensitized by electron acceptors in nonpolar solvents.[32–33] The involvement of the adiabatic pathway depends on charge-transfer contributions.

Electron Transfer Photochemistry

Aromatic nitriles and quinones (e.g., chloranil) are conventional electron transfer photosensitizers in polar solvents (e.g., acetonitrile). CIDNP analysis indicates that the radical cation of the 1,1-dimethylindene cyclobutane dimer is formed by electron transfer from the dimer to triplet chloranil.[34] The photosensitized reactions of the head-to-head *anti*-dimer of indene and 1,2-diphenylcyclobutane by the electron acceptors in acetonitrile give the corresponding olefins in relatively low quantum yields. No products are obtained from the free radical cations of the cyclobutanes and olefins.[35] This reaction path is attributed to rapid electron reversal from the sensitizer radical anion to the closed and ring-opened radical cations in a solvent cage prior to dissociation.[35] On the other hand, the photosensitized monomerization of the head-to-head anethole cyclodimer proceeds through a chain mechanism involving electron exchange between the olefin radical cation and a neutral dimer molecule.[35] A similar chain mechanism operates in the photosensitized cycloreversion of aryl-substituted cage compounds by aromatic nitriles,[36] semiconductors,[37] and pyrylium salts,[38] as well as by the excitation of charge transfer complexes with tetracyanoethene.[39] Polymethylbicyclo[2.2.0]-hexanes undergo the stereospecific Cope rearrangement through boat cyclohexane-1,4-diyl radical cations formed by photochemical electron transfer to the electron acceptors.[40]

A novel type of photosensitization called "redox photosensitization" has been applied to ring-splitting reactions of the indene cyclobutane dimers and related arylcyclobutanes using the aromatic hydrocarbon (ArH) and *p*-dicyanobenzene (DCNB) as the photosensitizer system.[41,42] The mechanism is shown in Figure 29.4. The cycloreversion reactions are catalyzed by the photogenerated ArH[+] through the π-complex (ArH·cyclobutane)[+] in cases where the oxidation potential of ArH is lower than those of cyclobutanes. Reactivities increase with the decrease in oxidation potential

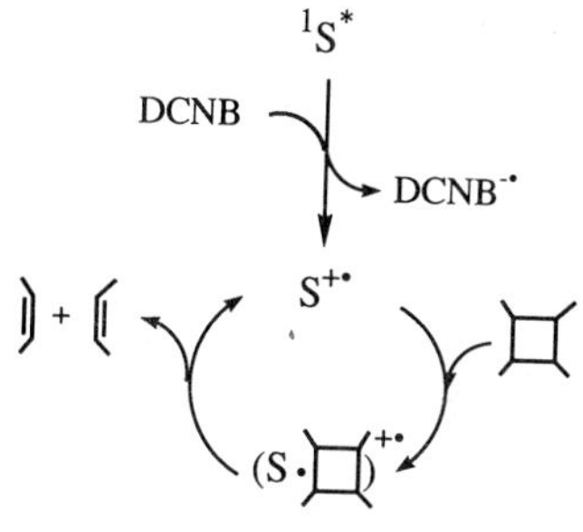

FIGURE 29.4   Redox-photosensitized splitting of the cyclobutane ring.

differences because of a higher population of the positive charge on the cyclobutane. Structure-reactivity relationships were discussed in terms of through-bond coupling between the aryl substituents (see Section 29.1). The redox-photosensitized cycloreversion proceeds with complete stereoretention.[42] Kinetic analysis demonstrated that the ring cleavage occurs at 10^9 s^{-1}, a rate constant compatible with a concerted process.[42] A similar mechanism was proposed for γ-radiolysis of the *syn* and *anti* head-to-head indene dimers and related cyclobutanes.[43]

The photoreactions of r-1,c-2-di(2-naphthyl)-t-3,t-4-di(methoxycarbonyl)cyclobutane (Figure 29.2) in the presence of triethylamine give only methyl naphthylacrylate in solvents of low polarity ($ε < 10$) through an exciplex, but also yield dinaphthylethene in increasing amounts with increasing solvent polarity as the consequence of participation of electron transfer from triethylamine to the excited singlet cyclobutane.[44] Indeed, it was shown that the cyclobutane radical anion generated by γ-radiolysis gives selectively dinaphthylethene and the radical anion of dimethyl fumarate. Spontaneous ring splitting of r-1,c-2,t-3,t-4-tetraphenylcyclobutane occurs to give *trans*-stilbene and the radical anion without intervention of the cyclobutane radical anion upon γ-radiolysis, even at 77K.[45,46]

Photorepair: Cycloreversion of Pyrimidine Cyclobutane Dimer

Exposure of biological systems to UV light causes lethal and/or mutagenic effects that are mainly due to formation of *cis,syn*-cyclobutane dimers (T◊T, T◊C, and C◊C) between adjacent pyrimidine bases in DNA (Equation 29.3).[47,48] Such harmful effects can be removed upon irradiation with near-UV and/or VIS light to UV-damaged biological systems.[49] This is called "photorepair" or "photo-reactivation," a rare photobiological phenomenon involving cycloreversion of the cyclobutane ring. Splitting of Py ◊ Pr′ to give the original pyrimidine bases occurs with nearly unit or relatively high quantum yields.[47,50–52] The system requires an enzyme (called "photolyase") that contains 1,5-dihydroflavin-adenine dinucleotide,[51] probably as the principal chromophore, and a tetrahydrofolate[52] or 8-hydroxy-5-deazaflavin[53] that can function as a photoantenna.[51,54]

$$
\text{T◊T, T◊C, C◊C} \quad
\begin{array}{c}
\text{photorepair} \\
\text{hv} > 300/\text{photolyase} \\
\overrightarrow{} \\
\underleftarrow{} \\
\text{hv} < 300 \text{ nm} \\
\text{UV damage}
\end{array}
\quad \text{T and/or C}
$$

(T = thymine; C = cytosine)   (29.3)

Since energy transfer from the photoexcited chromophores to pyrimidine dimers is too endo-thermic to occur, other mechanisms should operate in photorepair. In order to explore the molecular mechanisms for the initiation of photorepair, extensive studies have been performed with such dimer models (Py ◊ Py) as thymine and uracil dimers (for the structures of the isomers, see Figure 29.5).[55,56] Lamola reported a pioneering study in which it was reported that anthraquinone

cis , syn   *trans , syn*

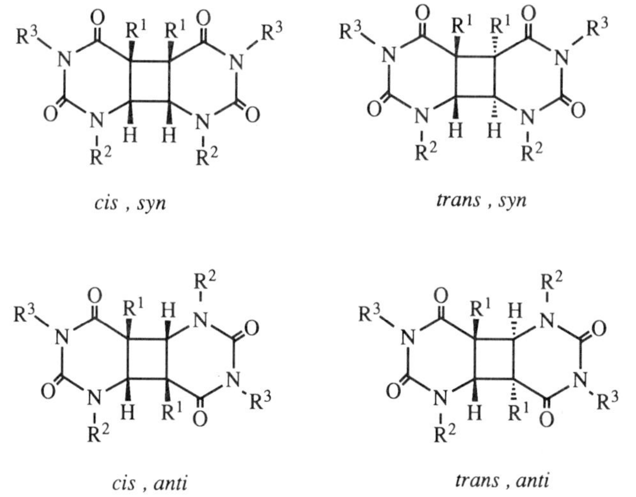

cis , anti   *trans , anti*

$(R^1, R^2, R^3 = H \text{ and/or } Me)$

FIGURE 29.5   The structures of the four stereoisomers of thymine ($R' = Me$) and uracil ($R' = H$) cyclobutane dimers.

sulfonate photosensitizes the monomerization of the *cis,syn*-dimer of 1,3-dimethylthymine through electron transfer from the dimer to the triplet quinone (Equation 29.4).[55] Photosensitized splitting of dimer models is also effected by quinones[57–59] and uranyl ion.[60] Additional evidence for an electron transfer mechanism (Equation 29.4) was again presented.[59,61] The quinones can monomerize the *cis,syn*- and *trans,syn*-dimers with similar efficiency but the *anti*-dimers either do not monomerize or do so very inefficiently.[57–59] The different reactivities were interpreted in terms of through-bond coupling between the n-orbitals of N(1) and N(1′), which is efficient in the *syn*-dimers but inefficient in the *anti*-isomers (Figure 29.5).[59,62] In most cases, however, quantum yields are significantly lower than unity.

$$\text{Sens} + \text{Py}\lozenge\text{Py} \longrightarrow {}^3\text{Sens}^* + \text{Py}\lozenge\text{Py} \longrightarrow \text{Sens}^{-\bullet} + \text{Py}\lozenge\text{Py}^{+\bullet} \longrightarrow \text{Sens} + 2\text{Py}$$

(Sens = quinones or uranil ion; Py$\lozenge$Py = dimer models)   (29.4)

Another type of model photosensitization is redox photosensitization using the phenanthrene/*p*-dicyanobenzene pair,[62] that can monomerize the *cis,syn*-dimer of 1,3-dimethylthymine much more efficiently than the *trans,syn*-isomer. The quantum yields for splitting are 0.6 in degassed solution but 100! in aerated solution. The effect of aeration was attributed to formation of a complex between the *cis,syn*-dimer and O_2, which is much more reactive toward catalysis by the photogenerated radical cation of phenanthrene than the uncomplexed dimer (Equations 29.5 to 29.7).

$$\text{Py}\lozenge\text{Py} + O_2 \rightleftharpoons \text{Py}\lozenge\text{Py} \cdot O_2 \tag{29.5}$$

$$\text{ArH} + \text{DCNB} \longrightarrow \text{ArH}^{+\bullet} + \text{DCNB}^{-\bullet} \tag{29.6}$$

$$\text{Py}\lozenge\text{Py} \cdot O_2 \xrightarrow{\text{ArH}^{+\bullet}} 2\text{Py} + O_2 \tag{29.7}$$

Splitting of dimer models can also be photosensitized by electron donors such as indole derivatives[63,64] and aromatic amines.[65] The monomerization proceeds through electron transfer from the excited photosensitizer to the dimer models (Equation 29.8). However, the quantum yields are very low in these cases.

$$\text{Sens} + \text{Py} \diamond \text{Py} \longrightarrow \text{Sens}^{+\bullet} + \text{Py} \diamond \text{Py}^{\bullet} \longrightarrow \text{Sens} + 2\text{Py} \qquad (29.8)$$

The model reactions clearly indicate that pyrimidine dimers can be split via electron transfer. An electron transfer mechanism is strongly suggested by laser flash[66] and ESR[67] spectroscopic studies on the photolysis of uracil and thymine dimers by *Escherichia coli* photolyase, though the direction of electron transfer from or to the excited-state chromophore is still in doubt. In most model reactions, the photosensitizers are not flavins but, with a few exceptions are nonbiological molecules. However, model reactions attempted using the oxidized and reduced forms of various flavins are very inefficient under neutral and weakly acidic or basic conditions.[68,69] It was found recently that riboflavin tetraacetate efficiently photosensitizes the monomerization of the dimers of 1,3-dimethylthymine and 1,3-dimethyluracil in the presence of perchloric acid.[70,71] The limiting quantum yields are unity in degassed acetonitrile, but much higher than unity with air contamination.[70] Electron transfer from the dimer to the protonated flavins in both the excited singlet and excited triplet states occurs to give radical-ion pairs that undergo the net monomerization without geminate recombination to the precursors in the ground state. In the presence of air, a chain carrier is produced from the protonated flavin to catalyze the monomerization of Py $\diamond$ Py · O$_2$.[70] Moreover, it was reported that 1,5-dihydroriboflavin tetraacetate generated *in situ* can photosensitize the monomerization of the dimethyluracil dimer in highly basic conditions by a chain mechanism.[72] Electron transfer from the excited-state deprotonated flavin to the dimer is the initiation process resulting in splitting of the dimer radical anion followed by electron transfer from the monomer radical anion to the dimer.

9.3 Other Photoreactions

Cyclobutanes undergo isomerization reactions through 1,4-biradicals generated by direct photolysis and triplet photosensitization, as well as through 1,4-radical ions produced by excited-state electron transfer. Irradiation of 5,6-disubstituted bicyclo[4.2.0]octa-2,4-dienes results in *cis,trans*-isomerization accompanied by the ring cleavage.[22] Similarly, the isomerization of *cis*-1,2-diphenylcyclobutane to the *trans*-isomer occurs along with formation of styrene on both direct photolysis and photosensitization by acetone.[23] The isomerization/splitting ratio under the latter condition is greater than that in the former. The isomerization of 1,2-diacylcyclobutanes to 2-acyl-2,3-dihydro-4*H*-pyrans occurs either via the excited singlet or the triplet state.[73] The photochemical transformation of truxones to C-nor-D-homo steroid systems has been achieved,[74] proceeding through 1,4-biradicals generated by photolytic β-fission of the ketones.

The photosensitized reactions of *cis*- and *trans*-1,2-diphenylcyclobutanes by the phenanthrene/*m*-dicyanobenzene give 1-phenyltetrahydronaphthalene in acetonitrile and 1-methoxy-1,4-diphenylbutane in acetonitrile/methanol via the ring-opened 1,4-radical cation.[75] On the other hand, these reactions are not photosensitized by electron acceptors in acetonitrile because of exclusive back-electron transfer of radical-ion pairs without dissociation into the free radical ions, while the photosensitized isomerization to the tetrahydronaphthalene occurs even to minor extents in nonpolar solvents via polar exciplexes (vide supra).[7,15] Photosensitization by 9,10-dicyanoanthracene effects the isomerization of *cis*-1,2-diphenoxycyclobutane to the *trans*-isomer in acetonitrile via the ring-opened 1,4-radical cation, but does not bring about the *trans*-to-*cis* isomerization.[76]

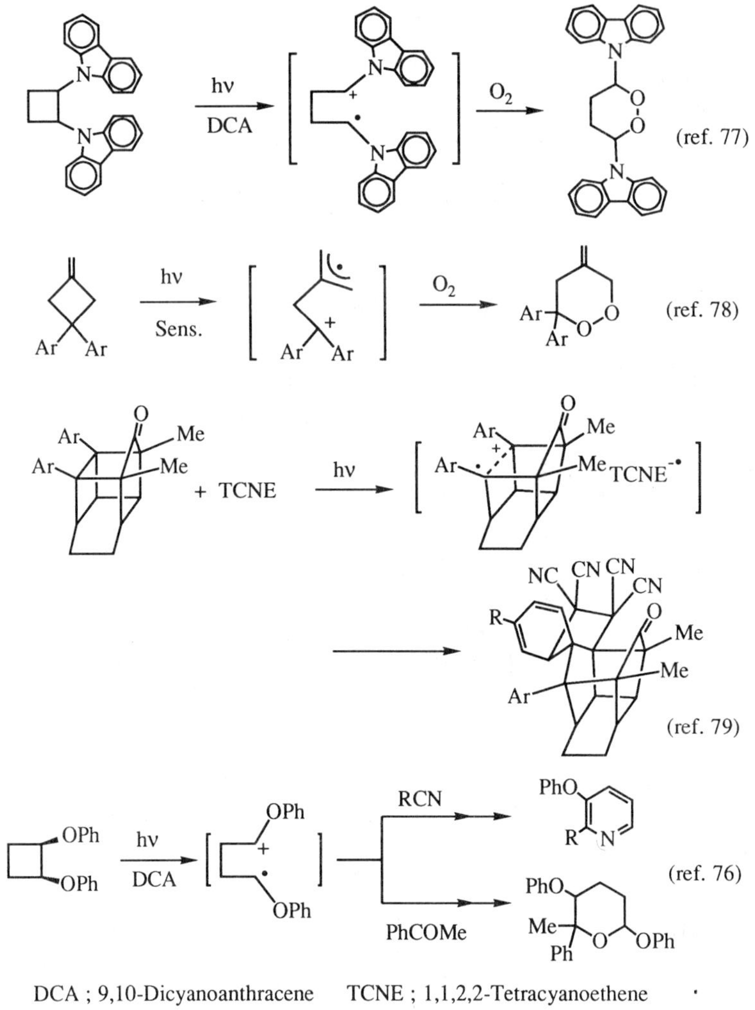

DCA ; 9,10-Dicyanoanthracene   TCNE ; 1,1,2,2-Tetracyanoethene

FIGURE 29.6   Photocycloaddition reactions of some cyclobutanes with unsaturated compounds.

Excited singlet 9,10-dicyanoanthracene abstracts an electron from *trans*-1,2-di(carbazol-9-yl)cyclobutane[77] and 3,3-diarylmethylenecyclobutanes[78] to generate the ring-opened radical cations that are trapped by O_2 to give the 1,2-dioxane compounds (Figure 29.6). The radical cation of 1,2-diphenoxycyclobutane can also be trapped by C≡N group of nitriles and C=O group of ketones to give pyridine and oxane compounds, respectively.[76] Similarly, cycloaddition of 1,1,2,2-tetracyanoethene to a cage ketone occurs through the radical cation of the cage ketone upon irradiation of a charge-transfer complex.[79]

29.4 Applications

Synthetic applications of photochemical [2+2]-cycloreversion reactions have been made, particularly for the preparation of unstable or unusual compounds, e.g., aza[14]- and aza[18]annulenes,[80,81] oxazepines,[82] azetinones,[83] Dewar furans,[84] bullvalenes,[85] pentalenes,[86] oxirenes,[87] and cyclobutadienes (Figure 29.7).[88] The photolytic ring cleavage of a tricyclic cyclobutanone provides a route for the

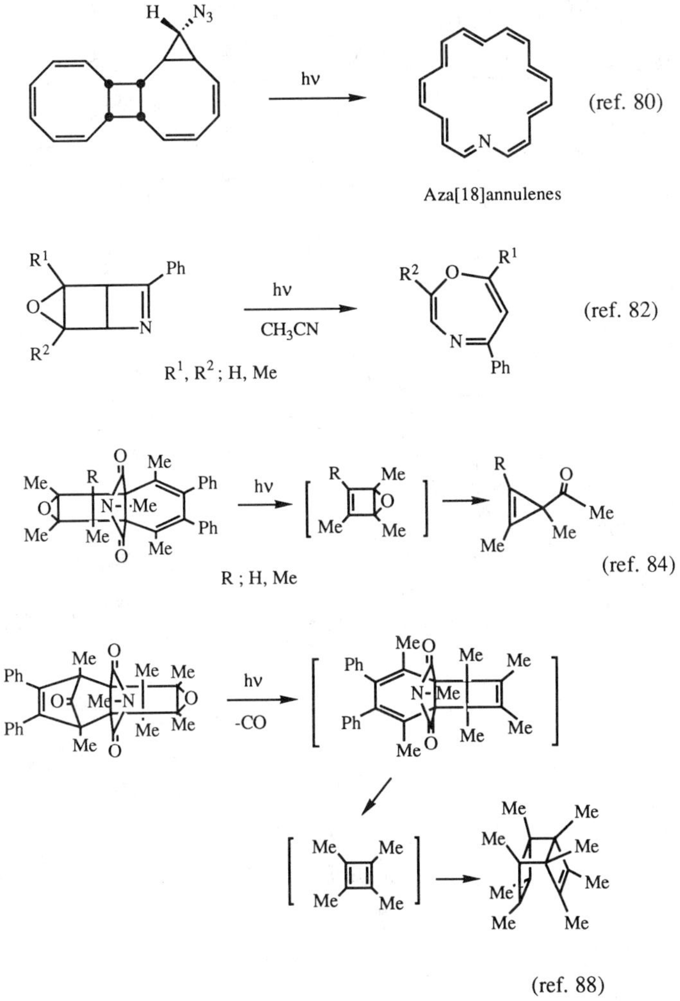

(ref. 80)

Aza[18]annulenes

(ref. 82)

R^1, R^2 ; H, Me

R ; H, Me

(ref. 84)

(ref. 88)

FIGURE 29.7 Synthetic applications of photochemical [2+2]-cycloreversion.

preparation of a 3,5-functionalized cyclopentenes.[89] Photoresponsible cyclobutano-crown ethers, synthesized by intramolecular [2+2]-photocycloaddition at $\lambda > 300$ nm, are reversibly transformed to the starting dienes at 220 nm.[90,91]

Ring expansion of fused cyclobutanols has been achieved by irradiation in the presence of HgO/ I_2 to provide a useful synthetic method for a variety of ring-expanded compounds (Figure 29.8).[92–94] Dicarbonyl(η^5-cyclopentadienyl)[cyclobutyl]irons undergo photolytic ring-expansion reactions to give metallacyclopentene complexes.[95,96]

Current applications using photoreactive polymers include photosensitive, photochromic, and photodegradative materials.[97–99] Polymers and polyesters containing cyclobutane moieties are readily derived by photocycloaddition of the cinnamic and coumarin chromophores, followed by photochemical cleavage in direction **b** to give new polymers containing double bonds in the polymer chain, as shown in Figure 29.9.

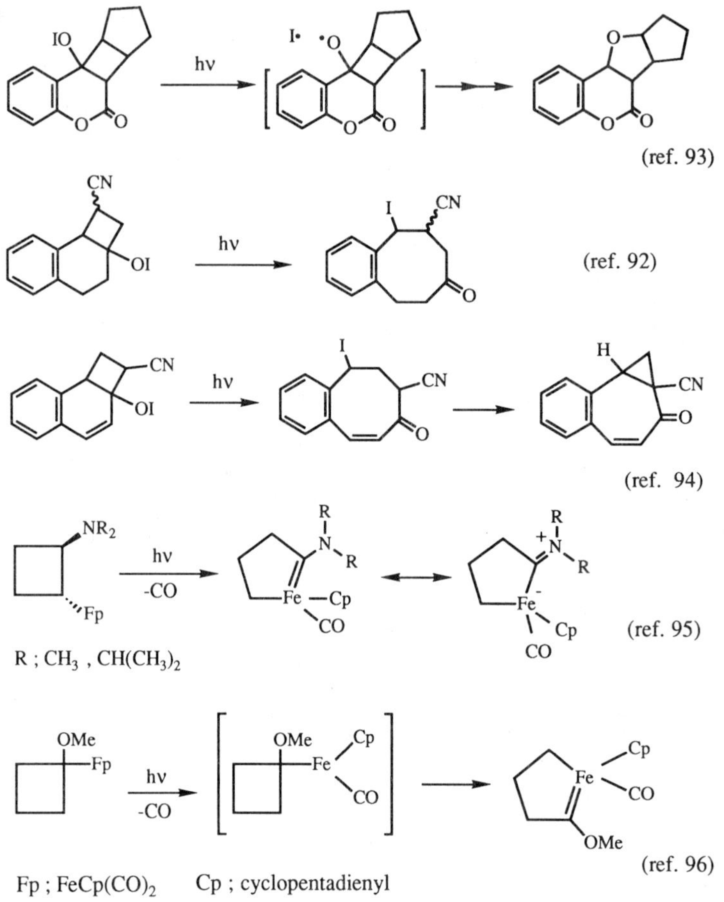

hν
(ref. 93)
hν
(ref. 92)
hν
(ref. 94)
hν
-CO
(ref. 95)
R ; CH3 , CH(CH3)2
hν
-CO
(ref. 96)
Fp ; FeCp(CO)2
Cp ; cyclopentadienyl

FIGURE 29.8 Photochemical ring expansion reactions of some cyclobutanes.

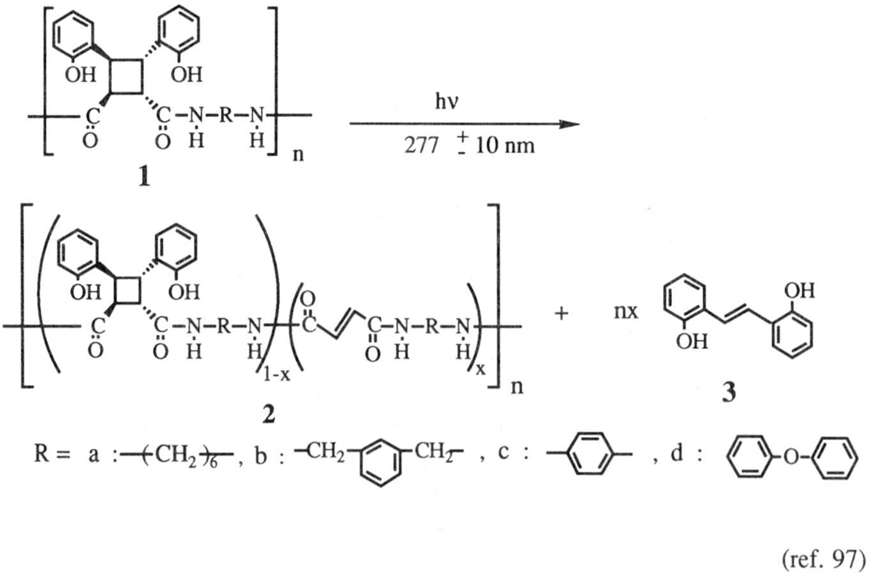

hν
277 ± 10 nm
1
2
3
nx
R = a : —(CH2)6—, b : —CH2—C6H4—CH2—, c : —C6H4—, d : C6H4—O—C6H4
(ref. 97)

FIGURE 29.9 Photocleavage of cyclobutanes in the polymer chain.

erences

1. Hoffman, R. and Davidson, R. B., The valence orbitals of cyclobutane, *J. Am. Chem. Soc.*, 93, 5699, 1971.
2. Moriarty, R. R., Stereochemistry of cyclobutane and heterocyclic analogs, *Top. Stereochem.*, 8, 271, 1974.
3. Schaumann, E. and Ketcham, R., [2+2]-Cycloreversion, *Angew. Chem. Int. Ed. Engl.*, 21, 225, 1982.
4. Kossanyi, J., Photochemical approach to the synthesis of natural products, *Pure Appl. Chem.*, 51, 181, 1979.
5. Hautala, R. R., King, R. B., and Kutal, C., *Solar Energy: Chemical Conversion and Storage*, Humana Press, Clifton, NJ, 1979.
6. (a) Steinmets, M. G., Photochemistry with short UV light, in *Organic Photochemistry*, Vol. 8, Padwa, A., Ed., Marcel Dekker, New York, 1987, chap. 2; (b) Adam, W. and Oppenlader, T., Direct photochemical cleavage of the cyclobutane ring in bicyclo[4.2.0]octane on 185 nm irradiation in solution, *Angew. Chem. Int. Ed. Engl.*, 23, 641, 1984; (c) Cate, R. C. and Hinkson, T. C., 2537-A Mercury-sensitized photochemical decomposition of perfluorocyclobutane, *J. Phys. Chem.*, 78, 2071, 1974.
7. Pac, C., Mechanism and structure-reactivity relationship in photosensitized reactions of some diarylcyclobutanes and quadricyclane by organic electron acceptors, *Pure Appl. Chem.*, 58, 1249, 1986.
8. Salem, L. and Wright, J. S., Vibrational modes, orbital symmetries, and unimolecular reaction paths, *J. Am. Chem. Soc.*, 91, 5947, 1969.
9. Hoffman, R., Interaction of orbitals through space and through bonds, *Acc. Chem. Res.*, 1, 1, 1971.
10. Harano, K., Ban, T., Yasuda, M., Osawa, E., and Kanematsu, K., Substituent effects in $[2\sigma + 2\sigma + 2\sigma]$ thermal decarbonylation of cage ketones. Remarkably effective elongation of strained C-C bond by through-bond coupling, *J. Am. Chem. Soc.*, 103, 2310, 1981.
11. Shima, K., Kimura, J., Yoshida, K., Yasuda, M., Imada, K., and Pac, C., Molecular structure of diarylcyclobutanes associated with reactivities in ring-cleavage reactions: Implications of conformationally controlled through-bond coupling, *Bull. Chem. Soc. Jpn.*, 62, 1934, 1989.
12. Woodward, R. B. and Hoffman, R., in *The Conservation of Orbital Symmetry*, Verlag Chemie, Weinheim/Bergstr., 1970.
13. Benson, W. S., *Thermochemical Kinetics, Methods for the Estimation of Thermochemical Data and Rate Parameters*, 2nd ed., John Wiley & Sons, New York, 1976.
14. Yasuda, M., Yoshida, K., Shima, K., Pac., C., and Yanagida, S., Thermal ring splitting reactions of diarylcyclobutanes: Significance of steric effects on orbital interaction states and biradical intermediates, *Bull. Chem. Soc. Jpn.*, 62, 1943, 1989.
15. Pac, C., Go-an, K., and Yanagida, S., Photochemical reactions of aromatic compounds. XLVI. Stereochemical reaction cources in the photolytic ring-cleavage reactions of diarylcyclobutanes: Implications of conformation-controlled orbital interactions in the excited-state chemistry, *Bull. Chem. Soc. Jpn.*, 62, 1951, 1989.
16. (a) Caldwell, R. A. and Creed, D., Exciplex intermediates in [2+2]photocycloadditions, *Acc. Chem. Res.*, 13, 45, 1980; (b) Mizuno, K., Caldwell, R. A., Tachibana, A., and Otsuji, Y., Stereospecific photocycloaddition of electron-deficient arylalkenes to 9-cyanophenanthrene via exciplex, *Tetrahedron Lett.*, 33, 5779, 1992.
17. Paquette, L. A. and Thompson, G. L., Stereochemical features of the thermal and photochemical fragmentations of [4.4.2]propella-2,4-dienes, *J. Am. Chem. Soc.*, 94, 7127, 1972.
18. Pac, C. Mizuno, K., and Sakurai, H., Photochemical reactions of aromatic compounds. XXIX. Photochemical and thermal cycloreversion of 1,2,2a,8b-tetrahydrocyclobuta[a]naphthalenes, *Bull. Chem. Soc. Jpn.*, 51, 329, 1978.
19. (a) Kaupp, G. and Stark, M., The *cis*-effect in photochemical cleavage of cyclobutanes, *Chem. Ber.*, 110, 3084, 1977; (b) Kaupp, G., Orientation in photochemical cyclobutane cleavage: *cis*-effect, *Angew. Chem.*, 84, 741, 1974; *Angew. Chem. Int. Ed. Eng.*, 13, 817, 1974.

20. Yonezawa, N., Yoshida, T., and Hasegawa, M., Symmetric and asymmetric photocleavage of the cyclobutane rings in head-to-head coumarin dimers and their lactone-opened derivatives, *J. Chem. Soc., Perkin Trans. I*, 1083, 1983.

21. Shizuka, H., Seki, I., Morita, T., and Iizuka, T., Photolyses of tetraphenylcyclobutanes at 254 nm, *Bull. Chem. Soc. Jpn.*, 52, 2074, 1979.

22. Caldwell, R. A., Photosensitized fragmentation of a bicyclo[4.2.0]octa-2,4-diene derivative. Preparation of the isomeric 1,2-diacetoxyethylenes, *J. Org. Chem.*, 34, 1886, 1969.

23. Jones, G., II and Chow, V. L., Geometric isomerization and cycloreversion in 1,2-diphenylcyclobutane. Photochemical vs. thermal activation, *J. Org. Chem.*, 39, 1447, 1974.

24. (a) Mizuno, K., Ichinose, N., Otsuji, Y., Caldwell, R. A., and Sink, R. M., unpublished results; Ichinose, N., Ph. D. Dissertation (University of Osaka Prefecture), 1989, 128.

25. Honda, K., Yabe, A., and Tanaka, H., Photochemistry of cyclobutanes. I. Photolysis of naphthalene *peri*-fused cyclobutanes in rigid media. Photocleavage from an upper triplet state, *Bull. Chem. Soc. Jpn.*, 49, 2384, 1976.

26. Kobashi, H., Ikawa, H., Kondo, R., and Morita, T., Effect of conformation on ring opening of *cis* and *trans* dimers of acenaphthylene in the triplet state. Direct detection of hydrocarbon biradical, 1,1'-biacenaphthene-2,2'-diyl, *Bull. Chem. Soc. Jpn.*, 55, 3031, 1982.

27. Takamuku, S. and Schnabel, W., Triplet sensitized cycloreversion of dinaphthyldimethoxycarbonylcyclobutane as studied by laser flash photolysis, *Chem. Phys. Lett.*, 69, 399, 1980.

28. Barton, D. H. R., Charpiot, B., Johnson, L. J., Motherwell, W. B., Scaiano, J. C., and Stanforth, S., Direct observation and chemistry of biradicals from photochemical decarbonylation of α-perphenylated cycloalkanones, *J. Am. Chem. Soc.*, 107, 3607, 1985.

29. Pac, C., Ohtsuki, T., Shiota, Y., Yanagida, S., and Sakurai, H., Photochemical reactions of aromatic compounds. XLII. Photosensitized reactions of some selected diarylcyclobutanes by aromatic nitriles and chloranil, implications of charge-transfer contributions on exciplex reactivities, *Bull. Chem. Soc. Jpn.*, 59, 1133, 1986.

30. Pac, C., Fukunaga, T., Ohtsuki, T., and Sakurai, H., Singlet-photosensitized ring-splitting and isomerization reactions of 1,2-diphenylcyclobutane by aromatic nitriles. A possible probe for relationship between reactivities and electronic nature of exciplex intermediates, *Chem. Lett.*, 1847, 1984.

31. Pac, C., Ohtsuki, T., Fukunaga, T., Yanagida, S., and Sakurai, H., Divergent solvent effects on photosensitized reactions of *cis*-1,2-diphenylcyclobutane by aromatic nitriles, *Chem. Lett.*, 1855, 1985.

32. Hasegawa, E., Okada, K., and Mukai, T., Exciplex isomerization in photosensitized cycloreversion reactions of cage compounds, *J. Am. Chem. Soc.*, 106, 6852, 1984.

33. Hasegawa, E., Okada, K., Ikeda, H., Yamashita, Y., and Mukai, T., Photosensitized [2+2] cycloreversion reactions of arylated cage compounds in nonpolar solvents. Highly efficient adiabatic exciplex isomerization, *J. Org. Chem.*, 56, 2170, 1991.

34. Roth, H. D. and Schilling, M. L. M., Nuclear spin polarization effects in radical ion pair reactions. A comparison between triplet state and radical ion reactivity, *J. Am. Chem. Soc.*, 103, 7210, 1981.

35. Pac, C., Fukunaga, T., Go-an, Y., Sakae, T., and Yanagida, S., Photochemical reactions of aromatic compounds. XLV. Controlling factors for reactivities and mechanisms in photoelectron transfer reactions of some selected diarylcyclobutanes with electron acceptors, *J. Photochem. Photobiol., A. Chemistry*, 41, 37, 1987.

36. Yamashita, Y., Ikeda, H., and Mukai, T., Photoinduced electron-transfer reactions of cage compounds. Novel pericyclic reactions involving a chain process, *J. Am. Chem. Soc.*, 109, 6682, 1987.

37. (a) Okada, K., Hisamitsu, K., and Mukai, T., Semiconductor-catalyzed [2+2]photocycloreversion of a strained cage molecule, *J. Chem. Soc., Chem. Commun.*, 941, 1980; (b) Okada, K., Hisamitsu, K., Takahashi, Y., Hanaoka, T., Miyashi, T., and Mukai, T., Semiconductor-catalyzed photocycloreversion, valence isomerization and [1,3]-sigmatoropic rearrangement, *Tetrahedron Lett.*, 25, 5311, 1984.

38. (a) Okada, K., Hisamitsu, K., and Mukai, T., Photocycloreversion-reaction of a cage molecule and related cyclobutanes with cationic sensitizers, *Tetrahedron Lett.*, 22, 1251, 1981; (b) Okada, K., Hisamitsu, K., Miyashi, T., and Mukai, T., Reaction mechanism of a pyrylium salt-sensitized photocycloreversion of a cage molecule, *J. Chem. Soc., Chem. Commun.*, 974, 1982.

39. Mukai, T., Sato, K., and Yamashita, Y., Cycloreversion reactions of phenylated cage compounds induced by electron transfer, *J. Am. Chem. Soc.*, 103, 670, 1981.

40. Tsuji, T., Miura, T., Sugiura, K., and Nishida, Photoinduced electron transfer reactions of 1,4-dimethyl- and 1,2,3,4,5,6-hexamethylbicyclo[2.2.0]hexanes. Evidence for boat cyclohexane-1,4-diyl radical cation and its stereospecific cleavage, *J. Am. Chem. Soc.*, 112, 1998, 1990.

41. Majima, T., Pac, C., and Sakurai, H., Redox-photosensitized reactions. 6. Stereospecific ring cleavage of 1,1a,2,2a-tetrahydro-7H-cyclobut[a]indene derivatives: Mechanism and structure-reactivity relationship, *J. Am. Chem. Soc.*, 102, 5265, 1980.

42. Majima, T., Pac, C., and Sakurai, H., Redox-photosensitized reactions. 5. Redox-photosensitized ring cleavage of 1-phenoxy-1,2,2a,3,4,8b-hexacyclobuta[a]naphthalene-8b-carbonitrile and its 2-methyl derivatives by redox photosensitization, *J. Chem. Soc., Perkin Trans. I*, 2795, 1980.

43. Majima, T., Pac, C., Takamuku, S., and Sakurai, H., Radiation-induced cycloreversion of indene cyclobutane dimers in *n*-butyl chloride by a chain reaction mechanism, *Chem. Lett.*, 1149, 1979.

44. Takamuku, S., Kuroda, T., and Sakurai, H., Cycloreversion of *cis*-1,2-dinaphthylcyclobutane via exciplex with triethylamine. Solvent effect on the mode of cycloreversion, *Chem. Lett.*, 377, 1982.

45. Takamuku, S., Kigawa, H., Miki, S., and Sakurai, H., Cycloreversion of tetraphenylcyclobutane induced by electron capture, *Chem. Lett.*, 797, 1979.

46. Takamuku, S., Dinh-Ngoc, B., and Schnabel, W., Pulse radiolysis of tetraphenylcyclobutane in hexamethylphosphoric triamide solution, *Z. Naturforschung.*, 339, 1281, 1978.

47. Cadet, J. and Vigny, P., The photochemistry of nucleic acids, in *Bioorganic Photochemistry*, Vol. 1, Morrison, H., Ed., Wiley Interscience, New York, 1990, 1.

48. Jagger, J., Ultraviolet inactivation of biological systems, in *Photochemistry and Photobiology of Nucleic Acids*, Vol. 2, Wang, S. H., Ed., Academic Press, New York, 1976, 147.

49. (a) Keiner, A., Effect of visible light on the recovery of streptomyces conidia from ultra-violet irradiation injury, *Proc. Natl. Acad. Sci. U.S.A.*, 35, 73, 1949; (b) Dulbecco, R., Reactivation of ultra-violet inactivated bacteriophage by visible light, *Nature*, 163, 949, 1949.

50. Harm, H., Repair of UV-irradiated biological systems: Photoreactivation, in *Photochemistry and Photobiology of Nucleic Acids*, Vol. 1, Wang, S. H., Ed., Academic Press, New York, 1976, 219.

51. Sancar, A. and Sancar, G. B., DNA repair enzymes, *Annu. Rev. Biochem.*, 57, 29, 1988.

52. (a) Eker, A. P. M., Hessels, J. K. C., and Dekker, R. H., Photoreactivating enzyme from Streptomyces griseus. VI. Action spectrum and kinetics of photoreactivation, *Photochem. Photobiol.*, 44, 197, 1986; (b) Sancar, G. B., Jorns, M. S., Payne, G., Fluke, D. J., Rupert, C. S., and Sancar, A., Action mechanism of *Escherichia coli* DNA photolyase, *J. Biol. Chem.*, 262, 492, 1987.

53. Payne, G. and Sancar, A., Absolute action spectrum of E-FADH$_2$ and E-FADH$_2$-MTHF forms of *Escherichia coli* DNA photolyase, *Biochemistry*, 29, 7715, 1990.

54. Malhotra, K., Kim, S.-T., Walsh, C., and Sancar, A., Roles of FAD and 8-hydroxy-5-deazaflavin chromophores in photoreactivation by *Anacystis nidulans* DNA photolyase, *J. Biol. Chem.*, 267, 15406, 1992.

55. Lamola, A. A., Photosensitization in biological systems and the mechanism of photoreactivation, *Mol. Photochem.*, 4, 107, 1972.

56. Pac, C. and Ishitani, O., Electron-transfer organic and bioorganic photochemistry, *Photochem. Photobiol.*, 48, 767, 1988.

57. Ben-Hur, E. and Rosenthal, I., Photosensitized splitting of pyrimidine dimers, *Photochem. Photobiol.*, 11, 163, 1970.

58. Sasson, S. and Dlad, D., Photosensitized monomerization of 1,3-dimethylthymine photodimers, *J. Org. Chem.*, 37, 3164, 1972.

59. Pac, C., Miyamoto, I., Masaki, Y., Furusho, S., Yanagida, Y., Ohno, T., and Yoshimura, A., Chloranil-photosensitized monomerization of dimethylthymine cyclobutane dimers and effect of magnesium perchlorate, *Photochem. Photobiol.*, 52, 973, 1990.

60. Rosenthal, I., Rao, M. M., and Salomon, J., Transition metal-ion photosensitized monomerization of pyrimidine dimers, *Biochem. Biophys. Acta*, 378, 165, 1975.

61. Kemmink, J., Eker, A. P. M., and Kaptein, R., CIDNP detected flash photolysis of *cis,syn*-1,3-dimethylthymine dimer, *Photochem. Photobiol.*, 44, 137, 1986.

62. (a) Pac, C., Kubo, J., Majima, T., and Sakurai, H., Structure-reactivity relationship in redox-photosensitized splitting of pyrimidine dimers and unusual enhancing effect of molecular oxygen, *Photochem. Photobiol.*, 36, 273, 1982; (b) Majima, T., Pac, C., Kubo, J., and Sakurai, H., Redox-photosensitized chain monomerization of *cis,syn*-dimer of dimethylthymine: Unusual effect of molecular oxygen, *Tetrahedron Lett.*, 21, 377, 1980.

63. Helence, C. and Charlier, M., Photosensitized splitting of pyrimidine dimers by indole derivatives and by triptophane-containing oligopeptides and proteins, *Photochem. Photobiol.*, 25, 429, 1977.

64. Van Champ, J. R., Young, T., Hartman, R. F., and Rose, S. D., Photosensitization of pyrimidine dimer splitting by a covelently bound indole, *Photochem. Photobiol.*, 45, 365, 1987.

65. Yeh, S.-R. and Falvey, D. E., Model studies of DNA photorepair: Radical anion cleavage of thymine dimers probed by nanosecond laser spectroscopy, *J. Am. Chem. Soc.*, 113, 8557, 1991.

66. Okamura, T., Sancar, A., Heelis, P. F., Begley, T. P., Hirata, Y., and Mataga, N., Picosecond laser photolysis studies on the photorepair of pyrimidine dimers by DNA photolyase. 1. Laser photolysis of photolyase-2-deoxyuridine dinucleotide photodimer complex, *J. Am. Chem. Soc.*, 113, 3143, 1991.

67. Kim, S.-T., Sancar, A., Essenmacher, C., and Babcock, G. T., Evidence from photoinduced EPR for a radical intermediate during photolysis of cyclobutane thymine dimer by DNA photolyase, *J. Am. Chem. Soc.*, 114, 4442, 1992.

68. Rokita, S. E. and Walsh, C. T., Flavin and 5-deazaflavin photosensitized cleavage of thymine dimer. A model of *in vivo* light-requiring DNA photorepair, *J. Am. Chem. Soc.*, 106, 4589, 1984.

69. Jorns, M. S., Photosensitized cleavage of thymine dimer with reduced flavin: A model for enzymic photorepair of DNA, *J. Am. Chem. Soc.*, 109, 3133, 1987.

70. Pac, C., Miyake, K., Masaki, Y., Yanagida, S., Ohno, T., and Yoshimura, A., Flavin-photosensitized monomerization of dimethylthymine cyclobutane dimer: Remarkable effects of perchloric acid and participation of excited-singlet, triplet, and chain-reaction pathways, *J. Am. Chem. Soc.*, in press.

71. Hartman, R. F. and Rosa, S. D., A possible chain reaction in photosensitized splitting of pyrimidine dimers by a protonated, oxidized flavin, *J. Org. Chem.*, 57, 2302, 1992.

72. Hartman, R. F. and Rosa, S. D., Efficient photosensitized pyrimidine dimer splitting by a reduced flavin requires the deprotonated flavin, *J. Am. Chem. Soc.*, 114, 3559, 1992.

73. Chaquin, P. and Kossanyi, J., Photochemistry of 1,2-diacylcyclobutanes, *Tetrahedron Lett.*, 3413, 1979.

74. Ceustermans, R. A. E., Martens, H. J., and Hoornaert, G. J., Photochemical transformation of truxones to C-nor-D-homo steroid systems, *J. Org. Chem.*, 44, 1388, 1979.

75. Gotoh, T., Kato, M., Yamamoto, M., and Nishijima, Y., Cation-radical transfer: Transfer efficiency in photosensitized isomerization reactions, *J. Chem. Soc., Chem. Commun.*, 90, 1981.

76. Evans, T. R., Wake, R. W., and Jaenicke, O., Singlet quenching mechanism — Sensitized dimerization of phenyl vinyl ether, in *The Exciplex*, Academic Press, New York, 1975, 345.

77. Mizuno, K., Murakami, K., Kamiyama, N., and Otsuji, Y., 9,10-Dicyanoanthracene-sensitized photo-oxygenation of *trans*-1,2-di(carbazol-9-yl)cyclobutane via electron transfer, *J. Chem. Soc., Chem. Commun.*, 462, 1983.

78. Miyashi, T., Takahashi, Y., Yokogawa, K., and Mukai, T., The single electron transfer induced degenerate methylenecyclobutane rearrangement through an allylically stabilized 1,4-cation radical, *J. Chem. Soc., Chem. Commun.*, 175, 1987.

79. Ikeda, K., Yamashita, Y., Kabuto, C., and Miyashi, T., Unusual cycloaddition of a cage ketone with tetracyanoethylene by irradiation of the electron donor-acceptor complex, *Tetrahedron Lett.*, 29, 5779, 1988.

80. Gilb, W. and Schroder, G., Aza[18]annulenes, *Angew. Chem. Int. Ed. Engl.*, 18, 312, 1979.

81. Rottele, H. and Schroder, G., Aza[14]annulenes, *Angew. Chem. Int. Ed. Engl.*, 19, 207, 1980.

82. Kurita, J., Iwata, K., and Tsuchiya, T., Synthesis of the first examples of fully unsaturated monocyclic 1,4-oxazepines, *J. Chem. Soc., Chem. Commun.*, 1188, 1986.

83. Kretschmer, G. and Warrener, R. N., A photochemical route to the unsaturated β-lactam, *N*-methyl-Azetione: A thermally labile ring-system, *Tetrahedron Lett.*, 1335, 1975.

84. R. N. Warrener, I. G. P. and R. A. Russel, A photochemical approach to methyl-substituted 5-oxabicyclo[2.1.0]pent-2-enes (Dewer furans), *J. Chem. Soc., Chem. Commun.*, 1464, 1984.

85. Schroder, G., Synthesis and properties of tricyclo[3.3.2.0^{4,6}]deca-2,7,9-triene (Bullvalene), *Chem. Ber.*, 97, 3140, 1964.

86. Hafner, K., Donges, R., Goedecke, E., and Kaiser, R., Concerning pentalene, 2-methylpentalene, and 1,3-dimethylpentalene, *Angew. Chem., Int. Ed. Engl.*, 12, 337, 1973.

87. Maier, G., Sayrac, T., and Reisenauer, H. P., Small rings. 42. Attempts to prepare oxirenes via photochemical cycloreversions, *Chem. Ber.*, 115, 2202, 1982.

88. (a) Masamune, S., Suda, M., Ona, H., and Leichter, L. M., Cyclobutadiene, *J. Chem. Soc., Chem. Commun.*, 1268, 1972; (b) Maier, G., Schneider, M., Kreiling G., and Mayer, W., Small rings. 35. Attempts to synthesize tetramethyltetrahedrane from heterocyclic precursors, *Chem. Ber.*, 114, 3922, 1981; (c) Maier, G., The cyclobutadiene problem, *Angew. Chem., Int. Ed. Engl.*, 13, 425, 1974.

89. Miller, R. D. and Abraitys, V. Y., Stereoselectivity in the photochemical cycloelimination of some polycyclic cyclobutanones, *J. Am. Chem. Soc.*, 94, 663, 1972.

90. Akabori, S., Kumagai, T., Habata, Y., and Sato, S., The preparation of photoresponsive cyclobutanocrown ethers by means of intramolecular [2+2]photocycloaddition, *Bull. Chem. Soc. Jpn.*, 61, 2459, 1988.

91. Akabori, S., Kumagai, T., Habata, Y., and Sato, S., Preparation of photoresponsive cyclobutane-1,2-dicarbonyl-capped 2.n diazacrown ethers by intramolecular [2+2] photocycloaddition, and their highly selective complexation with lithium cation, *J. Chem. Soc., Chem. Commun.*, 661, 1988.

92. Suginome, H., Itoh, M., and Kobayashi, K., Ring expansion through [2+2] photocycloaddition-β-scission sequence. Two carbon ring expansions of β-indanone and β-tetralone, *Chem. Lett.*, 1527, 1987.

93. Suginome, H., Liu, C. F., Seko, S., Kobayashi, K., and Furusaki, A., Photoinduced molecular transformations. 100. Formation of furocoumarins and furochromones via a β-scission of cyclobutanoxyl radicals generated from [2+2] photoadducts from 4-hydroxycoumarin and acrylic and cyclic alkenes. X-ray crystal structures of (6aα,6bα,9aα,9bα)-(±)-6b,7,8,9,9a,9b-hexahydro-9b-hydroxybenzo[*b*]cyclopenta[3,4]cyclobuta[1,2-*d*]pyran-6(6aH)-one, *cis*-(±)-6b,8,9,9a-tetrahydro-6H,7H-cyclopenta[4,5]furo[3,2-c][1]benzopyran-6-one, and *cis*-1,2,2a,8b-tetrahydro-8b-hydroxy-1,1,2,2-tetramethyl-3H-benzo[*b*]cyclobuta[*d*]pyran-3-one, *J. Org. Chem.*, 53, 5952, 1988.

94. Suginome, H., Liu, C. F., Tokuda, M., and Furusaki, A., Photoinduced transformation. Part 76. Ring expansion through a [2+2] photocycloaddition-β-scission sequence; the photorearrangement of *endo*-4-cyanotricyclo[6.4.0.0^{2,5}]dodeca-1(12),6,8,10-tetraen-5-yl hypoiodite to 4-cyanotricyclo-[6.4.0.0^{2,4}]dodeca-1(12),6,8,10-tetraen-5-one. X-ray crystal structure of 4-cyanotricyclo[6.4.0.0^{2,4}]-dodeca-1(12),6,8,10-tetraen-5-one, *J. Chem. Soc., Perkin Trans. I*, 327, 1985.

95. Stenstrom, Y. and Jones, W. M., Photoinduced ring expansliron of a cyclobutyliron σ-complex: An example of rearrangement of an alkyl group from saturated carbon to a transition metal to give a carbene complex, *Organometallics*, 5, 178, 1986.

96. Stenstrom, Y., Koziol, A. E., Palenik, G. J., and Jones, W. M., Photoinduced ring expansion of cycloalkyl iron σ-complexes to cyclic iron complexes, *Organometallics*, 6, 2079, 1987.

97. Chen, Y., K. S., Yonezawa, N., and Hasegawa, M., Linear dichroism induced by the photocleavage of polyamides containing cyclobutane rings, *J. Polym. Sci. Polym. Chem. Ed.,* 26, 3397, 1988.

98. Caccamese, S., Maravigna, P., Montaudo, G., Recca, A., and Scamporrino, E., Photolytic behavior of copolyamides from adipic and truxillic acids, *J. Poly. Sci. Pol. Chem. Ed.,* 17, 1463, 1979.

99. Hasegawa, M., Yonezawa, N., Kanoe, T., and Ikebe, Y., A new approach to fully conjugated polyamides: Preparation and photolysis of poly[p-phenylene *trans*-3,*cis*-4-*bis*(2-hydroxyphenyl)-1,*trans*-2-cyclobutane dicarboxylamide], *J. Polym. Sci. Polym. Lett. Ed.,* 20, 309, 1982.

30

Photochemistry of Oxiranes — Photoreactions of Epoxynaphthoquinones

uhiro Maruyama
o Institute of Technology

uo Kubo
ane University

).1 Definition of the Reaction

The photochemistry of oxiranes has been extensively explored and the potential synthetic utility of the photoreactions has received considerable attention. A number of reviews appeared in the 1970s.[1–3] The photochemistry of oxiranes is dramatically dependent upon the nature of the substituents. Considerable effort has been devoted to the elucidation of the photochemical properties of aryl-substituted oxiranes and α,β-epoxyketones.

Aryl-substituted oxiranes 1 have been shown to undergo photofragmentation in solution to give carbonyl compounds 4 and arylcarbenes 3, which can be trapped by alkenes and alcohols (Scheme 30.1). Carbonyl ylides (1,3-dipoles) 2 appeared to be the principal intermediates in the carbene formation. Their involvement has been established not only by inference through chemical trapping (e.g., as cycloadducts such as 8 from reactions with dipolarophiles (alkenes) and as products such as 9 from reaction with alcohols), but also by direct spectroscopic observations of 2 in low-temperature rigid matrices or in condensed media at room temperature by time-resolved laser flash photolysis.[4,5]

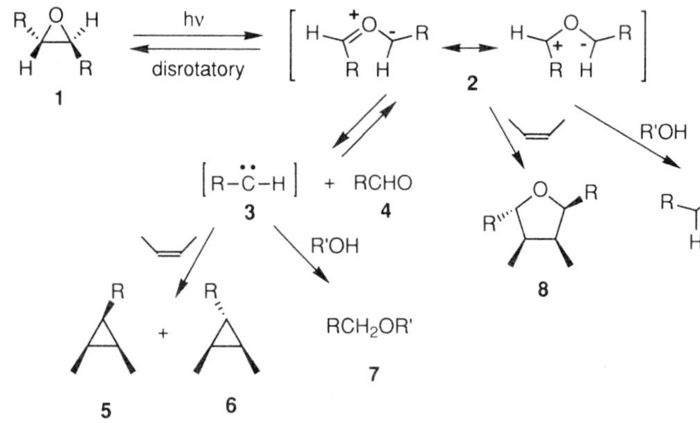

SCHEME 30.1

In the photochemistry of α,β-epoxyketones, it is generally accepted that the nπ*-state is the chemically significant excited state and two types of reactions may be distinguished: (1) photoisomerization to β-diketones accompanying migration of one of the substituent groups (Scheme 30.2),[6] and (2) photochromic valence isomerization between arylcyclopentenone oxides, such as 13, and pyrylium oxides, such as 14, that can be trapped by dipolarophiles (Equation 30.1).[7]

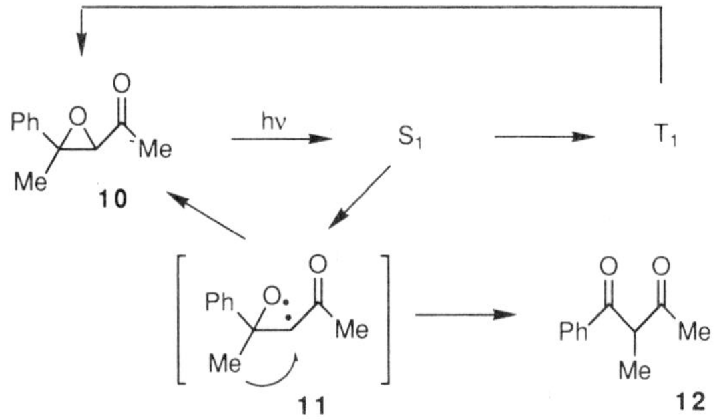

SCHEME 30.2

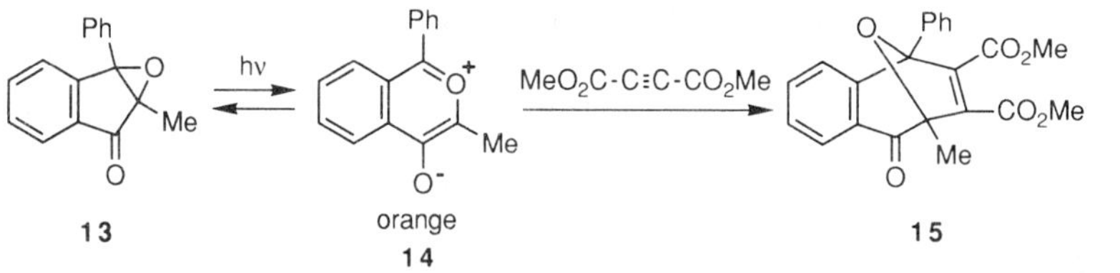

(30.1)

In this chapter, we concentrate on the photochemistry of epoxynaphthoquinones 16 that are oxiranes with vicinal dicarbonyl substituent groups. Epoxyquinones are metabolites of biologically important quinones; thus, their photochemical reactivity may be related to some photodamaging effect of quinonoid compounds.

30.2 Mechanism

Since the intersystem crossing of 16 is very efficient and the lowest triplet state (T_1) is nπ* in type, all photoreactions of 16 may be considered to originate from T_1.[8,9] The T_1 state of 16 is capable of the following types of reaction (Scheme 30.3): (1) C-C bond fission of the oxirane ring and formation of carbonyl ylides (17) or 1,3-biradicals (18); (2) C-O bond fission of the oxirane ring; (3) α-cleavage (the Norrish Type I reaction); (4) oxetane formation with alkenes; and (5) intramolecular or intermolecular hydrogen abstraction reactions with hydrogen donors.

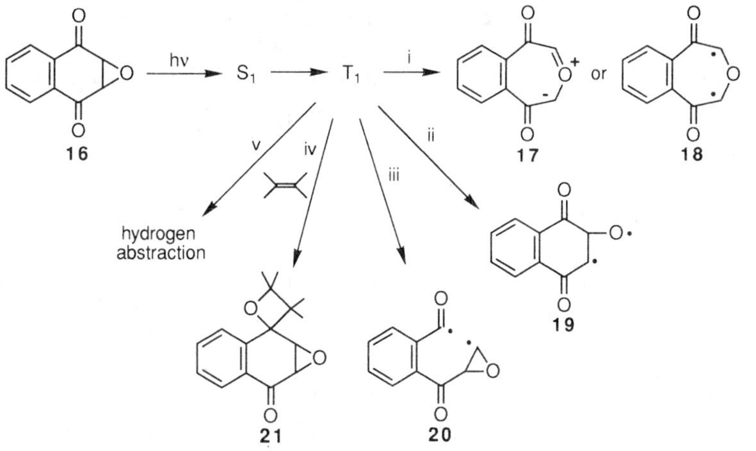

SCHEME 30.3

Photoreactions of 16 are highly dependent upon the substitution pattern at the C2 and C3 positions of 16. Nonsubstituted or 2-alkyl-substituted compounds 16a,b undergo photoreactions characteristic of the carbonyl chromophore; for instance, photoreactions with norbornene (22) gave oxetanes 23 and 24 (Equation 30.2). 2-Aryl or 2,3-disubstituted compounds 16c-e, however, react as carbonyl ylides (17) or 1,3-biradicals (18) via C-C bond cleavage of the oxirane ring; for example, irradiation in the presence of norbornene (22) afforded cycloadducts 25 and 26 (Equation 30.3).[9,10] The dialkyl, aryl, and diaryl substituents in 16c-e appear to provide stabilization that facilitates the formation of 17 or 18.

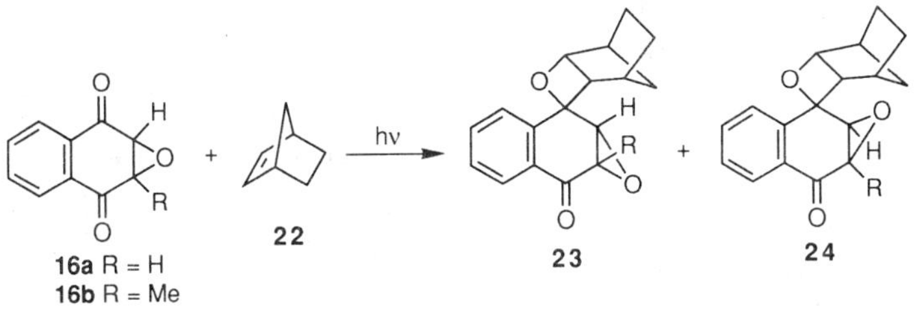

$$(30.2)$$

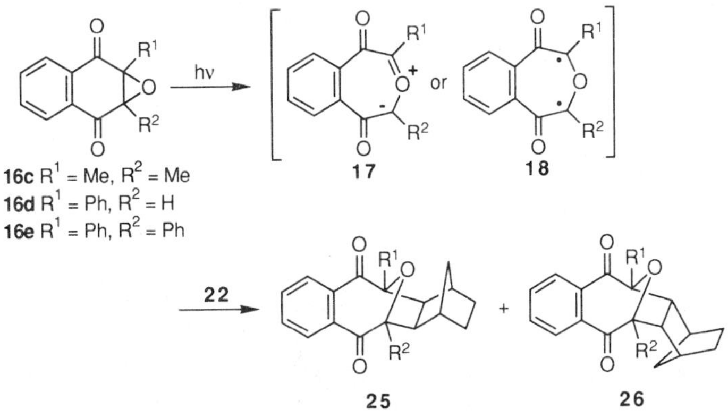

$$(30.3)$$

30.3 Synthetic Applications

Epoxynaphthoquinones **16** are photolabile and undergo photorearrangement or photodimerization. Photolysis of an extremely dilute solution of **16c** (<0.1 m*M* in benzene) gave triketone **28** and a mixture of alkylidenephthalides **32** and **33** (Scheme 30.4).[11] The triketone **28** is also photolabile and was converted on irradiation into **32** and **33** almost quantitatively. This photorearrangement was explained in terms of two possible mechanisms starting with C–O bond fission of the oxirane ring (path A) and α-cleavage (path B) (Scheme 30.4). 2-Acetyl-3-methyl-2,3-epoxy-2,3-dihydro-1,4-naphthoquinone[12] and **16e**[13] were also reported to undergo similar photorearrangement to give a mixture of alkylidenephthalides.

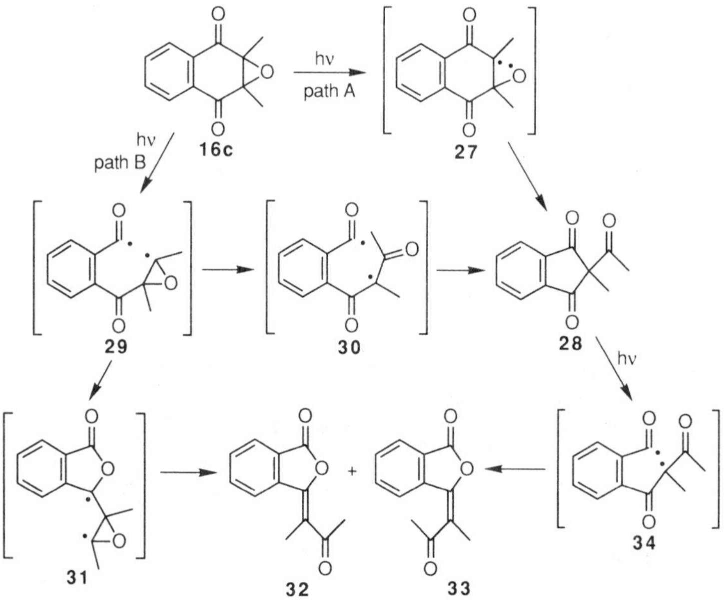

SCHEME 30.4

Photoreaction of **16c** was found to depend largely upon the concentration of **16c**. Photolysis of a concentrated solution of **16c** (0.1 *M* in benzene) thus gave the dimers **35** (65%) and **36** (20%) with complete quenching of the formation of **28**, **32**, and **33** (Equation 30.4). This dimerization was explained in terms of C–C bond fission of the oxirane ring of **16c** to form carbonyl ylide **17** or 1,3-biradical **18**, and trapping of **17** or **18** with the carbonyl group of ground-state **16c**.

le 30.1   Photocycloaddition of 2,3-Dimethyl-2,3-epoxy-2,3-dihydro-1,4-naphthoquinone (16c) to Alkenes 37a-k[a]

	Conc.[b] (M)	Irrad. time	Conv.[c] (h) (%)	38			39	40	
ene 37				endo	exo	endo/exo		Z	E
$_2$=CMe$_2$ 37a	0.2	2	100	—			41	13	13
	0.2	0.25	65		42	~	23	7	7
lopentene 37b	0.11	2	90	50	—	7.4	2	15	15
CH=CH$_2$ 37c	0.05	4	69	16	19	0.28	15	23	23
$_2$=CHCH$_2$OH 37d	0.1	0.5	82	84	—	4.5	—	6	5
$_2$=CHCO$_2$Me 37e	0.15	0.5	87	17	44	0.35	17	12	8
$_2$=CHCN 37f	0.2	0.5	72	86	—		4	2	2
)CH=CH$_2$ 37g	0.07	0.5	53	34	32	1.4	—	—	—
$_2$=CHCONHPh 37h	0.04	3	61	55	0		25	12	
$_2$=CHCONMePh 37i	0.04	2	72	1	18		29		41
$_2$=CHCH$_2$NHCOMe 37j	0.04	3	58	52	6	4	15	24	
$_2$=CHCH$_2$NMeCOMe 37k	0.04	2.5	88	1	17	0.3	15		57

Photoreactions were carried out in benzene.
Concentrations of alkenes 37a-k.
Conversion of 16c.

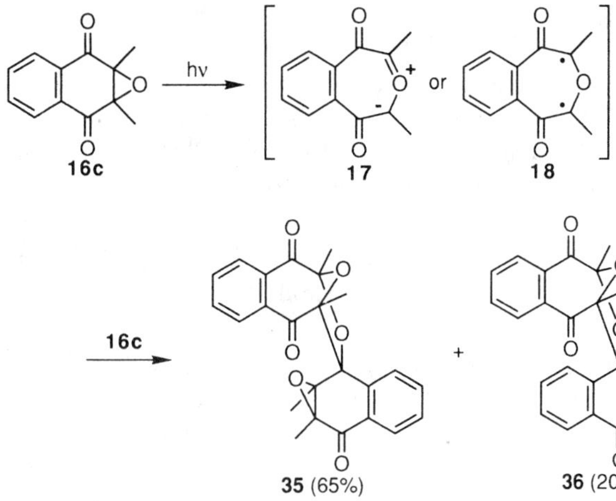

(30.4)

A wide variety of alkenes 37, including simple alkenes, cycloalkenes, aryl-substituted alkenes, electron-deficient alkenes, and electron-rich alkenes, can be used to trap 17 or 18 to give the primary adducts 38 with complete quenching of the dimer formation of 16c (Equation 30.5 and Table 30.1).[8–10,13–15] Cycloaddition of 17 or 18 to *trans*- and *cis*-2-butene proceeded stereospecifically. The excited triplet state of 16c was quenched by electron-rich alkenes via exciplex deactivation; this quenching competed with the path leading to 17 or 18.

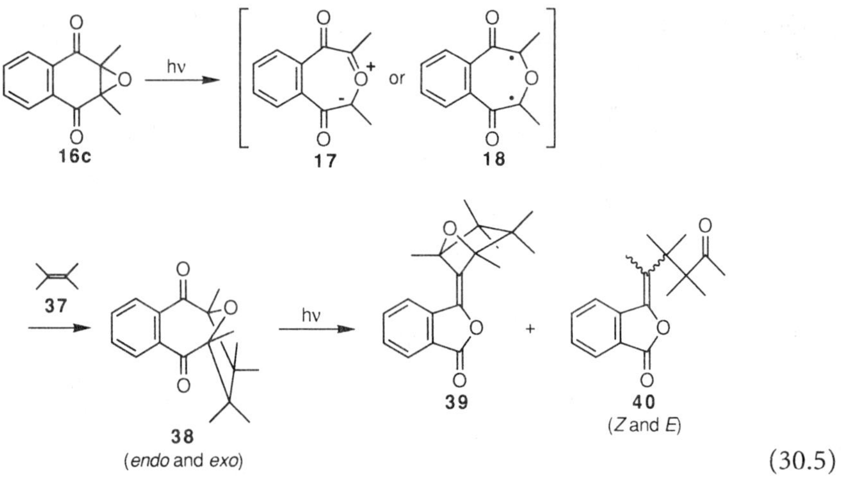

(30.5)

The *endo-exo* ratios of the primary adducts **38** at low conversions (e.g., <5%) are included in Table 30.1. Generally, *endo-exo* stereoselectivity is poor. However, there is a tendency toward *endo*-orientation for electron-releasing substituents such as alkyl, hydroxymethyl, and ethoxy groups, and toward *exo*-orientation for electron-attracting substituents such as methoxycarbonyl and phenyl groups, with the exception of the cyano group which showed a prominent *endo*-orientating tendency. These orientations were accounted for by considering the dipole-dipole interactions rather than secondary orbital overlap or steric factors. However, with acrylonitrile (**37f**), some other interactions, presumably secondary π-orbital overlap, would predominate over the dipole-dipole interaction. On the other hand, remarkably high *endo*-stereoselectivity was observed with *N*-monosubstituted acrylamide **37h**, or *N*-allylacetamide (**37j**), and inverted *exo*-stereoselectivity with *N,N*-disubstituted acrylamide **37i** or *N*-allyl-*N*-methylacetamide (**37k**) were explained by the favorable hydrogen-bonding interactions by the amide N-H bond in the transition state of the cycloaddition illustrated in **41** and **42**.[15]

Further irradiation of the primary adducts **38** gave spirophthalides **39** and a *Z-E* mixture of alkylidenephthalides **40** simultaneously. The photorearrangement is likely to begin with preferential α-cleavage at the more crowded side leading to a biradical **43**, followed by lactonization to give a 1,4-biradical **44** that cyclizes to give **39** and a *Z-E* mixture of **40** (Scheme 30.5). On the basis of the evidence of quantum yields and triplet lifetime, this preference is believed to arise from the different efficiencies of the radical recombination of biradicals such as **43** and **45**.[8] The distribution of products **38**, **39**, and **40** could be controlled by the length of irradiation and wavelength of the light source used.[8,10]

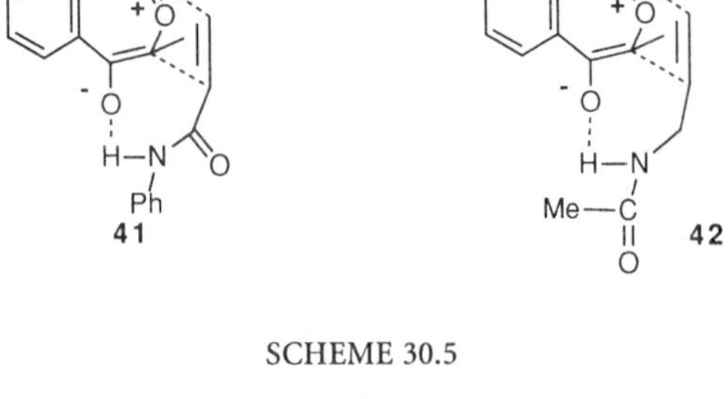

SCHEME 30.5

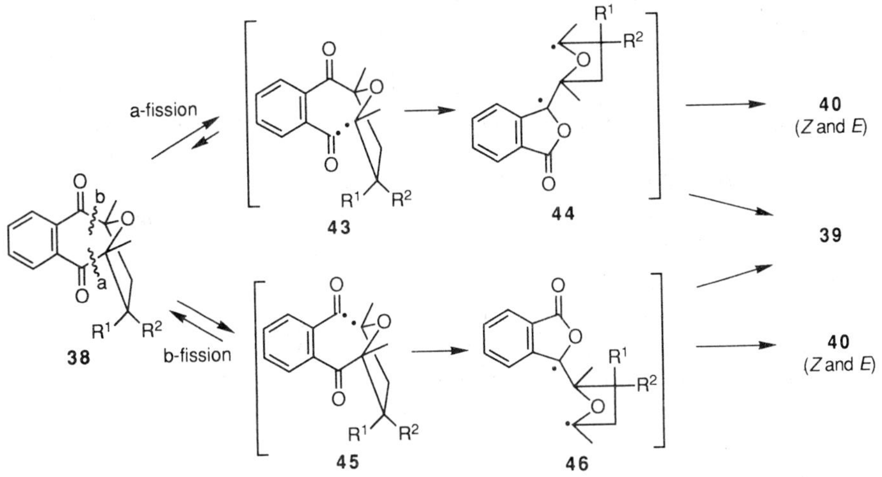

SCHEME 30.5 (continued)

A variety of aldehydes 47 and ketones 50 can also be used to trap 17 or 18 to give cycloadducts 48 + 49 and 51 in high yields, respectively (Scheme 30.6).[16] Further irradiation of 51 gave a Z-E mixture of alkylidenephthalides 54 via α-cleavage. Singlet oxygen[17] and alcohols[18] are also reported to trap 17 or 18.

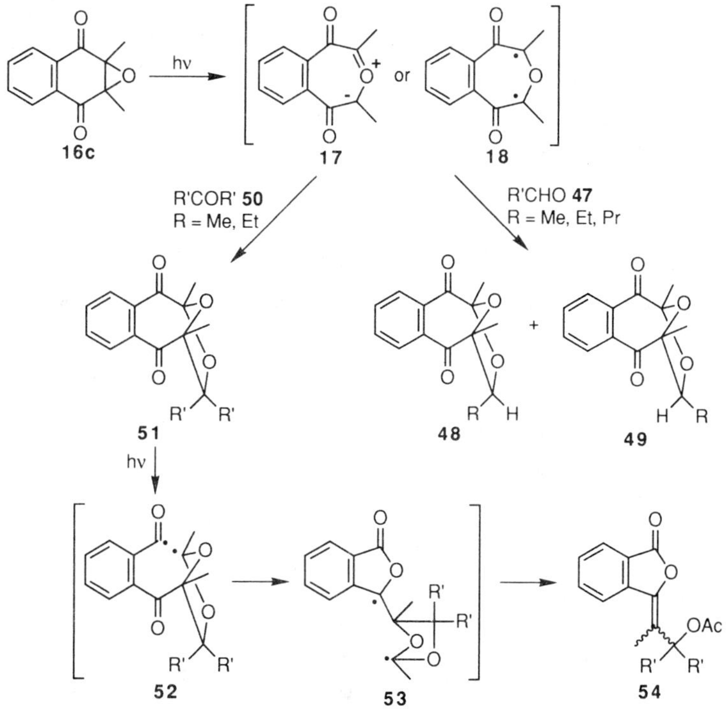

SCHEME 30.6

As for the hydrogen abstraction of epoxyquinones, the Norrish Type II photoreaction (intramolecular hydrogen abstraction) of epoxynaphthoquinones was reported to take place without oxirane ring opening to give a cyclobutanol that, on further irradiation, was converted into a new class of polycondensed ring system.[19] Photoreactions of 2-alkyl-substituted epoxynaphthoquinones with good hydrogen donors, such as 2-propanol or xanthene, have been reported and the reaction mechanism was examined by means of the CIDNP technique.[20]

The photochemistry of 2,3-imino-2,3-dihydro-1,4-naphthoquinones **55**, which are electronically analogous to epoxyquinones, was also investigated.[21] On irradiation with alkenes (e.g., **22**), the iminoquinone **55** gave cycloadducts such as **58** via C–C bond cleavage of the aziridine ring to afford 1,3-dipole **56** or 1,3-biradical **57** (Equation 30.6). Cycloadducts from **55** were inert photochemically, whereas the corresponding cycloadducts from epoxyquinones undergo further photorearrangement. The limiting quantum yields of the cycloaddition of **55** with alkenes ($\approx$0.01) were about 50 times smaller than those of epoxyquinones ($\approx$0.5). This difference in photoreactivity was explained by the difference of the excited-state character; absorption and emission spectra revealed that **55** had a large CT character.

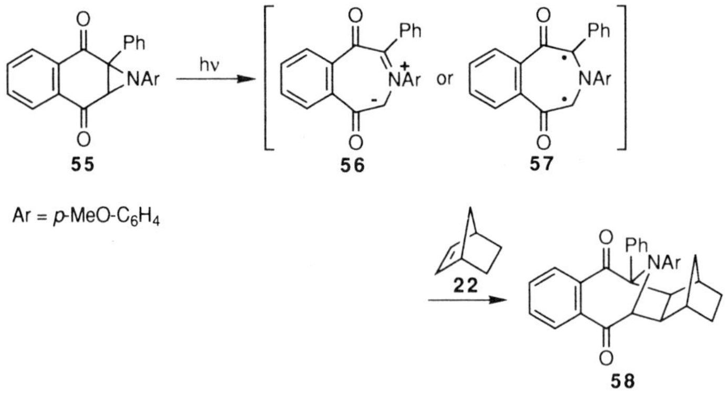

$$(30.6)$$

References

1. Bertoniere, N. R. and Griffin, G. W., *Organic Photochemistry*, Vol. 3, Marcel Dekker, New York, 1973, chap. 2.
2. Griffin, G. W. and Padwa, A., *Photochemistry of Heterocyclic Compounds*, John Wiley & Sons, New York, 1976, chap. 2.
3. Armarego, W. L. F., *Stereochemistry of Heterocyclic Compounds*, Part 2, John Wiley & Sons, New York, 1977, 12.
4. Das, P. K., *Handbook of Organic Photochemistry*, Vol. 2, CRC Press, Boca Raton, FL, 1987, 35.
5. Clark, K. B., Bhattacharyya, K., Das, P. K., Scaiano, J. C., and Schaap, A. P., Photochemistry of 2,3-Di(1′-naphthyl)oxiranes. Spectral and kinetic behavior of carbonyl ylides in condensed media, *J. Org. Chem.*, 57, 3706, 1992.
6. Markos, C. S. and Reusch, W., Photochemical Rearrangement of α,β-epoxy ketones. An elaboration of the mechanism, *J. Am. Chem. Soc.*, 89, 3363, 1967.
7. Zimmerman, H. E. and Simkin, R. D., Photochemical reactions of 2,3-epoxy-2-methyl-3-phenylindanone, *Tetrahedron Lett.*, 1847, 1964.
8. Osuka, A., Suzuki, H., and Maruyama, K., Photochemistry of epoxyquinones. 7. Photoinduced cycloaddition of 2,3-dimethyl-2,3-dihydro-2,3-epoxy-1,4-naphthoquinone to olefins, *J. Chem. Soc., Perkin Trans. I*, 2671, 1982.
9. Maruyama, K., Arakawa, S., and Otsuki, T., Photochemistry of epoxyquinones: Photoinduced cycloaddition of 2,3-epoxy-1,4-naphthoquinones with olefins, *Tetrahedron Lett.*, 2433, 1975.

10. Arakawa, S., Photochemistry of epoxyquinones. 2. Photoinduced cycloaddition reactions of aryl- or alkyl-substituted 2,3-epoxy-2,3-dihydro-1,4-naphthoquinones with olefins, *J. Org. Chem.*, 42, 3800, 1977.

11. Maruyama, K. and Osuka, A., Photochemistry of epoxyquinone. 4. Primary dimers in the photochemical reaction of 2,3-dimethyl-2,3-epoxy-2,3-dihydro-1,4-naphthoquinone, *J. Org. Chem.*, 45, 1898, 1980.

12. Maruyama, K. and Arakawa, S., Photo-induced isomerization of 2-acyl-3-methyl-1,4-naphthoquinone-2,3-epoxide, *Chem. Lett.*, 1974, 719.

13. Kato, H., Tezuka, H., Yamaguchi, K., Nowada, K., and Nakamura, Y., Photochemical and thermal reactions of heterocycles. 2. Photolysis of 2,3-diphenylnaphthoquinone 2,3-epoxide: Trapping of a cyclic carbonyl ylide and photoisomerization, *J. Chem. Soc., Perkin Trans. I*, 1029, 1978.

14. Giles, R. G. F., Green, I. R., and Mitchell, P. R. K., Epoxidation of 1,4-dihydro-1,4-methanoanthraquinones. Photochemistry and crystal structure determination of the products, *J. Chem. Soc., Perkin Trans. I*, 719, 1979.

15. Maruyama, K., Osuka, A., and Nakagawa, K., Photochemistry of epoxynaphthoquinones. 8. Endo-stereoselective photocycloaddition of 2,3-epoxy-2,3-dihydro-2,3-dimethyl-1,4-naphthoquinone to olefins containing amide group, *Bull. Chem. Soc. Jpn.*, 60, 1021, 1987.

16. Maruyama, K. and Osuka, A., Photo-induced cycloadditions of epoxyquinone to aldehydes and ketones, *Chem. Lett.*, 77, 1979.

17. Maruyama, K., Osuka, A., and Suzuki, H., Photoinduced oxygenation of an epoxynaphthoquinone, *J. Chem. Soc., Chem. Commun.*, 723, 1980.

18. Osuka, A., Suzuki, H., and Maruyama, K., Photochemical reaction of epoxynaphthoquinones with alcohols. An ionic trapping of carbonyl ylides, *Chem. Lett.*, 201, 1981.

19. Maruyama, K., Osuka, A., and Suzuki, H., Type II photoreaction of epoxynaphthoquinones, *J. Chem. Soc., Chem. Commun.*, 323, 1980.

20. Maruyama, K. and Arakawa, S., Photochemistry of epoxyquinones. 1. Photochemical reactions of 2-alkyl-2,3-epoxy-2,3-dihydro-1,4-naphthoquinones with hydrogen donors, *J. Org. Chem.*, 42, 3793, 1977.

21. Maruyama, K. and Ogawa, T., Nitrogen effects in photoreactions. Photochemistry of iminoquinones with olefins, *J. Org. Chem.*, 48, 4968, 1983.

31

Photodecarboxylation of Acids and Lactones

Peter Wan
University of Victoria

David Budac
University of Victoria

31.1 Introduction

The reactions to be discussed here involve the loss of carbon dioxide from photoexcited aliphatic and aromatic carboxylic acids and lactones, as shown by Equations 31.1 and 31.2, respectively. In addition, a summary of methods for inducing the photodecarboxylation of acids that are otherwise photoinert, by structural modification, will be discussed since these methods show great promise in organic synthesis.

$$R\text{-}\overset{\overset{\displaystyle O}{\|}}{C}\text{-}OH \xrightarrow{\ h\nu\ } RH \ + \ CO_2 \tag{31.1}$$

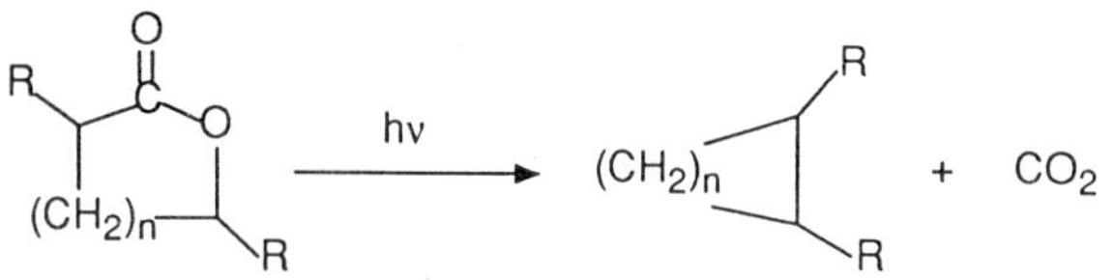

$$\tag{31.2}$$

Thermal decarboxylations are common in organic chemistry and are used frequently in organic synthesis, particularly with malonic acid derivatives.[1] Decarboxylation steps are also important in a number of biological processes such as the thiamine-mediated decarboxylation of pyruvic acid to acetaldehyde and CO_2, and the pyridoxal phosphate-mediated decarboxylation of amino acids.[2] Thermal decarboxylations occur commonly from the carboxylate ion and are accelerated by the presence of electron-withdrawing groups that can stabilize the incipient carbanion intermediate. Mechanistically, these decarboxylations may be viewed as the reverse of the addition of a carbanion to carbon dioxide. Thermal decarboxylations via radical intermediates are also common. However, as will be shown below, there are more mechanistic possibilities for *photo*decarboxylation.

Many of the earlier studies (prior to 1966) of photodecarboxylation of simple aliphatic carboxylic acids (e.g., formic, acetic, malonic, and pyruvic acids) and lactones (β-propiolactone, γ-butyrolactone, and δ-valerolactone) have been summarized by Calvert and Pitts in their classic work.[3] Most of these studies were carried out in the gas phase and, in general, photodecarboxylation

0-8493-8634-9/95/$0.00+$.50
© 1995 by CRC Press, Inc.

is accompanied by several competing reaction channels. For this reason and along with the fact that the parent carboxyl group absorbs at wavelengths shorter than 250 nm, the photochemistry of the carboxyl group has received considerably less attention than aldehydes and ketones. In recent times, however, photodecarboxylation has gained prominence as a reaction of some utility.

In the late 1970s, three reviews[4-6] were published that address some aspect of photodecarboxylation of organic compounds. However, apart from what appears in the appropriate section of the annual reviews of photochemistry published by the Royal Chemical Society (Specialist Periodical Report), an explicit review of photodecarboxylation did not appear until 1992.[7] It reviews work on photodecarboxylation of acids, esters, anhydrides, and lactones that were published during the period 1970–1992, with emphasis on mechanism and synthetic utility of aromatic substrates.

31.2 Mechanism

The observed mechanisms of photodecarboxylation of carboxylic acids may be broadly classified into three general categories: (1) heterolytic, (2) homolytic, or (3) mesolytic. They are defined according to the nature of their product-forming steps, as shown in Equation 31.3. A mesolytic mechanism involves decarboxylation from a radical cation precursor. The ejected electron in mesolytic cleavage may reside on the same molecule, solvated or transferred to an electron transfer sensitizer. Although carboxylate anions can photoionize and subsequently decarboxylate,[7] this overall mechanism is classified as homolytic since the product-forming step still involves formal homolysis of a radical precursor.

$$RCO_2H \xrightarrow{hv} \begin{cases} R^- + CO_2 + H^+ & \text{heterolytic} \\ R^{\cdot} + CO_2 + H^{\cdot} & \text{homolytic} \\ \left[RCO_2H\right]^{+\cdot} \xrightarrow{-e^-} R^{\cdot} + CO_2 + H^+ & \text{mesolytic} \end{cases}$$

$$(31.3)$$

Simple aliphatic carboxylic acids tend to photodecarboxylate via the homolytic mechanism.[3,8] A flash photolysis study of the photodecarboxylation of a number of simple aliphatic carboxylic acids (acetic, malonic, and pyruvic acids) in aqueous solution was carried out by Mittal et al.[8] In all cases, the major photolytic process observed for the protonated acids is homolysis of the C–C bond α to the carbonyl, resulting in radical intermediates [λ_{max} ($\cdot CO_2H$) = 235 nm; ε_{max} = 3.0 × 10^3 M^{-1} cm^{-1}]. Photolysis of the dissociated acids (at higher pH) failed to give any observable radicals. Since photodecarboxylation is also observed at these pHs, a heterolytic mechanism for photodecarboxylation cannot be ruled out.

In general, mechanistic studies of photodecarboxylation of simple arylacetic acids are encumbered by their relatively low quantum yields ($\Phi \leq 0.05$) for reaction. For example, although the photodecarboxylation of phenylacetic acid has been subjected to much study,[9-11] the details of the mechanism are not completely understood. The undissociated acid is believed to react via a homolytic mechanism (benzyl radical λ_{max} = 302, 314 nm),[11] while the carboxylate gives rise to a carbanion intermediate via a formal heterolytic mechanism (Equation 31.4). The possibility that the carbanion is formed in a stepwise manner (initial electron photoejection followed by decarboxylation and subsequent recombination of the electron with the benzyl radical) was not ruled out by Epling and Lopes[9] since the hydrated electron was observed by Meiggs et al.[10] in a flash photolysis study ($\lambda_{max} \approx$ 720 nm; $\tau <$ 40 μs). A number of α-hydroxy-substituted arylacetic acids (including the parent mandelic acid) and related substrates photodecarboxylate in aqueous solution via S_1 (pH 5 to 12) with high quantum yields (Φ = 0.2 to 0.7) via a heterolytic mechanism, giving the corresponding benzyl alcohols in high yields (Equation 31.5).[7] The substrates are unreactive at pH < pK_a. At pH > 12, an additional pathway is believed to operate since hydrobenzoins are also isolated. The proposed α-hydroxybenzyl carbanion intermediates (Equation 31.5) are novel in the sense that they are not readily available from traditional acid-base chemistry. The photodecarboxylation of *m*- and *p*-nitrophenylacetic acids and related compounds[12–14] has been studied by indirect and direct methods (nanosecond and picosecond laser and conventional microsecond flash photolysis) and is believed to react via a heterolytic (possibly adiabatic)[13] mechanism from T_1 (Φ = 0.6 to 0.7). The reaction of the *para*-isomer gives rise to a long-lived (on the order of minutes) *p*-nitrobenzyl carbanion intermediate (λ_{max} = 350 nm).[12,13] Stermitz and Huang[15] found that 2-, 3-, and 4-pyridylacetic acids photodecarboxylate from the zwitterion in aqueous solution via a formal heterolytic mechanism, with high quantum yields (Φ = 0.19 to 0.45). The optimum pH for the reaction is at the isoelectric point. The intermediates are believed to be the corresponding methylpyridine tautomers (1). Wan and co-workers[16,17] studied the reaction for a series of diarylacetic acids (e.g., 2 to 4) and found dramatic differences in relative photodecarboxylation efficiency. All of the substrates reacted via a heterolytic mechanism (Equation 31.6) from S_1, as demonstrated by incorporation of deuterium into the product from D_2O, lack of radical-derived products on direct photolysis, and low reactivity on triplet sensitization. The relative reactivity observed was opposite to that found in thermal decarboxylation; that is, 4 was the most reactive substrate (Φ = 0.60; k_d [primary photodecarboxylation rate constant from S_1] = 6 × 10^9 s^{-1}) and 2 the least reactive (Φ = 0.042; k_d = 9 × 10^6 s^{-1}). The authors argued that the relative reactivity may be rationalized by the notion that 4n cyclically conjugated carbanion intermediates have inherent stability (i.e., "aromatic") on the excited-state reaction surface.

$$PhCH_2CO_2H \xrightarrow{h\nu} PhCH_2 \cdot \quad \cdot CO_2H$$

$$PhCH_2CO_2^- \xrightarrow{h\nu} PhCH_2^- + CO_2 \tag{31.4}$$

$$\underset{\underset{OH}{|}}{Ar\overset{}{C}HCO_2^-} \xrightarrow[\substack{H_2O\text{-}CH_3CN \\ (\text{-}CO_2)}]{h\nu} Ar\overline{C}HOH \longrightarrow ArCH_2OH \tag{31.5}$$

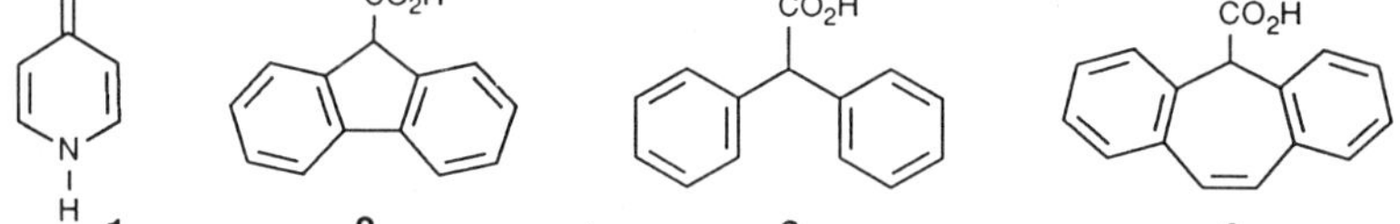

1   2   3   4

$$Ar_2CHCO_2^- \xrightarrow[\substack{H_2O\text{-}CH_3CN \\ (\text{-}CO_2)}]{h\nu} Ar_2CH^- \longrightarrow Ar_2CH_2 \tag{31.6}$$

$$CH_3COCO_2H \qquad PhCOCO_2H$$

5   6

The photodecarboxylation of pyruvic (5) and benzoylformic (phenylglyoxylic) (6) acids has been studied by several groups.[18–23] The quantum yields are high (Φ > 0.6) if carried out in aqueous solution. Benzoylformic acid (6) gave only benzaldehyde and CO_2, whereas 5 gave acetoin

$(CH_3COCH(OH)CH_3)$, CO_2, and a trace of acetaldehyde. Acetoin most likely arises via a secondary reaction of acetaldehyde. Photolysis of **5** in the gas phase is cleaner ($\Phi \approx 1$), giving only acetaldehyde and CO_2.[18,23] Photoreduction processes are observed when both **5** and **6** [λ_{max} (Ph–$\dot{C}$(OH)–CO_2H) = 313 nm; $\tau > 100\ \mu s$ in 2-PrOH][22] are photolyzed in organic solvents. Because of the high quantum efficiency and clean process for **6** in aqueous solution, use as a chemical actinometer over the excitation range 250 to 400 nm has been suggested.[21] With respect to mechanism, there is consensus that reaction is via T_1 (λ_{max} = 322 nm),[22] but both heterolytic[22] (via a benzoyl anion) and homolytic[18–22] (via either a benzoyl radical or α-hydroxyphenyl methyl carbene) mechanisms have been proposed. A concerted mechanism is suggested for the reaction of **5** in the gas phase.[23] Interestingly, the carboxylates of these acids (in solution) are much less reactive, supporting a concerted mechanism of decarboxylation from the acid form in which the acid proton is transferred directly to the α-carbonyl on loss of CO_2.

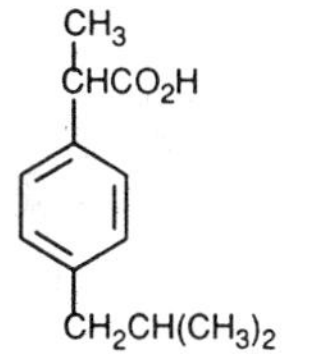

7 (Ibuprofen) 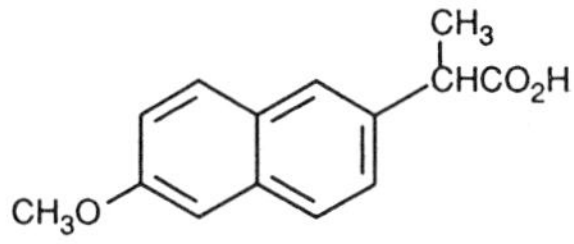   **8** (Naproxen) 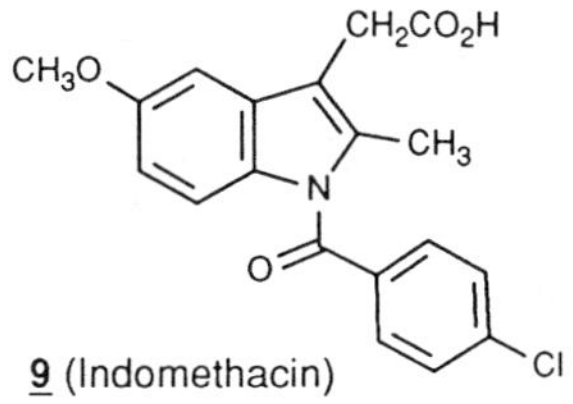   **9** (Indomethacin)

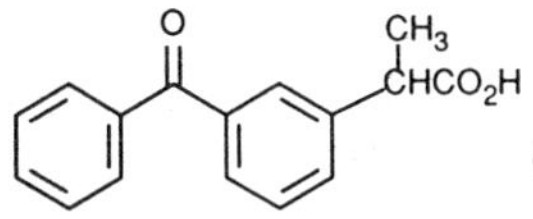

10 (Ketoprofen) 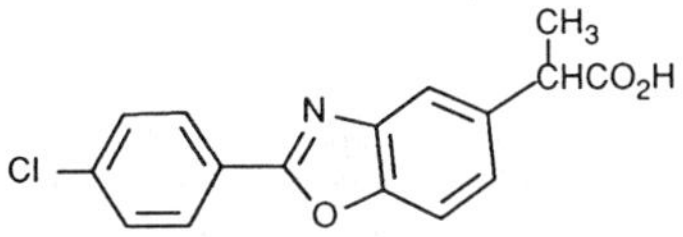   **11** (Benoxaprofen) 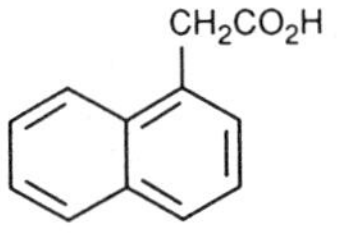   **12** (1-Naphthalene acetic acid)

A variety of carboxylic acids containing an arylacetic acid moiety are used as nonsteroidal antiinflammatory drugs (NSAID) (**7** to **11**) and as a plant growth regulator (**12**). The photochemistry of these substances has been studied to gauge their stability toward light, especially since some NSAIDs are suspected to be phototoxic. All of **7** to **12** photodecarboxylate with a range of quantum efficiencies and mechanistic possibilities.[7] Evidence available[7] indicate that **7** reacts via a homolytic mechanism; **9** via a heterolytic pathway; and all of **8**, **10**, and **11** via a mesolytic pathway. Lamp irradiation of **12** results in a homolytic pathway[7] for reaction, whereas laser flash photolysis induces a biphotonic mesolytic pathway [λ_{max} (1-naphthylmethyl radical) = 330 nm].[7,24] The latter pathway was also shown to operate for 4-methyl- and 4-methoxyphenylacetic acids.[24]

The mechanism for photodecarboxylation of lactones (photoextrusion)[6] usually involves initial homolysis of the O–C bond β to the carbonyl, followed by loss of CO_2, to generate a biradical. Heterolytic and mesolytic pathways can also be envisaged for lactones, but have not been shown to operate in general.[6]

31.3 Synthetic Applications

Apart from the well-known thermal decarboxylation of malonic acid derivatives and other β-ketoacids, where the carboxyl group is replaced with a hydrogen, the Hundsdiecker reaction[1] is a useful one-carbon degradation of a carboxylic acid, in which the carboxyl group is replaced with a halogen. The efficiency and mechanism of photodecarboxylation of acids depend critically on structure and are thus less useful in organic synthesis. To address this deficiency, several methods[7] have been developed to induce the general photodecarboxylation of carboxylic acids, either by converting them to activated esters or imides, which are then irradiated directly or by sensitization. The goal of these methodologies is to induce *homolytic* photodecarboxylation of otherwise unreactive compounds, giving radical intermediates that may then be used for further synthetic manipulation. Since carboxylic acids are widely available, this methodology has gained prominence recently as a useful synthetic methodology.

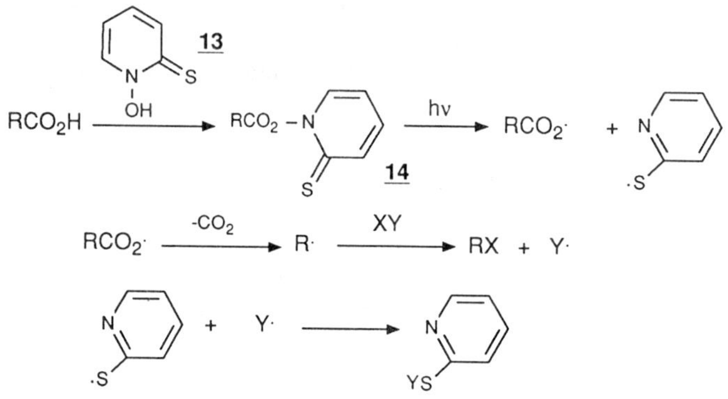

SCHEME 31.1

Perhaps the most well known of these methods is that developed by Barton and co-workers (Scheme 31.1).[7,25–32] The carboxylic acid is first converted to an acyl derivative of *N*-hydroxy-2-thiopyridone (13), to give ester 14, that on photolysis (tungsten lamp) yields the carboxyl radical; the carboxyl radical, on decarboxylation, gives the carbon-centered radical. This "tamed radical" can be trapped with a variety of reagents, including Bu_3SnH, alkenes, and halogens. Other activating groups related to 13 have also been used[26] successfully with quantum yields in the range 10 to 60. A radical chain mechanism has been proposed for the overall process.[26]

Related methods based on conversion of the acid to benzophenone oxime esters[7] or *N*-acyloxyphthalimides [use of sensitized irradiation with 1,6-*bis*(dimethylamino)pyrene or 1-benzyl-1,4-dihydronicotinamide with $Ru(bpy)_3Cl_2$][33,34] have been reported. In addition, sensitized irradiation of carboxylic acids with acridine has been reported to result in decarboxylation of the acid.[7,35]

The photodecarboxylation of lactones has been shown to be a useful synthetic reaction as a method for making ring systems (such as cyclopropanes, cyclobutenes, and cyclophanes) via the coupling of the biradical formed from loss of CO_2.[6,7] The synthetic utility is enhanced by the fact that lactones are generally stable compounds with well-established methodologies for their preparation. However, not all lactones photodecarboxylate cleanly, thus limiting the generality of this reaction. The classic example where photodecarboxylation of a lactone is used for synthesis is in the preparation of cyclobutadiene (Equation 31.7).[6,36] An electrocyclic ring closure from photolysis of α-pyrone 15 gives a strained bicyclic intermediate that, on further irradiation, decarboxylates to give the classic antiaromatic molecule cyclobutadiene (Equation 31.7). Cyclobutadiene generated in this way is stable at 8 K in an argon matrix[6] or at ambient or higher temperatures when trapped in a hemicarcerand.[36] Photolysis of *bis*lactone precursors related to 16 resulted in decarboxylation to give *o*-quinonedimethide intermediates that react further to give cyclobutanes and dimerization products.[37–39] The synthesis of the natural product aubergenone from commercially available α-santonin required the photodecarboxylation of γ-lactone 17 to give the cyclopropane 18 in 68% yield, which was subsequently taken on to aubergenone.[40] The widest utility of photodecarboxylation of lactones in synthesis is for the preparation of cyclophanes.[41–44] This is illustrated by the elegant preparation of [2.2]paracyclophane (20) in 70% yield from photolysis of the precursor bislactone 19 (Equation 31.8).

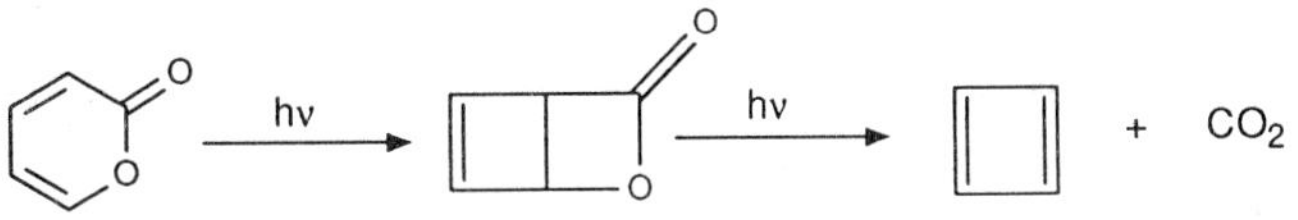

$$(31.7)$$

15

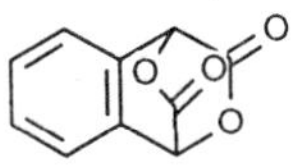

16 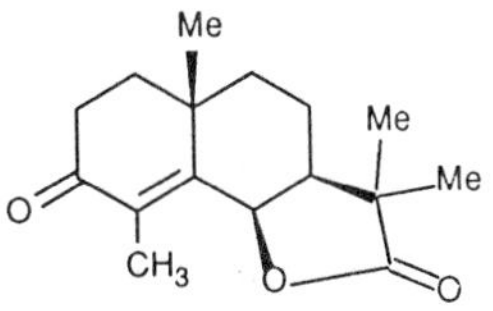   **17** 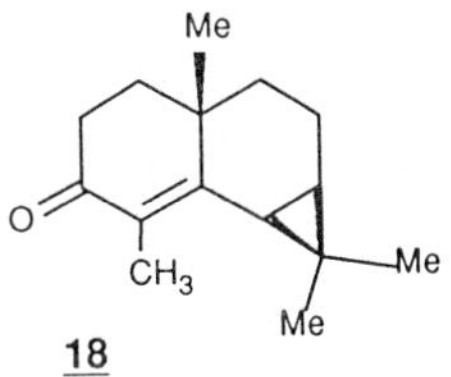   **18**

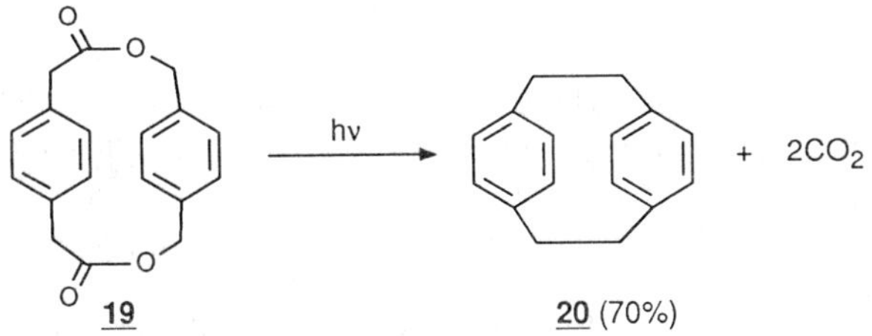

19   **20** (70%)

1.4 Industrial Applications

Two areas where photodecarboxylation reactions are believed to have possible industrial importance are (1) photodecomposition of commercially used organic acids and (2) solar energy conversion. In the former, methods for the efficient photodecomposition (via *photo*decarboxylation) of carboxylic acids used in industry (as herbicides, detergents, reagents, etc.) have been investigated. Use of photosensitizers[7] and transition metal ions as catalysts have been considered.[45,46] With respect to solar energy conversion, the photoelectrochemical conversion of simple carboxylic acids to alkanes (photo-Kolbe reaction) to be used as fuels has been investigated by several groups.[47–49] Use of the semiconductor catalyst TiO_2 in the presence of Pt results in the decarboxylation of acetic acid to methane and CO_2. Other simple carboxylic acids were also converted to the corresponding alkanes.[49]

Other applications may also be envisaged. Since heterolytic photodecarboxylation of alkali salts of carboxylic acids (vide supra) in aqueous solution gives rise to alkali metal hydroxides and hence an irreversible photoinduced increase in basicity, applications to photoresist and imaging technology is a possibility. The same reaction may also prove useful for laser-induced changes in pH in microenvironments where an irreversible increase in pH is required on short time scales.

Spectroscopic Data (λ_{max} of transients)

$\cdot CO_2H$	235 nm
Benzyl radical	302, 314 nm
$e^-(aq)$	720 nm
p-Nitrobenzyl carbanion	350 nm
$Ph-\dot{C}(OH)-CO_2H$	313 nm
$PhCOCO_2H$ (T_1 absorption)	322 nm
1-Naphthylmethyl radical	330 nm

Sensitizers Employed

1,6-*bis*(Dimethylamino)pyrene
1-Benzyl-1,4-dihydronicotinamide
$Ru(bpy)_3Cl_2$
Acridine

References

1. March, J., *Advanced Organic Chemistry,* 3rd ed., Wiley-Interscience, New York, 1985.
2. Fersht, A., *Enzyme Structure and Mechanism,* Freeman, Reading, UK, 1977.
3. Calvert, J. G. and Pitts, J. N., Jr., *Photochemistry,* John Wiley & Sons, New York, 1966.
4. Coyle, J. D., Photochemistry of carboxylic acid derivatives, *Chem. Rev.,* 78, 97, 1978.
5. Fox, M. A., The photoexcited states of organic anions, *Chem. Rev.,* 79, 253, 1979.
6. Givens, R. S., Photoextrusion of small molecules, *Org. Photochem.,* 5, 227, 1979.
7. Budac, D. and Wan, P., Photodecarboxylation: Mechanism and synthetic utility, *J. Photochem. Photobiol. A.,* 67, 135, 1992.
8. Mittal, L. J., Mittal, J. P., and Hayon, E., Photo-induced decarboxylation of aliphatic acids and esters in solution. Dependence upon state of protonation of the carboxyl group, *J. Phys. Chem.,* 77, 1482, 1973.
9. Epling, G. A. and Lopes, A., Fragmentation pathways in the photolysis of phenylacetic acid, *J. Am. Chem. Soc.,* 99, 2700, 1977.
10. Meiggs, T. O., Grossweiner, L. I., and Miller, S. I., Extinction coefficient and recombination rate of benzyl radicals. I. Photolysis of sodium phenylacetate, *J. Am. Chem. Soc.,* 94, 7981, 1972.
11. Meiggs, T. O. and Miller, S. I., Photolysis of phenylacetic acid and methyl phenylacetate in methanol, *J. Am. Chem. Soc.,* 94, 1989, 1972.
12. Margerum, J. D. and Petrusis, C. T., The photodecarboxylation of nitrophenylacetate ions, *J. Am. Chem. Soc.,* 91, 2467, 1969.
13. Craig, B. B., Weiss, R. G., and Atherton, S. J., Picosecond and nanosecond laser photolyses of *p*-nitrophenylacetate in aqueous media. A photoadiabatic decarboxylation process?, *J. Phys. Chem.,* 91, 5906, 1987.
14. Wan, P. and Muralidharan, S., Structure and mechanism in the photo-retro-Aldol type reactions of nitrobenzyl derivatives. Photochemical heterolytic cleavage of C–C bonds, *J. Am. Chem. Soc.,* 110, 4336, 1988.
15. Stermitz, F. R. and Huang, W. H., Thermal and photodecarboxylation of 2-, 3-, and 4-pyridylacetic acid, *J. Am. Chem. Soc.,* 93, 3427, 1971.
16. McAuley, I., Krogh, E., and Wan, P., Carbanion intermediates in the photodecarboxylation of benzannelated acetic acids in aqueous solution, *J. Am. Chem. Soc.,* 110, 600, 1988.
17. Krogh, E. and Wan, P., Photodecarboxylation of diarylacetic acids in aqueous solution: Enhanced photogeneration of cyclically conjugated eight π electron carbanions, *J. Am. Chem. Soc.,* 114, 705, 1992.
18. Leermakers, P. and Vesley, G. F., The photochemistry of α-keto acids and esters. I. Photolysis of pyruvic acid and benzoylformic acid, *J. Am. Chem. Soc.,* 85, 3776, 1963.

19. Closs, G. and Miller, R. J., Photoreduction and photodecarboxylation of pyruvic acid. Application of CIDNP to mechanistic photochemistry, *J. Am. Chem. Soc.*, 100, 3483, 1978.

20. Sawaki, Y. and Ogata, Y., Mechanism of the photoepoxidation with and photodecarboxylation of α-keto acids, *J. Am. Chem. Soc.*, 103, 6455, 1981.

21. Defoin, A., Defoin-Straatmann, R., Hildenbrand, K., Bittersmann, E., Kreft, D., and Kuhn, H. J., A new liquid phase actinometer: Quantum yield and photo-CIDNP study of phenylglyoxylic acid in aqueous solution, *J. Photochem.*, 33, 237, 1986.

22. Kuhn, H. K. and Görner, H., Triplet state and photodecarboxylation of phenylglyoxylic acid in the presence of water, *J. Phys. Chem.*, 92, 6208, 1988.

23. Hall, G. E., Muckerman, J. T., Preses, J. M., Weston, R. E., Jr., and Flynn, G. W., Time-resolved FTIR studies of the photodissociation of pyruvic acid at 193 nm, *Chem. Phys. Lett.*, 193, 77, 1992.

24. Steenken, S., Warren, C. J., and Gilbert, B. C., Generation of radical-cations from naphthalene and some derivatives, both by photoionization and reaction with SO_4^-: Formation and reactions studied by laser flash photolysis, *J. Chem. Soc., Perkin Trans.* II, 335, 1990.

25. Barton, D. H. R., The invention of chemical reactions: The last five years, *Tetrahedron*, 48, 2529, 1992.

26. Barton, D. H. R., Blundell, P., and Jaszberenyi, J. C., Quantum yields in the photochemically induced radical chemistry of acyl derivatives of thiohydroxamic acids, *J. Am. Chem. Soc.*, 113, 6937, 1991.

27. Barton, D. H. R., Chern, C.-Y., and Jaszberenyi, J. C., Homologation of carboxylic acids by improved methods based on radical chain chemistry of acyl derivatives of N-hydroxy-2-thiopyridone, *Tetrahedron Lett.*, 5013, 1992.

28. Barton, D. H. R., Chern, C.-Y., and Jaszberenyi, J. C., Two carbon homologation of carboxylic acids via carbon radicals generated from acyl derivatives of N-hydroxy-2-thiopyridone: synthesis of C_{n+2} α-keto-acids from C_n acids. (The "three carbon" problem), *Tetrahedron Lett.*, 5017, 1992.

29. Barton, D. H. R., Jaszberenyi, J. C., and Theodorakis, E. A., The invention of radical reactions. XXII. New reactions: Nitrile and thiocyanate transfer to carbon radicals from sulfonyl cyanides and sulfonyl isothiocyanates, *Tetrahedron*, 48, 2613, 1992.

30. Barton, D. H. R., Blundell, P., and Jaszberenyi, J. C., The invention of radical reactions. XXVIII. A new very photolabile O-acyl thiohydroxamic acid derivative as precursor of carbon radicals, *Tetrahedron*, 48, 7121, 1992.

31. Crich, D. and Quintero, L., Radical chemistry associated with the thiocarbonyl group, *Chem. Rev.*, 89, 1413, 1989.

32. Lipczynska-Kochany, E., Photochemistry of hydroxamic acids and derivatives, *Chem. Rev.*, 91, 477, 1991.

33. Okada, K., Okamoto, K., and Oda, M., A new and practical method of decarboxylation: Photosensitized decarboxylation of N-acyloxyphthalimides via electron transfer mechanism, *J. Am. Chem. Soc.*, 110, 8736, 1988.

34. Okada, K., Okamoto, K., Morita, N., Okubo, K., and Oda, M., Photosensitized decarboxylative Michael addition through N-(acyloxy)phthalimides via an electron-transfer mechanism, *J. Am. Chem. Soc.*, 113, 9401, 1991.

35. Okada, K., Okubo, K., and Oda, M., Decarboxylative photooxygenation of carboxylic acids by the use of acridine, *Tetrahedron Lett.*, 83, 1992.

36. Cram, D. J., Tanner, M. E., and Thomas, R., The taming of cyclobutadiene, *Angew. Chem. Int. Ed.*, 30, 1024, 1991.

37. Bleasdale, D. A., Jones, D. W., Maier, G., and Reisenauer, H. P., 2-Benzopyran-3-one, *J. Chem. Soc., Chem. Commun.*, 1095, 1983.

38. Jones, D. W. and Kneen, G., o-Quinonoid compounds. IX. Photodecarboxylation of 2-benzopyran-3-one adducts and photoreactions of the derived o-quinonedimethanes, *J. Chem. Soc., Perkin Trans 1*, 175, 1975.

39. Jones, D. W. and Kneen, G., Photodecarboxylation of 2-benzopyran-3-one adducts and photoreaction of the derived *o*-quinodimethanes, *J. Chem. Soc., Chem. Commun.,* 1038, 1972.

40. Murai, A., Abiko, A., Ono, M., and Masamune, T., Synthesis of aubergenone, a sesquiterpenoid phytoalexin from diseased eggplants, *Bull. Chem. Soc. Jpn.,* 55, 1191, 1982.

41. Filler, R., Cantrell, G. L., Wolanin, D., and Naqvi, S. M., Synthesis of polyfluoroaryl [2.2] cyclophanes, *J. Fluorine Chem.,* 30, 399, 1986.

42. Hilbert, M. and Solladie, G., Substituent effect during the synthesis of substituted [2.2] paracyclophane by photoextrusion of carbon dioxide from a cyclic diester, *J. Org. Chem.,* 45, 4496, 1980.

43. Truesdale, E. A., [2.2.2.2] Orthoparacyclophane, *Tetrahedron Lett.,* 3777, 1978.

44. Kaplan, M. L. and Truesdale, E. A., [2.2] Paracyclophane by photoextrusion of carbon dioxide from a cyclic diester, *Tetrahedron Lett.,* 3665, 1976.

45. Langford, C. H. and Quance, G. W., Photochemical decomposition of EDTA coordinated to cobalt(III): Products, thermal reactions, and relevance for outer sphere alcohol oxidation by the excited state, *Can. J. Chem.,* 55, 3132, 1977.

46. Carey, J. H. and Langford, C. H., Photodecomposition of Fe(III) aminopolycarboxylates, *Can. J. Chem.,* 51, 3665, 1973.

47. Sato, S., Photo-Kolbe reaction at gas-solid interfaces, *J. Phys. Chem.,* 87, 3531, 1983.

48. Muzyka, J. L. and Fox, M. A., Selective photoelectrochemical oxidation of vicinal cyclohexanedicarboxylic acids: A mechanistic study, *J. Org. Chem.,* 55, 209, 1990.

49. Kraeutler, B. and Bard, A. J., Heterogeneous photocatalytic decomposition of saturated carboxylic acids on TiO_2 powder. Decarboxylative route to alkanes, *J. Am. Chem. Soc.,* 100, 5985, 1978.

32

The Photochemistry of
Esters of Carboxylic Acids

J. Pincock
Dalhousie University

32.1 Introduction

This chapter deals only with the photochemistry of esters of carboxylic acids. The photochemistry of other esters such as phosphates[1] and sulfonates[2] has been recently reviewed. An extensive review from 1979 of the photochemistry of carboxylic acid esters[3] and a more recent one on photodecarboxylation,[4] which includes considerable discussion of ester photochemistry, can be used as the foundation for this chapter. The Chemical Society Reports[5] are also an excellent source of information. Reactions where the ester functional group is a substituent, but not directly involved in the photochemical step, will not be included. In Section 32.3, material will be organized into nine major reaction types. For the first five of these, the ester atoms (CO-O) remain bonded but, in some cases, modified; for the second four of these reaction types, fragmentation of the ester group occurs. Eight of these reaction types will be illustrated by current examples chosen to be as heavily referenced as possible, so that the earlier work can be found by the interested reader. This clearly does not represent a comprehensive review but was necessitated by the restriction on length. The ninth type, photocleavage of benzylic esters, the author's current interest, will be covered in more detail.

32.2 Chromophores

As discussed in detail,[3] aliphatic esters have only a weakly absorbing n,π^*-transition, at short wavelengths; for ethyl acetate, λ_{max} 212 nm ($\varepsilon = 48$), in the gas phase. Because of this, these simple esters have received little attention from the organic photochemist since the common light sources only provide wavelengths longer than 250 nm. Conjugation of the ester functional group with either aromatic rings or double bonds provides more easily accessible π,π^*-chromophores. Selected examples are given in Table 32.1. Substitution on the aromatic ring changes the properties of the excited state in the expected normal way.

0-8493-8634-9/95/$0.00+$.50
© 1995 by CRC Press, Inc.

Table 32.1   Chromophores and Excited-State Properties of Selected Esters

Compound	λ_{max} (nm)	ε (mol^{-1} cm^{-1})	E_S (kcal mol^{-1})	τ_S (ns)	Φ_F	E_T (kcal mol^{-1})	Ref.
Ph-(CO)-OCH$_3$	271	880	100[a]	—	0.0	77.9	6
PhCH$_2$O-(CO)-CH$_3$	257	214	105.8	14	0.05	—	7
NaphCH$_2$O-(CO)-CH$_3$	275	7300	89.3	41	0.14	61.3	8
NaphCH$_2$O-(CO)-CH$_2$Ph	275	7400	91.9	49	0.14	60.0	9

[a] The position of the o,o band cannot be determined because $\Phi_F = 0$.

32.3 Reaction Types

Cycloadditions

Two types of cycloadditions are possible — cyclobutane formation as in Equation 32.1[10] and oxetane formation as in Equation 32.2.[11] For the first example, the reaction is well established to occur by exciplex formation from the singlet state; therefore, the ester is the electron donor and the alkene must be a good electron acceptor. For Equation 32.1, the ester fluorescence is quenched by the alkene at the diffusional rate and weak exciplex emission can be observed at long wavelength. For Equation 32.2, the ester is too short-lived to be detected by nanosecond fluorescence, and reaction may occur by irradiation of a ground state charge-transfer complex.[12] In polar solvents like acetonitrile, dissociation to radical ions occurs and, for the substrates in Equation 32.2, can dominate the observed chemistry at high 1,1-diphenylethene concentrations, Equation 32.3[11] Now the ester is serving only as an electron-accepting photosensitizer to generate the radical cation of the alkene. The competition between cyclobutane and oxetane formation can be controlled by geometric factors in intramolecular cases as shown in Equation 32.4.[10]

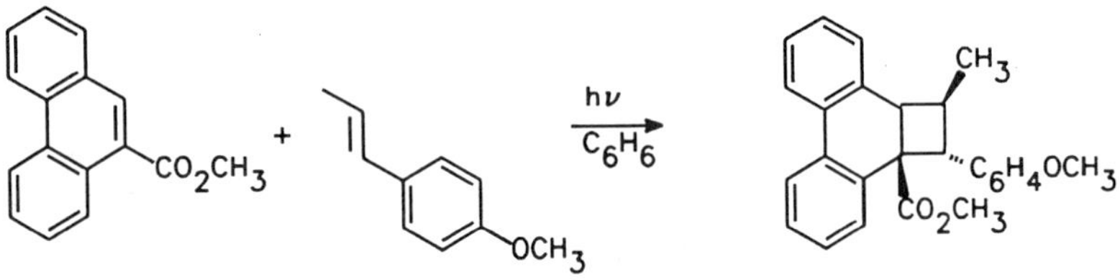

(32.1)

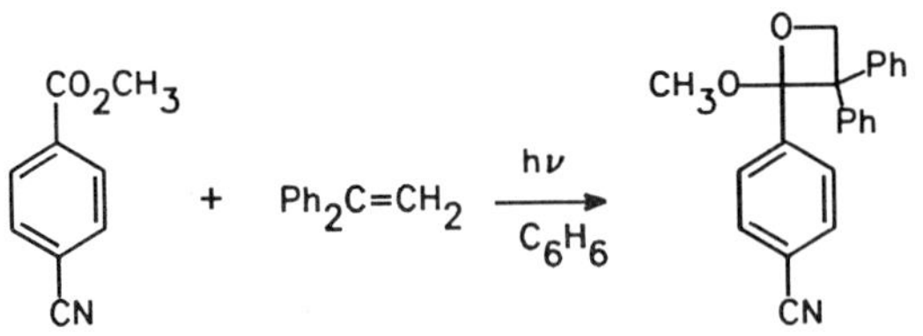

(32.2)

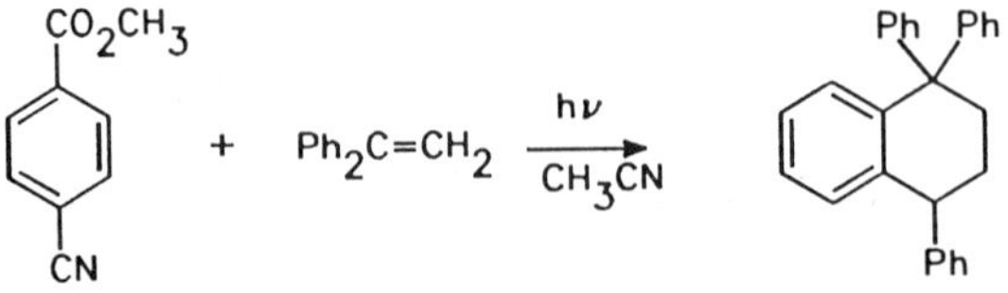

(32.3)

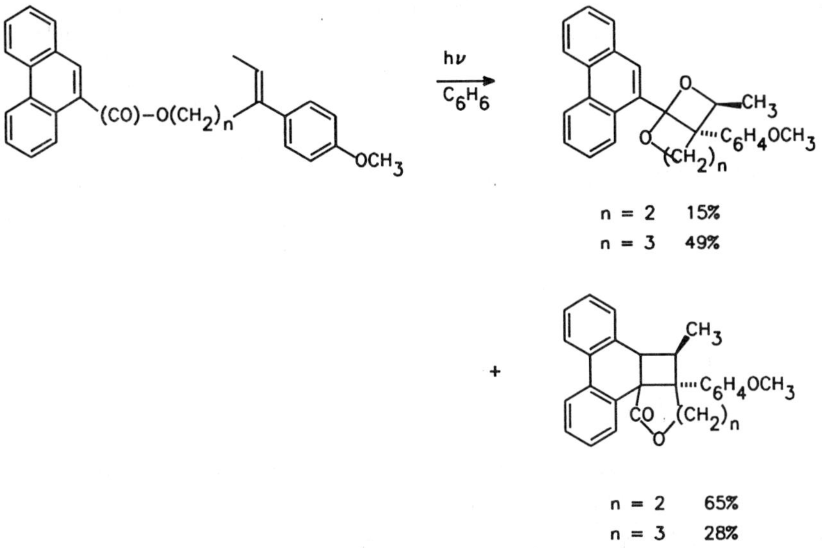

(CO)–O(CH2)n
OCH3
hν
C6H6
CH3
C6H4OCH3
(CH2)n
n = 2 15%
n = 3 49%
+
CH3
C6H4OCH3
CO
(CH2)n
n = 2 65%
n = 3 28%

(32.4)

Hydrogen Abstraction

Hydrogen atom abstraction reactions are very common for the n,π*-triplet of ketone carbonyl groups. For esters, this process is much less efficient because the lowest state is a π,π*-triplet, certainly for esters of aromatic acids and probably for many others. The hydrogen atom abstraction reactions that are observed are mainly from the n,π*-singlet state. Since this state is short-lived, intermolecular reactions cannot usually compete with rapid intersystem crossing. However, intramolecular reactions are more efficient and two general cases have been studied extensively.

The first is the Norrish Type II reaction involving γ-hydrogen abstraction, Equation 32.5.[13] In contrast to ketone chemistry, the quantum yields are low (~0.01). By studying the *erythro*- and *threo*-isomers, return of the biradical to starting material was also shown to be inefficient.

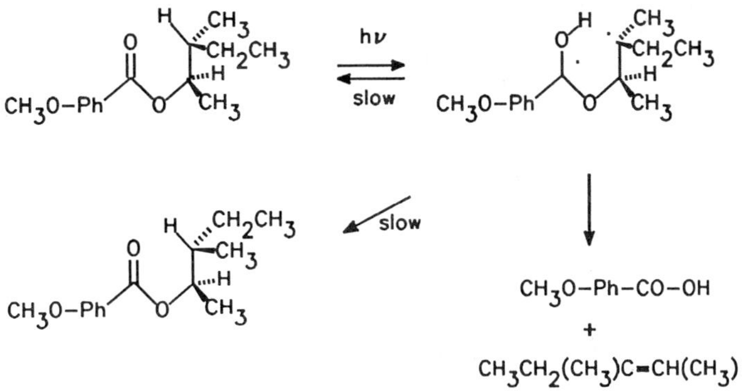

CH3O–Ph
hν
slow
slow
CH3O–Ph–CO–OH
+
CH3CH2(CH3)C=CH(CH3)

(32.5)

The second important intramolecular pathway is the photochemical deconjugation of α,β-unsaturated esters, Equation 32.6.[14] By using catalytic amounts of weak bases, the photoenol formed by hydrogen abstraction in the excited singlet state is converted to the unconjugated isomer. Many examples have been reported and, in general, the yields are good. A recent report on the use of chiral bases demonstrates that both high chemical yield (65%) and high enantiomeric excess (70%) in the deconjugated isomer are possible, Equation 32.7,[15] for B = (–)-Ph-CH(OH)-CH(CH_3)-NH_2.

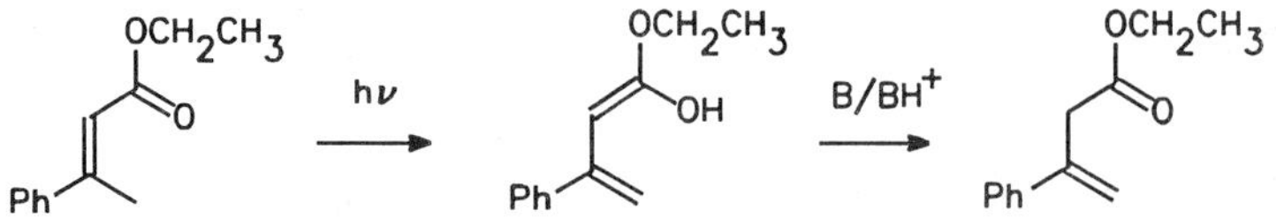

$$(32.6)$$

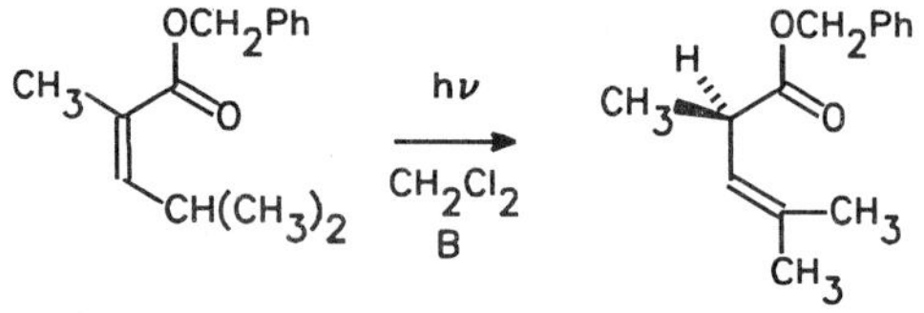

$$(32.7)$$

Photoisomerization

Like many functionalized alkenes, α,β-unsaturated esters can be photoisomerized by both direct and triplet-sensitized irradiations. The example of methyl cinnamate is shown in Equation 32.8.[16] The excited state involved in direct irradiations is still in doubt, although the weak fluorescence and lack of oxygen quenching suggests the singlet. The photostationary state is 46% Z for direct irradiation but can be increased to 88% Z by irradiation of the Lewis acid/ester adduct. The effect of the Lewis acid is explained by the increased equilibrium constant for complexation by the E-isomer that increases the preferential absorption of light by that isomer. Therefore, E is converted to Z more efficiently than Z to E. This latter method also works for alkyl substituted α,β-unsaturated esters[17] and, moreover, suppresses the deconjugation reaction described above in Section 32.3.

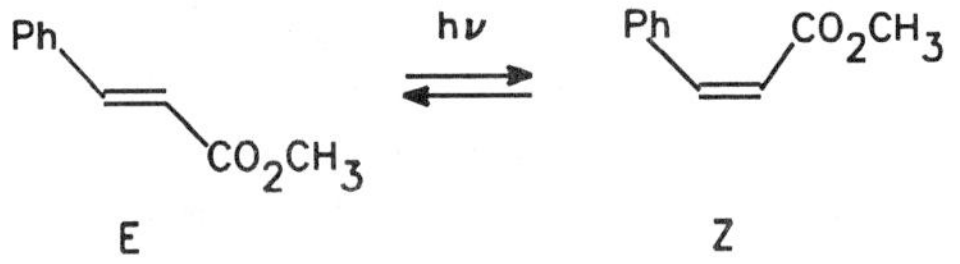

$$(32.8)$$

Rearrangements (Other than the Photo-Fries)

Examples of rearrangements of ester containing substrates are shown in Equations 32.9,[18] 32.10,[19] and 32.11.[20]

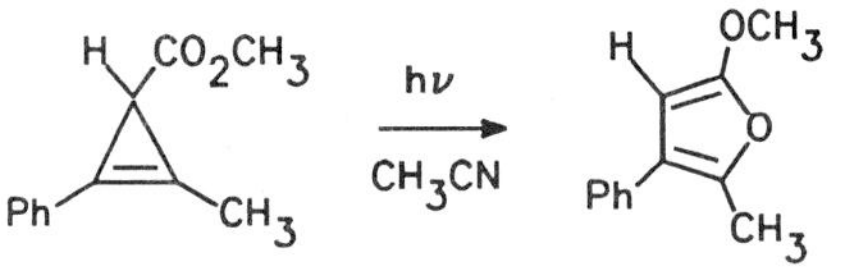

$$(32.9)$$

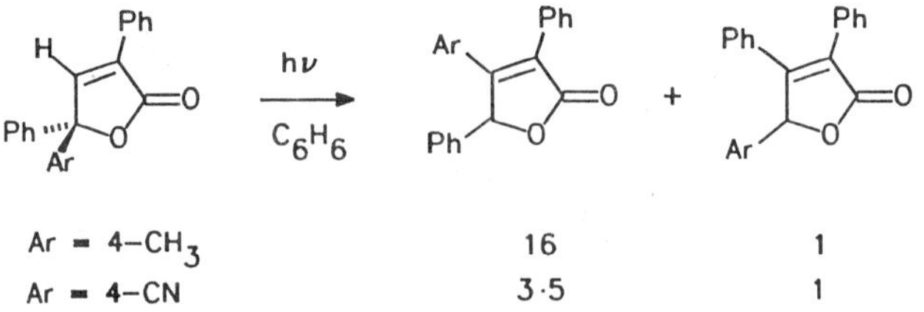

| Ar = 4-CH$_3$ | 16 | 1 |
| Ar = 4-CN | 3·5 | 1 |

$$(32.10)$$

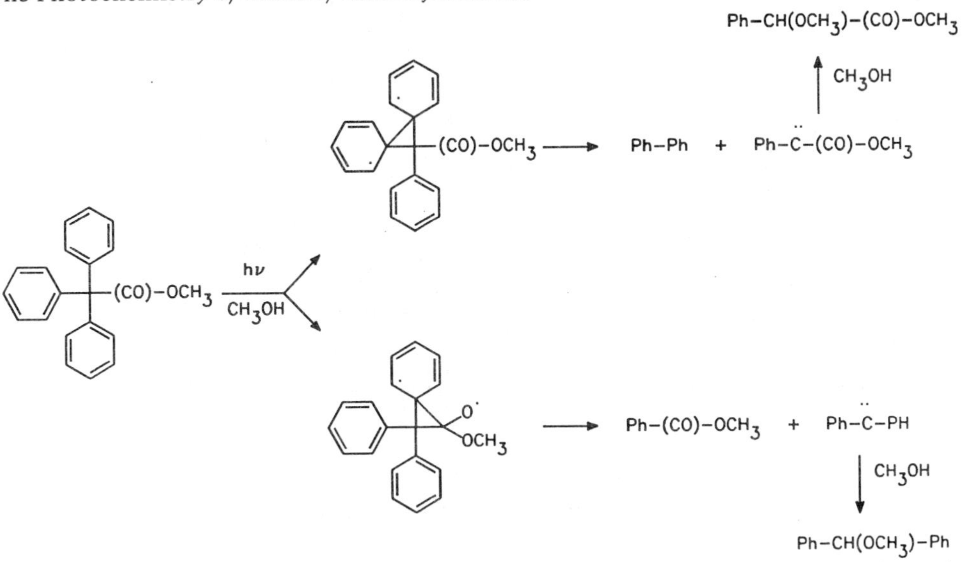

$$(32.11)$$

None of these reactions are inherently photochemistry of the ester functional group. In Equation 32.9,[18] the cyclopropene double bond is the chromophore inducing excited singlet state cyclopropene ring cleavage to the biradical/vinylmethylene, which either returns to cyclopropene (as determined by racemization at C3) or rearranges to the furan. The unexpected regiochemistry of cleavage of the methyl rather than the phenyl substituted σ-bond is of interest. A recent report on 3-acylcyclopropene photochemistry has appeared.[21] In Equation 32.10,[19] the ester remains intact but is conjugated through the double bond and has an effect on the polarity of the rearrangement of the excited state. The oxa-di-π-methane reaction, the oxygen analogue of the extensively studied di-π-methane reaction,[22,23] is known, but rare, for esters. An interesting variant of it is shown in Equation 32.11[23] along with the proposed mechanism through intermediate carbenes. The yields of the products indicate that aryl-aryl and aryl-carbonyl carbon bridging are equally efficient. Quenching studies suggest that aryl-aryl bridging occurs via the triplet state, whereas the other bridging occurs from a much shorter lived state that may be the singlet.

Excited-State Basicity

Many organic functional groups have enhanced basicity or acidity in the excited state.[24] Esters have received little attention in this respect, and no examples that actually lead to net chemical change are known. However, the basicity of the ethyl esters of both 1- and 2-naphthoic acid are considerably enhanced in the excited state.[25] Values for the pK_a of the protonated form are 1-ester, $pK_a(S_0)$ = −8.6, $pK_a(S_1)$ = −1.0, 0.9; 2-ester, $pK_a(S_0)$ = −8.2, $pK_a(S_1)$ = −2.5, 0.9.

Photo-Fries Rearrangement

This reaction has been reviewed recently[26] as well as in the past.[27,28] A typical example is shown in Equation 32.12[29] for photolysis of 1-naphthyl acetate. The ratio of the products is 1:0.45:0.12 in acetonitrile and 1:0.044: < 0.001 in Freon 113. Mechanistic evidence demonstrates that the reaction occurs from the excited singlet state and that the acetyl naphthols are formed by in-cage coupling. The triplet state is unreactive, presumably because the lower excited state energy is insufficient to cleave the carbon-oxygen bond.

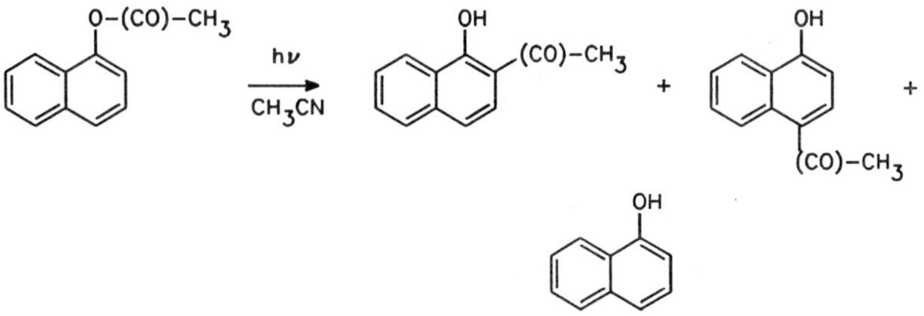

(32.12)

The photo-Fries rearrangement is also known for vinyl esters as in Equation 32.13.[30]

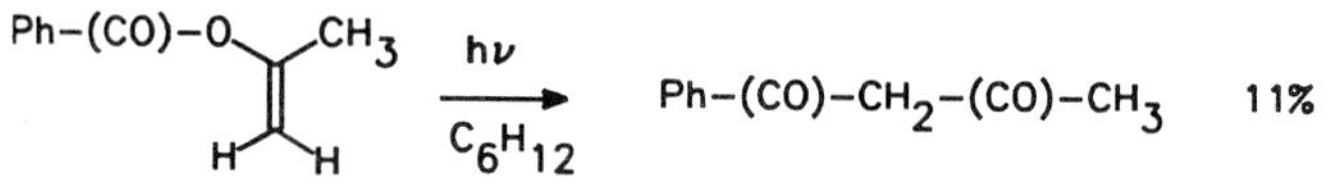

(32.13)

The cleavage of the carbonyl carbon to oxygen bond can lead to other products if the photo-Fries reaction is precluded for geometric reasons. An example is shown in Equation 32.14[31] where intramolecular hydrogen abstraction in the first formed biradical results in ketene formation. The ketene then reacts by addition with the nucleophilic solvent, methanol.

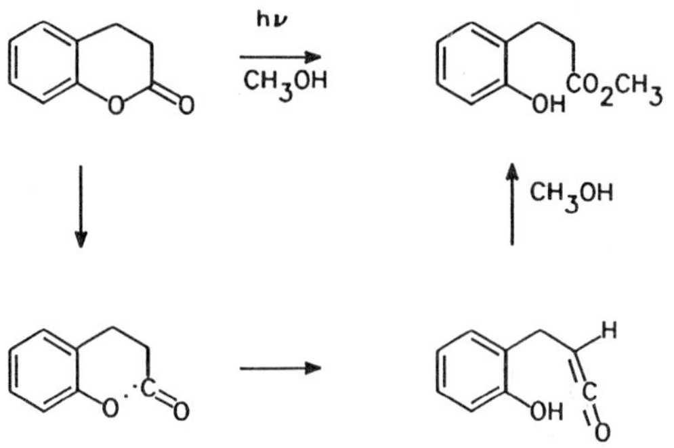

(32.14)

Fragmentation of Ester Radical Ions

The radical ions of esters often fragment efficiently. This has been studied in detail most recently by pulse radiolysis of esters of benzoic acids, Equation 32.15.[32] The benzoate ester serves to lower the reduction potential of the ester carbonyl group. The rate of this cleavage process has been found to correlate well with the BDE of the carbon-oxygen bond. Only one example is known for the photochemical electron transfer generation of radical anions, Equation 32.16.[33] The fact that electron transfer is occurring and results in the formation of radical ions is shown by the observation that the fluorescence of 1-methoxynaphthalene in methanol is quenched by methyl benzoate with a rate constant of 1.1×10^9 s^{-1}. This is an area that requires more examples to clarify the mechanism and the efficiency of the process.

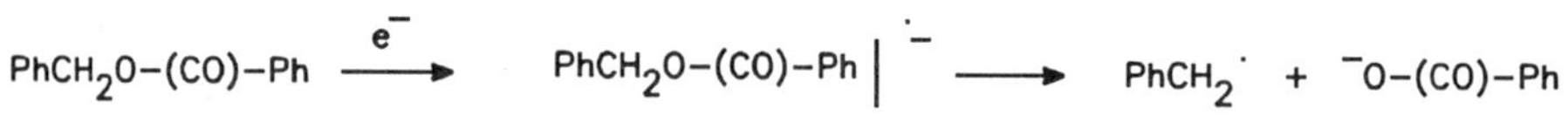

(32.15)

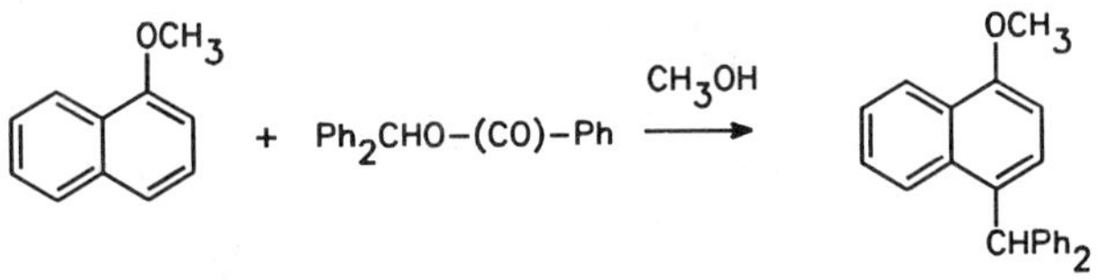

$$(32.16)$$

Decarbonylation Reactions

Most decarbonylation reactions reported are a side reaction in the photo-Fries rearrangement, Equation 32.12. However, in other cases where the chromophore is conjugated to the ester O-(CO) bond and promotes bond cleavage, decarbonylations can be efficient. For instance, the biradical formed initially can decarbonylate to nonaromatic enones, Equation 32.17.[34] Reaction, in a conjugate addition, of the very reactive enone with the nucleophilic solvent gives the observed photoproduct. Whether this reaction proceeds by initial (CO)-O (analogous to the first step of the photo-Fries reaction) or (CO)-C(Ph)$_2$ bond cleavage is not known. Another example is the photochemical conversion of 2 (3H)-furanones to α,β-unsaturated ketones, Equation 32.18.[35]

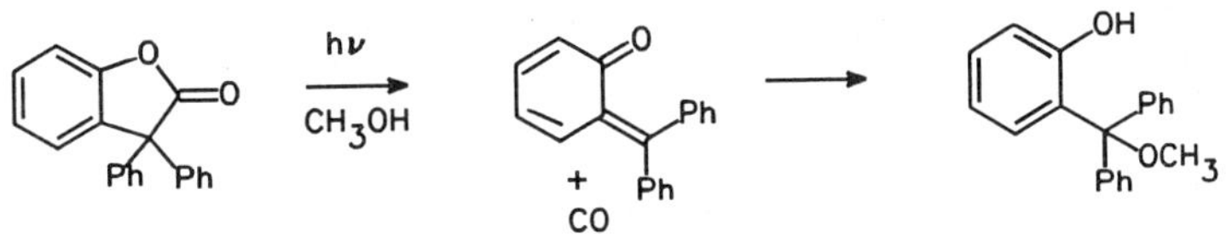

$$(32.17)$$

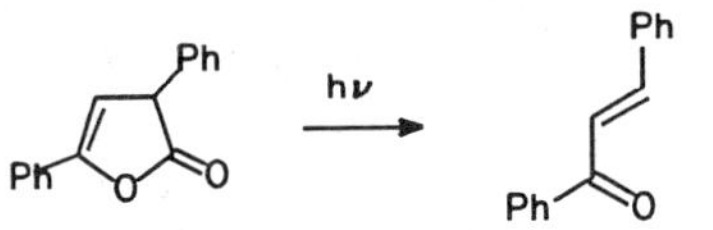

$$(32.18)$$

Photosolvolyis and Photodecarboxylation of Arylmethyl Esters

On photolysis of arylmethyl esters in nucleophilic solvents, the products of two quite different types are usually formed, Equation 32.19.

$$(32.19)$$

The first two products are derived from the arylmethyl cation and carboxylate anion, respectively, (photosolvolysis); the last product is derived from in-cage coupling of the arylmethyl radical with the alkyl radical that results from loss of CO$_2$ (photodecarboxylation). Since the ion-pair and the radical-pair chemistry usually occur in competition with one another, these two reactions will be discussed together. The general topics of photosolvolysis[36] and photodecarboxylation[4] have been reviewed. There is good evidence for both benzylic and 1-naphthylmethyl esters that the triplet state is unreactive.[8,37] Therefore, only the excited singlet state needs to be considered. A general mechanism is shown in Scheme 32.1.

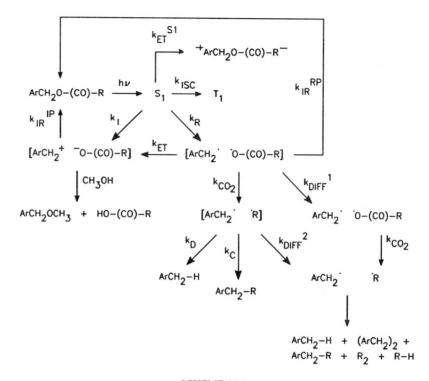

SCHEME 32.1

For simplicity, the intermediate ion-pair is shown as a contact pair although solvent separated pairs may play a role. The rate constants in Scheme 32.1 are k_{ISC}, intersystem crossing to the unreactive triplet; k_{ET}^{S1}, intramolecular electron transfer converting S_1 to an exciplex; k_1, heterolytic cleavage in S_1 of the carbon-oxygen bond; k_R, homolytic cleavage in S_1 of the carbon-oxygen bond; k_{ET}, electron transfer converting the radical-pair to the ion-pair (in polar solvents, oxidation potentials[38] show that the ion-pair will be more stable than the radical-pair); k_{IR}^{IP} and k_{IR}^{RP}, internal return of the ion-pair and the radical-pair, respectively, to starting ester; k_{CO_2}, decarboxylation of the acyloxy radical; k_{DIFF}^1 and k_{DIFF}^2, diffusion of the radical pair from the solvent cage either before or after, respectively, decarboxylation; k_c, in-cage radical coupling; and k_D, in-cage radical disproportionation.

Since product composition is dependent on the relative magnitude of all of these rate constants that will, in turn, be dependent on the substrate and the reaction conditions, predictions of product structures and yields may be very difficult. Fortunately, for many cases, one pathway often predominates. For instance, in nonpolar solvents, ion-pair formation is not likely, i.e., $k_R \gg k_1$ and k_{ET} is negligibly small. Therefore, only radical-pair chemistry results and decarboxylation becomes the major pathway. Many synthetically useful examples of this are catalogued in the reviews by Givens and Levi[3] and Wan.[4] Two examples used to clarify mechanisms are found in the work of Givens et al., Equations 32.20[39] and 32.21.[37] The higher yield of the mixed coupling product indicates that in-cage decarboxylation (k_{CO_2}) is somewhat faster than diffusional escape (k_{DIFF}^1) in agreement with current estimates[40] of $k_{CO_2} = 4.6 \times 10^9$ s^{-1} for the phenylacetyloxy radical (Table 32.2). The diffusional escape seems less important for the larger naphthylmethyl radical.

Table 32.2 Rates of Decarboxylation for the Alkyl Acyloxy Radicals Formed in the Photolysis in Methanol of 1-Naphthylmethyl Alkanoates[40]

R	$k_{CO_2} \times 10^{-9}$ (s^{-1})
CH_3	1.3
CH_3CH_2	2.0
$(CH_3)_2CH$	6.5
$(CH_3)_3C$	11
$PhCH_2$	4.6

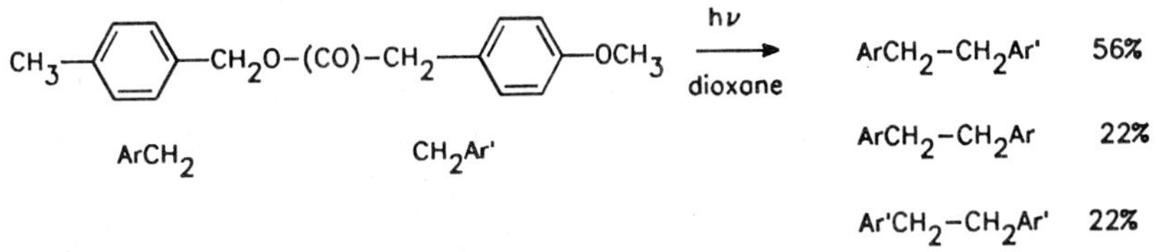

CH3
CH2O-(CO)-CH2
OCH3
hν
dioxane
ArCH2-CH2Ar' 56%
ArCH2-CH2Ar 22%
Ar'CH2-CH2Ar' 22%
ArCH2
CH2Ar'

(32.20)

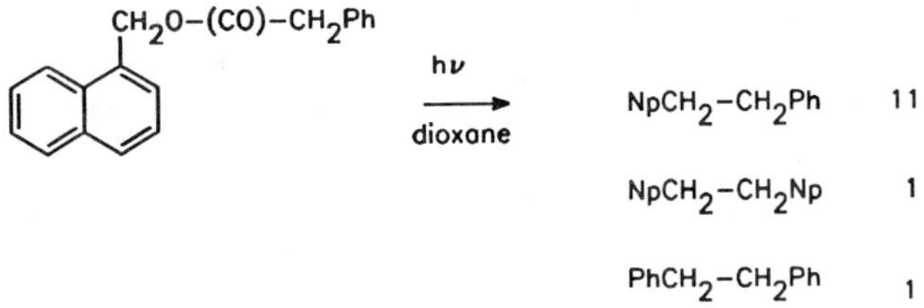

CH2O-(CO)-CH2Ph
hν
dioxane
NpCH2-CH2Ph 11
NpCH2-CH2Np 1
PhCH2-CH2Ph 1

(32.21)

In a pioneering study on benzylic ester photochemistry, Zimmerman and Sandel[41] reported on the photolysis of 3- and 4-methoxybenzyl acetate in aqueous dioxane. As shown in Equations 32.22 and 32.23, the yield of the benzyl alcohol derived from the benzyl cation was higher for the 3-methoxy than for the 4-methoxy isomer. This observation, which is contrary to ground-state expectations, was rationalized by SHMO calculations of charge distribution in the excited singlet state. An electron-donating group was predicted to be a better electron donor, in S_1, from the *meta*-position than from the *para*-position. The argument developed was that substituents controlled electron distribution in the excited state and this had an effect on the magnitude of k_I. For 3-methoxybenzyl acetate, k_I was greater than for the 4-methoxy isomer. Since that time (1963), this idea, called the "meta effect", has dominated mechanistic considerations of substituent effects for photolysis of benzylic esters.

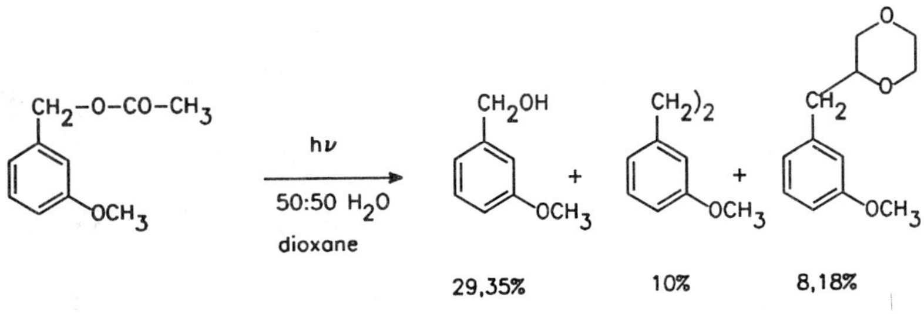

CH2-O-CO-CH3
OCH3
hν
50:50 H2O
dioxane
CH2OH
OCH3
+
CH2)2
OCH3
+
CH2
OCH3
29,35%
10%
8,18%

(32.22)

$$\text{"minor amount"} \quad 23.31\% \quad 31.36\% \qquad (32.23)$$

In 1989, the observation[9,42] was made, for a set consisting of eleven ring-substituted 1-naphthylmethyl phenylacetates, that product yields for photolysis in methanol could not be explained by this idea. The products obtained are shown in Equation 32.24 ($N = C_{10}H_{7-n}X_n$). Their yields could be rationalized by a mechanism of exclusive homolytic cleavage, i.e., $k_R \gg k_I$. Since the observed yields of out-of-cage products were low, the diffusional rate constants (k_{DIFF}^1 and k_{DIFF}^2) must be small. Therefore, product composition is being controlled by the ratio of k_{ET}/k_{CO_2}. Since R is a constant (PhCH$_2$), k_{CO_2} is a "radical clock" for k_{ET}. A good estimate[40] of this rate of decarboxylation (4.6×10^9 s^{-1}) allowed determination of k_{ET} values as a function of changes in substituent on the naphthalene ring. Since, and according to Marcus theory[43] of electron transfer, these rate constant should be dependent on the oxidation potential of the 1-naphthylmethyl radical, values for these were also measured.[38] Not surprisingly, they were correlated well by σ^+ ($\rho^+ = -7.1$).

$$NCH_2O\text{-}(CO)\text{-}CH_2Ph \xrightarrow[CH_3OH]{h\nu} NCH_2\text{-}OCH_3 \;+\; HO\text{-}(CO)\text{-}CH_2Ph \;+\; NCH_2\text{-}CH_2Ph$$

$$(32.24)$$

A plot of log k_{ET} vs. E_{OX} (Figure 32.1) shows the parabolic behavior predicted by Marcus theory. The rate of electron transfer converting the radical-pair to the ion-pair increases as the substituent becomes a better electron donor, passes through a maximum for 4-CH$_3$ and then decreases as the process becomes prohibitively exothermal. The value of the reorganization energy, $\lambda = 0.52$ eV, is reasonable for electron transfer in a contact pair over a short distance.[44,45]

This explanation suggests that product composition in these 1-naphthylmethyl cleavage reactions is controlled by the competition between k_{ET} and k_{CO_2}. This is quite different from the idea of the "meta effect" where the critical competition was thought to be between k_R and k_I.

A simple extension of this k_{ET}/k_{CO_2} competition idea is that rates of decarboxylation as a function of R can also be determined.[40] If the naphthalene ring is kept constant so that k_{ET} remains constant, then variation in R which varies the rate of decarboxylation should also change the product distribution. Analysis of products for the photolysis in methanol of the esters, C$_{10}$H$_7$CH$_2$O-(CO)-R where R = CH$_3$, CH$_3$CH$_2$, (CH$_3$)$_2$CH and (CH$_3$)$_3$C gave the values shown in Table 32.2 for k_{CO_2}. These values, which are the first obtained reasonable estimates for k_{CO_2} of alkylacyloxy radicals, should prove useful in discussions of reactions involving loss of carbon dioxide from acyloxy radicals.[46] Older estimates of $k_{CO_2} \approx 10^9$ s^{-1} for R = CH$_3$ by ESR[47] and CIDNP[48] studies are consistent with the value obtained.

The implications that these results have on other reactions of arylmethyl esters and arylmethyl substrates in general are not yet clear. A reinvestigation of substituent effects, using a wider range of electron-donating and electron-withdrawing groups, for benzyl acetates and benzyl pivalates has been completed and confirms the proposed mechanism. Photolysis of these esters in methanol leads to six major products, Equation 32.25.[7] The variation in product ratios as a function of X and R suggests that electron transfer is again a major contributor to the mechanism for formation of ionic intermediates. These systems are more complicated than the related naphthalene cases because diffusional escape is a dominant process. Moreover, for the highly electron-rich aromatic rings (f,g), a process other than electron transfer seems to be responsible for formation of the benzylic

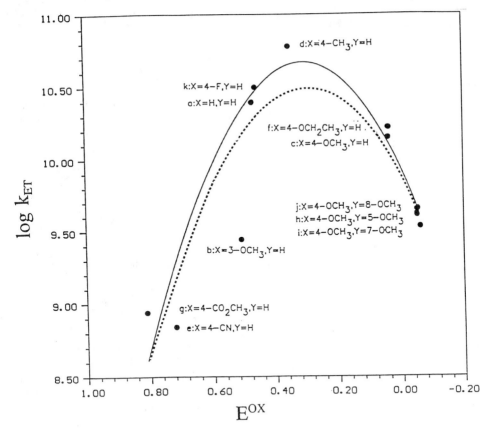

FIGURE 32.1 Plot of the log of the rate of electron transfer for converting the radical-pair to the ion-pair as a function of the oxidation potential (volts) of the 1-naphthylmethyl radical. The dashed line includes the point for the 3-methoxy compound; the solid line does not. A least squares fit gives a value for the reorganization energy of 0.52 eV for the solid line.

cation. This may be k_i or perhaps intramolecular electron transfer in the excited singlet state, i.e., k_{ET}^{S1} in Scheme 32.1. Work is in progress to try to understand these possibilities.

$$
\begin{array}{c}
\text{(Scheme 32.25)}
\end{array}
$$

a, X=H
b, X=4-CH$_3$
c, X=3-OCH$_3$
d, X=4-OCH$_3$
e, X=4-CN
f, X=3,5-(OCH$_3$)$_2$
g, X=3,4,5,-(OCH$_3$)$_3$

1· R=CH$_3$
2· R=C(CH$_3$)$_3$

(32.25)

Intramolecular electron transfer in the excited singlet state of esters (k_{ET}^{S1}) to form an exciplex has been well documented for benzoate esters, $ArCH_2O\text{-}CO\text{-}Ph$. The conjugation of the aromatic ring to the carbonyl group lowers the reduction potential. This process has been studied quantitatively by fluorescence quenching.[49] The rate constant can be as low as 10^7 s^{-1} if the aryl ring is a poor donor (4-CN) and the benzoate a poor acceptor (4-OCH$_3$). In this case, normal arylmethyl carbon-oxygen homolytic bond cleavage dominates. In contrast, $k_{ET}^{S1} = 10^{10}$ s^{-1} completely dominates benzylic bond cleavage if the aryl ring is a good donor (4-OCH$_3$) and the benzoate a good acceptor (4-CN).

Finally, the possibility of internal return must be mentioned. This process would obviously decrease the efficiency of product formation but, more importantly, could alter the ratio of products resulting from the radical and ionic pathways. For instance, if internal return in the ion-pair (k_{IR}^{IP} in Scheme 32.1) varies with the substituent on the aromatic ring then the yield of the products derived from the ion will not be dependent solely on the rate of formation of the ion-pair (k_{ET} and k_I). Mechanistic arguments developed about these rate constants based on product yields would then be incorrect.

Internal return in esters has been monitored by selectively labeling the alcohol oxygen of esters with ^{18}O. Measurement of the scrambling of this label between the original site and the carbonyl oxygen gives a measure of the efficiency of this process. Three examples are given below.

In Equation 32.26,[50] after 50% conversion of the ester to products, the label was 35% scrambled but the chiral center had maintained its configuration. In Equation 32.27[50] after 53% conversion to products, there was neither ^{18}O scrambling nor *trans-to-cis* isomerization of the lactone. In Equation 32.28,[51] after 48% conversion, the label was 68% scrambled. Based on these results, Givens has suggested that the observed ^{18}O scrambling does not occur through a fragmentation/recombination process. The evidence for this is the lack of scrambling or *trans-to-cis* isomerization for the lactone in Equation 32.27 and the lack of racemization of the ester in Equation 32.26. After benzylic carbon-oxygen bond cleavage, decarboxylation must be faster than the bond rotation that would lead to these processes. The conclusion reached is that the ^{18}O scrambling results from a concerted 1,3-migration of the benzylic carbon from the alcohol oxygen to the carbonyl oxygen of the ester. For the lactone, Equation 32.27, this process is sterically impossible so neither ^{18}O exchange nor isomerization results. For the ester in Equation 32.26, exchange is possible, but not racemization.

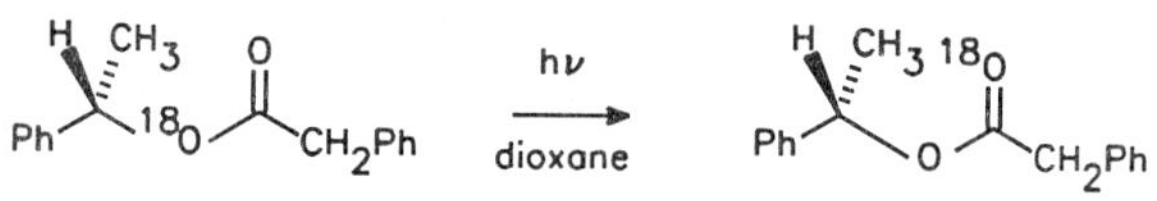

(32.26)

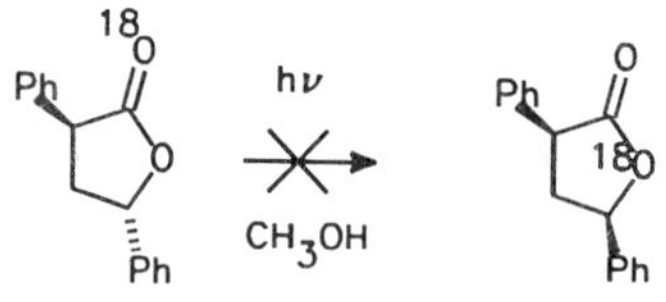

(32.27)

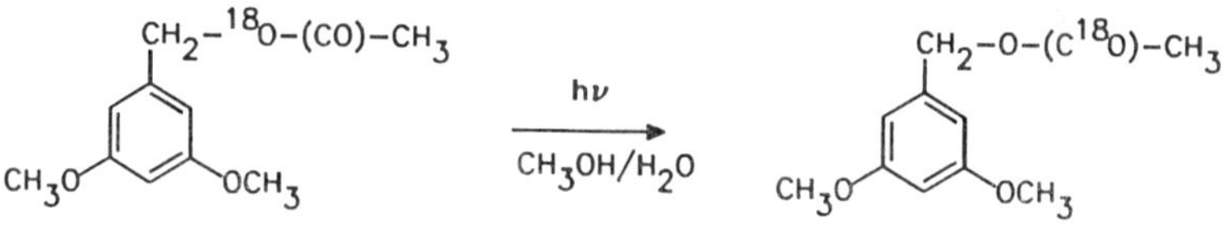

(32.28)

A study of the effect that aromatic substituents have on internal return is in progress.[52]

erences

1. Givens, R. S. and Kueper, L. W., III, Photochemistry of phosphate esters, *Chem. Revs.*, 93, 55, 1993.
2. Horspool, W. M., Photochemistry and radiation chemistry, *The Chemistry of Sulfonic Acids, Esters and Their Derivatives*, Patai, S. and Rappaport, Z., Eds., J. Wiley & Sons, New York, 1991, 501.
3. Givens, R. S. and Levi, N., The photochemistry of organic acids, esters, anhydrides, lactones and imides, *The Chemistry of Functional Groups, Supplement B, The Chemistry of Acid Derivatives*, Patai, S., Ed., 1979, 641.
4. Budac, D. and Wan, P., Photodecarboxylation: Mechanism and synthetic utility, *J. Photochem. Photobiol. A: Chem.*, 67, 135, 1992.
5. *Photochemistry, A Specialist Periodical Report*, The Royal Society of Chemistry, Vol. 1–23, 1970–1992.
6. Yamasaki, N., Inoue, Y., Yokoyama, T., Tai, A., Ishida, A., and Takamuku, S., Steric hindrance-induced dual fluorescence of congested benzenehexacarboxylates, *J. Am. Chem. Soc.*, 113, 1993, 1991.
7. Hilborn, J. W., MacKnight, E., Pincock, J. A., and Wedge, P. J., Photosolvolysis of substituted benzyl acetates and pivalates. A re-examination of substituent effects, *J. Am. Chem. Soc.*, 116, 3337, 1994.
8. Arnold, B., Donald, L., Jurgens, A., and Pincock, J. A., Homolytic versus heterolytic cleavage for the photochemistry of 1-naphthylmethyl derivatives, *Can. J. Chem.*, 63, 3140, 1985.
9. DeCosta, D. P. and Pincock, J. A., Photochemistry of substituted 1-naphthylmethyl esters of phenylacetic and 3-phenylpropanoic acid: Radical pairs, ion pairs and Marcus electron transfer, *J. Am. Chem. Soc.*, 115, 2180, 1993.
10. Sakuragi, H., Tokumaru, K., Itoh, H., Terakawa, K., Kikuchi, K., Caldwell, R. A., and Hsu, C.-C., Photochemical and photophysical behavior of *p*-methoxyphenylalkenyl phenanthrenecarboxylates. I. Structure and competitive formation of intramolecular cycloaddition products, *Bull. Chem. Soc. Jpn.*, 63, 1049, 1990.
11. Neunteufel, R. A. and Arnold, D. R., Radical ions in photochemistry. I. The 1,1-diphenylethylene cation radical, *J. Am. Chem. Soc.*, 95, 4080, 1973.
12. Arnold, D. R. and McMahon, K., Dalhousie University, personal communication.
13. Gano, J., Short lived intermediates. I. The Type II photofragmentation of *threo*- and *erythro*-1,2-dimethylbutyl acetate, *Tetrahedron Lett.*, 2549, 1969.
14. Duhaime, R. M., Lombardo, D. A., Skinner, I. A., and Weedon, A. C., Conversion of α,β-unsaturated esters to their β,γ-unsaturated isomers by photochemical deconjugation, *J. Org. Chem.*, 50, 873, 1985.
15. Piva, O., Mortezaie, R., Henin, F., Muzart, J., and Pete, J.-P., Highly enantioselective photodeconjugation of α,β-unsaturated esters. Origin of the chiral discrimination, *J. Am. Chem. Soc.*, 112, 9263, 1990.
16. Lewis, F. D., Oxman, J. D., Gibson, L. L., Hampsch, H. L., and Quillen, S. L., Lewis acid catalysis of photochemical reactions. 4. Selective isomerization of cinnamic esters, *J. Am. Chem. Soc.*, 108, 3005, 1986.
17. Lewis, F. D., Howard, D. K., Barancyk, S. V., and Oxman, J. D., Lewis acid catalysis of photochemical reactions. 5. Selective isomerization of conjugated butenoic and dienoic esters, *J. Am. Chem. Soc.*, 108, 3016, 1986.
18. Pincock, J. A. and Moutsokapas, A. A., An optically active cyclopropene as a mechanistic probe in cyclopropene photochemistry, *Can. J. Chem.*, 55, 979, 1977.
19. Padwa, A., Brookhart, T., Dehm, D., West, G., and Wubbels, G., Solvent control of migratory aptitudes in the photochemical rearrangement of 2(5H)-furanones, *J. Am. Chem. Soc.*, 99, 2347, 1977.
20. Shi, M., Okamoto, Y., and Takamuku, S., Photolysis of triphenylacetic acid and its methyl ester: A novel photochemical generation of carbene intermediates, *Bull. Chem. Soc. Jpn.*, 63, 3345, 1990.

21. Zimmerman, H. E. and Wright, C. W., Triplet photochemistry of acyl and imino cyclopropenes. A rearrangement to afford furans and pyrroles: Reaction and mechanism, *J. Am. Chem. Soc.*, 114, 6603, 1992.

22. Hixson, S. S., Mariano, P. S., and Zimmerman, H. E., The di-π-methane and oxa-di-π-methane rearrangements, *Chem. Rev.*, 73, 531, 1973.

23. Zimmerman, H. E., The di-π-methane (Zimmerman) rearrangement, in *Rearrangements in Ground and Excited States*, deMayo, P., Ed., Academic Press, New York, 1980, 131.

24. Ireland, J. F. and Wyatt, P. A. H., Acid-base properties of electronically excited states of organic molecules, *Adv. Phys. Org. Chem.*, 12, 131, 1976.

25. Kovi, P. J. and Schulman, S. G., Intermolecular hydrogen bonding and proton transfer in the ground and lowest excited singlet states of the ethyl esters of 1- and 2-naphthoic acid, *Spectrosc. Lett.*, 5, 443, 1972.

26. Martin, R., Photo-Fries rearrangement, *Org. Prep. Proc. Int.*, 24, 369, 1992.

27. Bellus, D., Photo-Fries rearrangement and related photochemical [1,j]-shifts (j = 3,5,7) of carbonyl and sulfonyl groups, *Adv. Photochem.*, 8, 109, 1971.

28. Kaupp, G., Photochemical rearrangements and fragmentations of benzene derivatives and annelated arenes, *Angew. Chem. Int. Ed.*, 19, 243, 1980.

29. Nakagaki, R., Hiramatsu, M., Watanabe, T., Tanimoto, Y., and Nagakura, S., Magnetic isotope and external magnetic field effects upon the photo-Fries rearrangements of 1-naphthyl acetate, *J. Phys. Chem.*, 89, 3222, 1985.

30. Libman, J., Sprecher, M., and Mazur, Y., Photochemistry of enolic systems. IV. Irradiation of enol trichloroacetates and a dienol trichloroacetate, *J. Am. Chem. Soc.*, 91, 2062, 1969.

31. Gutsche, C. D. and Oude-Alink, B. A. M., The photoinduced alcoholysis of 3,4-dihydrocoumarin and related compounds, *J. Am. Chem. Soc.*, 90, 5855, 1968.

32. Masnovi, J., Radical anions of esters of carboxylic acids. Effects of structure and solvent on unimolecular fragmentations, *J. Am. Chem. Soc.*, 111, 9081, 1989.

33. Decosta, D. P. and Pincock, J. A., unpublished results.

34. Padwa, A., Dehm, D., Oine, T., and Lee, G. A., Competitive keto-enolate photochemistry in the 3-phenylisocoumaranone system, *J. Am. Chem. Soc.*, 97, 1837, 1975.

35. Fillol, L., Miranda, M. A., Morera, I. M., and Sheikh, H., Photochemistry of 2(3H)-furanones and 2 (5H)-furanones, *Heterocycles*, 31, 751, 1990.

36. Cristol, S. J. and Bindel, T. H., Photosolvolysis, *Org. Photochem.*, 6, 327, 1983.

37. Givens, R. S., Matuszewski, B., and Neywick, C. V., Photodecarboxylation of esters. Photolysis of α- and β-naphthalenemethyl derivatives, *J. Am. Chem. Soc.*, 96, 5547, 1974.

38. Milne, P. H., Wayner, D. M., DeCosta, D. P., and Pincock, J. A., Substituent and charge distribution effects on the redox potentials of radicals. Thermodynamics for homolytic versus heterolytic cleavage in the 1-naphthylmethyl system, *Can. J. Chem.*, 70, 121, 1992.

39. Givens, R. S. and Oettle, W. F., Mechanistic studies in organic photochemistry. VI. Photodecarboxylation of benzyl esters, *J. Org. Chem.*, 37, 4325, 1972.

40. Hilborn, J. W. and Pincock, J. A., Rates of decarboxylation of acyloxy radicals formed in the photocleavage of substituted 1-naphthylmethyl alkanoates, *J. Am. Chem. Soc.*, 113, 2683, 1991.

41. Zimmerman, H. E. and Sandel, V. R., Mechanistic organic photochemistry. II. Solvolytic photochemical reactions, *J. Am. Chem. Soc.*, 85, 915, 1963.

42. DeCosta, D. P. and Pincock, J. A., Control of product distribution by Marcus type electron-transfer rates for the radical pair generated in benzylic ester photochemistry, *J. Am. Chem. Soc.*, 111, 8948, 1989.

43. Eberson, L., Ed., *Electron Transfer Reactions in Organic Chemistry*, Springer-Verlag, Berlin, 1987, 32.

44. Gould, I. R., Moody, R., and Farid, S., Electron transfer reactions in the Marcus inverted region: Differences in solvation and electronic coupling between excited charge-transfer complexes and geminate radical ion pairs, *J. Am. Chem. Soc.*, 110, 7243, 1988.

45. Closs, G. L., Calcaterra, L. T., Green, N. J., Penfield, K. W., and Miller, J. R., Distance, stereoelectronic effects and the Marcus inverted region in intramolecular electron transfer in organic radical ions, *J. Phys. Chem.*, 90, 3673, 1986.

46. Penn, J. H. and Owens, W. H., The photochemistry of mixed anhydrides: A search for selectivity in photochemically initiated bond cleavage reactions, *J. Am. Chem. Soc.*, 115, 82, 1993.

47. Braun, W., Rajbenbach, L., and Eirich, F. R., Peroxide decomposition and the cage effect, *J. Phys. Chem.*, 66, 1591, 1962.

48. Kaptein, R., Brokken-Zijp, J., and de Kanter, F. J. J., Chemically induced dynamic nuclear polarization. XI. Thermal decomposition of acetyl peroxide, *J. Am. Chem. Soc.*, 94, 6280, 1972.

49. DeCosta, D. P. and Pincock, J. A., Intramolecular electron transfer in the photochemistry of substituted 1-naphthylmethyl esters of benzoic acids, *Can. J. Chem.*, 70, 1879, 1991.

50. Givens, R. S., Matuszewski, B., Levi, N., and Leung, D., Photodecarboxylation. A labelling study. Mechanistic studies in photochemistry. 15, *J. Am. Chem. Soc.*, 99, 1896, 1977.

51. Jaeger, D. A., Photochemistry of aromatic compounds. Internal return in the photosolvolysis of 3,5-dimethoxybenzyl acetate, *J. Am. Chem. Soc.*, 97, 902, 1975.

52. Kim, J., Ngai, B., and Pincock, J. A., unpublished results.

33

Carbene Formation in the Photochemistry of Cyclic Ketones

S. M. Roberts

Exeter University

33.1 Introduction

The photon-catalyzed ring expansion of cyclic ketones to give oxacycloalkanes was featured in the early work of Yates[1] and further developed by Yates, Turro, and Hostettler. These initial investigations have been reviewed[2] and, together with later studies, showed that cyclobutanone derivatives have a proclivity to undergo photochemical ring expansion reactions, forming oxacarbenes as high-energy intermediates (Scheme 33.1). The oxacarbenes could be trapped by water or alcohols to give γ-lactols or 2-alkoxytetrahydrofurans respectively *(vide infra)*. A second mode of reaction of cyclobutanones is via a retro [2+2]-reaction to give ketene(s) and alkene(s) (Scheme 33.1). In contrast, cyclopentanone derivatives often undergo photolysis to form an alkene and an aldehyde. If this pathway is prohibited, a ketene and alkene are formed: only if both these pathways are frustrated is oxacarbene formation observed (Scheme 33.2), except in special cases e.g., when the five-membered ring is part of a strained structure. In a small number of cases, decarbonylation is observed as another reaction manifold for strained cycloalkanones.

Thus, 2,2-dimethylcyclobutanone (1) produces the acetal 2 on photolysis in methanol: the intermediate oxacarbene is trapped by the solvent (Scheme 33.2).[3] Oxygen insertion toward the more substituted center is a common feature of this type of reaction. The photogenerated oxacarbenes have also been trapped by alkenes in relatively low yield; for example, benzocyclobutenedione (3) yields the polycyclic product 4 on photolysis in the presence of 2-methylpropene (Scheme 33.3).[4]

Oxacarbenes derived from cyclobutanones[5] and cyclopentanones[6] can be trapped intramolecularly by adjacent hydroxyl groups (Scheme 33.4).

33.2 Mechanistic Considerations

The mechanism of the photoconversion of a cyclobutanone to the corresponding oxacarbene has been the subject of detailed investigation and some debate. The intermediacy of an oxacarbene is not in doubt following trapping experiments along the lines described above and also low temperature work.[7] Certainly, oxacarbene formation from cycloalkanones is a reversible process; as such, Agosta has shown that a cyclic oxacarbene (produced by a nonphotolytic mechanism) forms, among other things, the corresponding cycloalkanone (Scheme 33.5).[8]

0-8493-8634-9/95/$0.00+$.50

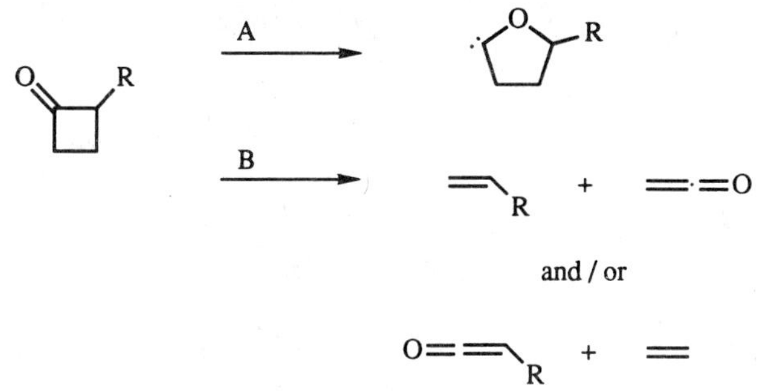

For cyclobutanones pathway A is generally preferred to pathway B.

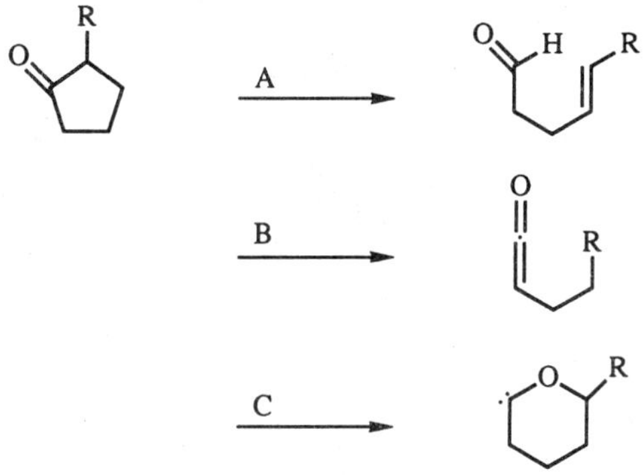

For cyclopentanones pathway A is generally preferred to pathway B which is generally preferred to pathway C.

SCHEME 33.1 Reaction pathways followed on photolysis of cyclobutanones and cyclopentanones

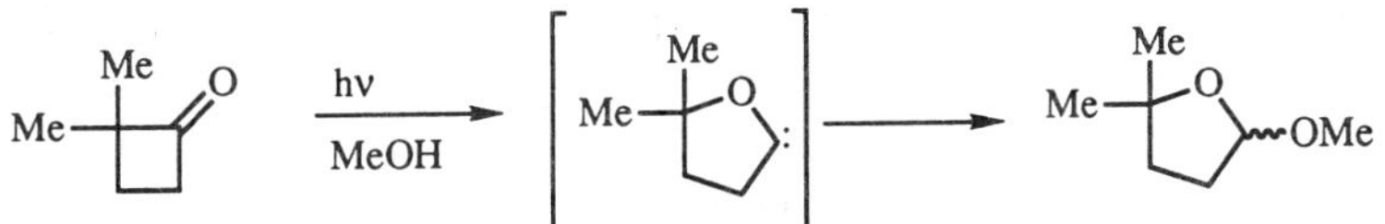

SCHEME 33.2 Photolysis of 2,2-dimethylcyclobutanone

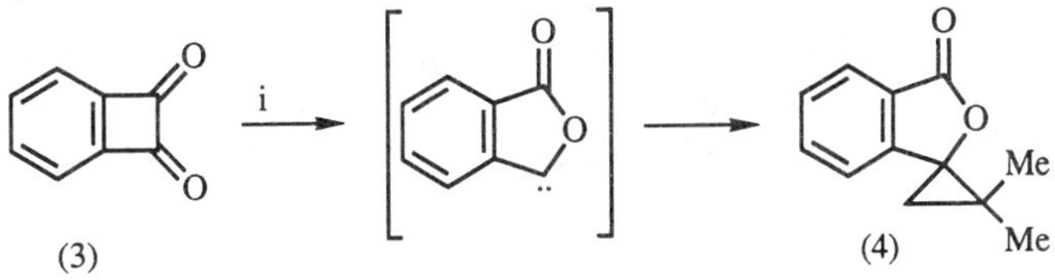

Reagents and conditions: i) hv, $=\!\!\!<^{Me}_{Me}$

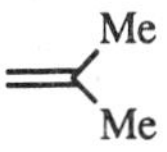

SCHEME 33.3 Photolysis of benzocyclobutenedione

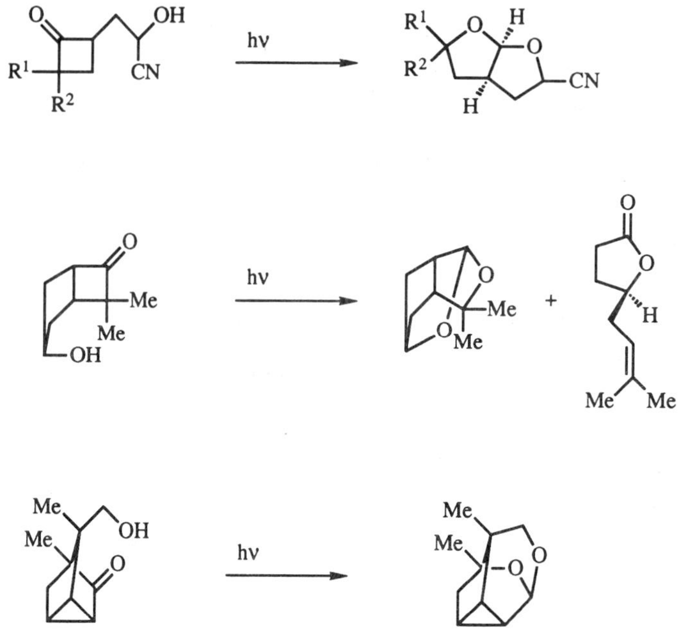

SCHEME 33.4   Intramolecular trapping of photogenerated oxacarbenes

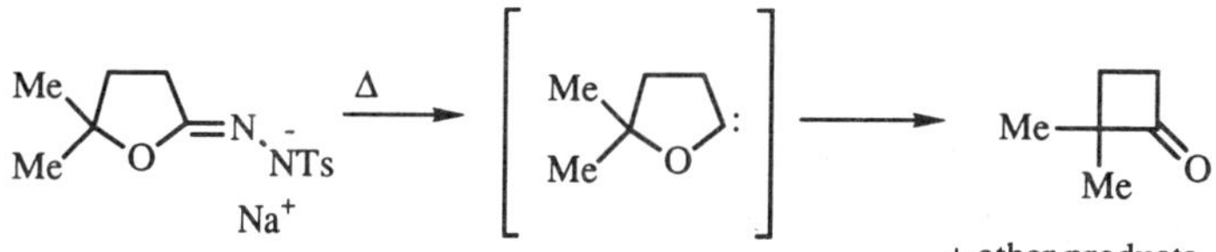

SCHEME 33.5   Rearrangement of a thermally generated oxacarbene into a cyclobutanone

In the photolyses of selected cyclobutanones, the retention of stereochemical integrity of the migrating carbon center militates against the intermediacy of a long-lived alkyl-acyl biradical (with the consequent rotation about the carbon-carbon bonds), but the possibility of having a very short-lived species (resulting from initial Norrish cleavage to give the more stable alkyl-acyl radical pair) cannot be ruled out (Scheme 33.6). In contrast, it is likely that the cyclopentanone photo-ring expansion reaction does proceed through an alkyl-acyl biradical with a discrete lifetime.[9] Turro and Morton have shown that it is a triplet excited state that is responsible for the ring expansion in cyclopentanones by the inhibition of acetal formation in the presence of known triplet quenchers.[10] Analogous quenching experiments with cyclobutanones failed, as did attempts to sensitize the ring expansion reaction using xanthone, acetophenone, and *m*-xylene.[11]

Structural components that appear to enhance the formation of the oxacarbene include α-methylene and α-(spiro)cyclopropyl units adjacent to the ketone moiety (Scheme 33.7).

The ring-expansion reaction can be observed for some six-membered ring ketones. For example, the silacycloalkanone **5** forms the corresponding oxacarbene, which can be trapped with diethyl fumarate to give the product **6** in 78% yield (Scheme 33.8).[14]

33.3 Synthetic Applications

One of the first synthetic applications of the cyclobutanone-to-alkoxytetrahydrofuran photocatalyzed ring expansion reaction was in the area of prostaglandin synthesis. The cyclobutanone derivative **7**

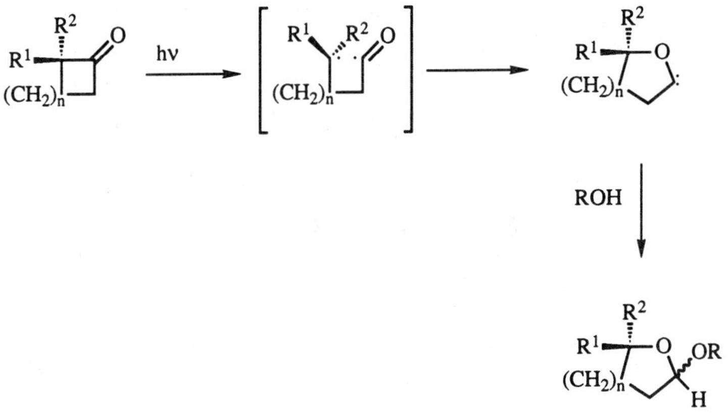

SCHEME 33.6   Detailed reaction pathway for photo-ring expansion of cyclic ketones

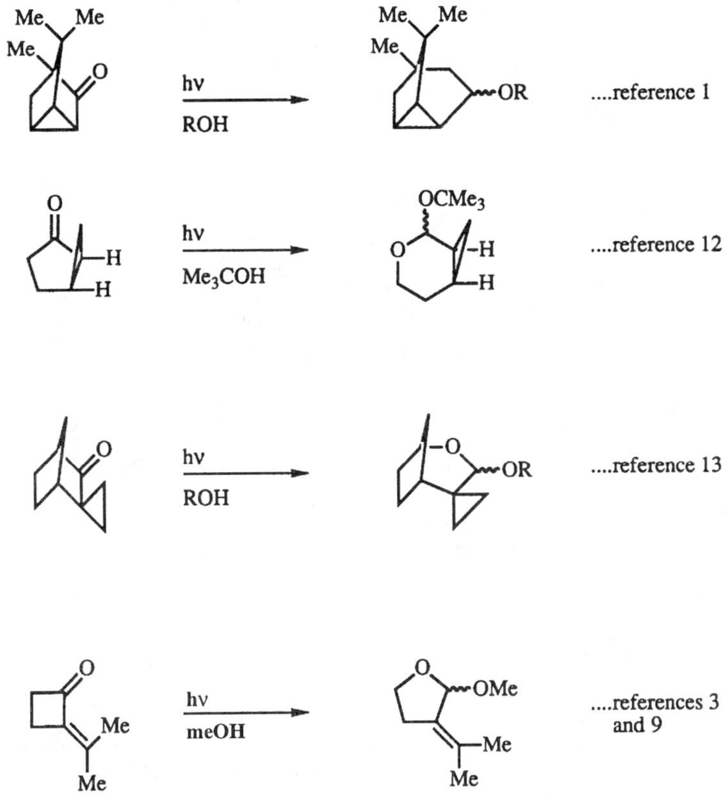

SCHEME 33.7   Ring expansion of cyclic ketones having an adjacent methylene or cyclopropyl unit

is readily prepared from cyclopentadiene in four steps. Photolysis in methanol gave the acetal **8**, an established intermediate in prostaglandin synthesis (Scheme 33.9).[15] A more important example of the ring expansion reaction in prostaglandin synthesis is illustrated in Scheme 33.10. Thus, the late stage prostaglandin intermediate **9** was photolyzed in aqueous acetonitrile to give a good yield (ca. 70%) of the lactol (**10**). The latter compound was not purified, but was subjected to a Wittig reaction to give prostaglandin-F$_2\alpha$ (**11**) directly (Scheme 33.10).[16]

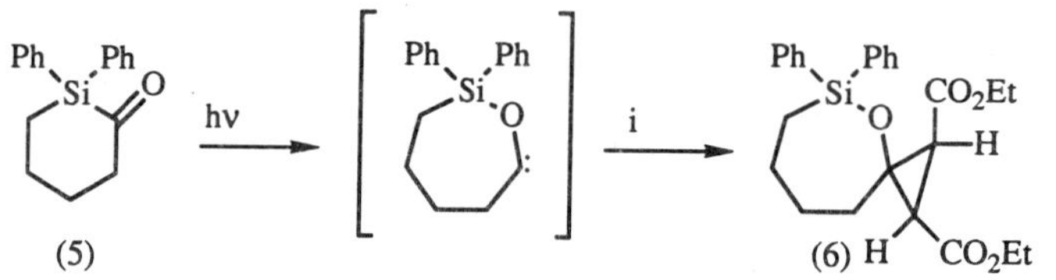

Reagents and conditions: i) $EtO_2CCH=CHCO_2Et$

SCHEME 33.8   Photolysis of a silacycloalkanone

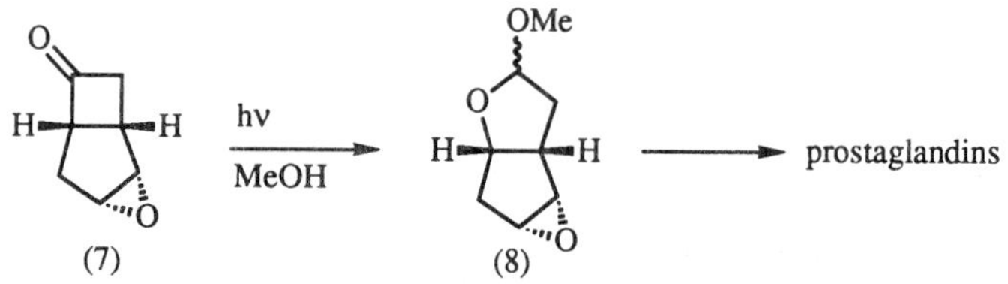

SCHEME 33.9   Photolysis of an epoxybicycloheptanone

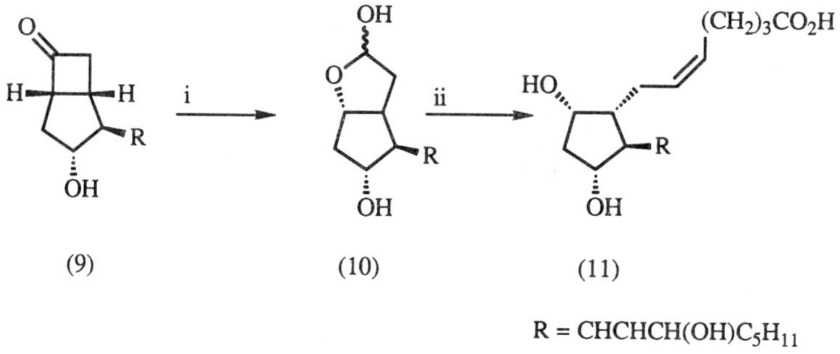

$$R = CHCHCH(OH)C_5H_{11}$$

Reagents and conditions: i) hv, MeCN, H_2O, quartz
ii) $Ph_3PCH(CH_2)_3CO_2^-$ then H^+, yield for two steps
48%.

SCHEME 33.10   Synthesis of prostaglandins and analogues

A later application of the photo-ring expansion reaction to the preparation of analogues of thromboxanes (Scheme 33.11) hinged on the fact that α-spiro-cyclopropylcyclopentanones undergo highly controlled photon-catalyzed reactions. The functionalized cyclopentanone 13 was converted into the oxacyclohexane derivative 14 on photolysis in methanol.[17]

An elegant example of the use of the photo-catalyzed ring expansion reaction in the synthesis of a natural product came from the Pirrung laboratory. The American authors photolysed the ketone 15 to produce the acetal 16, which was subsequently converted into muscarine (17) (Scheme 33.12).[18]

The photo-ring expansion reaction can also be applied to the synthesis of novel sulphur-containing heterocyclic systems; for example, the conversion of the β-dithiolactone (18) into the dithiaacetal 19 (Scheme 33.13).[19]

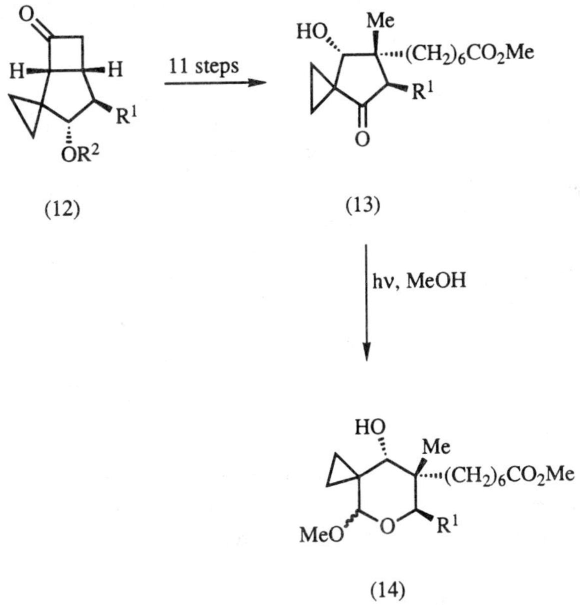
11 steps
(12)
(13)
hν, MeOH
(14)

SCHEME 33.11 Synthesis of a thromboxane analogue

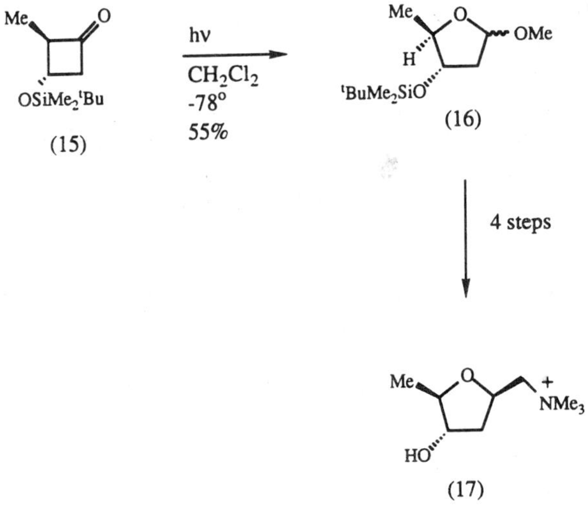
hν
CH_2Cl_2
-78°
55%
(15)
(16)
4 steps
(17)

SCHEME 33.12 Synthesis of muscarine

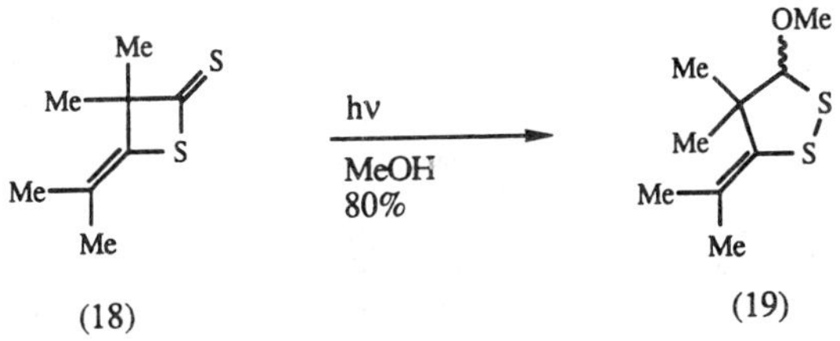
hν
MeOH
80%
(18)
(19)

SCHEME 33.13 Ring expansion of a thiabutanone

References

1. (a) Yates, P. and Kilmurry, L., Two photochemical reactions of cyclocamphanone, *Tetrahedron Lett.*, 1739, 1964; (b) Yates, P. and Kilmurry, L., An oxacarbene as an intermediate in the photolysis of cyclocamphanone, *J. Am. Chem. Soc.*, 88, 1563, 1966.

2. (a) Yates, P., The photochemical ring expansion of cyclic ketones, *J. Photochem.*, 5, 91, 1976; (b) Turro, N. J., Dalton, J. C., Dawes, K., Farrington, G., Hautala, R., Morton, D., Niemczyk, M., and Schore, N., Molecular photochemistry of alkanones in solution: α-Cleavage, hydrogen abstraction, cycloaddition and sensitization reactions, *Acc. Chem. Res.*, 5, 92, 1972; (c) Stohrer, W. D., Jacobs, P., Kaiser, K. H., Wieche, G., and Quinkert, G., *Top. Curr. Chem.*, 46, 181, 1974.

3. (a) Turro, N. J. and Southam, R. M., Molecular photochemistry. IV. Solution photochemistry of cyclobutanone and some derivatives, *Tetrahedron Lett.*, 545, 1967; (b) Morton, D. R. and Turro, N. J., Photochemical ring expansion of cyclic aliphatic ketones. Cyclobutanones and cyclopentanones, *J. Am. Chem. Soc.*, 95, 3947, 1973.

4. (a) Staab, H. A. and Ipaktschi, J., Photochemische Reaktionen des Benzocyclobuten-1,2-dions, *Tetrahedron Lett.*, 583, 1966; (b) Staab, H. A. and Ipaktschi, J., Photochemische Reaktionen des Benzocyclobuten-1,2-Dions, *Chem. Ber.*, 101, 1457, 1968.

5. (a) Butt, S., Davies, H. G., Dawson, M. J., Lawrence, G. C., Leaver, J., Roberts, S. M., Turner, M. J., Wakefield, B. J., Wall, W. F., and Winders, J. A., Resolution of 7,7-dimethylbicyclo[3.2.0]hept-2-en-6-one using Mortierella ramanniana and 3α,20β-hydroxy-steroid dehydrogenase. Photochemistry of 3-hydroxy-7,7-dimethylbicyclo[3.2.0]heptan-6-ones and the synthesis of (+)-eldanolide, *J. Chem. Soc., Perkin Trans. I*, 903, 1987; (b) Pirrung, M., Bicyclic acetals from oxacarbenes, *Angew. Chem. Int. Ed.*, 24, 1043, 1985.

6. Switlak, K., He, D., and Yates, P., Intramolecular trapping of an oxacarbene intermediate in the photochemical ring expansion of a cyclopentanone, *J. Chem. Soc., Perkin Trans. I*, 2579, 1992.

7. (a) Quinkert, G., Kaiser, K. H., and Stohrer, W. D., Low temperature electronic spectroscopic observations on light-induced cyclobutanone-tetrahydrofurylidene rearrangement, *Angew. Chem. Int. Ed.*, 13, 198, 1974; (b) Kesselmayer, M. A. and Sheridan, R. S., Direct observation of photochemical cleavage of a cyclopropylalkoxycarbene to an alkyne, *J. Am. Chem. Soc.*, 109, 5029, 1987; (c) Matsumura, M., Ammann, J. R., and Sheridan, R. S., Photochemical ring expansion of tetramethyl cyclobutanone revisited: Angular dependence of electronic absorption of singlet carbenes, *Tetrahedron Lett.*, 33, 1843, 1992; (d) see also Miller, R. D., Gölitz, P., Janssen, J., and Lemmens, J., Alternative precursors to 1,4-acyl alkyl biradicals: Cyclic N-acyl-1,1-diazenes, *J. Am. Chem. Soc.*, 106, 7277, 1984.

8. (a) Foster, A. M. and Agosta, W. C., Pyrolysis of lactone tosylhydrazone sodium salts, *J. Am. Chem. Soc.*, 94, 5777, 1972; (b) Smith, A. B., III, Foster, A. M., and Agosta, W. C., Two novel routes to spiro[3,4]octan-1-ones, *J. Am. Chem. Soc.*, 94, 5100, 1972; (c) Foster, A. M. and Agosta, W. C., Stereochemical consequences in the pyrolysis of lactone tosylhydrazone salts, *J. Am. Chem. Soc.*, 95, 608, 1973.

9. (a) Turro, N. J. and McDaniel, D. M., Stereospecific photoreactions of cyclobutanones, *J. Am. Chem. Soc.*, 92, 5727, 1970; (b) McDaniel, D. M. and Turro, N. J., Photochemistry of bicyclo[6.2.0]decan-9-ones in solution, *Tetrahedron Lett.*, 3035, 1972; (c) Quinkert, G., Kaiser, K. H., and Stohrer, W. D., Low temperature electronic spectroscopic observations on light-induced cyclobutanone-tetrahydrofurylidene rearrangement, *Angew. Chem. Int. Ed.*, 13, 198, 1974; (d) Quinkert, G., Jacobs, P., and Stohrer, W. D., Stereospecificity as an argument against the intermediacy of 1,4-alkyl/acyl diradicals in the light induced cyclobutanone-tetrahydrofurylidene isomerisation in condensed phase, *Angew. Chem. Int. Ed.*, 13, 197, 1974; (e) Quinkert, G. and Jacobs, P., Lichtinduzierte Reacktionen. IX. Die Stereospezifität der Lichtinduzierten Cyclobutanon/Tetrahydrofuryliden-Isomerisierung und Imre Mechanistische Konsequenz, *Chem. Ber.*, 107, 2473, 1974; (f) Grewal, R. S., Burnell, D. J., and Yates, P., Stereochemistry of oxacarbene formation in the photolysis of oxacyclopentan-3-ones, *J. Chem. Soc., Chem. Commun.*, 759, 1984.

10. (a) Morton, D. R., Lee-Ruff, E., Southam, R. M., and Turro, N. J., Molecular photochemistry. XXVII. Photochemical ring expansion of cyclobutanone, substituted cyclobutanones, and related cyclic ketones, *J. Am. Chem. Soc.,* 92, 4349, 1970; (b) Turro, N. J. and Morton, D. R., Mechanism of photochemical ring expansion reactions of cyclic ketones, *J. Am. Chem. Soc.,* 93, 2569, 1971.

11. (a) Turro, N. J. and McDaniel, D. M., The photoreduction of cyclic ketones by tri-*N*-butylstannane, *Mol. Photochem.,* 2, 39, 1970; (b) Jones, G. and Zalk, S. A., Comparative photo- and thermal chemistry of bicyclo[3.2.0]hept-2-en-6-one, *Tetrahedron Lett.,* 4095, 1973.

12. Dauben, W. G., Schutte, L., Schaffer, G. W., and Gagosion, R. B., Photoisomerisation of bicyclo[3.1.0]hexan-2-ones, *J. Am. Chem. Soc.,* 95, 468, 1973.

13. Yates, P. and Loutfy, R. O., Photochemical ring expansion of cyclic ketones via cyclic oxacarbenes, *Acc. Chem. Res.,* 8, 209, 1975.

14. (a) Brook, A. G., Kucera, H. W., and Pearce, R., Photolysis of 1,1-diphenylsilacyclohexanone in dimethylfumarate. The trapping of a siloxycarbene by an electron deficient olefin, *Can. J. Chem.,* 49, 1618, 1971; (b) see also, Stereospecificity in the photoisomerisation of steroidal α-ketols to lactones, *Tetrahedron Lett.,* 25, 3289, 1984.

15. (a) Crossland, N. M., Kelly, D. R., Roberts, S. M., Reynolds, D. P., and Newton, R. F., Photosynthetic routes to prostanoids. Ring expansions of some substituted cyclobutanones to form cyclic acetals or γ-lactols, *J. Chem. Soc., Chem. Commun.,* 681, 1979; (b) Crossland, N. M., Kelly, D. R., Roberts, S. M., Reynolds, D. P., and Newton, R. F., Synthesis of (±)-prostaglandin F$_2$α involving photolytic conversion of a cyclobutanone to a γ-lactol, *J. Chem. Soc., Chem. Commun.,* 683, 1979.

16. Howard, C. C., Newton, R. F., Reynolds, D. P., Wadsworth, A. H., Kelly, D. R., and Roberts, S. M., Total synthesis of prostaglandin-F$_2$α involving stereocontrolled and photo-induced reactions of bicyclo[3.2.0]heptanones, *J. Chem. Soc., Perkin Trans. I,* 852, 1980.

17. (a) Azadi-Ardakani, M., Loftus, G. C., Mjalli, A. M. M., Newton, R. F., and Roberts, S. M., Photochemistry of cyclopentanone derivatives in the synthesis of prostaglandin and thromboxane analogues, *J. Chem. Soc., Chem. Commun.,* 1709, 1989; (b) For a related transformation *en route* to pederol, see Pirrung, M. C. and Kenney, P. M., Synthesis of pederol, *J. Org. Chem.,* 52, 2335, 1987.

18. Pirrung, M. C. and De Amicis, C. V., Total synthesis of the muscarines, *Tetrahedron Lett.,* 29, 159, 1988.

19. Muthuramu, K., Sundari, B., and Ramamurthy, V., Photofragmentation reactions of dithiolactones, *Tetrahedron,* 39, 2719, 1983.

34

Norrish Type I Processes of Ketones: Basic Concepts

Cornelia Bohne
University of Victoria

The Norrish type I is the process in which the excitation of ketones leads to the homolytic cleavage of the α-bond of the carbonyl moiety. The reactivity toward α-cleavage of excited ketones is analogous to that observed for alkoxy radicals; that is, within a particular molecule, the weakest α-bond is cleaved, thus leading to the most stable radical. For different ketones, the cleavage is faster for the ketone in which the most stable radicals are formed. Additional increase in reactivity is obtained when strain is released, as in the case of cyclic ketones.[1,2] These general principles are exemplified in Table 34.1 (entries A-E).

The rate constants for Norrish type I processes are much faster for n,π* as compared to π,π* excited states. The n,π* reactivity is due to the weakening of the α-bond by overlap of this bond with the half vacant n-orbital on oxygen. This overlap is not possible for π,π* excited states. Substitution with electron-releasing groups on the aromatic ring of aryl alkyl ketones leads to stabilization of the π,π* excited state. For example, *para*-substitution on the phenyl ring of phenyl *t*-butyl ketone by methoxy or phenyl groups leads to a dramatic decrease of the reactivity (Table 34.1, entries F-H), as measured by the reaction quantum yield and rate constant of the α-cleavage reaction (Equation 34.1). Both of these substituents stabilize the triplet π,π* state in relation to the n, π* state.

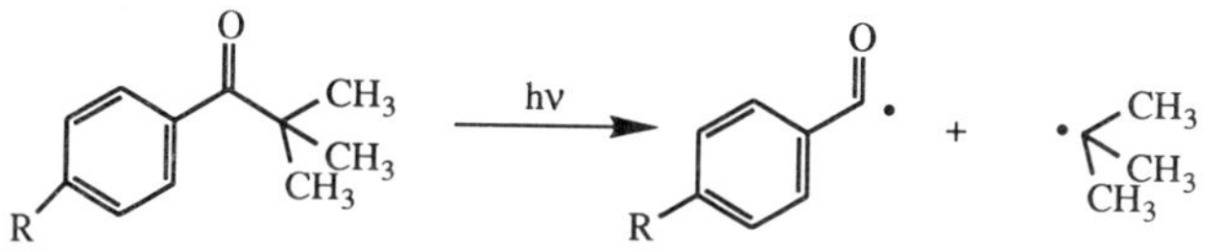

$$(34.1)$$

α-Cleavage reactivity of triplet n,π* states are generally about 100 times greater than from singlet states with the same configuration. This difference is rationalized by the use of state correlation diagrams[8,9] that relate the multiplicity, symmetry (electronic correlation), and energy of the different states of the reactants and products. In the α-cleavage reaction, the acyl radical formed can be bent or linear. For acyclic and unstrained cyclic ketones, the linear radical is too high in energy to be accessible from the first excited states. A correlation of the different excited states of the ketone and the possible radical pairs (or biradicals in the case of cyclic compounds) show that neither the singlet nor the triplet n,π* excited states have an electronic correlation with ground-state photoproducts. A triplet product can be obtained from the crossing of two triplet energy surfaces, whereas no such crossing is observed for the singlet state. Thus, the triplet reaction involves an internal conversion process, whereas for the singlet, a spin-forbidden intersystem crossing process is required.

0-8493-8634-9/95/$0.00+$.50
© 1995 by CRC Press, Inc.

Table 34.1 Rate Constants and Quantum Yields for α-Cleavage from the Triplet States of Ketones

Entry	Ketone	Quantum yield	Rate constant (s^{-1})	Ref.
A	$PhCOCH_2Ph$	~0.4	2×10^6	5
B	$PhCOCH(CH_3)Ph$	~0.4	3×10^7	5
C	$PhCOC(CH_3)_2Ph$	~0.3	1×10^8	5
D	Cyclopentanone	~0.2	2×10^8	4, 6
E	Cyclohexanone	~0.2	2×10^7	4, 6
F	$PhCOC(CH_3)_3$	~0.3	1×10^7	3
G	$4\text{-}CH_3O\text{-}C_6H_4COC(CH_3)_3$	~0.1	7×10^5	3
H	$4\text{-}C_6H_5\text{-}C_6H_4COC(CH_3)_3$	<0.001	Not measured	3
I	$CH_3COC(CH_3)_3$	0.33	$>10^9$	7

The Norrish type I process of aliphatic ketones is faster than for aromatic compounds (Table 34.1 entries F and I). This difference is based on energetic grounds. Triplet states of aromatic ketones are stabilized compared to those of aliphatic compounds and the α-cleavage reaction is approximately thermoneutral for the aromatic compounds.

The radicals formed in the Norrish type I process of acyclic ketones can undergo (1) recombination; (2) two acyl radicals can react forming diketones; or (3) the acyl radical can decarbonylate to form a second carbon-centered radical. All carbon-centered radicals can recombine or undergo disproportionation.

Recombination of the radicals formed by α-cleavage leads to product quantum yields that are less than unity. The participation of recombination reactions was demonstrated in the photolysis of α-phenylacetophenone, a chiral ketone (Scheme 34.1). Cleavage of the excited ketone occurs with unit quantum efficiency. The quantum yields of disproportionation products, recombination with retention of configuration, and recombination with racemization are 0.44, 0.27, and 0.33, respectively.[10] Configuration retention suggests that intersystem crossing from the triplet radical-pair necessary for recombination to occur in the cage is competitive with rotation of the radical-pair and diffusion into the solvent. The last two processes lead to racemization.

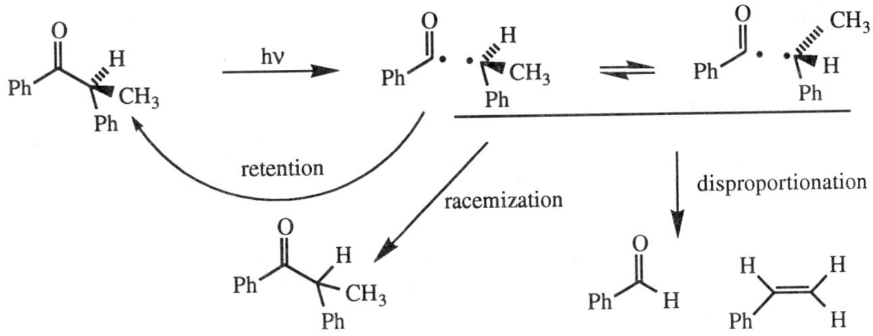

SCHEME 34.1

Decarbonylation of the acyl radical following α-cleavage occurs only if a very stable radical (e.g., *t*-alkyl or benzyl) is formed. This is seen, for example, in the case of dibenzylketones (Scheme 34.2)[11,12] or tetraphenylacetone[13] for which two benzyl radicals and two diphenylmethyl radicals are formed, respectively. For both reactions, the decarbonylation process is fast enough to avoid the formation of products involving the acyl radical. In the case of 4-methyldibenzylketone, only the product from pathway *a* (Scheme 34.2) is formed from geminate processes (due to recombination of the radical-pair initially formed), whereas all three pathways occur for random processes after separation of the radical-pair.

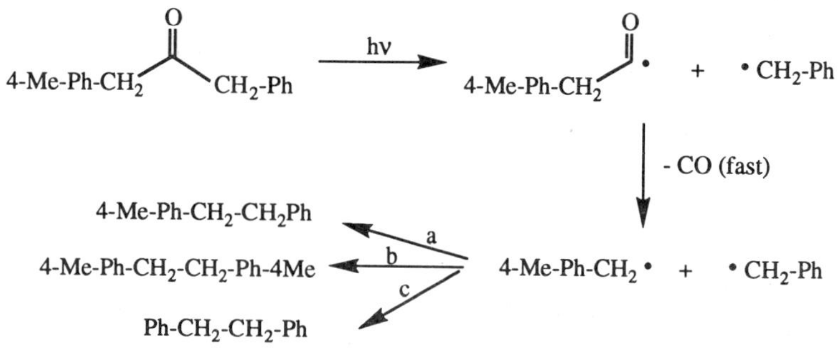

SCHEME 34.2

Norrish type I processes of cycloalkanones have a more diverse mechanistic scheme compared to acyclic ketones due to the possibility of intramolecular reactions (Scheme 34.3). In line with the observation made for acyclic compounds, the triplet reactivity exceeds that of the singlet. In addition, both radical stabilization and ring strain increase the rate constant for cleavage. For example, the rate constant for α-cleavage of cyclopentanone is 1 order of magnitude higher than for cyclohexanone (Table 34.1, entries D and E), and the cleavage rate constant for 2-methylcyclohexanone and 2,2-dimethylcyclohexanone are about 10 and 100 times larger, respectively, than for cyclohexanone.[4,6] Some singlet reactivity is observed when the rate constant for cleavage is fast enough to compete with intersystem crossing. Singlet reactivity is relatively important for molecules with considerable ring strain. Indeed, α-cleavage for cyclobutanone occurs predominantly from the singlet state (vide infra).

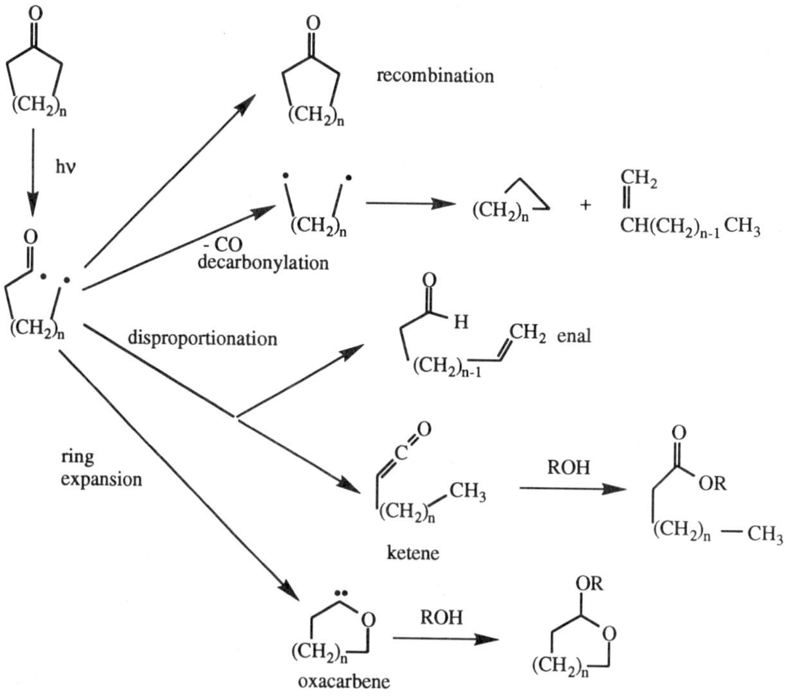

SCHEME 34.3

Direct evidence for biradical formation in Norrish type I processes was obtained from radical trapping reactions and direct methods such as chemically induced dynamic nuclear polarization (CIDNP) and laser flash photolysis. Alkyl acyl biradicals were trapped by nitric oxide and the adduct was characterized by ESR.[14] CIDNP is an NMR technique in which an enhanced absorption or emission signal is evidence for radical intermediates. Studies of cyclic ketones (cycloheptanone up to cycloundecanone) have established that aldehyde formation and product regeneration involve the intermediacy of a triplet acyl alkyl biradical.[15,16] In the case of benzocyclooctanone, CIDNP studies established the triplet biradical lifetime to be 189 ns,[17] one of the longest lifetimes reported for biradicals formed in Norrish type I processes.[18] Laser flash photolysis was employed to determine the lifetimes of a variety of biradicals formed after α-cleavage.[18] For example, the lifetimes of 1,5-biradicals such as 2,2-dimethyl and 2,2,5,5-tetramethylpentanone are appreciably shorter (11 to 14 ns) than those observed for the 1,6-biradicals formed in the homolysis of 2,2,6-trimethyl and 2,2,6,6-tetramethylcyclohexanone (31 to 45 ns).[19]

Product formation from acyl alkyl biradicals will be determined by the relative magnitudes of the rate constants of the reaction pathways shown in Scheme 34.3. Competition between the different pathways depends on spin multiplicity of the biradical and population of the molecules in the conformation suitable for reaction. Singlet biradicals have short lifetimes because recombination to regenerate the starting material is favored, whereas a triplet biradical must intersystem cross prior to such a reaction. Thus, conformational equilibration can occur for long-lived triplet biradicals.

When the α-carbon of the ketone is a chiral center, recombination of the biradical leads to epimerization at this center.[20] Photoepimerization and isolation of identical racemic mixtures (Equation 34.2) from epimeric starting materials were the initial mechanistic evidence for Norrish type I processes.

$$(34.2)$$

Decarbonylation of acyl alkyl biradicals has a significant activation barrier and only occurs when one or more of the following are satisfied: (1) radical-stabilizing substituents are present at the α- and α'-positions; (2) no hydrogens are readily accessible for disproportionation reactions; and (3) when significant ring strain is released by the loss of CO (Equation 34.3).[21,22] For example, decarbonylation can compete with disproportionation when either tertiary or benzylic radicals are formed. The competition between these processes are temperature dependent and decarbonylation can be favored at higher temperatures.[21]

$$(34.3)$$

Disproportionation products are the most commonly isolated products from the decay of acyl alkyl biradicals. The intramolecular nature of the reaction was established by deuterium labeling experiments. The partition between the two disproportionation reactions — enal and ketene formation — depends on the relative population of the conformers leading to each of these compounds. This relative population can be rationalized by analysis of transition state strain energies. This is exemplified by the enal:ketene product ratio for cyclopentanones and cyclohexanones derivatives shown in Table 34.2. For cyclopentanone and 2-substituted derivatives, the formation of the aldehyde is highly favored, but substituents in the 3-position diminish the

Table 34.2   Quantum Yield Ratio for Enal and Ketene Formation
in the Photolysis of Cyclic Ketones

Ketone	$\phi_{enal}/\phi_{ketene}$	Ref.
Cyclopentanone (CP)	23	23
2-Methyl CP	~40	24
2-Isopropyl CP	~30	24
3-Methyl CP 1,2 cleavage	2.0	23
3-Methyl CP 5,6 cleavage	2.4	23
3,3 Dimethyl CP	0.5	23
Cyclohexanone (CH)	1.6	24
cis- and trans-2-methyl-5-t-butyl CH	9.1	25
2-Methyl CH	2.3	25
cis-2,5-Dimethyl CH	2.2	25
4-Methyl CH	1.0	24
4-t-Butyl CH	0.2	24
3,5-Dimethyl CH	0.2	24

enal:ketene ratio. These results can be explained from the fact that for aldehyde formation, the number of hydrogens that can be abstracted diminishes and the transition state strain energy increases upon substitution at the 3-position. Cyclohexanones have enal:ketene ratios around 2.0. These ratios decrease upon substitution at both the 3- and 4-positions, but increase with bulky substituents in the 5-position. Substitution at the 4-position leads to a gauche interaction in the transition state for enal formation that is not present in the transition state for the ketene (Scheme 34.4, cleavage at 1–2 bond). This gauche interaction leads to a decrease of the yield of enal formation. The reduction observed for 3,5-dimethylcyclohexanone is due to the fact that the number of hydrogens available for enal formation is halved. The effect of substituents at the 5-position suggests that the transition state leading to ketene formation is destabilized. This is particularly noticeable when the substituent is the bulky t-butyl group.

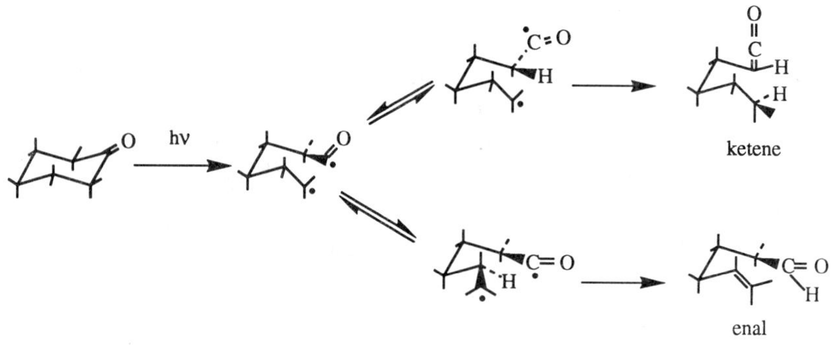

SCHEME 34.4

Ring expansion leading to the formation of oxacarbenes has been observed predominantly in the photochemistry of cyclobutanones (vide infra). Oxacarbene formation has also been observed in strained molecules where biradicals formed in the α-cleavage process cannot undergo disproportionation reactions due to structural constraints, i.e., no hydrogens to abstract or conformational inhibition of the transition state for ketene or enal formation.

The photochemistry of cyclobutanones is markedly different from that observed for cyclopentanones and cyclohexanones. Sensitization and quenching experiments have demonstrated that the α-cleavage process is predominantly an excited singlet state reaction.[26] Product yields lower than unity suggest a recombination pathway. Additional reactions of the singlet biradical are (1) β-cleavage (cycloelimination) to form an olefin and a ketene; (2) decarbonylation and cyclization of the carbon-carbon biradical; or (3) ring expansion to form the oxacarbene. Further, product formation in cyclobutanone photochemistry is stereoselective (Scheme 34.5),[2] a feature not encountered in the photochemistry of cyclopentanones and cyclohexanones.

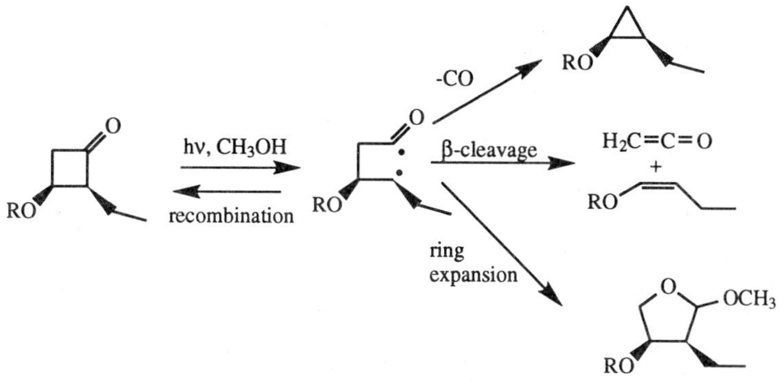

SCHEME 34.5

The unusual reactivity of cyclobutanones is ascribed to strain released upon α-cleavage. This lowers the energy of both the bent and linear acyl radicals relative to the n,π* state of the excited cyclobutanone, the effect being more pronounced for the linear biradical. The linear biradical energy is below the energy of the excited singlet state as there is no thermodynamic barrier for α-cleavage from the excited singlet state. Cyclobutanone α-cleavage from the singlet state competes favorably with intersystem crossing. In the case of cycloalkanones with larger rings, the energy of the linear biradical is higher than the excited state singlet energy.[27] Furthermore, the different reactivity observed for the biradicals formed from cyclobutanone and larger cycloalkanones can be rationalized by the kind of biradical (linear or bent) formed. For example, decarbonylation is rarely observed for cyclopentanones and cyclohexanones because this process is endothermic for the bent radical. The linear radical formed in the α-cleavage of cyclobutanones is close structurally to the transition state for decarbonylation and, in this case, the reaction occurs even at 77 K.[2]

In this chapter, the general and basic concepts involved in the Norrish type I reaction of ketones have been presented. No attempt was made to cover exceptions and to describe complexities when additional functional groups are present. In Chapter 35, some specific examples and synthetic applications involving Norrish type I processes of ketones are presented.

eferences

1. Gilbert, A. and Baggott, J., *Essentials of Molecular Photochemistry,* Blackwell Scientific Publications, Oxford, 1991.
2. Turro, N. J., *Modern Molecular Photochemistry,* Benjamin/Cummings, Menlo Park, CA, 1978.
3. Lewis, F. F. and Magyar, J. G., Photoreduction and α-cleavage of aryl alkyl ketones, *J. Org. Chem.,* 37, 2102, 1972.
4. Wagner, P. J. and Speorke, R. W., Triplet lifetimes of cyclic ketones, *J. Am. Chem. Soc.,* 91, 4437, 1969.
5. Heine, H. G., Hartmann, W., Kory, D. R., Magyar, J. G., Hoyle, C. E., McVey, J. K., and Lewis, F. D., Photochemical α-cleavage and free radical reactions of some deoxybenzoins, *J. Org. Chem.,* 39, 691, 1974.
6. Dalton, J. C., Dawes, K., Turro, N. J., Weiss, D. S., Barltrop, J. A., and Coyle, J. D., Type I and type II photochemical reactions of some five and six-membered cycloalkanones, *J. Am. Chem. Soc.,* 93, 7213, 1971.

7. Yang, N. C., Feit, E. D., Hui, M. H., Turro, N. J., and Dalton, J. C., Photochemistry of di-*tert*-butyl ketone and structural effects on the rate and efficiency of intersystem crossing of aliphatic ketones, *J. Am. Chem. Soc.*, 92, 6974, 1970.

8. Dauben, W. G., Salem, L., and Turro, N. J., A classification of photochemical reactions, *Acc. Chem. Res.*, 8, 41, 1975.

9. Salem, L., Surface crossings and surface touchings in photochemistry, *J. Am. Chem. Soc.*, 96, 3486, 1974.

10. Lewis, F. D. and Magyar, J. G., Cage effects in the photochemistry of *(S)*-(+)-2-phenylpropiophenone, *J. Am. Chem. Soc.*, 95, 5973, 1973.

11. Turro, N. J. and Weed, G. C., Micellar systems as "supercages" for reactions of geminate radical pairs. Magnetic effects, *J. Am. Chem. Soc.*, 105, 1861, 1983.

12. Turro, N. J. and Cherry, W. R., Photoreactions in detergent solutions. Enhancement of regioselectivity resulting from the reduced dimensionality of substrates sequestered in a micelle, *J. Am. Chem. Soc.*, 100, 7431, 1978.

13. Gould, I. R., Zimmt, M. B., Turro, N. J., Baretz, B. H., and Lehr, G. F., Dynamics of radical pair reactions in micelles, *J. Am. Chem. Soc.*, 107, 4607, 1985.

14. Maruthamuthu, P. and Scaiano, J. C., Biradical doublet trapping by nitric oxide. An electron spin resonance study, *J. Phys. Chem.*, 82, 1588, 1978.

15. Closs, G. L. and Doubleday, C. E., Determination of the average singlet-triplet splitting in biradicals by measurement of the magnetic field dependence of CIDNP, *J. Am. Chem. Soc.*, 95, 2735, 1973.

16. Closs, G. L. and Doubleday, C. E., Chemically induced nuclear spin polarization derived from biradicals generated by photochemical cleavage of cyclic ketones, and the observation of a solvent effect on signal intensities, *J. Am. Chem. Soc.*, 94, 9248, 1972.

17. Closs, G. L. and Miller, R. J., Laser flash photolysis with NMR detection. Submicrosecond time-resolved CIDNP. Kinetics of triplet states and biradicals, *J. Am. Chem. Soc.*, 103, 3586, 1981.

18. Johnston, L. J. and Scaiano, J. C., Time resolved studies of biradical reactions in solution, *Chem. Rev.*, 89, 521, 1989.

19. Weir, D. and Scaiano, J. C., Lifetimes of the biradicals produced in the Norrish type I reaction of cycloalkanones, *Chem. Phys. Lett.*, 118, 526, 1985.

20. Weiss, D. S., The Norrish type I reaction in cycloalkanone photochemistry, in *Organic Photochemistry*, Padwa, A., Ed., Marcel Dekker, New York, 1981.

21. Givens, R. S., Photoextrusion of small molecules, in *Organic Photochemistry*, Padwa, A., Ed., Marcel Dekker, New York, 1981.

22. Kurabayashi, K. and Mukai, T., Organic photochemistry. XXI. Photochemical and thermal behavior of bicyclo[4.2.1]nona-2,4,7-trien-9-one, *Tetrahedron Lett.*, 1049, 1972.

23. Badcock, C. C., Rickborn, B., and Pritchard, G. O., Photolysis of some cyclopentanones, *Chem. Ind.*, 1053, 1970.

24. Coyle, J. D., Product ratios in the ring-opening photoreactions of cycloalkanones, *J. Chem. Soc. B*, 1736, 1971.

25. Hammond, W. B. and Yeung, T. S., Photochemistry of 5-alkyl-2-methylcyclohexanones, *Tetrahedron Lett.*, 1169, 1975.

26. Morton, D. R., Turro, N. J., Solution phase photochemistry of cyclobutanones, in *Advances in Photochemistry*, Pitts, J. N., Hammond, G. S., and Gollnick, K., Eds., John Wiley & Sons, New York, 1974.

27. Turro, N. J., Farneth, W. E., and Devaquet, A., Salem diagrams as a device for the elucidation of photochemical reaction mechanisms. Application to the cleavage of cyclic alkanones, *J. Am. Chem. Soc.*, 98, 7425, 1976.

Norrish Type I Processes of Ketones: Selected Examples and Synthetic Applications

Cornelia Bohne
University of Victoria

35.1 Introduction

In Chapter 34 the general mechanistic aspects of Norrish type I reactions of aliphatic and cyclic ketones were described. In this chapter, selected examples are illustrated to highlight the importance of this reaction in the generation of transient intermediates, the effect of additional functionalities on ketone reactivity, and synthetic applications.

35.2 Generation of Transient Intermediates

Norrish type I reactions have been used extensively in the field of reactive intermediates to generate transient species for mechanistic studies. One of the most prominent applications has been the generation of biradicals in the photolysis of cyclic ketones.[1] Aspects related to the reactivity of acyl alkyl radicals have been discussed briefly in Chapter 34. α-Cleavage followed by rapid decarbonylation of the acyl alkyl biradical has been employed to generate α,ω-1,n diyls, and the effect of the chain length on their lifetimes has been studied.[1–3] As well as biradicalar species, α-cleavage and subsequent decarbonylation of a suitable precursor molecule has been employed to generate a variety of intermediates, such as xylylenes (Scheme 35.1) or cyclobutadienes.[4,5] In the case of 1,1,3,3-tetramethyl-2-indanone, the 1,4-biradical and the xylylene can be detected by laser flash photolysis.[5]

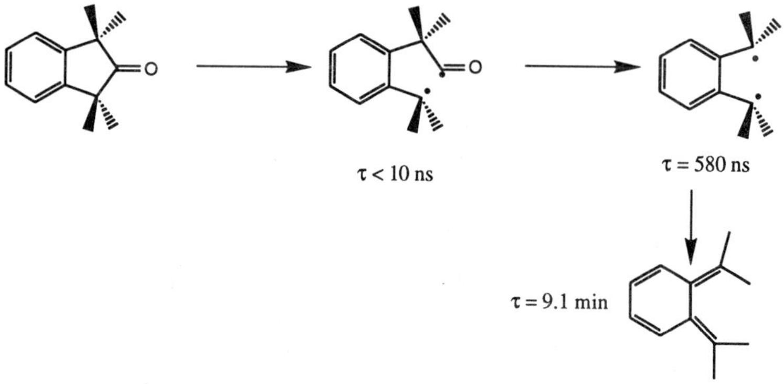

SCHEME 35.1   Photolysis of 1,1,3,3,-tetramethyl-2-indanone.

35.3 Effect of Additional Functionalities

Functionalities in key positions of the ketone can alter the mechanistic pathway for product formation. Rearrangement of biradicals formed after α-cleavage can in some instances compete with the typical disproportionation reactions that lead to enal or ketene formation. An example is the ring expansion shown in Scheme 35.2 that involves two sequential α-cleavage, ring-opening and cyclization reactions to form a ring-expanded ketone. Effective competition occurs due to kinetic factors, i.e., the rate constant for ring opening is fast. Alternatively, disproportionation can be inhibited by steric constraints (vide infra).

The formation of oxacarbenes is one of the most synthetically useful reactions involving Norrish type I processes, as this intermediate can, for example, react with alcohols or olefins (Scheme 35.3). Oxacarbenes are frequently formed in the photolysis of cyclobutanones, but are rare for cyclopentanones and cyclohexanones where other reactions such as ketene or enal formation compete. Introduction of spirocyclopropyl or *exo*-methylene groups in the α-position to the carbonyl leads to an increase of the quantum yield for oxacarbene formation. In these cases, the regiochemistry of the α-cleavage is reversed and bond breaking occurs at the least substituted α-position. This effect was proposed to be due to stabilization of the oxacarbene formed. For example, in the photolysis of α-spirocyclopropyl cyclopentanone in methanol, the acetal formed from the oxacarbene is one of the major products (Scheme 35.4). On the other hand, α-cleavage of cycloalkanones with spirocyclopropyl substitution in the β,γ or remote positions leads to an acyl-cyclopropylcarbinyl biradical. This biradical disproportionates or rearranges to a homoallylic radical.[6]

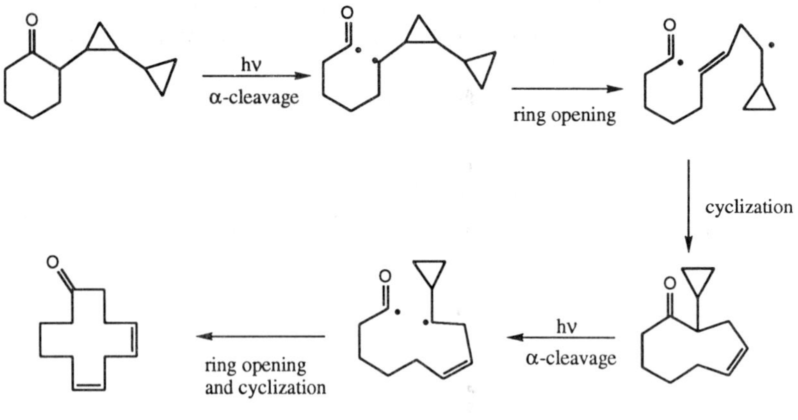

SCHEME 35.2

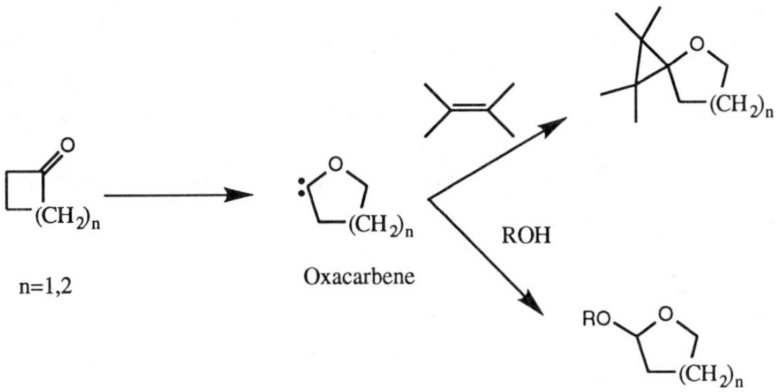

SCHEME 35.3

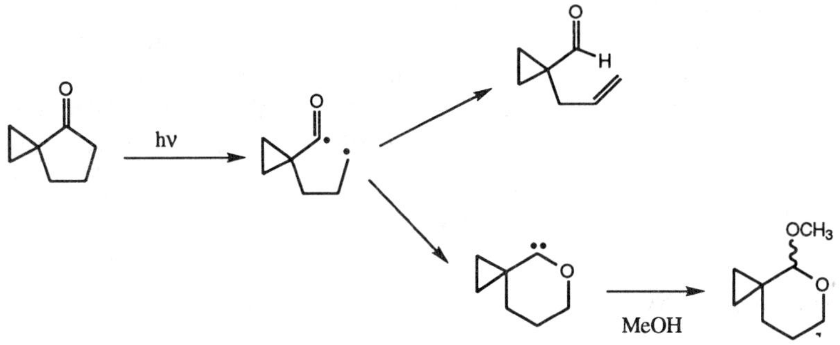

SCHEME 35.4 Photolysis of spirocyclopropyl cyclopentanone.

Ring expansion and oxacarbene formation has been observed in the photolysis of the cyclohexanone shown in Equation 35.1. In ethanol:diethyl ether (9:1, − 70°C), the yield of acetal formation is 72%.[7] Formation of the oxacarbene has been suggested to reflect the ability of the annular oxygen to stabilize the oxacarbene intermediate. However, this reaction is not general. The photolysis of a similar cyclohexanone (Equation 35.2) leads primarily to decarbonylation of the acyl alkyl biradical formed in the Norrish type I process.[6]

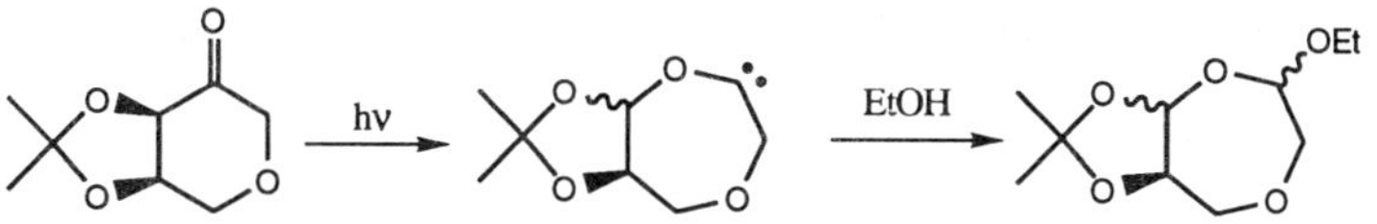

(35.1)

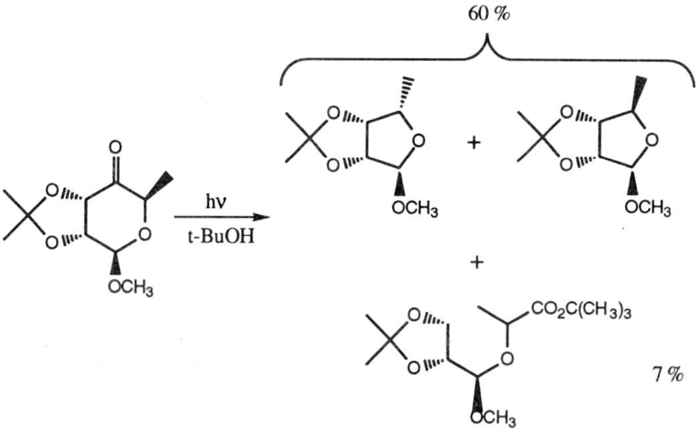

(35.2)

Oxacarbenes are also produced in compounds where the biradical formed in the Norrish type I process cannot attain the conformations necessary for disproportionation reactions. Some bicyclic compounds have this characteristic. In the case of the bicyclic cyclopentanone shown in Equation 35.3, the intermediate oxacarbene is trapped by an intramolecular reaction with the hydroxyl group.[8] This intramolecular trapping reaction is very efficient, and no trapping by solvent was observed when the reaction was carried out in methanol.

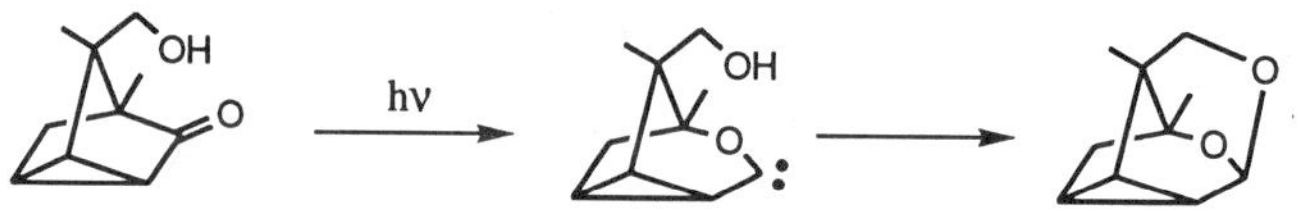

$$(35.3)$$

35.4 Synthetic Applications

Norrish type I reactions have not been employed extensively for synthetic purposes as, in most cases, competing radical reactions are involved and the reactions have low stereospecificity. The sensitivity of product distribution to subtle structural changes in the molecule to be photolyzed is an additional drawback. These characteristics lead to a poor predictability of the products formed for a particular starting compound. Nevertheless, α-cleavage reactions have been employed successfully for synthetic purposes in specific cases. The examples below are grouped with respect to the mechanistic pathway that follows α-cleavage.

Oxacarbene Intermediate

The intermediates of Norrish type I photolysis that have been most useful from the synthetic point of view are oxacarbenes. This reaction has been employed in the synthesis of prostaglandin $F_{2\alpha}$ (Scheme 35.5).[9] The reactions have an overall yield of 49% when performed without the purification of the lactol. Optimization experiments for the photochemical reaction of the cyclobutanone were performed with an epoxy ketone derivative that established the preferential formation of the *exo*-acetal. Further experiments, with a derivative of cyclobutanone I, determined that the lactol could be formed directly in aqueous media. Some cyclopentene, due to β-cleavage of the biradical, was formed as a side product. This reaction established an alternative synthetic route for prostaglandin $F_{2\alpha}$. The same compound can also be obtained in a non-photochemical reaction starting from the Corey lactone.

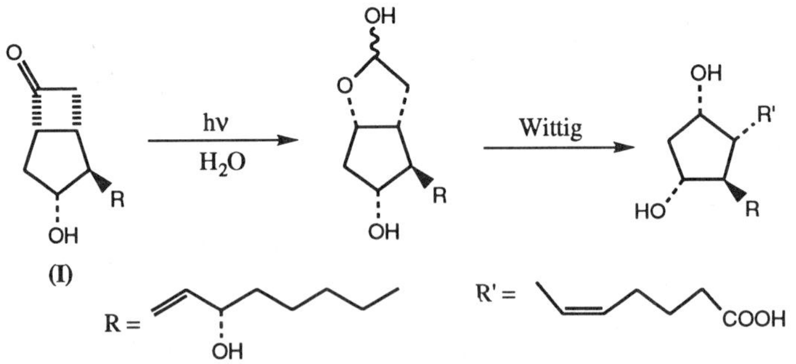

SCHEME 35.5   Synthesis of prostaglandin $F_2\alpha$.

Prostaglandin tromboxane-A_2 and B_2 analogues were synthesized in a reaction where the photochemically generated oxacarbene is intramolecularly trapped by a hydroxyl group. This reaction involves the α-cleavage of a cyclopentanone. To achieve a high yield of oxacarbene formation, compounds with spirocyclopropyl substitution at the α-position were employed.[9]

Disproportionation Reactions

Disproportionation reactions following α-cleavage have found some synthetic application. α-Cleavage of succinimides, followed by disproportionation, generates unsaturated amides that can undergo further photochemical transformation to yield azetidinediones (Equation 35.4).[10,11]

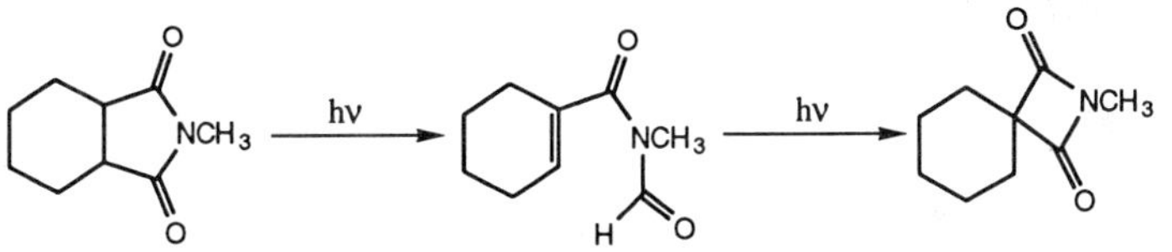

$$(35.4)$$

Prostaglandin C_2 can be synthesized from the hydroxyaldehyde II. This aldehyde was obtained in the intramolecular disproportionation reaction of the biradical formed in the Norrish type I reaction of a bicyclic cyclopentanone (Scheme 35.6).[12]

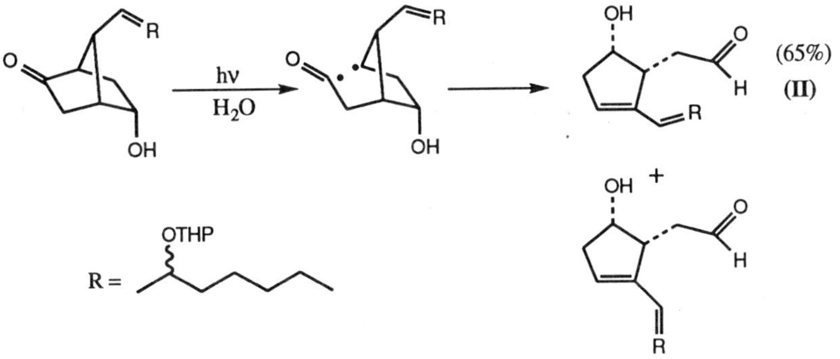

SCHEME 35.6   Synthesis of hydroxyaldehyde 2.

Intramolecular Trapping of Ketenes

The intramolecular trapping of photochemically generated ketenes has been exploited as a synthetic route for naturally occurring lactones and lactams. These reactions involve the selective trapping of the ketene by diol or amine groups (Equation 35.5) and are highly stereoselective.[13]

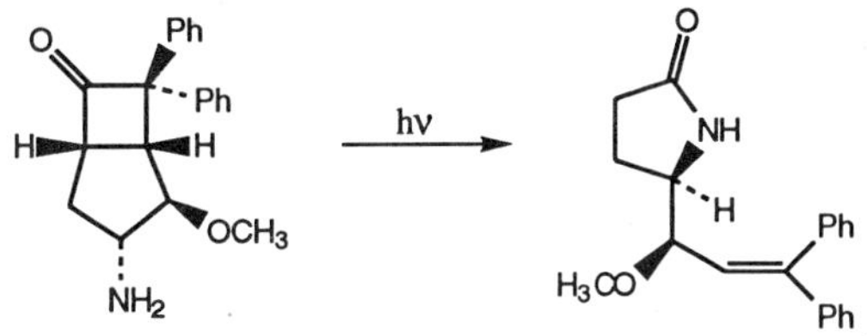

$$(35.5)$$

Decarbonylation

Decarbonylation of the acyl alkyl radical formed in α-cleavage processes can be achieved when the biradical is prevented from undergoing disproportionation reactions. This strategy has been applied in the synthesis of several polyspiro compounds (Scheme 35.7).[14]

35.5 Conclusion

Examples for the application of Norrish type I reactions for mechanistic studies, where this reaction provides one of the pathways for transient intermediate generation and for synthetic purposes, have

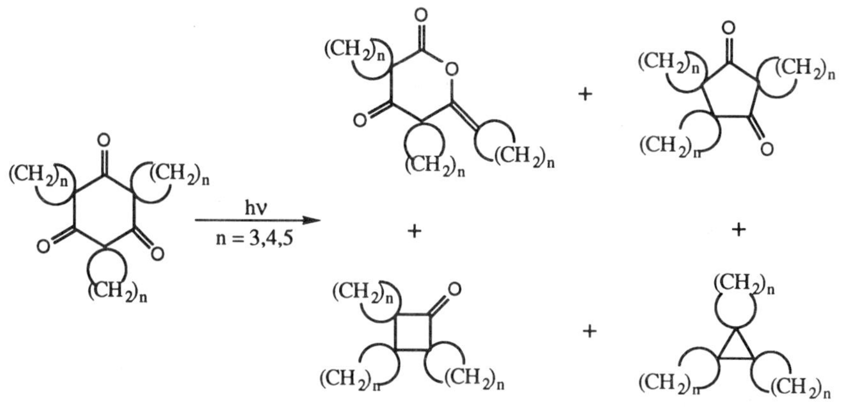

SCHEME 35.7

been presented. In addition, the effect of structural changes on the product distribution following α-cleavage has been addressed. The material described does not represent a complete literature survey, but does provide the reader with an overview of the potentialities of the Norrish type I reaction.

References

1. Johnston, L. J. and Scaiano, J. C., Time resolved studies of biradical reactions in solution, *Chem. Rev.*, 89, 521, 1989.
2. Zimmt, M. B., Doubleday, C., Jr., and Turro, N. J., Substituent and solvent effect on the lifetimes of hydrocarbon-based biradicals, *Chem. Phys. Lett.*, 134, 549, 1987.
3. Zimmt, M. B., Doubleday, C., Jr., Gould, I., and Turro, N. J., Nanosecond flash photolysis studies of intersystem crossing rate constants in biradicals: Structural effects brought about by spin-orbit coupling, *J. Am. Chem. Soc.*, 107, 6724, 1985.
4. Givens, R. S., Photoextrusion of small molecules, in *Organic Photochemistry*, Marcel Dekker, New York, 1981.
5. Scaiano, J. C., Wintgens, V., and Netto-Ferreira, J. C., Mechanistic studies of the photogeneration and photochemistry of *ortho*-xylylenes, *Pure Appl. Chem.*, 62, 1557, 1990.
6. Weiss, D. S., The Norrish type I reaction in cycloalkanone photochemistry, in *Organic Photochemistry*, Marcel Dekker, New York, 1981.
7. Collins, P. M., Oparaeche, N. N., and Whitton, B. R., Photochemical ring expansion of five- and six-membered ring ketones derived from pentoses, *J. Chem. Soc., Chem. Commun.*, 292, 1974.
8. Switlak, K., He, D., and Yates, P., Intramolecular trapping of an oxacarbene intermediate in the photochemical ring expansion of a cyclopentanone, *J. Chem. Soc., Perkin Trans. I*, 2579, 1992.
9. Newton, R. F., Carbonyl compounds: α-Cleavage, in *Photochemistry in Organic Synthesis*, The Royal Society of Chemistry, Special Publications, 1986.
10. Maruyama, K., Ishitoku, T., and Kubo, Y., Azetidine-2,4-diones via photoinduced ring contraction of succinimides, *J. Am. Chem. Soc.*, 101, 3670, 1979.
11. Maruyama, K., Ishitoku, T., and Kubo, Y., Photochemistry of aliphatic imides. Synthesis of azetidine-2,4-diones via photochemical isomerization of succinimides and *N*-formyl-*N*-methyl-α,β-unsaturated amides, *J. Org. Chem.*, 46, 27, 1981.
12. Crossland, N. M., Roberts, S. M., and Newton, R. F. Photosynthetic routes to prostaglandins. Synthesis of prostaglandin C_2 and analogues, *J. Chem. Soc., Perkin Trans. I*, 2397, 1979.

13. Rahman, S. S., Wakefield, B. J., Roberts, S. M., and Dowle, M. D., Intramolecular nucleophilic addition to photochemically generated ketenes as a versatile route to lactones and lactams; Synthesis of a mosquito pheromone, goniothalamin, argebtilactone, and the Streptomyces L-factor, *J. Chem. Soc., Chem. Commun.*, 303, 1989.

14. Krapcho, A. P. and Waller, F. J., Syntheses of trispirocyclopropanes via triplet photodecarbonylations of polymethyleneketene trimers, *J. Org. Chem.*, 37, 1079, 1972.

36

Photoinduced Intermolecular Hydrogen Abstraction Reactions of Ketones

Mordecai B. Rubin
Technion-Israel Institute of Technology

Fill a flask with 10 grams of benzophenone and 100 ml of isopropyl alcohol, add one drop of acetic acid, stopper and let stand near a window. In a week or less (depending on geography and the season of the year), filter off a nearly quantitative yield of lovely white crystals of analytically pure benzpinacol.[1] The other product is acetone. This is one of the simplest reactions a chemist can perform. It was correctly interpreted by Ciamician and Silber[2] in 1900 and formed the subject of the first mechanistic investigation in organic photochemistry carried out by Cohen[3] during the second decade of the 20th century. This reaction and the closely related photolysis of benzophenone (BP) with benzhydrol were the first excited-state reactions shown clearly to involve the triplet state as the reactive electronically excited species in one of the earliest flash photolysis investigations[4] and by other studies.[5] Subsequently, almost every possible technique has been applied to these reactions of benzophenone. A number of reviews of ketone photochemistry have appeared,[6,7] including a recent comprehensive review of hydrogen atom abstraction reactions.[8]

The early studies mentioned above, reinforced by many later results, established that the benzophenone ketyl radical (BPH) is formed from the triplet state of BP. The following simple mechanism was proposed:

$$BP + h\nu \rightarrow BP*^1 \xrightarrow{\text{isc}} BP*^3 \tag{36.1}$$

$$BP*^3 \rightarrow BP \tag{36.2}$$

$$BP*^3 + R_1R_2CHOH \rightarrow BPH\cdot + R_1R_2\dot{C}OH \tag{36.3}$$

$$BP + R_1R_2\dot{C}OH \rightarrow BPH\cdot + R_1R_2CO \tag{36.4}$$

$$2\,BPH\cdot \rightarrow (Ph)_2C(OH)C(OH)Ph_2 \tag{36.5}$$

Initial excitation of ketone to the singlet state is followed by fast intersystem crossing (isc) and relaxation to the lowest triplet state of the aromatic ketone. Aliphatic ketones undergo slower isc

0-8493-8634-9/95/$0.00+$.50
© 1995 by CRC Press, Inc.

Table 36.1   Rate Constants for Quenching of Ketones

Ketone	Substrate	$k \ (M^{-1} \ s^{-1})$
2-Pentanone	Cumene	8×10^4
Cyclopentanone	2-Propanol	1.1×10^7
Cyclohexanone	2-Propanol	1.8×10^6
Acetophenone	Cyclopentane	5×10^3
Benzophenone	2-Propanol	1.9×10^6
Benzophenone	Toluene	5×10^5
Benzophenone	Triethylamine	3×10^9

and may react from either singlet or triplet states. Reaction 36.2 summarizes all the physical deactivation processes available to triplet BP; high quantum yields indicate in many cases that these often do not compete with chemical reaction of the n,π^* triplet state.

Comparison of reactivities in hydrogen atom abstraction by thermally generated alkoxy radicals with photoreactivities of ketones showed considerable similarities,[9] suggesting that there is free radical character in the n,π^* triplet state of ketones. This provides a good analogy for the excited-state reaction in this system, the transfer of a hydrogen atom from the substrate to carbonyl oxygen resulting in formation of the benzophenone ketyl radical (BPH) and the alcohol-derived ketyl radical (Reaction 36.3). One of the striking successes of the early investigations was the good agreement between rate constants determined from flash photolysis and from quenching experiments. Considerable attention has focused on every detail of this hydrogen abstraction process; detailed discussion is beyond the scope of this short review. A few representative examples of rate constants are summarized in Table 36.1

Rate constants for ketones having π,π^* lowest triplet states (e.g., *p*-phenylbenzophenone and 2-acetylnaphthalene) or charge transfer states (e.g., Michler's ketone, or *p,p'*-*bis*-dimethylaminobenzophenone) are much smaller, resulting in low to negligible quantum yields for such compounds. Solvent polarity can affect the relative energies of these triplet states (n,π^* states favored in nonpolar solvents) and thereby have a pronounced influence on rate constants (and quantum yields) for hydrogen abstraction. To cite a particularly dramatic example, the rate constant for hydrogen abstraction by xanthone from isopropyl alcohol varies by a factor of about 500 as a function of solvent polarity.[10]

Reaction 36.4, formation of a second BPH radical by transfer of a hydrogen atom from hydroxypropyl radical to ground-state benzophenone, should really be written as an equilibrium. However, under ideal conditions, the quantum yield for disappearance of BP in isopropyl alcohol approaches a value of two so that this equilibrium must lie far to the right under these conditions. Subsequent combination of two BPH radicals (Reaction 36.5) affords the observed benzpinacol in a very rapid coupling reaction.

The simplicity of the mechanism presented above turned out to be the result of a fortuitous combination of high BP concentration and low light intensity used in early experiments. Subsequent investigation revealed a variety of additional free radical reactions that follow the initial hydrogen atom abstraction. The first of these became apparent when spectroscopic monitoring (UV/VIS) of the reaction was attempted. In theory, recording the progress of reaction by monitoring absorption spectra of a solution as a function of irradiation time should be a simple matter since BP absorbs at longer wavelengths than other reactants or products. Instead of the anticipated decrease in absorption with progress of reaction (degassed solution), an intense new absorption in the region of 300 nm was observed. This was attributed to the formation of one or more light-absorbing transients (LAT). Interruption of irradiation, admission of oxygen, and standing for several hours resulted in disappearance of the new absorption. The resulting spectrum could then be used for calculation of quantum yields. Eventually, it was established[11] that the LAT formation is a minor reaction path involving coupling of hydroxypropyl radicals at the *para*-position[12] of

BPH. The resulting enolic form of *para*-substituted dihydrobenzophenone undergoes air oxidation to substituted benzophenones, as shown by isolation of p-hydroxymethylbenzophenone (1) from the photoreaction of BP with methanol. A nuclear substitution product was also observed in a reaction of acetophenone. Similar spectroscopic behavior involving LAT-type intermediates has been observed in the reaction of BP with toluene.[13]

Further investigation established that the additional reactions (36.6–36.8) must be added to the five above to present a complete picture. Thus, Reaction 36.4 above, formation of a second BPH by reaction of alcohol-derived ketyl radical with ground-state BP, was originally suggested in order to account for the fact that the quantum yield for BP disappearance in isopropyl alcohol has a limiting value of 2 under ideal conditions (high BP concentration, low light intensity, degassed solution). This also explained the supposed absence of mixed pinacol 2 under these conditions. Subsequently, however, compounds of type 2 were detected in reactions of BP in isopropyl alcohol (up to 10%), ethyl alcohol (up to 32%) and methyl alcohol (up to 60%). Reaction 36.4 must be formulated as an equilibrium with steady-state concentrations of the various species dependent on structure and reaction conditions. The position of this equilibrium and the reactivities of the various ketyl radicals determine the overall quantum yield and the product composition. Factors which increase radical concentrations, such as high light intensity,[14] will lead to increasing importance of radical-radical reactions and vice versa. It should be noted that coupling products (1 and 2) other than benzpinacol can constitute as much as 70% of the total in the reaction of BP with methanol.

$$BPH \cdot + R_1R_2\dot{C}OH \rightarrow LAT \tag{36.6}$$

$$BPH \cdot + R_1R_2\dot{C}OH \rightarrow 2 \tag{36.7}$$

$$BPH \cdot + R_1R_2\dot{C}OH \rightarrow BP + R_1R_2CHOH \tag{36.8}$$

$$BPH \cdot + R_1R_2\dot{C}OH \rightarrow BPH_2 + R_1R_2CO \tag{36.9}$$

$$2\,BPH \cdot \rightarrow BP + Ph_2CHOH \tag{36.10}$$

$$2\,R_1R_2\dot{C}OH \rightarrow R_1R_2C(OH)C(OH)R_1R_2 \tag{36.11}$$

Reaction 36.8 involves reversal of the initial hydrogen atom abstraction step and has been experimentally established in a few cases. For example, appreciable racemization of optically active methyl 2-octyl ether occurred upon irradiation with benzophenone; about half of the originally formed ether radicals reverted to starting material in the course of the irradiation. This type of reaction is undoubtedly responsible for lowered quantum yields in many cases.

Reactions 36.9 and 36.10 involve disproportionation of ketyl radicals. They have not been observed in reactions of BP, but do occur with some aliphatic ketones (*vide infra*). These reactions reduce the quantum yield for disappearance of BP from the ideal value of 2. Reduction to alcohols is observed at high pH values. Formation of pinacol from the radical R_1R_2CHOH (Reaction 36.11) is a final possibility that has also not been observed with aromatic ketones.

The radicals BPH and R are formed initially as a triplet pair in a solvent cage and could, *a priori*, undergo isc to give a singlet pair that would be expected to undergo very fast coupling to unsymmetrical pinacols 2. The observation that 2 is indeed formed was originally interpreted as evidence that such in-cage reactions are important. However, the fact that the fraction of **2** present varies in a rational way with light intensity is inconsistent with its formation by a cage process. It appears that the events involved subsequent to hydrogen abstraction reaction generally involve separated free radicals.

$$BP^{*3} + RH \longrightarrow \overline{BPH\cdot + R\cdot} \quad \overset{3}{\underset{1}{\longrightarrow}} \quad \begin{array}{c} \textbf{BPH}\cdot + \textbf{R}\cdot \\[8pt] \overline{\textbf{BPH}\cdot + \textbf{R}\cdot} \end{array}$$

Mixtures of benzophenones show interesting behavior upon irradiation in isopropyl alcohol. For example, irradiation of a mixture of *p,p'*-dimethoxybenzophenone[3,14,15] (**3**) and BP results in exclusive reaction of **3** even when conditions are adjusted so that an equal number of photons are absorbed by each of the two ketones. This has been suggested[15] to be the result of energy transfer from BP*3 to **3** (Reaction 36.12, An = *p*-anisyl); it seems more probable[16] that it is the consequence of equilibration between aromatic ketones and their ketyl radicals (Reaction 36.13). Even more convincing evidence for equilibration between ketones and ketyl radicals comes from studies using labeled ketone or labeled alcohol.

$$BP^{*3} + An_2CO \rightleftarrows BP + An_2CO^{*3} \tag{36.12}$$

$$BPH\cdot + An_2CO \rightleftarrows BP + An_2\dot{C}OH \tag{36.13}$$

The preceding discussion has dealt in some detail with the photolysis of benzophenone in the presence of alcohols because of the historical importance of this reaction; the generalities that emerge are applicable to many hydrogen abstraction reactions. In fact, ketones with lowest n,π* triplet states abstract hydrogen from a considerable variety of C-H bonds including saturated hydrocarbons, toluene and substituted toluenes, ethers and thioethers, aldehydes, esters, amines, amides, phenols, etc. The free radical nature of intermediates in these reactions has made them of considerable interest from a practical point of view, particularly in polymer chemistry. Hydrogen abstraction from benzene by benzophenone has also been suggested.[17] An excellent hydrogen donor, even for reluctant ketones is tributyl stannane, Bu₃SnH, and rate constants for this hydrogen atom donor approach the diffusion-controlled limit in many cases.

The reaction of BP with toluene (Reaction 36.14) is typical. Hydrogen abstraction produces BPH and benzyl radicals that couple in all simple ways,[18] producing a mixture of benzpinacol, bibenzyl, and diphenyl benzyl carbinol with quantum yields of 0.2 to 0.5 for disappearance of BP. The mechanism of this reaction has been suggested to involve orientation-dependent complexation. The value of less than unity for the quantum yield has been attributed to this complex formation, but could also be the result of reversal of the initial hydrogen atom transfer step. This reaction was one of the first investigated by the CIDNP technique;[19] separated radicals and not a cage radical pair are involved as has been shown by chemical studies.[20]

$$BP + ArCH_3 \rightarrow Ph_2C(OH)C(OH)Ph_2 + ArCH_2CH_2Ar + Ph_2C(OH)CH_2Ar \tag{36.14}$$

The reaction of ketones with amines[8,21] can be fundamentally different from reactions described above and presents an alternative mechanism for hydrogen atom transfer. Photophysical studies demonstrated that: (1) the rate of quenching of triplet benzophenone by amines is several orders of magnitude larger than for other substrates and approaches the diffusion-controlled limit; (2) ketones whose triplet states are long lived and unreactive with hydrogen donors, such as isopropyl alcohol, are also quenched efficiently by amines; (3) the rate of triplet quenching can be correlated with ionization potential of the amine; and (4) the products of such quenching include ground-state reactants and radical ions. Extensive studies have led to acceptance of a mechanistic picture which spans the extremes of initial single electron transfer (SET) between amine and triplet ketone to conventional hydrogen atom transfer. The SET mechanism (Reaction 36.15) results in formation of benzophenone ketyl radical anion and amine radical cation as illustrated. Rapid back transfer of an electron regenerates starting materials and lowers the quantum yields. This energy-wasting process competes with transfer of a proton from cation radical to ketyl anion radical, resulting in overall hydrogen atom transfer, as illustrated.

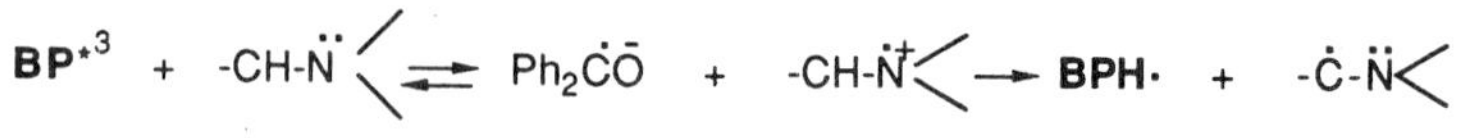

$$\textbf{BP}^{*3} \;+\; \text{-CH-}\ddot{\text{N}}\big< \;\rightleftharpoons\; \text{Ph}_2\dot{\text{C}}\bar{\text{O}} \;+\; \text{-CH-}\overset{+}{\text{N}}\big< \;\longrightarrow\; \textbf{BPH}\cdot \;+\; \text{-}\dot{\text{C}}\text{-}\ddot{\text{N}}\big< \qquad (36.15)$$

Turning to compounds other than aromatic ketones, the aldehydes present an interesting case. They contain both a potentially reactive carbonyl group and the abstractable aldehydic hydrogen so that no additional hydrogen atom donor need be present. An example is photolysis of benzaldehyde in an inert solvent such as benzene where (Reaction 36.16) the various coupling products of the initially formed benzoyl and phenyl hydroxymethyl radicals are produced, namely hydrobenzoin, benzoin, and benzil. The experimental result is complicated by the fact that the latter two absorb in the same region of the spectrum as the reacting aldehyde and are themselves capable of hydrogen atom abstraction. Reaction in restricting media such as cyclodextrin afforded benzoin and a *para*-coupling product.[22]

$$\text{PhCHO} \rightarrow \text{Ph}\dot{\text{C}}\text{HOH} + \text{Ph}\dot{\text{C}}\text{O} \rightarrow (\text{PhCHOH})_2 + (\text{PhCO})_2 + \text{PhCH(OH)COPh} \quad (36.16)$$

Simple ketones such as acetone and acetophenone behave very much like benzophenone. For example, photolysis of acetone in toluene solution results in formation of pinacol, benzyl dimethyl carbinol, and bibenzyl in good analogy with the reaction of BP. However, differences in the chemistry of free radicals formed initially are often encountered. Even with such a simple reaction as irradiation of acetone in isopropyl alcohol, CIDNP studies[23] have established that acetone ketyl radicals disproportionate to give isopropyl alcohol and the enol of acetone in addition to the expected pinacol (Reaction 36.17). This type of disproportionation is normally masked by ketonization of the enol to regenerate starting material.

$$(36.17)$$

Very little structural complexity is required to bring one into the fascinating and useful realm of intramolecular photochemical reactions of ketones (discussed in detail in another section of this volume). The two principal reactions are α-cleavage (Norrish type I, Reaction 36.18) and γ-hydrogen abstraction (Norrish type II, Reaction 36.19a, and Yang cyclization, Reaction 36.19b). A very brief discussion on the competition between such reactions and intermolecular abstraction follows.

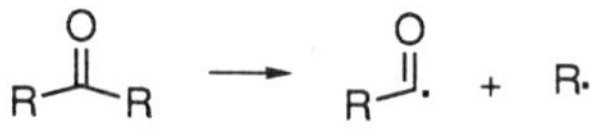

$$(36.18)$$

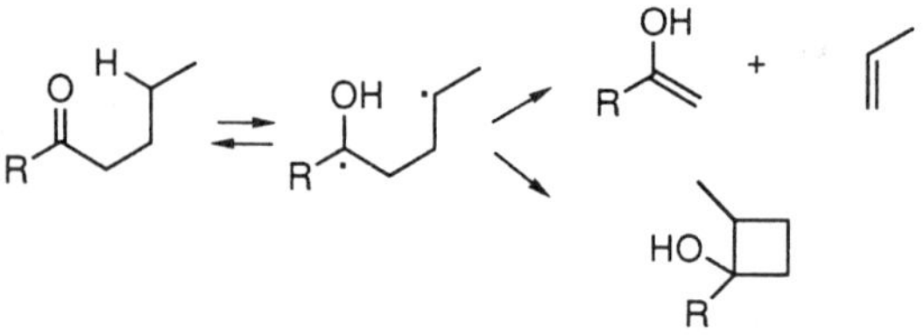

$$(36.19a)$$

$$(36.19b)$$

α-Cleavage is not observed with diaryl ketones. Its importance in reactions of other types of ketones can be related to the stabilization of the radicals formed. For example, irradiation of acetophenone in isopropyl alcohol produces the expected pinacol; products derived from benzoyl and methyl radicals are not observed. Benzoin ethers, on the other hand, undergo efficient α-cleavage (Reaction 36.20) to form benzoyl and stabilized phenyl alkoxymethyl radicals; these ethers are used commercially as initiators in photopolymerization. Another illustrative comparison is the reaction of 3-ketosteroids in isopropyl alcohol to produce pinacols (Reaction 36.21) as compared to that of 4,4-dimethyl-3-ketones. The latter undergo α-cleavage followed by intramolecular hydrogen atom transfer to form ketenes that are trapped as esters by the alcohol solvent (Reaction 36.22).

$$PhCOCH_3 \; \xrightarrow{\;X\;} \; Ph\dot{C}O + \dot{C}H_3 \qquad (20) \quad PhCO\overset{OR}{\underset{|}{C}}HPh \longrightarrow Ph\dot{C}O + Ph\dot{C}HOR \qquad (36.20)$$

(36.21)

(36.22)

Intramolecular hydrogen atom abstraction (Reaction 36.19) is almost invariably observed when a hydrogen atom is located, with proper orientation, sufficiently close to an electronically excited carbonyl group, particularly when a six-membered transition state (γ-hydrogen atom) is possible for the intramolecular reaction. As illustrated in Reaction 36.19, the resulting 1,4-biradical may undergo cleavage, bond formation to a cyclobutanol, or reversal to starting material. Classical examples are reactions of *o*-hydroxy- and *o*-alkylbenzophenones, butyrophenone and longer chain aromatic-aliphatic ketones, as well as alkanones of sufficient chain length. The thrust of investigations with compounds of this type has been to avoid intermolecular reaction by the use of solvents that are poor hydrogen atom donors. Hydrogen atom abstraction may be a minor process when irradiation is conducted in a good hydrogen-donating solvent.

Unsaturated ketones have an even more abundant array of reaction pathways than saturated ones. These include dimerization and other types of cycloaddition reactions (both inter- and intramolecular), rearrangements such as the di-π-methane and oxa-di-π-methane rearrangements, etc. Reduction products (pinacols, saturated ketones) originating in hydrogen atom abstraction are sometimes observed, for example, from irradiation of testosterone acetate in isopropyl alcohol. Again, the use of inert solvents suppresses the abstraction reaction. In some α,β-unsaturated cyclopentenones, hydrogen abstraction to the β-carbon atom may be observed.

Another complication in ketone photoreactions is electronic energy transfer. Ketones generally have high triplet energies and are often used as sensitizers for triplet state reactions. Diffusion-controlled transfer of triplet energy (Reaction 36.23) occurs whenever the triplet energy of the acceptor (A) is about 3 or more kcal mol^{-1} lower than that of the sensitizer; the resulting chemistry will be that of the triplet state of the acceptor. The rate of this bimolecular reaction will depend on the concentration of acceptor present. If a hydrogen-donating solvent is used in such an experiment, intermolecular hydrogen abstraction may compete with energy transfer at low acceptor concentrations. Evidence for such a process can sometimes be deduced from quantum yield behavior. The initial reaction products will be a ketyl radical such as BPH plus the solvent-derived radical (Reaction 36.4). These radicals may transfer hydrogen atoms to substrate so as to regenerate starting BP with concomitant formation of a radical derived from the acceptor reaction (Reaction

36.24), thus making it seem as if triplet energy transfer to the acceptor followed by hydrogen atom abstraction by the acceptor triplet had occurred. This process is sometimes referred to as "chemical sensitization".[24] To choose one example, the direct irradiation of *cis*-1,2-dibenzoyl ethylene (*cis*-DBE) results in rearrangement to a ketene that is trapped by the solvent alcohol as the corresponding ester. When the reaction was (supposedly) sensitized by BP in isopropyl alcohol solution, reduction to dibenzoylethane was reported. Subsequent examination[25] showed that the triplet state of *cis*-DBE is unreactive and that the reduction process involves hydrogen atom transfer from ketyl radicals to *cis*-DBE. Another example is presented in the section on hydrogen atom abstraction reactions of α-diketones.

$$BP^{*3} + A \rightarrow BP + A^{*3} \qquad\qquad (36.23)$$

$$BPH \cdot + A \rightarrow BP + AH \cdot \qquad\qquad (36.24)$$

References

1. Fieser, L. F., *Experiments in Organic Chemistry*, D. C. Heath and Co., New York, 2nd ed., 1941, 202.
2. Ciamician, G. and Silber, P., *Chem. Ber.*, 33, 2911, 1900.
3. Cohen, W. D., *Rec. Trav. Chim. Pays-Bas*, 39, 243, 1920.
4. Porter, G. and Wilkinson, F., *Trans. Faraday Soc.*, 59, 1686, 1963.
5. (a) Pitts, J. N., Letsinger, R. L., Taylor, R. P., Patterson, J. M., Recktenwald, G., and Martin, R. B., *J. Am. Chem. Soc.*, 80, 1068, 1959, (b) Bäckstrom, H. L. J. and Sandros, K., *Acta Chem. Scand.*, 14, 48, 1960; (c) Moore, W. M., Hammond, G. S., and Foss, R. P., *J. Am. Chem. Soc.*, 83, 2789, 1961.
6. Schönberg, A. and Mustafa, A., *Chem. Rev.*, 46, 181, 1947.
7. Photochemistry of cyclic ketones has been reviewed: Chapman, O. L. and Weiss, D. S., in *Organic Photochemistry*, Vol. 3, Chapman, O. L., Ed., Marcel Dekker, N.Y., 1973, chap. 3.
8. For a recent comprehensive review on hydrogen atom abstraction see: Wagner, P. and Park, B.-S., in *Organic Photochemistry*, Vol. 11, A. Padwa, Ed., Marcel Dekker, N.Y., 1991, chap. 4.
9. Walling, C. and Gibian, M. J., *J. Am. Chem. Soc.*, 87, 3361, 1965.
10. Scaiano, J. C., *J. Am. Chem. Soc.*, 102, 7747, 1980.
11. Chilton, J., Giering, L., and Steel, C., *J. Am. Chem. Soc.*, 98, 1865, 1976.
12. Coupling at the *ortho*-position of BPH is also possible but has not been demonstrated experimentally.
13. Rubin, M. B., unpublished results.
14. Rubin, M. B., *Tetrahedron Lett.*, 23, 4615, 1982.
15. Johnson, H. W., Pitts, J. N., Jr., and Burleigh, M., *Chem. Ind.*, 1493, 1964.
16. The triplet energy of BP is reported to be very slightly lower than that of 3; Leigh, W. J. and Arnold, D. R., *J. Chem. Soc., Chem. Commun.*, 406, 1980.
17. Buettner, A. V. and Dedinas, J., *J. Phys. Chem.*, 75, 187, 1971.
18. Coupling at *ortho*-positions is possible but has not been established.
19. (a) Closs, G. L. and Closs, L. E., *J. Am. Chem. Soc.*, 91, 4550, 1969; (b) Closs, G. L., Doubleday, C. E., and Paulson, D. R., *J. Am. Chem. Soc.*, 92, 2185, 1970.
20. Rubin, M. B., *Tetrahedron Lett.*, 3931, 1969.
21. Cohen, S. G., Parola, A., and Parsons, G. H, Jr., *Chem. Rev.*, 73, 141, 1973.
22. Rao, V. P. and Turro, N. J., *Tetrahedron Lett.*, 30, 4641, 1989.
23. Seifert, K.-G., *Chem. Ber.*, 107, 241, 1974.
24. For a discussion of complications in sensitized reactions see: Engel, P. S. and Monroe, B. M., *Advances in Photochemistry*, Vol. 8, Pitts, J. N., Jr., Hammond, G. S., and Noyes, W. A., Jr., Eds., Wiley-Interscience, New York, 1971, 245.
25. (a) Zimmerman, H. E. and Hull, V. J., *J. Am. Chem. Soc.*, 92, 6515, 1970; (b) Rubin, M. B. and Ben-Bassat, J. M., *Mol. Photochem.*, 3, 155, 1971.

37

Hydrogen Abstraction Reactions of α-Diketones

Mordecai B. Rubin
Technion-Israel Institute of
Technology

37.1 Introduction

α-Diketones (and the closely related *o*-quinones) are attractive substrates for photochemical investigation, offering many virtues to the photochemist. They are colored compounds with longest wavelength (n,π*) absorption maxima in the range 350 to 550 nm *(vide infra)*, making them amenable to irradiation with visible light sources. In almost all cases, the products of their photochemical reactions absorb at much shorter wavelengths than the starting diketones so that secondary photolysis can be avoided by the simple expedient of using Pyrex vessels or a cut-off filter to prevent light absorption by products. They undergo efficient intersystem crossing and their reactions almost invariably proceed from the lowest triplet state (assumed to have an *s-trans*-conformation whenever possible). They have weak fluorescence, together with strong phosphorescence, a combination than can provide considerable information in mechanistic investigations. Triplet energies of α-diketones lie in the relatively low range 50 to 55 kcal mol^{-1} so that their reactions can be sensitized by a variety of triplet sensitizers. They can also be quenched (usually for mechanistic purposes) by low-energy triplet quenchers, e.g., anthracene (E$_T$, 42 kcal mol^{-1}). A final virtue of these compounds is the fact that good methods for their synthesis are available. Several reviews of α-diketone photochemistry have appeared.[1-3]

The combination of these factors has made α-diketones the subject of many photochemical investigations going back to early photophysical studies of biacetyl. Even earlier, one of the first photochemical reactions reported in the chemical literature was that of 9,10-phenanthrenequinone (PQ) with diethyl ether, reported in 1886 as the result of leaving an ether solution of the quinone

0-8493-8634-9/95/$0.00+$.50
© 1995 by CRC Press, Inc.

Table 37.1   Long-Wavelength Absorption Maxima
of Selected α-Diketones

Compound	λ_{max}[a] (nm)
Biacetyl	447
Di-*t*-butyl diketone	362
Benzil	370
Mesitil	493
2-Furil	400
2-Pyridil	360
Cyclobutane-1,2-dione	489
Cyclobutenedione	340
[4.4.2]-Propella-11,12-dione	461
[4.4.2]-Propella-3,8-diene-11,12-dione	537
Cyclopentane-1,2-dione[b]	505
Cyclohexane-1,2-dione[b]	380
Cycloheptane-1,2-dione[b]	337
Cyclooctane-1,2-dione[b]	343
Cyclohexadecane-1,2-dione[b]	384
Tetrahydrofuran-3,4-dione[b]	540
Camphorquinone	470
Bicyclo[2.2.2]octane-2,3-dione	478

[a] Extinction coefficients in the range 20 to 50.
[b] α,α,α′,α′-tetramethyl.

near a window in Klinger's laboratory. The well-known fact that the intercarbonyl bond in an α-diketone is much weaker than other bonds (e.g., bond dissociation energies in biacetyl: CH_3-CO, ~80 kcal mol[-1]; CO-CO, 68 kcal mol[-1]) might prompt the assumption that a primary process of these compounds will be α-cleavage to give a pair of acyl radicals. This is true for biacetyl in the gas phase, but is rarely observed in solution photochemistry. Theoretical calculations suggest that the intercarbonyl bond is stronger in the excited state than in the ground state.[4] Even in cases where product structure is suggestive of α-cleavage of diketone, alternatives can be invoked to explain the results.[5]

37.2 Absorption Spectra

Longest-wavelength (n,π*) absorption maxima of representative α-diketones are collected in Table 37.1. While a number of factors are involved in the strong dependence of λ_{max} on structure, a major one is the torsion angle between adjacent carbonyl groups. This was established over 40 years ago and has been confirmed repeatedly since then; maximum values are observed for the two planar conformations with torsion angles of 0 or 180°. Extinction coefficients for these forbidden n,π* transitions are generally in the range 20 to 50; values in alcohol solutions are often lowered by partial conversion to hemiketals. The observation that the wide range of excitation energies is matched by a small range of triplet energies has been explained by postulating similar *s-trans*-conformations (whenever possible) for relaxed triplet states of all α-diketones. Additional maxima are observed around 300 nm.

37.3 Reaction with Oxygen

It should be emphasized that the following discussion deals with reactions of diketones *in the absence of oxygen*. In its presence, diketones are converted apparently to intermediate peroxy radicals, and the observed products are carboxylic acid anhydrides or the corresponding acids. When alkenes are present, the intermediate peroxy radicals react with the double bond, providing a mild procedure for epoxidation of olefins. In the absence of oxygen, the triplet dione undergoes a reaction for which the first step — addition of carbonyl oxygen to double bond — is analogous to that of the Paternò-Büchi reaction of monoketones. The products are ketooxetanes or dioxenes that may undergo further photochemical reactions.

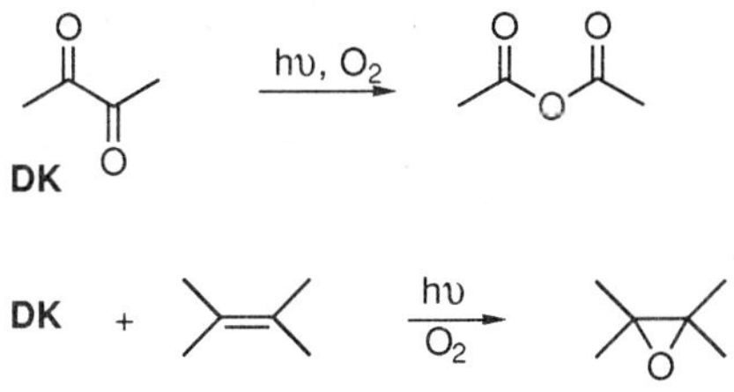

(Str.37.1)

⁊.4 The α-Diketones

The photochemistry of a wide variety of α-diketones has been investigated and includes both *inter-* and *intramolecular* reactions. The intramolecular pathway is preferred when a cyclic, six-membered transition state for hydrogen atom transfer is possible. Compounds investigated include dialkyl, alkyl-aryl, diaryl diketones, and many cyclic compounds such as those shown below. Compounds that exist as enolic tautomers (e.g., cyclohexane-1,2-dione) react as unsaturated ketones rather than as diones.[6] A minor caveat is the tendency of some diketones to form hydrates (and hemiketals); anhydrous conditions provide a simple remedy in such cases.

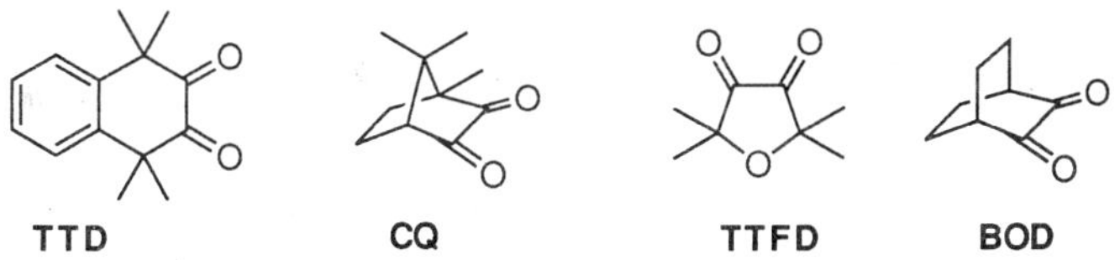

(Str.37.2)

The exceptions are cyclobutane-1,2-diones that *bis*decarbonylate to alkenes, cyclobutene-1,2-diones that may rearrange to *bis*ketenes and/or oxacarbenes, and homoallylically conjugated cyclic diones that undergo allylic rearrangement and *bis*decarbonylation. These are all intramolecular reactions that proceed from the singlet state; a critical factor is availability of a concerted pathway that can compete effectively with intersystem crossing to the triplet. Attempts to establish that the *bis*decarbonylation reactions involve concerted elimination of C_2O_2 have not been successful to date.[7]

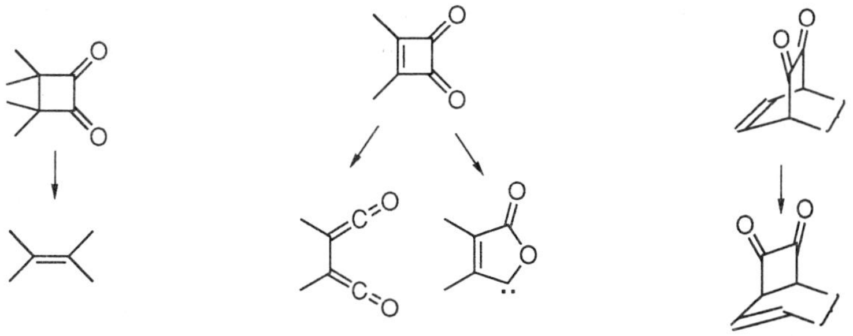

(Str.37.3)

The two major primary processes of triplet states of most α-diketones are hydrogen atom abstraction[8] and addition to multiple bonds in a process analogous to the Paternò-Büchi reaction, as mentioned earlier. Both of these reactions involve excitation of a diketone to its singlet state, followed by fast intersystem crossing to the triplet (reaction 1). This brief review will be limited to hydrogen atom abstraction; the Paternò-Büchi reaction of carbonyl compounds is discussed elsewhere in this volume. In contrast to monoketones (*cf.* benzophenone) with quantum yields as high as 2 in certain reactions, the quantum yields for intermolecular reactions of diones are in the range 0.01 to 0.5. This is partly due to competing deactivation of the excited state (reaction 2) and may also be the result of chemical reactions that regenerate starting materials from reaction intermediates (*vide infra*). Quantum yields of intramolecular reactions are often higher. Discussion of individual aspects of the reactions follows.

37.5 The Hydrogen Donor

The triplet state of an α-diketone can abstract hydrogen from almost any substrate (R-H, often present as the solvent) containing C-H bonds to yield (reaction #3) the neutral semidione radical DKH and the radical R. Substrates that have been shown to react with diketones in such reactions include primary and secondary alcohols, toluene and substituted toluenes, saturated hydrocarbons, alkenes having allylic hydrogens, aliphatic and aromatic aldehydes, ethers, sulfides, esters, and amines. Hydrogen atoms alpha to carbonyl or cyano groups are relatively inert.

Benzene has often been used as the solvent in such reactions. Reactions in benzene solution are slow when no intramolecular avenues are available, and produce mixtures of products. The formation of enols under these conditions is discussed later. Photoreduction in such solutions has been attributed to hydrogen abstraction from the ground state of the diketone present. Biphenyl has also been observed as a reaction product, leading to suggestions that phenyl radicals are formed somehow.

Reactions of amines deserve mention. Extensive studies of the photolysis of *mono*ketones in the presence of amines have demonstrated that hydrogen atom abstraction can be, in the extreme, a two-step process involving single electron transfer from amine to triplet ketone followed by a separate proton transfer step. α-Diketones might be expected to participate even more readily in electron transfer; this has been demonstrated to occur in photocycloaddition reactions with suitable alkenes. While hydrogen atom transfer definitely occurs in diketone-amine reactions, clear-cut evidence for an initial electron transfer step is lacking although it should be noted that high rate constants for quenching of biacetyl fluorescence and phosphorescence by amines are observed.[9] In the event, reactions of diones and amines generally result in complex product mixtures and are of minimal preparative value.

As noted above, the reactive excited state in H-atom abstraction reactions is the triplet n,π^* state of the diketone. The products of the primary photochemical reaction are the pair of radicals DKH and R, formed initially as a triplet pair in a solvent cage. In such cases, separation into free radicals may compete with intersystem crossing to an in-cage singlet pair that should undergo very rapid coupling. With the exception of the reaction of CQ with toluene, which has been shown (*vide infra*) to involve (in part) a cage process, little information is available on this point.

(Str.37.4)

7.6 The Semidione Radical

The resonance-stabilized semidione radical DKH exists as two rapidly equilibrating species.[10] These two are distinguishable in unsymmetrical diketones. Attempts to establish which of two nonidentical carbonyl groups in such compounds is responsible for hydrogen abstraction by examination of products are futile. The observed result for an unsymmetrical dione will be due to a combination of the factors affecting the equilibrium between the tautomeric semidione radicals and the relative rates of subsequent reactions.

(Str.37.5)

Equilibration between different semidione radicals has been demonstrated using tetramethyltetralindione (TTD, λ_{max} 388 nm) and camphorquinone (CQ, λ_{max} 470 nm). The considerable difference in absorption spectra of these two compounds made it possible to excite selectively either one or both simultaneously by appropriate choice of irradiating wavelength. When this was done in isopropyl alcohol solution, the only product obtained was the hydroxyketone TTDH$_2$, even when all of the light was initially absorbed by CQ (only one isomeric semidione radical shown); only the TTDH semidione radical could be detected in the esr spectrum of irradiated mixtures. Products derived from CQ appeared when most of the TTD had been consumed.

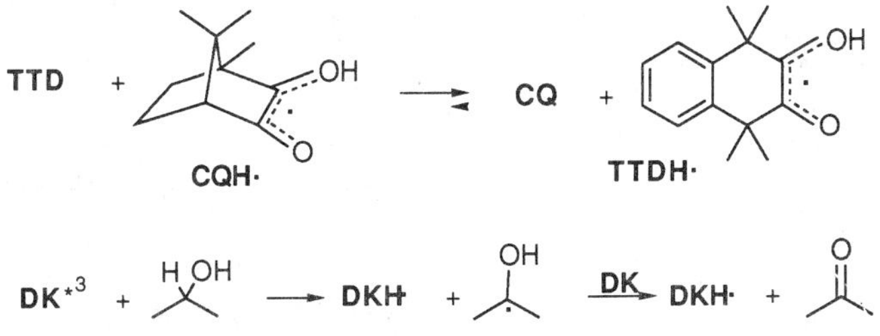

(Str.37.6)

A special feature of reactions with alcohols is the formation of DKH plus a ketyl radical; the latter may react with a molecule of ground-state diketone to form a monoketone and a second DKH. In the case of methanol as hydrogen donor, the initial hydroxymethyl radical reacts sufficiently rapidly with DKH to give coupling products in competition with hydrogen transfer to a second DKH plus formaldehyde. Other ketyl radicals, such as 2-hydroxypropyl (from isopropyl alcohol) and hydroxydiphenylmethyl radical (from benzhydrol), react exclusively with ground-state diketone leaving DKH as the only radical species in solution.

7.7 Intermolecular Reactions

A general scheme for these reactions is presented below. The major products observed in most cases are the result of coupling between DKH and R radicals (exceptions are noted below) and the overall result is addition of RH to starting diketone. As suggested by the two resonance forms shown for DKH, such coupling could occur, *a priori,* either at carbon or at oxygen with four coupling products theoretically possible for unsymmetrical diketones. Coupling at carbon produces an α-substituted-α-hydroxyketone (1) and coupling at oxygen would give a monosubstituted enediol (2) in equilibrium with its keto tautomer (3). The observed result for radical coupling reactions in all diketone photolyses is coupling at a carbon atom of DKH; that is, formation of 1. These reactions provide a convenient synthetic procedure for the synthesis of such compounds. A detailed study[11] of the reaction of CQ with *p*-xylene provided no evidence for formation of α-ketoether (3, R = *p*-CH$_3$C$_6$H$_4$CH$_2$); the only products, in addition to small amounts of CQH$_2$ and *p,p'*-dimethylbibenzyl, were compounds resulting from reaction of product (1, R = *p*-CH$_3$C$_6$H$_4$CH$_2$) with triplet CQ. The special case of *o*-quinones will be discussed at the end of this summary of diketone reactions.

$$\text{DK} \xrightarrow{h\upsilon} \text{DK}^{*1} \xrightarrow{isc} \text{DK}^{*3} \tag{37.1}$$

$$\text{DK}^{*3} \longrightarrow \text{DK} \tag{37.2}$$

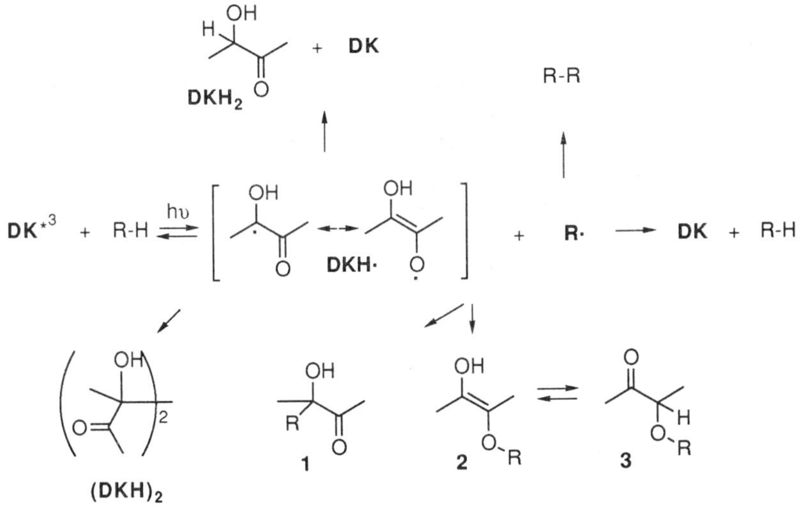

$$\tag{37.3}$$

Minor reactions are: coupling of two R radicals (e.g., to give bibenzyl), disproportionation of DKH to give reduced diketone (DKH$_2$)[12] plus starting diketone, or dimerization of DKH to thermally labile diketopinacol (DKH)$_2$. Pinacols have been observed only with simple open-chain diketones such as biacetyl and benzil. As noted above, hydrogen abstractions involving alcohols as the substrate can be followed by reaction to form a second DKH. In such cases, the only radical remaining will be DKH, and its bimolecular termination reactions will result. This provides a convenient method for reduction of cyclic α-diketones to α-hydroxyketones without further reduction to vicinal diol. As mentioned above, reaction in methanol leads to a mixture of coupling and disproportionation.

Apparent exceptions to the generalization that radical coupling occurs only at the carbon atom of DKH have been observed in reactions of olefins and of aldehydes with diketones. Detailed examination has established that these exceptions do not involve the coupling mechanism described above. Thus, formation of allylic ether 4 in the reaction of biacetyl with α-methylstyrene was shown to involve initial addition of diketone to the double bond to give biradical 5 followed by intramolecular transfer of a hydrogen atom, as illustrated. The alternative, abstraction of hydrogen to give allylic radical 6 plus DKH, followed by coupling at oxygen, was ruled out by deuterium labeling studies that showed none of the scrambling to be expected if 6 had been an intermediate. Products of allylic hydrogen atom abstraction have been observed in some cases; all of these are the result of coupling at a carbon atom of DKH, as illustrated for formation of 7 (two diastereomers) in the reaction of TTFD with cyclohexene.

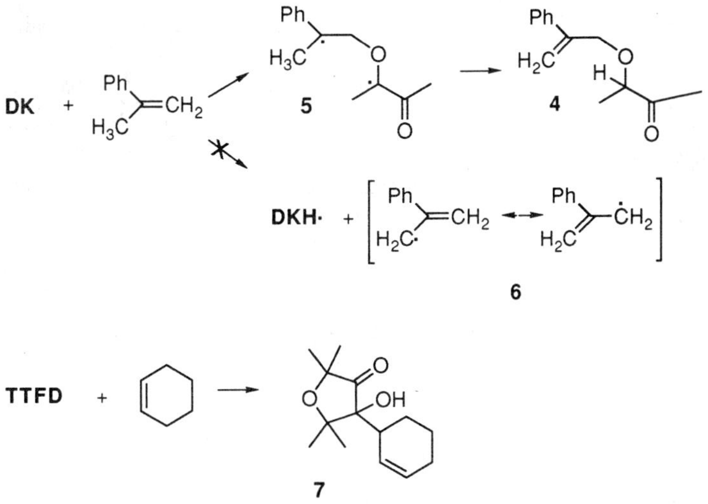

(Str.37.7)

The reaction of BOD with aromatic aldehydes produced mixtures of α-aroyl-α-hydroxyketone 8 and α-ketol-X-benzoate 9, which seemingly present an example of coupling of aroyl radicals either at carbon or at oxygen of the intermediate semidione radical BODH. A detailed study of the reaction with *p*-chlorobenzaldehyde showed that the ratio 8:9 changed markedly with light intensity (and other experimental variables), varying from 0.06 to 2.5. It was established that 9 is formed by a radical-molecule mechanism in which aroyl radicals add to ground-state diketone to form an α-keto radical that accepts a hydrogen atom from DKH to give the enolic form of 9, followed by ketonization. The result observed experimentally represents the outcome of competition between the radical-molecule reaction and the radical-radical reaction with the result determined by the experimental conditions used.

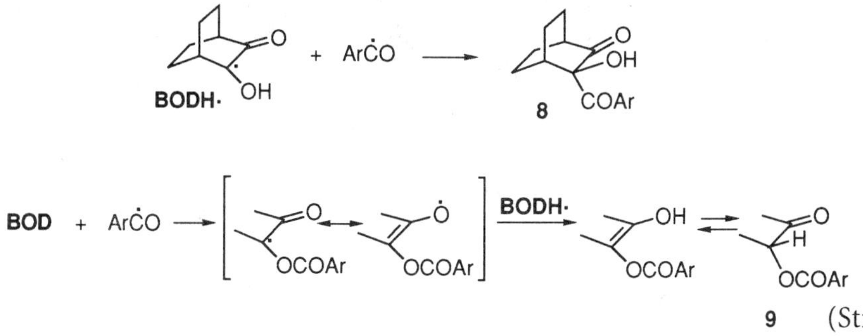

9   (Str.37.8)

7.8 Intramolecular Reactions

A variety of interesting and useful intramolecular H-abstraction reactions of α-diketones have been reported. These produce the biradical 10, which undergoes intramolecular coupling to form a new carbocyclic ring. The first example was the reaction of simple dialkyl diketones to give exclusively cyclobutanolones of type 11 and not acylcylobutanols. Cyclodecanedione reacted similarly to produce a bicyclic hydroxyketone as illustrated. Other examples include cycloalkyl diones as shown below. These reactions are similar mechanistically to the Yang cyclization reaction of monoketones (γ-hydrogen abstraction), but they often proceed with high chemical and quantum yields and without the competing fragmentation observed in Norrish type II reactions. Competing formation of semidione radicals may be observed when irradiation is performed in the presence of a good hydrogen atom donor such as isopropyl alcohol. The regiospecificity observed with diketones has been attributed to the *s-trans* geometry of excited states of such α-diketones, as illustrated. Even with a compound such as 1,8-diphenyloctane-4,5-dione (**12**), which offers benzylic as an alternative to simple secondary hydrogen atoms, the cyclobutanolone is the major (90% yield) product. Relative reactivities for primary, secondary, and tertiary hydrogens, determined by quenching studies, were 1:80:1000, respectively; however, respectable yields were obtained even when abstraction of a primary hydrogen atom was involved. An α-bromoketone reacted without loss of bromine, emphasizing the potential of reactions of this type for synthesis of polyfunctional compounds.

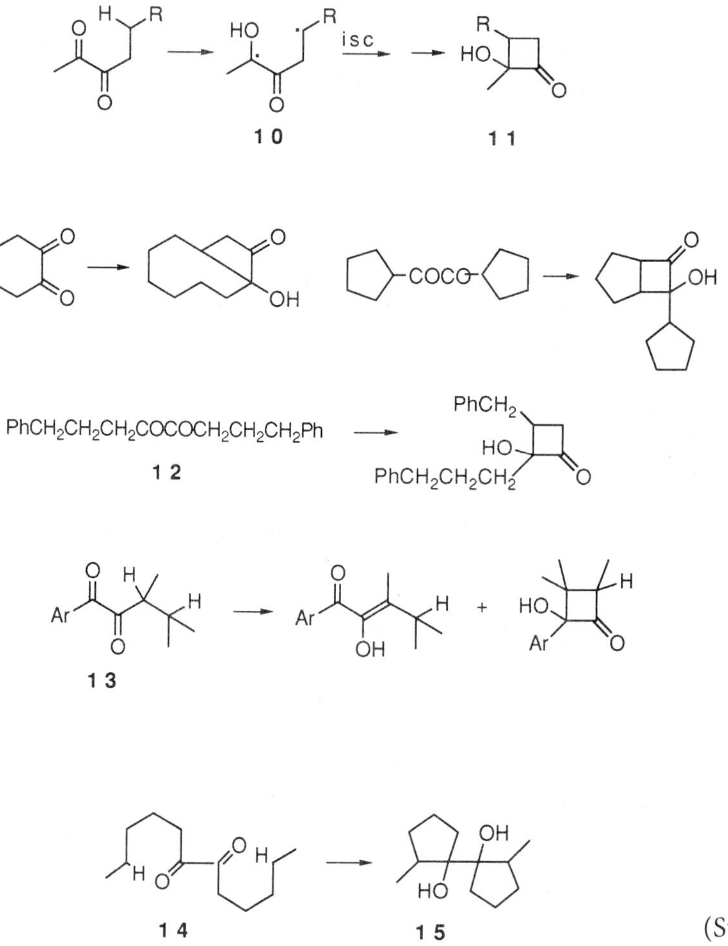

10   **11**

PhCH₂CH₂CH₂COCOCH₂CH₂CH₂Ph →

12

13

14   **15**   (Str.37.9)

Reactions of aryl alkyl diketones (**13**) also produced cyclobutanolones with relative rates for hydrogen atom abstraction of 1:60:390 (primary:secondary:tertiary). However, in many cases, quantum yields for disappearance of **13** were larger than for formation of cyclobutanolone, indicating occurrence of an additional photochemical reaction. This was shown to be concentration-dependent enolization of **13** with an even more marked dependence on structure (relative rates 1: 1×10^3:3×10^5 for primary, secondary, and tertiary hydrogens, respectively). Photoenolization of biacetyl had been observed earlier and occurs in certain other cases as well. The mechanism of this energy-wasting process is not clear; simple abstraction of an α-hydrogen atom, followed by coupling of a hydrogen at the oxygen atom of the resulting radical, does not provide a satisfactory explanation.

An interesting exception to cyclobutanolone formation is the final carbon-carbon bond-forming reaction in the synthesis of dimethyldodecahedrane. An α-diketone having the partial structure **14** (shown by X-ray analysis) afforded **15**, containing two new five-membered rings upon irradiation. This very useful, albeit "abnormal", result (abstraction of a δ-hydrogen), results from the restraint imposed by the rigid cage structure[13] of **13**, which enforces proximity of the normally more remote hydrogen atoms to the reactive carbonyl groups.

α,β-Unsaturated and *o*-alkylsubstituted aromatic diketones give products containing five-membered rings. This is illustrated for *o*-substituted aryl diketones **14**. The intermediate, equilibrating semidione radicals **17** and **18** could couple to form either four- or five-membered ring products; only the hydroxyindanone **19** was found. It should be noted that the alternative product, **20**, could undergo facile thermal rearrangement to **19** and not be detected. The presence, in fact, of equilibrating semidione radicals **17** and **18** was demonstrated nicely by trapping experiments with sulfur dioxide that gave two interconvertible sulfones (**21** and **22**), as illustrated. An example of similar cyclization of a conjugated alkenyl dione is shown below, **23** → **24**. More remote unsaturation can result in acyl migration, intramolecular cycloaddition to the double bond, etc.

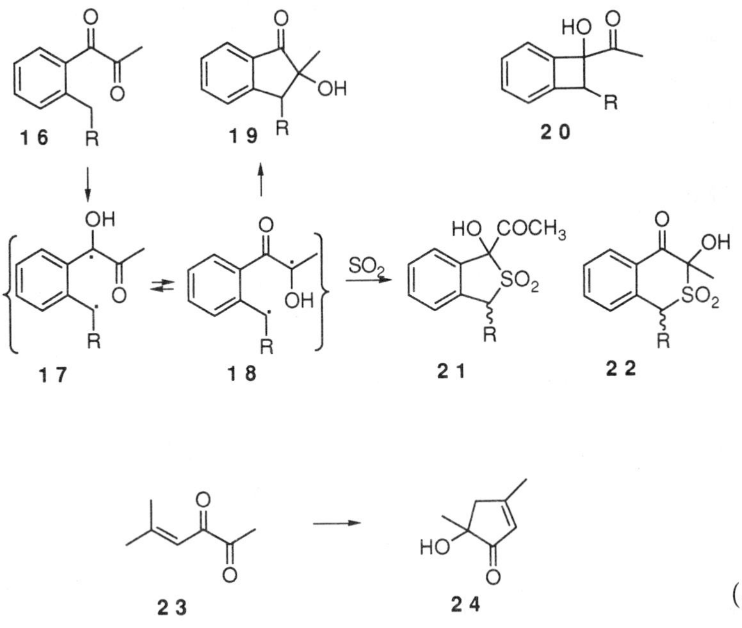

(Str.37.10)

37.9 Regeneration of Reactants from Radicals

In addition to the reactions described above, reversal of the hydrogen abstraction process to regenerate reactants (DKH + R → DK + RH) can occur after radicals in a cage undergo intersystem crossing to a singlet pair or after migration from the cage to give separated radicals. This has been demonstrated in the reaction of CQ with *p*-chlorobenzaldehyde by a study of the dependence of quantum yield on aldehyde concentration and by irradiating in the presence of D_2O where it was shown that, after partial conversion, aldehydic protium in recovered benzaldehyde had been replaced by deuterium. This must be due to exchange of protium in the semidione radical by deuterium, followed by transfer of deuterium from deuteriated semidione to *p*-chlorobenzoyl radical. The quantum yield studies indicated that about half of the radical pairs formed originally terminated by regeneration of dione and aldehyde, making this process a significant factor in reducing quantum yields. Its importance in other dione reactions has not been investigated.

37.10 *o*-Quinones

The photochemistry of *o*-quinones shows many parallels with that of α-diketones and is mentioned briefly here. In the presence of hydrogen atom donors, hydrogen abstraction to form semiquinone and donor radicals occurs. Reactions of PQ, the most extensively studied of these compounds, are shown below. With alcohols as hydrogen donors the products were dihydric phenols, some unusual quantum yield behavior has been reported in the reaction with ethanol. The addition products observed in all other cases were monosubstituted dihydric phenols (25), i.e., overall 1,4-addition, except with substituted toluenes. These gave 1,2-addition products (26) in addition to 25, the ratio 25:26 depended on *para*-substitution of the toluene. The preference for hydrogen atom abstraction alpha to an ether linkage over abstraction alpha to a carbonyl group is emphasized by the results obtained upon irradiation of PQ in ethyl propionate solution where only one product was detected as shown below. Nuclear substitution has been observed in reactions of *o*-naphthoquinones. The monosubstituted dihydroxy aromatics produced in most of these reactions can serve as starting materials for preparation of other monosubstituted dihydric phenols, as well as unsymmetrically disubstituted compounds.[14]

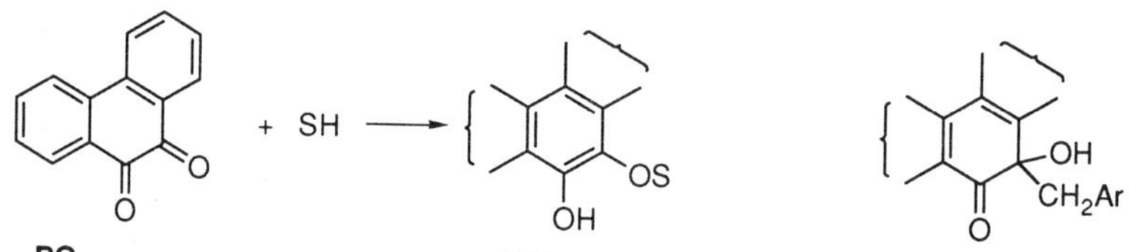

$$S = R\overset{|}{C}O, \ R\overset{|}{C}HOR, \ ArCH_2-, \ CH_3CH_2COO\overset{|}{C}HCH_3$$

(Str.37.11)

37.11 Sensitization

As noted in the Section 37.1, α-diketones have low triplet energies, making them amenable to sensitization via bimolecular electronic energy transfer from a variety of triplet sensitizers. When such energy transfer experiments were performed with CQ in isopropyl alcohol using benzophenone (BP) as sensitizer, the phenomenon known as "chemical sensitization" was observed. When the solvent is a good hydrogen atom donor, such as isopropyl alcohol, hydrogen atom abstraction by triplet BP can compete with triplet energy transfer to CQ. The resulting ketyl radicals then transfer hydrogen efficiently to ground state CQ. This process produces the same CQH radical, as obtained by reaction of triplet CQ (from direct irradiation) with isopropyl alcohol. However, since the quantum yield for abstraction from isopropyl alcohol by BP is significantly higher than that for abstraction by CQ, the quantum yield for photoreduction of CQ increased with decreasing CQ concentration; exactly the opposite of what would be expected for a bimolecular energy transfer process. Similar behavior was observed with the system BP, CQ, toluene. This method makes it possible to generate separately the two radicals formed in a solvent cage when triplet diketone abstracts hydrogen. Comparison of the results obtained by direct irradiation with those obtained in the "sensitized" experiment can provide information on similarities or differences between the two processes. It was inferred from such experiments that at least part of the reaction of toluene with CQ involves coupling of in-cage radicals. Attempted chiral sensitization using chiral CQ failed to give detectable enantiomeric excess.

37.12 Vicinal Polyketones

Finally, we note that vicinal tri- and tetraketones, known for over a century,[15] participate inefficiently, if at all, in hydrogen atom abstraction reactions. Triketones react to form complex product mixtures with overall quantum yields of less than 0.001; the mechanism may involve hydrogen abstraction.[16,17] Tetraketones undergo intramolecular rearrangement with loss of carbon monoxide to give reactive acyloxy ketenes,[18] which react with ground-state tetraketone forming tricyclic lactones.[19] Two exceptions are dimesityl and di-*t*-butyl tetraketones, which are photochemically inert. In a manner analogous to the behavior of *o*-methylbenzophenones, this lack of reactivity is actually a mask for intramolecular hydrogen abstraction resulting in formation of enols, as shown below, which revert to starting material by ketonization; deuterium was incorporated in the *o*-methyl groups when irradiation was performed in the presence of *t*-butyl-OD.[15]

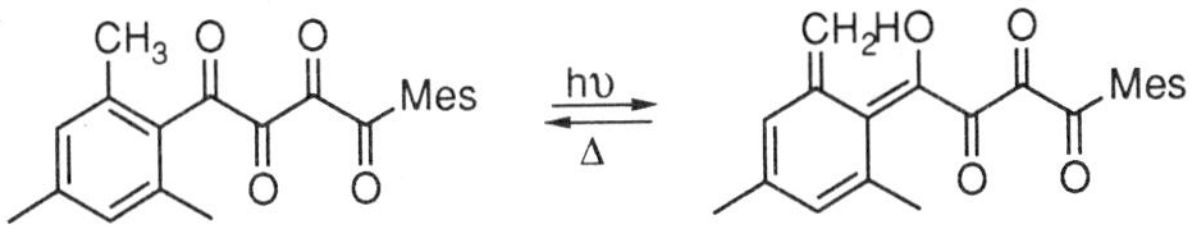

(Str.37.12)

7.13 Summary

Inter- or intramolecular hydrogen atom abstraction by photoexcited α-diketones is a very common reaction with many types of substrates and results in the formation of radical pairs or biradicals. These are of interest from the point of view of free radical chemistry. In addition, such reactions have considerable synthetic potential for reduction of diones and for formation of new carbon–carbon bonds to give a variety of multifunctional ring systems.

References

1. Rubin, M. B., Photochemistry of *o*-quinones and α-diketones, *Fortschr. Chem. Forsch.*, 13, 251, 1969.
2. Monroe, B. M., The Photochemistry of α-Dicarbonyl Compounds, *Adv. Photochem.*, 8, 77, 1971.
3. Rubin, M. B., Recent Photochemistry of α-Diketones, *Top. Curr. Chem.*, 129, 1, 1985.
4. Pawlikowski, M., Zgierski, M., and Orlando, M. Z., *Chem. Phys. Lett.*, 105, 612, 1984.
5. Rubin, M. B., Photochemistry of Vicinal Polyketones, in *Excited States in Organic Chemistry and Biology*, B. Pullman and N. Goldblum, D., Eds., Reidel, Dordrecht, Holland, 1977.
6. The considerable difference in absorption maxima between α-diketones and their enolic tautomers allows selective irradiation of the diketo form and concomitant shifting of the equilibrium.
7. Rubin, M. B., Patyk, A., and Sander, W., The absence of ethylenedione in photochemical *bis*decarbonylation reactions, *Tetrahedron Lett.*, 29, 6641, 1988.
8. For a recent comprehensive review of hydrogen atom abstraction by carbonyl compounds, see: Wagner, P. and Park, B.-S., in Organic Photochemistry, Padwa, A., Ed., Marcel Dekker, New York, 1991, Vol. 11, chap. 4.
9. Turro, N. J. and Engel, R., *J. Am. Chem. Soc.*, 91, 7113, 1969.
10. The suggestion of a single, symmetrical structure was withdrawn on the basis of esr spectra.
11. Rubin, M. B. and Gutman, A. L., Minor products in photoreactions of α-Diketones with arenes. Abstraction of hydroxylic hydrogen by triplet carbonyl. *J. Org. Chem.*, 51, 2511, 1986.
12. The initial product of disproportionation is the enediol which ketonizes to give hydroxyketone.
13. A similar result in a *p*-cyclophanedione has been observed by R. Krämer; unpublished result, this laboratory.
14. Rubin, M. B., *J. Org. Chem.*, 28, 1949, 1963.

15. For a general review of the chemistry of *vic*-polyketones, see: Rubin, M. B., *Chem. Rev.,* 75, 177, 1975; *Cf.* also Schönberg, A. and Singer, E., *Tetrahedron,* 34, 1285, 1978.
16. Urry, W. H., Pai, M. H., and Chen, C. Y., *J. Am. Chem. Soc.,* 86, 5342, 1964.
17. Rubin, M. B., Heller, M., Monisov, R., Gleiter, R., and Dörner, T., *J. Photochem. Photobiol. A: Chem.,* in press.
18. Rubin, M. B., Etinger, M., Sander, W., and Wierlacher, S., unpublished results.
19. Rubin, M. B., Krochmal, E. C., Jr., and Kaftory, M., *Rec. Trav. Chim. Pays-Bas,* 98, 85, 1979.

38

Norrish Type II Photoelimination of Ketones: Cleavage of 1,4-Biradicals Formed by γ-Hydrogen Abstraction

Peter J. Wagner
Michigan State University

38.1 The Norrish Type II Photoelimination Reaction

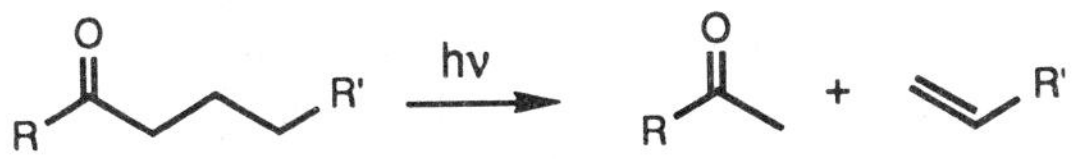

38.2 Mechanism

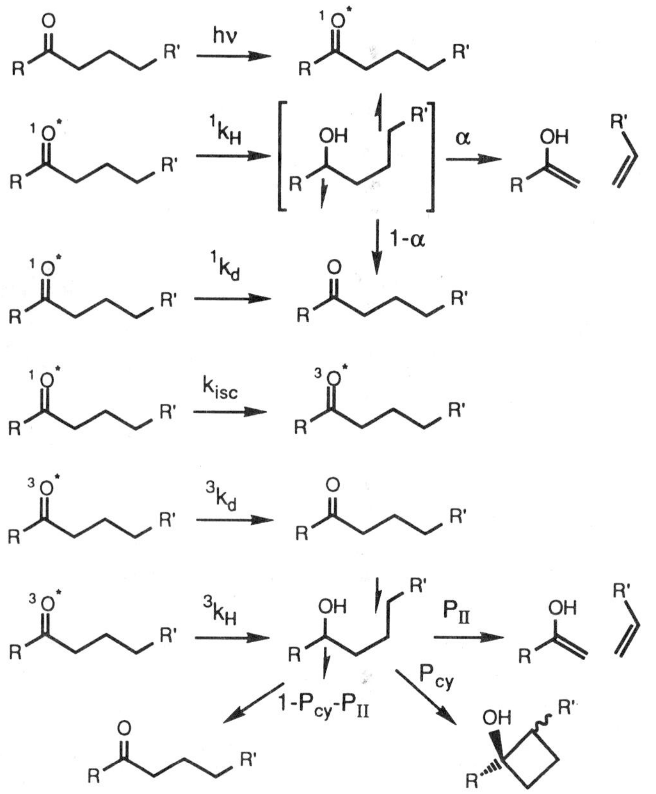

38.3 Byproducts

Cyclobutanols (usually)
Cyclopentanols (rarely)

38.4 History

A wide variety of carbonyl compounds undergo photoinduced intramolecular hydrogen atom abstraction to form 1-hydroxy-1,x-biradicals that then undergo two common competing reactions: coupling to produce cyclic alcohols, and disproportionation back to ketone or to various enols. The overall process parallels closely the well-known bimolecular photoreduction of ketones, the most common products of which are formed by radical coupling. These intramolecular hydrogen abstraction processes have been studied widely and several full reviews are available.[1–5]

Interestingly, the first and still best known example of such intramolecular hydrogen abstraction was the "type II" photoelimination discovered by Norrish,[6] who found that ketones with γ C-H bonds cleave to methyl ketones and alkenes rather than to acyl and alkyl radicals, the earlier discovered "type I" cleavage. Later workers found that both cleavage processes compete in certain ketones and that overall quantum yields are particularly low whenever the type II process occurs. Yang first found the cyclobutanols that accompany cleavage and suggested that both processes arise from a common 1,4-biradical intermediate.[7] As realization of this process grew, other workers investigated a large variety of ketones that form different kinds of biradicals and cyclic alcohols. Although cleavage and cyclization of 1,4-biradicals are linked mechanistically, the cleavage process is so easy to measure that the Norrish type II cleavage has been studied widely to gain basic mechanistic information about biradicals and about hydrogen abstraction reactions of excited ketones. This chapter is devoted only to the cleavage process discovered by and properly named after Norrish. Other sections of this handbook are devoted to cyclic alcohol formation initiated by hydrogen abstraction.

For years, the "type II" reaction was considered to be concerted; a neat six-atom transition state would lead to the alkene and the enol tautomer of the product ketone. This process would now be called a "retro-ene" reaction. In a key experiment, Calvert and Pitts verified by IR that the enol is indeed formed first and is converted rapidly to ketone.[8] However, Yang had already discovered competing cyclobutanol formation and pointed out that cleavage of a 1,4-biradical would give both types of products.[7] Before the concerted vs. two-step question was elucidated further another basic mechanistic puzzle was raised. One group found that type II cleavage of 2-pentanone was quenched by biacetyl,[9] which was known to quench excited triplets rapidly. Another group found that the reaction of 2-hexanone was not quenched under the same conditions.[10] The two groups obviously differed as to which excited state undergoes the reaction. The apparent conflict was solved neatly by the revelation that both ketones react from both states, with 2-hexanone undergoing more unquenchable singlet reaction than 2-pentanone.[11]

Before the mid 1960s, most studies were performed on aliphatic ketones. Pitts then showed that phenyl alkyl ketones also undergo the reaction;[12] Wagner and Hammond showed that the reaction is completely triplet derived.[13] Using phenyl ketones to isolate triplet reactivity, Wagner et al. suggested that the low quantum yields for type II reaction are caused by disproportionation of Yang's 1,4-biradical intermediate back to ketone. Soon, it was discovered that added Lewis bases markedly increase product quantum yields of triplet reactions, often to unity.[14,15] This behavior was attributed to hydrogen bonding by the hydroxy group of the biradical to the Lewis base; the H-bond must be broken during disproportionation but not during cleavage or cyclization. In 1972, the biradicals were successfully trapped with mercaptans[16] and racemization at the γ-carbon was equated quantitatively with "radiationless decay", thus proving that it involves reversion of a biradical intermediate to ketone.[17] At the same time, Yang showed that the triplet, but not the singlet, component of the photoreactivity of 5-methyl-2-hexanone also produces extensive racemization at the γ-carbon.[18] All of this work firmly established the 1,4-biradical as an intermediate in the triplet reaction but left open how much it is involved in the singlet reaction. Some unquenchable cyclization reactions indicate that singlet biradicals are indeed formed, even if they may not account for all the cleavage reaction.[19]

38.5 Quantum Efficiency

As the above description indicates, the Norrish type II reaction involves two consecutive intermediates: first an excited state (that can be singlet and/or triplet) and then a biradical. The quantum efficiency for cleavage, as described by Equation 38.1, depends on how many competitive reactions both intermediates undergo. Consequently it is necessary to understand how structure and environment affect the behavior of both species separately.

$$\Phi_{II} = \alpha^1 k_H \tau_S + P_{II} \Phi_{isc}^3 k_H \tau_T \qquad (38.1)$$

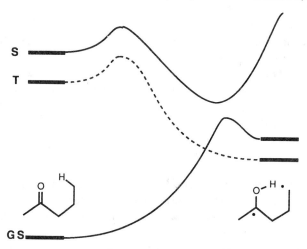

FIGURE 38.1   Potential energy diagram for excited-state hydrogen abstraction.

38.6 Excited-State Reactivity

Nature of Excited State

Photoexcitation of carbonyls effectively breaks the double bond to form a 1,2-biradical, thus conferring on the more electronegative oxygen atom radical reactivity that parallels closely that of alkoxy radicals. Two multiplicities — singlet and triplet — and two electronic configurations — n,π^* and π,π^* — have been shown to display reactivity.

Singlet vs. Triplet

Aliphatic ketones have n,π^* lowest singlets that undergo intersystem crossing to their n,π^* lowest triplets with rates ca. 10^8 s^{-1}, slow enough that some singlet reaction occurs. The reaction of a given ketone can be partitioned into its singlet or triplet components by the technique of differential quenching with conjugated dienes. All triplet reaction can be quenched and singlet reaction isolated by performing irradiations in dienes or substituted naphthalenes as solvent.[20] Figure 38.1 depicts the well-known different potential energy surfaces followed by the two states.[3] The low quantum efficiency always observed for singlet reactions indicates that partial hydrogen abstraction promotes radiationless decay directly back to reactant at the point of a forbidden crossing between ground- and excited-state surfaces.[21] The characteristic near-zero yields of cyclobutanol from the singlet state, even in alcohol solvents, suggest that some of the singlet reaction may actually occur in a concerted fashion; but triplet reaction occurs exclusively via biradicals that both cleave and cyclize.

The rapid ($>10^{10}$ s^{-1}) intersystem crossing in most phenyl ketones usually produces exclusive triplet reactivity. A few intramolecular hydrogen abstraction reactions of aryl ketones have been shown to involve excited singlets at least partially.[25,26] This pattern occurs when n,π^* singlet reaction is fast enough to compete with intersystem crossing; exclusive singlet reaction can occur if triplet reaction is so slow that it is totally suppressed by trace quenchers.

Table 38.1   Excited-State Behavior of $CH_3COCH_2CH_2R$[1,22–24]

R	Φ_{II}^S	Φ_{isc}	Φ_{II}^T	$^1k_r - 10^8$ s$^{-1} - {}^3k_r$	
CH_3	0.025	0.81	>0.36	1.0	0.13
CH_3CH_2	0.10	0.37	0.37	8.8	1.0
$(CH_3)_2CH$	0.07	0.18	0.17	20	3.8

Substituent Effects — n,π* vs. π,π*

Aryl ketones can have either n,π* or π,π* lowest triplets; the former are far more reactive than the latter. Ketones with π,π* lowest triplets do undergo hydrogen abstraction reactions. They display two quite different forms of triplet reactivity. In systems with π,π* triplets only a few kJ mol^{-1} below their n,π* triplets, most of the measured reactivity arises from low concentrations of the n,π* triplet in thermal equilibrium with the lower π,π* triplet.[27,28] Equation 38.2 describes the observed rate constants for hydrogen abstraction, with the n,π* state providing most or all of the reactivity when ΔE ($E_{n,\pi} - E_{\pi,\pi}$) < 21 kJ mol^{-1}.

$$k_{obs}^{H} = \chi_{n,\pi} k_{H}^{n,\pi} + \chi_{\pi,\pi} k_{H}^{\pi,\pi} \tag{38.2}$$

$$\chi_{n,\pi} = \left(1 - \chi_{\pi,\pi}\right) = e^{-\Delta E/RT} \Big/ \left[1 + e^{-\Delta E/RT}\right] \tag{38.3}$$

Table 38.2 compares the effects of ring substituents on rate constants for triplet state hydrogen abstraction by valerophenone, together with intrinsic $k_{n,\pi}$ values that reflect independently measured substituent effects on rate constants for bimolecular hydrogen abstraction ($\rho = 0.5$)[29] and

Table 38.2 Rate Constants for Triplet γ-Hydrogen Abstraction by Ring-Substituted Valerophenones in Benzene

Substituent	σ	k_{obs} 10^7 s^{-1}	% n,π*	$k_{n,\pi}$ 10^7 s^{-1}
m-(NH$^+$)	—	200	>99	200
p-(NH$^+$)	—	330	>99	330
o-(N)	—	19	>99	19
m-(N)	—	31	>99	31
p-(N)	—	68	>99	68
H	0	13	99	13
p-OCF$_3$	0.28	13	95	14
o-CF$_3$	—	13	>99	13
m-CF$_3$	0.43	32	>99	32
p-CF$_3$	0.54	28	>99	28
o-alkyl	—	3.0	25	12
m-alkyl	−0.07	3.9	35	12
p-alkyl	−0.17	1.8	18	10
o-OMe	—	0.30	3	10
m-OMe	0.12	0.02	0.15	15
p-OMe	−0.27	0.06	1	6
p-OAc	0.31	4.4	25	18
o-F	—	14	99	14
m-F	0.34	18	99	18
p-F	0.06	15	99	15
o-Cl	—	3.5	?	—
m-Cl	0.37	16	99	20
p-Cl	0.23	3.0	16	18
o-CO$_2$Me	—	3.6	?	—
m-CO$_2$Me	0.37	28	99	28
p-CO$_2$Me	0.45	12	40	30
o-CN	—	23	99	23
m-CN	0.56	30	99	30
p-CN	0.66	6.8	21	32
m-COR	0.38	14	50	28
p-COR	0.50	2.7	9	30
p-SMe	0	<0.001	<0.01	14
p-SCF$_3$	0.64	3.0	10	32

derived values for the percentage of n,π* triplets in equilibrium with π,π* triplets. The energy gaps between the two triplets represent the experimentally measured increases in activation energies for reaction produced by a substituent that causes the π,π* triplet to be lowest.[28,30] Only very strong electron donors produce a large enough energy gap that reactivity becomes imperceptible.

The second type of π,π* reactivity occurs in polynuclear aryl ketones with large $E_{n,\pi} - E_{\pi,\pi}$ values. It is often assumed incorrectly that π,π* triplets are totally unreactive at hydrogen abstraction. This notion is contradicted by the observed triplet reactivity of some naphthyl ketones, which is only 0.01 to 0.001% that of an n,π* triplet.[31,32] The original explanation for this large difference still seems reasonable, namely, the distinct spin localization in an n,π* triplet as opposed to the delocalized spin in a π,π* triplet.[31] This fact is usually depicted in terms of the n,π* triplet resembling a 1,2-biradical[33] and thus manifesting the chemical reactivity of an alkoxy radical,[1,34] while the π,π* triplets have little spin density on oxygen. Despite the intrinsic, if greatly diminished, reactivity of acylnaphthalene triplets, most reported type II cleavages of naphthyl ketones are very low quantum yield reactions from an unquenchable singlet.[25]

Substituent Effects — Inductive

Substituents on the α-carbon of ketones influence excited-state reactivity, although no extensive study has been done. In particular, electron-withdrawing groups seem to increase reactivity. For example, fluorination markedly enhances n,π* reactivity; however, it also promotes a π,π* lowest triplet.[35] α-Ketohexanoic acid shows enhanced reactivity compared to 2-hexanone.[36] α-Alkoxy ketones also show unusually high reactivity,[37] although much of the increase is due to conformational effects as discussed below.

Solvent Effects

For ketones with n,π* lowest triplets, there do not appear to be any significant solvent effects on k_H values. However, when the π,π* triplet is of lower or comparable energy, polar solvents reduce overall reactivity by lowering the excitation energy of the π,π* triplet and raising that of the n,π* triplet, thus decreasing $\chi_{n,\pi}$ values.[27]

Energetics

The n,π* triplet excitation energies of aliphatic ketones have not been measured exactly but generally are considered to be ~326 kJ mol^{-1} those of phenyl alkyl ketones are ~301 kJ mol^{-1}. Abstraction of even primary hydrogens is exothermic; both types of ketones display similar rate constants for attack on a given type of C-H bond,[1,38] benzylic stabilization of the radical product compensating for the lower energy of the phenyl ketone triplet. The activation energy for hydrogen abstraction from an unactivated methylene group is only ~16.6 kJ mol^{-1}.[28,39] The higher reactivity of n,π* singlets relative to triplets most likely reflects the increased exothermicity of reaction.

Nature of Hydrogen Atom

C-H bond Energy Effects

Table 38.3 lists measured rate constants for triplet state γ-hydrogen abstraction by γ-substituted butyrophenones. The relative values should hold for any type of hydrogen abstraction by n,π* triplets. The 1:30:200 per-bond primary:secondary:tertiary ratio is characteristic of electron-deficient species such as alkoxy radicals. γ-Substituents affect both C-H bond energies and electron density of the C-H bonds, both of which affect k_H values. The fourth column extracts a resonance factor by which conjugation of the γ-substituent with the developing γ-radical site enhances the rate constant. This factor was obtained by correcting the observed substituent effect on k_H for inductive deactivation, with a ρ_I value of -4.3 extrapolated from that for δ-substituents (see next section).

Inductive Effects

Table 38.4 lists measured rate constants for triplet state γ-hydrogen abstraction by δ-substituted valerophenones. Here, the substituents are not conjugated with the developing γ-radical site so that their effects are entirely inductive. They describe a ρ_I value of −1.85, which demonstrates the highly electron-deficient nature of the carbonyl n,π* state. It should be noted that I, Br, and RSO substituents enhance reactivity by what has been suggested to be anchimeric assistance.[40]

Geometric Effects

While two functional groups in a bimolecular reaction are free to rotate into all possible relative orientations and diffuse to van der Waals contact distance, the intervening bonds limit the number of geometries available to two separate functional groups involved in an intramolecular reaction. The resulting conformational limitations affect rate constants for hydrogen abstraction and thus determine regioselectivity.

For some time, there has been a simplistic belief that the hydrogen atom being abstracted should be restricted to lie along the long axis of the carbonyl *n*-orbital. This notion recognizes the directionality of *p*-orbitals and the fact that n,π* radical reactivity is centered on this particular *p* (or sp^2) orbital. However, many examples are known in which efficient hydrogen abstraction takes place with the developing H-O bond making a fairly large angle with respect to the nodal plane of the carbonyl π-system.[41–44] Wagner has pointed out that such reactivity probably demonstrates a $\cos^2$ dependence on this angle,[3] that being the electron density function for a *p*-orbital.

Scheffer has studied a variety of ketones that undergo γ-hydrogen abstraction in their crystalline states.[45,46] In analyzing their reactivity, he considered the *ground-state* parameters depicted in Scheme 38.1 to be the most important in determining reactivity: d, the distance between O and H; θ, the O–H–C angle; Δ, the C=O–H angle; and ω, the dihedral angle that the O–H vector makes

Table 38.3 Rate Constants in Benzene for Triplet γ-Hydrogen Abstraction by γ-Substituted Butyrophenones

γ-Substituent	σ_I	k_H, 10^7 s^{-1}	Resonance factor
H	0.0	0.7	—
Alkyl	−0.05	14–20	1.0
Dimethyl	−0.10	50	—
Phenyl	0.10	40	8.4
Vinyl	0.09	50	8.8
RS	0.23	64	54
PhS	0.30	45	60
RSO	0.52	1.2	13
R$_2$N	0.06	80	8
CH$_3$O	0.30	62	100
HO	0.25	40	40
PhO	0.38	22	60
OC(=O)Me	0.39	1.2	5
C(=O)OMe	0.30	1.0	1.6
Chloro	0.47	1.0	8
N$_3$	0.44	0.5	4
Fluoro	0.52	0.8	10
CN	0.59	0.4	11
RSO$_2$	0.60	0.04	1
NHR2$^+$	0.80	0.01	1
NR$_3^+$	0.90	0.001	1

Table 38.4 Rate Constants in Benzene for Triplet γ-Hydrogen Abstraction by δ-Substituted Valerophenones

δ-Substituent	σ_I	k_H, 10^7 s^{-1}
H	0.0	14
Alkyl	−0.05	18
R$_2$N	0.06	7.0
Phenyl	0.10	8.4
RS	0.23	4.8
PhS	0.30	4.7
CH$_3$O	0.30	2.6
C(=O)OMe	0.30	3.8
COOH	0.33	2.6
PhO	0.38	2.1
I	0.39	17
Br	0.44	5.6
Cl	0.47	2.2
CH$_2$Cl	0.20	5.6
N$_3$	0.44	1.8
RSO	0.52	12
CN	0.56	1.0
CH$_2$CN	0.24	4.5
CH$_2$CH$_2$CN	0.10	8.0
RSO$_2$	0.60	1.2
SCN		1.4
NHR$_2^+$	0.80	0.50

with respect to the nodal plane of the carbonyl π-system. Such a comparison of triplet reactivity with ground-state geometries is valid because n,π* excitation is known to be so highly localized on the carbonyl group that geometric changes in the rest of the molecule are negligible.

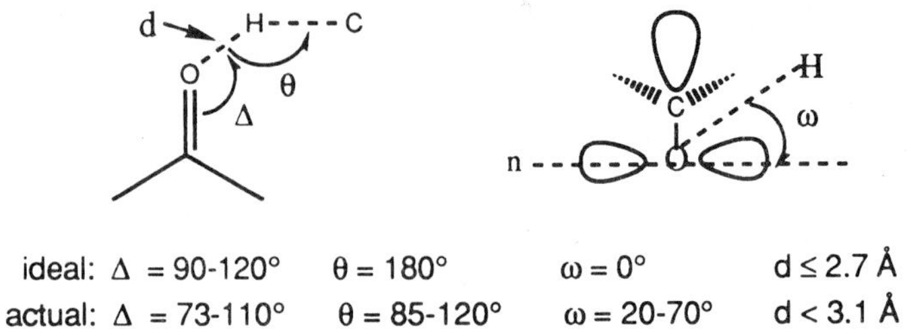

ideal: Δ = 90-120°   θ = 180°   ω = 0°   d ≤ 2.7 Å

actual: Δ = 73-110°   θ = 85-120°   ω = 20-70°   d < 3.1 Å

SCHEME 38.1

Scheffer suggested the theoretically "ideal" values for these parameters noted in Scheme 38.1. In over two dozen examples in which X-ray crystal structures were obtained for reactive ketones, the actual values observed are summarized in the scheme. The most important parameter appears to be d, since whenever two hydrogens at different distances <3.1 Å are available, the closer one is abstracted. The value of θ can obviously vary greatly from the linear arrangement thought to be preferable. Likewise, the value of ω can depart significantly from the "ideal" 0°. Both of these "deviations" from ideality have long been known from the reactivity of many steroidal ketones.[3,41] The meaning of Δ is the least clear, since n,π* excitation lengthens carbonyl bonds.[47]

Theoretical approaches to answering these orientational questions about hydrogen atom abstraction by n,π* states have reproduced the "ideal" values of Scheme 38.1.[48,49] There have been several experimental observations that suggest zero reactivity when ω = 90°.[50,51] All of these involve rigid polycyclic ketones or crystalline media, where the reactants are constrained to maintain their ground-state geometry. This lack of conformational mobility maximizes the possibility that a biradical formed by either α-cleavage or hydrogen abstraction might revert to starting ketone with 100% efficiency. This very problem affects cyclodecanone in solution. Yang found originally that it provides only the product of ε-hydrogen abstraction in benzene.[52] Sauers has now reported that products of both γ- and ε-hydrogen abstraction are formed in *t*-butyl alcohol, despite the fact that the ε-hydrogen lies at an angle ω of 90°, whereas ω is only 56° for the γ-hydrogen (Scheme 38.2).[53]

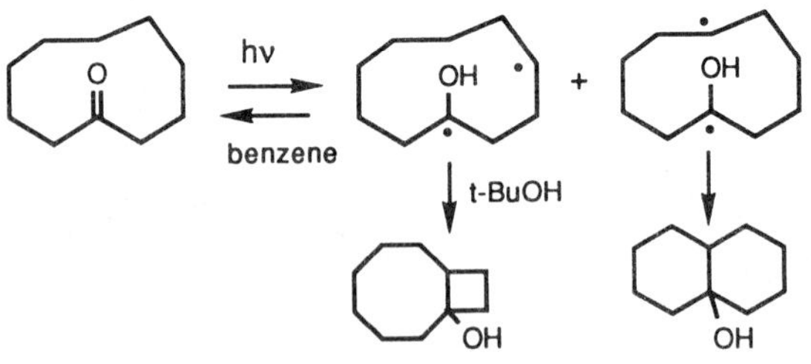

SCHEME 38.2

There remains the strong suspicion that hydrogen abstraction becomes slow as ω approaches 90°; but without systematic measurements of rate constants as a function of ω, no generalization is possible. Wagner et al. have reported one example for some sterically congested ketones whose triplets undergo δ-hydrogen abstraction. For hydrogens held at nearly identical distances from the oxygen, ones with ω = 20 to 30° react 3 to 4 times faster that ones with ω = 65 to 70°, the difference predicted by a cos² relationship.[43] Many more measurements of this type are needed for compounds that cannot undergo revertible competing photoreactions. In particular, a way to measure rate constants for solid-state reactions needs to be developed.

Conformational Effects

Rate constants for intramolecular hydrogen abstraction reflect conformational effects,[3,54] as depicted in Scheme 38.3. It is assumed that interconversion among the various conformers of a given excited state competes with reaction and decay. One or more conformations may have geometries close to that required for reaction and are dubbed "favorable" or "reactive". Also there are usually "unreactive" geometries in which the two functional groups are too far apart for reaction; these must be able to rotate into reactive conformations if they are to react rather than decay to ground state. Molecules that react as solids must have lowest energy conformations that are reactive. Most acyclic ketones exist preferentially in unreactive conformations, such that they do not react in the solid state. The lowest energy conformations of cyclic systems may have either reactive or unreactive geometries.

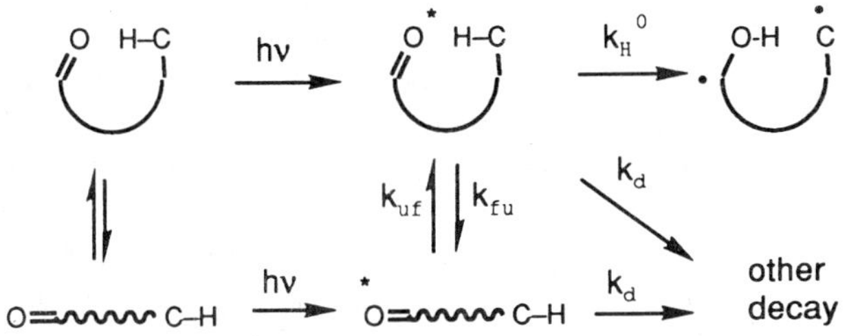

SCHEME 38.3

Dynamically, there are three classes of excited states: (1) those in which interconversion among conformers is much faster than decay reactions of the individual conformers, such that conformational equilibrium is established before reaction; (2) those in which conformational interconversions are all slower than decay, such that ground-state conformational preferences control photoreactivity; and (3) those in which unreactive conformers rotate into the reactive conformer, which reacts faster than it rotates back into the unreactive geometry. In this third case, reaction of the unreactive excited state is subject to rotational control, the intramolecular equivalent of intermolecular diffusion control.

The Norrish type II reaction can be subject to any of these different dynamic boundary conditions.[54] The first dramatic example of ground-state control of triplet reactivity was provided by Lewis' study of the Norrish type II reaction of benzoylcyclohexanes.[55] Photoenolization of *ortho*-alkyl ketones was the first reaction recognized to show rotational control of triplet reactivity.[56] Conformational equilibrium is much more common; an early example is the type II reaction of benzoylcyclobutanes.[57] Scheme 38.4 lists what determines observed k_H values for each of the three cases. For systems in which the excited state establishes conformational equilibrium before decay, there is only one excited-state lifetime, which is determined by the weighted decay rates of all conformers. If more than one is reactive, then the observed rate constant for reaction is the sum of possibly different rate constants k_r weighted by their individual equilibrium populations χ_r ($\Sigma\chi_r + \Sigma\chi_u = 1$). When ground-state or rotational control obtains, different excited states have different lifetimes whether they undergo the same or different reactions.

conformational equilibrium:	$1/\tau = \Sigma \chi_u \tau_u + \Sigma \chi_r \tau_r$	[4a]
	$k_H^{obs} = \Sigma \chi_r k_H$	[4b]
ground state control	$k_H^{obs} = k_H^0$	[5]
rotational control	$k_H^{obs} = k_{uf} + k_H^0(F^*)$	[6]

SCHEME 38.4

Several general comments can be made about different classes of structure. In acyclic ketones, the regioselectivity of hydrogen abstraction reflects different rate constants for cyclic transition states of different sizes, which are closely related to χ_r values. Normally, the largest α-substituent eclipses the carbonyl in ketones, with the lowest energy conformation being fully staggered. The α-β C-C bond must become gauche for γ-hydrogen abstraction to occur. Extra β-substituents increase the fraction of reactive conformers.[41]

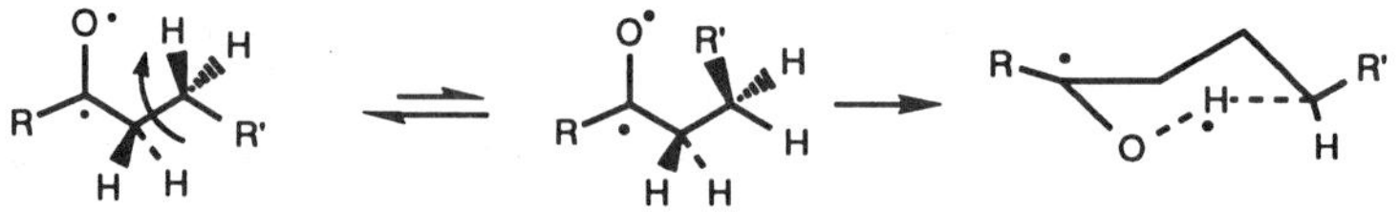

When the carbonyl and C-H bond are part of a ring or substituents on a ring, the conformational limitations of the ring may either increase or decrease the population χ_r of favorable conformations. Lewis demonstrated how incorporation of rings between the carbonyl and the γ-carbon can increase rate constants for γ-hydrogen abstraction.[58] Scheme 38.5 shows some ketones in which the activation energy for abstraction of a secondary hydrogen remains the 14.6 kJ mol^{-1} that it is for bimolecular reaction[28,30] but the entropy of activation becomes less negative as rings freeze intervening C-C bond rotations. Each such frozen rotation increases the rate constant by a factor of ~8, which corresponds to 4 eu. In all three cases, the major conformation is either a reactive one or of comparable energy and the rings limit the number of *unreactive* conformers.

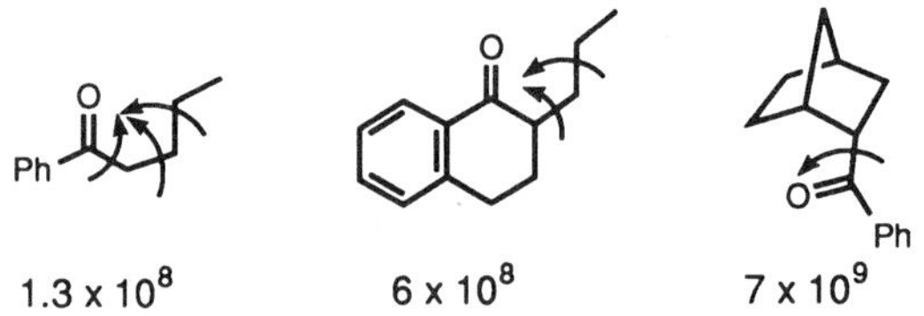

$$1.3 \times 10^8 \qquad 6 \times 10^8 \qquad 7 \times 10^9$$

SCHEME 38.5

In general, however, *there is no constant factor by which rings change intramolecular reactivity.* They may decrease population of a reactive conformation, as Alexander showed for benzoylcyclobutane;[57] limit reaction to one site, as the selective δ-hydrogen abstractions in Paquette's synthesis of dodecahedrane exemplified;[59] or prevent population of a reactive conformer, as Turro reported for cyclohexanones with axial α-alkyl groups.[60]

Regioselectivity

Competing intramolecular reaction rates are determined by the energy required for the molecule to rotate and/or twist into the geometries that bring the excited carbonyl and a remote C-H bond into the proper distance and orientation for reaction. Regioselectivities are determined by such energy and rate differences.

Acyclic Systems. The predominance of γ-hydrogen abstraction by excited straight chain ketones is so pronounced that for years it was thought that no ketone containing γ-hydrogens would react at any other position. A few notable exceptions were eventually found; δ-methoxyvalerophenone[61] and γ-benzoylbutyraldehyde[62] undergo both γ and δ-hydrogen abstraction with comparable rate constants, because both possess a highly reactive δ-C-H bond and an inductively deactivated γ-C-H bond. From these results, it was deduced that the intrinsic γ/δ selectivity in triplet ketones is 20:1.[41] No reaction occurs at any other position.

Houk has published several theoretical studies of intramolecular hydrogen transfer that reproduce accurately the preference for 1,5- vs. 1,6-hydrogen transfers in triplet ketones.[49] Much of the preference is entropic, as expected, but there is little enthalpy difference between the two modes. A chair-like transition state was suggested originally for γ-hydrogen transfers.[41] The calculations predict a strong preference for a linear C-H-O arrangement (θ = 180°), which is accommodated more easily in a seven-atom transition state than in a six-atom one. Thus, relative enthalpies of activation reflect both ring size strain and orbital orientation in the transition state. It must be reiterated that actual ketones react even when θ approaches 90° in the ground state and may adopt a chair-like transition state.

In summary, ketones react preferentially with γ-C-H bonds. Reducing their reactivity can allow more remote hydrogen abstractions to compete, but the only way to eliminate γ-hydrogen abstraction is to eliminate γ-hydrogens. This constraint does not apply to similar hydrogen abstraction reactions of thioketones and *N*-alkylimides.[5]

Cyclic Systems.   The interposition of even one ring between the carbonyl and any γ-C-H bonds imposes conformational restrictions that limit the molecular geometries possible. Consequently, regioselectivity becomes dominated by frozen rotations such that only certain C-H bonds are close enough to the carbonyl to react. Such structures can enhance or suppress the reactivity of γ-C-H bonds and also enhance the reactivity of more remote sites.

Charge Transfer Followed by Proton Transfer

Bimolecular photoreduction of ketones by good electron donors often proceeds by either electron transfer or charge transfer (CT) complexation, followed by a proton transfer. Intramolecular hydrogen transfers do not seem to follow this pathway as often. There is good evidence from the behavior of aminoketones that tight overlap of the donor lone pair with the half-empty *n*-orbital is necessary for triplet CT reaction. Rate constants for internal quenching in ω-dimethylaminoalkyl phenyl ketones are maximum for β- and γ-amino ketones, in which five- or six-atom rings can be formed by overlap of the nitrogen lone pair with the half-occupied carbonyl n orbital.[63] Interestingly, triplet α-dialkylaminoacetophenones do *not* undergo rapid internal charge transfer; they do, however, undergo normal γ-hydrogen abstraction.[26] The rapid internal charge transfer in triplet γ-dialkylaminobutyrophenones does not lead to efficient product formation;[63,64] the required orbital overlap seems to enforce a restricted orientation in the exciplex that prevents proton transfer to oxygen, such that CT becomes a purely competitive quenching process.

8.7 Biradical Behavior

Direct Study of Triplet Biradicals

In 1977, Scaiano and Small reported the first flash spectroscopic detection of a 1,4-biradical intermediate, from γ-methylvalerophenone.[65] They rapidly developed two independent methods to monitor triplet-generated hydroxy biradicals: (1) direct detection of their transient absorption; or (2) indirect detection by following the growth of the strongly absorbing paraquat radical-cation, which is produced when added paraquat oxidizes the hydroxy radical site.[66,67] Scaiano has summarized thoroughly several basic features of triplet 1,4-biradical lifetimes.[68] They are not very sensitive to substitution on the benzene ring or to steric features at the γ-carbon of the ketone, but they are increased by conjugating substituents such as γ-phenyls.[69] They are lengthened ca. threefold in Lewis base solvents, having values in the 25 to 50-ns range in hydrocarbons and 75 to 160 nsec in alcohols.[70] They are almost independent of temperature, having activation energies for decay of 0 to 4 kJ mol^{-1}.[65] They are shortened by bimolecular interaction with paramagnetic species such as oxygen[71] and nitroxide radicals.[72] The most significant structurally induced change is produced by the internal oxygen atom in the biradicals formed from α-alkoxyacetophenones, which have maximum lifetimes of only a few nanoseconds.[69]

Multiplicity

Overwhelming evidence has been found for the intermediacy of triplet 1,4-biradicals following internal hydrogen abstraction by triplet ketones, especially phenyl ketones. Some unquenchable cyclization reactions suggest strongly that singlet biradicals can be formed; unfortunately, they appear to be too short-lived to be detected by trapping or nanosecond spectroscopy. This evanescent character of singlet biradicals helped generate one popular view of how triplet biradicals proceed to singlet products.

Based entirely on his study of some 1,4-biradicals formed by γ-hydrogen abstraction in triplet phenyl ketones, Scaiano postulated that biradical lifetimes are determined by rates of intersystem crossing (isc) to the corresponding singlet biradicals. Moreover, he suggested further that product ratios reflect relative isc rates for different conformations of the triplet biradicals, the singlet biradical being too short-lived to convert into any other product-forming conformation.[73]

More recently, other workers have suggested an alternative view. From magnetic resonance studies of biradicals, Closs suggested that triplet biradicals cyclize whenever their two ends collide, but with a low probability that reflects the amount of singlet character mixed into the biradical by spin-orbit coupling.[74] Michl has suggested a comparable picture, purely from a theoretical viewpoint, with the added feature that incipient bonding of the two ends of the biradical induces additional spin-orbit coupling through a highly stabilizing covalent interaction.[75] Finally, from a comparison of the behavior of several types of 1,5-biradicals with 1,4-biradicals, Wagner has suggested that attainment of product-forming geometries does indeed induce isc so that product ratios reflect normal (small) barriers to bond rotation and product formation.[76] It is important to emphasize that these ideas refer only to the decay of biradicals unaffected by outside agents. Scaiano has provided several dramatic illustrations of how paramagnetic species can change the partitioning of 1,4-biradicals, presumably by inducing isc in conformations that normally decay relatively slowly.

Biradical Disproportionation Back to Ketone

It cannot be overemphasized that the quantum yields for two-step triplet reactions such as those initiated by hydrogen abstraction often reflect partitioning of the intermediate biradicals more than relative rate constants for triplet reaction. This is especially true for the very rapid intramolecular hydrogen abstractions of n,π* triplets, such as recorded in Tables 38.1 and 38.3. This fact was demonstrated first by the behavior of 2-hexanone, in which γ-deuteration *increases* the quantum yield of type II cleavage,[77] and of γ-substituted butyrophenones, in which quantum yields for type II cleavage decrease as rate constants for γ-hydrogen abstraction increase.[78]

Generally, it is true that almost all the hydroxybiradicals formed by internal hydrogen abstraction revert partially to starting ketone by an internal radical disproportionation reaction in which the carbon-centered radical site abstracts the hydrogen from the hydroxyl group. Often, this process is the major reaction of the intermediate biradical, so that the overall quantum yield of product formation is low even when hydrogen abstraction by the excited state is 100% efficient. Should the efficiency of hydrogen abstraction be low because of electronic or conformational problems, then biradical disproportionation lowers quantum efficiency even further.

The addition of Lewis bases inhibits disproportionation back to ketone and thus enhances quantum yields of product formation.[14,15] In many cases, reversion to ketone is totally suppressed and quantum yields approach unity, as discussed above. The ability of different additives to solvate the biradicals, as determined by the concentration needed to maximize product quantum yields, matches their basicities closely.[15] Alcohols, water, ethers, and pyridines are used commonly for this purpose, but all Lewis bases that are not also strong electron donors work to some extent. Not every ketone reacts with 100% efficiency even in the presence of strong Lewis acids. The equilibrium between the solvated and the unsolvated biradical obviously varies with biradical structure. Sometimes, only a fraction of the biradical is solvated; the rest still reverts to ketone.

Can a biradical undergo only disproportionation, so that no product is formed no matter how fast the excited state reacts? A recent example is provided by the behavior of cyclodecanone; its triplet undergoes both γ- and ε-hydrogen abstraction, the former faster than the latter, but none of the 1,4-biradical gives any product except in *t*-butyl alcohol.[53] In general, cyclic and polycyclic systems with little conformational freedom, such that the two radical sites might not be able to get away from each other, would seem the most likely to have this problem.

The only real difference between the reversion to ketone that occurs in both singlet and triplet hydrogen abstractions is the slowness of T→S isc, such that the triplet forms a detectable biradical intermediate. In the singlet process, the molecule can slip onto the pathway for reversion to ground-state ketone directly. This singlet state decay is not suppressed by added Lewis bases,[18,22-24] further evidence that it can proceed without first forming a transient intermediate.

Factors Affecting Biradical Cleavage Efficiency

The cleavage efficiency of triplet generated 1,4-biradicals varies widely with structure. Fortunately, the "shortness" of these 1,4-biradicals prevents several forms of disproportionation that 1,5-biradicals undergo.[4] Several factors affect the partitioning among cleavage, cyclization, and reversion to starting ketones.[80,81] Four geometries of the 1,4-biradical must be considered: two stretched-out staggered conformations and two coiled gauche conformations. Cleavage can occur from either gauche or staggered, whereas cyclization and disproportionation require the latter.

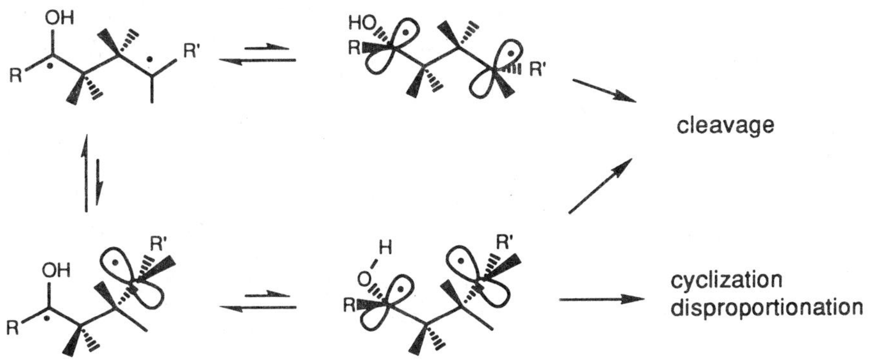

Cleavage appears to be governed by the stereoelectronic necessity for overlap of the breaking bond with both half-occupied *p*-orbitals.[78,82] Anything that prevents such molecular alignment retards or suppresses cleavage. In straight-chain ketones, cleavage/cyclization ratios tend to be at least 5:1. However, α-substitution by alkyl groups,[83] fluorines,[84] or rings[85] increases cyclization, primarily by destabilizing the geometry required for cleavage. α-Alkoxy ketones cyclize as much as they cleave;[37] in this case cleavage is not retarded as much as cyclization is enhanced, presumably by the oxygen relieving ring strain. Diketones do not undergo any cleavage, most likely because resonance holds the would-be breaking bond perpendicular to the hydroxy-radical site.[86] In contrast, environmental effects that prevent coiling or enforce stretched-out conformations make cleavage highly efficient.

Often distinction is made between coiled and stretched conformations in order to rationalize cyclization/cleavage ratios. However, disproportionation also requires a coiled conformation; it is not at all clear what affects the competition between cyclization and disproportionation, other than that hydrogen bonding to Lewis bases dramatically inhibits the latter and changes the stereoselectivity of the former. As an example, compare valerophenone and α,α-difluorovalerophenone. The cyclization/disproportionation ratio in benzene is 3:2 for the latter but only 1:9 for the former.

38.8 Synthetic Uses

Simple type II elimination has been put to use in several ways. Glycosides made from γ-hydroxyketones undergo photoelimination to give *O*-vinyl glycosides, which can undergo a Claisen rearrangement in certain deoxy sugars.[87] Properly alkylated bicyclo[4.2.0]octan-2-ones formed by [2+2]-photocycloaddition of cyclohexenones to alkenes cleave to δ,ε-unsaturated ketones.[88] Scharf has removed the C-17 side chain in steroids by photolysis.[89] Perhaps the most unusual use of type II elimination was the preparation of *m*-xylylene by Goodman and Berson.[90]

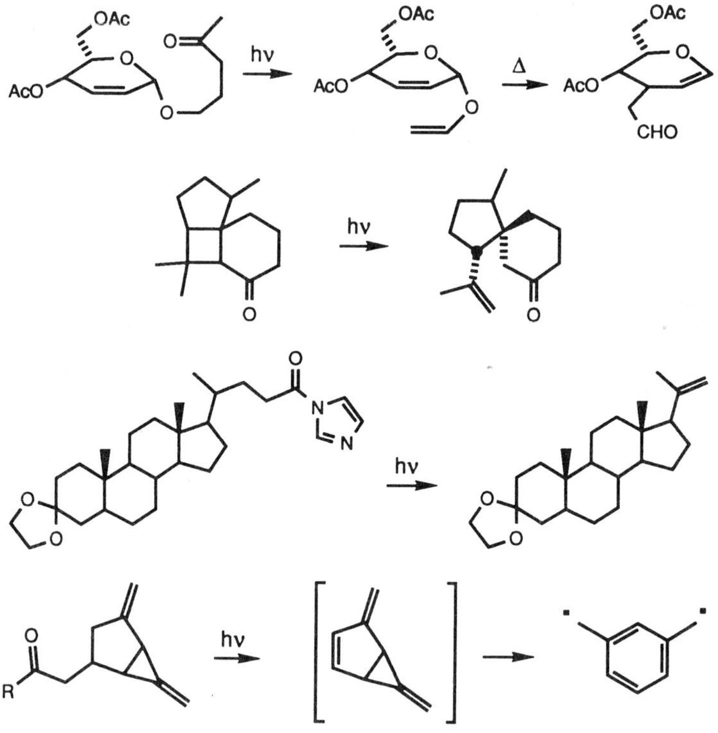

Finally, Wirz and Kresge have exploited brilliantly the type II elimination of various ketones induced by flash photolysis to allow kinetic studies of the transient enol photoproducts.[91]

38.9 Environmental Effects

Both the rate constants for competing triplet reactions and the partitioning of biradical intermediates may be affected by the environment. The major effects on triplet reactivity involve conformational changes induced by highly ordered media and decreases in the efficiency of α-cleavage of the radical. The major effects on biradicals involve conformational restrictions that impede or promote coiling and solvation that inhibits disproportionation back to ketone.

Several laboratories have found high quantum yields and product ratios that suggest a fairly polar environment for ketones irradiated in aqueous surfactant solution.[92] It is well accepted now that the polar end of the ketones and particularly the biradicals reside mostly near the micelle-solvent interface, called the Stern layer, which has Lewis base character. Although micelles produce negligible effects on triplet rate constants, they can improve type II/type I ratios by their "super-cage" effect that enhances the recoupling of radical pairs.[93]

The benzoin ethers were a mechanistic puzzle for years, since they were reported first to undergo only α-cleavage to radicals and *no* type II reaction in solution,[94] despite the high rate constants for γ-hydrogen abstraction in simple α-alkoxy ketones.[37] The use of methanol as solvent provides 5 to 10% type II reaction. Cyclization and type II elimination occur also when benzoin ethers dissolved in micelles[95] or adsorbed on silica are irradiated.[96] Irradiation of solid complexes of three common benzoin ethers with β-cyclodextrin produces 90% type II reaction.[97] All the "organized" media apparently force the ketone into a more reactive geometry than obtains in solution and, just as important, decrease translational mobility, thus making type I cleavage mostly revertible. There are several other examples of ketones that undergo mainly type I cleavage in solution but mostly or entirely type II reaction when complexed with β-cyclodextrin.[98] These high product selectivities are achieved only when the solid complex is irradiated. Aqueous solutions of ketone and cyclodextrin give more type II reaction than occurs in benzene, but radical cleavage still competes strongly.

Turro has looked at the effects of several zeolites on the photochemistry of α,α-dimethylvalerophenone,[99] which undergoes competitive type I and type II reaction in solution.[100] Depending on the cavity size of the zeolite, product ratios vary substantially. When mainly type I cleavage is observed, the cavities must be too small to allow the conformational changes required for type II reaction.

Weiss has explored the effects of liquid crystals on the type II reaction. Their effects on short ketones are variable, whereas their effects on long ketones can be dramatic. His study of various *p*-alkylalkanophenones in *n*-butyl stearate reveals key aspects of how guest molecules orient themselves in liquid crystals.[101] In short, when the carbonyl group is in the middle of the ketone molecule, very little γ-hydrogen abstraction occurs, apparently because most of the carbonyls are in the middle of a liquid crystal network and the alkyl chains are fully stretched alongside the stearate molecules. What little reaction does occur arises from random misaligned ketones. When the carbonyl group is at one end of the ketone, much more hydrogen abstraction occurs, presumably because of lower rigidity in the stearate framework. Whenever the γ-radical site is part of a long alkyl group that is held in a stretched geometry, cleavage dominates cyclization.

Whitten studied a ketone that is also a surfactant, ω-(*p*-toluyl)pentadecanoic acid.[102] When prepared as a monolayer film in arachidic acid, this ketone undergoes only a trace of type II reaction. In benzene, it cleaves to *p*-methylacetophenone with a normal $\Phi = 0.20$; in an aqueous SDS solution, $\Phi = 0.80$. The enhanced quantum efficiency in the micelle was discussed above; it is interesting that the molecule must be looped so that both carbonyls are in the Stern layer. The lack of γ-hydrogen abstraction in the monolayer is ascribed to the linear rigidity of the monolayer environment. The monolayer would appear to allow even less molecular flexibility than does the liquid crystal.

The photochemistry of polymers containing keto groups has been studied extensively and reviewed by Guillet.[103] The type II reaction has been of particular interest inasmuch as it is partially responsible for the photodegradation of polymers. In this regard, temperature effects on the quantum efficiency for cleavage of ethylene-CO copolymers have been associated with glass transitions that reduce conformational mobility and the eventual freezing out of key rotations.[104] At temperatures above the glass transition, the type II cleavage of both amorphous host polymers and guest ketones proceeds with efficiencies similar to those in solution. Whatever constraints the polymer places on molecular motion, they are not large enough to prevent the relatively small motions needed for γ-hydrogen abstraction and cleavage. Hrdlovic and Guillet have studied carefully the behavior of polyacrylophenones and copolymers of styrene and phenyl vinyl ketone.[105] Guillet has even produced a purposely photodegradable polystyrene copolymer based on type II cleavage of the backbone.

Wamser has studied the photochemistry of cyclohexylacetyl groups bonded by Friedel-Crafts acylation to polystyrene beads, which provides an α-cyclohexyl acetophenone structure.[106] The polymer-bound ketone reacts with almost the same quantum efficiency (0.5) as the free ketone in pentane, but is unreactive in ethanol, which does not swell the polymer and allow the molecular motion necessary for reaction.

The crystalline state also provides some unique behavior. Scheffer has studied five α-cycloalkyl-*p*-chloroacetophenones, all of which react both in solution and in the crystal.[46,107] They give comparable product ratios in both phases. With the knowledge that organic molecules crystallize normally in their most stable conformations, examination of solution rate constants for hydrogen abstraction revealed no correlation with the geometric parameters for reaction. The rate constants do correlate well with those for bimolecular attack of various radicals on the different sized cycloalkanes, so the intrinsic reactivities of the different C-H bonds appear to determine relative reactivities. Scheffer concluded that Equation 38.4 applies, with comparable values of χ for all five. He suggests that these ketones react not from their most stable conformations but rather from higher energy conformations that provide a more favorable geometry for reaction. The value of θ averages only 114° for the crystals. It is possible also that they all react from their crystal geometries, in which case χ_r values are all close to 1. All of the ketones except the cyclobutyl example are considerably more reactive than *p*-chlorovalerophenone despite their awkward geometries. The rate enhancements probably represent the normal entropic gain associated with the ring.[58] The high observed k values cannot contain a really low value of χ_r; thus, any common geometric adjustment in solution must involve a relatively low energy conformational change.

n = 1-5

Ring size	d (Å)	ω(°)	Δ(°)	k, 10^8 s^{-1}	% Cleavage
4	3.1	23	78	0.3	92
5	2.8	31	96	1.2	92
6	2.6	42	90	1.2	38
7	2.7	42	82	5.7	59
8	2.7	48	77	6.7	18

These results also provide unique insight into biradical behavior. The ketones with the smallest two rings undergo primarily cleavage; the next two split nearly 50:50; the largest ring gives mainly cyclobutanol. In the last case, the 18% elimination gives mainly *cis*-cyclooctene, the *cis/trans* ratio being 9:1 in benzene and 2.5:1 in the crystal. However, over 85% of the cyclization gives the *trans* ring junction. Scheffer provides an intriguing explanation by noting that the *p*-orbital on the ring in the biradical must twist 90 ± 45° to become parallel to the *p*-orbital at the hydroxy radical site and achieve the geometry required for both cleavage and cyclization. Rotation in the direction that gives *cis*-products is favored for the two smallest rings; rotation in the opposite direction, which favors formation of *trans*-products, is favored for the largest ring; the two rotations are equally likely for the six- and seven-member rings. Because the three largest ketones yield mostly *trans*-fused cyclization products but *cis*-cycloalkene cleavage product, Scheffer concludes that the pre-*cis*-biradical undergoes mainly cleavage and the pre-*trans*-biradical undergoes mainly cyclization. He proposes further that the phase-independence of the results indicates that these preferences are intrinsic to the structure of the molecules. Scheme 38.6 depicts the proposed biradical geometries and indicates the product ratios as a function of ring size.

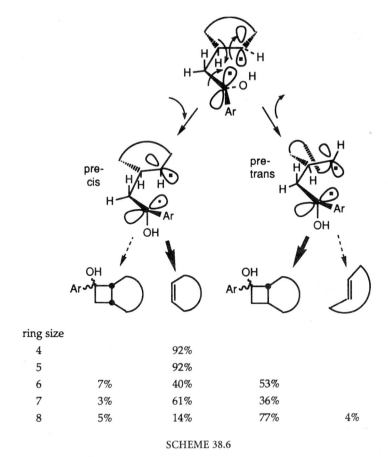

ring size			
4		92%	
5		92%	
6	7%	40%	53%
7	3%	61%	36%
8	5%	14%	77%	4%

SCHEME 38.6

References

1. Wagner, P. J., Type II photoelimination and photocyclization of ketones, *Acc. Chem. Res.*, 4, 168, 1971.
2. Dalton, J. C. and Turro, N. J., Photoreactivity of n,π* excited states of alkyl ketones, *Annu. Rev. Phys. Chem.*, 21, 499, 1970.
3. Wagner, P. J., Chemistry of excited triplet organic carbonyl compounds, *Topics Curr. Chem.*, 66, 1, 1976.
4. Wagner, P. J., 1,5-Biradicals and five-membered rings generated by δ-hydrogen abstraction in photoexcited ketones, *Acc. Chem. Res.*, 22, 83, 1989.
5. Wagner, P. and Park, B.-S., Photoinduced hydrogen atom abstraction by carbonyl compounds, *Org. Photochem.*, 11, 227, 1991.
6. Norrish, R. G. W. and Appleyard, M. E. S., Primary photochemical reactions. IV. Decomposition of methyl ethyl ketone and methyl butyl ketone, *J. Chem. Soc.*, 874, 1934.
7. Yang, N. C. and Yang, D.-H., Cyclobutanol formation from irradiation of ketones, *J. Am. Chem. Soc.*, 80, 2913, 1958.
8. McMillan, G. R., Calvert, J. G., and Pitts, J. N., Jr., Detection and lifetime of enol-acetone in the photolysis of 2-pentanone vapor, *J. Am. Chem. Soc.*, 86, 3602, 1964.
9. Ausloos, P. and Rebbert, R. E., Photoelimination of ethylene from 2-pentanone, *J. Am. Chem. Soc.*, 86, 4512, 1964.

10. Michael, J. L. and Noyes, W. A., Jr., Photochemistry of 2-pentanone with biacetyl, *J. Am. Chem. Soc.*, 85, 1027, 1963.

11. (a) Wagner, P. J. and Hammond, G. S., Mechanism of type II photoelimination, *J. Am. Chem. Soc.*, 87, 4009, 1965; (b) Dougherty, T. J., *J. Am. Chem. Soc.*, 87, 4011, 1965.

12. Baum, E. J., Wan, J. K. S., and Pitts, J. N., Jr., Reactivity of excited states. Intramolecular hydrogen atom abstraction in substituted butyrophenones, *J. Am. Chem. Soc.*, 88, 2652, 1966.

13. Wagner, P. J. and Hammond, G. S., Quenching of type II photoelimination reaction, *J. Am. Chem. Soc.*, 88, 1245, 1966.

14. Wagner, P. J., Solvent effects on type II photoelimination of phenyl ketones, *J. Am. Chem. Soc.*, 89, 5898, 1967.

15. Wagner, P. J., Kochevar, I. E., and Kemppainen, A. E., Type II photoprocesses of phenyl ketones. Procedures for determining meaningful quantum yields and triplet lifetimes, *J. Am. Chem. Soc.*, 94, 7489, 1972.

16. Wagner, P. J. and Zepp, R. G., Trapping by mercaptans of the biradical intermediates in type II photoelimination, *J. Am. Chem. Soc.*, 94, 287, 1972.

17. Wagner, P. J., Kelso, P. A., and Zepp, R. G., Type II photoprocesses of phenyl ketones. Evidence for a biradical intermediate, *J. Am. Chem. Soc.*, 94, 7480, 1972.

18. Yang, N. C. and Elliot, S. P., Photochemistry of *(S)*-(+)-5-methyl-2-heptanone, *J. Am. Chem. Soc.*, 91, 7550, 1969.

19. Yang, N. C., Morduchowitz, A., and Yang, D.-H., On the mechanism of photochemical formation of cyclobutanols, *J. Am. Chem. Soc.*, 85, 1017, 1963.

20. Wagner, P. J., 1-Methylnaphthalene-sensitized singlet state reactions of aliphatic ketones, *Mol. Photochem.*, 3, 169, 1971.

21. Salem, L. and Rowland, C., The electronic properties of diradicals, *Angew. Chem., Int. Ed.*, 11, 92, 1972.

22. Barltrop, J. A. and Coyle, J. C., Singlet and triplet participation in the photochemistry of simple alkanones, *Tetrahedron Lett.*, 3235, 1968.

23. Wagner, P. J., Differences between singlet and triplet state type II photoelimination of aliphatic ketones, *Tetrahedron Lett.*, 5385, 1968.

24. Yang, N. C., Elliot, S. P., and Kim, B., The mechanism of photochemistry of alkanones with γ hydrogens, *J. Am. Chem. Soc.*, 91, 7551, 1969.

25. Yang, N. C. and Shani, A., Photochemistry of β-naphthyl alkyl ketones in solution, *J. Chem. Soc., Chem. Commun.*, 815, 1971; Coyle, J. C., A type 2 photoelimination reaction of 1-naphthyl ketones, *J. Chem. Soc., Perkin Trans. 2*, 233, 1973.

26. Wagner, P. J. and Jellinek, T., Intramolecular quenching of the excited singlets of ω-dialkylamino alkyl ketones. Singlet state type II photoelimination of α-dimethylaminoacetophenone, *J. Am. Chem. Soc.*, 93, 7328, 1971.

27. Wagner, P. J., Kemppainen, A. E., and Schott, H. N., Effect of ring substituents on type II photoreactions of phenyl ketones, *J. Am. Chem. Soc.*, 95, 5604, 1973.

28. Encina, M. V., Lissi, E. A., Lemp, E., Zanocco, A., and Scaiano, J. C., Temperature dependence of the photochemistry of aryl alkyl ketones, *J. Am. Chem. Soc.*, 105, 1856, 1983.

29. Wagner, P. J., Truman, R. J., and Scaiano, J. C., Substituent effects on hydrogen abstraction by phenyl ketone triplets, *J. Am. Chem. Soc.*, 107, 7093, 1985.

30. Berger, M., McAlpine, E., and Steel, C., Substituted acetophenones. Importance of activation energies in mixed state models of photoreactivity, *J. Am. Chem. Soc.*, 100, 5147, 1978.

31. Hammond, G. S. and Leermakers, P. A., Photoreduction of 1-naphthaldehyde and 2-acetonaphthone, *J. Am. Chem. Soc.*, 84, 207, 1962.

32. deBoer, C. D., Herkstroeter, W. G., Marchetti, A. P., Schultz, A. P., and Schlessinger, R. H., Norrish type II rearrangement from π,π^* triplet states, *J. Am. Chem. Soc.*, 95, 3963, 1973.

33. Zimmerman, H. E., A new approach to mechanistic organic photochemistry, *Adv. Photochem.*, 1, 183, 1963.

34. Walling, C. and Gibian, M. J., Hydrogen abstraction reactions by the triplet states of ketones, *J. Am. Chem. Soc.*, 87, 3361, 1965.

35. Wagner, P. J., Thomas, M. J., and Puchalski, A. E., Photoreactivity of α-fluorinated phenyl alkyl ketones, *J. Am. Chem. Soc.*, 108, 7739, 1986.

36. Evans, T. R. and Leermakers, P. A., The Norrish type II process in α-keto acids. Photolysis of α-keto decanoic acid in benzene, *J. Am. Chem. Soc.*, 90, 1840, 1968.

37. Turro, N. J. and Lewis, F. D., Type II photoelimination and 3-oxetanol formation from α-alkoxyacetophenones and related compounds, *J. Am. Chem. Soc.*, 92, 311, 1970.

38. Previtali, C. M. and Scaiano, J. C., The kinetics of photochemical reactions. 1. Application of a modified bond-energy-bond-order method to the atom abstraction reactions of excited carbonyl compounds, *J. Chem. Soc., Perkin Trans.*, 2, 1667, 1972.

39. Giering, L., Berger, M., and Steel, C., Rate studies of aromatic triplet carbonyls with hydrocarbons, *J. Am. Chem. Soc.*, 96, 953, 1974.

40. Wagner, P. J., Lindstrom, M. J., Sedon, J. H., and Ward, D. R., Photochemistry of δ-halo ketones: Anchimeric assistance in triplet-state γ-hydrogen abstraction and β-elimination of halogen atoms from the resulting diradicals, *J. Am. Chem. Soc.*, 103, 3842, 1981.

41. Wagner, P. J., Kelso, P. A., Kemppainen, A. E., and Zepp, R. G., Type II photoprocesses of phenyl ketones. Competitive δ-hydrogen abstraction and the geometry of intramolecular hydrogen atom transfers, *J. Am. Chem. Soc.*, 94, 7500, 1972.

42. Chang, H. C., Popovitz-Biro, R., Lahav, M., and Leiserowitz, L., Mapping the molecular pathway during photoaddition of guest acetophenone and *p*-fluoroacetapheone to host deoxycholic acid as studied by X-ray defraction in systems undergoing single-crystal-to-single-crystal transformation, *J. Am. Chem. Soc.*, 109, 3883, 1987.

43. Wagner, P. J., Zhou, B., Hasegawa, T., and Ward, D. L., Diverse photochemistry of sterically congested α-arylacetophenones: Ground-state conformational control of reactivity, *J. Am. Chem. Soc.*, 113, 9640, 1991.

44. Ito, Y., Matsuura, T., and Fukuyama, K., Efficiency for solid-state photocyclization of 2,4,6-triisopropylbenzophenones, *Tetrahedron Lett.*, 29, 3087, 1988.

45. Ariel, S., Ramamurthy, V., Scheffer, J. R., and Trotter, J., Norrish type II reaction in the solid state: Involvement of a boat-like reactant conformation, *J. Am. Chem. Soc.*, 105, 6959, 1983.

46. Scheffer, J. R., The influence of the molecular crystalline environment on organic photorearrangements, *Org. Photochem.*, 8, 249, 1987.

47. Chandler, W. D. and Goodman, L., Allowed and forbidden character in $\pi^* \leftarrow$ n spectra of cycloalkanones, *J. Mol. Spectrosc.*, 35, 232, 1970.

48. Severance, D., Pandey, B., and Morrison, H., Reaction path analysis of hydrogen abstraction by the formaldehyde triplet state, *J. Am. Chem. Soc.*, 109, 3231, 1987.

49. Dorigo, A. E., McCarrick, M. A., Loncharich, R. J., and Houk, K. N., Transition structures for hydrogen atom transfers to oxygen. Comparisons of intermolecular and intramolecular processes and open- and closed-shell systems, *J. Am. Chem. Soc.*, 112, 7508, 1990.

50. (a) Sugiyama, N., Nishio, T., Yamada, K., and Aoyama, H., Photochemical reactions of bridged polycyclic ketones, *Bull. Chem. Soc. Jpn.*, 43, 1879, 1970.

51. (a) Sauers, R. R., Scimone, A., and Shams, H., Synthesis and photochemistry of some new pentacycloundecan-8-ones. Probes for hydrogen abstraction in the π-plane, *J. Org. Chem.*, 53, 6084, 1988; (b) Sauers, R. R. and Krogh-Jesperson, K., Analysis of Norrish type II reactions by molecular mechanics methodology, *Tetrahedron Lett.*, 30, 527, 1989.

52. Barnard, M. and Yang, N. C., Proximity effect in photochemical reactions, *Proc. Chem. Soc., London*, 302, 1958.

53. Sauers, R. R. and Huang, S.-Y., Analysis of Norrish type II reactions by molecular mechanics methodology: Cyclodecanone, *Tetrahedron Lett.*, 31, 5709, 1990.

54. Wagner, P. J. Conformational flexibility and photochemistry, *Acc. Chem. Res.*, 16, 461, 1983.

55. Lewis, F. D., Johnson, R. W., and Johnson, D. E., Conformational control of photochemical behavior. Competitive α cleavage and γ-hydrogen abstraction of alkyl phenyl ketones, *J. Am. Chem. Soc.*, 96, 6090, 1974.

56. Wagner, P. J. and Chen, C.-P., A rotation-controlled excited state reaction: The photoenolization of *ortho*-alkyl phenyl ketones, *J. Am. Chem. Soc.*, 98, 239, 1976.

57. Alexander, E. C. and Uliana, J. A., Photolysis of cyclobutyl aryl ketones. Evidence for the involvement of an excited state conformational equilibrium in their photoconversion to aryl bicyclo[1.1.1]pentanols, *J. Am. Chem. Soc.*, 96, 5644, 1974.

58. Lewis, F. D., Johnson, R. W., and Kory, D. R., Transition state for γ-hydrogen abstraction of alkyl phenyl ketones, *J. Am. Chem. Soc.*, 96, 6100, 1974.

59. Paquette, L. A. and Balogh, D. W., An experdient synthesis of 1,16-dimethyl-dodecahedrane, *J. Am. Chem. Soc.*, 104, 774, 1982.

60. Turro, N. J. and Weiss, D. S., Stereoelectronic requirements for the type II cleavage of *cis*- and *trans*-4-*t*-butylcyclohexanones, *J. Am. Chem. Soc.*, 90, 2185, 1968.

61. Wagner, P. J. and Zepp, R. G., γ vs. δ-Hydrogen abstraction in the photochemistry of β-alkoxyketones. An overlooked reaction of hydroxybiradicals, *J. Am. Chem. Soc.*, 93, 4958, 1971.

62. Ounsworth, J. and Scheffer, J. R., Intramolecular abstraction of aldehydic hydrogen by ketone triplets: Formation of 2-hydroxy-2-phenylcycloalkanones, *J. Chem. Soc., Chem. Commun.*, 232, 1986.

63. Wagner, P. J., Kemppainen, A. E., and Jellinek, T., Type II photoreactions of phenyl ketones. Competitive charge transfer in α-, γ-, and δ-dialkylamino ketones, *J. Am. Chem. Soc.*, 94, 7512, 1972.

64. Wagner, P. J. and Ersfeld, D. A., Solvent specific photochemistry involving an intramolecular amino ketone triplet exciplex, *J. Am. Chem. Soc.*, 98, 4515, 1976.

65. Small, R. D. Jr. and Scaiano, J. C., Direct detection of the biradicals generated in the Norrish type II reaction, *Chem. Phys. Lett.*, 50, 431, 1977.

66. Small, R. D. Jr. and Scaiano, J. C., Photochemistry of phenyl alkyl ketones. The lifetime of the intermediate biradicals, *J. Phys. Chem.*, 81, 2126, 1977.

67. Small, R. D. Jr., and Scaiano, J. C., One electron reduction of paraquat dication by photogenerated biradicals, *J. Photochem.*, 6, 453, 1976/77.

68. Scaiano, J. C., Laser flash photolysis studies of the reactions of some 1,4-biradicals, *Acc. Chem. Res.*, 15, 252, 1982.

69. Caldwell, R. A., Majima, T., and Pac, C., Some structural effects on triplet biradical lifetimes. Norrish II and Paternò-Büchi biradicals, *J. Am. Chem. Soc.*, 104, 629, 1982.

70. Small, R. D. Jr. and Scaiano, J. C., Solvent effects on the lifetimes of photogenerated biradicals, *Chem. Phys. Lett.*, 59, 246, 1978.

71. Small, R. D. Jr. and Scaiano, J. C., Interaction of oxygen with transient biradicals photogenerated from γ-methyl valerophenone, *Chem. Phys. Lett.*, 48, 354, 1977; Small, R. D., Jr. and Scaiano, J. C., Differentiation of excited-state and biradical processes. Photochemistry of phenyl alkyl ketones in the presence of oxygen, *J. Am. Chem. Soc.*, 100, 4512, 1978.

72. Encinas, M. V. and Scaiano, J. C., Interaction between photogenerated biradicals and free radicals: di-*tert*-butylnitroxide, *J. Photochem.*, 11, 241, 1979.

73. Scaiano, J. C., Does intersystem crossing in triplet biradicals generate singlets with conformational memory?, *Tetrahedron*, 38, 819, 1982.

74. Closs, G. and Redwine, O. D., Cyclization and disproportionation kinetics of triplet generated, medium chain length, localized biradicals measured by time-resolved CIDNP, *J. Am. Chem. Soc.*, 107, 4543, 1985.

75. Michl, J. and Bonacic-Koutecky, V., *Electronic Aspects of Organic Photochemistry*, Wiley, New York, 1990.

76. Wagner, P. J. and Park, B.-S., High diastereoselectivity in the cyclization of 1,5-biradicals: What causes such sizeable steric barriers to biradical coupling?, *Tetrahedron Lett.*, 32, 165, 1991; Wagner, P. J., Meador, M. A., Zhou, B., and Park, B.-S., Photocyclization of a-(*o*-tolyl)acetophenones: Triplet and 1,5-biradical reactivity, *J. Am. Chem. Soc.*, 113, 9630, 1991.

77. Coulson, D. R. and Yang, N. C., Deuterium isotope effects in the photochemistry of 2-hexanone, *J. Am. Chem. Soc.*, 88, 4511, 1966.

78. Wagner, P. J. and Kemppainen, A. E., Is there any correlation between quantum yields and triplet-state reactivity in type II photoelimination?, *J. Am. Chem. Soc.*, 90, 5896, 1968.

79. Caldwell, R. A., Dhawan, S. N., and Moore, D. E., pH Dependence of the lifetime of a Norrish II biradical, *J. Am. Chem. Soc.*, 107, 5163, 1985.

80. Scaiano, J. C., Lissi, E. A., and Encina, M. V., Chemistry of the biradicals produced in the Norrish type II reaction, *Rev. Chem. Intermediates*, 2, 139, 1978.

81. Wagner, P. J., Kelso, P. A., Kemppainen, A. E., McGrath, J. M., Schott, H. N., and Zepp, R. G., Type II photoprocesses of phenyl ketones. A glimpse at the behavior of 1,4 biradicals, *J. Am. Chem. Soc.*, 94, 7506, 1972.

82. Hoffman, R., Swaminathau, S., Odell, B. G., and Gleiter, R., A potential surface for a nonconcerted reaction. Tetramethylene, *J. Am. Chem. Soc.*, 92, 7091, 1970.

83. Lewis, F. D. and Hilliard, T. A., Photochemistry of methyl-substituted butyrophenones, *J. Am. Chem. Soc.*, 94, 3852, 1972.

84. Wagner, P. J. and Thomas, M. J., Enhanced photocyclization of α-fluoroketones, *J. Am. Chem. Soc.*, 98, 241, 1976.

85. Lewis, F. D., Johnson, R. W., and Ruden, R. A., Photochemistry of bicycloalkylphenyl ketones, *J. Am. Chem. Soc.*, 94, 4292, 1972.

86. Wagner, P. J., Zepp, R. G., Liu, K.-C., Thomas, M., Lee, T.-J., and Turro, N. J., Competing photocyclization and photoenolization of phenyl α-diketones, *J. Am. Chem. Soc.*, 98, 8125, 1976.

87. Descotes, G. et al., Photochemical synthesis of *o*-vinyl glycosides and their transformation into C-branched sugars, *Synthesis*, 711, 1979.

88. Manh, D. D. K., Ecoto, J., Fetizon, M., Colin, H., and Diez-Masa, J.-C., A new approach to spirosesquiterpenes of the acorane family, *J. Chem. Soc., Chem. Commun.*, 953, 1981.

89. Hilgeis, G. and Scharf, H.-D., Synthese von 5β-Androstan-3,17-dion aus Cholsäure, *Liebigs Ann. Chem.*, 1498, 1985.

90. Goodman, J. L. and Berson, J. A., Formation and intermolecular capture of *m*-quinodimethane, *J. Am. Chem. Soc.*, 106, 1867, 1984.

91. Haspra, P., Sutter, A., and Wirz, J., Acidity of acetophenone enol in aqueous solution, *Angew. Chem., Int. Ed.*, 18, 617, 1979; Chiang, Y., Kresge, A. J., Tang, Y. S., and Wirz, J., pK$_a$ and keto-enol equilibrium constant of acetone in aquious solution, *J. Am. Chem. Soc.*, 106, 460, 1984.

92. Turro, N. J., Liu, K.-C., and Chow, M.-F., Solvent sensitivity of type II photoreaction of ketones as a device to probe solute location in micelles, *Photochem. Photobiol.*, 26, 413, 1977.

93. Turro, N. J. and Weed, G. C., Micellar systems as "supercages" for reactions of geminate radical pairs. Magnetic effects, *J. Am. Chem. Soc.*, 105, 1861, 1983.

94. Pappas, S. P. and Chattopadhyay, A., Photochemistry of benzoin ethers. Type I cleavage by low energy sensitization, *J. Am. Chem. Soc.*, 95, 6484, 1973.

95. Devanathan, S. and Ramamurthy, V., Intramolecular orientation at the micellar interface: control of Norrish type I and type II reactivity of benzoin alkyl ethers via conformational effects, *J. Phys. Org. Chem.*, 1, 91, 1988.

96. deMayo, P., Nakamura, A., Tsang, P. W. K., and Wong, S. K., Surface photochemistry: deviation of the course of reaction in benzoin ether photolysis by absorption on silica gel, *J. Am. Chem. Soc.*, 104, 6824, 1982.

97. Dasarathu Reddy, G., Ramanathan, K. V., and Ramamurthy, V., Modification of photochemistry by cyclodextrin complexation: Competitive Norrish type I and type II reactions of benzoin alkyl ethers, *J. Org. Chem.,* 51, 3085, 1986.

98. Dasarathu Reddy, G. and Ramamurthy, V., Modification of photochemical reactivity by cyclodextrin complexation: Alteration of photochemical behavior via restriction of translational and rotational motions. Alkyldeoxybenzoins, *J. Org. Chem.,* 52, 5521, 1987.

99. Turro, N. J. and Wan, P., Photochemistry of phenyl alkyl ketones adsorbed on zeolite molecular sieves. Observation of pronounced effects on type I/type II photochemistry, *Tetrahedron Lett.,* 25, 3655, 1984.

100. Wagner, P. J. and McGrath, J. M., Competitive types I and II photocleavage of α,α-dimethylvalerophenone, *J. Am. Chem. Soc.,* 94, 3849, 1972.

101. He, Z. and Weiss, R. G., Length and direction-specific solute-solvent interactions as determined from Norrish II reactions of *p*-alkylalkanophenones in ordered phases of *n*-butyl stearate, *J. Am. Chem. Soc.,* 112, 5535, 1990.

102. Worsham, P. R., Eaker, D. W., and Whitten, D. G., Ketone photoreactivity as a probe of the microenvironment: Photochemistry of the surfactant ketone 16-oxo-16-*p*-tolylhexadecanoic acid in monolayers, micelles, and solution, *J. Am. Chem. Soc.,* 100, 7091, 1978.

103. Guillet, J. E., *Polymer Photophysics and Photochemistry,* Cambridge University Press, Cambridge, 1985, chap. 10; Guillet, J. E., Photochemistry and molecular motion in solid amorphous polymers, *Adv. Photochem.,* 14, 91, 1988.

104. Hartley, G. H. and Guillet, J. E., Photochemistry of ketone polymers. I. Studies of ethylene-carbon monoxide copolymers, *Macromolecules,* 1, 165, 1968.

105. Golemba, F. J. and Guillet, J. E., Photochemistry of ketone polymers. VII. Polymers and copolymers of phenyl vinyl ketone, *Macromolecules,* 5, 212, 1972.

106. Wamser, C. C. and Wagner, W. R., Type II photoelimination from α-cycloalkylacetophenones and a polystyrene-bound analogue, *J. Am. Chem. Soc.,* 103, 7232, 1981.

107. Ariel, S., Evans, S. V., Garcia-Garibay, M., Harkness, B. R., Omkaram, N., Scheffer, J. R., and Trotter, J., The generation of 1,4-biradicals in rigid media: Crystal structure-solid state reactivity correlation, *J. Am. Chem. Soc.,* 110, 5591, 1988.

Norrish Type II Processes of Ketones: Influence of Environment

Richard G. Weiss

Georgetown University

39.1 Introduction

Chapters 38, 39, 42–46, and 52 describe several important photochemical reactions of carbonyl-bearing molecules. Aspects of another important photochemical reaction of ketones, the Norrish type II processes (Equation 39.1) are treated in Chapters 47 and 49–51. In the present discussion, both fragmentation (E) and cyclization (C) processes of the transient hydroxy-1,4-biradicals (BR)[1] are included under the same rubric. Note that the *trans-* and *cis*-cyclobutanols (*t* and *c*) are defined as the isomers with their two alkyl or aryl substituents on the cyclobutane ring in *trans-* and *cis*-orientations, respectively.

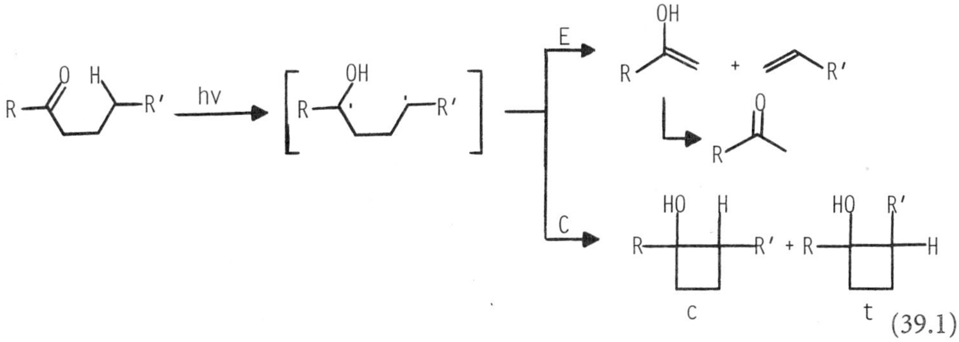

$$(39.1)$$

A more detailed mechanism for the reaction that includes the species considered important to product formation is presented in Scheme 39.1. In flexible ketones that contain few intramolecular steric constraints, the vast majority of the families of *cisoid*-1,4-biradicals (c-BR$_1$ and c-BR$_2$) appear to collapse to cyclization products, while the *transoid*-biradicals (t-BR) fragment. The environment in which the electronic excitation of a ketone molecule takes place can have important consequences on the eventual mixture of products and the quantum yields of their formation. This is due

0-8493-8634-9/95/$0.00+$.50
© 1995 by CRC Press, Inc.

to the influences of medium on (1) the efficiency of initial hydrogen abstraction (to form *i*-BR),
(2) the fraction of BR that returns to ketone, (3) the rates at which *i*-BR transforms to *c*-BR and
t-BR, (4) the equilibrium among the *c*-BR and *t*-BR, and (5) the lifetime(s) of the BR conformers.
Since alkanones undergo reaction in both their excited singlet and triplet manifolds,[2] each of the
factors above must be considered separately for ¹BR and ³BR species; aromatic ketones, which
undergo reaction almost exclusively from the triplet manifold, are conceptually simpler to treat
mechanistically. Additionally, alternative photochemical processes, to the extent that they occur,
will diminish the Norrish type II quantum yields; there is less likelihood that they can influence the
ratios of Norrish type II products.

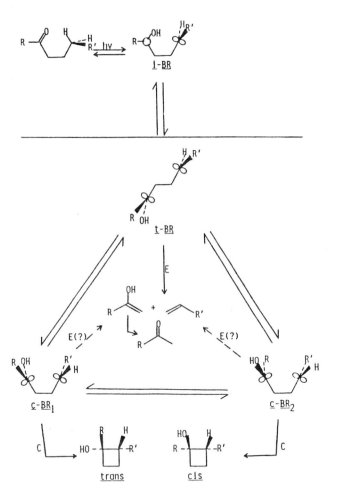

SCHEME 39.1.   A detailed mechanism of the Norrish II reaction which emphasizes the intermediates germane to product selection.

In spite of this complexity, clear manifestations of environmental control over specific aspects
of the processes outlined in Scheme 39.1 have been demonstrated. In this chapter, examples will be
cited that illustrate these points. More comprehensive accounts of this subject have appeared
elsewhere,[3] and interested readers should consult other sources for detailed descriptions of the
various media discussed here.[4] Although anisotropic environments will be emphasized, some
mention of isotropic solvent effects will be included also.

Table 39.1 Influence of Solvent on the Quantum
Yields for Loss of Valerophenone (Φ_-) and for
Appearance of the Fragmentation Product,
Acetophenone (Φ_E)

Solvent	Φ_-	Φ_E
Benzene	0.44	0.37
Hexane	0.46	0.40
t-BuOH	1.0	0.90
i-PrOH	1.0	0.68
Ethanol	1.0	0.70
Acetonitrile	1.0	0.85

From Reference 5.

9.2 Quantum Yields

Norrish type II quantum yields of both aliphatic and aromatic ketones are known to be sensitive to solvent polarity.[5] For instance, as shown in Table 39.1, valerophenone is converted to photoproducts much more efficiently in polar solvents than in nonpolar ones. Since activation energies for γ-hydrogen abstraction from excited singlet and triplet states of ketones, resulting in formation of i-BR, are usually less than 10 kcal mol⁻¹ in isotropic solvents,[6] quantum yields of the Norrish type II processes do not vary dramatically with temperature unless competing reactions like Norrish type I processes become important.

In a qualitative study, Slivinskas and Guillet[7] found that a symmetrical n-alkanone in its neat crystalline phase at room temperature was almost totally unreactive to UV radiation. Subsequent investigations on other *sym-n*-alkanones confirmed this observation and led to the conclusion that the very small amounts of photoproducts that do form after protracted irradiations arise from molecules at defect sites.[8]

The crystalline packing arrangement of n-alkanones has molecules in extended, rod-like conformations and packed in layers.[9] The long axes of the molecules are parallel and are frequently normal to the layer planes. Since the carbonyl groups of *sym-n*-alkanones reside at the middle of a layer, where conformational motions are attenuated most, it is not unreasonable that the quantum yields of their Norrish type II reactions are almost zero; such molecules cannot attain a shape conducive to γ-hydrogen abstraction (and i-BR formation) during their excited-state lifetimes. 2-*n*-Alkanones, by virtue of their carbonyl groups being near a layer interface (where conformational changes are more facile), do undergo the Norrish type II reactions with substantial (but unquantified) quantum yields.[10]

An even clearer demonstration of how the stiffness of the walls of a reaction cavity in a solvent matrix comprised of rod-like molecules packed in layers can influence the photoreactivity of guest ketones is provided by a study of the Norrish type II reactions of *p*-alkyl alkanophenones (AAP) in the hexatic B and solid layered phases of *n*-butyl stearate (BS).[11] The relative quantum yields for conversion of 2 wt% of isomers of AAP as a function of BS phase and benzoyl group position along a polymethylene chain of 21 carbon atoms is shown in Figure 39.1. The compounds g through i are the least reactive isomers when reaction is conducted in either of the BS ordered phases. Assuming nearly isomorphous substitution by extended molecules of AAP into a BS matrix, the benzoyl groups of the g, h, and i isomers are located in the vicinity of a layer middle, where it is stiffest.

The influence of the stiffness (and, to a certain extent, polarity) of the environment experienced by AAA on the quantum yields of its Norrish type II *fragmentation* component has been examined by Whitten and co-workers in various microheterogeneous media.[12] Their results (Table 39.2) indicate that the greater the stiffness of the polymethylene chains of surfactant molecules in various assemblies, the more difficult is the penetration of water molecules to the vicinity of the benzoyl groups of AAA. In each of the assemblies, the carboxylate group of AAA is anchored near the interface between the surfactant molecules and the aqueous part. This permits the average position of the benzoyl group of AAA to be removed from the vicinity of the interfaces. Since the quantum yields decrease as the solvent polarity is decreased (*N.B.*, benzene and *t*-butyl alcohol as isotropic media) and as the reaction cavity becomes less flexible and less able to support γ-hydrogen abstraction by the excited triplet state of AAA,[3b] solvent polarity and reaction cavity stiffness act synergistically.

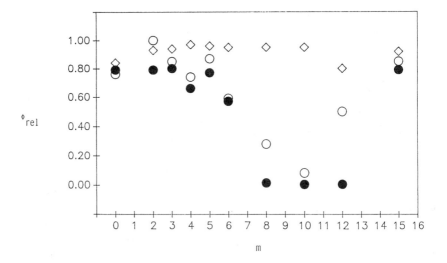

FIGURE 39.1   Relative quantum yields from loss of AAP isomers in BS as a function of solvent phase and chromophore position along the AAP polymethylene chain. Note that m+n = 21 in all cases. ◊, 30 °C (isotropic); ○, 20 °C (hexatic B); •, 10 °C (solid). (*Source:* He, Z. and Weiss, R. G., *J. Am. Chem. Soc.*, 112, 5535, 1990.)

Table 39.2   Relative Quantum Yields for Fragmentation of AAA in Various Media

AAA	Medium	Φ_{II}(fragmentation)[a]
a or b	Carboxymethylamylose (CMA)/ H$_2$O/NaCl	0.14
c	t-Butyl alcohol	1.02
c	Sodium dodecylsulfate (SDS) micelles/H$_2$O	0.81
c	Cetyltrimethylammonium chloride (CTAC) micelles/H$_2$O	0.72
c	Benzene	0.27
c	Dioctadecyldimethylammonium chloride (DODAC) vesicles/H$_2$O	0.22
c	Arachidic acid assemblies	<0.001[b]

[a] Relative to Φ_{II} (fragmentation) of butyrophenone in t-butyl alcohol.
[b] Φ for disappearance of AAA-c was ≤0.06 (no type I), suggesting some intermolecular reactions.
From Reference 12.

Table 39.3   Quantum Yields for Loss of Alkanophenones (Φ_-) and *trans/cis*-Cyclobutanol Ratios in Various Solvents

Ketone	Solvent	Φ_-[a]	t/c
Valerophenone	t-BuOH	1.00	1.5
	Benzene	0.33	3.6
	HTAC micelles	1.06	1.9
Octanophenone	t-BuOH	1.00	1.1
	Benzene	0.29	4.7
	HTAC micelles	0.71	1.2

[a] Relative to valerophenone in t-butyl alcohol whose absolute value is assumed to be 1.
From Reference 13.

In principle, removal of the carboxylate anchor and shortening of the polymethylene chain in AAA allow benzoyl groups in molecules like valerophenone and octanophenone to explore a greater fraction of the hydrophobic part of a microheterogeneous medium. Thus, aqueous micellar solutions of hexadecyl trimethyl ammonium chloride (HTAC) provide an environment that is very conducive to Norrish type II reactions of valerophenone and somewhat less so to octanophenone (Table 39.3).[13] Although both alkanophenones experience a rather polar environment, indicating that their benzoyl groups are near a micellar surface, the longer tail of octanophenone seems to force it to reside, on average, closer to the hydrocarbon tails of HTAC molecules.

9.3 Photoproduct Ratios

From Scheme 39.1, it is clear that the intermediate BR can follow several courses leading to fragmentation (E) and cyclization (C) products. Additionally, there are two families of diastereomeric c-BR: c-BR$_1$, the precursor of *trans*-cyclobutanol products *(t)*, and c-BR$_2$, the precursor of the *cis*-cyclobutanols *(c)*. Here, the focus is on the empirical *consequences* of various environments as well as on the shape, size, and functionality relationships between a reaction cavity and a reactive ketone that are responsible for variations in E/C and t/c photoproduct ratios. Examples will be limited to reactions involving simple alkanones and alkanophenones.

In addition to the aforementioned variations in quantum yield for Norrish type II reactions of ketones in nonpolar and polar media, the photoproduct ratios are also affected. Thus, the t/c ratio from irradiation of valerophenone in t-butyl alcohol doubles in the nonpolar solvent, benzene.[14] As seen in Table 39.3,[13] the t/c ratios from irradiation of valerophenone or octanophenone in HTAC micelles resemble closely the value from the more polar isotropic solvent, consistent with the

Table 39.4   Product Distributions from Irradiation of BOE in Methanol and
β-CD Complexes

BOE, R =	Solvent	Norrish type I	E	C
-CH₃	MeOH	98.7	1.3	0
	β-CD (solid)	8.0	69.3	22.7
	β-CD (aqueous)	81	14.8	4.2
-CH(CH₃)₂	MeOH	100	0	0
	β-CD (solid)	7.0	78.0	15.0
	β-CD (aqueous)	89.7	5.1	5.2
-(CH₂)₈H	β-CD (solid)	81	5	
	β-CD (aqueous)	88	12	

From Reference 15.

ketones and their BR intermediates being in the proximity of some water molecules even when imbedded in the micelles.

Aqueous solutions of cyclodextrins (CD) are somewhat analogous to micellar solutions in the environments which they offer to complexed ketones. However, the reaction cavity of a cyclodextrin torus, being smaller and stiffer, should be more selective with respect to the ketone molecules it accepts and the conformations they may adopt. The consequences of forcing reactions to occur within a torus, and allowing the guests to sample the outside environment, is exemplified by the results in Table 39.4 for irradiation of the benzoin ethers (BOE) shown in Equation 39.2.[15] Not only does entrapment of BOE molecules in a torus alter the E/C ratios, but it also drastically increases the fraction of reacting species that follows the Norrish type II course. Not all alkanophenone photoreactions are affected as severely by β-CD complexation as is BOE. Alkanophenones with alkyl groups varying from 3 to 13 methylene units yield E/C ratios that are very similar when the complexing agent is solid or aqueous β-CD.[16]

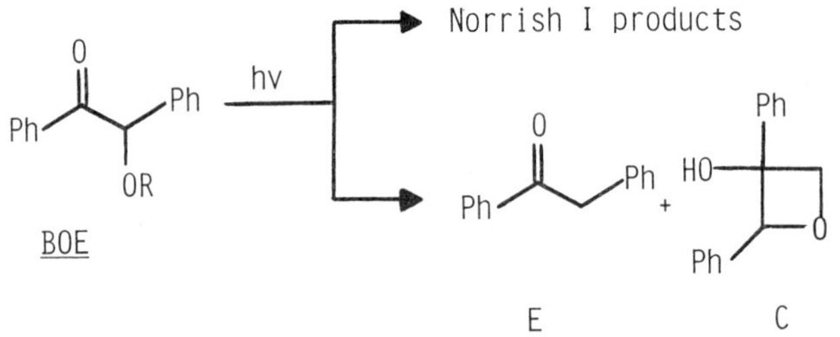

$$(39.2)$$

In very ordered interdigitated or normal bilayers made from 50 wt% of water and 50 wt% of either potassium stearate (KS), potassium palmitate (KP), or 1:1 potassium stearate:1-octadecanol (KSO),[17] the Norrish type II reactions of a homologous series of *sym-n-*alkanones have been shown to be very sensitive to the length of the ketones relative to the surfactant hosts.[18] The corresponding 2-alkanones yield photoproduct ratios that are similar to those found in highly polar media, regardless of the overall alkanone length.[18] Thus, as in the micellar solutions, the carbonyl groups of the 2-alkanones and the loci of reaction in their BR intermediates reside near a bilayer-aqueous interface. From the ratios shown in Figure 39.2, it is clear that the anisotropy of the bilayer matrices can influence the decay paths of the *sym-n-*alkanone-derived BR only when the lengths of the ketone guests and the lipophilic portion of the host molecules are nearly equal (i.e., when the substitution is isomorphous). A combination of photoproduct ratios and spectroscopic data supports a model of incorporation for *sym-n-*alkanones much longer than the host molecules in which they adopt a hairpin conformation and the carbonyl groups are at or near a bilayer-aqueous interface.

n-Butyl stearate (BS), in which the carboxyl anion head group has been "neutralized", forms a thermotropic hexatic B phase with host molecules again arranged hexagonally and in layers (like

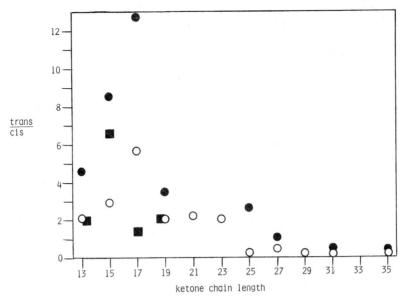

FIGURE 39.2 *trans/cis*-Ratios from irradiation of *sym-n*-alkanones in various gel phases of KP (15 °C, ■), KS (38 °C, ○), and KSO (38 °C, •). (*Source:* Treanor, R. L. and Weiss, R. G., *J. Am. Chem. Soc.,* 110, 2170, 1988.)

Table 39.5 Photoproduct Ratios from *2-* and *sym-n*-Alkanones in the Isotropic (30 °C), Hexatic B (20 °C), and Solid (0 °C) Phases of BS

Carbon atom chain length	T (°C)	2-n-Alkanone E/C	t/c	sym-n-Alkanone E/C	t/c
15	30	3.6	1.3	1.9	2.4
	20	11.7	1.3	2.3	2.8
	0	20.5	1.6	3.6	3.5
17	30	3.7	1.5	2.2	2.2
	20	15.1	1.9	3.2	10.8
	0	40.0	2.0	10.0	6.8
18	30	3.7	1.1		
	20	14.8	1.6		
	0	42.3	1.7		
19	30	4.1	1.3	1.8	
	20	16.5	2.4	6.2	13.3
	0	29.4	2.5	15.9	4.2
20	30	3.9	1.0		
	20	10.4	1.5		
	0	18.5	1.5		
21	30			2.0	2.5
	20			5.5	14.1
	0			>15	6.0

From Reference 8.

in the lyotropic KS phase), and two layered solid phases with orthorhombic packing.[9a,19] Although the *t/c* ratios from homologous series of *2-n*-alkanones do not appear to be influenced by any of the BS ordered phases, the E/C ratios are influenced and are largest when the total length is 17 to 19 carbon atoms (Table 39.5).[8] This is significantly shorter than a BS molecule and suggests a model in which there is significant H-bonding between the hydroxy of a BR of appropriate length and the carboxyl group of an adjacent BS molecule. By contrast, both the E/C and *t/c* ratios from irradiation

Table 39.6   Photoproduct Ratios from 2- and *sym-n*-Alkanones
in the Isotropic (45 °C), Phase II Solid (35 °C), and Phase I
Solid (25 °C) phases of C21

Carbon atom chain length	T (°C)	2-*n*-Alkanone		*sym-n*-Alkanone	
		E/C	t/c	E/C	t/c
18	45	4.0	1.5		
	35	3.9	4.6		
	25	14	2.4		
20	45	4.9	1.6		
	35	18	4.7		
	25	165	18		
21	45	2.4	1.0	1.8	2.5
	35	6.0	2.5	3.5	25
	25	46	3.6	>69	0.9
22	45	4.0	1.7		
	35	8.4	2.8		
	25	29	4.0		

[a] Error limits, which are large in the case of high ratios, have been
deleted for the sake of clarity.
From Reference 21.

of the *sym-n*-alkanone homologues in the ordered phases of BS change with ketone length. The homologue yielding the largest ratios has not been determined exactly, but it contains at least 19 to 21 carbon atoms and no more than 25.[8]

Similar studies on the Norrish type II reactions of AAP in BS support both the need for near isomorphous substitution by the guest ketones in the BS matrices and the importance of H-bonding interactions between BR intermediates and BS molecules.[11]

Removal of the carboxyl group of a BS molecule leaves heneicosane (C21) that conceptually should provide an even more easily defined reaction cavity for guest molecules undergoing Norrish type II reactions. Heneicosane (C21) forms two solid phases in which individual molecules are in fully extended conformations, are packed either hexagonally (phase II) or orthorhombically (phase I) within a layer, and are normal to the layer planes.[20] As evidenced from the results of several spectroscopic and calorimetric experiments, the solid phases of C21 do provide very restrictive reaction cages when *n*-alkanones are isomorphously substituted into a solid solvent matrix.[21] However, the solid *n*-alkane matrices appear able to incorporate only those alkanones whose overall length almost matches that of the host. When this occurs, very high E/C photoproduct ratios can be recorded (Table 39.6). Otherwise, the guest molecules are either rejected (i.e., phase separated) or placed in highly disturbed solid environments which are not common to the bulk. In both cases, the photoproduct ratios approach those from irradiations conducted in nonpolar *isotropic* media.

Unlike the reaction cavity afforded to ketones and their BR intermediates by the anisotropic phases of BS and C21, the reaction cavities of inclusion complexes and zeolites have "inflexible" walls, predetermined sizes, and functional groups that project from the walls at set orientations and may interact with some guest molecules. Thus, the course of Norrish type II reactions followed by ketones that are guests in these media is expected to be controlled by somewhat different factors than those imposed by the anisotropic phases of BS.

In evidence of this, 5-nonanone molecules irradiated within the narrow channels of an urea inclusion complex (5 Å diameter) form elimination products and a *trans*-cyclobutanol, but no more than 3% of the *cis*-isomer.[22] 2-Hexanone and 2-undecanone, irradiated in urea channels, favor the *trans*-cyclobutanol isomers over the *cis* in a similar fashion.[22] By contrast, a series of alkanophenones that are either unsubstituted or contain an alkyl group along the alkane chain show very little differences between the Norrish type II photoproduct ratios found in benzene solutions

Table 39.7 *trans/cis*-Ratios from Irradiation of *n*-Alkanones in Zeolites and in Hexane Solutions

Alkanone	Hexane	NaX	NaY	ZSM-5	ZSM-11
4-Nonanone	1.8	0.6	1.3	60	60
4-Undecanone	1.8	0.4	0.7	60	60
4-Dodecanone	1.7	0.7	0.9	65	70
4-Tridecanone	1.7	0.7	1.1	70	68
4-Tetradecanone	1.7	0.8	1.1	72	66
4-Decanone	1.8	0.4	0.7	60	60
3-Decanone	1.8	0.4	0.9	16	14
2-Decanone	1.5	0.6	1.0	6.0	6.5
3-Octanone	1.8	0.7	0.8	20	18
4-Octanone	1.8	0.7	1.3	15	18
2-Octanone	1.4	0.8	1.2	8.0	7.1
2-Heptanone	1.7	0.8	0.9	3.8	4.1
3-Heptanone	1.5	0.6	0.9	2.8	2.6
2-Hexanone	1.5	0.8	1.3	2.4	2.7

From Reference 25a.

and in complexes with Dianin's compound[23] (4-*p*-hydroxyphenyl-2,2,4-trimethylchroman, a non-polar host whose channels are effectively truncated at each 11 Å of length by a 2.8 Å constriction from six hydrogen-bonding hydroxyl groups[24]). Apparently, the volume (and its shape) allocated to each guest ketone molecule in these channels is adequate to allow it to explore virtually all of the conformations it would in an isotropic environment.

The necessity to limit the shape and volume of the reaction cavity if photoproduct ratios are to be controlled is demonstrated clearly when alkanones and alkanophenones are irradiated as guests in two types of zeolites, the x and y faujasites and ZSM-5 and ZSM-11 pentasils (Tables 39.7 and 39.8).[25] Whereas the alkanophenones and alkanones yield E/C ratios in the faujasites that are expected from a polar environment which imposes little spatial constraints upon the BR intermediates, only the bulkier alkanophenones in the narrower pentasil channels produce E/C ratios that indicate significant inhibition to formation of the more globular *c*-BR conformations. Although the *t/c* ratios from irradiation of the alkanones in the faujasites are similarly unaffected, they show a clear preference for the less-bulky *trans*-isomer in the narrower pentasil channels and when both alkyl chains on the cyclobutanol ring are "long" (i.e., when the carbonyl group of a reactant alkanone is near the center of the molecule).

As mentioned previously, *sym-n*-alkanones in their neat solid phases are nearly inert to UV/VIS radiation.[7,8] The very small amounts of photoproducts isolated after protracted irradiation are formed in E/C ratios reminiscent of reaction in a nonpolar isotropic medium. It is reasonable to ascribe these photoproducts to the reaction of molecules at defect sites within the crystalline lattices. By contrast, the much more photoreactive solid 2-*n*-alkanones with 14 to 19 carbon atoms yield E/C ratios that exceed 60 in optimal cases and at very low percentages of conversion.[10] At the same time, the *t/c* ratios remain close to the values found upon irradiation of the melt. The E/C ratios are a sensitive function of percent of conversion, and decrease to melt values after loss of 20 to 30% of starting material. Apparently, the photoproducts disrupt the solid phase order, allowing the BR intermediates to follow the motions that are normally attenuated. An example is 2-nonadecanone (mp 53–54 °C) whose E/C ratios at 23 °C at various percentages of conversion are 32.5 (10%), 16.1 (17%), and 3.8 (46%);[10] the E/C ratio from a hexane solution of the same ketone at 23 °C is 4.5.[10]

An extremely interesting example of how neat solid phases can alter Norrish type II photoreactions has been found with the three macrocylic diones (MCD).[26] As shown in Table 39.9, both the E/C and *t/c* ratios are dramatically different in the solid and solution phases. Dimorphs of the n = 7 and n = 12 homologues of MCD have been irradiated and result in very different E/C ratios that are partially explicable on the basis of crystal packing considerations.

Table 39.8 E/C and t/c Ratios from Alkanophenones (n = length of carbon atom chain attached to benzoyl group) in Zeolites and in Benzene Solutions

Zeolite	Pore dia (Å)	Cation radius (Å)[a]	Cage free volume (Å³)	n = 3 E/C	n = 4 E/C	n = 7 E/C	n = 7 t/c	n = 11 E/C	n = 11 t/c	n = 13 E/C	n = 13 t/c	n = 17 E/C	n = 17 t/c
(Benzene solution)				6.2	3.0	1.9	2.6	1.2	2.4	2.7	2.4	2.7	2.7
Na-A	4				2.7	2.3	2.2						
Li-X	8	0.6	873	3.9	1.3	1.6	2.6	0.6	1.4	0.4	2.1	2.5	1.8
Na-X	8	0.95	852	2.7	1.1	1.6	2.6	0.6	1.4	1.4	0.8	2.7	1.3
K-X	8	1.33	800	3.3	1.1	1.9	1.1	0.5	1.1	1.5	1.3	4.3	2.0
Rb-X	8	1.48	770	1.9	1.6	1.9	1.6	0.7	4.2	1.9	2.2	6.2	3.4
Cs-X	8	1.69	732	2.3	1.3	1.9	3.2	1.2	6.7	1.7	5.1	6.8	4.9
Na-Mordenite	7				2.5	2.2	2.3						
silicalite	6				>50	>50							
Na-ZSM-5	~5.5			73	>100	>100			>100		>100		
Na-ZSM-8	~5.5			82	>100	>100			>100		>100		
Na-ZSM-11	~5.5			56	>100	>100			>100		>100		
Na-Zeolite-β	~7.5–8.0			0.54	1.3	0.62		0.68		0.31		0.48	

[a] Ward, R.J. J. Catal., 10, 34, 1968.

From References 25b-d.

Table 39.9 Photoproduct ratios from irradiation of *MCD* in hexane solutions and in neat solid phases.

MCD	Phase	T (°C)	E/C	t/c
n = 7	k_1	20	0.01	0.11
	k_2	40	1.7	1.18
	Hexane solution	20	0.75	1.59
	Hexane solution	40	1.0	1.58
n = 8	k	20	0.15	28
	Hexane solution	20	0.69	2.47
n = 10	k	20	0.05	22.8
	Hexane solution	20	1.3	3.4
n = 12	Plate-k	RT	<0.01	<0.05
	Needle-k	RT	<0.01	>9

[a] k = crystal.
[b] RT = room temperature.
From Reference 26.

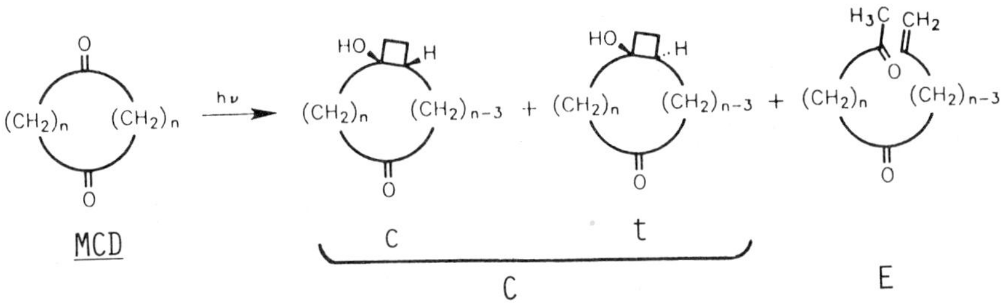

For the most part, solid phase irradiations of neat α-cycloalkyl acetophenones yield E/C ratios that are somewhat higher than the values found from acetonitrile solutions, but the *t/c* ratios in the two media are very similar.

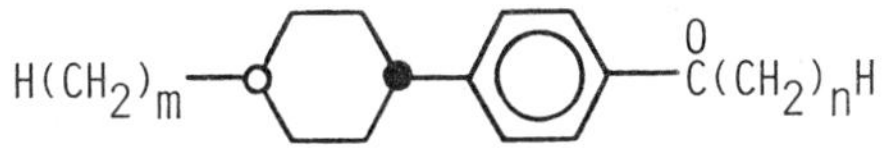

	m	n
a	5	4
b	7	4
c	7	6
d	7	8

Very different E/C ratios are found when the mesomorphic aromatic ketones LCK are irradiated in their neat solid phases or in solution.[27] An example of the behavior observed is presented in Figure 39.3. Note that the E/C ratios from irradiation of LCK-*d* in its smectic B phase are essentially the same as those found in the isotropic melt. LCK-*a* exhibits analogous behavior in that its E/C ratios from the nematic and isotropic melt phases are virtually the same, while the solid phase ratios are much higher. Arguments involving large free volume about the biradicaloid centers of the BR intermediates of LCK in their mesophases have been advanced to rationalize these results.[27]

The lack of a clear difference between the E/C ratios from the LCK in their mesophases and melt or solution phases should be contrasted with the results discussed previously for the AAP in the hexatic B phase of BS.[11] Similar contrasts between the solid phase E/C ratios from α-cycloalkyl acetophenones and the LCK should also be noted.

What these results and the others presented here indicate is that there are many factors that influence reactivity and selectivity in the Norrish type II reactions of ketones in ordered environments.

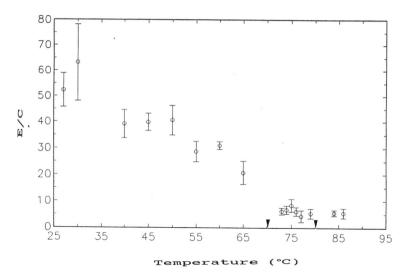

FIGURE 39.3   E/C ratios from irradiation of LCK-*d* in its neat isotropic, smectic B, and solid phases. Phase transition temperatures are indicated with arrows. (*Source:* Furman, I., Butcher, R. J., Catchings, R. M., and Weiss, R. G., *J. Am. Chem. Soc.*, 114, 6023, 1992.)

To understand how and why the photoreactions occur as they do, all of these factors must be considered. Unfortunately, the parameters associated with them are not known with reasonable precision in most systems; the state of the art dictates that empiricism rather that theory be the experimentalist's guide.

39.4 Industrial Applications

There are few industrial applications that exploit Norrish type II reactions of ketones in unconventional environments that the author is aware of.[28] However, there are possibilities. For example, cold stretching of films of poly(ethylene-co-carbon monoxide) results in a decrease in the fraction of photoreaction that proceeds via a Norrish type II course and an increase in the amount of Norrish type I reaction.[29] Consistent with this is the additional observation that film stretching increases the E/C ratios in the polymer. Thus, simply by altering extrusion and draw procedures, it should be possible either to control the rate of polymer degradation of ketone-containing films in sunlight or to vary the functional group content of the irradiated films.

Acknowledgments

The National Science Foundation is thanked for its support of the research described here that emanated from the laboratory of the author.

References

1. (a) Yang, N. C. and Yang, D. H., *J. Am. Chem. Soc.*, 80, 2913, 1958. (b) Wagner, P. J., Kelso, P. A., and Zepp, R. G., *J. Am. Chem. Soc.*, 94, 7480, 1972.
2. Yang, N. C. and Elliott, S. P., *J. Am. Chem. Soc.*, 91, 7550, 1969.
3. See, for instance, various chapters in: (a) Ramamurthy, V., Ed., *Photochemistry in Organized and Constrained Media*, VCH Publishers, New York, 1991; (b) Ramamurthy, V., Weiss, R. G., and Hammond, G. S., in *Advances in Photochemistry*, Vol. 18, Volman, D., Hammond, G. S., and Neckers, D. C., Eds., Wiley, New York, 1967; (c) Kalyanasundaram, K., *Photochemistry in*

Microheterogeneous Systems, Academic, Orlando, FL, 1987; (d) Thomas, J. K., *Chemistry of Excitation at Interfaces,* American Chemical Society, Washington, D.C., 1984.

4. Weiss, R. G., *Tetrahedron,* 44, 3413, 1988.

5. (a) Wagner, P. J., *J. Am. Chem. Soc.,* 89, 5898, 1967; (b) Wagner, P. J., *Acc. Chem. Res.,* 4, 168, 1971.

6. (a) Lissi, E. A. and Encina, M. V., in *Handbook of Organic Photochemistry,* Vol. II, Scaiano, J. C., Ed., CRC Press, Boca Raton, FL, 1989, chap. 7; (b) Encina, M. V., Nogales, A., and Lissi, E. A., *J. Photochem.,* 4, 75, 1975.

7. Slivinskas, J. A. and Guillet, J. E., *J. Polym. Sci., Polym. Chem. Ed.,* 11, 3043, 1973.

8. Treanor, R. L. and Weiss, R. G., *Tetrahedron,* 43, 1371, 1987.

9. (a) Sullivan, P. K., *J. Research (NBS),* 78A, 129, 1974; (b) Bailey, A. V., Mitcham, D., and Skau, E. L., *J. Chem. Eng. Data,* 15, 542, 1970; (c) Saville, W. B. and Shearer, G., *J. Chem. Soc.,* 127, 591, 1925.

10. Weiss, R. G., Chandrasekhar, S., and Vilalta, P. M., *Collect. Czech. Chem. Commun.,* 12, 531, 1992.

11. He, Z. and Weiss, R. G., *J. Am. Chem. Soc.,* 112, 5535, 1990.

12. (a) Hui, Y., Winkle, J. R., and Whitten, D. G., *J. Phys. Chem.,* 87, 23, 1983; (b) Winkle, J. R., Worsham, R. R., Schanze, K. S., and Whitten, D. G., *J. Am. Chem. Soc.,* 105, 3951, 1983.

13. Turro, N. J., Liu, K.-C., and Chow, M.-F., *Photochem. Photobiol.,* 26, 413, 1977.

14. Wagner, P. J., Kelso, P. A., Kemppainen, A. E., McGrath, J. M., Schott, H. N., and Zepp, R. G., *J. Am. Chem. Soc.,* 84, 7506, 1972.

15. (a) Dasaratha Reddy, G., Usha, G., Ramanathan, K. V., and Ramamurthy, V., *J. Org. Chem.,* 51, 3085, 1986; (b) Dasaratha Reddy, G. and Ramamurthy, V., *J. Org. Chem.,* 52, 5521, 1987.

16. Dasaratha Reddy, G., Jayasree, B., and Ramamurthy, V., *J. Org. Chem.,* 52, 3107, 1987.

17. Vincent, J. M. and Skoulios, A., *Acta Crystallogr.,* 20, 432, 441, 447, 1966.

18. Treanor, R. L. and Weiss, R. G., *J. Am. Chem. Soc.,* 110, 2170, 1988.

19. (a) Dryden, J. S., *J. Chem. Phys.,* 26, 604, 1957; (b) Krishnamurti, D., Krishnamurthy, K. S., and Shashidhar, R., *Mol. Cryst. Liq. Cryst.,* 8, 339, 1969; (c) Krishnamurthy, K. S. and Krishnamurti, D., *Mol. Cryst. Liq. Cryst.,* 6, 407, 1970; (d) Krishnamurthy, K. S., *Mol. Cryst. Liq. Cryst.,* 132, 255, 1986.

20. (a) Broadhurst, M. G., *J. Research (NBS),* 62A, 241, 1962; (b) Maroncelli, M., Qi, S. P., Strauss, H. L., and Snyder, R. G., *J. Am. Chem. Soc.,* 104, 6237, 1982; (c) Maroncelli, M., Strauss, H. L., and Snyder, R. G., *J. Chem. Phys.,* 82, 2811, 1985; (d) Schaerer, A. A., Busso, C. J., Smith, A. E., and Skinner, L. B., *J. Am. Chem. Soc.,* 77, 2017, 1955; (e) Brown, M. S., Grant, D. M., Horton, W. J., Mayne, C. L., and Evans, G. T., *J. Am. Chem. Soc.,* 107, 6698, 1985.

21. (a) Nunez, A. and Weiss, R. G., *J. Am. Chem. Soc.,* 109, 6215, 1987; (b) Nunez, A. and Weiss, R. G., *Bol. Soc. Chilena Quim.,* 35, 3, 1990; (c) Nunez, A., Hammond, G. S., and Weiss, R. G., *J. Am. Chem. Soc.,* 114, 10258, 1992; (d) Vilalta, P. M., Hammond, G. S., Weiss, R. G., *Langmuir,* 9, 1910, 1993.

22. Casal, H. L., de Mayo, P., Miranda, J. F., and Scaiano, J. C., *J. Am. Chem. Soc.,* 105, 5154, 1983.

23. Goswami, P. C., de Mayo, P., Ramnath, N., Bernard, G., Omkaram, N., Scheffer, J. R., and Wong, W. F., *Can. J. Chem.,* 63, 2719, 1985.

24. MacNicol, D. D., *Inclusion Compounds,* Vol. 2, Atwood, J. L., Davies, J. E., and MacNicol, D. D., Eds., Academic Press, New York, 1984, 1.

25. (a) Ramamurthy, V. and Sanderson, D. R., *Tetrahedron Lett.,* 33, 2757, 1992; (b) Turro, N. J. and Wan, P., *Tetrahedron Lett.,* 25, 3655, 1984; (c) Ramamurthy, V., Corbin, D. R., and Eaton, D. F., *J. Chem. Soc. Chem. Commun.,* 1213, 1989; (d) Ramamurthy, V., Corbin, D. R., and Johnston, L. J., *J. Am. Chem. Soc.,* 114, 3870, 1992.

26. (a) Lewis, T. J., Rettig, S. J., Scheffer, J. R., Trotter, J., and Wireko, F., *J. Am. Chem. Soc.,* 112, 3679, 1990; (b) Lewis, T. J., Rettig, S. J., Scheffer, J. R., and Trotter, J., *J. Am. Chem. Soc.,* 113, 8180, 1991.

27. (a) Furman, I. and Weiss, R. G., *J. Am. Chem. Soc.,* 114, 1381, 1992; (b) Furman, I., Butcher, R. J., Catchings, R. M., and Weiss, R. G., *J. Am. Chem. Soc.,* 114, 6023, 1992.

28. See for instance: Guillet, J., *Polymer Photophysics and Photochemistry,* Cambridge University Press, Cambridge, 1985.

29. (a) Bovey, F. A., Gooden, R., Schilling, F. C., Winslow, F. H., *Polym. Preprints,* 28, 238, 1987; (b) Gooden, R., Davis, D. D., Hellman, M. Y., Lovinger, A. J., and Winslow, F. H., *Macromolecules,* 21, 1212, 1988.

40

Regio- and Stereoselective Syntheses of Cyclopropanols, Azetidinols, and Pyrrolidinols

Hans-Georg Henning
Humboldt-University Berlin

40.1 Introduction

The carbonyl oxygen atom of n,π* excited ketones can add a hydrogen atom, generating a hydroxy C-radical. Generally, this hydrogen transfer is governed predominantly by steric factor. As a rule, the nearest hydrogen atom is transferred almost independently of the electronic properties of the neighboring atoms in the H-donating species.

This selection principle may be useful in photochemical syntheses of cyclic organic compounds from ketones with an aliphatic side chain. Favoring one of the transferrable hydrogen atoms by the choice of suitable experimental conditions should allow accomplishment of regio-, diastereo-, or even enantioselective syntheses of carbocyclic or heterocyclic hydroxy compounds.

Generally, most important for a selective synthesis is a relatively stable preferential conformation of the educt ketone with differentiation of the transferable hydrogen atoms of the aliphatic side chain by inter- or intramolecular effects, respectively. Therefore, the actual problem of the synthesis is the search for suitable experimental conditions leading to that conformation.

In this chapter, the significance of this idea is demonstrated with examples of photochemical syntheses of hydroxysubstituted cyclopropanes, azetidines, and pyrrolidines.

40.2 Cyclopropanols

On the pre-condition that the intramolecular atomic distances of the ketone are not markedly changed by an electronic excitation, the conformation I (Figure 40.1) would be unfavorable for an intramolecular cyclization. There is no doubt that the transfer of the γ-hydrogen atom (conformation II) is favored; that is why the first prerequisite for photochemical formation of a three-

0-8493-8634-9/95/$0.00+$.50
© 1995 by CRC Press, Inc.

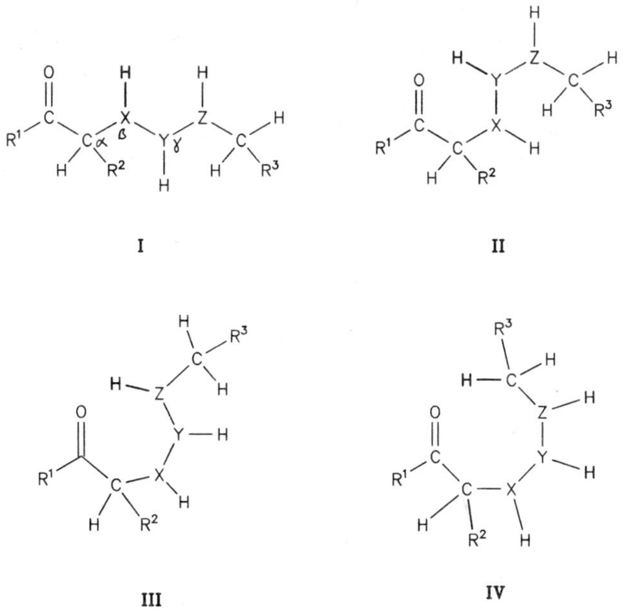

I

II

III

IV

FIGURE 40.1

membered ring is the absence of γ-hydrogen, for instance by the inclusion of a heteroatom like O or N-R at that site.

The photopreparation of 2-hydroxy-2-phenyl-cyclopropan-1-(N,N-dialkyl)carbamides via n,π* excited-state benzoylpropioamides may serve as an example.[1] The morpholide 1 probably adopts the conformations I, III, and perhaps IV. A mixture of the products 3 + 4 was obtained in 6.5 and 20% yield, (Equation 40.1). Similarly, the N,N-dibenzylamide of β-benzoylpropionic acid yielded 35% of the corresponding cyclopropanols. Five- or six-membered ring compounds were not observed.

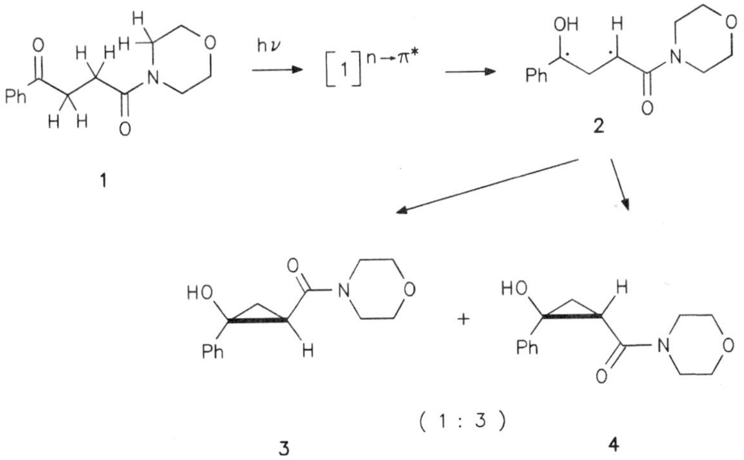

(1 : 3)

3   4   (40.1)

Besides regioselectivity, the distribution of 3 and 4 reveals a diastereoselectivity; this shows that both hydrogen atoms in the enantiotopic α-position are unequal. Obviously, the selectivity of the photocyclization 1 → 3/4 is not caused by intramolecular effects alone. There are signs that the formation of the preferred conformation for the cyclization product is also influenced by the solvent. An interaction between the nucleophilic oxygen atom of the ether and the oxygen atom of the ketocarbonyl group after n,π*-excitation (Equation 40.2) may favor hydrogen transfer from the β-position (conformation I). In the following radical recombination step, this arrangement is responsible for the shielding of one side of the preformed cyclopropane molecule, and for directing the morpholide group to the position *trans* to the hydroxy group. Regio- and diastereoselectivity are the consequences.

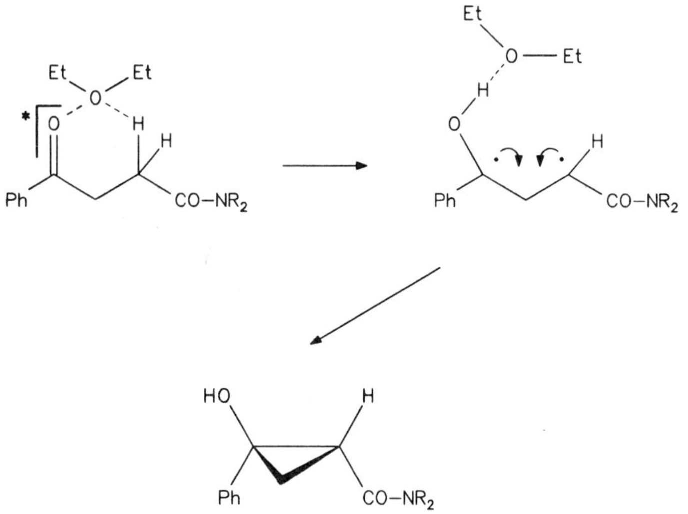

(40.2)

Irradiation of β-dialkylaminopropiophenones in dioxane or tetrahydrofuran bring about pinacolizations, fragmentations, isomerizations, and cyclizations.[2a] However, β-morpholino-propiophenones with α-methyl, α-phenyl, and β-phenyl substituents formed exclusively 2-morpholinocyclopropanols when irradiated in dioxane solution.[2b,c] Substituted β-piperazino-propiophenones reacted similarly.[2a] Reasons for the observed regioselectivity were not proposed by the authors.

By comparison with the regioselective photocyclization 1 → 3/4, the use of dioxane and tetrahydrofuran as solvents is of some interest because these solvents may have favored conformation I in the photoreaction of β-dialkylamino propiophenones. A careful investigation of the photochemical behavior of n,π* excited α-phenyl-β-morpholinopropiophenone 5 showed surprising results.[3] In ether, irradiation of racemic 5 formed 1,3-diphenyl-2-morpholinocyclopropanol as expected. After resolution by chiral-phase HPLC, irradiation of one of the pure enantiomers of 5 yielded only one stereoisomeric product instead of eight isomers. The vicinal coupling constant J_{2H-3H} 5,7 indicated that this product was one of the set of compounds 6–9 (Equation 40.3). Further investigations will examine in detail the enantiospecifity of this photocyclization.

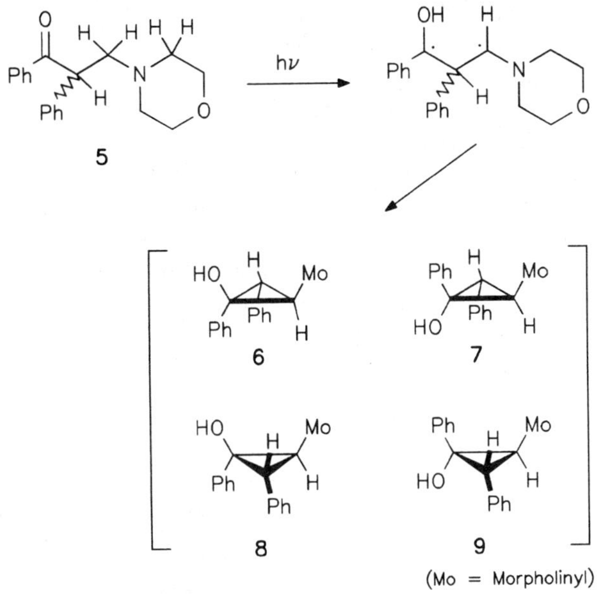

$$(40.3)$$

(Mo = Morpholinyl)

The influence that solvent has on the regioselective photocyclization of aminoketones was clearly proved in the case of n,π* excited *N*-(β-benzoylethyl)-*N*-acylaminoacetates **10** (Equation 40.4).[4] In ether solution, cyclopropanols **11** and pyrrolidinols **12** were formed. In benzene or benzene/cyclohexane, only pyrrolidinols **12** were generated (*cf.* Chapter 4). Obviously, the ether favored conformation **I** (Figure 40.1) of the electronically excited compounds.

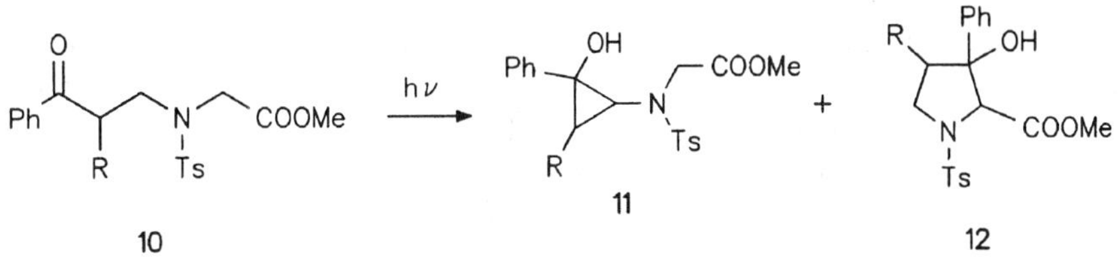

Solvent	R	yield (%)	
		11	12
Ether	H	15	12
	Me	5	35
	Ph	6	58
Benzene	H	0	59
	Me	0	70
	Ph	0	36

$$(40.4)$$

40.3 Azetidinols

The photochemical conversion of alkylketones to hydroxy-substituted four-membered rings demands the efficient transfer of a hydrogen atom from the γ-position of the side chain to the n,π* excited carbonyl group, generating a hydroxy biradical as an intermediate. This is easily possible if the molecule adopts conformation **II** (Figure 40.1).

There are a number of papers referring to the formation and the chemical reactions of those biradicals.[5,6] Depending on the conformation, the biradicals may undergo intramolecular recombination, yielding four-membered rings or they may suffer β-cleavage (Norrish type II reaction).[7] Thus, the achievement of effective photocyclization requires control of conformation II and the avoidance of β-cleavage (stereo-electronic control).

There are many examples showing that unhindered rotation around single bonds in the side-chain favors β-cleavage and give very low yields of cyclic products.

In 2-phenacylpiperidine 13, there are several transferable hydrogen atoms (Equation 40.5). Irradiation produces acetophenone in high yield, indicating the preference for γ-hydrogen abstraction and the involvement of Norrish type II cleavage. Thus, the main reaction pathway is the same as in the case of 2-(3,5-diphenyl-1-phenylsulfonyl-2-pyrazolin-5-yl)-acetophenone where only γ-hydrogen atoms are available for transfer.[9]

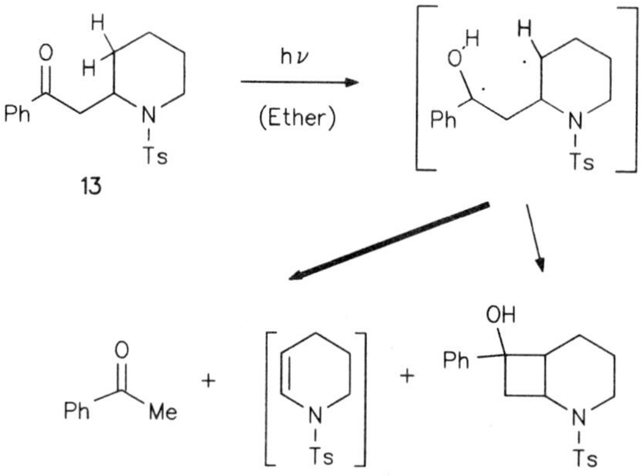

(40.5)

In α-aminoacetophenones (n,π* excitation), both β-cleavage and intramolecular cyclization compete.[10-13] Gold found that a change from *N,N*-dialkylamino- to *N*-acyl-*N*-alkyl-aminoacetophenones causes a shift in reactivity in favor of azetidinol formation.[11] In the case of aryl-α-amidoalkyl-ketones 14, utilization of this effect brings about an increase in the yield of azetidinols (to 66%).[14a]

It seems reasonable that stabilization of a special intermediate conformation by the *N*-acylgroup must have an effect on the steric result of the reaction. Example 14 (Equation 40.6) shows that this really is the case.[14b] ¹H NMR/NOE experiments proved that in this photoreaction, the diastereomeric products 15–17 are formed.[14c] The different yields show that this reaction is a diastereoselective photosynthesis of 3-aryl-3-azetidinols, with poor yields due to the competing β-cleavage.

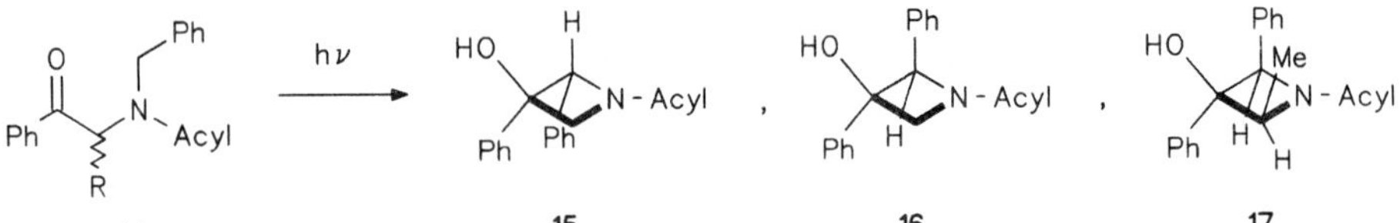

Acyl	R	yield (%)		
		15	16	17
Ts	H	23,4	4,8	—
SO₂–R'	H	0	24	—
CO–Ph	Me	—	—	65,7

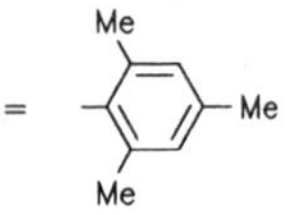

(40.6)

An attempt to obtain a pharmacological interesting azetidinol by photocyclization of the α-aminoacetophenone **18** failed.[15] Even utilization of techniques such as acylation of the product and working in an oxygen-free atmosphere produced very low yields of **19** (Equation 40.7).

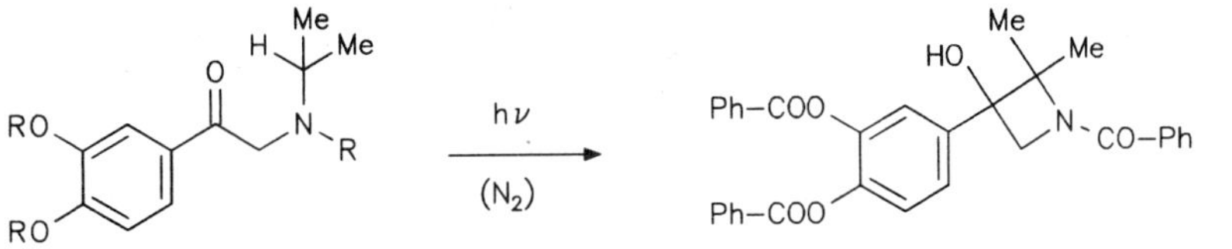

18

(a: R=H; b: R= CO–Ph)

19

(8,2% yield)

(40.7)

0.4 Pyrrolidinols

The pyrrolidine moiety is found in many natural products. Some examples are the amino acids proline and hydroxyproline as well as the pyrrolizidine alkaloid Retronecin **20**. The development of special and, as far as possible, stereoselective syntheses of pyrrolidine derivatives is most important because of their interesting biological properties (cf. **20** and the nootropicum Oxiracetam **21**; Figure 40.2).[16,17]

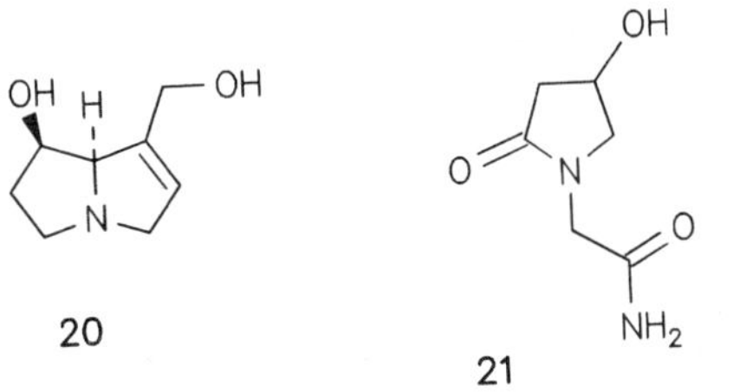

20

21

FIGURE 40.2

Numerous interesting papers have been published dealing with thermal methods of synthesis.[18–20] However, corresponding photochemical approaches are very rare. Since intramolecular photocyclization to yield five-membered rings involves an intermediate conformation **III** (Figure 40.1) it is essential that the transfer of hydrogen atoms from the β- and γ-positions (X and Y) must be avoided or suppressed. This can be achieved using *N,N*-disubstituted β-aminoketones for the pyrrolidine synthesis. γ-H transfer is impossible and if a carbonyl group occupies the β-portion, only δ-H transfer is possible.

In this respect, compound **22** satisfies all these conditions (Equation 40.8). H-abstraction could take place from the methylene group or from the *N*-methyl group, alternatively.

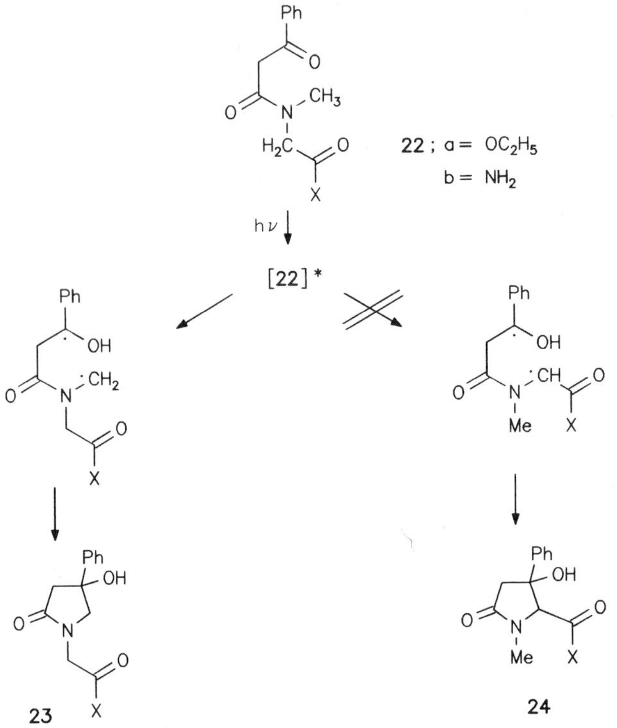

22 ; a = OC_2H_5
b = NH_2

(40.8)

23   24

The irradiation of **22** in diethylether shows that H-abstraction for the N-methyl group is predominant and yields the hydroxy pyrrolidinones **23** in 78 to 80% yield.[21a] This shows the important role that steric factors can play in hydrogen transfer processes. (The comparison of Oxiracetam **21** with the photoproduct **23b** is of some practical interest.) Under the same conditions, irradiation of the pyrrolidide **25** in diethylether proceeds diastereoselectively to yield the pyrrolizidine **26** in 60% yield (Equation 40.9).[21b]

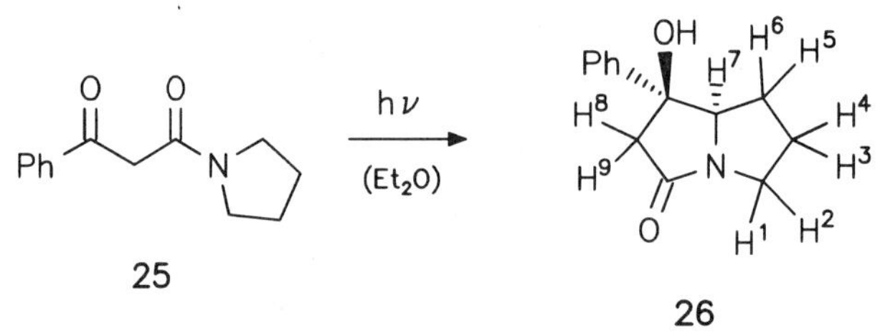

25

26   (40.9)

The photochemical behavior of 2-benzoyl-1,4-*bis*(tosyl)-piperazine **27** shows the great dependence on the outcome of the reaction of the stereochemistry in the **27** material and intermediates (Equation 40.10). In the absence of substituents in the 5-position (**27a**), irradiation in benzene brings about exclusively Norrish type II cleavage, affording the product **28**. Under the same condition, irradiation of 5,5-dimethyl-piperazine **27b** results in the formation of the diazabicyclo compounds **29** and **30**, respectively. The ratio of products is dependent on the proportions of the ring conformations A and B.

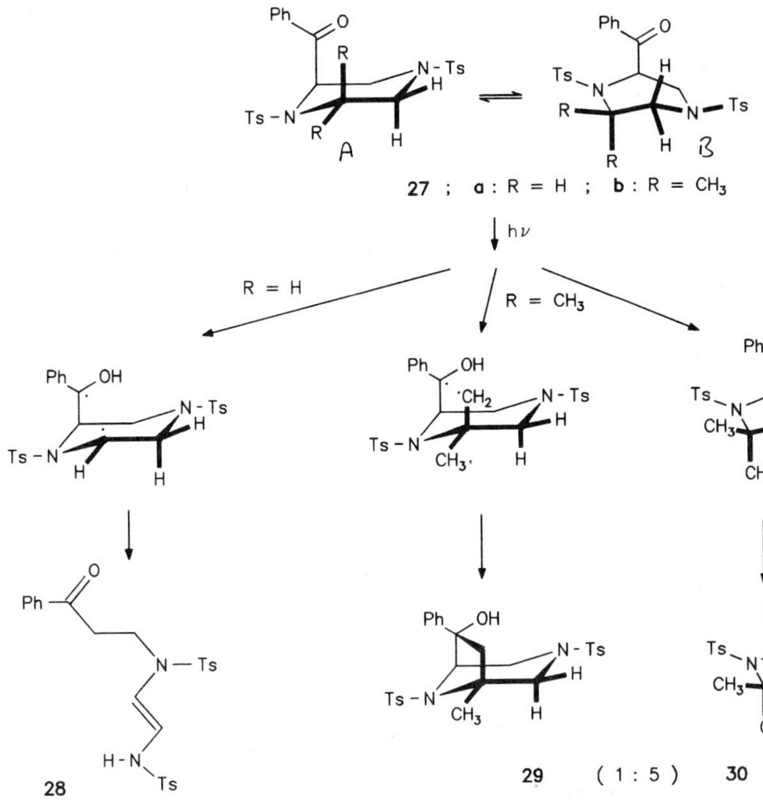

It is worth noting that this result contrasts with the reaction of compound **31** which, under comparable conditions but with diethylether as a solvent, yields 2-piperazino-cyclopropanol.[2a] These results are not predictable by simple examination of structural formulae (Figure 40.3), and it is important to take the conformations of the aminoketones into account. This comparison may lead to the conclusion that the intramolecular photocyclization of the β-aminopropiophenones **32** generally gives 2-aminocyclopropanols but no pyrrolidinols. As already shown (*cf.* reaction 10 → 11/12), this is not correct. If **32** in its n,π* state adopts conformation III (Figure 40.1), n,π* under influence of solvent or substituents R¹-R³ then a 1,5-biradical will be produced and pyrrolidinols will be formed. By analogy to the azetidinol synthesis 14 → 15–17 (Equation 40.6) pyrrolidinols can be favored the use of *N*-acyl substituent.[23] This acyl effect (Equation 40.11) may even be more dominant than the influence of solvent (ether). In addition, the methyl groups in the α-position increase the yield of pyrrolidinol to 90%. Obviously, one cannot explain this increase by an electronic effect and steric effects are most important.

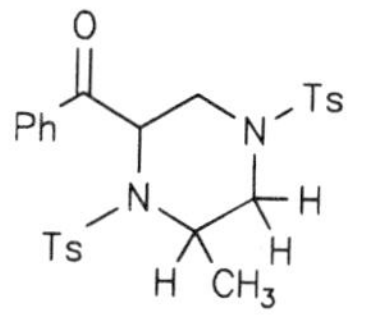

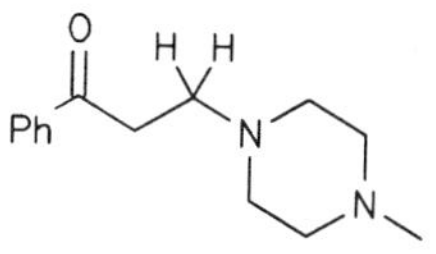

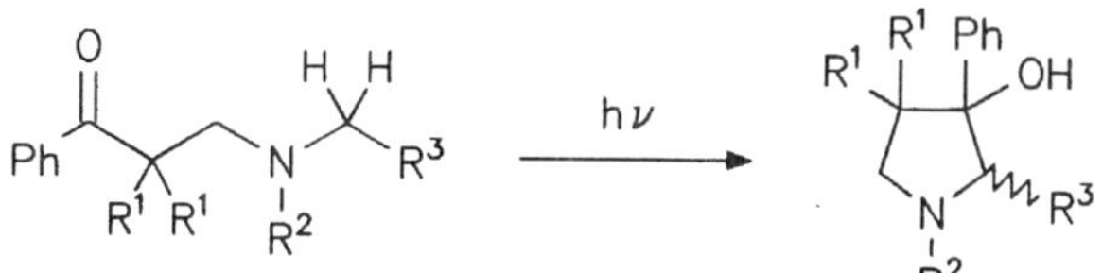

R^1	R^2	R^3	Solvent	33−yield (%) [a]	Ref.
H	CO−Ph	Ph	Ether	50,5	18a
Me	CO−Ph	Ph	Ether	90	18b
H	CO−NH$_2$	Ph	[b]	37	18c
H	Ts	Py[c]	[b]	38,3	18d

[a] Mixture of 2 diastereomers.

[b] Benzene : Cyclohexane = 70 : 30

[c] 3−Pyridyl

(40.11)

It is interesting that this reaction gave two diastereoisomers **33** although the diastereoselectivity was not established.[23]

A discussion of the problem of how an *N*-acyl group works in the formation of the transition state conformations of the hydrogen transfer and the biradical recombination is hitherto limited to consideration of models. This is further complicated by the fact that the result of the reaction also reflects an influence of the substituents, R^3, which are apparently remote from the excited carbonyl group. Another example shows the efficiency of the reaction and an n,π*-excitation of the glycinamides **34** in benzene provides either **35** or **36** in diastereoselective reactions (Equation 40.12).[24]

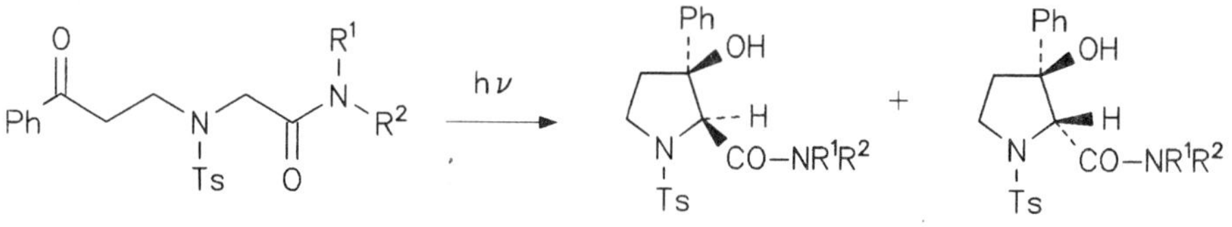

R^1	R^2	35 −yield (%)	36 −yield (%)
H	CH$_3$	68	−
H	CH$_2$−Ph	65	−
CH$_3$	CH$_3$	−	53
CH$_3$	CH$_2$−Ph	−	46

(40.12)

By considering all the known effects that influence the intermediate conformations, a photochemical diastereoselective synthesis of 3-hydroxy-prolines **38** has been developed.[4b] This involves the irradiation of substituted glycines 37 in benzene/cyclohexane, with λ ≥300 nm, and yields pure 38 with a relative *trans*-configuration (OH/COOR³) in the yields shown (Equation 40.13).

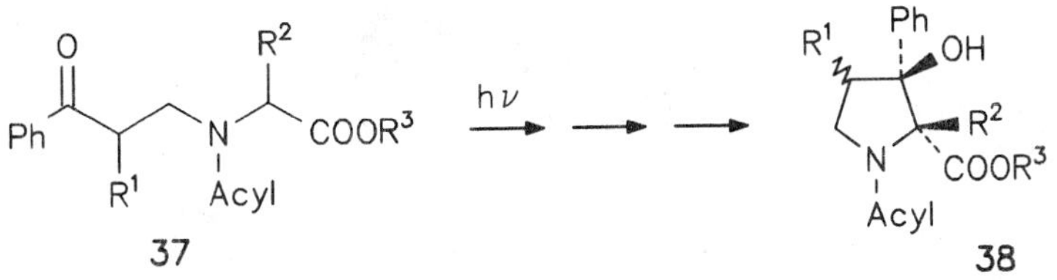

	Acyl	R¹	R²	R³	yield(%)	rel. conf. OH/R¹
a	Ts	H	H	Me	76	—
b	Ts	H	H	Et	85	—
c	Ts	Me	H	Me	70	trans
d	Ts	H	Me	Me	<5	
e	Ts	Ph	H	Me	40	cis
f	Ts	H	Ph	Me	0	
g	CO–Ph	H	H	Me	65	—
h	CO–Ph	Me	H	Me	62	trans
i	CO–Ph	Ph	H	Me	70	cis

(40.13)

Irradiation of the substituted β-alanine derivatives 39 confirmed the influence of conformation (Figure 40.1) and the influence of solvents and substituents. In ether, the 2-aminocyclopropanol 40 was formed. In benzene/cyclohexane, the primary product 41 gave 1-phenyl-2-oxa-6-aza-bicyclo [3.3.0]octan-3-one 42 in 35% yield (Equation 40.14).[4b] Obviously, conformation IV (Figure 40.1) of 39 does not play a role in the different steps of the reaction.

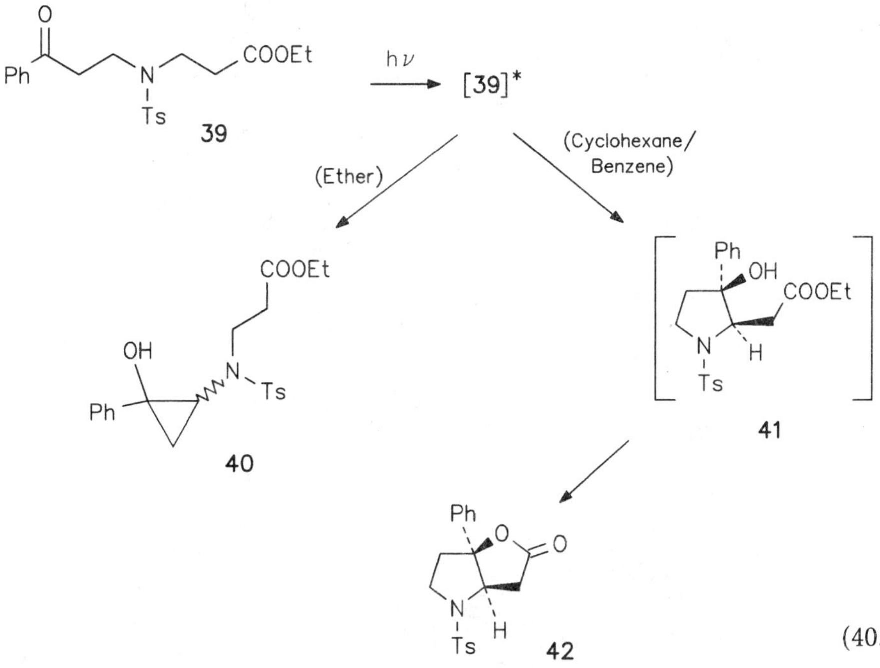

(40.14)

Table 40.1   Correlation Between the Barriers of
Amide Rotation in 37 and the 38-Yields

	37[a]		38-yield	
	Acyl (CO-R) R	$\Delta G_c^{\neq}$ (kJ mol^{-1})	(b) (%)	$\sigma^*(R)^c$
k	Me	79.8	25	0.0
l	i-Bu	78.4	33	−0.125
m	i-Pr	75.4	35	−0.200
n	t-Bu	61.5	43	−0.320

[a] $R^1 = R^2 = H; R^3 = Me$.
[b] Measured by quant. HPLC.
[c] Taft constant [26].

As already mentioned, there exists no proven explanation for the influence of an *N*-acyl group on the selectivity of the intramolecular photocyclization of aminoketones. A contribution to this problem is an investigation into the dependence of the energy barrier of the internal amide rotation and the chemical yields of pyrrolidinol in amino ketones **37**.[25] An unambiguous correlation was found for the compounds **37** with aliphatic *N*-acyl substituents. The lower the barrier $\Delta G_c^{\neq}$ the higher the yield of pyrrolidinol (Table 40.1). In the case of aromatic *N*-acyl groups, the separated rotamers **43a/b** ($\Delta G_c^{\neq} \geq 100$ kJ mol^{-1}) showed completely different photochemical behavior under identical conditions (irradiation $\lambda \geq 300$ nm; benzene/cyclohexane 20/80); **43a** decomposed to a mixture of different products while **43b** provided the hydroxyproline ester **44** in 60% yield (Equation 40.15).

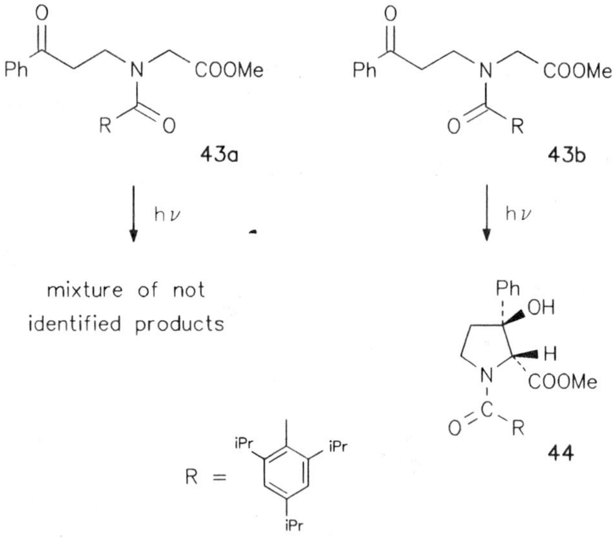

(40.15)

References

1. Henning, H. G., Berlinghoff, R., Mahlow, A., Köppel, H., and Schleinitz, K.-D., *N,N*-Dialkylamides of 2-hydroxy-2-phenyl-cyclopropane-1-carboxylic acid — photochemical synthesis and determination of the structure, *J. Prakt. Chem.*, 323, 914, 1981; *Chem. Abstr.*, 96, 141939, 1982.

2. (a) Roth, H. J., El Raie, M. H., and Schrauth, T., Photocyclization of 3-amino ketones to 2-amino-1-cyclopropanols and their isomerization, *Arch. Pharm.*, 307, 584, 1974; *Chem. Abstr.*, 81, 135596, 1974; (b) Roth, H. J. and El Raie, M. H., Photocyclization of 3-amino-ketones to 2-aminocyclopropanols, *Tetrahedron Lett.*, 2445, 1970; *Chem. Abstr.*, 73, 56042, 1970; (c) Roth, H. J. and El Raie, M. H., Photocyclization of 3-morpholinopropiophenones, *Arch. Pharm.*, 305, 213, 1972; *Chem. Abstr.*, 77, 19587, 1972.

3. Weigel, W. and Henning, H. G., unpublished results.

4. (a) Henning, H. G., Haber, H., and Buchholz, H., Photocyclization of *N*-tosyl-*N*-(β-benzoylethyl)glycine esters, *Pharmazie* 36, 160, 1981; *Chem. Abstr.*, 95, 81459, 1981; (b) Haber, H., Buchholz, H., Sukale, R., and Henning, H. G., Diastereoselective synthesis of 3-hydroxyprolines, *J. Prakt. Chem.*, 327, 51, 1985; *Chem. Abstr.*, 104, 130228, 1986.

5. Heinrich, P., in Methoden der Organischen Chemie (HOUBEN-WEYL), Herausgeber E. Mueller, Band IV/5b, 1975, 891.

6. Gilbert, A. and Baggott, J., *Essentials in Molecular Photochemistry*, Blackwell Scientific, London, 1991, 310ff.

7. Sengupta, D., Sumathi, R., and Chandra, A. K., Role of barriers to conformational changes in ketones undergoing the Norrish type II process, *J. Photochem. Photobiol.*, 60, 149, 1991; *Chem. Abstr.*, 115, 255469, 1991.

8. Henning, H. G. and Bartels, H., Norrish-II reactions of *N*-tosylpiperidinyl ketones, *Z. Chem.*, 23, 455, 1983; *Chem. Abstr.*, 100, 174630, 1984.

9. Lempert-Sréter, M., Tamás, J., and Lempert, K., Photolysis and fragmentation upon electron impact of 2-(3,5-diphenyl-1-phenyl-sulfonyl-2-pyrazolin-5-yl)-acetophenone, *Acta Chim. Acad. Sci. Hung.*, 80, 455, 1974.

10. Wagner, P. J., Chemistry of excited triplet organic carbonyl compounds, *Top. Curr. Chem.*, 66, 1, 1976.

11. Gold, E. H., Photolysis of α-*N*-alkylamidoacetophenones, a direct route to 3-azetidinols, *J. Am. Chem. Soc.*, 93, 2793, 1971.

12. Padwa, A., Eisenhardt, W., Gruber, R., and Pashayan, D., Electron transfer in the type II photoelimination of α-amino-acetophenones, *J. Am. Chem. Soc.*, 93, 6998, 1971.

13. (a) Hill, J. and Townend, J., Light-induced reactions of 2-(*N*-alkyl-*N*-arylamino)-cyclohexanones, *J. Chem. Soc., Chem. Commun.*, 1108, 1972; (b) Allworth, K. L., El-Hamamy, A. A., Hesabi, M. M., and Hill, J., Light-induced reactions of 2-(*N*-Alkyl-*N*-arylamino)acetophenones and related amino-ketones: Formation of 1,3-diarylazetidin-3-ols, *J. Chem. Soc., Perkin Trans. 1*, 1671, 1980; (c) Hesabi, M. M., Hill, J., and El-Hamamy, A. A., Light-induced reactions of heteroaryl-*N*-methylanilinomethyl-ketones: Formation of 3-heteroaryl-1-phenylazetidin-3-ols, *J. Chem. Soc., Perkin Trans. 1*, 2371, 1980.

14. (a) Henning, H. G., Fuhrmann, J., and Krippendorf, U., cis-2,3-Diaryl-azetidin-3-ols from n,π^*-excited *N*-benzyl-α-amino-acetophenones, *Z. Chem.*, 21, 36, 1981; *Chem. Abstr.*, 94, 208618, 1981; (b) Fuhrmann, J., Haupt, M., and Henning, H. G., Diastereoselective synthesis of 3-arylazetidin-3-ols via photocyclization of aryl α-amidoalkyl ketones, *J. Prakt. Chem.*, 326, 177, 1984; *Chem. Abstr.*, 101, 38283, 1984; (c) Fuhrmann, J., Köppel, H., Schleinitz, K.-D., and Henning, H. G., Studies of the structure of Z/E isomeric 2,3-diarylazetidin-3-oles, *J. Prakt. Chem.*, 324, 664, 1982; *Chem. Abstr.*, 97, 181590, 1982.

15. Henning, H. G., Fuhrmann, J., Haupt, M., Schöder, H., Knoll, A., and Bartels, H., Preparation and UV-stability of *O,O,N*-triacylisoprenalone, *Pharmazie*, 37, 224, 1982; *Chem. Abstr.*, 97, 182073, 1982.

16. Nishimura, Y., Kondo, S., and Umezawa, H., Synthetic studies on pyrrolizidine alkaloid antitumor agents. Enantioselective synthesis of retronecine and its enantiomer from D-glucose, *J. Org. Chem.*, 50, 5210, 1985.

17. Pellegata, R., Pinza, M., Pifferi, G., Gaiti, A., Mozzi, R., Tirillini, R., and Porcellati, G., Synthesis and biochemical activity of new alkyl and acyl derivatives of 4-hydroxy-2-pyrrolidinone, *Farmaco [Pavia] Ed. Sci.*, 36, 845, 1981; *Chem. Abstr.*, 96, 19902, 1982.

18. Flanagan, D. M. and Joullie, M. M., Synthetic strategies for the construction of 3-pyrrolidinol, a versatile nitrogen heterocycle, *Heterocycles*, 26, 2247, 1987.

19. Tamaru, Y., Kawamura, S., Bando, T., Tanaka, K., Hojo, M., and Yoshida, Z., Stereoselective intramolecular haloamidation of N-protected 3-hydroxy-4-pentenylamines and 4-hydroxy-5-hexenylamines, *J. Org. Chem.*, 53, 5491, 1988.

20. Hartwig, W. and Born, L., Diastereoselective and enantioselective total synthesis of the hepatoprotective agent Clausenamide, *J. Org. Chem.*, 52, 4352, 1987.

21. (a) Henning, H. G., Gelbin, A., and Schoeder, H., Preparation and regioselective photocyclization of N-benzoylethyl-sarcosine derivatives, *Pharmazie*, 29, 433, 1984; *Chem. Abstr.*, 102, 6096, 1985; (b) Wessig, P. and Henning, H. G., Diastereoselective photochemical synthesis of 1-hydroxy-1-phenyl-hexahydro-pyrrolizin-3-one — A structural relative of retronecine, publication in preparation.

22. Wessig, P., Legart, F., Hoffmann, B., and Henning, H. G., Synthesis and transannular photocyclization of 2-benzoyl-1,4-*bis*(tosyl)piperazines, *Liebigs Ann. Chem.*, 979, 1991.

23. (a) Henning, H. G., Dietzsch, T., and Fuhrmann, J., Synthesis of 3-aryl-3-pyrrolidinols via photocyclization of β-amino-propiophenones, *J. Prakt. Chem.*, 323, 435, 1981; *Chem. Abstr.*, 95, 186310, 1981; (b) Henning, H. G., Haupt, M., and Schoeder, H., Effect of substituents on the photocyclization of β-amidopropiophenones, *Z. Chem.*, 24, 60, 1984; *Chem. Abstr.*, 101, 90076, 1984; (c) Henning, H. G., Haber, H., and Knoll, A., Intramolecular cyclizations of β-ureido-propiophenones, *Z. Chem.*, 22, 58, 1982; *Chem. Abstr.*, 97, 23728, 1982; (d) Henning, H. G. and Walther, K., Photochemical preparation of 3'-hydroxy-3'-phenyl-N'-tosylnornicotine, *Pharmazie*, 37, 810, 1982; *Chem. Abstr.*, 98, 143698, 1983.

24. Walther, K., Kranz, U., and Henning, H. G., Preparation and diastereoselective photocyclization of N-(β-benzoylethyl)-N-tosyl-glycinamides, *J. Prakt. Chem.*, 329, 859, 1987; *Chem. Abstr.*, 109, 93557, 1988.

25. (a) Kernchen, F. and Henning, H. G., Preparation and photochemical behavior of both rotamers of methyl N-(β-benzoylethyl)-N-(2',4',6'-triisopropylbenzoyl) glycinates, *Z. Chem.*, 28, 219, 1988; *Chem. Abstr.*, 110, 154817, 1989; (b) Kernchen, F. and Henning, H. G., The behavior of N-acylglycinate rotamers in the photochemical glycine→proline conversion, *Monatsh. Chem.*, 120, 253, 1989.

26. Taft, R. W., The general nature of the proportionality of polar effects of substituent groups in organic chemistry, *J. Am. Chem. Soc.*, 75, 4231, 1953.

Analytical data:

1: ^{1}H-NMR (CDCl$_3$, 80 MHz): δ = 2.72 (t, 2H, CH$_2$-CONR$_2$), 3.34 (t, 2H, Ph-CO-CH$_2$), 3.60 (m, 8H, morph), 7.30–8.06 (m, 5H, arom); IR (KBr): ν = 1690, 1645cm^{-1}; UV (Ether): λ_{max} = 315 (1.9), 236 (4.2) nm.

3: ^{1}H-NMR (CDCl$_3$, 80 MHz): δ = 1.0–2.0 (m, 3H, cProp), 3.35 (m, 8H, morph), 4.98 (s, 1H, OH), 7.3–7.4 (m, 5H, arom); IR (KBr): ν = 3260, 1615 cm^{-1}.

4: ^{1}H-NMR (CDCl$_3$, 80 MHz): δ = 1.2–2.1 (m, 3H, cProp), 3.41 (s, 8H, morph), 5.09 (s, 1H, OH), 7.1 (m, 5H, arom); IR (KBr): ν = 3420, 1640 cm^{-1}.

5: ^{1}H-NMR (CDCl$_3$, 250 MHz): δ = 2.45 (m, 4H, morph), 2.65 (ABX-A, 1H, CH-N; J_{AX} = 5.1 Hz, J_{AB} = 12.2 Hz), 3.40 (ABX-B, 1H,CH-N; J_{BX} = 9.2 Hz, J_{AB} = 12.2 Hz), 3.60 (t, 4H, morph), 4.85 (ABX-X, 1H, CO-CH; J_{AX} = 5.1 Hz, J_{BX} = 9.2 Hz), 7.35 (m, 8H, arom), 7.95 (d, 2H, arom). IR (KBr): ν = 1677 cm^{-1}. UV (MeOH): λ_{max} = 318, 248 nm.

Product from 5 (6, 7, 8, or 9): ^{1}H-NMR (CDCl$_3$, 250 MHz): δ = 1.95 (s, 1H, OH), 2.35 (m, 2H, morph), 2.60 (m, 2H, morph), 2.70 (d, 1H, Ph-CH; J = 5.7 Hz), 2.75 (d, 1H, CH-N; J = 5.7 Hz), 3.50 (m, 4H, morph), 7.3 (m, 8H, arom), 7.70 (d, 2H, arom). IR (KBr): ν = 3287 cm^{-1}. UV (MeOH): λ_{max} = 236 nm.

10 (R = H): ^{1}H-NMR (CDCl$_3$, 80 MHz): δ = 2.31 (s, 3H, Ts-Me), 3.19–3.69 (m, 4H, CH$_2$-CH$_2$), 3.49 (s, 3H, OMe), 4.08 (s, 2H, N-CH$_2$-CO), 7.12–7.90 (m, 9H, arom). IR (CHCl$_3$): ν = 1754, 1683 cm^{-1}. UV (Ether): λ_{max} = 315 (1.78), 236 (4.35) nm.

10 (R = Me): ^{1}H-NMR (CDCl$_3$, 80 MHz): δ = 1.09 (d, 3H, α-C-Me), 2.30 (s, 3H, Ts-Me), 3.40 (s, 3H, OMe), 3.46 (d, 2H, β-CH$_2$), 3.85–4.08 (m, 1H, α-CH), 4.02 (s, 2H, N-CH$_2$-CO), 7.21–7.93 (m, 9H, arom). IR (CHCl$_3$): ν = 1750, 1677 cm^{-1}. UV (Ether): λ_{max} = 320 (1.85), 237 (4.36) nm.

10 (R = Ph): ^{1}H-NMR (CDCl$_3$, 80 MHz): δ = 2.33 (s, 3H, Ts-Me), 3.38 (s, 3H, OMe), 3.48–3.93 (m, 2H, β-CH$_2$), 3.78/4.15 (d, 2H, N-CH$_2$-CO), 5.05–5.25 (m, 1H, α-CH), 7.12–7.95 (m, 14H, arom). IR (CHCl$_3$): ν = 1752, 1677 cm^{-1}. UV (Ether): λ_{max} = 329 (2.33), 239 (4.36) nm.

11 (R = H): ^{1}H-NMR (CDCl$_3$, 80 MHz): δ = 1.88–2.13 (m, 1H, cProp), 2.30 (s, 3H, Ts-Me), 2.62–3.0 (m, 2H, cProp), 3.08 (s, 1H, OH), 3.51 (s, 3H, OMe), 3.79 (s, 2H, N-CH$_2$-CO), 6.90–7.52 (m, 9H, arom). IR (KBr): ν = 3500, 1750 cm^{-1}.

12 (R = H): ^{1}H-NMR (CDCl$_3$, 80 MHz): δ = 1.93–2.03 (m, 2H, C(4)H$_2$), 3.25 (s, 1H, OH), 3.38–3.75 (m, 2H, C(5)H$_2$), 3.64 (s, 3H, OMe), 4.45 (s, 1H, C(2)H), 7.12–7.38 (m, 7H, arom), 7.60–7.82 (m, 2H, arom). IR (CHCl$_3$): ν = 3575, 1735 cm^{-1}.

12 (R = Me): ^{1}H-NMR (CDCl$_3$, 80 MHz): δ = 0.68 (d, 3H, C(1)Me), 2.04–2.16 (m, 1H, C(1)H), 3.10 (s, 1H, OH), 3.25–3.68 (m, 2H, C(5)H$_2$), 3.58 (s, 3H, OMe), 4.39 (s, 1H, C(2)H), 7.19–7.35 (m, 7H, arom), 7.65–7.83 (m, 2H, arom). IR (CHCl$_3$): ν = 3567, 1734 cm^{-1}.

12 (R = Ph): ^{1}H-NMR (CDCl$_3$, 80 MHz): δ = 2.75 (s, 1H, OH), 3.25–3.90 (m, 2H, C(5)H$_2$), 3.57 (s, 3H, OMe), 4.02 (m, 1H, C(4)H), 4.70 (s, 1H, C(2)H), 6.70–6.87 (m, 2H, arom), 6.97–7.37 (m, 10 H, arom), 7.72–7.90 (m, 2H, arom). IR (CHCl$_3$): ν = 3562, 1751 cm^{-1}.

14 (Acyl = Ts, R = H): ^{1}H-NMR (CDCl$_3$, 80 MHz): δ = 2.35 (s, 3H, Ts-Me), 4.45 (s, 2H, CH$_2$-Ph), 4.53 (s, 2H, CO-CH$_2$), 7.1–7.8 (14H, arom). IR (CCl$_4$): ν = 1703 cm^{-1}. UV (Ether): λ_{max} = 324 (1.85), 238.5 (4.31) nm.

14 (Acyl = SO$_2$-R′, R = H): ^{1}H-NMR (CDCl$_3$, 80 MHz): δ = 2.20 (s, 3H, *p*-Me), 4.46 (s, 2H, N-CH$_2$-Ph), 4.52 (s, 2H, COCH$_2$), 6.62 (s, 4H, *o*-Me), 6.90–7.80 (m, 12H, arom). IR (CCl$_4$): ν = 1704 cm^{-1}. UV (Ether): λ_{max} = 324.2 (1.85), 240.6 (4.43) nm.

14 (Acyl = CO-Ph, R = Me): ^{1}H-NMR (CDCl$_3$, 80 MHz): δ = 1.32 (d, 3H, Me), 4.41 (s, 2H, N-CH$_2$-Ph), 5.81 (q, 1H, CH), 6.88–7.94 (m, 15H, arom). IR (CCl$_4$): ν = 1695, 1643 cm^{-1}. UV (Ether): λ_{max} = 328 (2.20), 235 (4.33) nm.

15: ^{1}H-NMR (CDCl$_3$/DMSO 2:1, 80 MHz): δ = 2.46 (s, 3H, Ts-Me), 3.80 (d, 1H, H$_a$), 4.22 (d, 1H, H$_b$), 4.95 (s, 1H, H$_c$), 6.28 (s, 1H, OH), 7.0–7.7 (m, 14H, arom). IR (KBr): ν = 3432 cm^{-1}.

16 (Acyl = Ts, R = H): ^{1}H-NMR (CDCl$_3$/DMSO 2:1, 80 MHz): δ = 2.49 (s, 3H, Ts-Me), 4.00 (s, 2H, H$_a$,H$_b$), 4.90 (s 1H, H$_c$), 5.78 (s, 1H, OH), 6.8–7.8 (m, 14H, arom). IR (KBr): ν = 3510 cm^{-1}.

16 (Acyl = SO$_2$-R′, R = H): ^{1}H-NMR (CDCl$_3$/DMSO 2:1, 80 MHz): δ = 2.55 (s, 3H, *p*-Me), 2.85 (s, 6H, *o*-Me), 4.15 (s, 2H, H$_a$,H$_b$), 5.35 (s, 1H, H$_c$), 6.49 (s, 1H, OH), 6.8–7.35 (m, 12H, arom). IR (KBr): ν = 3420 cm^{-1}.

17: ^{1}H-NMR (CDCl$_3$, 80 MHz): δ = 1.48 (d, 3H, Me), 1.98 (s, 1H, OH), 4.88 (q, 1H, C(4)H), 5.57 (s, 1H, C(2)H), 7.0–7.56 (m, 15H, arom). IR (KBr): ν = 3400 cm^{-1}.

18b: ^{1}H-NMR (CDCl$_3$, 80 MHz): δ = 1.2 (d, 6H, Me), 4.1 (m, 1H, CH), 4.7 (s, 2H, CH$_2$), 7.2–8.1 (m, 18H, arom). IR (KBr): ν = 1745, 1705, 1625 cm^{-1}.

19: ^{1}H-NMR (CDCl$_3$, 80 MHz): δ = 1.20 (s, 3H, Me), 1.71 (s, 3H, Me), 3.85 (s, 1H, OH), 4.16 (d, 1H, CH$_2$; J = 10 Hz), 4.58 (d, 1H, CH$_2$; J = 10 Hz), 7.2–8.1 (m, 18H, arom). IR (KBr): ν = 3400, 1750, 1630 cm^{-1}.

22a: ^{1}H-NMR (CDCl$_3$, 80 MHz): δ = 1.21 (t, 3H, ester-CH$_3$), 3.03 (s, 3H, Me), 4.07 (m, <4H, C(α)H$_2$ + ester-CH$_2$), 5.66 (s, <1H, CH-enol), 7.2–7.7 (m, 5H, arom), 16.20 (s, >1H, OH-enol). IR (KBr): ν = 1745, 1700, 1630 cm^{-1}. UV (Ether): λ_{max} = 300 (4.15), 227 (3.80) nm.

22b: ^{1}H-NMR (DMSO-d$_6$, 80 MHz): δ = 2.87 (m, 3H, Me), 3.92 (m, <4H, CH$_2$), 5.92 (s, <1H, CH-enol), 7.1–7.9 (m, 7H, arom + NH$_2$), 15.5 (s, >1H; OH-enol). IR (KBr): ν = 1690, 1650, 1620 cm^{-1}. UV (Ether): λ_{max} = 300 (4.06), 237 (4.02) nm.

23a: ^{1}H-NMR (DMSO-d_6, 80 MHz): δ = 1.12 (t, 3H, ester-CH$_3$), 2.45/2.77 (dd, 2H, C(3)H$_2$), 3.47/3.68 (dd, 2H, C(5)H$_2$), 4.12 (m, 4H, N-CH$_2$ + ester-CH$_2$), 5.69 (s, 1H, OH), 7.2–7.5 (m, 5H, arom). IR (KBr): ν = 3225, 1745, 1690 cm^{-1}.

23b: ^{1}H-NMR (DMSO-d_6, 80 MHz): δ = 2.37/2.82 (dd, 2H, C(3)H$_2$), 3.37/3.69 (dd, 2H, C(5)H$_2$), 3.60/3.91 (dd, 2H, N-CH$_2$), 5.75 (s, 1H, OH), 7.1–7.5 (m, 7H, arom + NH$_2$). IR (KBr): ν = 3310, 1680, 1650, 1625 cm^{-1}.

25 (keto-enol mixture): ^{1}H-NMR (CDCl$_3$, 300 MHz): δ = 1.85–2.0 (m), 3.4–4.6 (m), 4.04 (s), 5.64 (s) 7.29–8.03 (m, 5H, arom), 15. (s, 1H, OH). IR (KBr): ν = 2977, 2881, 1628, 1599 cm^{-1}. UV (MeCN): λ_{max} = 295 nm.

26: ^{1}H-NMR (CDCl$_3$, 300 MHz): δ = 1.69–1.80/2.0–2.28 (m, 4H, H^3-H^6), 2.69 (d, 1H, H^8), 3.01–3.09 (m, 1H, H^1), 3.30 (d, 1H, H^9), 3.61–3.69 (m, 1H, H^2), 4.12–4.20 (m, 2H, H^7 + OH), 7.26–7.50 (m, 5H, arom). IR (KBr): ν = 3203, 2977, 1676 cm^{-1}.

27a: ^{1}H-NMR (CDCl$_3$, 250 MHz): δ = 2.38 (s, 3H, Ts-Me), 2.42 (s, 3H, Ts-Me), 5.55 (t, 1H, N-CH-CO), 2.45–4.0 (m, 6H, pip), 7.2–7.8 (m, 13H, arom). IR (KBr): ν = 1692, 1345, 1162 cm^{-1}. UV (MeCN): λ_{max} = 331 (2.14), 236 (4.43) nm.

27b: ^{1}H-NMR (CDCl$_3$, 250 MHz): δ = 1.48 (s, 3H, Me), 1.60 (s, 3H, Me), 2.34 (s, 3H, Ts-Me), 2.39 (s, 3H, Ts-Me), 2.85–3.69 (m, 4H, pip), 5.55 (t, 1H, N-CH-CO), 7.16–7.78 (m, 13H, arom). IR (KBr): ν = 1702, 1345, 1164 cm^{-1}. UV (MeCN): λ_{max} = 322 (2.03), 235 (4.44) nm.

28: ^{1}H-NMR (CDCl$_3$, 250 MHz): δ = 2.3 (s, 3H, Ts-Me), 2.4 (s, 3H, Ts-Me), 3.0–4.2 (m, 6H, CH$_2$ + CH), 5.9 (s, 1H, NH), 7.0–7.8 (m, 13H, arom). IR (KBr): ν = 3226, 1677, 1348, 1166 cm^{-1}.

29: ^{1}H-NMR (CDCl$_3$, 250 MHz): δ = 1.41 (s, 3H, Me), 2.36 (d, 1H), 2.41 (s, 3H, Ts-Me), 2.47 (s, 3H, Ts-Me), 2.61 (d, 1H), 2.95 (d, 1H), 3.27 (d, 1H), 3.34 (s, 1H), 3.54 (d, 1H), 4.11 (d, 1H), 4.12 (s, 1H), 7.18–7.77 (m, 13H, arom). IR (KBr): ν = 3545, 1348, 1167 cm^{-1}.

30: ^{1}H-NMR (CDCl$_3$, 250 MHz): δ = 1.59 (s, 3H, Me), 1.77 (s, 3H, Me), 2.24 (s, 1H, OH), 2.33 (s, 3H, Ts-Me), 2.40 (s, 3H, Ts-Me), 3.13–3.23 (m, 2H), 4.29 (s, 1H), 4.47 (m, 1H), 7.0–7.8 (m, 13H, arom). IR (KBr): ν = 3480, 1344, 1170 cm^{-1}.

32 (R^1 = H, R^2 = CO-Ph, R^3 = Ph): ^{1}H-NMR (CDCl$_3$, 80 MHz): δ = 3.25 (t, 2H, α-CH$_2$), 3.65 (t, 2H, β-CH$_2$), 4.60 (s, 2H, N-CH$_2$-Ph), 7.83–7.95 (m, 15H, arom). IR (Nujol): ν = 1672, 1638 cm^{-1}. UV (Ether): λ_{max} = 241 (4.55) nm.

32 (R^1 = Me, R^2 = CO-Ph, R^3 = Ph): ^{1}H-NMR (CDCl$_3$, 80 MHz): δ = 1.31 (s, 6H, Me), 3.79 (s, 2H, β-CH$_2$), 4.39 (s, 2H, N-CH$_2$-Ph), 6.77–7.8 m, 15H, arom). IR (KBr): ν = 1670, 1640 cm^{-1}.

32 (R^1 = H, R^2 = CO-NH$_2$, R^3 = Ph): ^{1}H-NMR (CDCl$_3$, 80 MHz): δ = 3.2 (t, 2H, CH$_2$), 3.6 (t, 2H, CH$_2$), 4.4 (s, 2H, N-CH$_2$-Ph), 5.3 (s, 2H, NH$_2$), 7.3–7.9 (m, 10 H, arom). IR (KBr): ν = 1685, 1665 cm^{-1}.

32 (R^1 = H, R^2 = Ts, R^3 = 3-pyridyl): ^{1}H-NMR (CDCl$_3$, 80 MHz): δ = 2.34 (s, 3H, Ts-Me), 3.04 (t, 2H, α-CH$_2$), 3.74 (t, 2H, β-CH$_2$), 4.00 (s, 2H, N-CH$_2$-Py), 7.04–7.8 (m, 11H, arom), 8.41 (d, 2H, arom). IR (KBr): ν = 1680, 1330, 1170 cm^{-1}.

33 (R^1 = H, R^2 = CO-Ph, R^3 = Ph): ^{1}H-NMR (CDBr$_3$, 80 MHz): δ = 2.2 (m, 2H, C(4)H$_2$), 2.35 (s, 1H, OH), 4.0 (m, 2H, C(5)H$_2$), 5.5 (s, 1H, C(2)H), 7.7 (m, 2H, *ortho*-Ph). IR (KBr): ν = 3290, 1610 cm^{-1}.
^{1}H-NMR (CDCl$_3$, 80 MHz): δ = 2.0 (s, 1H, OH), 2.6 (m, 2H, C(4)H$_2$), 4.05 (m, 2H, C(5)H$_2$), 4.7 (s, 1H, C(2)H), 7.4 (m, 2H, *ortho*-Ph). IR (KBr): ν = 3360, 1610 cm^{-1}.

33 (R^1 = Me, R^2 = CO-Ph, R^3 = Ph):
^{1}H-NMR (CDCl$_3$, 80 MHz): δ = 0.73 (s, 3H, Me), 0.84 (s, 3H, Me), 1.46 (s, 1H, OH), 3.36/4.07 (dd, C(5)H$_2$), 6.03 (s, 1H, CH).
IR (KBr): ν = 3400, 1645 cm^{-1}.
^{1}H-NMR (CDCl$_3$, 80 MHz): δ = 0.39 (s, 3H, Me), 1.08 (s, 3H, Me), 3.04 (s, 1H, OH), 3.36/3.75 (dd, 2H, C(5)H$_2$), 5.58 (s, CH).
IR (KBr): ν = 3400, 1638 cm^{-1}.

33 (R^1 = H, R^2 = CO-NH$_2$, R^3 = Ph): ^{1}H-NMR (CDCl$_3$/DMSO 3:1): δ = 1.75–2.90 (m, 2H, C(4)H$_2$), 3.84 (q, 2H, C(5)H$_2$), 4.70 (s, 1H, CH), 4.76 (s, 2H, NH$_2$), 5.35 (s, 1H, OH), 6.55–7.25 (m, 10 H, Ph). IR (KBr): ν = 3400, 3350–3230, 1670 cm^{-1}.

33 (R^1 = H, R^2 = Ts, R^3 = 3-pyridyl):
  ^{1}H-NMR (CDCl$_3$, 80 MHz): δ = 1.63–2.00 (m, 2H, C(4)H$_2$), 2.35 (s, 3H, Ts-Me), 3.05 (s, 1H, OH), 3.45–3.91 (m, 2H, C(C5)H$_2$), 4.65 (s, 1H, CH), 6.70–8.13 (m, 13H, arom). IR (KBr): ν = 3400, 1320, 1150 cm^{-1}.
  ^{1}H-NMR (CDCl$_3$, 80 MHz): δ = 1.83–2.08 (m, 2H, C(4)H$_2$), 2.39 (s, 3H, Ts-Me), 3.70–4.03 (m, 3H, OH + C(5)H$_2$), 4.76 (s, 1H, CH), 6.75–8.00 (m, 13H, arom). IR (KBr): ν = 3480, 1330, 1160 cm^{-1}.

34 (R^1 = H, R^2 = Me): ^{1}H-NMR (CDCl$_3$, 80 MHz): δ = 2.32 (s, 3H, Ts-Me), 2.65 (d, 3H, N-Me), 3.04–3.64 (m, 6H, CH$_2$), 6.68 (m, 1H, NH), 7.06–7.88 (m, 9H, arom). IR (KBr): ν = 3405, 1684, 1657, 1348, 1160 cm^{-1}. UV (EtOH): λ$_{max}$ = 317 (1.90), 236 (4.41) nm.

34 (R^1 = H, R^2 = CH$_2$-(Ph): ^{1}H-NMR (CDCl$_3$, 80 MHz): δ = 2.36 (s, 3H, Ts-Me), 3.04–3.74 (m, 6H, CH$_2$), 4.34 (d, 2H, CH$_2$), 6.94–7.83 (m, 15H, OH + arom). IR (KBr): ν = 3380, 1682, 1670, 1345, 1160 cm^{-1}. UV (EtOH): λ$_{max}$ = 323 (1.83), 234 (4.41) nm.

34 (R^1 = R^2 = Me): ^{1}H-NMR (CDCl$_3$, 80 MHz): δ = 2.26 (s, 3H, Me), 2.74 (s, 3H, Me), 2.89 (s, 3H, Me), 3.12–3.62 (m, 4H, CH$_2$), 4.08 (s, 2H, CH$_2$), 7.05–7.89 (m, 9H, arom). IR (KBr): ν = 1680, 1663, 1330, 1150 cm^{-1}. UV (EtOH): λ$_{max}$ = 321 (1.85), 236 (4.40) nm.

34 (R^1 = Me, R^2 = CH$_2$-Ph): ^{1}H-NMR (CDCl$_3$, 80 MHz): δ = 2.26 (s, 3H, Me), 2.79 (s, 3H, Me), 3.12–3.68 (m, 4H, CH$_2$), 4.15 (s, 2H, CH$_2$), 4.34 (s, 2H, CH$_2$), 6.94–7.87 (m, 14H, arom). IR (KBr): ν = 1675, 1665, 1325, 1155 cm^{-1}. UV (EtOH): λ$_{max}$ = 318 (2.10), 233 (4.43) nm.

35 (R^1 = H, R^2 = Me): ^{1}H-NMR (CDCl$_3$, 80 MHz): δ = 1.45–2.08 (m, 2H, C(4)H$_2$), 2.39 (s, 3H, Ts-Me), 2.75 (d, 3H, N-Me), 3.59–3.84 (m, 2H, C(5)H$_2$), 4.25 (s, 1H, CH), 4.55 (s, 1H, OH), 6.78–7.78 (m, 10 H, NH + arom). IR (KBr): ν = 3390, 3260, 1662 cm^{-1}.

35 (R^1 = H, R^2 = CH$_2$-Ph): ^{1}H-NMR (CDCl$_3$, 80 MHz): δ = 1.50–2.03 (m, 2H, C(4)H$_2$), 2.36 (s, 3H, Ts-Me), 3.55–3.81 (m, 2H, C(5)H$_2$), 4.32 (s, 1H, CH), 4.32 (s, 1H, OH), 4.40 (d, 2H, CH$_2$), 6.79–7.75 (m, 15H, NH + arom). IR (KBr): ν = 3490, 3410, 1684, 1662 cm^{-1}.

36 (R^1 = R^2 = Me): ^{1}H-NMR (CDCl$_3$, 80 MHz): δ = 1.92–2.25 (m, 2H, C(4)H$_2$), 2.32 (s, 3H, Ts-Me), 2.54 (s, 3H, Me), 2.76 (s, 2H, Me), 3.32–3.82 (m, 2H, C(5)H$_2$), 4.62 (s, 1H, CH), 5.71 (s, 1H, OH), 7.12–7.75 (m, 9H, arom). IR (KBr): ν = 3420, 1630 cm^{-1}.

36 (R^1 = Me, R^2 = CH$_2$-Ph): ^{1}H-NMR (CDCl$_3$, 80 MHz): δ = 1.88–2.20 (m, 2H, C(4)H$_2$), 2.30 (s, 3H, Ts-Me), 2.46/2.62 (2s, 3H, Me), 3.32–3.82 (m, 2H, C(5)H$_2$), 4.05–4.66 (m, 2H, CH$_2$), 4.78 (s, 1H, CH), 5.74–5.96 (2s, 1H, OH), 6.80–7.75 (m, 14H, arom). IR (KBr): ν = 3320, 1635 cm^{-1}.

37a: ^{1}H-NMR (CDCl$_3$, 80 MHz): δ = 2.31 (s, 3H, Ts-Me), 3.19–3.69 (m, 4H, α-CH$_2$ + β-CH$_2$), 3.49 (s, 3H, OMe), 4.08 (s, 2H, N-CH$_2$-CO), 7.12–7.90 (m, 9H, arom). IR (CHCl$_3$): ν = 1754, 1683 cm^{-1}. UV (Ether): λ$_{max}$ = 315 (1.78), 236 (4.35) nm.

37b: ^{1}H-NMR (CDCl$_3$, 80 MHz): δ = 1.09 (t, 3H, ester-Me), 2.33 (s, 3H, Ts-Me), 3.30–3.68 (m, 4H, α-CH$_2$ + β-CH$_2$), 3.96 (q, 2H, ester-CH$_2$), 4.08 (s, 2H, N-CH$_2$-CO), 7.13–7.91 (m, 9H, arom). IR (CHCl$_3$): ν = 1747, 1682 cm^{-1}. UV (Ether): λ$_{max}$ = 318 (1.84), 238 (4.38) nm.

37c: ^{1}H-NMR (CDCl$_3$, 80 MHz): δ = 1.09 (d, 3H, α-C-Me), 2.30 (s, 3H, Ts-Me), 3.40 (s, 3H, OMe), 3.46 (d, 2H, β-CH$_2$), 3.85–4.08 (m, 1H, α-CH), 7.21–7.93 (m, 9H, arom). IR (CHCl$_3$): ν = 1750, 1677 cm^{-1}. UV (Ether): λ$_{max}$ = 320 (1.85), 237 (4.36) nm.

37d: ^{1}H-NMR (CDCl$_3$, 80 MHz): δ = 1.31 (d, 3H, CH-Me), 2.31 (s, 3H, Ts-Me), 3.40–3.71 (m, 4H, α-CH$_2$ + β-CH$_2$), 3.42 (s, 3H, OMe), 4.69 (q, 1H, N-CH-CO), 7.14–7.98 (m, 9H, arom). IR (CHCl$_3$): ν = 1744, 1680 cm^{-1}. UV (Ether): λ$_{max}$ = 319 (1.74), 237 (4.38) nm.

37e: ^{1}H-NMR (CDCl$_3$, 80 MHz): δ = 2.33 (s, 3H, Ts-Me), 3.38 (s, 3H, OMe), 3.48–3.93 (m, 2H, β-CH$_2$), 3.78/4.15 (d, 2H, N-CH$_2$-CO), 5.05–5.25 (m, 1H, α-CH$_2$), 7.12–7.95 (m, 14H, arom). IR (CHCl$_3$): ν = 1752, 1677 cm^{-1}. UV (Ether): λ$_{max}$ = 329 (2.33), 239 (4.36) nm.

37f: ^{1}H-NMR (CDCl$_3$, 80 MHz): δ = 2.34 (s, 3H, Ts-Me), 3.07–3.73 (m, 4H, α-CH$_2$ + β-CH$_2$), 3.48 (s, 3H, OMe), 5.79 (s, 1H, N-CH-CO), 7.18–7.77 (m, 14H, arom). IR (CHCL$_3$): ν = 1747, 1681 cm^{-1}. UV (Ether): λ$_{max}$ = 319 (1.76), 237 (4.39) nm.

37g: ^{1}H-NMR (CDCl$_3$, 80 MHz): δ = 3.18–3.98 (m, 4H, α-CH$_2$ + β-CH$_2$), 3.65 (s, 3H, OMe), 4.15 (s, 2H, N-CH$_2$-CO), 7.25–8.03 (m, 10 H, arom). IR (CHCl$_3$): ν = 1749, 1683, 1637 cm^{-1}. UV (Ether): λ$_{max}$ = 318 (1.89), 239 (4.19) nm.

37h: ^{1}H-NMR (CDCl$_3$, 80 MHz): δ = 1.14 (d, 3H, α-C-Me), 3.25–3.75 (m, 2H, β-CH$_2$), 3.51 (s, 3H, OMe), 3.92 (s, 2H, N-CH$_2$-CO), 4.06–4.20 (m, 1H, α-CH), 7.27–8.0 (m, 10 H, arom). IR (CHCl$_3$): ν = 1747, 1676, 1639 cm^{-1}. UV (Ether): λ$_{max}$ = 321 (1.81), 240 (4.23) nm.

37i: ^{1}H-NMR (CDCl$_3$, 80 MHz): δ = 3.39 (s, 3H, OMe), 3.63–4.25 (m, 4H, β-CH$_2$ + N-CH$_2$-CO), 5.27–5.48 (m, 1H, α-CH), 7.07–7.81 (m, 15H, arom). IR (CHCl$_3$): ν = 1747, 1677, 1636 cm^{-1}. UV (Ether): λ$_{max}$ = 328 (2.28), 240 (4.22) nm.

38a: ^{1}H-NMR (CDCl$_3$, 80 MHz): δ = 1.93–2.03 (m, 2H, C(4)H$_2$), 3.25 (s, 1H OH), 3.38–3.75 (m, 2H, C(5)H$_2$), 3.64 (s, 3H, OMe), 4.45 (s, 1H, C(2)H), 7.12–7.82 (m, 9H, arom). IR (CHCl$_3$): ν = 3575, 1735 cm^{-1}.

38b: ^{1}H-NMR (CDCl$_3$, 80 MHz): δ = 1.08 (t, 3H, ester-Me), 1.90–2.25 (m, 2H, C(4)H$_2$), 3.29 (s, 1H, OH), 3.40–3.75 (m, 2H, C(5)H$_2$), 4.08 (q, 2H, ester-CH$_2$), 4.40 (s, 1H, C(2)H), 7.13–7.80 (m, 9H, arom). IR (CHCl$_3$): ν = 3579, 1733 cm^{-1}.

38c: ^{1}H-NMR (CDCl$_3$, 80 MHz): δ = 0.68 (d, 3H, C(4)-Me), 2.04–2.16 (m, 1H, C(4)H), 3.10 (s, 1H, OH), 3.25–3.68 (m, 2H, C(5)H$_2$), 3.58 (s, 3H, OMe), 4.39 (s, 1H, C(2)H), 7.19–7.83 (m, 9H, arom). IR (CHCl$_3$): ν = 3567, 1734 cm^{-1}.

38e: ^{1}H-NMR (CDCl$_3$, 80 MHz): δ = 2.75 (s, 1H, OH), 3.25–3.90 (m, 2H, C(5)H$_2$), 3.57 (s, 3H, OMe), 4.02 (d, 1H, C(4)H), 4.70 (s, 1H, C(2)H), 6.70–7.90 (m, 14H, arom). IR (CHCl$_3$): ν = 3562, 1751 cm^{-1}.

38g: ^{1}H-NMR (CDCl$_3$, 80 MHz): δ = 2.08–2.35 (m, 2H, C(4)H$_2$), 3.41 (s, 1H, OH), 3.48–4.00 (m, 2H, C(5)H$_2$), 3.61 (s, 3H, OMe), 4.85 (s, 1H, C(2)H), 7.05–7.57 (m, 10 H, arom). IR (CHCl$_3$): ν = 3580, 1738, 1631 cm^{-1}.

38h: ^{1}H-NMR (CDCl$_3$, 80 MHz): δ = 0.71 (d, 3H, C(4)-Me), 2.25–2.72 (m, 1H, C(4)H), 3.35 (s, 1H, OH), 3.35–3.77 (m, 2H, C(5)H$_2$), 3.55 (s, 3H, OMe), 4.83 (s, 1H, C(2)H), 7.18–7.68 (m, 10H, arom). IR (CHCl$_3$): ν = 3590, 1738, 1628 cm^{-1}.

38i: ^{1}H-NMR (CDCl$_3$, 80 MHz): δ = 2.75 (s, 1H, OH), 3.61 (s, 3H, OMe), 3.79–4.00 (m, 2H, C(5)H$_2$), 4.30–4.55 (m, 1H, C(4)H), 5.13 (s, 1H, C(2)H), 6.98–7.75 (m, 15H, arom). IR (CHCl$_3$): ν = 3565, 1747, 1635 cm^{-1}.

39: ^{1}H-NMR (CDCl$_3$, 80 MHz): δ = 1.18 (t, 3H, ester-Me), 2.44 (s, 3H, Ts-Me), 2.55–2.82 (t, 2H, CH$_2$-COOEt), 3.30–3.82 (m, 6H, α-CH$_2$ + β-CH$_2$ + N-CH$_2$), 4.23 (q, 2H, ester-CH$_2$), 7.50–8.32 (m, 9H, arom). IR (CHCl$_3$): ν = 1727, 1683 cm^{-1}. UV (Ether): λ$_{max}$ = 319 (1.76), 238 (4.43) nm.

42: ^{1}H-NMR (CDCl$_3$, 80 MHz): δ = 1.73–2.28 (m, 2H, C(8)H$_2$), 2.30 (s, 3H, Ts-Me), 2.71 (d, 1H, C(4)H$_2$), 2.75 (s, 1H, C(4)H$_2$), 3.40–3.95 (m, 2H, C(7)H$_2$), 4.30 (d, 1H, C(5)H), 6.88–7.79 (m, 9H, arom). IR (CHCl$_3$): ν = 1790 cm^{-1}.

43a: ^{1}H-NMR (CDCl$_3$, 80 MHz): δ = 0.90–1.30 (m, 18H, *i*-Pr-Me), 2.66–3.06 (m, 3H, *i*-Pr-CH), 3.10–3.30 (m, 2H, α-CH$_2$), 3.35–3.60 (m, 2H, β-CH$_2$), 3.61 (s, 3H, OMe), 4.28 (s, 2H, N-CH$_2$-CO), 7.00–7.85 (m, 7H, arom). IR (CHCl$_3$): ν = 1747, 1681, 1640 cm^{-1}. UV (MeOH): λ$_{max}$ = 316 (1.95), 243 (4.13) nm.

43b: ^{1}H-NMR (CDCl$_3$, 80 MHz): δ = 0.70–1.30 (m, 18H, *i*-Pr-Me), 2.50–3.00 (m, 3H, *i*-Pr-CH), 3.35–3.55 (m, 2H, α-CH$_2$), 3.56 (s, 3H, OMe), 3.68–3.90 (m, 2H, β-CH$_2$), 3.99 (s, 2H, N-CH$_2$-CO), 6.96–8.04 (m, 7H, arom). IR (CHCl$_3$): ν = 1750, 1681, 1634 cm^{-1}. UV (MeOH): λ$_{max}$ = 315 (1.90), 243 (4.15) nm.

44: ^{1}H-NMR (CDCl$_3$, 80 MHz): δ = 1.05–1.35 (m, 18H, *i*-Pr-Me), 2.10–2.34 (m, 2H, C(4)H$_2$), 2.65–3.50 (m, 5H, C(5)H$_2$ + 3 *i*-Pr-CH), 3.02 (s, 1H, OH), 3.71 (s, 3H, OMe), 4.96 (s, 1H, C(2)H), 6.88–7.57 (m, 7H, arom). IR (KBr): ν = 3425, 1740, 1625 cm^{-1}.

41

Norrish Type II Processes Involving 1,6- and Greater Hydrogen Transfer Reactions

‟ard L. Descotes
‟ersity Lyon

41.1 Introduction

Intramolecular hydrogen abstraction by the excited carbonyl group is a very well-known primary photochemical process and generally involves a predominating six-membered cyclic transition state in the Norrish type II reaction. This γ-hydrogen abstraction (1,5-hydrogen shift) is greatly facilitated by favorable stereoelectronic or geometric requirements.[1] Hydrogen abstraction from other positions such as δ- (1,6-hydrogen shift) is disfavored for seven-membered cyclic transition states both statistically and energetically. Efficient δ-hydrogen abstraction occurs when there are no γ-hydrogens or when structural rigidity prevents access to these γ-hydrogens.[2-4] The applications involving δ-hydrogen abstraction has been put out to intensive use in spectacular syntheses of dodecahedrane[5] or in the series of carbohydrates[6] for example.

Remote hydrogen abstraction through eight-membered transition states remains extremely rare even in the photochemistry of carbonyl compounds, while medium and large cyclic transition states are mainly evoked in the photochemistry of nitrogen functional groups.

Synthetic applications for such long-distance Norrish type II processes will be summarized for the preparation of oxygen and nitrogen heterocyclic compounds that result from different supposed mechanisms.

41.2 Mechanisms of Long-Distance Hydrogen Phototransfers

1,6- and 1,7-Biradicals

Very few mechanistic studies are described on the exotic photochemical Norrish type II reactions involving 1,6- or 1,7-biradicals. The reported phototransformation of cyclodecanone 1 into

0-8493-8634-9/95/$0.00+$.50

10-decalols 2 proceeds via preferential abstraction of hydrogen from the ε- rather than the γ-carbon[4] in cyclohexane. Irradiation of cyclodecanone in *t*-butanol produced a 1-hydroxybicyclo [6.2.0] octane 3 as the major product, in agreement with theoretical calculations and with the known ability of hydroxylic solvents to promote reactions of triplet hydroxy biradicals (Scheme 41.1).

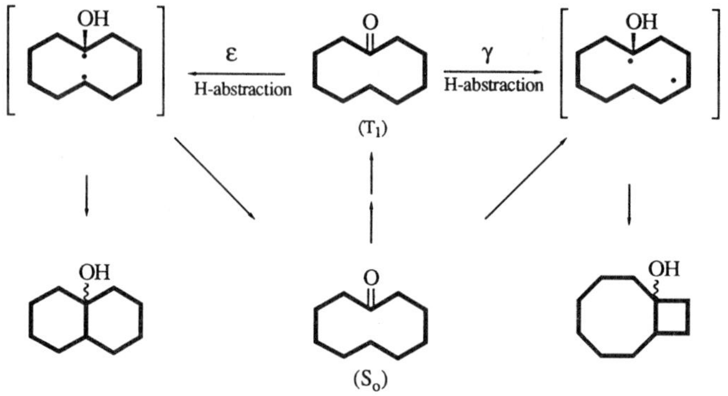

SCHEME 41.1   Photocyclization of cyclodecanone 1.

The photocyclization of α-[*o*-(benzyloxy)phenyl] acetophenone 4 (n = 1) due to ε-hydrogen abstraction by a triplet ketone is assumed to proceed via a 1,6-biradical intermediate with a comparable rate of hydrogen abstraction to that for δ-hydrogen abstraction by triplet *o*-(benzyloxy)benzophenone[8,9] (4, n = 0) (Scheme 41.2). The presumed 1,5- or 1,6-biradical can disproportionate to enol and can compete with photocyclization processes into indanols 5 (n = 0) or benzopyranols 5 (n = 1). This triplet reaction is quenched by dienes, and its low quantum yield may be due to competitive enolization reactions.

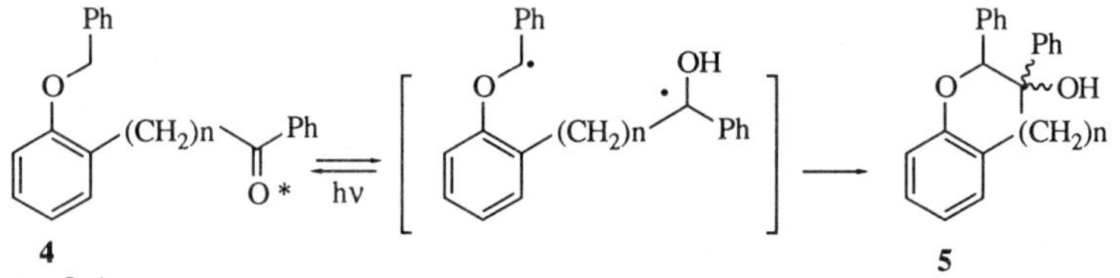

SCHEME 41.2   Photocyclization of α[*o*-(benzyloxy) phenyl] acetophenones 4.

1,6-Biradicals can also result from a rearrangement of 1,4-biradicals in a similar process to that proposed for biradicals derived from α-diketones.[10] The photocyclization of *o*-methylphenyl 1,3-diketones (6) into 7 is indicated to proceed through a six-membered transition state to give a 1,4-biradical (Scheme 41.3).

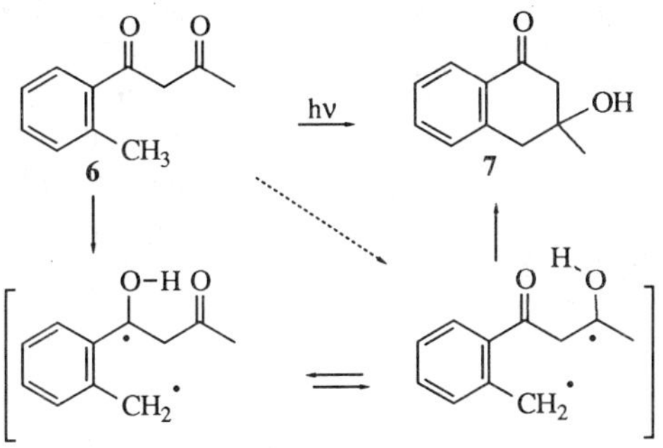

SCHEME 41.3 Photocyclization of *o*-methylphenyl 1,3-diketone 6.

The formation of 1,7-biradicals via a most unusual nine-membered transition state is evoked in the photochemistry of the *o*-benzyl substituted ketone (8) to yield tetrahydrobenzoxepinols (9).[11] The ζ-hydrogen abstraction reactions occur with rather low quantum efficiencies, probably resulting from competing intermolecular hydrogen abstraction to yield polymers (Scheme 41.4).

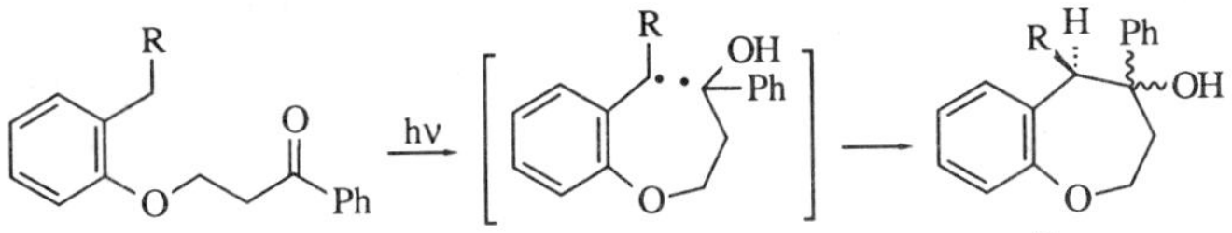

R = Phenyl, vinyl

SCHEME 41.4 Photocyclization of *o*-benzyl substituted ketones 8.

Intramolecular competition between δ- and ζ-hydrogen abstraction also depends on the position and on the nature of the substituents of the corresponding 1,5- and 1,7-biradicals.[12] The formation of spiroacetal 11 by UV irradiation of *o*-carbonylphenoxyisochroman (10) and δ-hydrogen photoabstraction is limited by the production of a crystalline [4.3.1]-bicyclo derivative 12 after abstraction of ζ-benzylic hydrogens (Scheme 41.5). The yield of the bicyclic product 12 decreases when the substituents R are bulky silyl or esters groups.

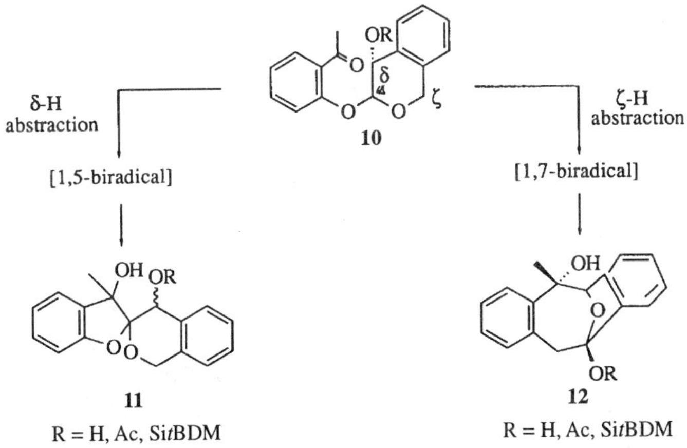

SCHEME 41.5 Competitive δ-H acetalic and ζ-H benzylic photoabstractions of 10.

In the case of γ,δ- and δ,ε-unsaturated amines **13**, hydrogen transformations proceed via 1,6- and 1,7-hydrogen shifts involving, respectively, seven- or eight-membered transition states. These photoreactions were found to be singlet reactions because they could not be sensitized or quenched.[13] Intramolecular charge exciplexes were suggested to explain the 1,6- and 1,7-electron-proton transfer (Scheme 41.6). Limited yields of pyrrolidines **14** (n = 2) and piperidines **14** (n = 3) are obtained by this type of hydrogen shift.

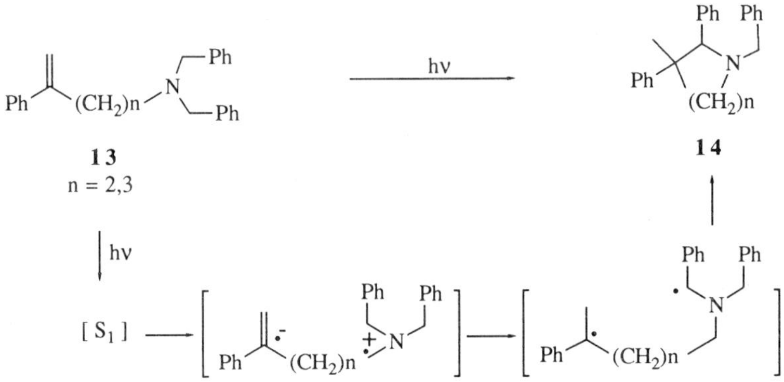

SCHEME 41.6   Photocyclization of γ,δ- and δ, ε-unsaturated amines **13**.

Long-Distance Hydrogen Abstraction

α-Ketoesters have been employed in a mild photoxidation reaction of alcohols by a classical Norrish type II reaction through a 1,4-diradical. However, with appropriate substitution, such compounds can provide a path to eight-membered ring lactones by a direct 1,9-hydrogen photoabstraction. Both 1,9-hydrogen abstraction and 1,5-hydrogen transfer are possible for this photocyclization,[14] but the use of deuterated compounds demonstrates clearly that in the case of **15**, only the 1,9-hydrogen abstraction reaction is involved in the formation of **16** (Scheme 41.7).

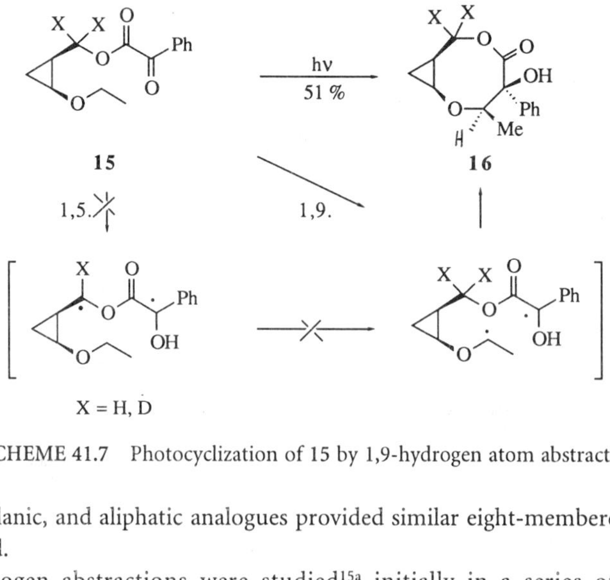

SCHEME 41.7   Photocyclization of **15** by 1,9-hydrogen atom abstraction.

Aromatic, cyclanic, and aliphatic analogues provided similar eight-membered ring lactones in rather good yield.

Remote hydrogen abstractions were studied[15a] initially in a series of *n*-alkylesters of 4-benzoylbenzoic acid (**17**) in order to bring about a remote functionalization (**18**) of a linear chain.

Hydrogen abstraction in this system does not occur below C-9 due to the inaccessibility of hydrogen. The rate constant of hydrogen abstraction increases by 0.5×10^4 sec^{-1} for each extra methylene group included in the ester alkyl group. In the case of a C_{16} chain, there is a great selectivity for hydrogen abstraction at C-14. These first observations and many others[15] indicate for remote H-abstraction, an increasing probability for cyclic conformation as carbon chains get larger (Scheme 41.8).

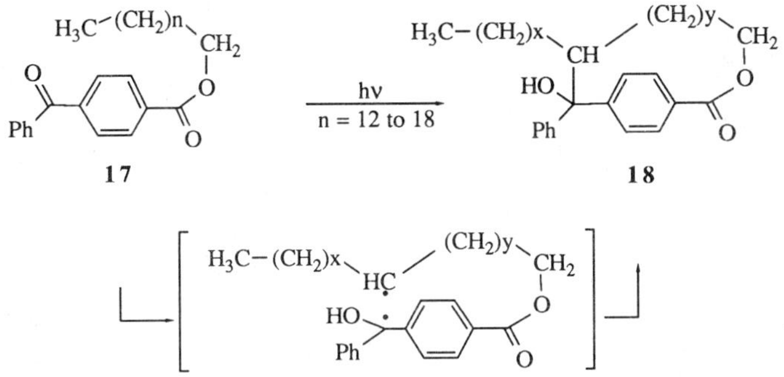

SCHEME 41.8 Remote oxidation of unactivated methylene groups of 17.

In the field of steroids,[15b] long-range oxidation by attack of photoexcited benzophenones on hydrogens at different positions of cholestanol (carbons 7, 12, 14) have been observed. For example, in 19 (n = 1), the oxygen atom of the benzophenone excited triplet state abstracts the hydrogen at C-14 of the steroidal backbone. The radical then undergoes inversion of configuration and the hydrogen at C-15 is transferred to the carbon atom of the ketyl radical affording the olefinic product 20. With a larger chain in 19 (n = 2) attack occurs at carbons 7 and 12 and leads to the coupling products 21 and 22 (Scheme 41.9). These long-distance hydrogen abstractions have low quantum yields and are not attractive for large-scale syntheses.[15d] More recent studies[16] have shown that triplet energy transfer in conformationally flexible bichromophoric molecules (cinnamyl esters of ω-benzoylcarboxylic acid,[16a] and diketones[16b]) can proceed by "through-space" rather than "through bonds" mechanisms.

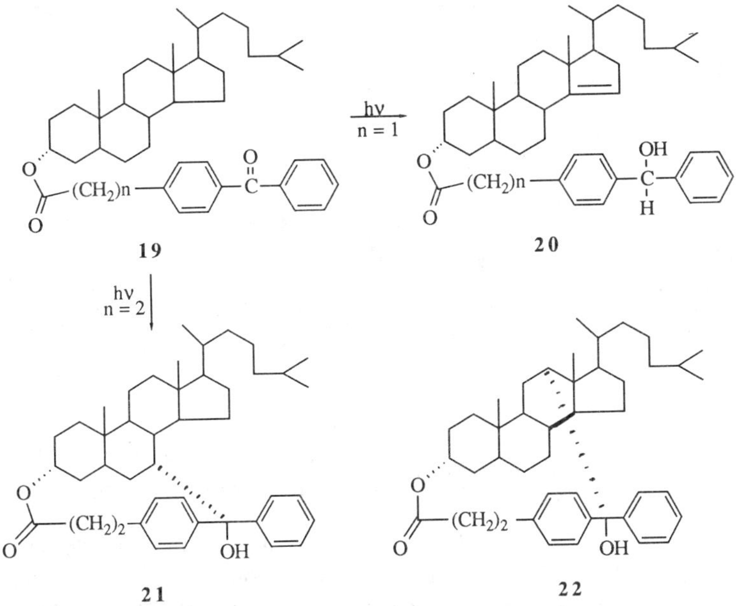

SCHEME 41.9 Remote oxidation and photocyclization of steroids 19.

The most useful remote hydrogen abstractions have been exemplified with imides that undergo cyclization via an electron transfer (SET) photochemical route. C–C bond formation between the imide carbonyl group and the terminal methyl mercapto group of 23 afforded mainly medium to large ring azathiacyclic systems 24.[17] The sulfur methyl group obviously facilitates the macrocyclic transition state since the oxygen methyl analog fails to undergo the reaction. Quantum yields of the photoreactions leading to 24 were comparable or better than those of aromatic imides where a smaller transition states is involved. A charge transfer complex was proposed whereby the biradical is formed prior to phototransfer and bond formation in the biradical. (Scheme 41.10).

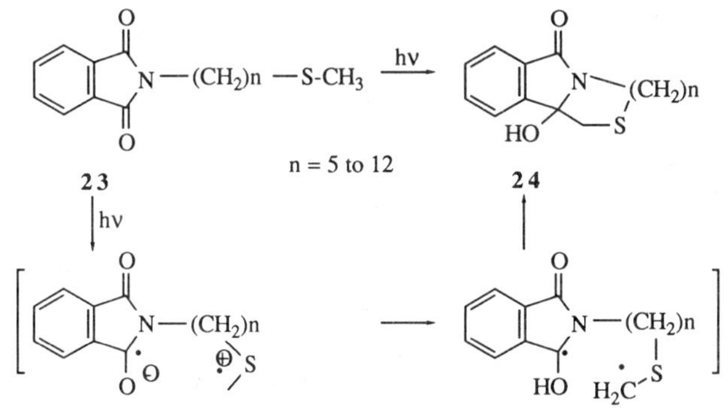

SCHEME 41.10 Remote photocyclization of sulfide-containing phtalimides 23.

A similar mechanism is also operative in the phthalimides 25 containing a 1,3-dithiolanyl group that were easily converted to macrocyclic compounds 26 and 27[18] (Scheme 41.11). The S-methine group of 25 ($x = 1$) was more reactive than the S-methylene group in this regioselective carbocyclization reaction. However, 1,3-dithiolanyl derivatives 25 ($x = 2$) failed to photocyclize.

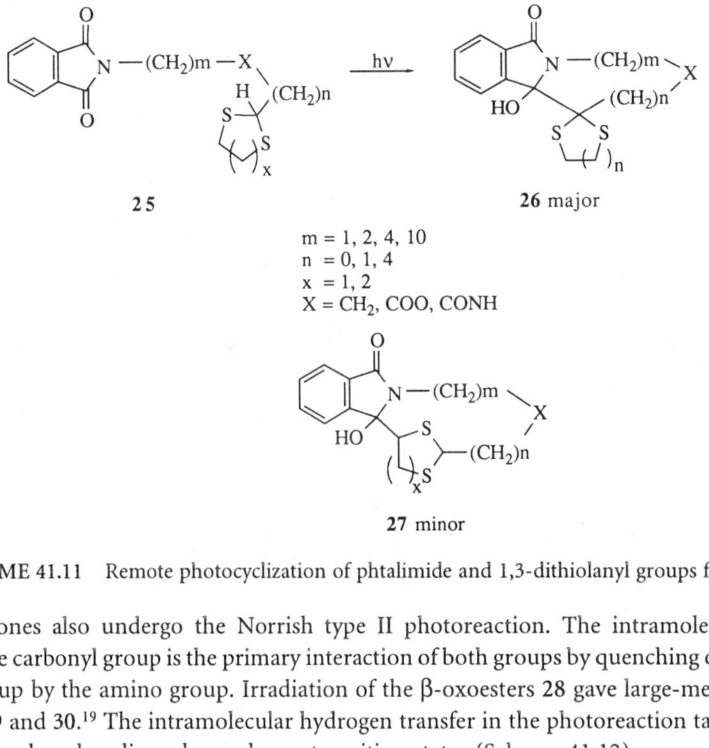

SCHEME 41.11 Remote photocyclization of phtalimide and 1,3-dithiolanyl groups for 25.

Aminoketones also undergo the Norrish type II photoreaction. The intramolecular charge transfer of the carbonyl group is the primary interaction of both groups by quenching of the excited carbonyl group by the amino group. Irradiation of the β-oxoesters 28 gave large-membered ring azlactones 29 and 30.[19] The intramolecular hydrogen transfer in the photoreaction takes place via a 9- to 12-membered cyclic and even larger transition states (Scheme 41.12).

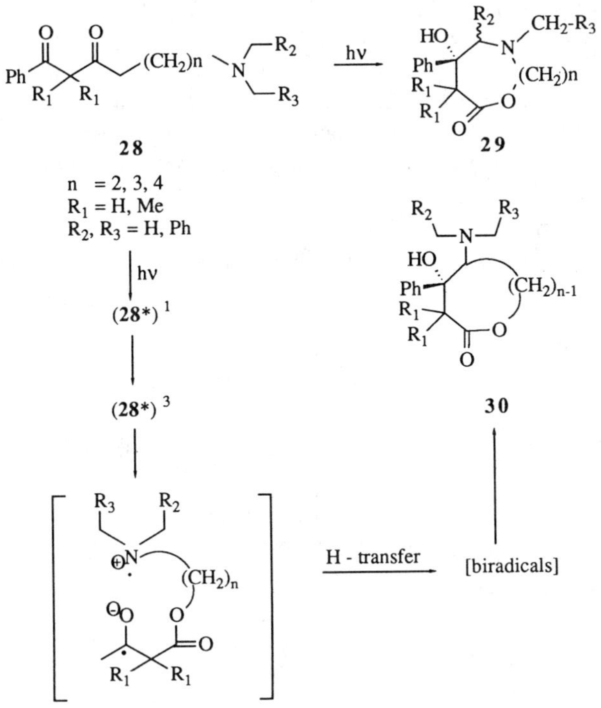

SCHEME 41.12 Remote photocyclization of (ω-dialkylamino) alkyl β-oxoesters 28.

.3 Synthetic Applications

To suppress the normal γ-hydrogen abstraction by the Norrish type II reaction, the introduction of heteroatoms at this position facilitates remote H-abstractions and syntheses of heterocyclic compounds. The main synthetic applications are in the field of oxygen, sulfur and nitrogen heterocycles.

Syntheses of Six- and Seven-Membered Oxygen Heterocycles

A photochemical synthesis of diastereoisomeric tetrahydropyran-3-ols 32 involves UV irradiation of the ketone 31. In the absence of γ- or δ-hydrogens, abstraction takes place at the ε-site[21] (Scheme 41.13).

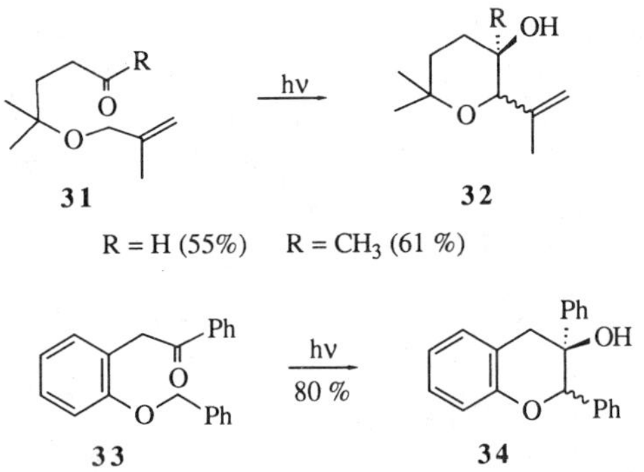

SCHEME 41.13 Photochemical syntheses of tetrahydropyran-3-ols 32 and benzopyranols 34.

The activated methallyloxy hydrogens of 31 can be replaced by benzyloxy hydrogens to extend the photocyclization to the synthesis of diastereoisomeric benzopyranols 34 via triplet ε-hydrogen abstraction in 33.[8]

The ζ-hydrogen benzylic transfer in 10 described previously confirms again (Scheme 41.5) the influence of the lability of benzylic hydrogens and the preference for such H-abstraction rather than acetal hydrogens from the δ-positions in the photocyclization of 10 to yield 12 as major seven-membered ring bicyclic compound.[12]

More functionalized aliphatic, cyclanic and aromatic molecules containing α-ketoesters groups (35, 37, 39) are readily photocyclized into eight-membered ring lactonic compounds (36, 38, 40) (Scheme 41.14).[14]

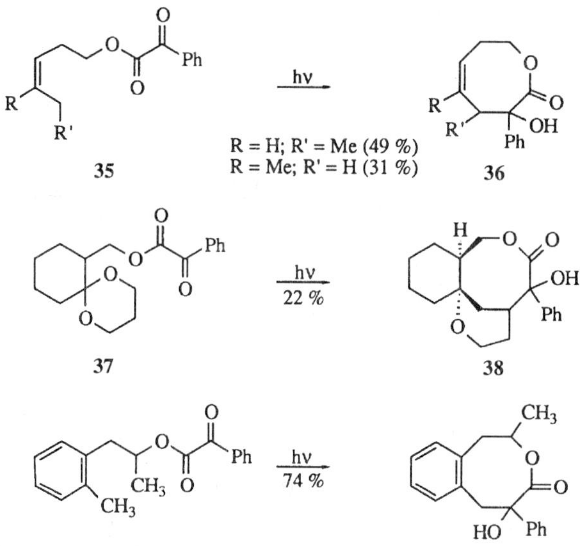

$$R = H; R' = Me\ (49\ \%)$$
$$R = Me; R' = H\ (31\ \%)$$

SCHEME 41.14   Photochemical syntheses of eight-membered ring lactonic derivatives 36, 38, and 40.

Syntheses of Nitrogen Heterocycles

From N-Substituted Alicyclic and Aromatic Imides: Syntheses of Diaza Heterocyclic Compounds

A range of common-, medium-, and large-ring photoproducts may be accessible by photocyclization of succinimide and maleimide derivatives in a reaction path related to that of phthalimides. Ring-expanded products (42, 43, 44) are obtained in low yields by way of a γ-hydrogen abstraction and acetidinol formation from 41. The opening of the intermediate affords the final products (Scheme 41.15.)[21] More reactive maleimides (45) or phthalimide derivatives (47) gave diazaheterocyclic photoproducts 46 and 48.[22] In contrast, photocyclization can take place in 49 by transfer of a benzylic hydrogen that is adjacent to nitrogen and is in the ε- or ζ-position relative to the excited carbonyl group to give 50 (n = 1, 2)[23] (Scheme 41.16).

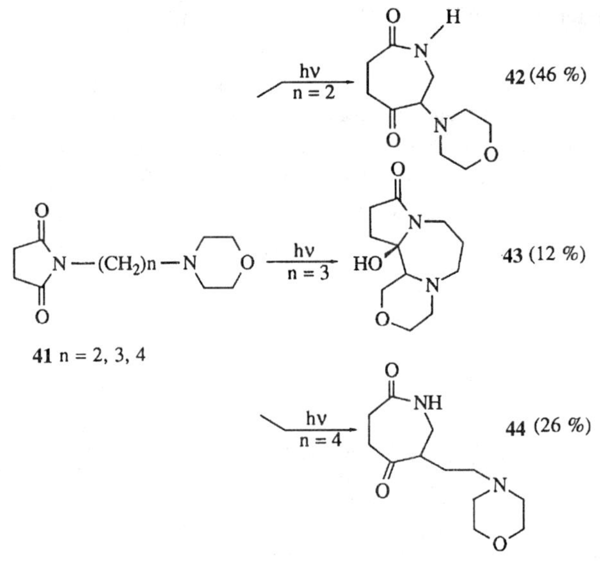

SCHEME 41.15   Photochemical syntheses of diazaheterocycles 42 and 43.

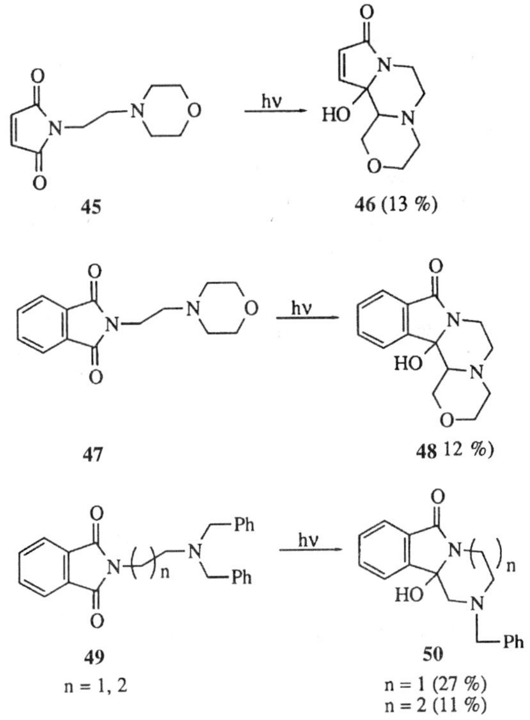

SCHEME 41.16   Photochemical syntheses of fused diazaheterocycles 46, 48, and 50.

From Sulfide-Containing Phthalimides and Thiophthalimides: Applications to Macrocyclic Peptide Syntheses

Phthalimides 51 possessing a terminal sulfide group in their N-alkylside chain, undergo photocyclizations to yield azathiacyclols 52 and 53 by way of Norrish type II reactions.[17a] This reaction has been extended to an easy synthesis of medium- to large-sized ring systems, of up to a 16-membered ring, by a C-C bond forming a path between the phthalimido carbonyl group and the terminal methyl-mercapto group. By the same methodology, macrocyclic peptides analogs 55 have been synthesized starting from the phthalimides 54 containing one or two peptide links.[17b] (Scheme 41.17). By this approach, macrocycles of up to 21-membered rings, equivalent in size to that of cyclic heptapeptides, were obtained in moderate yields. Similarly, the more reactive cyclic thioimides 56 with N-ω-phenyl alkyl substituents undergo the Norrish type II cyclization when benzylic hydrogens are available at the δ- or ε-position to the thiocarbonyl group[20] to afford compounds 57.

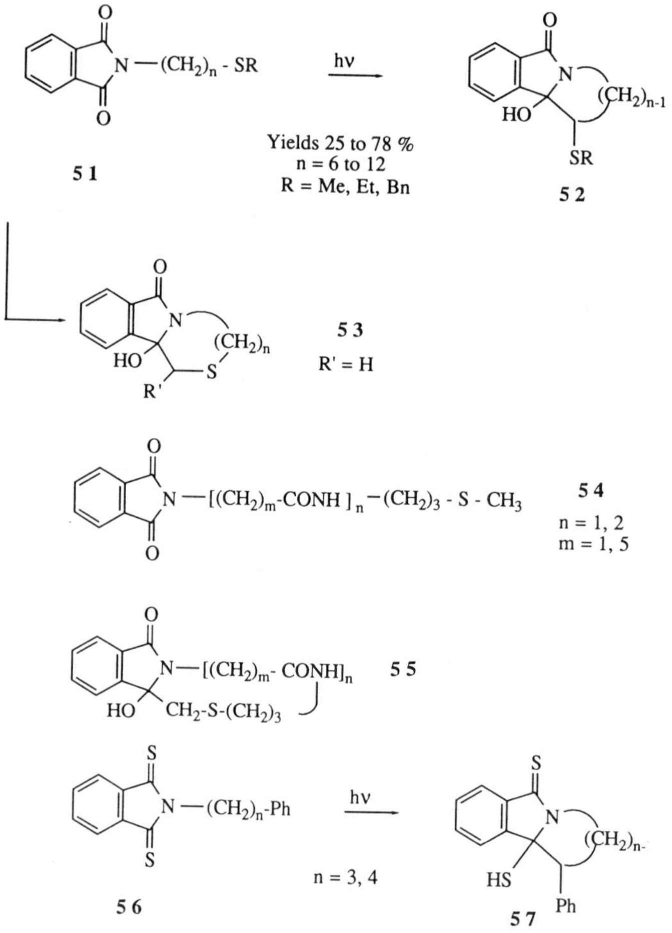

SCHEME 41.17 Photocyclization of sulfide-containing phtalimides 51 and 54 and thiomides 56.

From ω-N-Disubstituted Aminoalkyl β-Oxoesters: Syntheses of Nitrogen and Oxygen Heterocyclic Compounds

The (ω-dialkylamino) alkyl β-oxoesters 28 undergo photocyclization via remote hydrogen transfer to yield medium-sized azalactones 29 and/or aminolactones 30 in moderate yields (20 to 40%)[19] (Scheme 41.12).

Remote photocyclization of similar (dibenzylamine) ethyl benzoylacetate 58 proceeds through a 10-membered cyclic transition state to afford in 52% yield a mixture of *cis* and *trans* eight-membered azalactones 59.[24]

In contrast, the irradiation of unsaturated cyclic amino β-ketoesters such as 60 does not form a C-C bond at the ketonic group but instead undergoes an intramolecular SET followed by cyclization to the double bond to afford the bridged bicyclic amine 61 (Scheme 41.18).[25]

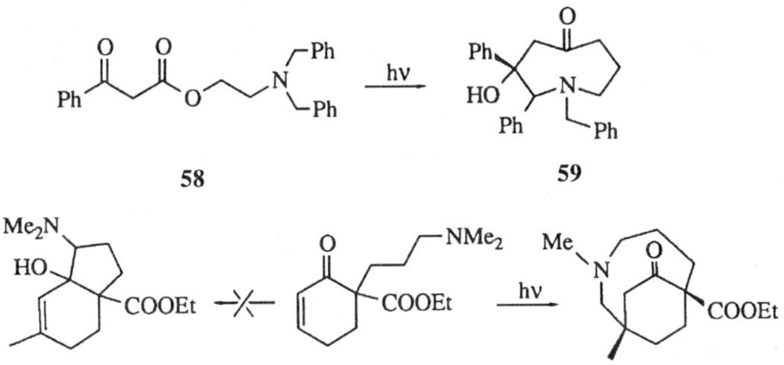

SCHEME 41.18   Photocyclization of N-substituted β-oxoesters 58 and 60.

eferences

1. Wagner, P. J., Type II photoelimination and photocyclization of ketones, *Acc. Chem. Res.*, 4, 168, 1971.
2. Wagner, P. J., Conformational flexibility and photochemistry, *Acc. Chem. Res.*, 16, 461, 1983.
3. Wagner, P. J., 1,5-Biradicals and five-membered rings generated by δ-hydrogen abstraction in photoexcited ketones, *Acc. Chem. Res.*, 22, 83, 1989.
4. Sauers, R. R. and Huang, S.-H., Analysis of Norrish type II reactions by molecular mechanics methodology: Cyclododecanone, *Tetrahedron Lett.*, 31, 5709, 1990.
5. Paquette, L. A., Ternansky, R. J., Balogh, D. W., and Kentgen, G., Total synthesis of dodecahedrane, *J. Am. Chem. Soc.*, 105, 5446, 1983.
6. Descotes, G., Synthetic saccharide photochemistry, *Top. Curr. Chem.*, 154, 39, 1990.
7. Schulte-Elte, K. H., Willhalm, B., Thomas, A. F., Stoll, M., and Ohloff, G., The photolysis and mass spectra of medium and large-ring ketones from C9 to C16, *Helv. Chim. Acta*, 54, 1759, 1971.
8. Meador, M. A. and Wagner, P. J., Photocyclization of α-[o-(benzyloxy) phenyl]acetophenone: Triplet state ε-hydrogen abstraction, *J. Org. Chem.*, 50, 419, 1985.
9. Wagner, P. J., Meador, M. A., and Schiano, J. C., Photocyclizations of o-(benzyloxy)acetophenone and benzophenone: Effects on biradical behavior, *J. Am. Chem. Soc.*, 106, 7988, 1984.
10. Hornback, J. M., Poundstone, M. L., Vadlamani, B., Graham, S. M., Gabay, J., and Patton, S. T., Photochemistry cyclization of o-methylphenyl 1,3-diketones, *J. Org. Chem.*, 53, 5597, 1988.
11. Carless, H. A. J. and Mwesigye-Kibende, S., Intramolecular hydrogen abstraction in ketone photochemistry: The first examples of ζ-hydrogen abstraction, *J. Chem. Soc., Chem. Commun.*, 1673, 1987.
12. Cottet, F., Cottier, L., and Descotes, G., Photocyclisations de cétoacétals aromatiques en vue d'une approche synthétique de la Crombénine, *Can. J. Chem.*, 68, 1251, 1990.
13. Aoyama, H., Arata, Y., and Omote, Y., Photocyclization of γ,δ- and δ,ε-unsaturated amines. Hydrogen transfer reactions via eight-membered cyclic transition states, *J. Chem. Soc., Chem. Commun.*, 1381, 1985.
14. Kraus, G. A. and Wu, Y., 1,5- and 1,9-Hydrogen atom abstractions. Photochemical strategies for radical cyclizations, *J. Am. Chem. Soc.*, 114, 8705, 1992.
15. (a) Breslow, R. and Winnik, M. A., Remote oxidation of unactivated methylenic groups, *J. Am. Chem. Soc.*, 91, 3083, 1969; (b) Breslow, R., Baldwin, S., Flechtner, T., Kalicky, P., Liu, S., and

Washburn, W., Remote oxidation of steroids by photolysis of attached benzophenone groups, *J. Am. Chem. Soc.*, 95, 3251, 1973; (c) Winnik, M. A., Lee, C. K., Basu, S., and Saunders, D. S., Conformational analysis of hydrocarbon chains in solution — Carbon tetrachloride, *J. Am. Chem. Soc.*, 96, 6182, 1974; (d) Breslow, R., Biomimetic control of chemical selectivity, *Acc. Chem. Res.*, 13, 170, 1980.

16. (a) Wagner, P. J. and El-Taliawi, G. M., Through-space intramolecular triplet energy transfer: The cinnamyl esters of ω-benzoyl carboxylic acid, *J. Am. Chem. Soc.*, 114, 8325, 1992; (b) Wagner, P. J. Giri, B. P., Frerking, H. W., Jr., and De Francesco, J., Spacer independent intramolecular triplet energy transfer in diketones, *J. Am. Chem. Soc.*, 114, 8326, 1992.

17. (a) Sato, Y., Nakai, H., Mizoguchi, T., Hatanaka, Y., and Kanaoka, Y., Regioselective remote photocyclization. Examples of a photochemical macrocyclic synthesis with sulfide-containing phtalimides, *J. Am. Chem. Soc.*, 98, 2349, 1976; (b) Sato, Y., Nakai, H., and Mizoguchi, T., A synthetic approach to cyclic peptide models by regioselective remote photocyclization of sulfide-containing phthalimide, *Tetrahedron Lett.*, 1889, 1976.

18. Wada, A., Nakai, H., Sato, Y., Hatanaka, Y., and Kanaoka, Y., A removable functional group in a photochemical macrocyclic synthesis, *Tetrahedron*, 39, 2691, 1983.

19. Hasegawa, T., Ogawa, T., Miyata, K., Karakizawa, A., Komiyama, M., Nishizawa, K., and Yoshioka, M., Photocyclization of (ω-dialkylamino)alkyl β-oxoesters via remote hydrogen transfer, *J. Chem. Soc., Perkin Trans. 1*, 901, 1990.

20. Machida, M., Oda, K., and Kanaoka, Y., Photochemistry of the nitrogen-thiocarbonyl systems. The Norrish type II reaction of the cyclic thioimides with a benzylic hydrogen, *Tetrahedron Lett.*, 26, 5173, 1985.

21. Carless, H. A. J. and Fekarurhobo, G. K., Photochemical ε-hydrogen abstraction as a route to tetrahydropyran-3-ols, *Tetrahedron Lett.*, 25, 5943, 1984.

22. Coyle, J. D. and Bryant, L. R. B., Synthesis of diazaheterocycles with a bridgehead nitrogen by photocyclization of N-substituted alicyclic imides, *J. Chem. Soc., Perkin Trans. 1*, 2857, 1983.

23. Coyle, J. D. and Newport, G. L., Fused imidazolidines, hexahydropyrazines and hexahydro-1,4-diazepines, *Synthesis*, 381, 1979.

24. Hasegawa, T., Miyata, K., Ogawa, T., Yoshihara, N., and Yoshioka, M., Remote photocyclization of (dibenzylamino)ethyl benzoylacetate. Intramolecular hydrogen abstraction through a ten-membered cyclic transition state, *J. Chem. Soc., Chem. Commun.*, 363, 1985.

25. Kraus, G. A. and Chen, L., Intramolecular photocyclization of aminoketones to form fused and bridged bicyclic amines, *Tetrahedron Lett.*, 32, 7151, 1991.

42

Electron Transfer Processes in Phthalimide Systems

Harald Mauder
*The Weizmann Institute of
Science*

Axel G. Griesbeck
*Institut für Organische Chemie,
Am Hubland*

42.1 Introduction and General Features

During the last 2 decades, the photochemistry of phthalimide systems has become a subject of widespread interest. The phthalimide chromophore shows a broad spectrum of reactivity leading mainly to cycloaddition (benzazepinedione) and photoreduction (isoindole formation) products by inter- as well as intramolecular processes. In contrast to succinimides, there are only a few examples of Paternò-Büchi reactions ([2+2]-cycloadditions) to the carbonyl group of the phthalimide system.[1] In the presence of electron donors (either intra- or intermolecular), the electronically excited phthalimide could also undergo electron transfer (SET) processes and act as an electron acceptor. This chapter will describe the transformations in the phthalimide system. Spectroscopic investigations and triplet quenching as well as sensitizing experiments demonstrated clearly that the SET process occurs via the S_1 (π,π^*) excited state[2,3] which has an excitation energy ΔE_{00} of ca. 80 kcal mol^{-1}.[4] Electrochemical measurements of N-substituted phthalimides in DMF gave reversible reduction potentials of -1.37 V (to its radical anion) and of -2.16 V (to its dianion).[5]

42.2 Single Electron Transfer in Phthalimide Systems

Alkenes

The application of the classical form of the Rehm-Weller equation[6] and available values of oxidation potentials of alkenes leads to three major reaction pathways that depend on the value of ΔG_{ET}. In these cases where ΔG_{ET} is positive (>5 kcal mol^{-1}) a formal $[\pi^2+\sigma^2]$-addition leads to benzazepinediones (1) as the sole products. With large negative values of ΔG_{ET}, on the other hand, exclusively single electron transfer (SET) processes are observed. In the absence of nucleophiles or other trapping reagents, a reversible electron transfer quenching process can take place.[7] The radical ion pair formed after electron transfer could be trapped by nucleophiles, e.g., alcohols or deactivated to reduction products. In the former cases, the alcohol adds to the alkene radical cation with *anti*-Markownikov regioselectivity leading to the more stabilized biradical intermediates.[8] Recombination of the radical sites gives the isoindolinone anion, which is subsequently protonated. If the electron transfer is only slightly endothermic, the [$\pi^2+\sigma^2$]-addition process can compete effectively with the SET process. In these cases, complex product mixtures are obtained.

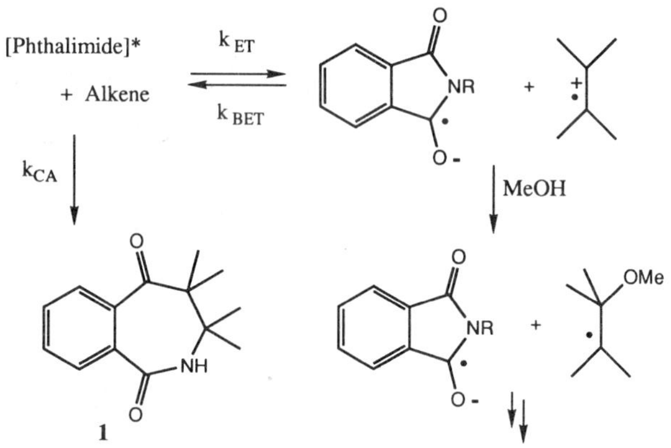

Intermolecular Examples

According to the Rehm-Weller equation, only olefins with oxidation potentials below 2.1 V (IP ≤ 9 eV) are able to operate as electron donors toward the phthalimide system. Thus, in nonpolar solvents like acetonitrile (neglectible Coulomb term), cyclic as well as acyclic tri- and tetrasubstituted olefins do not react in [$\pi^2+\sigma^2$]-additions.[7] These olefins operate effectively as electron transfer quenchers of the exited *N*-methylphthalimide singlet state. Dienes, however, with ionization potentials as low as about 8.6 eV, are still reactive in the cycloaddition reaction, demonstrating clearly how the two reaction pathways are influenced by the ratio of the reaction constants k_{CA}:k_{ET}.[9,10] Photoreduction products were first described by Kanaoka and Hatanaka[11] for *N*-methylphthalimide (NMP) and cyclohexene. At least in the case of 2,3-dimethyl-2-butene (2) as electron donor, the photoreduction is a SET-mediated process involving the singlet exited state of NMP.[12] The photoreduction products 3 and 4 probably arise from different radical ion pairs (contact and solvent separated).[13] Nucleophilic solvents are able to trap the intermediate radical ion pair, leading to products of solvent incorporation. With styrene (5), NMP gives in alcoholic solutions radical ion trapped products in high yield.[14] Depending on the structure of the solvent (= nucleophile), there are drastic differences in the product ratio between SET-products (A and B) and the formal [$\pi^2+\sigma^2$]-addition product C. Whereas methanol traps the styrene radical cation completely, the addition of *t*-butanol is noticeably sluggish. The increase in addition product (C) with increasing steric demand and decreasing polarity of the solvent indicates again the competition between these two pathways. The influence of aryl substituents at the phthaloyl chromophore on the regiochemistry of the photoreduction and the trapping reaction was investigated by Mazzocchi.[15] The results of these experiments prove unambiguously the radical anion character of the phthalimidoyl group.

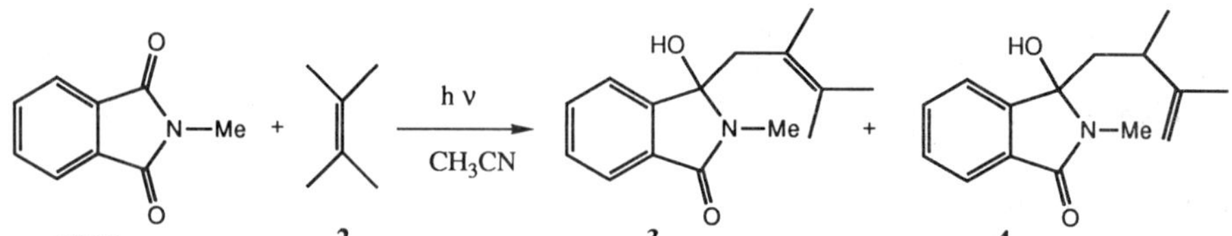

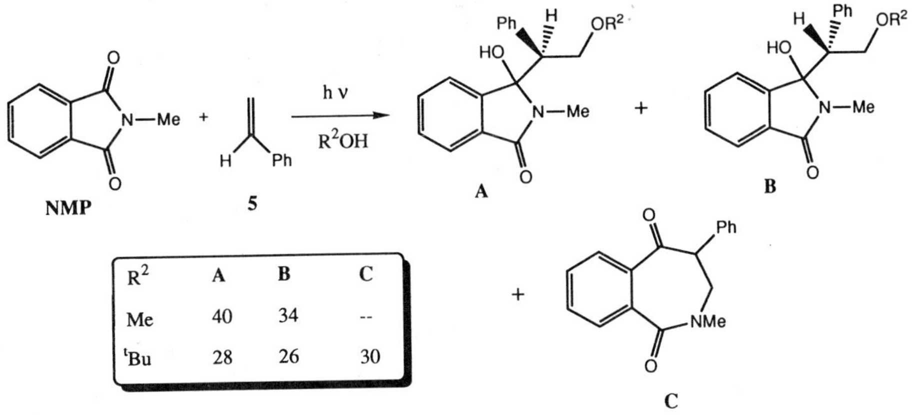

R^2	A	B	C
Me	40	34	--
^tBu	28	26	30

Intramolecular Examples

N-2-Propenyl phthalimide (6) does not react during irradiation in methanol.[16] Alkyl substitution of the terminal position of the C-C double bond activates the substrate and solvent-incorporated products are formed. After intramolecular SET, a nucleophilic solvent molecule adds with *anti*-Markownikov regioselectivity leading to the more stabilized biradical intermediate which, after recombination and protonation, gives the tricyclic products 9 and 10. The phenyl-substituted substrate 8, with a low oxidation potential, gives solvent-incorporated products even in less polar solvents as *t*-butanol in contrast to substrates with C-C double bonds of higher oxidation potential. This demonstrates again that less polar solvents decrease the reactivity in SET reactions.[3]

The 2-methyl substituted *N*-2-propenyl phthalimide (11) undergoes radical/radical recombination after the initial SET, which is at least one order of magnitude faster than the nucleophilic attack. Two different products 12 and 13 were observed after trapping by methanol or ring-enlargement reactions.[17]

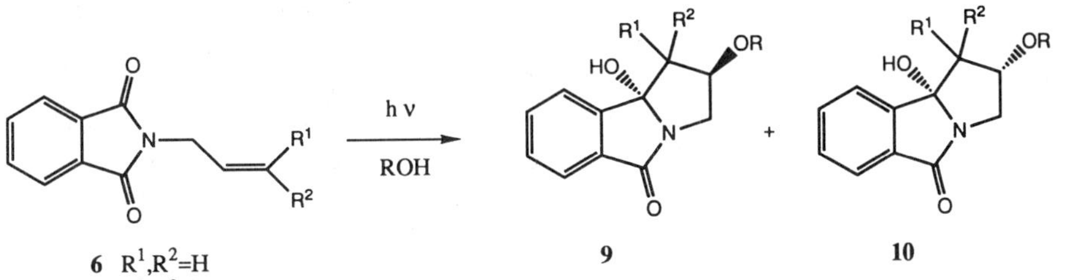

6 $R^1,R^2=H$
7 $R^1,R^2=Me$
8 $R^1=H, R^2=Ph$

N-3-alkenylphthalimides 14 cyclize either to five-membered (15) or six-membered (16) ring products, depending on the stability of the intermediate biradicals.[18] With remote styryl moieties, cyclization products with macrocyclic rings (up to 15 ring atoms) are easily attainable in reasonable yields.[19] The cyclization also operates efficiently with remote indoyl- and cyclopentenyl-substituents exclusively in an intramolecular fashion, which can be rationalized by the assumption of close contact between the donor and acceptor functionalities present in the ground state of the starting materials.[20]

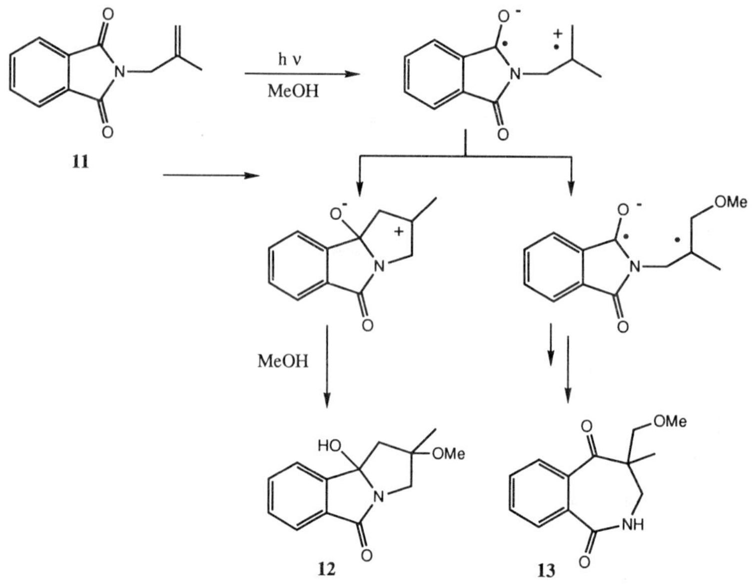

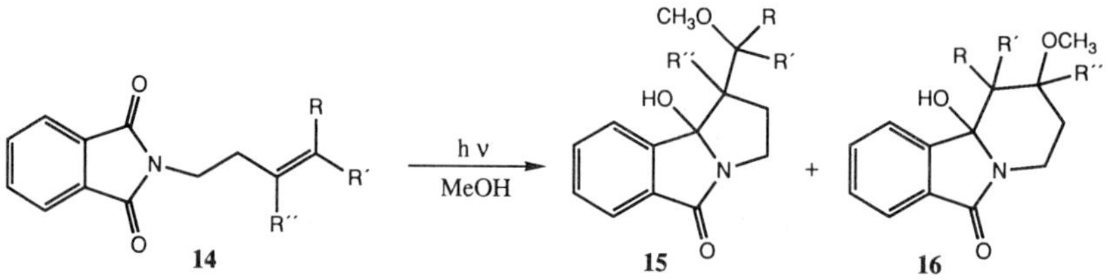

Ethers

Depending on the position of the side chain oxygen atom relative to the carbonyl group alkoxy-substituted phthalimides give rise to different heterocyclic products after photochemical excitation. In cases where oxygen is in the γ-position (as in **17**, the N atom is regarded as α with respect to the imide carbonyl carbon), usually the reactive position in N-alkyl-substituted phthalimides, H-abstraction takes place from the neighbouring δ-position leading to tricyclic hydroxy isoindolinones **18** in moderate yields. Even in substrates where the ether oxygen is located at the ε-position exclusively, δ-H-atoms are reactive. These products, which also are tricyclic hydroxy isoindolinones, now have an alkoxy substituent instead of a ring oxygen.[21] The exclusive H-abstraction from the position α to the oxygen is also observed in intermolecular reactions with ethers leading to reductive coupling products.[22] It was shown by Wagner[23] that an oxygen atom activates an adjacent C-H bond.

Amines

In an early report concerning the photochemistry of phthalimide systems containing a nitrogen atom in the side chain, Roth and Schwarz[24] examined several *bis*-phthalimido methylamines **19**. All these compounds cyclize via radical/radical recombination, with the methyl or methylene group adjacent to nitrogen, to give the corresponding phthalimido imidazolidines **20**. These products could undergo subsequent secondary photocyclization. When the same compounds (19) were irradiated in methanol, photoreduction and photorecombination products formed preferentially.[25]

The spectroscopy and photochemistry of *N*-(dibenzylaminomethyl)phthalimide (21) has been investigated intensively by Coyle and Newport.[26] They concluded that electron transfer takes place as the first reaction step in these systems and is followed by proton transfer and radical recombination (22). With alkyl-substituted amino side-chains, the reactive positions for secondary transformations are the corresponding methyl or methylene groups neighboring the heteroatom.[27,28] Methyl groups attached to nitrogen are about three times more reactive than methylene groups in these reactions. Despite the low yield (≤18%), the possibility for synthesis of macrocyclic ring systems (up to 16 ring atoms) makes this starting material interesting for synthetic purposes.

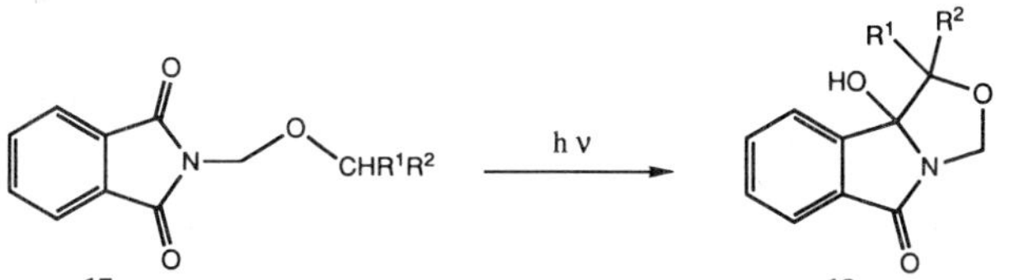

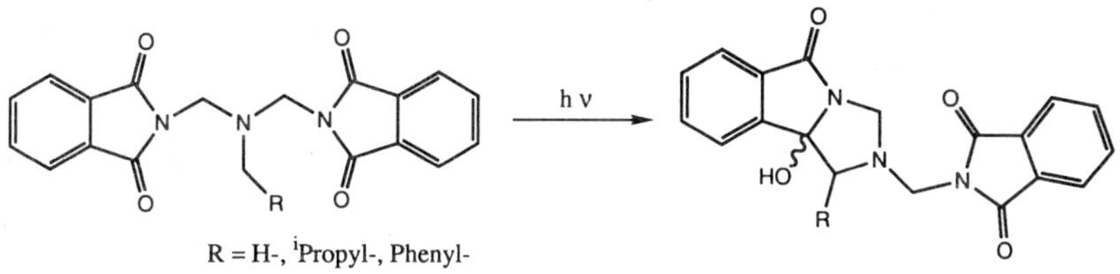

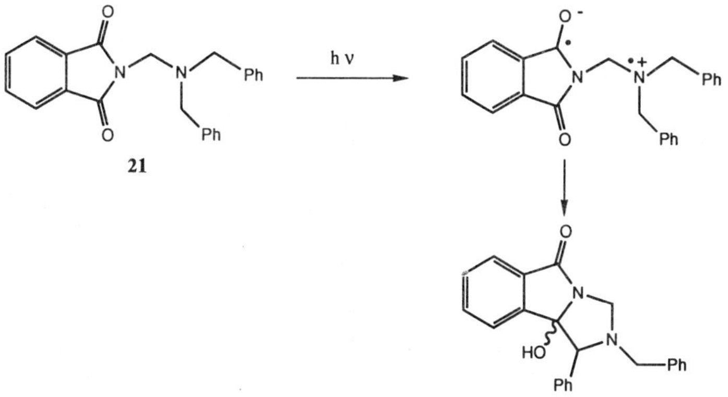

Thioethers

Thioethers show a pronounced reactivity in SET processes with phthalimide systems. Depending on the chain length, *N*-phthaloyl methylthioethers (23) are transformed into six- and seven-membered heterocycles 24 with high regioselectivity after proton transfer from the *S*-methyl group (ε- or ζ-position).[29] After electronic excitation of the phthalimide chromophore, electron transfer from the sulfur atom takes place to give a radical ion pair with close contact between oxygen and sulfur. This close geometry and the highly increased acidity of the α-CH groups explains the exclusive removal of hydrogen from the terminal *S*-methyl substituent. Even remote methylthio-ethers with amide or ester functions as components of the side chain can be transformed into macrocyclic ring systems with ring sizes up to 38 carbons.[30,31]

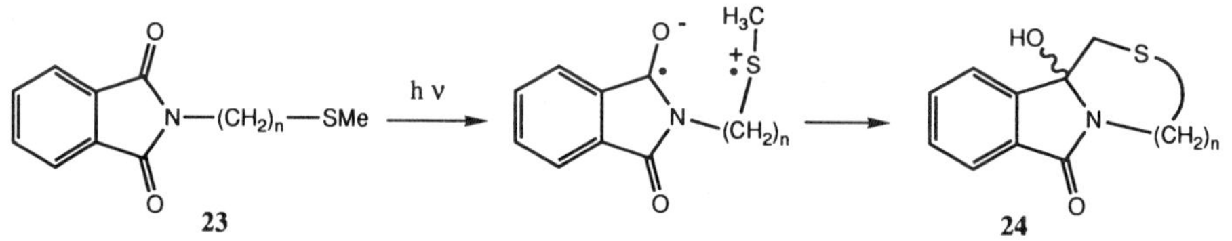

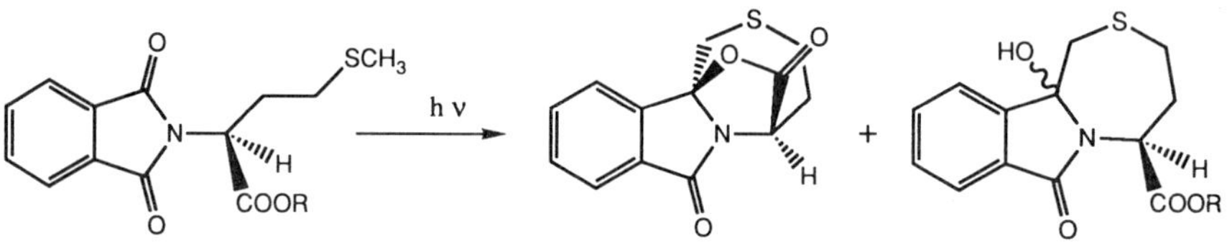

Analogously, irradiation of *N*-phthaloyl methionine methyl ester **25a** leads to a 45:55 mixture of two diastereomeric cyclization products **26a** in 84% yield.[32] *N*-Phthaloyl methionine **25b** itself gives, in acetone solution or in acetonitrile via triplet sensitization, the tetracyclic compound **27** in 75 to 85% yield besides minor amounts of the decarboxylation/cyclization product **26b**.[33] The formation of the novel product **27** is rationalized by the following reaction sequence. After SET from the sulfur atom to the phthalimide chromophore, the imidyl radical anion forms the lactone ring via nucleophilic attack at the carboxy group. Subsequent loss of a proton from the thioether radical cation affords product **27**.

When the corresponding sulfone of *N*-phthaloyl methionine methyl ester is irradiated, even after prolonged irradiation, no reaction could be observed. This is to be expected since there is no electron-donor function to participate in intramolecular SET.

42.3 Miscellaneous Reactions

Intermolecular reactions between phthalimides and α-silylsubstituted heteroatom compounds have been described by Yoon and Mariano.[34] α-Trimethylsilyl (TMS) substituents in electron donor systems are used widely to generate free radical intermediates following a reaction protocol that operates via SET processes.[35] α-Silyl substituted ethers, thioethers, or amines operate as electron donors in phthalimide photoreactions that proceed with high regioselectivity at the α-silyl-substituted site, first leading to a radical ion pair which after TMS-shift and radical coupling recombines to give an OTMS-substituted hydroxyisoindolinone **28**. Hydrolysis leads finally to **29** in reasonable to high yields.

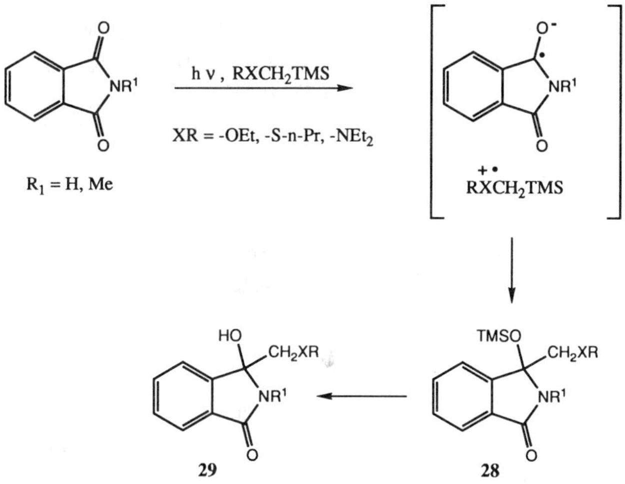

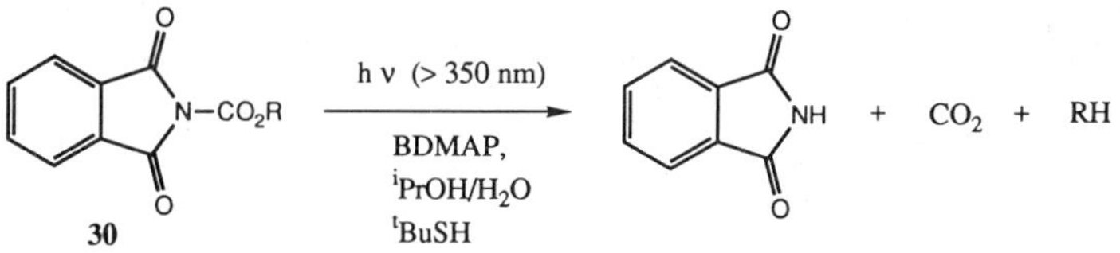

A methodology for decarboxylation of unactivated carboxylic acids via *N*-acyloxyphthalimides with use of a sensitized SET reaction was described by Okada and Oda.[36] Photolysis of *N*-acyloxyphthalimides 30 in solution with 1,6-*bis*(dimethylamino)pyrene (BDMAP) as sensitizer and *t*-BuSH as a hydrogen source gives the corresponding alkanes and phthalimide in excellent yields.

eferences

1. Takechi, H., Machida, M., and Kanaoka, Y., Intramolecular photoreactions of phthalimide-alkene systems. Oxetane formation of *N*-(ω-indol-3-ylalkyl)phthalimides, *Chem. Pharm. Bull.*, 36, 2853, 1988.
2. Hayashi, H., Nagakura, S., Kubo, Y., and Maruyama, K., Mechanistic studies of photochemical reactions of *N*-ethylphthalimides with olefins, *Chem. Phys. Lett.*, 72, 291, 1980.
3. Maruyama, K. and Kubo, Y., Photochemistry of *N*-(2-alkenyl)phthalimides. Photoinduced cyclization and elimination reactions, *J. Org. Chem.*, 46, 3612, 1981.
4. Coyle, J. D., Newport, G. L., and Harriman, A., Nitrogen-substituted phthalimides: Fluorescence, phosphorescence, and the mechanism of photocyclization, *J. Chem. Soc. Perkin Trans. 2*, 133, 1978.
5. Leedy, D. W. and Muck, D. L., Cathodic reduction of phthalimide systems in nonaqueous solutions, *J. Am. Chem. Soc.*, 93, 4264, 1971.
6. Rehm, D. and Weller, A., Kinetics of fluorescence quenching by electron and H-atom transfer, *Isr. J. Chem.*, 8, 259, 1970.
7. Mazzocchi, P. H., Minamikawa, S., and Wilson, P., The photochemical addition of alkenes to *N*-methylphthalimide. Evidence for electron transfer quenching, *Tetrahedron Lett.*, 45, 4361, 1978.
8. Urry, W. H., Stacey, F. W., Huyser, E. S., and Juveland, O. O., The peroxide- and light-induced additions of alcohols to olefins, *J. Am. Chem. Soc.*, 76, 450, 1954.

9. Mazzocchi, P. H., Minamikawa, S., Wilson, P., Bowen, M., and Narian, N., Photochemical additions of alkenes to phthalimides to form benzazepinediones. Addition of dienes, alkenes, vinyl ethers, vinyl esters, and an allene, *J. Org. Chem.*, 46, 4846, 1981.

10. Mazzocchi, P. H., Bowen, M. J., and Narian, N. K., Photochemical addition of dienes to *N*-alkylphthalimides, *J. Am. Chem. Soc.*, 99, 7063, 1977.

11. Kanaoka, Y. and Hatanaka, Y., Photoaddition of ethers and olefins to *N*-methylphthalimide, *Chem. Pharm. Bull.*, 22, 2205, 1974.

12. Mazzocchi, P. H. and Klinger, L., Photoreduction of *N*-methylphthalimide with 2,3-dimethyl-2-butene. Evidence for reaction through an electron transfer generated ion pair, *J. Am. Chem. Soc.*, 106, 7567, 1984.

13. Mazzocchi, P. H., Minamikawa, S., and Wilson, P., Competitive photochemical $\sigma^2+\pi^2$ addition and electron transfer in the *N*-methylphthalimide-alkene system, *J. Org. Chem.*, 50, 2681, 1985.

14. Maruyama, K. and Kubo, Y., Photo-induced solvent-incorporated addition of *N*-methylphthalimide to olefins. Reaction promoted by initial one electron transfer, *Chem. Lett.*, 851, 1978.

15. Mazzocchi, P. H., Wilson, P., Khachik, F., Klingler, L., and Minamikawa, S., Photochemical additions of alkenes to phthalimides. Mechanistic investigations on the stereochemistry of alkene additions and the effect of aryl substituents on the regiochemistry of alkene additions, *J. Org. Chem.*, 48, 2981, 1983.

16. Maruyama, K., Kubo, Y., Machida, M., Oda, K., Kanaoka, Y., and Fukuyama, K., Photochemical cyclization of *N*-2-alkenyl- and *N*-3-alkenylphthalimides, *J. Org. Chem.*, 43, 2303, 1978.

17. Mazzocchi, P. H. and Fritz, G., Photolysis of *N*-(2-methyl-2-propenyl)phthalimide in methanol. Evidence supporting radical-radical coupling of a photochemically generated radical ion pair, *J. Am. Chem. Soc.*, 108, 5362, 1986.

18. Machida, M., Oda, K., Maruyama, K., and Kubo, Y., Photocyclization of *N*-3-alkenylphthalimides. Effect of alkyl substitution on the formation of pyrroloisoindoles and pyridoisoindoles, *Heterocycles*, 14, 779, 1980.

19. Maruyama, K. and Kubo, Y., Solvent incorporated medium to macrocyclic compounds by the photochemical cyclization of *N*-alkenylphthalimides, *J. Am. Chem. Soc.*, 100, 7772, 1978.

20. Machida, M., Oda, K., and Kanaoka, Y., Photocyclization of N-[ω-(cycloalken-1-yl)alkyl]- and N-[ω-(inden-3-yl)alkyl]phthalimides. Synthesis of spiro-nitrogen multicyclic systems, *Tetrahedron*, 41, 4995, 1985.

21. Sato, Y., Nakai, H., Wada, M., Ogiwara, H., Mizoguchi, T., Migata, Y., Hatanaka, Y., and Kanaoka, Y., Photocyclization of *N*-alkoxylphthalimides with favored δ-hydrogen abstraction: Syntheses of oxazolol[4,3a]isoindoles and oxazolo[4,3a]-isoindole-1-spiro-1'-cycloalkane ring systems, *Chem. Pharm. Bull.*, 30, 1639, 1982.

22. Kanaoka, Y. and Hatanaka, Y., Photoaddition of ethers and olefins to *N*-methylphthalimide, *Chem. Pharm. Bull.*, 22, 2205, 1974.

23. (a) Wagner, P. J., Type II photoelimination and photocyclization of ketones, *Acc. Chem. Res.*, 4, 168, 1971; (b) Wagner, P. J., Kelso, P. A., Kemppainen, A. E., and Zepp, R. G., Type II photoprocesses of phenyl ketones. Competitive δ-hydrogen abstraction and the geometry of intramolecular hydrogen atom transfer, *J. Am. Chem. Soc.*, 94, 7500, 1972; (c) Wagner, P. J., Kelso, P. A., Kemppainen, A. E., McGrath, J. M., Schott, H. N., and Zepp, R. G., Type II photoprocesses of phenyl ketones. A glimpse at the behavior of 1,4-biradicals, *J. Am. Chem. Soc.*, 94, 7506, 1972.

24. Roth, H. J. and Schwarz, D., Photoreaktionen von *bis*-phthalimidomethyl-alkylaminen, *Arch. Pharm.*, 308, 218, 1975.

25. Roth, H. J., Schwarz, D., and Hundeshagen, G., Photoreaktionen von Phthalimid-Mannichbasen in Methanol, *Arch. Pharm.*, 309, 48, 1976.

26. Coyle, J. D. and Newport, G. L., Photochemical cyclization of *N*-(dibenzylaminomethyl)phthalimide, *Tetrahedron Lett.*, 899, 1977.

27. Machida, M., Takechi, H., and Kanaoka, Y., Remote photocyclization. Photochemical macrocyclic synthesis with *N*-(ω-methylanilino)alkyl-phthalimides, *Heterocycles*, 7, 273, 1977.

28. Machida, M., Takechi, H., and Kanaoka, Y., Photocyclization of *N*-aminoalkylphthalimides. Synthesis of multicyclic fused hexahydropyrazines and hexahydro-1,4-diazepines, *Heterocycles*, 14, 1255, 1980.

29. Sato, Y., Nakai, H., Ogiwara, O., Mizoguchi, T., Migita, Y., and Kanaoka, Y., Photochemistry of the phthalimide system. Photocyclization of the phthalimides with a sulfide chain: Synthesis of azacyclols by δ, ε, and ζ-hydrogen abstraction, *Tetrahedron Lett.*, 46, 4565, 1973.

30. (a) Sato, Y., Nakai, H., Mizoguchi, T., Hatanaka, Y., and Kanaoka, Y., Regioselective remote photocyclization. Examples of a photochemical macrocyclic synthesis with sulfide-containing phthalimides, *J. Am. Chem. Soc.*, 98, 2349, 1976; (b) Sato, Y., Nakai, H., Mizoguchi, T., and Kanaoka, Y., A synthetic approach to cyclic peptide models by regioselective remote photocyclization of sulfide-containing phthalimides, *Tetrahedron Lett.*, 22, 1889, 1976.

31. (a) Wada, M., Nakai, H., Aoe, K., Kotera, K., Sato, Y., Hatanaka, Y., and Kanaoka, Y., Application of the remote photocyclization with a pair system of phthalimide and methylthio groups, *Tetrahedron*, 39, 1273, 1983; (b) Wada, M., Nakai, H., Sato, Y., Hatanaka, Y., and Kanaoka, Y., A removable functional group in a photochemical synthesis. Remote photocyclization with a pair system of phthalimide and 1,3-dithiolanyl groups, *Tetrahedron*, 39, 2691, 1983.

32. Sato, Y., Nakai, H., Mizoguchi, T., Kawanishi, M., Hatanaka, Y., and Kanaoka, Y., Photodecarboxylation of *N*-phthaloyl-α-amino acids, *Chem. Pharm. Bull.*, 30, 1262, 1982.

33. Griesbeck, A. G., Mauder, H., Müller, I., Peters, E.-M., Peters, K., and von Schnering, H. G., Photochemistry of *N*-phthaloyl derivatives of methionine, *Tetrahedron Lett.*, 34, 453, 1993.

34. (a) Yoon, U.-C. and Mariano, P. S., Mechanistic and synthetic aspects of amine-enone single electron transfer photochemistry, *Acc. Chem. Res.*, 25, 233, 1992; (b) Yoon, U.-C., Kim, H.-J., and Mariano, P. S., Electron transfer-induced photochemical reactions in imide-RXCH$_2$TMS systems. Photoaddition of α-trimethylsilyl substituted heteroatom containing compounds to phthalimides, *Heterocycles*, 29, 1041, 1989.

35. Ohga, K. and Mariano, P. S., Electron transfer initiated photoaddition of allylsilanes to 1-methyl-2-phenylpyrrolinium perchlorate. A novel allylation methodology, *J. Am. Chem. Soc.*, 104, 617, 1982.

36. Okada, K., Ogamoto, K., and Oda, M., A new and practical method of decarboxylation: Photosensitized decarboxylation of *N*-acyloxyphthalimides via electron transfer mechanism, *J. Am. Chem. Soc.*, 110, 8736, 1988.

43

Oxetane Formation: Intermolecular Additions

Axel G. Griesbeck
Institut für Organische Chemie,
University of Cologne

43.1 Introduction and Definition of the Reaction

The first intermolecular photocycloaddition of an aromatic carbonyl compound (benzaldehyde) to an alkene (2-methyl-2-butene) was reported by Paternò and Chieffi: "In una prima esperienza abbiamo esposto un miscuglio equimolecolare di amilene (gr. 43) e di aldeide benzoica (gr. 67) in un tubo chiuso, dal 5 dicembre 1907 al 20 marzo 1908, cioè per circa tre mesi e mezzo della stagione invernale".[1] This long-term experiment (104 days!) was repeated by Büchi and co-workers in the mid-1950s and confirmed the oxetane *structure* of the main product.[2] The *regioselectivity* of this specific transformation, however, was not correctly established until a publication by Yang et al.[3] But as yet, no exact analysis of the *stereoselectivity* has been reported. There are several special features about this reaction that are worthy of discussion: the influence of the properties of the state of the electronically excited species (which normally is the carbonyl compound) and of the properties of the alkene on rate, efficiency, and selectivity of the oxetane formation. Furthermore, a detailed discussion of *diastereo- and enantioselective modifications* and of *intramolecular* variants will be given elsewhere. This chapter describes applications of intermolecular Paternò-Büchi reactions. After discussion of the features of the carbonyl addends, the olefinic counterparts are discussed. A number of extensive reviews about special features of this reaction type have appeared in the last 15 years and are recommended for further information.[4]

43.2 Mechanism of the Paternò-Büchi Reaction

Theoretical Models

Turro has classified two possible primary trajectories in considering the orbital interactions between alkenes and n,π^* excited carbonyls: (1) the nucleophilic attack of the alkene toward the carbonyl half-filled n-orbital, characterized as the "perpendicular approach", and (2) the nucleophilic attack of the carbonyl by its half-filled π^*-orbital toward the alkenes empty π^*-orbital, characterized as the "parallel approach".[5] First-order orbital correlation diagrams are in line with

0-8493-8634-9/95/$0.00+$.50
© 1995 by CRC Press, Inc.

this model and predict the formation of a carbon-carbon bonded 1,4-biradical for the parallel and a carbon-oxygen bonded biradical for the perpendicular approach.[6] Which approach dominates is controlled by the relative positions of the alkene HOMO and LUMO. For interactions between electron-rich alkenes and carbonyl compounds, the perpendicular approach is favored while, for electron-deficient alkenes, the parallel approach is involved. Several *ab initio* calculations dealing with these have been published.[7]

Experimental Results

The biradical model has been in vogue for at least 2 decades in describing chemo- and regioselectivity phenomena. At least for triplet 1,4-biradicals (2-oxatetramethylenes or preoxetanes), this assumption has been confirmed by experimental fact. Trapping experiments (triplet oxygen and sulfur dioxide)[8] as well as the application of radical clocks[9] have demonstrated that short-lived (1 to 10 ns) triplet 1,4-biradicals are formed when triplet excited carbonyl compounds interact with alkenes. Spectroscopic evidence for these species came from laser flash photolysis of electron-rich olefins with benzophenone.[10] Also, in this case, the radical anion of the ketone was detected, demonstrating that electron transfer processes can interfere with the formation of oxetanes. The existence of an exciplex as a precursor of the 1,4-biradicals as well as of the radical ion pair was deduced from a comparison of oxetane formation with the fluorescence quenching of singlet excited ketones by electron-rich and electron-deficient alkenes.[11] A series of experiments on substrate diastereoselectivity was undertaken to differentiate between reactive states (n,π^* vs. π,π^*) and multiplicities (singlet vs. triplet). Thus, simple rules have been deduced that could be used as first approximations: triplet n,π^* carbonyls show much less stereoselectivity than singlet n,π^* carbonyls.[12] More details are given in Chapter 45, which covers stereoselectivity in oxetane formation.

3.3 Substrates for the Paternò-Büchi Reaction

Carbonyl Compounds

Aromatic Ketones and Aldehydes

There are many Paternò-Büchi reactions involving substituted benzophenones or benzaldehydes. Some of these reactions are discussed in context with their olefinic reaction partners. The acyl derivatives of furan and thiophene are a special group of aromatic carbonyl reagents that could, in principle, undergo [2+2]-cycloaddition toward the C=O group as well as toward one of the ring C=C groups. The reaction periselectivity is controlled by the heteroatom and by the substitution pattern that influences the nature of the lowest excited triplet state.[13] Cantrell[14] has published the results of a series of experiments with 2- and 3-substituted benzoyl-, acetyl-, and formylthiophenes and furans. The aldehydes 1 and 3 (X = S,O) gave the oxetanes 2,4 highly selectively with 2,3-dimethyl-2-butene as alkene addend.[14] Similar results were reported for 2-benzoylfuran and 2-benzoylthiophene 5 (X = S,O), whereas the corresponding 2-acetyl substrates 6 gave mixtures of [2+2]- and [4+2]-photoproducts 7–9.[15]

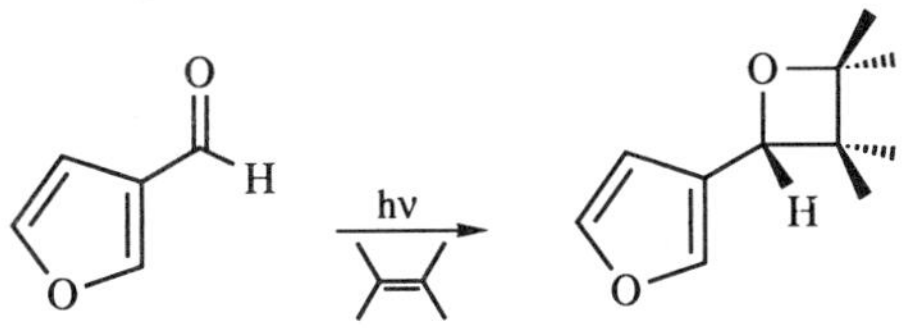

1   2 (54%)

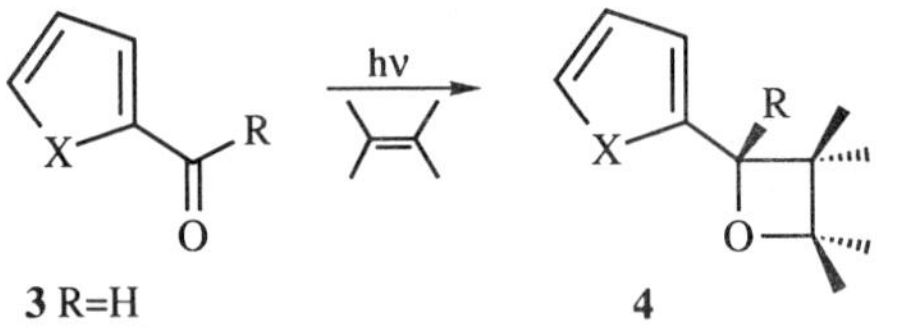

R=H, X=S 46%
R=Ph, X=O 27%
R=Ph, X=S 76%

3 R=H
5 R=Ph

4

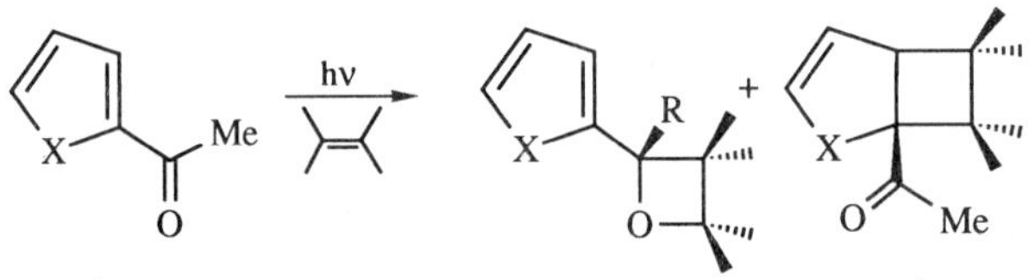

+ [4+2]

6

	7	**8**	**9**
X=O	8%	33%	0%
X=S	11%	10%	38%

Carboxylic Acid Derivatives and Nitriles

The photochemistry of carboxylic acid derivatives has been summarized by Coyle.[16] For arene carboxylic acid esters, it has been shown that [2+2]-cycloaddition competes with hydrogen abstraction by the excited ester from an allylic position of the alkene. The addition of methyl benzoate 10 to 2-methyl-2-butene gave a 1:1 mixture of the Paternò-Büchi adduct 11 and the coupling product 12.[17] Less electron-rich alkenes, e.g., cyclopentene, did add preferentially to the benzene ring of 10 affording the *ortho*- and *meta*-cycloaddition products. Furans could also be added photochemically to methyl benzoate and other arenecarboxylic acid esters. The resulting bicyclic oxetanes could be transformed into a series of synthetically valuable products.[18] [2+2]-Cycloadducts and/or their cleavage or rearrangement products have also been described for photoreactions of alkenes with diethyl oxalate,[19] benzoic acid,[20] and carbamates.[21] The site selectivity of alkene addition to benzonitrile 13 has been studied with a series of cyclic and acyclic olefins. Again, two reaction modes could be observed: [2+2]-cycloaddition to the nitrile group leading to 2-azabutadienes (deriving from the azetines formed initially, e.g., 14 from 2,3-dimethyl-2-butene[22]) and *ortho*-cycloaddition to the benzene ring (e.g., 15 from 2,3-dihydrofuran[23]).

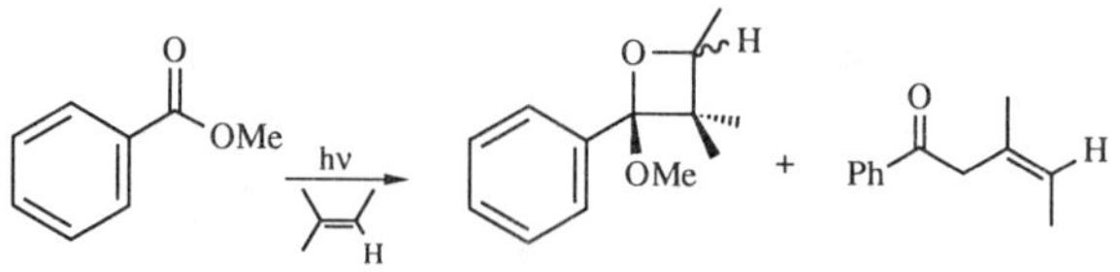

10　　　　**11** (33%)　　　　**12** (34%)

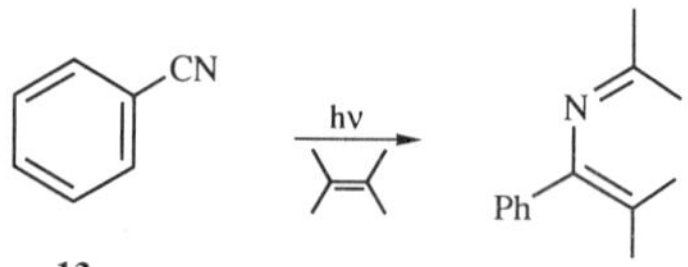

13　　　　**14** (66%)

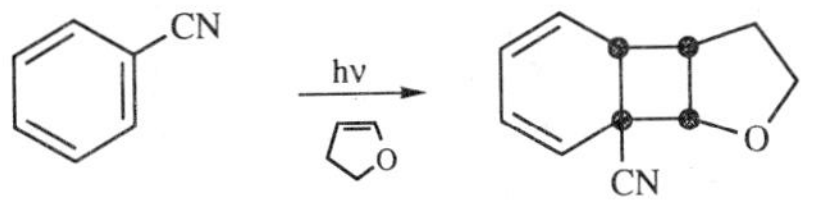

α-Ketocarbonyl Compounds, Acyl Cyanides

α-Diketones undergo primary photochemical addition to olefins to form [4+2]- and [2+2]-cycloadducts in competition with hydrogen abstraction, α-cleavage, and enol formation. The product ratio 17:18:19:20 from the biacetyl 16/2,3-dimethyl-2-butene photoreaction depends strongly on the solvent polarity and reaction temperature, indicating that an exciplex intermediate with pronounced charge separation and possibly free radical ions from a photo-induced single electron transfer (SET) step is involved.[24] Similar product ratios were reported for the methyl pyruvate/2,3-dimethyl-2-butene photoreaction. In this case, however, a state selectivity effect is responsible for the formation of the different ether and alcohol products.[25] Obviously, the existence of allylic hydrogens favors the formation of unsaturated acyclic products via hydrogen migration steps at the triplet biradical level. More electron-rich alkenes without (or with unfavorable) allylic hydrogens do give oxetanes with excited α-dicarbonyl compounds. Furan, indene,[26] as well as isopropenyl ethyl ether[27] were converted to the corresponding [2+2]-cycloadducts (e.g., 21) with biacetyl 16 and methyl pyruvate, respectively. Chiral phenyl pyruvates have been investigated intensively as carbonyl addends with medium to remarkably high diastereoselectivities (*vide infra*). Benzoyl cyanide 22 gives mixtures of cycloaddition and coupling products with 2,3-dimethyl-2-butene,[28] whereas the addition of 22, and a series of other acyl cyanides, to furan is chemoselective and leads to bicyclic oxetanes 23 with variable *endo/exo* ratios.[29]

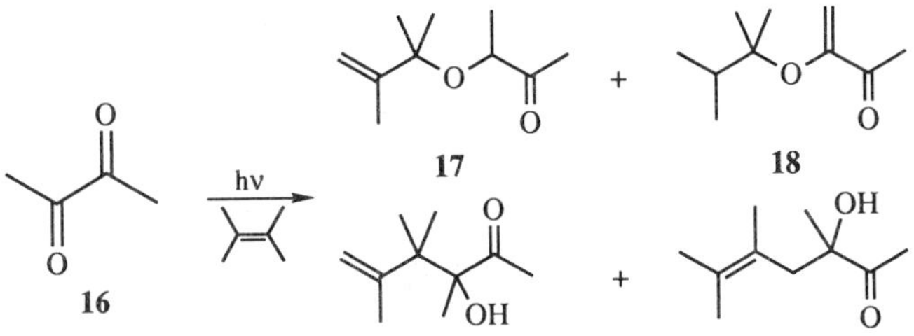

	17 : 18 : 19 : 20
in CH₃CN/r.t.:	56 : 35 : 3 : 6
in n-hexane/r.t.:	33 : 3 : 23 : 41

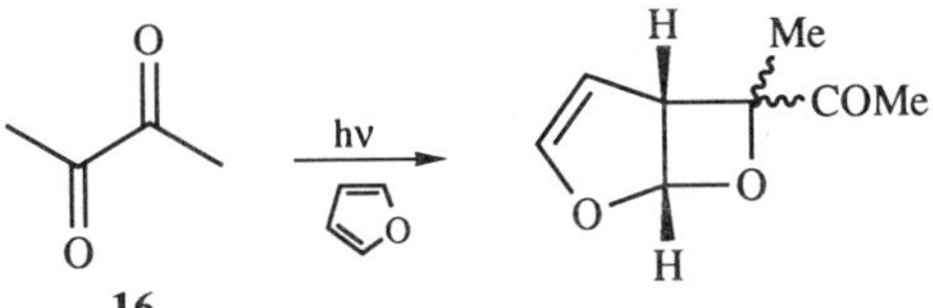

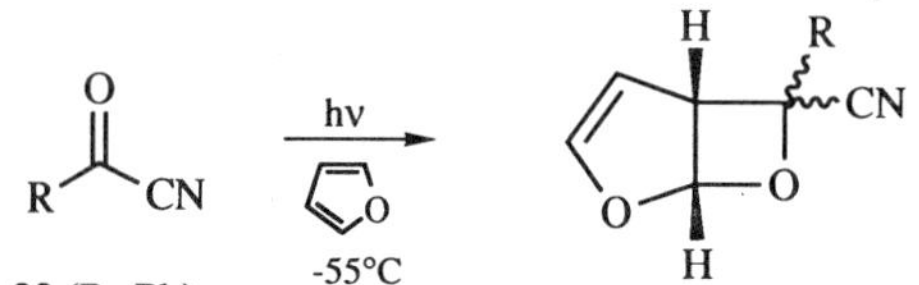

22 (R=Ph)

23 (95% for R=Ph; endo/exo = 5.3 : 1)

Enones and Ynones

There is a striking difference between the photochemical reactivity of α,β-unsaturated enones and the corresponding ynones. Whereas many cyclic enones undergo [2+2]-cycloaddition to alkenes at the C=C double bond of the enone (probably from the triplet π,π* state) to yield cyclobutanes, acyclic enones could easily deactivate by rotation about the central C-C single bond. Ynones, on the other hand, behave much more like alkyl-substituted carbonyl compounds and add to (sterically less emcumberd) alkenes to yield oxetanes.[30,31] The *regioselectivity* of this Paternò-Büchi reaction is similar to that of aliphatic or aromatic carbonyl compounds with a preference for primary attack at the less-substituted carbon atom (e.g., **25** and **26** from the reaction of but-3-in-2-one **24** with isobutylene). A serious drawback is the low *chemoselectivity*. For most substrates, a (formal) [3+2]-cycloaddition is the major reaction path that constitutes a possible rearrangement at the preoxetane biradical level.[32] A detailed kinetic and spectroscopic investigation of the reaction showed that the [2+2]-adducts **27** are formed from the singlet biradical precursors, whereas the [3+2]-adducts **28** derive from the corresponding triplet biradicals.[33]

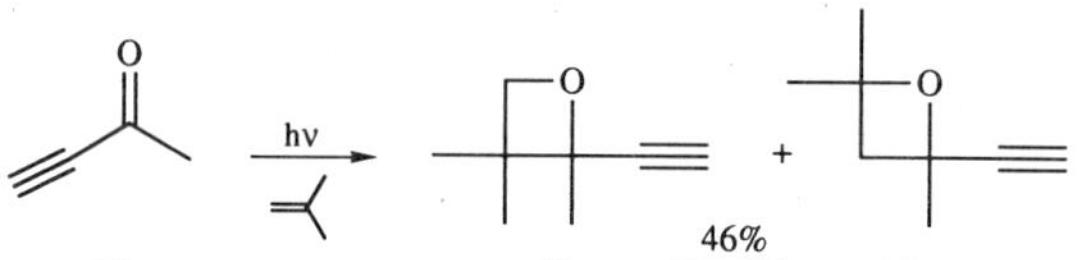

24   **25**   14 : 86   **26**
                              46%

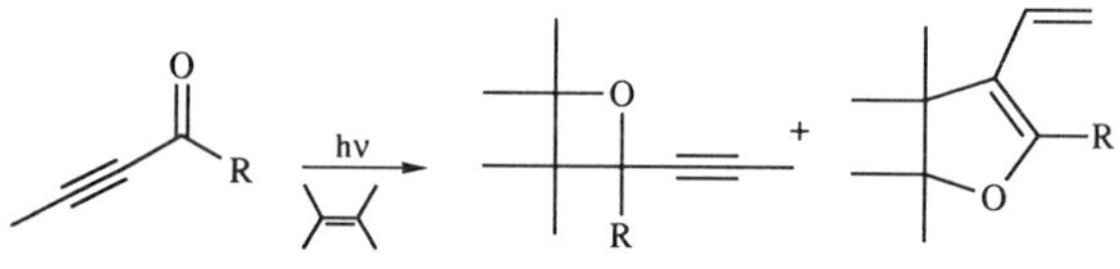

27   **28**

Quinones

The photoaddition of 1,4-benzoquinone **29** to electron-donor substituted alkenes is an efficient process that leads to spiro-oxetanes (e.g., **30a**) in high yield.[34] The use of quinones as carbonyl compounds is advantageous because their n,π*-transitions (430–480 nm) are shifted to larger wavelengths. Strained alkenes such as norbornene or norbornadiene are most reactive and large amounts of rearrangement products are formed.[35] Due to their convenient absorption behavior, benzoquinones have been used to study trapping reactions of intermediates. Wilson published the first oxygen trapping experiments using the 1,4-benzoquinone **29**/*t*-butylethene system.[36] Whereas a triplet 1,4-biradical was formulated as the most probable intermediate at that time, later work on intramolecular trapping reaction favored the formation of radical ion pairs.[37] Efficient lactonization reaction to form **32** during irradiation of pent-4-enoic acid **31** and **29** accounts for an olefin radical cation that undergoes electrophilic addition to the carboxyl group. Another type of rearrangement has been reported in the photoreaction of tetramethylallene and 1,4-benzoquinone **29**,[38] where 5-hydroxy-indan-2-ones **33** are formed in high yields probably via unstable spiro-oxetanes **30b** as intermediates.

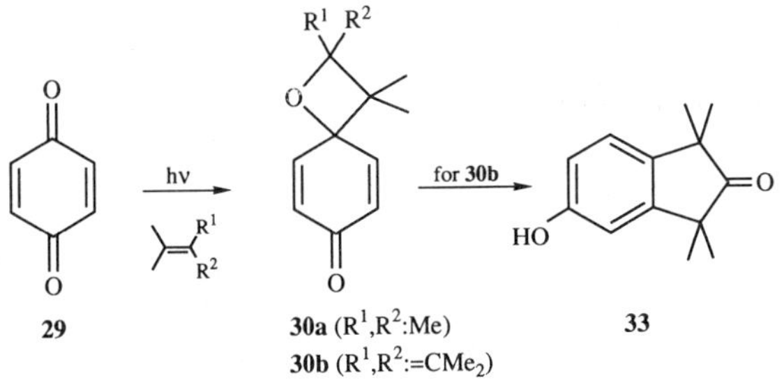

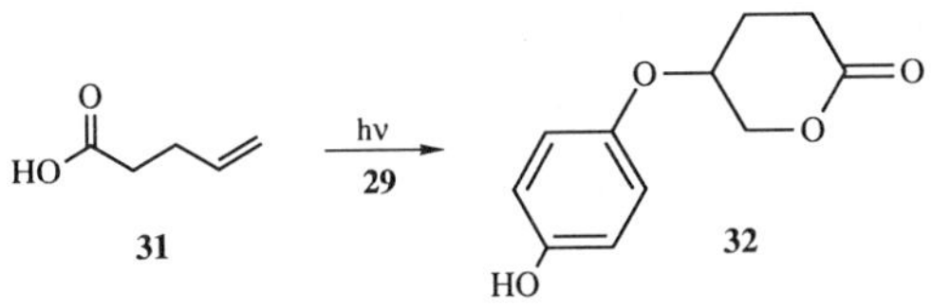

29   **30a** (R^1,R^2:Me)   **33**
          30b (R^1,R^2:=CMe$_2$)

31   **32**

Olefin Compounds

Alkenes, Alkyl- and Aryl-Substituted

Arnold reported a series of reactions between benzophenones and monoalkenes. In most cases, the oxetanes were the major products and could be isolated in good yields.[39] Benzaldehyde 34, in which the triplet n,π* is the reactive state,[40] is less chemoselective and with cyclohexene gives the oxetane 35 as well as several hydrogen-abstraction and radical coupling products 36–38.[41] Clear evidence for a sufficient long-lived intermediate came from investigations of the stereoselectivity of the Paternò-Büchi reaction with *cis*- and *trans*-2-butene as substrates. When acetone[42] or benzaldehyde[43] was used as the carbonyl addend, complete stereo-randomization was observed. Acetaldehyde and 2-naphthaldehyde showed stereoselective addition reactions that confirmed the singlet n,π* as the reactive state.[43]

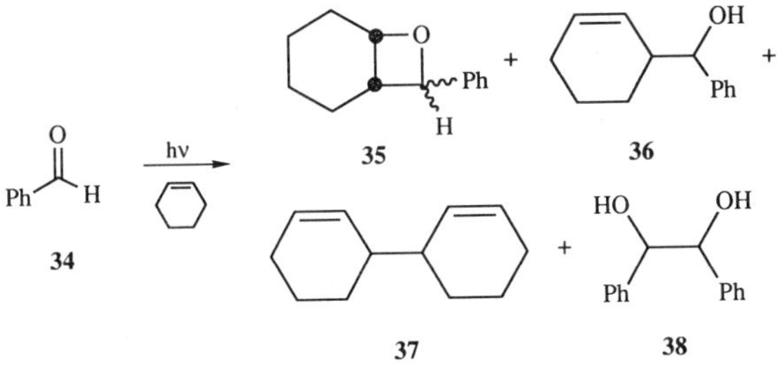

35   **36**

34

37   **38**

Alkenes, Electron Donor-Substituted

The regioselectivity of the Paternò-Büchi reaction with acyclic enol ethers is substantially higher than with the corresponding unsymmetrically alkyl-substituted olefins. This effect was used for the synthesis of a variety of 3-alkoxyoxetanes and a series of derivatives.[44] The diastereomeric *cis*- and *trans*-1-methoxy-1-butenes were used as substrates for the investigation of the spin state influence on reactivity, regio- and stereoselectivity.[45] The use of trimethylsilyloxyethene 40 as electron-rich alkene is advantageous, and several 1,3-anhydro-apiitol-derivatives such as 41 could be synthesized via photocycloaddition with 1,3-diacetoxy-2-propanone 39.[46] Branched-chain erythrono-1,4-lactones are accessible from an oxetane 44 that was derived (thermally) from diethyl mesoxalate 42 and 2,2-diisopropyl-1,3-dioxole 43.[47] An impressive improvement in the regioselectivity of oxetane formation was discovered with 2,3-dihydrofuran 45 as the alkene addend. With acetone/2,3-dihydrofuran 45 the cycloadduct 46 was obtained in a >200:1 ratio of the two possible regioisomers.[48]

The photocycloaddition of cyclic thioenol ethers to benzophenone has also been studied. In contrast to acyclic enol ethers, these substrates exhibit high regioselectivity with almost exclusive formation of the 3-alkylthio oxetanes.[49]

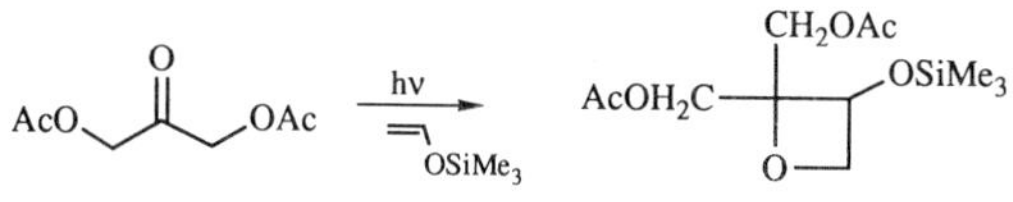

Alkynes

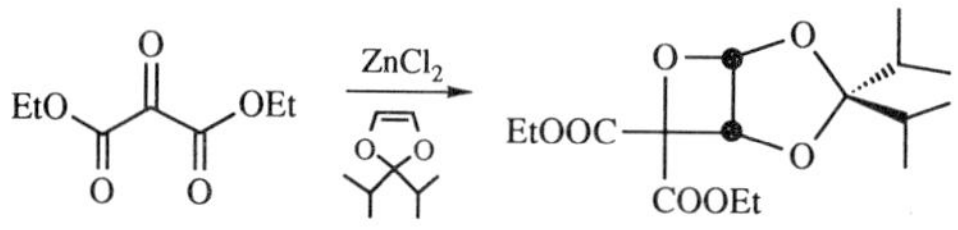

Oxetenes (oxets) have been postulated as primary photoadducts from carbonyl compounds and alkyl- and aryl-substituted alkynes and alkylthioacetylenes.[50] The first evidence for such an unstable intermediate came from the work of Friedrich and Bower, who reported a lifetime of several hours at −35 °C for the benzaldehyde 34/2-butyne photoproduct 47.[51] On further irradiation in the presence of excess benzaldehyde, a *bis*-oxetane 48 was formed. At elevated temperatures, rapid ring-opening to give α,β-unsaturated ketones occurs, the major products for photocycloaddition of alkynes with carbonyl addends at room temperature. The parent oxetene has been prepared from 3-hydroxyoxetane.[52]

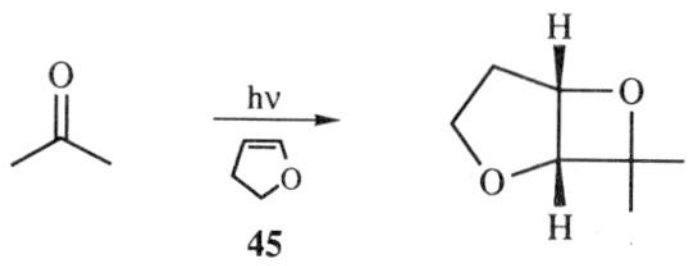

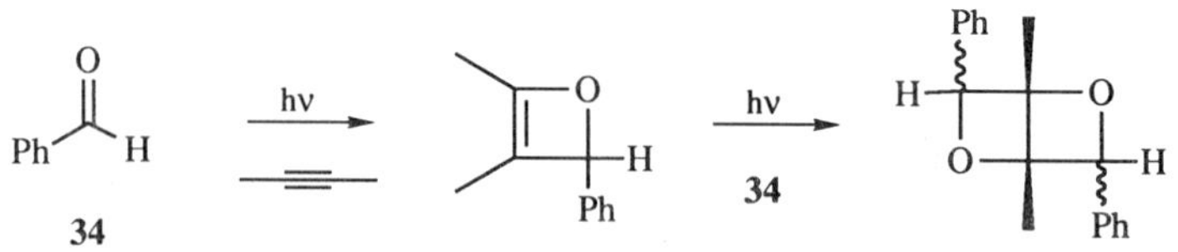

Allenes and Ketenimines

The photocycloaddition of a variety of carbonyl compounds toward methyl-substituted allenes has been reported to proceed with high quantum yields (0.59 for acetophenone/tetramethylallene)[53] to give 1:1 and 1:2 adducts.[54] The 2-alkylideneoxetanes are useful precursors for the synthesis of cyclobutanones; e.g., 51a,b from the benzophenone 49/tetramethylallene-cycloadduct 50.[55] Upon prolonged irradiation in the presence of an excess of ketone, the monoadducts 50 are converted into 1,5- and 1,6-dioxaspiro[3.3]heptanes (e.g., 52a,b).[53] 2- and 3-Iminooxetanes could be prepared by photolysis of ketenimines in the presence of aliphatic or aromatic ketones.[56]

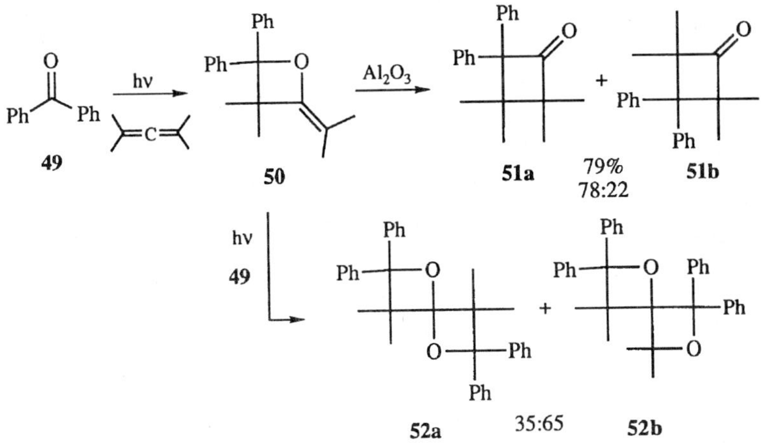

Dienes and Enynes

Because of their low triplet energies (55 to 60 kcal mol^{-1}), 1,3-dienes are often used as quenchers for the excited triplet states of carbonyl compounds. Besides physical quenching, however, cycloaddition to form oxetanes can occur as a side reaction, as shown for benzophenone.[57] Chemical yields are low because of competing diene dimerization and hydrogen abstraction reactions. The corresponding photoreactions with aliphatic ketones[58] or aldehydes[59] are much more effective in the sense that oxetanes are formed with high quantum yields and good chemical yields by a mechanism involving the singlet excited carbonyl addend. Carless and Maitra[60] have shown that the photocycloaddition of acetaldehyde **53** to the diastereomeric *E*- and *Z*-penta-1,3-dienes is highly regio- and stereoselective (e.g., oxetanes **54a-d** from the *Z*-isomer). This is also evidence for a singlet mechanism. A site-selective reaction has been reported for the addition of benzophenone to an 1,3-enyne with exclusive addition to the C=C double bond.[61]

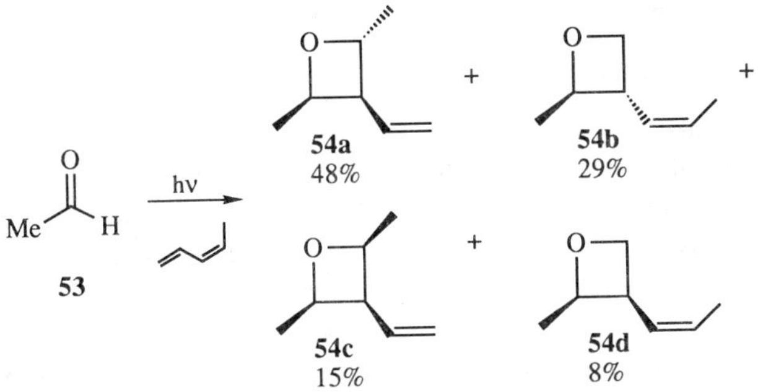

Furans

The photocycloaddition of benzophenone **49** to furan **55a** was originally described by Schenck et al.[62] Besides the 1:1 adduct **56**, two regioisomeric 2:1 adducts **57a,b** were also isolated;[63] the structure of **57a** was revised by Toki and Evanega.[64] All prostereogenic carbonyl addends when added photochemically to furan showed regioselectivities >99:1 in favor of the bicyclic acetal product. This is also the case for 2-substituted furans, however, mixtures of acetal- and ketal-type oxetanes (e.g., for **55b**) were obtained.[65] The use of furans with steric-demanding substituents (e.g., **55c**)[66a] or an acetyl substituent (e.g., **55d**)[66b] at the 2-position improves largely the site selectivity. A large number of carbonyl compounds has been investigated in the last 10 years by Zamojski[67] and especially by Schreiber,[68] who used furan-carbonyl adducts as intermediates in total syntheses of natural products. Acid-catalyzed rearrangement of these adducts is a useful method for the synthesis of 3-substituted furans.[69]

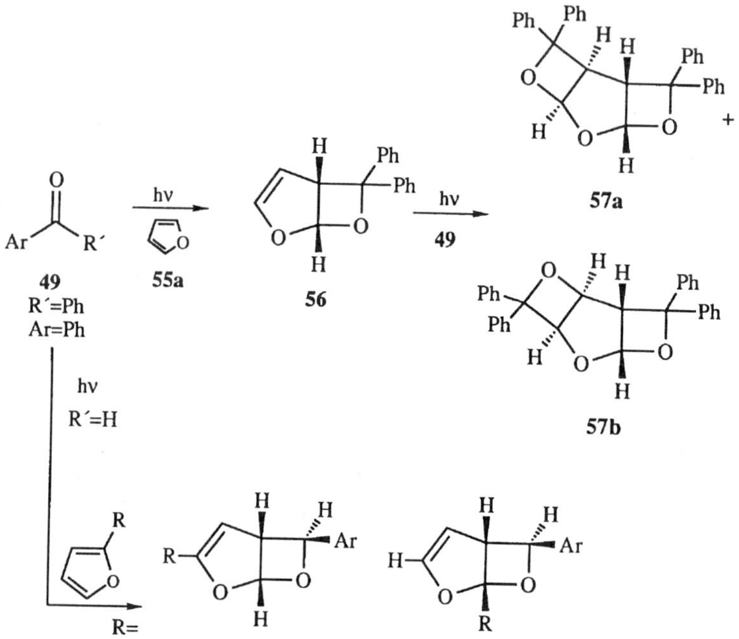

	R=		:		
55b	Me	43	:	57	Ar = Ph
55c	SiPr$^i_3$	>20	:	1	Ar = Ph
55d	Ac	1	:	>20	Ar = 4-MeOPh

Other Heteroaromatic Substrates

Methyl-substituted thiophenes afford oxetanes in high regioselectivity when they are reacted with excited benzophenone.[70] Pyrroles, imidazoles, and indoles behave similarly if the nitrogen atom is substituted with electron-accepting groups.[71] Pyrroles, when substituted at the nitrogen atom by alkyl groups do, however, give rise to rearranged pyrrroles, probably via an oxetane intermediate.[72]

Strained Hydrocarbons

The photocycloaddition of triplet benzophenone **49** to norbornene was originally reported by Scharf and Korte.[73] The photoproduct **58**, which is formed in high exo-selectivity, could be cleaved thermally to yield the δ,ε-unsaturated ketone **59**. This is an application of the "carbonyl-olefin-metathesis" (COM) concept.[74] The 1,4-biradical formed in the interaction of norbornene with *o*-dibenzoyl-benzene was trapped intramolecularly by the second carbonyl moiety.[75] A highly regioselective reaction of triplet benzophenone was reported with 5-methylenenorborn-2-ene, with preferential attack toward the *exo* C=C-double bond.[76] A number of publications have appeared that discuss photocycloaddition reactions of triplet carbonyl compounds to norbornadiene and quadricyclane and the competition between the Paternò-Büchi reaction and the sensitized norbornadiene/quadricyclane interconversion.[77] Oxetane formation has been reported for the photoreaction of biacetyl and *p*-quinones with benzvalene.[78]

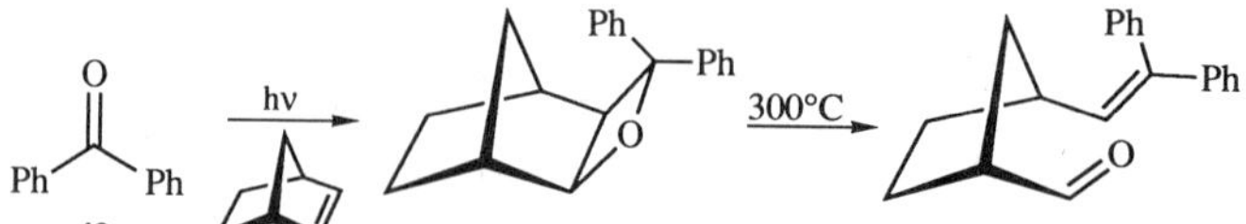

Alkenes, Electron Acceptor-Substituted

In contrast to photocycloaddition reactions of carbonyl compounds to electron-rich alkenes (which shows a low degree of stereoselectivity in the case of triplet excited carbonyls), the corresponding reactions with electron-deficient alkenes, such as cyanoalkenes, are, although rather inefficient, highly stereoselective.[79] Kinetic analysis showed that these reactions involve the attack by the singlet excited carbonyl via a parallel approach.[80] An important side reaction is the photosensitized geometrical isomerization of the alkene C=C double bond, e.g., *cis*-1,2-dicyanoethylene and acetone gives the *cis*-oxetane **60** and *trans*-1,2-dicyanoethylene. 2-Norbornanone was used as a model reagent for investigation of the influence of steric hindrance on the face-selectivity of oxetane formation with electron-donor and electron acceptor-substituted alkenes.[81]

60

·ferences

1. Paternò, E. and Chieffi, G., Sintesi in chimica organica per mezzo della luce. Nota II. Composti degli idrocarburi non saturi con aldeidi e chetoni, *Gazz. Chim. Ital.*, 341, 1909.
2. Büchi, G., Inman, C. G., and Lipinsky, E. S., Light-catalyzed organic reactions. I. The reaction of carbonyl compounds with 2-methyl-2-butene in the presence of ultraviolet light, *J. Am. Chem. Soc.*, 76, 4327, 1954.
3. Yang, N. C., Nussim, M., Jorgenson, M. J., and Murov, S., Photochemical reactions of carbonyl compounds in solution. The Paternò-Büchi reaction, *Tetrahedron Lett.*, 3657, 1964.
4. (a) Inoue, Y., Asymmetric photochemical reactions in solution, *Chem. Rev.*, 92, 741, 1992; (b) Porco, J. A., Jr. and Schreiber, S. L., The Paternò-Büchi reaction, *Comprehensive Organic Synthesis*, Trost, B. M., Fleming, I., and Paquette, L. A., Eds., 5, 151, 1991; (c) Demuth, M. and Mikhail, G., New developments in the field of photochemical synthesis, *Synthesis*, 145, 1989; (d) Carless, H. A. J., *Photochemistry in Organic Chemistry*, Coyle, J. D., Ed., The Royal Society of Chemistry, Special Publication, 57, 95, 1986; (e) Carless, H. A. J., Photochemical synthesis of oxetanes, *Synthetic Organic Photochemistry*, Horspool, W. M., Ed., Plenum Press, New York, 425, 1984; (f) Jones, G., II, Synthetic applications of the Paternò-Büchi reaction, *Org. Photochem.*, 5, 1, 1981.
5. Turro, N. J., Dalton, J. C., Dawes, K., Farrington, G., Hautala, R., Morton, D., Niemczyk, M., and Schore, N., Molecular photochemistry of alkanones in solution: α-cleavage, hydrogen abstraction, cycloaddition, and sensitization reactions, *Acc. Chem. Res.*, 5, 92, 1972.
6. Dauben, W. G., Salem, L., and Turro, N. J., A classification of photochemical reactions, *Acc. Chem. Res.*, 8, 41, 1975.
7. (a) Salem, L., Surface crossings and surface touchings in photochemistry, *J. Am. Chem. Soc.*, 96, 3486, 1974; (b) Bigot, B., Devaquet, A., and Turro, N. J., Natural correlation diagrams. A unifying theoretical basis for analysis of n orbital initiated ketone photoreactions, *J. Am. Chem. Soc.*, 103, 6, 1981.
8. Wilson, R. M., The trapping of biradicals and related photochemical intermediates, *Org. Photochem.*, 7, 339, 1985.
9. Shimizu, N., Ishikawa, M., Ishikura, K., and Nishida, S., Photocycloaddition of aromatic carbonyl compounds to vinylcyclopropene and its derivatives, *J. Am. Chem. Soc.*, 96, 6456, 1974.
10. Freilich, S. C. and Peters, K. S., Observation of the 1,4-biradical in the Paternò-Büchi reaction, *J. Am. Chem. Soc.*, 103, 6255, 1981.
11. (a) Barltrop, J. A. and Carless, H. A. J., Photocycloaddition of aliphatic ketones to α,β-unsaturated nitriles, *J. Am. Chem. Soc.*, 94, 1951, 1972; (b) Dalton, J. C., Wriede, P. A., and Turro, N. J., Photocycloaddition of acetone to 1,2-dicyanoethylene, *J. Am. Chem. Soc.*, 92, 1318, 1970.
12. Turro, N. J. and Wriede, P. A., The photocycloaddition of acetone to 1-methoxy-1-butene. A comparison of singlet and triplet mechanisms and biradical intermediates, *J. Am. Chem. Soc.*, 92, 320, 1970.

13. (a) Arnold, D. R. and Clarke, B. M., The effect of an adjacent methyl group on the excited state reactivity of 2-benzoylthiophene, *Can. J. Chem.*, 53, 1, 1975; (b) Arnold, D. R. and Hadjiantoniou, The effect of an adjacent methyl group on the excited state reactivity of 3-benzoylthiophene, *Can. J. Chem.*, 56, 1970, 1978.

14. Cantrell, T. S., Reactivity of photochemically excited 3-acylthiophenes, 3-acylfurans, and the formylthiophenes and -furans, *J. Org. Chem.*, 42, 3774, 1977.

15. Cantrell, T. S., Photochemical reactions of 2-acylthiophenes, -furans, and -pyrroles with alkenes, *J. Org. Chem.*, 39, 2242, 1974.

16. Coyle, J. D., Photochemistry of carboxylic acid derivatives, *Chem. Rev.*, 78, 97, 1978.

17. Cantrell, T. S. and Allen, A. C., Photochemical reactions of arenecarboxylic acid esters with electron-rich alkenes: 2+2 Cycloaddition, hydrogen abstraction, and cycloreversion, *J. Org. Chem.*, 54, 135, 1989.

18. Cantrell, T. S., Allen, A. C. and Ziffer, H., Photochemical 2+2 cycloaddition of arenecarboxylic acid esters to furans and 1,3-dienes. 2+2 Cycloreversion of oxetanes to dienol esters and ketones, *J. Org. Chem.*, 54, 140, 1989.

19. Tominaga, T., Odaira, Y., and Tsutsumi, S., Photochemical synthesis of oxetanes from diethyl oxalate, *Bull. Chem. Soc. Jpn.*, 40, 2451, 1967.

20. Cantrell, T. S., Photochemical reactions of benzoic acid. Cycloaddition, hydrogen abstraction, and reverse type II elimination, *J. Am. Chem. Soc.*, 95, 2714, 1973.

21. Tominaga, T. and Tsutsumi, S., Photocycloaddition of carbamates to 1,1-diphenylethylene, *Tetrahedron Lett.*, 3175, 1969.

22. Cantrell, T. S., Photochemical cycloadditions of benzonitrile to alkenes, factors controlling the site of the addition, *J. Org. Chem.*, 42, 4238, 1977.

23. Mattay, J., Runsink, J., Heckendorn, R., and Winkler, T., Photoreactions of benzonitrile with cyclic enol ethers, *Tetrahedron*, 43, 5781, 1987.

24. Turro, N. J., Shima, K., Chung, C.-J., Tanielian, C., and Kanfer, S., Photoreactions of biacetyl and tetramethylethylene. Solvent and temperature effects, *Tetrahedron Lett.*, 2775, 1980.

25. Shima, K., Sawada, T., and Yoshinaga, H., Photoaddition of singlet and triplet methyl pyruvate with 2,3-dimethyl-2-butene, *Bull. Chem. Soc. Jpn.*, 51, 608, 1978.

26. Ryang, H.-S., Shima, K., and Sakurai, H., Photoaddition reaction of biacetyl, *J. Org. Chem.*, 38, 2860, 1973.

27. Shima, K., Kawamura, T., and Tanabe, K., The photoaddition of methyl pyruvate to methyl-substituted olefins, *Bull. Chem. Soc. Jpn.*, 47, 2347, 1974.

28. Cantrell, T. S., Photochemical reactions of benzoyl cyanide and benzoyl halides with alkenes, *J. Chem. Soc. Chem. Commun.*, 637, 1975.

29. Zagar, C. and Scharf, H.-D., The Paternò-Büchi reaction of achiral and chiral acyl cyanides with furan, *Chem. Ber.*, 124, 967, 1991.

30. Jorgenson, M. J., Photoaddition of acetylenic ketones to olefines, *Tetrahedron Lett.*, 5811, 1966.

31. Kwiatkowski, G. T. and Selley, D. B., The photochemistry of conjugated acetylenic olefins and carbonyl derivatives, *Tetrahedron Lett.*, 3471, 1968.

32. Hussain, S. and Agosta, W. C., Photochemical [3+2] cycloaddition of α,β-acetylenic ketones with simple olefins, *Tetrahedron*, 37, 3301, 1981.

33. Saba, S., Wolff, S., Schröder, C., Margaretha, P., and Agosta, W. C., Studies on the photochemical reactions of α,β-acetylenic ketones with tetramethylethylene, *J. Am. Chem. Soc.*, 105, 6902, 1983.

34. Bryce-Smith, D., Gilbert, A., and Johnson, M. G., Formation of spiro-oxetanes by photoaddition of olefins to *p*-benzoquinone, *J. Chem. Soc. (C)*, 383, 1967.

35. Bunce, N. J. and Hadley, M., Mechanism of oxetane formation in the photocycloaddition of *p*-benzoquinone to alkenes, *Can. J. Chem.*, 53, 3240, 1975.

36. Wilson, R. M. and Musser, A. K., Photocyclizations involving quinone-olefin charge-transfer exciplexes, *J. Am. Chem. Soc.*, 102, 1720, 1980.

37. Fehnel, E. A. and Brokaw, F. C., Photocycloaddition reactions of norbornadiene and quadricyclane with *p*-benzoquinone, *J. Org. Chem.*, 45, 578, 1980.

38. Ishibe, N., Hashimoto, K., and Yamaguchi, Y., Photochemical addition of allenes to *p*-quinones, *J. Chem. Soc. Perkin Trans. 1*, 318, 1975.

39. Arnold, D. R., Hinman, R. L., and Glick, A. H., Chemical properties of the carbonyl nπ* state. The photochemical preparation of oxetanes, *Tetrahedron Lett.*, 1425, 1964.

40. Yang, N. C., Loeschen, R., and Mitchell, D., On the mechanism of the Paternò-Büchi reaction, *J. Am. Chem. Soc.*, 89, 5465, 1967.

41. Bradshaw, J. S., Ultraviolet irradiation of carbonyl compounds in cyclohexene and 1-hexene, *J. Org. Chem.*, 31, 237, 1966.

42. Carless, H. A. J., Photocycloaddition of acetone to acyclic olefins, *Tetrahedron Lett.*, 3173, 1973.

43. Yang, N. C., Kimura, M., and Eisenhardt, W., Paternò-Büchi reactions of aromatic aldehydes with 2-butenes and their implication on the rate of intersystem crossing of aromatic aldehydes, *J. Am. Chem. Soc.*, 95, 5058, 1973.

44. (a) Schroeter, S. H. and Orlando, C. M., The photocycloaddition of various ketones and aldehydes to vinyl ethers and ketene diethyl acetal, *J. Org. Chem.*, 34, 1181, 1969; (b) Schroeter, S. H., The synthesis of 3-hydroxyaldehydes, 3-hydroxy acetals, and 3-hydroxy ethers from 2-alkoxyoxetanes, *J. Org. Chem.*, 34, 1188, 1969.

45. (a) Turro, N. J. and Wriede, P. A., The photocycloaddition of acetone to 1-methoxy-1-butene. A comparison of singlet and triplet mechanisms and biradical intermediates, *J. Am. Chem. Soc.*, 90, 6863, 1968; (b) Turro, N. J. and Wriede, P. A., The photocycloaddition of acetone to 1-methoxy-1-butene. A comparison of singlet and triplet mechanisms and biradical intermediates, *J. Am. Chem. Soc.*, 92, 320, 1970.

46. Araki, Y., Nagasawa, J.-I., and Ishido, Y., Photochemical cycloaddition of 1,3-diacetoxy-2-propanone to (trimethylsilyloxy)ethylene, *Carbohydrate Res.*, 91, 77, 1981.

47. Mattay, J. and Buchkremer, K., Thermal and photochemical oxetane formation. A contribution to the synthesis of branched-chain aldonolactones, *Helv. Chim. Acta*, 71, 981, 1988.

48. Carless, H. A. J. and Haywood, D. J., Photochemical synthesis of 2,6-dioxabicyclo[3.2.0]heptanes, *J. Chem. Soc., Chem. Commun.*, 1067, 1980.

49. Morris, T. H., Smith, E. H., and Walsh, R., Oxetane synthesis: Methyl vinyl sulphides as new traps of excited benzophenone in a stereoselective and regiospecific Paternò-Büchi reaction, *J. Chem. Soc., Chem. Commun.*, 964, 1987.

50. (a) Büchi, G., Kofron, J. T., Koller, E., and Rosenthal, D., Light catalyzed organic reactions. V. The addition of aromatic carbonyl compounds to a disubstituted acetylene, *J. Am. Chem. Soc.*, 78, 876, 1956; (b) Miyamoto, T., Shigemitsu, Y., and Odaira, Y., Photoaddition of carboxylate esters to diphenylacetylene, *J. Chem. Soc., Chem. Commun.*, 1410, 1969; (c) Bradshaw, J. S., Knudsen, R. D., and Parish, W. W., Irradiation of benzaldehyde in 1-hexyne, *J. Org. Chem.*, 40, 529, 1975; (d) Mosterd, A., Matser, H. J., and Bos, H. J. T., Photoaddition of non-cisoid 1,2-diketones and phenylglyoxal to alkylthioacetylenes; preparation of 3-alkylthiofurans, *Tetrahedron Lett.*, 4179, 1974.

51. Friedrich, L. E. and Bower, J. D., Detection of an oxetene intermediate in the photoreaction of benzaldehyde with 2-butyne, *J. Am. Chem. Soc.*, 95, 6869, 1973.

52. Friedrich, L. E. and Lam, P. Y.-S., Syntheses and reactions of 3-phenyloxete and the parent unsubstituted oxete, *J. Org. Chem.*, 46, 306, 1981.

53. Gotthardt, H., Steinmetz, R., and Hammond, G. S., Cycloaddition of carbonyl compounds to allenes, *J. Org. Chem.*, 33, 2774, 1968.

54. Arnold, D. R. and Glick, A. H., The photocycloaddition of carbonyl compounds to allenes, *J. Chem. Soc., Chem. Commun.*, 813, 1966.

55. Gotthardt, H. and Hammond, G. S., Some interesting rearrangements in the 2-isopropylidene oxetane series, *Chem. Ber.*, 107, 3922, 1974.

56. (a) Singer, L. A., Davis, G. A., and Knutsen, R. L., Photocycloaddition of acetone to ketenimines. *Syn-anti* exchange barriers in β-imino oxetanes, *J. Am. Chem. Soc.*, 94, 1188, 1972; (b) Singer, L. A. and Davis, G. A., Photokinetic studies on benzophenone. Photocycloaddition to ketenimines and self-quenching of the benzophenone triplet, *J. Am. Chem. Soc.*, 95, 8638, 1973.

57. Barltrop, J. A. and Carless, H. A. J., Organic photochemistry. XI. The photocycloaddition of benzophenone to conjugated dienes, *J. Am. Chem. Soc.*, 93, 4794, 1971.

58. (a) Hautala, R. R., Dawes, K., and Turro, N. J., Stereoselectivity and regioselectivity in the photocycloaddition of 2-methyl-2,4-hexadiene and acetone, *Tetrahedron Lett.*, 1229, 1972; (b) Barltrop, J. A. and Carless, H. A. J., Organic photochemistry. XIV. Photocycloaddition of alkyl ketones to conjugated dienes, *J. Am. Chem. Soc.*, 94, 8761, 1972.

59. (a) Kubota, T., Shima, K., Toki, S., and Sakurai, H., The photoaddition of propionaldehyde to cyclohexa-1,3-diene. Oxetane formation by a singlet mechanism, *J. Chem. Soc., Chem. Commun.*, 1462, 1969; (b) Shima, K., Kubota, T., and Sakurai, H., Organic photochemical reactions. XXIV. Photocycloaddition of propanal to 1,3-cyclohexadiene, *Bull. Chem. Soc. Jpn.*, 49, 2567, 1976.

60. Carless, H. A. J. and Maitra, A. K., Photocycloaddition of acetaldehyde to *(E)-* and *(Z)*-penta-1,3-diene, *Tetrahedron Lett.*, 1411, 1977.

61. Carless, H. A. J., Photocycloaddition of benzophenone to a conjugated enyne, *Tetrahedron Lett.*, 2265, 1972.

62. Schenck, G. O., Hartmann, W., and Steinmetz, R., Four-membered ring synthesis by the photosensitized cycloaddition of dimethyl maleic anhydride to olefins, *Chem. Ber.*, 96, 498, 1963.

63. (a) Ogata, M., Watanabe, H., and Kano, H., Photochemical cycloaddition of benzophenone to furans, *Tetrahedron Lett.*, 533, 1967.

64. (a) Toki, S. and Sakurai, H., On the structure of the 2:1 adduct of benzophenone with furan, *Tetrahedron Lett.*, 4119, 1967; (b) Evanega, G. R. and Whipple, E. B., The photochemical addition of benzophenone to furan, *Tetrahedron Lett.*, 2163, 1967.

65. (a) Toki, S., Shima, K., and Sakurai, H., Organic photochemical reactions. I. The synthesis of substituted oxetanes by the photoaddition of aldehydes to furans, *Bull. Chem. Soc. Jpn.*, 38, 760, 1965; (b) Sekretár, S., Rudá, J., and Stibrányi, L., Photochemical reactions of 2-substituted furans with some carbonyl compounds, *Coll. Czech. Chem. Commun.*, 49, 71, 1984.

66. (a) Schreiber, S. L., Desmaele, D., and Porco, J. A., Jr., On the use of unsymmetrically substituted furans in the furan-carbonyl photocycloaddition reaction: Synthesis of a kadsurenone-ginkgolide hybrid, *Tetrahedron Lett.*, 29, 6689, 1988; (b) Carless, H. A. J. and Halfhide, A. F. E., Highly regioselective [2+2] photocycloaddition of aromatic aldehydes to acetylfurans, *J. Chem. Soc., Perkin Trans. 1*, 1081, 1992.

67. (a) Jarosz, S. and Zamojski, A., Asymmetric photocycloaddition between furan and optically active ketones, *Tetrahedron*, 38, 1453, 1982; (b) Kozluk, T. and Zamojski, A., The synthesis of 3-deoxy-DL-streptose, *Tetrahedron*, 39, 805, 1983.

68. (a) Schreiber, S. L. and Satake, K., Application of the furan-carbonyl photocycloaddition reaction to the synthesis of the *bis*(tetrahydrofuran) moiety of asteltoxin, *J. Am. Chem. Soc.*, 105, 6723, 1983; (b) Schreiber, S. L. and Satake, K., Total synthesis of (±)-asteltoxin, *J. Am. Chem. Soc.*, 106, 4186, 1984; (c) Schreiber, S. L. and Satake, K., Studies of the furan-carbonyl photocycloaddition reaction: The determination of the absolute stereostructure of asteltoxin, *Tetrahedron Lett.*, 27, 2575, 1986.

69. (a) Kitamura, T., Kawakami, Y., Imegawa, T., and Kawanishi, M., One-pot synthesis of 3-substituted furan: A synthesis of perillaketone, *Synth. Commun.*, 7, 521, 1977; (b) Jarosz, S. and Zamojski, A., Rearrangement of 6-substituted 2,7-dioxabicyclo[3.2.0]hept-3-enes to furans, *J. Org. Chem.*, 44, 3720, 1979.

70. Rivas, C., Pacheco, D., Vargas, F., and Ascanio, J., Synthesis of oxetanes by photoaddition of carbonyl compounds to thiophene derivatives, *J. Heterocyclic Chem.*, 18, 1065, 1981.

71. (a) Rivas, C. and Bolivar, R. A., Synthesis of oxetanes by photoaddition of carbonyl compounds to pyrrole derivatives, *J. Heterocyclic Chem.*, 13, 1037, 1976; (b) Nakano, T., Rodriquez, W., de Roche, S. Z., Larrauri, J. M., Rivas, C., and Pérez, C., Photoaddition of ketones to imidazoles, thiazoles,

isothiazoles and isoxazoles. Synthesis of their oxetanes, *J. Heterocyclic Chem.*, 17, 1777, 1980; (c) Julian, D. R. and Tringham, G. D., Photoaddition of ketones to indoles: Synthesis of oxeto[2,3-b] indoles, *J. Chem. Soc., Chem. Commun.*, 13, 1973.

72. (a) Matsuura, T., Banba, A., and Ogura, K., Photoinduced reactions. XLV. Photoaddition of ketones to methylimidazoles, *Tetrahedron*, 27, 1211, 1971; (b) Jones, G., II, Gilow, H. M., and Low, J., Regioselective photoaddition of pyrrols and aliphatic carbonyl compounds. A new synthesis of 3(4)-substituted pyrrols, *J. Org. Chem.*, 44, 2949, 1979.

73. Scharf, D. and Korte, F., Photosensitized cyclodimerization of norbornene, *Tetrahedron Lett.*, 821, 1963.

74. Jones, G., II, Schwartz, S. B., and Marton, M. T., Regiospecific thermal cleavage of some oxetane photoadducts: Carbonyl-olefin metathesis in sequential photochemical and thermal steps, *J. Chem. Soc., Chem. Commun.*, 374, 1973.

75. Shigemitsu, Y., Yamamoto, S., Miyamoto, T., and Odaira, Y., A novel photocycloaddition of dibenzoylbenzene to olefins, *Tetrahedron Lett.*, 2819, 1975.

76. Gorman, A. A., Leyland, R. L., Parekh, C. T., and Rodgers, M. A. J., The reaction of triplet benzophenone with 5-methylenenorborn-2-ene: Regiospecific oxetane formation, *Tetrahedron Lett.*, 1391, 1976.

77. (a) Barwise, A. J. G., Gorman, A. A., Leyland, R. L., Parekh, C. T., and Smith, P. G., The inter- and intramolecular addition of aromatic ketone triplets to quadricyclane, *Tetrahedron*, 36, 397, 1980; (b) Gorman, A. A., Leyland, R. L., Rodgers, M. A. J., and Smith, P. G., Concerning the mechanism of interactions of triplet benzophenone with norbornadienes and quadricyclanes, *Tetrahedron Lett.*, 5085, 1973; (c) Kubota, T., Shima, K., and Sakurai, H., The photocycloaddition of benzophenone to norbornadiene, *Chem. Lett.*, 343, 1972.

78. Christl, M. and Braun, M., Photocycloadditions of benzvalene, *Angew. Chem. Int. Ed.*, 601, 28, 1989.

79. (a) Beereboom, J. J. and von Wittenau, M. S., The photochemical conversion of fumaronitrile and acetone to oxetanes, *J. Org. Chem.*, 30, 1231, 1965; (b) Turro, N. J., Wriede, P., Dalton, J. C., Arnold, D., and Glick, A., Photocycloaddition of alkyl ketones to electron-deficient double bonds, *J. Am. Chem. Soc.*, 89, 3950, 1967.

80. (a) Turro, N. J., Wriede, P., and Dalton, J. C., Evidence for a singlet-state complex in the photocycloaddition of acetone to *trans*-1,2-dicyanoethylene, *J. Am. Chem. Soc.*, 90, 3274, 1968; (b) Dalton, J. C., Wriede, P. A., and Turro, N. J., Photocycloaddition of acetone to 1,2-dicyanoethylene, *J. Am. Chem. Soc.*, 92, 1318, 1970.

81. (a) Turro, N. J. and Farrington, G. L., Quenching of the fluorescence of 2-norbornanone and derivatives by electron-rich and electron-poor ethylenes, *J. Am. Chem. Soc.*, 102, 6051, 1980; (b) Turro, N. J. and Farrington, G. L., Photoinduced oxetane formation between 2-norbornanone and derivatives with electron-poor ethylenes, *J. Am. Chem. Soc.*, 102, 6056, 1980.

Oxetane Formation:
Addition to Heterocycles

Carlos Rivas

Franklin Vargas
Instituto Venezolano de
Investigaciones Científicas

44.1 Introduction

The photoaddition of carbonyl compounds to alkenes to generate the four-membered oxetane ring was first reported by Paternò and Chieffi[1a] in Italy in 1909. They proposed the formation of a novel four-membered oxygen heterocycle, an oxacyclobutane or oxetane in a single photochemical step when benzophenone was irradiated by sunlight in the presence of 2-methyl-2-butene (Scheme 44.1).

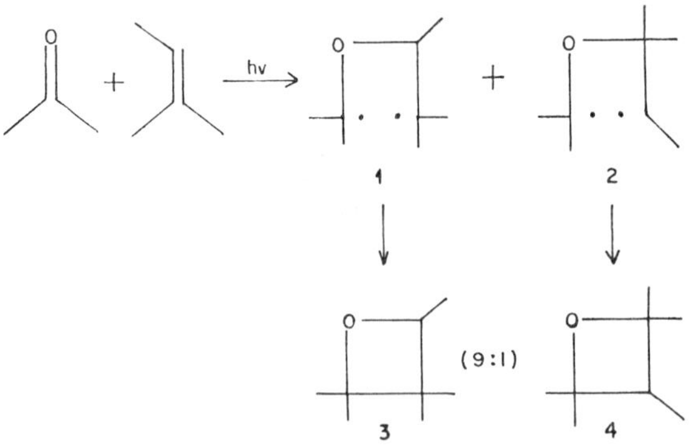

SCHEME 44.1

Although Paternò and Chieffi suggested the correct structure for the photoproduct, it was not until 1954 that Büchi and collaborators[1b] reinvestigated the reaction and confirmed unambiguously the structure proposed originally by the Italian workers. In the 1960s, the usefulness of the photocycloaddition was recognized, and many examples of synthetic value started to appear in the literature.[2] This reaction has been known since 1964 as the Paternò-Büchi reaction.[3]

0-8493-8634-9/95/$0.00+$.50
© 1995 by CRC Press, Inc.

Although a variety of alternative classical synthetic procedures[4,5] for oxetane formation are known, the Paternò-Büchi photocycloaddition became the method of choice to prepare oxetanes. On the other hand, examples of thermal addition of carbonyl compounds to alkenes remain rare, as may be inferred from the restrictions imposed by orbital symmetry on $[2+2]\pi$-cycloaddition reactions.[6]

The usefulness of this photocycloaddition in synthesis lies not only on the facility to produce the oxetane photoadducts that may have different potential applications as suggested by the number of patents registered,[7] but also by the variety of methods developed to open the oxetane ring as a means of producing other functional groups. In heterocyclic chemistry, it has been a useful method for the synthesis of 3-substituted heterocycles.[8,9]

4.2 Mechanistic Studies

In general, with rare exceptions, the light-absorbing and reacting species in these cycloadditions is the carbonyl compound, since the reaction can be brought about by irradiation in a region where only the carbonyl compound absorbs. On the other hand, absorption by the oxetane photoproduct is negligible and the reactions can go therefore to completion without the interference of side photoreactions.[9] In some instances, in heterocyclic systems it is possible that a second excited carbonyl molecule can undergo photocycloaddition to the remaining double bond in the ring and a 2:1 adduct is obtained.[10,11] In general, carbonyl compounds that undergo the photocycloaddition reaction are also photoreduced upon irradiation in isopropyl alcohol. Subject to the limitation of triplet-triplet energy transfer to the olefin or to the diene system in heterocyclic compounds, the converse is also true. Namely, carbonyl compounds that undergo photoreduction in isopropyl alcohol can afford oxetanes unless they have a triplet energy high enough for the olefin or the conjugated system in the heterocycle to behave as a quencher.[2] Thus, both reactions take place from the same excited state, namely the lowest n,π^* state.

Aliphatic carbonyl compounds often undergo oxetane formation from both the singlet and the triplet states, whereas aromatic carbonyl compounds that undergo very rapid intersystem crossing to the triplet form oxetanes exclusively from the triplet state.[9]

The fact that heterocycles may act either as efficient or inefficient quenchers can be a very important property to explain their reactivity toward excited carbonyl compounds in oxetane-forming reactions.[12] In addition to energy transfer, there are other paths for deactivation available to excited carbonyl compounds.[2] It is fairly well accepted that the immediate precursors of the oxetanes are biradicals[13] whose existence has been evidenced by purely chemical means[9] as well as by the application of picosecond spectrocopy.[14] Two mechanisms[13,15] have been proposed to explain the low-energy pathways leading to the biradical intermediate. They are (1) nucleophilic attack initiated by the half-occupied π^*-orbital of the carbonyl oxygen atom on the unoccupied π^*-orbital of an electron-deficient olefin in the plane of the molecule. This LUMO-LUMO interaction is called a "parallel approach"; and (2) an electrophilic attack initiated by the half occupied n-orbital of the carbonyl oxygen atom on the unoccupied π^*-orbital of an electron-rich olefin in a perpendicular direction to the plane of the molecule. This HOMO-HOMO interaction is called a "perpendicular approach". Thus, the immediate precursors of these intermediates are $^3\pi,\pi^*$ or n,π^* states. From their interaction with a suitable ground-state substrate, three types of biradicals can possibly be formed:[15] (1) an exciplex with the excitation localized on one of the two partners, (2) a neutral conventional biradical, or (3) an ionic biradical or radical ion pair. Scheme 44.2, adapted from Turro,[15] may serve to illustrate the above for the reaction between an n,π^* state and a heterocycle.

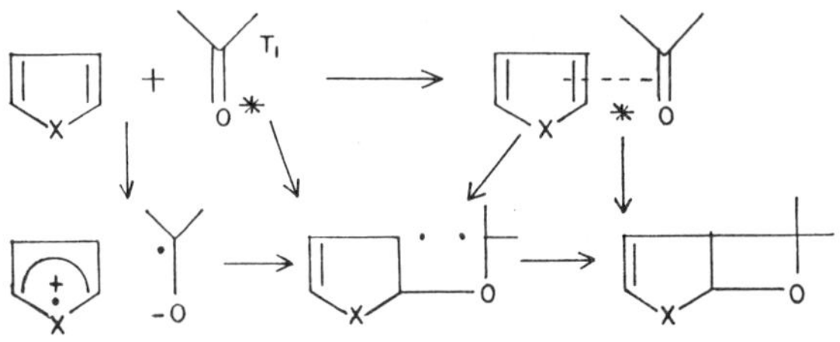

SCHEME 44.2

Even though the existence of exciplex intermediates in the Paternò-Büchi cycloaddition reaction is a controversial matter, Turro's proposal (Scheme 44.2) continues to be very useful in explaining stereochemical and mechanistic situations that are difficult to explain otherwise.

44.3 Orientation and Stereochemistry

The factors controlling the regio- and stereochemistry of triplet state photocycloadditions are not well understood. In contrast, the corresponding photoreactions via singlet states are more predictable, occurring with high stereospecificity. For example, the reaction of n,π^* states of acetone with enol ethers gives photoadducts where there is more stereospecificity in S_1 than in T_1, and there is less competing *cis-trans*-isomerizations from S_1 than from T_1. These observations are consistent with the fact that S_1 in the presence of an enol ether directly gives a singlet biradical, while a T_1 and an enol ether yield a triple biradical that has to undergo intersystem crossing (isc) before bond formation. The lack of stereospecificity in the case of the triplet biradical may be explained in terms of a long-lived intermediate that allows for the loss of stereospecificity before complete bonding is achieved.

Recently, it has been shown that the product stereoselectivity in photocycloaddition reactions of aromatic aldehydes and mono-olefins such as cyclopentene, cyclohexene, 2,3-dihydrofuran, and 2,3-dihydropyran can be attributed to the reactive conformations of the intermediate triplet 1,4-biradicals.[16] On the other hand, great importance has been given to the stability of the biradical intermediate to explain the regiochemical outcome of the reaction. One of the classical examples, where this generalization works as expected, is in the photocycloaddition of benzophenone to 2-methyl-2-butene which gives a mixture of two oxetanes 3 and 4 in the ratio of 9:1 (Scheme 44.1). This result is explained in terms of the stability of the biradical intermediate, where 1, the precursor of 3, is more stable than 2, the precursor of 4. Nevertheless, this generalization is not always applicable and, in many cases of low regiocontrol, the exciplex model can be useful to explain the course of the reaction. In the photocycloaddition reactions of excited carbonyl compounds to five-membered heterocycles, the product structure does not always conform to that expected from the most stable biradical intermediate. Carbonyl compounds, either ketones or aldehydes, usually add in such a way that the oxygen atom of the four-membered ring ends up bonded to one of the two carbon atoms next to the heteroatom in the five-membered ring (position-2) or to the carbon atom next to the heteroatom numbered lowest in heterocycles with two heteroatoms. The photocycloaddition of benzophenone to 1,2-dimethylimidazole[17] can illustrate a situation where there is ambiguity as to the relative stability of the two possible biradical intermediates that can be formed. However, calculations can be used to explain the experimental results. Thus, the energies of the two possible biradical intermediates 5 and 6 were calculated by a localization method using simple Hückel calculations. The results obtained indicated that 5 is more stable than 6. This is with the experimental results from the photocycloaddition which yields 4-α-hydroxy,diphenyl-methylimidazole) 7, a product of ring-opening of an unstable oxetane (Scheme 44.3).

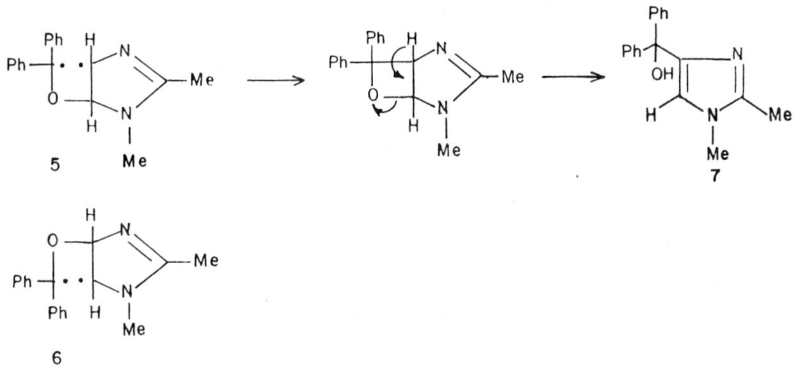

SCHEME 44.3

Griesbeck and Stadtmüller[18] have devoted a great deal of attention to the problem of the formation of regioisomers where the predominant product does not conform to that expected from the most stable biradical intermediate. They suggest that regioselectivity in these systems could be determined by the preorientation of the substrates rather than by the difference in biradical stability. This idea is in agreement with several studies regarding the regiochemistry of the Paternò-Büchi reaction[19] and with spectroscopic evidence for the involvement exciplex intermediates.[20] It is likely that the regioselectivity in heterocyclic systems, such as furans, thiophenes, selenophenes, etc., undergoing Paternò-Büchi reactions, could be fitted into this scheme whenever the products obtained cannot conform to those expected from the biradical intermediate stability rule. Stern-Volmer analysis of the ability of heterocycle, such as pyrroles and selenophenes, to quench benzophenone photoreduction in isopropyl alcohol indicates that these heterocycles have a tendency to form exciplexes.[21]

4.4 Photocycloadditions of Carbonyl Compounds to Heterocyclic Systems

The heterocyclic compound that has received more attention by far as a substrate for excited carbonyl compounds has been furan and its derivatives. The reason for this preference may be the fact that it readily reacts with the excited species to give oxetanes in good yields and in a clean photochemical process that facilitates the isolation of products and, most important of all, because of the potential applications of some of the oxetanes derived from it as the starting point in the synthesis of certain natural products.

The direct [2+2]-photocycloaddition to benzophenone to furan was first achieved by two groups of workers: Schenck, Hartman, and Steimetz[22a] in Germany and Hammond and Turro[22b] in the U.S. in 1963. The product was later unequivocally identified by ^{1}H-NMR as 6,6-diphenyl-2,7-dioxabicyclo[3,2,0]-Δ^3 heptene.[23] Thereafter, this reaction was extended to several methyl-substituted furans[24,25] and to benzofuran derivatives.[26] These in turn were used as substrates for a variety of aliphatic and aromatic carbonyl compounds.[24]

In contrast to the product mixture of two isomeric oxetanes, obtained by Toki, Shima, and Sakurai[24] from the photolysis of aldehydes such as benzaldehyde, acetaldehyde, and propionaldehyde in 2-methylfuran, which result from the addition of the carbonyl compound to either double bond of the furan, irradiations of benzophenone in 2-methyl-, 3-methyl-,[25] 3-hydroxymethyl-,[27] and 2,3-dimethyl furan[28] resulted in cycloaddition to the more substituted double bond in the ring as the only product of the reaction. All these reactions afford regiospecifically the head-to-head product, as demonstrated by NOE measurements.[28] The stereospecificity of products obtained by Sakurai according to the NMR evidence, corroborated by Whipple and Evanega,[29] was shown to be the result of an *exo* cycloaddition. Ogata et al.[10] isolated 2:1 adducts by prolonged irradiation of benzophenone in furan and 2-methylfuran, as well as by first isolating the 1:1 adduct and then photolyzing it in benzene solution in the presence of benzophenone. Soon thereafter, the *anti*-stereochemistry of the 2:1 adduct was demonstrated by other workers.[30]

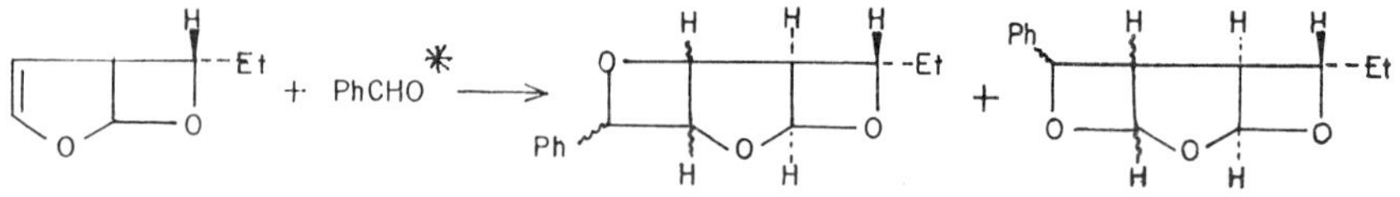

SCHEME 44.4

Mixed 2:1 adducts have been obtained by irradiating a carbonyl compound in the presence of a 1:1 adduct already prepared from furan and a different carbonyl compound.[11]

In contrast to the accepted role of each of the two reactants in Paternò-Büchi reactions, DeBoer[26a] reported evidence that the triplet state of methyl coumarilate is capable of reacting with ground-state benzophenone to yield an oxetane. A series of oxetanes have been synthesized by Kawase et al.[26b] from derivatives of benzofuran and carbonyl compounds such as benzophenone and benzaldehyde.

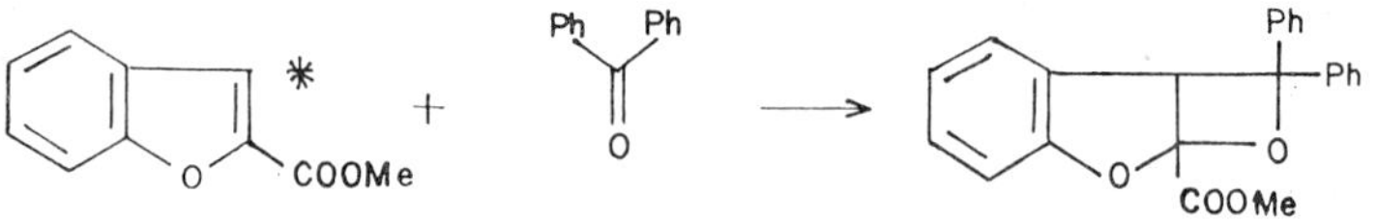

SCHEME 44.5

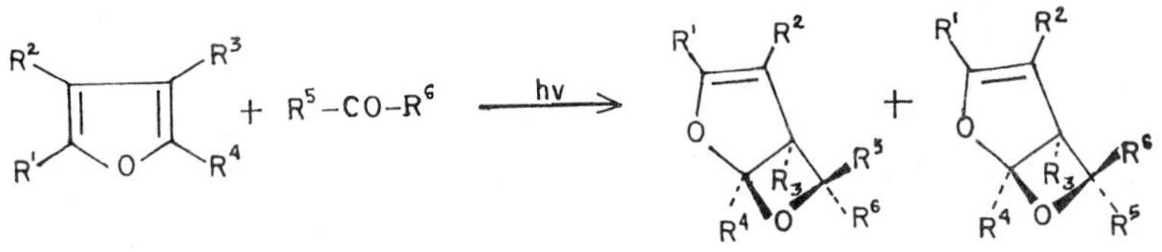

		Ref.
1)	$R^1=R^2=R^3=R^4=R^5=R^6=Ph$	22
2)	$R^1=R^2=R^3=R^4=H$; $R^5=Ph$, $R^6=4$-Me-Ph	25c
3)	$R^1=R^2=R^3=R^4=H$; $R^5=Ph$ $R^6=4MeO$-Ph	25c
4)	$R^1=R^2=R^3=R^4=H$; $R^5=Ph$, $R^6=4$-Cl-Ph	25c
5)	$R^1=R^2=R^3=R^4=H$; $R^5=Ph$, $R^6=2$-Thienyl	25c
6)	$R^1=R^2=R^3=R^4=H$; $R^5=Ph$, $R^6=Ph$	8
7)	$R^1=R^2=R^3=R^4=H$; $R^5=H$, $R^6=Me$	8
8)	$R^1=R^2=R^3=R^4=H$; $R^5=H$, $R^6=CO_2Bu$	8
9)	$R^1=R^2=R^3=R^4=H$; $R^5=H$, $R^6=CH_2OCOMe$	8
10)	$R^1=R^2=R^3=R^4=H$; $R^5=H$, $R^6=CH_2Pr^i$	8
11)	$R^1=R^2=R^3=R^4=H$; $R^5=H$, $R^6=CO_2Et$	8
12)	$R^1=R^2=R^3=R^4=H$; $R^5=H$, $R^6=2$-Furyl	8
13)	$R^1=R^2=R^3=R^4=H$; $R^5=H$, $R^6=CH=CHMe$	36
14)	$R^1=R^2=R^3=R^4=H$; $R^5=H$, $R^6=1$-Naphthyl	25c
15)	$R^1=R^2=R^3=R^4=H$; $R^5=H$, $R^6=2$-Naphthyl	25c
16)	$R^1=R^2=R^3=H$; $R^4=Me$; $R^5=R^6=Ph$	25a
17)	$R^1=R^2=R^3=H$; $R^4=Me$; $R^5=H$, $R^6=Ph$	24
18)	$R^2=R^3=R^4=H$; $R^1=Me$; $R^5=H$, $R^6=Ph$	24
19)	$R^1=R^2=R^3=H$; $R^4=Me$; $R^5=H$, $R^6=Me$	24
20)	$R^2=R^3=R^4=H$; $R^1=Me$; $R^5=H$, $R^6=Me$	24
21)	$R^1=R^2=R^3=H$; $R^4=Me$; $R^5=H$, $R^6=4$-MePh	25c
22)	$R^1=R^2=R^3=H$; $R^4=Me$; $R^5=H$, $R^6=4$-MeOPh	25c
23)	$R^1=R^2=R^3=H$; $R^4=CH_2OH$; $R^5=H$, $R^6=Ph$	25a
24)	$R^1=R^2=R^4=H$; $R^3=CH_3$; $R^5=R^6=Ph$	28
25)	$R^1=R^2=R^4=H$; $R^3=CH_2OH$; $R^5=R^6=Ph$	27
26)	$R^1=R^2=R^3=H$; $R^4=Me$; $R^5=Ph$; $R^6=4$-Cl-Ph	25c
27)	$R^1=R^2=R^3=H$; $R^4=Me$; $R^5=Ph$, $R^6=2$-Thienyl	25c
28)	$R^1=R^2=R^3=H$; $R^4=Me$; $R^5=H$, $R^6=1$-Naphthyl	25c
29)	$R^1=R^2=R^3=H$; $R^4=Me$; $R^5=H$, $R^6=2$-Naphthyl	25c
30)	$R^2=R^3=H$; $R^1=R^4=Me$; $R^5=H$, $R^6=Ph$	23
31)	$R^2=R^3=H$; $R^1=R^4=Me$; $R^5=H$, $R^6=Et$ or Ph	36
32)	$R^1=R^3=H$; $R^2=R^4=Me$; $R^5=H$, $R^6=Ph$	25b
33)	$R^2=R^4=H$; $R^1=R^3=Me$; $R^5=H$, $R^6=Ph$	25b
34)	$R^1=R^2=H$; $R^3=R^4=Me$; $R^5=H$, $R^6=Ph$	28
35)	$R^1=R^2=R^3=R^4=R^5=H$, $R^6=Et$	8
36)	$R^1=R^2=R^3=R^4=H$; $R^5=H$, $R^6=CH_2-CH_2$ CHME$_2$	8
37)	$R^1=R^2=R^3=R^4=H$; $R^5 =$ furyl-CO, $R^6=$furyl	35
38)	$R^1=R^2=R^3=R^4=H$; $R^5C_2H_{50}$-,H, $R^6=CO_2$ C_2H_5	35
39)	$R^2=R^3=R^4=H$; $R^1=CH(OAc)_2$; $R^5=H$; $R^6=Ph$	37
40)	$R^1=R^2=R^3=H$; $R^4=CH(OAc)_2$; $R^5=H$; $R^6=Ph$	37
41)	$R^1=R^2=R^3=R^4=H$; $R^5=R^6=$Cyclopentyl	8d

SCHEME 44.6

More interesting than the photocycloaddition of aromatic ketones to furan has been the photocycloaddition of aldehydes to this heterocycle.[24] The importance of the oxetanes from aldehyde-furan photocycloadditions is illustrated by a number of natural product syntheses in which these compounds represent very valuable intermediates.[31] Among the natural products syntheses involving oxetane precursors are those of avenaciolide[32] and asteltoxin.[33]

The inspiration for the application of the Paternò-Büchi photocycloaddition of aldehydes to furan derivatives in this elegant synthesis of natural products originated with the work of Sakurai et al.,[24] who showed that these reactions give rise to *exo*-substituted photoadducts with a high degree of regio- and stereoselectivity. It was recognized by Schreiber that in this reaction, the furan can be considered an equivalent to a *Z*-enolate and that the photoadduct can be viewed as a protected aldol. The outcome of the overall reactions can be a stereoselective aldol condensation, a very important reaction in organic synthesis. Hydrolysis of the product yields an acyclic chain with a *threo*-configuration.

Hambelek and Just[32a] have developed methods for the chemical conversion of furan-aldehyde adducts (R-CHO: R = Ph, Pr^i, CH$_2$OTB DMS or CH$_2$OB$_z$) into trisubstituted monocyclic oxetanes that are potential intermediates in natural product synthesis. These authors have synthesized an oxetane from 2-methylfuran and the aldehyde R-COO-CH$_2$CHO that was used as the starting point in the synthesis of racemic oxetanocin.[32b] It is interesting that when ketones of the type RCOCN (R = C$_6$H$_4$(CO)$_2$N) photo-add to furan, the ratios of *endo-exo*-isomers obtained are temperature dependent.[33] The photocycloaddition of phenyl glyoxalates of different sugars to furan has been used to induce asymmetric induction in the Paternò-Büchi reaction.[34]

44.5 The Paternò-Büchi Reaction in Heterocyclic Systems Other than Furan

In contrast to furan and its methyl derivatives that react readily with excited carbonyl compounds to yield oxetanes, thiophene, pyrrole, selenophene, and other heterocycles with more than one heteroatom in the ring were until the late 1960s thought to be inert to oxetane forming reactions. This view was reflected in the following statement, "Furan is unique in its photochemical reactivity with benzophenone in that other heterocyclic compounds (thiophene, pyrrole, isoxazole or oxazole) do not undergo analogous oxetane formation".[30b] Attempts to synthesize oxetanes from thiophene, pyrrole, and *N*-methyl-pyrrole were unsuccessful.

In the course of an investigation on the photoaddition of 2,3-dimethylmaleic anhydride to several thiophene derivatives sensitized by benzophenone, it was found that when 2,5-dimethylthiophene was used as the substrate, in addition to a small amount of the expected cycloaddition product, another crystalline compound was obtained in good yield.[39] The latter was identified by spectroscopic methods as an oxetane. In order to study the scope of the cycloaddition reaction, carbonyl compounds such as 1- and 2-naphthaldehyde, benzaldehyde, 2-, 3-, and 4-benzoylpyridine, 1-benzoylpyrrole, acetophenone, and 2-benzoylthiophene, respectively, were irradiated in the presence of 2,5-dimethylthiophene. 2,3-Dimethyl,[40] 3,4-dimethyl-, and 2,3,5-trimethylthiophene were also used as substrates for excited state additions (?) by benzophenone. Of the carbonyl compounds irradiated, only 1-naphthaldehyde, 2-, 3-, and 4-benzoylpirydine and 2-benzoylthiophene were shown to react with 2,5-dimethylthiophene to yield oxetanes.[41]

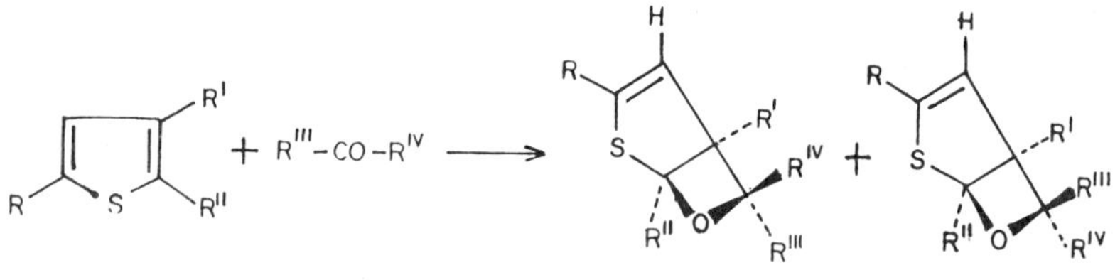

a) R = R^{II} = Me ; R^I = H; R^{III} = R^{IV} = Ph

b) R = R^{II} = Me ; R^I = H; R^{III} = H ; R^{IV} = Ph

c) R = R^{II} = Me ; R^I = H; R^{III} = 2-PYRIDYL; R^{IV} = Ph

d) R = R^{II} = Me ; R^I = H; R^{III} = 3-PYRIDYL; R^{IV} = Ph

e) R = R^{II} = Me ; R^I = H; R^{III} = 4-PYRIDYL; R^{IV} = Ph

f) R = R^{II} = Me ; R^I = H; R^{III} = 2-THYENYL; R^{IV} = Ph

g) R = H ; R^I = R^{II} = Me; R^{III} = R^{IV} = Ph

SCHEME 44.7

It is interesting to note that all unsymmetrical (R^I-CO-R) carbonyl compounds that yield oxetanes give a mixture of two geometrical isomers as demonstrated by the NMR spectra of the compounds obtained.

Furthermore, of the di- and trimethylthiophene derivatives used as substrates for excited benzophenone, only 2,3-dimethylthiophene affords an oxetane in fairly good yield.[40] In this case, as with the other heterocycles studied, benzophenone adds to the more substituted double bond in the ring.

Selenophene itself does not give oxetanes in the presence of excited benzophenone. However, substitution of one methyl group in the ring suffices to make it reactive toward benzophenone upon irradiation. Thus, 2-methylselenophene, under these conditions affords an oxetane in good yield.[42] 3-Methyl-[43] and 2,5-dimethyl-[43] selenophene are also reactive toward excited benzophenone, yielding the corresponding oxetanes.

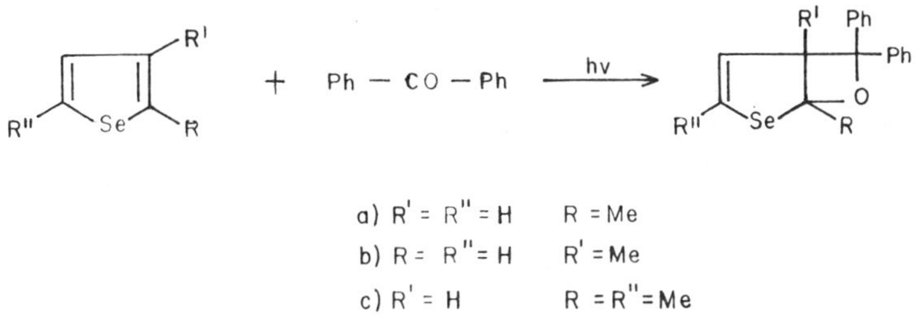

a) R^I = R^{II} = H   R = Me
b) R = R^{II} = H   R^I = Me
c) R^I = H   R = R^{II} = Me

SCHEME 44.8

Tellurophene[43,44] decomposes when attempts have been made to use it as a substrate for excited benzophenone. Perhaps, the tellurium atom is too large and it is extruded from the ring before it has a chance to react. Nevertheless, tellurophene has been used in [3+4]-cycloaddition reactions with α,α-dibromoketones in the presence of iron nonacarbonyl, thermally and photochemically affording the corresponding stable [3+4] adduct.[44]

Silacyclopentadienes (siloles) have recently attracted wide attention from a photochemical standpoint because they undergo dimerizations and cycloadditions with 1,1-dimethoxyethene. Irradiation of benzophenone and 1,1-dimethyl-2,5-diphenylsilacyclopentadiene yields two oxetanes by a SET mechanism. It was shown that the silole acts as an effective electron donor.[45]

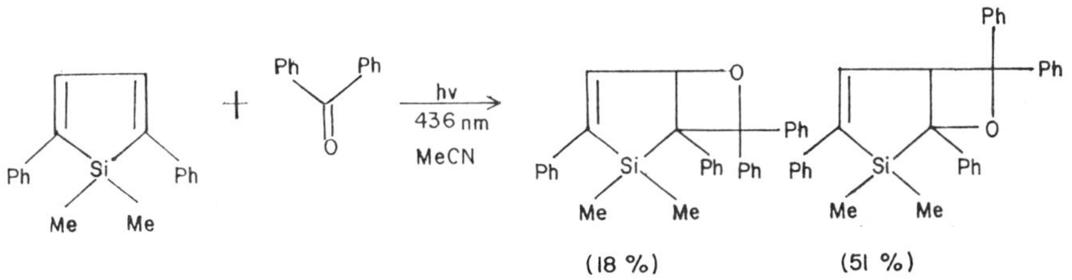

SCHEME 44.9

Pyrrole and *N*-methylpyrrole do not react with excited benzophenone. Nevertheless, *N*-phenylpyrrole yields a hydroxy compound that could have an unstable oxetane as its precursor.[46] The photocycloaddition of benzophenone to pyrrole to yield an oxetane was only possible when an electron-attracting group (benzoyl) was bonded to the nitrogen atom in the ring.[46,47] As shown in Scheme 44.10 the product obtained is the 2:1 adduct. Other electron-attracting groups such as acetyl, benzenesulfonyl, and carbomethoxy were also tested, but no success was attained as far as deactivating the aromatic system of pyrrole. A benzyl group was also observed to have no effect.

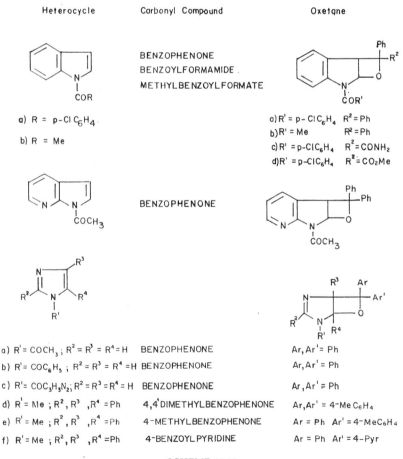

SCHEME 44.10

In general, the oxetanes obtained from nitrogen heterocycles, not substituted with electron-attracting groups, are unstable and the only products isolated are alcohols resulting from ring opening. A typical example of this came from the work of Matsura et al.[17] who attempted the photocycloaddition of benzophenone and acetophenone to 1,2-dimethylimidazole. This results in the formation of products in low yield. The precursor of these ring-opened compounds is probably the unstable oxetane (Scheme 44.3).

By protecting the N-atom with electron-withdrawing groups, oxetanes from various nitrogen heterocycles such as indole,[49] azaindole,[50] and imidazole[51,52] were prepared and examples of the synthetic utility are shown in Scheme 44.11.

Heterocycle	Carbonyl Compound	Oxetane

(indole structure, R = COR)

BENZOPHENONE
BENZOYLFORMAMIDE
METHYLBENZOYLFORMATE

a) R = p-Cl C$_6$H$_4$

b) R = Me

a) R' = p-Cl C$_6$H$_4$ R^2 = Ph
b) R' = Me R^2 = Ph
c) R' = p-ClC$_6$H$_4$ R^2 = CONH$_2$
d) R' = p-ClC$_6$H$_4$ R^2 = CO$_2$Me

(azaindole structure, COCH$_3$)

BENZOPHENONE

(oxetane structure, COCH$_3$)

(imidazole structure R^3, R^2, R^4, R')

a) R' = COCH$_3$; R^2 = R^3 = R^4 = H BENZOPHENONE Ar, Ar' = Ph

b) R' = COC$_6$H$_5$; R^2 = R^3 = R^4 = H BENZOPHENONE Ar, Ar' = Ph

c) R' = COC$_3$H$_3$N$_2$; R^2 = R^3 = R^4 = H BENZOPHENONE Ar, Ar' = Ph

d) R' = Me ; R^2, R^3, R^4 = Ph 4,4'DIMETHYLBENZOPHENONE Ar, Ar' = 4-Me C$_6$H$_4$

e) R' = Me ; R^2, R^3, R^4 = Ph 4-METHYLBENZOPHENONE Ar = Ph Ar' = 4-MeC$_6$H$_4$

f) R' = Me ; R^2, R^3, R^4 = Ph 4-BENZOYLPYRIDINE Ar = Ph Ar' = 4-Pyr

SCHEME 44.11

In addition to imidazoles, other heterocycles with two heteroatoms have been tested for oxetane formation: 2,4-dimethylthiazole 3,5- and 4,5-dimethylisoxazole afforded oxetanes[47,48] in good yield when irradiated in the presence of excited benzophenone. 4-Methylisothiazole does not yield an oxetane when irradiated with benzophenone but instead yields a hydroxyphenyl derivative presumably via a hydrogen abstraction, radical combination pathway.

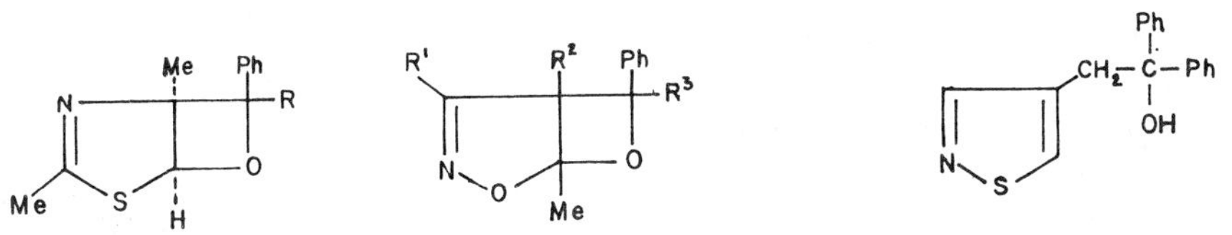

a) R = Ph

b) R = C_5H_4N (Pyr)

a) R' = Me ; R² = H ; R³ = Ph

b) R' = Me ; R² = H ; R³ = Pyr

c) R' = H ; R² = Me ; R³ = Ph

SCHEME 44.12

Photoadditions of chiral ketoesters to oxazole afford oxetanes that are cleaved with methanol to generate precursors of *erythro* sugar derivatives.[54]

Vinylene carbonate[55] has also been used successfully in photocycloaddition reactions to carbonyl compounds as illustrated by the synthesis of the branched-chain sugar, D,L-apiose, in which the first step is the formation of an oxetane by reacting vinylene carbonate with diacetoxyacetone. Treatment of the oxetane with base brings about deacetylation and ring opening via a hemiacetal.

There are very few heterocyclic compounds with three heteroatoms that undergo the Paternò-Büchi reaction. Bentrade and Darnall[56] showed that acetone and perdeuteroacetone add photochemically to phospholen. The initial adduct is thought to be an oxetane that rearranges to the isolated product.

The contrast between furan and its methyl derivatives as good substrates for Paternò-Büchi reactions with the other five-membered heterocycles investigated thus far that are either inert or poor substrates has been emphasized throughout this discussion. An attempt has been made to correlate inertness with their quenching activity on the photoreduction of benzophenone since the excited state of the compound involved is the same for both types of reaction. In fact, Stern-Volmer plots[21] (ϕ_o/ϕ_q vs. Q) for thiophene and its monomethylderivatives give linear plots, which indicate that thiophene is a good quencher and, furthermore, its monomethyl derivatives are even more efficient than the parent compound. However, 2,5-dimethylthiophene turns out to be a very poor quencher. These results could answer the question, since substitution of electron-donating groups to a conjugated diene system should raise its lowest triplet energy level to a higher value above that of the ground state of the unsubstituted diene. Thus, substitution of one methyl group in the thiophene nucleus raises its lowest triplet energy level T_1 at the thiophene closer to that of benzophenone, but still not high enough to change it from a quencher to an acceptor of the excited ketone for cycloaddition to take place. Substitution of two methyl groups as in the case of 2,5-dimethylthiophene raises T_1 of the thiophene to a higher level than that of T_1 of benzophenone; energy transfer then becomes inefficient and the reaction takes place; 2,3-dimethylthiophene also gives oxetane as expected. The Stern-Volmer plots for pyrroles and selenophenes as quenchers of photoreduction of benzophenone are curved and have the shape of a segment of a parabola.[21] This result indicates that, in addition to the triplet state of benzophenone, the heterocycle is quenching another excited state that may presumably be an exciplex between T_1 of benzophenone and the heterocycle itself. In the case of linear Stern-Volmer plots the poorest quenchers are the best substrates for excited ketones and the opposite is observed for compounds such as the selenophenes which give parabola-like Stern-Volmer plots.[21]

44.6 Conclusions

1. Furan and its methyl derivatives behave as a system of two double bonds capable of accepting one or two molecules of carbonyl compound to form a single or a double oxetane according to the conditions of the reaction.

2. In methyl-substituted furans, thiophenes, and selenophenes the excited carbonyl compounds react preferentially at the double bond with the higher substitution when carbonyl compounds are used with triplet energy (T_1) similar to benzophenone. With carbonyl compounds of higher triplet energy, such as benzaldehyde or propionaldehyde, the reaction may occur on either double bond.

3. Thiophene and its monomethyl-derivatives, pyrrole and its monomethyl derivatives, selenophene, and tellurophene are inert to the reaction. Substitution of one methyl group in selenophene, two methyl groups in thiophene, and protection of the N-atom with electron-attracting groups in pyrrole and other nitrogen heterocycles make these heteroaromatics reactive to the [2+2]-cycloaddition.

4. Reactivity of these heterocycles can be correlated to their ability as quenchers of photoreduction of benzophenone.

5. Yields of the oxetane-forming reaction follow the rule pyrrole < thiophene < furan (20, 50, 80% average yield).

6. Addition of the carbonyl compound to double bonds in the ring always takes place in a *cis*-manner, and the oxygen of the oxetane ring is always located next to the carbon atom adjacent to the heteroatom in the five-membered ring. Addition of a second carbonyl molecule to form a 2:1 adduct is not regioselective.

7. There is still considerable scope for research on heterocycles in general and especially on those different from furan such as thiophenes, pyrroles, selenophenes, phospholes, arsoles, siloles, etc.

References

1. (a) Paterno, E. and Chieffi, G., Synthesis in organic chemistry by means of light. II. Compounds of the unsaturated hydrocarbons with aldehydes and ketones, *Gazz. Chim. Ital.*, 39, 341, 1909; *Chem Abstr.*, 5, 681, 1911; (b) Büchi, G., Inman, C. G., and Lipinski, E. S., Light catalyzed organic reactions. I. The reaction of carbonyl compounds with 2-methyl-2-butene in the presence of ultraviolet light, *J. Am. Chem. Soc.*, 76, 4327, 1954.

2. Arnold, D. R., The photocycloaddition of carbonyl compounds to unsaturated systems: The syntheses of oxetanes, in *Advances in Photochemistry*, Vol. 6, Noyes, W. A., Jr., Hammond, G. S., and Pitts, J. N., Jr., Eds., Interscience, New York, 1968.

3. Yang, N. C., Photochemical Reactions of Carbonyl Compounds in Solution: Paternò-Büchi Reaction, *Pure and Appl. Chem.*, 6, 591, 1964.

4. Searles, S., Jr., Compounds with three and four-membered rings, in *The Chemistry of Heterocyclic Compounds*, Weissberger, A., Ed., Vol. 19, Part II, Interscience, New York, 1964, chap. 9.

5. (a) Still, W. C., Allyloxycarbanions, cyclization to vinyloxetanes, *Tetrahedron Lett.*, 2115, 1976; (b) Biggs, J. B., Three new convenient preparations of oxetane, *Tetrahedron Lett.*, 4285, 1975.

6. Zimmerman, H. E., The Möbius-Hückel concept in organic chemistry. Application to organic molecules and reactions, *Acc. Chem. Res.*, 4, 272, 1971.

7. Jones, G., II, Synthetic applications of the Paternò-Büchi reaction, in *Organic Photochemistry*, Vol. 5, Padwa, A., Ed., Marcel Decker, New York, 1981, 1.

8. (a) Zamojski, A. and Kosluk, T., Synthesis of 3-substituted furans, *J. Org. Chem.*, 42, 1089, 1977; (b) Kitamura, T., Kawakami, Y., Imagagua, T., and Kawanisi, N., One pot synthesis of 3-substituted furan: A synthesis of perillaketone, *Syn. Commun.*, 7, 521, 1977; (c) Jones, G., II, Gilow, H. M., and

Low, J., Regioselective photoaddition of pyrroles and aliphatic carbonyl compounds. A new synthesis of 3(4)-substituted pyrroles, *J. Org. Chem.*, 44, 2949, 1979; (d) Cantrell, T. S., Allen, A. C., and Ziffer, H., Photochemical 2+2 cycloaddition of arenecarboxylic acid esters to furans and 1,3-dienes. 2+2 Cycloreversion of oxetanes to dienol esters and ketones, *J. Org. Chem.*, 54, 140, 1989; (e) Liu, W. H. and Wu, H. J., Synthesis of furanoid natural products, *J. Chin. Chem. Soc., (Taipei)*, 35, 241, 1988; *Chem. Abstr.*, 110, 94846, 1989; (f) Bolivar, R. A., Tasayco, M. L., Rivas, C., and Leon, V., Photocycloaddition of benzophenone to α-Angelica lactone and thermal rearrangement of the product, *J. Heterocyclic Chem.*, 20, 205, 1983.

9. Carless, H. A. J., Carbonyl compounds: Cycloaddition, in *Photochemistry in Organic Synthesis*, Coyle, J. D., Ed., The Royal Society of Chemistry, London, 1986, 96.

10. Ogata, M., Watanabe, H., and Kano, H., Photochemical cycloaddition of benzophenone to furan, *Tetrahedron Lett.*, 533, 1967.

11. Itokawa, H., Matsumoto, H., Oshima, T., and Mihashi, S., Photocycloaddition of benzaldehyde to 2,3-dihydrofuran derivatives, *Yukugaku Zasshi*, 107, 767, 1987; *Chem. Abst.*, 108, 186609, 1988.

12. Bolivar, R. A. and Rivas, C., Quencher effect of thiophene and its monomethylderivatives on photoreduction and photocycloaddition reactions of ketones, *J. Photochem.*, 19, 95, 1982.

13. Kopecky, J., *Organic Photochemistry. A Visual Approach*, CCH Publishers, New York, 1992, 126.

14. Freilich, S. C. and Peters, K. S., Picosecond dynamics of the Paternò-Büchi reaction, *J. Am. Chem. Soc.*, 107, 3819, 1985.

15. Turro, N. J., "Modern Molecular Photochemistry", Benjamin-Cummings, Menlo Park, Ca, 1978.

16. Griesbeck, A. G. and Stadtmüller, S., Photocycloaddition of benzaldehyde to cyclic olefins: electronic control of endo stereoselectivity, *J. Am. Chem. Soc.*, 112, 1281, 1990.

17. Matsuura, T., Banba, A., and Ogura, K., Photoinduced reactions-XLV, photoaddition of ketones to imidazoles, *Tetrahedron*, 27, 1211, 1971.

18. Griesbeck, A. G. and Stadtmüller, S., Electronic control of stereoselectivity in photocycloaddition reactions. 4. Effects of methyl substituents at the donor olefin. *J. Am. Chem. Soc.*, 113, 6923, 1991.

19. (a) Yang, N. C. and Eisenhardt, W. J., On the mechanism of Paternò-Büchi reaction of alkanals, *J. Am. Chem. Soc.*, 93, 1277, 1971; (b) Carless, H. A. J., Photocycloaddition of acetone to acylic olefins, *Tetrahedron Lett.*, 3173, 1973; (c) Carless, H. A. J., Maitra, A. K., and Trivedi, H. S., Photochemical cycloadditions of aldehydes to styrenes, *J. Chem. Soc., Chem. Commun.*, 984, 1979; (d) Schore, N. E. and Turro, N. J., Mechanism of the interaction of n,π* excited alkanones with electron rich ethylenes, *J. Am. Chem. Soc.*, 97, 2482, 1975.

20. (a) Kochevar, I. E. and Wagner, P. J., Quenching of triplet phenylketones by olefins, *J. Am. Chem. Soc.*, 94, 3859, 1972; (b) Caldwell, R. A., Sovocool, G. W., and Gajewski, R. P., Primary interaction between diaryl ketone triplets and simple alkenes isotope effects, *J. Am. Chem. Soc.*, 95, 2549, 1973; (c) Caldwell, R. A. and Creed, D., Exciplex intermediates in [2+2] photocycloadditions, *Acc. Chem. Res.*, 13, 45, 1980.

21. Bolivar, R. A., Machado, R., Montero, L., Vargas, F., and Rivas, C., Quencher effect of five-membered ring heterocycles on the photoreduction and photocycloaddition reactions of benzophenone, *J. Photochem.*, 22, 91, 1983.

22. (a) Schenck, G. O., Hartmann, W., and Steinmetz, R., Vierringsynthesen durch photosensibilisierte cycloaddition von dimethylmaleinsäure an olefine, *Chem. Ber.*, 96, 498, 1963; (b) Hammond, G. S. and Turro, N. J., Organic photochemistry. The study of photochemical reactions provides new information on the excited states of molecules, *Science*, 142, 1541, 1963.

23. Gagnaire, D. and Payo-Subiza, E., Couplages des protons d'un methyle a travers cinq liaisons en resonance magnetique nucleaire: Ethers vinyliques heterocycliques, *Bull. Soc. Chim., France*, 2623, 1963.

24. (a) Toki, S., Shima, K., and Sakurai, H., Organic photochemical reactions. I. The synthesis of substituted oxetanes by photoaddition of aldehydes to furans, *Bull. Chem. Soc. Jpn.*, 38, 760, 1965; (b) Shima, K. and Sakurai, H., Organic photochemical reactions. IV. Photoaddition reactions of various carbonyl compounds to furan, *Bull. Chem. Soc., Jpn.*, 39, 1806, 1966.

25. (a) Rivas, C. and Payo, E., Synthesis of oxetanes by photoaddition of benzophenone to furans, *J. Org. Chem.*, 32, 2918, 1967; (b) Rivas, C., Payo, E., Mantecon, J., and Cortés, L., Photochemical cycloadditions of organic compounds *Photochem. Photobiol.*, 7, 807, 1968; (c) Rivas, C., Bolivar, R. A., and Cucarella, M., Photoadditions of carbonyl compounds to five-membered ring heterocycles, *J. Heterocyclic Chem.*, 19, 529, 1981.

26. (a) De Boer, C., Oxetane formation from excited state olefin and ground state ketone, *Tetrahedron Lett.*, 4977, 1971; (b) Kawase, Y., Yamaguchi, S., Ochimi, H., and Horita, H., The photochemical reaction of benzofuran derivatives with benzophenone or benzaldehyde, *Bull. Chem. Soc. Jpn.*, 47, 2660, 1974.

27. Bolivar, R. A., Döppert, K., and Rivas, C., A carbon-13 NMR assignment study of a series of oxetanes derived from carbonyl compounds and furan and thiophene derivatives, *J. Heterocyclic Chem.*, 19, 317, 1982.

28. Nakano, T., Rivas, C., Perez, C., and Tori, K., Configuration and stereochemistry of photochemical products by application of nuclear overhauser effect. Adducts of benzophenone to methyl-substituted furans and 2,5-dimethylthiophene, and of methyl-substituted maleic anhydrides to thiophene and its methyl derivatives and benzo b-thiophene, *J. Chem., Soc., Perkin Trans.*, 2322, 1973.

29. Whipple, E. B. and Evanega, G. R., The assignment of configuration to the photoaddition products of unsymmetrical carbonyls to furan using pseudo-contact shifts, *Tetrahedron*, 24, 1299, 1968.

30. (a) Leitich, J., Comments on the paper, Photochemical cycloadditions of benzophenone to furans, *Tetrahedron Lett.*, 1937, 1967; (b) Evanega, G. R. and Whipple, E. B., The photochemical addition of benzophenone to furan, *Tetrahedron Lett.*, 2163, 1967.

31. (a) Schreiber, S. L., [2+2]-Photocycloadditions in the synthesis of chiral molecules, *Science*, 227, 857, 1985; (b) Schreiber, S. L. and Hoveida, H. A., Synthetic studies of the furan-carbonyl photocycloaddition reaction. A total synthesis of (±)-avenaciolide, *J. Am. Chem. Soc.*, 106, 7200, 1984; (c) Schreiber, S. L. and Satake, K., Total synthesis of (±)-asteltoxin, *J. Am. Chem. Soc.*, 106, 4186, 1984.

32. (a) Hambalek, R. and Just, G., Trisubstituted oxetanes from 2,7-dioxabicyclo-[3,2,0] hept-3-enes, *Tetrahedron Lett.*, 31, 4693, 1990; (b) Hambalek, R. and Just, C., A short synthesis of (±)oxetanocin, *Tetrahedron Lett.*, 31, 5445, 1990.

33. Zagar, C. and Scharf, H. D., The Paternò-Büchi reaction of achiral and chiral acylcyanides with furan (chiral induction in photochemical reactions XIII), *Chem. Ber.*, 124, 967, 1991.

34. Pelzer, P., Julten, P., and Scharf, H. D., (Chirale Induction bei Photochemischen Reaktionen X) Isoselectivität bei der Asymetrischen Paternò-Büchi — Reaktion unter Verwendung von Kohlenhydraten als Chirale Auxiliare, *Chem. Ber*, 122, 487, 1989.

35. Al-Jalal, N. A., Ijam, M. J., Al-Omran, F., and Gopalakrishnan, B., Photoaddition of furil and diethyloxalate to electron-rich olefins, *J. Photochem. Photobiol, A. Chem.*, 55, 339, 1991.

36. (a) Schreiber, S. L., Hoveyda, A. H., and Wu, H. J., A photochemical route to the formation of *threo* aldols, *J. Am. Chem. Soc.*, 105, 660, 1983. (b) Paleta, O., Suobada, J., and Deket, V., UV-light induced cycloaddition of 3,4-dichloro-3,4,4-trifluoro-2-butanone to 2-methyl-2-butene, *Collect. Czech., Chem. Commun.*, 43, 2932, 1978.

37. Sekretar, S., Ruda, J., and Stibranyi, L., Photochemical Reactions of 2-Substituted Furans with Carbonyl Compounds, *Top Furan Chem., Proc. Symp Furan Chem.*, 4th., 1983, 98, *Chem. Abstr.*, 101, 129950, 1984.

38. Rivas, C., Vélez, M., and Crescente, O., Synthesis of an oxetane by photoaddition of benzophenone to a thiophene derivative, *J. Chem. Soc., Chem. Commun.*, 1474, 1970.

39. Rivas, C., Pacheco, D., Vargas, F., and Ascanio, J., Synthesis of oxetanes by photoaddition of carbonyl compounds to thiophene derivatives, *J. Heterocyclic Chem.*, 18, 1065, 1981.

40. Rivas, C. and Bolivar, R. A., Synthesis of oxetanes by photoaddition of carbonyl compounds to 2,5-dimethylthiophene, *J. Heterocyclic Chem.*, 10, 967, 1973.

41. Rivas, C., Pacheco, D., and Vargas, F., Synthesis of an oxetane by photoaddition of benzophenone to a selenophene derivative, *Acta SudAmer. de Quim.*, 2, 1, 1982; *Chem. Abstr.*, 99, 38394, 1983.

42. Rivas, C. and Vargas, F., unpublished results.

43. Pacheco, D., Vargas, F., and Rivas, C., Thermal and photochemical reactions between, α-dibromoketones and five-membered heterocycles in the presence of iron nonacarbonyl, *Phosphorus and Sulfur*, 25, 245, 1985.

44. Kyushin, S., Ohkura, Y., Nakadaira, Y., and Ohashi, M., Novel photo-induced reactions of 1,1-dimethyl-2,5-diphenylsilacyclopentadiene with benzophenone derivatives, *J. Chem. Soc., Chem. Commun.*, 718, 1990.

45. Rivas, C., Vélez, M., Cucarella, M., Bolivar, R. A., and Flores, S. E., Síntesis de un oxetano por fotoadición de la benzofenona a un derivado Pirrólico, *Acta Cient. Venezolana*, 22, 145, 1971; *Chem. Abstr.*, 114279t, 1972.

46. Rivas, C. and Bolivar, R. A., Synthesis of oxetanes by photoaddition of carbonyl compounds to pyrrole derivatives, *J. Heterocyclic Chem.*, 13, 1037, 1976.

47. Julian, D. R. and Tringham, G. D., Photoaddition of ketones to indoles synthesis of oxeto 2–3b indoles, *J. Chem. Soc., Chem. Commun.*, 13, 1973.

48. Nakano, T. and Santana, M., Photoaddition of benzophenone to azaindole: Synthesis of the oxetane of 7-azaindole, *J. Heterocyclic Chem.*, 13, 585, 1976.

49. Nakano, T., Rivas, C., Perez, C., and Larrauri, J. M., Photoaddition of ketones to imidazoles: Synthesis of oxetanes, *J. Heterocyclic Chem.*, 13, 175, 1976.

50. Ito, Y., Ji-Ben, M., Suzuki, S., Kusumaga, Y., and Matsuura, T., Efficient photochemical oxetane formation from 1-methyl-2,4,5-triphenylimidazole and benzophenones, *Tetrahedron Lett.*, 26, 2093, 1985.

51. Nakano, T., Rodriguez, W., de Roche, S. Z., Larrauri, J. M., Rivas, C., and Perez, C., Photoaddition of ketones to imidazoles, thiazoles, isothiazoles and isooxazoles: Synthesis of their oxetanes, *J. Heterocyclic Chem.*, 17, 1777, 1980.

52. Weuthen, M., Scharf, H. D., Runsink, J., and Vassen, R., Chiral induction in photochemical reactions. VIII. Diastereoselective photo-aldol reaktion von chiralen ketoestern mit 2,3-dihydrooxazolen, *Chem. Ber.*, 121, 971, 1988.

53. Araki, Y., Nagasawa, J., and Ishido, Y., Synthetic studies of carbohydrate derivatives by photochemical reactions. 16. Synthesis of DL-apiose derivatives by photochemical cycloaddition of 1,3-dihydroxypropan-2-one derivatives with ethanediol or ethanol derivatives, *J. Chem. Soc., Perkin Trans. 1*, 12, 1981.

54. Bentrade, W. C. and Darnall, K. R., Oxetane intermediate in the photocondensation of acetone with 2,2,2-trimethoxy-4 5-dimethyl-1,3,2-dioxaphospholene, *J. Chem. Soc., Chem. Commun.*, 862, 1969.

45

Oxetane Formation: Stereocontrol

Axel G. Griesbeck
Institut für Organische Chemie,
University of Cologne

45.1 Introduction

The combination of two prostereogenic substrate molecules, i.e., the carbonyl and the alkene, in the course of a Paternò-Büchi reaction leads to a photoadduct with three new stereogenic centers. Control of the relative and absolute configuration is a challenge for synthetic chemistry in that many interesting products could be derived in principle from oxetane precursors. A detailed knowledge of the photophysical properties of the electronically excited compound (which is in most cases the carbonyl addend) is necessary to understand (and predict) the stereochemical result of such a [2+2]-cycloaddition reaction. Therefore, the configuration of the excited state, its lifetime, and rate of internal conversion as well as intersystem crossing capabilities should be known. For clean transformations, the carbonyl group should be the only absorbing chromophore in the reaction mixture (i.e., the product should not absorb at the wavelength used), the solvent should not interfere with the cycloaddition step by competing reactions (e.g., hydrogen abstraction), and the polarity influence of the reaction medium on biradical or photoinduced single electron transfer (SET) steps ought to be carefully investigated.

0-8493-8634-9/95/$0.00+$.50
© 1995 by CRC Press, Inc.

5.2 The Carbonyl Site

Parallel and Perpendicular Approach

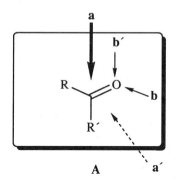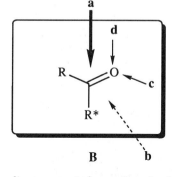

enantiotopic faces a,a′ and b,b′   *diastereotopic* faces a,b and c,d

The simple model for the spatial distribution of electron density in the n,π* state shows that a non-prostereogenic n,π* excited carbonyl compound (A) has two pairs of enantiotopic faces for inter-action with an alkene addend. Analogously to a ground-state nucleophilic addition reaction, considering the Umpolung effect, electron-deficient alkenes should interact preferentially with the nucleophilic π*-orbital. Such an orientation has been named the "parallel approach." On the other hand, electron-rich alkenes should interact preferentially with the electrophilic n-orbital perpen-dicular to the π-plane. Consequently, such an orientation has been named the "perpendicular approach." Analyses of the product stereochemistry cannot uncover these primary orientation phenomena. A study of Stern-Volmer kinetics of several 2-norbornanones with the electron-rich cis-diethoxyethylene (c-DEE) and the electron-poor trans-dicyanoethylene (t-DCE) as quenchers did corroborate this model.[1] In these cases, the carbonyl compounds are chiral (B) and exhibit two pairs of diastereotopic faces for interaction with the alkene addends. The 2-norbornanone skeleton could be attacked by t-DCE via a parallel approach from the exo- or from the endo-face. From the fluorescence quenching data in Table 45.1, it can be seen that the exo-blocked compound 2 is quenched five times more slowly than the parent compound 1. The quenching by c-DEE, however, is not affected by groups blocking the π-plane of the carbonyl group. If the two diastereotopic faces of the perpendicular n-plane are blocked additionally by methyl groups (e.g., in compound 3) the rate constant for c-DEE quenching drops by a factor of >50, whereas the rate constant for t-DCE is now not affected. In an additional paper, it was shown by Turro and Farrington that the t-DCE addition is a highly stereoselective process, indicating an exclusive interaction with the first excited singlet state of the carbonyl compound.[2] An increase in steric hindrance toward the approach from the exo-side reduced the rate of fluorescence quenching; however, the efficiency of oxetane forma-tion was increased. Therefore, it was interpreted that for sterically more hindered ketones, a possible exciplex intermediate is much more effectively transformed into the photoproduct. This may be due to a puckering effect of the n,π*-excited carbonyl group that leads to enhanced efficiency of photoproduct formation from the exciplex intermediate.

Non-Prostereogenic Carbonyls

Ketones with homotopic faces have been used widely to determine the stereoselectivity in Paternò-Büchi reactions with E- and Z-isomers of electron-rich and electron-poor olefins. The

Table 45.1   Fluorescence Quenching of 2-Norbornanone
Singlets by *trans*-DCE and *cis*-DEE [1]

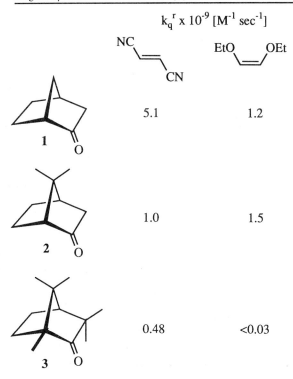

$$k_q^r \times 10^{-9} \ [M^{-1} \ sec^{-1}]$$

1	5.1	1.2
2	1.0	1.5
3	0.48	<0.03

photocycloaddition of acetone (**4**) to *cis*- or *trans*-2-butene (**5**) leads to a mixture of diastereoisomeric oxetanes **6** and **7** in a constant ratio of 64:36 (±2%) independent from the substrate configuration.[3] No isomerization of the alkene substrates could be detected at low to medium conversion. The photocycloaddition of acetone (**4**) to *cis*- and *trans*-1-methoxy-1-butene (**8,9**) has been studied in order to show the divergent stereoselectivity of singlet and triplet excited carbonyl states.[4] By use of piperylene as the triplet quencher and investigation of the concentration (quencher as well as substrate) dependency of the stereoselectivity of the Paternò-Büchi reaction, a consistent mechanism was established. At high concentrations of the alkene or of the triplet quencher, maximum d.r. (diastereomeric ratio) values of 82:18 and 27:73 were found for the oxetanes from **8** and **9**, respectively. Therefore, it was concluded that the stereochemistry of the initial butene is retained in the oxetanes **10** and **11** (regioisomers ignored) when singlet excited acetone attacks the alkene, whereas, a scrambling effect is obtained for the triplet case, indicating a long-lived triplet biradical intermediate.

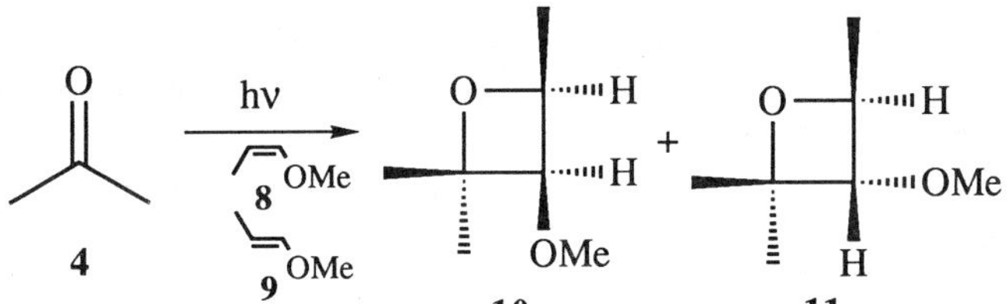

In some cases, however, a high degree of stereoselectivity could be obtained even with "pure" triplet excited carbonyl compounds. In these cases, e.g., the photocycloaddition of benzophenone to several methyl vinyl sulphides 12, the intermediary triplet 1,4-biradical undergoes preferentially one of two possible cyclization modes after intersystem crossing.[5] The nonstereospecific (using Zimmerman's definition[6]) nature of this reaction has been demonstrated by the use of stereoisomeric substrates that lead to oxetanes 13 in identical diastereoisomeric ratios. Another possibility for obtaining stereoisomerically pure oxetanes is the use of alkenes that simultaneously serve as quenchers for the carbonyl triplets, e.g., dienes. The photocycloaddition of acetone (4) and 2-methyl-2,4-hexadiene (14) represents such a process, leading to two regioisomeric oxetanes 15 and 16 where the substrate configuration is retained in 16.[7]

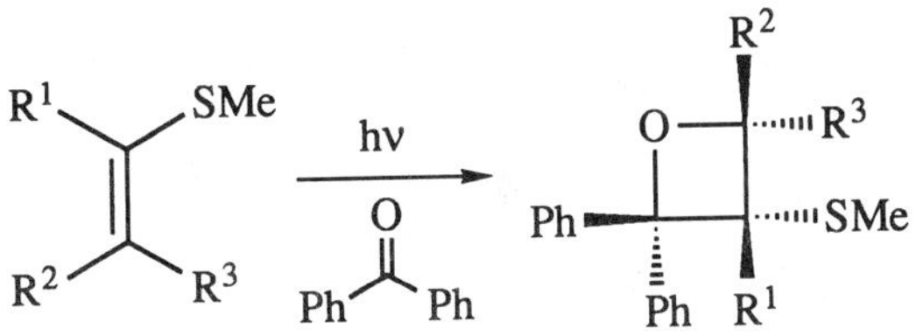

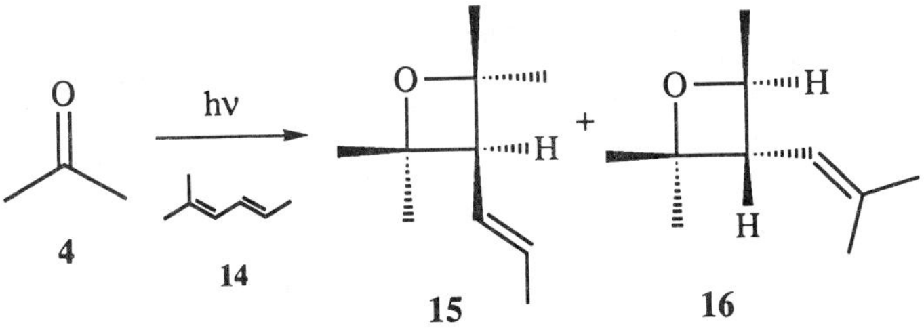

Prostereogenic Carbonyls: Product Stereoselectivity

Inherent Diastereoselectivity

In contrast to the results with acetone *(vide supra)*, the photoaddition of acetaldehyde (17) to the 2-butene 5 is highly stereoselective (or stereospecific[6]).[8] Aromatic carbonyl compounds such as benzaldehyde (18) show low stereoselectivities independent of their concentrations (in contrast to aliphatic ketones) due to fast intersystem crossing into the triplet manifold of the carbonyl substrate. If, however, the configuration of the triplet excited state switches from n,π* to π,π*, the product stereoselectivity rises again. In these cases (e.g., 2-naphthaldehyde 19), the carbonyl singlet is the reactive state for photocycloaddition reactions. Obviously, the quantum yields for such a process are low (about 5% of triplet reactivity) because of efficient deactivation via the triplet π,π^* states. Another concentration study has been published by Jones et al.[9] where they describe the photocycloaddition of aliphatic aldehydes to medium-ring cycloalkenes. The variation in photoadduct distribution is due to stereospecific addition of the aldehyde singlet (dominant at high alkene concentrations) accompanied by a more stereorandom triplet pathway at lower concentrations of the cycloalkenes.[9]

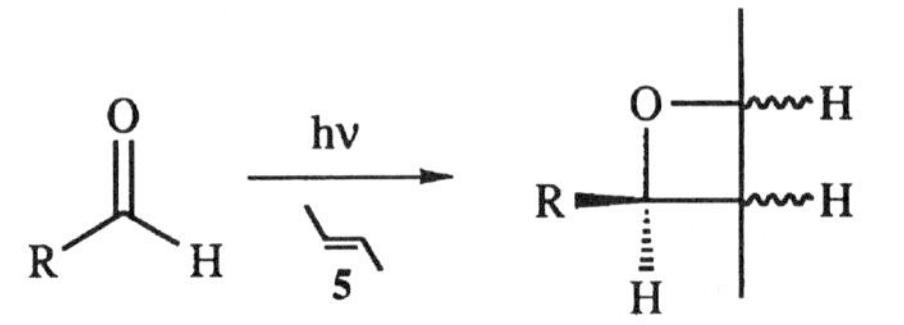

Selectivity ratio (E/Z):

17 (R=Me) 95 : 5
18 (R=Ph) 62 : 38
19 (R=Nphth) 94 : 6

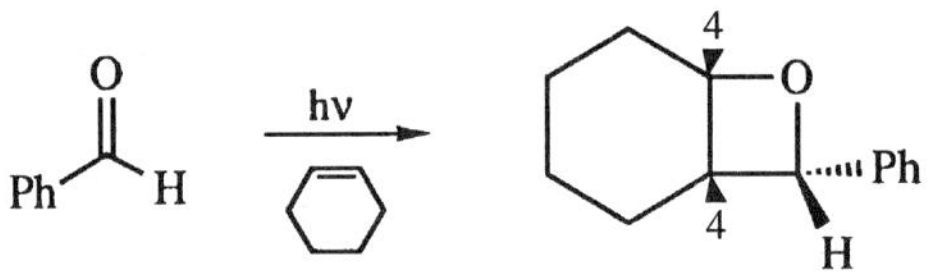

18   **20**   **21**

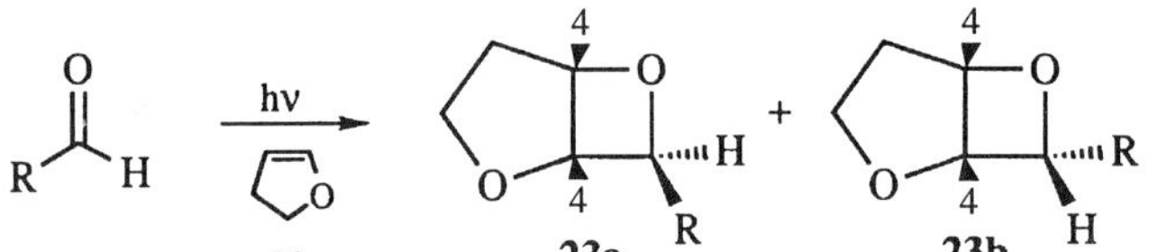

22   **23a**   **23b**

18 (R=Ph) 12 : 88
24 (R=Mes) < 2 : 98
25 (R=Nphth) >98 : 2

For the Paternò-Büchi reaction of cycloalkenes with prostereogenic aromatic carbonyl compounds (which show rapid intersystem crosssing to the triplet excited carbonyl), the results were not so straightforward. The reaction of cyclohexene (20) with benzaldehyde (18) was reported in the literature,[10] and the spectral data of the main product 21 (35%) described as "consistent with the unusual assignment of *endo*-stereochemistry".[11] A detailed study of the photocycloaddition of (triplet) benzaldehyde with several cycloalkenes revealed that *endo*-stereoselectivity is an inherent property for these processes.[12] The addition of 18 to 2,3-dihydrofuran (22) was particularly significant: only one regioisomer 23 (analogous to the reaction with acetone)[13] is formed in a diastereoisomeric ratio of 88:12. Increasing steric demand of the carbonyl addend leads to an increase in diastereoselectivity, e.g., for mesityl aldehyde 24, only the *endo*-diastereoisomer 23b could be detected.[14] Substituent effects have also been described for methyl-substituted cycloalkenes that were in accord with the postulated principle for control of stereoselectivity in triplet reactions.[15] These selectivity effects could be explained by the assumption of certain 1,4-biradical conformers that fulfill the prerequisites (Salem rules)[16] for rapid intersystem crossing to form the 1,4-singlet biradicals. These can interconvert without spin barrier into the products and thus exhibit a "conformational memory".[12,15]

In contrast to the triplet pathway, singlet excited carbonyl compounds such as the 1- or 2-naphthaldehydes 25 do add with high *exo*-stereoselectivity to 2,3-dihydrofuran.[17] Both singlet *and* triplet excited carbonyl compounds underwent photocycloaddition to furan 26 with high *exo*-selectivity to give adducts 27. Additionally, only one regioisomer is formed in these highly efficient reactions.[18] The oxetanes formed in these reactions are acid labile compounds and can be converted into aldol products. Schreiber et al.[19] used this property for the synthesis of several key compounds as substrates for the synthesis of natural product and created the name "photo-Aldol reaction". Besides the addition of aldehydes to the unsubstituted furan, ketones and esters as well as substituted furans were used as starting materials.[20,21] In most cases, the site selectivity (for furans with unsymmetric substitution pattern) is low,[22] but the regioselectivity (exclusive formation of the acetal product) and the stereoselectivity stays extraordinarily high.[23] The same degree of regio- and stereoselectivity has been observed for other heterocyclic substrates such as thiophenes, oxazoles, pyrazoles, and many more.[24] A somewhat lower degree of stereoselectivity was reported for the photocycloaddition of carbonyl compounds to 1,3-cyclopentadiene (with propanal: *exo/endo* = 80:20)[25] and 1,3-cyclohexadiene (with acetaldehyde: *exo/endo* = 88:12).[26] Another group of highly efficient carbonyl addends are α-keto esters such a alkyl pyruvates, alkyl glyoxylates, and esters of phenyl glyoxylic acid. Photocycloaddition with electron-rich cycloalkenes such as dioxoles and furans have been reported to proceed with high regio- and diastereoselectively. The Paternò-Büchi reaction of 2,2-diisopropyl-1,3-dioxole 28 with ethyl pyruvate 29 leads to a 80:20 mixture of diastereoisomeric oxetanes 30.[27] Zamojski has reported a highly selective reaction of triplet excited *n*-butyl glyoxylate 31 with furan 26. Similar to the reactions with aromatic aldehydes,[18] only the *exo*-photoproduct 32 is formed.

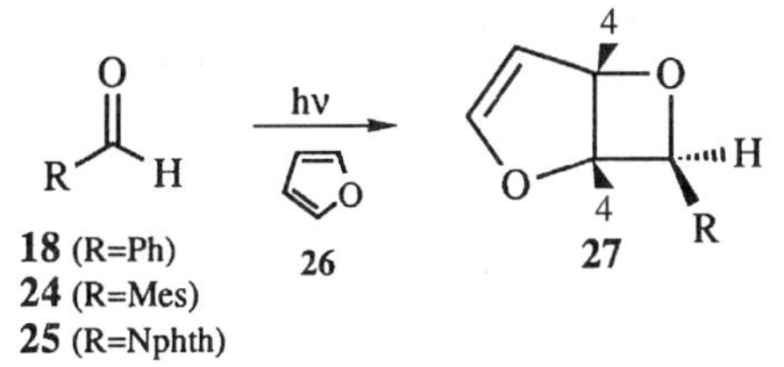

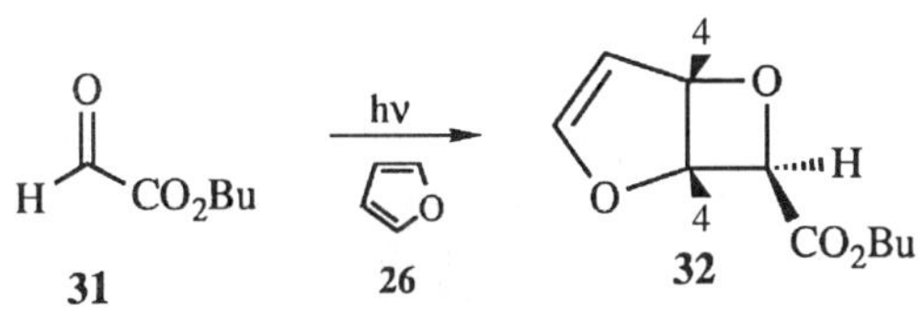

18 (R=Ph)
24 (R=Mes)
25 (R=Nphth)

26

27

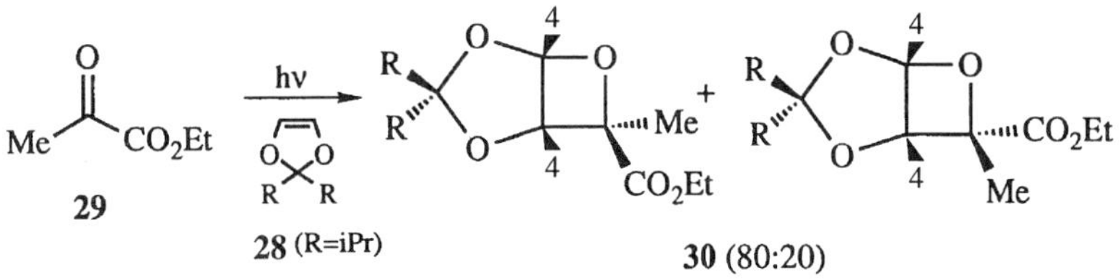

29

28 (R=iPr)

30 (80:20)

31

26

32

Induced Diastereoselectivity

The first report concerning an "asymmetric" Paternò-Büchi reaction with a chiral carbonyl component was reported in 1979 by Gotthardt and Lenz.[29] The photocycloaddition of the enantiomerically pure menthyl ester of phenylglyoxylic acid **33** with 2,3-dimethyl-2-butene **34** gave the oxetane **35** with a diastereomeric excess of only 37%. The corresponding chiral glyoxylates in addition to furan **26** gave even lower diastereomeric excesses of 5 ± 2%.[30] The unique behavior of chiral phenyl glyoxylates was demonstrated later by Scharf and co-workers. Despite the fact that in all cases the stereogenic centers are localized in the alcohol part of the α-keto ester and therefore remarkably far away from the reactive (triplet excited) carbonyl group, the (induced) diastereoselectivities were exceedingly high (>96% in many cases) with 8-phenylmenthol as the chiral auxiliary.[31] The temperature dependence of the auxiliary-induced diastereoselectivity was studied by the same group.[32] The photoaddition of phenyl glyoxylates to several electron-rich cycloalkenes also used as substrates in reactions with α-keto esters, was studied.[33] Most of the additives exhibited the normal isoselectivity effects but an *inversion* of the induced diastereoselectivity was found at certain temperatures. Consequently, this effect was named the *isoinversion principle* and it was demonstrated to be a general phenomenon for many two-step reactions.[34] The inherent (noninduced) diastereoselectivity for these photoadditions (e.g., oxetane **38** from the dioxole **36** and the 8-phenylmenthyl phenyl glyoxylate **37**) was >96% with the phenyl group being directed into the *endo*-position. The application of Cram-like model compounds such as isopropylidene glyceraldehyde[35] or acyclic α-chiral ketones[36] did not lead to high stereoselectivities.

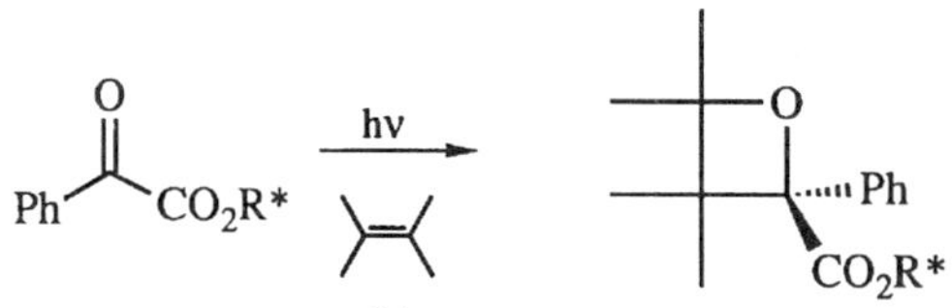

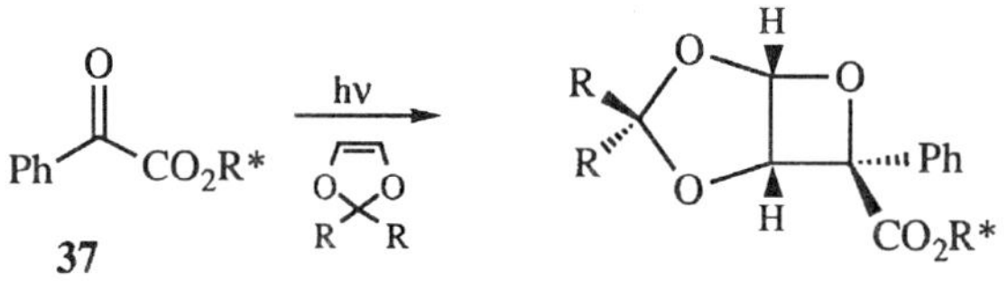

45.3 The Alkene Site

Inherent Diastereoselectivity

A great number of acyclic and cyclic alkenes and alkadienes have been used as addends for Paternò-Büchi reactions with carbonyl compounds and many of them are mentioned elsewhere. Dioxoles[37] and 2,3-dihydrooxazoles[38] have been found to be remarkably effective addends. These alkenes are highly electron-rich compounds with low-lying ionization potentials and also serve as potent electron-donor substrates. This property makes the application of electron-rich alkenes sometimes critical because SET (single electron-transfer) reactions can interfere with the "normal" photocycloadditions via triplet 1,4-biradicals. The use of nonpolar solvents is therefore recommended. However, an analysis of the energetics (Rehm-Weller relationship) should always be included.[39] Intensive investigations of the diastereofacial selectivity of the photocycloaddition of ketones to norbornene and norbornadiene have been reported. The addition of biacetyl (39) to norbornene 40a is highly (>24:1) *exo*-selective, whereas the *syn*-7-*t*-butyl derivative 40b showed inverted (<1:30) *exo*-selectivity.[40] Introduction of a hydroxy group at the 7-*syn*-position of norbornene (40c) re-inverts the the diastereoselectivity: in this case, the *exo*-adduct 41c is formed with d.e. >97%. The latter effect could be due to hydrogen bonding that pre-complexes the excited carbonyl species. A similar photoreaction has been reported for norbornadiene 42.[41] As was shown by Gorman et al.,[42] ketone triplets do not add to 42 but interconvert it to quadricyclane that serves as a substrate for the subsequent Paternò-Büchi reaction.

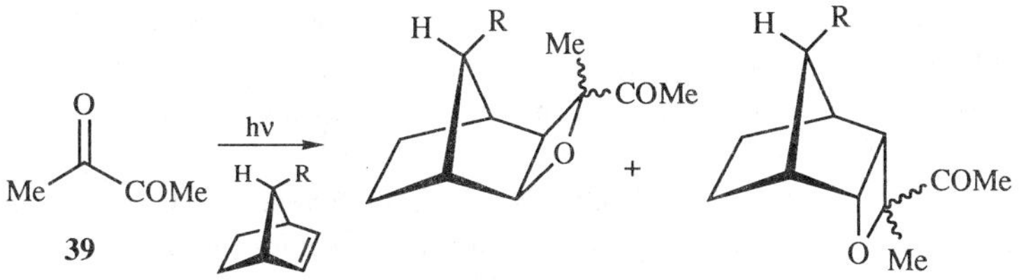

40a (R=H)   **41a** (> 24 : 1)
40b (R=tBu)   **41b** (< 1 : 30)
40c (R=OH)   **41c** (> 30 :1)

Induced Diastereoselectivity

The use of chiral alkenes is a less often used possibility for "asymmetric" photocycloadditions. An optically active acetal was used as substrate in the synthesis of prostaglandins. The induced (*anti*) diastereoselectivity was high (>90%), and the inherent diastereoselectivity was not reported.[43] An interesting solvent dependency of the face-selectivity was reported for 3,4,6-tri-*O*-acetyl-D-glucal as a acarbohydrate substrate.[44]

ferences

1. Turro, N. J. and Farrington, G. L., Quenching of the fluorescence of 2-norbornanone and derivatives by electron-rich and electron-poor ethylenes, *J. Am. Chem. Soc.*, 102, 6051, 1980.
2. Turro, N. J. and Farrington, G. L., Photoinduced oxetane formation between 2-norbornanone and derivatives with electron-poor ethylenes, *J. Am. Chem. Soc.*, 102, 6056, 1980.
3. Carless, H. A. J., Photocycloaddition of acetone to acyclic olefins, *Tetrahedron Lett.*, 3173, 1973.
4. Turro, N. J. and Wriede, P. A., The photocycloaddition of acetone to 1-methoxy-1-butene. A comparison of singlet and triplet mechanisms and singlet and triplet biradical intermediates, *J. Am. Chem. Soc.*, 92, 320, 1970.
5. Morris, T. H., Smith, E. H., and Walsh, R., Oxetane synthesis: Methyl vinyl sulphides as new traps of excited benzophenone in a stereoselective and regiospecific Paternò-Büchi reaction, *J. Chem. Soc., Chem. Commun.*, 964, 1987.
6. Zimmerman, H. E., Singer, L., and Thyagarajan, B. S., *J. Am. Chem. Soc.*, 81, 108, 1959, footnote 16.
7. Hautala, R. R., Dawes, K., and Turro, N. J., Stereoselectivity and regioselectivity of the photocycloaddition of 2-methyl-2,4-hexadiene and acetone, *Tetrahedron Lett.*, 1229, 1972.
8. Yang, N. C., Kimura, M., and Eisenhardt, W., Paternò-Büchi reaction of aromatic aldehydes with 2-butenes and their implication on the rate of intersystem crossing of aromatic aldehydes, *J. Am. Chem. Soc.*, 95, 5058, 1973.

9. Jones, G., II, Khalil, Z. H., Phan, X. T., Chen, T.-J., and Welankiwar, S., Divergent stereoselectivity in the photoaddition of alkanals and medium ring cycloalkenes, *Tetrahedron Lett.,* 22, 3823, 1981.

10. Bradshaw, J. S., Ultraviolet irradiation of carbonyl compounds in cyclohexene and 1-hexene, *J. Org. Chem.,* 31, 237, 1966.

11. Jones, G., II, Synthetic applications of the Paternò-Büchi reaction, *Organic Photochemistry,* Padwa, A., Ed., 5, 1, 1981.

12. Griesbeck, A. G. and Stadtmüller, S., Photocycloaddition of benzaldehyde to cyclic olefins: Electronic control of *endo* stereoselectivity, *J. Am. Chem. Soc.,* 112, 1281, 1990.

13. Carless, H. A. J. and Haywood, D. J., Photochemical synthesis of 2,6-dioxabicyclo[3.2.0]-heptanes, *J. Chem. Soc., Chem. Commun.,* 1067, 1980.

14. Griesbeck, A. G. and Stadtmüller, S., Regio- and stereoselective photocycloaddition reactions of aromatic aldehydes to furan and 2,3-dihydrofuran, *Chem. Ber.,* 123, 357, 1990.

15. Griesbeck, A. G. and Stadtmüller, S., Electronic control of stereoselectivity in photocycloaddition reactions. 4. Effects of methyl substituents at the donor olefin, *J. Am. Chem. Soc.,* 113, 6923, 1991.

16. Salem, L. and Rowland, C., The electronic properties of diradicals, *Angew. Chem. Int. Ed.,* 11, 92, 1972.

17. Griesbeck, A. G., Mauder, H., Peters, K., Peters, E.-M., and von Schnering, H. G., Photocycloadditions with α- and β-naphthaldehyde: Complete inversion of diastereoselectivity as a consequence of differently configured electronic states, *Chem. Ber.,* 124, 407, 1991.

18. (a) Schenck, G. O., Hartmann, W., and Steinmetz, R., Four-membered ring synthesis by the photosensitized cycloaddition of dimethyl maleic anhydride to olefins, *Chem. Ber.,* 96, 498, 1963; (b) Toki, S., Shima, K., and Sakurai, H., Organic photochemical reactions. I. The synthesis of substituted oxetanes by the photoaddition of aldehydes to furans, *Bull. Chem. Soc. Jpn.,* 38, 760, 1965; (c) Shima, K. and Sakurai, H., Photoaddition reactions of various carbonyl compounds to furan, *Bull. Chem. Soc. Jpn.,* 39, 1806, 1966.

19. (a) Schreiber, S. L. and Satake, K., Application of the furan-carbonyl photocycloaddition reaction to the synthesis of the *bis*(tetrahydrofuran) moiety of asteltoxin, *J. Am. Chem. Soc.,* 105, 6723, 1983; (b) Schreiber, S. L., Hoveyda, A. H., and Wu, H.-J., A photochemical route to the formation of threo aldols, *J. Am. Chem. Soc.,* 105, 660, 1983; (c) Schreiber, S. L. and Satake, K., Total synthesis of (±)-asteltoxin, *J. Am. Chem. Soc.,* 106, 4186, 1984; (d) Schreiber, S. L., [2+2] Photocycloadditions in the synthesis of chiral molecules, *Science,* 227, 857, 1985.

20. Feigenbaum, A., Pete, J.-P., and Poquet-Dhimane, A.-L., Stereoselective synthesis of unsaturated oxetanes, *Heterocycles,* 27, 125, 1988.

21. Cantrell, T. S., Allen, A. C., and Ziffer, H., Photochemical 2+2 cycloaddition of arenecarboxylic acid esters to furans and 1,3-dienes. 2+2 cycloreversion of oxetanes to dienol esters and ketones, *J. Org. Chem.,* 54, 140, 1989.

22. Exceptions due to steric and electronic reasons are known: (a) Carless, H. A. J. and Halfhide, A. F. E., Highly regioselective [2+2] photocycloaddition of aromatic aldehydes to acylfurans, *J. Chem. Soc., Perkin Trans. 1,* 1081, 1992; (b) Schreiber, S. L., Desmaele, D., and Porco, J. A., Jr., On the use of unsymmetrically substituted furans in the furan-carbonyl photocycloaddition reaction: Synthesis of a kadsurenone-ginkgolide hybrid, *Tetrahedron Lett.,* 29, 6689, 1988.

23. Jarosz, S. and Zamojski, A., Asymmetric photocycloaddition between furan and optically active ketones, *Tetrahedron,* 38, 1453, 1982.

24. (a) Rivas, C. and Bolivar, R. A., Synthesis of oxetanes by photoaddition of carbonyl compounds to pyrrole derivatives, *J. Heterocyclic Chem.,* 13, 1037, 1976; (b) Julian, D. R. and Tringham, G. D., Photoaddition of ketones to indoles: Synthesis of oxeto[2,3-b]indoles, *J. Chem. Soc., Chem. Commun.,* 13, 1973; (c) Rivas, C., Pacheco, D., Vargas, F., and Ascanio, J., Synthesis of oxetanes by photoaddition of carbonyl compounds to thiophene derivatives, *J. Heterocyclic Chem.,* 18, 1065, 1981; (d) Nakano, T., Rodriguez, W., de Roche, S. Z., Larrauri, J. M., Rivas, and Pérez, C., Photoaddition of ketones to imidazoles, thiazoles, isothiazoles and isoxazoles. Synthesis of their oxetanes, *J. Heterocyclic Chem.,* 17, 1777, 1980.

25. Shima, K., Kubota, T., and Sakurai, H., Organic photochemical reactions. XXIV. Photocycloaddition of propanal to 1,3-cyclohexadiene, *Bull. Chem. Soc. Jpn.*, 49, 2567, 1976.

26. Hoye, T. R. and Richardson, W. S., A short, oxetane-based synthesis of (±)-sarracenin, *J. Org. Chem.*, 54, 688, 1989.

27. Mattay, J. and Buchkremer, K., Thermal and photochemical oxetane formation with α-ketoesters, *Heterocycles*, 27, 2153, 1988.

28. Zamojski, A. and Kozluk, T., Synthesis of 3-substituted furans, *J. Org. Chem.*, 42, 1089, 1977.

29. Gotthardt, H. and Lenz, W., Unusually high asymmetric induction in the photochemical formation of oxetanes, *Angew. Chem. Int. Ed.*, 18, 868, 1979.

30. Jarosz, S. and Zamojski, A., Asymmetric photocycloaddition between furan and chiral alkyl glyoxylates, *Tetrahedron*, 38, 1447, 1982.

31. Nehrings, A., Scharf, H.-D., and Runsink, J., Photochemische Darstellung eines L-Erythrose-Bausteins und sein Einsatz bei der Synthese von Methyl-2,3-O-isopropyliden-β-L-apio-L-furanosid, *Angew. Chem.*, 97, 882, 1985; *Angew. Chem. Int. Ed.*, 25, 877, 1985.

32. Buschmann, H., Scharf, H.-D., Hoffmann, N., Plath, M., and Runsink, J., Chiral induction in photochemical reactions. 10. The principle of isoinversion: A model of stereoselection developed from the diastereoselectivity of the Paternò-Büchi reaction, *J. Am. Chem. Soc.*, 111, 5367, 1989.

33. (a) Koch, H., Runsink, J., and Scharf, H.-D., Investigation of chiral induction in photochemical oxetane formation, *Tetrahedron Lett.*, 24, 3217, 1983; (b) Koch, H., Scharf, H.-D., Runsink, J., and Leismann, H., Regio- and diastereoselectivity in the oxetane formation of chiral phenylglyoxylates with electron rich olefins, *Chem. Ber.*, 118, 1485, 1985; (c) Pelzer, R., Jütten, P., and Scharf, H.-D., Chiral induction in photochemical reactions. IX. Isoselectivity in the asymmetric Paternò-Büchi reaction using carbohydrates as chiral auxiliaries, *Chem. Ber.*, 122, 487, 1989.

34. Buschmann, H., Scharf, H.-D., Hoffmann, N., and Esser, P., Das Isoinversionsprinzip — ein allgemeines Selektionsmodell in der Chemie, *Angew. Chem.*, 103, 480, 1991.

35. Schreiber, S. L. and Satake, K., Studies of the furan-carbonyl photocycloaddition reaction: The determination of the absolute stereostructure of asteltoxin, *Tetrahedron Lett.*, 27, 2575, 1986.

36. Jarosz, S. and Zamojski, A., Asymmetric photocycloaddition between furan and optically active ketones, *Tetrahedron*, 38, 1453, 1982.

37. Meier, L. and Scharf, H.-D., A new synthesis of 4,5-unsubstituted 1,3-dioxoles, *Synthesis*, 517, 1987.

38. Weuthen, M., Scharf, H.-D., and Runsink, J., 3-Acetyl-2,3-dihydro-2,2-dimethyloxazole: synthesis, properties, and its application as olefinic partner in the Paternò-Büchi reaction, *Chem. Ber.*, 120, 1023, 1987.

39. Mattay, J., Ladungstransfer und Radikalionen in der Photochemie, *Angew. Chem.*, 99, 849, 1987.

40. Sauers, R. R., Valenti, P. C., and Tavss, E., The importance of steric effects on the photocycloadditions of biacetyl to norbornenes, *Tetrahedron Lett.*, 3129, 1975.

41. Kubota, T., Shima, K., and Sakurai, H., The photocycloaddition of benzophenone to norbornadiene, *Tetrahedron Lett.*, 343, 1972.

42. (a) Barwise, A. J. G., Gorman, A. A., Leyland, R. L., Parekh, C. T., and Smith, P. G., The inter- and intramolecular addition of aromatic ketone triplets to quadricyclane, *Tetrahedron*, 36, 397, 1980; (b) Gorman, A. A. and Leyland, R. L., The reaction of benzophenone triplets with norbornadiene and quadricyclane, *Tetrahedron Lett.*, 5345, 1972; (c) Gorman, A. A., Leyland, R. L., Rodgers, M. A. J., and Smith, P. G., Concerning the mechanism of interactions of triplet benzophenone with norbornadienes and quadricyclanes, *Tetrahedron Lett.*, 5085, 1973.

43. Morton, D. R. and Morge, R. A., Total synthesis of 3-oxa-4,5,6-trinor-3,7-inter-*m*-phenylene prostaglandins. 1. Photochemical approach, *J. Org. Chem.*, 43, 2093, 1978.

44. Araki, Y., Senna, K., Matsuura, K., and Ishido, Y., Supplementary aspects in the photochemical addition of acetone to 3,4,6-tri-O-acetyl-D-glucal, *Carbohydr. Res.*, 60, 389, 1978.

46

Oxetane Formation:
Intramolecular Addition

Howard A. J. Carless
Birkbeck College

Intramolecular oxetane formation can occur on photolysis of a range of unsaturated carbonyl compounds, and is subject only to the limitation that other photochemical processes (e.g., intramolecular hydrogen abstraction, Norrish type II reaction, or oxa-di-π-methane rearrangement) do not dominate. The highly ordered structure of the intramolecular adducts, combined with good chemical reactivity of the oxetane ring, have led to several applications of the reaction in synthesis.[1,2]

The earliest report of intramolecular Paternò-Büchi reaction came from gas-phase studies by Srinivasan in 1960,[3] in which photolysis of the parent γ,δ-unsaturated ketone hex-5-en-2-one (**1a**) was suggested to give the 2-oxabicyclo[2.2.0]hexane (**2a**). Later solution-phase work by Yang[4] extended this example to those in which there was methyl substitution at the double bond; ketone **1b** led to a 2:5 ratio of oxetanes **2b** and **3b,** where the latter cross-addition product has the orientation favored typically via a mechanism involving alkyl substitution of a 1,4-biradical intermediate. Yang's work suggested also that **3a** was a product from the photolysis of **1a** [ratio **2a:3a** = 6:4], and showed that in the case of photolysis of less substituted examples such as **1a**, the oxetane **3a** could not be isolated but decomposed thermally to cyclopentenol (**4**) on attempted separation. When the double bond of the unsaturated ketones is tetrasubstituted, oxetanes having the orientation of **3** are the dominant or exclusive photoproducts.[5] In other examples where the alkene component is capable of showing *cis-trans* isomerization (such as hept-5-en-2-one), Morrison[6] has shown that the reaction mechanism involves a singlet exciplex leading to an oxetane having the orientation of **2**. The formation of the exciplex provides a pathway for excited-state deactivation, lowering the yield of *cis-trans* isomerization of the reactant. Thus, it is often possible to obtain stereoselectivity in intramolecular photocycloaddition without dominant loss of double bond geometry in the reactant.

0-8493-8634-9/95/$0.00+$.50

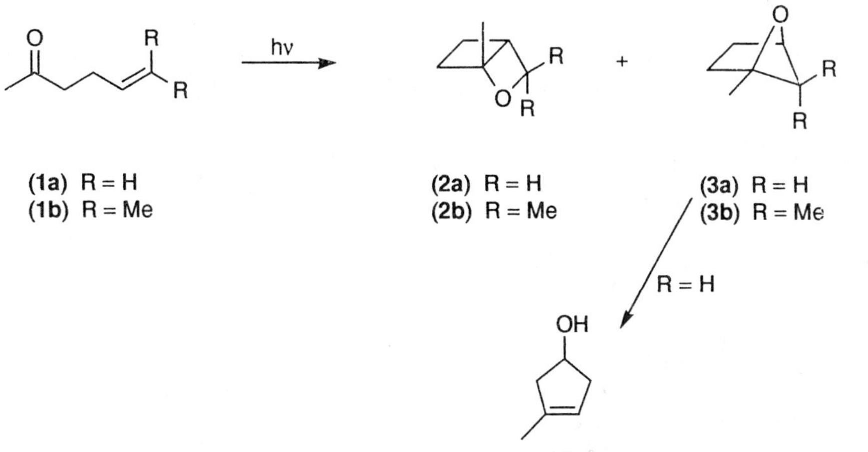

(1a) R = H
(1b) R = Me

(2a) R = H
(2b) R = Me

(3a) R = H
(3b) R = Me

R = H

(4)

A similar intramolecular Paternò-Büchi reaction may occur at double bonds that are either more remote or closer to the carbonyl group. Kossanyi and co-workers[7] have investigated the photochemical reactions of δ,ε-unsaturated aldehydes. For **5**, both orientations of bicyclic adduct **6** and **7** are produced, although in general only adducts having the 6-oxabicyclo[3.2.0] heptane ring system (**6**) are formed reliably.

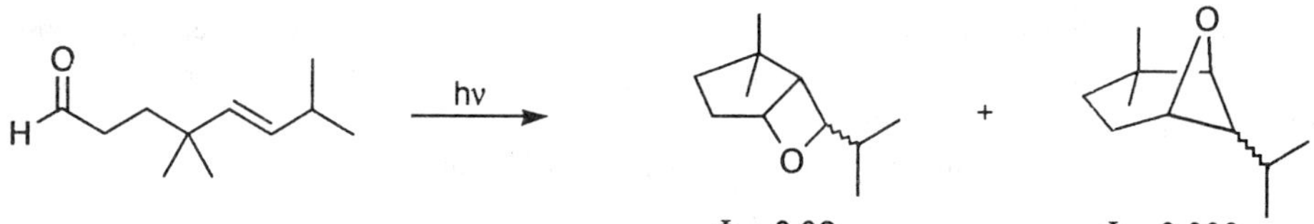

$\Phi = 0.02$

$\Phi = 0.006$

(5)

(6)

(7)

The allenic ketone **8** prefers to undergo intramolecular cycloaddition to the γ,δ- rather than the δ,ε-double bond, to yield the bicyclic methyleneoxetane **9**.[8]

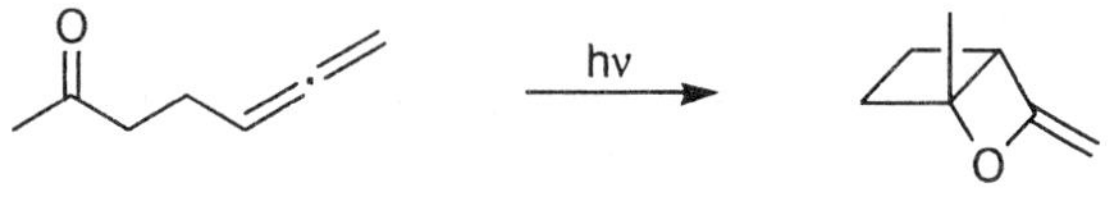

(8)

(9)

The βγ,γδ-unsaturated allenic ketone **10** undergoes a remarkably efficient cycloaddition on acetone-sensitized irradiation to yield the acetone-trapped tricyclic adduct **11**, without competing oxa-di-π-methane rearrangement or 1,3-acyl shift.[9]

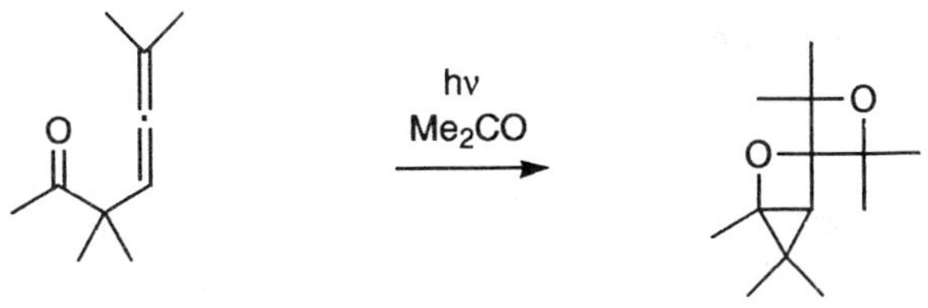

(10)

(11) 70%

Among the most efficient examples known of oxetane formation are those involving unsaturated α-dicarbonyls; thus, 12 gives cycloaddition in high chemical and quantum yields, behaving as if it were a γ,δ-unsaturated carbonyl compound undergoing cross-addition.[10]

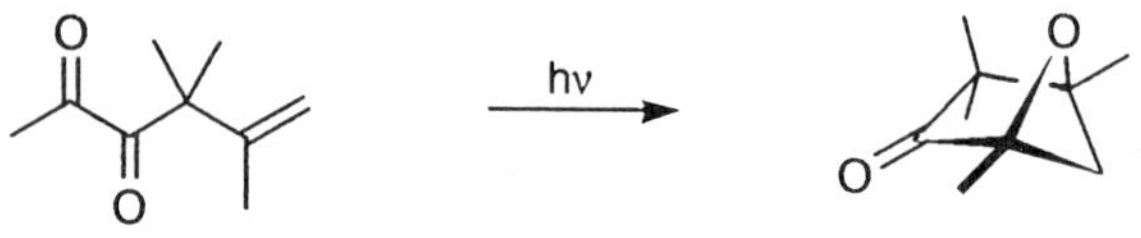

(12)

94%, Φ = 0.6

Photochemical oxetane formation and cleavage provides the opportunity for an interesting functional group interconversion, as shown by irradiation of the αβ,γδ-unsaturated ketone 13. The unisolated oxetane decomposed thermally by ring fission in the alternative sense to its formation, to yield the dienyl aldehyde 14 in 54% overall yield.[11]

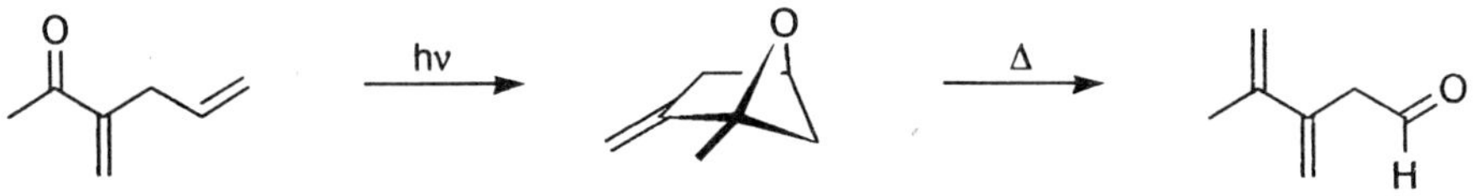

(13)   (14)

A similar photochemical oxetane formation/cleavage sequence has allowed the preparation of a paracyclophane: thus, irradiation of the long-chain unsaturated ketone 15 gave carbonyl cycloaddition at the remote double bond to afford oxetane (16) in 83% yield. Treatment of 16 with silica gel in hexane brought about elimination of formaldehyde, leading to the paracyclophane alkene 17 in 90% yield.[12]

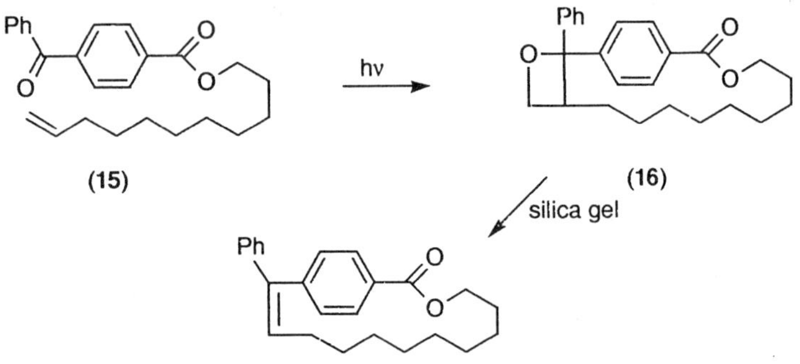

(15)   (16)

silica gel

(17)

Although conjugated dienes usually act as efficient triplet quenchers of carbonyl excited states, the *(E)-retro*-γ-ionone 18 exhibits exclusive oxetane (19) formation on irradiation in ethanol, possibly via a singlet excited state.[13]

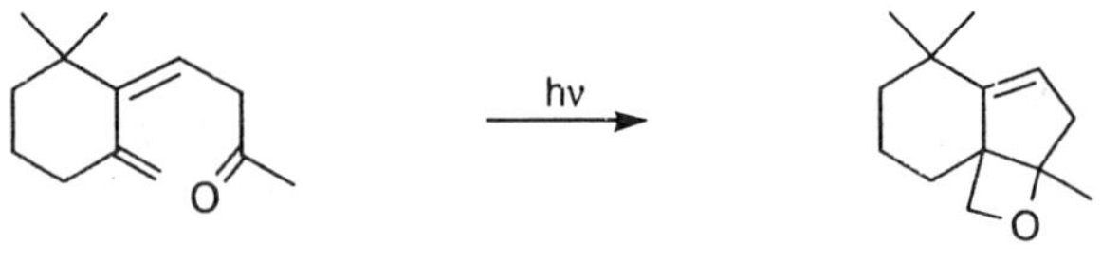

(18)   (19)

The easily prepared 2-allylcycloalkanones give intramolecular photocycloaddition, with Norrish type I reaction being competitive only in the cyclopentanone series. Irradiation of 2-allylcyclohexanone, for example, gave 62% of an oxetane mixture in which addition/cross-addition products analogous to 2 and 3 were formed in a 1:2 ratio.[14] The thermal instability of the cross-adducts has led to synthetic applications; 2-allylcycloheptanone, for instance, gives cross-addition oxetane 20 as the dominant (82%) isomer. Heating the oxetanes above 100 °C converted 20 to a mixture of unsaturated alcohols (cf. 4) and dienes, which was dehydrated and dehydrogenated on stronger heating to azulene 21.[15]

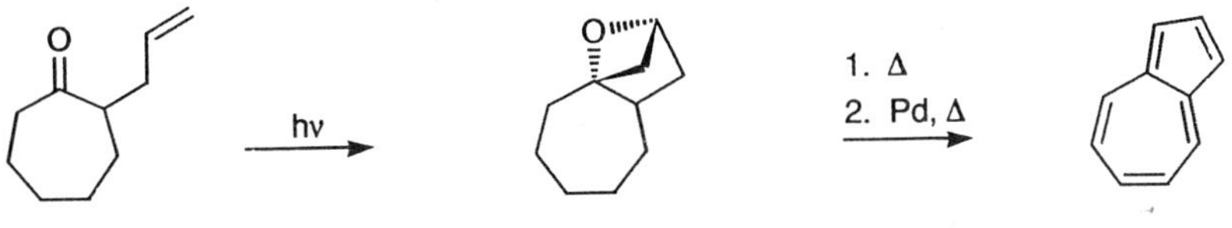

(20) (21)

Occasionally, the unsaturated carbonyl compounds that are reactants are themselves derived from the common Norrish type I photochemical cleavage of cyclic ketones, as in the production of tricyclic oxetanes by photolysis of norbornanone 22[16] or of camphor.[17]

(22)

Intramolecular carbonyl/alkene photocycloaddition has been observed in transannular situations, and the resultant oxetanes have been cleaved to give useful access to both saturated and unsaturated alcohols. Irradiation of *trans*-cyclodec-5-enone (23) gave the tricyclic oxetane 24 in 55% yield, and subsequent hydride reduction led to the *trans*-decalol 25.[18] Mihailovic[19] has demonstrated such transannular chemistry in the photolysis of the enone 26, easily available from cholesterol. Irradiation of 26 in acetone solution led to the oxetane 27 (52%), and acidic cleavage of the oxetane ring converted 27 to the 1α-hydroxycholesterol 3-acetate derivative 28 (82%).[19] The latter compound is an important intermediate in the synthesis of biologically active 1α-hydroxy-vitamin D_3.

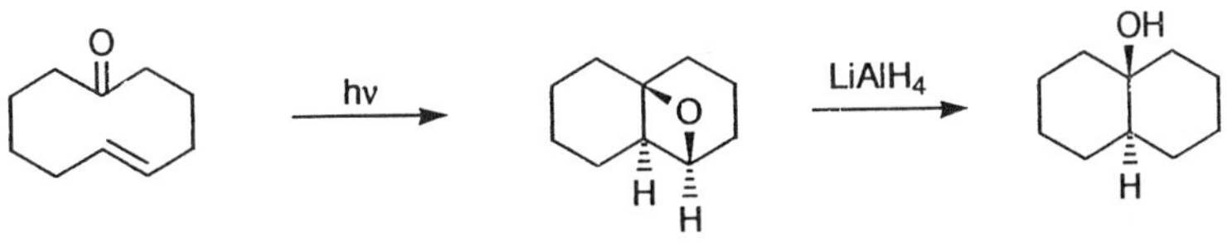

(23) (24) (25)

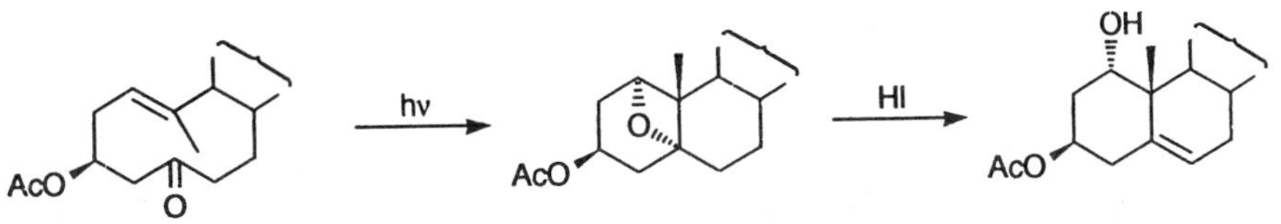

(26) (27) (28)

In the photoreactions of β,γ-unsaturated carbonyl compounds having the alkene constrained in four- to six-membered rings, the oxa-di-π-methane rearrangement tends to dominate on sensitized irradiation, or a 1,3-acyl shift on direct irradiation.[20] However, for seven- and eight-membered cyclic alkenes, where the olefin can twist in the excited state, oxetane formation is paramount on direct irradiation,[21] as illustrated for the cycloheptenyl ketone 29 which yielded the unusual tricyclic oxetane 30 on irradiation in cyclohexane.[22] In constrained systems such as the equatorially substituted 2-acetyl methylenecyclohexane 31, an oxetane 32 has been reported as the sole photoproduct, although it was only isolated in low yield (13%).[23]

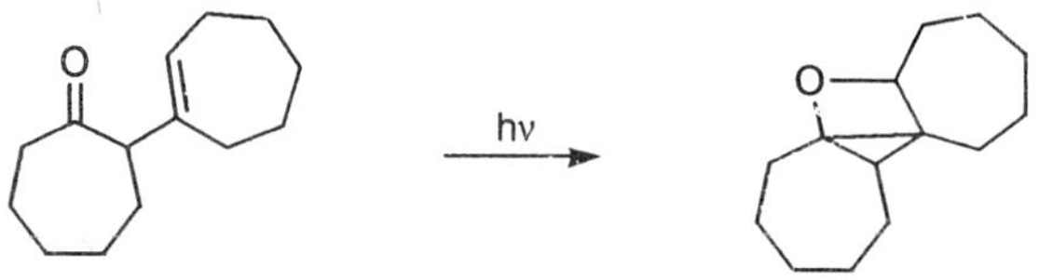

(29)   (30)

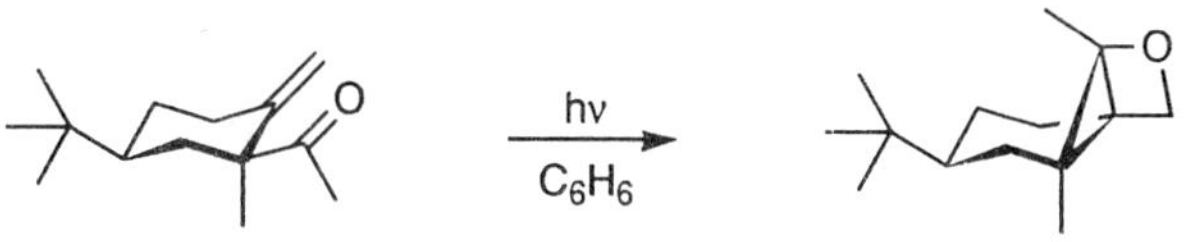

(31)   (32)

Over a period of almost 20 years, Sauers and co-workers have investigated thoroughly the photochemical reactions of *endo*-5-acylnorbornenes and their derivatives.[24] Intramolecular [2+2]-cycloaddition, leading to caged oxetanes (e.g., 33 → 34)[25] has been shown to be efficient both in terms of quantum yield[26] and chemical yield. The synthetic utility of the reaction is greatly extended by the reductive cleavage of the oxetane ring to give a 2,5-methylene-bridged norbornanol 35, using lithium aluminium hydride[25] or lithium/liquid ammonia.[27]

(33)   (34)   (35)

The reaction has been applied to the production, from norbornenyl ketones, of a wide range of caged oxetanes, such as polychlorinated examples,[28] and is compatible with a number of substituents at the α-position of the acyl ketone (e.g., OH, Cl[29] or even pyrrolyl or thienyl ketones.[30] Such oxetane-forming reactions have been exploited in the synthesis of several interesting molecules having unsaturation within the rigid "stellane" ring system of 36. The optically active enone 36, for example, has been prepared from (−)-5-*endo*-acetylnorbornene via photocycloaddition and lithium diethylamide-induced ring cleavage of the oxetane 37 to afford an unsaturated alcohol, followed by CrO₃/pyridine oxidation.[31] The use of lithium diisopropylamide has been found to improve the yields of the unsaturated alcohols from oxetane cleavage and was applied to the route from the homologous oxetane 38 to the diene 39.[32]

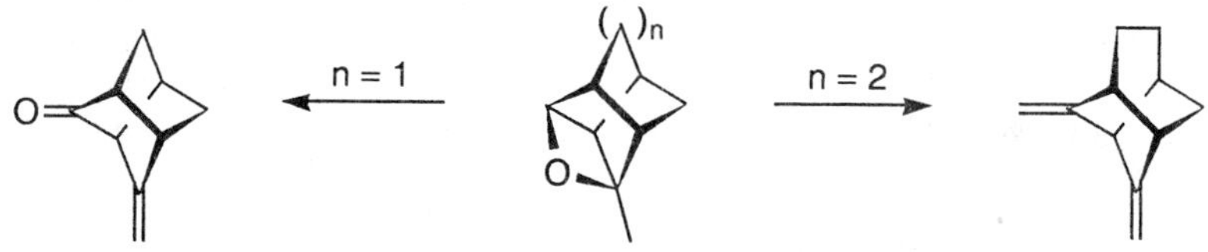

The methylenenorbornenyl ketone 40 undergoes chemoselective attack by the excited carbonyl group at the endocyclic double bond to give the oxetane 41 (60%).[33] Lithium diisopropylamide-induced oxetane ring cleavage led to the unsaturated alcohol 42 (68%), which was followed by oxidation and Wittig reaction to yield the remarkable stellatriene 43.

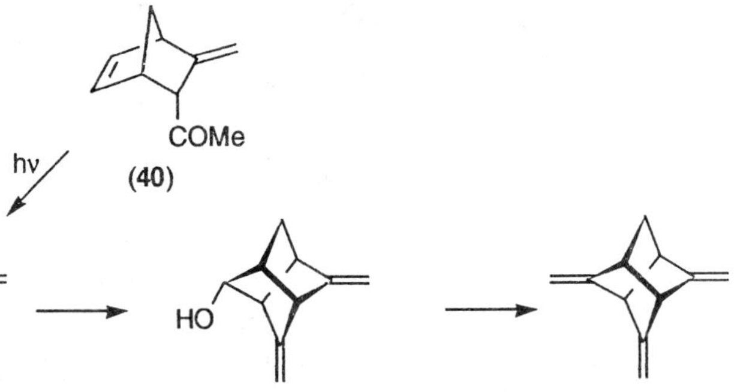

Imides are also capable of acting as the carbonyl components in intramolecular Paternò-Büchi reactions. Photolysis of the N-2-methylallyl imide 44, for example, gave an oxetane almost quantitatively.[34] Subsequent treatment with aqueous acid produced the ring-expanded amide adduct 45 in high yield. A similar intramolecular photocycloaddition occurs on irradiation of an N-alkenyl phthalimide:[35] the resultant oxetane has been found to undergo facile retro-[2+2]-cleavage to give a phenylcycloalkene, in a reaction analogous to that outlined in 16 → 17.

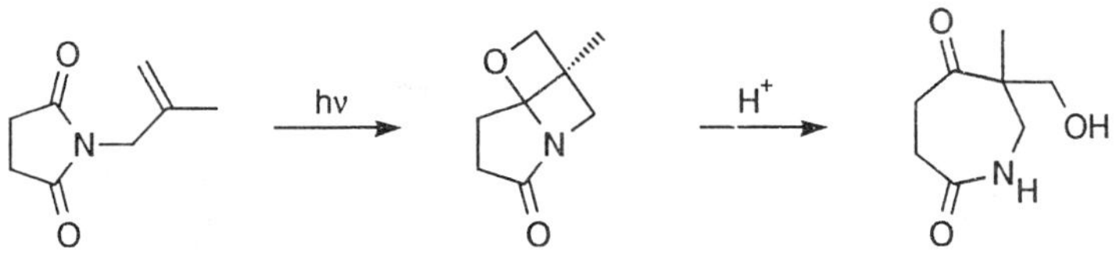

Photocycloaddition of aliphatic carbonyl compounds to vinyl ethers is a very efficient photoreaction that has been exploited in an intramolecular manner. Thus, irradiation of the vinyloxyketone 46, formally a γ,δ-unsaturated ketone, led to the 2,5-dioxabicyclo[2.2.0]hexane 47 in 40% yield.[36] In the related case of photolysis of 1-cyclopentenyloxypropan-2-one, besides oxetane, a diketone was also formed as the result of a 1,3-acyl shift in the reactant.[37]

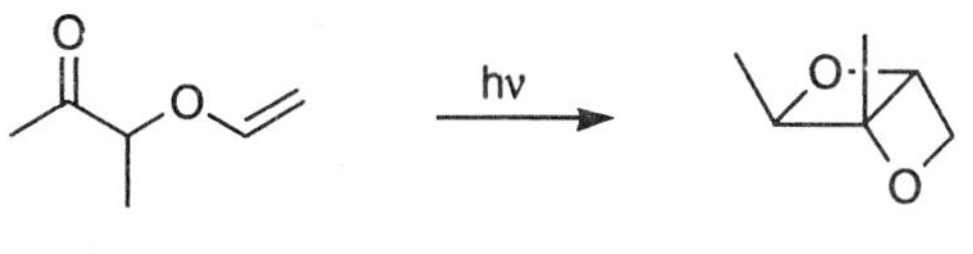

The homologous vinyloxy-carbonyl compounds **48** undergo a facile intramolecular photocycloaddition on irradiation in benzene solution to yield the bicyclic adducts **49**,[38] analogous in orientation to the cycloaddition of a δ,ε-unsaturated aldehyde to give oxetane **6**.

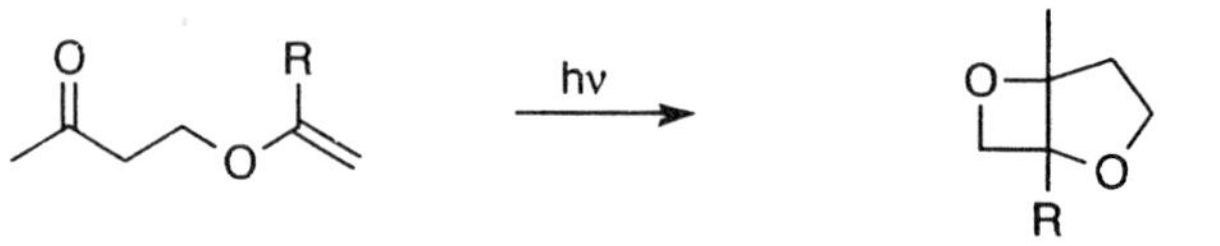

(48)

(49) R = H, 63%
R = Me, 72%

This area of photochemistry has received attention because of the occurrence of the 2,6-dioxabicyclo[3.1.1]heptane ring system in thromboxane A_2 (TXA$_2$, **50**), the unstable

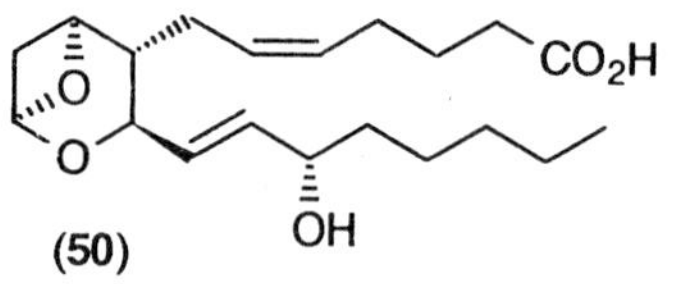

(50)

biologically active counterpart of the prostacyclins. The strained ring system and very reactive acetal functionality have combined to make the synthesis of TXA$_2$ highly challenging. Analogues having halo-substituents at the CH$_2$-position of the oxetane ring are more stable toward hydrolysis and have been prepared.[39] Photochemical syntheses of the core bicyclic ring of TXA$_2$ are potentially attractive, as they can be carried out under neutral and aprotic conditions. Although photolysis of β-vinyloxyketones **48** does not appear to produce an oxetane having the required orientation of addition (i.e., **50**), Carless and Fekarurhobo[40] have prepared homologues of the TXA$_2$ ring system by irradiation of the vinyloxyketone **51**, which gave the bicyclic oxetanes **52** and **53** in 11 and 54% yields, respectively.

(51)   **(52)**   **(53)**

Fried and co-workers[41] have examined the intramolecular [2+2]-addition of a difluorovinyl ether aldehyde **54** that gave two isomers of the tricyclic oxetanes **55** as well as a dioxepene **56** (ratios 5:4:1), the latter probably being formed via the Paternò-Büchi 1,4-biradical. There was no evidence for formation of the desired fluorinated TXA$_2$ ring system (**57**).

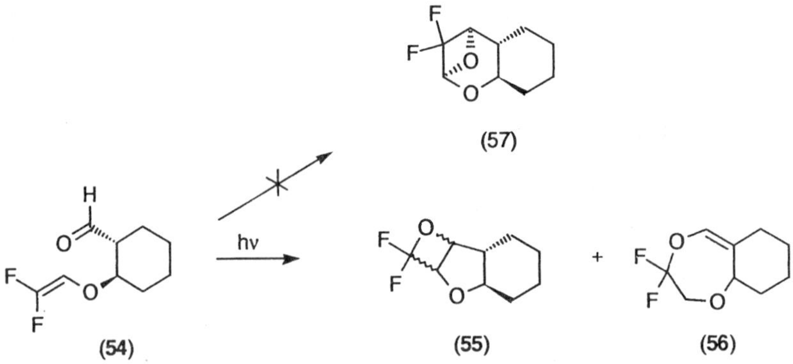

(57)

(54)   **(55)**   **(56)**

The remarkably efficient photocycloaddition of excited carbonyl compounds to vinyl ethers has allowed their reaction in situations where the functional groups would normally be considered as remote. Photochemical reaction of the vinyloxyketones 58, for example, gave regioselective intramolecular cycloaddition to afford the oxetane 59.[42] Trapping of the reactive acetal 59 as its methanol adduct gave a synthesis of medium-ring acetals such as 60, which was applicable in lower yield (45 to 80%) to systems having both shorter and longer interconnecting chains.

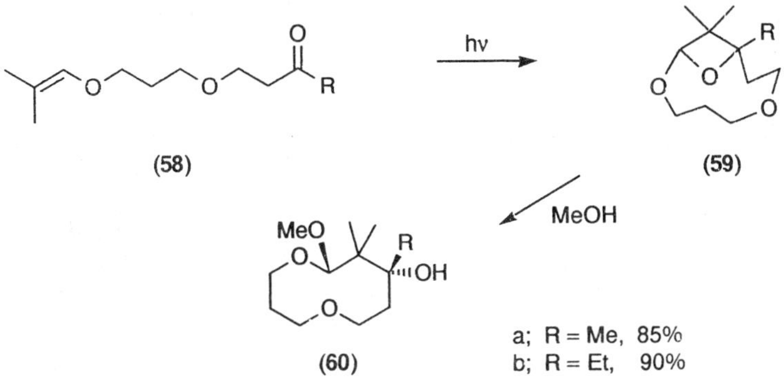

(58)

hv

(59)

MeOH

(60)

a; R = Me, 85%
b; R = Et, 90%

Schreiber and Porco have reported the efficient photocycloaddition of carbonyl compounds to furans in an intramolecular sense in the synthesis of fused polycyclic oxygen heterocycles.[1] Irradiation of the 3-substituted furan 61 gave the oxetanes 62 and 63 (ratio 2:3) as major adducts, with minor competition from the Norrish type II reaction.

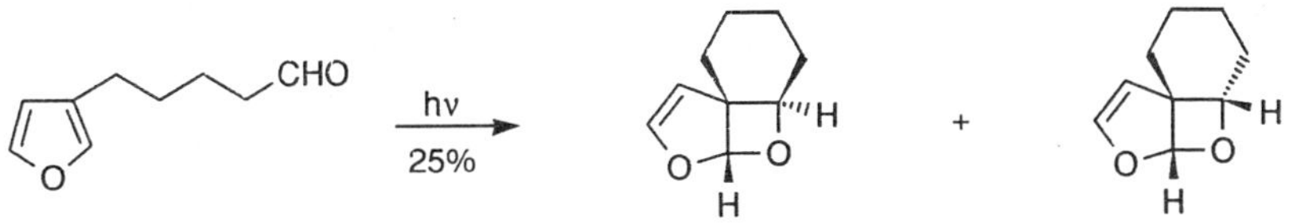

(61)

hv
25%

(62)

+

(63)

However, on irradiation of the shorter chain analogue 64, the chiral stereocenter leads to asymmetric induction in which only a single stereoisomer of intramolecular adduct 65 is produced. Subsequent steps of reduction and hydrolysis provide a well-functionalized spirocycle 66 in nearly quantitative yield.

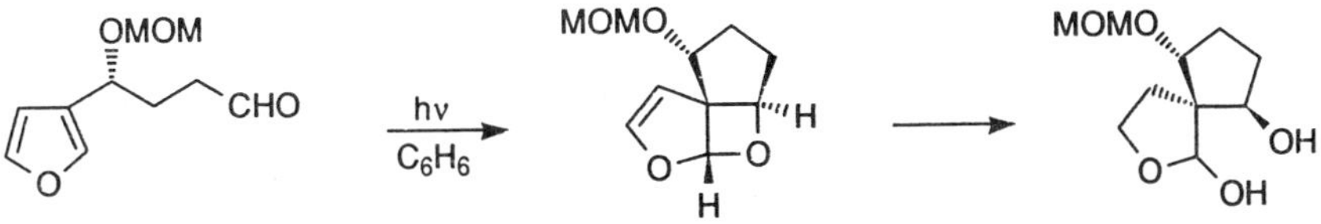

(64)

hv
C₆H₆

(65)

(66)

eferences

1. Porco, J. A., Jr. and Schreiber, S. L., The Paternò-Büchi reaction, in *Comprehensive Organic Synthesis*, Vol. 5, Trost, B. M. and Fleming, I., Eds., Pergamon Press, Oxford, 1991, 151.
2. Carless, H. A. J., Photochemical syntheses of oxetans, in *Synthetic Organic Photochemistry*, Horspool, W. M., Ed., Plenum Press, New York, 1984, 425.
3. Srinivasan, R., Photoisomerization of 5-hexen-2-one, *J. Am. Chem. Soc.*, 82, 775, 1960.
4. Yang, N. C., Nussim, M., and Coulson, D. R., Photochemistry of δγ-unsaturated ketones in solution, *Tetrahedron Lett.*, 1525, 1965.

5. Berger, J., Yoshioka, M., Zink, M. P., Wolf, H. R., and Jeger, O., Zur photochemie tetraalkyl substituierter γ-keto-olefine, *Helv. Chim. Acta*, 63, 154, 1980.

6. Morrison, H., Photochemistry of organic bichromophoric molecules, *Acc. Chem. Res.*, 12, 383, 1979; Kurowsky, S. R. and Morrison, H., Photochemistry of bichromophoric molecules. The solution phase photoisomerization of 5-hepten-2-one, *J. Am. Chem. Soc.*, 94, 507, 1972.

7. Guiard, B., Furth, B., and Kossanyi, J., Reactivity of δε-unsaturated aldehydes, *Bull. Soc. Chim. France*, 1552, 1976.

8. Crandall, J. K. and Mayer, C. F., The photochemistry of 5,6-heptadiene-2-one, *J. Org. Chem.*, 34, 2814, 1969.

9. Kudrawcew, W., Frei, B., Wolf, H. R., and Jeger, O., Photolysis of homoconjugated allene ketones, *Heterocycles*, 17, 139, 1982.

10. Bishop, R. and Hamer, N. K., Photochemistry of some non-conjugated unsaturated 1,2-diketones, *J. Chem. Soc. (C)*, 1197, 1970.

11. Cormier, R. A. and Agosta, W. C., Abstraction of allylic hydrogens vs. other processes in the photochemistry of three doubly unsaturated ketones, *J. Am. Chem. Soc.*, 96, 1867, 1974.

12. Bichan, D. and Winnik, M., A photochemical paracyclophane synthesis: Intramolecular oxetane formation, *Tetrahedron Lett.*, 3857, 1974.

13. van Wageningen, A. and Cerfontain, H., Photolysis of retro-γ-ionones, *Tetrahedron Lett.*, 3679, 1972.

14. Kossanyi, J., Jost, P., Furth, B., Daccord, G., and Chaquin, P., Intramolecular photoaddition of the excited carbonyl group of cycloalkanones to a non-conjugated ethylenic double bond, *J. Chem. Res. (S)*, 368, 1980; *J. Chem. Res. (M)*, 4601, 1980.

15. Jost, P., Chaquin, P., and Kossanyi, J., Une nouvelle voie d'acces au squelette de l'azulene, *Tetrahedron Lett.*, 465, 1980.

16. Meinwald, J. and Chapman, R. A., The solution photochemistry of some bicyclic ketones, *J. Am. Chem. Soc.*, 90, 3218, 1968.

17. Agosta, W. C. and Herron, D. K., The solution photolysis of camphor, *J. Am. Chem. Soc.*, 90, 7025, 1968.

18. Lange, G. L. and Bosch, M., Photochemistry of *trans*-5-cyclodecenone: A transannular Paternò-Büchi reaction, *Tetrahedron Lett.*, 315, 1971.

19. Mihailovic, M. L., Lorenc, L., Pavlovic, V., and Kalvoda, J., A convenient synthesis of 1α- and 1β-hydroxycholesterol, *Tetrahedron*, 33, 441, 1977.

20. Carless, H. A. J., Enone and dienone rearrangements, in *Photochemistry in Organic Synthesis*, Coyle, J. D., Ed., Royal Society of Chemistry, London, 1986, 118.

21. Engel, P. S. and Schexnayder, M. A., Systematic structural modification in the photochemistry of β,γ-unsaturated ketones. I. Cyclic olefins, *J. Am. Chem. Soc.*, 97, 145, 1975.

22. Cookson, R. C. and Rogers, N. R., Photochemistry of 2-(cycloalkyl-1-enyl)cycloalkanones. A new photochemical reaction of β,γ-unsaturated ketones, *J. Chem. Soc., Perkin Trans. 1*, 1037, 1974.

23. Dalton, J. C. and Chan, H.-F., Photochemistry of 2-aceto-2-methyl methylenecyclohexanes, *Tetrahedron Lett.*, 3351, 1974.

24. Sauers, R. R., Rousseau, A. D., and Byrne, B., Mechanistic studies on the photocycloaddition of γ,δ-unsaturated ketones, *J. Am. Chem. Soc.*, 97, 4947, 1975.

25. Sauers, R. R., Kelly, K. W., and Sickles, B. R., Synthesis and chemistry of some 2-substituted tricyclo[3.3.0.0^{3,7}]octanes, *J. Org. Chem.*, 37, 537, 1972.

26. Sauers, R. R. and Rousseau, A. D., Intramolecular photocycloadditions of naphthyl ketones, *J. Am. Chem. Soc.*, 94, 1776, 1972.

27. Sauers, R. R., Schinski, W., Mason, M. M., O'Hara, E., and Byrne, B., Reductive cleavage of polycyclic oxetanes, *J. Org. Chem.*, 38, 642, 1973.

28. Sauers, R. R., Bierenbaum, R., Johnson, R. J., Thich, J. A., Potenza, J., and Schugar, H. J., Synthesis and chemistry of some polychlorinated oxetanes, *J. Org. Chem.*, 41, 2943, 1976.

29. Sauers, R. R. and Lynch, D. C., Photochemistry of some heteroatom-substituted 5-acylnorbornenes, *J. Org. Chem.*, 45, 1286, 1980.

30. Sauers, R. R., Hagedorn, A. A., III, Van Arnum, S. D., Gomez, R. P., and Moquin, R. V., Synthesis and photochemistry of heterocyclic norbornenyl ketones, *J. Org. Chem.*, 52, 5501, 1987.

31. Nakazaki, M., Naemura, K., and Kondo, Y., Synthesis and chiroptical properties of optically active derivatives of tricyclo[3.3.0.0^{3,7}]octane and oxatricyclononanes, *J. Org. Chem.*, 41, 1229, 1976.

32. Gleiter, R. and Kissler, B., 2,4-Dimethylenetricyclo[3.3.0.0^{3,7}]octane and 2,9-dimethylene tricyclo[4.3.0.0^{3,8}]nonane. Synthesis and through-space interactions, *Tetrahedron Lett.*, 28, 6151, 1987.

33. Gleiter, R., Sigwart, C., and Kissler, B., 4,6-Dimethylenetricyclo[3.3.0.0^{3,7}]octane-2-one and 2,4,6-trimethylenetricyclo[3.3.0.0^{3,7}]octane, *Angew. Chem. Int. Ed.*, 28, 1525, 1989.

34. Maruyama, K. and Kubo, Y., Formation of intramolecular oxetanes in the photolysis of *N*-2-alkenyl alicyclic imides, *J. Org. Chem.*, 42, 3215, 1977.

35. Mazzocchi, P. H., Klingler, L., Edwards, M., Wilson, P., and Shook, D., Intra- and intermolecular Paternò-Büchi reactions on phthalimides. Isolation of the oxetane, *Tetrahedron Lett.*, 24, 143, 1983.

36. Dalton, J. C. and Tremont, S. J., Photochemical synthesis of a 2,5-dioxabicyclo[2.2.0]hexane, *Tetrahedron Lett.*, 4025, 1973; Discovery of a photo-Cope reaction of β-oxa-γ,δ-enones and its implications for the mechanisms of intramolecular photocycloaddition, *J. Am. Chem. Soc.*, 97, 6916, 1975.

37. Mattay, J., Rearrangement and intramolecular cycloaddition of 1-cyclopentenyloxy-2-propanone, *Tetrahedron Lett.*, 21, 2309, 1980.

38. Carless, H. A. J. and Haywood, D. J., Photochemical syntheses of 2,6-dioxabicyclo[3.2.0]heptanes, *J. Chem. Soc., Chem. Commun.*, 1067, 1980.

39. Witkowski, S., Rao, Y. K., Premchandran, R. H., Halushka, P. V., and Fried, J., Total synthesis of (+)-10,10-difluorothromboxane A$_2$ and its 9,11 and 15 stereoisomers, *J. Am. Chem. Soc.*, 114, 8464, 1992.

40. Carless, H. A. J. and Fekarurhobo, G. K., A photochemical route to a thromboxane A$_2$ ring analogue, *J. Chem. Soc., Chem. Commun.*, 667, 1984.

41. Fried, J., Kittisopikul, S., and Hallinan, E. A., Intramolecular 2+2 addition of a difluorovinyl ether aldehyde yields [3.2.0] and no [3.1.1] products, *Tetrahedron Lett.*, 25, 4329, 1984.

42. Carless, H. A. J., Beanland, J., and Mwesigye-Kibende, S., The Paternò-Büchi reaction as a route to medium-ring ethers and acetals, *Tetrahedron Lett.*, 28, 5933, 1987.

47

Photo-Fries Reaction and Related Processes

Miguel A. Miranda
*Universidad Politcnica de
Valencia*

47.1 Definition of the Reaction

The photo-Fries rearrangement, analogous to the classical Lewis acid-catalyzed Fries counterpart, was first reported by Anderson and Reese in 1960.[1] The prototype for this reaction is the photo-chemical transformation of phenyl acetate (1) into *o*- and *p*-hydroxyacetophenone (2 and 3). Since the initial discovery, a number of variations have been devised and the process has been extended to systems (4) such as aryl carbonates, carbamates, sulphonates, and sulphamates, as well as anilides, sulphonanilides, and sulphenanilides. Analogous 1,3-migrations in the corresponding enol derivatives (7) have been observed. Much of this work was done during the first decade from the discovery of the reaction and is summarized in two excellent review articles.[2,3] In this chapter, we shall limit ourselves to the most recent achievements on the rearrangement of aryl esters, paying special attention to the key mechanistic aspects and the most relevant synthetic applications.

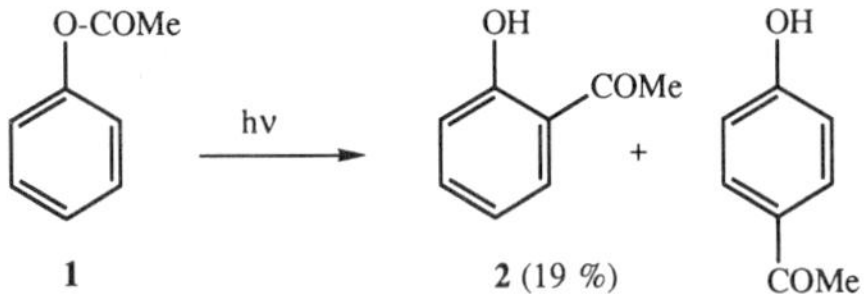

0-8493-8634-9/95/$0.00+$.50
© 1995 by CRC Press, Inc.

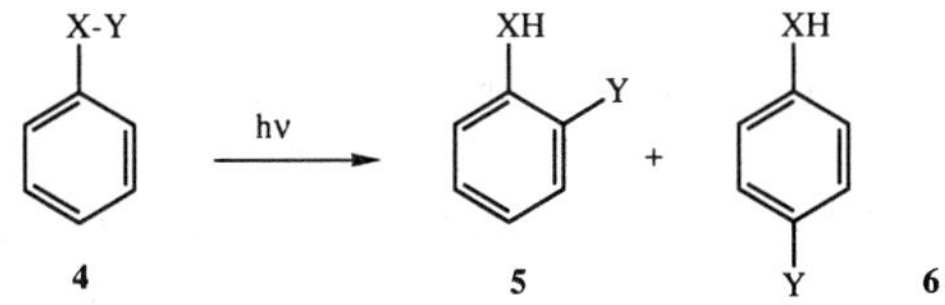

X = O, NH, NR
Y = COR, COOR, CONHR, SO$_2$R, SO$_3$H, etc.

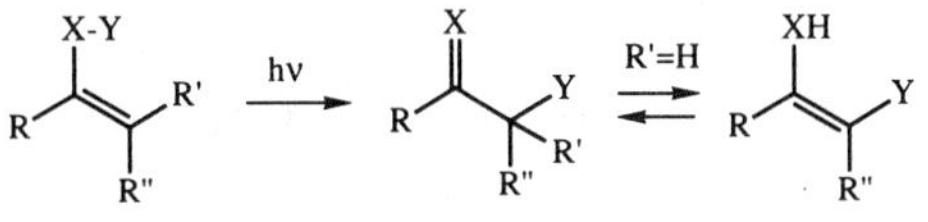

.2 Mechanism

Many reports have appeared during the last few years in connection with the mechanism of the photo-Fries reaction. These contributions have confirmed that photo-Fries rearrangement is a singlet reaction that occurs through homolytic cleavage of the carbonyl-oxygen bond, to give a caged radical pair. In-cage recombination affords the acyl migration products, while hydrogen abstraction by the aryloxy radical leads to the formation of phenols, which are the most common byproducts. This is summarized below for phenyl acetate (1).

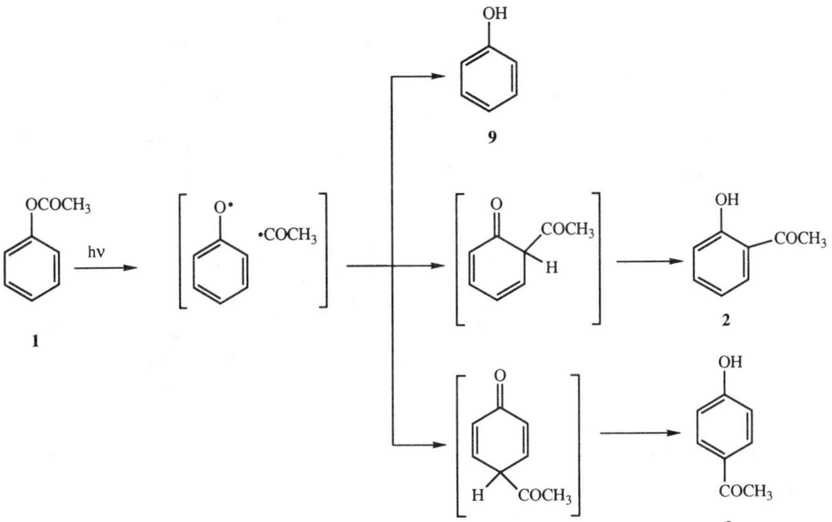

The most convincing arguments in support of this rationalization include: (1) detection of the phenoxy radical by spontaneous Raman spectroscopy in the photolysis of phenyl acetate;[4] (2) spin trapping of acetyl or benzoyl radicals in the photolysis of phenyl acetate or phenyl benzoate, respectively, using 2-methyl-2-nitrosopropane;[5] (3) measurement of magnetic isotope effects and external magnetic field effects upon the photo-Fries rearrangement of 1-naphthyl acetate and its ^{13}C-carbonyl-labeled analogue;[6] and (4) measurement of kinetic isotope effects (^{18}O at the phenolic oxygen, ^{14}C at the α-carbon and ^{14}C at the *ortho*-carbon) as well as magnetic isotope effects (^{13}C at the α-carbon) in the photo-Fries rearrangement of p-methoxyphenyl acetate.[7]

47.3 Synthetic Applications

Exploitation of the photo-Fries rearrangement as a general synthetic method depends on the degree of variability in the phenolic and the acyl substructures, as well as the possibility of controlling the outcome of the process through manipulation of the experimental conditions. These aspects have been extensively investigated during recent years.

Variations in the Phenolic Moiety

Soon after its discovery, the photo-Fries reaction was applied to esters of a wide variety of phenols.[2,3] However, the presence of certain electron-withdrawing substituents (for instance acyl groups) in the phenolic ring was found to inhibit the rearrangement. This limitation is also inherent to the classical Fries rearrangement, due to the electrophilic nature of electron-withdrawing substituents. The fact that the photochemical version of the reaction proceeds in neutral media and at room temperature has allowed the polyacylation of phenols, using enol acetates or cyclic acetals as carbonyl blocking groups.[8-11] This is shown below for the conversion of *p*-hydroxyacetophenone (3) into the hydroxydiketone (13).

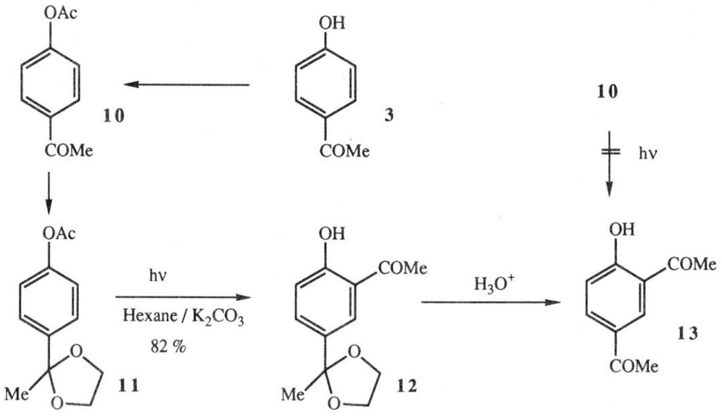

In addition to the photoreactivity of esters of simple phenols, the condensed polynuclear analogues may form esters which are also prone to undergo the photo-Fries rearrangement. In this context, the studies on naphthyl esters[12-14] can be cited among the most significant work. Likewise, this photoreaction has been employed for the oxygen-to-carbon migration in condensed heterocyclic compounds. Some examples are the ring acylations achieved by photolysis of the acyloxy derivatives of coumarins,[15] benzopyrans,[16] indoles,[17] and dihydrobenzofurans.[18] Thus, the absolute configuration of rutaretin methyl ether (20) has been established by transformation into a derivative, whose independent synthesis was achieved from S-marmesin, through a sequence that involves photo-Fries rearrangement of the diacetate 14.[18]

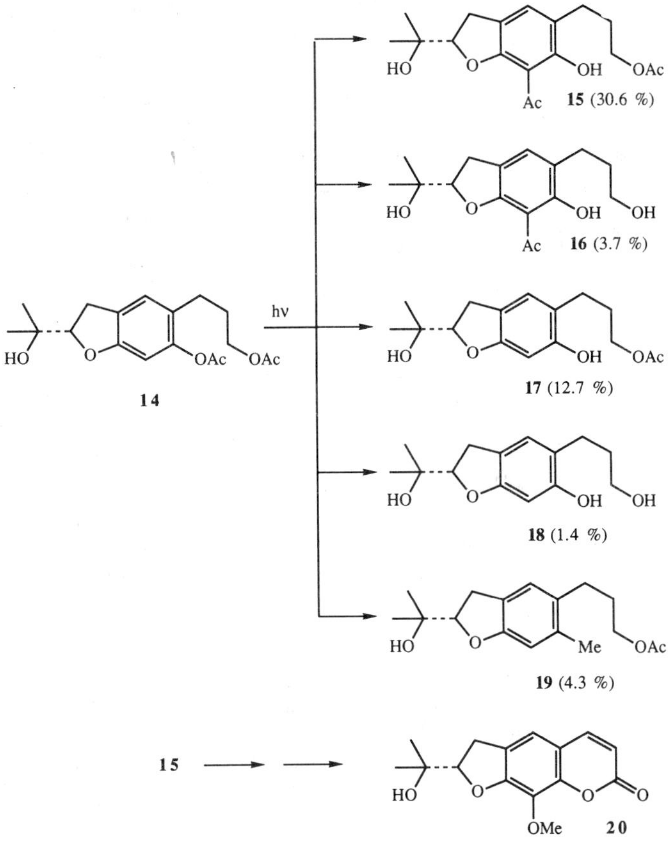

Variations in the Acyl Moiety

When the acyl group carries a second functionality (double or triple bond, hydroxy group, carbonyl group, etc.), interaction of the latter with the *ortho*-hydroxy group, subsequent to the rearrangement step, may be a general principle for the construction of heterocyclic ring systems. This has been applied to the synthesis of furanones,[19,20] chromones,[21–23] chromenes,[24–26] flavones,[27–29] aurones,[29] xanthones,[30–33] thioxanthones,[34] or acridones.[35]

For instance, photolysis of aryl esters of phenylpropiolic acid (21) gives rise to *o*-hydroxyaryl phenylethynyl ketones (22). The latter compounds can undergo cyclization in two different ways, depending upon the reaction conditions: when potassium carbonate in acetone is used, flavones (23) are formed by a 6-*endo*-dig process;[36] by contrast, the use of sodium ethoxide in ethanol favors the 5-*exo*-dig process,[36] to give aurones (24) as the major products.[29]

The side-chain functionality can be masked, as in the case of the protected ketoesters (25), whose irradiation and subsequent deprotection provide an alternative entry to the flavone system.[28]

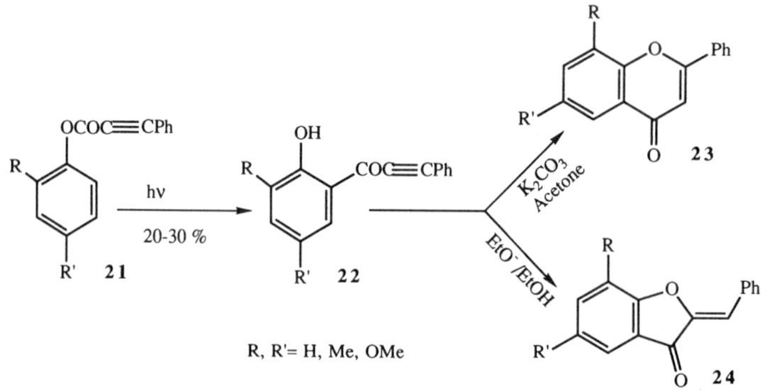

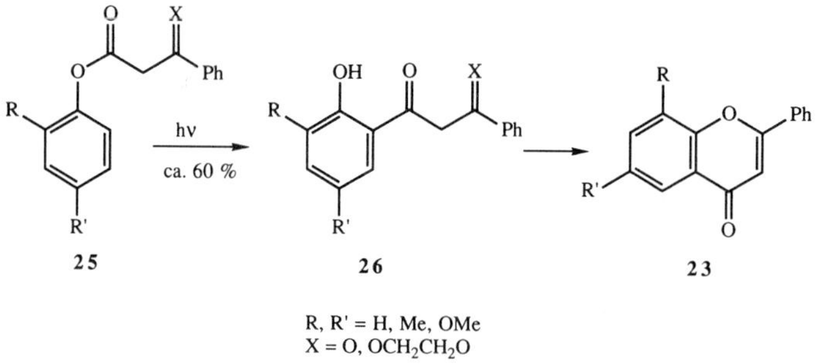

R, R'= H, Me, OMe

R, R' = H, Me, OMe
X = O, OCH₂CH₂O

In another application, the total synthesis of bikaverin (29) has been achieved using the photo-Fries rearrangement of the protected everninic acid ester (27) as a key step for the construction of the benzo[b]xanthene skeleton.[31]

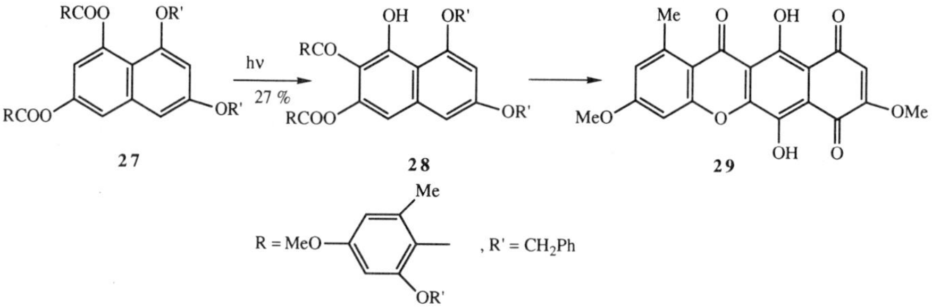

R = MeO— , R' = CH₂Ph

Use of Heterogeneous Media

In view of its paradigmatic nature as a process involving in-cage recombination vs. diffusion of organic radical pairs, the photo-Fries rearrangement has been chosen as a model reaction to study the influence of heterogeneous media as modifiers of reactivity and selectivity. In an early study, the photo-Fries rearrangement of aryl esters was attempted on the surface of silica gel.[37] Subsequently, cyclodextrins have been employed to carry out the photo-Fries rearrangement of aryl esters in a restricted environment.[38–41] The initial observations[38,39] appeared to indicate cyclodextrin-enhanced *para*-selectivity; however, more recent works[40,41] report an increase in the *ortho*-product attributable to cyclodextrin complexation. Other modifications involve the use of starch,[39] amylose,[39] sodium dodecyl sulfate,[42] or potassium carbonate[10,43] to create the heterogeneous irradiation system.

More conventional liquid-liquid biphasic conditions have been also conveniently employed to increase the efficiency of the photo-Fries reaction. For instance, *p*-methoxyphenyl esters of α,β-unsaturated carboxylic acids (30) undergo photochemical rearrangement, followed by basic cyclization, to give 4-chromanones (32) with moderate overall yields. Using a two-phase system, benzene/aqueous sodium hydroxide, chromanones are obtained directly in nearly quantitative yields. Subsequent reduction/dehydration affords 2*H*-chromenes (33), interesting because of their antijuvenile hormone activity.[25]

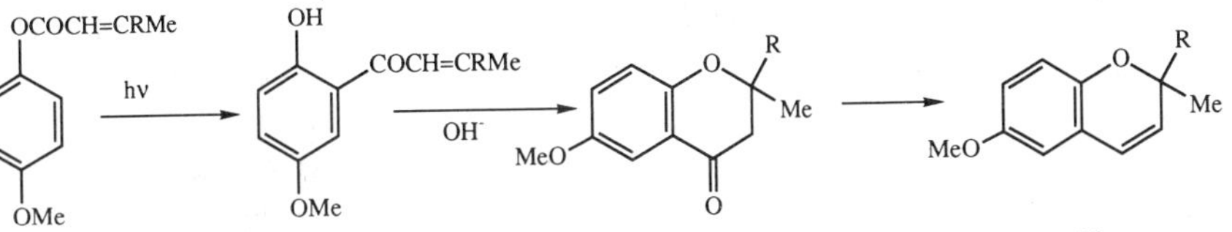

7.4 Industrial Applicability

Much attention has been paid to the photo-Fries rearrangement of aromatic polyesters due to the remarkable photostabilization produced by the resulting polymer-bound *o*-hydroxycarbonyl chromophores. This property has been attributed to the high extinction coefficient of the photo-Fries products (internal filters) and to their ability to dissipate the absorbed energy by nonphotochemical pathways, such as excited-state proton transfer.[44,45] Bisphenol A polycarbonates,[46,47] poly(aryl cinnamates),[47,48] aromatic diisocyanate-based polyurethanes,[50,51] fluorene-based polyesters,[52] or 1- and 2-naphthyl methacrylate copolymers[53,54] are among the polymeric materials capable of undergoing this type of photochemical transformation. In some cases, a wavelength-dependent photochemistry of the polymer has been observed,[46,48,49] due to competition between the photo-Fries reaction and other photoprocesses (crosslinking, oxidation, etc.).

Another field of application is the design of polymeric imaging systems.[55] The lithographic potential of a photochemical reaction is based on the possibility of dissolving selectively either the exposed or the unexposed areas of a polymer film. Since all the photo-Fries products are phenols, the irradiated polyester should be easily dissolved in aqueous base, while the unchanged starting material should remain undissolved. The key photochemical step of this sequence of processes is shown below for the conversion of poly(*p*-acetoxystyrene) (34) into (35).[56]

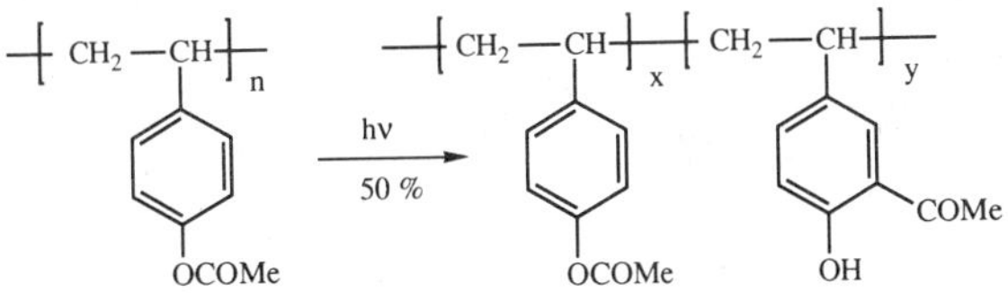

References

1. Anderson, J. C. and Reese, C. B., Photo-induced Fries rearrangement, *Proc. Chem. Soc. London*, 217, 1960.
2. Stenberg, V. I., Photo-Fries reaction and related arrangements, *Org. Photochem.*, 1, 127, 1967.
3. Bellus, D., Photo-Fries rearrangement and related photochemical [1,j]-shifts (j = 3, 5, 7) of carbonyl and sulfonyl groups, *Adv. Photochem.*, 8, 109, 1971.
4. Beck, S. M. and Brus, L. E., Transient intermediates in the photo-Fries isomerization of phenyl acetate via spontaneous Raman spectroscopy, *J. Am. Chem. Soc.*, 104, 1805, 1982.
5. Rosenthal, I., Mosoba, M. M., and Riesz, P., Spin trapping with 2-methyl-2-nitrosopropane: Photochemistry of carbonyl-containing compounds. Methyl radical formation from dimethyl sulfoxide, *Can. J. Chem.*, 60, 1486, 1982.

6. Nakagaki, P., Hiramatsu, M., Watanabe, T., Tanimoto, Y., and Nagakura, S., Magnetic isotope and external magnetic field effects upon the photo-Fries rearrangement of 1-naphthyl acetate, *J. Phys. Chem.*, 89, 3222, 1985.

7. Shine, H. J. and Subotkowski, W., Kinetic (^{18}O and ^{14}C) and magnetic (^{13}C) isotope effects in the photo-Fries rearrangement of 4-methoxyphenyl acetate, *J. Org. Chem.*, 52, 3815, 1987.

8. García, H., Miranda, M. A., Roquet-Jalmar, M. F., and Martínez-Utrilla, R., Influence of enol acetylation on the photo-Fries rearrangement of an *ortho*-acylaryl benzoate, *Liebigs Ann. Chem.*, 2238, 1982.

9. García, H., Martínez-Utrilla, R., Miranda, M. A., and Roquet-Jalmar, M. F., Intra- and intermolecular photoreactions of *o*-benzoyloxyacetophenone derivatives, *J. Chem. Res. (S)*, 350, 1982.

10. García, H., Miranda, M. A., and Primo, J., The photo-Fries rearrangement of acetoxyacetophenones using cyclic acetals as carbonyl blocking groups in the presence of potassium carbonate. An improved procedure for the synthesis of diacylphenols, *J. Chem. Res., (S)*, 100, 1986.

11. García, H., Martínez-Utrilla, R., and Miranda, M. A., Cyclic acetals as carbonyl blocking groups in the photo-Fries rearrangement of acyl substituted aryl esters, *Tetrahedron*, 41, 3131, 1985.

12. Crouse, D. J., Hurlbut, S. L., and Wheeler, D. M. S., Photo-Fries rearrangements of 1-naphthyl esters in the synthesis of 2-acylnaphthoquinones, *J. Org. Chem.*, 46, 374, 1981.

13. Fariña, F., Martínez-Utrilla, R., and Paredes, M. C., Polycyclic hydroxyquinones. VIII. Preparation of acetylhydroxynaphthazarines by photo-Fries rearrangement. A convenient synthesis of spinochrome A, *Tetrahedron*, 38, 1531, 1982.

14. Greenland, H., Pinhey, J. T., and Sternhell, S., The photochemistry of 2-acetoxynaphthalen-1(2H)-ones, *J. Chem. Soc., Perkin Trans. 1*, 1789, 1986.

15. Kulshrestha, S. K., Dureja, P., and Mukerjee, S. K., Photo-induced reactions. IV. Studies on photo-Fries migration of some coumarins, *Ind. J. Chem.*, 23B, 1064, 1984.

16. Miranda, M. A., Primo, J., and Tormos, R., Photochemistry of 7-acetoxybenzopyran derivatives. Synthesis of eupatoriochromene and encecalin, *Tetrahedron*, 45, 7593, 1989.

17. Chan, A. C. and Hilliard, P. R., Regioselectivity of the photo-Fries rearrangement in acetoxyindoles, *Tetrahedron Lett.*, 30, 6483, 1989.

18. Ishii, H., Sekiguchi, F., and Ishikawa, T., Studies on the chemical constituents of rutaceous plants. XLI. Absolute configuration of rutaretin methyl ether, *Tetrahedron*, 37, 285, 1981.

19. Fillol, L., Martínez-Utrilla, R., Miranda, M. A., and Morera, I., Photochemical versus aluminium chloride-catalyzed Fries rearrangement of aryl hydrogen succinates. Synthesis of 2(3H)-furanones, *Monatsh. Chem.*, 120, 863, 1989.

20. Martínez-Utrilla, R. and Miranda, M. A., Indirect hydroquinone succinoylation via a photo-Fries rearrangement. Application to the synthesis of enol lactones, *Tetrahedron Lett.*, 21, 2281, 1980.

21. Alvaro, M., García, H., Miranda, M. A., and Primo, J., Neighbouring group participation in the photolysis of aryl esters of unsaturated 1,4-dicarboxylic acids, *Recl. Trav. Chim., Pays-Bas*, 105, 233, 1986.

22. Alvaro, M., García, H., Iborra, S., Miranda, M. A., and Primo, J., New photochemical approaches to the synthesis of chromones, *Tetrahedron*, 43, 143, 1987.

23. Miranda, M. A., Primo, J., and Tormos, R., Influence of the stereochemistry on the rate of cyclization of *cis* and *trans* o-hydroxyaryl alkenyl ketones. Mechanistic implications, *Tetrahedron*, 43, 2323, 1987.

24. Miranda, M. A., Primo, J., and Tormos, R., A new synthesis of precocene II and precocene III based on the photo-Fries rearrangement of a sesamol ester, *Heterocycles*, 32, 1159, 1991.

25. Miranda, M. A., Primo, J., and Tormos, R., A new synthesis of 4-chromanones, *Heterocycles*, 19, 1819, 1982.

26. Miranda, M. A., Primo, J., and Tormos, R., Studies on the synthesis of precocenes. The photo-Fries rearrangement of esters of α,β-unsaturated carboxylic acids and meta-oxygenated phenols, *Heterocycles*, 27, 673, 1988.

27. García, H., Iborra, S., Miranda, M. A., and Primo, J., Application of the photo-Fries rearrangement of aryl dihydrocinnamates to the synthesis of flavonoids, *Heterocycles*, 23, 1983, 1985.

28. García, H., Iborra, S., Miranda, M. A., and Primo, J., Photolysis of cyclic acetals of aryl benzoylacetates as the key step in a new synthesis of flavones, *Heterocycles*, 24, 2511, 1986.

29. García, H., Iborra, S., Miranda, M. A., and Primo, J., 6-*Endo*-dig versus 5-*exo*-dig ring closure in *o*-hydroxyaryl phenylethynyl ketones. A new approach to the synthesis of flavones and aurones, *J. Org. Chem.*, 51, 4432, 1986.

30. Lewis, J. R. and Paul, J. G., Oxidative coupling. 11. Approaches to the synthesis of bikaverin, *J. Chem. Soc., Perkin Trans. 1*, 770, 1981.

31. Katagiri, N., Nakano, J., and Kato, T., Synthesis of bikaverin, *J. Chem. Soc., Perkin Trans. 1*, 2710, 1981.

32. Díaz-Mondéjar, M. R. and Miranda, M. A., 2'-Acetoxy-2-hydroxy-5-methoxybenzophenone. Photochemical synthesis, transacylation and cyclization to 2-methoxyxanthone, *Tetrahedron*, 38, 1523, 1982.

33. Díaz-Mondéjar, M. R. and Miranda, M. A., Photolysis of 2-aryloxy- or 2-arylthio-1,3-benzodioxan-4-ones, *Heterocycles*, 22, 1125, 1984.

34. Belled, C., Miranda, M. A., and Simón-Fuentes, A., Fototransposición de Fries en derivados de tiosalicilato de fenilo. Una nueva aproximación a la síntesis de tioxantonas, *An. Quím.*, 85C, 39, 1989.

35. Belled, C., Miranda, M. A., and Simón-Fuentes, A., Fototransposición de Fries de esters fenólicos de los ácidos antranílico y *N*-acetilantranílico. Aplicaciones a la síntesis de heterociclos, *An. Quím.*, 86, 431, 1990.

36. Baldwin, J. E., Rules for ring closure, *J. Chem. Soc., Chem. Commun.*, 734, 1976.

37. Avnir, D., de Mayo, P., and Ono, I., Biphasic photochemistry: The photo-Fries rearrangement on silica gel, *J. Chem. Soc., Chem. Commun.*, 1109, 1978.

38. Ohara, M. and Watanabe, K., Selective photochemical Fries rearrangement of phenyl acetate in the presence of β-cyclodextrin, *Angew, Chem., Int. Ed.*, 14, 820, 1975.

39. Chênevert, R. and Voyer, N., Photochemical rearrangement of phenyl benzoate in the presence of cyclodextrins and amilose, *Tetrahedron Lett.*, 25, 5007, 1984.

40. Syamala, M. S., Rao, B. N., and Ramamurthy, V., Modification of photochemical reactivity by cyclodextrin complexation: Product selectivity in photo-Fries rearrangement, *Tetrahedron*, 44, 7234, 1988.

41. Veglia, A. V., Sánchez, A. M., and de Rossi, R. H., Change of selectivity in the photo-Fries rearrangement of phenyl acetate induced by β-cyclodextrin, *J. Org. Chem.*, 55, 4083, 1990.

42. Singh, A. K. and Sonar, S. N., Photorearrangement of aryl esters in micellar medium, *Synth. Commun.*, 15, 1113, 1985.

43. García, H., Primo, J., and Miranda, M. A., The photo-Fries rearrangement in the presence of potassium carbonate: A convenient synthesis of *ortho*-hydroxyacetophenones, *Synthesis*, 901, 1985.

44. Allen, N. S., Photostabilizing action of *ortho*-hydroxy aromatic compounds: A critical review, *Polym. Photochem.*, 3, 167, 1983.

45. Ranby, B. and Rabek, J. F., *Photodegradation, Photo-oxidation and Photostabilization of Polymers*, Wiley, New York, 1975.

46. Rivaton, A., Sallet, D., and Lemaire, J., The photochemistry of bisphenol A polycarbonate reconsidered, *Polym. Photochem.*, 3, 463, 1983.

47. Torikai, A., Murata, T., and Fueki, K., Photo-induced reactions of polycarbonate studied by ESR, viscosity and optical absorption measurements, *Polym. Photochem.*, 4, 255, 1984.

48. David, M., Creed, D., Griffin, A. C., Hoyle, C. E., and Venkataran, K., Photochemical crosslinking of main-chain liquid-crystalline polymers containing cinnamoyl groups, *Makromol. Chem. Rapid Commun.*, 10, 391, 1989.

49. Creed, D., Griffin, A. C., Hoyle, C. E., and Venkataran, K., Chromophore aggregation and concomitant wavelength-dependent photochemistry of a main-chain liquid-crystalline poly(aryl cinnamate), *J. Am. Chem. Soc.*, 112, 4049, 1990.

50. Hoyle, C. E. and Kim, K. J., Photolysis of aromatic diisocyanate-based polyurethanes in solution, *J. Polym. Sci. Part A Polym. Chem.*, 24, 1879, 1986.
51. Hoyle, C. E., Chawla, C. P., and Kim, K. J., The effect of flexibility on the photodegradation of aromatic diisocyanate-based polyurethanes, *J. Polym. Sci. Part A Polym. Chem.*, 26, 1295, 1988.
52. Lo, J., Lee, S. N., and Pearce, E. M., Photo-Fries rearrangement of fluorene-based polyarylates, *J. Appl. Polym. Sci.*, 29, 35, 1984.
53. Holden, D. A., Jordan, K., and Safarzadeh-Amiri, A., Studies of polymer photostabilization using fluorescence spectroscopy: Photochemistry of naphthyl methacrylate copolymers, *Macromolecules*, 19, 895, 1986.
54. Wang, Z., Holden, D. A., and McCourt, F. R. W., Generation of nonrandom chromophore distributions by the photo-Fries reaction of 2-naphthyl acetate in poly(methyl methacrylate), *Macromolecules*, 23, 3773, 1990.
55. Tessier, T. G., Frechet, J. M. J., Willson, C. G., and Ito, H., The photo-Fries rearrangement and its use in polymeric imaging systems, *Am. Chem. Soc. Symp. Ser.*, 266, 269, 1984.
56. Frechet, J. M. J., Tessier, T. G., Willson, C. G., and Ito, H., Poly[*p*-(formyloxy)styrene]: Synthesis and radiation-induced decarbonylation, *Macromolecules*, 18, 317, 1985.

48

Photorearrangement Reactions of Cyclohexenones

David I. Schuster
New York University

48.1 Introduction

The molecular rearrangements of photoexcited cyclohexenones have received a great deal of attention for over 30 years. The subject has been reviewed several times,[1,2] most recently by this author in 1989.[3] The present chapter will attempt to summarize briefly the main features of these classic photochemical rearrangement reactions.

Type A photorearrangements in either polar or nonpolar solvents involve conversion of 4,4-disubstituted cyclohexenones into bicyclo[3.1.0]hexanones (lumiketones), a reaction that is often accompanied by ring contraction to 3-substituted cyclopentenones.[2] Classic examples of these transformations are given in Equations 48.1 and 48.2 for 4,4-dimethylcyclohexenone (1) and testosterone acetate (4).[4,5] Thus, in *t*-butyl alcohol, 1 is transformed into lumiketone 2 as well as 3-isopropylcyclopent-2-en-1-one (3), while in acetic acid two additional cyclopentanone derivatives are isolated. In a completely analogous manner, 4 is converted into 5 and 6 on irradiation in *t*-BuOH. Chemical yields of the lumiketones are usually optimal in *t*-BuOH in which competitive photoreactions of the enones (e.g., photoreduction and photodeconjugation) are minimized. Photodimerization of cyclohexenones is a general competing reaction of these compounds,[2] but can be virtually eliminated by using low concentrations of the enones. The quantum efficiencies for rearrangement are quite low, generally less than 0.01,[2] so that long irradiation times (many hours, even days) are required to effect substantial conversion of starting material. Possible reasons for this will be discussed later.

0-8493-8634-9/95/$0.00+$.50
© 1995 by CRC Press, Inc.

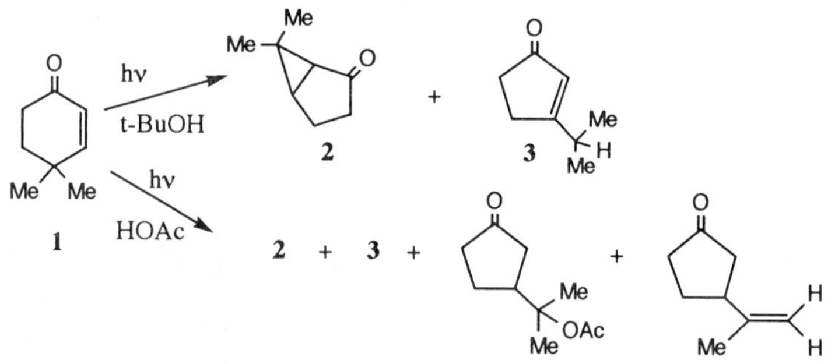

$$(48.1)$$

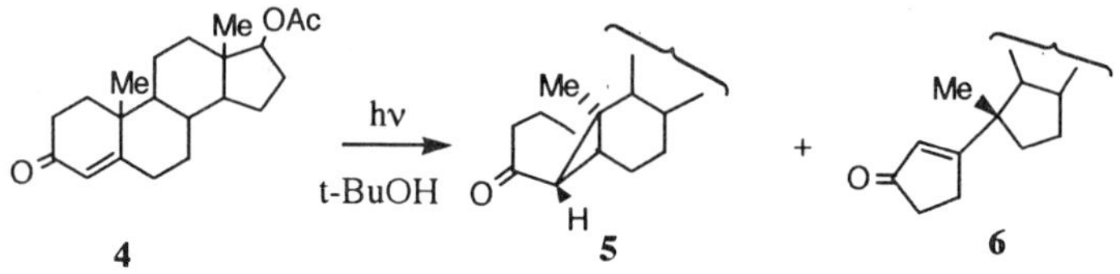

$$(48.2)$$

Dauben and co-workers established some time ago that the photorearrangement of cyclohexenones to lumiketones only occurs when two substituents are present at C-4 of the enone, and that at least one of them must be an alkyl group.[6] In the case of 4-alkyl-4-aryl-2-cyclohexenones such as 7, the lumiketone photorearrangement competes with migration of the phenyl group, as shown in Equation 48.3.[7] The former process is more prominent in polar solvents such as acetonitrile, dimethylformamide, and methanol, but only phenyl migration (type B reaction) is seen in benzene and ether. Only aryl migration is seen in 4,4-diarylcyclohexenones such as 8 (Equation 48.4).[8] A totally analogous situation exists with 4-alkyl-4-vinylcyclohexenones; only vinyl migration is observed with dienone 9 (Equation 48.5).[9] This type of photorearrangement, which is structurally analogous to the well-known di-π-methane photorearrangement,[10] will be discussed later.

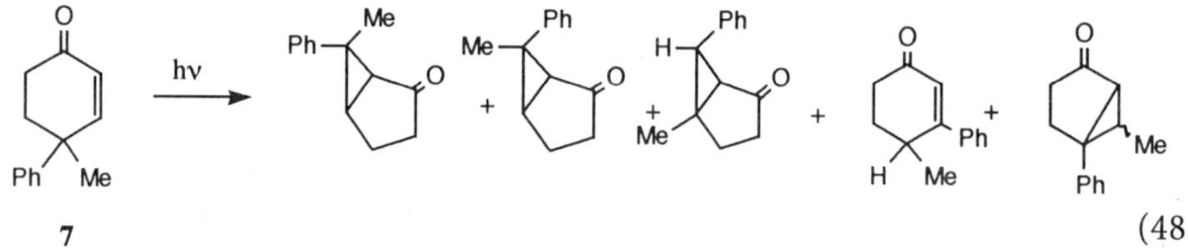

$$(48.3)$$

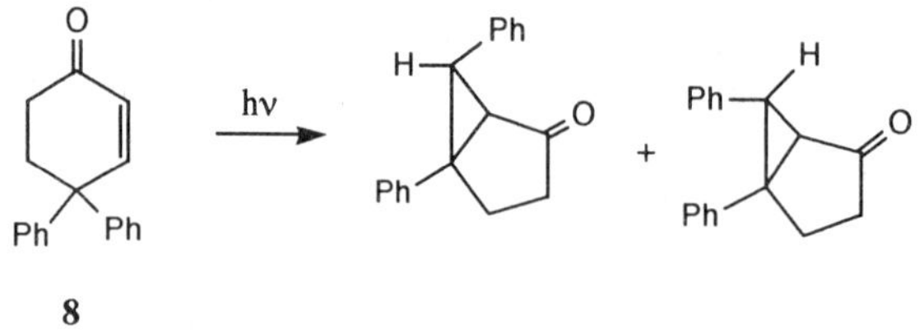

$$(48.4)$$

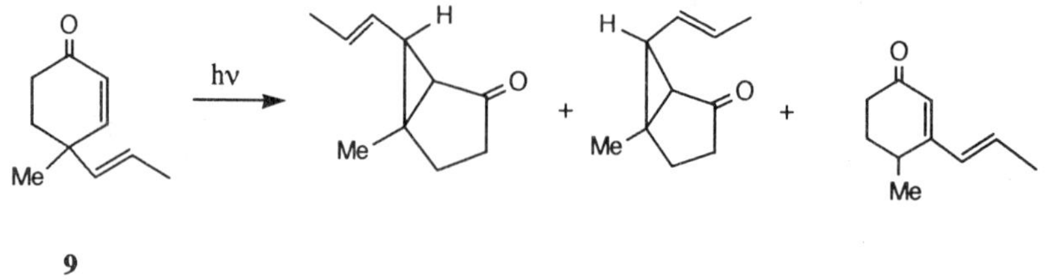

$$(48.5)$$

8.2 Mechanism and Stereochemistry of the Lumiketone Photorearrangement of Cyclohexenones

Schaffner and co-workers demonstrated, using a deuterium-labeled testosterone derivative, that the photorearrangement to the lumiketone (Equation 48.2) occurs with retention of configuration at C-1 and inversion at C-10 (analogous to C-5 and C-4 of a simple cyclohexenone).[11] Chapman reported that photorearrangement of the optically active phenanthrone 10 proceeded with inversion at C-10 and no detectable loss of optical purity (Equation 48.6).[12] These findings suggested that the rearrangement involves more or less concerted bond switching and does not proceed stepwise via biradical intermediates leading to loss of stereochemical integrity at the stereogenic centers. This was confirmed by Schuster and co-workers using the simple chiral cyclohexenones 11 and 12 in which any bias induced by the fused ring systems above was absent (see Equations 48.7 and 48.8).[13]

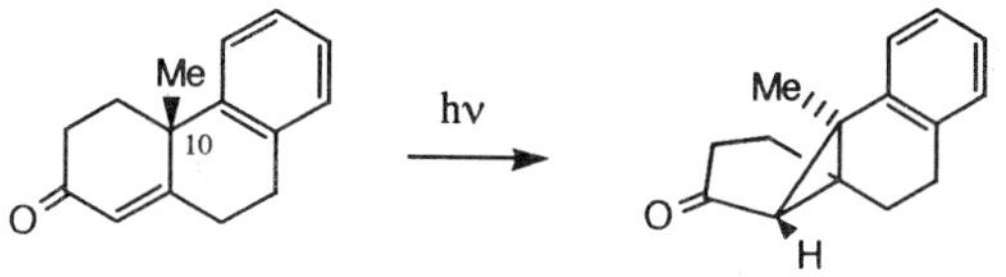

(48.6)

(48.7)

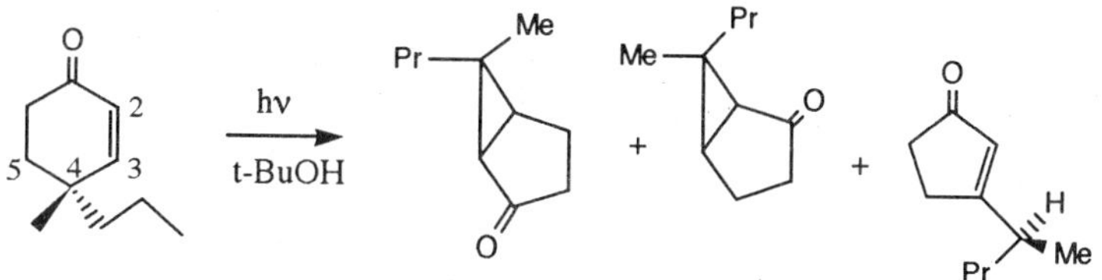

(48.8)

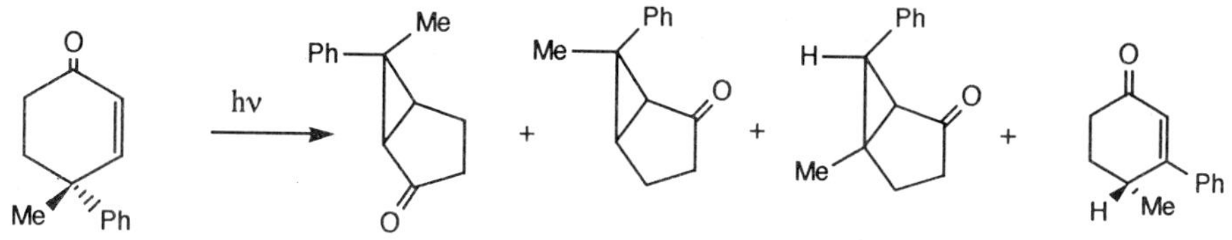

The optical purity of the lumiketones was identical to that of the starting enones, and there was no loss of optical purity in recovered 11 even after 325 h of continuous irradiation. This establishes that cleavage of the bond between C-4 and C-5 of the enone is concerted with formation of the bonds between C-3 and C-5 and between C-2 and C-4. In a formal sense, the reaction occurs with inversion of configuration at C-4 and retention at C-5. Using Woodward-Hoffmann terminology,[14] this corresponds to a $_{\pi}2_a + _{\sigma}2_a$ process, with antarafacial addition to both reactant bonds, the C2-C3 π-bond and the C4-C5 σ-bond. This is depicted for monocyclic cyclohexenones in Scheme 48.1. In fused ring systems such as 4 and 10, the rearrangement occurs on only one face of the enone because of steric constraints (i.e., the necessity of *cis*-fusion of the three-membered ring to both the five- and six-membered rings), affording only one lumiketone product. Shaik[15] has suggested that intersystem crossing might be facilitated by the twisting necessary to achieve the geometry corresponding to the $_{\pi}2_a + _{\sigma}2_a$ transition state, in which case spin inversion might well occur along the pathway leading directly from the enone triplet to the lumiketone in its ground electronic state. This hypothesis is consistent with the finding that the rigid bicyclononenone 13a, which is a fused ring analogue of enone 1, does not undergo the analogous photorearrangement to lumiketone 14, which itself is not strained prohibitively (Equation 48.9).[16] In fact, the photochemical behavior of enones

13a and **13b**, which are severely constrained with respect to twisting around the C=C bond, is atypical with respect to that of most cyclohexenones, in that hydrogen abstraction reactions predominate, even from normally unreactive solvents such as *t*-butyl alcohol and acetone.[16,17]

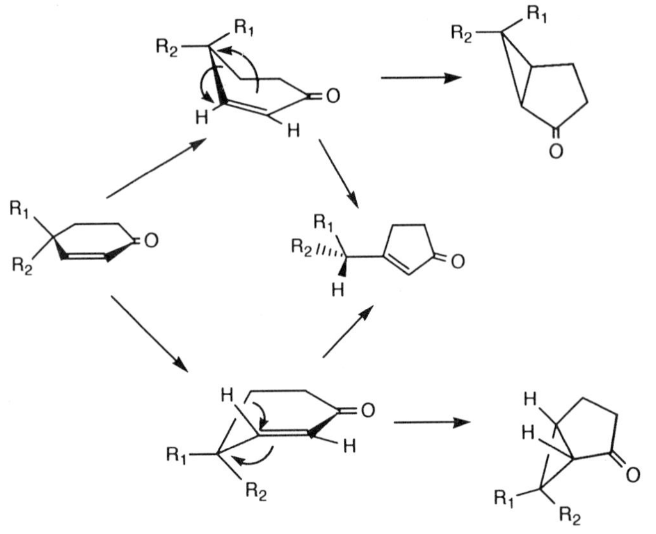

SCHEME 48.1

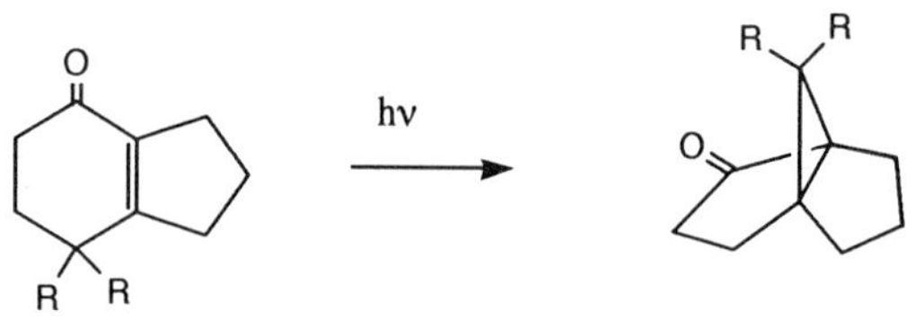

13

a) R = Me
b) R = H

(48.9)

Recent studies using transient absorption spectroscopy (laser flash photolysis) and time-resolved photoacoustic calorimetry confirm that relaxed triplet excited states of simple cyclohexenones are indeed highly twisted.[18,19] Thus, cyclohexenone itself has a triplet energy of 63 kcal/mol[-1], about 10 kcal mol[-1] lower than that of the conformationally rigid analogy **13b**. Substituted cyclohexenones have intermediate energies. Based upon the effect of substituents at C-3 on enone triplet energy, it was proposed that cyclohexenone π,π^* triplets relax toward the geometry shown in **15**, in which the β-carbon of the enone

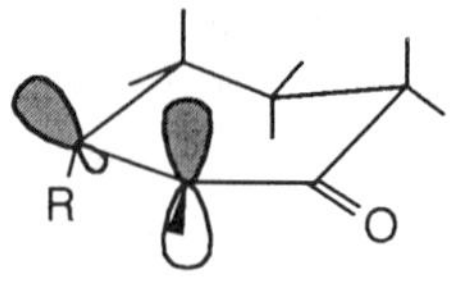

15

becomes pyramidal (sp^3) while the α-carbon remains trigonal (sp^2). By analogy with studies of the effect of geometry on intersystem crossing in alkenes,[20] spin-orbit coupling leading to radiationless decay to the ground-state potential surface is predicted to be optimal at this geometry. To the extent that twisting around the C=C bond occurs in the triplet excited state, the triplet- and ground-state potential surfaces approach each other in energy, leading to increasingly rapid radiationless decay from the upper to the lower electronic state. Thus, it is not surprising that the triplet lifetimes of cyclohexenone and 4,4-disubstituted analogues are only ca. 25 ns in both polar and nonpolar solvents, while that of BNEN (13b) is 1500 ns. All 3-substituted cyclohexenones studied thus far (3-methyl-, 3-ethyl-, 3-isopropyl-, and 3-t-butyl-cyclohexenone, as well as 3,5,5-trimethyl-cyclohexenone) all have triplet lifetimes between 70 and 120 ns.[19,21]

Schuster proposed that the low quantum efficiency for the lumiketone photorearrangement is due to the competition between reversion to ground state and intramolecular $_\sigma2_a + _\pi2_a$-cycloaddition at the twisted enone triplet geometry after decay to the ground-state potential surface.[2,13] This scheme by no means precludes the possibility suggested by Shaik[15] that spin inversion occurs concomitantly with formation of lumiketones. According to this picture, the quantum efficiency of lumiketone formation ought to increase with temperature, but no data to this effect have yet been published.

The Type A photorearrangement of cyclohexenones is structurally analogous to the well-known photorearrangement of 2,5-cyclohexadienones to bicyclo[3.1.0]hexenones.[22,23] There is good evidence that the latter reaction proceeds via a triplet-derived zwitterion intermediate, as shown in Equation 48.10 for the classic example of 4,4-diphenyl-2,5-cyclohexadienone.[22]

$$\text{(48.10)}$$

The sequence of intermediates shown in Equation 48.10 was confirmed by Schuster and Patel using 4-methyl-4-trichloromethyl-2,5-cyclohexadienone,[24] in which case the zwitterions could be intercepted by nucleophiles.[24,25] In an attempt to mechanistically link the lumiketone photorearrangements of cyclohexenones and cyclohexadienones, Chapman proposed that the cyclohexenone photorearrangement proceeds via a "polar state" depicted as 16 (see Scheme 48.2),[1a,5] although it was never specified clearly whether 16 was meant to represent a ground- or excited-state species. There is a fundamental distinction between the oxyallyl species shown in Equation 48.10 and structure 16, which could represent a resonance contributor to the ground state or conceivably a polarized singlet excited state species. While one could formally account for the course of rearrangement by ring contraction to 17 followed by ring closure to 2, such a mechanism does not account for the stereospecificity of the reaction nor the fact that twisting of the reactive enone species seems to be mandatory for the reaction to take place. Further evidence against the polar state hypothesis was provided by the results of a study of enone 18 by Schuster and Brizzolara.[27] It was observed that two competing reactions occurred in this system — rearrangement to lumiketone 19 and H-abstraction leading to loss of CH_2OH radicals (Equation 48.11). Quenching experiments showed that these reactions proceeded from a common triplet excited state. No products attributable to a polar state intermediate were observed. In cases where products of nucleophilic attack are observed, it is likely that these are not formed directly from cyclohexenone excited states, but represent secondary reactions of the lumiketones.

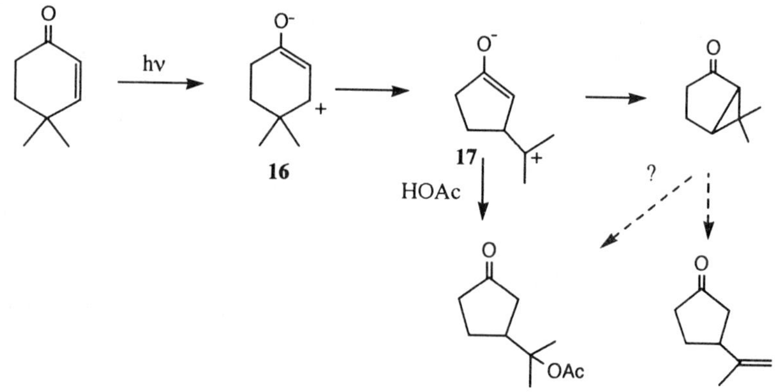

SCHEME 48.2

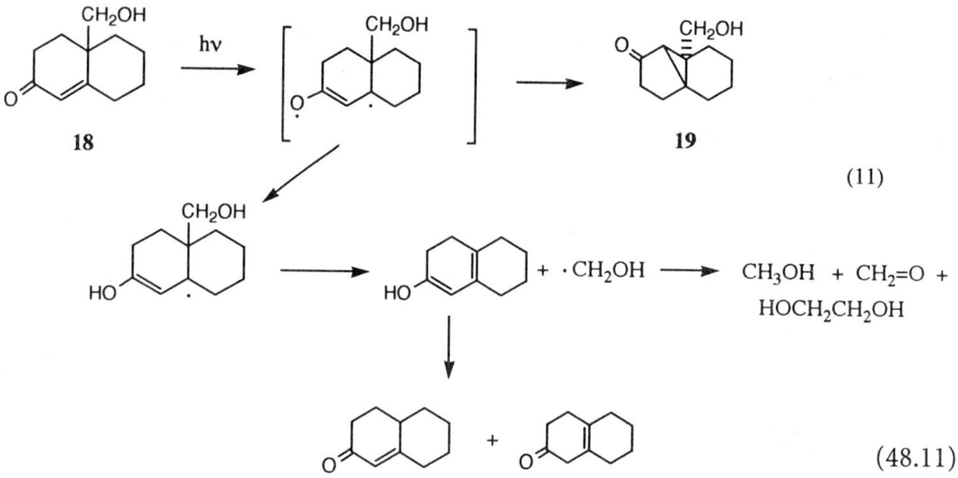

(11)

(48.11)

The nature of the triplet excited state responsible for photorearrangement to lumiketones has been the subject of considerable speculation.[2] It is now generally agreed that the lowest triplet state of most cyclohexenones is a π,π^* state, and it is this state that undergoes energetic relaxation by twisting around the C=C bond, as described earlier.[19] Spectroscopic studies of steroid enones at low temperatures also suggest strongly that the lowest triplet is a π,π^* state.[28] The fact that lumiketone formation from **7** predominates in more polar solvents is also consistent with this assignment. In the case of phenanthrone **10**, five products are seen upon irradiation in 2-propanol, the lumiketone **20** and four reduction products, **21** to **24** (Equation 48.12).[29] The Stern-Volmer slopes for quenching by naphthalene are distinctly different for **20** and **21** vs. **22** to **24**. Clearly the pathway to formation of **23** and **24** involves H-abstraction on the carbonyl oxygen atom, a characteristic reaction of $^3n,\pi^*$ states of aldehydes and ketones. Thus, by default, the triplet state leading to **20** must be the $^3\pi,\pi^*$ state. The fact that formation of the *cis*-fused dihydroketone **21** arises stereoselectively from the $^3\pi,\pi^*$ state can be rationalized in terms of the constraints associated with approach of the H-donor (2-propanol) to a twisted enone moiety. No such constraints operate for reduction of the $^3n,\pi^*$ state.[29]

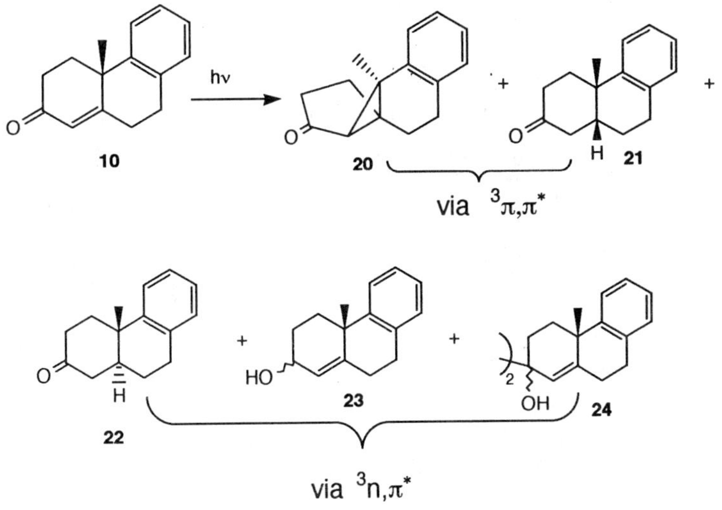

$$\text{via } ^3\pi,\pi^*$$

$$\text{via } ^3n,\pi^* \tag{48.12}$$

8.3 Mechanism and Stereochemistry of the Type B Photorearrangement of Cyclohexenones: Aryl and Vinyl Migrations

As indicated earlier, 1,2-aryl migration competes with formation of the lumiketone when an aryl group is present at C-4 of the cyclohexenone. Based upon the dependence of product ratios on the polarity of the solvent, Dauben and co-workers proposed that the lowest $^3n,\pi^*$ state was responsible for the aryl migration and the $^3\pi,\pi^*$ state for formation of the two lumiketones,[7] consistent with the triplet assignment made above. Zimmerman established the migratory aptitudes of substituted phenyl groups,[30] and have shown they are consistent with migration to a carbon atom with odd-electron (i.e., radical) character, and not to a carbocationic center (as in Chapman's polar state **16**).

Although they proceed via $^3n,\pi^*$ and not via $^3\pi,\pi^*$ states, these reactions, in many ways, are quite analogous to the di-π-methane photorearrangements that have been so extensively studied by Zimmerman and co-workers, and by others.[10] Thus, on direct or triplet-sensitized excitation of 4,4-diphenylcyclohexenone **8**, the major product is the 6-*endo*-phenyl bicyclic enone **25** whose formation can be rationalized in terms of the bridged intermediate **26** (Scheme 48.3). Formation of minor amounts of the 6-*exo*-phenyl stereoisomer **27** was explained in terms of a competitive route via the open biradical **28** shown in Scheme 48.3. Quantum yields vary as a function of the electronic character of the migrating and nonmigrating groups and can be as large as 0.18, considerably larger than for lumiketone formation.

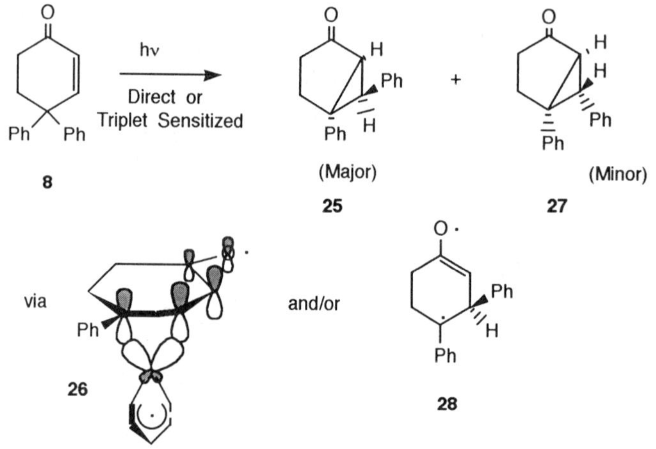

SCHEME 48.3

Study of the chiral enone 12 showed that both phenyl migration products are formed stereospecifically without any loss of optical purity.[13] One can account for this result in terms of initial
formation of a bridged intermediate 29 analogous to that proposed by Zimmerman, as shown in
Scheme 48.4. This species then undergoes either H-migration from C-3 to C-4 (to give 30) or
bonding between C-2 and C-4 (to give 31), concomitant with cleavage of the C4-phenyl bond. It
is perhaps significant that no 6-*exo*-phenyl product (32) is formed, indicating that ring opening of
29 to give the open biradical precursor to 32 does not take place.

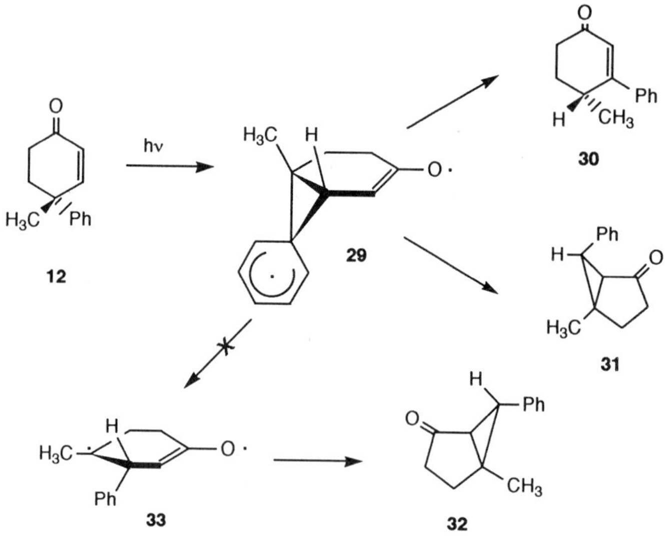

SCHEME 48.4

Zimmerman has shown recently that the allenic analogue of the diphenylcyclohexenone 8,
namely compound 34, undergoes a structurally analogous photorearrangement to give only 35,
which is the analogue of the *endo*-phenyl ketone 25 (see Equation 48.13).[32]

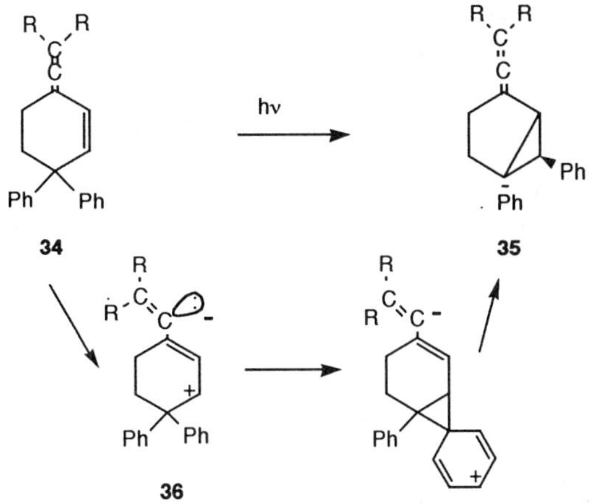

(48.13)

The quantum yield for the **34** to **35** conversion is 0.106. The reactive excited state in this reaction is the lowest singlet, while the corresponding triplet is unreactive (presumably due to a free rotor effect), in complete contrast to the mechanism of the enone photorearrangement. The proposed pathway involves transfer of electron density from the π-MO of the ring to the approximately in-plane sp^2 hybrid orbital on the central carbon of the allenic moiety, to give an intermediate zwitterion depicted as **36**. This then undergoes phenyl bridging followed by ring closure. It is interesting to note that the mode of excitation in allene **34** is electronically the exact opposite of the $n \rightarrow \pi^*$ excitation that initiates the photochemistry of cyclohexenones.

Zimmerman determined the effect of replacing the phenyl group in enone **25** by an aryl group whose triplet energy is lower than that of the enone moiety. The photorearrangements of the aryl-substituted cyclohexenones **37**, where the aryl group is either *p*-biphenylyl, α-naphthyl, or β-naphthyl, are shown in Equation 48.14, along with the quantum yields measured in benzene and *t*-BuOH.[33,34]

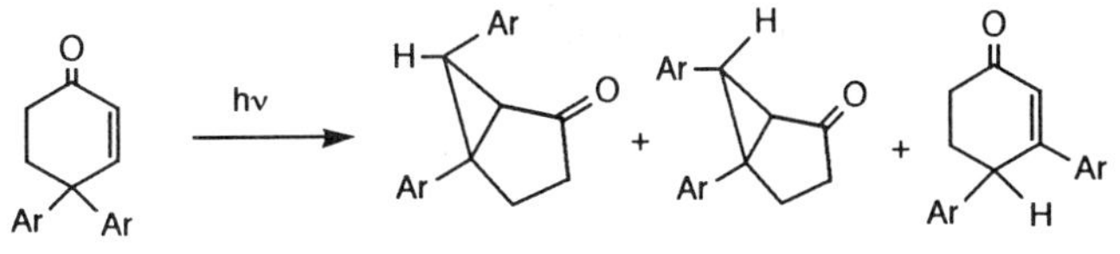

Ar	Solvent	Quantum Yields		
p-biphenyl	t-BuOH	0.26	0.024	0.020
	Benzene	0.33	0.019	0.013
α-naphthyl	t-BuOH	0.46		0.54
	Benzene	0.43		0.57
β-naphthyl	t-BuOH	0.38	0.02	
	Benzene	0.40	0.02	

(48.14)

It can be seen that the reaction course is similar, giving the rearranged *endo*-aryl bicyclohexanone **38** as the major product. The isomeric *exo*-aryl ketones **39** and the rearranged cyclohexenones **40** are minor products. Triplet quenching and sensitization studies indicate that these photorearrangements all proceed via triplet states, but there is a significant difference between the nature of that triplet as a function of the aryl substituent. In the parent system **25**, the reactive triplet is an n,π* state. In the case of **37** with Ar = biphenylyl, Zimmerman concludes that the equilibration of excitation between the enone and biphenyl moieties is faster than the rate of rearrangement.[33] The increase in quantum efficiency for rearrangement by a factor of 10 compared to **25** is attributed to better delocalization of the odd electron density in the bridged intermediate analogous to **26**. For **37** with Ar = α- or β-naphthyl, Zimmerman and Solomon propose that intramolecular triplet energy transfer from the higher energy enone triplet T_2 ($E_T \simeq 69$ kcal mol^{-1}) to the lower energy naphthyl triplet T_1 ($E_T \simeq 61$ kcal mol^{-1}) is faster than any competitive process.[34] The rearrangement then takes place from the naphthalene-like π,π* triplet by a classic di-π-methane rearrangement. The net result is a large increase in the quantum efficiency for the rearrangement.

48.4 Photoisomerization of *cis*-2-Cyclohexenones to *trans*-2-Cyclohexenones

The photoisomerization of conjugated cyclic enones to their ground-state *trans*-isomers is a well known process in seven- and eight-membered ring systems.[35] In these systems, *trans*-cycloalkenones can be directly detected spectroscopically (UV, IR) at low temperatures, and are trapped by reactive dienes with formation of *trans*-fused Diels-Alder adducts (see Equation 48.15). Bonneau has confirmed the formation of *trans*-cycloheptenone upon irradiation of the *cis*-isomer using transient absorption spectroscopy.[36] This species, $\lambda_{max} = 265$ nm, has a lifetime of 45 s in cyclohexane at room temperature, but is much shorter lived in alcohol solvents (33 ms in MeOH, 74 ms in EtOH) due to rapid nucleophilic addition across the strained C=C bond, a reaction that is probably acid-catalyzed (see below). Using CH$_3$OD, Hart has shown that nucleophilic photoaddition to cycloheptenones and cyclooctenones is stereospecific, positioning the methoxy and deuterium moieties exclusively *trans*.[37] This has been rationalized in terms of a mechanism involving *syn*-addition of MeOH(D) to the ground-state *trans*-cycloalkenones, as depicted schematically in Equation 48.16. The activation energy for decay of *trans*-cycloheptenone in cyclohexane, which is assumed to be the barrier for thermal isomerization to the *cis*-isomer, is 15.2 kcal mol^{-1}.[36]

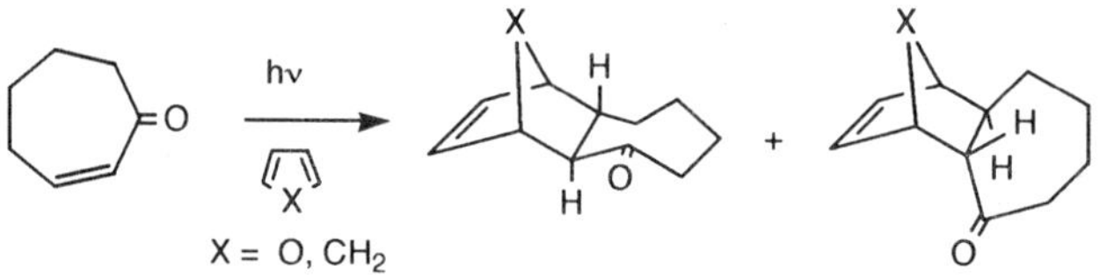

$$(48.15)$$

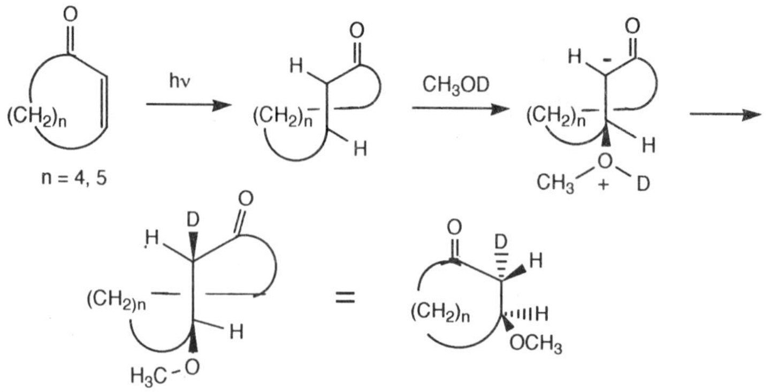

$$(48.16)$$

trans-Cyclohexenones have occasionally been postulated as intermediates in photochemical reactions of cyclohexenones, but have yet to be directly detected in a manner that has proved successful for its seven- and eight-membered ring analogues. *trans*-Cyclohexenes have been conclusively detected and trapped on direct excitation of 1-acetylcyclohexene[38] and on triplet-sensitized excitation of 1-phenylcyclohexene[39] and a number of other substituted cyclohexenes.[40] In all cases, the UV absorption of the *trans*-isomer is considerably red-shifted relative to that of the *cis*-isomer, and the barriers to thermal isomerization of *trans* to *cis* are on the order of 10 kcal mol⁻¹. Verbeek and co-workers[41] have calculated that *trans*-cyclohexene lies in a potential minimum 56 kcal mol⁻¹ above the ground state, and that the barrier for isomerization to the *cis*-isomer is ≈15 kcal mol⁻¹. Because of the increased strain associated with insertion of an additional trigonal center into a *trans*-cyclohexene molecule, the energy of *trans*-cyclohexenone relative to its *cis*-isomer should be larger than 56 kcal mol⁻¹, and the barrier for isomerization should be ≤10 kcal mol⁻¹. Using TCSCG-3–21G calculations, Johnson[42] has found two minima for *trans*-cyclohexenone, one with a twist boat and a second with a chair-like geometry, both lying ≈60 kcal mol⁻¹ above the ground state. The barrier for conversion of the twist boat *trans*-cyclohexenone to the *cis*-isomer is calculated to be ≈13 kcal mol⁻¹.

Perhaps the best experimental evidence for formation of a *trans*-cyclohexenone comes from studies of Pummerer's ketone **41**. Hart has shown that photoaddition of MeOD to **41** occurs to give the *trans*-adduct **42**, as shown in Equation 48.17.[43] The reaction course is completely analogous to that of the medium-ring enones, and Hart therefore concludes that excitation of **41** results "in an excited state or intermediate in which the carbon-carbon double bond is twisted by more than 90°," followed by *syn*-addition of methanol. The kinetic isotope effect of 4.3 ± 0.5, measured using mixtures of MeOH and MeOD, is similar to that found for photoadditions of methanol to medium-ring enones.[37] Further support for formation of a *trans*-isomer of **41** comes from the observations by Schuster and Mintas[44] that only *trans*-fused Diels-Alder adducts are formed on photoexcitation of **41** in neat furan (Equation 48.18). The fact that 1-methylnaphthalene does not quench formation of these adducts is consistent with a triplet lifetime of **41** of <15 ns,[45] suggesting that twisting around the C=C bond is extremely facile in this system.[19] Irradiation of **41** in the presence of alkenes leads to preferential formation of *trans*-fused [2+2]-cycloadducts,[44] but this in itself is not conclusively diagnostic for the intermediacy of ground-state *trans*-cyclohexenones.[46,47] Rodriguez-Hahn and co-workers have observed a similar stereospecific photoaddition of methanol to decompostin (**43**) (see Equation 48.19).

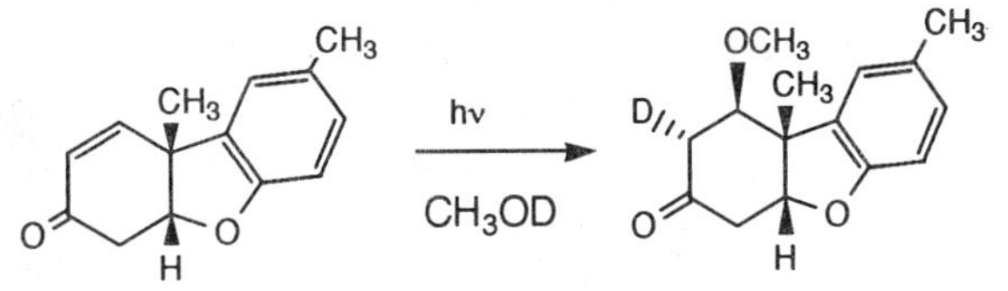

(48.17)

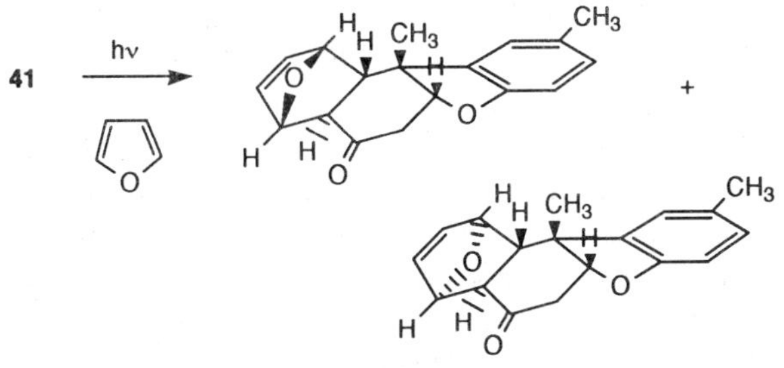

(48.18)

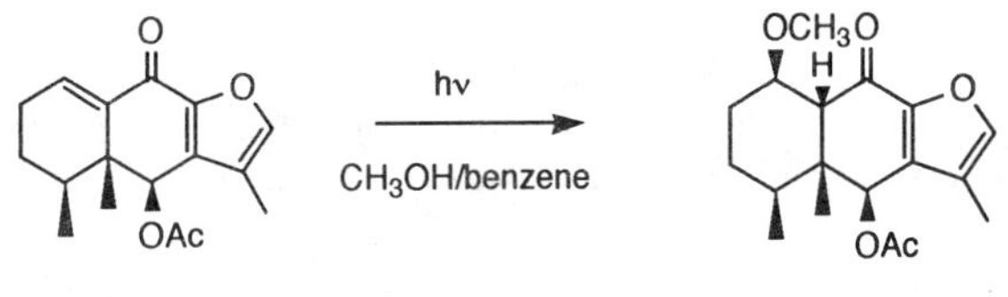

(48.19)

Photoaddition of alcohols to cyclohexenones is not a common reaction. Only traces of methanol adducts can be detected after extensive irradiation of cyclohexenone in methanol.[48] Rudolph and Weedon[49] have recently reported that photochemical deconjugation of 3-alkylcyclohexenones such as isophorone 44 (3,5,5-trimethyl-2-cyclohexenone) in benzene to give 45, as shown in Equation 48.20, requires the presence of an acid (typically acetic acid).

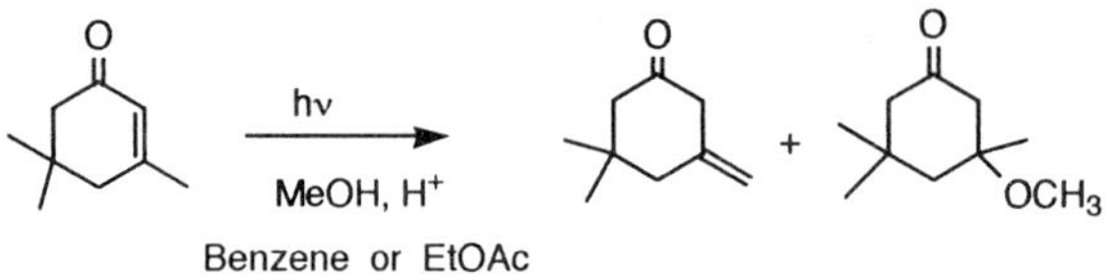

44   **45**   **46**

$$(48.20)$$

If methanol is present, photodeconjugation is accompanied by formation of the methanol adduct 46. Using a kinetic analysis based on the dependency of quantum efficiencies on acid concentration, the authors argue that the species undergoing protonation must have a lifetime >1 μs, and therefore cannot be the enone triplet (whose lifetime they estimate to be ≈10 ns); they suggest that the ground-state *trans*-cyclohexenone is the species undergoing protonation to give a tertiary carbocation, which either loses a proton or is captured by methanol.[49] However, several observations by Schuster and co-workers do not support this mechanism.[50] First, the triplet lifetime of isophorone is 79 ns, considerably longer than that estimated by Rudolph and Weedon.[49] More importantly, laser flash studies demonstrate that for 3-methylcyclohexenone (3MCH), which undergoes the same acid-catalyzed reactions as isophorone, the enone triplet is the species undergoing protonation under the reaction conditions. The rate constant for protonation of MCH triplets by sulfuric acid in ethyl acetate is 1.7×10^9 M^{-1} s^{-1}, while that for protonation by acetic acid is estimated to be somewhat smaller. From the data of Rudolph and Weedon,[49] a rate constant of 1×10^8 M^{-1} s^{-1} for protonation of the isophorone triplet by acetic acid can be estimated, sufficient to account for the measured quantum efficiency of the reactions (maximum 0.04). Thus, a *trans*-cyclohexenone is not a *requisite* intermediate in these reactions. Furthermore, the results of studies using photoacoustic calorimetry[18] argue against the formation of long-lived strained ground-state intermediates in appreciable yield from simple cyclohexenones, since formation and decay of the enone triplet excited states account for >98% of the excitation energy. This is not the case for 1-acetylcyclohexene, where a substantial heat discrepancy is observed,[18] consistent with the formation of a ground-state *trans*-isomer.[38]

Schuster, Scaiano, and co-workers[51] reported kinetic data that indicate that during photocycloaddition of 4,4-dimethylcyclohexenone to alkenes, the species being intercepted by alkenes is not the enone triplet, but rather some species derived from the enone triplet, one possibility being a *trans*-cyclohexenone. Since there is abundant data that demonstrate that alkenes typically intercept enone triplets directly en route to [2+2]-cycloadducts, this observation remains a distinct anomaly.

In summary, *trans*-cyclohexenones formed via triplet excited states of *cis*-2-cyclohexenone have been suggested as intermediates in a number of photochemical reactions, but these elusive species have yet to be firmly pinned down. Theoretical calculations[42] indicate that although such species represent minima on the ground-state potential surface, they are kinetically labile. Nonetheless, *trans*-cyclohexenones should be detectable using fast reaction techniques at low temperatures.

References

1. (a) Chapman, O. L., in *Advances in Photochemistry*, Vol. 1, Noyes, W. A., Jr., Hammond, G. S., and Pitts, J. N., Jr., Eds., Wiley-Interscience, New York, 1963, 323–420; (b) Schaffner, K., in *Advances in Photochemistry*, Vol. 4, 1966, 81–112.

2. Schuster, D. I., in *Rearrangements in Ground and Excited States*, Vol. 3, de Mayo, P., Ed., Academic Press, New York, 1980, 167–279.

3. Schuster, D. I., in *The Chemistry of Enones*, Patai, S. and Rappoport, Z., Eds., Wiley, Chichester, U.K., 1989; 623–756.

4. Kwie, W. W., Shoulders, B. A., and Gardner, P. D., *J. Am. Chem. Soc.*, 84, 2268, 1962. Shoulders, B. A., Kwie, W. W., Klyne, W., and Gardner, P. D., *Tetrahedron*, 21, 2973, 1965.

5. Chapman, O. L., Rettig, T. A., Griswold, A. A., Dutton, A. I., and Fitton, P., *Tetrahedron Lett.*, 2049, 1963.

6. Dauben, W. G., Shaffer, G. W., and Vietmeyer, N. D., *J. Org. Chem.*, 33, 4060, 1968.

7. Dauben, W. G., Spitzer, W. A., and Kellogg, M. S., *J. Am. Chem. Soc.*, 93, 3674, 1971.

8. Zimmerman, H. E. and Wilson, J. W., *J. Am. Chem. Soc.*, 86, 4036, 1964; Zimmerman, H. E., Rieke, R. D., and Scheffer, J. R., *J. Am. Chem. Soc.*, 89, 2033, 1967; Zimmerman, H. E. and Morse, R. L., *J. Am. Chem. Soc.*, 90, 954, 1968; Zimmerman, H. E. and Hancock, K. G., *J. Am. Chem. Soc.*, 90, 3749, 1968; Zimmerman, H. E. and Elser, W. R., Jr., *J. Am. Chem. Soc.*, 91, 887, 1969.

9. Nobs, F., Burger, U., and Schaffner, K., *Helv. Chim. Acta*, 60, 1607, 1977; Swenton, J. S., Blankenship, R. M., and Sanitra, R., *J. Am. Chem. Soc.*, 97, 4941, 1975.

10. For a comprehensive review, see Hixson, S. S., Mariano, P. S., and Zimmerman, H. E., *Chem. Rev.*, 73, 531, 1973.

11. Bellus, D., Kearns, D. R., and Schaffner, K., *Helv. Chim. Acta*, 52, 971, 1969.

12. Chapman, O. L., Sieja, J. B., and Welstead, W. J., Jr., *J. Am. Chem. Soc.*, 88, 161, 1966.

13. Schuster, D. I., Brown, R. H., and Resnick, B. M., *J. Am. Chem. Soc.*, 100, 4504, 1978.

14. Woodward, R. B. and Hoffmann, R., *The Conservation of Orbital Symmetry*, Verlag Chemie/ Academic Press, Weinheim, Germany, 1970, 89–100.

15. Shaik, S. S., *J. Am. Chem. Soc.*, 101, 2736, 1979; Shaik, S. S. and Epiotis, N. D. ibid. 1978, 100, 18.

16. Schuster, D. I. and Hussain, S., *J. Am. Chem. Soc.*, 102, 409, 1980; Hussain, S. Ph.D. Dissertation, New York University, 1979.

17. Schuster, D. I., Woning, J., Kaprinidis, N. A., Pan, Y., Cai, B., Barra, M., and Rhodes, C. A., *J. Am. Chem. Soc.*, 114, 7029, 1992.

18. Schuster, D. I., Heibel, G. E., Caldwell, R. A., and Tang, W., *Photochem. Photobiol.*, 52, 645, 1990.

19. Schuster, D. I., Dunn, D. A., Heibel, G. E., Brown, P. B., Rao, J. M., and Woning, J., *J. Am. Chem. Soc.*, 113, 6245, 1991.

20. Ni, T., Caldwell, R. A., and Melton, L. A., *J. Am. Chem. Soc.*, 111, 457, 1989.

21. Schuster, D. I. and Woning, J., unpublished results.

22. Zimmerman, H. E., in *Advances in Photochemistry*, Vol. 1, Noyes, W. A., Jr., Hammond, G. S., and Pitts, J. N., Jr., Eds, Wiley-Interscience, New York, 1963, 183–208; Zimmerman, H. E. and Schuster, D. I. *J. Am. Chem. Soc.*, 1962, 84, 4527; Zimmerman, H. E. and Swenton, J. S., *J. Am. Chem. Soc.*, 1967, 89, 906.

23. For a review, see Schaffner, K. and Demuth, M., in *Rearrangements in Ground and Excited States*, Vol. 3, de Mayo, P., Ed., Academic Press, New York, 1980, 281–348.

24. Patel, D. J. and Schuster, D. I., *J. Am. Chem. Soc.*, 90, 5137, 1968; Schuster, D. I. and Patel, D. J., *J. Am. Chem. Soc.*, 90, 5145, 1968.

25. Schuster, D. I. and Liu, K., *Tetrahedron*, 37, 3329, 1981.

26. Samuel, C. J., *J. Chem. Soc. Perkin Trans. 2*, 736, 1981.

27. Schuster, D. I. and Brizzolara, D. F., *J. Am. Chem. Soc.*, 92, 4357, 1970.

28. Kearns, D. R., Marsh, G., and Schaffner, K., *J. Chem. Phys.* 49, 3316, 1968; Marsh, G., Kearns, D. R., and Schaffner, K., *Helv. Chim. Acta*, 51, 1890, 1968; Marsh, G., Kearns, D. R., and Schaffner, K., *J. Am. Chem. Soc.*, 93, 3129, 1971.

29. Chan, A. C. and Schuster, D. I., *J. Am. Chem. Soc.*, 108, 4561, 1986.

30. Zimmerman, H. E. and Lewin, N. *J. Am. Chem. Soc.*, 91, 879, 1969.

31. Zimmerman, H. E. *Tetrahedron*, 30, 1617, 1974.

32. Zimmerman, H. E., Baker, M. R., Bottner, R. C., Morrissey, M. M., and Murphy, S., *J. Am. Chem. Soc.*, 115, 459, 1993.

33. Zimmerman, H. E., Jian-hua, X., King, R. B., and Caufield, C. E., *J. Am. Chem. Soc.*, 107, 7724, 1985.

34. Zimmerman, H. E., Caufield, C. E., and King, R. B., *J. Am. Chem. Soc.*, 107, 7732, 1985.

35. Eaton, P. E. and Lin, K., *J. Am. Chem. Soc.*, 86, 2087, 1964; *J. Am. Chem. Soc.*, 87, 2052, 1965; Corey, E. J., Tada, M., LeMahieu, R., and Libit, L., *J. Am. Chem. Soc.*, 87, 2051, 1965.

36. Bonneau, R., Fornier de Violet, P., and Joussot-Dubien, J., *Nouv. J. Chim.*, 1, 31, 1977.

37. Dunkelblum, E. and Hart, H., *J. Am. Chem. Soc.*, 99, 644, 1977; Hart, H. and Dunkelblum, E., *J. Am. Chem. Soc.*, 100, 5141, 1978.

38. Bonneau, R. and Fornier de Violet, P., *C. R. Acad. Sci. Paris, Ser. C*, 284, 631, 1977.

39. Goodman, J. L., Peters, K. S., Misawa, H., and Caldwell, R. A., *J. Am. Chem. Soc.*, 108, 6803, 1986.

40. Bonneau, R., *J. Photochem.*, 36, 311, 1987.

41. Verbeek, J., van Lenthe, J. H., Timmermans, P. J. J. A., Mackor, A., and Budzelaar, P. H. M., *J. Org. Chem.*, 52, 2955, 1987.

42. Johnson, R. P., Univ. of New Hampshire, private communication.

43. Dunkelblum, E., Hart, H., and Jeffares, M., *J. Org. Chem.*, 43, 3409, 1978.

44. Mintas, M., Schuster, D. I., and Williard, P. G., *J. Am. Chem. Soc.*, 110, 2305, 1988. *Tetrahedron*, 44, 6001, 1988.

45. Scaiano, J. C. and Schuster, D. I., unpublished results.

46. Schuster, D. I., Lem, G., and Kaprinidis, N. A., *Chem. Rev.*, 93, 3, 1993.

47. Schuster, D. I., this volume, pp. ?

48. Rodriguez-Hahn, L., Esquivel, B., Ortega, A., Garcia, J., Diaz, E., Cardena, J., Soriano-Garcia, M., and Toscano, A., *J. Org. Chem.*, 50, 2865, 1985.

49. Rudolph, A. and Weedon, A. C., *J. Am. Chem. Soc.*, 111, 8756, 1989.

50. Schuster, D. I., Yang, J.-M., and Woning, J., unpublished results.

51. Schuster, D. I., Brown, P. B., Capponi, L., Rhodes, C. A., Scaiano, J. C., and Weir, D., *J. Am. Chem. Soc.*, 109, 2533, 1987.

49

Photodeconjugation of Unsaturated Enones and Acid Derivatives

ı-Pierre Pete
ersité de Reims
npagne-Ardenne

).1 Introduction

The photochemical deconjugation of α,β-unsaturated carbonyl compounds to their β,γ-unsaturated isomers, which was uncovered more than 30 years ago,[1–3] has been shown to be a general process. The reaction is of great theoretical and synthetic interest and can be used, in principle, as a simple procedure for the conversion of conjugated enones or esters into their thermodynamically less stable β,γ-unsaturated isomers. However, from the early reports in the literature, various systems having this structure were apparently inert under the irradiation conditions and molecules such as mesityl oxide were recovered unchanged. All these reactions were shown to involve photoenol intermediates that isomerize either into the starting conjugated carbonyl derivative or the deconjugated isomer (Scheme 49.1).

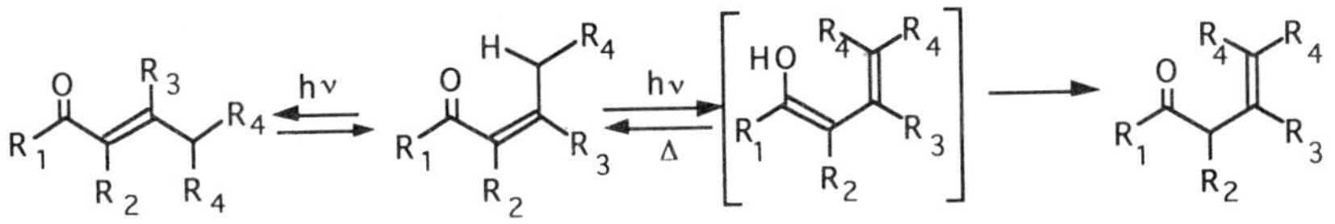

SCHEME 49.1

During the last past 10 years, the reaction has attracted the interest of the scientific community, and a better comprehension of the deconjugation mechanism and of the deactivation pathways available to the photoenol intermediate has revealed new synthetic applications even in the field of asymmetric synthesis.

0-8493-8634-9/95/$0.00+$.50

Table 49.1   Typical UV Absorption Data

Compound	Solvent	π,π^* absorption λ_{max}, nm (ε_{max})	n,π^* absorption λ_{max}, nm (ε_{max})
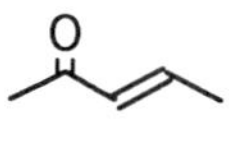	Water Ethanol n heptane	226 (14000) 224 (9750) 215 (11500)	300 (62) 330 (27)
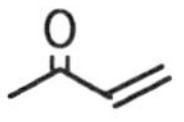	Water n heptane	231 (9650) 222,5 (8700)	303 (60) 330 (27)
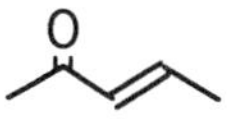	Ethanol Hexane	237 (12600) 229,5 (12600)	315 (78) 327 (97,5)
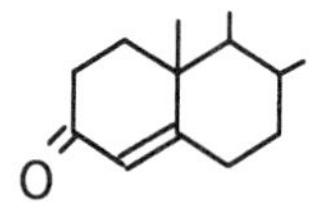	t BuOH Benzène	240 (> 10000)	308 (77) 333 (55)
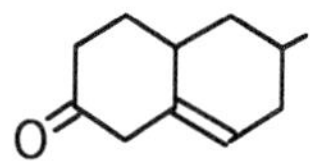	Ethanol	213 (2040)	293 (170)
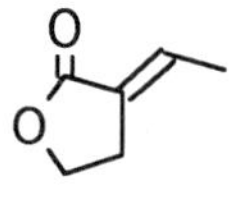	Méthanol Cyclohexane	219 (11500) 215 (11500)	- -
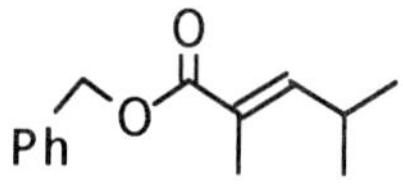	Ethanol	212 (12900)	
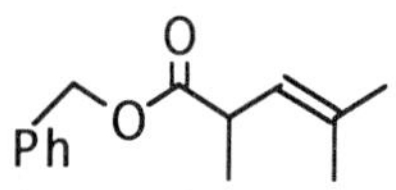	Ethanol	220 (16100) 255 (490)	

49.2 Photoreactivity of Aliphatic α,β-Unsaturated Esters and Ketones

α,β-Unsaturated ketones and the corresponding acid derivatives show distinct absorption bands in the UV due to n,π* and π,π* transitions, respectively. For enones, the n,π* transition occurs at wavelengths higher than 280 nm, with a maximum appearing at about 320 nm depending slightly on the solvent (Table 49.1). The π,π* transition is observed at shorter wavelengths, and the position of the maximum absorption has been correlated with the number and the nature of the substituents on the ethylenic bond,[4] The π,π* transition is subject to a very important solvent effect and a bathochromic shift is observed for the absorption maximum when more polar solvents are used. Conjugated acid derivatives absorb at shorter wavelengths than the corresponding enones and the n,π* absorption is masked usually by the far stronger π,π* absorption band.

When aliphatic or medium-ring enones or the corresponding acid derivatives are irradiated either in the n,π* or π,π* absorption band, an efficient Z,E isomerization is observed and a photostationary mixture of Z,E diastereoisomers is rapidly attained.[5] The geometric isomerization can be sensitized by triplet sensitizers of triplet energies higher than 300 kJ mol⁻¹, and acetone can be used conveniently as the solvent and the sensitizer. For longer irradiation times, the formation of a photodienol and then of the deconjugated carbonyl compound can be observed with a quantum yield of the order of 0.05 to 0.10. From kinetic measurements, it appears that the photodienol intermediate is formed only from *Z*-stereoisomers and that neither sensitization by triplet sensitizers, nor quenching of the dienol formation by triplet quenchers can be detected. For these reasons, it can be concluded that the deconjugation process is a singlet reaction.

Deconjugation is a very general process and has been observed for the unsaturated esters,[5c,6–29] acids,[7,9,30–35] lactones,[36,37] amides,[20] aldehydes,[38] aliphatic enones,[14,39–68] medium-ring unsaturated ketones,[70–74] and even the cyclohexenones[75–84,109–113] whose structures are given in Scheme 49.2.

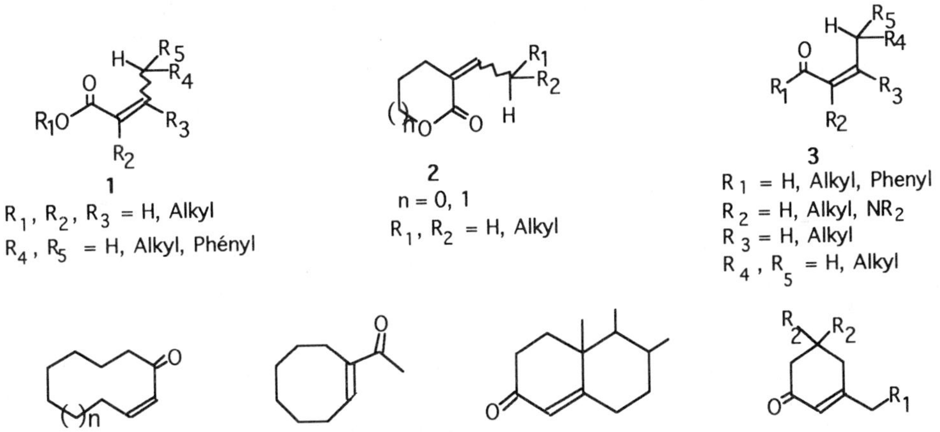

1

R$_1$, R$_2$, R$_3$ = H, Alkyl
R$_4$, R$_5$ = H, Alkyl, Phényl

2
n = 0, 1
R$_1$, R$_2$ = H, Alkyl

3
R$_1$ = H, Alkyl, Phenyl
R$_2$ = H, Alkyl, NR$_2$
R$_3$ = H, Alkyl
R$_4$, R$_5$ = H, Alkyl

4

5

6
R$_1$, R$_2$ = H, Alkyl

7
R$_1$, R$_2$ = H, Alkyl

SCHEME 49.2

In the absence of γ-substituents the starting α,β-unsaturated ester or enones can have very low reactivity.[47,48,85] However, even in these cases, the corresponding deconjugated isomers can be isolated in good yields when the irradiation is carried out in the presence of weak bases.[22–24,53,58,62]

The starting material absorbs at a longer wavelength than the deconjugated derivative. By irradiating α,β-unsaturated enones at 366 nm or α,β-unsaturated esters at 254 nm, most of the light is absorbed by the starting material and irradiations can be carried out until completion. The deconjugated isomer can be formed in high chemical yields from esters, acids, and lactones. For enones, the chemical yield depends on the photochemical stability of the deconjugated isomer.

9.3 Photodienols as Intermediates

The photodienol is produced from the singlet excited state of the corresponding *cis*-unsaturated compound; its formation has been proposed to involve a concerted and antarafacial 1,5-sigmatropic hydrogen migration according to the Woodward-Hoffmann rules.[13,67] As a result of the cyclic transition state, the stereochemistry of the 1,2-double bond should be unique and *Z* in the absence of α-substituents, at least for low conversion yields (Scheme 49.3). As will be seen later, the stereospecificity of the photoenolization process has been verified indirectly by trapping of the dienol intermediate.

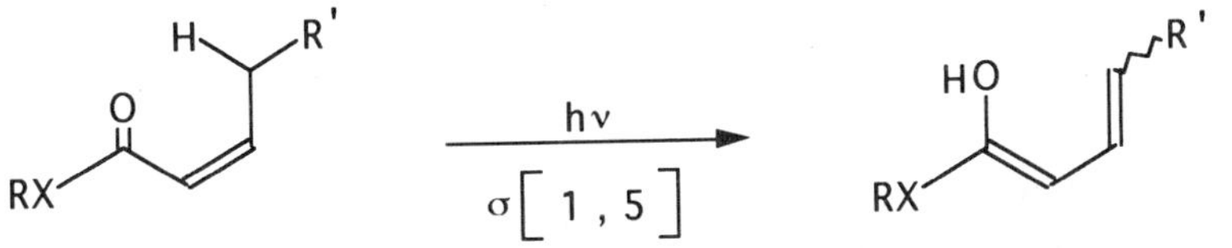

SCHEME 49.3

Lifetimes of dienols depend strongly on their substitution and on the solvents used. Dienols derived from acid derivatives have relatively short lifetimes and their concentration is kept at a low level when unsaturated acid derivatives are irradiated at 254 nm using low-pressure mercury lamps. Under these conditions, quenching of the starting excited molecule by the dienol is not important and the *Z,E*-isomerization of the enol double bond is a very minor process.[22,24] In contrast, lifetime of the dienols derived from conjugated ketones can be quite long.[73,86,87] For high conversion yields, the dienols can play the role of triplet quenchers and some geometric isomerization is observed.

Dienol intermediates can be detected directly in the reaction mixture by IR spectroscopy. A strong O-H stretch at 3450 cm[-1], olefinic absorption at 1656 cm[-1], and a C-O stretch at 1191 cm[-1] have been observed for the very stable dienols formed from acetylcyclooctene[73] in acetonitrile. Dienols have also been detected by IR in the gas phase (49) (Scheme 49.4).

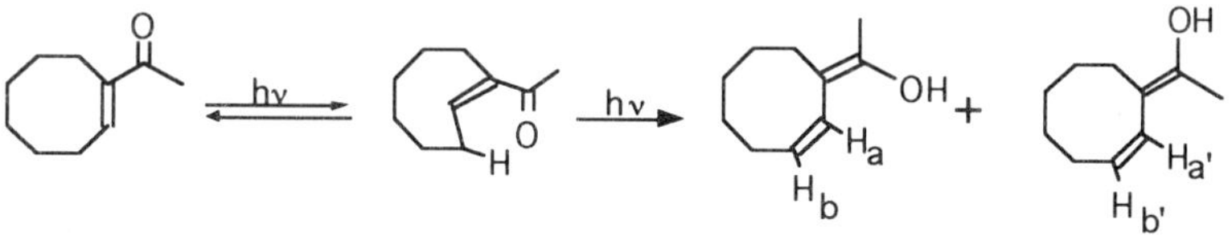

SCHEME 49.4

The ^{1}H/NMR spectra show new signals at 6.48, 6.14, and 5.22 ppm, which can be attributed to the vinylic protons Ha, Ha' and Hb, Hb'. Similar studies have been carried out on mesityl oxide and related enones.[71,86] When the photolysis is carried out in water in the presence of potassium hydroxide, a transient species with an absorption at 290 nm is assigned to the dienolate chromophore.[66]

49.4 Deactivation of Photodienols

In the absence of other reagents, two main processes are available to the intermediate: return to the starting unsaturated molecule and tautomerism to the β,γ-unsaturated isomers. In pure nonpolar solvents, the dienol leads mainly or only to the starting material through a thermal 1,5-sigmatropic hydrogen shift, and the activation parameters of the rearrangements have been evaluated from the NMR data[86] (Scheme 49.5).

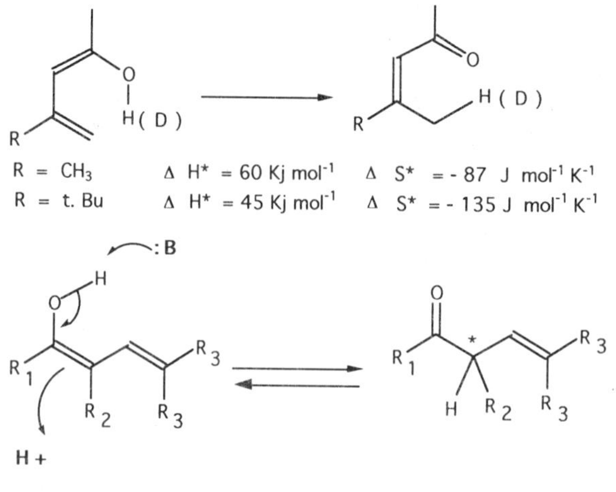

R = CH$_3$	Δ H* = 60 Kj mol[-1]	Δ S* = - 87 J mol[-1] K[-1]
R = t. Bu	Δ H* = 45 Kj mol[-1]	Δ S* = - 135 J mol[-1] K[-1]

SCHEME 49.5

The deconjugated isomer is formed by prototropy which is either catalyzed by acids[32,87,88] or bases or even impurities.[85] In polar solvents containing small quantities of base, the formation of the β,γ-unsaturated ester or ketone involves the intermediacy of a dienolate that is protonated in the α-position.

In the presence of methanol-O-d, incorporation of deuterium onto the α-carbon of the deconjugated isomer is observed. Exchange of the enolic proton in the deuterated solvent is a very fast process that competes with the 1,5-sigmatropic shift. For this reason, irradiation of an α,β-unsaturated carbonyl derivative results also in incorporation of deuterium at the γ-position of the starting material. This incorporation was recognized early on in the study as an indication that dienol formation is an important path for energy dissipation for the excited molecules. Molecules such as mesityl oxide, which are recovered almost quantitatively from the reaction mixture after very long irradiation times, also incorporate deuterium onto the γ-position.[47,57]

Decay of dienols to α,β-unsaturated esters, lactones, or ketones by a 1,5-sigmatropic shift is an efficient process and very sensitive to steric hindrance. γ-Substituents hinder the adoption of the conformation needed for the intramolecular 1,5-hydrogen shift and formation of the deconjugated isomer in a catalyzed process becomes competitive. Addition of a weak base such as imidazole or dimethylimidazole can help the transformation of the dienol into the β,γ-unsaturated isomer[23,67] (Scheme 49.6)

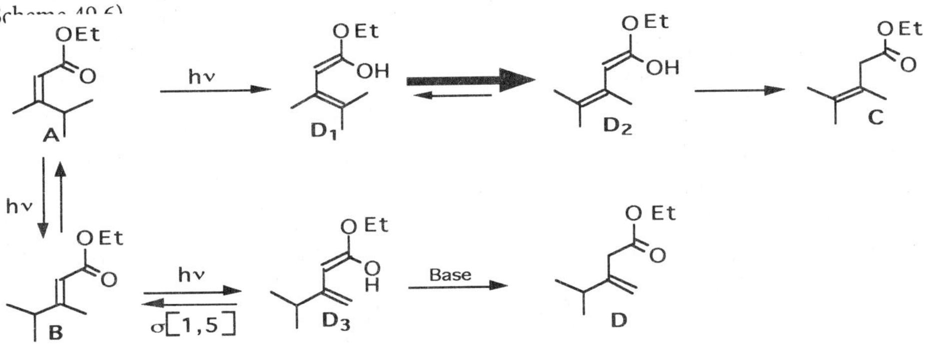

SCHEME 49.6

For example, continued irradiation of a mixture of A and B at the photostationary state in methanol leads to a mixture of deconjugated esters C and D, where isomer C is predominant. When small amounts of base are added to the reaction mixture, a considerable increase in the concentration of D is observed. The results can be easily rationalized if we assume that the dienol formed by γ-H-abstraction from A is more stable in the less-hindered D_2 conformation, which does not allow the 1,5-sigmatropic shift. Decay of D_2 to the deconjugated ester C is the normal process. Excitation of stereoisomer B and γ-H-abstraction from the methyl group produces the dienol D_3 directly in its least-hindered conformation. In the absence of base, dienol D_3 decays preferentially by a thermal hydrogen shift to the conjugated ester. Introduction of a base in the reaction mixture allows the interception of D_3 and makes the formation of D competitive with the intramolecular process.[23]

The dienols derived from enones can be trapped by acetic anhydride or trimethylsilylchloride.[60,67,68] In the latter case, though the initial Z-stereochemistry of the enol intermediate is preserved in the siloxydene, the presence of (E)-siloxydiene can also be observed at high conversion (Scheme 49.7).

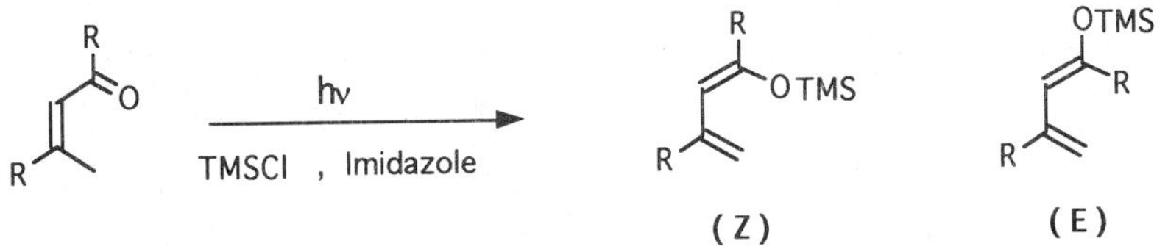

SCHEME 49.7

The formation of the *E*-stereoisomer is presumed to arise from triplet-sensitized isomerization of the *Z*-isomer. Diels-Alder reaction of conformationally rigid dienols is a very useful reaction.[89] However, trapping of these linear photodienols in Diels-Alder reactions has not yet been reported, which is indicative that the dienol has too short a lifetime in its *S-cis*-conformation to be trapped by dienophiles in a bimolecular reaction.

49.5 Asymmetric Photodeconjugation

If α-substituents are present in the starting α,β-unsaturated esters, conjugated lactones, or enones, photodeconjugation creates a new chiral center in the products. Under normal conditions, protonation of the dienol in the α-position produces a racemic mixture. When the irradiation is carried out in aprotic solvents in the presence of small quantities of chiral catalyst I*, an asymmetric protonation takes place and one enantiomer of the deconjugated isomer is now favored[90–97] (Scheme 49.8).

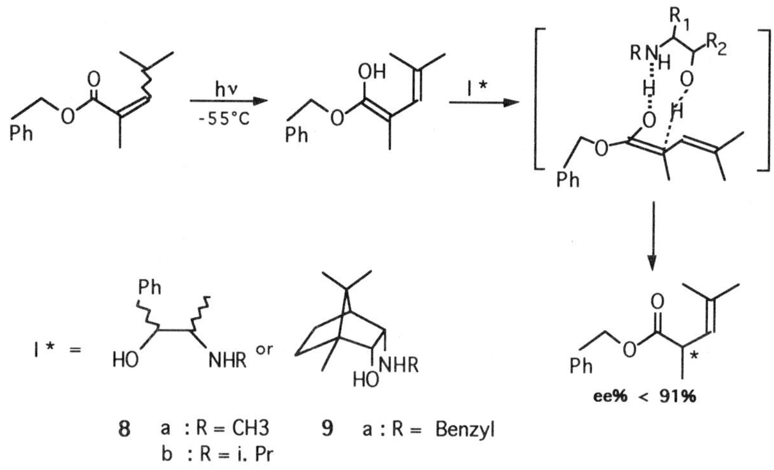

SCHEME 49.8

The ee (enantiomeric excess) % is considerably increased in the presence of chiral β-aminoalcohols rather than isolated chiral alcohols or amines.[97] The synergic effect of amino and hydroxyl groups implies that cyclic transition states can be involved in the catalyzed tautomerism of the photodienol. A model with a nine-membered transition state has been proposed to rationalize the chirality of the new chiral center. Deprotonation of the enol by the amino group and, at the same time, protonation on the α-carbon by the hydroxyl group of the chiral inductor explains the observed results.[93,94,97] The effect of temperature on the selectivity has been studied and high ee% have been reported for temperatures lower than –40 °C. For example, photodeconjugation of benzyl 2,4-dimethylpent-2-enoate in the presence of ephedrine or bornyl derivatives led to ee% values up to 70 and 90%, respectively (Tables 49.2 and 49.3).

These selectivities, the highest ever observed for a photochemical reaction in solution, are also the best described for the enantioselective protonation of enols or enolates.[98,99] The high ee% values are consistent with a stereospecific photoenolization and with the proposal of an antarafacial 1,5-sigmatropic hydrogen shift from the γ-carbon to the oxygen of the singlet excited carbonyl in the photochemical step.

When the photodeconjugation reaction is applied to unsaturated esters derived from chiral alcohols, protonation of the dienol leads to a mixture of diastereoisomers. To observe high diastereoselectivities, the chiral alkoxy group has to mask selectively one face of the enol. Only very

Table 49.2 Asymmetric Photodeconjugation of 1 in CH_2Cl_2 in the Presence of a Chiral β-Aminoalcohol[a]

R_1	R_2	R_3	R_4	R_5	Inductor	Θ(°C)	Chemical yield %	ee%	Configuration
CH_2-C_6H_5	CH_3	H	CH_3	CH_3	(+) 9	−55	71	91	R
					(+) 8a	−40	64	37	S
					(−) 8b	−40	65	70	R
C_2H_5	CH_3	H	C_2H_5	C_2H_5	(+) 8a	−40	70	70	S
C_2H_5	CH_3	H	CH_3	CH_3	(+) 8a	−40	68	16	S
					(−) 8b	−40	84	40	R
CH_3	C_2H_5	H	CH_3	CH_3	(−) 8a	−78	90	10	R
C_2H_5	CH_3	H	$(CH_2)_4$		(+) 8a	−40	68	10	S
					(−) 8b	−40	67	32	R
C_2H_5	CH_3	H	$(CH_2)_5$		(+) 8a	−40	72	21	S
					(−) 8b	−40	59	63	R
$C_6H_5CH_2$	CH_3	H	$(CH_2)_5$		(+) 8a	−40	60	31	S
					(−) 8b	−40	60	52	R
$C_6H_{11}CH_2$	CH_3	H	CH_3	CH_3	(+) 8a	−40	73	17	S
					(−) 8b	−40	74	43	R
$C_6H_5CH_2$	CH_3	H	H	C_6H_{13}	(+) 8a	−40	75	9	S
					(−) 8b	−40	70	38	R

[a] 1 (10^{-2} mol l^{-1}); λ = 254 nm; Inductor (10^{-3} mol l^{-1})

Table 49.3 Discrimination Parameters $\Delta\Delta H^\#$ and $\Delta\Delta S^\#$ Observed in the Photodeconjugation of 1 for $R_2 = R_4 = R_5 = CH_3$, $R_3 = H$

R_1	Inductor	Solvent	$\Delta\Delta H^\#$ KJ mol^{-1}	$\Delta\Delta S^\#$ J mol^{-1}K^{-1}	$\Delta\Delta G^\#$ at 229K KJ mol^{-1}
CH_2-C_6H_5	(+) 8a	CH_2Cl_2	−2.9	−6	−1.5
	(+) 8a	*n*-Hexane	−3.9	−10	−1.7
	(+) 8a	*n*-Pentane	−3	−8	−1.2
	(+) 8a	*n*-Octane	−5.7	−16	−2
	(+) 8b	CH_2Cl_2	−10	−29	−3.4
CH_2-C_6H_{11}	(+) 8a	CH_2Cl_2	−3.7	−13	−0.7
	(+) 8b	CH_2Cl_2	−5.8	−17	−1.9

From Reference 97.

large alkoxy groups give satisfactory results,[100–103] and a diastereoselectivity of higher than 97% has been reported for some diacetonylglucose derivatives[103] (Scheme 49.9).

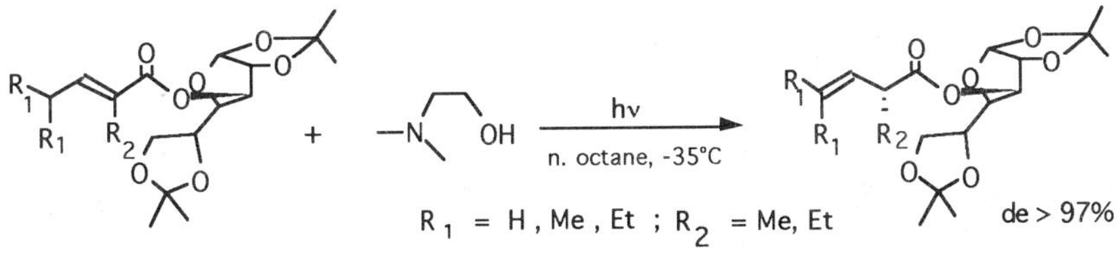

R_1 = H, Me, Et ; R_2 = Me, Et   de > 97%

SCHEME 49.9

Enantioselective deconjugation of enones is less satisfactory for several reasons: the corresponding photoenols can be stable enough, even in the presence of very small quantities of a chiral aminoalcohol, to be able to quench the triplet excited state of the starting enone and lead to some amounts of the stereoisomeric enol; consequently, the selectivity of the protonation step is lowered.[104] The deconjugated enones can also be photoreactive,[43,44,55,74] and electron transfer processes can become competitive as soon as chiral amines are introduced into the reaction mixture.[104] The chemical yields of the deconjugated enones are usually lower than for the corresponding esters.

Photodeconjugation can be very useful in multistep synthesis. Trapping of the photodienol by trimethylsilylchloride leads to activated dienes for Diels-Alder reactions that might be difficult to prepare by other routes. Furthermore, regioselective photodeconjugation in the presence of a base has already been applied to the synthesis of natural products.[27,85]

49.6 Photodeconjugation of Cyclic Enones

Z,E Photoisomerization of enones 4 and 5 having no γ-H available in the favorable configuration of the ethylenic double bond is an efficient process and a strained *trans*-stereoisomer is produced rapidly. In the *trans*-stereoisomer a γ-hydrogen can be abstracted as from linear enones and photodeconjugation is observed as described in Scheme 49.4.

trans-Cyclohexenones cannot be obtained as stable isomers[114–118] and their deconjugation involving a photochemical γ-H-abstraction is highly improbable. However, β,γ-unsaturated enones are observed when cyclohexenones or acetylcyclohexene are irradiated in *t*-butanol or in nonpolar solvents,[2,75–84,109–113] the reaction can be sensitized by triplet sensitizers such as acetophenone[79] in contrast with the singlet photodeconjugation of other enones and unsaturated acid derivatives. The presence of an enol intermediate is detected by deuterium incorporation on the α-carbon when the reaction is carried out in *t*-butanol-O-d. See Scheme 49.10. Some deuterium incorporation in the deconjugated isomer is also observed when hexadeuteriobenzene is used as the solvent. The complexity of the reactivity of cyclohexenone derivatives is illustrated by surprising solvent and wavelength effects, and some conflicting results have been reported.[84] An important contribution to the understanding of photodeconjugation of cyclohexenones was recently provided by the observation that the reaction, as determined for isophorone, needs the presence of impurities in order to proceed. Furthermore, the maximum quantum yield is obtained as soon as 0.01 molar equivalent acetic acid is introduced into the benzene solution (Scheme 49.11). The observations are consistent with the formation of a twisted and very strained *trans*-cyclohexenone. Protonation of this labile intermediate, as already described for excited cyclohexenes,[119] gives a carbocation that can be trapped by alcohols or deprotonated to the deconjugated enone.

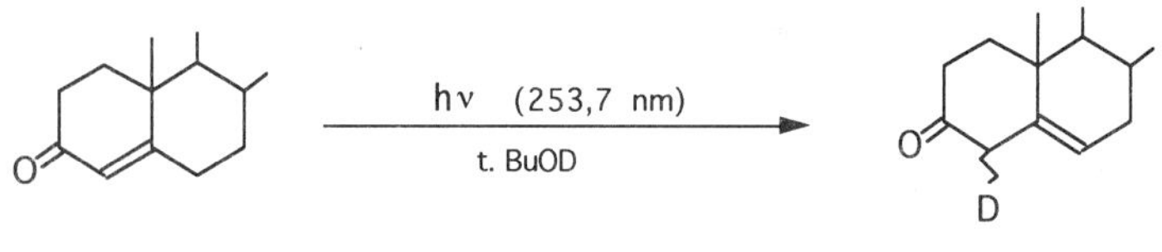

SCHEME 49.10

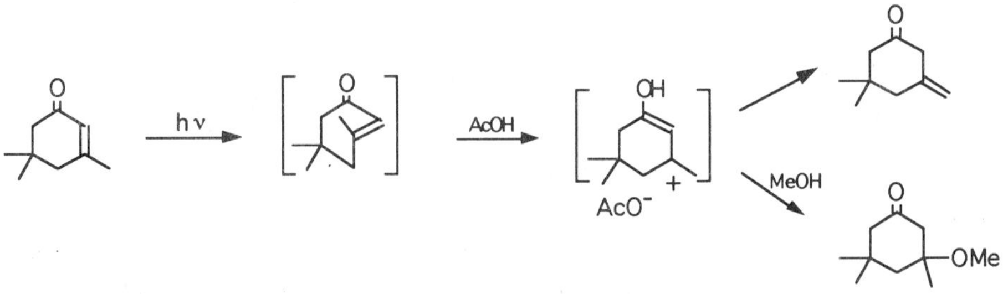

SCHEME 49.11

If *Z,E*-photoisomerisation is the main side reaction for α,β-unsaturated esters, lactones, and noncyclic enones [2+2]-cycloadditions, lumiketone rearrangements, and other cyclization processes can be competitive for cyclohexenone derivatives.

The photodeconjugation reaction involves a γ-hydrogen abstraction and formation of conjugated dienols. Photodienols cannot only be obtained from α,β-unsaturated ketones and acid derivatives but also from O-alkyl benzo- and acetophenones excited in their triplet states. The dienols are formed from these molecules as a mixture of stereoisomers and can be characterized by spectroscopy or trapped by chemical means. However, for aromatic ketones, the energy resonance of the benzene ring is too high to allow the formation of the deconjugated isomers.[89] The photodeconjugation process is limited to ketones and acid derivatives linked to an aliphatic ethylenic bond.

Due to the mild conditions used, to the high chemical yields obtained, and to the stereospecific formation of dienolic intermediates, the photodeconjugation process is very attractive for synthetic applications and even for asymmetric purposes.

ferences

1. Lutz, R. E., Bailey, P. S., Dien, C. K., and Rinker, J. W., The *cis*- and *trans*-β-aroyl-α- and β-methylacrylic acids and β-aroyl-α-methylenepropionic acids, *J. Am. Chem. Soc.*, 75, 5039, 1953.
2. Levina, R. Y., Kostin, V. N., and Gembitskii, P. A., Photochemical isomerization of vinylketones into allylketones, *Zhur. Obshchei Khim.*, 29, 2456, 1959.
3. Mousseron-Canet, M., Mousseron, M., and Legendre, P., Isomérisation photochimique dans la série de l'α-ionone, *Bull. Soc. Chim. Fr.*, 1509, 1961.
4. Jaffe, H. H. and Orchin, M., Theory and applications of UV spectroscopy, *Theory and Applications of UV Spectroscopy*, John Wiley & Sons, New York, 1966, 204, and references therein.
5. (a) Borrell, P. and Holmes, J. D., Photochemical *cis-trans* isomerization of 2-carbomethoxy-2-butene, *J. Photochem.*, 1, 433, 1972/73; (b) Morrison, H. and Rodriguez, O., Organic photochemistry. XXIX. *Z-E* photoisomerization of 3-methyl-3-penten-2-one. Evidence for non radiative decay, *J. Photochem.*, 3, 471, 1974/75; (c) Barltrop, J. A. and Wills, J., The mechanism of the photoisomerisation of α,β-unsaturated esters, *Tetrahedron Lett.*, 32, 4987, 1968.
6. Jorgenson, M. J., Photochemical hydrogen abstraction in unsaturated esters, *J. Chem. Soc., Chem. Commun.*, 137, 1965.
7. Kropp, P. J., and Krauss, H. J., Photochemistry of cycloalkenes. IV. Comparison with crotonic acid, *J. Org. Chem.*, 32, 3222, 1967.
8. Jorgenson, M. J. and Gundel, L., On the mechanism of the photochemical reaction of conjugated esters. The importance of geometrically isomeric excited states, *Tetrahedron Lett.*, 4991, 1968.
9. Rando, R. R. and Doering, W. E., β,γ-Unsaturated acids and esters by photochemical isomerization of α,β-congeners, *J. Org. Chem.*, 33, 1671, 1968.
10. Jorgenson, M. J., A large deuterium solvent isotope effect on a photochemical reaction, *J. Am. Chem. Soc.*, 91, 198, 1969.
11. Jorgenson, M. J., Photochemistry of α,β-unsaturated esters. VII. The photolytic behavior of vinylcyclopropanecarboxylates, *J. Am. Chem. Soc.*, 91, 6432, 1969.
12. Buchi, G. and Feairheller, S. H., Photochemical reactions. XIV. Additions to ethyl propiolate, *J. Org. Chem.*, 34, 609, 1969.
13. Jorgenson, M. J. and Patumtevapibal, S., Ring size and conformational effects on photodeconjugation of cycloalkylidene esters, *Tetrahedron Lett.*, 489, 1970.
14. Crandall, J. K. and Mayer, C. F., Photochemistry of some carbonyl conjugated 1,5-hexadienes, *J. Org. Chem.*, 35, 3049, 1970.
15. Rhoads, S. J., Chattopadhyay, J. K., and Waali, E., Double bond isomerizations in unsaturated esters and enol ethers. I. Equilibrium studies in cyclic and acyclic systems *J. Org. Chem.*, 35, 3352, 1970.

16. Scheffer, J. R. and Boire, B. A., Solution photochemistry. V. Differences in singlet and triplet state reactivities of some acyclic 1,7-dienes, *J. Am. Chem. Soc.,* 93, 5490, 1971.

17. Sundberg, R. J., Lin, L. S., and Smith, F. X., Photochemical deconjugation as a synthetic route to 1,2,3,6-tetrahydropyridine-4-acetic acid esters from $\Delta^{4,\alpha}$-piperidine-4-acetic acid esters, *J. Org. Chem.,* 38, 2558, 1973.

18. Majeti, S. and Gibson, T. W., Photochemistry of esters. Contrast in reactivity between acyclic and cyclic conjugated esters, *Tetrahedron Lett.,* 4889, 1973.

19. Gibson, T. W., Majeti, S., and Barnett, B. L., Photochemistry of esters. II. Effects of ring size on the photochemical behavior of α,β-unsaturated esters, *Tetrahedron Lett.,* 4801, 1976.

20. Mazzochi, P. H., Bowen, M. W., and Kachinsky, J., Generation of an aliphatic methyleneketen. Photolysis of *N*-methyl-4,4-dimethyl-3-isopropylideneazetidin-2-one, *J. Chem. Soc., Chem. Commun.,* 53, 1977.

21. Leitich, J., Partale H., and Polansky, O. E., Über die Photochemie cyclischer Acylale von Alkyliden-, Alkenyliden- und Benzylidenmalonsäuren, *Chem. Ber.,* 112, 3293, 1979.

22. Skinner, I. A. and Weedon, A. C., Photoenolization of α,β-unsaturated esters; effect of base upon product distribution and reaction rate in photochemical deconjugation, *Tetrahedron Lett.,* 4299, 1983.

23. Weedon, A. C., Base catalysis as an alternative explanation for an apparent deuterium solvent isotope effect in the photochemical reactions of α,β-unsaturated esters, *Can. J. Chem.,* 62, 1933, 1984.

24. Duhaime, R., Lombardo, D., Skinner, I. A., and Weedon, A. C., Conversion of α,β-unsaturated esters to their β,γ-unsaturated isomers by photochemical deconjugation, *J. Org. Chem.,* 50, 873, 1985.

25. Marjerrison, M. and Weedon, A. C., The importance of photoenolization as a route for radiation-less decay in β-alkyl-α,β-unsaturated carbonyl compounds, *J. Photochem.,* 33, 113, 1986.

26. Lombardo, D. A. and Weedon, A. C., Photoenolization of conjugated esters. Synthesis of a San Jose scale pheromone by partially regio-controlled photochemical deconjugation, *Tetrahedron Lett.,* 27, 5555, 1986.

27. Freeman, P. K., Siggel, L., Chamberlain, P. H., and Clapp, G. E., The photochemistry of methyl geranate, a model chromophore for insect juvenile hormone analogs, *Tetrahedron,* 44, 5051, 1988.

28. Freeman, P. K. and Siggel, L., The kinetics of photochemical deconjugation reactions of methyl geranate, *Tetrahedron,* 44, 5065, 1988.

29. *Cf.* also References 90 to 99.

30. Mousseron, M., Isomérisation photochimique de systèmes polyéniques, *Pure Appl. Chem.,* 9, 481, 1964.

31. Crowley, K. J., Photochemical formation of allenes in solution, *J. Am. Chem. Soc.,* 85, 1210, 1973.

32. Crowley, K. J., Schneider, R. A., and Meinwald, J., Photoisomerization of phorone, *J. Chem. Soc. (C),* 571, 1966.

33. Itoh, M., Tokuda, M., Hataya M., and Suzuki, A., Photochemical reactions of α,β-unsaturated acids and esters, *Hokkaido Daigaku Kogakubu Kenkyu Kokoku,* 60, 37, 1971.

34. Biot, J. M., de Keukeleire, D., and Verzele, M., Photoisomerization in α,β-unsaturated carboxylic acids, *Bull. Soc. Chim. Belg.,* 86, 973, 1977.

35. *Cf.* also Reference 96.

36. Ohga, K. and Matsuo, T., A study on the photochemistry of α,β-unsaturated γ-lactones. II. Photoisomerization of 3-ethylidene-4,5-dihydro-2(3H)-furanone, *Bull. Soc. Chim. Jpn.,* 46, 2181, 1973.

37. Henin, F., Mortezaei, R., and Pete, J. P., A convenient synthesis of α-vinyllactones: The photodeconjugation of α,β-unsaturated γ or δ-lactones, *Synthesis,* 1019, 1983.

38. Mc.Dowell, C. A. and Sifniades, S., Isomerization as a primary process in the photolysis of crotonaldehyde, *J. Am. Chem. Soc.,* 84, 4606, 1962.

39. *Cf.* Reference 31.

40. Henin, F., Muzart, J., Pete, J. P., M'Boungou M'Passi, A., and Rau, H., Enantioselective protonation of a simple enol: Aminoalcohol-catalyzed ketonization of a photochemically produced 2-methylinden-3-ol, *Angew. Chem. Int. Ed.*, 30, 416, 1991.

41. De Mayo, P., Stothers, J. B., and Yip, R. W., Irradiation of β-ionone, *Can J. Chem.*, 39, 2135, 1961.

42. Van Wageningen, A. and de Boer, Th., Irradiation of *trans*-α-ionone, *Rec. Trav. Chim. Pays-Bas*, 89, 797, 1970.

43. Visser, C. P. and Cerfontain, H., Photochemistry of dienones. Part IX. Photochemistry of *(E)* and *(Z)*-α-ionone, *Rec. Trav. Chim. Pays-Bas*, 100, 153, 1981.

44. Visser, C. P., Van der Wel, H., and Cerfontain, H., Photochemistry of dienones. XI. Photochemistry of the pivaloyl analogues of *(E)* -α- and β-ionone, *Rec. Trav. Chim. Pays-Bas*, 102, 302, 1983.

45. Lutz, R. E., Bailey, P. S., Dien, C. K., and Rinker, J. W., The *cis*- and *trans*-aroyl-α and β-methylacrylic acids and β-aroyl-α-methylenepropionic acids, *J. Am. Chem. Soc.*, 75, 5039, 1953.

46. Jorgenson, M. J. and Yang, N. C., A novel photochemical reaction. Conversion of α,β-unsaturated ketones to acetonylcyclopropanes, *J. Am. Chem. Soc.*, 85, 1698, 1963.

47. Yang, N. C. and Jorgenson, M. J. Photochemical isomerization of simple α,β-unsaturated ketones, *Tetrahedron Lett.*, 1203, 1964.

48. Schneider, R. A. and Meinwald, J., Photochemical reactions of α,β-unsaturated carbonyl compounds with olefins, *J. Am. Chem. Soc.*, 89, 2023, 1967.

49. Coomber, J. W. and Pitts, J. N., Jr., Molecular structure and photochemical reactivity. XIV. The vapor phase photochemistry of *trans*-crotonaldehyde, *J. Am. Chem. Soc.*, 91, 4955, 1969.

50. Friedrich, L. E. and Schuster, G. B., Irradiation of α,β-unsaturated ketones. Search for intermediate oxabicyclobutanes, *J. Am. Chem. Soc.*, 94, 1193, 1972.

51. Kawai, M. and Naya, K., Photochemistry of fukinone and pulegone, *Chem. Lett.*, 389, 1972.

52. Hatanaka, A. and Ohgi, T., Leaf alcohol. XIX. Photochemical isomerization of leaf aldehyde, *Agrik. Biol. Chem.*, 36, 1263, 1972.

53. Bonnet, F. and Lemaire, J., Transitions non radiatives de l'oxyde de mésityle, *Bull. Soc. Chim. Fr.*, 1973, 1185.

54. Smith, A. B. and Agosta, W. C., Photochemical reactions of 1-cyclopentenyl and 1-cyclohexenyl ketones, *J. Am. Chem. Soc.*, 95, 1961, 1973.

55. Cookson, R. C. and Rogers, N. R., Photochemistry of 2-(cycloalk-1-enyl) cycloalkanones. New photochemical reaction of β,γ-unsaturated ketones, *J. Chem. Soc., Perkin Trans 1*, 1037, 1974.

56. Arnould, J. C. and Pete, J. P., Photochemistry of α-dialkylamino enones. I. A new oxidative cyclisation of chalcone derivatives, *Tetrahedron Lett.*, 2459, 1975.

57. Tada, K. and Miura, K., The photo-enolization of β-disubstituted α,β-unsaturated ketone, *Bull. Soc. Chim. Jpn.*, 49, 713, 1976.

58. Deflandre, A., Lheureux, A., Rioual, A., and Lemaire, J., Comportement photochimique de cétones α,β-insaturées sous excitation directe, *Can. J. Chem.*, 54, 2127, 1976.

59. Arnould, J. C., Enger, A., Feigenbaum, A., and Pete, J. P., Photolyse de cétones conjuguées hétérosubstituées linéaires, *Tetrahedron*, 35, 2501, 1979.

60. Wan, C. S. K. and Weedon, A. C. The photochemical enolization of an aliphatic α,β-unsaturated ketone, *J. Chem. Soc., Chem. Commun.*, 1235, 1981.

61. Prasad, G., Giri, B. P. and Mehrota, K. N., Photochemistry of benzyl monobenzhydrylimine, *J. Org. Chem.*, 47, 2353, 1982.

62. Eng, S. L., Ricard, R., Wan, C. S. K. and Weedon, A. C., Photochemical deconjugation of α,β-unsaturated ketones, *J. Chem. Soc., Chem. Commun.*, 236, 1983.

63. Stevens, R. V. and Pruit, J. R., On the annulation of Δ²-tetrahydropyridines. An expeditious total synthesis of the protoberberine alkaloid karachine, *J. Chem. Soc., Chem. Commun.*, 1425, 1983.

64. Henderson, W. A. Jr. and Ullman, E. F., Photochemistry of 2-benzyl- and 2-benzhydryl-3-benzoylchromones, *J. Am. Chem. Soc.*, 87, 5424, 1965.

65. Ferrier, R. J. and Tyler, P. C., Synthesis and photolysis of some carbohydrate 1,6-dienes, *Carbohydr. Res.*, 136, 249, 1985.

66. (a) Duhaime, R. M. and Weedon, A. C., Direct observation of a photochemically produced dienol: Evidence for a non-catalyzed reketonization pathway unavailable to simple enols, *J. Am. Chem. Soc.*, 107, 6723, 1985. (b) Duhaime, R. M. and Weedon, A. C., Direct measurement of the rates of reketonization of dienolates produced by photochemical enolization of β-alkyl α,β-unsaturated ketones in aqueous basic solution, *J. Am. Chem. Soc.*, 109, 2479, 1987.

67. Ricard, R., Sauvage, P., Wan, C. S. K., Weedon, A. C., and Wong, D. F., Photochemical enolization of β-alkyl-α,β-unsaturated ketones, *J. Org. Chem.*, 51, 62, 1986.

68. Wan, C. S. K., Weedon, A. C., and Wong, D. F., Stereoselective and regioselective thermal and photochemical preparation of siloxy dienes, *J. Org. Chem.*, 51, 3335, 1986.

69. Nozaki, H., Mori, T., and Noyori, R., Preparation and photochemical isomerization of 2-cyclodecenones, *Tetrahedron*, 22, 1207, 1966.

70. Carlson, R. G. and Bateman, J. H., The photochemical rearrangement of cis-2-cyclodecenone, *Tetrahedron Lett.*, 4151, 1967.

71. Noyori, R., Inoue, H., and Kato, M., Photolysis of A-acetylcyclooctene. Direct observation of dienol intermediate in photochemical deconjugation of α,β-unsaturated ketone, *J. Am. Chem. Soc.*, 92, 6699, 1970.

72. Marchesini, A., Pagani, G., and Pagnoni, U. M., The stereochemical fate of the biradical intermediate in the photolysis of cyclododecen-2-one, *Tetrahedron Lett.*, 1041, 1973.

73. Noyori, R., Inoue, H., and Kato, M., Direct observation of dienol intermediates in photochemical deconjugation of an α,β-unsaturated ketone. Photoisomerization of 1-acetylcyclooctene, *Bull. Soc. Chim. Jpn.*, 49, 3673, 1976.

74. Gleiter, R., Sander, W., Irngartinger, H., and Lenz, A., A simple path to tricyclo [5.3.0.0^{2,8}]decane and its derivatives, *Tetrahedron Lett.*, 23, 2647, 1982.

75. Wehrli, H., Wenger, R., Schaffner, K., and Jeger, O., Zur photochemischen Isomerisierung von 10α-Testosteron, *Helv. Chim. Acta*, 46, 678, 1963.

76. Dauben, W. G., Shaffer, G. W., and Wietmeyer, N. D., Alkyl-substitution effects in the photochemistry of 2-cyclohexenones, *J. Org. Chem.*, 33, 4061, 1968.

77. Furutachi, N., Wakadaira, Y., and Nakanishi, K., Photoirradiation of steroid 5-en-7-one systems and a mutual exchange of C$_4$ and C$_6$ in 3β-acetoxycholest-5-en-7-one, *J. Am. Chem. Soc.*, 91, 1028, 1969.

78. Kuwata, S. and Schaffner, K., Spezifisch π → π*- induzierte Keton-Photoreaktionen: Die doppelbindungsverschiebung von O-Acetyl-10α-testosteron. Protonisierung der Doppelbindung seines Δ^5 Isomeren, *Helv. Chim. Acta*, 52, 173, 1969.

79. Bellus, D., Kearns, D. R., and Schaffner, K., Zur Photochemie von α,β-ungesättigten cyclischen Ketonen: Spezifische Reactionen der n,π*- und π,π*-Triplettzustände von O-Acetyl-testosteron und 10-Methyl-Δ^{1,9}-octalon-2, *Helv. Chim. Acta*, 52, 971, 1969.

80. Gloor, J., Schaffner, K., and Jeger, O., Spezifisch π → π*- induzierte Photoisomerisierungen von 10-Dimethoxymethyl-Δ^{1,9}-octal-2-one. Ein Photochemischer Zugang zu [4.4.3]-12-oxapropellan-Derivaten, *Helv. Chim. Acta*, 54, 1864, 1971.

81. Margaretha, P. and Schaffner, K., Zur Photochemie von Δ^{1,9}-10-Methyl-2-octalon, *Helv. Chim. Acta*, 56, 2884, 1973.

82. Gloor, J. and Schaffner, K., Spezifisch π → π*- induzierte Reactionen von γ-Dimethoxymethylcyclohexen-2-onen: 1,3-Umlagerung und Wasserstoffabstraktion durch das α-Kohlenstoffatom, *Helv. Chim. Acta*, 57, 1815, 1974.

83. Gioia, B., Ballabio, M., Beccali, E. M., Cecchi, R., and Marchesini, A., Photochemical behaviour of bicyclo[5.3.1]undec-1(10)-en-9-one, *J. Chem. Soc., Perkin Trans. 1*, 560, 1981.

84. Rudolph, A. and Weedon, A. C., Acid catalysis of the photochemical deconjugation reaction of 3-alkyl-2-cyclohexenones, *J. Am. Chem. Soc.*, 111, 8756, 1989.

85. Weedon, A. C., *The Chemistry of Enols*, S. Pataï, Ed., John Wiley & Sons, New York, 1990, 591.

86. Duhaime, R. M. and Weedon, A. C., Direct observation of dienols produced by photochemical enolisation of α,β-unsaturated ketones: Rates and activation parameters for dienol reketonisation via a 1,5-hydrogen shift, *Can. J. Chem.*, 65, 1867, 1987.

87. Lewis, F. D. and Oxman, J. D., Lewis acid enhancement of photochemical *trans → cis* isomerization of α,β-unsaturated esters, *J. Am. Chem. Soc.*, 103, 7345, 1981.

88. Lewis, F. D., Howard, D. K., Barancyk, S. V., and Oxman, J. D., Lewis acid catalysis of photochemical reactions 5. Selective isomerization of conjugated butenoic and dienoic esters, *J. Am. Chem. Soc.*, 108, 3016, 1986.

89. Sammes, P. G., Photoenolisation, *Tetrahedron*, 32, 405, 1976.

90. Henin, F., Mortezaei, R., and Pete, J. P., Enantioselective photodeconjugation of conjugated lactones induced by small amounts of a chiral inductor, *Tetrahedron Lett.*, 26, 4945, 1985.

91. Mortezaei, R., Henin, F., Muzart, J., and Pete, J. P., Enantioselective photodeconjugation of α,β-unsaturated esters in the presence of catalytic amounts of a chiral-inducing entity, *Tetrahedron Lett.*, 26, 6079, 1985.

92. Mortezaei, R., Piva, O., Henin, F., Muzart, J., and Pete, J. P., Evaluation of the steric interactions responsible for the enantioselective photodeconjugation of α,β-unsaturated esters, *Tetrahedron Lett.*, 27, 2997, 1986.

93. Piva, O., Henin, F., Muzart, J., and Pete, J. P., Enantioselective photodeconjugation of α,β-unsaturated esters: Effect of the nature of the chiral agent, *Tetrahedron Lett.*, 27, 3001, 1986.

94. Pete, J. P., Henin, F., Mortezaei, R., Muzart J., and Piva, O., Enantioselective photodeconjugation of conjugated esters and lactones, *Pure Appl. Chem.*, 58, 1257, 1986.

95. Piva, O., Henin, F., Muzart, J., and Pete, J. P., A very enantioselective photodeconjugation of α,β-unsaturated esters, *Tetrahedron Lett.*, 28, 4825, 1987.

96. Henin, F., Mortezaei, R., Pete, J. P., and Piva, O., Photodéconjugaison énantiosélective d'esters et de lactones conjugées en présence d'éphédrine, *Tetrahedron*, 45, 6171, 1989.

97. Piva, O., Mortezaei, R., Henin, F., Muzart, J., and Pete, J. P., Highly enantioselective photodeconjugation of α,β-unsaturated esters. Origin of the chiral discrimination, *J. Am. Chem. Soc.*, 112, 9263, 1990.

98. Duhamel, L., Duhamel, P., Launay, J. C., and Plaquevent, J. C., Asymmetric protonations, *Bull. Soc. Chim. Fr.*, II, 421, 1984.

99. (a) Fehr C. and Galindo, J., Synthesis of (R)-(+)- and (S)-α-damascone by tandem Grignard reaction. Enantioselective protonation: Evidence for the intermediacy of a chiral complex, *J. Am. Chem. Soc.*, 110, 6909, 1988; (b) Fehr, C., Enantioselective protonation of enolates in natural product synthesis, *Chimia*, 253, 1991.

100. Mortezaei, R., Awandi, D., Henin F., Muzart J., and Pete, J. P., Diastereoselective photodeconjugation of α,β-unsaturated esters, *J. Am. Chem. Soc.*, 110, 4824.F, 1988.

101. Awandi, D., Henin, F., Muzart, J., and Pete, J. P., Reversal of diastereoselectivity in protonation of chiral photodienols, *Tetrahedron: Asymmetry*, 2, 1101, 1991.

102. Piva, O. and Pete, J. P., Diacetone and D-glucose: efficient chiral building block for asymmetric photodeconjugation, *Tetrahedron: Asymmetry*, 3, 759, 1992.

103. Charlton, J. L., Pham, V. C., and Pete, J. P., Lactate as chiral auxiliary in asymmetric photodeconjugation of unsaturated esters, *Tetrahedron Lett.*, 33, 6073, 1992.

104. Henin, F., Muzart, J., Pete, J. P., and Piva, O., Enantioselective photodeconjugation of α-substituted α,β-unsaturated ketones, *New J. Chem.*, 15, 611, 1991.

105. Gardner, P. D. and Hamil, H. F., A photochemical ester rearrangement induced by homoconjugation excitation, *J. Am. Chem. Soc.*, 83, 3531, 1961.

106. Yamada, Y., Uda, H., and Nakanishi, K., Synthesis of 6-methylbicyclo[4.2.0] octan-2-one by photochemical additions, *J. Chem. Soc., Chem. Commun.*, 423, 1966.

107. Hayashi, J., Furutachi, N., Nakadaira, Y., and Nakanishi, K., The structures of two dimers obtained upon irradiation of cholest-5-en-7-one, *Tetrahedron Lett.*, 4589, 1969.

108. Schuster, D. I. and Brizzolara, D. F., The mechanism of photoisomerization of cyclohexenones. 10-hydroxymethyl-$\Delta^{1,9}$-2-octalone. The question of hydrogen abstraction from benzene by ketone triplets, *J. Am. Chem. Soc.*, 92, 4357, 1970.

109. Jeger, O. and Schaffner, K., On photochemical transformations of steroids, *Pure Appl. Chem.*, 21, 247, 1970.

110. Nobs, F., Burger, U., and Schaffner, K., Specifically (π,π^*)-induced cyclohexenone reactions. 4a-(Z-1-propenyl)-bicyclo[4.4.0]dec-1-(8a)-en-2-one and 4a-(Z-1-propenyl bicyclo[4.4.0]deca-1 (8a) 7-dien-2-one, *Helv. Chim. Acta*, 60, 1607, 1977.

111. Shiloff, J. D. and Hunter, N. R., Solvent effects on the photocycloaddition and photoenolisation reactions of isophorone, *Can. J. Chem.*, 57, 3301, 1979.

112. Williams, J. R., Mattei, P. L., Abdel-Magid, A., and Blount, J. F., Photochemistry of estr-4-en-3-ones: 17β-hydroxy-4-estren-3-one, 17β-acetoxy-4,9-estradien-3-one, 17β-hydroxy-4,9,11-estratrien-3-one, and norgestrel. Photochemistry of 5α-estran-3-ones, *J. Org. Chem.*, 51, 769, 1986.

113. Wintgens, V., Guérin, B., Lenholm, H., Brisson, J. R., and Scaiano, J. C., Photochemistry of the inclusion complex of (R)-$\Delta^{1,9}$-10-methyl-2-octalone with α-cyclodextrin, *J. Photochem. Photobiol. A. Chemistry*, 44, 1968.

114. Bonneau, R. and Fornier De Violet, P., Evidence for the strained *trans*-acetyl-l-cyclohexene and a transient species assigned to the orthogonal triplet state, *C.R. Acad. Sc. Ser. C*, 284, 631, 1977.

115. Goldfarb, T. D., Kinetic studies of transient photochemical isomers of 2-cycloheptenone, 1-acetylcyclohexene and 2-cyclohexenone, *J. Photochem.*, 8, 39, 1978.

116. Bonneau, R., Transient species in photochemistry of enones. The orthogonal triplet state revealed by laser photolysis, *J. Am. Chem. Soc.*, 102, 3816, 1980.

117. Pienta, N. J., Photoreactivity of α,β-unsaturated carbonyl compounds. 2. Fast transients from irradiation of 2-cyclohexenones and amines, *J. Am. Chem. Soc.*, 106, 2704, 1984.

118. Schuster, D. I., Dum, D. A., Heibel, G. E., Brown, P. B., Rao, J. M., Woning J., and Bonneau, R., Enone photochemistry. Dynamic properties of triplet excited states of cyclic conjugated enones as revealed by transient absorption spectroscopy, *J. Am. Chem. Soc.*, 113, 6245, 1991.

119. Kropp, P. J., *Photochemistry of Alkenes in Solution Organic Photochemistry*, Vol. 4, Marcel Dekker, 1979, 1.

50

Phase Effects on the Competition Between Hydrogen Abstraction and Photocycloaddition in Bicyclic Enones

ɪn R. Scheffer
iversity of British Columbia

ɪes Trotter
iversity of British Columbia

ɪna D.
ɪdmundsdottir
iversity of British Columbia

Depending on their structure, cyclic α,β-unsaturated ketones display a wide range of photochemical reactivity that includes rearrangement, [2+2]-cycloaddition, *cis,trans*-isomerization, and hydrogen abstraction.[1] Although relatively rare, the hydrogen abstraction photoreactions of cyclic enones are nevertheless well documented, and it is this reaction that forms the main focus of this chapter.

Hydrogen atom transfer to enones possessing lowest (π,π*) triplet states, particularly those that are conformationally rigid, occurs preferentially to the β-carbon atom of the α,β-unsaturated double bond, a reactivity pattern that reflects the electronic nature of the excited state as well as the energetic advantage associated with locating the radical center in the product at its more favorable position α to the carbonyl group. Both inter- and intramolecular versions of this reaction are known; with regard to the former process, the work of Schuster and co-workers is most definitive. Schuster, Woning, Kaprinidis, Pan, Cai, Barra, and Rhodes[2] showed, for example, that photolysis of the bicyclic enone 1 (Scheme 50.1) in hydrogen atom-donating solvents gave products derived from the radical species 2. This intermediate, which could be detected spectroscopically, was estimated to have a lifetime of 4.1 μs.

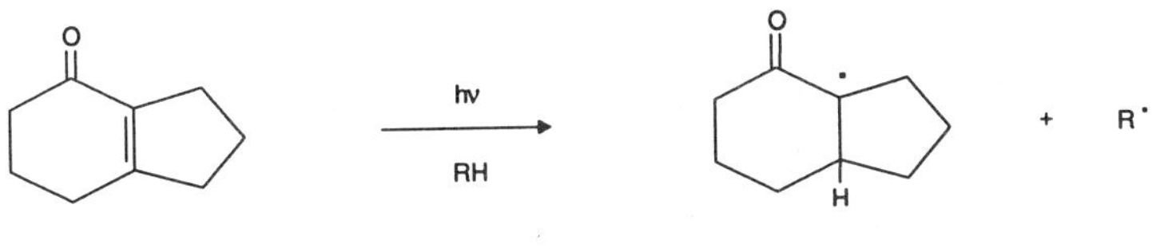

SCHEME 50.1

Intramolecular hydrogen atom abstraction by the β-carbon of enones leads to biradical intermediates that can either disproportionate or couple to form novel photoproducts. An example taken from the work of Wolff, Schreiber, Smith, and Agosta[3] is depicted in Scheme 50.2. Thus, irradiation of the cyclopentenone derivative 3 was shown to lead to photoproducts 5 (59%), 6 (24%), and 7

(11%), a reaction that was interpreted as occurring through biradical 4 as the result of six-membered transition state transfer of the methine hydrogen atom of the side chain to the β-carbon of the enone.

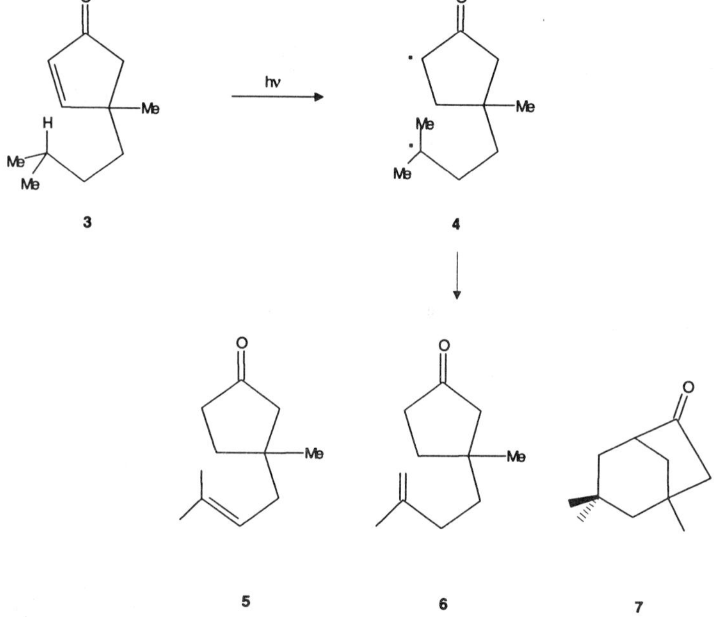

SCHEME 50.2

Our interest in this type of hydrogen atom abstraction came about as a result of our work with *cis*-fused bicyclic enones of general structure 8 (Scheme 50.3).[4] Compounds of this type are readily prepared through Diels-Alder addition reactions of *p*-benzoquinones to 1,3-dienes, followed by chemical modification of one of the two carbonyl groups in the adduct. The enones so produced, epimers 8A and 8B, underwent intramolecular [2+2]-cycloaddition when photolyzed in solution to form the interesting cage molecules 9A and 9B. In marked contrast, when the same enones were photolyzed *in their neat crystalline form*, completely different results were obtained. Under these conditions, the photoproducts were the novel tricyclic ketones of general structure 10 and 11. Not only are these photoproducts different from those formed in solution, but they differ fundamentally from one another as well: compound 10 possesses the tricyclo[4.4.0.0^{3,7}]decane carbon skeleton, whereas structure 11 has the tricyclo[4.4.0.0^{3,10}]decane ring system. How can two reactants that differ only in their relative stereochemistry at a remote center lead to such different photochemical results? The answer to this question was worked out with the help of extensive X-ray crystallographic analysis of the enones 8A and 8B.[5] As one will see, the key lies in the fact that epimers 8A and 8B adopt different conformations in the solid state and that the conformers react photochemically through unique, conformation-specific processes involving the transfer of *different hydrogen atoms* to the β-carbon of the enone moiety.

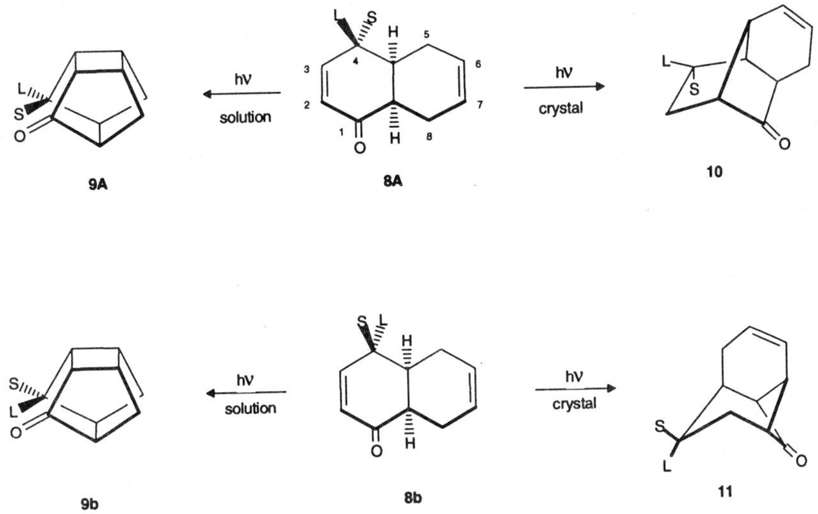

S = small group, L = large group; other substituents omitted for clarity

SCHEME 50.3

The structure-reactivity relationships to be outlined in the following pages have been deduced from the study of a large number of compounds during a period of over 10 years. Space limitations do not permit a full exposition of the results, and we have chosen therefore to illustrate the main principles involved through a discussion of the photochemistry of five representative enones, compounds **12** to **16** (Scheme 50.4). The interested reader who desires more detail and further examples is referred to the original literature cited. Each enone is the result of a Diels-Alder reaction between duroquinone and the appropriate diene, followed by sodium borohydride reduction of one of the two carbonyl groups in the adduct. The larger group at C-4 is therefore hydroxyl and the smaller is hydrogen. In most cases, a mixture of epimeric alcohols was produced, and these were separated and studied individually; acetylation of the alcohols led to the corresponding acetates.

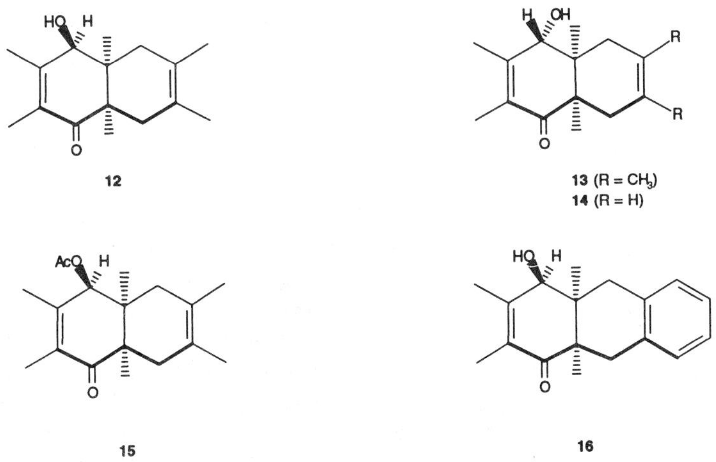

12

13 (R = CH₃)
14 (R = H)

15

16

SCHEME 50.4

Photolysis ($\lambda > 330$ nm) of enone **12** (*cf.* **8A**, Scheme 50.3) in benzene solution afforded a 96% isolated yield of the intramolecular [2+2]-cycloadduct **17** (Scheme 50.5).[6] In contrast, when crystals of enone **12** were irradiated, the nearly exclusive product proved to be the ketol **18**. The solid state photolyses were conducted by coating, through slow evaporation, the inside of a flask with a thin, polycrystalline film of the sample and then irradiating it from within using the standard immersion well configuration. This allowed a nitrogen atmosphere to be maintained over the sample, and cooling could be carried out by immersing the entire apparatus in an appropriate cooling bath. In the case of enone **12**, which is typical, 70 mg of compound was irradiated for 2.1 h at -74 °C to afford, after column chromatography, 39 mg of recovered starting material and 25 mg of photoproduct **18**.

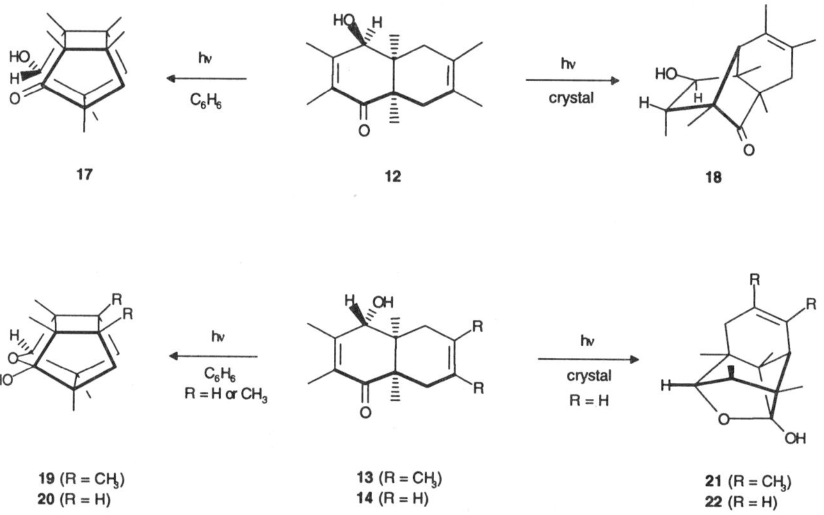

SCHEME 50.5

Enone **13** (*cf.* **8B**, Scheme 50.3), which is the C-4 epimer of enone **12**, also underwent facile [2+2]-photocycloaddition in solution (Scheme 50.5).[6] This leads to a cage compound in which the hydroxyl group is directed toward the transannular carbonyl group and, as a result, the photoproduct is isolated in its internal hemiacetal form, **19**. In analogous fashion, photolysis of benzene solutions of enone **14** gave hemiacetal **20**. Interestingly, in this case, significant amounts of the solid-state photoproduct **22** were also formed.[6] In the solid state, enone **13** proved to be photoinert, whereas its analogue, **14**, reacted smoothly to form the novel photoproduct **22**. The structure of this latter compound, which was determined by X-ray crystallography, can be seen to be the internal hemiacetal of photoproduct type **11** (Scheme 50.3) in which L = OH.

As mentioned earlier, the X-ray crystal structures of the reactant enones provided the key to understanding these complex transformations. The crystal structures showed that enones of type **8A** (e.g., **12**) invariably crystallize in a conformation that is very different from the conformation adopted by enones of type **8B** (e.g., **13** and **14**). These conformations are depicted in Scheme 50.6. Both are characterized by flattened cyclohexenone rings *cis*-fused to half-chair cyclohexene rings; and in each case, the larger substituent at C-4 (the hydroxyl group) occupies the more favorable pseudoequatorial position. This is consistent with the general observation that conformationally flexible molecules crystallize in or near their lowest energy conformations, since the alternative half-chair conformers in each case would have the larger substituent in the pseudoaxial position.

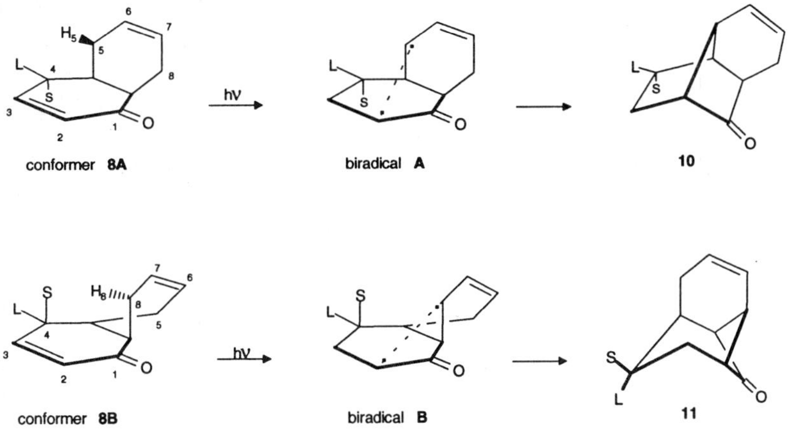

SCHEME 50.6

The solid-state reaction mechanisms now stand revealed. Type A enones react via intramolecular transfer of H-5 to C-3, the β-carbon atom of the α,β-unsaturated ketone. After hydrogen transfer (which, by the way, is a rare example of a five-membered transition state process), the biradical produced (biradical A, Scheme 50.6) undergoes C2-C5 bond formation resulting in photoproducts of type 10 (e.g., 18). In type B enones, it is the H-8 allylic hydrogen atom that is abstracted through a six-membered transition state, thus resulting in biradical B; closure of this species by C2-C8 bonding then affords photoproducts of type 11 (e.g., 22). Two further points deserve mention at this juncture. First, both conformers A and B are unfavorable for intramolecular [2+2]-photocycloaddition. The double bonds are neither close together nor parallel and, as a result, no cage products are formed in the solid state. The second point is that the proposed hydrogen abstraction reactions and carbon-carbon bond formations *are* geometrically feasible from conformers A and B. In eight separate examples,[4] the C ··· H abstraction distances, as determined by X-ray crystallography, ranged from 2.72 to 2.85 Å and are all less than the sum of the van der Waals radii for carbon and hydrogen (2.90 Å).[7] Similarly, the C-2 ··· C-5 and C-2 ··· C-8 bonding distances fell in the range of 3.17 to 3.42 Å, once again very close to the van der Waals radii sum of 3.40 Å. Of course, these distances are only a rough indication of hydrogen abstractability and carbon-carbon bond-forming ability, since they refer to ground-state, not excited-state, geometry. For a discussion of this point that takes into consideration the likelihood that the β-carbon atom of enones becomes pyramidalized in the excited state, see Reference 8.

Having accounted for the "conformation-specific" solid-state photochemistry of enones 12 and 14 (the lack of reactivity of enone 13 will be discussed later), the question arises next as to the reasons for the preference displayed by these compounds for intramolecular [2+2]-photocycloaddition in solution. After all, the conformations shown in Scheme 50.6 are undoubtedly the major conformers present in solution, so why are more of photoproduct types 10 and 11 not seen in this medium as well? We propose that this stems from a very fast [2+2]-photoreaction from a minor, high-energy conformer in solution; specifically, conformer C (below), in which the enone double bond and the cyclohexene double bond are parallel and close together. This is an eclipsed conformer, so it would be predicted to have a higher conformational energy than either A or B, in which the dihedral angles about the ring junction carbon-carbon bond are all close to 60°.

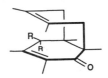

conformer C

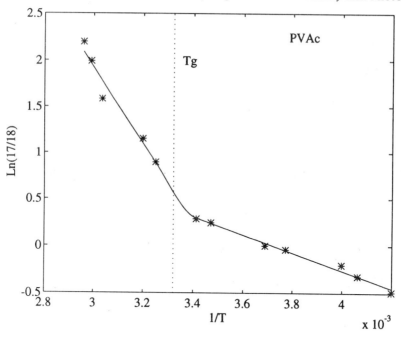

FIGURE 50.1

It seems likely that, in solution, equilibrium is established among the enone conformers in the *excited state* since, as pointed out by Lewis et al.,[9] if this were not the case, the photoproduct ratios would reflect the ground-state conformer distribution; i.e., hydrogen abstraction from conformers 8A and 8B would predominate. In the one case where it was measured (enone 14), the excited-state lifetime was found to be sufficiently long (*ca.* 0.01 s) to be consistent with this picture. In addition, the weak phosphorescence observed in the case of enone 14 was consistent with a $(\pi,\pi^*)^3$ excited state, the same excited state identified by others as being responsible for hydrogen abstraction by an enone β-carbon.[10]

The mechanistic and kinetic picture presented above predicts that if one were to retard the rate of isomerization among the various enone conformers by, for example, increasing the solvent viscosity, one should get more of the hydrogen abstraction-type photoproducts 10 and 11. This was found to be the case.[11] Enone 12 was embedded in a poly(vinyl acetate) matrix by dissolving the two components in chloroform, casting a film, and pumping off the solvent. Photolysis of such films led to mixtures of photoproducts 17 and 18 whose ratio was temperature dependent. The results of these measurements, plotted as ln (17/18) vs. 1/T, are shown in Figure 50.1. Interestingly, the data can be fit by two reasonably good straight lines of different slope that meet at the glass transition temperature. This indicates that the 17/18 ratio increases more rapidly above T_g than below it. This can be ascribed to a matrix-derived restriction of the motions required for conformational isomerization below T_g; above T_g, the free volume of the polymer matrix increases more rapidly with temperature and such restrictions are eased.

The photochemistry of enone 13 which, it will be recalled, was photoinert in the crystalline state, was also investigated in a polymer matrix.[11] We hoped that significant amounts of the "missing" solid-state photoproduct 21 would be obtained under these conditions, and these hopes were realized. Irradiation of enone 13 in a poly(methyl methacrylate) matrix at room temperature led to a mixture of [2+2]-cycloadduct 19 (20%) and hemiacetal 21 (80%). These results suggest that the lack of photoreactivity of enone 13 in the solid state is most likely due to a crystal lattice packing effect that is not present in the polymer matrix. Our original suggestion[6] that the unreactivity of enone 13 in the solid state might be due to a prohibitively long C-3 ⋯ H-8 abstraction distance (2.92 Å) is probably not correct.

We turn now to a discussion of the solid-state and solution-phase photochemistry of enone 15. This compound, like all the other enones studied, underwent facile intramolecular [2+2]-photocycloaddition in benzene solution (Scheme 50.7), but its solid-state photoreactivity proved to be startlingly different. Irradiation of crystals of enone 15 afforded the novel keto-acetate 27 as the sole volatile product.[12,13] This compound is the result of abstraction of H-5 by C-2, *the α-carbon* of the α,β-unsaturated double bond, followed by C3-C5 bonding.

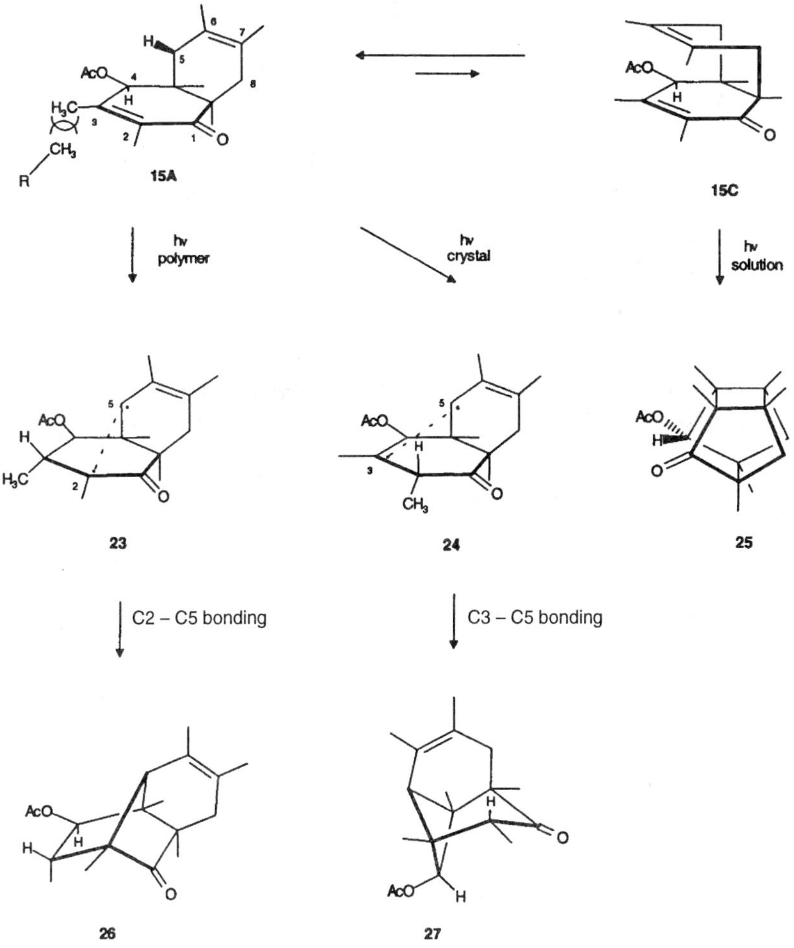

SCHEME 50.7

Why should hydrogen transfer in the case of enone 15 be to the α-position, rather than to the β-position as in all the other enones studied? The crystal structure reveals that enone 15 has a normal type A conformation and indicates that there is nothing unusual about the C-2 ⋯ H-5 or C-3 ⋯ H-5 distances. The packing diagram for 15 did, however, indicate a particularly close contact between the methyl group at C-3 and a methyl group of a neighboring molecule below it in the lattice (15A(?) is A, Scheme 50.7). This, we suggest, is the feature responsible for the abnormal solid-state photobehavior of enone **15**. The argument proceeds as follows: the activation energy of the normally favored hydrogen transfer to C-3 is raised through steric hindrance to the downward motion of the methyl group that accompanies pyramidalization at this center. In contrast, there is no such contact involving the methyl group attached to C-2, and hydrogen transfer to this site is unimpeded. Computer-assisted molecular modeling studies support this hypothesis.[12,13]

If this explanation is correct, then immobilizing enone **15** in conformation **A** in a matrix that lacks the specific close contact to the C-3 methyl group and that slows the conformational isomerization to conformer **C** might well lead to the "normal" solid-state photoproduct expected in this case, namely compound **26** (Scheme 50.7). This is exactly what was found.[14] When microscope slides coated with thin films of poly(methyl methacrylate) containing dissolved **15** were irradiated at −50 °C, keto-acetate **26** was the major (65%) product formed. The finding that small amounts (*ca.* 10%) of photoproduct **27** were also formed under these conditions may indicate the presence of microcrystallites of enone **15** in PMMA. It is more likely, however, that this simply reflects a close competition between the two hydrogen transfer pathways.

The final compound to be discussed in this review is enone **16** (Scheme 50.8). This material was prepared in order to see what photochemistry would ensue in solution when the normally observed [2+2]-photocycloaddition reaction is "blocked" by making the 6,7-double bond part of an aromatic ring. Although intramolecular [2+2]-photocycloaddition reactions between aromatic double bonds and enones are known,[15] it was expected that this process would be sufficiently slow in solution to allow alternative hydrogen abstraction pathways to be competitive. Such was indeed found to be the case.[8] Thus, irradiation of enone **16** in acetonitrile through Pyrex afforded a 90% yield of photoproduct **31**, the result of intramolecular transfer of benzylic hydrogen atom H8 to the β-carbon atom of the enone. On the other hand, photolysis of crystals of compound **16** afforded predominantly keto-alcohol **30**, the product arising from transfer of hydrogen atom H-5 to the β-carbon atom.

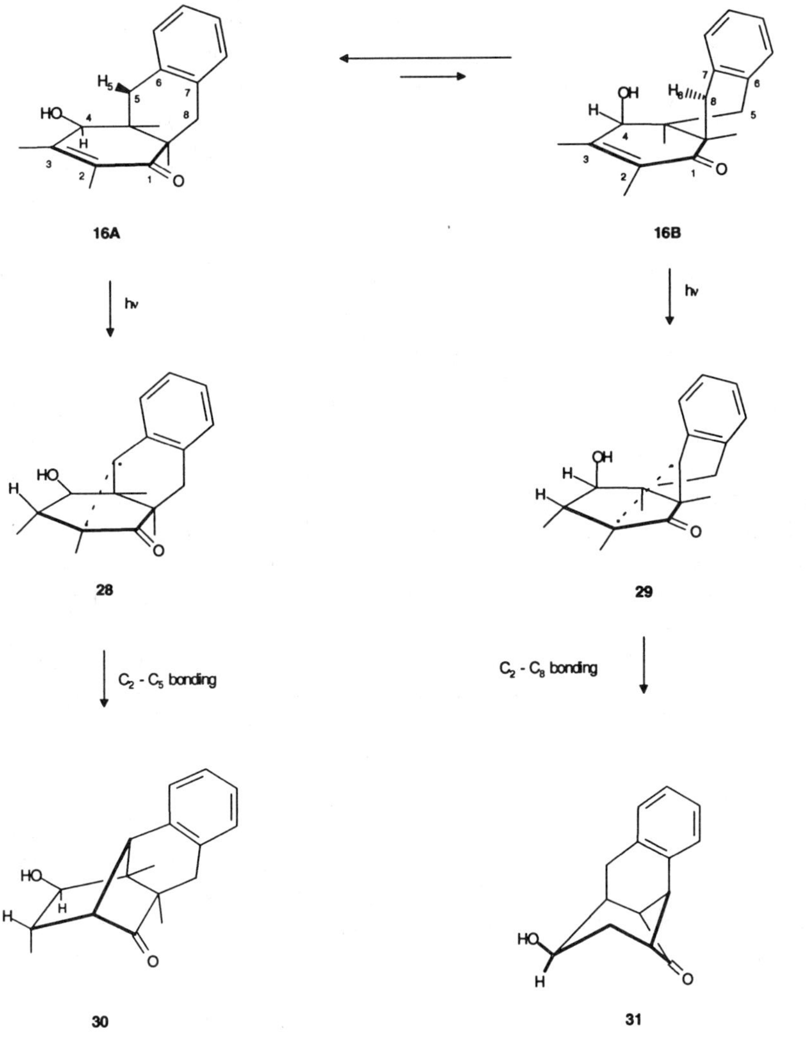

SCHEME 50.8

These results can be analyzed as shown in Scheme 50.8. The conformation of compound **16** in the solid state is **16A**, the one that places the larger substituent at C-4 (the hydroxyl group) in a pseudoequatorial position. This situates H-5 over C-3 (d = 2.85 Å) and leads to normal type A solid-state photobehavior that is exactly analogous to that observed in the case of enone **12** and a number of other similar compounds. Conformer **16A** is undoubtedly the major species present in solution as well, but in this medium it exists in rapid equilibrium with a number of other conformers, including **16B**. This is a minor conformer owing to the presence of a pseudoaxial hydroxyl group at C-4, but it is evidently much more reactive than conformer **16A**, and like other type B conformers, undergoes photoinduced C-3 ⋯ H-8 hydrogen atom abstraction. Eclipsed conformer **16C** (not shown) is also certainly present in small amounts in solution, but is photochemically unreactive owing to the reluctance of the molecule to involve the aromatic ring in a [2+2]-photocycloaddition process.

Like the other enones studied, equilibrium must be established between conformers **16A** and **16B** in the excited state; otherwise, the photoproduct ratio would reflect the ground-state conformer distribution, which is not the case. As pointed out by Lewis et al.,[9] under these circumstances, the Curtin-Hammett principle applies, and the product ratio should depend only on the relative magnitude of the two rate-limiting photochemical steps. This analysis predicts, therefore, that the six-membered transition state process of C-3 ⋯ H-8 hydrogen abstraction from conformer **16B** should be favored by a substantial amount over the five-membered transition state process of C-3 ⋯ H-5 abstraction from conformer **16A**. This conclusion is entirely consistent with the well-known competitive advantage that γ-hydrogen atom abstraction enjoys in the Norrish type II photoreaction.[16]

To summarize, the results described above demonstrate that bicyclic enones of general structure **8** possess what may be termed *latent* reactivity that is revealed when a photochemical stimulus is applied in the solid state. This situation arises when the liquid phase reaction is particularly rapid and occurs through a minor, high-energy conformer. In the crystal, however, organic molecules nearly always adopt (and are restricted to) their more stable, lower-energy conformations.[17] If this conformer reacts differently from, and more slowly than, its higher-energy (solution) isomer, new products will be observed in the solid. Thus, different conformers afford different photoproducts in the two media, a circumstance we have termed "conformation-specific" photochemistry.

The property of the crystalline state that is of primary importance in establishing conformation-specific photobehavior is its ability to restrict the interconversion of the reactive conformers (viscosity effect). We have demonstrated that other viscous media, such as polymer matrices, are capable of exerting a similar effect. The most interesting situation arises when, in addition to its viscosity effect, the crystal lattice exerts a highly specific intermolecular steric effect that alters reactivity. When this occurs, as in the case of enone **15**, unique and characteristic photoreactivity is observed in each of *three* different media: the isotropic liquid phase, the pure crystalline phase, and the polymeric matrix environment. Medium effects of this complexity are, to the best of our knowledge, unprecedented.

References

1. Schuster, D. I., Photochemical rearrangements of enones, in *Rearrangements in Ground and Excited States*, de Mayo, P., Ed., Academic Press, New York, 1980, chap. 17.
2. Schuster, D. I., Woning, J., Kaprinidis, N. A., Pan, Y., Cai, B., Barra, M., and Rhodes, C. A., Photochemical and photophysical studies of bicyclo[4.3.0]non-1(6)-en-2-one, *J. Am. Chem. Soc.*, 114, 7029, 1992.
3. Wolff, S., Schreiber, W. L., Smith, A. B., III, and Agosta, W. C., Photochemistry of cyclopentenones. Hydrogen abstraction by the β-carbon, *J. Am. Chem. Soc.*, 94, 7797, 1972.
4. (a) Scheffer, J. R., Trotter, J., Appel, W. K., Greenhough, T. J., Jiang, Z. Q., Secco, A. S., and Walsh, L., Crystal lattice control of unimolecular photorearrangements. Medium-dependent photochemistry of cyclohexenones, *Mol. Cryst. Liq. Cryst.*, 93, 1, 1983.
5. Trotter, J., Structural aspects of the solid-state photochemistry of tetrahydronaphthoquinones, *Acta Cryst.*, B39, 373, 1983.
6. Appel, W. K., Jiang, Z. Q., Scheffer, J. R., and Walsh, L., Crystal lattice control of unimolecular photorearrangements. Medium-dependent photochemistry of cyclohexenones, *J. Am. Chem. Soc.*, 105, 5354, 1983.
7. Bondi, A., Van der Waals volumes and radii, *J. Phys. Chem.*, 68, 441, 1964.
8. Ariel, S., Askari, S., Scheffer, J. R., and Trotter, J., Latent photochemical hydrogen abstractions realized in crystalline media, *J. Org. Chem.*, 54, 4324, 1989.
9. Lewis, F. D., Johnson, R. W., and Johnson, D. E., Conformational control of photochemical behavior. Competitive α-cleavage and γ-hydrogen abstraction of alkyl phenyl ketones, *J. Am. Chem. Soc.*, 96, 6090, 1974.
10. Chan, A. C. and Schuster, D. I., Stereospecific photoreduction of polycyclic α,β-unsaturated ketones, *J. Am. Chem. Soc.*, 108, 4561, 1986.

11. Gudmundsdottir, A. D. and Scheffer, J. R., A comparative study of an organic photorearrangement in solution, in the pure crystalline phase and in a polymer film, *Tetrahedron Lett.,* 30, 419, 1989.

12. Ariel, S., Askari, S., Scheffer, J. R., Trotter, J., and Walsh, L., Steric compression control of photochemical reactions in the solid state, *J. Am. Chem. Soc.,* 106, 5726, 1984.

13. Ariel, S., Askari, S., Evans, S. V., Hwang, C., Jay, J., Scheffer, J. R., Trotter, J., Walsh, L., and Wong, Y.-F., Reaction selectivity in solid state chemistry, *Tetrahedron,* 43, 1253, 1987.

14. Gudmundsdottir, A. D. and Scheffer, J. R., Observation of a photorearrangement unique to the polymer matrix, *Tetrahedron Lett.,* 30, 423, 1989.

15. Kushner, A. S., Photoisomerization of the 1,4-naphthoquinone-cyclopentadiene adduct. A novel intramolecular [6+2] cycloaddition, *Tetrahedron Lett.,* 3275, 1971.

16. Wagner, P. and Park, B.-S., Photoinduced hydrogen atom abstraction by carbonyl compounds, in *Organic Photochemistry,* Vol. 11, Padwa, A., Ed., Marcel Dekker, New York, 1991, chap. 4.

17. Dunitz, J. D., *X-Ray Analysis of the Structure of Organic Molecules,* Cornell University, Ithaca, 1979, chap. 7.

51

Intramolecular Photocycloadditions to Enones: Influence of the Chain Length

Jochen Mattay
*Organisch-Chemisches Institut
der Universität Münster*

51.1 Introduction

The intramolecular photocycloaddition of olefins to enones was first reported by Ciamician in 1908 when he observed the conversion of carvone to carvone camphor (Scheme 51.1).[1] Since the 1960s, the reaction has gained considerable attention, a fact that is reflected in a great number of publications including numerous detailed reviews.[2] Enone-alkene photocycloaddition has become one of the most frequently utilized photochemical reactions since it provides the key step for the synthesis of cyclobutane systems. This modus operandi recognized early on by Corey,[3] Eaton,[4] and de Mayo[5] and nowadays often applied by others.[2] Although many of the early examples were intramolecular, most of the applications have been devoted to intermolecular cycloadditions. Only 1 decade ago, the former version again found attention.[2a,b]

The mechanism used today, following the pioneering work of de Mayo[5] and recent investigations of Schuster[2c] and Weedon,[7] is shown in Scheme 51.2. Excitation and intersystem crossing yields the triplet excited state of the enone E, which interacts with the olefin O in different ways. All these possibilities involve 1,4-biradicals that control the regiochemistry of the cyclobutane formation. This was demonstrated clearly by Weedon et al.[7] for the cyclopentenone-ethyl vinyl ether system in which the regioisomer ratio is determined by the relative efficiency of the cyclobutane formation and the cleavage of the 1,4-biradical to the starting materials. Whether an exciplex is involved in the formation of the 1,4-biradical is an even more open question after these results.

51.2 Mechanism and Regioselectivity

In general, the mechanism shown in Scheme 51.2 can also be applied to the intramolecular version of the enone photocycloaddition. The special feature that, in addition, controls the selectivity of this

0-8493-8634-9/95/$0.00+$.50
© 1995 by CRC Press, Inc.

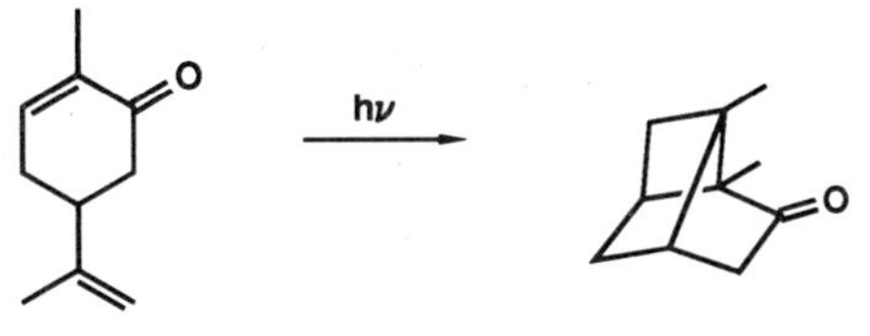

SCHEME 51.1   Conversion of carvone to carvone camphor.

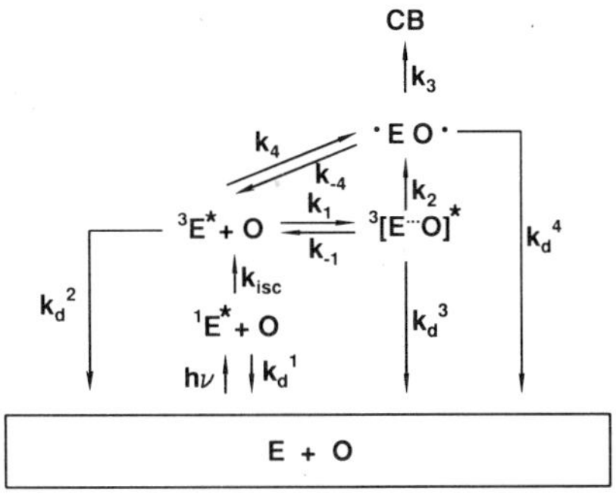

SCHEME 51.2   Triplet mechanism for enone photocycloaddition[a].

reaction is determined by the chain length. This had been recognized already by Srinivasan[8] and Hammond[9] in the 1960s and was rationalized in terms of the "rule of five" by postulating a biradical mechanism.[2,6,10–12] The work of Agosta and co-workers[12d] is noteworthy. They suggested an initial bonding at $C(\alpha)$ of 1- and 2-acyl-1,5-hexadienes leading to biradicals of the type 4a (Scheme 51.3). In addition, other intramolecular enone-olefin systems resemble the cyclization of 5-hexenyl radicals.[13] In connection with the new results published by Weedon et al.,[7] this rationalization has received additional support.

An alternative view of looking at these reactions has been suggested by Gleiter and co-workers.[14] According to these authors, the regioselectivity is strongly influenced by the symmetry of the frontier orbitals and by steric effects.[14a] "Through-space" and "through-bond" effects either act in combination as in 5 or oppose each other as in 4. Consequently, both π-bonds of 5 interact similarly to isolated ethenes leading to the head-to-head cycloadduct 5b. On the contrary, the stronger "through-bond" interaction in 4 results in an inversion of the regioselectivity with the formation of the head-to-tail cycloadduct.[4b]

Gleiter has also supported this model by many experiments[14b] (Schemes 51.4–51.6). If the conformations of unsaturated cyclohexenones are as shown in Scheme 51.6, the C_2-type arrangement A should be favored for 7, whereas the alternative C_s-type geometry B should be preferred in the case of 13. The formation of the corresponding cycloadducts 7b,c and 13a, respectively, agrees with this view. Moreover, all the other unsaturated cyclohexenones with propano links and even those with butano links behave similarly (Scheme 51.5). On the other hand, cyclohexenones with an ethano link show a remarkable substituent effect (Scheme 51.4). A methoxy group at the α-position of the double bond does not alter the regioselectivity as expected from the polarization of electronically excited cyclohexenone (Corey's model); on the contrary, it increases the formation

---

[a] Simplified mechanism (e.g., without consideration of E-E interaction etc.): E = enone, O = olefin, [E···O] = exciplex, ·EO· = 1,4-biradical, CB = cyclobutane, k_d = deactivation prosses, k_{isc} = intersystem crossing

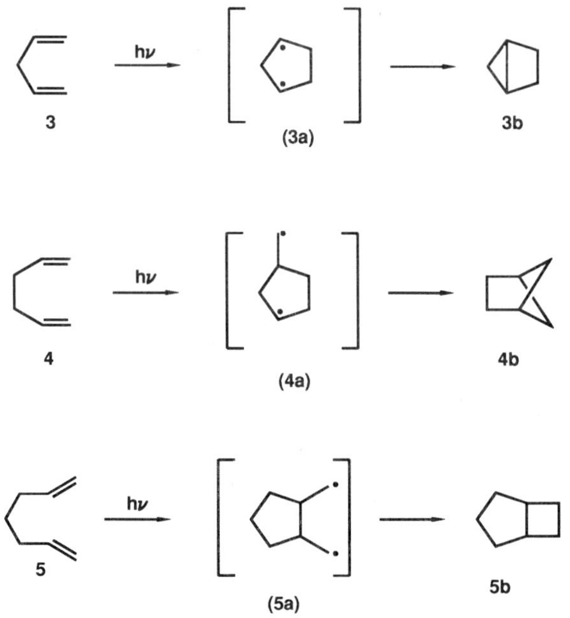

SCHEME 51.3   Regiochemistry of the intramolecular [2+2]-photocycloaddition of 1,4-, 1,5-, and 1,6-dienes according to Gleiter and Sander.[13]

of the HT-product (*cf.* **6a-c** and **7b,c**). However, introduction of an electron-releasing group in the β-position as in **8, 9,** and **10** leads to an increased predominance of the HH-product. Although this effect may be rationalized on the basis of Corey's assumption of reversed polarization in electronically excited cyclohexenone,[3] calculations by means of the SINDO/1 procedure do not support this view.[14b] Assuming a stepwise mechanism for the ring closure via biradicals of the type **4a** and **5a** (*cf.* Scheme 51.3), the authors[14b] can explain the exclusive formation of the HT-products **6a-c** from **6** and the HH-products **11a,b** from **11** (Schemes 51.7 and 51.8).[14b] Unfavorable steric effects in the biradical (**6a-d**) result in the formation of the stereoisomers **6a-c** exclusively. Product **6d** is not formed during the photoreaction, as shown by an independent synthesis of **6d**.

The preferred formation of the HH-products from **8, 9,** and **10** requires further comment. On the one hand, Corey's model of an interaction between polarized cyclohexenone in the excited state and the enol ether (Scheme 51.9) agrees with the observed regioselectivity. An argument based on the most stable biradicals also confirms the observation provided that the initial C-C-bond formation takes place at C(α) leading to the cyclohexane biradical **17**. Both views are limited in their applications to the other systems (e.g., **11** → **11a,b** in comparison **6** → **6a-c**). Although the general validity of the through-bond interaction model remains to be clarified, Gleiter's rationalization — which also considers electronic effects resulting from the polarization of the double bonds and the conformation of the reactive triplet states — at least seems to be more applicable. Further details may be found in the literature.[14,17]

Agosta and co-workers have drawn attention to the biradical reversion in the intramolecular photochemistry of carbonyl-substituted 1,5-hexadienes.[12d] Whereas 1-acyl and 2-acylhexadienes behave quite similarly up to the formation of the biradical intermediate, the 3-keto hexadienes do not. Initial bonding at C(α) of the enone in the former systems (**18, 19**) yields the freely reverting biradicals **18a** and **19a**. The 3-keto derivative (**20**), however, does not revert to the starting compound because of initial bonding at C(β) (*cf.* Scheme 51.10). The influence of the substituent R on the rate of reversion and the rate of conversion to the products supports Agosta's model.

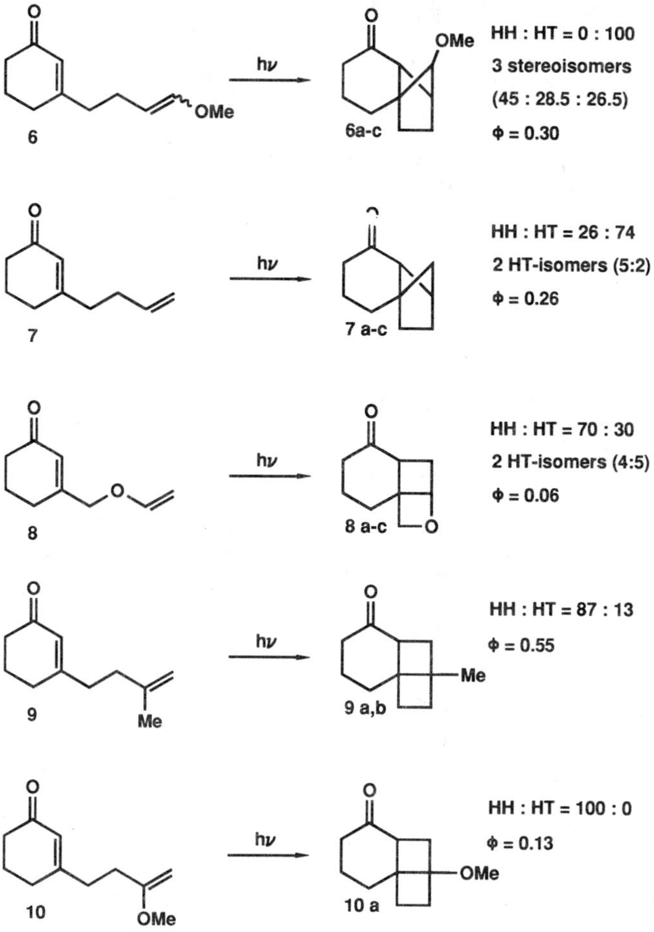

SCHEME 51.4 Ratio of the head-to-head (HH) to the head-to-tail (HT) regioisomer for unsaturated enones 6–10[a].

1.3 Selected Examples

Following the organizational scheme of Crimmins,[6b] selected examples are discussed, starting with 2-alkenylenones and concluding with the de Mayo reaction (Scheme 51.11).

2-Alkenylenones

Intramolecular photocycloadditions of 2-alkenylenones comprise reactions of enones of structure 21 in Scheme 51.11. According to the mechanistic conditions ("rule of five", through-bond inter-action etc., *cf.* Section 51.2), the initial attack of the enone on the alkene double bond gives rise to the formation of 1,4-biradical intermediates of type **3a, 4a,** or **5a** depending on the connecting chain length (n = 1–3, *cf.* Scheme 51.3).

A representative example for n = 1 is the synthesis of the hirsutane carbon skeleton (Scheme 51.12).[18] Irradiation of the dicyclopentenylmethane **29** in methanol resulted in the formation of the *cis-cisoid-cis*-tricyclo[6.3.0.0^{2,6}]undecanone derivative **31**. The reaction proceeds presumably via

---

[a] Only the main regioisomer is shown; ϕ = quantum yield of product formation. For further details see the following references: **6, 8,** and **10** (Ref. 14a), **7** (Ref. 12b, c), **9** (Ref. 12b).

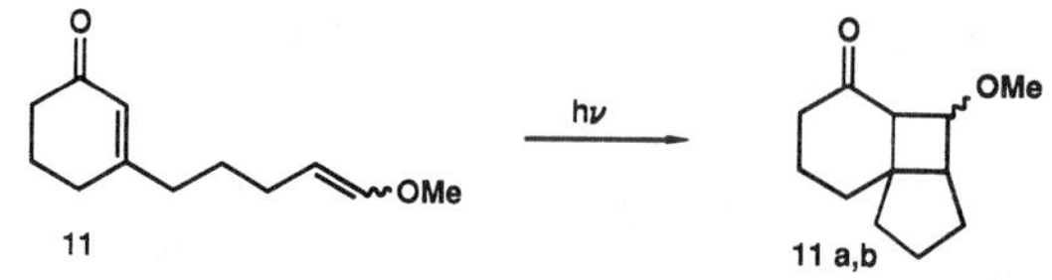

11

11 a,b

2 stereoisomers
(1:1)
Φ = 0.46

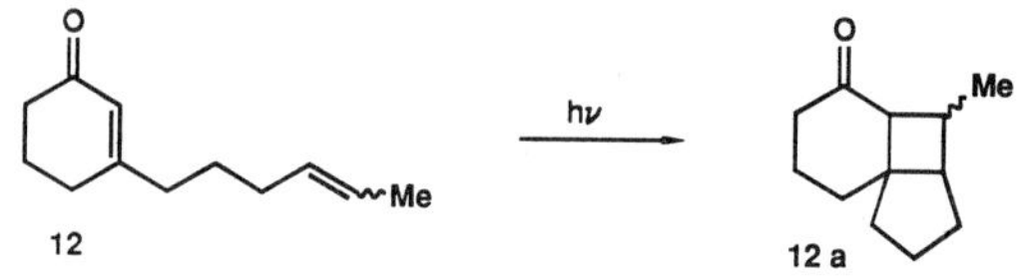

12

12 a

2 stereoisomers
(1:1)
Φ = 0.43 (E)
Φ = 0.55 (Z)

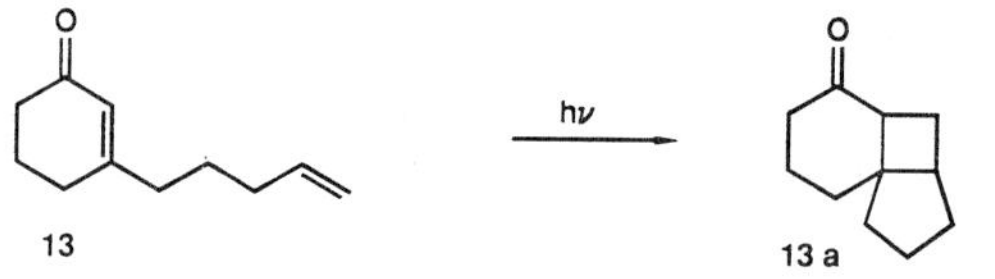

13

13 a

Φ = 0.5
92%

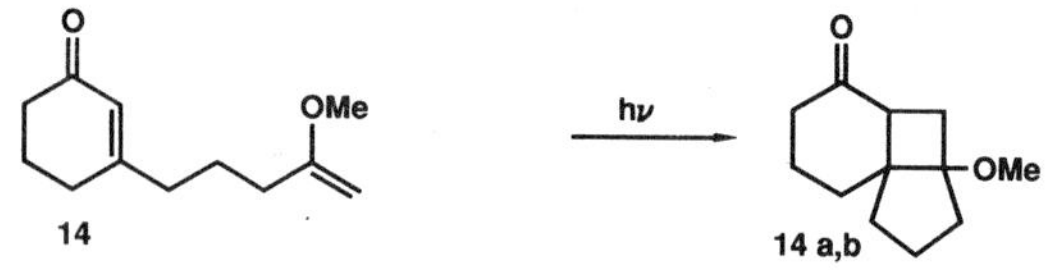

14

14 a,b

2 stereoisomers
(47:53)
Φ = 0.71

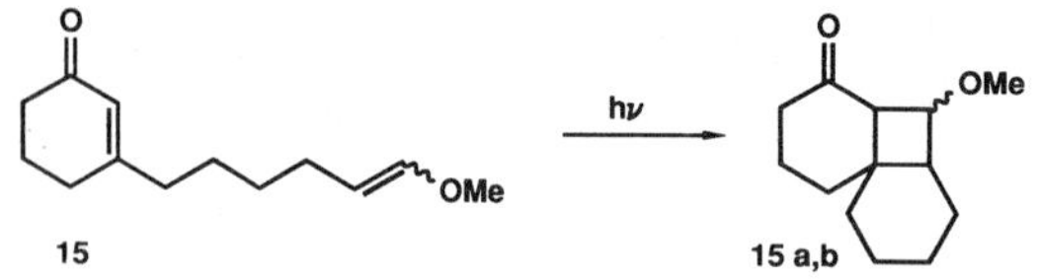

15

15 a,b

2 stereoisomers
(1:1)
Φ = 0.31

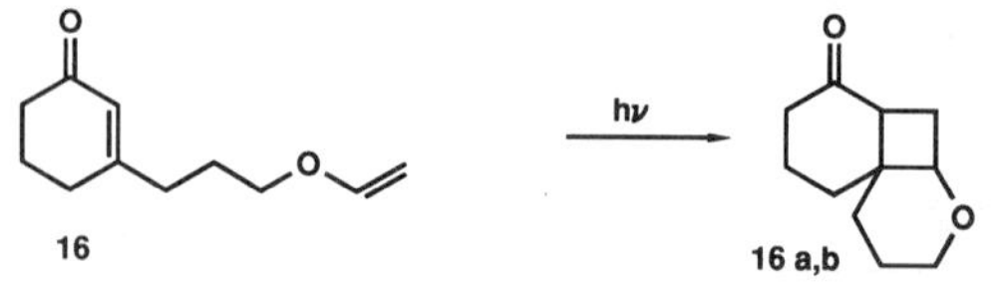

16

16 a,b

2 stereoisomers
(88:12)
Φ = 0.50

SCHEME 51.5   Regiospecific formation of head-to-head (HH) isomers from unsaturated enones 11–16[a].

[a]See footnote a of Scheme 51.4. Further details may be taken from the following references: 11, 14–15 (Ref. 14a), 12 (Ref. 15), 13 (Ref. 15, 16, 27).

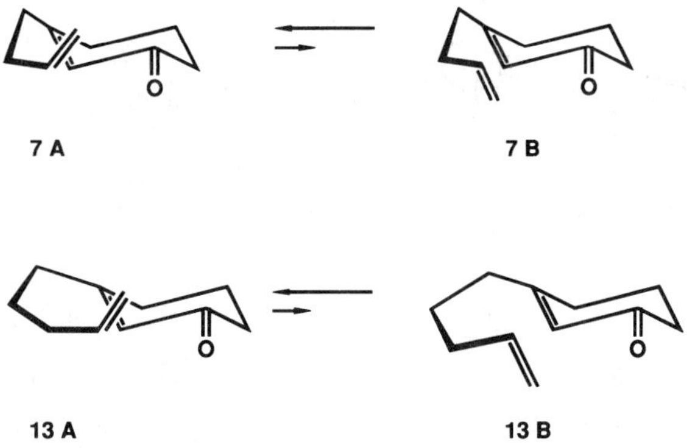

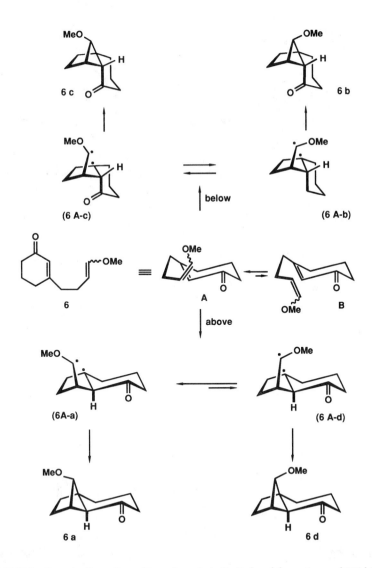

SCHEME 51.6   Conformations of unsaturated cyclohexenones according to Gleiter et al.[14]

SCHEME 51.7   Stepwise formation of 6 a-c from 6 via C_α-C_β-bond formation and HT-biradicals.

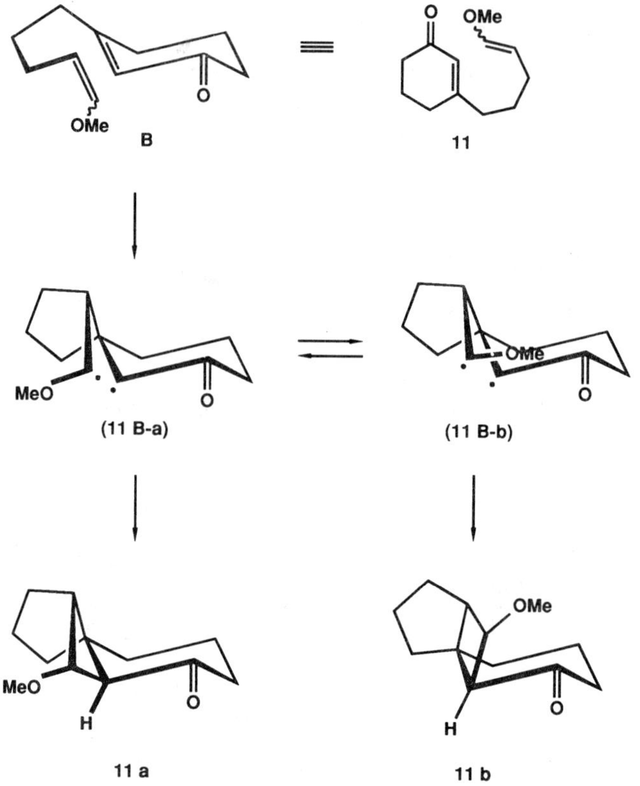

SCHEME 51.8 Stepwise formation of 11a and b from 11 via C_β-C_β-bond formation and HH-biradicals.

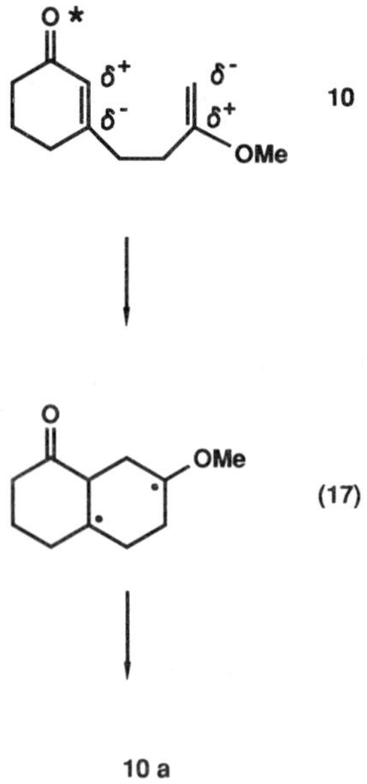

SCHEME 51.9 Polarization of cyclohexenone in the excited state according to Corey.

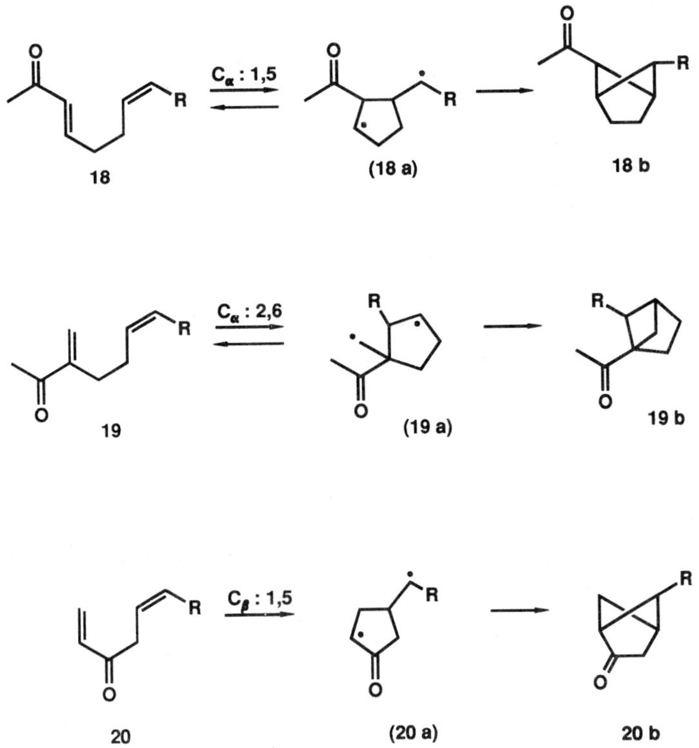

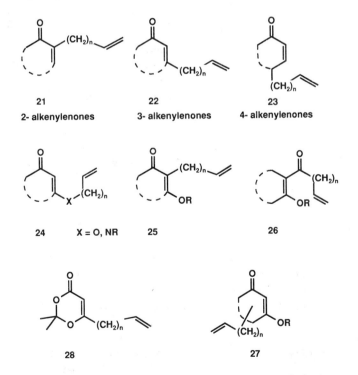

SCHEME 51.10 Biradical reversion in the intramolecular photochemistry of carbonyl-substituted 1,5-hexadienes according to Agosta et al.[12d]

21	22	23
2- alkenylenones	3- alkenylenones	4- alkenylenones

24 X = O, NR	25	26

28	27

1,3- Dicarbonyl Derivates (intramolecular de Mayo Reaction)

SCHEME 51.11 Selected alkenylenones and 1,3-dicarbonyl derivatives.

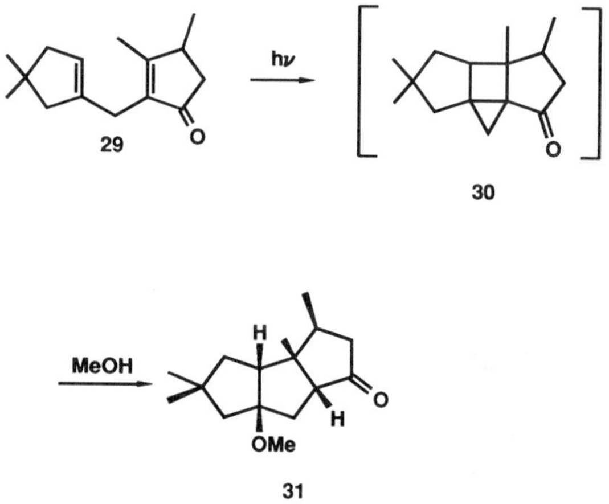

SCHEME 51.12   Synthesis of the hirsutane carbon skeleton.[18]

the highly strained *head-to-head* adduct **30**, which implies the intermediacy of a 1,4-biradical of type **3a** shown in Scheme 51.3. The intermediate **30** rearranges rapidly by a nucleophilic ring opening by the solvent. Further examples are given in Scheme 51.13, which not only show the influence of the "rule of five", but also that of polar and steric effects.

3-Alkenylenones

3-Alkenylenones also have a high potential in organic synthesis. One representative example was reported by Pirrung who utilized the high regio- and stereoselectivity in the photoreaction of **41** for the synthesis of isocumene **44** (Scheme 51.14).[24]

The reaction of a dienone with a spacer of two atoms between the enone and the alkene unit (n = 2) should give preferentially a bicyclo[2.1.1.]hexane skeleton. In many cases, however, the bicyclo[2.2.0]hexane skeleton is also formed concurrently. Two effects can change the regioselectivity toward the *straight* addition mode (head-to-head): placement of a substituent on the C-5 position of the 1,5-hexadiene system and incorporation of the conjugate double bond in a five- or six-membered ring. Representative examples with cyclopentenones **45** to **47** are shown in Scheme 51.15. Analogous substituted cyclohexenones behave similarly.[12b,25] Agosta and co-workers have taken advantage of the influence of alkyl substitution to prepare some unusual fenestranes.[25]

In view of the mechanistic features discussed in Section 51.2, the results of Becker et al.[16] are noteworthy. The authors presented evidence that no radical reversion is operative since the quantum yields of conversion *E*-**48** and *Z*-**48** are quite high, and no alkene isomerization could be detected in recovered starting material (compare, however, the results of Agosta et al.[12d]).

4-Alkenylenones

4-Alkenylenones have not been applied as widely as the 2- and 3-alkenylenones. Dauben has reported the photocycloadditions of various allenes attached to cyclohexenones and cyclopentenones at the 4-position. Two examples are shown in Scheme 51.16. Irradiation of the cyclopentenone **49** resulted in the formation of the *straight* adduct **49a** and product **49b** in a ratio of 1:1.[26a] Compound **49b** was formed by reaction across the 1,2-double bond of the allene function. Photocycloaddition of the cyclohexenone derivative **50**,[26b] as a 3:2 mixture of *anti*- and *syn*-diastereoisomers at −70 °C, resulted in quantitative cycloaddition of the *anti*-diastereoisomer to yield mainly the *cis*-decalin system **50a**. The *syn*-diastereomer gave three minor photoproducts that were not characterized.

Other examples of photocycloaddition reactions of cyclohexenones with alkenyl and alkynyl substituents at the 4-position have been reviewed by Crimmins.[6b]

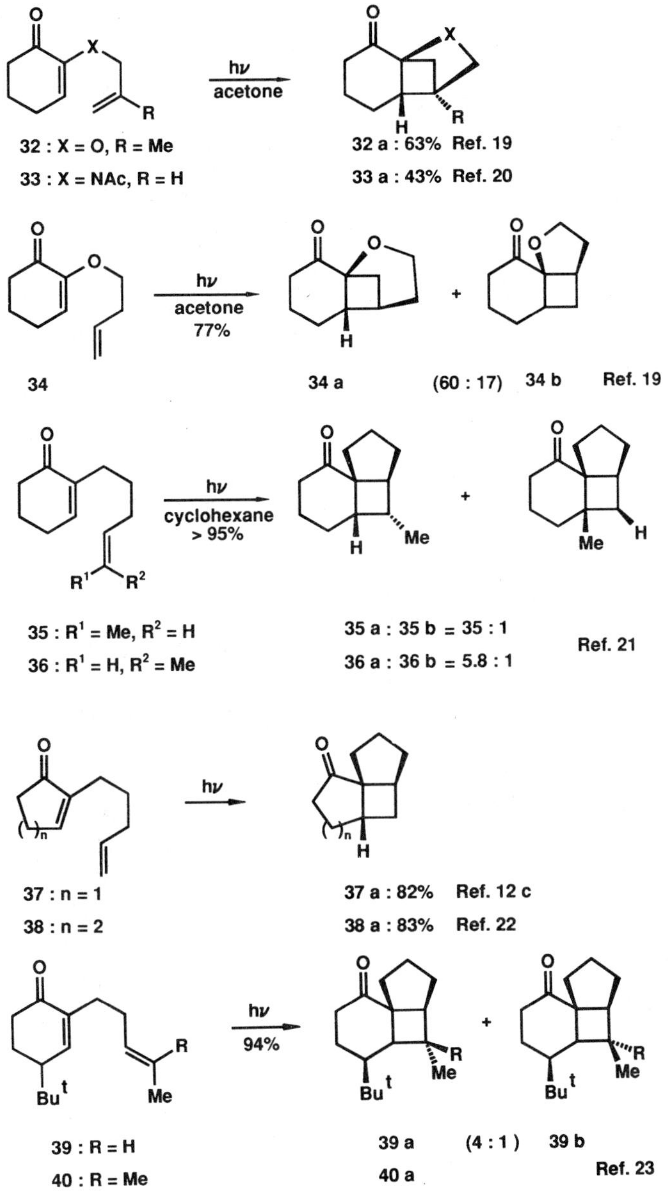

SCHEME 51.13 Representative examples of intramolecular photocycloadditions of 2-alkenylenones.

Intramolecular de Mayo Reaction

The majority of the synthetic interest in the de Mayo reactions stems from the ready formation of 1,5-dicarbonyl compounds and their subsequent rich chemistry, e.g., the formation of cyclohexenones on exposure of the products to aldol condensation. As an extension of the de Mayo reaction, which is the photocycloaddition of enolated 1,3-diketones to double bonds, enol esters, vinylogous esters and amides, and dioxolenones have been employed as enone components. In the following, the focus is on intramolecular cycloadditions of enol esters, dioxolenones and vinylogous esters and amides as β-diketone equivalents. The examples summarized in Scheme 51.17 belong to the de Mayo reaction in the broader sense as mentioned above. For more details, see Crimmins,[6b] Oppolzer,[6c] and Conrads and Mattay.[2a]

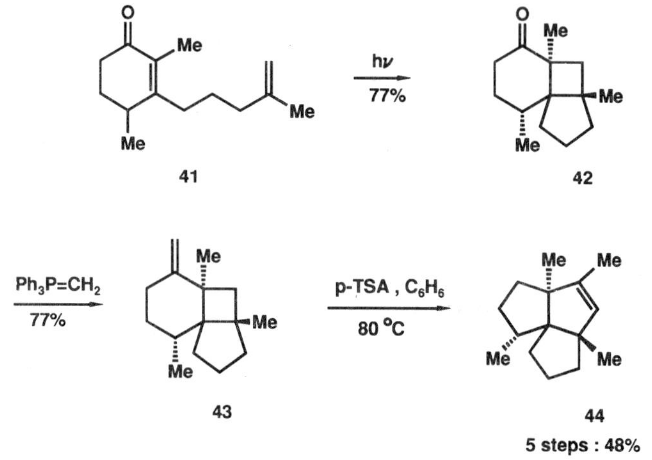

SCHEME 51.14   Synthesis of isocumene 44.[24]

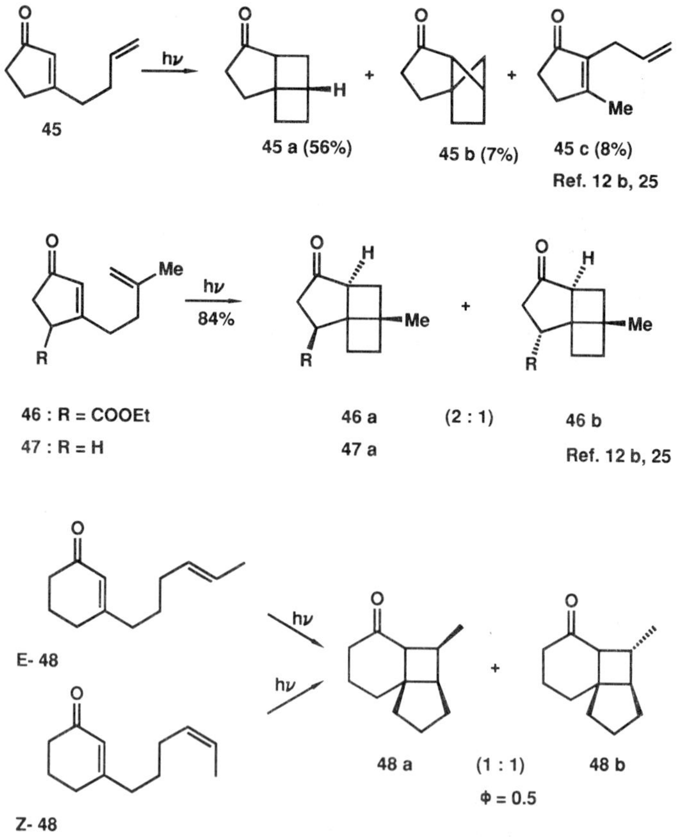

SCHEME 51.15   Representative examples of intramolecular photocycloadditions of 3-alkenylenones.

In Scheme 51.18, the synthesis of the taxane skeleton is illustrated. Irradiation of the mixture of diastereomers 63 resulted in cycloaddition of one diastereoisomer to yield 64 and recovery of the other. Oxidation with RuO_4 followed by base hydrolysis provides the diketone 66 in good yield. The product 66 shows the basic skeleton of taxol, currently a molecule of interest due to its anticancer activity.[34]

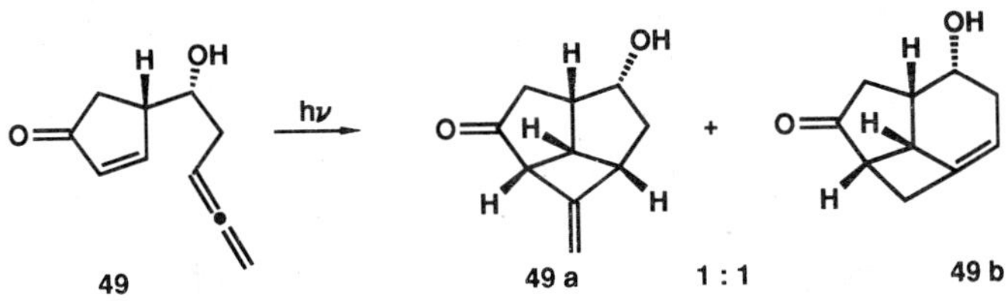

49 49 a 1 : 1 49 b

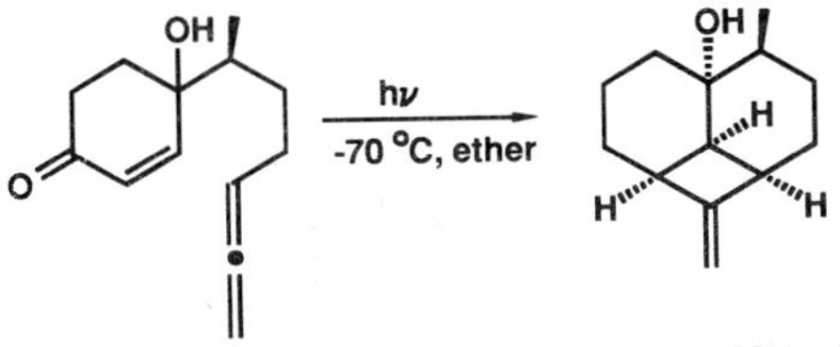

50 anti : syn = 3 : 2 50 a 60% + 3 minor products

SCHEME 51.16 Intramolecular photocycloadditions of alkene-containing cycloalkenones.

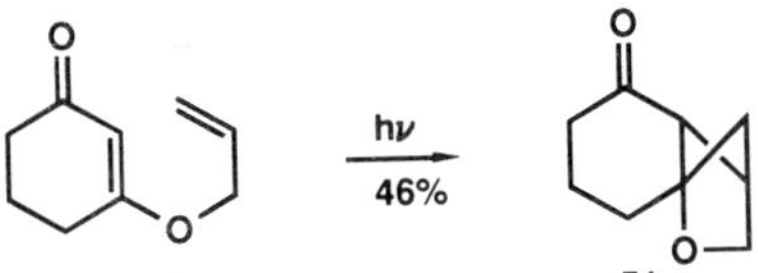

51 51 a Ref. 27

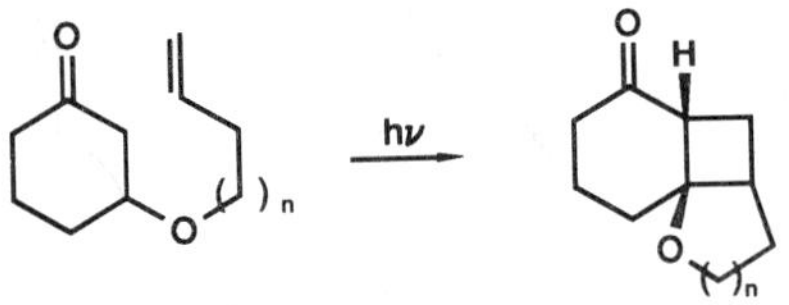

52 : n = 1 52 a : 82%

53 : n = 2 53 a : 28% Ref. 27

54 : n = 3 54 a : 25%

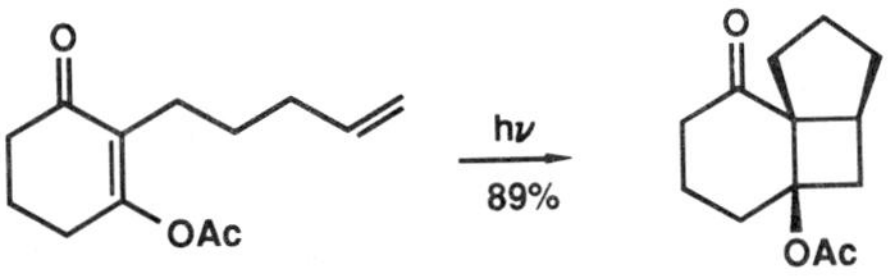

55 55 a Ref. 28

SCHEME 51.17 Selected intramolecular de Mayo reactions.

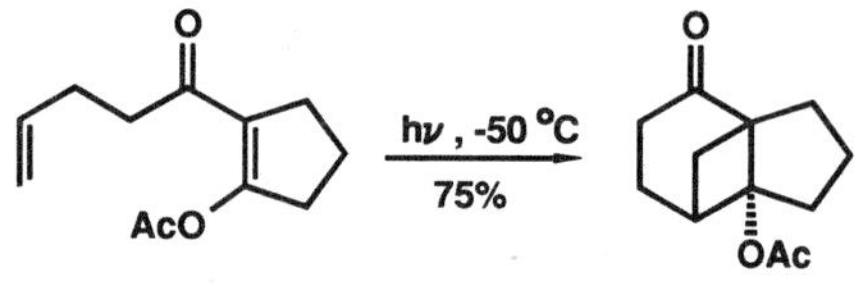

56   **56 a**   **Ref. 29**

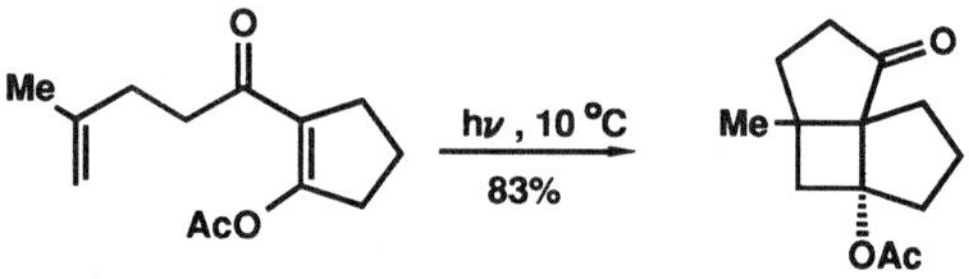

57   **57 a**   **Ref. 30**

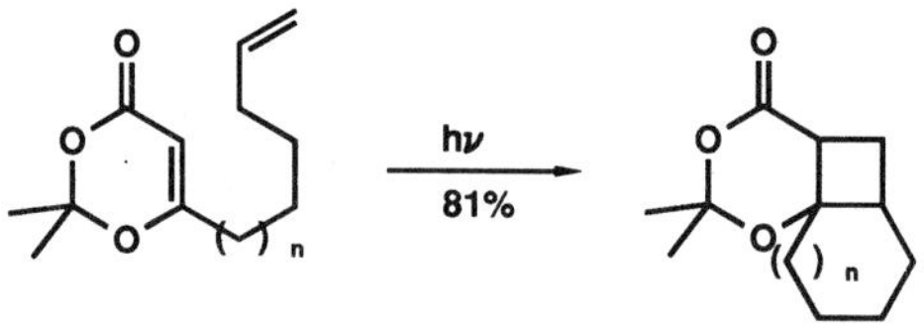

58 (n = 1, 2)   **58 a**   **Ref. 31**

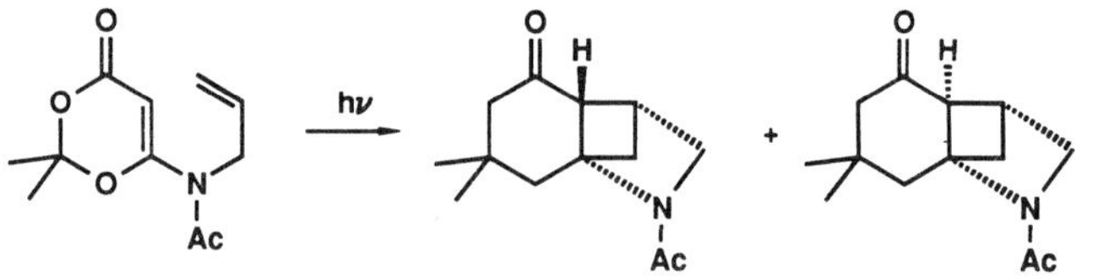

59   **59 a : 31%**   **59 b : 25%   Ref. 32**

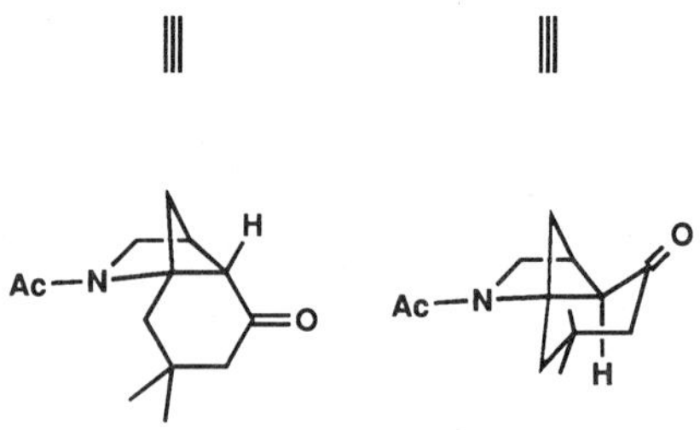

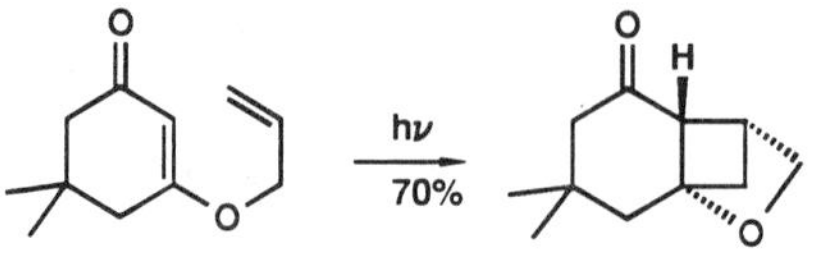

60   **60 a**   **Ref. 31**

SCHEME 51.17   (continued)

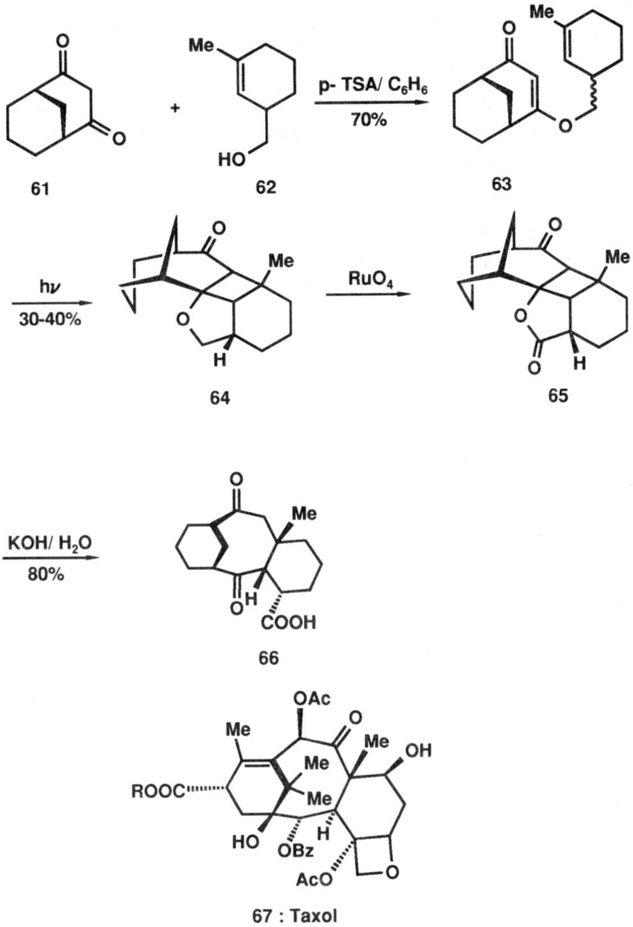

SCHEME 51.18   Synthesis of the taxane skeleton.[33]

References

1. Ciamician, G. and Silber, P., Chemische Lichtwirkungen, XIII Mitteilung, *Ber. Dtsch. Chem. Ges.,* 41, 1928, 1908.
2. Selected general reviews: (a) Conrads, R. and Mattay, J., [2+2] Photocycloadditions, in *Stereoselective Synthesis of Organic Compounds,* Vol. 3, Helmchen, G., Hoffmann, R. W., and Mulzer, J., Eds., Houben-Weyl E22, Thieme Verlag, Stuttgart, 1994, chap. 1.6.1.4; (b) Crimmins, M. T., Photochemical cycloadditions, in *Comprehensive Organic Synthesis,* Vol. 5, Trost, B. M., Ed., Pergamon, Oxford, 1991, chap. 2.3; (c) Schuster, D. I., The chemistry of enones, Part 2, in *The Chemistry of Functional Groups,* Patai, S. and Rappoport, Z., Eds., Wiley, Chichester, 1989; (d) Weedon, A. C., Enone photochemical cycloaddition in organic synthesis, in *Synthetic Organic Photochemistry,* Horspool, W. M., Ed., Plenum Press, New York, 1984, 1981, chap. 2.
3. Corey, E. J., Bass, J. D., LeMahieu, R., and Mita, R. B., A study of the photochemical reactions of 2-cyclohexenones with substituted olefins, *J. Am. Chem. Soc.,* 86, 5570, 1964.
4. Eaton, P. E., Photochemical reactions of simple alicyclic enones, *Acc. Chem. Res.,* 1, 50, 1968.
5. de Mayo, P., Enone photoannelation, *Acc. Chem. Res.,* 4, 41, 1971.
6. Selected recent reviews on intramolecular cycloadditions to enones: (a) Becker, D. and Haddad, N., Applications of intramolecular 2 + 2 photocycloadditions in organic synthesis, in *Organic Photochemistry,* Vol. 10, Padwa, A., Ed., Marcel Dekker, New York, 1989, chap. 1; (b) Crimmins, M. T., Synthetic applications of intramolecular enone-olefin photocycloadditions, *Chem. Rev.,* 88, 1453,

1988; (c) Oppolzer, W., Intramolecular [2+2]photoaddition/cyclobutane-fragmentation sequence in organic synthesis, *Acc. Chem. Res.,* 15, 135, 1982.

7. Andrew, D., Hastings, D. J., Oldroyd, D. L., Rudolph, A., Weedon, A. C., Wong, D. F., and Zhang, B., Triplet 1,4-biradical intermediates in the photocycloaddition reactions of enones and *N*-acylindoles with alkenes, *Pure Appl. Chem.,* 64, 1327, 1992.

8. Srinivasan, R. and Carlough, K. H., Mercury (^{3}P$_1$) photosensitized internal cycloaddition reactions to 1,4- 1,5- and 1,6-dienes, *J. Am. Chem. Soc.,* 89, 4932, 1967.

9. Liu, R. S. H. and Hammond, G. S., Photosensitized internal addition of dienes to olefins, *J. Am. Chem. Soc.,* 89, 4936, 1967.

10. Dilling, W. L., Intramolecular photochemical cycloaddition reactions of nonconjugated olefins, *Chem. Rev.,* 66, 373, 1966.

11. Gilbert, A. and Baggott, J., in *Essentials of Molecular Photochemistry,* Blackwell Scientific, London, 1991, chap. 6.7.

12. (a) Agosta, W. C. and Wolff, S., Photochemistry of carbonyl-substituted hexadienes, *Pure Appl. Chem.,* 54, 1579, 1982; (b) Wolff, S. and Agosta, W. C., Regiochemical control in intramolecular photochemical reactions of 1,5-hexadien-3-ones and 1-acyl-1,5-hexadienes and intramolecular photochemical reactions of 2-acyl-1,5-hexadienes, *J. Am. Chem. Soc.,* 105, 1292 and 1299, 1983; (c) Matlin, A. R., George, C. F., Wolff, S., and Agosta, W. C., Regiochemical control in intramolecular photochemical reactions of 1,6-heptadienes: Carbonyl-substituted 1-(4-alkenyl)-1-cyclopentenes, *J. Am. Chem. Soc.,* 108, 3385, 1986; (d) Schröder, C., Wolff, S., and Agosta, W. C., Biradical reversion in the intramolecular photochemistry of carbonyl-substituted 1,5-hexadienes, *J. Am. Chem. Soc.,* 109, 5491, 1987.

13. Beckwith, A. L. J., Regio-selectivity and stereo-selectivity in radical reactions, *Tetrahedron,* 37, 3073, 1981.

14. (a) Gleiter, R. and Sander, W., Light-induced [2+2]-cycloadditions of non-conjugated dienes — influence of through-bond interaction, *Angew. Chem.,* 97, 575, 1985, *Angew. Chem. Int. Ed.,* 24, 566, 1985; (b) Gleiter, R. and Fischer, E., Regiochemistry of the intramolecular [2+2] photocycloaddition of enones to vinyl ethers as a function of chain length, *Chem. Ber.,* 125, 1899, 1992.

15. Cargill, R. L., Dalton, J. R., O'Connor, S., and Michels, D. G., Tricyclo[6.3.0.0^{1,6}]undecan-2-one and its isomerization to tricyclo[3.3.3.0]undecan-2-one, *Tetrahedron Lett.,* 1978, 4465.

16. Becker, D., Nagler, M., Hirsh, S., and Ramun, J., Intramolecular [2+2] Photocycloadditions of *E* and *Z* olefins to cyclohex-2-enone, *J. Chem. Soc., Chem. Commun.,* 1983, 371.

17. Fischer, E., Lichtinduzierte [2+2] Cycloadditionen nicht konjugierter Diene, Ph.D. Thesis, Heidelberg, 1988, chap. 4.

18. Kueh, J. S. H., Mellor, M., and Pattenden, G., Photocyclisations of dicyclopent-1-enyl methanes to tricyclo[6.3.0.0^{2,6}]undecanes: A synthesis of the hirsutane carbon skeleton, *J. Chem. Soc., Chem. Commun.,* 1978, 5.

19. Ikeda, M., Takahashi, M., Uchino, T., Ohno, K., Tamura, Y., and Kido, M., Intramolecular [2+2] photocycloaddition of 2-(alkenyloxy)cyclohex-2-enones, *J. Org. Chem.,* 48, 4241, 1983.

20. Ikeda, M., Uchino, T., Takahashi, M., Ishibashi, H., Tamura, Y., and Kido, M., Intramolecular [2+2] photocycloaddition of 2-[*N*-acyl-*N*-(2-propenyl)amino]cyclohex-2-enones: Synthesis of 2-azabicyclo[2.1.1]hexanes, *Pharm. Bull.,* 33, 3279, 1985.

21. Becker, D., Haddad, N., and Sahali, Y., Topological and steric effects in mechanism of intramolecular [2+2] photocycloadditions, *Tetrahedron Lett.,* 30, 4429, 1989.

22. Cargill, R. L., Dalton, J. R., O'Connor, S., and Michels, D. G., Tricyclo[6.3.0.0^{1,6}]undecan-2-one and its isomerization to tricyclo[3.3.3.0]undecan-2-one, *Tetrahedron Lett.,* 1978, 4465.

23. Becker, D. and Haddad, N., About the stereochemistry of intramolecular [2+2] photocycloadditions, *Tetrahedron Lett.,* 27, 6393, 1986.

24. Pirrung, M. C., Total synthesis of (±)-isocomene, *J. Am. Chem. Soc.,* 101, 7130, 1979; and Total synthesis of (±)-isocomene and related studies, *J. Am. Chem. Soc.,* 103, 82, 1981.

25. Venepalli, B. R. and Agosta, W. C., Fenestranes and the flattening of tetrahedral carbon, *Chem. Rev.,* 87, 399, 1987.

26. (a) Dauben, W. G., Rocco, V. P., and Shapiro, G., Intramolecular [2+2] photocycloaddition of 4-substituted cyclopent-2-en-1-ones, *J. Org. Chem.,* 50, 3155, 1985; (b) Dauben, W. G. and Shapiro, G., Stereochemistry of intramolecular [2+2] photocycloadditions of 4-(4,5-hexadienyl)-2-cyclohexen-1-ones, *Tetrahedron Lett.,* 26, 989, 1985.

27. Mattay, J., Banning, A., Bischof, E. W., Heidbreder, A., and Runsink, J., Photoreactions of enones with amines — Cyclization of unsaturated enones and reductive ring opening by photoinduced electron transfer (PET), *Chem. Ber.,* 124, 2119, 1992.

28. (a) Oppolzer, W. and Bird, T. G. C., Intramolecular de Mayo reactions of 3-acetoxy-2-alkenyl-2-cyclohexenones, *Helv. Chim. Acta,* 62, 1199, 1979; (b) Begley, M. J., Mellor, M., and Pattenden, G., A new approach to fused carbocycles. Intramolecular photocycloadditions of 1,3-dione enol acetates, *J. Chem. Soc., Chem. Commun.,* 1979, 235.

29. (a) Begley, M. J., Mellor, M., and Pattenden, G., New synthetic approaches to fused-ring carbocycles based on intramolecular photocycloadditions of 1,3-diene enol esters, *J. Chem. Soc., Perkin Trans. I,* 1983, 1905; (b) Seto, H., Hirokawa, S., Fujimoto, Y., and Tatsuno, T., Synthesis of bicyclo [4.3.1]dec-2-en-7-one via intramolecular [2+2] photocycloaddition, *Chem. Lett.,* 1983, 989.

30. Seto, H., Fujimoto, Y., Tatsuno, T., and Yoshioka, H., Synthetic studies on carotane and dolastane type terpenes: A new entry to the total synthesis of (±)-daucene via intramolecular [2+2] photocycloaddition, *Synth. Commun.,* 15, 1217, 1985.

31. Winkler, J. D., Hey, J. P., and Hannon, F. J., Intramolecular photocycloadditions on dioxolenones: An efficient method for the synthesis of medium-sized rings, *Heterocycles,* 25, 55, 1987.

32. Tamura, Y., Ishibashi, H., Jirai, M., Kita, Y., and Ikeda, M., Photochemical syntheses of 2-aza- and 2-oxabicyclo[2.1.1]hexane ring systems, *J. Org. Chem.,* 40, 2702, 1975.

33. Kojima, T., Inoue, Y., and Kakisawa, H., Synthesis of a (±)-3β-trinortaxane derivative, *Chem. Lett.,* 1985, 323.

34. Wender, P. A. and Mucciaro, T. P., A new and practical approach to the synthesis of taxol and taxol analogues: The pinene path, *J. Am. Chem. Soc.,* 114, 5878 1992 and references therein.

52

Photocycloaddition Reactions of Cyclopentenones with Alkenes

Alan C. Weedon
University of Western Ontario

52.1 Introduction

The focus of this chapter is the photocycloaddition reaction of alkenes with 2-cyclopentenones to yield cyclobutane adducts. The reaction is shown in Scheme 52.1 and was first reported by Eaton in 1962.[1] The mechanistic aspects of the reaction have been reviewed frequently, as have the synthetic applications;[2–8] consequently, in this short chapter, selected examples are used to highlight the synthetic potential and to define the factors governing the stereochemical and regiochemical outcome. In addition, recent work that has led to a revision of the accepted mechanism is discussed.

52.2 The Reaction Mechanism

Much of the early work on the mechanism of the photocycloaddition reaction between alkenes and cyclic α,β-unsaturated ketones was performed by de Mayo[3] who built on the qualitative results obtained by Corey[9] for the photoaddition of alkenes to 2-cyclohexenones. The sequence shown in Scheme 52.2 was formulated by de Mayo[10] and has become known as the Corey-de Mayo mechanism. In this sequence, the triplet excited state of the enone is intercepted by the alkene to form a triplet exciplex that decays to one or more isomeric ground-state triplet 1,4-biradical species. The biradicals are formed by bonding of one terminus of the alkene π-system to either the 2-position or the 3-position of the cyclopentenone. Following spin inversion, the biradicals collapse to ground-state starting materials in competition with closure to the cyclobutane products. For a given alkene-enone combination, there may be several isomeric biradical species that differ according to which termini of the alkene and cyclopentenone double bonds have become bonded.

0-8493-8634-9/95/$0.00+$.50
© 1995 by CRC Press, Inc.

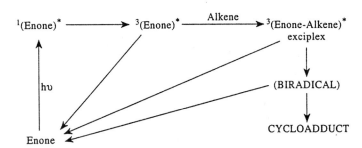

SCHEME 52.1

SCHEME 52.2

Quantitative information is available for many of the steps in this sequence. Intersystem crossing of the singlet excited state of the enone occurs so rapidly that the triplet excited state is formed with a quantum yield of unity.[11] In the absence of alkene, the cyclopentenone triplet excited state has a lifetime of 185 ns in cyclohexane, which is shortened by self quenching if the enone concentration is sufficiently high.[12] Interaction of the triplet excited state of the enone with alkenes occurs with a rate constant of the order of 10^8 M^{-1} s^{-1} and this varies relatively little with alkene structure.[13] Recent work has suggested that the exciplex species shown in Scheme 52.2 is not, in fact, a necessary intermediate in the reaction, but rather that the triplet excited state of the enone reacts with the alkene to form the triplet 1,4-biradicals directly.[13–15] This will be discussed further below. The triplet 1,4-biradical intermediates are estimated to have a lifetime of approximately 50 ns.[16] This estimate is derived from a study of the products of the photocycloaddition reaction between 2-cyclopentenone and vinylcyclopropane (Scheme 52.3). The biradical intermediates **1** and **2** in this reaction rearrange in competition with closure to cycloadducts **3** and **4**; in the rearrangement, the cyclopropyl carbinyl radical present in **1** and **2** opens to a homoallylic radical, which then disproportionates to give the isolated products **5** and **6**. The rate of the rearrangement reaction in closely related structures is known and can be used to clock the rate of the closure reaction of the biradicals **1** and **2** and hence estimate their lifetimes.[16]

The structures of the biradicals formed in the reaction of 2-cyclopentenones with alkenes have been determined for some systems by chemical trapping experiments.[14–16] For example, the reaction between 2-cyclopentenone and ethyl vinyl ether gives a mixture of stereoisomers of the regioisomers **7** and **8** (Scheme 52.4) in the ratio 24:76. The regiochemical arrangement in **7** and **8** are normally described as "head-to-head" and "head-to-tail", respectively. When the reaction is performed with hydrogen selenide present to act as a hydrogen atom donor, quantitative trapping of the biradical intermediates occurs to give compounds **9** to **12** only, in the ratio 43:7:24:26.[14,15] The formation of these compounds as the only trapped species indicates that biradicals **13** and **14** are the sole intermediates in the reaction and that the alternative biradicals **15** and **16** are not formed. This result appears to be general; with few exceptions, in the photocycloaddition reaction of alkenes with 2-cyclopentenone the biradicals are found to be derived from bonding of both the 2-position and the 3-position of the enone to the less-substituted end of the alkene.[14,15]

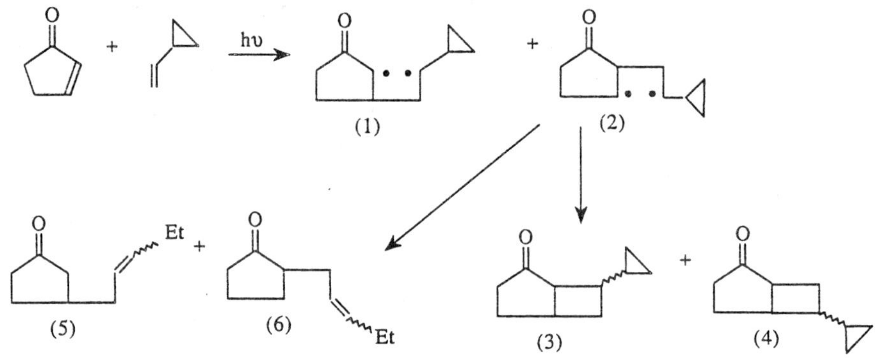

SCHEME 52.3

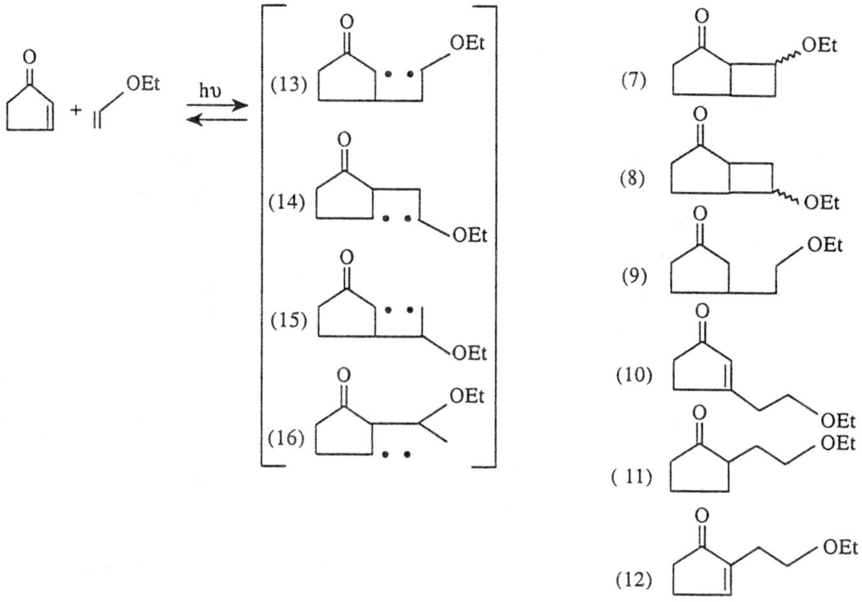

SCHEME 52.4

For the reaction of 2-cyclopentenone with ethyl vinyl ether, comparison of the cycloadduct regiochemical ratio (7:8 = 24:76) with the trapped product ratio (9:10:11:12 = 43:7:24:26) indicates that the biradicals 13 and 14 are formed in a 1:1 ratio and that the cycloadduct regiochemical ratio is determined by the greater preference of biradical 14 (as compared with biradical 13) to close to product rather than revert to starting materials. This result also appears to be general; for the systems that have been examined to date,[14,15] the amount of each product regioisomer formed is not determined by the relative amounts of their biradical precursors produced in the reaction but rather by the relative efficiencies with which the biradicals close to products in competition with reversion to the starting materials.

The regioselectivity of the photocycloaddition reaction between cyclopentenones and alkenes is in general not very high. Monosubstituted ethylenes possessing a good electron-donating substituent such as an alkoxy group favor formation of the head-to-tail regioisomer, as seen above in the

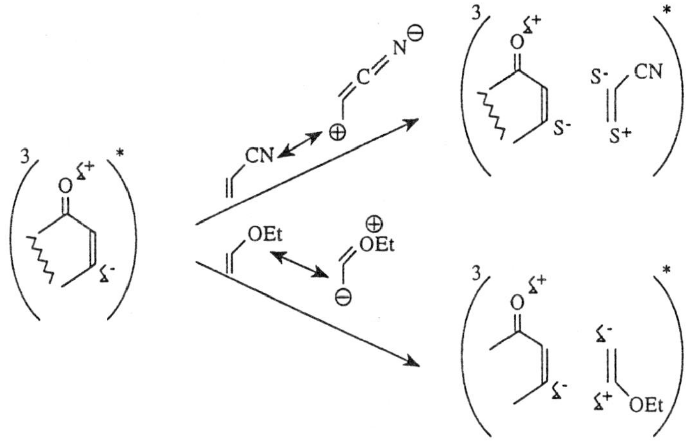

SCHEME 52.5

formation of 7 and 8 in the ratio 24:76;[14,15,17,18] however, the regioselectivity rarely exceeds this and usually falls far below. For example, adducts of 2-cyclopentenone with alkyl-substituted ethylenes are obtained with a regioisomeric ratio of close to 1:1, while 1,1-dialkylethylenes yield the head-to-tail and head-to-head isomers in a 60:40 ratio.[6] Prior to the recent evidence described above that the reaction regiochemistry is apparently governed by the fates of the partitioning biradicals, any regioselectivity in the reaction was explained by the assumption that the putative exciplex intermediate favored a particular alignment of the dipole of the enone excited state with that of the ground-state alkene. It was proposed that this aligned exciplex dictated the preferential formation of some biradicals over others.[3,6] This treatment assumed that the enone triplet excited state possessed a dipole moment reversed from that in the ground state, and that this dipole interacted with electron-rich or electron-deficient alkenes in the sense shown in Scheme 52.5. The concept of an exciplex directing the regiochemistry of the reaction was developed originally to explain the preferential formation of head-to-head cycloadducts in the reaction of 2-cyclohexenone with alkenes possessing electron-withdrawing groups such as acrylonitrile.[3,9] Since that time, exceptions to this regiochemical preference have been reported[19–21] for cyclohexenone, and the number of examples where electron-poor alkenes yield mainly the head-to-tail regioisomers now exceeds the number of earlier reports that the head-to-head product dominated. The situation is similar for 2-cyclopentenone; methyl cyclobutene-1-carboxylate[22] and acrylonitrile[15] both add to 2-cyclopentenones to yield a mixture of regioisomeric adducts in which the head-to-tail isomer dominates. Similarly, methyl acrylate adds to the cyclopentenone 17 to give the head-to-tail cycloadduct 18 and the head-to-head isomer 19 in a 7:1 ratio (Scheme 52.6).[23] In the reaction of 2-cyclopentenone with acrylonitrile, the head-to-head and head-to-tail regioisomers 20 and 21 are obtained in a ratio of 38:62; the biradical intermediates 22 and 23 have been generated by an alternative route (Norrish type I cleavage and decarbonylation of diketones 24 and 25), and their ratio for partitioning between cycloadducts and starting material determined (Scheme 52.7). The result indicates that for this reaction, the product regiochemistry results from more efficient closure of biradical 23 relative to biradical 22, rather than preferential formation of one or the other biradical.[15]

The photocycloaddition reaction of allene with 2-cyclopentenone is the only reaction that seems to favor head-to-head adduct formation consistently.[24–6] The biradicals that are intermediates in this reaction have been trapped with hydrogen selenide,[27] and it has been shown that here also the regiochemistry is determined by the fates of the biradicals rather than the relative proportions generated. Thus, the major regioisomer 26 is formed from the biradicals 27 and 28, which are produced in least amount, while the minor regioisomer 29 is formed from the biradical 30, which is formed in greatest amount (Scheme 52.8). Reversion to the starting materials is therefore more favorable for 30, while closure to products occurs more efficiently for 27 and 28.

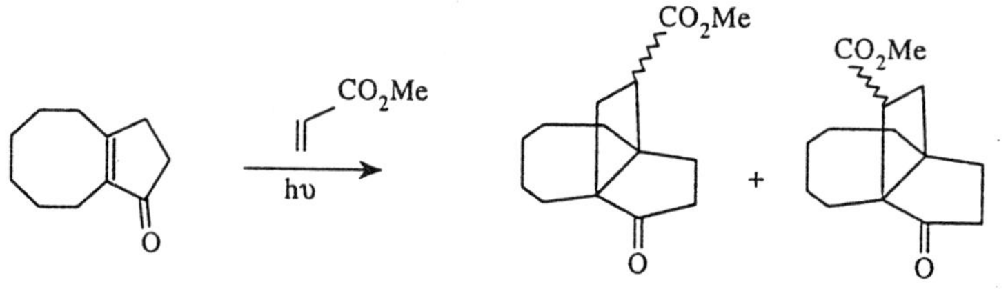

(17)   (18)   (19)

SCHEME 52.6

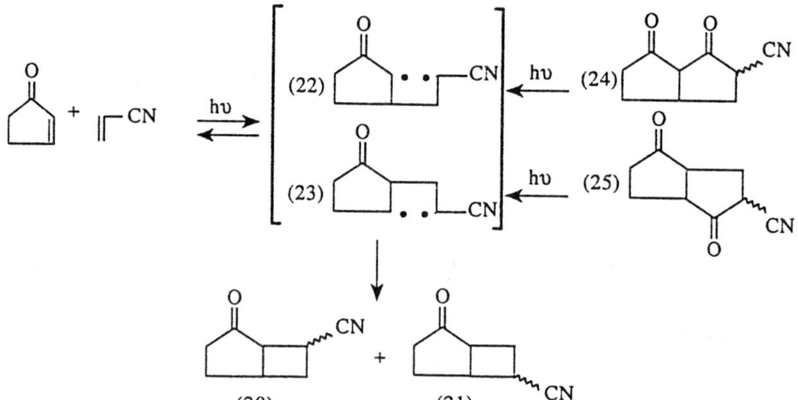

(20)   (21)

SCHEME 52.7

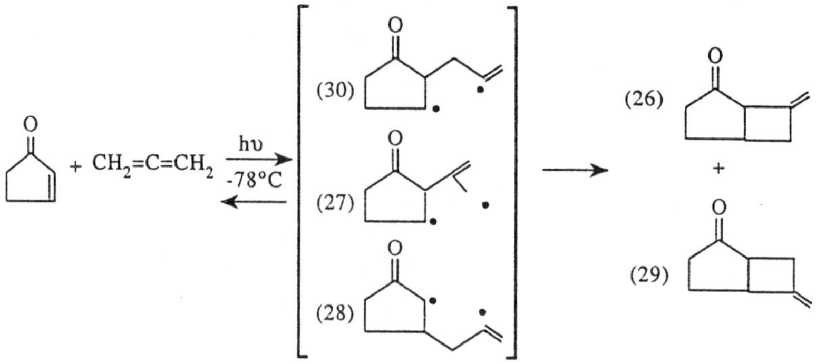

SCHEME 52.8

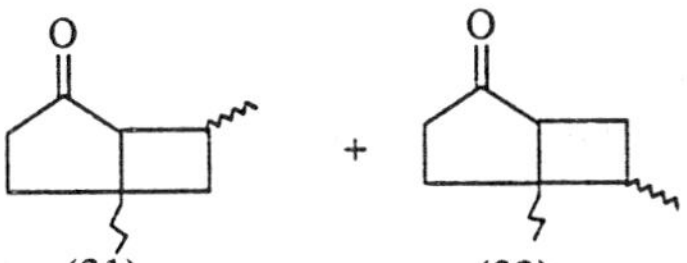

(31)   (32)

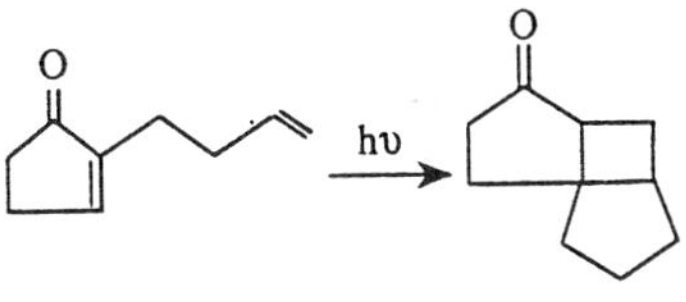

SCHEME 52.9

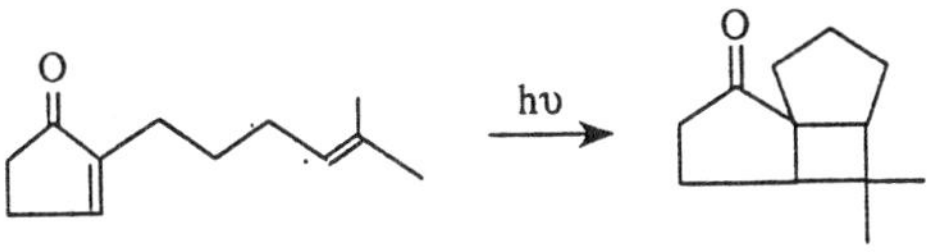

SCHEME 52.10

For synthetic applications, the maximum control of regiochemistry that can be achieved in the intermolecular photocycloaddition of alkenes to 2-cyclopentenone is of the order of 80% in favor of the head-to-tail regioisomer if an alkoxy-substituted ethylene is used, and ~90% in favor of the head-to-head regioisomer if allene is used. Attempts have been made to overcome the almost complete lack of regiocontrol in the addition of alkyl-substituted ethylenes by performing the reaction in micelles.[28] For example, addition of 1-octene to 3-butyl-2-cyclopentenone yields the head-to-head and head-to-tail products **31** and **32** in a 53:47 ratio when the reaction is carried out in cyclohexane as solvent. When the reactants are solubilized in aqueous potassium dodecanoate micelles, the same reaction yields **31** and **32** in a 12:88 ratio. This selectivity is ascribed to alignment of the alkene and the enone such that their nonpolar substituents are both directed toward the nonpolar interior of the micelle.[28]

The most effective control of regiochemistry is achieved by tethering the alkene to the enone.[5,6,29] This has been extensively investigated by Agosta and co-workers and is illustrated in Schemes 52.9 and 52.10 for 3-alkenyl-2-cyclopentenones and 2-alkenyl-2-cyclopentenones, respectively.[30] The regioselectivity in the intramolecular reaction is frequently governed by the "rule of five", which predicts for this application that the major product is formed from a biradical in which a five-membered ring has been produced.[31–33] This is illustrated in Scheme 52.11; the major product can arise from biradical **33**, whereas the minor product must be formed from a biradical such as **34** in which a larger ring has been produced.[30] A full discussion of the effects of tether length and point of attachment, as well as the effect of different substituents on the regiochemical outcome of the intramolecular reaction has been presented by Crimmins in a recent review.[29]

The biradical species formed as intermediates in the photocycloaddition reaction of alkenes with enones are sufficiently long-lived that conformational relaxation can be expected to occur.[16] Consequently, any stereochemistry present in the alkene is lost in the cyclobutane products. This does not mean necessarily that the composition of the mixture of stereoisomeric cycloadducts obtained from each geometrical isomer of the alkene will be the same since the amount of each of the isomeric biradical intermediates produced from each alkene stereoisomer may be different. This

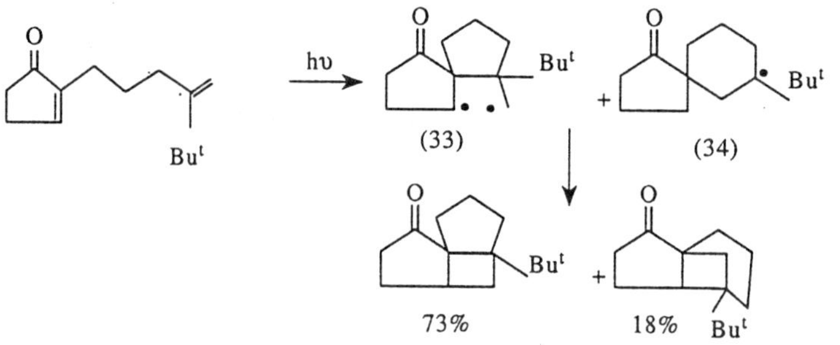

(33)   (34)

73%   18%

SCHEME 52.11

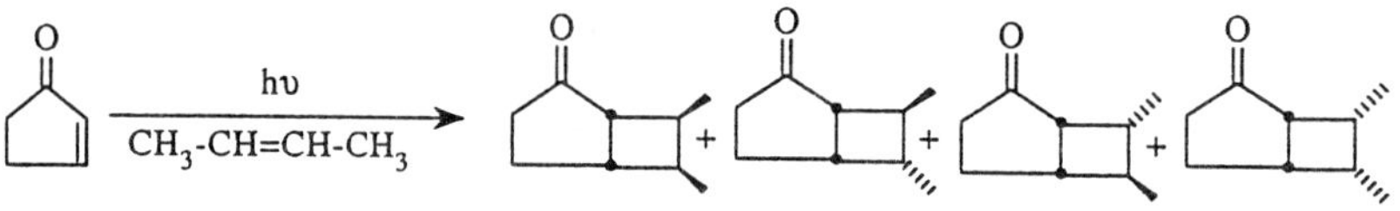

relative yields:
cis-alkene   42.6 43.5 11.7 2.2
trans-alkene 50.2 28.5 19.2 2.1

SCHEME 52.12

is illustrated in Scheme 52.12 for the addition of *cis*- and *trans*-2-butene to 2-cyclopentenone.[15] Closure of the 1,4-biradical intermediates always yields *cis*-stereochemistry at the ring fusion positions of the cyclopentane-cyclobutane ring junction, presumably because formation of the *trans*-fused isomers would result in the development of a prohibitive level of ring strain.

A remarkable number of substituents on both the alkene and the cyclopentenone are compatible with cycloadduct formation. Successful reaction has been reported for cyclopentenones possessing alkyl, aryl, nitrile, carboxyl, silyl, and alkoxyl functions at the 2-position or 3-position of the enone ring.[6] Cyclopentenones possessing oxygen functions at the 2-position or the 3-position are enols of the corresponding 1,2- or 1,3-diketones; photocycloaddition of alkenes to these enones is often referred to as the de Mayo reaction and this is reviewed in Chapter 54. The reaction is also successful for alkenes bearing alkyl, alkoxyl, carbonyl, or chloro substituents directly on the double bond.[6] One mechanistic puzzle, which has yet to be resolved, is an explanation for the ability of cyclopentenones to form cycloadducts with dienes[34,35] and with cyano-substituted alkenes such as acrylonitrile, and fumaro- and maleonitrile.[36] These alkenes all possess triplet energies below that of cyclopentenone and would be expected to quench the enone triplet excited state rather than react with it.[15]

52.3 Synthetic Applications of Intermolecular Photocycloaddition Reactions of 2-Cyclopentenone with Alkenes

The photocycloaddition reaction of alkenes to cyclopentenones has been applied to the synthesis of a number of natural products containing cyclobutane rings. One of the earliest examples of this application is shown in Scheme 52.13 for the preparation of α-bourbonene, 35, and β-bourbonene, 36.[37] This sequence utilizes the fact that the cycloaddition always yields *cis*-fusion of the cyclobutane-cyclopentane ring junction. It also illustrates a general phenomenon which is that for the addition

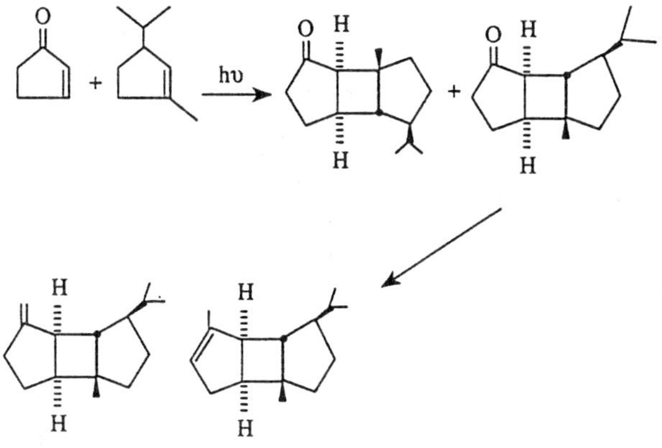

(35) (36)

SCHEME 52.13

of a cyclopentene to a cyclopentenone the formation of products possessing *cis-anti-cis*-geometry around the cyclobutane ring is favored. This corresponds to an *exo*-orientation of approach of the enone to the alkene rather than an *endo*-orientation of approach, which would yield a product with *cis-syn-cis*-geometry. The preferential formation of products apparently arising from *exo*-approach suggests that steric interactions are an important factor in determining the product geometry. The configuration of the isopropyl substituent in the cycloadducts shown in Scheme 52.13 is also that expected for approach of the enone to the less-hindered face of the alkene.

The *cis-anti-cis*-geometry of the carbon skeleton of the bourbonenes is also found in the natural product spatol, 37. A recent synthesis, shown in Scheme 52.14, uses a similar approach to that applied for the preparation of the bourbonenes for the generation of the spatol degradation product (and synthetic precursor), compound 38.[38] The structure of the cycloadduct suggests that the approach of the enone again occurs to the less-hindered face of the alkene 39. Quite fortuitously, since no precedent would have predicted it, the reaction is completely regioselective.

Other cyclobutane-containing natural products that have been prepared by photocycloaddition of alkenes to cyclopentenones are lineatin, 40,[39] and grandisol, 41.[40-42]

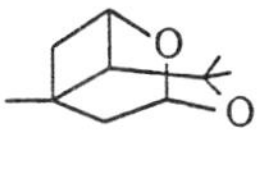

(40) 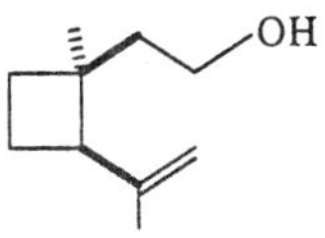 (41)

Many synthetic applications of the intermolecular photocycloaddition reaction between cyclopentenones and alkenes have involved ring opening or ring expansion of the cyclobutane ring of the cycloadducts to generate the carbon skeleton of natural products. For example, the head-to-tail regioselectivity attainable by addition of 1,1-dialkoxyethylenes to cyclopentenone has been utilized to generate the adduct 42a which, following elaboration of the cyclopentanone ring and hydrolysis of the ketal function, undergoes Baeyer-Villiger oxidation to the prostoglandin precursor 43 (Scheme 52.15).[43] Capnellene, 44, has been prepared from the homologue of 42a, compound 42b, by expansion of the cyclobutane ring using ethyl diazoacetate (Scheme 52.15).[44] The photocycloaddition of 1,1-diethoxyethylene is also regioselective when esters of cyclopentenone-3-carboxylic acid are the enone; in addition, if chiral alcohols are used to prepare the esters, then some diastereoselectivity is obtained.[45] Diastereoselectivity has also been reported for the photocycloaddition of the chiral ketal 45 to the methyl ester of cyclopentanone-3-carboxylic acid.[46]

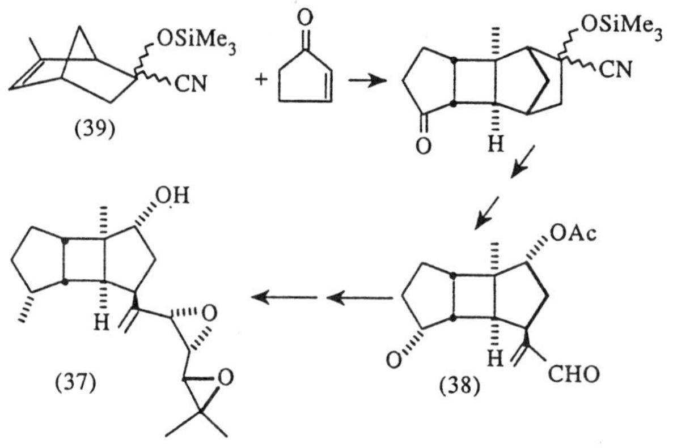

SCHEME 52.14

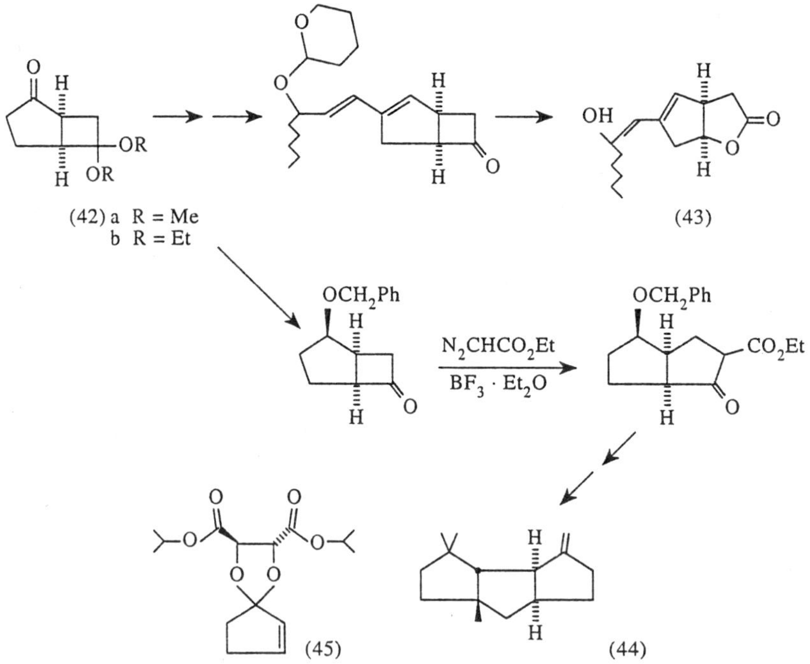

SCHEME 52.15

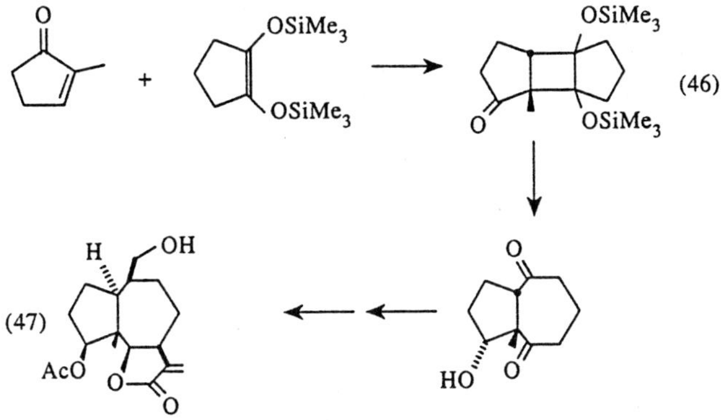

SCHEME 52.16

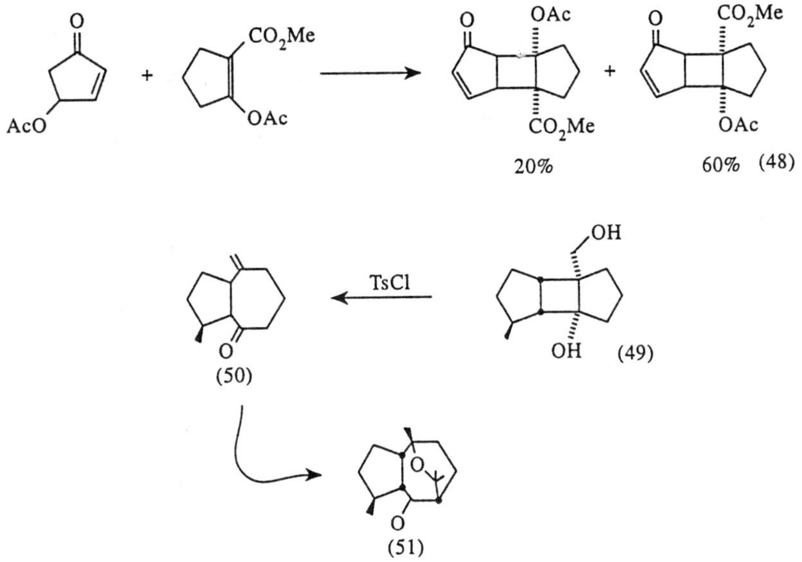

SCHEME 52.17

Several methods have been developed for the generation of perhydroazulene ring systems by ring opening of adducts of cyclopentenones with cyclopentenes. For example, oxidative cleavage of cycloadduct **46** with lead tetra-acetate yields a perhydroazulene, which can subsequently be converted into hysterin, **47**, (Scheme 52.16),[47] and the cycloadduct **48** (Scheme 52.17) undergoes base catalyzed ring expansion following conversion to **49** and tosylation. The product **50** was then converted to epikessane, **51**.[48] The carbon skeleton present in **50** has also been generated by a radical initiated pathway as shown in Scheme 52.18.[49]

The regioselective addition of allene to cyclopentenone to produce head-to-head adducts has been utilized for the preparation of a series of diterpenoids from the synthetic intermediate **52** (Scheme 52.19).[50] Interestingly, both of the major photoadducts yielded **52** upon ozonolysis followed by treatment with methoxide. The necessary isomerization probably occurs by equilibration

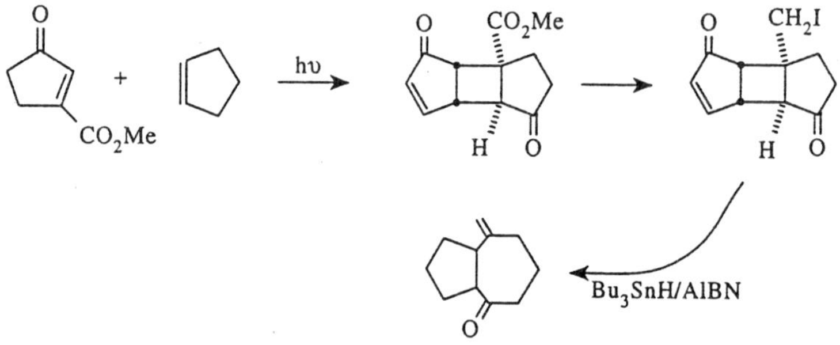

SCHEME 52.18

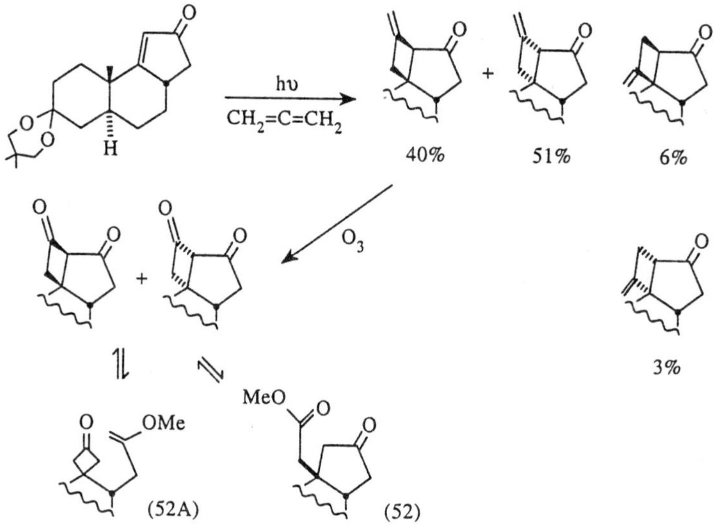

SCHEME 52.19

with the cyclobutane **52A**. Acid-catalyzed contraction of the cyclobutane ring of a derivative of an allene adduct has been used to prepare a model compound possessing many of the structural features necessary for the preparation of marasmic acid, **53** (Scheme 52.20).[51]

52.4 Synthetic Applications of Intramolecular Photocycloaddition Reactions of 2-Cyclopentenone with Alkenes

The intramolecular photocycloaddition of an alkene to a 2-cyclopentenone has been applied to the development of routes leading to fenestranes (i.e., compounds containing four rings fused to a central quaternary carbon). The fenestrane **54**, which contains two cyclobutane rings fused to two

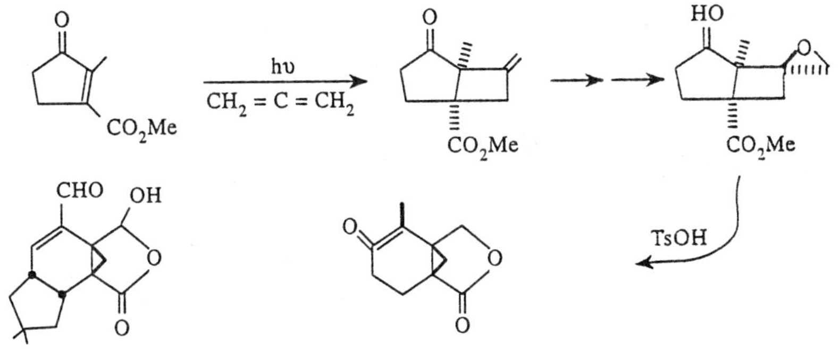

SCHEME 52.20

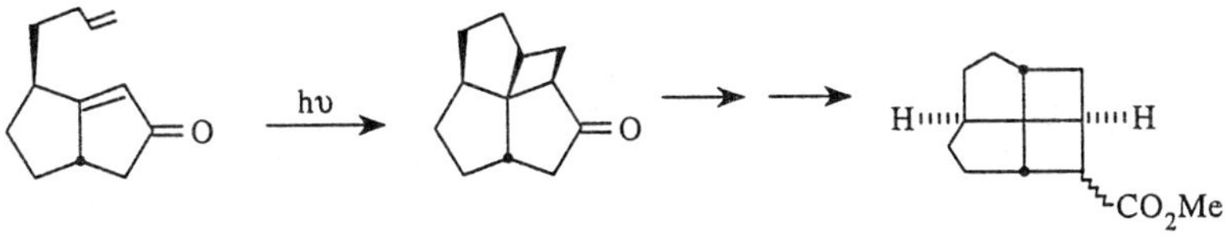

SCHEME 52.21

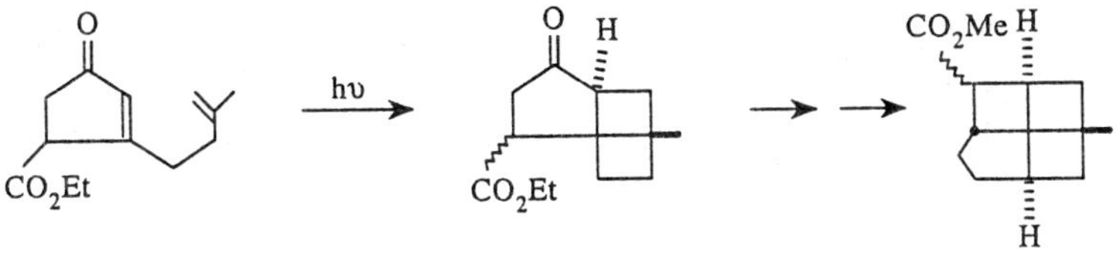

SCHEME 52.22

cyclopentane rings, has been prepared by ring contraction of the cycloadduct 55 obtained from the intramolecular photocycloaddition of enone 56 (Scheme 52.21).[52] The regiochemistry of this reaction is that predicted by the "rule of five." Agosta and Wolff have found that 3-(3′-butenyl)-2-cyclopentenones can be induced to yield "non-rule-of-five"-derived products if the butenyl side chain possesses an alkyl substituent at the 3′-position; this has been used to generate fenestrane 57 by cyclization and ring contraction of the cycloadduct 58 (Scheme 52.22).[53,54]

The intramolecular photocycloaddition reaction of 3-(4′-pentenyl)-2-cyclopentenones can normally be relied on to yield cycloadducts with the regiochemistry predicted by the "rule of five," and these systems have been used for the preparation of polycyclic natural products such as pentalene, 59, and pentalenic acid, 60, (Scheme 52.23),[55] siliphinene, 61, (Scheme 52.24)[56] and laurenene, 62, (Scheme 52.25).[57] A route to the methyl ester of pentalenolactone G, 63, has been reported that utilizes intramolecular photocycloaddition of an allene, as shown in Scheme 52.26.[58] The regiochemistry of the addition is the same as that observed in intermolecular allene additions. However, it has been found that if an allene is attached to the 4-position of a cyclopentenone by a two-atom tether, then a 1:1 mixture of regioisomers such as 64 and 65 is obtained (Scheme 52.27).[59] Cycloadduct 65 is of special interest in that it contains a bridgehead olefin and also because

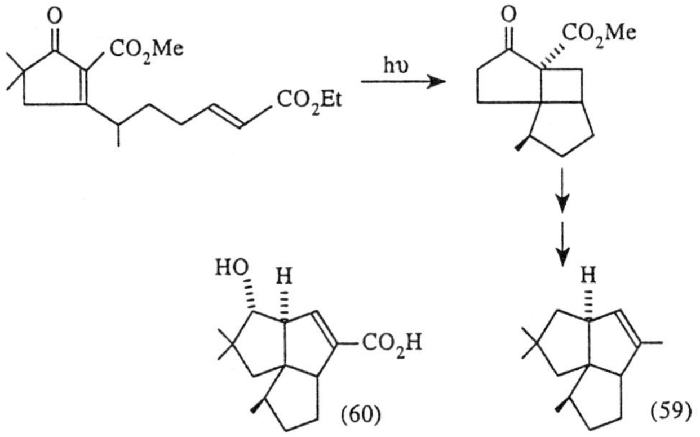

SCHEME 52.23

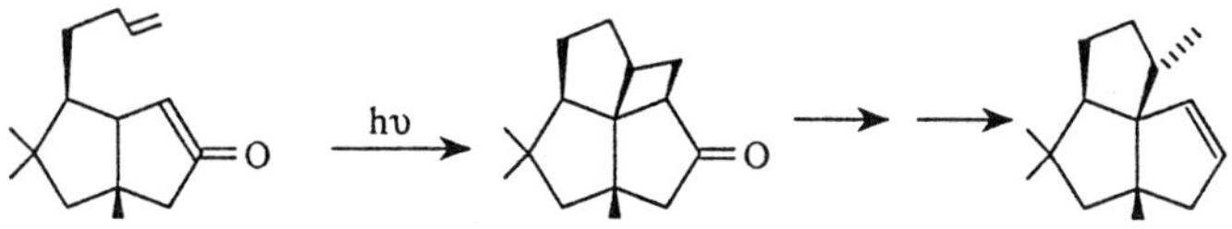

SCHEME 52.24

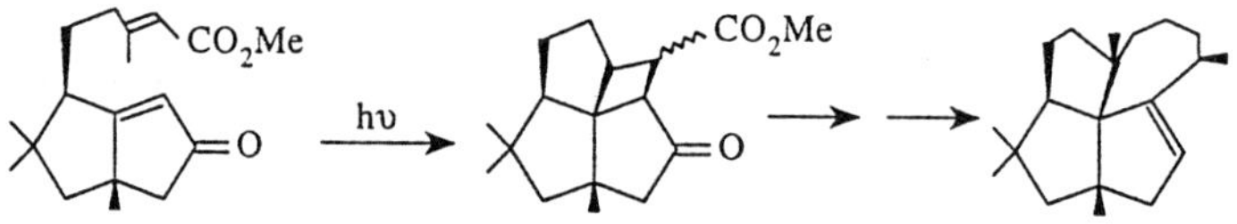

SCHEME 52.25

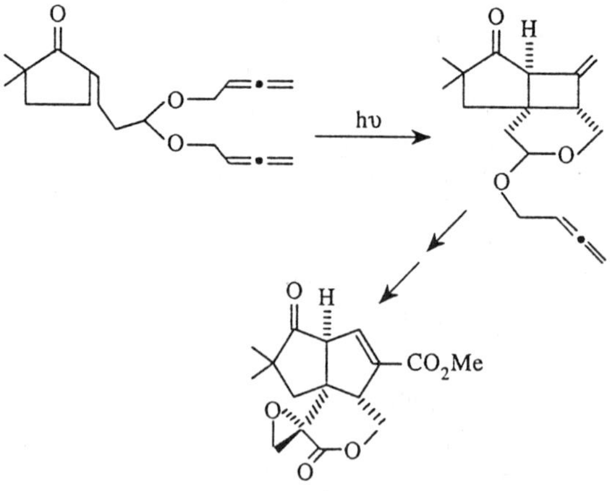

SCHEME 52.26

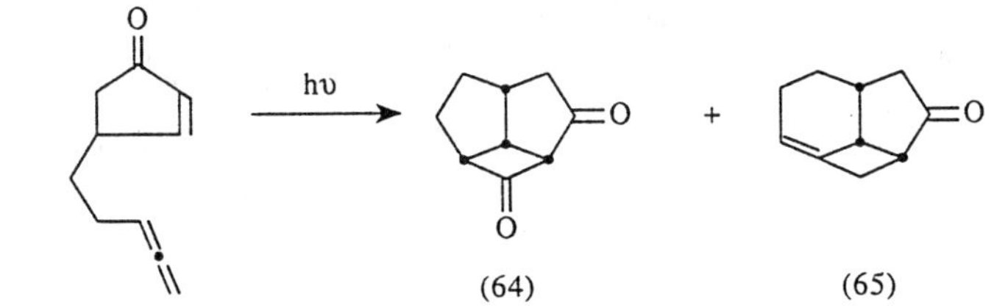

(64)   (65)

SCHEME 52.27

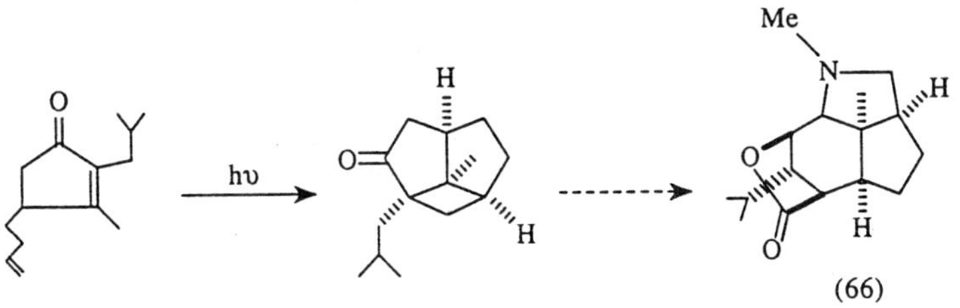

(66)

SCHEME 52.28

(67)

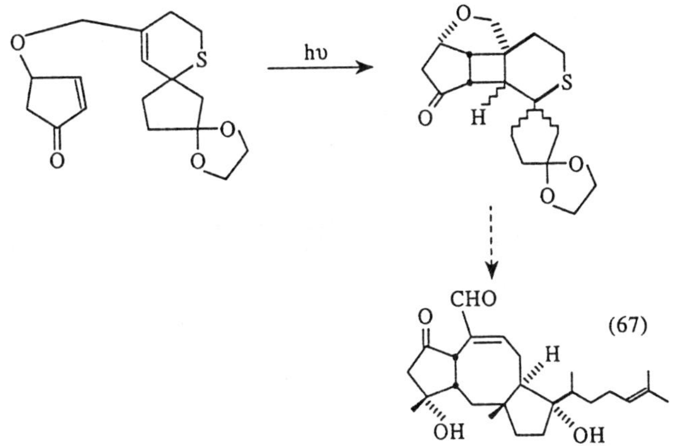

SCHEME 52.29

it is derived from a "non-rule-of-five" biradical. If instead of an allene, an alkene is attached to the enone 4-position by a two-atom tether, then the normal "rule-of-five" product is observed; this has been used for the synthesis of a possible precursor of the alkaloid dendrobine, **66**, (Scheme 52.28)[60] and as an approach for the preparation of natural products containing the dicyclopenta-cyclooctane ring system such as ophiobolin B, **67**, (Scheme 52.29).[61]

There are few examples of the synthesis of natural products using intramolecular photocycloaddition reactions of 2-cyclopentenones possessing alkenes attached to the 2-position. The carbon skeleton of hirsutene, **68**, has been prepared by intramolecular photocycloaddition of the enone **69** in methanol solution (Scheme 52.30).[62] It is proposed that the isolated product, **70**, is formed by solvent attack on the cycloadduct **71**. Intramolecular photocycloaddition has been attempted with compound **72** (Scheme 52.31) as a possible route for the preparation of pleuromutilin, **73**. However, the reaction fails and, instead, intramolecular hydrogen abstraction occurs followed by coupling of the radical pair to give **74** as the isolated product.[63]

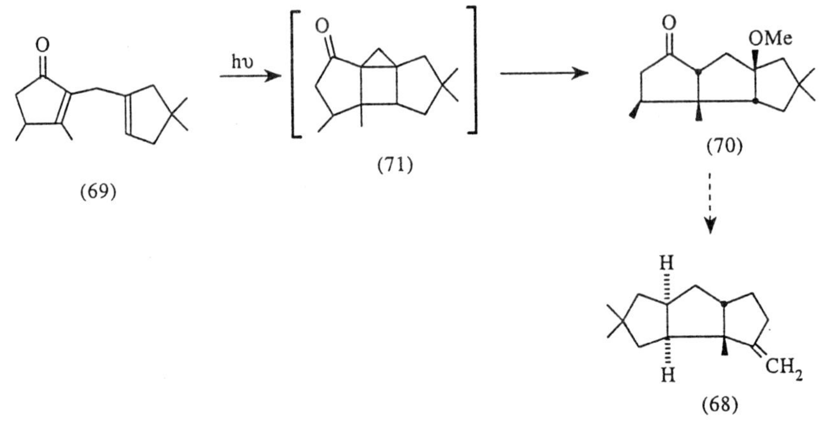

SCHEME 52.30

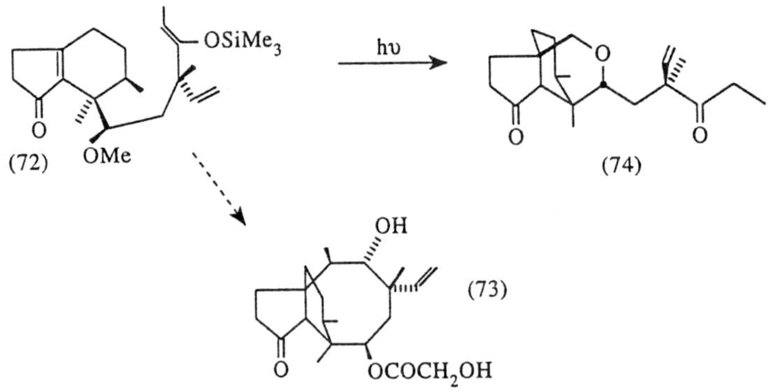

SCHEME 52.31

References

1. Eaton, P. E., On the mechanism of the photodimerization of cyclopentenone, *J. Am. Chem. Soc.*, 84, 2454, 1962.
2. Eaton, P. E., Photochemical reactions of simple alicyclic enones, *Acc. Chem. Res.*, 1, 50, 1967.
3. de Mayo, P., Enone photoannelation, *Acc. Chem. Res.*, 4, 41, 1970.
4. Bauslaugh, P. G., Photochemical cycloaddition reactions of enones to alkenes; synthetic application, *Synthesis*, 287, 1970.
5. Baldwin, S. W., Synthetic aspects of 2 + 2 cycloadditions of α,β-unsaturated carbonyl compounds, *Org. Photochem.*, 5, 123, 1981.
6. Weedon, A. C., Enone photochemical cycloaddition in organic synthesis, in *Synthetic Organic Photochemistry*, Horspool, W. M., Ed., Plenum, New York, 1984, chap. 2.
7. Schuster, D. I., The photochemistry of enones, in *The Chemistry of Enones*, Rappoport, Z., Ed., Wiley, Chichester, 1989, chap. 15.
8. Schuster, D. I., New insights into an old mechanism: [2+2]Photocycloaddition of enones to alkenes, *Chem. Rev.*, 93, 3, 1993.
9. Corey, E. J., Bass, J. D., LeMahieu, R., and Mitra, R. B., A study of the photochemical reactions of 2-cyclohexenones with substituted olefins, *J. Am. Chem. Soc.*, 86, 5570, 1964.

10. Loutfy, R. O. and de Mayo, P., On the mechanism of enone photoannelation: Activation energies and the role of the exciplexes, *J. Am. Chem. Soc.*, 99, 3559, 1977.

11. Wagner, P. J. and Bucheck, D. J., A comparison of the photodimerizations of 2-cyclopentenone and 2-cyclohexenone in acetonitrile, *J. Am. Chem. Soc.*, 91, 5090, 1969.

12. Caldwell, R. A., Tang, W., Schuster, D. I., and Heibel, G. E., Nanosecond kinetic absorption and calorimetric studies of cyclopentenone: The triplet, self-quenching, and the predimerization biradicals, *Photochem. Photobiol.*, 53, 159, 1991.

13. Schuster, D. I., Heibel, G. E., and Brown, P. B., Are triplet exciplexes involved in the [2+2] photocycloaddition of cyclic enones to alkenes?, *J. Am. Chem. Soc.*, 110, 8261, 1988.

14. Hastings, D. J. and Weedon, A. C., Origin of the regioselectivity in the photochemical cycloaddition reactions of cyclic enones with alkenes: Chemical trapping evidence for the structures, mechanism of formation, and fates of 1,4-biradical intermediates, *J. Am. Chem. Soc.*, 113, 8525, 1991.

15. Andrew, D., Hastings, D. J., Oldroyd, D. L., Rudolph, A., Weedon, A. C., Wong, D. F., and Zhang, B., Triplet 1,4-biradical intermediates in the photocycloaddition reactions of enones and *N*-acylindoles with alkenes, *Pure Appl. Chem.*, 64, 1327, 1992.

16. Rudolph, A. and Weedon, A. C., Radical clocks as probes of 1,4-biradical intermediates in the photochemical cycloaddition reactions of 2-cyclopentenone with alkenes, *Can. J. Chem.*, 68, 1590, 1990.

17. Termont, D., De Keukeleire, D., and Vandewalle, M., Regio- and stereo-selectivity in [π2+π2] photocycloaddition reactions between cyclopent-2-enone and electron-rich alkenes, *J. Chem. Soc., Perkin Trans. I*, 2349, 1977.

18. Griesbeck, A. G., Stadtmüller, S., Busse, H., Bringmann, G., and Buddrus, J., Photoreaktionen zwischen 2-cyclopenten-1-on und enolethern, *Chem. Ber.*, 125, 933, 1992.

19. Tada, M. and Nieda, Y., Ring size effect on [2+2] photochemical cycloaddition of enones with cyclic olefins, *Bull. Chem. Soc. Jpn.*, 61, 1416, 1988.

20. Lange, G. L., Organ, M. G., and Lee, M., Reversal of regioselectivity with increasing ring size of alkene component in [2+2] photoadditions, *Tetrahedron Lett.*, 31, 4689, 1990.

21. Swapna, G. V. T., Lakshmi, A. B., Rao, J. M., and Kunwar, A. C., Mechanistic implications of photoannelation reaction of 4,4-dimethyl-cyclohex-2-ene-1-one and acrylonitrile — regio- and stereochemistry of the major photoadduct by ¹H and ¹³C NMR spectroscopy, *Tetrahedron*, 45, 1777, 1989.

22. Wender, P. A. and Lechleiter, J. C., A photochemically mediated (4C+2C) annelation. Synthesis of (±)-10-epijunenol, *J. Am. Chem. Soc.*, 100, 4321, 1978.

23. Tobe, Y., Nakayama, A., Kakiuchi, K., Odaira, Y., Kai, Y., and Kasai, N., Synthesis, conformation, and structure of 8,11-*bis*(methoxycarbonyl)[6]-paracyclophane, *J. Org. Chem.*, 52, 2639, 1987.

24. Eaton, P. E., Photocondensation reactions of unsaturated ketones, *Tetrahedron Lett.*, 3695, 1964.

25. Sydnes, L. K. and Stensen, W., Regioselective photocycloaddition of 2-cyclopentenone to some allenes, *Acta Chem. Scand.*, 40, 1986, 657.

26. Stensen, W., Svendsen, J. S., Hofer, O., and Sydnes, L. K., Photochemical [2+2] cycloadditions. III. Addition of 4-substituted 2-cyclopentenones to allene; configuration determination by lanthanide-induced shift studies, *Acta Chem. Scand.*, 42, 259, 1988.

27. Maradyn, D. J., Sydnes, L. K., and Weedon, A. C., Origin of the regiochemistry in the photochemical cycloaddition reaction of 2-cyclopentenone with allene: Trapping of triplet 1,4-biradical intermediates with hydrogen selenide, *Tetrahedron Lett.*, 34, 2413, 1993.

28. Berenjian, N., de Mayo, P., Sturgeon, M.-E., Sydnes, L. K., and Weedon, A. C., Biphasic photochemistry: Micelle solutions as media for photochemical cycloadditions of enones, *Can. J. Chem.*, 60, 425, 1982.

29. Crimmins, M. T., Synthetic applications of intramolecular enone-olefin photocycloadditions, *Chem. Rev.*, 88, 1453, 1988.

30. Matlin, A. R., George, C. F., Wolff, S., and Agosta, W. C., Regiochemical control in intramolecular photochemical reactions of 1,6-heptadienes: Carbonyl-substituted 1-(4-alkenyl)-1-cyclopentenes, *J. Am. Chem. Soc.*, 1986, 108, 3385.

31. Srinivasan, R. and Carlough, K. H., Mercury ($^3P_1$) photosensitized internal cycloaddition reactions in 1,4-, 1,5-, and 1,6-dienes, *J. Am. Chem. Soc.*, 89, 4932, 1967.

32. Liu, R. S. and Hammond, G. S., Photosensitized internal addition of dienes to olefins, *J. Am. Chem. Soc.*, 89, 4936, 1967.

33. Beckwith, A. L. J., Regio-selectivity and stereo-selectivity in radical reactions, *Tetrahedron*, 37, 3073, 1981.

34. Cantrell, T. S., Photochemical cycloaddition of cyclohexenone and cyclopentenone to conjugated dienes, *J. Org. Chem.*, 39, 3063, 1974.

35. Demuth, D., Pandey, B., Wietfeld, B., Said, H., and Vaider, J., Photocycloadditions of 2-(trimethylsilyloxy)-1,3-butadiene to 2-cycloalkenones, *Helv. Chim. Acta*, 71, 1392, 1988.

36. Schuster, D. I., Heibel, G. E., and Woning, J., The mechanism of interaction of triplet 3-methylcyclohex-2-en-1-one with maleo- and fumarodinitrile: Evidence for direct formation of triplet 1,4-biradicals in [2+2] photocycloaddition without the intermediacy of exciplexes, *Angew. Chem. Int. Ed.*, 30, 1345, 1991.

37. White, J. D. and Gupta, D. N., The total synthesis of α- and β-bourbonene, *J. Am. Chem. Soc.*, 90, 6171, 1968.

38. Salomon, R. G., Sachinvala, N. D., Roy, S., Basu, B., Raychaudhuri, S. R., Miller, D. B., and Sharma, R. B., Total synthesis of spatane diterpenes: The tricyclic nucleus, *J. Am. Chem. Soc.*, 113, 3085, 1991.

39. Mori, K. and Sasaki, M., Synthesis of (±)-lineatin, the unique tricyclic pheremone of trypodendron lineatum (Oliver), *Tetrahedron Lett.*, 1329, 1979.

40. Cargill, R. L. and Wright, B. W., A new fragmentation reaction and its application to the synthesis of (±)-grandisol, *J. Org. Chem.*, 40, 120, 1975.

41. Rosini, G., Salomini, A., and Squarcia, F., A new convenient preparation of *cis*-(±)-2-acetyl-1-methyl cyclobutane acetic acid, *Synthesis*, 942, 1979.

42. Mori, K., Synthesis of both enantiomers of grandisol, the boll weavil pheremone, *Tetrahedron*, 34, 915, 1978.

43. De Keukeleire, D., de Wilde, H., Verzele, F., Wyffels, W., and Vandewalle, M., 6,6-dimethoxybicyclo(3,2,0)heptan-2-one: An intermediate for prostanoid synthons, *Bull. Chem. Soc. Belg.*, 88, 79, 1979.

44. Liu, H. J. and Kulkarni, M. G., Total synthesis of (±)-$\Delta^{9(12)}$-capnellene, *Tetrahedron Lett.*, 26, 4847, 1985.

45. Herzog, H., Koch, H., Scharf, H.-D., and Runsink, J., Chiral induction in photochemical reactions. V. Regio- and diastereoselectivity in the photochemical [2+2] cycloaddition of chiral cyclenone-3-carboxylates with 1,1-diethoxyethene, *Tetrahedron*, 42, 3547, 1986.

46. Lange, G. L. and Decicco, C. P., Asymmetric induction in mixed photoadducts employing α,β-unsaturated homochiral ketals, *Tetrahedron Lett.*, 29, 2613, 1988.

47. Demuynck, M., De Clercq, P., and Vanderwalle, M., (±)-Hysterin: Revised structure and total synthesis, *J. Org. Chem.*, 44, 4863, 1979.

48. Liu, H. J. and Lee, S. P., Total synthesis of S-epikessane and dehydrokessane, *Tetrahedron Lett.*, 3699, 1977.

49. Lange, G. L. and Gottardo, C., Free radical fragmentation of derivatives of [2+2] photoadducts, *Tetrahedron Lett.*, 31, 5985, 1990.

50. Piers, E., Abeysekera, B. F., Herbert, D. J., and Suckling, I. D., Total synthesis of the stemodane-type diterpenoids (±)-stemodin and (±)-maritimol. Formal total synthesis of (±)-stemodinone and (±)-2-desoxystemodinone, *Can. J. Chem.*, 63, 3418, 1985.

51. Tobe, Y., Sato, J., Sorori, T., Kakiuchi, K., and Odaira, Y., Cyclobutylcyclopropylcarbinyl type rearrangement of 1-oxaspirohexane derivatives. A new entry into functionalized norcaranes, *Tetrahedron Lett.*, 27, 2905, 1986.

52. Dauben, W. G. and Walker, D. M., Synthesis of [4.5.5.5]fenestrane and a [4.4.5.5]fenestrane derivative, *Tetrahedron Lett.*, 23, 711, 1982.

53. Bhaskar Rao, V., George, C. F., Wolff, S., and Agosta, W. C., Synthesis and structural studies in the [4.4.4.5]fenestrane series, *J. Am. Chem. Soc.*, 107, 5732, 1985.

54. Venepalli, B. R. and Agosta, W. C., Fenestranes and the flattening of tetrahedral carbon, *Chem. Rev.*, 87, 399, 1987.

55. Crimmins, M. T. and DeLoach, J. A., Intramolecular photocycloadditions-cyclobutane fragmentation: Total synthesis of (±)-pentalenene, (±)-pentalenic acid, and (±)-deoxypentalenic acid, *J. Am. Chem. Soc.*, 108, 800, 1986.

56. Crimmins, M. T. and Mascarella, S. W., Radical cleavage of cyclobutanes: Alternative routes to (±)-siliphinene, *Tetrahedron Lett.*, 28, 5063, 1987.

57. Crimmins, M. T. and Gould, L. D., Intramolecular photocycloaddition. Cyclobutane fragmentation: Total synthesis of (±)-laurenine, *J. Am. Chem. Soc.*, 109, 6199, 1987.

58. Pirrung, M. C. and Thomson, S. A., Total synthesis of pentalenolactone G methyl ester, *J. Org. Chem.*, 53, 227, 1988.

59. Dauben, W. G., Rocco, V. P., and Shapiro, G., Intramolecular [2+2] photocycloaddition of 4-substituted cyclopent-2-en-1-ones, *J. Org. Chem.*, 50, 3155, 1985.

60. Connelly, P. J. and Heathcock, C. H., An approach to the total synthesis of dendrobine, *J. Org. Chem.*, 50, 4135, 1985.

61. De Gregori, A., Jommi, G., Sisti, M., Gariboldi, P., and Merati, F., Studies directed towards the total synthesis of dicyclopenta[a,d]cyclooctane terpenoids, *Tetrahedron*, 44, 2549, 1988.

62. Kueh, J. S. H., Mellor, M., and Pattenden, Photocyclisations of dicyclopent-1-enyl methanes to tricyclo[6.3.0.0^{2,6}]undecanes: A synthesis of the hirsutane skeleton, *J. Chem. Soc., Chem. Commun.*, 5, 1978.

63. Paquette, L. A., Wiedeman, P. E., and Bulman-Page, P. C., A relay approach to (±)-pleuromutilin. III. Direct degradation of the natural product to the key diketone intermediate and its chemospecific functionalisation, *Tetrahedron Lett.*, 26, 1611, 1985.

53

[2+2]-Photocycloadditions of Cyclohexenones to Alkenes

David I. Schuster
New York University

53.1 Introduction

Following Eaton's original discovery of the photocyclodimerization of 2-cyclopentenone[1] and of photocycloaddition of this enone to cyclopentene,[2] Corey and co-workers reported [2+2]-photocycloadditions of 2-cyclohexenone to a variety of alkenes[3] and established many of the characteristic features of this reaction. The potential of this type of reaction as a key step in the synthesis of natural products was first shown by Corey in the landmark synthesis of caryophyllene.[4] Subsequently, inter- as well as intramolecular [2+2]-photocycloadditions have become part of the standard repertoire of synthetic organic chemists, and this process is now probably the most widely used photochemical reaction in synthetic organic chemistry.[5] Since the synthetic utility of this reaction has been recently reviewed,[5] this article will focus on recent insights into its mechanism.[6]

Photodimerization of cyclic enones in solution to give mixtures of *cis-anti* head-to-head (HH) and head-to-tail (HT) [2+2]-photodimers occurs via the lowest enone triplet, established using triplet sensitization and quenching techniques.[7-11] In 1969, Wagner[10,11] suggested that the inefficiency in photodimerization of cyclopentenone (CP) and cyclohexenone (CH) was associated principally with reversion of initially formed dimeric intermediates, possibly 1,4-biradicals, to two ground-state molecules, and that the extent of this reversion was probably different for HH as opposed to HT dimerization. If so, there would be no direct relationship between quantum

0-8493-8634-9/95/$0.00+$.50
© 1995 by CRC Press, Inc.

efficiencies of dimer formation and triplet-state reactivities, a phenomenon that was common in nonconcerted reactions involving triplet excited states.[10] de Mayo came to similar conclusions at the same time regarding addition of cyclopentenone triplets to alkenes.[12]

Based upon earlier spectroscopic studies of steroid enones[13] and calculations of energies of relaxed n,π* and π,π* states,[14] de Mayo concluded that the lowest energy enone triplet state and the one responsible for photoaddition reactions was probably the π,π* state.[15] This conclusion has been amply confirmed by recent studies by Schuster and co-workers using transient absorption spectroscopy[16] and time-resolved photoacoustic calorimetry.[17] However, in some cases (specifically rigid enones), the lowest n,π* and π,π* triplets may be close in energy.[5a,16,18] Estimates by de Mayo[12,15] of the rate constants k_d and k_a for enone triplet decay and for additions of enone triplets to alkenes, respectively, were made, as in Wagner's photodimerization study,[11] from quenching kinetics using piperylene and acenaphthene as triplet quenchers, assuming that enone triplet quenching occurred at the diffusion-controlled rate. Schuster and co-workers measured triplet lifetimes in solution of a number of cyclic enones using flash photolysis[16] and demonstrated that alkenes directly quench these triplets.[19,20] Trends in these values will be discussed later, but it should be noted here that in overlapping systems these rate constants are about an order of magnitude smaller than de Mayo's estimates[12,15] because of the overly generous value assumed for the rate constant for triplet energy transfer.

.2 General Characteristics of [2+2]-Photocycloadditions of Cyclohexenones

Stereochemistry of Ring Fusions of Cycloadducts

Corey established early on that *trans*-fused as well as *cis*-fused cycloadducts are formed on photoaddition of 2-cyclohexenone to simple alkenes, and the former are often the major products.[3] This is illustrated in Equation 53.1 for photoaddition to 1,1-dimethoxyethene.

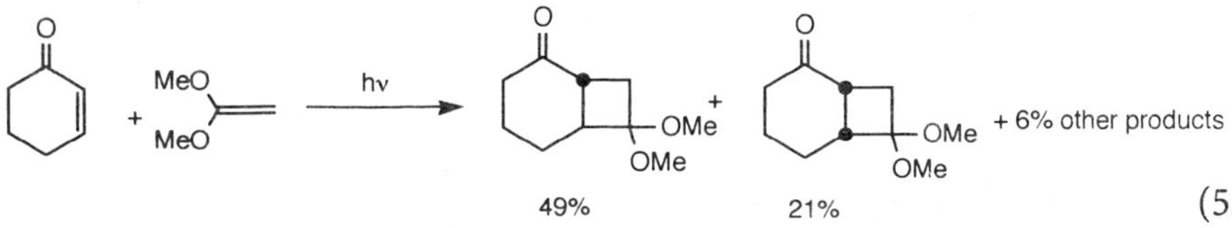

(53.1)

This observation is not confined to simple monocyclic enones; e.g., testosterone acetate **1b** gives both *cis*- and *trans*-fused photoadducts (**2b** and **3b**) with cyclopentene in an initial ratio of 2:1 (Equation 53.2).[21]

1a R = $OCOCH_2CH_3$
1b R = OAc

2a R = $OCOCH_2CH_3$
2b R = OAc

3a R = $OCOCH_2CH_3$
3b R = OAc

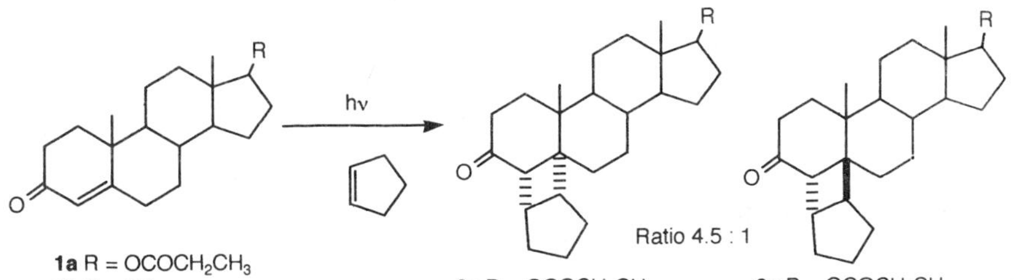

(53.2)

A gradual change in favor of the *cis*-adduct **2b** is due to secondary phototransformations of the adducts that preferentially consume **3b**.[22] This stereoselectivity must have a kinetic rather than thermodynamic basis, since the *cis*-fused structure with a boat cyclohexanone ring and a relatively flat cyclobutane ring is thermodynamically more stable than the corresponding *trans*-fused structure that has diequatorial linkage of a twisted cyclobutane ring to a relatively undistorted half-chair cyclohexanone ring.[21] The difference in energies of the stereoisomers from addition of 2-cyclohexenone to 2,3-dimethyl-2-butene from MM2 calculations is *ca.* 3.0 kcal mol^{-1}, but increases to 7.4 kcal mol^{-1} for cyclohexenone-cyclopentene adducts corresponding to 2 and 3.[21] In some cases, e.g., addition to 2-methylpropene, formation of cycloadducts is accompanied by formation of addition products attributed to disproportionation of 1,4-biradical intermediates (Equation 53.3).[3] Wilson and coworkers have shown that both *cis*- and *trans*-fused [2+2] cycloadducts are formed from photoaddition of cyclohexenones to *buckminsterfullerene* (C_{60}).[71]

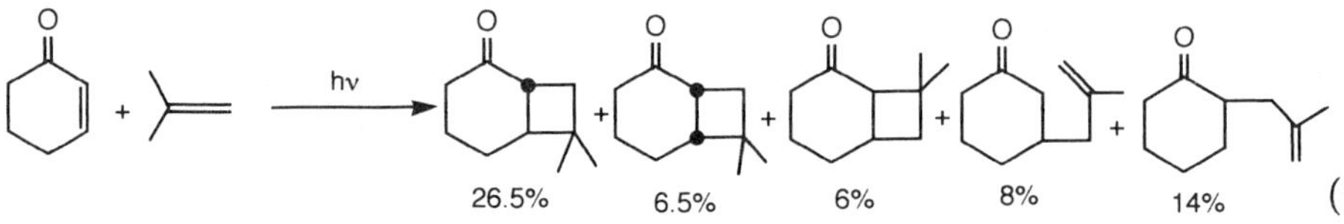

$$26.5\% \quad 6.5\% \quad 6\% \quad 8\% \quad 14\% \qquad (53.3)$$

Stereochemical Integrity of the Alkene Component in Photocycloadditions

Intermolecular Photocycloadditions

Corey reported that an identical mixture of cycloadducts was obtained from photoaddition of cyclohexenone to either *Z*- or *E*-2-butene, indicating that the stereochemistry of the alkene component is lost in the course of the reaction.[3] Recovery of the alkene component at various reaction times indicated <1% isomerization of the starting material had occurred. This finding demonstrated that stereomutation occurred at some intermediate reaction stage, and that the two new sigma bonds in the cycloadduct must be formed sequentially. It was not clear until recently (see below) whether initial bonding occurred at the α- or β-carbon of the enone, or both. The most reasonable mechanistic interpretation involves rotationally equilibrated triplet 1,4-biradicals as precyclization intermediates. The lack of isomerization of the starting materials indicated, at least in this system, that reversion of these 1,4-biradicals to starting material by fragmentation is not competitive with cyclization.

Cargill and co-workers[23] found that bicyclo[4.3.0]non-1(6)-en-2-one (BNEN, 4) gives the same four products on photoaddition to either *Z*- or *E*-2-butene, again indicating that the stereochemical integrity of the alkene component is lost during the reaction (see Equation 53.4).

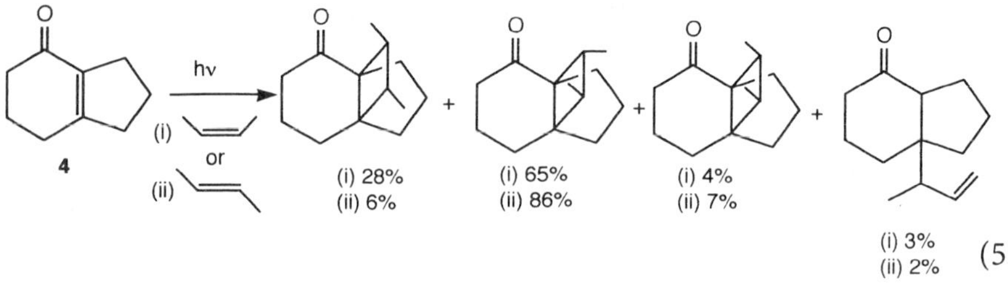

$$
\begin{array}{cccc}
\text{(i) } 28\% & \text{(i) } 65\% & \text{(i) } 4\% & \\
\text{(ii) } 6\% & \text{(ii) } 86\% & \text{(ii) } 7\% & \\
& & & \text{(i) } 3\% \\
& & & \text{(ii) } 2\% \qquad (53.4)
\end{array}
$$

A similar study involving photoaddition of cyclopentenone to *Z*- and *E*-1,2-dichloroethene was reported by Dilling and co-workers.[24] In both these cases, the possibility of reversion of the biradicals to starting materials resulting in *Z,E*-isomerization of the alkene was not explicitly considered or investigated. McCullough found that at 2% conversion in the photoaddition of 3-phenylcyclohexenone to *Z*-2-butene, recovered alkene contained 9% of the *E*-alkene,[25] demonstrating that biradical reversion to ground-state starting materials can be a major source of inefficiency in these reactions.

Biradical reversion appears to be a much more important process in enone additions to electron deficient alkenes. Thus, Schuster and co-workers found that in the reaction of 3-methylcyclohexenone with *Z*- and *E*-1,2-dicyanoethene (maleo- and fumaronitrile) isomerization of the alkenes accompanies formation of cycloadducts.[26] Based upon quantum yields for all processes and the rate constants for quenching of the enone triplet by these alkenes, the authors concluded that alkene isomerization occurred by reversion of 1,4-biradical intermediates (i.e., a Schenck-type mechanism)[27] rather than by triplet energy transfer from the enone to the alkenes, which was a distinct possibility because of the relatively low triplet energies of these particular alkenes.[28] The full significance of biradical reversion in affecting the course of enone photocycloadditions has only emerged recently, and will be discussed later.

Intramolecular Photocycloadditions

Stereochemical scrambling in intramolecular photocycloadditions of cyclohexenones with tethered alkene moieties has also been investigated. Becker found that the isomeric β-linked 1-acyl-1,6-heptadienes 5 and 6 (R = CH$_3$) give a 1:1 mixture of stereoisomeric cycloadducts 7 and 8 but the dienes do not equilibrate during the irradiation (Equation 53.4).[29] Similar results were obtained later with 5 (R = isopropyl) and with a *cis*-dideuterio analogue, demonstrating that steric effects did not play an important role. The results indicate that initial bonding occurs between C-2 (the β-carbon of the enone) and C-6 of the heptadiene moiety in accord with the famous "rule of five"[30] to give triplet 1,4-biradicals whose lifetimes are sufficiently long to allow complete rotational equilibration prior to ring closure. Thus, the reaction proceeds via a completely equilibrated mixture of 1,4-biradicals 9 and 10, which do not revert to starting materials (Equation 53.5).

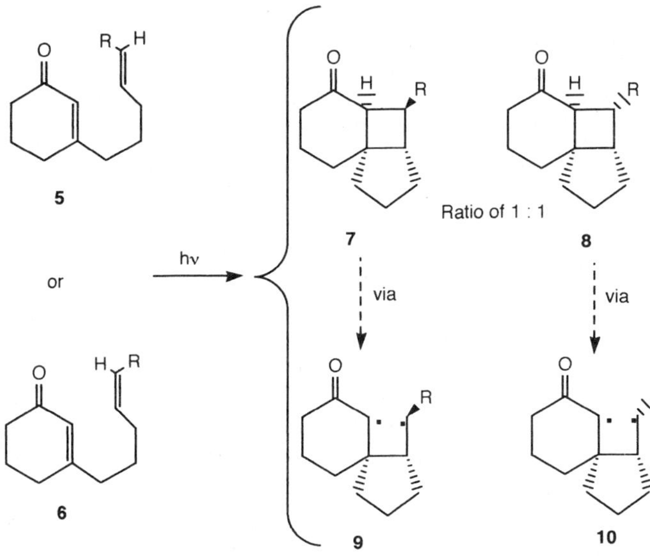

(53.5)

In contrast, Agosta found that the 1-acyl-hexadienes 11 undergoes scrambling of the label (D or CH$_3$) on the C=C bond competitive with formation of photoproducts.[31] Similar results were found for an acyclic analogue. For 11, reaction most likely occurs via 1,4-biradicals 12 formed by initial 1,5-bonding to the α-carbon of the enone. Inversion and rotation at the radical center then occurs prior to fragmentation or cyclization. Based upon quantum yields, the ratio of rates of biradical reversion (k$_r$) to product formation (k$_p$) are 0.75 for 11a and 1.81 for 11b. It is obvious that the competition between the various pathways for biradical formation and decay depends critically on the substitution pattern in these systems.

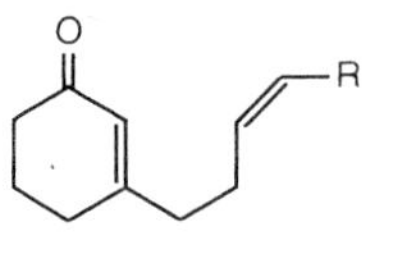

11a R = D
11b R = Me

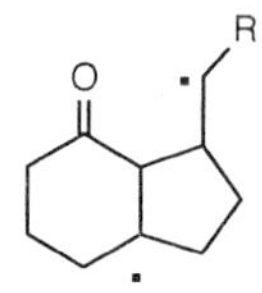

12

Becker[29] also investigated the photoreactions of the α-linked cyclohexenones **13** to **17**, where again no Z,E-isomerization of the starting materials was observed (Equation 53.6).

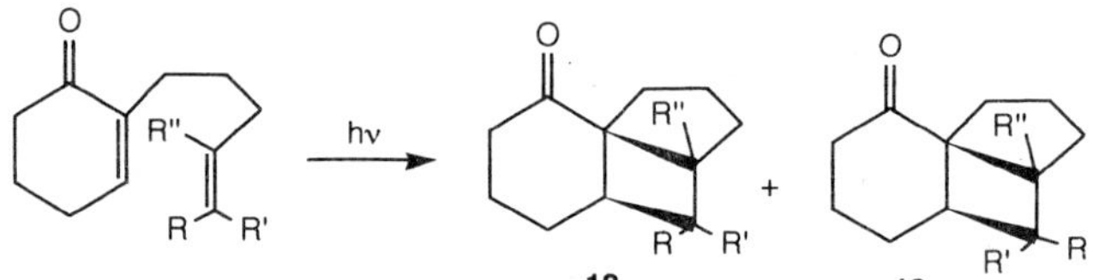

13 R = Me, R' = R'' = H
14 R' = Me, R = R'' = H
15 R = i-Pr, R = R'' = H
16 R' = i-Pr, R = R'' = H
17 R = R'' = D, R' = H

18

19

Ratio of **18** : **19** = 35 : 1 for **13**
Ratio of **18** : **19** = 5.8 : 1 for **14**
Ratio of **18** : **19** = 3.8 : 1 for **15**
Ratio of **18** : **19** = 1.4 : 1 for **16**
Ratio of **18** : **19** = 2 : 1 for **17**

$$(53.6)$$

While mixtures of stereoisomeric cycloadducts were formed in each case, no simple pattern of reactivity emerges. While both **13** and **14** afford preferentially diastereomer **18**, the ratio of **18** to **19** is quite different in each case, indicating the reaction is more complicated than that of the analogous β-linked dienones. With the propyl-substituted dienones **15** and **16**, cycloadducts **18** are again formed in preference to **19** but to different extents. Even with the dideuterodienone **22**, diastereoselectivity is observed on ring closure. Clearly, these results require the participation of several competing reaction pathways that have yet to be fully elucidated.

Regiochemistry in Enone-Alkene [2+2]-Photocycloadditions

One of the most significant findings in Corey's pioneering study was that photocycloadditions of cyclohexenones to unsymmetrical alkenes were invariably regioselective.[3] Thus, in addition of cyclohexenone to 2-methylpropene (Equation 53.3) and 1,1-dimethoxyethene (Equation 53.1), a clear preference for formation of head-to-tail (HT) vs. head-to-head (HH) adducts is observed. Similar selectivity was also found for photoaddition of cyclohexenone to allene, vinyl acetate, methyl vinyl ether, and benzyl vinyl ether, by Cantrell[32] for addition of 3-methylcyclohexenone (3-MCH) to 1,1-dimethoxyethene (DME), and by Lenz[33] for photoaddition of a steroid enone to 1,1-dialkoxyethene and 2-methylpropene. Corey claimed that, in contrast, HH adducts are preferentially formed between cyclohexenone and acrylonitrile although the adduct structures (including stereochemistry) were not firmly established.[3] Cantrell[32] reported that photoaddition of 3-MCH to acrylonitrile also gave mainly two HH adducts; the structure of a third adduct was not firmly established. On the other hand, Rao and co-workers recently established that the major photoadduct from 4,4-dimethylcyclohexenone and acrylonitrile had an HT structure,[34] while Weedon and co-workers established that the ratio of HH:HT adducts formed from 2-cyclopentenone and acrylonitrile was 3:4.8.[35] Thus, the suggestion[3] that the regioselectivity in [2+2]-photocycloadditions is reversed in enone additions to electron-deficient, vis a vis electron-rich alkenes, appears not to be correct.

Strong evidence that regioselectivity in [2+2]-photocycloadditions does not follow a simple rule comes from a study by Lange on addition of enones **20a** and **20b** to a series of methyl 1-cycloalkene-1-carboxylates (Equation 53.7).[36]

(53.7)

While additions to the cyclobutenyl ester gave exclusively HH adducts in line with Corey's generalization, increasing proportions of HT adducts were obtained as the ring size of the alkene was increased, resulting eventually in a reversal of regioselectivity. Thus, the HH:HT adduct ratio was 1:1 for 20a and 60:40 for 20b with the cyclopentenyl ester, and 1:9 and 1:20 for reaction, respectively, of 20a and 20b with the cyclohexenyl ester. Thus, the regiochemistry associated with enone-alkene photocycloadditions is clearly more complex than originally envisioned by Corey.[3] A mechanistic rationalization of these findings will be deferred for later discussion.

Reactivity of Alkenes Toward Photoexcited Enones

Corey and co-workers determined "relative reactivities" for additions of alkenes to photoexcited 2-cyclohexenone by measuring the ratio of adducts derived competitively from 1:1 mixtures of alkenes.[3] The "relative rate factors" found, correcting for statistical factors, were 1,1-dimethxyethene, 4.66; methoxyethene, 1.57; cyclopentene, 1.00 (reference compound); isobutene, 0.40; allene, 0.234; and acrylonitrile was much less "reactive" than any of the other alkenes listed. Based on these data, Corey concluded that photoexcited cyclohexenone was a moderately electrophilic species toward alkenes.

These data, which were critical elements in the formulation of Corey's exciplex mechanism for enone photocycloadditions, were generally accepted for a long time as proper measures of alkene reactivity in photocycloadditions.[5b-d] However, since product ratios in photochemical processes reflect relative quantum efficiencies for disappearance of starting materials and/or formation of products, they cannot be equated with relative rates of reaction.[5a,10,12,37] The lack of a relationship between product quantum yields and rate constants of a single reaction step was demonstrated many years ago by Wagner in the case of Norrish type II reactions of aromatic ketones. Here, quantum efficiencies are determined entirely by the competition between reversion of 1,4-biradical intermediates to starting material and progress on pathways leading to products, and do not correlate at all with the rate constants for formation of the biradicals from enone triplet states.[38] Since 1,4-biradicals also play a crucial role in enone photocycloaddition to alkenes, one should not expect relative yields of enone-alkene photoadducts to reflect directly the rates of the initial interaction of alkenes with enone triplet excited states.[5a,15,19,37,39]

Rate constants for interaction (quenching) of triplet excited states of cyclic enones with alkenes were first reported by Schuster et al. using transient absorption spectroscopy (nanosecond flash photolysis).[16,19,20] The rate constants k_q are obtained from the relationship $(\tau_T)^{-1} = (\tau_o)^{-1} + k_q[\text{alkene}]$, where τ_o is the limiting triplet lifetime of the enone at the enone concentration utilized in the absence of alkene. The decay of enone triplet absorption at 280 nm could be conveniently followed subsequent to excitation of the enones (cyclopentenone [CP], 3-methyl-cyclohexenone [3-MCH], testosterone acetate [TA], and BNEN, 4]) in acetonitrile and cyclohexane at 355 nm using the third harmonic of a Nd:YAG laser. In all cases, decays were cleanly first order. Quantum efficiencies for capture of enone triplets by alkenes, ϕ_{tc}, are given by $k_q \tau_T[\text{alkene}]$ using experimentally determined values of k_q and τ_T.

There is absolutely no correlation between the rate constants k_q for interaction of the enone triplets with the alkenes and the overall quantum efficiency for formation of products derived from this interaction.[6] In general, electron-rich alkenes give higher quantum yields for adduct formation, while the highest rate constants for triplet quenching are found with electron-deficient alkenes, such as acrylonitrile and fumaronitrile. Values of ϕ_{tc} at the alkene concentrations at which product quantum yields were measured (0.5 M in most cases) are always much higher than the efficiency of product formation. This is perhaps the clearest evidence for efficient formation of intermediates that revert to starting materials in competition with progress to adducts. There is no correlation between the rate constants k_q and ionization potentials of the alkenes, as would be expected if quenching involved formation of some kind of donor-acceptor complex with the enone acting as the electron acceptor. In fact, the data suggest that the enone π,π^* triplet is not electrophilic. There is also no significant difference between rate constants measured in polar (acetonitrile) and non-polar (cyclohexane) solvents, arguing against charge polarization in the transition state for the initial triplet enone-alkene interaction.

As mentioned, two different mechanisms for quenching of enone triplets could be operating with electron-deficient alkenes: a Schenck-like addition mechanism with formation of 1,4-biradicals[27] and triplet energy transfer. It was shown that quenching of 3-methylcyclohexenone by 1,2-dicyanoethenes involves a Schenck mechanism,[26] while interaction of higher energy triplets of the rigid enone BNEN with the same alkenes involves transfer of triplet excitation.[18] The difference can be attributed to inhibition of triplet energy transfer to alkenes from 3-MCH and similar conformationally flexible enones by the Dexter exchange mechanism due to poor π-overlap between the twisted chromophore of the twisted triplets and the alkenes.[40] No such inhibition would be involved with the conformationally constrained enone BNEN.

53.3 Mechanistic Proposals

The Corey-de Mayo Exciplex Mechanism

On the basis of the regiochemistry and the "relative rate factors" observed in studies of alkene additions to photoexcited cyclohexenone (see above), Corey[3] proposed in 1964 that the first step of the [2+2]-photocycloaddition of enones to alkenes involved interaction of a polarized enone triplet excited state with the ground-state alkene to give an "oriented π-complex", illustrated in 21 for addition of cyclohexenone to methoxyethene. Assuming (incorrectly, as it turned out) that the reactive excited state of the enone was a $^3n,\pi^*$ state, the charge polarization from extended Hückel calculations was predicted to be as shown, with electron density higher at β-carbon than at α-carbon.[41] In the donor-acceptor π-complex, the alkene ground state was proposed to act as the electron donor and the enone excited state as the electron acceptor, the two moieties being held together by coulombic attraction. Corey noted[3] that it was likely that the π-complex model could not be extended to enone photodimerization nor to reactions of enones with alkenes possessing strong electron-withdrawing substituents such as CN or COOR. Steric factors remained to be assessed. The alternative hypothesis that the regiospecificity in enone photocycloadditions might be governed by preferences in formation of 1,4-biradicals was rejected, since this did not explain the observed selectivity in photoaddition of cyclohexenone to DME as well as the "relative rate factors" for alkenes. However, 1,4-biradicals were invoked to rationalize the occasional formation of disproportionation products (see Equation 53.3) as well as the loss of stereochemistry on photoaddition of cyclohexenone to the isomeric 2-butenes.

21

de Mayo suggested that Corey's proposed intermediates might decay to regenerate ground-state reactants competitive with formation of products.[9,12,42] He noted that quantum yields for adduct formation from CP and a variety of alkenes were in no case greater than 0.50, even in neat alkene. de Mayo initially concluded that the intermediate undergoing revision was the exciplex, rather than a biradical derived from the exciplex, since little if any alkene isomerization occurred ($\Phi < 0.033$) on irradiation of CP in the presence of 3-hexene.[42] He also did not exclude the possibility that adducts might arise, at least in part, directly from the exciplex, bypassing the biradical. The tetramethylene 1,4-biradical was considered later to be the key intermediate, in part because variations in the quantum yield for photoadditions with temperature could be explained most easily in terms of changes in the partitioning of the 1,4-biradical intermediate.[43] Loutfy and de Mayo[44] continued to maintain that a triplet exciplex is first formed, irreversibly, from the enone triplet and the alkene ground state and that the exciplex collapses to one or more 1,4-biradicals, which either cyclize or revert to ground-state starting materials. Insufficient evidence was available to indicate whether the first bond in the adduct is formed α or β to the enone carbonyl group. However, the kinetic data[44] of Loutfy and de Mayo in no way *requires* formation of an exciplex precursor to biradicals in enone [2+2]-photocycloadditions. Caldwell concluded that a complex, most likely a π-complex, was an obligatory precursor to triplet 1,4-biradicals in the analogous process of oxetane formation from benzophenone and alkenes, based on secondary kinetic isotope effects associated with initial quenching of the ketone triplet, formation of oxetanes, and *cis,trans*-isomerization of the alkenes.[45]

The final version of the Corey-de Mayo mechanism is shown in Scheme 53.1.[44]

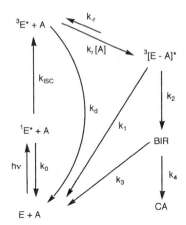

E = enone, A = alkene, BIR = 1,4-biradical, CA = cycloadducts

SCHEME 53.1

Identification of the Reactive Enone Intermediate

As already noted, it had been concluded that the reactive excited state of the enone component in [2+2]-photocycloadditions to alkenes is a $^3\pi,\pi^*$ state.[15,44] Additional corroboratory evidence comes from studies on 4,4-dimethylcyclohexenone 22, which gives the products shown in Equation 53.8.

22

23   (53.8)

upon irradiation in the presence of 2,3-dimethyl-2-butene.[37,46,47] Kinetic studies demonstrated that adduct formation is competitive with unimolecular photorearrangement of **22**, and that these products arise from the same enone excited state ($^3\pi,\pi^*$). Irradiation of **22** in neat alkene affords oxetane **23** in addition to the other products. The Stern-Volmer slope for quenching by naphthalene of oxetane formation, a reaction attributed to $^3n,\pi^*$ states,[48] is indeed different from that for quenching of all the other products in Equation 53.7. There is convincing evidence from transient absorption spectroscopy[16] and photoacoustic calorimetry[17] that relaxed $^3\pi,\pi^*$ states of simple cyclohexenones are highly twisted. The intriguing possibility that triplets of flexible cyclohexenones might give ground state *trans*-cyclohexenones en route to [2+2]-cycloadducts and other photoproducts has been frequently considered,[49-56] but remains controversial.

Bauslaugh's Biradical Proposal: Trapping and Detection of Biradical Intermediates in Enone [2+2]-Photocycloadditions

In 1970, Bauslaugh[57] proposed that the regiochemistry in [2+2]-photocycloadditions of enones could be explained without invoking exciplexes. It was merely necessary to consider the most likely conformations of various possible HH and HT biradicals and whether they were more likely to cyclize to give either *cis*- or *trans*-fused adducts or to undergo fragmentation. This proposal has recently received support from Weedon and co-workers who succeeded in trapping biradicals in [2+2]-photocycloaddition reactions of cyclopentenone using hydrogen selenide.[35,58] Thus, photoaddition of cyclopentenone to cyclopentene in benzene gives a mixture of *cis-syn-cis*- and *cis-anti-cis*-adducts; but in the presence of 0.3 M H$_2$Se, no cycloadducts were formed. The new products observed were derived by H-transfer to the putative 1,4-biradical intermediates, followed by reduction and disproportionation. The ratio of adducts suggests that the HH and HT biradicals are formed in a 1:1 ratio; that is, there is essentially no difference in reactivity toward alkenes at the α- and β-carbons of the enone triplet.

An even more revealing finding involved the photoaddition of cyclopentenone to ethyl vinyl ether (EVE),[35,58] which gives a 3:1 mixture of HT and HH cycloadducts. Adduct formation is totally suppressed by H$_2$Se, yielding instead a mixture of four new products attributable to trapping of the HT and HH biradicals (Scheme 53.2). The yields of product indicate that, once again, HH and HT biradicals are formed in a 1:1 ratio, and that only the more substituted biradicals are formed (i.e., no detectable products were found derived from biradicals containing a primary radical center). Regioselectivity in adduct formation must therefore originate in differences in the extent to which the HH and HT biradicals revert to starting material in competition with ring closure. Why the HT biradical shows a much greater preference to undergo cyclization vs. fragmentation, *vis a vis* the HH biradical, is not yet known, but may relate to differing populations of extended vs. closed conformations of these isomeric biradicals. A similar preference for cyclization of HT vs. HH biradicals was shown in other photoaddition reactions of cyclopentenone.[35] No analogous biradical trapping data has been reported as yet for cyclohexenone photoadditions.

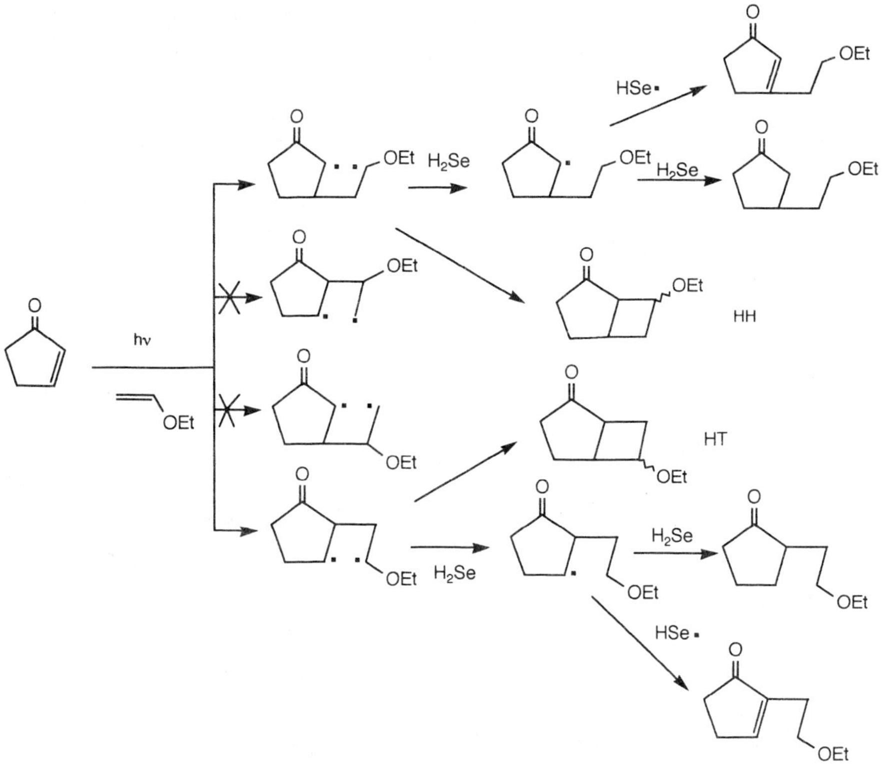

SCHEME 53.2

One approach to determination of triplet 1,4-biradical lifetimes is via generation of biradicals that are capable of undergoing competitive rearrangements at a known rate. The ring opening of cyclopropylcarbinyl to 3-butenyl radicals[59] has been used as a "radical clock" for estimation of biradical lifetimes. Thus, Becker and co-workers generated diradicals from dienones analogous to 5 with a cyclopropyl substituent on the C=C bond in the sidechain.[60] Rearrangement products as well as normal intramolecular [2+2]-cycloadducts were isolated in a ratio of 45:55, showing that the ring opening of the intermediate biradical occurred at roughly the same rate as ring closure (Equation 53.9). The biradical lifetimes were therefore estimated to be on the order of 50 ns. In a related study, photoaddition of cyclopentenone to vinylcyclopropane gave HH and HT cycloadducts as well as products derived from ring opening of the intermediate biradicals, whose lifetimes were estimated to be *ca.* 50 ns (Equation 53.10).[61] However, photoaddition of cyclopentenone to 1,6-heptadiene gave only the expected [2+2]-cycloadducts, and no products derived from rearrangement of the initial 1,4-biradicals to 1,6-biradicals were observed. Since the rate of rearrangement of 1-hexenyl to cyclopentylmethyl radicals is less than 10^5 s^{-1},[62] the initial triplet 1,4-biradicals must have lifetimes substantially less than 10 μs. A similar observation was made in intramolecular photoadditions of cyclohexenones with a tethered diene moiety.

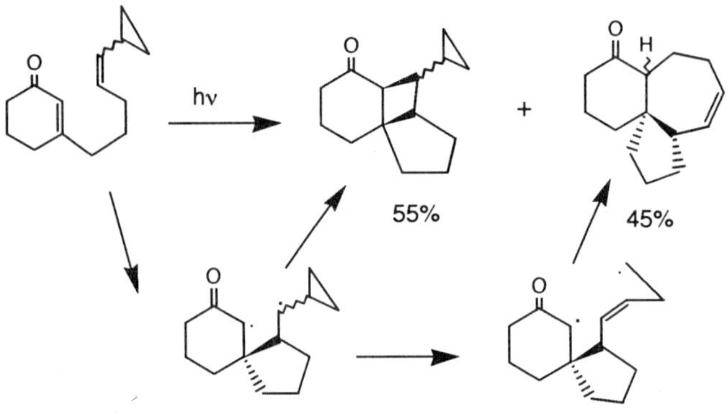

$$(53.9)$$

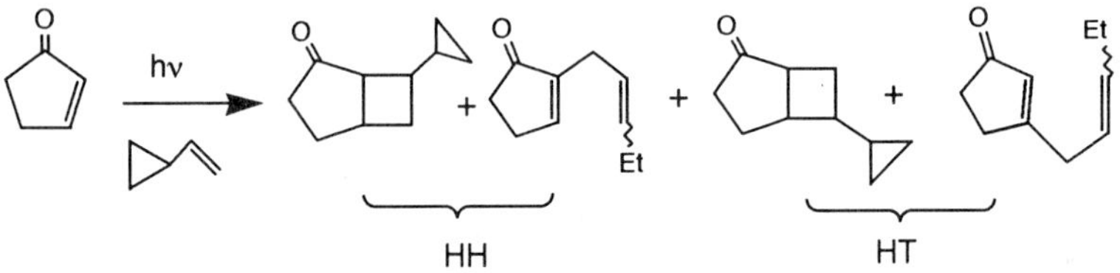

HH   HT (53.10)

Schuster, Caldwell, and co-workers[63] recently succeeded in directly detecting triplet 1,4-biradicals in enone photodimerizations. Nanosecond flash excitation of cyclopentenone (CP) in acetonitrile gives complex transient decay profiles at 280 nm that can be resolved into two first-order decays, one dependent on the concentration of the enone and a second that is not. The former is concluded to be the CP triplet,[6,11,12,51,63] whose limiting CP triplet lifetime in acetonitrile is 380 ± 75 ns, much higher than Wagner's estimate of 3 ns[11] as well as the value of 30 ns reported by Bonneau.[51] Bonneau's value is consistent with the value of 37 ns for the concentration-independent lifetime of the second transient in the laser flash studies, which corresponds to the (weighted) average lifetime of the mixture of HH and HT triplet 1,4-biradicals formed by CP self-quenching (Equation 53.11).

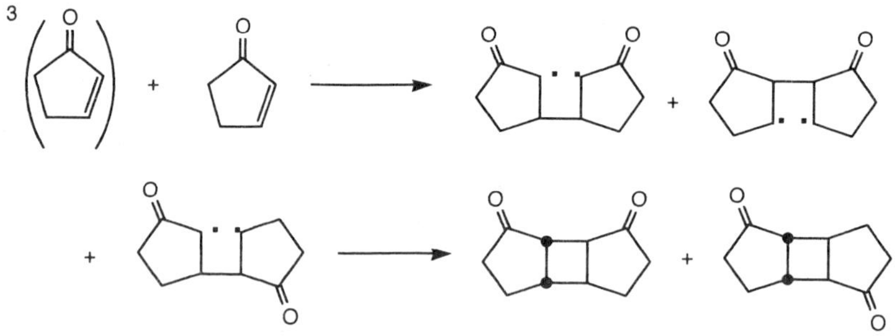

$$(53.11)$$

(Note that two stereogenic centers are generated on formation of the first C-C bond, so that each of the three biradicals shown is a mixture of two diastereomers; thus, six possible 1,4-biradicals are formed in this reaction.) Using time-resolved photoacoustic calorimetry (PAC) at moderate CP concentrations, a short-lived transient species was detected, concluded to be the mixture of dimeric CP 1,4-biradicals, whose lifetime was in perfect agreement with that from the flash study.[63] The average energy of the biradicals (relative to a pair of CP ground-state molecules) was 47.4 ± 1.7 kcal mol^{-1}. Using the Benson group additivity technique,[64] the energies of the HH and HT biradicals (A and B) were estimated to be 44 and 51 kcal mol^{-1}, respectively. These are the first such data determined for triplet 1,4-biradicals derived from cyclic enones.

More recently, the PAC technique has been used to measure lifetimes and energies of triplet 1,4-biradicals derived from some model enones and alkenes, as the first step in a broad study of such species.[65] The biradical lifetimes vary from 20 to 900 ns. The shortest lifetimes are for biradicals derived from 3-methylcyclohexenone and cyanoalkenes, while the longest lifetime is for this enone with 2,3-dimethyl-2-butene. A typical value is 59 ± 5 ns for the biradical(s) derived from testosterone acetate and cyclopentene. Again, it must be emphasized that these represent average lifetimes of the biradicals formed in the reaction, some of which may not even yield products to a significant extent. The average energies of these biradicals vary over a considerable range, from 36 to 60 kcal mol^{-1}, relative to the ground states of the reactants. These values are in good agreement with estimates based on Benson's additivity rules.[64] Further discussion of these data will be presented in due course. It is clear, however, that the triplet biradicals derived from enone triplets and alkenes have lifetimes allowing complete conformational equilibration, as has been implied by the stereo-chemical observations discussed previously.

Weedon failed to intercept biradicals using H_2Se from photoaddition of cyclopentenone (CP) to acrylonitrile (AN).[35] Xanthone-sensitized irradiation of CP in the presence of AN yields CP dimers and not CP-AN cycloadducts. Furthermore, upon direct irradiation, the quantum yield of enone dimer formation increased with increasing CP concentration, while the quantum yield for CP-AN adduct formation was constant. Since the two reactions occur via different CP excited states, Weedon suggested that the CP singlet excited state may be the species interacting with AN.[35] However, Schuster and co-workers[66] found that CP photodimerization is quenched by AN, and 1-methylnaphthalene quenches formation of both CP dimers and CP-AN adducts, although the Stern-Volmer quenching slopes are not precisely identical. It has been established[19] that CP-AN adducts do not arise from a route involving triplet energy transfer to AN followed by attack of AN triplets on ground-state enone. These observations suggests that the photocycloaddition reaction occurs via a CP triplet which is different from that leading to CP dimers. Clearly, more work is needed to resolve this mechanistic conundrum. There is evidence that photodimerization of cyclohexenone may also arise from a triplet state different from that responsible for [2+2]-photocycloaddition.[37]

The Bauslaugh-Schuster-Weedon Biradical Mechanism for Enone-Alkene Photocycloadditions

While the Corey-de Mayo exciplex mechanism[3,15,44] for photocycloaddition of enones to alkenes has provided a stimulus for workers in this field, and undoubtedly has had heuristic value in accounting for experimental observations in a vast number of reactions,[5] there seems to be no experimental basis for the proposal that the initial interaction of enone excited states and alkenes involves formation of an enone-alkene donor-acceptor π-complex. There is no reason why exciplexes should continue to be invoked in mechanistic discussions of enone [2+2]-photocycloaddition reactions. The mechanism shown in Scheme 53.3, essentially that postulated by Bauslaugh, is sufficient to explain the course of these reactions. In this mechanism, the only intermediates invoked are enone triplet excited states and triplet and singlet adduct biradicals. It is possible that examples may yet be uncovered where evidence *requiring* initial formation of exciplexes will be obtained. Until such time, it is proposed that the mechanism of Scheme III should be used as the framework for discussion of [2+2]-photocycloadditions of enones to alkenes.

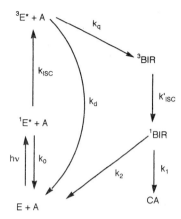

E = enone, A = alkene, BIR = 1,4-biradical, CA = cycloadducts

SCHEME 53.3

One possible benefit of this revised mechanism is that more attention may be paid to photoadditions of cyclohexenones to electron-deficient alkenes, now that the mechanistic bias against such reactions has been removed. While these are sometimes accompanied by unwanted side reactions, such as polymer formation in the case of AN,[19,32] these reactions oftentimes proceed quite cleanly, as in Lange's examples.[36] Just as [2+2]-photoaddition of enones to electron-rich alkenes have been profitably used by synthetic chemists,[5] there is no reason why analogous reactions using electron-poor alkenes should not also find synthetic utility.

53.4 Future Directions in Mechanistic Research on Cyclohexenone Photocycloadditions

The focus of attention in future mechanistic work on [2+2]-photocycloadditions must necessarily be centered on the triplet 1,4-biradical intermediates derived from enone triplets and alkenes. The physical and chemical properties of such species need to be determined in a systematic fashion, and the electronic and steric factors that control their rates of fragmentation and cyclization need to be elucidated. Theoretical calculations will be of some assistance in this connection, as will increased understanding of the dynamic properties of flexible triplet biradicals.[67] It is not at all obvious why certain cyclohexenone photoadditions (e.g., to *Z*- and *E*-2-butene) are not accompanied by at least partial isomerization of the recovered alkene, and why there is no loss of alkene stereochemistry in recovered starting material in the intramolecular photocycloadditions reported by Becker.[29] An obvious goal is to attempt to correlate the extent of isomerization in the reactant alkene with independently determined biradical partitioning factors.

The possibility of altering the course of enone [2+2]-photocycloadditions by modification of reaction conditions needs to be investigated. One approach would be to carry out such reactions in nonhomogenous media.[68] Thus, photodimerization of coumarin in cyclodextrins,[69] as well as of cyclopentenone and cyclohexenone in dry zeolites,[70] give product distributions quite different from those in fluid solution. While stereochemical and regiochemical control of [2+2]-photocycloadditions in solution does not appear possible, they are achievable in restricted environments. Thus, [2+2] photodimerization of cyclopentenone and cyclohexenone in certain zeolites affords predominantly HH dimers, whereas HT dimers are the major products in solution.[70] Also, *trans*-fused dimers are major products on photodimerization of cyclohexenone in zeolites. Thus, both the regio- and stereoselectivity of cyclohexenone [2+2] photocycloadditions are significantly changed by using zeolites as the reaction medium. Finally, the factors that control the stereochemistry of the 6/4 ring fusion in cyclohexenone [2+2]-photoadditions still need to be elucidated completely.

References

1. Eaton, P. E. and Hurt, W. S., Photodimerization of cyclopentenone. Singlet or triplet?, *J. Am. Chem. Soc.*, 88, 5038, 1966.

2. Eaton, P. E., On the mechanism of the photodimerization of cyclopentenone, *J. Am. Chem. Soc.*, 84, 2454, 1962.

3. Corey, E. J., Bass, J. D., LeMahieu, R., and Mitra, R. B., A study of the photochemical reactions of 2-cyclohexenones with substituted olefins, *J. Am. Chem. Soc.*, 86, 5570, 1964.

4. Corey, E. J., Mitra, R. B., and Uda, H., Total synthesis of the *d,l*-caryophyllene and *d,l*-isocaryophyllene, *J. Am. Chem. Soc.*, 86, 485, 1964.

5. (a) Weedon, A. C., Enone photochemical cycloaddition in organic synthesis, in *Synthetic Organic Chemistry*, Horspool, W. M., Ed., Plenum Press, New York, 1984, 61; (b) Baldwin, S. W., Synthetic aspects of 2+2 cycloadditions of α,β-unsaturated carbonyl compounds, in *Organic Photochemistry*, Vol. 5, Padwa, A., Ed., Marcel Dekker, New York, 1981, 123; (c) Carless, H. A. J., Enone and dienone rearrangements, in *Photochemistry in Organic Synthesis*, Coyle, J. D., Ed., Royal Soc. Chem., London, 1986, 118; (d) Wender, P. A., Alkenes: Cycloaddition, in *Photochemistry in Organic Synthesis*, Coyle, J. D., Ed., Royal Soc. Chem., London, 1986, 163; (e) Carless, H. A. J., Terpenoids, in *Photochemistry in Organic Synthesis;* Coyle, J. D., Ed., Royal Soc. Chem., London, 1986, 210; (f) Oppolzer, W., Intramolecular [2+2] photoaddition/cyclobutane-fragmentation sequence in organic chemistry, *Acc. Chem. Res.*, 15, 135, 1982.

6. Schuster, D. I., Lem, G., and Kaprinidis, N. A., New insights into an old mechanism: [2+2] photocycloaddition of enones to alkenes, *Chem. Rev.*, 93, 3, 1993.

7. Lam, E. Y. Y., Valentine, D., and Hammond, G. S., Mechanisms of photochemical reactions in solution. XLIV. Photodimerization of cyclohexenone, *J. Am. Chem. Soc.*, 89, 3482, 1967; Ruhlen, J. L. and Leermakers, P. A., Photochemistry of cyclopentenone in various media, *J. Am. Chem. Soc.*, 89, 4944, 1967.

8. Chapman, O. L., Koch, T. H., Klein, F., Nelson, P. J., and Brown, E. L., Two triplet mechanisms in photochemical addition of 2-cyclohexenones to 1,1-dimethoxyethylene, *J. Am. Chem. Soc.*, 90, 1657, 1968.

9. de Mayo, P., Pete, J. P., and Tchir, M., Cyclopentenone photocycloaddition. A reaction from a higher triplet state, *J. Am. Chem. Soc.*, 89, 5712, 1967; de Mayo, P., Pete, J. P., and Tchir, M. F., Photochemical synthesis. 22. On photochemical cycloaddition: Cyclopentenone, *Can. J. Chem.*, 46, 2535, 1968.

10. Wagner, P. J. and Bucheck, D. J., Inefficiency and reversibility in the photodimerization of 2-cyclopentenone, *Can. J. Chem.*, 47, 713, 1969.

11. Wagner, P. J. and Bucheck, D. J., A comparison of the photodimerizations of 2-cyclopentenone and of 2-cyclohexenone in acetonitrile, *J. Am. Chem. Soc.*, 91, 5090, 1969.

12. de Mayo, P., Nicholson, A. A., and Tchir, M. F., Evidence for reversible intermediate formation in cyclopentenone cycloaddition, *Can. J. Chem.*, 47, 711, 1969.

13. Kearns, D. R., Marsh, G., and Schaffner, K., Excited singlet and triplet states of a cyclic conjugated enone, *J. Chem. Phys.*, 49, 3316, 1968; Marsh, G., Kearns, D. R., and Schaffner, K., Spektroskopische Untersuchung einiger α,β-ungesättigter cyclischer Ketone, *Helv. Chim. Acta*, 51, 1890, 1968; Marsh, G., Kearns, D. R., and Schaffner, K., Investigation of singlet-triplet transitions by phosphorescence excitation spectroscopy. IX. Conjugated enones, *J. Am. Chem. Soc.*, 93, 3129, 1971.

14. Devaquet, A., Potential energy sheets for the n-π^* and π-π^* triplet states of α,β-unsaturated ketones, *J. Am. Chem. Soc.*, 94, 5160, 1972.

15. de Mayo, P., Enone photoannelation, *Acc. Chem. Res.*, 4, 41, 1971.

16. Schuster, D. I., Dunn, D. A., Heibel, G. E., Brown, P. B., Rao, J. M., Woning, J., and Bonneau, R., Enone photochemistry. Dynamic properties of triplet excited states of cyclic conjugated enones as revealed by transient absorption spectroscopy, *J. Am. Chem. Soc.*, 113, 6245, 1991, and earlier papers cited.

17. Schuster, D. I., Heibel, G. E., Caldwell, R. A., and Tang, W., Determination of triplet excitation energies of cyclic enones by time-resolved photoacoustic calorimetry, *Photochem. Photobiol.*, 52, 645, 1990.

18. Schuster, D. I., Woning, J., Kaprinidis, N. A., Pan, Y., Cai, B., Barra, M., and Rhodes, C. A., Photochemical and photophysical studies of bicyclo[4.3.0]non-1(6)-en-2-one, *J. Am. Chem. Soc.*, 114, 7029, 1992.

19. Schuster, D. I., Heibel, G. E., Brown, P. B., Turro, N. J., and Kumar, C. V., Are triplet exciplexes involved in [2+2] photocycloaddition of cyclic enones to alkenes?, *J. Am. Chem. Soc.*, 110, 8261, 1988.

20. Heibel, G. E., Energetics and reactivity of excited states of cyclic α,β-unsaturated ketones, Ph. D. Dissertation, New York University, 1990; Brown, P. B., Photocycloaddition reactions of alkenes and 2-cyclohexenone: A mechanistic study, Ph. D. Dissertation, New York University, 1988.

21. Schuster, D. I., Kaprinidis, N. A., Wink, D. J., and Dewan, J. C., Stereochemistry of [2+2] photocycloaddition of cyclic enones to alkenes: Structural and mechanistic considerations in formation of *trans*-fused cycloadducts, *J. Org. Chem.*, 56, 561, 1991.

22. Kaprinidis, N. A., Woning, J., Schuster, D. I., and Ghatlia, N. D., Photochemistry of steroidal enones: Formation of an exceptionally stable ketene by an α-cleavage reaction, *J. Org. Chem.*, 57, 755, 1992.

23. Peet, N. P., Cargill, R. L., and Bushey, D. F., Synthesis and acid-catalyzed rearrangements of tricyclo[4.3.2.0] undecanones, *J. Org. Chem.*, 38, 1218, 1973.

24. Dilling, W. L., Tabor, T. E., Boer, F. P., and North, P. P., Organic photochemistry. IX. The photocycloaddition of 2-cyclopentenone to *cis*- and *trans*-dichloroethylene. Evidence for initial attack at carbon-3 and rotational equillibration of biradical intermediates, *J. Am. Chem. Soc.*, 92, 1399, 1970.

25. McCullough, J. J., Ramachandran, B. R., Snyder, F. F., and Taylor, G. N., Kinetics of photochemical addition of 3-phenyl-2-cyclohexenone to tetramethylethylene, *J. Am. Chem. Soc.*, 97, 6767, 1975.

26. Schuster, D. I., Heibel, G. E., and Woning, J., The mechanism of interaction of triplet 3-methylcyclohex-2-en-1-one with maleo- and fumarodinitrile: Evidence for direct formation of triplet 1,4-biradicals in [2+2] photocycloadditions without the intermediacy of exciplexes, *Angew. Chem., Int. Ed.*, 30, 1345, 1991.

27. Schenck, G. O. and Steinmetz, R., Neuratige durch Benzophenon photosensibilisierte Additionen von Maleinsäureanhydrid an Benzol und andere Aromaten, *Tetrahedron Lett.*, 1, 1960; Gollnick, K. and Schenck, G. O., Mechanism and stereoselectivity of photosensitized oxygen transfer reactions, *Pure Appl. Chem.*, 9, 507, 1964.

28. (a) Wong, P. C., Triplet energies of fumaronitrile and maleonitrile, *Can. J. Chem.*, 60, 339, 1982; (b) Lavilla, J. A. and Goodman, J. L., The energetics and kinetics of relaxed alkene triplet states as determined by pulsed time-resolved photoacoustic calorimetry, *Chem. Phys. Lett.*, 141, 149, 1987.

29. (a) Becker, D. and Haddad, Y. S., Topological and steric effects in mechanism of intramolecular [2+2] photocycloadditions, *Tetrahedron Lett.*, 30, 4429, 1989; (b) Becker, D., Nagler, M., Sahali, Y., and Haddad, N., Regiochemistry and stereochemistry of intramolecular [2+2] photocycloaddition of carbon-carbon double bonds to cyclohexenones, *J. Org. Chem.*, 56, 4537, 1991.

30. (a) Srinivasan, R. and Carlough, K. H., Mercury ($^3P_1$) photosensitized internal cycloaddition reactions in 1,4-, 1,5-, and 1,6-dienes, *J. Am. Chem. Soc.*, 89, 4932, 1967; (b) Liu, R. S. H. and Hammond, G. S., Photosensitized internal additions of dienes to olefins, *J. Am. Chem. Soc.*, 89, 4936, 1967; (c) Gleiter, R. and Sander, W., Light-induced [2+2] cycloaddition reactions of non-conjugated dienes — the effect of through-bond interaction, *Angew. Chem., Int. Ed.*, 24, 566, 1985; (d) Fischer, E. and Gleiter, R., Regiochemistry of the intramolecular [2+2]-photocycloaddition of cyclohexenone to vinyl ethers, *Angew. Chem. Int. Ed.*, 28, 925, 1989.

31. Schroder, C., Wolff, S., and Agosta, W. C., Biradical reversion in the intramolecular photochemistry of carbonyl-substituted 1,5-hexadienes, *J. Am. Chem. Soc.*, 109, 5491, 1987.

32. Cantrell, T. S., Haller, W. S., and Williams, J. C., Photocycloaddition reactions of some 3-substituted cyclohexenones, *J. Org. Chem.*, 34, 509, 1969.

33. (a) Lenz, G. R., Photocycloaddition reactions of conjugated enones, *Rev. Chem. Intermed.*, 4, 369, 1981; (b) Lenz, G. R., The isolation and characterization of a steroidal *trans*-fused cyclobutanone from an enone [2+2] photoadduct, *J. Chem. Soc., Chem. Commun.*, 803, 1982; (c) Lenz, G. R., The photocycloaddition of an 5α-androst-1-en-3-one to olefins, *J. Chem. Soc., Perkin Trans. 1*, 2397, 1984.

34. Swapna, G. V. T., Lakshmi, A. B., Rao, J. M., and Kunwar, A. C., Mechanistic implications of photoannelation reaction of 4,4-dimethyl-cyclohex-2-en-1-one and acrylonitrile; regio- and stereochemistry of the major photoadduct by ^{1}H and ^{13}C NMR spectroscopy, *Tetrahedron*, 45, 1777, 1989.

35. Andrew, D., Hastings, D. J., Oldroyd, D. L., Rudolph, A., Weedon, A. C., Wong, D. F., and Zhang, B., Triplet 1,4-biradical intermediates in the reactions of enones and *N*-acylindoles with alkenes, *Pure Appl. Chem.*, 64, 1327, 1992.

36. Lange, G. L., Organ, M. G., and Lee, M., Reversal of regioselectivity with increasing ring size of alkene component in [2+2] photoadditions, *Tetrahedron Lett.*, 31, 4689, 1990.

37. Schuster, D. I., The photochemistry of enones, in *The Chemistry of Enones*, Part 2, Patai. S. and Rappoport, Z., Eds., John Wiley & Sons, Chichester, U.K., 1989, 623.

38. Wagner, P. J., Type II photoelimination and photocyclization of ketones, *Acc. Chem. Res.*, 4, 168, 1971.

39. Cargill, R. L., Morton, G. H., and Bordner, J., Stereochemistry of photochemical cycloaddition: 4-*tert*-Butylcyclohex-2-en-1-one and ethylene, *J. Org. Chem.*, 45, 3929, 1980.

40. Scaiano, J. C., Leigh, W. J., Meador, M. A., and Wagner, P. J., Sterically hindered triplet energy transfer, *J. Am. Chem. Soc.*, 107, 5806, 1985.

41. (a) Hoffman, R., An extended Hückel theory. I. Hydrocarbons, *J. Chem. Phys.*, 39, 1397, 1963; (b) Hoffman, R., Extended Hückel theory. IV. Carbonium ions, *J. Chem. Phys.*, 40, 2480, 1964; (c) Zimmerman, H. E. and Swenton, J. S., Mechanistic organic photochemistry. VIII. Identification of the n-π* triplet excited state in the rearrangement of 4,4-diphenyl cyclohexadienone, *J. Am. Chem. Soc.*, 86, 1436, 1964.

42. de Mayo, P., Nicholson, A. A., and Tchir, M. F., Photochemical synthesis. 22. On photochemical cycloaddition: Cyclopentenone, *Can. J. Chem.*, 48, 225, 1970.

43. (a) Montgomery, L. K., Schueller, K., and Bartlett, P. D., Cycloaddition. II. Evidence of a biradical intermediate in the thermal addition of 1,1-dichloro-2,2-difluoroethylene to the geometrical isomer of 2,4-hexadiene, *J. Am. Chem. Soc.*, 86, 622, 1964; (b) Bartlett, P. D. and Wallbillich, G. E. H., Cycloaddition. XI. Evidence for reversible biradical formation in the addition of 1,1-dichloro-2,2-difluoroethylene to the stereoisomers of 1,4-dichloro-1,3-butadiene, *J. Am. Chem. Soc.*, 91, 409, 1969; (c) Bartlett, P. D. and Porter, N. A., The stereochemical course of cyclic azo decompositions, *J. Am. Chem. Soc.*, 5317, 1968; (d) Wagner, P. J. and Schott, H. N., Polar substituent effects in the reactions of 1,4-biradicals, *J. Am. Chem. Soc.*, 91, 5383, 1969.

44. Loutfy, R. D. and de Mayo, P., On the mechanism of enone photoannelation: Activation energies and the role of exciplexes, *J. Am. Chem. Soc.*, 99, 3559, 1977.

45. Caldwell, R. A., Sovocool, G. W., and Gajewski, R. J., Primary interaction between diaryl ketone triplets and simple alkenes. Isotope effects, *J. Am. Chem. Soc.*, 95, 2549, 1973.

46. Schuster, D. I., Greenberg, M. M., Nuñez, I. M., and Tucker, P. C., Identification of the reactive electronic excited state in the photocycloaddition of alkenes to cyclic enones, *J. Org. Chem.*, 48, 2615, 1983.

47. Tucker, P. C., The photochemistry of 2-cyclohexenones. The [2+2] photocycloaddition of 2-cyclohexenones to alkenes, Ph. D. Dissertation, New York University, 1988.

48. (a) Turro, N. J., *Modern Molecular Photochemistry*, Benjamin/Cummings, Menlo Park, CA, 1978, 432 (b) Cowan, D. O. and Drisko, R. L., *Elements of Organic Photochemistry*, Plenum Press, New York, 1976, 181.

49. Schuster, D. I., Brown, P. B., Capponi, L. J., Rhodes, C. A., Scaiano, J. C., Tucker, P. C., and Weir, D., Mechanistic alternatives in photocycloaddition of cyclohexenones to alkenes, *J. Am. Chem. Soc.,* 109, 2533, 1987.

50. Bonneau, R. and Fornier de Violet, P., Mise en évidence d'un isomère trans tendu de l'acétyl-1-cyclohèxene et d'une forme métastable attribueé à un état triplet orthogonal, *C. R. Acad. Sci. Paris, Ser. C.,* 284, 631, 1977; Goldfarb, T. J., Kinetic studies of transient photochemical isomers of 2-cycloheptenone, 1-acetyl cyclohexene and 2-cyclohexenone, *J. Photochem.,* 8, 29, 1978.

51. Bonneau, R., Transient species in photochemistry of enones. The orthogonal triplet state revealed by laser photolysis, *J. Am. Chem. Soc.,* 102, 3816, 1980.

52. Eaton, P. E., Photochemical reactions of simple alicyclic enones, *Acc. Chem. Res.,* 1, 50, 1968.

53. Bowman, R. M., Calvo, C., McCullough, J. J., Rasmussen, P. W., and Snyder, F. F., Photoaddition of 2-cyclohexenone derivatives to cyclopentene — an investigation of stereochemistry, *J. Org. Chem.,* 37, 2084, 1972.

54. Verbeek, J., van Lenthe, J. H., Timmermans, P. J. J. A., Mackor, A., and Budzelaar, P. H. M., On the existence of *trans*-cyclohexene, *J. Org. Chem.,* 52, 2955, 1987.

55. Dunkelblum, E., Hart, H., and Jeffares, M., Stereochemistry of the photoinduced addition of methanol to Pummerer's ketone, a 2-cyclohexenone, *J. Org. Chem.,* 43, 3409, 1978; Mintas, M., Schuster, D. I., and Williard, P. G., Stereochemistry and mechanism of [4+2] photocycloaddition of Pummerer's ketone to furan, *J. Am. Chem. Soc.,* 110, 2305, 1988; Mintas, M., Schuster, D. I., and Williard, P. G., Stereochemistry and mechanism of the [2+2] and [4+2] photocycloaddition of alkenes and dienes to Pummerer's ketone, *Tetrahedron,* 44, 6001, 1988.

56. Rudolph, A. and Weedon, A. C., Acid catalysis of the photochemical deconjugation reaction of 3-alkyl-2-cyclohexenones, *J. Am. Chem. Soc.,* 111, 8756, 1989.

57. Bauslaugh, P. G., Photochemical cycloaddition reactions of enones to alkenes: Synthetic applications, *Synthesis,* 287, 1970.

58. Hastings, D. J. and Weedon, A. C., Origin of the regioselectivity in the photochemical cycloaddition reactions of cyclic enones with alkenes. Chemical trapping. Evidence for the structures, mechanism of formation and fates of 1,4-biradical intermediates, *J. Am. Chem. Soc.,* 113, 8525, 1991.

59. (a) Mathew, L. and Warkentin, J., The cyclopropylmethyl free radical clock. Calibration for the range 30–89 °C, *J. Am. Chem. Soc.,* 108, 8525, 1991; (b) Castellino, J. and Bruice, T. C., Intermediates in the epoxidation of alkynes by cytochrome P-450 models. 2. Use of the *trans*-2-*trans*-3-diphenylcyclopropyl substitute in a search for radical intermediates, *J. Am. Chem. Soc.,* 110, 7512, 1988.

60. Becker, D., Haddad, N., and Sahali, Y., Trapping of 1,4-diradical intermediate in intramolecular [2+2] photocycloaddition, *Tetrahedron Lett.,* 30, 2661, 1989.

61. Rudolph, A. and Weedon, A. C., Radical clocks as probes of 1,4-biradical intermediates in the photochemical cycloaddition reactions of 2-cyclopentenone with alkenes, *Can. J. Chem.,* 68, 1590, 1990.

62. Lusztyk, J., Maillard, B., Deyard, S., Lindsay, D. A., and Ingold, K. U., Kinetics for the reaction of a secondary alkyl radical with tri-*n*-butylgermanium hydride and calibration of a secondary alkyl radical clock reaction, *J. Org. Chem.,* 52, 3509, 1987.

63. Caldwell, R. A., Tang, W., Schuster, D. I., and Heibel, G. E., Nanosecond kinetic absorption and calorimetric studies of cyclopentenone: The triplet, self quenching, and the predimerization biradicals, *Photochem. Photobiol.,* 53, 159, 1991.

64. Benson, S. W., *Thermochemical Kinetics,* 2nd ed., Wiley, New York, 1976.

65. Kaprinidis, N. A., Lem, G., Courtney, S. H., and Schuster, D. I., Determination of the energies and lifetimes of triplet 1,4-biradicals derived in [2+2] photocycloaddition reactions of enones with alkenes using photoacoustic calorimetry, *J. Am. Chem. Soc.,* 115, 3324, 1993.

66. Schuster, D. I., Lem, G., and Kaprinidis, N. A., unpublished results.

67. For a recent review, see: Doubleday, C., Jr., Turro, N. J., and Wong, V. F., Dynamics of flexible triplet biradicals, *Acc. Chem. Res.,* 22, 199, 1989.

68. Fox, M. A., Ed., *Organic Transformations in Nonhomogeneous Media,* ACS Symposium Series 278, American Chemical Society, Washington, D. C., 1985.

69. Moorthy, J. M., Venkatesan, K., and Weiss, R. G., Photodimerizations of coumarins in solid cyclodextrin inclusion complexes, *J. Org. Chem.,* 57, 3292, 1992.

70. Lem, G., Kaprinidis, N. A., Schuster, D. I., Ghatlia, N. D., and Turro, N. J., Regioselective photodimerization of enones in zeolites, *J. Am. Chem. Soc.,* 115, 7009, 1993.

71. Wilson, S. R., Kaprinidis, N., Wu, Y., and Schuster, D. I., A new reaction of fullerenes: [2+2] photocycloaddition of enones, *J. Am. Chem. Soc.,* 115, 8495, 1993; Wilson, S. R., Wu, Y., Kaprinidis, N. A., Schuster, D. I., and Welch, C. J., Resolution of enantiomers of *cis-* and *trans-*fused C_{60}-enone [2+2] photoadducts, *J. Org. Chem.,* 58, 6548, 1993.

[2+2]-Photocycloaddition Reactions of Enolized 1,3-Diketones and 1,2-Diketones with Alkenes: The de Mayo Reaction

Alan C. Weedon
University of Western Ontario

54.1 Introduction

The photocycloaddition reaction of alkenes with the enol forms of 1,3-diketones and 1,2-diketones is commonly referred to as the de Mayo reaction. The primary products are cyclobutanes formed by bonding of the alkene termini to the carbon-carbon double bond of the enol, as shown in Schemes 54.1 and 54.2. With 1,3-diketones, there is generally enough of the enol form **1a** present in equilibrium with the parent diketone **2** to allow direct or sensitized excitation of the enol. The cyclobutane adducts **3a** formed are not isolable since they undergo a retro-aldol ring opening to a 1,5-diketone **4** under the reaction conditions; however, if the enol is trapped as the ether (i.e., **1b**) or the acetate (i.e., **1c**), then the cyclobutane adducts **3b** or **3c** can be isolated if desired. With 1,2-diketones, the enol form of the diketone has to be trapped as the enol ether **5a** or the enol acetate **5b** in order to raise the concentration of the enolized form to a high enough level for it to absorb light or be sensitized.

0-8493-8634-9/95/$0.00+$.50

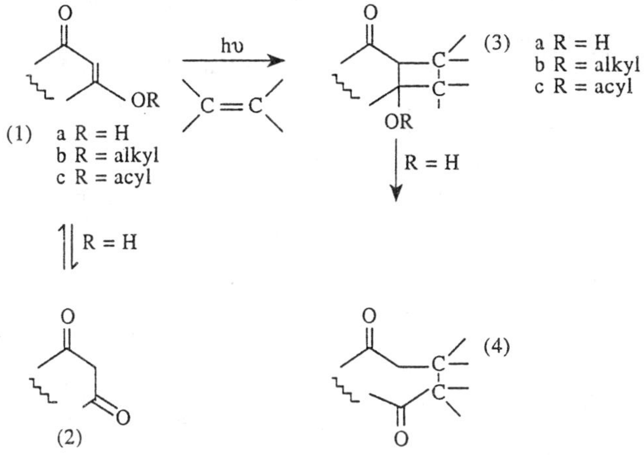

SCHEME 54.1

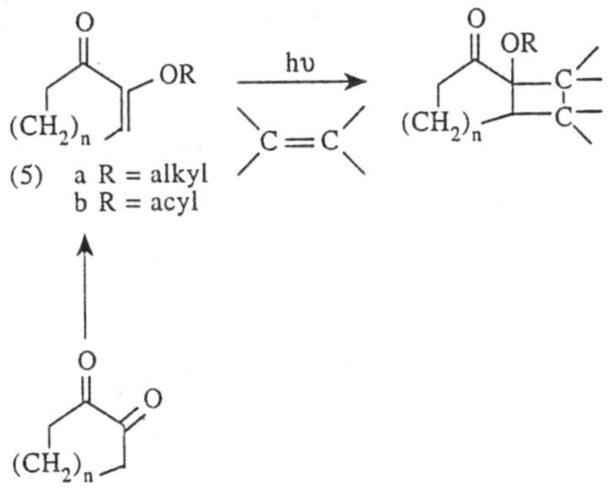

SCHEME 54.2

The reaction shown in Scheme 54.1 is general for the enol forms of cyclopentane- and cyclohex-ane-1,3-diones but fails for larger ring diketones unless the alkene is tethered to the diketone so that it can add in an intramolecular fashion. The reaction also proceeds if the enol oxygen is replaced by a nitrogen function (e.g., 6) and for vinyligous lactones (e.g., 7). The enols of acyclic 1,3-dicarbonyl compounds such as that of acetylacetone also react photochemically with alkenes to give acylcyclobutanols that subsequently ring open to acyclic 1,5-dicarbonyl compounds.

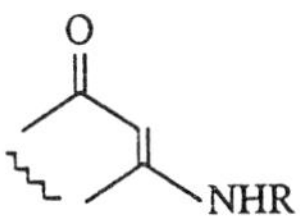

(6) 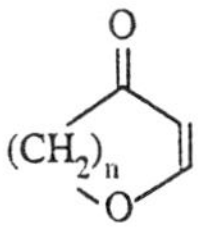   (7)

The photocycloaddition of alkenes to the enols of 1,2-diketones has received less attention but appears to be general for the trapped enols of cyclopentane-1,2-diones. The reaction is reported to fail for the trapped enols of cyclohexane-1,2-diones unless the alkene is tethered so as to allow intramolecular addition.

The photocycloaddition reaction of enolized diketones with alkenes has been the subject of many reports. No comprehensive review has yet been published, although some aspects of the reaction have been discussed in the secondary literature.[1-4] Since an exhaustive review of the reaction is not possible in the space available here, this chapter will attempt to describe the capabilities and limitations of the reaction for synthetic applications using a limited number of selected examples.

54.2 The Reaction Mechanism

The mechanism of the photocycloaddition reaction of enolized diketones with alkenes has not been investigated extensively. However, it is generally assumed to proceed by a mechanism similar to that operating in the cyclobutane-forming reaction that occurs when 2-cyclopentenones or 2-cyclohexenones are irradiated with UV light in the presence of alkenes. For many years,[3,4] the accepted model for the mechanism of the latter reactions was that shown in Scheme 54.3. In this model, the triplet excited state of the enone is intercepted by the alkene to yield an exciplex. This occurs in competition with relaxation of the triplet to the ground state, which is facilitated by rotation around the carbon-carbon double bond of the enone. The triplet exciplex is proposed to react to form a triplet 1,4-biradical by bonding of either the 2-position or the 3-position of the enone to a terminus of the alkene π-system. Following spin inversion, the 1,4-biradical then either cleaves to regenerate the ground-state enone and alkene, or ring closes to give the cyclobutane adduct. Recent results[5,6] have suggested that the inclusion of an exciplex intermediate in this mechanism is unnecessary. In addition, the structures of the intermediate triplet 1,4-biradicals have been inferred from the results of chemical trapping experiments.[5-7] Most commonly, the 1,4-biradicals are formed as isomeric mixtures derived from bonding of either the 2-position or the 3-position of the enone to the terminus of the alkene less able to support a radical center. This is consistent with a preference for the triplet enone to react with the alkene to produce the more stable biradical intermediates if it is assumed that a carbon radical center adjacent to a carbonyl group is of similar stability to a secondary carbon radical. If the reaction of enolized diketones proceeds by a similar mechanism, then the reaction with monosubstituted alkenes would be expected to yield predominantly the triplet 1,4-biradical **8** rather than the less stable biradicals **9** to **11**.

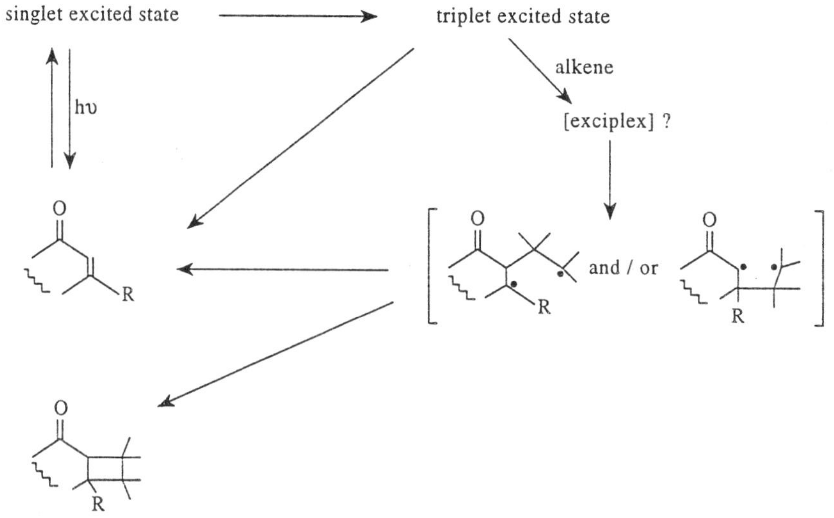

SCHEME 54.3

The intermediacy of the triplet excited state in the photocycloaddition reaction of enolized diketones has been confirmed by a study of the reaction between acetylacetone and cyclohexene.[8] It was found that the reaction could be sensitized by acetophenone (triplet energy 74 kcal mol[-1]) and that it was quenched by piperylene (triplet energy 59 kcal mol[-1]). Benzophenone (triplet energy 68 kcal mol[-1]) did not sensitize the reaction, which suggests that the enolized diketone triplet excited state energy is ~70 kcal mol[-1].

The intermediacy of the triplet 1,4-biradical is supported by the occasional observation of byproducts resulting from intramolecular disproportionation of the biradical.[9,10] For example,[9] photoaddition of cyclopentene to the enol of dimedone yields mainly the cyclooctanedione derived from retro-aldol ring opening of the primary cycloadduct 12; however, 2-cyclopentyldimedone 13 is also isolated as a minor product and is most likely formed from intramolecular disproportionation of the 1,4-biradical 14.

4.3 Consequences of the Mechanism for Synthetic Applications

The finding that the cycloaddition reaction proceeds, at least in the cases examined, from the triplet excited state of the enolized diketones means that the reaction can be sensitized if desired, but also that the presence of energy-transfer quenchers may inhibit the reaction. Therefore, oxygen should be excluded from the reaction, and it would be expected that cycloaddition partners with triplet energies less than that of the enolized diketone would quench the excited state rather than participate in cycloaddition. Nonconjugated alkenes normally possess vertical triplet energies higher than those estimated for simple enolized diketones and therefore can be expected to be suitable cycloaddition partners. Dienes, styrenes, and stilbenes possess low triplet energies and would be expected to act as quenchers rather than as cycloaddition partners. In general, this appears to be true; for example, it was noted above that piperylene (*trans*-1,3-pentadiene) quenches the cycloaddition reaction between acetylacetone and cyclohexene. However, exceptions do exist. Thus, stilbene will add to the enol tautomers of dibenzoylmethane and 1-phenyl-1,3-butanedione,[11] and the enol of methyl 2,4-dioxopentanoate adds to styrene[12] and dienes.[13–16] Similarly, adducts arising from intramolecular photocycloaddition of the trapped enols 15 and 16 have been reported.[17] These exceptions may originate from a lowering of the triplet excited-state energy of the enols so that they fall below those of the cycloaddition partners. Conjugation could achieve this lowering in the enols of dibenzoylmethane, 1-phenylbutane-1,3-dione, and methyl-2,4-dioxopentanoate, while it has been suggested that partial rotation around the carbon-carbon double bonds of 15 and 16 yields relaxed triplets with energies lower than the vertical triplet energies of the side-chain alkenes.[20]

Since triplet excited-state relaxation by partial rotation around a double bond brings the excited-state energy surface closer to the ground-state energy surface, deactivation of the triplet to the ground state can become rapid and make cycloaddition noncompetitive. This is often used to explain why intermolecular cycloaddition is generally observed only for addition of alkenes to the enols of smaller-ring cyclic diones (i.e., cyclopentanediones and cyclohexanediones), in which the ring prevents rotation about the enol double bond in the excited state. This limitation can be overcome if the alkene is tethered and available for intramolecular cycloaddition, as in the eight-membered ring derivative 17, although the reaction is still inefficient.[17] In the case of the enol tautomers of acyclic 1,3-diketones, it is thought that intramolecular hydrogen bonding in the enol form (e.g., as in 18) slows twisting long enough for the triplet excited state to react with alkenes. In the absence of the hydrogen bond, the photocycloaddition reaction fails unless the alkene is tethered and available for intramolecular addition. Such an intramolecular reaction was first reported[18,19] for the system 19, which is the trapped enol form of the 2-acylcyclopentanone 20; the adduct obtained had structure 21. Many more examples of intramolecular addition of the trapped enols of acyclic 1,3-dicarbonyl compounds have since appeared.[2]

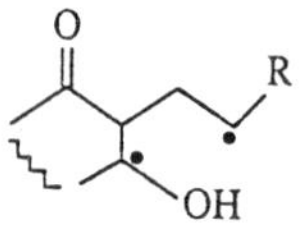

(8)

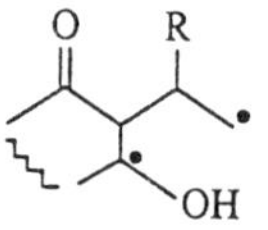

(9)

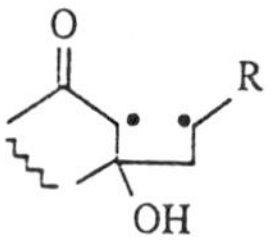

(10)

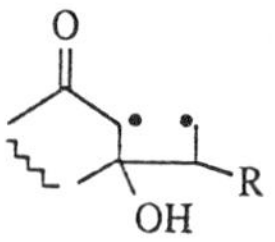

(11)

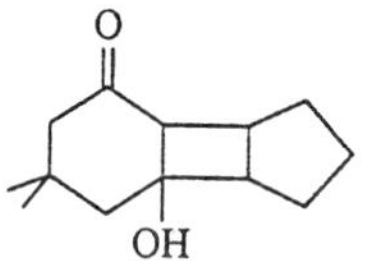

(12)

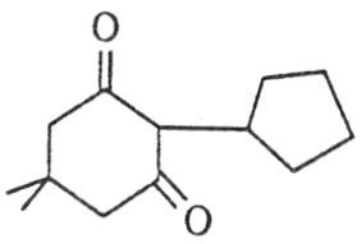

(13)

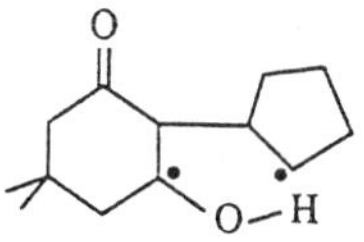

(14)

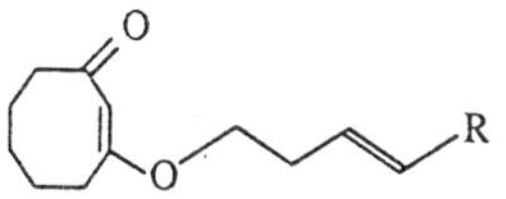

(15) R = CH = CH$_2$
(16) R = Ph
(17) R = nC$_3$H$_7$

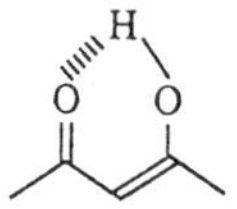

(18)

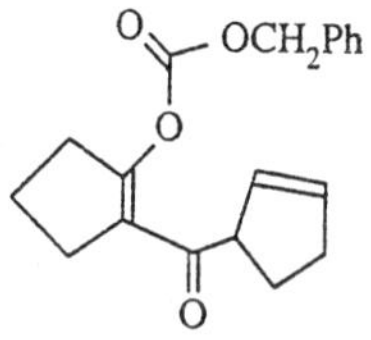

(19)

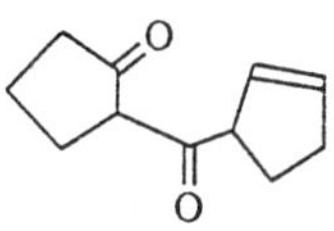

(20)

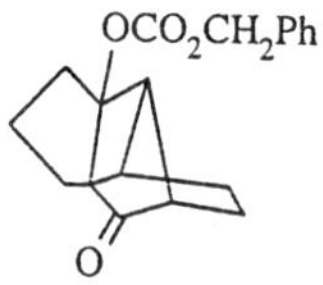

(21)

Two orientations of addition are possible when nonsymmetrically substituted alkenes add to enolized diketones. This is shown for enolized 1,3-diketones in Scheme 54.4. Attempts have been made to use the exciplex model for the mechanism of enone-alkene photocycloadditions to predict the regiochemical outcome of this reaction; this approach uses the assumption that the product regiochemistry results from favorable alignment of the dipole of the triplet excited enol with the dipole of the ground-state alkene in the exciplex intermediate. As indicated in Section 54.2, recent results suggest that an exciplex intermediate may not be involved and that the reaction may in fact proceed via the more stable triplet 1,4-biradical intermediates. If this is so, then it would be expected that for enols of 1,3-diketones, the reaction regiochemistry would be determined by the preferential formation of biradical 8 which would lead to product 22 rather than 23. Neither of these approaches are particularly successful for the prediction of the regiochemistry of the photocycloaddition reaction between enolized diketones and alkenes since, in many cases, it is difficult to estimate the likely dipole of the enolized diketone triplet state or the relative stabilities of the possible biradical intermediates. In addition, the reaction is not always very regioselective. For example, alkenes bearing a good electron-donating substituent, such as an alkoxy group, invariably favor the formation of the regioisomer 22,[20–23] but isopropenyl acetate yields mainly 23[24] while alkyl substituted alkenes generally yield mixtures of both 22 and 23 with little regiochemical preference.[20,24–28] Addition of allene to trapped enols of 1,3-diketones is reported to be regioselective, but the actual orientation obtained has been found to depend upon the structure of the enol; regioisomer 22 dominates if the enol is trapped as an alkyl ether,[29] whereas regioisomer 23 is formed preferentially if the enol is trapped as the acetate ester[29] or as a carbonate ester.[30] Reversal of the regiochemical preference with structure is also seen for the addition of electron-deficient alkenes; cycloaddition between the enol of 1,3-cyclohexadiene and methyl acrylate[28] yields the product of retro-aldol ring opening of 23; whereas, in the addition of methylcyclohexene-1-carboxylate to the methyl ether of the enol of 1,3-cyclohexanedione, the opposite regioisomer (i.e., 22) is obtained.[31]

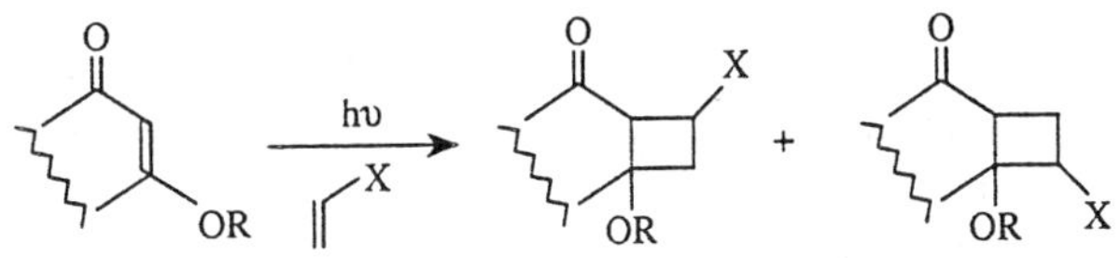

SCHEME 54.4

Choice of solvent has also been found to exert a large effect on the reaction regiochemistry. For example, the reaction between the trapped enol 24 and the alkene 25 yields a mixture of the regioisomers 26 and 27 in close to a 1:1 ratio when the reaction is performed in a polar solvent; when a nonpolar solvent is used the regioisomeric ratio becomes 95:5 in favor of 27.[26] However, the generality of this solvent effect has not been demonstrated.

The low and somewhat unpredictable regioselectivity of the reaction can be overcome by tethering the alkene to the enolized diketone so that an intramolecular photocycloaddition can take place.[2] The alkene can be attached in such a manner that only one orientation of addition is feasible; alternatively, use can be made of the so-called "rule of five" that allows prediction of the reaction regiochemistry in cases where both orientations of addition are possible. The "rule of five" reflects the propensity for cyclizations yielding a radical center by addition to a double bond to produce a five-membered ring.[32–35] The control of regiochemistry that can be obtained is seen in the intramolecular photocyclization of 28 and 29; compound 28 yields the product of closure of the biradical 30;[36] whereas with 29, the opposite orientation of alkene addition occurs and the product of closure of biradical 31 is obtained.[37]

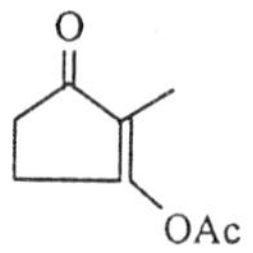

(24)

(25)

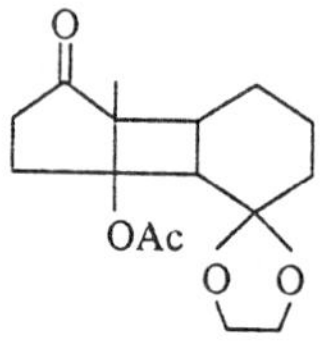

(26)

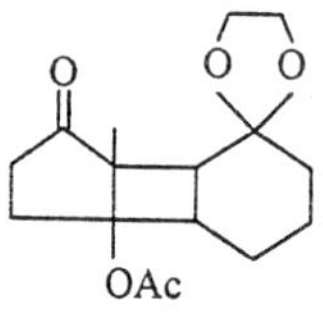

(27)

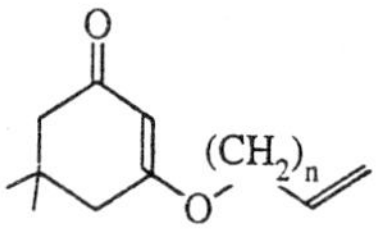

OAc

OAc

OAc

O O

O O

O

(28) (n = 1)
(29) (n = 2)

$(CH_2)_n$

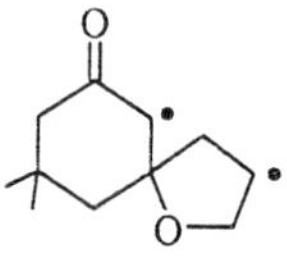

(30)

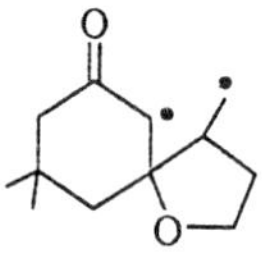

(31)

When a cyclobutane ring is formed by addition of an alkene to an enolized diketone, four sp^2 carbons are converted to sp^3 carbons so that as many as four new chiral centers can be produced. With enolized 1,3-diketones, two of these chiral carbons can be lost when the primary adduct undergoes retro-aldol opening to a 1,5-diketone. Because of the intermediacy of a triplet 1,4-biradical, mixtures of stereoisomers about the alkene-derived centers are normally obtained and any stereochemistry present in the original alkene is also lost. The only exception to this is when a four- or five-membered cyclic alkene is used, in which case *cis*-fusion of the alkene-derived ring to the newly formed cyclobutane ring is obtained. As would be expected, some stereocontrol can be achieved in intramolecular cycloadditions and examples of this will be seen in the next section.

54.4 Synthetic Applications

The first reported example of the photocycloaddition of an enolized 1,3-dicarbonyl compound to an alkene was for the addition of acetylacetone to cyclohexene, and it was shown that the 1,5-diketone obtained as the isolated product can be used for the synthesis of six-membered rings by aldol cyclization.[24,38] This sequence has been applied in a synthesis of the natural products valerane 32 and isovalerane 33 using the enol of formyl acetone as shown in Scheme 54.5.[39] The addition was reported to yield the *cis*-ring fusion stereochemistry preferentially; this was fortuitous since addition of cyclohexenes to enones normally gives mixtures of *cis*- and *trans*-isomers.[40] As noted above, with smaller-ring cycloalkenes, the stereochemistry of the ring junction in the primary cycloadduct must be *cis*; this has been utilized in a synthesis of the natural product hirsutene 34 as shown in Scheme 54.6.[41,42]

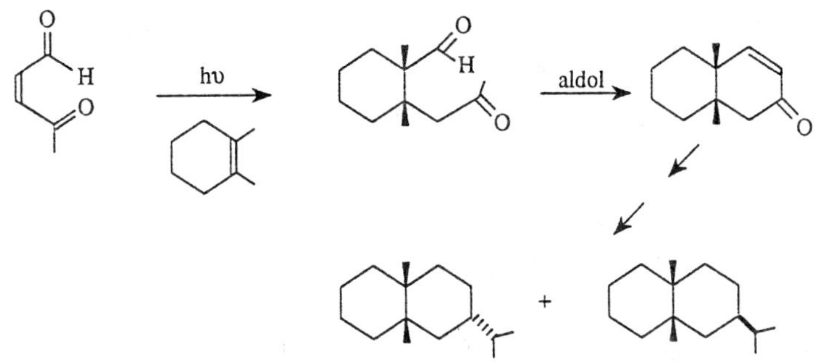

SCHEME 54.5

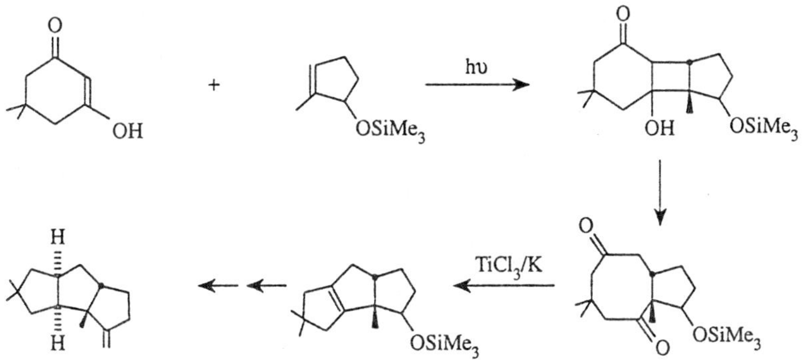

SCHEME 54.6

Nonsymmetrical 1,3-dicarbonyl compounds such as formyl acetone, and tricarbonyl compounds such as diformyl acetic acid and the ester of 2,4-dioxopentanoic acid, can enolize in more than one direction. Photocycloaddition is generally observed with only one of the possible enols (an example with formyl acetone was seen in Scheme 54.5), and the observed patterns of reaction for such compounds have been summarized.[1,3,4]

Most applications of the intermolecular photocycloaddition reaction between enolized cyclic 1,3-dicarbonyl compounds and alkenes have used enols trapped as the alkyl, silyl, or ester derivative.[2–4] This ensures a high concentration of the desired enol and overcomes the solubility problems sometimes encountered with the parent compound, as well as avoiding the potential competing thermal or photochemical reactivity of the 1,3-dicarbonyl function. It also allows selective manipulation of one of the two carbonyl functions before the other is regenerated in the retro-aldolization step. An example of this is seen in the synthesis of β-himachalene 35 in which the cycloadduct 27 is sequentially reduced with borohydride and mesylated to give 36. Treatment of 36 with base removes the acetyl group originally present to trap the enol and induces ring opening and elimination to yield 37, a precursor of 35.[28]

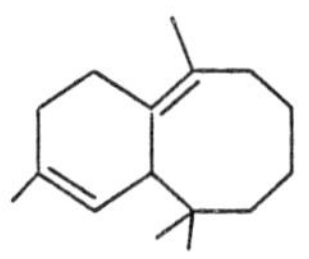

(35) 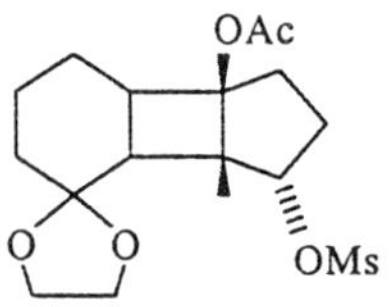   (36) 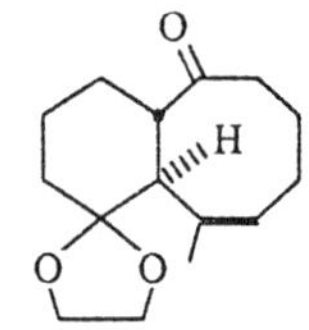   (37)

In the case of cyclopentane-1,2-diones, the amount of the enol tautomer present in equilibrium with the diketone is normally insufficient for photocycloaddition and so it must be trapped. If the enol acetate ester is used, then the cycloadduct obtained can be rearranged with base to give a hydroxy-substituted bicycloheptanone; treatment of this with lead tetraacetate yields a 4-oxocylohexane carboxylic acid ester as shown in Scheme 54.7. This sequence has been used for the synthesis of acorenone 38[43] and methyl isomarasmate 39.[44] The intermolecular photoaddition of alkenes to trapped enols of cyclohexane-1,2-diones has been reported to fail;[21] however, the reaction does proceed in an intramolecular fashion.[45,46] Thus, photocyclization of 40 and 41 yields the adducts 42 and 43, respectively. The change in product regiochemistry is consistent with the preferential formation of the "rule of five" biradical intermediate. However, with 44, both orientations of addition are observed with the non-"rule of five" product 45 preferred over the expected product 46.

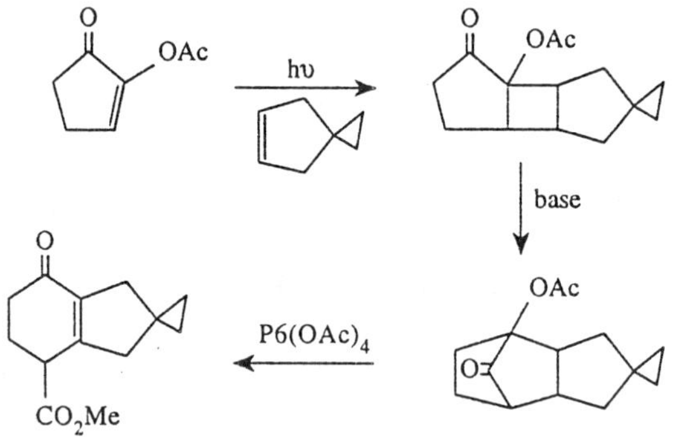

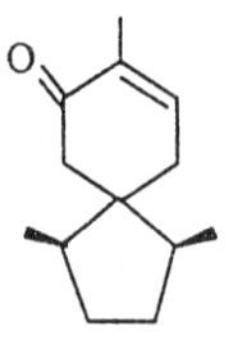

(38) 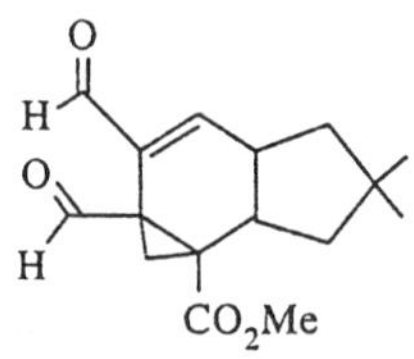   (39)

SCHEME 54.7

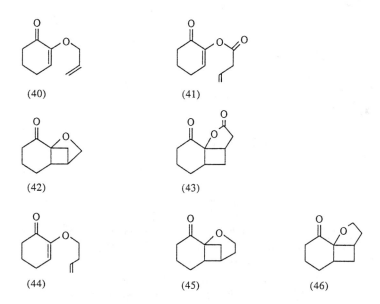

(40)   (41)

(42)   (43)

(44)   (45)   (46)

Intramolecular addition of tethered alkenes to trapped enols of 1,3-diketones has been used extensively for the preparation of natural products and their analogues.[2] Several groups have explored the use of the reaction for the synthesis of the carbon skeleton present in Taxol, **47**. One of these approaches is shown in Scheme 54.8 and uses a model in which an alkene tether traps the enol form of homocamphorquinone;[47] this sequence also demonstrates the control of stereochemistry and regiochemistry that can be attained in the intramolecular application of the de Mayo reaction.

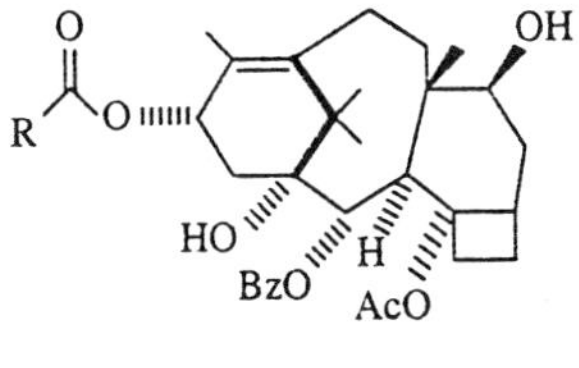

(47)

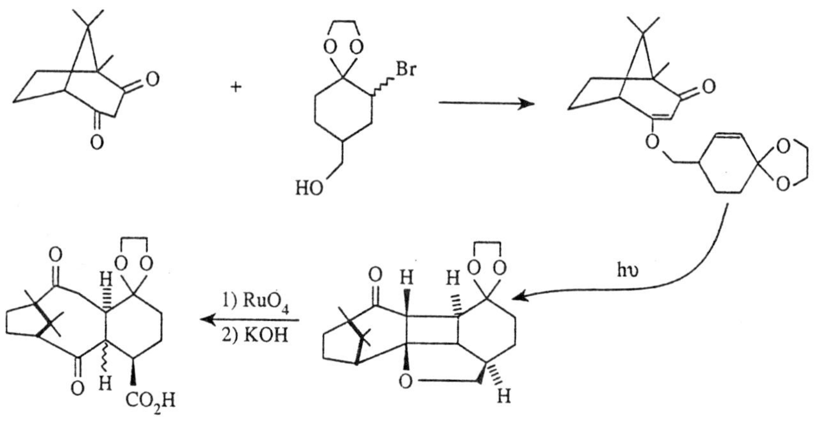

SCHEME 54.8

When the enol oxygen of the enolized 1,2- or 1,3-diketone is replaced by a nitrogen function, the photocycloaddition reaction with alkenes still proceeds. For example, ethyl vinyl ether adds to compound **48** to give cycloadduct **49**, which is not isolated since it spontaneously opens and hydrolyzes to give diketone **50**.[23] When alkenes are tethered to the nitrogen atom of **48**, cycloadducts are also formed and the influence of the length of the tether upon the orientation of addition has been examined in detail.[2,48] The reaction is also successful when the enol oxygen of enolized 1,2-diketones is replaced by a nitrogen atom bearing a tethered alkene. For example, compound **51** yields the cycloadduct **52**; the orientation of addition is that expected for a "rule of five" biradical intermediate.[49] Amino analogues of acyclic enolized 1,3-diketones will react with alkenes in an intermolecular fashion[50,51] and also if the alkene is tethered and available for intramolecular addition; thus, the product of photocycloaddition of compound **53** is adduct **54**, which spontaneously opens to **55**, a synthetic precursor of the alkaloid mesembrine.[52]

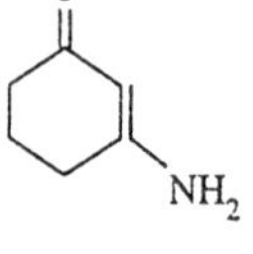

(48)

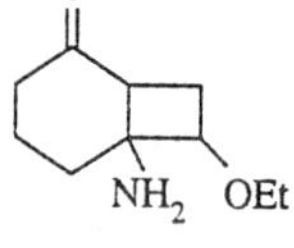

(49)

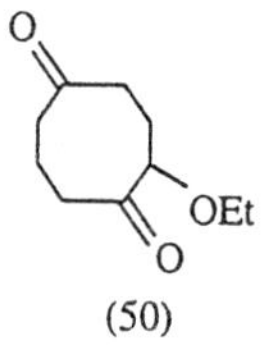

(50)

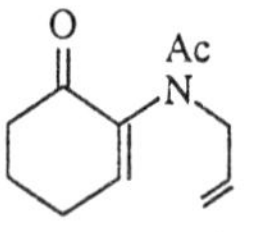

(51)

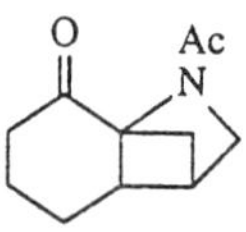

(52)

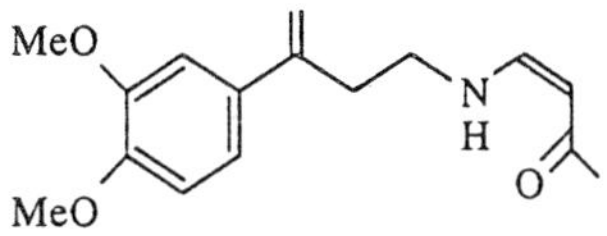

(53)

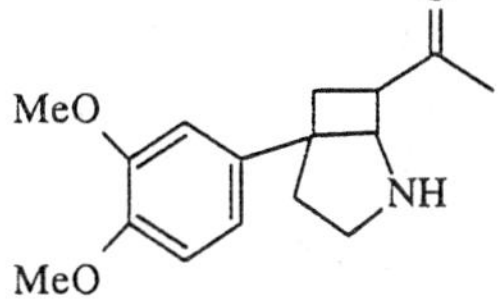

(54)

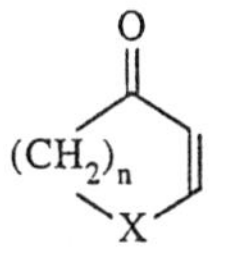

(7) X = 0
(56) X = NR

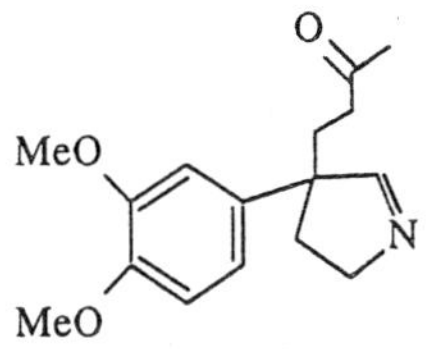

(55)

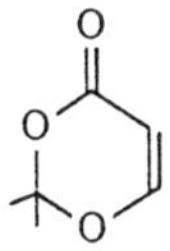

(57)

The vinyligous lactone 7 described in the introduction to this chapter and the nitrogen analogue 56 (a vinylogous lactam) may be regarded as derivatives of enolized 1,3-diketones, and both undergo photochemical cycloaddition with alkenes.[3,53-55]

The photocycloaddition reaction with alkenes is reported to fail for β-ketoesters[56,57] which instead give products of carbonyl group photochemistry (e.g., Paternò-Büchi addition products and hydrogen abstraction products). Cyclobutane formation can, however, be achieved using dioxolone derivatives such as 57.[58]

eferences

1. Weedon, A. C., Photochemical reactions involving enols, in *The Chemistry of Enols*, Rappoport, Z., Ed., Wiley, Chichester, 1990, chap. 9.
2. Crimmins, M. T., Synthetic applications of intramolecular enone-olefin photocycloadditions, *Chem. Rev.*, 88, 1453, 1988.
3. Weedon, A. C., Enone photochemical cycloaddition in organic synthesis, in *Synthetic Organic Photochemistry*, Horspool, W. M., Ed., Plenum, New York, 1984, chap. 2.
4. Baldwin, S. W., Synthetic aspects of 2 + 2 cycloadditions of α,β-unsaturated carbonyl compounds, *Org. Photochem.*, 5, 123, 1981.
5. Hastings, D. J. and Weedon, A. C., Origin of the regioselectivity in the photochemical cycloaddition reactions of cyclic enones with alkenes: Chemical trapping evidence for the structures, mechanism of formation, and fates of 1,4-biradical intermediates, *J. Am. Chem. Soc.*, 113, 8525, 1991.
6. Andrew, D., Hastings, D. J., Oldroyd, D. L., Rudolph, A., Weedon, A. C., Wong, D. F., and Zhang, B., Triplet 1,4-biradical intermediates in the photocycloaddition reactions of enones and N-acylindoles with alkenes, *Pure Appl. Chem.*, 64, 1327, 1992.
7. Rudolph, A. and Weedon, A. C., Radical clocks as probes of 1,4-biradical intermediates in the photochemical cycloaddition reactions of 2-cyclopentenone with alkenes, *Can. J. Chem.*, 68, 1590, 1990.

8. Nozaki, H., Murita, M., Mori, T., and Noyori, R., Photochemical behaviour of enolic β-diketones towards cycloolefins, *Tetrahedron*, 24, 1821, 1968.

9. Pauw, J. E. and Weedon, A. C., A synthesis of the tricyclo[6.3.0.0^{2,6}]undecane system, *Tetrahedron Lett.*, 23, 5485, 1982.

10. Berry, N. M., Darey, M. C. P., and Harwood, L. M., Photochemical 2-alkylation of cyclohexane-1,3-diones with enol ethers, *Tetrahedron Lett.*, 27, 2319, 1986.

11. Casals, P.-F., Ferard, J., and Ropert, R., Photoaddition de dicetones-1,3 aromatiques sur divers carbures styreniques: Orientation et stereospecificite de l'addition, *Tetrahedron Lett.*, 3077, 1976.

12. Takeshita, H., Mori, A., and Nakamura, S., Synthetic photochemistry. XXVIII. A photochemical C$_5$-homologation of 4-isopropenyltoluene with methyl 2,4-dioxopentanoate to isolaurene and a formal synthesis of cuparene, *Bull. Chem. Soc. Jpn.*, 57, 3152, 1984.

13. Kato, N. and Takeshita, H., Synthetic photochemistry. XXIX. A convenient preparation of 1,2,3-substituted cyclopentenes from the photoadducts of methyl 2,4-dioxopentanoate-isoprene, *Bull. Chem. Soc. Jpn.*, 58, 1574, 1985.

14. Takeshita, H., Komiyama, K., and Okaishi, K., Synthetic photochemistry. XXXII. Structure of by-products formed in the photocycloaddition of myrcene to methyl 2,4-dioxoalkanoates, *Bull. Chem. Soc. Jpn.*, 58, 2725, 1985.

15. Kato, N., Nakanishi, K., and Takeshita, H., Synthetic photochemistry. XXXIV. Synthetic strategy of 5–8–5-membered tricyclo higher terpenoids based on the condensation of two optically-active iridoids, C$_{10}$-synthons obtained from photocycloadduct of methyl 2,4-dioxopentanoate-isoprene, and its application to a synthesis of the basic carbon skeleton of fusicoccane, *Bull. Chem. Soc. Jpn.*, 59, 1109, 1986.

16. Hatsui, T., Nojima, C., and Takeshita, H., Synthetic photochemistry. LIV. The photoaddition reaction of methyl 2,4-dioxopentanoate with 2,5-dimethyl-2,4-hexadiene, sterically crowded conjugated diene, *Bull. Chem. Soc. Jpn.*, 63, 1611, 1990.

17. Pirrung, M. C. and Webster, N. J. G., Mechanism of intramolecular photocycloadditions of cyclooctenones, *J. Org. Chem.*, 52, 3603, 1987.

18. Oppolzer, W. and Godel, T. A new and efficient total synthesis of (±)-longifolene, *J. Am. Chem. Soc.*, 100, 2583, 1978.

19. Oppolzer, W. and Godel, T., Synthesis of (±)- and enantiomerically pure (+)-longifolene and of (±)- and enantiomerically pure (+)-sativene by an intramolecular de Mayo reaction, *Helv. Chim. Acta*, 67, 1154, 1984.

20. Cantrell, T. S., Haller, W. S., and Williams, J. C., Photocycloaddition reactions of some 3-substituted cyclohexenones, *J. Org. Chem.*, 34, 509, 1969.

21. Matsumoto, T., Shirahama, H., Ichihara, A., Kagawa, S., and Matsumoto, S., Photochemical cycloaddition of 2-acetoxy-2-en-1-ones to 1,1-diethoxyethylene and formation of bicyclo(2,2,1)-heptane from the adduct, *Tetrahedron Lett.*, 4103, 1969.

22. Cantrell, T. S., Photochemical transformations of some β-substituted enones, *Tetrahedron*, 27, 1227, 1971.

23. Tsuda, Y., Kiuchi, F., Umeda, I., Iwasa, E., and Sakai, Y., Photocycloaddition of 3-methoxycyclohexenone and 3-aminocyclohexenone to ethoxyethylene: Stereochemistry of the cycloadducts, *Chem. Pharm. Bull.*, 34, 3614, 1986.

24. de Mayo, P. and Takeshita, H., The formation of heptanediones from acetylacetone and alkenes, *Can. J. Chem.*, 41, 440, 1963.

25. Cadogan, J. I. G., Hey, D. H., and Sharp, J. T., The reaction of acetylacetone with 1-octene, *Proc. Chem. Soc.*, 142, 1964.

26. Challand, B. D. and de Mayo, P., Solvent effects in the stereochemistry of photocycloaddition, *J. Chem. Soc., Chem. Commun.*, 982, 1968.

27. Lange, G. L., Campbell, H. M., and Neidert, E., Preparation of substituted spiro (4,5)decan-7-ones. An approach to the synthesis of acorenones, *J. Org. Chem.*, 38, 2117, 1973.

28. Challand, B. D., Hikino, H., Kornis, G., Lange, G. L., and de Mayo, P., Photochemical cycloaddition. Some applications of the use of enolized β-diketones, *J. Org. Chem.*, 34, 794, 1969.

29. Benchikh-le-Hocine, M., Khac, D. D., Cervantes, H., and Fetizon, M., Model studies in Taxane diterpene synthesis, *Nouv. J. Chim.,* 10, 715, 1986.

30. Kaczmarck, and Blechert, S., de Mayo Reaktionen mit Allen. Ein kurzer Weg zum Bicyclo[5.3.1]undecansystem der Taxane, *Tetrahedron Lett.,* 27, 2845, 1986.

31. Tada, M. and Nieda, Y., Ring size effect on [2+2] photochemical cycloaddition of enones with cyclic olefins, *Bull. Chem. Soc. Jpn.,* 61, 1416, 1988.

32. Srinivasan, R. and Carlough, K. H., Mercury (^{3}P$_1$) photosensitized internal cycloaddition reactions in 1,4-, 1,5-, and 1,6-dienes, *J. Am. Chem. Soc.,* 89, 4932, 1967.

33. Liu, R. S. and Hammond, G. S., Photosensitized internal addition of dienes to olefins, *J. Am. Chem. Soc.,* 89, 4936, 1967.

34. Beckwith, A. L. J., Regio-selectivity and stereo-selectivity in radical reactions, *Tetrahedron,* 37, 3073, 1981.

35. Matlin, A. R., Turk, B. E., McGarvey, D. J., and Manevich, A. A., Intramolecular photocycloaddition reactions of 3-(2-propenoxy)cyclopent-2-en-1-ones and 3-(2-propenoxy)cyclohex-2-en-1-ones, *J. Org. Chem.,* 57, 4632, 1992.

36. Tamura, Y., Ishibashi, H., Hirai, M., Kita, Y., and Ikeda, M., Photochemical syntheses of 2-aza- and 2-oxabicyclo[2.1.1]hexane ring systems, *J. Org. Chem.,* 40, 2702, 1975.

37. Tamura, Y., Ishibashi, H., Kita, Y., and Ikeda, M., Intramolecular photocycloaddition of 3-(but-3-enyl)oxy- and 3-(pent-4-enyl)oxy-cyclohex-2-enones: Stereospecific formation of 2-oxabicyclo-[3,2,0]heptane and -[4,2,0]octane systems, *J. Chem. Soc., Chem. Commun.,* 101, 1973.

38. de Mayo, P., Takeshita, H., and Sattar, A. B. M. A., The photochemical synthesis of 1,5-diketones and their cyclisation: A new annulation process, *Proc. Chem. Soc.,* 119, 1962.

39. Baldwin, S. W. and Gawley, R. E., Photoannelations with α-formyl ketones. A facile synthesis of the *cis*-9,10-dimethyldecal ring system of the valerane sesquiterpenes, *Tetrahedron Lett.,* 3969, 1975.

40. Agami, C., Levisalles, J., and Rizk, T., Stereochemical aspects of the photochemical reaction of cyclohexene with acetylacetone, *J. Chem. Res. Synop.,* 166, 1988.

41. Disanayaka, B. W. and Weedon, A. C., A short synthesis of hirsutene using the de Mayo reaction, *J. Chem. Soc., Chem. Commun.,* 1282, 1985.

42. Disanayaka, B. W. and Weedon, A. C., Application of the de Mayo reaction to the preparation of tricyclo[6.3.0.0^{2,6}]undecanes. A photochemical synthesis of (±)-hirsutene, *J. Org. Chem.,* 52, 2905, 1987.

43. Lange, G. L., Neidert, E. E., Orram, W. J., and Wallace, D. J., *Can. J. Chem.,* 56, 1628, 1978.

44. Helmlinger, D., de Mayo, P., Nye, M., Westfelt, L., and Yeats, R. B., The synthesis of methyl isomarasmate, *Tetrahedron Lett.,* 349, 1970.

45. Ikeda, M., Takahashi, M., Uchino, T., Ohno, K., and Tamura, Y., Intramolecular [2+2] photocycloaddition of 2-(alkenyloxy)cyclohex-2-enones, *J. Org. Chem.,* 48, 4241, 1983.

46. Ikeda, M., Takahashi, M., Uchino, T., and Tamura, Y., Intramolecular photo[2+2] cycloaddition of 2-alkenoyloxycyclohex-2-enones, *Chem. Pharm. Bull.,* 32, 538, 1984.

47. Berkowitz, W. F., Amarasekara, A. S., and Perumattam, J. J., A photochemical approach to the Taxanes, *J. Org. Chem.,* 52, 1119, 1987.

48. Vogler, B., Bayer, R., Meller, M., Kraus, W., and Schell, F. M., Photo-aza-Claisen rearrangements of cyclic enaminones, *J. Org. Chem.,* 54, 4165, 1989.

49. Ikeda, M., Uchino, T., Takahashi, M., Ishibashi, H., Tamura, M., and Kido, M., Intramolecular [2+2] photocycloaddition of 2-[N-acyl-N-(2-propenyl)amino]cyclohex-2-enones: Synthesis of 2-azabicyclo[2.1.1]hexanes, *Chem. Pharm. Bull.,* 33, 3279, 1985.

50. Tietze, L. F., Bergmann, A., and Bruggemann, K., Regioselective photochemical cycloaddition of enamine-carbaldehydes and alkenes: Synthesis of 1,4-dihydropyridines and 2-hydroxy-1,2,3,4-tetrahydropyridines, *Tetrahedron Lett.,* 24, 3579, 1983.

51. Tietze, L. F. and Bergmann, A., Synthesis of 1,4-dihydropyridine nucleosides by photochemical cycloaddition, *Angew. Chem., Int. Ed.,* 24, 127, 1985.

52. Winkler, J. D., Muller, C. L., and Scott, R. D., A new method for the formation of nitrogen-containing ring systems via the intramolecular photocycloaddition of vinylogous amides. A synthesis of mesembrine, *J. Am. Chem. Soc.*, 110, 4831, 1988.

53. Matsui, T., Kawano, Y., and Nakayama, M., Preparation and structure of acetylene adduct of 4,6-di-*O*-acetyl-1,5-anhydro-2-deoxy-*d*-erythro-hex-1-en-3-ulose, *Chem. Express*, 5, 697, 1990.

54. Gebel, R.-C. and Margaretha, P., Photochemistry of 2-methyl-2-trifluoromethyl- and 2,2-*bis*(trifluoromethyl)-3(2H)-furanone, *Chem. Ber.*, 123, 855, 1990.

55. Guerry, P., Blanco, P., Brodbeck, H., Pasteris, O., and Neier, R., 1-Methoxycarbonyl-substituiertes 2,3-dihydropyridin-4(1H)-on (= methyl-1,2,3,4-tetrahydro-4-oxopyridin-1-carboxylat) als chromophor für die photochemische [2+2]-cycloaddition, *Helv. Chim. Acta*, 74, 163, 1991.

56. Tada, M., Kokubo, T., and Sato, T., Photoreaction of acetoacetate, *Bull. Chem. Soc. Jpn.*, 43, 2162, 1970.

57. Tada, M., Harada, H., and Miura, K., The photocycloaddition of 3-cyano-2,4-pentandione with cyclohexene and its quenching by enol ether, *Bull. Chem. Soc. Jpn.*, 51, 839, 1978.

58. Baldwin, S. W., Martin, G. F., and Nunn, D. S., Total synthesis of (+)-elemol by photoannelation, *J. Org. Chem.*, 50, 5720, 1985.

Photorearrangement Reactions of Cross-Conjugated Cyclohexadienones

Arthur G. Schultz
Rensselaer Polytechnic Institute

55.1 Definition of the Reaction

Cross-conjugated dienones 1 undergo two types of primary photorearrangement. The type A photorearrangement[1] produces a bicyclo[3.1.0]hex-3-en-2-one 3 via oxyallyl Zwitterion 2. In polar protic solvents, it is sometimes possible to trap Zwitterion 2 to give a bicyclo[3.1.0]hexanone 4. More often, rearrangement with trapping occurs to give a 4-substituted cyclopent-2-en-1-one 5.

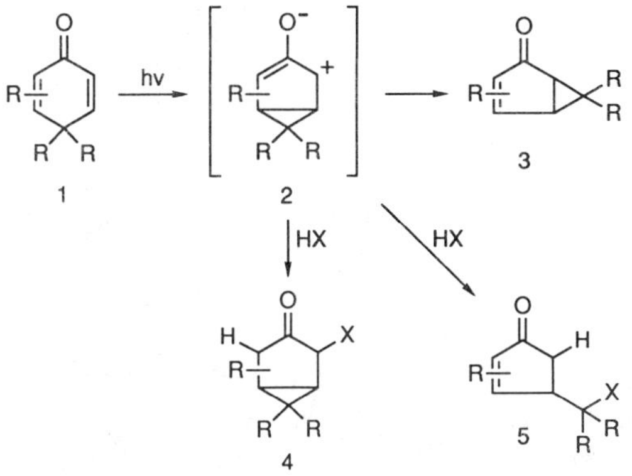

The second type of rearrangement begins by homolytic cleavage of an exocyclic bond at C-4 to give the radical pair 6 (or a biradical from spirodienones). Hydrogen-atom transfer pathways from 6 provide phenols or quinone methides.

0-8493-8634-9/95/$0.00+$.50
© 1995 by CRC Press, Inc.

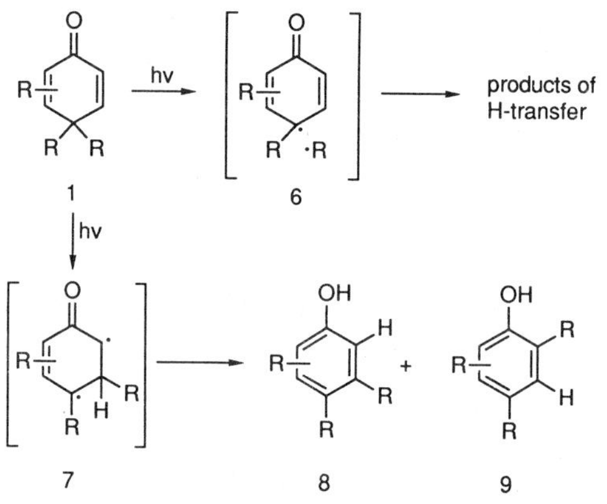

Alternatively, a di-π-methane type of rearrangement[2] can occur when one of the C-4 substituents is an aromatic ring or a vinyl group. This process involves 1,2-migration of the aryl or vinyl substituent to give the hypothetical biradical 7, from which competing 1,2-migrations of the C-3 hydrogen atom and the C-3-substituent R produce phenols 8 and 9.

Products of secondary photorearrangement are often encountered. The type B photo-rearrangement of bicyclo[3.1.0]hexenone 3 occurs by way of oxyallyl Zwitterion 10; 1,2-rearrangements from 10 give phenols 11 and 12. Bicyclo[3.1.0] hexenones can also undergo photorearrangement to 2,4-cyclohexadien-1-ones 13. The linearly conjugated dienones undergo reversible photochemical ring opening to diene ketenes 14, which give diene carboxylic acid derivatives in the presence of appropriate nucleophiles, HX.

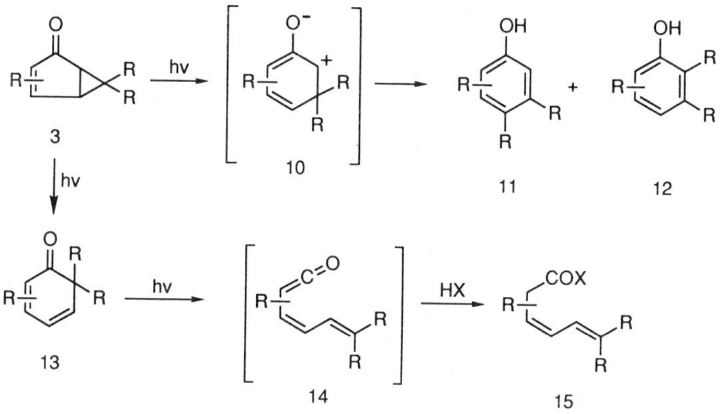

55.2 Mechanism

The complex photochemical and thermal reaction pathways characteristic of 2,5-cyclohexadien-1-ones have been summarized in several reviews.[4–10] The generally accepted mechanism for the type A photorearrangement[1] involves photoexcitation of 1 to the n,π* singlet state, followed by intersystem crossing to the triplet n,π* state 16. Bridging from C-3 to C-5 in 16 gives 17 initially in the n,π* excited state; electron demotion provides the ground-state oxyallyl Zwitterion 2, from which migration of the bridging carbon atom gives 3. It has been demonstrated that this 1,4-sigmatropic rearrangement occurs with inversion of configuration of the migrating group.[11]

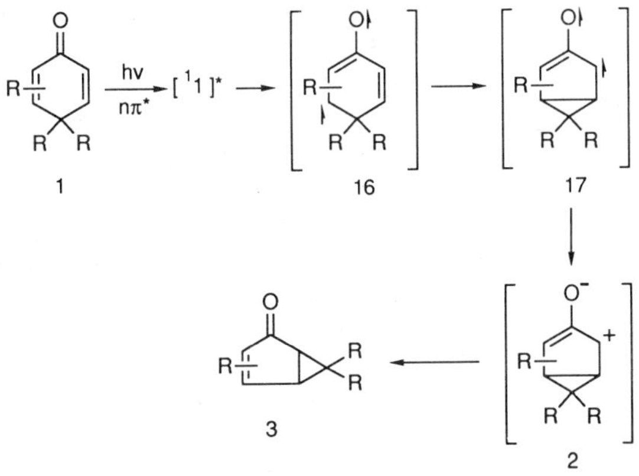

Gas-phase and solution photochemical studies with 4,4-dimethyl-2,5-cyclohexadien-1-one sug-gest that biradicals are involved in gas-phase photolyses, while Zwitterion 2 is the reactive interme-diate in aqueous dioxane.[12] On the other hand, photolysis of spirodienone 18 in ether solution was found to give rise to phenolic products characteristic of biradical 19.[13]

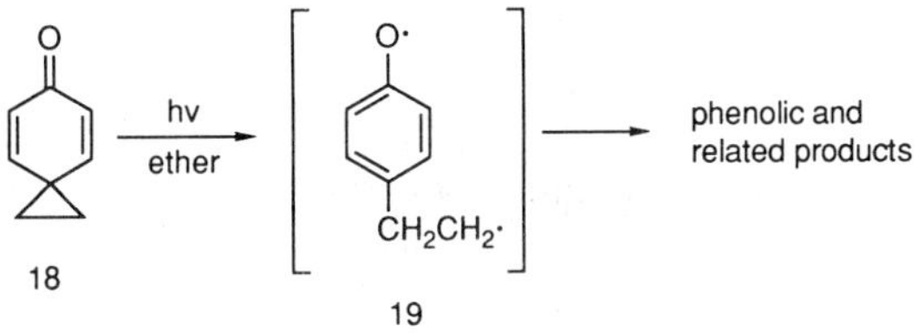

The possibility that ring strain in 18 makes this a special case was considered; studies with the 4-trichloromethyl derivative 20 provide compelling evidence that both biradical and Zwitterionic species are involved in the solution photochemistry of this 2,5-cyclohexadienone.[14] A simplified schematic for the photorearrangement chemistry of 20 is shown below. It was found that the radical-like and Zwitterion-like reactivities are sensitized by benzophenone (30-fold excess; 95% incident light absorbed by benzophenone) and both reactivities are quenched by piperylene.[14b]

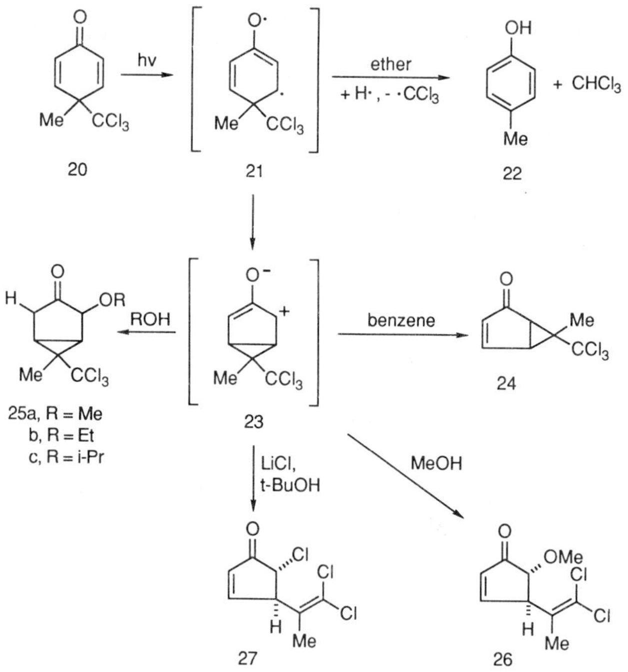

The formation of nucleophilic trapping products **25**, **26**, and **27** at the expense of bicyclo[3.1.0]hexenone **24** is suggestive of a common intermediate along this branch of the scheme; e.g., oxyallyl Zwitterion **23**.[14e] In subsequent studies it was found that photolysis of **20** in the presence of cyclopentadiene gave two 1:1 cycloadducts of zwitterion **23**.[14f] Additional evidence for the intermediacy of Zwitterions in dienone photochemistry comes from ground-state studies, wherein it has been shown that examples of **2** generated by Favorskii rearrangement isomerize to bicyclo[3.1.0]hexenones.[11,15]

The presence of LiCl (*cf.*, **20** → **27**) has no effect on the formation of bicyclohexenones from dienones that lack the electron-withdrawing 4-trichloromethyl substituent.[14e] The rather efficient generation of trapping products from photolyses of **20** may be a result of an unusually long lifetime of Zwitterion **23**, although ring-fused dienone **28**, that is constrained with respect to rearrangement to a bicyclohexenone but should have no difficulty in forming Zwitterion **29**, also failed to give adducts with appropriate trapping reagents.[16]

The triplet lifetime of **28** is much greater than excited-state lifetimes of type A reactive dienones, suggesting that rearrangement and decay of the excited state back to starting material are coupled processes.

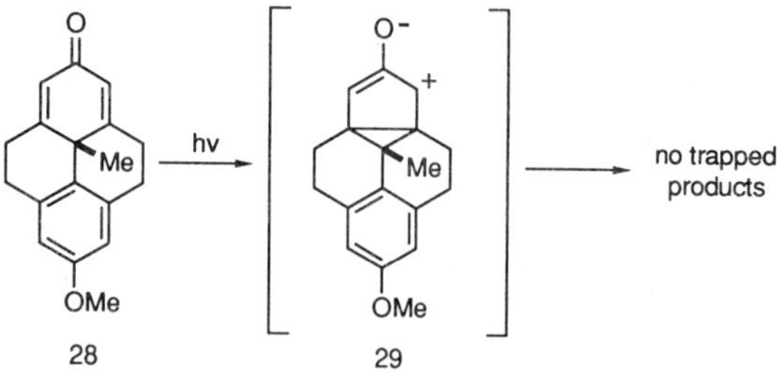

The absence of trapping products from photoreactions of **28** coupled with the absence of an allowed thermal pathway for the return of Zwitterion to dienone has forced the conclusion that **29**

"as a discrete intermediate" is not involved in the photochemistry of 28.[16] It has been shown that enantiomerically pure 4-carbomethoxy-3-methoxy-4-methyl-2,5-cyclohexadien-1-one (30) undergoes photoracemization along with type A photorearrangement to bicyclo[3.1.0]hexenones 31 and 32.[17] This result requires that a bond to C-4 in 30 is cleaved by some process that does not always result in formation of a bicyclohexenone. It is unlikely that an exocyclic bond to C-4 is cleaved because 4-substituted phenols are absent from the product mixtures (*cf.*, 18 → 19); additional mechanistic speculation is available.[17]

The excited state responsible for type A photoreactivity can also be intercepted by alkenes to give [2+2]-cycloaddition products. Thus, irradiation of 30 in THF:cyclopentene (1:1) at 25 °C produces four [2+2]-cycloadducts along with products of type A photorearrangement.[18] Type A photoreactivity of 30 is almost negligible at −80 °C in the presence of cyclopentene; the four cycloadducts can be obtained in ~96% yield under these conditions. A discussion of the highly regio- and stereoselective intramolecular [2+2]-photocycloadditions of 4-(3′-alkenyl)-2,5-cyclohexadien-1-ones[19,20] is presented in the section of this handbook concerned with trapping reactions of cyclohexadienones.

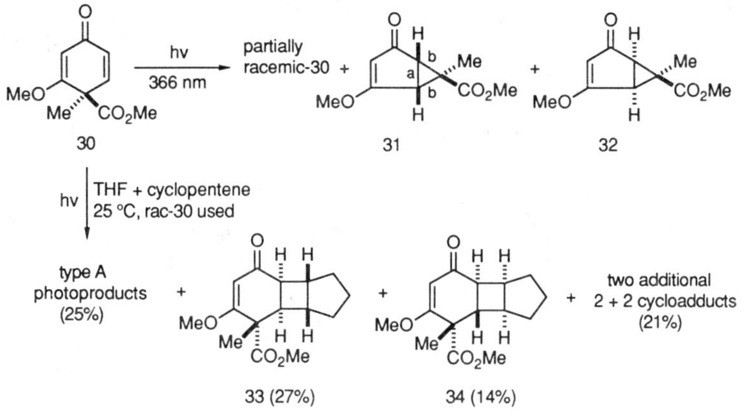

The regioselectivity of the type A photorearrangement is highly dependent on the skeletal structure within which the dienone resides and on the type and position of functionality on the dienone ring. The situation is complicated further by the potential for secondary photoreactivity of type A reaction products, the previously indicated effects of solvents and other additives on product composition, and probably other factors that remain to be elucidated. Despite this high degree of complexity, much progress has been made in defining the opportunities for regiocontrol and it is now possible to predict the regioselectivity of the photorearrangement of a new 2,5-cyclohexadienone with a reasonable degree of confidence.

The photorearrangement of 3-methoxy-4,4-diphenyl-2,5-cyclohexadien-1-one (35) has proven to be of major importance in defining a relationship between the timing of the 1,4-rearrangement step in bicyclo[3.1.0]hexenone formation and the regioselectivity of the rearrangement process. Irradiation of 35 in benzene solution produced bicyclohexenones 38 and 39 in a ratio of 1.4:1.[21] Zwitterion 37, prepared by the Favorskii rearrangement, was found to rearrange exclusively to 38. To explain the concurrent formation of 38 and 39 from the photolysis of 35, it was suggested that rearrangement to 39 occurs from an electronically excited bridged species 36, the photochemical precursor of Zwitterion 37. An approximately 20% yield of 3-methoxy-4,5-diphenylphenol also was obtained from irradiation of 35 (*cf.*, 1 → 8).

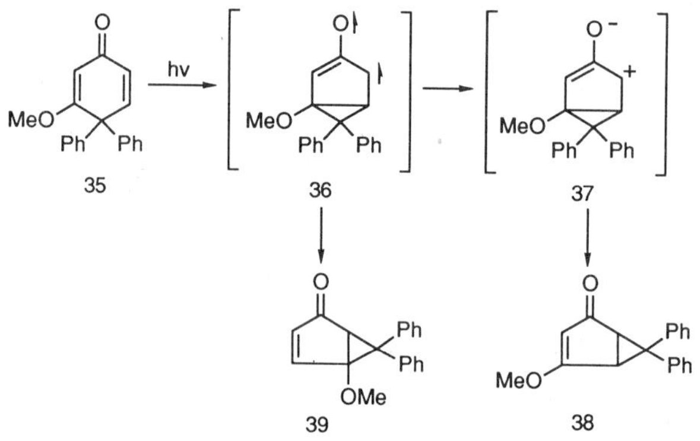

The quantum yields for photoproduct formation in benzene are $\phi = 0.27$ for **38**, 0.19 for **39**, and 0.12 for the formation of 3-methoxy-4,5-diphenylphenol.[21] Acetophenone sensitization gave the same product distribution with the same quantum efficiencies. In methanol, the type A efficiencies increased without an appreciable change in the regioselectivity, but the efficiency of formation of 3-methoxy-4,5-diphenylphenol decreased.

The photorearrangement of 3-cyano-2,5-cyclohexadienone **40** and the rearrangement of Zwitterion **41**, prepared by ground-state chemistry, are completely regioselective.[22] It is unfortunate for preparative purposes that bicyclohexenone **42** is consumed almost as rapidly as it is generated photochemically to give the secondary photoproducts **43** and **44**. The situation is somewhat improved by the utilization of *m*-methoxyacetophenone as sensitizer (large quantities because of a relatively small extinction coefficient at ~317 nm), but a combination of fractional crystallization and chromatography are required to obtain **42** by this procedure.

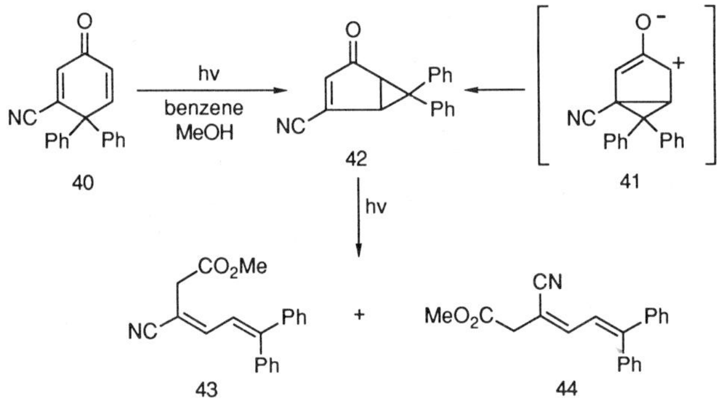

4-Methoxybicyclo[3.1.0]hexenones such as **31**, **32**, and **38** do not readily rearrange to ring-opened products or phenols under the photolysis conditions required to prepare these materials from their respective cyclohexadienones. With light of wavelength ~366 nm, **31** converts to **32** (the enantiomer of that shown) by reversible cleavage of a "b" cyclopropane bond, but at shorter wavelength (irradiation through Pyrex) the compounds **31**, **32**, and related bicyclohexenones photorearrange to phenolic products via cleavage of the cyclopropane bond "a" (*cf.*, **3** → **10** → **11** + **12**).[17]

Inter- and intramolecular trapping studies with furan have provided indirect evidence for the intermediacy of regioisomeric bicyclohexenones related to **39** in the photorearrangements of 4-carbomethoxy- and 4-cyano-3-methoxy-2,5-cyclohexadien-1-ones.[23] Thus, the outstanding regioselectivity observed for the type A photorearrangement of **30** and other 4-alkyl-substituted derivatives is the result of differential photoreactivity of the derived bicyclohexenones; the 4-methoxy regioisomers are "photostable" at 366 nm, while the 5-methoxy isomers photoconvert to the 4-methoxybicyclohexenones. The reader should turn to the section of this handbook describing the trapping reactions of cyclohexadienones for more detail.

The regioselectivity of type A photorearrangement of C-2-substituted 2,5-cyclohexadien-1-ones is governed by electronic rather than steric effects to give C1- rather than C3-substituted bicyclo[3.1.0]hex-3-en-2-ones. Brief irradiation of the 2-carbomethoxydienone **45** at 254 nm gave **46** (31%), while extended irradiation gave phenol **47**.[24] The exclusive formation of **46** is thought to be a result of stabilization by the carbomethoxy group of the transition state for the 1,4-sigmatropic rearrangement.

45 46 47 (43%, Pyrex)

48a, X = CO_2Me
b, X = CN
c, X = OMe
d, X = Me

49

50a, X = CO_2Me (88%)
b, X = CN (77%)
c, X = OMe (low yield)
d, X = Me (41%)

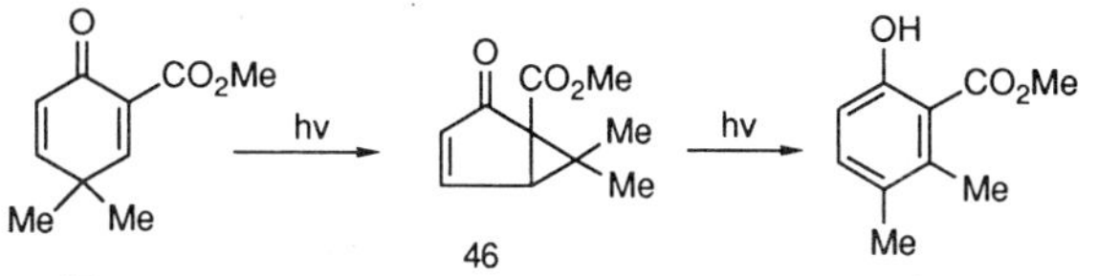

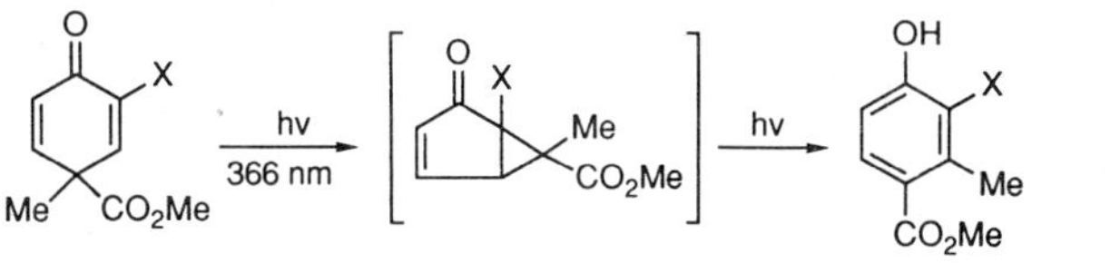

Another study showed that both electron-withdrawing and electron-donating groups at C-2 provide a high level of regiocontrol; e.g., **48** → **49** → **50**.[25] Type B photoreactivity of the derived bicyclohexenones currently imposes a practical limitation to the synthetic utility of these findings, although the attendant methodology does offer a potentially useful preparation of certain highly substituted phenols. Examples of regioselectivity resulting from other less clearly defined effects, are available in the literature.[26–29]

Pioneering studies of the stereoselectivity of the type A photorearrangement focused on steric effects of substituents at C-3 and C-5 of the dienone.[30] It was found that as the substituents at C-3 and C-5 increase in size, there is greater stereoselectivity in bicyclohexenone formation; e.g., **51** → **52** + **53**. The trend in selectivities is consistent with a minimization of steric interactions between the larger *n*-propyl substituent at C-4 and R_1 and R_2 in the transition state leading from a planar dienone to the favored *endo*-*n*-propyl-substituted Zwitterion.

51a, $R_1 = R_2 = H$
b, $R_1 = H$; $R_2 = Me$
c, $R_1 = R_2 = Me$

52a (56%)
b (32%)
c (9%)

53a (44%)
b (68%)
c (91%)

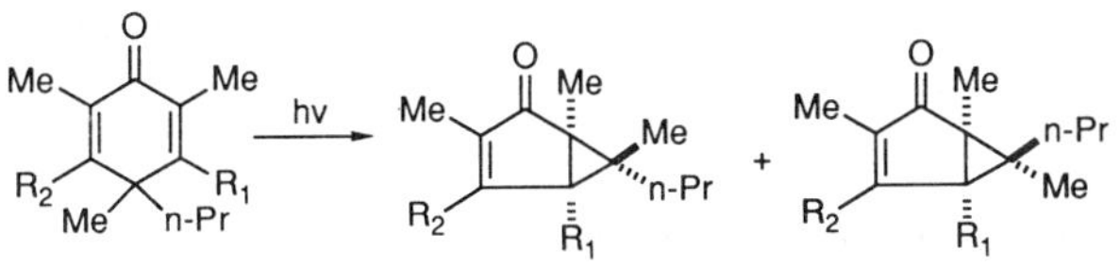

The selective formation of the *exo*-6-trichloromethylbicyclo[3.1.0]hexenone (*exo:endo*, 15:1) from photolysis of 3,4-dimethyl-4-trichloromethyl-2,5-cyclohexadien-1-one, which is contrary to expectations based on the steric rationale, has been explained by an argument involving repulsion of the C=O and C-CCl_3 dipoles in the transition state leading to the *endo*-trichloromethyl-substituted Zwitterion.[31] Steric effects of C4-substitution on the photochemistry of Δ^1-4-alkyltestosterones has been examined, but no orderly dependence of reaction kinetics on the size of the alkyl group was observed.[32]

Both the steric and dipole arguments offer insufficient explanations for the completely diastereoselective photorearrangement of dienone 54 to bicyclohexenone 55.[33] Conformational analysis supported by X-ray crystallographic studies of 54 and 55 suggested that the type A photorearrangement of 54 occurs in accord with the principle of least motion in the transition state leading from 54 to the favored Zwitterion. Bicyclohexenone 55 is photostable at 366 nm (*cf.*, 31 and 32), but rearranges to its diastereomer and type B photoproducts on photolysis through Pyrex.

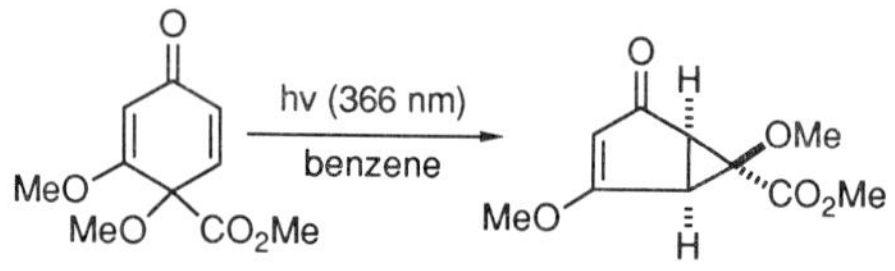

(Str.M)

It was noted earlier that 3-methoxy-4,5-diphenylphenol is formed as a minor product from photolysis of 3-methoxy-4,4-diphenyl-2,5-cyclohexadien-1-one (35). The aryl migration is unusual in that the type A rearrangement is considered to be much more efficient in terms of quantum yields and rates than migration of an aryl group from C-4 to the β-carbon atom of an enone triplet state.[21] It is believed that the 3-methoxy-substituent lowers the energy of the π,π* triplet state of the dienone to such an extent that 4-aryl migrations (and [2+2]-cycloadditions in the presence of alkenes; e.g., 30 → 33, etc.) are able to compete with the type A rearrangement characteristic of the n,π* triplet state. Exclusive 1,2-aryl migration has been observed for certain 4,4-diphenyl-2,5-cyclohexadien-1-ones,[34] but these rearrangements were proposed to occur from excited states of the migrating aryl groups; e.g., 56 → 57.

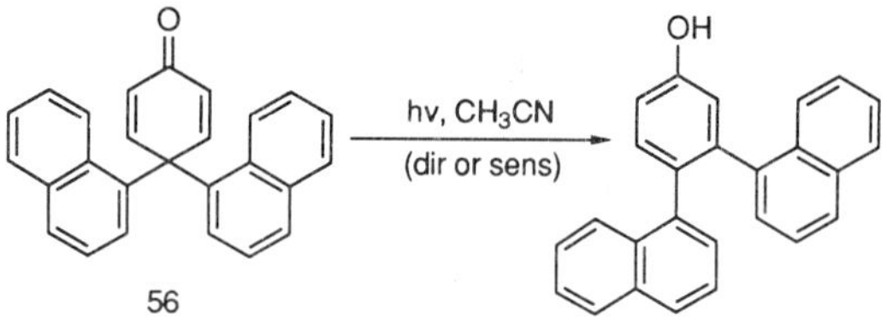

(Str.N)

Dienone-phenol rearrangements also have been observed with the 4-vinyl dienones 58 and 61.[35] These rearrangements must occur by way of the excited state of the dienone and were proposed to follow the di-π-methane mechanism; e.g., 1 → 7 → 8 + 9. It is of certain, but as yet undefined, mechanistic interest that type A photochemistry is observed when photolyses of 58 are carried out in the presence of trifluoroacetic acid.

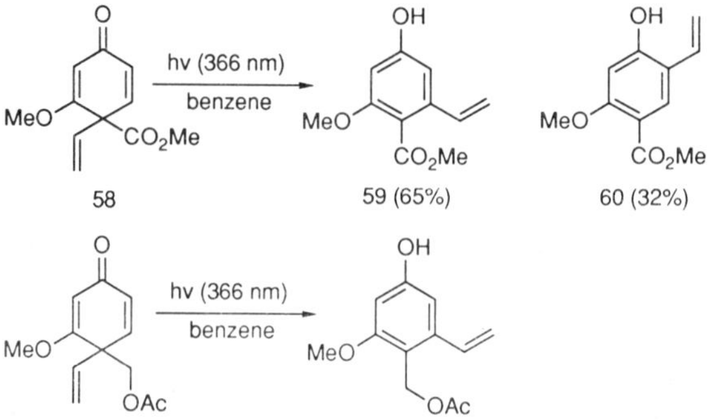

(Str.O)

.3 Synthetic Applications

The type A photorearrangement manifold has provided the greatest opportunities for utilization of cross-conjugated cyclohexadienones in organic photochemical synthesis. Primary photoproduct distributions depend on the type and position of substituents on the dienone and on the nature of the photolysis medium. The importance of secondary photoreactivity is dependent on substituent effects as well as the wavelength and duration of the photolysis. These variables are considered in detail sufficient for most synthetic purposes in the preceding mechanistic discussion. A representative selection of the most chemically efficient photoconversions of 2,5-cyclohexadien-1-ones is presented in this section. The reader should consult review articles for a more comprehensive summary of synthetic applications.[6,7,36]

Type A Photorearrangement of Bicyclic Dienones Related to α-Santonin

The prototype considered here is the photoconversion of α-santonin (63) to the annelated bicyclohexenone lumisantonin (64). Although 64 is photoreactive and will convert to the linearly conjugated dienone 65, lumisantonin can be obtained in 42% yield from photolysis of 63 in dioxane if the irradiation is terminated at the appropriate time.[26]

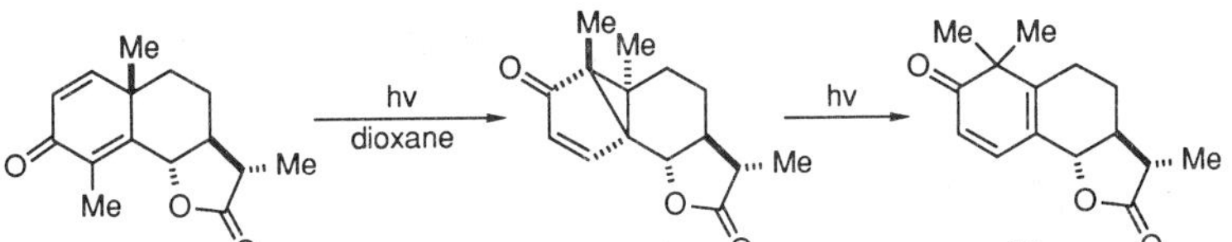

63   64 (42%)   65

In an analogous fashion, 3-keto-10-methyl-$\Delta^{1,4}$-hexahydronaphthalene derivatives 66 give bicyclohexenones 67[37,38] and the derivatives of 17β-hydroxyandrosta-1,4-dien-3-one (68) give 69[39,40].

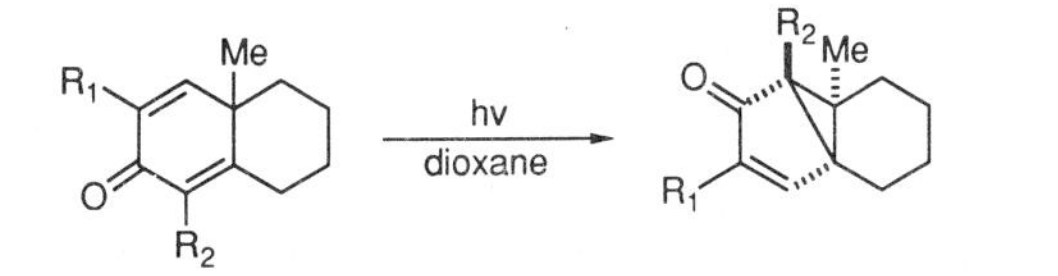

66a, R_1 = Me; R_2 = H   67a, R_1 = Me; R_2 = H (67%)
  b, R_1 = H; R_2 = Me   b, R_1 = H; R_2 = Me (70%)

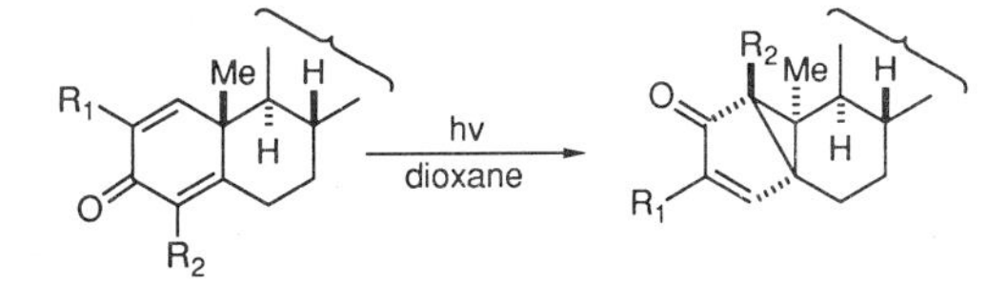

68a, R_1 = R_2 = H   69a, R_1 = R_2 = H (62%)
  b, R_1 = H; R_2 = Me   b, R_1 = H; R_2 = Me (60-70%)
  c, R_1 = R_2 = Me   c, R_1 = R_2 = Me (60%)

Photolysis of dienone 70, prepared by annelation of (+)-dihydrocarvone with methyl vinyl ketone, followed by catalytic hydrogenation gave bicyclohexanone 71. This substance was converted to the sesquiterpene (−)-axisonitrile-3 (72) via reductive ring opening of an intermediate vinylcyclopropane.[41]

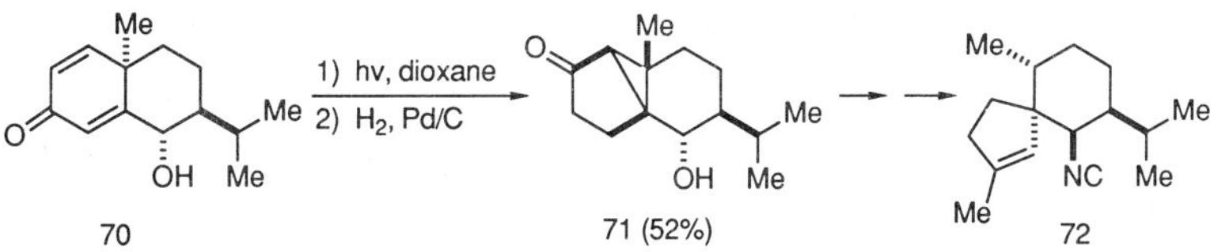

70   71 (52%)   72

Photolyses of the 2-methoxy-2,5-cyclohexadienones **73a** and **73b** are poorly regioselective to give **74a/74b** as major products and **75a/75b** as minor products (*cf.*, 48c → 50c).[29]

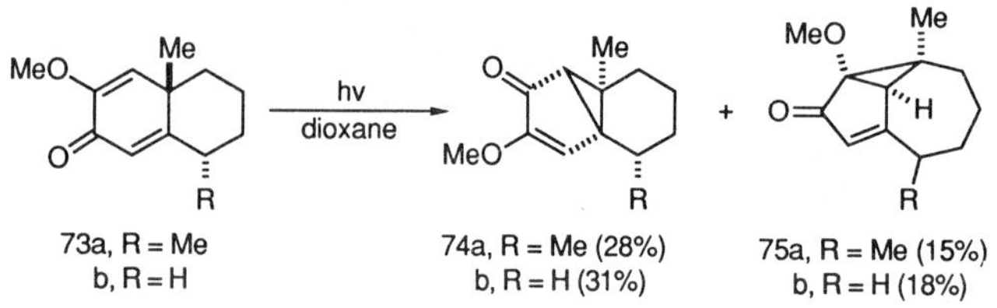

| 73a, R = Me | 74a, R = Me (28%) | 75a, R = Me (15%) |
| b, R = H | b, R = H (31%) | b, R = H (18%) |

Photorearrangements in Protic Solvents

Irradiation of α-santonin (**63**) in aqueous acetic acid provides isophotosantonic lactone **76** in good yields.[26,42] Analogous photochemistry has been observed for several other 3-keto-10-methyl-$\Delta^{1,4}$-hexahydronaphthalenes. For example, hydroazulenone **77** is obtained on photolysis of **66b** in aqueous acetic acid.[43,44] On the other hand, it has been shown that the 2-methyl dienone **66a** gives mainly the spiro-hydroxyketone **78** on irradiation in aqueous acetic acid.[37]

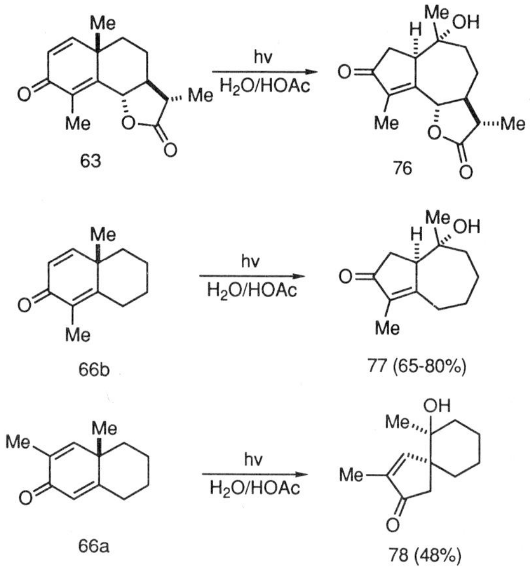

Irradiation of the 4-methoxy analogue of **66b** in glacial acetic acid gives the corresponding acetoxy analogue of fused bicycle **78**.[45] Similar behavior was exhibited by the 6/5-fused dienone **79**, which gave **80** in the key step of an efficient construction of the sesquiterpene *dl*-oplopanone (**81**).[46] Photolysis of the 2-methoxydienone **73a** gave **82**, an intermediate in a total synthesis of *dl*-α-vetispirene (**83**).[45]

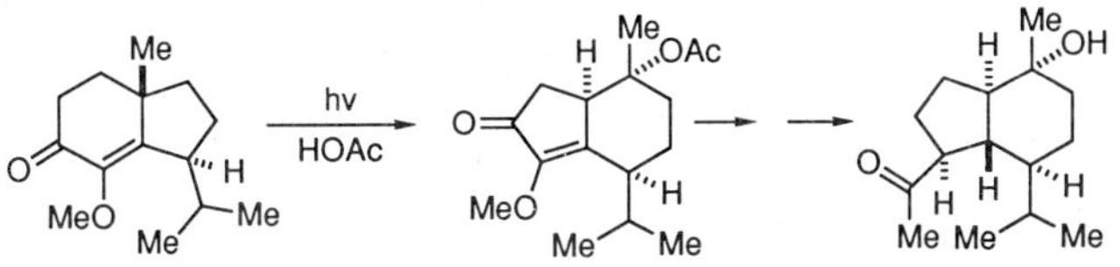

79   80 (91%)   81

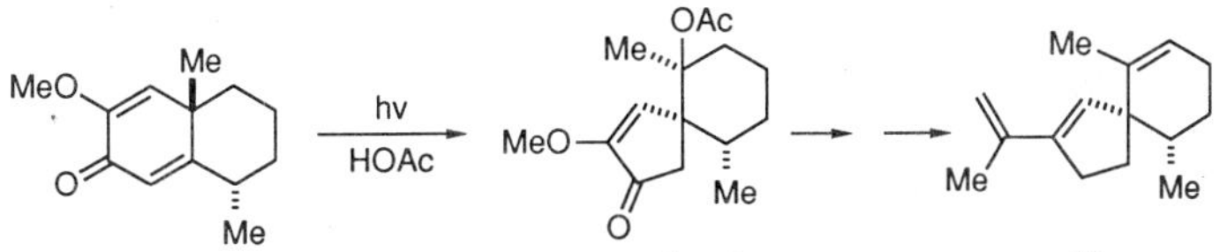

73a   82 (89%)   83

Photoreactions of Monocyclic Dienones

Photolysis of quinone monoketal **84** in methanol gives 2-(trimethoxymethyl)-2-cyclopentenone (**85**), although this result is said not to be generalizable.[47] By contrast, six examples of the conversion **86** to **87** were reported to occur in excellent yields with regioselectivity typical of other 2,5-cyclohexadien-1-ones. Since quinone monoketals can be prepared from hydroquinones by efficient electrochemical procedures[48] this methodology might be useful for the synthesis of natural products containing five-membered ring systems.

Photorearrangements of a series of 4-alkyl-4-methoxy-2,5-cyclohexadien-1-ones **88a–88d** gave 4-(alkyldimethoxymethyl)cyclopent-2-en-1-ones (**89a–89d**) in good yields.[49] It is noteworthy that these cyclopentenones have a ketone carbonyl group in protected form and that they serve as substrates for conjugate addition reactions. The dienones were prepared from the corresponding 4-alkylphenol by well-developed procedures.[50]

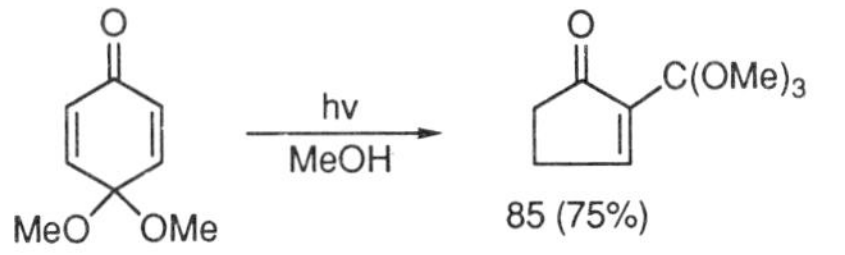

84   85 (75%)

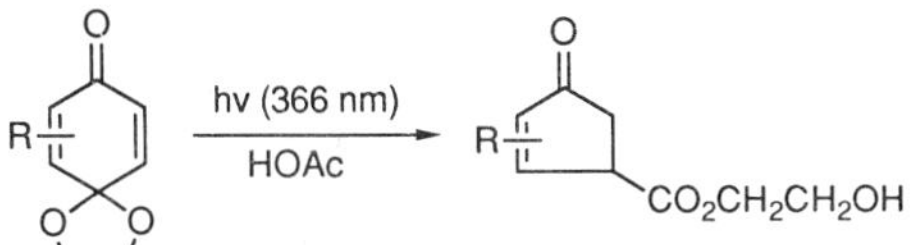

86   87 (R = H; 82%)

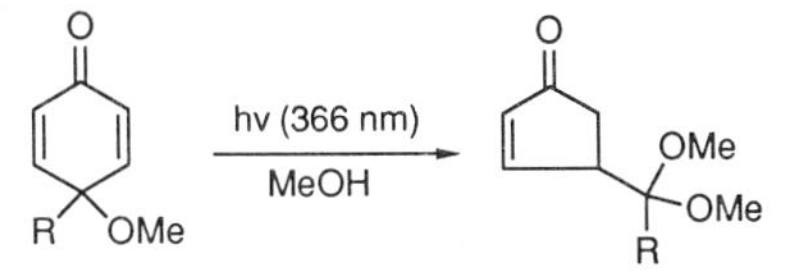

88a, R = Me
b, R = Et
c, R = i-Pr
d, R = t-Bu

88a, R = Me (67%)
b, R = Et (74%)
c, R = i-Pr (51%)
d, R = t-Bu (37%)

Irradiation of 3-methoxy-2,5-cyclohexadien-1-ones **90a–90e** produced mixtures of diastereomeric 4-methoxybicyclo[3.1.0]hex-3-en-2-ones **91** and **92**.[17] Continued irradiation of the mixtures of **91** and **92** with 366-nm light gave predominately the diastereomeric series **91a–91e** (~9:1 for the distribution of **91** and **92**). The attendant conversion of **91b** to lactone **93** suggests that this photochemistry might be useful for prostanoid and other cyclopentanoid natural product syntheses. Dienones **90** are available by Birch reduction-alkylation of benzoic acid derivatives.[51]

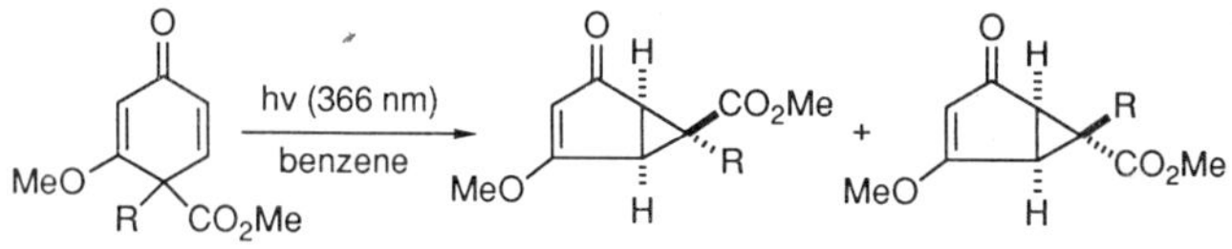

90a, R = Me
b, R = Et
c, R = $CH_2CH{=}CH_2$
d, R = CH_2CH_2OAc
e, R = $CH_2CH_2CH_2Cl$

91

92

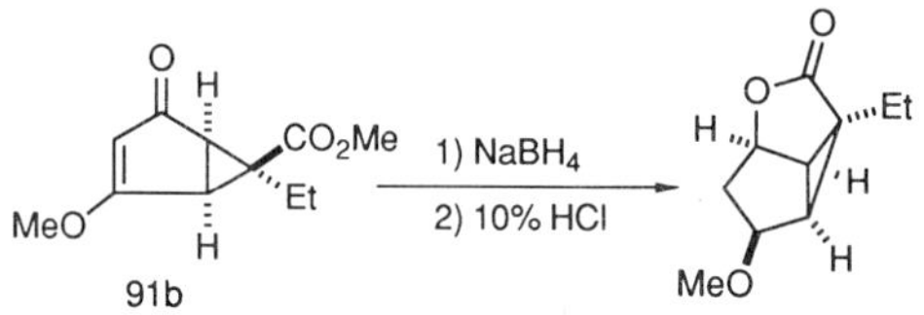

91b

93 (~100%)

Dienone to Phenol Photorearrangements

The dienone-phenol photorearrangement is useful for the preparation of highly substituted phenols. Photolyses of 4-alkyl-4-carbomethoxy-2,5-cyclohexadienones give mixtures of 2- and 4-carbomethoxyphenols (**94** → **95** + **96**).[17] The 4-benzyl derivatives **97a** and **97b** undergo regioselective photorearrangements to give **98a** and **98b**, respectively. Although bicyclohexenones derived from 3-methoxy-2,5-cyclohexadienones are relatively photostable at 366 nm, dienone irradiations through Pyrex provide the corresponding phenols in good to excellent yields (**90e** → **99**). Irradiation of the benzo-annelated dienone **100** gives naphthol **101**.[52]

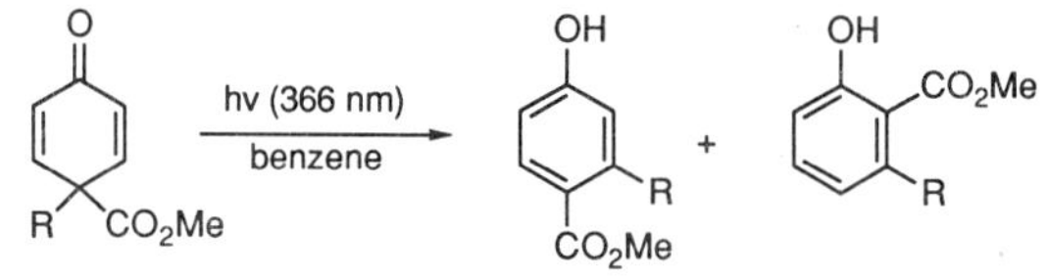

94, R = $CH_2CH_2CH_2Cl$

95 (46%)

96 (13%)

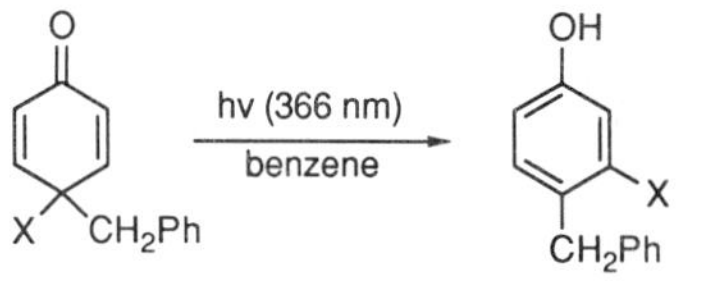

97a, X = CO_2Me
b, X = CN

98a, X = CO_2Me (67%)
b, X = CN (81%)

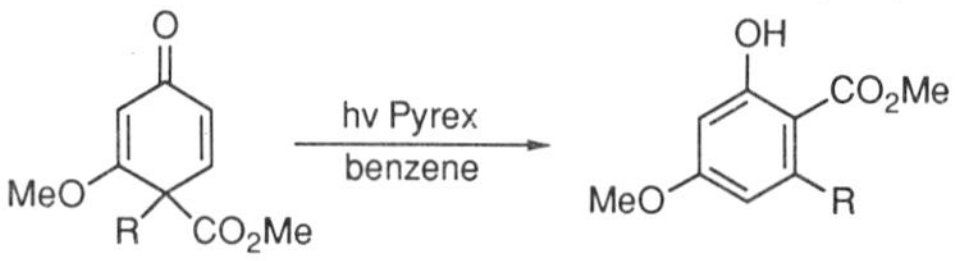

90e, R = $CH_2CH_2CH_2Cl$

99 (80%)

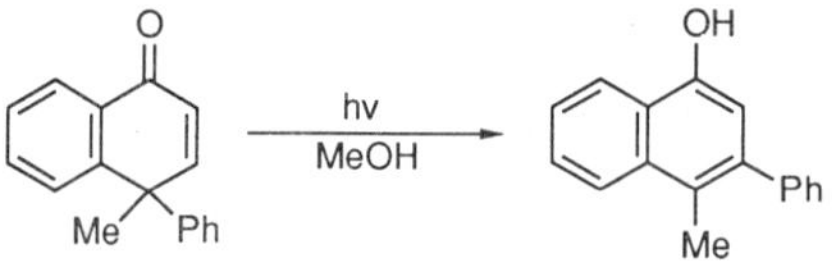

100

101 (55%)

1,3-Photorearrangements of Vinyl Ethers of Spiroquinol Ketals

A new photorearrangement of vinyl ethers of spiroquinol ketals, which occurs by the biradical pathway 1 → 6, provides a high yielding synthesis of spiro-fused 2,5-cyclohexadien-1-ones; e.g., 102 → 103.[53,54] This singlet-state process promises to be valuable for the synthesis of amaryllidacae alkaloids[54] and other spiro-fused cyclohexane ring systems.[55]

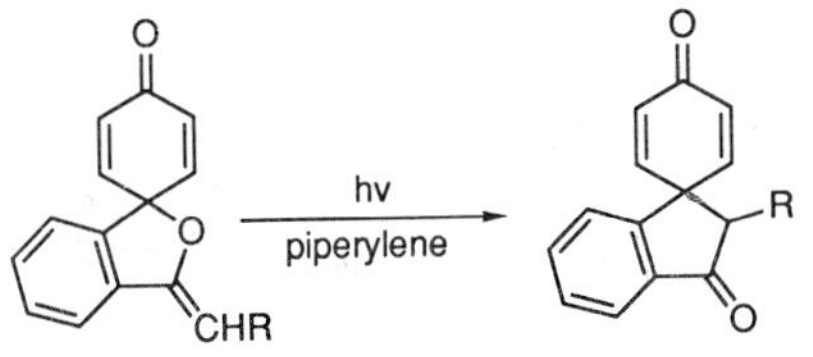

102a, R = Me
b, R = Ph

103a, R = Me (80%)
b, R = Ph (67%, Ø = 0.98)

Methods for Preparation of 2,5-Cyclohexadienones

Many of the synthetic methods available in the literature have been noted in recent reports;[17,54] additional methods are available.[56-63]

References

1. Zimmerman, H. E. and Schuster, D. I., A new approach to mechanistic organic photochemistry. IV. Photochemical rearrangements of 4,4-diphenylcyclohexadienone, *J. Am. Chem. Soc.*, 84, 4527, 1962.
2. Zimmerman, H. E., The di-π-methane rearrangement, *Org. Photochem.*, 11, 1, 1991.
3. Zimmerman, H. E. and Epling, G. A., Generation of photochemical intermediates without light. The type B Zwitterion. Mechanistic origins, *J. Am. Chem. Soc.*, 94, 7806, 1972.
4. Zimmerman, H. E., A new approach to mechanistic organic photochemistry, *Adv. Photochem.*, 1, 183, 1963.
5. Chapman, O. L., Photochemical rearrangements of organic molecules, *Adv. Photochem.*, 1, 323, 1963.
6. Schaffner, K., Photochemical rearrangements of conjugated cyclic ketones: The present state of investigations, *Adv. Photochem.*, 4, 81, 1966.
7. Kropp, P., Photochemical transformations of cyclohexadienones and related compounds, *Org. Photochem.*, 1, 1, 1967.
8. Chapman, O. L. and Weiss, D. S., Photochemistry of cyclic ketones, *Org. Photochem.*, 3, 197, 1973.
9. Schuster, D. I., Mechanisms of photochemical transformations of cross-conjugated cyclohexadienones, *Acc. Chem. Res.*, 11, 65, 1978.
10. Schaffner, K. and Demuth, M. M., *Rearrangements in Ground and Excited States*, Vol. 3, de Mayo, P., Ed., Academic Press, New York, 1980.
11. (a) Zimmerman, H. E. and Crumrine, D. S., Photochemistry without light and the stereochemistry of the type A dienone rearrangement. Organic photochemistry. XXXVI, *J. Am. Chem. Soc.*, 90, 5612, 1968; (b) Brennan, T. H. and Hill, R. K., Stereochemistry of a 1,4-sigmatropic rearrangement, *J. Am. Chem. Soc.*, 90, 5614, 1968.
12. Swenton, J. S., Saurborn, E., Srinivasan, R., and Sonntag, F. I., A criterion for Zwitterionic intermediates in the photochemistry of 2,5-cyclohexadienones, *J. Am. Chem. Soc.*, 90, 2990, 1968.
13. (a) Schuster, D. I. and Polowczyk, C. J., The photolysis of spiro[2.5]octa-1,4-dien-3-one in ethyl ether. A note on the mechanism of photolysis of 2,5-cyclohexadienones, *J. Am. Chem. Soc.*, 86, 4502, 1964; (b) Schuster, D. I. and Krull, I. S., Evidence for a 1,2-hydrogen-atom migration in a photochemically generated diradical, *J. Am. Chem. Soc.*, 88, 3456, 1966.

14. (a) Schuster, D. I. and Patel, D. J., The radical fragmentation route on photolysis of 2,5-cyclohexadienones. 4-Methyl-4-trichloromethyl-2,5-cyclohexadienone, *J. Am. Chem. Soc.,* 87, 2515, 1965; (b) Schuster, D. I. and Patel, D. J., Stereospecificity and solvent control of lumiproduct formation vs. fragmentation in dienone photochemistry, *J. Am. Chem. Soc.,* 88, 1825, 1966; (c) Patel, D. J. and Schuster, D. I., The photochemistry of 4-methyl-4-trichloromethyl-2,5-cyclohexadienone. I. The nature of the products, *J. Am. Chem. Soc.,* 90, 5137, 1968; (d) Schuster, D. I. and Patel, D. J., The photochemistry of 4-methyl-4-trichloromethyl-2,5-cyclohexadienone. II. Mechanistic studies and characterization of the excited state, *J. Am. Chem. Soc.,* 90, 5145, 1968; (e) Schuster, D. I. and Liu, K.-C., Nucleophilic trapping of photochemically-generated Zwitterions. Evidence for the intermediacy of Zwitterions in the photochemical rearrangement of cyclohexadienones, *J. Am. Chem. Soc.,* 93, 6711, 1971; (f) Samuel, C. J., The type A Zwitterion in cyclohexadienone photochemistry: chemical trapping and stereochemistry, *J. Chem. Soc., Perkin II,* 736, 1981.

15. Zimmerman, H. E., Crumrine, D. S., Döpp, D., and Huyffer, P. S., Photochemistry without light and the stereochemistry of the type A dienone rearrangement. Organic photochemistry. XXXVIII, *J. Am. Chem. Soc.,* 91, 434, 1969.

16. Zimmerman, H. E. and Jones, G., II, The photochemistry of a cyclohexadienone structurally incapable of rearrangement. Exploratory and mechanistic organic photochemistry. XLVII, *J. Am. Chem. Soc.,* 92, 2753, 1970.

17. Schultz, A. G., Lavieri, F. P., Macielag, M., and Plummer, M., 2,5-Cyclohexadien-1-one to bicyclo[3.1.0]hexenone photorearrangement. Development of the reaction for use in organic synthesis, *J. Am. Chem. Soc.,* 109, 3991, 1987.

18. Schultz, A. G. and Taveras, A. G., The first intermolecular 2 + 2 photocycloaddition of 2,5-cyclohexadien-1-ones to alkenes, *Tetrahedron Lett.,* 29, 6881, 1988.

19. Schultz, A. G., Plummer, M., Taveras, A. G., and Kullnig, R. K., Intramolecular 2 + 2 photocycloadditions of 4-(3'-alkenyl)- and 4-(3'-pentynyl)-2,5-cyclohexadien-1-ones, *J. Am. Chem. Soc.,* 110, 5547, 1988.

20. Schultz, A. G., Geiss, W., and Kullnig, R. K., The regio- and diastereoselectivity of the intramolecular 2 + 2 photocycloadditions of 4-(2'-isopropyl-3'-butenyl)-2,5-cyclohexadien-1-ones, *J. Org. Chem.,* 54, 3158, 1989.

21. Zimmerman, H. E. and Pasteris, R. J., Type A Zwitterions and cyclohexadienone photochemical rearrangements. Mechanistic and exploratory organic photochemistry, *J. Org. Chem.,* 45, 4876, 1980.

22. Zimmerman, H. E. and Pasteris, R. J., Regioselectivity in cyclohexadienone photochemistry. Role of Zwitterions in type A photochemical and dark reactions. Mechanistic and exploratory organic photochemistry, *J. Org. Chem.,* 45, 4864, 1980.

23. Schultz, A. G. and Reilly, J., 2,5-Cyclohexadien-1-one photochemistry. Reversible type B oxyallyl Zwitterion formation from photorearrangements of bicyclo[3.1.0]hex-3-en-2-ones, *J. Am. Chem. Soc.,* 114, 5068, 1992.

24. Broka, C. A., Photorearrangements of carbomethoxy-substituted cyclohexadienones in neutral media, *J. Org. Chem.,* 53, 575, 1988.

25. Schultz, A. G. and Hardinger, S. A., Photochemical and acid-catalyzed dienone-phenol rearrangements. The effect of substituents on the regioselectivity of 1,4-sigmatropic rearrangements of the type A intermediate, *J. Org. Chem.,* 56, 1105, 1991.

26. Arigoni, D., Bosshard, H., Bruderer, H., Büchi, G., Jeger, O., and Krebaum, L. J., Photochemische reaktionen. Über gegenseitige beziehungen und umwandlungen bei bestrahlungsprodukten des santonins, *Helv. Chim. Acta,* 40, 1732, 1957.

27. Barton, D. H. R., de Mayo, P., and Shafiq, M., The mechanism of the light-catalyzed transformation of santonin into 10-hydroxy-3-oxoguai-4-en-6:12-olide, *Proc. Chem. Soc. (London),* 205, 1957.

28. Kropp, P. J., Photochemical rearrangements of cross-conjugated cyclohexadienones. VII. 6,9-Dimethylspiro[4.5]deca-6,9-dien-8-one, *Tetrahedron,* 21, 2183, 1965.

29. Caine, D., Deutsch, H., Chao, S. T., Van Derveer, D. G., and Bertrand, J. A., Photochemical conversion of methoxy-substituted 6/6-fused cross-conjugated cyclohexadienones into isomeric tricyclodecenones, *J. Org. Chem.*, 43, 1114, 1978.

30. Rodgers, T. R. and Hart, H., Control of the stereochemistry of 2,5-cyclohexadienone photoisomerization by steric factors, *Tetrahedron Lett.*, 4845, 1969.

31. Schuster, D. I., Prabhu, K. V., Adcock, S., van der Veen, J., and Fujiwara, H., Stereoselectivity in the photochemical rearrangement of 3,4-dimethyl-4-trichloromethyl-2,5-cyclohexadienone, *J. Am. Chem. Soc.*, 93, 1557, 1971.

32. Schuster, D. I. and Barringer, W. C., Synthesis and photochemistry of Δ^1-4-alkyltestosterones, *J. Am. Chem. Soc.*, 93, 731, 1971.

33. Schultz, A. G. and Harrington, R. E., Substituent effects on the photorearrangements of 4-alkoxy-4-carbomethoxy-3-methoxy-2,5-cyclohexadien-1-ones, *J. Org. Chem.*, 56, 6391, 1991.

34. Zimmerman, H. E. and Lynch, D. C., Rapidly rearranging excited states of bichromophoric molecules-mechanistic and exploratory organic photochemistry, *J. Am. Chem. Soc.*, 107, 7745, 1985.

35. Schultz, A. G. and Green, N. J., Photochemistry of 4-vinyl-2,5-cyclohexadien-1-ones. A remarkable effect of substitution on the type A and dienone-phenol photorearrangements, *J. Am. Chem. Soc.*, 114, 1824, 1992.

36. Krapcho, A. P., Synthesis of carbocyclic spiro compounds via rearrangement routes, *Synthesis*, 425, 1976.

37. Kropp, P. J., The photochemical properties of a 2-methyl-1,4-dien-3-one, *J. Am. Chem. Soc.*, 86, 4053, 1964.

38. Kropp, P. J., The acid-catalyzed cleavage of cyclopropyl ketones related to lumisantonin, *J. Am. Chem. Soc.*, 87, 3914, 1965.

39. Dutler, H., Ganter, C., Ryf, H., Utzinger, E. C., Weinberg, K., Schaffner, K., Arigoni, D., and Jeger, O., Photochemische reaktionen. Photochemische umwandlungen von O-acetyl-1-dihydrotesteron I, *Helv. Chim. Acta*, 45, 2346, 1962.

40. Weinberg, K., Utzinger, E. C., Arigoni, D., and Jeger, O., Photochemische reaktionen. Beeinflussung der photochemischen isomerisierung gekreuzter dienone durch substituenten am chromophor, *Helv. Chim. Acta*, 43, 236, 1960.

41. Caine, D. and Deutsch, H., Total synthesis of (−)-axisonitrile-3. An application of the reductive ring opening of vinylcyclopropanes, *J. Am. Chem. Soc.*, 100, 8030, 1978.

42. Barton, D. H. R., de Mayo, P., and Shafiq, M., The constitution of photosantonic acid, *Proc. Chem. Soc. (London)*, 345, 1957.

43. Caine, D. and Dawson, J. B., The synthesis of a model perhydroazulene derivative, *J. Org. Chem.*, 29, 3108, 1964.

44. Kropp, P. J., Photochemical rearrangements of cross-conjugated cyclohexadienones. V. A model for the santonin series, *J. Org. Chem.*, 29, 3110, 1964.

45. Caine, D., Boucugnani, A. A., Chao, S. T., Dawson, J. B., and Ingwalson, P. F., Stereospecific synthesis of 6,C-10-dimethyl (r-5-C¹)spiro[4.5]dec-6-en-2-one and its conversion into (±)-α-vetispirene, *J. Org. Chem.*, 41, 1539, 1976.

46. Caine, D. and Tuller, F. N., Photochemical rearrangements of 6/5-fused cross-conjugated cyclohexadienones. Application to the total synthesis of *dl*-oplopanone, *J. Am. Chem. Soc.*, 93, 6311, 1971.

47. Pirrung, M. C. and Nunn, D. S., Photochemical rearrangements of quinone monoketals. Synthesis of substituted cyclopentenones, *Tetrahedron Lett.*, 29, 163, 1988.

48. Swenton, J. S., Quinone *bis*- and monoketals via electrochemical oxidation. Versatile intermediates for organic synthesis, *Acc. Chem. Res.*, 16, 74, 1983.

49. Taveras, A. G., Jr., Photorearrangement of 4-alkyl-4-alkoxy-2,5-cyclohexadienones: Synthesis of 4-(alkyldimethoxymethyl)cyclopent-2-en-1-ones, *Tetrahedron Lett.*, 29, 1103, 1988.

50. Corey, E. J., Barcza, S., and Klotmann, G., A new method for the directed conversion of the phenoxy grouping into a variety of cyclic polyfunctional systems, *J. Am. Chem. Soc.*, 91, 4782, 1969.

51. Schultz, A. G., Enantioselective methods for chiral cyclohexane ring synthesis, *Acc. Chem. Res.*, 23, 207, 1990.

52. Zimmerman, H. E., Hahn, R. C., Morrison, H., and Wani, M. C., Mechanistic and exploratory photochemistry. XII. Characterization of the reactivity of the β-carbon in the triplet excited state of α,β-unsaturated ketones, *J. Am. Chem. Soc.*, 87, 1138, 1965.

53. Wang, S., Callinan, A., and Swenton, J. S., Photochemical 1,3-oxygen-to-carbon migrations. An efficient, high-yield route to protected spirodienones, *J. Org. Chem.*, 55, 2272, 1990.

54. Swenton, J. S., Callinan, A., and Wang, S., Efficient syntheses of vinyl ethers of spiroquinol ketals and their high-yield photochemical oxygen-to-carbon[1,3]-shift to spiro-fused 2,5-cyclohexadienones, *J. Org. Chem.*, 57, 78, 1992.

55. Biggs, T. N. and Swenton, J. S., Thermal and photochemical rearrangements of cyclopropyl ethers of *p*-quinols. Competing reaction pathways leading to five- and six-membered ring spirocyclic ketones, *J. Org. Chem.*, 57, 5568, 1992.

56. Zaidi, J. H., Waring, A. J., Synthesis of 3,4-dihydro-3,3,8a-trimethylnaphthalene-1,6(2*H*, 8a*H*)-dione, a 4-acylcyclohexa-2,5-dienone, *J. Chem. Soc., Chem. Commun.*, 618, 1980.

57. Danishefsky, S., Morris, J., Mullen, G., and Gammill, R., Total synthesis of *dl*-tazettine and 6a-epipretazettine: a formal total synthesis of *dl*-pretazettine. Some observations on the relationship of 6a-epipretazettine and tazettine, *J. Am. Chem. Soc.*, 104, 7591, 1982.

58. Ponpipom, M. M., Yue, B. Z., Bugianesi, R. L., Brooker, D. R., Chang, M. N., and Shen, T. Y., Total synthesis of kadsurenone and its analogs, *Tetrahedron Lett.*, 27, 309, 1986.

59. Tamura, Y., Yakura, T., Haruta, J., and Kita, Y., Hypervalent iodine oxidation of *p*-alkoxyphenols and related compounds: a general route to *p*-benzoquinone monoacetals and spiro lactones, *J. Org. Chem.*, 52, 3927, 1987.

60. Haack, R. A., and Beck, K. R., Synthesis of substituted spiro[4.5]deca-3,6,9-triene-2,8-diones: an expeditious route to the spiro[4.5]decane terpene skeleton, *Tetrahedron Lett.*, 30, 1605, 1989.

61. Callinan, A., Chen, Y., Morrow, G. W., and Swenton, J. S., Spiro-annulated 2,5-cyclohexadienones via oxidation of 2′-alkenyl-*p*-phenyl phenols with iodobenzene diacetate, *Tetrahedron Lett.*, 31, 4551, 1990.

62. Kita, Y., Tohma, H., Kikuchi, K., Inagaki, M., and Yakura, T., Hypervalent iodine oxidation of *N*-acyltyramines: synthesis of quinol ethers, spirohexadienones, and hexahydroindol-6-ones, *J. Org. Chem.*, 56, 435, 1991.

63. Pirrung, M. C. and Nunn, D. S., Synthesis of quinone monoketals by diol exchange, *Tetrahedron Lett.*, 33, 6591, 1992.

56

Photochemical Rearrangements of 6/6- and 6/5-Fused Cross-Conjugated Cyclohexadienones

rury Caine
niversity of Alabama

56.1 Introduction

Interest in the photochemical behavior of cross-conjugated cyclohexadienones began in the 19th century with reports of the light-induced reactions of the naturally occurring sesquiterpene α-santonin (1).[1] However, it was not until the mid-1950s that the modern era of photochemical investigation of these systems began with the elucidation of the structures of isophotosantonic lactone (2),[2,3] lumisantonin (3),[3–6] photosantonin (5a), and photosantonic acid (5b) (Scheme 56.1).[7,8] Madzasantonin (4), an intermediate in the pathway between 3 and 5, was isolated somewhat later.[9,10] Simultaneously, with the research on the α-santonin photoproducts, reports of the photochemical behavior of steroidal systems [e.g., prednisone acetate which contains a 1,4-diene-3-one chromophore in ring A] began to appear in the literature.[11,12]

0-8493-8634-9/95/$0.00+$.50
© 1995 by CRC Press, Inc.

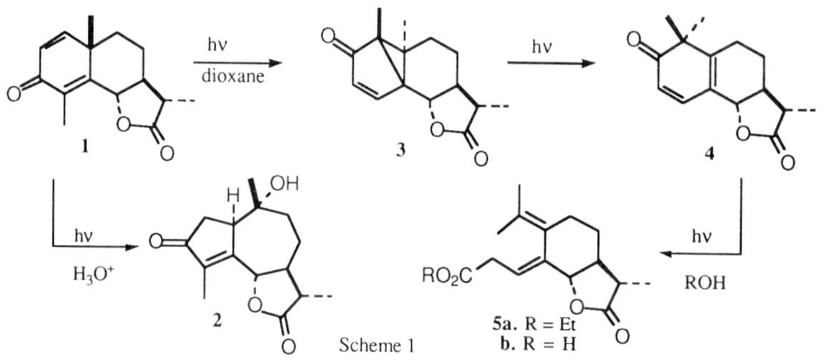

SCHEME 56.1

The fascinating rearrangements exhibited by α-santonin and related steroidal systems and the recognition of the potential synthetic value of these types of reactions have led to investigation of the photochemistry of a variety of monocyclic, fused-ring, and spirocyclic compounds containing cross-conjugated cyclohexadienone ring systems. Reactions of bicyclic 6/6- and 6/5-fused dienones will be the major focus of this chapter. The first study of a simple 6/6-fused cyclohexadienone was reported in 1963;[13] and, although the photochemistry of steroidal B-nor-1,4-diene-3-one had been reported earlier,[14] the reactions of simple bicyclic 6/5-fused systems were first published in 1968.[15] Several excellent reviews of cyclohexadienone photochemistry have been published.[16-18]

56.2 Mechanism

Excitation of intense π,π^* transitions around 240 nm or weaker n,π^* transitions around 300 nm can effect rearrangements of cross-conjugated cyclohexadienones. These reactions are generally believed to occur via triplet n,π^* excited states that may be populated by intersystem crossing from the singlet, or by both intersystem crossing and internal conversion in cases where π,π^* excited states are produced by the initial irradiation.[16,18] On the basis of studies involving monocyclic 4,4-disubstituted-2,5-dienones, Zimmerman and Schuster[19] have proposed a unifying mechanism that accounts generally for the photochemical behavior of most cross-conjugated cyclohexadienones in nonnucleophilic and nucleophilic media.

This mechanism, as applied to ring A unsubstituted and 2- and 4-methyl-substituted 6/6- or 6/5-fused bicyclic dienone systems such as 6 (n = 1 or 2), is illustrated in Scheme 56.2. A key intermediate, the zwitterionic species 9 (n = 1 or 2), is produced by: (1) electronic excitation and intersystem crossing to give the n,π^* triplet intermediate 7 (n = 1 or 2); (2) 1,5-bonding to give the excited-state cyclopropyl intermediate 8 (n = 1 or 2); and (3) electron demotion to the ground-state species. Bicyclo[3.1.0]hex-2-en-3-ones derivatives (lumiproducts) 10 (n = 1 or 2), which are obtained in good yields on irradiation of most bicyclic dienones in nonnucleophilic solvents such as dioxane,[16-18] may arise from 9 (n = 1 or 2) via symmetry-allowed 1,4-sigmatropic rearrangements (path a). However, in protic nucleophilic solvents such as aqueous acetic acid, 9 (n = 1 or 2) is protonated on oxygen to give the mesoionic species 11 (n = 1 or 2) which may undergo a variety of reactions such as: (1) nucleophilic attack of water at C-9 or C-10 with concomitant cleavage of the internal bond of the cyclopropane ring to give bicyclic hydroxy ketones 12 (n = 1 or 2) (path b); (2) nucleophilic attack of the solvent at C-10 with cleavage of the external bond of the cyclopropane ring to give spirocyclic hydroxy ketones 13 (path c); or (3) proton loss from C-1 with concomitant cleavage of the 5,9-bond to give 5/6-fused linearly conjugated dienones 14 (path d). Direct collapse of 9 (n = 1) to 14 may also occur in aprotic media (path e).

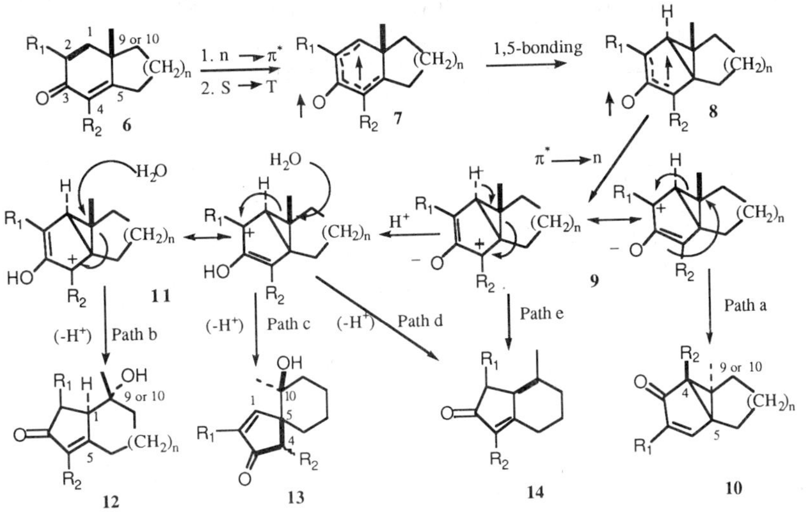

SCHEME 56.2

It has been pointed out that, theoretically, bicyclohexenones such as **10** may arise directly from the starting dienones via photochemically allowed [σ2a+π2a]-cycloaddition processes.[20a,b] However, several lines of evidence provide support for the intermediacy of zwitterionic species in these reactions.[17,21–26] Among the most convincing of these are the irradiations of 4-methyl-4-trichloromethyl-2,5-cyclohexadienone (**15**) in acidic methanol and in benzene containing cyclopentadiene which led to products **17**[17,22] and **18**,[24] respectively. These compounds, which contain an intact cyclopropane ring, arose from trapping of the zwitterionic intermediate **16** by nucleophilic attack of the solvent[17,22] and [3+4]-cycloaddition,[24] respectively (Scheme 56.3). Apparently, 1,4-sigmatropic rearrangement or solvent attack involving C-4 of intermediate **16** is relatively slow because the trichloromethyl group induces positive character at that position. As will be discussed in a later section, tricyclic α-acetoxy ketones containing cyclopropane rings have been isolated recently from the irradiation of a bicyclic 6/5-fused dienone in acetic acid.[26]

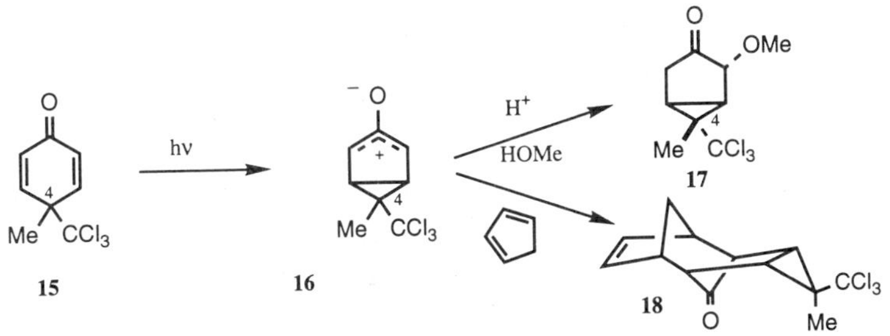

SCHEME 56.3

The ability of primary photoproducts such as the lumiproducts **10** to yield secondary, and even tertiary and quaternary photolysis products, adds to the complexity of cyclohexadienone photochemical reactions.[16,18] As shown in Scheme 56.4, electronic excitation of **10** leads to a zwitterionic intermediate such as **19** which can rearrange by three possible pathways (paths a-c) to give new cyclohexadienones **6**, **20** or **21**, or the phenolic tautomer of **20** in the cases where $R_2 = H$. The nature of the substituents R_1 and R_2, and particularly, the size of ring B and its substitution pattern can influence profoundly the course of these rearrangements. Space does not permit a detailed discussion of lumiproduct rearrangements here, but the subject has been covered thoroughly in earlier reviews.[16,18]

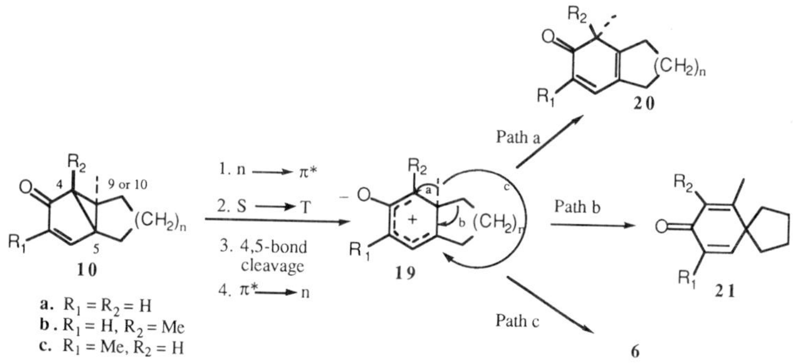

SCHEME 56.4

56.3 Scope and Limitations

Normally, lumiproducts (e.g., 10) are best prepared by irradiations of dilute solutions of dienones in dioxane or toluene with a low-pressure mercury lamp that emits most of its radiation at 254 nm.[16] Best yields of hydroxy ketone products (e.g., 12 or 13) are obtained by irradiations of dilute aqueous acetic acid solutions of dienones with a medium- or high-pressure mercury lamp screened by a Pyrex filter. Acetoxy or alkoxy ketones corresponding to 12 or 13 may be obtained using glacial acetic acid or alcohols (or alcoholic acetic acid) as the solvent.

Influence of Substituents at C-2 and C-4 on the Dienone Chromophore

Like α-santonin, steroidal and simple 6/6-fused dienones such as 6 (n = 2) that are unsubstituted at C-2 and C-4, or contain methyl groups at one of these positions, rearrange to the corresponding lumiproducts in good yields (59 to 70%) upon irradiation in dioxane.[16] Similar results were obtained for the corresponding 6/5-fused systems 6 (n = 1) except that in addition to 10 (n = 1), 6a (n = 1), and to a lesser extent 6b (n = 1), also gave the corresponding linearly conjugated dienones 14 in low yields.[15] Interestingly, irradiation of a B-nor-steroidal dienone related to 6a (n = 1) gave exclusively the corresponding 5/6-fused linearly conjugated dienone under the same conditions.[14] Apparently, the added strain due to the presence of the *trans*-fused C ring of the steroid caused collapse of the intermediate related to 9a (n = 1) to the dienone (*cf.* path e Scheme 56.2) in place of the normal 1,4-sigmatropic rearrangement.

When 6/6-fused dienones such as 6 (n = 2) are irradiated in protic media, the mode of cleavage of the mesoionic intermediate 11 is controlled by the electronic effects of substituents (Scheme 56.2). For example, on irradiation in aqueous acetic acid, the 4-methyl dienone 6b (n = 2) yielded exclusively the 5/7-fused hydroxy ketone 12b (n = 2),[27] while the 2-methyl compound 6c (n = 2) yielded exclusively the spiro hydroxy ketone 13c.[28] Under similar conditions, the 6α-methyl derivative of 6a (n = 2), i.e., a ring A unsubstituted system, yielded an approximately 1:1 mixture of the corresponding 6α-methyl derivatives of 12a (n = 2) and 13a.[29] [In each case, the hydroxy ketone products were accompanied by varying amounts of phenols, which arose as secondary photoproducts derived from the related lumiproducts 10 (n = 2).[29]] Thus, the location of an electron-releasing alkyl group at C-2 or C-4 leads to stabilization of the resonance structure of 11 with a positive charge at that position and cleavage via path c or path b, respectively, takes place; when ring A is unsubstituted, the two resonance forms of **11** are approximately equal in energy so that products derived from both cleavage pathways are obtained.

The 4-methyl substituted 6/5-fused dienone **6b** (n = 1) showed similar behavior to **6b** (n = 2) and gave the 5/6-fused hydroxy ketone **12b** (n = 1) in 59% yield upon irradiation in aqueous acetic acid,[15b] and the related 5/6-fused ethoxy ketone upon irradiation in ethanol.[15a] However, the reactions of dienones **6a** (n = 1) and **6c** (n = 1) differed significantly from those of the corresponding 6/6-fused systems. For example, upon irradiation in methanolic acetic acid or methanol alone, **6a** (n = 1) gave a 3:3:2 mixture of the 6/5-fused methoxy ketone **22**, the tricyclic methoxy ketone **23**, and the linearly conjugated dienone **14a** in about 65% yield (Scheme 56.5).[30] Under similar conditions, dienone **6c** gave an 8:5 mixture of the 2-methyl derivatives of **22** and **23**. Compounds such as **12b** (n = 1) (Scheme 56.2), **22**, and **14a** are easily rationalized as resulting from cleavage of the internal bond of the cyclopropane ring in the mesoionic intermediate **11** (n = 1) (Scheme 56.2). No spiro products were isolated from any of the irradiations of the 6/5-fused dienones. In **11** (n = 1), the fusion of the cyclopropane ring to the five-membered B ring appears to increase the strain in the internal bond to the point that its cleavage is favored greatly even when a methyl substituent, which should favor cleavage of the external bond, is present at C-2.

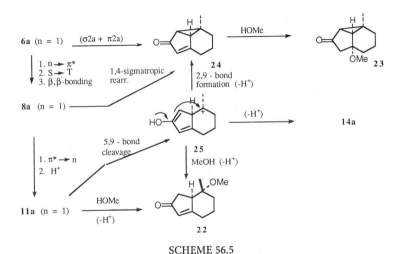

SCHEME 56.5

The mechanism of formation of tricyclic ketones such as **23** is interesting. Such products possibly arise via photochemical or thermal 1,4-addition of methanol to a strained tricyclic enone such as **24** (Scheme 56.5). Enone **24** may arise by one of several possible pathways. Firstly, a photochemically-allowed [σ2a+π2a]-cycloaddition involving the 5,9-σ-bond and the 1,2-π-bond could yield **24** directly from dienone **6a** (n = 1).[20] Secondly, there may be competition between electron demotion and a 1,4-sigmatropic rearrangement with retention of configuration at C-9 in a triplet intermediate such as **8** (n = 1), which would lead to **24**.[30] Finally, a ground-state reaction in which intermediate **11a** (n = 1) undergoes cleavage of the 5,9-bond of the cyclopropane ring to give an enolic carbocation **25** which produces enone **24** by closure of the 2,9-bond is a possibility.[18] Intermediate **25** is also a possible precursor of methoxy enone **22** and dienone **14a**. Present evidence does not allow the exclusion of any of these possibilities.

In addition to the lumiproduct **27a**, irradiation of the 6/6-fused 2-methoxy dienone **26a** in dioxane at 254 nm gave smaller amounts of the tricyclic enone **28a** and the phenol **29a** as primary photoproducts; similar results were obtained for the 6α-methyl derivative of **26a**, i.e., **26b** (Scheme 56.6).[31] The phenol **29a** is probably produced by a photochemically induced 1,2-methyl shift.[28] The abnormal lumiproduct **28a** may be formed by a pathway similar to one of the three described above for the conversion of **6a** (n = 1) into **24**. The possibility of a 1,4-sigmatropic rearrangement with retention of configuration at C-10 prior to electron demotion to a zwitterionic species seems most reasonable, but the role of the methoxy group in promoting this process is unclear.[31] Irradiation of dienone **26b** in glacial acetic acid gave the spiro acetoxy ketone **30b** in 89% yield; the use of aqueous acetic acid provided the corresponding spiro hydroxy ketone but in much lower yield because of hydrolysis of the enol ether function.[32]

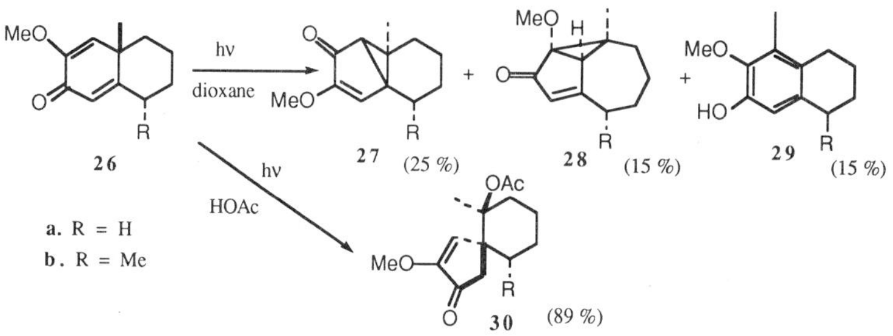

SCHEME 56.6

There have been no reports of the irradiations of the 4-methoxy dienones such as **31** in nonnucleophilic solvents. However, photolysis of the related steroidal compound 4-methoxy-2,4-cholestadien-3-one in dioxane gave the normal lumiproduct exclusively.[33] The 4-methoxy 6/6- and 6/5-fused dienones **31a** and **31b** gave the 5/7-and 5/6-fused acetoxy enones **32a**[32] and **32b**,[34] respectively, in excellent yields upon irradiation in glacial acetic acid. Thus, the photochemical reactions of dienones **26** and **31a** in protic solvents appear to proceed via mesoionic intermediates analogous to **11** (n = 2) with the electron-releasing methoxy groups favoring solvolytic cleavage of the external and internal bonds of the cyclopropane rings, respectively. Both the strain in the internal bond of the cyclopropane ring and the location of the methoxy group in a species analogous to **11** (n = 1) would favor conversion of **31b** into **32b**.

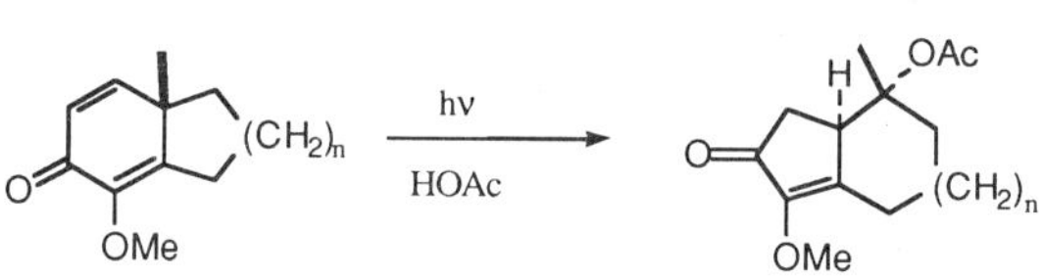

6/6-Fused cross-conjugated cyclohexadienones containing electron-withdrawing carboxy and carbomethoxy groups at C-2 and C-4 have been irradiated in nonnucleophilic and nucleophilic solvents (Scheme 56.7).[35,36] The 2-carboxy and 2-carbomethoxy dienones **33** and **41** did not yield normal lumiproducts such as **11** when irradiated in anhydrous dioxane. Rather, dienone **33** gave the 5/7-fused dienone **36** and the 5/7-fused enone lactone **37** in 67 and 16% yields, respectively. Upon excitation of **33**, a cyclopropyl intermediate such as **34** in which the carbonyl group is protonated internally by the carboxyl group should be produced. Here, the presence of the electron-withdrawing group would favor build-up of more positive charge at C-4 than at C-2. Intermediate **34** could undergo loss of a proton from the methyl group and concomitant cleavage of the 5,10-bond (path a) to yield the β-keto acid **35** which would decarboxylate to give **36**. On the other hand, intramolecular nucleophilic attack of the carboxylate oxygen at C-10 of **34** with cleavage of the 5,10-bond of the cyclopropane ring (path b) would give enone lactone **37** directly.

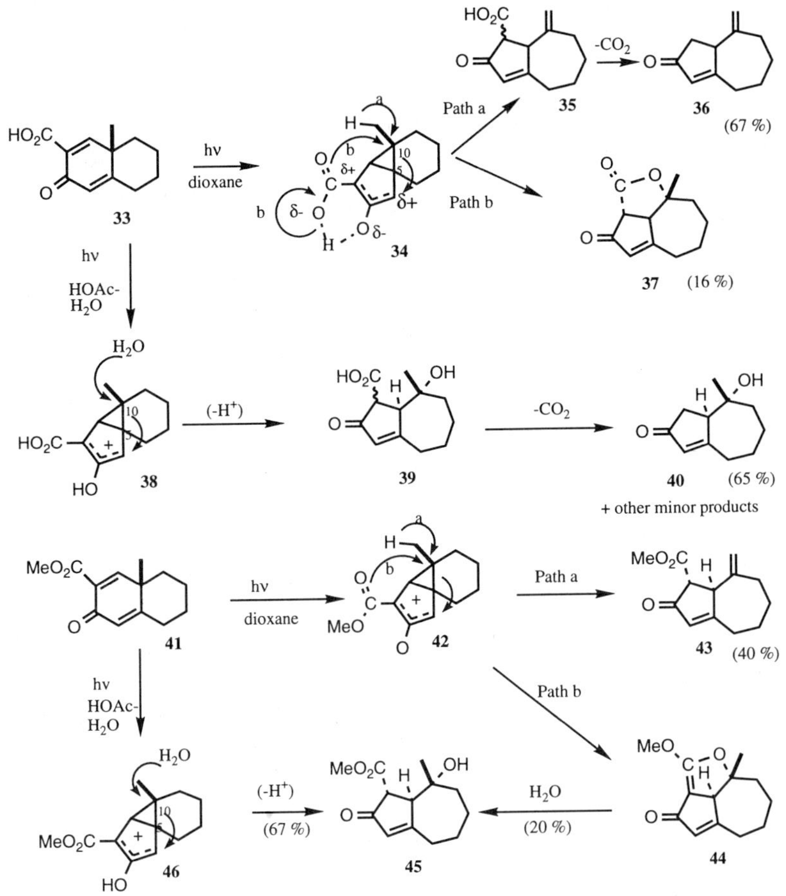

SCHEME 56.7

Irradiation of 33 in aqueous acetic acid gave the 5/7-fused hydroxy ketone 40 in 65% yield along with small quantities of other products. In this medium, the cyclopropyl intermediate 38, obtained by β,β′-bonding and protonation, should undergo solvolytic cleavage of the 5,10-bond as shown to give the β-keto acid 39, which would decarboxylate to 40. Thus, 2-carboxy dienones such as 33 are excellent precursors of ring A unsubstituted 5/7-fused hydroxy ketones.

When dienone 41 was irradiated in dioxane, the carbomethoxy group suppressed the normal 1,4-sigmatropic rearrangement of the zwitterionic intermediate 42. Instead, 42 underwent competitive proton loss from the methyl group and 5,10-bond cleavage to give dienone ester 43 (path a) in 40% yield and intramolecular nucleophilic attack of the carbomethoxyl group at the electron deficient C-10 (path b) to give the unstable intermediate 44 (path b). This was hydrolyzed when exposed to water to give the hydroxy ester 45 in 20% yield.[36] Dienone 41 was converted into 45 in high yield on irradiation in aqueous acetic acid.[35] Although the intermediacy of a species such as 44 cannot be ruled out in this reaction, the usual intermolecular attack of water on the mesoionic intermediate 46 would produce this product. Irradiation of 2-formyl 6/6-fused dienones in aqueous acetic acid leads to 5/7-fused hydroxymethylene compounds that can be deformylated to the corresponding ring A unsubstituted 5/7-fused systems.[35,37]

The dienone 47, containing a carboxyl group at C-4, was smoothly converted into products having a spiro[5.4]decane system when irradiated in dioxane or aqueous acetic acid (Scheme 56.8). Use of the former medium produced the spiro dienone 48 in 62% yield while the latter led to the keto lactone **49** in 40% yield. Dienone **48** apparently arose via decarboxylation of the corresponding β-keto acid **50** which would be obtained by proton loss from C-9 and concomitant cleavage of the 1,10-bond in the chelated cyclopropyl intermediate **51**. Protonation of **51** in acetic acid would give the corresponding mesoionic intermediate which could undergo solvolytic attack at C-10 to produce a γ-hydroxy β-keto acid which could lactonize to **49**.

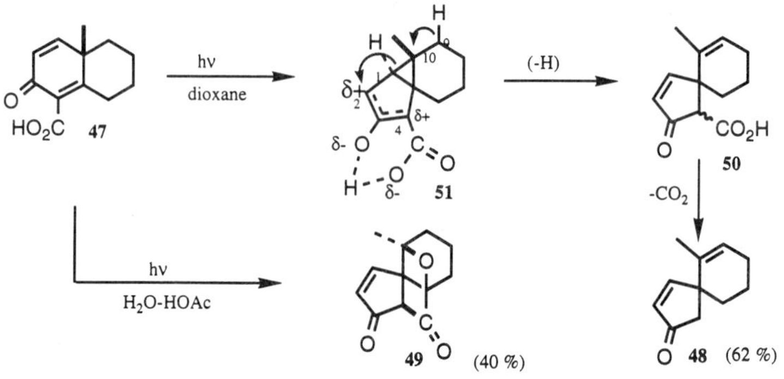

SCHEME 56.8

The irradiation of 4-formyl and 4-carbomethoxy dienones related to **47** failed to yield photo-products with rearranged carbon skeletons.[35] It has been suggested that in these cases, photochemically promoted deconjugation of the 4,5-bond competes with the normal β,β'-bonding process.[35] In the corresponding 4-carbomethoxy 6/5-fused dienone where double-bond deconjugation would be less favorable for steric reasons, skeletal rearrangement occurred to give a 5/6-fused dienone analogous to **14** upon irradiation in toluene solution.[36] The lumiproduct, containing a carbomethoxy group at C-4, was formed reversibly in this reaction.[36]

Other Ring A Substituent Effects

The presence of electron-withdrawing groups at the angular positions of bicyclic dienones can influence profoundly their photochemical reactivity and modes of rearrangement. For example, irradiation of the 6/6-fused dienone **52a** containing a carboethoxy substituent at C-10 in both anhydrous dioxane and aqueous acetic acid gave largely the lumiproduct **53a** along with small amounts of the linearly conjugated dienone **54a**, the tricyclic enones ester **55a**, and other products (Scheme 56.9).[38]

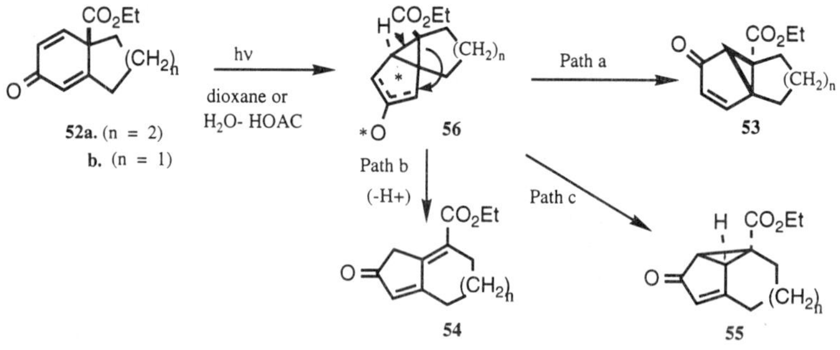

SCHEME 56.9

Apparently, the electron-withdrawing ester group in the zwitterionic intermediate **56a** (* = –/+) stabilizes the cyclopropane ring toward solvolytic cleavage, although 1,4-sigmatropic rearrangement of C-10 from C-5 to C-2 (path a) still occurs to give 53a. Dienone 54a could arise via collapse of 56a (* = −/+) (path b). The tricyclic compound 55a, which has the same carbon skeleton as the tricyclic ketone 28, possibly is obtained via a similar pathway, i.e., a rare 1,4-sigmatropic rearrangement of C-10 from C-5 to C-2 with retention of configuration in the excited-state species 56a (* = ·/·) (path c).[39] Dienone 52b, which contains a five-membered ring B, gave exclusively the 5/6-fused dienone 54b under similar conditions.[39] This reaction provides a further demonstration of the strong propensity of 6/5-fused dienones to yield 5/6-fused systems.

The 10-hydroxy dienone 57a and its 4-methyl derivative 57b have been shown to rearrange photochemically to the 5/7-fused enediones 58a and 58b in low yields on irradiation in benzene (Scheme 56.10).[40] This behavior is anticipated because the zwitterionic cyclopropanol intermediates 59a,b could collapse with cleavage of the 5,10-bond as shown. Irradiation of the 2-methyl hydroxy dienone 57c under similar conditions gave the butenolide 60. This photoproduct is considered to be formed via the spiro enedione 61, produced from cleavage of the 1,10-bond in 59c, by a complex photochemical rearrangement pathway.[40]

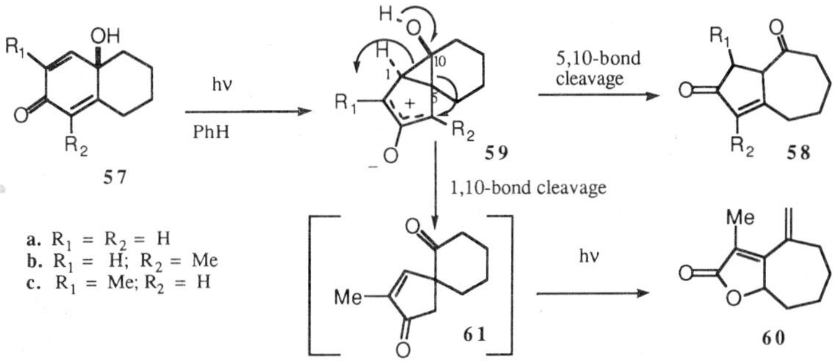

a. $R_1 = R_2 = H$
b. $R_1 = H$; $R_2 = Me$
c. $R_1 = Me$; $R_2 = H$

SCHEME 56.10

1-Methyl-substituted 6/6-fused dienones such as those derived from nootkatone and α-vetivone yielded the expected lumiproducts upon irradiation in dioxane.[41] Dienones 62[42] and 63,[43] in which C-1 is located at a ring fusion, failed to undergo photochemical rearrangements under a variety of irradiation conditions. Presumably, steric factors prevent β,β′-bonding in the excited states of these dienones. In the hydrogen donor solvent 2-propanol, dienone 62 underwent a hydrogen abstraction reaction that led to the loss of the methyl group and aromatization of the A ring.[42] In the same solvent, photoreduction of one of the double bonds of dienone 63 occurred.[43]

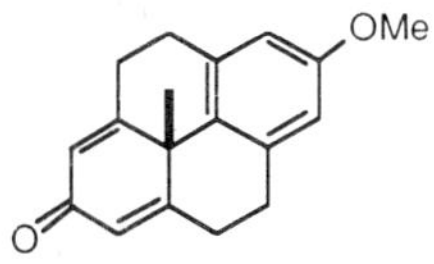

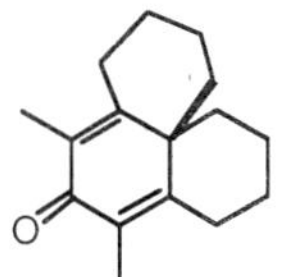

Effects of Ring B Substituents

Various structural features associated with ring B substituents may have an important effect on the reactivity and modes of rearrangement of bicyclic dienones. For example, irradiation of dienone 64 in aqueous acetic acid gave the 5/7-fused hydroxy ketone 65 as the major product, along with the spirocyclic dienone 66 and the phenol 67.[44] Apparently, the 5/7-fused system 65 was formed in higher yield than usual because the β-methyl group at C-6 hindered solvent attack from the topside of the mesoionic cyclopropyl intermediate derived from 64. Proton loss from C-9 and cleavage of the external bond of the cyclopropane ring in this intermediate would yield the spiro dienone 66. Phenol 67 probably arose via rearrangement of the undetected lumiproduct derived from 64.

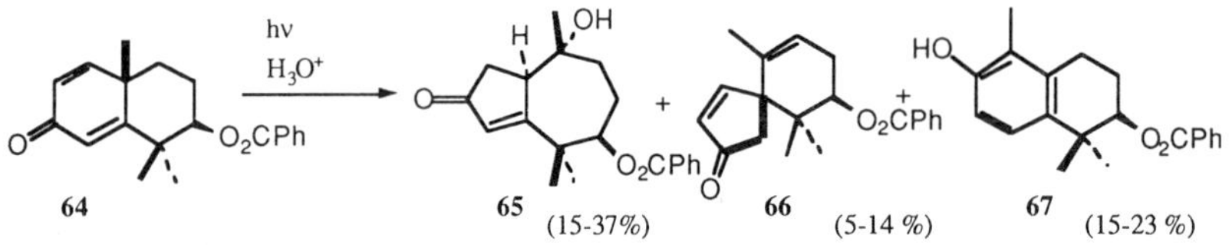

The tricyclic dienone **68a,** containing a 10α-methyl group and a β-oriented dimethylcyclopropane ring fused to the B ring, has been shown to be inert toward UV irradiation in aqueous acetic acid,[45a] benzene, or ether.[45b] The nonreactivity of **68a** apparently results from the fact that β,β'-bonding in an excited-state intermediate would lead to a highly strained tetracyclic system with the adjacent cyclopropane rings *cis*-fused on the six-membered ring. Interestingly, compound **68b,** which contains a carboxy group at C-2, was easily converted into the tricyclic dienone **69** upon irradiation in dioxane.[46] The manner in which the carboxy group renders **68b** photolabile is unclear. It may affect the energy and electronic distribution of the excited-state species sufficiently to allow the formation of an intramolecularly hydrogen bonded tetracyclic intermediate analogous to **34** or it may cause the intervention of a different reaction pathway.

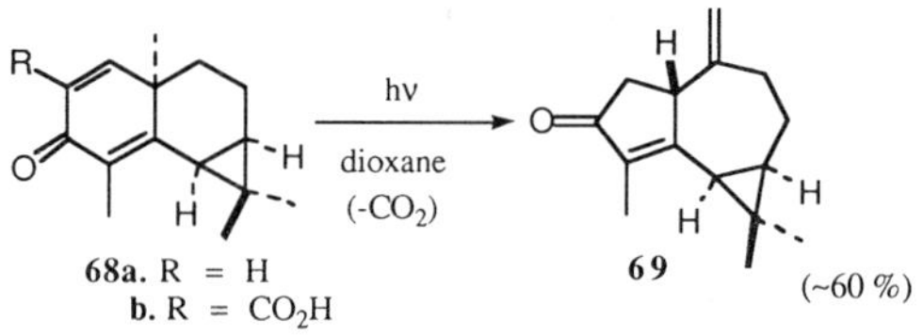

In contrast to **68a,** its 10-epimer **70a** readily rearranged to the corresponding lumiproduct **71**[47] or 5/7-fused hydroxy ketone **72**[48] upon irradiation in dioxane or aqueous acetic acid, respectively (Scheme 56.11). Apparently, β,β'-bonding to give intermediate **73** occurred readily in this case. The related ring A unsubstituted dienone **70b** reacted similarly to **70a** when irradiated in dioxane, but it was prone to undergo thermal opening of the cyclopropane ring in aqueous acetic acid. However, it gave a ring A unsubstituted 5/7-fused acetoxy ketone related to **72** in high yield upon irradiation in glacial acetic acid. The dimethylcyclopropane ring apparently prevents β-attack of the solvent at C-10 of intermediate **73b,** which would yield a spiro product.

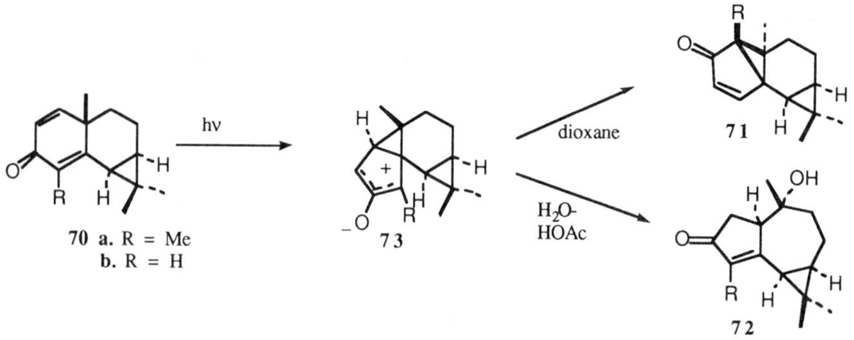

SCHEME 56.11

Bicyclic 6/6- and 6/5-fused dienones containing oxy substituents at the δ-position (C-9 or C-8) on ring B exhibit interesting photochemical behavior.[26] The 6/6-fused acetoxy dienone **74** underwent photochemical rearrangement to give largely the 5/7-fused hydroxy ketone **75** upon irradiation in aqueous acetic acid. This result is possibly attributable to participation of the acetoxyl group in the selective opening of the 5,10-bond of the cyclopropane ring in the mesoionic cyclopropyl intermediate **76.**[26]

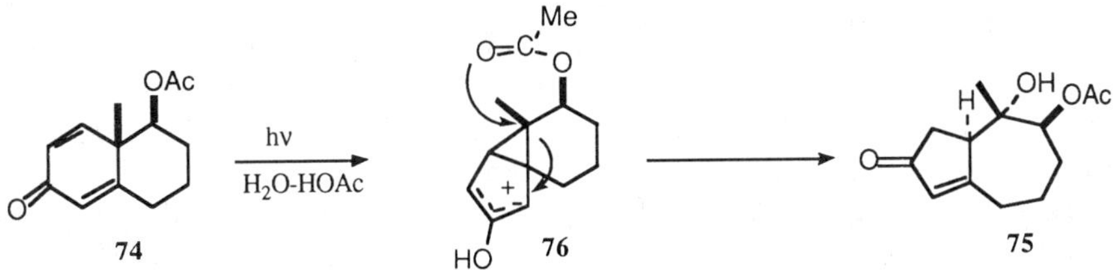

The 5/6-fused dienone **77** in which the δ-oxy substituent is protected as a *t*-butyldiphenylsilyl derivative gave the four photoproducts shown upon irradiation in glacial acetic acid.[26] Compounds **78** and **81** have structures similar to the products obtained from irradiation of the ring B unsubstituted, 6/5-fused dienone **6a** (n = 1) and probably arise by similar pathways (Scheme 56.5). The tricyclic products **79** and **80** represent the first examples of solvent trapping of cyclopropyl intermediates during photolysis of fused-ring dienones. In dienone **77**, it appears that the δ-oxy substituent reduces sufficiently the rate of cleavage of the cyclopropyl intermediate analogous to **11** (n = 1) to permit solvent attack at C-2 and C-4. This effect may be due to steric factors or be electronic in origin. The bulky C-8 substituent would reduce the rate of solvent attack at C-9, but, also, the electron-withdrawing effect of the oxy substituent may destabilize the transition state for opening of the cyclopropane ring by increasing the amount of positive charge at C-9. It should be noted that both types of dienones, i.e., **15** and **77**, in which solvent trapping of intermediates with the cyclopropane ring intact has been observed, contain electron-withdrawing δ-substituents.

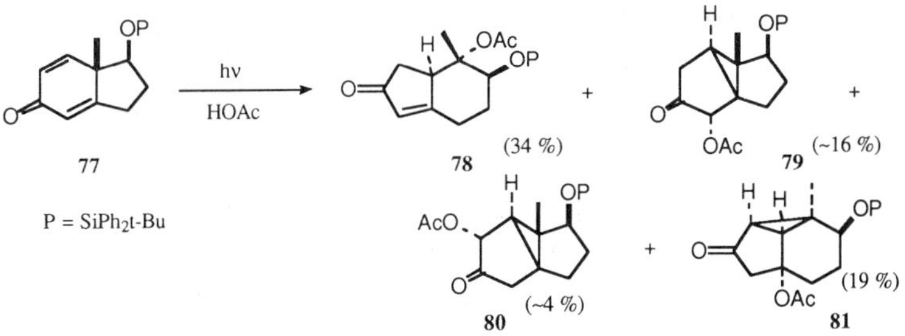

56.4 Synthetic Applications

Since photochemical rearrangements of cross-conjugated cyclohexadienones are highly stereoselective and subject to considerable amount of control depending upon the wavelength of the UV light and the solvent employed as well as substituent effects, these reactions have been used extensively as key steps in the synthesis of natural products — particularly, sesquiterpenes.

Several spirocyclic sesquiterpenes have been prepared from lumiproducts of the type **10a** (n = 2). Acid treatment of the 6α-methyl derivative of the lumiproduct **10a** (n = 2) gave a spiro dienone which was utilized in the first synthesis of racemic β-vetivone.[49] A similar reaction of the 6β-methyl compound also gave a spiro dienone precursor to the spirovetivane sesquiterpenes.[50] The lumiproduct prepared from irradiation of dehydro-α-vetivone in dioxane at 254 nm was converted into racemic dehydro-β-vetivone on irradiation in aqueous acetic acid with >300-nm light.[41] Catalytic reduction of this spirocyclic trienone gave a 3:1 mixture of racemic β-vetivone and racemic 10-epi-β-vetivone.[41] The lumiproduct **83**, derived from the 6/6-fused dienone **82**, was converted into the spirocyclic marine natural product (−)-axisonitrile-3 (**84**) in six steps.[51] The 2-methoxy acetoxy enone **30b** has also been employed as a key intermediate for the synthesis of racemic α-vetispirene.[32]

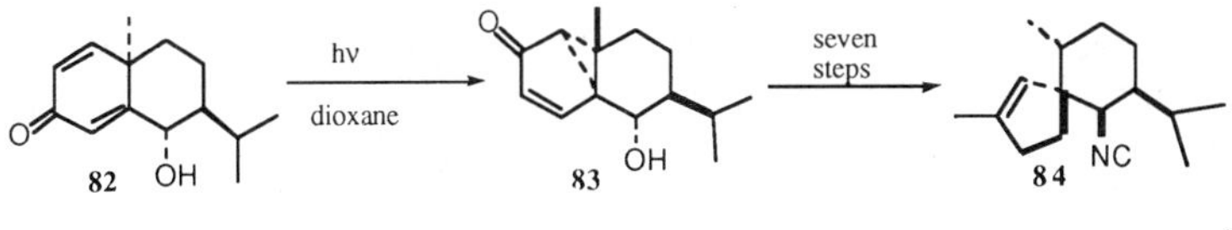

Photochemical rearrangements in protic media of sesquiterpenes such as α-santonin, with a eudesmane carbon skeleton, provide excellent routes to compounds such as 2, which have a guaiane skeleton. For example, the *O*-acetyl derivative of 2 has been converted into arborescin,[52] 1-epicyclocolorenone,[53] desacetoxymatricarin,[54] achillin,[55] dihydroarbiglovin,[56] and (–)-estafiatin.[57] Also, 8-epiartemisin has been transformed into geigerin acetate in very low yield via its photoproduct *O*-acetyl-8-epi-isophotoartemisic lactone.[58] 5-epi-α-Bulnesene, 4-epi-α-bulnesene, and α-bulnesene itself have been prepared via pathways involving the photochemical conversions of various cyclohexadienones containing a eudesmane skeleton into the corresponding 5/7-fused hydroxy ketones.[59]

Regio- and stereoselective hydrogenation of the *exo*-methylene group in the tricyclic dienone 69, produced from irradiation of dienone 68b, provided a facile approach to the sesquiterpene (–)-cyclocolorenone.[46] Likewise, the tricyclic hydroazulene derivative 72, derived from (–)-dehydroepimaalienone (70a), was a key intermediate in the synthesis of (–)-4-epiglobilol and (+)-4-epiaromadendrene.[48]

Upon irradiation in glacial acetic acid, the 6β-isopropyl derivative of the 4-methoxy 6/5-fused dienone 31b yielded the corresponding isopropyl derivative of the 6/5-fused acetoxy ketone 32b, which was converted into racemic oplopanone in nine steps.[34] 6/5-Fused dienones such as 77 and its C-8 epimer may be potentially useful intermediates for the synthesis of the highly functionalized oplopane sesquiterpene tussilagone.[26,60] The 6β-isopropyl derivative of dienone 6a (n = 1) gave the corresponding 5/6-fused acetoxy ketone on irradiation in glacial acetic acid. Expansion of ring A in this compound gave the racemic cadinane derivatives 3-oxo-α-cadinol and α-cadinol.[61] A 5/7/6-fused tricyclic acetoxy ketone has been prepared from the corresponding 6/6/6-fused dienone and used in a relay total synthesis of grayanotoxin II.[62]

References

1. Simonsen, J. and Barton, D. H. R., *The Terpenes,* Volume III, Cambridge University Press, New York, 292, 1952.
2. Barton, D. H. R., de Mayo, P., and Shafiq, M., Photochemical transformations. I. Some preliminary investigations, *J. Chem. Soc.,* 929, 1957.
3. Arigoni, D., Bosshard, H., Bruderer, H., Büchi, G., Jeger, O., and Krebaum, L. J., Photochemical reactions (2). Reciprocal relations and transformations in irradiation products of the santonins, *Helv. Chim. Acta,* 40, 1732, 1957.
4. Barton, D. H. R., de Mayo, P., and Shafiq, M., Photochemical transformations. II. The constitution of lumisantonin, *J. Chem. Soc.,* 140, 1958.
5. Barton, D. H. R. and Gilham, P. T., Photochemical transformations. IX. The stereochemistry of lumisantonin, *J. Chem. Soc.,* 4596, 1960.
6. Cocker, W., Crowley, K., Edward, J. T., McMurry, T. B. H., and Stuart, E. R., The chemistry of santonin. IV. Some irradiation products of the santonins, *J. Chem. Soc.,* 3416, 1957.
7. Barton, D. H. R., de Mayo, P., and Shafiq, M., Photochemical transformations. V. The constitution of photosantonic acid and derivatives, *J. Chem. Soc.,* 3314, 1958.
8. van Tamelen, E. E., Levin, S. H., Brenner, G., Wolinsky, J., and Aldrich, P. E., The structure of photosantonic acid, *J. Am. Chem. Soc.,* 81, 1666, 1959.
9. Chapman, O. L. and Englert, L. F., A mechanistically significant intermediate in the lumisantonin to photosantonic acid conversion, *J. Am. Chem. Soc.,* 85, 3028, 1963.
10. (a) Fisch, M. H. and Richards, J. H., The mechanism of photoconversion of santonin, *J. Am. Chem. Soc.,* 85, 3029, 1963; (b) The mechanism of the photochemical rearrangement of lumisantonin, *J. Am. Chem. Soc.,* 90, 1547, 1968.
11. Dutler, H., Bosshard, H., and Jeger, O., Photochemical reactions. Light-catalyzed dienone-phenol rearrangement, *Helv. Chim. Acta,* 40, 494, 1957.

12. Barton, D. H. R. and Taylor, W. C., Photochemical transformation. IV. The photochemistry of prednisone acetate, *J. Chem. Soc.*, 2500, 1958.

13. Kropp, P. J. and Erman, W. F., Photochemical rearrangements of cross-conjugated cyclohexadienones. II. Unsubstituted 1,4-diene-3-ones, *J. Am. Chem. Soc.*, 85, 2456, 1963.

14. Bozzato, G., Throndsen, H. P., Schaffner, K., and Jeger, O., The photochemical isomerization of B-nor-1-dehydrotestosterone acetate in dioxane solution, *J. Am. Chem. Soc.*, 86, 2073, 1964.

15. (a) Caine, D., Alexander, A. M., Ming, K., and Powers, W. J., III, Photochemical rearrangements of bicyclic 6/5-fused cross-conjugated cyclohexadienones and related compounds, *J. Org. Chem.*, 37, 706, 1972; (b) Powers, W. J., III, Photochemical Rearrangements of Cross-Conjugated Cyclohexadienones Related to Indane, Ph.D. Dissertation, Georgia Institute of Technology, Atlanta, GA, 1968.

16. Kropp, P. J., Photochemical transformations of cyclohexadienones and related compound, *Org. Photochem.*, 1, 1, 1967.

17. Schuster, D. I., Mechanisms of photochemical transformations of cross-conjugated cyclohexadienones, *Acc. Chem. Res.*, 11, 65, 1978.

18. Schaffner, K. and Demuth, M., Photochemical rearrangements of conjugated cyclic dienones, in *Rearrangements in Ground and Excited States*, Vol. 3, de Mayo, P. Ed., Academic Press, New York, 1980, 281.

19. Zimmerman, H. E. and Schuster, D. I., A new approach to mechanistic organic photochemistry. II. Photochemical rearrangements of 4,4-diphenylcyclohexadienone, *J. Am. Chem. Soc.*, 84, 4527, 1962.

20. (a) Woodward, R. B. and Hoffman, R., The conservation of orbital symmetry, *Angew. Chem. Int. Ed.*, 8, 781, 1969; (b) Coxon, J. M. and Halton, B. *Organic Photochemistry*, Cambridge University Press, New York, 1974, 92.

21. Zimmerman, H. E., Crumrine, D. S., Döpp, D., and Huyffer, P. S., Photochemistry without light and the stereochemistry of the type A dienone rearrangement. Organic photochemistry. XXXVIII, *J. Am. Chem. Soc.*, 91, 434, 1969.

22. Schuster, D. I. and Abraitys, V. Y., Isolation of a bicyclo[3.1.0]hexan-3-one on photolysis of a cyclohexan-2,5-dienone: Direct evidence for a zwitterion intermediate, *J. Chem. Soc., Chem. Commun.*, 419, 1969.

23. Schuster, D. I. and Liu, K.-C., Experimental proof for the intermediacy of zwitterions in the pathway leading to lumiketones upon irradiation of 2,5-cyclohexadienones, *Tetrahedron*, 37, 3329, 1981.

24. Samuel, C. J., The type A zwitterion in cyclohexadienone photochemistry: Chemical tapping and stereochemistry, *J. Chem. Soc., Perkin Trans. II*, 736, 1981.

25. Schultz, A. G., Myong, S. O., and Puig, S., Intramolecular azide cycloaddition to a photochemically generated Zwitterion, *Tetrahedron Lett.*, 25, 1011, 1984.

26. Caine, D., Kotian, P. L., and McGuiness, M. D., Synthesis and photochemical rearrangements of bicyclic cross-conjugated cyclohexadienones containing δ-oxy substituents, *J. Org. Chem.*, 56, 6307, 1991.

27. (a) Kropp, P. J., Photochemical rearrangements of cross-conjugated cyclohexadienones. V. A model for the santonin series, *J. Org. Chem.*, 29, 3110, 1964; (b) Caine, D. and Dawson, J. B. The synthesis of a model perhydroazulene derivative, *J. Org. Chem.*, 29, 3108, 1964.

28. Kropp, P. J., The photochemical properties of a 2-methyl-1,4-dien-3-one, *J. Am. Chem. Soc.*, 86, 4053, 1964.

29. Kropp, P. J. and Erman, W. F., Photochemical rearrangements of cross-conjugated cyclohexadienones. II. Unsubstituted 1,4-dien-3-ones, *J. Am. Chem. Soc.*, 85, 2456, 1963.

30. Caine, D., Gupton, J. T., III, Ming, K., and Powers, W. J., III, Photochemical rearrangements of 6/5-fused cross-conjugated cyclohexadienones in protic solvents, *J. Chem. Soc., Chem. Commun.*, 469, 1973.

31. Caine, D., Deutsch, H., Chao, S. T., Van Derveer, D. G., and Bertrand, J. A., Photochemical conversion of methoxy-substituted 6/6-fused cross-conjugated cyclohexadienones into isomeric tricyclodecenones, *J. Org. Chem.*, 43, 1114, 1978.

32. Caine, D., Boucugnani, A. A., Chao, S. T., Dawson, J. B., and Ingwalson, P. F., Stereospecific synthesis of 6,c-10-dimethyl(r-5-C′)spiro[4.5]dec-6-en-2-one and its conversion into (±)-α vetispirene, *J. Org. Chem.*, 41, 1539, 1976.

33. Pète, J. P. and Wolfhugel, J. L., Photochemical transformations of enones: Influence of the environment on the reactivity of α-methoxy cyclohexenones, *Tetrahedron Lett.*, 4637, 1973.

34. Caine, D. and Tuller, F. N., The total synthesis of *dl*-oplopanone, *J. Org. Chem.*, 38, 3663, 1973.

35. Caine, D., Brake, P. F., de Bardeleben, J. F., Jr., and Dawson, J. B., The influence of electron-withdrawing substituents on the photochemical behavior of bicyclic 6/6-fused cross-conjugated cyclohexadienones, *J. Org. Chem.*, 38, 967, 1973.

36. Broka, C. A., Photochemical rearrangements of carbomethoxy-substituted cyclohexadienones in neutral media, *J. Org. Chem.*, 53, 575, 1988.

37. Shiozaki, M., Mori, K., Matsui, M., and Hiraoka, T., Photochemical synthesis of compounds with a grayanotoxin skeleton, *Tetrahedron Lett.*, 657, 1972.

38. (a) Kropp, P. J., The photochemical behavior of 3-keto-9-carboethoxy $\Delta^{1,4}$-hexahydronaphthalene, *Tetrahedron Lett.*, 3647, 1964; (b) Kropp, P. J., personal communication.

39. Ming, K., Substituent Effect on Photochemical Rearrangement of Cross-Conjugated Cyclohexadienones Related to Indanone, Ph.D. Dissertation, Georgia Institute of Technology, 1972.

40. Burkinshaw, G. F., Davis, B. R., and Woodgate, P. D., The photochemistry of some 4-hydroxycyclohexa-2,5-dienones, *J. Chem. Soc. C*, 1607, 1970.

41. Caine, D., Chu, C.-Y., and Graham, S. L., Photochemical pathways for the interconversion of nootkatane and spirovetivane sesquiterpenes, *J. Org. Chem.*, 45, 3790, 1980.

42. Zimmerman, H. E. and Jones, G., II, The photochemistry of a cyclohexadienone structurally incapable of rearrangement. Exploratory and mechanistic organic photochemistry. XLVII, *J. Am. Chem. Soc.*, 92, 2753, 1970.

43. Mock, W. L. and Rumon, K. A., Photoreduction of 2,4-dimethyl-3-oxo-3,5,6,7,8,8a-hexahydro-1,8a-butanonaphthalene, a nonphotorearranging cross-conjugated cyclohexadienone, *J. Org. Chem.*, 37, 400, 1972.

44. Kropp, P. J., Photochemical rearrangements of cross-conjugated cyclohexadienones. III. An example of steric control, *J. Am. Chem. Soc.*, 85, 3779, 1963.

45. (a) Streith, J. and Blind, A., Stereospecific photochemical synthesis of aromadendrane derivatives, *Bull. Soc. Chem. Fr.*, 2133, 1968; (b) Kropp, P. J. and Krauss, H. J., Photochemical rearrangement of a γ, δ-cyclopropyl-α,β-unsaturated ketone, *J. Org. Chem.*, 32, 4118, 1967.

46. Caine, D. and Ingwalson, P. F., The influence of substituents on the photochemical behavior of cross-conjugated cyclohexadienones. A facile total synthesis of (−)-cyclocolorenone, *J. Org. Chem.*, 37, 3751, 1972.

47. Caine, D., Deutsch, H., and Gupton, J. T., III, Photochemical rearrangements of cross-conjugated cyclohexadienones related to epimaalienone, *J. Org. Chem.*, 43, 343, 1978.

48. Caine, D., and Gupton, J. T., III, Photochemical rearrangements of cross-conjugated cyclohexadienones. Application to the synthesis of (−)-4-epiglobulol and (+)-4-epiaromadendrene, *J. Org. Chem.*, 40, 809, 1975.

49. Marshall, J. A. and Johnson, P. C., The structure and synthesis of β-vetivone, *J. Org. Chem.*, 35, 192, 1970.

50. Caine, D., Boucugnani, A. A., and Pennington, W. R., A convenient photochemical synthesis of a precursor to the spirovetivanes, *J. Org. Chem.*, 41, 3632, 1976.

51. Caine, D. and Deutsch, H., Total synthesis of (−)-axisonitrile-3. An application of the reductive ring opening of vinylcyclopropanes, *J. Am. Chem. Soc.*, 100, 8030, 1978.

52. Suchy, M., Herout, V., and Sorm, F., Terpenes. CLXV. Structure of guainolides artabsin and arborescin, *Coll. Czech. Chem. Commun.*, 29, 1829, 1964.

53. Büchi, G., Kauffman, J. M. and Loewenthal, H. J. E., Synthesis of 1-epicyclocolorenone and the stereochemistry of cyclocolorenone, *J. Am. Chem. Soc.*, 88, 3403, 1966.

54. White, E. H., Eguchi, S., and Marx, J. N., The synthesis and stereochemistry of desacetoxymatricarin and the stereochemistry of matricarin, *Tetrahedron*, 25, 2099, 1969.

55. Marx, J. N. and White, E. H., The stereochemistry and synthesis of achillin, *Tetrahedron*, 25, 2117, 1969.

56. Marx, J. N. and McGaughey, S. M., Synthesis of dihydroarbiglovin and the stereochemistry of arbiglovin, *Tetrahedron*, 28, 3583, 1972.

57. Edgar, M. T., Greene, A. E., and Crabbé, P., Stereoselective synthesis of (−)-estafiatin, *J. Org. Chem.*, 44, 159, 1979.

58. Barton, D. H. R., Pinhey, J. T., and Wells, R. J., Photochemical transformations. XV. Synthetic studies on giegerin and its derivatives, *J. Chem. Soc.*, 2518, 1964.

59. Piers, E. and Cheng, K. F., Stereoselective synthesis of α-bulnesene, 4-epi-α-bulnesene, and 5-epi-α-bulnesene, *Can. J. Chem.*, 48, 2234, 1970.

60. Caine, D. and Kotian, P. L., Synthesis and photochemical rearrangement of (1R,7aS)-1-(*tert*-butyldiphenylsiloxy)-7a-methyl-5(7aH)-indanone, *J. Org. Chem.*, 57, 6587, 1992.

61. Caine, D. and Frobese, A. S., The photochemical total synthesis of (±)-3-oxo-α-cadinol and (±)-α-cadinol, *Tetrahedron Lett.*, 3107, 1977.

62. Gasa, S., Hamanaka, N., Matsunaga, S., Okuno, T., Takeda, N., and Matsumoto, T., Relay total synthesis of grayanotoxin II, *Tetrahedron Lett.*, 553, 1976.

Cyclohexadienone Photochemistry: Trapping Reactions

Arthur G. Schultz
Rensselaer Polytechnic Institute

57.1 Definition of the Reaction

In this chapter, trapping reactions of intermediates generated from photolysis of 2,5-cyclohexadien-1-ones will be considered. Both inter- and intramolecular reactions have been reported; but because the intermolecular photoreactions of 2,5- (and 2,4-) cyclohexadienones have been presented elsewhere in this handbook, the focus of discussion in this chapter will be on the intramolecular process. Developments in this area have been made only within the last few years; consequently, the full potential in organic synthesis largely remains to be established.

As shown below, irradiation of a 4,4-disubstituted 2,5-cyclohexadien-1-one **1**, usually in the region of 300–366 nm, produces an excited species [1]* that can undergo the type A photorearrangement[1] to a bicyclo[3.1.0]hex-3-en-2-one **2**. Alternatively, [1]* can react with a trapping agent (X) tethered to the cyclohexadienone ring.

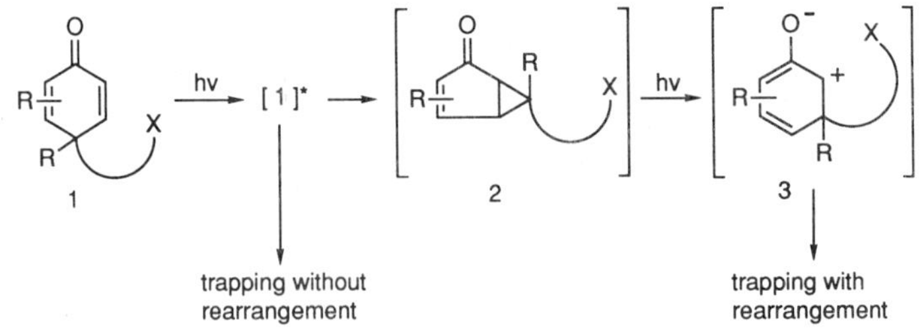

When **2** is produced, then it may be converted to the oxyallyl Zwitterion **3** in a second photochemical step (the type B photorearrangement).[2] Substituted phenols are usually obtained from **3** by rearrangement; but with suitably tethered reagents (X), the oxyallyl Zwitterion can be intercepted to give an interesting assortment of carbocyclic and heterocyclic systems. Although it is possible to envision substrates with attachments at C-2 through C-6, only examples with tethers at C-4 have been reported.

0-8493-8634-9/95/$0.00+$.50

The photoexcited state [4]* of a 2,4-cyclohexadien-1-one can rearrange to a diene ketene intermediate **5**. Most of the reported trapping reactions of **5** involve external nucleophiles such as water, alcohols, and amines; however, macrocyclizations are possible when the substrate contains an appropriate trapping agent (X). Outstanding progress has been made in the synthesis of macrolides by this strategy.[3] Direct trapping of [4]* prior to rearrangement has not as yet been reported.

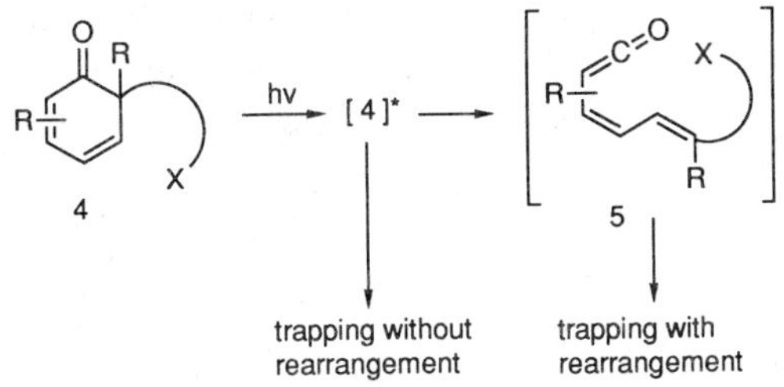

The reader should note that products arising from diene ketenes are occasionally observed from photoreactions of bicyclo[3.1.0]hex-3-en-2-ones.[4a] Furthermore, highly substituted 2,4-cyclohexadien-1-ones have been found to photorearrange to bicyclo[3.1.0]hexenones.[4b] Although these features of dienone photochemistry do not seem to have played a role in the trapping reactions to be considered in this chapter, they should be kept in mind when designing a synthetic plan that involves substantial modification of the substitution pattern of the prototype systems.

7.2 Mechanism

Trapping with Rearrangement

Photorearrangements of 4-alkyl-4-carbomethoxy-2,5-cyclohexadien-1-ones **6a** at 366 nm result in the formation of phenols **9** and **10**.[5] Phenol formation occurs by type A photorearrangement[1] of **6a** to bicyclo[3.1.0]hexenone **7**, type B photorearrangement[2] of **7** to oxyallyl Zwitterion **8**, and competing carbomethoxy group rearrangements to C-2 and C-4 of oxyallyl Zwitterion **8**. The exclusive migration of the carbomethoxy group in Zwitterion **8** is noteworthy. An analogous migration tendency has been determined for the carbethoxy group in the acid-catalyzed dienone-phenol rearrangement.[6]

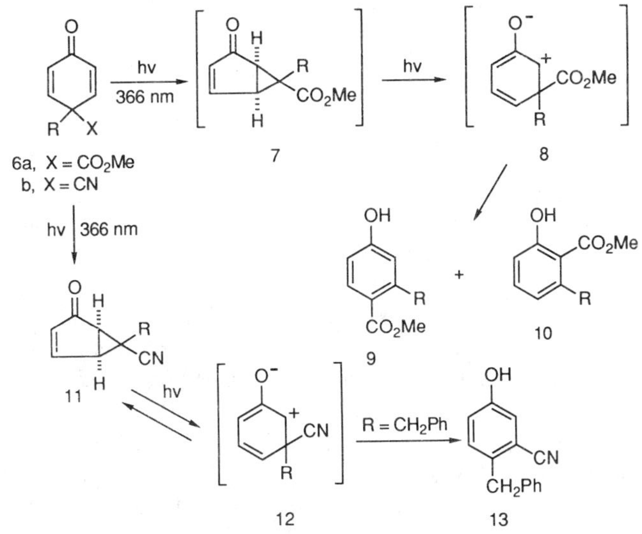

The corresponding photorearrangements of 4-alkyl-4-cyano-2,5-cyclohexadien-1-ones **6b** were found to give isolable bicyclo[3.1.0]hexenones **11** that do not undergo further photorearrangement to phenols.[5] However, irradiation of the 4-benzyl derivative of **6b** gave **11** (R = CH_2Ph), which photorearranged to 4-benzyl-3-cyanophenol (**13**) on continued irradiation.

These observations are consistent with reversible formation of the type B Zwitterion. Zwitterion **12** does not rearrange to phenolic products when groups with poor migratory aptitude are at C-4 of the starting 2,5-cyclohexadienone (e.g., cyano and primary alkyl); rather, reversion of **12** to **11** occurs presumably by a symmetry-allowed photochemical disrotatory electrocyclization of the pentadienyl cation located in **12**.[7] It has been noted that a thermal disrotatory closure of the type B Zwitterion is forbidden.[2]

Cycloaddition studies have confirmed that photorearrangements of 6-alkyl-6-cyanobicyclo[3.1.0]-hex-3-en-2-ones to the type B Zwitterion are reversible (e.g., **16** → **17**).[7] Irradiation of the cyclohexadienone **14b** gives a mixture of diastereomeric bicyclohexenones **16**. Continued irradiation of **16** results in a change in the distribution of bicyclohexenone diastereomers but does not produce type B photorearrangement products. However, irradiation of **14b** in the presence of 3 equivalents of furan afforded the furan-oxyallyl Zwitterion cycloadduct **18**, and irradiation of **16** under the same reaction conditions also gave **18**. Such intermolecular trapping reactions with furan have precedent in the earlier literature.[8a] The intramolecular furan trapping reaction (**14a** → **15**) has also been reported.[7]

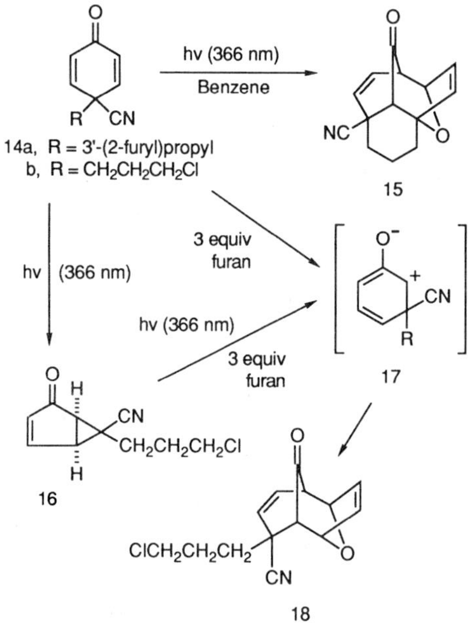

hv (366 nm)
Benzene
R
CN
14a, R = 3'-(2-furyl)propyl
b, R = CH2CH2CH2Cl
NC
O
15
3 equiv
furan
hv (366 nm)
hv (366 nm)
3 equiv
furan
O⁻
+
CN
R
17
H
CN
CH2CH2CH2Cl
H
16
ClCH2CH2CH2
CN
18

It is noteworthy that the type A Zwitterion generated photochemically from 4-methyl-4-trichloromethyl-2,5,cyclohexadien-1-one has been trapped with cyclopentadiene.[86] This trapping experiment is successful apparently because the type A intermediate in this case has a relatively long lifetime.

4-Carbomethoxy- and 4-cyano-3-methoxy-2,5-cyclohexadien-1-ones undergo photorearrangement at 366 nm to give mixtures of diastereomeric 6-carbomethoxy- and 6-cyano-4-methoxybicyclo[3.1.0]hex-3-en-2-ones; none of the corresponding 5-methoxy regioisomers were detected.[5] Continued irradiation of the bicyclohexenone diastereomers at 366 nm results in photoequilibration to give mainly the 6-*endo*-carbomethoxy- and 6-*endo*-cyanobicyclohexenones. Trace amounts of phenolic products are formed after extended irradiation of the 6-carbomethoxybicyclohexenones at 366 nm, while irradiation through Pyrex (≥300 nm) affords phenols as the sole reaction products. A wavelength effect on the quantum yields for photorearrangements of bicyclo[3.1.0]hex-3-en-2-ones has confirmed that the formation of phenols is relatively inefficient at 366 nm.[9]

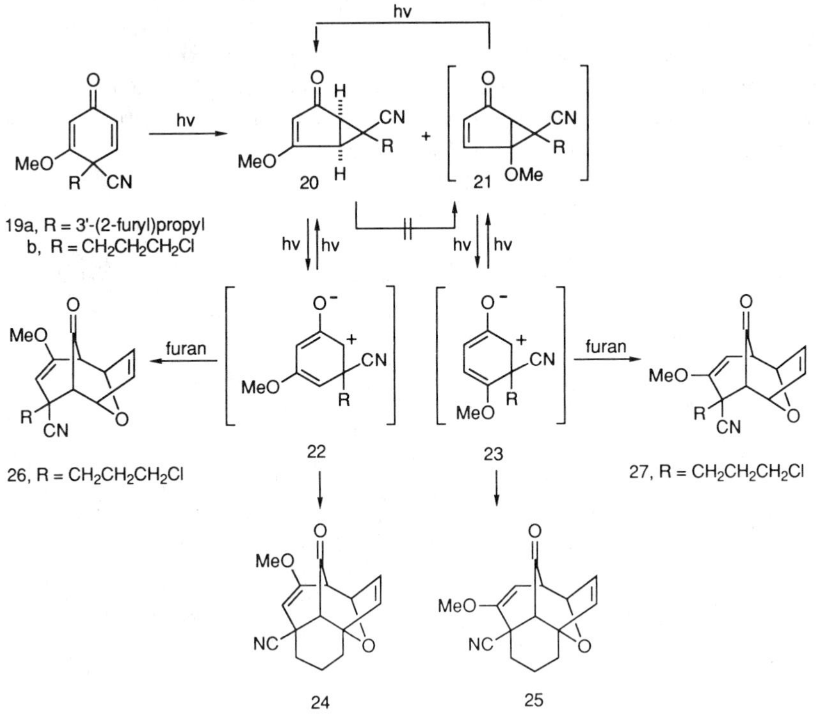

Irradiation of **19a** at 366 nm provided bicyclohexenone diastereomers **20** along with an intramolecular oxyallyl Zwitterion-furan cycloadduct **25** possessing an unexpected substitution pattern.[7] Bicyclohexenone **20** was found to be photostable at 366 nm but photorearranged to the intramolecular oxyallyl Zwitterion-furan cycloadduct **24** when irradiated through Pyrex.

The intermolecular oxyallyl Zwitterion-furan cycloaddition was examined with 3-methoxy-2,5-cyclohexadien-1-one **19b**. On irradiation at 366 nm, **19b** photorearranges to bicyclohexenone **20**, but Pyrex-filtered irradiation of **19b** in the presence of 3 equivalents of furan gave the 6-*endo*-cyanobicyclohexenone corresponding to structure **20** and cycloadducts **26** and **27**. At 366 nm, **19b** gave a mixture of bicyclohexenone diastereomers **20** and only the rearranged cycloadduct **27**.

A mechanistic rationale involving reversible formation of the oxyallyl Zwitterion **22** from **19a** and **19b** was used to explain the distributions of photoproducts **20**, **24**, and **26**. The isolation of the unexpected oxyallyl Zwitterion-furan cycloadducts **25** and **27** necessitated the formulation of a second type B Zwitterion, namely **23**. Thus, type A photorearrangements of **19a** and **19b** give not only the isolable 4-methoxybicyclo[3.1.0]hexenone **20**, but also the 5-methoxy regioisomer **21**. Bicyclohexenone **21** has not been detected from irradiations of **19b** in the absence of furan, which implies that **21** must undergo photorearrangement to the 4-methoxybicyclohexenone **20**; the mechanism for the conversion of **21** to **20** has not been elucidated.

Trapping Without Rearrangement

It has been found that the excited state of 4-(3′-alkenyl) and 4-(3′-pentynyl)-2,5-cyclohexadien-1-ones can be intercepted by the tethered unsaturation to give products of [2+2]-cycloaddition.[10] As illustrated below for **28a**, this process is in competition with normal type A/B photochemistry. There is some evidence to show that the type A photorearrangement can be somewhat selectively inhibited by irradiation of the 4-(3′-alkenyl)-2,5-cyclohexadien-1-one in pentane solution at −78 °C.[11]

The type A photorearrangement can be avoided completely by substitution at C-3 with a methoxy group. Irradiation of **28b** at 366 nm in benzene solution deaerated with N_2 gave the tricyclo[5.2.1.0^{5,10}]dec-2-en-4-one **31** in >95% yield with attendant formation of <5% of the regioisomeric tricyclodecenone. Thus, while the 3-methoxy substituent in **31** facilitates intramolecular [2+2]-cycloaddition, cyclobutane formation occurs primarily at the unsubstituted double bond.[10] Stereochemical studies indicate that 1,4-biradicals precede cyclobutane formation. The 4-allyl and 4-(4'-pentenyl)-2,5-cyclohexadien-1-ones do not undergo [2+2]-photocycloadditions under conditions utilized for **31** and related substrates.[10]

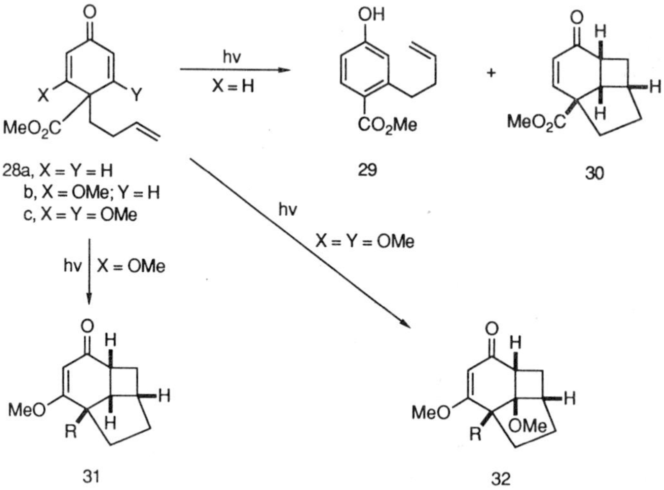

28a, X = Y = H
b, X = OMe; Y = H
c, X = Y = OMe

29

30

31

32

It is believed that the 3-methoxy group promotes [2+2]-cycloaddition by a mechanism that involves mixing of the π,π^* triplet state with the n,π^* triplet state normally associated with type A reactivity. Whereas 4-carbomethoxy-3,5-dimethoxy-4-methyl-2,5-cyclohexadien-1-one has been found to be unreactive in the type A manifold, it does undergo [2+2]-photocycloaddition to cyclopentene,[12] and 4-(3'-butenyl)-4-carbomethoxy-3,5-dimethoxy-2,5-cyclohexadien-1-one (**28c**) was found to give tricyclodecenone **32** in quantitative yield.[10]

The development of the diastereoselective Birch reduction-alkylation[13] has provided the wherewithal to examine the enantioselectivity of 2,5-cyclohexadienone photoreactions. In contrast to (4R)-4-carbomethoxy-3-methoxy-4-methyl-2,5-cyclohexadien-1-one, which was found to undergo partial photoracemization under type A conditions, enantiomerically pure **28b** underwent intramolecular [2+2]-photocycloaddition without racemization.[10] The latter result, if general, will be of importance to workers considering the utilization of the [2+2]-photocycloaddition in asymmetric organic synthesis.

Acetoxymethyl derivatives **33a** and **33b** have been converted to tricyclodecenones **34a** and **34b** in essentially quantitative yield.[10] Thus, the presence of a C-4 strongly electron-withdrawing substituent is not essential for the diversion of the photochemical reactivity of 4-(3'-butenyl)-2,5-cyclohexadienones from type A behavior to intramolecular cycloaddition. This is noteworthy because it has been found that electronegative substituents at C-4 tend to lower the triplet-state energy of the 2,5-cyclohexadienone.[14]

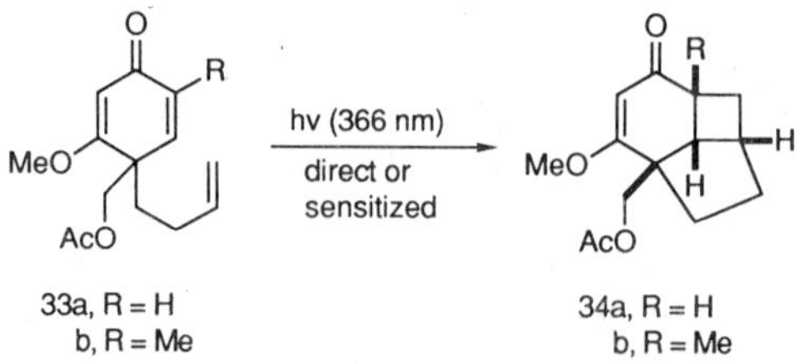

33a, R = H
b, R = Me

34a, R = H
b, R = Me

The photoconversion of **33a** to **34a** can be sensitized by benzophenone (E_T, 69 kcal mol^{-1})[15] and xanthone (E_T, 74 kcal mol^{-1});[15] however, Michler's ketone (E_T, ~62 kcal mol^{-1})[16] did not sensitize the photoconversion of **28b** to **31**. The experimentally determined triplet-state energies for 2,5-cyclohexadienones without substituents at C-3 have been reported to be 67 to 71 kcal mol^{-1}.[14,17] Xanthone may be more efficient than benzophenone in triplet-energy transfer to 2,5-cyclohexadienone **33a**, and energy transfer from Michler's ketone to **33a** appears to be prohibitively endothermic.[10] It also has been reported that Michler's ketone does not sensitize the photoconversion of santonin to lumisantonin.[18]

The photoreactivity of **33a** is not significantly depressed by the presence of up to 1.5 M piperylene. Generally, triplet quenching studies of 2,5-cyclohexadienones have used extremely high concentrations of quenchers.[17–20] It has been suggested that triplet quenching is inefficient because the triplet state is short lived.[20]

Photocyclization by way of the triplet state of the 4-(4′-phenyl-3′-butenyl) substituent has been reported for the 3-methoxy-2,5-cyclohexadienone **35a**.[21] The conversion to **36a** can be carried out by direct irradiation at 366 nm or by sensitization with Michler's ketone. Inasmuch as the photoconversion of **28b** to **31** is not sensitized by Michler's ketone, it has been reasonably concluded that the sensitized conversion of **35a** to **35b** occurs by energy transfer from Michler's ketone to the 4-(4′-phenyl-3′-butenyl)chromophore in **35a** (E_T styrene, 61.7 kcal mol^{-1}).[16]

Direct irradiation of **35b** gave **36b** in 84% isolated yield. The absence of significant quantities of type A photoproducts from **35b** is compatible with a mechanism involving efficient quenching of the photoexcited 2,5-cyclohexadienone by the styrene chromophore. However, an alternative mechanism involving faster rates of cyclization of the excited 2,5-cyclohexadienone to the 4′-phenyl-3′-butenyl group compared to the 3′-butenyl substituent cannot be excluded. Although mechanistic details are insufficient at present, it appears that this modification should be synthetically useful, especially with 2,5-cyclohexadienones that are prone to undergo type A photorearrangements rather than [2+2]-photocycloaddition (*cf.*, **28a** → **29** + **30**).

35a, X= OMe
b, X = H

36a, X = OMe
b, X = H

7.3 Synthetic Applications

Trapping with Rearrangement

The type B oxyallyl Zwitterion **3**, generated from a 2,5-cyclohexadien-1-one by two successive photorearrangements, undergoes intramolecular cycloadditions with furan, azide, and alkene subunits. The trapping agent has been attached to C-4 of the dienone ring by way of a two- or three-atom tether. This combination results in the formation of a five- or six-membered ring that is common to both rings of the cycloadduct nucleus. Ketone and alcohol groups at C-11 of steroidal cross-conjugated dienones also have been shown to be effective trapping reagents for the type B Zwitterion.

Irradiation of 4-(furfuryloxy)-2,4,6-trimethylcyclohexa-2,5-dien-1-one (**37**) in dry, degassed benzene solution (6×10^{-3} M) at 366 nm for 1 h, followed by flash chromatography on silica gel, provided the bridged furan adduct **38** in 80% isolated yield.[22] In an analogous fashion, photolysis of **39a** gave **40a** in nearly quantitative yield.[23]

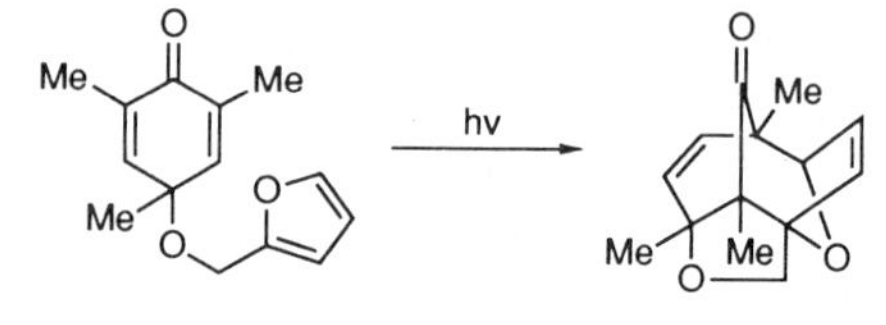

37 → 38

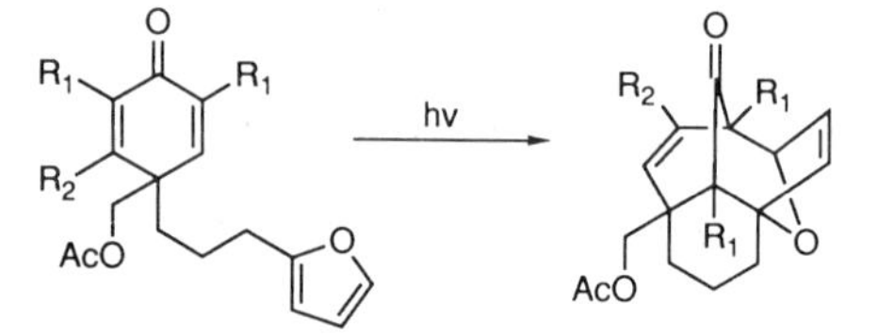

39a, R$_1$ = Me; R$_2$ = H
  b, R$_1$ = H; R$_2$ = OMe

40a, R$_1$ = Me; R$_2$ = H
  b, R$_1$ = H; R$_2$ = OMe

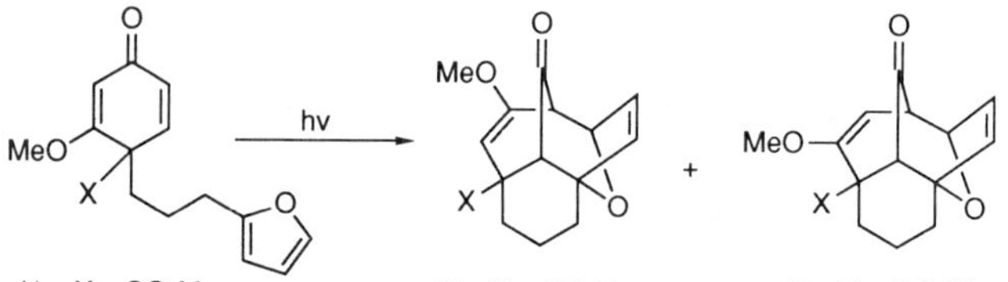

41a, X = CO$_2$Me
  b, X = CN

42a, X = CO$_2$Me
  b, X = CN

43a, X = CO$_2$Me
  b, X = CN

It has been found that Zwitterion addition to the furan ring in the conversions 37 → 38, 39a → 40a, and all other cases examined occur from an *endo*-orientation. *endo*-Additions also have been reported for the intermolecular process.[8a] Also of interest is the observation that the vinyl-substituted Zwitterionic unit in the type B intermediate reacts exclusively as a two-electron component in cycloadditions to the furan ring.

Irradiation of 39b taken to partial conversion gives a mixture of intermediate bicyclo[3.1.0]hex-3-en-2-ones and the bridged furan adduct 40b. After further irradiation (30 h) of the mixture, the bicyclohexenones were completely converted to 40b. The reluctance of 4-methoxybicyclo[3.1.0]-hexenones to undergo the type B photorearrangement at 366 nm requires that Pyrex-filtered light be used in experiments involving 3-methoxy-2,5-cyclohexadien-1-ones.

Additional complications arise when the 3-methoxy group is present along with the 4-carbomethoxy or 4-cyano substituent. Mixtures of regioisomeric cycloadducts are produced from photolyses of 41a and 41b, and migration of the carbomethoxy group is competitive with cycloaddition in the type B Zwitterion generated from 41a.

Dipolar cycloadditions to the type B Zwitterion represent a potentially rich area for synthetic investigation. To date, only the relatively stable alkyl azide group has been examined. Irradiations of 44a and 44b in THF solutions with 366-nm light give bicyclic triazenes 45a and 45b in 64 and 75% isolated yields, respectively.[24]

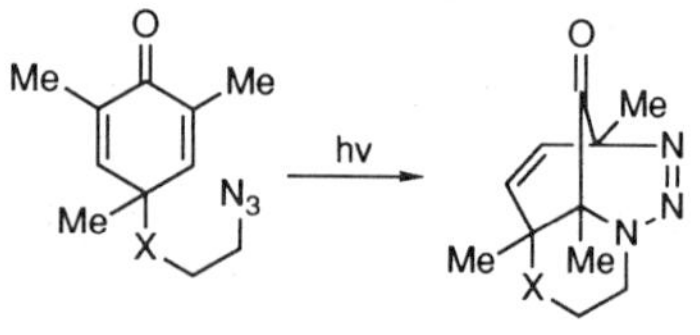

44a, X = CH$_2$
  b, X = O

45a, X = CH$_2$
  b, X = O

Photolysis of the 4-(3'-azidopropyl)-4-carbomethoxy-2,5-cyclohexadien-1-one (46a) gave 3-(3'-azidopropyl)phenols. Only 6-(3'-azidopropyl)-6-*endo*-carbomethoxy-4-methoxybicyclo[3.1.0]-

hex-3-en-2-one (80%) was obtained from **46b**; **46c** did provide a bridged triazene **47**, albeit in only ~10% yield. These data suggest that the azide is somewhat less reactive than the furan group in oxyallyl Zwitterion trapping reactions.

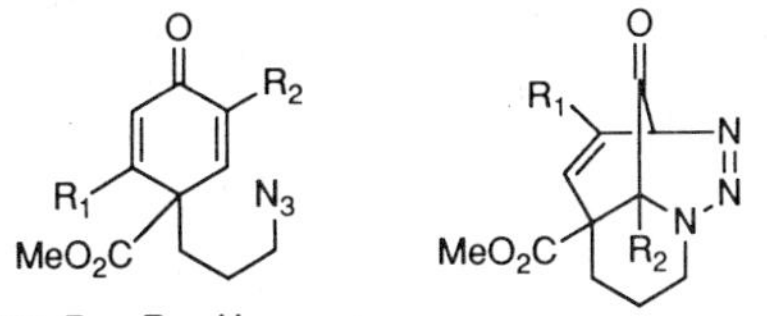

46a, $R_1 = R_2 = H$
b, $R_1 = OMe$; $R_2 = H$
c, $R_1 = R_2 = OMe$

47, $R_1 = R_2 = OMe$

Intramolecular trapping reactions of the type B Zwitterion with tethered alkenyl groups have provided examples of every anticipated mode of cycloaddition. The Zwitterionic unit in **48** behaves as a four-electron component in the two [5+2]-atom cycloadditions resulting in product types **49** and **50**, while [3+2]-atom cycloadditions give rise to either dienol ether **51** or bridged cyclopentanone **52**.[25]

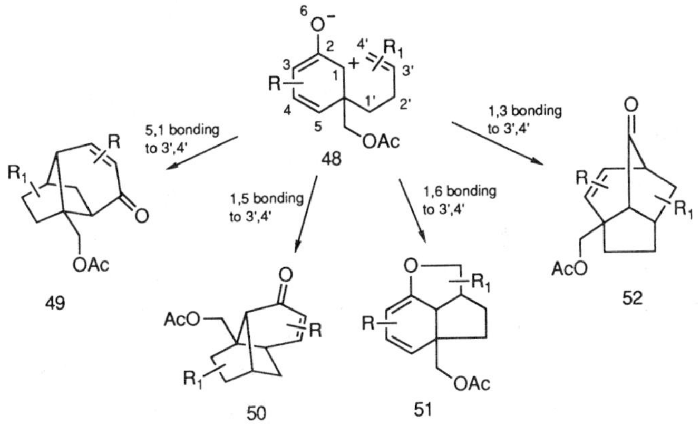

Irradiation of **53a** at 366 nm gave an inseparable 3:1 mixture of **55a** and **56a**; acetate cleavage and chromatographic separation of the alcohols gave **55b** (61%) and **56b** (10%). Similarly, **53b** gave **55d** and **56d**.

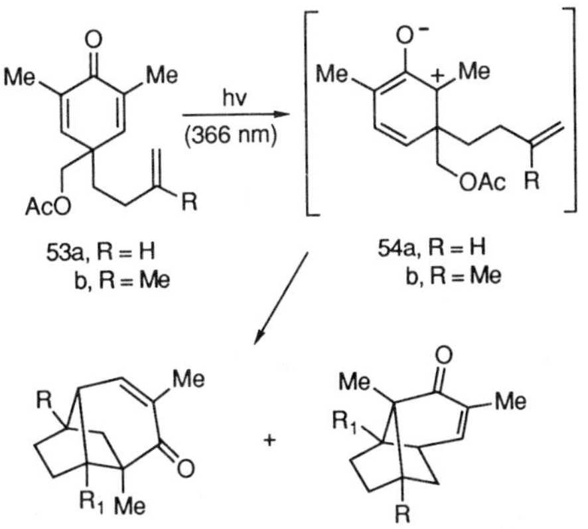

53a, R = H
b, R = Me

54a, R = H
b, R = Me

55a, R = H; R_1 = CH_2OAc
b, R = H; R_1 = CH_2OH
c, R = Me; R_1 = CH_2OAc
d, R = Me; R_1 = CH_2OH

56a, R = H; R_1 = CH_2OAc
b, R = H; R_1 = CH_2OH
c, R = Me; R_1 = CH_2OAc
d, R = Me; R_1 = CH_2OH

The effect of substituents at C-4′ of the tether has been examined by photolyses of dienones **57a–c**. The major products obtained from **57a** and **57c** are dienol ethers **58a** and **58c**. The photoreaction mixtures are complicated; in the case of vinyl aldehyde **57c**, oligomerization predominates. By contrast, photolysis of the *trans*-allylic alcohol derivative **57b** gave dienol ether **58b** in 71% isolated yield.

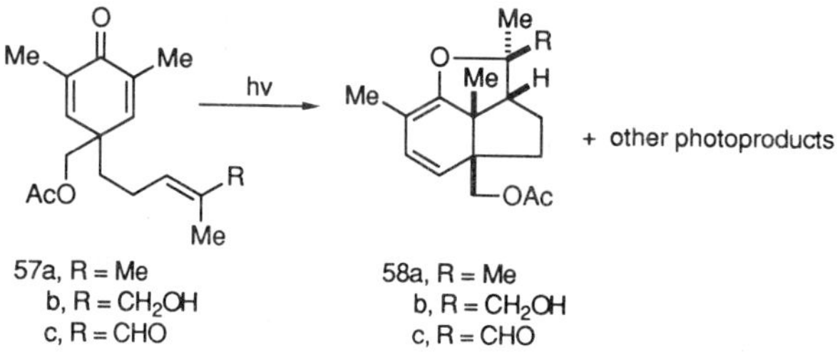

57a, R = Me
 b, R = CH₂OH
 c, R = CHO

58a, R = Me
 b, R = CH₂OH
 c, R = CHO

A study of the photorearrangement chemistry of 11-keto steroid **58** has uncovered some intramolecular trapping reactions of potential preparative value.[26] It has been found that photolysis of **58** in dioxane first generates the bicyclo[3.1.0]hexenone **59**, which rearranges to a 4-methylphenol on continued irradiation. The formation of 4-methylphenol occurs by methyl migration in the intermediate type B Zwitterion **60**; in 45% aqueous acetic acid, **59** gives the internal hemiketal **61a** in 47% yield. In a subsequent study,[27] it was found that the analogous 11β-hydroxy analogue of **59** gave **61b** in 86% yield, whereas the epimer gave the expected 4-methylphenol because the 11α-hydroxy group is too far removed to be an effective trapping agent.

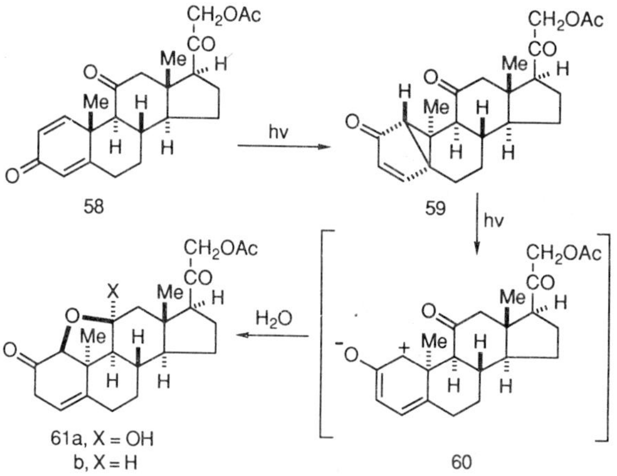

58

59

61a, X = OH
 b, X = H

60

Trapping Without Rearrangement

The 3-methoxy-2,5-cyclohexadien-1-ones **62a–e** undergo efficient [2+2]-photocycloaddition to give cyclobutanes with a synthetically useful range of positional substitution. Photocycloadditions of both *cis*- and *trans*-3′-pentenyl derivatives give a 9:1 mixture of cyclobutanes **63a** and **63b**. Examples **62c–e** demonstrate that it is possible to generate quaternary centers on the cyclobutane ring, and with **62e** adjacent quaternary centers. It is remarkable that type A photoreactivity is not encountered even in these examples that involve substantial steric crowding in the derived cycloadduct.[10]

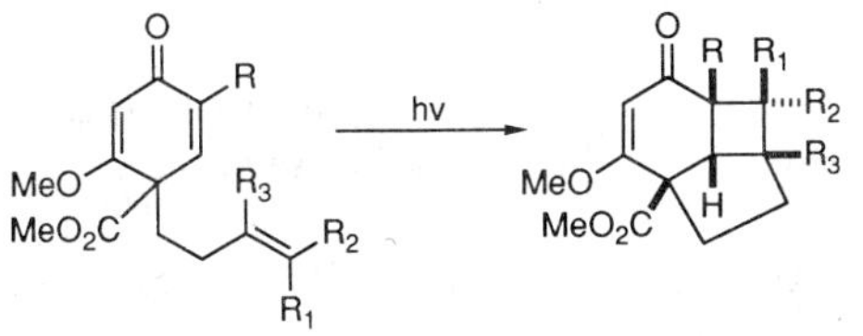

62a, R_1 = Me; R = R_2 = R_3 = H
b, R_2 = Me; R = R_1 = R_3 = H
c, R_3 = Me; R = R_1 = R_2 = H
d, R_1 = R_2 = Me; R = R_3 = H
e, R = R_1 = R_2 = Me; R_3 = H

63a, R_1 = Me; R = R_2 = R_3 = H
b, R_2 = Me; R = R_1 = R_3 = H
c, R_3 = Me; R = R_1 = R_2 = H
d, R_1 = R_2 = Me; R = R_3 = H
e, R = R_1 = R_2 = Me; R_3 = H

In the single example of acetylene participation, it has been shown that **64** gave cyclobutene **65**. Hydrogenation of **65** provides **63b** in a stereospecific complement to the photocyclization involving the alkenyl derivative **62a**.[10]

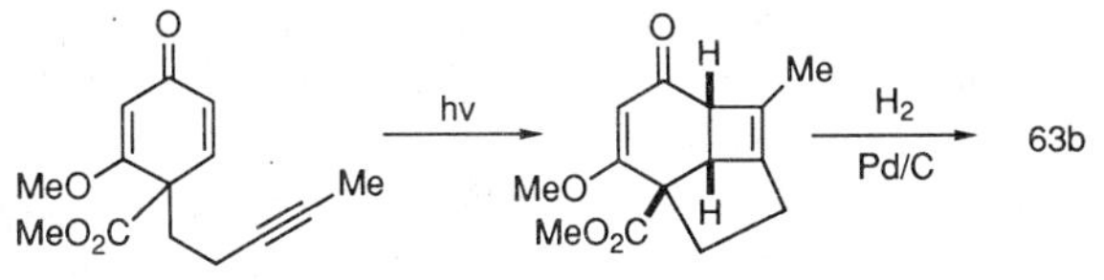

The regio- and diastereoselectivity of the intramolecular [2+2]-photocycloaddition of 4-(2'-isopropyl-3'-butenyl)-2,5-cyclohexadienones has been examined.[11] Solid-state studies have revealed some interesting and perhaps synthetically useful effects of the packing of molecules in crystals on the diastereoselectivity of a 4-(3'-butenyl)-2,5-cyclohexadienone.[28] The intramolecular [2+2]-photocycloadditions of 4-carbomethoxy-3-methoxy-4-(1'-oxa-3'-butenyl)-2,5-cyclohexadien-1-one has been found to be less regioselective than the corresponding 4-(3'-butenyl) analogue (e.g., **28b**).[29]

Very recently, it has been reported that α-silyamino- and α-silylamido-2,5-cyclohexadien-1-ones undergo 9,10-dicyanoanthracene (DCA)-SET-sensitized radical cyclization to form *cis*-ring-fused hydroisoquinolines in modest to good yields; e.g., **66** → **67**. The major direct irradiation reaction pathway for the α-silylamido derivatives involves type A photochemistry, whereas direct irradiations of the α-silyamino analogues give near-exclusive conversion to the *cis*-hydroisoquinoline. These photocyclizations are expected to have value in yohimbane and related alkaloid syntheses.[30]

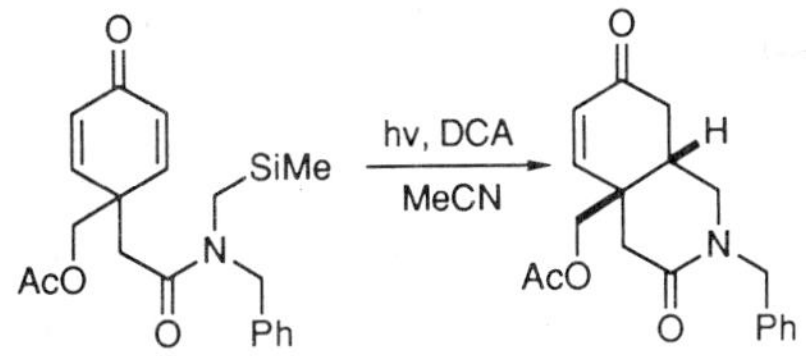

References

1. Zimmerman, H. E. and Schuster, D. I., A new approach to mechanistic organic photochemistry. IV. Photochemical rearrangements of 4,4-diphenylcyclohexadienone, *J. Am. Chem. Soc.*, 84, 4527, 1962.

2. Zimmerman, H. E. and Epling, G. A., Generation of photochemical intermediates without light. The type B Zwitterion. Mechanistic origins, *J. Am. Chem. Soc.*, 94, 7806, 1972.

3. Quinkert, G., Billhardt, U.-M., Jakob, H., Fischer, G., Glenneberg, J., Nagler, P., Autze, V., Heim, N., Wacker, M., Schwalbe, T., Kurth, Y., Bats, J. W., Dürner, G., Zimmermann, G., and Kessler, H., Photolactonisierung: ein neuer synthetischer zugang zu makroliden, *Helv. Chim. Acta*, 70, 771, 1987.

4. (a) Zimmerman, H. E., Keese, R., Nasielski, J., and Swenton, J. E., Mechanistic organic photochemistry. XVI. The photochemistry of 6,6-diphenylbicyclo-[3.1.0]hex-3-en-2-one, *J. Am. Chem. Soc.*, 88, 4895, 1966; (b) Griffiths, J. and Hart, H., A new general photochemical reaction of 2,4-cyclohexadienones, *J. Am. Chem. Soc.*, 90, 5296, 1968.

5. Schultz, A. G., Lavieri, F. P., Macielag, M., and Plummer, M., 2,5-Cyclohexadien-1-one to bicyclo[3.1.0]hexenone photorearrangement. Development of the reaction for use in organic synthesis, *J. Am. Chem. Soc.*, 109, 3991, 1987.

6. Marx, J. N., Argyle, J. C., and Norman, L. R., Migration of electronegative substituents. I. Relative migratory aptitude and migration tendency of the carbethoxy group in the dienone-phenol rearrangement, *J. Am. Chem. Soc.*, 96, 2121, 1974.

7. Schultz, A. G. and Reilly, J., 2,5-Cyclohexadien-1-one photochemistry. Reversible type B oxyallyl Zwitterion formation from photorearrangements of bicyclo[3.1.0]hex-3-en-2-ones, *J. Am. Chem. Soc.*, 114, 5068, 1992.

8. Chapman, O. L., Clardy, J. C., McDowell, T. L., and Wright, H. E., Photoisomerization of 5,5-dimethylbicyclo[4.1.0]hept-3-en-2-one and 4,6,6-trimethylbicyclo[3.1.0]hex-3-en-2-one, *J. Am. Chem. Soc.*, 95, 5086, 1973; (b) Samuel, C. J., The type A Zwitterion in cyclohexadienone photochemistry: Chemical trapping and stereochemistry, *J. Chem. Soc., Perkin II*, 736, 1981.

9. Dauben, W. G., Cogen, J. M., Behar, V., Schultz, A. G., Geiss, W., and Taveras, A. G., Wavelength dependent photoisomerization of bicyclo[3.1.0]-hexenones, *Tetrahedron Lett.*, 33, 1713, 1992.

10. Schultz, A. G., Plummer, M., Taveras, A. G., and Kullnig, R. K., Intramolecular 2+2 photocycloadditions of 4-(3′-alkenyl)- and 4-(3′-pentynyl)-2,5-cyclohexadien-1-ones, *J. Am. Chem. Soc.*, 110, 5547, 1988.

11. Schultz, A. G., Geiss, W., and Kullnig, R. K., The regio- and diastereoselectivity of the intramolecular 2+2 photocycloadditions of 4-(2′-isopropyl-3′-butenyl)-2,5-cyclohexadien-1-ones, *J. Org. Chem.*, 54, 3158, 1989.

12. Schultz, A. G. and Taveras, A. G., The first intramolecular 2 + 2 photocycloadditions of 2,5-cyclohexadien-1-ones to alkenes, *Tetrahedron Lett.*, 29, 6881, 1988.

13. Schultz, A. G., Enantioselective methods for chiral cyclohexane ring synthesis, *Acc. Chem. Res.*, 23, 207, 1990.

14. Zimmerman, H. E., Binkley, R. W., McCullough, J. J., and Zimmerman, G. A., Organic photochemistry. XXVII. Electronically excited states, *J. Am. Chem. Soc.*, 89, 6589, 1967.

15. Herkstroeter, W. G., Lamola, A. A., and Hammond, G. S., Mechanisms of photochemical reactions in solution. XXVII. Values of triplet excitation energies of selected sensitizers, *J. Am. Chem. Soc.*, 86, 4537, 1964.

16. Murov, S. L., *Handbook of Photochemistry*, Marcel Dekker, New York, 1973, Section 1.

17. Zimmerman, H. E. and Swenton, J. S., Mechanistic organic photochemistry. VIII. Identification of the n-π* triplet excited state in the rearrangement of 4,4-diphenylcyclohexadienone, *J. Am. Chem. Soc.*, 86, 1436, 1964.

18. Fisch, M. H. and Richards, J. H., The mechanism of the photoconversion of santonin, *J. Am. Chem. Soc.*, 85, 3029, 1963.

19. Zimmerman, H. E. and Swenton, J. J., Mechanistic organic photochemistry. XXI. Electronic details of the 2,5-cyclohexadienone rearrangement, *J. Am. Chem. Soc.*, 89, 906, 1967.

20. Schuster, D. I., Fabian, A. C., Kong, N. P., Barringer, W. C., Curran, W. V., and Sussman, D. H., Efficiencies of quenching of short-lived excited triplet states of ketones with dienes, *J. Am. Chem. Soc.*, 90, 5027, 1968.

21. Schultz, A. G. and Geiss, W., Effect of triplet state sensitization on inter- and intramolecular 2 + 2 photocycloadditions of 2,5-cyclohexadien-1-ones, *J. Am. Chem. Soc.*, 113, 3490, 1991.

22. Schultz, A. G., Puig, S., and Wang, Y., The photochemistry of cyclohexa-2,5-dien-1-ones. Intramolecular cycloaddition of a furan to an intermediate oxyallyl Zwitterion, *J. Chem. Soc., Chem. Commun.*, 785, 1985.

23. Schultz, A. G., Macielag, M., and Plummer, M., Intramolecular cycloadditions to oxyallyl Zwitterions generated from photorearrangements of 2,5-cyclohexadien-1-ones, *J. Org. Chem.*, 53, 391, 1988.

24. Schultz, A. G., Myong, S. O., and Puig, S., Intramolecular azide cycloaddition to a photochemically generated Zwitterion, *Tetrahedron Lett.*, 25, 1011, 1984.

25. Schultz, A. G. and Plummer, M., Intramolecular cycloadditions of alkenes to oxyallyl Zwitterions generated from photorearrangements of 2,5-cyclohexadien-1-ones, *J. Org. Chem.*, 54, 2112, 1989.

26. Williams, J. R., Moore, R. H., Li, R., and Blount, J. F., Structure and photochemistry of lumiprednisone and lumiprednisone acetate, *J. Am. Chem. Soc.*, 101, 5019, 1979.

27. Williams, J. R., Moore, R. H., Li, R., and Weeks, C. M., Photochemistry of 11α- and 11β-hydroxy steroidal bicyclo[3.1.0]hex-3-en-2-ones in neutral and acidic media, *J. Org. Chem.*, 45, 2324, 1980.

28. Schultz, A. G., Taveras, A. G., Taylor, R. E., Tham, F. S., and Kullnig, R. K., Solid-state photochemistry. Remarkable effects of the packing of molecules in crystals on the diastereoselectivity of the intramolecular 2 + 2 photocycloaddition of a 4-(3′-butenyl)-2,5-cyclohexadien-1-one, *J. Am. Chem. Soc.*, 114, 8725, 1992.

29. Schultz, A. G. and Harrington, R. E., Substituent effects on the photorearrangements of 4-alkoxy-4-carbomethoxy-3-methoxy-2,5-cyclohexadien-1-ones, *J. Org. Chem.*, 56, 6391, 1991.

30. Jung, Y. S., Swartz, W. H., Xu, W., Mariano, P. S., Green, N. J., and Schultz, A. G., Exploratory studies of α-silylamino and α-silylamido 2,5-cyclohexadien-1-one SET photochemistry. Methodology for synthesis of functionalized hydroisoquinolines, *J. Org. Chem.*, 57, 6037, 1992.

58

Photorearrangement Reactions of Linearly Conjugated Cyclohexadienones

Arthur G. Schultz
Rensselaer Polytechnic Institute

58.1 Definition of the Reaction

Linearly conjugated dienones 1 undergo reversible photochemical ring opening to diene ketenes 2, which afford diene carboxylic acid derivatives 3 by reaction with protic nucleophilic reagents.[1] With certain heavily substituted dienones, the intermediate diene ketene can undergo intramolecular [4+2]-cycloaddition to give bicyclo[3.1.0]hexenones 4; direct photorearrangement of 1 to 4 is also possible.[2] Photorearrangements of 2-acyloxy substituted dienones to phenolic products have been observed; e.g., 1 → 5.[1]

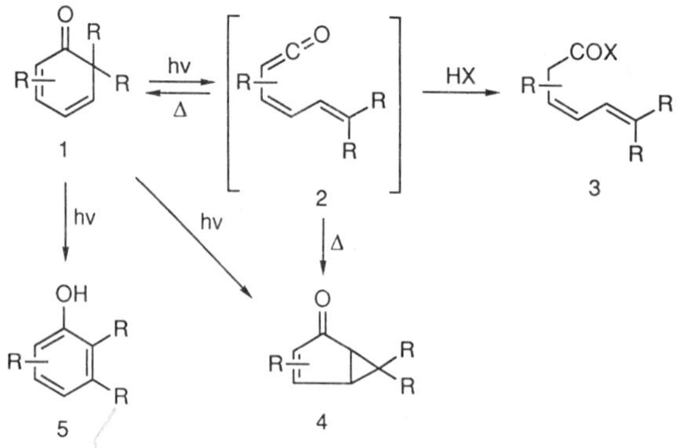

0-8493-8634-9/95/$0.00+$.50

8.2 Mechanism

Early developments in the photochemistry of 2,4-cyclohexadienones have been reviewed in detail.[3-6] Diene ketenes 2 generally are assumed to be reactive intermediates in the photocleavage of 1; however, direct evidence for ketene 7 has been obtained from the IR spectrum of the mixture produced by irradiation of 6-methyl-6-dichloromethyl-2,4-cyclohexadienone (6) as a liquid nitrogen cooled glass.[7] A detailed study of the photochemistry of 6,6-dimethyl-2,4-cyclohexadienone has provided spectroscopic evidence for reversible formation of a diene ketene, a quantum yield that is independent of the wavelength of irradiation (313 and 365 nm), and the observation that the photoreaction is not quenched by piperylene.[8a] An analogous study of diene ketenes photogenerated from 2-acetoxy-2,4-cyclohexadienones (*o*-quinolacetates) is available.[8b]

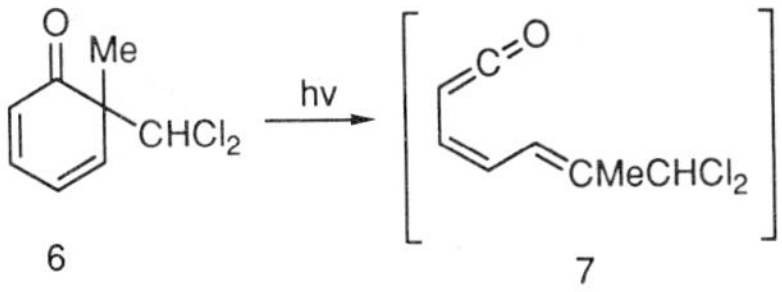

Hexamethyl-2,4-cyclohexadienone (8) photoisomerizes to diene ketene 9 which, in the absence of a strong nucleophile, reverts back to 8 by a thermal electrocyclization or undergoes intramolecular cycloaddition to give bicyclo[3.1.0]hexenone 10.[2] It is noteworthy that irradiation of 8 in methanol (or hexane) solution gives only 10, while incorporation of dimethylamine in the solvent leads to the corresponding diene amide in >90% yield. Spectroscopic evidence for the intermediacy of a diene ketene in both reactions was obtained. A related study of the partitioning of an intermediate diene ketene to bicyclo[3.1.0]-hexenone and diene amide has been published.[9]

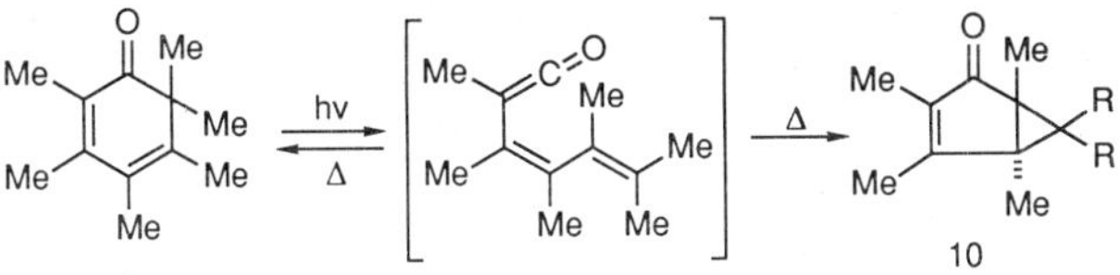

A study of substituent effects on the photochemistry of 2,4-cyclohexadienones has revealed that diene ketenes may be formed by selective irradiation of the dienone n,π* band (λ_{max} ~354 nm), while bicyclo[3.1.0]-hexenones may be obtained directly from the dienone π,π* band (λ_{max} ~302 nm).[2] Addition of methyl substituents to the dienone ring and/or performing the photoreaction in strong acid (CF$_3$CO$_2$H or silica gel) shifts the position of the π,π* band to longer wavelength, resulting in the formation of increased amounts of bicyclohexenone.

Constitutional isomers of 2,4-cyclohexadienones may be assigned by consideration of UV and ^{13}C NMR spectral data.[10a] Although the majority of diene carboxylic acid derivatives obtained from ring-opening reactions of dienones are 1,2-adducts, occasionally 1,4- and/or 1,6-adducts are produced. Spectroscopic and mechanistic information regarding the constitution of adducts is available,[8b,10b] and a two-dimensional NMR technique for the determination of atom connectivities has been reported.[10c]

A diene ketene has been shown to be an intermediate in the photointerconversion of steroidal diastereoisomers 11 and 12.[11] The contribution of 11 and 12 to the photostationary state depends on the wavelength of the light and the solvent. The diene ketene was observed by low-temperature spectroscopic studies and was trapped by reaction with cyclohexylamine. It has been demonstrated that a diene ketene generated by an unrelated photochemical reaction undergoes thermal and photochemical isomerization to a 2,4-cyclohexadienone.[12]

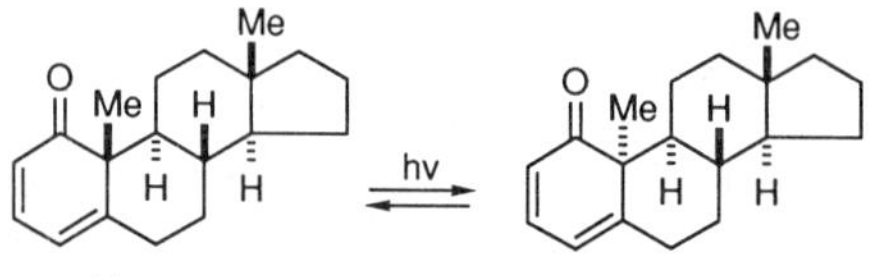

Diene carboxylic acid derivatives produced by photocleavage of 2,4-cyclohexadienones tend to be isomerically pure at the C3-C4 double bond (*Z*-configuration),[6,13] but a mixture of *E*- and *Z*-configurations at the C5-C6 double bond. Only the 6-acetoxy-2,4-cyclohexadienones,[14–17] 6-(benzoyloxy)-2,4-cyclohexadienones,[17] and 6-(2-oxopropyl)-2,4-cyclohexadienones[18,19] have been reported to undergo highly diastereoselective photoisomerization to diene ketenes. For example, photolysis of 6-acetoxy-2,4,6-trimethyl-2,4-cyclohexadienone (13a) in methanol gives 14a in 93% yield along with only 2% of the corresponding C5-C6 isomer.[15] By contrast, the unsaturated hydrocarbon analogue 13b gives 14b (~70%) and 30% of the C5-C6 isomer. These product distributions are a result of reaction kinetics rather than photoequilibration; the factors that may control diene stereoselectivity have been discussed.[14,17]

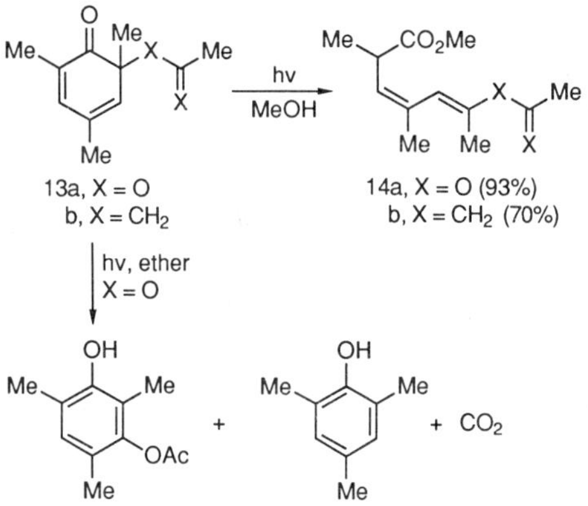

Dienone 13a undergoes relatively slow photochemical conversion to phenols 15 and 16 in the presence of weak nucleophiles or on irradiation in ether solution. Acetoxyl migration has been suggested to occur via the intermediate Zwitterion 17. Dienone-phenol photorearrangements that involve intermediate bicyclo[3.1.0]-hexenones have been reported also.[20]

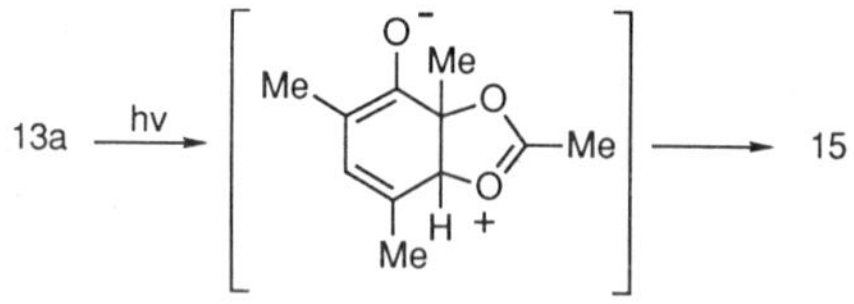

Certain benzodihydrofurans photorearrange to phenolic olefins.[21] For example, irradiation of 2-carbomethoxybenzodihydrofuran 18 gives phenol 20 by a pathway that involves the intermediacy of spirocyclopropane-2,4-cyclohexadienone 19. Benzodihydrofurans that would be expected to photorearrange to spirocyclopropane-2,4-cyclohexadienones incapable of rearrangement to a phenolic olefin are stable to irradiation in ether, methanol, or THF-pyrrolidine; e.g., 21. It is assumed that 22 is being formed and, remarkably, this dienone undergoes reversion to 21 rather than ring opening to diene ketene 23. Dienone 22 should strongly absorb the incident light (certainly compared to 21), but angle strain in the *exo*-methylenecyclopropane 23 may cause dienone ring opening to be inefficient relative to 1,3-rearrangement back to **21**.[19] Dienones also have been detected[22] in photo-Fries rearrangements[23] of phenyl esters to *ortho*- and *para*-substituted phenols.

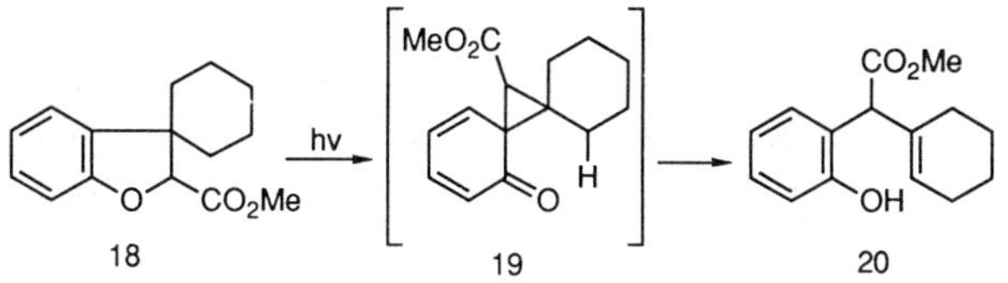

3.3 Synthetic Applications

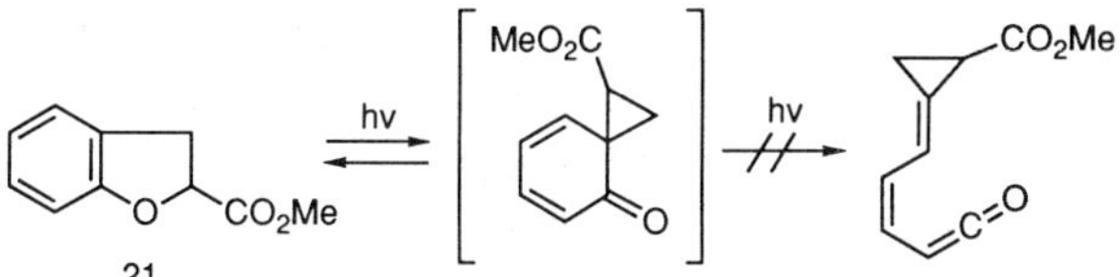

The focus of this discussion is restricted to representative synthetic applications of the ring opening-trapping reactions of 2,4-cyclohexadienones. This process, which provides convenient access to a wide range of derivatives of diene carboxylic acids, is not limited to monocyclic dienones.

Photocleavage of Ring-Fused Dienones

Dienone **24** has been shown to be an intermediate in the photoconversion of lumisantonin to photosantonic acid (**25**).[24] It is noteworthy that **24** is not detected by UV spectroscopy during the conversion of lumisantonin to **25** in aqueous organic solvents because the rate of disappearance of the intermediate is greater than the rate of formation. Dienone **24** is obtained by irradiation of lumisantonin in anhydrous ether in a Pyrex vessel.

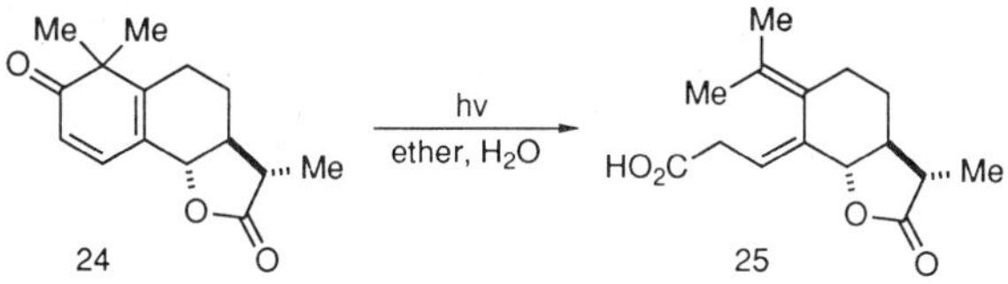

The diene ketene intermediate detected in the photointerconversion of steroidal dienones **11** and **12** is unreactive toward weak nucleophiles such as 2,2,2-trifluoroethanol, but is reactive with cyclohexylamine to give products of 1,6- as well as 1,2-addition.[11] Subsequently, it was found that irradiation in the presence of both trifluoroethanol and the strong aprotic nucleophile 1,4-diazabicyclo[2.2.2]-octane (DABCO) does result in the formation of a trifluoroethyl ester (1,2-adduct).[25] Low-temperature UV spectra have provided support for the intermediacy of the DABCO-diene ketene addition complex **26**. This modification should be generally useful for dienone reactions with other weak protic nucleophiles and, in fact, has been used with great success in examples involving intramolecular ketene trapping; *vide infra*.

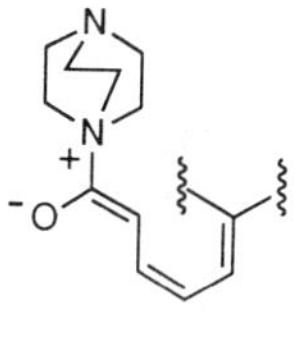

26

Photocleavage of Monocyclic Dienones

Diene carboxylic acid derivatives prepared by photocleavage reactions of 2-acetoxy- and 2-allyl-2,4-cyclohexadienones can be obtained usually in good to excellent yields.[1,26] The configurational integrity of the diene structural unit is highly dependent on the wavelength of the light utilized in the photolysis. Many of the early reports describe dienes with *E*-configuration at the C3-C4 double bond; however, it has been shown that the *Z*-isomers formed initially undergo photoisomerization to the *E*-isomers if light of sufficiently low wavelength is used.[27]

The products of dienone photocleavage have found some utility in organic and natural products synthesis. For example, photolysis (sun lamp) of dienone **27**, obtained by Wessely oxidation of *o*-cresol, in ether-methanol solution gave an intermediate acetoxydienic ester, which was subsequently converted to keto ester **28** (~50% yield from *o*-cresol). This substance was used to construct the *trans*-hydrindan system **29**, an intermediate in the synthesis of 11-keto steroids.[28]

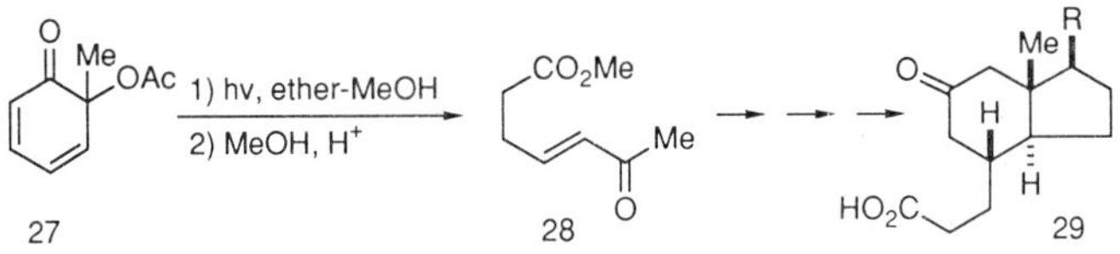

Dienone **30** is prepared from methyl 2-methoxybenzoate by a Birch reduction-alkylation procedure.[29] Irradiation of **30** in methanol solution with 366-nm light gave the diene dicarboxylic ester **31** as a mixture of C5-C6 olefin isomers in 98% isolated yield.[30] It is noteworthy that this ring cleavage is not complicated by a photo-Fries-related[23] rearrangement to phenolic ester **33**. The azidopropyl side chain in **31** enabled the construction of the C(2)-substituted proline derivative **32** and related heterocyclic materials.[30]

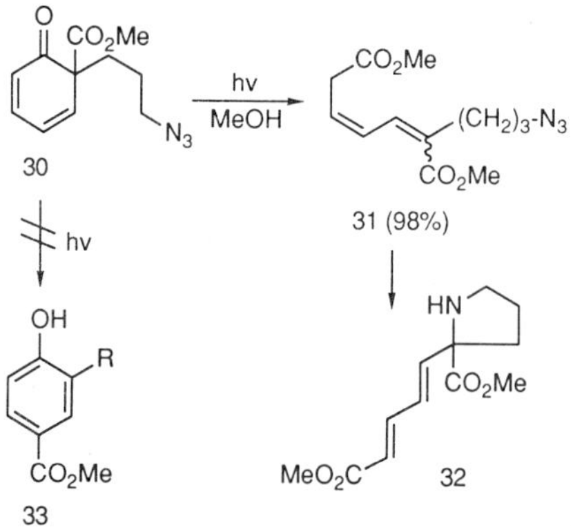

The diastereoselectivities of chiral amine addition to ketenes generated from photocleavage of 6-(2-oxopropyl)-2,6-dimethyl-2,4-cyclohexadienones **34a** and **34b** are in the low to moderate range. Product distributions of ~70:30 were obtained when photolyses were carried out in the presence of one equivalent of *d*- or *l*-ephedrine in THF solution. The major products obtained from *d*-ephedrine were determined to be **34a** and **34b**.[19]

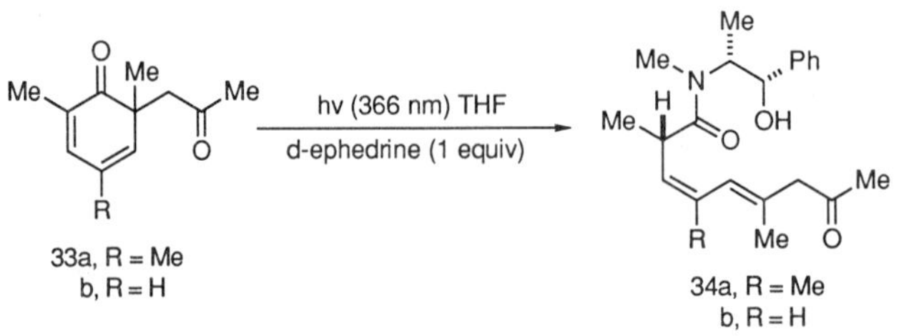

The pivitol reaction in a remarkable synthesis of crocetin dimethyl ester **35** is the photocleavage of *bis*-dienone **36** in methanol to give **37**. Dienone **36** was assembled by alkylation of 2,6-dimethylphenol with *(E)*-1,4-dibromobut-2-ene followed by *bis*-Claisen rearrangement.[31]

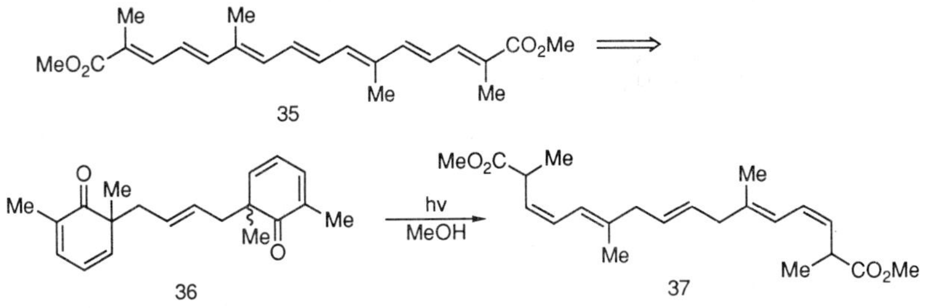

Macrolide Constructions: Aspicilin Synthesis

Perhaps the most impressive applications of dienone photocleavage reactions have been in the area of macrolide construction.[25,32,33] The essence of photolactonization is shown in the conversion of the 6-acetoxy-2,4-cyclohexadienone **38** to diene ketene **39**. Cyclization of **39** would provide macrolide **40**, while dimerization would give the diolide **41**. This process has been studied in great detail.[25,32] A specific application, the total synthesis of the lichen macrolide (+)-aspicilin (**48**), will illustrate the simplicity of this strategy for the assembly of certain types of macrolactones.

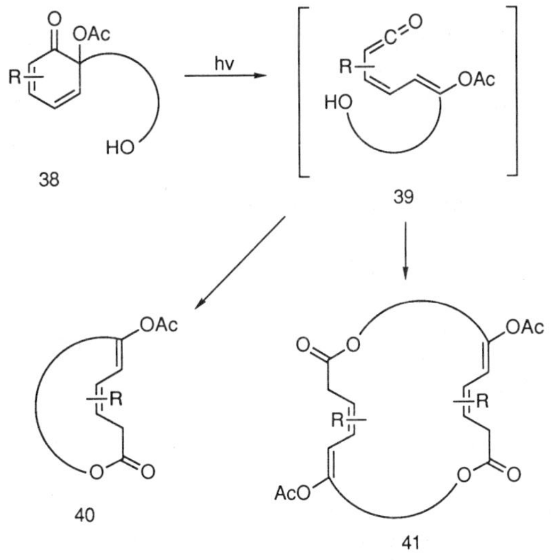

The 6-acetoxy-2,4-cyclohexadienone **42** was constructed from phenol derivative **43**, the protected 1-bromo-9-nonanol **44**, and *(S)-(–)*-methyloxirane **45** (prepared from *(S)-(–)*-methyl lactate).[33] The phenylsulfonyl group in **42** was introduced for purposes of regiocontrol in the Wessely acetoxylation and also to provide functionality for conversion of the photolactone **46** into the dienone lactone **47**. The key photolactonization affords a 2.5:1 mixture of diastereomers **46** in 69% yield.

Intraannular chirality transfer resulting from the C-17 methyl substituent enabled the C-6 carbonyl group to be stereoselectively reduced in 96% yield. The final transformations set the stage for hydroxylation of the C4-C5 double bond with OsO_4 to give (+)-aspicilin (**48**) in 40% yield.

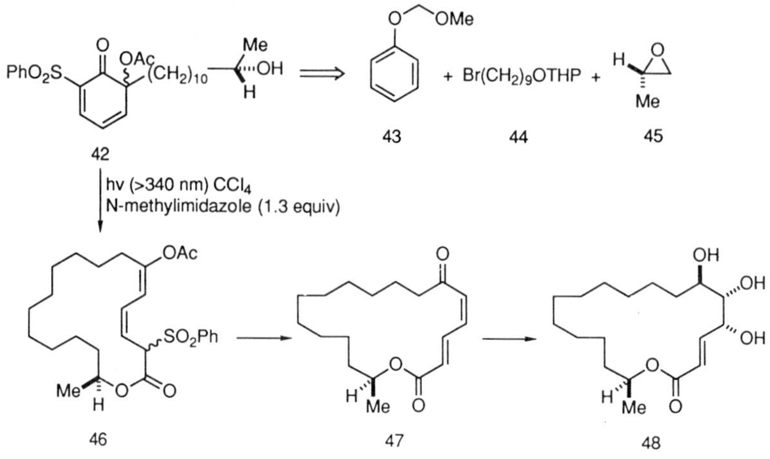

Methods for Preparation of 2,4-Cyclohexadienones

The most important methods for the preparation of 2,4-cyclohexadienones have already been noted. Additional methods are available.[34-40]

References

1. Barton, D. H. R. and Quinkert, G., Photochemical transformations. VI. Photochemical cleavage of cyclohexadienones, *J. Chem. Soc.*, 1, 1960.
2. (a) Griffiths, J. and Hart, H., On the mechanism of 2,4-cyclohexadienone photoisomerization, *J. Am. Chem. Soc.*, 90, 3297, 1968; (b) Griffiths, J. and Hart, H., A new general photochemical reaction of 2,4-cyclohexadienones, *J. Am. Chem. Soc.*, 90, 5296, 1968.
3. Quinkert, G., Light-induced formation of acids from cyclic ketones, *Angew. Chem., Int. Ed.*, 4, 211, 1965.
4. Kropp, P., Photochemical transformations of cyclohexadienones and related compounds, *Org. Photochem.*, 1, 1, 1967.
5. Quinkert, G., Thermally reversible photoisomerizations, *Angew. Chem., Int. Ed.*, 11, 1072, 1972.
6. Quinkert, G., Photochemistry of linearly conjugated cyclohexadienones in solution, *Pure Appl. Chem.*, 33, 285, 1973.
7. Chapman, O. L. and Lassila, J. D., Direct observation of ketene intermediates in photochemical reactions, *J. Am. Chem. Soc.*, 90, 2449, 1968.
8. (a) Quinkert, G., Bronstert, B., Egert, D., Michaelis, P., Jürges, P., Prescher, G., Syldatk, A., and Perkampus, H.-H., Photochemische öffnung linear-konjugierter cyclohexadienone zu den *seco*-isomeren dienylketenen, *Chem. Ber.*, 109, 1332, 1976; (b) Quinkert, G., Kleiner, E., Freitag, B.-J., Glenneberg, J., Billhardt, U.-M., Cech, F., Schmieder, K. R., Schudok, C., Steinmetzer, H.-C., Bats, J. W., Zimmerman, G., Dürner, G., Rehm, D., and Paulus, E. F., Über Dienketen aus *o*-chinolacetaten, *Helv. Chim. Acta*, 69, 469, 1986.
9. Lemmer, D. and Perst, H., 2,4-Cyclohexadienon-bicyclo[3.1.0]hexenon-valenzisomerisierung nachweis von keten-zwischenstufen, *Tetrahedron Lett.*, 2735, 1972.
10. (a) Quinkert, G., Dürner, G., Kleiner, E., Adam, F., Haupt, E., and Leibfritz, D., Regeln zur bestimmung von spektren bei 2,4-cyclohexadien-1-one, *Chem. Ber.*, 113, 2227, 1980; (b) Quinkert, G., Billhardt, U.-M., Paulus, E. F., Bats, J. W., and Fuess, H., Adducts of the ketene (1E,3Z)-1,2,3,4,5-pentamethyl-6-oxo-1,3,5-hexatrienyl acetate, *Angew. Chem. Int. Ed.*, 23, 442, 1984; (c) Kessler, H., Griesinger, C., and Lautz, J., Determination of connectivities via small proton-carbon couplings with a new two-dimensional NMR technique, *Angew. Chem. Int. Ed.*, 23, 444, 1984.

11. Quinkert, G., Englert, H., Cech, F., Stegk, A., Haupt, E., Leibfritz, D., and Rehm, D., Über das *seco*-isomere dienylketen aus 2,4-androstadien-1-on, *Chem. Ber.*, 112, 310, 1979.

12. Hobson, J. D., Al Holly, M. M., and Malpass, J. R., Cyclohexa-2,4-dienone-dienketen valence isomerisations: Thermal and photochemical interconversions, *J. Chem. Soc., Chem. Commun.*, 764, 1968.

13. Quinkert, G., Hintzmann, M., Michaelis, P., Jürges, P., Appelt, H., and Krüger, U., Darstellung und konfigurationsbestimmung von derivaten sämtlicher 6-phenylhepta-3,5-diensäuren, *Liebigs Ann. Chem.*, 748, 38, 1971.

14. Baldwin, J. E. and McDaniel, M. C., Stereochemical selectivities in the electrocyclic valence isomerizations of cyclobutenones and 2,4-cyclohexadienones, *J. Am. Chem. Soc.*, 90, 6118, 1968.

15. Quinkert, G., Bronstert, B., and Schmieder, K. R., Three isomerization routes originating from different electronic states of a linearly conjugated cyclohexadienone, *Angew. Chem. Int. Ed.*, 11, 637, 1972.

16. (a) Morris, M. R. and Waring, A. J., Cyclohexadienones: Stereospecific photochemical rearrangements of *o*-quinol acetates, *J. Chem. Soc., Chem. Commun.*, 526, 1969; (b) Morris, M. R. and Waring, A. J., Stereospecific photoisomerizations of 6-acetoxy-2,3,4,5,6-pentamethylcyclohexa-2,4-dienone, *J. Chem. Soc. (C)*, 3269, 1971.

17. Waring, A. J., Morris, M. R., and Islam, M. M., Stereoselective photochemical ring-opening of cyclohexa-2,4-dienones, *J. Chem. Soc. (C)*, 3274, 1971.

18. Schultz, A. G., Ranganathan, R., and Kulkarni, Y. S., On the photoreactivity of benzodihydrofurans and 2,4-cyclohexadien-1-ones, *Tetrahedron Lett.*, 23, 4527, 1982.

19. Schultz, A. G. and Kulkarni, Y. S., Enantioselective processes. Reaction of optically active amines with photochemically generated ketenes, *J. Org. Chem.*, 49, 5202, 1984.

20. Perst, H., Photo-umlagerung von 2,4,6-triphenyl-*o*-chinolacetat, *Tetrahedron Lett.*, 3601, 1970.

21. Schultz, A. G., Napier, J. J., and Sundararaman, P., Stereochemistry of benzodihydrofuran-2-carboxylic acid ester photorearrangement, *J. Am. Chem. Soc.*, 106, 3590, 1984.

22. Kalmus, C. E. and Hercules, D. M., A mechanistic study of the photo-Fries rearrangement, *Tetrahedron Lett.*, 1575, 1972.

23. Stenberg, V. I., Photo-Fries and related rearrangements, *Org. Photochem.*, 1, 127, 1967.

24. (a) Chapman, O. L. and Englert, L. F., A mechanistically significant intermediate in the lumisantonin to photosantonic acid conversion, *J. Am. Chem. Soc.*, 85, 3028, 1963; (b) Fisch, M. H. and Richards, J. H., The mechanism of the photoconversion of santonin, *J. Am. Chem. Soc.*, 85, 3029, 1963.

25. Quinkert, G., Fischer, G., Billhardt, U.-M., Glenneberg, J., Hertz, U., Dürner, G., Paulus, E. F., and Bats, J. W., A photochemical entry to macrocyclic mono- and dilactones, *Angew. Chem. Int. Ed.*, 23, 440, 1984.

26. Barton, D. H. R., some photochemical rearrangements, *Helv. Chim. Acta*, 42, 2604, 1959.

27. (a) Collins, P. M. and Hart, H., *J. Chem. Soc. (C)*, 1197, 1967; (b) Quinkert, G., Bronstert, B., Michaelis, P., and Krüger, U., *Angew. Chem. Int. Ed.*, 9, 240, 1970.

28. Stork, G. and Sherman, D. H., Efficient *de novo* construction of the indanpropionic acid precursor of 11-keto steroids. An improved internal Diels-Alder sequence, *J. Am. Chem. Soc.*, 104, 3758, 1982.

29. Schultz, A. G., Dittami, J. P., Lavieri, F. P., Salowey, C., Sundararaman, P., and Szymula, M. B., The synthesis and selected chemistry of 6-alkyl-6-carbalkoxy- and 6-alkyl-6-(aminocarbonyl)-2,4-cyclohexadien-1-ones and cyclohexadienone ketals, *J. Org. Chem.*, 49, 4429, 1984.

30. Schultz, A. G., Staib, R. R., and Eng, K. K., 2,4-Cyclohexadien-1-ones in organic synthesis. Further studies of molecular rearrangements occurring from products of intramolecular azide-olefin cycloadditions, *J. Org. Chem.*, 52, 2968, 1987.

31. Quinkert, G., Schmieder, K. R., Dürner, G., Hache, K., Stegk, A., and Barton, D. H. R., Eine einfache synthese von dimethylcrocetin, *Chem. Ber.*, 110, 3582, 1977.

32. Quinkert, G., Billhardt, U.-M., Jakob, H., Fischer, G., Glenneberg, J., Nagler, P., Autze, V., Heim, N., Wacker, M., Schwalbe, T., Kurth, Y., Bats, J. W., Dürner, G., Zimmermann, G., and Kessler, H., Photolactonisierung: ein neuer synthetischer zugang zu makroliden, *Helv. Chim. Acta*, 70, 771, 1987.

33. Quinkert, G., Heim, N., Glenneberg, J., Billhardt, U.-M., Autze, V., Bats, J. W., and Dürner, G., Total synthesis of the enantiomerically pure lichen macrolide (+)-aspicilin, *Angew. Chem. Int. Ed.*, 26, 362, 1987.

34. Andersson, G., Periodate oxidation of phenols. XIX. Nondimerizing *o*-quinols, *o*-quinol ethers, and *o*-quinone ketals, *Acta Chem. Scand. B*, 30, 64, 1976.

35. Belanger, A., Berney, D. J. F., Borschberg, H.-J., Brousseau, R., Doutheau, A., Durand, R., Katayama, H., Lapalme, R., Leture, D. M., Liao, C.-C., MacLachlan, F. N., Maffrand, J.-P., Marazza, F., Martino, R., Moreau, C., Saint-Laurent, L., Saintonge, R., Soucy, P., Ruest, L., and Deslongchamps, P., Total synthesis of ryanodol, *Can. J. Chem.*, 57, 3348, 1979.

36. Topgi, R. S., Novel synthesis of cyclohexa-2,4-dien-1-ones. Its use in a partial synthesis of the chromophore portion of phomenoic acid, *J. Org. Chem.*, 54, 6125, 1989.

37. Bauta, W. E., Wulff, W. D., Pavkovic, S. F., and Zaluzec, E. J., Cyclohexadienone annulations of aryl carbene complexes of chromium: New strategies for the synthesis of indole alkaloids, *J. Org. Chem.*, 54, 3249, 1989.

38. Khodabocus, A., Shing, T. K. M., Sutherland, J. K., and Williams, J. G., Substituent control of stereochemistry in the divinylketene-cyclohexadienone cyclization, *J. Chem. Soc., Chem. Commun.*, 783, 1989.

39. Soukup, M., Lukác, T., Zell, R., Roessler, F., Steiner, K., and Widmer, E., Ein neuer Zugang zu 2,6,6-trimethylcyclohexa-2,4-dienon aus 4-oxoisophoron, *Helv. Chim. Acta*, 72, 365, 1989.

40. Schultz, A. G., Harrington, R. E., and Tham, F. S., Regio- and stereoselective epoxidation of chiral 1,4-cyclohexadienes, *Tetrahedron Lett.*, 33, 6097, 1992.

59

1,4-Quinone Cycloaddition Reactions with Alkenes, Alkynes, and Related Compounds

David Creed
University of Southern Mississippi

59.1 Introduction

This chapter covers those reactions of photochemically excited 1,4-quinones (*p*-quinones), Q, that lead to cyclic adducts. The archetypal reactions are illustrated in Scheme 59.1 in which formation of a spiro-oxetane, **1**, and/or a cyclobutane, **2**, by reaction of Q with an alkene is depicted. Reaction of Q with an alkyne can lead to formation of a metastable spiro-oxete, **3**, which undergoes facile thermal rearrangement to a *p*-quinone methide, **5**, and/or a cyclobutene, **4**. Photoadditions of 1,4-quinones to alkenes were first reported by Bryce-Smith and Gilbert[1] in 1964 and, for additions to an alkyne, by Bryce-Smith et al.[2] and Zimmerman and Craft,[3] also in 1964. Hydrogen atom abstraction reactions of excited Q are not included in this chapter, even if they result in cyclic products, but electron transfer reactions leading to cycloadducts are included.

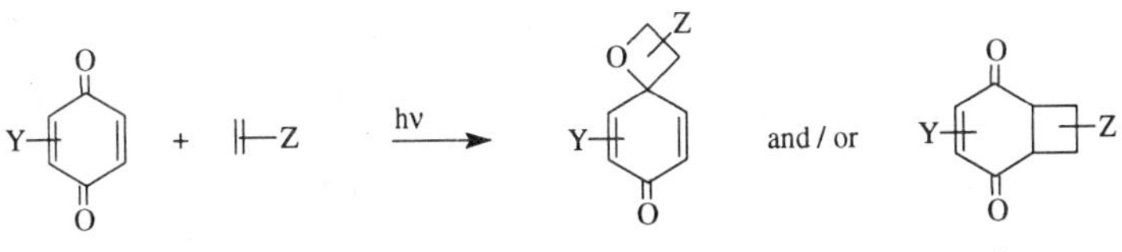

1 2

SCHEME 59.1 General reactions of 1,4-quinones with alkenes and alkynes.

0-8493-8634-9/95/$0.00+$.50

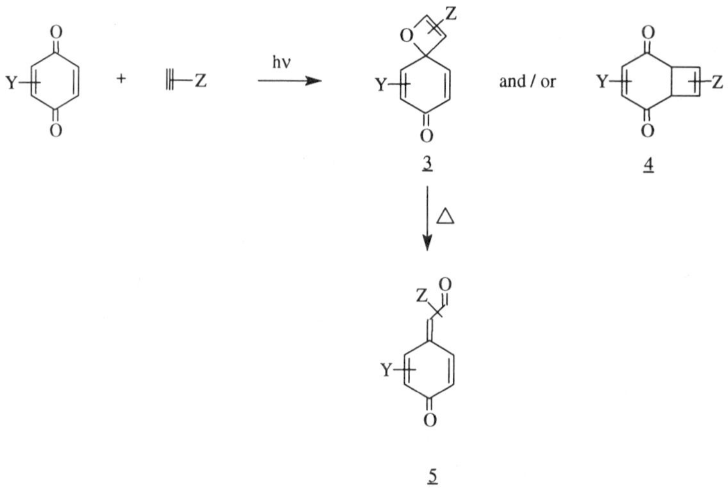

SCHEME 59.1 (continued)

Photochemical reactions of quinones have been known since the early days of the science of photochemistry. There are a number of excellent past reviews[4] of the general topic of quinone photochemistry, the most recent by Maruyama and Osuka.[5] These reviews cover the literature published prior to about 1985 on photocycloaddition reactions of 1,4-quinones. A comprehensive survey of this entire topic has not been attempted in the present chapter which emphasizes basic synthetic and mechanistic aspects of the topic and synthetically interesting results, especially those published since 1985, involving 1,4-quinone additions to alkenes, alkynes, and related compounds.

59.2 Mechanisms

Almost all the photochemical reactions of quinones (Q) studied from a mechanistic standpoint have been shown to occur from the lowest triplet state $[Q(T_1)]$ of the quinone. Singlet lifetimes, in general, are sufficiently short to preclude involvement of the singlet state in bimolecular photochemical reactions of 1,4-quinones. Charge transfer and exciplex and/or radical ion formation are postulated in all recent mechanistic studies. Once triplet-state formation has occurred, adduct formation with an alkene or alkyne (A) occurs either by an exciplex to biradical pathway or via radical ion formation and subsequent coupling reactions. The latter processes are most likely in polar solvents such as acetonitrile. The exciplex/biradical pathway (Scheme 59.2) involves formation of a triplet exciplex (^3E) that collapses to a triplet biradical (^3BR), which is the immediate precursor of the products (P) or reverts to starting materials. Since most of these types of reactions are not regio- or stereospecific, undoubtedly more than one biradical is involved. There is also the possibility of more than one exciplex being involved, especially in the cases where cycloaddition occurs to both the carbonyl group and nucleus of Q. Even in the case where an n,π^* excited state interacts with a simple alkene, CNDO calculations[6] lead to the suggestion of two "exciplexes" or "CT complexes", an n-complex, 6, and a π-complex, 7. However, there is not yet any direct experimental evidence for more than one exciplex. The exciplex or exciplexes can be stabilized, presumably by charge transfer (CT) interactions, but free ions are not formed. Exciplex reversion to $Q(T_1)$ + A and/or decay back to Q + A competes, presumably, with biradical formation. Within this mechanism, there is a clear trend for n,π^* triplet states of Q to afford products from addition of A to the quinone carbonyl group, whereas π,π^* states afford products from addition of A to the quinone nucleus.

$$
\begin{array}{ccccccccc}
Q & \xrightarrow{h\nu} & Q(S_1) & \rightarrow & Q(T_1) + A & \rightleftharpoons & {}^3E & \rightarrow & {}^3BR \rightarrow P \\
 & & & & \downarrow & & \downarrow & & \downarrow \\
 & & & & Q & & Q + A & & Q + A
\end{array}
$$

SCHEME 59.2 The Exciplex/Biradical Mechanism for Photocycloadditions of Unsaturated Compounds to Quinones.

Several pieces of evidence support the exciplex/biradical mechanism. Thus, both *cis*- and *trans*-2-butene add photochemically[7] to 1,4-benzoquinone (BQ) to give the same ratio of the isomeric spiro-oxetanes **8** and **9**. This indicates a common triplet biradical intermediate. The presence of oxygen in many quinone/alkene photocycloaddition systems leads to 1,2,4-trioxanes such as **10** that are formed in competition with spiro-oxetanes.[8] Intramolecular examples of trioxane formation are also known[9] in quinones substituted with alkenyl groups. The ratio of regioisomeric spiro-oxetanes **11** and **12** from photoaddition of BQ to 3,3-dimethylbutene is unchanged even in the presence of sufficient oxygen (11 atm) to divert 43% of product formation to the trioxane **10**, suggesting[8] that a biradical *precursor* such as a triplet exciplex is the species trapped by O_2. Transients believed to be triplet exciplexes with nanosecond lifetimes have been observed[10] in the laser flash photolysis of BQ in the presence of 3,3-dimethylbutene, tetramethylallene, and tetraphenyl allene. Most recently, Bryce Smith et al.[11] have studied the formation of spiro-oxetanes from photoaddition of a large number of electron-donor and electron-acceptor alkenes to BQ. The reactions were not stereospecific and product formation increased with increasing electron donating ability of the alkene. Initial rates of product formation were enhanced in acetonitrile. In all cases, except that of isopropenyl acetate, the regiochemistry of spiro-oxetane formation was the opposite of that expected if addition of the alkene had formed the more stable biradical. For example, additions of ethyl vinyl ether and phenyl vinyl ether afforded only spiro-oxetanes **13** and **14**, albeit in very low yields (4 and 5%, respectively). Formation of these oxetanes was rationalized on the basis of the intermediacy of a charge transfer-stabilized exciplex. In contrast to BQ, where spiro-oxetanes are usually the only photoadducts, alkene additions to 1,4-naphthoquinone (NQ) often give both spiro-oxetanes and cyclobutanes.[12] This effect in NQ may be due to the close proximity of n,π^* and π,π^* triplet states, with the former only about 25 kJ mol^{-1} below the latter.[13] The comparable energy gap in BQ is 76 kJ mol^{-1}. Thus, photoaddition of 2-methylpropene to NQ affords[12] a 1:1 ratio of regioisomeric spiro-oxetanes **15** and **16** and cyclobutane **17**, in a total yield of 70%. In general, the formation of spiro-oxetanes became more favored with increased electron-donating ability of the alkene and therefore lower ΔG^0 values for electron transfer. The electron-accepting alkenes, acrylonitrile and methyl acrylate, gave only cyclobutane adducts.

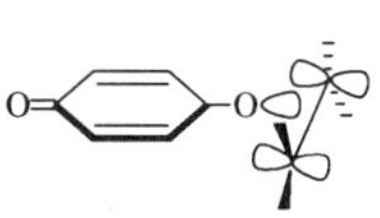

n-Complex

6

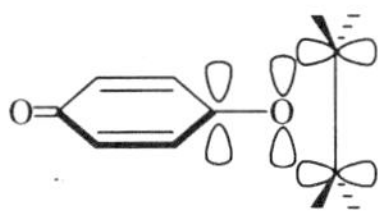

π-Complex

7

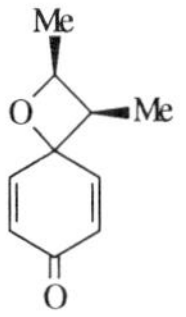

8

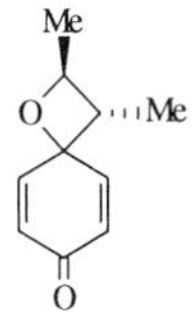

9

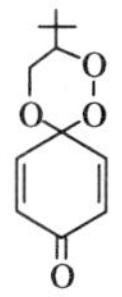

10

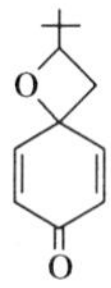

11

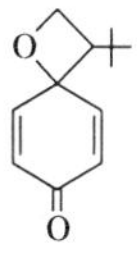

12

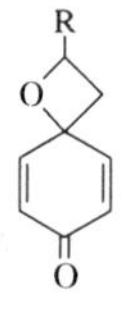

13 R = OEt
14 R = OPh

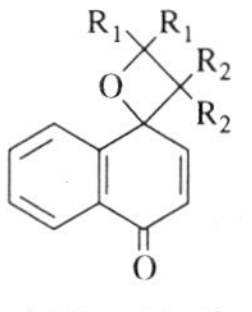

15 R_1 = Me; R_2 = H
16 R_1 = H; R_2 = Me

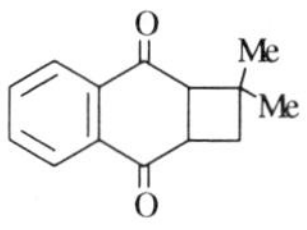

17

The radical ion mechanism (Scheme 59.3) involves one electron transfer from A to $Q(T_1)$ and either coupling or escape of the radical ions Q^- and A^+ from the encounter. Products and starting materials arise from subsequent coupling and electron transfer reactions both within the radical ion pair and between the free ions. The most important factors determining the likelihood of this mechanism are presumably the electron-accepting ability of $Q(T_1)$, rather than its specific n,π* or π,π* character, and the electron-donating ability of A. Electron donation from $Q(T_1)$ to A does not seem to have been documented to date. Solvent stabilization of the ions is also crucial to the occurrence of this mechanism.

$$Q \xrightarrow{h\nu} Q(S_1) \quad \rightarrow \quad Q(T_1) + A \quad \rightarrow \quad (Q^-\,A^+) \quad \rightleftharpoons \quad Q^-_{solv.} + A^+_{solv.}$$

$$\downarrow \qquad\qquad\quad \downarrow \qquad\qquad\quad \downarrow$$

$$Q \qquad\qquad\quad Q + A, P \qquad\quad Q + A, P$$

SCHEME 59.3   The Radical Ion Mechanism for Photoaddition of Quinones to Unsaturated Compounds.

There have been a number of studies in which evidence has been presented for the radical ion mechanism. Recently, Maruyama and Imahori[14] have made a detailed mechanistic study of a fairly typical reaction of this type: that of 2,3-dichloro-1,4-naphthoquinone, 18, with 1,1-diarylethylenes, e.g., 19. This reaction affords not only the "normal" cyclobutane adduct, 20, but also ethylene adducts (e.g., 21) that can be photocyclized to 5-phenylbenz[a]anthracene-7,12-diones (e.g., 22) Extensions of this latter reaction have led to synthesis of a large number of polycyclic aromatic hydrocarbons. Formation of 20 can be quenched by the triplet quenchers O_2 and fluoranthene. Quantum yields for reaction are higher in benzene than in acetonitrile, suggesting that back electron transfer is more effective in acetonitrile. In previous work,[15] strong CIDNP signals attributed to back electron transfer were observed for both Q and A starting materials. In the present work,[14] the triplet state of 18 was observed by picosecond flash photolysis but an exciplex could not be observed in nonpolar solvents. A transient decaying on a microsecond time scale and assigned to a contact radical ion pair was observed in acetonitrile. It is presumably formed by quenching of T_1 by A. Solvent-separated ion pairs, as evidenced by the formation of coupling products derived from attack of A^+ on A, were only formed when the radical cation could be stabilized by electron-donating groups, as in the case of 1,1-*bis*(4-methoxyphenyl)ethylene.

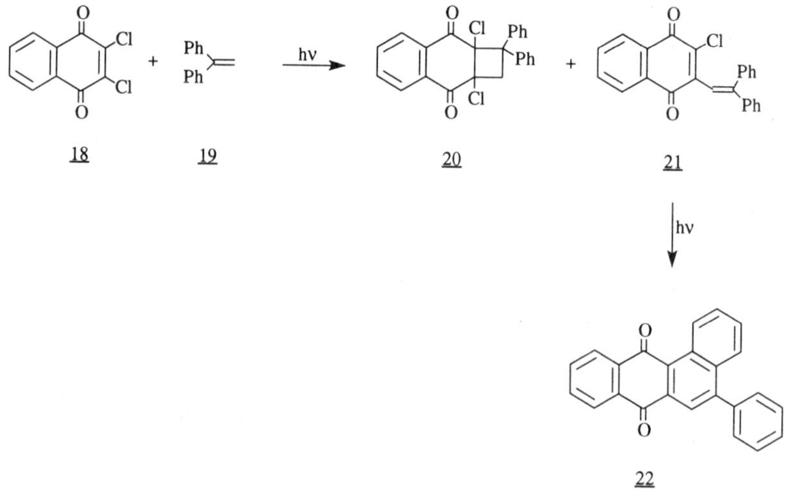

).3 Synthetic Applications

1,4-Quinone cycloadditions to alkenes and alkynes can afford several potentially useful types of products. However, it is worth pointing out that yields are extremely variable and reactions, since they are from the triplet state, do not usually occur either regio- or stereospecifically. In the author's experience, one of the most important factors affecting the yields of product(s) is the wavelength of irradiation. In general, the longer the wavelength of irradiation, the higher will be the yield. This is because the cycloadducts formed as primary products in many reactions are unsaturated ketones with strongly UV-absorbing conjugated chromophores. Wilson et al.[8] have recommended the Argon-ion CW laser as an excellent source for specific excitation of the longest wavelength transitions of 1,4-quinones, thereby avoiding secondary photolysis of primary products. Similar results can be obtained with the strong visible (e.g., 436 and 478 nm) lines of the mercury arc lamp or with white light sources (e.g., quartz halogen lamps) with suitable cut-off filters. These latter sources are cheaper and safer than a powerful CW laser. The alkene structure is undoubtedly an important factor. Bryce-Smith et al.[11] found that good electron-donating alkenes such as ethyl vinyl ether gave much lower (5%) yields of oxetane product than "simple" alkenes in which yields range up to 40%, for 2,3-dimethyl-2-butene and Z-cyclooctene. In singlet exciplex cycloadditions, there is a tendency for the more thermodynamically stable exciplexes to give lower quantum yields of cycloadducts. A similar effect may be operating here. Hydrogen atom abstraction by photoexcited Q may also contribute to low yields of cycloadducts. Thus, Wilson et al.[8] obtained near-quantitative yields of trioxane 10 and oxetanes 11 and 12 from 3,3-dimethyl-1-butene (t-butyl ethylene), a "simple" alkene with no readily abstractable hydrogen atoms. The choice of solvent is also important, especially for reactions involving n,π* triplet states of 1,4-quinones, since *(vide supra)* these states have a pronounced tendency to abstract hydrogen atoms in competition with the desired reaction. Therefore, the use of solvents such as carbon tetrachloride and the more environmentally benign substitute, 1,1,2-trifluoro-1,2,2-trichloroethane (in which the BQ triplet lifetime is several microseconds[16]), both of which lack abstractable hydrogen atoms, is strongly recommended. Benzene is a commonly used solvent for studies of quinone photochemistry. However, despite the absence of readily abstractable hydrogen atoms, benzene quenches the triplet state of BQ. The lifetime of $BQ(T_1)$ is less than 200 ns in benzene, an effect attributable to electron transfer quenching since the quenching of $BQ(T_1)$ by a series of substituted benzenes correlates with the electron-donating ability[16] of these compounds.

A number of recent examples of cycloaddition reactions of quinones to alkenes, alkynes, and related compounds are given below. Recent examples of intramolecular cycloadditions and trioxane-forming photooxidative cycloadditions to alkenes are also briefly discussed.

Cycloadditions to Alkenes and Their Derivatives

Alkenes can add to 1,4-quinones at the nucleus to afford cyclobutanes and/or at the carbonyl group to give spiro-oxetanes as primary photoproducts (Scheme 59.1). As noted above, both of these types of adducts contain chromophores that absorb strongly in the UV. Therefore, light sources and filters must be chosen so as to suppress secondary photochemical reactions. The oxetane photoproducts formed are often quite sensitive to the presence of acid, which can induce dienone-phenol rearrangement reactions. Additions of 1,4-quinones to dienes, trienes, etc., are also well known.[4,5] A recent example of addition to a triene (cycloheptatriene) is discussed below.

The most thorough recent study of formation of spiro-oxetanes between BQ and "simple" alkenes is that of Bryce-Smith et al.[11] This study was stimulated by the discovery that spiro-oxetanes are cytotoxic *in vitro* and neurotoxic *in vivo*. In general, irradiations were conducted in benzene using the Pyrex-filtered light from a mercury arc. The highest isolated yields (40%) were from simple alkenes such as 2,3-dimethyl-2-butene and Z-cyclooctene, the lowest from cyclopentene (1%) and "electron-rich" alkenes such as ethyl vinyl ether (4%). It seems possible that these relatively low yields, in at least some cases, may have to do with the conditions of irradiation. For comparison, Wilson et al.[8] obtained 78% of oxetanes and 13% of a trioxane upon (long wavelength) argon ion CW laser irradiation of BQ and cyclooctene under 11 atm oxygen in an "inert" solvent (CCl_4). The mechanistic aspects of the study by Bryce-Smith et al.[11] are discussed above. Bryce-Smith et al.[12] have also studied the photoadditions of alkenes to NQ under the same irradiation conditions as for the reactions of BQ. Ethene, cyclopentene, styrene, methyl acrylate, and acrylonitrile gave only cyclobutane photoadducts, whereas Z-cyclooctene and 2,3-dimethyl-2-butene gave only spiro-oxetanes. In several cases (oct-1-ene, 2-methylpropene, 3,3-dimethylbut-1-ene, vinyl acetate, and ethyl vinyl ether), both spiro-oxetane and cyclobutane adducts are isolated. There was a general trend for spiro-oxetane formation to be favored with ethenes of low oxidation potential. Total yields were variable, ranging from 5% for acrylonitrile, which gives only a cyclobutane, to 90% for Z-cyclooctene, which gives only a spiro-oxetane.

Kraus et al.[17] have observed that 2-ethoxy-1,4-naphthoquinone, which most likely has a lowest π,π^* triplet state, reacts photochemically (Hg lamp through quartz) in benzene under nitrogen with 3-butenol to afford a 53% yield of the adduct 23. The final product was assumed to arise from a secondary photochemical reaction of an initially formed [2+2]-cycloadduct 24. Presumably, no attempt was made to isolate this initial cycloadduct by excitation at a longer wavelength.

Miyashi et al.[18] have studied the photochemical addition of the well-known fluxional molecule barbaralone, 25, with tetrahalogenated 1,4-benzoquinones such as *p*-chloranil (CA), *p*-bromanil, and *p*-fluoranil. Irradiation of CA with 25 in nonpolar solvents such as benzene afforded adducts 26 (major) and 27 (minor) in up to ~40% yield. In solvent mixtures containing water or methanol, the adducts 28 and 29, respectively, were primary products. An exciplex to biradical to rearranged biradical to adduct mechanism was suggested for nonpolar solvents; whereas, in polar solvents, reaction is believed to occur via the trapping by water or methanol of a contact radical ion pair.

High-pressure mercury lamp irradiation (no filter specified) of BQ and the complex alkene 30 in deuteriochloroform gave[19] a single oxetane adduct 31, with *exo*-stereochemistry about the oxetane ring, in 78% yield. 1,4-Naphthoquinone gave an analogous adduct in 44% yield. Thermolysis of adducts such as 31 to cyclobutenes and/or oxetes (or their derivatives) was not reported, although 30 was described as a potentially useful "transfer reagent" for acetylene.

Mori and Takeshita[20] have reported the isolation of 50–70% yields of a spirocyclic ether 32 from Pyrex-filtered mercury arc irradiation of BQ in benzene in the presence of cycloheptatriene. In the case of NQ, adducts 33 and 34 from additions to the quinone nucleus were formed, together with the two possible isomeric spiro-oxetanes, an observation consistent with the increased π,π^* nature of the lowest triplet state of NQ. Within this context, it is not surprising that tetramethyl-1,4-benzoquinone (duroquinone) gives[21] only photoproducts from addition to the Q ring when irradiated in the presence of cycloheptatriene since methyl substituents on the BQ ring tend to increase the π,π^* nature of the lowest triplet state.[13]

Several other cycloadditions of alkene derivatives to benzoquinone in the presence of oxygen to form spiro-oxetanes in competition with trioxane formation are described in the section on trioxane-forming reactions.

Cycloadditions to Alkynes

Photochemical [2+2]-cycloadditions of 1,4-quinones to alkynes afford, in principle, cyclobutenes from addition to the quinone nucleus and spiro-oxetes from addition to the carbonyl group (Scheme 59.1). The isolation of cyclobutene adducts from this reaction is well known but spiro-oxetes are never isolated. They presumably undergo thermal rearrangement. In the only recently reported example of this type of reaction, Kim et al.[22] isolated the quinodimethanes 35 and 36 from addition of 1,4-diphenylbuta-1,3-diyne to BQ and CA. The primary product of this reaction is presumably the spiro-oxete 37.

Cycloadditions to Allenes

Irradiation of BQ in the presence of tetraphenylallene in CCl_4 or 10% acetic acid in CCl_4 affords[10] the indene 38. In the presence of methanol, 2-methoxy-1,1,3-triphenylindene and 3-methoxy-1,1,3,3-tetraphenylpropene are also obtained. Protonation of an exciplex is proposed as a key step in the reaction mechanism. Transients observed by laser flash photolysis of BQ in the presence of tetraphenylallene, tetramethylallene, and 3,3-dimethylbutene are identified tentatively as exciplexes. Maruyama and Imahori[23,24] have observed that halo-substituted 1,4-naphthoquinones form spiropyran 39 and/or cyclobutane 40 adducts upon irradiation with phenylallenes in benzene above 320 nm where only the quinone is directly excited. Monophenylallene affords 41 exclusively in 74% yield, but those diphenylallenes that react at all give spiropyrans exclusively in up to 89% yield. On the basis of the energetics of electron transfer, substituent effects, solvent effects, and CIDNP experiments, it is concluded that the reaction mechanism involves initial electron transfer forming an ion pair (or exciplex), followed by conversion to a biradical that can cyclize and rearrange to the spiropyran product.

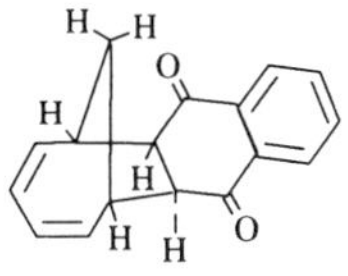

33	34	35 R = H
		36 R = Cl

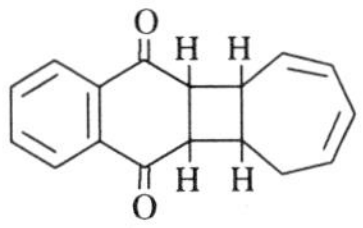

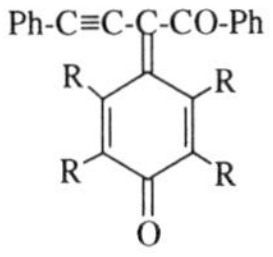

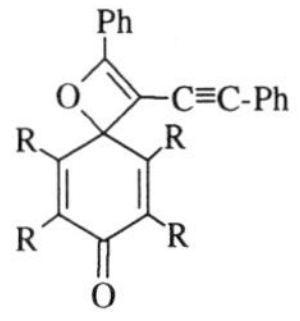

37

R = H or Cl

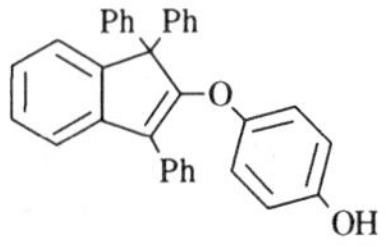

38

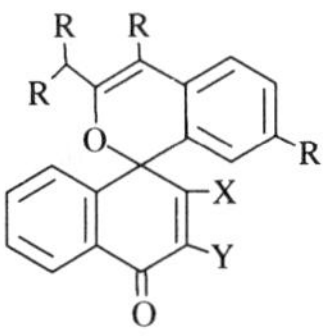

39
41, R = H
   X = Y = Cl

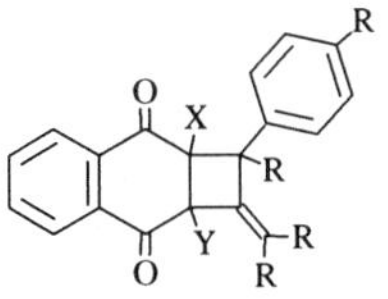

40

Intramolecular Cycloadditions

There have been a few recent interesting examples of intramolecular cycloadditions of alkenyl-substituted 1,4-quinones affording complex fused-ring systems. Miyake et al.[25] irradiated ($\lambda >$

330 nm) rotational isomers **42** and **43** of 2,5-(*trans*-4-octeno)-*p*-benzoquinones in THF-methanol. In the case of **42** (R = Me), a single product, **44**, was obtained in 62% yield ($\Phi = 0.080$), presumably via ketene **45**. The use of Michler's ketone as a triplet sensitizer did not lead to the formation of **44**, suggesting this intramolecular reaction is a (rare) example of a singlet-state reaction of a benzoquinone. Participation of singlet states of benzoquinones in bimolecular reactions is improbable because of the short lifetimes of such states. In contrast to the behavior of **42**, its rotational isomer, **43**, yielded a different product, the methanol adduct **46** in 90% yield ($\Phi = 0.070$). A frontier molecular orbital explanation is given to account for the preferred reaction at the 2- and 5-positions of the quinone ring in **43** compared to reaction at the 2- and 4-positions in **42**.

The photochemistry of the "norbornadienyl quinone", **47**, has been investigated[26] as a possible solar energy storage system. Several quinones of this type undergo high chemical yield ("nearly quantitative" for $R^1 = OMe$, $R^2 = H$), reasonable quantum yield (ca. 0.2), intramolecular cycloaddition reactions, giving the quadricyclane derivatives **48** upon irradiation with visible light ($\lambda > 410$ nm) in dichloromethane. These photochemical [2+2]-photocycloadditions can be reversed thermally to regenerate the original quinone in an exothermic process.

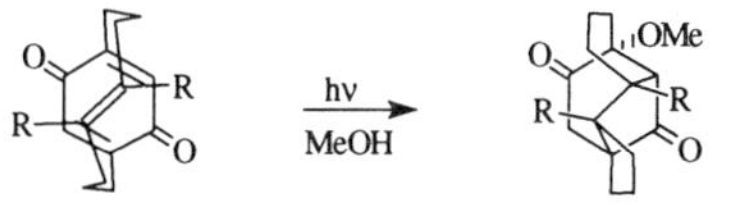

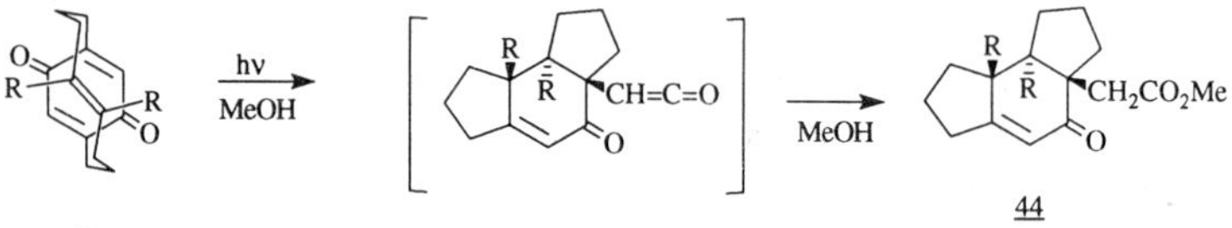

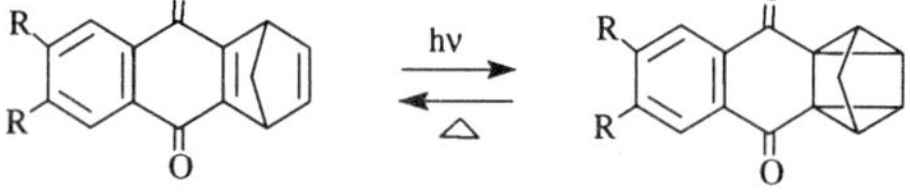

Iwamoto and Takuwa[27] have used 1,1-disubstituted alkenes to trap intermediates in the photochemical reactions of allyl-1,4-benzoquinones. These reactions lead to novel tricyclic products. Thus, **49**, upon irradiation ($\lambda > 410$ nm) in benzene in the presence of alkenes affords stereoisomeric products **50** and **51** in yields of up to 89% (for **49**, R = Me, reacting with methyl methacrylate). Yields were greatest for electron-deficient alkenes, suggesting that the species trapped (presumed to be a biradical) has nucleophilic character at the position where trapping occurs. Steric effects were also concluded to be important since 1,2-disubstituted alkenes, even electron-deficient alkenes such as dimethyl maleate, were unreactive, as were 1,1-disubstituted alkenes with bulky substituents (e.g., 1,1-diphenylethene).

Trioxane-Forming Reactions

Interception by oxygen of one of the intermediates (either ^3E or ^3BR) in the photocycloaddition of alkenes to 1,4-quinones can lead to novel trioxanes.[8,9] Adam et al.[28,29] have exploited this reaction in photochemical approaches to the anti-malarial drug, Qinghaosu (Artemisinin), 52. Thus, BQ undergoes[28] regiospecific photoaddition (Ar ion laser at 415–515 nm) to enol lactone 53 in carbon tetrachloride to afford, under an argon atmosphere, the oxetane stereoisomers 54 and 55. In the presence of oxygen, the regioisomeric trioxanes 56 and 57 were also formed, together with the oxetanes and a singlet oxygen-derived oxidation product of the enol lactone. Similarly, BQ undergoes photoaddition[29] (455–515 nm argon ion laser illumination) to the pyranone 58 in CCl$_4$ to afford mainly the oxetane 59 in *ca.* 40% yield under argon. The other regioisomer, 60, was not detected. Under ca. 10 atm oxygen, 59 and three isomeric trioxanes, 61 (*cis-* and *trans-*) and 62 were obtained in a total yield of 62% with a ratio of oxetane to trioxane of 45:55. The regioisomeric oxetane, 60, was detected by nmr but not isolated in the experiments carried out under oxygen.

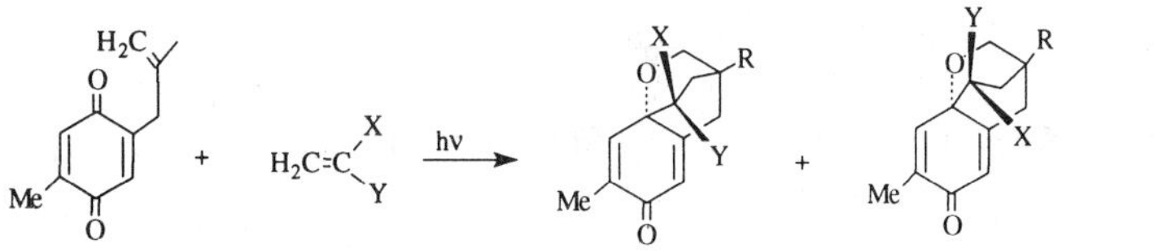

49
R = H or Me

50
R = H, Me
X = H, Me

51
Y = Ph, 4-ClC$_6$H$_4$, 4MeOC$_6$H$_4$, CO$_2$Me, CN

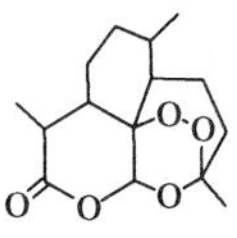

52

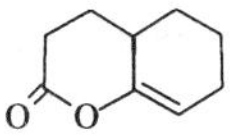

53

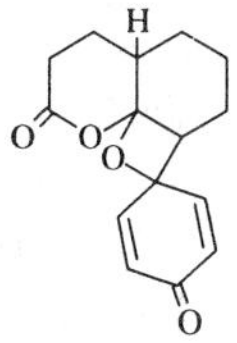

54 : cis
55 : trans

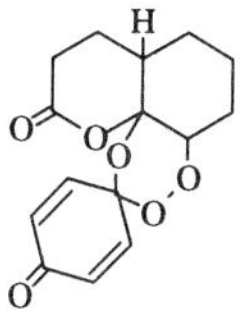

56

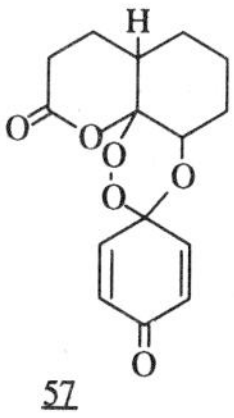

57

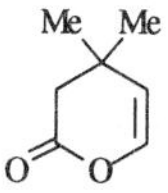

58

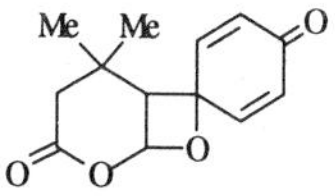

59

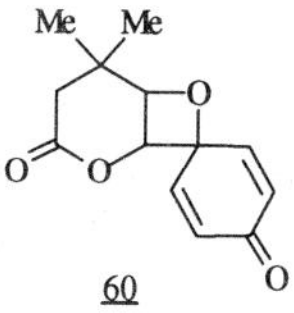

60

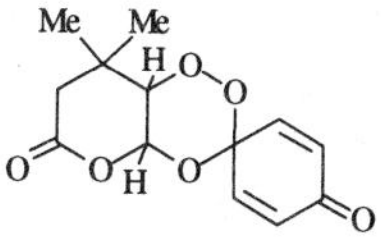

61

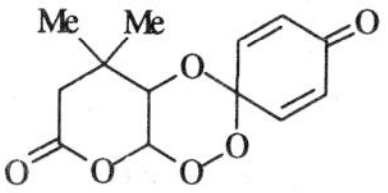

62

Acknowledgments

I thank the National Science Foundation (EPSCoR program), the State of Mississippi, and the University of Southern Mississippi for financial support.

References

1. (a) Bryce-Smith, D. and Gilbert, A., Photoaddition of *p*-quinones to olefins: New syntheses of oxetans and phenols, *Proc. Chem. Soc.*, 87, 1964; (b) Bryce-Smith, D. and Gilbert, A., 1:1 and 2:1 Photoaddition of cyclooctene and cycloocta-1,5-diene to chloranil, *Tetrahedron Lett.*, 3471, 1964.

2. Bryce-Smith, D., Fray, G. I., and Gilbert, A., Photo adduct of *p*-benzoquinone and diphenylacetylene, *Tetrahedron Lett.*, 2137, 1964.

3. Zimmerman, H. E. and Craft, L., Photochemical reaction of benzoquinone with tolan, *Tetrahedron Lett.*, 2131, 1964.

4. (a) Schonberg, A. and Mustafa, A., Photoreactions of non-enolizable ketones in sunlight, *Chem. Rev.*, 40, 181, 1947; (b) Bruce, J. M., Light-induced reactions of quinones, *Quart. Rev.*, 21, 405, 1967; (c) Rubin, M. B., *Fortsch. Chem. Forsch.*, 13, 251, 1969; (d) Bruce, J. M., Photochemistry of quinones, in *The Chemistry of the Quinonoid Compounds*, Patai, S., Ed., John Wiley & Sons, New York, 1974, chap. 9.

5. Maruyama, K. and Osuka, A., Recent advances in the photochemistry of quinones, in *The Chemistry of Quinonoid Compounds*, Vol. 2, Patai, S. and Rappoport, Z., Eds., John Wiley & Sons, New York, 1988, chap. 13.

6. Wilson, R. M., Outcalt, R., and Jaffé, H. H., Triplet carbonyl-olefin charge-transfer complexes. A CNDO treatment, *J. Am. Chem. Soc.*, 100, 301, 1978.

7. Bunce, N. J. and M. Hadley, On the mechanism of oxetane formation in the photocycloaddition of *p*-benzoquinone to alkenes, *Can. J. Chem.*, 53, 3240, 1975.

8. Wilson, R. M., Wunderly, S. W., Walsh, T. F., Musser, A. K., Outcault, R., Geiser, F., Gee, S. K., Brabender, W., Yerino, Jr., D., Conrad, T. T., and Tharp, G. A., Laser photochemistry: Trapping of quinone-olefin preoxetane intermediates with molecular oxygen and chemistry of the resulting 1,2,4-trioxanes, *J. Am. Chem. Soc.*, 104, 4429, 1982.

9. (a) Creed, D., Werbin, H., and Strom, E. T., Photochemistry of electron transport quinones. II. Model studies with plastoquinone-1 [2,3-dimethyl-5-(3-methylbut-2-enyl)-1,4-benzoquinone], *J. Am. Chem. Soc.*, 93, 502, 1971; (b) Wilson, R. M., Walsh, T. F., and Gee, S. K., Laser photochemistry: The wavelength dependent oxidative photodegradation of Vitamin K analogs, *Tetrahedron Lett.*, 21, 3459, 1980; (c) Maruyama, K., Muraoka, M., and Naruta, Y., Photo-oxygenation of alkenoyl-1,4-quinones by atmospheric oxygen. Formation of stable cyclic products, *J. Chem. Soc., Chem. Commun.*, 1282, 1980.

10. Schnapp, K. A., Wilson, R. M., Ho, D. M., Caldwell, R. A., and Creed, D., Benzoquinone-olefin exciplexes: The observation and chemistry of the *p*-benzoquinone-tetraphenylallene exciplex, *J. Am. Chem. Soc.*, 112, 3700, 1990.

11. Bryce-Smith, D., Evans, E. E., Gilbert, A., and McNeil, H. S., Factors influencing the regiochemistry of spiro-oxetane formation from the photocycloaddition of ethenes to 1,4-benzoquinone, *J. Chem. Soc., Perkin Trans. 2*, 1587, 1991.

12. Bryce-Smith, D., Evans, E. E., Gilbert, A., and McNeil, H. S., Photoaddition of ethenes to 1,4-naphthoquinone: Factors influencing the site of reaction, *J. Chem. Soc., Perkin Trans. 1*, 485, 1992.

13. Bunce, N. J., Ridley, J. E., and Zerner, M. C., On the excited states of *p*-quinones and an interpretation of the photocycloaddition of *p*-quinones to alkenes, *Theor. Chim. Acta*, 45, 283, 1977.

14. Maruyama, K. and Imahori, H., Photoreactions of halogeno-1,4-naphthoquinones with electron-rich alkenes, *J. Chem. Soc., Perkin Trans. 2*, 257, 1990.

15. Maruyama, K., Otsuki, T., and Tai, S., Photoinduced electron-transfer-initiated aromatic cyclization, *J. Org. Chem.*, 50, 52, 1985.

16. Creed, D., Caldwell, R. A., and Rodgers, M. A. J., unpublished results, 1987.

17. Kraus, G. A., Shi, J., and Reynolds, D., A reinvestigation of the photochemistry of 2-alkoxy-1,4-naphthoquinones, *Synth. Commun.*, 20, 1837, 1990.

18. Miyashi, T., Konno, A., Takahashi, Y., Kaneko, A., Suzuki, T., and Mukai, T., Photocycloaddition reactions of tricyclo[3.3.1.0^{2,8}]nona-3,6-dien-9-one (Barbaralone) and carbonyl compounds, *Tetrahedron Lett.*, 30, 5297, 1989.

19. Tian, G. R., Mori, A., Kato, N., and Takeshita, H., Synthetic photochemistry. XLVI. Cycloaddition of *exo, endo*-2,7-*bis*(methoxycarbonyl)-11,12-dioxatetracyclo[6.2.1^{3,6}.0^{2,7}]dodeca-4,9-diene and conjugated enones and quinones, *Bull. Chem. Soc. Jpn.*, 62, 506, 1989.

20. Mori, A. and Takeshita, H., Synthetic photochemistry. XXX. The addition reactions of cycloheptatriene with some aromatic *p*-quinones, *Bull. Chem. Soc., Jpn.*, 58, 1581, 1985.

21. Ogino, K., Minami, T., Kozuka, S., and Kinshita, T., Photochemical reactions of duroquinone with cyclic polyenes. Synthesis of new cage compounds, *J. Org. Chem.*, 45, 4694, 1980.

22. Kim, S. S., Yoo, D. Y., Kim, A. R., Cho, I., and Shim, S. C., Photoaddition reactions of *p*-quinones to conjugated diynes, *Bull. Korean Chem. Soc.*, 10, 66, 1989.

23. Maruyama, K. and Imahori, H., A novel (2+4) photocyclization reaction between quinone and allene, *Chem. Lett.*, 725, 1988.

24. Maruyama, K. and Imahori, H., (2+4) Photocyclization between quinones and allenes via photo-induced electron transfer, *J. Org. Chem.*, 54, 2692, 1989.

25. Miyake, M., Tsuji, T., Furusaki, A., and Nishida, S., Distinctive photochemical behavior of rotationally isomeric 2,5-(*trans*-4-octeno)-*p*-benzoquinones, *Chem. Lett.*, 47, 1988.

26. Suzuki, T., Yamashita, Y., Mukai, T., and Miyashi, T., Photocycloaddition of norbornadienes fused with quinone units, *Tetrahedron Lett.*, 29, 1405, 1988.

27. Iwamoto, H. and Takuwa, A., Photochemical reactions of allyl-1,4-benzoquinones. Trapping of biradical intermediates with olefins, *Chem. Lett.*, 5, 1993.

28. Adam, W., Kliem, U., Mosandl, T., Peters, E.-M., Peters, K., and von Schnering, H. G., Preparative visible laser photochemistry: Quinghaosu-type 1,2,4-trioxanes by molecular oxygen trapping of Paternò-Büchi triplet 1,4-diradicals derived from 3,4-dihydro-4,4-dimethyl-2H-pyran-2-one and quinones, *J. Org. Chem.*, 53, 4986, 1988.

29. Adam, W., Kliem, U., Peters, E.-M., Peters, K., and von Schnering, H. G., Quinghaosu-type 1,2,4-trioxanes by molecular oxygen trapping of Paternò-Büchi triplet 1,4-diradicals derived from the bicyclic enol lactones $\Delta^{1,6}$- and $\Delta^{1,10}$-2-oxabicyclo[4.4.0]decen-2-one and *p*-benzoquinone, *J. Prakt. Chem.*, 330, 391, 1988.

60

Photochemical Hydrogen Abstraction Reactions of Quinones

Kazuhiro Maruyama
Kyoto Institute of Technology

Yasuo Kubo
Shimane University

60.1 Definition of the Reaction

Photochemistry of quinones, *p*- and *o*-quinones, has been the subject of extensive investigation and a number of excellent reviews on the photochemistry of quinones are now available.[1–5] Hydrogen abstraction (H-abstraction) is one of the most typical photoreactions of quinones and a wide variety of substrates, including alcohols, ethers, aldehydes, alkyl aromatics, and others, has been identified as potential H-donors.

Considering the H-abstraction from alcohols, generally the quinones (Q) are reduced to the corresponding hydroquinones (QH_2) and the alcohols (RR′CHOH) are dehydrogenated to the corresponding aldehydes or ketones (RCOR′).[2,4] The first step in the process involves H-abstraction by the triplet excited states of Q from the α-position of the alcohols to give a radical pair composed of the semiquinone radicals (QH·) and hydroxyalkyl radicals [RR′C(OH)·] (Equation 60.1).

$$^3Q^* + RR'CHOH \rightarrow QH\cdot + RR'C(OH)\cdot \qquad (60.1)$$

$$QH\cdot \rightarrow Q^{-} + H^+ \qquad (60.2)$$

$$RR'C(OH)\cdot + Q \rightarrow RCOR' + QH\cdot \qquad (60.3)$$

$$2QH\cdot \rightarrow QH_2 + Q \qquad (60.4)$$

In the presence of base, QH· dissociates to the semiquinone radical anion (Q^{-}), whereas the dissociation is suppressed in acidic or nonpolar media (Equation 60.2). Aldehydes or ketones (RCOR′) arise from the oxidation of RR′C(OH)· by ground-state Q rather than from disproportionation of RR′C(OH)· (Equation 60.3). Disproportionation of QH· results in the formation of QH_2 (Equation 60.4).

0-8493-8634-9/95/$0.00+$.50
© 1995 by CRC Press, Inc.

Ethers can also act as the H-donor, and frequently, the H-abstraction from ethers was reported to give 1:1 adducts; photoreaction of 9,10-phenanthraquinone (**1**) and ethers **2** gave hydroquinone monoethers **3** as the major product (Equation 60.5).[6,7]

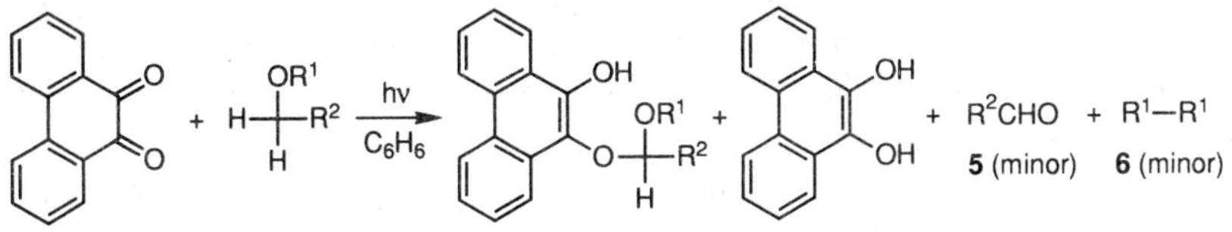

1 **2** **3** (major) **4** (minor) **5** (minor) **6** (minor)

R^1 = alkyl or aralkyl, R^2 = alkyl or aryl (60.5)

Irradiation of *p*-quinones, such as **7**, in the presence of aldehydes **8** normally gives acylhydroquinones **9** as the predominant products; only a trace of the hydroquinone monoesters **10** is formed (Equation 60.6).[8] On the contrary, formation of hydroquinone monoesters, such as **11**, is more common in the photoreactions of the *o*-quinones, such as **1** on reaction with **8** (Equation 60.7).[7,9,10]

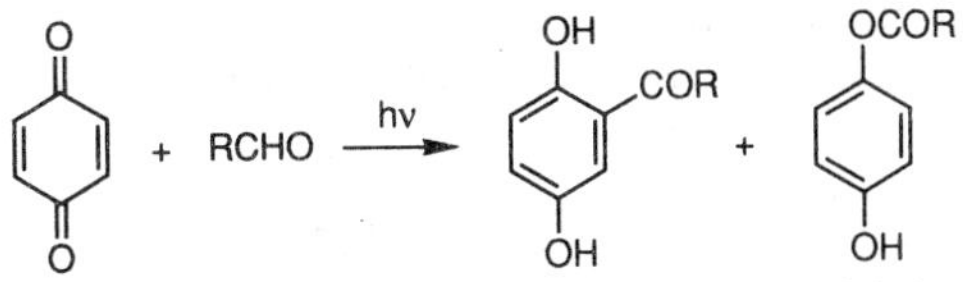

7 **8** **9** (major) **10** (minor) (60.6)

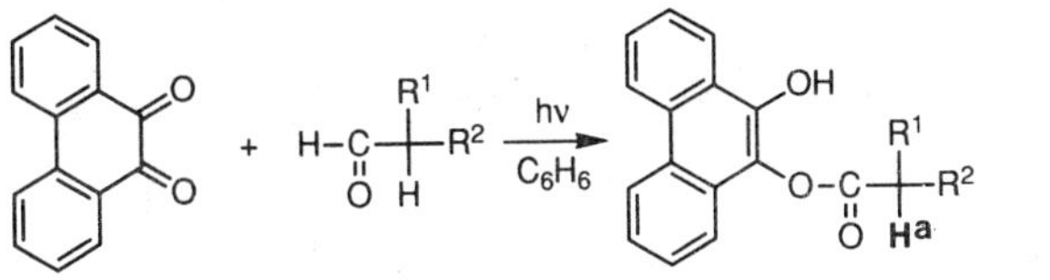

1 **8** **11** (nearly quantitative)

R^1 = alkyl or aryl, R^2 = alkyl or aryl (60.7)

The photoinduced H-abstraction of *p*-quinones from alkyl aromatics results in formation of the hydroquinones together with the dehydrogenated hydrocarbons such as dehydrodimers or alkenes.[11] Whereas, ketols (such as **13**) are formed as the major product in the photoinduced H-abstraction of *o*-quinones (such as **1**) from alkyl aromatics **12** (Equation 60.8).[7,12]

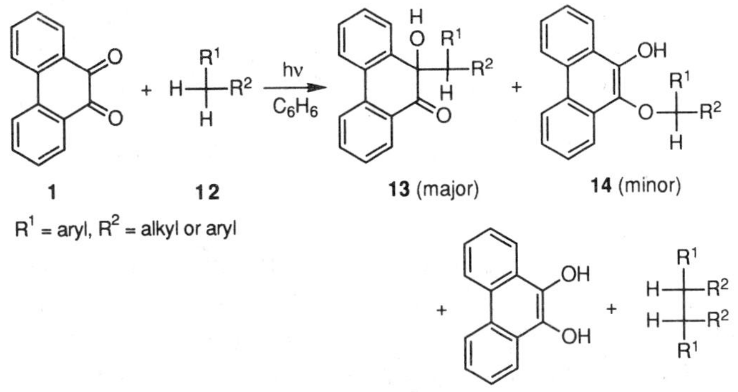

1 **12** **13** (major) **14** (minor)

R^1 = aryl, R^2 = alkyl or aryl

4 (minor) **15** (minor) (60.8)

60.2 Mechanism

Undoubtedly, most photoreactions of quinones proceed via the triplet excited state, since the intersystem crossing of the singlet excited states of quinones is a rapid process.[5] Exceptionally, anthraquinones bearing amino and hydroxy groups have the lowest singlet (CT) excited states whose lifetimes are sometimes long enough for chemical reactions.[13] When the triplet excited state of the substrate molecule lies below the triplet state of the quinones, electronic energy transfer has been shown to occur in many cases.[5]

The photoinduced H-abstraction of quinone can occur primarily by one of two mechanisms. One is the direct H-atom transfer from the H-donor (HD) to the n,π^* triplet states of quinones to give a radical pair composed of QH· and a substrate radical (·D) (path A in Scheme 60.1). Quantum yields of 0.5 to 1.0 and even higher values for the H-abstraction have been reported for quinones with low-lying n,π^* triplet states. Substitution of the quinones by electron-donating groups gives rise to low-lying intramolecular charge transfer (CT) or π,π^*-states in these molecules with a concomitant decrease in reactivity toward H-abstraction. Another possible mechanism is the two-step part of electron transfer, followed by proton transfer via the radical ion pair state (path B). There is still controversy about whether the primary reaction of quinone triplet states is the direct H-transfer or the electron-transfer reaction mode, although the photoreaction of quinones has long been studied. Nanosecond and picosecond time-resolved spectroscopy has revealed the coexistence of these two mechanisms for the tetrachloro-1,4-benzoquinone (p-chloranil)/durene system.[14]

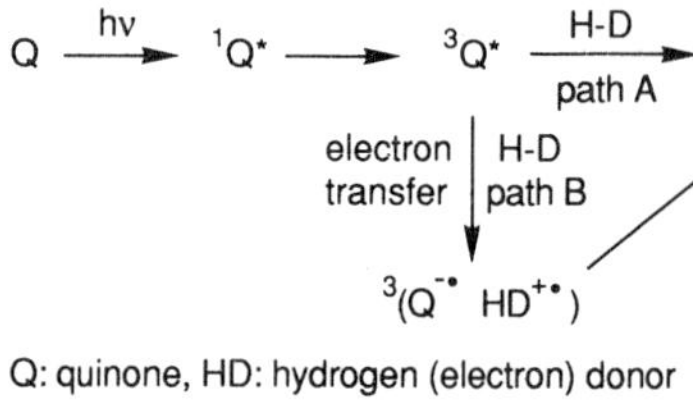

Q: quinone, HD: hydrogen (electron) donor

SCHEME 60.1

Final products of the H-abstraction of quinones arise from the chemical reactions of a radical pair of QH· and ·D depending largely on the chemical nature of the both species and on the dynamics of the radical pair. Although it was previously proposed that adducts of quinones and aldehydes formed by the H-abstraction path arose via scavenging of the acyl radicals by the ground-state quinones,[2,4] an in-cage recombination of the acyl radical and the semiquinone radical is now well recognized as a major pathway for their formation. The in-cage mechanism was proved by the results from the reaction of thermally generated acyl radical (18) and ground-state 1 (Equation 60.9), which gave a quite different product 19 from that of the photochemical reaction (Equation 60.7).[15]

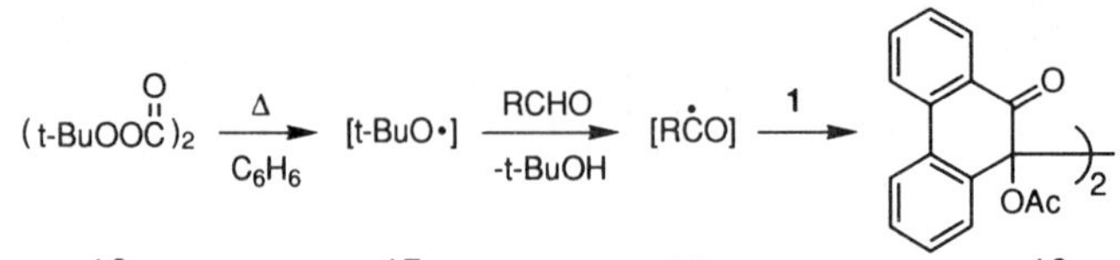

$$(60.9)$$

The chemically induced dynamic nuclear polarization (CIDNP) technique also provided definitive evidence for the in-cage mechanism.[16,17] Thus, irradiation of a solution of 1 and aldehyde 8 in the cavity of an ^{1}H NMR spectrometer caused enhancement of the resonance due to the proton "H^a" in the product 11 (Equation 60.7), confirming the intervention of a solvent-caged triplet radical pair of the semiquinone radical of 1 and the acyl radical for the formation of adduct 11.[7,18]

The CIDNP effects observed during the irradiation also indicate the intermediate formation of unstable adducts of the semiquinone radicals of various quinones and the substrate radicals generated by the H-abstraction. For example, 9-xanthyl radical (Xan·) was shown to add to semiquinone radicals at almost all the possible sites, as shown in Figure 60.1.[19-21] Since the methine H of 13 and 14 (Equation 60.8 or Scheme 60.2) were strongly polarized but in opposite directions to each other during the irradiation of 1 with alkyl aromatics 12, it was concluded that 14 was formed by in-cage recombination of the singlet radical pair of 20 and 21, which arose via homolysis of the vibrationally excited primary adduct 22 (Scheme 60.2).[7]

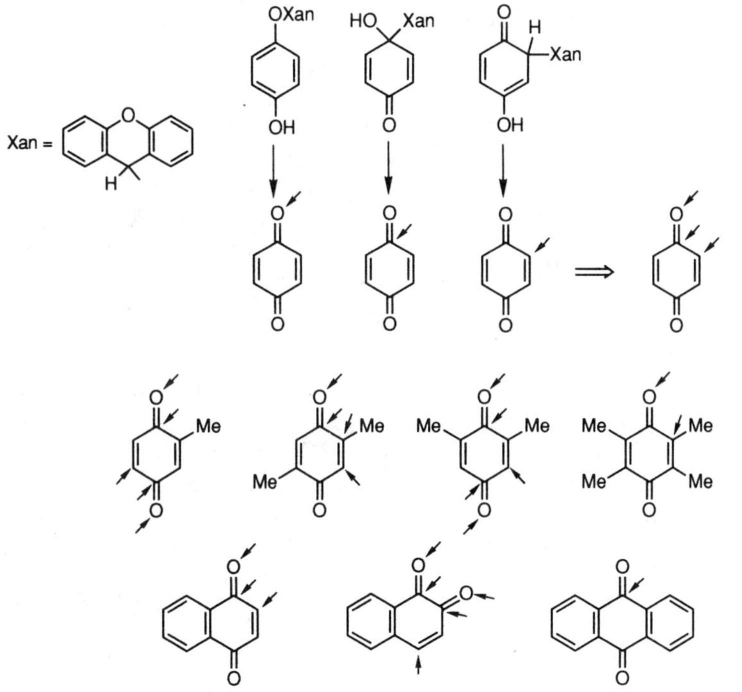

FIGURE 60.1

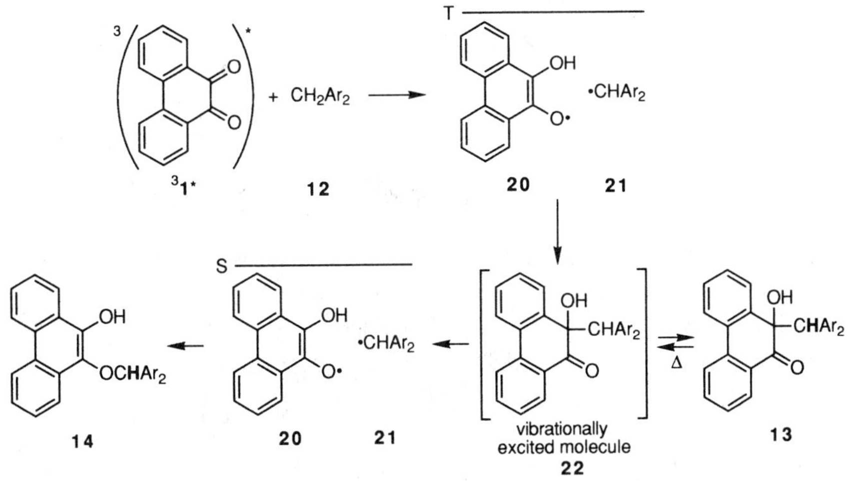

SCHEME 60.2

60.3 Synthetic Applications

The formation of adducts by this H-abstraction process implies same potential synthetic utility of the photoinduced intermolecular H-abstraction of quinones from a variety of the H-donors. For example, photoreactions of naphthoquinone (23) with α,β-unsaturated aldehydes (24), giving adducts 25, was reported to be useful for the synthesis of analogues of α- and β-lapachones (26 and 27) which were known for their antimicrobial and antitumor activity (Equation 60.10).[22]

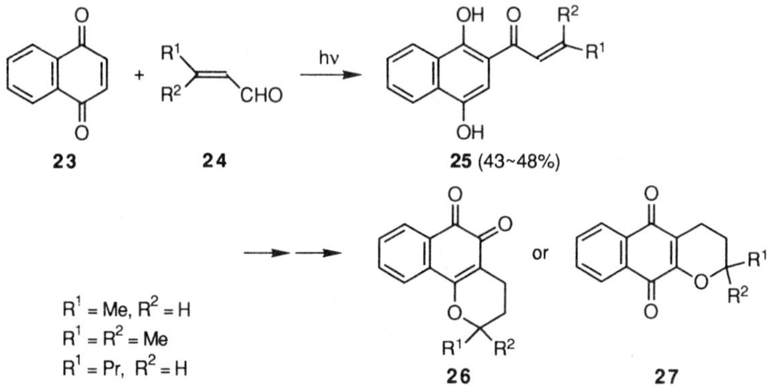

$$(60.10)$$

On the other hand, photoinduced intramolecular H-abstraction of quinones bearing side chains provides useful synthetic routes to a variety of ring systems. Normally, intramolecular H-abstraction from the γ-position proceeds via a six-membered transition state, similar to the Norrish type II reaction of aromatic ketones, to give arene-fused dihydrofuran derivatives 32 in high yields. The formation of a cyclobutanol such as 33 is rare (Scheme 60.3).[5]

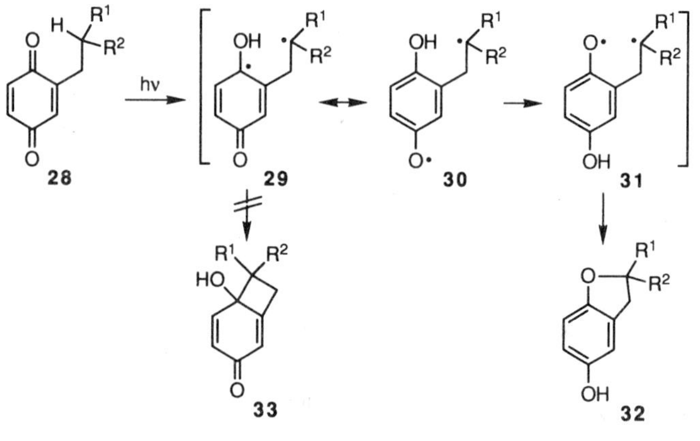

SCHEME 60.3

The photoinduced intramolecular H-abstraction of aminoquinones have received considerable attention due to their potential use in synthetic routes toward mitomycin antibiotics. For example, the photoinduced intramolecular H-abstraction of aminobenzoquinone 34 gave 35, which was oxidized to pyrrolidone-quinone 36 (Equation 60.11).[23]

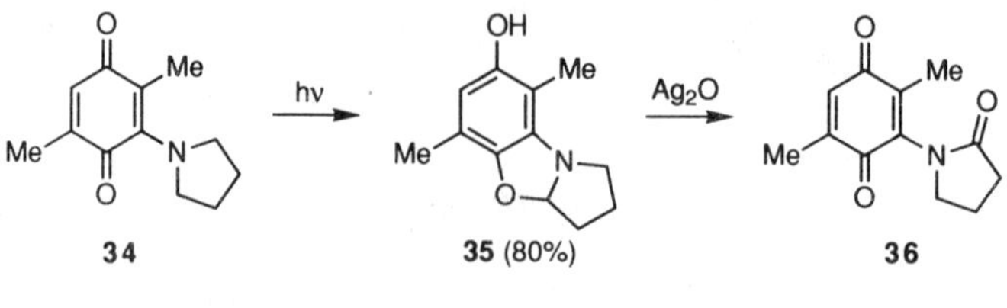

$$(60.11)$$

Introduction of an active methine or methylene group at the neighboring position to the amino group in aminoquinones provided a preparative route to new heterocyclic quinones (Equation 60.12).[24]

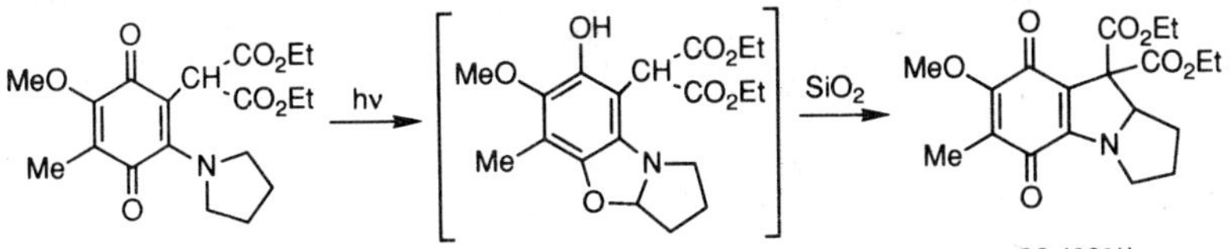

$$(60.12)$$

Similar reactions also occur in the photoreactions of amino-1,4-naphthoquinones.[25] On the other hand, irradiation of a solution of naphthoquinone 40 bearing a 3-pyrrolin-1-yl group and oxidation of the resulting solution with air gave quantitatively a naphthoquinone 41 bearing a 1-pyrrolyl group 41 by a mechanism involving intramolecular H-abstraction (Equation 60.13).[26] An isoindole derivative 43 was formed transiently by the photoinduced intramolecular H-abstraction of the isoindoline-substituted naphthoquinone 42.[27] The intermediacy of 43 was confirmed by obtaining, good yields of Diels-Alder adducts, such as 44 obtained by addition of N-phenylmaleimide (Equation 60.14).

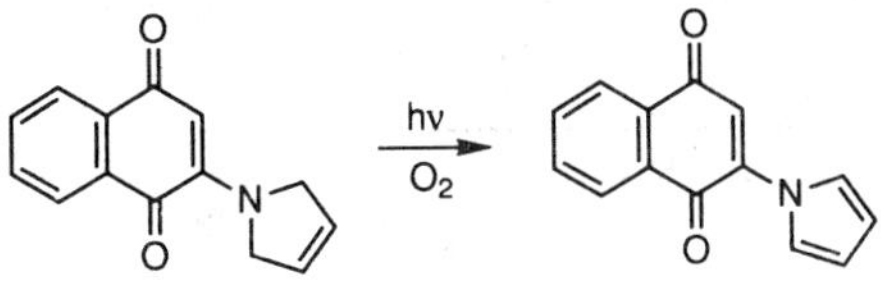

$$(60.13)$$

$$(60.14)$$

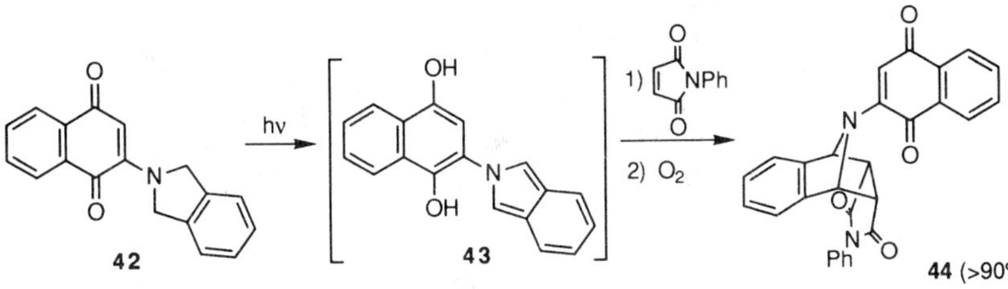

Besides dihydrofuran derivatives (e.g., 32), the photoinduced intramolecular H-abstraction of quinones bearing side chains can yield many other ring systems. Photoreaction of 2-alkoxy-3-(2-oxoalkyl)-1,4-naphthoquinones (e.g., 45) gave (1,3)-dioxepino-1,4-naphthoquinones, such as 49 (Equation 60.15).[28] The photocyclization was explained by a mechanism that involved the intramolecular H-abstraction in 45 to give a biradical 46, cyclization of 46 to 47, and intramolecular disproportionation of 47 to afford 48. Oxidation by O_2 yields the final product. An intramolecular H-abstraction from a remote position in a highly site-selective manner was reported in the photoreactions of acetylglycine-anthraquinone molecules 50 (n = 1~3) (Equation 60.16).[29] The site-selective reactions of the two carbonyl groups of the anthraquinone moiety depended upon the length of the spacer between the acetylglycyl and anthraquinone moiety; the cyclization site changed from the carbonyl group at the 9-position (51) to that of the 10-position (52) as the spacer length increased. This simple system may be helpful for the conformational analysis of a flexible molecule in solution and the selective functionalization of a molecule with many reactive sites.

Table 60.1   UV Characteristics of the Substrates

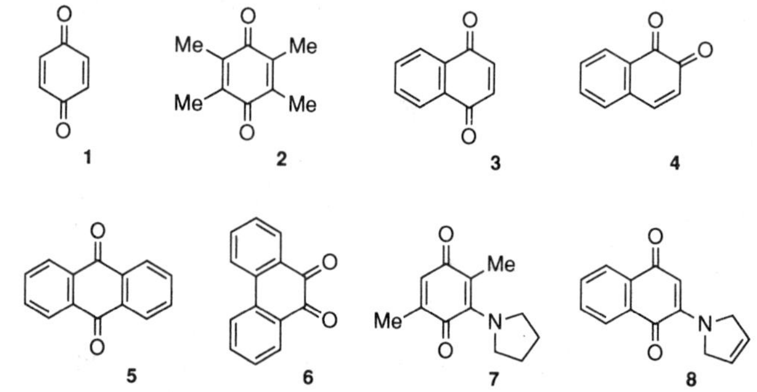

Substrate	Solvent	λ_{max}/nm (log ε/mol^{-1} dm^{-1} cm^{-1})
1	CHCl$_3$	430 (1.5), 300 (3.0)[a]
2	Cyclohexane	455 (1.4), 430 (1.3), 328 (2.4), 263 (4.4), 256 (4.4)[a]
3	Et$_2$O	403 (1.7), 331 (3.5)[a]
4	Et$_2$O	538 (1.6), 389 (3.6), 331 (3.5), 252 (4.5), 248 (4.5)[a]
5	O(CH$_2$CH$_2$)$_2$O	400 (1.9), 323 (3.7), 265 (4.6), 256 (4.5)[a]
6	O(CH$_2$CH$_2$)$_2$O	490 (1.7), 395 (3.3), 313 (3.7), 260 (4.3)[a]
7	EtOH	245 (4.4)[b]
8	C$_6$H$_6$	460, 276[c]

[a] *Kagakubinran kisohen*, 3, Maruzen, Tokyo, 1984.
[b] Ref. 23.
[c] Ref. 26.

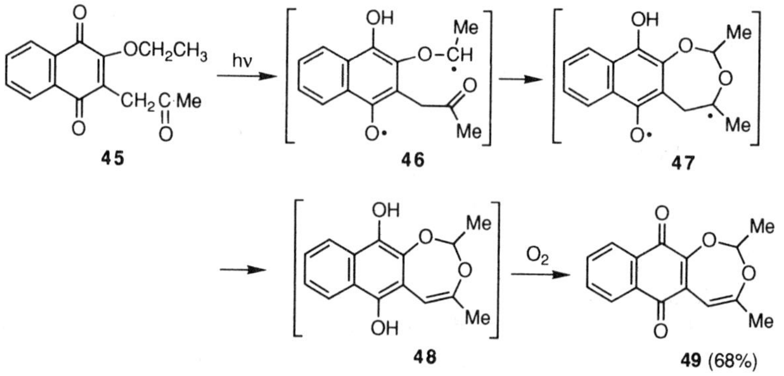

(60.15)

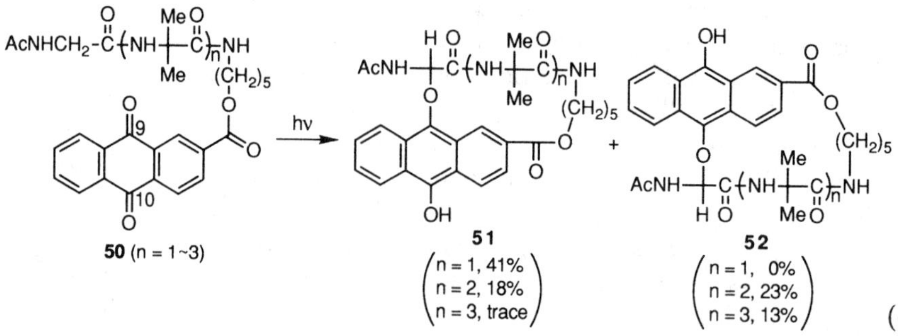

(60.16)

ferences

1. Schonberg, A. and Mustafa, A., Reactions of nonenolizable ketones in sunlight, *Chem. Rev.*, 40, 181, 1947.
2. Bruce, J. M., Light-induced reactions of quinones, *Quart. Rev.*, 21, 405, 1967.
3. Rubin, M. B., Photochemistry of *o*-quinones and α-diketones, *Fortsch. Chem. Forsch.*, 13, 251, 1969.
4. Bruce, J. M., *The Chemistry of the Quinonoid Compounds*, Part 1, John Wiley & Sons, New York, 1974, chap. 9.
5. Maruyama, K. and Osuka, A., *The Chemistry of the Quinonoid Compounds*, Vol. 2, Part 1, John Wiley & Sons, New York, 1988, chap. 13.
6. Rubin, M. B., Photochemical reactions of diketones. The 1,4-addition of ethers to 9,10-phenanthrenequinone, *J. Org. Chem.*, 28, 1949, 1963.
7. Maruyama, K., Otsuki, T., and Naruta, Y., Photochemical reaction of 9,10-phenanthrenequinone with hydrogen donors. Behavior of radicals in solution as studied by CIDNP, *Bull. Chem. Soc. Jpn.*, 49, 791, 1976.
8. Bruce, J. M. and Cutts, E., Light-induced and related reactions of quinones. I. The mechanism of formation of acetylquinol from 1,4-benzoquinone and acetaldehyde, *J. Chem. Soc. (C)*, 449, 1966.
9. Takuwa, A., The photochemical reaction of 1,2-naphthoquinones with aldehydes. III. The reactions with aromatic aldehydes and α,β-unsaturated aliphatic aldehydes, *Bull. Chem. Soc. Jpn.*, 50, 2973, 1977.
10. Takuwa, A., Iwamoto, H., Soga, O., and Maruyama, K., The formation of acetylcatechol in the photochemical reaction of *o*-benzoquinones with acetaldehyde, *Bull. Chem. Soc. Jpn.*, 55, 3657, 1082.
11. Maruyama, K. and Arakawa, S., The photochemical reaction of 1,4-naphthoquinone derivatives with hydrogen donors, *Bull. Chem. Soc. Jpn.*, 47, 1960, 1974.
12. Rubin, M. B. and Zwitkowits, P., Photochemical reactions of diketones. II. The 1,2-addition of substituted toluenes to 9,10-phenanthrenequinone, *J. Org. Chem.*, 29, 2362, 1964.
13. Davies, A. K., Gee, G. A., McKellar, J. F., and Phillips, G. O., Fluorescence quenching of 2-piperidinoanthraquinone, *J. Chem. Soc., Perkin Trans. 2*, 1742, 1973.
14. Kobashi, H., Funabashi, M., Kondo, T., Morita, T., Okada, T., and Mataga, N., Coexistence of hydrogen atom transfer reactions through and not through triplet ion pair between *p*-chloranil and durene, *Bull. Chem. Soc. Jpn.*, 57, 3557, 1984.
15. Maruyama, K., Sakurai, H., and Otsuki, T., The addition reaction of acyl radicals to 9,10-phenanthrenequinone in the presence of the corresponding aldehydes. A support for the in-cage mechanism of the photochemical reaction of 9,10-phenanthrenequinone with aldehydes, *Bull. Chem. Soc. Jpn.*, 50, 2777, 1977.
16. Maruyama, K., Shindo, H., Otsuki, T., and Maruyama, T., Chemically induced dynamic nuclear polarization in the photochemical reaction of phenanthraquinone with hydrogen donors. I. Kinetics of nuclear spin polarization, *Bull. Chem. Soc. Jpn.*, 44, 2756, 1971.
17. Shindo, H., Maruyama, K., Otsuki, T., and Maruyama, T., Chemically induced dynamic nuclear polarization in the photochemical reaction of phenanthraquinone with hydrogen donors. II. A consideration of unusual nuclear spin polarization, *Bull. Chem. Soc. Jpn.*, 44, 2789, 1971.
18. Maruyama, K. and Miyagi, Y., Photo-induced condensation reaction of *p*-quinones with aldehydes, *Bull. Chem. Soc. Jpn.*, 47, 1303, 1974.
19. Maruyama, K., Otsuki, T., and Takuwa, A., Photochemical reduction of *p*-benzoquinones studied by CIDNP technique, *Chem. Lett.*, 1972, 131.
20. Maruyama, K., Otsuki, T., Takuwa, A., and Arakawa, S., Photochemical reduction of *p*-quinones with hydrogen donors studied by CIDNP technique, *Bull. Chem. Soc. Jpn.*, 46, 2470, 1973.
21. Maruyama, K., Takuwa, A., and Soga, O., Photochemical reduction of 1,2-naphthoquinones with xanthen. Investigation by means of a photo-CIDNP technique, *J. Chem. Soc., Perkin Trans. 2*, 255, 1979.

22. Maruyama, K. and Naruta, Y., Syntheses of α- and β-lapachones and their homologues by way of photochemical side chain introduction to quinone, *Chem. Lett.*, 847, 1977.

23. Falci, K. J., Franck, R. W., and Smith, G. P., Approaches to the mitomycins. Photochemistry of aminoquinones, *J. Org. Chem.*, 42, 3317, 1977.

24. Akiba, M., Ikuta, S., and Takada, T., A convenient synthesis of 7-methoxymitosene by the photolysis of aminobenzoquinones, *J. Chem. Soc., Chem. Commun.*, 817, 1983.

25. Akiba, M., Kosugi, Y., Okuyama, M., and Takada, T., A convenient photosynthesis of aziridinopyrrolo[1,2-a]benz[f]indoloquinone and heterocyclic quinones as model compounds of mitomycins by a one-pot reaction, *J. Org. Chem.*, 43, 181, 1978.

26. Maruyama, K., Kozuka, T., and Otsuki, T., The intramolecular hydrogen abstraction reaction in the photolysis of aminated 1,4-naphthoquinones, *Bull. Chem. Soc. Jpn.*, 50, 2170, 1977.

27. Maruyama, K., Kozuka, T., Otsuki, T., and Naruta, Y., A transient formation of N-substituted isoindole by means of photochemical intramolecular hydrogen abstraction, *Chem. Lett.*, 1125, 1977.

28. Maruyama, K., Osuka, A., Nakagawa, K., Jinsenji, T., and Tabuchi, K., Novel photoinduced cyclization of 2-alkoxy-3-(2-oxoalkyl)-1,4-naphthoquinone, *Chem. Lett.*, 1505, 1988.

29. Maruyama, K., Hashimoto, M., and Tamiaki, H., Site-selective photocyclization of acetylglycine-anthraquinone molecules, *Chem. Lett.*, 1455, 1991.

61

The Photochemistry of Sulfonamides and Sulfenamides

. Pincock
ousie University

.1 Sulfonamides

Introduction

The first rule of photochemistry is that excited-state reactions can occur only if the functional group of interest absorbs photons in the accessible region of the UV or VIS spectrum. Although the subject of deeper UV photochemistry has attracted considerable interest,[1] for most photochemists this, in practice, means wavelengths greater than 250 nm. Because of this, a variety of functionalized aliphatic compounds such as ethers, alcohols, esters, phosphates, nitriles, and amines, which are transparent above 250 nm, have received little attention from photochemists. Sulfonamides fall into this class. Even the task of finding UV absorption spectra of saturated sulfonamides proved to be unsuccessful.[2] However, the fact that $PhCH_2SO_2NH_2$ has a UV spectrum (λ_{max} 259 nm (ε, 203), CH_3OH solvent),[4] which is not significantly different from the B band of toluene,[5] demonstrates that the sulfonamide functional group cannot be contributing significantly to absorption in that region of the spectrum.

Despite this fact, there is a reasonably long history of sulfonamide photochemistry, as has been outlined in a recent excellent review[6] of the general topic of $C-SO_2-X$ (X = Cl, Br, I, O, or N) photochemistry. In this chapter, rather than simply recataloging these results, specific examples will be emphasized from a mechanistic point of view. Most of this photochemistry is of compounds that contain chromophores (usually aromatic rings) that allow reactivity of the sulfonamide functional group, i.e., $ArSO_2NR_2$, $ArCH_2SO_2NR_2$, or RSO_2NHAr. For the first two of these examples, the major interest has been the design of photolabile protecting groups for amines; that will be the topic of the next section. The last example usually leads to "photo-Fries" chemistry as described later.

0-8493-8634-9/95/$0.00+$.50
© 1995 by CRC Press, Inc.

Sulfonamides as Photolabile Protecting Groups for Amines

The N-sulfonyl group has often been recommended as a useful protecting group for amines because of its stability in a variety of chemical conditions such as acid or base hydrolysis, catalytic hydrogenation, and oxidation. This stability, which is a positive feature, makes removal difficult. Usually, sodium in liquid ammonia is necessary although both electrochemical[7] and chemical reduction[8] have been proposed as useful alternatives. Obviously, these methods may not be compatible with other reactive functional groups present in the molecule. Because of these facts, photochemical methods are appealing; three quite different approaches have been proposed, as described below. The general topic of photolabile protecting groups in organic chemistry has also been reviewed.[9]

Direct Photolysis of Arylsulfonamides ($ArSO_2NR_1R_2$)

The work of Pète et al.,[10,11] which followed an earlier, short report by D'Souza and Day,[12] demonstrated that direct photolysis at 254 nm of arylsulfonamides of aliphatic amines and amino acids gives reasonable to good yields of the free amine (Equation 61.1).

$$Ar-SO_2-NR_1R_2 \xrightarrow{h\nu} H-NR_1R_2$$

$$(61.1)$$

Some selected examples are given in Table 61.1. Although the multiplicity of the reactive excited state is not known for these reactions, both the excited state singlet ($\sim$418 kJ mol^{-1}) and the triplet ($\sim$334 kJ mol^{-1}) for simple benzene derivatives have enough energy to break the S-N bond in the sulfonamide, which should be weaker than the C-S bond in dimethyl sulfone (280 kJ mol^{-1}).[13] Although low yields (3 to 6%) of the aromatic hydrocarbon (ArH), indicating C-S bond cleavage, were detected, the major products can be rationalized by homolytic S-N bond cleavage. The amino radical presumably abstracts hydrogen from the solvent (at least in the case of ether) to form the amine. The sulfonyl radical, $ArSO_2\cdot$, undergoes a bimolecular redox reaction to sulfonyl (RSO$_3$) and sulfenyl (R-S) species. The major disadvantage of this procedure is the requirement for irradiation at short wavelengths (250 to 280 nm) where many other reactive functional groups absorb.

Direct Photolysis of Arylmethylsulfonamides ($ArCH_2SO_2NR_1R_2$)

The direct photolysis of arylmethylsulfonyl compounds leads to efficient C-S bond cleavage, Equation 61.2. A recent review is available,[14] and the reactivity of the resulting sulfonyl radicals has been studied by EPR spectroscopy.[15] Because of the low bond dissociation energy (221 kJ mol^{-1})[13] of the C-S bond, this homolytic bond cleavage process occurs from both the singlet and triplet excited state for benzylic[16] and 1-naphthylmethyl[16,17] sulfones. The fate of the resulting sulfonyl radical depends on X[18] in Equation 61.2; for X equals NH$_2$, the loss of sulfur dioxide appears to be very rapid. An S-N bond dissociation energy for $\cdot SO_2NH_2$ of only 75 kJ mol^{-1} has been estimated by *ab initio* calculations.[19]

Table 61.1   Yield of Amine on Photolysis of Aryl Sulfonamides

	ArSO$_2$NR$_1$R$_2$			Amine	
Ar	R$_1$	R$_2$	Solvent	Yield (%)	Ref.
4-MeC$_6$H$_4$-	-CH$_2$CH$_3$	-CH$_2$CH$_3$	Ether	84^a	11
C$_6$H$_5$-	-CH$_2$CH$_3$	-CH$_2$CH$_3$	Ether	77^a	11
4-MeC$_6$H$_4$-	H	-C$_6$H$_{11}$	Ether	80^a	11
4-MeC$_6$H$_4$-	H	Leucine	H$_2$O	22–48^b	12
4-MeC$_6$H$_4$-	H	Histidine	H$_2$O	46–57^b	12

^{a} The sum of the free amine plus the ArSO$_3^{-+}$ NH$_2$R$_1$R$_2$ salt.
^{b} Depends on temperature, the low yields at 25 °C, and the high values at 90 °C.

Table 61.2   Yield of Amine on Photolysis of Arylmethylsulfonamides

$ArCH_2SO_2NR_1R_2$

Ar	R_1	R_2	Solvent	Amine Yield (%)	Ref.
C_6H_5-	$-C_4H_9$	$-C_4H_9$	2-Propanol	78	20
C_6H_5-	-H	$-C_6H_{11}$	2-Propanol: 3% NaOH	96	20
C_6H_5-	-H	1-Adenine	H_2O	Quantitative	21
$3,5$-$diCH_3O$-C_6H_3-	-H	3-Adenine	H_2O	Quantitative	21
$3,5$-$diCH_3O$-C_6H_3-	-H	7-Adenine	H_2O	Quantitative	21
Q^a	$-C_6H_{13}$	$-C_6H_{13}$	2-Propanol	96	24
Q^a	-H	$-C_6H_5$	2-Propanol	58	24
Q^a	-H	1-Adenine	2-Propanol	95	24

$^a Q =$

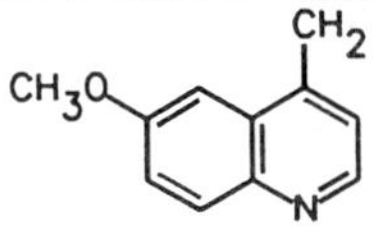

$$ArCH_2-SO_2-X \xrightarrow{h\nu} ArCH_2^{\cdot} \quad {}^{\cdot}SO_2-X \qquad (61.2)$$

These considerations have lead to the development of excellent protecting groups for amines, as shown in Equation 61.3 for the irradiation of benzylic sulfonamides in 2-propanol.[20] The products are easily separated and yields of isolated amine are high, as the few selected examples in Table 61.2 indicate. For aliphatic amines, the chromophore can be phenyl since there is no competing absorption. Surprisingly, this is still true for irradiation of adenine derivatives[21] at 254 nm where competitive absorption by the heterocyclic nucleotide must be occurring. This last reaction is an example of the well-known photosolvolysis[22] through cation/anion pairs, since the arylmethyl derivative is converted to the arylmethanol in water. Whether the ions are produced through direct heterolytic cleavage or by homolytic cleavage followed by electron transfer that converts the radical pair to the ion pair is a subject of much current interest.[23]

$$ArCH_2-SO_2-NR_1R_2 \xrightarrow[(CH_3)_2CHOH]{h\nu} H-NR_1R_2 \;+\; ArCH_3 \;+\; SO_2 \qquad (61.3)$$

The major advantage of the arylmethyl derivatives as protecting groups is exemplified by the development[24] of the quinoline derivative (Q in Table 61.2). Now, the strong π,π absorption of the quinoline (λ_{max}, 360 nm (ε, 50,000)) allows for irradiation at longer wavelengths where other reactive functional groups are transparent. No evidence was obtained for ionic intermediates in these reactions.

Photochemical Electron Transfer Reduction of Arylsulfonamides

The development of this procedure was based on an initial observation by Umezawa et al.[25] who reported that irradiation of 6,7-dimethoxy-1,2,3,4-tetrahydroisoquinoline N-tosylates, 1, in the presence of $NaBH_4$ resulted in photochemical cleavage and high yields of the corresponding free amines, 2, Equation 61.4. Since compound 1 also lacked the characteristic strong fluorescence of 1,2-dimethoxybenzene, this process was explained by intramolecular electron transfer in the excited singlet state, from the methoxy substituted aromatic donor to the sulfonyl substituted aromatic acceptor. The observation that electron transfer was favorable for the tosylate, 1, but not the corresponding mesylate, indicated the importance of the more easily reduced aromatic sulfonyl group.

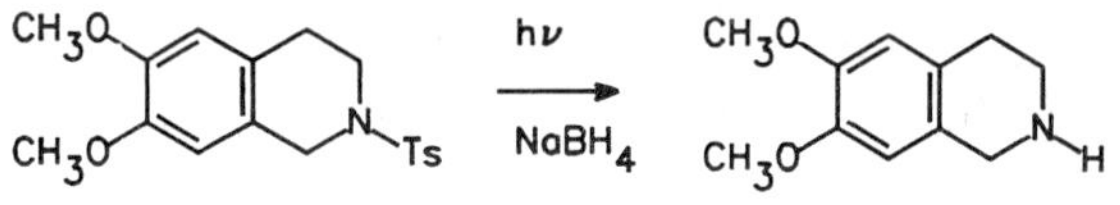

$$(61.4)$$

This idea was then extended[26,27] to intermolecular sensitization as in Equation 61.5 (1,2-DMB is 1,2-dimethoxybenzene), where the yield of the liberated amine was 87% for all three cases. A variety of methoxy aromatic donors was tested as sensitizers, and the following mechanism, Equations 61.6 to 61.11, was proposed.

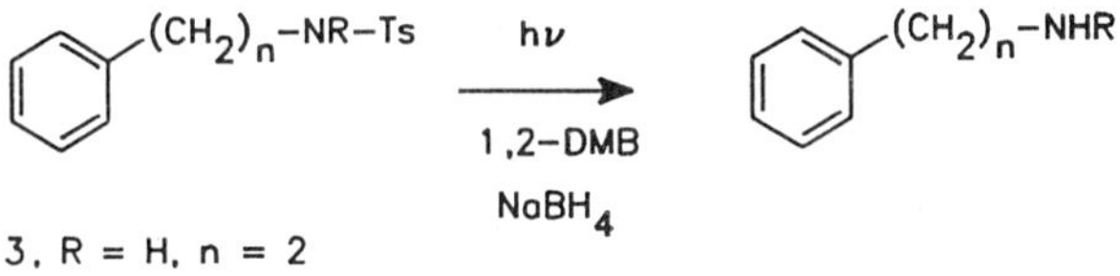

3, R = H, n = 2
4, R = CH_3, n = 2
5, R = CH_3, n = 1

$$(61.5)$$

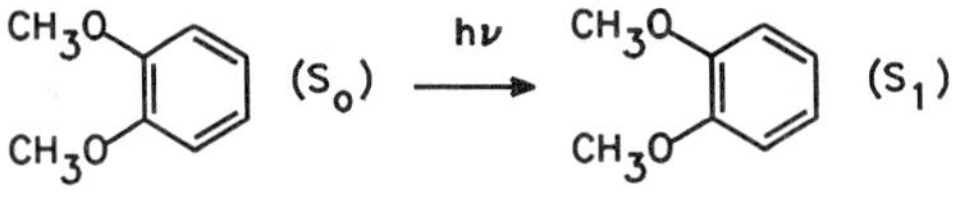

$$(61.6)$$

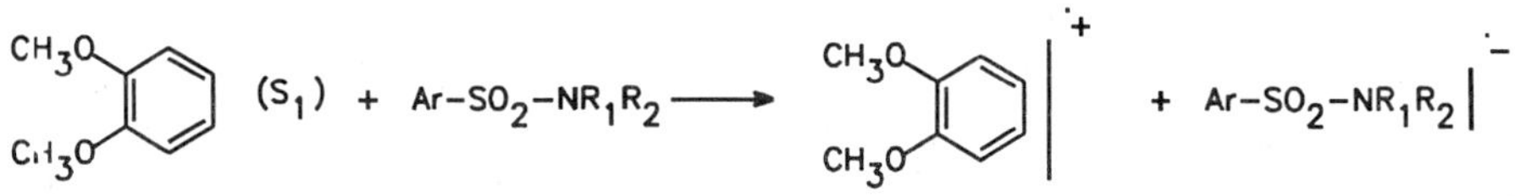

$$(61.7)$$

$$Ar-SO_2-NR_1R_2 \big|^{\cdot -} \longrightarrow Ar-SO_2^{\cdot} + {}^-NR_1R_2 (8)$$

$$(61.8)$$

$$(61.9)$$

$${}^-NR_1R_2 + H_2O \longrightarrow H-NR_1R_2 + {}^-OH$$

$$Ar-SO_2^{\cdot} \xrightarrow{\text{NaBH}_4} Ar-SO_2H \big|^{\cdot -}$$

$$(61.10)$$

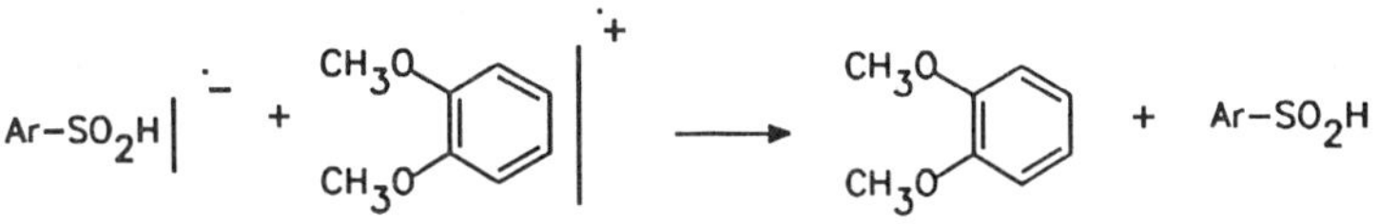

$$(61.11)$$

Equation 61.6 shows the original photochemical excitation with 1,2-DMB as an example of a donor. Intermolecular electron transfer followed by diffusional separation gives the radical ions, Equation 61.7. (Back electron transfer to regenerate starting materials, the major step leading to inefficiency in these reactions is not shown.) The electron-transfer step follows the Rehm-Weller[28] equation and becomes diffusional as measured by fluorescence quenching, as it becomes exothermal because of the lower oxidation potential of the donor. The radical anion of the sulfonamide then cleaves, Equation 61.8, in analogy to the well-established electrochemical reduction of sulfonamides.[7] The sodium borohydride (other reductants like ascorbic acid, NH_3BH_3, and hydrazine were also successful) is apparently required to reduce the sulfonyl radical to a sulfinate radical anion, Equation 61.10. Oxidation to the sulfinate anion then regenerates the photo sensitizer, Equation 61.11.

Table 61.3 Yield of Amine on Photochemical Electron Transfer Reduction of Toluenesulfonyl Amides

$TsNR_1R_2$				
R_1	R_2	Donor	Amine Yield (%)	Ref.
CH_3-	-CH_2CH_2Ph	1,5-DMN	88	27
CH_3-	Ph	1,5-DMN	75	27
H	2-hydroxycyclohexyl	1,5-DMN	77	27
Boc-gly-L-lys-OCH_3	H	1,5-DMN	82	30
H	-CH_2CH_2Ph	NONa	100	30
CH_3-	-CH_2Ph	NONa	98	30

This process works extremely well for a variety of amines, including small peptides, and irradiation can be through Pyrex ($\lambda > 290$ nm) provided 1,5-dimethoxynaphthalene (1,5-DMN) is used as the donor. Table 61.3 and Equation 61.5 show selected examples. The observation that this procedure works well even for aromatic amines (75% yield, $R_1 = CH_3$, $R_2 = Ph$) is of interest since normally photo-Fries reactions are a major competing pathway (see below). A water-soluble sensitizing donor, **6**, has also been designed.[27]

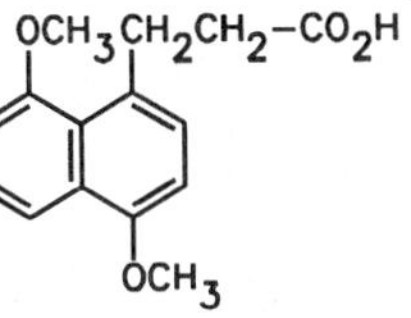

6

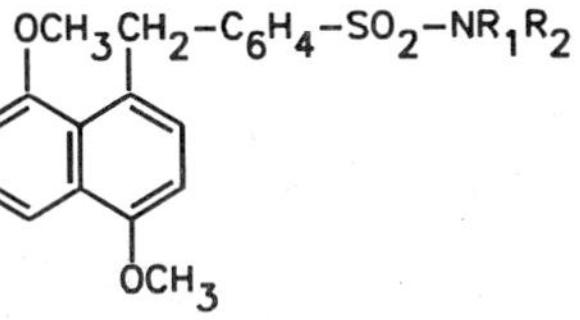

7 ($DMNBS-NR_1R_2$)

Recent extensions to this idea incorporate the donor and the acceptor into the same molecule, **7** (DMNBS), so that the electron transfer again becomes intramolecular.[29] The selectivity of this procedure is demonstrated by Equation 61.12 where only the toluene sulfonamide group close to the donor is removed. Other protecting groups, like BOC, Cbz, Ms, and Bz, are also unaffected by the deprotection.

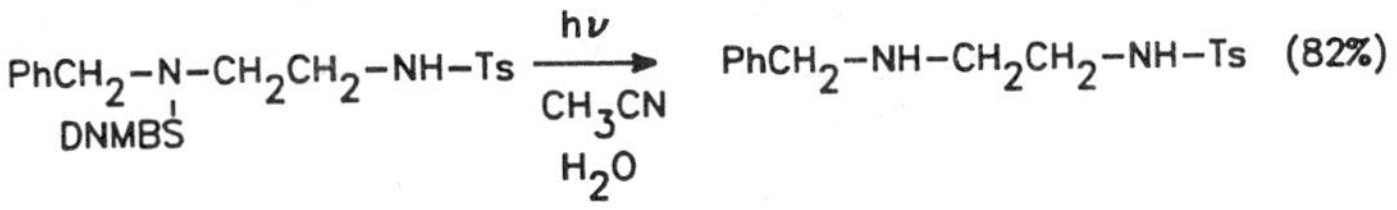

$$PhCH_2-\underset{DNMBS}{N}-CH_2CH_2-NH-Ts \xrightarrow[\substack{CH_3CN \\ H_2O}]{h\nu} PhCH_2-NH-CH_2CH_2-NH-Ts \quad (82\%)$$

$$(61.12)$$

Finally, absorption at wavelengths as long as 350 nm and quantum efficiencies as high as 0.6 can be obtained if naphthoxide anions, NONa, are used as donors[30] as shown in Equation 61.13. Examples are given in Table 6.3 and yields of liberated amine are obviously excellent.

The conclusion from this section on photolabile protecting groups for amines is that procedures with very high yields of deprotection are now available. The last one, using naphthoxide anion as an electron-transfer reductant, satisfies all the requirements of a synthetic photochemist: mild conditions, long wavelengths (350 nm), and high chemical and quantum yields. It remains to be seen if the demands of the practitioners of pure synthetic organic chemistry will also be met by these promising procedures.

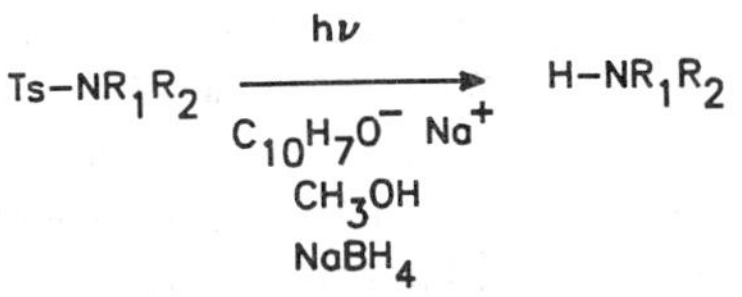

$$Ts-NR_1R_2 \xrightarrow[\substack{C_{10}H_7O^- Na^+ \\ CH_3OH \\ NaBH_4}]{h\nu} H-NR_1R_2$$

$$(61.13)$$

Photo-Fries Rearrangement of Sulfonamides of Aromatic Amines (RSO₂NR₁Ar)

The photo-Fries rearrangement occurs for many functionalized aromatic rings, including aryl sulfonamides. The reaction has been reviewed.[31,32] A classic example is given in Equation 61.14.[33] The generally accepted mechanism is homolytic cleavage of the S-N bond of the sulfonamide followed by in-cage coupling of the radical pair. The absence (or at least low yield) of the *meta*-isomer is rationalized by the low spin density at that position in the aryl aminyl radical. The amine results from hydrogen atom abstraction from the solvent. Recently, it has been reported[34] that, for benzenesulfonanilide, the yield of aniline can be reduced and the yield of *ortho*-product increased by photolysis in water with β-cyclodextrin. Photolysis of the β-cyclodextrin:benzenesulfonamide adduct in the solid state gave the *ortho*-photo-Fries product exclusively.

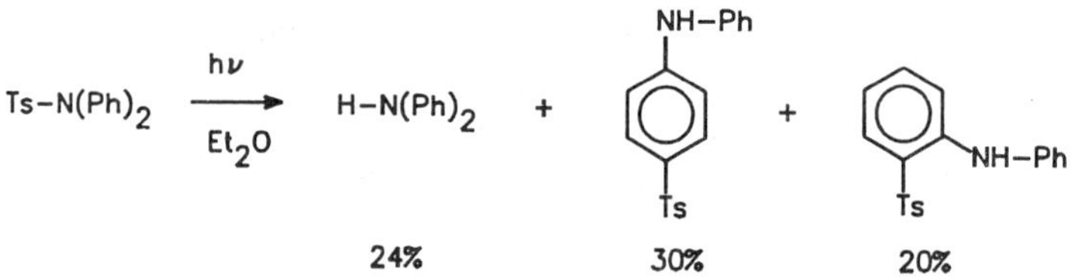

$$\text{Ts-N(Ph)}_2 \xrightarrow[\text{Et}_2\text{O}]{h\nu} \text{H-N(Ph)}_2 \ + \ \text{(structure, 30\%)} \ + \ \text{(structure, 20\%)}$$

24%   30%   20%   (61.14)

From a photochemist's point of view, the best studied example of these reactions is by Lally and Spillane[35,36] who photolyzed the sodium salt of sulfamic acid in methanol solvent, Equation 61.15. The mass balance is very good for these highly efficient reactions and the ratio of *o:m:p* isomers agrees well with spin density calculations. The yield of aniline can be increased by adding radical scavengers (BuSH) to the reaction mixture. Stern-Volmer quenching studies with 1,3-pentadiene indicate that these reactions are occurring from the triplet state; the curvature in the plot for aniline formation implies two excited states are involved. The proposed mechanism is by a triplet radical pair formed by homolytic cleavage of the N-S bond.

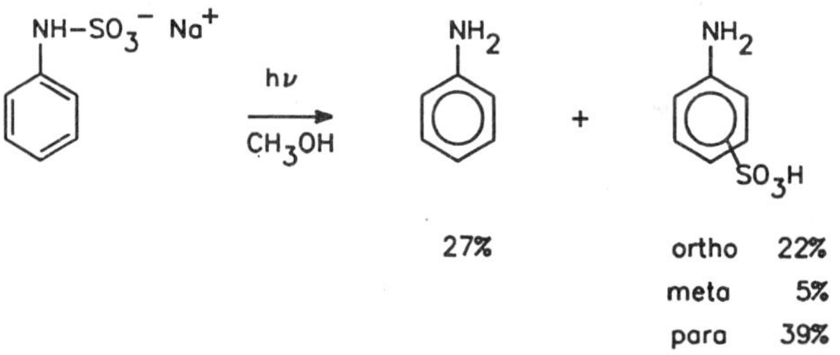

$$\xrightarrow[\text{CH}_3\text{OH}]{h\nu}$$

27%   ortho 22%
             meta   5%   (61.15)
             para   39%

Miscellaneous Reactions Involving Sulfur-Nitrogen Bond Cleavage in the Excited State

As implied by the above discussions, much of sulfonamide photochemistry can be rationalized by an initial S-N bond cleavage of the excited state to form a radical pair. The fate of the radical pair depends on the structure of the two radicals and the reaction conditions. These reactions are well-catalogued in Horspool's review.[6] Some cases involve loss of SO₂ and some do not; clearly, the strength of the C-S bond in the sulfonyl radical[18,19] and the lifetime of the radical pair are critical. Two selected examples are given in Equations 61.16[37] and 61.17.[38] Studies of the latter reaction by laser flash photolysis suggest that both the singlet and triplet excited states are photoreactive. As well, a transient is detected that is either the biradical resulting from S-N bond cleavage or the hydrogen migration precursor tautomer of the photoproduct.

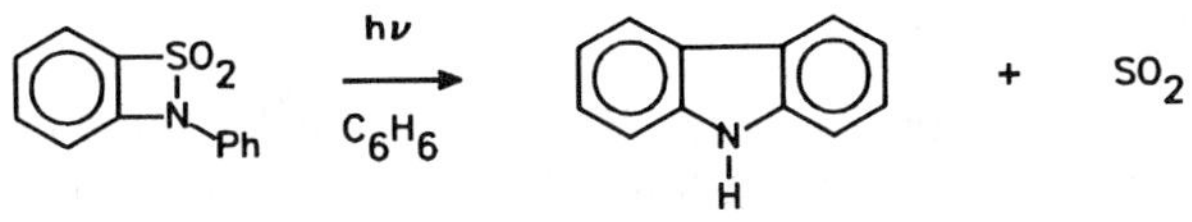

$$(61.16)$$

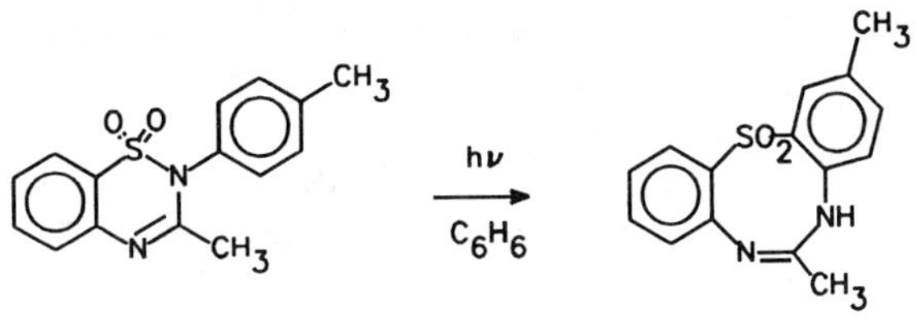

$$(61.17)$$

Miscellaneous Reactions Not Involving Sulfur-Nitrogen Bond Cleavage in the Excited State

Only a few examples of this class of reaction are known. The best studied case is by Pète et al.[39] One example of many is given in Equation 61.18. Although SO_2 is lost during the process, the initial photochemical step is thought to be aryl-vinyl bonding from the excited triplet state of the enone.

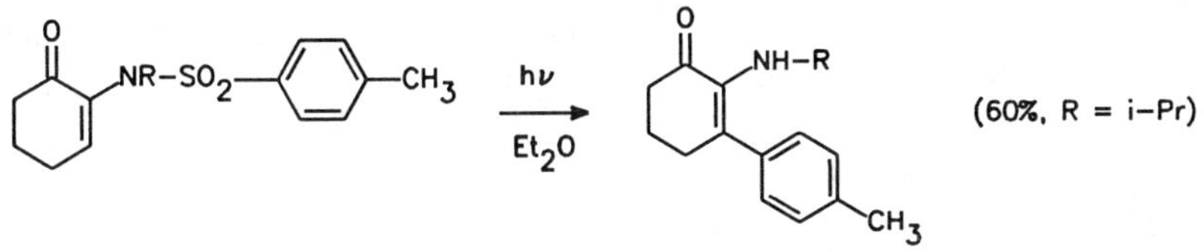

$$(60\%,\ R = i\text{-}Pr)$$

$$(61.18)$$

Another unusual example where the sulfonamide functional group remains intact is given by Equation 61.19.

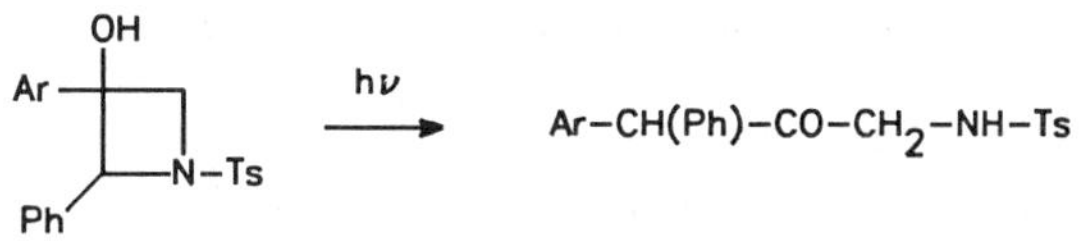

$$(61.19)$$

1.2 Sulfenamides

The photochemistry of sulfenamides is a relatively unstudied topic. A recent review[41] is available in which no references more recent than 1978 are reported. A search[2] for UV spectral data of alkyl sulfenamides again proved to be unsuccessful. As in sulfonamide photochemistry, chromophores are created by aromatic rings bound to the sulfur or nitrogen and sulfur-nitrogen homolytic bond cleavage is (so far) the only observed process. Photo-Fries, as well as cage-escape, products result,[42] Equation 61.20. These processes have also been used to generate and study EPR spectra of R(RS)N· radicals.[43] No quantitative or physical organic photochemical studies have been reported. Clearly, this is a topic where more needs to be done.

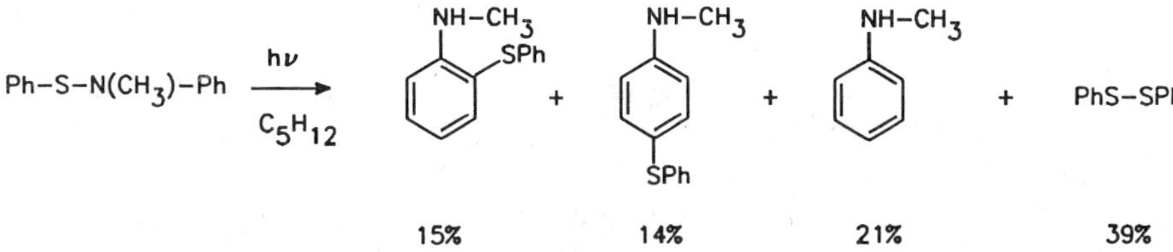

$$(61.20)$$

References

1. For a review of cyclobutene photochemistry, see Leigh, W.J., Techniques and applications of far-UV photochemistry in solution. The photochemistry of C_3H_4 and C_4H_6 hydrocarbons, *Chem. Rev.*, 93, 487, 1993.

2. An early review by Block[3] catalogues UV data for many organic sulfur compounds, but not sulfonamides or sulfenamides. A Chemical Abstracts search under "sulfonamides, ultraviolet" and "sulfenamides, ultraviolet" was unsuccessful.

3. Block, E., The photochemistry of organic sulfur compounds, *Quart. Rep. Sulfur Chem.*, 4, 237, 1969.

4. Grasselli, J. G. and Ritchey, W. M., Eds., *Atlas of Spectral Data and Physical Constants of Organic Compounds*, Vol. IV, CRC Press, Boca Raton, FL, 1975, 685.

5. Silverstein, R. M., Bassler, G. C., and Morrill, T. C., in *Spectrometric Identification of Organic Compounds*, 5th ed., J. Wiley & Sons, New York, 1991, 306.

6. Horspool, W. M., Photochemistry and radiation chemistry, *The Chemistry of Sulfonic Acids, Esters and Their Derivatives*, Patai, S. and Rappaport, Z., Eds., J. Wiley & Sons, New York, 1991, 501.

7. Cottrell, P. T. and Mann, C. K., *Electrochemical reduction of arylsulfonamides*, *J. Am. Chem. Soc.*, 93, 3579, 1971.

8. Closson, W. D. and Schulenberg, J. S., On the mechanism of sulfonamide cleavage by arene anion radicals, *J. Am. Chem. Soc.*, 92, 650, 1970.

9. Pillai, V. N. R., Photoremovable protecting groups in organic synthesis, *Synthesis*, 1, 1980.

10. Abad, A., Mellier, D., Pete, J. P., and Portella, C., Photolysis of alkylsulfonates and sulfonamides. Regeneration of alcohols and amines, *Tetrahedron Lett.*, 4555, 1971.

11. Pete, J. P. and Portella, C., Photolysis of N-alkylarenesulfonamides, *J. Chem. Res. (S)*, 20, 1979.

12. D'Souza, L. and Day, R. A., Photolytic cleavage of sulfonamide bonds, *Science*, 160, 882, 1968.

13. McMillen, D. F. and Golden, D. M., Hydrocarbon bond dissociation energies, *Annu. Rev. Phys. Chem.*, 33, 493, 1982.

14. Still, I. W. J., The photochemistry of sulfoxides and sulfones, *The Chemistry of Sulfones and Sulfoxides*, Patai, S., Rappaport, Z., and Stirling, C., Eds., John Wiley & Sons, 1988, 873.

15. Chatgilialoglu, C., Griller, D., and Rossini, S., Amino and alkoxysulfonyl radicals, *J. Org. Chem.*, 54, 2734, 1989.

16. Givens, R. S., Hrinczenko, B., Liu, H.-S., Maluszewski, B., and Tholen-Collison, J., Photoextrusion of SO_2 from arylmethyl sulfones: Exploration of the mechanism by chemical trapping, chiral and CIDNP probes, *J. Am. Chem. Soc.*, 106, 1779, 1984.

17. Arnold, B., Donald, L., Jurgens, A., and Pincock, J. A., Homolytic versus heterolytic cleavage for the photochemistry of 1-naphthylmethyl derivatives, *Can. J. Chem.*, 63, 3140, 1985.

18. Langler, R. F., Marini, Z. A., and Pincock, J. A., The photochemistry of benzylic sulfonyl compounds: The preparation of sulfones and of sulfinic acids, *Can. J. Chem.*, 56, 903, 1978.

19. Boyd, R. J., Gupta, A., Langler, R. F., Lounie, S. P., and Pincock, J. A., Sulfonyl radicals, sulfinic acid, and related species: An *ab initio* molecular orbital study, *Can. J. Chem.*, 58, 331, 1980.

20. Pincock, J. A. and Jurgens, A., Sulfonamides as photolabile protecting groups for amines, *Tetrahedron Lett.*, 1029, 1979.

21. Er-Rhaimini, A., Mohsinaly, N., and Mornet, R., The photosolvolysis N-arylmethyladenines. Photoremovable N-arylmethyl protective groups or N-containing compounds, *Tetrahedron Lett.*, 31, 5757, 1990.

22. Cristol, S. J. and Bindel, T. H., Photosolvolysis, in *Organic Photochemistry*, Vol. 6, Marcel Dekker, New York, 1983, 327.

23. DeCosta, D. P. and Pincock, J. A., Photochemistry of substituted 1-naphthylmethyl esters of phenylacetic and 3-phenylpropanoic acid: Radical pairs, ion pairs and Marcus electron transfer, *J. Am. Chem. Soc.*, 115, 2180, 1993.

24. Epling, G. A. and Walker, M. E., A new photochemically removable protecting group for amines, *Tetrahedron Lett.*, 23, 3843, 1982.

25. (a) Umezawa, B., Hoshino, O., and Sawaki, S., Organic photochemistry. II. Tetrahydroquinolines from their *N*-tosylates, *Chem. Pharm. Bull.*, 17, 1120, 1969; (b) Hoshino, O., Sawaki, S., and Umezawa, B., Organic photochemistry. III. Photoreductive cleavage of *p*-toluenesulfonamides, *Chem. Pharm. Bull.*, 18, 182, 1970.

26. Hamada, T., Nishida, A., Matsumoto, Y., and Yonemitsu, O., Photohydrolysis of sulfonamides via donor-acceptor ion pairs with electron-donating aromatics and its application to the selective detosylation of lysine peptides, *J. Am. Chem. Soc.*, 102, 3978, 1980.

27. Hamada, T., Nishida, A., Matsumoto, Y., and Yonemitsu, O., Selective removal of electron-accepting *p*-toluene and naphthalenesulfonyl protecting groups for amino function via photoinduced donor-acceptor ion pairs with electron-donating aromatics, *J. Am. Chem. Soc.*, 108, 140, 1986.

28. Rehm, D. and Weller, A., Kinetics of fluorescence by electron and H-atom transfer, *Isr. J. Chem.*, 8, 259, 1970.

29. Hamada, T., Nishida, A., and Yonemitsu, O., A new amino protecting group readily removable with near ultraviolet light as an application of electron-transfer photochemistry, *Tetrahedron Lett.*, 28, 4241, 1989.

30. Art, J. F., Kestemont, J. D., and Soumillion, J. Ph., Photodetosylation of sulfonamides initiated by electron transfer from an aromatic sensitizer, *Tetrahedron Lett.*, 32, 1425, 1991.

31. Bellus, D., Photo-Fries rearrangement and related photochemical [1,j]-shifts (j = 3,5,7) of carbonyl and sulfonyl groups, *Adv. Photochem.*, 8, 109, 1971.

32. Kaupp, G., Photochemical rearrangements and fragmentations of benzene derivatives and annelated arenes, *Angew. Chem. Int. Ed.*, 19, 243, 1980.

33. Weiss, B., Durr, H., and Haas, H. J., Photochemistry of sulfonamides and sulfonylureas: A contribution to the problem of light induced dermatoses, *Angew. Chem. Int. Ed.*, 19, 648, 1980.

34. Pitchumani, K., Manickam, M. C. D., and Srinivasan, C., Effect of cyclodextrin encapsulation on photo-Fries rearrangement of benzenesulfoanilide, *Tetrahedron Lett.*, 32, 2975, 1991.

35. Lally, J. M. and Spillane, W. J., The photochemistry of phenylsulfamic acid: Photorearrangement and photodegredation, *J. Chem. Soc., Chem. Commun.*, 8, 1987.

36. Lally, J. M. and Spillane, W. J., The photochemistry of *para*-substituted phenylsulfamates — photo-Fries rearrangements, *J. Chem. Soc., Perkin Trans. 2*, 803, 1991.

37. Ao, M. S. and Burgess, E. M., Benzothiazete 1,1-dioxides, *J. Am. Chem. Soc.*, 93, 5298, 1971.

38. Kumar, C. V., Gopidas, K. R., Bhattacharyya, K., Das, P. K., and George, M. V., Photoinduced ring enlargement reactions of 2H-1,2,4-benzothiadiazine 1,1-dioxides. Steady state and laser flash photolysis studies, *J. Org. Chem.*, 51, 1967, 1986.

39. Cossy, J. and Pete, J.-P., A study of the mechanism of the photochemical transformation of *N*-alkyl 2-arenesulfonamido-2-cyclohexenone to *N*-alkyl 2-amino-3-aryl-2-cyclohexenone, *Tetrahedron*, 37, 2287, 1981 and preceding references.

40. Fuhrmann, J., Dietzsch, T., and Henning, H. G., Rearrangement of *N*-tosyl-2,3-diarylazetidin-3-ols to tosylaminomethyl benzhydryl ketones, *Z. Chem.*, 23, 52, 1990.

41. Horspool, W. M., Photochemistry and radiation chemistry, *The Chemistry of Sulfenic Acids and Their Derivatives*, S. Patai, Ed., John Wiley & Sons, New York, 1990, 517.

42. Ando, T., Nijima, M., and Tokura, N., Oxidative, photoinduced, and thermal sulfur-nitrogen bond fission in benzenesulfenanilides, *J. Chem. Soc., Perkin Trans. 1*, 2227, 1977.

43. Miura, Y., Asada, H., and Kinoshita, M., ESR studies of *N*-(arylthio)-*t*-butylaminyls, *Bull. Chem. Soc. Jpn.*, 50, 1855, 1977.

62

Photochemical Methods for Protection and Deprotection of Sulfur-Containing Compounds

V. N. Rajasekharan Pillai
Mahatma Gandhi University &
Jawaharlal Nehru Centre for
Advanced Scientific Research
Kottayam, Kerala, India

62.1 Introduction

Organic sulfur-containing compounds undergo a wide variety of light-induced reactions that are of immense synthetic and technological significance. Sulfur-containing functional groups form part of many complex synthetic and biological molecules. Effective manipulation of these groups is important in synthetic design involving such systems. Selective and specific methods for the photolytic protection and activation of sulfur-containing functional groups and the photochemical protection and cleavage of sulfur containing blocking groups are discussed in this chapter. The photochemical blocking methods for the thiol group, light-induced deprotection of arene sulfonamides, photolytic cleavage of 2,4-dinitrobenzenesulfenyl derivatives of carboxylic acids, photochemistry of 2-thionothiazolidines, and photolytic dithioacetalization reactions are covered specifically.

62.2 Photoprotection of Thiol Groups

The light-induced reactions involving the thiol groups are very significant in the phototransformations of biological macromolecules. The synthetic processes involving the amino acid cysteine and its derivatives also demand effective manipulation and control on the reactivity of the thiol groups. The conventional thiol protection by the benzyl group and its removal by catalytic hydrogenolysis is complicated by the poisoning due to the sulfur and consequent reduction in yields. In this context, photochemical methods that involve neutral working conditions at room temperature are attractive alternatives. The most commonly used photochemically removable protective function

0-8493-8634-9/95/$0.00+$.50
© 1995 by CRC Press, Inc.

for the thiol group in the amino acid cysteine is the 2-nitrobenzyl group.[1] The 2-nitrobenzyl group can be introduced into cysteine (1) by reacting it with 2-nitrobenzyl chloride in the presence of a base like triethylamine. *S*-2-nitrobenzyl cysteine (2) thus produced on irradiation underwent quantitative deprotection yielding cysteine, cystine (3), and cysteic acid (4), along with 2-nitrosobenzaldehyde and its photolytic products (Scheme 62.1).

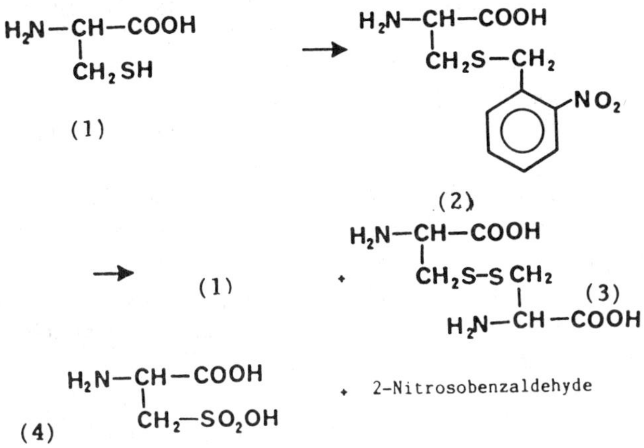

SCHEME 62.1 2-Nitrobenzyl protection of thiol group in cysteine.

The mechanism of the photochemical process responsible for the deprotection of the thiol group in cysteine involves an intramolecular hydrogen abstraction by the excited nitro group. The light-induced internal oxidation-reduction reactions of aromatic nitro compounds containing an *ortho*-nitro group have been the subject of much investigation.[2-4] In this intramolecular rearrangement, the nitro function is reduced to a nitroso group and an oxygen is inserted into the carbon-hydrogen bond located at the 2-position. The intramolecular hydrogen abstraction by the excited nitro group is followed by an electron redistribution to the *aci*-nitro form, which rearranges to the nitroso derivative (Scheme 62.2).

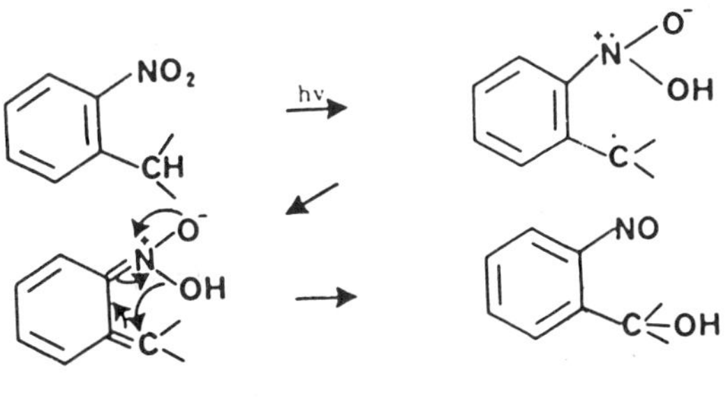

SCHEME 62.2 Photochemical oxygen transfer reactions in ortho-substituted aromatic nitro compounds.

The presence of aldehyde trapping reagents and quenchers of photooxidation in the reaction mixture determines the molar ratio of the three amino acids formed in the photolysis. Irradiation of 2-nitrobenzyl cysteine without additives gives only 62% free cysteine and 35% cysteic acid; whereas, irradiation in the presence of either an aldehyde reagent or a photooxidation quencher leads to a decrease in the formation of cysteic acid. Irradiation in the presence of a mixture of semicarbazide hydrochloride and dimethylsulfide reduced the formation of cysteic acid to about 3%, while cysteine and cystine were obtained in 55 and 42% yields, respectively. This enhancement of yields by the addition of aldehyde reagents results from the trapping of the 2-nitrosobenzaldehyde and subsequent minimization of its reaction with the newly exposed thiol to form a thiohemiacetal.[5] The effect of the photooxidation quencher is to prevent the oxidation of the thiol to the sulphonate. The 2-nitrobenzyl group has been found to be stable under all the conditions of peptide synthesis, and it has been used in the synthesis of peptides involving cysteine.[6]

62.3 Photochemistry of Sulfonyl Protecting Groups

Sulfonamides have been used as excellent protective groups for amines because they lower the nucleophilicity of the nitrogen as well as allow conversion of primary amines to secondary amines by alkylation.[7] The *p*-toluenesulfonyl (tosyl) group is the most commonly employed sulfonyl protecting group for the amino function. The tosyl group is usually removed by treatment with strong acids or bases. As an alternative to these methods, D'Souza and Day observed that sulfonamides can be deprotected photolytically to liberate the free amino group.[8] In addition to tosyl, α- and β-naphthalenesulfonyl and methane-sulfonyl derivatives were also found to undergo photolytic cleavage to liberate the amino function. A variety of tosylated amino acids and peptide derivatives were found to undergo deblocking by UV irradiation under conditions that neither affect the peptide amide bonds nor cause the destruction of the amino acid residues.[8]

The 5-dimethylamino-1-naphthalenesulfonyl (dansyl) group is one of the most important amino protecting groups finding application in amino acid characterization and peptide analysis. This has also been used as a photolytically removable sulfonyl protecting function for the amino groups in amino acids and peptides.[9,10] The major photolysis products of the dansylic derivatives (5) identified from neutral or basic solutions were dansylic acid (6), the corresponding amino compound (7), and ammonia (Scheme 62.3). The quantum yields for the photocleavage reactions vary with the medium and the highest quantum yields for the cleavage of the sulfonamide bonds were observed in acid medium.

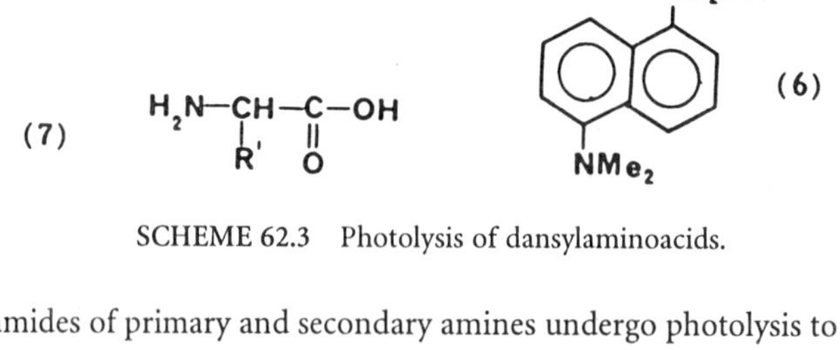

SCHEME 62.3   Photolysis of dansylaminoacids.

Benzylsulfonamides of primary and secondary amines undergo photolysis to liberate amines in excellent yields.[11] Solvents of increased hydrogen atom-donating ability were found to enhance the yields of photolytic deprotection. The method was found unsuitable for aromatic amines, presumably because the more strongly absorbing aromatic amino chromophore changes the nature of the excitation process. The attempt to circumvent this problem using the 4-nitrobenzenesulfonamide chromophore was also unsuccessful. There was no change even after prolonged irradiation, suggesting that the lower excitation energy does not lead to carbon-sulfur bond cleavage. The photochemistry of the benzylsulfonyl system is characterized by its homolytic cleavage to benzyl and sulphonyl radicals.[12] The final products are dependent on the stability of the sulfonyl radicals and the solvent used (Scheme 62.4).

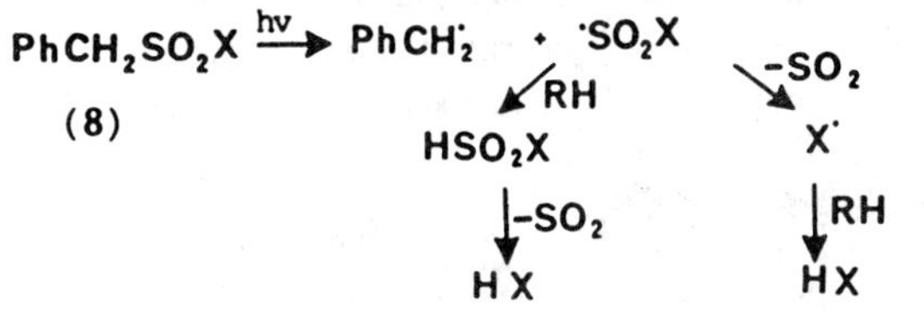

SCHEME 62.4 Photolytic cleavage of benzylsulfonyl derivatives.

The quinolinemethylsulfonamide group was found to undergo efficient photolytic removal even in the case of aromatic amines and other complex amines.[13] 2-Aryl-4-quinoline derivatives on irradiation with light of wavelength 350 nm underwent facile fragmentation.[14,15] The sulfonylchlorides of the type (9) produce sulfonamides (11) in excellent yields on reaction with amines (10) (Scheme 62.5). These sulfonamides on irradiation with a Rayonet photochemical reactor using 350-nm lamps undergo facile deprotection to liberate the free amines in yields ranging from 32 to 96%. The use of this light-sensitive sulfonylchloride is particularly advantageous for the protection of biologically important amines like guanine, adenine, and cytosine.[13]

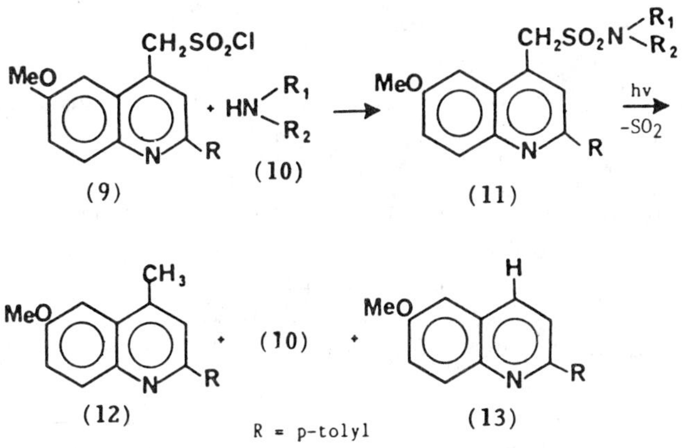

SCHEME 62.5 Photolytic deprotection of sulfonamides.

A free-radical mechanism has been postulated for the photolytic cleavage of arenesulfonamides.[16,17] Ring substituents significantly affect the photolysis reaction. Arenesulfonyl radicals, which are known to be produced in the photolysis of arenesulfonamides, are intermediates in the photolytic rearrangement of arenesulfonamides.[18,19] An unusual desulfonation reaction has been reported in the photolysis of 2-(N-alkylaryl-sulfonylamido)cyclohexane (14).[20,21]

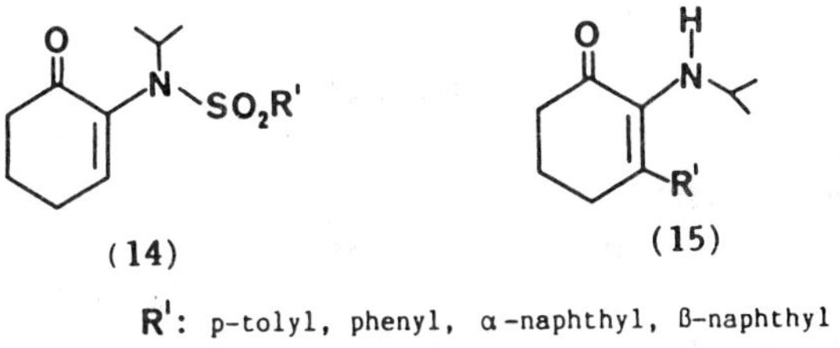

R': p-tolyl, phenyl, α-naphthyl, β-naphthyl

The mechanism of the photodetosylation reaction has been investigated in the case of N-tosyl-1,2,3,4-tetrahydroisoquinolines (16).[22] The reaction in this case was initiated by the formation of an exciplex or electron transfer between the excited dimethoxybenzene and tosyl moieties, followed by the formation of a biradical (17) and a nitrogen centred radical (18), both of which were converted to (19). Deuterium labeling studies showed that the routes through (17) and (18) contributed almost equally. The photolytic deprotection of tosyl groups from tetrahydroisoquinolines and other *N*-heterocyclic compounds has also been investigated extensively.[23–28]

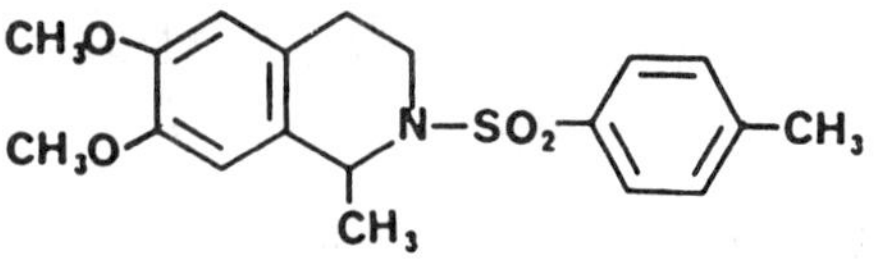

(16)

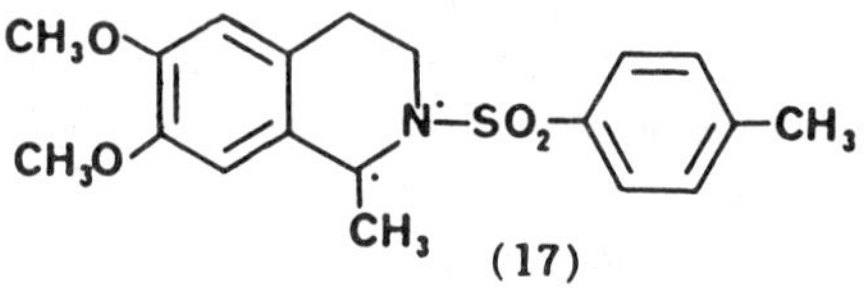

(17)

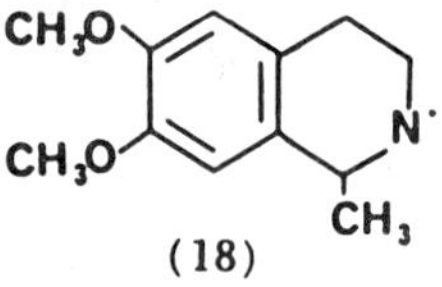

(18) 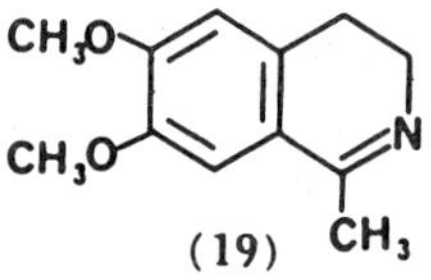 (19)

Selective removal of electron-accepting *p*-toluene- and naphthalenesulfonyl protecting groups for amino function via photoinduced donor-acceptor ion pairs with electron-donating aromatics has been reported.[29,30] When *N*-tosylamines were irradiated in the presence of an electron-donating aromatic compound such as 1,2- and 1,4-dimethoxybenzenes and 1,5-dimethoxy naphthalene and a reductant like sodium borohydride, ascorbic acid, ammonia, borane, and hydrazine, detosylation proceeded quite easily to give the corresponding amines. Mechanistic studies based on fluorescence quenching, quantum yield measurement, and free energy calculation have shown that this photo-reaction involves an electron transfer from an electron-donating aromatic to an electron-accepting sulfonamide. This photochemical hydrolysis of tosylamides by a sensitized electron transfer reaction has been observed to be useful in peptide chemistry. The *N*-tosyl group is one of the most stable protecting groups for amino function, and only the reduction with Na/NH$_3$ is practical as a useful deprotection method.[31] Many complicating side reactions have been reported,[32] and the selective cleavage of the *N*-tosyl protecting group has been reported as an attractive alternative. The photoenhanced reduction of carbonyl compounds by NaBH$_4$ reported recently is a mechanistically related reaction.[33]

Tosyl esters were hydrolyzed by >300-nm wavelength light in the presence of dimethoxy-naphthalene and hydrazine.[34] Quenching experiments indicated that the reaction proceeds via electron transfer from dimethoxynaphthalene to the tosyl group. The reaction was successfully applied to the deprotection of the tosyl esters of sugar and nucleoside derivatives.[34]

62.4 Photolysis of the Benzenesulfenyl Group

The use of 2,4-dinitrobenzenesulfenyl esters as photocleavable protecting groups for the carboxylic function was investigated by Barton et al.[35–37] These esters are formed by the reaction of 2,4-dinitrosulfenylchloride with the carboxylic acids. The photolysis proceeds in excellent yields to liberate the carboxyl group (Scheme 62.6). The formation of 2,4-dinitrophenyl phenylsulphide as a major photoproduct in benzene solvent suggests that the deprotection reaction is a nucleophilic displacement of the electrophilic excited state by the aromatic solvent. The formation of the 2-aminobenzenesulfonic acid derivative follows from the internal photoredox reaction between the 2-nitro group and the sulfur side chain.[38–40]

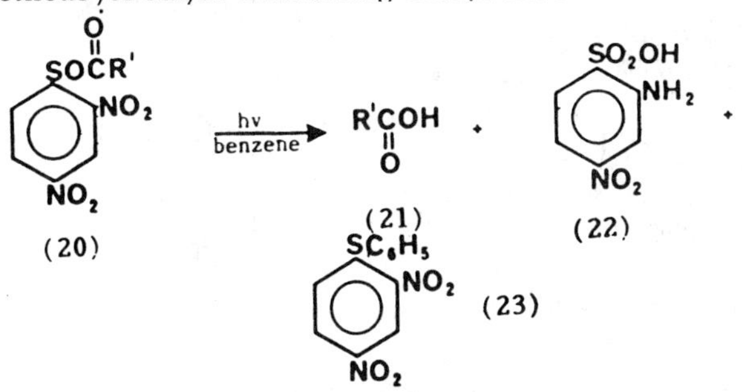

SCHEME 62.6   Photolytic deprotection of 2,4-dinitrobenzenesulphenyl group.

2.5 Cleavage of *N*-Acyl-2-thionothiazolidines

The photochemical activation of the carboxyl group via *N*-acyl-2-thionothiazolidines has been reported by Burton and White.[41] Photolysis of the *N*-acyl derivatives (26) of 2-thionothiazolidine (24) in the presence of ethanol resulted in the formation of the corresponding ethyl esters. The *N*-acyl derivatives (26) are prepared conveniently by the treatment of the sodium salt of (24) with the corresponding acyl chloride in benzene. The kinetic product from this reaction is the *S*-acyl derivative (25); these thioesters undergo a facile S → N thermal rearrangement to yield the 3-acylated product (26) (Scheme 62.7).[42,43]

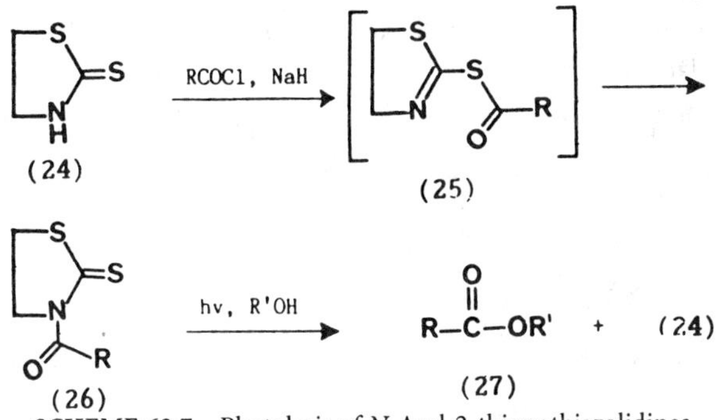

SCHEME 62.7   Photolysis of *N*-Acyl-2-thionothiazolidines.

Acylated 2-thionothiazolidines are relatively stable toward both acidic and basic reagents and can be stored indefinitely in the dark. However, upon irradiation, they undergo cleavage of the acyl substituent and the parent 2-thionothiazolidine can be recovered in quantitative yield. If irradiation is carried out in ethanol, the corresponding ethyl ester is obtained, with an efficiency that is markedly dependent on the nature of the R substituent. Thus, aryl derivatives afford a generally good yield of esters, whereas aliphatic systems give poor yields. Irradiation of *N*-benzoyl-2-thionothiazolidine resulted in a quantitative recovery of the starting material. A mechanism involving γ-hydrogen abstraction by the sulfur atom is postulated for this photochemical activation process.[41] This mechanism (Scheme 62.8) is analogous to the mechanistic pathway for the photolysis of *o*-phenylethylthiobenzoates investigated by Barton and co-workers.[44] This photochemical activation approach has been extended to the synthesis of amides and peptides.[45–47]

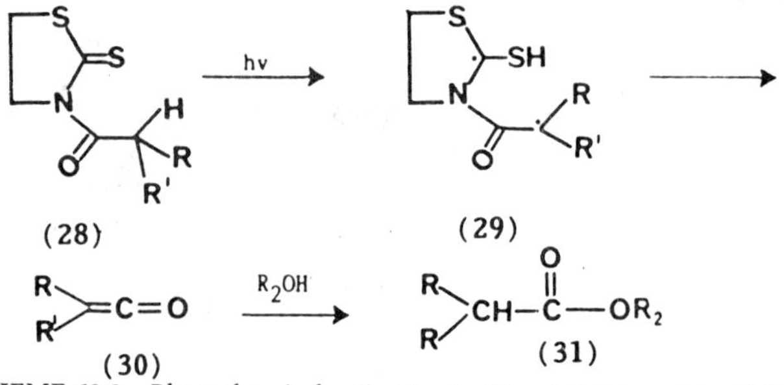

(28)   (29)

(30)   (31)

SCHEME 62.8   Photochemical activation in *N*-acyl-2-thionothiazolidine.

62.6 Photolytic Dethioacetalization

In a study of the photolysis of sulfur-containing steroids, Takahashi et al. observed that dethioacetalization occurred upon photolysis, resulting in the formation of the parent ketone.[48] It was also found that this method was applicable to ethylene- and dibenzylidine-dithioacetal derivatives. Because of the relative stability of the thioacetals to base and acid, it is difficult to regenerate the parent carbonyl compounds.[49] Thus, the photochemical methods offer a simple operation for dethioacetalization (Scheme 62.9). It differs from the conventional methods in that neutral conditions are used. In a typical experiment, a mixture of 5α-cholestane-3-one-ethylene dithioacetal (32) and benzophenone in hexane was irradiated for 3.5 h at room temperature to give 5α-cholestane-3-one (33). A number of other simple cyclic ketones (34) and steroidal ketones have also been deprotected in a similar manner.[49]

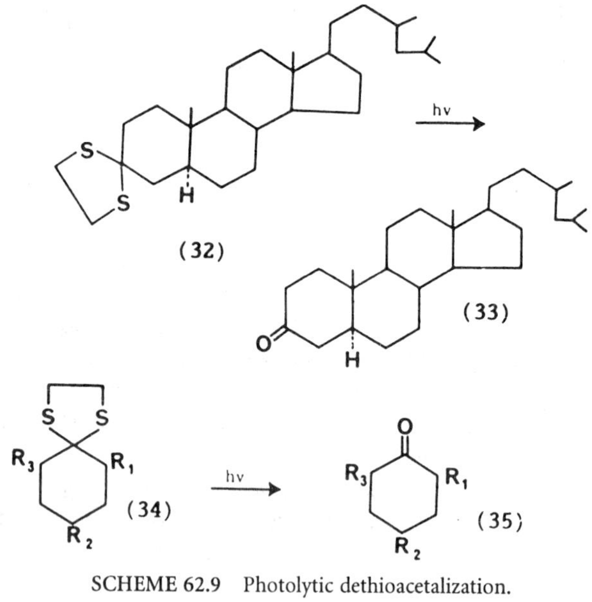

(32)

(33)

(34)   (35)

SCHEME 62.9   Photolytic dethioacetalization.

62.7 Conclusion

Photolytic reactions involving the sulfenyl, sulfonyl, and thioacetal functions provide a facile method of specific protection of functional groups. These light-sensitive groups can be cleaved by light of suitable wavelength under neutral conditions at room temperature. The protection and photochemical release of sulfur-containing compounds like thiols and thio-containing amino acids also involve light-induced reactions of synthetic and biological significance. The photocleavage of *N*-acyl-2-thionothiazolidines can be used for the activation of carboxyl groups. Mechanistically, the photochemical formation of thiyl radicals and their subsequent reaction under different conditions form the basis of almost all the methods for the photochemical protection and deprotection of sulfur-containing compounds.

References

1. Hazum, E., Gottlieb, P., Amit, B., Patchornik, A., and Fridkin, M., *Peptides: Proc. Eur. Pep. Symp., 16th.,* K. Brunfeldt Ed., Scriptor Pub., Copenhagen, 1981, 105.
2. de Mayo, P., Ultraviolet photochemistry of simple unsaturated systems. *Adv. Org. Chem.,* 2, 367, 1960.
3. de Mayo, P. and Reid, S. T., *Quart. Rev. Chem. Soc.,* 15, 393, 1961.
4. Morrison, H. A., *The Chemistry of the Nitro and Nitroso Groups,* Part I, Feuer, H., Ed., Wiley, New York, 1970, 185.
5. Wolman, Y., *Chemistry of the Thio Group,* Part II, Patai, S., Ed., John Wiley, New York, 1974, 105.
6. Amit, B., Hazum, E., Fridkin, and M., Patchornik, A., A photolabile protecting group for the phenolic hydroxyl function of tyrosine *Int. J. Pept. Protein Res.* 9, 91, 1977.
7. Hendrickson, J. B. and Bergeron, R., Protection of and monoalkylation of amines *Tetrahedron Lett.,* 345, 1970.
8. D'Souza, L. and Day, R. A., Photolytic cleavage of sulfonamide bonds *Science,* 160, 882, 1968.
9. D'Souza, L., Bhat, K., Madiah, M., and Day, R. A., Photolysis of dansyl amino acids and dansyl peptide, *Arch. Biochim. Biophys.,* 141, 690, 1970.
10. Carpenter, D. C., Splitting of the CONH linkage by ultraviolet light, *J. Am. Chem. Soc.,* 62, 289, 1940.
11. Pincock, J. A. and Jurgens, A., Sulfonamides as photolabile protecting groups for amines, *Tetrahedron Lett.,* 1029, 1979.
12. Langler, R. F., Marini, Z., and Pincock, J. A., The photochemistry of benzylic sulfonyl compounds: the preparation of sulfones and sulfonic acids, *Can. J. Chem.,* 56, 903, 1978.
13. Epling, G. A. and Walker, M. E., A new photochemically removable protecting group for amines, *Tetrahedron Lett.,* 23, 3843, 1982.
14. Epling, G. A. and Ayengar, N. K., Photochemical fragmentation of phototoxic 2-aryl quinoline methanols, *Tetrahedron Lett.,* 3007, 1976.
15. Epling, G. A., Ayengar, N. K., Lopes, A., and Yoon, U. C., Photochemical reduction and decarboxylation of 2-phenyl quinoline-4-carboxylic acids, *J. Org. Chem.,* 43, 2928, 1978.
16. Abad, A., Mellier, D., Pete, J. D., and Portella, C., Photolysis of alkyl sulfonates and sulfonamides. Regeneration of alcohols and amines, *Tetrahedron Lett.,* 4555, 1971.
17. Mellier, D., Pete, J. P., and Portella, C., Mechanism of the photolysis of esters and amides. Influence of the nature of the acid group during the reaction, *Tetrahedron Lett.,* 4559, 1971.
18. Nozaki, H., Okada, T., Noyori, R., and Kawanishi, M., Photochemical rearrangement of arenesulfonamilides to p-amino ditosyl sulfones, *Tetrahedron,* 22, 2177, 1966.
19. Arnould, J. C. and Pete, J. P., Photochemistry of α-dialkyl amino enones. I. New oxidative cyclization of chalcone derivatives, *Tetrahedron Lett.,* 2459, 1975.
20. Arnould, J. C., Cossy, J., and Pete, J. P., Photolysis of 2(N-alkyl-arylsulfonamide)cyclohexenone, An unusual and useful desulfonation reaction. *Tetrahedron Lett.,* 3919, 1976.
21. Arnould, J. C., Cossy, J., and Pete, J. P., Photochemical reactivity of α-amino-enones: cyclization and new type of reaction of α-sulfonamidocyclohexenone *Tetrahedron,* 36, 1585, 1980.
22. Hamada, T., Nishida, A., and Yonemitsu, O., Mechanism of photodetosylation of N-tosyl-1,2,3,4-tetrahydroisoquinolines involving electron transfer in the excited state, *Heterocycles,* 12, 647, 1979.
23. Umezawa, B., Hoshino, O., and Sawaki, S., Organic Photochemistry. I. 3,4-Dihydroisoquinolines from tetrahydroisoquinoline N-tosylate *Chem. Pharm. Bull.,* 17, 1115, 1969.
24. Hoshino, O., Sawaki, S., and Umezawa, B., Organic Photochemistry. III. Photo-reductive cleavage of p-toluene sulphonamides, *Chem. Pharm. Bull.,* 18, 182, 1970.

25. Umezawa, B., Hoshino, O., and Sawaki, S., Organic Photochemistry. II. Tetrahydroisoquinolines from their N-tosylates, *Chem. Pharm. Bull.*, 17, 1120, 1969.

26. Somei, M. and Natsume, M., Photochemical rearrangements for the synthesis of 3-, 4- and 6-substituted indoles, *Tetrahedron Lett.*, 2451, 1973.

27. Umezawa, B., Hoshino, O., and Yamanashi, Y., Photolysis of N-methyl-4,7-diacetoxy-6-methoxy-1,2,3,4-tetrahydroisoquinoline, *Tetrahedron Lett.*, 933, 1969.

28. Hoshino, O., Yamanashi, Y., and Umezawa, B., Reaction of N-methyl-4,7-diacetoxy-6-methoxy-1,2,3,4-tetrahydroisoquinoline with amines or thiols, *Tetrahedron Lett.*, 337, 1969.

29. Hamada, T., Nishida, A., Matsumoto, Y., and Yonemitsu, O., Photohydrolysis of sulfonamides via donor-acceptor ion pairs with electron donating aromatic and its application to the selective detosylation of lysine peptides. *J. Am. Chem. Soc.*, 102, 3978, 1980.

30. Hamada, T., Nishida, A., and Matsumoto, Y., Selective removal of electron-acceptors p-toluene sulfonyl and naphthalene sulphonyl protecting groups for amine function via photoinduced donor-acceptor ion pairs with electron donating aromatics. *J. Am. Chem. Soc.*, 108, 140, 1986.

31. Greene, T. W., *Protective Groups in Organic Synthesis,* Wiley-Interscience, Chichester, 1981, 285.

32. Schon, I., Sodium liquid-ammonia reduction in peptide chemistry *Chem. Rev.*, 84, 287, 1984.

33. Choi, J. H., Kim, D. W., and Shim, S. C., Photoenhanced reduction of carbonyl compounds by sodium borohydride *Tetrahedron Lett.*, 27, 1157, 1986.

34. Nishida, A., Hamada, and T., Yonemitsu, O., Hydrolysis of tosyl esters initiated by an electron-transfer, *J. Org. Chem.*, 53, 3386, 1988.

35. Barton, D. H. R., Chow, Y. L., Cox, A., and Kirby, G. W., *Tetrahedron Lett.*, 1055, 1962.

36. Barton, D. H. R., Chow, Y. L., Cox, A., and Kirby, G. W., *J. Chem. Soc.*, 3571, 1965.

37. Barton, D. H. R., Nakano, T., and Sammes, P. G., *J. Chem. Soc. (C)*, 322, 1968.

38. Pillai, V. N. R., *Chem. Ind. (London)*, 456, 1976.

39. Pillai, V. N. R. and Padmanabhan, P. V., *J. Ind. Chem. Soc.*, 55, 279, 1978.

40. Pillai, V. N. R., *Chem. Ind. (London)*, 665, 1977.

41. Burton, L. P. G. and White, J. D. *Tetrahedron Lett.*, 21, 3147, 1980.

42. Nagao, Y., Kawabata, K., and Fujita, E., *J. Chem. Soc. Chem. Commun.*, 330, 1978.

43. Nagao, Y., Kawabata, K., and Fujita, E., *J. Chem. Soc. Perkin Trans. I*, 2470, 1980.

44. Barton, D. H. R., Balton, M., Magnus, P. D., and West, P. J., *J. Chem. Soc. Perkin Trans. I*, 1580, 1973.

45. Nagao, Y., Seno, K., and Kawabata, K., *Tetrahedron Lett.*, 21, 841, 1980.

46. Pillai, V. N. R., *Proc. IUPAC Symp. Photochem.*, 9th, Pau, France, 1982, 274.

47. Pillai, V. N. R. and Purushothaman, E., *Int. Conf. Org. Synth. 5th*, Freiburg, W. Germany, 1984, 92.

48. Takahashi, T., Nakamura, C. Y., and Satoh, Y., *J. Chem. Soc., Chem. Commun.*, 680, 1977.

49. Seebach, D., *Synthesis*, 17, 1969.

63

Solution Photochemistry
of Thioketones

Ramamurthy
rtment of Chemistry
ne University
Orleans, Louisiana

Nageshwer Rao
Corporation,
chnology Division
cott, New York

Pushkara Rao
hem America Inc.,
arch & Development
enter
mouth Junction,
Jersey

3.1 Introduction

The comparatively softer character and the lower electronegativity of sulfur, coupled with the high polarizability of the $C=S$ bonding, give rise to important differences in reactivity between thiocarbonyl and carbonyl compounds. While there are certain similarities in the primary photoprocesses between carbonyl and thiocarbonyl compounds, significant differences in the final products obtained through these primary photoreactions exist; also, there are reactions that are specific to the thiocarbonyl chromophore.[1] A brief outline of the excited-state reactions (excluding hydrogen abstraction that forms an independent chapter) of thioketones is provided here. Thioesters, thioamides, and other related compounds do not form part of this presentation. For comprehensive coverage of the literature, readers should refer to several recent reviews.[2-4] Photochemical investigation of thiones is laden with certain unique problems specific to this chromophore: air oxidation, self-quenching, and reaction with solvents. These aspects have been investigated in detail in recent years with the hope that these problems can be alleviated. Results of such studies are briefly touched upon.

3.2 Photophysical Aspects

Thioketones, in general, are highly colored compounds and possess electronic absorption in the region 200 to 700 nm.[5,6] Absorption spectra of three typical thiones, belonging to different classes, are shown in Figure 63.1. The longest wavelength absorption in each one of these is attributed to the spin-forbidden S_0 to T_1 (n,π*) transition. The intensity of the S_0 to T_1 transition (ε ~ 1–5) results from enhanced spin-orbit coupling due to mixing of the n,π* triplet with the nearby π,π* state and from a modest heavy-atom effect due to the sulfur atom. A stronger band (ε ~ 100) in the visible region, with small differences in band maxima between aromatic, aliphatic, and arylalkyl thiones,

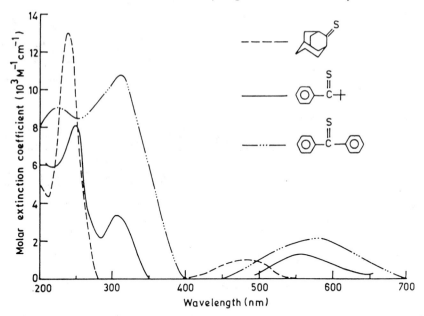

FIGURE 63.1 Electronic absorption spectra of three model thiones belonging to aliphatic, aromatic, and aryl alkyl series, recorded in hydrocarbon solvents. Note the shift in the maxima of n,π* and π,π* bands depending on the conjugation provided by the substituent.

corresponds to the S_0 to S_1 (π,π*) transition. Additional bands attributed to π,π*, n,σ*, and Rydberg transitions are present below 350 nm. Thiocarbonyl compounds are strikingly different from carbonyl compounds in that they possess a large energy gap between the n,π* (S_1) and the π,π*(S_2) singlet states.

In the case of thioketones, the decay rates from S_2 (i.e., the reciprocal of the S_2 lifetime) have been established to be inversely related to the energy gap between the S_2 and the S_1 states.[7] The lifetime of S_2 is also dependent on the nature of the solvent: for example, the S_2 lifetimes of xanthione in benzene, hexane, and perfluorohexane are 11, 25, and 162 ps respectively.[8] The relatively long lifetime results in observable fluorescence and reactions from the higher excited S_2 state of thioketones.[9,10]

The dynamics of the triplet state (T_1) have been explored using phosphorescence lifetime,[11] triplet-triplet absorption,[12] intermolecular triplet quenching, and energy transfer techniques.[13] The general picture is that thione triplets are formed in high yields by intersystem crossing from S_1 have an unquenched triplet lifetime in the range of microseconds, and phosphoresce moderately in fluid media.[7] The lowest excited singlet states are very short lived, exhibit no measurable prompt emission, and have never been observed by techniques designed to detect transients. The existence of an equilibrium between S_1 and T_1, a result of a small energy gap and strong spin-orbit coupling between these states, has been established in a number of aromatic and alicyclic thiones by monitoring the delayed fluorescence from S_1.[14]

63.3 Photooxidation

Thiones, in general, are unstable under aerated conditions, especially in presence of light. The oxidation of thiobenzophenone to benzophenone has been known for nearly a century.[15] While most thiones are oxidized to the corresponding carbonyl compounds, a few exceptions have been noted recently. For example, di-t-butylthione is oxidized in an aerated atmosphere by light to the corresponding ketone and sulfine.[16] This observation has led to detailed mechanistic studies and the following conclusions have resulted:[17,18]

1. The ratio of products, ketone:sulfine, is dependent on the steric and electronic features of the molecule carrying the thiocarbonyl functionality (Schemes 63.1–63.3).

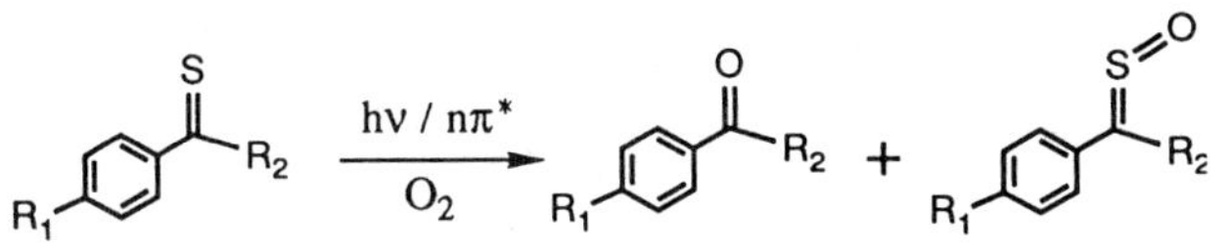

R_1 = OCH$_3$; R_2 = t-butyl	80%	~2%
R_1 = Cl ; R_2 = t-butyl	52%	25%
R_1 = OCH$_3$; R_2 = Phenyl	57%	0%
R_1 = Cl ; R_2 = Phenyl	83%	~5%

SCHEME 63.1

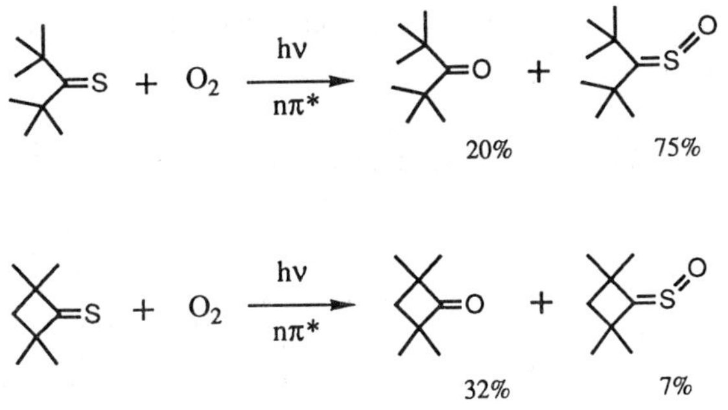

20%   75%

32%   7%

SCHEME 63.2

30%   15%

70%

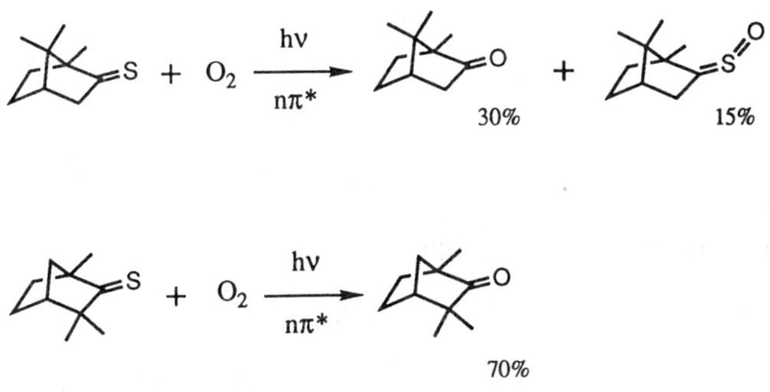

SCHEME 63.3

2. Singlet oxygen, generated by energy transfer from the triplet state of thione to the ground state of oxygen, is the oxidizing species (Scheme 63.4).[18]

$$T^0 \xrightarrow{\ h\nu\ } T^{*1} \tag{1}$$

$$T^{*1} \xrightarrow{\ ISC\ } T^{*3} \tag{2}$$

$$T^{*3} + {}^3O_2 \longrightarrow T^0 + {}^1O_2 \tag{3}$$

$$T^0 + {}^1O_2 \longrightarrow K + S \tag{4}$$

T = Thione ;   K = Ketone ;   S = Sulfine

SCHEME 63.4

3. The proposed mechanism, illustrated in Scheme 63.5, involves attack of singlet oxygen on the *n*-lobe of the thiocarbonyl chromophore. Intermediates 1,2,3-thiadioxetane and zwitterions/biradicals are suggested to precede the formation of the isolated products. Steric influence on the decay paths of these intermediates has been invoked to account for the differences in the product distribution (ketone:sulfine) between sterically congested and sterically free thiones (e.g., thiofenchone vs. thiocamphor, Scheme 63.1).[17a] For example, the presence of C-7 methyl groups in the bicyclo [2.2.1]heptane series is suggested to hinder the closure of the zwitterionic/biradical intermediate to the ketone precursor thiadioxetane. Under such conditions, sulfine formation dominates (Scheme 63.6).

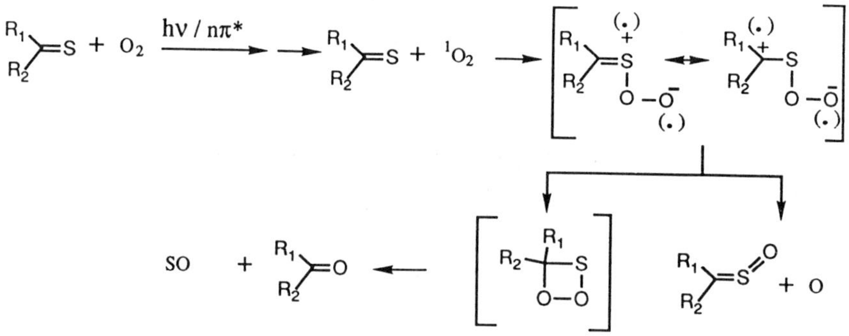

SCHEME 63.5

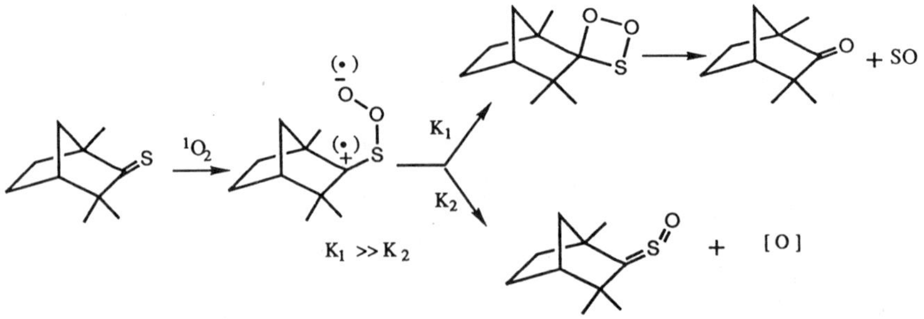

SCHEME 63.6

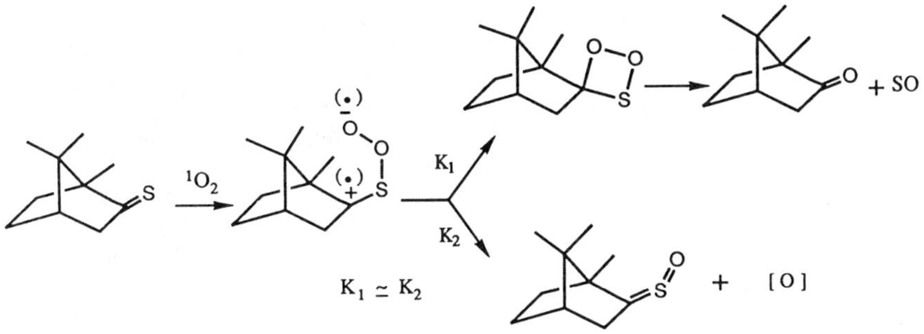

SCHEME 63.6 (continued)

4. The influence of the substituents on the ratio of products (ketone:sulfine) in the case of diaryl and aryl alkyl thiones has been interpreted on the basis of their ability to stabilize the two forms of the zwitterionic intermediate.[17b] As illustrated in Scheme 63.7, of the two forms, only one (i.e., **B**) can serve as a precursor for both sulfine and ketone; form **A**, unsuited for cyclization, is expected to yield only sulfine. While electron-donating substituents favor form **B**, electron-releasing substituents stabilize form **A**; accordingly, the ratios of sulfine:ketone vary as shown in Scheme 63.3.

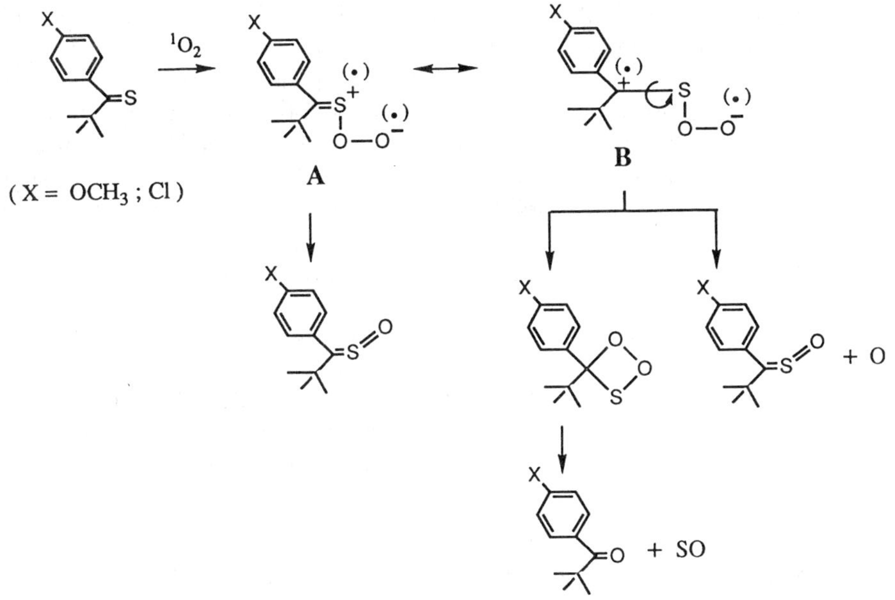

SCHEME 63.7

It should be clear from the above summary that it is advisable to carry out photochemical and photophysical studies of thiones under thoroughly degassed conditions.

3.4 Self-Quenching and Photodimerization

Triplets of all thiones thus far investigated undergo self-quenching at near diffusion-controlled rates (Table 63.1).[19] Self-quenching in the triplet state of carbonyl compounds is known,[20] but, it occurs at rates at least three orders of magnitude lower than that in the corresponding thiocarbonyl

Table 63.1   Rate of Self-Quenching in Thiocarbonyl Triplets at Room Temperature

Thione	k_{sq} $(M^{-1} s^{-1})$	Solvent
Thiobenzophenone[a]	4.2×10^9	Cyclohexane
Michler's thione[a]	2.9×10^9	Cyclohexane
4,4'-Dimethoxythiobenzophenone[b]	2.6×10^9	Benzene
Thioxanthione[b]	7.1×10^9	Benzene
Pivalothiophenone (PT)[c]	1.5×10^9	Benzene
p-Methoxy PT[c]	2.7×10^9	Benzene
p-Fluoro PT[c]	1.3×10^9	Benzene
p-Chloro PT[c]	0.58×10^9	Benzene
Admantanethione[d]	9.9×10^9	n-Hexane
2,2,4-Tetramethyl cyclobutanethione[e]	6.9×10^9	Benzene
Di-t-butylthione[e]	0.13×10^9	Benzene

[a] Kemp, D. R. and de Mayo, P., *J. Chem. Soc., Chem. Commun.*, 233, 1978.

[b] Kumar, C. V., Quinn, L., and Das, P. K., *J. Chem. Soc., Faraday Trans. 2*, 80, 783, 1984.

[c] Bhattacharyya, K., Ramamurthy, V., and Das, P. K., *J. Phys. Chem.*, 91, 5626, 1987.

[d] Lawrence, A. H., de Mayo, P., Bonneau, R., and Jousset-Dubien, J., *Mol. Photochem.*, 5, 361, 1973.

[e] Bhattacharyya, K., Kumar, C. V., Das, P. K., Jayasree, B., and Ramamurthy, V., *J. Chem. Soc., Faraday Trans. 2*, 81, 1383, 1985.

compounds. Self-quenching that occurs in triplet thiones can, to some extent, be controlled by steric and electronic features of the molecular framework carrying the thiocarbonyl functionality.[21] The donor-acceptor nature of this quenching process has been established by investigating the quenching of triplet xanthione by a variety of structurally different thiones possessing triplet energies above that of xanthione. Self-quenching does depend, although to a small degree, on the electron-donor and electron-acceptor ability of the *para*-substituent in the case of pivalothiophenone.[22] Further, steric crowding around the thiocarbonyl chromophore decreases the rate of self-quenching.[23] These observations have led to the conclusion that the interaction between an excited triplet and a ground-state thione occurs through n-orbitals. Such an interaction, stabilized by donor-acceptor-type features, results in physical decay to the ground state via an excimer. No emission from the proposed triplet excimer has been observed.

Triplet self-quenching leads occasionally to stable products (1,3-dithietanes) and such self-quenching has been characterized in a few thiones (Scheme 63.8).[24] Dimerization of adamantanethione occurring upon excitation to S_1, has been investigated in detail.[25] The quantum yield for the formation of the dimer 1,3-dithietane, upon S_1 irradiation is very low ($<10^{-4}$) owing to the high reversibility of the proposed intermediates — triplet excimer and biradical. Only rarely are 1,2-dithietanes formed.[26] Such examples are shown in Scheme 63.9. Although photodimerization is not a general reaction, this reaction is unique since no ketones display this behavior.

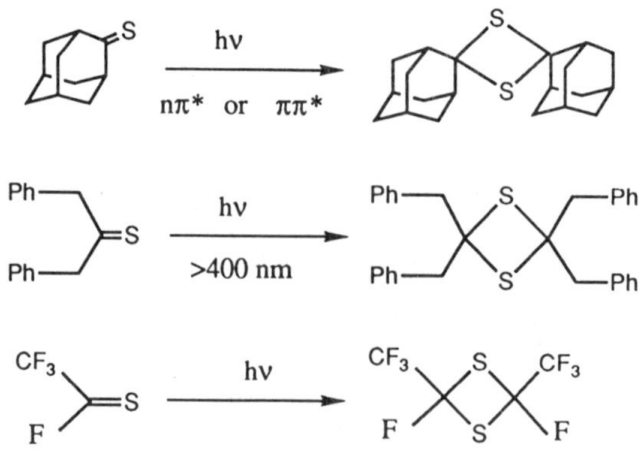

SCHEME 63.8

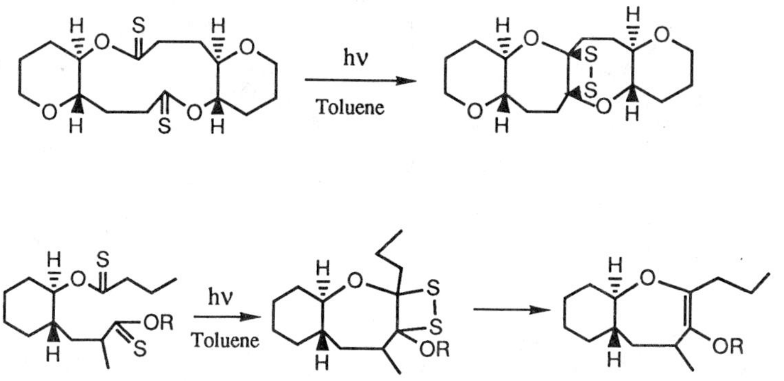

SCHEME 63.9

The initial observation with respect to self-quenching in the S_2 state was made in the case of thiobenzophenone.[27a] This was attributed to quenching by energy transfer ($S_2 + S_0 \rightarrow S_0 + S_1$). Since then, self-quenching in the S_2 state has been observed with a number of thiones.[27b] Energy transfer to a lower lying S_1 state has been shown not to be the major process by which the self-quenching occurs. High rates of S_2 self-quenching remain unexplained.[24] Only in the case of the formation of the adamantanethione dimer (1,3-dithietane) has the S_2 state been implicated via a nonemitting excimeric state.[28]

3.5 Addition to Unsaturated Systems

The photoaddition of a number of unsaturated organic molecules (olefins,[25,28–38] acetylenes,[39] allenes,[40] imines,[41] and nitriles[42]) to thiones has been extensively studied. Product studies, mechanistic investigations, and theoretical analyses have occupied the interest of a number of groups for over 2 decades. Aliphatic and aromatic thiones as well as enethiones (α,β-unsaturated thiones and thiocoumarins) have been shown to undergo addition to unsaturated systems from both the lower and the higher excited states.[37,38] The characteristics and the products of these additions from the two states are different, and these are briefly presented below.

Addition from the Lower Excited State

The parent and substituted thiobenzophenones and xanthione have been utilized as models to investigate the photoaddition reactions of diarylthiones.[1–4] Since all these systems show identical behavior, the results of only one system, thiobenzophenone, are presented below. Thiobenzophenone adds to both electron-rich and electron-poor olefins upon irradiation with visible light (S_0 to S_1 excitation).[30] Generally, thietane is the resultant addition product, although 1,4-dithianes have been isolated at higher concentration of thiones with certain olefins (Scheme 63.10). The addition yielding the thietane is nonstereospecific but is regioselective.

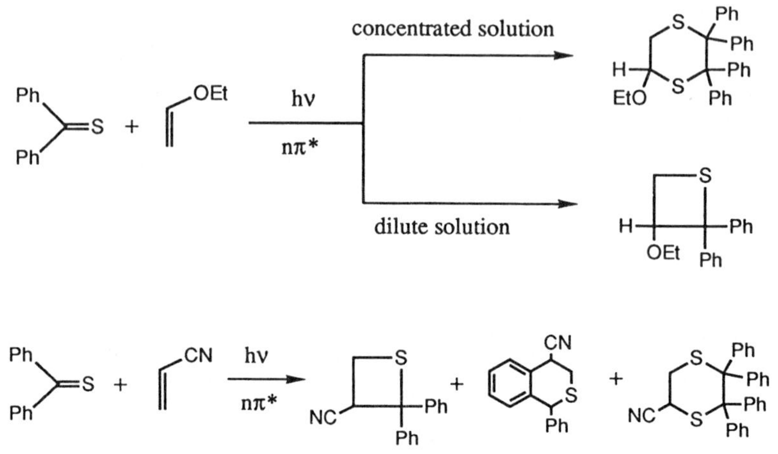

SCHEME 63.10

Biradicals and triplet exciplexes (thione-olefin complex) are suggested as precursors to the stable product thietane and/or 1,4-dithiane.[30,33] Involvement of a biradical intermediate is supported by the following observations (Scheme 63.11): (1) upon irradiation of thiobenzophenone in presence of olefins (whose triplet energies are higher than that of the thione), geometric isomerization of the olefins occurs; (2) the regioselective nature of the final thietane can be predicted on the basis of the stability of possible biradicals that would result from the addition of the thione to olefins; (3) 1,4-dithiane, a trapping product of the biradical by ground-state thione, is formed at high concentration of thiones; and (4) methyl viologen radical cation, an electron-transfer product from methyl viologen to the biradical, has been identified spectroscopically during the flash photolysis of xanthione in the presence of allenes and methyl viologen.[35] The intermediacy of an exciplex has its support from the linear relationship observed between the quenching rate of triplet xanthione by olefins and the ionization potential of the quencher.[33] No direct evidence for the involvement of a biradical and an exciplex is available in these addition reactions.

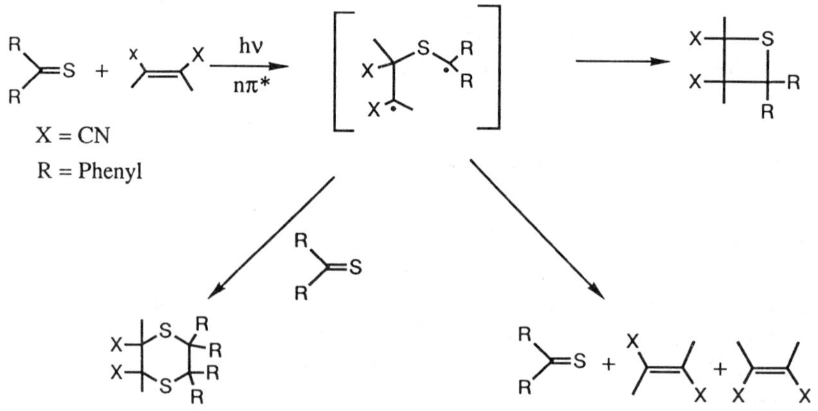

SCHEME 63.11

α,β-Unsaturated thiones behave very similar to diarylthiones upon excitation to the S_1 level in presence of electron-rich olefins.[37c] Thietanes are the primary products; however, 1,4-dithianes are also formed at high concentration of thiones (Scheme 63.12). Unlike the case of α,β-unsaturated enones, no addition occurs to the α,β C-C double bond. Thiocoumarin behaves in an analogous fashion.[38]

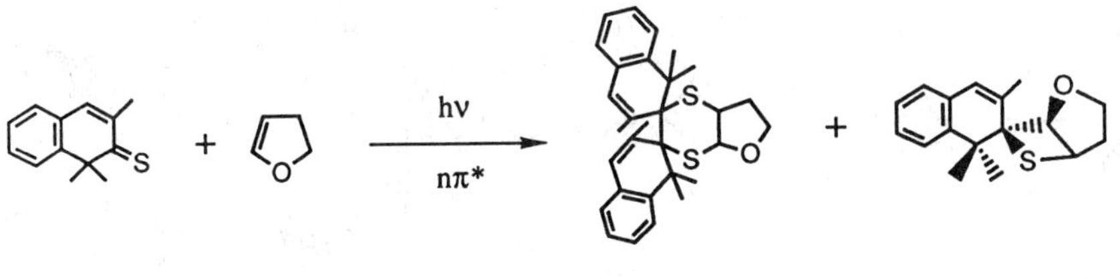

SCHEME 63.12

Adamantanethione and di-*t*-butylthione have been investigated as models for alkylthiones.[25,34] Final products obtained from these two systems differ and, therefore, results from both thiones are presented below. Adamantanethione adds to electron-rich and electron-poor olefins upon excitation to the S_1 level. The addition, as in the case of diarylthiones, is nonstereospecific but regioselective (Scheme 63.13). Unlike the case of thiobenzophenone, no 1,4-dithianes are formed. Evidence in favor of 1,4-biradical intermediate has accumulated through product studies. Sterically conjested thiones such as di-*t*-butylthione, do not yield thietane in the presence of olefins; instead, they yield open-chain products (Scheme 63.14). The primary intermediate of addition from both thiones is believed to be the same: namely a biradical that results from the attack of olefin on the partially occupied *n*-orbital of thiocarbonyl chromophore. The difference in the final products obtained between the two systems (adamantanethione and di-*t*-butylthione) is attributed to steric factors; in the case of di-*t*-butylthione, the steric interference by methyl groups prevents the closure of 1,4-biradical to the thietane.

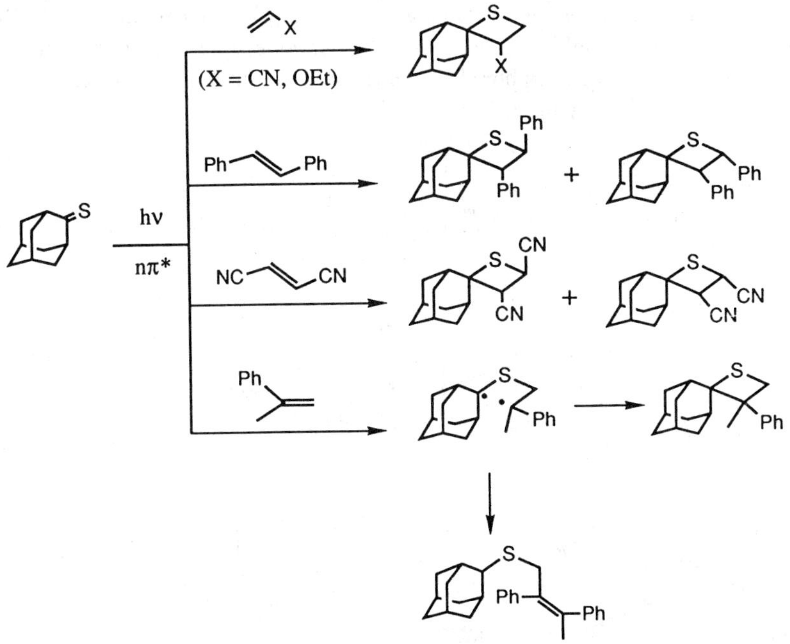

SCHEME 63.13

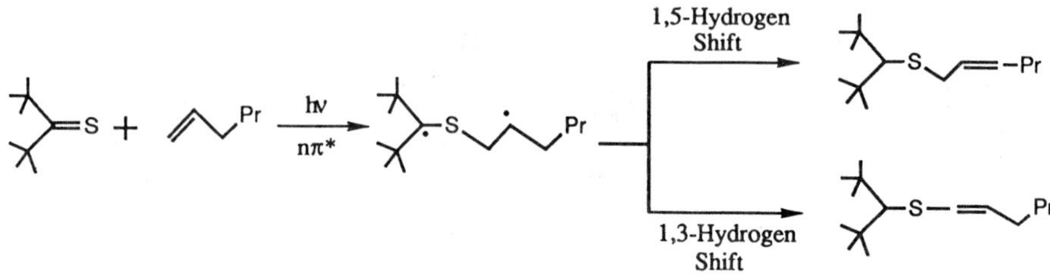

SCHEME 63.14

Based on sensitization and quenching studies, the addition reactions described above have been established to occur via the triplet state. The quantum yield of the isolated adduct is low in most cases (possibly due to the reversibility of biradical and exciplex intermediates). Since the S_0 to S_1 absorption coefficient is low, high concentrations of thiones have to be used in order to achieve absorption of a significant portion of the incident light. Such high concentration leads to a very high rate of self-quenching that reduces significantly the efficiency of the addition. Strategies to circumvent these problems need to be devised.

Addition from the Higher Excited State

Upon excitation to S_2, thiobenzophenone adds to electron-poor olefins to yield thietanes (Scheme 63.15).[31,32] The addition is stereospecific and regioselective. Similar behavior is observed with α,β-unsaturated thiones, with no addition occurring to the α,β C-C double bond (Scheme 63.16).[37a,b] Based on sensitization and quenching studies, the reactive state has been identified as the S_2 state in both systems. In the case of thiobenzophenone, 1,3-dithiane, isolated at low temperature, is suggested as the precursor for the thietane, at least during low temperature (-70 °C) reaction (Scheme 63.17). Adamantanethione as well as di-t-butylthione, upon excitation to the S_2 state, add to both electron-rich and electron-poor olefins to yield thietanes.[28,34] The addition, while stereospecific, is not regioselective (Scheme 63.18).

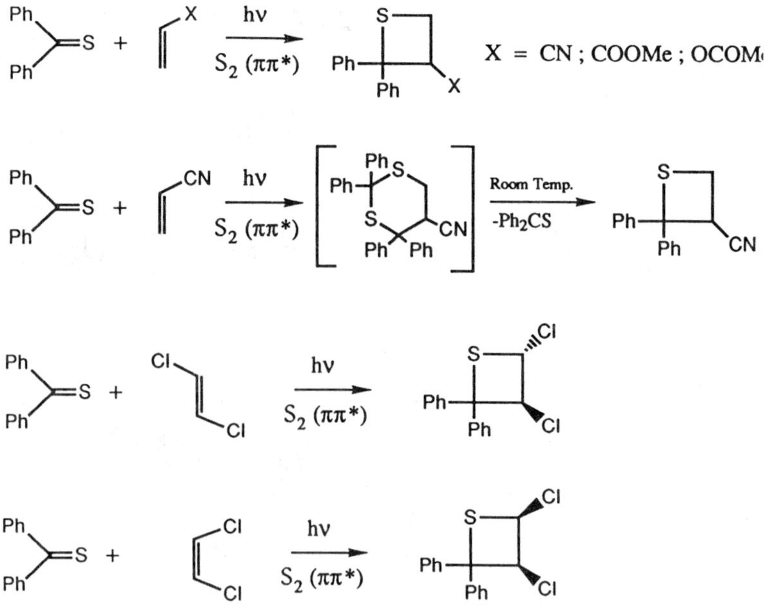

SCHEME 63.15

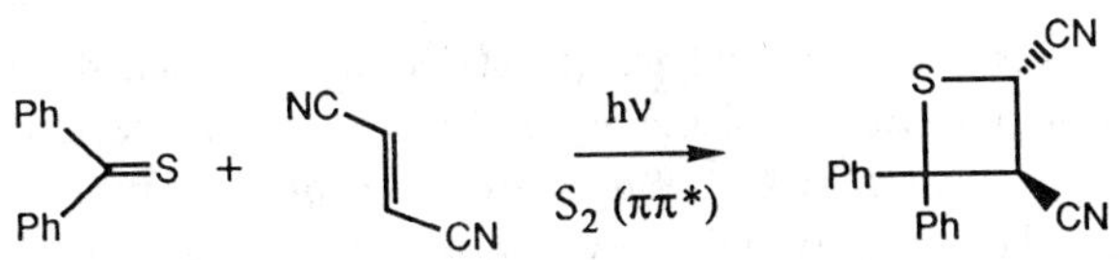

SCHEME 63.15 (continued)

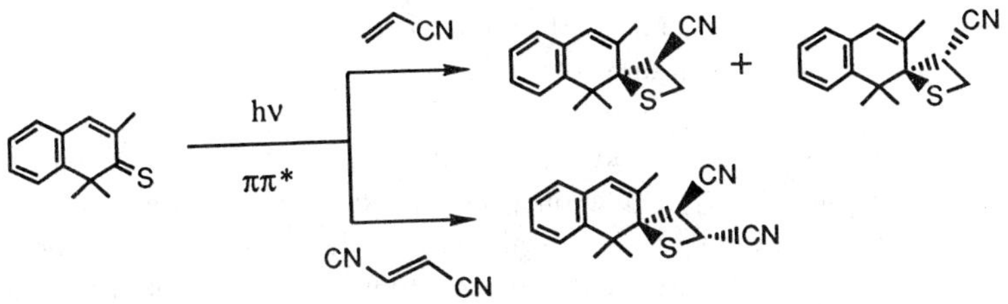

SCHEME 63.16

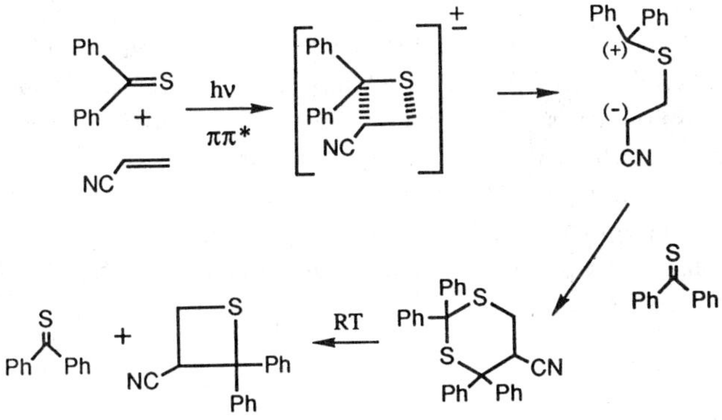

SCHEME 63.17

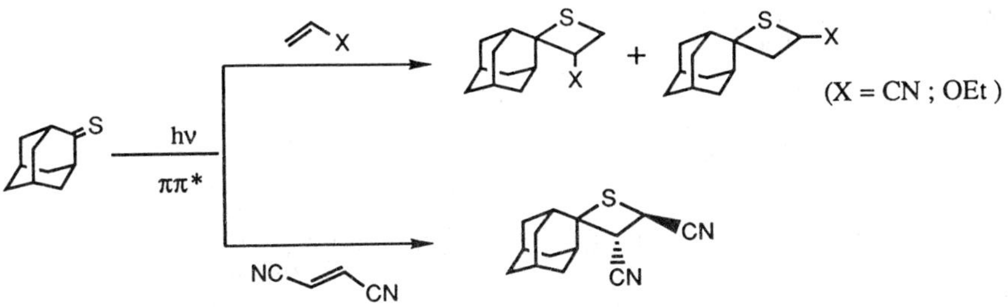

$(X = CN ; OEt)$

SCHEME 63.18

An exciplex intermediate is proposed to be involved in the addition process. In general, the addition that occurs from the S_2 state has a high quantum yield (>0.2 compared to ~10^{-4} from T_1 addition). An important difference between aryl (as well as α,β-unsaturated thiones) and alkylthiones in reactivity toward olefin exists; while both electron-poor and electron-rich olefins undergo addition to the S_2 state of alkylthiones, only electron-poor olefins add to diaryl and α,β-unsaturated thiones. This, as well as regio- and stereoselectivities obtained during the photoaddition of thiones to olefins, has been rationalized on the basis of a model based on the PMO approach.[36]

63.6 α-Cleavage

Norrish type I α-cleavage is a fundamental reaction of electronically excited carbonyl compounds.[43] Aliphatic and acyclic ketones yield alkyl and acyl radicals on α-cleavage. Cyclic ketones can yield oxacarbene intermediates as well as these radicals. The rate of α-cleavage depends on the triplet energy of the ketone and the stability of the resulting primary products intermediates: radicals and oxacarbenes. A few of the well-known acyclic ketones that undergo α-cleavage are di-*t*-butyl ketone, dibenzyl ketone, deoxybenzoins, and benzoin ethers. Based on what is known with ketones, one would predict that the triplet state (T_1) of the corresponding thioketones are too low in energy to induce α-cleavage. Indeed, none of the corresponding thioketones of the above systems undergo α-cleavage upon excitation to the S_1 or S_2 states. In fact, no acyclic thioketones have been shown thus far to undergo an α-cleavage process.

A large number of cyclic ketones are known to α-cleave from the lowest excited states. Classic examples are cyclopropanones, cyclobutanones, cyclopentanones, cyclohexanones, and bicyclic ketones such as fenchone and camphor.[43] Of the thioketones derived from these cyclic ketones, only cyclobutanethioketone and its derivatives have been established to undergo α-cleavage.[44–46] In addition to cyclobutanethioketones, cyclopropenethioketones undergo α-cleavage.[47,48] In the following paragraphs, results of the photolysis of cyclopropenethioketones and cyclobutanethioketones are briefly presented.

Three typical systems that have been subjected to photolysis are 2,2,4,4-tetramethyl-cyclobutanethione; 2,2,4,4-tetramethylcyclobutane-1,3-dithione; and a number of 3-substituted-2,2,4,4-tetramethylcyclobutane-4-thiones. In addition, 2,2,4,4-tetramethyl substituents have been replaced by several spirocyclic systems. All these molecules undergo α-cleavage from the lowest triplet state (as established by sensitization and quenching studies) upon excitation to either the S_1 or S_2 state. Products of photolysis in benzene and in methanol are shown in Schemes 63.19–63.22. In methanol, all molecules except 2,2,4,4-tetramethylcyclobutanethione yield a product at solvent addition. Apart from this, the products formed in benzene and in methanol are the same for all the cases shown in the above schemes.

SCHEME 63.19

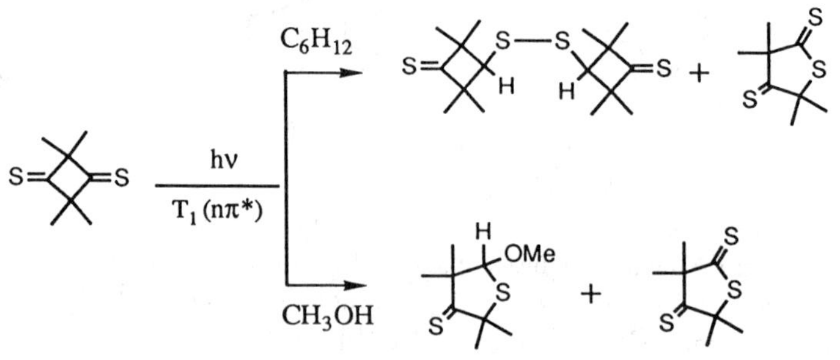

SCHEME 63.20

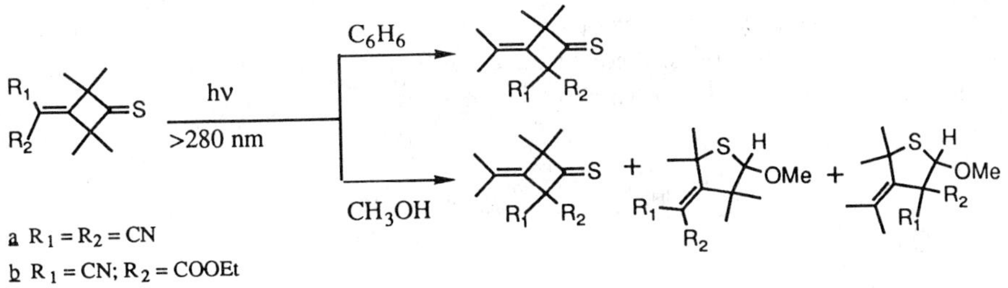

SCHEME 63.21

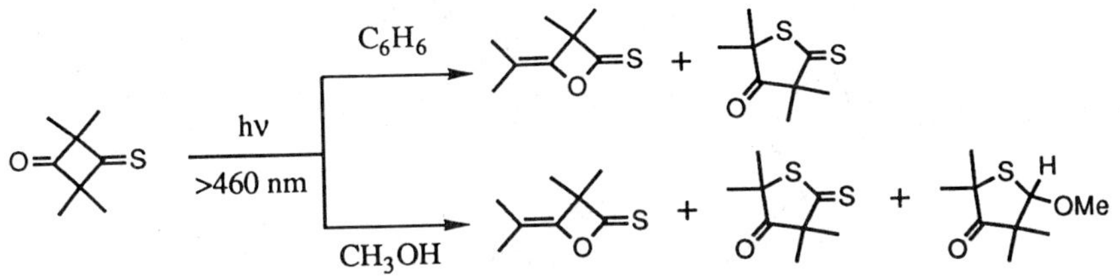

SCHEME 63.22

Products of photolysis have been rationalized on the basis of a mechanism involving α-cleavage as the primary process and biradicals and thiacarbenes as intermediates (Scheme 63.23).[46] MINDO/3 calculations were utilized to address a number of mechanistic questions concerning the α-cleavage process discussed above. Conclusions reached on the basis of these calculations are as follows: (1) Unsubstituted cyclobutanethione cannot cleave since the product biradical is much higher in energy than the triplet state. *gem*-Dimethyl substitution and/or conjugation at the 3-position is required to lower the energy of the biradical to near that of the triplet state. (b) Concerted ring expansion to a thiacarbene is electronically allowed both in the monothione and in the dithione systems; however, such a process is not thermodynamically favored in the absence of conjugation at the 3-position. In general, these predictions are consistent with the observations described in Schemes 19 through 22.

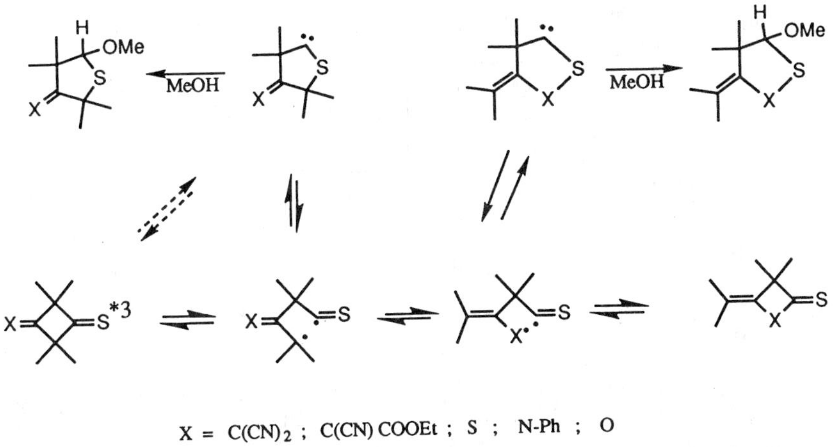

X = C(CN)$_2$; C(CN)COOEt ; S ; N-Ph ; O

SCHEME 63.23

Cyclopropenethiones are another class of molecules that undergo α-cleavage from the lowest excited triplet state.[47] Excitation of diphenylcyclopropenethione into the n,π* or π,π* excited states in a deaerated solution gives 2,3,5,6-tetraphenylthieno[3,2-b]thiophene. Irradiation in methanol gives two products — methyl-1-methoxy-phenyl-thiocinnamate and 2-methoxy-3,4-diphenylthiete (Scheme 63.24). The formation of these products has been rationalized by the mechanism illustrated in Scheme 24 and involves a Norrish type I α-cleavage as the primary photoprocess.

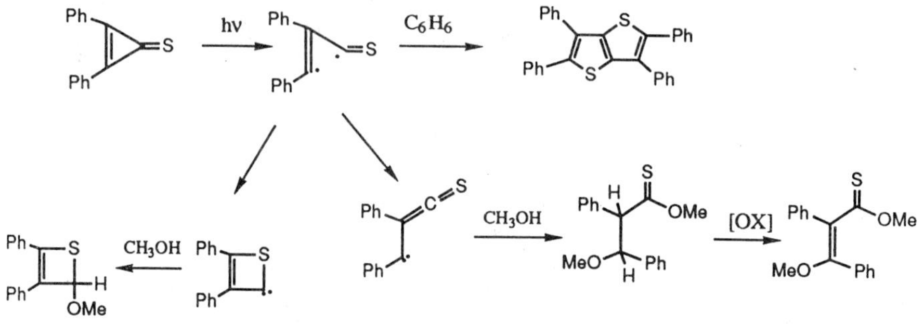

SCHEME 63.24

A remarkable regioselectivity is observed during the α-cleavage process of arylalkylcyclopropenethiones.[48] Photolysis of phenylalkylcyclopropenethiones (Scheme 63.25) in methanol gives rise to a single product — 1-methoxy-2-phenylalkylthio-2-butenoate. However, irradiation in methanol under oxygen-saturated conditions produces 2-phenyl-3-keto-alkyl-thiobutenoate. Formation of these products has been rationalized on the basis of an α-cleavage. According to the mechanism shown in Scheme 63.24, preferential formation of the more stabilized carbene (phenyl substituted) would be expected. However, the isolated products are not compatible with this prediction. The current explanation involves a preferred fast radiationless decay of the reactive triplet to a higher-energy intermediate (a less-stable carbene), probably due to better overlap of the energy surfaces. Such thinking has been invoked earlier in the case of cyclopropene ring opening.

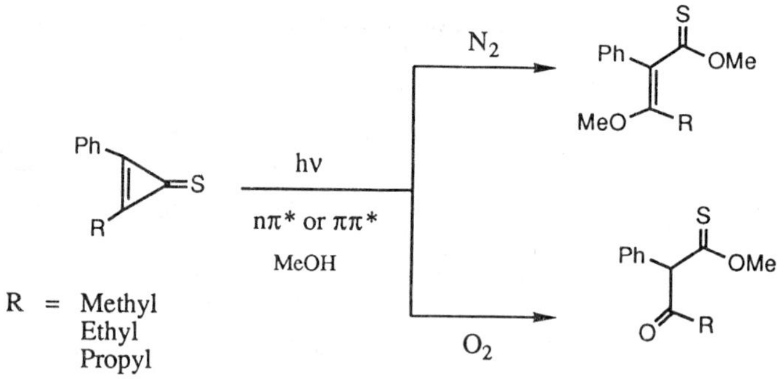

R = Methyl
    Ethyl
    Propyl

SCHEME 63.25

63.7 Final Remarks

We have shown above, with selected examples, that thioketones behave photochemically differently from the corresponding ketones. They show activity from both lower and higher excited states. Most of the reactions observed thus far have been traced to either the T_1 or S_2 state. Photoaddition to unsaturated systems and photooxidation are common reactions and are generally observed with most thiones. The α-cleavage reaction is less common and occurs only with thiones that are structurally activated. Photodimerization is also a less-common reaction, although self-quenching occurs in every thione at a near diffusion-controlled rate. Factors that control the extent of decay of triplet and singlet (S_2) excimers via physical and chemical channels need to be understood.

Acknowledgments

VPR gratefully acknowledges the support of Drs. Robert Mininni and Alex Jen. BNR thanks Ed Frankoski and Dr. Steve Schubert for their continuous support and encouragement.

ferences

1. de Mayo, P., Thione photochemistry and the chemistry of the S_2 state, *Acc. Chem. Res.*, 9, 52, 1976.
2. Turro, N. J., Ramamurthy, V., Cherry, W., and Farneth, W., The effect of wavelength on organic photoreactions in solution. Reactions from upper excited states, *Chem. Rev.*, 78, 125, 1978.
3. (a) Ramamurthy, V., Thiocarbonyl photochemistry, *Org. Photochem.*, 7, 231, 1985; (b) Coyle, J. D., The photochemistry of thiocarbonyl compounds, *Tetrahedron*, 41, 5393, 1985; (c) Rao, V. P., The photoreactivity of thiocarbonyl compounds, *Sulfur Reports*, 12, 359, 1992.
4. (a) Steer, R. P., Structure and decay dynamics of electronic excited states of thiocarbonyl compounds, *Rev. Chem. Intermediates*, 4, 1, 1981; (b) Steer, R. P. and Ramamurthy, V., Photophysics and intramolecular photochemistry of thiones in solution, *Acc. Chem. Res.*, 21, 380, 1988; (c) Maciejewski, A., and Steer, R. P., Photophysics, physical photochemistry and related spectroscopy of thiocarbonyls, *Chem. Rev.*, 93, 67, 1993.
5. (a) Gupta, S. D., Chowdhury, M., and Bera, S. C., 6500Å Transition of thiobenzophenone, *J. Chem. Phys.*, 53, 1293, 1970; (b) Emis, C. A., and Oosterhoff, L. J., The n-π* absorption and emission of optically active *trans*-β-hydrindinone and *trans*-β-thiohydrindanone, *J. Chem. Phys.*, 54, 4809, 1971; (c) Lees, W. A. and Burawoy, A., Electronic spectra of organic molecules and their interpretation. XII. The effect of solvents on the electronic spectra of thiobenzophenone and its derivatives, *Tetrahedron*, 20, 2229, 1964.
6. (a) Blackwell, D. S. L., Liao, C. C., Loutfy, R. O., de Mayo, P., and Pasyc, S., On the luminescence of thiocarbonyl compounds, *Mol. Photochem.*, 4, 171, 1972.
7. Maciejewski, A., Safarzadeh-Amiri, A., Verrall, R. E., and Steer, R. P., Radiationless decay of the second excited singlet states of aromatic thiones: Experimental verification of the energy gap law, *Chem. Phys.*, 87, 295, 1984.
8. (a) Maciejewski, A. and Steer, R. P., Effect of solvent on the subnanosecond decay of the second excited singlet state of tetramethylindanethione, *Chem. Phys. Lett.*, 100, 540, 1983; (b) Maciejewski, A. and Steer, R. P., Photophysics of the second excited singlet states of xanthione and related thiones in perfluoroalkane solvents, *J. Am. Chem. Soc.*, 105, 6738, 1983.
9. (a) Hui, M. H., de Mayo, P., Suau, R., and Ware, W. R., Thione photochemistry: Fluorescence from higher excited states, *Chem. Phys. Lett.*, 31, 257, 1975; (b) Mahaney, M. and Huber, H. R., Fluorescence from the second excited singlet of aromatic thioketones in solution, *Chem. Phys.*, 9, 371, 1975.
10. Falk, K. J. and Steer, R. P., Photophysics and intramolecular photochemistry of adamantanethione, thiocamphor and thiofenchone excited to their second excited singlet states: Evidence for subpicosecond photoprocesses, *J. Am. Chem. Soc.*, 111, 6518, 1989.
11. (a) Safarzadehi-Amiri, A., Verrall, R. E., and Steer, R. P., Decay dynamics of aromatic thione triplet states in fluid solution, *Can. J. Chem.*, 61, 894, 1983; (b) M. Mahaney and J. R. Huber, The nature of the red emission of aromatic thioketones, *J. Photochem.*, 5, 333, 1976.
12. Cox, A., Kemp, D. R., Lapouyade, R., de Mayo, P., Joussot-Dubien, J., and Bonneau, R., Thione Photochemistry: The *peri* cyclization of some polycyclic aromatic thiones, *Can. J. Chem.*, 53, 2386, 1975.
13. (a) Kumar, C. V., Quin, L., and Das, P. K., Aromatic thioketone triplets and their quenching behaviour towards oxygen and di-*tert*-butylnitroxy radical. A laser-flash-photolysis study, *J. Chem. Soc., Faraday Trans. II*, 80, 783, 1984; (b) Kumar, C. V., Davis, H. F., and Das, P. K., Photophysical behaviors of *N*-methylthioacridone triplet and efficiency of singlet oxygen generation from its quenching by Oxygen, *Chem. Phys. Lett.*, 109, 184, 1984.

14. Szymanski, M. and Steer, R. P., Vibrational activation in the radiationless decay of the S_2, S_1, T_1, and T_2 states of aromatic thiones in solution: Red edge effects, *J. Phys. Chem.*, 96, 8719, 1992.

15. Gattermann, L. and Schulze, H., Ueber thiobenzophenon, *Chem. Ber.*, 29, 2944, 1896.

16. (a) Tamagaki, S., Akatsuka, R., Nakamura, M., and Kozuka, S., Photooxidation of di-*tert*-butylthioketone, *Tetrahedron Lett.*, 3665, 1979; (b) Tamagaki, S. and Hotta, K., A sulphine mechanism for the singlet oxygenation of thiones, *J. Chem. Soc., Chem. Commun.*, 598, 1980; (c) Rao, V. J. and Ramamurthy, V., Photooxidation of di-*tert*-butylthioketone, *Curr. Sci.*, 49, 199, 1980; (d) Rao, V. J. and Ramamurthy, V., Photochemical oxidation of thiones: Di-*tert*-butylthioketone, *Ind. J. Chem., B*, 19, 143, 1980.

17. (a) Rao, V. J., Muthuramu, K., and Ramamurthy, V., Oxidation of thioketones by singlet and triplet oxygen, *J. Org. Chem.*, 47, 127, 1982; (b) Ramanath, N., Ramesh, V., and Ramamurthy, V., Photochemical oxidation of thioketones: Steric and electronic aspects, *J. Org. Chem.*, 48, 214, 1983.

18. (a) Ramesh, V., Ramanath, N., and Ramamurthy, V., Efficiency of singlet oxygen production by thiocarbonyls, *J. Photochem.*, 18, 293, 1982; (b) Ramesh, V., Ramanath, N., Rao, V. J., and Ramamurthy, V., Rates of oxidation of thioketones by singlet oxygen, *J. Photochem.*, 18, 109, 1982.

19. (a) Kemp, D. R. and de Mayo, P., The detection of transients in thiocarbonyl by flash photolysis, *J. Chem. Soc., Chem. Commun.*, 233, 1972; (b) Lawrence, A. H., de Mayo, P., Bonneau, R., and Jousset-Dubien, J., Thione Photochemistry: The detection of a saturated thione triplet by laser flash photolysis, *Mol. Photochem.*, 5, 361, 1973; (c) Bruhlmann, V. and Huber, J. R., Triplet state quenching by ground state molecules of the same kind: Xanthione in solution, *Chem. Phys. Lett.*, 54, 606, 1978.

20. Schuster, D. I., Energy wastage processes in ketone photochemistry, *Pure Appl. Chem.*, 41, 601, 1975.

21. (a) Ramesh, V., Ramanath, N., and Ramamurthy, V., The problem of triplet self-quenching in thioketone photochemistry, *J. Photochem.*, 23, 141, 1983; (b) Rajee, R. and Ramamurthy, V., Energy wastage in organic photochemistry: Self-quenching in thiones, *J. Photochem.*, 11, 135, 1979.

22. Bhattacharyya, K., Ramamurthy, V., and Das, P. K., A laser flash photolysis study of pivalothiophenone triplets. Steric and electronic effects in thione photoreaction kinetics, *J. Phys. Chem.*, 91, 5626, 1987.

23. Bhattacharyya, K., Kumar, C. V., Das, P. K., Jayasree, B., and Ramamurthy, V., Laser flash photolysis study of aliphatic thioketones, *J. Chem. Soc., Faraday Trans. 2*, 81, 1383, 1985.

24. Kozlowski, J., Maciejewski, A., Szymanski, M., and Steer, R. P., Photochemistry of aromatic thiones in their S_2 and T_1 states, *J. Chem. Soc., Faraday Trans. 2*, 88, 557, 1992.

25. (a) Liao, C. C. and de Mayo, P., Photoadditions of an alicyclic thioketone, *J. Chem. Soc., Chem. Commun.*, 1525, 1971; (b) Lawrence, A. H., Liao, C. C., de Mayo, P., and Ramamurthy, V., Thione photochemistry: Cycloaddition in a saturated alicyclic system, *J. Am. Chem. Soc.*, 98, 2219, 1976.

26. Nicolaou, K. C., Defrees, S. A., Hwang, C. K., Stylianides, N., Carrel, P. J., and Snyder, J. P., Dithiatopazine and related systems. Synthesis, chemistry, X-ray crystallographic analysis and calculations, *J. Am. Chem. Soc.*, 112, 3029, 1990.

27. (a) Liu, R. S. H. and Ramamurthy, V., Self-quenching in photocycloaddition of thiobenzophenone to crotononitrile: A case of energy transfer from S_2, *Mol. Photochem.*, 3, 261–265, 1971; (b) Maciejewski, A. and Steer, R. P., Bimolecular quenching of the excited singlet state of tetramethylindanethione, *J. Photochem.*, 24, 303, 1984.

28. Lawrence, A. H., Liao, C. C., de Mayo, P., and Ramamurthy, V., Thione photochemistry. Mechanism of the short wavelength cycloaddition of adamantanethione: Evidence for an excimer derived from S_2, *J. Am. Chem. Soc.*, 98, 3572, 1976.

29. Ohno, A., Photocycloaddition of thiocarbonyl compounds to olefins and related reactions, *Int. J. Sulfur Chem., B*, 6, 183, 1971.

30. (a) Ohno, A., Ohnishi, Y., Fukuyama, M., and Tsuchihashi, G., Photocycloaddition of thiocarbonyl compounds to olefins. The reaction of thiobenzophenone with styrene and substituted styrenes, *J. Am. Chem. Soc.*, 90, 7038, 1968; (b) Tsuchihashi, G., Yamauchi, M., and Fukuyama, M., Photocycloaddition of thiobenzophenone to olefins, *Tetrahedron Lett.*, 1971, 1967; (c) Ohno, A.,

Ohnishi, Y., and Tsuchihashi, G., Photocycloaddition of thiocarbonyl compounds to olefins. The reactions of thiobenzophenone with case III olefins, *Tetrahedron Lett.*, 283, 1969.

31. (a) Ohno, A., Ohnishi, Y., and Tsuchihashi, G., Photocycloaddition of thiocarbonyl compounds to olefins. The reactions of thiobenzophenone with electron-deficient olefins, *Tetrahedron Lett.*, 161, 1969; (b) Ohno, A., Ohnishi, Y., and Tsuchihashi, G., Photocycloaddition of thiocarbonyl compounds to olefins. The reaction of thiobenzophenone with various types of olefins, *J. Am. Chem. Soc.*, 91, 5038, 1969; (c) Gotthardt, H. and Lenz, W., Dependence of the optical induction on the mechanism of the photochemical thietane formation, *Tetrahedron Lett.*, 2879, 1979.

32. de Mayo, P. and Nicholson, A., Thione photochemistry: On thiobenzophenone photoannelation, *Isr. J. Chem.*, 10, 341, 1972; de Mayo, P. and Shizuka, H., On the mechanism of photocycloaddition of thiobenzophenone at 366 nm, *J. Am. Chem. Soc.*, 95, 3942, 1973.

33. Turro, N. J. and Ramamurthy, V., On the mechanism of photocycloaddition of aromatic thiones (n,π* triplet) to multiple bonds, *Tetrahedron Lett.*, 2423, 1976.

34. (a) Rajee, R. and Ramamurthy, V., Wavelength dependent photochemical reactions: Photocycloaddition and hydrogen abstraction reactions of di-*tert*-butylthione, *Tetrahedron Lett.*, 37, 3463, 1978; (b) Ohno, A., Uohama, M., Nakamura, K., and Oka, S., Photocycloaddition of thiocarbonyl compounds to multiple bonds. XI. Photoaddition of di-*tert*-butyl thioketone with olefins, *Tetrahedron Lett.*, 1905, 1977.

35. Kamphuis, J., Bos, H. J. T., Visser, R. J., Huizer, B. H., and Varma, C. A. G. O., An extremely short lived 1,4-biradicals as intermediates in the photocycloaddition reactions of triplet state aromatic thiones with allenes, *J. Chem. Soc., Perkin Trans. II*, 1867, 1986.

36. Pushkara Rao, V., Chandrasekar, J., and Ramamurthy, V., Thermal and photochemical cycloaddition reactions of thiocarbomyls: A qualitative MO analysis, *J. Chem. Soc., Perkin Trans. II*, 647, 1988.

37. (a) Pushkara Rao, V., and Ramamurthy, V., Upper excited state reactions of α,β-unsaturated thiones: Photocycloaddition to electron-deficient olefins, *J. Org. Chem.*, 50, 5009, 1985; (b) Pushkara Rao, V. and Ramamurthy, V., Photochemistry of α,β-unsaturated thiones: Cycloaddition to electron-deficient olefins from higher excited states, *J. Org. Chem.*, 53, 332, 1988; (c) Pushkara Rao, V. and Ramamurthy, V., Photochemistry of α,β-unsaturated thiones: Addition to electron-rich olefins from T$_1$, *J. Org. Chem.*, 50, 327, 1988.

38. Devanathan, S. and Ramamurthy, V., Photochemistry of α,β-unsaturated thiones: Cycloaddition of thiocoumarin to electron-rich and electron-deficient olefins from T$_1$, *J. Org. Chem.*, 53, 741, 1988.

39. (a) Gotthardt, H. and Nieberl, S., Photochemical and thermal reactions of thiones with substituted alkynes, *Tetrahedron Lett.*, 3563, 1976; (b) Ohno, A., Koizumi, T., Ohnishi, Y., and Tsuchihashi, G., Photocycloaddition of thiocarbonyl compounds to multiple bonds. VI. The reaction of thiobenzophenone with acetylenic compounds, *Tetrahedron Lett.*, 2025, 1970; (c) Brouwer, A. C. and Bos, H. J. T., Photochemical and thermal (2+2) addition of thiones to hetero-substituted acetylenes. Rearrangement of intermediate theites to α,β-unsaturated dithioesters and thioamides, *Tetrahedron Lett.*, 209, 1976; (d) Gotthardt, H. and Huss, O. M., Light induced formation of stable thietes, *Tetrahedron Lett.*, 3617, 1978; (e) Bouwer, A. C., George, A. V. E., and Bos, H. J. T., Photochemical reactions of some aromatic thiones with bis (*tert*-butylthio)ethyne. Isolation and properties of thietes, the photochemical (2+2)-cycloadducts and their isomeric α,β-unsaturated dithioesters, *Recl. Trav. Chim. Pays. Bas.*, 102, 83, 1983.

40. (a) Bos, H. J. T., Schinkel, H., and Wijsman, Th. C. M., Photocycloadditions of aromatic thiones to allenes, *Tetrahedron Lett.*, 3905, 1971; (b) Hostra, G., Kamphius, J., and Bos, H. J. T., Photocycloaddition reactions of the first stable thioaldehyde: 2,4,6-Tri(*tert*-butyl) thiobenzaldehyde with substituted allenes, *Tetrahedron Lett.*, 873, 1984.

41. (a) Ohno, A., Kito, N., and Koizumi, T., Photocycloaddition of thiocarbonyl compounds to multiple bonds. VIII. The reaction of thiobenzophenone with imines, *Tetrahedron Lett.*, 2421, 1971; (b) Visser, R. G., Baaij, J. P. B., Bouwer, A. C., and Bos, H. J. T., Photochemical (2+2) cycloaddition of aromatic thiones to keteneimines. Acid catalyzed rearrangement of iminothietanes, *Tetrahedron Lett.*, 4343, 1977.

42. Blackwell, D. S. L., de Mayo, P., and Suau, R., Thione photochemistry: Cycloaddition to the nitrile function, *Tetrahedron Lett.,* 91, 1974.

43. Turro, N. J., *Modern Molecular Photochemistry,* University Science Books, Mill Valley, CA, 1991, chap. 13.

44. (a) Muthuramu, K. and Ramamurthy, V., Photochemistry of dimethylthioketene dimers, *J. Org. Chem.,* 45, 4532, 1980; (b) Muthuramu, K., Sundari, B., and Ramamurthy, V., Strain assisted α-cleavage reactions of cyclobutanethiones, *Ind. J. Chem., B,* 20, 797, 1981; (c) Muthuramu, K. and Ramamurthy, V., Strain assisted α-cleavage reactions of thioketones: Cyclobutanethiones, *J. Org. Chem.,* 48, 4482, 1983.

45. (a) Muthuramu, K., and Ramamurthy, V., Photofragmentation reactions of thiocarbonyl compounds, *Chem. Lett.,* 1261, 1981; (b) Photolysis of dispiro-substituted 3-thioxo-1-cylobutanones, Kimura, K., Fuikuda, Y., Negoro, T., and Odaira, Y., *Bull. Chem. Soc. Jpn.,* 54, 1901, 1980.

46. Rao, B. N., Chandrasekhar, J., and Ramamurthy, V., Photochemical, photophysical and theoretical studies on cyclobutanethiones: α-Cleavage reactions, *J. Org. Chem.,* 53, 745, 1988.

47. Singh, S., Bhadbhade, M., Venkatesan, K., and Ramamurthy, V., Strain assisted α-cleavage reactions of thioketones: Diphenylcyclopropenethiones, *J. Org. Chem.,* 47, 3550, 1982.

48. (a) Singh, S. and Ramamurthy, V., Regioselectivity in α-cleavage reactions: Arylalkyl-cyclopropenethiones, *J. Org. Chem.,* 50, 3732, 1985; (b) Usha, G., Nageshwer Rao, B., Chandrsasekhar, J., and Ramamurthy, V., The origin of regioselectivity in α-cleavage reactions of cyclopropenethiones: Potential role of pseudo Jahn-Teller effect in substituted cyclopropenyl systems, *J. Org. Chem.,* 51, 3630, 1986.

64

Thiocarbonyls: Photochemical Hydrogen Abstraction Reactions

Pushkara Rao
...hem America Inc.,
...arch & Development
...nter
...mouth Junction,
... Jersey

Nageshwer Rao
... Corporation, Technology
...vision
...cott, New York

... Ramamurthy
...rtment of Chemistry
...ne University
... Orleans, Louisiana

64.1 Introduction

Hydrogen abstraction reaction of carbonyl compounds is one of the most extensively investigated organic photochemical reactions.[1-3] Thiocarbonyl compounds, similar to carbonyl compounds, have been established to undergo hydrogen abstraction processes. Thiocarbonyls differ substantially from carbonyl compounds both in terms of electronic configuration of the reactive excited states and the mode of attack of the thiocarbonyl chromophore on hydrogen-bearing substrates. In carbonyls, hydrogen abstraction generally occurs from the n,π^* state, and the partially filled electrophilic n-orbital initiates this process.[3] In contrast, thiocarbonyls abstract hydrogen from both the upper π,π^* state and the lower n,π^* state, and the abstracted hydrogen can add at either the C or the S atom of the thiocarbonyl chromophore.[4] A brief summary of the electronic absorption characteristics and general photoreactivity[5] of thiocarbonyl compounds is presented in a companion chapter. This chapter outlines the photochemical hydrogen abstraction reactions of organic molecules containing the thiocarbonyl chromophore.

64.2 Intramolecular Hydrogen Abstraction

When their structures permit, thiocarbonyls undergo distinct wavelength-dependent intramolecular hydrogen abstraction reaction. The course of hydrogen abstraction and the subsequent product formation are related to the substituents present on the thiocarbonyl function and the wavelength of excitation.

Aryl Alkyl Thiones

Aryl alkyl thiones containing saturated hydrocarbon alkyl chains, upon excitation to the $S_2(\pi,\pi^*)$ state, yield cyclopentane thiols via a δ-hydrogen abstraction from the alkyl chain (Scheme 64.1).[6]

This reaction does not occur upon irradiation of the $S_1(n,\pi^*)$ band. Successful quenching of the reaction by singlet quenchers and unsuccessful triplet sensitization are taken to support $S_2(\pi,\pi^*)$ as the reactive state. A mechanism involving a short-lived 1,5-biradical has been proposed for this δ-hydrogen abstraction process.[7] The reaction pathway pursued by thiocarbonyl compounds is clearly different from that undertaken by the analogous compounds bearing a carbonyl chromophore.[8] The synthetic value of δ-hydrogen abstraction by aryl alkyl thiones has been demonstrated by employing this process as a key step in the synthesis of a sesquiterpenoid, ($\pm$)-Cuparene (Scheme 64.2).[9]

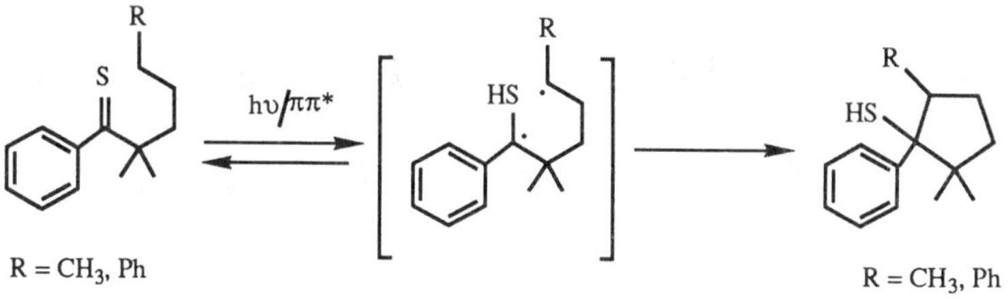

SCHEME 64.1

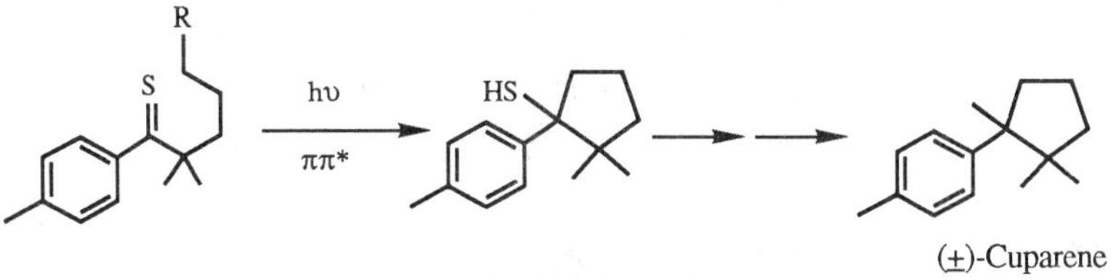

SCHEME 64.2

Aryl alkyl thiones lacking δ-hydrogens (e.g., 1-phenyl-2,2,5,5-tetramethyl-thiohexanone) do not undergo any photoreaction, irrespective of the wavelength of excitation.[7] However, when the δ-carbon is replaced by an ethereal oxygen atom, photochemical hydrogen abstraction occurs from carbons adjacent (ε and γ) to the oxygen atom (Scheme 64.3) as a result of activation of the neighboring hydrogens by the heteroatom.[10] ε-Hydrogen abstraction leads to six-membered ring thiols, while γ-hydrogen abstraction leads to cyclobutane thiols and type-II β-cleavage to olefins. Whereas the γ-hydrogen abstraction can be initiated by light of short and long wavelength, ε-hydrogen abstraction is specific and occurs only upon excitation to the S_2 state.

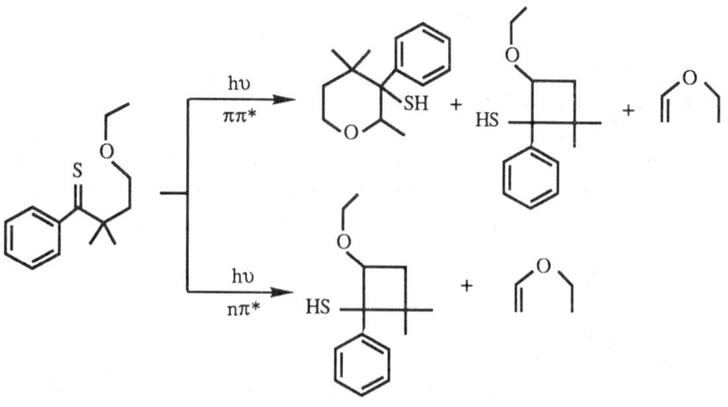

SCHEME 64.3

A β-hydrogen abstraction process leading to cyclopropane thiols has also been observed, but this is restricted to a few classes of thiones. One such class is the aryl alkyl thiones having β-hydrogens activated by the functional groups such as phenyl and S-alkyl groups (Scheme 64.4).[11] Excitation of such thiones brings about wavelength-independent cyclization [$S_2(\pi,\pi^*)$ or $S_1(n,\pi^*)$] that is presumed to proceed via a 1,3-biradical intermediate. Quenching and sensitization studies suggest the lowest n,π* triplet as the reactive state. Structurally similar ketones are generally photostable; in a few cases, only slow type-I fission is observed.[12]

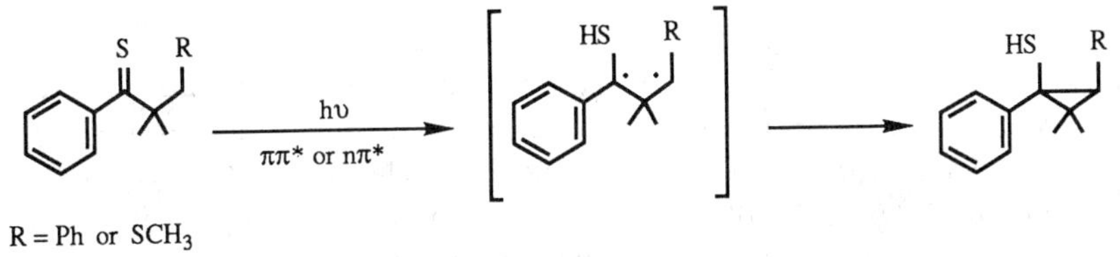

R = Ph or SCH₃

SCHEME 64.4

Cyclization leading to a six-membered thiolane derivative has been observed during the irradiation of a thermally stable thioaldehyde, 2,4,6-tri-*t*-butylthiobenzaldehyde (Scheme 64.5).[13] This excitation wavelength-independent cyclization is the result of transfer of a hydrogen atom to the carbon of the thiocarbonyl chromophore. Possible involvement of a radical intermediate in this photocyclization process is indicated by the formation of the same thiolane derivative as one of the products during the reaction of 2,4,6-tri-*t*-butylthiobenzaldehyde with free radicals such as 1-cyano-1-methylethyl and *t*-butyl radicals.

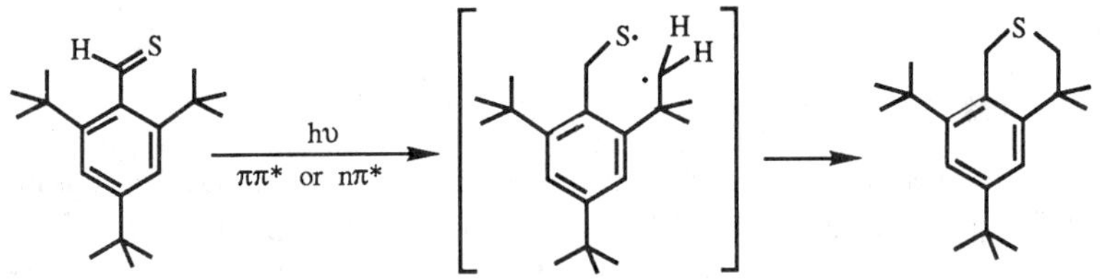

SCHEME 64.5

Dialkyl Thiones

Bridged bicyclic thiones (e.g., bicycloheptane thiones and bicyclooctane thiones) on irradiation yield β-insertion products.[14] This reaction, which occurs only upon excitation to the S_2 (π,π*) state, is illustrated in Scheme 64.6 using the bicyclic thiones, thiofenchone and 2,2-dipropyl thiofenchone, as examples. In both these thiones, hydrogen abstraction leading to tricyclic thiols takes place from the carbon β to the thiocarbonyl group. Availability of β, γ, and δ-hydrogens in alkyl side chain of 2,2-dipropyl thiofenchone does not change the course of the reaction. Participation of a β-*endo*-hydrogen during the cyclization has been confirmed by deuterium labeling studies. This reaction differs in at least two respects from the β-insertion process occurring in the aryl alkyl thiones (Scheme 4) discussed above: (1) the reaction from bridged bicyclic thiones occurs only from the S_2 state, whereas that from aryl alkyl thiones originates from the S_1 state; and (2) the abstraction in the case of aryl alkyl thiones requires activation of the hydrogens, whereas no such activation is required in bridged bicyclic thiones. No analogous reaction is known in structurally similar bicyclic ketones.

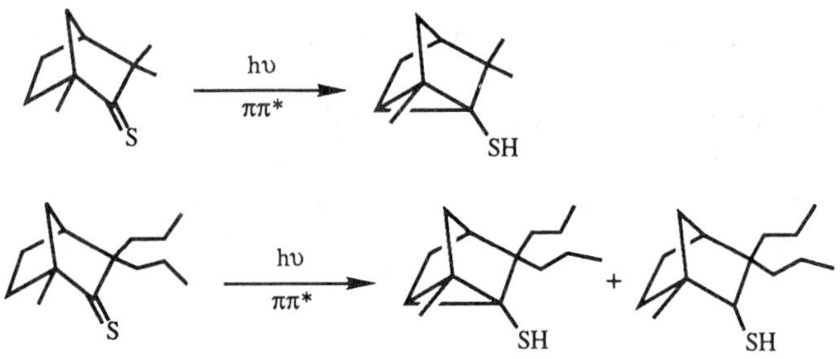

SCHEME 64.6

Monocyclic thiones such as 2,2,5,5-tetraalkylthiocyclopentanones and 2,2,6,6-tetraalkylthio-cyclohexanones also react from the $S_2(\pi,\pi^*)$ state similar to bicyclic thiones, but the mode of hydrogen abstraction is dependent on the alkyl side chain (Scheme 64.7).[15] When the alkyl groups are methyl groups, the products of hydrogen abstraction are cyclopropyl thiols (β-insertion). However, irradiation of thiones with a longer alkyl chain leads to γ-insertion as the major and β-insertion as the minor products; a δ-insertion pathway is virtually absent.

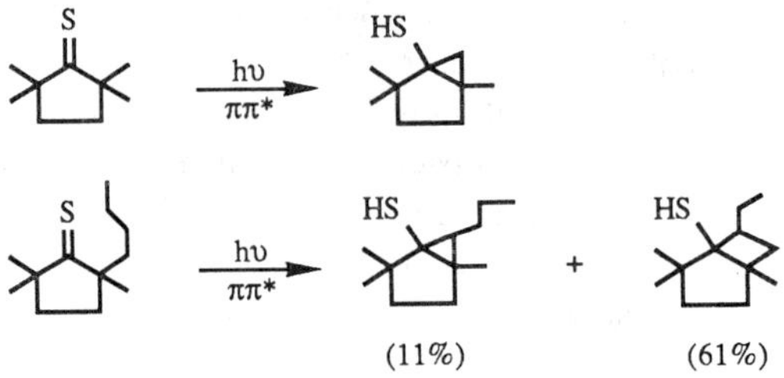

(11%)   (61%)

SCHEME 64.7

Diaryl Thiones

Certain aromatic thioketones, having a *peri*-hydrogen adjacent to the thiobenzoyl group, cyclize on irradiation.[16] This photocyclization produces a product bearing a five-membered ring containing a sulfur atom (Scheme 64.8). Mechanistic investigations suggest that the lowest singlet excited state may be responsible for this process. Incorporation of deuterium from co-solvent D_2O suggests that a dipolar intermediate may be involved in the cyclization process.

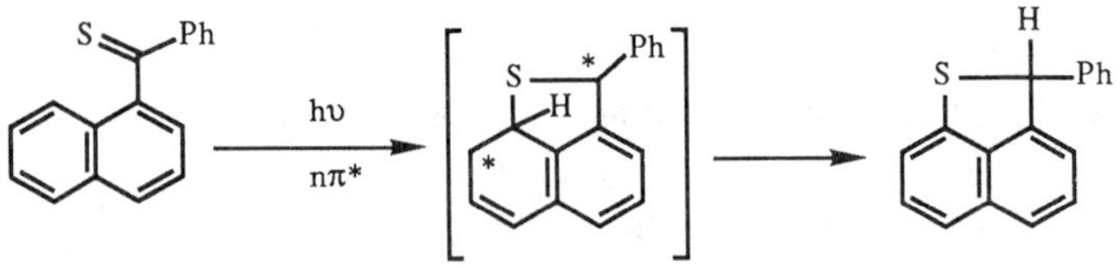

(* may represent 0, 1 or 2 electrons)

SCHEME 64.8

Photothioenolization via intramolecular hydrogen abstraction has been observed during the photolysis (S_1; n,π*) of *O*-benzylthiobenzophenone (Scheme 64.9).[17] The enethiol is stable at low temperature, but rapidly reverts back to thioketone at room temperature. A short-lived enethiol has been trapped with acetylenedicarboxylates to yield benzothiapyran.

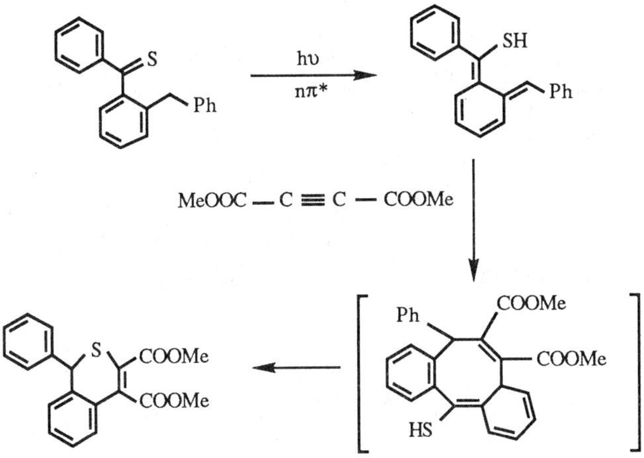

SCHEME 64.9

Thioesters and Thioimides

O-Alkyl thiobenzoates, possessing activated γ-hydrogen atoms in the O-alkyl chain, upon excitation to the lowest n,π* state, undergo a γ-hydrogen abstraction reaction (Scheme 64.10).[18] The 1,4-biradical resulting from this reaction collapses via elimination and cyclization processes; while the elimination pathway (path E) leads to an olefin and thiobenzoic acid, the cyclization pathway (path C) leads to an oxetane. Substituents at the β-carbon of the O-alkyl chain influence the extent of participation of these two pathways. Bulky alkyl groups at the β-carbon favor path C over path E. Elimination leading to olefins and thiobenzoic acids is overwhelmingly preferred in systems possessing no β-substituents. However, when the substituents at the β- and the γ-carbons are parts of a ring system, only elimination products are observed. The stereospecific nature of this elimination has been demonstrated by deuterium incorporation studies in cholesteryl thiobenzoate (Scheme 64.11).[19] This photoelimination offers an extremely mild synthetic method for dehydrating many homoallylic alcohols.[20]

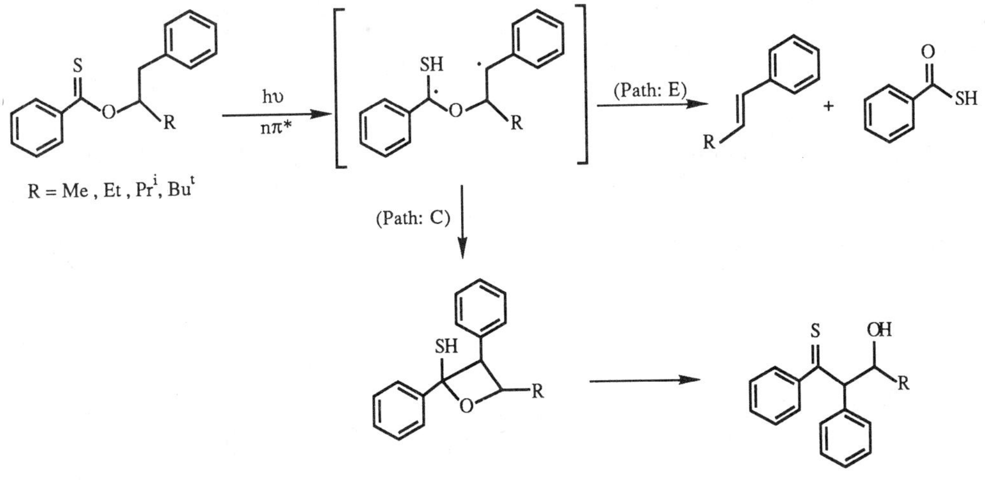

SCHEME 64.10

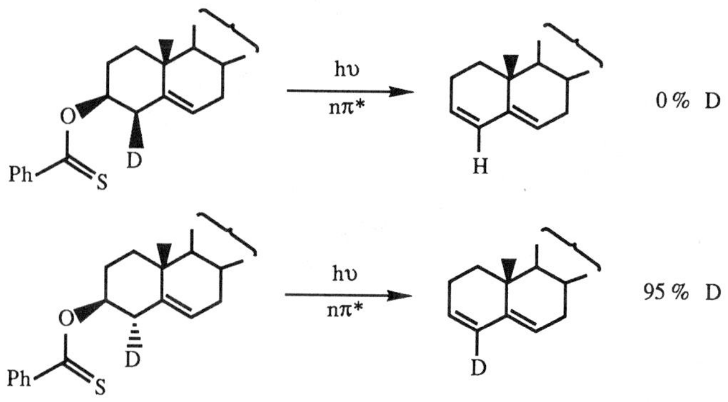

SCHEME 64.11

Similar to thioesters, acyclic monothioimides (Scheme 64.12) undergo photochemical γ-hydrogen abstraction to produce β-lactams (type-II cyclization products) and thioamides (type-II cleavage products).[21] This reaction, which has been generalized using a number of structurally related acyclic and semicyclic monothioimides, provides an attractive method to synthesize β-lactams possessing sulfur atoms at the C-4 position.[22]

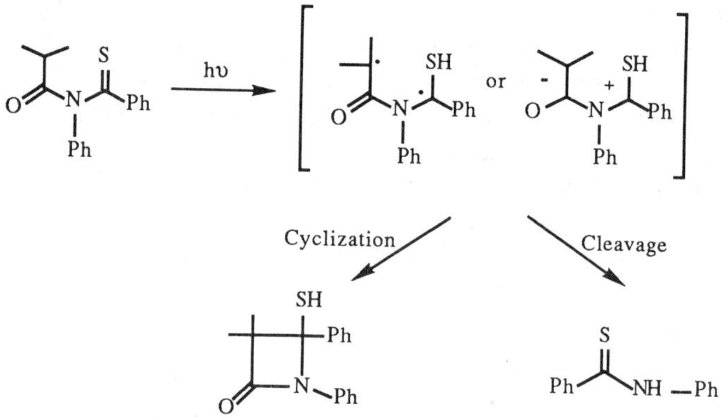

SCHEME 64.12

An interesting photorearrangement *via* an initial β-hydrogen abstraction has been reported for *N*-acylthiobenzamides.[23] As illustrated in Scheme 64.13, *N*-acylthiobenzamides upon irradiation (>400 nm) in benzene rearrange to give α-acylaminothioketones. The formation of α-acylaminothioketone is attributed to the ring opening of an aziridine intermediate that is produced by β-cyclization of *N*-acylthiobenzamide.

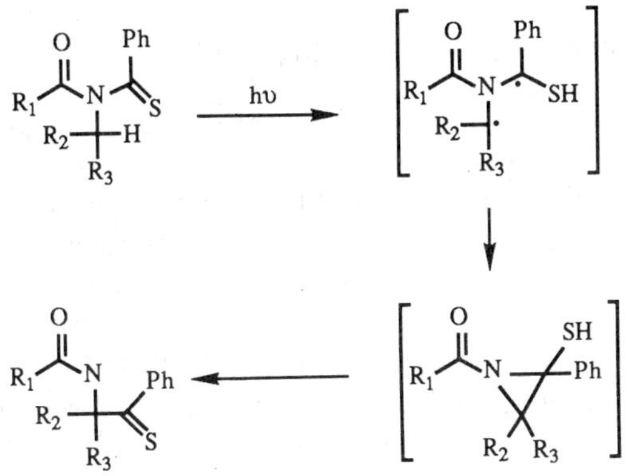

$$R_1 = R_2 = Me; \ R_3 = Me \ or \ Ph$$

$$R_1 = Ph; \ R_2 = Me; \ R_3 = Me, Et \ or \ Ph$$

SCHEME 64.13

δ-Hydrogen abstraction resulting in imidazolidinones and imidazolidinethiones has been observed during the irradiation of *N*-thioaroylureas and *N*-thioaroylthioureas, respectively (Scheme 64.14).[24] The products of irradiation have been converted smoothly to imidazolones and imidazolthiones by the acid-catalyzed elimination of hydrogen sulfide. Since *N*-thioaroylureas and *N*-thioaroylthioureas can be prepared easily, this photochemical hydrogen abstraction process provides a simple approach for the synthesis of imidazolones and imidazolthiones.

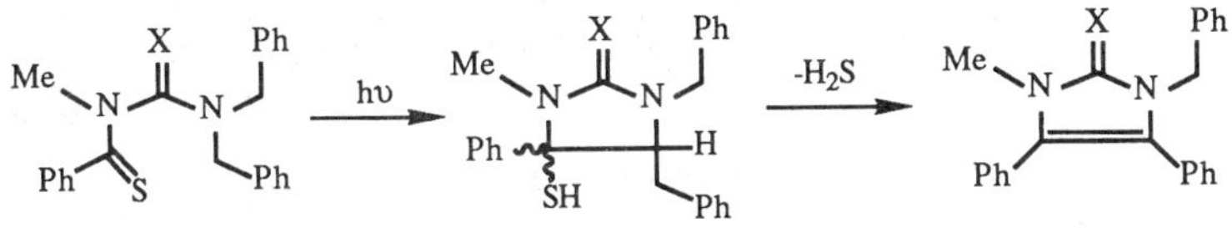

X = O or S

SCHEME 64.14

Conjugated thiocarbonyls, such as 4-thiouracils substituted at the 5-position by hydroxyalkyl chains (Scheme 64.15), upon excitation either to the $S_2(\pi,\pi^*)$ or $S_1(n,\pi^*)$ state, undergo intramolecular hydrogen abstraction to yield a 1,4- or a 1,5-biradical. These biradicals do not undergo the usual cyclization or elimination; instead, they take an alternative pathway that involves the delocalization of the adjacent carbon-carbon double bond.[25] It has been suggested that the photoreactive state is of n,π^* character.

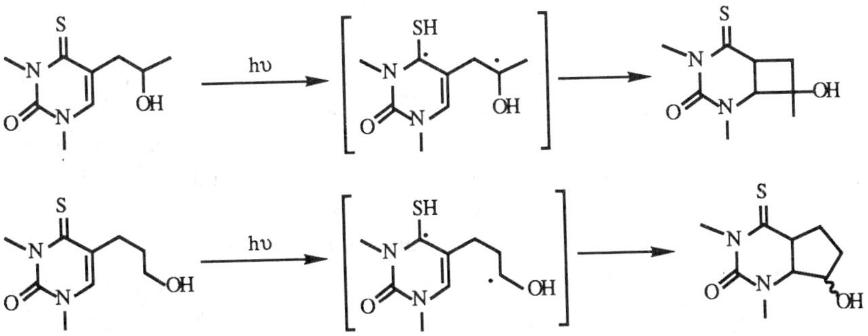

SCHEME 64.15

64.3 Intermolecular Hydrogen Abstraction

Thiocarbonyl compounds in hydrogen atom-containing solvents undergo intermolecular hydrogen abstraction when irradiated in either the $S_2(\pi,\pi^*)$ or the $S_1(n,\pi^*)$ bands. The ease of photoreduction is governed by the strength of the C-H bond in the hydrogen donor, as well as the structure and electronic configuration of the excited thiocarbonyl function. Two classes of thiocarbonyls, diaryl and dialkyl thiones, have been studied in some detail.

Upon excitation to S_2 or S_1, thiobenzophenone reacts with various hydrogen-donating solvents.[26] When the hydrogen donor is ethanol, the products of excitation are diphenylmethanethiol and the corresponding sulfide and disulfide derivatives (Scheme 64.16). The reactive excited state is the n,π* triplet of thiobenzophenone. Mechanistic studies indicate that initial hydrogen abstraction leads to a thioketyl radical similar to the ketyl radical involved in benzophenone photoreduction. The reduction products observed are believed to originate from trapping of the ketyl radical by the ground-state thiobenzophenone. This reduction behavior is quite different from that of benzophenone, which gives benzpinacols under similar conditions. The difference in the end result between thione and ketone reductions has been attributed to the better trapping ability of ground-state thiobenzophenone over benzophenone.

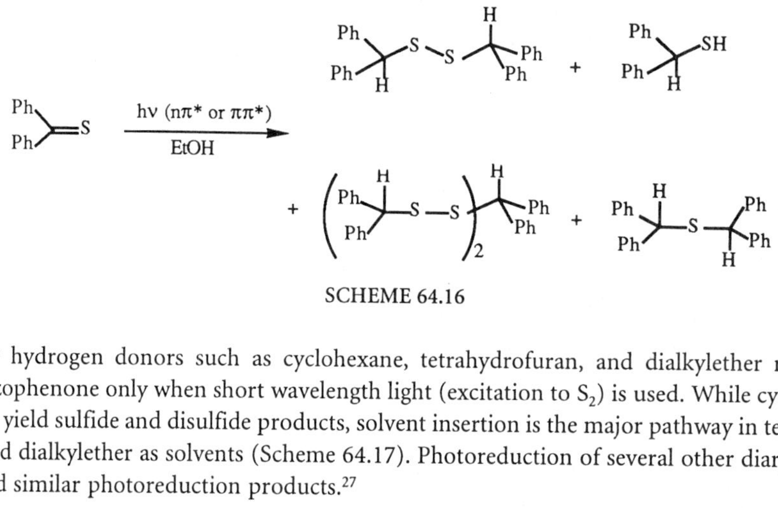

SCHEME 64.16

Other hydrogen donors such as cyclohexane, tetrahydrofuran, and dialkylether react with thiobenzophenone only when short wavelength light (excitation to S_2) is used. While cyclohexane reacts to yield sulfide and disulfide products, solvent insertion is the major pathway in tetrahydrofuran and dialkylether as solvents (Scheme 64.17). Photoreduction of several other diaryl thiones also yield similar photoreduction products.[27]

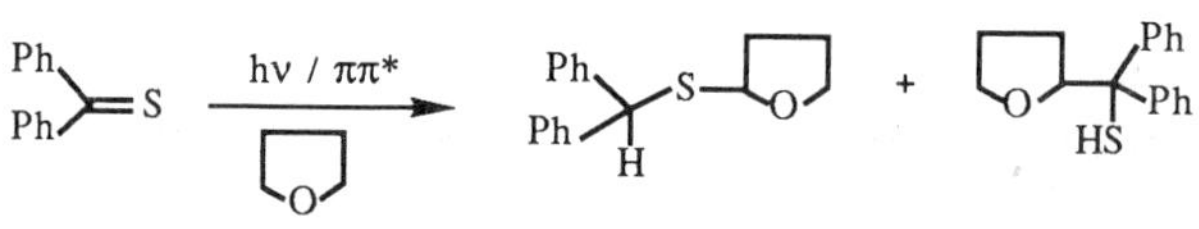

SCHEME 64.17

Photoreduction of adamantanethione (as a model for dialkyl thiones) has been investigated. Irradiation of adamantanethione in the n,π* band in cyclohexane gives only the 1,3-dithietane dimer and no reduction products are observed. However, under the same conditions when 2-adamantanethiol is used as the hydrogen donor, di(2-adamantyl) disulfide is obtained.[28] A mechanism involving a number of radical intermediates has been proposed.

In contrast to the n,π* state, the π,π* state of adamantanethione is a fairly indiscriminate abstracter of hydrogen atoms.[29] In cyclohexane, solvent insertion products, sulfide and thiol (Scheme 64.18) together with 1,3-dithietane dimer are formed. This reduction has been shown to involve hydrogen abstraction and radical pairs. Based on the dependence of solvent viscosity on the product distribution, it has been proposed that the insertion products result from the recombination of radicals within the solvent cage. Similar to adamantanethione, di-*t*-butyl thione undergoes photoreduction also in many hydrogen-donating solvents and addends.[30]

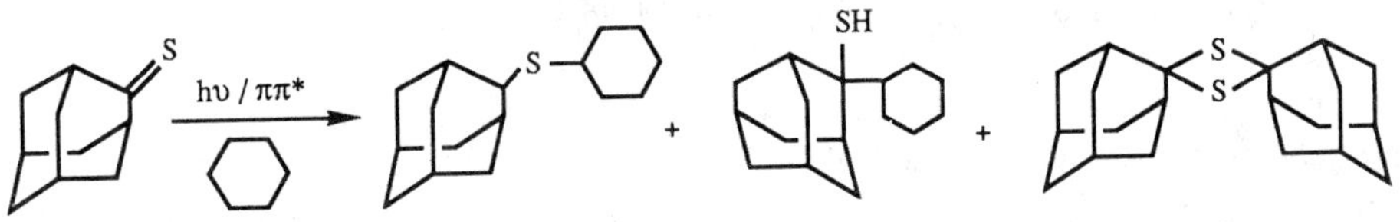

SCHEME 64.18

4.4 Conclusion

Hydrogen abstraction occurs in thiocarbonyls from both the upper (π,π^*) and the lower (n,π^*) excited states. Thiocarbonyls yield both cyclic ring and open-chain products resulting from hydrogen abstraction. The synthetic application of some of these reactions (five-membered ring products from δ-hydrogen abstraction in aryl alkyl thiones, cleavage products from γ-abstraction in *O*-alkyl thiobenzoates, β-lactam derivatives from γ-hydrogen abstraction in thioimides, and imidazolones from δ-insertion in *N*-thioaroylureas) have already been recognized. Based on these synthetic applications and on the feasibility of selectively generating carbocyclic or heterocyclic rings of various sizes, photochemical hydrogen abstraction reactions of thiocarbonyls may become a useful and routine synthetic strategy. Room still exists for detailed mechanistic studies of a number of novel hydrogen abstraction processes, especially considering the sophisticated tools that have become available since the original discovery of these processes.

Acknowledgments

VPR gratefully acknowledges the support of Drs. Robert Mininni and Alex Jen. BNR thanks Ed Frankoski and Dr. Steve Schubert for their continuous support and encouragement.

References

1. Turro, N. J., *Modern Molecular Photochemistry*, Benjamin/Cummins, Menlo Park, CA, 1978.
2. For a review of the synthetic applications of photoreductions, see: (a) Schoenberg, A., *Preparative Organic Chemistry*, Springer-Verlag, New York, 1968; (b) Cowan, D. O. and Drisco, R. L., *Elements of Organic Photochemistry*, Plenum Press, New York, 1976, 75.
3. (a) Wagner, P. J., *Top. Curr. Chem.*, 66, 1, 1976; (b) Wagner, P. J., in *Rearrangements of Ground and Excited States*, Vol. 3, de Mayo, P., Ed., Academic, Orlando, FL, 1980, chap. 20; (c) Wagner, P. J., *Acc. Chem. Res.*, 16, 461, 1983.
4. (a) Bigot, B., *Isr. J. Chem.*, 23, 116, 1983; (b) Sumathi, K. and Chandra, A. K., *J. Photochem. Photobiol. A*, 43, 313, 1988.
5. For major reviews on the photochemistry and photophysics of thiocarbonyl compounds, see: (a) de Mayo, P., *Acc. Chem. Res.*, 9, 52, 1976; (b) Steer, R. P., *Rev. Chem. Intermed.*, 4, 1, 1981; (c) Ramamurthy, V., *Org. Photochem.*, 7, 231, 1985; (d) Coyle, J. D., *Tetrahedron*, 41, 5393, 1985; (e) Ramamurthy, V., and Steer, R. P., *Acc. Chem. Res.*, 21, 380, 1988; (f) Pushkara Rao, V., *Sulfur Reports*, 12, 359, 1992; (g) Maciejewski, A. and Steer, R. P., *Chem. Rev.*, 93, 67, 1993.
6. (a) de Mayo, P. and Suau, R., *J. Am. Chem. Soc.*, 96, 6807, 1974; (b) Couture, A., Ho, K. W., Hoshino, M., de Mayo, P., Suau, R., and Ware, W. R., *J. Am. Chem. Soc.*, 98, 6218, 1976.
7. Ho, K. W. and de Mayo, P., *J. Am. Chem. Soc.*, 101, 5725, 1979.

8. (a) Wagner, P. J. and Hammond, G. S., *J. Am. Chem. Soc.*, 88, 1245, 1966; (b) Lewis, F. D. *J. Am. Chem. Soc.*, 92, 5602, 1970; (c) Wagner, P. J. and McGrath, J. M., *J. Am. Chem. Soc.*, 94, 3849, 1972; (d) Lewis, F. D. and Hillard, T. A. *J. Am. Chem. Soc.*, 94, 3852, 1972.

9. de Mayo, P. and Suau, R., *J. Chem. Soc., Perkin Trans. I*, 2559, 1974.

10. Basu, S., Couture, A., Ho, K. W., Hoshino, M., de Mayo, P., and Suau, R., *Can. J. Chem.*, 59, 246, 1981.

11. (a) Couture, A., Hoshino, M., and de Mayo, P., *J. Chem. Soc. Chem. Commun.*, 131, 1976; (b) Couture, A., Gomez, J., and de Mayo, P., *J. Org. Chem.*, 46, 2010, 1981.

12. (a) Stermitz, F. R., Nicodem, D. E., Muralidharan, V. P., and O'Donnell, C. M., *Mol. Photochem.*, 2, 87, 1970; (b) Wagner, P. J., Kelso, P. A., Kemppainen, A. Haug and Graber, D. R., *Mol. Photochem.*, 2, 81, 1970.

13. Okazaki, R., Ishii, A., Fukuda, N., Oyama, H., and Inamoto, N., *Tetrahedron Lett.*, 849, 1984.

14. Blackwell, D. S. L. and de Mayo, P., *J. Chem. Soc., Chem. Commun.*, 130, 1973.

15. Blackwell, D. S. L., Lee, K. H., de Mayo, P., Petrasiunas, G. L. R., and Reverdy, G., *Nouv. J. Chim.*, 3, 123, 1979.

16. (a) Lapouyade, R. and de Mayo, P., *Can. J. Chem.*, 50, 4068, 1972; (b) Cox, A., Kemp, D. R., Lapouyade, R., de Mayo, P., Joussot-Dubien, J., and Bonneau, R., *Can. J. Chem.*, 53, 2386, 1975.

17. (a) Kito, N. and Ohno, A., *J. Chem. Soc., Chem. Commun.*, 1338, 1971; (b) Kito, N. and Ohno, A., *Int. J. Sulfur Chem.*, 8, 427, 1973.

18. (a) Achmatowicz, S., Barton, D. H. R., Magnus, P. D., Poulton, G. A., and West, P. J., *J. Chem. Soc., Chem. Commun.*, 1014, 1971; (b) Barton, D. H. R., Bolton, M., Magnus, P. D., and West, P. J., *J. Chem. Soc., Chem. Commun.*, 632, 1972; (c) Wirtz, J., *J. Chem. Soc., Perkin Trans. II*, 1307, 1973; (d) Barton, D. H. R., Bolton, M., Magnus, P. D., and West, P. J., *J. Chem. Soc., Perkin Trans. II*, 1580, 1973; (e) Ogata, Y., Takagi, K., and Ihda, S., *J. Chem. Soc., Perkin Trans. II*, 1725, 1975.

19. Achmatowicz, S., Barton, D. H. R., Magnus, P. D., Poulton, G. A., and West, P. J., *J. Chem. Soc., Perkin Trans. I*, 1567, 1973.

20. Barton, D. H. R., Bolton, M., Magnus, P. D., and West, P. J., *J. Chem. Soc., Perkin Trans. I*, 1574, 1973.

21. Sakamoto, M., Aoyama, H., and Omote, Y., *Tetrahedron Lett.*, 4475, 1985.

22. (a) Sakamoto, M., Aoyama, H., and Omote, Y., *Tetrahedron Lett.*, 1335, 1986; (b) Sakamoto, M., Tanaka, M., Fukuda, A., Aoyama, H., and Omote, Y., *J. Chem. Soc., Perkin. Trans. I*, 1353, 1988; (c) Sakamoto, M., Watanabe, S., Fujita, T., and Tohnishi, M., *J. Chem. Soc., Perkin. Trans. I*, 2203, 1988.

23. Sakamoto, M., Aoyama, H., and Omote, Y., *J. Org. Chem.*, 49, 1837, 1984.

24. Aoyama, H., Sakamoto, M., and Omote, Y., *Chem. Lett.*, 1397, 1983.

25. Fourrey, J. L., Henry, G., and Jouin, P., *Tetrahedron Lett.*, 2375, 1976.

26. (a) Ohno, A. and Kito, N., *Int. J. Sulfur. Chem. A*, 1, 26, 1971; (b) Oster, G., Citarel, L., and Goodman, M., *J. Am. Chem. Soc.*, 84, 7036, 1962; (c) Kito, N. and Ohno, A., *Bull. Chem. Soc. Jpn.*, 46, 2487, 1973; (d) Ohnishi, Y. and Ohno, A., *Bull. Chem. Soc. Jpn.*, 46, 3868, 1973; (e) Formosinho, S. J., *J. Chem. Soc., Faraday Trans. II*, 72, 1332, 1976.

27. (a) Capitanio, D. A., Pownall, H. J., and Huber, J. R., *J. Photochem.*, 3, 225, 1974; (b) Bruhlmann, U. and Huber, J. R., *J. Photochem.*, 10, 205, 1979.

28. Bolton, J. R., Chen, K. S., Lawrence, A. H., and de Mayo, *J. Am. Chem. Soc.*, 97, 1832, 1975.

29. (a) Lawrence, A. H., Liao, C. C., de Mayo, P., and Ramamurthy, V., *J. Am. Chem. Soc.*, 98, 3572, 1976; (b) Law, K. Y. and de Mayo, P., *J. Am. Chem. Soc.*, 101, 3251, 1979.

30. (a) Rao, V. J. and Ramamurthy, V., *Ind. J. Chem. B.*, 18, 265, 1979; (b) Rajee, R. and Ramamurthy, V., *Tetrahedron Lett.*, 3463, 1978; (c) Ohno, A., Uohama, M., Nakamura, K., and Oka, S., *Bull. Chem. Soc. Jpn.*, 52, 1521, 1979.

Photorearrangement of Thioarenes

ain Lablache-Combier
iversité des Sciences et
echniques de Lille

65.1 Introduction

This chapter deals with the ring rearrangement of the thioarenes: thiophenes, thiazoles, and isothiazoles. Photoreactions of compounds in which the side chain is the reactive part of the molecule (stilbene-type cyclization, photo-Fries rearrangement, reactions of NO_2 groups linked to the ring or to a side chain) are excluded.

The photochemistry of sulfur-containing aromatic rings has already been reviewed.[1-6] A general introduction to the photoisomerization of heteroaromatic compounds is given at the start of Chapter 82 in this book. To explain the photorearrangement of these arenes, five mechanistic paths were described by Padwa.[4] These routes are:

1. The ring contraction-ring expansion route.
2. The internal cyclization-isomerization route. (Mechanisms 1, 2, and 5 are outlined in Chapter 82.)
3. The Van Tamelen-Whitesides general mechanism. This involves initial cleavage of the weakest single bond to form a cyclopropene or its heterocyclic analog in equilibrium with a bicyclic isomer, as shown below in Scheme 65.1, thus accounting for the interchange of the 2- and 4-positions of the ring.

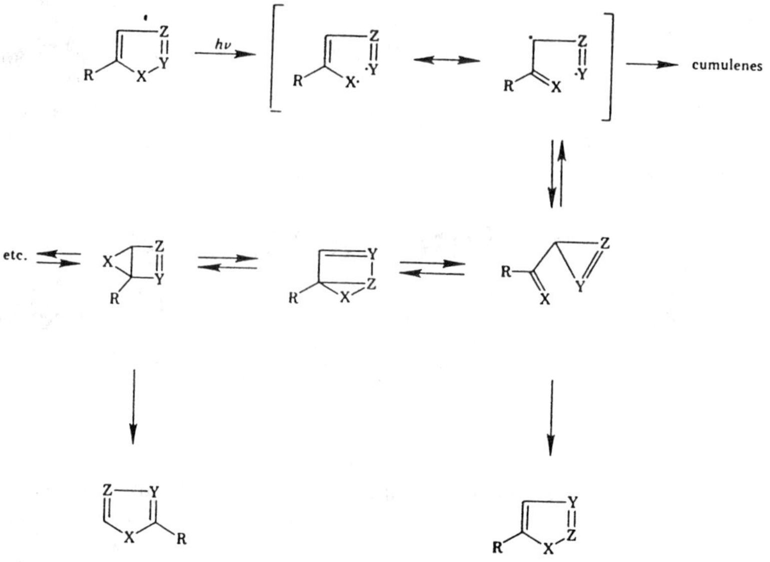

SCHEME 65.1

4. The Zwitterion-Tricyclic Route. This involves valence shell expansion of a sulfur atom. According to this mechanism, the sulfur *3d* orbitals are allowed to interact with the neighboring double bond to give a tricyclic Zwitterion, which subsequently collapses to the rearranged products (Scheme 65.2).

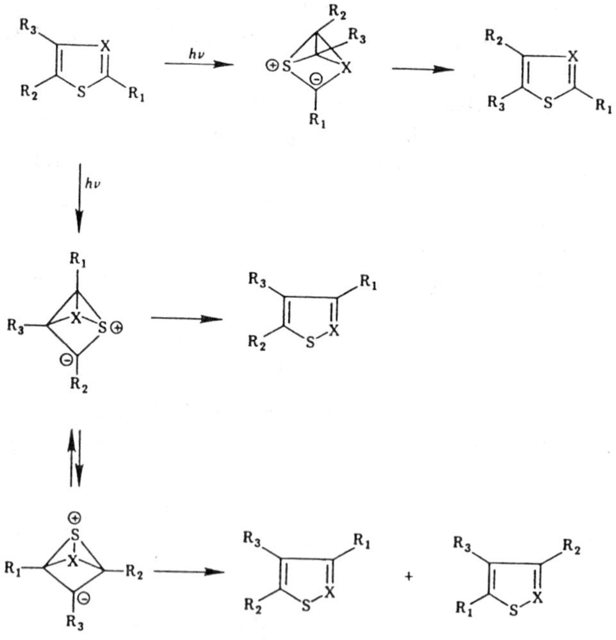

SCHEME 65.2

5. The fragmentation-readdition route

5.2 Thiophene, Alkylthiophenes, and Arylthiophenes

The pioneering and the most extensive studies in this field have been performed by Wynberg's group.[7] The mechanism of the photoisomerization of some compounds (e.g., some furans or isoxazoles[4]) has been proved unambiguously; but in the thiophene case, only Dewar thiophene itself and the Dewar derivative of some thiophenes substituted by electron-withdrawing groups have been characterized. Thus, the mechanism of photoisomerization of most thiophenes described remains speculative, even if intermediates have been trapped.[8-11]

Experimental Data

The photoisomerization of 2-arylthiophene to 3-arylthiophene seemed *a priori* to be a rather simple reaction,[12] and the following observations have been made.

1. The photorearrangement occurs within the thiophene ring and not the aryl ring.[13]
2. The aryl group remains linked to the same carbon atom of the thiophene ring during the photorearrangement. This has been proved by a ^{14}C labeling experiment on 2-phenylthiophene[14] (Scheme 65.3).

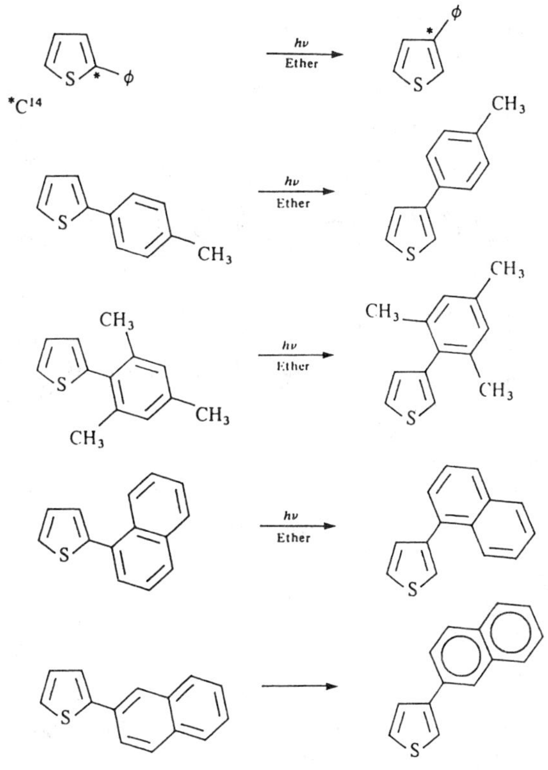

SCHEME 65.3

3. The data obtained from the photoisomerization of deuterium-labeled 2-phenylthiophene[15] and from methylphenylthiophenes[16] show that the thiophene ring photoisomerization is a rather complex reaction. Scrambling occurs at all possible ring positions of the isomerized 3-phenylthiophene, and the recovered starting material is also found to have undergone deuterium scrambling.[15,17]

4. The major path of photorearrangement of phenyl-, deuterio-, and methyl-substituted 2-phenylthiophenes involves an interchange of the C2-C3 atoms, without concomitant inversion of the C4-C5 carbon atoms.

5. Phenyl-, deuterio-, and methyl-substituted 3-phenylthiophenes exhibit considerable specificity of rearrangement, as shown in Scheme 65.4.

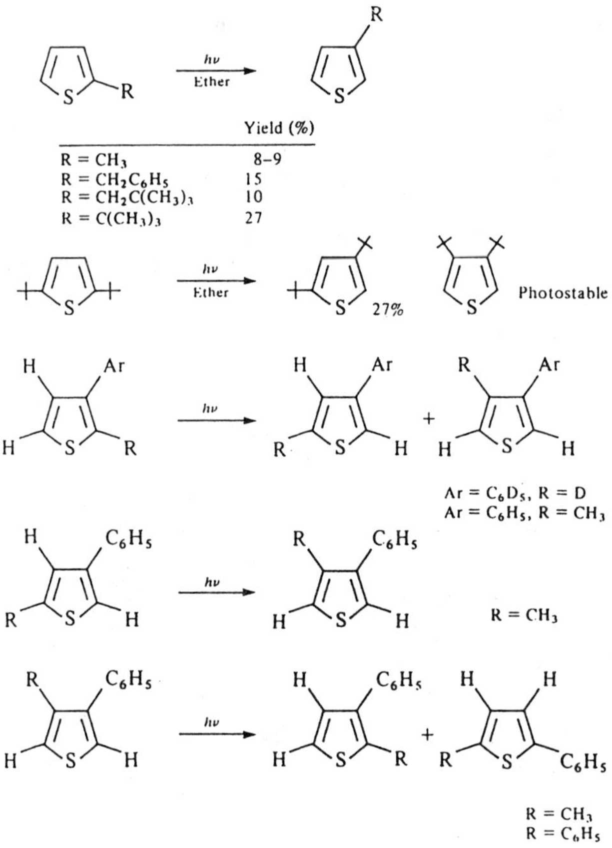

SCHEME 65.4

6. Some 2-alkylthiophenes also photoisomerize.[18]

Mechanisms

Numerous mechanisms have been postulated to rationalize the rearrangement patterns. The reactive excited state is the singlet in the case of 2-phenylthiophene.[19] The proposed mechanisms include the following.

1. Initial formation of a thiophene valence bond isomer,[20] analogous Dewar (1–1'), prismane (Ladenburg) (2–3'), or benzvalene (3–2') bond isomers of benzene.[21] This is not in accord with experimental data (Scheme 65.5).

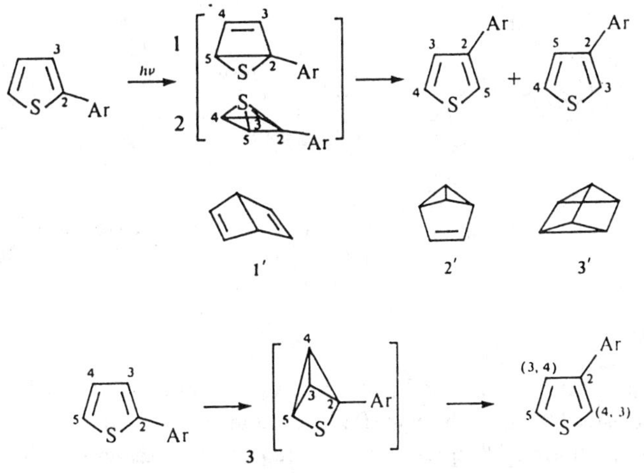

SCHEME 65.5

2. A ring contraction-ring expansion mechanism analogous to the one by which 3,5-diphenylisoxazole photoisomerizes to 2,5-diphenyloxazole[22] (Scheme 65.6). This mechanism can explain satisfactorily the major rearrangement paths of phenyl-, deuterio-, and methyl-substituted arylthiophenes. However, Wynberg rejected this mechanism on the assumption that, in the photorearrangement of 2-phenylthiophene to 3-phenylthiophene, the intermediate would be thioaldehyde 4 and not thioketone 5, which was, *a priori*, expected to be more stable than 4.[20] π-Bond order calculations lead, in fact, to the opposite conclusion.[11]

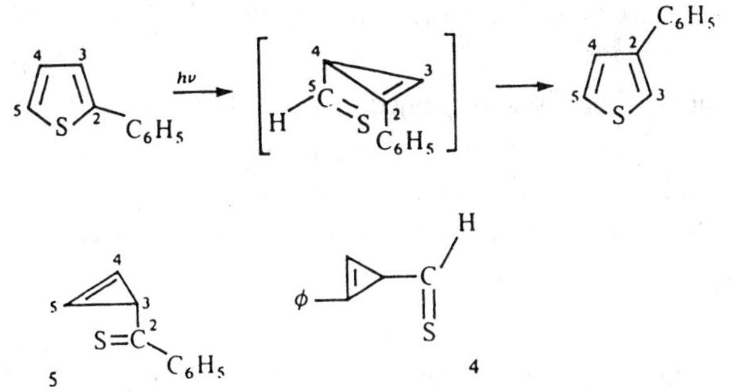

SCHEME 65.6

3. A Zwitterionic intermediate[20] (Scheme 65.7a). This mechanism, supported by Wynberg, fits well with the observed data on 2-arylthiophenes, but not for 3-arylthiophenes. 3-Phenylthiophene itself does not photoisomerize, but slowly decomposes. Wynberg proposed the intermediate shown in Scheme 65.7b to account for the experimental data on 3-arylthiophenes.

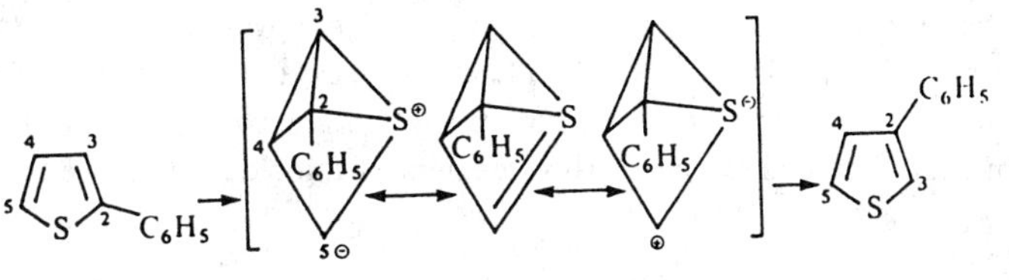

SCHEME 65.7a

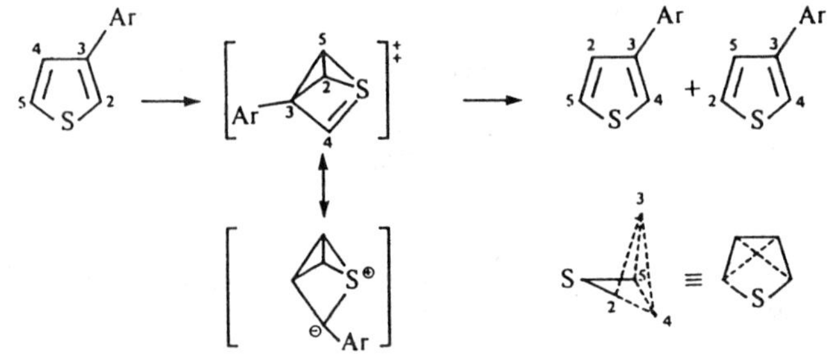

SCHEME 65.7b

4. Kellogg proposed another mechanism in which, within the intermediate, a two-atom fragment is twisted 90° out of the plane formed by the remaining three atoms 6 shown in scheme 7b.[23] Indeed, this might represent a stable geometrical situation on the singlet excited-state potential surface, from which isomerization is supposed to take place.[24] This mechanism also explains the observed phenylmethyl-photorearrangements, but it has no experimental or theoretical support.

5. It has been suggested that the difference in photochemical behavior between 2-phenylthiophene and 3-phenylthiophene is related to the electronic structure of the initial photoreactive states.[25] For 2-phenylthiophene, the lowest singlet excited state is delocalized over the whole π-system; whereas, in 3-phenylthiophene, the singlet excited state is mainly localized on the phenyl moiety. For 2-phenylthiophene, there is a crossing of π,π^* electronic states and the photoisomerization becomes irreversible; whereas, for 3-phenylthiophene, no crossing occurs and a "no-reaction reaction" is preferred.

6. Van Tamelen and Whitesides[26] proposed the mechanism in Scheme 65.8. This explains the majority of the results encountered, including those for 3-arylthiophenes, if the following assumptions are made:

  • Dewar structures and cyclopropenylthiocarbonyls are in equilibrium.
  • In the formation of a given cyclopropene, the ring contractions occur in such a fashion as to place the phenyl group on the double bond of the cyclopropene. Although the requirements of the second assumption are not always adhered to, it does account for the major products of the photoreactions.

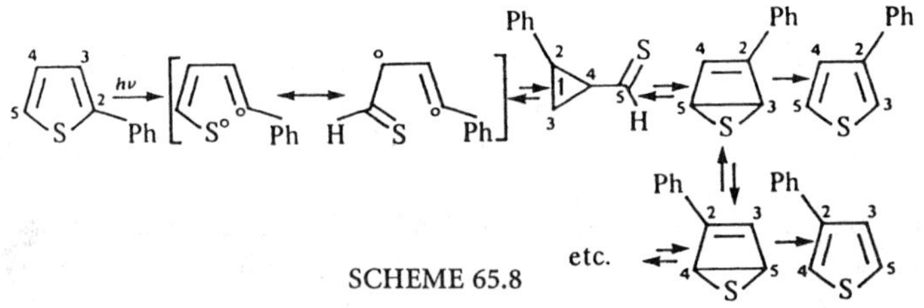

SCHEME 65.8

Ab initio SCF and CI calculations lead to the conclusion that the 3–5 and 2–4 transposition of carbon atoms of thiophene may proceed from the lowest singlet state $^1B_2$ of thiophene via Dewar thiophene.[27] The $^1B_2$ state of thiophene would easily convert to the biradical intermediate 7 in one step. The internal conversion of this species to the S_0 state would cause the transposition of the carbon atoms. 2-Phenylthiophene would photoisomerize by the same mechanism, whereas 3-phenylthiophene would prefer the ring contraction-ring expansion route in which 2–3 or 4–5 carbon atom transpositions can occur (involvement of thioaldehyde 4, see Scheme 6).

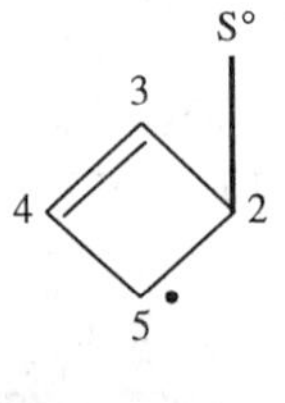

SINDO 1 calculations led to the conclusion that in the case of 2-cyanothiophene, the internal cyclization route is the more favorable. This has been shown experimentally to be exact, and that in the case of 2-methylthiophene, no clear decision in favor of the Zwitterion tricyclic mechanism can be made, and therefore that this mechanism cannot be excluded.[28]

Dewar thiophenes has been isolated and or trapped in the following cases:

1. **Thiophene.** When thiophene is irradiated at 229 nm in solution in furan, the two 1:1 adducts are formed (Scheme 65.9).[29] Dewar thiophene itself is obtained when the reaction is performed a −170 °C in a glassy matrix. When the photolysis is performed at 10K in an argon matrix, the compounds shown in Scheme 65.10 are obtained.[30] The formation of vinylacetylene, CS_2, and propyne correlates well with the previously reported gas-phase photolysis of thiophene.[31] Irradiation of the thiophene photolysate at longer wavelengths (254, 280, and even 320 nm) caused the disappearance of the Dewar thiophene and of the cyclopropene-3-thioaldehyde (note the formation of this compound) with a concomitant increase of the thiophene.

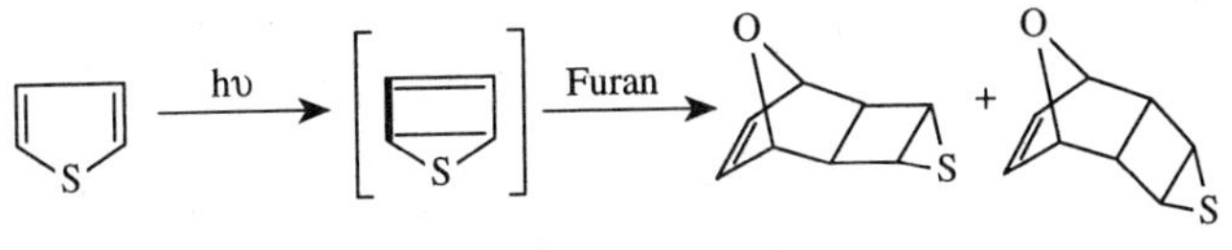

SCHEME 65.9

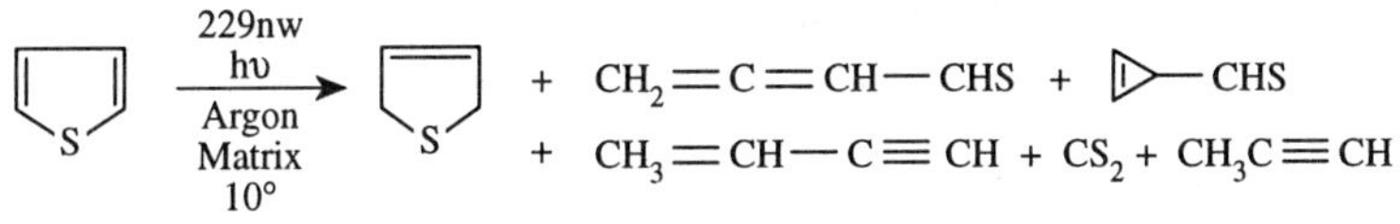

SCHEME 65.10

2. **2,3,5-Tetrakis(trifluoromethyl) thiophene.**[32–34] In general, electron-withdrawing groups stabilize the intermediate(s) in the photoisomerization of aromatic compounds.[5,35] Irradiation of 2,3,5-tetrakis(trifluoromethyl) thiophene yields the Dewar derivative 8 which has a half-life of 5.1 h in benzene at 160 °C. Furthermore, it undergoes Diels-Alder cycloaddition with a variety of dienes.[33–38]

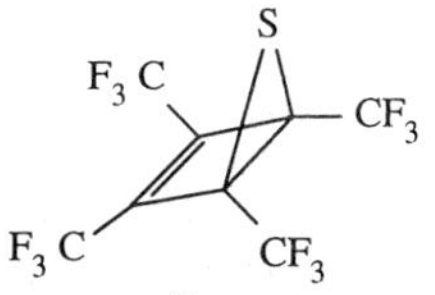

8

3. **Cyanothiophenes.**[39–40] In the case of 2- and 3-cyanothiophene, the Dewar thiophenes are not isolated, but have been trapped by furan. The isomerization of these cyanothiophenes is thought to proceed by the internal cyclization route[39] that arises from the S_1 state of the thiophene. Dewar thiophenes are isolated when methylcyanothiophenes are irradiated.[40] Their interconversions have been studied in considerable detail.

Irradiation of either 3-methyl-4-cyano- or 2-methyl-3-cyano-thiophene and furan in cyclohexane at room temperature afford the same Diels-Alder adducts. This result means that the same mixture of Dewar derivatives is formed from either starting material. Low-temperature studies have shown that the Dewar derivative 9 is detected when the reaction is performed at −68 °C in [$^2H_8$]-THF (see Scheme 65.11). Its half-life is 2 min at −35 °C and it converts into the isomeric species 10. A similar isomerization is observed for the conversion of 11 into 12.

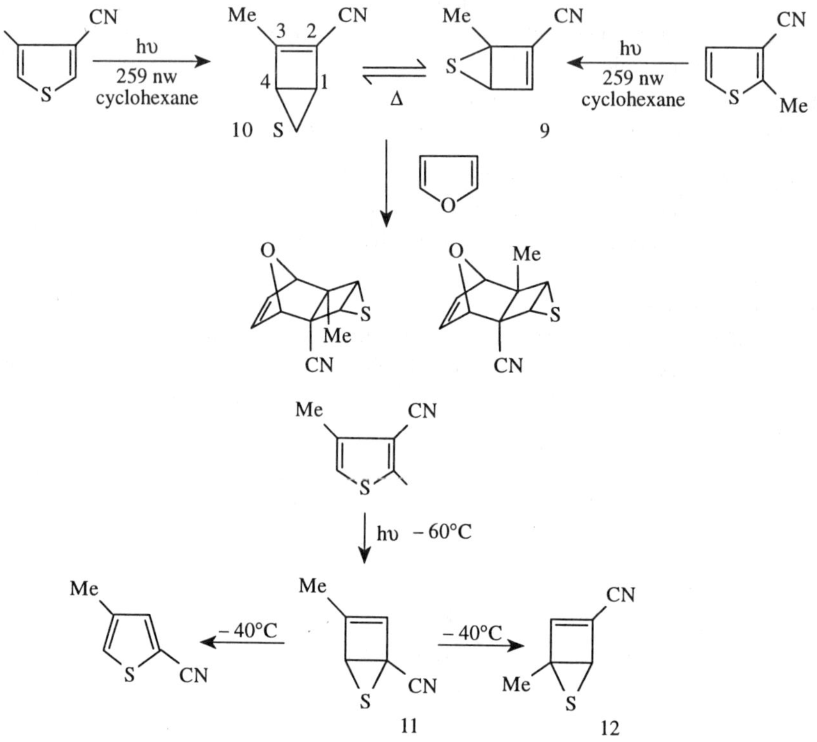

SCHEME 65.11

4. **2,3,4-Trifluoromethylthiophene.**[41] In this example, only one Dewar benzene is trapped by furan as the Diels-Alder adduct (Scheme 65.12).

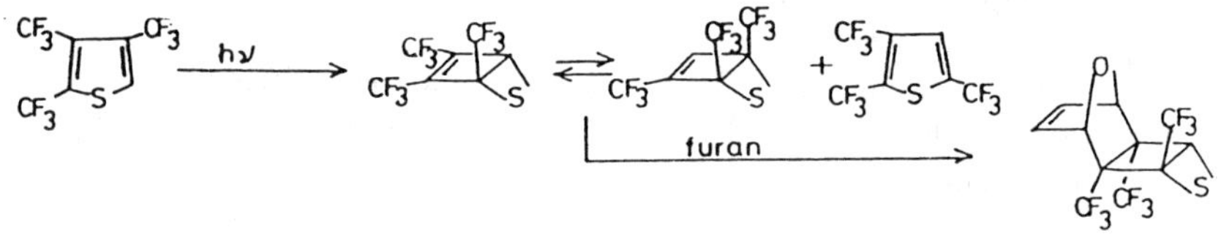

SCHEME 65.12

5. **2,3-Difluoromethylthiophene.**[42] This thiophene gives an equilibrium mixture of the 2,3- and the 3,4-*bis*(trifluoromethyl) Dewar thiophenes, together with isomerized thiophenes (Scheme 65.13).

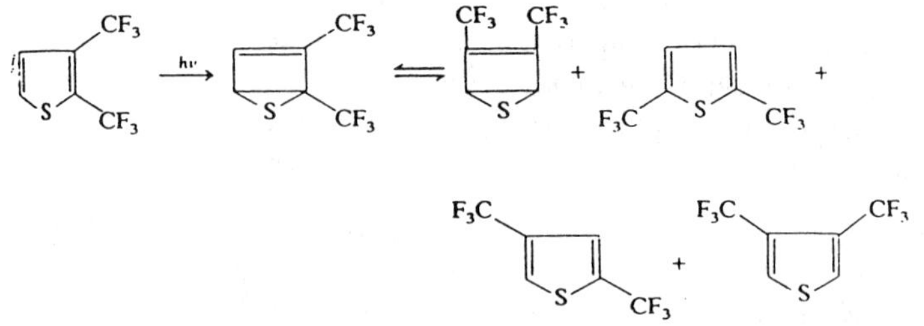

SCHEME 65.13

6. **2,5-*Bis*(trifluoromethyl)thiophene.** In contrast to the previous example, irradiation of 2,5-*bis*(trifluoromethyl)thiophene gave only 2,4-*bis*(trifluoromethyl)thiophene with no evidence for the formation of Dewar thiophene. There is no obvious pathway via a Dewar thiophene to the isomer isolated, and the alternative species **13** or **14** have been proposed as intermediates in this transformation (Scheme 65.14).

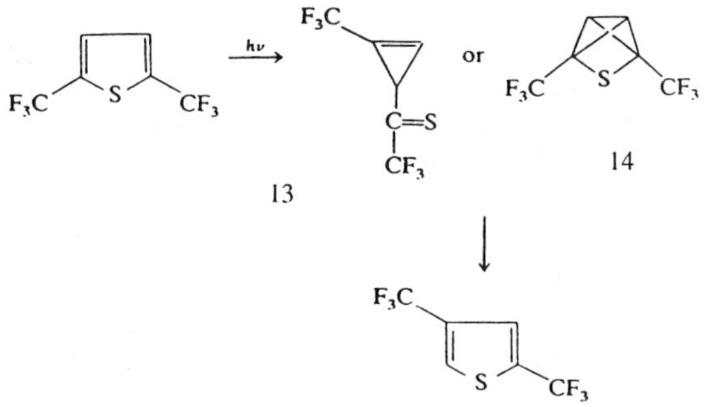

SCHEME 65.14

Trapping of Intermediates by Primary Amines

The intermediates formed by photorearrangement of thiophenes have been trapped by primary amines. Thus, irradiation of thiophene, the methylthiophenes, 2-phenylthiophene, and the dimethylthiophenes in amines brings about the conversion into pyrroles.[8,9] The corresponding furans are also photoconverted into the same pyrrole(s), suggesting that the intermediates have a similar structure in both cases.[10,11]

The structure of the pyrroles obtained from the irradiation of thiophenes in propylamine cannot be accounted for if it is assumed that the amine reacts on some Zwitterionic intermediate.[11] On the other hand, cyclopropenylthioketone or cyclopropenylthioaldehyde intermediates and Dewar thiophene intermediates can explain some of the experimental results. (Scheme 65.15). Thus, if it is assumed that the structure of the intermediate is either a cyclopropenylthioketone or cyclopropenylthioaldehyde, or a Dewar thiophene, depending upon the substituent(s) on the thiophene ring, all the experimental data can be explained. During their investigation of the reactivity of *tetrakis*(trifluoromethyl) Dewar thiophene **8**,[35,43] Kobayashi and co-workers showed that the Dewar thiophene **8**, upon heating, is converted to *N*-phenyl-2,3,4,5-*tetrakis*(trifluoromethyl)pyrrole on heating with aniline. Such a result implies that there is interconversion between this pyrrole and its Dewar isomer. These data support the hypotheses of Couture and Lablache-Combier about a possible reaction between a Dewar thiophene and a primary amine, leading to a pyrrole.[10,11] However, *N*-phenyl-cyclopropenylimine (**16**) is not converted either thermally or photochemically to a pyrrole.[43] This failure does not mean that a cyclopropenylimines cannot be intermediates in the phototransformation of thiophenes into pyrroles for the following reasons.

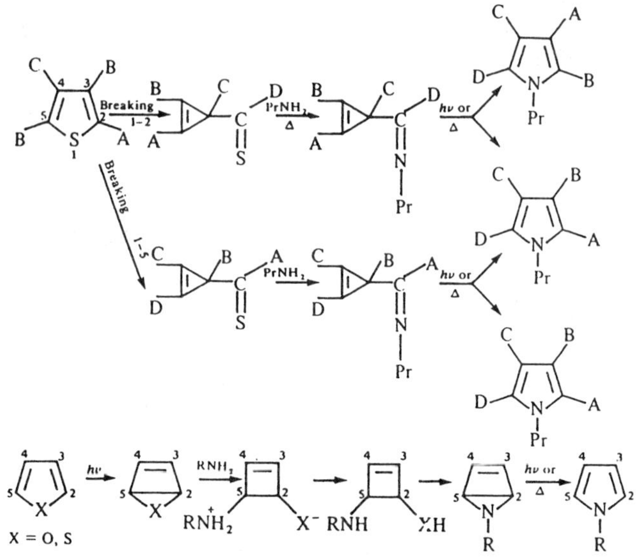

SCHEME 65.15

1. Substitution of a hydrogen by a trifluoromethyl group drastically changes the behavior and the stability of a molecule.[35]
2. Cyclopropenylketone 17 irradiated in *n*-butylamine is converted to a 2:1 mixture of *N*-butyl-2,5-dimethylpyrrole and *N*-butyl-2,4-dimethylpyrrole. The first step of this reaction is certainly imine formation. Some other cyclopropenylketones behave similarly.[44–46]
3. Imine 15 is converted to a pyrazole by heating or to an imidazole by UV irradiation.[47]

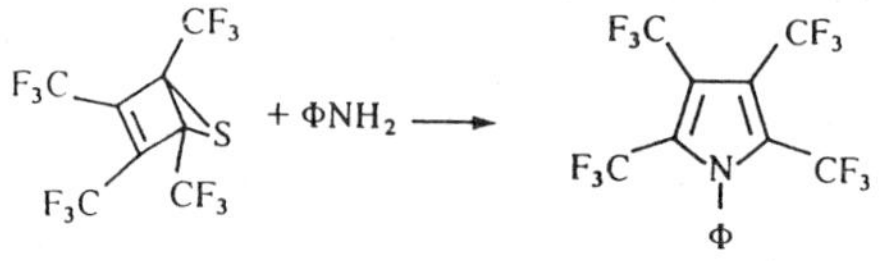

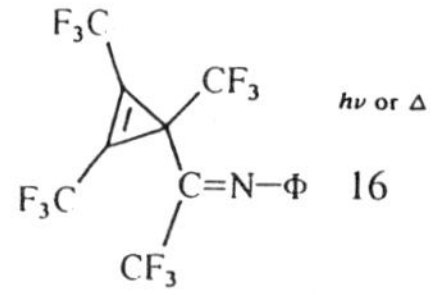

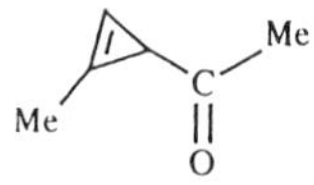

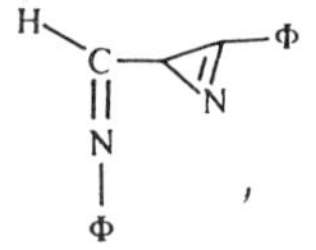

Irradiation of 2,5 diphenylisoxazole in propylamine leads to the formation of a pyrazole and an imine[48] (Scheme 65.16).

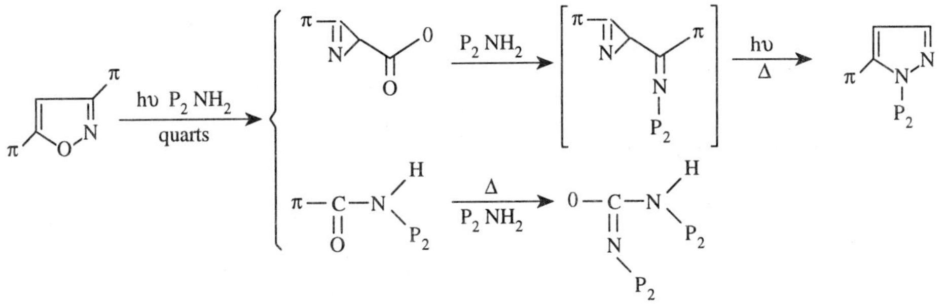

SCHEME 65.16

Heteroarylthiophenes

Photoisomerization of dithienyl compounds has been reported (see Scheme 65.17).[18] The mechanism suggested by Wynberg to take into account the structure of the product formed is completely different from those suggested for alkyl- or aryl-thiophene photorearrangements (see Scheme 65.18). However, 2(2-thienyl)furan rearranges analogously to 2-phenylthiophene. This reaction is dominated by the transformation of the thiophene ring since this incorporates the weakest C-X bond.[49] During a study of the photocyclization of terthiophenes, it was shown that a photorearrangement occurred prior to photocyclization. The likely mechanism is illustrated in Scheme 65.19.[50]

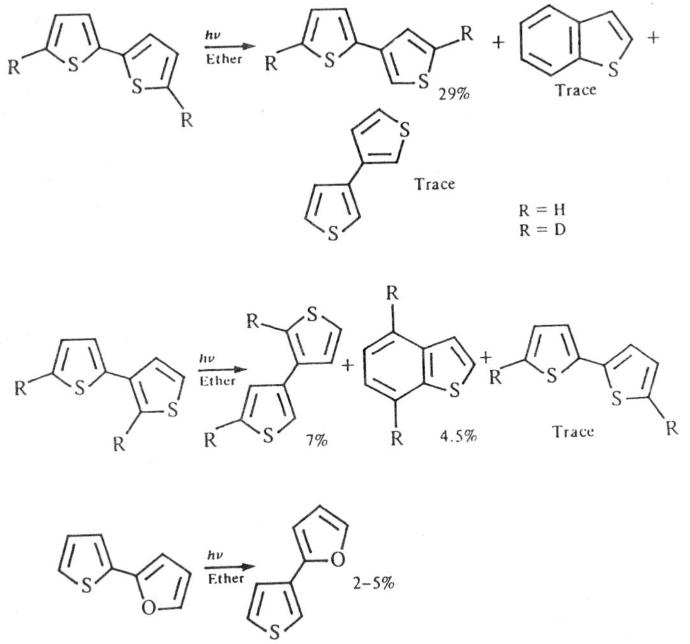

SCHEME 65.17

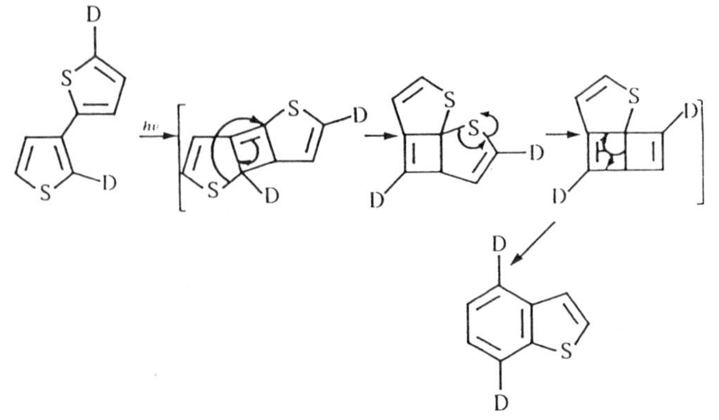

SCHEME 65.18

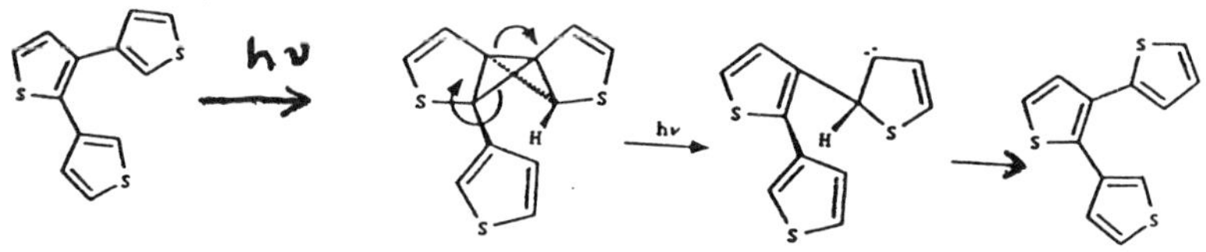

SCHEME 65.19

No reports on the photorearrangement of benzo[*b*]thiophene or other condensed thiophenes have appeared.

Conclusion

To conclude this discussion on the photoisomerization of thiophene and its alkyl or aryl derivatives, it can be said that the mechanism of this reaction is not yet proven clearly and may differ from one compound to the other. For certain derivatives, a ground-state cyclopropene-thioaldehyde or cyclopropene-thioketone, or a Dewar thiophene may be the correct intermediate. On the other hand, for some other derivatives no ground-state intermediate is involved and it has been suggested that a cyclopropene-thioketone or cyclopropene-thioaldehyde and a Dewar thiophene may be two equivalent graphical representations of the same excited-state hypersurface. (See Section 65.1 of this Chapter 82) A similar question has been asked regarding the photoisomerization of six-membered aromatic *N*-oxide derivatives where a ground-state oxaziridine intermediate may or may not be involved.[51]

65.3 Isothiazoles and Thiazoles

Isothiazole and thiazole photochemistry is an intriguing field. The main photoreaction is isomerization, and several suggestions closely related to those presented earlier in this chapter for thiophene photochemistry have been made regarding the mechanisms involved have been put forward. Isothiazole[52] and some methylisothiazoles have been irradiated. The product from isothiazole in either the presence or absence of propylamine was thiazole, and the methylisothiazoles rearranged as shown in Scheme 65.20. The mechanism is assumed to involve Zwitterionic intermediates analogous to those proposed by Wynberg for 2-phenylthiophene.[20] It was observed that the yield of isomerization products increased with increasing solvent polarity, a fact that supports the proposed mechanism.[53] Involvement of a Dewar-type intermediate does not provide a satisfactory explanation for the formation of **18** and **19** from **20**.

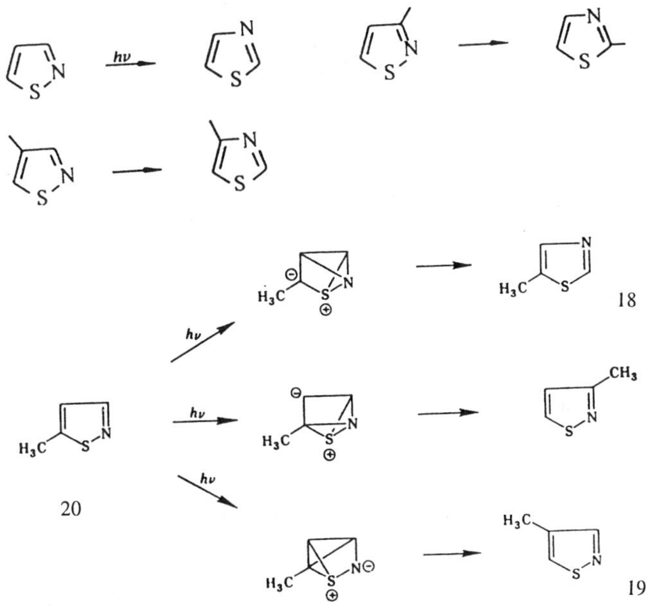

SCHEME 65.20

The photochemistry of phenylisothiazoles and phenylthiazoles has been intensively studied by Vernin et al.[54–60] and by others.[61–63] Some typical examples of the rearrangements encountered are given in Scheme 65.21. The results indicate the following order of ease of photoisomerization: 2-phenylthiazole, 5-phenylisothiazole > 5-phenylthiazole, 3-phenylisothiazole > 4-phenylthiazole, 4-phenylisothiazole. The rearrangement has been shown to take place only in the thiazole ring and, for example, 2-methylphenylthiazole derivatives rearrange without carbon transpositions in the phenyl groups.[54] Sensitizing and quenching experiments indicate that the reactive state in the case of 2-phenylthiazole is a singlet, whereas a triplet state is involved in the case of 3-phenylisothiazole.[60] The mechanism illustrated in Scheme 65.22 explains the majority of the experimental results obtained from thiazoles and isothiazole.[56,57]

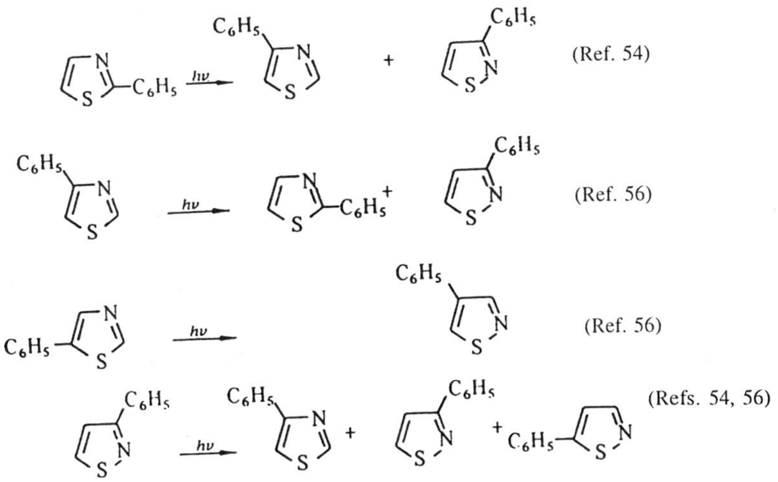

SCHEME 65.21

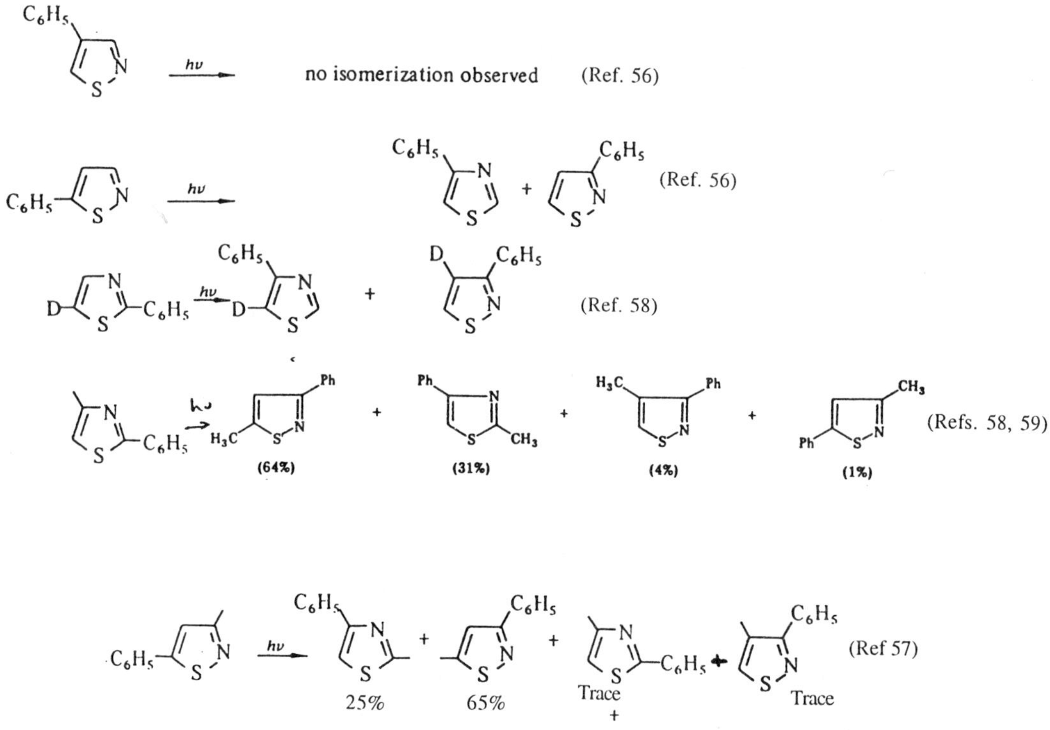

SCHEME 65.21 (continued)

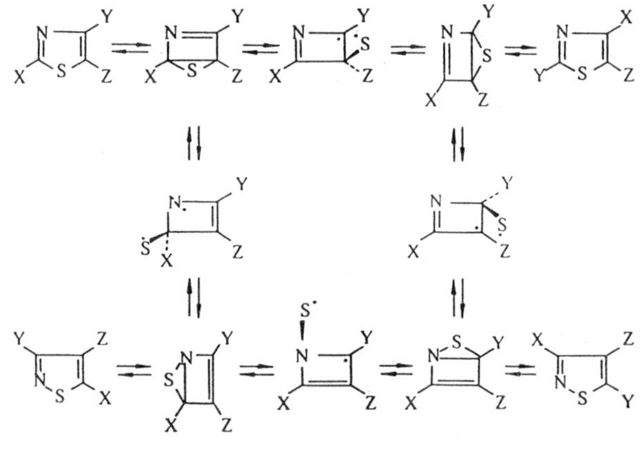

SCHEME 65.22

It is noteworthy that the major products are those derived from bicyclic intermediates in which the phenyl groups are conjugated with the double bond.[55,56] π-Bond-order Hückel calculations confirm that such intermediates are more probable.[56] The bond that is easiest to break is presumably the one for which the π-bond order in the excited state is lowest. Further mechanistic information comes from incorporation of deuterium during the photorearrangement of 2-phenylthiazole and 4-phenylthiazole in benzene containing 2% D_2O.[61,62,64] These observations appear to favor a tricyclic Zwitterionic intermediate in which deuterium incorporation is possible rather than a valence-bond isomerization mechanism. Thus, the following two mechanisms have been proposed in the case of thiazole (Scheme 65.23).[65] One (type A) involves exchange of positions 2 and 4 or 3 and 5; the other (type B) involves interchange of positions 2 and 3 together with exchange of 4 and 5. Tricyclic sulphonium ions and bicyclic species are postulated as intermediates in these transformations (Scheme 65.24).

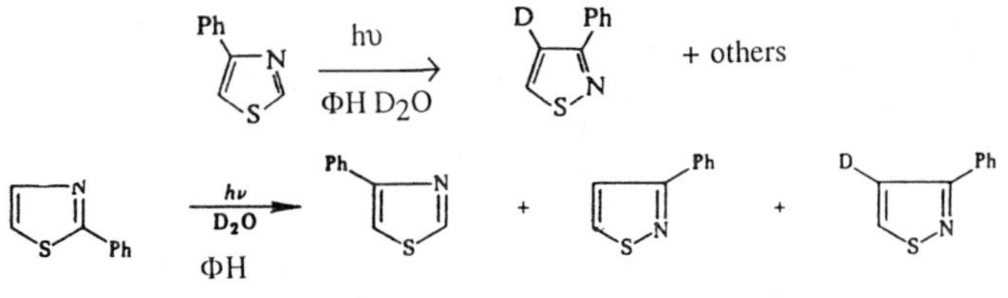

SCHEME 65.23

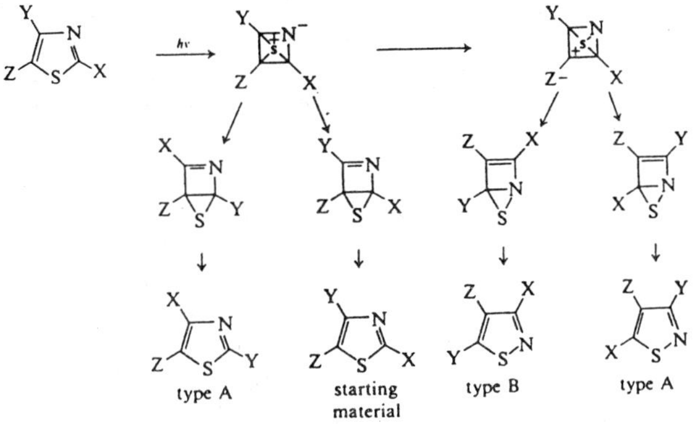

SCHEME 65.24

Irradiation of a 1:1 complex of peplomycin (PEM) (21) and Cu(II) results in isomerization of 2,4′-bithiazole ring to produce photo-PEM 22 accompanied by a small amount of 24.[66,67] Irradiation (302 nm) of 1:5 complex of PEM and Cu(II) enhanced the formation of 24 (30% yield) (Scheme 65.25).[68]

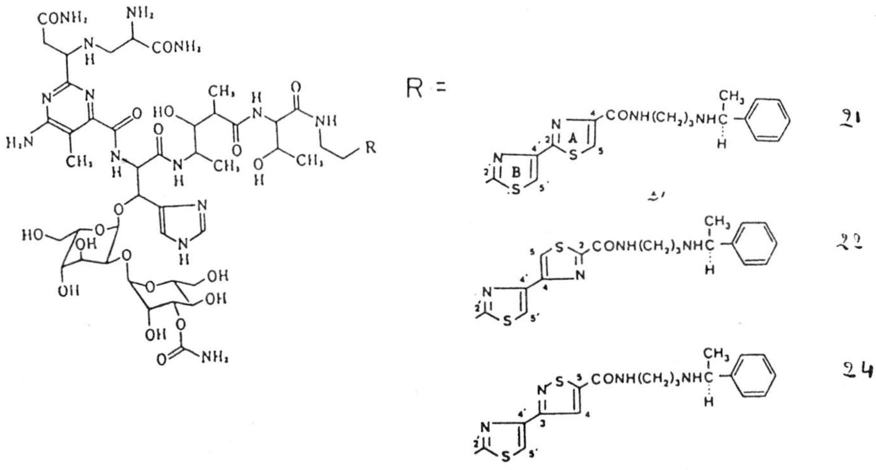

SCHEME 65.25

The photochemistry of the model compounds 25 and 26, in which the former contains the photoactive bithiazole chromophore, has been examined.[66,68] With the bithiazole 25, two rearrangement products are formed, and the isomerization in both cases is confined to the thiazole ring bearing the carboethoxy group. The major of the two products was assigned structure **27** and arises by the same rearrangement as occurs in the bleomycin antibiotics, while the minor product, assigned structure **28**, has undergone a skeletal rearrangement to produce an isothiazole ring. On the basis of the failure to quench the reaction with piperylene or to sensitize it with acetophenone or benzophenone, it was concluded that the rearrangement arises from the singlet excited state. On direct irradiation, the quantum yields of formation of **27** and **28** were 0.023 and 0.0005, respectively. The phenyl-substituted analogue **26** reacted similarly.

In the trithiazole **29**, in contrast to dithiazole **25** and phenylthiazole **26**, the thiazole ring B undergoes regiospecific photoisomerization while ring A is unaffected.[69] Dewar isomers are proposed as intermediates to explain the photochemical formation of **30** from **29**.[69] The authors[69] argue that the rearrangement proceeds by migration of sulfur in the Dewar intermediates **31** and **32** and made use of the calculated HOMO and LUMO charge densities in the three thiazole rings of **29** to explain why the rearrangement is selective for the central ring. The authors[69] acknowledge, however, that the rearrangement may also proceed via the ring contraction-ring expansion mechanism described previously, although this would not explain why it is so selective for the central ring of **29**.

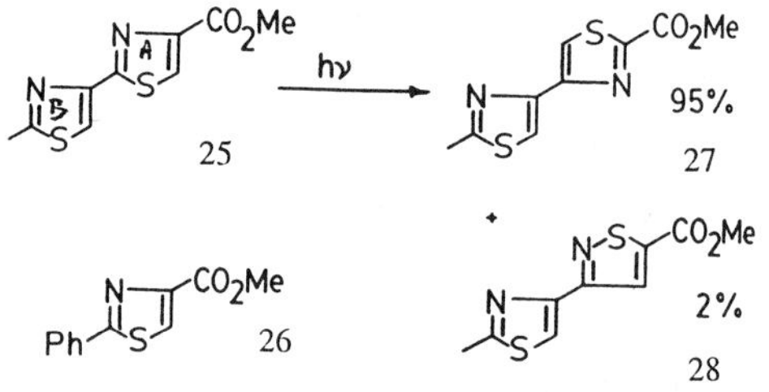

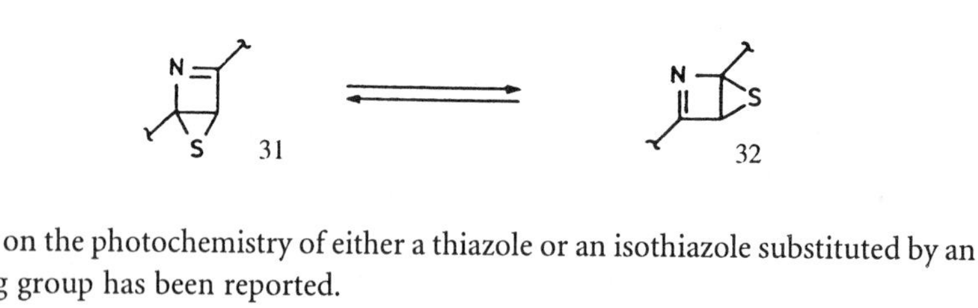

No study on the photochemistry of either a thiazole or an isothiazole substituted by an electron-withdrawing group has been reported.

65.4 Aminothiazolium Salts

The first step of the photodegradation reaction of 4-aminothiazolium salts is suggested to be the formation (Scheme 65.26) of a Dewar-type intermediate which then is desulfurized leading to a 2-imino-1,2-dihydroazete intermediate.[70] Ring opening also occurs during the irradiation of the salt shown in Scheme 65.27.[71]

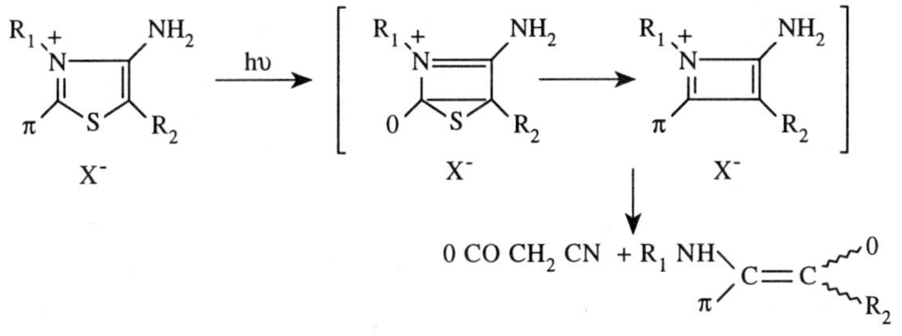

SCHEME 65.26

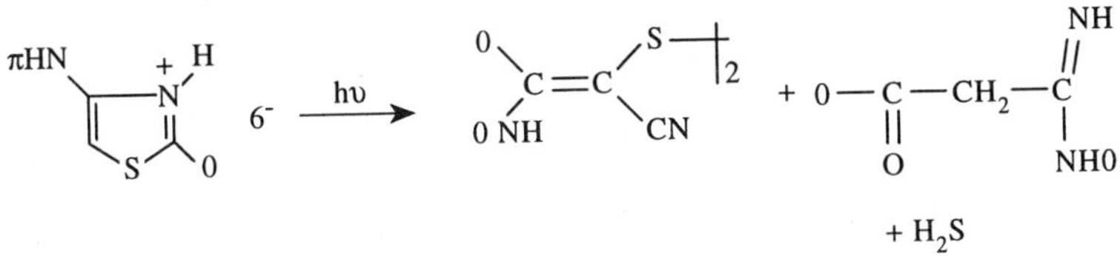

SCHEME 65.27

5 Benzoisothiazoles

The S-N bond undergoes photocleavage during the irradiation of benzo[*d*]isothiazole. Benzo[*c*]isothiazole undergoes similar fission and yields gave benzophenone, presumably via the corresponding thione, and a dibenzodiazocine (Scheme 65.28).[72] A flash photolytic study[73] has shown that the photochemical additions of benzo[*c*]isothiazole to alkenes and alkynes proceed by direct 1,2-addition of the alkene or alkyne to the N-S bond of the photoexcited isothiazole. A biradical species is not an intermediate as had been proposed initially.[74] N-S bond cleavage is responsible[75] for the major product formed in the photochemical reactions of 3-chloro- or 3-methoxy-2,1-benzoisothiazole. The hydrogens that appear in the amino group in the product are probably abstracted from solvent (Scheme 65.29).

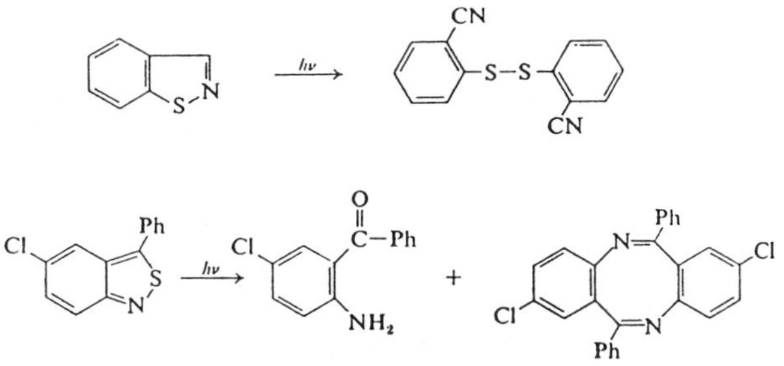

SCHEME 65.28

(X = Cl, OMe)

SCHEME 65.29

Sunlight-induced ring cleavage of thieno[2,3-*c*]isothiazoles to give thiophenes with accompanying loss of elemental sulfur is believed to involve initial N-S bond homolysis (Scheme 65.30).[76] The first step of the photoreactions of 1,2,3-thiadiazoles or 1,2,3-benzothiadiazoles is the loss of nitrogen and the formation of thiirenes.[77–79] In the case of 5-phenyl-1,2,3,4-thiatriazole, the stable phenylthiazirine is formed as the primary product obtained after loss of N_2.[80]

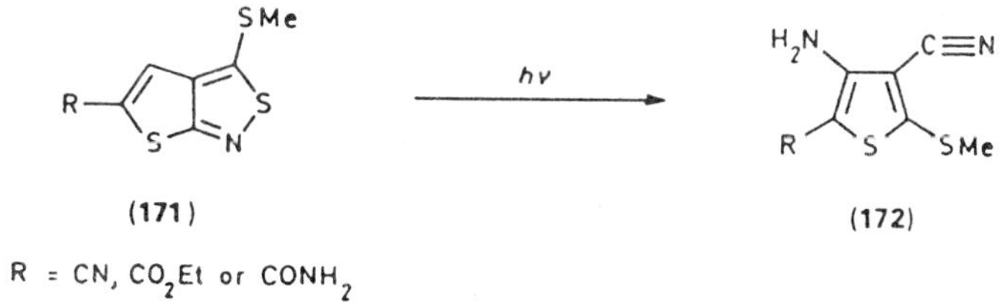

(171)   (172)

R = CN, CO$_2$Et or CONH$_2$

SCHEME 65.30

References

1. Lablache-Combier, A. and Remy, M. A., Photoisomerisations des dérivés hétérocycliques aromatiques à cinq chainons, *Bull. Soc. Chim. Fr.,* 679, 1971.
2. Lablache-Combier, A., Photoisomérisation de composés aromatiques B. Composés à cinq chainons, *L'Actualité Chimique,* 9, Déc. 1973.
3. Lablache-Combier, A., Photoisomerization of five membered heterocyclic compounds, in *Photochemistry of Heterocyclic Compounds,* Buchardt O., Ed., John Wiley & Sons, New York, 1976, 123.
4. Padwa, A., Photochemical rearrangements of five membered ring heterocycles, in *Rearrangements in Ground and Excited States,* Vol. 3, de Mayo, P., Ed., Academic Press, New York, 1980, 501.
5. Zupan, N. and Sket, B., Photochemistry of fluoro substituted aromatic and heteroaromatic molecules, *Isr. J. Chem.,* 17, 92, 1978.
6. Lablache-Combier, A., Photochemical reactions of thiophenes, in *Chemistry of Heterocyclic Compounds Volume 44 Part 1: Thiophene and its Derivatives,* Gronowitz, Ed., John Wiley & Sons, New York, 1985, 745.
7. Wynberg, H., Some observations on the chemical, photochemical and spectral properties of thiophenes, *Acc. Chem. Res.,* 4, 65, 1971.
8. Couture, A. and Lablache-Combier, A., Thiophene photochemistry, *J. Chem. Soc., Chem. Commun.,* 524, 1969.
9. Couture, A. and Lablache-Combier, A., Photochimie du thiophene, *Tetrahedron,* 27, 1059, 1971.
10. Couture, A. and Lablache-Combier, A., Analogy between furan and thiophene photochemistry, *J. Chem. Soc., Chem. Commun.,* 891, 1971.
11. Couture, A., Delevallée, A., Lablache-Combier, A., and Parkanyi, C., Etude comparative des photoréactions de thiophènes et de furannes avec la propylamine, *Tetrahedron,* 31, 785, 1975.
12. Wynberg, H. and Van Driel, H., The photochemical rearrangement or arylthiophenes, *J. Am. Chem. Soc.,* 87, 3998, 1965.
13. Wynberg, H., Van Driel, H., Kellogg, R. M., and Buter, J., Observation on the scope of arylthiophene rearrangements, *J. Am. Chem. Soc.,* 89, 3487, 1967.
14. Wynberg, H. and Van Driel, H., Further evidence for scrambling in the photochemical rearrangement of 2-phenylthiophene, *J. Chem. Soc., Chem. Commun.,* 203, 1966.

15. Kellogg, R. M. and Wynberg, H., Investigation of phenylthiophene photorearrangements by deuterium labeling techniques, *J. Am. Chem. Soc.*, 89, 3495, 1967.

16. Wynberg, H., Beekhuis, G. E., Van Driel, H., and Kellogg, R. M., Photorearrangements of phenylmethylthiophenes, *J. Am. Chem. Soc.*, 89, 3498, 1967.

17. Wynberg, H., Kellogg, R. M., Van Driel, H., and Beekhuis, G. E., Photochemical photorearrangements of arylthiophenes, *J. Am. Chem. Soc.*, 88, 5047, 1966.

18. Kellogg, R. M., Dik, J. K., Van Driel, H., and Wynberg, H., Rearrangements of alkylthiophenes and the dithienyls, *J. Org. Chem.*, 35, 2737, 1970.

19. Kellogg, R. M. and Wynberg, H., Quenching of an excited singlet state of 2-phenylthiophene, *Tetrahedron Lett.*, 5895, 1968.

20. Wynberg, H., Kellogg, R. M., Van Driel, H., and Beekhuis, G. E., Observation on the mechanism of arylthiophene rearrangements, *J. Am. Chem. Soc.*, 89, 3501, 1967.

21. Bryce-Smith, D. and Gilbert, A., The organic photochemistry of benzene, *Tetrahedron*, 32, 1309, 1976.

22. Ullman, E. F. and Singh, B., Photochemical transposition of ring atoms in five membered heterocycles. The photorearrangement of 3,5-diphenylisoxazole, *J. Am. Chem. Soc.*, 88, 1844, 1966.

23. Kellogg, R. M., A model for the photochemically induced valence bond isomerization of thiophenes, *Tetrahedron Lett.*, 1429, 1972.

24. Epiotis, N. D., *The Theory of Organic Reactions*, Springer Verlag, Berlin, 1978.

25. Mehlhorn, A., Fratev, F., and Monev, V., Low energy excited singlet states of some monophenyl substituted 5-membered heterocycles and their photoisomerization, *Tetrahedron*, 37, 3627, 1981.

26. Van Tamelen, E. E. and Whitesides, T. H., Valence tautomers of heterocyclic aromatic species, *J. Am. Chem. Soc.*, 93, 6129, 1971.

27. Matsushita, T., Tanaka, H., and Nishimoto, K., A theoretical study on the photoisomerization of thiophene, *Theor. Chim. Acta*, 63, 55, 1983.

28. Jug, K. and Schluff, H. P., SINDO 1, Study of the photoisomerization mechanisms of thiophenes, *J. Org. Chem.*, 56, 129, 1991.

29. Rendall, W. A., Torres, M., and Strausz, O. P., Dewar thiophene: Its generation and trapping with furan, *J. Am. Chem. Soc.*, 107, 723, 1985.

30. Rendall, W. A., Clement, A., Torres, M., and Strausz, O. P., Dewar furan and Dewar thiophene: Low-temperature matrix photolysis of furan and thiophene, *J. Am. Chem. Soc.*, 108, 1691, 1986.

31. Wiebe, H. A. and Heicklen, J., Photolysis of thiophene vapor, *Can. J. Chem.*, 47, 2965, 1969.

32. Wiebe, H. A., Braslavsky, S., and Heicklen, J., The photolysis of tetra (trifluoromethyl)thiophene vapor, *Can. J. Chem.*, 50, 2721, 1972.

33. Kobayashi, Y., Kumadaki, I., Ohsawa, A., and Sekine, Y., Derivation of oxahomocubanes from Dewar thiophene, *Tetrahedron Lett.*, 2841, 1974.

34. Kobayashi, Y., Kumadaki, I., Ohsawa, A., Sekine, Y., and Mochizuki, H., Synthesis and reactions of Dewar thiophene, *Chem. Pharm. Bull.*, 23, 2773, 1975.

35. Kobayashi, Y., and Itumaro, K., Valence bond isomers of aromatic compounds stabilized by trifluoromethyl groups, *Acc. Chem. Res.*, 14, 76, 1981.

36. Kobayashi, Y., Kumadaki, I., Ohsawa, A., and Sekine, Y., Isomerization of Dewar thiophene catalysed by phosphorus compounds, *Tetrahedron Lett.*, 1639, 1975.

37. Kobayashi, Y., Kumadaki, I., Ohsawa, A., Sekine, Y., and Ando, A., Reactions of *tetrakis*(trifluoromethyl) Dewar thiophene, *Heterocycles*, 6, 1587, 1977.

38. Kobayashi, Y., Kumadaki, I., Ohsawa, A., Sekine, Y., and Ando, A., Diels Alder reactions of 1,2,3,4-*tetrakis*(trifluoromethyl)-5-thiabicyclopent-2-ene, *J. Chem. Soc., Perkin Trans. 1*, 2355, 1977.

39. Barltrop, J. A. and Day, A. C., Phototransformations of cyanothiophenes: Permutation Pattern Analysis and the chemical trapping of an intermediate 5-thiobicyclo-pent-2-ene, *J. Chem. Soc., Chem. Commun.*, 881, 1979.

40. Barltrop, J. A., Day, A. C., and Irving, E., Cyano substituted 5-thiabicyclopent-2-enes: Reactions and relevance to cyanothiophene phototranspositions, *J. Chem. Soc., Chem. Commun.*, 966, 1979.

41. Kobayashi, Y., Kawada, K., Ando, A., and Kumadaki, I., Photoisomerization of trifluoromethylated thiophenes, *Heterocycles,* 20, 174, 1983.

42. Kobayashi, Y., Kawada, K., Ando, A., and Kumadaki, I., Photoisomerization of *bis*(trifluoromethyl)-thiophenes, *Tetrahedron Lett.,* 25, 1917, 1984.

43. Kobayashi, Y., Ando, A., Kawada, K., and Kumadaki, I., Some reactions of valence-bond isomers of *tetrakis*(trifluoromethyl)pyrroles, *J. Org. Chem.,* 45, 2968, 1980.

44. Tsuchiya, T., Arai, H., and Igeta, H., Photolysis of pyridazine N-oxides: Formation of cyclopropenylketones, *J. Chem. Soc., Chem. Commun.,* 550, 1972.

45. Tsuchiya, T., Arai, H., and Igeta, H., A photochemical preparation of N-substituted pyrroles from pyridazine N-oxides, *Chem. Pharm. Bull.,* 29, 1516, 1973.

46. Tsuchiya, T., Arai, H., and Igeta, H., Formation of cyclopropenylketones and furans from pyridazine N-oxides by irradiation, *Tetrahedron,* 29, 2747, 1973.

47. Padwa, A., Smolanoff, J., Tremper, A., Intramolecular photochemical and thermal cyclization reactions of 2-vinyl substituted 2H-azirines, *Tetrahedron Lett.,* 29, 1974.

48. Delevallée, A. and Lablache-Combier, A., unpublished results.

49. Wynberg, H., Sinningc, H. J., and Creemers, H. M. J. C., The thienyl furans, *J. Org. Chem.,* 36, 1011, 1971.

50. Jayasuriya, N., Kagan, J., Owens, J. E., Kornak, E. P., and Perrine, D. M., Photocyclization of terthiophenes, *J. Org. Chem.,* 54, 4203, 1989.

51. Lablache-Combier, A., Photoisomerization of six-membered heterocyclic compounds, in *Photochemistry of Heterocyclic Compounds,* Buchardt, O., Ed., John Wiley & Sons, New York, 1976, 207.

52. Catteau, J. P., Lablache-Combier, A., and Pollet, A., Isothiazole photoisomerization, *J. Chem. Soc., Chem. Commun.,* 1018, 1969.

53. Lablache-Combier, A. and Pollet, A., Photoreactions des methylisothiazoles, *Tetrahedron,* 28, 3141, 1972.

54. (a) Vernin, G., Dou, H. J. M., and Metzger, J., Photoisomérisation des aryl-2 thiazoles, *Compt. Rend. Acad. Sci. Paris,* 271, 1616, 1970; (b) Vernin, G., Poite, J. C., Metzger, J., Aune, J. P., and Dou, H. J. M., Isomérization photochimique des phénylthiazoles et phénylisothiazoles, *Bull. Soc. Chim. Fr.,* 1103, 1971.

55. Vernin, G., Jauffred, R., Richard, C., Dou, H. J. M., and Metzger, J., Reaction of thiazoyl-2-yl and benzothiazol-2-yl radicals with aromatic compounds, *J. Chem. Soc., Perkin Trans. 2,* 1145, 1972.

56. Vernin, G., Riou, C., Dou, H. J. M., Bouscasse, L., Metzger, J., and Loridan, G., Transposition photochimique en série hétérocyclique. Arylthiazoles et isothiazoles isomères, *Bull. Soc. Chim. Fr.,* 1743, 1973.

57. Riou, C., Poite, J. C., Vernin, G., and Metzger, J., Les réactions de transposition photochimique en série hétérocyclique. IV. Photoisomérisation des phénylméthylthiazoles et isothiazoles isomères, *Tetrahedron,* 30, 879, 1974.

58. Riou, C., Vernin, G., Dou, H. J. M., and Metzger, J., Les réactions de transposition photochimique en série hétérocyclique. II. Photoisomérisation du phényl-2-deutéro-5-thiazole et du phényl-2-méthyl-4-thiazole. Influence des solvants et d'additifs divers sur la photoisomérisation du phényl-2-thiazole. Résultats expérimentaux, *Bull. Soc. Chim. Fr.,* 2673, 1972.

59. Vernin, G., Poite, J. C., Dou, H. J. M., Metzger, J., and Vernin, G., Etude des radicaux methyl-3-isothiazolyl-5 en série aromatique, *Bull. Soc. Chim. Fr.,* 3157, 1972.

60. Vernin, G., Treppendahl, S., and Metzger, J., Etudes photochimiques en série hétérocyclique. V. Mise en évidence d'oxazoles par photooxydation de thiazoles, *Helv. Chim. Acta,* 60, 284, 1977.

61. Kojima, M. and Maeda, M., Photorearrangements of 2,5- and 2,4-diphenylthiazole, *J. Chem. Soc., Chem. Commun.,* 386, 1970.

62. Maeda, M. and Kojima, M., Mechanism of the photorearrangements of phenylthiazoles, *Tetrahedron Lett.,* 3523, 1973.

63. Ohashi, N., Io, A., and Yonezawa, T., *J. Chem. Soc., Chem. Commun.,* 1148, 1970.

64. Maeda, M., Kawahara, A., Kai, M., and Kojima, M., Reaction pathway in the photorearrangements of phenylisothiazoles, *Heterocycles*, 3, 389, 1975.

65. Maeda, M. and Kojima, M., Mechanism of the photorearrangements of phenylisothiazoles, *J. Chem. Soc., Perkin Trans. 1*, 685, 1978.

66. Morii, T., Matsuura, T., Saito, I., Suzuki, T., Kuwahara, J., and Sugiura, Y., Phototransformed bleomycin antibiotics. Structure and DNA cleavage activity, *J. Am. Chem. Soc.*, 108, 7089, 1986.

67. Morii, T., Saito, I., Matsuura, T., Kuwahara, J., and Sugiura, Y., New lumi bleomycin containing thiazolylisothiazole ring, *J. Am. Chem. Soc.*, 109, 938, 1987.

68. Saito, I., Morii, T., Okumura, Y., Mori, S., Yamaguchi, K., and Matsuura, T., Ring selective photorearrangement of bithiazoles, *Tetrahedron Lett.*, 27, 6385, 1986.

69. Saito, I., Morii, T., Yamaguchi, K., and Matsuura, T., Remarkably high selectivity in photoisomerization of trithiazoles, *Tetrahedron Lett.*, 29, 3963, 1988.

70. Kato, H., Wakao, K., Yamada, A., and Mutoh, Y., Products via transient quasi antiaromatic azetine intermediates generated by desulfurisation of photochemical intermediates from a thiazolium-4-olate and 4-aminothiazolium salts, *J. Chem. Soc., Perkin Trans. 1*, 189, 1988.

71. Chinone, A., Huseya, Y., and Ohta, M., The photochemical reactions of several mesoionic compounds, *Bull. Chem. Soc. Jpn.*, 43, 2650, 1970.

72. Ohashi, M., Ezaki, A., and Yonezawa, T., Photodecomposition of benzoisothiazoles, *J. Chem. Soc., Chem. Commun.*, 617, 1974.

73. Neckers, D. C., Sindler-Kulyk, M., Scaiano, J. C., Stewart, L. C., and Weiz, D., Excited properties of 2-phenylbenzothiazole and 3-phenyl-1,2-benzothiazole, *J. Photochem.*, 39, 59, 1987.

74. Sindler-Kulyk, M., Neckers, D. C., and Blount, J. R., Photocycloaddition of 2-phenyl-1, 2-benzoisothiazole, *Tetrahedron*, 37, 3377, 1981.

75. Jackson, B., Schmid, H., and Hansen, H. J., Einige Bestrahlungs experiments mit 2, 1-benzoisothiazoles, *Helv. Chim. Acta*, 62, 391, 1979.

76. Corsaro, A., Guerrera, F., Sarva, M. C., and Siracusa, M. A., The light induced conversion of thieno[2,3-*c*]isothiazoles into thiophenes, *Heterocycles*, 27, 2539, 1988.

77. White, R. C., Scoby, J., and Roberts, T. D., Reduction as a probe for benzothiirene intermediates. Thermal and photochemical decomposition of 1,2,3-benzothiadiazoles, *Tetrahedron Lett.*, 2785, 1979.

78. Font, J., Torres, M., Gunning, H. E., and Strausz, O. P., Gas phase photolysis of 1,2,3-thiadiazoles: Evidence for thiirene intermediates, *J. Org. Chem.*, 43, 2487, 1978.

79. Krantz, A. and Laureni, J., Characterization of matrix-isolated antiaromatic three membered heterocycles. Preparation of the elusive thiirene molecule, *J. Am. Chem. Soc.*, 103, 486, 1981.

80. Holm, A., Harrit, N., and Trabjerg, I. B., Strong evidence for thiazirines as stable intermediates at cryogenic temperatures in the photolytic formation of nitrile sulphides from aryl-substituted 1,2,3,4-thiatriazole, thiatriazole-3-oxide, and 1,3,4-oxathiazol-2-one, *J. Chem. Soc., Perkin Trans. 1*, 746, 1978.

66

E,Z-Isomerization of Imines, Oximes, and Azo Compounds

Hiroshi Suginome
Hokkaido University

66.1 A Short Historical Background

Imines, oximes, and azo compounds having either the nitrogen-carbon or nitrogen-nitrogen double bond in the Z- or E-configuration exhibit a wide variety of photochemical reactions. The photochemistry of azobenzene and its derivatives is especially of great technological significance. Among the diverse photochemical reactions exhibited by these molecules, this chapter overviews briefly the photointerconversion between E- and Z-isomers, which is the principal photoreaction of these groups of compounds.

The first recorded example of the E,Z-isomerization of imines was that of oximes by Hantzsch in Zürich, who reported in 1890 that E,Z-isomerization took place when the tolylphenyl ketone oxime or α-anisylphenyl ketone oxime was irradiated.[1] Subsequent to this study, Ciamician and Silver at Bologna reported that when E-o- and p-nitrobenzaldehyde oximes, suspended in benzene, were exposed to sunlight, they were converted into the corresponding Z-nitrobenzaldehyde oximes.[2] This report was confirmed and extended to O-methyloximes by Brady and Dunn in London (Scheme 66.1).[3] Stoermer also found a similar photochemical E,Z-photoisomerization with certain ketoximes.[4]

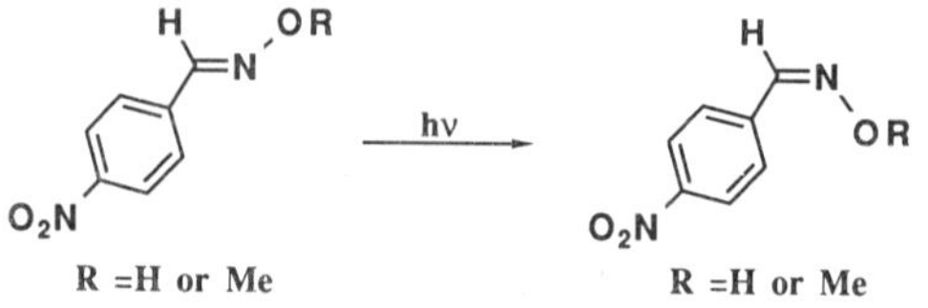

SCHEME 66.1

0-8493-8634-9/95/$0.00+$.50
© 1995 by CRC Press, Inc.

Subsequent to Thiele's preparation of azomethane — the first azoalkane — and the observation of its thermal decomposition to give ethane and nitrogen in 1909,[5] Ramsperger first reported in 1928 that azomethane decomposed similarly to ethane and nitrogen upon irradiation with monochromatic light.[6] Since then, studies on the photochemistry of azoalkanes, especially those by Ansloos,[7] Bartlett,[8] and Steel and their colleagues during the 1950s and 1960s, have been concerned mainly with the mechanistic aspects of elimination of nitrogen in connection with the behavior of the generated biradicals.[17]

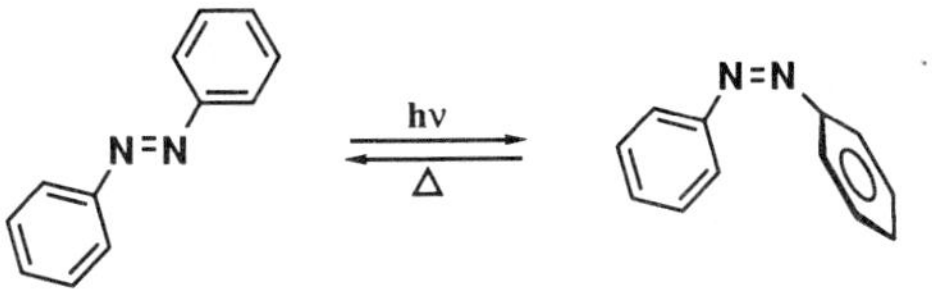

SCHEME 66.2

The first report concerning the photochemical *E,Z*-isomerization of the nitrogen-nitrogen double bond dealt with an azobenzene and its derivatives that have a greater stability than azoalkanes by Hartley in 1937.[9] The possibility of isomerism in azobenzenes had been suggested for many years before this report. Thus, Hartley succeeded in isolating *Z*-azobenzene, which thermally reverted to *E*-azobenzene, and measured its physical and chemical properties (Scheme 66.2). In the following years, the photoisomerization was found to be general for azobenzenes and their naphthalene or heterocyclic derivatives. Some 25 years later, Hutton and Steel found that *E,Z*-photoisomerization was a general reaction even for azoalkanes.[10]

The photochemical *E,Z*-isomerization of imines carrying alkyl, aryl, or acyl substituents at the nitrogen of their C=N group, on the other hand, was reported much later. This was due to the fact that the energy barrier to interconversion was normally small for imines and that, at ambient temperatures, a thermal reversion from the *Z*-form to the more stable *E*-form takes place readily. Thus, Fischer and Frei reported in 1957 that the irradiation of some diarylimines in solution at −100 °C induced a change in the UV spectrum and led to a photoequilibrium.[11] The spectrum of the starting solution was restored upon allowing the solutions to warm up. They correctly interpreted this change in the spectra as being due to *E,Z*-interconversions of the imines (Scheme 66.3).

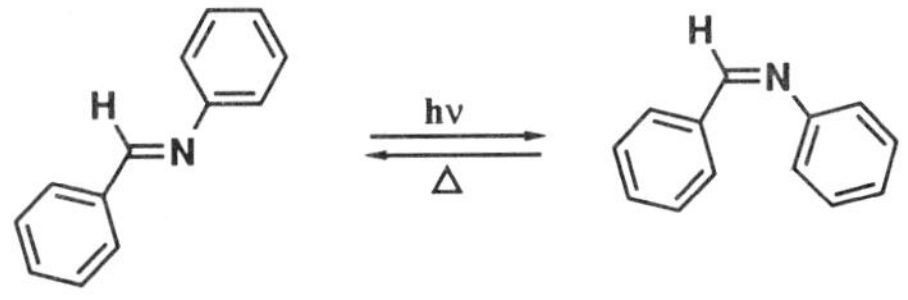

SCHEME 66.3

Since the 1960s, numerous investigations have been carried out on the nature of the C-N and N-N double-bond photoisomerization and competitive nature of the *E,Z*-isomerization with other processes, such as photocyclization, photoreduction, and photorearrangement. The variety of substrates studied included oxime, oxime ethers, aliphatic and aromatic azo compounds, hydrazones, and Schiff bases. The accumulated results are the subjects of several comprehensive and thorough review articles by Wettermark,[12] Padwa,[13] Pratt,[14] Griffiths,[15] Engel,[16,17] and Dürr.[18]

A notable application of *E,Z*-photoisomerization of azobenzene since the late 1970s involves the construction of photoresponsive molecules, such as those containing both crown ether and azobenzene units. Irradiation of these complex molecules carrying an azobenzene function is accompanied by profound conformational changes; these photoresponsive molecules have great potential for application to the problem of acceleration of ion transport, photocontrolled solvent extractions, etc. This topic has been the subject of a recent review.[19]

66.2 *E,Z*-Photoisomerization of Oximes and Nitrones

An isolated -C=N-OH group in saturated oximes in a hydrocarbon solvent gives rise to a band at 192 to 200 nm (ε; 7800–9440),[20] which has been assigned to the π,π^* transition band.[20] The corresponding *O*-methyl oxime also revealed absorption at a similar wavelength with a similar intensity (cyclohexanone *O*-methyl oxime; 201 nm; ε; 7000).[20] No distinct low-intensity band or shoulder attributable to a n,π^* transition has been observed, and any weak n,π^* band occurring near 190 nm should be submerged in the π,π^* band. Conjugation of the hydroxyimino group with a carbon-carbon double bond results in the appearance of highly intense absorption near 236–260 nm (ε; 1.1–2.6 × 10⁴),[20,21] probably assignable to π,π^* transitions.

In 1963, de Mayo and colleagues found that arylaldoximes, such as benzaldoxime in protic solvents, undergo a Beckmann-type photorearrangement to give the corresponding amide upon irradiation through quartz.[22a] The major reaction for energy dissipation was, however, *E,Z*-photoisomerization. They suggested, based on a variety of experimental evidence, that the amides are produced from an oxaziridine intermediate formed from an excited oxime as outlined in Scheme 66.4. The intermediacy of the oxaziridine in this rearrangement was confirmed subsequently by de Mayo[22b] and by Oine and Mukai.[23]

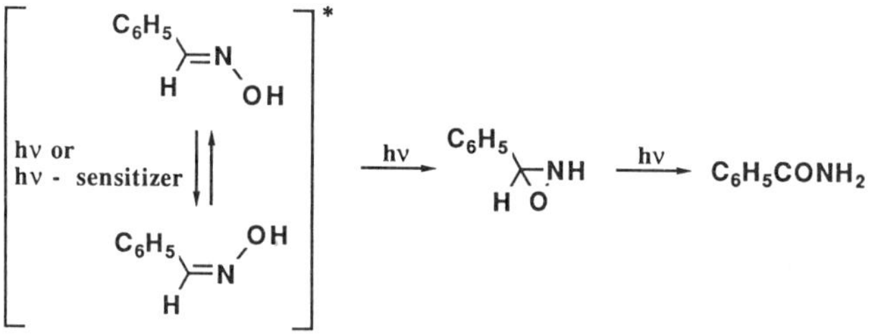

SCHEME 66.4

Subsequent to this discovery, the formation of amides and lactams was found to be the major competing photoreaction of oximes; Just and colleagues found that the irradiation of a methanolic solution of cyclohexanone oxime gave caprolactam.[24] Suginome et al. found that the chirality of the migrating carbon centre α to the hydroxyimino group in the formation of lactams in the photo-Beckmann rearrangement is retained;[25] typically, 5α-cholestan-6-one oxime gives a 1:1 ratio of the two corresponding 5α-lactam isomers, while 5β-cholestan-6-one oxime gives a 1:1 ratio of two 5β-lactam isomers in the photo-Beckmann rearrangement. On the basis of these data and other evidence, Suginome et al. proposed the path of photo-Beckmann rearrangement in terms of the simple scheme (Scheme 66.5) involving a photochemical *E,Z*-isomerization, followed by transformations of the excited singlet *E*- and *Z*-oximes into oxaziridine intermediates and a reorganization of the resulting singlet excited oxaziridines to the lactams in a fully concerted manner.[25]

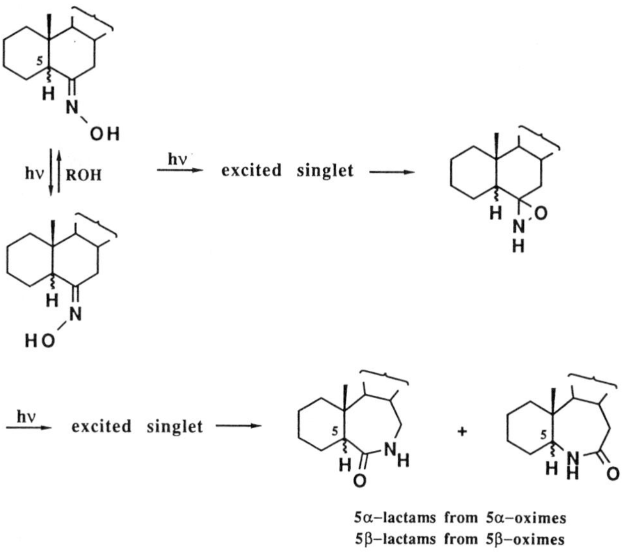

SCHEME 66.5

Koyano and Tanaka found that the E-to-Z and Z-to-E-photoisomerization takes place upon the irradiation of a solution of α-cyano-α, N-diphenylnitrone in the presence of a triplet sensitizer, while irradiation of E- or Z-forms of the nitrone in the absence of a sensitizer resulted in the formation of the corresponding oxaziridine.[26] These results indicated that photochemical E,Z and Z,E-isomerization takes place from the triplet excited nitrone, while the photochemical formation of the oxaziridine takes place from a singlet excited nitrone (Scheme 66.6).

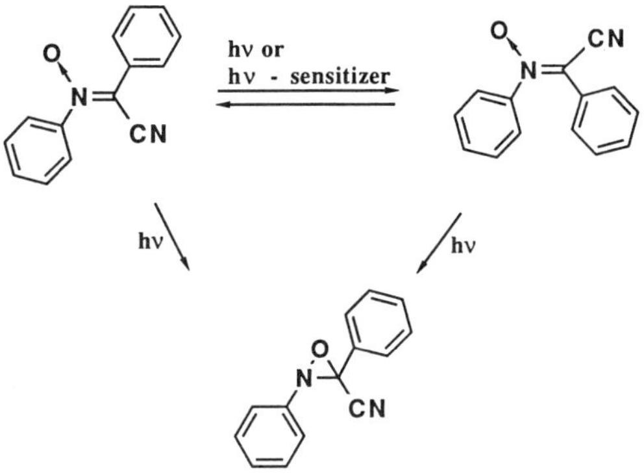

SCHEME 66.6

The oxime ethers (such as outlined in Scheme 66.7) are attractive candidates for mechanistic studies concerning the E,Z-photoisomerization of imines (*vide infra*), since the presence of the alkoxyl group drastically reduces the rates of thermal interconversion ($k < 10^{-3}$ at 60 °C) and allows mechanistic studies to be carried out at ambient temperatures.[27] Thus, Padwa and Albrecht studied the photochemical E,Z-isomerization about the carbon-nitrogen bond of O-methyl oxime ethers in some detail.[27] They found that upon direct irradiation of the O-methyl oximes of acetophenone, a photostationary state Z:E ratio of 2.20 can be achieved. Photoisomerization about the C-N double bond of the oxime ethers could also be induced by triplet excitation. A photostationary state vs. the triplet energy of the sensitizer plot indicates: (1) a high-energy region in which the stationary state ratio is ca. 1.5; (2) a gradual increase in the Z:E ratio from 301 down to 267 kJ of triplet energy; and (3) a sharp decrease from 267 to 226 kJ. The latter two observations are explained by a nonvertical excitation of the acceptor, as in the case of stilbene photochemistry. Since the direct irradiation could not be quenched with high concentration of piperylene, an electronically excited singlet appears to be the reactive state in the direct irradiation. The O-methyl oxime ether of 2-acetophenone has also been reported to undergo facile *Z,E*-photoisomerization.[27b] The composition of the photostationary state appears to be concentration-dependent, with the *Z*-isomer predominating at low concentration. Fluorescence quenching studies and photosensitization experiments gave evidence for the involvement of the singlet state.[27b] Photoisomerization of some α-oxo-oxime ethers was also investigated later in some detail.[28]

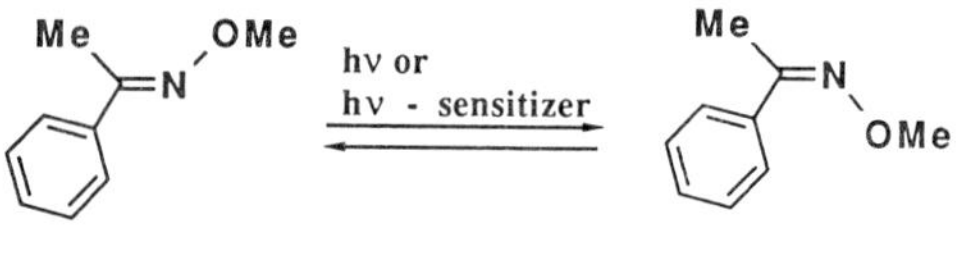

SCHEME 66.7

It has been found recently that a triplet-sensitized photoisomerization of *N*-methyl-1-(2-anthryl)ethanimine in the presence of benzil results in one-way *Z*-to-*E*-isomerization with a quantum yield of 11, but no reverse *E*-to-*Z*-photoisomerization (Scheme 66.8).[29] An explanation on this one-way isomerization involving a quantum chain process was given on the basis of some spectroscopic results.[28]

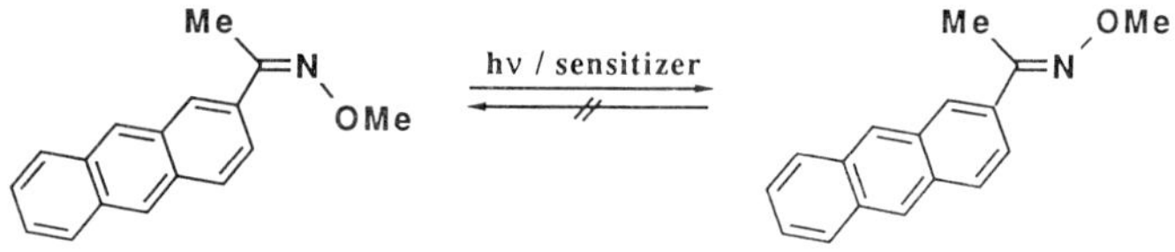

SCHEME 66.8

66.3 *E,Z*-Photoisomerization of Imines

An isolated azomethine chromophore in alkylimines gives rise to two absorption bands in the UV region of the spectrum at about 230–260 nm (ε; ca. 2×10^2) and at 170–180 nm (ε; 10^4). The former was assigned to the n,π* transition band and the latter to the π,π* transition.[12] Aryl-substituted imines absorb at longer wavelength.[30] They exhibit a number of bands, and their positions depend on the substituents on the aromatic ring. These bands have been interpreted as being of the π,π* and charge-transfer type;[30] benzylideneaniline exhibits an intense absorption maximum at 252 nm and a shoulder at 315 nm. The absorption attributable to the n,π* transition has not been assigned unambiguously, and may be submerged in the π,π* band. It has been known that the absorption spectrum of benzylideneaniline distinctly differs with those of stilbene and azobenzene. It has been proved that *E*-benzylideneaniline exists in a conformation in which the PhC=N moiety is planar, but the plane of the *N*-phenyl ring makes an angle with the rest of the molecule. This allows conjugation with the lone-pair electron on the nitrogen atom, but reduces conjugation between the phenyl rings through the C=N bond. The angle of rotation in the *E*-isomer in the gas, solution, and solid phases was estimated to be in the 30 to 55° range by various physical techniques.[31] It seems to be general agreement that the aniline ring of the *Z*-form is also rotated.

In 1957, Fischer and Frei found that the irradiation of some diarylamines in solution at −100 °C induced distinct changes in the UV spectra of the solutions, with the changes leading to an equilibrium.[11] The spectrum of the original isomer was restored by allowing the solutions to warm up. This report was the first spectroscopic observation of the *E,Z*-photoisomerization of imines. They estimated the thermal relaxation to have an activation energy of 67–71 kcal mol^{-1} which was appreciably lower than those of azobenzene and stilbene (96 and 176) kJ mol^{-1}. Later, Wettermark and colleagues[32] demonstrated that the *E,Z*-isomerism of *N*-benzylideneaniline and the related compounds can take place even at room temperature (Scheme 66.9). At this temperature, thermal relaxation proceeds with a half-life of approximately 1s. They obtained the range of activation energies of *Z,E*-thermal relaxation for a number of derivatives of *p*-substituted *N*-benzilideneaniline.

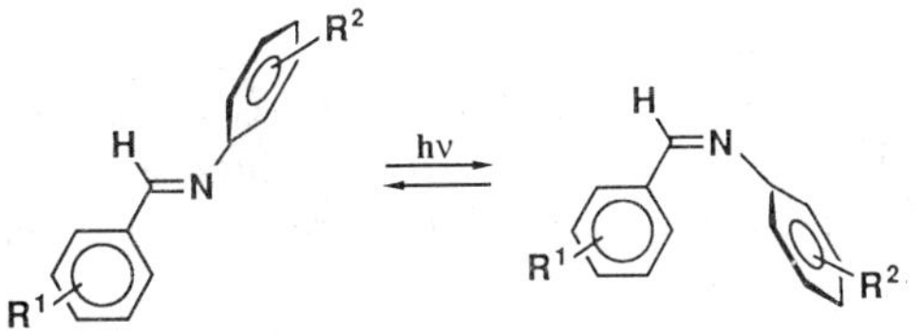

SCHEME 66.9

The formation of *Z*-forms of benzilidene aniline and its derivatives was also observed in matrices at 77K by ^{1}H-NMR,[33a] ^{13}C-NMR,[33b] and UV spectroscopy,[33c] in Nujol at 77K by IR spectroscopy,[33c] and in solution at room temperature by flash photolysis.[33d]

The photochemistry of the salicylideneanilines suffers from competing reversible photoreactions involving the transfer of the phenolic proton to the imino nitrogen, along with the formation of the corresponding *o*-quinoid forms (photochromism)[32a,34a-d] or a zwitterion.[34e]

Many simple imines are susceptible to facile hydrolysis to give their parent ketones in the presence of even a trace of moisture. The small amount of ketones generated affects the results of the photoreaction of imines. Thus, a major activity in the investigation of the photochemical behavior of imines has been on imine derivatives such as *O*-alkyl oximes and hydrazones, which are less susceptible to hydrolysis. Extensive studies thus carried out on the *E,Z*-photoisomerization of the configurationally stable oxime ethers have already been described (*vide supra*).[27-29]

Studies on the photochemical *E,Z*-isomerization of phenylhydrazones have also been carried out.[35-39] It was reported that the introduction of a nitro group to the *N*-aryl group of pyridine-2-aldehyde phenylhydrazone[35] enhances the quantum yield of the photoisomerization of the *E*-isomer to the *Z*-isomer in benzene, but decreases the quantum yield of the *Z*-isomer to the *E*-isomer, reaching zero.[37] Thus, a direct irradiation of the *E*-isomer of pyridine-2-aldehyde 4-nitrophenylhydrozone in benzene resulted in a complete isomerization to the *Z*-form, the stability of which apparently arises from an intramolecular hydrogen bond between the nitrogen of the pyridine ring and NH (Scheme 66.10).[37]

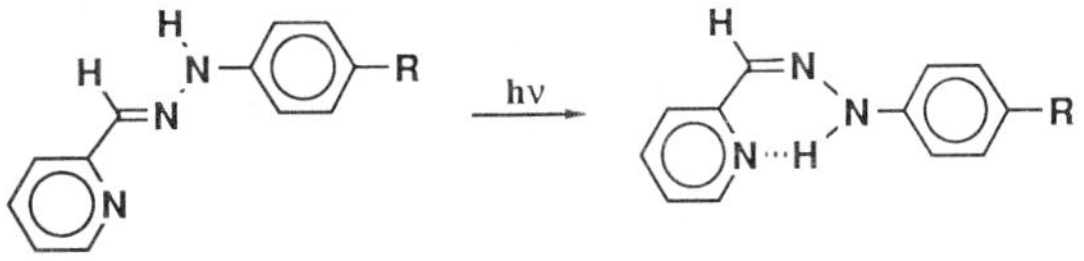

R = H or NO$_2$

SCHEME 66.10

The readers are advised to consult comprehensive reviews[13,14] and the references cited in the original papers for more details concerning the photochemistry of the carbon-nitrogen double bond.

66.4 *E,Z*-Isomerization of Azo Compounds

Azoalkanes exhibit absorption bands in two regions: one is in the 320–380 nm (ε, 10 ~ 150) region, and the other is in the 200-nm region (ε ~ 1000).[16,40] The weak longer-wavelength band has been assigned to the n,π^* transition and the stronger shorter wavelength band to the n,σ^* transition. Bathochromic and hyperchromic effects of the n,π^* band are observed in going from *E*-azoalkanes to the *Z*-isomer; *E*-azoisopropane exhibits the n,π^* band at 360 nm with ε ~ 14.5, while the *Z*-isomer exhibits the corresponding absorption at 380 nm with ε ~ 140 in nonpolar solvents.[16,41]

Azobenzene exhibits three absorptions at longer wavelength.[15,40,42] The lowest-energy transitions occur at approximately 440 nm (ε ~ 500) and 430 nm (ε ~ 1500) in *E*- and *Z*-azobenzenes, respectively, which are assignable to the n,π^* transition. The second absorption band occurs at 280 nm (ε = 5100) for the *Z*- and *E*-forms at 314 nm (ε ~ 17,000). The marked difference between the two isomers is due to the nonplanar conformation of the *Z*-isomer in solution. The highest-energy absorption band occurs in the 230–240-nm region for both isomers. An introduction of an electron-donating substituent to the azobenzene system results in the appearance of the absorption bands in the visible region (azo dyes).

The principal reactions of azo compounds upon irradiation with UV light are either elimination of nitrogen to give alkyl radicals or *E,Z*-isomerization. Until the 1950s, studies on the photoreaction of aliphatic azo compounds largely focused on the elimination of nitrogen in gas phase.[7a,43] For example, the photolysis of azoisopropane in the gas phase generates isopropyl radicals and nitrogen with unit quantum efficiency (Scheme 66.11).[44] Many of the aliphatic azo compounds (such as 2,2'-azoisobutyronitrile[45]), which generate alkyl radicals readily upon irradiation or pyrolysis, have been used as initiators of such radical reactions as polymerization and radical cyclization in organic synthesis.

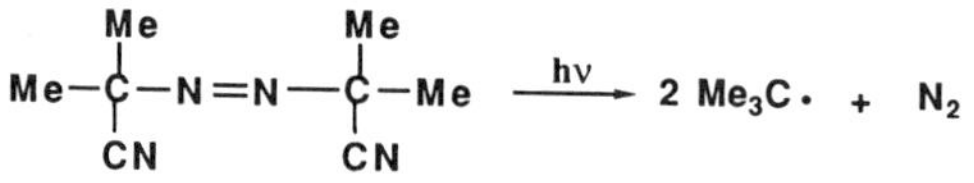

SCHEME 66.11

However, in 1964, Hutton and Steel found that irradiation into the n,π^* band of *E*-azomethane in solution at room temperature caused a decrease in the quantum yield of nitrogen formation to 0.17 and led to *E,Z*-photoisomerization, as detected by UV and ^{1}H-NMR spectroscopy (Scheme 66.12).[10] More energetic irradiation into the n,σ^* band of aliphatic *E*-azoalkanes (such as azoisopropane) using UV light in the 200 nm region leads to the elimination of nitrogen along with the formation of vibrationally excited alkyl radicals.[46]

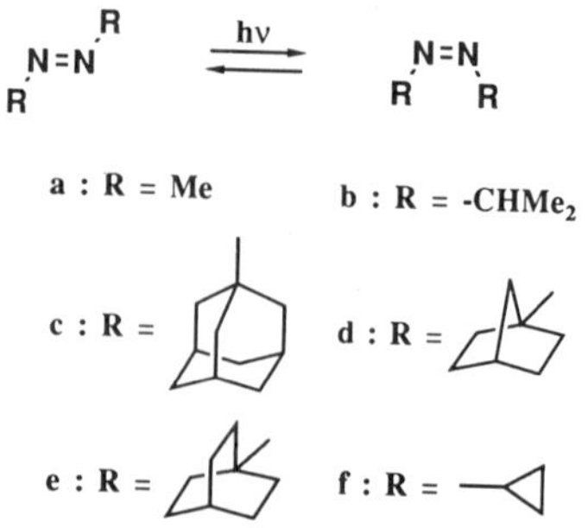

SCHEME 66.12

Subsequently, Steel and colleagues as well as Mill and Stringham found that *E,Z*-isomerization is an important general process competing with dissociation and many of the photogenerated

Z-isomers undergo remarkably facile thermal decomposition to give nitrogen and alkyl radicals.[44,47] Irradiation of *E*-azomethane at 25 °C in benzene gives the *Z*-isomer with a quantum yield $\Phi_{E \to Z}$ = 0.42; the quantum yield $\Phi_{Z \to E}$ of the reverse reaction was found to be 0.45.[48] The steric effect and the nature of the incipient radical largely govern the stability of *Z*-azo alkenes. *Z*-azo-2-methyl-2-propane decomposes with an activation energy of 96 kJ, some 84 kJ less than that of the *E*-isomers.[47a] An activation energy of the thermal decomposition of *Z*-azoisopropane[47b] is 171 kJ, which is only about 33 kJ less than that of the *E*-isomer, while the *Z*-isomer of azo*bis*(1,1-dimethyl-2-propene) decomposes at −120 °C.[16,47c] This mechanism, *E* → *Z*-isomerization followed by thermal decomposition of the unstable *Z*-isomer to give alkyl radicals, was found to be general for symmetric[49] and unsymmetric azo compounds.[50]

The *Z*-isomer, having considerable steric bulk and high energy of decomposition due to the generation of high-energy radicals, is expected to be stable and can be isomerized thermally to the *E*-form without the loss of nitrogen. Thus, Engel and colleagues synthesized *Z*-di-1-adamantyldiazene and *Z*-di-1-norbornyldiazene by UV irradiation of their *E*-isomers in toluene at 0 °C (Scheme 12).[51] The quantum yields of the elimination of nitrogen in these *E* → *Z*-isomerization were extremely low. Engel and colleague also synthesized *Z*-azacyclopropane by irradiation of the corresponding *E*-isomer with 313 nm light (Scheme 12). It exhibited extraordinary thermal stability for the elimination of nitrogen.[52] Similarly, eight-membered cyclic *E*-azoalkane was synthesized by Overberger and colleagues, as outlined in Scheme 66.13.[53] The formation of a seven-membered cyclic *E*-azoalkane has been reported;[54] a direct irradiation (403 nm, n,π* band) or a triplet-sensitized irradiation of 1,4-dihydronaphtho[1,8-de][1,2]diazepine resulted in a smooth isomerization to the *E*-isomer, while irradiation of either the *E*- or *Z*-isomer with 313 nm light (π,π* band) induced a clear removal of nitrogen to give acenaphthene via a biradical, as outlined in Scheme 66.14.

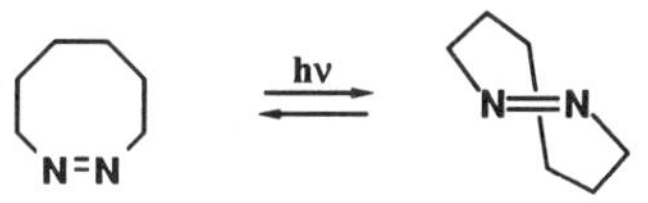

SCHEME 66.13

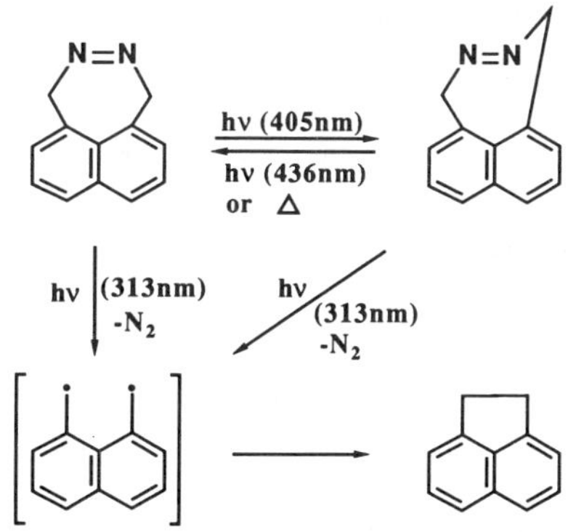

SCHEME 66.14

The mechanism by which *E*- and *Z*-azo compounds are interconverted has remained unclarified. $\phi_{E,Z}$ for *E*-azopropane and *E*-azocumene over the 20 to −190 °C temperature range indicated no significant barriers to isomerization.[16] Theoretical studies indicated that the mechanism involving rotation about the N-N bond is the preferred pathway for the n,π* state, whereas that involving in-plane motion of the substituent group is the preferred one for the ^{3}π,π* state.[55]

E,Z-isomerization of azo compounds by a triplet sensitization has received some attention and has been summarized in a review.[17]

Although aromatic azo compounds also undergo photochemical *E,Z*-isomerization, the elimination of nitrogen to give aryl radicals is negligible. This is due to the stronger C-N bond and the lower excited energy. A full understanding of the nature of the excited state associated with the photoisomerization of azobenzene is more difficult than the hydrocarbon analogue, such as stilbene, because of the involvement of an additional n,π* state. (The readers should refer to a review article with regard to this aspect.[15])

1

Since the first discovery of the *Z*-isomer of azobenzene by Hartley (Scheme 2),[9] the generality of the photoisomerization reaction has been demonstrated with a variety of azobenzenes and their derivatives.[56] A convenient isolation procedure of *Z*-azobenzene, *Z*-2,2'-azopyridine, and azonaphthalenes using column chromatography has been described.[57] The benzene rings in the *Z*-azobenzene are not planar.[15,40]

The proportion of *Z*-isomer at the photostationary state should be increased if the wavelength of light applied is not absorbed by the *Z*-isomer. Fischer and colleagues showed that the proportion of *Z*-azobenzene at the photostationary state increased to 91% with 365 nm light, while it was 37% with unfiltered UV light, since the *E*-isomer has a much higher extinction coefficient than does the *Z*-isomer.[58]

Although the mechanism of the photoisomerization may involve either twisting around the central N=N bond in the excited state (rotation mechanism) or inversion at one of the nitrogen atoms (inversion mechanism), this has been a subject of debate.[59] Evidence has recently been reported for the involvement of an inversion mechanism for n,π* excitations.[60] An *E* → *Z*-photoisomerization took place when *E*-2,19-dithia[3,3](4,4')-*trans*-diphenyldiazeno(4)phane 1 in benzene was irradiated. This was taken as unequivocal proof for an inversion mechanism of azobenzene, since a planar inversion in this molecule is possible and rotation should be severely hindered. The quantum yields were found to be wavelength independent, in contrast to those of the parent azobenzene. Further evidence was reported[60b] that normal azobenzenes isomerize via an inversion from the n,π* state; but that upon π,π* excitation, a rotational feature becomes important. Evidence for the involvement of an inversion mechanism has also been obtained from a study of an substituted azobenzene.[60c]

The activation energy of the thermal reversion of *Z*-azobenzenes to their *E*-forms has been shown to be *ca.* 100 kJ mol^{-1} for a wide range of derivatives. The value is some 54 kJ lower than that observed for the isomerization of stilbene, thus suggesting the inversion mechanism.[61]

A fundamental study on the sensitized photochemical *E,Z*-isomerization of azobenzene in solution was again carried out recently.[62]

More recently, the photochemistry of 4-diethylamino- and 4-diethylamino-4'-methoxyazobenzene was studied as reasonable models for commercially useful monoazo-dyes.[63] It was concluded that the lowest singlet and triplet states of these azobenzene derivatives are only capable of geometrical isomerization, whereas hydrogen abstraction takes place from the high-lying triplet state of both the *E*- and *Z*-forms of these dyes.[63]

66.5 Applications

Synthesis

The ground-state reaction to imines, oximes, and azo-compounds invariably produces the more stable E-isomers, and photochemical isomerization is practically the only means for preparing their less-stable isomers,[57] as in the case of olefins, such as stilbene. For example, the pharmacologically active Z-isomer of isonicotinaldehyde was prepared by photoisomerization (Scheme 66.15).[64] A less stable E-isomer of methyl (Z)-O-methyl-hydroxymates can be isolated in 38% yield by UV irradiation of the Z-isomer in benzene, followed by preparative TLC (Scheme 66.16).[65]

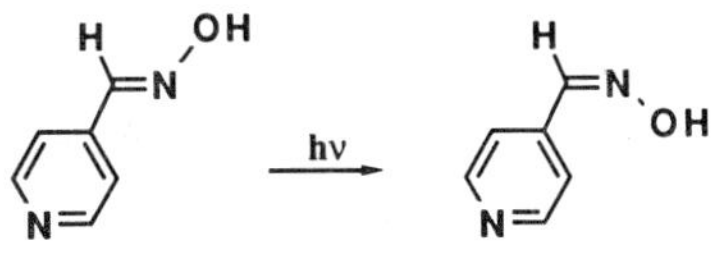

SCHEME 66.15

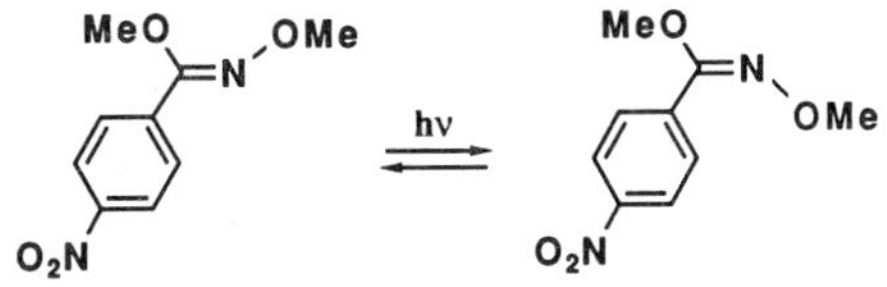

SCHEME 66.16

The E:Z-photostationary ratio is largely governed by the relative extinction coefficient of the absorptions of the two isomers at the wavelength used for direct irradiation.[27,58] Isomerization can also be achieved by triplet sensitization in which the relative values of the triplet energies are important in governing the ratio of the isomers.[27]

Photoresponsive Molecules Containing Azobenzene Unit

The photoinduced E,Z-isomerization of azobenzene has attracted considerable attention of chemists since the late 1970s when several groups of investigators recognized the E,Z-isomerization of azobenzene to be useful as a new tool to enforce reversible changes in the conformation of a lamellar multibilayer,[66] a synthetic bilayer membrane,[67] polymers,[68] cyclodextrins,[69] and crown ethers.[70] Considerable efforts concerning a variety of molecular systems have been devoted to this aspect of the photoisomerism of azobenzene during this decade. Extensive work by Shinkai and colleagues on the synthesis of a number of photoresponsive azobenzene-bridged crown ethers and studies of their functions is especially impressive. In general, 70 to 80% of the photoresponsive molecules containing the E-azobenzene unit can be converted to the Z-isomer upon irradiation with UV light ($330 < \lambda < 380$ nm). The E-isomer is quantitatively regenerated, either thermally or upon irradiation with visible light ($\lambda > 420$ nm). Because of space limitations, only a few of the results in this area are highlighted in this chapter.

Thus, Shinkai and colleagues prepared an azobenzene-bridged aza-crown ether 2 and found that the binding efficiency of 2 for alkali metal ions can be modified by an E,Z-photoisomerization (Scheme 66.17). The result was rationalized in terms of photoinduced expansion of the aza-crown ether size.[70a] Similarly, the E-isomer of a new photoresponsive cryptand 3 containing a 2, 2'-azopyridine bridge prepared by them can bind such heavy metal ions as Cu^{2+}, Ni^{2+}, Co^{2+}, and Hg^{2+} from aqueous solution into the organic phase, whereas the photogenerated Z-isomer has been found to be unable to bind (Scheme 66.17).[71]

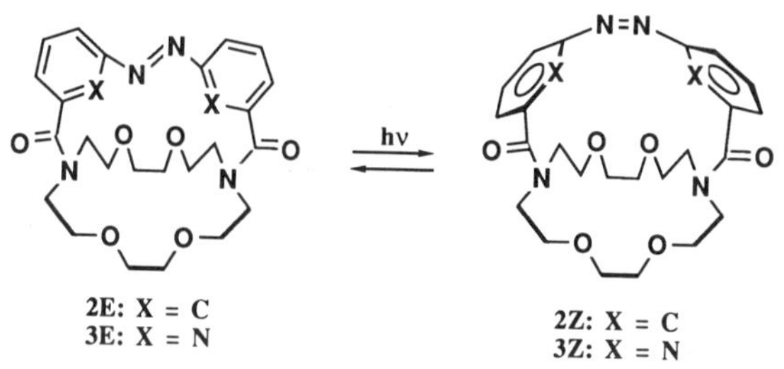

SCHEME 66.17

In contrast, the *E*-isomer of azobenzenophane crown ether **4** lacks totally the ability to extract metal ions, whereas the *Z*-isomer is able to bind alkali metal cations (Scheme 66.18).[72]

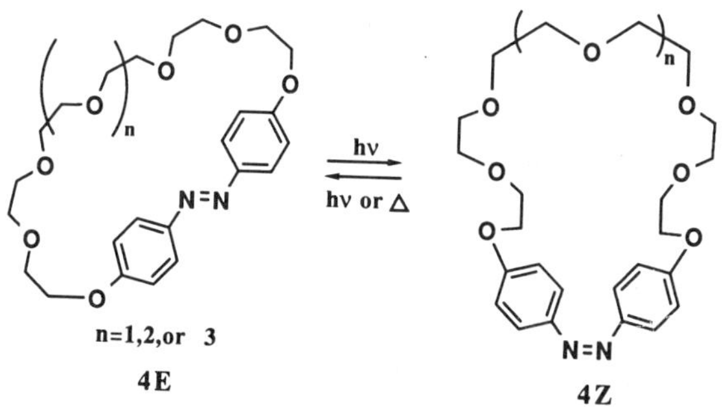

SCHEME 66.18

The *Z*-form of the photoresponsive crown ether **5**, prepared by Shiga and colleagues, lacks affinity for an alkali metal ion, whereas the *E*-form is able to do so.[70b] The *E*-isomer of a *bis*crown ether **6** and the photoisomerized *Z*-isomer (Scheme 66.19) exhibit a contrasting ion extraction ability;[73] the ratio *(E:Z)* of extractability for Na+ against that for K+ is 238-fold. Rb+ and Cs+ are also extracted by **6Z** more efficiently that by **6E**. Moreover, the rate of the K+ transport across a liquid (*o*-dichlorobenzene) membrane was found to be suppressed by light when a hydrophobic counterion was used, whereas it was accelerated by light when a relatively hydrophilic counteranion was used. The ion-binding ability of the *E*-form of the photoresponsive crown ether **7** changes on irradiation as a consequence of an intramolecular interaction between the ammonium group and the crown ether in the *Z*-isomer, as outlined in Scheme 66.20.[74] Similarly, the ion-binding properties of the crown ether **8** are appreciably altered upon *E* → *Z*-photoisomerization.[75]

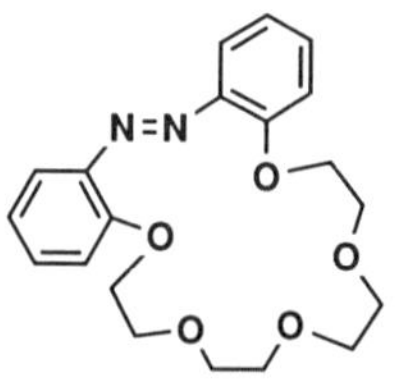

5

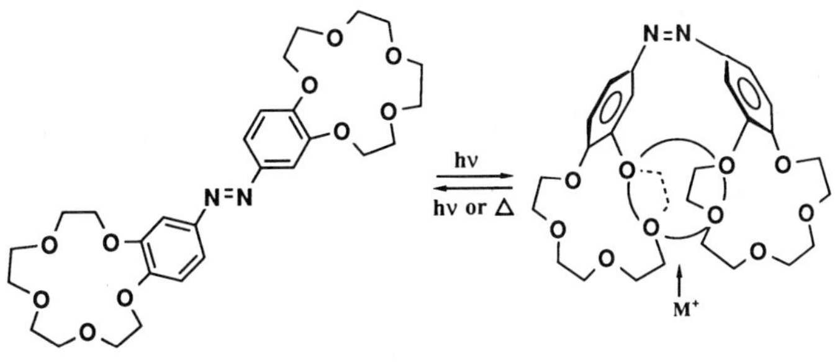

SCHEME 66.19

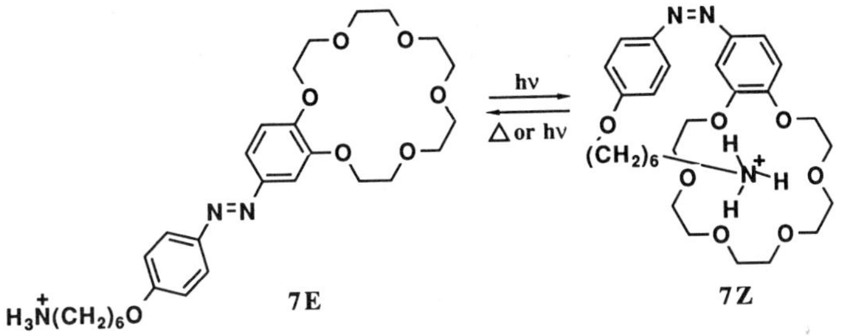

SCHEME 66.20

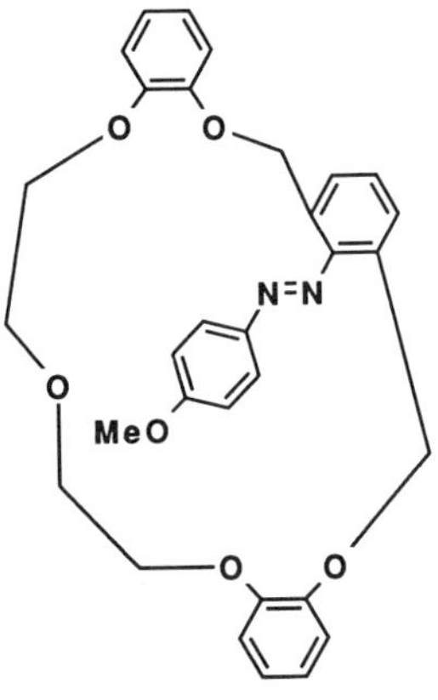

8

In another attempt, azobenzene-appended γ-cyclodextrin 9 was prepared. On irradiation with 320 to 390 nm light, 70% of the *E*-isomer in water isomerized to the *Z*-isomer, which exhibited an enhanced binding ability for (−)-borneol and (+)-fenchone.[76] Benzo-18-crown-6 **10** linked to a phenol was synthesized. Upon irradiation of UV light, the azo linkage isomerized to the *Z*-isomer. It slowly isomerized back to the *E*-form in the dark and rapidly upon irradiation of visible light. An enhanced extraction of Ca^{2+} by the *Z*-isomer has been observed. The *Z*-structure constitutes a "photoresponsive anion cap".[77]

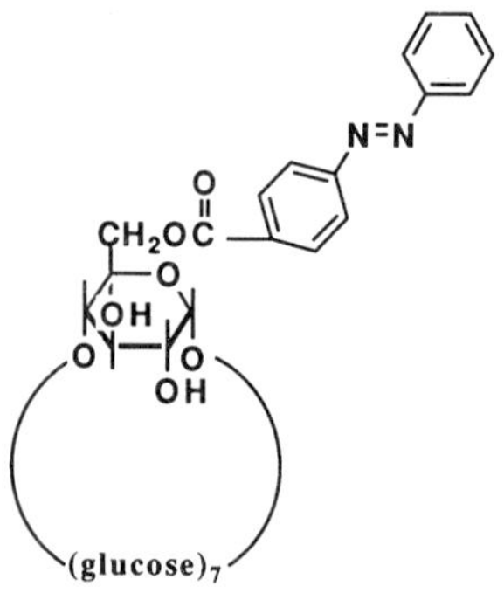

9

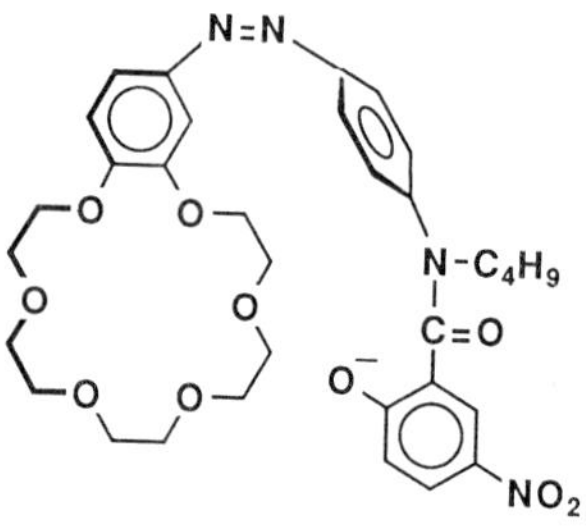

10

Any readers interested in this important application of the photoisomerization of azo compounds in this fascinating field should refer to the review article[18] and the original papers cited.[78]

References

1. Hantzsch, A., Die stereochemisch-isomerin Oximes des *p*-Tolyl-phenylketones, *Ber.*, 23, 2325, 1890; Hantzsch, A., Üeber stereoisomere Ketoxime, *Ber.*, 24, 51, 1891.
2. Ciamician, G. and Silber, P., Chemische Lichtwirkungen, *Ber.*, 36, 4266, 1903.
3. Brady, O. L. and Dunn, F. P., The isomerism of the oximes. II. The nitrobenzaldoximes, *J. Chem. Soc.*, 103, 1619, 1913; Brady, O. L. and McHugh, G. P., The isomerism of the oximes. XVI. The action of ultraviolet light on aldoximes and their derivatives, *J. Chem. Soc.*, 125, 547, 1924.
4. Stoermer, R., Über die Umlagerung der stabilen Stereoisomeren in labile durch ultraviolettes Light (II), *Ber.*, 44, 637, 1911.
5. Thiele, J., Über Hydrazo- und Azomethan, *Ber.*, 42, 2575, 1909.
6. Ramsperger, H., The photochemical decomposition of azomethane, *J. Am. Chem. Soc.*, 50, 123, 1928.
7. See, for example: (a) Steacie, E. W. R., *Atomic and Free Radical Reactions*, Reinhold, New York, 1954, 376; (b) Rebbert, R. W. and Ausloos, P., The photolysis and radiolysis of $CH_3N_2CH_3$ and $CH_3N_2CH_3$-$CD_3N_2CD_3$ mixtures, *J. Phys. Chem.*, 66, 2253, 1962.
8. See, for example: Nelson, S. F. and Bartlett, P. D., Azocumene. I. Preparation and decomposition of azocumene. Unsymmetrical coupling products of the cumyl radical, *J. Am. Chem. Soc.*, 88, 137, 1966; Bartlett, P. D. and McBride, J. M., Configuration, conformation and spin in radical pairs, *Pure Appl. Chem.*, 15, 89, 1967.
9. Hartley, G. S., The *cis*-form of azobenzene, *Nature*, 140, 281, 1937.
10. Hutton, R. F. and Steel, C., Photoisomerization of azomethane, *J. Am. Chem. Soc.*, 86, 745, 1964.
11. Fischer, E. and Frei, Y., Photoisomerization equilibria involving the C=N double bond, *J. Chem. Phys.*, 27, 808, 1957.

12. Wettermark, G., Photochemistry of the carbon-nitrogen double bond, in *The Chemistry of the Carbon-Nitrogen Double Bond*, Patai, S., Ed., Interscience, New York, 1969, 565.

13. Padwa, A., Photochemistry of the carbon-nitrogen double bond, *Chem. Rev.*, 77, 37, 1977.

14. Pratt, A. C., The chemistry of imines, *Chem. Soc. Rev.*, 6, 63, 1977.

15. Griffiths, J., Photochemistry of azobenzene and its derivatives, *Chem. Soc. Rev.*, 1, 481, 1972.

16. Engel, P. S. and Steel, C., Photochemistry of aliphatic azo compounds in solution, *Acc. Chem. Res.*, 5, 242, 1972.

17. Engel, P. S., Mechanism of the thermal and photochemical decomposition of azoalkanes, *Chem. Rev.*, 80, 99, 1980.

18. Dürr, H. and Ruge, B., Triplet states from azo compounds, *Top. Curr. Chem.*, 66, 55, 1976.

19. (a) Shinkai, S. and Manabe, O., Photoresponsive crown ethers, *Yukigoseikagaku* (in Japanese), 40, 92, 1982; (b) Shinkai, S. and Manabe, O., Host guest complex chemistry. 3. Photocontrol of ion extraction and ion transport by photofunctional crown ethers, *Top. Curr. Chem.*, 121, 67, 1984; (c) Ueno, A. and Osa, T., Photocontrol of molecular functions, *Yukigoseikagaku* (in Japanese), 38, 207, 1980.

20. (a) Orenski, P. J. and Clossen, W. D., The ultraviolet absorption spectra of oximes, *Tetrahedron Lett.*, 3629, 1967; (b) Suginome, H., Takahashi, H., and Masamune, T., The photo-Beckmann rearrangement of 3α,5-cyclo-5α-cholestan-6-one oxime, *Bull. Chem. Soc. Jpn.*, 45, 1836, 1972.

21. Suginome, H., Kaji, M., Ohtsuka, T., Yamada, S., Ohki, T., Senboku, H., and Furusaki, A., Photo-induced molecular transformations. 130. Novel stereospecific photorearrangement and stereospecific addition of methanol in steroidal α,β-unsaturated cyclic ketone oximes, *J. Chem. Soc., Perkin Trans. 1*, 427, 1992.

22. (a) Amin, J. H. and de Mayo, P., The irradiation of aryl aldoximes, *Tetrahedron Lett.*, 1585, 1963; (b) Izawa, H., de Mayo, P., and Tabata, T., The photochemical Beckmann rearrangement, *Can. J. Chem.*, 47, 51, 1969.

23. Oine, T. and Mukai, T., Evidence for the formation of oxiziridines during the irradiation of oximes, *Tetrahedron Lett.*, 157, 1969.

24. (a) Taylor, R. T., Douek, M., and Just, G., Photolysis of oximes, *Tetrahedron Lett.*, 4143, 1966; (b) Cunningham, M., Ng Lim, L. S., and Just, G., Photochemistry of oximes. III. Photochemical Beckmann rearrangement, *Can. J. Chem.*, 49, 2891, 1971.

25. (a) Suginome, H. and Takahashi, H., Stereochemical aspects of the photo-Beckmann rearrangement. Stereochemical integrity of the terminus of the migrating carbon in the photo-Beckmann rearrangements of 5α- and 5β-cholestan-6-one oximes, *Bull. Chem. Soc. Jpn.*, 48, 576, 1975; (b) Suginome, H. and Yagihashi, F., Photoinduced transformations. 36. Stereochemical integrity of the terminus of the migrating carbon in the photo-Beckmann rearrangements of some cholestanone oximes, *J. Chem. Soc., Perkin Trans. 1*, 2488, 1977.

26. Koyano, K. and Tanaka, I., The photochemical and thermal isomerization of *trans*- and *cis*-α-cyano-α-phenyl-N-phenylnitrones, *J. Phys. Chem.*, 69, 2545, 1965.

27. (a) Padwa, A. and Albrecht, F., Photochemical *syn-anti* isomerization about the carbon-nitrogen double bond, *J. Am. Chem. Soc.*, 96, 4849, 1974; (b) Padwa, A. and Albrecht, F., *J. Org. Chem.*, 39, 2361, 1974.

28. Baas, P. and Cerfontain, H., Photochemistry of α-oxo-oximes. 3. Photoisomerization of some α-oxo-oxime ethyl ethers, *J. Chem. Soc., Perkin Trans. 2*, 151, 1979.

29. Furuuchi, H., Arai, T., Kuriyama, Y., Sakuragi, H., and Tokumaru, K., One-way photoisomerization of the C=N double bond. Isomerization of (Z)-N-methoxy-1-(2-anthryl)ethanimine, *Chem. Lett.*, 847, 1990.

30. (a) El-Bayoumi, M. A., El-Aasser, M., and Abdel-Halim, F., Electronic spectra and structures of Schiff's bases. I. Benzanils, *J. Am. Chem. Soc.*, 93, 586, 1971; (b) El-Aasser, M., Abdel-Halim, F., and El-Bayoumi, M. A., *J. Am. Chem. Soc.*, 93, 590, 1971.

31. (a) Bürgi, H. B. and Dunitz, J. D., Crystal and molecular structures of benzylideneaniline, benzylideneaniline-p-carboxylic acid and p-methoxybenzilidene-p-nitroaniline, *Helv. Chim. Acta*, 53, 1747, 1970; (b) Bally, T., Haselbach, E., Lanyiova, J., Marschner, E., and Rossi, M., Concerning

the conformation of isolated benzilideneaniline, *Helv. Chim. Acta,* 59, 486, 1976; (c) Traetteberg, M., Hilmo, I., Abraham, R. J., and Ljunggren, S., The molecular structure of *N*-benzylidene-aniline, *J. Mol. Struct.,* 48, 395, 1978; (d) Akaba, R., Tokumaru, K., and Kobayashi, T., Electronic structures and conformations of *N*-benzylideneanilines. I. Electronic absorption spectral study combined with CNDO/S CI calculations, *Bull. Chem. Soc. Jpn.,* 53, 1993, 1980; Akaba, R., Tokumaru, K., Kobayashi, T., and Utsunomiya, C., Electronic structures and conformations of *N*-benzylideneanilines. II. Photoelectron spectral study, *Bull. Chem. Soc. Jpn.,* 53, 2002, 1980.

32. (a) Anderson, D. G. and Wettermark, G., Photoinduced isomerization of anils, *J. Am. Chem. Soc.,* 87, 1433, 1965; (b) Wettermark, G., Weinstein, J. Sousa, J., and Doglioti, L., Kinetics of *cis-trans* isomerization of *para*-substituted *N*-benzylidene anilines, *J. Phys. Chem.,* 69, 1584, 1965.

33. (a) Kobayashi, M., Yoshida, M., and Minato, H., Configuration of the photoisomers of benzylideneanilines, *J. Org. Chem.,* 41, 3322, 1976; (b) Yoshida, M. and Kobayashi, M., Configuration and conformation of the photoisomers of *N*-[*p*-(dimethylamino) benzilidene]anilines, *Bull. Chem. Soc. Jpn.,* 54, 2395, 1981; (c) Lewis, J. W. and Sandorfy, C., An infrared study of the photoisomerization of *N*-benzilideneaniline, *Can. J. Chem.,* 60, 1720, 1982; (d) Kanamaru, N. and Kimura, K., Photoinduced isomerization of benzalaniline, *Mol. Photochem.,* 5, 427, 1973.

34. (a) Cohen, M. D. and Schmidt, G. M. T., *J. Phys. Chem.,* 66, 2442, 1962; (b) Wettermark, G. and Dogliotti, L., Transient species in the photolysis of anils, *J. Chem. Phys.,* 40, 1486, 1964; (c) Potashnik, R. and Ottolenghi, M., Photoisomerization of photochromic anils, *J. Chem. Phys.,* 51, 3671, 1969; (d) Rossenfeld, T., Ottolenghi, M., and Meyer, A. Y., Photochromic anils. Structure of photoisomers and thermal relaxation processes, *Mol. Photochem.,* 5, 39, 1973; (e) Lewis, J. W. and Sandorfy, C., A spectroscopic study of proton transfer and photochromism in *N*-(2-hydroxybenzylidene)aniline, *Can. J. Chem.,* 60, 1738, 1982 and papers cited therein.

35. Schulte-Frohlinde, D., Quantenausbeuten bei der photochemischen syn ⇄ *anti*-umlagerung der pyridin-aldehyd-(2)-phenylhydrazone, *Liebigs Ann. Chem.,* 615, 114, 1958.

36. Condorelli, G., Costanzo, L. L., Pistara, S., and Giuffrida, S., The photochemical isomerization of 2-naphthaldehydephenylhydrazone, *Zeitsch. Phys. Chem. Neue Folge,* 90, 58, 1974.

37. Condorelli, G., Contanzo, L. L., Alicata, L., and Giuffrida, A., The photochemical isomerization of the pyridine-2-aldehyde 4-nitrophenylhydrazone, *Chem. Lett.,* 227, 1975.

38. Condorelli, G., Costanzo, L. L., Giuffrida, A., and Pistara, S., On the mechanism of the photoisomerization of the benzaldehyde and its derivative 4-nitrophenylhydrazones, *Zeitsch. Phys. Chem. Neue Folge,* 96, 97, 1975.

39. Courtot, P., Pichon, R., and Le Saint, J., Photochromise par isomerization *syn-anti* de phenylhydrazone-2 de tricetones-1,2,3 et de dicetones-1,2 substitutes, *Tetrahedron Lett.,* 1181, 1976.

40. (a) Rubin, M. B., Hart, R. R., and Keubler, N. A., Electronic states of the azoalkanes, *J. Am. Chem. Soc.,* 89, 1564, 1967; (b) Rau, H., Spectroskopische eigenshaften organischer azoverbindungen, *Angew. Chem.,* 85, 248, 1973.

41. Calvert, J. G. and Pitts, J. N., Jr., *Photochemistry,* Wiley, New York, 1967, 450.

42. Beveridge, D. L. and Jaffe, H. H., The electronic structure and spectra of *cis-* and *trans-*azobenzene, *J. Am. Chem. Soc.,* 88, 1948, 1966.

43. For a review, see Reference 41, p. 462.

44. Abram, I. I., Milne, G. S., Solomon, B. S., and Steel, C., The photochemistry of aliphatic azo compounds. The role of triplets and singlets in their photochemistry, *J. Am. Chem. Soc.,* 91, 1220, 1969.

45. For example, see: (a) Back, R. and Sivertz, C., The photolysis of 2,2′-azo-*bis*-isobutyronitrile, *Can. J. Chem.,* 32, 1061, 1954; (b) Roy, J.-C., Nash, J. R., Williams, R. R., Jr., and Hamill, W. H., Diffusion kinetics: The photolysis of azo-*bis*-isobutyronitrile, *J. Am. Chem. Soc.,* 78, 519, 1956; (c) Smith, P. and Rosenberg, A. M., The kinetics of the photolysis of 2,2′-azo-*bis*-isobutyronitrile, *J. Am. Chem. Soc.,* 81, 2037, 1957.

46. Arin, M. L. and Steel, C., Photochemistry of azoisopropane in the 2000-Å region, *J. Phys. Chem.,* 76, 1685, 1972.

47. (a) Mill, T. and Stringham, R. S., Photoisomerization of azoalkanes, *Tetrahedron, Lett.,* 1853, 1969; (b) Fogel, L. D., Rennert, A. M., and Steel, C., Thermal decomposition and isomerization of *cis*-azoisopropane, *J. Chem. Soc., Chem. Commun.,* 536, 1975; (c) Engel, P. S. and Bishop, D. J., Thermolysis of *cis* and *trans* azoalkanes, *J. Am. Chem. Soc.,* 97, 6754, 1975.

48. For example, see: Baird, N. C. and Swenson, J. R., Quantum organic photochemistry. IV. The photoisomerization of diimide and azoalkanes, *Can. J. Chem.,* 51, 3097, 1973.

49. Engel, P. S. and Bartlett, P. D., The sensitized photolysis of acyclic azo compounds. Singlet energy transfer, *J. Am. Chem. Soc.,* 92, 5883, 1970.

50. Porter, N. A., Marnett, L. J., Lochmüller, C. H., Closs, G. L., and Shobataki, M., Application of chemically induced dynamic nuclear polarization to a study of the decomposition of unsymmetric azo compounds, *J. Am. Chem. Soc.,* 94, 3664, 1972.

51. Engel, P. S., Melaugh, R. A., Page, M. A., Szilagyi, S., and Timberlake, J. W., Stable *cis* dialkyldiazenes (azoalkanes): *cis*-di-1-adamantyldiazene and *cis*-di-1-norbornyldiazene, *J. Am. Chem. Soc.,* 98, 1976; Chae, W.-K., Baughman, S. A., Engel, P. S., Bruch, M., Özmeral, C., Szilagyi, S., and Timberlake, J. W., Decomposition and isomerization of bridgehead *cis*-1,2-diazenes (azoalkanes), *J. Am. Chem. Soc.,* 103, 4824, 1981.

52. Engel, P. S. and Gerth, D. B., Azocyclopropane, *J. Am. Chem. Soc.,* 103, 7689, 1981.

53. Overberger, C. G., Chi, M.-S., Pucci, D. G., and Barry, J. A., *trans*-Azo linkages in eight-, nine-, and ten-membered cyclic azo compounds, *Tetrahedron Lett.,* 4565, 1972; Overberger, C. G. and Chi, M.-S., Photochemical isomerization of eight-membered azo compounds, *J. Org. Chem.,* 46, 303, 1981.

54. (a) Gisin, M. and Wirz, J., *Helv. Chim. Acta,* 59, 2273, 1976; (b) Pagin, R. M., Burnett, M. N., and Dodd, J. R., *J. Am. Chem. Soc.,* 99, 1972, 1977.

55. Thompson, A. M., Goswami, P. C., and Zimmerman, G. L., Kinetic analysis of the photochemistry of alkyldiazenes in hydrocarbon solution. The quasi-steady state, *J. Phys. Chem.,* 83, 314, 1979.

56. For example, see: Brod, W. R., Gould, J. H., and Wyman, G. M., The relation between the absorption spectra and the chemical constitution of dyes. XXV. Phototropism and *cis-trans* isomerism in aromatic azo compounds, *J. Am. Chem. Soc.,* 74, 4641, 1952.

57. (a) Cook, A. H., The preparation of some *cis*-azo-compounds, *J. Chem. Soc.,* 876, 1938; (b) Campbell, N., Henderson, A. W., and Taylor, D., Geometrical isomerism of azo-compounds, *J. Chem. Soc.,* 1281, 1953; (c) Frankel, M., Wolovsky, R., and Fischer, E., Geometrical isomerism of the azonaphthalenes, *J. Chem. Soc.,* 3441, 1955.

58. Fischer, E., Frankel, M., and Wolovsky, R., Wavelength dependence of photoisomerization equilibria in azocompounds, *J. Chem. Phys.,* 23, 1367, 1955.

59. For example, see: Gegiou, D., Muszkat, K. A., and Fischer, E., Temperature dependence of photoisomerization. V. The effect of substituents on the photoisomerization of stilbenes and azobenzenes, *J. Am. Chem. Soc.,* 90, 3907, 1968.

60. (a) Rau, H. and Lüddecke, E., On the rotation-inversion controversy on photoisomerization of azobenzenes. Experimental proof of inversion, *J. Am. Chem. Soc.,* 104, 1616, 1982; (b) Rau, H., Further evidence for rotation in the π,π^* and inversion in the n,π^* photoisomerization of azobenzenes, *J. Photochem.,* 26, 221, 1984; (c) Tanaka, T., Sueishi, Y., Yamamoto, S., and Nishimura, N., Pressure dependence of the photostationary *trans/cis* concentration ratio of 4-dimethylamino-4'-niroazobenzene. A new method of evaluating the reaction volume, *Chem. Lett.,* 1203, 1985.

61. Talaty, E. R. and Fargo, J. C., Thermal *cis-trans*-isomerization of substituted azobenzenes: A correction of the literature, *J. Chem. Soc., Chem. Commun.,* 65, 1967.

62. Ronayette, J., Arnaud, R., Lebourgeois, P., and Lemaire, J., Isomerisation photochimique de lázobenzene en solution. I, *Can. J. Chem.,* 52, 1848, 1974; Ronayette, J., Arnaud, R., and Lemaire, J., Isomerisation photosensibilisee par des colorants et photoreduction de l'azobenzene en solution. II. *Can. J. Chem.,* 52, 1858, 1974.

63. Albini, A., Fasani, E., and Pietra, S., The photochemistry of azodyes. Photoisomerization versus photoreduction from 4-diethylaminoazobenzene and 4-diethylamino-4'-methoxyazobenzene, *J. Chem. Soc., Perkin Trans. 2,* 1021, 1983.

64. Posiomek, E. J., Photochemical isomerization. Synthesis of *anti*-isonicotinaldehyde oxime derivatives, *J. Pharm. Sci.*, 54, 333, 1965.

65. Ogino, K., Matsumoto, T., and Kozuka, S., Photoisomerization of substituted *O*-methyl-*p*-nitrobenzohydroxymates, *Me. Fac. Eng., Osaka Univ.*, 16, 1545, 1979 (*Chem. Abstr.*, 92, 58910, 1980).

66. Balasubramanian, D., Subramani, S., and Kumar, C., Modification of a model membrane structure by embedded photochrome, *Nature (London)*, 254, 252, 1975.

67. (a) Kano, K., Tanaka, Y., Ogawa, T., Shimomura, M., Okahata, Y., and Kunitake, T., Photoresponsive membranes., Regulation of membrane properties by photoreversible *cis-trans* isomerization of azobenzenes, *Chem. Lett.*, 421, 1980; (b) Kunitake, T., Nakashima, N., Shimomura, M., Okahata, Y., Kano, Y., and Ogawa, T., Unique properties of chromophore-containing bilayer aggregates: Enhanced chirality and photochemically induced morphological change, *J. Am. Chem. Soc.*, 102, 6642, 1980.

68. (a) Pieroni, O., Houben, J. L., Fissi, A., Costantino, P., and Ciardelli, F., Reversible conformational changes induced by light in poly(L-glutamic acid) with photochromic side chains, *J. Am. Chem. Soc.*, 102, 5913, 1980; (b) Ueno, A., Takahashi, K., Anzai, J., and Osa, T., Photocontrol of polypeptide helix sense by *cis-trans* isomerism of side-chain azobenzene moieties, *J. Am. Chem. Soc.*, 103, 6410, 1981.

69. Ueno, A., Tomita, Y., and Osa, T., Photoresponsive binding ability of azobenzene-appended γ-cyclodextrin, *Tetrahedron Lett.*, 24, 5245, 1983 and references cited therein.

70. (a) Shinkai, S., Nakaji, T., Nishida, Y., Ogawa, T., and Manabe, O., Photoresponsive crown ethers. 1. *cis-trans* Isomerism of azobenzene as a tool to enforce conformational changes of crown ethers and polymers, *J. Am. Chem. Soc.*, 102, 5860, 1980; (b) Shiga, M., Takagi, M., and Ueno, K., Azo-crown ethers. The dyes with azo group directly involved in the crown ether skeleton, *Chem. Lett.*, 1021, 1980.

71. Shinkai, S., Kouno, T., Kusano, Y., and Manabe, O., Photoresponsive crown ethers. 7. Proton and metal ion catalyses in the *cis-trans* isomerization of azopyridines and an azopyridine-bridged cryptand, *J. Chem. Soc., Perkin Trans. 1*, 2741, 1982.

72. Shinkai, S., Minami, T., Kusano, Y., and Manabe, O., Photoresponsive crown ethers. 8. Azobenzenophane-type "switched-on" crown ethers which exhibit an all-or-nothing change in ion-binding ability, *J. Am. Chem. Soc.*, 105, 1851, 1983.

73. Shinkai, S., Nakaji, T., Ogawa, T., Shigematsi, K., and Manabe, O., Photoresponsive crown ethers. 2. Photocontrol of ion extraction and ion transport by a *bis*(crown ether) with a butterfly-like motion, *J. Am. Chem. Soc.*, 103, 111, 1981.

74. Shinkai, S., Ishihara, M., Ueda, K., and Manabe, O., On-off-switched crown ether-metal ion complexation by photoinduced intramolecular ammonium group 'tail-biting', *J. Chem. Soc., Chem. Commun.*, 727, 1984.

75. Shinkai, S., Miyazaki, K., and Manabe, O., A photochemically "switched-on" crown ether containing an intramolecular 4-methoxyphenyl azo substituent, *Angew. Chem. Int. Ed.*, 24, 866, 1985.

76. Ueno, A., Tomita, Y., and Osa, T., Photoresponsive binding ability of azobenzene-appended γ-cyclodextrin, *Tetrahedron Lett.*, 24, 5245, 1983.

77. Shinkai, S., Minami, T., Kusano, Y., and Manabe, O., Photoresponsive crown ethers. 5. Light-driven ion transport by crown ethers with a photoresponsive anion cap, *J. Am. Chem. Soc.*, 104, 1967, 1982.

78. Shinkai, S., Yoshida, T., Manabe, O., and Fuchita, Y., Photoresponsive crown ethers. 20. Reversible photocontrol of association-dissociation equilibria between azo*bis*(benzo-18-crown-6) and diammonium cation, *J. Chem. Soc., Perkin Trans. 1*, 1431, 1988, their earlier papers and the references cited therein.

67

Photocycloaddition Reactions To Imines

kehiko Nishio
ersity of Tsukuba

While [2+2]-photocycloadditions of olefins to carbon-carbon, carbon-oxygen, and carbon-sulfur double bonds to yield the expected four-membered ring compounds are well characterized and often employed in organic synthesis, similar cycloadditions to imines are less frequently encountered.[1] The reason invoked generally is poorer reactivity of the excited imino group due to the rapid radiationless decay that results by twisting around the carbon-nitrogen double bond (*syn,anti*-isomerization). In this chapter, photocycloaddition reactions of imines to C=C and C=N bonds are summarized.

The first example of such a reaction was reported by Searles and Clasen.[2] A 1,2-diazetidine (2) has been proposed as the unstable intermediate formed in the photoreaction of N-(4-dimethylamino-benzylidene)aniline (1) which produces the stilbene (3) and azobenzene (4). This reaction has not been repeated following a thorough reinvestigation by Ohta and Tokumaru.[3] The photodimer of benzaldehyde cyclohexylimine (5) was postulated originally to be the 1,3-azetidine (6) by Kan and Furey[4] but subsequently was identified as the reductive dimer (7) by Padwa et al.[5]

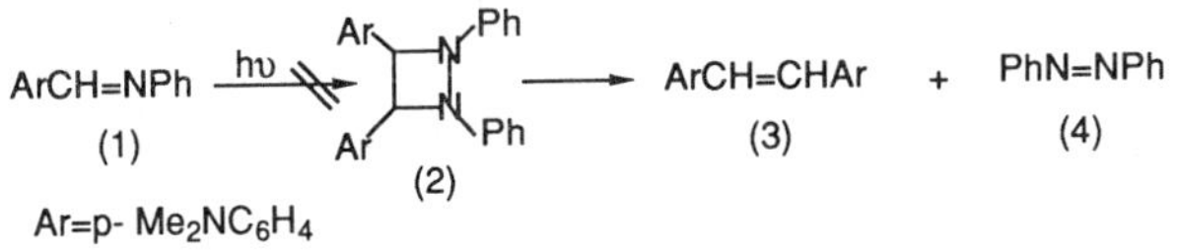

$$\text{ArCH=NPh} \xrightarrow{h\upsilon} (2) \longrightarrow \text{ArCH=CHAr} + \text{PhN=NPh}$$

(1)   (2)   (3)   (4)

Ar=p- Me$_2$NC$_6$H$_4$

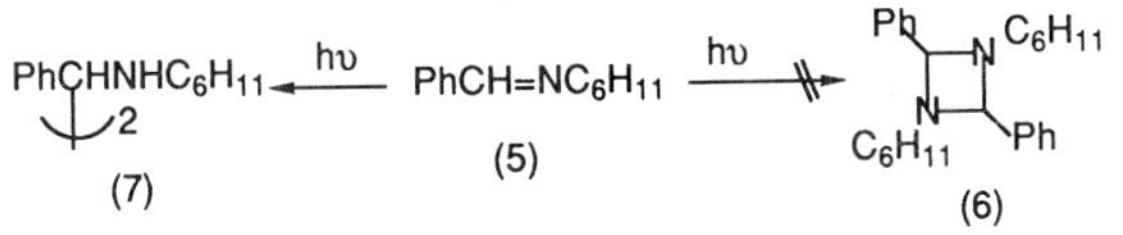

$$\text{PhCHNHC}_6\text{H}_{11} \xleftarrow{h\upsilon} \text{PhCH=NC}_6\text{H}_{11} \xrightarrow{h\upsilon} (6)$$

(7)   (5)   (6)

The first documented example of the photocycloaddition of olefins to the C=N bond was reported by Tsuge et al.[6] Irradiation of the oxadiazole (8) and indene in the presence of iodine leads to the formation of the [2+2]-adduct (9). A different pathway is followed on direct irradiation in benzene leading to the formation of the oxadiazepine (10). Analogous [2+2]-photocycloaddition reactions of the oxadiazole (8) to furan or benzothiophen in the presence or absence of a triplet sensitizer have been reported.[7] In the presence of iodine, an unexpected photoaddition occurs leading to the formation of 3-acetylhydrazone (11) and acetylfuran (12).[7a]

0-8493-8634-9/95/$0.00+$.50
© 1995 by CRC Press, Inc.

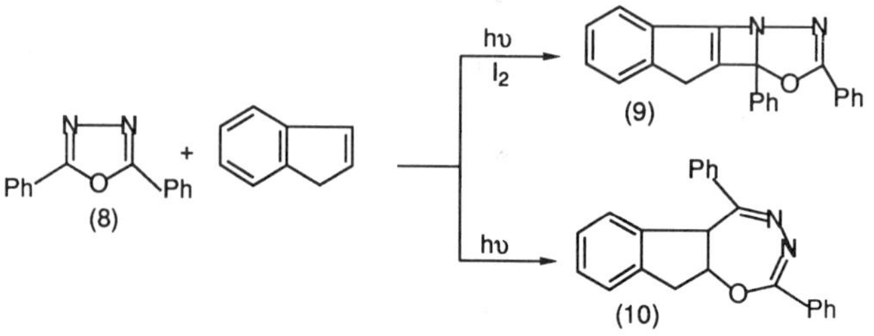

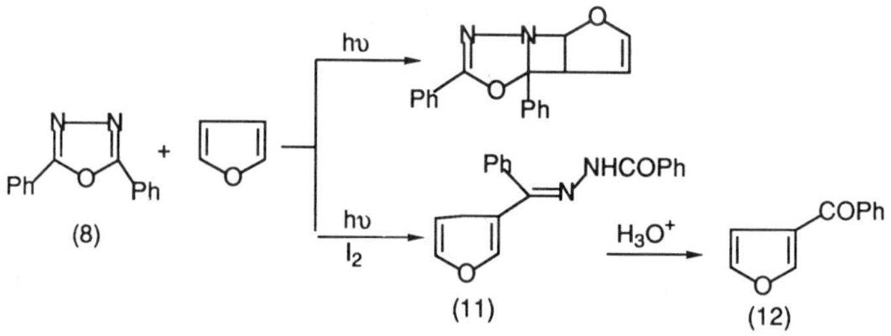

Photoaddition to the carbon-nitrogen double bond is less easily effected, particularly in the case of acyclic imines. The fluorinated imine (13), however, is reported to give the 1,3-diazetidine (14) on irradiation in acetone.[8] The *syn,anti*-isomerization exhibited by the fluorinated imine (15) and the cycloaddition reaction to yield the azetidine (16) are proposed to involve the T_1 state.[9] The cycloaddition reaction is not observed for nonfluorinated analogues.

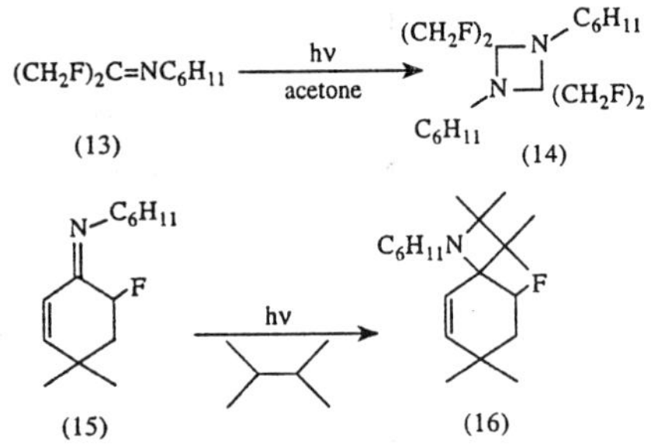

Surprisingly, a concentrated solution of 2-phenylbenzoxazole (17) in cyclohexane is reported to undergo dimerization to the 1,3-diazetidine (18) on irradiation.[10] In dilute solution, this reaction is no longer efficient. 2-(4-Fluorophenyl)benzoxazole has been shown to form the head-to-tail dimer on irradiation in a two-phase system.[11] Photodehalogenation competes with photodimerization in the corresponding 2-(4-chlorophenyl)derivative and dehydrobromination is observed exclusively with 2-(4-bromophenyl)benzoxazole.[11b] 1,3-diazetidine is claimed to be a product of the solid-state irradiation of cinnoline.[12]

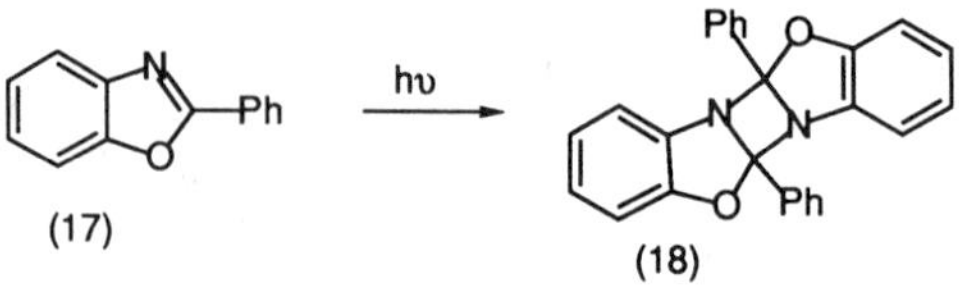

Various 4-cycloalkylideneoxazole-5(4H)-ones (19) have been observed to undergo photodimerization in the solid state;[13] the cyclohexylidene derivative (19), for example, is converted in this way to the 1,3-diazetidine (20) in almost quantitative yield.

(19) (20)

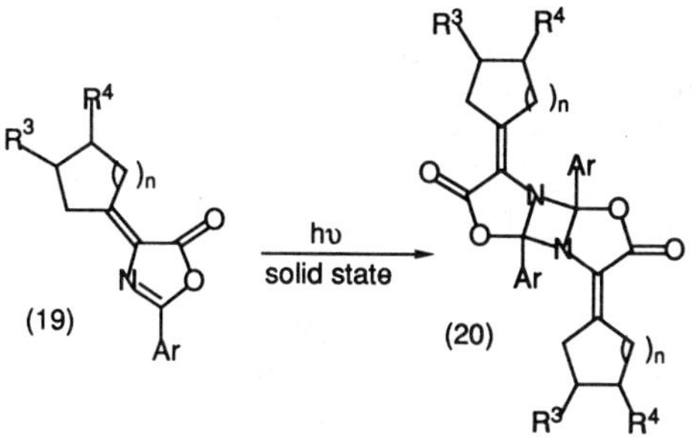

Unlike simple alkyl and aryl substituted systems, imines conjugated with electron-withdrawing carbonyl or aryl functions at nitrogen or carbon appear to participate more readily in excited state [2+2]-cycloaddition processes.

A full report of the photoaddition of olefins to the imine linkages of azauracil and azathymine has appeared. 1,3-Dimethyl-6-azauracil and 1,3-dimethyl-6-azathymine undergo high-yield cycloaddition reactions to a variety of olefins including ethylene, tetramethylethylene, isobutene, ethyl vinyl ether, vinyl acetate, and isopropenyl acetate to produce bicyclic azetidines.[14] For example, acetone-sensitized irradiation of azauracil (21) in the presence of isobutene yields the regioisomeric azetidines (22) and (23) in a 92:8 ratio.[14b] With ethyl vinyl ether or vinyl acetate and 1,3-dimethyl-6-azauracil (21), the epimeric azetidines (24) and (25) are obtained in comparable yields.[14a,b] The corresponding photoaddition of 6-azauracils to isopropenyl or cyclohexenyl acetate forms the basis of a versatile high-yield synthesis of 5-substituted 6-azauracils.[14c,d] In an analogous fashion, the 6-azathymine photoadds to both tetramethylethylene and vinyl acetate to yield azetidines.[14b] The regiochemical outcome of these processes can be explained in terms of either orientation in a triplet olefin-azauracil complex or stabilization of biradicals formed by stepwise addition of olefin to triplet azauracil. The adducts (26) are also formed readily by the sensitized photoaddition of the azauracil (21) to maleimides.[15]

(21) (22) (23)

(21) (24) (25)

R^1=H R^2=OEt
R^1=H R^2=$OCOCH_3$
R^1=Me R^2=$OCOCH_3$

(21) (26)

(27) (28)

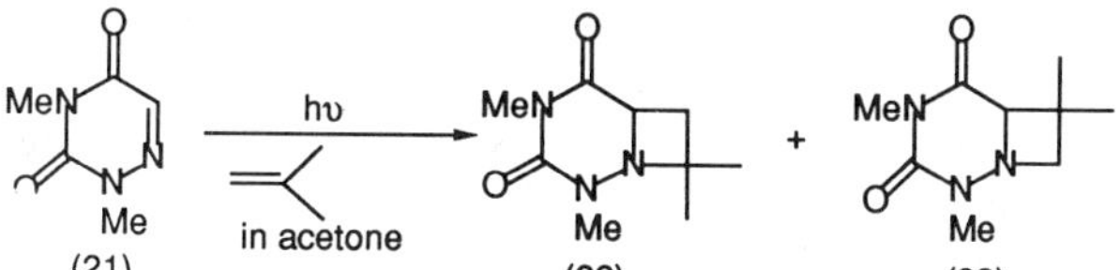

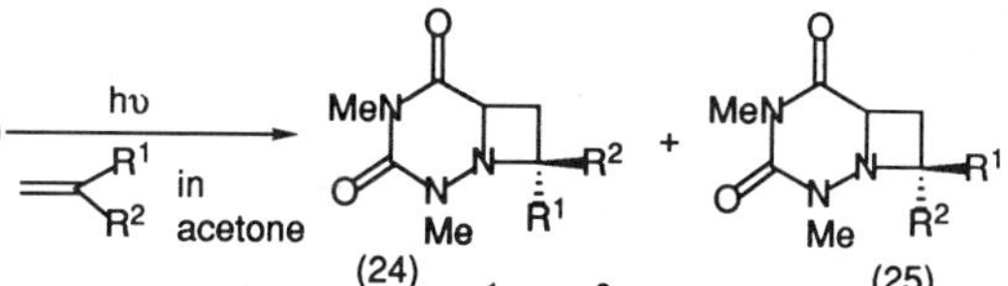

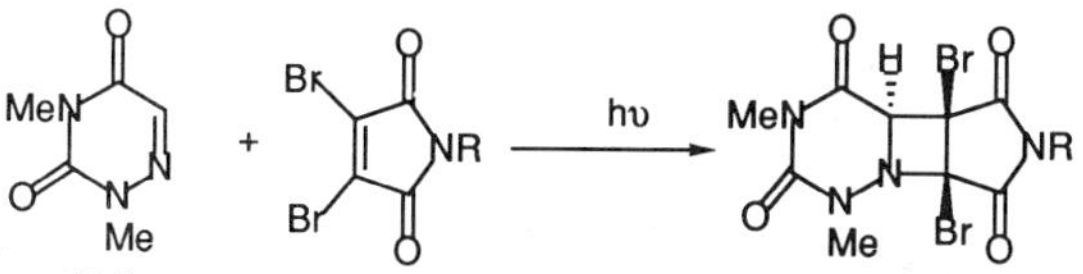

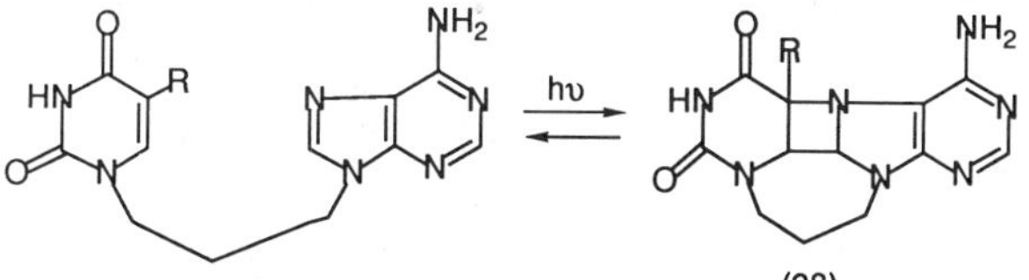

The first example of an intramolecular photocycloaddition was claimed for the pyrimidine-purine dinucleotide analogues (27); the major products of irradiation in aqueous solution are believed to be the thermally and photochemically unstable azetidine (28).[16] Acetone-sensitized irradiation of pyrimidine-6-azapyrimidine dinucleotide analogues also yields azetidines.[17]

The photoaddition of olefins to the isoindolone (29) and oxazolinones (30) have been described.[18] Irradiation of 3-ethylisoindolone (29) with electron-rich olefins such as 1,1-dimethoxyethylene, 2-butene, cyclohexene, and furan results in regiospecific, but nonstereospecific cycloaddition and affords the azetidines (31–33, 36–37).[18a,b] With an electron-deficient olefin such as fumaronitrile, 3-ethoxyisoindolone (29) shows no tendency to undergo [2+2]-photocyclo-addition.[18a] In certain instances, competing reactions are observed. With isobutene and 2-butene, the major products (34, 38) are those derived by a photochemical "ene-type" reaction, whereas with tetramethylethylene a unique reaction occurs leading to the formation of the azepinone (35). This reaction could be analogous to the ring expansion processes encountered with phthalimides.[19] The photoaddition of 3-ethoxyisoindolone (29) to tetramethylethylene and *cis*-2-butene has been shown to be quenched by di-*t*-butyl nitroxide, *cis*-piperylene, and biacetyl. A mechanism that accounts for the formation of all products in terms of a triplet exciplex has been proposed.[18d] The 2-aryl-2-oxazolin-4-ones (30) also undergo [2+2]-photocycloaddition with electron-rich olefins to give cycloadducts (39, 40).[18c] The photocycloaddition proceeds with high regiospecificity. The bicyclic azetidine (42) is prepared by irradiation of the 1,3-oxazin-4-one (41) in the presence of 1,1-dimethoxyethylene.[20] This product is used as the precursor for the formation of the azetidine (43), which is prepared by the pyrolysis of (42) at 225 °C.

As depicted above, olefin-imine photocycloadditions are efficient excited-state reactions when the imine function has ring constraint and contains electron-withdrawing groups on the carbon or nitrogen atoms. In addition, electron-rich olefins are generally required for these reactions. However, this requirement is not rigorous, and the photoaddition of imines and electron-poor olefins has been observed. 1,3-Dimethyl-6-azauracil (21) photoadds to the electron-poor olefin dibromomaleimide to give the [2+2]-adduct (26).[15]

Irradiation of both the quinoxalin-2-ones (44) and the benzoxazin-2-ones (45) with electron-deficient olefins affords the [2+2]-adducts (46, 47) as a mixture of stereoisomers with high degrees of regioselectivity.[21a-c] A triplet excited state is involved. In a similar manner, the imines (44, 45) are converted into the azetidines by photoaddition to aryl alkenes.[21d] Tetra-hydroquinoxalinones undergo regioselective photoaddition to electron-poor and to aryl olefins to give azetidines.[21e] A triplet excited state of the pteridine-2,4,7-triones (48) is photoreactive and adds to olefins to afford reasonable yields of the azetidines (49).[21f]

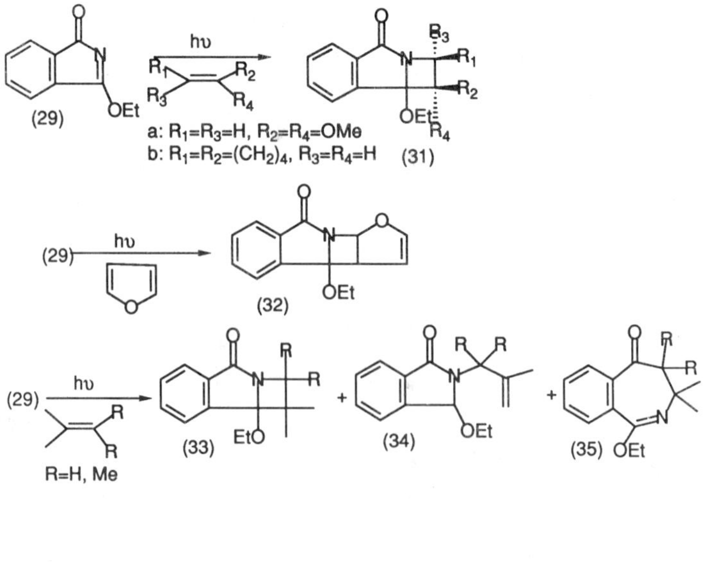

a: R$_1$=R$_3$=H, R$_2$=R$_4$=OMe
b: R$_1$=R$_2$=(CH$_2$)$_4$, R$_3$=R$_4$=H (31)

(32)

R=H, Me

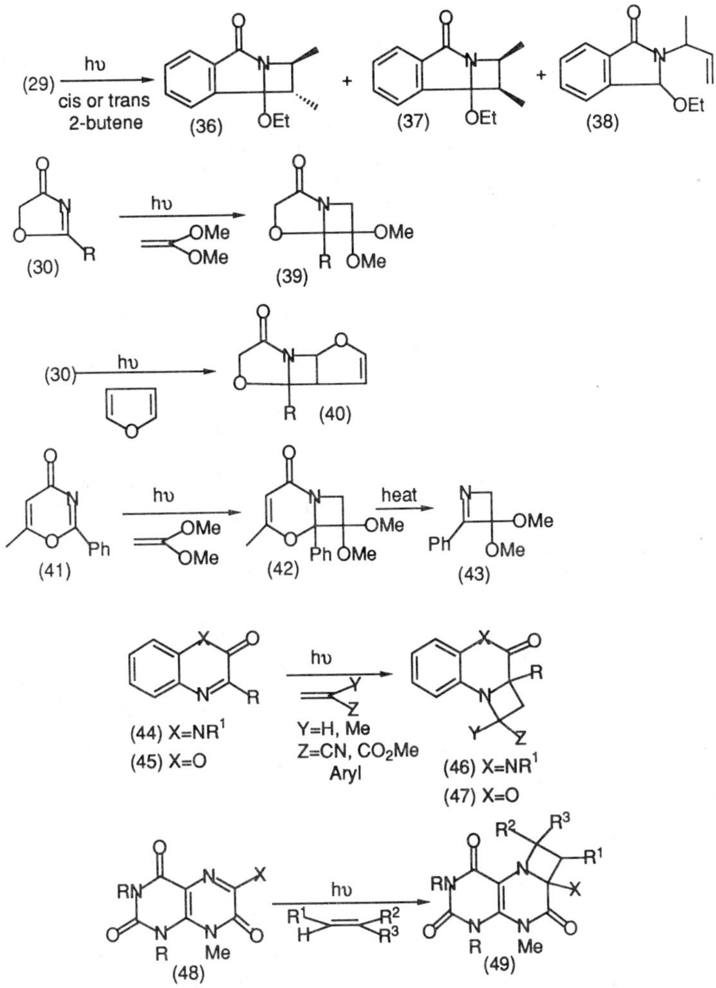

The activation of C=N bonds in heteroaromatic compounds by the trifluoromethyl group toward photocycloaddition has been demonstrated.[21] Both electron-rich and electron-poor olefins add photochemically to 3-trifluoromethylquinoxalin-2-one (50) and to 3-trifluoro-methylbenzoxazin-2-one (51) to afford the azetidines (52–55).[22a] The intra- and intermolecular [2+2]-cycloadducts (58, 59) are obtained in good yields by irradiation of the 2-trifluoromethyl-4(3H)-quinoxalinones (56, 57). The reaction is dependent on the nature of the substituent on C-2 and when this is H or Cl, no photocycloaddition reactions take place.[22b] The photoassisted cycload-dition reaction of 3-trifluoromethyl-quinoxalinones and -benzoxazinone with ketene has also been reported.[22c] The azetidines having a trifluoromethyl group are synthesized by the photochemical [2+2]-cycloaddition of the trifluoromethyl substituted derivative of triazin-5-ones and -3,5-diones with olefins.[22d]

3-Aryl-2-isoxazolines (60) with electron-withdrawing substituents on the 3-aryl moiety react regiospecifically with furan, thiophene, and indene to give the cycloadducts (61–63).[23a-c] The *p*-methoxy and unsubstituted phenyl derivatives of (60) fail to undergo cycloaddition with these olefins. This novel substituent effect suggests that a donor-acceptor interaction (exciplex or charge transfer complex) between the singlet excited state of the 2-isoxazolines and olefinic substances is important. Similarly, the excited-state reactivity of *p*-cyano- and *p*-methoxycarbonyl-analogues of (60) with aromatic substrates has been studied.[23d,e] Irradiation of the isoxazolines (60) in benzene leads to the formation of the tricyclic azetidines (64). In contrast, irradiation of the isoxazolines (60) in methylated benzene derivatives such as toluene, xylene, and mesitylene gives benzyl-adducts via a hydrogen atom abstraction radical combination path. The hydrogen abstraction takes place at the carbon-nitrogen double bond.[23e]

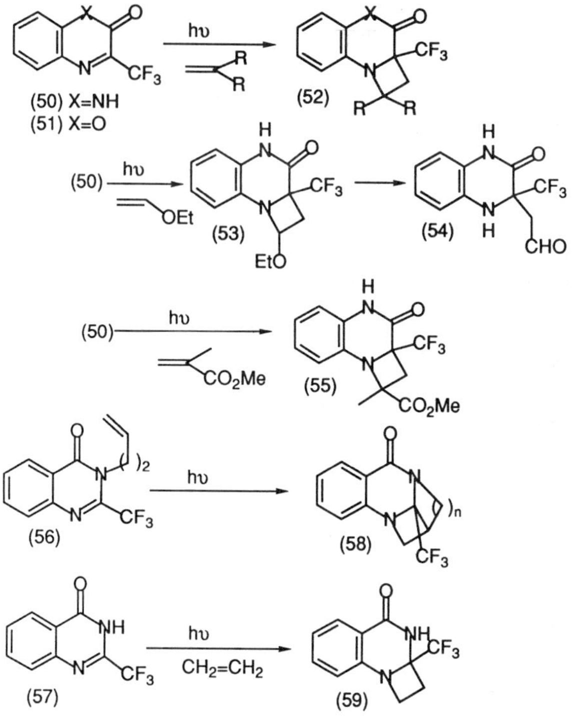

(50) X=NH
(51) X=O
(52)

(53)
(54)

(55)

(56)
(58)

(57)
(59)

The photoaddition of olefins to the 3-aryl-1,2-benzothiazoles (65) is formally a [π2+σ2]-cycloaddition. The reaction is regiospecific with respect to the direction of addition of ethyl vinyl ether and stereospecific with respect to the reaction with *cis*- and *trans*-2-butene.[24a,b] Photoaddition of 2-phenylbenzothiazole (67) to olefins is reported to give the 1,5-benzothiazepines (68) in a regiospecific and stereospecific manner.[24c] Photoaddition reactions of (67) with electron-rich alkynes give substituted 1,5-benzothiazepines in a one-step process.[24d] Two mechanisms are possible in theory, one involves [2+2]-cycloaddition to the C=N bond, followed by thermal ring opening of the resultant azetidine and a second involving C-S bond homolysis and subsequent radical addition to the olefins. On the evidence available, the latter pathway seems more reasonable. Photoaddition of the electron-rich acetylenes to 3-phenyl-1,2-benzoisothiazole (65) leads to the formation of 3,4-benzo-2,6-thiazabicyclo[3,2,0]hepta-3,6-dienes (69, 70).[24e] The route to these products may involve [2+2]-cycloaddition to yield an azetidine followed by S-N bond homolysis and then rearrangement to afford the final product. Electronically excited stilbene is trapped by the 8-substituted caffeines (71) to yield the propellanes (72) and the addition products (73–75).[25] The formation of azetidines is proposed to account for the photoaddition products. Photoaddition of stilbene with the benzothiazole (76) has been reported to afford the azetidine (77) and the insertion product (78).[26] The azetidinone (79) is formed by the photocycloaddition between an excited stilbene and phenyl isocyanate.[27]

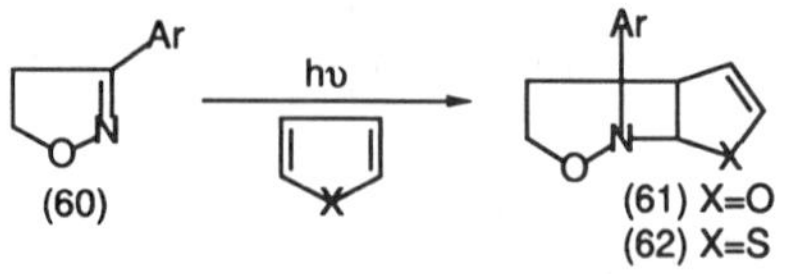

(60)
(61) X=O
(62) X=S

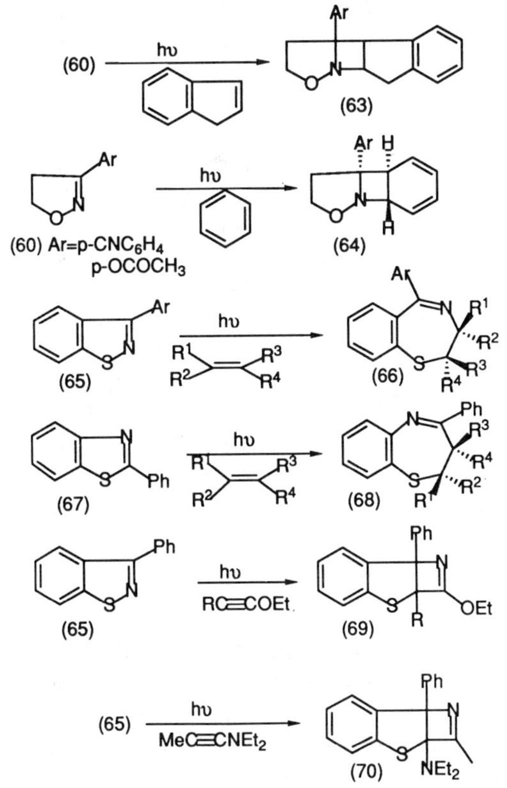

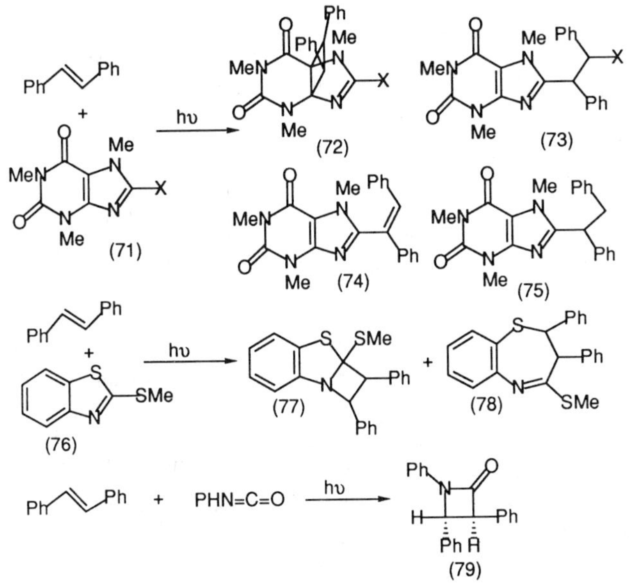

The regiospecific addition of the 6-substituted phenanthridines (80) to electron-rich olefins to give the azetidines (81) and/or the azocines (82) in ethanol has been described.[28] The latter products are shown to be secondary photoproducts arising by ring opening and photosolvolysis of the primary adducts (81). The reaction proceeds via an exciplex formed from the π,π* singlet state of the phenanthridine. An equivalent intramolecular cycloaddition has been accomplished in high yield by direct or acetone-sensitized irradiation of the imine (83) to give the azetidine (84).[29] An intramolecular cycloaddition of this type has been used in the construction of the azapropellane (86) from the oxime ether (85).[30]

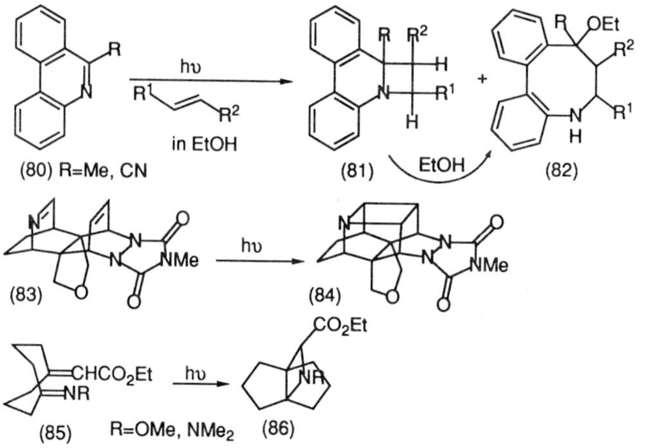

(80) R=Me, CN   in EtOH   (81)   EtOH   (82)

(83)   hυ   (84)

(85)   R=OMe, NMe₂   (86)

Irradiation of the 1-aza diene (87) in the solid state produces the bicyclo[3,1,0]hexene derivatives (88) and the intermolecular [2+2]-cycloadduct (89); whereas in solution, the former (88) are produced.[31] This primary photoproduct proves to be extremely unstable, undergoing rapid hydrolysis to the imide (89). Inspection of the X-ray crystal structure of the azadiene (87) suggests that the formation of [2+2]-cycloadduct in the solid state is primarily the result of topochemical control.

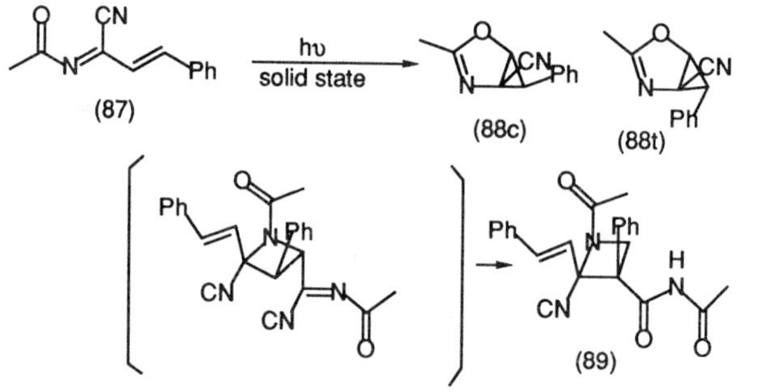

(87)   hυ   (88c)   (88t)
            solid state

(89)

A variety of the substituted β-lactams (91–93), including a cepham analog, are synthesized by the photochemical reactions of the chromium carbene complex (90) with the substituted imines.[32a,b] Photolysis of the chromium alkoxycarbene complex (90) with the N-(benzoloxycarbonyl)-imidazolines produces the protected azapemams (94).[32c] Hydrogenolysis of (94) gives the free azapenams (95). Hydro-1,4-diazaepin-5-ones (96) are obtained by hydrogenolysis under acidic conditions. Although the oxazines (98) and oxazolines are inert toward the chromium carbene complex (90), they are converted to the bicyclic β-lactams (99) by the photolytic reaction of the molybdenum carbene complexes (97).[32d] Photolytic reaction of chromium carbene complexes containing amino group [$(CO)_5Cr=C(H)NR_2$] with imine, oxazine, oxazole, imidates, thiazines, and thiazolines produces β-lactams in fair to good yields.[32e] For example, the oxazine (98) is converted to the β-lactam (100) by photolytic reaction with the aminocarbene complex (90). The dibenzyl-β-lactam (100) is cleanly debenzylated by hydrogenolysis to give the amino-β-lactam (101), which is of biological interest.

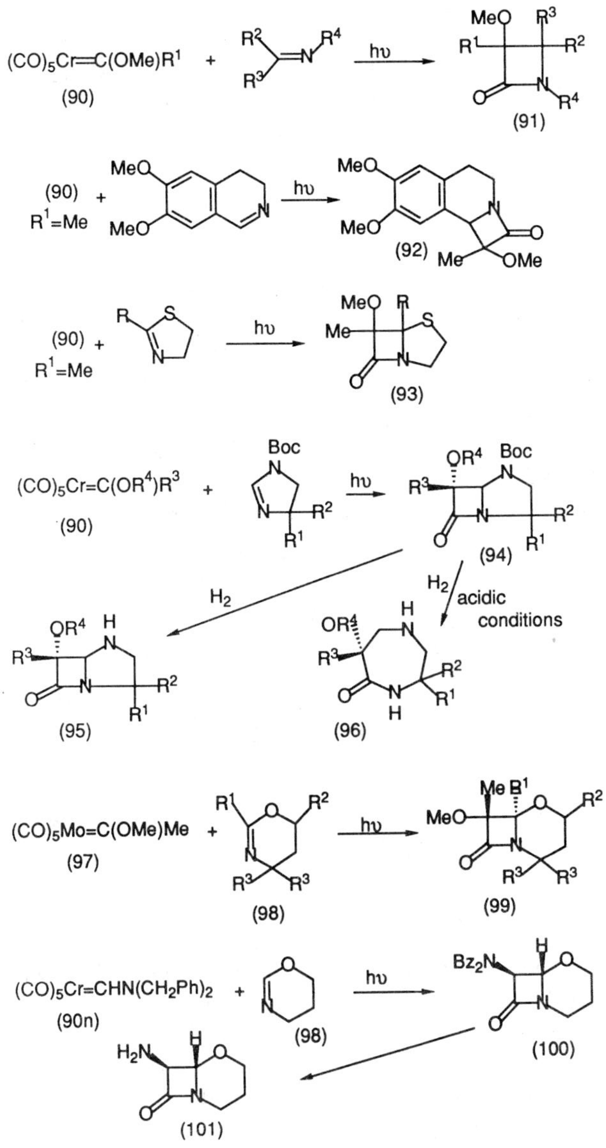

A reasonable mechanism for the β-lactam-forming reaction is shown in the scheme, in which photolytic generation of the chromium-ketene complex (102) from the chromium carbene complex (90n) and subsequent reaction of this complexed ketene with the substrate produces the observed products (103).[32f] Photolysis of chromium carbene complexes containing a chiral, optically active alcohol group produces lactones in high yield and high diastereoselectivity which are converted to optically active amino acids.[32f] Monocyclic-β-lactams are prepared by the photolytic reaction of a chromium carbene complex with the s-1,3,5-triazines (104).[32g] The stereoselectivity of the reaction of imine with ketene generated by the reaction of acid chloride with triethylamine and that for the complexed ketene generated by photolysis of chromium carbene complexes in the presence and absence of added triethylamine have been compared.[32h] Triethylamine addition to reactions of carbene complexes affords results that closely parallel those of ketene generated from acid chlorides. Photochemical reaction of chromium carbene complexes and iminodithiocarbanates affords the β-lactams (105) that upon oxidation, afford the 4-oxo-β-lactams (106).[33]

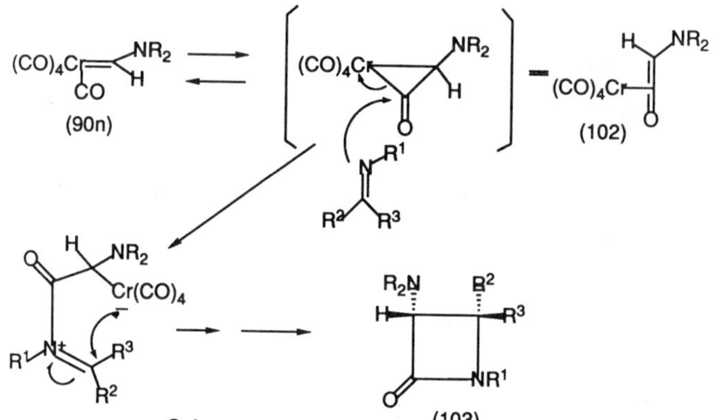

$(CO)_4C$ ... NR_2 ... $\longrightarrow$ ... $(CO)_4Cr$ ... NR_2 ... $(CO)_4Cr$ ... NR_2

CO

(90n) (102)

R^1

R^2 R^3

$Cr(CO)_4$

R^1 R^3

R^2

R_2N R^2

R^3

NR^1

Scheme (103)

R

$(CO)_5Cr$ NBz_2 $+$ RN NR $\xrightarrow{h\upsilon}$ Bz_2N

H

(90n) (104) NR

(103)

R^2 SMe

$(CO)_5Cr$ R^2 $+$ NR^1 $\xrightarrow{h\upsilon}$ R^1 SMe

R^3 MeS SMe

NR^1

(105)

R^2 O

R^1 $Ox,$

O NR^1

(106)

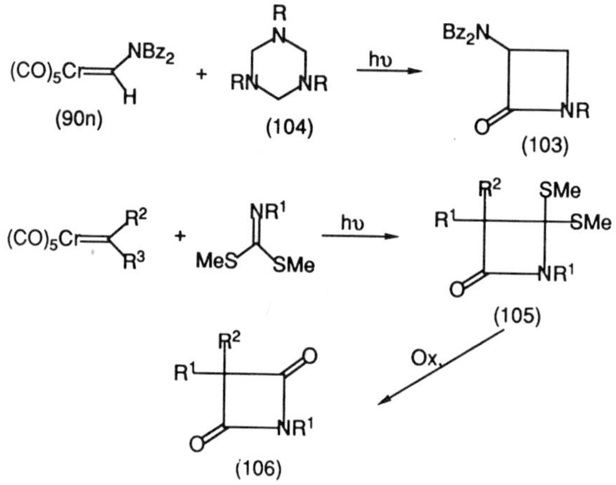

References

1. (a) Padwa, A., Photochemistry of the carbon-nitrogen double bond, *Chem. Rev.*, 77, 37, 1977; (b) Pratt, A. C., The photochemistry of imines, *Chem. Soc. Rev.*, 6, 63, 1977; (c) Mariano, P. S., The photochemistry of substances containing the C=N moiety with emphasis on electron transfer processes, in *Organic Photochemistry*, Vol. 9, Padwa, A., Ed., Marcel Dekker, New York, 1987, 1.
2. Searles, S., Jr. and Clasen, R. A., A 1,2-diazetidine intermediate from photocyclization of a Schiff base, *Tetrahedron Lett.*, 1627, 1965.
3. Ohta, H. and Tokumaru, K., Photochemistry of N-(4-dimethyl-aminobenzylidene)aniline, *Bull. Chem. Soc. Jpn.*, 48, 1669, 1975.
4. Kan, R. O. and Furey, R. L., Photochemical formation of 1,3-diazetidines, *J. Am. Chem. Soc.*, 90, 1666, 1968.
5. Padwa, A., Bergmark, W., and Pashayan, D., On the photoreduction of benzaldehyde N-alkylimines, *J. Am. Chem. Soc.*, 90, 4458, 1968.
6. (a) Tsuge, O., Tashiro, M., and Oe, K., Photochemical reaction of 2,5-diphenyl-1,3,4-oxadiazole with indene, *Tetrahedron Lett.*, 3971, 1968; (b) Oe, K., Tashiro, M., and Tsuge, O., Photochemistry of heterocyclic compounds. 5. Photochemical reaction of 2,5-diaryl-1,3,4-oxadiazoles with indene, *J. Org. Chem.*, 42, 1496, 1977.
7. (a) Tsuge, O., Oe, K., and Tashiro, M., Photochemistry of heterocyclic compounds. II. The photochemical reaction of 2,5-disubstituted 1,3,4-oxadiazoles with furan, *Tetrahedron*, 29, 41, 1973; (b) Oe, K., Tashiro, M., and Tsuge, O., Photochemistry of heterocyclic compounds. VII. Photochemical reaction of 2,5-diphenyl-1,3,4-oxadiazole with benzo[b]thiophenes, *Bull. Chem. Soc. Jpn.*, 50, 3281, 1977.

8. Margaretha, P., Photochemistry of aliphatic imines. The photochemical behaviour of fluorinated N-isopropylidene-cyclohexylamines, *Helv. Chim. Acta*, 65, 290, 1982.

9. Margaretha, P., Photochemistry of 2-cyclohexene-imine and 2,3,4,4a,5,6-hexahydroquinolines, *Helv. Chim. Acta*, 61, 1025, 1978.

10. (a) Roussilhe, J., Despax, B., Lopez, A., and Paillous, N., Photodimerization of 2-phenylbenzoxazole and its acid-catalysed reversion as a new system for light energy conversion, *J. Chem. Soc., Chem. Commun.*, 380, 1982; (b) Roussilhe, J., Fargin, E., Lopez, A., Despax, B., and Paillous, N., Photochemical behavior of 2-phenylbenzoxazole. Synthesis of 1,3-diazetidine via intermolecular [2π + 2π]cycloaddition of two carbon-nitrogen double bonds, *J. Org. Chem.*, 48, 3736, 1983.

11. (a) Paillous, N., Forgues, S. F., Jaud, J., and Devillers, J., [2 + 2]-Cycloaddition of two C=N double bonds. First structural evidence for head-to-tail photodimerization in the 2-phenylbenzoxazole series, *J. Chem. Soc., Chem. Commun.*, 578, 1987; (b) Forgues, S. F. and Paillous, N., Photodehalogenation and photodimerization of 2-(4-halophenyl)benzoxazoles. Dependence of the mechanism on the nature of the halogen atom, *J. Org. Chem.*, 51, 672, 1986.

12. Marshell, P. A., Mooney, B. A., Prager, R. H., and Ward, A. D., Central nervous system active compounds. XI. Cinnolinyl-isobenzofuranones [1-(3-phtalidyl)cinnolin-4(1H)-ones], *Aust. J. Chem.*, 34, 2619, 1981.

13. Lawrenz, D., Mohr, S., and Wendlander, B., Formation of 1,3-diazetidines via C-N dimerization of 4-cycloalkylidene-oxazol-5(4H)-ones in the solid state, *J. Chem. Soc., Chem. Commun.*, 863, 1984.

14. (a) Hyatt, J. A. and Swenton, J. S., Photochemical reactivity of 2,4-dimethyl-1,2,4-triazine-3,5(2H)-dione (1,3-dimethyl-6-azauracil), *J. Chem. Soc., Chem. Commun.*, 1144, 1972; (b) Swenton, J. S. and Hyatt, J. A., Photosensitized cycloaddition to 1,3-dimethyl-6-azauracil and 1,3-dimethyl-6-azathymine. An imine linkage unusually reactive toward photocycloaddition, *J. Am. Chem. Soc.*, 96, 4879, 1974; (c) Swenton, J. S. and Balchunis, R. J., A versatile high yield photochemical synthesis of 5-substituted 6-azauracils, *J. Heterocycl. Chem.*, 11, 453, 1974; (d) Swenton, J. S. and Balchunis, R. J., Photochemical functionalization of 6-azauracils to 5-substituted-6-azauracils, *J. Heterocycl. Chem.*, 11, 917, 1974.

15. Szilágyi, G. and Wanhoff, H., Novel photocycloaddition of dibromomaleimides to 1,3-dimethyl-6-azauracil, *Angew. Chem. Int. Ed.*, 19, 1026, 1980.

16. Paszyc, S., Skalski, B., and Wenska, G., Photochemical reactions of some pyrimidine-purine dinucleotides analogs, *Tetrahedron Lett.*, 449, 1976.

17. Zasada-Parzyńska, A., Celewicz, L., and Golankiewicz, K., Synthesis and photochemical properties of pyrimidine-6-azapyrimidine analogs of dinucleotides with propanone bridge, *Synth. Commun.*, 16, 1177, 1986.

18. (a) Koch, T. H. and Howard, K. A., 2 + 2 Photocycloaddition to a carbon nitrogen double bond. I. 3-Ethoxyisoindolone, *Tetrahedron Lett.*, 4035, 1972; (b) Haward, K. A. and Koch, T. H., Photochemical reactivity of keto imino ethers. V. (2 + 2)Photocycloaddition to the carbon nitrogen double bond of 3-ethoxyisoindolone, *J. Am. Chem. Soc.*, 97, 7288, 1975; (c) Rodehorst, R. M. and Koch, T. H., Photochemical reactivity of keto imino ethers. VI. Type I rearrangement and (2 + 2)photocycloaddition to carbon-nitrogen double bond of 2-oxazolin-4-ones, *J. Am. Chem. Soc.*, 97, 7298, 1975; (d) Anderson, D. R., Keute, J. S., Koch, T. H., and Moseley, R. H., Di-*tert*-butyl nitroxide quenching of the photoaddition of olefins to carbon-nitrogen double bond of 3-ethoxyisoindolenone, *J. Am. Chem. Soc.*, 99, 6332, 1977.

19. Kanaoka, Y., Photoreactions of cyclic imides. Examples of synthetic organic photochemistry, *Acc. Chem. Res.*, 11, 407, 1978; Mazzocchi, P. H., *The photochemistry of imides*, in Organic Photochemistry, Vol. 5, Padwa, A., Ed., Marcel Dekker, New York, 1981, 421; Coyle, J. D., *Phthalimide and Its Derivatives*, in Synthetic Organic Photochemistry, Horspool, W. M., Ed., Plenum Press, New York and London, 1984, 259.

20. Koch, T. H., Higgins, R. H., and Schuster, H. F., An azetidine from a photocycloaddition reaction followed by a retro Diels-Alder fragmentation, *Tetrahedron Lett.*, 431, 1977.

21. (a) Nishio, T., (2 + 2)Photocycloaddition of the carbon-nitrogen double bond of quinoxalin-2(1H)-ones to electron-deficient olefins, *J. Org. Chem.*, 49, 827, 1984; (b) Nishio, T. and Omote, Y., Photocycloaddition reactions of 1,4-benzoxazin-2-ones and electron-poor olefins, *J. Org. Chem.*, 50, 1370, 1985; (c) Nishio, T., Photochemical reactions of quinoxalin-2-ones and related compounds, *J. Chem. Soc., Perkin Trans. 1*, 565, 1990; (d) Nishio, T. and Omote, Y., Photocycloaddition of quinoxalin-2-ones and benzoxazin-2-ones to aryl alkenes, *J. Chem. Soc., Perkin Trans. 1*, 2611, 1987; (e) Nishio, T., Kondo, M., and Omote, Y., Photochemical reactions of tetrahydroquinoxalin-2(1H)-ones and related compound, *Helv. Chim. Acta*, 74, 225, 1991; (f) Nishio, T., Nishiyama, T., and Omote, Y., Photochemical [2 + 2]-cycloaddition of the C≠ N bond of pteridin-2,4,7-triones to alkenes, *Liebigs Ann. Chem.*, 441, 1988.

22. (a) Kaneko, C., Kasai, K., Watanabe, H., and Katagiri, N., Cycloadditions in synthesis. XXXI. 2 + 2 Photocycloaddition of 3-trifluoromethyl derivatives of 2-quinoxalin-2(1H)-one and 1,4-benzoxazin-2-one to olefins: Effects of the trifluoromethyl group, *Chem. Pharm. Bull.*, 34, 4955, 1986; (b) Kaneko, C., Kasai, K., Katagiri, N., and Chiba, T., Photoaddition of 4(3H)-quinazolone derivatives to olefins: Effects of the 2-substituent, *Chem. Pharm. Bull.*, 34, 3672, 1986; (c) Katagiri, N., Kasai, K., and Kaneko, C., Synthesis of 4-trifluoromethylazetidin-2-ones by a novel 2 + 2 photocycloaddition of 3-trifluoromethyl-quinoxalin-2-one or 1,4-benzoxazin-2-one to ketene, *Chem. Pharm. Bull.*, 34, 4429, 1986; (d) Katagiri, N., Watanabe, H., and Kanekeo, C., Cycloaddition in synthesis. XXXVII. Synthesis of 6-trifluoromethyl-1,2,4-triazines and -1,2,4-triazin-5-ones and their pericyclic reactions with olefins, *Chem. Pharm. Bull.*, 36, 3354, 1988.

23. (a) Kumagai, T., Kawamura, Y., and Mukai, T., Photocycloaddition of 3-aryl-2-isoxazolines with five-membered heterocycles, *Chem. Lett.*, 1357, 1983; (b) Mukai, T., Kumagai, T., Saiki, H., and Kawamura, Y., Photochemical behavior of cyclic imino ethers: The N-O bond fission, *syn-anti* isomerization and cycloaddition reactions in the C≠ N-O chromophore, *J. Photochem.*, 17, 365, 1981; (c) Kawamura, Y. Kumagai, T., and Mukai, T., Photocycloaddition reaction of 3-aryl-2-isoxazolines with indene. Generation of [2 + 2]cycloadduct stereoisomers, *Chem. Lett.*, 1937, 1985; (d) Kumagai, T., Shimizu, K., Kawamura, Y., and Mukai, T., Photochemistry of 3-aryl-2-isoxazoline, *Tetrahedron*, 37, 3365, 1981; (e) Kumagai, T., Kawamura, Y., and Mukai, T., Photochemical reaction of 3-aryl-2-isoxazolines with methylated benzenes, *Tetrahedron Lett.*, 24, 2279, 1983.

24. (a) Sindler-Kulyk, M. and Neckers, D. C., Photocycloadditions to 3-phenyl-1,2-benzoisothiazole, *Tetrahedron Lett.*, 22, 529, 1981; (b) Sindler-Kulyk, M. and Neckers, D. C., Photocycloaddition to 3-phenyl-1,2-benzoisothiazole, *Tetrahedron*, 37, 3377, 1981; (c) Sindler-Kulyk, M. and Neckers, D. C., Photocycloadditions to 2-phenylbenzothiazole, *Tetrahedron Lett.*, 22, 2081, 1981; (d) Sindler-Kulyk, M. and Neckers, D. C., Photochemistry of 2-phenyl-benzothiazole with ethoxyacetylene and ethoxypropyne. Synthesis of 1,5-benzothiazepines, *J. Org. Chem.*, 47, 4914, 1982; (e) Sindler-Kulyk, M. and Neckers, D. C., Photocycloaddition reactions of 3-phenyl-1,2-benzoisothiazole and alkynes, *J. Org. Chem.*, 48, 1275, 1983.

25. Kaupp, G. and Ringer, E., Multifunktionelle photoaddition von stilben an coffeinderivate und benzothiazole, *Chem. Ber.*, 119, 1525, 1986.

26. Kaupp, G. and Grüter, H. W., Bekannte und neue reaktionstypen bei der photoreaktion von stilbenen mit cyclischen iminen, *Chem. Ber.*, 114, 2844, 1981.

27. Kubota, T. and Sakurai, H., The photocycloaddition of stilbene to phenyl isocyanate, *J. Chem. Soc., Chem. Commun.*, 362, 1972.

28. (a) Futamura, S., Ohta, H., and Kamiya, Y., Photocycloaddition of 6-cyanophenanthridine to electron-rich olefins, *Chem. Lett.*, 655, 1980; (b) Futamura, S., Ohta, H., and Kamiya, Y., The photocycloaddition of 6-substituted phenanthridines to electron-rich olefins, *Bull. Chem. Soc. Jpn.*, 55, 2190, 1982.

29. Fischer, G., Fritz, H., and Prinzbach, H., An intramolecular imine/ene-photo-[2 + 2]-cycloaddition reaction, *Tetrahedron Lett.*, 27, 1269, 1986.

30. Malamidou-Xenikaki, E., and Nicolaides, D. N., Synthesis of heterocyclic propellanes. Preparation and transannular reactions of 5-ethoxycarbonylmethylene-cyclooctanone and the corresponding oximes and hydrazones, *Tetrahedron,* 42, 5081, 1981.

31. Teng, M., Lauher, J. W., and Fowler, F. W., Solid-state and solution photochemistry of a 1-aza diene, *J. Org. Chem.,* 56, 6840, 1991.

32. (a) McGuire, M. A. and Hegedus, L. S., Synthesis of β-lactams by the photolytic reaction of chromium carbene complexes with imines, *J. Am. Chem. Soc.,* 104, 5538, 1982; (b) Hegedus, L. S., McGuire, M. A., Schultze, L. M., Yijun, C., and Anderson, O. P., Reaction of chromium carbene complexes with imines. Synthesis of β-lactams, *J. Am. Chem. Soc.,* 106, 2680, 1984; (c) Betschart, C. and Hegedus, L. S., Synthesis of azapenams, diazepinones, and dioxocyclams via the photolytic reaction of chromium alkoxycarbene complexes with imidazolines, *J. Am. Chem. Soc.,* 114, 5010, 1992; (d) Hegedus, L. S., Schultze, L. M., Toro, J., and Yijun, C., Photolytic reaction of chromium and molybdenum carbene complexes with imines, *Tetrahedron,* 41, 5833, 1985; (e) Borel, C., Hegedus, L. S., Krebs, J., and Satoh, Y., Synthesis of amino-β-lactams by the photolytic reaction of imines with penta-carbonyl[(dibenzylamino)carbene]chromium(0), *J. Am. Chem. Soc.,* 109, 1101, 1987; (f) Hegedus, L. S., de Weck, G., and D'Andrea, S., Evidence for the intermediacy of chromium-ketene complexes in the synthesis of β-lactams by the photolytic reaction of chromium-carbene complexes with imines. Use in amino acid synthesis, *J. Am. Chem. Soc.,* 110, 2122, 1988; (g) Hegedus, L. S. and D'Andrea, S., Synthesis of monocyclic β-lactams by the photolytic reaction of chromium carbene complexes with s-1,3,5-triazines, *J. Org. Chem.,* 53, 3113, 1988; (h) Hegedus, L. S., Montgomery, J., Nakamura, Y., and Snustad, D. C., A contribution to the confusion surrounding the reaction of ketenes with imines to produce β-lactams. A comparison of stereoselectivity dependence on the method of ketene generation: Acid chloride/triethylamine vs. photolysis of chromium-carbene complexes, *J. Am. Chem. Soc.,* 113, 5784, 1991.

33. Alcaide, B., Dominguez, G., Plamet, J., and Sierra, M. A., Chromium-carbene-mediated synthesis of 4-oxo β-lactam (malonimides) and malonic acid derivatives, *J. Org. Chem.,* 57, 447, 1992.

68

Photocycloaddition Reactions of Indoles

Alan C. Weedon
University of Western Ontario

68.1 Introduction

Since the isolation and characterization of indole by Baeyer[1] in 1866, the chemistry of the indole ring system and its derivatives has been extensively investigated and a very large literature has accumulated.[2] Somewhat surprisingly, during the first century of the study of indole chemistry only a handful of articles appeared in which the photochemical properties of the indole ring system were discussed. In part, this paucity arises from the relative efficiency of photophysical routes for decay of indole excited states that detract from the potential excited-state chemistry. However, in 1973 two reports appeared[3,4] that described how the placement of an acyl substituent on the indole nitrogen activated the indole ring toward photochemical cycloaddition with alkenes and carbonyl compounds to give cyclobutanes and oxetanes, respectively. Subsequently, it was reported[5] that irradiation of indoles with alkynes yields cyclobutene adducts; very recently it was shown that indole will undergo an electron transfer-sensitized Diels-Alder reaction with dienes.[6,7] This chapter reviews the published work describing the photochemical cycloaddition of the π-bonds of alkenes, dienes, alkynes, and carbonyl compounds across the 2,3-position of the indole ring. The basic reaction for each cycloaddition partner is shown in Scheme 68.1.

0-8493-8634-9/95/$0.00+$.50
© 1995 by CRC Press, Inc.

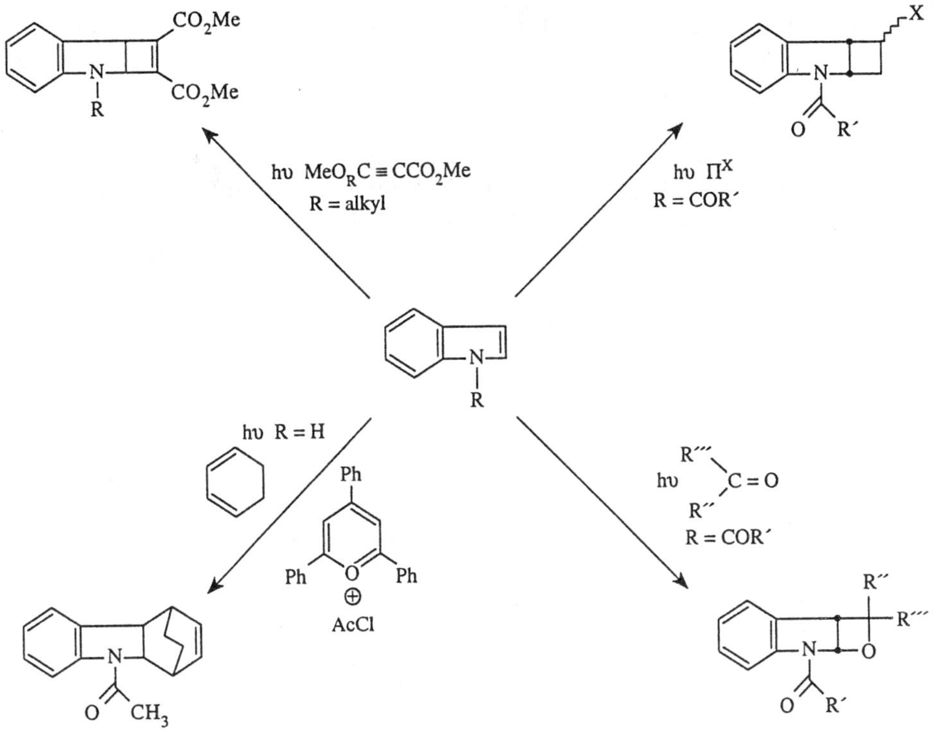

SCHEME 68.1

3.2 The Reaction Mechanism

As with thermal cycloaddition reactions, the successful synthetic application of photochemical cycloaddition reactions requires an appropriate knowledge of the reaction mechanism in order to predict and control the reaction regiochemistry and stereochemistry. In this section, the mechanistic models that have been developed to explain the dependence of product distribution on the structures of the reaction partners are summarized.

Cycloaddition of Alkenes to N-Acylindoles

In the first report of the photochemical cycloaddition of an alkene to an indole, Julian and Foster noted[4] that for the photoaddition to proceed it was necessary for an acyl group to be present on the indole nitrogen atom. In addition, the reaction was found to be regioselective but not necessarily stereoselective. Thus, various combinations of methyl acrylate, acrylonitrile, acrylamide, ethyl vinyl ether, acrylic acid, methyl vinyl ketone, isobutene, or vinyl acetate with N-benzoylindole, N-(p-chlorobenzoyl)indole, 4-methoxy-N-(p-chlorobenzoyl)-indole, or N-acetylindole gave the head-to-tail regioisomers 1a and 1b, and little or none of the head-to-head regioisomer 2 when the indole derivative was irradiated in benzene solution using Pyrex-filtered UV light. Indole and N-methylindole were unreactive, although the formation of substitution products derived via an electron-transfer route has been observed when N-alkyl or N-unsubstituted indoles are irradiated with acrylonitrile in acetonitrile.[8] In most cases, the cycloaddition reaction gave the *exo*-stereoisomer 1a as the major or sole product, although with vinyl acetate substantial quantities of the *endo*-stereoisomer 1b were also obtained. This result was confirmed by Ikeda and co-workers[9] who also examined the reaction for N-benzoylindoles possessing a variety of substituents at the 2- and 3-positions of the indole ring. In order to rationalize the observed stereoselectivity and regioselectivity, they suggested that the reaction proceeds via the intermediacy of the biradical **3** in which the 2-position of the indole ring has become bonded to the less-substituted terminus of the alkene. Julian and Foster in their original report[4] noted that the cycloaddition reaction was sensitized by acetophenone and quenched by naphthalene, although to an unspecified degree. This suggests that the reaction proceeds via the *N*-acylindole triplet excited state, although it does not exclude concurrent reaction of the singlet excited state. The involvement of the triplet excited state is consistent with the intermediacy of a triplet 1,4-biradical such as **3**, which could then provide an opportunity for the system to undergo spin inversion and attain the singlet ground-state surface.

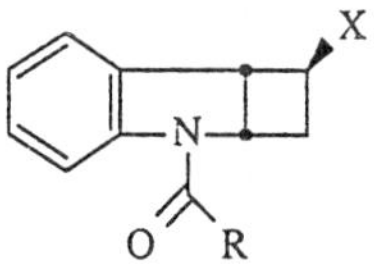

(1a) 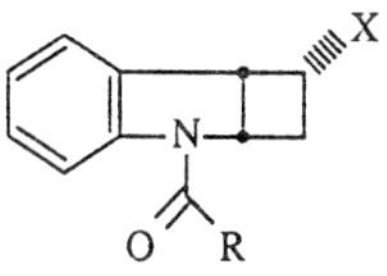   (1b) 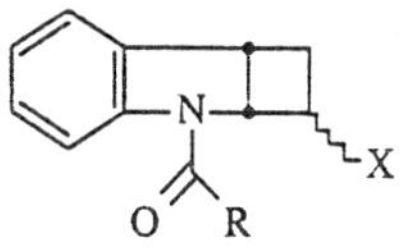   (2) 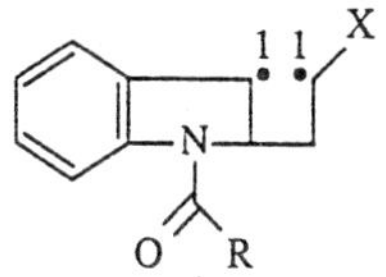   (3)

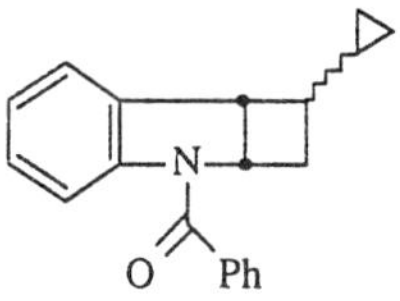

(4) 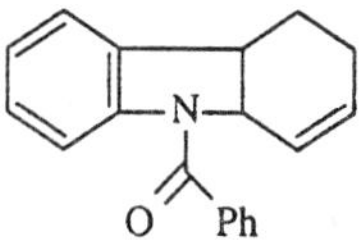   (5) 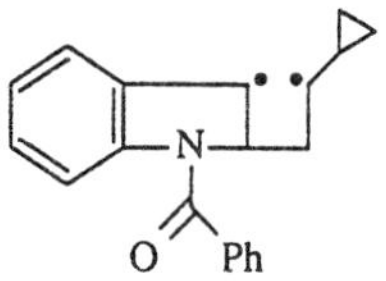   (6) 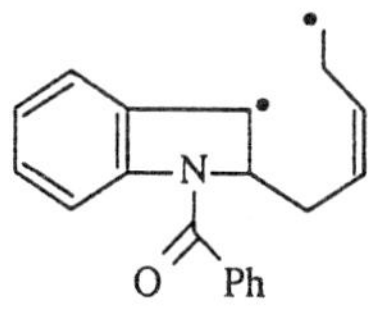   (7)

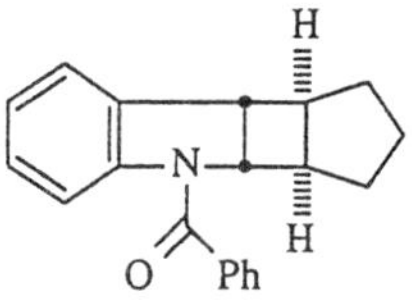

(8) 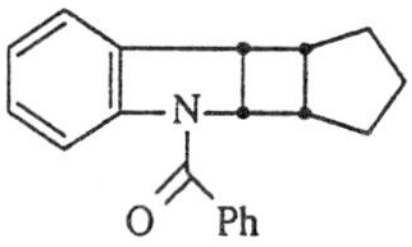   (9) 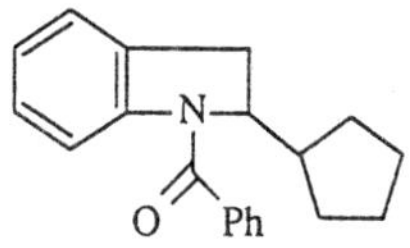   (10)

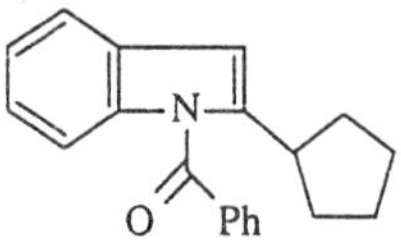

(11) 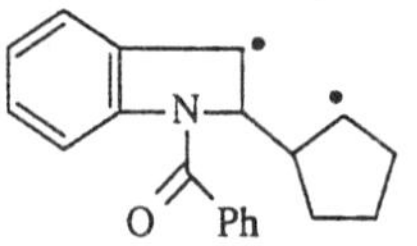   (12) 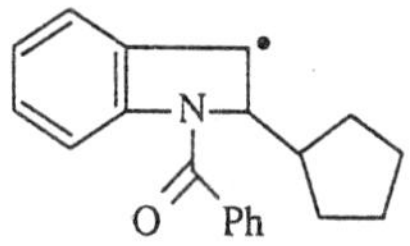   (13)

A mechanistic pathway involving the formation of such a triplet 1,4-biradical from interaction of the acylindole triplet excited state has been confirmed by a number of studies.[10-16] The *N*-benzoylindole chromophore is weakly fluorescent,[10,11] and the fluorescence is not quenched by added cyclopentene;[12] therefore, the singlet excited state is not involved in the cycloaddition with this alkene. The variation of cycloaddition quantum yield with alkene concentration for the reaction of *N*-benzoylindole with cyclopentene has been determined.[12] The data are consistent with a mechanistic model in which the singlet excited state of the *N*-acylindole intersystem crosses to the triplet excited state (lifetime 28 ns) with an efficiency of 0.39; the latter reacts with cyclopentene (rate constant $\sim 10^7 M^{-1} s^{-1}$) to give one or more triplet 1,4-biradical intermediates which, following spin inversion, either close to product (16%) or revert to ground-state starting materials (84%). The intermediacy of triplet 1,4-biradical intermediates allows memory of any alkene stereochemistry to be lost; thus, similar cycloadduct product mixtures are obtained with both the *cis*- and the *trans*-isomers of 2-butene or 4-octene.[13] In addition, the formation of isomerized alkene is observed in these reactions. Since the triplet excited-state energy of the alkenes is higher than that of the acylindoles, this is indicative of reversion of the biradical to starting materials in competition with closure to product. The photochemical cycloaddition reaction between *N*-benzoylindole and vinylcyclopropane yields the cyclobutane adducts 4 and the cycloheptene adduct 5;[14] formation of the latter is indicative of the intermediacy of the biradical 6 formed by bonding of the 2-position of the *N*-acylindole to the less-substituted end of the alkene. In 6, the cyclopropylcarbinyl radical can rearrange to the homoallylic radical 7, which ring closes to 5; the rate constant for the rearrangement of a cyclopropylcarbinyl radical to a homoallylic radical is known and this allows estimation of a value for the lifetime of the biradical 6 of approximately 100 ns. The intermediacy and structures of the 1,4-biradicals have also been confirmed by chemical trapping with hydrogen selenide.[15] For the reaction of *N*-benzoylindole with cyclopentene, the adducts 8 and 9 are produced; however, in the presence of hydrogen selenide compounds, 10 and 11 are obtained instead. These are the products of reduction of stereoisomers of biradical 12 as well as of partial reduction of 12 to give 13 followed by disproportionation with hydrogen selenyl radical. The overall mechanism for the photocycloaddition reaction between *N*-acylindoles and alkenes is summarized in Scheme 68.2.

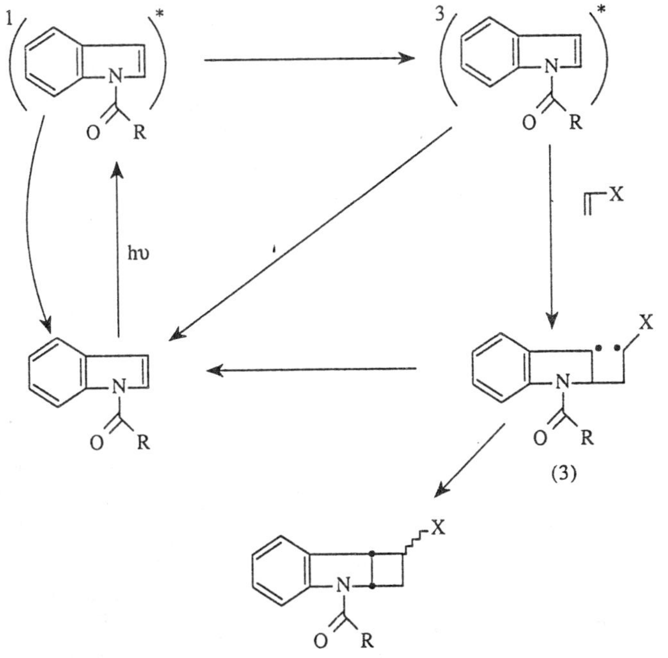

SCHEME 68.2

Photocycloaddition of Dienes to Indoles

The triplet excited states of indole and *N*-acylindoles are higher in energy than those of conjugated dienes and, consequently, energy transfer quenching rather than photocycloaddition occurs when the indoles are irradiated in the presence of conjugated dienes.[12] However, indole is a good electron donor in both the ground and excited states so that electron transfer can become exothermal in the presence of a suitable electron acceptor. Thus, irradiation of methanolic solutions of indoles in the presence of acrylonitrile leads to the formation of (1′-cyanoethyl)indoles that are produced by electron transfer from the singlet excited indole to the acrylonitrile.[8] Similarly, the formation of Diels-Alder cycloadducts when pyrilium salts are irradiated in the presence of indoles and 1,3-cyclohexadienes is thought to be initiated by electron transfer from the ground-state indole to the excited state of the pyrilium salt as shown in Scheme 68.3.[6,7] The indole radical cation produced may then attack the diene by bonding of the indole 3-position to one terminus of the diene; back-electron transfer and ring closure subsequently yields the product as a mixture of *exo*- and *endo*-isomers. With substituted dienes, the initial bond appears to be formed so as to maximize stabilization of the allylic radical by the substituent; consequently, the cycloaddition is regioselective. For example, 1-substituted dienes yield **14** rather than **15**, while 2-substituted dienes give **15** rather than **14**. The indoline nitrogen present in the products of these reactions makes them more basic and better electron donors than indole itself. As a result, the products inhibit the reaction by quenching the excited state of the pyrilium salt. This can be overcome by performing the reaction in the presence of an acylating agent such as acetyl chloride, methyl chloroformate, or toluene sulfonyl chloride; these acylate the indoline and make it a poor electron donor.

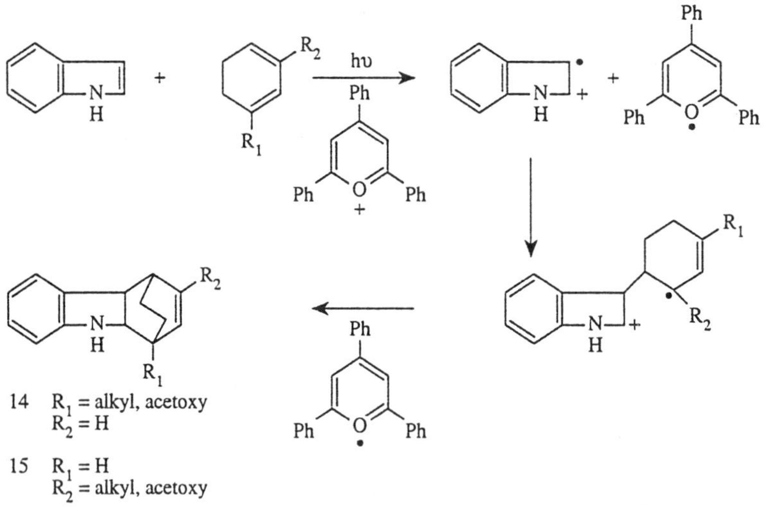

14   R_1 = alkyl, acetoxy
     R_2 = H

15   R_1 = H
     R_2 = alkyl, acetoxy

SCHEME 68.3

Photocycloaddition of Alkynes to Indoles

The cycloaddition of dimethyl acetylenedicarboxylate to *N*-alkylindoles to give cyclobutenes is sensitized by acetophenone,[5,17,18] and it has been proposed that the reaction proceeds by interaction of the acetylene diester with the indole triplet excited state to form a triplet 1,4-biradical intermediate such as **16**.[18] When the reaction is performed with 1,3-dimethylindole in methanol rather than benzene or acetonitrile, cyclobutene adduct formation is suppressed in favor of formation of geometrical isomers of the substitution product **17**; this is consistent with trapping of the biradical intermediate **16** by abstraction of a hydrogen atom. Similarly, when the reaction is performed using high concentrations of the indole, cyclobutene formation is suppressed in favor of formation of geometrical isomers of **18** and **19**; this is also consistent with trapping of biradical **16**, but this time by addition to the ground-state indole precursor.

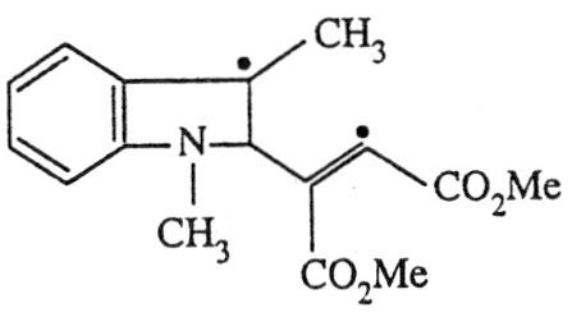

(16)

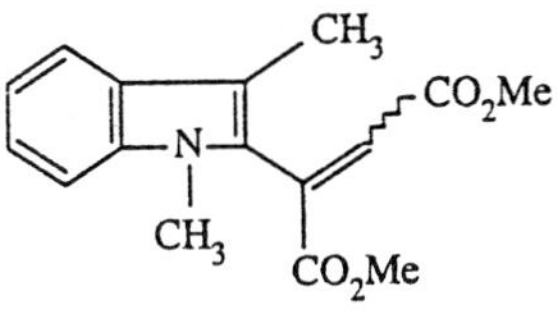

(17)

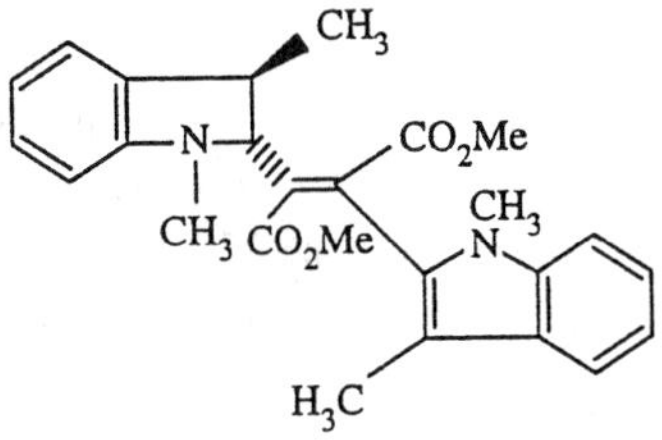

(18)

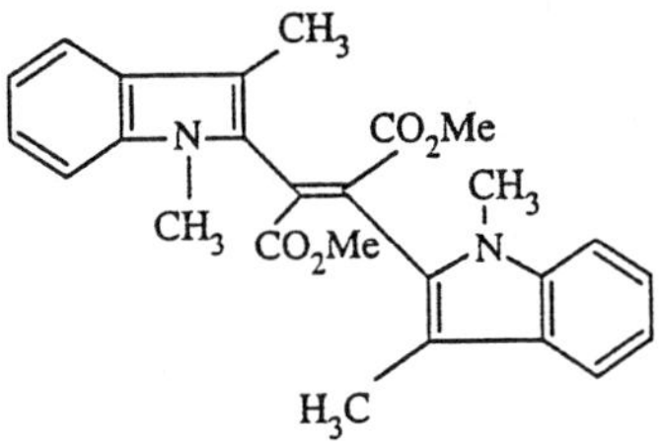

(19)

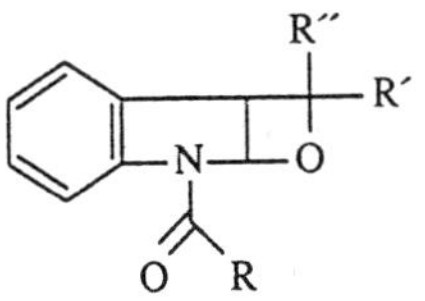

(20)

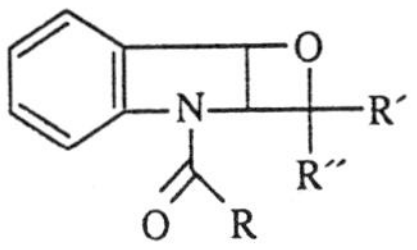

(21)

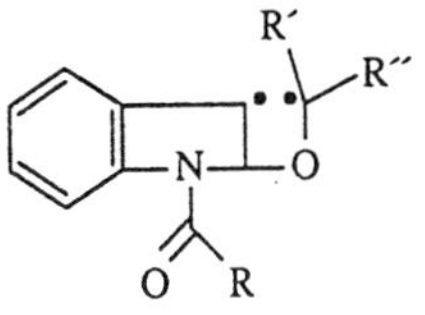

(22)

Photocycloaddition of Carbonyl Compounds to *N*-Acylindoles

In the first report of this reaction, Julian and Tringham describe how the photocycloaddition of carbonyl groups to *N*-acetylindole and *N*-(*p*-chlorobenzoyl)indole is only successful for those carbonyl compounds that possess triplet excited-state energies below 68 kcal mol^{-1}. Thus, the reaction proceeds for benzophenone but fails for acetone, benzaldehyde, and propionaldehyde. The authors proposed that the triplet excited-state energy of the *N*-acylindoles are of this order (this has subsequently been confirmed[10,12]) and that light absorption by the *N*-acylindole is followed by intersystem crossing and energy transfer to the carbonyl compound. It is then the triplet excited state of the carbonyl compound that cycloadds to the ground-state *N*-acylindole so that the latter serves as the alkene partner in a Paternò-Büchi reaction. The reaction is regioselective and yields cycloadducts with structure **20** rather than **21**. By analogy with other Paternò-Büchi reactions and with the mechanism for alkene cycloaddition to *N*-acylindoles described above, it is reasonable to assume that the regiochemistry is determined by the preferential formation of a triplet 1,4-biradical intermediate in which the indole 2-position has become attached to the carbonyl oxygen, as shown in structure **22**.

As with the alkene photocycloaddition reaction, the reaction does not proceed for indole itself, which implies that the function of the acyl group in these reactions is to reduce the aromaticity of the indole ring by localizing the electron pair on nitrogen in the amide bond and so convert the indole to an indene-like system.

68.3 Synthetic Applications

Photocycloaddition of Alkenes to *N*-Acylindoles

Selective cleavage of one of the bonds of the cyclobutane ring formed in the photocycloaddition of alkenes to *N*-acylindoles offers a route to 2-substituted, 3-substituted, or 2,3-disubstituted indoles as shown in Scheme 68.4. Alternatively, oxidation of the indoline ring of the cycloadducts could yield an indolo-cyclobutene, which would be expected to ring open and thus provide a new entry into synthetically useful[19] indolo-2,3-quinodimethanes. This is also shown in Scheme 68.4. Some of these opportunities have been examined. Ikeda and co-workers have converted the vinyl acetate cycloadducts 23 to the cyclobutanones 24; Baeyer-Villiger oxidation of these gives lactones 25. Alternatively, the cyclobutanones 24 can be converted to the corresponding oximes, which allows access to 2-cyanomethylindoles 26 if they are subjected to Beckmann rearrangement conditions.[20] Hydrolysis and oxidative cleavage of the methyl acrylate cycloadduct 27 yields the cyclobutene 28.[21,22] When this is heated briefly at ~300°C, ring opening to the benzazepine 29 occurs in reasonable yield. If the thermolysis is performed in the presence of silver ions, the ring opening proceeds in refluxing xylene.[21,22]

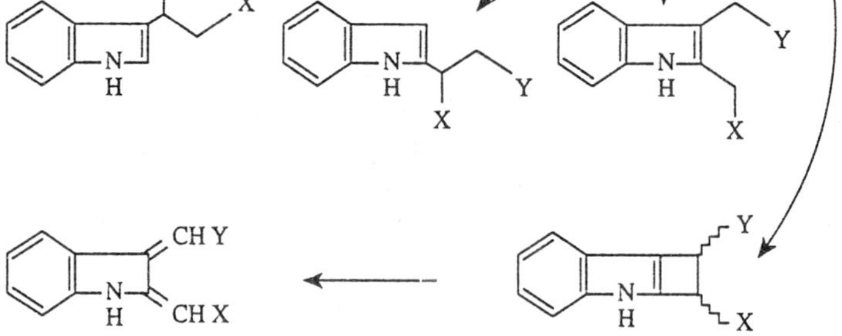

SCHEME 68.4

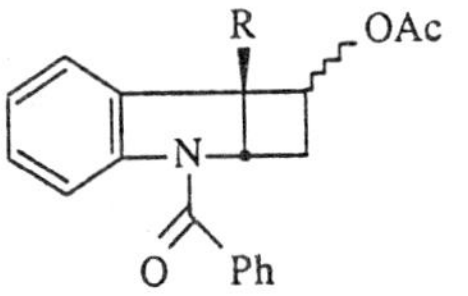

(23) R = H,Me

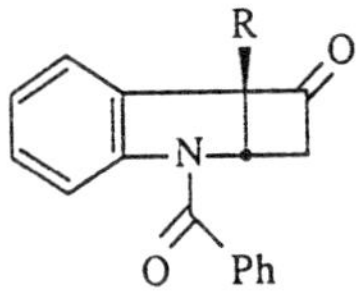

(24)

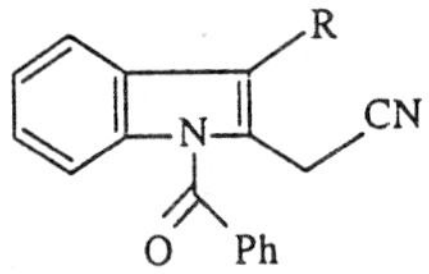

(26)

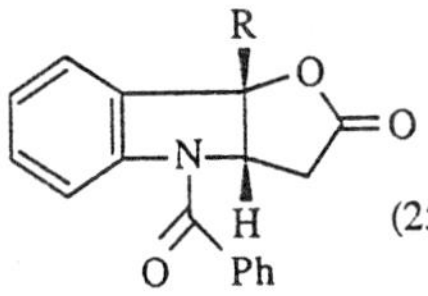

(25)

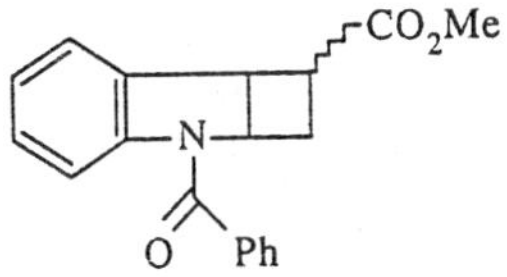

(27)

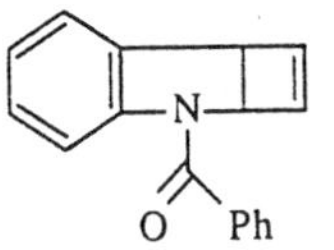

(28)

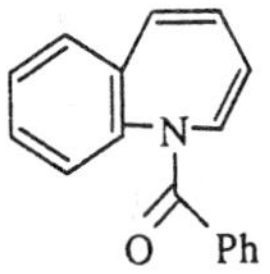

(29)

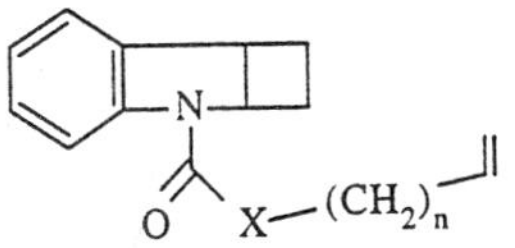

(30) X = CH$_2$ n = 0
(31) X = CH$_2$ n = 1
(32) X = CH$_2$ n = 2
(38) X = 0 n = 0
(39) X = 0 n = 1
(40) X = 0 n = 2
(41) X = 0 n = 3

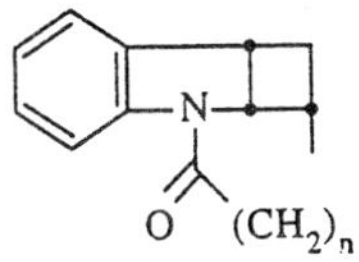

(33) n = 1
(34) n = 2
(35) n = 3

With monosubstituted alkenes, the mechanism of the photocycloaddition of alkenes to *N*-acylindoles dictates preferential formation of the head-to-tail regioisomers **1**. This selectivity is advantageous for synthetic application as long as the head-to-head isomer **2** is not desired. It has recently been demonstrated that the normal regioselectivity can be reversed and the head-to-head regioisomer accessed if the alkene is appropriately tethered to the indole.[23] Thus, the *N*-alkenoylindoles **30** to **32** undergo intramolecular photocycloaddition to yield adducts **33** to **35**; sequential ethanolysis and acetylation then give a single stereoisomer of the opposite regioisomer (e.g., **36**) to that obtained in the intermolecular photocycloaddition (e.g., **37** from addition of ethyl 4-pentenoate to *N*-acetylindole). The intramolecular addition does not proceed for the *N*-alkenyloxycarbonylindoles **38** to **41**, apparently because the alkenyloxy group is frozen in an unsuitable conformation for the duration of the excited-state lifetime.[23]

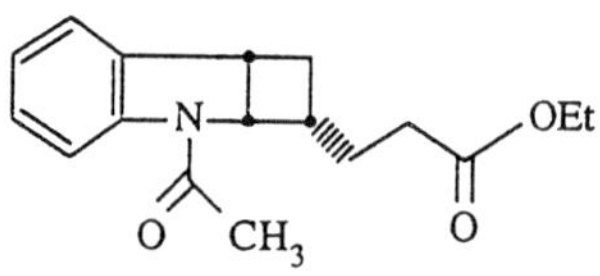

(36)

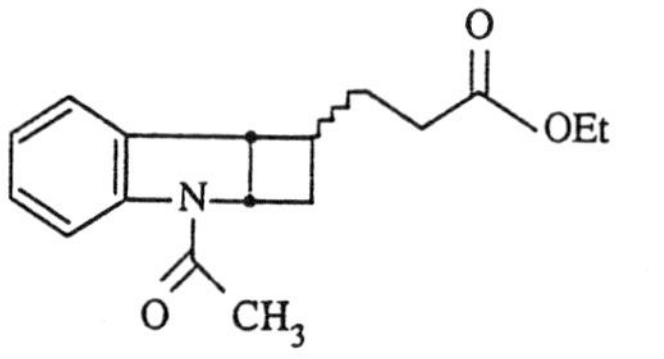

(37)

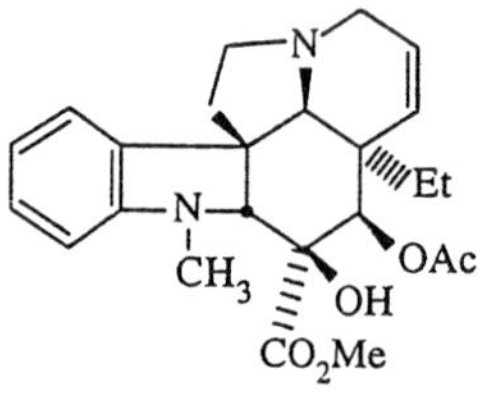

(42)

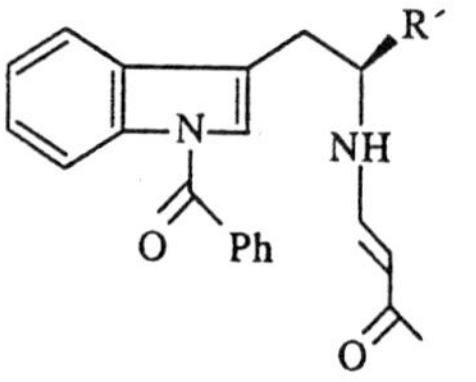

(43) R=OBu^t R´= CO$_2$Me

(47) R=OCH$_2$Ph   R´= C

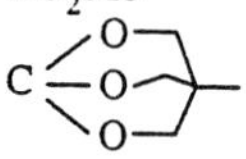

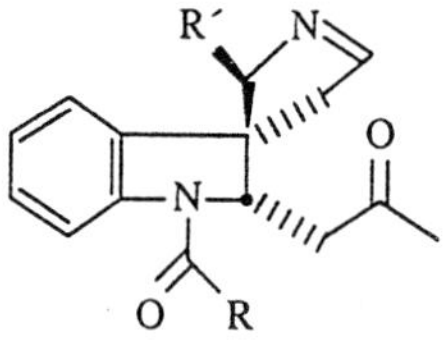

(44) R=OBu^t R´= CO$_2$Me

(48) R=OCH$_2$Ph   R´= C

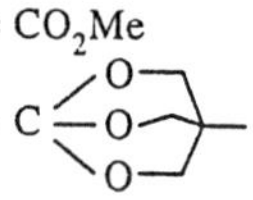

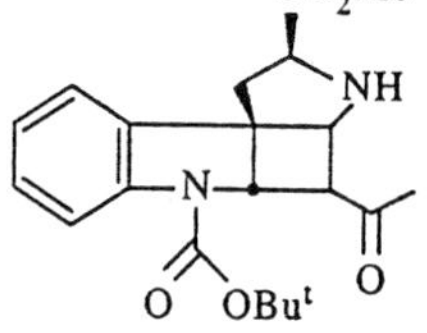

(45)

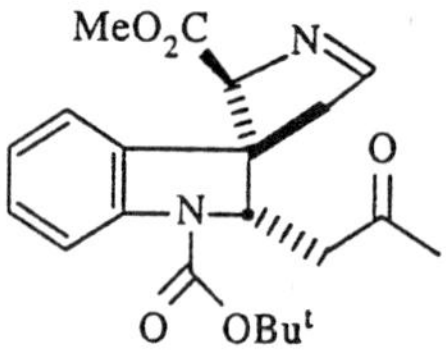

(46)

As expected from consideration of the reaction mechanism, the photocycloaddition of alkenes that possess nonidentical substituents at each end of the π-bond is not regioselective. For example, addition of methyl 3-methyl-2-butenoate to N-acylindoles yields mixtures of regioisomers.[15] The regiochemistry in such cases can also be controlled by tethering of the alkene to the indole and this has been used in a formal synthesis of vindorosine 42.[24] In this route, UV light irradiation of 43 gave mainly 44, which is presumably formed by spontaneous opening of the intramolecular photocycloadduct 45. The product was accompanied by a minor stereoisomer possessing structure 46; the stereoselectivity of the reaction was improved by the use of 47, instead of 43, to give 48. Compound 48 was then converted to a precursor of vindorosine described previously.

The utilization of the photocycloaddition reaction between N-acylindoles and alkenes for syn-thetic purposes requires that any competing photochemistry be avoided. It has been shown that many substituents on both the alkene and the indole are compatible with obtaining good chemical yields of photocycloadducts; an exception is an acetic acid substituent such as is present in the antiinflammatory drug Indomethacin, 49. Indomethacin undergoes decarboxylation rather than cycloaddition when irradiated with UV light in the presence of cyclopentene.[11] Another reaction that can interfere to a large extent is photo-Fries rearrangement of the *N*-acylindole to give ring acylated indoles.[25] However, this reaction apparently proceeds from an upper singlet or triplet excited state, or a vibrationally hot singlet excited state,[25] and thus can be suppressed completely by sensitizing the cycloaddition with a high-energy sensitizer such as acetophenone.[26] A wide variety of *N*-acyl substituents have been found to activate the indole ring toward photocycloaddition; these include acyl groups that can be readily removed under a variety of mild neutral, acidic, or basic conditions.[26] For example, compounds **50** to **53** all yield photocycloadducts with cyclopentene, and the acyl group can be removed from the adducts by treatment with fluoride ion (**50**), bicarbonate (**51**), trifluoroacetic acid (**52**), or by hydrogenolysis (**53**). It has also been shown that *N*-acylindoles can photodimerize from the triplet excited state;[27,28] for example, UV light irradiation of *N*-ethoxycarbonylindole yields a mixture of **54** and **55** in a 2:1 ratio.[28] This reaction is easily suppressed in favor of alkene cycloaddition by using alkene concentrations substantially higher than that of the indole derivative.

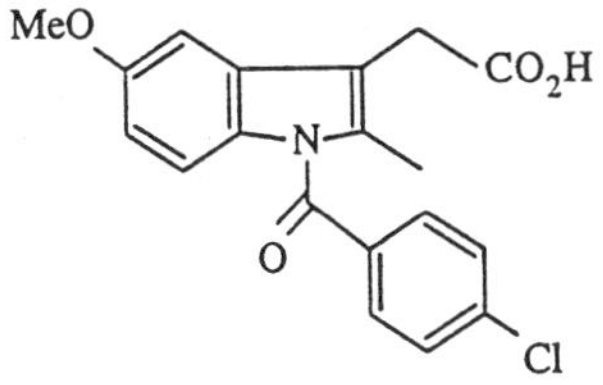

(49)

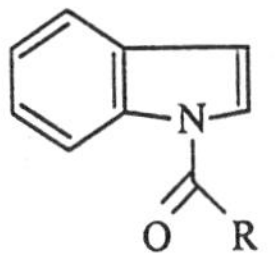

(50) R=OCH$_2$CH$_2$SiMe$_3$
(51) R=OCH$_2$CH$_2$CN
(52) R=OBu^t
(53) R=OCH$_2$Ph

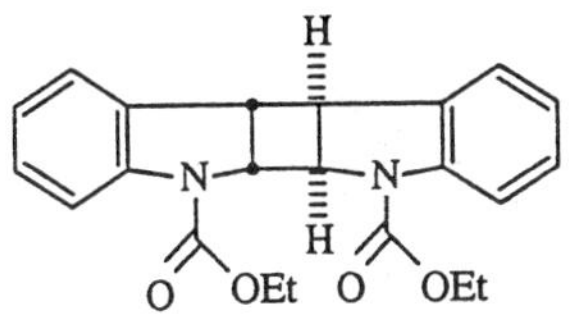

(54)

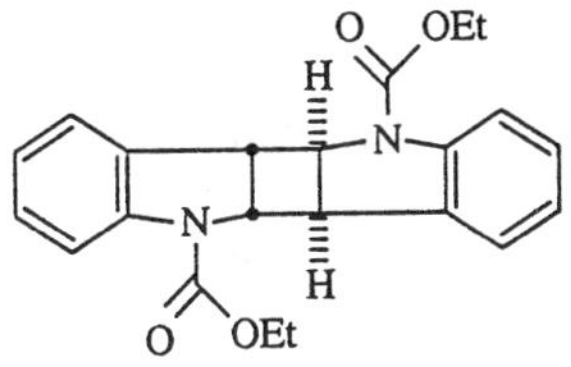

(55)

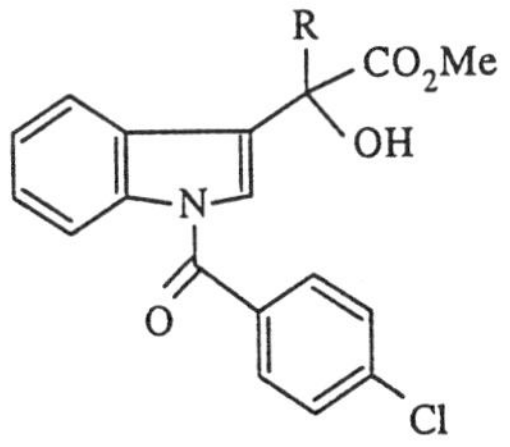

(56) R=Ph
(57) R=Me

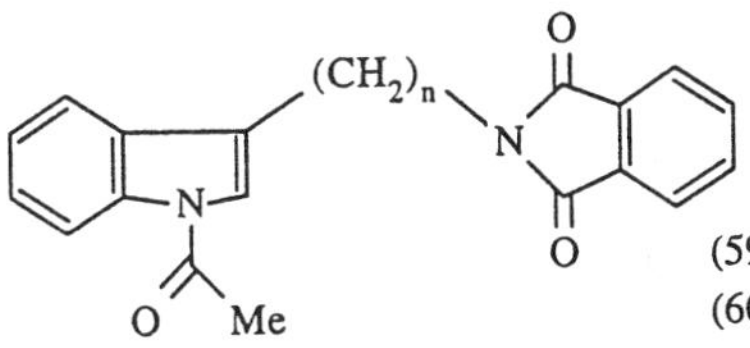

(58)

(59) n=2
(60) n=3
(61) n=4
(62) n=5

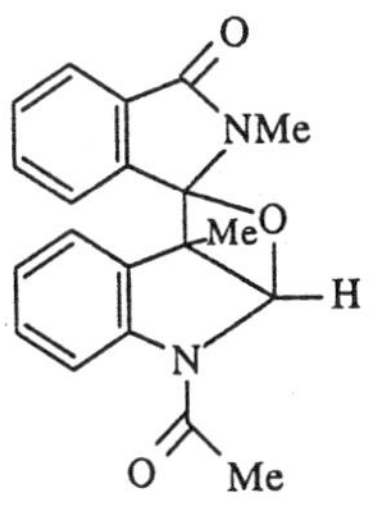

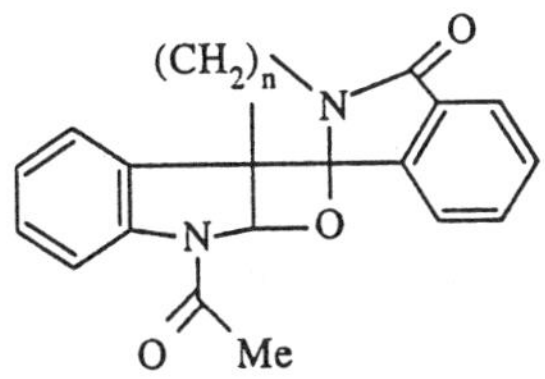

(63) n=2
(64) n=3
(65) n=4
(66) n=5

Photocycloaddition of Dienes to Indoles

The photocycloaddition of dienes to indoles has not yet been applied synthetically, although it has been noted that the adducts could be useful synthetic building blocks since the cyclohexene ring of the reaction products 14 and 15 could be oxidatively cleaved, and the enol acetate function present in 15 can be hydrolyzed to a ketone and could then be opened by Baeyer-Villiger oxidation.[6,7]

The reaction appears to be generally applicable for the addition of alkyl- or acetoxy-substituted conjugated cyclohexadienes to indole; however, methoxycyclohexadienes are reported to be unstable under the reaction conditions.[7] The reaction also proceeds poorly for noncyclic conjugated dienes and for substituted indoles.

Photocycloaddition of Alkynes to Indoles

It was noted above that oxidative decarboxylation of the adducts of N-acylindole with methyl acrylate yields cyclobutenes such as 28 that can ring-open to benzazepines such as 29. The photocycloaddition of dimethyl acetylenedicarboxylate to indoles provides an alternative and more direct route to indolo-cyclobutenes, and hence to benzazepines, and is applicable to indoles with a variety of substitution patterns as long as an alkyl group is present on the indole nitrogen atom.[5,17,18] In the case of N,3-dimethylindole, interference from the formation of products of trapping of a putative biradical intermediate by the alkyne (e.g., 17) or the indole (e.g., 18 and 19) can be avoided by appropriate choice of solvent or indole concentration.[18]

Photocycloaddition of Carbonyl Compounds to N-Acylindoles

The mechanism of photocycloaddition of carbonyl compounds to N-acylindoles to give oxetanes suggests that any carbonyl compound with a triplet energy of less than that of the N-acylindole should react. Compounds that have been reported to cycloadd to give oxetanes are benzophenone, benzoylformamide, methyl benzoyl formate, methyl pyruvate, and N-methyl phthalimide.[3,29,30] The oxetanes derived from addition of methyl benzoyl formate and methyl pyruvate to N-(p-chlorobenzoyl)indole are apparently unstable with respect to opening of the oxetane and yield alcohols 56 and 57, respectively. The question of the stereochemistry in the oxetane adducts of unsymmetrical carbonyl compounds such as benzoylformamide does not appear to have been addressed.[3] However, the addition of a variety of N-acylindoles to N-methylphthalimide is reported to be stereoselective and yields the more sterically hindered oxetanes such as 58.[29,30] This reaction is also unusual in that addition of double bonds to phthalimides normally occurs across the bond between the carbonyl group and the imide nitrogen.[31] Intramolecular photocycloaddition of N-acylindoles to a carbonyl group of N-alkylphthalimides has also been reported.[32,33] The reaction proceeds for compounds 59 to 62 to give adducts 63 to 66, but fails if the phthalimide is linked to the 2-position of the indole; instead, photo-Fries rearrangement of the indole N-acyl group occurs.

References

1. Baeyer, A., Ueber die Reduction aromatischer Verbindungen mittelst Zinkstaub, *Ann. Chem.*, 140, 295, 1866.
2. Weissberger, A. and Taylor, E. C., Eds., *The Chemistry of Heterocyclic Compounds*, Vol. 25, Wiley, New York, 1972 (Part 1), 1972 (Part 2), 1979 (Part 3), 1982 (Part 4).
3. Julian, D. R. and Tringham, G. D., Photoaddition of ketones to indoles: Synthesis of oxeto[2,3-b]indoles, *J. Chem. Soc., Chem. Commun.*, 13, 1973.
4. Julian, D. R. and Foster, R., Photoaddition of olefins to indoles: Synthesis of tetrahydro-1H-cyclobut[b]indoles, *J. Chem. Soc., Chem. Commun.*, 311, 1973.

5. Davis, P. D. and Neckers, D. C., Photocycloaddition of dimethylacetylene dicarboxylate to activated indoles, *Tetrahedron Lett.*, 2979, 1978.

6. Gieseler, A., Steckhan, E., and Wiest, O., Photoinduced, electron-transfer-catalyzed Diels-Alder reaction between indole and 1,3-cyclohexadienes, *Synlett*, 275, 1990.

7. Gieseler, A., Steckhan, E., Wiest, O., and Knoch, F., Photochemically induced radical cation Diels-Alder reaction of indole with electron rich dienes, *J. Org. Chem.*, 56, 1405, 1991.

8. Yamasaki, K., Matsuura, T., and Saito, I., Exciplex involvement in the photoaddition of acrylonitrile to indoles, *J. Chem. Soc., Chem. Commun.*, 944, 1974.

9. Ikeda, M., Ohno, K., Mohri, S., Takahashi, M., and Tamura, Y., Regio- and stereochemical aspects of [2+2] photocycloaddition between 1-benzoylindoles and olefins, *J. Chem. Soc., Perkin Trans. 1*, 405, 1984.

10. Disanayaka, B. W. and Weedon, A. C., Charge transfer fluorescence of some *N*-benzoylindoles, *Can. J. Chem.*, 65, 245, 1987.

11. Weedon, A. C. and Wong, D. F., The photochemistry of Indomethacin, *J. Photochem. Photobiol.*, 61, 27, 1991.

12. Disanayaka, B. W. and Weedon, A. C., The mechanism of the photochemical cycloaddition reaction of *N*-benzoylindole with cyclopentene, *Can. J. Chem.*, 68, 1685, 1990.

13. Hastings, D. J. and Weedon, A. C., Stereochemical studies of the photochemical cycloaddition reaction of alkenes with *N*-benzoylindole and *N*-carboethoxyindole; evidence for biradical intermediacy, *Can. J. Chem.*, 69, 1171, 1991.

14. Hastings, D. J. and Weedon, A. C., Structures and lifetimes of 1,4-biradical intermediates in the photochemical cycloaddition reactions of *N*-benzoylindole with alkenes, *J. Org. Chem.*, 56, 6326, 1991.

15. Hastings, D. J. and Weedon, A. C., The origin of the regioselectivity in the 2 + 2 photochemical cycloaddition reactions of *N*-benzoylindole with alkenes: Trapping of 1,4-biradical intermediates with hydrogen selenide, *Tetrahedron Lett.*, 32, 4107, 1991.

16. Andrew, D., Hastings, D. J., Oldroyd, D. L., Rudolph, A., Weedon, A. C., Wong, D. F., and Zhang, B., Triplet 1,4-biradical intermediates in the photocycloaddition reactions of enones and *N*-acylindoles with alkenes, *Pure Appl. Chem.*, 64, 1327, 1992.

17. Davis, P. D. and Neckers, D. C., Photocycloaddition of dimethyl acetylenedicarboxylate to activated indoles, *J. Org. Chem.*, 45, 456, 1980.

18. Davis, P. D., Neckers, D. C., and Blount, J. R., Photocycloaddition of dimethyl acetylenedicarboxylate to 1,3-dimethylindole, *J. Org. Chem.*, 45, 462, 1980.

19. Pindur, U. and Erfanian-Abdoust, H., Indolo-2,3-quinodimethanes and stable cyclic analogues for regio- and stereocontrolled syntheses of [b]-anelated indoles, *Chem. Rev.*, 89, 1681, 1989.

20. Ikeda, M., Uno, T., Homma, K., Ohno, K., and Tamura, Y., Beckman fission of some fused cyclobutanones: A new entry into indole-2-acetonitrile and benzo[b]thiophene-2-acetonitrile, *Synth. Commun.*, 10, 437, 1980.

21. Ikeda, M., Ohno, K., Uno, T., and Tamura, Y., Synthesis and some properties of 1*H*-1-benzazepines, *Tetrahedron Lett.*, 21, 3403, 1980.

22. Ikeda, M., Ohno, K., Takahashi, M., Uno, T., Tamura, Y., and Kido, M., *J. Chem. Soc., Perkin* Trans. 1, 741, 1982.

23. Oldroyd, D. L. and Weedon, A. C., Intramolecular photochemical cycloadditions of *N*-alkenyloxycarbonylindoles and *N*-alkenoylindoles, *J. Chem. Soc., Chem. Commun.*, 1491, 1992.

24. Winkler, J. D., Scott, R. D., and Williard, P. G., Asymmetric induction in the vinylogous amide photocycloaddition reaction. A formal synthesis of vindorosine, *J. Am. Chem. Soc.*, 112, 8971, 1990.

25. Oldroyd, D. L. and Weedon, A. C., Solvent- and wavelength-dependent photochemistry of *N*-benzoylindole and *N*-ethoxycarbonylindole, *J. Photochem. Photobiol. A*, 57, 207, 1991.

26. Weedon, A. C. and Zhang, B., Removable groups for the activation of indole photochemistry, *Synthesis*, 95, 1992.

27. Hino, T., Taniguchi, M., Date, T., and Iidaka, Y., Photodimerization of indole derivatives, *Heterocycles*, 7, 105, 1977.
28. Oldroyd, D. L., Payne, N. C., Vital, J. J., Weedon, A. C., and Zhang, B., *Tetrahedron Lett.*, 34, 1087, 1993.
29. Takechi, H., Machida, M., and Kanaoka, Y., Intermolecular photoaddition of *N*-methylphthalimide to indole derivatives: Regio- and stereoselective formation of oxeto[2,3-b]indoles, *Heterocycles*, 23, 1373, 1985.
30. Takechi, H., Machida, M., and Kanaoka, Y., Intermolecular photoreactions of phthalimide-alkene systems. Regio- and stereoselective oxetane formation from *N*-methylphthalimide and *N*-acetylindole derivatives, *Chem. Pharm. Bull.*, 36, 3770, 1988.
31. Coyle, J. D., Phthalimide and its derivatives, in *Synthetic Organic Photochemistry*, Horspool, W. M., Ed., Plenum, New York, 1984, chap. 4.
32. Machida, M., Takechi, H., and Kanaoka, Y., Photoreaction of *N*-(ω-indol-3-ylalkyl)phthalimides: Intramolecular oxetane formation of the aromatic imide system, *Tetrahedron Lett.*, 23, 4981, 1982.
33. Takechi, M., Machida, M., and Kanaoka, Y., Intramolecular photoreactions of phthalimide-alkene systems. Oxetane formation of *N*-(ω-indol-3-ylalkyl)phthalimides, *Chem. Pharm. Bull.*, 36, 2853, 1988.

SET-Induced Inter- and Intramolecular Additions to Iminium Cations

Patrick S. Mariano
University of Maryland -
College Park

69.1 Introduction

This chapter discusses the general concepts that serve as the foundation for a new class of single electron transfer (SET)-promoted photochemical reactions of substances containing the iminium cation grouping. In this review, the fundamentals of SET photochemistry are briefly outlined first, followed by a survey of the characteristics of iminium cations as electron acceptors. The general reaction pathways occurring in SET-induced excited-state processes of iminium cations are then summarized and recent examples of each reaction type are given. Finally, applications of these photochemical processes to complex molecule synthesis are presented to demonstrate the synthetic versatility of the chemistry.

69.2 SET Photochemistry of Iminium Cations

General Concepts of SET Photochemistry

The concepts serving as the basis for SET photochemistry originate from early investigations of excited-state complex (excimer and exciplex) formation.[1-7] It is now clear that charge transfer (CT) is an important stabilizing feature of excited-state complexes as demonstrated by the relationships between exciplex energies and the ionization potentials and electron affinities of the partners[8] and the medium polarity.[9,10] Another important aspect of CT in excited-state complexes is related to the decay of these species to neutral and/or charged radical intermediates. Exciplex collapse by this

0-8493-8634-9/95/$0.00+$.50
© 1995 by CRC Press, Inc.

mode leading to net SET is favored in highly polar solvents[11] and by electronic factors that lead to odd electron and charge stabilization. Thus, in polar solvents, ion radical formation can effectively compete with exciplex decay by other modes including emission or direct chemical reaction.

The ability to predict qualitatively when SET pathways will compete with other routes for excited-state decay is aided by considerations of empirically derived[12,13] and theoretically supported[14–16] relationships that exist between the SET free energy changes (ΔG_{SET}) and rate constants (k_{SET}), and between ΔG_{SET} and oxidation ($E_{1/2}(+)$) and reduction ($E_{1/2}(-)$) potentials and excited-state energies ($\Delta E_{0,0}$) of the donor-acceptor pairs. These relationships applied to bimolecular SET show that k_{SET} approaches the diffusion limit (ca. 10^{10} $M^{-1}s^{-1}$) when $0 > \Delta G_{SET} = E_{1/2}(+) - E_{1/2}(-) - \Delta E_{0,0}$.

The unique reaction pathways followed in SET photochemical processes are governed by the nature of the reactive intermediates involved. Unlike classical photochemistry where excited-state electronic characteristics control chemistry, SET-induced processes are driven by the chemical reactivity of neutral and charged radical intermediates. When viewed together, the ability to predict (1) when SET will be an important route followed in excited-state reactions, (2) when ion radical formation will be favored, and (3) the types of reactions that will occur based on ion and radical chemistry, can be useful in the design and mechanistic analysis of novel photochemical processes.

Iminium Cations As Acceptors in SET Processes

Iminium cations, $R_2 \overset{+}{N} = \overset{\cdot}{C} R_2$, and their *N*-heteroaromatic relatives play major roles in organic chemistry as both electronically interesting substances and reactive intermediates.[17] Pertinent to the current discussion are the electron-acceptor properties of these species as reflected by their participation in nucleophilic addition and reduction reactions. The low-lying LUMOs in iminium cations lead to ready reduction, as seen in the $E_{1/2}(-)$ values for alkyl- and aryl-substituted analogs that lie between −1.96 and −0.84 V.[18–21] SET reductions of iminium cations leads to the generation of stabilized[22] α-amino radicals, $R_2 N - \overset{\cdot}{C} R_2$, which undergo typical free-radical reactions (e.g., C-C bond formation to form 1,2-diamines).

69.3 General Features of Iminium Cation SET Photoreactions

As anticipated, iminium cations participate in a variety of SET-promoted photoreactions with neutral electron donors. These processes leading to C-C bond formation are driven by the often predictable properties of donor-derived cation radicals. As such, novel transformations have been uncovered and these have been applied to the construction of complex organic structures.

SET photoreactions between neutral donors and iminium cation acceptors can be promoted by excitation of either species in polar solvents (Scheme 69.1). The cation (2) and neutral α-amino (1) radicals arising in this way can participate in a host of secondary reactions, including (1) radical coupling, (2) nucleophile addition, and (3) α-CH deprotonation, which lead to precursors of adducts. Another general and highly selective reaction pathway followed by donor cation radicals that bare α-trialkylsilyl substituents is desilylation. The chemical efficiencies of intermolecular processes proceeding by these pathways (Scheme 69.1) can be limited by competing radical reactions (e.g., H-atom abstraction, dimerization) of radical intermediates that diffuse from initially formed solvent caged pairs. As expected, when occurring intramolecularly, these processes do not suffer from this limitation and, consequently, are often more efficient.

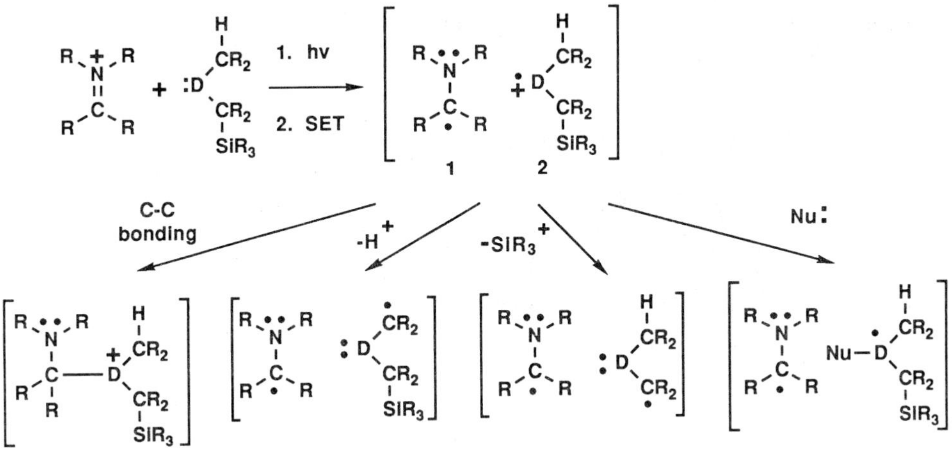

SCHEME 69.1

9.4 Photoaddition and Photocyclization Reactions of Donor-Iminium Cation Acceptor Systems

Olefin-Iminium Cation Reactions

Calculations of ΔG_{SET} show that SET from electron-rich olefins ($E_{1/2}(+) \cong +2.0$ V) to the singlet excited states of vinyl or aryl conjugated iminium cations ($E_{1/2}^{S1}(-) \cong +2.9$ V) is energetically (and thus kinetically) favorable.[23,24] The observations that several olefins of this type photoadd to 2-phenyl-1-pyrrolinium perchlorate (3) via the intermediacy of ion radical intermediates show the chemical consequences of SET in these systems. An instructive example is the highly regioselective addition of isobutylene 4 to 3, induced by $\lambda > 250$-nm irradiation in MeOH, which furnishes the adduct 5 in an 81% yield (Scheme 69.2).[23–26] The mechanism of this reaction involves MeOH addition to the isobutylene cation radical providing the radical pair 6, which undergoes C-C bonding to form 5.

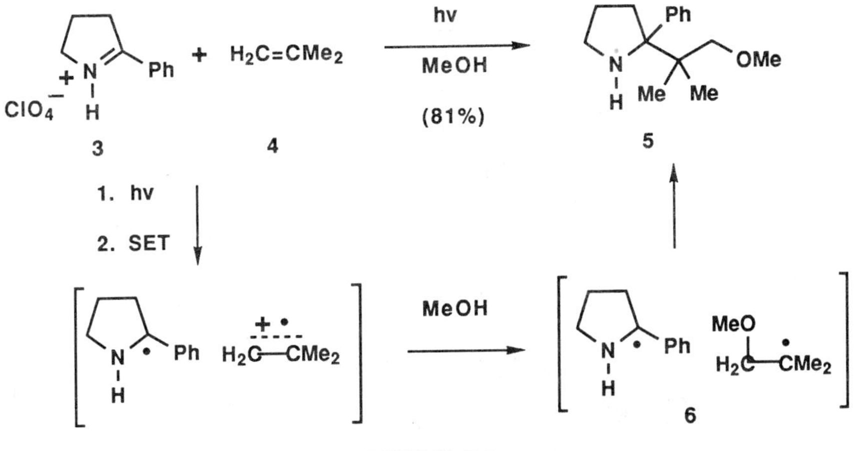

SCHEME 69.2

Other examples of this reaction type are the photocyclizations of the *N*-allyl and α′-C-vinyl iminium perchlorates 7 and 8 (Scheme 69.3).[27]

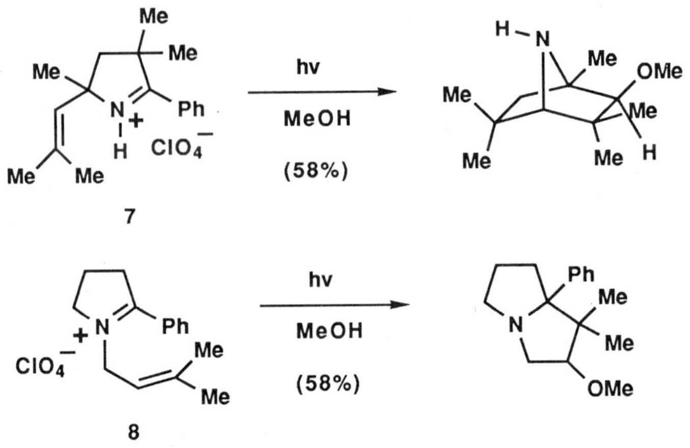

SCHEME 69.3

The efficiencies of the SET-promoted olefin-iminium cation photoadditions should depend on the olefin oxidation potentials. Consistent with this evaluation are the reports that electron-poor olefins 9 such as methyl acrylate and acrylonitrile react with the singlet excited state of 3 to give adducts 10 resulting from sequential [2+2]-cycloaddition-cationic rearrangement rather than SET pathways (Scheme 69.4).[26,28]

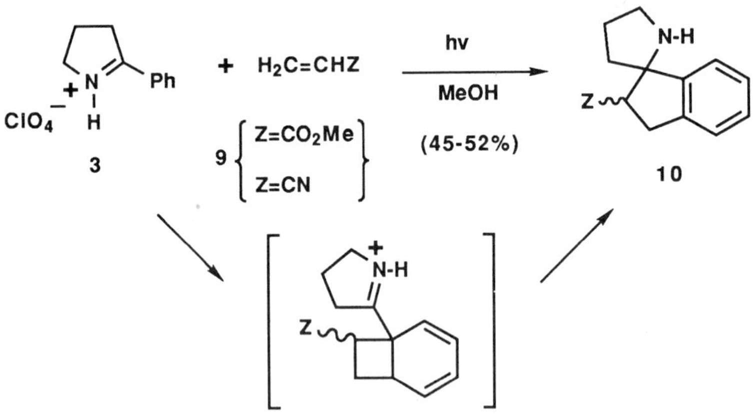

SCHEME 69.4

N-Heteroaromatic cations containing appended, electron-rich olefin groups undergo similar, SET-promoted photocyclization (e.g., 11 → 13, Scheme 69.5) reactions.[29,30] Analogous to intermolecular reactions with related iminium cations, when the $E_{1/2}(+)$ value of the olefin appendage is elevated by removal of the electron-donating methyl groups, as in the pyridinium salt 12, the efficiency of the SET pathway is reduced and the typical pyridinium cation excited-state ring contraction reaction to form aminocyclopentene 14 predominates.

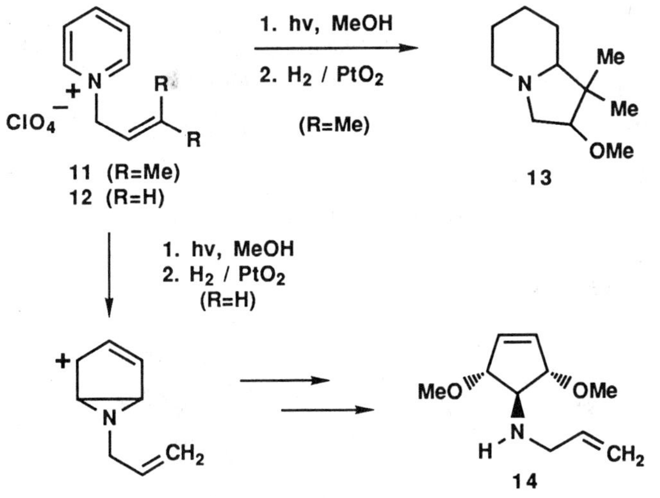

SCHEME 69.5

In some cases, olefin-derived cation radicals, arising by SET to excited, conjugated iminium cations, undergo allylic deprotonation in competition with nucleophilic addition. This behavior is seen in the photoadditions of cyclohexene **15** and methyl dimethylacrylate **17** to pyrrolinium perchlorate **3**, where adducts **16** and **18**, respectively, are formed by this alternate pathway (Scheme 69.6).[27]

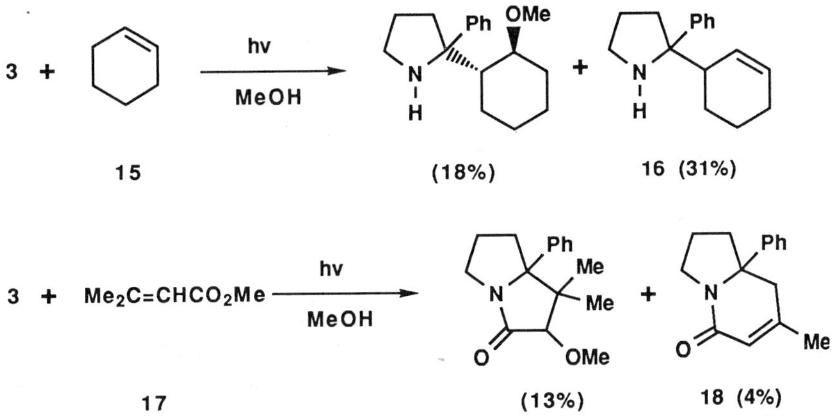

SCHEME 69.6

Olefin cation radical α-CH deprotonation is a specific example of a general process in which α-electrofugal groups adjacent to positive centers are eliminated. This view, together with knowledge about the fast and selective desilylation reactions of β-silyl carbocations, led to the design of highly efficient and selective SET-promoted photoaddition and photocyclization reactions of iminium cations with allylsilanes. Examples are found in the photoadditions of allylsilanes **20** to pyrrolinium salt **19** that generate adducts **21** (Scheme 69.7).[31,32] The sequential SET-desilylation mechanism (shown in Scheme 69.7) proposed for these processes is consistent with their regiochemical outcomes.

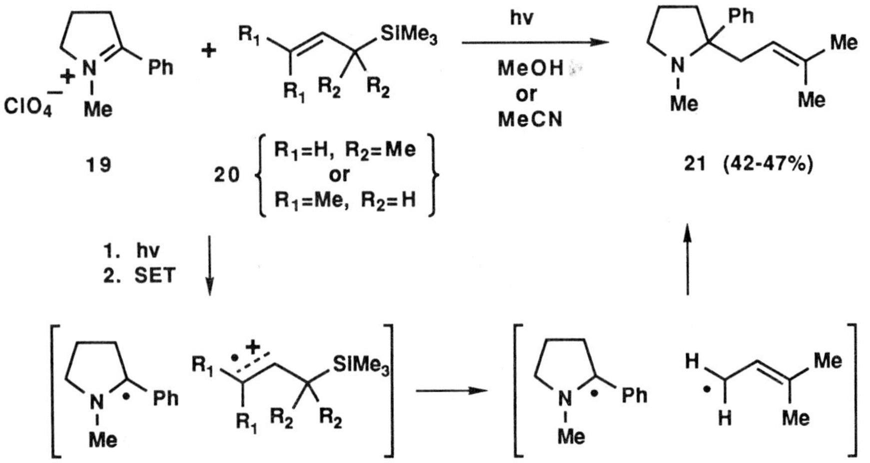

SCHEME 69.7

The synthetic potential of this SET photochemistry is shown by the highly efficient conversions of the silylmethallyl-iminium perchlorates 22 to the corresponding amine products 23, processes that create hindered, quaternary spirocyclic centers (Scheme 69.8).[33]

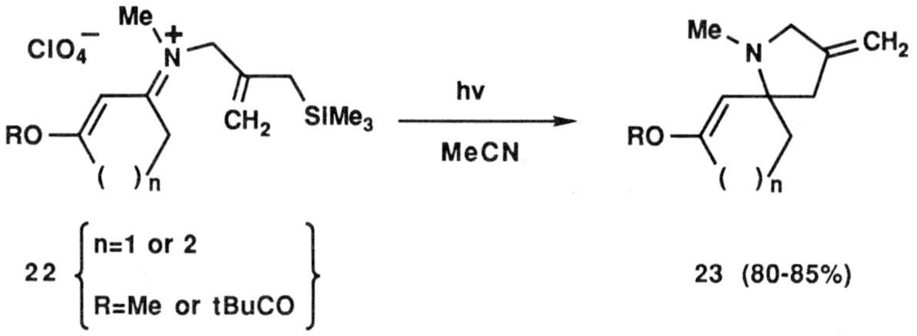

SCHEME 69.8

In investigations probing the SET photochemistry of the N-trimethylsilylmethylallyl-pyrrolinium salts 24, factors influencing competitive routes available to singlet 1,5-biradical cations 25 were delineated (Scheme 69.9).[34,35] Using deuterium labeling to distinguish between pyrrolizidine ring formation by sequential desilylation-biradical cyclization (giving 26:27 = 1) vs. cation biradical cyclization-desilylation (giving only 27), it was shown that cyclization (k_c) is favored over desilylation (k_d) of 25 for systems containing *p*-electron-donating groups (e.g., $k_d/k_c = 0.85$ for Y = OCH_3), while desilylation predominates for systems with *p*-electron-withdrawing groups (e.g., $k_d/k_c = 5.9$ for Y = CF_3). These effects are attributed to the control of k_c by α-amino radical SOMO (singly occupied molecular orbital) energies, which themselves are governed by aryl substituents (e.g., energy is higher with Y = OCH_3).

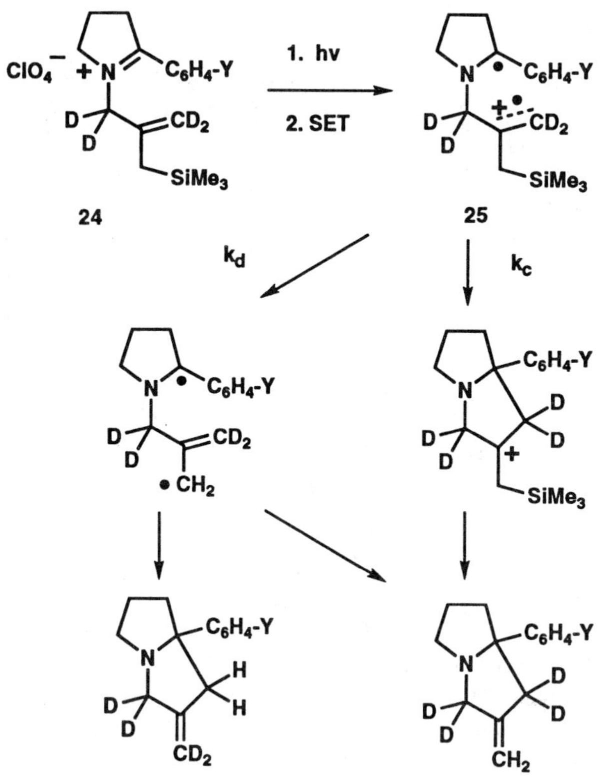

SCHEME 69.9

Arene-Iminium Cation Photoreactions

Electron-rich arenes are predicted to be good donors to excited states of conjugated iminium cations, as are their singlet excited states with ground-state iminium cation acceptors. These properties are reflected in the photoaddition reactions occurring between arenes 29 (containing benzylic H, SiMe$_3$, and SnBu$_3$ substitution) with the pyrrolinium salts 4 and 28 (Scheme 69.10) and induced by irradiation of either the cation or arene chromophores.[36] These processes lead to production of the pyrrolidine adducts 30 in yields that depend on the electrofugal group (E) present in the arene. These observations are rationalized by the competition between the rates of loss of electrofugal groups in the intermediate arene cation radical (Bu$_3$Sn > Me$_3$Si > H) vs. cage collapse of the cation radical pair.

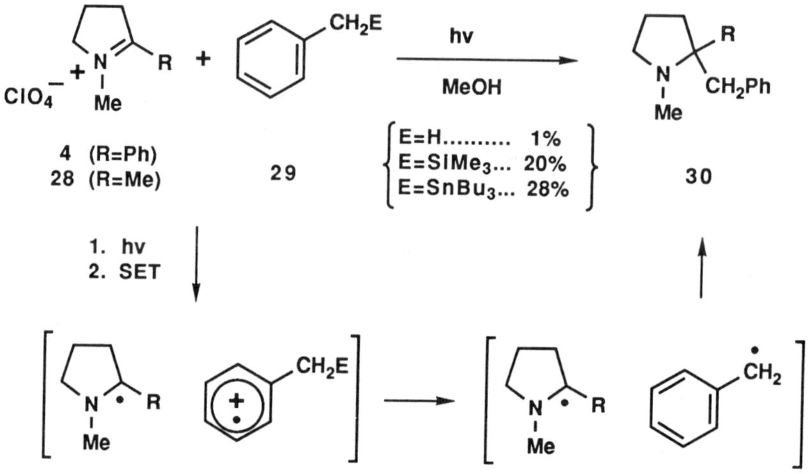

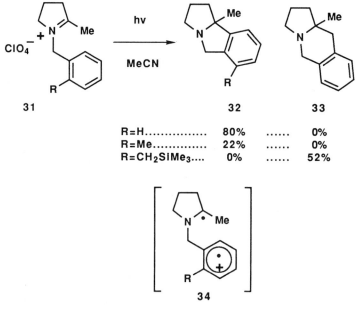

SCHEME 69.10

Studies[35,37] of linked arene-pyrrolinium cation systems **31** have uncovered another interesting facet of these photochemical processes. This is exemplified best by the dramatic dependence of the nature of the photoreactions of **31** (i.e., benzopyrrolizidines **32** or benzoindolizidine **33** produced) on the arene ring substituent. Thus, benzopyrrolizidines **32** are generated when biradical cation **34** cyclization occurs faster than loss of a benzylic electrofugal group; the latter route gives the biradical precursor of benzoindolizidine **33** (Scheme 69.11).

SCHEME 69.11

9.5 Synthetic Applications of Iminium Cation SET Photocyclization Reactions

The chemical efficiencies as well as the structural and functional outcomes of iminium cation SET-promoted photocyclization reactions are ideal for their employment in novel *N*-heterocycle or alkaloid synthetic methodologies. Investigations of the applications of strategies based on these methodologies have focused on synthetic approaches for preparing substances with key skeletal features of members of several alkaloid families. For example, SET-induced photocyclization of the *N*-trimethylsilylmethylallyl-iminium salt **35** leading to **36** is a useful step in the preparation of pentacyclic enol ester **37**,[38] a substance possessing the structure of members of the harringtonine alkaloid family (Scheme 69.12). In an erythrina alkaloid synthetic plan, photospirocyclization of the silylalkenyl-dihydroisoquinolinium perchlorate **38** is used to produce the spirocyclic amine **39** in a route (Scheme 69.13) to the erythrinane **40**.[39,40] Finally, the protoberberine alkaloid tetracyclic framework is accessed readily by SET-promoted photocyclization of *N*-silylxylyl-dihydroisoquinolinium perchlorates, exemplified by the transformation of **41** to stylopine (**42**) (Scheme 68.14).[41-43]

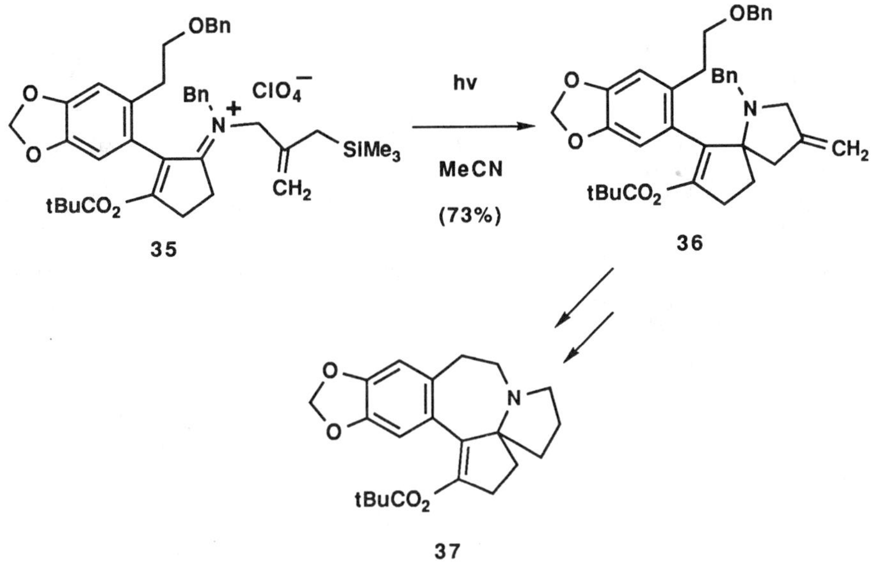

SCHEME 69.12

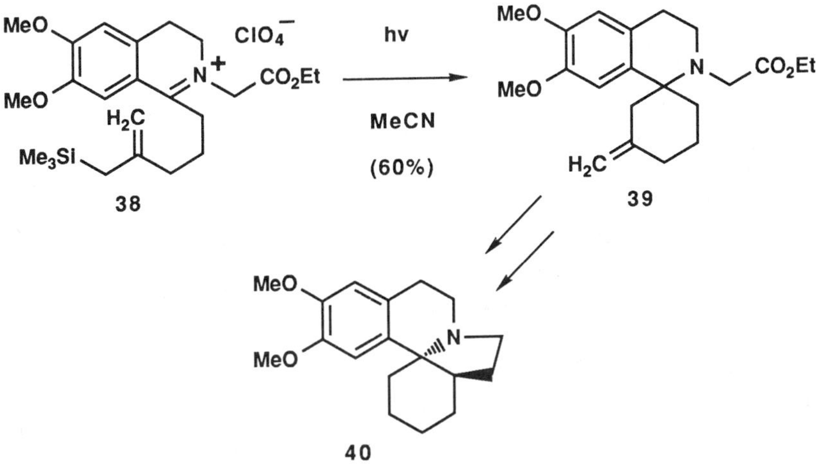

SCHEME 69.13

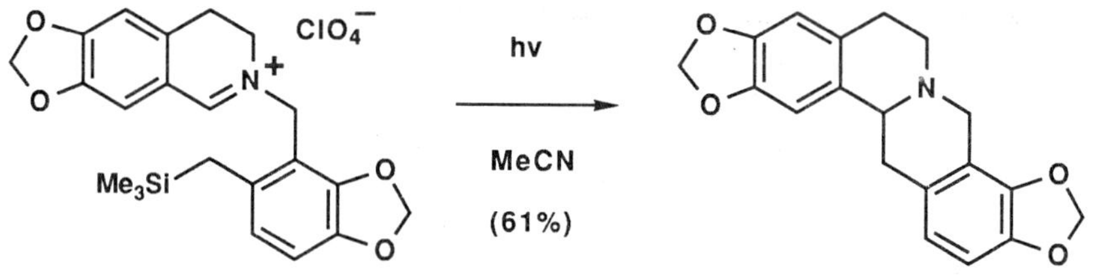

SCHEME 69.14

References

1. Forster, T. and Kasper, K., Ein konzentration sumschlag der fluoreszenz, *Z. Phys. Chem.*, 1, 275, 1954.
2. Birks, J. B., Excimer fluorescence of aromatic compounds, *Prog. React. Kinet.*, 5, 181, 1970.
3. Forster, T., Excimers, *Angew. Chem. Int. Ed.*, 8, 333, 1969.
4. Stevens, B. and Hutton, E., Radiative lifetime of the pyrene dimer and the possible role of excited dimers in energy transfer processes, *Nature*, 186, 1045, 1960.
5. Walker, M. S., Bednar, T. W., and Lumry, R., Exciplex formation in the excited state of indole, *J. Chem. Phys.*, 45, 3455, 1966.
6. Hochstrasser, R., Mixed dimer emission from pyrene crystals containing perylene, *J. Chem. Phys.*, 36, 1099, 1962.
7. Leonhardt, H. and Weller, A., Electronenubertagungsreaktioner des angeregten perylens, *Ber. Bunsenges. Physik. Chem.*, 67, 791, 1963.
8. Knibbe, H., Rehm, D., and Weller, A., Formation of molecular complexes in the excited state. Correlation between the emission maxima and the reduction potential of an electron acceptor, *Z. Phys. Chem.*, 56, 95, 1967.
9. Knibbe, H., Rehm, D., and Weller, A., Intermediates and kinetics of fluorescence quenching by electron transfer, *Ber. Bunsenges. Phys. Chem.*, 72, 257, 1968.
10. Ware, W. R. and Richter, H. P., Fluorescence quenching via charge transfer. The perylene-*N,N*-dimethylaniline system, *J. Chem. Phys.*, 48, 1595, 1968.
11. Knibbe, H., Rollig, K., Schafer, F. P., and Weller, A., Charge transfer complex and solvent shared ion pairs in fluorescence quenching, *J. Chem. Phys.*, 47, 1184, 1967.
12. Beens, H. and Weller, A., Triplet complex formation in the excited state, *Chem. Phys. Lett.*, 2, 140, 1968.
13. Rehm, D. and Weller, A., Kinetics of fluorescence quenching by electron and H-atom transfer, *Isr. J. Chem.*, 8, 259, 1970.
14. Indelli, M. T. and Scandola, F., Remarks on free-energy correlations of rate constants for electron transfer quenching of electronically excited states, *J. Am. Chem. Soc.*, 100, 7733, 1978.
15. Scandola, F. and Balzani, V., Free energy relationships for electron transfer processes, *J. Am. Chem. Soc.*, 101, 6140, 1979.
16. Scandola, F., Balzani, V., and Schuster, G. B., Free energy relationships for reversible and irreversible electron transfer processes, *J. Am. Chem. Soc.*, 103, 2519, 1981.
17. Bohme, H. and Viehe, H. G., *Iminium Salts in Organic Chemistry*, Vol. 9, John Wiley & Sons, New York, 1976.
18. Andrieux, C. P. and Saveant, J. M., Electrodimerization II. Reduction mechanism of immonium cations, *J. Electroanal. Chem.*, 26, 223, 1970.
19. Andrieux, C. P. and Saveant, J. M., Electrooxidation of 1, 2-enediamines, *J. Electroanal. Chem.*, 28, 339, 1970.

20. Andrieux, C. P. and Saveant, J. M., Electrodimerization. EPR detection of transient immonium radicals, *J. Electroanal. Chem.*, 28, 446, 1970.

21. Holy, A., Kupicka, J., and Arnold, Z., The polarographic behavior of polymethinium salts, *Coll. Czech. Chem. Commun.*, 30, 4127, 1965.

22. Griller, D. and Lossing, F. P., On the thermal chemistry of α-aminoalkyl radicals, *J. Am. Chem. Soc.*, 103, 1586, 1981.

23. Mariano, P. S., Electron-transfer mechanisms in photochemical transformations of iminium salts, *Acc. Chem. Res.*, 16, 130, 1983.

24. Mariano, P. S., The photochemistry of iminium salts and related heteroaromatic systems, *Tetrahedron Rep. 156*, 39, 3845, 1983.

25. Mariano, P. S., Stavinoha, J. L., Pepe, G., and Meyer, E. F., Novel photochemical addition reactions of iminium salts. Electron transfer initiated additions of olefins to 2-phenyl-1-pyrrolinium perchlorate, *J. Am. Chem. Soc.*, 100, 7114, 1978.

26. Stavinoha, J. L. and Mariano, P. S., Electron transfer photochemistry of iminium salts. Olefin photoadditions to 2-phenyl-1-pyrrolinium perchlorate, *J. Am. Chem. Soc.*, 103, 3136, 1981.

27. Stavinoha, J. L., Mariano, P. S., Leone-Bay, A., Swanson, R., and Bracken, C. J., Photocyclizations of *N*-allyliminium salts leading to production of substituted pyrrolidines, *J. Am. Chem. Soc.*, 103, 3148, 1981.

28. Mariano, P. S. and Leone, A., Photocycloadditions of electron poor olefins to 2-aryl-1-pyrrolinium salts, *Tetrahedron Lett.*, 4581, 1980.

29. Yoon, U. C., Quillen, S. L., Mariano, P. S., Swanson, R., Stavinoha, J. L., and Bay, E., Electron transfer initiated photocyclizations of *N*-allylpyridinium and quinolinium salts. Preparation of indolizidines, *Tetrahedron Lett.*, 919, 1982.

30. Yoon, U. C., Quillen, S. L., Mariano, P. S., Swanson, R., Stavinoha, J. L., and Bay, E., Exploratory and mechanistic aspects of the electron transfer photochemistry of olefin *N*-heteroaromatic cation systems, *J. Am. Chem. Soc.*, 105, 1204, 1983.

31. Ohga, K. and Mariano, P. S., Electron transfer initiated photoadditions of allylsilanes to 1-methyl-2-phenyl-1-pyrrolinium perchlorate. A new allylation methodology, *J. Am. Chem. Soc.*, 104, 617, 1982.

32. Ohga, K., Yoon, U. C., and Mariano, P. S., Exploratory and mechanistic studies of the electron transfer initiated photoaddition reactions of allylsilane-iminium salt systems, *J. Org. Chem.*, 49, 213, 1984.

33. Ullrich, J. W., Chiu, F. T., Tiner-Harding, T., and Mariano, P. S., Electron transfer initiated photospirocyclization reactions of β-enaminone derived allyliminium salts, *J. Org. Chem.*, 49, 220, 1984.

34. Chung, L. T. and Mariano, P. S., Dual diradical and cation diradical cyclization mechanisms for silylalkenyl iminium salt electron transfer induced photocyclization reactions, *J. Am. Chem. Soc.*, 109, 5287, 1987.

35. Cho, I. S., Chung, L. T., and Mariano, P. S., Electron transfer induced photochemical reactions of (silylallyl)-iminium and -benzylpyrrolinium salts by dual diradical and diradical cation cyclization pathways, *J. Am. Chem. Soc.*, 112, 3594, 1990.

36. Borg, R. M., Heuckeroth, R. O., Lan, J. Y., Quillen, S. L., and Mariano, P. S., Arene-iminium salt electron transfer photochemistry. Mechanistically interesting photoaddition processes, *J. Am. Chem. Soc.*, 109, 2728, 1987.

37. Lan, J. Y., Heuckeroth, R. O., and Mariano, P. S., Electron transfer induced photocyclization reactions of arene-iminium salt systems. Effects of cation diradical deprotonation and desilylation on the nature and efficiencies of reaction pathways followed, *J. Am. Chem. Soc.*, 109, 2738, 1987.

38. Kavash, R. W. and Mariano, P. S., Application of a SET-induced photospirocyclization methodology to Harringtonine ring construction, *Tetrahedron Lett.*, 30, 4185, 1989.

39. Ahmed-Schofield, R. and Mariano, P. S., Sequential electron transfer-desilylation methods for diradical photogeneration as part of synthetic routes for erythrina alkaloid synthesis, *J. Org. Chem.*, 50, 5667, 1985.

40. Ahmed-Schofield, R. and Mariano, P. S., A photochemical route for erythrinane ring construction, *J. Org. Chem.,* 52, 1478, 1987.

41. Ho, G. D. and Mariano, P. S., Novel photochemical diradical cyclization methods for protobererine alkaloid synthesis. Preparation of (±)-xylopinine and (±)-stylopine, *J. Org. Chem.,* 52, 704, 1987.

42. Ho, G. D. and Mariano, P. S., Exploratory, mechanistic and synthetic aspects of silylarene-iminium salt SET-photochemistry. Studies of diradical cyclization processes and applications to protoberberine alkaloid synthesis, *J. Org. Chem.,* 53, 5113, 1988.

43. Cho, I. S., Chang, S. S., Ho, C., Lee, C. P., Ammon, H. L., and Mariano, P. S., Stereochemical characteristics of SET-promoted photochemical reactions of dihydroisoquinolinium salts, *Heterocycles,* 31, 3910, 1991.

Photochemistry of
N-Oxides

Angelo Albini
University of Torino; currently
the University of Pavia

Elisa Fasani
University of Pavia

Maref M. Amer
University of Zagazig

70.1 Introduction

The N-oxide group, characterized by a donative bond between a nitrogen and an oxygen atom, is present in different compounds, such as tertiary amine N-oxides $R_3N \rightarrow O$ (nitrogen sp^3-hybridized), nitrones, azoxy compounds, and heterocyclic N-oxides $R-X=N \rightarrow O$ (nitrogen sp^2-hybridized) and nitrile N-oxides $R-C\equiv N \rightarrow O$ (nitrogen sp-hybridized). These compounds differ greatly in their chemical and physical properties.[1] Indeed, as far as the photochemical behavior is concerned, it looks as if these various molecules contain different chromophores, in the sense that the lowest excited state does not involve orbitals localized on the NO function in every case. Nevertheless, the overwhelming majority of these compounds are photoreactive, and usually the chemical reaction involves cleavage of the $N-C_\alpha$ or of the N–O bond or of both.[2-4]

Indeed, the N-oxide function is one of the few groups to which a monomolecular photoreactivity is associated. Quantum yields for reactions are often moderate to high, and, when they are colored, the high reactivity associated with these compounds is made apparent by the changes that occur when they are exposed to sunlight. Thus, the photoreactivity of azoxynaphthalene was first noted when the compound was first characterized,[5] and the liability of quinoxaline 1,4-dioxide was first inferred from the change of the UV spectrum of solution prepared for spectroscopy work and the loss of antibacteriological properties when exposed to light.[6] In the following discussion, the relevant photochemistry is discussed according to the chromophore present.

70.2 Tertiary Amine N-Oxides

N-Oxidation introduces no new absorptions in the spectrum of aliphatic tertiary amines. Such compounds do not absorb above 200 nm and nothing has been reported about their photochemistry. However, when a light-absorbing moiety is present — a typical example being the N-oxides of benzyl and β-phenylethylamine — the N-oxide function is the site of the photoreaction, even if it is not directly involved in the electronic excitation. The processes observed are Meisenheimer rearrangement to substituted hydroxylamines, e.g., with the N-oxides of N,N-dimethyl-αmethylbenzylamine[7] or of protopine alkaloids (Equation 70.1)[8] or fragmentation to amides as with berbine *N*-oxides.[9]

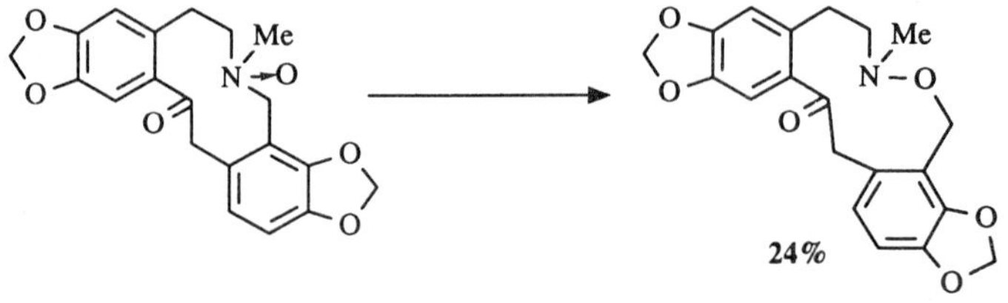

24%

(70.1)

With arylamines, *N*-oxidation causes a strong hypso- and hypochromic shift of the absorption.[10] The photochemical reaction is *N*-deoxygenation in this case,[11,12] with oxygen transfer to the solvent or intramolecularly to the α-position, e.g., Equation 70.2.

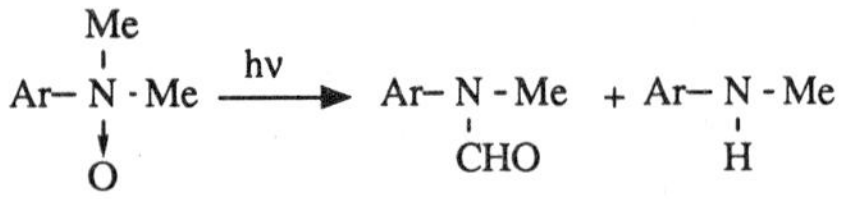

$$\text{Ar–}\overset{\overset{\displaystyle Me}{|}}{\underset{\underset{\displaystyle O}{\downarrow}}{N}}\text{·Me} \xrightarrow{h\nu} \text{Ar–}\underset{\underset{\displaystyle CHO}{|}}{N}\text{·Me} + \text{Ar–}\underset{\underset{\displaystyle H}{|}}{N}\text{·Me}$$

(70.2)

70.3 Nitrones and Azoxy Compounds

In *N*-oxides with a nitrogen of *sp²*-hybridization type, the lowest-energy absorption band corresponds generally to a π,π* transition with strong ICT character.[13,14] This band is often red-shifted with respect to the nonoxidized parent compound and has intensity depending widely on the structure. Furthermore, a strong hypsochromic shift is observed with polar and protic solvents (since it involves charge transfer from the oxygen atom). Although this description of the excited states may be appropriate for nitrones, azoxy derivatives, and heterocyclic *N*-oxides, the chemical outcome of the irradiation is different in these cases.

Both aliphatic[15] and aromatic nitrones[16–18] undergo cyclization to oxaziridines, a reaction that occurs with excellent yield, provided that the light used for the irradiation is filtered appropriately in order to avoid secondary photoconversion of the oxaziridines to amides, and offers a useful entry for the synthesis of these three-membered heterocycles (Equation 70.3).

91%

(70.3)

Ring-fused oxaziridines are obtained from cyclic nitrones[17,19–21] (in some cases, however, only ring-contracted amides,[17,22] lactams,[23,24] and products resulting from the insertion of the oxygen atom in the ring,[23,25] ring-opened isocyanates[25] or ketones are obtained,[23] presumably through further reaction of the oxaziridines). The photocyclization of nitrones occurs stereoselectively, at least at low temperature,[17] and asymmetric induction in chiral solvents has been observed.[26]

Aliphatic azoxy derivatives also photocyclize to oxadiaziridines, and examples of monocyclic[27,28] and bicyclic derivatives have been reported.[29–31] The reaction is generally thermally and/or photochemically reversible, and thus the steric or the positional isomerization of the azoxy group observed following irradiation occurs through reversible cyclization.[28,31] Some azine *N*-oxides and a bicyclic azoxy derivative have been found to cleave photochemically to diazo derivatives, probably via oxaziridines (Equation 70.4).[32]

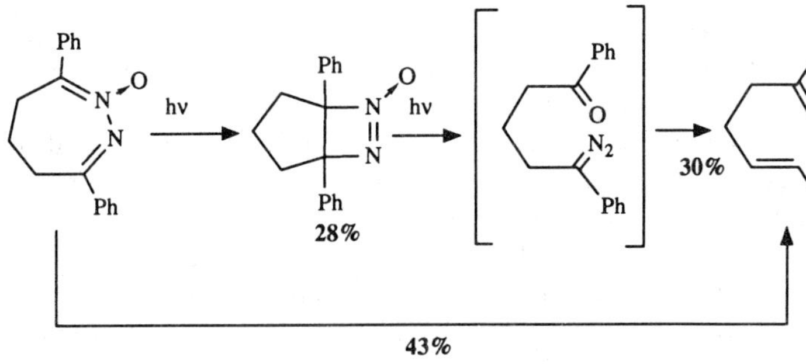

$$(70.4)$$

Aromatic azoxy derivatives photorearrange to *o*-hydroxyazo compounds,[33–38] a reaction that has been termed, rather improperly, a "photo-Wallach" rearrangement. The oxygen atom attacks the aryl ring distant from the N–O group (attached to the other nitrogen atom) and yields an oxadiazolidine intermediate. When there is no *ortho*-position free on this aryl group, the rearrangement fails.[37,38] This intermediate is converted to the final product, either directly by nucleophilic or acid-catalyzed proton shift or alternatively through cleavage to yield a diazonium-phenolate radical ion pair and recombination (trapping of the diazonium cation is possible). The latter path is more important in apolar solvents (Equation 70.5).[33,34] Substituents with a strong electron-withdrawing or electron-donating effect slow down this rearrangement, and then other photoprocesses, in particular a 1,2-oxygen shift (possibly via an oxadiaziridine), are observed.[33,35] The azoxy-to-hydroxyazo rearrangement has also been reported in the solid state as well as with mixed aliphatic-aromatic derivatives.[36]

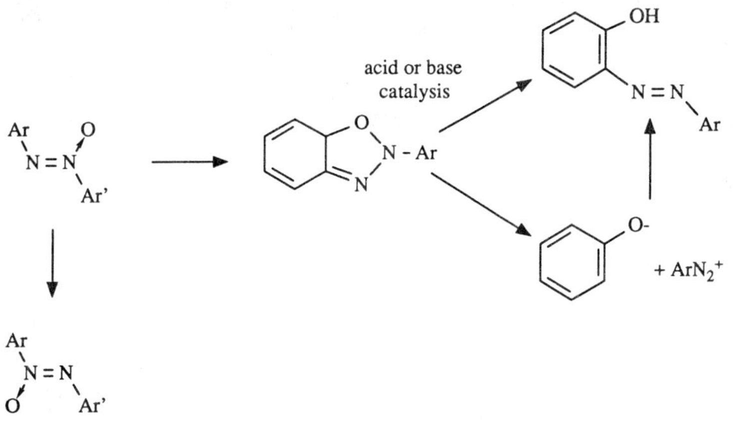

$$(70.5)$$

0.4 Heteroaromatic *N*-Oxides

Rearrangement

As with nitrones, the lowest excited singlet state of six-membered heteroaromatic *N*-oxides is an ICT π,π^* state.[40] With pyridine and other monocyclic *N*-oxides, the transition is weak and experiences strong solvatochromism; while with aza-naphthalene and aza-anthracene *N*-oxides, the transition is strong and less solvent dependent; in the latter series, fluorescence appears. Calculations show that the ICT character correspondingly diminishes,[1,40] and the photophysical behavior becomes more similar to that of carbocyclic aromatics. However, *N*-oxides with three (and four) aromatic rings are still photoreactive.

Irradiation usually leads to rearrangement (see next section, which concerns *N*-deoxygenation). There has been a long debate concerning whether or not the different reactions observed could be interpreted through a single mechanism.[2-4] Indeed, most of the reactions could be rationalized as involving a primary rearrangement to an oxaziridine, followed by further thermal or photochemical transformations. However, contrary to the case of nitrones — where oxaziridines have indeed been isolated in many cases — with heterocyclic *N*-oxides, there is no strong evidence for such an intermediate and the processes observed are better rationalized as arising via biradical or zwitterionic paths without involving an intermediate of any stability. In the following, the photochemical behavior of six-membered heterocycles is discussed first, and that of five-membered derivatives afterwards. The main reaction paths are gathered in Equation 70.6.

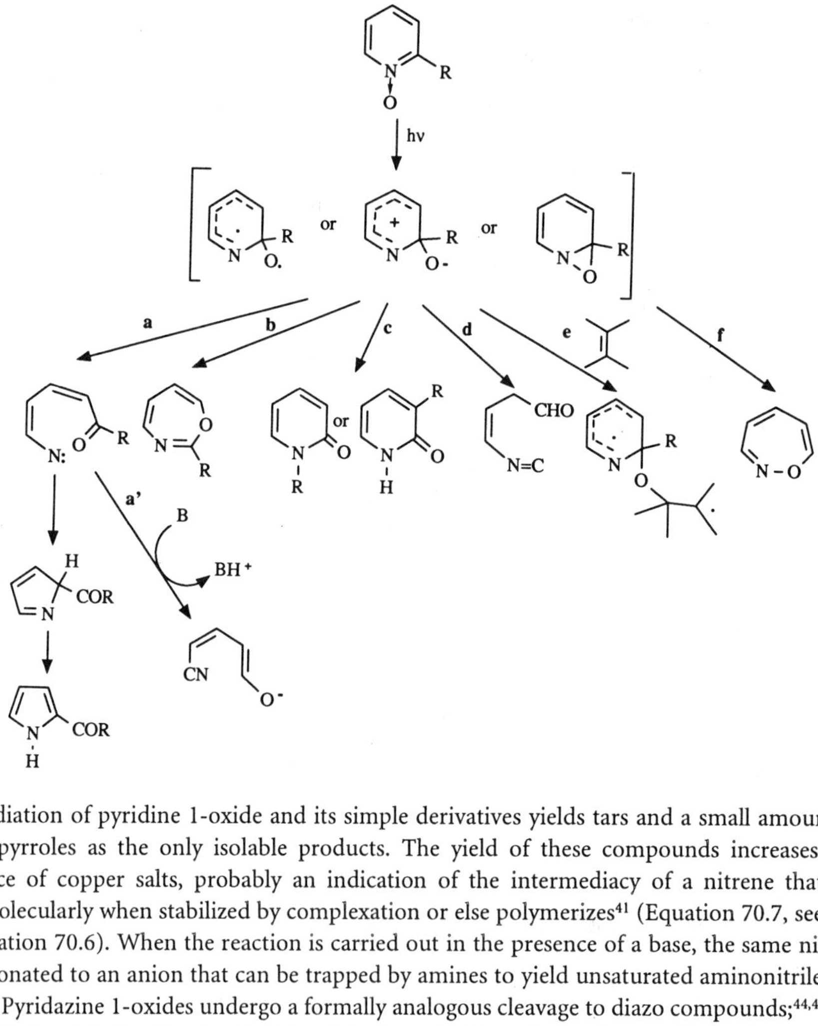

(70.6)

Irradiation of pyridine 1-oxide and its simple derivatives yields tars and a small amount of 2-formylpyrroles as the only isolable products. The yield of these compounds increases in the presence of copper salts, probably an indication of the intermediacy of a nitrene that reacts intramolecularly when stabilized by complexation or else polymerizes[41] (Equation 70.7, see path a in Equation 70.6). When the reaction is carried out in the presence of a base, the same nitrene is deprotonated to an anion that can be trapped by amines to yield unsaturated aminonitriles (path a').[42,43] Pyridazine 1-oxides undergo a formally analogous cleavage to diazo compounds;[44,45a] while in pyridazine 1,2-dioxide, the N–N bond is cleaved ultimately to give 3,4-dihydroisoxazolo[4,5-d]isoxazole.[45b] 2,6-Dicyanopyridine 1-oxide gives the 2-formylpyrrole in good yield,[46] while polyphenylpyridine oxides undergo efficient ring enlargement to 1,3-oxazepines (path b).[47] These seven-membered heterocycles rearrange easily to 1-acylpyrroles that are also obtained as minor products from various pyridine oxides. The ring contraction of some pyrimidine[48] and pyrazine 1-oxides,[49a] as well as of 1,2,4-triazine 4-oxides[49b] to imidazoles and respectively 1,2,4-triazoles, probably occurs analogously via photochemical ring enlargement followed by thermal contraction.

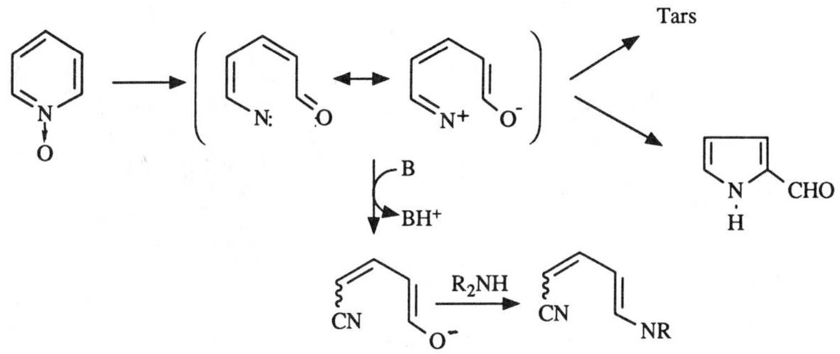

$$(70.7)$$

The photochemistry of azanaphthalene *N*-oxides is dominated by two reactions: ring enlargement to benzoxazepines (or aza analogues) in aprotic solvent (path b) and rearrangement to lactams (path c) in protic solvents. The first reaction has been observed with the *N*-oxides of quinolines (Equation 70.8),[50–55] isoquinolines,[53–57] several substituted quinoxalines,[58,59] and some quinazolines.[60] The products are moisture sensitive and care is required during the work-up, except when they are stabilized by substituents (such as Ph,[51–53,56] CN,[52,53,56] MeO,[54] and CF$_3$[55]). With these substituents present, ring enlargement occurs in protic medium and is followed by nucleophilic addition. The product from the parent isoquinoline 2-oxide cannot be isolated, but the open-chain adducts obtained with nucleophiles are indicative that its structure is 1,3-benzoxazepine.[50] In contrast, quinoxaline 1-oxide and its 3-methoxy derivative undergo cleavage to isocyanides (Equation 70.9, path d in Equation 70.6)[58] and 1,4-diphenylphthalazine 2-oxide[44] as well as benzo-1,2,3-triazine 3-oxide[61] are cleaved to diazoketones (path a). The latter *N*-oxide, as well as cinnoline 2-oxide, also undergoes different fragmentations with formal loss of NO, CO, or N$_2$.[62]

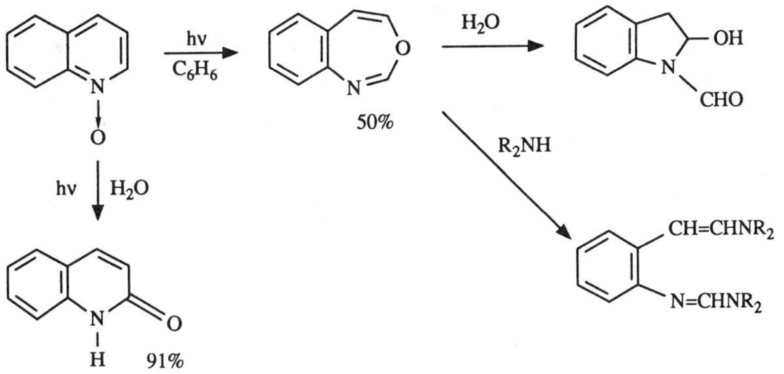

$$(70.8)$$

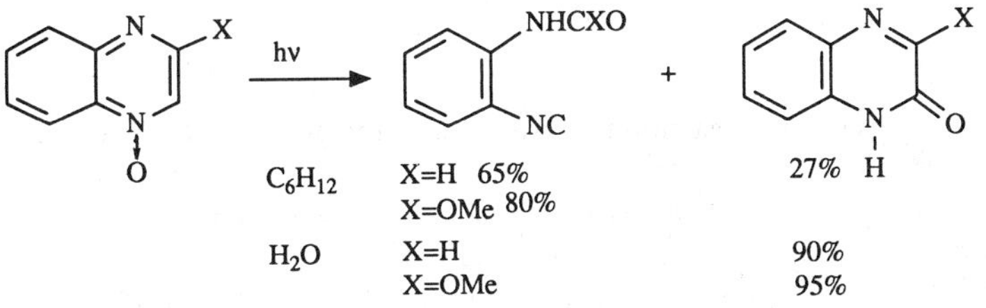

$$(70.9)$$

The rearrangement to lactams is quite general, with the exception of the substituted derivatives mentioned above. Substituents at the α-position may migrate either to the nitrogen or to the neighboring carbon atom (or both) during the rearrangement.[63–65] An application of this rearrangement is seen in the synthesis of the alkaloid ravenine (Equation 70.10).[66]

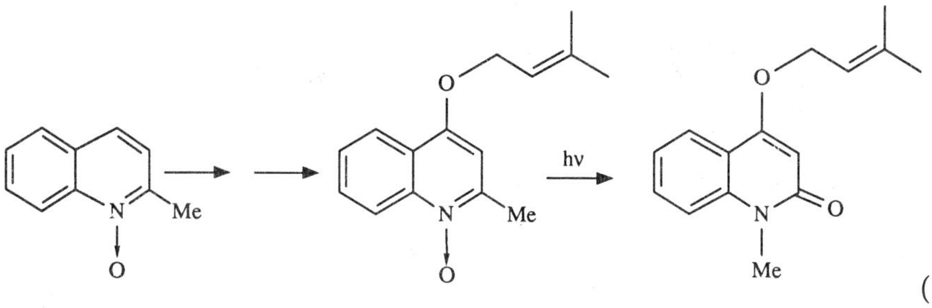

(70.10)

In apolar solvents, 1- and 4-azaphenanthrene N-oxides rearrange to naphthoxazepines,[67] and 6-aryl- and 6-cyanophenanthridine 5-oxides likewise undergo ring enlargement to dibenzoxazepines[68–70] while the parent compound yields only phenanthridone.[67] The photoreactivity of the cyano derivative shows a variety of reaction paths (Equation 70.11). Thus, in ethanol, N-ethoxyphenanthridone is formed;[69,70] and in the presence of 2,3-dimethylbutene, an addition process takes place (path e).[68,71] All the N-oxides mentioned above yield lactams on irradiation in protic solvents. With 6-alkylphenanthridine 5-oxides, migration of the substituent to the nitrogen atom takes place (a reaction applied in the synthesis of the alkaloid perlonine).[72] If, however, the substituent at the 6-position is capable of forming a stable free radical (e.g., a benzhydryl group), then the group undergoes elimination rather than migration.[68,73] This last reaction, as well as the alkene addition mentioned above, are indications of the particular importance of a biradical structure during the rearrangement.

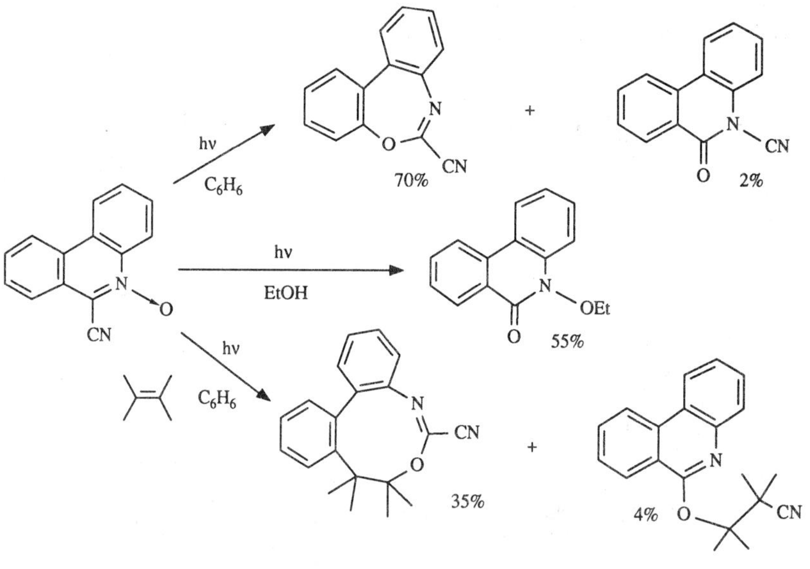

(70.11)

The reaction path changes again with aza-anthracene N-oxides. With 9-cyano- and chloroacridine 10-oxide, the main products are dibenzo-1,2-oxazepines (the valence tautomers of oxaziridines, path f),[74] and the same type of compound may be formed from related N-oxides, where products arising from further photochemical or thermal reactions are actually isolated. In general, the main products from acridine[75] and phenazine N-oxides[76–78] following irradiation in aprotic solvents are indole and benzimidazole derivatives, respectively (this is an example of path a; the benzo ring becomes perpendicular to the molecular plane during the rearrangement). In protic solvents, the products obtained are the oxepino[1,2-b]quinoline or quinoxaline, respectively (Equation 70.12). Cleavage to benzoxazolylpentadienenitriles is also observed with phenazine N-oxide.[76]

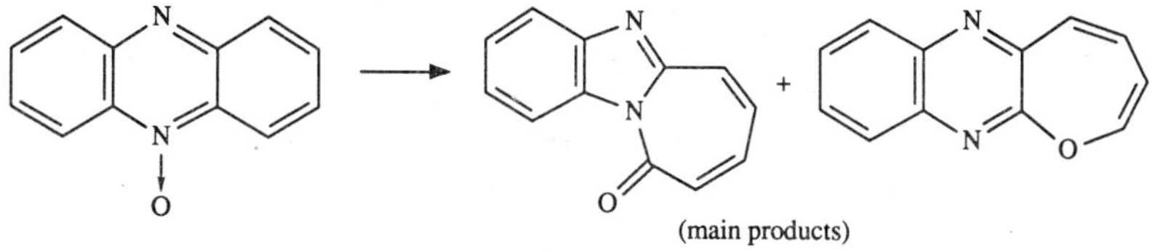

(main products) (70.12)

Benzo[*a*]acridine[79] and phenazine 7-oxide[80] undergo ring enlargement to an oxazepine, or an oxadiazepine, respectively, while with benzo[*a*]phenazine 12-oxide, rearrangement to benzimidazole predominates.[80] Purine[81–84] and pyrazolopyrimidine N-oxides[85] (usually studied in water, where N-hydroxy tautomers may predominate) generally undergo photodeoxygenation rather than rearrangement, though migration of the oxide function,[81] cleavage of the pyrimidine ring,[84,85] and rearrangement to lactams[82,83] have also been observed (the last reaction occurs also with some pteridine N-oxides).[86]

The photochemistry of five-membered heterocyclic N-oxides involves mainly rearrangement to lactams (e.g., for 1-alkyl-1,2,4-triazole 4-oxides[87] and 1-alkylbenzimidazole 3-oxides in protic medium[88]) or cleavage to open-chain amides, as with 1-alkylbenzimidazole 3-oxides in aprotic medium[88] or with 1-*H*-tetraarylimidazole 3-oxides.[89] Positional isomerization of the oxide function is also observed commonly (e.g., with (benzo)thiadiazoles,[90,91] benzofuroxans,[92] and benzotriazoles[93]) and at least in some cases it has been shown that the reaction involves a ring-opening/ring-closure mechanism.[91]

N-Deoxygenation

The previous discussion has shown that rearrangement is the common result of the irradiation of heterocyclic N-oxides. However, deoxygenation is an accompanying reaction and, as a rule, its importance increases when the substrates are more readily reduced, i.e., in polycyclic rather than monocyclic N-oxides, in α- and γ-diaza and obviously in polyaza N-oxides, and in substrates carrying strong electron-withdrawing substituents. Thus, deoxygenation is significant with pyridazine,[45a] pyrazine,[48] and cinnoline N-oxides[62] and, at least under suitable conditions, becomes the only path observed (although with a low quantum yield) e.g., with the N-oxides of benzoylpyridines,[94] benzo[*c*]cinnoline,[95] nitrophenazine,[96] alloxazine,[120] and purine.[81–84]

Two main mechanisms have been considered for this process:[2,3] either deoxygenation is a triplet-state process (and rearrangement a singlet-state reaction),[94] or an intermediate formed along the rearrangement path (e.g., an oxaziridine) transfers oxygen to suitable substrates.[70] There is probably no unitary mechanism;[97] thus, it appears likely that benzoylpyridine N-oxides[94] and probably some dioxides[98] are deoxygenated via the triplet through hydrogen abstraction (the effect of an external magnetic field suggests that dual paths may operate also with other N-oxides[99]). On the other hand, 2-nitrophenazine 10-oxide reacts via the singlet state through an electron transfer path,[96] and 6-cyanophenanthridine via a rearranged intermediate.[70]

From the preparative point of view, it should be noted that irradiation in the presence of suitable acceptors (boron trifluoride,[100] triphenylphosphine,[101] phosphite,[102] amines[96]) is a general method for the reduction of N-oxides, particularly useful for the selective N-deoxygenation in the presence of nitro groups.[96,102]

On the other hand, N-oxides can be considered as useful photochemical oxidants in reactions such as the hydroxylation of aromatics[103–106] or alkanes,[103,107-109] the epoxidation of alkenes,[108,110] the oxidation of epoxides,[111] the dealkylation of alkylamines,[112] the dehydrocyclization of nucleosides to cyclonucleosides,[113] the desulfurization of P(v) derivatives,[97] and the S-oxidation of alkyl sulfides.[114] The common oxygen donors are pyridazine 1-oxide[103,104] and some polycyclic derivatives, such as pyrimido[5,4-*g*]pteridine N-oxide.[105,112,113] Here again, the mechanism of the oxidation is not unambiguous. The fact that alkanes are also attacked has suggested that "activated" atomic oxygen is released,[107,108,110] although in more selective oxidations, as those obtained with pteridine derivatives, an electron transfer path is followed.[112,113] The oxidation of aromatic substrates through this photochemical method has been considered as a model for biochemical oxidation.[106,109,114]

Industrial Application

The favorable UV/VIS absorption properties of *N*-oxides, which can be extended to any desired range by choosing a suitable substrate, and their generalized photoreactivity have stimulated some application. Thus, polymers containing vinylpyridine (or quinoline, acridine, etc.) *N*-oxides have been found to undergo photo-cross-linking conveniently.[115] Furthermore, *N*-oxides have been used for photolithography,[116] and in photography for the hardening of films[117] and as additives for developers.[118,119]

70.5 Nitrile *N*-Oxides

Nitrile *N*-oxides are themselves reactive intermediates that are often generated *in situ*. Therefore, photochemical studies of such substrates are limited to derivatives stabilized by steric hindrance. In these cases,[121] it appears that on irradiation they rearrange to acylnitrenes (which insert intramolecularly) or to isocyanates (which undergo nucleophilic addition).

References

1. Albini, A. and Pietra, S., *Heterocyclic N-Oxides*, CRC Press, Boca Raton, FL, 1992.
2. Albini, A. and Alpegiani, M., Photochemistry of the *N*-oxide function, *Chem. Rev.*, 84, 43, 1984.
3. Spence, A. G., Taylor, E. C., and Buchardt, O., The photochemical reactions of azoxy compounds, nitrones, and aromatic amine *N*-oxides, *Chem. Rev.*, 70, 231, 1970; Buchardt, O., Amine and imine *N*-oxides, in *Houben Weil Handb. Org. Chem.*, G. Thieme Verlag, Stuttgart, Vol. 4/5b, 1975, 1282.
4. Bellamy, F. and Streith, J., The photochemistry of aromatic *N*-oxides: A critical review, *Heterocycles*, 4, 1931, 1976.
5. Wacker, L., On α-azoxynaphthalene, *Liebigs Ann. Chem.*, 317, 375, 1901.
6. Landquist, J. K., Photochemical decomposition of quinoxaline mono and di *N*-oxides, *J. Chem. Soc.*, 2830, 1953.
7. Schoellkopf, U., Patsch, M., and Schafer, M., Indication of the radicalic course of the Meisenheimer rearrangement, *Tetrahedron Lett.*, 2515, 1964.
8. Iwasa, K. and Takao, N., The pyrolysis and photolysis of the protopine type alkaloids, *Heterocycles*, 20, 1535, 1983.
9. Chinnasamy, P., Minard, R. D., and Schamma, M., The photolysis of berbine *N*-oxides, *Tetrahedron*, 36, 1515, 1980.
10. Colonna, M. and Risaliti, A., UV spectra and character of the N–O bond in 2- and 4-phenylpyridine *N*-oxide, *Ann. Chim. (Rome)*, 48, 1395, 1958.
11. Stenberg, W. I. and Schiller, J. E., Excited state predictions on amine oxides, *J. Am. Chem. Soc.*, 97, 424, 1975.
12. Albini, A., Fasani, E., Moroni, M., and Pietra, S., Photochemical decomposition of 4-arylazo- and 4-arylazoxy-*N*, *N*-dialkylaniline *N*-oxides, *J. Chem. Soc., Perkin Trans. 2*, 1439, 1986.
13. Ghersetti, I., Maccagnani, G., Mangini, A., and Montanari, F., Infrared and ultraviolet absorption spectra of pyridine *N*-oxides, *J. Heterocycl. Chem.*, 6, 859, 1969.
14. Kubota, T., Electronic spectra and electronic structure of some basic heterocyclic *N*-oxides, *Bull. Chem. Soc. Jpn.*, 35, 946, 1962.
15. Druelinger, M. L., Shelton, R. W., and Lammert, S. R., Photochemistry of methylenenitrones and related compounds, *J. Heterocycl. Chem.*, 13, 1001, 1976.

16. Splitter, J. S., Su, T. M., Ono, H., and Calvin, M., Orbital-symmetry control in the nitrone-oxaziridine system, *J. Am. Chem. Soc.,* 93, 4075, 1971.

17. Ning, R. Y., Field, G. F., and Sternbach, L. H., Quinoxalines and 1, 4-benzodiazepines. XLVI. Photochemistry of nitrones and oxaziridines, *J. Heterocycl. Chem.,* 7, 475, 1970.

18. Ono, H., Splitter, J. S., and Calvin, M., The effect of phenyl substituent in the nitrogen inversion barrier in a 2-phenyloxaziridine, *Tetrahedron Lett.,* 4107, 1973.

19. Black, D. S. C. and Watson, K. G., Photorearrangement of some bicyclic oxaziridines, *Aust. J. Chem.,* 26, 2502, 1973.

20. Suginome, H., Mizuguchi, T., and Masanume, T., Formation of oxaziridine by the irradiation and benzoylation of (22S, 25S)-N-acetyl-11a-aza-C-homo-5α-veratra-11a, 13(17)diene-3β, 11β, 23β-triol 11a-oxide, *Bull. Chem. Soc. Jpn.,* 50, 987, 1977.

21. Black, D. S. C. and Johnstone, L. M., Some reactions of bicyclic 1-pyrrole 1-oxides, *Aust. J. Chem.,* 37, 577, 1984.

22. Black, D. S. C. and Boscacci, A. B., Photorearrangement of 3-oxo-1-pyrroline 1-oxides, *Aust. J. Chem.,* 30, 1109, 1980; Black, D. S. C. and Boscacci, A. B., Synthesis and photochemistry of 2-acyl-1-pyrroline 1-oxides, *Aust. J. Chem.,* 30, 1353, 1980.

23. Aurich, H. G. and Grigo, U., Photochemistry of oxindolinilydenamine N-oxides, *Chem. Ber.,* 109, 3849, 1976.

24. Oliveros, E., Antoun, H., Rivière, M., and Lattes, A., Comparison between the thermal and photochemical rearrangement of spirooxaziridines, *J. Heterocycl. Chem.,* 13, 623, 1976.

25. Doepp, D., Photoisomerisation and acidolysis of 3, 3-dimethyl-3*H*-indole 1-oxides, *Chem. Ber.,* 109, 3849, 1976.

26. Boyd, D. R. and Neill, D. C., Asymmetric synthesis of a pyramidal nitrogen centre in a chiral medium, *J. Chem. Soc., Chem. Commun.,* 51, 1977.

27. Greene, F. D. and Hecht, S. S., Oxadiaziridines. Synthesis, valence isomerization and reactivity, *J. Org. Chem.,* 35, 2842, 1970.

28. Taylor, K. G., Isaac, S. R., and Swigert, J. L., Photolytic isomerization of azoxyalkanes, *J. Org. Chem.,* 41, 1146, 1976.

29. Dolbier, W. R., Matsui, K., Michl, J., and Horak, D. V., 2, 2-dimethylisoindene and 5, 5-dimethylbenzobicyclo[2.1.0]pent-2-ene, *J. Am. Chem. Soc.,* 99, 3876, 1977.

30. Olsen, H. and Pedersen, C. L., Photochemical isomerization of bicyclo *cis*-1, 2-diazene N-oxides to oxadiaziridines, *Acta Chem. Scand.,* B36, 701, 1982.

31. Squillacote, M. E., Bergman, A., De Felippis, J., and West, E. M., Thermal stability of *cis*-oxaziridines, *Tetrahedron Lett.,* 28, 275, 1987.

32. Williams, W. M. and Dolbier, W. R., Photochemical reaction of cyclic azine monoxides, *J. Am. Chem. Soc.,* 94, 3955, 1972.

33. Albini, A., Fasani, E., Moroni, M., and Pietra, S., Photochemistry of some methoxy and dimethylamino derivatives of azoxybenzene, *J. Org. Chem.,* 51, 88, 1986.

34. Bunce, N. J., Schoch, J. P., and Zerner, M. C., Photorearrangement of azoxybenzene to 2-hydroxyazobenzene. Evidence for electrophilic substitution by oxygen, *J. Am. Chem. Soc.,* 99, 7986, 1977.

35. Bunce, N. J., A new mode of photoreaction in the azoxybenzene series, *Can. J. Chem.,* 53, 3477, 1977.

36. Doepp, D. and Muller, D., Solid state photolysis: Photocleavage of a sterically hindered azoxybenzene into aryldiazonium phenoxide, *Tetrahedron Lett.,* 3863, 1978.

37. Goon, D. J. W., Murray, N. G., Schoch, J. P., and Bunce, N. J., Evidence against a hydrogen abstraction mechanism in the photorearrangement of azoxybenzene to 2-hydroxyazobenzene, *Can. J. Chem.,* 51, 3827, 1973.

38. Bunce, N. J., Goon, D. J. W., and Schoch, J. P., Synthesis and photolysis of some naphthylazoxy compounds, *J. Chem. Soc., Perkin Trans. 1,* 688, 1976.

39. Taylor, K. G. and Riehl, T., Synthesis of new azoxy compounds by photolytic isomerization, *J. Am. Chem. Soc.,* 94, 250, 1972.

40. Scholz, M., Electronic structure of pyridine N-oxide, *J. Prakt. Chem.*, 323, 571, 1981; Scholz, M., MO calculation of electronic spectra of pyridine N-oxide. A critical review, *J. Prakt. Chem.*, 324, 85, 1982.

41. Bellamy, F. and Streith, J., Effect of copper salts on the photochemistry of monosubstituted pyridine N-oxides, *J. Chem. Res. (S)*, 18, 1979; Bellamy, F., Streith, J., and Fritz, H., Photochemical behaviour of monosubstituted pyridine N-oxide in water, *Nouv. J. Chem.*, 3, 115, 1979.

42. Lohse, C., Hagedorn, L., Albini, A., and Fasani, E., Photochemistry of pyridine N-oxide, *Tetrahedron*, 44, 2591, 1988.

43. Buchardt, O., Christensen, J. J., Nielsen, P. E., Koganty, R. R., Finsen, L., Lohse, C., and Becher, J., Photochemical ring-opening of pyridine N-oxide to 5-oxo-2-pentenenitrile and/or 5-oxo-3-pentenenitrile, *Acta Chem. Scand.*, B34, 31, 1980.

44. Tomer, K. B., Harrit, N., Rosenthal, I., Buchardt, O., Kumler, P. L., and Creed, D., Photochemical behaviour of aromatic 1, 2-diazine N-oxides, *J. Am. Chem. Soc.*, 95, 7402, 1973.

45. (a) Tsuchiya, T., Arai, H., and Igeta, H., Formation of cyclopropenylketones and furans from pyridazine N-oxide by irradiation, *Tetrahedron*, 29, 2747, 1973; (b) Ohsawa, A., Arai, H., Igeta, H., Akimoto, T., and Tsuji, A., The photoisomerization of pyridazine 1, 2-dioxide; formation of 3a, 6a-dihydroisoxazolo[1,2-*d*]isoxazole, *Tetrahedron*, 35, 1267, 1979.

46. Ishikawa, M., Kaneko, C., Yokoe, L., and Yamada, S., Photolysis of 2, 6-dicyanopyridine 1-oxide, *Tetrahedron*, 25, 295, 1969.

47. Buchardt, O., Pedersen, C. L., and Harrit, N., Light-induced ring expansion of pyridine N-oxides, *J. Org. Chem.*, 37, 3592, 1972.

48. Roeterdink, F., van der Plas, H. C., and Kandijs, A., Photochemistry of pyrimidine N-oxides, *Recl. Trav. Chim. Pays-Bas*, 94, 16, 1975.

49. (a) Ikekawa, N., Honma, Y., and Kenkyusho, R., Photochemical reactions of pyrazine-N-oxides, *Tetrahedron Lett.*, 1197, 1967; (b) Neunhoeffer, H. and Boehnisch, V., Reactions of 1, 2, 4-triazine 4-oxides, *Liebigs Ann. Chem.*, 153, 1976.

50. Albini, A., Bettinetti, G., and Minoli, G., On 1, 3-benzoxazepine and 3, 1-benzoxazepine, *Tetrahedron Lett.*, 3761, 1979.

51. Buchardt, O., Kumler, P. L., and Lohse, C., Photolysis of phenylquinoline N-oxides in solution. Solvent influence on the product distribution, *Acta Chem. Scand.*, 23, 2149, 1969.

52. Buchardt, O., Jensen, B., and Larsen, I. K., The formation of benz[*d*]-1,3-oxazepines in the photolysis of quinoline N-oxides in solution, *Acta Chem. Scand.*, 21, 1841, 1967.

53. Kaneko, C., Yamada, S., and Ishikawa, M., Irradiation of N-oxides of α-cyanoazanaphthalenes in aprotic solvents, *Tetrahedron Lett.*, 2145, 1966.

54. Albini, A., Fasani, E., and Maggi Dacrema, L., Photochemistry of methoxy substituted quinoline and isoquinoline N-oxides, *J. Chem. Soc., Perkin Trans. 1*, 2738, 1980.

55. Kaneko, C., Hayashi, S., and Kobayashi, Y., Photolysis of 2-(trifluoromethyl)quinoline 1-oxides and 1-(trifluoromethyl)-isoquinoline 2-oxides, *Chem. Pharm. Bull.*, 22, 2147, 1974.

56. Simonsen, O., Lohse, C., and Buchardt, O., The photolysis of 1-phenyl- and 1-cyano-substituted isoquinoline N-oxides to benz[*f*][1,3]oxazepines, *Acta Chem. Scand.*, 24, 268, 1970.

57. Brenmer, J. B. and Wiryachitra, P., The photochemistry of papaverine N-oxide, *Aust. J. Chem.*, 26, 437, 1973.

58. Albini, A., Colombi, R., and Minoli, G., Photochemistry of quinoxaline 1-oxide and some of its derivatives, *J. Chem. Soc., Perkin Trans. 1*, 924, 1978.

59. Kaneko, C., Yamada, S., Yokoe, I., and Ishikawa, M., Study of the stable photoproducts derived from quinoline 1-oxide and quinoxaline 1-oxide, *Tetrahedron Lett.*, 1873, 1967.

60. Field, G. F. and Sternbach, L. H., Quinazoline and 1, 4-benzodiazepines. XLII. Photochemistry of some N-oxides, *J. Org. Chem.*, 33, 4438, 1968.

61. Horspool, W. M., Kerkshaw, J. R., Murray, A. W., and Stevenson, G. M., Photolysis of 4-substituted 1,2,3-benzotriazine 3-N-oxides, *J. Am. Chem. Soc.*, 85, 2390, 1973.

62. Horspool, W. M., Kerkshaw, J. R., and Murray, A. W., Photolysis of 4-methylcinnoline 1- and 2-N-oxides, *J. Chem. Soc., Chem. Commun.*, 345, 1973.

63. Buchardt, O., Tomer, K. B., and Madsen, V., A photochemical 1,2-deuterium shift. Irradiation of quinoline N-oxide and quinoline N-oxide-2-d₁, *Tetrahedron Lett.*, 1311, 1971.

64. Lohse, C., Primary photoprocesses in isoquinoline N-oxides, *J. Chem. Soc., Perkin Trans. 2*, 229, 1972.

65. Kaneko, C., Yamamoto, A., and Hashiba, M., Ring contraction reactions of methylquinoline 1-oxide 5-carboxylates via the corresponding benz[a]-1,3-oxazepines. A facile synthesis of methyl indole-4-carboxylate and its derivatives, *Chem. Pharm. Bull.*, 27, 946, 1979.

66. Kaneko, C., Naito, T., Hashiba, M., Fujii, H., and Somei, M., A new synthesis of ravenine and related alkaloids by means of a photo-rearrangement reaction of 4-alkoxy-2-methylquinoline 1-oxides, *Chem. Pharm. Bull.*, 27, 1813, 1979.

67. Albini, A., Bettinetti, G., and Minoli, G., Photochemistry of some azaphenanthrene N-oxides, *J. Chem. Soc., Perkin Trans. 1*, 1159, 1980.

68. Albini, A., Fasani, E., and Frattini, V., Medium and substituent effect on the photochemistry of phenanthridine N-oxides. Is an intermediate of diradical character involved in the photorearrangement of heterocyclic N-oxides?, *J. Chem. Soc., Perkin Trans. 2*, 235, 1988.

69. Kaneko, C., Hayashi, R., Yamamori, M., Tokumura, K., and Itoh, M., Photochemical reactions of 6-cyanophenanthridine 5-oxide, *Chem. Pharm. Bull.*, 26, 2508, 1978.

70. Tokumura, K., Goto, H., Kashibara, H., Kaneko, C., and Itoh, M., Formation and reaction of oxaziridine intermediate in the photochemical reaction of 6-cyanophenanthridine 5-oxide at low temperature, *J. Am. Chem. Soc.*, 102, 5643, 1980.

71. Albini, A., Fasani, E., and Buchardt, O., Radicaloid intermediates in the photochemistry of 6-cyanophenthridine N-oxide, *Tetrahedron Lett.*, 23, 4849, 1982.

72. Ridley, A. B. and Taylor, W. C., Synthesis of 6-(3,4-dimethoxyphenyl)-4-oxo-3, 4-dihydrobenzo[c][2,7]naphtyridin-6-ium chloride (perlonine chloride), *Aust. J. Chem.*, 40, 631, 1987.

73. Taylor, E. C. and Spence, G. G., Group migration in the photolysis of 6-substituted phenanthridine 5-oxide, *J. Chem. Soc., Chem. Commun.*, 767, 1966.

74. Yamada, S. and Kaneko, C., Photochemistry of acridine 10-oxide: Synthesis and reaction of dibenz[c, f]-1,2-oxazepines, *Tetrahedron*, 35, 1273, 1979.

75. Yamada, S., Ishikawa, M., and Kaneko, C., Photochemistry of acridine N-oxides, *Chem. Pharm. Bull.*, 23, 2818, 1973.

76. Albini, A., Bettinetti, G., and Pietra, S., Photoreactions of phenazine 5-oxide, *Tetrahedron Lett.*, 3657, 1972.

77. Albini, A., Bettinetti, G., and Pietra, S., Photoisomerization of substituted phenazine N-oxides, *Gazz. Chim. Ital.*, 105, 15, 1975.

78. Kawata, H., Niizuma, S., Kumagai, T., and Kokubun, H., Photochromism of phenazine N-dioxides in organic solvents, *Chem. Lett.*, 767, 1985.

79. Kaneko, C., Yamada, S., and Ishikawa, M., Synthesis of dibenzo[c,g]-2,5-diaza-1, 6-oxido[10]annulene and 12-methyldibenz[c,g]-2-aza-1,6-oxido[10]annulene, *Tetrahedron Lett.*, 2329, 1970.

80. Albini, A., Barinotti, A., Bettinetti, G., and Pietra, S., The photoisomerization of benzo[a]phenazine 7-oxide, *Gazz. Chim. Ital.*, 106, 871, 1976.

81. Lam, F. L., Brown, G. B., and Parham, J. C., Photoisomerization of 1-hydroxy to 3-hydroxyxanthine. Photochemistry of related 1-hydroxypurines, *J. Org. Chem.*, 39, 1391, 1974.

82. Lam, F. L. and Parham, J. C., Photochemistry of 1-hydroxy- and 1-methoxyhypoxanthines, *J. Org. Chem.*, 38, 2397, 1973.

83. Lam, F. L. and Parham, J. C., Photochemistry of purine 3-oxides in hydroxylic solvents, *Tetrahedron*, 38, 2371, 1982.

84. Lam, F. L. and Parham, J. C., The photoreactions of 6-methyl- and 6, 9-dimethylpurine 1-oxides, *J. Am. Chem. Soc.*, 97, 2839, 1975.

85. Bose, S. N., Kumar, S., Davies, J. M., Sethi, S. K., and McCloskey, J. A., Conversion of formycin into the fluorescent isoguanosine analogue 7-amino-3-(β-D-ribofuranosyl)-1H-pyrazolo[4,3-d]pyrimidin-5(4H)-one, *J. Chem. Soc., Perkin Trans. 1*, 2421, 1984.

86. Lam, F. L. and Lee, T. C., Photochemistry of some pteridine N-oxides, *J. Org. Chem.*, 43, 167, 1978.

87. Timpe, H. J. and Becher, H. G. O., The photochemistry of isoelectronic 1, 2, 4-triazole 4-N-oxide and 4-imino-1,2,4-triazolium ylide, *J. Prakt. Chem.*, 314, 324, 1972.

88. Ogata, M., Matsumoto, H., Takahashi, S., and Kano, H., Photolysis of 1-benzyl-2-ethylbenzimidazole 3-oxides, *Chem. Pharm. Bull.*, 18, 964, 1970.

89. Woolhouse, A. D., Photoisomerization of tetraaryl-1H-imidazole 3-oxides, *Aust. J. Chem.*, 32, 2059, 1979.

90. Pedersen, C. L., Lohse, C., and Poliakoff, M. Photolysis of benzo[c]-1,2,5-thiadiazole 2-oxide. Spectroscopic evidence for the reversible formation of 2-thionitrosonitrobenzene, *Acta Chem. Scand.*, B32, 625, 1978.

91. Braun, H. P., Zeller, K. P., and Meier, H., Photolysis of 1,2,3-thiadiazole 2-oxide, *Liebigs Ann. Chem.*, 1257, 1975.

92. Calzaferri, G., Gleiter, R., Knauer, K. H., Martin, H. D., and Schmidt, E., Photochromism of 4-substituted benzofuroxans, *Angew. Chem. Int. Ed.*, 13, 86, 1974.

93. Serve, M. P., Feld, W. N., Seybold, P. C., and Steppel, R. N., Synthesis of 1-methyl-1, 2, 3-benzotriazole 2-oxide, *J. Heterocycl. Chem.*, 12, 811, 1975.

94. Albini, A., Fasani, E., and Frattini, V., Photochemistry of 2- and 4-benzoylpyridine N-oxides, *J. Photochem.*, 37, 355, 1987.

95. Tanigaka, R., Photoreduction of azoxybenzene to azobenzene, *Bull. Chem. Soc. Jpn.*, 41, 1664, 1968.

96. Pietra, S., Bettinetti, G., Albini, A., Fasani, E., and Oberti, R., Photochemical reactions of nitrophenazine 10-oxide with amines, *J. Chem. Soc., Perkin Trans. 2*, 185, 1978.

97. Bharaway, R. K. and Davidson, R. S., The use of N-oxides to photoinduce the oxidative desulfurization and deselenation at pentacovalent phosphorous, *J. Chem. Res. (S)*, 406, 1987.

98. Lin, S. K. and Wang, H. Q., Radical mechanism in the photoreaction of organic N-oxides: Some paradiazine N,N-dioxides, *Heterocycles*, 24, 659, 1986.

99. Hata, N., Ono, Y., and Nakagawa, F., The effect of external magnetic field on the photochemical reaction of isoquinoline N-oxide in various alcohols, *Chem. Lett.*, 603, 1979.

100. Hata, N., Ono, Y., and Kawasaki, M., Photoinduced deoxygenation reaction of heterocyclic N-oxides, *Chem. Lett.*, 25, 1975.

101. Kaneko, C., Yamamori, M., Yamamoto, A., and Hayashi, R., Irradiation of aromatic amine oxides in dichloromethane in the presence of triphenylphosphine: A facile deoxygenation procedure of aromatic amine N-oxides, *Tetrahedron Lett.*, 2799, 1978.

102. Kaneko, C., Yamamoto, A., and Gomi, M., A facile method for the preparation of 4-nitropyridine and -quinoline derivatives: Reduction of aromatic amine N-oxides with triphenylphosphite under irradiation, *Heterocycles*, 12, 227, 1979.

103. Igeta, H., Tsuchiya, T., Yamada, M., and Arai, H., Photoinduced oxygenation of hydrocarbons by pyridazine N-oxide, *Chem. Pharm. Bull.*, 16, 767, 1968.

104. Tsuchiya, T., Arai, H., and Igeta, H., Photo-induced oxygenation by pyridazine N-oxides. II. Formation of epoxides from ethylenic compounds, *Tetrahedron Lett.*, 2213, 1970.

105. Sako, M., Shimada, K., Hirata, K., and Maki, Y., Photochemical hydroxylation of benzene derivatives by pyrimido[5,4-g]pteridine N-oxide, *Tetrahedron Lett.*, 26, 6493, 1985.

106. Maki, Y., Sako, M., Shimada, K., Murase, T., and Hiroto, K., Pyrimido[5,4-g]pteridine N-oxide as a simple chemical model of hepatic monooxygenase, *Stud. Org. Chem.*, 33, 465, 1988.

107. Strub, H., Strehler, C., and Streith, J., Photoinduced nitrene, carbene, and atomic oxygen transfer reactions starting from the corresponding pyridinium N-, C-, and O-ylides, *Chem. Ber.*, 120, 355, 1987.

108. Schneider, H. J. and Sanerbrey, R., Photo-oxidation of alkanes with *N*-oxides, *J. Chem. Res. (S)*, 14, 1987.

109. Akhatar, N. M., Boyd, D. R., Neill, J. D., and Jerina, D. M., Stereochemical and mechanistic aspects of sulfoxides, epoxides, arene oxides and phenol formation by photochemical oxygen atom transfer from aza-araomatic *N*-oxides, *J. Chem. Soc., Perkin Trans. 1*, 1693, 1980.

110. Ogawa, Y., Iwasaki, S., and Okuda, S., A study of the transition state in the photooxygenation by aromatic amine *N*-oxides, *Tetrahedron Lett.*, 2747, 1969.

111. Ito, Y. and Matsuura, T., The reaction of epoxides with oxygen-transfer reagents, *J. Chem. Soc., Perkin Trans. 1*, 1871, 1981.

112. Sako, S., Shimada, K., Hirota, K., and Maki, Y., Photochemical oxygen atom transfer reaction by heterocyclic *N*-oxides involving a single electron transfer process: Oxidative demethylation of *N,N*-dimethylaniline, *J. Am. Chem. Soc.*, 108, 6039, 1986.

113. Sako, C., Shimada, K., Hirota, K., and Maki, Y., Photochemical intramolecular cyclization of purine and pyrimidine nucleosides induced by an electron acceptor, *J. Chem. Soc., Chem. Commun.*, 1704, 1986.

114. Ogawa, Y., Iwasaki, S., and Okuda, S., Photochemical aromatic hydroxylation by aromatic amine *N*-oxides, remarkable solvent effect on NIH-shift, *Tetrahedron Lett.*, 22, 3637, 1981.

115. Decout, J. L., Lablache-Combier, A., and Loucheux, C., Photocrosslinking in polymers and copolymers containing pyridine *N*-oxide groups, *J. Polym. Sci., Polym. Chem. Ed.*, 18, 2371, 1980; Decout, J. L., Lablache-Combier, A., and Loucheux, C., Photocrosslinking in polymers and copolymers containing various amine *N*-oxide groups, *J. Polym. Sci., Polym. Chem. Ed.*, 18, 2391, 1980.

116. Hiraoka, H., Welsch, L. W., Deep UV photolitography with composite photoresists containing poly(olefin sulfones), *Org. Coat. Appl. Polym. Sci. Proc.*, 48, 48, 1983.

117. Booker, R. A., In situ film hardening with pyridine N-oxide and aldehyde precursors, U.S. Pat. 4504578 A, 1985, *Chem. Abstr.*, 102, 157941e, 1985.

118. Bourgeois, G. J., Gaudiana, R. A., Sahatjian, R. A., Photographic products comprising dye developers and *N*-oxides, U.S. Pat. 4203766, 1980., *Chem. Abstr.*, 93, 177209w, 1980.

119. Ciurca, S. J. and Brault, A. T., Photographic color materials with *N*-oxides as oxidant, French Pat. 2232777, 1975, *Chem. Abstr.*, 84, 67800p, 1975.

120. Gladys, M. and Knappe, W. R., Stepwise photoreduction of (iso)alloxazine *N*-oxides, *Z. Naturforsch.*, 29B, 549, 1974.

121. Just., G. and Zehetner, W., Photolysis of podocarponitrile oxide and mesitonitrile oxide, *Tetrahedron Lett.*, 3389, 1967.

Photoreactions of Pyridinium Ylides

Takashi Tsuchiya
Hokuriku University

71.1 Introduction

In contrast to the isoelectronic pyridine *N*-oxides, pyridinium ylides (1) and (2) appeared late on the photochemical scene. Since three research groups[1-3] almost simultaneously demonstrated the photoinduced rearrangement of pyridine *N*-imides (1) into 1,2-diazepines in 1968–1969, photoreaction of the imides has received considerable attention. Pyridine and related azine *N*-imides are readily prepared mainly by direct *N*-amination of parent azines with *O*-substituted hydroxylamines, followed by acylation. With the introduction of the powerful aminating reagent, *O*-mesitylenesulfonylhydroxylamine (MSH), by Tamura in 1972,[4] the scope of the photoreactions of azine *N*-imides has been expanded. Compared with the *N*-imides (1), only a few examples of photoreactions of pyridinium methylides (2) are known. (Scheme 71.1.)

(1)   **(2)**

SCHEME 71.1

The nucleophilicity of the imino nitrogen, the electrophilicity of the azine ring, and the dipolar character permit the *N*-imides (1) to react in a number of ways, depending upon the nature of the ring, the substituents, the reagents, and the conditions. Aspects of these reactions of azine *N*-imides, including photoreactions, have been reviewed extensively.[5-7] Several reviews concerning the photoinduced rearrangement to diazepines have also been published.[8-11]

0-8493-8634-9/95/$0.00+$.50

1.2 Photoreaction Processes and Mechanisms

Although the photolysis of trialkylamine *N*-acylimides leads either to no reaction or to N-N bond cleavage,[12] pyridine *N*-imides undergo photoreactions essentially similar to those observed for pyridine *N*-oxides.[13] The typical photoreaction processes of *N*-imides (3) are (1) triplet-derived fragmentation of the N-N bond leading to the parent pyridines (4) and nitrenes (5), and (2) singlet-derived electrocyclization of the 1,3-dipoles to the diaziridine intermediates (6), which undergo a thermal or photochemical ring opening to the 1,2-diazepines (7) or an N-N bond fission followed by aromatization giving rise to 2-aminopyridines (8), as shown in Scheme 71.2. In some cases, the diaziridines (6) undergo further rearrangement to yield the 1,3-diazepines (10) via the azirine intermediates (9).

In the presence of triplet photosensitizers such as eosine or 3,4-benzopyrene, a notable increase of the N-N bond cleavage has been observed,[14] thus supporting a singlet pathway for the rearrangement and a triplet pathway for the competing fragmentation. Evidence for the formation of triplet nitrenes in the fragmentation process is provided by several experiments.[15] Although attempts to detect or trap the diaziridine intermediates (6) by spectroscopic or chemical means have been unsuccessful, the intermediacy of (6) has been supported by studies on the deuterium isotope effects.[16] These rearrangements of the *N*-imides (3) to the diazepines (7) and aminopyridines (8) via the diaziridines (6) do not occur thermally.

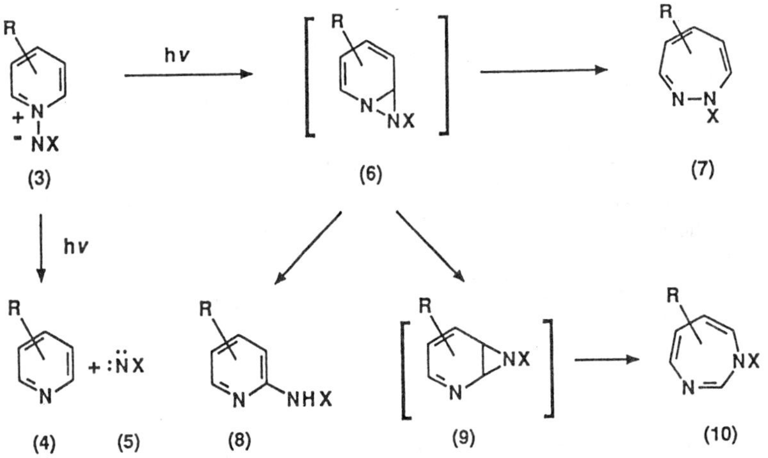

X = Ph, Me, COR', CO$_2$R', SO$_2$R', etc. (R' = alkyl or phenyl)
R = H, Me, Ph, Cl, Br, CN, OMe, SMe, N(Me)$_2$, CO$_2$R', etc.

SCHEME 71.2

71.3 Photoreactions and Synthetic Applications

Pyridine *N*-Imides

The type of substituents on the imino group and on the pyridine ring exerts an influence on the photoreaction of the *N*-imides (3). In the photolysis of pyridine *N*-phenylimides and *N*-alkylimides, the initial N-N bond fission is the exclusive reaction to yield only pyridines (4) and nitrenes (5).[15] In contrast, pyridine *N*-acylimides, *N*-alkoxycarbonylimides, and *N*-sulfonylimides give the corresponding 1,2-diazepines (7) in high to moderate yields.[8–11,17–19]

The reaction courses are also affected markedly by ring substituents.[17–19] Substituents at the 2-position lead to regiospecific cyclization exclusively toward the less-hindered 6-position to give 3-substituted [1H]-1,2-diazepines. With electron-donating groups at the 4-position, ring expansion to diazepines is observed, while electron-withdrawing substituents at the 4-position inhibit the ring expansion. In the cases of 3-substituted pyridines, electron-withdrawing groups show a strong directing effect to afford 4-substituted [1H]-1,2-diazepines, while electron-donating groups act indiscriminately giving both 4- and 6-substituted diazepines. These results have been rationalized in terms of Hückel MO calculations.[19]

The 3-dimethylamino compound (11) shows somewhat different photochemical behavior, affording the bicyclic compound (13), which may be formed from the 1,3-diazepine (12) derived by the pathway shown in Scheme 71.3.[20]

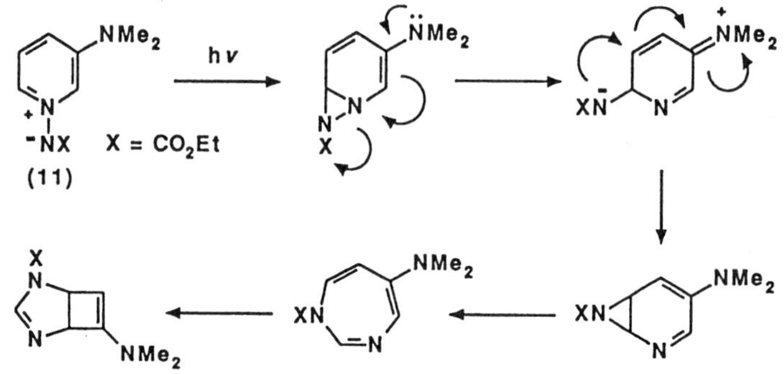

SCHEME 71.3

Photolysis of the *N*-vinylimides (14),[21] (15),[22] and (16)[23] results in no rearrangement and only fragmentation to give pyridine and various products derived from the nitrenes released by N-N bond fission. The cycloalkane-annulated pyridine *N*-imide (17: n = 2) is converted into the diazepine (18), while a competing 1,5-electrocyclization is preferred on irradiation of the imides (17: n = 3 or 4) yielding the tricyclic compounds (19).[24] (Scheme 71.4.)

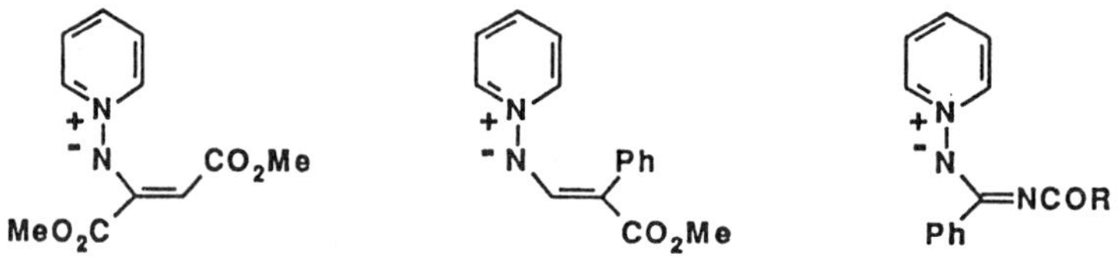

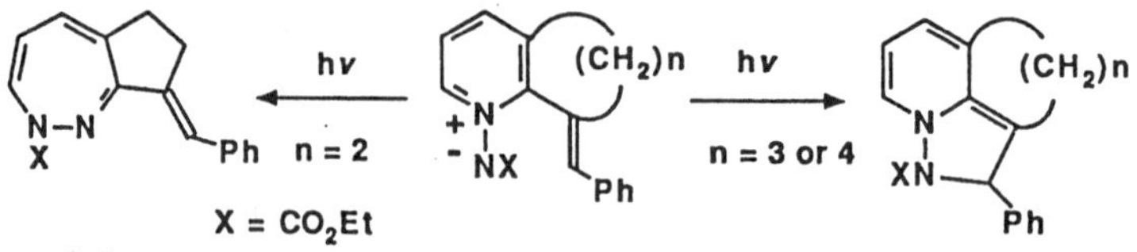

SCHEME 71.4

Quinoline and Related Fused Pyridine *N*-Imides

Photolysis of quinoline *N*-imide dimers (20) in methylene chloride containing acetic acid affords [1H]-1,2-benzodiazepines (23) in moderate yields, together with only small amounts of

2-aminoquinolines and the parent quinolines.[25] The equilibrium between the dimers (20) and the monomeric imides (21) is confirmed in the reaction solvent by NMR studies, suggesting that the diaziridines (22) are formed from (21). The effects of quinoline ring substituents on the photoreaction are also observed. Similarly, the fused pyridine *N*-imides (24) condensed with aromatic heterocyclic rings or their dimers afford the corresponding fused [1H]-1,2-diazepines (25).[26] (Scheme 71.5.)

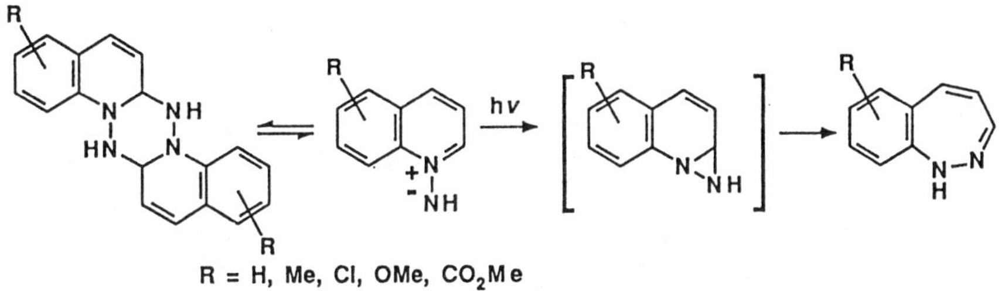
hν
R = H, Me, Cl, OMe, CO2Me

(20) (21) (22) (23)

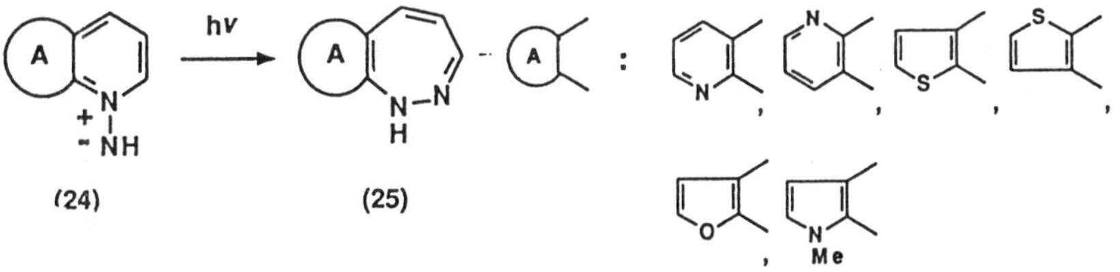
hν

(24) (25)

SCHEME 71.5

In contrast, irradiation of quinoline *N*-acylimides and *N*-ethoxycarbonylimides (26) gives only 2-aminoquinolines and quinoline, and no ring expansion products. However, in alcoholic solvents, the dihydrodiazepines (27) are produced.[27] (Scheme 71.6.)

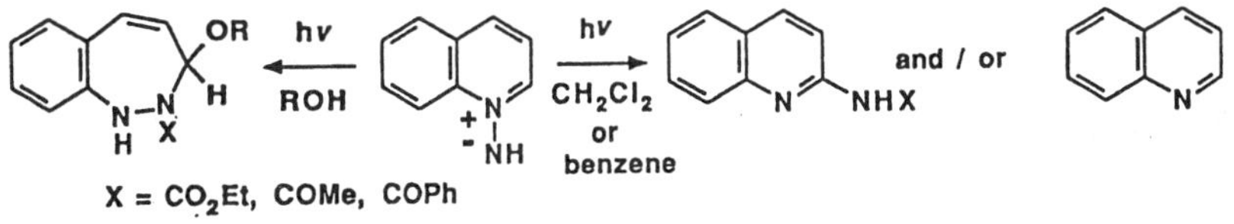
hν
ROH
hν
CH2Cl2
or
benzene
and / or
X = CO2Et, COMe, COPh

(27) (26)

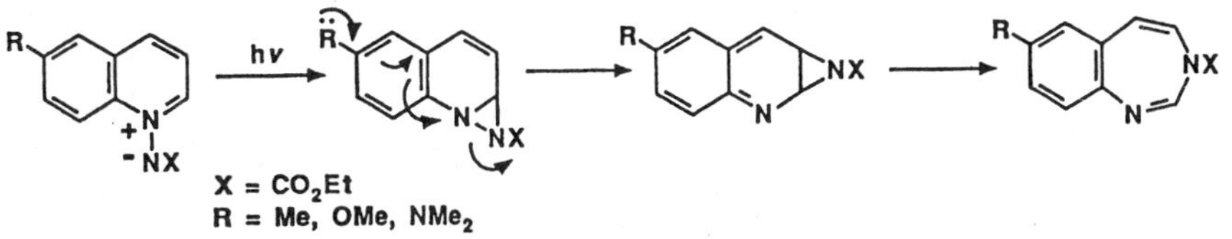
hν
X = CO2Et
R = Me, OMe, NMe2

(28) (29) (30) (31)

SCHEME 71.6

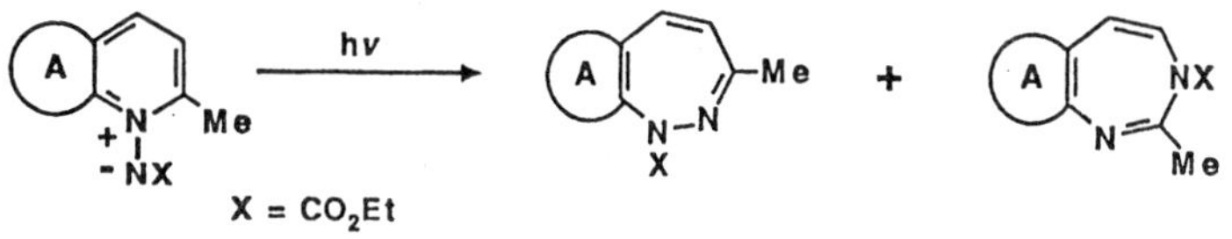
hv
A
+N
-NX
Me
X = CO$_2$Et
-Me
N-N
X
+
NX
N
Me

(32) (33) (34)

SCHEME 71.6 (continued)

It should be noted that the quinoline *N*-imides (28), with an electron-donating substituent at 6- or 8-position, are converted into [3H]-1,3-benzodiazepines (31), probably via the azirine intermediates (30).[28] The electron-donating group may provide assistance for rupture of the N-N bond in (29) and for further cyclization of the resulting dipolar intermediates into azirines (30). In the case of the fused pyridine *N*-imides (32), the initial cyclization takes place to either side of the pyridine ring, thus forming both 1,2- (33) and 1,3-diazepines (34).[29]

Isoquinoline and Related Fused Pyridine *N*-Imides

Irradiation of 1-unsubstituted isoquinoline *N*-acylimides (35) in methylene dichloride gives 1-aminoisoquinolines (36),[27,30] while, in ethanol as solvent, 4-ethoxyisoquinoline (38) is obtained via the adducts (37) on irradiation.[27] In contrast, 1-substituted isoquinoline *N*-imides (39) undergo a two-step rearrangement to give [1H]-1,3-benzodiazepines (42), presumably via the intermediates (40) and (41).[31] (Scheme 71.7.)

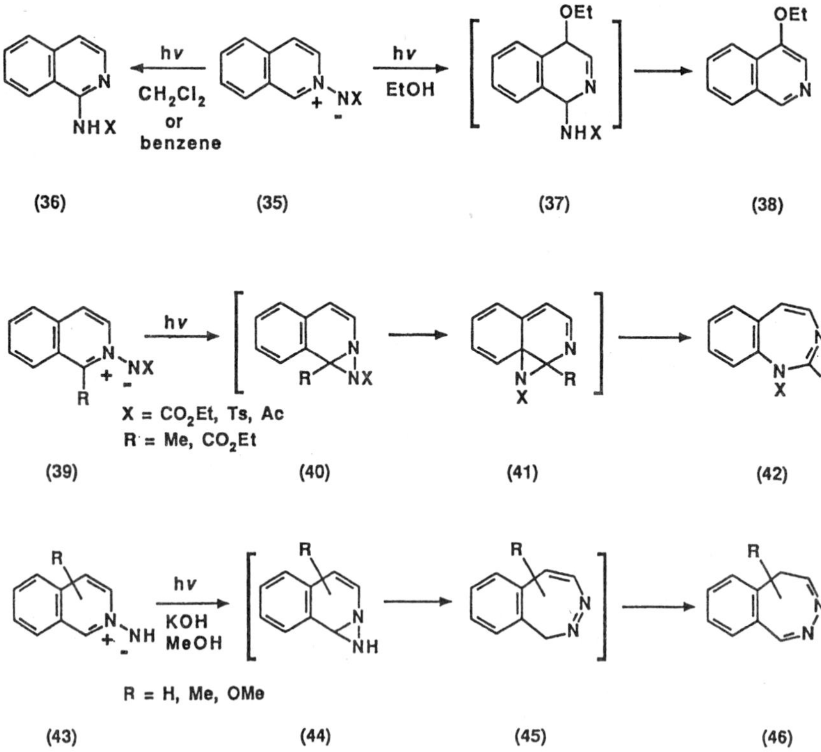
hv
CH$_2$Cl$_2$
or
benzene
NHX
EtOH
OEt
(36) (35) (37) (38)
X = CO$_2$Et, Ts, Ac
R = Me, CO$_2$Et
(39) (40) (41) (42)
KOH
MeOH
R = H, Me, OMe
(43) (44) (45) (46)

SCHEME 71.7

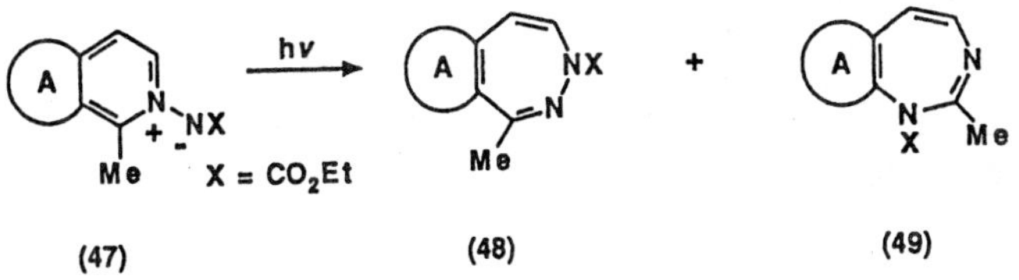

(47)   (48)   (49)

SCHEME 71.7 (continued)

Isoquinoline *N*-unsubstituted imides (43), on irradiation in the presence of a base such as potassium hydroxide, yield the [5H]-2,3-benzodiazepines (46).[32] The reaction occurs only in the presence of bases, the function of which may be to effect the final tautomerization of the unstable [1H]-diazepines (45) to the more stable [5H]-isomers (46).

The [c]-fused methylpyridine *N*-imides (47), condensed with thiophene, furan, or pyrrole, are converted on irradiation into both [3H]-2,3- (48) and [1H]-1,3-diazepines (49).[33] In the absence of the methyl group, only aminopyridine formation is observed.

Diazine *N*-Imides

Photolysis of pyrazine *N*-imides (50) results in the formation of the pyrazoles (53); in the reaction, the formation of hydrogen cyanide, acetonitrile, or benzonitrile is also observed.[34] The proposed pathway, via the triazepine intermediates (51), is outlined in Scheme 71.8. In a similar fashion, irradiation of the pyrimidine *N*-imide (54) and pyridazine *N*-imides (56) gives the pyrazole (55) and pyrroles (57), respectively.[35]

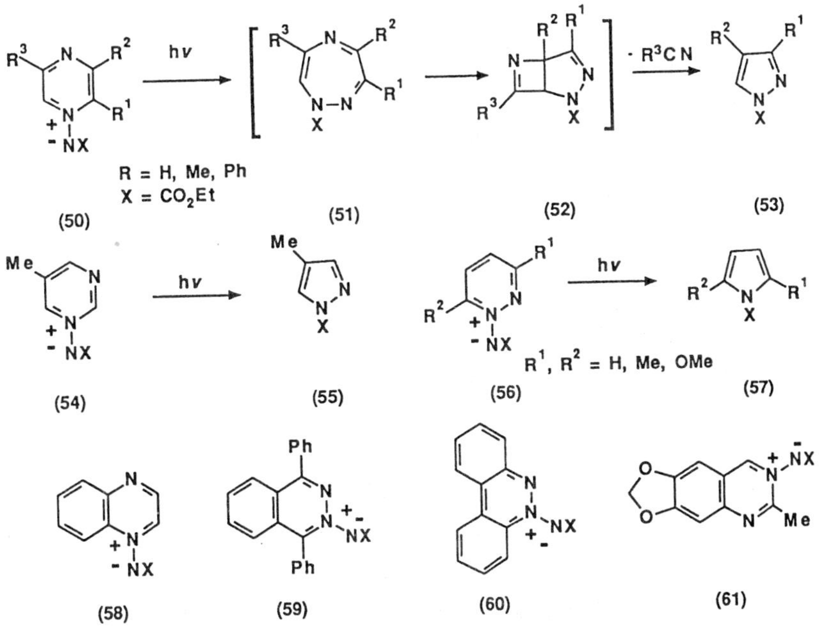

(50)   (51)   (52)   (53)

(54)   (55)   (56)   (57)

(58)   (59)   (60)   (61)

SCHEME 71.8

Irradiation of quinoxaline (58),[36] phthalazine (59),[37] and benzocinnoline *N*-imides (60)[38] results only in fragmentation or rearrangement to α-amino derivatives. The 4-unsubstituted quinoxaline 3-acylimides (61) produce dihydrotriazepines together with other three products on irradiation in ethanol.[39]

Pyridinium Methylides

Pyridinium dicyanomethylide (62) in benzene undergoes photoinduced rearrangement to yield the vinylpyrrole (63) and dicyanonorcaradiene (64);[40] the latter is formed by the reaction of dicyanomethylcarbene, extruded from (62), with the solvent. Photolysis of pyridazinium dicyanomethylides (65) affords the pyrazoles (66) and vinylcyclopropenes (67),[41] presumably via aziridine intermediates formed by cyclization of (65). Both ring-expansion to azepine (69) and fragmentation to isoquinoline and the carbene (70), which trimerizes to (71), occur on irradiation of the isoquinoline methylide (68).[42] (Scheme 71.9.)

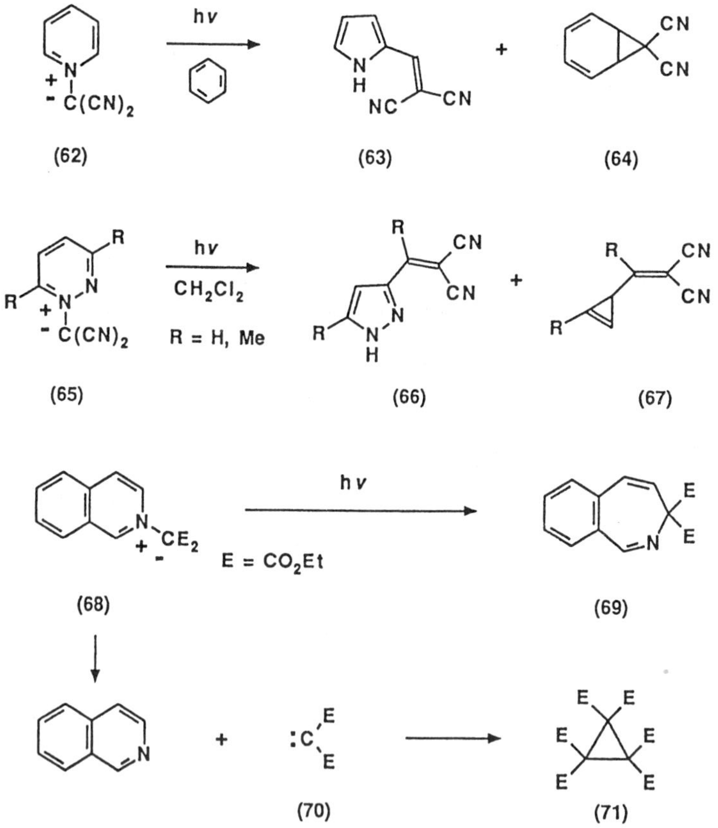

SCHEME 71.9

References

1. Streith, J. and Cassal, J. M., Photochemical synthesis of ethyl 1H-diazepine-1-carboxylate, *Angew. Chem. Int. Ed.*, 7, 129, 1968.
2. Sasaki, T., Kanematsu, K., and Kakehi, A., Structures of the photochemical isomerization products of pyridinium ylides. Diazepines and their Diels-Alder adducts, *J. Chem. Soc., Chem. Commun.*, 432, 1969.

3. Snieckus, V., The photolysis of *N*-acyliminopyridinium ylides, *J. Chem. Soc., Chem. Commun.*, 831, 1969.

4. Tamura, Y., Minamikawa, J., Miki, Y., Matsugashita, S., and Ikeda, M., A novel method for heteroaromatic *N*-imides, *Tetrahedron Lett.*, 4133, 1972; Tamura, Y., Minamikawa, J., and Ikeda, M., *O*-Mesitylenesulfonylhydroxylamine and related compounds — Powerful aminating reagents, *Synthesis*, 1, 1977.

5. McKillip, W. J., Sedor, E. A., Culbertson, B. M., and Wawzonek, S., The chemistry of aminimides, *Chem. Rev.*, 73, 255, 1973.

6. Timple, H.-J., Heteroaromatic *N*-imines, *Adv. Heterocycl. Chem.*, 17, 213, 1974.

7. Tamura, Y. and Ikeda, M., Advances in the chemistry of heteroaromatic *N*-imines and *N*-aminoazonium salts, *Adv. Heterocycl. Chem.*, 29, 71, 1981.

8. Nastasi, M., The chemistry of 1,2-diazepines, *Heterocycles*, 4, 1509, 1976.

9. Streith, J., Photochemistry as a tool in heterocyclic synthesis. From pyridinium *N*-ylides to diazepines and beyond, *Heterocycles*, 6, 2021, 1977.

10. Snieckus, V. and Streith, J., 1,2-Diazepines: A new vista in heterocyclic chemistry, *Acc. Chem. Res.*, 14, 348, 1981.

11. Tsuchiya, T., Ring transformation of cyclic amine *N*-imides, *Yakugaku Zasshi*, 103, 373, 1983; Tsuchiya, T., Chemistry of 1,2-benzodiazepines and related compounds, *J. Synth. Org. Chem., Jpn.*, 39, 99, 1981; Tsuchiya, T., Chemistry of 1,3-diazepines, *J. Synth. Org. Chem., Jpn.*, 41, 641, 1983.

12. Sedor, E. A., The thermolysis of trialkylamine carboethoxylamide, *Tetrahedron Lett.*, 323, 1971; Smith, S. R. and Briggs, P. C., The pyrolysis and photolysis of trialkylamine benzimide, *J. Chem. Soc., Chem. Commun.*, 120, 1965.

13. Spence, G. G., Taylor, E. C., and Buchardt, O., The photochemical reactions of azoxy compounds, nitrones, and aromatic amine *N*-oxides, *Chem. Rev.*, 70, 231, 1970; Albini, A. and Alpegiani, M., The photochemistry of the *N*-oxide function, *Chem. Rev.*, 84, 43, 1984.

14. Streith, J., Luttringer, J. P., and Nastasi, M., Synthesis and rearrangements of 1,2-diazepines, *J. Org. Chem.*, 36, 2962, 1971.

15. Snieckus, V. and Kan, G., The photolysis of Schneider's 1-phenyliminopyridinium ylides, *J. Chem. Soc., Chem. Commun.*, 172, 1970; Nastasi, M., Strub, H., and Streith, J., Formation of a triplet nitrene by the photolysis of *N*-aminopyridinium ylide, *Tetrahedron Lett.*, 4719, 1976.

16. Kwart, H., Benko, D. A., Streith, J., Harris, D. J., and Shuppiser, J. L., The significance of a secondary deuterium isotope effect in the photorearrangement of 1-iminopyridinium ylides, *J. Am. Chem. Soc.*, 100, 6501, 1978; Kwart, H., Benko, D. A., Streith, J., and Shuppiser, J. L., Identification by means of heavy-atom isotope effects of a photoactivated reaction as a ground-state process, *J. Am. Chem. Soc.*, 100, 6502, 1978.

17. Sasaki, T., Kanematsu, K., Kakehi, A., Ichikawa, I., and Hayakawa, K., The photochemical intramolecular 1,3-dipolar cycloaddition of substituted 1-ethoxycarbonyliminopyridinium ylides, *J. Org. Chem.*, 35, 426, 1970.

18. Balasubramanian, A., McIntosh, J. M., and Snieckus, V., The photoisomerization of 1-iminopyridinium ylides to 1(1H),2-diazepines, *J. Org. Chem.*, 35, 433, 1970.

19. Nastasi, M. and Streith, J., Photoinduced regiospecific electrocyclization of pyridinium *N*-imides, *Bull. Soc. Chim., France*, 630, 1973.

20. Kurita, J., Kojima, H., Enkaku, M., and Tsuchiya, T., Synthesis of monocyclic 1,3-diazepines. 2. Substituent effects on the thermal ring-conversion of 1,2-diazepines into 1,3-diazepines, *Chem. Pharm. Bull.*, 29, 3696, 1981.

21. Sasaki, T., Kanematsu, K., and Kakehi, A., Synthesis and characterization of *N*-vinyliminopyridinium ylides, *J. Org. Chem.*, 37, 3106, 1972.

22. Kakehi, A., Ito, S., Funahashi, T., and Ota, Y., Thermal and photochemical behavior of sterically hindered *N*-vinyliminopyridinium ylides, *J. Org. Chem.*, 41, 1570, 1976.

23. Kakehi, A., Ito, S., Uchiyama, K., Konno, Y., and Kondo, K., Thermolysis and photolysis of various *N*-imidoyliminopyridinium ylides, *J. Org. Chem.*, 42, 443, 1977.

24. Baum, G., Friderichs, A., Kümmell, A., Massa, W., and Seitz, G., A novel photoinduced synthesis of substituted 3-azaquinolizinone, *Chem. Ber.*, 121, 411, 1988.

25. Tsuchiya, T., Kurita, J., and Snieckus, V., General photochemical synthesis of 1H-1,2-benzodiazepines from N-iminoquinolinium ylide dimers, *J. Org. Chem.*, 42, 1856, 1977.

26. Tsuchiya, T., Enkaku, M., Kurita, J., and Sawanishi, H., Syntheses of novel 1H-1,2-diazepines condensed with aromatic heterocyclic rings, *Chem. Pharm. Bull.*, 27, 2183, 1979.

27. Tamura, Y., Ishibashi, H., Tsujimoto, N., and Ikeda, M., The photorearrangement and thermolyses of N-benzoyliminoisoquinolinium and quinolinium betaines, *Chem. Pharm. Bull.*, 19, 1285, 1971; Shiba, T., Yamane, K., and Kato, H., Photolysis of iminoquinolinium and iminothiazolinium ylides, *J. Chem. Soc., Chem. Commun.*, 1952, 1970; Tamura, Y., Matsugashita, S., Ishibashi, H., and Ikeda, M., Photochemistry of N-acyliminoisoquinolinium and -quinolinium betaines, *Tetrahedron*, 29, 2359, 1973.

28. Tsuchiya, T., Okajima, S., Enkaku, M., and Kurita, J., Photochemical Synthesis of 3H-1,3-benzodiazepines from quinoline N-acylimides, *Chem. Pharm. Bull.*, 30, 3757, 1982; Tsuchiya, T., Okajima, S., Enkaku, M., and Kurita, J., Formation of 3H-1,3-benzodiazepines from quinoline N-acylimides, *J. Chem. Soc., Chem. Commun.*, 211, 1981.

29. Tsuchiya, T., Enkaku, M., and Okajima, S., Photolysis of thieno-, furo-, and pyrrolo-*[b]*pyridine N-imides: Formation of 3H-1,3-diazepines, *J. Chem. Soc., Chem. Commun.*, 454, 1980; Tsuchiya, T., Enkaku, M., and Okajima, S., Photolysis of thieno-, furo-, and pyrrolo-[b]pyridine N-imides. Formation of fused 1H-1,2- and 3H-1,3-diazepines, *Chem. Pharm. Bull.*, 29, 3173, 1981.

30. Becher, J. and Lohse, C., The photochemical rearrangement of isoquinolinium N-acyl ylides to 1-acylaminoisoquinolines via the excited singlet state, *Acta Chem. Scand.*, 26, 4041, 1972.

31. Tsuchiya, T., Enkaku, M., Kurita, J., and Sawanishi, H., Formation of novel 1H-1,3-benzodiazepines in the photolysis of isoquinoline N-imides, *J. Chem. Soc., Chem. Commun.*, 534, 1979; Tsuchiya, T., Enkaku, M., and Okajima, S., Photochemical synthesis of novel 1H-1,3-benzodiazepines from isoquinoline N-imides, *Chem. Pharm. Bull.*, 28, 2602, 1980.

32. Kurita, J., Enkaku, M., and Tsuchiya, T., Photochemical synthesis of 2,3-benzodiazepines from isoquinoline N-imides, *Chem. Pharm. Bull.*, 30, 3764, 1982.

33. Tsuchiya, T., Sawanishi, H., Enkaku, M., and Hirai, T., Photolysis of thieno-, furo-, and pyrrolo-[c]pyridine N-imides: Formation of novel fused 1H-1,3- and 3H-2,3-diazepines, *Chem. Pharm. Bull.*, 29, 1539, 1981.

34. Tsuchiya, T., Kurita, J., and Ogawa, K., Photolysis of N-ethoxycarbonyliminopyrazinium ylides, *J. Chem. Soc., Chem. Commun.*, 250, 1976.

35. Tsuchiya, T., Kurita, J., and Takayama, K., Photochemical behavior of pyrazine, pyrimidine, and pyridazine N-imides, *Chem. Pharm. Bull.*, 28, 2676, 1980.

36. Agai, B., Lempert, K., and Møller, J., Synthesis and photoisomerization of N-ethoxycarbonyl-N-(1-quinoxalino)amide, *Acta Chim., Budapest*, 80, 465, 1974.

37. Lempert, K. and Zauer, K., The synthesis and some reactions of N-(2-phthalazino)- and N-(1,4-diphenyl-2-phthalazino)benzamidates, *Acta Chim. Acad. Sci., Hung.*, 88, 81, 1976.

38. Gait, S. F., Peek, M. E., Rees, C. W., and Storr, R. S., Benzo[c]cinnoline N-imides, *J. Chem. Soc., Perkin Trans. 1*, 19, 1975.

39. Fetter, J., Lempert, K., and Møller, J., The synthesis and photochemistry of ethyl N-(2-methyl-4-methylene-6,7-methylenedioxy-3,4-dihydro-3-quinazolinyl)-N-phenylcarbamate, *Tetrahedron*, 34, 2557, 1978.

40. Streith, J., Blind, A., Cassal, J.-M., and Sigwalt, O., Photochemistry of pyridinium ylides, *Bull. Soc. Chim., France*, 948, 1969.

41. Arai, H., Igeta, H., and Tsuchiya, T., Photolysis of pyridazinium dicyanomethylides, *J. Chem. Soc., Chem. Commun.*, 521, 1973.

42. Lablache-Combier, A. and Surpateanu, G., Formation of an azepine by photochemical reaction of N+-C− type cycloimmonium ylide, *Tetrahedron Lett.*, 3081, 1976.

72

Photochemistry of Aza-Substituted Dienes

Diego Armesto
Universidad Complutense

72.1 Introduction

The photochemistry of the C=N bond has not been studied as extensively as that of the C=O and C=C bonds. However, in the last 10 or 15 years, there has been an increasing interest in this class of compounds, particularly in areas such as the photochemistry of azirines, iminium salts, and azadienes. In some instances, the reactivity observed for the azadienes is similar to that described for alkenes and carbonyl compounds. However, this class of compounds often shows a very different reactivity. This chapter deals with the photochemical reactivity of aza-substituted dienes.

72.2 E,Z-Isomerization

There are many examples of E,Z-isomerization in imines and related compounds. Thus, the irradiation of the E-imines, both under direct or sensitized conditions, brings about the formation of the Z-isomers (Scheme 72.1).[1,2] However, the reaction is considerably less interesting than the analogous one in alkenes. The reason for this is that the energy barrier for the thermal conversion of the less stable diastereoisomer into the more stable one is very low (ca. 58 to 75 kJ mol^{-1}).[3] As a result, it is usually impossible to isolate the Z-isomer since it reverts to starting material during work-up. In many cases, the excited states of simple imines deactivate via isomerization around the C-N double bond. This accounts for the lack of photochemical reactivity of some acyclic compounds in which the chromophore is a C=N functional group. The geometrical photoisomerization of the C=N can take place by two different mechanisms. One of them is rotation around the C-N double bond, in a similar manner as the isomerization of the C=C. This mechanism has been postulated in cases for which the first excited state of the C=N moiety has π,π^* character. However, the C=N could also isomerize by an inversion mechanism consisting of a change in the hybridization of the nitrogen lone pair of electrons from sp^2 to sp. This mechanism probably occurs via the n,π^* excited state.[4]

0-8493-8634-9/95/$0.00+$.50
© 1995 by CRC Press, Inc.

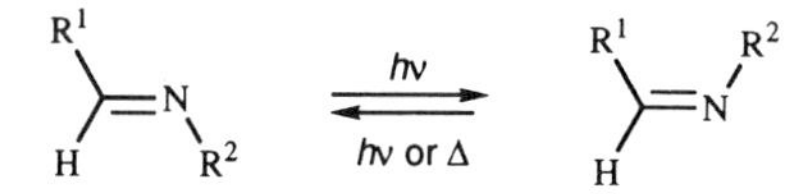

SCHEME 72.1

In the case of the oximes and oxime derivatives, the Z-stereoisomer is usually stable thermally. Therefore, the reaction has some synthetic utility; for instance, the pharmacologically active *anti*- or *Z*-isonicotinaldehyde oxime has been obtained by irradiation of the corresponding *syn*- or *E*-isomer.[5] Most of the cases of E,Z-isomerization reported for azadienes refer to thermally stable C=N derivatives. For instance, the irradiation of the *E,E*- or the *Z,E*-isomer of the 1-azadiene (1) brings about a photostationary mixture of the two diastereoisomers (Scheme 72.2).[6] This example allows a comparison between the photochemical reactivity of the C=C and the C=N toward geometric isomerization. The irradiation results in rapid C-N double bond isomerization accompanied by slower C-C double bond isomerization. A similar behavior is observed for the protonated 1-azadiene (2)[7] and for the 2-azadiene (3)[8] that, in addition to isomerization around the C=N, also show isomerization of the C=C moiety. As should be expected in cases in which the C=C bond is part of a six-membered ring (as in the hydrazone (4)[9] and the cyclohexenone derivatives (5)[10]), the isomerization takes place around the C=N bond exclusively. Related compounds such as the triphenylformazone (6)[11] (Scheme 72.3) and the 4,5-diaza-1,3,5,7-tetraene (7)[12] (Scheme 72.4) behave similarly.

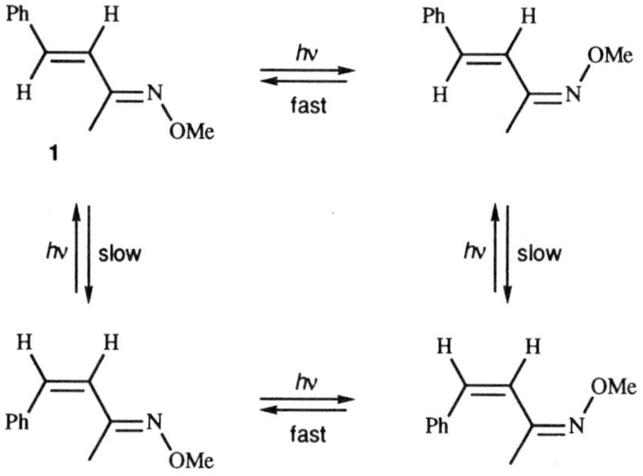

SCHEME 72.2

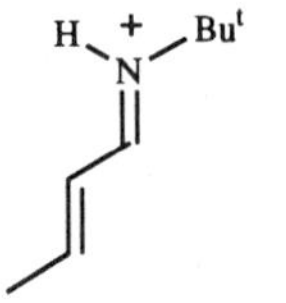

2

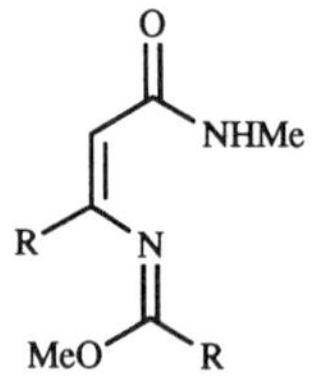

3

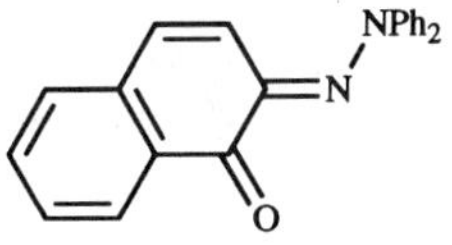

4

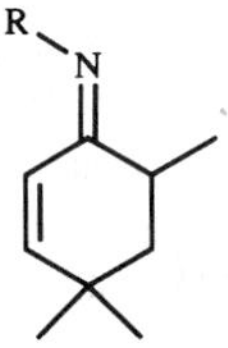

5a: R = C$_6$H$_{11}$
5b: R = OH

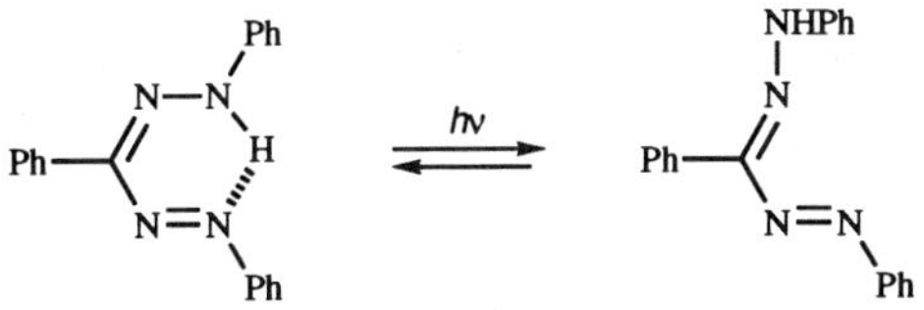

6

SCHEME 72.3

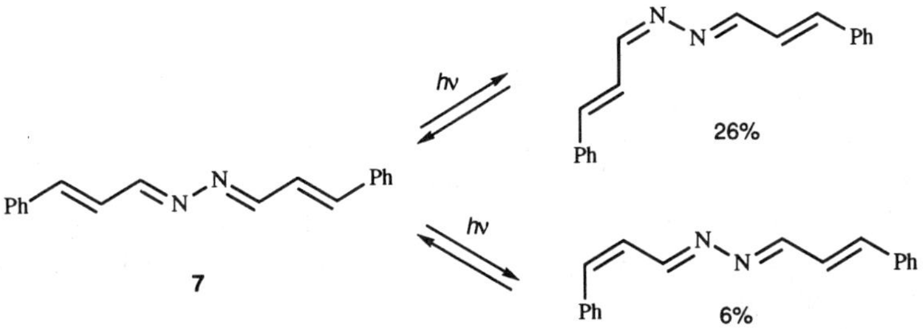

26%

7

6%

SCHEME 72.4

Apart from the synthetic potential of the *Z,E*-photoisomerization in unsaturated imines and related compounds, this photochemical process is of importance because of its implication in the vision process. Thus, one of the key steps in the transformation of visible light into electrical pulses transmitted to the brain by the optic nerve is the photoisomerization of the C-11 *cis*-iminium salt (**8**) to the all *trans*-isomer (Scheme 72.5).[13]

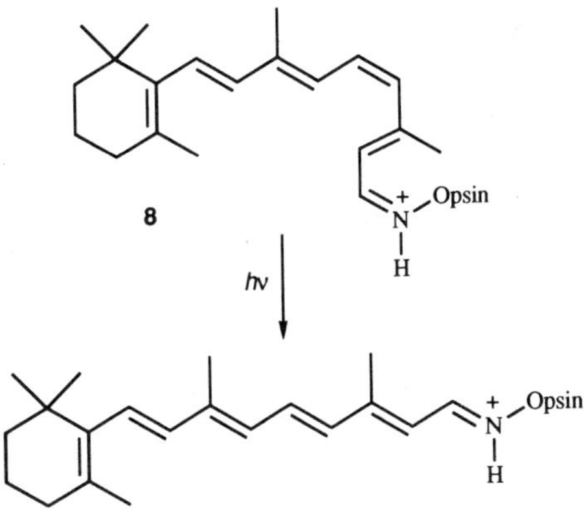

8

SCHEME 72.5

72.3 Photochemistry of Acyloxy-aza-1,3-dienes

The photochemistry of functionalized conjugated azadienes has not been studied to the same extent as that of related compounds such as 1,3-dienes and conjugated enones. However, acyloxyazadienes have been the subject of some systematic studies.

Influence of the Degree of Conjugation on the Photoreactivity of 4-Acyloxy-2-aza-1,3-dienes

Interesting photochemical reactivity has been observed in the direct irradiation of 4-acyloxy-2-azabuta-1,3-dienes (9). These compounds undergo different photochemical reactions depending on the degree of twist around the N-C3 bond of the 2-azadiene skeleton. Thus, direct irradiation of the *E*-dienes (10) gave the corresponding *Z*-isomers exclusively (Scheme 72.6).[14] However, under similar conditions, diene (11a) gave the 1,3-benzoyl-migrated compound (12a).[14] 1,3-Benzoyl migrations are common in enol esters and are the result of the excitation of the benzoyl ester moiety. The different reactivity observed for dienes (10) and diene (11a) can be due to conformational factors. Thus, dienes (10) are flat, as demonstrated by the high extinction coefficient ($\varepsilon \simeq$ 20,000) absorption in the 350 nm of the UV spectra, and also by X-ray diffraction analysis.[15] Azadienes (11) also show an absorption in this region; but in these cases, the extinction coefficient is lower ($\varepsilon \simeq$ 1300 to 2300) due to the presence of the 3-methyl substituent that reduces the overlap between the imine and the enol ester entities. With full conjugation of the diene, the absorption above 280 nm is dominated by the π,π^* absorption of the diene and, not surprisingly, *E,Z*-isomerization is the sole photochemical event as shown by dienes (10). The angle of twist around the C3-N bond in diene (11a) leads to a situation where the absorption of the benzoyl ester group tails above 280 nm and becomes the active chromophore. This excitation brings about the 1,3-benzoyl migration. The direct irradiation of the acetate (11b) fails to yield a photoproduct, in accord with the postulate that the principal absorbing moiety in diene (11a) is the benzoyl function. The pair of azadienes (11c) and (11d) also fits into this scheme. Thus, while the azadiene (11d) is unreactive on direct irradiation, the azadiene (11c) is converted on direct or COD-quenched irradiation into the diketone (12b) analogous to the diketone (12a). In contrast, acetophenone-sensitized irradiation of dienes (11a-d) brings about *E,Z*-isomerization to afford the azadienes (13a-d) in each case.

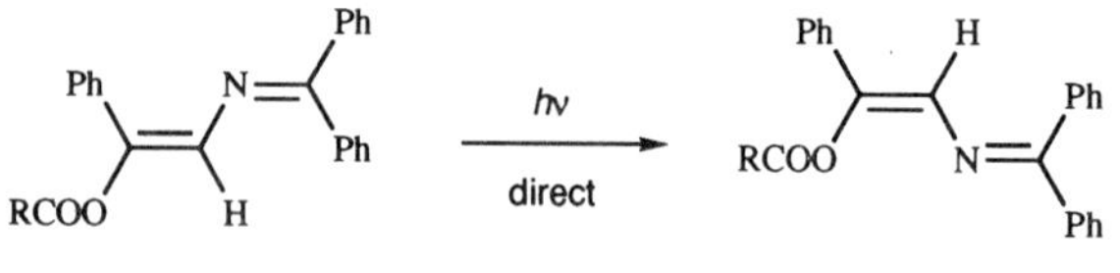

10: R = Ph or Me

SCHEME 72.6

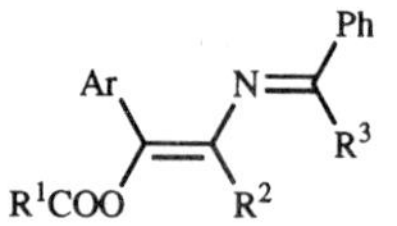

9

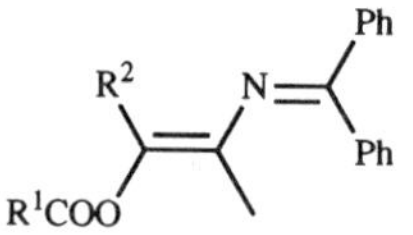

11a: R^1 = R^2 = Ph
11b: R^1 = Me, R^2 = Ph
11c: R^1 = Ph, R^2 = Me
11d: R^1 = R^2 = Me

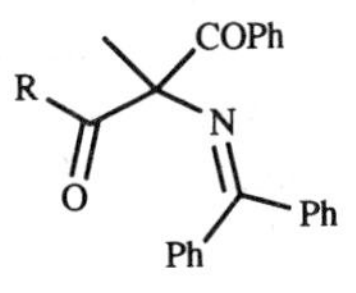

12a: R = Ph
12b: R = Me

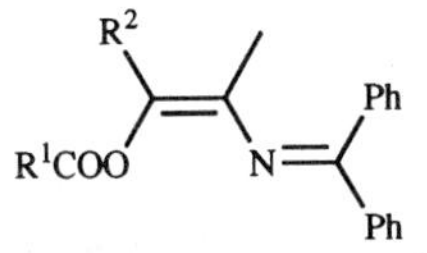

13a: R^1 = R^2 = Ph
13b: R^1 = Me, R^2 = Ph
13c: R^1 = Ph, R^2 = Me
13d: R^1 = R^2 = Me

Interestingly, azadienes (14) with 3-aryl substituents undergo an unexpected photochemical reaction to give dihydro-oxazoles (15) in a general reaction that is also very efficient both in chemical and light usage terms (Scheme 72.7).[16] The overall rearrangement involves formally a 1,2-aroyl migration followed by cyclization. This novel reactivity was interpreted as a consequence of the degree of conjugation between the imine and the enol ester moieties. An X-ray diffraction analysis[17] demonstrated that in azadienes (14), the C3-N twist is almost 40° in the crystal, which places the nitrogen lone pair of electrons in correct alignment for single electron transfer (SET) to occur. The molecules (14) have an ideal structural set-up with both electron-rich and electron-deficient moieties, namely the nitrogen-substituted C-C double bond and the ester, respectively. Thus, a mechanism involving SET was proposed as shown in Scheme 72.8.

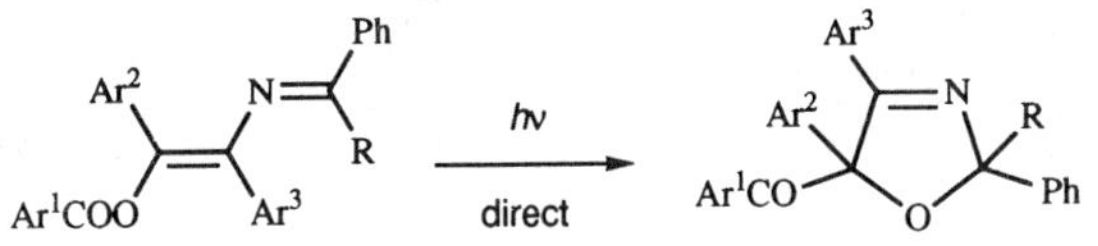

SCHEME 72.7

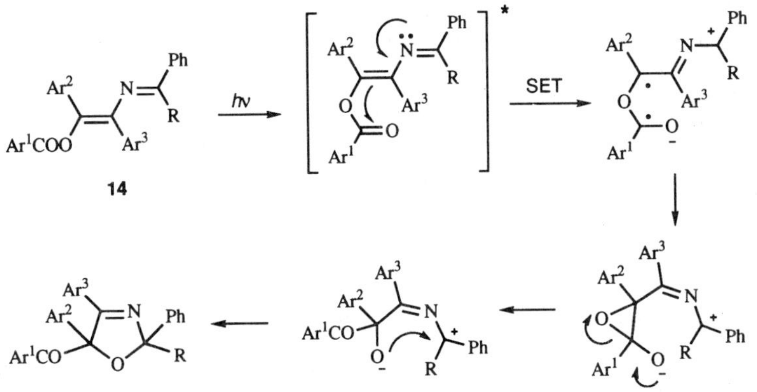

SCHEME 72.8

The applicability of this mechanism was tested by carrying out the irradiation of (14) in the presence of perchloric acid in order to suppress the SET by protonation of the nitrogen. This protonation alters the electron-donating properties of the nitrogen-substituted C-C double bond by the suppression of the availability of the nitrogen lone pair. Under these conditions the cyclization to the dihydro-oxazoles is completely suppressed giving clear evidence that a SET mechanism is responsible for the hitherto unknown 1,2-acyl migration in an enol ester. However, while protonation suppresses the 1,2-acyl migration reaction in these azadienes, irradiations in perchloric acid for a brief time leads to their conversion into the isoquinolinones (16) in addition to E,Z-isomerization (Scheme 72.9).[18,19] The cyclization can be carried out by replacing perchloric acid with BF_3 etherate and under these conditions azadienes (14) cyclize to afford isoquinolinones in yields greater than that for the perchloric acid reaction. The analysis of the influence of substituents on the quantum yields of the reaction is supportive of a SET mechanism from the aryl ring in C4 to the iminum group, as shown in Scheme 72.10.[20]

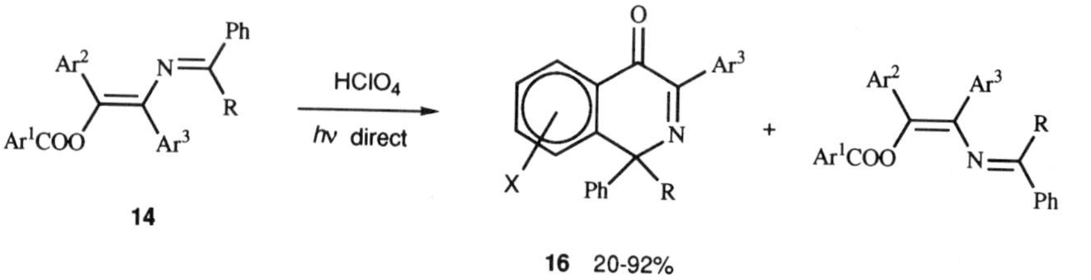

SCHEME 72.9

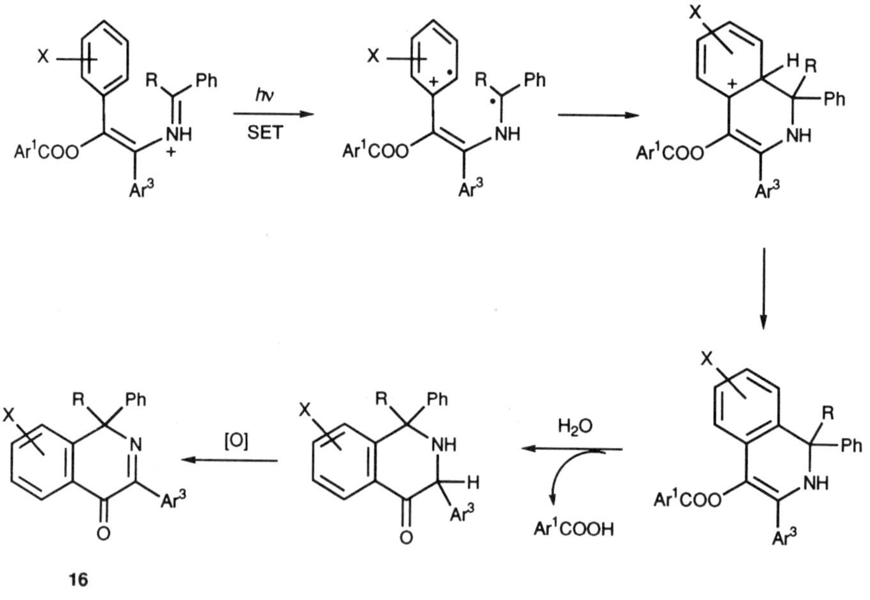

SCHEME 72.10

Cyclizations of this type can be considered formally as a photo-Mannich reaction. There are very few examples of such a cyclization path reported in the literature but one such is the conversion of the 1-styrylpyridinium salts (17) into benzo[a]quinolizinium salts (18) (Scheme 72.11).[21]

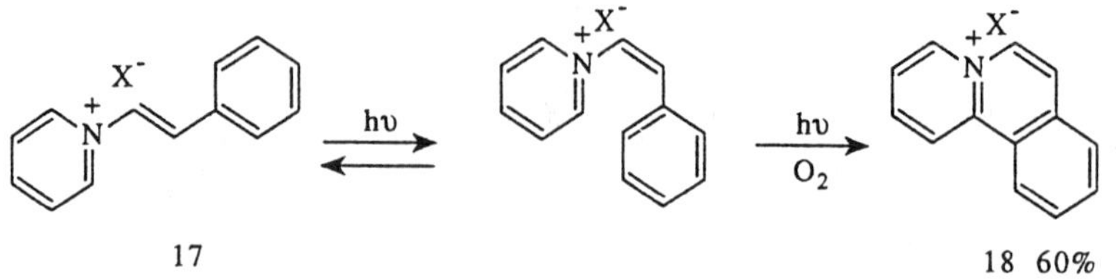

SCHEME 72.11

Photochemistry of 1-Acyloxy-1-aza-1,3-dienes

The position of the nitrogen in the skeleton of the azadiene is important in determining the photochemical behavior. This is illustrated by the reactions observed with 1-acyloxy-1-azadienes (19).[22]

Direct irradiation of azadienes (19) gives quinolines (20) in a reaction that can be rationalized in terms of a six-electron process involving the atoms of the diene skeleton and the aryl group at C-4. This cyclization is typical for six-electron systems and is similar to the transformation of 1,4-diphenylbuta-1,3-diene into 1-phenylnaphthalene[23] or to cyclizations of the *cis*-stilbene type.[24] With azadienes (19), cyclization affords a dihydro-intermediate (21) that readily undergoes elimination of benzoic acid, presumably by a thermal process, to afford the final product (Scheme 72.12). The influence of substituents on the cyclization was not quantitatively evaluated but it appears that there is little effect exerted by the substituted aryl group at C-4. The results obtained with the 2-azadienes (14) showed that electron transfer was important for the outcome of the photoreaction. However, the photochemistry of 1-azadienes (19) show that the nitrogen does not influence the outcome of the reaction. This approach to 7-substituted quinolines has some synthetic value since substitution at this site is normally difficult to carry out. This difficulty arises from the fact that electrophilic substitution of quinolines introduces substituents at positions C-5 and C-8, while nucleophilic attack takes place at C-2. Alternative approaches to quinolines substituted at C-7 suffer from the problem of using intermediates that, in some cases, are difficult to synthesize.

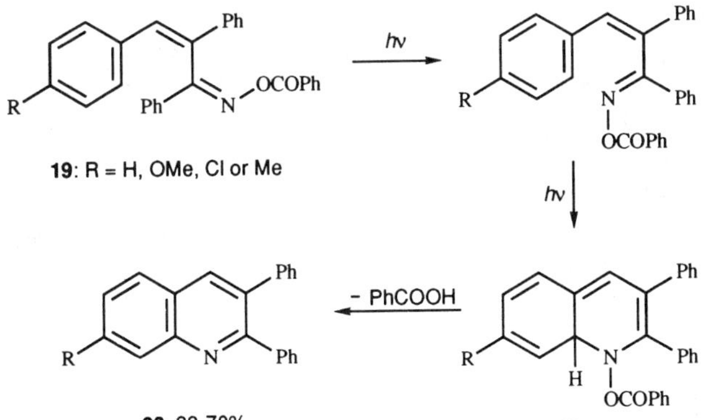

SCHEME 72.12

A similar photochemical behavior to that observed for (19) has been described for α,β-unsaturated oximes. Thus, irradiation of 1-aza-1,3-diene (22) brings about the formation of the quinoline derivative (23) (Scheme 72.13).[25] Another example of a six-electron cyclization in an 1-aza-1,3-diene has been reported in the irradiation of the 1-methoxy-1-aza-1,3-diene (24) that yields the corresponding pentacyclic compound (25) in 80% yield (Scheme 72.14).[26]

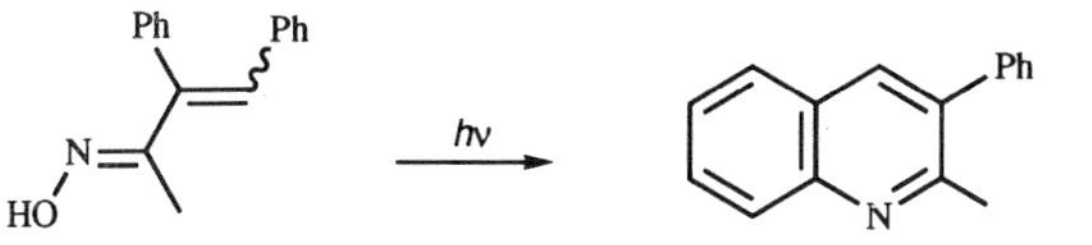

SCHEME 72.13

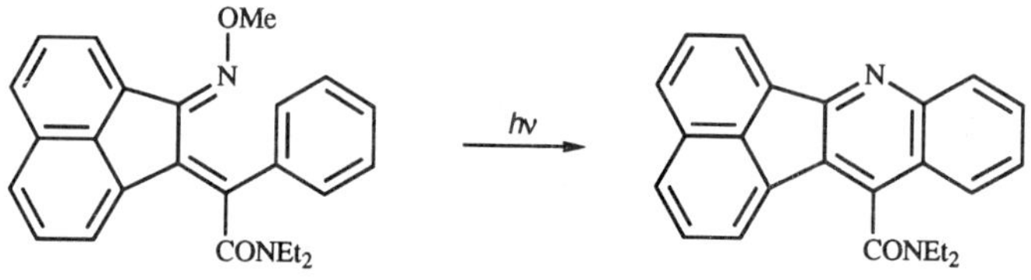

24

25 80%

SCHEME 72.14

Photochemistry of 1-Acyloxy-1,4-diaza-1,3-dienes

Studies carried out on the photochemistry of 1-acyloxy-1,4-diaza-1,3-dienes (26) have shown that these compounds undergo some interesting photochemical reactions. The photochemical reactivity of the diazadienes (26) is dependent on the type of substituent on N-4 and falls into two categories: (1) those with an abstractable hydrogen on the substituent at N-4 and (2) those where this substituent is a phenyl group. Irradiation of the iminooxime (27a) in acetone gives only (28) as the sole photochemical product arising by *E,Z*-isomerization of the oxime moiety. When the diazadiene (26a) was irradiated under identical conditions, a high yield (71%) of the imidazole (29a) was obtained.[27] The difference in reactivity between (26a) and (27a) could be due to abstraction of the oxime hydrogen in (27a) by acetone in its excited state, bringing about *E,Z*-isomerization; whereas, the diazadiene (26a) undergoes a different hydrogen abstraction to afford the radical (30). Cyclization of this followed by elimination of benzoic acid affords the imidazole (29a) (Scheme 72.15). The formation of the imidazole also occurs on irradiation in methylene dichloride, in the absence of acetone as the hydrogen abstracting reagent, although in lower yield (30%). This result is interpreted by the involvement of an intermolecular hydrogen abstraction with the excited benzoate ester as the hydrogen abstracting group. The failure of the acetyl derivative (26b) to afford the imidazole (29a) on irradiation in methylene dichloride, using a Pyrex filter, is supportive of the proposed mechanism. However, irradiation of (26b) in acetone follows the usual hydrogen abstraction/cyclization path to afford the imidazole (29a) as well as the corresponding isomer (31). The reaction has been extended to the diazadienes (26c) and (26d), affording the imidazoles (29b) and (32), respectively.

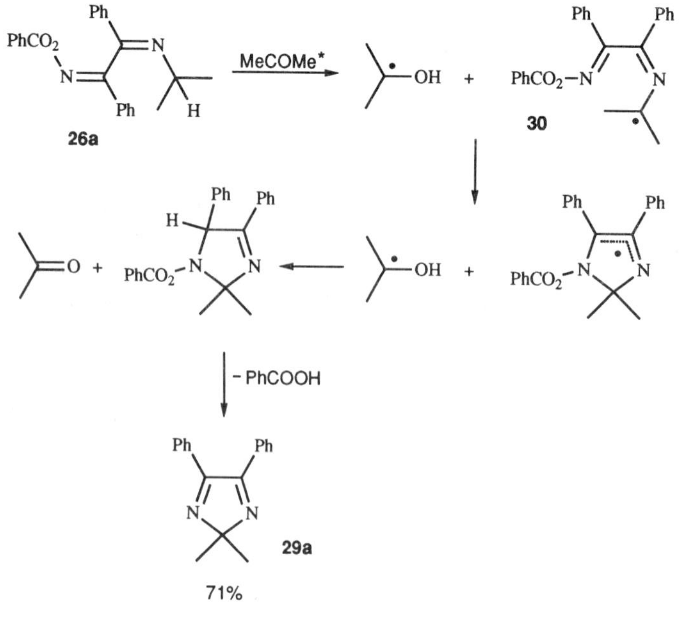

26a

30

- PhCOOH

29a

71%

SCHEME 72.15

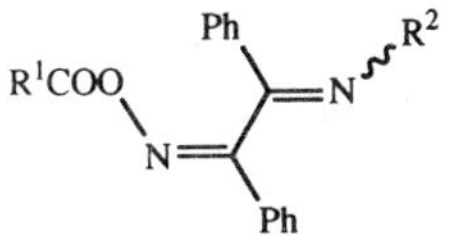

26a: R^1 = Ph, R^2 = Pr^j
26b: R^1 = Me, R^2 = Pr^j
26c: R^1 = Ph, R^2 = PhMeCH
26d: R^1 = Ph, R^2 = PhCH$_2$
26e: R^1 = R^2 = Ph
26f: R^1 = Me, R^2 = Ph

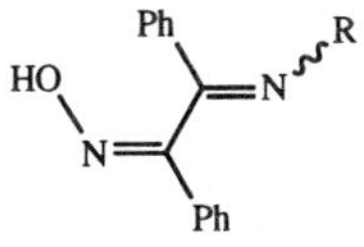

27a: R = Pr^j
27b: R = Ph

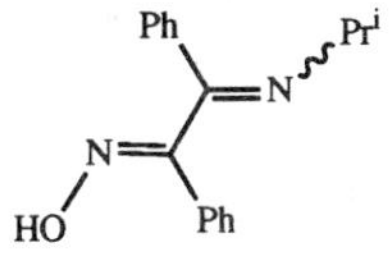

28

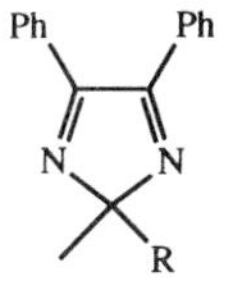

29a: R = Me
29b: R = Ph

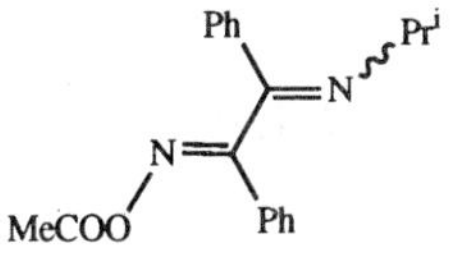

31

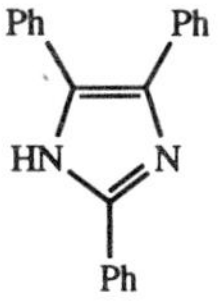

32

An alternative mode of reaction is seen with the diazadienes (**26e**) and (**26f**).[28] These compounds are incapable of undergoing the hydrogen abstraction process described above. Thus, irradiation in methylene dichloride affords the quinoxaline derivative (**33**). The formation of this product arises by a conventional six-electron cyclization similar to the formation of quinoline derivatives from 1-azadienes.[22] In this case, photocyclization affords the intermediate (**34**), which subsequently loses benzoic or acetic acid to give the final product (Scheme 72.16). This reaction can also be carried out in acetone and, under these conditions, the yield of quinoxaline increases. The use of acetone brings about the process in a shorter time and presumably shows that the triplet state is involved. Interestingly, a different reaction path is followed when the irradiation of diazadiene (**26e**) is carried out in the presence of BF$_3$ etherate. Under these conditions, the nitrogen lone pair coordinates to the boron and this complex is photochemically reactive and yields the phenanthridine derivative (**35a**) (Scheme 72.17). The change in reaction mode is in agreement with earlier results[28] that demonstrated the reluctance of Schiff's bases to undergo a *cis*-stilbene-type cyclization unless the nitrogen lone pair was protonated. More recent studies[29] have shown that the irradiation of boron trifluoride complexes of Schiff's bases brings about rapid ring closure to the corresponding phenanthridine derivatives in good yield. The diazadiene (**27b**) is also photochemically reactive, and irradiation in methylene dichloride gives both the quinoxaline (**33**) and a phenanthridine derivative (**35b**). These experiments with the 1,4-diazadienes (**26**) and (**27**) show that a variety of different reaction types (e.g., hydrogen abstraction and cyclization, *E,Z*-isomerization, and six-electron cyclizations leading to quinoxaline and phenanthridine skeleta) can take place. As far as can be judged, there is no evidence for electron transfer involvement in these systems.

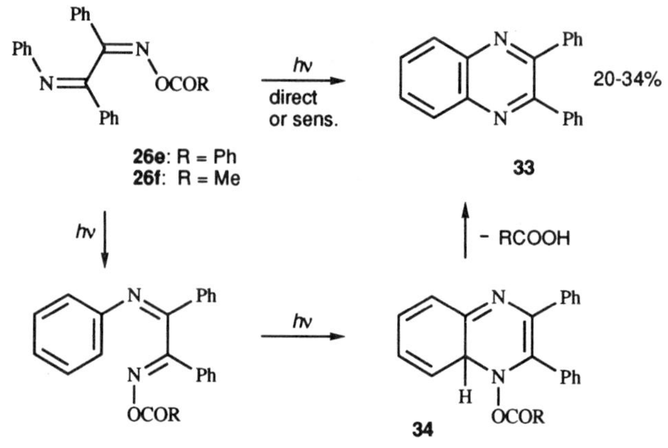

SCHEME 72.16

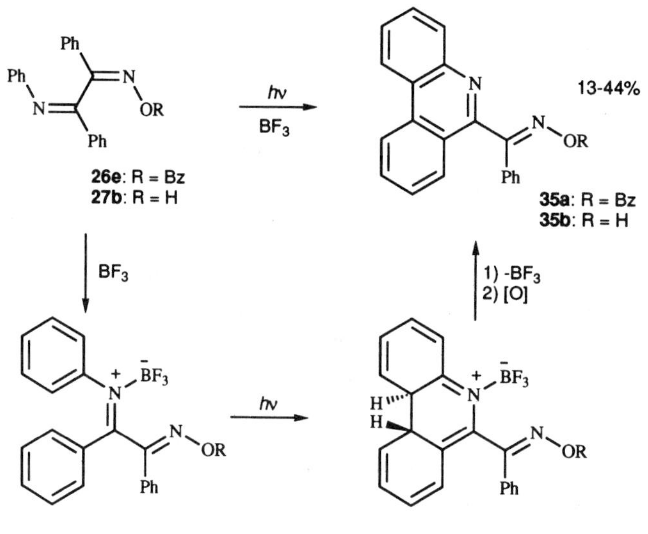

SCHEME 72.17

72.4 Miscellaneous Reactions of Azadienes

Apart from the six-electron cyclization discussed earlier, conjugated azadienes can also undergo other photochemical cyclizations. Thus, there are quite a few examples in the literature of 4π-electrocyclic ring closures in the photochemistry of cyclic conjugated 1-aza and 2-azadienes. Some selected reactions of this type are shown in Scheme 72.18.[30–32]

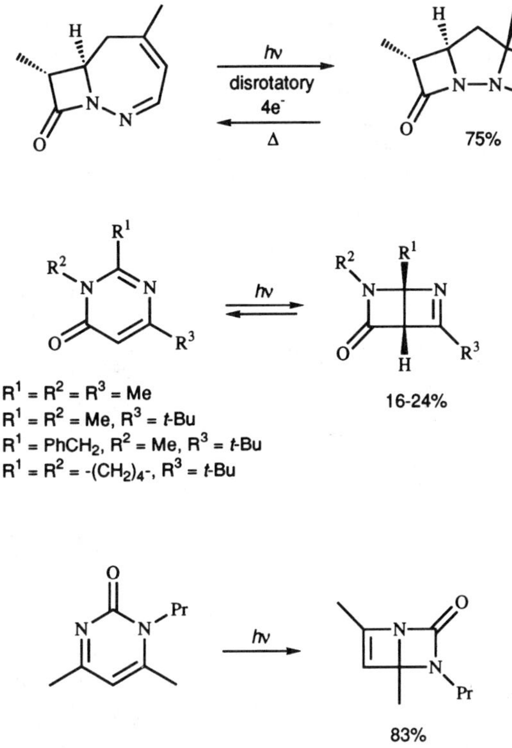

$R^1 = R^2 = R^3 = Me$
$R^1 = R^2 = Me, R^3 = t\text{-}Bu$
$R^1 = PhCH_2, R^2 = Me, R^3 = t\text{-}Bu$
$R^1 = R^2 = -(CH_2)_4\text{-}, R^3 = t\text{-}Bu$

16-24%

83%

SCHEME 72.18

Some reactions which are common in the photochemistry of dienes and enones have seldom been reported in azadienes. For instance, there are very few examples of hydrogen abstraction reactions by the excited C-N double bond in the photochemistry of azadienes, while these are very common in the photochemistry of ketones. A reaction of this type is the photocyclization of the substituted dihydroisoquinoline (36) to a spirobenzylisoquinoline (37) (Scheme 72.19).[33]

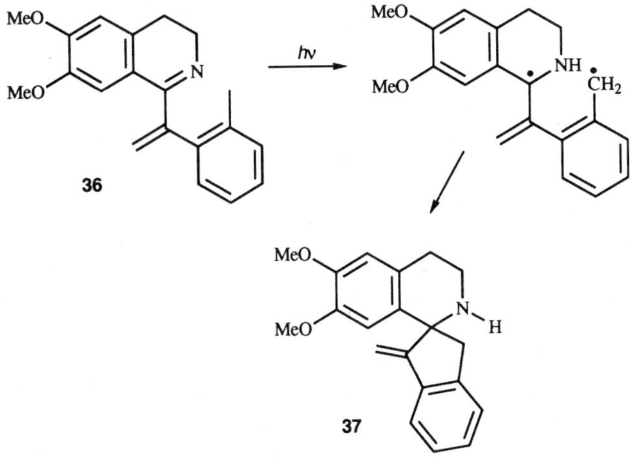

SCHEME 72.19

Photochemical [2+2]-cycloadditions are encountered very frequently in conjugated alkenes and ketones. However, there are only a few cases of such a reactivity in the photochemistry of C=N compounds. Most of the cases reported of this process refer to *intra-* or intermolecular cycloadditions of cyclic imines to alkenes. A reaction of this type in an azadiene has been described in the intramolecular cycloaddition of the oxime ether (38) that yields an azapropellane (39) in 71% yield (Scheme 72.20).[34] Cycloadditions between two C=N bonds are even less frequent. However, one example is the [2+2] dimerization of an azadiene. Thus, the irradiation in the solid state of the cyclohexylidene (40) yields diazetine (41) in almost quantitative yield (Scheme 72.21).[35]

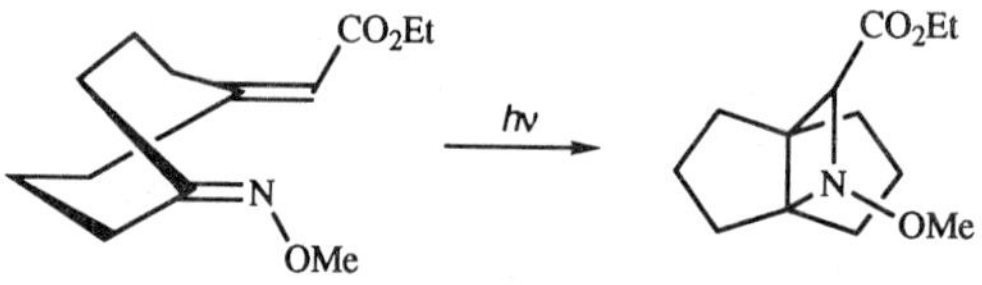

SCHEME 72.20

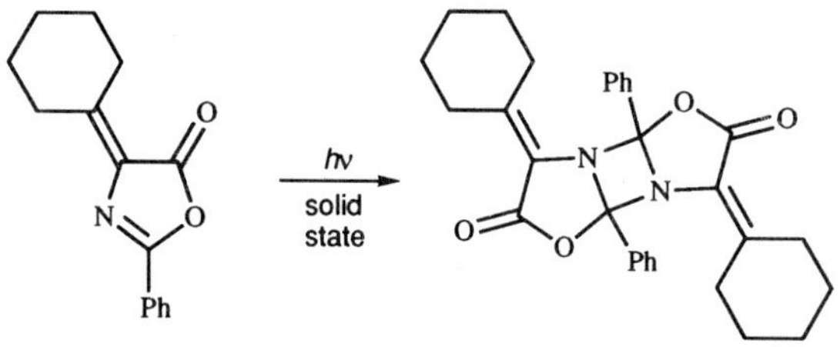

SCHEME 72.21

Other areas of interest in azadiene photochemistry are the aza-di-π-methane rearrangement of 1-aza-1,4-dienes and the intramolecular photoaddition of unsaturated iminium salts. Because of their interest, these two topics are discussed in detail in separate chapters of this book.

References

1. Padwa, A., Photochemistry of the carbon-nitrogen double bond, *Chem. Rev.*, 77, 37, 1977.
2. Pratt, A. C., The photochemistry of imines, *Chem. Soc. Rev.*, 6, 63, 1977.
3. Wettermark, G., *The Chemistry of the Carbon-Nitrogen Double Bond*, Patai, S., Ed., Wiley-Interscience, 1969, 580.
4. Rau, H. and Lüddecke, E., On the rotation-inversion controversy on photoisomerization of azobenzenes. Experimental proof of inversion, *J. Am. Chem. Soc.*, 104, 1616, 1982.
5. Poziomek, E. J. and Vaughan, L. G., Synthesis and configurational analysis of picolinaldehyde oximes, *J. Pharm. Sci.*, 54, 811, 1965.
6. Pratt, A. C. and Abdul-Majid, Q., Photochemistry of the carbon-nitrogen double bond. 1. Carbon-nitrogen vs. carbon-carbon double bond isomerisation in the photochemistry of α,β-unsaturated oxime ethers: The benzylideneacetone oxime O-methyl ether, *J. Chem. Soc., Perkin Trans. 1*, 1691, 1986.
7. Childs, R. F. and Dickie, B. D., The photoisomerizations of protonated Schiff base derivatives of some enals, *J. Chem. Soc., Chem. Commun.*, 1268, 1981; Childs, R. F. and Dickie, B. D., Photochemical and thermal stereomutations of 3-aryl-2-propenylideniminium salts, *J. Am. Chem. Soc.*, 105, 5041, 1983.

8. Takahashi, T., Hirokami, S., Kato, K., Nagata, M., and Yamazaki, T., Formation and reactions of Dewar 4-pyrimidones in the photochemistry of 4-pyrimidones at low temperature. 2, *J. Org. Chem.,* 48, 2914, 1983.

9. Wettermark, G. *The Chemistry of the Carbon-Nitrogen Double Bond,* Patai, S., Ed., Wiley-Interscience, 1969, 574.

10. Margaretha, P., Photochemistry of 2-cyclohexenimines and 2,3,4,4a,5,6-hexahydroquinolines. Preliminary communication, *Helv. Chim. Acta,* 61, 1025, 1978; Margaretha, P., Mechanistic aspects of the photochemical formation of α,β-unsaturated ketones, *Tetrahedron Lett.,* 4205, 1974.

11. Schulte-Frohlinde, D., Quantum yields in the photochemical *syn-anti* rearrangement of 2-pyridine carboxaldehyde phenylhydrazone, *Justus Liebigs Ann. Chem.,* 622, 47, 1959.

12. Dale, J. and Zechmeister, L., On the stereochemistry of azines: Cinnamalazine and phenylpentadienalazine, *J. Am. Chem. Soc.,* 75, 2379, 1953.

13. Kliger, D. S. and Menger, E. L., Vision: An overview, *Acc. Chem. Res.,* 8, 81, 1975.

14. Armesto, D., Gallego, M. G., and Horspool, W. M., Conformation dependent photochemistry of dienes. Substituent effects on the photochemical reactions of some 4-acyloxy-2-azabuta-1,3-dienes, *J. Chem. Soc. Perkin Trans. 1,* 2663, 1987.

15. Florencio, F., Mohedano, J., and Garcia-Blanco, S., Structure of benzoyloxytetraphenylazadiene, *Acta Crystallogr., Sect. C: Cryst. Struct. Commun.,* C43, 1631, 1987.

16. Armesto, D., Ortiz, M. J., Perez-Ossorio, R., and Horspool, W. M., A novel photochemical 1,2-acyl migration in an enol ester. The synthesis of 3-oxazoline derivatives, *Tetrahedron Lett.,* 24, 1197, 1983; Armesto, D., Horspool, W. M., Ortiz, M. J., and Perez-Ossorio R., Photochemistry of 4-acyloxy-2-azabuta-1,3-dienes. A novel photochemical 1,2-acyl migration in an enol ester. The synthesis of 2,5-dihydro-oxazole derivatives, *J. Chem. Soc., Perkin Trans. 1,* 623, 1986.

17. Mohedano, J. M., Florencio, F., and Garcia-Blanco, S., Crystal structure of 4-benzoyloxy-1,1,4-triphenyl-2-aza-1,3-diene, *Z. Kristallogr.,* 180, 131, 1987.

18. Armesto, D., Horspool, W. M., Langa, F., Ortiz, M. J., Perez-Ossorio, R., and Romano, S., A new synthesis of 1,1-diphenyl-3-arylisoquinolin-4-ones by the novel cyclization of 2-azabuta-1,3-dienes, *Tetrahedron Lett.,* 26, 5213, 1985.

19. Armesto, D., Gallego, M. G., Ortiz, M. J., Romano, S., and Horspool, W. M., A synthesis of isoquinolinones by the photochemical cyclization of 2-azabuta-1,3-dienes in the presence of acids, *J. Chem. Soc., Perkin Trans. 1,* 1343, 1989.

20. Armesto, D., Horspool, W. M., Ortiz, M. J., and Romano, S., A mechanistic study on the photochemical cyclization of protonated 2-aza-1,3-diene derivatives, unpublished results.

21. Arai, S., Takeuchi, T., Ishikawa, M., Takeuchi, T., Yamazari, M., and Hida, M., Syntheses of condensed polycyclic azonia aromatic compounds by photocyclization, *J. Chem. Soc., Perkin Trans. 1,* 481, 1987.

22. Armesto, D., Gallego, M. G., and Horspool, W. M., A photochemical synthesis of quinoline derivatives by the cyclization of N-benzoyloxy-4-aryl-2,3-diphenyl-1-azabuta-1,3-dienes, *J. Chem. Soc., Perkin Trans. 1,* 1623, 1989.

23. Fonken, G. J., Photochemical formation of polynuclear aromatic compounds from diaryl polyenes, *Chem. & Ind.,* 1327, 1962.

24. Laarhoven, W. H., Photochemical cyclization and intramolecular cycloaddition of conjugated aryl olefins. I. Photocyclization with dehydrogenation, *Rec. J. R. Neth. Chem. Soc.,* 102, 185, 1983; Laarhoven, W. H., Photochemical cyclization and intramolecular cycloaddition of conjugated aryl olefins. II. Photocyclization without dehydrogenation and photocycloadditions, *Rec. J. R. Neth. Chem. Soc.,* 102, 241, 1983.

25. Glinka, J., Photochemical reactions of α-phenylbenzylideneacetone oxime, *Pol. J. Chem.,* 53, 2143, 1979.

26. Elferink, V. H. M. and Bos, H. J. T., Novel photochemical and thermal electrocyclization to fused quinolines, *J. Chem. Soc., Chem. Commun.,* 882, 1985.

27. Armesto, D., Horspool, W. M., Apoita, M., Gallego, M. G., and Ramos, A., Photochemical reactivity of imines from benzil mono-oxime esters, *J. Chem. Soc., Perkin Trans. 1,* 2035, 1989.

28. Badger, G. M., Joshua, C. P., and Lewis, G. E., Photocatalysed cyclization of benzalaniline, *Tetrahedron Lett.*, 3711, 1964.

29. Prabhakar, S., Lobo, A. M., and Tavares, R. M., Boron complexes as control synthons in photocyclisations: An improved phenanthridine synthesis, *J. Chem. Soc., Chem. Commun.*, 884, 1978.

30. Tschamber, T., Fritz, H., and Streith, J., Stereospecific molecular design. Synthesis of a new heterocyclic system, *Helv. Chim. Acta*, 68, 1359, 1985.

31. Hirokami, S., Takahashi, T., Kurosawa, K., Nagata, M., and Yamazaki, T., Photochemistry of 4-pyrimidones: Isolation of Dewar isomers, *J. Org. Chem.*, 50, 166, 1985.

32. Nishio, T., Kameyama, S., and Omote, Y., Photochemistry of pyrimidin-2(1H)-ones. Intramolecular γ-hydrogen abstraction by the nitrogen of the imino group, *J. Chem. Soc., Perkin Trans. 1*, 1147, 1986.

33. Hirai, Y., Egawa, H., Wakui, Y., and Yamazaki, A photochemistry of 1-(1-phenylvinyl)-3,4-dihydroisoquinoline, *Heterocycles*, 25, 201, 1987.

34. Malamidou-Xenikaki, E. and Nicolaides, D. N., Synthesis of heterocyclic propellanes. Preparation and transannular reactions of 5-ethoxycarbonylmethylene-cyclooctanone and the corresponding oximes and hydrazones, *Tetrahedron*, 42, 5081, 1986.

35. Lawrenz, D., Mohr, S., and Wendländer, B., Formation of 1,3-diazetines via C-N dimerization of 4-cycloalkylidene-oxazol-5(4H)-ones in the solid state, *J. Chem. Soc., Chem. Commun.*, 863, 1984.

The Aza-di-π-methane
Rearrangement

·go Armesto
·ersidad Complutense

3.1 Introduction

The photochemistry of the C=N bond has not been studied as extensively as the photoreactivity of other chromophores such as the C=C bond and the C=O bond. However, in the last 15 years, there has been an increasing interest in this class of compounds. Among the new reactions uncovered in recent times in the study of the photochemistry of C=N derivatives, the aza-di-π-methane rearrangement has been one of the major achievements in this field.

3.2 The Aza-di-π-methane Rearrangement
of β,γ-Unsaturated Imines

The di-π-methane (DPM) reaction of 1,4-alkenes and its counterpart the oxa-di-π-methane (ODPM) rearrangement of β,γ-unsaturated ketones have been known for many years and a large number of studies have been devoted to these two reactions.[1] However, the extension of the rearrangement to 1-aza-1,4-dienes is relatively recent. The first two examples of aza-di-π-methane (ADPM) reactivity were described in the sensitized irradiation of the β,γ-unsaturated imine (1) that yielded exclusively the corresponding cyclopropyl imine (2),[2a] which hydrolyzed to the corresponding aldehyde on isolation (Scheme 73.1), and in the cyclization of the cyclic oximes (3) that gives the tricyclic derivatives (4) (Scheme 73.2).[3] However, while the aza-di-π-methane reaction of β,γ-unsaturated imines has proved to be very general, the corresponding rearrangement of the oximes has only been observed for the compounds (3) and closely related systems and acyclic β,γ-unsaturated oximes do not undergo the cyclization.

0-8493-8634-9/95/$0.00+$.50
© 1995 by CRC Press, Inc.

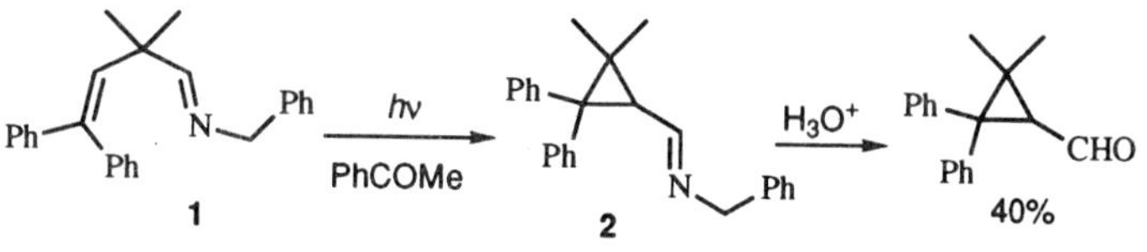

SCHEME 73.1

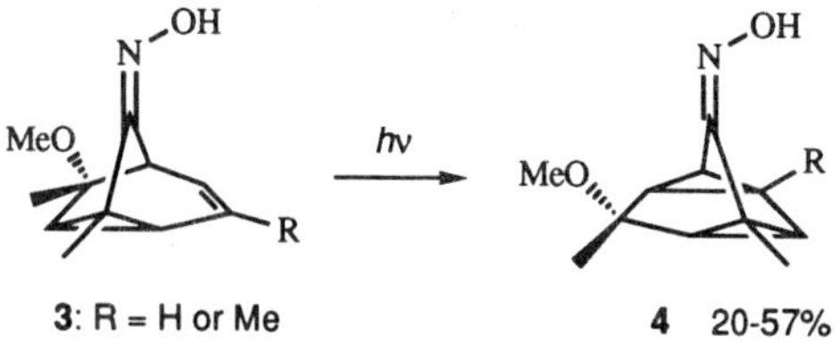

3: R = H or Me   4   20-57%

SCHEME 73.2

Early studies on the aza-di-π-methane reaction demonstrated that the moisture-sensitive imines (5), readily obtained by reaction of amines with the aldehyde (6), were photochemically reactive. Using acetophenone- or phenanthrene-sensitized irradiation, brief exposure brought about conversion into a cyclic product. As a result of the moisture sensitivity of the photoproduct isolation of the compound was not possible. However, hydrolysis of the photolysate provided a stable compound that was identified as the cyclopropane carbaldehyde (7) (Scheme 73.3).[2b]

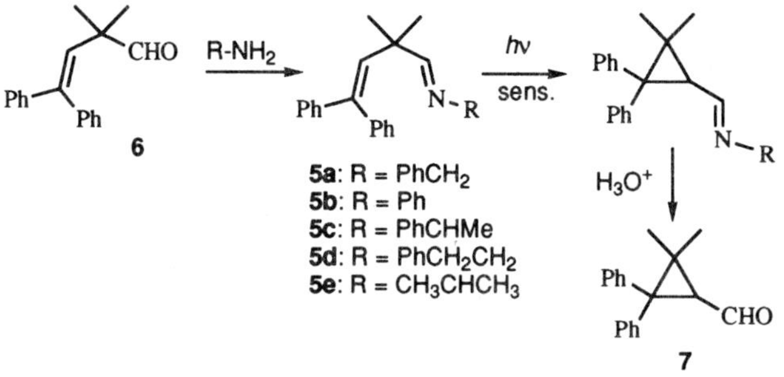

5a: R = PhCH$_2$
5b: R = Ph
5c: R = PhCHMe
5d: R = PhCH$_2$CH$_2$
5e: R = CH$_3$CHCH$_3$

SCHEME 73.3

Direct irradiation was also effective in this conversion but the process was much less efficient. This observation and the fact that the reaction does not occur in the presence of triplet quenchers supports the postulate that the reaction is taking place via the triplet excited state. The reaction was interpreted as the first example of the acyclic aza-di-π-methane rearrangement and the preliminary thoughts on the mechanism are shown in Scheme 73.4.

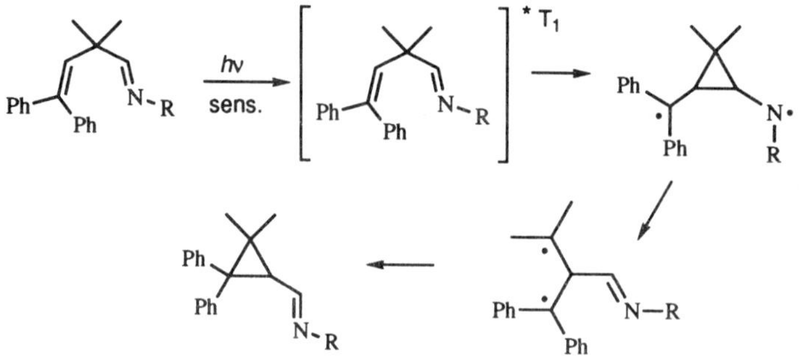

SCHEME 73.4

This route to a cyclopropyl aldehyde (7) has some synthetic value since irradiation of the starting aldehyde (6) does not bring about an oxa-di-π-methane rearrangement and, instead, decarbonylation occurs efficiently by a Norrish type I process.[4] Thus, the approach of converting the aldehyde to an imine followed by irradiation and hydrolysis provides a path for circumventing the normal photochemical reactivity of the aldehyde. As will be seen later, such a *modus operandi* is quite general. This early study showed, in qualitative terms, that the efficiency of the reaction was to some extent substituent dependent. The poorest yields were obtained with alkyl substituents attached to the imine nitrogen, while the best yields were recorded with a phenyl or benzyl group. The principal difference between the N-alkyl derivatives and the N-phenyl derivative of the same imine is the ability of the phenyl group to conjugate with the lone pair on the nitrogen rather than with the imine π-bond.[5] The phenyl ring of the N-benzyl substituent can also interact with the nitrogen lone pair by a homo conjugative effect. Thus, it seemed likely that the nitrogen lone pair was in some way involved in the cyclization step. It has been established that the lowest-energy band in the photoelectron spectrum of simple imines arises by removal of an electron from the nitrogen lone pair.[6] Consequently, based on this observation, one hypothesis is that single electron transfer (SET) from the nitrogen lone pair to the alkene moiety — 1,1-diphenyl vinyl is a known electron acceptor[7] — makes the cyclization less efficient. Thus, if it is possible to increase the ionization potential of the imine, then the cyclization should become more efficient. In order to test this hypothesis in a quantitative fashion, the efficiency of the cyclization of the series of 1-aryl derivatives of 3,3-dimethyl-5,5-diphenyl-1-azapenta-1,4-diene (8) was studied.[8] All of them undergo the acetophenone-sensitized aza-di-π-methane rearrangement described above and yield a cyclopropylimine. Again, the cyclopropylimine was not isolated and, as usual, hydrolysis afforded the aldehyde (7).

The photochemical reactivity of the imines was demonstrated qualitatively in irradiations where azadiene (8e) yielded the cyclopropyl derivative more efficiently than any of the other imines studied. The quantitative study, from which the quantum yield values shown in Table 73.1 were obtained, fall into a distinct pattern where the imine (8b) is the least efficient and the imine (8e) is the most efficient.

As mentioned before, it has been demonstrated that the N-aryl group overlaps more with the nitrogen lone pair than with the imine π-system. Clearly, the nature of the substituent on the aryl function will exercise some control over the availability of the nitrogen lone pair to undergo an electron transfer to the 1,1-diphenyl vinyl moiety. This transfer will be greatest in azadiene (8b) and poorest in imine (8e). In fact, the enhancement in yield between the p-methoxy-substituted imine and the p-cyano-substituted imine is 32. The other substituents, a p-chloro and an m-methoxy group in imines (8c) and (8d) have a very small effect on the quantum yield for the cyclization. This is borne out by the results shown in Table 73.1. Further substantiation for the interaction between the aryl group and the nitrogen lone pair comes from a Hammett plot of log Φ against the σ^+ substituent constants which gives a ρ value of 1.1 with a correlation coefficient of 0.998.

8a: Ar = Ph
8b: Ar = p-MeOC$_6$H$_4$
8c: Ar = p-ClC$_6$H$_4$
8d: Ar = m-MeOC$_6$H$_4$
8e: Ar = p-CNC$_6$H$_4$

8

9a: Ar = Ph
9b: Ar = p-MeC$_6$H$_4$
9c: Ar = p-ClC$_6$H$_4$
9d: Ar = m-FC$_6$H$_4$
9e: Ar = p-CF$_3$C$_6$H$_4$

9

Table 73.1   Quantum Yield of Cyclization of Azadienes (8)

Azadiene	Φ	Relative Φ
8a	0.0035	1.00
8b	0.0004	0.13
8c	0.0040	1.12
8d	0.0039	1.10
8e	0.0147	4.15

Table 73.2   Quantum Yield of Cyclization of Azadienes (9)

Azadiene	Φ	Relative Φ
9a	0.0092	1.00
9b	0.0066	0.72
9c	0.0111	1.21
9d	0.0144	1.56
9e	0.0167	1.82

Additional proof in favor of this proposed adverse electron transfer was obtained from the quantitative study of the acetophenone-sensitized photocyclization of the *N*-benzyl-substituted azadienes (9).[9] Once again, all of them undergo the aza-di-π-methane rearrangement, yielding a cyclopropylimine, and the quantum yields for the rearrangement show again a dependence on the nature of the substituents on the *N*-benzyl group. The reaction is most efficient with electron-withdrawing substituents (Table 73.2).

The linear relationship between log Φ and σ^+ suggests that there is a homoconjugative interaction between the benzyl group and the nitrogen lone pair. Both of these results confirm the conjugative effect of the aryl and benzyl substituents with the nitrogen lone pair. These results are supportive of the postulate that an electron transfer from the imine nitrogen to the alkene group is detrimental to the cyclization.

To take account of these results, a more accurate representation of the mechanism involving the electron transfer process that is seen to adversely affect the efficiency of the reaction was proposed and is shown in Scheme 73.5.[9] This mechanism gives a reasonable explanation for the failure of a variety of molecules to undergo the aza-di-π-methane rearrangement. Thus, if there is an efficient SET from the nitrogen lone pair to the alkene moiety due to the low ionization potential of the lone pair, then the rearrangement is either very inefficient or fails altogether. Thus, this could be the reason for the failure of the oxime (10)[10] and the oxime ether (11)[11] to undergo the aza-di-π-methane rearrangement.

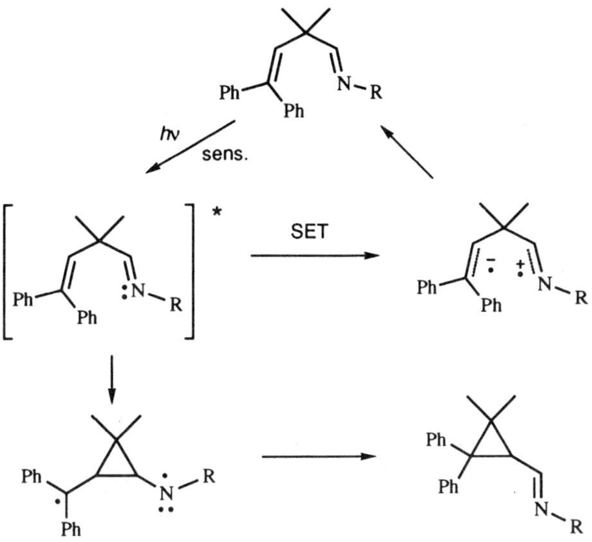

SCHEME 73.5

3.3 Extension of the Aza-di-π-methane Rearrangement to C-N Double Bond Stable Derivatives

The logical extension of the foregoing was to seek a way by which the ionization potential of the oxime could be raised. This was readily achieved by incorporating an electron-withdrawing group by simple acetylation of the oxime to afford (12). When this compound was irradiated using acetophenone sensitization, the ADPM rearrangement occurs to afford the product (13) in 86% yield.[12]

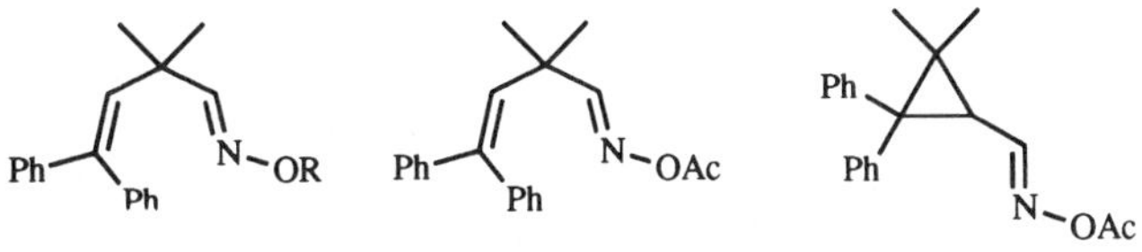

10: R = H
11: R = Me   **12**   **13**

The discovery of this facile rearrangement of a stable oxime acetate derivative of the aldehyde (6) was a major step forward and prompted a study of the scope of this particular system. Thus, a series of oxime acetate derivatives (14) was prepared and studied.[13] The success of this reaction mode was further demonstrated by the photochemical conversion of the oxime acetates (14a) and (14b) into the corresponding cyclopropane derivatives (15a) and (15b) in 90 and 18% yield, respectively (Scheme 73.6).

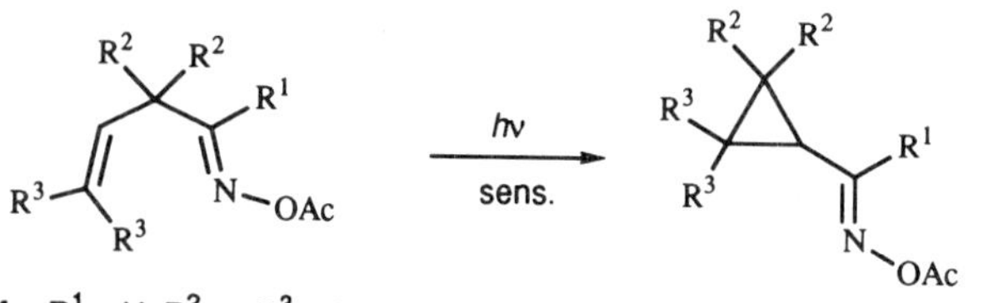

14a: $R^1 = H$, $R^2 = R^3 = Ph$
14b: $R^1 = R^2 = Me$, $R^3 = Ph$
14c: $R^1 = R^3 = Ph$, $R^2 = Me$
14d: $R^1 = H$, $R^2 = R^3 = Me$
14e: $R^1 = R^2 = R^3 = Me$

15a: $R^1 = H$, $R^2 = R^3 = Ph$
15b: $R^1 = R^2 = Me$, $R^3 = Ph$
15c: $R^1 = H$, $R^2 = R^3 = Me$

SCHEME 73.6

From these results and from the previous conversion of (12), it can be seen that the aldoxime acetates are more reactive than the ketoxime acetate (14b) and this is further substantiated by the failure of oxime acetate (14c) to afford a cyclized product after 20 h irradiation. Changing the substitution on the terminal carbon of the azadiene system to methyl groups in oxime acetate (14d) does not alter the reaction path and using acetone as sensitizer the cyclopropane (15c) was obtained in 32% yield. The success of the aza-di-π-methane rearrangement of this all-alkyl substituted oxime acetate (14d) is of great importance. A similar reaction has not been reported in the oxa-di-π-methane reaction[1] and there is only one example in the di-π-methane system reported by Baeckstrom[14] and by Bullivant and Pattenden.[15] The other oxime acetate (14e) was unreactive under acetone sensitization and is further confirmation that the ketoxime acetates are less reactive in qualitative terms. Earlier, it was pointed out that the efficiency of the cyclization was dependent on the nature of the substituent on the nitrogen. Further support for these mechanistic proposals was obtained by quantum yield measurements. The results from this quantitative work show that the quantum efficiency for the formation of (13) from the oxime acetate (12) by acetophenone sensitization is 0.12, which is 10-fold better than the best quantum yield for the cyclization of the aryl substituted imines (8). The aza-di-π-methane reaction can in fact be extremely efficient, as shown by the cyclization of oxime acetate (14a) which yields the corresponding cyclopropane (15a) with a quantum yield of 0.82. The high efficiency of this reaction could be due to the diphenyl substitution of the central carbon. Zimmerman et al.[16] have also observed similar changes in the di-π-methane reaction of (16), where the quantum yield of cyclization was measured as 0.42 for the sensitized cyclization. These results show that in some cases, the aza-di-π-methane reaction is more efficient than the di-π-methane counterpart.

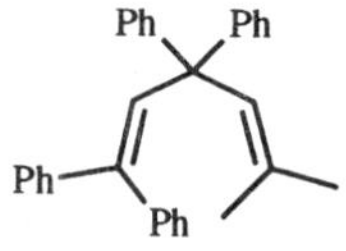

16

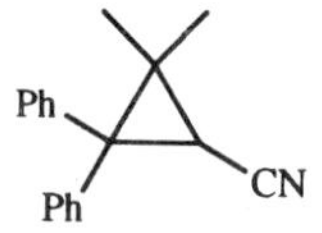

19

With the establishment of the need for an electron-withdrawing substituent on the oxime system, the next step was to seek other functional groups that might be equally efficient as well as giving stable crystalline derivatives. It was demonstrated readily that the derivatives (17) could be prepared easily by conventional methods.[17] Among all these derivatives, only the trifluoroacetates show thermal instability and care has to be taken in handling these compounds. The irradiation of compound (17a) using acetophenone as sensitizer afforded the semicarbazone derivative (18a) in 40% yield. Similar photochemical reactivity was shown by the other derivatives (17b)[17b] and (17c)[17a]

(Scheme 73.7). In the case of (**17b**), a different work-up procedure was used, that of hydrolysis of the photolysate, which allowed the isolation of aldehyde (**7**, 71%) rather than the acetylhydrazone derivative. The photocyclization of the trifluoroacetate derivative (**17d**) is very efficient, and irradiation for a mere 10 min affords the cyanocyclopropane (**19**) in 80% isolated yield.[17b] The photocyclization in this example is followed by an efficient thermal elimination of trifluoroacetic acid during the work-up procedure. This provides a new synthetic route to cyanocyclopropanes from derivatives of β,γ-unsaturated aldehydes. A direct path by the irradiation of β,γ-unsaturated nitriles fails because of their photochemical inertness.[10] The ADPM rearrangement of the benzoate derivative (**17e**)[17a] is also efficient and gives the corresponding cyclopropane (**18e**) in 90% yield after 20-min irradiation.

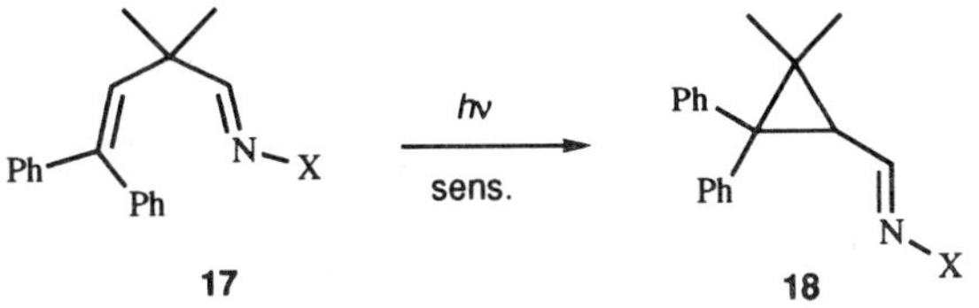

17a = 18a: X = NHCONH₂
17b = 18b: X = NHCOCH₃
17c = 18c: X = NHCOPh
17d = 18d: X = OCOCF₃
17e = 18e: X = OCOPh

SCHEME 73.7

Thus, in general, the hydrazine derivatives undergo cyclization less efficiently than the oxime derivatives. The poorer efficiency observed with the hydrazine derivatives could be explained by the fact that amide groups have poorer electron-withdrawing capacity than esters. The study has been extended to include the derivatives of the ketone (**20**). Although previous experiments have demonstrated that the imine and oxime acetate derivatives of ketone (**20**) undergo the ADPM, the derivatives (**21a**) and (**21b**) were shown to be unreactive in the aza-di-π-methane rearrangement. However, the trifluoroacetate (**21a**) does react photochemically and affords diene (**22**).[17b]

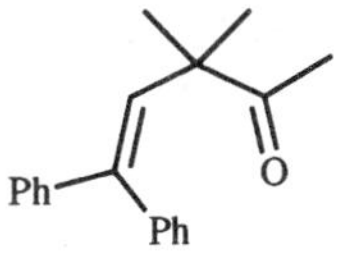

20

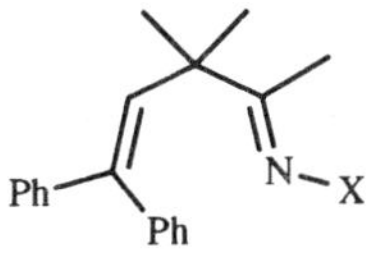

21a: X = OCOCF₃
21b: X = OCOPh

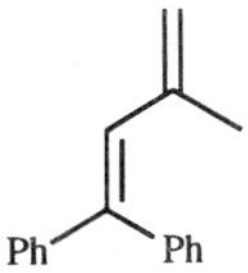

22

A fragmentation reaction of this type has not been reported previously in the photochemistry of 1-aza-1,4-diene derivatives. The mechanism shown in Scheme 73.8 was proposed to account for this result. The key step in the reaction is a SET to the trifluoroacetyl group from the 1,1-diphenyl vinyl moiety. The involvement of this may be due to the greater electron-accepting properties of the trifluoroacetyl group compared to the others used above. This is followed by hydrogen abstraction from the proximate methyl group. The resultant radical cation/radical anion undergoes fragmentation and hydrogen abstraction as shown in Scheme 8. That the aldehyde derivative (**17d**) does not undergo fragmentation is presumably due to the absence of an abstractable acidic hydrogen.

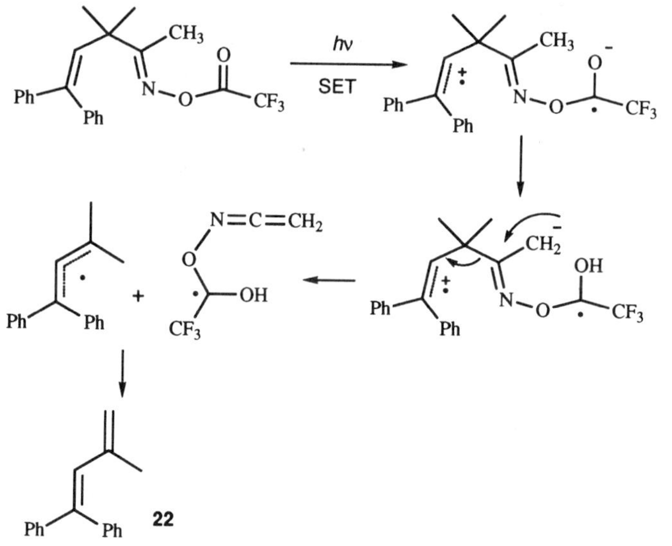

SCHEME 73.8

The foregoing demonstrates qualitatively the efficiency of the ADPM rearrangement for this series of derivatives. One of the most important advantages of the compounds used in this study is that the hydrazine derivatives and the cyclopropanes derived from them are stable and crystalline and are obtained in high yield. The success with these derivatives indicates that there is no adverse effect, apart from a slight decrease in efficiency, in incorporating a nitrogen adjacent to the imine nitrogen. It is clear that the ADPM rearrangement is not restricted to imines and oxime acetates, but can be readily extended to other common derivatives of carbonyl compounds. The study has also indicated some limits to the type of derivative that can be used. Thus, if the derivative has a functional group that can undergo SET from the triplet vinyl moiety, then a fragmentation path can be operative.

An alternative method for raising the ionization potential of the oxime lone pair electrons could be by coordination to a Lewis acid. This was demonstrated for the oxime derivatives (10) and (23) when they were irradiated in benzene solution in the presence of boron trifluoride etherate. Typically, oxime (10) affords a low yield (3%) of the cyclopropyl aldehyde (7) by the aza-di-π-methane rearrangement, but the principal product (48%) was identified by X-ray diffraction analysis as the dihydro-isoxazole (24a).[18] The other oximes (23a) and (23b) were cyclized successfully under similar conditions to yield the dihydro-isoxazoles (24b, 35%) and (24c, 30%) respectively.

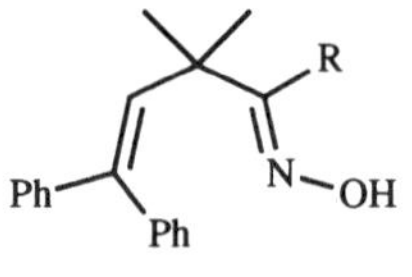

10: R = H	**24a**: R = H	**7**
23a: R = Ph	**24b**: R = Ph	
23b: R = Me	**24c**: R = Me	

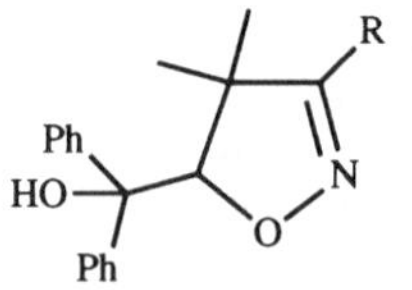

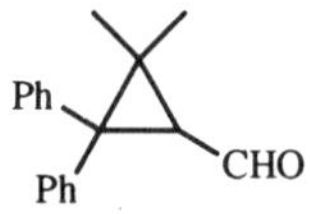

The preferred route to these products is shown in Scheme 73.9. Here, it is envisaged that the complex of the oxime and BF$_3$ is photochemically excited and undergoes electron transfer from the 1,1-diphenyl vinyl moiety to the iminium salt, a process with ample precedent. The resultant radical cation/radical anion reacts intramolecularly to afford the dihydro-isoxazole system. The conversion of (25) to the final product (24a) probably occurs during isolation. The formation of the cyclopropyl derivative (7), albeit in low yield, demonstrates that oxime (10) can be made to undergo the ADPM rearrangement if the electron transfer from the imine (oxime) nitrogen is suppressed. However, it is obvious that the ADPM rearrangement is not efficient and competes very poorly with the alternative process of electron transfer from the 1,1-diphenyl vinyl moiety to the iminium salt. Furthermore, the boron must play an important part in the reaction since the rearrangement observed under these conditions does not occur when the irradiation is carried out using perchloric acid to protonate the oxime nitrogen. The reaction described above provides a new route to the synthesis of dihydro-isoxazoles.

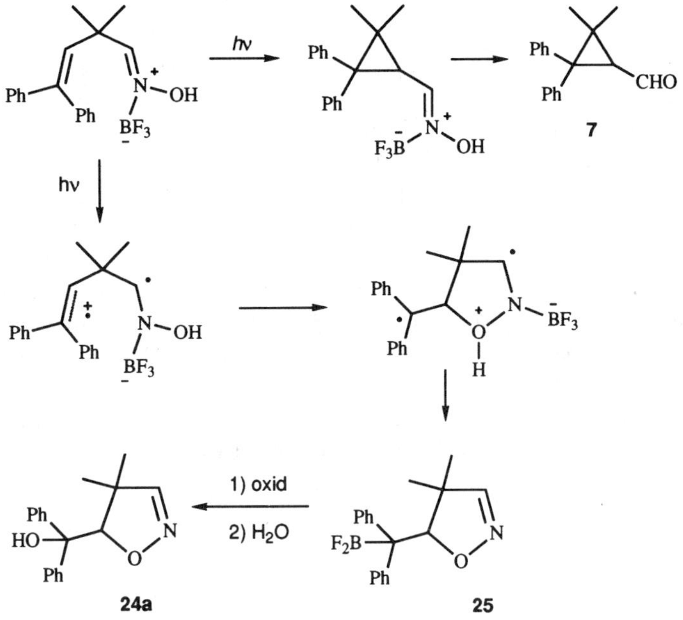

SCHEME 73.9

3.4 Steric and Electronic Effects on the Aza-di-π-methane Rearrangement

All the preceding illustrates clearly the power and the generality of the ADPM process but, in some respects, the question of the influence of substitution pattern on the rearrangement has not been addressed. This question is vital, of course, if the synthetic potential of the reaction is to be realized fully and if use can be made of it in the synthesis of naturally occurring compounds such as the pyrethroids and related systems. To this aim, a study on the influence of changes in substitution pattern was carried out. It was shown readily that on sensitized irradiation, the oxime acetate (26) undergoes conventional aza-di-π-methane rearrangement into the cyclopropyl oxime acetate (27) in 71% yield.[19] This reaction is stereospecific affording only the *trans*-isomer shown. It was particularly surprising that no evidence for deactivation of the excited state by the free rotor effect was observed. Such an effect is well documented in the triplet reactivity of acyclic 1,4-dienes.[20]

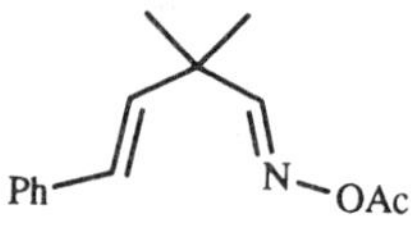

26

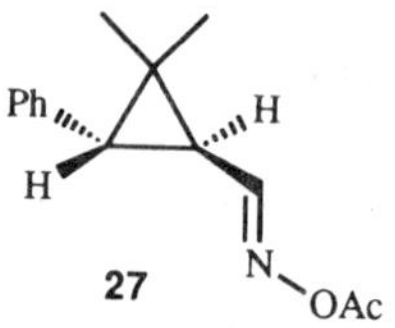

27

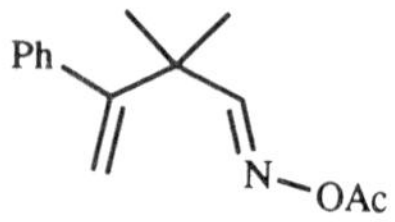

28

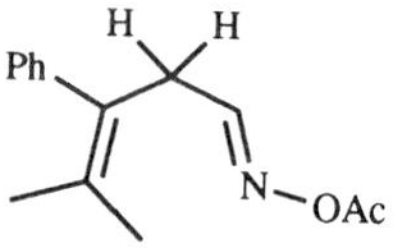

29

The successful cyclization of (26) prompted a study of the 1-aza-1,4-diene (28) in which the 5-position is unsubstituted. The absence of substituents on this carbon makes this a good example for free rotor activity in the triplet excited state, and the incorporation of a phenyl at the 4-position still ensures that triplet energy will be transferred to the alkene moiety. Surprisingly, the irradiation of the oxime acetate (28) follows a different reaction path and affords the oxime acetate (29) in 62% yield. This reaction is the first example of a photochemical 1,3-migration of a C=N group in a 1-aza-1,4-diene system. The failure of this diene to undergo the aza-di-π-methane is thought to be due to preferential formation of the stabilized biradical, which can be produced by bonding between the methylene end of the excited-state alkene and the carbon of the oxime acetate. This alternative path is outlined in Scheme 73.10. Here, it is shown that energy transfer from the sensitizer affords, as always, the triplet alkene. Conventional bridging would yield a biradical in which there was minimal stabilization. The molecule, therefore, follows a path that forms a better biradical. This route yields the cyclobutane intermediate that subsequently ring opens by rupture of bond "a" to yield the diene (29).

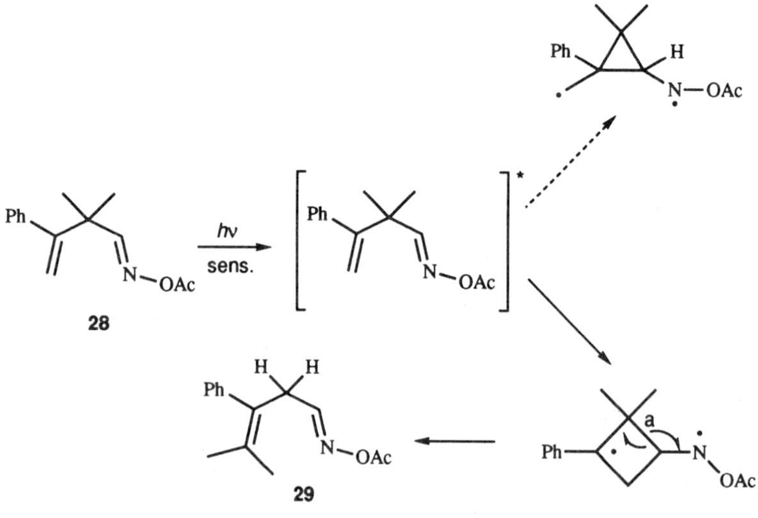

SCHEME 73.10

Free rotor deactivation was also a possible explanation of the failure of the azadienes (30) to undergo the sensitized ADPM process.[21] Thus, the irradiation of the E-oxime acetate (30a) for 1 h, using acetone as sensitizer, gave only a 2:1 mixture of the E- and Z-isomers. Prolonged irradiation of up to 7 h did not change the ratio indicating that this was the composition of the photostationary state. Even direct irradiation of the oxime acetate (30a) at 254 nm brought about *E,Z*-isomerization, affording a separable mixture of the isomers in a ratio of 1.25:1 (Scheme 73.11). A similar result was obtained from the acetone-sensitized irradiation of the *E*-oxime acetate (30b) where a 2.5:1 mixture of the *E*- and *Z*-isomers was formed. The failure of these two oxime acetates (30a) and (30b) to undergo cyclization was considered to be due to either a deactivation of the excited triplet state by a free rotor effect or an alternative deactivation path via SET from the nitrogen lone pair to the alkene moiety. However, irradiation of the oxime acetates (30c) and (30d), in which the electron-withdrawing groups have been replaced by a methoxymethyl or by an acetoxymethyl group, again brings about *E,Z*-isomerization exclusively. These results demonstrate clearly that SET is not responsible for the absence of aza-di-π-methane reactivity in these oxime acetates. However, the incorporation of a second substituent at the 5-position gave oxime acetates (31), which were again photoreactive in the ADPM rearrangement (Scheme 73.12).[21,22] Thus, irradiation of the *E*-oxime acetate (31a) by acetone sensitization for 1.5 h also brings about *E,Z*-isomerization in a ratio of 2.5:1 (72% yield) but also produces new products as an inseparable mixture (28%). This was a 1:1 mixture of compounds that were identified as the ADPM product (32a). The other oxime acetates (31b) and (31c) behave in a similar manner and acetone-sensitized irradiation yield photolysates composed of a mixture of *E,Z*-isomers and cyclopropanes (32b, 54%) and (32c, 21%), respectively.

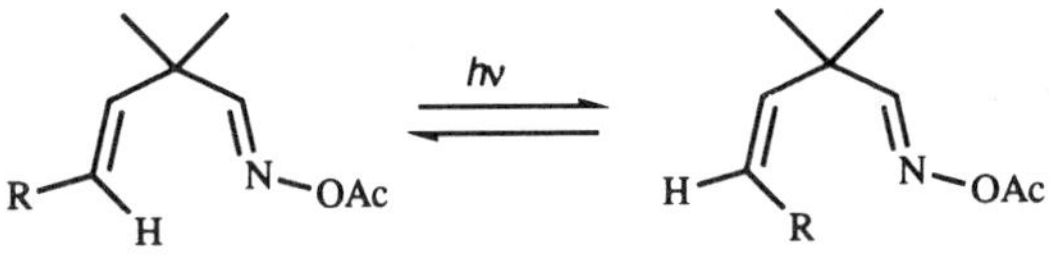

30a: R = CO$_2$Et
30b: R = CN
30c: R = CH$_2$OCOMe
30d: R = CH$_2$OMe

SCHEME 73.11

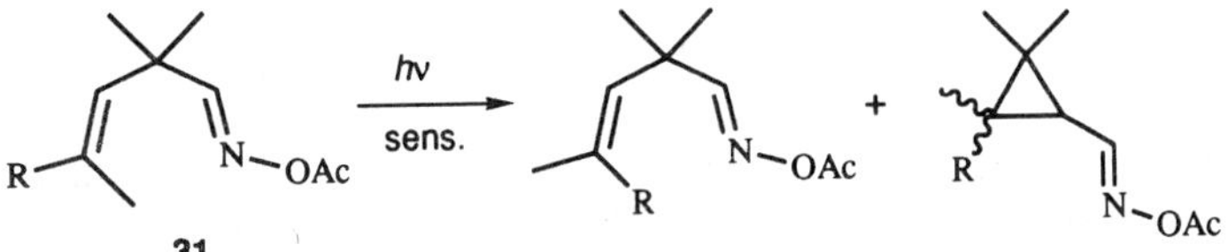

31a = 32a: R = CO$_2$Et
31b = 32b: R = CN
31c = 32c: R = CH$_2$OCOMe

SCHEME 73.12

However, it was surprising that the introduction of a methyl group should have such a profound effect on the outcome of the reaction. The possibility that free-rotor deactivation was not the controlling feature was considered. An alternative reason based on the stability of the biradical in the bridged intermediate could well determine whether or not cyclization will be successful.

In order to obtain additional evidence in favor of this proposal, a study on the influence of the size of the substituents at 5-position of the 1-aza-1,4-diene system was carried out.[23] Increasing the bulk of the ester function in (30a) by the use of the *t*-butyl ester (33a) did not change the outcome of the reaction and, again, only *E,Z*-isomerization took place. Irradiation of the bulky 5-cyclohexyl-substituted diene (33b) also fails to yield products by way of the ADPM path and, again, only *E,Z*-isomerization occurs on benzene sensitization. Even the 5,5-dicyclohexyl derivative (33c) is unreactive. This study demonstrates clearly that the lack of aza-di-π-methane reactivity observed in the monosubstituted aza-dienes (30) is not due to a free rotor effect.

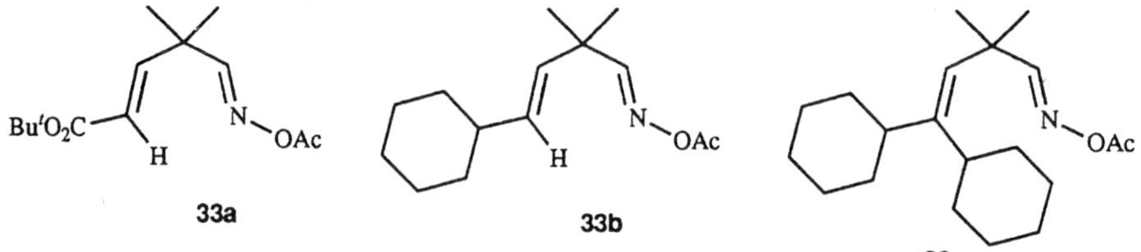

33a   **33b**   **33c**

73.5 Other Synthetic Applications of the Aza-di-π-methane Rearrangement

All the evidence gathered regarding the effect of substitution at C-5 on the outcome of the ADPM point to the fact that dienes which fail to undergo the rearrangement do so because of a poorly stable cyclopropyl biradical intermediate. Further support for this concept of radical stabilization was obtained in the highly efficient (up to 80% yield) photochemical cyclization of the 1-aza-1,4,6-trienes (34) to the corresponding cyclopropanes.[24] The resultant cyclopropylimines are hydrolyzed readily to yield the cyclopropyl aldehydes (35) (Scheme 73.13). This particular reaction provides a major route to the synthesis of cyclopropyl aldehydes and acids (36), which can be readily converted into molecules of commercial importance such as pyrethroid derivatives.[25] On the other hand, the photochemistry of dienes (31) and trienes (34) has opened a new photochemical route to cyclopropanes with two functional groups.

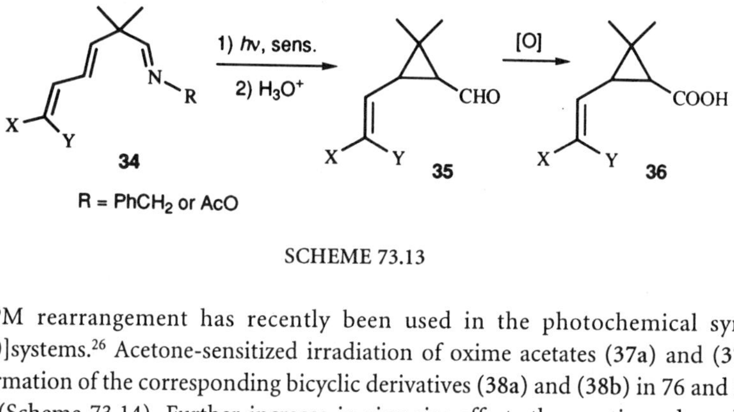

SCHEME 73.13

The ADPM rearrangement has recently been used in the photochemical synthesis of bicyclo[n.1.0]systems.[26] Acetone-sensitized irradiation of oxime acetates (37a) and (37b) brings about the formation of the corresponding bicyclic derivatives (38a) and (38b) in 76 and 24% yield, respectively (Scheme 73.14). Further increase in ring size affects the reaction adversely and the oxime acetate (37c) is unreactive by the ADPM path. This decrease of efficiency in the formation of product arising by the ADPM rearrangement as the ring size increases could be due to relaxation of the triplet state by twisting of the C=C bond. To test this possibility, the dihydronaphthalene derivative (39) was irradiated using acetophenone as sensitizer. Under these conditions (39) was transformed into the tricyclic product (40) in 90% yield after irradiation for 20 min. It is interesting to note that the photochemical reactivity of the methyl ketones (41) related to the aldehydes (42), used as precursors of the oxime acetates (37), has been studied.[27] This work established that the cyclohexenyl and cycloheptenyl ketones are unreactive by the oxa-di-π-methane process. Cyclobutenyl and cyclopentenyl ketones undergo the ODPM rearrangement on triplet-sensitized irradiation, although as a secondary reaction path. The results obtained in the study of oxime acetates (37) show that the ADPM rearrangement is much more general than the oxa-di-π-methane analogue, and avoids secondary reactions such as 1,3-acyl migration or decarbonylation. The ADPM conversion of the oxime acetates (37) into the bicyclic aldehyde derivatives (38) is the first photochemical synthesis of such compounds and opens a new synthetic approach to bicyclic naturally occurring compounds.

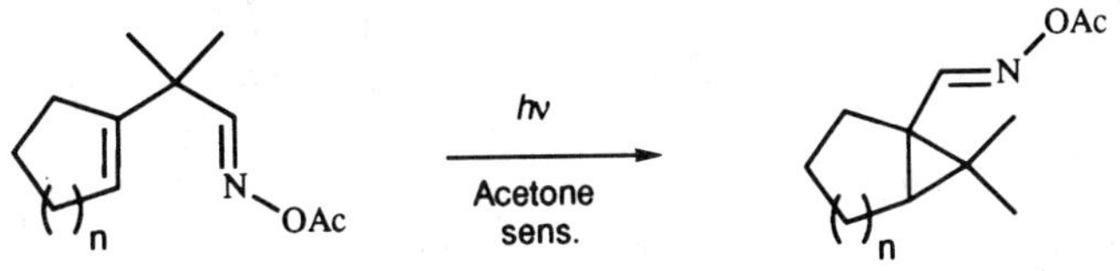

37a: n = 1
37b: n = 2
37c: n = 3

38a: n = 1
38b: n = 2

SCHEME 73.14

39

40

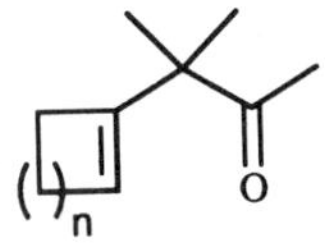

41a: n = 1
41b: n = 2
41c: n = 3
41d: n = 4
41e: n = 5

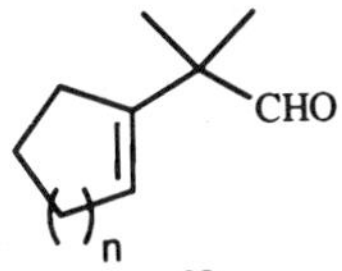

42

The ADPM rearrangement has also been observed in some heterocyclic systems. Thus, 4*H*-1,2-diazepines (**43**) rearrange to 6*H*-1,4-diazepines (**45**) via a 2,6-diazabicyclo[3.2.0]hepta-2,6-diene (**44**) in a process that can be considered as an example of an ADPM reaction (Scheme 73.15).[28] The photochemical rearrangement of the dihydro-benzocarbazoles (**46**) to the indenoquinolines (**47**) takes place via an ADPM mechanism, as shown in Scheme 73.16.[29]

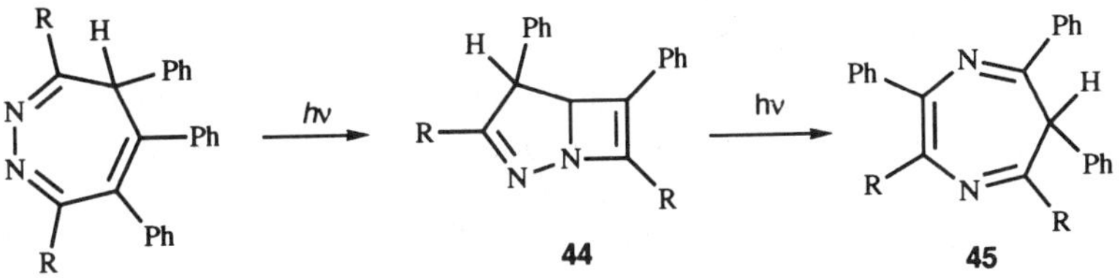

43 R = Me or Ph

44

45

SCHEME 73.15

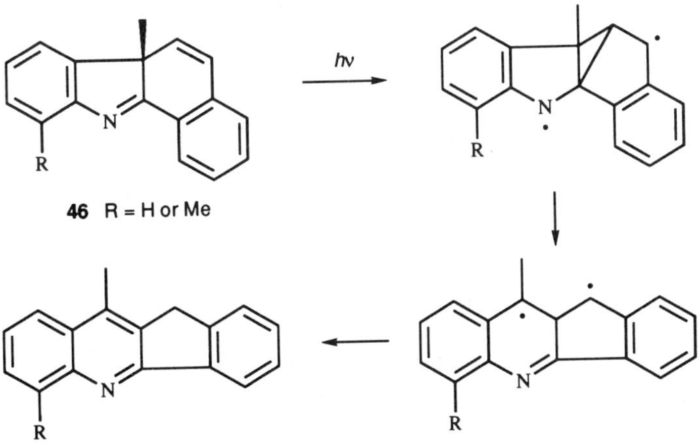

46 R = H or Me

47

SCHEME 73.16

The competition between the di-π-methane and the aza-di-π-methane rearrangements has been studied in the photochemistry of pyrazine derivatives of barrelene (48a–b). Irradiation of these brings about the formation of pyrazinosemibullvalenes in a typical di-π-methane process. Compound (48a) gives (49a) by a DPM path and (50a) and (51a) by an ADPM path. However, compound (48b) yields semibullvalenes (50b) and (51b) exclusively coming from an ADPM rearrangement. This result shows that the ADPM reaction could compete favorably with the di-π-methane rearrangement.[30] A similar study has been carried out using benzoquinoxalinobarrelenes that afford benzoquinoxalinosemibullvalenes from both the ADPM and the DPM paths.[31] Again, in these cases, the major product is formed by the ADPM rearrangement.

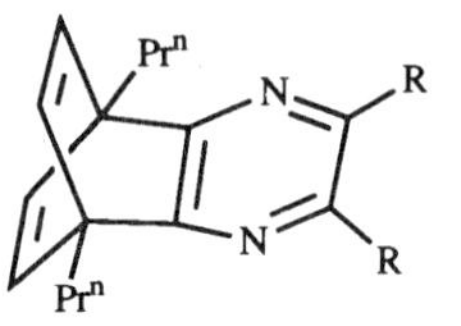

48a: R = H
48b: R = CN

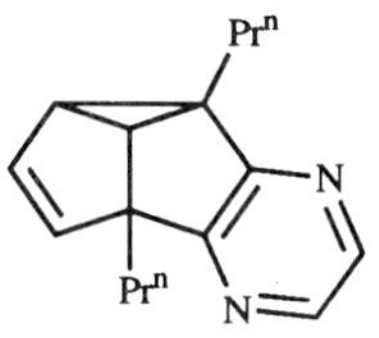

49a (53%)

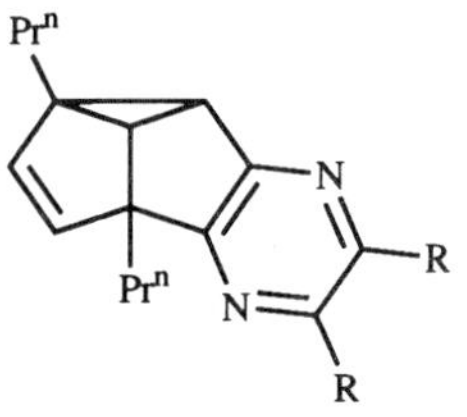

50a: R = H (25%)
50b: R = CN (57%)

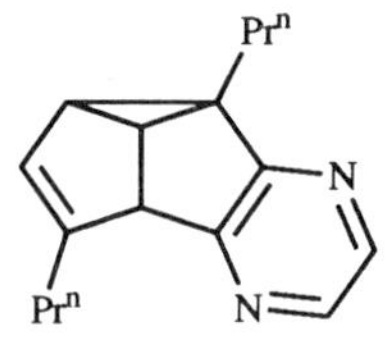

51a: R = H (22%)
51b: R = CN (43%)

The studies carried out thus far on the aza-di-π-methane rearrangement show that the reaction is very general for imines and other stable C-N double bond derivatives from β,γ-unsaturated aldehydes and ketones yielding cyclopropyl derivatives. In this rearrangement, the SET process is detrimental to the efficiency of the reaction; but if SET can be minimized, the ADPM process becomes efficient in both chemical and light-usage terms. From a synthetic point of view, the ADPM overcomes the usual decarbonylation path that is the normal photochemical behavior of β,γ-unsaturated aldehydes. This allows the transformation of these aldehydes and some ketones, those that do not undergo the oxa-di-π-rearrangement, into the corresponding cyclopropyl carbonyl compound. This synthetic route has been used in a novel and efficient path to molecules of commercial importance. However, recent studies in this field have shown that some β,γ-unsaturated aldehydes[32a] and oximes[32b] undergo efficient oxa- and aza-di-π-methane rearrangements, respectively. These results are in clear contrast to the general opinion that such compounds are inert in this reaction mode. These surprising observations will force further studies on the scope of these two novel reactions.

References

1. Zimmerman, H. E., The di-π-methane rearrangement, in *Organic Photochemistry*, Vol. 11, Padwa, A., Ed., Marcel Dekker, New York, 1991, 1; Demuth, M., Synthetic aspects of the oxa-di-π-methane rearrangement, in *Organic Photochemistry*, Vol 11, Padwa, A., Ed., Marcel Dekker, New York, 1991, 37.

2. (a) Armesto, D., Martin, J. F., Perez-Ossorio, R., and Horspool, W. M., A novel aza-di-π-methane rearrangement. The photoreaction of 4,4-dimethyl-1,6,6-triphenyl-2-azahexa-2,5-diene, *Tetrahedron Lett.*, 23, 2149, 1982; (b) Armesto, D., Horspool, W. M., Martin J. F., and Perez-Ossorio, R.,

The synthesis and photochemical reactivity of β,γ-unsaturated imines. An aza-di-π-methane rearrangement of 1-azapenta-1,4-dienes, *J. Chem. Res. (S)*, 46, 1986; *(M)*, 0631, 1986.

3. Nitta, M., Kasahara, I., and Kobayashi, T., Azadi-π-methane rearrangement involving an oxime group, *Bull. Chem. Soc. Jpn.*, 54, 1275, 1981.

4. Pratt, A. C., Photochemistry of β,γ-unsaturated carbonyl compounds. 3,3-Dimethyl-5,5-diphenylpent-4-en-2-one and 2,2-dimethyl-4,4-diphenylbut-3-enal, *J. Chem. Soc., Perkin Trans. 1*, 2496, 1973.

5. Pratt A. C., The photochemistry of imines, *Chem. Soc. Rev.*, 6, 63, 1977.

6. Sandorfy C., Spectroscopy of nonaromatic Schiff bases, *J. Photochem.*, 17, 297, 1981.

7. Arnold, D. R. and Maroulis, A. J., Radical ions in photochemistry. 4. The 1,1-diphenylethylene anion radical by photosensitization (electron transfer), *J. Am. Chem. Soc.*, 99, 7355, 1977.

8. Armesto, D., Horspool, W. M., Langa, F., and Perez-Ossorio, R., Substitution effects on the aza-di-π-methane rearrangement of imines, *J. Chem. Soc., Perkin Trans. 2*, 1039, 1987.

9. Armesto, D., Horspool, W. M., and Langa, F., The aza-di-π-methane rearrangement of 1-aryl-4,4-dimethyl-6,6-diphenyl-2-azahexa-2,5-dienes. The influence of substituents on the N-benzyl group, *J. Chem. Soc. Perkin Trans. 2*, 903, 1989.

10. Armesto, D., Horspool, W. M., Langa, F., Martin, J. F., and Perez-Ossorio, R., Studies on the scope of the aza-di-π-methane rearrangement of β,γ-unsaturated imines, *J. Chem. Soc., Perkin Trans. 1*, 743, 1987.

11. Pratt, A. C. and Q. Abdul-Majid, Photochemistry of the carbon-nitrogen double bond. 2. An investigation of the 3-methylenepropan-1-imine and 3-oxopropan-1-imine chromophores, *J. Chem. Soc., Perkin Trans. 1*, 359, 1987.

12. Armesto, D., Horspool, W. M., and Langa, F., The aza-di-π-methane rearrangement of O-acetyl-2,2-dimethyl-4,4-diphenylbut-3-enal oxime, *J. Chem. Soc., Chem. Commun.*, 1874, 1987.

13. Armesto, D., Horspool, W. M., Langa, F., and Ramos, A., Extension of the aza-di-π-methane reaction to stable derivatives. Photochemical cyclization of β,γ-unsaturated oxime acetates, *J. Chem. Soc., Perkin Trans. 1*, 223, 1991.

14. Baeckstrom, P., Photochemical formation of chrysanthemic acid and cyclopropylacrylic acid derivatives, *Tetrahedron*, 34, 3331, 1978; Baeckstrom, P., Multiplicity dependence of the di-π-methane photochemistry. Regiospecific formation of tetramethylvinylciclopropanes and chrysanthemic acid derivatives, *J. Chem. Soc., Chem. Commun.*, 476, 1976.

15. Bullivant, M. J. and Pattenden, G., A photochemical di-π-methane rearrangement leading to methyl chrysanthemate, *J. Chem. Soc., Perkin Trans. 1*, 256, 1976.

16. Zimmerman, H. E., Boettcher, R. J., and Braig, W., Accentuation of the di-π-methane reactivity by central carbon substitution. Mechanistic and exploratory organic photochemistry. LXXV, *J. Am. Chem. Soc.*, 95, 2155, 1973.

17. (a) Armesto, D., Horspool, W. M., Mancheño, M. J., and Ortiz, M. J., The aza-di-π-methane rearrangement of stable derivatives of 2,2-dimethyl-4,4-diphenylbut-3-enal, *J. Chem. Soc., Perkin Trans. 1*, 2348, 1990; (b) Armesto, D., Horspool, W. M., Mancheño, M. J., and Ortiz, M. J., Chemically efficient aza-di-π-methane photoreactivity with novel stable derivatives of β,γ-unsaturated carbonyl compounds, *J. Chem. Soc., Perkin Trans. 1*, 2325, 1992.

18. Armesto, D., Barnes, J. C., Horspool, W. M., and Langa, F., Intramolecular electron transfer in the novel photoreaction of some β,γ-unsaturated oxime-boron trifluoride complexes. A new synthetic path to dihydroisoxazoles, *J. Chem. Soc., Chem. Commun.*, 123, 1990.

19. Armesto, D., Agarrabeitia, A. R., Horspool, W. M., and Gallego, M. G., Unexpected influence of mono-phenyl substitution on the photochemistry of β,γ-unsaturated oxime acetates, *J. Chem. Soc., Chem. Commun.*, 934, 1990.

20. Zimmerman, H. E., Kamm, K. S., and Werthemann, D. W., Single photon counting and magic multipliers in direct measurements of single state di-π-methane rearrangement rates in the picosecond range. Mechanistic organic photochemistry. LXXXIII, *J. Am. Chem. Soc.*, 96, 439, 1974.

21. Armesto, D., Gallego, M. G., and Horspool, W. M., The photochemical synthesis of potential pyrethroid components by the aza-di-π-methane rearrangement of β,γ-unsaturated oxime acetates, *Tetrahedron Lett.,* 31, 2475, 1990.

22. Armesto, D., Gallego, M. G., and Horspool, W. M., Photochemistry of β,γ-unsaturated oxime acetates. Aza-di-π-methane reactivity of functionalised all-aliphatic systems. A photochemical approach to pyrethrin-like cyclopropane derivatives, *Tetrahedron,* 46, 6185, 1990.

23. Armesto, D., Horspool, W. M., Gallego, M. G., and Agarrabeitia, A. R., Steric and electronic effects on the photochemical reactivity of oxime acetates of β,γ-unsaturated aldehydes, *J. Chem. Soc., Perkin Trans. 1,* 163, 1992.

24. Armesto, D., Gallego, M. G., Horspool, W. M., and Bermejo, F., ES. P. 9100648, 13th March 1991; 92/16499, PCT/ES92/00017, 13th February 1992.

25. Naumann, K., *Chemistry of Plant Protection, Synthetic Pyrethroid Insecticides: Structure and Properties,* Vol. 4, Springer-Verlag, Berlin 1990.

26. Armesto, D. and Ramos, A., Photochemical synthesis of oxime acetates derivatives of 1-carbaldehydrobicyclo [n.1.0] alkanes by the aza-di-π,-methane rearrangement. *Tetrahedron,* 49, 7159, 1993.

27. Engel, P. S. and Schexnayder, M. A., Systematic structural modifications in the photochemistry of β,γ-unsaturated ketones. I. Cyclic olefins, *J. Am. Chem. Soc.,* 97, 145, 1975.

28. Demlehner, U. and Sauer, J., Aza-di-π-methan-umlagerungen?, *Tetrahedron Lett.,* 25, 5627, 1984; Reissenweber, G. and Sauer, J., Aza-di-π-methan-umlagerungen, *Tetrahedron Lett.,* 18, 4389, 1977.

29. Kulagowski, J. J., Mitchell, G., Moody, C. J., and Rees, C. W., Preparation and rearrangement of 6a-methyl-6aH-benzo[a]carbazole and 11b-methyl-11bH-benzo[c]carbazole, *J. Chem. Soc., Chem. Commun.,* 650, 1985.

30. Liao, C. C., Hsieh, H. P., and Lin, S. Y., Photorearrangement of some pyrazinobarrelenes, *J. Chem. Soc., Chem. Commun.,* 545, 1990.

31. Liao, C. C. and Yang, P. H., Photochemistry of benzoquinoxalino-barrelenes, *Tetrahedron Lett.,* 33, 5521, 1992.

32. (a) Armesto, D., Ortiz, M. J., and Romano, S., The oxa-di-π-methane rearrangement of β,γ-unsaturated aldehydes, unpublished results. (b) Armesto, D., Ramos, A., and Mayoral E. P., The aza-di-π-methane rearrangement of β,γ-unsaturated oximes, *Tetrahedron Lett.,* 35, 3785, 1994; Armesto, D., Ortiz, M. J., Ramos, A., Horspool, W. M., and Mayoral E. P., A study of the competition between the di-π-methane and the aza-di-π-methane rearrangement in 2-vinyl-β,γ-unsaturated oxime derivatives. The novel aza-di-π-methane reactivity of β,γ-unsaturated oximes, *J. Org. Chem.,* in press.

lix Müller
Goldschmidt AG

chen Mattay
ganisch-Chemisches Institut
der Universität Münster

Azirine Photochemistry

The 2*[H]*-azirine ring is, as are all other small ring heterocycles, a very useful synthon, especially for the construction of larger ring heterocycles.[1] The synthesis of 2*[H]*-azirines with a diverse choice of substituents is fairly simple, and there are many classic and modern synthetic pathways available.[1,2,3] Photochemists have focused their attention on this reaction and on subsequent additions to dipolarophiles over the last 2 decades.[4-6]

Irradiation with light of 280-nm wavelength into the n,π* band of an aryl-substituted azirine 1 leads to the opening of the ring and the formation of the nitrile ylide 2 (Scheme 74.1, path A), that can be added in a high yield to dipolarophiles such as acrylonitrile.[7] The 1,3-dipolar cycloaddition reacts by a concerted pathway and the diastereoselectivity is high.

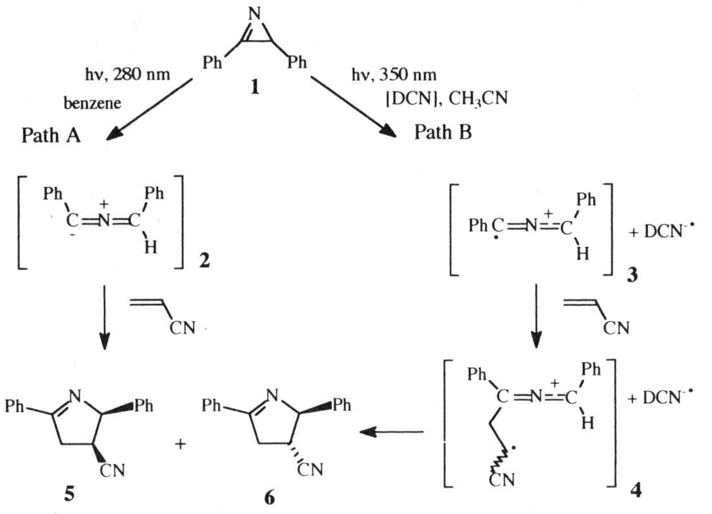

Path A: **5**: 10 %, **6**: 90 %

Path B: **5**: 50 %, **6**: 50 %

SCHEME 74.1   Mechanisms of photoinduced cycloadditions of azirines.

Very early on in the study of such systems, the mechanism was substantiated by low-temperature UV spectroscopy showing the presence of the ylide 2, which absorbs at 350 nm.[8] Laser spectroscopy allowed the measurement of lifetimes of ylides like 2 and these were found to be in the range of 10^{-3} s. The rate constant for the reaction of 2 with acrylonitrile has been determined to be k = 1.2×10^{6} M^{-1} s^{-1}.[9,10]

Fairly recently, a new way of photochemical azirine ring opening has been developed.[11,12] In a polar solvent such as acetonitrile, an electron acceptor [1,4-naphthalene dicarbonitrile (DCN)] is irradiated with light of 350-nm wavelength. The excited DCN accepts an electron from the azirine

0-8493-8634-9/95/$0.00+$.50

1, which opens spontaneously to the 2-azaallenyl radical cation 3 (Scheme 1, path B). The reaction with acrylonitrile leads to the same products as obtained under the conditions of direct irradiation, but the selectivity is nonexistent.[12] The cycloaddition involving a photoelectron transfer process (PET) is not concerted and it is possible to trap another intermediate such as 4. The spectroscopic investigation of the reaction has revealed the UV absorbance of 3 to be at 485 nm. The lifetime of 3 is 1.43 μs; the rate constant for a typical reaction has been determined as $7.8 \times 10^9\ M^{-1}\ s^{-1}$.[13] The different data obtained for both reactions show that the direct irradiation gives good selectivity but poor reactivity; whereas in the PET case, the opposite behavior is observed.

With this mechanistic knowledge, both reactions have been used extensively to synthesize five-membered heterocycles. Padwa first established the reaction of 2 with different olefins (See Scheme 74.2).[14]

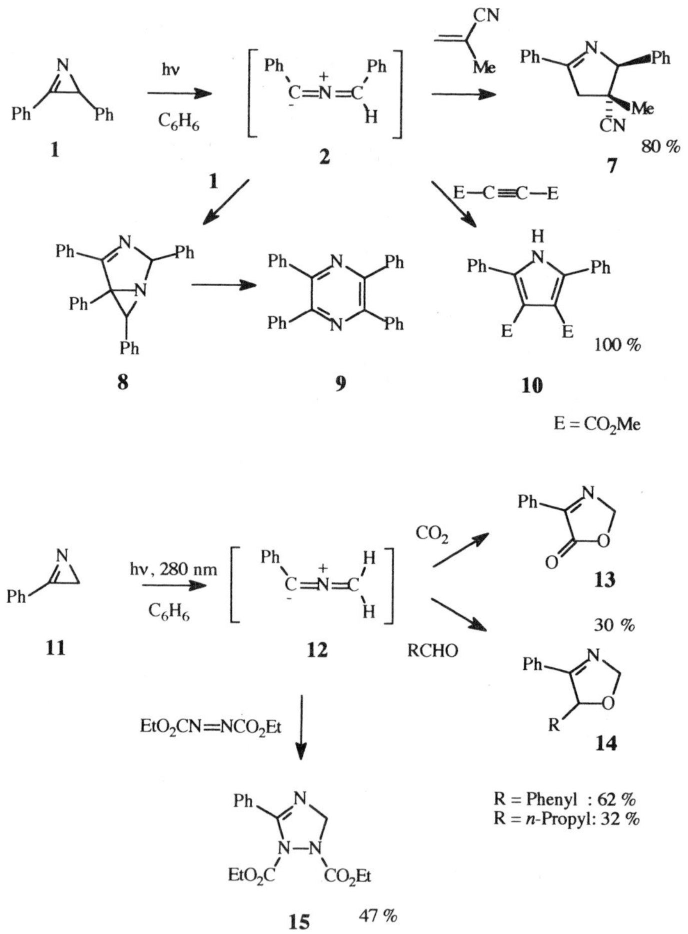

SCHEME 74.2   1,3-Dipolar cycloadditions of azirines by means of direct irradiation.

With electron-deficient olefins, pyrrolidines such as 5, 6, and 7 are formed. Electron-rich olefins do not react with the ylide 2. Under these conditions, the ylide reacts with another azirine molecule to form the bicycle 8, which on prolonged irradiation forms the pyrazine 9. These same compounds, 8 and 9, are formed if azirine 1 is irradiated in the absence of dipolarophile. The synthesis of pyrroles is possible with acceptor-substituted alkynes and compounds like 10 are formed in excellent yields.

Schmid and, later, Heimgartner investigated the reaction of ylides like 12 with carbonyl compounds.[14,15] With carbon dioxide, the oxazolinone 13 is formed regioselectively. The addition of aldehydes to 12 leads to oxazolines 14 in variable yields. 1,2,4-Triazolines such as 15 are formed by the addition of the ylide 12 to diazo compounds. The yields are reasonable.[16] This reaction has been demonstrated recently for aminoazirines.[17]

The scope of five-membered heterocycles that can be prepared by this procedure is quite large. Nevertheless, an important limitation is the requirement of the aryl substituent in the 3-position at the 2[H]-azirine ring, which shifts the n,π*-band to the red and allows irradiation at a wavelength at which products are not destroyed during irradiation. In addition, even some dipolarophiles, as imines, are not attacked at all by ylides like 2 and 12.[6]

The limitations of the 1,3-dipolar cycloaddition have been overcome by means of PET. For example, the 2-azaallenyl radical cation reacts with imines to form imidazoles 18[12,18] (Scheme 74.3).

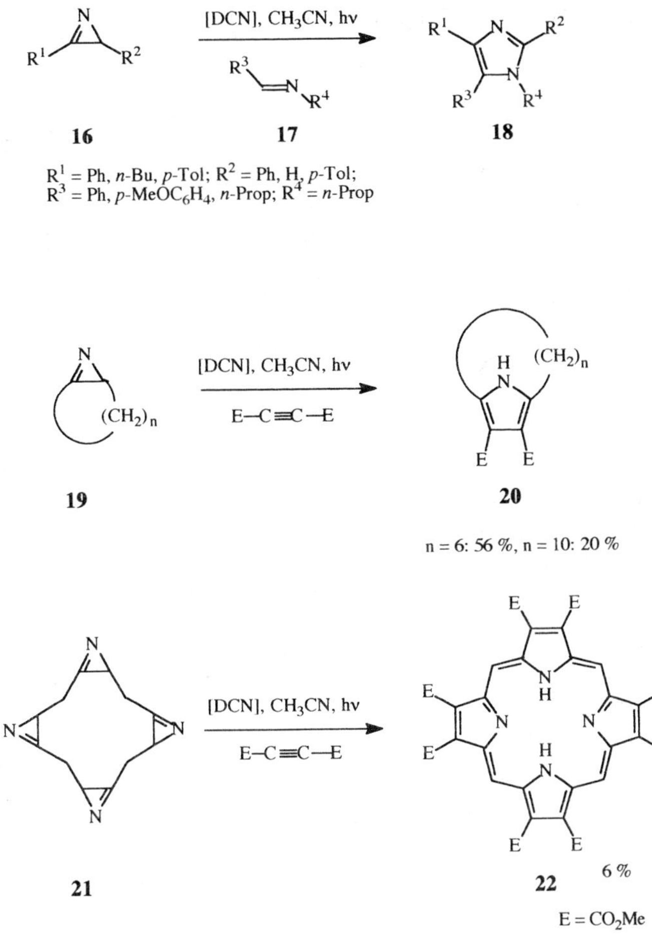

R^1 = Ph, *n*-Bu, *p*-Tol; R^2 = Ph, H, *p*-Tol;
R^3 = Ph, *p*-MeOC$_6$H$_4$, *n*-Prop; R^4 = *n*-Prop

n = 6: 56 %, n = 10: 20 %

6 %

E = CO$_2$Me

SCHEME 74.3 Reactions of azirines under the conditions of photoinduced electron transfer.

The yields of product depend on the substitution pattern of the starting materials 16 and 17. If R^1 to R^3 are phenyl substituents, the yield of imidazole is 87%; but for an imidazole with R^1 and R^3 as alkyl substituents and R^2 = hydrogen, the yield drops to 40%. As side products, bicyclic compounds analogous to 8 were identified.

As the last mentioned example indicates, alkyl-substituted azirines react as well under PET conditions to give the same product type as obtained from aryl-substituted azirines upon direct irradiation. Bicyclic azirines 19 can also be added to electron-poor alkynes to form pyrrolophanes 20 under PET conditions. Depending on the length of the methylene chain, the yields vary from

20% (n = 10) to 56% (n = 6).[19] Imidazolophanes can be prepared using this procedure, but the yields are not encouraging.

Although it was not possible yet to build up (2,2)pyrrolophanes or (2,2,2)pyrrolophanes, the reaction has been used to prepare a porphyrin 22 from a tetrakisazirine 21 in small amounts.[20] This was the first synthesis of a porphyrin system without a pyrrole derivative as starting material.

Thus, with the enhancement of reactivity through PET control, the photocycloadditions of azirines are an even more attractive approach to five-membered heterocycles.

Selected characteristic data for compounds mentioned in this chapter. Numbers refer to schemes.

DCN
UV (acetonitrile): λ_{max} = 239 nm (log ε = 4.503), 312 (3.950), 332 (3.879);
$E_{1/2}$red = −1.67 V vs. Ag/AgNO$_3$ (0.1 M) in acetonitrile.

1 Mp. 61 °C
^{1}H-NMR (300 MHz, CDCl$_3$): δ = 3.33 (s, 1H, CHPh), 7.1–7.9 (m, 10H, aromat. -H);-
^{13}C-NMR (75.5 MHz, CDCl$_3$): δ = 34.65 (C2), 124.60–141.11 (aromat. C), 163.85 (C3);-
MS (70 eV) m/z (%) = 193 (M$^+$, 100), 192 (21), 165 (22), 89 (60);-
IR (KBr): ν (C=N) = 1740 cm^{-1};-
UV (acetonitrile): λ_{max} = 245 nm (log ε = 4.182), 277 (3.176), 285 (3.017).

2 UV (cyclohexane): λ_{max} = 350 nm

3 UV (*n*-butyl chloride): λ_{max} = 485 nm

18 Mp. 92 °C.
^{1}H-NMR (300 MHz, CDCl$_3$): δ = 1.05 (tt ^{3}J = 7.3, ^{4}J = 2.1 Hz, 3H, CH$_3$), 1.45 (tq ^{3}J = 7.3, 7.7 Hz, 2H, CH$_2$), 3.95 (tq ^{3}J = 7.7, ^{4}J = 2.1 Hz, 2H, N-CH$_2$), 7.0–7.9 (m, 15 H,aromat. -H);-
^{13}C-NMR (75.5 MHz, CDCl$_3$): δ = 10.62 (CH$_3$), 23.64 (CH$_2$CH$_3$), 46.15 (N-CH$_2$), 134.42 (C5), 137.50 (C4), 147.49 (C2), 126.04–137.01 (aromat. C);-
MS (70 eV) m/z (%) = 338 (M$^+$, 100), 295 (28), 193 (8), 165 (34), 89 (38);-
IR (KBr): ν = 3000 cm^{-1}, 1590, 1550, 1480, 1435, 1385, 740, 680;-
UV (acetonitrile) λ_{max} = 232 nm (log ε = 4.519), 255 (3.146), 280 (3.230).

19 (n = 6) Bp. 76 °C / 20 Torr
^{1}H-NMR (300 MHz, CDCl$_3$): δ = 1.2–1.8 (m, 8 H, H-C3 - H-C6), 2.01 (m, 2 H, H-C7), 2.56 (m, 2 H, H-C2), 2.80 (m, 1 H, H-C8);-
^{13}C-NMR (75.5 MHz, CDCl$_3$): δ = 19.77, 20.07, 21.04, 21.18, 25.66, 25.80 (C2-C7), 29.01 (C8), 152.53 (Cl);-
MS (70 eV) m/z (%) = 123 (M$^+$, 6), 122 (56), 108 (10), 95 (36), 94 (52), 81 (32), 80 (78), 54 (76), 41 (100);-
IR (NaCl); ν (C=N) = 1765 cm^{-1}.

20 (n = 6) brown oil.
^{1}H-NMR (300 MHz, CDCl$_3$): δ = 1.3–1.5 (m, 8H, 4 CH$_2$), 2.8 (m, 4H, CH$_2$-C2), 3.50 (s, 6H, OCH$_3$);-
^{13}C-NMR (75.5 MHz, CDCl$_3$): δ = 19.48, 24.38, 32.65 (6 CH$_2$), 52.05 (2 OCH$_3$), 111.52 (C3), 116.51 (C2), 167.25 (C=O);-
MS (70 eV): m/z 265 (M$^+$, 6%), 234 (39), 216 (40), 206 (47), 174 (18), 166 (24), 159 (20), 148 (16), 147 (18), 122 (8), 94 (10), 91 (21), 79 (23), 59 (100);-
IR (cap.): ν = 3440 cm^{-1}, 3100, 3000, 1845, 1740, 1535;-
UV (acetonitrile): λ_{max} = 260 nm (log ε = 3.305), 196 (4.812).

21 Mp. 128 °C
^{1}H-NMR (300 MHz, CDCl$_3$): δ = 2.45 (d, 8H, ^{3}J = 6.8 Hz, CH$_2$), 2.91 (t, 4H, ^{3}J = 6.8 Hz, CH);-
^{13}C-NMR (75.5 MHz, CDCl$_3$): δ = 25.44 (CH$_2$), 30.98 (CH), 154.06 (C=N);-
MS (70 eV): m/z 212 (M$^+$, 0.7%), 198 (1), 170 (41), 156 (38), 107 (24), 91 (100);-
IR (KBr): ν = 2950 cm^{-1}, 1770, 1545, 1165, 1005;-
UV: (acetonitrile) λ$_{max}$ = 235 nm (log ε = 3.65).

22 Mp. 298 °C.
^{1}H-NMR (300 MHz, CDCl$_3$): δ = –3.40 (s, 2H, NH), 3.58 (s, 24H, OCH$_3$), 10.78 (s, 4H, CH);-
^{13}C-NMR (75.5 MHz, CDCl$_3$): δ = 53.30 (8 OCH$_3$), 107.61 (4 CH), 136.94 (8 C3), 139.10 (8 C2),
152.03 (8 C=O);-
MS (70 eV): m/z 774 (M$^+$, 1.5%), 745 (2), 605 (3), 387 (15), 203 (66), 118 (100);-
IR (KBr): ν = 3420 cm^{-1}, 2980, 1725, 1440, 1235, 1036;-
UV: (CH$_3$OH) λ$_{max}$ = 491 nm (log ε = 3.641), 409 (4.916), 332 (4.410), 205 (5.335).

References

1. Nair, V., Azirines, in *Small Ring Heterocycles*, Part 1, Hassner, A., Ed., Wiley, New York, 1983, 217.
2. Heimgartner, H., 3-Amino-2H-azirine, Bausteine für α,α-disubstituierte α-Aminosäuren in Heterocyclen- und Peptidsynthesen, *Angew. Chem. Int. Ed.*, 30, 238, 1991; *Angew. Chem.*, 103, 271, 1991.
3. Bannert, K., Azidobutatrien und Azidobutenine, *Chem. Ber.*, 122, 1175, 1989; Bannert, K., Synthese, Strukturzuordnung, Photolyse und Thermolyse von 2,3-Diazido-1,3-butadienen, *Chem. Ber.*, 120, 1891, 1987.
4. Padwa, A., Azirine photolysis and cycloaddition reactions, in *Synthetic Organic Photochemistry*, Horsepool, W., Ed., Plenum, New York, 1984, 313.
5. Hansen, H.-J. and Heimgartner, H., Nitriles ylides, in *1,3-Dipolar Chemistry*, Vol. I, Padwa, A., Ed., Wiley, New York, 1984, 177.
6. Gerber, U., Heimgartner, H., Schmid, H., and Hansen, H.-J., A review on the photochemistry of 2H-Azirines, *Heterocycles*, 6, 143, 1977.
7. Padwa, A., Dharan, M., Smolanoff, J., and Wetmore, S. I., Jr., Observation on the scope of the photoinduced 1,3-dipolar addition reactions of arylazirines, *J. Am. Chem. Soc.*, 95, 1945, 1973; Padwa, A., Dharan, M., Smolanoff, J., and Wetmore, S. I., Jr., Photochemical transformations of small ring heterocyclic compounds. XLVII. Electronic details of the photocycloaddition of arylazirines, *J. Am. Chem. Soc.*, 95, 1954, 1973.
8. Sieber, W., Gilgen, P., Chaloupka, S., Hansen, H. J., Schmid, H., Tieftemperatur Bestrahlungen von 3-Phenyl-2H-azirinen, *Helv. Chim. Acta*, 56, 1679, 1973.
9. Padwa, A., Rosenthal, R. J., Dent, W., Filho, P., Turro, N. J., Hrovat, D. A., and Gould, I. R., Steady-state and laser photolysis studies of substituted 2H-azirines. Spectroscopy, absolute rates, and Arrhenius behavior for the reaction of nitrile ylides with electron deficient olefins, *J. Org. Chem.*, 49, 3174, 1984.
10. Naito, I., Morihara, H., Ishida, A., Takamuku, S., Isomura, K., and Taniguchi, H., Photochemistry of 2H-azirine-formation of nitrile ylide evidenced by laser flash photolysis and pulse radiolysis, *Bull. Chem. Jpn.*, 64, 2757, 1991.
11. Müller, F. and Mattay, J., Photocycloadditions. Control by energy and electron transfer, *Chem. Rev.*, 93, 99, 1993.
12. Müller, F. and Mattay, J., [3+2] Cycloadditions with azirine radical cation: A new synthesis of N-substituted imidazoles, *Angew. Chem. Int. Ed.*, 30, 1336, 1991; *Angew. Chem.*, 103, 1352, 1991.
13. Müller, F., Mattay, J., and Steenken, S., Radical Cation [3+2] Cycloadditions of 2H-azirines. Mechanistic studies concerning the intermediate radical cation, *J. Org. Chem.*, 58, 4462, 1993.

14. Giesendanner, H., Heimgartner, H., Jackson, B., Winkler, T., Hansen, H.-J., and Schmid, H., Photochemische Cycloadditionen von 3-Phenyl-2H-azirinen mit Aldehyden, *Helv. Chim. Acta*, 56, 2611, 1973.

15. Dietliker, K. and Heimgartner, H., Photochemisch induzierte Reaktionen von 3-Amino-2H-azirinen, *Helv. Chim. Acta*, 66, 262, 1983.

16. Gilgen, P., Heimgartner, H., and Schmid, H., Photoinduzierte 1,3-dipolare Cycloadditionen von 3-Phenyl-2H-azirinen an Azodicarbonsäurediethylester, *Helv. Chim. Acta*, 57, 1382, 1974.

17. Villalgordo, J. M. and Heimgartner, H., Reaction of diphenyl phosphorochloridate with amide enolates: A new convenient synthesis of 2-monosubstituted 3-(*N*-Methyl-*N*-phenylamine)-2H-azirines, *Helv. Chim. Acta*, 75, 1866, 1992.

18. Müller, F. and Mattay, J., [3+2] Cycloadditions with azirines under the conditions of photoinduced electron transfer. A new method for the synthesis of imidazoles and heterophanes, *Chem. Ber.*, 126, 543, 1993.

19. Müller, F. and Mattay, J., A new synthesis for imidazolo- and pyrrolophanes by [3+2] cycloaddition with azaallenyl radical cations, *Angew. Chem. Int. Ed.*, 31, 209, 1992; *Angew. Chem.*, 104, 207, 1992.

20. Müller, F., Karwe, A., and Mattay, J., A new synthesis for porphyrin systems by four sequential [3+2] cycloadditions with azaallenyl radical cations, *J. Org. Chem.*, 57, 6080, 1992.

75

Photochemical Decomposition of Cyclic Azoalkanes

aldemar Adam
iversity of Wurzburg

oskun Sahin
iversity of Wurzburg

75.1 Introduction

The term "azoalkanes" is used extensively and refers to substances in which nitrogen is bonded to carbon; heteroatoms may be present elsewhere in the azoalkane structure. This chapter deals with cyclic azoalkanes because of the notable difference between their dissociative behavior from that of the acyclic derivatives; however, the latter may provide clues for uncovering key structural and energetic factors governing dissociation of azoalkanes. In the case of cyclic azoalkanes, the loss of molecular nitrogen provides an efficient method for producing biradicals, which are currently of great interest.[1]

Although azoalkanes have been known since 1909,[2] their rich chemistry through thermal or photochemical extrusion of nitrogen to form highly ring-strained, sterically crowded, fluxional, antiaromatic, and other unusual molecules has become apparent during the last 3 decades.[3] The importance of this synthetic approach is that the azo linkage serves as a precursor for introducing the critical (usually last) bond in a complex target molecule. Besides their synthetic potential for unusual and novel structures, azoalkanes constitute a convenient source of radicals and biradicals of nearly any desired structure. The available synthetic methodology, the improvements in analytical techniques, the development of efficient spectroscopic methods, and the availability of good commercial UV light sources all have contributed to enrich our knowledge of the physical properties and chemical processes of such short-lived, reactive intermediates.

Most azoalkanes exhibit a weak n,π^* band in the 300- to 400-nm region with the extinction coefficients for the *cis*-isomers about 10 times higher than that for the *trans*-isomers. Although the n,π^* transition is much weaker than the π,π^* transition ($\varepsilon \approx 10^4$ to 10^5), the former is suitable for the photoextrusion of molecular nitrogen from azoalkanes. Since the excitation energies for the triplet states (E_T) are ca. 50–60 kcal mol^{-1} triplet sensitization can be performed effectively with the usual carbonyl sensitizers (acetone, benzophenone, benzil, etc.).

0-8493-8634-9/95/$0.00+$.50
© 1995 by CRC Press, Inc.

Upon thermal or photochemical excitation, the majority of azoalkanes lose molecular nitrogen; only a few undergo transformations retaining the nitrogen. Both categories are, however, of synthetic utility and interest. Most frustrating is the third category of azoalkanes, namely the "reluctant" azoalkanes, which are essentially inert to denitrogenation, especially during photolysis. The term "reluctant" signifies that the quantum yields for denitrogenation are less than 5%. Thus, an elegant synthetic sequence may falter at its last step, when attempts to introduce the last carbon-carbon bond by nitrogen extrusion are thwarted by the resistance of the azoalkane to denitrogenation. Clearly, the electronically excited, "reluctant" azoalkane possesses an activation energy (E_a ca. 2–10 kcal mol^{-1}) toward nitrogen elimination. Consequently, denitrogenation no longer takes place on the lowest-energy path and photophysical or photochemical energy dissipation (e.g., fluorescence, radiationless decay, *Z,E*-isomerization, intramolecular cyclization, or phototautomerization) can become competitive.[4]

75.2 Mechanisms for Decomposition of Azoalkanes

The question of one-bond vs. two-bond cleavage has been one of the persistent controversial mechanistic issues ever since the denitrogenation of azoalkanes has been investigated. However, in the last few years, consensus of opinion has been emerging both on the theoretical and experimental fronts, which supports the stepwise elimination of N_2.

Thermal Decomposition

To distinguish between mechanisms a and b (Scheme 75.1), the E_a values for the decomposition of alkyl-substituted acyclic azoalkanes were compared as early as 1929.[5] Since the E_a values for the unsymmetrical derivatives fell between that of the two symmetrical ones, it was concluded that both C-N bonds break at once. Since this early work, much effort has been expended in trying to decide between mechanisms a and b for azoalkane thermolysis.[6]

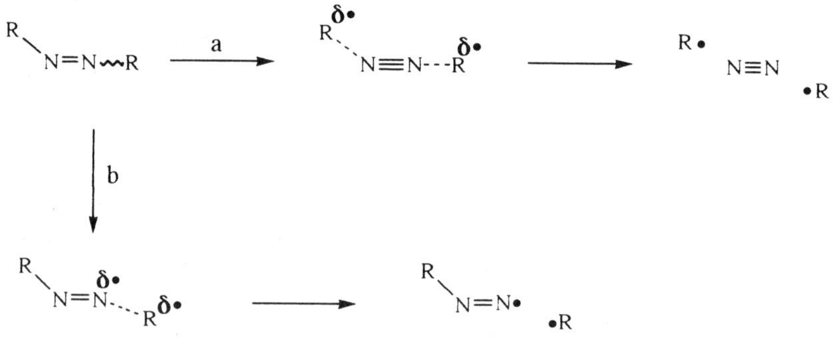

SCHEME 75.1

Thus, thermochemical kinetic studies as well as a variety of other techniques have been employed to distinguish one-bond from two-bond cleavage in azoalkanes. Isomerization of the highly unsymmetrical acyclic azoalkane 1 (Scheme 75.2) was observed during thermolysis, which suggests one-bond rupture.[7] Other isomerizations were reported for unsaturated bi- and tricyclic azoalkanes; e.g., the thermal isomerization of the azoalkane 3 (Scheme 75.3).[8,9] Recently, one-bond cleavage was also shown to be the preferred route in the thermolysis of simple, saturated, bicyclic azoalkanes.[10,11] In summary, thermal decomposition seems to proceed mainly by a stepwise cleavage of the C-N bonds; the more unsymmetrical the azo compound, the more unsymmetrically it cleaves.

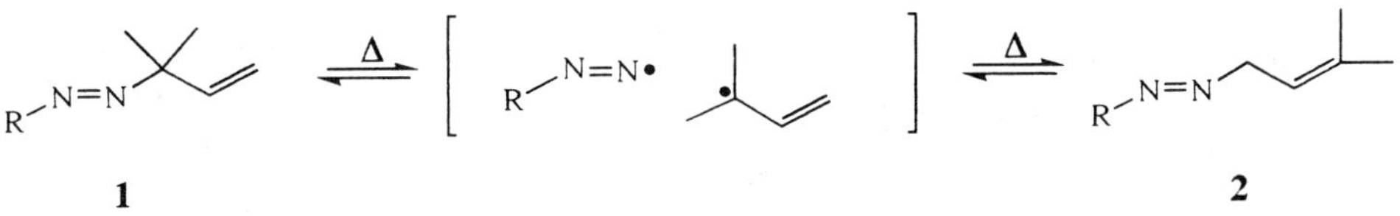

SCHEME 75.2

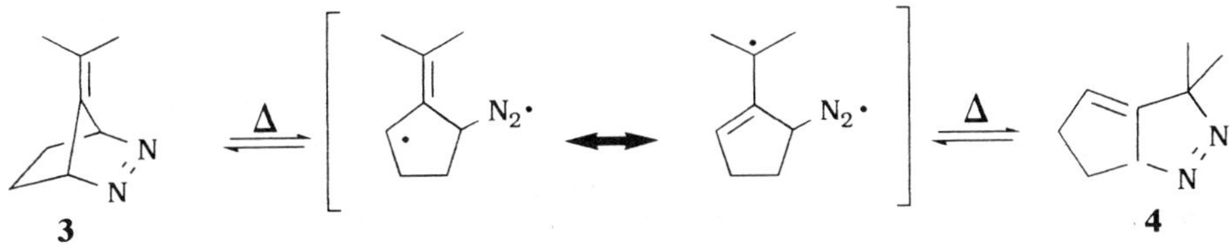

SCHEME 75.3

Photochemical Decomposition

Triplet biradicals, which result upon sensitized photolysis of bicyclic azoalkanes, can be trapped efficiently with radical scavengers (e.g., molecular oxygen[12] or nitroxides[13]); whereas, in the direct photolysis, usually no trapping products have been observed. In view of this spin-state effect,[2d,14] a clear differentiation between direct (^{1}n,π^*) and triplet-sensitized (^{3}n,π^*) excitation should be exercised.

Direct Photolysis

As in the thermal and also in the photochemical case the question arises if both C-N bonds cleave simultaneously or stepwise. Although quantum mechanical calculations favor stepwise homolysis in the photolytic decay of *cis*-diimide,[15] it has been verified experimentally only recently that this is also a feasible process in the direct photochemical decomposition of azoalkanes.[9,16,17] Thus, an intermediary diazoalkane 6 could be detected by IR and UV spectroscopy during denitrogenation of the triyclic azoalkane 5 (Scheme 75.4) and its further photolysis led to diene 7 through the corresponding carbene. The intervention of diazenyl biradicals has been postulated during direct photolysis and indeed has been observed to have lifetimes of between 5 to 20 ns.[18a,b] Furthermore, it was reported that vapor-phase photolysis of azoalkane 8 (Scheme 75.5) in a helium atmosphere led to triplet diazenyl biradicals with a lifetime of about 25 ns.[18c]

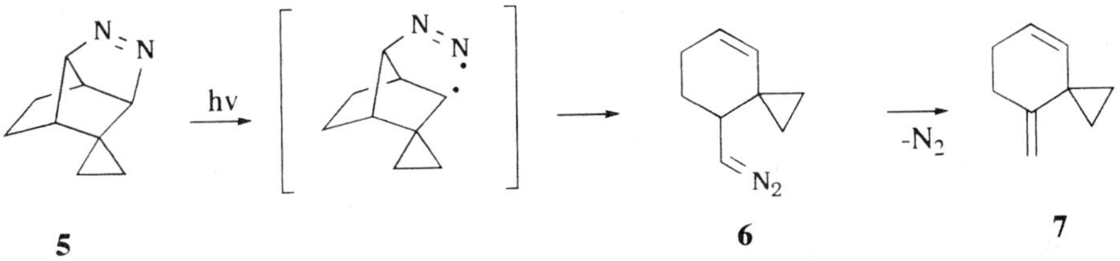

SCHEME 75.4

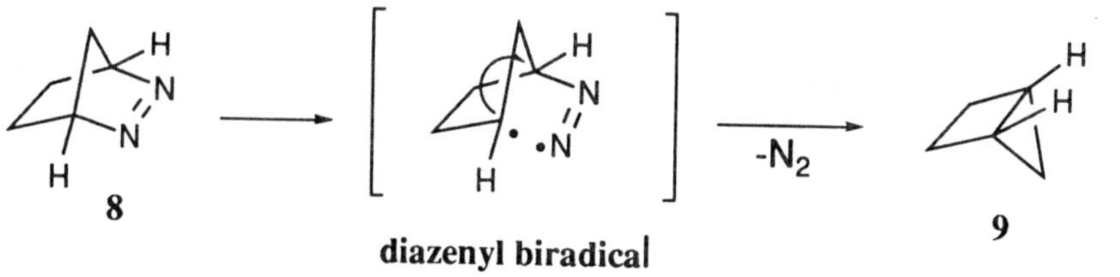

diazenyl biradical

SCHEME 75.5

Triplet-Sensitized Photolysis

Only a few examples exist in which triplet sensitization of azoalkanes in solution allows the observation of products derived from intermediary diazenyl biradicals, e.g., the retrocleavage of **10** (Scheme 75.6).[19] Oxygen-induced intersystem crossing from the triplet to the singlet state of the diazenyl biradical may be responsible in this special case. Usually, due to fast nitrogen loss, triplet diazenyl biradicals generated from bicyclic azoalkanes are too short-lived to affect the product ratios![1b]

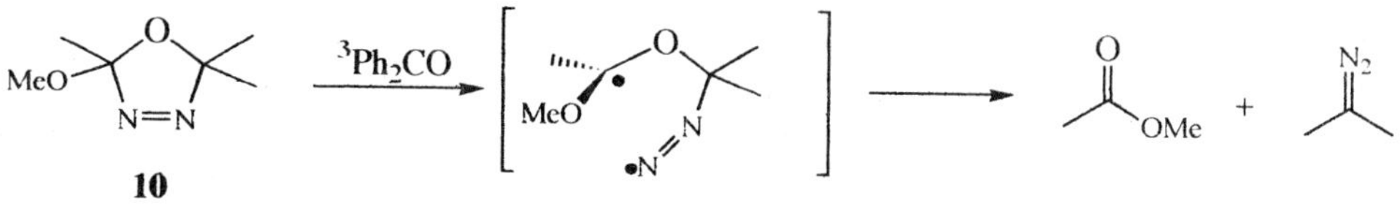

SCHEME 75.6

Stereoselectivity of the Denitrogenation Process

Thermal activation, as well as direct photochemical excitation of stereolabeled bicyclic azoalkanes, leads mainly to the double inverted denitrogenated products, as illustrated for azoalkane **8** (Scheme 75.5).[20] Nevertheless, few examples exist in which quantitative retention[21] or double inversion[22] takes place. The original S_H2 mechanism,[20a] in which the diazenyl biradical loses N_2 by backside attack of the carbon radical site on the nitrogen-bearing carbon atom, seems to account best for most of the experimental facts (Scheme 75.5). On the other hand, triplet-sensitized photolysis of such azoalkanes generates a planar triplet biradical that leads to completely stereorandomized products.[23] The influence of the reaction conditions on the product distribution is displayed in Table 75.1 for the stereolabeled azoalkane *exo*-d$_2$-**8**.

Photosensitized Electron Transfer-Induced Decomposition

A relatively new aspect of azoalkane chemistry is the induced denitrogenation by photosensitized single electron transfer (SET). The formation of chlorinated hydrocarbons in the direct photolysis of azoalkane **12** (Scheme 75.7) in CCl_4 was one of the first examples for SET reaction of azoalkanes.[24] Later, SET reactions were also observed in the photolysis of analogous azoalkanes in $BrCCl_3$.[25] In this case, the azoalkanes serve as suitable electron donors in SET reactions. There is a considerable body of evidence that indicates that the transient radical cations of azoalkanes can undergo fragmentation, rearrangement, or addition reactions. A particularly effective method to sensitize photochemical single electron transfer reactions of azoalkanes is to employ 2,4,6-triphenylpyrylium tetrafluoroborate ($TPP^+BF_4^-$) as sensitizer.[26] Recently, a remarkable stereochemical memory effect was observed in the SET reactions of stereolabeled azoalkanes and bicyclopentanes.[27]

Table 75.1 Product Studies of the Photolyses and Pyrolyses of Azoalkane *exo*-d$_2$-8[a]

| | Product distribution | | | Ratio | |
Conditions	*exo*-d$_2$-9:	*endo*-d$_2$-9:	11	*exo/endo*	Ref.
180 °C, 100–200 torr	74	25	1	3.0	20b
Gas phase, 1–2 torr	50	50		1.0	20a
Gas Phase, 350–760 torr	60	40		1.5	20a
333–364 nm, *n*-pentane	75	25	trace	3.0	20c
hv matrix, –70 °C	33	67		0.5	20a
350 nm, Ph$_2$CO, C$_6$H$_6$	50	50	trace	1.0	20b
185 nm, *n*-heptane	37	12	43[b]	3.0	20b
Laser-jet, Ph$_2$CO, C$_6$D$_6$	30	30	7[c]	1.0	20c

[a] For structures, cf. Scheme 75.5; 11 is

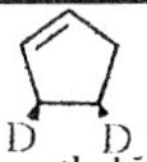

[b] In addition, 6% 1,4-pentadiene and 2% methylenecyclobutane as secondary photoproducts were formed.

[c] Also, 33% dimers were observed.

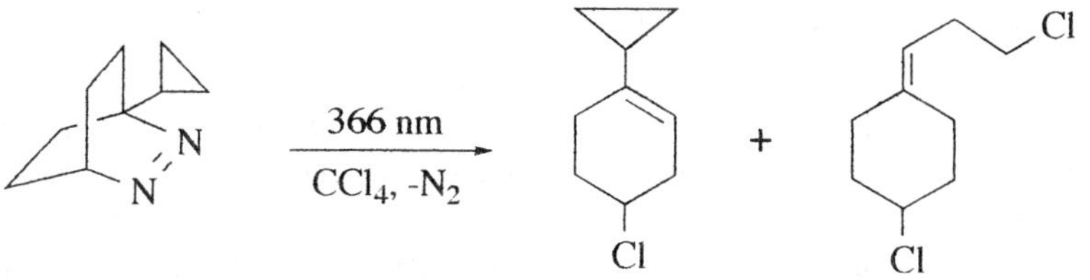

SCHEME 75.7

5.3 Photochemical Reactions

Cyclization

The smallest cyclic azo compound, diazirine **15**, suffers nitrogen loss on photolysis to give a carbene. This retrocyclic reaction is general for a variety of substituted diazirines.[28] The generation of carbenes by photodenitrogenation is more commonly achieved through the photolysis of readily accessible diazo compounds. The higher homologous diazetines **16** serve as precursors for introducing double bonds. Despite their structural simplicity, diazetines are synthetically not readily available, and the involvement of radiationless decay upon photochemical excitation is an important competitive pathway in their photochemistry.[29]

Photodeazetation of pyrazolines **17** provides a general synthesis of cyclopropanes **18**. With methylenepyrazolines and their derivatives, trimethylenemethane[30] biradicals are formed on photolysis and, depending on the photolysis mode, different ratios of rearrangement products were observed.[2b,28] The smallest and most strained fused bicycle **22** has been prepared by nitrogen extrusion from the fused azoalkanes **19**[31] and **20**,[32] as well as from the bridged azoalkane **21**.[10] It was shown that nitrogen loss is stepwise; first, **20** is formed. Diazoalkanes, generated by cycloreversion, intervene and are the cause for the low yields of **22**. The housane **9** can be prepared either from the fused azoalkane **23**[33] or from **8**. This constitutes the very first utilization of azoalkanes as precursors to highly strained molecules through thermal denitrogenation.[34] In a similar manner, the heterocyclic housanes **25** were prepared from **24**.[35] Fusion of a cyclopropane ring to an aryl moiety leads to the highly strained benzocyclopropene **27**.[36] This synthetic strategy is used in general for the preparation of the highly strained bi- and polycycles **28–43**. It is interesting to observe that in the cases of **29** and **31**, the rings are *anti* in the product, although in the azo precursors they were initially *syn* to the incipient cyclobutane ring; thus, N$_2$ extrusion has taken place with double inversion.[37] Furthermore, in these stereoisomers, denitrogenation is greatly enhanced by the anchimeric assistance of the cyclopropane ring in 28 and the cyclobutane ring in 30. This concerted elimination affords the cyclic 1,4- or 1,5-diene as the major product. In contrast, when the azo linkage and the cyclobutane ring are *syn* to one another as in 32, the thermal (140°C in CHCl$_3$) and photolytic denitrogenation give essentially quantitatively the fused *anti*-tricycle 33.[38] Depending on the photolysis conditions, 35 is formed in high yield (90%), along with the homotropilidene 36.[39] The isomeric azo compounds 37 and 38 lead on irradiation to the bicyclodiene 39 rather than the expected tetracycles.[40] Photolysis of the bridged azoalkane 40 affords the bridged hydrocarbons 41 and 42, whereas, on pyrolysis, a [4+2]-retrocyclization occurs to give a pyrazole quantitatively, which on 1,3 hydrogen shift tautomerizes to its hydrazone.[41]

One of the earliest target molecules in bridged bicyclization was the synthesis of quadricyclane 44. Upon thermolysis or triplet-sensitized photolysis of the azoalkane 43, quadricyclane was formed in up to 90% yield.[42] Another much studied and intriguing example is the synthesis of prismane 46. On thermolysis, the caged azoalkane 45 affords benzene quantitatively instead of the desired prismane. However, on direct photolysis, 10% of prismane was formed besides benzene, Dewar benzene, and benzvalene.[43] Thus, the synthesis of caged hydrocarbons from suitable azo precursors through bridged bicyclization on denitrogenation is a viable synthetic approach to such complex structures; this has also been shown in the gas-phase photolytic conversion of 47 to give 48.[44]

The azoalkane 49 did not produce the desired caged product, neither on thermolysis nor on photolysis. Instead, on heating, it affords barrelene 50 quantitatively, while triplet-sensitized photolysis gives only semibullvalene 51, the di-π-methane rearrangement product of barrelene. On direct photolysis, both barrelene (24%) and semibullvalene (73%) are formed. Analogous results are observed for the benzo- and naphtho-annelated derivatives.[45] Valuable information on the mechanism of the di-π-methane rearrangement was acquired from related systems.[46] Cyclizations to either spiroalkanes[47] or spiro-conjugated systems[48] are well-known denitrogenation reactions of suitable azoalkanes. Many other examples of highly strained molecules such as 53 and 55 have been prepared in this way.

Transformations Without Loss of Nitrogen

As mentioned before, not all azoalkanes extrude molecular nitrogen on thermal or photochemical activation; instead, they undergo transformations with retention of nitrogen. This behavior is, however, of mechanistic interest and synthetic utility. An early example was observed in the photolysis of the peralkylated 3H-pyrazoles. Thus, while photolysis of 56 at −60 °C in *n*-pentane brought about nitrogen loss and the formation of the expected cyclopropene as the major product, photolysis at −60 °C in methanol gave the valence tautomer 57 exclusively. The latter is unstable and reverts to 56 on warming to room temperature.[49] A related example is the rearrangement of azoalkane 58 into the aziridine derivative 59 on sensitized photolysis. On standing at 20 °C, the thermally labile 59 retrocyclizes into the diazo derivative 60.[50] A similar rearrangement was observed with the azoalkane 61. While direct photolysis afforded the expected tricycloalkane 62, sensitized photolysis gave the stable and isolable aziridine 63 quantitatively.[46] This type of azoalkane rearrangement is of synthetic value since such aziridines would be difficult to prepare by classical methods.

An interesting and unexpected rearrangement was observed for the attempted photodeazetation of the azoalkane 45. While on direct photolysis, 45 afforded the expected prismane in low yield (cf. Table 75.2), on sensitized photolysis the diazacyclooctatetraene 64 was the major product (67%).[43b] Also quite surprising was the intramolecular cycloaddition of the azo linkage to the juxtaposed C=C bond in the azoalkane 65 to give the caged product 66. This type of phototransformation is unique in azoalkane chemistry![51]

Photoreduction is another transformation pathway, as shown in the case of 67.[52] In fact, hydrogen abstraction by azo esters and aryl azo compounds are well-known.[2d] The photochemistry of cyclic azoalkanes with six-membered rings and larger resembles that of their acyclic counterparts

Table 75.2   Selected Decomposition Products of Cyclic Azoalkanes[a]

Azoalkano	Transformation product	Ref.

a) Cyclic

15   **16**

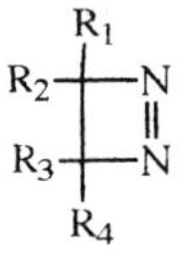

17   **18**

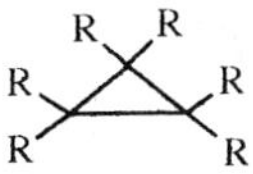

b) Bicyclic

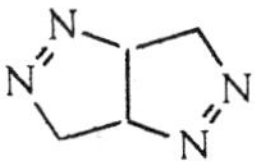

19

20   **22**

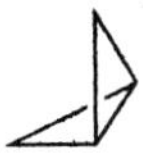

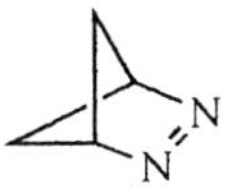

21

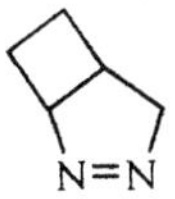

23   **9**

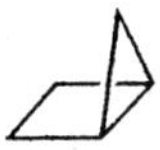

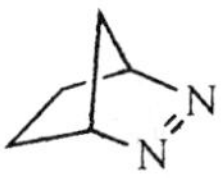

8

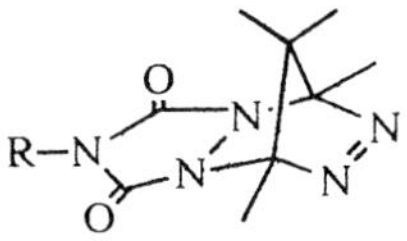

24   **25**

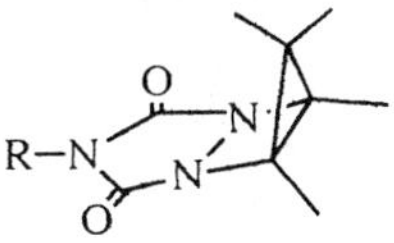

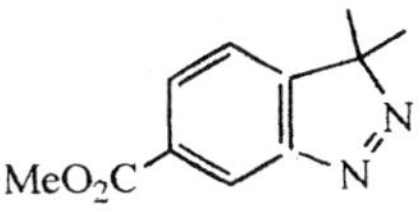

26   **27**

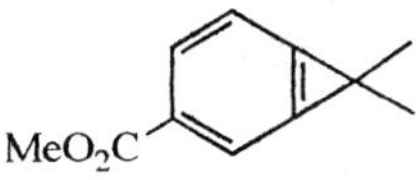

c) Polycyclic

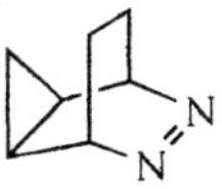

28   **29**

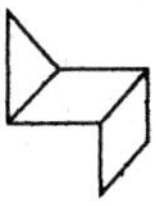

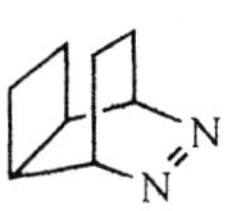

30   **31**

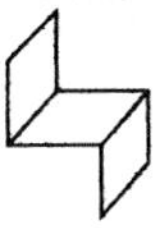

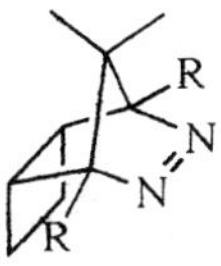

32   **33**

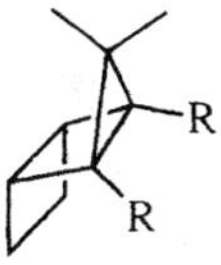

Table 75.2 (continued) Selected Decomposition Products of Cyclic Azoalkanes[a]

Azoalkano		Transformation product	Ref.
c) Polycyclic			
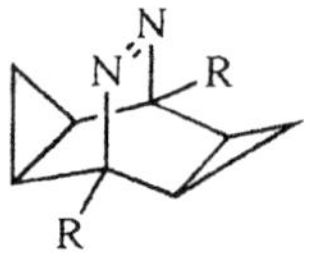	**34**	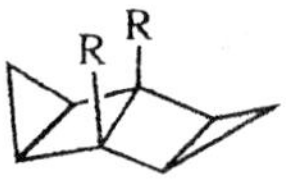 **35** 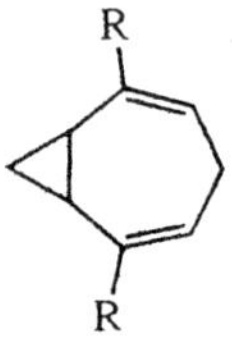	**36**
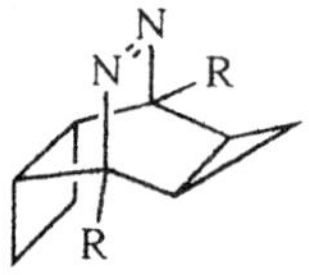	**37**	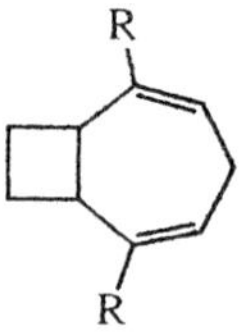	**39**
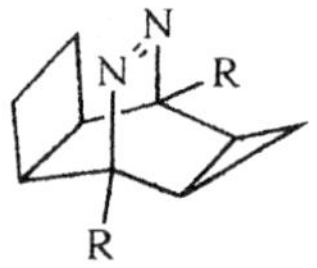	**38**		
d) Bridged			
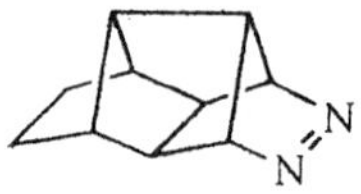	**40**	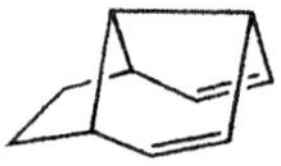 **41** 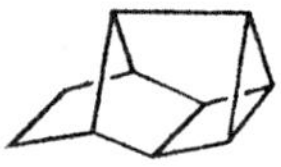	**42**
e) Cage			
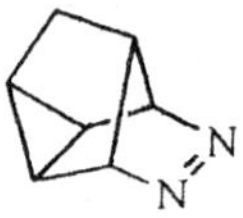	**43**	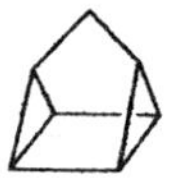	**44**
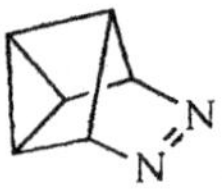	**45**	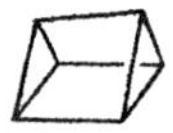	**46**
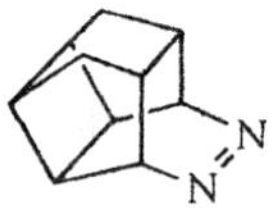	**47**	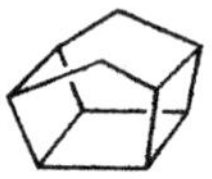	**48**
f) Fluxional			
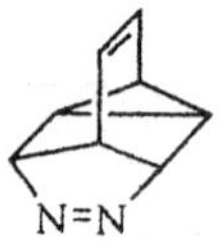	**49**	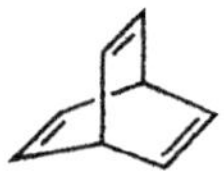 **50** 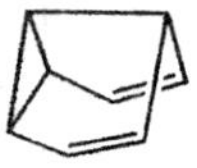	**51**
g) Spiro			
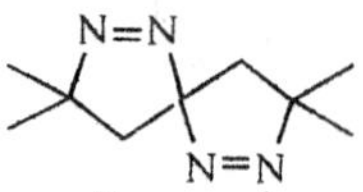	**52**	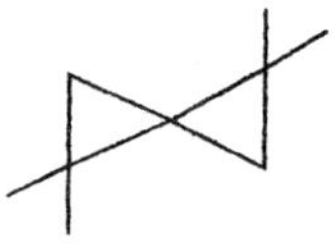	**52**
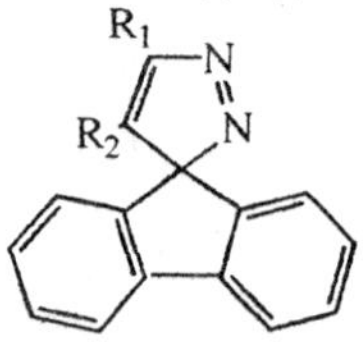	**54**	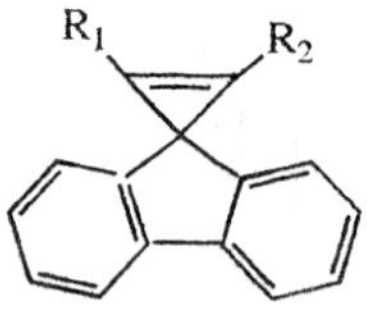	**55**

[a] References to the original literature are given in the text.

in that Z,E-isomerization is competitive with denitrogenation.[53,54] Thus, on irradiation of the eight-membered azoalkane 69, 60% of it was converted to the cyclic *trans*-isomer 70, which photoisomerized back to the *cis*-69.[53] That not all acyclic azoalkanes photodenitrogenate is demonstrated by the norbornyldiazene 71, which on photolysis affords the stable *cis*-isomer.[54]

Reluctant Azoalkanes

Although the majority of azoalkanes lose molecular nitrogen on either thermal activation or photochemical excitation (cf. Table 75.2) or undergo transformations with retention of the nitrogen (cf. Table 75.3), there exist azoalkanes that are essentially inert to photochemical denitrogenation. This frustrating third category of azoalkanes, namely the so-called "reluctant" azoalkanes, thwarted elegant synthetic sequences in its last step due to the fact that the azoalkane was resistant to denitrogenation. An early classical example is the attempt to prepare cubane from diazabasketene 76. Thus, prolonged photolysis afforded a low yield of cyclooctatetraene, while vacuum flash pyrolysis gave azacyclooctatetraene.[55] The common structural feature of these "reluctant" azoalkanes is that the azo linkage is part of a six-membered ring (except 69, 71, and 73).

Much mechanistic work has been expended in order to gain an understanding of why such azoalkanes lose nitrogen reluctantly and some progress has been made along these lines. For example, denitrogenation of 75 was enhanced by placing radical-stabilizing substituents such as cyano, vinyl, halogen, etc. at the α-carbon to the azo linkage, or by introducing strain into the ring through an exocyclic cyclopropyl or cyclobutyl ring.[2d] Alternatively, vapor-phase photolysis at elevated temperatures proved to be successful in denitrogenating 47.[44,57] Thus, there exists an activation energy for photodenitrogenation since azoalkanes lose nitrogen more efficiently at higher temperatures.

An alternative approach to photodenitrogenate "reluctant" azoalkanes is to employ 185-nm radiation.[58] Azoalkanes 69, 71, and 73–76 effectively denitrogenate on exposure to such "high-energy" (155 kcal) radiation, as shown by the quantum qields (Φ) in Table 75.4. Higher excited states (π,π^*, Rydberg) appear to be responsible for the enhanced quantum yields. To explain the wavelength-dependent products or product distributions, Zwitterions have been postulated as intermediates in the 185-nm photolysis. These intermediates are not accessible in conventional 350-nm (n,π^* excitation) photolysis. Another effective method to photoextrude N_2 is the formation of radical cations through electron transfer on photosensitization. For a detailed discussion, the interested reader is referred to the literature.[20c,58d]

5.4 Applications

The photochemical and/or thermal deazetation of azoalkanes has served quite generally an important role in the mechanistic elucidation of radical and biradical reactions. In addition, valuable information on the lifetimes and chemical reactivities was acquired by trapping such species (primarily triplet states) with appropriate scavengers (Table 75.5).[59–65] Fundamental information on the factors that control intersystem crossing (ISC) in triplet biradicals[66] was obtained through these studies.[1b,67] Azoalkane 8 is used as an actinometer in photochemical reactions. An industrial application of azoalkanes is their use as radical initiators, e.g., azoisobutyronitrile (AIBN).

The synthetic utilization of either photochemical or thermal denitrogenation of azoalkanes is so extensive that the usefulness and convenience of this synthetic method cannot be overemphasized. Besides the numerous cyclization reactions to prepare strained molecules, bicyclization that leads to fused, bridged, caged, fluxional, and spiro structures must be mentioned in this context. Very important applications concern the strategic introduction of double bonds, e.g., the preparation of antiaromatic thiirene[68] and cyclobutadiene,[69] strained cyclooctyne,[70] or sterically congested *bis*(2,2,5,5-tetramethylcyclopentylidene),[71] which demonstrate the value of azoalkanes as precursors for unusual and complex structures (Table 75.2).

Table 75.3 Transformations of Azoalkanes without Loss of Nitrogen

Azoalkane			Transformation Product		Ref.
a) Rearrangement					
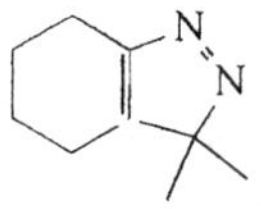	**56**	hv ⇌ Δ	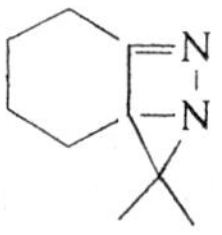	**57**	49
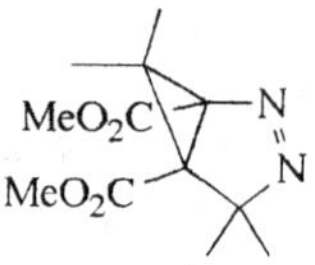	**58**	hv, Ph_2CO →	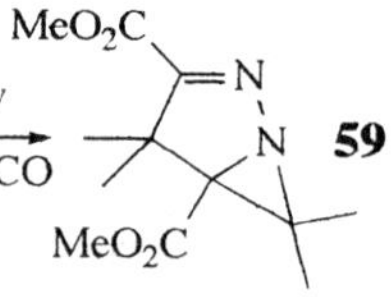 **59** → 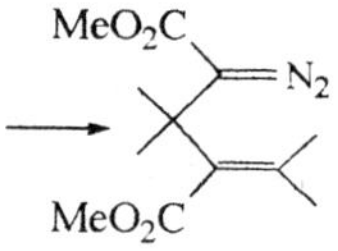	**60**	50
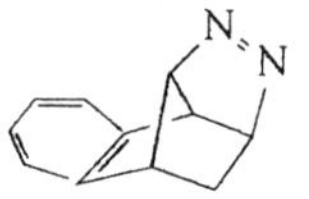	**61**	hv; Ph_2CO	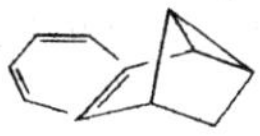	**62**	
			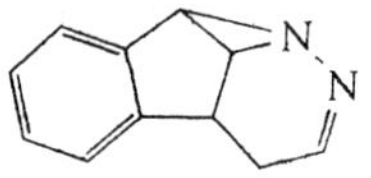	**63**	46
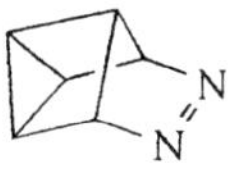	**45**	hv, Ph_2CO →	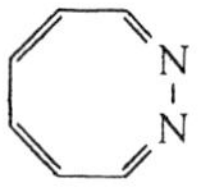	**64**	43
b) Intramolecular Cycloaddition					
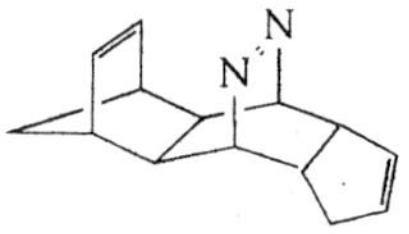	**65**	hv →	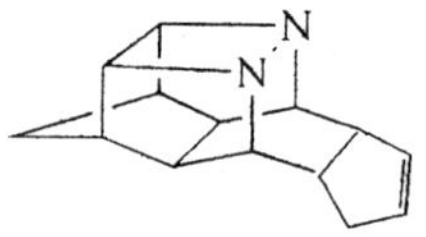	**66**	51
c) Photoreduction					
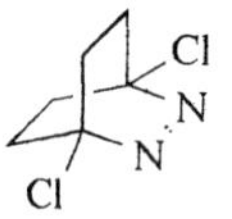	**67**	hv →	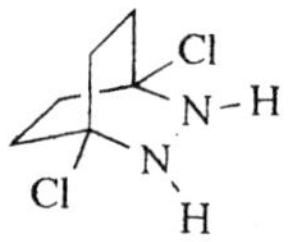	**68**	52
d) Z / E - Isomerization					
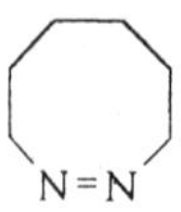	**69**	hv ⇌	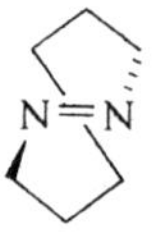	**70**	53
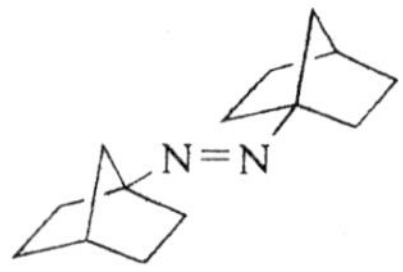	**71**	hv ⇌	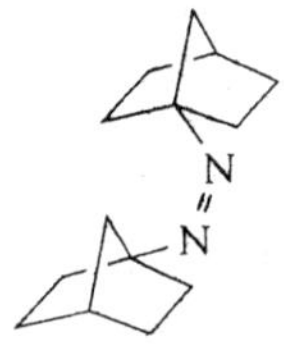	**72**	

Table 75.4   Reluctant Azoalkanes

Compound		Φ_{350}	Ref.	Φ_{185}	Ref.
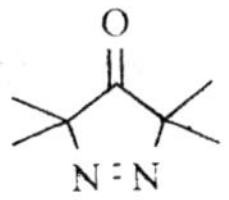	73	0.012	56	0.63	58b
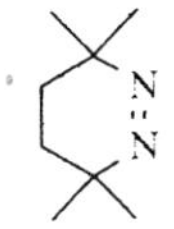	74	0.008	2d	0.22	58a
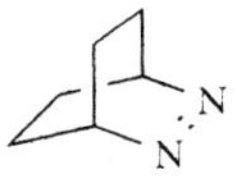	75	0.011	2d	0.50	58a
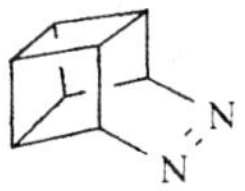	76	0.00006	55	0.20	58a
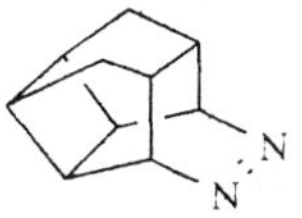	47	0.01	2d	–	
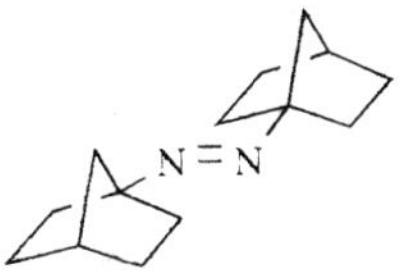	71	0.0008	54	0.13	58c
	69	0.015	53	0.69	58d

[a] Quantum yields were determined by using 2,3-diazabicyclo[2.2.1]heptene (8) as actinometer, for which the photodenitrogenation efficiency is 100% at 185 and 350 nm.[58a]

Table 75.5 Some Examples of Biradical Trapping Reactions

Compound	Biradical	Trapping product[a]		Ref.
	8	$^3O_2$ / $R_2NO\bullet$	77 / 78	59 / 13
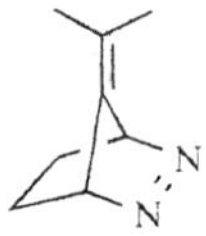	3	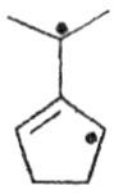 $^3O_2$	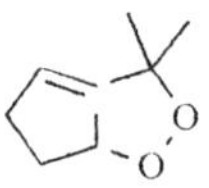 79	59
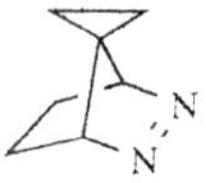	80	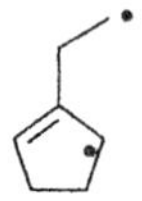 $^3O_2$	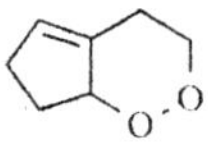 81	60
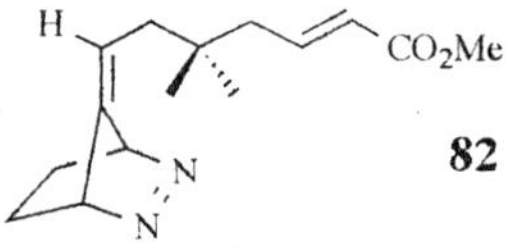	82		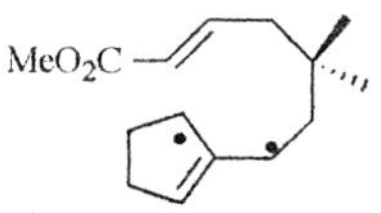 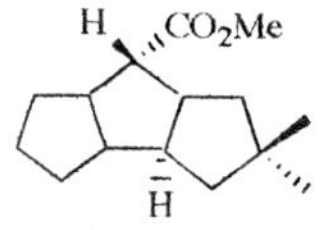 83	61
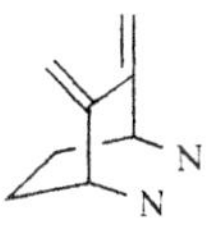	84	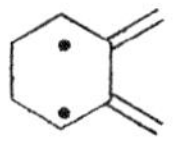 $^3O_2$	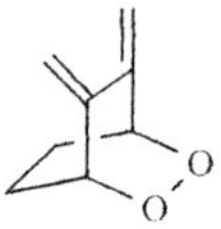 85 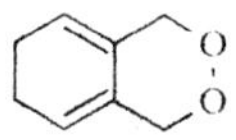 + 86	62
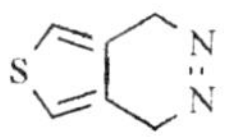	87	Ar	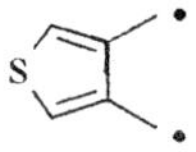  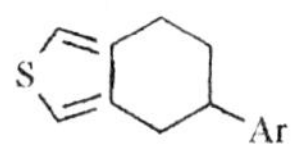 88	63
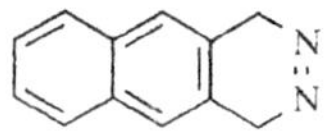	89	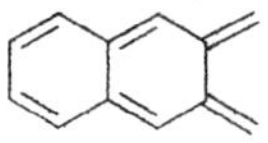 $^3O_2$	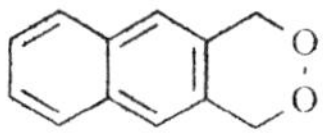 90	64
	91	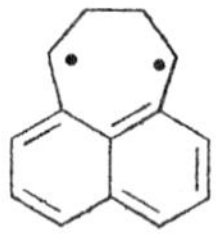 $^3O_2$	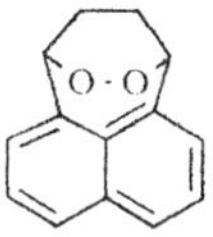 92 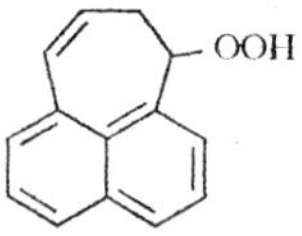 + 93	65

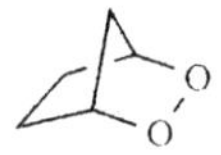

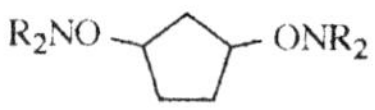

[a] The rate constants $[k(O_2)]$ for the reaction of biradicals with molecular oxygen is diffusion-controlled (ca. 10^{-10} M^{-1} s^{-1}).[1b]

ferences

1. (a) Borden, W. T., Ed., *Diradicals,* Wiley, New York, 1982; (b) Adam, W., Grabowski, S., and Wilson, R. M., Localized cyclic triplet diradicals. Lifetime determination by trapping with oxygen, *Acc. Chem. Res.,* 23, 165, 1990; (c) Johnston, L. J. and Scaiano, J. C., Time-resolved studies of biradical reactions in solution, *Chem. Rev.,* 89, 521, 1989.

2. Thiele, J., Über Hydrazo- und Azomethan, *Chem. Ber.,* 42, 2575, 1909.

3. (a) Patai, S., Ed., *The Chemistry of the Hydrazo, Azo and Azoxy Groups,* Vol. I, II, Wiley, London, 1975; (b) Meier, H. and Zeller, K. P., Thermische und photochemische Stickstoffeliminierungen, *Angew. Chem.,* 89, 876, 1977; (c) Adam, W. and De Lucchi, O., The synthesis of unusual organic molecules from azoalkanes, *Angew. Chem. Int. Ed.,* 19, 762, 1980; (d) Engel, P. S., Mechanism of the thermal and photochemical decomposition of azoalkanes, *Chem. Rev.,* 80, 99, 1980.

4. (a) Clark, W. D. K. and Steel, C., Photochemistry of 2,3-diazabicyclo[2.2.2]oct-2-ene, *J. Am. Chem. Soc.,* 93, 6347, 1971; (b) Engel, P. S. and Steel, C., Photochemistry of aliphatic azo compounds in solution, *Acc. Chem. Res.,* 6, 275, 1973.

5. Ramsperger, H. C., The thermal decomposition of methyl isopropyl di-imide: A homogeneous unimolecular reaction. The thermal decomposition of hydrazoic acid and methyl azide, *J. Am. Chem. Soc.,* 51, 2134, 1929.

6. (a) Hiberty, P. C., and Jean, V., Organic transition states. 6. Thermal decomposition of 1-pyrazolines, *J. Am. Chem. Soc.,* 101, 2538, 1979; (b) Dannenberg, J. J. and Tanaka, K., Theoretical studies of radical recombination reactions. 1. Allyl and azo allyl radicals, *J. Am. Chem. Soc.,* 107, 671, 1985.

7. Engel, P. S. and Gerth, D. B., Thermolysis of acyclic azoalkanes: Simultaneous or stepwise C-N homolysis?, *J. Am. Chem. Soc.,* 105, 6849, 1983.

8. Chichra, D., Platz, M., and Berson, J., Generation and capture of common intermediates from proto-planar and proto-bisected trimethylenemethane precursors. Thermal rearrangement of a methylene-pyrazoline, *J. Am. Chem. Soc.,* 99, 8507, 1977.

9. Reedich, D. E. and Sheridan, R. S., Direct evidence for both thermal and photochemical stepwise cleavage in a pair of isomeric azo compounds, *J. Am. Chem. Soc.,* 110, 3697, 1988.

10. Chang, M. H., Jain, R., and Dougherty, D. A., Chemical activation as a probe of reaction mechanism. Synthesis and thermal decomposition of 2,3-diazabicyclo[2.1.1]hex-2-enes, *J. Am. Chem. Soc.,* 106, 4211, 1984.

11. Simpson, C. J. S. M. and Adam, W., Dynamics of the thermal decomposition of 2,3-diazabicyclo[2.2.1]hept-2-ene, *J. Am. Chem. Soc.,* 113, 4728, 1991.

12. Wilson, R. M., The trapping of biradicals and related photochemical intermediates, in *Organic Photochemistry,* Vol. 7, Padwa, A., Ed., Marcel Dekker, New York, 1985, 339.

13. Adam, W. and Bottle, S., Benzophenone-sensitized photolysis of the azoalkane diazabicyclo-[2.2.1]hept-2-ene (DBH): Trapping of the 1,3-cyclopentadiyl triplet diradical by a nitroxide, *Tetrahedron Lett.,* 32, 1405, 1991.

14. Engel, P. S. and Culotta, A. M., Photolysis of azoalkanes. Reactions and kinetics of the 1-cyclopropylcyclopentane-1,3-diyl biradical and the 1-cyclopropylcyclopentyl radical, *J. Am. Chem. Soc.,* 113, 2686, 1991.

15. Bigot, B., Sevin, A., and Devaquet, A., Photochemical extrusion of nitrogen in azo compounds. An *ab initio* SCF-CI study, *J. Am. Chem. Soc.,* 100, 2639, 1978.

16. Edmunds, A. J. F. and Samuel, C. J., Photochemical deazetation of 2,3-diazabicyclo[2.2.2]oct-2-ene: Pseudorotation of the cyclohexanediyl biradical, *J. Chem. Soc., Perkin Trans. 1,* 1267, 1989.

17. Adam, W., Dörr, M., Hill, K., Peters, E.-M., Peters, K., and von Schnering, H. G., Synthesis, photolysis, and thermolysis of azoalkanes derived from spirocyclopropane-substituted norbornenes: Competitive nitrogen loss and diazoalkane formation, *J. Org. Chem.,* 50, 587, 1985.

18. (a) Burton, K. A. and Weisman, R. B., Stepwise photodissociation of vapor-phase azomethane, *J. Am. Chem. Soc.,* 112, 1804, 1990; (b) Boate, D. R. and Scaiano, J. C., Transient phenomena in the photochemistry of *trans*-azomethane, *Tetrahedron Lett.,* 30, 4633, 1989; (c) Adams, J. S., Weisman,

R. B., and Engel, P. S., Photodissociation of a bicyclic azoalkane: Time-resolved coherent anti-Stokes Raman spectroscopy studies of vapor-phase 2,3-diazabicyclo[2.2.1]hept-2-ene, *J. Am. Chem. Soc.*, 112, 9115, 1990.

19. (a) Majchrzak, M. W., Békhazi, M., Tse-Sheepy, I., and Warkentin, J., Photolysis of 2-alkoxy-Δ^3-1,3,4-oxadiazolines. A new route to diazoalkanes, *J. Org. Chem.*, 54, 1842, 1989; (b) Adam, W. and Finzel, R., UV-laser photochemistry: Retro-cleavage in the benzophenone-sensitized photolysis of Δ^3-1,3,4-oxadiazolines into diazoalkanes, *Tetrahedron Lett.*, 31, 863, 1990.

20. (a) Roth, W. R. and Martin, M., Zur Stereochemie des thermischen und photochemischen Zerfalls von 2,3-Diaza-bicyclo[2.2.1]hepten-(2), *Liebigs Ann. Chem.*, 702, 1, 1967; (b) Adam, W., Oppenländer, T., and Zang, G., Photochemistry of the azoalkanes 2,3-diazabicyclo[2.2.1]hept-2-ene and spiro[cyclopropane-1,7'-[2,3]diazabicyclo[2.2.1]hept-2-ene]: On the questions of one-bond vs. two-bond cleavage during the denitrogenation, cyclization vs. rearrangement of the 1,3-diradicals, and double inversion, *J. Org. Chem.*, 50, 3303, 1985; (c) Adam, W., Denninger, U., Finzel, R., Kita, F., Platsch, H., Walter, H., and Zang, G., Comparative study of the pyrolysis, photoinduced electron transfer (PET), and laser-jet and 185-nm photochemistry of alkyl-substituted bicyclic azoalkanes, *J. Am. Chem. Soc.*, 114, 5027, 1992.

21. Paquette, L. A. and Leichter, L. M., Synthesis and thermal rearrangement of tricyclo[3.2.0.0^{2,4}]-hept-6-enes. An analysis of structural requirements for effective intramolecular trapping of a 1,3-diradical by a remote cyclobutene-ring, *J. Am. Chem. Soc.*, 93, 5128, 1971.

22. Martin, H. D., Heiser, B., and Kunze, M., Neuartige Bildungsweise von *cis, trans*-1,5-Cyclooctadien, *Angew. Chem.*, 90, 735, 1978.

23. (a) Adam, W., Hannemann, K., and Wilson, R. M., Sensibilisierte UV-Laserphotolyse von Azoalkanen: Einflüsse der Konformation auf Spinumkehr und Lebensdauer von Triplettdiradikalen, *Angew. Chem.*, 97, 1072, 1985; (b) Adam, W., Platsch, H., Reinhard, G., and Wirz, J., Effect of 2,2-dimethyl substitution on the lifetimes of cyclic hydrocarbon triplet 1,3-biradicals, *J. Am. Chem. Soc.*, 112, 4571, 1990.

24. Engel, P. S., Keys, D. E., and Kitamura, A., "Spring-loaded" biradicals. The radical and electron-transfer photochemistry of bridgehead cyclopropyl-substituted 2,3-diazabicyclo[2.2.2]oct-2-enes (DBO's), *J. Am. Chem. Soc.*, 107, 4964, 1985.

25. Engel, P. S., Kitamura, A., and Keys, D. E., Fluorescence quenching and photoreactions of 2,3-diazabicyclo[2.2.2]oct-2-enes. A case of charge transfer and hydrogen atom transfer, *J. Org. Chem.*, 52, 5015, 1987.

26. Adam, W. and Dörr, M., Wagner-Meerwein rearrangements of radical cations generated by triphenylpyrryliumtetrafluoroborate photosensitized electron transfer of azoalkanes, *J. Am. Chem. Soc.*, 109, 1570, 1987.

27. Adam, W., Walter, H., Chen, G.-F., and Williams, F., Photochemical and ESR spectral evidence for a stereoselective rearrangement of radical cations derived from azoalkanes and bicyclopentanes, *J. Am. Chem. Soc.*, 114, 3007, 1992.

28. Braslavsky, S. and Heicklen, J., The gas-phase thermal and photochemical decomposition of heterocyclic compounds containing nitrogen, oxygen, or sulfur, *Chem. Rev.*, 77, 473, 1977.

29. Engel, P. S., Hayes, R. A., Keifer, L., Szilagyi, S., and Timberlake, J. W., Extrusion of nitrogen from cyclic and bicyclic azo compounds, *J. Am. Chem. Soc.*, 100, 1876, 1978.

30. Berson, J. A., The chemistry of trimethylenemethanes, a new class of biradical reactive intermediates, *Acc. Chem. Res.*, 11, 446, 1978.

31. Franck-Neumann, M. and Dietrich-Buchecker, C., Total stereospecific synthesis of *cis*-chrysanthemic methyl ester: The cyclopropenic way (1), *Tetrahedron Lett.*, 21, 671, 1980.

32. Schneider, M. and Csacsko, B., Bildung von Diazoverbindungen beim Zerfall von 2,3-Diazabicyclo[3.1.0]hex-2-enen: [3+2]-Cycloreversionen, *Angew. Chem.*, 89, 905, 1977.

33. Keppel, R. A., Multiple mechanism in the thermal and photochemical decomposition of 2,3-diazabicyclo[3.1.0]hex-2-enes, *J. Am. Chem. Soc.*, 94, 1350, 1972.

34. Criegee, R. and Rimmelin, A., Darstellung und Eigenschaften von Bicyclo[0.1.2]pentan, *Chem. Ber.*, 90, 414, 1957.
35. Wiley, P. F., *The Chemistry of Heterocyclic Compounds*, Vol. 33, Wiley, New York, 1978, 1073.
36. Anet, R. and Anet, F. A. L., Synthesis of benzocyclopropene derivative, *J. Am. Chem. Soc.*, 86, 525, 1964.
37. Tanida, H. and Teratake, S., Photolytic and thermal decompositions of *exo*-6,7-diaza[3.2.1.0^{2,4}]oct-6-ene. Evidence for unimportance of cyclopropyl participation in the former decomposition, *Tetrahedron Lett.*, 377, 1970.
38. Paquette, L. A. and Leichter, L. M., Pyrolysis of *anti*-tricyclo[3.2.0.0^{2,4}]heptanes. The role of 2,4-substitution in the operation of σ bond assisted cyclobutane fragmentations, *J. Am. Chem. Soc.*, 93, 4922, 1971.
39. (a) Fühlhuber, H. D., Gousetis, G., Troll, T., and Sauer, J., Photolysen von tetracyclischen Azoverbindungen (*exo, exo*-9,10-Diazatetracyclo[3.3.2.0^{2,4}.0^{6,8}]dec-9-ene). Ein leichter Zugang zu substituierten Homotropilidenen, *Tetrahedron Lett.*, 3903, 1978; (b) Gousetis, G. and Sauer, J., Ein Beitrag zum Mechanismus der Photolyse tetracyclischer Azoverbindungen, *Tetrahedron Lett.*, 1295, 1979.
40. (a) Allred, E. L. and Voorhees, K. J., Orientational effects on cyclopropyl participation in the thermolysis of azo compounds. Assessment of the *endo* configuration, *J. Am. Chem. Soc.*, 95, 620, 1973; (b) Paquette, L. A. and Epstein, M. J., Assessment of the relative efficiencies of *syn*-dispored cyclopropane, cyclobutane, and cyclobutane ring participation in the photolysis of azo compounds. Kinetic analysis of a *cis*-bicyclo[5.2.0]nona-2,5,8-triene to *cis* bicyclo[6.1.0]nona-2,4,6-triene rearrangement, *J. Am. Chem. Soc.*, 95, 6717, 1973.
41. (a) Paquette, L. A., Kramer, J. D., Lavrik, P. B., and Wyvratt, M. J., Photoisomerization of triquinacene congeners, *J. Org. Chem.*, 42, 503, 1977; (b) Paquette, L. A., Wyvratt, M. J., Berk, H. C., and Moerch, R. E., Domino Diels-Alder cycloadditions to 9,10-dihydrofulvalene and 11,12-dihydrosesquifulvalene. A synthetic tool for the elaboration of polycondensed acyclic systems, *J. Am. Chem. Soc.*, 100, 5845, 1978.
42. Turro, N. J., Cherry, W. R., Mirbach, M. F., and Mirbach, M. J., Energy acquisition, storage, and release. Photochemistry of cyclic azoalkanes as alternate entries to the energy surfaces interconnecting norbornadiene and quadricyclane, *J. Am. Chem. Soc.*, 99, 7388, 1977.
43. (a) Katz, T. J. and Acton, N., Synthesis of prismane, *J. Am. Chem. Soc.*, 95, 2738, 1973; (b) Trost, B. M., Scudder, P. H., Cory, R. M., Turro, N. J., Ramamurthy, V., and Katz, T. J., 1,2-Diaza-2,4,6,8-cyclooctatetraene, *J. Org. Chem.*, 44, 1264, 1979.
44. (a) Turro, N. J., Liu, K.-C., Cherry, W., Liu, J.-M., and Jacobson, B., Photoelimination of nitrogen from reluctant cyclic azo compounds. Preparation of some novel cyclopentadiene cyclodimers, *Tetrahedron Lett.*, 19, 555, 1978; (b) Quast, H., Fuss, A., and Heublein, A., Cyclopropanimine durch Photolyse von Pyrazoliniminen: Thermische Reaktion eines angeregten Zustandes, *Angew. Chem.*, 92, 55, 1980; (c) Turro, N. J., Liu, J.-M., Martin, H.-D., and Kunze, M., Photoelimination of nitrogen from cyclic azoalkanes: An exceptionally labile and an exceptionally reluctant diazabicyclo[2.2.2]octene, *Tetrahedron Lett.*, 21, 1299, 1980.
45. Zimmermann, H. E., Boetticher, R. J., Buehler, N. E., Keck, G. E., and Steinmetz, M. G., Independent generation of cyclopropyldicarbinyl diradical species of the di-π-methane rearrangement. Excited singlet, triplet, and ground-state hypersurfaces of barrelene photochemistry, *J. Am. Chem. Soc.*, 98, 7680, 1976.
46. Adam, W., Carballeira, N., and DeLucchi, O., Unusual thermal and photochemical transformations of the azoalkane 2,3-diaza-7,8-benzotricyclo[4.3.0.0^{4,9}]nona-2,7-diene, *J. Am. Chem. Soc.*, 102, 2107, 1980.
47. (a) Schneider, M., Schuster, O., and Rau, H., Addition von 2-Diazopropan an Allen. Asymmetrische Zerstörung von 1-Pyrazolen mit circular polarisiertem Licht, *Chem. Ber.*, 110, 2180, 1977; (b) Krapcho, A. P., Synthesis of carbocyclic spiro compounds via cycloaddition routes, *Synthesis*, 77, 1978.

48. (a) Dürr, H., Schmidt, W., and Sergio, R., Dibenzo[2.4]spirene und deren Umlagerungsprodukte durch Photofragmentierung und Thermolyse von Spiropyrazolen, *Liebigs Ann. Chem.*, 1132, 1974; (b) Dürr, H. and Gleiter, R., Spirokonjugation, *Angew. Chem.*, 90, 591, 1978.

49. Closs, G. L., Böll, W. A., Heyn, H., and Dev, V., Photochemical conversions of 3H-pyrazoles to cyclopropenes and 1,2-diazabicyclo[2.1.0]pent-2-enes, *J. Am. Chem. Soc.*, 90, 173, 1968.

50. Franck-Neumann, M., Martina, D., and Dietrich-Buchecker, C., Photolyses directes et sensibilisées d'homopyrazolenines et de *bis* Δ^1-pyrazolines, *Tetrahedron Lett.*, 16, 1763, 1975.

51. (a) Berning, W. and Hünig, S., Photochemische [2+2]-Cycloaddition zwischen parallelen CC- und NN-Doppelbindungen, *Angew. Chem.*, 89, 825, 1977; (b) Albert, B., Berning, W., Burschka, C., Hünig, S., and Prokschy, F., Intramolekulare [2+2]-Cycloaddition zwischen parallelen C=C- und N=N-Bindungen, *Chem. Ber.*, 117, 1465, 1984; (c) Hünig, S. and Prokschy, F., [3+2]-Cycloadditionen zwischen parallelen C=C- und N-alkylierten N=N-Bindungen, *Chem. Ber.*, 117, 2099, 1984; (d) Beck, K. and Hünig, S., Substituierte Isopyrazole als elektronenarme Diene zur Synthese von 2,3-Diazabicyclo[2.2.1]heptenen und deren Photoreaktionen, *Chem. Ber.*, 120, 477, 1987; (e) For related examples, cf. Klingler, O. and Prinzbach, H., Außergewöhnlich stabile *cis, cis*-Trialkyltriaziridine durch Azo/Nitren-Addition, *Angew. Chem.*, 99, 579, 1987.

52. Luttke, W. and Schabacker, V., 1,4-Dichlor-bicyclo[2.2.0]hexan durch Photolyse von 1,4-Dichlor-2,3-diaza-bicyclo[2.2.2]octen-(2), *Liebigs Ann. Chem.*, 698, 86, 1966.

53. Overberger, C. G., Chi, M. S., Pucci, D. G., and Barry, J. A., *Trans*-azo linkages in eight-, nine-, and ten-membered cyclic azo compounds, *Tetrahedron Lett.*, 13, 4565, 1972.

54. Engel, P. S., Melaugh, R. A., Page, M. A., Szilagyi, S., and Timberlake, J. W., Stable *cis*-dialkyldiazenes (azoalkanes): *cis*-Di-1-adamautyldiazene and *cis*-di-1-norbornyldiazene, *J. Am. Chem. Soc.*, 98, 1971, 1976.

55. McNeil, D. W., Kent, M. E., Hedaya, E., D'Angelo, P. F., and Schissel, P. O., Azocine. The flash vacuum pyrolsis of 7,8-diazapentacyclo[4.2.2.0^{2,5}.0^{3,9}.0^{4,10}]dec-7-ene (Diazabasketene), *J. Am. Chem. Soc.*, 93, 3817, 1971.

56. Engel, P. S. and Shen, L., Photochemical and thermal decomposition of 1-pyrazolines, *Can. J. Chem.*, 52, 4040, 1978.

57. Mirbach, M. J., Liu, K. C., Mirbach, M. F., Cherry, W. R., Turro, N. J., and Engel, P. S., Spectroscopic properties of cyclic and bicyclic azoalkanes, *J. Am. Chem. Soc.*, 100, 5122, 1978.

58. (a) Adam, W. and Mazenod, F., Facile, liquid-phase denitrogenation of "reluctant" azoalkanes on photolysis with 185-nm radiation, *J. Am. Chem. Soc.*, 102, 7131, 1980; (b) Adam, W., Fuss, A., Mazenod, F. P., and Quast, H., Reluctant azoalkanes: Short-wavelength (185 nm) liquid-phase photolysis and high-temperature (400–1000°C) gas-phase pyrolysis of 3,3,5,5-tetramethylpyrazolin-4-one, *J. Am. Chem. Soc.*, 103, 998, 1981; (c) Adam, W., Mazenod, F., Nishizawa, Y., Engel, P. S., Baughman, S. A., Chae, W.-K., Horsey, D. W., Quast, H., and Seiferling, B., Reluctant azoalkanes: The photochemical behaviour of acyclic, bridgehead-centered azoalkanes on 185-nm and 350-nm irradiation, *J. Am. Chem. Soc.*, 105, 6141, 1983; (d) Adam, W., and Oppenländer, T., 185-nm-Photochemie von Olefinen, gespannten Kohlenwasserstoffen und Azoalkanen in Lösung, *Angew. Chem.*, 98, 659, 1986.

59. Wilson, R. M. and Geiser, F., Biradical trapping: The formation of bicyclic peroxides via the thermal and photodecomposition of azo compounds in the presence of oxygen, *J. Am. Chem. Soc.*, 100, 2255, 1978.

60. Adam, W., Hannemann, K., and Hössel, P., Preparative UV-laser photochemistry of the azoalkane spiro(2,3-diazabicyclo[2.2.1]hept-2-ene-7′,1-cyclopropane), *Tetrahedron Lett.*, 25, 181, 1984.

61. (a) Little, R. D. and Muller, G. W., A regiospecific and highly stereoselective approach to the synthesis of linearly fused tricyclopentanoids. Intramolecular diyl trapping reactions, *J. Am. Chem. Soc.*, 101, 7129, 1979; (b) Little, R. D., The intramolecular diyl trapping reaction. A useful tool for organic synthesis, *Chem. Rev.*, 86, 875, 1986.

62. Roth, W. R. and Scholz, B. P., Das 2,3-Dimethylen-1,4-cyclohexadiyl, *Chem. Ber.*, 115, 1197, 1982.

63. Stone, K. J., Greenberg, M. M., Blackstock, J. C., and Berson, J. A., Heterocyclic aromatic non-Kekulé molecules. Synthesis and solution-phase chemistry of the singlet biradicals 3,4-dimethylenefurane and 3,4-dimethylenethiophene, *J. Am. Chem. Soc.,* 111, 3659, 1989.

64. Gisin, M. and Wirz, J., Photolysis of the azo-precursors of 2,3- and 1,8-naphthoquinodimethane, *Helv. Chim. Acta,* 59, 2273, 1976.

65. (a) Burnett, M. N., Boothe, R., Clark, E., Gisin, M., Hassaneen, H. M., Pagni, R. M., Persy, G., Smith, R. J., and Wirz, J., 1,4-Perinaphthadiyl. Singlet- and triplet-state reactivity of a conjugated hydrocarbon biradical, *J. Am. Chem. Soc.,* 110, 2527, 1988; (b) Adam, W., Platsch, H., and Wirz, J., Oxygen trapping and thermochemistry of a hydrocarbon singlet biradical: 1,3-Diphenylcyclopentane-1,3-diyl, *J. Am. Chem. Soc.,* 111, 6896, 1989.

66. Salem, L. and Rowland, C., Die elektronischen Eigenschaften von Diradikalen, *Angew. Chem.,* 84, 86, 1972.

67. Goldberg, A. H. and Dougherty, D. A., Effects of through-bond and through-space interactions on singlet-triplet energy gaps in localized biradicals, *J. Am. Chem. Soc.,* 105, 284, 1983.

68. Font, J., Torres, M., Gunning, H. E., and Strausz, O. P., Gas-phase photolysis of 1,2,3-thiadiazoles: Evidence for thiirene intermediates, *J. Org. Chem.,* 43, 2487, 1978.

69. Masamune, S., Nakamura, N., and Spadaro, J., 1,2-*Bis*(β-tosylethoxycarbonyl)diazene. Its application to the 2,3-diazabicyclo[2.2.0]hexene system, *J. Am. Chem. Soc.,* 97, 918, 1975.

70. Lalezari, I., Shafiee, A., and Yalpani, M., Selenium heterocycles, VIII. Pyrolysis of cycloalka-1,2,3-selenadiazoles (1), *J. Heterocycl. Chem.,* 9, 1411, 1972.

71. Krebs, A. and Rüger, W., Synthese und Eigenschaften des bi-2,2,5,5-Tetramethylcyclopentylidens, eines sterisch gehinderten Alkens, *Tetrahedron Lett.,* 20, 1305, 1979.

Carbene Formation by Extrusion of Nitrogen

Heinz Dürr
Universität des Saarlandes

Aboel-Magd A.
  Abdel-Wahab
Assiut University

76.1 Introduction

Reactive intermediates have always occupied a significant place in all branches of chemistry. Carbenes, as examples of such intermediates, have been the subject of many interesting studies concerning generation, reaction mechanism, selectivity, multiplicity, and even stability studies. Really, it is no exaggeration to claim a major role for carbenes in the modern chemist's armory for the synthesis of a large number of molecules that are of interest both for research and for applications.

The correct name of any carbene is obtained by attaching either the prefix carbene, or the suffix-ylidene to the name of the skeleton from which it is derived. This chapter focuses on the generation of divalent carbon intermediate by extrusion of nitrogen. Several other methods are possible, like formation from tosylhydrazone salts, oxiranes, aziridines, dioxolanes, and pyrazolenines, from cyclopropanes by photocycloelimination, from transition-metal carbene complexes or by α-elimination of α,α-dihalogen compounds.[1,2]

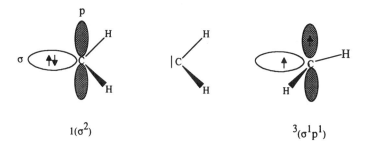

$1(\sigma^2)$ $\qquad\qquad\qquad\qquad$ $3(\sigma^1 p^1)$

A carbene (illustrated above) can exist, in principle, as a singlet $^1(\sigma^2)$ or as a triplet $^3(\sigma p)$ species.[3] Higher excited configurations are possible but do not play a major role in preparative experiments.

Carbenes, as reactive intermediates, are detected by flash spectroscopy. This method is also used to measure directly the rate constant of the addition reaction of free carbenes to olefins. UV

0-8493-8634-9/95/$0.00+$.50
© 1995 by CRC Press, Inc.

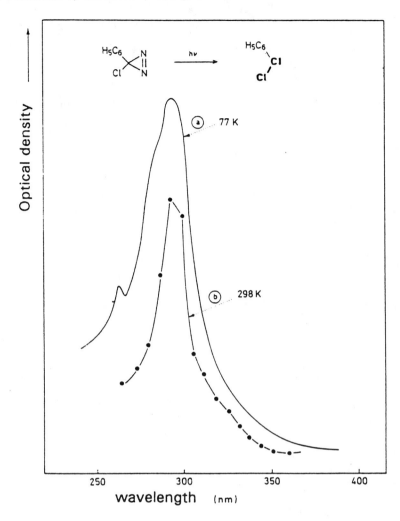

FIGURE 76.1 Ultraviolet absorption spectrum of chlorophenylcarbene in (a) 3-methylpentane at 77 K; (b) isooctane at 20°C. (Adapted from Houben Weyl, E 19b/1.)

absorption spectroscopy of chlorophenylcarbene (Figure 76.1), generated from 3-chloro-3-phenyldiazirine, and its rate of addition to different olefins (Table 76.1) are measured, for example, at 77 and at 293 K in isooctane.[4a]

A further method available for the study of triplet carbenes exclusively is ESR spectroscopy.[3] CIDNP experiments are used as well to investigate the spin state of carbenes in solution. Generation of a carbene from 3H-diazirine or diazocompounds by thermolysis or photolysis with evolution of nitrogen yields a free carbene.

Table 76.1   Addition Reaction Constants of Chlorophenylcarbene with Olefins

Alkene	$k_{abs}[l\ mol^{-1}\ s^{-1}]^a$	$k_{abs}^{rel\ b}$	...-cyclopropane
	2.8×10^8	1.00	1-Chlor-1-phenyl-tetramethyl-...
	1.3×10^8	0.46	3-Chlor-3-phenyl-1,1,2-trimethyl-...
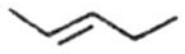	5.5×10^6	0.020	1-Chlor-3-ethyl-2-methyl-1-phenyl-...
	2.2×10^6	0.0079	2-Butyl-1-chlor-1-phenyl-...

^{a} Experimental error + 10%.
^{b} Relative equilibrium constants/absolute equilibrium constants.
Source: Adapted from Reference 39.

UV Absorptions and Quantum Yields of Photoinduced N_2 Extrusion from Diazo Compounds

Diazo compounds	λ_{max} (nm)	ε	Quantum yield φ_{N_2}	Ref.
CH_2N_2	~410	3	4	4b
$H_3C–CH{=}N_2$	440	3.5	—	4b
	470	3.5		
$H_5C_6–CH{=}N_2$	491	26		4c
	275	22000		
$(H_5C_6)_2C{=}N_2$	526	101		4d
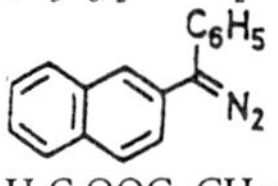	288	21300	0.78	4d
	513	100		
	287	20600	0.69	
$H_5C_2OOC–CH{=}N_2$	360	21		4d
	269	7110	0.66	
	247	7650		
$H_5C_6–CO–CH{=}N_2$	294	13500	0.46	4d
	250	12300		
$4\text{-}H_3C\text{-}C_6H_4–CO–CH{=}N_2$	297	16800	0.42	4d
	260	13200		
$4\text{-}H_3CO\text{-}C_6H_4–CO–CH{=}N_2$	304	24100	0.36	4d
$4\text{-}O_2N\text{-}C_6H_4–CO–CH{=}N_2$	307	13300	0.18	4d
	264	15500		
$4\text{-}Cl\text{-}C_6H_4–CO–CH{=}N_2$	299	14500	0.41	4d
	257	15200		
$3\text{-}O_2N\text{-}C_6H_4–CO–CH{=}N_2$	297	14200	0.22	4d
$2\text{-}H_3CO\text{-}C_6H_4–CO–CH{=}N_2$	288	13000	0.49	4d
	252	12300		
$(H_5C_6–CO)_2C{=}N_2$	275	16600	0.31	4d
	256	22200		
$H_5C_6–CO–CN_2–COOCH_3$	274	10100	0.35	4d
	253	10700		
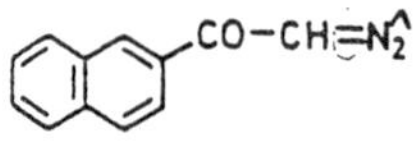	301	11500	0.31	4d
	309	19700	0.36	4d
	261			
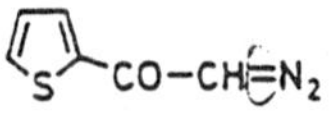				
$ClCH{=}N_2$	485	—		4e
	518	—		
	545	—		
	301	3310	0.24	4d

UV Absorptions and Quantum Yields of Photoinduced N_2 Extrusion from Diazo Compounds

Diazo compounds	λ_{max} (nm)	ε	Quantum yield φ_{N_2}	Ref.
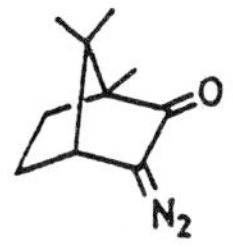	252	12000		
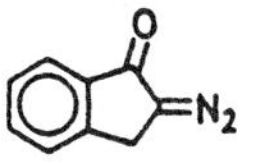	321 256	12150 19000	0.14	4d
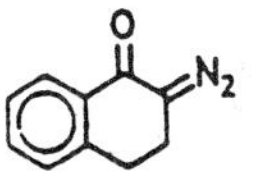	326 260	12000 9780	0.21	4d

Quantum yield φ_{N_2} determined at the wavelength given (in MeOH).

Diazo compounds	λ_{max} (nm)	ε	Quantum yield φ_{N_2}	Ref.
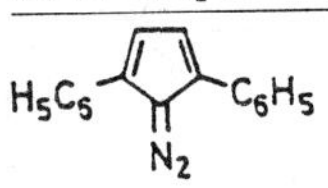	404 368 318 270	2800 4000 5900 10600	—	4f
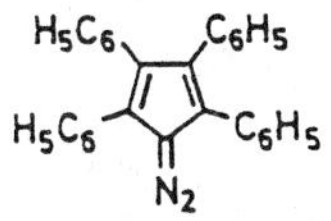	390 335 285 230	2800 13000 17700 26900	—	4f
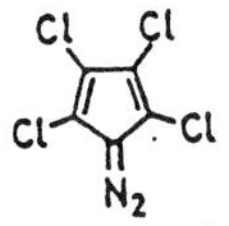	355 310 305	19500 20500 21400	—	4f
$N_2=CH-CO-(CH_2)_4-CO-CH=N_2$	270 248	17300 19000	0.34	4d
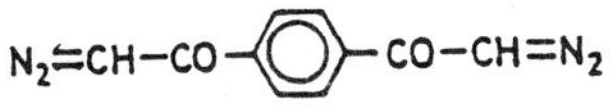 $N_2=CH-CO-\!\!\bigcirc\!\!-CO-CH=N_2$	311 264	22500 17500	0.15	4d
$N_2=CH-CO-CO-CH=N_2$	318 270	11700 15900	0.31	4d
$Hg\left[\begin{smallmatrix}N_2\\ \|\\ C-COOR\end{smallmatrix}\right]_2$	380 264	107 24900	—	4g

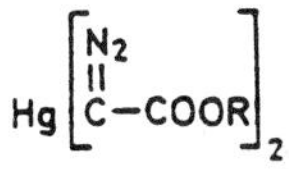

Quantum yields determined for N_2 production.

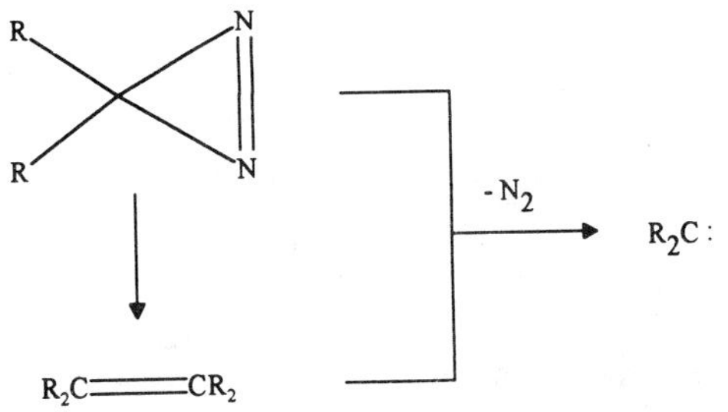

Table 76.2   Relative Reactivities of Alkenes with Bromophenylcarbene[5,6,7]

| Alkene | $H_5C_6-CHBr_2/(H_3C)_3C-OK/(H_3C)_3C-OH$; 25° | | |
		With 18-crown-6	Without 18-crown-6
(see structure)	4.4	4.1	1.6
	2.5	2.4	1.3
	1.00	1.00	1.00
	0.53	0.50	0.29
	0.25	0.24	0.15

Source: Adapted from Reference 39.

Other methods involving base-induced α-elimination from haloalkanes often include the formation of a carbenoid.

$$R_2C\begin{smallmatrix}H\\ \\X\end{smallmatrix} \;+\; M^+Base^- \;\underset{+H\,base}{\overset{-H\,base}{\rightleftharpoons}}\; R_2C\begin{smallmatrix}M\\ \\X\end{smallmatrix} \longrightarrow \begin{cases} \text{carbenoid} \\ \downarrow -MX \\ R_2C\,| \end{cases}$$

$-MX$

Relative reactivities of olefins with bromophenylcarbene generated either (1) by irradiation of 3-bromo-3-phenyldiazirene or (2) from base-induced α-elimination of α,α-dibromotoluene in the presence and in the absence of 18-crown-6 showed that addition of 18-crown-6 (via the complex of carbene with KCl–K–O-*t*-butylate) as well as photolysis of diazirine afforded the free carbene.[5] α-Elimination normally proceeds via the carbenoid.

The relative reactivities of alkenes with bromophenylcarbene are given in Table 76.2.[5–7]

Carbenes can react according to (1) intramolecular or (2) intermolecular. The basic reactions are addition and insertion. Triplet carbenes may also react via an abstraction mechanism.

As typical examples, the intermolecular and intramolecular reactions with a double bond are shown:

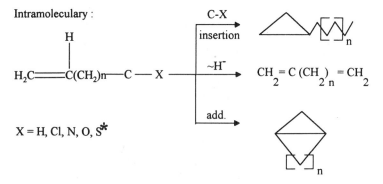

*C-X insertion is not included in this survey, because the emphasis is on the synthetic use of carbene chemistry

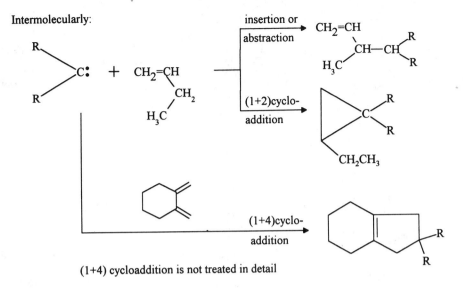

(1+4) cycloaddition is not treated in detail

As a matter of convenience, generation of each type of carbene will be treated separately.

6.2 Formation of Carbenes

Carbenes of Coordination Number 1

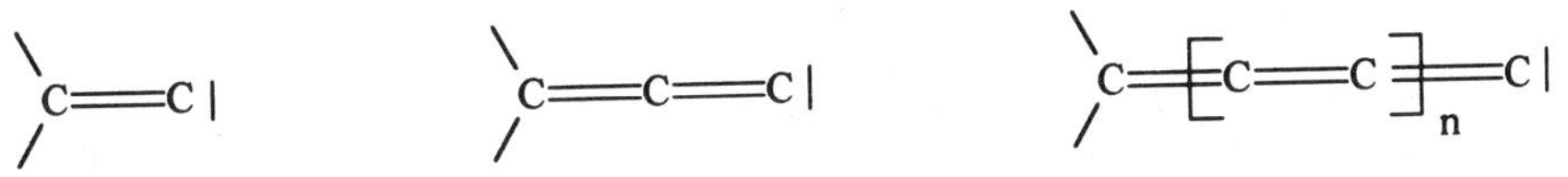

1

Alkylidene

2

Alkenylidene

3

Alkpolyenylidene

n>1

Type 1 of carbenes $>C(=C)_n=C:$ (n = 0 → ∞) were mainly formed by methods that include α-elimination, although thermal extrusion of nitrogen was reported from tetrazolones,[8] and 1-aryl-2-diazo-2-oxo-1-silylethane[9] or from 2,3-*bis*[N-nitrosoacylamino]-1,1-diphenylcyclopropane.[10]

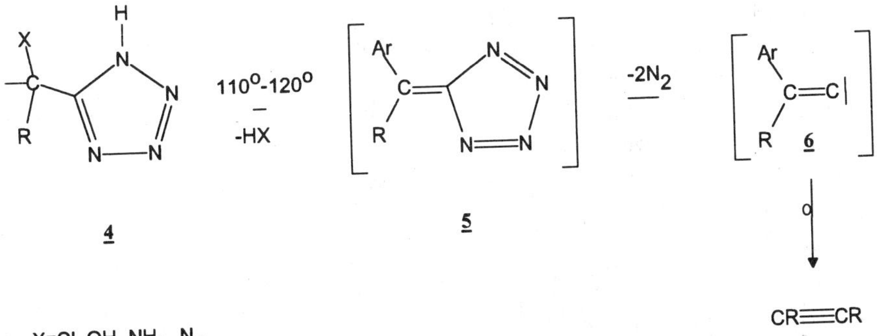

4

110°-120°
—
-HX

5

-2N$_2$

6

CR≡CR

7

X=Cl, OH, NH$_2$, N$_3$

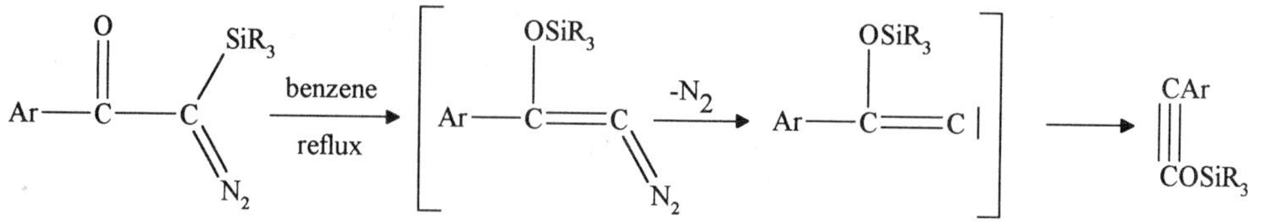

8   **9**   **10**   **11**

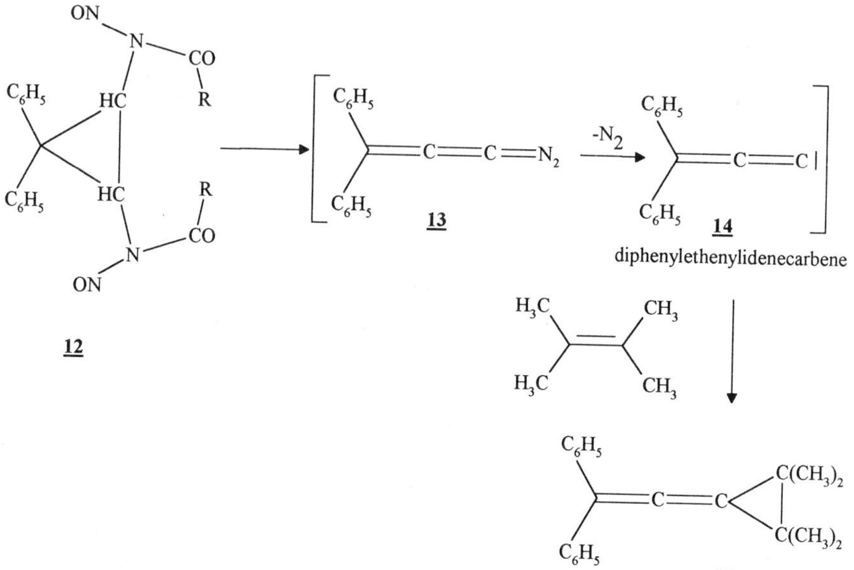

12   **13**   **14**

diphenylethenylidenecarbene

15

Carbenes of Coordination Number 2

Acyclic Homocarbenes (Methylene, Alkyl-, Dialkyl-, Alkenyl-, and Alkynylcarbenes)

From 3H-Diazirine. Herzberg and collaborators[11–13] provided the first experimental evidence that a metastable, bent singlet was initially produced from diazoalkane and decayed to a ground-state triplet.[14,15] The liquid-phase chemistry of methylene, for example, is simple in comparison with gas-phase chemistry. However, hot molecules produced by extremely exothermic reactions of methylene play a minor role in the gas phase. Carbenes have been used on countless occasions to make three-membered rings by [1+2]-cycloaddition.

Photolysis and thermolysis of suitably substituted diazirines could be considered as ideal method for generation of free methylene as well as for alkyl- and dialkylcarbenes. Diazirines are preferred precursors compared to acyclic diazo compounds because of their stability and ease of handling.[16] The parent 3H-diazirine reveals a weak absorption ($\varepsilon_{max} = 200$) in the range of 280 to 350 nm. Photolysis of 3H-diazirine in xenon matrix at 4 K gives triplet ground-state methylene; at 77 K, ethenyl radicals are formed.[17–19]

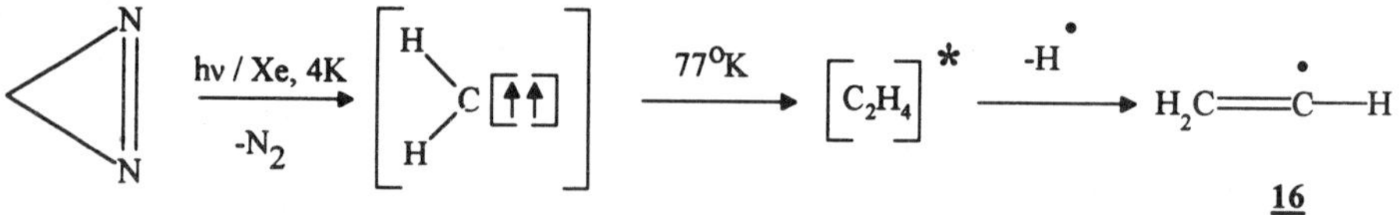

16

The singlet-triplet energy difference of methylene was found to be about 7.5 kcal mol^{-1}.[20] Direct photolysis of 3H-diazirine in chloroform gives 1,1,2-trichloroethane (92% yield) by carbon-chlorine insertion, while in the presence of tetradecane, the carbon-hydrogen insertion product, 1,1,1-trichloroethane, is formed. This fact is explained in the following scheme[21,22] employing CIDNP-NMR measurements to demonstrate the formation of singlet biradicals.

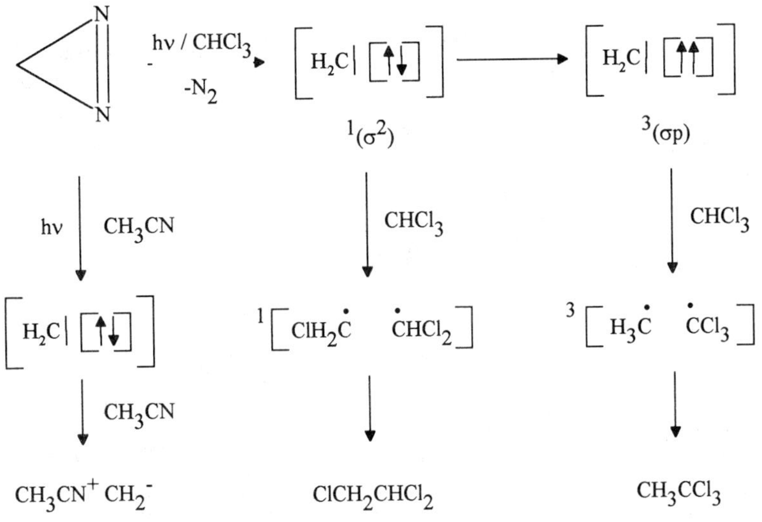

In acetonitrile as solvent, irradiation of diazirine gives a nitrile-ylide as a primary product having an absorption at 280 nm. This ylide was characterized by Turro[23,24] using the pulse-laser technique. It undergoes 1,3-cycloaddition reactions with diethylacetylenedicarboxylate.[25]

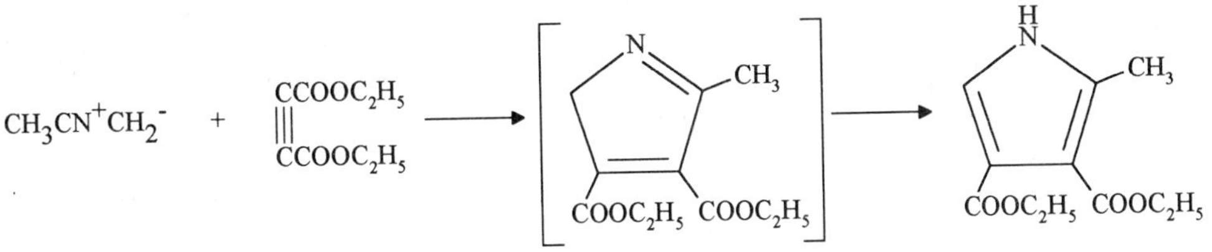

3,4-dicarboethoxy-2-methylpyrrol (45%)

An interesting find was obtained with 3-isopropyl-, 3-t-butyl-, and 3-(1-methyl-1-phenyl-ethyl)-2H-diazirine. Sensitized photolysis of **22** gives very similar results as thermolysis either in the gas phase or in decalin. However, direct photolysis of **22** in benzene leads to similar products but in different ratios. This can be explained by formation of an electronically excited, less selective carbene, formed upon direct photolysis,[2,26–29] a result supported by the work of Frey and Stevens.[28]

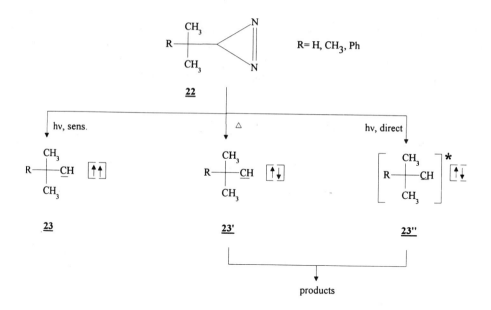

22

Condition	1,1-Dimethylcyclopropane	2-Methyl-2-butene
Gas phase, hv	51.0	49.0
Benzene, hv, 25°C	44.7	53.3
Thioxanthone, hv, 25°C	90.8	9.2
Gas phase, 160°C	92.0	8.0
Decaline, 180°C	88.4	11.6

Table 76.3   Photo- and Thermal Decomposition of 3-*t*-Butyl-3*H*-diazirine[2,26]

From Diazoalkanes. As early as 1901, Hantzsch and Lehman[30] and Staudinger and Kupfer[31] reported on the photochemical and thermal decomposition of diazomethane to afford of methylene. Subsequently, Hine[32] and Doering[33] started the era of carbene chemistry and excellent reviews were published.[34,35] Compared to other methods, the generation of carbenes from diazoalkanes finds wide application in spite of the toxicity and the tendency of low-molecular weight diazoalkanes to explode. Diazoalkanes show a weak UV absorption band between 300 and 500 nm, assigned to a forbidden n,π* transition. Photoexcitation of the long-wave bands of diazo compounds results in an excited singlet state, which may react as follows:

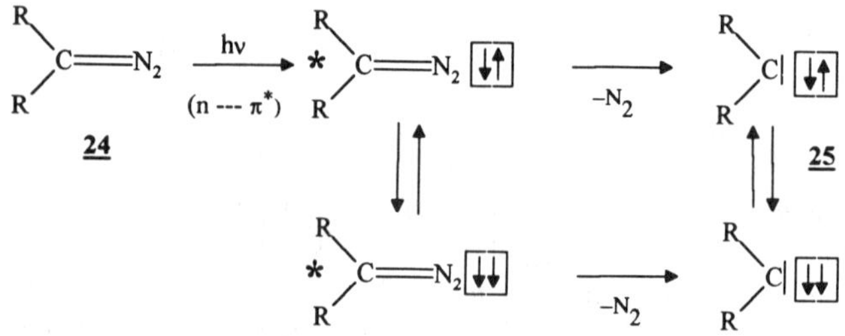

The rate of disappearance of the diazo compound, the quantum yield of the photolysis, should be independent of the nature and concentration of reactants. This is the kinetic criterion for the intermediacy of free radicals. Diazomethane shows a weak UV absorption band between 300 and 500 nm and a sharp band between 200 and 260 nm.[37,38] Photolysis at this band brings about elimination of nitrogen and produces methylene.

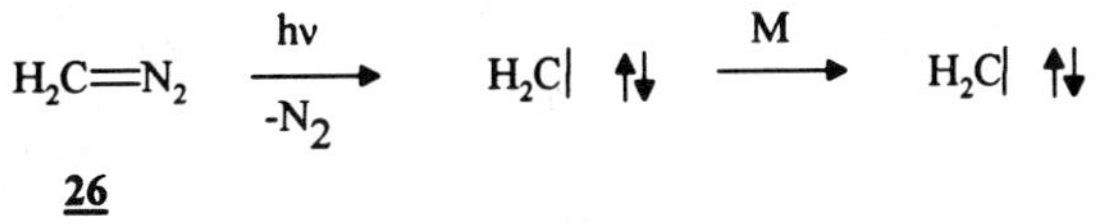

$$H_2C=N_2 \xrightarrow[-N_2]{h\nu} H_2C| \updownarrow \xrightarrow{M} H_2C| \updownarrow$$

26

Evolution of nitrogen from higher diazoalkanes gives the corresponding alkylcarbenes that may undergo (1) intramolecular insertion into β-C-H bonds, (2) 1,2-hydride, alkyl, or aryl shifts, or (3) intramolecular addition in the presence of a double bond; for example:

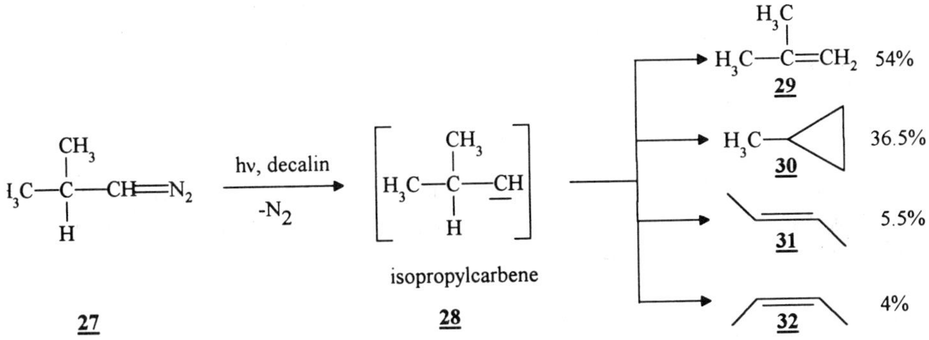

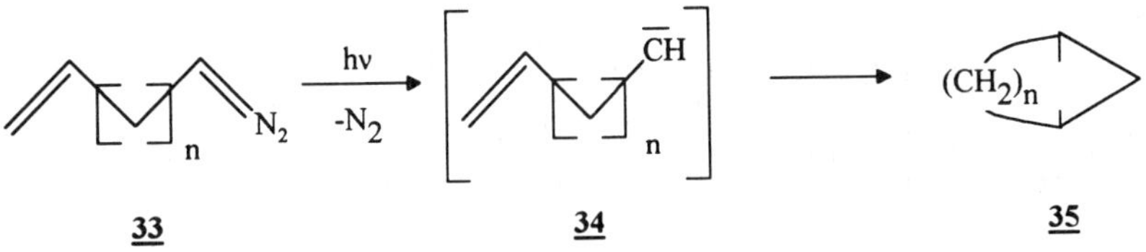

isopropylcarbene

27   **28**

29 54%

30 36.5%

31 5.5%

32 4%

33   **34**   **35**

By comparison, thermolysis of 1-diazo-2-methylpropane gave only methylcyclopropane (33%) and isobutene (67%). Many other carbenes were generated via thermolysis and/or photolysis of the corresponding diazoprecursors (see References 36 to 39). Kirmse and Horn[36] found, in a study of the catalytic, thermal, and photolytic decomposition of diazoalkanes in various solvents, that similarly to diazirines the photodecomposition gave a less selective intermediate than the thermal reaction. Broadly speaking, it may be said that free carbenes are formed from diazo compounds in photochemical and thermal reactions, whereas all other methods yield carbenoids.

Alkenylcarbenes: The situation completely changes when the intramolecular double bond is α,β to the carbene carbon atom or held in close proximity by steric or structural factors. With few exceptions, it can be stated that if a singlet carbene can undergo internal cycloaddition or insertion, it will not be possible to trap it efficiently with an external reagent.[40] Generally, the bicyclo [n.1.0] system is formed. Some other reaction pathways can be predominant, as for instance with 3-diazo-1-propene.[41]

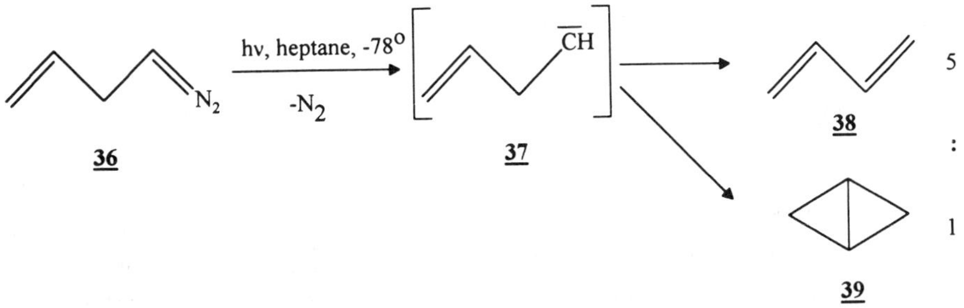

36   **37**   **38** 5

39 : 1

Also, generation of α-alkenyl carbenes in the absence of nucleophilic reagents leads to the formation of cyclopropenes in high yield.[42,43]

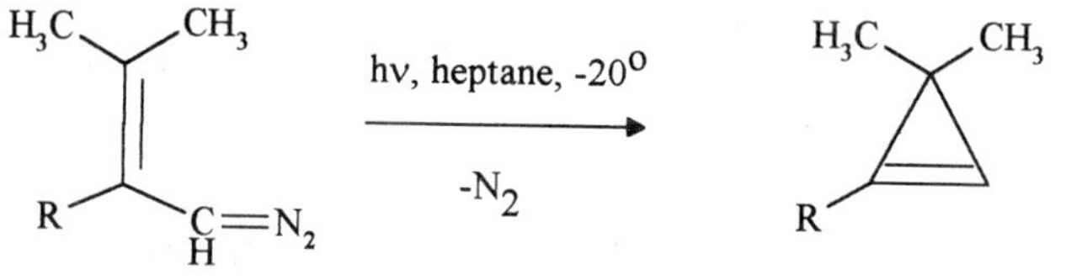

R= CH$_3$ 70%

    C$_6$H$_5$ 60 - 80%

    p-CH$_3$OC$_6$H$_4$ 90%

Alkynylcarbenes

Skell and Klebe reported in 1960 that photolysis of diazopropyne *42* in *E*- and *Z*-2-butene gives a carbene that is slightly nonstereospecific and a mixture of three cyclopropanes *44* to *46* in the ratio 4:2.5:1 with *Z*-2-butene being isolated.[44]

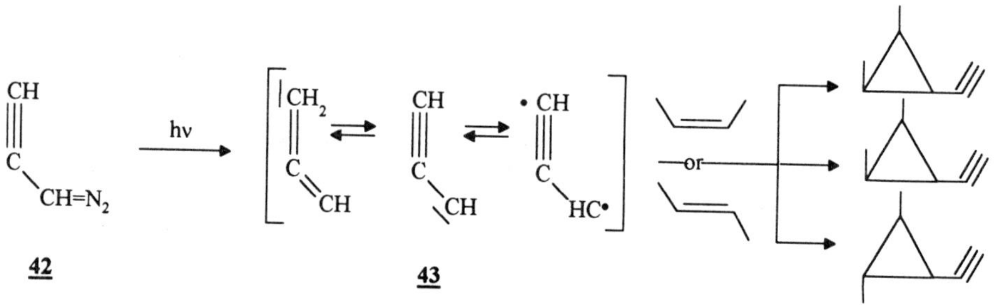

44 : 45 : 46   =   4 : 2.5 : 1   (Z-2-carbene)

                   1 : 2.3 : 63   (E-2-carbene)

Bamford-Stevens Reaction (Methylene-Alkyl-Dialkylcarbenes).   Thermolysis and/or photolysis of tosylhydrazone salts, accessible by condensation of *p*-toluenesulfonylhydrazone with aldehydes or ketones, gives rise to the elimination of the sulfinate salt and forms the intermediate diazo compound. Generation of free carbenes by this method is sometimes in doubt and as a consequence this method it is not included in this article except when generation of a carbene from the diazirine or diazo compound is not available.[2,45]

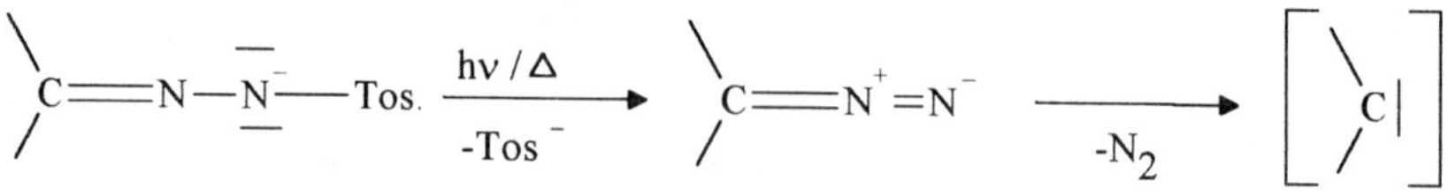

Isocyclic Carbenes

Different cyclopropyl-, cyclobutyl-, cyclopentyl-, cyclohexyl-, and cycloheptyl- (or higher) carbenes are known.[40] A survey is given in the literature.[46] Generally, the shift of an alkyl group results in ring expansion.[40]

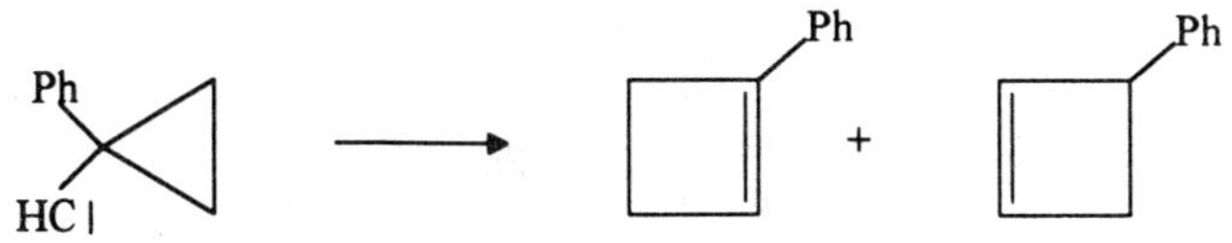

48 49 50

90% 10%

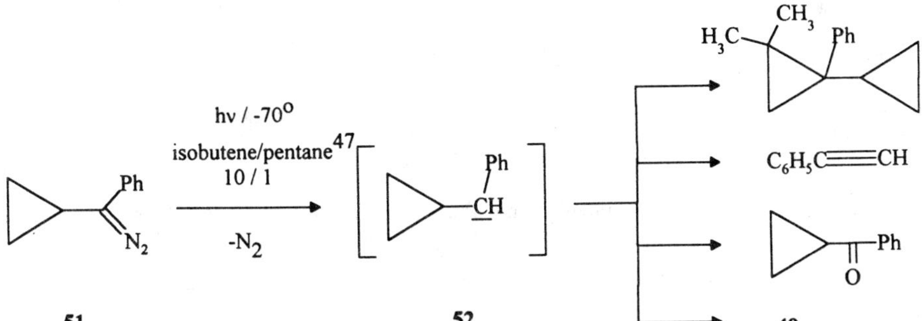

51 52 49

53

54

55

49 : 53 : (54 : 55) = 27% : 34% : 1.9%

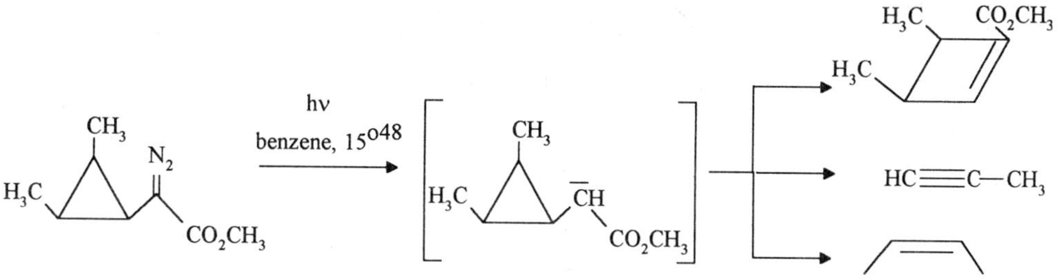

56 57

58 : (59 : 60) = 67% :3

Thermolysis or photolysis of cubylphenylcarbene 61 in the presence of ethanol affords two isomers of ethoxy homocubene 64 and 65.[49,50]

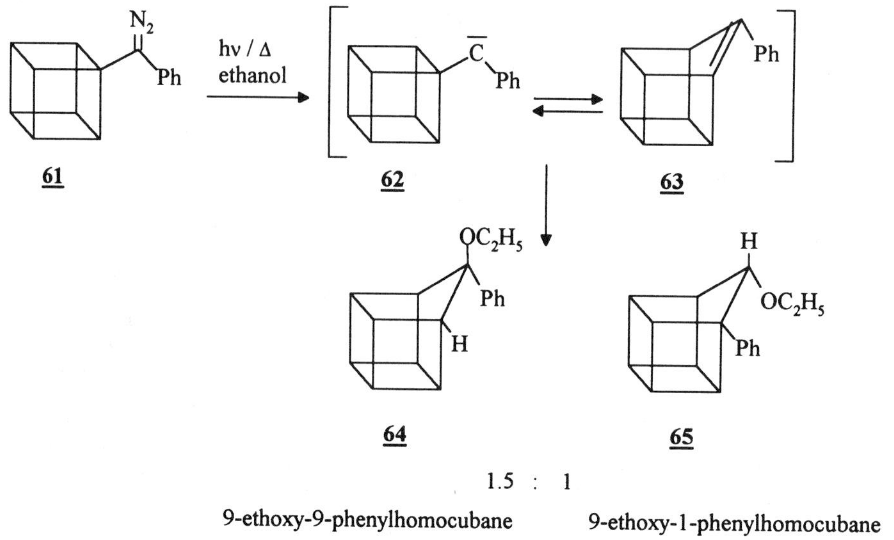
N2
Ph
hν / Δ
ethanol
C
Ph
Ph
61
62
63
OC2H5
Ph
H
H
OC2H5
Ph
64
65
1.5 : 1
9-ethoxy-9-phenylhomocubane
9-ethoxy-1-phenylhomocubane

Also, cyclopentyl-,[51] cyclohexyl-,[51] and cycloheptylcarbene[52] are generated mainly through a Bamford-Stevens reaction and a mixture of products is obtained.

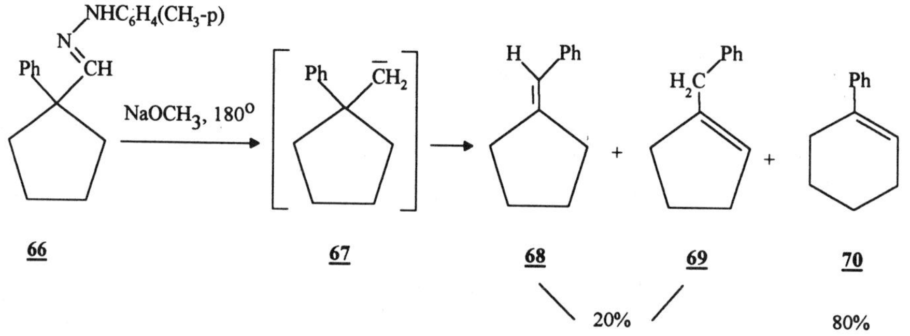
NHC6H4(CH3-p)
N
CH
Ph
NaOCH3, 180°
Ph
CH2
H
Ph
H2C
Ph
Ph
66
67
68
69
70
20%
80%

Cycloalkenylcarbenes

Cycloalkenylcarbenes are carbenes that contain a cyclic side group. Tolane is the major decomposition product from the decomposition of diphenylcyclopropene diazomethane and a minor product is formed via normal ring expansion.[52] Cycloheptatrienylcarbene yields only benzene, acetylene, and heptafulvene in addition to cyclooctatetraene.[53]

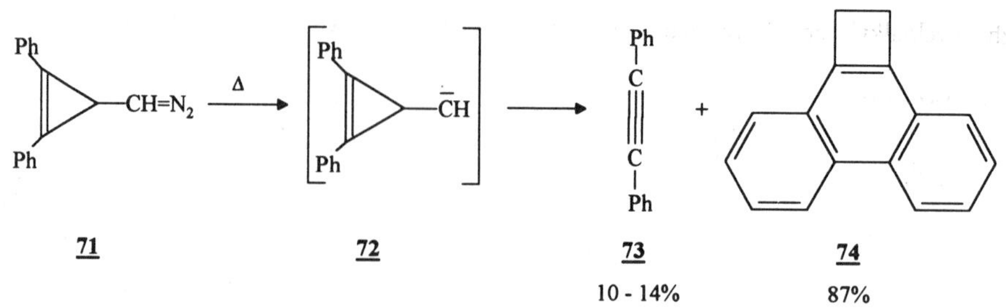
Ph
Ph
CH=N2
Δ
Ph
Ph
CH
Ph
C
C
Ph
71
72
73
74
10 - 14%
87%

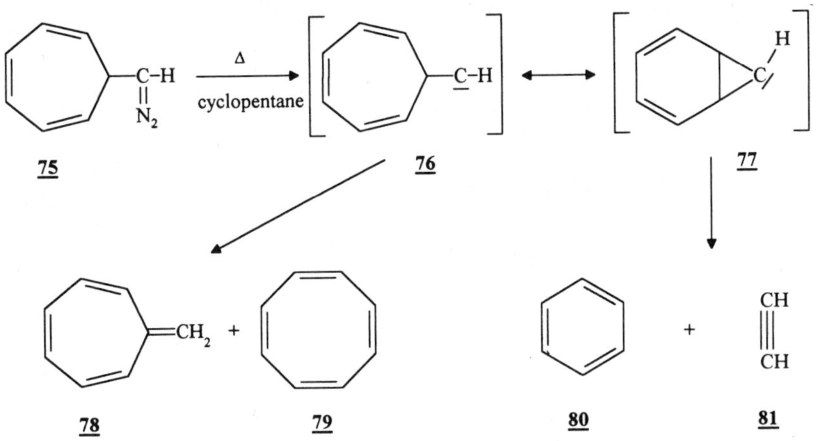

Cycloalkylidenecarbenes

Optically active cyclopropylidenes give optically active allenes.[54,55] Decomposition of diazocyclopentane and diazocyclohexane leads only to cyclopentene and cyclohexene, respectively, but their isomers, cyclopentyl- and cyclohexyldiazirine show different behavior depending on the method of generation. However, photolysis is less selective than thermal decomposition.[56,57]

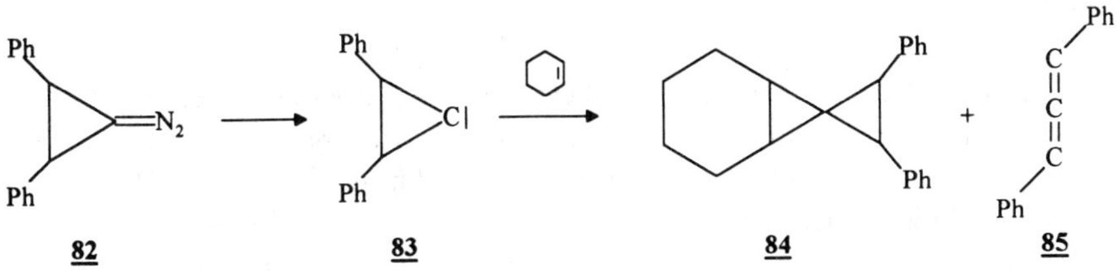

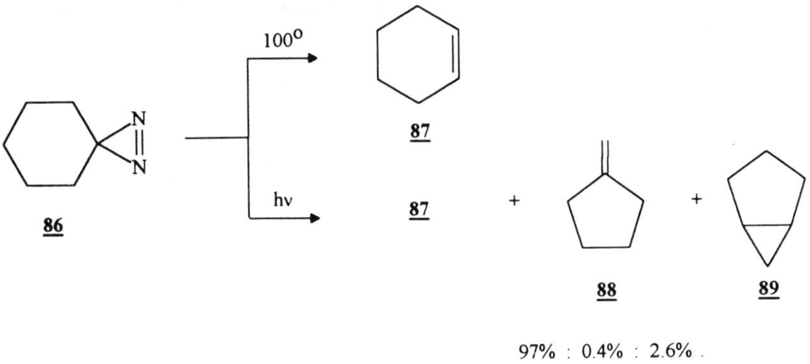

97% : 0.4% : 2.6%

Higher cycloalkylidenes have been generated through methods which do not involve free carbenes.[57]

Cycloalkenylidenes

This class of carbenes is of special interest as the carbene carbon is attached to a conjugated α,β-unsaturated system. Within this definition, a different series of cycloalkenylidenes can be constructed that contain a total of 4n or 4n + 2 electrons.[58] The resonance structures of these carbenes show that carbenes A and E have nucleophilic and carbenes B, C, and D have electrophilic centers.

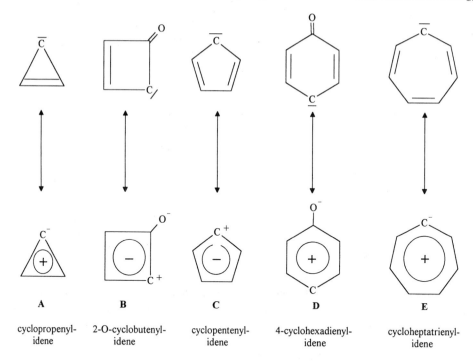

A	B	C	D	E
cyclopropenyl- idene	2-O-cyclobutenyl- idene	cyclopentenyl- idene	4-cyclohexadienyl- idene	cycloheptatrienyl- idene

A variety of phenyl-substituted cyclopentadienylidenes has been examined. Irradiation of 5-diazocyclopentadiene at 8 to 20 K in argon afforded a triplet carbene,[59,60] while irradiation at 293 K in methanol gives a mixture of 1- and 2-methoxy-1,3-cyclopentadiene quantitatively.[61] Irradiation of 5-diazo-1,4-diphenyl-1,3-cyclopentadiene in benzene produced in addition to the dimer 1,1',4,4'-tetraphenylpentafulvalene (6%), 1,4-diphenyl-7*H*-benzocycloheptatriene.[62] Tetrasubstituted 5-diazocyclopentadienes have received attention for generation of the spirocyclopropane adduct upon irradiation in alkenes.[63–65]

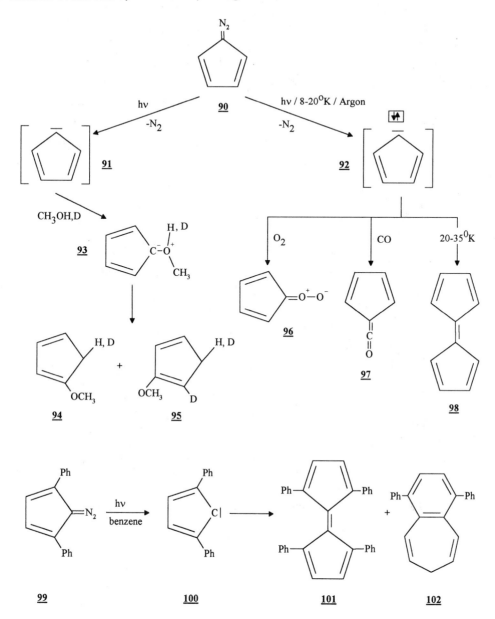

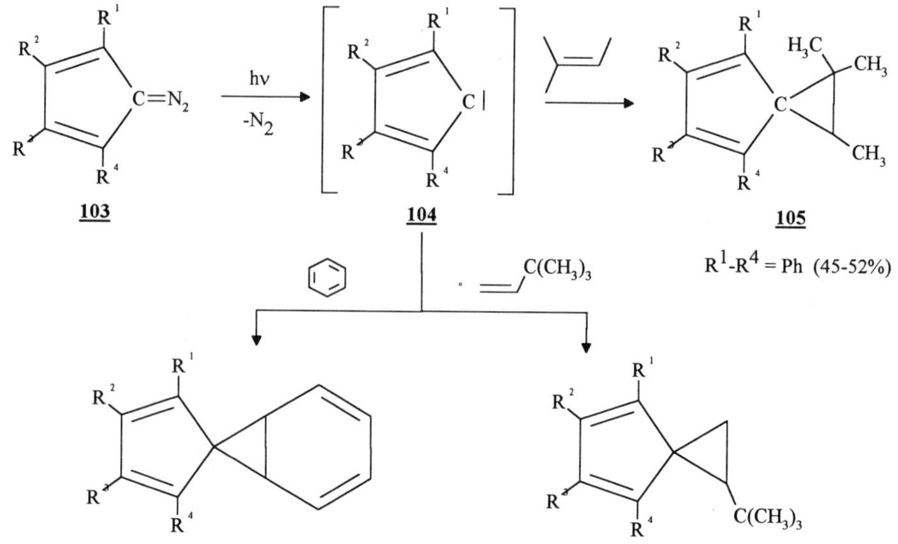

103   **104**   **105**

R^1-R^4 = Ph (45-52%)

106   **107**

c) R^1-R^4 = Ph
d) R^1,R^4 = Ph; R^2, R^3 = H   b) R^1-R^4 = Br (95%)
e) R^1-R^4 = Cl[66]
f) R^1-R^4 = o-phenylene

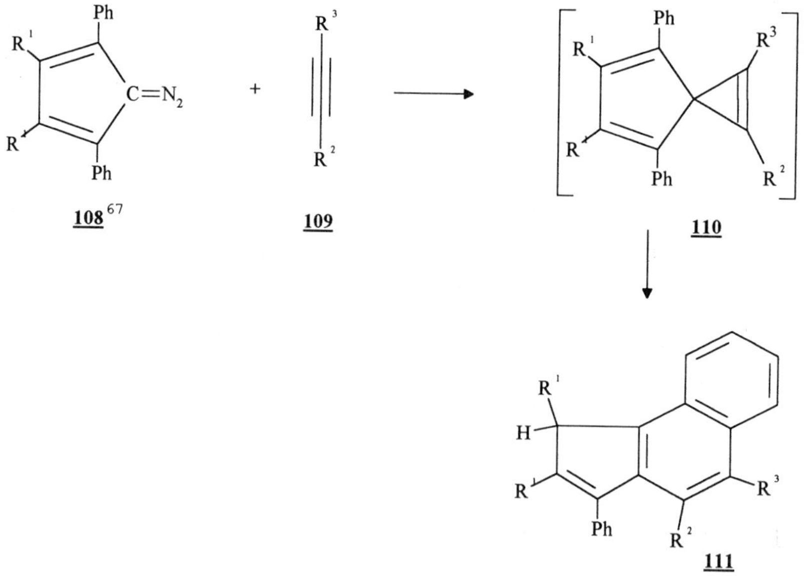

108[67]   **109**   **110**

111

R^1	R^2	R^3	R^1	R^2	R^3
H	H	$CO_2C_2H_5$	C_6H_5	H	$CO_2C_3H_9$
H	CO_2CH_3	CH_3	C_6H_5	CO_2CH_3	CO_2CH_3

Photolysis of 2,5-di-*t*-butyl-3-diazo-6-oxo-1,4-cyclohexadiene at λ = 480 nm (sun lamps) in 2,3-dimethyl-2-butene gave the spirocompound (**112**) in 80% yield.[68]

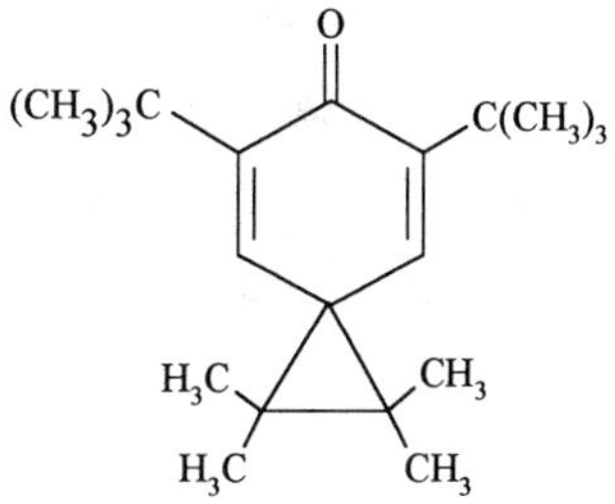

112

Arylcarbenes

Arylcarbenes are probably the most intensively studied carbenes. Normally, arylcarbenes have triplet ground states, as is indicated by ESR and/or CIDNP experiments.[69] Singlet ground-state arylcarbenes are also known.[70,71] Arylcarbenes can be subdivided into (1) monoarylcarbenes, (2) diarylcarbenes, and (3) carbenes in which the carbene carbon is part of the aromatic system.

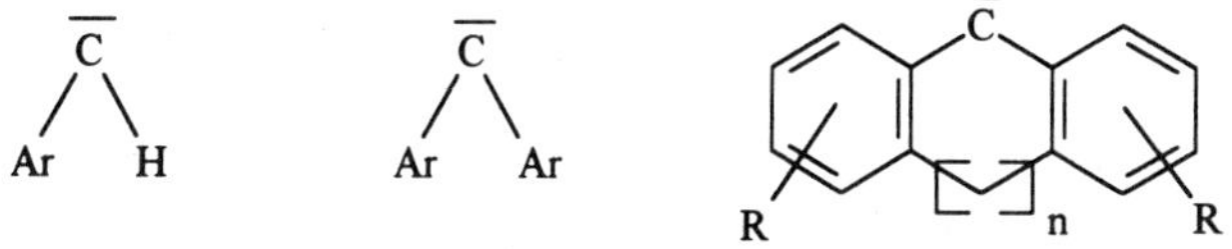

A   **B**   **C**

From Diazirines. Diazirines have been used by many authors for generation of arylcarbenes. Diazirines can directly be decomposed to give carbenes or to rearrange to their diazo isomers, which in turn affords the carbene. Several examples are known.[72,73]

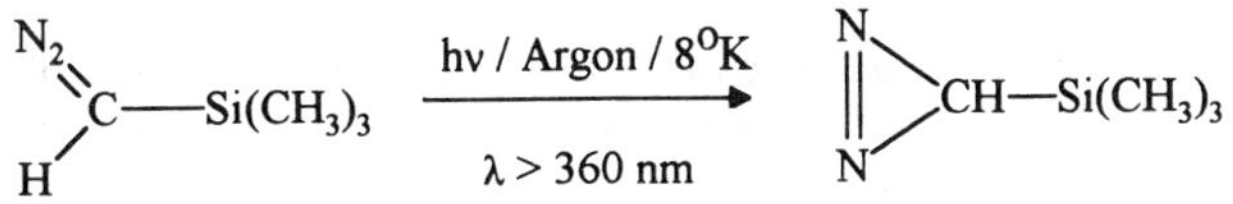

113   **114**

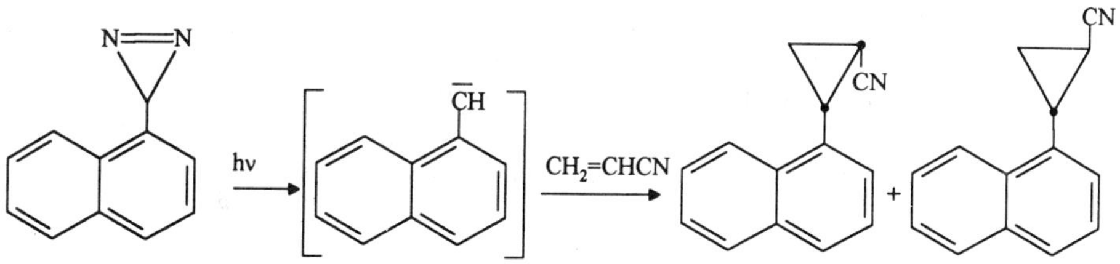

115   **116**   **117**   **118**

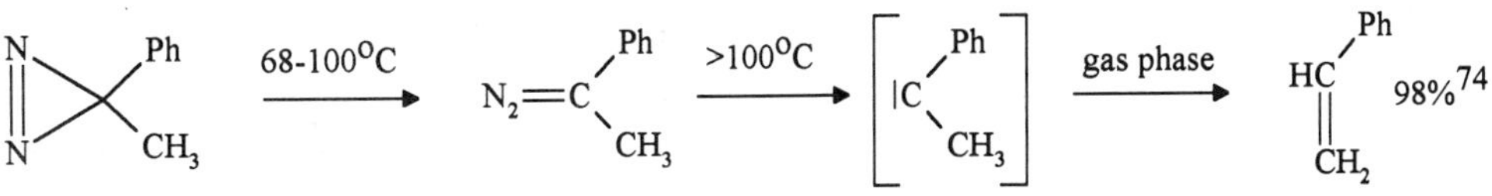

119   **120**   **121**   **122**

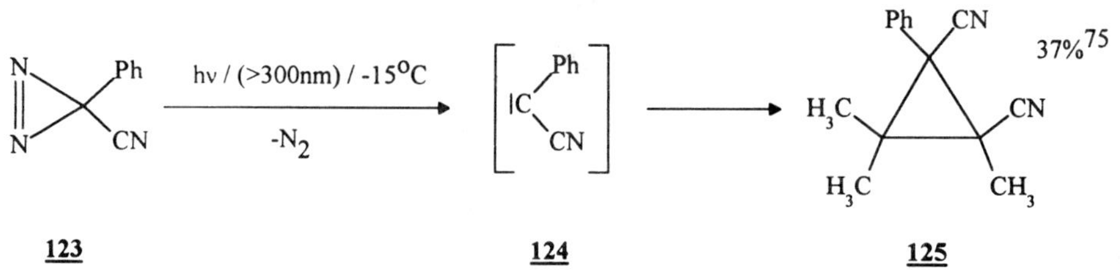

123　　　　　　　　　　**124**　　　　　　　　　　**125**

From Diazo Compounds. Diazo compounds decompose with extrusion of nitrogen under the effect of light or heat to produce carbenes. Phenylcarbene[76] as well as diphenylcarbene[77] possess triplet ground states that are in equilibrium with the singlet configurations.

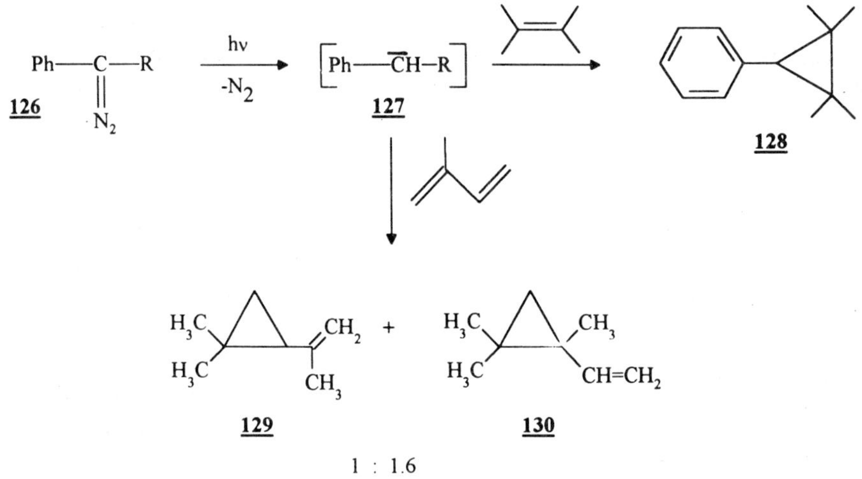

126　　　　　　　**127**　　　　　　　　　　　　　　　　　　**128**

129　　　　　　　**130**

1 : 1.6

Decomposition of phenyldiazomethane in olefins leads to the formation of cyclopropanes.[13,78] The reaction is largely stereospecific (>96%), and there is only a slight preference for the phenyl group to become *syn* to the groups attached to the olefins. Fluorenylidenes have found special interest in the field of carbene chemistry. A large number of fluorenylidene derivatives has been prepared and subjected to extensive studies. Diphenylcarbene represents a compound bridged by an infinite chain. Fluorenylidene has two aryl groups connected by a zero bridge. Fluorenylidene seems to be an exceptional carbene (unlike diphenylcarbene) in that it adds easily and without complications to a variety of olefins.[79,80]

Again, an equilibrium state exists between triplet and singlet states at ambient temperature. Neither fluorenylidene nor diphenylcarbene can be induced to add in a 1,4 fashion to dienes such as **136** to **142**.[81]

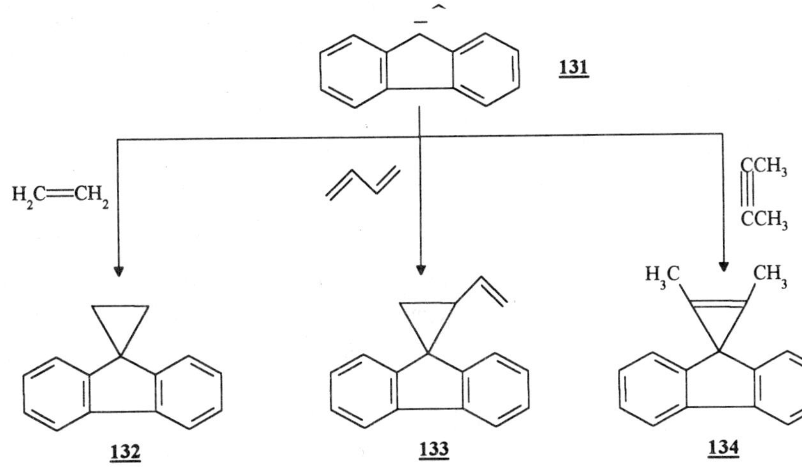

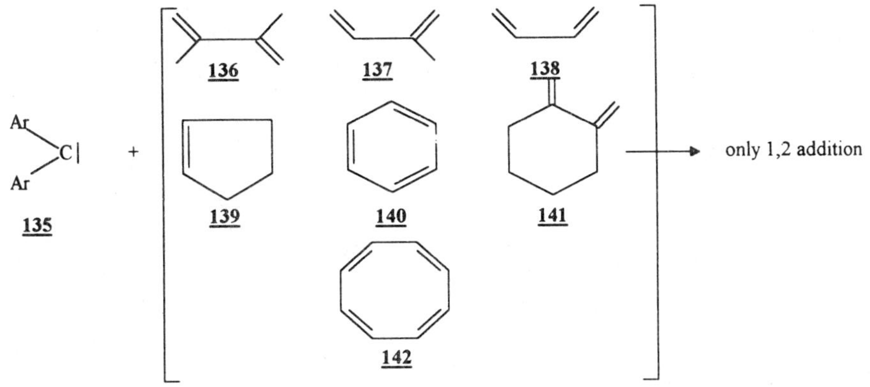

Fluorenylidene attacks benzene to give a mixture of the valence isomeric norcaradiene 143 and cycloheptatriene 144.[82a]

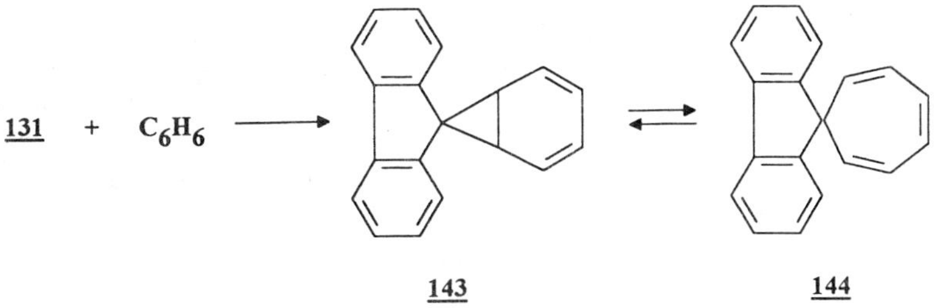

Other aryl- or diarylcarbenes are known, such as carbenes 145 to 152. The carbene 145 is similar to diphenylcarbene in its properties.

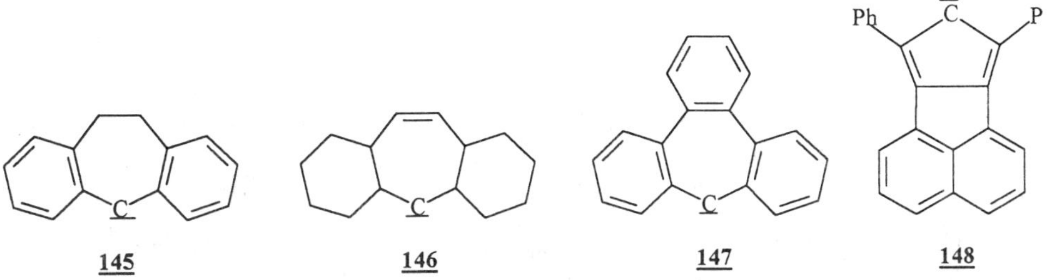

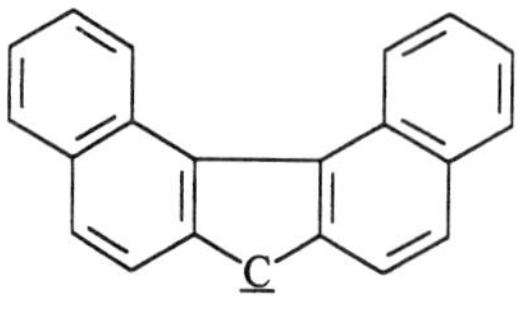

149 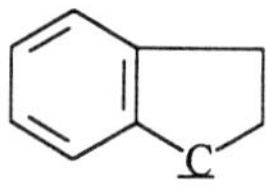   **150** 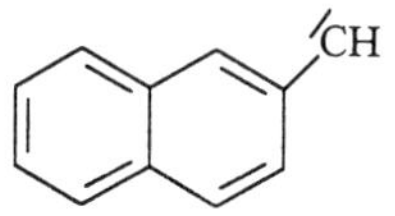   **151** 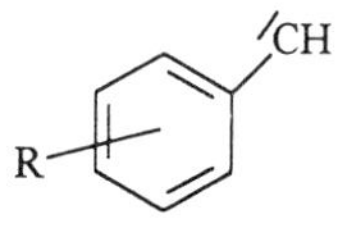   **152**

A few di- and polycarbenes are known and most were studied by spectroscopic techniques. The dicarbenes 153 and 155 are generated by irradiation of the corresponding bisdiazo precursors. Reaction of 153 in toluene gives 154, while 155 undergoes addition to benzene with ring expansion to 156.[82b,c]

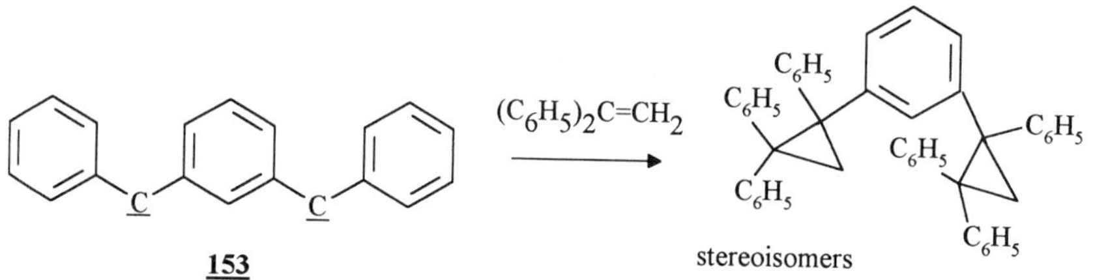

153   stereoisomers

                            154

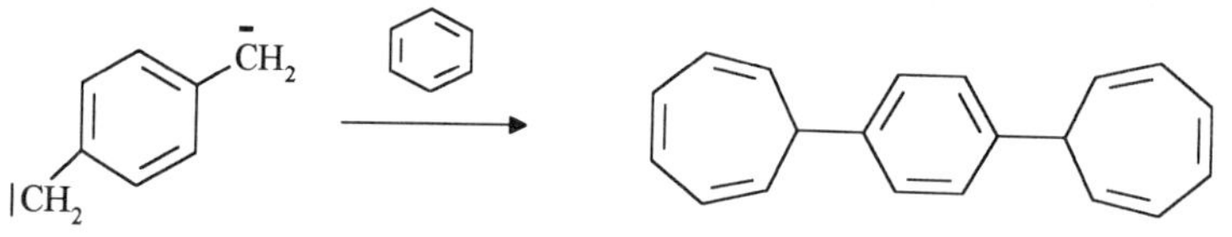

155   **156**

Oxygen-Containing Carbenes

This section includes carbalkoxy- and ketocarbenes. The Wolff rearrangement[2,83] is common to both keto- and carbalkoxycarbenes. This intramolecular reaction is of special significance in the Arndt Eistert synthesis of homologous acids and esters as well as in the synthesis of strained small rings.

Carbalkoxycarbenes undergo both insertion and cycloaddition. Carbethoxycarbene 157 inserts into the α- and β-position of diethylether in the ratio 4:1 to give 159 and 160 in addition to 161 which may be formed from an ylide.[84]

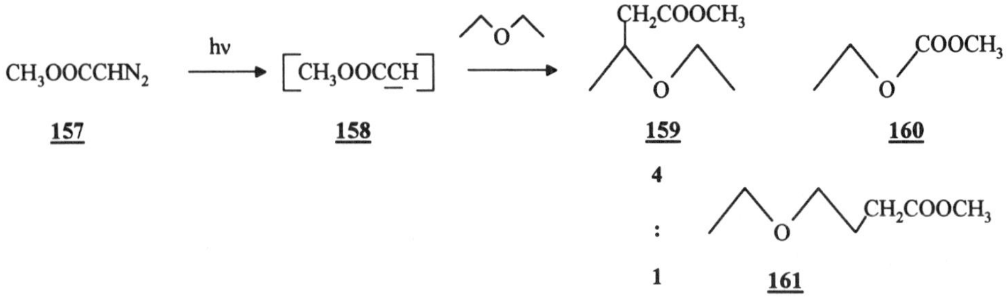

157   **158**   **159**   **160**

                                  4

                                  :

                                  1   **161**

Cycloaddition of 158 to olefins has long been known to be stereospecific with cyclic or asymmetric double bonds to give mainly the less hindered exoadduct 162.[85,86]

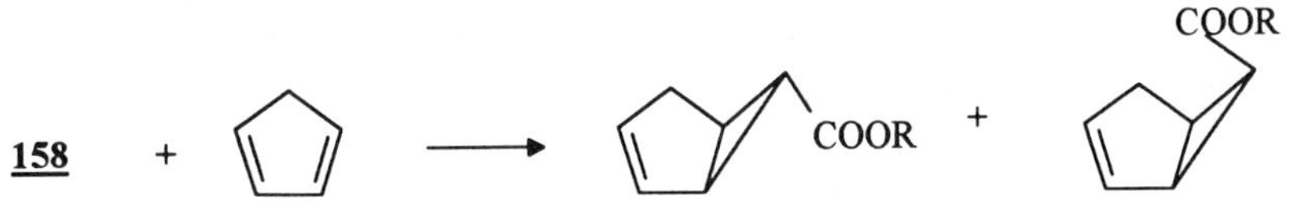

158 + ⟶ **162** + **163**

3 : 1

Ketocarbenes are most prone to Wolff rearrangement; for example, **167** is formed as a product of the rearrangement of **166** during photolysis of **164**.[87,88]

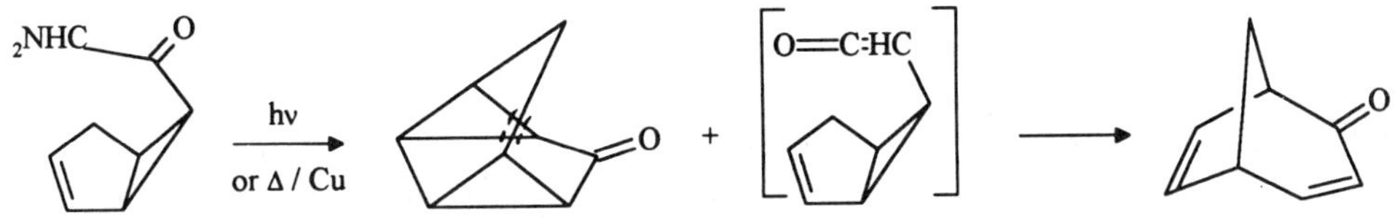

164   **165**   **166**   **167**

However, photolysis of diazoketone **168** in cyclohexene presumedly affords adduct **169** because the mobility of an amide in Wolff rearrangement is low.[89]

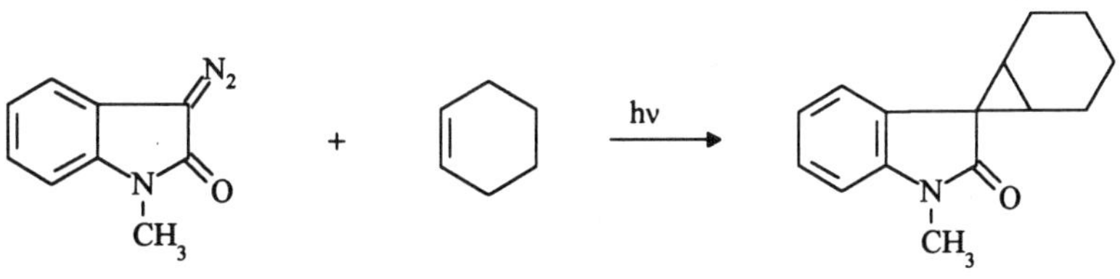

168   **169**

Several other oxygen-containing carbenes are also known, such as carbenes **170** through **175**.

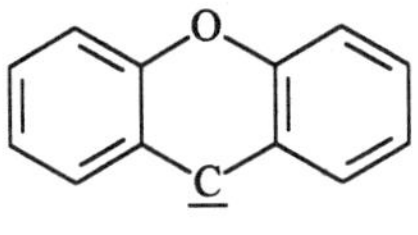

170 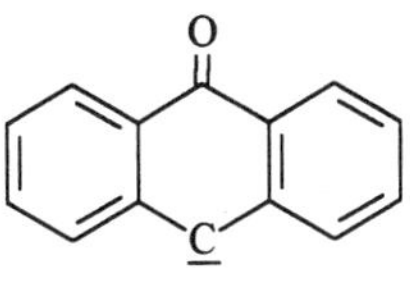   **171** 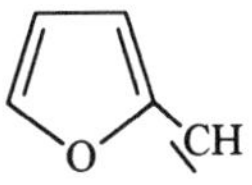   **172** 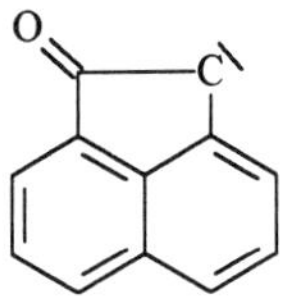   **173**

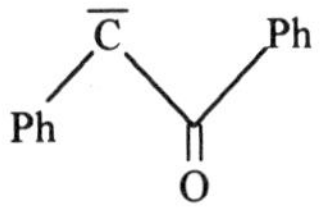

174 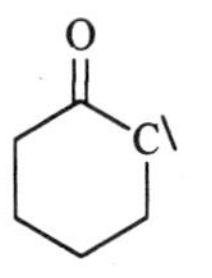   **175**

Halocarbenes

Most of the work on halocarbenes is concerned with carbenoids. Pyrolysis of chlorodiazomethane leads to a less selective intermediate than its carbenoid counterpart.[90] Addition to 2-butene is stereospecific, but the *syn/anti* ratios in various olefins were markedly different.[90]

Photochemically or thermally generated carbenes, unlike most carbenoids, undergo the insertion reaction easily. These results are consistent with the involvement of a free halocarbene in its singlet state.

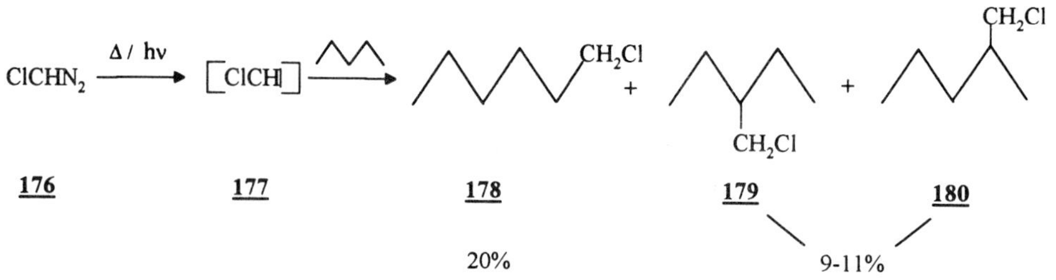

176   **177**   **178**   **179**   **180**

20%   9-11%

Difluorocarbene, prepared by thermolysis or photolysis of difluorodiazirine, yields 1,1-difluorocyclopropanes via stereospecific addition to olefins, consistent with a singlet ground state.[91] The relative reactivity of alkenes with bromophenylcarbene is indicated in Table 76.2.

Sulfur-Containing Carbenes

Sulfur-containing carbenes are known, and several review articles have been published.[92,93] The diazo precursors are prepared either by the action of weak bases on the appropriate *N*-nitroso compound[94] or by the use of diazotransfer reactions.[95,96] The sulfonylcarbene **182** undergoes an addition reaction to olefins and acetylenes to give **183** and **184**.[96,97]

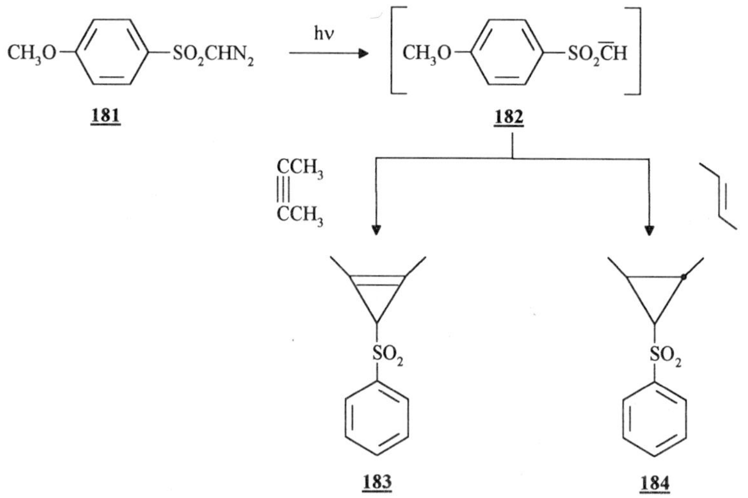

181   **182**

183   **184**

The chemistry of the polynuclear-ring sulfur-containing carbenes thioxanthenylidene-5,5-dioxide **185** indicates easy cycloaddition to styrenes.[98] It reacts from its singlet state and behaves as an electrophilic carbene.

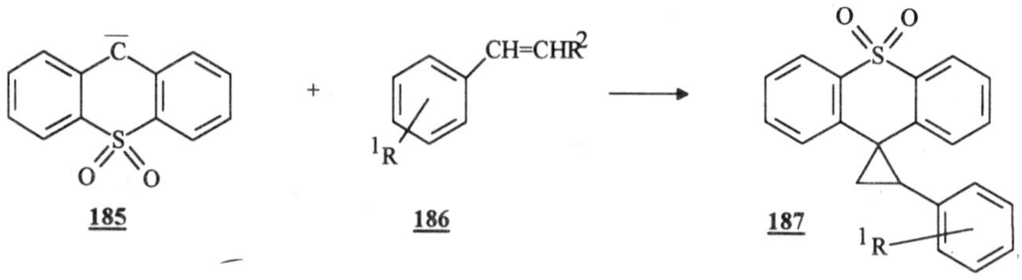

185   **186**   **187**

Nitrogen-Containing Carbenes

Cyanocarbenes, for example, have a triplet ground state as confirmed by ESR spectroscopy.[99] Their stereospecific[14] addition to olefins gives cyclopropanes. The benzene adduct is norcaradiene 190 and not cycloheptatriene.[100]

188 189 190

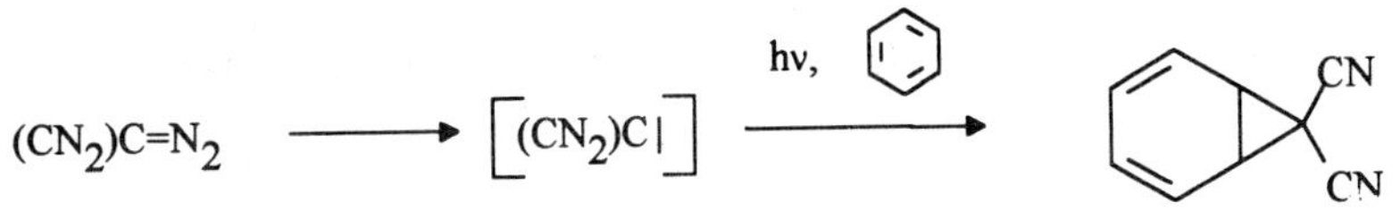

Mono- and dinitrodiazomethane are known.[101,102] Diazafluorenylidenes 191 to 193 add to styrenes and other olefins, mainly affording cyclopropanes 195.[103–105]

191 192 193

194 195

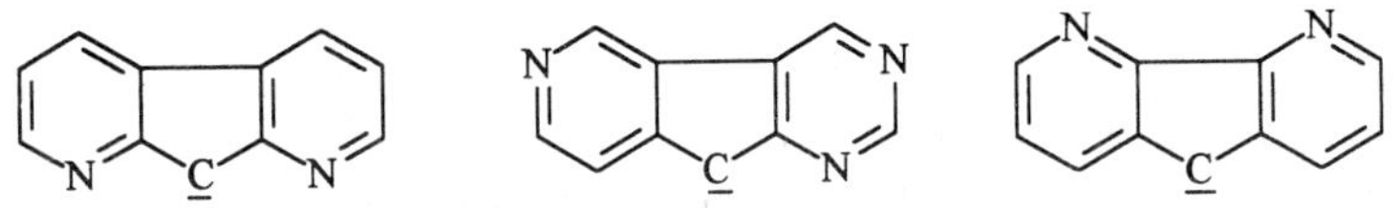

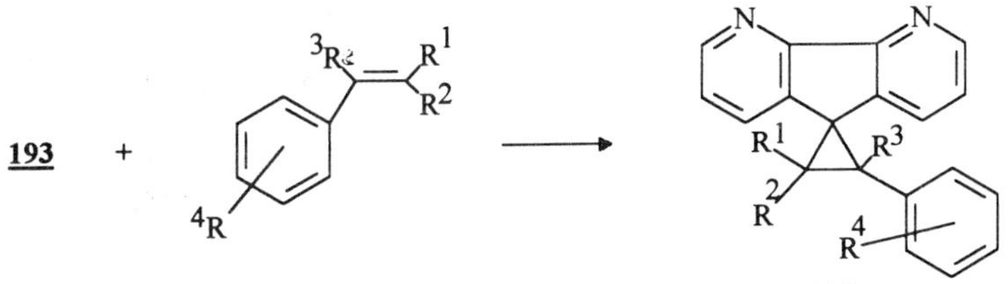

A recent review on nitrogen-containing carbenes is available.[106]

Although only few stable ylides have been made from keto- or carbalkoxycarbenes, carbenes 197 and 200 form stable ylides with alkyl sulfides[107] and isoquinoline,[108] 198 and 201, respectively.

196 197 198

199 200 201

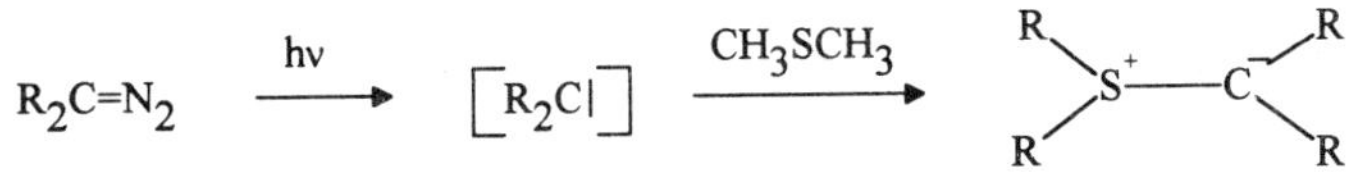

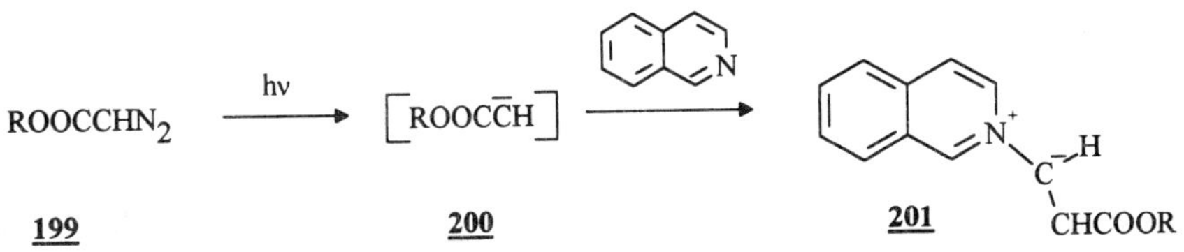

Phosphorus-Containing Carbenes[109]

Generation of phosphorus-containing carbenes from the corresponding diazo precursors is known. Addition to olefins and benzene leads to cyclopropanes 208, insertion products, and norcaradienes 207.[110,111]

202 203 204 205

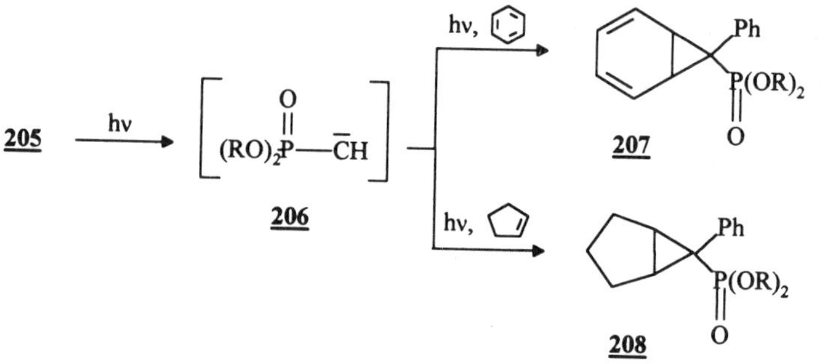

(Str.W2)

Phosphorus-containing carbenes insert into the O-H bond of methanol and undergo Wolff-type rearrangement. The ease of insertion and addition depends on the group attached to the carbene carbon atom.[110]

Carbenes Containing Other Elements

Several diazocompounds, carbene precursors, containing iron, silicon, cadmium, zinc, germanium, lead, tin, mercury, silver, and lithium, are known. Ferrocenylphenylcarbene and ferrocenylmethylcarbene are found to be similar to diphenylcarbene in their reactions with olefins.[112–114]

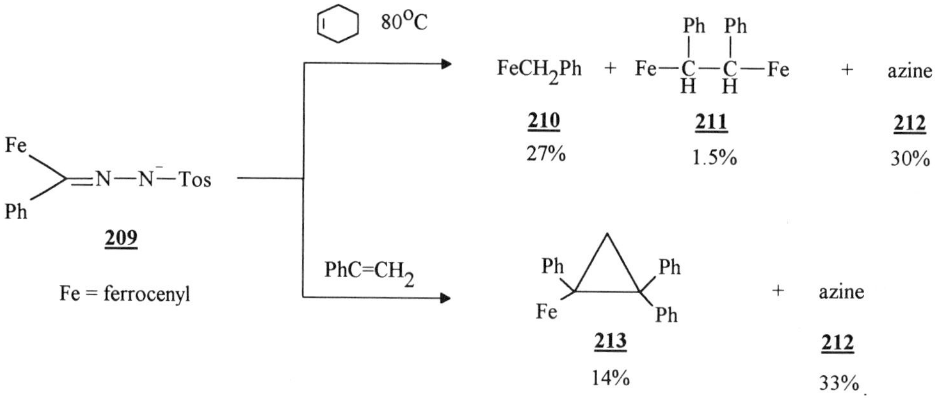

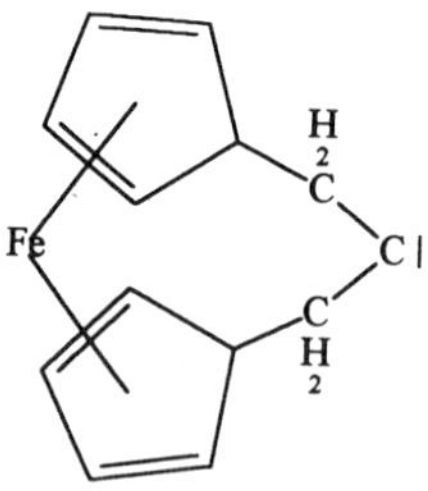

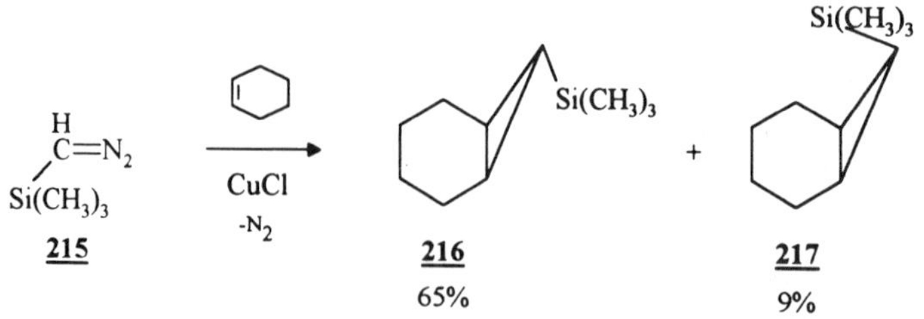

Silicon-containing carbenes, such as trimethylsilyldiazomethane **215**, afford cyclopropanes **216** and **217** on catalyzed decomposition.[115]

Acknowledgments

Our thanks are due the DAAD for financial help (A. A. Abdel-Wahab) and for assistance in typing by S. Kenziora and drawing of reaction schemes by C. Andreis.

eferences

1. Moss, R. A. and Jones, Jr., M., *Carbenes*, I and II, John Wiley & Sons, New York, 1973, 1975.
2. Kirmse, W., *Carbene Chemistry*, Academic Press, New York, 1971.
3. (a) Symous, M., *Chemical and Biochemical Aspects of Electron-Spin Resonance Spectroscopy*, Van Nostrand Reinhold, New York, 1978; (b) *Determination of Organic Structures by Physical Methods*, Vol. 3, Nachod, F. C., and Zuckermann, J. J., Eds., Academic Press, New York, 1971.
4. (a) Gould, I. R., Turro, N. J., Butcher, J., Jr., Doubleday, C., Jr., Hacker, N. P., Lehr, G. F., Moss, R. A., Cox, D. P., Guo, W., Munjal, R. C., Perez, L. A., and Fedorynski, M., Time resolved flash-spectroscopic investigations of the reactions of singlet arylhalocarbenes, *Tetrahedron*, 41, 1587, 1985; (b) Brinton, R. K. and Volman, D. H., The ultraviolet absorption spectra of gaseous diazomethane and diazoethane. Evidence for the existence of ethylidene radicals in diazoethane photolysis, *J. Chem. Phys.*, 19, 1394, 1951; (c) Closs, G. D. and Moss, R. A., Carbenoid formation of acrylcyclopropanes from olefins, benzal bromides and organolithium compounds and from photolysis of aryldiazomethanes, *J. Am. Chem. Soc.*, 86 4042, 1964; (d) Kirmse, W. and Horner L., Quantenausbeuten der Photolyse aliphatischer Diazoverbindungen, *Liebigs Ann. Chem.*, 625, 34, 1959; (e) Closs, G. L. and Coyle, J. J., The halogenation of diazomethane. Study of the reactivities of carbenes derived from halodiazomethanes, *J. Am. Chem. Soc.*, 87, 4270, 1965; (f) Dürr, H. and Scheppers, G., Reaktionen von Carbena-cyclopentadienen mit Cycloheptatrien, *Liebigs Ann. Chem.*, 740, 63, 1970; (g) Strausz, O. P., Do Minh, T., and Fout, J. J., The formation and reactions of monovalent carbon intermediates. II. Further studies on the decomposition of diethyl mercurybisdiazoacetate, *J. Am. Chem. Soc.*, 90, 1930, 1968.
5. Moss, R. A. and Pilkiewicz, F. G., Crown ethers in carbene chemistry. The generation of free phenylhalocarbenes, *J. Am. Chem. Soc*, 96, 5632, 1974.
6. Moss, R. A., Joyce, M. A., and Huselton, J. K., The olefinic selectivity of dibromocarbene, *Tetrahedron Lett.*, 4621, 1975.
7. Stang, P. I. and Magnum, M. G., Unsaturated carbenes from primary vinyl triflates. The nature of vinylidene carbene intermediates, *J. Am. Chem. Soc.*, 91, 6478, 1975.
8. Berhinger, H. and Mainer, M., Ein neuer Abbau des Tetrazol-rings — Acetylen aus substituierten 5-Methyl-1H-Tetrazolen, *Tetrahedron Lett.*, 1663, 1966.
9. Mass, G. and Brückmann, R., Preparation of 1-aryl-2-siloxylalkynes from silylated α-diazocarbonyl compounds, *J. Org. Chem.*, 50, 2801, 1985.
10. Northington, D. J. and Jones, W. M., Diphenyldiazoalkene and diphenylalkenyl diazotate, *Tetrahedron Lett.*, 317, 1971.
11. Herzberg, G. and Shoosmith, Spectrum and structure of the free methylene radical, *Nature*, 183, 1801, 1959.
12. Herzberg G., The spectra and structures of free methyl and free methylene, *Proc. R. Soc. London Ser. A*, 262, 291, 1961.
13. Herzberg, G. and John, J. W. C., The spectrum and structure of singlet CH_2, *Proc. R. Soc. London Ser. A*, 295, 107, 1966.
14. Braun, W. A., Bass, A. M., and Pilling, M., Flash photolysis of ketene and diazomethane: the production and reaction kinetics of triplet and singlet methylene, *J. Chem. Pys.*, 52, 5931, 1970.
15. Borodko, Y. G., Shilov, A. E., and Shteinmann, A. A., Molecular nitrogen activation, *Dokl. Akad. Nauk. SSR*, 168, 581, 1966; *Chem. Abstr.*, 65, 8198, 1966.
16. Liu, M. T., The thermolysis and photolysis of diazines, *Chem. Soc. Rev.*, 11, 127, 1981.

17. Bernheim, R. A., Berhard, H. W., Wang, P. S., Wood, K. S., and Skell, P. S., ^{13}C Hyperfine interaction in CD_2, *J. Chem. Phys.*, 54, 4120, 1971.

18. Wasserman, E., Kuck, V. J., Hutton, R. S., Anderson, E. P., and Yager, W. A., ^{13}C Hyperfine interaction and geometry of methylene, *J. Chem. Phys.*, 54, 4120, 1971.

19. Bernheim, R. A. and Chien, S. H., EPR of methylene — temperature dependence in a xenon matrix, *J. Chem. Phys.*, 66, 5703, 1977.

20. Shavitt, I., Geometry and singlet-triplet energy gap in methylene: a critical review of experimental and theoretical determinations, *Tetrahedron*, 41, 1531, 1985.

21. Roth, H. D., Reactions of methylene in solutions. Selective abstraction reactions of $^1CH_2$ and $^3CH_2$, *J. Am. Chem. Soc.*, 93, 4935, 1971; *Acc. Chem. Res.*, 10, 85, 1977.

22. Turro, N. J. and Cha, Y., Spectroscopic and chemical evidence for methylene singlet-triplet intersystem crossing in solution, *Tetrahedron Lett.*, 27, 6149, 1986.

23. Turro, N. J., Cha, Y., and Gould, I. R., Reactivity and intersystem crossing of singlet methylene in solution, *J. Am. Chem. Soc.*, 109, 2101, 1987.

24. Turro, N. J., Cha, Y., Gould, I. R., Padwa, A., Gasdaska, J. R., and Tomas, M., Carbene and silicone routes toward a simple nitrile ylide. Spectroscopic, kinetic and chemical characterization, *J. Org. Chem.*, 50, 4417, 1985.

25. Padwa, A., Gasdaska, I. R., Tomas, M., Turro, N. J., Cha, Y., and Gould, I. R., Carbene and silicone routes as methods for the generation and dipolar cycloaddition reactions of methyl nitrile ylide, *J. Am. Chem. Soc.*, 108, 6739, 1986.

26. Mansor, A. M. and Stevens, I. D. R., Hot radical effects in carbene reactions, *Tetrahedron Lett.*, 1733, 1966.

27. Shang, K.-T. and Shechter, U., Roles of multiplicity and electronic excitation on intramolecular reactions of alkyl-carbenes in condensed phase, *J. Am. Chem. Soc.*, 101, 5082, 1979.

28. Frey, H. M. and Stevens, I. D. R., The photolysis of 3-*t*-butyl-diazine, *J. Chem. Soc.*, 1301, 1965.

29. Kraska, A. R., Chang, K.-T., Chang, S.-T., Mosely, C. C., and Shechter, H., The effects of multiplicity and excitation on the behavior of 2-methyl-2-phenylpropylidene. The intramolecular chemistry of 2,2-diphenylpropylidene, *Tetrahedron Lett.*, 23, 1627, 1982.

30. Hantzsch, A. and Lehman, M., Über Derivate des Isodiazomethans, *Ber. Dtsch. Chem. Ges.*, 34, 2522, 1901.

31. Staudinger, H. and Kupfer, O., Über Reaktionen des Methylens. III. Diazomethan, *Ber. Dtsch. Chem. Ges.*, 45, 501, 1912.

32. Hine, J., Carbon dichloride as an intermediate in the basic hydrolysis of chloroform. A mechanism for substitution reactions at a saturated carbon atom, *J. Am. Chem. Soc.*, 72, 2438, 1950.

33. Doering, W. v. E. and Hoffmann, A. K., The addition of dichlorcarbene to olefins, *J. Am. Chem. Soc.*, 76, 6162, 1952.

34. Kirmse, W., Neues über Carbene, *Angew. Chem.*, 73, 161, 1961.

35. Dürr, H., Triplet-intermediates from diazo-compounds-carbenes, *Topics Curr. Chem.*, 55, 87, 1975.

36. Kirmse, W. and Horn, K., Vergleich der katalytischen, thermischen und photolytischen Zersetzung von Diazoalkanen, *Chem. Ber.*, 100, 2698, 1967.

37. Figuera, J. M., Fernandez, E., and Avila, J., Photolysis of diazo-*n*-propane. A route for the photochemical activation of propylene, *J. Phys. Chem.*, 78, 1348, 1974.

38. Figuera, J. M., Perez, J. M., and Wolf, A. P., Photolysis of diazo-*n*-butane, *J. Chem. Soc., Faraday Trans. 1*, 1, 1905, 1975.

39. Zeller, K.-P. and Guel, H., *Houben-Weyl*, Methoden der Org. Chemie, Band E 19b/1, Regitz, M., Georg Thieme Verlag, Stuttgart, New York, 1989, chap. 2.

40. Baron, W. J., De Camp, M. R., Hendrick, M. E., Jones, M., Jr., Levin, H., and Sohn, M. B., *Carbenes I*; Jones, M. and Moss, R. A., Eds., John Wiley & Sons, New York, 1973, chap. 1.

41. Lemal, D. M., Menger, F., and Clark, G. W., Bicyclobutane, *J. Am. Chem. Soc.*, 85, 2529, 1963.

42. Closs, G. L., Closs, L. E., and Böll, W. A., The base-induced pyrolysis of tosylhydrazones of α,β-unsaturated aldehydes and ketones. A convenient synthesis of some alkylcyclopropanes, *J. Am. Chem. Soc.*, 85, 3796, 1963.

43. Arnold, D. R., Humphreys, R. W., Leigh, W. J., and Palmer, G. E., Electronic excited state of small ring compounds — cyclopropene, vinylcarbene and vinylmethylene, *J. Am. Chem. Soc.*, 98, 6225, 1976.

44. Skell, P. S. and Klebe, J., Structure and properties of propargylene C₃H₂, *J. Am. Chem. Soc.*, 82, 241, 1960.

45. Shapiro, R. H., Alkenes from tosylhydrazones, *Org. React.*, 23, 405, 1976.

46. F. Bockes and U. H. Brinker, Carbene (oide) carbine, Houben Weyl, Vol. E 19b / part 1, Thieme Verlag, Stuttgart, 1989.

47. Moss, R. A. and Wetter, W. P., Cyclopropylphenylcarbene — thermal control of intermolecular addition, *Tetrahedron Lett.*, 22, 997, 1981.

48. Galluci, R. and Jones, M., Jr., The stereochemistry of intramolecular reactions of cyclopropylcarbenes, *J. Am. Chem. Soc.*, 98, 7704, 1976.

49. Eaton, P. E. and Hoffmann, K.-L., 9-Phenyl-1(9)-monocarbene — probably the most twisted olefin yet known and the carbene 1-phenyl-9-homocubylidene, its rearrangement product, *J. Am. Chem. Soc.*, 109, 5285, 1987.

50. Moss, R. A., Fantina, M. E., and Munjal, R. C., Intermolecular additions of cyclobutylchlorocarbene, *Tetrahedron Lett.*, 1277, 1979.

51. Wilt, J. W., Kostruik, J. M., and Orlowski, R. C., Ring-size effects in the neophyl rearrangement. V. The carbenoid decomposition of 1-phenylcycloalkane — carboxaldehyde tosylhydrazones, *J. Org. Chem.*, 30, 1052, 1965.

52. White, E. H., Meier, G. E., Graeve, R., Zirngibl, U., and Friend, E. W., The synthesis and properties of diphenyl-cyclopropenyldiazomethane and a structural reassignment of the so-called diphenyltetrahedrane, *J. Am. Chem. Soc.*, 88, 611, 1966.

53. Zimmerman, H. E. and Sousa, L. R., Cycloheptatrienyldiazomethane. Its synthesis and behavior; C₈H₈ Chemistry. Correlation diagrams and nodal properties, *J. Am. Chem. Soc.*, 94, 834, 1972.

54. Jones, W. M. and Walbrick, J. M., Effects of solvent on the cyclopropylidene-alkene conversion, *J. Org. Chem.*, 34, 2217, 1969.

55. Jones, W. M. and Wilson, J. W., Jr., The stereochemistry of the cyclopropane-alkene conversion, *Tetrahedron Lett.*, 1587, 1965.

56. Frey, H. M. and Scaplehorn, Thermal decomposition of the diazirines. Part II. 3,3-Tetramethylenediazirine, 3,3-pentamethylenediazirine and 3,3-diethyldiazine, *J. Chem. Soc. (A)*, 968, 1966.

57. Friedman, L. and Shechter, H., Transannular and hydrogen rearrangement reactions in carbenoid decomposition of diazocycloalkanes, *J. Am. Chem. Soc.*, 83, 3159, 1961.

58. Dürr, H., *Topics Curr. Chem.*, 40, 163, 1973.

59. Chapman, O. L., Photochemistry of diazocompounds and azides in argon, *Pure App. Chem.*, 51, 331, 1979.

60. Baird, M. S., Dunkin, I. R., Hacker, N., Poliakoff, M., and Turner, J. J., Cyclopentadienylidene — a matrix isolation study exploiting photolysis with unpolarized and plane-polarized light, *J. Am. Chem. Soc.*, 103, 5190, 1981.

61. Kirmse, W., Loosen, K., and Sluma, H. D., Carbenes and the O–H bond: cyclopentadienylidene and cycloheptatrienylidene, *J. Am. Chem. Soc.*, 103, 5935, 1981.

62. Dürr, H. and Scheppers, G., Addition von varbena-cyclopentadienen an Benzol, *Justus Liebigs Ann. Chem.*, 734, 141, 1970.

63. Dürr, H. and Scheppers, G., Reaktion des photochemisch erzeugten Carbens 1.2.3.4-Tetraphenyl-5-carbena-cyclopentadien-(1.3) mit Olefinen, *Chem. Ber.*, 100, 3236, 1967.

64. Dürr, H. and Bujnoch, W., Gezielte Erzeugung von Triplett-Carbena-Cyclopentadienen durch "Energy-Transfer", *Tetrahedron Lett.*, 1433, 1973.

65. Dürr, H. and Scheppers, G., Photochemische Synthese von substituierten 5H- und 7H-Benzocycloheptenen, *Angew. Chem.*, 9, 359, 1968.

66. Mc Bee, E. T. and Sienkowski, K. J., The preparation and photolytic decomposition of tetrabromodiazocyclopentadiene, *J. Org. Chem.*, 38, 1340, 1973.

67. Dürr, H. and Schrader, L., Eine neue Benzo-cyclopropen Synthese, *Angew. Chem.*, 11, 426, 1969.
68. Koser, G. F. and Pirkle, H. W., *J. Org. Chem.*, 32, 1992, 1967.
69. Trozzolo, A. M. and Wasserman, E., *Carbenes II*, Moss, R. A. and Jones, M., Jr., Eds., John Wiley & Sons, 1975, 185.
70. Chuang, C., Lapin, S. C., Schrok, A. K., and Schuster, G. B., Diemthoxyfluorenylidene — a ground state singlet carbene, *J. Am. Chem. Soc.*, 107, 3238, 1985.
71. Lapin, S. C. and Schuster, G. B., Chemical and physical properties of 9-xanthylidene — a ground state singlet aromatic carbene, *J. Am. Chem. Soc.*, 107, 424, 1985.
72. Chedekel, M. R., Skoglund, M., Kreeger, R. L., and Shechter, H., Solid state chemistry — discrete trimethylsilylmethylene, *J. Am. Chem. Soc.*, 98, 7846, 1976.
73. Meier, H., *Chemistry of Diazirines II*, Liu, M.T.H., Ed., CRC Press, Boca Raton, FL, 1987, chap. 6.1.
74. Liu, M. T. H. and Ramakrishnan, K., Thermal decomposition of phenylmethyldiazirine — effect of solvent on product distribution, *J. Org. Chem.*, 42, 3450, 1977.
75. Cox, D. P., Moss, R. A., and Terpenski, J., Exchange reactions of halodiazirines — synthesis of fluorodiazirines, *J. Am. Chem. Soc.*, 105, 6513, 1983.
76. Kuzai, M., Lüerssen, H., and Wentrup, C., ESR-spectroscopischer Nachweis thermisch erzeugter Triplett-Nitrene und photochemisch erzeugter Triplett-Cycloheptatrienylidene, *Angew. Chem.*, 98, 476, 1986.
77. Wentrup, C., Arylcarbene in Houben Weyl B E 19/1, 848, Thieme Verlag, Stuttgart, 1989, 848.
78. Gutsche, C. D., Bacham, G. L., and Coffey, R. S., Chemistry of bivalent carbon intermediates. IV. Comparative intermolecular and intramolecular reactivities of phenylcarbene to various bond types, *Tetrahedron*, 18, 617, 1962.
79. Jones, M., Jr. and Rettig, K. R., Some properties of triplett fluorenylidene — detection of the singlet state, *J. Am. Chem. Soc.*, 87, 4015, 1965.
80. D'yakanov, I. A., Dushina, V. P., and Goldonikov, Reaction of 9-diazofluorene with trimethylvinylsilane, *Zh. Obsch. Khim.*, 39, 923, 1969; *J. Gen. Chem. USSR*, 39, 887, 1969.
81. Jones, M., Jr., Ando, W., Hendrick, M. E., Kukzycki, A., Jr., Howley, P. M., Hummel, K. M., and Malament, D. S., Irradiation of methyl-diazomalonate in solution — reactions of singlet and triplet carbenes with carbon–carbon double bonds, *J. Am. Chem. Soc.*, 94, 7469, 1972.
82. (a) Dürr, H. and Kober, H., Photochemistry of small ring compounds — new spironorcaradienes. Reversible norcaradiene — cycloheptatriene valence-isomerization, *Angew. Chem. Int. Ed.*, 10, 342, 1971; (b) Murray, R. W. and Kaplan M. L., Sigmatropic reactions in the 1,4-*bis*(cycloheptatrienyl)benzene isomers, *J. Am. Chem. Soc.*, 88, 3527, 1966; (c) Muraharshi, S. I., Yoshimura, Y. Y., and Moritani, I., Quintet carbenes: *m*-phenylene-*bis*(phenylmethylene) and *m*-phenylene-*bis*(methylene), *Tetrahedron*, 28, 1485, 1972.
83. Weygand, F. and Bestman, H. J., Neue präparartive Methoden der Org. Chemie III. Synthesen unter Verwendung von Diazoketonen, *Angew. Chem.*, 72, 535, 1960.
84. Maas G., Organooxycarbonylcarbene, in Houben-Weyl, B.E 19b/Part 2, p. 1134, Thieme Verlag, Stuttgart.
85. Warkentin, J., Singleton, E., and Edgar, J. F., Reactions of ethyldiazoacetate with cyclopentadiene; synthesis of the epimeric ethyl-bicyclo[3.1.0]hex-2-ene-6-carboxylates, *Can. J. Chem.*, 43, 3456, 1965.
86. Skell, P. S. and Etter, R. M., Steric discrimination in reactions of ethoxycarbonylcarbene: norcarane-7-carboxylic acid, *Proc. Chem. Soc.*, 433, 1961.
87. Freeman, P. K. and Kuper, D. G., Synthesis of bicyclo[3.2.1]-octa-3,6-dien-2-one. An unusual valence isomerization, *Chem. Ind.*, 424, 1965.
88. Meinwald, J. and Wahl, G. H., Jr., Tetracyclo[3.3.0.0^{2,8}.0^{4,6}]-octan-3-one, *Chem. Ind.*, 425, 1965.
89. Moriconi, E. J. and Murray, J. J., Pyrolysis and photolysis of 1-methyl-3-diazooxindole. Base decomposition of isatin 2-tosylhydrazone, *J. Org. Chem.*, 29, 3577, 1964.
90. Closs, G. L. and Schwartz, G. M., Carbenes from alkyl halides and organolithium compounds. II. The reactivity of chlorocarbene in its addition to olefins, *J. Am. Chem. Soc.*, 82, 5729, 1960.

91. Mitsch, R. A., Difluorodiazirine. III. Synthesis of difluorocyclopropanes, *J. Am. Chem. Soc.*, 87, 758, 1965.

92. Van Leusen, A. M. and Strating, J., Chemistry of sulfonyldiazomethanes, *Q. Rep. Sulfur. Chem.*, 5, 67, 1970.

93. Schank, K., Organothio-, Sulfinyl- and Sulfonyl-carbene, *Houben Weyl*, E 19b / Teilband 2, 1989.

94. Van Leusen, A. M. and Strating, J., Chemistry of α-diazosulfones. V. The synthesis of arylsulfonyldiazomethanes and alkylsulfonyldiazomethanes, *Rec. Trav. Chim.*, 84, 151, 1965.

95. Regitz, M., *Transfer of Diazo Groups, Newer Methods of Preparative Organic Chemistry*, Vol. VI, Academic Press, New York, 1971.

96. Van Leusen, A. M., Mulder, R. J., and Strating, J., Chemistry of α-diazosulfones. IX. Synthesis of sulfonylcyclopropanes by addition of a sulfonylcarbene to alkenes, *Rec. Trav. Chim.*, 86, 225, 1967.

97. Abramovitch, R. A. and Roy, J., Reactions of benzenesulfonylcarbene, *J. Chem. Soc., Chem. Commun.*, 542, 1965.

98. Abdel-Wahab, A. A., Doss, S. H., Frühauf, E. M., Dürr, H., Gould, I. R., and Turro, N. J., Carbenadibenzocycloheptane — steady-state and time-resolved spectroscopic laser studies, *J. Org. Chem.*, 52, 434, 1987.

99. Ciganek, E., Dicyanocarbene, *J. Am. Chem. Soc.*, 88, 1974, 1966.

100. Ciganek, E., 7,7-Dicyanonorcaradienes, *J. Am. Chem. Soc.*, 87, 652, 1965.

101. Schöllkopf, U. and Markusch, P., Preparation of dinitrodiazomethane by nitration of nitrodiazomethane, *Angew. Chem. Int. Ed.*, 8, 612, 1969.

102. Schöllkopf, U., Tonne, P., Schafer, H., and Markusch, P., Synthesen von Nitro-diazo-essigsäureestern, Nitro-cyan- und Nitro-trifluormethyl-diazo-methan, *Ann.*, 722, 45, 1969.

103. Li, Y. Z. and Schuster, G. B., Photochemistry of 9-diazo-3,6-diazofluorene: through-space or through-bond transmission of electronic effects, *J. Org. Chem.*, 52, 3975, 1987.

104. Mohamed, O. S., Dürr, H., Ismail, M. T., and Abdel-Wahab, A. A., A route to 4,5-diazofluorenylidene: preparation, photo- and thermal rections of 9-diazo-4,5-diazofluorene, *Tetrahedron Lett.*, 30, 1935, 1989.

105. Abdel-Wahab, A. A., Ismail, M. T., Mohamed, O. S., and Dürr, H., *J. Org. Chem.*, submitted.

106. Heydt H. and Regitz M., Carbene mit einer N-Funktion am Carben C-Atom Houben-Weyl E19b/2, Thieme Verlag, Stuttgart, 1989, 1785.

107. Ando, W., Yagihara, T., Tozune, S., and Migita, T., Formation of stable sulfonium ylides via photodecomposition of diazocarbonyl compounds in dimethyl sulfide, *J. Am. Chem. Soc.*, 91, 2786, 1964.

108. Zugravescu, I., Rucinschi, E., and Surpateanu, G., The Action of carbenes on *N*-heterocycles (I) *Tetrahedron Lett.*, 941, 1970.

109. Heydt H. and Regitz M., Phosphorcarbene oder Azido-carbene, Houben Weyl E19b / 2, Thieme Verlag, Stuttgart, 1989, 1822.

110. Regitz, M., Scherer, H., and Anschütz, W., Über die Reaktivität von Phosphoro- und Phosphinylcarbenen, *Tetrahedron Lett.*, 753, 1970.

111. Günther, H., Tunggal, B. D., Regitz, M., Scherer, H., and Keller, T., Application of carbon 13-resonance spectroscopy 2. Two new norcaradiene-cycloheptatriene equilibria, *Angew. Chem. Int. Ed.*, 10, 563, 1971.

112. Sonoda, A., Moritani, I., Saraie, T., and Wada, T., Reactions of ferrocenyl carbenes. 3. Thermal decomposition of acylferrocene tosylhydrazone sodium salt, *Tetrahedron Lett.*, 2943, 1969.

113. Sonoda, A. and Moritani, I., Reactions of ferrocenyl-carbene. III. Additions of some α-ferrocenyl-carbenes to 1,1-diphenyl-ethylene, *Bull. Chem. Soc. Jpn.*, 43, 3522, 1970.

114. Sonoda, A. and Moritani, I., Reactions of ferrocenyl-carbene. IV. Synthesis of [3]ferrocenophan-2-one tosylhydrazone and the thermal decomposition of its sodium salt, *J. Organometal. Chem.*, 26, 133, 1971.

115. Seyferth, D., Dow, A. W., Menzel, H., and Flood, T. C., Trimethylsilyldiazomethane and trimethylsilylcarbene, *J. Am. Chem. Soc.*, 90, 1080, 1968.

Nitrene Formation by Photoextrusion of Nitrogen from Azides

Takashi Tsuchiya
Hokuriku University

77.1 Definition of the Reaction and Mechanism

Photolysis of organic azides leads almost without exception to the elimination of nitrogen, forming nitrenes; therefore, the photoreaction of the azides can be rationalized in terms of nitrene intermediates, which in turn undergo a variety of inter- and intramolecular reactions such as insertion, addition, hydrogen abstraction, and rearrangement reactions.

The photolysis proceeds generally by initial formation of a singlet nitrene that usually decay to the lower-energy triplet state by intersystem crossing in cases where the singlet nitrene does not react, as shown in Scheme 77.1. Therefore, nitrene intermediates can behave as triplet as well as singlet species, depending on the nature of the nitrene and on the photoreaction conditions. For triplet nitrene reactions, deactivation of the singlet formed initially is required and, for example, can be achieved via the "heavy atom" effect of the solvent containing halogen atoms such as bromobenzene or the presence of a triplet sensitizer. In either event, nitrenes have a short lifetime and thus undergo stabilization by the above-mentioned reactions.

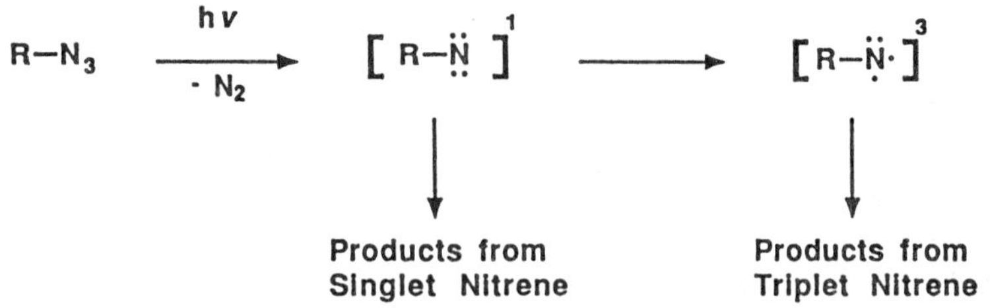

$$R-N_3 \xrightarrow[\;-N_2\;]{h\nu} \left[R-\ddot{\underset{\cdot\cdot}{N}} \right]^1 \longrightarrow \left[R-\ddot{\underset{\cdot}{N}}\cdot \right]^3$$

Products from Singlet Nitrene   **Products from Triplet Nitrene**

SCHEME 77.1

Thermolysis of organic azides also readily produces nitrene and, at first, the photolysis was viewed only as an additional process of formation of nitrenes. Recently, however, the photolysis has been noticed as an important process to the production of nitrenes under milder conditions. Therefore, low temperatures can be employed and also the use of sensitizers allows the generation

0-8493-8634-9/95/$0.00+$.50

of the nitrene specifically in the singlet or the triplet state. Furthermore, the photoreactions are useful for the synthesis of thermally unstable compounds.

Many example reaction of nitrenes generated by either thermolysis or photolysis of organic azides have been reported and extensively reviewed.[1-15] Thus, only the most common types of reaction are outlined, together with several limited examples in the following section. For more details, the readers should consult the references quoted.

7.2 Photoreactions and Synthetic Applications

Alkyl Azides

The typical reactions of the nitrenes generated photochemically from alkyl azides are (1) imine formation by either 1,2-hydrogen migration or 1,2-alkyl migration, and (2) intramolecular hydrogen abstraction with ring closure; the former is more synthetically useful. For example, the imines (2), generated from primary azides (1), give aldehydes (3) by hydrolysis. This conversion provides a valuable route for the generation of an aldehyde function in protected sugars.[16] In the photolysis of the tertiary alkyl azide (4), competing 1,2-alkyl migrations occur to produce the imines (5) and (6).[17] Analogous reactions are observed on irradiation of the cycloalkane azides (7)[18] and (10),[19] forming the strained bridgehead imines (8) and (11), which can be trapped with methanol or cyanide ion as the adducts (9) and (12), respectively, as shown in Scheme 77.2.

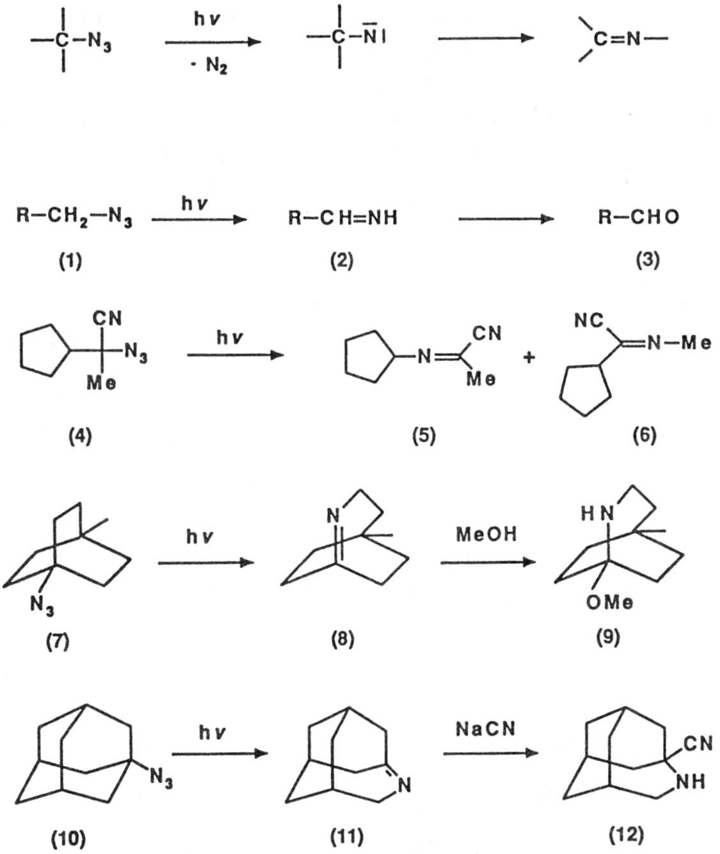

SCHEME 77.2

Acyl Azides

Photolysis of acyl azides results in either nitrene formation or the Curtius-type rearrangement to an isocyanate. The latter is considered generally to involve a concerted process and no nitrene intermediates are involved. The singlet acylnitrenes undergo mainly stereoselective addition to C=C bonds producing aziridines or insertion into C-H bonds, whereas triplet nitrenes undergo hydrogen abstraction to yield amides, as shown in Scheme 77.3.

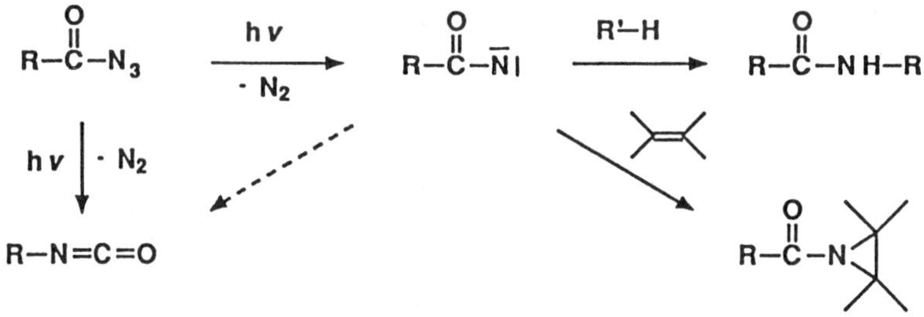

SCHEME 77.3

Phenylcarbamoyl azide (13), on irradiation in methanol, is readily converted into the hydrazinecarboxylate (15) via the isocyanate (14).[20] Alkoxycarbonyl azides (16)[21] and (18)[22] are efficiently transformed on irradiation into the cyclic amides (17) and (19), respectively, via the intramolecular C-H insertion of the nitrenes generated initially. Aziridines formed by addition of nitrenes to olefins are potential intermediates in synthesis. For example, the amino-ketal (22) is obtained by irradiation of 2-methylenepyran (20) in methanol in the presence of methyl azidoformate via the aziridine (21).[23] Benzoyl and ethoxycarbonyl nitrenes react with diketene (23), giving rise to the pyrrolinones (25) via the adducts (24).[24] Photolysis of benzoyl azide in acetonitrile produces the oxadiazole (26), whereas in acetone the dioxazole (27) is formed, probably via the addition of benzoyl nitrene to the solvent.[25] (See Scheme 77.4.)

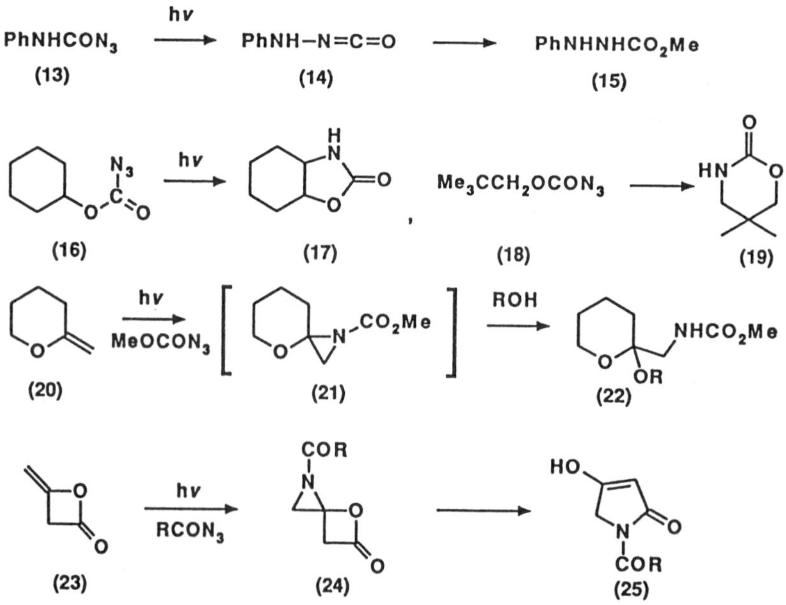

SCHEME 77.4

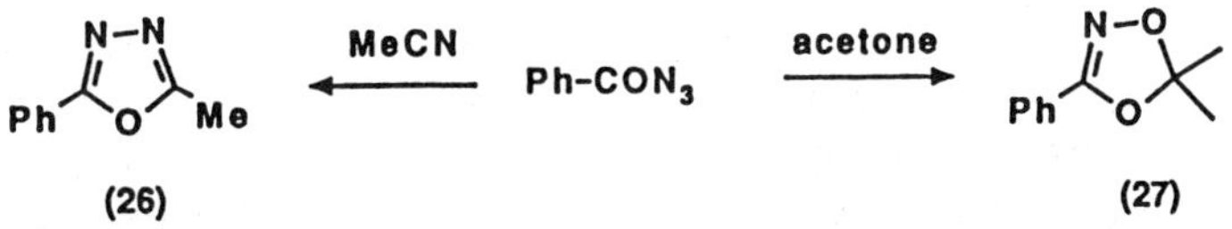

SCHEME 77.4 (continued)

Vinyl Azides

The most general photoreaction of vinyl azides is the formation of 2*H*-azirines. Even though the reaction is a well-established process, its mechanism still remains unresolved and the initial 2*H*-azirine formation from vinyl azides is, in some cases, believed not to involve vinylnitrene intermediates. However, 2*H*-azirine are considered to be in thermal equilibrium with vinylnitrenes, and thus the photoreactions of vinyl azides appear to involve vinylnitrene as the key intermediates as well as 2*H*-azirines. The reactions of 2*H*-azirines have been used extensively in synthesis.[26] (Scheme 77.5.)

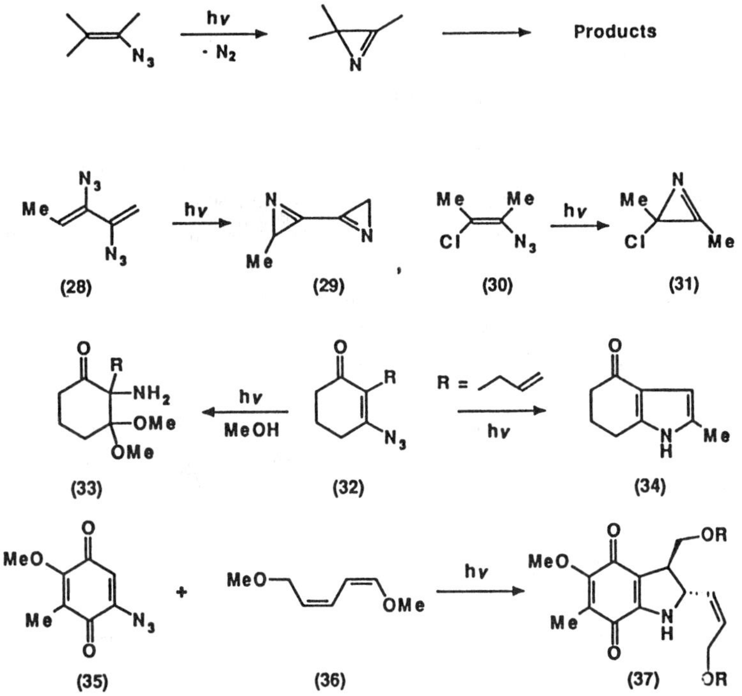

SCHEME 77.5

The vinyl azides (28)[27] and (30)[28] are converted by photoelimination of nitrogen into the 2*H*-azirines (29) and (31), respectively. Photolysis of 2-alkyl-3-azido-2-cyclohexen-1-ones (32) in methanol affords mainly the amino-ketal derivatives (33) via the ring opening of the azirine intermediates by methanol; in the case of the 2-allyl derivative, the tetrahydroindole (34) is also produced.[29] A mitomycin precursor (37) can be prepared with high stereoselectivity by photolysis of the azidoquinone (35) in the presence of the *cis, cis*-diene (36).[30]

Aryl Azides

The reaction of singlet arylnitrenes generated photochemically from aryl azides has been developed and many of these reactions have been described in several reviews.[8-11,14] Consequently, only major points of current interest are shown here. Intramolecular insertion reactions of aryl- and heteroaryl-nitrenes to form five- and six-membered *N*-heterocyclic rings are well-known general high-yield processes. Photolysis of *o*-azidobiphenyls (38) results in the formation of the carbazoles (39) in high yields.[9,31] Replacement of the phenyl ring of the biphenyls by various heterocyclic rings such as pyridine,[32] pyrimidine,[33] thiophene,[31] pyrazole,[34] and benzothiazole[35] rings also leads to cyclizations in good yield on irradiation. 2-Vinyl-3-azidothiophenes (40) also undergo cyclization to the thienopyrroles (41).[36] Biaryl azides of type (42) cyclize initially to the five-membered rings (43), which then undergo rearrangement to the six-membered compounds (44).[37] (Scheme 77.6.)

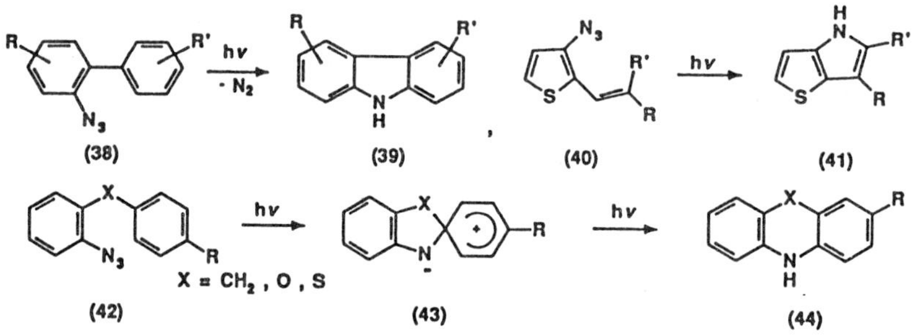

SCHEME 77.6

The singlet phenylnitrenes (46) generated from phenyl azides (45) undergo an intramolecular cyclization to form benzazirines (47), which are in equilibrium with the didehydroazepines (48). The nucleophilic trapping of the azirines with amines or alkoxide ions affords the 3*H*-azepines (51) via the aziridines (49) and the 1*H*-azepines (50),[14,38] as shown in Scheme 77.7. The effects of substituents and nucleophiles on the reaction and the direction of the azirine formation by substituted phenylnitrenes have been well documented. This ring expansion reaction is very useful for the synthesis of a variety of fully unsaturated seven-membered *N*-heterocycles such as azepines, diazepines, triazepines, and related condensed azepines. For example, photolysis of 2-,[39] 3-,[40] and 4-azidopyridines[41] (52) gives 5*H*-1,3-diazepines (53) and (54) and 5*H*-1,4-diazepines (55), respectively. Similarly, 4-azidoquinolines (56) afford 1,4-benzodiazepines (57),[42] and 2,4-benzodiazepines (59) are obtained from 4-azidoisoquinolines (58).[43] The diazine azides (60)[44] and (62)[39] also undergo the photoinduced ring expansion forming 1,2,5- (61) and 1,3,5-triazepines (63).

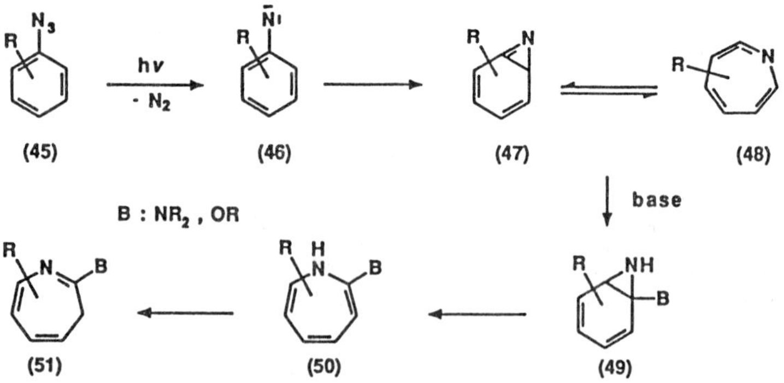

SCHEME 77.7

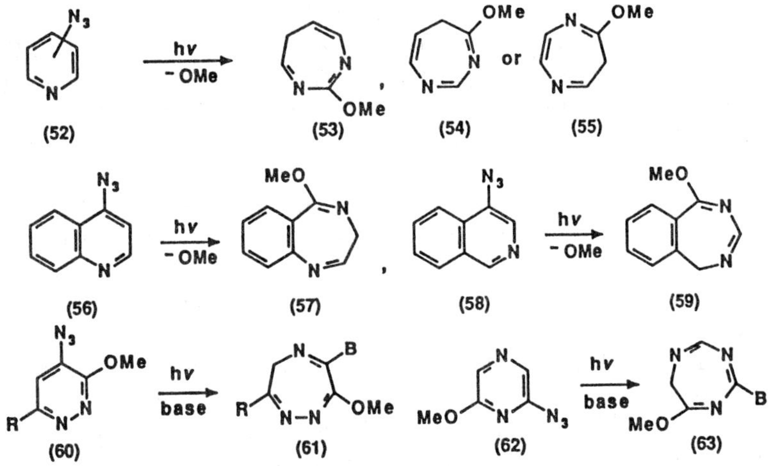

SCHEME 77.7 (continued)

Miscellaneous Azides

Photolysis of silyl, sulfonyl, phosphoryl, and phosphonyl azides also leads to the elimination of the nitrogen, producing the corresponding nitrenes that undergo essentially similar reactions to the above-mentioned C-azides.[14]

References

1. Horner, L. and Christmann, A., Nitrenes, *Angew. Chem., Int. Ed.*, 2, 599, 1963.
2. Abramovitch, R. A. and Davis, B. A., Preparation and properties of imido intermediates, *Chem. Rev.*, 64, 149, 1964.
3. Lwowski, W., Nitrene and the decomposition of carbonylazides, *Angew. Chem., Int. Ed.*, 6, 897, 1967.
4. L'abbe, G., Decomposition and addition reactions of organic azides, *Chem. Rev.*, 69, 343, 1969.
5. Lewis, R. D. and Saunders, W. H., Alkylnitrenes, in *Nitrenes*, Lwowski, W., Ed., John Wiley & Sons, New York, 1970, chap. 3.
6. Lwowski, W., Carbonylnitrenes, in *Nitrenes*, Lwowski, W., Ed., John Wiley & Sons, New York, 1970, chap. 6.
7. Edwards, O. E., Acylnitrene cyclizations, in *Nitrenes*, Lwowski, W., Ed., John Wiley & Sons, New York, 1970, chap. 7.
8. Smalley, R. K. and Suschitzky, H., Reactive intermediates in organic synthesis. III. Nitrenes, *Chem. Ind.*, 1338, 1970.
9. Iddon, B., Meth-Cohn, O., Scriven, E. F. V., Suschitzky, H., and Gallagher, P. T., Developments in arylnitrene chemistry: Syntheses and mechanisms, *Angew. Chem., Int. Ed.*, 18, 900, 1979.
10. Wentrup, C., Carbenes and nitrenes in heterocyclic chemistry: Intramolecular reactions, *Adv. Heterocycl. Chem.*, 28, 231, 1981.
11. Scriven, E. F. V., Current aspects of the solution chemistry of arylnitrenes, in *Reactive Intermediates*, Vol. 2, Abramovitch, R. A., Ed., Plenum, New York, 1982, chap. 1.
12. Reid, S. T., The photochemistry of nitrogen-containing heterocycles, *Adv. Heterocycl. Chem.*, 30, 239, 1982.
13. Moor, H. W. and Goldish, D. M., Vinyl, aryl, and acyl azides, in *The Chemistry of Functional Groups, Supplement D, The Chemistry of Halides, Pseudo-halides and Azides, Part 1*, Patai, S. and Rappoport, Z., Eds., John Wiley & Sons, New York, 1983, chap. 8.

14. Scriven, E. F. V. and Turnbull, K., Azides: Their preparation and synthetic uses, *Chem. Rev.*, 88, 297, 1988.

15. Schuster, G. B. and Platz, M. S., Photochemistry of phenyl azides, *Adv. Photochem.*, 17, 69, 1992.

16. Durette, P. L., Rosegay, A., Walsh, M. A. R., and Shen, T. Y., Synthesis of [6-³H]-*N*-acetylmuramyl-
L-alanyl-D-isoglutamine, *Tetrahedron Lett.*, 291, 1979; Horton, D., Luetzow, A. E., and Wease, J. C.,
Generation of an aldehyde group in a protected sugar by photolysis of a primary azide, *Carbohyd.
Res.*, 8, 366, 1968.

17. Barone, A. D. and Watt, D. S., The photolysis of α-azidonitriles, *Tetrahedron Lett.*, 3673, 1978.

18. Quest, H. and Seiferling, B., Photolysis of 1-azido-4-methylbicyclo[2.2.2]octanes, *Ann. Chem.*,
1553, 1982.

19. Sasaki, T., Eguchi, S., and Okano, T., Synthesis of novel 4-azahomoadamantano[3,4]-fused heterocycles via hydrocyanation of 4-azahomoadamant-3-ene, *J. Org. Chem.*, 46, 4474, 1981.

20. Lwowski, W., deMauriac, R. A., Tompson, M., Wilde, R. E., and Chen, S.-Y., Curtius and Lossen
rearrangement. III. Photolysis of certain carbamoyl azides, *J. Org. Chem.*, 40, 2608, 1975.

21. Wright, J. J. and Marton, J. B., Thermolysis and photolysis of 3β-lanostenyl azidocarbonate:
Functionalization of the 4α-methyl group, *J. Chem. Soc., Chem. Commun.*, 668, 1976.

22. Alewood, P. F., Benn, M., and Reinfried, R., Cyclization of azidoformates to tetrahydro-1,3-oxazin-
2-one and oxazolidin-2-one, *Can. J. Chem.*, 52, 4083, 1974.

23. Kozlowska-Gramsz, E. and Descotes, G., Photochemical addition of methyl azidoformate to 2-
methylenetetrahydropyran and 2-methyl-5,6-dihydropyran, *J. Heterocycl. Chem.*, 20, 671, 1983.

24. Kato, T., Suzuki, Y., and Sato, M., Reaction of diketene with acyl azides, *Chem. Pharm. Bull.*, 27,
1181, 1979.

25. Eibler, E. and Sauer, J., A Contribution to the isocyanate formation in the photolysis of acyl azides,
Tetrahedron Lett., 30, 2569, 1974.

26. Fowler, F. W., Syntheses and reactions of 1-azirines, *Adv. Heterocycl. Chem.*, 13, 45, 1971; Padwa,
A. and Carlsen, P. H. J., Nitrile ylides and nitrenes from 2*H*-azirines, in *Reactive Intermediates*, Vol.
2, Abramovitch, R. A., Ed., Plenum, New York, 1982, chap. 2.

27. Banert, K., Synthesis of new bi-2*H*-azirin-3-yl compounds from diazides, *Tetrahedron Lett.*, 26,
5261, 1985.

28. Ciabattoni, J. and Cabell, M., Jr., 3-Chloro-1-azirines. Photochemical formation and thermal
isomerization, *J. Am. Chem. Soc.*, 93, 1482, 1971.

29. Tamura, Y., Yoshimura, Y., Nishimura, T., Kato, S., and Kita, Y., Photolysis and thermolysis of 3-
azido-2-cyclohexen-1-ones, *Tetrahedron Lett.*, 351, 1973.

30. Naruta, Y., Nagai, N., Yokota, T., and Maruyama, K., Stereospecific photochemical cyclization of
azidoquinone with *E,E*- and *Z,Z*-dienes. Application to the synthesis of an important precursor
toward mitomycins, *Chem. Lett.*, 1185, 1986.

31. Lindley, J. M., McRobbie, I. M., Meth-Cohn, O., and Suschitzky, H., Competitive cyclisation of
singlet and triplet nitrenes. 5. Mechanism of cyclization of 2-nitrenobiphenyls and related systems,
J. Chem. Soc., Perkin Trans. 1, 2194, 1977.

32. Boyer, J. H. and Lai, C.-C., Fragmentation of 2-(2-azidophenyl)-pyridine and isomerisaton of
5λ⁵σ³-pyrido[1,2-*b*]indazole, *J. Chem. Soc., Perkin Trans. 1*, 74, 1977.

33. Hyatt, J. A. and Swenton, J. S., A facile synthesis of 9*H*-pyrimido[4,5-b]indole via photolysis of 4-
azido-5-phenylpyrimidine, *J. Heterocycl. Chem.*, 9, 409, 1972.

34. Albini, A., Bettinetti, G. F., and Minoli, G., Dimethylbenzopyrazolotriazole vs. methyl-
pyrazoloquinoxaline from 3,4-dimethyl-1-(2-nitrophenyl)pyrazole, *Chem. Lett.*, 331, 1981.

35. Hawkins, D., Lindley, J. M., McRobbie, I. M., and Meth-Cohn, O., Competitive cyclization of
singlet and triplet nitrenes. 9. 2-(2-Nitrenophenyl)benzothiazoles and -benzimidazoles, *J. Chem.
Soc., Perkin Trans. 1*, 2387, 1980.

36. Gairns, R. S., Moody, C. J., and Rees, C. W., Photochemical conversion of 3-azido-2-vinylthiophenes
into thienopyrazoles and of 2-azidostyrenes into indoles. High migratory aptitude of sulfur substituents, *J. Chem. Soc., Perkin Trans. 1*, 501, 1986.

37. Lindley, J. M., Meth-Cohn, O., and Suschitzky, H., Competitive cyclization of singlet and triplet nitrens. 6. The cyclisation of 2-azidophenyl thienyl sulfides, *J. Chem. Soc., Perkin Trans 1*, 1198, 1978.

38. Purvis, R., Smalley, R. K., Strachan, W. A., and Suschitzky, H., The photolysis of *o*-azidobenzoic acid derivatives: A practicable synthesis of 2-alkoxy-3-alkoxycarbonyl-3*H*-azepines, *J. Chem. Soc., Perkin Trans. 1*, 191, 1978; Schrock, A. K. and Schuster, G. B., Photochemistry of phenyl azide: Chemical properties of the transient intermediates, *J. Am. Chem. Soc.*, 106, 5228, 1984.

39. Sawanishi, H. and Tsuchiya, T., Ring expansion of α-azidoazines: Formation of the first examples of fully unsaturated monocyclic 1,3,5-triazepines, *J. Chem. Soc., Chem. Commun.*, 723, 1990.

40. Sawanishi, H., Tajima, K., and Tsuchiya, T., Syntheses of 5*H*-1,3-diazepines and 2*H*-1,4-diazepines from 3-azidopyridines, *Chem. Pharm. Bull.*, 35, 4101, 1987.

41. Sawanishi, H., Tajima, K., and Tsuchiya, T., Syntheses of 6*H*-1,4-diazepines and 1-acyl-1*H*-1,4-diazepines from 4-pyridylazides, *Chem. Pharm. Bull.*, 35, 3175, 1987.

42. Sashida, H., Fujii, A., and Tsuchiya, T., Syntheses of fully unsaturated 1*H*- and 3*H*-benzodiazepines from 4-quinolyl azides, *Chem. Pharm. Bull.*, 35, 3182, 1987; Sashida, H., Kaname, M., and Tsuchiya, T., Synthesis of *N*-unsubstituted 1*H*-1,4-benzodiazepines stabilized by intramolecular hydrogen bonding, *Chem. Pharm. Bull.*, 38, 2919, 1990.

43. Sawanishi, H., Sashida, H., and Tsuchiya, T., Photochemical synthesis of 1*H*-2,4-benzodiazepines from 4-azidoisoquinolines, *Chem. Pharm. Bull.*, 33, 4564, 1985.

44. Sawanishi, H., Saito, S., and Tsuchiya, T., Synthesis of fully unsaturated 1,2,5-triazepines by photochemical ring expansion of 4-azidopyridazines, *Chem. Pharm. Bull.*, 38, 2992, 1990.

Photochemistry of Silylldiazoalkanes

Toshio Shimizu
University of Tsukuba

Wataru Ando
University of Tsukuba

78.1 Definition and Mechanism of the Reaction

Since silyldiazoalkanes, trimethylsilyldiazomethane, and trimethylsilyl ethoxycarbonyl diazomethane had been first reported by Lappert et al.[1] and Schöllkopf et al.[2] in 1967, their reactivities have been a subject of great interest. Diazomethane is a well-known indispensable reagent in organic syntheses; however, the utility of diazomethane is often handicapped by its toxicity and explosive nature. On the other hand, trimethylsilyldiazomethane, a derivative of diazomethane where all of the hydrogen is replaced by a trimethylsilyl group, is thermally stable greenish-yellow liquid and even can be distilled at atmospheric pressure (bp 96 °C).[3] Because of its stability and structure, trimethylsilyldiazomethane has found great use as a stable reagent compared with diazomethane. Some applications for organic syntheses have been reviewed up to date.[4–7] This section deals with the photochemistry of silyldiazoalkanes.

After the first preparation[1,8] of trimethylsilyldiazomethane 1 from lithiated diazomethane and trimethylsilylchloride by Lappert et al. (5%) [Equation (1) in Scheme 78.1]. Some other synthetic methods for 1 have been reported. A reaction of diazomethane with trimethylsilyl triflate in the presence of diisopropylethylamine gives 1 in good yield (74%) [Equation (2) in Scheme 78.1].[9] Treatment of *N*-nitroso-*N*-trimethylsilylmethylurea with base also gives 1 along with hexamethyldisiloxane [Equation (3) in Scheme 78.1].[3,10] Other nitroso precursors are also available for the preparation of 1;[11–14] however, the synthesis of these *N*-nitroso compounds can often be multi-step reaction. Diazotization of trimethylsilylmethyllithium by tosyl azide gives 1 in a yield of 38% [Equation (4) in Scheme 78.1].[15] An effective synthetic method by Shioiri et al.[16] utilizes a diazo transfer step from diphenylphosphoryl azide (85%) [Equation (5) in Scheme 78.1].[16]

0-8493-8634-9/95/$0.00+$.50
© 1995 by CRC Press, Inc.

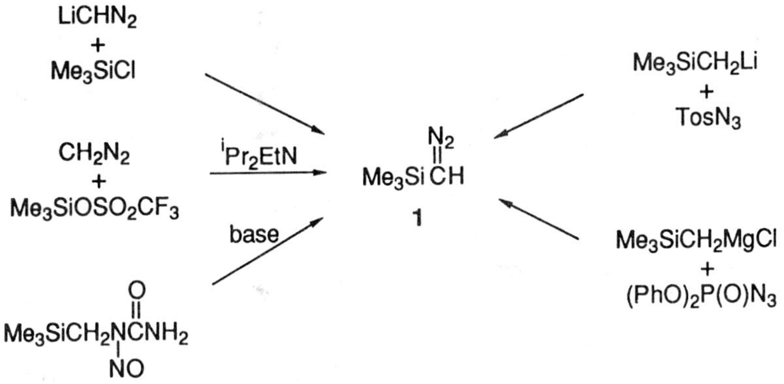

LiCHN$_2$
+
Me$_3$SiCl

CH$_2$N$_2$
+
Me$_3$SiOSO$_2$CF$_3$

$$\xrightarrow[\text{base}]{^i\text{Pr}_2\text{EtN}}$$

O
‖
Me$_3$SiCH$_2$NCNH$_2$
|
NO

Me$_3$SiCH$_2$Li
+
TosN$_3$

$$\underset{\textbf{1}}{\overset{\overset{\text{N}_2}{\|}}{\text{Me}_3\text{Si CH}}}$$

Me$_3$SiCH$_2$MgCl
+
(PhO)$_2$P(O)N$_3$

SCHEME 78.1

Alkyl-, aryl-2,[17,18] alkoxycarbonyl-3,[2,19] and silyl-4a[20] substituted silyldiazomethanes are also known and are the subjects of study. Furthermore, some diazoalkanes substituted with 14 group elements 5a,[21] 6,[17] 4b,[20] 4c,[20] 5b,[1] 7,[2] and 4d[22] have also been prepared (Scheme 78.2).

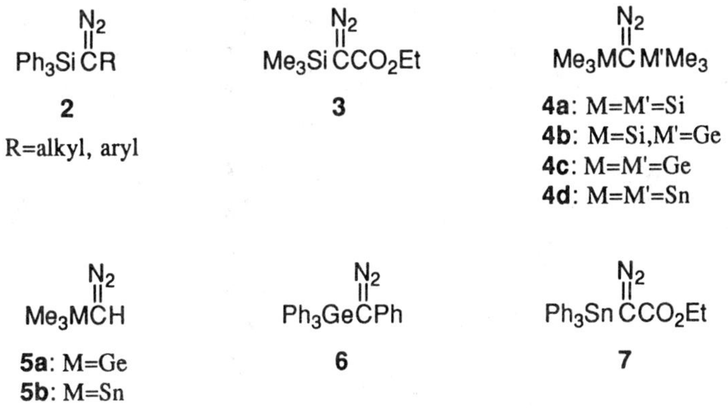

$$\underset{\textbf{2}}{\overset{\overset{\text{N}_2}{\|}}{\text{Ph}_3\text{Si CR}}}$$

R=alkyl, aryl

$$\underset{\textbf{3}}{\overset{\overset{\text{N}_2}{\|}}{\text{Me}_3\text{Si CCO}_2\text{Et}}}$$

$$\underset{}{\overset{\overset{\text{N}_2}{\|}}{\text{Me}_3\text{MC M'Me}_3}}$$

4a: M=M'=Si
4b: M=Si,M'=Ge
4c: M=M'=Ge
4d: M=M'=Sn

$$\underset{}{\overset{\overset{\text{N}_2}{\|}}{\text{Me}_3\text{MCH}}}$$

5a: M=Ge
5b: M=Sn

$$\underset{\textbf{6}}{\overset{\overset{\text{N}_2}{\|}}{\text{Ph}_3\text{GeCPh}}}$$

$$\underset{\textbf{7}}{\overset{\overset{\text{N}_2}{\|}}{\text{Ph}_3\text{Sn CCO}_2\text{Et}}}$$

SCHEME 78.2

Initial photochemical study of silyldiazoalkanes has been carried out by Schöllkopf et al.[2,19] They demonstrated that the irradiation of trimethylsilyl ethoxycarbonyl diazomethane 3 in isobutene afforded ethyl 2,2-dimethyl-1-(trimethylsilyl)cyclopropanecarboxylate 8 (Scheme 78.3). Similarly, irradiation of 1 in *trans*-2-butene furnished olefin 9 along with cyclopropane derivative 10 (Scheme 78.3).[23] The formation of these product implicates the intermediates of carbenes in the photolysis.

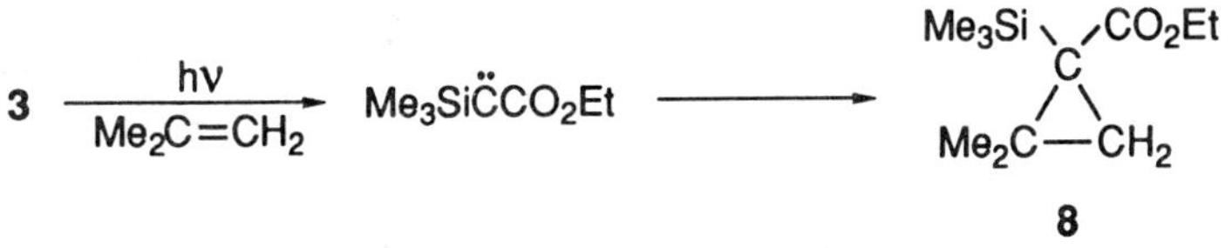

$$\textbf{3} \xrightarrow[\text{Me}_2\text{C}=\text{CH}_2]{h\nu} \text{Me}_3\text{Si}\ddot{\text{C}}\text{CO}_2\text{Et} \longrightarrow$$

Me$_3$Si   CO$_2$Et
＼C／
／ ＼
Me$_2$C—CH$_2$

8

SCHEME 78.3

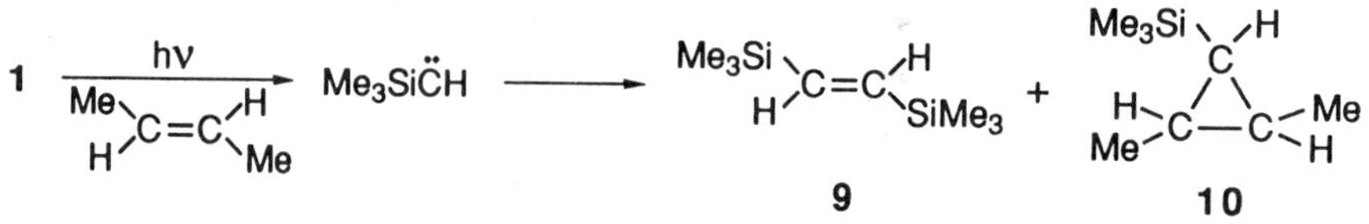

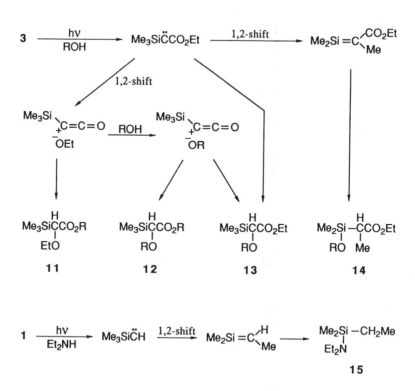

SCHEME 78.3 (continued)

Almost all photochemical and thermal reactions of silyldiazoalkanes involve silylcarbene chemistry. When **3** is photolyzed in alcohols, four products **11–14** are obtained in approximately 60 to 90% yield (Scheme 78.4).[24,25] The formation of **14** confirms that the carbene generated from **3** affords an unstable silene via 1,2-methyl migration. The formation of **11** and **12** also indicates the existence of trimethylsilyl ethoxy ketene via 1,2-ethoxy migration. The product **15** obtained by trapping of silene by amine is also formed upon irradiation of **1** with diethyl amine (Scheme 78.4).[26] The migration of silyl groups also occurs when polysilylated diazomethanes **16** are irradiated to give corresponding trapped products **17** and **18** (Scheme 78.5).[27] Silene itself dimerizes to give 1,3-disilacyclobutane **19** when R is hydrogen in **16**. The 1,2-migration reactions of groups resulting in the transformation of carbenes into silenes have also been described in many other reports.[20,28–31] Formation of silene via 1,2-migration has been demonstrated spectroscopically by the matrix isolation techniques.[32]

SCHEME 78.4

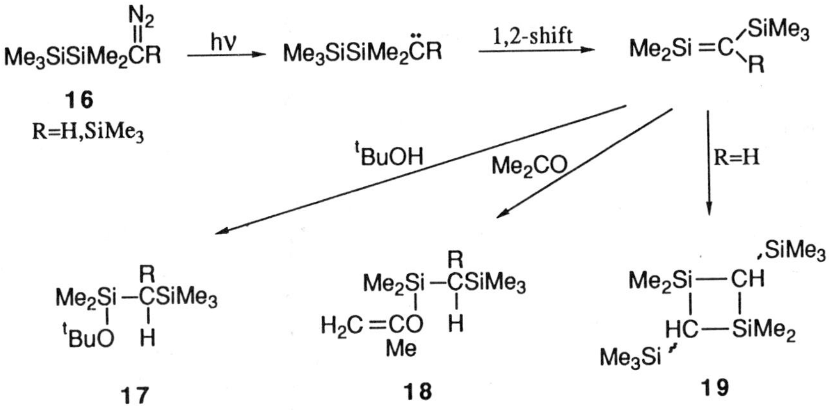

SCHEME 78.5

Interesting photochemical behavior of **1** is reported by two independent research groups.[33,34] Irradiation of **1** in an argon matrix at 8K produces a photostationary state involving a new species, identified as trimethylsilyldiazirine (**20**) (Scheme 78.6). Diazirine **20** is reasonably stable at room temperature, but further irradiation of the mixture yields the silene via 1,2-methyl migration within the carbene. Warming the matrix affords olefin **9** and disilacyclobutane **21**.

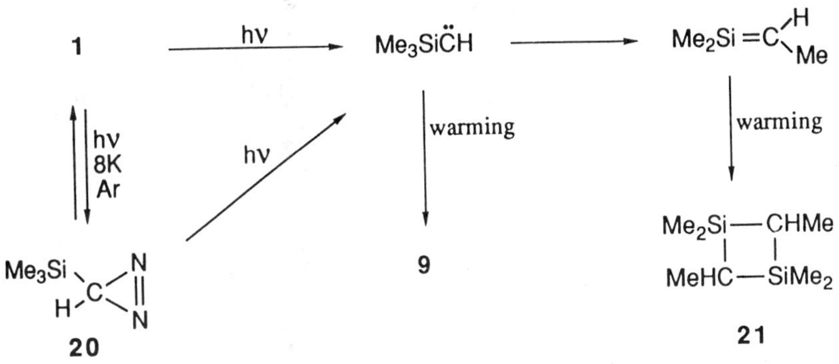

SCHEME 78.6

78.2 Applications

Photoirradiation of pentamethyldisilanyl adamantyl diazo ketone **22** also generates a carbene. Conjugated silenes **23**, formed via group migration in the carbene, undergo [2+2]-cycloaddition to give siloxetenes **24** when R is 1-adamantyl or *t*-butyl. When R is isopropyl or methyl group the conjugated silenes **23** yield the head-to-tail cyclodimers **25** (Scheme 78.7).[35,36]

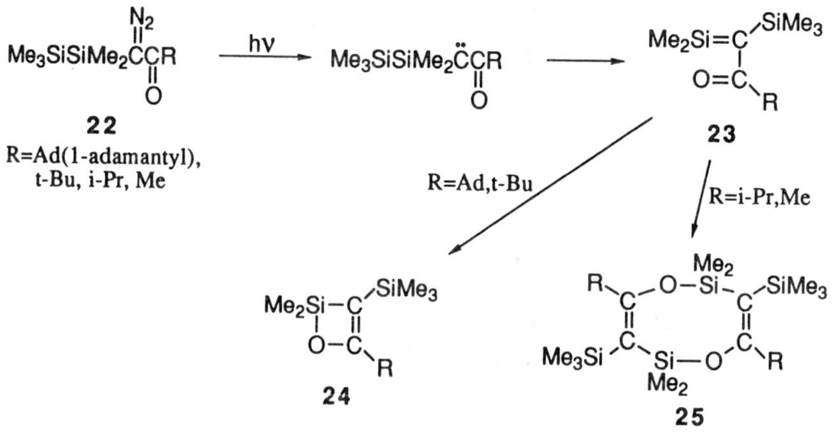

SCHEME 78.7

Interesting functionalized transient species silatoluene **26** and silafulvene **27** are generated when silacyclopentadienyl diazomethane **28** or the diazirine derivative **29** are irradiated (Scheme 78.8).[37,38] Silatoluene **26** and silafulvene **27** are generated from **30** via Si–C insertion and 1,2-methyl migration, respectively. The existence of **26** and **27** is confirmed by trapping with alcohol, butadiene, and benzophenone.

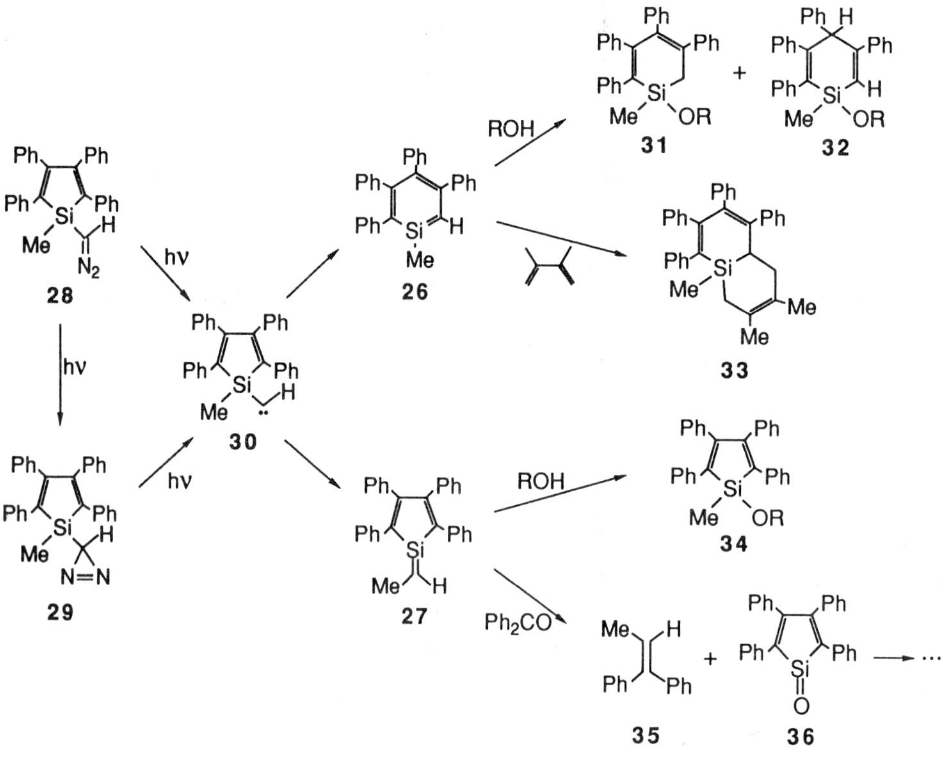

SCHEME 78.8

Recently, an examination of the photoreaction of the more complex silyldiazo system has been carried out.[39] Irradiation of a benzene solution of the *bis*(silyldiazomethyl) compound **37a** with excess *t*-butyl alcohol gives the adduct of bissilene **38a** (Scheme 78.9). The photolysis of **37a** and **37b** without trapping agents yields interesting compounds *trisilabicyclo*[1.1.1]pentanes **39a** and **39b** along with considerable quantities of polymeric material. Bicyclopentanes **39a** and **39b** are considered to be formed via the siladienes **41**. A low-temperature NMR study revealed one more product, **40b**, which is easily polymerized at 25 °C. Compound **40b** is also considered to be formed by parallel addition of **41**.

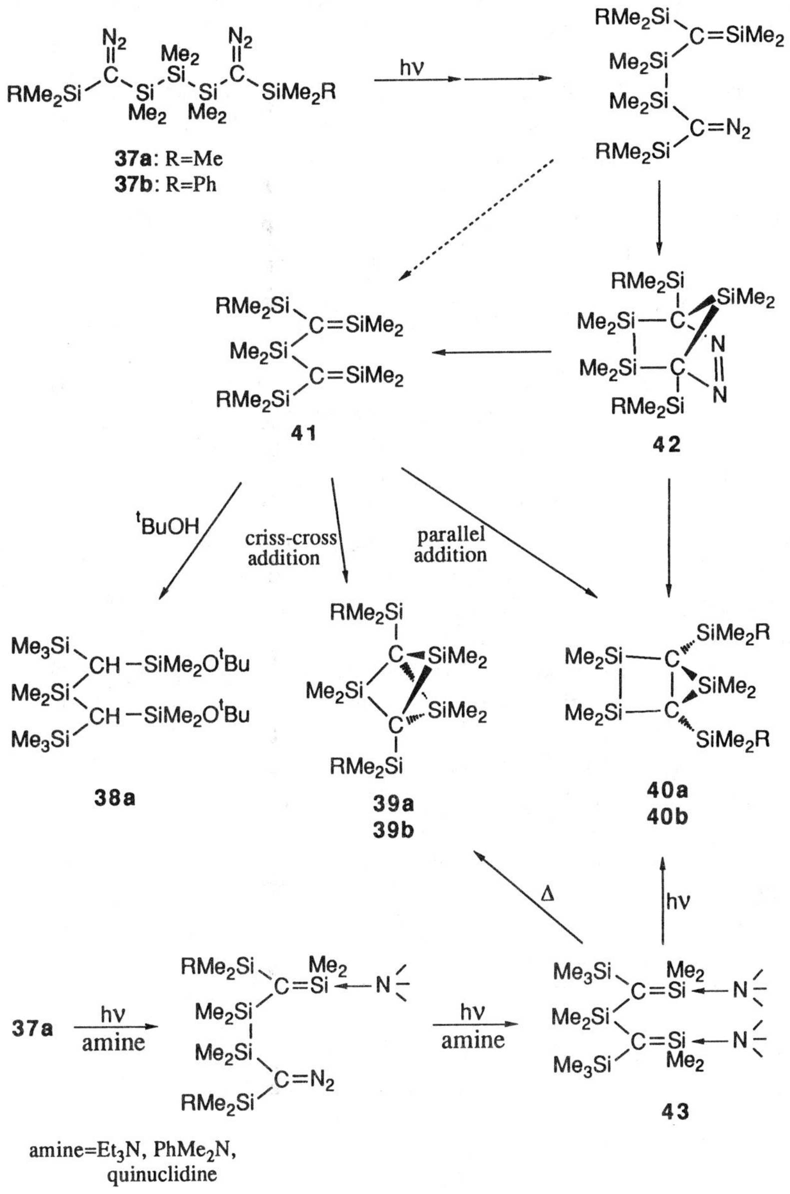

SCHEME 78.9

An alternative mechanism for the formation of **40b** is the path via bicyclic diazo compound **42**. The photolysis of **37a** with amines gives siladiene-amine complexes **43**.[40] The *bissilene* complexes **43** decompose thermally to give the bicyclopentane **39a**. The photolysis of **43** (amine = PhMe$_2$N) yields the parallel adduct **40a**.

The photoreaction of silyldiazoalkanes is a good route to silylcarbenes under mild conditions. The thermolysis also gives the silylcarbene, but subsequent thermal reaction is different from the photochemical process. Some of photochemical and thermal reactions have been discussed in some recent books.[7,41,42]

References

1. Lappert, M. F. and Lorberth, J., *J. Chem. Soc., Chem. Commun.,* 836, 1967.
2. Schöllkpf, U. and Rieber, N., *Angew. Chem.,* 79, 906, 1967; *Angew. Chem. Int. Ed.,* 6, 884, 1967.
3. Seyferth, D., Dow, A. W., Menzel, H., and Flood, T. C., *J. Am. Chem. Soc.,* 90, 1080, 1968.
4. Aoyama, T. and Shioiri, T., *Farumashia,* 20, 762, 1984 (Japanese).
5. Anderson, R., *Synthesis,* 717, 1985.
6. Shioiri, T. and Aoyama, T., *Yuki Gosei Kagaku Kyokaishi,* 44, 149, 1986 (Japanese).
7. Larson, G. L., Ed., *Advances in Silicon Chemistry,* Vol. 1, JAI Press, Connecticut, 1991.
8. Lappert, M. F., Lorberth, J., and Poland, J. S., *J. Chem. Soc., (A),* 2954, 1970.
9. Martin, M., *Synth. Commun.,* 13, 809, 1983.
10. Seyferth, D., Menzel, H., Dow, A. W., and Flood, T. C., *J. Organometal. Chem.,* 44, 279, 1972.
11. Sheludyakov, V. D., Khatuntsev, G. D., and Mironov, V. F., *Zh. Obshch. Khim.,* 39, 2785, 1969.
12. Crossman, J. M., Haszeldine, R. N., and Tipping, A. E., *J. Chem. Soc., Dalton,* 483, 1973.
13. Schöllkopf, U. and Scholz, H.-U., *Synthesis,* 271, 1976.
14. Aoyama, T. and Shioiri, T., *Chem. Pharm. Bull.,* 29, 3249, 1981.
15. Barton, T. J. and Hoekman, S. K., *Synth. React. Inorg. Met.-Org. Chem.,* 9, 297, 1979.
16. Mori, S., Sakai, I., Aoyama, T., and Shioiri, T., *Chem. Pharm. Bull.,* 30, 3380, 1982.
17. Brook, A. G. and Jones, P. F., *Can. J. Chem.,* 47, 4353, 1969.
18. Kaufmann, K. D., Auräth, B., Träger, P., and Rühlmann, K., *Tetrahedron Lett.,* 4973, 1968.
19. Schöllkopf, U., Hoppe, D., Rieber, N., and Jacobi, V., *Justus Liebigs Ann. Chem.,* 730, 1, 1969.
20. Barton, T. J. and Hoekman, S. K., *J. Am. Chem. Soc.,* 102, 1584, 1980.
21. Krommes, P. and Lorberth, J., *J. Organometal. Chem.,* 127, 19, 1977.
22. Lappert, M. F. and Poland, J. S., *J. Chem. Soc., Chem. Commun.,* 156, 1969.
23. Haszeldine, R. N., Scott, D. L., and Tipping, A. E., *J. Chem. Soc., Perkin Trans. 1,* 1440, 1974.
24. Ando, W., Hagiwara, T., and Migita, T., *J. Am. Chem. Soc.,* 95, 7518, 1973.
25. Ando, W., Sekiguchi, A., Hagiwara, T., Migita, T., Chowdhry, V., Westheimer, F. H., Kammula, S. L., Green, M., and Jones, M., Jr., *J. Am. Chem. Soc.,* 101, 6393, 1979.
26. Kreeger, R. L. and Shechter, H., *Tetrahedron Lett.,* 2061, 1975.
27. Sekiguchi, A. and Ando, W., *Chem. Lett.,* 871, 1983.
28. Ando, W., Sekiguchi, A., and Migita, T., *Chem. Lett.,* 779, 1976.
29. Ando, W., Sekiguchi, A., Migita, T., Kammula, S., Green, M., and Jones, M., Jr., *J. Am. Chem. Soc.,* 97, 3818, 1975.
30. Ando, W., Sekiguchi, A., and Sato, T., *J. Am. Chem. Soc.,* 103, 5573, 1981.
31. Sekiguchi, A., Sato, T., and Ando, W., *Organometallics,* 6, 2337, 1987.
32. Sekiguchi, A. and Ando, W., *Chem. Lett.,* p.2025, 1986.
33. Chapman, O. L., Chang, C.-C., Kolc, J., Jung, M. E., Lowe, J. A., Barton, T. J., and Tumey, M. L., *J. Am. Chem. Soc.,* 98, 7844, 1976.
34. Chedekel, M. R., Skoglund, M., Kreeger, R. L., and Shechter, H., *J. Am. Chem. Soc.,* 98, 7846, 1976.
35. Sekiguchi, A. and Ando, W., *J. Am. Chem. Soc.,* 106, 1486, 1984.
36. Maas, G., Schneider, K., and Ando, W., *J. Chem. Soc., Chem. Commun.,* 72, 1988.
37. Ando, W., Tanikawa, H., and Sekiguchi, A., *Tetrahedron Lett.,* 24, 4245, 1983.
38. Sekiguchi, A., Tanikawa, H., and Ando, W., *Organometallics,* 4, 584, 1985.
39. Ando, W., Yoshida, H., Kurishima, K., and Sugiyama, M., *J. Am. Chem. Soc.,* 113, 7790, 1991.
40. Yoshida, H., Kurishima, K., Nagase, S., and Ando, W., submitted for publication.
41. Patai, S. and Rappoport, Z., Ed., *The Chemistry of Organic Silicon Compounds,* John Wiley & Sons, Chichester, 1989.
42. Colvin, E., *Silicon in Organic Synthesis,* Butterworths, London, 1981.

79

Matrix Isolation Techniques in Diazirine Photochemistry

)bert S. Sheridan
ıiversity of Nevada

'9.1 Introduction

The ability of organic chemists to spectroscopically characterize reactive intermediates in a direct manner has played a major role in the advancement of current mechanistic understanding. Spectroscopic data, of course, inherently provide structural and electronic information. Observational interrogation of reactive intermediates during the course of reactions, on the other hand, also provides reactivity information. One powerful method for the "static" spectroscopic observation of reactive intermediates is matrix isolation. In this technique, (relatively) stable precursor molecules are frozen into low-temperature glasses in high dilution. Irradiation of the matrices liberates the intermediates, which because of the low temperatures and rigid surroundings are protected from subsequent reactions. The photoproducts are then probed spectroscopically, (usually with IR, EPR, or UV/VIS), degraded photochemically, trapped in warming experiments, etc. Besides the intrinsic structural information that such studies afford, the spectroscopic data provide a guide to time-resolved solution experiments.

The following will describe work using the matrix isolation technique to investigate the photochemistry of a variety of diazirines (1).[1] Generally, irradiations of these small-ring azo compounds produce the corresponding carbenes (2). The low-temperature methods permit the carbenes to be observed spectroscopically directly and studied, and their photochemical behavior can be probed in turn. The basic thrust of this chapter is more technique oriented than many others in this volume.

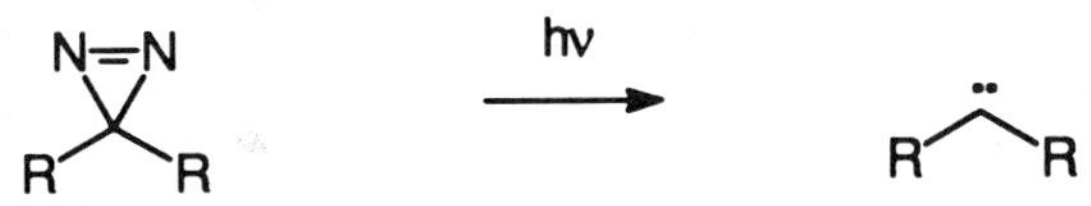

1 2

0-8493-8634-9/95/$0.00+$.50

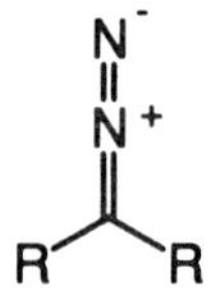

3

Diazirines (1) offer a useful alternative to the isomeric diazo compounds (3) for the photochemical generation of a number of carbenes.[1] In particular, we have been interested in the spectroscopy and photochemistry of ground-state singlet carbenes. One approach to stabilizing the singlet state in preference to the triplet state in these species hinges on substituents with lone-pair or other available electrons, which can interact selectively with the nominally empty *p*-orbital of the singlet carbenes. The necessary electron-donating character of the R groups in most cases renders the corresponding diazo compounds too unstable to be useful as precursors. On the other hand, synthetic routes to appropriate diazirines (1) are known[1] or have been worked out during the course of these studies.

79.2 Matrix Photochemistry of Diazirines

Matrix Preparation

Matrix isolation techniques have been well reviewed,[2] and excellent descriptions of the fine details appear in a number of original sources. Hence, we will concentrate only on some of the major issues. At the heart of the low-temperature apparatus is a closed-cycle helium refrigerator, capable of achieving temperatures of 6K. The refrigerator cools a 1-in. diameter window, which for most of our experiments is made of cesium iodide. The cold parts of the refrigerator head are surrounded by a vacuum shroud that is generally evacuated to ca. 10^{-6} torr. The shroud is fitted with KBr windows to permit optical throughput, and is rotatable so that matrix samples can be directed onto the window.

Sample preparation for matrix experiments is considerably more critical than for most other photochemical investigations. The matrices must be produced in such a fashion that substrate molecules are well isolated in the support material, and good optical quality must be maintained. Experimentally, the precursors are deposited from the gas phase onto the cooled windows along with large excesses of the desired matrix material. We use Ar or N_2 routinely for matrix media, and samples are deposited at window temperatures of 15 to 25K.

The method of matrix deposition depends on the volatility and thermal stability of the substrate. For example, chloromethoxydiazirine (4)[3] could be premixed in the gas phase with a ca. 1000-fold excess of argon or nitrogen before deposition. Such preparations generally give the highest degree of sample isolation. In contrast, methoxymethyldiazirine (5)[4] is unstable at room temperature, and has to be co-deposited onto the window along with matrix gas introduced from a separate port. Evaporating 5 from a tube maintained at $-110\,°C$ directly onto the 20K window, with simultaneous deposition of a large excess of N_2, protected the sample against thermal degradation, and gave a well-isolated matrix.

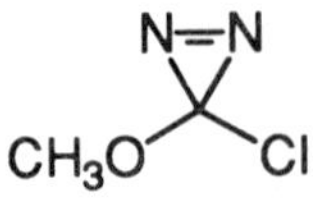

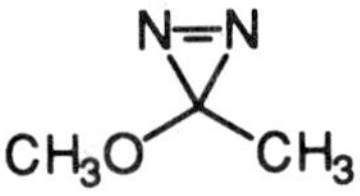

Following preparation of the matrices, samples are cooled to the minimum refrigerator temperature (6 to 12K, in general) and IR and UV/VIS spectra are recorded. The width of the IR absorptions gives some indication of the degree of sample isolation; well-isolated samples tend to exhibit narrow lines.

Matrix Photolyses

Photochemistry is induced in the matrix-isolated samples with either broad-band irradiation from a high-pressure Xe lamp, where wavelengths are selected by cut-off filters, or with monochromatic light from a high-pressure Hg lamp/monochromator combination. Careful control of irradiation wavelengths has proved to be critical in the study of many of the carbene systems described below. The carbenes are themselves photolabile and must be protected from absorbing the light used to photolyze the diazirines. The progress of the photoreaction is closely followed by IR and UV/VIS spectroscopy. The initial photoproducts are characterized by comparison with spectra of known compounds, by photodegredation, by comparison to calculated spectra, and by bimolecular trapping.

Typical of the experiments carried out, irradiation of diazirine 4 matrix isolated in N_2 at 10K at 370 nm produced methoxychlorocarbene (6), which could be characterized by IR and UV/VIS spectroscopy.[3] The identity of 6 was confirmed by generation in an Ar matrix containing HCl. Subsequent annealing of the matrix at 35K caused the IR bands assigned to the carbene to disappear, with concomitant growth of trapped product 7 bands. Carbene 6 exhibited a UV λ_{max} at 318 nm. Importantly, the "action spectrum" of photochemical destruction of the IR bands of 6 fit the UV spectrum. Thus, the two spectroscopic observations were linked to a common intermediate.

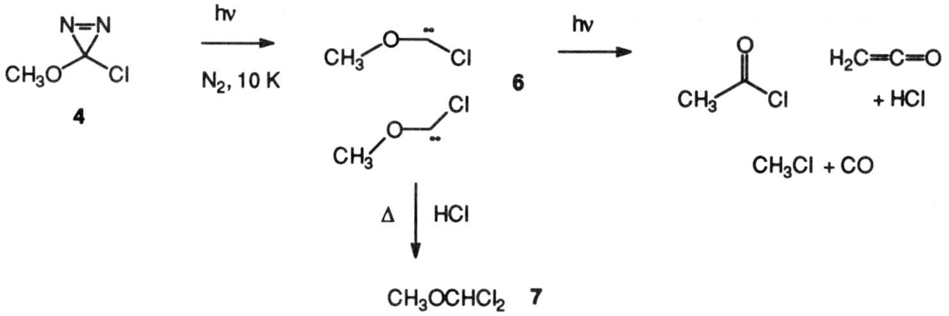

The IR spectrum reveals important information regarding the structure of the carbene.[3] Two sets of IR absorbances were observed that showed somewhat different photoreactivity; one set was selectively destroyed by irradiation on the short-wavelength side of the UV spectrum of 6, and the other was more labile when irradiation was performed on the long-wavelength side. These were assigned to the Z- and E-isomers of the carbene, as shown above. Carbene 6 also exhibited considerable C-O double-bond character, as revealed by a strong band in the IR at 1300 cm^{-1}, midway between carbonyl and ether stretching frequencies. Deuterium and ^{18}O labeling studies confirmed that this absorption corresponds to a C-O stretch.

Irradiation of carbene 6 at 320 nm gave an interesting array of products. Although acetyl chloride is, formally at least, a 1,2-shift product of the carbene, the origin of the other products is less straightforward. In particular, ketene and HCl require substantial skeletal reorganization. It is believed that irradiation of the carbene initially brings about C-O bond cleavage leading subsequently to a methyl radical, a chlorine atom, and CO. Various recombinations of these species lead to the observed products.

Irradiation of matrix-isolated phenoxychlorodiazirine (8) generated phenoxychlorocarbene (9) in a similar manner, which again could be characterized by IR, UV/Vis, and trapping.[5] The λ_{max} of 9, at 320 nm, is similar to that of 6. The photochemistry of 9 also resembles that of 6, with a 1,2-carbon shift (benzoyl chloride) and decarbonylation (chlorobenzene) products observed. Two geometric isomers of the carbene were again detected, with different IR and slightly different UV spectra. Interestingly, however, irradiation of the *Z*-isomer produced some of the *E*-isomer, and vice versa. The photointerconversion in this case, in contrast to results with methoxychlorocarbene, was attributed to the stronger C-O bond in 9, affording a carbene less susceptible to fragmentation.

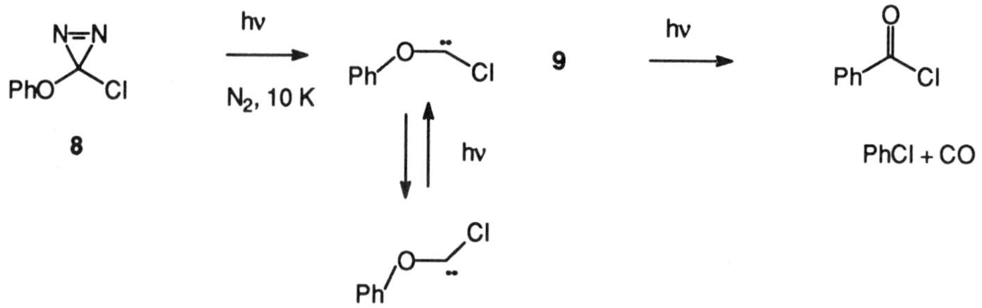

The photochemistry of methoxyfluorodiazirine (10)[6] parallels exactly that of the chloro carbene 6. It was most convenient, however, to characterize this carbene chemically by dimerization; warming to 35K a N_2 matrix containing 11 produced 12, as observed by IR spectroscopy. A significant point of difference from the oxychlorocarbenes is the UV absorption of methoxyfluorocarbene (11). The λ_{max} of 11 is shifted to considerably shorter wavelengths compared to 6 and 9. Stronger interaction between the fluorine substituent lone pair and the carbenic center leads to a shift to higher energy of the π^* molecular orbital, giving a λ_{max} of 240 nm for 11. One practical consequence of this blue shift is that it is easier to generate the fluorocarbene with broadband irradiation than it is the chlorocarbenes, since the carbene UV absorption is well removed from that of the diazirine starting material.

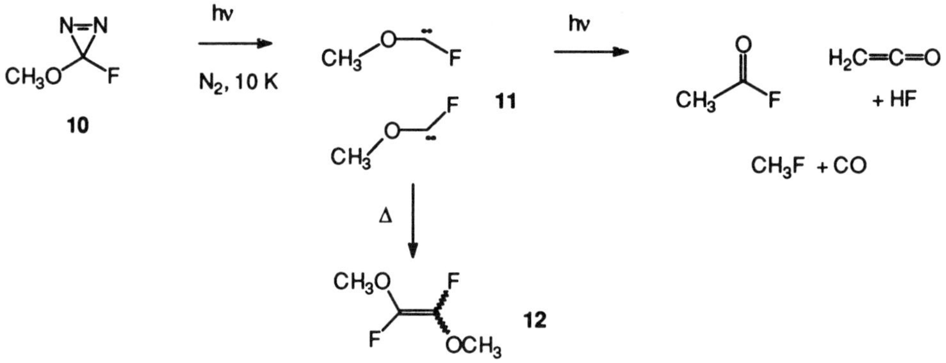

The carbene systems described previously have all benefited from the dual stabilizing effects of an alkoxy and a halo substituent. Alkyl-substituted carbenes with only one stabilizing group have been studied more recently. The first such system to be investigated was methylmethoxydiazirine (5).[4] These investigations represented the first direct observation of a singlet alkyl carbene. Irradiation at 312 nm of 5 isolate in a N_2 matrix gave methylmethoxycarbene (13), along with other products. In this case, careful control of the irradiation wavelength was particularly necessary to prevent subsequent photochemistry of the carbene 13, which absorbed at *longer* wavelengths than the alkoxyhalocarbenes and than the starting material. The carbene showed a broad absorbance in the UV/VIS λ_{max} = 390 nm. This observation fits both the trends seen in the alkoxyhalo systems and theoretical predictions.[7] The wavelength dependence of the photochemical disappearance of the carbene IR bands was found to correlate with the electronic spectrum. Separate sets of IR bands for 13 could again be attributed to cis- and trans-isomers. Here, the cis-isomer (13a) absorbed at longer wavelengths in the UV/VIS than the trans-isomer (13b). The assignments of the IR and electronic spectra were confirmed by *ab initio* calculations carried out by Sheridan's group[7] and by others.[4,8]

Trapping experiments with methanol also substantiated the structural assignment of 13. Carbene 13 was photochemically generated in a 3-methylpentane matrix containing $CH_3OH(4:1)$ at 20K. Warming the matrix to 80K caused the disappearance of the IR bands of 13 and growth of the IR spectrum of acetal 14.

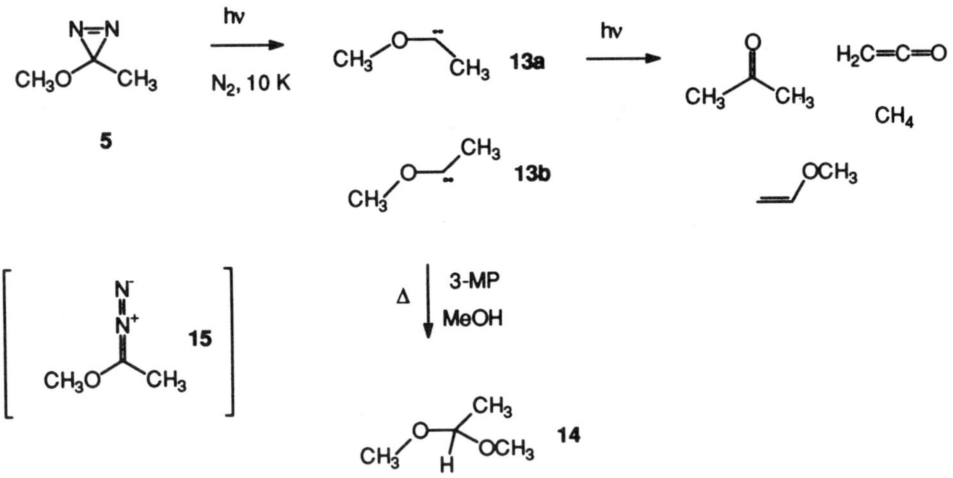

The photochemistry of carbene 13 is similar to that of the alkoxyhalocarbenes. Broad-band irradiation of 13 at >340 nm gave acetone, methane, and ketene. These products can be sensibly attributed to C-O cleavage of the carbene to a methyl radical acetyl radical pair, followed by recombination or disproportionation. However, the carbene also gave methyl vinyl ether. Clearly, this product arises from a photochemically induced 1,2-H shift in the carbene.[4]

The results obtained from matrix irradiation of 5 proved valuable in interpreting and guiding solution time-resolved spectroscopic studies on methylmethoxycarbene (13). Initial investigations by others[9] attributed a 390-nm absorption resulting from laser irradiation of 5 to the corresponding diazo compound 15. The results[4] from the matrix study showed that this band corresponded, in fact, to carbene 13. Minor amounts of diazo 15 could also be detected in the matrix experiments, and its spectra and chemistry were easily distinguished from those of the carbene.

Theoretical studies[10] indicate that halogen substituents can also stabilize singlet carbenes via lone-pair donation, although not as effectively as alkoxy groups. Irradiation of chloromethyldiazirine (16) or fluoromethyldiazirine (17) in low-temperature matrices, however, gave no spectroscopic evidence for the carbenes by IR or UV/VIS. Only the H-shifted vinyl halides were observed.[11]

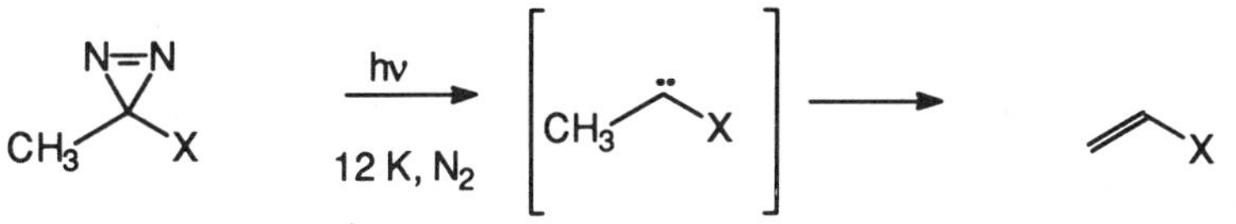

16 X = Cl
17 X = F

In an attempt to arrest the facile 1,2-hydrogen shift that prevented observation of the halocarbenes, the chlorocyclopropyl system was investigated.[12] Irradiation of N_2 matrix-isolated 3-chloro-3-cyclopropyldiazirine (18) at 334 nm gave chlorocyclopropylcarbene (19), along with small amounts of other products. The carbene (orange-colored) exhibited a broad UV/VIS spectrum with a λ_{max} at 460 nm, and a stronger band in the UV at 240 nm. This significantly longer wavelength absorption of 19, compared to those of alkoxy carbenes, reflects the weaker electron-donating ability of the chlorine substituent. The "action spectrum" for photodestruction of the carbene IR bands roughly matched the electronic spectra. Carbene 19 could be trapped by annealing in an HCl-containing matrix to give 20, verified by comparison with the IR spectrum of authentic material. In this case, it appeared that only one conformation of the carbene 19 was observed. Irradiation at either the long- or the short-wavelength sides of either the UV or VIS absorptions of the carbene resulted in simultaneous destruction of all of the IR bands of 19.

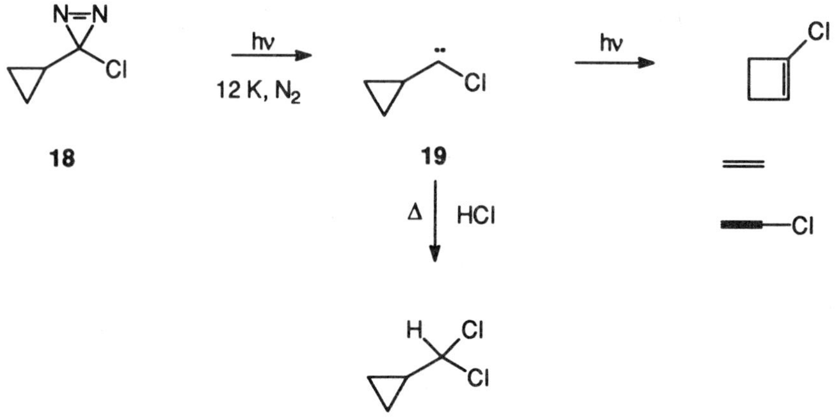

The chlorocyclopropylcarbene (19) was photolabile. Irradiation at 450 nm produced chlorocyclobutene via ring expansion, and chloroacetylene and ethylene via fragmentation. The strong carbene absorption at 246 nm could also be observed in solution by laser photolysis of diazirine 18.[12,13] A lifetime of 10^{-6} s for carbene 19 at room temperature was ascertained.

It is thought that cyclopropyl groups can stabilize considerably a singlet carbene via homoconjugative interactions with the nominally empty p-orbital, analogous to cyclopropylcarbinyl carbocations.[14] The cyclopropyl group, as seen above, also does not undergo C-H shifts readily. The possibility was thus considered that two cyclopropyl substituents might provide, for the first time, an observable dialkyl singlet ground-state carbene. In contrast to reports of earlier failures,[15] it was found that dicyclopropyldiazirine (21) could be synthesized from the cyclohexylimine of dicyclopropylketone.[16] Irradiation of an N_2 matrix containing 21 at 334 nm gave dicyclopropylcarbene (22) and dicyclopropyldiazomethane (23). Here, the identity of the carbene was confirmed by trapping with CO. Warming to 38K an Ar matrix doped with 2% CO containing carbene 22 resulted in a decrease of the bands of 22 and concurrent increase in the bands of dicyclopropylketene (24).

The UV/VIS spectrum of 22 showed a broad band at 400 to 600 nm (λ_{max} = 490 nm), along with a stronger absorbance at λ_{max} = 230 nm; both bands grew and disappeared simultaneously with the bands of the IR of the carbene. The "action spectrum" of photodestruction of the IR peaks assigned to 22 matched roughly the UV/VIS spectrum, substantiating that the same species was observed by both methods. As in the case of chlorocyclopropylcarbene,[12] irradiation of matrices containing 22 at either end of the long- or short-wavelength UV/VIS absorbances caused all of the IR bands of the carbene to disappear simultaneously (as did the UV/VIS spectra). This implies the presence of only one conformer of 22. Irradiation of carbene 22 converted it rapidly to cyclopropylcyclobutene (ring expansion), cyclopropylacetylene, and ethylene (fragmentation).

The only other dialkyl carbenes that have been directly observed previous to this work were triplet ground-state species.[17] In contrast, the above spectroscopic results, particularly the visible absorbance, indicate that 22 is a ground-state singlet. For example, triplet methylene only absorbs at wavelengths much shorter than 200 nm,[18] whereas the singlet shows broad absorption from 550 to 950 nm.[19]

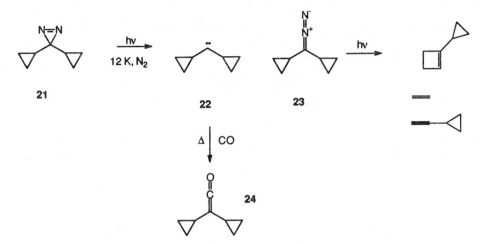

Ab initio calculations confirmed the conclusions of these studies on **22**.[16] Three conformations of **22** (**22a, b,** and **c**) can benefit maximally from interaction between the empty carbenic *p*-orbital and the Walsh orbitals of both cyclopropanes. Only **22a** and **b**, however, would be expected to exhibit low barriers for ring expansion. The geometries of **22a, b,** and **c** were optimized at the MP2/ 6–31G* level (relative energies: 0, 4.0, and 12.8 kcal mol⁻¹ for **a, b,** and **c**, respectively, uncorrected for zero-point energy). Vibrational calculations indicated that a C_{2v} structure for **22c** was not an energy minimum (in contrast to **22a** and **b**). Relaxation to a skewed C_2 geometry, however, produced a local minimum. The vibrational spectrum calculated for **22a** was in excellent agreement with the observed IR data. The calculated wavelengths for the lowest-energy electronic transition from the CIS method (configuration interaction with single excitations, 6–31G*) for **22a, b,** and **c**, respectively, were 475, 538, and 629 nm. The *ab initio* calculations thus support the contention that **22** is a ground-state singlet, and that only conformation **22a** is observed.

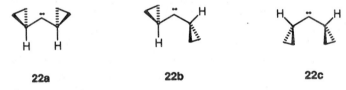

Spectroscopic Properties of Diazirines

Diazirine	X, Y	UV/VIS
4	CH_3O, Cl	350, 365, 382 nm
5	CH_3O, CH_3	318, 337, 351, 368 nm
8	PhO, Cl	328, 342, 360 nm
10	CH_3O, F	329, 344, 361 nm
18	*cis*-C_3H_5, Cl	329, 340, 357 nm
21	*cis*-C_3H_5, *cis*-C_3H_5	353 nm

eferences

1. For general references on the preparation, properties, and reactions of diazirines, see: Liu, M. T. H., Ed., *Chemistry of Diazirines*, Vol. I and II, CRC Press, Boca Raton, FL, 1987.
2. (a) Dunkin, I. R., The matrix isolation technique and its application to organic chemistry, *Chem. Soc. Rev.*, 9, 1, 1980; (b) Sheridan, R. S., Matrix isolation photochemistry, in *Organic Photochemistry*, Vol. 8, Marcel Dekker New York, 1987, 159, and references therein.
3. (a) Kesselmayer, M. A. and Sheridan, R. S., Infrared spectrum and photochemistry of methoxychlorocarbene, *J. Am. Chem. Soc.*, 106, 436, 1984; (b) Kesselmayer, M. A. and Sheridan, R. S., Methoxychlorocarbene. Matrix spectroscopy and photochemistry, *J. Am. Chem. Soc.*, 108, 99, 1986.

4. Sheridan, R. S., Moss, R. A., Wilk, B. K., Shen, S., Wlostowski, M., Kesselmayer, M. A., Subramanian, R., Kmiecik-Lawrynowicz, G., and Krogh-Jespersen, K., Direct observational studies of a singlet alkylcarbene: Methylmethoxycarbene, a remarkably selective nucleophile, *J. Am. Chem. Soc.*, 110, 7563, 1988.

5. Kesselmayer, M. A. and Sheridan, R. S., Phenoxychlorocarbene. Spectroscopy and photochemical interconversion of geometric isomers, *J. Am. Chem. Soc.*, 108, 844, 1986.

6. Du, X.-M., Fan, H., Goodman, J. L., Kesselmayer, M. A., Krogh-Jespersen, K., LaVilla, J. A., Moss, R. A., Shen, S., and Sheridan, R. S., Reactions of dimethoxycarbene and fluoromethoxycarbene with hydroxyl compounds: Absolute rate constants and the heat of formation of dimethoxycarbene, *J. Am. Chem. Soc.*, 112, 1920, 1990.

7. Matsumura, M., Ammann, J. R., and Sheridan, R. S., Photochemical ring expansion of tetramethylcyclobutanone revisited: Angular dependence of electronic absorption of singlet carbenes, *Tetrahedron Lett.*, 33, 1843, 1992.

8. Hess, B. A., Jr. and Smentek-Mielczarek, S., *Ab initio* computation of the infrared spectra of alkoxycarbenes, *J. Mol. Struct.*, 227, 265, 1991.

9. Moss, R. A., Shen, S., Hadel, L. M., Kmiecik-Lawrynowicz, G., Wlostowska, J., and Krogh-Jespersen, K., Absolute rate and philicity studies of methoxyphenylcarbene. An extended range for carbenic ambiphilicity, *J. Am. Chem. Soc.*, 109, 4341, 1987.

10. Evanseck, J. D. and Houk, K. N., Theoretical predictions of activation energies for 1,2-hydrogen shifts in singlet carbenes, *J. Phys. Chem.*, 94, 5518, 1990.

11. Green, I. and Sheridan, R. S., unpublished results.

12. Ho, G. J., Krogh-Jespersen, K., Moss, R. A., Shen, S., Sheridan, R. S., and Subramanian, R., Kinetics of a carbene rearrangement: The 1,2-carbon migration of cyclopropylchlorocarbene, *J. Am. Chem. Soc.*, 111, 6875, 1989.

13. (a) Moss, R. A., Ho, G.-J., Shen, S., and Krogh-Jespersen, K., Activation parameters for a 1,2 carbon shift in a carbene rearrangement, *J. Am. Chem. Soc.*, 112, 1638, 1990; (b) Bonneau, R., Liu, M. T. H., and Rayez, M. T., Rates for 1,2 migration in alkylchlorocarbenes, *J. Am. Chem. Soc.*, 111, 5973, 1989; (c) Liu, M. T. H. and Bonneau, R., Energy barrier for cyclopropyl-chlorocarbene rearrangement measured by direct observation of the carbene in laser flash spectroscopy, *J. Phys. Chem.*, 93, 7298, 1989.

14. (a) Hoffmann, R., Zeiss, G. D., and Van Dine, G. W., The electronic structure of methylenes, *J. Am. Chem. Soc.*, 90, 1485, 1968; (b) Schoeller, W. W., Ring expansion reaction of cyclopropylcarbene to cyclobutene, *J. Org. Chem.*, 45, 2161, 1980; (c) Shevlin, P. B. and McKee, M. L., A theoretical investigation of the intramolecular reactions of cyclopropylmethylene, *J. Am. Chem. Soc.*, 111, 519, 1989; (d) Wang, R. and Deng, C., Theoretical studies of carbenes and carbenoids. 5. Intramolecular reactions of cyclopropylcarbene, *Tetrahedron*, 44, 7355, 1988; (e) Chou, J.-H., McKee, M. L., De Felippis, J., Squillacote, M., and Shevlin, P. B., Experimental and computational evidence indicating that the initial carbene conformation is product determining in the reactions of cyclopropylmethylene, *J. Org. Chem.*, 55, 3291, 1990.

15. Church, R. R. R. and Weiss, M. J., Diazirines. II. Synthesis and properties of small functionalized diazirine molecules. Some observations on the reaction of a diaziridine with the iodine-iodide ion system, *J. Org. Chem.*, 35, 2465, 1970.

16. Ammann, J. R., Subramanian, R., and Sheridan, R. S., Dicyclopropylcarbene: Direct characterization of a singlet dialkylcarbene, *J. Am. Chem. Soc.*, 114, 7592, 1992.

17. (a) Gano, J. E., Wettach, R. H., Platz, M. S., and Senthilnathan, V. P., Di-*tert*-butylcarbene: The low-temperature photochemistry of di-*tert*-butyldiazomethane, *J. Am. Chem. Soc.*, 104, 2326, 1982; (b) Myers, D. R., Senthilnathan, V. P., Platz, M. S., and Jones, M., Jr., Diadamantylcarbene in solution, *J. Am. Chem. Soc.*, 108, 4232, 1986.

18. Herzberg, G., The spectra and structures of free methyl and free methylene, *Proc. Roy. Soc., London*, A262, 291, 1961.

19. Herzberg, G. and Johns, J. W. C., The spectrum and structure of singlet CH_2, *Proc. Roy. Soc., London*, 295, 107, 1964.

80

Remote Functionalization by Nitrites: The Barton Reaction

roshi Suginome
kaido University

0.1 Definition of the Reaction and A Short Historical Background

The photolysis of organic nitrites of appropriate constitution and conformation in solvents such as benzene, toluene, or acetonitrile transforms them into δ-nitroso alcohols via a homolytic fission of the O-N bond of their nitrosoxy group, followed by an intramolecular δ-hydrogen abstraction of the resulting alkoxyl radicals to generate a δ-carbon radical that combines with the generated nitric oxide. The nitroso alcohols are isolated as δ-hydroxyimino alcohols, formed by a spontaneous thermal isomerization, or as the nitroso dimers (Scheme 80.1).[1] This transformation has been named the "Barton reaction".[2]

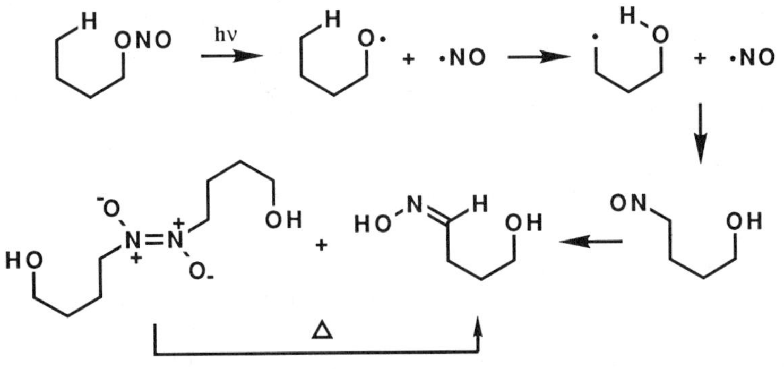

SCHEME 80.1

Spectroscopic as well as photochemical investigations of organic nitrites in the vapor phase were initiated during the 1930s at Oxford by Thompson and colleagues.[3] The interpretation of the primary mode of the photochemical decomposition of nitrite esters by Thompson, however,

0-8493-8634-9/95/$0.00+$.50

differed from what has become established now as the primary process for nitrite photolysis. Subsequent to Thompson's study, Coe and Doumani reinvestigated the vapor-phase photolysis of *t*-butyl nitrite, confirming that the products of photolysis were acetone and nitrosomethane (Scheme 80.2).[4] They interpreted the results in terms of a concerted photochemical molecular rearrangement of *t*-butyl nitrite, since no ethane was detected in this photolysis.

$$(CH_3)_3C-ONO \xrightarrow{\ h\nu\ } (CH_3)_2C=O \ + \ CH_3NO$$

SCHEME 80.2

Gray and Style proposed for the first time, based on their study on the photolysis of methyl nitrite,[5] that the primary step of nitrite photolysis is the homolysis of the ONO bond. The process was entirely analogous to the mechanism that had already been proposed by Steacie[6] for the gas-phase pyrolysis of organic nitrites. Tharte then undertook an extensive investigation of the photolysis of primary, secondary, and tertiary alkyl nitrites, believing that his results could be accommodated by either of the two possible mechanisms,[7] depending on the wavelength of the incident light and the structure of the alkyl nitrite.

Since these early studies on nitrite photolysis, a number of extensive studies (as summarized by Calvert and Pitts[8]) have established that the major primary reaction of the photolysis of nitrite esters in solution is the homolysis of the O-N bond of the nitrosoxy group to give alkoxyl radicals and nitric oxide (Scheme 80.3).[8]

$$RONO \xrightarrow{\ h\nu\ } RO\cdot \ + \ \cdot NO$$

SCHEME 80.3

The photochemistry of nitrite esters, however, became a subject of considerable attention when Barton showed its usefulness for organic synthesis at the beginning of the 1960s.[1] Barton and co-workers showed that the photolysis of suitably constituted steroidal alcohol nitrites brought about an intramolecular exchange of the NO of the nitrite group with a hydrogen atom attached to a carbon atom in the δ-position, as mentioned above. This reaction has been shown not only to be one of the most useful photochemical processes for the selective functionalization of an unactivated carbon atom in organic synthesis, but also has led to a renaissance of the concept of remote functionalization by an intramolecular reaction, such as the Hofmann-Löffler-Freytag reaction.[9] This led to a general recognition of the utility of radical reactions in organic synthesis.

The historical and mechanistic aspects and the synthetic applications of the reaction have been the subject of a number of comprehensive review articles by Barton[10] as well as co-workers.[2,11]

80.2 Mechanism of the Barton Reaction

Alkyl nitrites, which can readily be prepared by reactions of the corresponding alcohols with nitrosyl chloride in pyridine, exhibit UV absorptions in two regions: one at 210 to 230 nm (ε 1000–1700) and the other (with a fine structure) at ~360 to 380nm (ε ~ 80).[8,13b] The absorption at the longer wavelength is assigned to the n, π* transition; that at the shorter wavelength is assigned to an intramolecular charge-transfer band.[8,12] The irradiation of the weak n,π* band causes homolysis of the O-N bond to give the corresponding alkoxyl radicals and nitric oxide with a quantum yield 0.7, as has been proven by the photolysis of steroidal nitrite esters in methanol using 365-nm monochromatic light.[13b] A quantum yield of 0.76 was also reported for the photolysis of octyl nitrite in heptane.[14] A quantum yield less than unity in the solution photolysis of the nitrites may indicate the reversibility of the primary process of the dissociation.

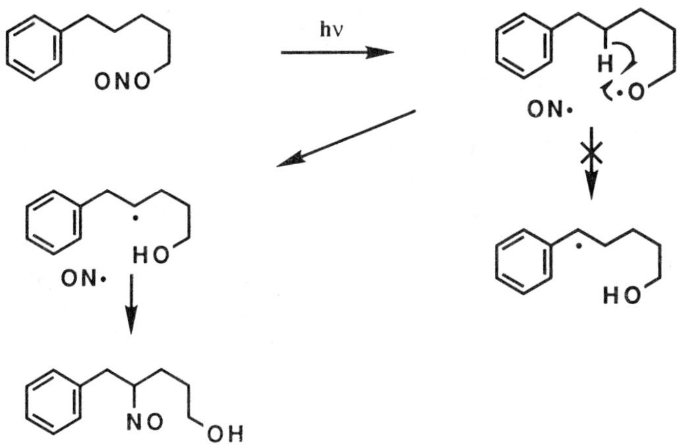

SCHEME 80.4

The thermolysis of nitrite esters in solution, on the other hand, resulted also in the decomposition of nitrite esters, but did not lead to any homolysis of the N-O bond. This was attributed to an ionic decomposition due to protonation of the ONO bond with adventitious proton as the impurity in the solution.[15]

Alkoxyl radicals, generated thus, abstract intramolecularly a hydrogen that is appropriately located. A number of experiments with a variety of cyclic and acyclic systems[2,11,16,17] has established that hydrogen abstraction involving a six-membered cyclic transition state is preferred over that involving five- or (and more) membered cyclic transition states; the alkoxyl radical generated by the photolysis of the nitrite of 5-phenylpentanol abstracts a hydrogen attached to the δ-carbon, but not a hydrogen attached to the ξ-carbon, to generate stabilized benzyl radicals, (Scheme 80.5).[16] On the basis of a number of results in the steroid field, together with an inspection of Dreiding models, the most appropriate distances between an O-radical and a carbon bearing a hydrogen atom to be abstracted have been estimated to be in the 2.5–2.7 Å range.[18]

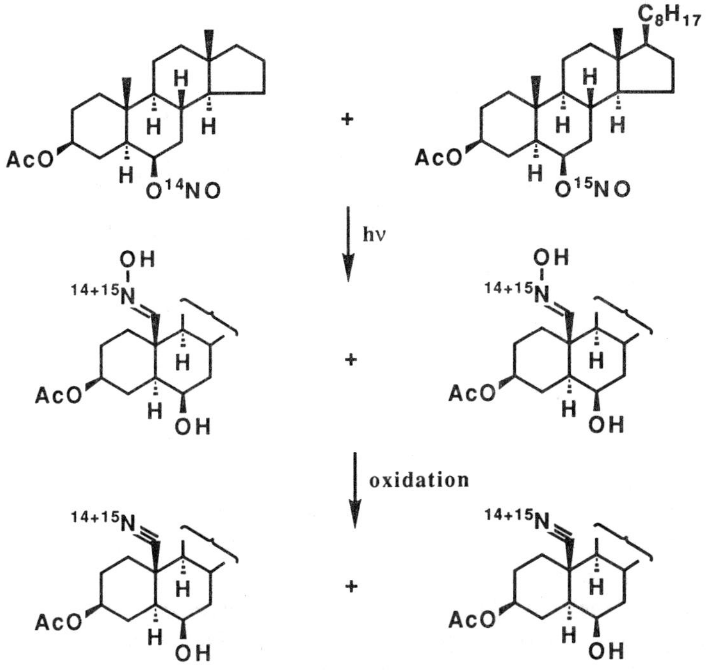

SCHEME 80.5

Carbon radicals, generated thus, then combine with nitric oxide to form nitroso compounds that readily isomerize to afford the corresponding oximes (or nitrosodimers). Quantum yield studies of the overall process cited above[8,13b] indicated that this free radical reaction is not a chain process. The behavior of the nitric oxide generated in the solution was studied using labeled nitrites;[19,20] the photolysis of a mixture of each equimolecule of ^{15}N-nitrite A and ^{14}N-nitrite B gave a mixture of two products (arising from intramolecular hydrogen abstraction) in which ^{15}N and ^{14}N were completely scrambled in the two products (Scheme 80.6).[19] Thus, it was concluded there is no solvent cage within which the NO is retained during the process. A mass spectrometric determination on the labeled nitrite recovered after a partial completion of the photolysis showed that no scrambling had taken place at this stage. These experiments indicated that non-cage combinations take place after hydrogen abstraction, although the primary homolysis step is a cage process, as outlined in Scheme 80.6.[19]

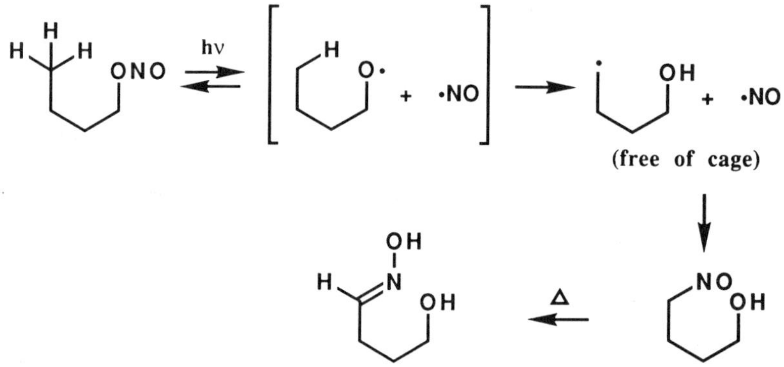

SCHEME 80.6

In agreement with this mechanism, 18-iodopredonisolone 21-acetate, but not the 18-oxime, is obtained when predonisolone 11β-nitrite 21-acetate in benzene is photolyzed in the presence of iodine (Scheme 80.7).[21]

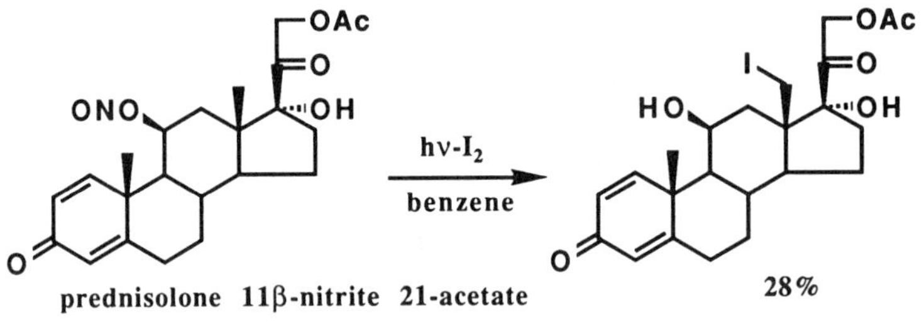

prednisolone 11β-nitrite 21-acetate   28%

SCHEME 80.7

80.3 Synthetic Applications

Since angular methyl groups and alkoxyl radicals with a 1,3-diaxial relationship in a rigid steroid and some terpenoid molecules ideally satisfy hydrogen abstraction through a 6-membered cyclic transition state, the Barton reaction has most successfully been applied to the field of steroids and some di- and triterpenoids. By this reaction, functional groups can, in principle, be introduced into the unactivated angular 10β- or 13β-methyl groups of steroids. As shown in Scheme 80.8, a hydrogen attached to the 13β-carbon can be abstracted by an alkoxyl radical at C-11, C-15, or C-20, and a hydrogen attached to the 10β-carbon from an alkoxyl radical at C-2, C-4, C-6, or C-11. Thus, a one-step introduction of functionality into an unactivated position can be achieved.

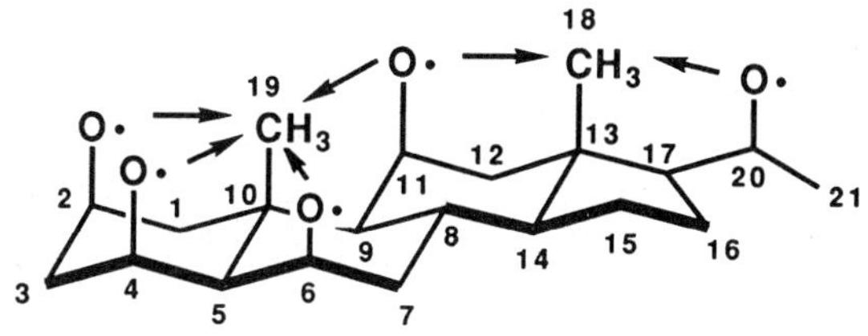

SCHEME 80.8

In the following, some representative examples among the numerous applications of this reaction found over 3 decades are presented. For more details, the reader should consult the review articles cited.[2,11]

1. Functionalization of 10β-Me of a steroid from the 2β-ol nitrite (Scheme 80.9).[22]

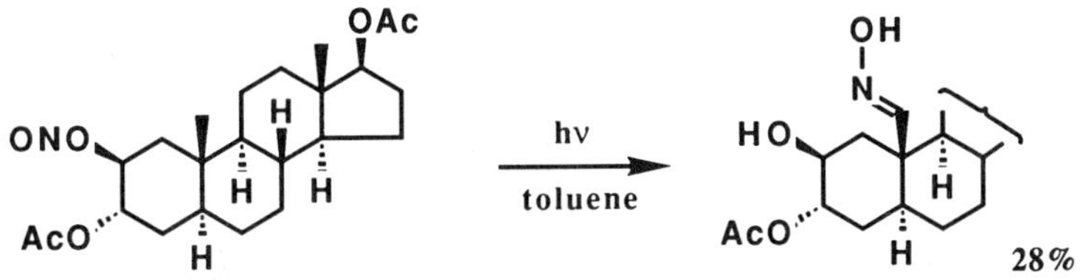

SCHEME 80.9

2. Functionalization of 10β-Me of a steroid from the 6β-ol nitrite (Scheme 80.10).[1]

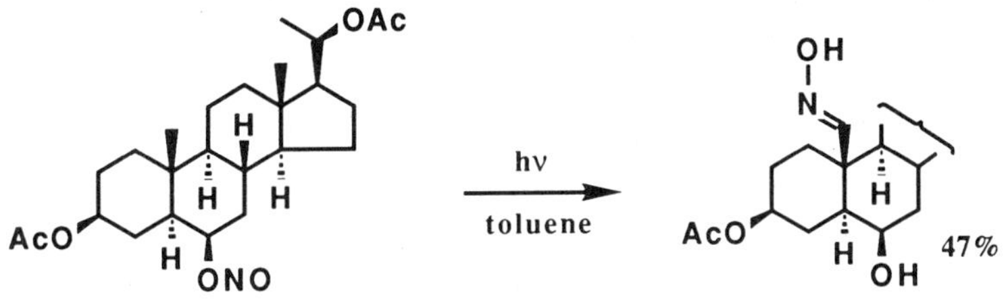

SCHEME 80.10

3. Functionalization of 10β- and 13β-Me of steroids from the 11β-ol nitrite: synthesis of aldosterone acetate (Schemes 80.11 and 80.12).[23,24] These examples are the most well-known applications of the reaction found by Barton, as well as his colleagues. The irradiation of corticosterone acetate nitrite affords aldosterone acetate oxime, which with nitrous acid gives aldosterone 12-acetate (Scheme 80.11). The overall yield, however, is rather low (15%), since the attack by the 11β-alkoxyl radical on C-19 competes with the desired attack on C-18. The incorporation of the 1,2-double bond into the nitrite could avoid the undesired C-19 attack by the radical; a far better yield (47%) of 19-oxime was achieved, as shown in Scheme 80.11. These examples of the functionalization may indicate that the Barton reaction is sensitive to structural changes, and that the distance and conformation requirements between the alkoxyl radical and the methyl group, which are in a 1,3-diaxial relationship, is rather strict (Scheme 80.12).

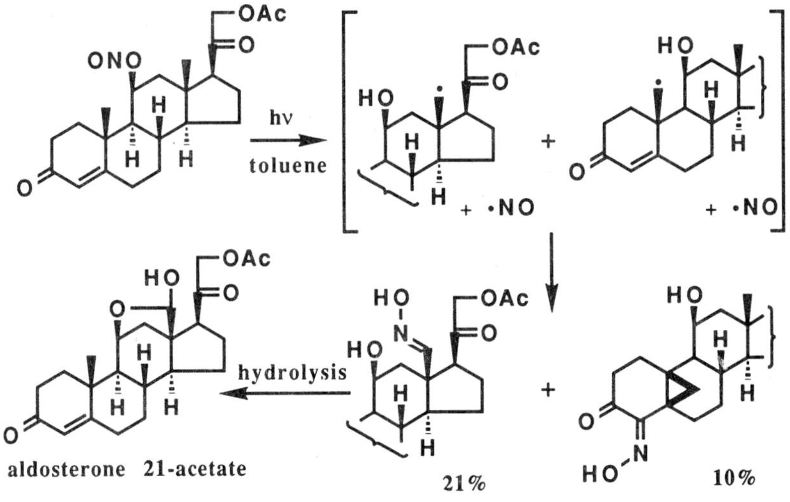

SCHEME 80.11

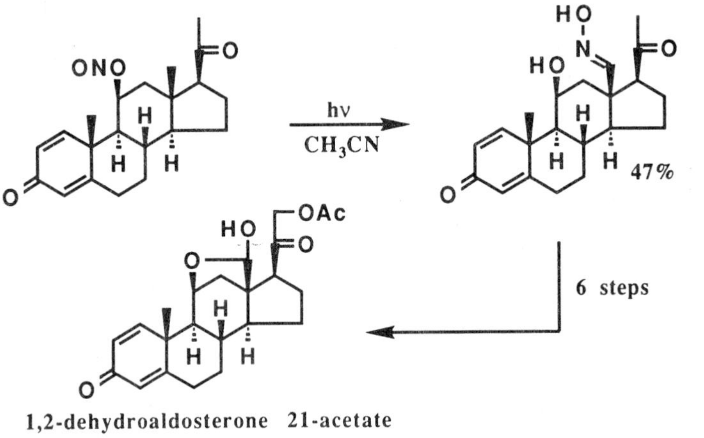

SCHEME 80.12

4. Functionalization of 14α-Me of a steroid from steroidal 7α-ol nitrite (Scheme 80.13).[25]

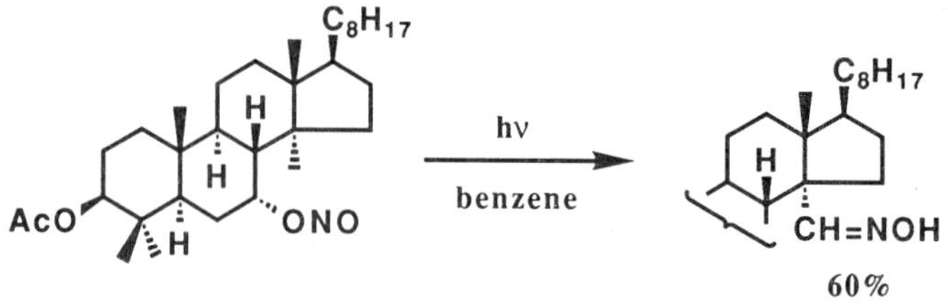

SCHEME 80.13

5. Functionalization of 13β-Me from steroidal 20α-ol nitrite:[26] synthesis of deoxofukujusonorone (Scheme 80.14).

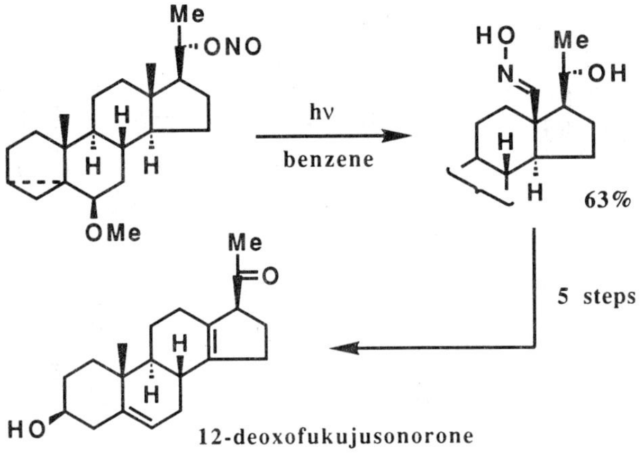

SCHEME 80.14

6. Functionalization of an angular methyl of terpenoid (Scheme 80.15).[27]

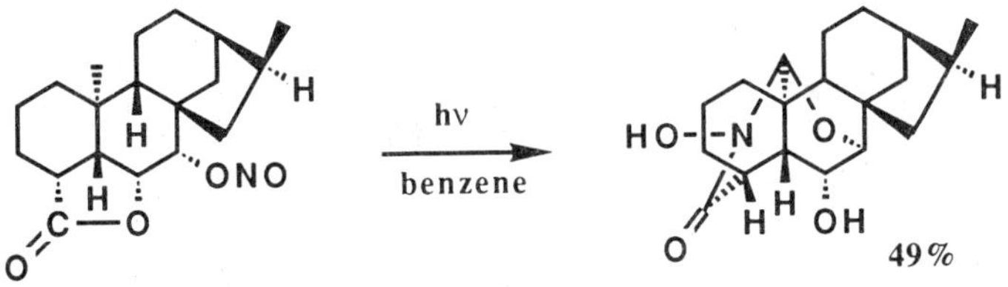

SCHEME 80.15

0.4 Accompanying Reactions in the Barton Reaction in the Photolysis of Nitrites

The following reactions compete with the Barton reaction in the photolysis of nitrites when the alkoxyl radical and the hydrogen to be abstracted is not ideally disposed for the Barton reaction. The fundamental nature of these competitive reactions in alkoxyl radicals generated from simple alcohols has been investigated extensively.[28] The competing reactions occasionally predominate over the Barton reaction (Scheme 80.19), or comprise an exclusive reaction (Schemes 80.16–80.18) that can be used for synthetic purposes. The following are representative examples. The mechanistic paths in each competing reaction are included.

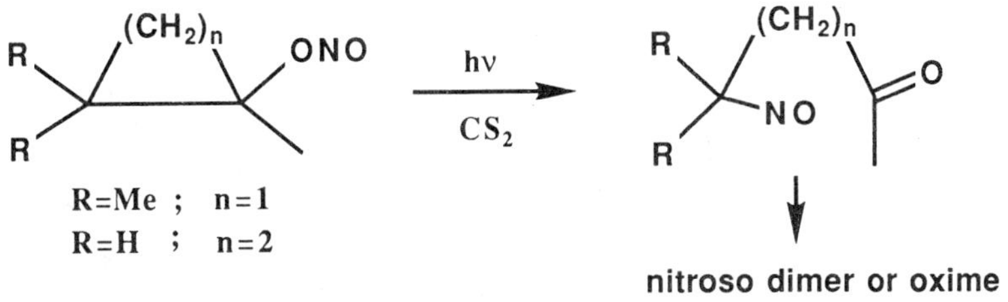

R=Me ; n=1
R=H ; n=2

nitroso dimer or oxime

SCHEME 80.16

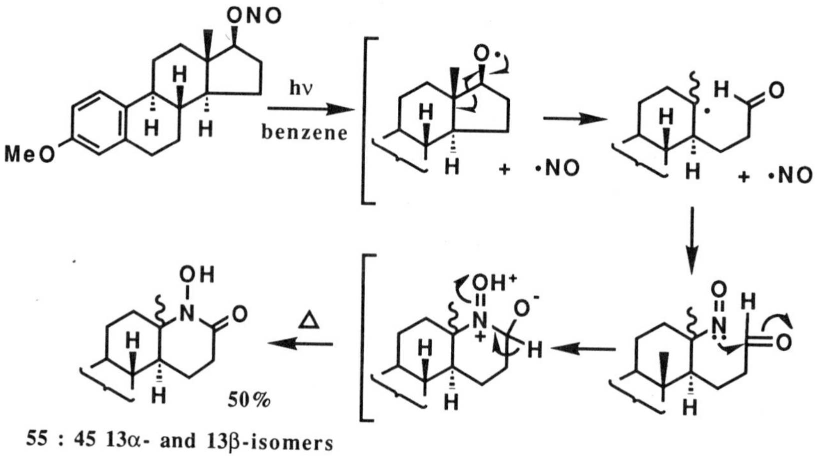

SCHEME 80.17

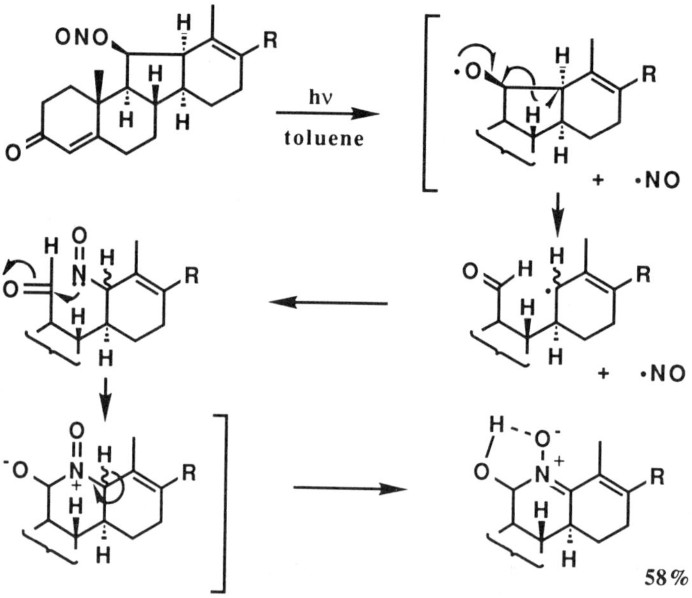

SCHEME 80.18

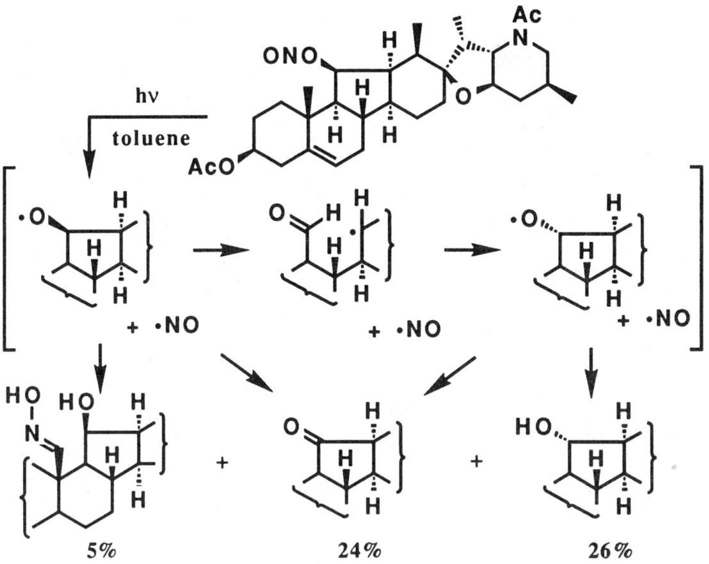

SCHEME 80.19

1. β-*Cleavage and epimerization of alkoxyl radicals:* β-scission is the most common and important competing reaction, as was first observed by Coe and Doumani[4] (Scheme 80.2). It takes place readily especially when the scission relieves some strain or provides stabilized carbon-centered radicals (such as tertiary and allyl radicals), as outlined in the following examples:

- β-Cleavage of cyclic alcohol nitrites[29,30] (Scheme 80.16).
- Formation of cyclic hydroxamic acids[31] (Scheme 80.17).
- Formation of cyclic nitrone[13,32] (Scheme 80.18).
- Epimerization of alkoxyl radicals[33,34] (Schemes 80.19 and 80.20).

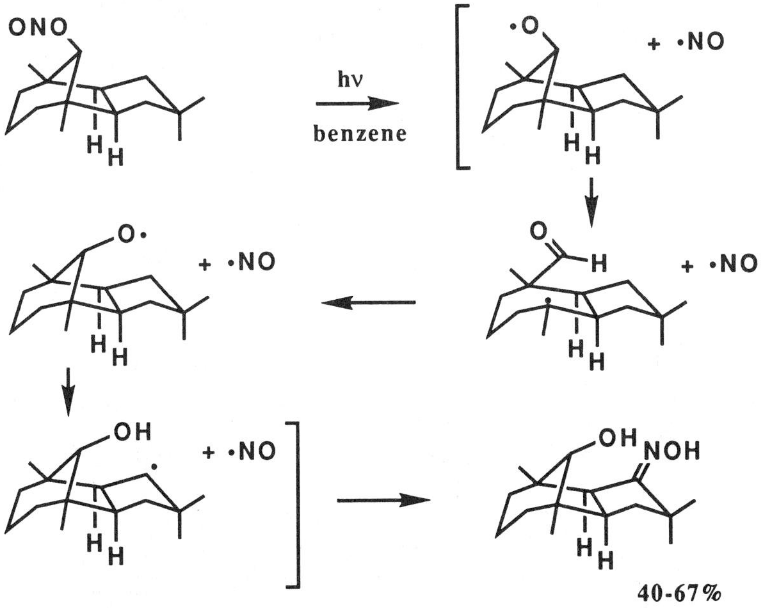

SCHEME 80.20

2. *Intramolecular addition of the alkoxyl radicals to the C-C double bond:*[35]

   • The formation of oxirane[36] (Scheme 80.21).

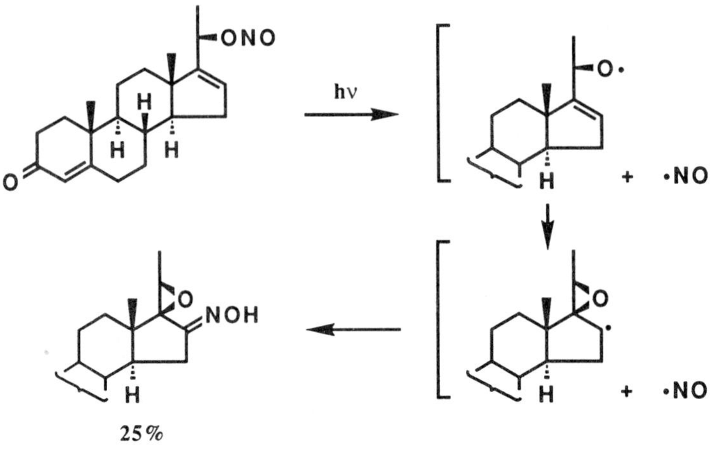

25%

SCHEME 80.21

   • The formation of tetrahydrofuran ring[37] (Scheme 80.22).

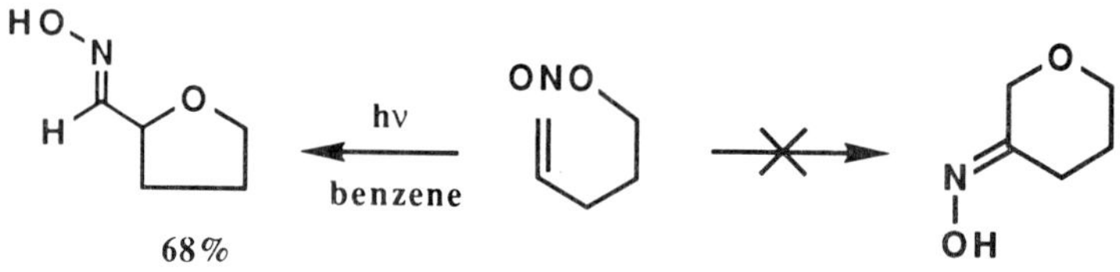

68%

SCHEME 80.22

3. *Disproportionation or α-hydrogen fission:* The photolysis of cyclic secondary alcohol nitrites is sometimes accompanied by the formation of the corresponding cyclic ketones by a loss of α-hydrogen of the alkoxyl radicals arising from either disproportionation or, less likely, α-hydrogen fission.

4. *Intermolecular hydrogen abstraction:* The parent alcohols are formed as a by-product by hydrogen abstraction from another molecule when the above-mentioned reactions are inefficient.

References

1. Barton, D. H. R., Beaton, J. M., Geller, L. E., and Pechet, M. M., A new photochemical reaction, *J. Am. Chem. Soc.,* 82, 2640, 1960; Barton, D. H. R., Beaton, J. M., Geller, J. E., and Pechet, M. M., A new photochemical reaction, *J. Am. Chem. Soc.,* 83, 4076, 1961.
2. Nussbaum, A. L. and Robinson, C. H., Some recent developments in the preparative photolysis of organic nitrites, *Tetrahedron,* 17, 35, 1962.
3. Purkis, C. H. and Thompson, H. W., The photochemistry of nitrates, nitrites and nitro-compounds. II, *J. Chem. Soc., Trans. Faraday Soc.,* 32, 1466, 1936; Thompson, H. W. and Dainton, F. S., The photochemistry of alkyl nitrites. III, *J. Chem. Soc., Trans. Faraday Soc.,* 33, 1546, 1937.
4. Coe, C. S. and Doumani, T. F., Photochemical decomposition of *t*-butyl nitrite, *J. Am. Chem. Soc.,* 70, 1516, 1948.

5. Gray, J. A. and Style, D. W. G., Photolysis of methyl nitrite, *J. Chem. Soc., Trans. Faraday Soc.*, 48, 1137, 1952.

6. Steacie, E. W. R. and Shaw, G. T., The homogeneous unimolecular decomposition of gaseous alkyl nitrites. II. The decomposition of ethyl nitrite, *J. Chem. Phys*, 2, 345, 1934.

7. Tarte, P., Recherches experimentales sur la decomposition photochimique des nitrites d'alkyles, *Bull. Soc. Roy Sci. Liege*, 22, 226, 1953.

8. Calvert, J. G. and Pitts, J. N., Jr., *Photochemistry*, Wiley, New York, 1966, 480.

9. (a) Hofmann, A. W., Ueber die einwirkung des broms in alkalisher lösung auf die amine, *Ber.*, 16, 558, 1883.; Hofmann, A. W., Zur kenntniss der coniin-gruppe, *Ber.*, 18, 5 and 109, 1985; (b) Löffler, K. and Freytag, C., Über eine neue bildungsweise von *N*-alkylierten Pyrrolidin, *Ber.*, 42, 3427, 1909; (c) Corey, E. J. and Hertler, W. R., A study of the formation of haloamines and cyclic amines by the free radical chain decomposition of *N*-haloammonium ions (Hofmann-Löffler reaction), *J. Am. Chem. Soc.*, 82, 1657, 1960; (d) Wolff, M. E., Cyclization of *N*-halogenated amines (The Hofmann-Löffler reaction), *Chem. Rev.*, 63, 55, 1963.

10. Barton, D. H. R., The use of photochemical reactions in organic synthesis, in *Organic Photochemistry*, Vol. 2, Butterworths, London, 1968, chap. 1.

11. (a) Akhtar, M., Some recent developments in the photochemistry of organic nitrites and hypohalites, in *Advances in Photochemistry*, Vol 2, Noyes, W. A., Hammond, G. S., and Pitts, J. N., Jr., Eds., Interscience, New York, 1964, 263; (b) Hesse, R. H., Barton reaction, in *Advances in Free Radical Chemistry*, 13, 83, 1969; (c) Chow, Y. L., Photochemistry of nitro and nitroso compounds, in *The Chemistry of Amino, Nitroso, Nitro Compounds and Their Derivatives*, Patai, S., Ed., Wiley, New York, 1982, 241.

12. Tanaka, M., Tanaka, J., and Nagakura, S., The electronic structures and electronic spectra of some aliphatic nitroso compounds, *Bull. Chem. Soc. Jpn.*, 39, 766, 1966.

13. (a) Suginome, H., Tsuneno, T. Sato, N., Maeda, N., Masamune, T., Shimanouchi, H., Tsuchida, Y., and Sasada, Y., Photoinduced transformations. XXX. Photorearrangement of (22*S*, 25*S*)-*N*-acetylveratra-5,8,13(17)-trienine-3β,11β,23β-triol 3,23-diacetate 11-nitrite, a fused cyclopentenyl nitrite, to two spiroisoxazolines, *J. Chem. Soc., Perkin Trans. 1*, 1297, 1976, (b) Suginome, H., Mizuguchi, T., Honda, S., and Masamune, T., Photoinduced transformations. 33. Mechanism of photoinduced rearrangement of a (22*S*, 25*S*)-*N*-acetylveratr-13(17)-enin-11β-yl nitrite, a fused cyclopentyl nitrite, to a nitrone, *J. Chem. Soc., Perkin Trans. 1*, 927, 1977; (c) Suginome, H., Maeda, N., and Masamune, T., Photoinduced transformations. XXXI. Photoinduced rearrangement of (22*S*, 25*S*)-*N*-acetyl-5α-veratra-8,13(17)-dienine-3β,11β,23β-triol 3,23-diacetate 11-nitrite to two spiroisoxazolines, *J. Chem. Soc., Perkin Trans. 1*, 1312, 1976.

14. Kabasakalian, P. and Townley, E. R., Photolysis of nitrite esters in solution. I. Photochemistry of *n*-octyl nitrite, *J. Am. Chem. Soc.*, 84, 2711, 1962.

15. Barton, D. H. R., Ramsay, G. C., and Wege, D., Photochemical transformations. XXI. A comparison of the photolysis and pyrolysis of organic nitrites, *J. Chem. Soc. (C)*, 1915, 1967.

16. Kabasakalian, P., Townley, E. R., and Yudis, M. D., Photolysis of nitrite esters in solution. II. Photochemistry of aromatic alkyl nitrites, *J. Am. Chem. Soc.*, 84, 2716, 1962.

17. Breslow, R., Biomimetic chemistry, *Chem. Soc. Rev.*, 1, 553, 1972.

18. Heusler, K. and Kalvoda, J., Intramolecular free-radical reactions, *Angew. Chem. Int. Ed.*, 3, 525, 1964.

19. Akhtar, M. and Pechet, M. M., The mechanism of the Barton reaction, *J. Am. Chem. Soc.*, 86, 265, 1964.

20. Suginome, H., Mizuguchi, T., and Masamune, T., Photoinduced transformations. 32. Scrambling of unlabelled and [^{15}N] nitrogen monoxide in the photo-induced rearrangement of (22*S*, 25*S*)-5α-veratr-13(17)-enin-11β-yl nitrites to nitrones, *J. Chem. Soc., Perkin Trans. 1*, 2365, 1976.

21. Akhtar, M., Barton, D. H. R., and Sammes, P. G., Some radical exchange reactions during nitrite ester photolysis, *J. Am. Chem. Soc.*, 87, 4601, 1965.

22. Wolff, M. E. and Morioka, T., C-19 Functional steroids. III. 2,19-Disubstituted androstane and cholestane derivatives, *J. Org. Chem.*, 30, 423, 1963.

23. Barton, D. H. R. and Beaton, J. M., A synthesis of aldosterone acetate, *J. Am. Chem. Soc.*, 82, 2641, 1960; *J. Am. Chem. Soc.*, 83, 4083, 1961.

24. Barton, D. H. R., Basu, N. K. J., Day, M., Hesse, R. H., Pechet, M. P., and Starratt, A. N., Improved synthesis of aldosterone, *J. Chem. Soc., Perkin Trans. 1*, 2243, 1975.

25. Bentley, T. J., McGhie, J. F., and Barton, D. H. R., The synthesis of 32-oxygenated lanostane derivatives, *Tetrahedron Lett.*, 2497, 1965.

26. Suginome, H., Nakayama, Y., and Senboku, H., Photoinduced molecular transformations. 131. Synthesis of 18-norsteroids, deoxofukujusonorone and the related steroids based on a selective β-scission of alkoxyl radicals as the key step, *J. Chem. Soc.*, 1837, 1992.

27. Hanson, J. R., The chemistry of the tetracyclic diterpenoids. II. The shape of ring B of the kaurenolides, *Tetrahedron*, 22, 1701, 1966.

28. Gray, P. and Williams, A., The thermochemistry and reactivity of alkoxyl radicals, *Chem. Rev.*, 59, 239, 1959; Bacha, J. D. and Kochi, J. K., Polar and solvent effects in the cleavage of *t*-alkoxyl radicals, *J. Org. Chem.*, 30, 3272, 1965; Walling, C., Padwa, A., Positive halogen compounds. VI. Effects of structure and medium on the β-scission of alkoxyl radicals, *J. Am. Chem. Soc.*, 85, 1593, 1963, and the references cited in these papers.

29. Depuy, C. H., Jones, H. L., Gibson, D. H., Cyclopropanols. VIII. Low-temperature thermolysis of cyclopropyl nitrites, *J. Am. Chem. Soc.*, 90, 5306, 1968.

30. Depuy, C. H., The chemistry of cyclopropanols, *Acc. Chem. Res.*, 1, 41, 1968.

31. (a) Robinson, C. H., Gnoj, O., Mitchell, A., Wayne, R., Townley, E., Kabasakalian, P., Oliveto, E. P., and Barton, D. H. R., The photolysis of organic nitrites. II. Synthesis of steroidal hydroxamic acids, *J. Am. Chem. Soc.*, 83, 1771, 1961; (b) Robinson, C. H., Gnoj, O., Mitchell, A., Oliveto, E. P., and Barton, D. H. R., The photochemical rearrangement of steroidal 17-nitrites, *Tetrahedron*, 21, 743, 1965; (c) Suginome, H., Yonekura, N., Mizuguchi, T., and Masamune, T., On the mechanism of the formation of steroidal cyclic hydroxamic acid in the photolysis of steroidal 17β-ol nitrite, *Bull. Chem. Soc. Jpn.*, 50, 3010, 1977; (d) Suginome, H., Maeda, N., and Kaji, M., Photoinduced transformations. 60. Photoinduced rearrangements of cholesteryl nitrites with monochromatic light, *J. Chem. Soc., Perkin Trans. 1*, 111, 1982.

32. Suginome, H., Sato, N., and Masamune, T., Photochemical formation of cyclic nitrone from nitrite of a fused 5-membered ring alcohol, *Tetrahedron*, 27, 4863, 1971.

33. Nickon, A., Iwadare, T., Mcguire, F. J., Mahajan, J. R., Narang, S. A., and Umezawa, B., The structure, stereochemistry, and genesis of α-caryophyllene alcohol (apollan-11-ol), *J. Am. Chem. Soc.*, 92, 1688, 1970.

34. Suginome, H., Sato, N., and Masamune, T., Photolysis of nitrites of 3O, N- diacetyl-22,27-imino-17,23-oxidojervan-5-ene-3β-11α- and 3β, 11β-diols, *Tetrahedron Lett.*, 1557, 1967; Suginome, H., Sato, N., and Masamune, T., Some observation on photolysis of fused cyclopentyl nitrites, *Bull. Chem. Soc. Jpn.*, 42, 215, 1969.

35. For a review, see Surzur, J.-M., Radical cyclizations by intramolecular additions, in *Reactive Intermediates*, Vol. 2, Abramovitch, Ed., Plenum, New York, 1982, chap. 3.

36. Rieke, R. D. and Moore, N. A., The cyclic addition of hetero radicals. II. Cyclic additions of alkoxyl radicals in alkenes, *J. Org. Chem.*, 37, 413, 1972.

37. Nussbaum, A. L., Wayne, R., Yuan, E., Sarre, O. Z., and Oliveto, E. P., Photolysis of organic nitrites. VIII. Intramolecular addition to a double bond, *J. Am. Chem. Soc.*, 87, 2451, 1965.

81

Photochemical Reactivity of the Nitro Group

etrich Döpp
versität Duisburg

1.1 Introduction

As evidenced by the first publication[1] on the solar photoreduction of nitrobenzene more than a century ago, nitro compounds, especially aromatic ones, are among the oldest study subjects of photochemists. Before the early 1960s, however, research in this field was not very active; but since that time, a general and steady increase in activity is to be noted.

Previous reviews treat nitro compound photochemistry in general[2,3] or specialize on nitroalkanes[4] or nitroarenes.[5] In a recent book,[6] photophysical properties and light-induced reactions of aromatic nitrocompounds have been treated in the general context of the photochemistry of aromatic compounds.

This chapter will review the photoreactions of organic nitro compounds with special emphasis on new developments and avoid, for limitations of space, duplication of earlier reviews wherever justified.

Most of the research on nitro compound photochemistry is done with nitroarenes, and thus the transformations of these will be given more coverage than nitroalkanes and nitroalkenes.

Occasionally, it is tempting to compare nitro compound photochemistry with that of carbonyl compounds. There are important differences, however. First, the nitro group itself may be trans-formed into a larger variety of other functional groups. Second, while the carbonyl group in ketones is integrated into the skeleton of a molecule, the nitro group never is; third, the relative ease with which the C-N bond is broken or contributes by predissociation[7] to radiationless decay has no equivalent in carbonyl photochemistry.

It is generally accepted that a nitro group attached to a molecule provides low-lying excited states, especially low-lying triplet states, since it enhances the $S_1 \rightarrow T_1$ transition rate without appreciably changing the $T_1 \rightarrow S_0$ transition rate[8] (but see below for a refinement of that statement). Therefore, many but by no means all light-induced reactions of nitro compounds originate from triplet excited states. It has been proposed that these may be grouped into states of n,π* (easily photoreducible by either H-atom or electron abstraction, e.g., nitrobenzene), π,π* (comparable reluctant to photoreduction, e.g., 1-nitronaphthalene, but subject to photosubstitution), and charge transfer (CT) character (widely unreactive, e.g., 4-nitroaniline).[9] The characterization of lower excited states must be regarded as difficult since the longest wavelength absorption band may represent transitions into three different electronically excited states, termed n,π*, L_b (short axis polarized) and L_a (long axis polarized).[10] The relative energies of these states depend on the nature of substituents and the solvent used.[10] For *p*-nitroaniline, the CT character of its lowest triplet and its dissoziation to the 4-nitroanilino radical have been demonstrated.[11] Unlike *p*-cyano- or *p*-acylsubstituted *N,N*-dimethylanilines, *p*-nitro-*N,N*-dimethylaniline does not exhibit dual fluorescence[12,13] and is regarded as unable to relax to a TICT state.[13] It is noteworthy that 4-nitro-*N,N*-dimethylaniline, 2,6-dimethyl-4-nitro-*N,N*-dimethylaniline, and 1-methyl-5-nitroindoline do show luminescence above 20,000 cm^{-1},[13] while it had been stated that due to the availability of the predissociation mode for deactivation, luminescence from nitrocompounds cannot be observed above that critical value.[7] The above result may reflect a possible increase of bond order in the emitting excited state.

It has been pointed out that a nitro group may introduce efficient promotion of radiationless decay by local vibrations modulating the overlap of lone pairs on adjacent oxygen atoms, and the distortion in its localized n,π* excited state relative to the ground state may cause strong phonon coupling in the radiationless transitions $S_1 \rightarrow S_0$ and $T_1 \rightarrow S_0$.[14]

Thus, a nitro group may influence typically the photophysics and the photochemistry of a system (dichotomy of pathways, observability of transients) as exemplified in the behavior of nitrostilbenes.[15-17] The properties of an arylnitrene may also be modified by a nitro group.[18] There are further examples of light-induced reactions in which the nitro group itself is not transformed because more favorable channels exist, as in the photoreduction of nitrophenazines[19] and in many nucleophilic aromatic photosubstitutions.

The past 2 decades have also seen highly valuable additions to the techniques of investigation already available. Besides nanosecond excitation, picosecond techniques have become widely used, and in addition to time-resolved absorption and emission, time-resolved resonance Raman scattering and time-resolved microwave conductivity have been introduced now (for one example each of either techniques, see References 16 and 17). Mention of techniques used will be made in the following sections wherever appropriate.

81.2 Nitroalkanes

The lower nitroalkanes are photolabile in the gas and liquid phase as well as in frozen solutions.[2-4] UV absorption of nitromethane in the gas phase and in solution has been thoroughly investigated[20] and the lack of fluorescence and phosphorescence in solution has been noted. The electronic structure and the photophysical properties of nitroalkanes[21] as well as their photochemical reactions in the liquid phase and in solution[4,22] have been reviewed previously.

Mononitroalkanes and Nitrocycloalkanes

Primary Processes

Gas-phase photolysis of simple nitroalkanes leads to complex product mixtures.[4,22] Generally, C-N bond homolysis is the dominating initial step (Equation 81.1) and the following events reflect the options open to the homolysis products (among others, Equations 81.2 and 81.3). Inevitably, the photochemistry characteristic for nitrites[3] is involved. Pathways of minor importance (Equations 81.4 and 81.5) have also been suggested.[4]

$$-\underset{|}{\overset{|}{C}}-NO_2 \quad \xrightarrow{h\nu} \quad -\underset{|}{\overset{|}{C}}\cdot \quad + \quad NO_2 \qquad (81.1)$$

$$-\underset{|}{\overset{|}{C}}\cdot \; + \; NO_2 \quad \xrightarrow{\hspace{2cm}} \quad -\underset{|}{\overset{|}{C}}-O-NO \qquad (81.2)$$

$$-\underset{|}{\overset{|}{C}}-O-N=O \quad \xrightarrow{h\nu/\Delta} \quad -\underset{|}{\overset{|}{C}}-O\cdot \; + \; NO \qquad (81.3)$$

$$CH_3 - NO_2 \quad \xrightarrow{h\nu} \quad CH_2O + HONO \qquad (81.4)$$

$$CH_3-CH(R)NO_2 \quad \xrightarrow{\hspace{2cm}} \quad H_2C{=}CHR + HNO_2 \qquad (81.5)$$

Alkyl radicals and NO_2 have been identified by ESR and have been shown to emanate in irradiations at low temperature.[23] When the energy can be dissipated, recombination to nitrites is observable; otherwise, the latter decompose rapidly to alkoxy and NO radicals. Vibrationally excited NO_2 is liberated from singlet excited nitromethane and 2-nitropropane by laser photolysis.[24] Polarized emissions (200 and 218 nm) of photodissociating nitromethane have been recorded.[25] The quantum yield of nitromethane photolysis has also been determined.[26] UV absorption bands in gas-phase and solution spectra have been interpreted to be of n,π^* nature on the basis of CNDO/S and INDO/S calculations.[20] *Ab initio* multiple reference double-excitation configuration-interaction ground- and excited-state (singlet and triplet) potential curves for nitromethane decomposition have been worked out.[27] Also, the action of nitromethane as quencher of polycyclic aromatic hydrocarbon fluorescence has been investigated.[28,29]

Irradiation of Mononitroalkanes in Solution

Due to secondary photolyses, the spectrum of products obtained upon irradiation of cyclohexane solutions of nitromethane[30] and nitroethane[31] is dependent markedly on the irradiation wavelength chosen. All wavelengths applied seem to excite the $(n,\pi^*)S_1$ state, which then crosses efficiently to T_1 and the latter either is homolyzed or (but to a much lesser extent) may abstract an H-atom, since $CH_3\dot{N}O_2H$ radicals may be detected.[30]

Upon 254-nm irradiation in deoxygenated cyclohexane solution, the following products (quantum yields in brackets) have been found from nitromethane:[30] dicyclohexyldiazene-*N,N'*-dioxide (= nitrosocyclohexane dimer, ~0.023), cyclohexanol (~0.003), cyclohexanone (~0.004), cyclohexyl nitrate (~0.001), and nitrocyclohexane (~0.001). With 313-nm irradiation, no nitrosocyclohexane dimer was obtained, but cyclohexanol (~0.002), cyclohexanone (~0.001), cyclohexyl nitrate (~0.003), nitrocyclohexane ($\sim 10^{-4}$), and cyclohexanone oxime ($\sim 4 \times 10^{-4}$). Nitrosocyclohexane absorbs more efficiently at 313 nm than at 254 nm; thus, the quantum yield of disappearance of nitrosocyclohexane dimer is 0.17 at 313 nm and ~0.03 at 254 nm.[30] An obvious rationale is that nitromethane undergoes homolysis to NO_2 and $\cdot CH_3$; the latter would attack cyclohexane to generate cyclohexyl. In turn, NO_2 may trap cyclohexyl or cyclohexyloxyl, which may be formed by the process in Equation 81.6:

$$C_6H_{11}^{\cdot} + CH_3NO_2 \rightarrow CH_3 - \dot{N}(O) - OC_6H_{11} \rightarrow CH_3NO + C_6H_{11}O^{\cdot} \qquad (81.6)$$

NO is liberated from methyl nitrite (Equation 81.3) or formed as follows

$$CH_3-NO+NO_2 \rightarrow CH_3NO_2+NO \qquad (81.7)$$

and nitrosocyclohexane and nitrocyclohexane are formed by radical combination:

$$C_6H_{11}{}^{\cdot} +NO_x \rightarrow C_6H_{11}NO_x\,(x=1\,\text{or}\,2) \qquad (81.8)$$

Irradiation of nitroethane in cyclohexane with $\lambda > 290$ nm gave 28% cyclohexanone oxime, 7% cyclohexyl nitrate, and 4% nitrocyclohexane;[31a] 254-nm irradiation afforded mainly nitrosocyclohexane. The latter, upon further irradiation at $\lambda > 290$ nm, gave the three aforementioned cyclohexane derivatives and additionally N-cyclohexyl-ε-caprolactam (1),[31a] the formation of which may be explained as shown in Scheme 81.1. There is no direct deoxygenation of the nitroalkane and hydrogen transfer to the excited nitroethane need not to be envisaged to explain the products.[31b] Photoreduction of primary and secondary nitroalkanes to oximes has been achieved by electron transfer.[32]

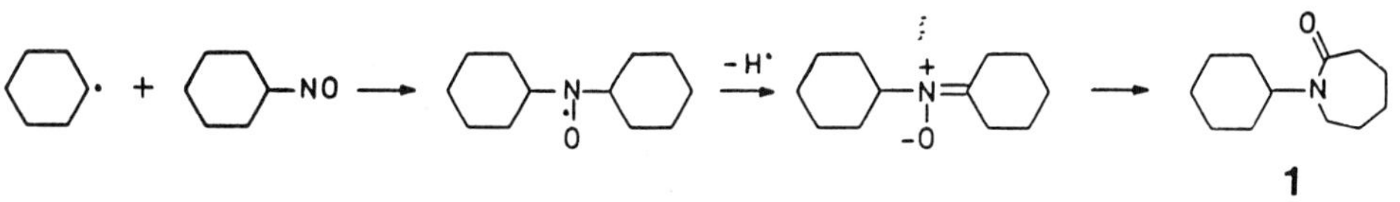

SCHEME 81.1

Irradiation (254 nm) of nitromethane in 2-propanol gave methane, methyl nitrite, isopropyl nitrite, acetone, acetone oxime, pinacol, and 1-methyl-2-(1-hydroxy-1-methylethyl)diazene-N,N'-dioxide.[33] The following processes are likely to be effective:

$$CH_3NO_2 \rightarrow CH_3{}^{\cdot} + NO_2 \rightarrow CH_3-ONO \rightarrow CH_3O^{\cdot} + NO \qquad (81.9)$$

$$CH_3{}^{\cdot} +(CH_3)_2CHOH \rightarrow CH_4 +(CH_3)_2\dot{C}OH \qquad (81.10)$$

$$2(CH_3)_2\dot{C}OH \rightarrow (CH_3)_2C(OH)(CH_3)_2 \qquad (81.11)$$

$$(CH_3)_2\dot{C}OH+NO \rightarrow (CH_3)_2C(OH)NO \qquad (81.12)$$

$$CH_3{}^{\cdot} +NO \rightarrow CH_3-NO \qquad (81.13)$$

$$2\,CH_3NO \rightarrow CH_3-N(O)\uparrow N(O)CH_3 \qquad (81.14)$$

$$CH_3NO+(CH_3)_2C(OH)NO \rightarrow CH_3-N(O)\uparrow N(O)-C(OH)(CH_3)_2 \qquad (81.15)$$

$$CH_3NO_2 +R^{\cdot} \rightarrow R-O-\dot{N}(O)-CH_3 \rightarrow CH_3NO+RO^{\cdot} \qquad (81.16)$$

Again, H-atom abstraction by (triplet) excited nitromethane does not contribute, and the observed solvent-derived radicals originate most likely from H-abstraction by $CH_3^{\cdot}$ only.

From irradiation of nitromethane in benzene, anisole, nitrobenzene, *o*- and *p*-nitrophenol, biphenyl, phenol, and traces of toluene have been identified.[34]

Photochemistry of Nitroalkanes and Nitrocycloalkanes Bearing Additional Functional Groups

The cyclic bisallylic nitrocompound **2** gave the corresponding hydroxycompound **3** with retention of configuration upon photolysis in methanol[35] via C-N homolysis to **4**, nitrite formation, O-N homolysis, and hydrogen capture (Scheme 81.2). The high stereoselectivity observed points to a close contact within the radical pair **4**.

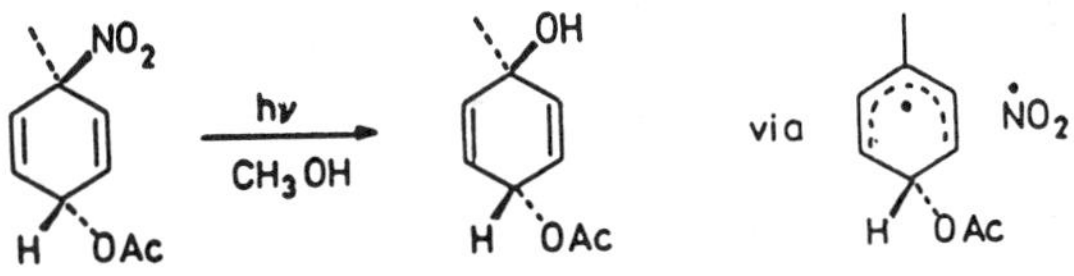

SCHEME 81.2

6-β-Nitrocholest-4-ene (**5**) was converted into compounds **8–11** by 254-nm irradiation in methanol.[36] A highly plausible rationale is C-N homolysis of **5** and recombination to generate the nitrites **6** and **7**, which in turn undergo O-N homolysis and transformation into the final products observed, the oximes **10** and **11** being products of the well-known Barton reaction.[3] 4-β-Nitrocholest-6-ene underwent a completely analogous sequence.[36]

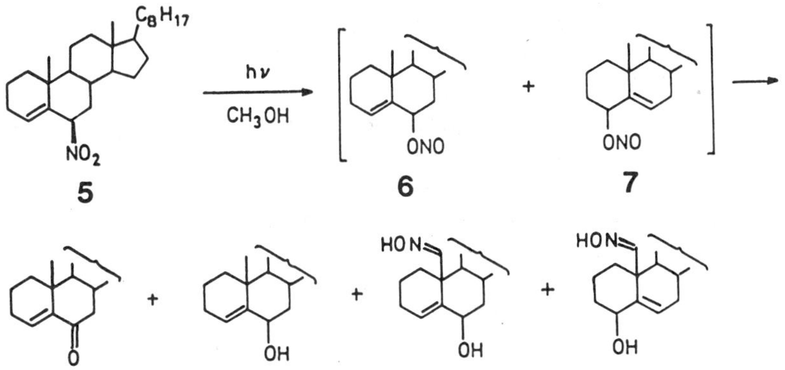

SCHEME 81.3

Photolysis of certain α-nitroepoxides, such as **12**, is observed in methanol or 2-propanol but not in *t*-butanol, acetonitrile, or in hydrocarbons.[37] The products observed (**13–15**) may originate from three parallel pathways: (a) light-induced reduction of the nitro group and ring opening, (b) light-induced ring opening and 1,2-shift of the nitro group, followed by reduction, and (c) loss of HNO$_2$ after uptake of ROH. The corresponding nitro alkene is not an intermediate.

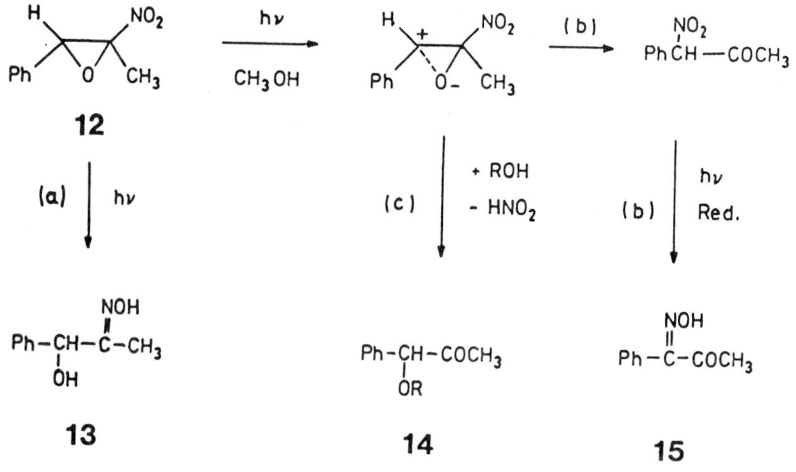

SCHEME 81.4

2-Nitrocyclohexanone and 2-nitrocycloheptanone tend to undergo transformation into cyclic *N*-hydroxyimides.[38]

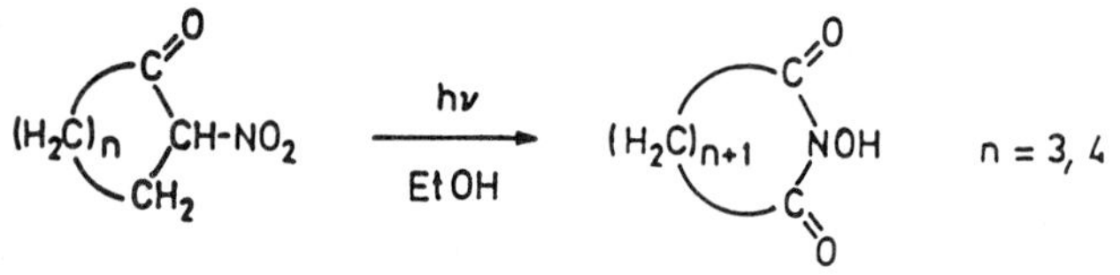

SCHEME 81.5

Monocyclic α-nitroketones do not resemble polycyclic α-nitroketones, however, since a more complex product pattern is found for 2-nitro-α-camphor.[38] Steroidal α-nitroketones have been transformed successfully by irradiation in ethanol solution into the corresponding α-diketones and α-hydroxyiminoketones. Cyclic ring-enlarged *N*-hydroxyimides are formed from five-membered ring steroidal ketones only.[39–42] The results depend to some extent on whether or not the enol form predominates in the solvent ethanol and on ring size.[42] A rationale for the conversion of a cyclic α-nitroketone into the corresponding *N*-hydroxyimide 17 has also been given, as demonstrated for 16-nitro-5-androstan-17-one (16) via the singlet excited state of its enol form.[42]

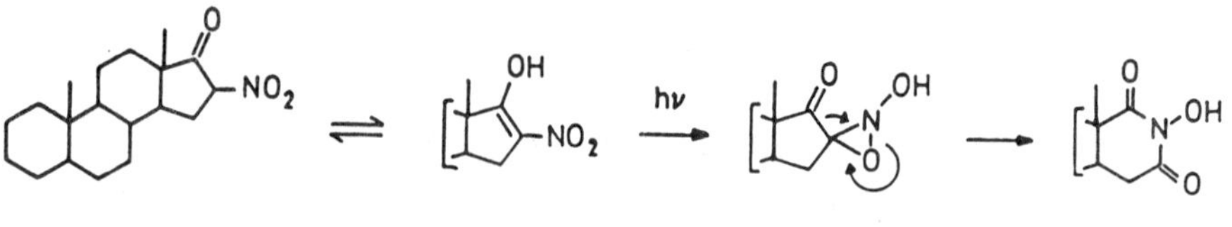

SCHEME 81.6

Efficient displacement of both nitro groups is observed upon photolysis of steroidal α,α′-dinitroketones to yield the corresponding α-diketones 19 and 20, as exemplified for 18.[40]

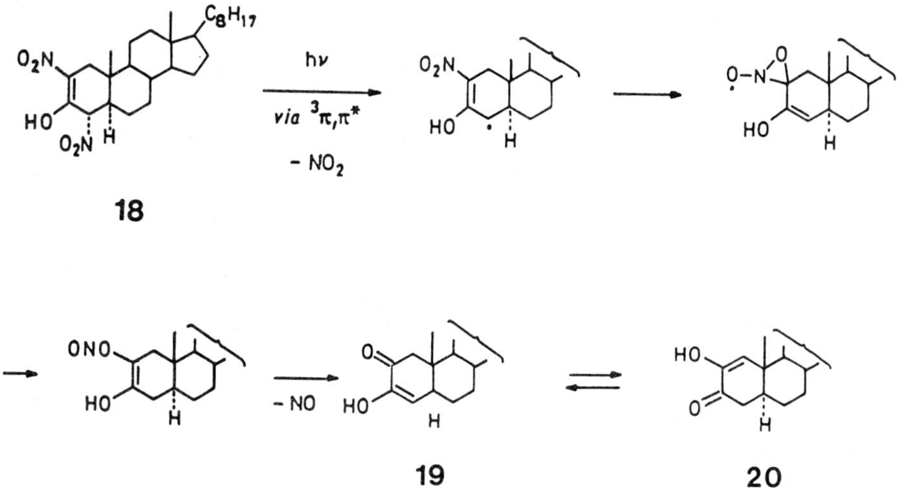

SCHEME 81.7

Light-induced reactions of α-halogenated nitroalkanes have been reviewed,[3] as well as nucleophilic substitutions at the α-carbon of tertiary nitroalkanes and 4-nitrobenzyl compounds, which take place by a light-accelerated radical anion chain reaction (S_{RN}^1).[43]

Tetranitromethane

When illuminated in a frozen matrix at −77 °C, tetranitromethane (TNM) shows homolysis to generate NO_2 and $(O_2N)_3C$ radicals[23] (Equation 81.17), while trinitromethanide anion ($\lambda_{max} \approx$ nm) is formed in aqueous alcohols[44] with quantum yields of 0.5 (methanol) or 0.35 (ethanol)[45] (Equation 81.18). TNM is useful as a nitrating agent for aromatics, and irradiation enhances its efficiency markedly (Equation 81.19). An isomerization to trinitromethylnitrite, which is envisaged to be a more efficient nitronium ion transfer agent, has been suggested.[46]

$$C(NO_2)_4 \xrightarrow{h\nu} NO_2 + C(NO_2)_3 \rightarrow ONO-C(NO_2)_3 \tag{81.17}$$

$$ONO-C(NO_2)_3 + ArH \rightarrow Ar(H)NO_2^+ + :C(NO_2)_3^- \tag{81.18}$$

$$Ar(H)NO_2^+ + :C(NO_2)_3^- \rightarrow Ar-NO_2 + HC(NO_2)_3 \tag{81.19}$$

Charge-transfer excitation of a binary ArH-TNM complex as the initiating process of light-induced nitration of aromatics by TNM had been suggested earlier[47] and worked out more recently by Kochi[48] and Eberson and Radner.[49,50] Due to rapid bond rupture in the TNM radical anion after excitation, a triad is formed (Equation 81.20) that exercises several options for its further very complex chemistry, the main one being nitro/trinitromethyl addition (Equation 81.21) and the minor one direct nitration of the aryl radical cation by NO_2 (Equation 81.22).[48–50] Ar-NO_2 may also be formed by elimination of nitroform from the addition product of Equation 81.21.

$$ArH-----C(NO_2)_4 \xrightarrow[CH_2Cl_2]{h\nu} [ArH^+ ----NO_2 ---^- :C(NO_2)_3](21) \tag{81.20}$$

$$21 \rightarrow ArH(NO_2)[C(NO_2)_3] \tag{81.21}$$

$$21 \rightarrow ArHNO_2^+ + :C(NO_2)_3^- \rightarrow ArNO_2 + HO(NO_2)_3 \tag{81.22}$$

Nitronic Acids and Nitronate Anions

These species are in a ground-state equilibrium with the corresponding nitroalkanes or nitrocycloalkanes and are included here.

1-Nitrooctane is deprotonated in aqueous ethanolic sodium hydroxide solution, and the anion strongly (ε 10,200) absorbs at 233 nm. 2-Nitrobornane nitronate has a strong UV maximum at 240 nm.[51a]

Irradiation of 1-nitrooctane nitronate in aqueous ethanolic sodium hydroxide solution at 254 nm gave caprylhydroxamic acid in 30% yield, which could be improved to 85% using a methanolic methylamine medium.[51a] Likewise, using 254-nm light and either alkaline water/ethanol mixtures or alkoxide containing alcohols as solvent, the nitronates of phenylnitromethane, 2-nitronorbornane, 1-phenyl-2-nitropropane, various 2-nitrocycloalkanones of different ring size and several steroidal nitrocompounds were converted into the corresponding hydroxamic acids or (ring-enlarged) *N*-hydroxylactams,[51,52] as illustrated in Scheme 81.8. The course of the reaction is depicted, using **21** as the representative, in scheme 81.9.[52]

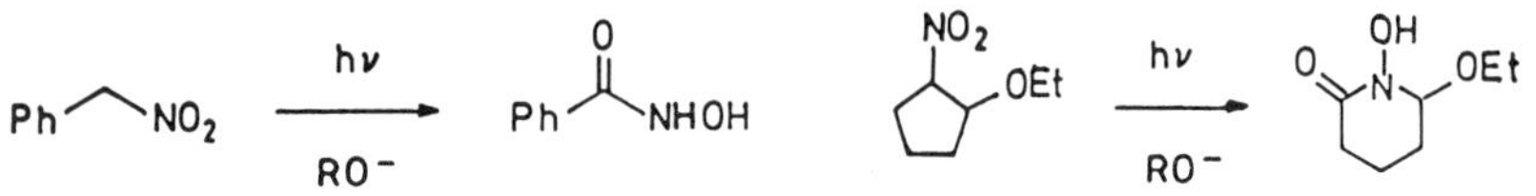

SCHEME 81.8

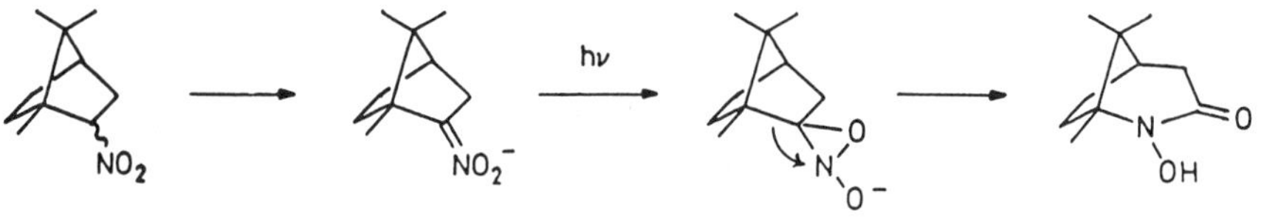

21

SCHEME 81.9

Five representative nitronates were found to have excited-state lifetimes between 0.54×10^{-5} and 1.26×10^{-4} s, which are regarded as too long for a singlet excited state; since oxygen decreased the quantum yields (e.g., for 2-nitrobornane $10^3 \times \phi = 5.69$ under argon and 2.07 in aerated solution), it is proposed that the reactive excited state is a π,π^* triplet.[51,52]

For some nitro compounds, relatively stable *aci*-forms (nitronic acids) exist. Their behavior on irradiation, however, seems to be generally different from that of the nitronate anions. 4-*t*-Butyl-*aci*-nitrocyclohexane, when irradiated in dry benzene at room temperature, gave 14% *Z*- and 35% *E*-isomer of the parent nitro compound, 14% 4-*t*-butylcyclohexanone and 33% of the oxime of that ketone.[53] *aci*-9-Nitrofluorene was converted likewise into a mixture of 60 to 62% 9-fluorenone, 22 to 24% 9-hydroximinofluorene, 4% 9,9′-bifluorenylidene, and 6 to 8% 9,9′-dinitro-9,9′-bifluorene by photolysis in a methanol/*t*-butanol mixture.[53]

81.3 Nitroalkenes

α,β-Unsaturated nitro compounds may undergo one or several of the following processes under illumination.

1. *E,Z-isomerization:*[54–59] Such equilibrations often precede other processes. Using sensitizers of varied triplet energy, different *E/Z* ratios of 2-nitro-2-butene may be reached.[54]

2. *Deconjugation to allylic nitro compounds:*[53,56,60–62] This reaction is typical for compounds bearing a secondary or tertiary hydrogen atom at the γ-carbon; for example, **22** → **23** and **24** → **25**,[56] (see Scheme 81.10).

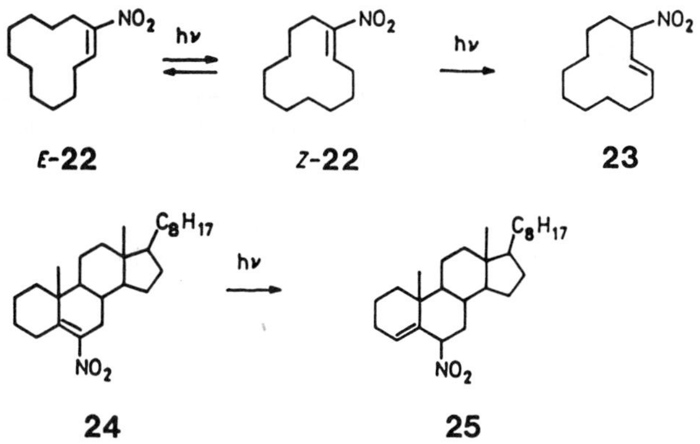

SCHEME 81.10

3. *Conversion into* α-*hydroximinoketones and other reactions via isomerization to labile nitrites:*[3,55,56,62-68] For the first example[63] known, the sequence of steps may be as follows (26 → 31, Scheme 81.11). Nitrites such as 28 can have a fleeting existence at best, but an oxy-radical (probably 29) has been detected by ESR albeit not at room temperature but in frozen benzene solution. Similarities to the nitrite formation in nitroalkanes are limited, since C-N bond homolysis can be bypassed via intermediates such as 27. Unlike a normal alkoxy radical, 29 has the option to recombine within the solvent cage with NO at the vinylogous position.

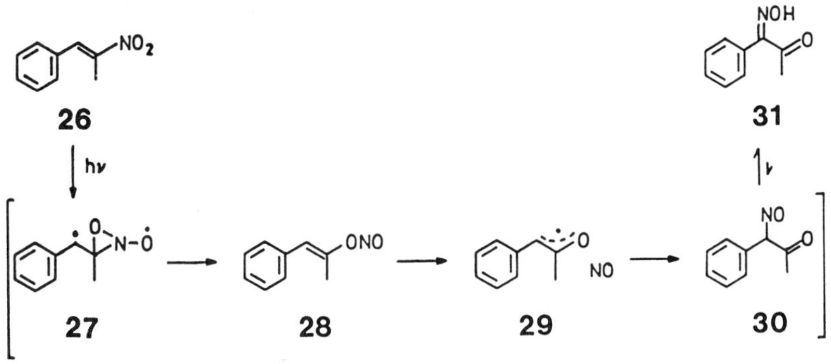

SCHEME 81.11

When no hydrogen is available at the β-terminus of the double bond, other reaction channels are followed, as exemplified for 32, prior to or after the C-N recombination (Scheme 81.12).[68]

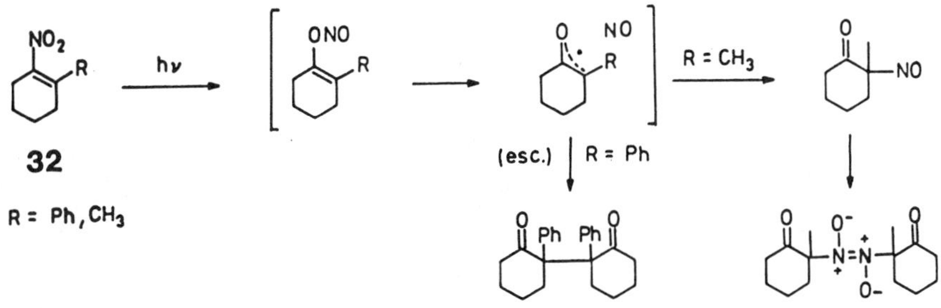

SCHEME 81.12

4. *Processes initiated by an intramolecular [2+2]-addition of one N-O bond to the C=C bond:*[55,65,67,68,70]
In this novel reaction,[67] an unstable oxazete-*N*-oxide (**33, 35**) is presumed as the primary product, which in turn either fragments into a carbonyl compound and a nitrile oxide[67,68] (to be intercepted with a dipolarophile) or is attacked by nucleophiles[67] (Scheme 81.13). It should be noted that **34** has been transformed into the dimethylacetal **36** of an α-oximino ketone, in which the former nitro-substituted carbon has retained the nitrogen function.

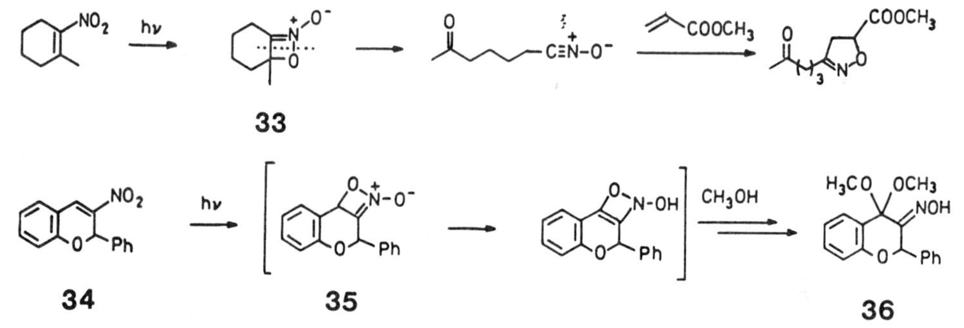

33

34   **35**   **36**

SCHEME 81.13

There are, however, also cases that do not fall under the aforementioned modes. Triplet excited **37** is, albeit in low efficiency, isomerized into **38** and thus is more reminiscent of the photoreactivity of the cyclohexenone **39** (Scheme 81.14). Extended conjugation opens pathways of formal oxygen atom transfer to the δ-carbon atom in α,β,γ,δ-unsaturated nitro compounds such as **40**[72,74] and **41**[73,74] (Scheme 81.15). Formally, one may regard these transformations as vinylene homologous cases of that of **35**.

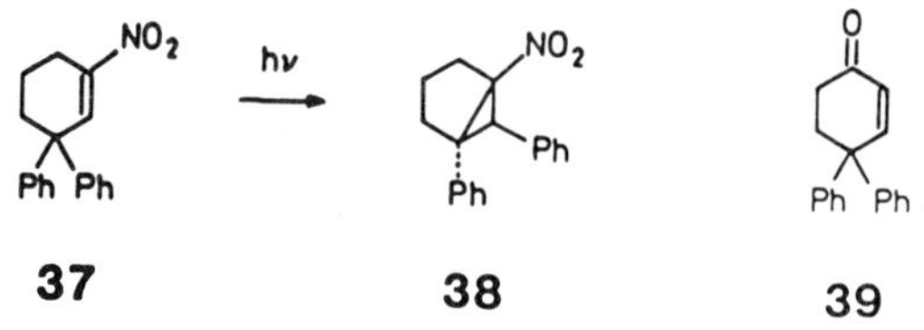

37   **38**   **39**

SCHEME 81.14

40

41

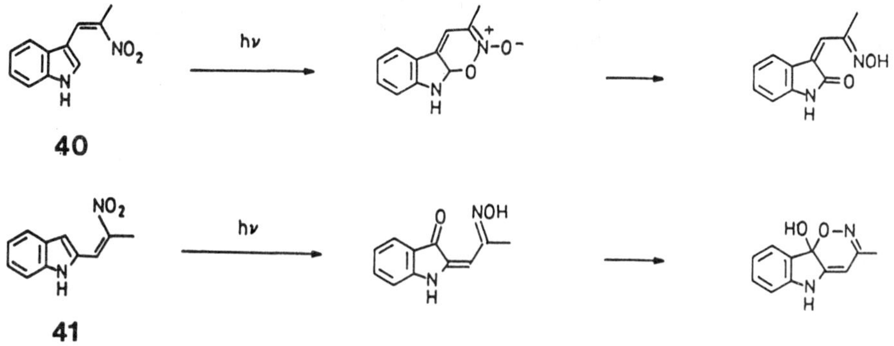

SCHEME 81.15

1.4 Nitroarenes

Nitro → Nitrite Rearrangement and Subsequent Reactions

Nitroanthracenes

Since Chapman and co-workers proposed a detailed mechanistic scheme for the photoconversions of 9-nitroanthracene,[75] this compound and related systems have been of considerable interest to photophysicists and photochemists. 9-Nitroanthracene (NA) and 9-cyano-10-nitroanthracene (CNA) and 9-benzoyl-10-nitroanthracene (BNA) are nonfluorescent,[76–81] and efforts to detect genuine phosphorescence[76,77,79–81] have been a matter of dispute for some time. Hamanoue and co-workers present a phosphorescence spectrum for NA with band maxima at 685 and 760 nm and with a lifetime of 14 ms at 77K in EPA,[79] and for BNA (700 and 775 nm)[76,77] and CNA (~740 and 820 nm)[76] under the same conditions. Testa[80,81] offers a massive warning that all luminescence observed from NA might arise from impurities. Generation of anthryloxy radicals by whatever mechanism is claimed to be fast even at 4.2K[81] and formation of highly fluorescent or phosphorescent products therefrom may well hamper the search for genuine emissions. Scrupulous purification of the samples may pose another problem.

Time-resolved absorption studies for NA, BNA, and CNA[78] after nanosecond and picosecond excitation revealed long (72 to 86 ps) build-up times for triplet-triplet absorption. It was concluded that indirect intersystem crossing $S_1(\pi,\pi^*) \rightarrow T_n(n,\pi^*) \rightarrow T_1(\pi,\pi^*)$ is the most important process to populate T_1 in accord with El-Sayed's rule[83] and that the above-mentioned build-up times represent the rate of $T_n \rightarrow T_1$ interconversion.[78]

An upper triplet state T_n (n,π^*) was assigned as responsible for the rearrangement of NA.[77,78] This involves C-O bond formation accompanied by C-N bond rupture. An isomerization to the nitrite **43** has been proposed,[75] and although it has never been identified directly, it is widely regarded as a logical intermediate.[3] Under photolysis conditions, O-N bond rupture in **43** is anticipated to generate the anthryloxy radical **44**. Its presence in solid photolysates at 77K has been demonstrated by ESR (in frozen benzene solutions).[69] Also, the absorption spectra of the corresponding anthryloxy radicals are superimposed on the T_1-T_n time-resolved absorptions of CNA and BNA.[76] The latter tend to give anthrols as the final products by abstraction of a H-atom by the respective anthryloxy radicals out of the matrix.[77] **44** may as well abstract hydrogen to form **45** or recombine with NO at C 10. If this carbon bears hydrogen, tautomerization of **47** to **46** follows immediately; otherwise, as demonstrated for 9-chloro-10-nitro- and 9,10-dinitroanthracene, the nitrosocompounds persist.[84] Escape from the pair with NO is another possibility for **44**, ultimately leading to symmetric coupling to generate bianthrone **48**.[75] Also, recombination on oxygen to regenerate **43** has been demonstrated to occur in a polymeric film parallel to hydrogen capture.[82] The scheme summarizes the current knowledge, including the oxidation of **44** to anthraquinone (**49**).

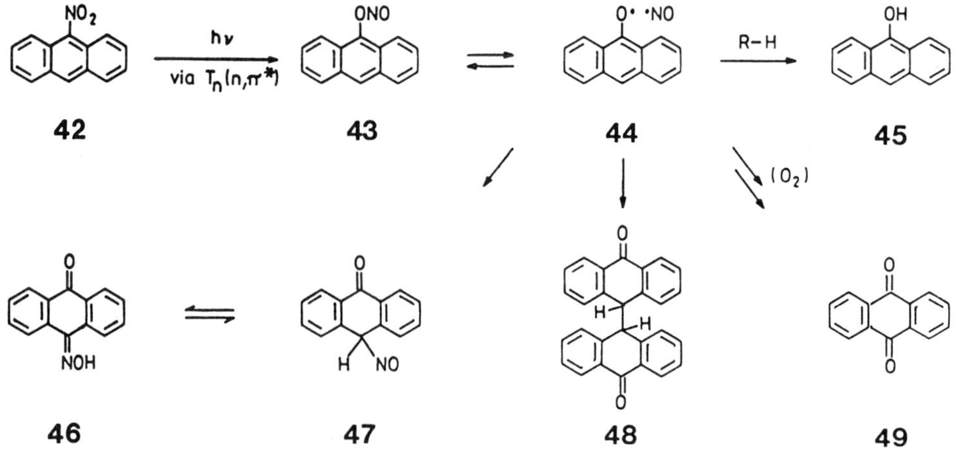

SCHEME 81.16

Other Nitroaromatic Compounds

Besides the nitroanthracenes, various other polycyclic nitroarenes were found to undergo this rearrangement, albeit with formation of final products of different types.[3,75] Among the recent examples are 1-nitro-2-phenylnaphthalene (50) giving 51–53[85] and 2,5-di-*t*-butyl-1-nitronaphthalene (54) yielding 55,[86] a behavior in marked contrast to that of *o*-nitro-*t*-butylbenzenes (see Scheme 81.17 and Section 81.4).

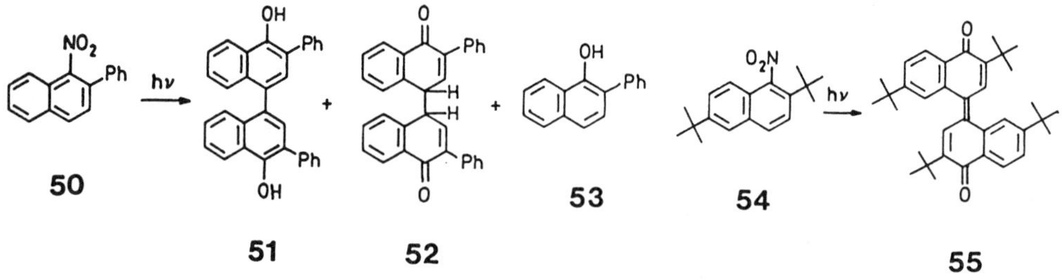

SCHEME 81.17

Spectroscopic and photochemical properties for the three mononitropyrenes (1-, 2-, and 4-) have been compared.[87] 2-Nitropyrene is remarkably stable to light, showing practically no conversion under illumination over 300 nm. 1-Nitropyrene is converted into 1-hydroxy-2-nitropyrene via a sequence of steps analogous to 42 → 44 and combination with NO at C-2 and subsequent oxidation of the nitroso derivative.[87,88] 4-Nitropyrene undergoes phototransformation into a mixture of hitherto uncompletely characterized products.[87]

Photoaddition of the Nitro Group to π-Systems

Irradiation (254 nm) of a 1:4 mixture of nitrobenzene and 2-methyl-2-butene gives rise to acetone, azobenzene, and perhydro-1,4-dioxa-2,5-diazines.[89] The most logical precursor would be the 1,3,2-dioxazolidine; i.e., the adduct of nitrobenzene by means of its oxygen atoms to the double bond. This adduct may fragment into acetone and a reactive species capable of dimerization.

Triplet n,π*-excited nitrobenzene adds stepwise to cyclohexene; the cycloadduct 56, analogous to a primary ozonide, could be isolated at −70 °C. Catalytic hydrogenation of 56 gave aniline and a mixture of the cyclohexane-1,2-diol diastereomers[90,91] (Scheme 81.18).

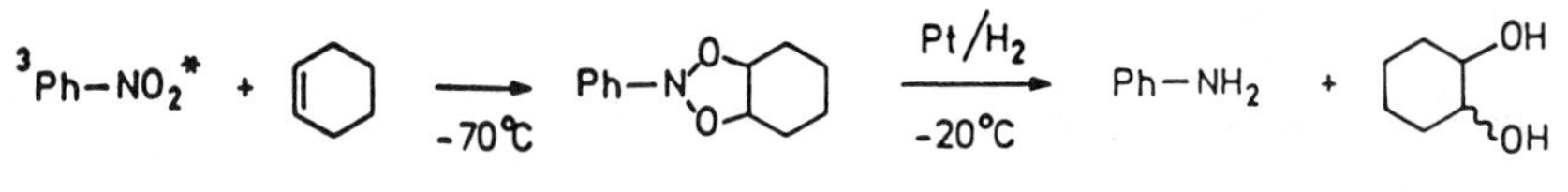

56

SCHEME 81.18

Irradiation of 1,3- or 1,4-dinitrobenzene (57a,b), as well as 1,3,5-trinitrobenzene (57c), in dichloromethane in the presence of adamantylideneadamantane (58) gave dioxazolidines (59a-c), which are stable at room temperature. The structure of 59c was unambiguously proven by an X-ray crystal structure determination.[92] The compound 59c is stable in refluxing benzene but not in refluxing toluene. It readily decomposes to adamantanone, 3,5-dinitrophenylhydroxylamine, and 3,3′,5,5′-tetranitroazoxybenzene at room temperature in dichloromethane containing *p*-toluene sulfonic acid.[92]

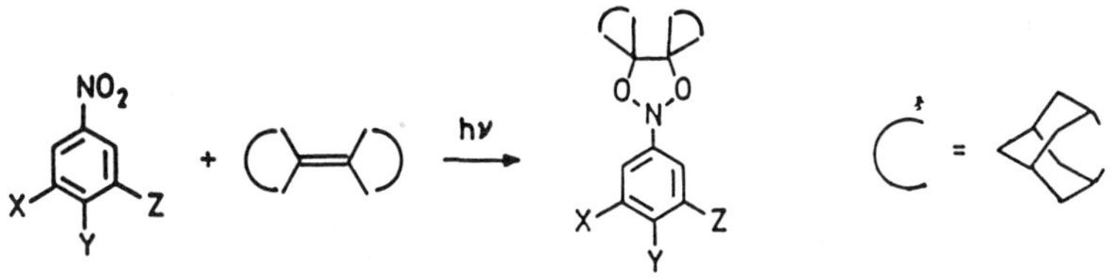

57 a-c   **58**   **59 a-c**

a. X = NO$_2$, Y = Z = H; **b:** X = Z = H Y = NO$_2$; **c:** X = Z = NO$_2$, Y = H

SCHEME 81.19

1-Nitronaphthalene with its lowest π,π^*-triplet state cannot be added in the same way to simple olefins.[91] On the other hand, excited 1-nitronaphthalene and 3-nitrochlorobenzene were found to induce the cleavage of the substituted ring in mono- or dimethoxynaphthalenes in appreciable yields,[93,94] and photoexcited nitro- and 3-chloronitrobenzene were able to induce 2 to 8% ring cleavage.[94,95] The most logical explanation would be an addition of the NO$_2$ group to one of the formal double bonds and subsequent rearrangement, as outlined for 1,4-dimethoxynaphthalene in Scheme 81.20.

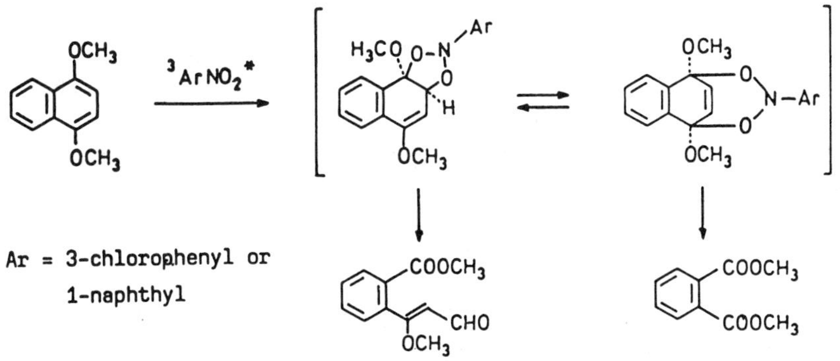

Ar = 3-chlorophenyl or
1-naphthyl

SCHEME 81.20

An early product study is available for the reaction of excited nitrobenzene with diphenylacetylene.[96] Intramolecular attacks of both nitro oxygen atoms on π-systems attached to the *ortho*-carbon of a nitrobenzene are inferred to explain the respective reaction products in some deep-seated, multistep, light-induced rearrangements.[3]

Photoreduction

Photoreduction Under Neutral Conditions

Nitrobenzene undergoes efficient reduction (disappearance quantum yield $\phi_d \approx 0.011$ in degassed solution) when excited with 366-nm light in 2-propanol.[97]

Following n,π*-excitation, H-abstraction from the solvent initiates a sequence of steps, ultimately affording phenylhydroxylamine and acetone.[97] Since the triplet yield of nitrobenzene had been determined to be 0.67 ± 0.10 using Lamola and Hammond's method;[98] the low efficiency is attributed to fast (10^9 s^{-1}) radiationless decay.[99] A transient absorbing at 440 and 625–650 nm was observed upon picosecond excitation of nitrobenzene, *p*-nitrotoluene, and *o*-nitrotoluene in THF or acetonitrile.[100] The similarity in absorption characteristics and kinetic behavior strongly suggests that a single transient is produced from all three compounds. For nitrobenzene, a lifetime of 800 ps was found; this agrees well with the earlier estimate of 1 ns for the n,π* triplet.[99] Quenchability of this transient by *trans*-piperylene ($E_T = 247$ kJ mol^{-101a}) points to its triplet nature ($k_q = 10^9$ M^{-1}s^{-1} for nitrobenzene and 0.2×10^9 M^{-1}s^{-1} for *o*-nitrotoluene); and by comparison of the observed rate constant with the diffusion-controlled limit in THF, an E_T of 242 kJ mol^{-1} is delineated for nitrobenzene.[100]

Slow hydrogen abstraction from 2-propanol upon 313-nm irradiation is observed from the π,π* triplet state of 2-nitronaphthalene,[102] whereas the longer-lived (2.5×10^{-6} s)[99] ^{3}(π,π*)1-nitronaphthalene (triplet yield 0.63 ± 0.10)[99] is practically unreactive toward 2-propanol (hydrogen abstraction rate constant <10^2 M^{-1}s^{-1})[99] and thus requires tri-*n*-butyl-stannane in benzene as reductant.[99] Photoreduction by 2-propanol to the corresponding arylhydroxylamines is also found for a variety of *m*- and *p*-acceptor-substituted nitrobenzenes[103–105] and 4-nitropyridine-1-oxides.[106–109]

There seems to be general agreement[3] that in neutral solution the photoreduction is initiated by H-atom abstraction out of the solvent by the triplet excited nitroaromatic (Equation 81.23). This step is likely followed by one more H-atom transfer to generate nitrosobenzene hydrate and acetone (Equation 81.24). The overall process leading to phenylhydroxylamine still remains subject to some speculation, and even Equation 81.24 is certainly only one of several options open to both initially formed radicals.[110,111]

$$^{3}(ArNO_{2})^{*} + (CH_{3})_{2}CHOH \rightarrow Ar\dot{N}O_{2}H + (CH_{3})_{2}\dot{C}-OH \tag{81.23}$$

$$Ar\dot{N}O_{2}H + (CH_{3})_{2}\dot{C}OH \rightarrow ArN(OH)_{2} + (CH_{3})_{2}CO \tag{81.24}$$

By necessity, the entire process has to be complex[110] since the products of two-electron, four-electron, and six-electron reduction (nitrosobenzene, phenylhydroxylamine, and aniline, respectively) are formed sequentially and may be present at the same time in the undergoing mixture photolysis. Also, it should be pointed out that $Ar\dot{N}O_2H$ is in equilibrium with the radical anion (Equation 81.25).

$$Ar\dot{N}O_{2}H \rightleftarrows ArNO_{2}^{-} + H^{+} \tag{81.25}$$

It thus may be formed as well by electron transfer to the excited nitro compound followed by protonation. Its rate of disproportionation (in acidic solution) has been determined[112] to be 3×10^8 M^{-1}s^{-1} (Equation 81.26).

$$2 Ar\dot{N}O_{2}H \rightarrow ArN(OH)_{2} + ArNO_{2} \tag{81.26}$$

Additional information can be obtained from pulse radiolysis experiments and absorbancy and *in situ* ESR monitoring.[111] A variety of *p*-substituted nitrobenzenes was found to oxidize the solvent derived α-hydroxyalkyl radicals [·CH$_2$OH, CH$_3$ĊHOH, (CH$_3$)$_2$ĊOH, generated by H-abstraction from the alcohol with OH radicals] to the corresponding carbonyl compounds with liberation of a proton and the nitrobenzene radical anion.[111] Whereas ·CH$_2$OH solely adds to the nitro group, addition and electron transfer to the nitro compound compete in case of the higher α-hydroxyalkyl radicals. The adducts, in turn, undergo fragmentation to the products mentioned.

Some effort has been spent on ESR detection of intermediate radicals during photolysis.[113] Whereas the primary product of hydrogen transfer, (namely, ArṄO$_2$H) is never detected, which is attributed to its rapid further reactions, spin-trapped products derived from addition of solvent-derived radicals to ground-state nitroaromatic are seen. It should be pointed out that secondary products of coupling of solvent-derived radicals to ArṄO$_2$H have been isolated.[114]

The photoreduction of 4-nitrobenzonitrile in 2-propanol has been shown by CIDNP studies[105] to proceed through an initial Ar-ṄO$_2$H/(CH$_3$)$_2$ĊOH pair, which exercises the following options: (1) back transfer of H to give nuclear polarized reactants, (2) second transfer of H to yield the nitroso hydrate, or (3) escape reactions. The escaped ArṄO$_2$H radical may encounter other radicals, dissociate to the radical anion in presence of a base,[112] or serve as reductant for other species, as suggested earlier.[110] Likewise, the solvent-derived radical may transfer a hydrogen atom to a ground-state nitro compound. The addition of α-hydroxyalkyl radicals generated by pulse radiolysis to ground-state nitrobenzene and the fate of the adduct radicals have been investigated in detail,[111] the net effect being a one-electron reduction of nitrobenzene. The overall process of photoreduction by 2-propanol via initial hydrogen atom transfer therefore should be envisaged as outlined in Equations 81.27 through 81.38 († denotes observability of nuclear polarization).

$$^3\text{ArNO}_2\,{}^* + (\text{CH}_3)_2\text{CHOH} \rightarrow \text{Ar}\dot{\text{N}}\text{O}_2\text{H}/(\text{CH}_3)_2\dot{\text{C}}\text{OH} \tag{81.27}$$

$$\xrightarrow{\;-\text{H}\;} \text{ArN(OH)}_2 + (\text{CH}_3)_2\text{CO} \tag{81.28}$$

$$\longrightarrow \text{ArNO}_2^\dagger + (\text{CH}_3)_2\text{CHOH}^\dagger \tag{81.29}$$

$$\xrightarrow{\;\text{esc.}\;} \text{Ar}\dot{\text{N}}\text{O}_2\text{H} + (\text{CH}_3)_2\dot{\text{C}}\text{OH} \tag{81.30}$$

$$\text{Ar}\dot{\text{N}}\text{O}_2\text{N} \rightleftarrows \text{Ar}\dot{\text{N}}\text{O}_2^- + \text{H}^+ \tag{81.31}$$

$$(\text{CH}_3)_2\dot{\text{C}}\text{OH} + \text{ArNO}_2 \rightarrow \text{Ar}\dot{\text{N}}\text{O}_2\text{H} + (\text{CH}_3)_2\text{CO} \tag{81.32}$$

$$(\text{CH}_3)_2\dot{\text{C}}\text{OH} + \text{ArNO}_2 \rightarrow \text{Ar}\dot{\text{N}}(\text{O}) - \text{O} - \text{C(OH)}(\text{CH}_3)_2 \tag{81.33}$$

$$\text{ArN(OH)}_2 \rightarrow \text{ArNO} + \text{H}_2\text{O} \tag{81.34}$$

$$\text{ArNO} + (\text{CH}_3)_2\text{CHOH} \xrightarrow{\;h\nu\;} \text{Ar}\dot{\text{N}}\text{HO} + (\text{CH}_3)_2\dot{\text{C}}\text{OH} \tag{81.35}$$

$$\text{Ar}\dot{\text{N}}\text{O}_2\text{H} + \text{Ar}\dot{\text{N}}\text{HO} \rightarrow \text{Ar}\dot{\text{N}}\text{O}_2\text{H}/\text{Ar}\dot{\text{N}}\text{HO} \tag{81.36}$$

$$\text{Ar}\dot{\text{N}}\text{O}_2\text{H}/\text{Ar}\dot{\text{N}}\text{HO} \rightarrow \text{ArNO}_2^\dagger\text{ArNHOH}^\dagger \; \text{(in neutral medium)} \tag{81.37}$$

$$\rightarrow \text{ArN(OH)}_2 + \text{ArNO}^\dagger \; \text{(in acidic medium)} \tag{81.38}$$

Photoreduction in the Presence of Acid

Nitrobenzene and various substituted nitrobenzenes, when irradiated in 12 M hydrochloric acid, give mixtures of oligo- and polychlorinated anilines.[115] The quantum efficiency of nitrobenzene disappearance (0.11 in 12 M HCl at 313 nm) drops markedly with decreasing concentration of HCl. An electron transfer from chloride ion to $^3(\text{n},\pi^*)$nitrobenzene is regarded as the initial step, since electron-donating substituents tend to decrease the efficiency. As with the quantum yield, the rate of nitrobenzene photoreduction in 2-propanol is markedly enhanced by the presence of hydrochloric acid,[116–118] and chlorinated anilines again dominate among the products. Again, phenylhydroxylamine is regarded as a key intermediate product. After initial electron transfer,[118,119] the radical anion is rapidly protonated to generate $PhNO_2H$.

Electron transfer has also been demonstrated from dihydroacridine to $^3(n,\pi^*)$nitrobenzene in the presence of perchloric acid in aqueous acetonitrile.[120] The overall six-electron photoreduction of nitrobenzene to aniline in this system is interpreted by the steps outlined below (Equations 81.39 to 81.44).[120]

$$^3PhNO_2{}^* + AcrH_2 \rightarrow PhNO_2^- + AcrH_2^+ \tag{81.39}$$

$$PhNO_2^- + H^+ \rightarrow PhNO_2H \tag{81.40}$$

$$PhNO_2H + AcrH_2^+ \rightarrow PhN(OH)_2\, AcrH^+ \tag{81.41}$$

$$PhN(OH)_2 \rightarrow PhNO + H_2O \tag{81.42}$$

$$PhNO + AcrH_2 + H^+ \rightarrow PhNHOH + AcrH^+ \tag{81.43}$$

$$PhNHOH + AcrH_2 + H^+ \rightarrow PhNH_2 + AcrH^+ \tag{81.44}$$

The process described by Equation 81.44 is regarded to be truly photochemical; both excited phenylhydroxylamine and excited dihydroacridinium may be the active species.[120]

The $^3(n,\pi^*)$4-nitropyridinium ion is the active species in the analogous reductions of this heterocycle by electron transfer from chloride ion.[121,122] 1-Nitronaphthalene, being photochemically inactive in neutral 2-propanol, is easily reduced in the presence of large amounts of hydrochloric acid,[123] as is 5-nitroquinoline.[124] Again, electron transfer from chloride ion, followed by rapid protonation, is a more likely rationale[124] than protonation of the excited state of the nitro compound[123] as anticipated earlier.

Lewis acid complexation may alter significantly the reaction pattern. Irradiation of the 1:1 boron trichloride/nitrobenzene complex in cyclohexane causes deoxygenation to form nitrosobenzene, boric acid, and chlorocyclohexane.[125]

Photoreduction in Basic Media

Photoreduction of nitrobenzenes is carried out in alkaline or buffered alcoholic solutions[126–128] to give anilines, azo, and azoxy compounds more efficiently than in neutral aqueous-alcoholic solutions. The latter two classes are typical products of reduction under basic conditions. Buffered aqueous methanolic formate solutions were also successfully applied.[127] The nature of the alcohol applied as solvent and reductant is also important: the disappearance quantum yield for m-dinitrobenzene increases in the order methanol < ethanol < 2-propanol.[128] For p-dinitrobenzene, replacement of CH_3OH by CD_3OD causes a decrease in quantum yield of disappearance from 0.011 to 0.0035, which is tantamount to $k_H/k_D \approx 2.6$.[128] This isotope effect drops to unity (for the m-isomer) and to 1.4 (for p-dinitrobenzene, respectively) in the presence of alkoxide and OH$^-$, indicating a change of mechanism. A diffusion-controlled electron transfer from the alkoxide ion,[129] followed by a rapid proton transfer, is proposed.[128] Nitrobenzene may also be reduced by cyanide ion.[130]

Photoreduction of nitrobenzenes with secondary and/or tertiary amines[131–137] has been found to be efficient; at the same time, these reactions are oxidative dealkylations of the amines used. With primary amines, photosubstitution of suitable nucleofugal groups may compete with photoreduction.[138,139] The oxidative N-demethylation of $PhN(CH_3)_2/PhN(CD_3)_2$ mixtures by photoexcited 3-chloronitrobenzene[135] shows isotope effects near unity; whereas, N-methyl-N-(trideutero-methyl)aniline $PhN(CH_3)CD_3$ experiences a significant discrimination between the "light" and "heavy" methyl group in N-demethylation, and $k_{CH3}/k_{CD3} = 3.5$ in benzene and 2.5 in acetonitrile. This effect is interpreted as follows: $^3(n,\pi^*)$-excited 3-chloronitrobenzene and N,N-dimethylaniline form a tight ion pair (especially in benzene), which internally transfers a proton from the aminium cation to the nitro radical anion before the two ions have a chance to be separated. Discrimination between H and D in this step is not possible, when either CH_3 or CD_3 are available only, i.e., in the mixture of isotopomers. This points to rate-limiting ion pair formation. The same conclusion can be reached for the photoreduction of various o-alkylnitrobenzenes with secondary or tertiary aliphatic amines.[134]

In this context, it should be pointed out that a number of o-alkyl- and o,o'-dialkylnitrobenzenes fail to undergo photoreduction by the H-atom donor 2-propanol and experience intramolecular H-abstractions instead (see Section 81.4). In the presence of secondary or tertiary aliphatic amines, optimally as solvents, however, electron transfer reduction by-passes the hindered H-atom transfer and successfully competes with intramolecular hydrogen abstraction.[131,132,134] Irradiation of 60a-c in neat triethyl amine affords the nitrones 61a-c, most likely via the processes outlined generally in Equations 81.45 to 81.53. Compound 60d, on the other hand, is just photoreduced to the corresponding amine 62 and hydroxylamine, the latter giving the nitroso compound 63 on work-up with admission of air[134] (Scheme 81.21).

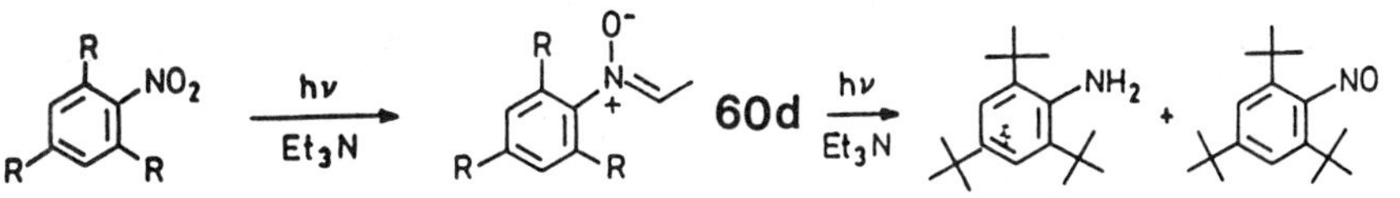

60a-c   **61a-c**   **60d**   **62**   **63**

a: R = CH_3; b: R = C_2H_5; c: R = i-C_3H_7; d: R = t-C_4H_9

SCHEME 81.21

$(Ar = C_6H_2R_3$ as in 60, $R' = CH_3, C_2H_5, C_3H_7)$

$$ArNO_2{}^* + (R'-CH_2)_3N \rightarrow [ArNO_2^- \dots (R'-CH_2)_3N^{+\cdot}] \quad (64) \qquad (81.45)$$

$$64 \rightarrow [Ar\dot{N}O_2H\dots(R'CH_2)_2N-CHR'] \quad (65) \qquad (81.46)$$

$$65 \rightarrow ArN(OH)-O-CHR'-N(CH_2R')_2 \quad (66) \qquad (81.47)$$

$$66 \rightarrow ArNO + R'CHOH - N(CH_2R')_2 \quad (67) \qquad (81.48)$$

$$67 \rightarrow HN(CH_2R')_2 + R'-CHO \qquad (81.49)$$

$$Ar-NO + Ar\dot{N}O_2H \rightarrow Ar\dot{N}HO + ArNO_2 \qquad (81.50)$$

$$Ar-\dot{N}HO + Ar\dot{N}O_2H \rightarrow ArNHOH + ArNO_2 \qquad (81.51)$$

$$ArNHOH + R'CHO \rightarrow ArN(O)\uparrow CHR' + H_2O \qquad (81.52)$$

$$ArNHOH + R'CHOH - N(CH_2R')_2 \rightarrow ArN(O)\uparrow CHR' + HN(CH_2R')_2 \quad (81.53)$$

This list of reactions is not necessarily comprehensive. Diethylamine (IP = 8.01 ± 0.01 eV[101b]) may also be used, but the quantum yield of starting material (60a,c,d) conversion does not exceed 0.08 ± 0.01, while with triethylamine (IP = 7.50 ± 0.2 eV[101b]) values of 0.47 ± 0.01 (60a), 0.25 ± 0.01 (60c), and 0.39 ± 0.02 (60d) are reached.[134]

Relatively little information is available about photoreductions of higher annellated nitroarenes by amines. For 1-nitronaphthalene and 2,7-dinitronaphthalene, absorption spectra and decay kinetics of triplet exciplexes with N,N,N',N'-tetramethyl-p-phenylendiamine, N,N-dimethyl-p-anisidine, and triphenylamine have been investigated[139] after nanosecond excitation at 400 nm. Electronically excited 1-nitronaphthalene dealkylates tertiary amines and, at the same time, undergoes reduction.[140]

Heterogeneous Photoreductions

6-Nitrocoumarin (64), 5-nitro-8-methoxypsoralen, 1-nitronaphthalene, 1,4-dinitrobenzene, 3- and 4-nitroacetophenone, 4-nitroanisole, and 4-nitrotoluene have been shown to undergo photo-induced reduction in 45 to 92% yield to the corresponding amines in ethanolic solution on photoexcited titania (a wide band-gap semiconductor) particles using 350-nm irradiation.[141] Even 9-nitroanthracene, which never undergoes photoreduction when directly excited,[75] is reduced to 9-aminoanthracene in 85% yield[141] (Scheme 81.22).

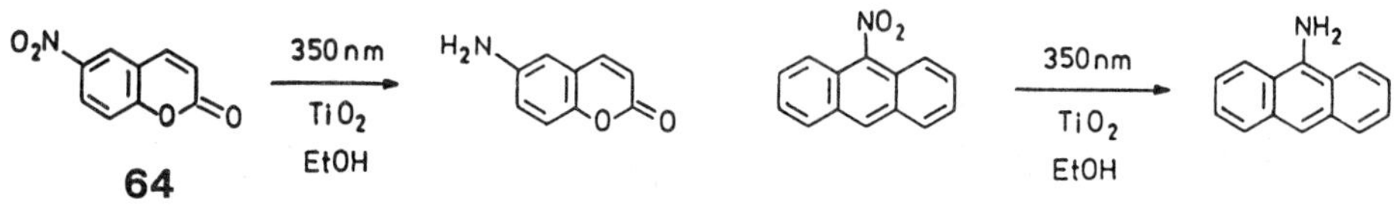

SCHEME 81.22

A surface-induced reduction of 4-nitrobenzoic acid on an irradiated roughened silver surface to the corresponding amino and azo compounds has been reported recently.[142]

Photoredox Reactions

Intramolecular Interactions Between Amino and Nitro Groups

Upon photoexcitation, a nitro group may also be reduced by an amino function contained in the same molecule (but not in conjugation). Depending on the length of the oligomethylene chain linking the two chromophores, N-[ω-(4-nitrophenoxy)alkyl]-anilines (65)[143–145] and their 4-nitro-1-naphthoxy analogues[146] in benzene solution either undergo a light-induced cleavage into aniline and the nitroso aldehyde 66 or the so-called photo-Smiles rearrangement to 67 (see section on photosubstitutions below) (Scheme 81.23). If the anilino H atom in 65 is replaced by benzyl, either the chain α-methylene or the benzylic position may be oxidized.[147] This photoredox reaction, but not the photo-Smiles rearrangement, is influenced markedly by a magnetic field; this indicates the involvement of biradical intermediates.[143–147] These are formed by a net hydrogen transfer (probably in the nonconcerted way) from a N-methylene group to the nitro group. The primary biradical 68, being most likely a triplet initially, partitions between "cage" and "escape" products (the latter promoted by the applied magnetic field).

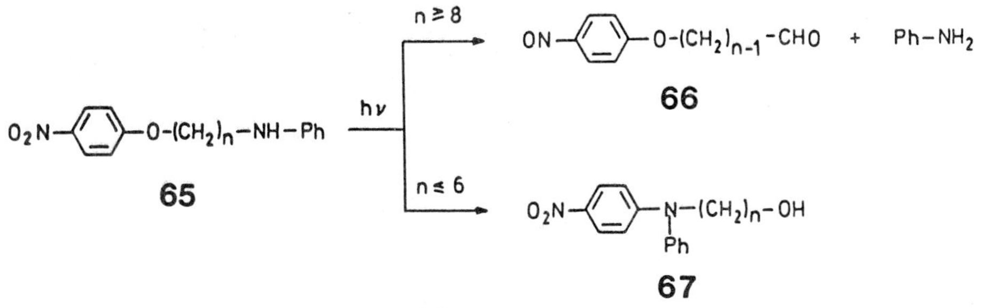

SCHEME 81.23

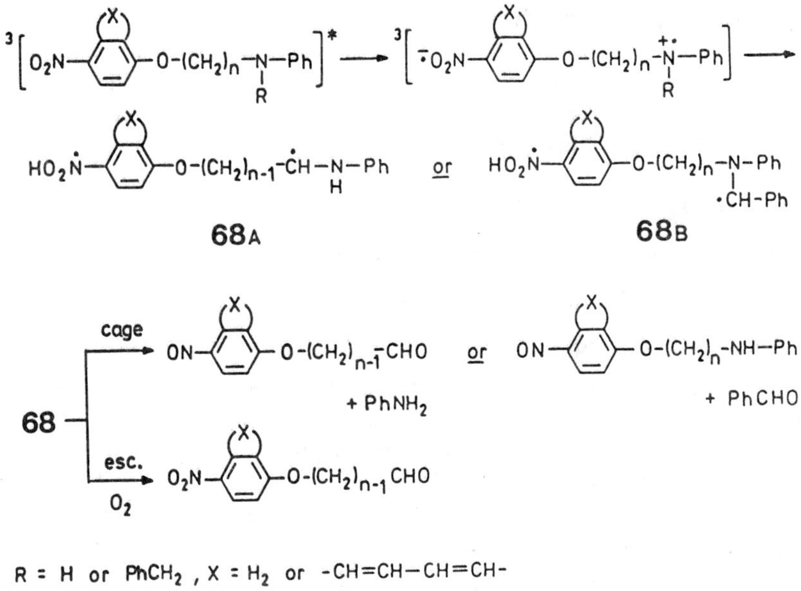

R = H or PhCH$_2$, X = H$_2$ or -CH=CH-CH=CH-

SCHEME 81.24

Competitive inter- and intramolecular photoredox reactions are also observed in the homologous series of compounds 4-O$_2$N-C$_6$H$_4$-O(CH$_2$)$_n$-C$_6$H$_4$-3-NMe$_2$ (69) with n = 2–12, where the nitro group is photoreduced to nitroso and the dimethylamino group is oxydatively mono-demethylated.[148] This requires first a net hydrogen transfer from CH$_3$N to NO$_2$ and subsequently an oxygen atom transfer from the resulting hydroxynitroxide to ·CH$_2$N (Scheme 81.25). This sequence may be intra- or intermolecular, since the nitroso-N-methyl product 70 is accompanied by 71 and 72, thus suggesting a competitive intermolecular pathway. Only small reactivity differences are encountered through the entire homologous series[148] (Scheme 81.26).

SCHEME 81.25

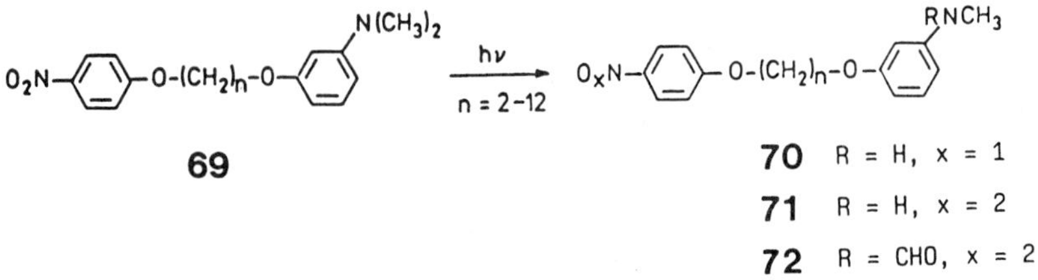

SCHEME 81.26

An effective monodealkylation of 4-nitro-*N,N*-dialkylanilines (Scheme 81.27) has been observed upon irradiation (≥290 nm) in 1 *M* sodium methoxide solution in methanol.[149] The presence of base is essential for the reaction to be efficient.[150] The reaction is fastest with R = CH$_3$, and 4-nitroso-*N,N*-dimethylaniline has been observed in almost equal amounts to the product of demethylation.[150] This suggests an overall photoredox process involving two molecules of starting material, but the details still await clarification.

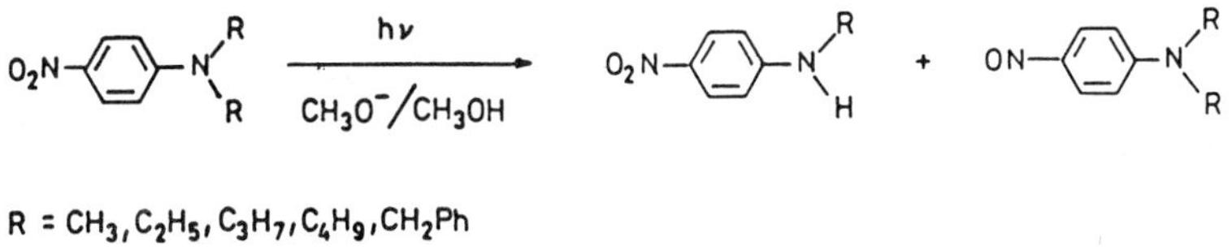

R = CH$_3$, C$_2$H$_5$, C$_3$H$_7$, C$_4$H$_9$, CH$_2$Ph

SCHEME 81.27

Solvent-Mediated Photoredox Reactions.

The presence of water seems to be required for certain light-induced redox reactions involving the nitro group and other functionalities residing at the benzenoid ring.

The "photoisomerization" of 4-nitrobenzaldehyde to 4-nitrosobenzoic acid, as originally demonstrated by Wubbels[151–152] and Reisch,[153] has been reinvestigated.[154] The quantum yield at 254-nm irradiation is 0.034 in 99% water/1% acetonitrile and is dependent on [H$_2$O], but independent of pH in the range 0 to 10 and concentration of starting material. Further, it is unaffected by radical trapping agents such as dioxane, methanol, or oxygen. In aqueous ammonia, *p*-nitrosobenzamide is obtained; this suggests a ketene-type intermediate[150,151,153] (**74**). While an electron transfer from water to ^{3}(n,π*)-73 had been proposed earlier,[152] the results of a nanosecond flash study[154] now point to a deprotonation of the formyl group of triplet excited **73**. With this key step, everything is en route to the final product **75**, since all further steps from **74** onward are to be regarded as conventional ground-state processes (Scheme 81.28).

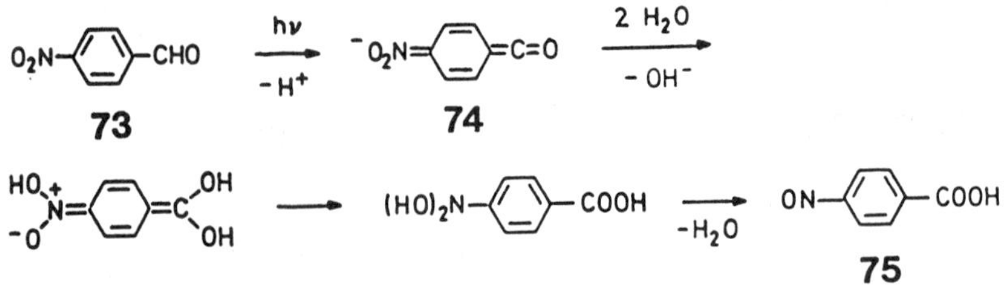

SCHEME 81.28

The same intermediate (**74**) is proposed in the photodecarboxylation of *p*-nitrophenylglyoxylic acid to 4-nitrosobenzoic acid[155] (Scheme 81.29). Decarboxylation of the triplet excited carboxylate is inferred from the pH dependence of the quantum yield. A transient absorption at 350 nm is assigned to 74 on the basis of its long lifetime (1.3 to 1.6 µs in water between pH 5 and 12).[155]

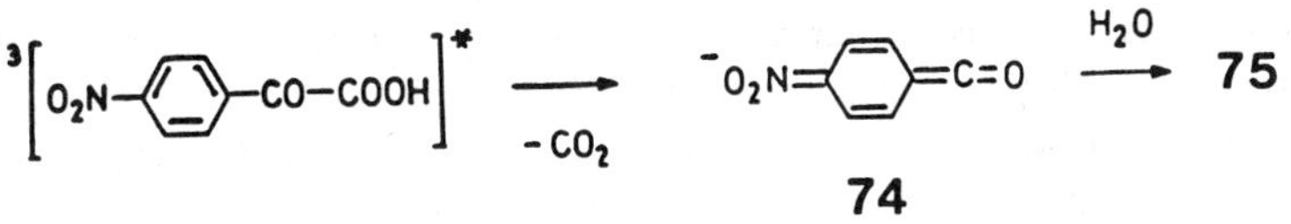

SCHEME 81.29

m-Nitrophenylacetaldehyde and *m*-nitrophenylacetic acid,[156] as well as *m*- and *p*-nitrobenzylalcohol[157,158] undergo a light-induced oxidation at the benzylic carbon and at the same time a reduction of the nitro function; the reduction products, however, have not always been characterized.[156] Formation of *m*-nitrotoluene from 3-nitrobenzylic precursors[156] points to liberation of the 3-nitrobenzyl anion from the excited starting material (Scheme 81.30). Water is, in any case, essential for the process.

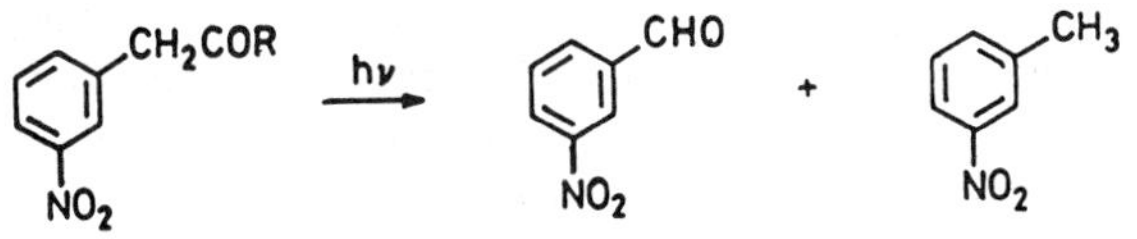

SCHEME 81.30

A peculiar pH dependence is found for the light-induced redox reactions of *p*- and *m*-nitrobenzyl alcohol in water.[157] Whereas the *p*-isomer undergoes an OH^--promoted (pH > 11) process (Scheme 31), the *m*-isomer shows the inverse dependence on pH. At pH = 7, the quantum yield of *m*-nitrobenzaldehyde formation is 0.055, the efficiency is increased with increasing acidity and peaks around $H_0 = -1$. For *p*-nitrobenzylalcohol (76), the transformation can be triplet sensitized and also quenched by 3,5-cyclohexadiene-1,2-dicarboxylic acid anion and may be outlined as shown, but details await clarification. A marked isotope effect on the quantum yield is observed when the benzylic hydrogen atoms are replaced by deuterium.[157,158]

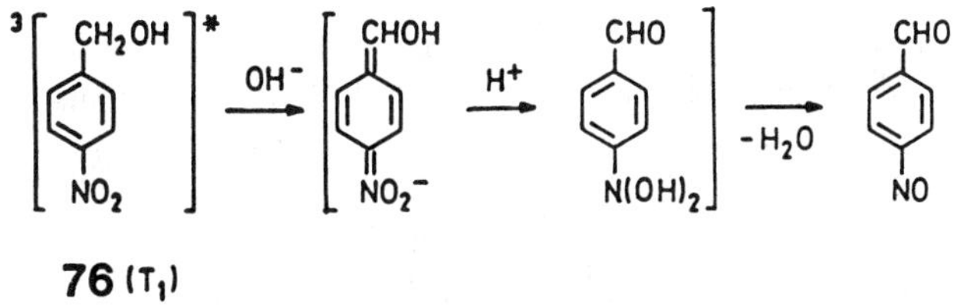

76 (T₁)

SCHEME 81.31

It should be noted, though, that nonphotochemical pathways for such redox reactions exist also: 4-nitroacetophenone is transformed into 4-nitrobenzoic acid and 4-aminoacetophenone by 0.5 *M* NaOH.[159]

Liberation, Protonation, and Oxidation of *m*- and *p*-Nitrobenzyl Anions

m- and *p*-Nitro groups may facilitate the light-induced hydroperoxygenation at benzylic positions in benzyl alcohols[160] and ethers.[161,162] The reaction probably involves photogenerated nitrobenzyl carbanions, which then may be trapped by oxygen. The photohydration (in the anti-Markownikow sense) of *m*-nitrostyrenes[163–165] may be coupled with oxygen uptake in a tandem addition of water and oxygen.[166] The proposed mechanism involves addition of water to the β-carbon in the excited triplet state and thus formation of the *m*-nitrobenzylic carbanion, which is subject to either protonation or trapping by oxygen. Identical quantum yields (0.32) for the tandem addition and photohydration (in absence of oxygen) of *m*-nitrostyrene suggest a common primary intermediate in both reactions.[166] A variety of triplet excited *m*-nitrobenzyl alcohols, ethers, and esters undergo photooxidation to *m*-nitrobenzaldehyde or -acetophenone, respectively, in 20 to 50% aqueous sulfuric acid with product formation quantum yields of 0.3 to 0.4.[167]

p-Nitrobenzylphosphonate dianions[168,169] and electron-rich 4-nitrobenzyl arenesulfonates[170–172] undergo moderate to efficient carbon-heteroatom bond rupture upon irradiation. The liberated sulfonic acids may be used to initiate polymerizations. An intramolecular electron transfer from the phosphonate or arenesulfonic acid moiety into the 4-nitrobenzyl system is envisaged as the initiating process.[169–171] Upon irradiation in ethanol, the above-mentioned phosphonate ($\phi_{dec} = 0.7$) gives 1,2-*bis*-(4-nitrophenyl)ethane, 4-nitrotoluene, *ortho*-phosphate, and alkyldihydrogen-phosphate.[168,169]

Suitable β(3- and 4-nitrophenyl)ethanols (77) liberate, when being irradiated in water or water/solvent mixtures, the corresponding nitrobenzylanions 78; the hydroxylated β-carbon ends up as a carbonyl group or an equivalent thereof.[173,174] The term "photo-retro-aldol" type reaction has been coined for this process[173,174] (Scheme 81.32). The same anions (78) may also be generated by a dioxolenium ion from 79. The anions may be protonated, trapped by O_2 to give hydroperoxides or be oxidized to the radical that ultimately dimerizes.[173,174]

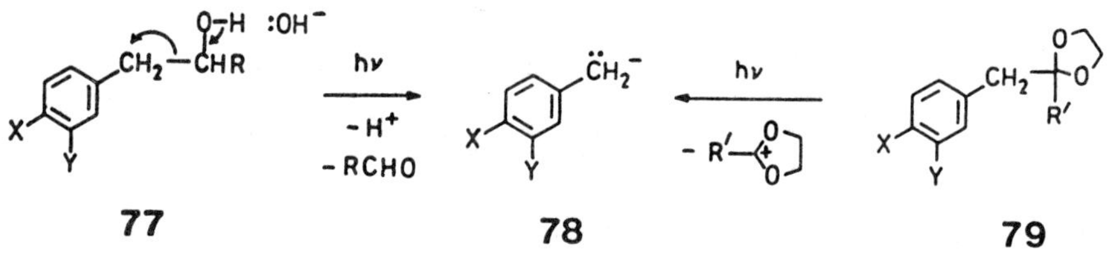

R = H or Ph; X = H, Y = NO₂ or Y = H, X = NO₂; R´ = Ph or H

SCHEME 81.32

Photochemical decarboxylation of *m*-nitrophenylacetic acids in aqueous solution[175–177] has been followed by millisecond,[176] nanosecond, and picosecond[177] flash experiments and been shown to liberate the nitrobenzyl anions, which ultimately are protonated or oxidized to the symmetric dimer. The anion may also transfer an electron to the starting material, whereby the long-lived radical anion of the latter becomes detectable by ESR.[178]

There are exceptions, too. Whereas the retro-aldol-like photocleavage may be more easily understood for *p*-nitrophenyl-β-(oxyalkyl)-derivatives, it is less plausibly rationalized for the isomeric *m*-derivatives 80. Participation of non-Kekulé-structures (81, 82), as earlier suggested for the photosolvolysis of (*m*-nitrophenyl)tritylether,[179] may well serve as an explanation.[180] The main product is 83, and no trace of *m*-nitrotoluene (indicative of the participation of *m*-nitrobenzyl anion) was detected (see Scheme 81.33).

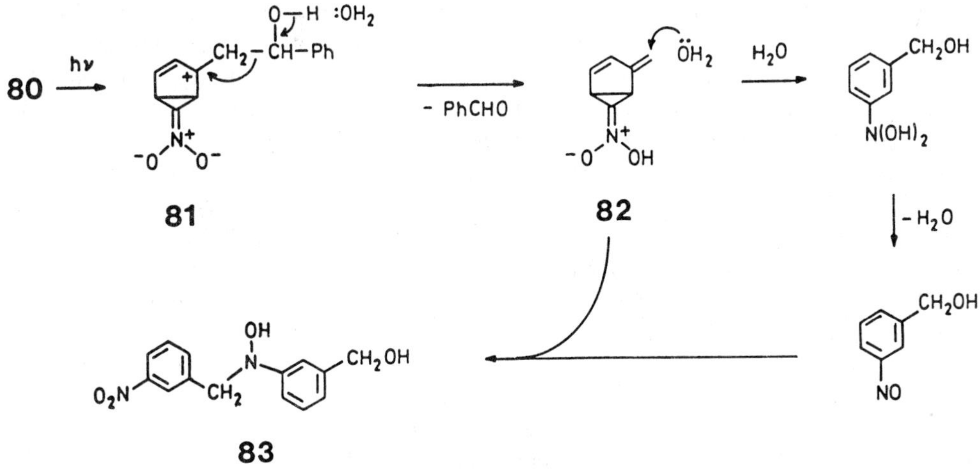

SCHEME 81.33

Intramolecular Hydrogen Abstractions and Subsequent Reactions

Benzylic Hydrogen Atoms

Following the classical paper of Ciamician and Silber[181] on the photoisomerization of *o*-nitrobenzaldehyde into *o*-nitrosobenzoic acid, the experiments revealed that, in general, compounds with benzylic hydrogen atoms in an *ortho*-position to a nitro group are light sensitive ("Sachs' rule").[182] Numerous investigations since then have confirmed this statement; for reviews see References 2, 3, 56, 183.

The intramolecular abstraction of an α-hydrogen atom in **84** is usually followed by relaxation to the *aci*-nitro form **86**, which in a basic environment dissociates (pK$_a$ 1.1–3.7)[184] to the *aci*-nitro anion (**87**). The latter two processes are responsible for various cases of photochromism observed with this type of alkylnitrobenzenes.[185] The following scheme summarizes the options in a prototype.

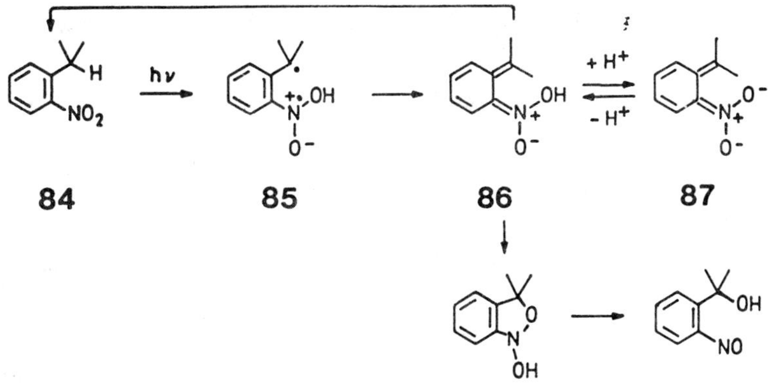

SCHEME 81.34

ESR-detectable radicals are most likely aminyl oxides[186] and thus due to secondary photolysis. The *aci*-nitro form of 2,4,6-trinitrotoluene and its anion have been indirectly proven by conventional flash photolysis[187] or steady illumination in solution.[188]

Nanosecond[189b–193] and picosecond[189a,190] laser excitation of o-nitrobenzyl alcohols,[189a] ethers,[189] and esters,[190–192,194] as well as of cyclic acetals of o-nitrobenzaldehyde[194] and some o-nitrotoluenes[193] has allowed monitoring of the build-up of the nitronic acid (86)[189–193] and its deprotonation to 87 (equal to the o-nitrobenzyl anion!).[190–193] Direct observation of nitronic acids in solution from 308-nm laser (15 ns) excitation of various o-nitrobenzylesters was also made by time-resolved resonance Raman spectroscopy using a 420-nm excitation (15 ns) of the photolysate.[192] In acetonitrile, only the corresponding nitronic acid is observed (lifetime 80 ms); whereas in methanol, the nitronic acid is demonstrated to be in equilibrium with the nitronate (lifetime 100 ms).[192] The picture may be complicated, though, due to the possible existence of up to four isomeric forms of the nitronic acid.[191] Early transients have also been assigned to the triplet excited state[189a] and the subsequent triplet biradical[189b] formed by intramolecular hydrogen abstraction within the excited triplet. Nitronic acid formation does occur competitively, however, from both the singlet and the triplet excited states,[189,190] if not preferentially from the singlet.[189a]

In rigid systems such as 88,89 with bridgehead benzylic hydrogen atom, where nitronic acid formation is severely hindered, the direct biradical (90,91) route to isolable products may be verified.[194] After picosecond excitation, transients with lifetimes of 770 and 410 ps have been detected and assigned to triplet excited 88 and 89, respectively. Biradicals (90,91) following the triplet have not been observed.[194] The accessible O...H....C arrangement in 88,89 deviates by 53° from linearity, but the O...H distance is more critical (88: r_{OH} = 1.6 Å, 89: r_{OH} = 1.3 Å), and the transient (^{3}88*), which experiences the longer O...H distance, also has the longer lifetime.

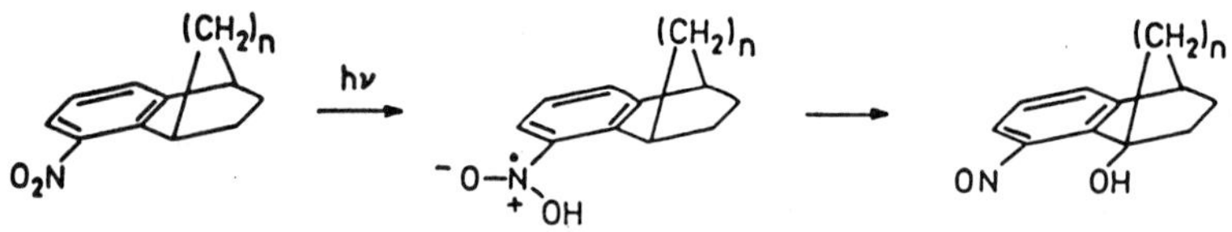

SCHEME 81.35

The well-known photorearrangement of 2-nitrobenzaldehyde[2,3,181,195] may be depicted as in the following scheme. The quantum yield for nitrosobenzoic acid formation has been determined to be close to 0.5 in solution, and the triplet lifetime was determined to 0.6 ns (after 350-nm excitation).[196]

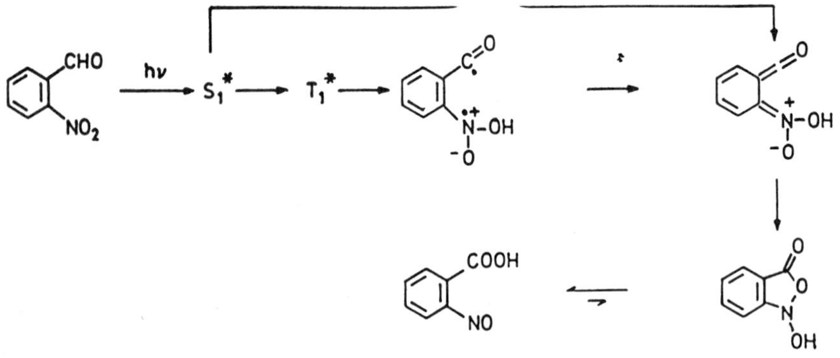

SCHEME 81.36

Picosecond excitation of o-nitrobenzaldehyde in benzene[196] or acetonitrile or THF solution[197] revealed a transient absorbing at 440 nm that decayed within nanoseconds and was assigned a ketene structure.[196,197] It decayed more rapidly in aqueous or alcoholic acetonitrile (or THF) than in dry acetonitrile. A decrease in yield of that transient by added *trans*-piperylene was interpreted in terms of a triplet precursor,[196] whereas other investigators did not find any influence of added *cis*-piperylene on the yield of transient.[197] Since no absorption between 625 and 650 nm assignable to a triplet was observed, it has been concluded[197] that in direct irradiation, hydrogen transfer will occur directly from the singlet excited state.

Derivatives of *o*-nitrobenzaldehyde show analogous behavior: crystalline *o*-nitrobenzalaniline undergoes color changes during irradiation and ultimately transformation into *o*-nitrobenzanilide[198,199] and 2-phenylazobenzoic acid.[199] *o*-Nitrobenzaldehyde *N*-acylhydrazones form 5-nitrophthalazines upon irradiation or may liberate benzyne.[200]

Side-Chain β-Hydrogen Atoms

In *o*-nitroalkylbenzenes, light-induced intramolecular abstraction of a β-hydrogen atom in the presence of abstractable α-hydrogen atoms is rare. 2-(2-Nitrophenyl)ethanol is photocyclized to 1-hydroxyoxindole,[201] and formation of 5,7-diethyloxindole has been claimed upon excitation of 2,4,6-triethylnitrobenzene, even in 2-propanol.[202] No α-hydrogens are available in *o*-nitro-*t*-alkylbenzenes. Like the interactions with benzylic positions, the photocyclizations observed with *o*-nitro-*t*-butylbenzenes 92[202–213] are at the same time photoredox processes and cyclisative dehydrations leading first to (often not isolable) 3*H*-indole-1-oxides 93, and most of the other products found are derived therefrom.

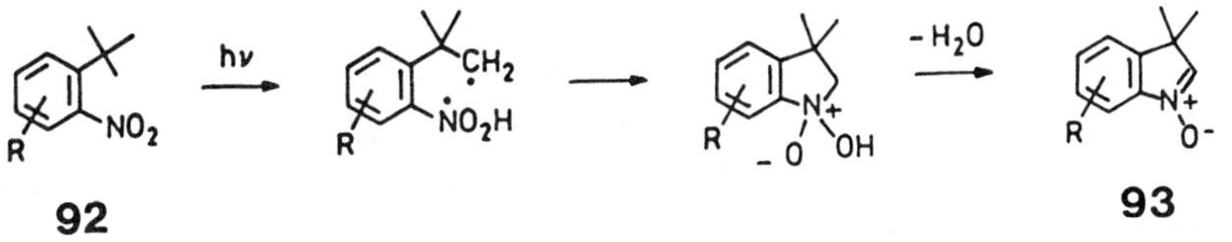

SCHEME 81.37

The nature of these compounds strongly depends on photolysis and work-up conditions: hydration followed by air oxidation of 93 yields 1-hydroxy-2-indolinones 94.[202,203,205–213] Various products of thermal or light-induced isomerization 95–97 and deoxygenation (98) of 93 are also generally found.[202,204,207–211] The parent nitrone 93 itself is in some cases accessible by photolysis of crystalline 92 and rapid chromatographic separation.[207,208,211]

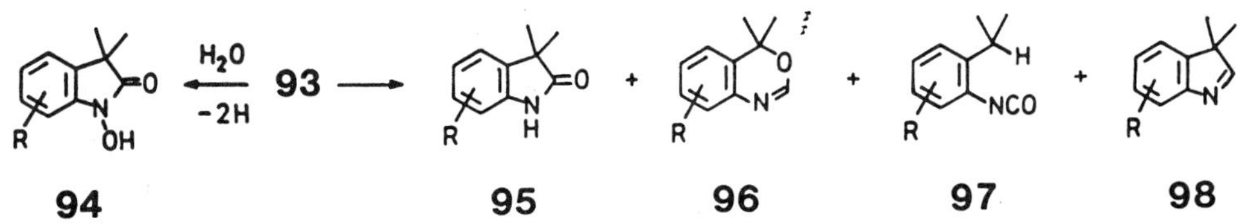

SCHEME 81.38

Since a constant fraction of 93 can always be transformed into 94, quantum yields for formation of 94 being close to 10^{-2} are a meaningful measure of reactivity;[206,212] *m*- and *p*-electron-donating groups R retard, electron acceptors R facilitate the process.[212] Both, the n,π* singlet and triplet states are active,[206,213] since the process may be sensitized by benzophenone (irradiation at 254 nm is necessary), but is only partially quenchable by octafluoronaphthalene and piperylene.[213] Hydrogen abstraction by the nitro group is demonstrated to be rate limiting and irreversible by the occurrence of significant deuterium isotope effects.[209]

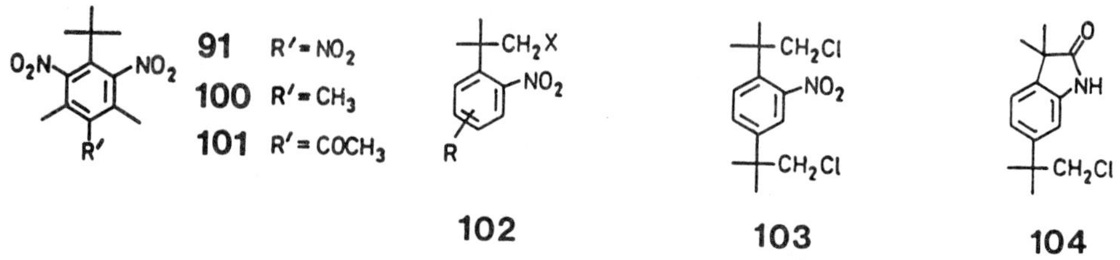

SCHEME 81.39

When benzylic hydrogens *ortho* to the nitro group are available in addition to the β-hydrogens of a *t*-butyl group, as in compounds 99–101, attack at *t*-butyl occurs almost exclusively.[211] This may be due largely to the considerable torsion of the nitro group relative to the benzene ring (77 to 90°),[214] which establishes an unfavorable geometry for hydrogen abstraction since ideally the N-O...H angle should be 90 to 120° (depending on the degree of hybridization between pure p or sp^2 assigned to the lone pair orbital) and the C-H...O angle 180°.[194,214] On this basis, the number of hydrogens actually available in the *t*-butyl group of 99 and 100 must be reduced to just two; and since the efficiency of H-abstraction is strongly decreased by increasing distance r_{OH}, one hydrogen atom each seems to be in an optimal situation with the following parameters: 99: C-H...O 136°, H...O-N 100°, r_{OH} 2.57 Å; 100: C-H...O 140°, H...O-N = 99°, r_{OH} 2.51 Å.[214] Whether ground-state geometries in the crystal resemble the geometry of the n,π* excited state, however, must remain open.

With apparently extreme selectivity, one hydrogen atom is abstracted out of the CH_2-X group in compounds 102 upon excitation.[215,216] Whether a comparable smaller C-H bond dissociation energy contributes or just favorable geometries are solely responsible cannot be answered yet conclusively. For crystalline 103, indeed, one of the methylene hydrogens has optimal parameters: C-H...O 131.9°, H...O-N 117.5°, r_{OH} 211 Å.[215] During solid-state photolysis of 103, hydrogen chloride is evolved and the indolinone 104 is readily formed.

It should be noted, though, that the excited-state hydrogen atom transfer does not only require an optimal N-O...H angle (see above), but also should ideally occur within the plane defined by the NO_2 group, which means the dihedral angle O-N-O...H should be close to 180°. Severe steric hindrance and torsion of the NO_2 group relative to the benzenoid ring (e.g., brought about by an *o*-methyl or an *o'-t*-butyl) as in 99–101 may well prevent intramolecular hydrogen abstraction even out of benzylic positions.[217] Probably along these lines, 2,6-di-*t*-butyl-1-nitronaphthalene (54), although structurally related to 92, does not show any sign of hydrogen abstraction out of the 2-*t*-butyl group but undergoes the nitro → nitrite rearrangement instead (see Section IV 81.4, Nitro → Nitrite Rearrangement and Subsequent Reactions).[86]

Miscellaneous Compounds

α-*N*-(2,4-Dinitrophenyl)α-amino acids, after light-induced intramolecular hydrogen abstraction from the amino acid α-position, undergo cyclization to imidazolones or cleavage into readily analyzable products.[3] Some α-*N*-(2,4,6-trinitrophenyl)amino acids show fragmentation into 2-nitroso-4,6-dinitroaniline, aldehyde, and carbon dioxide upon irradiation in aqueous solution at pH 8.[218]

o-Nitrophenylmethyl- and benzyl ethers may be photocyclized to benzoxazoles,[219,220] or undergo rearrangement to 2-hydroxybenzanilides together with formation of minor amount of *o*-nitrophenols and other products.[220] (2,4-Dinitrophenylthio)acetic acids, on the other hand, decarboxylate upon excitation and liberate (2,4-dinitrophenylthio)methanide.[222] Anion stabilization by sulfur may well contribute to this reaction.

Photosubstitutions

A number of light-induced polar substitutions exist, in which the nitro group may not be involved specifically, but acts through its electron-withdrawing properties. Among these are the light-induced solvolysis of *m*-nitrophenyl trityl ether[179] and the displacement of acetoxy against phenyl in 2,4-dinitrobenzenesulfenylacetate upon irradiation in benzene solution,[223,224] which may be rationalized by an electrophilic attack of the sulfur atom in the excited sulfenyl derivative on benzene. The most prominent reaction, in this context, is the nucleophilic aromatic photosubstitution of suitably substituted nitroarenes.[3,5,225–228]

Nucleophilic Aromatic Photosubstitution

A comprehensive treatment is beyond the scope of this chapter, and only a few salient features will be treated in the light of selected examples.

In the attack of a nucleophile on the $^3(\pi,\pi^*)$-excited nitroarene, the nitro group exerts a strong *meta*-directing effect; displacements of substituents *para* to the nitro group are comparatively sluggish.[225–228]

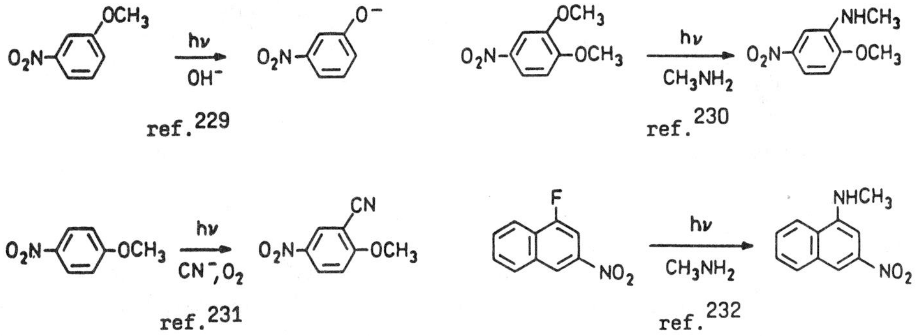

SCHEME 81.40

3,5-Dinitroanisole shows a very clean substitution of methoxy by OH^-, starting from the $^3(\pi,\pi^*)$ state[233] with a lifetime of 5×10^{-8} s or shorter, depending on nucleophile concentration. A transient (λ_{max} 410 nm, τ 5×10^7 s) is interpreted to be an exciplex of the reactants. The envelope of the first UV absorption band covers a weak n,π^* and a strong π,π^* transition.[14] The ordering of states both in the singlet and triplet manifold is dependent on the solvent: in aprotic media such as acetonitrile, S_1 is of n,π^* nature and has a lifetime of 10 ps, T_1 (n,π^*) of 780 ps. In hydrogen-bonding solvents, both S_1 and T_1 are delocalized hydrogen-bonded π,π^* states, and the triplet lifetime may vary over three orders of magnitude.[14]

The wealth of results accumulated[2,3,225–228] has led to attempts to rationalize the orientation, and two rules based on FMO arguments,[234] on a more advanced treatment,[235] and on earlier considerations[236] have been proposed:[237]

1. *Nucleophilic photosubstitutions that involve a one-step formation of a σ-complex via direct interaction between an excited aromatic substrate and a nucleophile are (substrate-)HOMO controlled:*

$$\text{ArH}^* + \text{Nu}: \ \rightarrow \ \sigma-\text{complex} \quad (S_N 2\text{Ar}^* \text{ mechanism}) \qquad (81.54)$$

2. *Nucleophilic photosubstitutions that involve electron transfer from a nucleophile to a nitroaromatic substrate and subsequent recombination of the resultant radical ions are (substrate-)LUMO controlled:*

$$\text{ArH}^* + \text{Nu}: \rightarrow \text{ArH}^{\cdot -} + \text{Nu}^{\cdot} + \rightarrow \sigma-\text{complex} \qquad (81.55)$$

It is assumed that the lowest excited state can be approximated by a one-electron configuration in which an electron is promoted from the (substrate-)HOMO to the (substrate-)LUMO (each level now being singly occupied).

The reactions of *m*-nitroanisole with OH[-229] or 3,4-dimethoxynitrobenzene (4-nitroveratrole) with methylamine[230] (see above) fall under rule (1). Rule (2) applies to cases where a substrate of high electron affinity is attacked by a nucleophile with low ionization potential. After initial electron transfer, a strong interaction develops between the (substrate-)SOMO (being the LUMO in the ground state) and the (nucleophile-)SOMO, and thus the orientation is LUMO-controlled.

The following intramolecular examples fall under rule (2)[237] since an aromatic amine serves well as an electron donor (see also the next section).

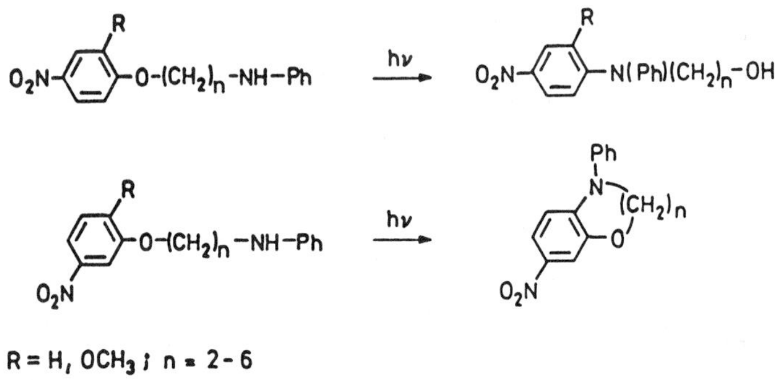

R = H, OCH₃ ; n = 2 - 6

SCHEME 81.41

In reality, things may be more complex though. A recent study[238] demonstrates that electron transfer may also occur from OH⁻ and aliphatic amines. The reactive triplet of 4-nitroveratrole was followed by the Meisenheimer complex with piperidine, as revealed by time-resolved resonance Raman scattering.

Depending on the ionization potential of the nucleophile offered, suitable nitroarenes should show reactions under rules (1) and (2) in parallel with the same nucleophile.[239,240] S_1^*- and T_1^*-4-nitroveratrole react with hexylamine via the S_N2Ar^* pathway to give *N*-hexyl-2-methoxy-5-nitroaniline and via electron transfer from amine to T_1^* and ultimate formation of *N*-hexyl-2-methoxy-4-nitroaniline.[239]

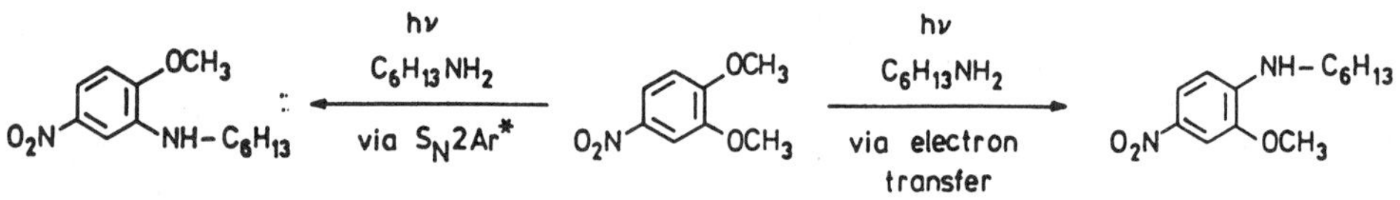

SCHEME 81.42

Via S_N2Ar^*, NO_2 group displacement has been demonstrated for the reactions of primary amines with triplet excited 4-nitroanisole[241] and 1-methoxy-4-nitronaphthalene.[242] The latter undergoes clean substitution of methoxy via the electron transfer route with secondary amines,[242] the former with primary ones.[241] The picture may further be complicated by nitro group photoreduction, as demonstrated for 4,5-dinitroveratrole.[243]

For the family of compounds 1-methoxy-4-nitronaphthalene, 3- and 4-nitroanisole, as well as 4-nitrovertrole, Bunce and Scaiano[242] give a compilation of photosubstitutions under rules (1) and (2) and this is summarized as follows: (1) S_N2Ar^* competes with electron transfer, especially when amines as nucleophiles are involved. This mode is effective as long as electron transfer is not energetically favored. (2) S_N2Ar^* leads to displacement of a substituent *meta* to nitro; or, if this is not possible, to *ipso*-substitution at the nitro group. (3) Electron transfer leads to replacement of substituents *para* to nitro; thereby, the NO_2 group stabilizes the radical anion. This substitution mode competes with photoreduction.

There are objections[241] concerning a too simple HOMO/LUMO treatment (see above). It was pointed out that interactions of charged species (radical ions) may not be suitably described by this approach. It was suggested to take σ-complex stabilities into account also.[241]

The lowest excited triplet of 2-nitrothiophene is subject to photosubstitution of the nitro group by cyanide, cyanate, hydroxide, and sulfite,[244] but to an (outer-sphere) electron transfer quenching by chloride, bromide, iodide, thiocyanate, and thiosulfate[244] as well as from hexacyanoferrate(II) and Tl(I)ions.[245]

Photo-Smiles Rearrangement

A closely related but intramolecular photosubstitution is the light-induced Smiles rearrangement, for which a duality of mechanisms has also been substantiated.[246,247]

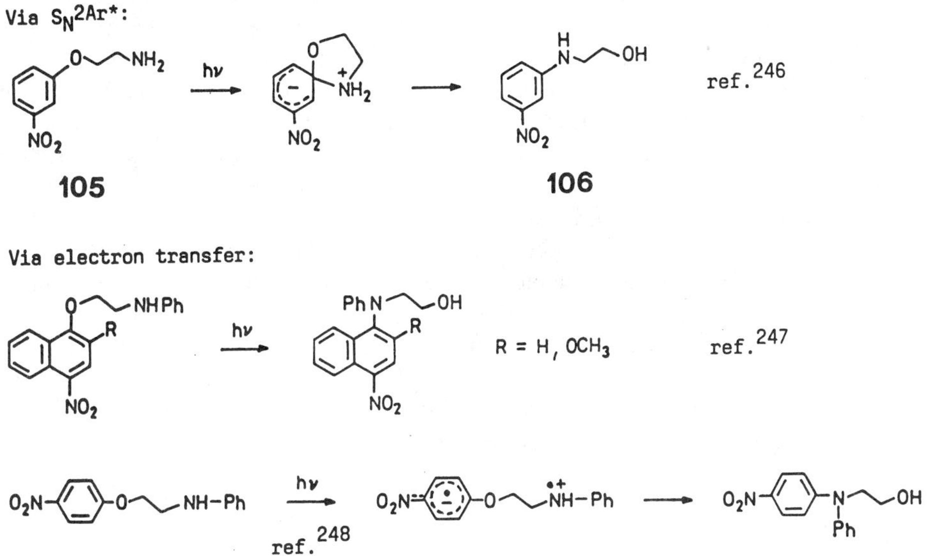

SCHEME 81.43

It is plausible that the anilino moiety may function better as an electron donor toward the excited nitroarene entity than a primary NH$_2$ group. The *o*- and *p*-isomers of **105** undergo thermal isomerization to the corresponding *N*-(β-hydroxyethyl)anilines **106**[249] in aqueous alkali. For more details of the light-induced rearrangement, especially the promotion by base and about the species on which the base acts, see References 151 and 249 to 252; for competition with other reactions dependent on the methylene linking nitroaryl and anilino moieties, see References 143, 146, 253, and 254.

1.5 Applications

The *o*-Nitrobenzyl Group in Photochemical Deprotection and Polymer Photochemistry

An *o*-nitrobenzyl (or 2-nitro-4,5-dimethoxybenzyl) group attached to O or N of an OH- or NH-containing functionality to be protected is stable toward removal by acid hydrolysis. It is easily transformed, however, into a readily hydrolyzable *o*-nitrosobenzaldehyde derivative upon irradiation (see Section 81.4, Benzylic Hydrogen Atoms).[255]

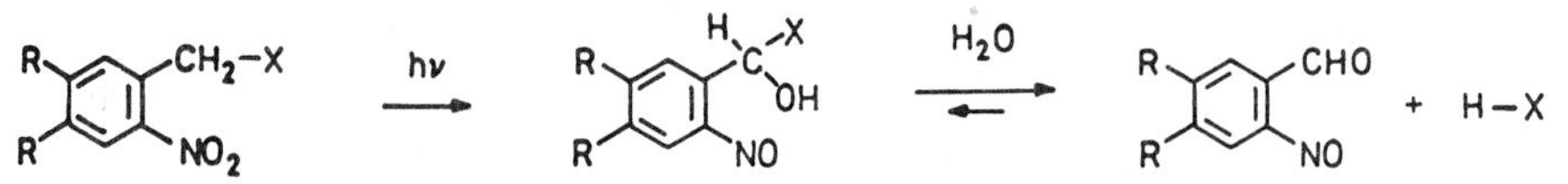

H-X = H-OOCNHR′, H-OR′, H-NHCOR′, H-OOCCH(NH₂)R′, sugars, phosphates

R = H or OCH₃

SCHEME 81.44

Good overviews of the applications are given in References 3 and 256. Among the more recent suggestions for using this principle are: liberation of fluorophores[257] and phosphate anion,[258] as well as the disintegration of liposomes[259] or the deprotection of α-amino acid carboxyl groups.[260] *o*-Nitrobenzyl alcohol has also been suggested as a protecting agent for aldehydes and ketones,[261] for which 1-(2-nitrobenzyl)ethane-1,2-diol has already been introduced.[262] *N*-(2-Nitrobenzyl)-1-naphthamide is photolytically cleaved into 1-naphthamide and 2-nitrosobenzaldehyde at −78 °C.[263]

Poly(*o*-nitrobenzylacrylate) films assume partial solubility in alkali when irradiated,[264] and co-polymers with such *o*-nitrobenzylesters are suggested as positive resists of high sensitivity and high thermal stability.[264,265] *o*-Nitrobenzylcholate derivatives have been introduced as light-sensitive compounds in poly(methyl-methacrylate-methacrylic acid) matrices,[266] and a deep UV resist containing *o*-nitrobenzylsilylether groups in the main chain has been described.[267] 4-(2-Nitrophenyl)-1,4-dihydropyridine has also been proposed as a light-sensitive dissolution inhibitor.[268] Liberation of carboxylic acids[269] and *p*-toluenesulfonic acid[270] from their *o*-nitrobenzylesters within polymer matrices has been studied.

Related to this in its original intention is the photochemical liberation of carboxylic acids out of *o*-nitroanilides. Originally suggested as a method of deprotection,[271] it has been used to cleave aromatic amide linkages in a poly(*N*-alkyl-*o*-nitroanilide) and demonstrated the usefulness of the latter as thermally stable, photosensitive polymer.[272]

Miscellaneous Applications

Nucleophilic aromatic photosubstitution (see Section 81.4, Photosubstitutions, Nucleophilic Aromatic Photosubstitutions) has been suggested for photoaffinity labeling using 4-nitroanisole, 2-fluoro-4-nitroanisole, and 4-nitroveratrole at primary and secondary amino groups.[273,274]

Excited nitrobenzene or derivatives thereof are able to oxidize hydrocarbons by hydrogen atom abstraction and radical coupling;[275] this principle may be used to hydroxylate benzylic positions[276] or to introduce hydroxyl functions or double bonds into certain steroids.[277] Nitrobenzene moieties have also been attached to steroids via chains of varied length and used, by excitation, for "remote oxidation".[278]

There are some successful light-induced syntheses of heterocycles by photocyclization of suitable nitroaromatic compounds, as in the preparation of phenazine-*N*-oxides from *N*-acetyl-*o*-nitrodiphenylamines[279,280] or of acridones, dibenzo[*c,f*][1,2]diazepin-11-one-5-oxides and -5,6-dioxides, together with 2,1-benzisoxazoles from 2,2′-dinitrodiphenylmethanes in ethanol containing sulfuric acid.[281–284]

References

1. Ciamician, G. and Silber, P., Über die Einwirkung des Lichtes auf eine alkoholische Nitrobenzollösung, *Ber. Dtsch. Chem. Ges.*, 19, 2899, 1886.
2. Morrison, H. A., The photochemistry of the nitro and nitroso groups, in *The Chemistry of Nitro and Nitroso Groups*, Part I, Feuer H; Ed., Interscience, New York, 1969, chap. 4, 165.

3. Chow, Y. L., Photochemistry of nitro and nitroso compounds, in *The Chemistry of Functional Groups, Supplement F (The Chemistry of Amino, Nitroso and Nitro Compounds and Their Derivatives)*, Patai, S., Ed., John Wiley & Sons, Chichester, 1982, chap. 6.

4. Paszyc, S., Photochemistry of nitroalkanes, *J. Photochem.*, 2, 183, 1973.

5. Döpp, D., Reactions of aromatic nitro compounds via excited triplet states, *Topics Curr. Chem.*, 55, 49, 1975.

6. Malkin, J., *Photophysical and Photochemical Properties of Aromatic Compounds*, CRC Press, Boca Raton, FL 1992.

7. Lippert, E. and Kelm, J., Spektroskopische Untersuchungen Über die Rolle des Käfig-Effektes bei der Prädissoziation aromatischer Nitroverbindungen, *Helv. Chim. Acta*, 61, 279, 1978.

8. McGlynn, S. P., Azumi, T., and Kinoshita, M., *Molecular Spectroscopy of the Triplet State*, Prentice-Hall, Englewood Cliffs, NJ, 1969, 252.

9. Frolov, A. N., El'tsov, A. V., Sosonkin, I. M., and Kuznetsova N. A., Photoreduction of nitro compounds. III. Relation between energy of lower triplet level and polarographic potentials for one-electron reduction of nitrocompounds, *Zh. Org. Khim.*, 9, 973, 1973.

10. Nakagaki, R., Photoreduction of nitroaromatic species. Properties and reactivity of excited states, *Yuki Gosei Kagaku Kyokaishi (J. Synthet. Org. Chem. Jpn.)* 48, 65, 1990.

11. Wolleben, J. and Testa, A. C., Charge transfer triplet state of *p*-nitroaniline, *J. Phys. Chem.*, 81, 429, 1977.

12. Rettig, W., Wermuth, G., and Lippert, E., Photophysical primary processes in solutions of *p*-substituted dialkylanilines, *Ber. Bunsenges. Phys. Chem.*, 83, 692, 1979.

13. Dobkowski, J., Herbich, J., Waluk, J., Koput, J., and Kühnle, W., The nature of the excited states of *p*-nitro-*N,N*-dimethylaniline, *J. Lumin.*, 44, 149, 1989. See also: Plotnikov, V. G. and Komarov, V. M., Theoretical interpretation of luminescent properties of nitroanilines, *Spectrosc. Lett.*, 9, 265, 1976.

14. Varma, C. A. G. O., Plantenga, F. L., Huizer, A. H., Zwart, J. P., Bergwerf, P., and Van der Ploeg, J. P. M., Picosecond and nanosecond kinetic spectroscopic investigations of the relaxation and the solute-solvent reaction of electronically excited 3,5-dinitroanisole, *J. Photochem.*, 24, 133, 1984.

15. (a) Schulte-Frohlinde, D. and Görner, H., *cis-trans* Photoisomerization of 4-nitrostilbenes, *Pure Appl. Chem.*, 51, 279, 1979 and references cited therein; (b) Görner, H. and Schulte-Frohlinde, D., Study of the *trans* → *cis* photoisomerization of 4-nitro-4'-dimethylaminostilbene in toluene solutions, *J. Mol. Struct.*, 84, 227, 1982; (c) Görner, H., *cis-trans*-Photoisomerization of nitrostilbenes. 12. Analysis of the pathways for *cis* ↔ *trans* photoisomerization of 4-nitro-, 4,4'-dinitro-, and 4-nitro-4'-methoxystilbene in solution, *Ber. Bunsenges. Phys. Chem.*, 88, 1199, 1984; (d) Görner, H., Schulte-Frohlinde, D., *cis-trans*-Photoisomerization of Nitrostilbenes. 13. Laser study of the triplet state of 4-nitrostilbenes in solution; Estimation of the equilibrium constant (^{3}t* ↔ ^{3}p*) and the rate constant for intersystem crossing (^{3}p* → ^{1}p), *Ber. Bunsenges. Phys. Chem.*, 88, 1208, 1984; (e) Görner, H. and Schulte Frohlinde, D., The role of triplet states in the *trans* → *cis* photoisomerization of quaternary salts of 4-nitro-4'-azastilbene and their quinolinium analogues. 6, *J. Phys. Chem.*, 89, 4105, 1985; (f) Görner, H., Photoinduced electron transfer vs. *trans* → *cis* photoisomerization for quaternary salts of 4-nitro-4'-azastilbene and their quinolinium analogues. 7, *J. Phys. Chem.*, 89, 4112, 1985; (g) Görner, H., The *cis-trans* isomerization of nitrostilbenes. XV. Mixed singlet and triplet mechanism for *trans* → *cis* photoisomerization of 4-nitro-4'-dialkylaminostilbenes in non-polar solvents, *J. Photochem. Photobiol., A: Chem.*, 40, 325, 1987; (h) Gruen, H. and Görner, H., *trans* → *cis* Photoisomerization, fluorescence, and relaxation phenomena of *trans*-4-nitro-4'-(dialkylamino)stilbenes and analogues with a nonrotatable amino group, *J. Phys. Chem.*, 93, 7144, 1989; (i) Schanze, K. S., Shin, D. M., and Whitten, D. G., Photochemical reactions in organized assemblies. 43. Micelle and vesicle solubilization sites. Determination of micropolarity and microviscosity using photophysics of a dipolar olefin, *J. Am. Chem. Soc.*, 107, 507, 1985.

16. Warman, J. M., de Haas, M. P., Hummel, A., Varma, C. A. G. O., and van Zeyl, P. H. M., Dipole moment changes in the singlet and triplet excited states of 4-Dimethylamino-4'-nitrostilbene detected by nanosecond time resolved microwave conductivity, *Chem. Phys. Lett.*, 87, 83, 1982.

17. Richter, C. and Hub, W., *Springer Proc. Phys., 4 (Time Resolved Vibrational Spectroscopy)*, 179, 1985.

18. Liang, T.-Y. and Schuster, G. B., Photochemistry of 3- and 4-nitrophenyl azides: Detection and characterization of reactive intermediates, *J. Am. Chem. Soc.*, 109, 7803, 1987.

19. (a) Albini, A., Bettinetti, G. F., Fasani, E., and Minoli, G., Photochemical reactions of primary amines on nitrophenazines, *J. Chem. Soc., Perkin Trans. 1*, 299, 1978; (b) Albini, A., Bettinetti, G. F., and Minoli, G., Photochemical reactions of amines on nitrophenazines. 2. Photoreduction versus photosubstitution, *J. Chem. Soc., Perkin Trans. 1*, 1980, 1980.

20. Marciniak, B., Koput, J., and Paszyc, S., Photochemical investigation of nitromethane in solution. I. Some photophysical properties of nitromethane, *Bull. Acad. Polon. Sci., Ser. Sci. Chim.*, 27, 843, 1979.

21. Marciniak, B., Koput, J., and Paszyc, S., Electronic structure and photophysical properties of nitroalkanes, *Wiadomósci Chemizne*, 36, 291, 1982.

22. Kozubek, H., Marciniak, B., and Paszyc, S., Photochemistry of nitroalkanes in the liquid phase and in solutions. *Wiadomósci Chemizne*, 33, 583, 1979.

23. Bielski, B. J. H. and Timmons, R. B., Electron paramagnetic resonance study of the photolysis of nitromethane, methyl nitrite and tetranitromethane at 77K, *J. Phys. Chem.*, 68, 347, 1964.

24. Spears, K. G. and Brugge, S. P., Vibrationally excited NO_2 from CH_3NO_2 and $2\text{-}C_3H_7NO_2$ photodissociation, *Chem. Phys. Lett.*, 54, 373, 1978.

25. Lao, K. Q., Jensen, E., Kash, P. W., and Butler, L. J., Polarized emission spectroscopy of photodissociating nitromethane at 200 and 218 nm, *J. Chem. Phys.*, 93, 3958, 1990.

26. Jarosiewicz, M., Szychlinski, J., and Piszczek, L., Rendement Quantique de la Photolyse du Nitrométhane, *J. Photochem.*, 29, 343, 1985.

27. Roszak, S., Kaufman, J. K., *ab initio* Multiple reference double-excitation configuration-interaction ground and excited state potential curves for nitromethane decomposition, *J. Chem. Phys.*, 94, 6030, 1991.

28. Dreeskamp, H., Koch, E., and Zander, M., On the fluorescence quenching of polycyclic aromatic hydrocarbons by nitromethane, *Z. Naturforsch., A.*, 30A, 1311, 1975.

29. Amszi, V. L., Cordero, Y., Smith, B., Tucker, S. A., Acree, W. E., Jr., Yang, C., Abu Shagara, E., and Harvey, R. G., Spectroscopic investigation of fluorescence quenching agents: Effect of nitromethane on the fluorescence emission behavior of selected cyclopenta-PAH, aceanthrylene and fluorene derivatives, *Appl. Spectrosc.*, 46, 1156, 1992.

30. Marciniak, B. and Paszyc, S., Photochemical studies of nitromethane in solution. II. Nitromethane in cyclohexane, *Bull. Acad. Polon. Sci., Ser. Sci. Chim.*, 28, 473, 1980.

31. (a) Reid, S. T. and Wilcox, E. J., Photochemistry of nitroalkanes in cyclohexane. Formation of *trans*-azocyclohexane di-N-oxide, *J. Chem. Soc. Chem. Commun.*, 646, 1975, (b) Reid, S. T., Tucker, J. N., and Wilcox, E. J., Photochemical transformations. VII. Solution photochemistry of nitroalkanes: The reaction products, *J. Chem. Soc., Perkin Trans. 1*, 1359, 1974.

32. Takechi, H. and Machida, M., Photochemical conversion of aliphatic nitro compounds into oximes, *Synthesis*, 206, 1989.

33. Kozubek, H. and Paszyc, S., Photochemical studies of nitromethane in solution. III. Nitromethane in isopropyl alcohol, *Bull. Acad. Polon. Sci., Ser. Sci. Chim.*, 28, 481, 1980.

34. Paszyc, S., Kozubek, H., and Marciniak, B., Photochemical studies on the nitromethane-benzene system in liquid phase, *Bull. Acad. Polon. Sci., Ser. Sci. Chim.*, 25, 951, 1977.

35. Fischer, A. and Ibrahim, P. N., Photolysis of nitro-β-γ-alkenes: Nitro-2,5-cyclohexadienes, *Tetrahedron*, 46, 2737, 1990.

36. Suginome, H., Takakuwa, K., and Orito, K., Functionalization of unactivated carbon involving photochemical intramolecular rearrangement of nitro group attached to tetrahedral carbon to nitrosooxy group, *Chem. Lett.*, 1357, 1982.

37. Saito, I., Takami, M., Konoike, T., and Matsuura, T., Photoinduced reactions. LXXIII. Solvent dependence in the photochemical reaction of α-nitroepoxides, *Bull. Chem. Soc. Jpn.*, 46, 3198, 1973.

38. Reid, S. T. and Tucker, J. N., The photochemistry of α-nitroketones, *Chem. Commun.*, 1609, 1971.

39. Suginome, H., Kurokawa, Y., and Orito, K., The photochemistry of steroidal 6-membered α-nitro ketones, *Bull. Chem. Soc. Jpn.*, 61, 4005, 1988.

40. Suginome, H. and Kurokawa, Y., The photoinduced removal of nitro groups from steroidal 6-membered α,α'-dinitro cyclic ketones, *Bull. Chem. Soc. Jpn.*, 62, 1107, 1989.

41. Suginome, H. and Kurokawa, Y., The photochemistry of steroidal 6-membered cyclic α-nitro enones, *Bull. Chem. Soc. Jpn.*, 62, 1343, 1989.

42. Suginome, H. and Kurokawa, Y., Photoinduced molecular transformations. 104. Pathways of the photorearrangements of five-membered cyclic steroidal α-nitro ketones to *N*-hydroxy cyclic imides, cyclic hydroxamic acid, and cyclic imide, *J. Org. Chem.*, 54, 5945, 1989.

43. Kornblum, N., Radical anion reactions of nitro compounds. *The Chemistry of Functional Groups, Supplement F: The Chemistry of Amino, Nitroso and Nitro Compounds and Their Derivatives, Part 1*, Patal, S., Ed., John Wiley & Sons, Chichester, 1982, 361.

44. (a) Slovetskii, V. I., Balykin, V. P., and Fainzilberg, A. A., Photochemistry of aliphatic nitro compounds. I. Flash photolysis of tetranitromethane solutions, *Izv. Akad. Nauk SSSR, Ser. Khim.*, 2181, 1975; (b) Slovetskii, V. I. and Balykin, V. P., Photochemistry of aliphatic nitro compounds. 2. Mechanism of the photochemical decomposition of tetranitromethane with respect to solvent nature, *Izv. Akad. Nauk SSSR, Ser. Khim.*, 2186, 1975.

45. Kerimov, O. M., Maksyutov, E. M., Milanich, A. I., and Slovetskii, V. I., Photochemistry of aliphatic nitro compounds and pulsed photolysis of tetranitromethane using a UV laser, *Izv. Akad. Nauk SSSR, Ser. Khim.*, 623, 1979.

46. Isaacs, N. S. and Abed, O. H., The mechanism of aromatic nitration by tetranitromethane, *Tetrahedron Lett.*, 23, 2799, 1982.

47. Kholmogorov, V. E. and Gorodyskii, V. A., Photochemistry of complexes of benzene derivatives with tetranitromethane, *Zh. Fiz. Chem.*, 46, 63, 1972.

48. Kochi, J. K., Inner sphere electron transfer in organic chemistry. Relevance to electrophilic aromatic nitration, *Acc. Chem. Res.*, 25, 39, 1992 and references cited therein.

49. (a) Eberson, L. and Radner, F., Light initiated and thermal nitration reactions during photolysis of naphthalene/tetranitromethane or 1-methoxynaphthalene/tetranitromethane in dichloromethane, *J. Am. Chem. Soc.*, 113, 5825, 1991 and references cited therein; (b) Eberson, L., Hartshorn, M. P., and Radner, F., Photochemical nitration by tetranitromethane. VI. Predominant nitro/trinitromethyl addition to naphthalene in dichloromethane and acetonitrile, *J. Chem. Soc., Perkin Trans. 2*, 1793, 1992, and references cited therein.

50. Eberson, L., Hartshorn, M. P., and Radner, F., Photochemical nitration by tetranitromethane. VII. Mode of formation of the nitro substitution products from 1,4-dimethylnaphthalene in dichloromethane and acetonitrile, *J. Chem. Soc., Perkin Trans 2*, 1799, 1992.

51. (a) Yamada, K., Tanaka, S., Naruchi, K., and Yamamoto, M., Novel intramolecular photorearrangement of nitronate anions, *J. Org. Chem.*, 47, 5283, 1982; (b) Yamada, K., Kanekiyo, T., Tanaka, S., Naruchi, K., and Yamamoto, M., Novel intramolecular photorearrangement of alkane nitronate anions, *J. Am. Chem. Soc.*, 103, 7003, 1981; (c) Yamada, K., Kishikawa, K., and Yamamoto, M., Stereospecificity of the photorearrangement of nitronate anions and its utilization for stereospecific cleavage of cyclic compounds, *J. Org. Chem.*, 52, 2327, 1987.

52. Edge, G. J., Imam, S. H., and Marples, B. A., Steroids, 21. Photorearrangement of steroidal nitronate salts and a *n*-butyl-spirooxaziridine, *J. Chem. Soc., Perkin Trans. 1*, 2319, 1984.

53. Grant, R. D., Pinhey, J. T., Rizzardo, E., and Smith, G. C., The photochemistry of α,β-unsaturated nitro compounds and nitronic acids. Concerning deconjugation and α,β-unsaturated ketone formation, *Aust. J. Chem.*, 38, 1505, 1985.

54. Descotes, G., Bahurel, Y., Bourrillot, M., Pingeon, G., and Rostaing, R., Nitrooléfines. II. Reactions d'isomerization thermique et photochimique des nitrooléfines, *Bull. Soc. Chim. Fr.*, 290, 1970.

55. Crosby, P. M., Salisbury, K., and Wood, G. P., Photochemical reactions of an α,β-unsaturated nitro compound, *J. Chem. Soc., Chem. Commun.*, 312, 1975.

56. Cridland, J. S., Moles, P. J., Reid, S. T., and Taylor, K. T., The photochemistry of cyclic and acyclic nitroalkenes, *Tetrahedron Lett.*, 4497, 1976.

57. Yokoyama, K., Kato, M., and Noyori, R., Photo-induced cycloaddition of 1-nitrocyclooctene and cyclopentadiene, *Bull. Chem. Soc. Jpn.*, 50, 2201, 1977.

58. Humphrey-Baker, R. A., Salisbury, K., and Wood, G. P., Photochemical reactions of an α,β-unsaturated nitro compound, *J. Chem. Soc., Perkin Trans. 2*, 659, 1978.

59. Becker, H. D., Soerensen, H., and Sandros, K., Photochemical isomerization and dimerization of 1-(9-anthryl)-2-nitroethylene, *J. Org. Chem.*, 51, 3223, 1986.

60. Descotes, G., Bahurel, Y., Bourrillot, M., Pingeon, G., and Rostaing, R., Nitrooléfines. III. Condensations thermique et photochimique des nitrooléfines avec les aldéhydes α-β éthyléniques, *Bull. Soc. Chim. Fr.*, 295, 1970.

61. Chapman, O. L., Griswold, A. A., Hoganson, E., Lenz, G., and Reasoner, J., Photochemistry of unsaturated nitro compounds, *Pure Appl. Chem.*, 9, 585, 1964.

62. Pinhey, J. T., Rizzardo, E., and Smith, G. C., The photochemistry of 3β-acetoxy-6-nitrocholest-5-ene and 6-nitrocholest-5-ene. The mechanism of photodeconjugation of α,β-unsaturated nitro compounds, *Aust. J. Chem.*, 31, 97, 1978.

63. Chapman, O. L., Cleveland, P. G., and Hoganson, E. D., A Photochemical rearrangement of nitro olefins, *J. Chem. Soc., Chem. Commun.*, 101, 1966.

64. Shaffer, G. W., Photochemistry of β-bromo-β-nitrostyrene, *Can. J. Chem.*, 48, 1948, 1970.

65. Saito, I., Takami, M., and Matsuura, T., Nitro-nitrite rearrangement and intramolecular cycloaddition in the photochemistry of nitro-olefins, *Tetrahedron Lett.*, 3155, 1975.

66. Hunt, R. G. and Reid, S. T., Photochemical transformations. 11. Photorearrangement of 1-(2-pyridyl)- and 1-(3-pyridyl)-2-nitropropenes, *J. Chem. Soc., Perkin Trans. 1*, 2462, 1977.

67. Reid, S. T., Thompson, J. K., and Mushambi, C. F., Photochemical transformations. 15. A novel photoreaction of 3-nitro-2-phenyl-2*H*-chromene in methanol, *Tetrahedron Lett.*, 24, 2209, 1983.

68. Grant, R. D. and Pinhey, J. T., The photochemistry of aliphatic and alicyclic α,β-unsaturated nitro compounds. A study of double bond cleavage following intramolecular cyclization and nitro-nitrite rearrangement, *Aust. J. Chem.*, 37, 1231, 1984.

69. Ioki, Y., Aryloxyl radicals by photorearrangement of nitro-compounds, *J. Chem. Soc., Perkin Trans. 2*, 1240, 1977.

70. Grant, R. D., Pinhey, J., and Rizzardo, E., The photochemistry of α-nitrostilbenes and related compounds. A study of double bond cleavage following intramolecular cyclization and nitro-nitrite rearrangement, *Aust. J. Chem.*, 37, 1217, 1984.

71. Zimmerman, H. E., Roberts, L. C., and Arnold, R., Photochemical rearrangements of an unsaturated nitro compound. Mechanistic and exploratory organic photochemistry, *J. Org. Chem.*, 42, 621, 1977.

72. Cridland, J. S. and Reid, S. T., Photochemical rearrangement of 3-(2-nitro-1-propenyl)indole, *J. Chem. Soc., Chem. Commun.*, 125, 1969.

73. Hunt, R. G., Reid, S. T., and Taylor, K. T., Further studies in the photorearrangement of heterocyclic nitroalkenes: A facile synthesis of 6-hydroxy-1,2-oxazines, *Tetrahedron Lett.*, 2861, 1972.

74. Velezheva, V. S., Yaroslavskii, I. S., and Suvorov, N. N., Photoisomerization of α,β-unsaturated nitro compounds of the indole series into alkylidene derivatives of 2- and 3-oxoindolines, *Zh. Org. Khim.*, 18, 2403, 1982.

75. Chapman, O. L., Heckert, D. C., Reasoner, J. W., and Thackaberry, S. P., Photochemical studies on 9-nitroanthracene, *J. Am. Chem. Soc.*, 88, 5550, 1966.

76. Hamanoue, K., Nakayama, T., Ushida, K., Kajiwara, K., and Shigenobu, Y., Photophysics and photochemistry of nitroanthracenes. 1. Primary processes in the photochemical reactions of 9-benzoyl-10-nitroanthracene and 9-cyano-10-nitroanthracene studied by steady state photolysis and nanosecond laser photolysis, *J. Chem. Soc., Faraday Trans.*, 87, 3365, 1991.

77. Hamanoue, K., Amano, M., Kimoto, M., Kajiwara, Y., Nakayama, T., and Teranishi, H., Photochemical reactions of nitroanthracene derivatives in fluid solutions, *J. Am. Chem. Soc.*, 106, 5993, 1984.

78. Hamanoue, K., Hirayama, S., Nakayama, T., and Teranishi, H., Nonradiative relaxation process of the higher excited states of *meso*-substituted anthracenes, *J. Phys. Chem.*, 84, 2074, 1980.

79. Hirayama, S., Kajiwara, Y., Nakayama, T., Hamanoue, K., and Teranishi, H., Correct assignment of the low-temperature luminescence from 9-nitroanthracene, *J. Phys. Chem.*, 89, 1945, 1985.

80. Snyder, R. and Testa, A. C., Nature of the low-temperature emission from 9-nitroanthracene, *J. Phys. Chem.*, 85, 1871, 1981.

81. Testa, A. C., Luminescence and photochemistry of 9-nitroanthracene at 4.2K, *J. Photochem. Photobiol. A: Chem.*, 47, 309, 189.

82. (a) Testa, A. C. and Wild, U. P., Holographic photochemical study of 9-nitroanthracene, *J. Phys. Chem.*, 90, 4302, 1986; (b) See also: Hamanoue, K., Hirayama, S., Hidaka, T., Ohaya, H., Nakayama, T., and Teranishi, H., Photochemical reaction of nitroanthracene derivatives in polymethyl methacrylate, *Polym. Photochem.*, 1, 57, 1981.

83. El-Sayed, M. A., Spin-orbit coupling and the radiationless processes in nitrogen heterocyclics, *J. Chem. Phys.*, 38, 2834, 1963.

84. Galliani, G. and Rindone, B., The photolysis of 9-nitroanthracene derivatives, *Gazz. Chim. Ital.*, 107, 435, 1977.

85. Boyer, J. H. and Patel, J., 1,1′-Dihydroxy-2,2′-diphenyl-4,4′-binaphthyl, *Org. Prep. Proc. Int.*, 14, 418, 1982.

86. Döpp, D. and Wong, C. C., Photochemie aromatischer Nitroverbindungen, XVIII. Photolyse von 2,6-Di-*tert*-butyl-1-nitronaphthalin, *Chem. Ber.*, 121, 2045, 1988.

87. van den Braken-van Leersum, A. M., Tintel, C., van t'Zelfde, M., Cornelisse, J., and Lugtenburg, J., Spectroscopic and photochemical properties of mononitropyrenes, *Recl. Trav. Chim.*, 106, 120, 1987.

88. Yasuhara, A. and Fuwa, K., Formation of 1-nitro-2-hydroxypyrene from 1-nitropyrene by photolysis, *Chem. Lett.*, 347, 1983.

89. Büchi, G. and Ayer, D. E., Light-catalyzed organic reactions. IV. The oxidation of olefins with nitrobenzene, *J. Am. Chem. Soc.*, 78, 1689, 1956.

90. Charlton, J. L. and de Mayo, P., The photocycloaddition of nitrobenzene to alkenes, *Can. J. Chem.*, 46, 1041, 1968.

91. Charlton, J. L., Liao, C. C., and de Mayo, P., Photochemical synthesis. The addition of aromatic nitro compounds to alkenes, *J. Am. Chem. Soc.*, 93, 2463, 1971.

92. Okada, K., Saito, Y., and Oda, M., Photochemical reaction of polynitrobenzenes with adamantylideneadamantane: The X-ray structure analysis and chemical properties of the dispiro N-(2,4,6-Trinitrophenyl)-1,3,2-dioxazolidine product, *J. Chem. Soc., Chem. Commun.*, 1731, 1992.

93. Saito, I., Takami, M., and Matsuura, T., Oxidative ring cleavage of methoxynaphthalenes with photo-excited aromatic nitro compounds, *Tetrahedron Lett.*, 659, 1974.

94. Saito, I., Takami, M., and Matsuura, T., Oxidative cleavage of aromatic methoxy compounds with photo-excited aromatic nitro compounds, *Bull. Chem. Soc. Jpn.*, 48, 2865, 1975.

95. Saito, I., Takami, M., and Matsuura, T., Photoinduced reactions. LXVI. Photochemical reaction of *m*-chloronitrobenzene with methoxybenzenes: Oxidative ring cleavage of aromatics, *Chem. Lett.*, 1195, 1972.

96. Scheinbaum, M. L., The photochemical reaction of nitrobenzene and tolane, *J. Org. Chem.*, 29, 2200, 1964.

97. Hurley, R. and Testa, A. C., Photochemical n→π* Excitation of nitrobenzene, *J. Am. Chem. Soc.*, 88, 4330, 1966.

98. Lamola, A. A. and Hammond, G. S., Mechanisms of photochemical reactions in solution. XXXIII. Intersystem crossing efficiencies, *J. Chem. Phys.*, 43, 2129, 1965.

99. Hurley, R. and Testa, A. C., Triplet state yield of aromatic nitro compounds, *J. Am. Chem. Soc.*, 90, 1949, 1968.

100. Yip, R. W., Sharma, D. K., Giasson, R., and Gravel, D., Picosecond excited state absorption of alkyl nitrobenzenes in solution, *J. Phys. Chem.*, 88, 5770, 1984.

101. Murov, St. L., *Handbook of Photochemistry*, Marcel Dekker, New York, 1973, 5 and 197.

102. Obi, K., Bottenheim, J. W., and Tanaka, I., The photoreduction of 2-nitronaphthalene in 2-propanol, *Bull. Chem. Soc. Jpn.*, 46, 1060, 1973.

103. Trotter, W. and Testa, A. C., Photoreduction of aromatic nitro compounds by tri-*n*-butylstannane, *J. Am. Chem. Soc.*, 90, 7044, 1968.

104. Hashimoto, S. and Kano, K., The photochemical reduction of nitrobenzene and its reduction intermediates. X. The photochemical reduction of the monosubstituted nitrobenzenes in 2-propanol, *Bull. Chem. Soc., Jpn.*, 45, 549, 1972.

105. Levy, N. and Cohen, M. D., Photoreduction of 4-cyano-1-nitrobenzene in propan-2-ol, *J. Chem. Soc., Perkin Trans. 2*, 553, 1979.

106. Kaneko, C., Yamada, S., Yokoe, I., Hata, N., and Ubukata, Y., Photochemical reduction of 4-nitropyridine-1-oxide and its methyl substituted derivatives in ethanol, *Tetrahedron Lett.*, 4729, 1966.

107. Hata, N., Okutsu, E., and Tanaka, I., The primary photochemical process of 4-nitropyridine *N*-oxide, *Bull. Chem. Soc. Jpn.*, 41, 1769, 1968.

108. Hata, N., Ono, I., and Tsuchiya, T., The primary photochemical process of 4-nitropyridine *N*-oxide, II. Decay kinetics of the photochemical intermediates, *Bull. Chem. Soc. Jpn.*, 45, 2386, 1972.

109. Ono, I. and Hata, N., The photosensitized reduction of 4-nitropyridine *N*-oxides, *Bull. Chem. Soc. Jpn.*, 45, 2951, 1972.

110. Barltrop, J. A. and Bunce, N. J., Organic photochemistry. VIII. The photochemical reduction of nitro-compounds, *J. Chem. Soc. (C)*, 1467, 1968.

111. Jagannadham, V. and Steenken, S., One-electron reduction of nitrobenzenes by α-hydroxyalkyl radicals via addition/elimination. An example of an organic inner-sphere electron-transfer reaction, *J. Am. Chem. Soc.*, 106, 6452, 1984.

112. Asmus, K.-D., Wigger, A., and Henglein, A., Pulsradiolytische Untersuchung einiger Elementarprozesse der Nitrobenzolreduktion, *Ber. Bunsenges. Phys. Chem.*, 70, 862, 1966.

113. (a) Ward, R. L., Photoinduced paramagnetism in solutions of nitrobenzene in tetrahedrofuran, *J. Chem. Phys.*, 38, 2588, 1963; (b) McMillan, M. and Norman, R. O. C., Electron spin resonance studies. XVII. Reactions of some free radicals with nitroalkanes, *J. Chem. Soc. (B)*, 590, 1968; (c) Chachaty, C. and Forchioni, A., Radicaux nitroxydes en phase liquide par photolyse des composes nitres. I. Solutions de nitrobenzène dans les alcools, *Tetrahedron Lett.*, 307, 1968; (d) Janzen, E. G. and Gerlock, J. L., Substituent effects on the photochemistry and nitroxide radical formation of nitro aromatic compounds as studied by electron spin resonance spin trapping techniques, *J. Am. Chem. Soc.*, 91, 3108, 1969; (e) Cowley, D. J. and Sutcliffe, L. H., Electron spin resonance studies of free-radical reactions in photolysed solutions of chloronitrobenzenes, *J. Chem. Soc. (B)*, 569, 1970; (f) Sleight, R. B. and Sutcliffe, L. H., E.S.R Investigation of alkoxynitroxide free radicals from the photolysis of nitrobenzenes and of heterocyclic nitro compounds, *Trans. Faraday Soc.*, 67, 2195, 1971; (g) Wong, S. K. and Wan, J. K. S., Some electron spin resonance observations in the photoreduction of nitrobenzenes, *Can. J. Chem.*, 51, 753, 1973; (h) Sancier, K. M., ESR of photoproduced nitroaromatic radicals: Correlation between photolysis and radiolysis, *Radiat. Res.*, 81, 487, 1980; (i) Alberti, A., Dellonte, S., Paradisi, C., Roffia, S., and Pedulli, G. F., Reactions of triplet carbonyl compounds and nitro derivatives with silanes, *J. Am. Chem. Soc.*, 112, 1123, 1990.

114. Hart, H. and Link, J. W., Radical coupling in the photoreduction of an aromatic nitro compound in ether, *J. Org. Chem.*, 34, 758, 1969.

115. (a) Letsinger, R. L. and Wubbels, G. G., Photoinduced substitution. IV. Reaction of aromatic nitro compounds with hydrochloric acid, *J. Am. Chem. Soc.*, 88, 5041, 1966; (b) Wubbels, G. G. and Letsinger, R. L., Photoreactions of nitrobenzene and monosubstituted nitrobenzenes with hydrochloric acid. Evidence concerning the reaction mechanism, *J. Am. Chem. Soc.*, 96, 6698, 1974.

116. Hurley, R. and Testa, A. C., Nitrobenzene photochemistry. II. Protonation in the excited state, *J. Am. Chem. Soc.*, 89, 6917, 1967.

117. Hashimoto, S., Sunamoto, J., Fujii, H., and Kano, K., Photochemical reduction of nitrobenzene and its reduction intermediates. III. The photochemical reduction of nitrobenzene, *Bull. Chem. Soc. Jpn.*, 41, 1249, 1968.

118. Wubbels, G. G., Jordan, J. W., and Mills, N. S., Hydrochloric acid catalyzed photoreduction of nitrobenzene by 2-propanol. The question of protonation in the excited state, *J. Am. Chem. Soc.*, 95, 1281, 1973.

119. Cu, A. and Testa, A. C., Evidence for electron transfer in the photoreduction of aromatic nitro compounds, *J. Am. Chem. Soc.*, 96, 1963, 1974.

120. Fukuzumi, S. and Tokuda, Y., Efficient six-electron photoreduction of nitrobenzene derivatives by 10-methyl-9,10-dihydroacridine in the presence of perchloric acid, *Bull. Chem. Soc. Jpn.*, 65, 831, 1992.

121. Hashimoto, S., Kano, K., and Ueda, K., Photochemical reduction of nitrobenzene and its reduction intermediates. IX. The photochemical reduction of 4-nitropyridine in a hydrochloric acid-isopropyl alcohol solution, *Bull. Chem. Soc. Jpn.*, 44, 1102, 1971, and earlier papers.

122. Cu, A. and Testa, A. C., Photochemistry of 4-nitropyridine in acid solutions, *J. Phys. Chem.*, 77, 1487, 1973.

123. Trotter, W. and Testa, A. C., Photoreduction of 1-nitronaphthalene by protonation in the excited state, *J. Phys. Chem.*, 74, 845, 1970.

124. Cu, A. and Testa, A. C., Photochemistry of the nitro group in aromatic heterocyclic molecules, *J. Phys. Chem.*, 79, 644, 1975.

125. Trotter, W. and Testa, A. C., Modification of nitrobenzene photochemistry by molecular complexation, *J. Phys. Chem.*, 75, 2415, 1971.

126. El'tsov, A. V., Kuznetsova, N. A., and Frolov, A. N., Photochemical reduction of nitro compounds in an alkaline medium, *Zh. Org. Khim.*, 7, 817, 1971.

127. Frolov, A. N., Kuznetsova, N. A., El'tsov, A. V., and Rtishev, N. I., Transfer of electron (hydride ion) during photoreduction of nitro compounds in alkaline and neutral alcoholic buffer solutions, *Zh. Org. Khim.*, 9, 963, 1973.

128. Kuznetsova, N. A., El'tsov, A. V., Fomin, G. V., and Frolov, A. N., Isotopic effect during the photoreduction of aromatic nitro compounds in aqueous alcohol and alkaline aqueous alcohol media, *Zh. Fiz. Khim.*, 49, 115, 1975.

129. Frolov, A. N., Kuznetsova, N. A., Rtishev, N. I., and El'tsov, A. V., Participation of alkoxyl ions in the photoreduction of nitro compounds in alkaline alcohol media, *Zh. Org. Khim.*, 10, 2562, 1974.

130. Vink, J. A. J., Cornelisse, J., and Havinga, E., Photoreactions of aromatic compounds — Photoreduction of nitrobenzene by cyanide ion, *Rec. Trav. Chim.*, 90, 1333, 1971.

131. Döpp, D., Die Produkte der Bestrahlung des 2-Nitro-1,4-di-*tert*-butylbenzols in aliphatischen Aminen, *Chem. Ber.*, 104, 1058, 1971.

132. Döpp, D., Müller, D., and Sailer, K.-H., Photochemie aromatischer Nitroverbindungen, VIII. Photoreduktion sterisch gehinderter Nitrobenzole in Triethylamin, *Tetrahedron Lett.*, 2137, 1974.

133. Takami, M., Matsuura, T., Saito, I., Photochemical Reaction of Aromatic Nitro Compounds with Aromatic Amines, *Tetrahedron Lett.*, 661, 1974.

134. Döpp, D., Müller, D., Photochemistry of aromatic nitro compounds. XIII. Photoreduction of sterically hindered nitrobenzenes in aliphatic amines, *Rec. Trav. Chim.*, 98, 297, 1979.

135. Döpp, D., Heufer, J., *N*-Demethylation of *N,N*-dimethylanilines by photoexcited 3-nitrochlorobenzene, *Tetrahedron Lett.*, 23, 1553, 1982.

136. (a) Sundarajan, K., Ramakrishnan, V., and Kuriacose, J. C., Photoreaction of nitrobenzene with aliphatic amines, *Ind. J. Chem.*, 22B, 257, 1983; (b) Sundarajan, K., Ramakrishnan, V., and Kuriacose, J. C., Photoreduction of nitrobenzene by aliphatic amines — A mechanistic study, *Ind. J. Chem.*, 23B, 1068, 1984.

137. Sundarajan, K., Ramakrishnan, V., and Kuriacose, J. C., Electron spin resonance studies on the radicals produced in the photoreduction of nitrobenzene by aliphatic amines, *J. Photochem.*, 27, 61, 1984.

138. (a) Mir, M., Marquet and J., Cayon, E., Solid state photochemistry of ternary cyclodextrin complexes: Total selectivity in the photoreduction of nitrophenyl ethers by 1-phenylethylamine, *Tetrahedron Lett.*, 33, 7053, 1992; (b) Kuzmič, P., Pavličková, L., Souček, M., Photoreaction of *N*-butyl-3,4-dimethoxy-6-nitrobenzamide with butylamine. A model study for lysine directed photoaffinity labeling, *Coll. Czech. Chem. Commun.*, 52, 1780, 1987.

139. Levin, P. P. and Kuz'min, V. A., Laser photolysis study of triplet exciplexes of nitro-naphthalenes with tertiary aromatic amines, *Izv. Akad. Nauk SSSR, Ser. Khim.*, 2367, 1987.

140. Döpp, D., Lin, J., and Abdel-Latif, F., unpublished.Pavlíčková,

141. Mahdavi, F., Bruton, T. C., and Li, Y., Photoinduced reduction of nitro compounds on semiconductor particles, *J. Org. Chem.*, 58, 744, 1993.

142. Sun, S., Birke, R. L., Lombardi, J. R., Leung, K. P., and Genack, A. Z., Photolysis of *p*-nitrobenzoic acid on roughened silver surfaces, *J. Phys. Chem.*, 92, 5965, 1988.

143. Nakagaki, R., Hiramatsu, M., Mutai, K., and Nagakura, S., Photochemistry of bichromophoric chain molecules containing electron donor and acceptor moieties. Dependence of reaction pathways on the chain length and mechanism of photoredox reaction of *N*-[ω-(*p*-nitrophenoxy)-alkyl]anilines, *Chem. Phys. Lett.*, 121, 262, 1985.

144. Nakagaki, R., Hiramatsu, M., Mutai, K., Tanimoto, Y., and Nagakura, S., Photochemistry of bichromophoric chain molecules containing electron donor and acceptor moieties. External magnetic field effects upon the photochemistry of *N*-[ϖ-(*p*-nitrophenoxy)-alkyl]anilines, *Chem. Phys. Lett.*, 134, 171, 1987.

145. Nakagaki, R., Mutai, K., Hiramatsu, M., Tukada, H., and Nagakura, S., Magnetic field effects upon photochemistry of bichromophoric chain molecules containing nitroaromatic and arylamino moieties: Elucidation of reaction mechanism and control of reaction yields, *Can. J. Chem.*, 66, 1989, 1988.

146. Nakagaki, R., Mutai, K., and Nagakura, S., Photochemistry of chain molecules containing 4-nitro-1-naphthoxyl and anilino chromophores. Switching of reaction pathways due to methylene chain length and magnetic field effects, *Chem. Phys. Lett.*, 154, 581, 1989.

147. Nakagaki, R., Mutai, K., and Nagakura, S., Magnetic Field Effects upon Photoredox Reactions of Bifunctional Chain Molecules Containing Anilino and Nitro-aromatic Chromophores, *Chem. Phys. Lett.*, 167, 439, 1990.

148. Ishii, Y., Tukada, H., Nakagaki, R., and Mutai, K., Magnetic field effects and reaction mechanism in photoredox reaction of nitro group, *Chem. Lett.*, 1559, 1990.

149. Döpp, D., Gerding, B., and Weiler, H., unpublished results.

150. Döpp, D. and Gerding, B., Base promoted photodealkylation of 4-nitro-*N,N*-dialkylnilines, *IX IUPAC Symposium on Photochemistry, Abstracts of Contributed Papers*, P 100, Pau, France, 1982.

151. Wubbels, G. G., Hautala, R. R., Letsinger, R. L., Photoinduced Substitution. Photoisomerization of *p*-nitrobenzaldehyde, *Tetrahedron Lett.*, 1689, 1970, and references cited therein.

152. Wubbels, G. G., Kalhorn, T. F., Johnson, D. E., and Campbell, D., Mechanism of the water-catalyzed photoisomerization of *p*-nitrobenzaldehyde, *J. Org. Chem.*, 47, 4664, 1982.

153. Reisch, J. and Weidmann, K. G., Photochemie des *p*-Nitrobenzaldehyds, *Arch. Pharm. (Weinheim)*, 304, 906, 1971.

154. Görner, H., submitted.

155. Görner, H., Currell, L. J., and Kuhn, H. J., Photoreaction of (*p*-nitrophenyl)glyoxylic acid to *p*-nitrosobenzoic acid in aqueous solution, *J. Phys. Chem.*, 95, 5518, 1991.

156. Wan, P. and Yates, K., Photochemistry of α-aryl carbonyl compounds in aqueous solution, *J. Chem. Soc., Chem. Commun.*, 275, 1982.

157. Wan, P. and Yates, K., Photoredox chemistry of *m*- and *p*-nitrobenzyl alcohols in aqueous solution. Observation of novel catalysis by the hydronium and hydroxide ions in these photoreactions, *J. Org. Chem.*, 48, 136, 1983.

158. Wan, P. and Yates, K., Photoredox chemistry of nitrobenzyl alcohols in aqueous solution. Acid and base catalysis of reaction, *Can. J. Chem.*, 64, 2076, 1986.

159. Wan, P. and Xigen, Xu, Disproportionation of 4-nitroacetophenone to 4-aminocetophenone and 4-nitrobenzoic acid, *J. Org. Chem.*, 54, 4473, 1989.

160. Wan, P. and Yates, K., Photochemical oxidation of nitrobenzyl alcohols in aqueous solution, *J. Chem. Soc., Chem. Commun.*, 1023, 1981.

161. Wan, P., Photooxygenation of nitroaromatic compounds. Catalytic effects in the reaction of molecular oxygen with nitrobenzyl derivatives, *Tetrahedron Lett.*, 26, 2387, 1985.

162. Wan, P., Muralidharan, S., McAuley, I., and Babbage, C. A., Photooxygenation of nitrobenzyl derivatives. Mechanism of photogeneration and hydrolysis of α-hydroperoxynitrobenzyl ethers, *Can. J. Chem.*, 65, 1775, 1987.

163. Wan, P., Culshaw, S., and Yates, K., Photohydration of aromatic alkenes and alkynes, *J. Am. Chem. Soc.*, 104, 2509, 1982.

164. Kalanderopoulos, P. and Yates, K., Intramolecular proton transfer in photohydration reactions, *J. Am. Chem. Soc.*, 108, 6290, 1986.

165. Wan, P., Davis, M. J., and Teo, M. A., Photoaddition of water and alcohols to 3-nitrostyrenes. Structure-reactivity and solvent effects, *J. Org. Chem.*, 54, 1354, 1989.

166. Wan, P. and Davis, M. J., Photochemical tandem addition of water and oxygen to the alkene moiety of *m*-nitrostyrenes, *J. Photochem. Photobiol. A: Chem.*, 48, 387, 1989.

167. Rafizadeh, K. and Yates, K., Acid-catalyzed photooxidation of *m*-nitrobenzyl derivatives in aqueous solution, *J. Org. Chem.*, 51, 2777, 1986.

168. Okamoto, Y., Iwamoto, N., and Takamuku, S., Photochemical carbon-phosphorus bond cleavage of nitro-substituted benzylphosphonic acids, *J. Chem. Soc., Chem. Commun.*, 1516, 1986.

169. Okamoto, Y., Iwamoto, N., Toki, S., and Takamuku, S., Photochemical carbon-phosphorus bond cleavage of (nitrobenzyl) phosphonate ions, *Bull Chem. Soc. Jpn.*, 60, 277, 1987.

170. Yamaoka, T., Adachi, H., Matsumoto, K., Watanabe, H., and Shirosaki, T., Photochemical dissociation of *p*-nitrobenzyl-9,10-dimethoxy-anthracene-2-sulfonate via intramolecular electron transfer, *J. Chem. Soc., Perkin Trans. 2*, 1709, 1990.

171. Naitoh, K. and Yamaoka, T., Photochemical dissociation of *p*-nitrobenzylsulfonate esters via an intramolecular electron transfer, *J. Chem. Soc., Perkin Trans. 2*, p. 663, 1992.

172. Yamaoka, T., Omote, T., Adachi, H., Kikuchi, N., Watanabe, Y., and Shirosaki, T., Photochemical dissociation of *p*-nitrobenzyl aromatic sulfonate and its application to chemical amplification resists, *J. Photopolym. Sci. Technol.*, 3, 275, 1990.

173. Wan, P. and Muralidharan, S., Photochemical retro-aldol type reactions of nitrobenzyl derivatives. Mechanistic variations in the elimination of nitrobenzyl carbanions from nitrobenzyl derivatives on photolysis, *Can. J. Chem.*, 64, 1949, 1986.

174. Wan, P. and Muralidharan, S., Structure and mechanism in the photo-retro-aldol type reactions of nitrobenzyl derivatives. Photochemical heterolytic cleavage of C-C bonds, *J. Am. Chem. Soc.*, 110, 4336, 1988; see also: Wan, P. and Muralidharan, S., *J. Am. Chem. Soc.*, 112, 4611, 1990, for error correction.

175. (a) Margerum, J. D., Transient photodecarboxylation intermediates, *J. Am. Chem. Soc.*, 87, 3772, 1965; (b) Margerum, J. D. and Petrusis, C. T., Photodecarboxylation of nitrophenylacetate ions, *J. Am. Chem. Soc.*, 91, 2467, 1969.

176. Craig, B. B. and Atherton, S. J., Kinetic and spectral properties of the photogenerated *p*-nitrobenzyl carbanion in aqueous media, *J. Chem. Soc., Perkin Trans. 2*, 1929, 1988.

177. Craig, B. B., Weiss, R. G., and Atherton, S. J., Picosecond and nanosecond laser photolyses of *p*-nitrophenylacetate in aqueous media. A photoadiabatic decarboxylation process?, *J. Phys. Chem.*, 91, 5906, 1987.

178. Muralidharan, S. and Wan, P., ESR studies of photogenerated nitrobenzyl carbanions in aqueous solution, *J. Photochem. Photobiol. A: Chem.*, 57, 191, 1991.

179. Zimmerman, H. E. and Somasekhara, S., Mechanistic organic photochemistry. III. Excited state solvolyses, *J. Am. Chem. Soc.*, 85, 922, 1963.

180. Muralidharan, S., Beveridge, K. A., and Wan, P., Evidence for the trapping of a non-Kekulé intermediate in the photo-retro-aldol type reaction of *m*-nitrobenzyl derivatives, *J. Chem. Soc., Chem. Commun.*, 1426, 1989.

181. Ciamician, G. and Silber, P., Chemische Lichtwirkungen. 2. Mittheilung, *Ber. Dtsch. Chem. Ges.*, 34, 2040, 1901.

182. Sachs, F. and Hilpert, S., Chemische Lichtwirkungen (Vorläufige Mittheilung), *Ber. Dtsch. Chem. Ges.*, 37, 3425, 1904.

183. Nurmukhametov, R. N. and Sergeev, A. M., Photochemistry of *o*-nitrobenzyl compounds, *Zh. Fiz. Khim.*, 64, 308, 1990.

184. Wettermark, G., Black, E., Dogliotti, L., Reactions of photochemically formed transients from 2-nitrotoluene, *Photochem. Photobiol.*, 4, 229, 1965.

185. See, for example: (a) Wettermark, G. and Ricci, R., General acid catalysis of the fading of photoisomerized 2,4-dinitrotoluene, *J. Chem. Phys.*, 39, 1218, 1963; (b) Wettermark, G., Photochromism of *o*-nitrotoluenes, *Nature*, 677, 1962; (c) Wettermark, G., A flash photolysis study of 2-(2,4-dinitrobenzyl)pyridine in water, *J. Am. Chem. Soc.*, 84, 3658, 1962; (d) Klemm, E., Klemm, D., Kleinschmidt, J., and Graness, A., Untersuchungen zum Substituenteneinfluß auf die Lebensdauer der Polymethin-Farbform photochromer Dinitrobenzylpyridine, *Z. Phys. Chem. (Leipzig)*, 262, 621, 1981, and both earlier and subsequent papers; (e) Dessauer, R. and Paris, J. P., Photochromism, *Adv. Photochem.*, 1, 275, 1963 and references cited therein.

186. (a) Strom, E. T. and Weinstein, J., Free radicals from the irradiation of *o*-nitrobenzyl compounds, *J. Org. Chem.*, 22, 3705, 1967; (b) Filby, W. G. and Günther, K., Nitrogen centered free radicals in the solution phase photolysis of *o*-nitrobenzaldehyde. Their nature and origin, *Z. Naturforsch.*, 28b, 810, 1973.

187. Suryanarayaman, K. and Capellos, C., Flash photolysis of 2,4,6-trinitrotoluene solutions, *Int. J. Chem. Kinet.*, 6, 89, 1974.

188. Burlinson, N. E., Sitzman, M. E., Kaplan, L. A., and Kayser, E., Photochemical generation of the 2,4,6-trinitrobenzyl anion, *J. Org. Chem.*, 44, 3695, 1979.

189. (a) Yip, R. W., Sharma, D. K., Giasson, R., and Gravel, D., Photochemistry of the *o*-nitrobenzyl system in solution: Evidence for singlet state intramolecular hydrogen abstraction, *J. Phys. Chem.*, 89, 5328, 1985; (b) Yip, R. W., Wen, Y. X., Gravel, D., Giasson, R., and Sharma, D. K., Photochemistry of the *o*-nitrobenzyl system in solution: Identification of the biradical intermediate in the intramolecular rearrangement, *J. Phys. Chem.*, 95, 6078, 1991.

190. Schupp, H., Wong, W. K., and Schnabel, W., Mechanistic studies of the photorearrangement of *o*-nitrobenzyl esters, *J. Photochem.*, 36, 85, 1987.

191. Zhu, Q. Q., Schnabel, W., and Schupp, H., Formation and decay of nitronic acid in the photorearrangement of *o*-nitrobenzyl esters, *J. Photochem.*, 39, 317, 1987.

192. Schneider, S., Fink, M., Bug, R., and Schupp, H., Investigation of the photorearrangement of *o*-nitrobenzylesters by time-resolved resonance Raman spectroscopy, *J. Photochem. Photobiol. A: Chem.*, 55, 329, 1991.

193. Craig, B. B. and Atherton, S. J., Laser photolysis of nitro aromatics, *Proc. SPIE-Int. Soc. Opt. Eng.*, 482, 96, 1984.

194. Gravel, D., Giasson, R., Blanchet, D., Yip, R. W., and Sharma, D. K., Photochemistry of the *o*-nitrobenzyl system in solution: Effects of O....H distance and geometrical constraint on the hydrogen transfer mechanism in the excited state, *Can. J. Chem.*, 69, 1193, 1991.

195. de Mayo, P. and Reid, S. T., Photochemical rearrangements and related transformations, *Quart. Rev.*, 15, 393, 1961.

196. George, M. V. and Scaiano, J. C., Photochemistry of *o*-nitrobenzaldehyde and related studies, *J. Phys. Chem.*, 84, 492, 1980.

197. Yip, R. W. and Sharma, D. K., The reactive state in the photorearrangement of *o*-nitrobenzaldehyde, *Res. Chem. Intermed.*, 11, 109, 1989.

198. Hadjoudis, E. and Hayon, E., Solid state photochemical rearrangement of *o*-nitrobenzylideneaniline, *J. Phys. Chem.*, 74, 2224, 1970.

199. Ried, W. and Wilk, M., Zur Photochemie ungesättigter Nitroverbindungen. I. Über den Einfluß der Substitution auf die Lichtumlagerung von *o*-Nitroazomethinen, *Liebigs Ann. Chem.*, 590, 91, 1954.

200. Maki, Y., Furuta, T., and Suzuki, M., Photochemical reactions. 20. Photolysis of *o*-nitrobenzaldehyde N-acylhydrazones leading to benzyne and 5-nitrophthalazines, *J. Chem. Soc., Perkin Trans. 1*, 553, 1979, and earlier papers.

201. Bakke, J., The Photocyclization of 2-(*o*-nitrophenyl)-ethanol to N-hydroxyoxindole, *Acta Chem. Scand.*, 24, 2650, 1970.

202. Kitaura, Y. and Matsuura, T., Photoinduced reactions. XLVI. Photochemistry of hindered nitrobenzene derivatives, *Tetrahedron*, 27, 1583, 1971.

203. Döpp, D., Photochemically induced reaction between a *t*-butyl side chain and an *ortho*-nitro-group, *Chem. Commun.*, 1248, 1968.

204. Barclay, L. R. C. and McMaster, I. T., Photolysis of 2,4,6-tri-*t*-butylnitrobenzene, *Can. J. Chem.*, 49, 676, 1971.

205. Döpp, D., Lichtinduzierter intramolekularer Ringschluß von *o*-nitro-*tert*-butylbenzol, *Chem. Ber.*, 104, 1035, 1971.

206. Döpp, D., Über den Mechanismus der lichtinduzierten Bildung von 1-Hydroxy-3,3-dimethyl-3*H*-indolonen-(2) aus *o*-Nitro-*tert*-butylbenzolen, *Chem. Ber.*, 104, 1043, 1971.

207. Döpp, D., Photochemie aromatischer Nitroverbindungen. V. Die Produkte der Photolyse von kristallinem 2-Nitro-1,4-di-*tert*-butylbenzol, *Tetrahedron Lett.*, 2757, 1971.

208. Döpp, D. and Sailer, K.-H., Photochemie aromatischer Nitroverbindungen, VI. Photolyse von kristallinem 2-Nitro-1,3,5-tri-*tert*-butylbenzol, *Tetrahedron Lett.*, 2761, 1971.

209. Döpp, D. and Brugger, E., Photochemie aromatischer Nitroverbindungen. VII. Photolyse von β-deuterierten 1-*tert*-Butyl-2-nitrobenzolen, *Chem. Ber.*, 106, 2166, 1973.

210. Döpp, D. and Sailer, K.-H., Photochemie aromatischer Nitroverbindungen. IX. Photolyseprodukte des 1,3,5-Tri-*tert*-butyl-2-nitrobenzols, *Chem. Ber.*, 108, 301, 1975.

211. (a) Döpp, D. and Sailer, K.-H., Photochemie aromatischer Nitroverbindungen, X. Photolyse von Keton-Moschus, Xylol-Moschus und Tibeten-Moschus, *Tetrahedron Lett.*, 1129, 1975, (b) Döpp, D. and Sailer, K.-H., Photochemie aromatischer Nitroverbindungen. XI. Vergilbung von kristallinem Keton-Moschus, Xylol-Moschus und Tibeten-Moschus, *Chem. Ber.*, 108, 3483, 1975.

212. Döpp, D. and Brugger, E., Photochemie aromatischer Nitroverbindungen. XII. Substituenteneinflüsse auf die Photocyclisierung von 1-*tert*-Butyl-2-nitrobenzolen, *Liebigs Ann. Chem.*, 554, 1979.

213. Döpp, D. and Brugger, E., Photochemie aromatischer Nitroverbindungen. XIV. Triplettsen-sibilisierung und -löschung bei der Photocyclisierung von 1,4-Di-*tert*-butyl-2-nitrobenzol, *Liebigs Ann. Chem.*, 1965, 1979.

214. Padmanabhan, K., Döpp, D., Venkatesan, K., and Ramamurthy, V., Solid state photochemistry of nitro compounds: Structure-reactivity correlations, *J. Chem. Soc., Perkin Trans. 2*, 897, 1986.

215. Padmanabhan, K., Venkatesan, K., Ramamurthy, V., Schmidt, R., and Döpp, D., Structure-reactivity correlation of photochemical reactions in organic crystals: Intramolecular hydrogen abstraction in an aromatic nitro compound, *J. Chem. Soc., Perkin Trans. 2*, 1153, 1987.

216. Döpp, D. and Schmidt, R., unpublished results; Schmidt, R., Ph.D. thesis, University Duisburg, 1989.

217. Döpp, D., Arfsten-Romberg, U., Bolz, W., van Hoof, W., and Kosfeld, H., Photo-Vergilbung von Ambrette-Moschus (4-*tert*-Butyl-3-methoxy-2,6-dinitrotoluol) in methanolischer Alkalilauge, *Chem. Ber.*, 112, 3946, 1979.

218. Frederiksen, J., Larsen, B. D., and Harrit, N., Photolysis of *N*-2,4,6-trinitrophenyl-substituted amino acids, *Tetrahedron Lett.*, 32, 5823, 1991.

219. Oguchi, S., Torizuka, H., Photocyclization of *o*-Nitrophenyl Alkyl Ethers, *Bull. Chem. Soc. Jpn.*, 53, 2425, 1980.

220. Jacob, E. D. and Joshua, C. P., Photolysis of 2-nitrophenyl benzyl ethers in neutral and acidic media, *Ind. J. Chem., Sect. B*, 23 B, 808, 1984.

221. Obara, H., Onodera, J., and Hattori, M., The formation of 2-acylamino-4-*t*-butyl-6-nitrophenols by the photoirradiation of 1-alkoxy-4-*t*-butyl-2,6-dinitro-benzenes, *Chem. Lett.*, 421, 1982.

222. Goudie, R. S. and Preston, P. N., Photolysis of compounds containing an *o*-nitroarylthio-substitu-ent, *J. Chem. Soc. (C)*, 3081, 1971.

223. Barton, D. H. R., Chow, Y. L., Cox, A., and Kirby, G. W., Photochemical transformations. XIX. Photosensitive protecting groups, *J. Chem. Soc.*, 3571, 1965.

224. Barton, D. H. R., Nakano, T., and Sammes, P. G., Photochemical transformations. XXII. Reactions of 2,4-dinitrobenzenesulfenyl derivatives, *J. Chem. Soc. (C)*, 322, 1968.

225. Cornelisse, J. and Havinga, E., Photosubstitution reaction of aromatic compounds, *Chem. Rev.*, 75, 353, 1975, and earlier reviews.

226. Havinga, E. and Cornelisse, J., Aromatic photosubstitution reactions, *Pure Appl. Chem.*, 47, 1, 1976.

227. Cornelisse, J., Lodder, G., and Havinga, E., Pathways and intermediates of nucleophilic aromatic photosubstitutions, *Rev. Chem. Intermed.*, 2, 1979, 231.

228. Parkanyi, C., Aromatic photosubstitutions, *Pure Appl. Chem.*, 55, 331, 1983.

229. de Jongh, R. O. and Havinga, E., Photoreactions of aromatic compounds. VI. The mechanism of the photohydrolysis of *m*-nitroanisole, *Rec. Trav. Chim.*, 85, 275, 1966.

230. Kronenberg, M. E., van der Heyden, A., and Havinga, E., Photoreactions of aromatic compounds. XIII. Photosubstitution of 4-nitroveratrole with methylamine; a Convenient Synthesis of *N*-methyl-4-nitro-*o*-anisidine, *Rec. Trav. Chim.*, 86, 254, 1967.

231. Letsinger, R. L., McCain, J. H., Photoinduced substitution. III. Replacement of aromatic hydrogen by cyanide, *J. Am. Chem. Soc.*, 88, 2884, 1966.

232. Lammers, J. G. and Lugtenberg, J., Photoreactions of aromatic compounds. XXXII. Fluorine as a leaving group in nucleophilic photosubstitution reactions of some nitronaphthalene derivatives, *Tetrahedron Lett.*, 1777, 1973.

233. de Gunst, G. P. and Havinga, E., Photoreactions of aromatic compounds. XXIX. Pathway and intermediates of the photoreactions of 3,5-dinitroanisole with nucleophiles, *Tetrahedron*, 29, 2167, 1973.

234. Fleming, I., Frontier orbitals and organic chemical reactions, John Wiley & Sons, New York, 1967.

235. Epiotis, N. D. and Shaik, S., Qualitative potential energy surfaces. 4. Aromatic substitution, *J. Am. Chem. Soc.*, 100, 29, 1978.

236. van Riel, H. C. H. A., Lodder, G., and Havinga, E., Photochemical methoxide exchange in some nitromethoxybenzenes. Role of the nitro group in S_N2Ar reactions, *J. Am. Chem. Soc.*, 103, 7257, 1981.

237. Mutai, K., Nakagaki, R., and Tukada, H., A rationalization of orientation in nucleophilic aromatic photosubstitution, *Bull. Chem. Soc. Jpn.*, 58, 2066, 1985.

238. van Eijk, A. M. J., Huizer, A. H., and Varma, C. A. G. O., Radical anion and Meisenheimer complexes arising from interaction of the hydroxyl ion or aliphatic amines with photoexcited 4-nitroveratrole studied by electron spin resonance and time-resolved resonance Raman scattering, *J. Photochem. Photobiol., A: Chem.*, 56, 183, 1991.

239. (a) Cantos, A., Marquet, J., Moreno-Mañas, M., The regioselectivity of nucleophilic aromatic photosubstitution. The photoreaction of 4-nitroveratrole with *n*-hexylamine, *Tetrahedron Lett.*, 28, 4191, 1987; (b) Cantos, A., Marquet, J., Moreno-Mañas, M., and Castello, A., On the regioselectivity of the nucleophilic aromatic photosubstitutions of 4-nitroveratrole. A threefold mechanistic pathway, *Tetrahedron*, 44, 2607, 1988.

240. (a) Kuzmič, P., Pavličková, L., Velek, J., and Souček, M., Photoreactions of 3,4-dimethoxy-1-nitrobenzene. A pH-dependence of regioselectivity, *Coll. Czech. Chem. Commun.*, 51, 1665, 1986; (b) Van Eijk, A. M. J., Huizer, A. H., Varma, C. A. G. O., and Marquet, J., Dynamic behavior of photoexcited solutions of 4-nitroveratrole containing OH⁻ or amines, *J. Am. Chem. Soc.*, 111, 88, 1989.

241. Cantos, A., Marquet, J., Moreno-Mañas, M., Gonzalez-Lafont, A., Lluch, J. M., and Bertran, J., On the regioselectivity of 4-nitroanisole photosubstitution with primary amines. A mechanistic and theoretical study, *J. Org. Chem.*, 55, 3303, 1990.

242. Bunce, N. J., Cater, S. R., Scaiano, J. C., and Johnston, L. J., Photosubstitution of 1-methoxy-4-nitronaphthalene with amine nucleophiles: Dual pathways, *J. Org. Chem.*, 52, 4214, 1987.

243. Marquet, J., Moreno-Mañas, M., Vallribera, A., Virgili, A., Bertran, J., Gonzalez-Lafont, A., and Lluch, J. M., The nucleophilic aromatic photosubstitutions of 4,5-dinitroveratrole with amines. Photoreductions of aromatic dinitrocompounds, *Tetrahedron*, 43, 351, 1987.

244. Berci Filho, P., Neumann, M. G., and Quina, F. H., The photocyanation of 2-nitrothiophene and the mechanism of nucleophilic aromatic photosubstitution, *J. Chem. Res. Synop.*, 70, 1991.

245. Martins, L. J. A., Electron transfer reactions of the 2-nitrothiophen triplet state by laser flash photolysis, *J. Chem. Soc., Faraday Trans. 1*, 78, 533, 1982.

246. Wubbels, G. G., Halverson, A. M., Oxman, J. D., and De Bruyn, V. H., Regioselectivity of photochemical and thermal Smiles rearrangements and related reactions of β-(nitrophenoxy)ethylamines, *J. Org. Chem.,* 50, 4499, 1985.

247. Mutai, K. and Nakagaki, R., Further evidence for the regioselectivity rules in nucleophilic aromatic photosubstitution, *Bull. Chem. Soc. Jpn.,* 58, 3663, 1985.

248. Yokoyama, K., Nakagaki, R., Nakamura, J., Mutai, K., and Nagakura, S., Spectroscopic and kinetic study of an intramolecular aromatic nucleophilic substitution. Reaction mechanism of a photo-Smiles rearrangement, *Bull. Chem. Soc. Jpn.,* 53, 2472, 1980.

249. Wubbels, G. G., Halverson, A. M., and Oxman, J. D., Thermal and photochemical Smiles rearrangements of β-(nitrophenoxy)ethylamines, *J. Am. Chem. Soc.,* 102, 4849, 1980.

250. (a) Wubbels, G. G. and Celander, D. W., Base catalysis in a photochemical Smiles rearrangement. A case of general base catalysis of a photoreaction, *J. Am. Chem. Soc.,* 103, 7669, 1981; (b) Wubbels, G. G. and Sevetson, B. R., General base catalysis of photo-Smiles rearrangement of 1-(4-nitrophenoxy)-2-anilinoethane in aqueous solution. Identity of the proton donating intermediate, *J. Phys. Org. Chem.,* 2, 177, 1989.

251. Wubbels, G. G., Sevetson, B. R., and Sanders, H., Competitive catalysis and quenching by amines of photo-Smiles rearrangement as evidence for a zwitterionic triplet as the proton-donating intermediate, *J. Am. Chem. Soc.,* 111, 1018, 1989.

252. Wubbels, G. G. and Cotter, W. D., α-Cyclodextrin complexation as a probe of heterolytic general base-catalyzed photo-Smiles rearrangements, *Tetrahedron Lett.,* 30, 6477, 1989.

253. Mutai, K. and Kobayashi, K., Photoinduced intramolecular aromatic nucleophilic substitution (the photo Smiles rearrangement) in amino ethers, *Bull. Chem. Soc. Jpn.,* 54, 462, 1981.

254. Mutai, K., Nakagaki, R., and Tukada, H., Magnetic field effect as a probe for radical pair intermediates in the photoreactions of nitroaromatic ethers, *Chem. Lett.,* 2261, 1987.

255. Patchornik, A., Amit, B., and Woodward, R. B., Photosensitive protecting groups, *J. Am. Chem. Soc.,* 92, 6333, 1970.

256. Binkley, R. W. and Flechtner, T. W., Photoremovable protecting groups, in *Synthetic Organic Photochemistry,* Horspool W. M., Ed., Plenum, New York, 1984, 375, and pertinent references cited therein.

257. Cummings, R. T. and Krafft, G. A., Photoactivable fluorophores. 1. Synthesis and photoactivation of *o*-nitrobenzyl-quenched fluorescent carbamates, *Tetrahedron Lett.,* 29, 65, 1988.

258. Baldwin, J. E., McConnaughie, A. W., Moloney, M. G., Pratt, A. J., and Shim, S. B., New photolabile phosphate protecting groups, *Tetrahedron,* 46, 6879, 1990.

259. Kusumi, A., Nakahama, S., and Yamaguchi, K., Liposomes that can be disintegrated by photo-irradiation, *Chem. Lett.,* 433, 1989.

260. Ajayaghosh, A. and Pillai, V. N. R., 2'-Nitrobenzhydryl polystyrene resin: A new photolabile polymeric support for carboxyl protection in amino acids, *Natl. Acad. Sci. Lett. (India),* 149, 1988.

261. Gravel, D., Murray, S., and Ladouceur, G., *o*-Nitrobenzyl alcohol, a simple and efficient reagent for the photoreversible protection of aldehydes and ketones, *J. Chem. Soc., Chem. Commun.,* 1828, 1985.

262. Hebert, J. and Gravel, D., *o*-Nitrophenylethylene glycol. Photosensitive protecting group for aldehydes and ketones, *Can. J. Chem.,* 52, 187, 1971.

263. Peyser, J. R. and Flechtner, T. W., *N*-(α-Hydroxy-2-nitrosobenzyl)-1-naphthamide: A photochemical intermediate, *J. Org. Chem.,* 52, 4645, 1987.

264. Barzynski, H. and Sänger, D., Zur Photolyse von makromolekularen *o*-Nitrobenzylderivaten, *Angew. Makromol. Chem.,* 93, 131, 1981.

265. Reichmanis, E., Wilkins, C. W., and Chandross, E. A., A novel approach to *o*-nitrobenzyl photochemistry for resists, *Vac. Sci. Technol.,* 19, 1338, 1981.

266. Reichmanis, E., Gooden, R., Wilkins, C. W., Jr., and Schonhorn, H., A study of the photochemical response of *o*-nitrobenzyl cholate derivatives in P(MMA-MAA) matrices, *J. Polym. Sci., Polym. Chem. Ed.,* 21, 1075, 1983.

267. Hayase, S., Onishi, Y., and Yasunobu, H., Deep-UV photoresist containing *ortho*-Nitrobenzylsilyl ether structure in the main chain, *J. Electrochem. Soc.*, 134, 2275, 1987.

268. Yamaoka, T., Watanabe, H., Kosaki, K., and Asano, T., Photochemical behavior of 2,6-dimethyl-3,5-dicarboxyalkyl-4(2'-nitrophenyl)-1,4-dihydropyridines and application to a positive resist, *J. Imaging Sci.*, 34, 50, 1990.

269. Reichmanis, E., Smith, B. C., and Gooden, R., *o*-Nitrobenzyl Photochemistry: Solution vs. solid state behavior, *Polym. Sci., Polym. Chem. Ed.*, 23, 1, 1985.

270. Houlihan, F. M., Shugard, A., Gooden, R., and Reichmanis, E., The photochemistry of nitrobenzyl tosylates in polymer matrices, *Polym. Prepr. (Am. Chem. Soc. Div. Polym. Chem.)*, 29, 543, 1988.

271. Amit, B. and Patchornik, A., Photorearrangement of *N*-substituted *o*-nitroanilides and nitroveratramides. Potential photosensitive protecting group, *Tetrahedron Lett.*, 2205, 1973.

272. MacDonald, S. A. and Willson, C. G., *Poly(N-alkylnitroamides), Polymer Materials for Electronic Applications*, Feit, E. D., Wilkins, and C. W., Jr., Eds., ACS Symp. Ser., 184, ACS, Washington D. C., 1982, 73.

273. Cervello, J., Figueredo, M., Marquet, J., Moreno-Mañas, M., Bertran, J., and Lluch, J. M., Nitrophenyl ethers as possible photoaffinity labels. The nucleophilic aromatic photosubstitution revisited, *Tetrahedron Lett.*, 25, 4147, 1984.

274. Figueredo, M., Marquet, J., Moreno-Mañas, M., and Pleixats, R., The photoreactions of 2-fluoro-4-nitroanisole with amines. The search for new biochemical photoprobes, *Tetrahedron Lett.*, 30, 2427, 1989.

275. Weller, J. W. and Hamilton, G. A., The photooxidation of alkanes by nitrobenzene, *J. Chem. Soc., Chem. Commun.*, 1390, 1970.

276. Libman, J., Nitroaromatic compounds for selective photochemical hydroxylation, *J. Chem. Soc., Chem. Commun.*, 868, 1977.

277. Libman, J. and Bermann, E., Photoexcited nitrobenzene for benzylic hydroxylation: The synthesis of 17β-acetoxy-9α-hydroxy-3-methoxyestra-1,3,5(10)-triene, *Tetrahedron Lett.*, 2191, 1977.

278. Scholl, P. C. and van de Mark, M. R., Remote oxidation with photoexcited nitrobenzene derivatives, *J. Org. Chem.*, 38, 2367, 1973.

279. Maki, Y., Suzuki, M., Hosokami, T., and Furuta, T., Photochemistry of *N*-acyl-2-nitrodiphenylamines. Novel photochemical synthesis of phenazine *N*-oxides, *J. Chem. Soc., Perkin Trans. 1*, 1354, 1974.

280. (a) Fasani, E., Pietra, S., and Albini, A., The photocyclization of *N*-acyl-2-nitrodiphenylamines to phenazine *N*-oxides: Scope and mechanism, *Heterocycles*, 33, 573, 1992; (b) Fasani, E., Mella, M., and Albini, A., Photochemical reactions of 2,4-dinitrodiphenylacetamide and related compounds: Spectroscopic and chemical identification of intermediates, *J. Chem. Soc., Perkin Trans. 1*, 2689, 1992.

281. Christudhas, M. and Joshua, C. P., Photochemistry of 2,2'-dinitrodiphenylmethanes. II. Irradiation of 5,5'-dimethyl-2,2'-dinitrodiphenylmethanes in neutral, acidic and alkaline media, *Aust. J. Chem.*, 35, 2377, 1982, and earlier papers.

282. Joshua, C. P. and Christudhas, M., Photochemistry of 2,2'-Dinitrodiphenylmethanes. 3. Irradiations of 5,3'-dimethyl-2,2'-dinitrodiphenylmethanes in neutral, acidic and alkaline media, *Ind. J. Chem. Sect. B*, 22 B, 432, 1983.

283. Christudhas, M., Jacob, E. D., and Joshua, C. P., Photolysis of 2-nitrodiphenylmethanes in isopropanol and acidified aqueous ethanol, *Ind. J. Chem. Sect B.*, 23 B, 815, 1984.

284. Mathew, T. and Joshua, C. P., Photolysis of benzoyl esters of 2,2'-dinitrodiphenylcarbinols in neutral and basic media, *J. Photochem. Photobiol., A: Chem.*, 60, 319, 1991.

Photorearrangement of Nitrogen-Containing Arenes

Alain Lablache-Combier
Université des Sciences et
Techniques de Lille

82.1 Introduction

This chapter deals only with the ring rearrangement reactions of nitrogen-containing aromatic compounds, and excludes aromatic derivatives containing a sulfur atom. The reactions of this latter group will be described in another chapter of this book.

Photoisomerization, especially of aromatic compounds, is a very general process.[1] In some cases, thermodynamically unstable ground-state intermediates have been isolated, trapped, or detected by spectroscopic methods as shown schematically in Scheme 82.1a. In other cases, indirect proof of the existence of ground-state intermediate was obtained. No intermediate may be formed in the ground state but rationalization of the structure of the photoproducts formed can be obtained from a "speculative" intermediate. This process is illustrated in Scheme 82.1b for an adiabatic process (very rare) or in Scheme 82.1c for a diabatic process.[2]

0-8493-8634-9/95/$0.00+$.50
© 1995 by CRC Press, Inc.

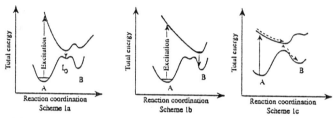

SCHEME 82.1 (a) Cross section of the hypothetical hypersurface illustrating the thermal (nonobserved) rearrangement of A via I to B. There is an intermediate I_0 in the photoisomerization of A to B. (b) Cross section of the excited state leading to the reaction. There is no intermediate in the photoisomerization of A to B. (c) There is no intermediate neither to the thermal or the photochemical isomerization.

In cases where the aromatic ring is substituted by electron-withdrawing groups, such as cyano or perfluoroalkyl, intermediates have been identified most often. Some intermediates have been characterized nevertheless for unsubstituted aromatic compounds. Most of the work dealing with aromatic compounds photorearrangements has been carried out in the period 1964–1982. There are obviously two types of N-containing aromatic compounds, the five-membered system (pyrroles, pyrazoles, imidazoles, oxazoles, isoxazoles, 1, 2, 4 oxadiazoles for the mono cyclic compounds) and the six-membered system (pyridine, pyrazine, pyrimidine, pyridazine for the monocyclic compounds). The photoreactions of the five-membered rings have been reviewed previously.[3-8] Fewer reviews have been written on the photochemistry of six-membered azaaromatic compounds.[3-4,9-12] There is a good analogy between the mechanism of the photoisomerization of these compounds with that for the rearrangement of benzene and its derivatives, an area that has been widely studied and reviewed.[9,11,13,14]

82.2 Five-Membered Azaaromatic Compounds

In the introduction of a review article, Padwa[8] rationalized the different mechanisms by which five-membered azaaromatic compounds photoisomerize. There are five.

 1. The Ring Contraction-Ring Expansion Route
It is the weakest single bond that cleaves. It is by such a mechanism that compounds such as isoxazoles photoisomerize.

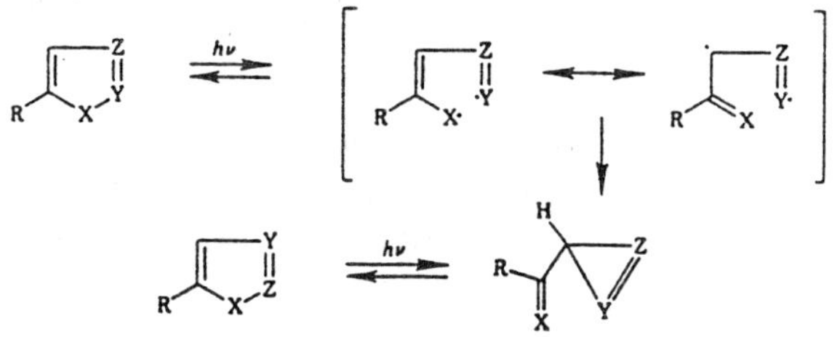

SCHEME 82.2

 2. The Internal Cyclization-Isomerization Route
This mechanism involves an initial disrotatory formation of a bicyclic isomer, followed by a [1,3]-sigmatropic shift to a second bicyclic isomer, which then undergoes a subsequent disrotatory ring opening to the rearranged product.

 These transformations are allowed as concerted photochemical processes in systems having appropriate symmetry according to the Woodward-Hoffmann rules.[15] The [1,3]-sigmatropic shift may terminate at either position-3 (Z atom) or -4 (C atom) to give either bicyclic isomer 1 or 2. Termination at both positions is possible to give rearranged heterocycles 3 and/or 4. The observed regioselectivity can be attributed to the preferred formation of the more stable bicyclic isomer. It is by such a mechanism that some pyrroles and imidazoles photoisomerize.

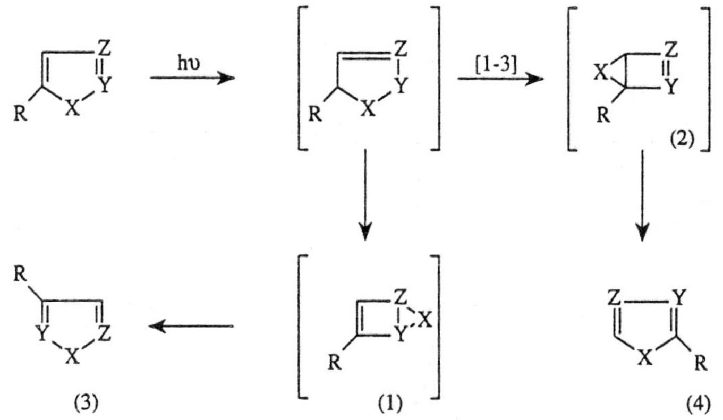

SCHEME 82.3

3. The Van Tamelen-Whitesides General Mechanism

This mechanism is a combination of the two previous mechanisms. It was postulated to explain data of some furans[16] or thiophenes (see Chapter 65 on thioarenes rearrangement). It will not be invoked in the reactions described in this chapter.

4. The Zwitterion-Tricycle Route

It has been proposed to rationalize the dates of some thiophenes and thiazoles. It will be discussed in the chapter dealing with thioarene rearrangements.

5. The Fragmentation-Readdition Route

1,2,5-Oxadiazoles react by this mechanism. Fragmentation compounds are more often obtained than readdition products.

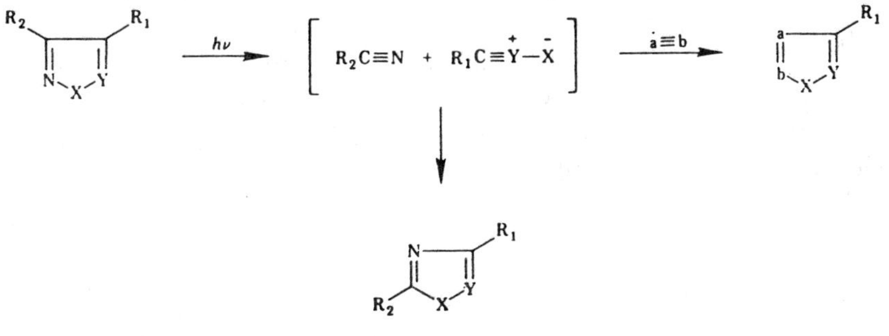

SCHEME 82.4

Pyrroles

Pyrrole, 2,4- and 2,5-dimethylpyrrole have been irradiated in the vapor phase and only degradation products were found.[17,18] In the condensed phase, pyrroles isomerize in several ways and the reactions most frequently observed lead to 1,2- and 1,3-isomerization. Thus, *N*-benzylpyrrole on irradiation in methanol as well as in the vapor phase gives a mixture of 2-benzyl- and 3-benzylpyrrole.[19] Optically active (+)-*N*-(1-phenylethyl)pyrrole undergoes a similar rearrangement to the 2- and 3-substituted derivatives in 12.5 and 2.7% yield, respectively, with 54% retention of configuration when irradiated as a neat liquid. In methanol solution, the conversion was greater but only 32% retention of configuration was found for the two isomers. In both cases, the optical activity of the starting material was virtually unchanged.[19]

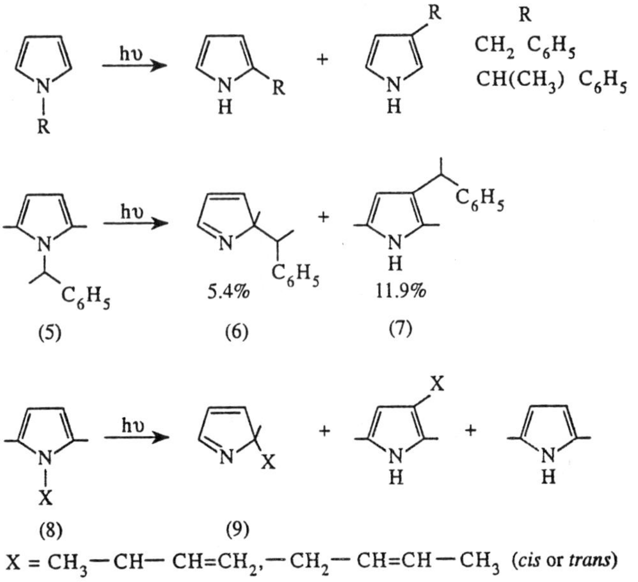

SCHEME 82.5

If both C-2 and C-5 are blocked as in compound 5, a mixture of the 2*H*-pyrrole 6 and compound 7 is formed. Again in this case, recovered substrate has lost no optical activity.[19] These results were taken as evidence against a free radical mechanism and it has been found that *N*-allylpyrroles rearrange photochemically by a concerted mechanism. It turns out that no 1-substituted isomers can be isolated.[20] The compound 9 does not photoisomerize further. The irradiation of the *cis*-crotyl substrate leads only to the formation of *cis*-crotyl and 1-methyllallyl products, and similarly the *trans*-substrate leads only to the *trans*-product. Oxygen has no effect on the product ratio. *N*-Allylpyrrole itself reacts in a similar way, leading to a mixture of 2- and 3-allylpyrrole. Other allyl migrations are shown in Scheme 82.6.[21]

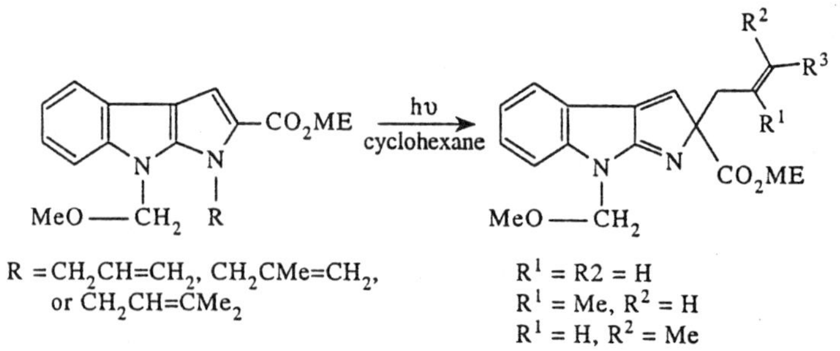

SCHEME 82.6

Contrary to the preceding examples, it appears that irradiation of some *N*-substituted 2,4-diphenylpyrroles leads to products via a free radical mechanism, a fact corroborated by the nonreactivity of *N*-cyclohexyl- and *N*-(2-phenylethyl)-2,4-diphenylpyrrole. *N*-*t*-butyl-2,4-diphenylpyrrole photocleaves under identical conditions.[22]

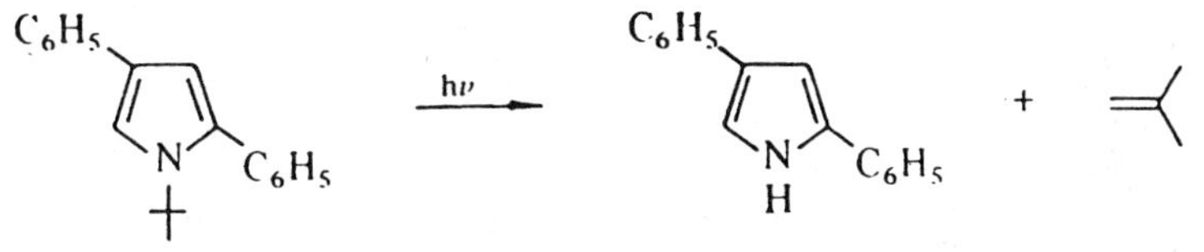

SCHEME 82.7

The photoisomerization of *N*-acetylpyrroles to a mixture of 2-acetyl- and 3-acetylpyrrole[23] is believed to be concerted[24] and not like the usual photo-Fries reaction that occurs by a free radical mechanism.[25] A study of the product distribution as a function of time indicates that both the 2- and 3-isomers are primary photoproducts in the reaction. If the 2-position is blocked by a methyl group, some of the 3-isomer arises and the 2*H*-2-acetyl compound formed initially then thermally gives the 3-isomer.[23] The isomer **10** is the major product of irradiation of *bis*(2-acetylpyrrol-1-yl)-methane in dioxane.[26]

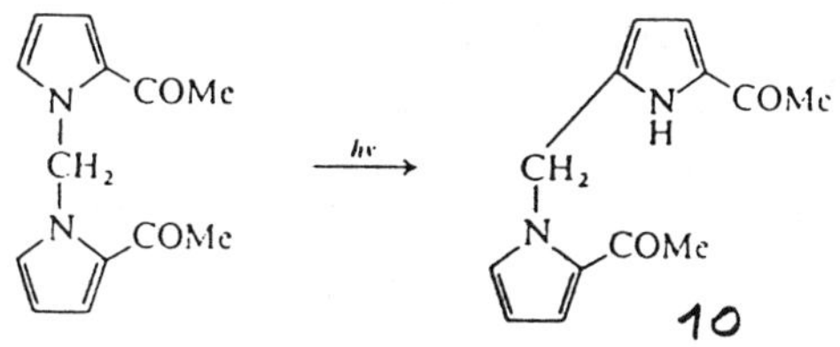

SCHEME 82.8

A photochemically induced 1,3-acyl shift is implicated in the rearrangement of 1-acylindoles to 3-acylindolenines.[27,28] Spontaneous ring expansion of these photoisomers is observed when R = $CH_2CH_2NH_2$, leading to the lactams **11** and thus providing a general synthetic route to *Strychnos* and *Aspidosperma* alkaloids.

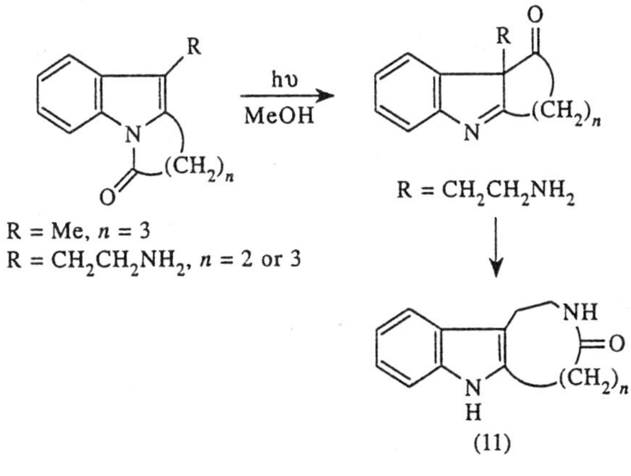

SCHEME 82.9

When irradiated in methanol, 2-cyanopyrrole photoisomerizes to 3-cyanopyrrole (55%).[29] Pyrrole 2-carboxaldehyde is also formed in 5% yield in this reaction. 3-Cyanopyrrole is photostable. The photoisomerization of 3-, 4-, and 5-methyl-2-cyanopyrrole in acetonitrile can be rationalized by the following scheme.[30]

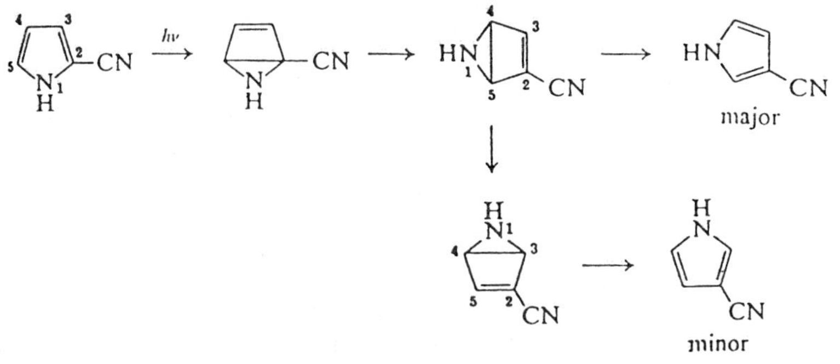

SCHEME 82.10

The suggestion that 2,5-bonding in the excited (π,π^*) singlet state initiates the transposition of 2-cyanopyrroles is supported by a correlation diagram[31,32] and has additional experimental support from the trapping of azabicyclopentene intermediates with methanol and furan.[33]

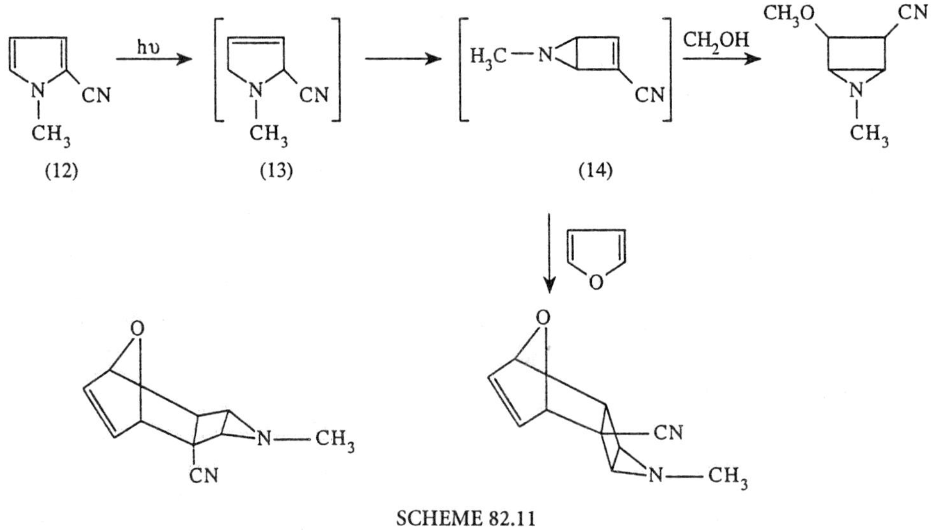

SCHEME 82.11

The formation of azabicyclopentene 14 involves apparently a thermally activated step since irradiation of 12 at –68°C yielded no detectable rearrangement product. If, as seems probable, the 2,5-bonded species 13 is the precursor of 14, the temperature effect suggests that the thermal reversion of 13 to 12 has a lower activation energy than the isomerization to 14. Further evidence regarding this point was obtained by studying the thermolysis of the furan photoadduct 15 derived from 2-cyano-5-methyl-pyrrole; cyanopyrroles 16, 17, 18, and 19 were obtained in a ratio of 4:6:2:1. These results have been explained by assuming that the thermal retrogression of 15 yields 20, which in turn gives access to the three isomeric azabicyclopentenes 21, 22, and 23 that can be derived from 20 by one or more nitrogen "walk" steps. As the temperature is lowered, "walking" of the aziridine ring becomes progressively less favorable relative to aromatization. At 30°C, the formation of a photoproduct requiring more than a single "walk" step is rare; and when such scrambling is observed (i.e., 16–18), it is a relatively minor pathway.[33]

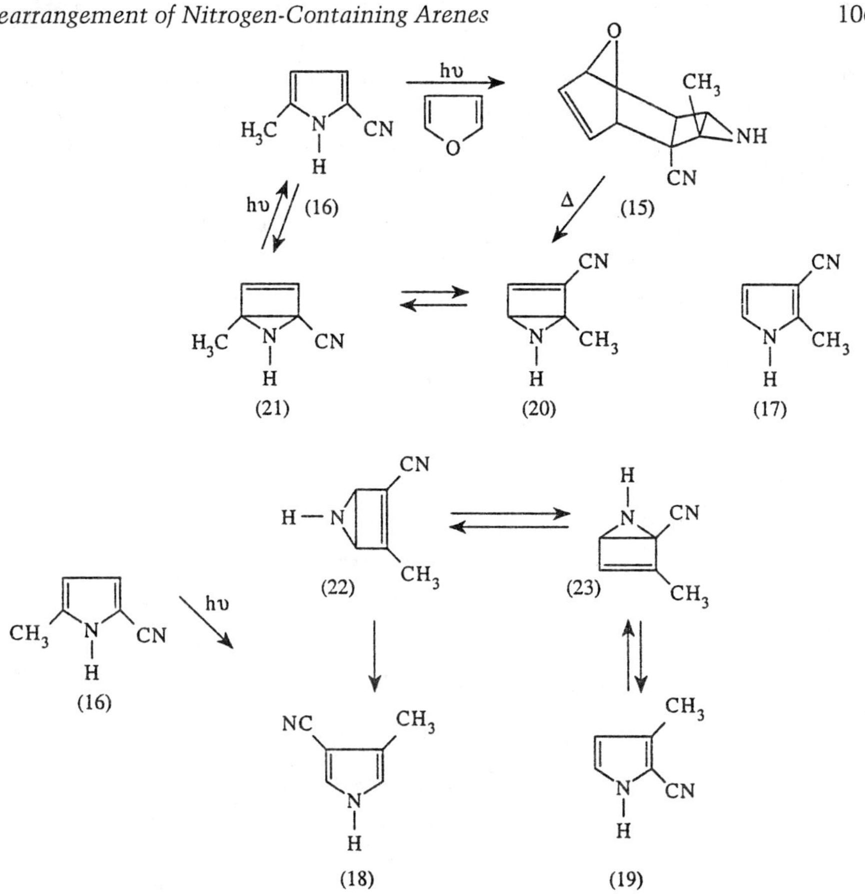

SCHEME 82.12

Although neither *N*- nor 3-(trimethylsilyl)pyrroles undergoes photorearrangement reactions, irradiation of the 2-isomer in degassed pentane gives an 84% yield of the 3-isomer.[34] The process is interpreted reasonably by the path used to explain the rearrangement of 2-cyanopyrrole. The possible role of "Dewar" pyrroles or cyclopropenylimines as intermediates in the photorearrangement of tetrakis(trifluoromethyl)pyrroles has also been investigated,[35] but the valence-bond isomer 25 does seem to be involved as an intermediate in the conversion of the *N*-phenylpyrrole 24 into the cyclobutindole 26.

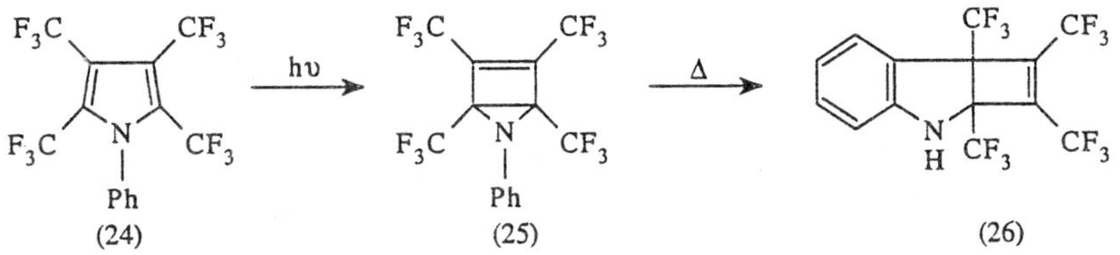

SCHEME 82.13

No isomerization involving a change in the structure of the aromatic nucleus of indoles has been reported.

Pyrazoles

Pyrazoles are photoisomerized to imidazoles. Schmid proposed that these photoreactions proceed by a ring-expanson/ring-contraction mechanism involving an intermediate such as 27.[36] However,

Labhart, by CNDO calculation, showed that the Zwitterionic intermediate **28** was preferred in the case of the pyrazole.[37] Another path must be involved in the photorearrangement of 1,3,5-trimethylpyrazole to 1,2,4-trimethylimidazole. Here, an internal cyclization isomerization route is proposed.[38,39]

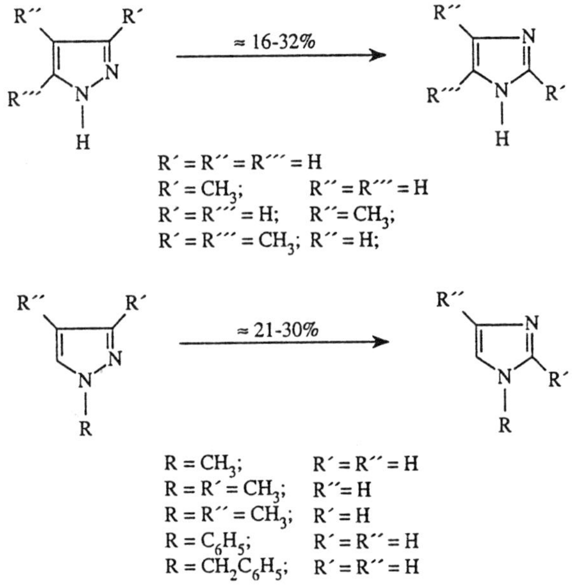

$$R' = R'' = R''' = H$$
$$R' = CH_3; \qquad R'' = R''' = H$$
$$R' = R''' = H; \quad R'' = CH_3;$$
$$R' = R''' = CH_3; \ R'' = H;$$

$$R = CH_3; \qquad R' = R'' = H$$
$$R = R' = CH_3; \quad R'' = H$$
$$R = R'' = CH_3; \ R' = H$$
$$R = C_6H_5; \qquad R' = R'' = H$$
$$R = CH_2C_6H_5; \ R' = R'' = H$$

SCHEME 82.14

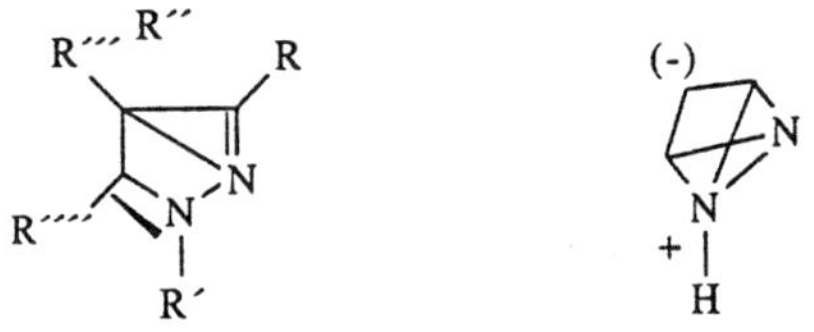

(27)   (28)

SCHEME 82.15

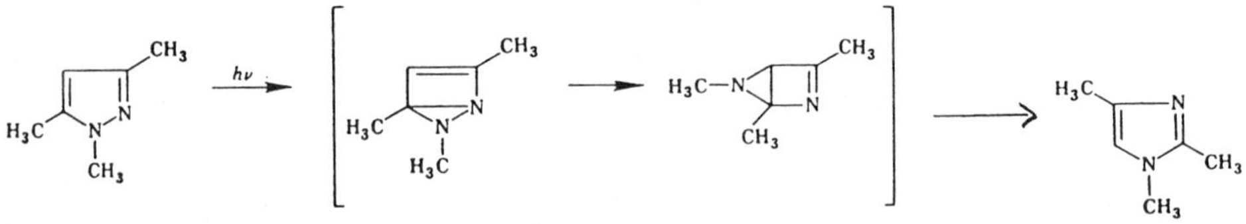

SCHEME 82.16

The isolation of 3-anilino-2,3-diphenylacrylonitrile as a product from the irradiation of 1,4,5-triphenylpyrazole[40] allows one to speculate on the existence of a third route for the pyrazole-imidazole rearrangement. Aminoacrylonitriles such as **31** give imidazoles **32** on irradiation, probably via four-membered ring intermediates.[41,42] The possibility exists, therefore, that pyrazoles with an unsubstituted 3-position may rearrange by the sequence pyrazole aminoacrylonitrile imidazole. Some of the rearrangements of pyrazoles to imidazoles that have been described[36] could occur via an aminoacrylonitrile; and in the cases of pyrazole and 3-methylpyrazole, it has been shown[42–44] that the appropriate aminoacrylonitrile does undergo photoconversion to the imidazole. Thus, photorearrangement of 5-methylpyrazole gives 2-methylimidazole as well as 4-methylimidazole; the latter may arise via the aminocrotonitrile 31 (R = CH$_3$). It should be noted, however, that aminoacrylonitrile 29 did not afford imidazole 30 under either sensitized or unsensitized conditions.[40] This could possibly be because compound 29 exists predominantly in the form with a *trans*-arrangement of the nitrile and amino groups; molecular models indicate considerable strain in the *cis*-form.

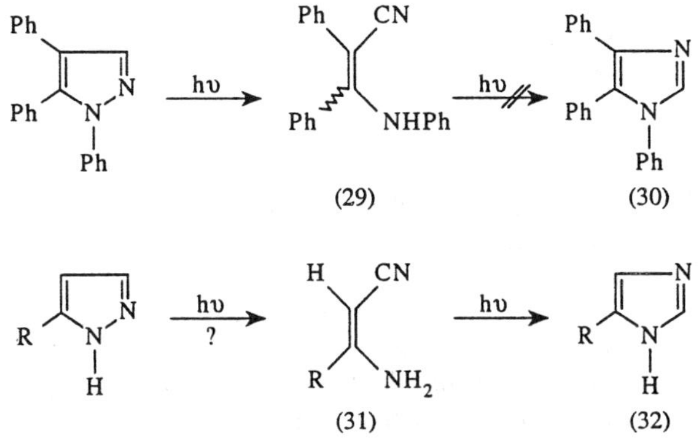

SCHEME 82.17

The photochemical conversion of diaminomaleonitrile into 4-amino-5-cyanoimidazole 33 supports the previous hypothesis. Such paths are of interest to chemists concerned with prebiotic mechanisms, and Kagan[45] has tested the hypothesis that 4-amino-3-cyanopyrazole 34 is an intermediate by irradiation at 300 nm. This converts 34 into 33 in low yield, but all attempts to detect 34 in the reaction of diaminomaleonitrile were unsuccessful. A pyrazolopyrimidine is also formed on photolysis of 34, presumably as a result of HCN formation and subsequent addition to ground-state 34.

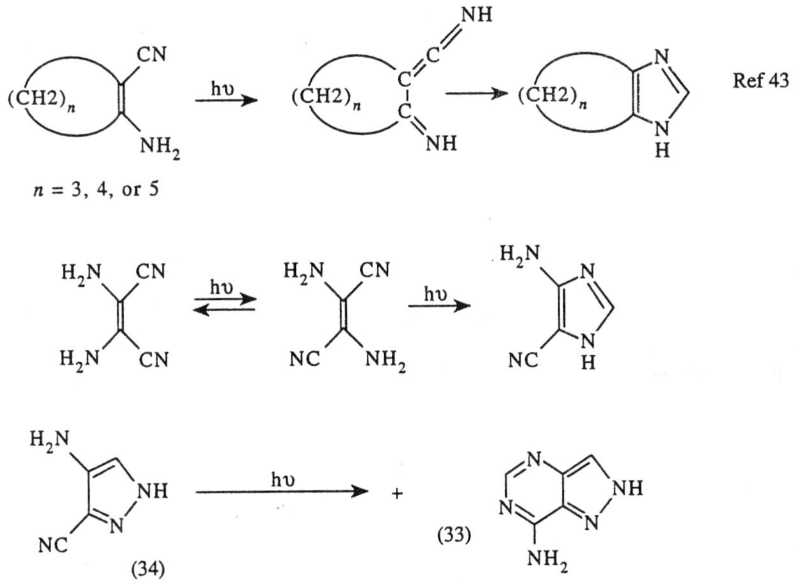

SCHEME 82.18

To explain the photoisomerization of *N*-methyl-3-cyanopyrazoles, it has been suggested that the reaction follows two different pathways:[46] (A) 1,5-interchange, proposed to arise by 2,5-bonding, to give a diazabicyclopentene that then isomerizes by a nitrogen "walk" and aromatization; and (b)

2,3-interchange probably via an azirine as intermediate.[46] In contrast to this, however, 1,5-dimethyl-3-trifluoromethylpyrazole undergoes the phototransposition process exclusively via the former pathway.

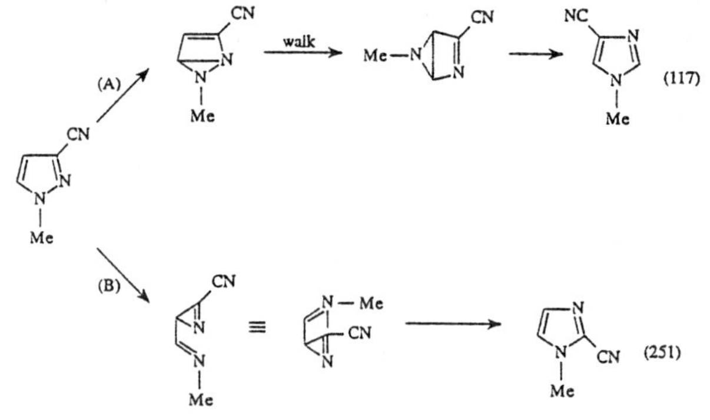

SCHEME 82.19

Another reaction path is suggested in which both the enaminonitrile and imidazole are formed.[47] This is exemplified by the photoisomerization of pyrazole, 1-methylpyrazole and 4-cyano-1-methylpyrazole to form 3-aminoacrylonitrile, 3-(methylamino)acrylonitrile, and 2-cyano-3-(methylamino)acrylonitrile, respectively, together with the corresponding imidazole.

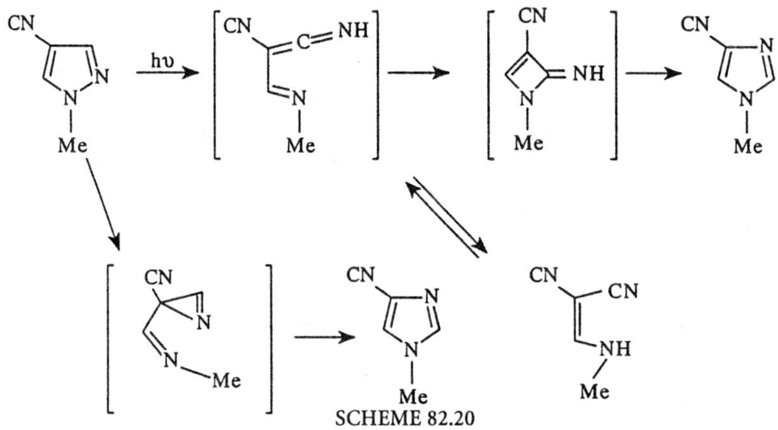

SCHEME 82.20

Indazoles

Indazole and 3-substituted indazoles photoisomerize to benzimidazole[48] by paths similar to that for the conversion of pyrazole into imidazole. Interestingly when R = phenyl, no reactions occurs.

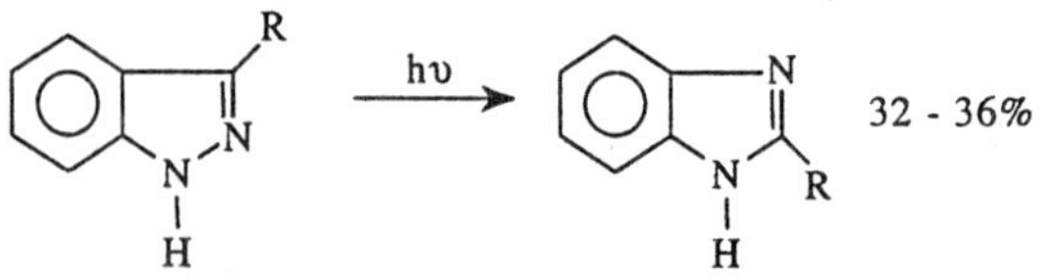

R = CH₃ or C₂H₅

$R = CH_3$ or C_2H_5

SCHEME 82.21

On photosensitization, 1-substituted indazoles lead to open-chain compounds.[36]

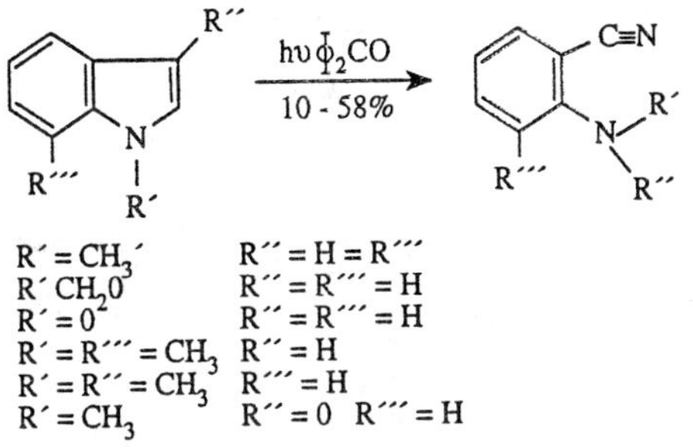
R″
hυ ϕ2CO
10 - 58%
C≡N
N
R‴
R′
R‴
N
R′
R″
R′ = CH3′
R′ CH2O
R′ = O
R′ = R‴ = CH3
R′ = R″ = CH3
R′ = CH3
R″ = H = R‴
R″ = R‴ = H
R″ = R‴ = H
R″ = H
R‴ = H
R″ = O R‴ = H

SCHEME 82.22

A mixture of both these paths is observed with some indazoles with substituents on the six-membered ring when irradiation affords benzimidazoles principally and also an open-chain nitrile.

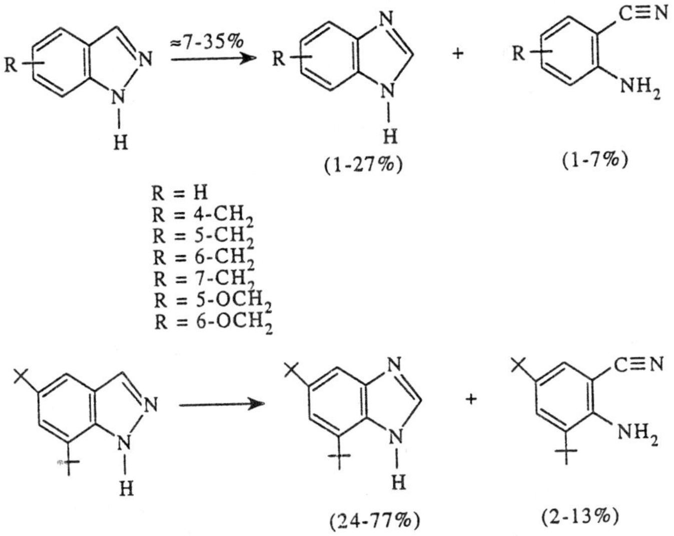
≈7-35%
R
N
N
H
R
N
N
H
+
R
C≡N
NH2
(1-27%)
(1-7%)
R = H
R = 4-CH2
R = 5-CH2
R = 6-CH2
R = 7-CH2
R = 5-OCH2
R = 6-OCH2
N
N
H
N
N
H
+
C≡N
NH2
(24-77%)
(2-13%)

SCHEME 82.23

2-*t*-Butyl-4-methylindazole and 1,3-dimethylindazole do not photoisomerize at low temperature. For these two compounds in the absence of photochemical reactions from S_1, the sum ϕ_F + ϕ_T equals unity within the limit of error.[49] Interestingly, some 2-alkylindazoles are converted to 1-alkylbenzimidazoles upon irradiation. This is a singlet-state reaction. It was proposed that the reaction proceeds via the internal cyclization-isomerization route[37,38,40,50] and involves intermediates such as **35**. Such tricyclic intermediates are formed nearly quantitatively when the reaction is performed at –60°C.[52,53]

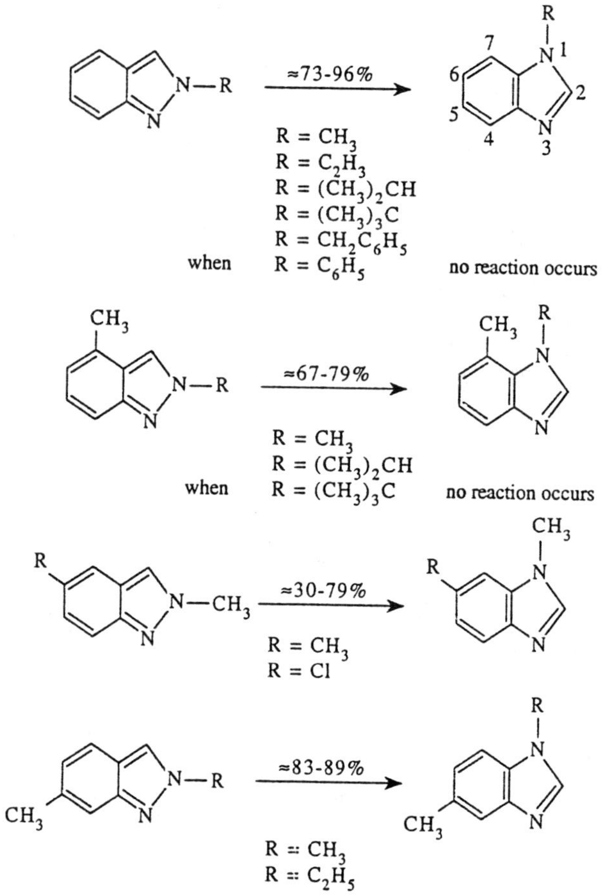

SCHEME 82.24

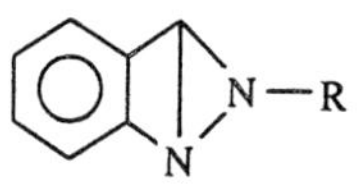

(35)

SCHEME 82.25

The photorearrangement of 1-methylindazole to *N*-methylanthranilonitrile, and of 2-methylindazole to 1-methylbenzimidazole, must proceed from the singlet excited states. Their triplet states absorbs at 420 nm (E_T 279 kJ mol^{-1}) and at 405 nm (E_T 281 kJ mol^{-1}), respectively, and the first-order rate constant for decay is approximately 10^4 s^{-1}. These states are quenched by oxygen or piperylene, but such treatment does not affect the photoisomerization. UV irradiation at $-160°C$ of these compounds permits the detection of radicals that may be formed by cleavage of the N-N bond.[51]

In dilute sulphuric acid (pH 2–4), photorearrangement to 1-alkylbenzimidazoles is suppressed and the dihydroazepinones **36** and **37** are formed; whereas in strongly acid solution, the *o*-aminoacetophenones are obtained.[54] The relative quantum yields for the formation of photoproducts as a function of pH were measured, and the results suggest that benzimidazoles and **36** are products of the photorearrangement of neutral indazole, whereas **37** and aminobenzophenone arise from the protonated indazole.

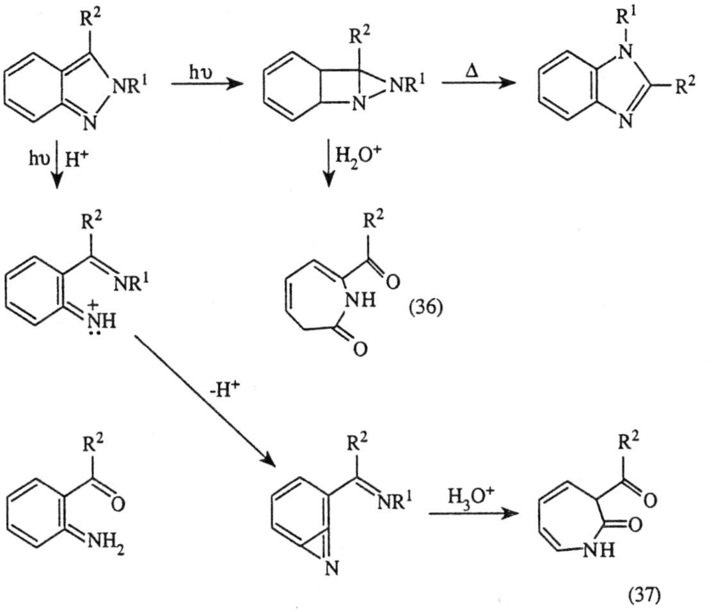

SCHEME 82.26

In aqueous ethanol, below 120K, 2-aminoacylbenzenes are formed by irradiation of indazoles and involve the intermediacy of a nitrene and an imine.[55]

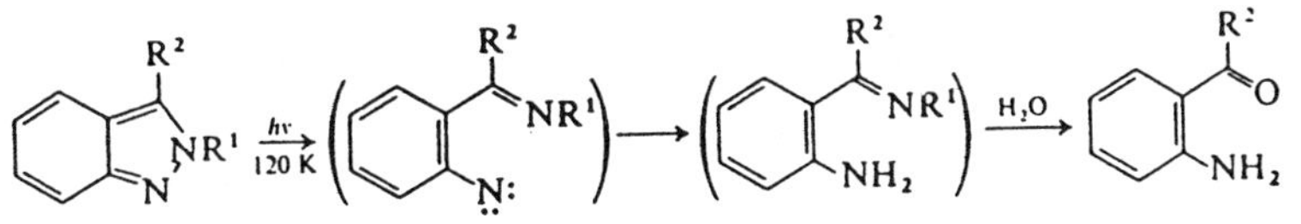

SCHEME 82.27

Imidazoles

Irradiation of a number of methyl-substituted imidazoles has been reported to result in the scrambling of the substituent groups.[38,39] These photoreactions involve a formal interchange of the 1,5- or 2,4-carbons of the imidazole ring. They proceed probably by the formation of a Dewar-type intermediate rather than by a ring-contraction ring-expansive sequence. This latter path allows interchange only of the 2,3- or 4,5-position.

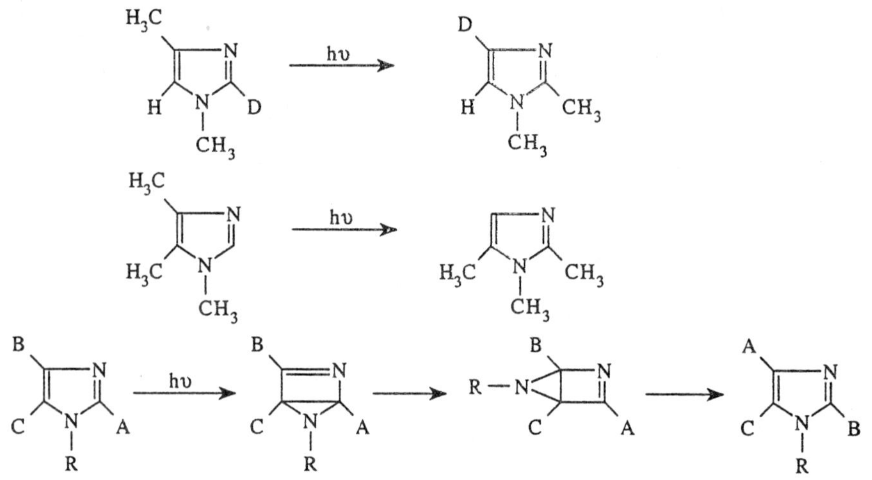

SCHEME 82.28

Isoxazoles

Many examples of the photorearrangement of isoxazoles have been reported in the literature.[56–62] The first step in the rearrangement is a ring-contraction ring-expansion route leading to azirine intermediates that have been characterized in many cases. These are converted photochemically either to the starting isoxazole or to an oxazole. The nature of the excited states involved in these reactions have been determined mainly in the case of 3,5-diphenylisoxazole.[55,56] The photoreaction is wavelength dependent.

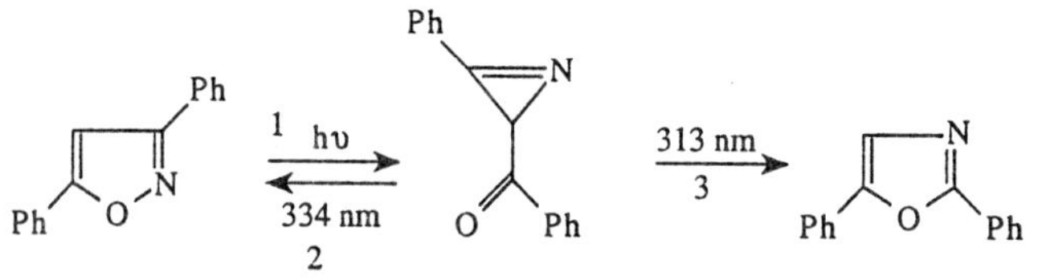

SCHEME 82.29

The mechanism of the photoisomerization of isoxazole can be summarized as follows.[62] Reaction (1): Isoxazole in the lowest $^1(n,\pi^*)$ state interconverts to azirine intermediate in the ground state through the N2-C3-C4-C5 torsional deformation, which leads to O1-N2 bond scission and N2-C4 bond formation simultaneously. Reaction (2): The S_1 state of the azirine intermediate is an (n,π^*) state of the carbonyl chromophore. The deformation of the N2-C3-C4 bond angle in the T_1 state produced through ISC from the S_1 state causes the N2-C4 bond rupture. Subsequently, odd electrons on the O1 and N2 atoms combine to form the O1-N2 bond, leading to the formation of isoxazole. This reaction does not involve a nitrene intermediate. Reaction (3): The azirine intermediate in the S_2 state, which is an (n,π^*) state of the ketimine chromophore, causes the C3-C4 bond rupture by the deformation of the C3-N2-C4 bond angle. It is possible to follow the conrotatory motion in such a ring-opening process. The transformation to oxazole proceeds via ISC from the S_2 to T_1 states. The doubly and triply excited configurations will exercise small modifications on these reactions. This reaction does not involve any intermediates. The possible reaction path of photoisomerization reaction of isoxazoles to oxazoles through azirine intermediates can be shown pictorially in Figure 82.1.

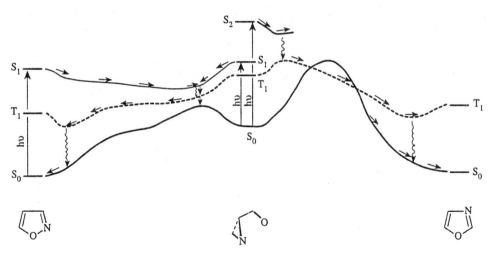

FIGURE 82.1 The possible mechanism of the photoisomerization of isoxazole to oxazole.[62]

The fact that it is the S_1 excited state of isoxazoles that is involved in the phototransformation is supported by the failure of 38 to photoisomerize.[59] In these compounds, as in benzophenone, ϕ_{ic} is nearly unity. Similar wavelength-dependent photoreactions of isoxazole azirine pairs were reported in the cases of 3-p-methoxyphenyl-5-phenyl-,[57] 3-phenyl-5-α-naphthyl-,[58] 3-methyl-5-phenyl-,[58] and 3-phenyl-5-methylisoxazoles.[58] 3-Acetyl-2-methyl-1-azirine is formed when 3,5-dimethylisoxazole is irradiated.[63]

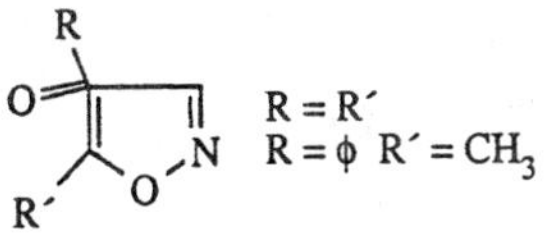

R = R'
R = ϕ R' = CH_3

(38)

SCHEME 82.30

The isocyanide 40 has been established as an intermediate in the photorearrangement at –77°C of isoxazole 39 to oxazole 43.[64,65] Spectroscopic evidence for the azirine 41 as a precursor to 42 has been reported, and the azirine itself is believed to be formed.

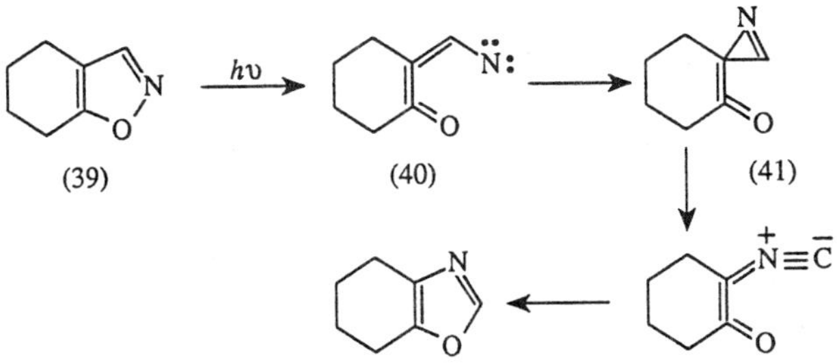

(39) $h\upsilon$ (40)   (41)

(43)   (42)

SCHEME 82.31

No spectroscopic evidence for an isonitrile has been detected in the rearrangement of the corresponding 3-methylated compound, although on irradiation it affords the corresponding oxazole.[64,65] Another example of isoxazole-oxazole photorearrangement is reported[66] in which the acylazirine intermediate 44 can be isolated by choosing an appropriate exciting wavelength. The azirine gives the oxazole on 254-nm irradiation, but with longer wavelengths it forms an acylketenimine.

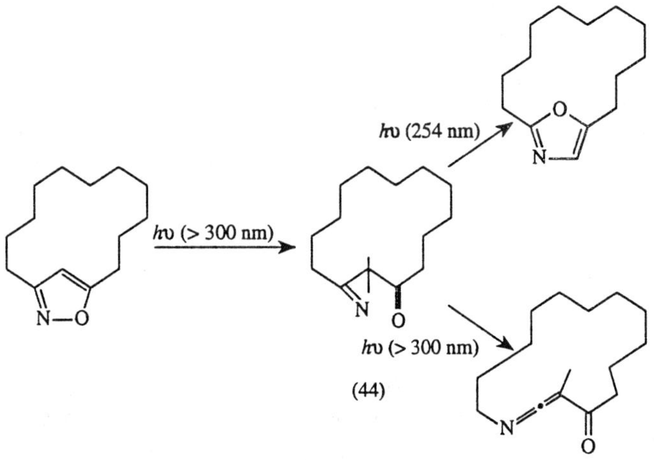

hυ (> 300 nm)

hυ (254 nm)

(44)

hυ (> 300 nm)

SCHEME 82.32

The bicyclic isoxazoles 45 have been described as a synthetic route to bridged oxazoles and imidazoles.[67]

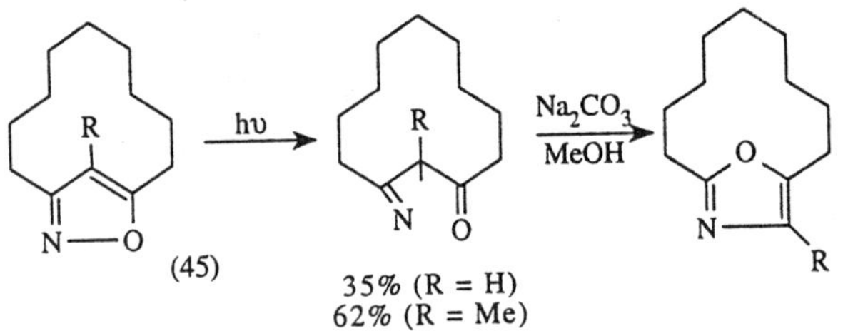

hυ

Na₂CO₃ / MeOH

$$Na_2CO_3 / MeOH$$

(45)

35% (R = H)
62% (R = Me)

SCHEME 82.33

Irradiation of 3-aryl-5-amino-isoxazoles also affords azirines.[68,69] In the transformation of 46 to 47, better yields are obtained when the reaction is performed thermally.[70]

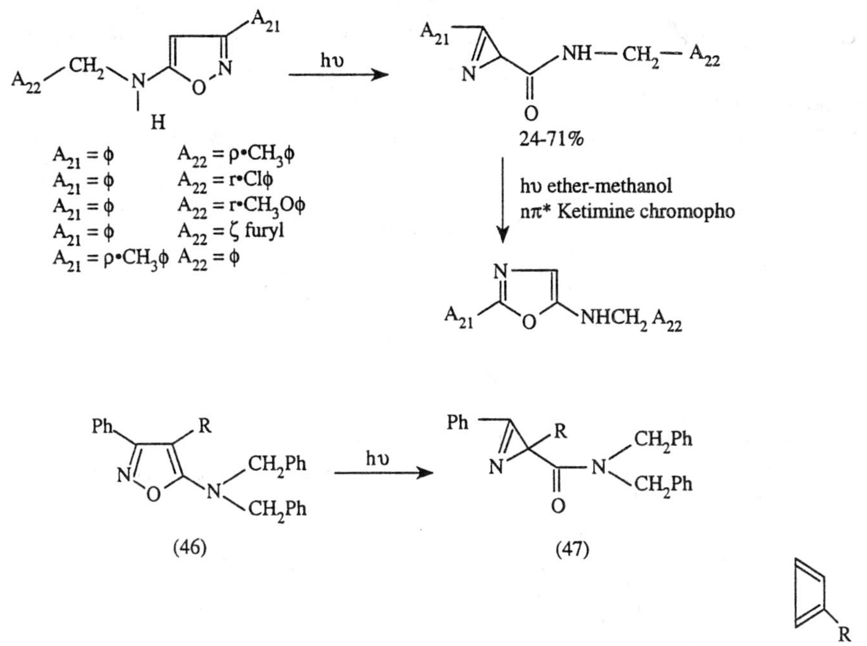

$A_{21} = \phi$ $A_{22} = p \cdot CH_3 \phi$
$A_{21} = \phi$ $A_{22} = r \cdot Cl \phi$
$A_{21} = \phi$ $A_{22} = r \cdot CH_3 O \phi$
$A_{21} = \phi$ $A_{22} = \zeta$ furyl
$A_{21} = p \cdot CH_3 \phi$ $A_{22} = \phi$

24-71%

$h\upsilon$ ether-methanol
$n\pi^*$ Ketimine chromopho

(46)   (47)

SCHEME 82.34

3,4-Dimethyl-5-aminoisoxazole irradiated in diethylether leads to an azirine and a ketinimine.[71]

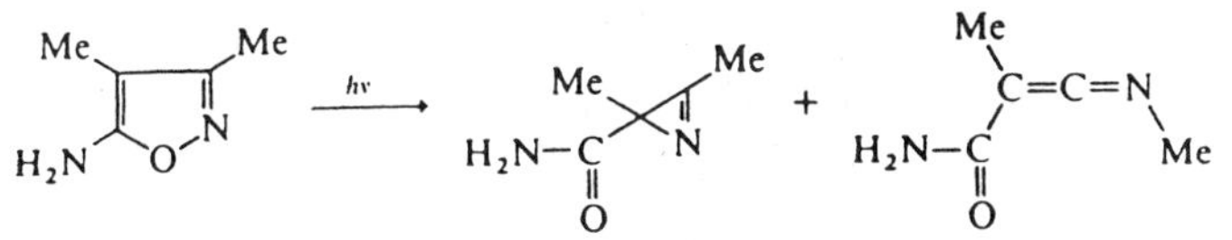

SCHEME 82.35

Products **48 to 50** were obtained on irradiation of the isoxazol-5-ylhydrazines (**51**).[72] These products were also formed in the corresponding thermal reaction and, in both cases, are believed to arise via intermediate 2*H*-azirine-2-carbohydrazides. An alternative pathway involving the ketenimine **52** has been proposed to account for the formation of the pyrazolone **53**.

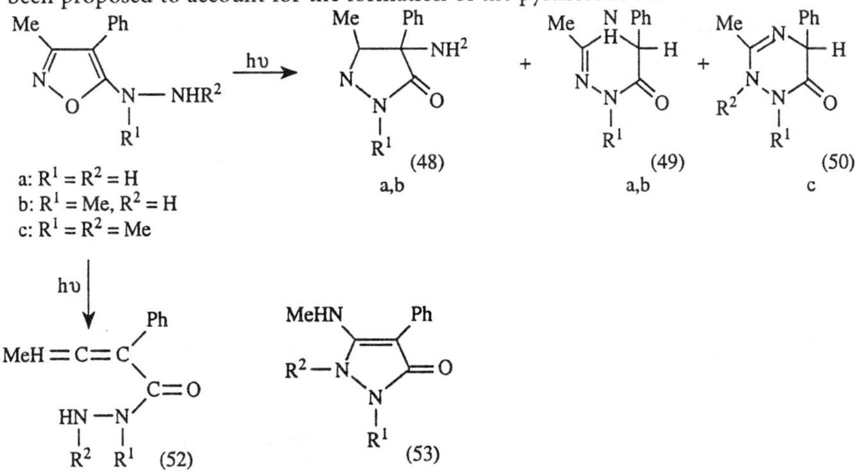

a: $R^1 = R^2 = H$
b: $R^1 = Me, R^2 = H$
c: $R^1 = R^2 = Me$

(48) a,b   (49) a,b   (50) c

(52)   (53)

SCHEME 82.36

In the following case, where the oxazole formed is thermally unstable, an amide is isolated.[68] The same type of reaction occurs when the aryl substituent is *p*-tolyl and the corresponding azirines can be isolated (30–36%).

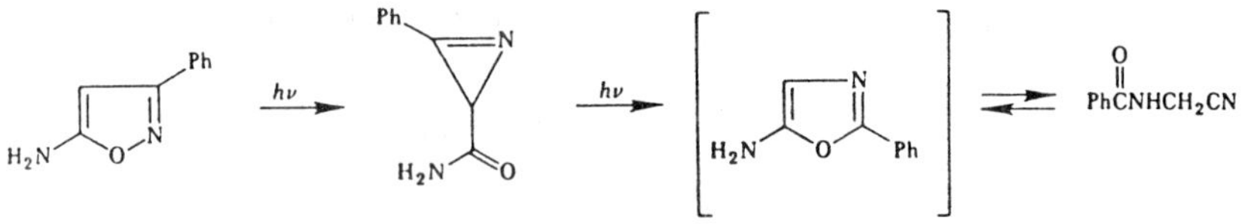

SCHEME 82.37

The mechanism of the thermal isomerization of the amino oxazole into benzoylaminoacetonitrile has been described[73] and involves the intermediacy of a mesoionic 5-imino-3*H*-oxazole.

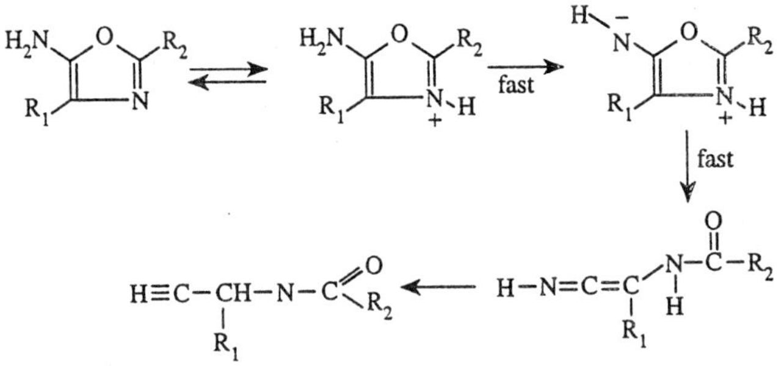

SCHEME 82.38

2-Amino-3-ethoxycarbonyl-3-phenylisoxazole isomerizes to the corresponding 2-phenyloxazole, while 5-amino-4-ethoxycarbonylisoxazole undergoes a ring cleavage. In both cases, the azirine is the postulated intermediate, but it has not been isolated.[74]

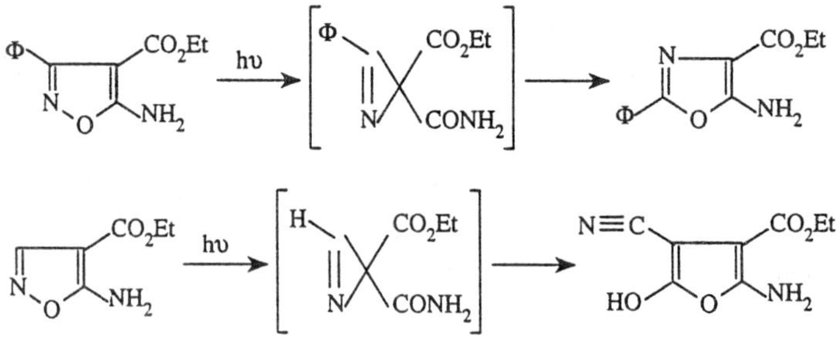

SCHEME 82.39

Open-chain compounds have also been obtained from 3,4,5-triphenyl isoxazole.[75]

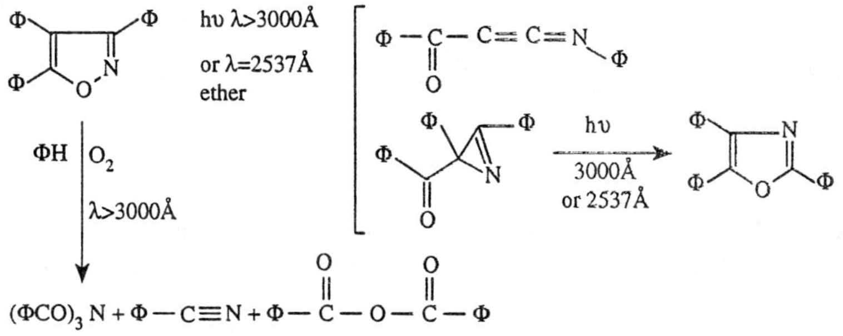

SCHEME 82.40

Ketoketinimines such as **54** and azirines have been trapped by methanol in the following cases.[76] When **55a** was irradiated in ethanol, the corresponding ethoxy derivative was formed, but the diethoxy-amide could not be isolated. 2,5-Dimethyloxazole leads to the same products, but more slowly than 3,5-dimethylisoxazole. 2,5-Diphenylisoxazole cannot be converted to open-chain compounds in methanol.[76]

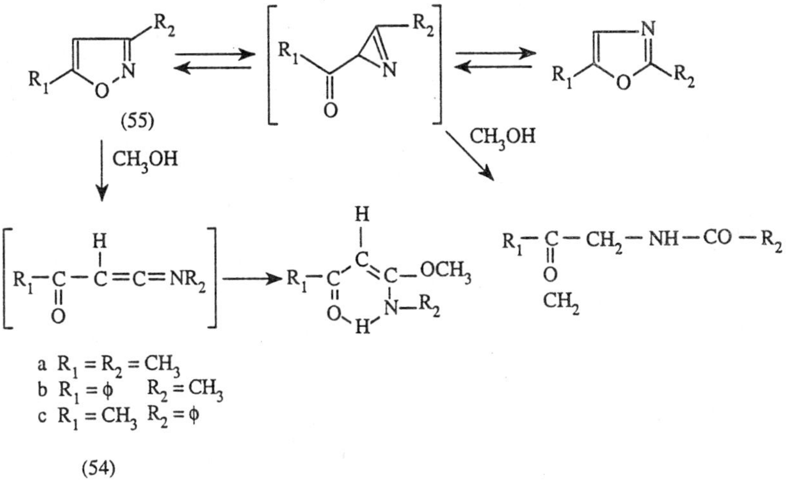

$$a \quad R_1 = R_2 = CH_3$$
$$b \quad R_1 = \phi \quad R_2 = CH_3$$
$$c \quad R_1 = CH_3 \quad R_2 = \phi$$

(54)

SCHEME 82.41

In some cases, the azirines are not isolated.[77]

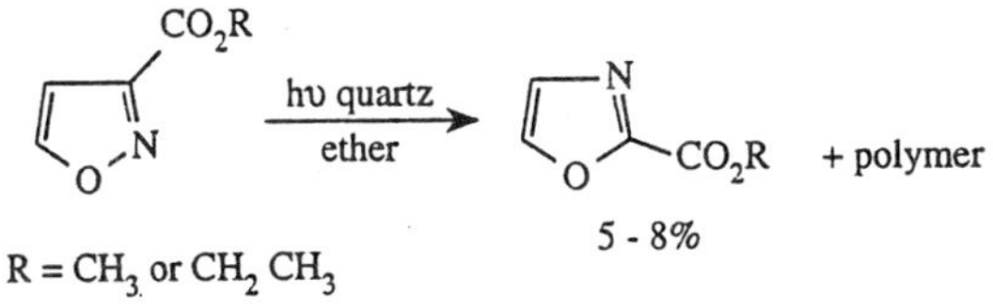

$$R = CH_3 \text{ or } CH_2 CH_3$$

5 - 8%

SCHEME 82.42

Ethyl-2-phenyl-1-azirine-1-carboxylate cyclizes upon irradiation to an oxazole.[78]

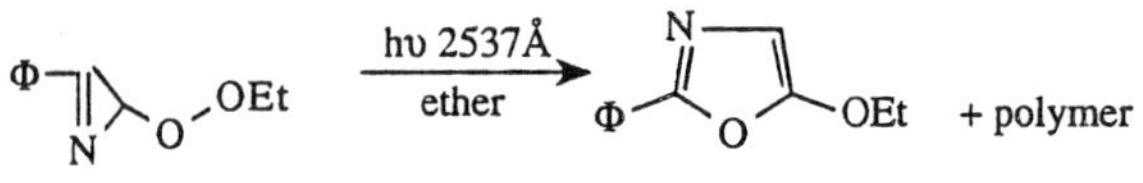

SCHEME 82.43

The isoxazole-3-carbonitrile gives no simple photoproduct, but is converted rapidly into a black polymeric material. 3-Acetyl-5-methylisoxazole is photostable when irradiated at $\lambda > 300$ nm. At 254 nm, it photoisomerizes via a triplet-state process and disappears with a quantum yield of 0.20 $\pm \times 0.02$. Further irradiation of **56** leads to the oxazole. The 1,3-diketone **57** is formed by a singlet-state reaction.[79]

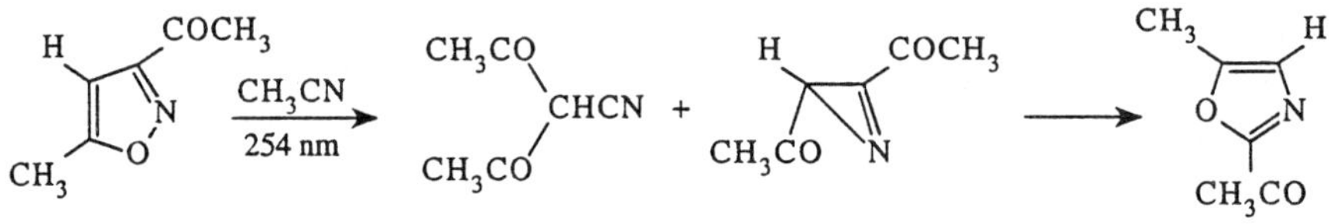

SCHEME 82.44

3-Hydroxyisoxazoles are converted to oxazolones upon irradiation, probably through intermediates such as **58**.[80,81] The oxazolone obtained from 3 hydroxy-4,5-dimethylisoxazole is not photostable.[80]

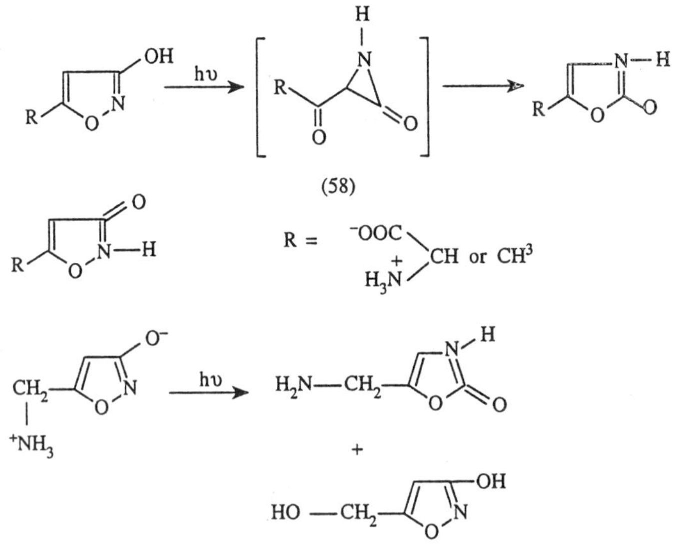

SCHEME 82.45

3-Methyl- (or 3-phenyl-) 5-methyl-4-acetylisoxazoles are photoisomerized to 3-methyl- (or 3-phenyl)-4-acetyl-5-methyloxazole.[82,83] The various processes involved can be rationalized by deuterium labeling.[82]

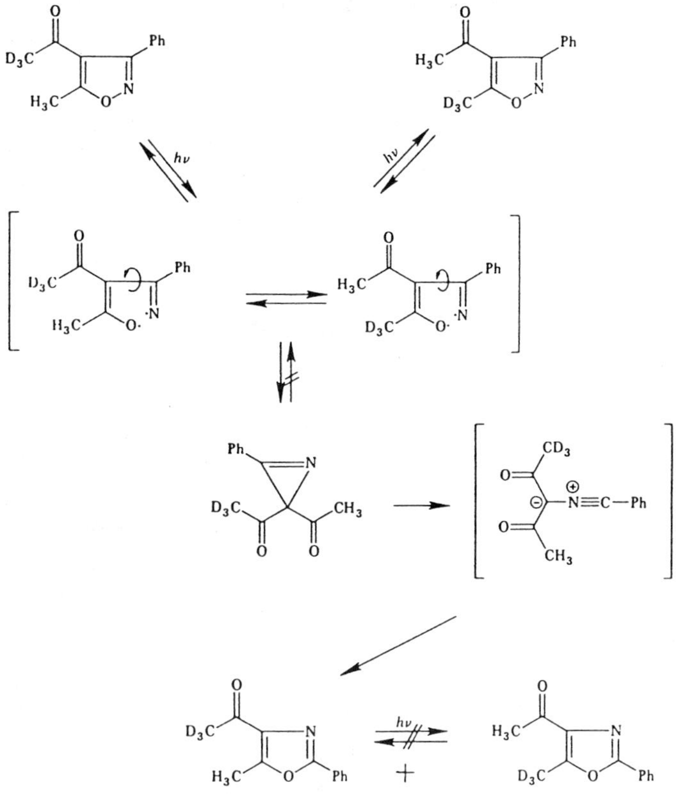

SCHEME 82.46

Attempts to rationalize the selectivity observed in the case of 3,5-diphenyl-4-acetylisoxazole have been made.[83,84]

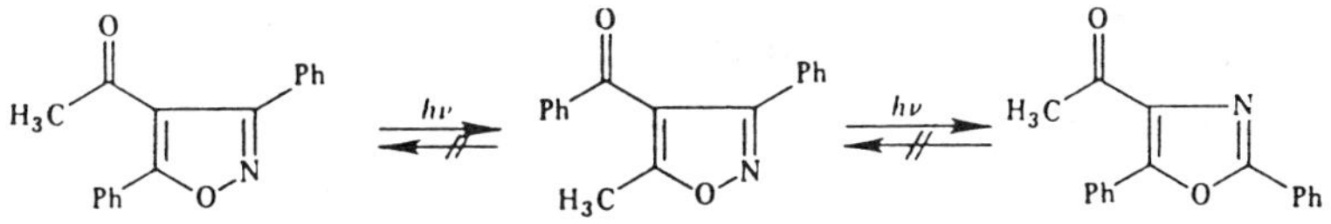

SCHEME 82.47

In the study of the photochemical reactivity of a series of 4-acylisoxazoles that react from the triplet state, the 2*H*-azirine intermediates have been isolated. They are converted to the corresponding oxazole on heating. The very low quantum yields measured for the rearrangement can be explained by assuming that the major fate of initial homolysis to the biradical by N–O cleavage is reclosure to the starting isoxazole rather than formation of the 2*H*-azirine. There is also some evidence for ketene imine formation.[85]

Compounds **59** and **60** form the corresponding oxetane by intramolecular cyclization upon irradiation with long wavelength UV light ($\lambda > 300$ nm). At 254 nm, however, **59** leads to a mixture of diastereoisomeric 2*H*-azirines **61** and oxetanes **62**.[86] At 300 nm, the quantum yield of formation of **62** from **59** is 0.11 ± 0.01; while at 254 nm, the quantum yield of the formation of **62** is 0.024 ± 0.003 and the quantum yield for the formation of the azirines is 0.024 ± 0.003. Two excited states of **59** are involved. If it is assumed that the oxetane is formed from the same reactive stated in both cases, it can be calculated that about 22% of the upper states decay to the oxetane-forming state.

In other words, the inefficiency associated with the reaction leading to 2*H*-azirines is not limited by oxetane formation.

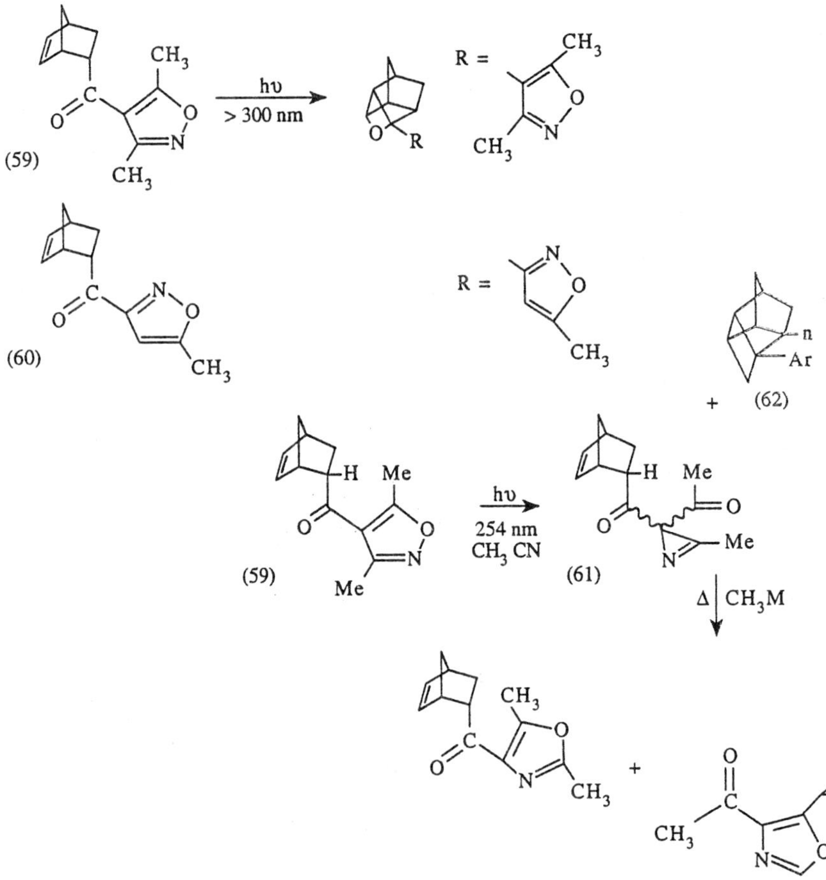

SCHEME 82.48

The photoisomerization of 4-benzalamino-3-methyl-5-styrylisoxazole to an imidazole[87] can be explained by an homolytic rupture of the N-O bond.

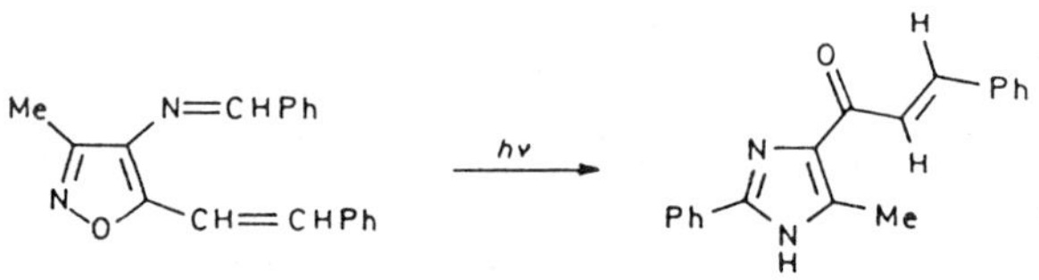

SCHEME 82.49

Carbon-oxygen bond homolysis followed by radical recombination appears to be responsible for the photorearrangement of the 2-allyloxyoxazoles to the oxazolin-2-ones.[88] 2-Benzyloxy-4,5-diphenyloxazole is reported to undergo an analogous rearrangement to give the corresponding oxazolin-2-one and bibenzyl.

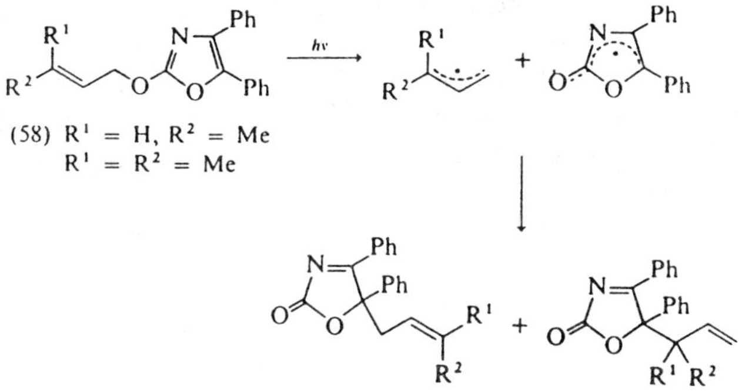

SCHEME 82.50

Benzisoxazoles (Indoxazenes)

An isonitrile is the intermediate of the photoisomerization of benzoisoxazole to benzoxazole.[89,90] This has been detected by IR at 77K in a KBr matrix.[64] This intermediate is converted to benzoxazole by heating. Quenching and sensitization experiments have shown that benzoxazole is formed from a π,π^* singlet state of isoxazole ($E_s = 393$ kJ mol^{-1}), whereas 2-cyanophenol results from the reaction of an n,π^* triplet state (347 kJ mol^{-1}). Similarly, the photoisomerization of 3-methylbenzisoxazole to 2-methylbenzoxazole is a singlet-state reaction.[64] (Interestingly, 2-cyanophenol also undergoes photoisomerization to benzoxazole[91–93] and involves the isonitrile as the intermediate.[89])

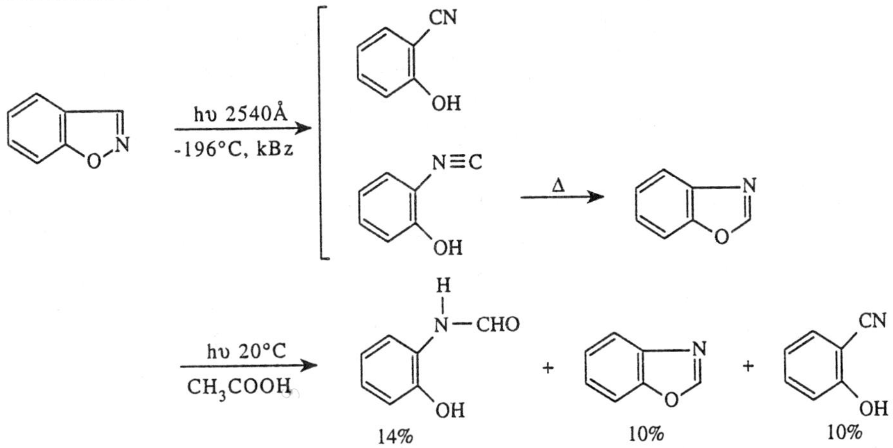

SCHEME 82.51

The irradiation of benzoisoxazoles in concentrated sulfuric acid is a synthetic path to carbonyl compounds.[94]

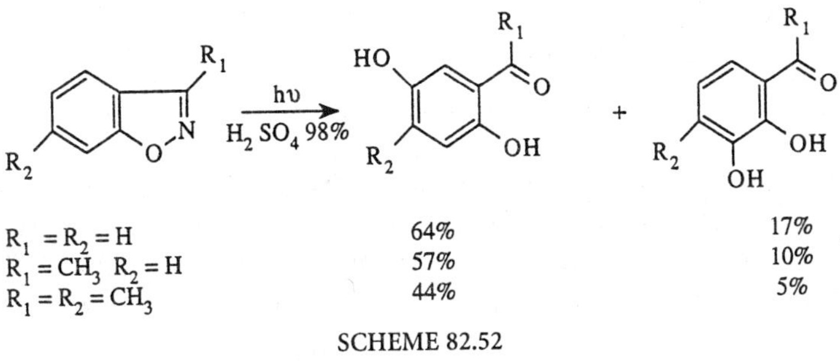

$R_1 = R_2 = H$	64%	17%
$R_1 = CH_3$, $R_2 = H$	57%	10%
$R_1 = R_2 = CH_3$	44%	5%

SCHEME 82.52

Photolysis of 2,1-benzoisoxazoles under these conditions leads to products by way of a nitrenium ion,[95,96] while 1,2-benzoisoxazoles under similar conditions react via an aryloxenium ion.[97] In the case of benzimidazole, its 3-methyl derivative, and the case of two naphthisoxazoles, the involvement of an azirine in their photoisomerization has been proven by spectroscopic studies.[98] The spiroazirines from these reactions are thermally much more unstable than those involved in the isomerization of isothiazoles.

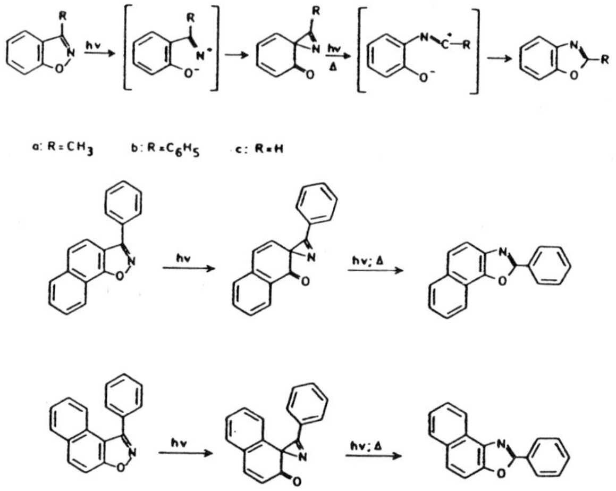

a: R = CH₃   b: R = C₆H₅   c: R = H

SCHEME 82.53

Isoxazolo[4,5-c], [5,4-b], [4,5-b], and [5,4-c]-pyridines are photoisomerized to the corresponding oxazopyridines.[99–103] In each case, a 2H-azirine is supposed to be an intermediate. In the case of the 3-methyl derivatives, ketinimines that can be trapped by water are also obtained. Compounds 64 (R_1 = Me, R_2 = R_3 = H) and 65 were characterized by flash photolysis as transients when 63 (R_1 = Me, R_2 = R_3 = H) is irradiated in an alcohol (R′ OH). Their lifetimes are in the millisecond range.[104]

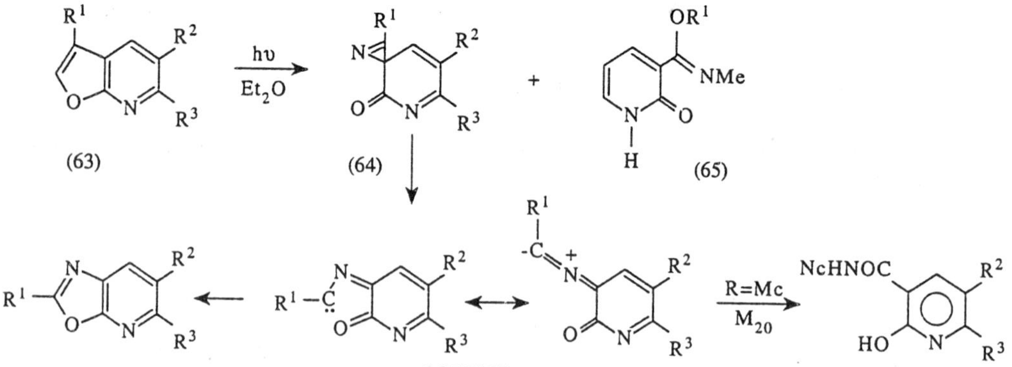

SCHEME 82.54

Compound 66, bearing a hydrazino group in the 4-position, is converted only to a 1-aminopyrazolopyridine.[103] The reaction proceeds probably by the intermediary of an azirine that was not characterized. This isomeric oxazole is one of the products of irradiation of 3-methylisoxazolo[4,5-d]pyridine in CCl₄. In the presence of HO⁻, a degradation occurs.[105]

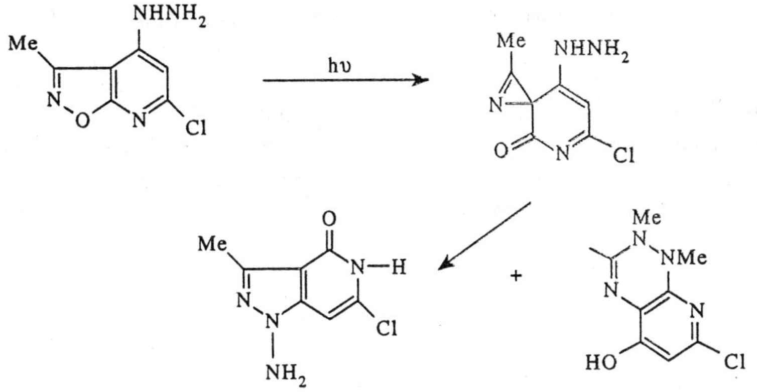

SCHEME 82.55

As in the case of compound **66**, compound **67** is converted to a pyrazolopyrimidine when irradiated.[105]

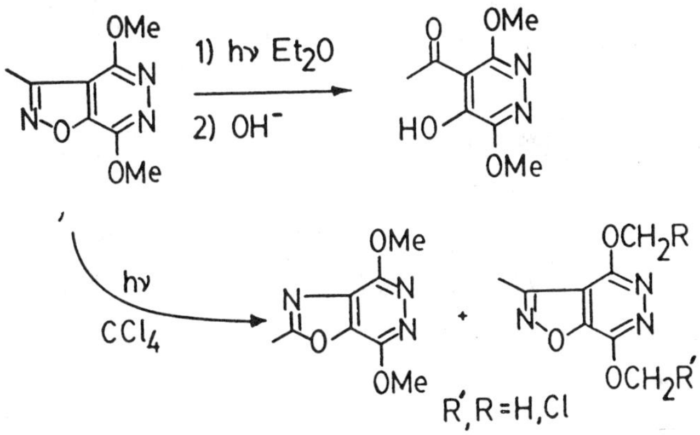

R´, R = H, Cl

SCHEME 82.56

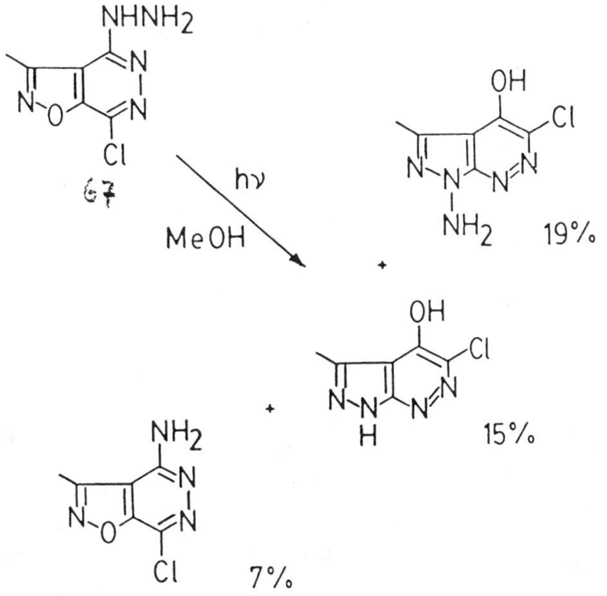

SCHEME 82.57

It is worthwhile noticing that with the isoxazolopyridine or pyrimidine, the five-membered ring undergoes photorearrangement at the expense of the six-membered ring.

Isomerization also occurs on irradiation of 3-hydroxyl-1,2-benzixoxale, which yields benzisoxazolinone.[106] Similar compounds are obtained from the irradiation of *N*-alkyl-1,2-benzisoxazolinones. However, the 3-methoxy derivative **68** reacts extremely slowly and this could suggest that the tautomeric form of the 3-hydroxy derivative is the reactive species in the conversion of the 3-hydroxy derivative discussed above.

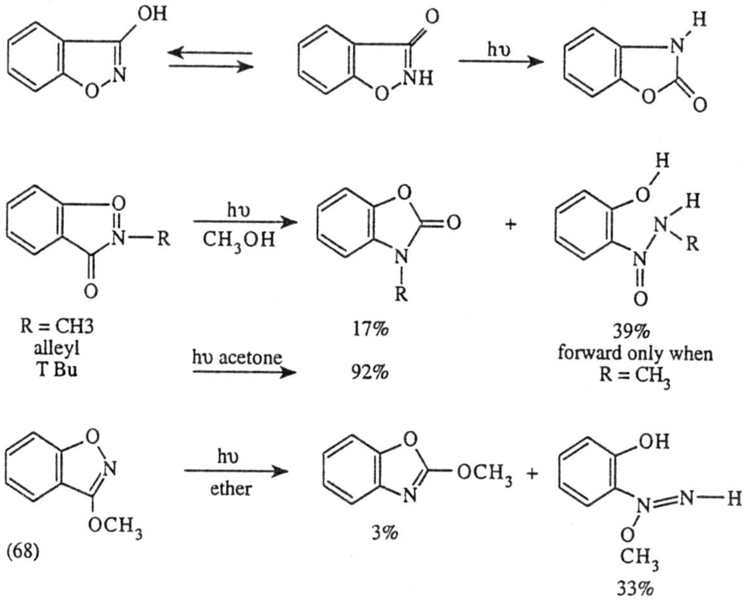

SCHEME 82.58

A low-temperature IR study of the photolysis excluded a reaction path via an isocyanate or a spiro α-lactam and the suggested mechanism[106] is shown in Scheme 82.59. Sensitization experiments demonstrate that the hydroxy as well as the methoxy derivatives react predominantly from the triplet state.[106]

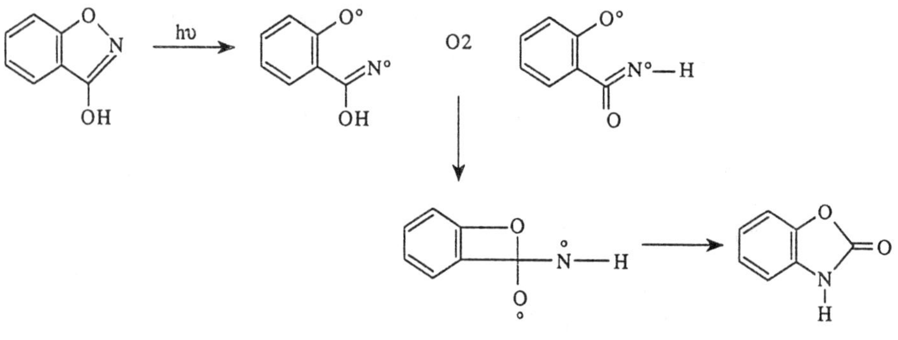

SCHEME 82.59

Polycyclic Isoxazoles

Some polycyclic isoxazoles are isomerized to open-chain compounds.[107]

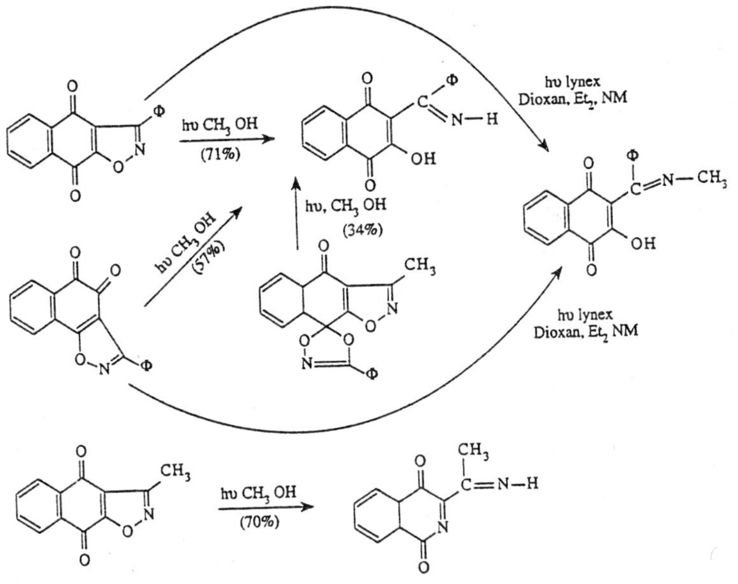

SCHEME 82.60

Anthranils or 2–1 Benzoisoxazoles

The photochemistry of anthranils is very interesting. It has been found that they can ring-expand and undergo additions of a nucleophilic reagent. However, this ring-expansion reaction takes place only with anthranils that are unsubstituted in the 7-position.[108–110] The mechanism shown in Scheme 82.61 has been proposed for the reaction, although a nitrene has not been observed with certainty. However, it has been shown that the photolysis of 2-azidoacetophenone leads to the same products as those formed from 3-methylanthranil, when irradiated in neutral methanol.[109,110] In some cases, benzoxazoles are also generated. Ring expansion reactions have been described, whereby the isoxazole ring opening undergoes N-O bond fission. Radical attack on the 3-aryl-substituent group affords the products shown in Scheme 82.62.[109,111,112]

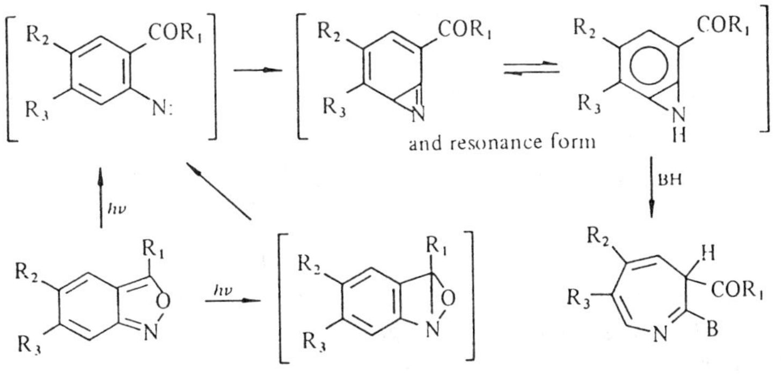

SCHEME 82.61

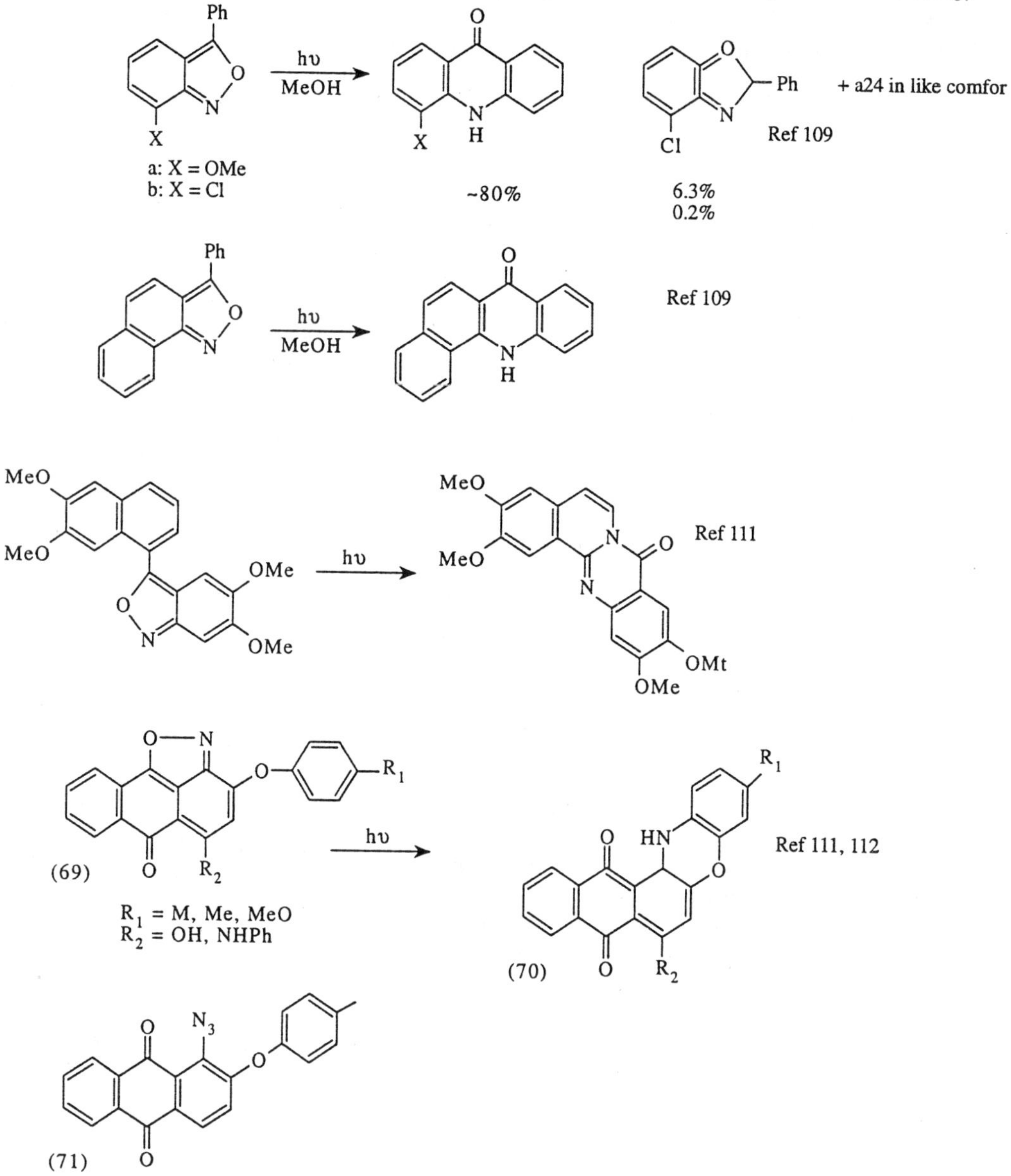

a: X = OMe
b: X = Cl

~80%

6.3%
0.2%

+ a24 in like comfor

Ref 109

Ref 109

Ref 111

(69)

R_1 = M, Me, MeO
R_2 = OH, NHPh

(70)

Ref 111, 112

(71)

SCHEME 82.62

In the case where R_1 = H_1, R_2 = OH,[112] it is reported that formation of 70 from 69 can be sensitized, thus indicating triplet excited-state reactivity. The addition of bromoethane to the reaction mixture increases the quantum yield of product formation significantly, although it does not change the fluorescent lifetime of 69. However, it is likely that the enhancement of the quantum yield is not due to heavy-atom acceleration of intersystem crossing from the singlet excited state to the triplet of 69. To explain these results, it is suggested that 70 is also formed via the singlet excited state of 69 and involves a singlet nitrene. The influence of the heavy-atom additive is to increase the rate of conversion of the singlet nitrene to its triplet. The triplet nitrene then proceeds to 70, while the singlet nitrene either proceeds to 70 also, or reverts to 69. This explanation is supported by the results of experiments in which the nitrene was generated independently by sensitized photolysis of the azide 71. This procedure gave 70 while direct irradiation gave both 69 and 70.

Open-chain compounds are formed when anthranils are irradiated in acidic medium.[94-96,114-115] The argument in favor of the involvement of a protonated nitrene in the course of these reactions is supported by the fact that the same compounds are formed from the thermal decomposition of 2-azido-5-methylacetophenone in sulfuric acid.[114]

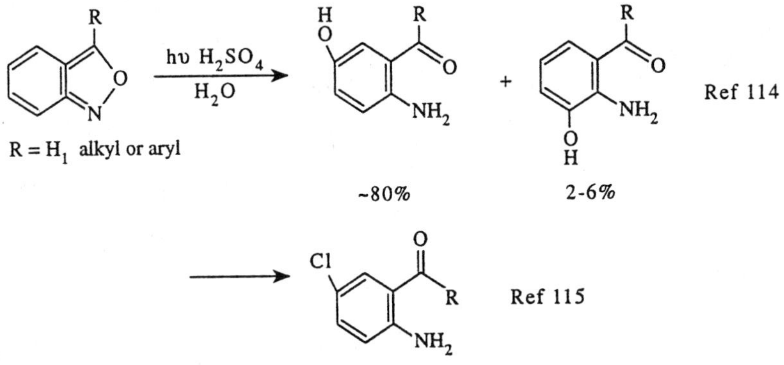

SCHEME 82.63

Oxazoles

Oxazoles, being the products in photochemical rearrangement of isoxazoles, also undergo a molecular reorganization under the influence of UV-VIS light. The reaction is complex and, in some cases, solvent dependent. For example, irradiation of 2,5-diphenyloxazole in ethanol gives 4,5-diphenyloxazole (transposition of atoms 2 and 4) and 3,5-diphenyloxazole (transposition of atoms 2 and 3). When irradiated in benzene, 2,4-diphenyloxazole (transposition of atoms 4 and 5) was obtained. Further irradiation of 2,4-diphenyloxazole in benzene produced 3,4-diphenylisoxazole in good yield.[116–118] (In the case of 2-methyl-5-phenyloxazole, the transposition of atoms 3 and 5 was also observed.)

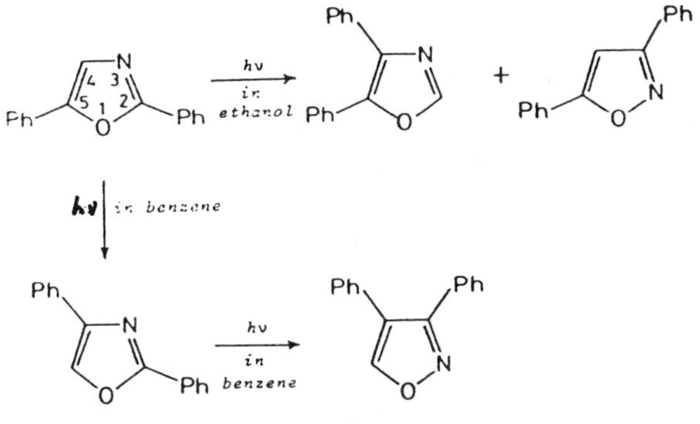

SCHEME 82.64

Calculations[119] show that a 1-azirine intermediate in its S_1 state is formed from the lowest $^1(n,\pi^*)$ state of oxazole in which there is an out-of-plane distortion. Intersystem crossing (ISC) to the T_1 state gives rise to the interchange of positions 2 and 3 to yield isoxazole (type A process in figure below). On the other hand, the lowest $^1(\pi,\pi^*)$ state of oxazole brings about two types of transpositions, i.e., the 4,5-transposition and the 2,4- and 3,5-transpositions. Positions 4 and 5 are interchanged via a 2-azirine intermediate when the out-of-plane distortion at this state and the internal conversion to the S_1 state occur (type A process). On the other hand, the 2,4- or 3,5-transposition is caused by the sigmatropic shift of an oxygen atom when a bicyclic intermediate is formed by disrotatory ring closure and subsequent ISC to the T_1 state (type B process in Scheme 82.65).

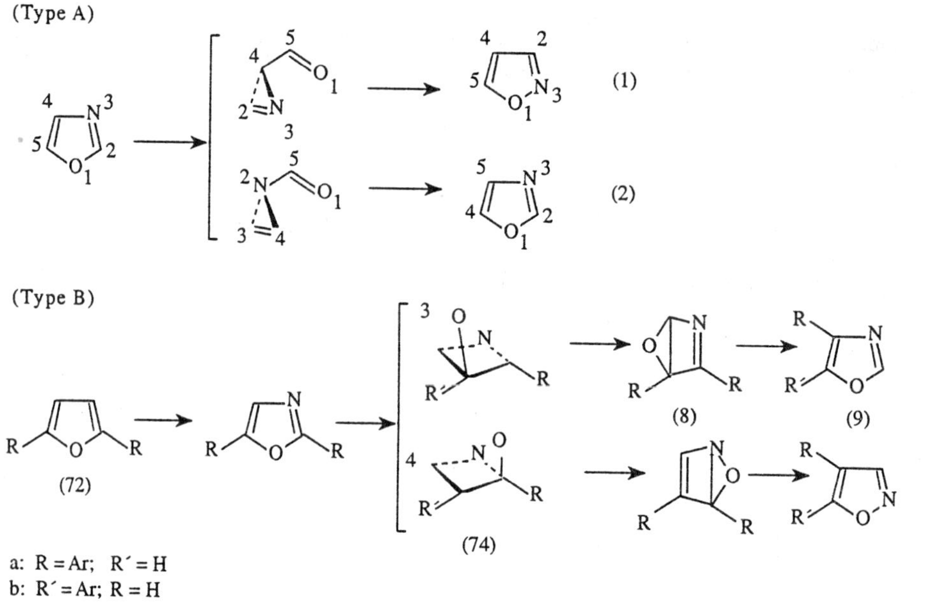

(Type A)

(Type B)

a: R = Ar; R′ = H
b: R′ = Ar; R = H

SCHEME 82.65

The difference observed between 2-aryl **72a** and 5-aryloxazoles **72b** can also be rationalized.[119] Thus, when an aromatic substituent is on the C2 or C5 atom, the amplitudes of $p\pi$ AOs of the LUMO on these atoms increase. Therefore, the bicyclic intermediate formation is favorable in the π,π^* state of these two compounds. In addition, the stabilization due to the conjugation effect with a substituent might play an important role in the determination of the reaction course. In intermediate **73a** (or **74b**), there is conjugation between an aromatic substituent and the C2-N3-C4 (or N3-C4-C5) part; whereas, in **73b** (or **74a**), there is not. That is, one can expect that the transition state is stabilized considerably due to conjugation and, as a result, the energy barrier is lowered. Therefore, the 2,4-transposition is favorable in **72a** (path 3) and the 3,5-transposition in **72b** (path 4 in Scheme 82.65).

Azirines have been characterized[117,118] in the photoreaction of 2-phenyloxazole, 2-phenyl-5-methyloxazole, 2-phenyl-4-methyloxazole, and 2-aryl-5-phenyloxazole. The structures are **75, 76, 77,** and **78,** respectively, and are shown in Figure 82.2.

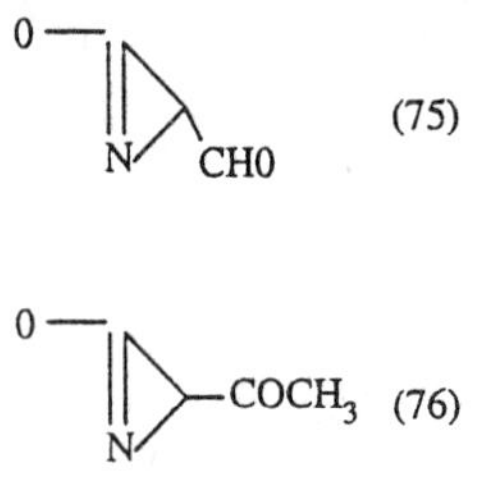

FIGURE 82.2   The structures of the azirines formed from 2-phenyloxazole (**75**), 2-phenyl-5-methyloxazole (**76**), 2-phenyl-4-methyloxazole (**77**), and 2-aryl-5-phenyloxazole (**78**).

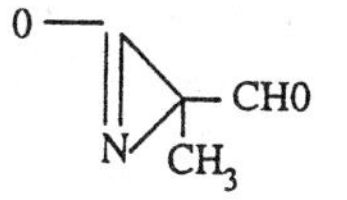

(77)

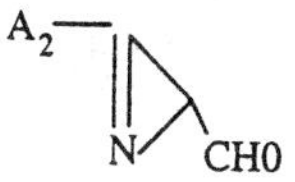

(78)

FIGURE 82.2 (continued)

2,5-Dimethyloxazole[76] also transforms into an azirine that has not been isolated but can be trapped by reaction with solvent as shown in Scheme 82.66.

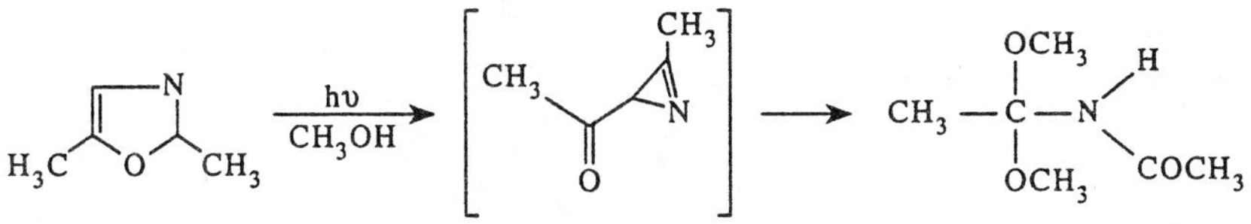

SCHEME 82.66

3,5-Dimethylisoxazole, on irradiation in an alcohol, leads to similar compounds,[76b] perhaps involving an azirine.[120] This reaction has been described in the following scheme.

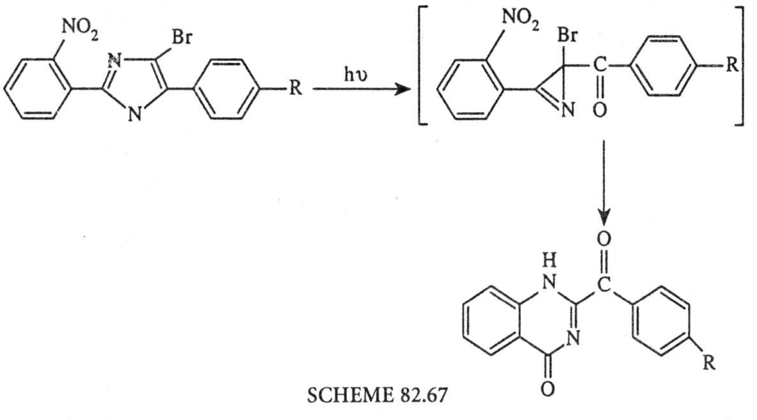

SCHEME 82.67

No ring photoisomerization has been described in the case of benzoxazole.

1,2,4-Oxadiazoles

The ring-opening process is proposed to rationalize the rearrangement of the 1,2,4-oxadiazoles 79 to give the corresponding 1,3,4-oxadiazoles.[121] Thus, irradiation of 79 in methanol results in ring opening to give a species represented by the resonance forms 80 and 81; closure of this to 82 and subsequent rearrangement leads to the isolated product. Changes in substitution on 79 to R_2 = phenyl or dimethylamino instead of amino or methylamino influences the reaction and in these cases the intermediates 80–81 afford 83 as the isolated photochemical product. The intermediacy of 80 and 81 has been tested by comparing the photochemistry of the isoxazole 56a with that of the azide **84**.[79] Photolysis of the latter would be expected to lead to nitrogen expulsion and formation of a nitrene analogous to **81**. If this is in fact an intermediate in the photorearrangement of **56a**, then the product obtained from **84** should be identical with that obtained from irradiation of **56a**. In the event irradiation of **56a** and **84** did indeed give the same product, the acylazirine **56**. However, photolysis of **56a** gave an additional major product, the diacetylacetonitrile **57**, which was not formed from **84**. The formation of **56** from **84** was inhibited by triplet quenchers, but sensitization of the reaction of **56a** resulted in complete suppression of the formation of **56**. Under these conditions, the oxazole **56b** was obtained as a secondary photoproduct produced by sensitized isomerization. The observations can all be accounted for if a nitrene is assumed to be the reactive intermediate, with **56** being derived from the triplet nitrene and **57** from the singlet nitrene. In the case of **79** where R_2 = OH or NHR, the reacting compound may be the tautomer **85** (Scheme 82.68).

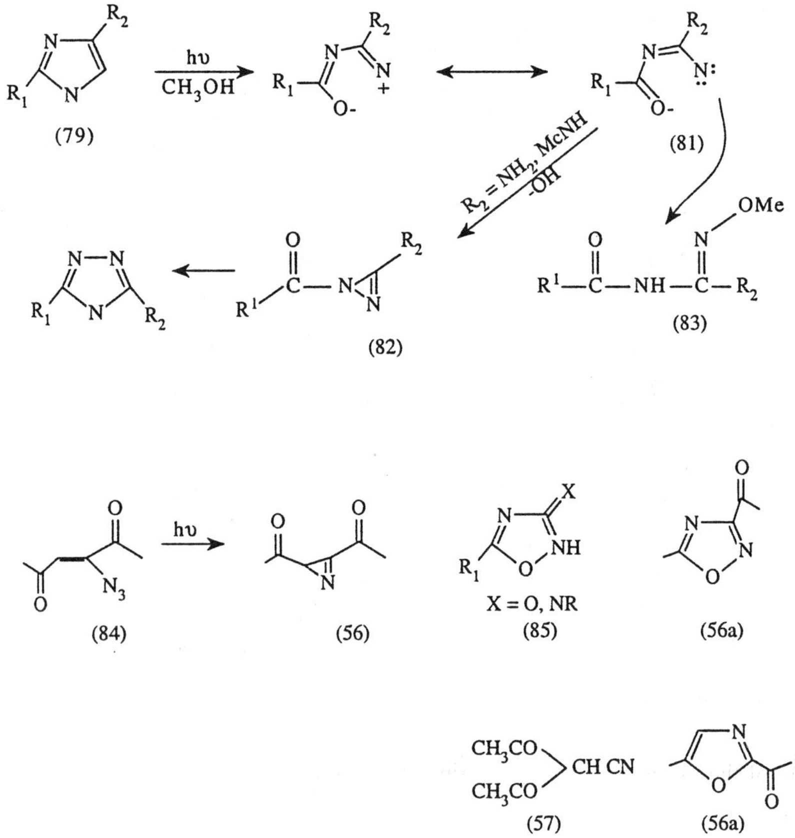

SCHEME 82.68

When the R_2 substituent in **79** is a 3-phenoxy or a 3-enaminoketone group, an intermediate species derived from photolysis of the ring N-O bond and characterized by a continuous 6π-electron system involving the side chain, was suggested to account for the ring closure encountered.[122,123]

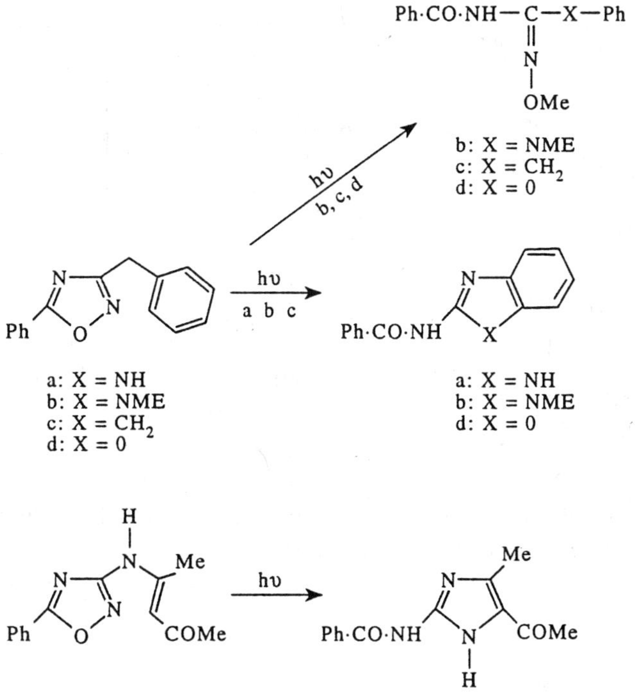

SCHEME 82.69

The proposed mechanism in this case is shown in Scheme 82.70.

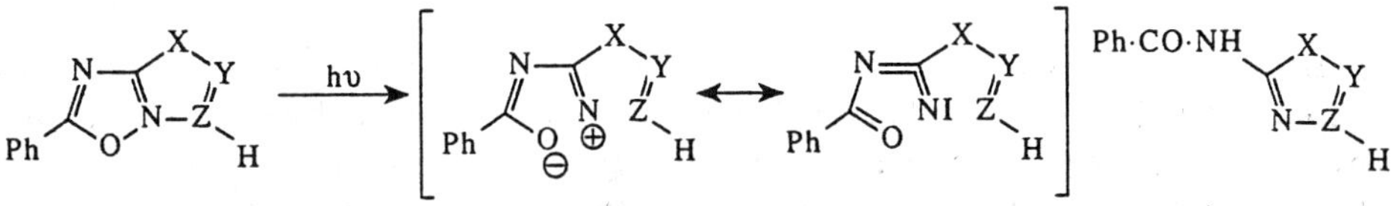

SCHEME 82.70

Primary products of irradiation of the 5-aryl-3-acetylamino-1,2,4-oxadiazoles **86** are the iso-meric oxadiazoles **87**, which are presumed to be formed from the biradical intermediates **88**.[124] The primary products **87** are also photochemically reactive and are transformed to the quinazolones **89**.[124,125] This secondary rearrangement may proceed by N-O bond fission in **87** to give **90**, followed by cyclization of the nitrogen onto the aryl ring. An alternative could involve electrocyclic ring closure in **87** by a process commonly observed for *N*-aryl benzamides and could involve the intermediate **91**.

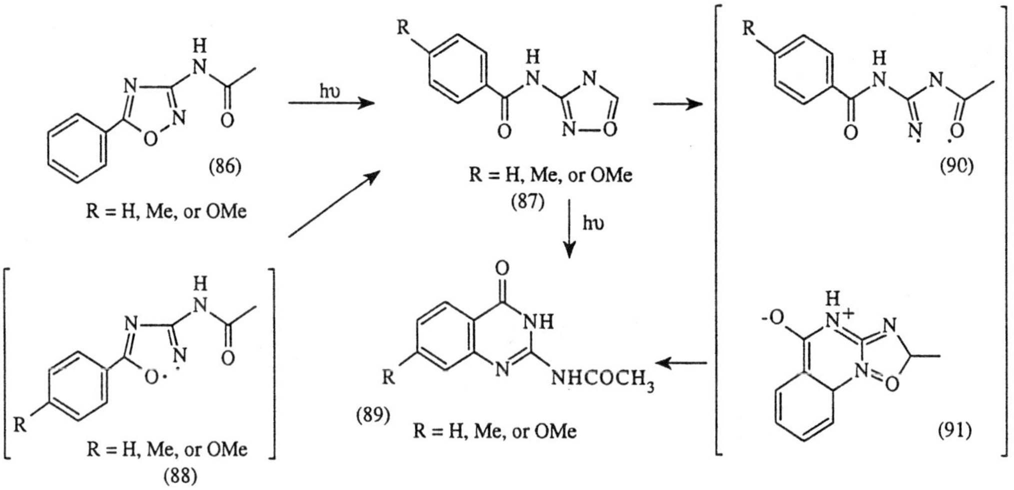

SCHEME 82.71

Similarly, 3-aroylamino-5-methyl-1,2,4-oxadiazoles are converted into the 2-acetylamino-quinazolin-4-ones when irradiated in methanol.[125,126]

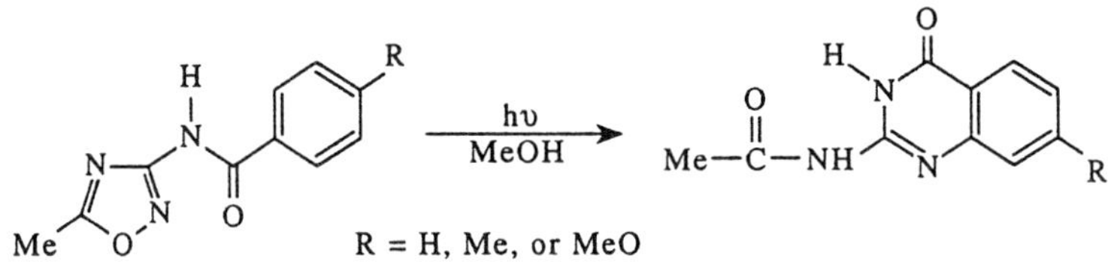

SCHEME 82.72

A similar choice of mechanisms is possible[127] in the photochemical rearrangement observed for the styryloxadiazoles 92. Here, irradiation of *E*-92, R = Ph, results in rapid *E,Z*-isomerization of the side chain double bond and slower formation of the quinoline 93, which could result from either N-O bond cleavage to 94, followed by cyclization of the nitrogen onto the benzene ring, or from electrocyclic closure of 92 to 95 followed by thermal rearrangement to 93. The isomerization to the quinoline 93 is much less efficient for *E*-92, R = Me, although *E,Z*-isomerization of the styryl double bond is still fast. The authors[127] argue that the phenyl group of 92, R = Ph, causes the excitation to be localized in the oxadiazole ring; whereas, in 92, R = Me, excitation is localized in the styryl chromophore; this implies that isomerization to the quinoline proceeds therefore by N-O cleavage via 94 rather than electrocyclic closure to 95.[127]

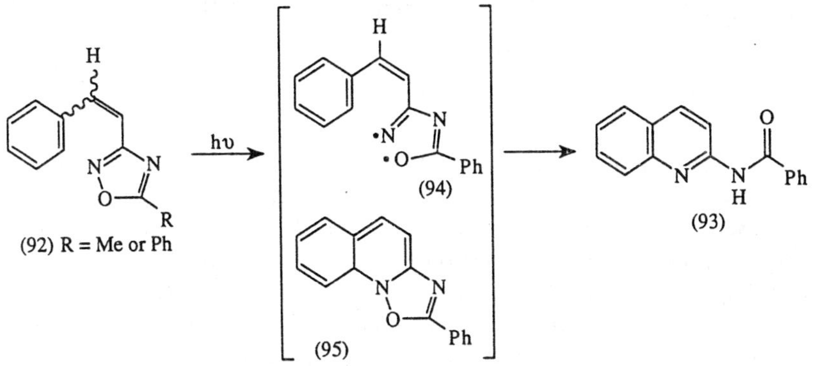

SCHEME 82.73

1,2,5 Oxadiazoles

Irradiation of oxadiazoles leads to fragmentation and the formation of compounds of low molecular weight.[131,132]

Benzofurazan

Intermediates **97** and **99** have been characterized by UV and IR spectroscopy in the conversion of benzofurazan **96** into the carbamate **101** and the azepine **102** by irradiation in methanol and benzene, respectively.[128] The nitrile oxide **99** can be trapped as the isoxazole **103** with dimethyl acetylenedicarboxylate.[129] In the presence of triethyl phosphite, the *cis,cis*-dinitrile **98** is obtained from benzofurazan in high yield.[130]

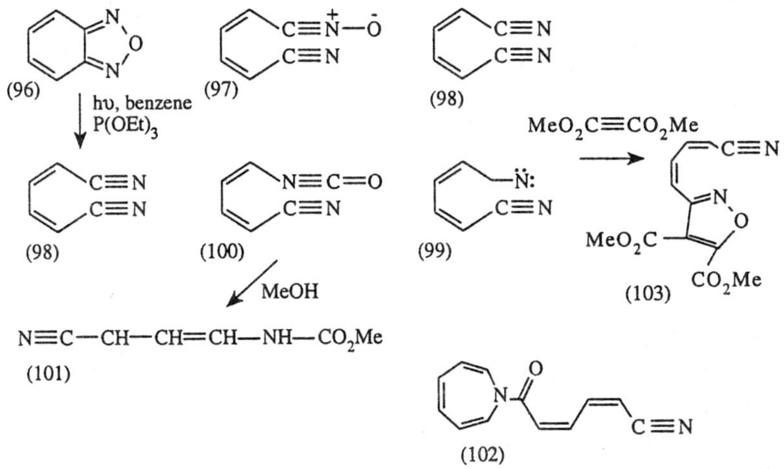

SCHEME 82.74

82.3 Six-Membered Azaaromatic Compounds

Analogous to the well-known scrambling and valence isomerization reactions of benzenes,[13,14] it has been found that many azabenzenes can undergo scrambling reactions.[4,6,9,11,12] In a few cases, valence isomers have been isolated. However, the formation of valence isomers has been indicated in many cases by trapping experiments, and valence isomers have been suggested as intermediates to explain many reactions.

Pyridines

Electronically excited pyridine has never been observed to undergo any radiative relaxation; thus, the efficiency of the radiationless transition is 100%. However, substituted pyridines do undergo radiative deactivations.[133–136] The lowest excited singlet state of pyridine has been classified as n,π^* with an energy of 4.3 eV (414 kJ mol^{-1}).[137,138] The nature of the lowest triplet state of pyridine is less certain. According to Hoover and Kasha,[133] it should be the forbidden $^3A_2$ (n,π^*) state (334 kJ mol^{-1}), whereas Evans[139] has characterized it as $^3A_1$ (π,π^*) on the basis of UV spectroscopy under very high oxygen pressure (100 atm). By this technique, a band at 3.68 eV (354 kJ mol^{-1}) was observed. In addition, ion and electron impact studies in the vapor phase indicate a triplet level of 3.8 eV (366 kJ mol^{-1}).[140] More detailed discussions of this point are available in the literature.[141,142]

When pyridine is irradiated in butane at $-15°C$, it is converted to Dewar pyridine, which has been characterized[143] and trapped by sodium borohydride or water.[144,145] Irradiation of pyridine in an argon matrix at 8K affords the Dewar pyridine initially, but extended irradiation brings about decomposition to acetylene and HCN.[146] Similarly, irradiation of pyridine in the vapor phase yields the same products.[147]

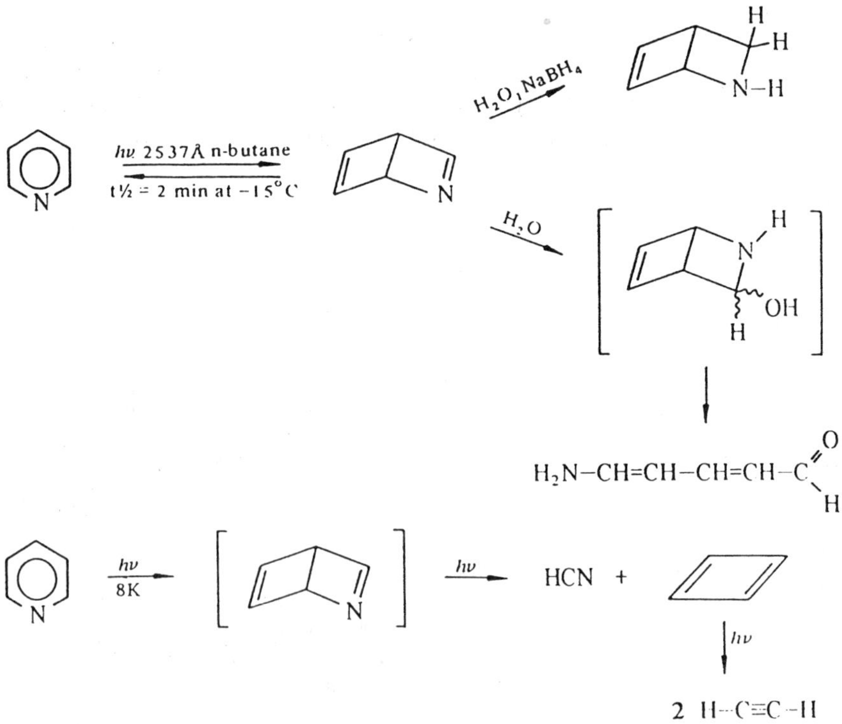

SCHEME 82.75

Alkylpyridines are isomerized upon irradiation in the vapor phase. Thus, irradiation of 2-picoline (2-methylpyridine) in the vapor phase at 1 to 1.5 torr leads to a mixture of 3- and 4-picoline.[148] The irradiation at 10 torr of a series of picolines and lutidines (dimethylpyridines) resulted in various isomerizations, which are believed to occur through intermediate azaprismanes and are illustrated in Scheme 82.76.[149]

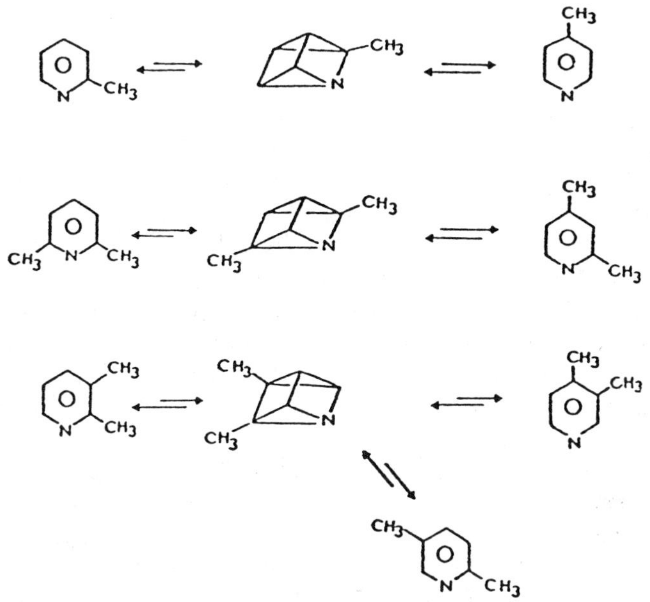

SCHEME 82.76

Similar ring carbon transpositions occur with 2-chloromethylpyridine, which is photoizomerized to 4-chloromethylpyridine.[150] In the case of 3,4-lutidine, two Dewar pyridines have been characterized.[151]

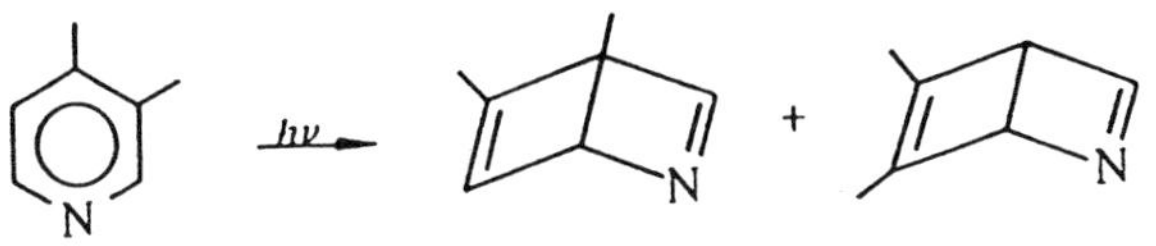

SCHEME 82.77

A Dewar pyridine has been trapped by $NaBH_4$ during irradiation of 3,5 lutidine; but in the case of 2,4,6-collidine, the corresponding piperidine, formed by reduction, is the only reaction product.[151]

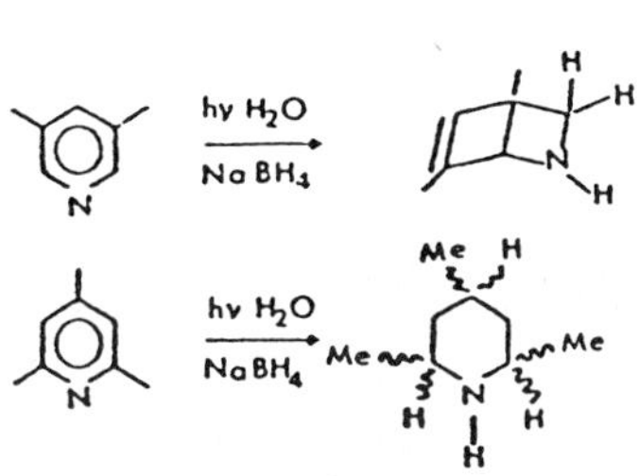

SCHEME 82.78

Irradiation of 2- and 4-picoline, or 3,4-lutidine in cyclohexane, results in isomerization and substitution reactions;[152–154] whereas, 2,3-, 2,4-, or 2,5-lutidine undergo substitution reactions only.

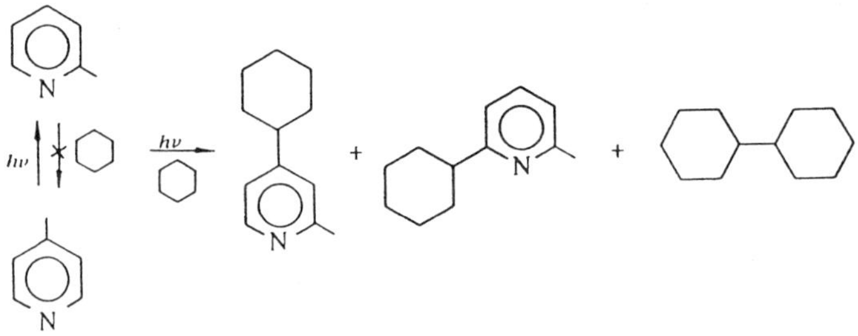

SCHEME 82.79

Wilzbach and Rausch[151] have irradiated a series of alkylpyridines in acetonitrile. This resulted in rather complex scrambling of the alkyl groups, and the observation of a temperature effect.

It is worthy of note, in contrast to the foregoing, that methyl derivatives of quinoline, isoquinoline, pyrazine, pyrimidine, pyridazine, quinoxaline, cinnoline, and acridine, as well as 2-, 3-, and 4-methylpyridine or 2- and 4-ethylpyridine are all subject to photodealkylation. Methyl or ethyl radicals have been detected by ESR during UV irradiation of the methylpyridines or ethylpyridines at 93K in CD_3OD.[153] There is probably no relation between photodealkylation of these compounds and their photoisomerization.

By comparison with benzene, it has not been possible to trap any pyridine or picoline or lutidine photorearrangement intermediate by reaction with an olefin.[13,14] Irradiation at 254 nm of the pyridine derivatives **104 a,b** in aqueous alkali yields quantitatively the 3-substituted methylene-2-azabicyclo[2,2,0]hex-5-ene compounds **105 a,b**.[155,156] Irradiation of these nonaromatic isomers in diethylether using a high-pressure mercury lamp results in the formation of the *ortho*-substituted anilines **106**, and hence it is deduced that the overall reaction involves two photons. The reactions proceed as shown in Scheme 82.80.[155,156]

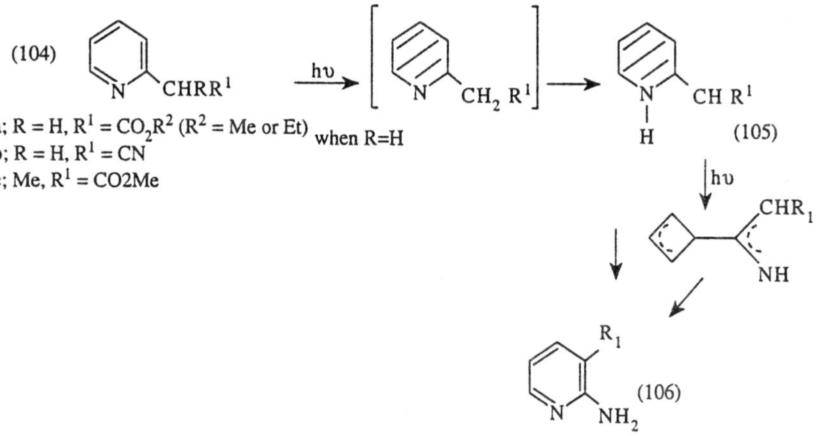

SCHEME 82.80 $R^1 = CN$ or CO_2R^2 ($R^2 = Me$ or Et)

On the other hand, irradiation of **104c** gives no volatile products. Further investigation of the details of the reaction have been made with the methyl[6-*H*]-2-pyridyl acetate **107**. Irradiation of this in diethylether:butanol (1:1) as solvent affords the ring-deuteriated methyl anthranilates in 20% yield.[157] Examination of the deuterium distribution in the product and starting material revealed that in the former approximatively equal amounts of the [4-^2H]- and [6-^2H]-isomers **108** and 109 (ca. 40% of each) are produced with 10 to 20% of the other two deuteriated positional isomers, whereas the latter contained 90% of the original isomer 107. In order to account for these distributions, it is suggested that scisson of the N1-C6 bond of the pyridine occurs to give the biradical intermediate 110, which then undergoes bond formation between C4 or C6 and C7 as shown in Scheme 82.81.

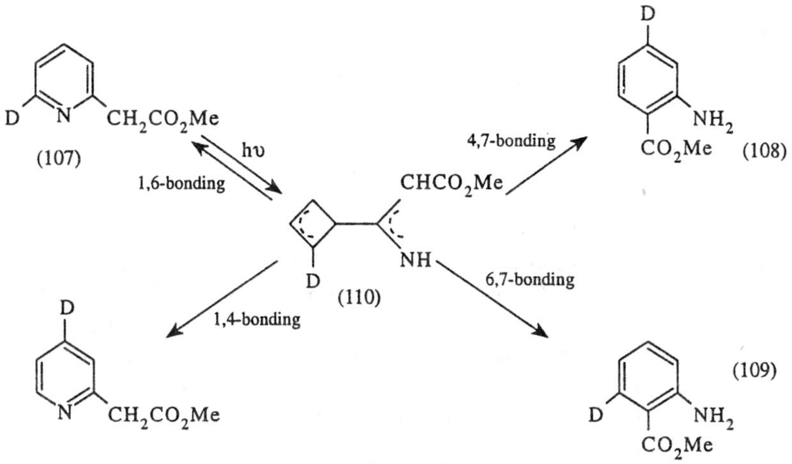

SCHEME 82.81

The reactivity of ethyl methyl-2-pyridyl acetates has been reported to be very dependent upon the position of the methyl substituent.[158] Thus, while the 3-, 4-, and 5-methyl-isomers all yield the corresponding anthranilates ($\phi_{disappearance}$ = 0.09 to 0.25), the 6-isomer has a very low photoreactivity and the quantum yield for disappearance is only 0.006. Reasons for this large difference in reactivity are not obvious from the suggested reaction mechanism.

Dewar isomers 111 are also reported on irradiation of the salt of 2-aminopyridine in methanol[15] or on irradiation of 2-fluoropyridine.[151]

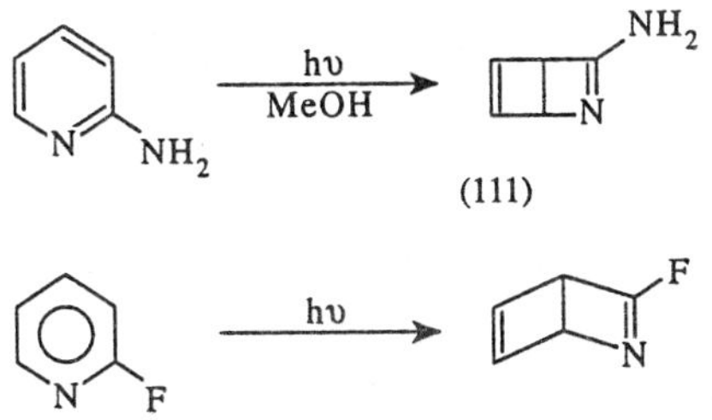

SCHEME 82.82

A pentafluoro-Dewar-pyridine, tentatively assigned the structure 112, has been isolated in very low yield from the complex mixture obtained from about 400 individual photolyses of pentafluoropyridine.[160] The half-life of this isomer is about 5 days at room temperature. Flash photolysis studies of pentafluoropyridine provide evidence for two azafulvene isomers 113 and 114 with half-lifes in the range of a millisecond (3×10^{-3} and 22×10^{-3} s), and it is suggested that an azabenzvalene is the precursor to both the fulvenes and the Dewar isomer, although this mechanistic argument is quite speculative.[160] The reactive excited state of pentafluoropyridine is proposed to be the S_1 state.[160]

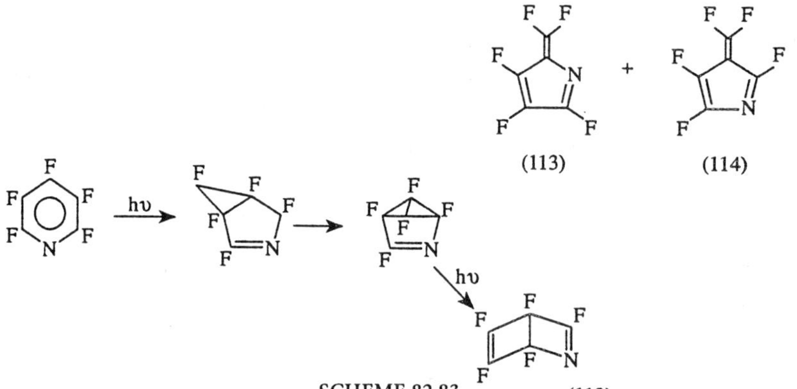

(113)   (114)

SCHEME 82.83   (112)

In pyridine, as well as in pyridazine, pyrimidine, and pyrazine, the photoisomerization proceeds from the lowest singlet state. The mechanism of the formation of the corresponding Dewar isomer has been analyzed theoretically.[161] The perfluoropyridine photoisomerization is the only case where an azabenzvalene is proposed to be an intermediate in the course of the photoisomerization of a pyridine compound. Such isomers have been proposed as intermediates in the formation of a pyridine from cyclopropenyl-2*H*-azirine.[162]

Pentachloropyridine is subject to C1-C3 homolytic bond cleavage upon UV irradiation. The radical thus formed reacts with solvent.[163–165] Similar C-C1 bond cleavage occurs in the case of compounds such as 115 or 116. In these two cases, tricyclic compounds are formed by cyclization of the intermediate radical.[164] The behavior of chloropyridines is therefore similar to that of various substituted chlorobenzenes.[166]

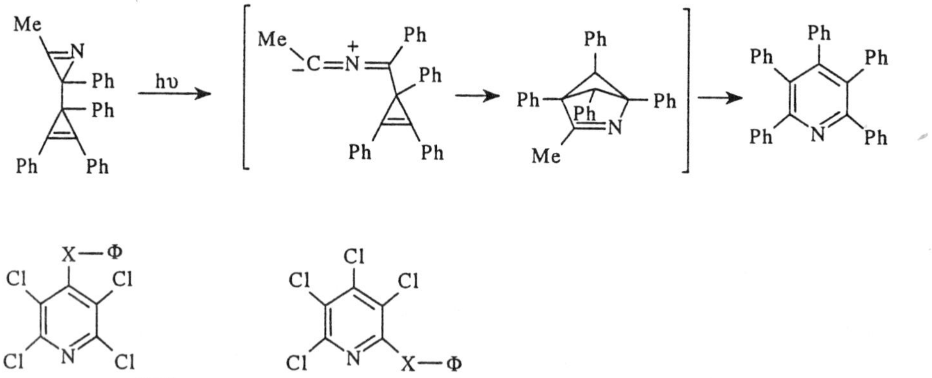

(115)

(116)

X = S, O, NH

SCHEME 82.84

When the pyridine ring is substituted by perfluoroalkyl groups, a Dewar pyridine is formed by UV irradiation. It is converted by further irradiation to azaprismane.[167–170] In each case, a 1-azabicyclo[2,2,0]hexadiene derivative [e.g., 117] is formed together with the azaprismanes, the structures of which are deduced from those of the pyridines [e.g., 120, 121, and 122] formed on rearomatization. To account for the structures of the azaprismanes, it is proposed that rearrangement of the 2-aza-bicyclo[2,2,0]hexadiene derivative formed initially takes place to yield the 1-aza isomer. Thus, whereas 117 can yield 119 and thence 122, it would appear that the formation of the azaprismane 118 requires the 1-aza-compound 123, which may be produced by rearrangement of the 2-aza-isomer 124 as shown in Scheme 82.85.[177]

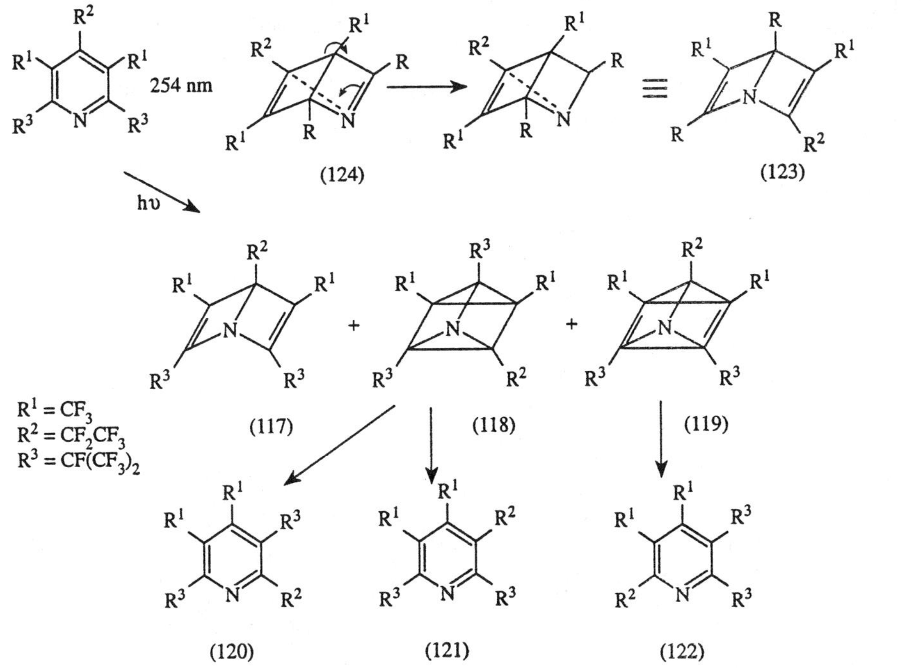

$R^1 = CF_3$
$R^2 = CF_2CF_3$
$R^3 = CF(CF_3)_2$

SCHEME 82.85

Some other Dewar pyridines have been isolated[171,173] by the use of a slow transference technique.[172] The aza-Dewar benzenes revert to the heteroarenes at 160°C, but are generally unchanged at room temperature after 6 months.

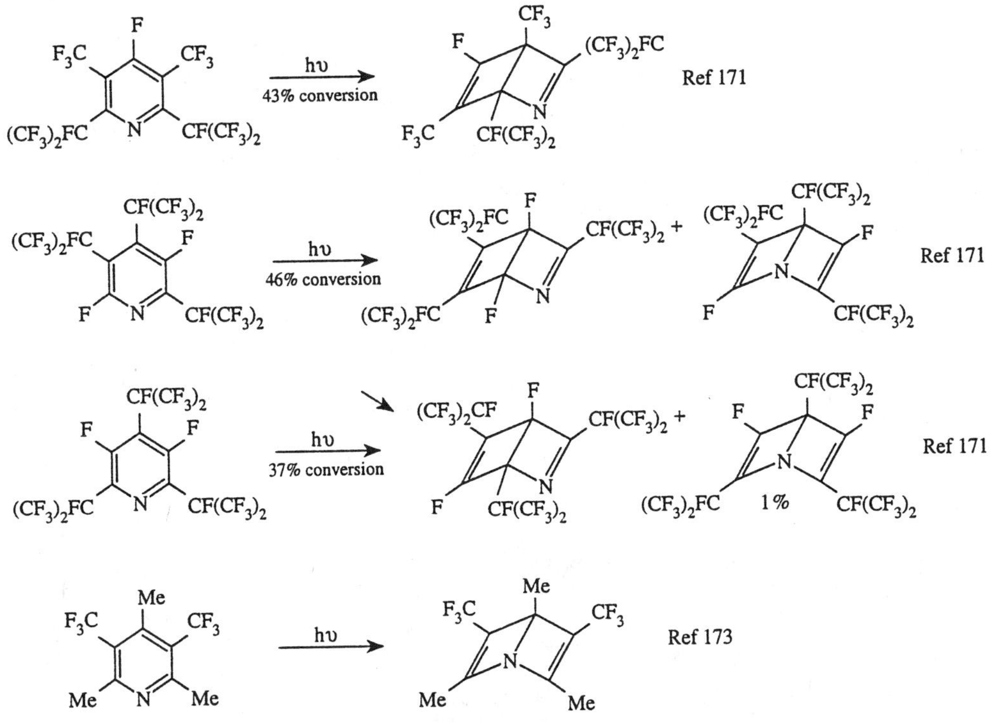

SCHEME 82.86

The tri-*t*-butylpyridines 125 are converted similarly into 2-azabicyclo[2,2,0]hexa-2,5-dienes 126, a process that can easily be reversed thermally by heating to 140°C; whereas, the 1-azabicyclo[2,2,0]-hexa-2,5-dienes 127 undergo intramolecular photocycloaddition to give the azaprismanes 128.[174]

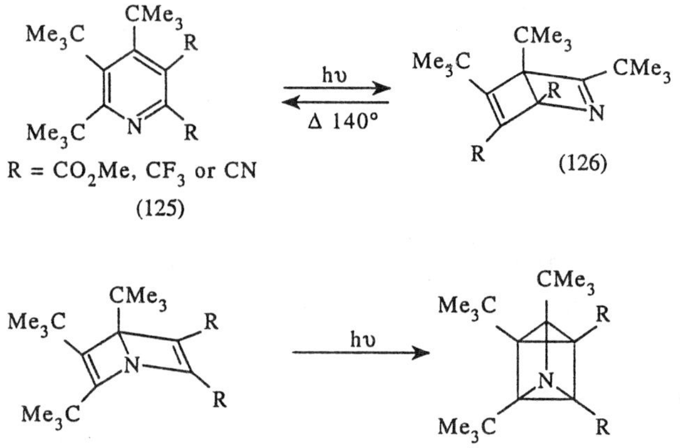

R = CO_2Me, CF_3 or CN
(125)
(126)
R = CF_3 or, CO_2Me
(127)
(128)

SCHEME 82.87

Photorearrangement of 3,5-dicarboalkoxypyridines proceeds probably by the intermediacy of a Dewar pyridine and an azaprismane, but none of the postulated intermediates have been isolated or characterized.[175,176]

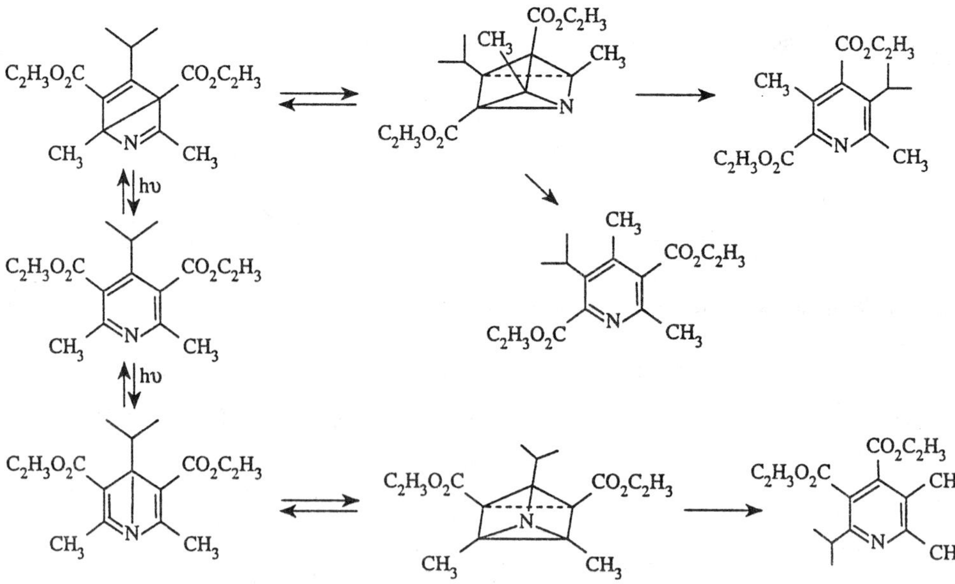

SCHEME 82.88

A rearrangement to a pyrrole occurs during the photoreaction of 3,5 dicarboethoxy-2,4,6-trimethylpyridine with ethanol or methanol. It can take place as shown in Scheme 82.89.[175,176]

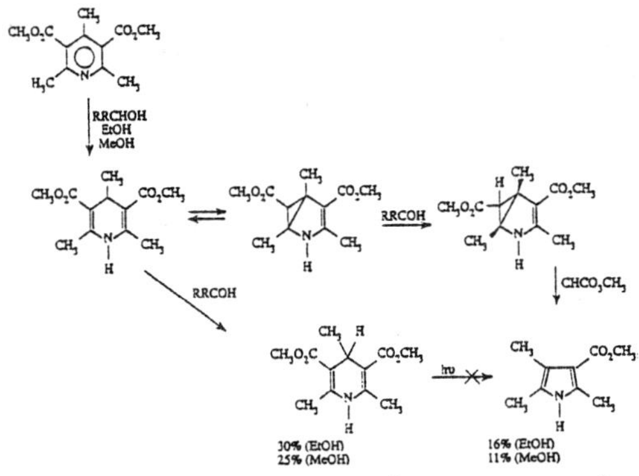

SCHEME 82.89

Polycyclic Monoaza Six-Membered Aromatic Compounds

By comparison with some naphthalene derivatives,[177] quinoline, isoquinoline and their methyl derivatives are not isomerized by UV irradiation. Acridine, phenanthridine, and their superior homologs are like anthracene and phenanthrene, and nearly all the aromatic hydrocarbons with more than three rings are not photoisomerized by UV irradiation. An intramolecular double proton transfer occurs in the lowest excited singlet state of 7-azaindole hydrogen-bonded dimers.[183]

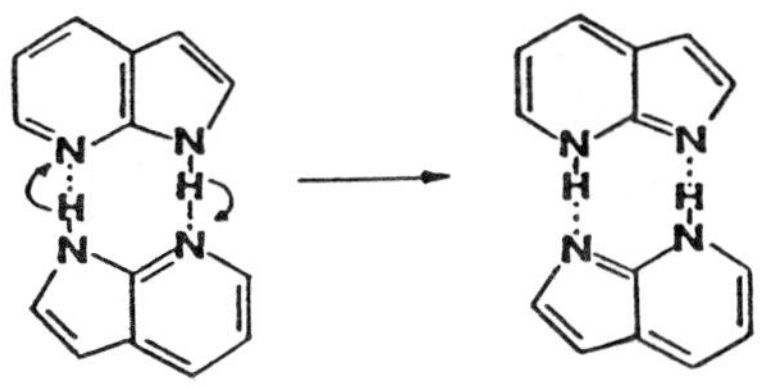

SCHEME 82.90

Methylpyridinium Chloride

When the nitrogen is quaternized by a methyl group, the photochemical behavior of the pyridine ring is changed.[178] Thus, methylpyridinium chloride, when photoexcited to its π,π^* state, leads to 1-methyl-ozaniabenzvalene. This can be trapped by water and methanol. The photoreactions of 3,4,5-trideuteropyridine and of some methyl picolinium chlorides or lutidinium chlorides show that the 1-methyl-ozaniabenzvalene leads to two differents cations **129** or **130**, which react further with the nucleophile. In products derived from cation **129** (path a, Scheme 82.91), there is no isomerization of the atoms of the ring; whereas rearrangement takes place in products derived from the cations **130** (path b, Scheme 82.91).

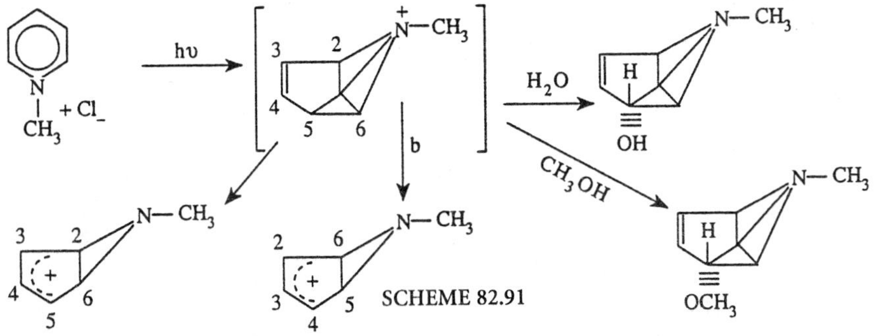

SCHEME 82.91

The methyl group can exert a strong directive influence on the nature of the cations formed.[178]

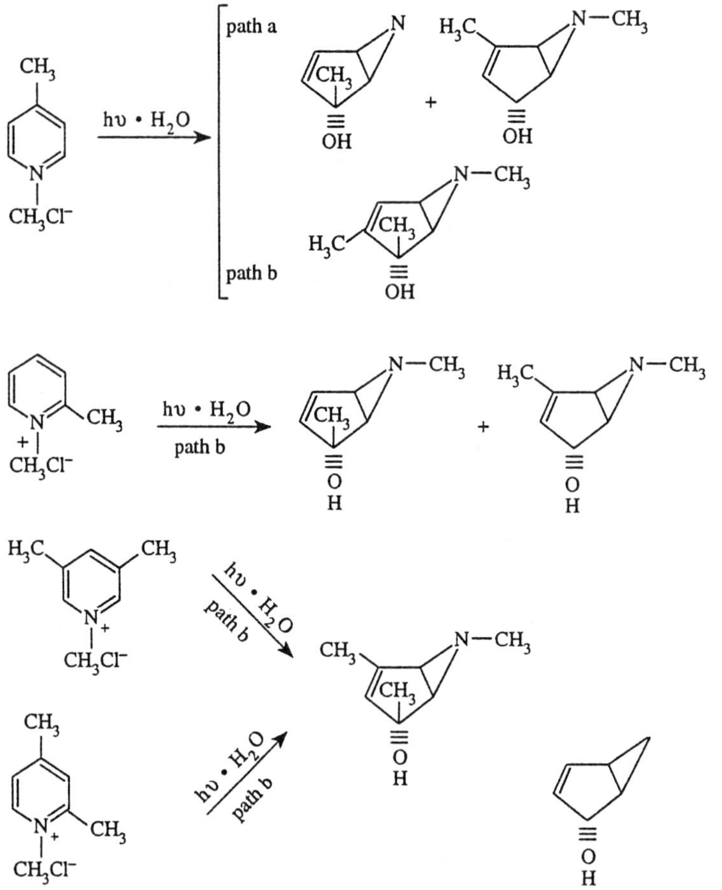

SCHEME 82.92

These reactions are similar to those encountered with benzene, which, when irradiated in acidified water, gives **131**.[179] A corresponding photohydration occurs in the case of protonated pyridines.[178] Thus, when 3,5-lutidine is irradiated in acidic D_2O, the compound disappears with a quantum yield of 0.1 and 2,4-lutidine-D_2 is formed subsequently. These reactions suggest that Dewar pyridine, which is formed only when the nitrogen of the pyridine is not substituted, arises from an n,π^* excited state.

The photobehavior of pyridinium iodide (Scheme 82.93) is different from pyridinium chloride. The iodine ion has a low ionization potential and an electron transfer occurs in the excited state.[180] Methyl-1-pyridinium, methyl-1-collidinium, and methyl-1-quinolinium iodides are subject to similar reactions.[181]

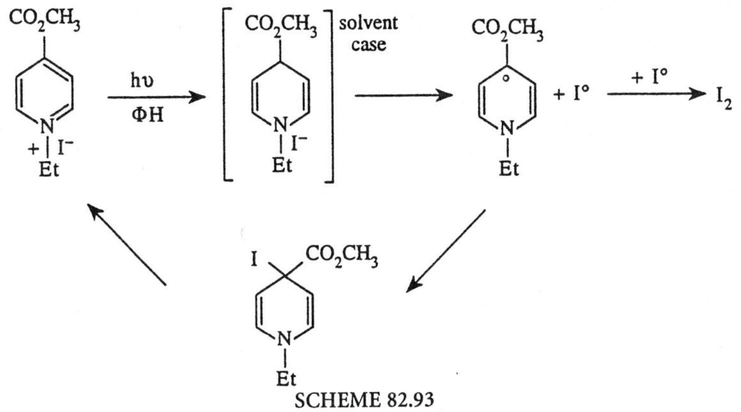

SCHEME 82.93

A photochemically induced 1,4-aryl shift occurs when 1-aryl-2-(2′-benzimidazolyl)-4,6-diphenylpyridinium salts are irradiated in methanol and sodium methoxide. They are converted (Scheme 82.94) into 2-[2′-(1′-arylbenzimidazolyl)]4,6-diphenylpyridines.[182]

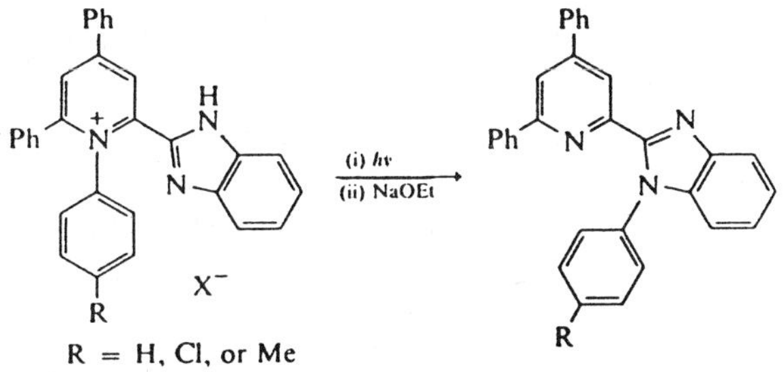

R = H, Cl, or Me

SCHEME 82.94

Diazines

Pyrazines-Pyrimidines

While the photochemical isomerization of pyrazine to pyrimidine in the vapor phase at 254 nm is reversible, this is not the case in isooctane solution.[184,185] Almost no reaction appears to take place at 313 nm.[185,186] Furthermore, it was observed that irradiation of 2-methylpyrazine in the vapor phase leads to a mixture of 4- and 5-methylpyrimidine and an unidentified compound. The structure of the pyrimidines obtained from the irradiation of 2,5- and 2,6-dimethylpyrazine was taken as evidence for the intermediacy of diazabenzvalenes.[185] An attempt to trap a diaza-Dewar benzene with sodium borohydride from the irradiation of pyrazine failed. Instead, the hexahydro derivative, piperazine, was formed.[151] The quantum yield for isomerization of pyrimidine or pyrazine in the vapor phase was found to be wavelength dependent. It was observed to be about 100 times higher at 254 nm than at 313 nm.[186] It was also found that the intersystem crossing efficiency from S_2 (π,π^*) was 0.1.[187] From these studies and from the study of the $Hg(^3P_1)$-sensitized isomerization of some pyrimidines and pyrazines,[188] it was inferred that a high-energy triplet or vibrationally excited T_1 state was involved in the photoisomerization of these diazines (Scheme 82.95).

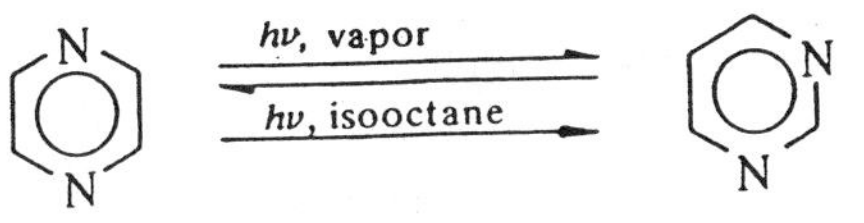

SCHEME 82.95

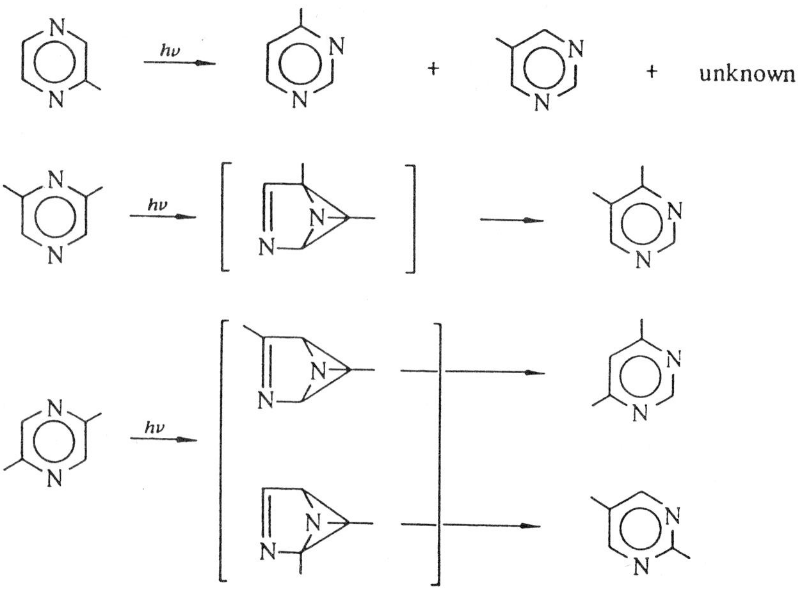

SCHEME 82.95 (continued)

The ring opening of 2,6-dimethyl-4-aminopyrimidine (Scheme 82.96) under alkaline aqueous conditions is believed to proceed through a Dewar intermediate.[189,190]

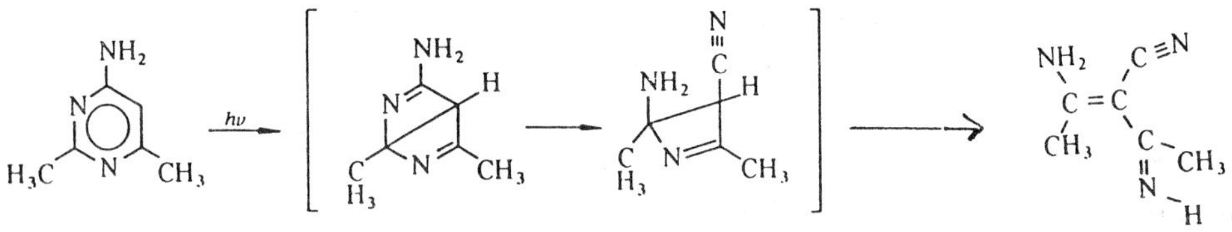

SCHEME 82.96

No photochemical isomerization of quinoxaline, phenazine, benzopyrimidine, or higher homologs has been detected. The phototransformation of benzo(3,4)cyclobuta(1,2b)quinoxaline 132 is presumed to involve initial ring opening to a biradical, which abstracts hydrogen from the solvent and forms the product 133.[191]

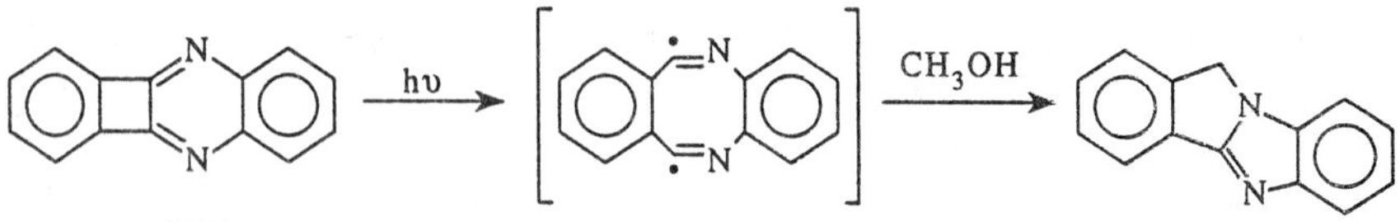

(132)   (133)

SCHEME 82.97

Pyridazines

Pyridazine is photoisomerized in the vapor phase to give a mixture of pyrimidine and pyrazine.[186]

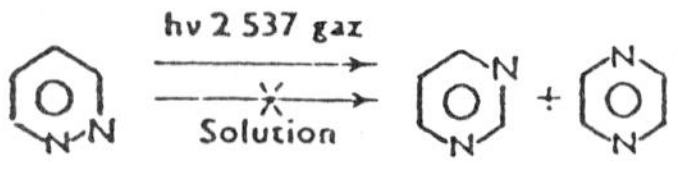

SCHEME 82.98

When irradiated in the gas phase ($\lambda = 365$–366 nm, 100–160°C), pyrazine is photodecomposed to N_2 and vinylacetylene.[186b] Chlorinated pyridazines have been shown to undergo photoisomerization to chlorinated pyrazines.[192] Tetrachloropyridazine, for example, affords tetrachloropyrazine on irradiation in perhalogenated solvents. Prolonged irradiation leads to the formation of tetrachloropyrimidine. The reactive excited state is S_1 (n,π^*); an activation barrier of 16 kJ mol^{-1} is encountered as the excited state of tetrachloropyrazine rearranges to tetrachloropyrazine.

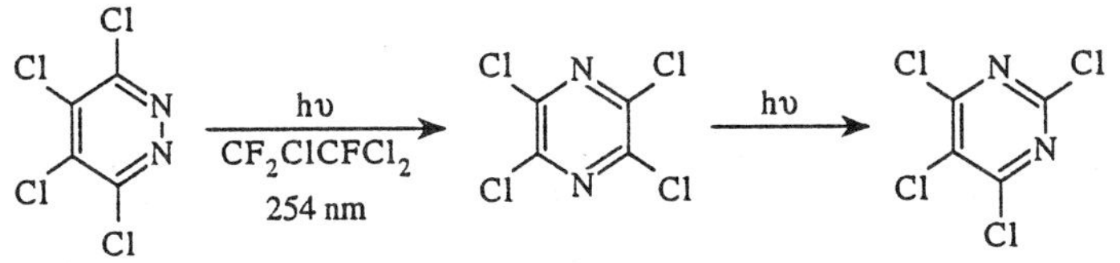

SCHEME 82.99

No intermediates have been detected, but the mechanism of these reactions can be as shown in Scheme 82.100.[192]

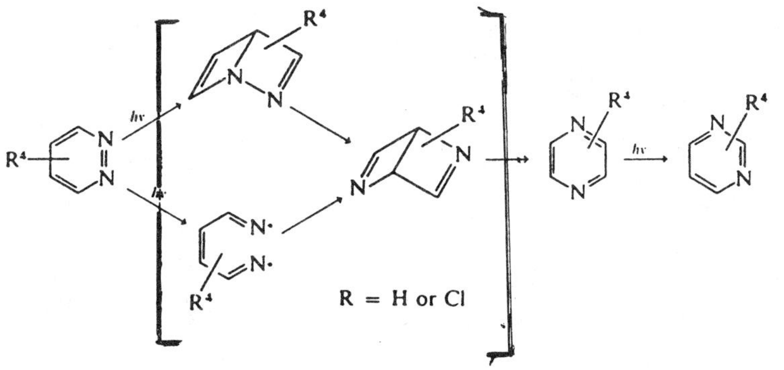

SCHEME 82.100

Perfluoropyridazine is photoisomerized to perfluoropyrazine, which in turn can be photoisomerized to perfluoropyrimidine in a much less efficient reaction.[193]

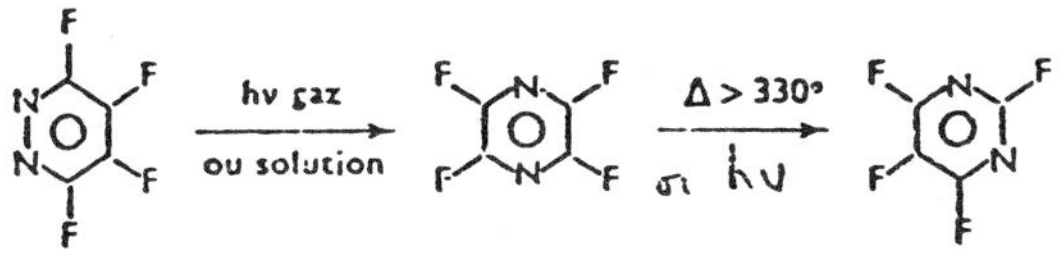

SCHEME 82.101

The compound 134, however, has been detected as a transient, formed on flash photolysis of tetrafluoropyridiazine.[194] Analogous ring contractions were observed in perfluorinated pyrimidine and pyrazine.

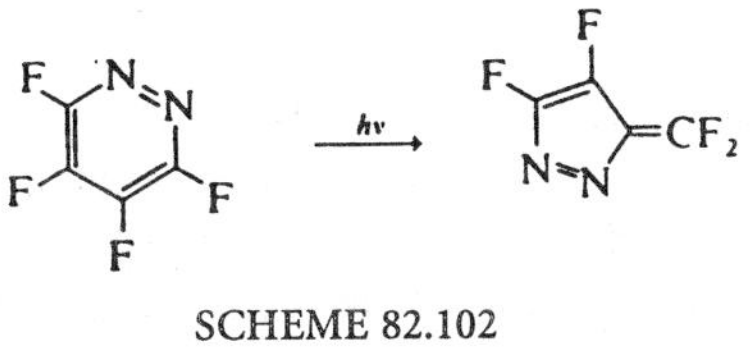

SCHEME 82.102

Diaza-Dewar benzenes have been isolated in some cases,[169,172,195,196] as illustrated below.

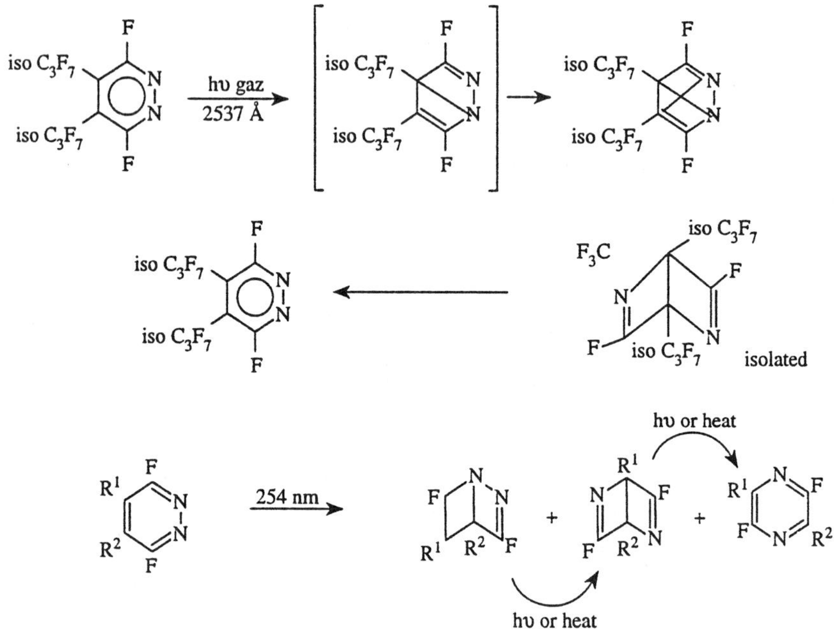

SCHEME 82.103

The photoreactivity shown by the perfluoroalkyl compounds can be rationalized as follows.[196] Intermediates **135** and **136** have been isolated, and the reaction path does not involve a diazaprismane as was suggested initially.[197]

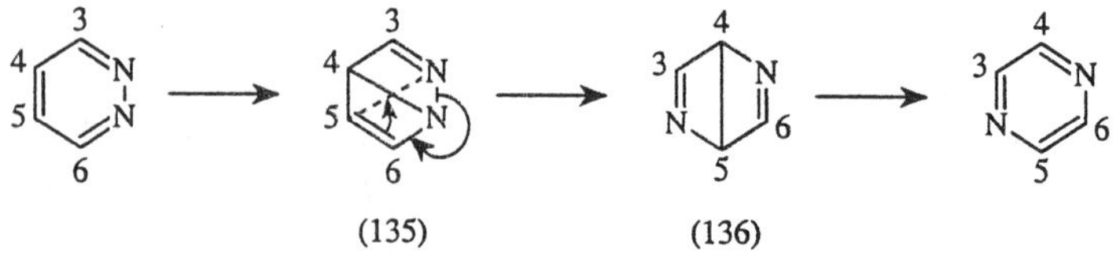

SCHEME 82.104

Irradiation of the fluorinated pyridazine **137** in a flow system gave products that are dimers of a possible intermediate azete **138**, formed, presumably along with the nitrile **139**, by thermal fragmentation of the *para*-bonded isomer **140**.[198] The corresponding pyrazine **141** and the diazabicyclo[2,2,0]hexadiene **142** were also obtained, as well as dimers of the azete **138**.

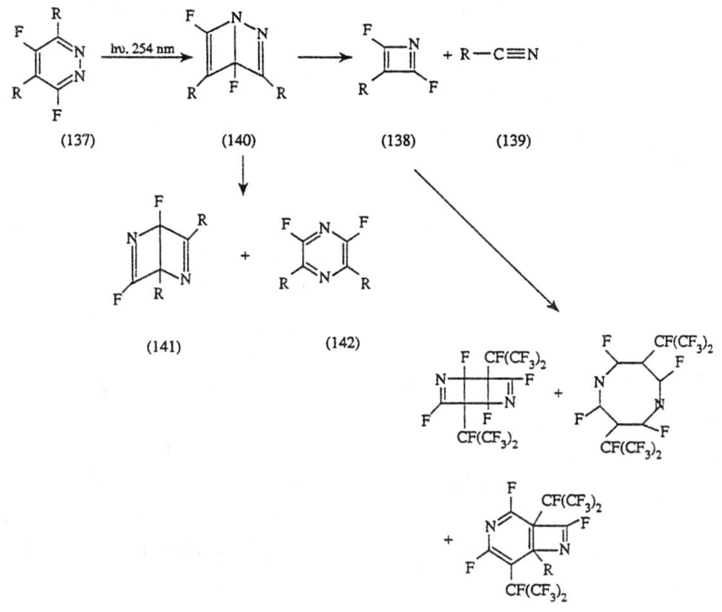

SCHEME 82.105

The thermal rearrangements of some perfluoroalkylpyridazines and of perfluoropyridazines to pyrimidines have been reported. They involve the intermediacy of diazabenzvalene[199] or diazabicyclo propenyls.[200] With one exception, all the benzo-fused derivatives of the diazines studied do not photoisomerize.[201] The exception to this rule is perfluorocinnoline, which photoisomerizes to perfluoroquinazoline.[201]

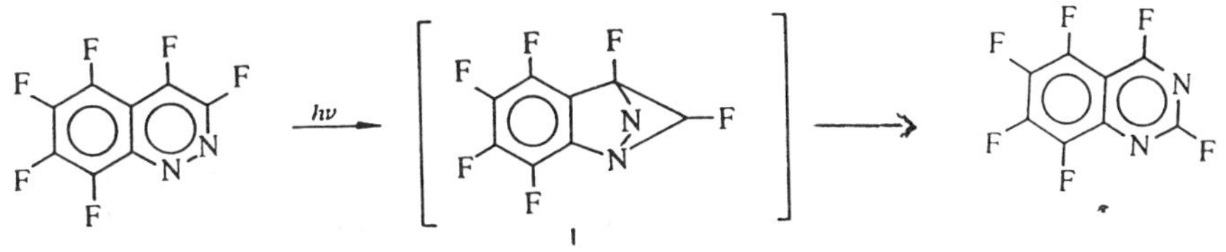

SCHEME 82.106

Triazines and Tetrazines

No photoisomerization has been observed upon irradiation of triazines and tetrazines. Irradiation as a rule leads to dissociation.[202] In the case of *sym*-tetrazines, nitrogen elimination is a two-photon process.[203,204] The involvement of isomeric forms in the photodecomposition of 1,4-*sym*-tetrazines is postulated.[205]

eferences

1. Phillips, D., Lemaire, J., Burton, C. S., and Noyes, W. A., Jr., Isomerization as a route for radiation-less transitions, in *Advances in Photochemistry*, Vol. 5, Noyes, W. A., Jr., Hammond, G. S., and Pitts, J. N., Jr., Eds., Interscience, New York, 1968, 329.
2. Forster, Th., Diabatic and adiabatic processes in photochemistry, *Pure Appl. Chem.*, 24, 443, 1970.
3. Zupan, N., and Sket, B., Photochemistry of Fluorosubstituted aromatic and heteroaromatic molecules, *Isr. J. Chem.*, 17, 92, 1978.

4. Beak, P., and Messer, R. W., The photoisomerization of heteroaromatic nitrogen compounds, in *Organic Photochemistry*, Vol. 2, Chapman, O. L., Ed., Marcel Dekker, New York, 1969, 117.

5. Lablache-Combier, A. and Remy, M. A., Photoisomerisation des dérivés hétérocycliques à cinq chainons, *Bull. Soc. Chim. Fr.*, 679, 1971.

6. Lablache-Combier, A., Photoisomerisation de composés aromatiques. B. Composés à 5 chainons, *L'Actualité Chimique*, 9, n° 7, Dec. 1973.

7. Lablache-Combier, A., Photoisomerization of five membered heterocyclic compounds, in *Photochemistry of Heterocyclic Compounds*, Buchardt, O., Ed., John Wiley & Sons, New York, 1976, 123.

8. Padwa, A., Photochemical rearrangements of five membered ring heterocycles, in *Rearrangements in Ground and Excited States*, Vol. 3, de Mayo, P., Ed., Academic Press, New York, 1980, 501.

9. Lablache-Combier, A., Photoisomerisation de composés aromatiques. A. Composés à 6 chainons, *L'Actualité Chimique*, 6, n° 6, Nov. 1973.

10. Lablache-Combier, A., Réactions de photoaddition et de photosubstitution de dérivés aromatiques azotés à 6 chainons, *Eléments de Photochimie Avancée*, Courtot, P., Ed., Hermann, Paris, 1972, 229.

11. Ivanoff, N. and Lahmani, F., Isomérisations photochimiques du noyau aromatique, in *Eléments de Photochimie Avancée*, Courtot, P., Ed., Hermann, Paris, 1972, 131.

12. Lablache-Combier, A., Photoisomerization of six membered heterocyclic compounds, *Photochemistry of Heterocyclic Compounds*, Buchardt, O., Ed., John Wiley & Sons, New York, 1976, 207.

13. Bryce-Smith, D. and Gilbert, A., The organic photochemistry of benzene, *Tetrahedron Lett.*, 32, 1309, 1976.

14. Bryce-Smith, D. and Gilbert, A., Rearrangement of benzene ring, *Rearrangements in Ground and Excited States*, Vol. 3, de Mayo P., Ed., Academic Press, New York, 1980, 349.

15. Hoffmann, R. and Woodward, R. B., The conservation of orbital symmetry, *Acc. Chem. Res.*, 1, 17, 1968.

16. Van Tamelen, E. E. and Whitesides, T. H., Valence tautomers of heterocyclic aromatic species, *J. Am. Chem. Soc.*, 93, 6129, 1971.

17. Chung Wu, E. and Heicklen, J., Photolysis of pyrrole vapor at 2139 Å and room temperature, *J. Am. Chem. Soc.*, 93, 3432, 1971.

18. Chung Wu, E. and Heicklen, J., Photolysis of 2,5- and 2,4-dimethylpyrrole vapors at room temperature, *Can. J. Chem.*, 50, 1678, 1972.

19. Patterson, J. M. and Burka, L. T., The photoisomerization of benzylpyrroles, *Tetrahedron Lett.*, 2215, 1969.

20. Patterson, J. M., Ferry, J. D., and Boyd, M. R., Photoisomerization of (substituted allyl)dialkylpyrroles, *J. Am. Chem. Soc.*, 95, 4356, 1973.

21. Moody, C. J. and Ward, J. G., Synthesis and photochemical rearrangement of 1-allyl-1,8-dihydropyrrole [2,3-*b*] indoles, *J. Chem. Soc., Chem. Commun.*, 646, 1984.

22. Padwa, A., Gruber, R., and Pashayan, D., Contrast of photochemical and pyrolytic models for mass spectral fragmentation. Dealkylation of *N*-alkyl-2,4-diphenylpyrroles, *Tetrahedron Lett.*, 3659, 1968.

23. Patterson, J. and Bruser, D. M., Photoisomerization of some substituted *N*-acetylpyrroles, *Tetrahedron Lett.*, 2959, 1973.

24. Shizuka, H., Ono, S., Morita, T., and Tanaka, I., A theoretical consideration of the isomerization of *N*-acetylopyrrole, *Mol. Photochem.*, 3, 203, 1971.

25. Bellus, D., Photo-Fries rearrangement and related photochemical [1,j] shifts (j = 3,5,7) of carbonyl and sulfonyl groups, in *Advances in Photochemistry*, Vol. 8, Pitts, J. N., Jr., Hammond, G. S., and Noyes, W. A., Jr., Eds. Wiley Interscience, New York, 1971, 109.

26. Houwen, O. H. and Tavares, D. F., Mass spectral rearrangement. Photoisomerization and photosolvolysis of *bis*(2-acetylpyrrol-1-yl) methane, *Tetrahedron Lett.*, 4995, 1978.

27. Ban, Y., Yoshida, K., Goto, J., and Oishi, T., Novel photoisomerization of 1-acylindoles to 3-acylindolenines: General entry to the total synthesis of Strychnos and Aspidosperma Alkaloids, *J. Am. Chem. Soc.*, 103, 6990, 1981.

28. Ban, Y., Yoshida, K., Goto, J., Oishi, T., and Takeda, E., A synthetic road to the forest of Strychnos, Aspidosperma, Schizozygane and Eburnamine alkaloids by way of the novel photoisomerization,

Tetrahedron, 39, 3657, 1984.

29. Hiraoka, H., Photoisomerization of 2-cyanopyrroles, *J. Chem. Soc., Chem. Commun.,* 1306, 1970.

30. Barltrop, J. A., Day, A. C., Moxon, P. D., and Ward, R. R., Permutation patterns and the phototranspositions of 2-cyanopyrroles, *J. Chem. Soc., Chem. Commun.,* 786, 1975.

31. Behrens, S. and Jug, K., Sindo 1 study of the photoisomerization of 2-cyanopyrroles to 3-cyanopyrroles, *J. Org. Chem.,* 55, 2288, 1990.

32. Behrens, S. and Jug, K., Coupling of electronic and nuclear motion in the photochemical internal cyclization of pyrrole, *Chem. Phys. Lett.,* 170, 377, 1990.

33. Barltrop, J. A., Day, A. C., and Ward, R. W., Phototransposition of 2-cyanopyrroles: Evidence for the intermediacy of 5-azabicyclo[2.1.0.]pen-2-enes, *J. Chem. Soc., Chem. Commun.,* 131, 1978.

34. Barton, T. J. and Hussmann G. P., Photoisomerization of 2-(trimethylsilyl)pyrroles, *J. Org. Chem.,* 50, 5881, 1985.

35. Kobayashi, Y., Ando, A., Kawada, K., and Kumadaki, I., Derivation of tetrakis(trifluoromethyl)(Dewar pyrroles) from tetrakis (trifluoromethyl)(Dewar thiophene), *J. Org. Chem.,* 45, 2966, 1980.

36. Tiefenthaler, H., Dorschelen, W., Goth, H., and Schmid, H., Photoisomerisierung von Pyrazolen und Indazolen zu Imidazolen — bzw. Benzimidazolen und 2-Aminobenzonitrilen, *Helv. Chim. Acta,* 50, 2244, 1967.

37. Labhart, H., Heinzelmann, W., and Dubois, J. P., On the search for the mechanism of photoreactions of som heterocyclic compounds, *Pure Appl. Chem.,* 24, 495, 1970.

38. Beak, P., Miesel, J. L., and Messer, W. R., The photoisomerization of 1,4,5-trimethylimidazole and 1,3,5 trimethylpyrazole, *Tetrahedron Lett.,* 5315, 1967.

39. Beak, P. and Messer, W., Photorearrangements of some *N*-methyldiazoles, *Tetrahedron,* 25, 3287, 1969.

40. Grimshaw, J. and Mannus, D., Photocyclization and photoisomerisation of 1,3,4- and 1,4,5-triphenylpyrazole, *J. Chem. Soc., Perkin Trans. 1,* 2096, 1977.

41. Koch, T. H. and Rodehorst, R. M., A quantitative investigation of the photochemical conversion of diaminomaleonitrile to diaminofumaronitrile and 4-amino-5-cyanoimidazole, *J. Am. Chem. Soc.,* 96, 6707, 1974.

42. Ferris, J. P., Sanchez, R. A., and Orgel, L. E., Studies in prebiotic synthesis. Synthesis of pyrimidines from cyanoacetylene and cyanate, *J. Mol. Biol.,* 33, 693, 1968.

43. Ferris, J. P. and Kuder, K. E., The photochemical conversion of enaminonitriles to imidazoles, *J. Am. Chem. Soc.,* 92, 2527, 1970.

44. Ferris, J. P. and Trimmer, R. W., Photochemical conversion of enaminonitriles to imidazoles, *J. Org. Chem.,* 41, 19, 1976.

45. Kagan, J. and Melnick, B., The synthesis and photochemistry of 4-amino-3-cyanopyrazoles, *J. Heterocycl. Chem.,* 16, 1113, 1979.

46. Barltrop. J. A., Day, A. C., Mack, A. G., Shahrisa, A., and Wakamatsu, S., Competing pathways in the phototransposition of pyrazoles, *J. Chem. Soc., Chem. Commun.,* 604, 1981.

47. Wakamatsu, S., Barltrop, J. A., and Day, A. C., Photochemical conversion of pyrazoles to enaminotriles and imidazoles, *Chem. Lett.,* 667, 1982.

48. Tiefenthaler, H., Dorscheln, W., Goth, H., and Schmid, H., Photoisomerisation von Indazolen zu Benzimidazolen, *Tetrahedron Lett.,* 2999, 1964.

49. Bircher, P., Pantke, E. R., and Labhart, H., Temperature dependence of the deactivation of electronically excited indazoles in solution, *Chem. Phys. Lett.,* 11, 347, 1971.

50. Dubois, J. P. and Labhardt, H., Untersuchungen zur Photochemie von 2-methylindazol, *Chimia,* 23, 109, 1969.

51. Ferris, J. P., Prabhu, K. V., and Strong, R. L., Photochemical rearrangements of indazoles. Investigation of the triplet excited states of 1- and 2-methylindazole, *J. Am. Chem. Soc.,* 97, 2835, 1975.

52. Heinzelmann, W., Marky, M., and Gilgen, P., Zum Mechanismus der Photochemischen Umwandlung von 2-Alkyl-indazolen in 1-Alkyl-benzimidazole. I Struktur un Reaktivität eines Zwischenproduktes, *Helv. Chim. Acta,* 59, 1512, 1976.

53. Heinzelmann, W., Marky, M., and Gilgen, P., Zum Mechanismus der Photochemischen Umwandlung von 2-Alkylindazolen in 1-Alkylbenzimidazolen, *Helv. Chim. Acta,* 59, 1528, 1976.

54. Heinzelmann, W., Marky, M., and Gilgen, P., Zum Mechanismus der Photochemie von 2-Alkylindazolen in Wasserigen Lösungen, *Helv. Chim. Acta,* 59, 2362, 1976.

55. Heinzelmann, W., Tieftemperatur-Photochemie von 2-Alkylindazolen, *Helv. Chem. Acta,* 61, 618, 1978.

56. Ullman, E. F. and Singh, B., Photochemical transposition of ring atoms in five-membered heterocycles. The photorearrangement of 3,5-diphenylisoxazole, *J. Am. Chem. Soc.,* 88, 1844, 1966.

57. Singh, B. and Ullman, E. F., Photochemical transposition of ring atoms in 3,5-diarylisoxazoles, an unusual example of wavelength control in a photochemical reaction of azirines, *J. Am. Chem. Soc.,* 89, 6911, 1967.

58. Ullman, E. F., Excited intermediates in solution photochemistry, *Acc. Chem. Res.,* 1, 353, 1968.

59. Singh, B., Zweig, A., and Gallivan, J. B., Wavelength dependent photochemistry of 2-aroyl-3-aryl-2*H*-azirines — Mechanistic studies, *J. Am. Chem. Soc.,* 94, 1199, 1972.

60. Padwa, A., Smolanoff, J., and Tremper, A., Intramolecular cycloaddition reactions of vinylsubstituted 2*H*-azirines, *J. Am. Chem. Soc.,* 97, 4682, 1975.

61. Tanaka, H., Matsushita, T., Osamura, Y., and Nishimoto, K., M.O. study on the photochemical isomerization of isoxazole, *Int. J. Quantum Chem.,* 18, 463, 1980.

62. Tanaka, H., Osamura, Y., and Matsushita, T., An M.O. study of the reaction mechanism of photoisomerization from isoxazole via azirine intermediate to oxazole, *Bull. Chem. Soc. Jpn.,* 54, 1293, 1981.

63. Murature, D. A., Perez, J. D., De Bertorello, M. M., and Bertorello, H. E., 3-Acetyl-2-methyl-1-azirine as photoproduct and the intermediate in the thermal isomerization of 3,5-dimethylisoxazole into 2,5-dimethyloxazole, *An. Assoc. Quim. Argentina,* 64, 337, 1976.

64. Ferris, J. P., Antonucci, F. R., and Trimmer, R. W., Mechanism of the photoisomerization of isoxales and 2-cyanophenol to oxazoles, *J. Am. Chem. Soc.,* 95, 919, 1973.

65. Ferris, J. P. and Trimmer, R. W., Mechanistic studies on the photochemical reactions of isoxazoles, *J. Org. Chem.,* 41, 13, 1976.

66. Albanesi, S. and Marchesini, A., Photochemical behaviour of the 16-methyl(3,5)[11] isoxazolophane and of the related 14-methyl-15-azabicyclo[12.1.0] pentadec-15(1)-en-13-one, *Tetrahedron Lett.,* 1875, 1979.

67. Beccalli, E. M., Majori, L., Marchesini, A., and Torricelli, C., Synthesis of oxazolo- and imidazolophanes, *Chem. Lett.,* 659, 1980.

68. Nishiwaki, T., Nakano, A., and Matsuoka, M., Thermally induced dimerization of 5-aminoisoxazoles and 2*H*-azirines and photochemistry of 5-amino isoxazoles, *J. Chem. Soc. C,* 1825, 1970.

69. Nishiwaki, T. and Fujiyama, F., Cleavage of 5-benzylaminooxazoles. Photoproducts of *N*-benzyl-2*H*-azirine-2 carboxamides by dialkyl phosphite, *J. Chem. Soc., Perkin Trans. 1,* 1456, 1972.

70. Lipshutz, B. H. and Reuter, D. C., Cyclopeptide alkaloid model studies, *Tetrahedron Lett.,* 29, 6067, 1988.

71. Murature, D. A. and De Bertorello, M. M., Estudio fotoquimico en solucion del 5-amino-3,4-dimethylisoxazol, *An. Assoc. Quim. Argentina,* 69, 177, 1981.

72. Adembri, G., Camparini, A., Donati, D., and Ponticelli, F., Photochemical rearrangement of isoxazol-5-ylhydrazines and *N*-(3-methyl-4-phenyl isoxazol-5-yl) acetamide, *Tetrahedron Lett.,* 4439, 1978.

73. Kille, G. and Fleury, J. P., Propriétés des amino-5 oxazoles, *Bull. Soc. Chim. Fr.,* 4631, 1968.

74. Wamhoff, H., Zur Photosensibilisierten Isomerisierung von 2,5-Dihydrofuranen und Isoxazolen, *Chem. Ber.,* 105, 748, 1972.

75. Kurtz, D. W. and Schechter, H., Photolysis of 2-Diazo-2-phenylacetophenone.0-benzyloxine and its subsequent products, *J. Chem. Soc., Chem. Commun.,* 689, 1966.

76. Sato, T., Yamamoto, K., and Fukui, K., The photochemical reaction of 3,5 disubstituted isoxazoles, *Chem. Lett.,* 111, 1973.

76b. Sato, T., Yamamoto, K., Fukui, K., Saito, K., Hayakawa, K., and Yoshiie, S., Metal catalysed organic photoreactions. Photoreactions of 3,5-dimethylisoxazole with and without catalytic assistance by Copper II salts, *J. Chem. Soc., Perkin Trans. 1*, 783, 1976.

77. Good, H. R. and Jones, G., Synthesis with isoxazoles: The production of an isoxazolo[2,3-*a*]pyridinimine salt and the photochemical conversion of isoxazole-3-carboxylates into oxazole-2-carboxylates, *J. Chem. Soc. C*, 1196, 1971.

78. Nishiwaki, T., Kitamura, T., and Nakano, A., A novel synthesis of 1-azirine having an ester function and observation of their mass spectra, *Tetrahedron*, 26, 453, 1970.

79. Sauers, R. R. and Van Arnum, S. D., Some novel isoxazole photochemistry: A comparison with vinylazide chemistry, *Tetrahedron Lett.*, 28, 5797, 1987.

80. Goth, H., Gagneux, A. R., Eugster, C. H., and Schmid, H., 2(3*H*)-oxazolone durch Photoumlagerung von 3-Hydraoxyisoxazolen-Synthesis von Muscazon, *Helv. Chim. Acta*, 50, 137, 1967.

81. Nakagawa, M., Nakamura, T., and Tomita, K., Photolysis of 3-hydroxyisoxazoles, *Agric. Biol. Chem.*, 38, 2205, 1974.

82. Padwa, A., Chen, E., and Ku, A., Thermal and photochemical valence isomerizations of 4-carbonyl-substituted isoxazoles, *J. Am. Chem. Soc.*, 97, 6484, 1975.

83. Dietliker, K., Gilgen, P., Heimgartner, H., and Schmid, H., Photochemie von in 4-stellung substituierten 5-methyl-3 phenylisoxazolen, *Helv. Chim. Acta*, 59, 2074, 1976.

84. Reference 8, p. 521.

85. Sauers, R. R., Hadel, L. M., Scimone, A. A., and Stevenson, T. A., Photochemistry of 4-acylisoxazoles, *J. Org. Chem.*, 55, 4011, 1990.

86. Sauers, R. R., Hagedorn, A. A., III, Van Arnum, S. A., Gomez, R. P., and Moquin, R. V., Synthesis and photochemistry of heterocyclic norbornenyl ketones, *J. Org. Chem.*, 52, 5501, 1987.

87. Keshava Raro, V., Rajanarendar, E., and Krishna Murty, A., Photoisomerisation of 4-benzalamino-3-methyl-5 styrylisoxazole, *Ind. J. Chem.*, 27B, 80, 1988.

88. Padwa, A. and Cohen, L. A., Aza-claisen rearrangements in the 2-allyloxy substituted oxazole ring, *Tetrahedron Lett.*, 915, 1982.

89. Goth, H. and Schmid, H., Photoisomerisierung von Benzisoxazolen zu Benzoxazolen, *Chimia*, 20, 148, 1966.

90. Heinzelmann, W. and Marky, M., Photochemie von Benzisoxazolen, *Helv. Chim. Acta*, 57, 376, 1974.

91. Ferris, J. P. and Antonucci, F. R., Synthesis of heterocycles by photochemical cyclization of *ortho*-substituted benzene derivatives, *J. Chem. Soc., Chem. Commun.*, 126, 1972.

92. Ferris, J. P. and Antonucci, F. R., Photochemistry of *ortho*-substituted benzene derivatives and related heterocycles, *J. Am. Chem. Soc.*, 96, 2010, 1974.

93. Ferris, J. P. and Antonucci, F. R., Mechanisms of the photochemical rearrangements of *ortho*-substituted benzene derivatives and related heterocycles, *J. Am. Chem. Soc.*, 96, 2014, 1974.

94. Georgarakis, M., Doppler, Th., Marky, M., Hansen, H. J., and Schmid, H., Photolyse von Indazolen, Benzisoxazolen und Anthranilen in Saurer Losung, *Helv. Chim. Acta*, 54, 2916, 1971.

95. Doppler, T., Schmid, H., and Hansen, H. J., Photolyse von 3-Methyl-2,1 benzisoxazol (3-methylanthranil) und 2-Azido-acetophenon in Gegenwart von Schwefel Säure und Benz olderivaten, *Helv. Chim. Acta*, 62, 304, 1979.

96. Giovannini, E. and De Soussa, B. F. S. E., Photolyse des 3-Phenyl-2,1-benzisoxazol und Einiger Seiner Derivate in Bromwasserstoffsaüre, *Helv. Chim. Acta*, 62, 198, 1979.

97. Doppler, T., Schmid, H., and Hansen, H. J., Zur Photochemie von 1,2 Benzisoxazolen in Stark Saurer Lösung, *Helv. Chim. Acta*, 62, 314, 1979.

98. Grellman, K. H. and Tauer, E., Photolysis of benz- and naph-isoxazoles: Evidence for the intermediate formation of azirines, *J. Photochem.*, 6, 365, 1977.

99. Adembri, G., Camparani, A., Donati, D., Ponticelli, F. and Tedeschi, P., Photochemical rearrangements of 3 methylisoxazoles [4,5-*c*] pyridines, *Tetrahedron Lett.*, 22, 2121, 1981.

100. Skötsch, C. and Breitmaier, E., Oxazolo [5,4-*b*] pyridine durch Photoumlagerung von Isoxazolo [5,4-*b*] pyridinen, *Chem. Ber.*, 112, 3282, 1979.

101. Chimichi, S., Tedeschi, P., Nesi, R., and Ponticelli, F., Carbon-13 NMR studies on azolopyridines. The oxazolopyridine systems, *Magn. Res. Chem.*, 23, 86, 1985.

102. Campari, A., Chimichi, S., Ponticelli, F., and Tedeschi, P., Azabenzoxazoles: Synthesis of 3-methylisoxazole [5,4-*c*] and 2-methyloxazole [5,4-*c*] pyridines, *Heterocycles*, 19, 1511, 1982.

103. Donati, D., Fusi, S., and Ponticelli, F., Photochemical rearrangements of 3-methylisoxazolopyridines, *Heterocycles*, 27, 1899, 1988.

104. Donati, D., Ponticelli, F., Bicchi, P. and Meucci, N., Flash photolysis study of 3-methylisoxazolo [5,4-*b*] pyridine, *J. Phys. Chem.*, 94, 5271, 1990.

105. Campari, A., Ponticelli, F., and Tedeschi, P., Synthesis and photochemical reactivity of 3-methylisoxazolo [4,5-*d*] pyridazines, *J. Heterocycl. Chem.*, 22, 1561, 1985.

106. Darlage, L. J., Kinstle, T. H., and MacIntosh, C. L., Photochemical rearrangements of 1,2 benzisoxazolinones, *J. Org. Chem.*, 36, 1088, 1971.

107. Ogata, M., Matsumoto, H., and Kano, H., Photochemical synthesis of 2-benzimidoyl-3-hydroxy-1,4-naphtoquinone and its analogues — A new type of growth regulator, *J. Chem. Soc., Chem. Commun.*, 218, 1973.

108. Ogata, M., Kano, H., and Matsumoto, H., Photorearangement of anthranils into azepines, *J. Chem. Soc., Chem. Commun.*, 397, 1968.

109. Ogata, M., Matsumoto, H., and Kano, H., Photorearrangement of anthranils into azepines, *Tetrahedron.*, 25, 5205, 1969.

110. Berwick, M. A., A comparative study of the photolytic decompositions of 2-azidoacetophenone and 3-methylanthranil, *J. Am. Chem. Soc.*, 93, 5780, 1971.

111. Ahmad, Y., Begum, T., Hussain Qureshi, I., Atta-ur-Rahman, Zaman, K., Changfu, X., and Clardy, J., A novel rearrangement of a papaverine derivative ito isoquino [1,2-*b*] quinazoline derivative, *Heterocycles*, 26, 1841, 1987.

112. Dmitriev, F. M., Gornostaev, L. M., Gritsan, N. P., and El'tsov, A. V., Fotoperegruppirovka 3-ariloksiantra [1,9-*cd*]-6-izoksazolonov v nafto [2,3-*a*] fenoksazin-8,13-diony, *Zh. Org. Khim.*, 21, 2452, 1985.

113. El'tsov, A. V., Dmitrev, F. M., Gornostaev, L. M., and Rtishchev, N. J., Fotoperegruppirorka 3-ariloksiantra [1,9-*cd*]-6-izoksazolonov v nafto [2,3-*a*] fenoksazin-8,13-diony. Fotoliz 1-azido-2-ariloksi-4-oxiantrakhinonov, *Zh. Org. Khim.*, 22, 2361, 1986.

114. Doppler, T., Jansen, H. J., and Schmid, H., Die Photolyse von Anthranilen in Saurer Losung: Verfleich mit der Photochemischen und Thermischen Zersetzung Entsprechender 2-Azido-acylbenzole in Saurer Losung, *Helv. Chim. Acta*, 55, 1730, 1972.

115. Giovannini, E., Rosalez, J., and De Souza, B., Photoinduzierte Nucleophile Substitution von 3-Substituerten 2,1 benzisoxazolen (unter Ringöffnung), *Helv. Chim. Acta*, 54, 2111, 1971.

116. Masaharu, K. and Maeda, M., The photochemical rearrangement of 2,5-diphenyloxazole, *Tetrahedron Lett.*, 2379, 1969.

117. Maeda, M. and Kojima, M., Photorearrangement of 2-phenyloxazole, *J. Chem. Soc., Chem. Commun.*, 539, 1973.

118. Maeda, M. and Kojima, M., Photorearrangements of phenyloxazoles, *J. Chem. Soc., Perkin Trans. 1*, 239, 1977.

119. Tanaka, H., Matsushita, T., and Nishimoto, K., A theoretical study on the photochemical transposition reaction of oxazole, *J. Am. Chem. Soc.*, 105, 1753, 1983.

120. Silberg, I. A., Macarovici, R., and Palibroda, N., Photolysis of 2-orthonitrophenyl-5 aryloxazoles. A new route to 2-aroyl-1*H*-4 quinazolinones, *Tetrahedron Lett.*, 1321, 1976.

121. Buscemi, S., Cicero, M. G., Vivona, N., and Caronna, T., Photochemical behaviour of some 1,2,4-oxadiazole derivatives, *J. Chem. Soc., Perkin Trans. 1*, 1313, 1988.

122. Buscemi, S., Cicero, M. G., and Vivona, N., Heterocyclic photorearrangement. Photochemical behaviour of some 3,5-disubstituted 1,2,4-oxadiazoles in methanol at 254 nm, *J. Heterocyclic Chem.*, 25, 931, 1988.

123. Buscemi, S. and Vivona, N., Heterocyclic photorearrangements. Photoinduced rearrangements of 1,2,4-oxadiazoles substituted by an XYZ side chain sequence, *J. Heterocyclic Chem.*, 25, 1551, 1988.

124. Buscemi, S., Macaluso, G., and Vivona, N., Herterocyclic photorearrangements. Photochemical behaviour of some 3-acetylamino-5-aryl-1,2,4 oxadiazoles. A photo induced isoheterocyclic rearrangement, *Heterocycles*, 29, 1301, 1989.

125. Buscemi, S. and Vivona, N., Heterocyclic photorearrangements — Photoinduced rearrangements of some 3-aroylamino-5 methyl-1,2,4 oxadiazoles, *Heterocycles*, 29, 737, 1989.

126. Buscemi, S. and Vivona, N., Heterocyclic photorearrangements. Some investigations of the photochemical behaviour of 3-acylamino-1,2,4 oxadiazoles. Synthesis of the quinazolin-4-one system, *J. Chem. Soc., Perkin Trans. 2*, 187, 1991.

127. Buscemi, S., Cusmano, G., Gruttadauria, M., Heterocyclic photorearrangements. Photoinduced rearrangement of 3- styril 1,2,4 oxadiazoles, *J. Heterocyclic Chem.*, 27, 861, 1990.

128. Heinzelmann, W., Gilgen, P., Zum Mechanismus der Photochemie des Benzfurazans, *Helv. Chim. Acta*, 59, 2727, 1976.

129. Yavari, I., Esfandiari, S., Mostashari, A. J., and Hunter, P. W. W., Photoreaction of benzofurazan and dimethylacetylenedicarboxylate. Synthesis of isomeric isoxazoles. Carbon-13 nuclear magnetic resonance spectra of isoxazoles and oxazoles, *J. Org. Chem.*, 40, 2880, 1975.

130. Mukai, T., Nittu, S., and Ohine, T., Cyano compound by photochemical reaction, Japan P. 7428176, *Chem. Abstr.*, 83, 43086, 1975.

131. Cantrell, T. S. and Haller, W. S., The photolysis of 1,2,5 oxadiazoles, 1,2,5 thiadiazoles and 2*H*-1,2,3-triazoles, *J. Chem. Soc., Chem. Commun.*, 977, 1968.

132. Mukai, T., Oine, T., and Matsubara, A., Photoreaction of 3,4-diphenyl-1,2,5 oxadiazole, *Bull. Chem. Soc. Jpn.*, 42, 581, 1969.

133. Hoover, R. J. and Kasha, M., Observation of phosphorescence in pyridines, *J. Am. Chem. Soc.*, 91, 6508, 1969.

134. Weisstuch, A. and Testa, A. C., A fluorescence study of aminopyridines, *J. Phys. Chem.*, 72, 1982, 1968.

135. Hotchandani, S. and Testa, A. C., Evidence for inversion of triplet levels in 4-*N* dimethyl-aminopyridine, *J. Chem. Phys.*, 54, 4508, 1971.

136. Schulman, S. G., Capomacchia, A. C., and Rietta, M. S., Lowest excited singlet state pKa* values of the isomeric aminopyridines, *Anal. Chem. Acta*, 56, 91, 1971.

137. Mac Glynn, S. P., Azumi, T., and Kinoshita, M., *Molecular Spectroscopy of Triplet State*, Prentice Hall, Englewood Cliffs, NJ, 1969, 86.

138. Hochstrasser, R. M., and Michaluk, J. W., Excited state dipole moment of pyridine, *J. Chem. Phys.*, 55, 4668, 1971.

139. Evans, D. F., Magnetic perturbation of singlet-triplet transitions, *J. Chem. Soc.*, 3885, 1957.

140. Doering, J. P. and Moore, J. H., Observation of a singlet-triplet transition in gas phase pyridine by ion and electron impact, *J. Chem. Phys.*, 56, 2176, 1972.

141. Noyes, W. A., Jr., and Al-Ani, K. E., The photochemistry of some simple aromatic molecules in the gaseous state, *Chem. Rev.*, 74, 29, 1974.

142. Ross, G., Ultraviolet spectroscopy and excited states of heterocyclic molecules, in *Photochemistry of Heterocyclic Compounds*, Buchardt, O., Ed., John Wiley & Sons, New York, 1976, 1.

143. Wilzbach, K. E., and Rausch, D. J., Photochemistry of nitrogen heterocycles. Dewar pyridine and its intermediacy in photoreduction and photohydratation of pyridine, *J. Am. Chem. Soc.*, 92, 2178, 1970.

144. Joussot-Dubien, J. and Houdard, J., Reversible photolysis of pyridine in aqueous solution, *Tetrahedron Lett.*, 4389, 1967.

145. Joussot-Dubien, J. and Houdard-Peyrere, J., Photolyse de la pyridine en solution aqueuse, *Bull. Soc. Chim. Fr.*, 2619, 1969.

146. Chapman, O. L., Mac Intosh, C. L., and Pacansky, J., Cyclobutadiene, *J. Am. Chem. Soc.*, 95, 614, 1973.

147. Mathias, E. and Heicklen, J., The gase phase photolysis of pyridine, *Mol. Photochem.*, 4, 483, 1972.

148. Roebke, W., Gas phase photolysis of 2-picoline, *J. Phys. Chem.*, 74, 4198, 1970.

149. Pascual, O. S. and Tuazon, L. O., Photolysis of 2- and 4-picolines, *Philippines Nucl. J.*, 1, 49, 1966 (*Chem. Abstr.*, 66, 115127b, 1967).

150. Shim, S. C. and Kim, S. S., Photochemistry and thermochemistry of picolylchlorides, *Bull. Korean Chem. Soc.*, 3, 110, 1982 (*Chem. Abstr.*, 98, 72072, 1983).

151. Wilzbach, K. E. and Rausch, D. J., unpublished results.

152. Caplain, S. and Lablache-Combier, A., Gas phase photochemistry of picolines and lutidines, *J. Chem. Soc., Chem. Commun.*, 1018, 1969.

153. Caplain, S., Castellano, A., Catteau, J. P., Lablache-Combier, A., and Vermeersch, G., unpublished results.

154. Caplain, S., Castellano, A., Catteau, J. P., and Lablache-Combier, A., Mécanisme de la photosubstitution de la pyridine en solution, *Tetrahedron*, 27, 3541, 1971.

155. Ogata, Y. and Takagi, K., Photoisomerisation of 2-pyridylacetonitrile to anthranilonitrile, *J. Am. Chem. Soc.*, 96, 5933, 1974.

156. Ogata, Y. and Takagi, K., Photochemistry of 2-picolines in alkaline media. Intermediacy of Dewar pyridines and their methides, *J. Org. Chem.*, 43, 944, 1978.

157. Takagi, K. and Ogata, Y., Photolysis of methyl [6–2H]2-pyridylacetate. Selective distribution of deuterium labeling, *J. Chem. Soc., Perkin Trans. 2*, 1980, 1977.

158. Takagi, K. and Ogata, Y., Photoisomerization of substituted 2-methyl pyridines to *ortho*-substituted anilines, *J. Chem. Soc., Perkin Trans. 2*, 1148, 1977.

159. Krzeczek, J., Szurgot, T., and Tomasik, P., Photochemical transformations of 2-aminopyridinium salts with some mineral acids, *Pol. J. Chem.*, 59, 1259, 1985.

160. Ratajckaz, E. and Sztuba, B., The gas phase photolysis of pentafluoropyridine, *J. Photochem.*, 13, 233, 1980.

161. Vysotskii, Y. B. and Sivyakova, L. N., Quantum chemical analysis of recyclisation reactions. Photoisomerization of six membered heterocycles, *Khim. Geterotsikl. Soedin.*, 357, 1986, (*Chem. Abstr.*, 105, 5962, 1986).

162. Padwa, A., Akiba, M., and Cohen, L. A., Small ring heterocycles. Role of azabenzvalene in the thermolysis of 3-cyclopropenyl substituted oxazolinones, *J. Am. Chem. Soc.*, 104, 286, 1982.

163. Ager, E., Chivers, G. E., and Suschitzki, H., Photolysis of pentachloropyridine and pentachloropyridine 1-oxyde, *J. Chem. Soc., Chem. Commun.*, 505, 1972.

164. Bratt, J. and Suschitzky, H., Photochemical cyclisation of substituted polyhalogenopyridines, *J. Chem. Soc., Chem. Commun.*, 949, 1972.

165. Ager, E., Chivers, G. E., and Suschitzky, H., Photochemistry of pentachloropyridine and some derivations, *J. Chem. Soc., Perkin Trans. 1*, 1125, 1973.

166. Sharman, R. K. and Karasch, N., The photolysis of iodoaromatic compounds, *Angew. Chem. Int. Ed.*, 7, 36, 1968.

167. Barlow, M. G., Dingwall, J. G., and Haszeldine, R. N., Valence bond isomers of heterocyclic compounds. Isomers of pentakis(pentafluoroethyl)pyridine, *J. Chem. Soc., Chem. Commun.*, 1580, 1970.

168. Barlow, M. G., Haszeldine, R. N., and Dingwall, J. G., The valence-bond isomers of pentakis(pentafluoroethyl)pyridine, *J. Chem. Soc., Perkin Trans. 1*, 1542, 1973.

169. Chambers, R. D., Middleton, R., and Corbally, R. P., Transpositions in aromatic rings, *J. Chem. Soc., Chem. Commun.*, 731, 1975.

170. Chambers, R. D. and Middleton, R., Rearrangements involving azaprismanes, *J. Chem. Soc., Perkin Trans. 1*, 1500, 1977.

171. Chambers, R. D. and Middleton, R., Stable 2-azabicyclo[2,2,0]hexa-2,5-dienes derivatives, *J. Chem. Soc., Chem. Commun.*, 154, 1977.

172. Chambers, R. D., Maslakiewicz, J. R., and Srivastava, K. G., Valence isomers of fluorinated pyridazines, *J. Chem. Soc., Perkin Trans. 1*, 1130, 1975.

173. Kobayashi, Y., Ohsawa, A., Baba, M., Sato, T., and Kumadaki, I., Synthesis and reactions of a substituted Dewar pyridine, *Chem. Pharm. Bull.*, 24, 2219, 1976.

174. Hees, V., Volgelbacher, U. J., Michels, G., and Regitz, M., Steric effects on valence isomerizations in the Dewar pyridine/azaprismane/pyridine system, *Tetrahedron*, 45, 3115, 1989.

175. Kellog, R. M., Van Bergen, T. J., and Wynberg, H., Photochemical ring contraction, reduction and solvent addition in pyridines, *Tetrahedron Lett.*, 5211, 1969.

176. Van Bergen, T. J. and Kellog, R. M., Photochemistry of 3,5-dicarboalkoxypyridines. Reduction and rearrangement, *J. Am. Chem. Soc.*, 94, 8451, 1972.

177. Mandella, W. L. and Franck, R. W., Photoisomerization of peri di-*t*-butylnaphtalenes, *J. Am. Chem. Soc.*, 95, 971, 1973.

178. Kaplan, L., Pavlik, J. W., and Wilzbach, K. E., Photohydratation of pyridium ions, *J. Am. Chem. Soc.*, 94, 3283, 1972.

179. Katz, T. J., Wang, E. J., and Acton, N., A benzvalene synthesis, *J. Am. Chem. Soc.*, 93, 3782, 1971.

180. Kosower, E. M. and Lindquist, L., Flash photolysis of a pyridinium iodide through the charge transfer band, *Tetrahedron Lett.*, 4481, 1965.

181. Cozzens, R. F. and Gover, T. A., Flash photolysis of the charge-transfer band of 1-methylpyridinium, 1-methylcollidinium and 1-methylquinolinium iodides, *J. Phys. Chem.*, 74, 3003, 1970.

182. Katritzky, A. R., de Ville, G., Patel, R. C., and Harlow, R., A 1,4-photochemical aryl shift, *Tetrahedron Lett.*, 1241, 1982.

183. Ingham, K. C., Abu-Elgheit, M., and El-Bayoumi, M. A., Confirmation of biprotonic phototautomerism in 7-azaindole hydrogen-bonded dimers, *J. Am. Chem. Soc.*, 93, 5023, 1971.

184. Lahmani, F., Ivanoff, N., and Magat, M., Sur une isomérisation photochimique de la pyrazine, *C. R. Acad. Sc. Paris*, 263, 1005, 1966.

185. Lahmani, F. and Ivanoff, N., Photoisomerization of pyrazine and of its methylderivatives, *Tetrahedron Lett.*, 3913, 1967.

186. Magat, M., Ivanoff, N., Lahmani, F., and Pileni, M. P., Transition non radiative dans les molécules: Compte-rendu de la 20ème réunion annuelle de la Société de Chimie Physique, Paris, 1969, *J. Chim. Phys.*, 67, 212, 1970.

186b. Fraser, J. R., How, L. H., and Weir, N. A., Photolysis of pyridazine in the gas phase, *Can. J. Chem.*, 53, 1456, 1975.

187. Ivanoff, N., Lahmani, F., Delouis, J. F., and Le Gouill, J. L., Photosensibilisation du 2-butène et du bicyclo[2.2.1.]heptadiène par la pyrazine en phase gazeuse, *J. Photochem.*, 2, 199, 1973/74.

188. Lahmani, F. and Ivanoff, N., Mercury sensitization of the isomerization of diazines, *J. Phys. Chem.*, 76, 2245, 1972.

189. Wierzchowski, K. L., Shugar, D., and Kratzitzky, A. R., Primary photoproduct of 2,6-dimethyl-4-aminopyrimidine, *J. Am. Chem. Soc.*, 85, 827, 1963.

190. Wierzchowski, K. L. and Shugar, D., Photochemistry of 4-aminopyrimidines: 2,6-Dimethyl-4-aminopyrimidine, *Photochem. Photobiol.*, 2, 377, 1963.

191. Sarkisian, J. I. and Binkley, R. W., The photolysis of benzo[3,4]cyclobuta[1,2]quinoxaline, *J. Org. Chem.*, 35, 1228, 1970.

192. Fox, M. A., Lemal, D. M., Johnson, D. W., and Hohman, J. R., Photolysis of some chlorinated pyridazines, *J. Org. Chem.*, 47, 398, 1982.

193. Johnson, D. W., Anstel, V., Feld, R. S., and Lemal, D. M., The pyridazine-pyrazine photorearrangement, *J. Am. Chem. Soc.*, 92, 7505, 1970.

194. Ratajczak, E. and Price, D., Detection of a transient after flash photolysis of perfluorinated diazines in the gas phase, *Bull. Acad. Pol. Sci. Ser. Sci. Chim.*, 29, 315, 1981.

195. Chambers, R. D., Musgrave, W. K. R., and Srivastava, K. C., A *para*-bonded isomer of an aromatic diazine, *J. Chem. Soc., Chem. Commun.*, 264, 1971.

196. Chambers, R. D., Mac Bride, J. A. H., Maslakiewicz, J. R., and Srivastava K. C., Photochemistry of halogenocarbon compounds. Rearrangement of pyridazines to pyrazines, *J. Chem. Soc., Perkin Trans. 1*, 396, 1975.

197. Allison, G. G., Chambers, R. D., Chebukov, Y. A., Mac Bride, J. A. H., and Musgrave, W. K. R., The isomerization of perfluoropyridazines to perfluoropyrimidines and to perfluoropyrazines, *J. Chem. Soc., Chem. Commun.*, 1200, 1969.

198. Chambers, R. D. and Maslakiewicz, J. R., Possible generation of a fluorinated azacyclobutadiene, *J. Chem. Soc., Chem. Commun.*, 1005, 1976.

199. Chambers, R. D., Mac Bride, J. A. H., and Musgrave, W. K. R., Thermal rearrangement of perfluoropyridazine and perfluoroalkylpyridazines to pyrimidines, *J. Chem. Soc. C*, 3384, 1971.

200. Chambers, R. D., Clark, M., Maslakiewicz, J. R., and Musgrave, W. K. R., Pyridazine rearrangements, *Tetrahedron Lett.*, 2405, 1973.

201. Chambers, R. D., Mac Bride, J. A. H., and Musgrave, W. K. R., Hexafluorocinnoline: synthesis and the photochemical isomerization to hexafluoroquinazoline, *J. Chem. Soc., Chem. Commun.*, 739, 1970.

202. Scheiner, P. and Dinda, J. F., Jr., Product formation in tetrazole photolysis, *Tetrahedron*, 26, 2619, 1970.

203. Burland, D., Carmona, F., and Pacansky, J., The photodissociation of *s*-tetrazine and dimethyl-*s*-tetrazine, *Chem. Phys. Lett.*, 56, 221, 1978.

204. Dellinger, B., Paczkowski, M. A., Hochstrasser, R. N., and Smith, A. B., Observation of transient intermediates in the photochemical decomposition of substituted *s*-tetrazines, *J. Am. Chem. Soc.*, 100, 3242, 1978.

205. King, D. S., Denny, C. T., Hochstrasser, R. M., and Smith, A. B., The photochemical decomposition of 1.4-*s*-tetrazine is N2, *J. Am. Chem. Soc.*, 99, 271, 1977.

Photochromic Nitrogen-Containing Compounds

Heinz Dürr
Universität des Saarlandes

83.1 Introduction — Definition

Photochromism[1] is a reversible transformation of a single chemical species between two states, whose absorption spectra are recognizably different, brought about, in at least one direction, by electromagnetic radiation (Equation 83.1).

$$A(\lambda_1) \underset{h\nu_2,\Delta}{\overset{h\nu_1}{\rightleftarrows}} B(\lambda_2) \qquad (83.1)$$

Species A is transformed into higher-energy form B as a result of irradiation with light. The reverse reaction to A usually occurs as a spontaneous thermal process, but may also be light-induced. Equation 83.1 holds only for unimolecular reactions. More recently, bimolecular photochromic systems, based on a cycloaddition or cycloreversion, have also been discovered (Equation 83.2).

$$A+B \underset{h\nu_2,\Delta}{\overset{h\nu_1}{\rightleftarrows}} C \qquad (83.2)$$

The photochromic systems[1,2] illustrated in Equation 83.1 can be classified into several groups on the basis of the photochemically induced primary step:

1. *Z,E-(cis,trans)*-Isomerization
2. Pericyclic reactions

0-8493-8634-9/95/$0.00+$.50
© 1995 by CRC Press, Inc.

3. Tautomerization
4. Homolytic bond cleavage
5. Electron transfer reaction

In this chapter, only N-containing photochromic molecules are selected. However, this covers many classes of photochromic systems. Therefore, photochromism based on triplet-triplet absorption of stilbenes, thioindigo, bianthrylidenes, fulgides, and salicylates is not discussed. Nitrogen-containing photochromic systems are dealt with if the reacting fragment of the molecule contains N. One exception is made for the very important spiropyrans (vide infra).

83.2 *Z,E-(cis,trans)*-Isomerization

Azobenzenes

When the double bond of azo compounds is excited either directly or via sensitizers, a reversible Z,E-isomerization is observed.[3,4,5]

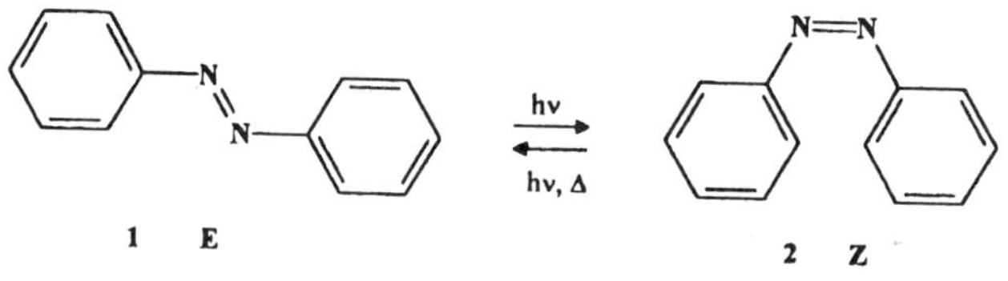

SCHEME 83.1

Azobenzene isomerization normally is a very clean reaction. In the presence of oxygen, slow photooxidation to azobenzene occurs.[3]

The E,Z-isomerization can be induced by the direct (singlet) or sensitized (triplet) route from the n,π^* or π,π^* state. The thermal back-reaction from the Z-isomer is slow in azobenzene and fast in amino- or aminonitro-substituted azobenzenes (pseudo-stilbenes). The color change is more in intensity than in wavelength.[3] In geometrically locked azobenzene, i.e., 3, the Z-form might be stable and cannot isomerize.[3]

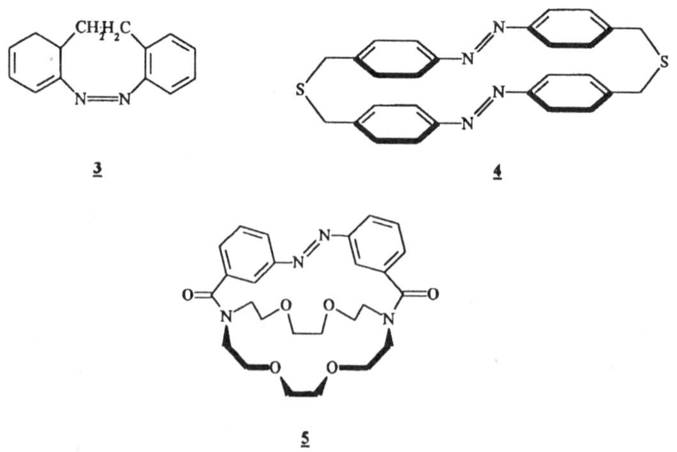

SCHEME 83.2

Compounds of type **4** show equal quantum yields $\phi_{E,Z}$ when excited either in the n,π^* or π,π^* state.

More recently, so-called photoresponsive crown ethers have also been prepared and, in these too, the light-sensitive behavior is based on *Z,E*-isomerization.[4b,6] *Z,E*-isomerization also plays an important role in simpler compounds such as benzylidene anilines[4a] and azomethines. The latter are of importance in biological systems such as rhodopsin in the human eye.[1]

83.3 Pericyclic Reactions

A large number of photochromic organic molecules utilize pericyclic reactions as the basis for photochromic materials.

Electrocyclization

Electrocyclic reactions have proved to be especially suitable as a structural unit for photochromic systems. Scheme 83.3 shows three different types of ring-opening reactions. However, only a few of the many possible electrocyclizations are important for photochromism.

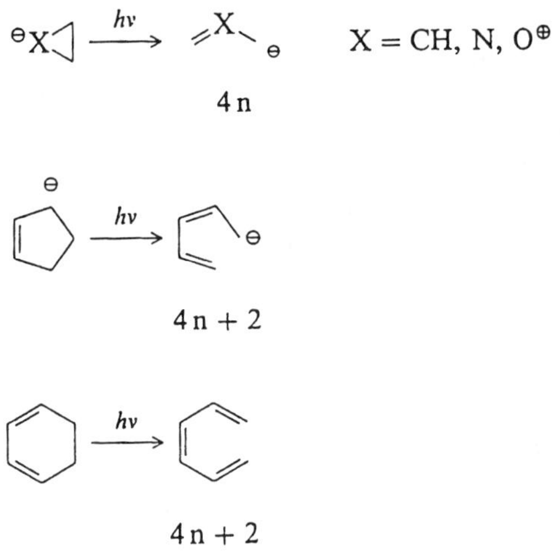

$$X = CH, N, O^{\oplus}$$

4 n

4 n + 2

4 n + 2

SCHEME 83.3

4n-Systems

Here, the three-membered heterocycles have an outstanding role. Griffin and co-workers prepared a new class of stable carbonyl ylides from epoxydiphenylsuccinimide **6** and diphenylepoxymaleic anhydride **7**.[7] Colors are generated when the bicyclic oxides **6** and **7** are irradiated at 77 K. In rigid matrices (2-methyltetrahydrofuran) ring openings afford the ylides **8** and **9**. The colors of **8** and **9** persist after the matrix is warmed up, softens and appears to become fluid (130 to 140 K).

Color is not visible at ambient temperature in fluid 2-methyltetrahydrofuran; however, **8** and **9** are stable in the solid state, even at room temperature.

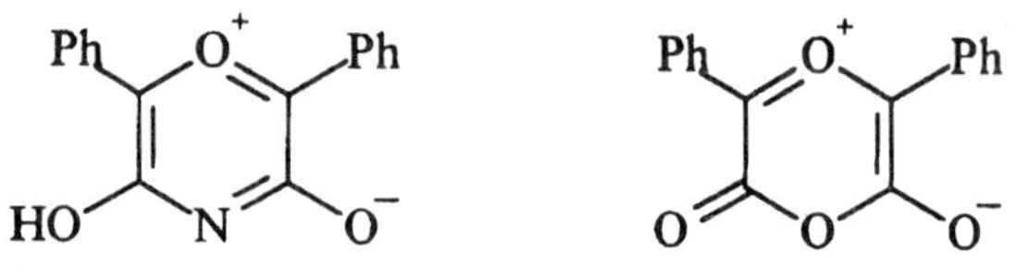

SCHEME 83.4

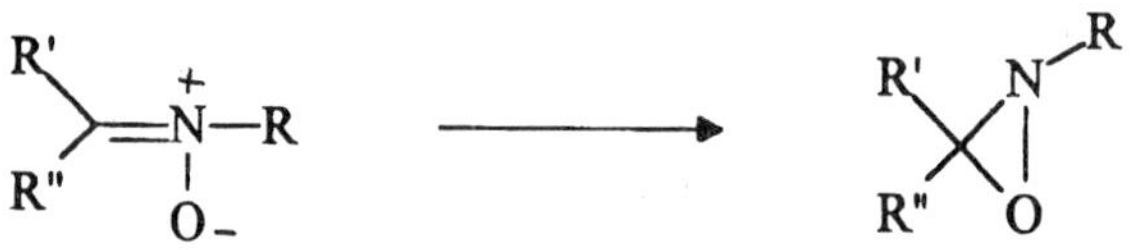

In an analogous way, the nitrones 10 can be converted by irradiation into the oxaziridines 11 (Scheme 83.4). The latter are powerful oxidizing agents, highly reactive, and in most cases they cannot be isolated. The reverse reaction takes place upon heating. Here, the open form 10 is more stable. Interesting photochromic molecules of the aziridine type have been found by Cromwell[8] and Tap Do Minh and Trozzolo et al.[9-11] Aziridines show photochromic behavior in the crystalline state although some systems are photochromic in solution. An especially efficient photochromic system is found in the bicyclic aziridine 12, which undergoes a light-induced reaction according to Scheme 83.5 to give monocyclic 13.[9-11]

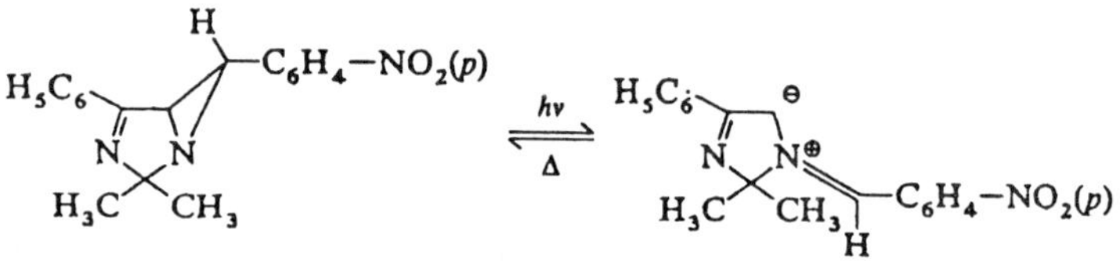

SCHEME 83.5

4n + 2 Systems: (5 Atoms, 6 Electrons).   A very efficient photochromic system was found in 1979.[12,13] It is based on the electrocyclic ring closure of azapentadienes to azacyclopentenes.

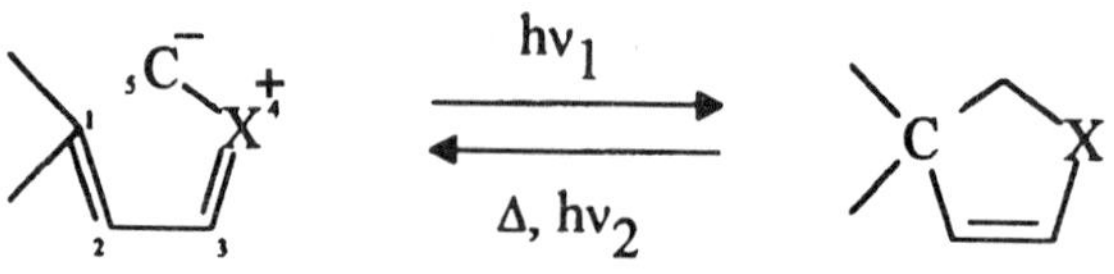

SCHEME 83.6

Nitrogen or oxygen can be in the different positions (1 to 5) of the azapentadiene systems. Thus, a great versatility is available as far as different classes of five-membered heterocycles are concerned. The different photochromic compounds based on this interconversion are given in Scheme 83.7.

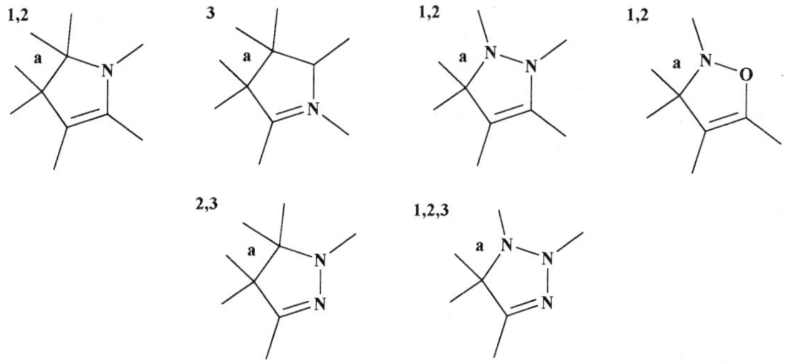

SCHEME 83.7

This scheme shows the classification (types) and characteristic structural features of the new photochromic systems (a: σ-bond formed upon 1,5-electrocyclization). According to the position of the heteroatom in the aza-pentadiene, fragments are classified as type 2, etc.[14]

Type 2 Systems: Systems of type 2 can be regarded as "1,3-dipoles of an allylic type".[14] The organic chemist's entire armory has been brought to bear on the study of 1,3-dipolar cycloadditions;[15] but despite this, it is only quite recently that the reaction has been successfully used in an effective photochromic system.

A new photochromic system whose structural basis is the spiro-dihydroindolizines[17] was found and studied intensively.[12–14] It can be prepared easily by different routes,[12–14] a typical one is shown in Scheme 83.8. When spirocyclopropenes with the general structure 16[12] are allowed to react with pyridine or pyridazine derivatives, photochromic spiro[1,8-a]dihydroindolizines 19, Y = CH or Y = N, respectively, are formed.[16–18a]

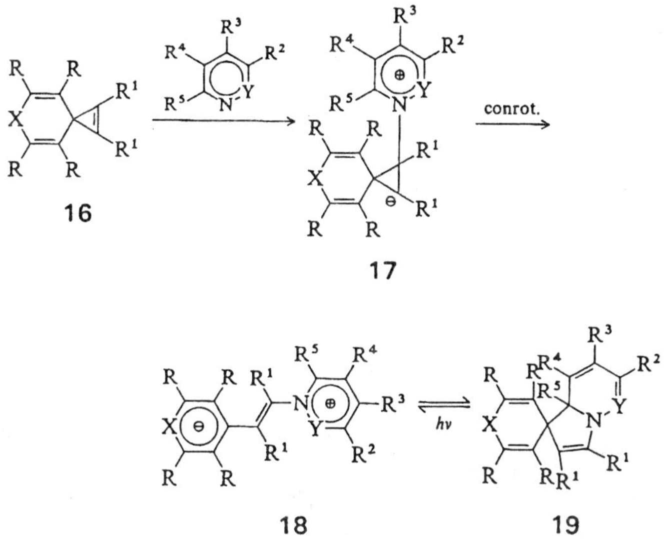

SCHEME 83.8   Cyclopropene route to 18 and 19. Solvent: CH_2Cl_2/ether; temperature: 20 to 25°C; yield: 12 to 89%.

Compounds with a very wide range of absorption wavelengths can be synthesized using this "cyclopropene route". The photochromic properties are not linked, however, to spiro-dihydroindolizines; simpler systems (diphenyl derivatives) can also be prepared easily. Thus, by introducing a variety of different substituents in all positions of **19**, molecules with tailor-made properties are accessible. Crown ether-linked dihydroindolizines can be made that allow special effects due to host, guest, or supramolecular assemblies.[18b]

Type 3 Systems: Systems of the type such as **20** on irradiation in the presence of oxygen give indoles. However, owing to subsequent thermal reactions, such as sigmatropic H-shifts (and possibly oxidations) at moderate or high temperatures, irreversible secondary reactions occur to yield di- and tetrahydroindoles and the system therefore cannot be regarded as being reversibly photochromic.[19]

Mixed Systems (1,2-; 2,3-; 1,2,3-): Mixed photochromic systems (**20** to **24**) can be synthesized as well and a survey of typical examples is given below:

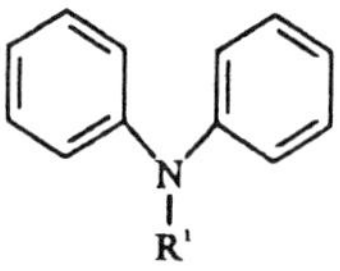

20

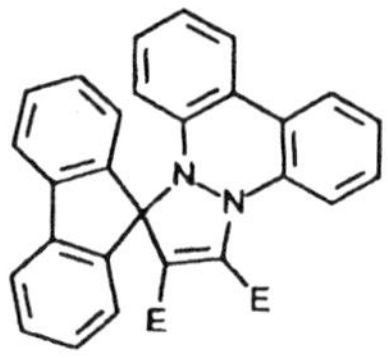

21[20] 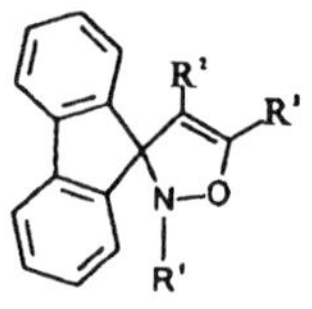   **22**[21] 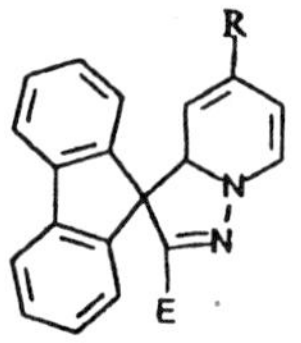   **23**[22]

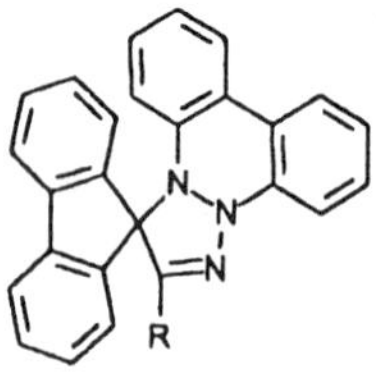

24[20]

SCHEME 83.9

Biphotochromic systems containing (1) two DHI-molecules[23] and (2) a DHI- and a salicylideneanil[24] have also been prepared and studied.

1,5-Electrocyclization: The UV spectra of the dihydroindolizines **19**, X = CH, and dihydropyrrolo[1,2-*b*]pyridazine **19**, X = N, usually have absorption maxima in the near-UV between 360 and 410 nm. The betaines **18** produced by photochemical ring opening absorb at about 505 to 726 nm.

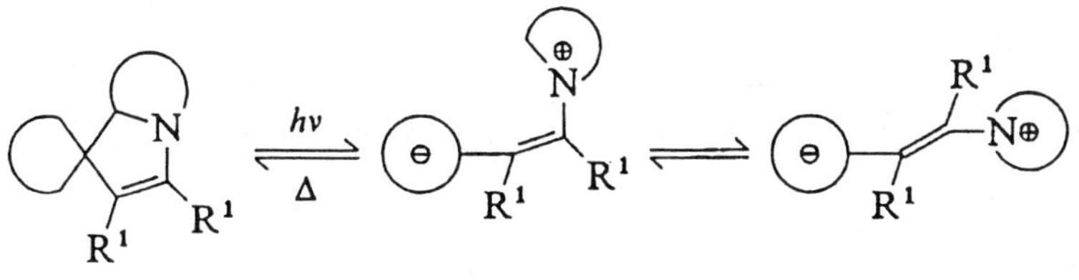

19   *E*-**18**   *Z*-**18**

SCHEME 83.10

The σ-bonds in the various azoheterocycles that undergo photochemical cleavage are marked in Scheme 83.7. These cases differ significantly from the ring opening of the cyclopentenyl anion since in the previous molecules, no anion is formed. The photoproduced compounds from these electroneutral five-membered heterocycles are zwitterions or betaines in all cases.

The ring opening of **19** occurs easily in solution, in polymers, and in rigid matrices at room temperature. In some cases, ring opening was observed also in the crystalline state.

The Multiplicity of the Photochemical Reaction **19** → **18**: The photochemical reaction for the ring opening of **19** to **18** was demonstrated to proceed via an excited singlet or a fast-reacting triplet state that does not undergo bimolecular quenching (for details see Reference 25). The determination of the quantum yield for ring opening **19** → **18** was carried out using nonlinear optimization methods (for details see Reference 26). The values for ϕ_R are of the order of 0.4 to 0.8 for the forward reaction and 0.1 for the backward reaction. The quantum yields are also wavelength dependent.

An important system belonging to the class of five-membered heterocyclic interconversions are the phytochromes (P_R), which have been studied by Schaffner and Braslavsky et al. These systems play an important biochemical role in regulating plant growth.[27]

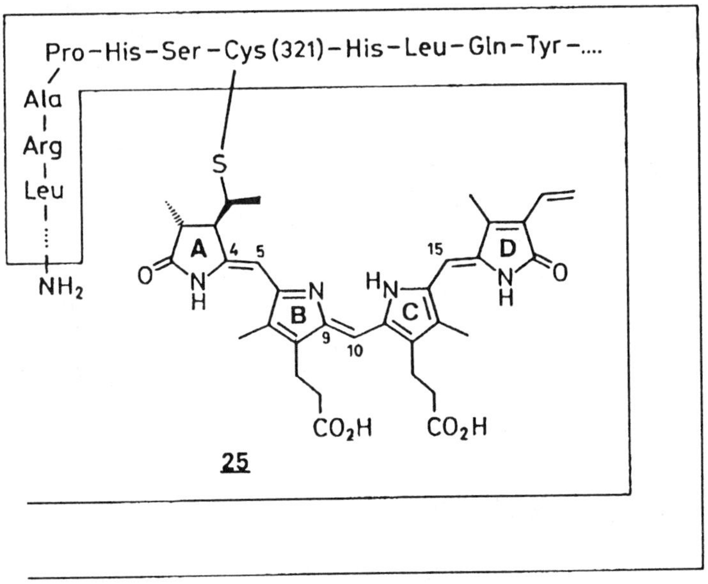

SCHEME 83.11   Structure of the P_r chromophore of phytochrome.

4n + 2 Systems: (6 Atoms, 6 Electrons).   An electrocyclic (4n + 2)-reaction (n = 1) is also found in the spiropyrans **26**[28–31] as shown in Scheme 83.12. UV irradiation of the colorless or slightly yellow spiro compounds **26** leads to cleavage of the C-O bond to give the open-chain molecule **27** that absorbs strongly in the visible region. These highly interesting photochromic spiropyrans have been studied very intensively.[1,32] This system has also led to quite a number of practical applications. Bräuchle et al., among others, have shown that an unstable intermediate with an orthogonally bridged structure is formed during ring opening of nitrobenzo spiran at 4.2 K.[33]

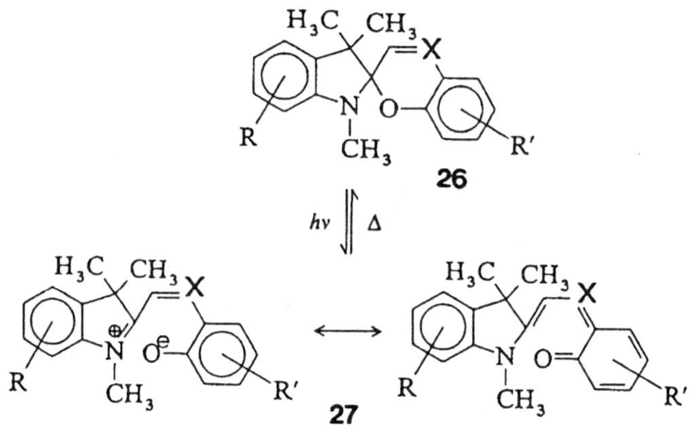

SCHEME 83.12   R and R' can be a large number of organic groups; each ring can also contain several substituents.

If in scheme 83.12 X is equal to N, the molecules **26** belong to the class of spiroxazines.[34–40] This photochromic system undergoes very fast ring opening to the colored form **27** with a good quantum yield ($\phi_R = 0.2$ to 0.8) in solution as in films. The color fades thermally very rapidly. The photochemical back-reaction is, however, very slow. No photochromism is observed in the crystalline state. In connection with the high long-term photostability, spiroxazines are of great commercial value. They belong to the best photochromic systems used today.[39] Recently, a series of spiroxazines has been investigated and their photochemical[40] properties have been discussed in detail. Related electrocyclizations were also studied with 1-nitro-2-aryl-alkenes.[41]

Cycloadditions

[2 + 2]-Cycloadditions

The photodimerization of 2-phenylbenzoxazole **28** produces a head-tail photodimer **29** involving two carbon-nitrogen double bonds.[42]

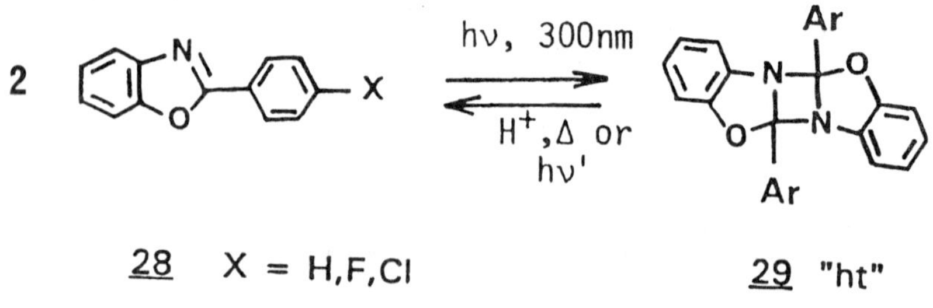

SCHEME 83.13

The photodimer **29** is labile in fluid solution but thermally stable in the solid state at room temperature. This system has been proposed as a system for light-energy conversion, since the thermal reversion of the photodimer to the starting material **28** releases 116 kJ mol^{-1}. A similar reversible [2 + 2]-cycloaddition has been reported for 1,1'-polymethylene-*bisthymines* (n = 2 to 6). The dimerization ($\lambda = 300$ nm) can be reverted by irradiation with $\lambda = 254$-nm light.[43,44]

[4 + 4]-Cycloaddition

2-Aminopyridine as the HCl-salt can undergo a [4 + 4]-photodimerization.[45–47] The diphosphate or dimethanesulfate of the dimer **31** can be cleaved by short wavelength irradiation. Chandross has used this reversible photoreaction, employing crystals of the dimer to write holographic gratings.[48] In a [4 + 4]-cycloaddition, acrizidiniumbromide **32** gave **33**.[49,50] Related studies were carried out with polymethylene-linked acrizidinium salts.[51]

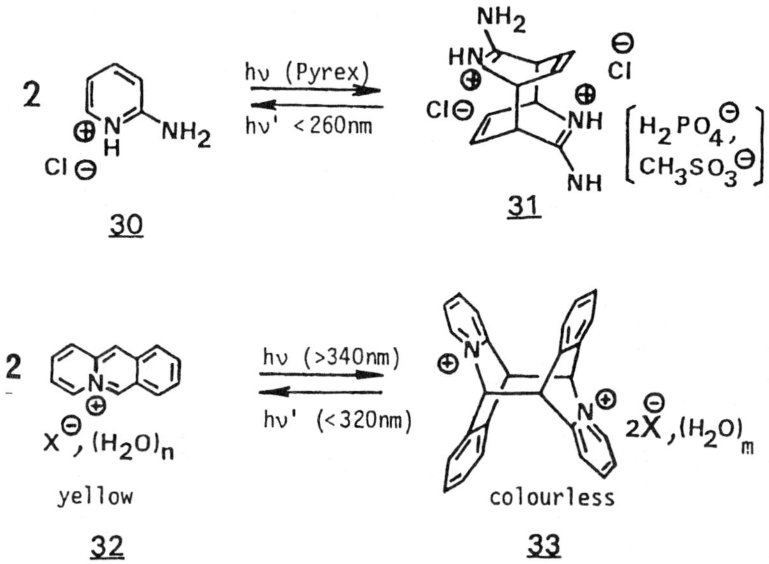

SCHEME 83.14

83.4 Tautomerization

Light-induced tautomerizations can occur within the salicylidene-anilines in the form of prototropic rearrangements such as **34** → **35**.[52–57]

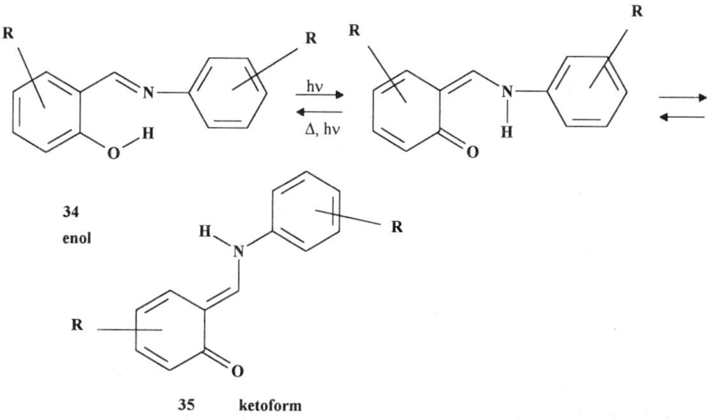

SCHEME 83.15

These molecules are photochromic in the crystalline state (α and β modification) and in rigid glasses. The large Stokes shift (F-spectrum) is explained by proton transfer in the excited state.

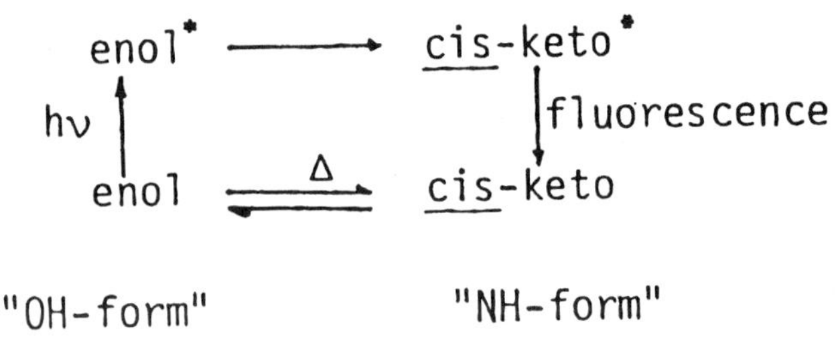

SCHEME 83.16

The mechanism has been studied in great detail.[58]

Similar behavior is found in 2-, 3-, or 4-(salicylidenamino)pyridines.[58,59] However, compounds 36 a to c show only thermochromism in the crystalline state. In rigid glasses at low temperature, salicylideneanilines 36 are also photochromic.

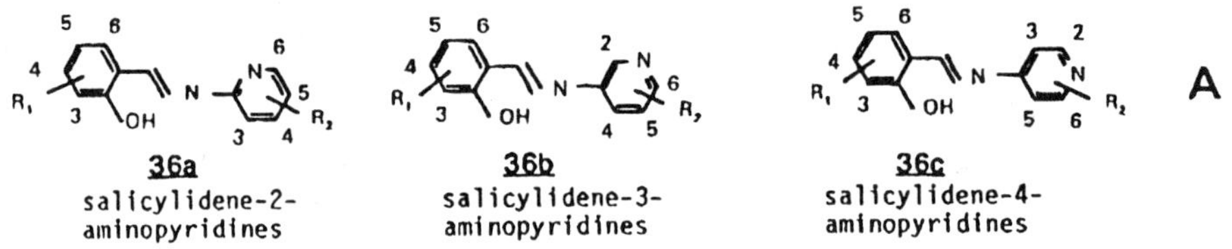

SCHEME 83.17

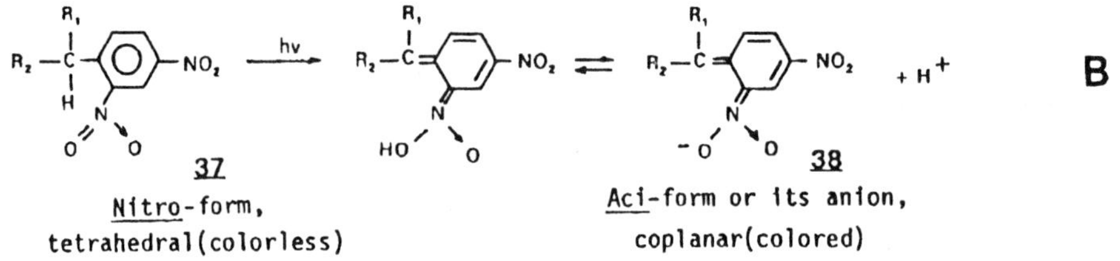

SCHEME 83.18

A related tautomerization occurs in aromatic nitro compounds 37. It is designated as an aci-nitro-phototautomerization (Scheme 18).[58]

An interesting tautomerization occurs in benzotriazoles (TINUVIN) that are used as light filters in dye protection.[60] Tautomerization processes in porphyrins are the bases of a photochromic reaction used in hole burning.[61] This process has been used for making a molecular computer. Dithizonates were used as photochromic materials in ophthalmic lenses.[62]

83.5 Dissociation/Homolytic Cleavage of Bonds

Typical examples of this type are the cleavage reactions of octaphenyl-1,1′-bipyrrolyl and of hexaphenyl-1,1′-biimidazolyl, which on exposure to light yield the colored free radicals 39 and 40, respectively. These reactions form the basis of a photochromic system.[63]

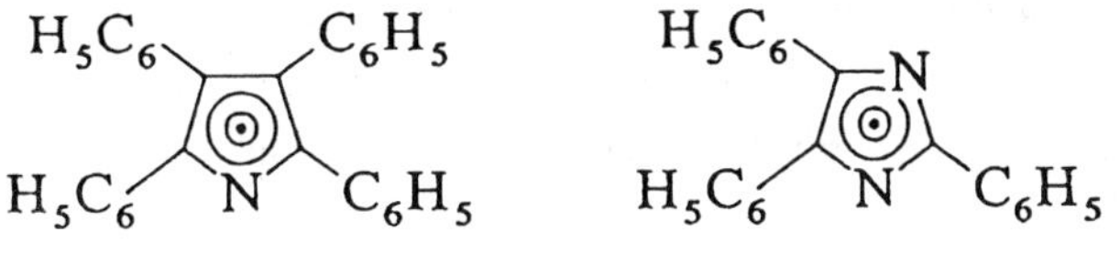

39   **40**

SCHEME 83.19

Photochromism is observed both in the crystalline state and in solution.

83.6 Electron Transfer/Redox Photochromism

The best-known photochromic system based on an electron transfer reaction is undoubtedly that of AgCl, which is decomposed by light according to Equation 83.3.

$$AgCl \xrightarrow{h\nu} Ag + \dot{Cl} \qquad (83.3)$$

This electron transfer reaction is of great practical importance in photochromic spectacle lenses based on silicate glass.[64] An example of a photochromic electron transfer reaction in organic chemistry is the methylene blue system, which has been investigated by Parker (Scheme 83.20).[65]

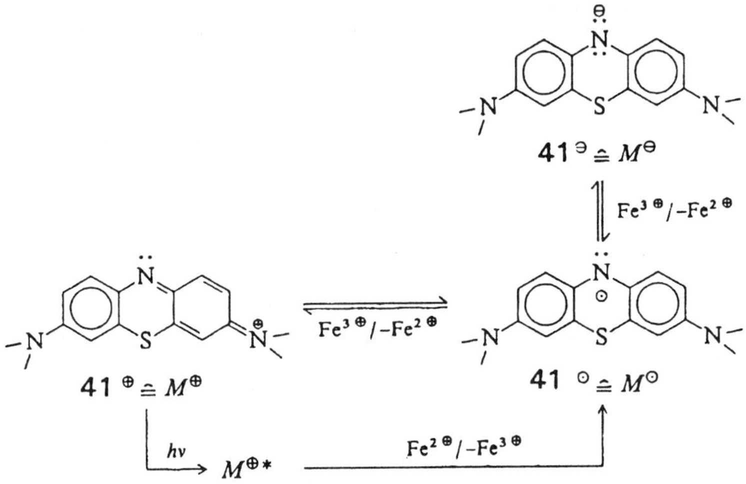

SCHEME 83.20

83.7 Applications

The inherent properties of photochromic systems make them suitable for use in many different areas. Basically, there are four main groups of possible application, each making use of different special properties:

1. The control and measurement of radiation intensity
2. Visual contrast effects
3. Solar energy conversion
4. Information imaging and storage

In the following, representative examples are given in Table 83.1. The physical (and chemical) characteristics that are important in applications of photochromic systems include:

1. The spectral regions where absorption occurs, or the operating wavelengths
2. Sensitivity
3. Reversibility and useful life (number of cycles)
4. Properties as a storage medium

For a more detailed treatment of application aspects, see Reference 1.

Table 83.1   Applications of Photochromic Materials

Control and measurement of radiation intensities	Contrast effects	Recording and storage of information
a) Optical filters	Masks for contrast equalization in photography	a) Imaging
— Lenses	— Film materials	— Instantaneous imaging materials
— Glasses		— Office copying materials
— Cuvettes	Amplifying contrast in projected images	
	— Projection screens	— Microfilm
Radiation protection equipment	— Projection walls	— Microfilm copying materials
— Glasses		— Sprays
— Spectacles	Production of printing plates and circuit boards	b) Digital information
— Films	— Photoresists	— Slow storage
— Containers	— Photopolymers	— Bulk data storage (archives, etc.)
Protection from sun	Solar energy conversion	Holography
Protection from flash exposure, lasers, etc.	Energy storage	— Storage disks
		— Storage films
Document preservation		— Crystals
Temperature indicators		Data display systems
		— Display screens (image screens)
b) Actinometry, dosimetry		Molecular computers
— Actinometer		Q-switches (for lasers)
— Dosimeter		Molecular electronics

Adapted from Reference 14.

Table 83.2   Some Data for the Betaines 18 and the Spiro[1,8a-dihydroindolizines] (and their *N*-analogues) 19. –R–R– = –(CH=CH)$_2$– (benzoannelation). $R^1 = CO_2CH_3$. Of the betaines, 18k and 18s could be isolated. UV data determined at 20°C in CHCl$_3$.[a]

						19			18	
	X	Y	R^2	R^3	R^4	R^5	m.p. (°C)	Yield (%)	λ_{max} (nm)(ε)	λ_{max} (nm)	$k_{38\rightarrow39}$ (s^{-1})
a	—[b]	CH	H	H	H	H	135	63	384 (10500)	586	4.88×10^{-3}
b	—	CD	D	D	D	D	138	73	383	586	3.4×10^{-3}
c	—	CH	H	OCH$_3$	H	H	145–147	73	376 (7730)	552	11.0×10^{-3}
d	H; H	CH	H	(CH=CH)$_2$	H		162	52	385 (13380)	572	1.2×10^{-3}
e	—	CH	H	(CH=CH)$_2$	H		162	78	378 (14390)	600	40[c]
f	—	CCH$_3$	H	H	H	H	130	31	383 (11100)	570	2.21×10^{-3}
g	—	N	H	H	H	H	146	74	389 (9480)	505	14×10^{-3}
h	—	C(CH=CH)$_2$	H	H	H		192	84	360 (15400)	724	2.7[c]
i	—	C(CH=CH)$_2$	H	H	CH$_3$		36	40	363 (12250)	726	0.1[c]
j[e]	—	CH	H	H	H	H	184	12	376 (10600)	694	832[c]
k[f]	—	CH	H	H	H	H	194	89	392 (11300)	629	27.7[c,d]
l	C=O	CH	H	H	H	H	183	61	388 (8260)		

[a] Adapted from Reference 14.
[b] A dash signifies a zero bridge.
[c] Determined flash photolytically by Dr. H. Hermann, Mülheim a. d. Ruhr (FRG).
[d] Cyclization in two-part steps; the value quoted refers to the rapid step, which is neither first order nor second order.
[e] $R^1 = H$.
[f] $R = C_6H_5$.

ble 83.3 Fluorescence Lifetime τ_s, Fluorescence Quantum Yields Φ_R, and Rate Constants k_{ic} and k_R for the otochemical Ring-Opening $19 \rightarrow 18$ at 25°C[a]

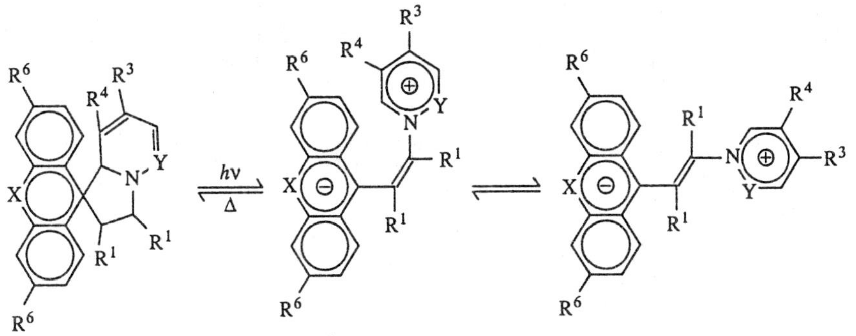

						τ_s	Φ_F	k_F		k_{ic}		$k_R{}^b$
X	Y	R¹	R³	R⁴	R⁶	(ns)	×10³	(s⁻¹)	Φ_{ic}	(s⁻¹)	$\Phi_R{}^b$	(s⁻¹)
—	CH	E	H	H	H	1.94	—	—	0.67	2.20×10^8	0.43	3.41×10^8
H, H	CH	E	(CH=CH)₂	H	0.50	—	—	0.20	1.63×10^9	0.80	4.04×10^8	
Cl₂)	H, H	CH	E	(CH=CH)₂	Cl	0.15	1.0	3.12×10^6	0.29	4.67×10^9	0.70	1.93×10^9
—	N	E	H	H	H	0.68	—	—	0.65	5.12×10^8	0.35	9.61×10^8
—	N	CN	H	H	H	0.21	3.5	1.74×10^7	0.36	3.00×10^9	0.63	1.74×10^9
C=O	N	E	H	H	H	0.42	2.2	5.24×10^6	0.41	1.38×10^9	0.58	9.76×10^8
H, H	N	E	H	H	H	0.43	2.3	5.40×10^6	0.43	1.30×10^9	0.56	1.00×10^9

[a] Adapted from Reference 14.
[b] Determined at 77 K.

Table 83.4 Spectrokinetic Values for Some Characteristic Substituted Compounds Through Nitrogen-Containing Systems[a]

		$k\Delta$ (s⁻¹) 25°C (O.F.→C.F.) toluene	λ_{max} O.F. (nm) toluene
H = (*)Indoline (A)	R³		
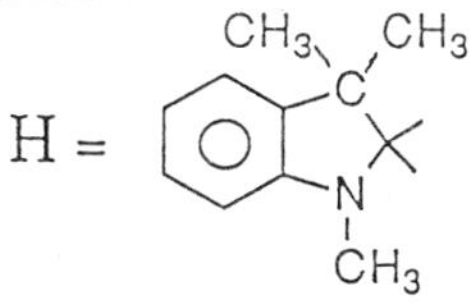	H	2×10^{-2}	610
	C_6H_5	242	625
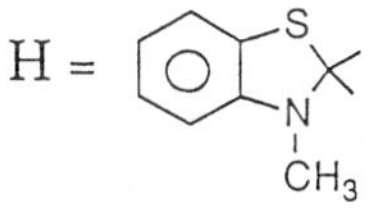 H = Benzothiazoline(B)	CH₃	2.6	600
	C₂H₅	23.2	635
	i-C₃H₇	680	610
	C₆H₁₁	612	610
	C₆H₅	6.4	635
	OCH₃	9.5×10^{-3}	640
	OC₆H₅	16.2×10^{-3}	625
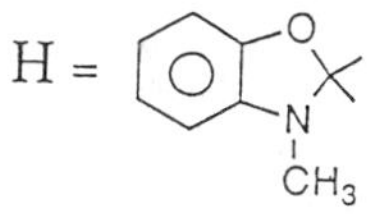 H = Benzoxazoline(C)	CH₃	0.82	600
	OCH₃	6×10^{-2}	620
	C₆H₅	3.30	600

Table 83.4 (continued)   Spectrokinetic Values for Some Characteristic Substituted Compounds Through Nitrogen-Containing Systems[a]

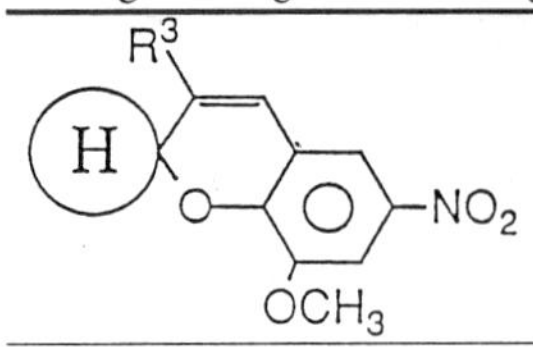	$k\Delta$ (s^{-1}) 25°C (O.F.→C.F.) toluene	λ_{max} O.F. (nm) toluene
H = Thiazolidine (D)		
CH$_3$	17	610
OCH$_3$	1.8×10^{-2}	600
H = Oxazolidine (E)		
CH$_3$	9×10^{-2}	545
OCH$_3$	9.2×10^{-4}	550
H = Pyrrolidine (F)		
CH$_3$	3.6	410, 580
C$_2$H$_5$	15	410, 585
H = CH$_3$ 1,3-Thiazine (G)		
CH$_3$	138	420, 600
C$_6$H$_5$	13.7	420, 590
H = CH$_3$ 1,3-Oxazine (H)		
CH$_3$	1.3×10^{-2}	420, 530
OCH$_3$	2.3×10^{-3}	412, 538
H = CH$_3$ 1,4-Thiazine (I)		
CH$_3$	50	590
H = (*) Piperidine (J)		
H	0.40	420, 575
CH$_3$	0.13	440

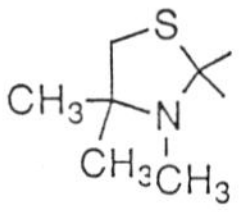

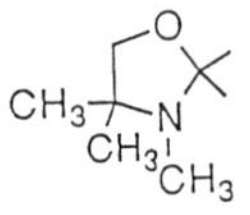

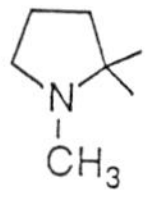

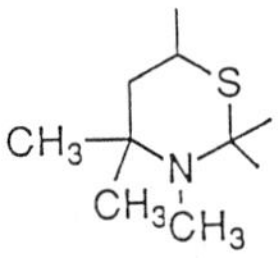

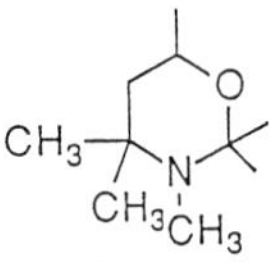

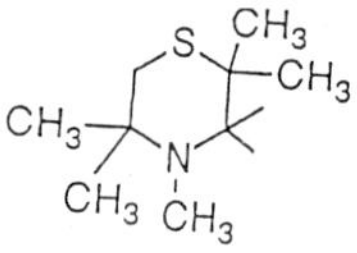

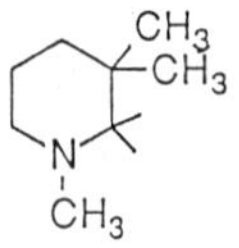

[a] Adapted from Reference 32.

Table 83.5   Compared Kinetic Bleaching Values in Toluene at 25°C[32]

| 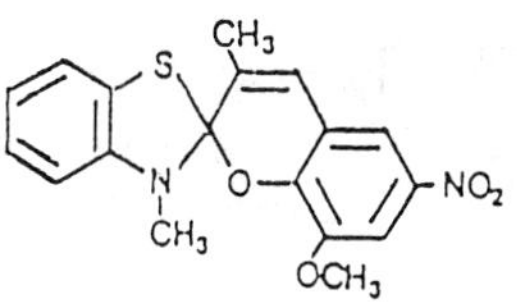 Benzothiazoline | 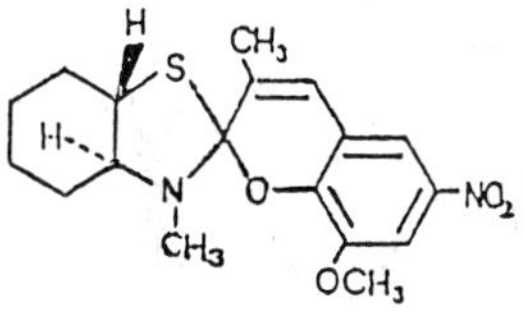 4',5'-Tetramethylenethiazolidine |
| $k\Delta = 2.6\ s^{-1}$ | $k\Delta = 98.8\ s^{-1}$ |

Table 83.6   Influence of Structural Features and Solvent on the Fading-Rate of the Open Form (O.F.) and the λ_{max} of both Closed Form (C.F.) and O.F.[32]

Compound	$k\Delta(s^{-1})$	λ_{max} C.F. (nm)	λ_{max} O.F. (nm)	Solvent
	17 2.4×10^{-4}	355.3 356	610 450–460	Toluene Ethanol
	14.5 1.15×10^{-4}	356 356	590 465	Toluene Ethanol
	9×10^{-2} 5.3×10^{-5}	335 349	545 450	Toluene Ethanol
	5.35×10^{-2} 1.6×10^{-6}	352.5 352	540 442	Toluene Ethanol

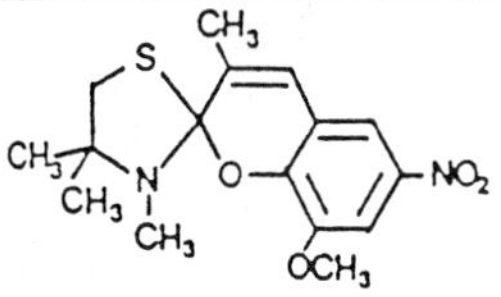

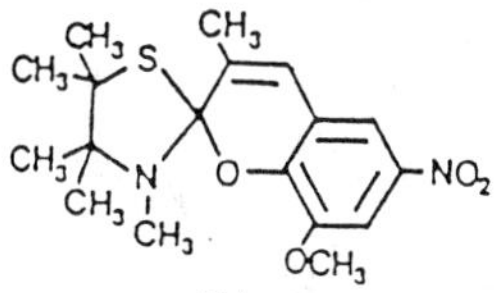

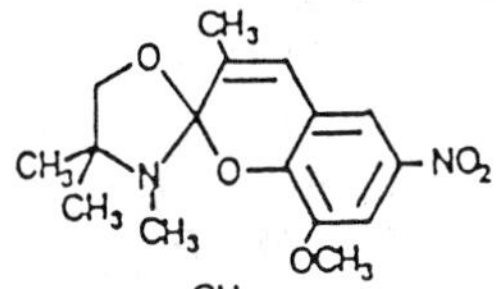

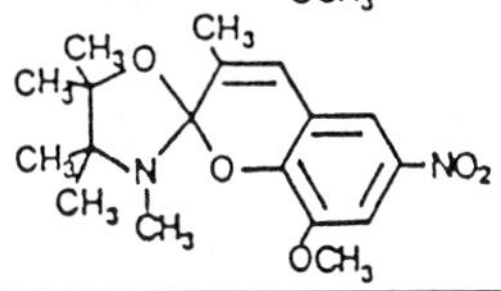

Table 83.7 Parent Ring Systems of Known Spirooxazines

Adapted from Reference 39.

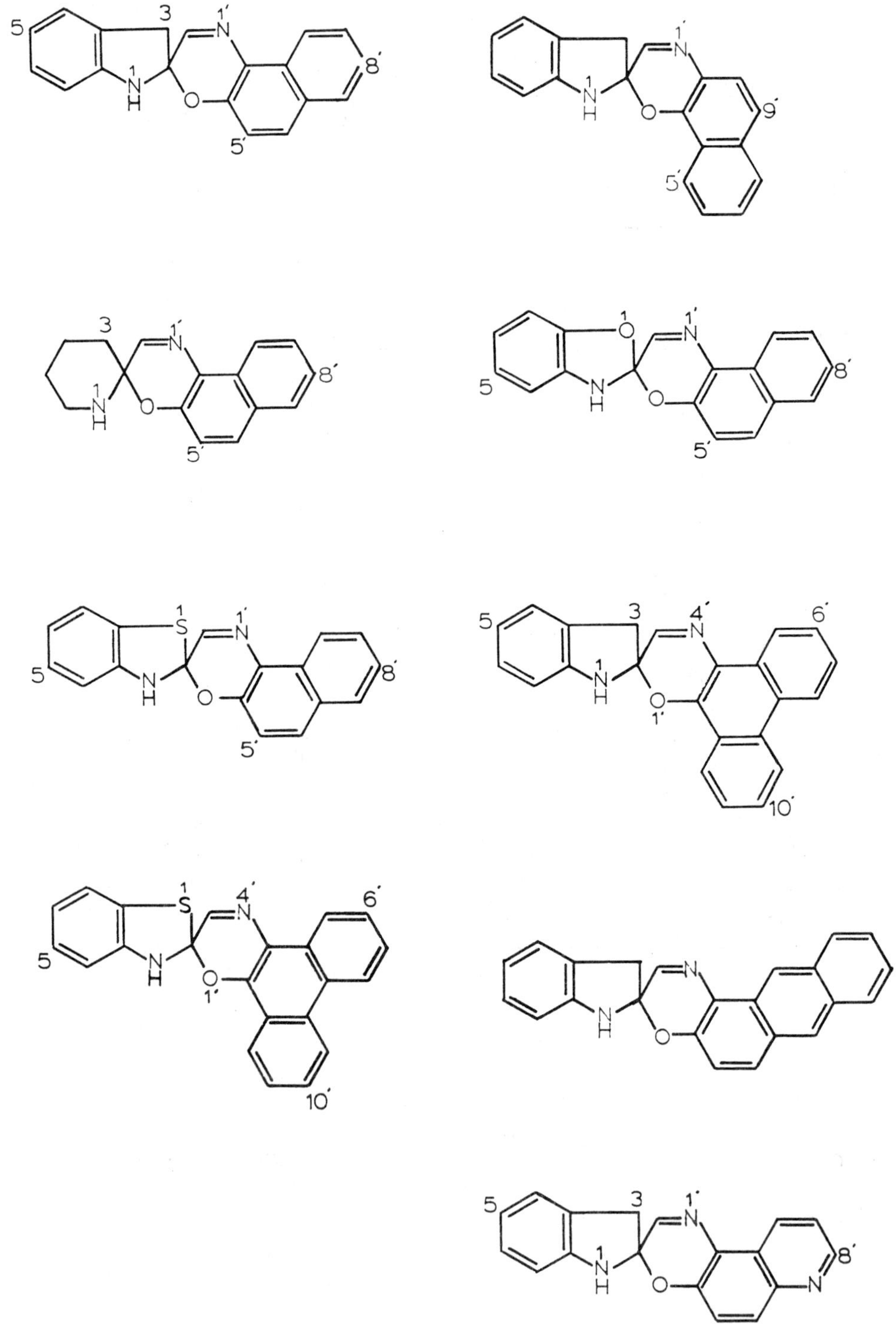

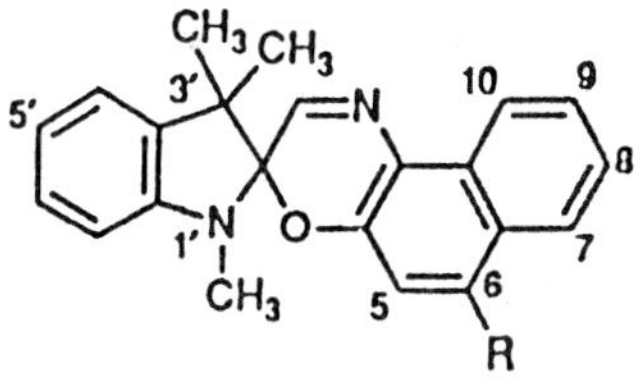

NOSI1   H

NOSI2

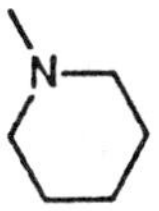

NOSI3

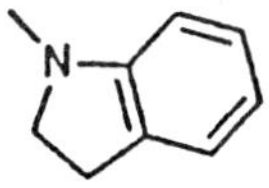

NOSI4

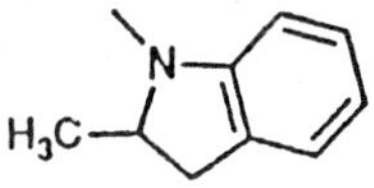

Table 83.8a   Spectral Parameters of the Colored Form of NOSI Compounds in Toluene at 294 K

Compound	λ_{max}/nm	$\varepsilon_{max}/10^{-3}$ dm^{-3} mol^{-1} cm^{-1} Fischer's Method	Intensity Variation
NOSI1	592 ± 1	NA[a]	31 ± 3
NOSI2	560 ± 1	32 ± 2	32 ± 3
NOSI3	583 ± 1	48 ± 3	45 ± 4
NOSI4	588 ± 1	43 ± 2	43 ± 4

[a] NA: not applicable.

Table 83.8b   Kinetic and Photochemical Parameters of NOSI Compounds at 352 nm in Toluene at 294 K

Compound	ϕ_B/ϕ_A	$\phi_A + \phi_B \approx \phi_A$	k/s^{-1}
NOSI1	NA	0.23 ± 0.10	0.270 ± 0.010
NOSI2	0.03 ± 0.01	0.42 ± 0.20	0.044 ± 0.002
NOSI3	<0.04[b]	0.20 ± 0.10	0.077 ± 0.004
NOSI4	0.07 ± 0.05	0.85 ± 0.20	0.099 ± 0.004

[b] Measured in a polyurethane matrix.
Adapted from Reference 40.

eferences

1. Brown, G. H., *Photochromism*, Wiley Interscience, New York, 1971; Dürr, H. and Bouas-Laurent H., Eds., *Photochromism — Molecules and Systems*, Elsevier, Amsterdam, 1990; El'tsov, A. V., *Organic Photochromes*, Plenum, New York, 1990.
2. Review articles and books: Murray, R. D., *Silverless Imaging Systems, Neblette's Handbook Photogr. Reprogr.*, 7th ed., 1977, 397; Barachevskii, A. S., Lashkov, G. I., and Tsekhomskii, V. A., Photochromism and Its Use, *Izd. Khimiya, Moscow*, 1977; *Chem. Abstr.*, 89, 1978; Heller, H. G., The development of photochromic compounds for use in optical information stores, *Chem. Ind. (London)*, 193, 1978; Kholmanskii, A. S., Zubkov, A. V., and Dyumaev, K. M., The nature of the primary photochemical step in spiropyrans, *Russ. Chem. Rev.*, 50, 305, 1981; El'tsov, A. V., Bren, V. A., and Gerasimenko, Y. E., *Organic Photochromic Substances*, P. P. Khimiya, Leningradskoe, Otdelenie, Leningrad, 1982, 285; *Chem. Abstr.*, 99, 1983, B: 4775y; Smets, G., Photochromic phenomena in the solid phase, *Adv. Polym. Sci.*, 50, 17, 1983; Fischer, E., Photochromism and other reversible photoreactions in the dianthrylidenes and implications regarding environmental control of photoreactions, color-structure correlations and other points, *Rev. Chem. Intermed.*, 5, 393, 1984; Marcus, L. H., Photochromic materials. June 1976 - May 1980 (citations from the NTIS data base), Gov. Rep. Announce. Index (U.S.) 80, 5760 (1980); Photochromic materials June 1970 - June 1980 (citation from the Engineering Index data base), *Chem. Abstr.*, 94, 130192r, Photochromic materials June 1976 - May 1980 (citations from the NTIS data base), *Chem. Abstr.*, 94, 130193s, Photochromic materials. June 1974 - May 1980 (citations from the International Aerospace Abstracts data base), *Chem. Abstr.*, 94, 130196v (1981); Workshop on Photochromics, Xth IUPAC Symposium on Photochemistry, Interlaken, 1984.

3. Rau, H., Azo compounds, in *Photochromism — Molecules and Systems*, Dürr, H. and Bouas-Laurent, H., Eds., Elsevier, Amsterdam, 1990, chap. 4.

4. Ball, P. and Nichols, C. H., *Dyes Pigm.*, 6, 13, 1985, Photochromism of the azo tautomer of 4-phenylazo-1-naphthol and its *o*-methylether in solvents and polymer substrates, *Chem. Abstr.*, 102, 96938k, 1985; Shinkai, S., Okawa, T., Kusano, Y., Manabe, O., Kikuwa, K., Goto, T., and Matsuda, T., Photoresponsive crown ethers. 4. Influence of alkali metal cations on photoisomerization and thermal isomerization of azo*bis*(benzocrownether)s, *J. Am. Chem. Soc.*, 104, 1960, 1982.

5. Labsky, J., Mikes, F., and Kalal, J., The photochromism of spiropyrans bound on side chains of soluble polymers, *Polym. Bull.*, 4 (12), 711, 1981.

6. Shinkai, E. and Manabe, O., *Top. Curr. Chem.*, 121, 67, 1984.

7. Griffin, G. W., Nishiyama, K., and Ishikawa, K., A potential precursor for 2,3-diphenyloxirene, *J. Org. Chem.*, 42, 180, 1977.

8. Cromwell, N. H. and Cauglhilan, J. A., Ethylene imine ketones, *J. Am. Chem. Soc.*, 67, 2235, 1945; Cromwell, N. H. and Hocksma, H., Ethylene imine ketones. IV. Isomerism and absorption spectra, *J. Am. Chem. Soc.*, 71, 708, 1949.

9. Trozzolo, A. M., Yager, W., Griffin, G., Kristinsson, H., and Sarkar, I. Direct evidence for the formation of diphenymethylene in the photolysis of triphenyl- and tetraphenyloxirane, *J. Am. Chem. Soc.*, 89, 3357, 1967; Do Minh, T. and Trozzolo, A. M., Photochromic aziridines. I. The mechanism of photochromism in 1,3-diazabicyclo[3.1.0]hex-3-enes and related aziridines, *J. Am. Chem. Soc.*, 94, 4046, 1972.

10. Trozzolo, A. M., Leslie, T. M., Sarpotdar, A. S., Small, A. D., Ferraudi, G. J., Do Minh, T., and Hartless, R. L., Photochemistry of some three-membered heterocycles, *Pure Appl. Chem.*, 51, 261, 1979.

11. Trozzolo, A. M., Sarpotdar, A. S., Leslie, T. M., Hartless, R. L., and Do Minh, T., The photochemistry of aryl-substituted three-membered heterocycles, *Mol. Cryst. Liq. Cryst.*, 50, 201, 1979.

12. Hauck, G. and Dürr, H., 1,8*a*-Dihydroindolizine als Komponenten neuer photochromer Systeme, *Angew. Chem.*, 91, 1010, 1979; *Angew. Chem. Int. Ed.*, 18, 945, 1979.

13. Dürr, H., 4n + 2 Systems based on 1,5-electrocyclization, in *Photochromism — Molecules and Systems*, Dürr, H. and Bouas-Laurent, H., Eds., Elsevier, Amsterdam, 1990, chap. 6.

14. Dürr, H., Perspektiven auf dem Gebiet der Photochromie: 1,5-Elektrozyklisierung von heteroanalogen Pentadienyl-Anionen als Basis eines neuartigen Systems, *Angew. Chem.*, 101, 427, 1989; *Angew. Chem. Int. Ed.*, 28, 413, 1989.

15. Huisgen, R., *Angew. Chem.*, 92, 979, 1980; 1,5-Elektrozyklisierungen - ein wichtiges Prinzip der Heterozyklenchemie, *Angew. Chem. Int. Ed.*, 19, 947, 1980; For a review, see Taylor E. C. and Turchi I. J., 1,5-dipolar cyclizations, *Chem. Rev.*, 79, 181, 1979.

16. Gross, H. and Dürr, H., Neue Synthese von Spiro[1,8*a*]dihydroindolizinen; ein neues photochemisch schaltbares System, *Angew. Chem.*, 94, 204, 1982; *Angew. Chem. Int. Ed.*, 21, 216, 1982; *Angew. Chem. Suppl.*, 559, 1982.

17. Dürr, H. and Hauck, G., Photochrome Spiro[1,8*a*-dihydroindolizine]-ein Verfahren zu ihrer Herstellung und ihre Verwendung in strahlungsempfindlichen Materialien, G. Pat. 2906193, 1991.

18. (a) Dürr, H., Groß, H., and Zils, K. D., Photochrome Spiro(1,8*a*-tetra-hydroindolizine), G. Pat. 3220257, 1994; (b) Dürr, H., Thome, A., Kranz, C., Kilburg, H., Bossmann, S., Braun, B., Franzen, K. P., and Blasius, E., Supramolecular effects on photochromism. Properties of crown ether-modified dihydroindolizines, *J. Phys. Org. Chem.*, 5, 689, 1992.

19. Grellmann, K. H., Kühnle, W., Weller, H., and Wolff, T., The photochemical formation of dihydrocarbazoles from diphenylamines and their thermal rearrangement and disproportionation reactions, *J. Am. Chem. Soc.*, 103, 6889, 1981; Grellmann, K. H., Schmitt, M., and Weller, H., The photoinduced stereospecific formation of 9-ethyl-1,4*a*-dimethyl-4,4*a*-dihydrocarbazole from *N*-ethyl-2,6-dimethyl-diphenylamine and its photoreactions, *J. Chem. Soc., Chem. Commun.*, 591, 1982; Schultz, A. G., Photochemical six-electron heterocyclization reactions, *Acc. Chem. Res.*, 16, 210, 1983.

20. Dürr, H., Thome, A., Steiner, U., Ulrich, T., Rabe, E., and Krüger, C., 1-H-Benzo[c]pyrazolo-[1,2a]cinnolines: A novel photochromic system, *J. Chem. Soc., Chem. Commun.*, 338, 1988.

21. Bach, V., Diploma thesis, University of Saarbrücken, 1984; Gilchrist, T. L. and Rees, C. W., Thermolysis of salts of 2-substituted acrylic acids. Novel reductions of a vinyl bromide, *J. Chem. Soc., C*, 779, 1968.

22. Schommer, C., Thesis, University of Saarbrücken 1987; Dorweiler, C., Münzmay, T., Spang, P., Holderbaum, M., Dürr, H., Raabe, E., and Krüger, C., Photochromie einfacher 1,8a-Dihydroindolizine bzw. 1,8a-Dihydro-5-azaindolizine, *Chem. Ber.*, 121, 843, 1988; Dürr, H., Schommer, C., and Münzmay, T., Dihydropyrazolopyridine und *Bis*(dihydroindolizine) — neuartige mono- und difunktionelle photochrome Systeme, *Angew. Chem.*, 98, 565 (1986) and 25, 572, 1986.

23. Dürr, H. and Spang, P., Photochrome Systeme nach Maß: Erste bichromophore Spiro-1,8a-dihydroindolizine, *Angew. Chem.*, 96, 277, 1984; *Angew. Chem. Int. Ed.*, 25, 572, 1986; 23, 241, 1984.

24. Hadjoudis, E., Holderbaum, M., and Dürr, H., Novel biphotochromic systems: Dihydroindolizines with Shiff base moiety, *J. Photochem. Photobiol. A: Chem.*, 58, 37, 1991.

25. Gross, H., Dürr, H., and Rettig, W., Photochromic systems. 8. Emission spectra of photochromic spiro[1,8a]dihydroindolizines and mechanism of the electrocyclic ring opening reaction, *J. Photochem.*, 26, 165, 1984.

26. Bär, R., Gauglitz, G., Benz, R., Poster, J., Spang, P., and Dürr, H., Photokinetische Untersuchungen an photochromen Systemen der Dihydroindolizine, *Z. Naturforsch.*, A39, 662, 1984.

27. Braslavsky, S., Holzwarth, A. R., and Schaffner, K., Konformationsanalyse, Photophysik und Photochemie der Gallenpigmente; Bilirubin- und Biliverdindimethylester und verwandte lineare Tetrapyrrole, *Angew. Chem.*, 95, 670, 1983; *Angew. Chem., Int. Ed.*, 22, 656, 1983.

28. Bergmann, E. D., Weizmann, A., and Fischer, E., Structure and polarity of some polycyclic spirans, *J. Am. Chem. Soc.*, 72, 5009, 1950.

29. Fischer, E. and Hirshberg, Y., Formation of colored forms of spirans by low-temperature irradiation, *J. Chem. Soc.*, 4522, 1952.

30. Hirshberg, Y. and Fischer, E., Photochromism and reversible multiple internal transitions in some spiropyrans at low temperature. II., *J. Chem. Soc.*, 297, 3129, 1954.

31. Bercovici, T., Heiligman-Rim, R., and Fischer, E., Photochromism in spiropyrans. VIII. Photochromism in acidified solutions, *Mol. Photochem.*, 23, 189, 1969.

32. Guglielmetti, R., 4n + 2 Systems: Spiropyrans, in *Photochromism — Molecules and Systems*, Dürr, H. and Bouas-Laurent, H., Eds., Elsevier, Amsterdam, 1991, chap. 8.

33. Gehrtz, M., Bräuchle, C., and Voitländer, J., Photochromic forms of 6-nitrobenzospiropyran. Emission spectroscopic and ODMR investigations, *J. Am. Chem. Soc.*, 104, 2094, 1982; Lenoble, C. and Becker, R. S., Photophysics, photochemistry, kinetics and mechanism of the photochromism of 6'-nitro-indolinospiropyran, *J. Phys. Chem.*, 90, 62, 1986.

34. Fox, R. E., Res.Rep. and Test; Final Report Contr. AF 41, 657, 1961, AD 444226; Chu, N. Y., Photochromism of spiroindolinonaphthoxazine. I. Photophysical properties, *Can. J. Chem.*, 61, 300, 1983.

35. Ono, H. and Osada, T., Photochromic compound and compos., U.S. Patent 3562 172, 1971.

36. Pottier, E., Du Best, R., Guglielmetti, R., Taridieu, P., Kellmann, A., Tfibel, F., Lenoir, P., and Aubard, J., Effets de substituant d'heteroatome et de solvant sur le cinetique de decoloration thermique et les spectres d'absorption de photomerocyanines en serie spiro[indoline-oxazine], *Helv. Chim. Acta*, 73, 303, 1990.

37. Hovey, R. J., Chu, N. Y., Piusz, P. G., and Fudsmann, C. H., Photochromic comp., U.S. Patent 4342 668, 1982.

38. Melzig, M. and Martinuzzi, G., Photochromic subst., PCT Int. Appl. WO 85 02 619, 1985.

39. Chu, N. Y., 4n + 2 Systems: Spiroxazines, in *Photochromism — Molecules and Systems*, Dürr, H. and Bouas-Laurent, H., Eds., Elsevier, Amsterdam, 1990, chap. 10.

40. Wilkinson, F., Hobley, J., and Naftaly, M., Photochromism of spironaphthoxazines: Molar absorption coefficients and quantum efficiencies, *J. Chem. Soc., Faraday Trans.*, 88, 1511, 1992.

41. Sousa, J. A., Weinstein, and Bluhm, A. L., The photochromism of 1-aryl-2-nitroalkenes, *J. Org. Chem.*, 34, 3320, 1969; Humphry-Baker, R. A., Salisbury, K., and Wood, G. P., Photochemical reactions of an α,β-unsaturated nitro compound, *J. Chem. Soc. Perkin Trans. 2*, 659, 1978.

42. Yano, E., Tatsura, K., and Ikegami, K., *Proc. of the XXth IUPAC Symp. on Photochem.*, 232, Abstracts, Bologna, 1988.

43. Leonard, N. J., Mc Credie, R. S., Logne, M. W., and Cundall, R., Solid state ultraviolet irradiation of 1,1'-trimethylenebisthymine and photosensitized irradiation of 1,1'-polymethylenebisthymines, *J. Am. Chem. Soc.*, 95, 2320, 1973.

44. Beukers, R. and Berends, W., Effects of ultraviolet irradiation on nucleic acid and their components, *Biochim. Biophys. Acta*, 41, 550, 1960; 49, 181, 1961.

45. Ayer, W. A., Hayatsu, R., de Mayo, P., Reid, S. T., and Stothers, J. B., The photodimers of α-pyridones, *Tetrahedron Lett.*, 648, 1961.

46. Paquette, L. A. and Slomp, G., Derivatives of *trans*-3,7-diazatricyclo[4.2.2.2^{2,5}]dodecane, *J. Am. Chem. Soc.*, 85, 765, 1963.

47. Taylor, E. and Kan, R. O., Photochemical dimerization of 2-amino-pyridines and 2-pyridines, *J. Am. Chem. Soc.*, 85, 776, 1963.

48. Tomlinson, W. J., Chandross, E. A., Fork, R. L., Pryde, C. A., and Lamola, A. A., *Appl. Optics*, 11, 543, 1972.

49. Bradscher, C. R., Beavers, L. E., and Jones, J. H., Acridinium salts, *J. Org. Chem.*, 22, 1740, 1957.

50. Chandross, E. A., Fork, R. L., Lamola, A. A., and Tomlinson, N. J., Optical storage devices, U.S. Patent 3668 663, 1972.

51. Wagner, J., Bending, J., and Kreysig, D., Mechanismus der reversiblen intramolekularen [π4s + π4s]-Photocycloaddition von α,ω-*Bis*-(9-acridiziniumyl)-alkanen, *J. Prakt. Chem.*, 326, 747 and 757, 1984.

52. Senier, A. and Shepheard, F. G., Studies in phototropy and thermotropy. I. Arylidene- and naphthylidene-amines, *J. Chem. Soc.*, 95, 1943, 1909; 101, 1950, 1912.

53. De Gaouck, V. and Le Fevre, R. J. W., The phototropy of anils and a note on the phototropy of solutions of the leuco-cyanides of malachite- and brilliant-greens, *J. Chem. Soc.*, 1457, 1939.

54. Hadjoudis, E. and Hayon, E., Flash photolysis of some photochromic *N*-benzylideneanilines, *J. Phys. Chem.*, 74, 3184, 1970.

55. Rosenfeld, T., Ottolenghi, M., and Meyer, A. Y., Photochromic anils. Structure of photoisomers and thermal relaxation process, *Mol. Photochem.*, 5, 39, 1973.

56. Miller, L. J. and Margerum, D. J., *Tech. Chem. (NY)*, 3, 557, 1971; Jaques, P., Biava, J. P., Goursot, A., and Faure, J., Investigation of tautomerism and photochromism exhibited by a series of hydroxyazodyes. 1. Spectroscopic study and conformational analysis of fundamental states, *Chim. Phys. Phys. Chim. Biol.*, 76, 56, 1979; and Jaques, P., Investigations of tautomerism and photochromism exhibited by a series of hydroxyazoic cationic dyes. II. Kinetic behavior, *Chim. Phys. Phys. Chim. Biol.*, 79, 352, 1982; Substituent effects on the tautomerism and photochromism exhibited by a series of hydroxyazo cationic dyes for polyester fibers, *Dyes Pigm.*, 5, 351, 1984.

57. Lewis, J. W. and Sandorfy, J. W., A spectroscopic study of proton transfer and photochromism in *N*-(2-hydroxybenzylidene)aniline, *Can. J. Chem.*, 60, 1738, 1982; Becker, R. S., Lenoble, C., and Zein, A., A comprehensive investigation of the photophysics and photochemistry of salicylideneaniline and derivatives of phenylbenzothiazole including solvent effects; Photophysics and photochemistry of the nitro-derivatives of salicylideneaniline and 2-(2'-hydroxyphenyl)-benzothiazole and solvent effects, *J. Phys. Chem.*, 91, 3509, 3517, 1987.

58. Hadjoudis, E., Tautomerism by hydrogen transfer in anils, aci-nitro and related compounds, in *Photochromism — Molecules and Systems*, Dürr, H. and Bouas-Lambert, H., Eds., Elsevier, Amsterdam, 1990, chap. 17.

59. Hadjoudis, E., Moustakali-Mavridis, I., and Xexakis, J., Effect of crystal and molecular structure on the thermochromism and photochromism of some salicylidene-2-aminopyridines, *Isr. J. Chem.*, 18, 202, 1979; Moustakali-Mavridis, I., Hadjoudis, E., and Mavridis, A., Structure of thermochromic

Schiff bases. II. Structures of *N*-salicylidene-3-aminopyridine and *N*-(5-methoxysalicylidene)-3-amino-pyridine, *Acta Crystallogr.*, B36, 1126, 1980; Hadjoudis, E., Photochromism and thermochromism of *N*-salicylideneanilines and *N*-salicylideneaminopyridines, *J. Photochem.*, 17, 355, 1981.

60. Kramer, H. E. A., Tautomerism by hydrogen transfer in salicylate, triazoles and oxazoles, in *Photochromism — Molecules and Systems*, Dürr, H. and Bouas-Laurent, H., Eds., Elsevier, Amsterdam, 1990, chap. 16.

61. Wild, U., Spectral hole-burning, in *Photochromism — Molecules and Systems*, Dürr, H. and Bouas-Laurent, H., Eds., Amsterdam, 1990, chap. 28.

62. Meriwether, L. S. and Breitner, E. C., Kinetic and infrared study of photochromism of metal dithizonates, *J. Am. Chem. Soc.*, 87, 4448, 1965.

63. Blinder, S. M. and Lord, N. W., Electron spin resonance of tetraphenylpyrryl radical, *J. Chem. Phys.*, 36, 540, 1961; Maeda, K., Chinone, A., and Hayashi, T., The photochromism, thermochromism and piezochromism of dimers of tetraphenylpyrryl, *Bull. Chem. Soc. Jpn.*, 43, 1431, 1970.

64. Gliemeroth, G. and Mader, K. H., Phototropes Glas, *Angew. Chem.*, 82, 421, 1970; *Angew. Chem., Int. Ed.*, 9, 434, 1970.

65. Parker, C. A., Photoreduction of methyleneblue. Some preliminary experiments by flash photolysis, *J. Phys. Chem.*, 63, 26, 1959.

84

Photobehavior of Alkyl Halides

Paul J. Kropp
*University of North Carolina
at Chapel Hill*

84.1 Photochemical Behavior

Principal Pathways

Alkyl iodides afford mixtures of radical- and ion-derived photoproducts in solution, with the latter predominating usually. Indeed, this is a powerful method for generating carbocations, including many that cannot be prepared readily by other methods. Alkyl bromides display similar photobehavior, but with a lower proportion of ionic products. Analogous behavior has been observed also for phenyl thioethers and selenoethers,[1] as well as some organosilicon iodides.[2] In a process related to the formation of ionic intermediates, irradiation of dihalomethanes in the presence of alkenes results in cyclopropanation, a synthetically useful procedure that complements traditional methods. This chapter, which is concerned with alkyl halides, is a major expansion of an earlier summary.[3] The solution-phase photobehavior of aryl, benzylic, and homobenzylic halides has been reviewed, along with brief summaries of alkyl systems.[4] The photobehavior of alkyl halides in the gas phase has also been reviewed.[5]

Spectroscopic Properties

Alkyl iodides have a long wavelength absorption arising from an $n\sigma^*$ transition that is red-shifted with increasing substitution about the iodine-bearing carbon atom, ranging in λ_{max} in hydrocarbon solvents from 258 nm for CH_3I to 269 nm for tertiary iodides (ε 375–675).[6] It is slightly blue-shifted with increasing solvent polarity, ranging in λ_{max} from 254 nm for CH_3I to 267 nm for tertiary iodides in CH_3OH (ε 380–735). There is also a strong shorter wavelength band at 194 nm arising from an nR(6s) transition. The $N\sigma^*$ absorption in bromides is less intense and occurs at much shorter wavelengths, ranging in λ_{max} from 202 nm for CH_3Br to 215 nm for tertiary bromides in hydrocarbon solvents (ε 200–300). Alkyl chlorides have little or no absorption above 200 nm, rendering them difficult to study photochemically. The absorption spectra of geminal dihalides are more complex and extend to longer wavelengths due to mutual interaction of the two halogen

0-8493-8634-9/95/$0.00+$.50
© 1995 by CRC Press, Inc.

atoms. For example, CH_2I_2 has absorptions at 212 (ε 1580), 240 (ε 600), and 290 (ε 1300) nm. These arise from four $n\sigma^*$ transitions, with the highest wavelength absorption consisting of two unresolved peaks.

Irradiation Procedures

Since alkyl iodides have maximum absorption near the principal mercury line at 254 nm, they can be irradiated easily with a low-pressure mercury lamp and quartz or Vycor optics. Bromides, on the other hand, require light of shorter wavelengths, such as that emitted by a medium-pressure mercury arc.

The by-product HX can frequently influence the course of reaction, by both re-adding to elimination products and serving as an efficient hydrogen atom donor, and is best removed from the irradiation mixture as it is formed. Triethylamine can be used as a scavenger in nonpolar solvents. However, in polar solvents, it leads to increased formation of the reduction product RH because of competing reaction via the amine-halide exciplex 1 (Scheme 84.1). The poorer electron donor HO⁻ has been found to be a suitable HX scavenger in alcoholic solvents.[7]

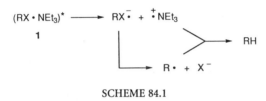

SCHEME 84.1

84.2 Mechanisms

It has long been recognized that absorption of light by the carbon-halogen chromophore results in homolytic cleavage of the bond. Not only is the antibonding σ-orbital occupied in the lowest-lying excited singlet state, but more than sufficient energy is available for bond cleavage. For example, the 0,0 energy levels of a typical tertiary iodide and bromide, estimated from the onsets of absorption at 315 and 250 nm, are 380 and 476 kJ mol⁻¹, respectively — substantially higher than the bond dissociation energies of 226 and 284 kJ mol⁻¹, respectively. It should be noted, however, that although the excited state is dissociative, it does not cleave with unit efficiency. The quantum yield for dissociation of ethyl iodide, for example, has been determined to be 0.30 to 0.32 in the gas phase, in the pure liquid, and in hexane.[8]

Early studies on the photobehavior of alkyl halides were conducted usually in the gas phase and interpreted in terms of radical intermediates.[5] Similar thinking carried over to early solution-phase studies.[4a] However, several "anomalous" products appeared in the solution phase that required skeletal rearrangements appropriate for carbocations but not typical of radical intermediates. These include the formation of (*E*)-2-butene (3) from iodide 2[9] and the rearranged alkene 5 from iodide 4,[10] which could readily arise via the ionic intermediates shown. In view of the formation of such products, an extensive study of the solution-phase photobehavior of alkyl iodides was undertaken, which provided firm evidence for the involvement of cationic intermediates and led to the proposal that they arise via initial homolytic cleavage followed by electron transfer within the resulting radical pair cage (6), as outlined in Scheme 84.2.[11] As discussed below, the resulting carbocations display behavior typical of other "free" cations generated by high-energy processes with little or no solvent participation.[12]

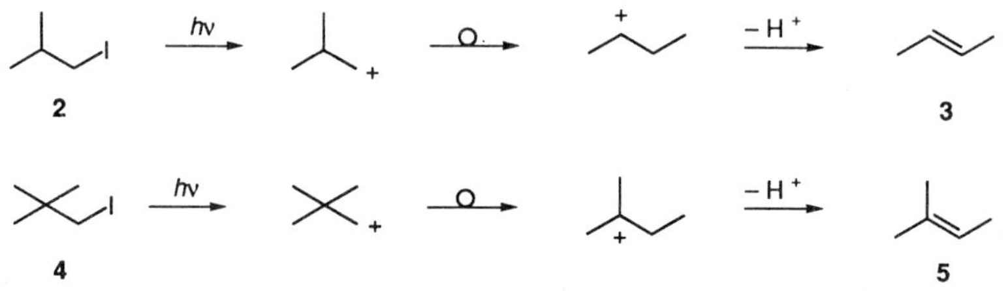

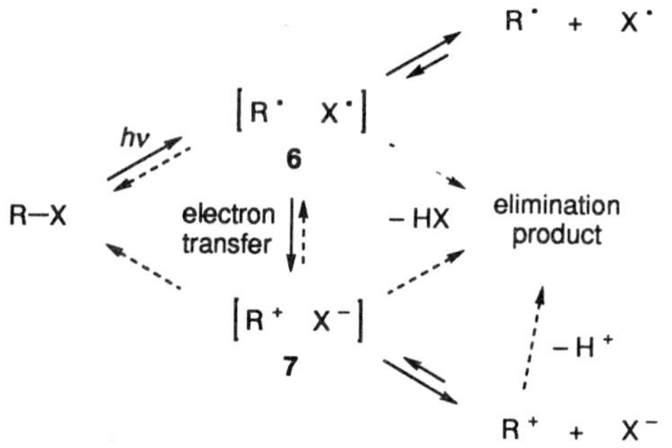

SCHEME 84.2

The initial studies centered on the 1-norbornyl halides **8**. On irradiation in CH_3OH, they afford a mixture of the reduction product norbornane (**9**) and the ether **10**, with the former predominating from bromide **8b** and the latter from iodide **8a**.[11,13] The reduction product **9** arises via abstraction of a hydrogen atom by the 1-norbornyl radical from the medium and ether **10** via nucleophilic trapping of the 1-norbornyl cation. Irradiation of either halide in CH_3OD afforded ether **10** with no detectable incorporation of deuterium, indicating that it does not arise via acid-catalyzed addition of the alcohol to an unsaturated intermediate such as the bridgehead alkene **12** or the propellane **13**.

The higher ratio of ionic to radical products from iodide **8a** has proven to be general for iodides compared with the corresponding bromides. It is surprising since the more electronegative bromine atom might be expected to undergo electron transfer to the ion pair more readily than iodine atom (Scheme 84.2). The observed trend apparently reflects a higher rate of diffusion from the radical pair cage relative to the rate of electron transfer for the lighter bromine atom.

As expected, the ratio of ionic to radical products is also dependent on the ionization potential of the alkyl radical R·. For example, in contrast with 1-iodonorbornane (**8a**), the 3-hydroxy derivative **14** afforded more of the reduction product **15** than ionic product **16** in CH_3OH.[14] Apparently, the electron-withdrawing OH group impedes electron transfer. It has been proposed that the ionization potential of R· must be ≤ 836 kJ mol[-1] for electron transfer to occur.[15]

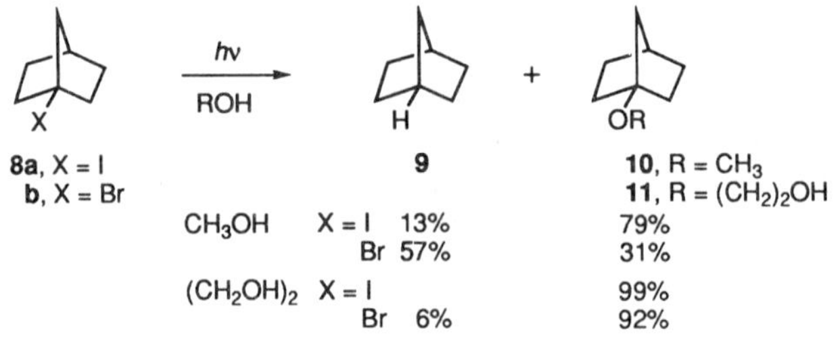

8a, X = I		**9**	**10**, R = CH_3
b, X = Br			**11**, R = $(CH_2)_2OH$
	CH_3OH	X = I 13%	79%
		Br 57%	31%
	$(CH_2OH)_2$	X = I	99%
		Br 6%	92%

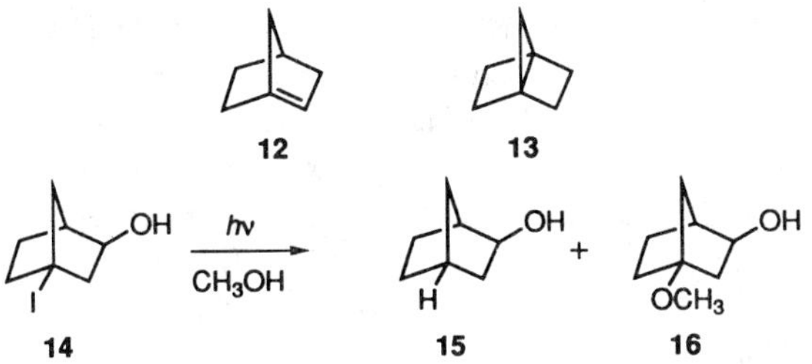

Consistent with the proposed mechanism, the ratio of ionic to radical product is dependent on the viscosity of the medium. Thus, irradiation of halides 8 in $(CH_2OH)_2$, which is only slightly more polar than CH_3OH but substantially more viscous, resulted in almost exclusive formation of the nucleophilic trapping product 11 from both the bromide and iodide.[11,13] The extended lifetime of the radical pair in the more viscous medium permits electron transfer to compete more effectively with diffusion of the radical components from the cage.

The quantum yield for disappearance of iodide 8a in CH_3OH at 254 nm is 0.10.[13] The low value is probably due, at least in part, to recombination of the radical and ion pairs 6 and 7. The quantum yield was somewhat higher in CH_3OH saturated with O_2, with formation of the radical product 9 quenched almost totally and that of the ionic product 10 reduced substantially.[13] This was accompanied by the formation of a new product, hydroperoxide 17, arising from trapping of R·. Bromide 8b displayed similar behavior except that formation of norbornane (9) was not totally quenched. Quenching of the ionic product 10 by O_2 is consistent with the proposed coupling of the radical and ionic pathways by electron transfer (Scheme 84.2).

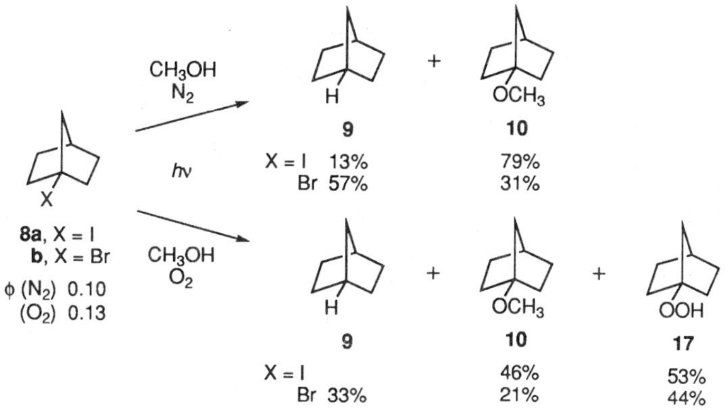

84.3 Examples

Monohalides

The photobehavior of alkyl halides varies with structure, depending on the ease of β-elimination.

Bridgehead and Related Halides: Nucleophilic Trapping vs. Reduction

Irradiation of bridgehead halides is a convenient method for generating bridgehead carbocations, which are readily trapped nucleophilically. Even though tertiary, these cations are difficult to form by conventional methods because bridgehead carbon atoms are inhibited structurally from relaxing to a planar geometry to provide the sp^2 hybridization preferred by carbocations. Thus, iodide 8a was recovered quantitatively from treatment with CH_3OH in the dark for 24 h — even in the presence of silver ion, which normally facilitates solvolysis of alkyl iodides.[13] By contrast, a methanolic solution of 8a afforded ether 10 readily on exposure to UV light.

Being difficult to form, bridgehead cations are highly reactive. Irradiation of halides **8** in a variety of media provided a convenient means of exploring the remarkable reactivity of the 1-norbornyl cation (**18**).[13] It was trapped efficiently by diethyl ether and THF to afford the ethers **19** and **20**. Even in aqueous THF, a substantial portion of the butenyl ether **20** was obtained. Trapping by *t*-butyl alcohol afforded some of the ditertiary ether **21**, but was accompanied by formation of 1-norbornanol (**22**) and 2-methylpropene. Cation **18** was trapped by CH_2Cl_2 to afford chloride **23**; in aqueous CH_3CN, the Ritter product **24** was formed. It should be noted that radical abstraction from CH_2Cl_2 occurs at H and ionic abstraction at Cl.

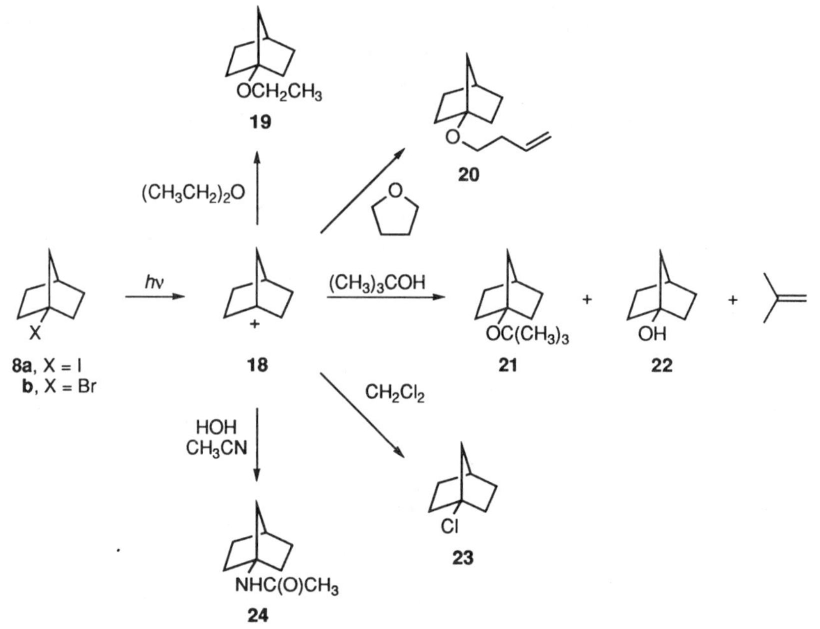

Interestingly, the 1,4-diiodide **25a** afforded mainly the acetal **28** rather than the dimethoxy product **27**.[16] Acetal **28** apparently arises from fragmentation of the 4-methoxy cation **29** prior to trapping with CH_3OH since it was also the principal product from separate irradiation of the 4-methoxy iodide **26a**. In contrast with diiodide **25a**, the 1,4-dibromide **25b** afforded predominantly the reduction product **8b**.

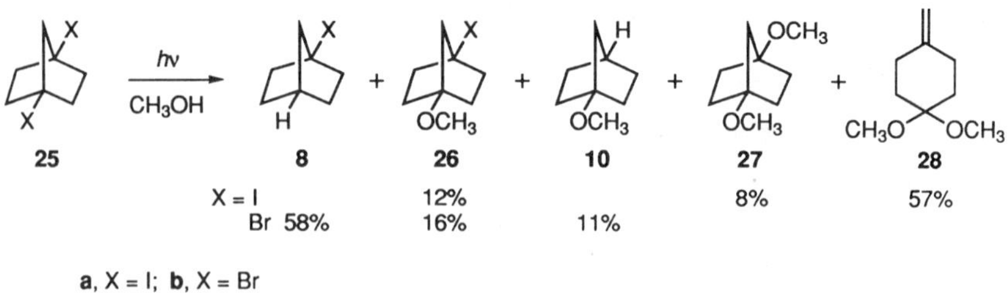

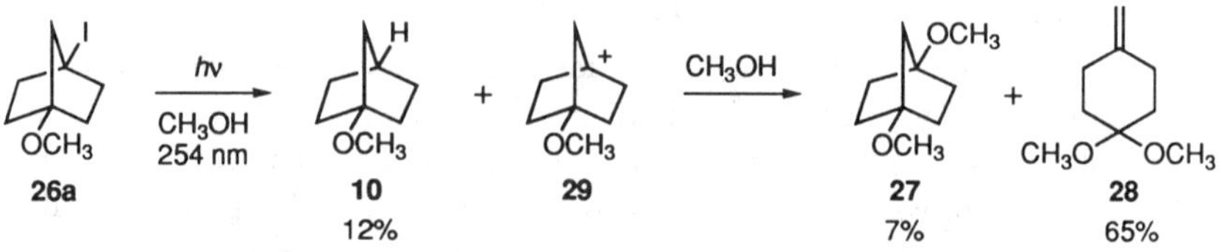

By analogy with 1-iodonorbornane (**8a**), the bridgehead iodides **30** afforded the corresponding ethers **32**, along with small amounts of the reduction products **31**.[17] Once again, irradiation in

CH₃OD afforded ethers **32** with no detectable deuterium incorporation. The bicyclo[2.2.2]- and bicyclo[3.2.1]octyl cations were also obtained by irradiation of 1-(iodomethyl)norbornane (**33**) in CH₃OH to afford ethers **32a** and **32b** in a ratio approaching the kinetic ratio of 1:2.[13] By contrast, silver ion-assisted methanolysis of iodide **33** afforded ether **32a** predominantly, arising from the thermodynamically more stable bicyclo[2.2.2]octyl cation, along with small amounts of the bicyclo[3.2.1]octyl ether **32b** and unrearranged ether **34**.

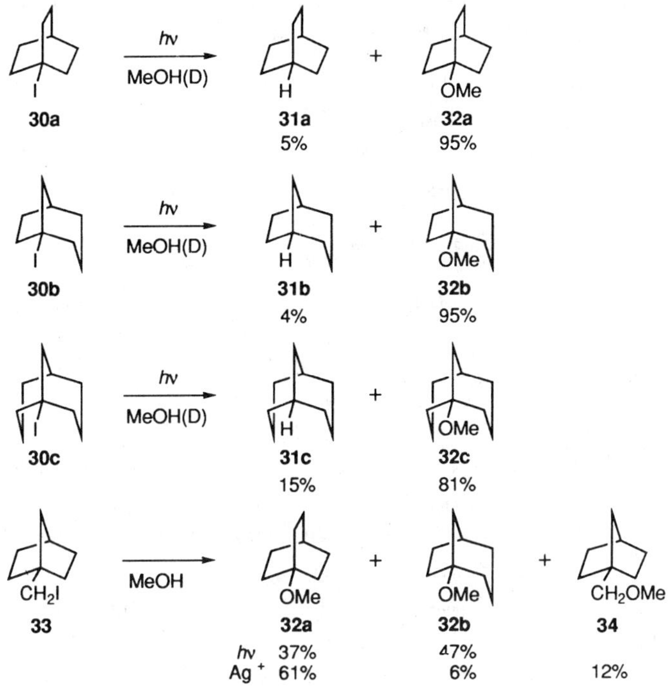

Irradiation of iodocubane (**35**) afforded the methyl ether **37**, accompanied by a small amount of cubane **36**, the reduction product.[18] The high ratio of ionic-to-radical product indicates that the cubyl cation is easier to form than 1-norbornyl cation, in agreement with solvolysis data and *ab initio* calculations showing stabilization of cubyl cation by cross-ring bonding to the β carbon atoms.[19] The 1,4-diiodide **38** afforded the mono- and disubstituted derivatives **39** and **40** in CH₃OH and wet CH₃CN.

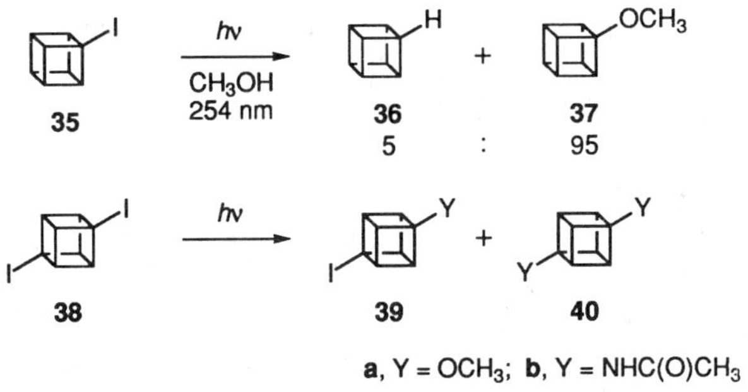

a, Y = OCH₃; **b**, Y = NHC(O)CH₃

Irradiation of the 1-adamantyl halides **41** in a variety of alcohols afforded the ethers **43** along with a small amount of adamantane (**42**), the reduction product.[13,20] The greater ease of forming the 1-adamantyl, compared with the 1-norbornyl, cation is reflected in the higher ratio of ionic-to-radical product from both the iodide and bromide. Once again, use of the viscous solvent (CH₂OH)₂ increased greatly the amount of ionic product 43 from the bromide 41b compared with CH_3OH.[13] Adamantyl bromides are significantly more sensitive to the presence of electron-withdrawing substituents than are the corresponding iodides. Thus, the 1,3-diiodides 44a and 45a afforded the corresponding diethers 44c and 45c in high yield, whereas the dibromides 44b and 45b afforded principally the corresponding monoether 43 (R = OCH_3) and its 3,5-dimethyl derivative.[20] Similarly, the tribromide 46 afforded mainly the diether 44c in CH_3OH,[20] and the tetrabromide 47b afforded only 1-acetamido-3,5,7-tribromoadamantane in wet CH_3CN.[21] By contrast, the tetraiodide 47a afforded the tetraamide 47c in 51 to 60% yield under similar conditions.[21]

41a, X = I
b, X = Br

42 **43**

		42	43
CH_3OH	X = I		99%
	Br	14%	53%
$(CH_2OH)_2$	X = I		98%
	Br	6%	83%

44a, X = I
b, X = Br
c, X = OCH_3

45a, X = I
b, X = Br
c, X = OCH_3

46

47a, X = I
b, X = Br
c, X = NHC(O)CH_3

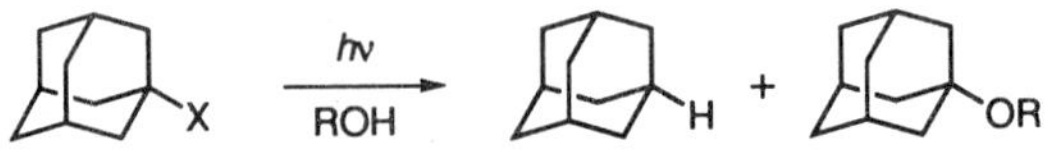

Non-bridgehead Halides: Elimination vs. Nucleophilic Trapping and Reduction

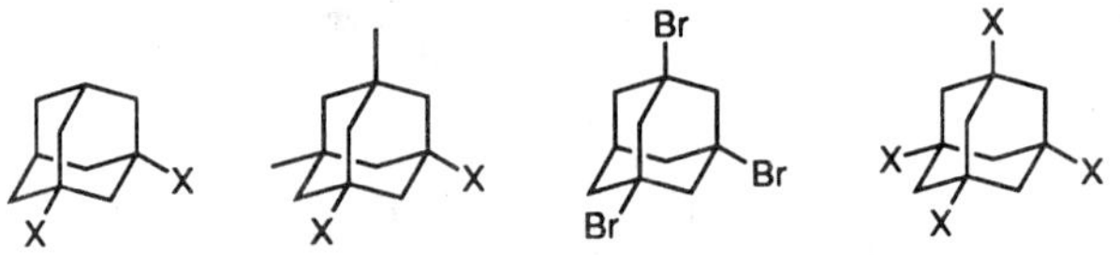

Except for systems like the preceding bridgehead halides that are strongly inhibited from undergoing elimination, alkyl halides undergo competing elimination and trapping in nucleophilic solvents. The product ratio depends on the degree of structural inhibition to elimination and the nucleophilicity of the solvent, but there is a strong propensity for undergoing elimination that is characteristic of "free" cations formed by high-energy processes without substantial solvent participation.[12] Frequently, cationic rearrangements precede elimination and nucleophilic trapping. Competing radical behavior is usually observed also.

1,2- and 1,3-Elimination. Irradiation of the 2-adamantyl halides 48 in CH_3OH afforded principally the 2-adamantyl ether 51, along with the reduction product adamantane (42).[13,20] However, when the irradiation was conducted in the presence of NH_4OH as an HX scavenger, the 1,3-elimination product 49 and rearranged elimination product 50 were also obtained.[7,22] In the absence of scavenging, these latter products undergo addition of HX to reform the starting halides 48. Irradiation of iodide 48a in the nonnucleophilic solvent, ether, containing triethylamine as a scavenger afforded the 1,3-elimination product 49 principally.[22]

48a, X = I
b, X = Br

		42	49	50	51
	X = I	1%			96%
	Br	19%			48%
NH_4OH	X = I	1%	20%	4%	75%
	Br	22%	23%	6%	40%

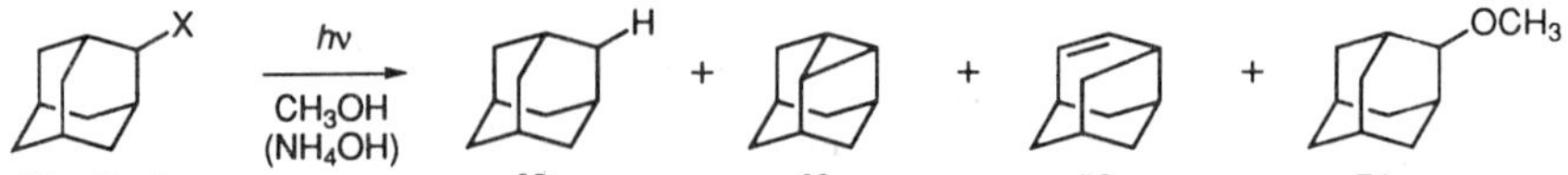

Another cation, like bridgehead cations, that is difficult to form by conventional methods is the 7-norbornyl cation, the problem in this case being angle strain. Thus, extended treatment of 7-iodonorbornane (52) with methanolic $AgNO_3$ afforded only a trace of ether 53 (R = CH_3).[23] By contrast, irradiation in CH_3OH readily afforded ether 53 (R = CH_3), along with smaller amounts of the reduction product norbornane (9), the unprecedented 1,3-elimination product 54, and the rearrangement products 55 to 57. The 2-chloro derivatives 58 underwent *syn, anti*-interconversion in competition with formation of the reduction product 60, the ethers 61 and 63, and the 1,3-elimination product 62. By contrast, treatment with methanolic AgNO$_3$ resulted in ionization of the more reactive chlorine substituent to afford ethers 59. Thus, irradiation at 254 nm is a simple method for inducing selective reaction by the iodine substituent.

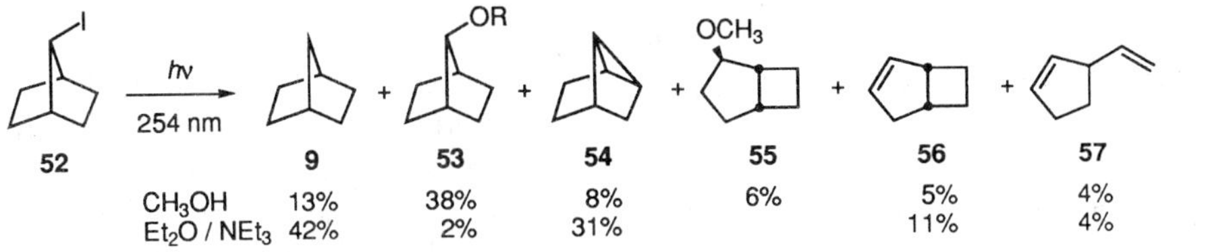

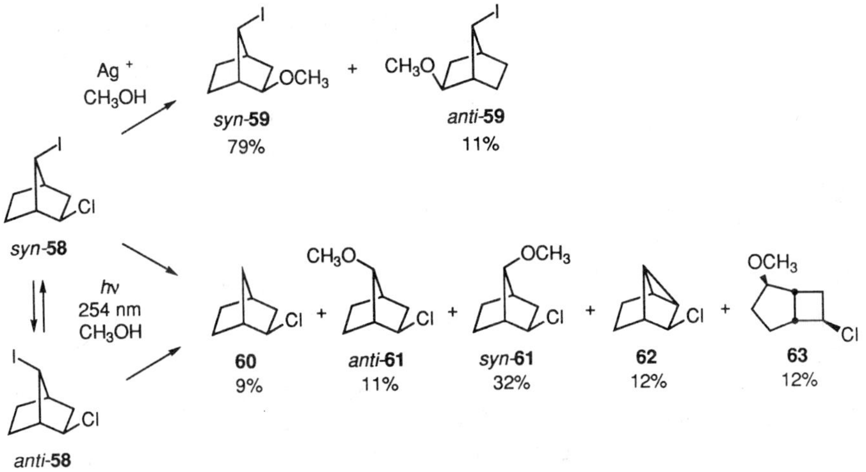

By contrast, the 2-norbornyl cation is easily formed. Moreover, the 2-norbornyl system can undergo elimination readily. On irradiation in CH$_3$OH or (CH$_3$)$_3$COH, the 2-norbornyl iodides 64a underwent *exo, endo*-equilibration and afforded the 1,2- and 1,3-elimination products 66 and 67, along with the nucleophilic trapping product 68.[7,24] The corresponding bromides 64b each afforded the same products, along with small amounts of the radical products 65 and 9, but underwent no detectable interconversion.

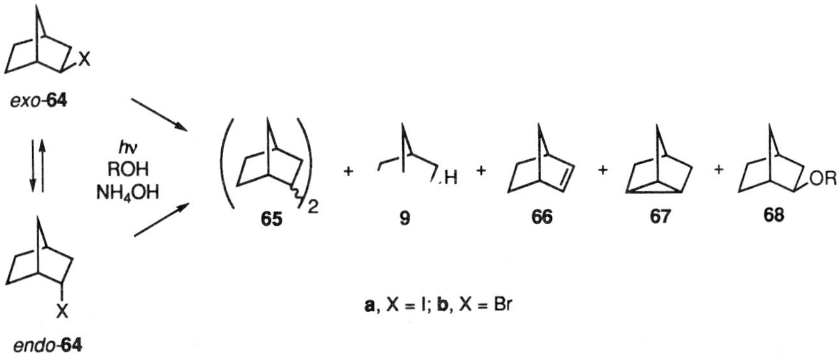

Clearly, the ethers 68 and the 1,3-elimination product nortricyclene (67) are ionic in origin. However, the 1,2-elimination product 2-norbornene (66) might arise in principle via either hydrogen atom transfer within the radical pair 6 or proton transfer within the ion pair 7 (Scheme 84.2). Deuterium labeling studies revealed that it arises by both pathways, with the percentage of ionic pathway being 36% for iodides **64a** and 64 to 69% for bromides **64b**.[7] These distributions, combined with the yields of the ionic products nortricyclene (**67**) and the ether **68** (R = CH$_3$), afford a ratio of in-cage radical to total ionic products of 1:3 for iodides **64a** but only 1:8 for bromides **64b**. Thus, of the material that is not lost by diffusion from the radical pair cage (**6**), a higher percentage is converted to products via ionic intermediates when X = Br, as expected based on the relative electronegativities of bromine and iodine.

Once out of the cage, the alkyl radical can abstract a hydrogen atom from the solvent or undergo combination/disproportionation with a second alkyl radical. Another available pathway in the case of iodides is transfer of an iodine atom from a molecule of unreacted starting material. Halogen atom transfer occurs readily for alkyl iodides ($k \approx 2 \times 10^5\ M^{-1}\ s^{-1}$)[25] and accounts for the observed interconversion of iodides **64a** (Scheme 84.3). In the more viscous solvent (CH$_3$)$_3$COH, in which escape from the radical cage (**6**) occurs more slowly, there is substantially less epimerization. In contrast with iodides, halogen atom transfer is much slower for alkyl bromides ($k \approx 6 \times 10^2\ M^{-1}\ s^{-1}$),[25] permitting combination/disproportionation to compete. Hence, bromides **64b** undergo no detectable epimerization.

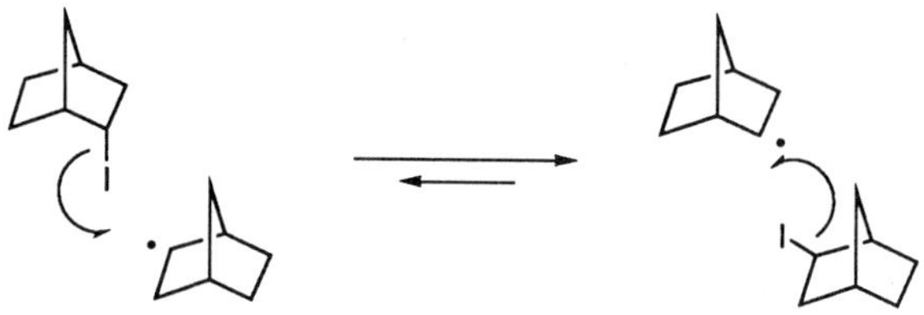

SCHEME 84.3

Analogously with the parent 2-norbornyl iodides **64a**, the iodo ethers **69a** underwent *exo,endo*-equilibration, accompanied by formation of the reduction product **70a**, the rearranged elimination product **73a**, the 1,3-elimination product **74a**, and the ethers **75a** and **76a**.[17] Also formed was the acetal **72**. Surprisingly, deuterium labeling studies showed that this arises via elimination to the bridgehead alkene **71** followed by addition of CH$_3$OH. In the presence of furan, the bridgehead alkene **71** was trapped by cycloaddition. Formation of the highly strained intermediate **71** underscores the strong driving force for elimination exhibited by alkyl halides on irradiation.

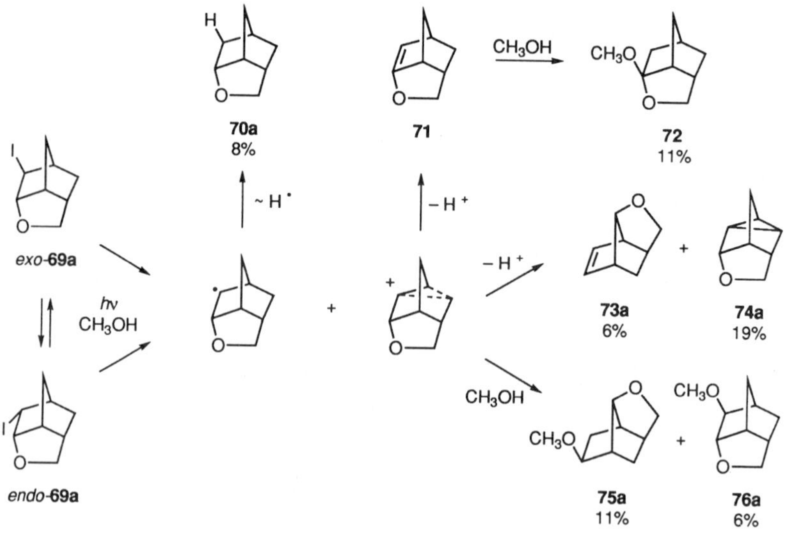

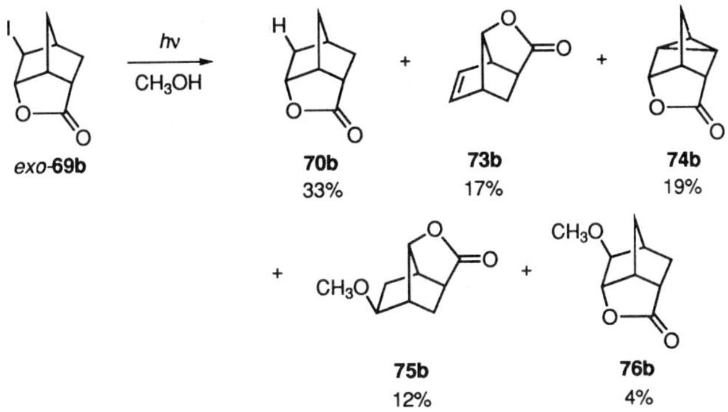

exo-69b

70b
33%

73b
17%

74b
19%

75b
12%

76b
4%

By contrast, the corresponding lactone 69b underwent no detectable epimerization and afforded principally the reduction product 70b.[17] The added inductive effect of the carbonyl oxygen atom apparently both retards electron transfer and renders the radical intermediate more reactive toward hydrogen abstraction from the medium, decreasing its lifetime and competing with the iodine exchange that leads to epimerization of iodo ether *exo*-69a (Scheme 84.3). The analogous elimination and trapping products 73b to 76b were formed, but there was no detectable formation of a product corresponding to acetal 72, presumably because the increased strain introduced by the sp^2-hybridized carbonyl carbon atom precludes elimination to a bridgehead alkene.

The analogous iodo ethers 77 underwent *exo,endo*-equilibration similarly, along with formation of the elimination product 78 and rearranged elimination product 79.[26] In CH_3OH, the rearranged acetal 80 was formed also.

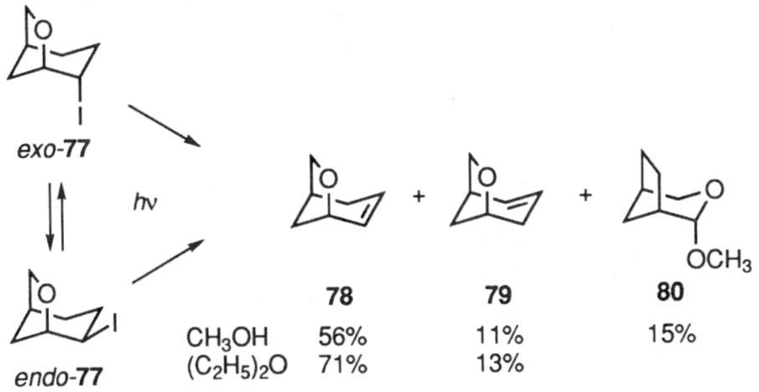

exo-77

hv

78

79

80

endo-77

CH_3OH	56%	11%	15%
$(C_2H_5)_2O$	71%	13%	

Epimerization was observed also in the interconversion of the deoxy iodo sugars 81 and 82.[27] Interconversion occurred from either isomer with a quantum yield greater than unity, consistent with a mechanism analogous to that shown in Scheme 84.3. Also formed was the reduction product 83. Similarly, the deoxy iodo sugar 84 afforded the reduction product 85.[27] Irradiation of deoxy halo sugars is a useful synthetic application of the radical photobehavior of alkyl halides, affording the corresponding deoxy sugars in high yield under mild conditions.[28] Ionic products are not formed, even in nucleophilic solvents. Except at the anomeric position, carbohydrates form carbocations with difficulty because of the large number of electron-withdrawing oxygen atoms.

81 82 83

84 $\phi = 0.31$

85 95%

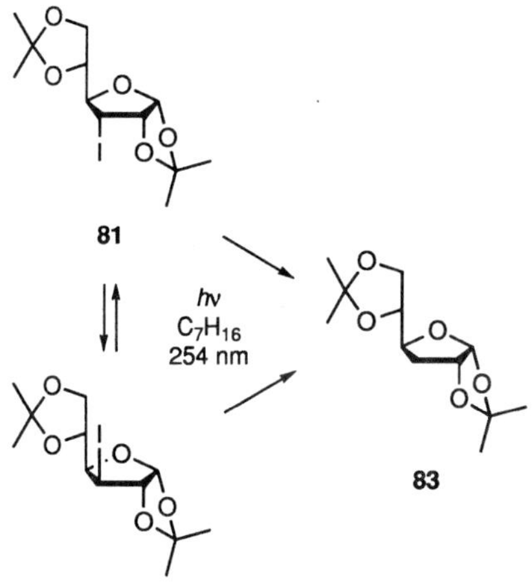

The *endo*-2-norbornyl analog longibornyl iodide (86) afforded principally the elimination products longifolene (87) and longicyclene (88), accompanied by a small amount of alkene 89 resulting from a transannular hydride shift.[29] Solvolysis of iodide 86 afforded only longifolene (87) and longicyclene (88), with none of the transannular elimination product 89. Formation of this product on irradiation illustrates once again the high reactivity of "free" cationic intermediates generated photolytically from alkyl halides. Another example is the formation of the strained alkene 91 from the 2-(iodomethyl)norbornane 90.[30]

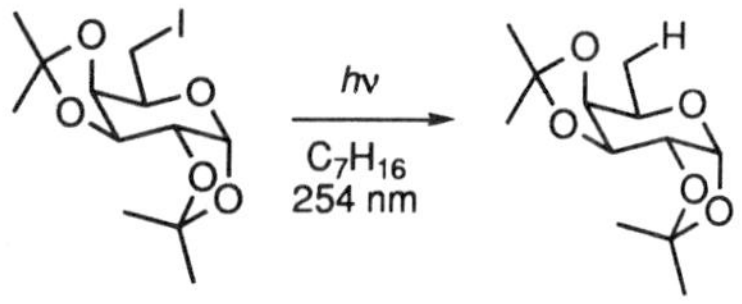

Irradiation of 9-iodocamphor (92) in CH_3OH afforded the ring-expanded elimination products 93 and 94, accompanied by a mixture of ethers 95.[31] Thus, the C-I chromophore can be irradiated selectively at 254 nm in the presence of a ketone carbonyl group. Under similar conditions, (iodomethyl)trimethylsilane (96) afforded principally the silyl ether 99, resulting from rearrangement of the α-silyl cation 97.[32] By contrast, silver ion-assisted methanolysis of iodide 96 afforded only the unrearranged ether 98 via S_N2 displacement, avoiding involvement of the α-silyl cation 97.

86 87 34% 88 37% 89 12%

90 91

92 93 17% 94 32% 95 46%

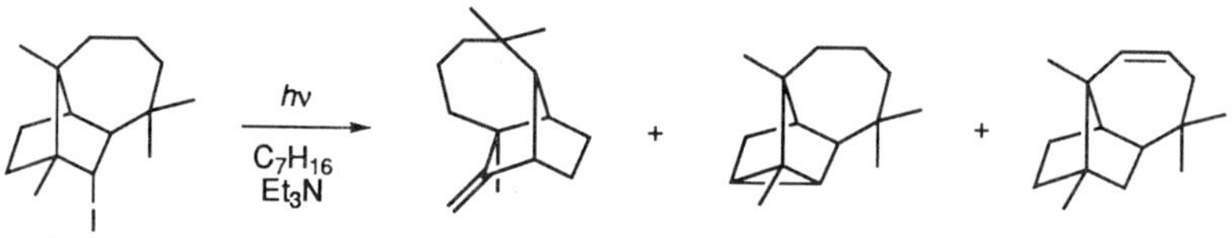

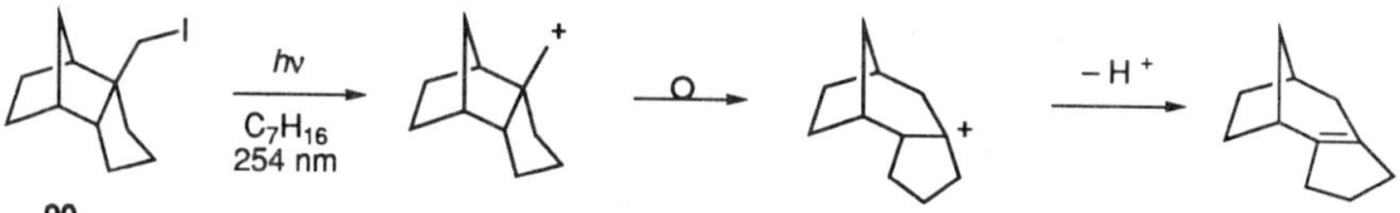

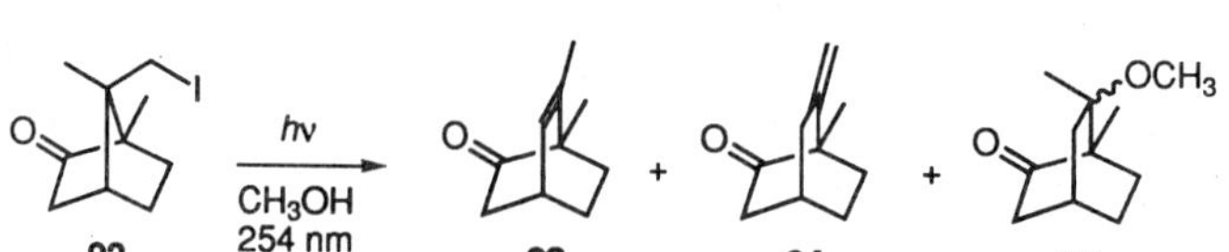

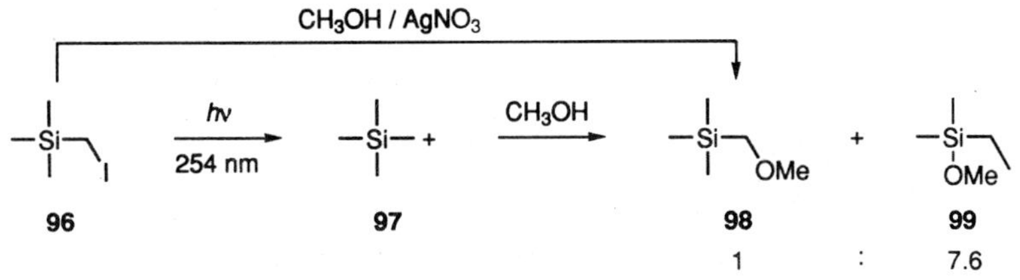

Even highly stabilized allylic cations undergo substantial elimination in a nucleophilic solvent when generated photochemically. Thus, the allylic iodide 100a and bromide 100b afforded principally the elimination product 102, along with the methyl ethers 103 and 104 on irradiation in CH₃OH.[33] Similar behavior was exhibited by the tetramethyl derivative 106. By contrast, the chloride 100c simply underwent equilibration with the allylic isomer 105, presumably via the radical pair.

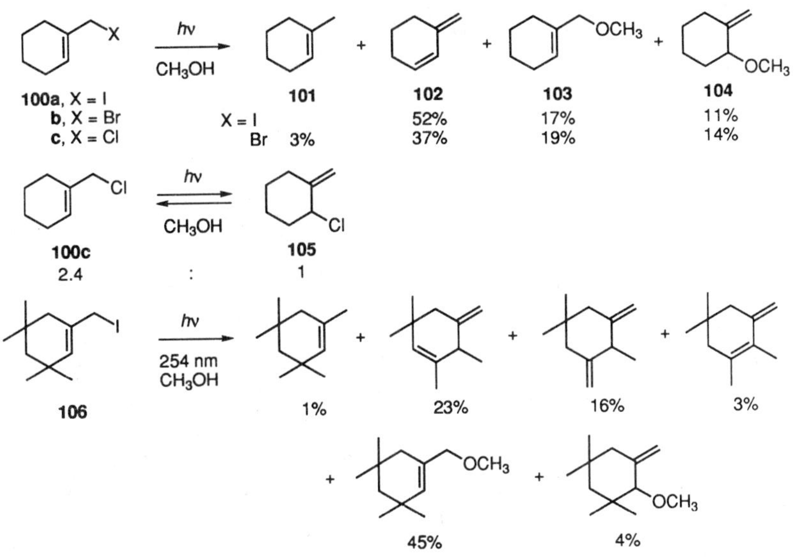

The 1-halooctanes 107 and iodocyclohexane (114) show once again the propensity of alkyl halides for photoelimination, giving primarily the elimination products 1-octene (110) and cyclohexene (116) in CH₃OH and only small amounts of the ethers 113 and 117.[11,13,34] As is typical, bromide 107b gave appreciable amounts of the out-of-cage radical products 108 and 109. The *t*-butylcyclohexyl iodide 118 afforded substantial amounts of the rearranged elimination product 121.[26]

In a dramatic application of the propensity of alkyl halides to undergo photoelimination, irradiation of the diiodosulfone 122a proved to be the method of choice for effecting elimination to diene 123, which was formed in good yield.[35] By contrast, the dihalosulfones 122 were either inert toward conventional elimination conditions or gave mainly methyl ethers on treatment with methanolic KOH. Irradiation of the dibromosulfone 122b gave mainly reduction and no detectable formation of the elimination product 123.

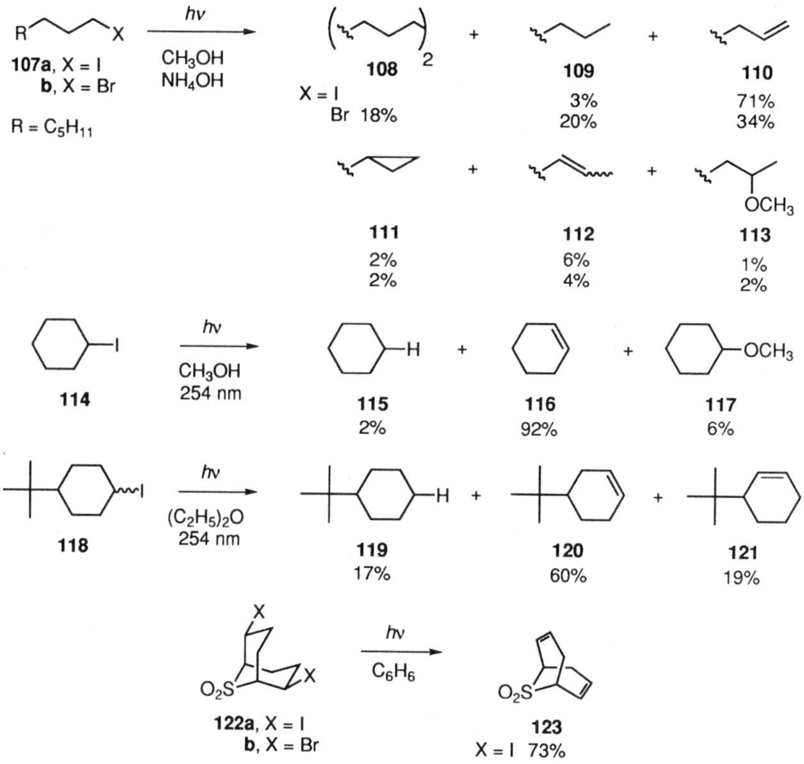

α-Elimination: A Secondary Pathway to Unsaturated Products. In the preceding discussion it was assumed that elimination products such as 1-octene (110) are formed via transfer of a β-hydrogen atom or a β-proton from the corresponding radical or carbocation, respectively. However, deuterium labeling studies with 1-iodooctane (107a) showed that 17 to 29% of the 1-octene (110) is formed via α-elimination to the carbene intermediate 124, depending on the solvent and irradiation conditions (Scheme 84.4).[34] Similarly, 27 to 30% of the cyclohexene (116) from 1-iodocyclohexane (114) arises via α-elimination. However, only 7% of the 1-octene (110) from bromide 107b arises via α-elimination.[7]

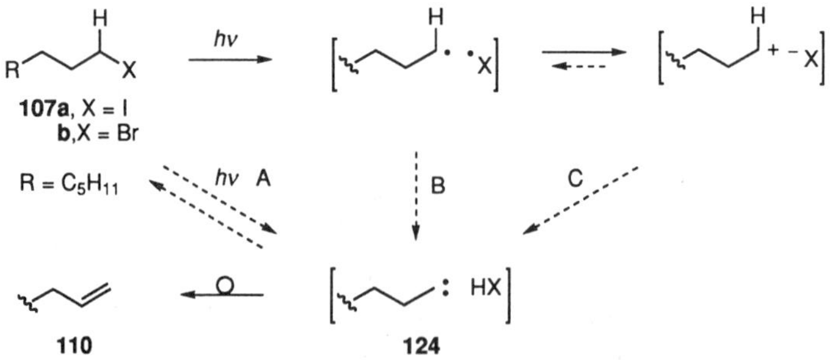

SCHEME 84.4

As shown in Scheme 84.4 with the 1-halooctane system 107 as a model, the caged carbene-HX pair 124 may undergo undetected insertion to regenerate halide 107 in competition with rearrangement of the carbene to 1-octene (110). Thus, α-elimination occur more extensively than reflected by the deuterium labeling results. Three possible routes to the carbene intermediate are outlined. The most obvious, direct formation via concerted loss of HX (path A), is inconsistent with the observed lack of a deuterium isotope effect on the relative involvement of α-elimination in the formation of 1-octene (110). It is not clear which of the other two paths is involved in carbene formation. There is, however, precedent for the formation of carbene intermediates via transfer of α-protons from carbocations.[36]

Trapping by Aromatic Solvents. Typically, alkyl halides undergo slow elimination on irradiation in aromatic solvents, accompanied by the formation of alkylated aromatic products in low-to-moderate yields (Scheme 84.5).[37] These latter products were obtained in higher yields with methyl,[38] trifluoromethyl,[39] and allyl iodides,[40] which are incapable of undergoing competing elimination. Similarly, irradiation of 1-iodonorbornane (8a) in toluene afforded a mixture of 1-tolylnorbornanes (8, X = $C_6H_4CH_3$) in good yield.[13] Imidazole and its derivatives underwent facile trifluoromethylation or perfluoroalkylation on irradiation of perfluoroalkyl iodides in their presence in CH_3OH solution.[41] However, when the photolyses of simple alkyl halides were performed in CH_3OH or CH_3CN solutions containing aromatic substrates, nucleophilic trapping products (alkyl methyl ethers or *N*-alkylacetamides) were formed at the expense of aromatic substitution products.[37]

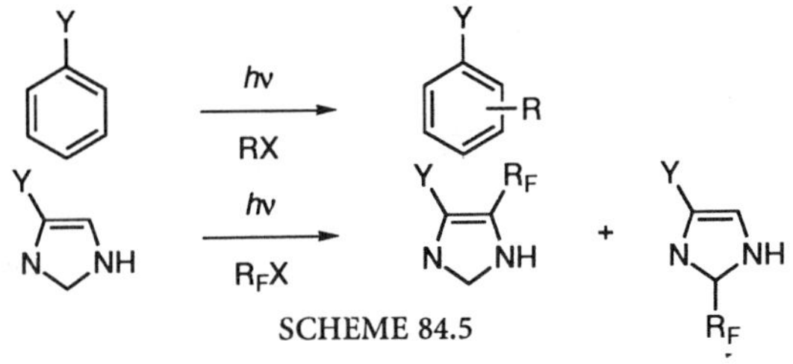

SCHEME 84.5

Since the absorption spectra of the C–I and aryl chromophores have extensive overlap, it is not clear whether photoreaction involves direct absorption by the alkyl halide or whether the light is being absorbed by the aromatic substrate, followed by either energy or electron transfer with the alkyl halide. It is also not clear whether the aromatic substrates are capturing radical or cationic intermediates. Simple alkyl iodides afforded mixtures of toluene adducts having *o:m:p* ratios consistent with cationic attack.[37] This was reinforced by the formation of ethers or acetamides at the expense of aryl substitution products in the presence of CH_3OH or CH_3CN. On the other hand, the 1-tolylnorbornanes (8, X = $C_6H_4CH_3$) from 1-iodonorbornane (8a) were formed in a ratio typical of that for homolytic substitution.[13] It would be expected that methyl and trifluoromethyl iodides would afford only radical intermediates. However, they exhibited a lower reactivity at the *ortho*-position than is generally typical of homolytic substitution reactions.[38,39] Similar results were afforded by allyl iodide.[40] Either competing trapping of radical and cationic intermediates is involved, or the mechanism is more complex than is apparent.

Cyclization. Trapping of the radical or cationic intermediates resulting from irradiation of alkyl halides can occur intramolecularly also. The 4-phenyl-1-butyl halides 125 afford the cyclized product 126, along with the additional products 127 to 130.[7,42] As usual, elimination predominates over aryl substitution. The allylic analog 131, which is structurally incapable of undergoing elimination, afforded the cyclization product 132, along with the amide 133, in CH_3CN.[42a] By contrast, the 2-phenyl-1-ethyl iodide 134a gave only the ether 136, whereas the bromide 134b gave a mixture of ether 136 and the out-of-cage radical product 135.[42b] Analogous behavior was exhibited in other alcoholic solvents.

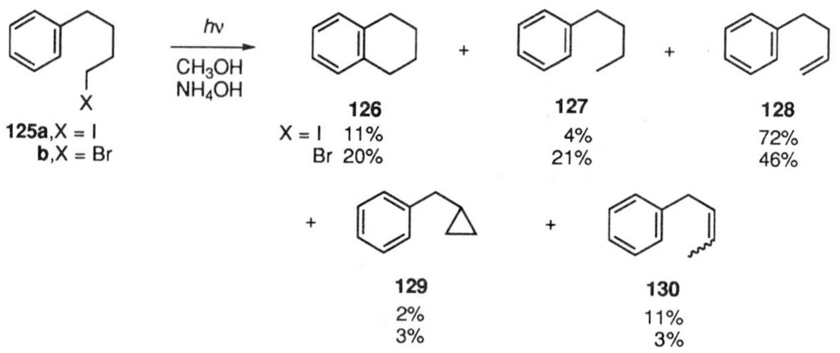
hv
CH3OH
NH4OH
X
125a,X = I
b,X = Br
126
X = I 11%
Br 20%
127
4%
21%
128
72%
46%
129
2%
3%
130
11%
3%

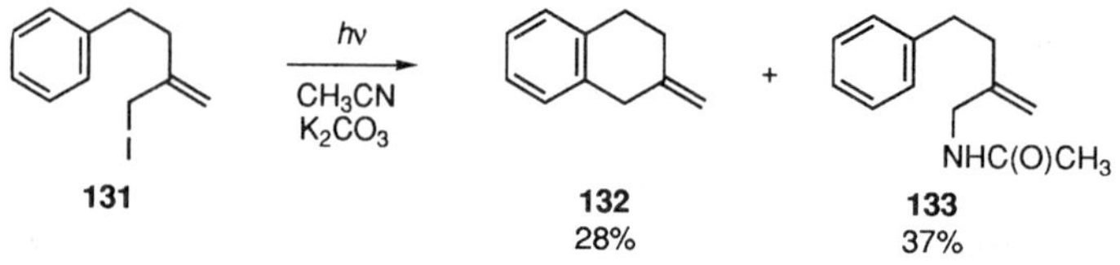
hv
CH3CN
K2CO3
I
131
132
28%
NHC(O)CH3
133
37%

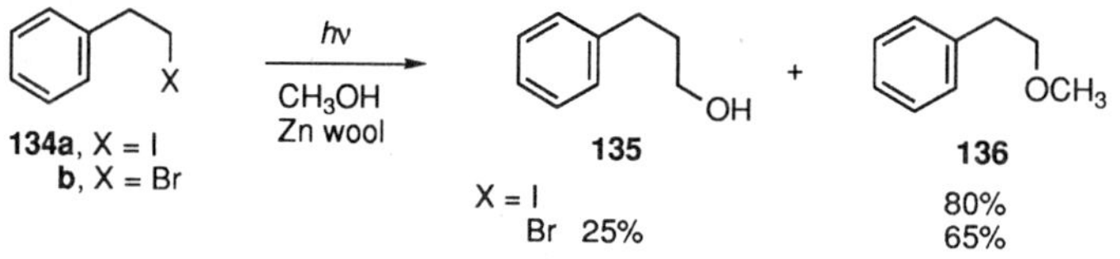
hv
CH3OH
Zn wool
X
134a, X = I
b, X = Br
OH
135
X = I
Br 25%
OCH3
136
80%
65%

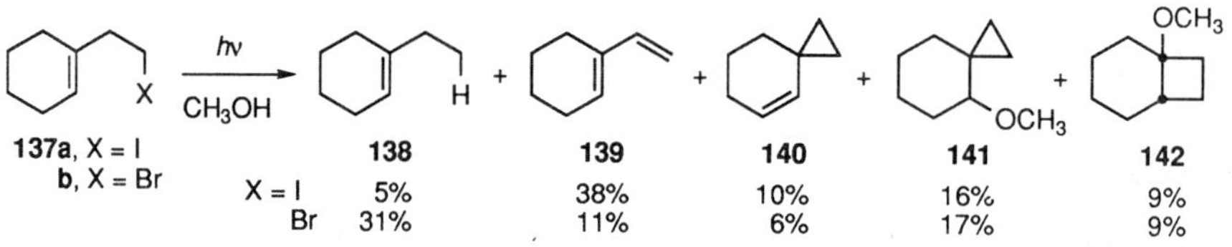
hv
CH3OH
X
137a, X = I
b, X = Br
H
138
X = I 5%
Br 31%
139
38%
11%
140
10%
6%
OCH3
141
16%
17%
OCH3
142
9%
9%

Cyclization can occur also with halo alkenes. The homoallylic halides 137 afforded the cyclized products 140 to 142, along with the reduction product 138 and elimination product 139.[33] Citronellyl iodide (143) afforded a mixture of the elimination product 145 and cyclization products 146 to 148, accompanied by a small amount of the reduction product 144.[29,43] For reasons not clear, cyclization was somewhat more favored relative to elimination with increasing temperature and on going to the more polar solvent THF. Analogous behavior was exhibited by the 2,3-dihydrofarnesyl iodides 149, which afforded the elimination products 150 and cyclization products 151.[43] Neryl iodide (*Z*)-152 afforded similarly a substantial amount of the cyclization product limonene (154), along with the elimination product myrcene (153).[44] The isomeric geranyl iodide (*E*-152), which is structurally inhibited from undergoing cyclization, afforded a small amount of limonene (154), apparently via initial *E*,*Z*-isomerization. Likewise, the farnesyl iodide (2*Z*)-155 afforded the cyclization products 157 principally, whereas (2*E*)-155 underwent elimination to 156 predominantly.[44]

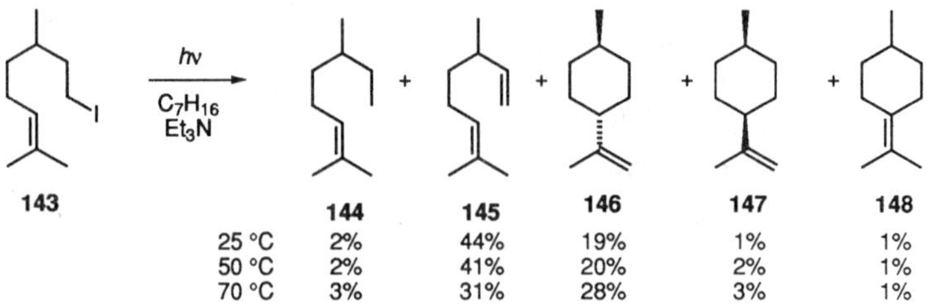
hv
C7H16
Et3N
I
143
144
145
146
147
148
25 °C 2% 44% 19% 1% 1%
50 °C 2% 41% 20% 2% 1%
70 °C 3% 31% 28% 3% 1%

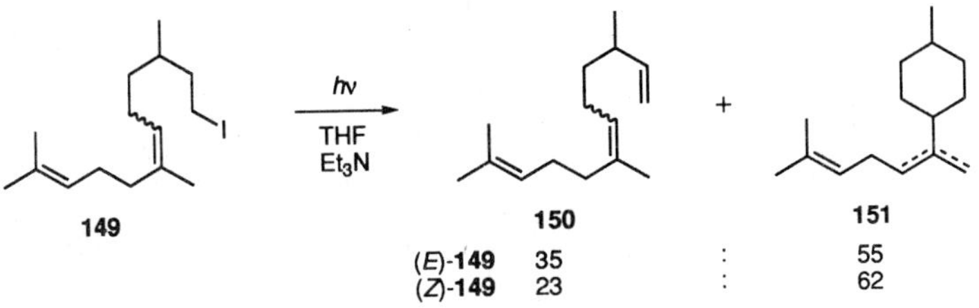

hν
THF
Et_3N
149
150
151
(E)-149 35 : 55
(Z)-149 23 : 62

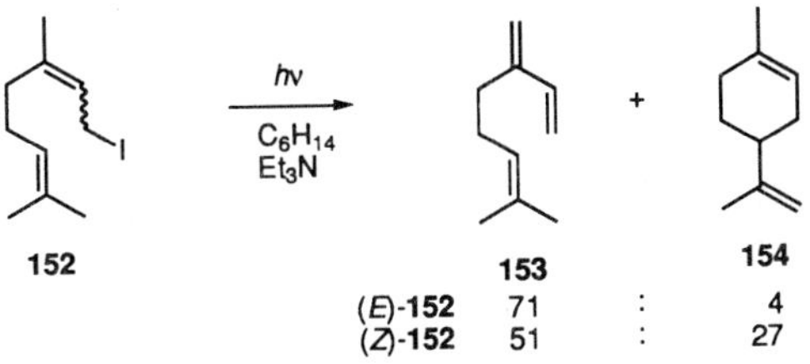

hν
C_6H_{14}
Et_3N
152
153
154
(E)-152 71 : 4
(Z)-152 51 : 27

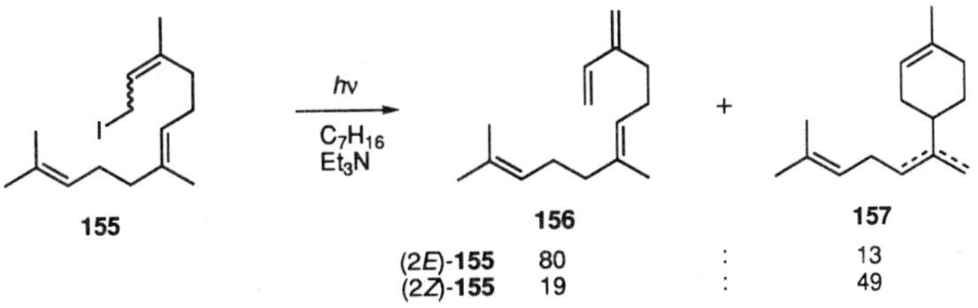

hν
C_7H_{16}
Et_3N
155
156
157
(2E)-155 80 : 13
(2Z)-155 19 : 49

Vinyl Halides

The photobehavior of vinyl iodides closely parallels that of their saturated alkyl analogs in solution, serving as a convenient and powerful method for the generation of vinyl cations that can be effected even at low temperature and in solvents of widely varying polarity. Once again, however, there is a strong propensity for elimination.

As with that of their saturated analogs, the photobehavior of vinyl iodides was initially interpreted exclusively in terms of radical intermediates. One of the first to be studied in solution, 1-iodopropene (158), was observed to undergo *E,Z*-isomerization accompanied by elimination to propyne, reduction to propene, and fragmentation to ethyne and CH_3I.[45,46] The radical pair 159 was proposed as the sole intermediate, even though elimination of HI could involve the corresponding ion pair equally well. The behavior of 4-iodo-3-heptene (160), which afforded the additional elimination product, allene 163, besides the reduction products 161 and the alkyne 162, was interpreted similarly as radical in character.[47]

(E)-158
hν
C_6H_{12}
(Z)-158
159
~H·
+ HI
H
+ CH_3I

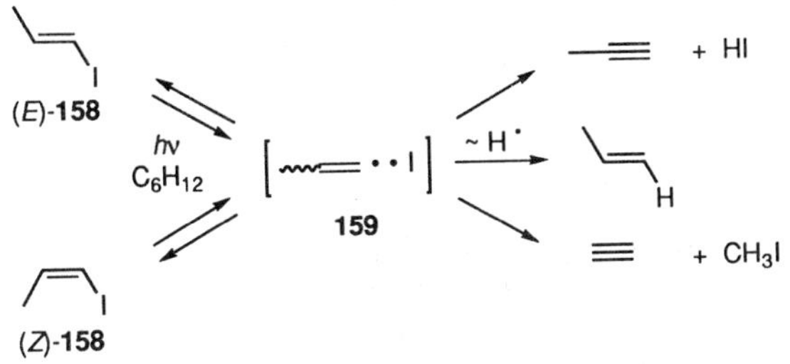

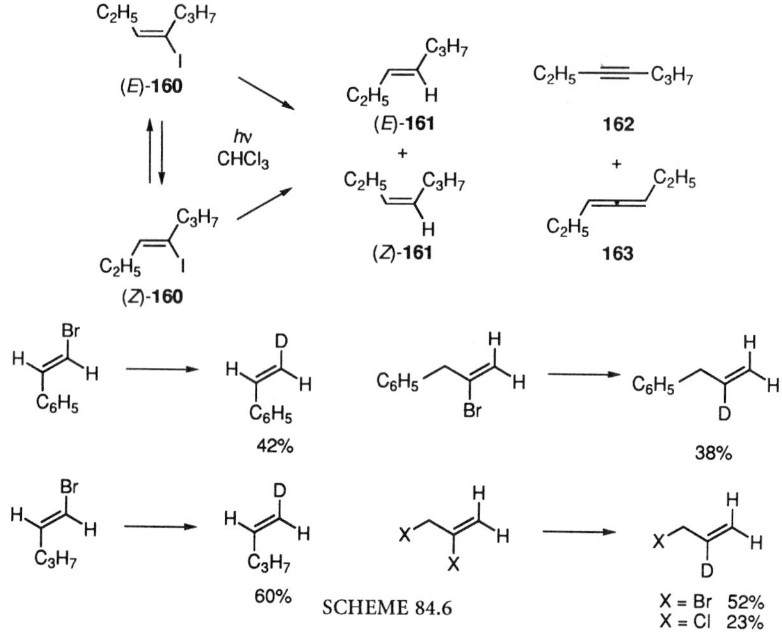

SCHEME 84.6

Advantage was taken of the formation of reduction products, which are undoubtedly of radical origin, to prepare specifically deuterated alkenes (Scheme 84.6).[48] It is interesting that substitution of a vinyl halogen atom by deuterium can be effected selectively in the presence of allylic bromides or chlorides, which normally are much more reactive, because they do not absorb strongly under the irradiation conditions. This was seen also in the irradiation of the insecticide, bromodan (164), which afforded a mixture of the reduction products 165 and 167 in CH_3OH and 166 and 167 in C_6H_{14}.[49] It is not clear why the chlorine substituent at C-2 or C-3 is more reactive in CH_3OH or C_6H_{14}, respectively.

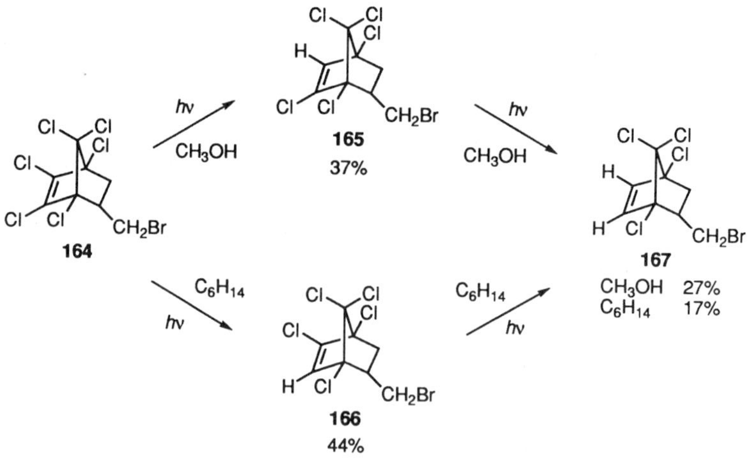

Evidence for the competing involvement of ionic intermediates in the photobehavior of vinyl halides was obtained by incorporating the chromophore in a cyclic environment, in which elimination is structurally inhibited. Thus, 1-iodocyclohexene (168a) afforded the nucleophilic trapping products 172 to 174 in high yields, clearly showing involvement of the vinyl cation 170.[50] Irradiation in CH_3OH containing Zn as an acid scavenger afforded the vinyl ether 171 (65% yield) instead of the acetal 172. As is typical, irradiation of bromide 168b in CH_3OH afforded principally cyclohexene (116), the reduction product, accompanied by smaller amounts of the acetal 172.

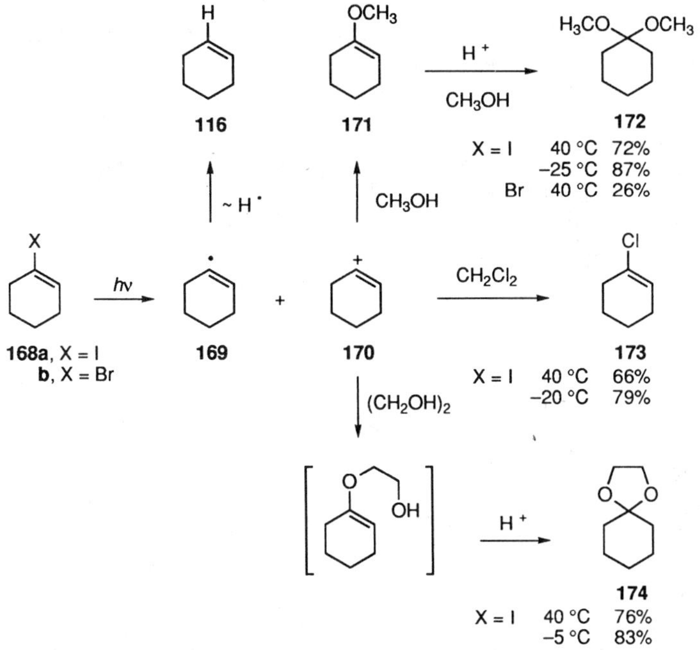

There was a moderate increase in the formation of the nucleophilic trapping products **172** to **174** relative to the reduction product cyclohexene (**116**) with decreasing temperature. The origin of this effect, which is general for vinyl iodides, is not clear. It does not appear to be due principally to the increased viscosity of the medium at lower temperatures since vinyl halides show no significant increase in the yield of ionic products on going from CH_3OH to the highly viscous solvent $(CH_2OH)_2$. Saturated alkyl halides, which exhibit increased ionic behavior in more viscous solvents, do not exhibit a similar temperature effect.[33]

The cyclopentenyl analog **175** afforded principally the reduction product **176**, accompanied by only a small amount of the trapping product **177** or **178** in CH_3OH or CH_2Cl_2, respectively, presumably because of the increased strain associated with incorporating an *sp*-hybridized center in a five-membered ring.[50] Nonetheless, these were the first unequivocal examples of the generation of a 1-cyclopentenyl cation from a cyclopentenyl precursor. The even more highly constrained cyclohexenyl halides **179** and **180** afforded exclusively isolongifolene (**179c**), the reduction product, or neoisolongifolene (**180c**), respectively, on irradiation in CH_3OH.[51]

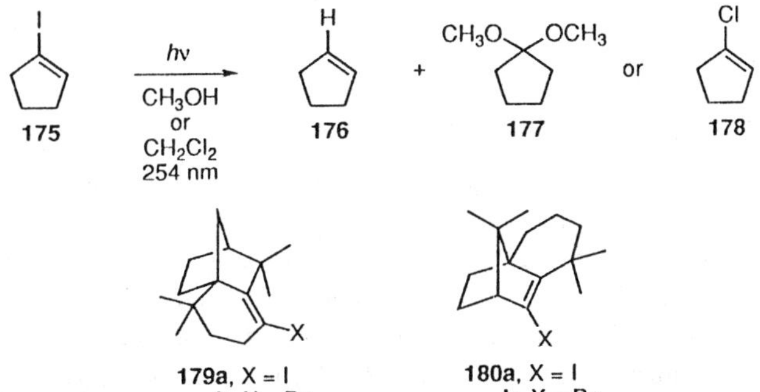

In the higher homolog, 1-iodocycloheptene (**181a**), the nucleophilic trapping product **184a** (Y = OCH_3 or Cl) predominated but was accompanied by the secondary products **185a** and **187a** arising from competing elimination to the strained allene **183a**.[50] In 1-iodocyclooctene (**181b**), elimination to allene **183b**, leading to the secondary products **185b** to **187b**, predominated over nucleophilic trapping in CH$_3$OH and occurred exclusively in CH$_2$Cl$_2$. In each case, some of the reduction product 182 was obtained also.

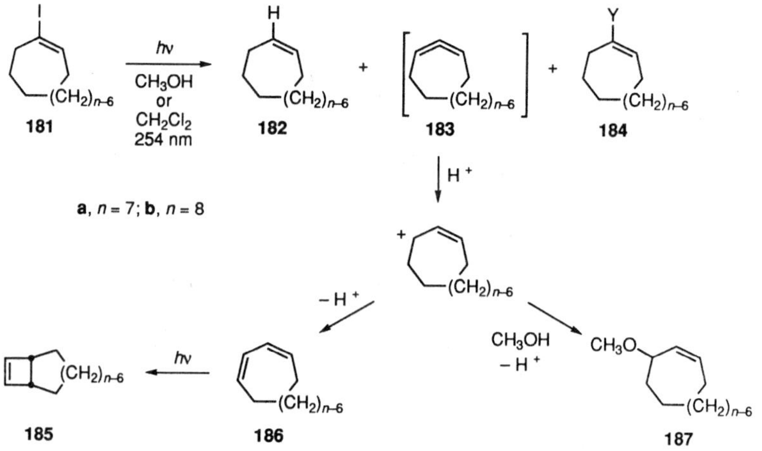

a, *n* = 7; **b**, *n* = 8

In the acyclic analog 188, elimination occurred exclusively even in CH$_3$OH to give the secondary products 191 and 192, along with the reduction product 189.[50] In CH$_3$OD, the allylic ether 192 was formed with deuterium incorporation as expected from protonation of the allene intermediate 190. In marked contrast with its facile photoelimination, iodide 188 is totally inert toward methanolic AgNO$_3$.[50]

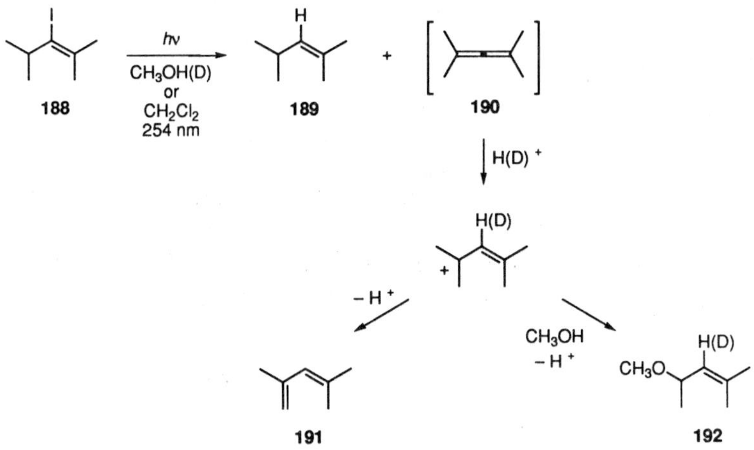

The exocyclic analog 193 similarly displays competing radical and ionic photobehavior. This is significant since vinyl cations generally require stabilization by an α-substituent. Except for (halomethylene)cyclopropanes and β-aryl systems, which gain special stabilization from the β-substituents, α-unsubstituted vinyl cations have not been generated solvolytically. Interestingly, the resulting vinyl cation 194, which can assume the preferred linear geometry of an *sp*-hybridized center, undergoes rearrangement to the cyclohexenyl cation 170, which is α-substituted but cannot be linear. Thus, irradiation of iodide 193 in CH$_2$Cl$_2$ or (CH$_2$OH)$_2$ afforded the reduction product 195 and either the vinyl chloride 173 or acetal 174, respectively.[50]

193 194 195 170 173 174

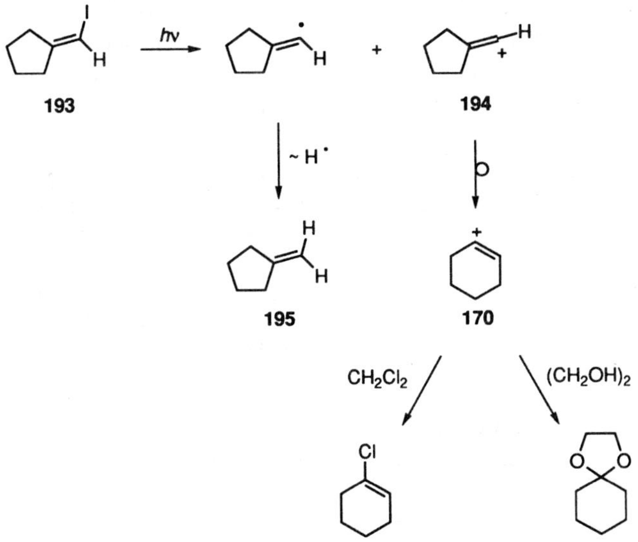

Analogous behavior was exhibited by the higher homolog (iodomethylene)cyclohexane.[50] The smaller homolog 196, however, proved to be a poor precursor to the highly strained cyclopentenyl cation 201. Irradiation in CH_2Cl_2 afforded principally the enyne 202, along with small amounts of the reduction product 197 and vinyl chlorides 198 and 178.[50] It is not clear whether enyne 202 arises from fragmentation of the radical 199 or cation 200.

197 2%, 198 2%, 196, 199, 200, 201, 178 2%, 202 61%

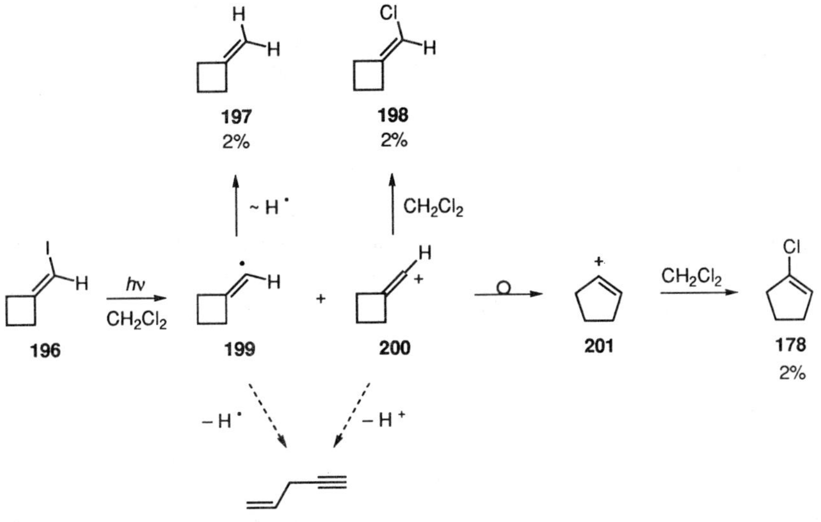

The exocyclic vinyl halides 203 show clearly the strong effect of α-substituents on vinyl cations. The α-unsubstituted derivatives 203 (R = H), which are inert toward methanolic $AgNO_3$, afforded only the reduction product camphene [204 (R = H)] on irradiation in CH_3OH.[52] Ring expansion, as exhibited by the exocyclic iodide 193, would afford a highly strained bicyclo[3.2.1]octenyl cation. By contrast, the methyl derivatives 203 (R = CH_3) afforded a mixture of the reduction products 204 (R = CH_3) and the allene 205, and the phenyl derivatives 203 (R = C_6H_5) afforded principally the ethers 206 (R = CH_3) in CH_3OH.[52] The ratio of ethers 206 to reduction products 204 (R = C_6H_5)

from bromide **203b** (R = C₆H₅) was 3.2:1.5:0.1:0.9 for the solvent series CH₃OH, CH₃CH₂OH, (CH₃)₂CHOH, and (CH₃)₃COH, respectively.[53] This follows the decreasing polarity of these solvents except for (CH₃)₂CHOH, which is an excellent hydrogen atom donor.

203a, X = I   **204**   **205**   **206**
   b, X = Br

The photobehavior of aryl-substituted vinyl halides has been extensively studied by the research groups of Taniguchi and Lodder.[54] Numerous examples of competing radical and ionic photobehavior have been observed, but are beyond the scope of this report to review in detail. Two of the earliest studies involved the 2,2-diarylvinyl halides **207**, which underwent predominant rearrangement-elimination to the tolans **208** along with competing formation of several radical products.[55,56]

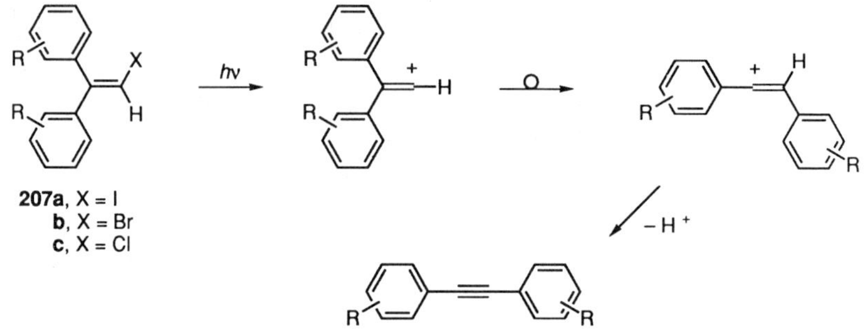

207a, X = I
   b, X = Br
   c, X = Cl

208

Alkynyl Halides

Early studies concentrated mostly on phenylethynyl halides and were interpreted in terms of radical intermediates.[57] Specific attempts to induce ionic photobehavior from the 1-halo-1-hexynes **209** in polar media, including the use of low temperature and viscous solvents, afforded almost exclusively radical-derived products.[15] This is consistent with the high ionization potentials of 1-alkynyl radicals.

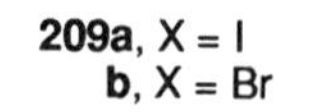

$$C_4H_9\!-\!\!\equiv\!\!-X$$

209a, X = I
   b, X = Br

Dihalides

Dihalides undergo photocleavage of a single carbon-halogen bond and, like monohalides, exhibit competing radical and ionic behavior. In the case of geminal diiodides, the α-iodo substituent stabilizes the adjacent cation, facilitating ionic behavior.

Vicinal Dihalides

Laser flash photolysis of 1,2-dibromoethane (**210**) afforded two bromine atoms per photon absorbed.[58] Irradiation of dihaloketones **211** in CH₃OH afforded the (*E*)-enones **212**.[59] Competing ionic behavior resulted in accompanying formation of the dimethoxy derivatives **213**. The corresponding dibromoesters **214** gave a mixture of the (*E*)- and (*Z*)-unsaturated esters **215**.[59] Apparently absorption by the phenyl group was critical under the reaction conditions since the aliphatic analog 216 afforded bromides 217 and 218 instead of the enones 219.[59] This problem was avoided by irradiation in the presence of $(C_2H_5)_3N$, which presumably involves radical anion formation (Scheme 84.1). Under these conditions, a wide variety of vicinal dihalides undergo dehalogenation in moderate to excellent yields.[59,60]

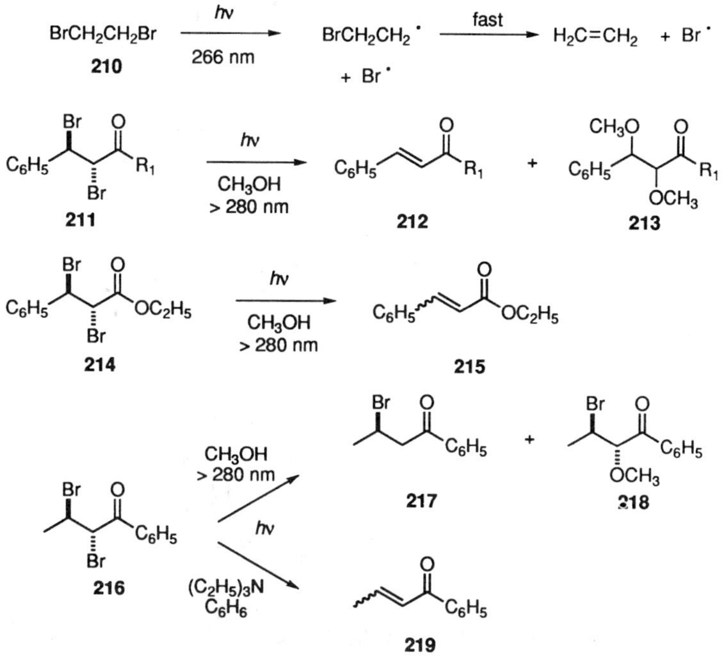

Geminal Dihalides

Appropriate dihalomethanes effect cyclopropanation of alkenes whereas their higher homologs, which are capable of undergoing elimination, exhibit intramolecular behavior.

Dihalomethanes: Cyclopropanation of Alkenes.   Simmons, co-discoverer of the cyclopropanation of alkenes with CH_2I_2 and Zn(Cu) couple, and co-workers observed early on that irradiation of CH_2I_2 in the presence of cyclohexene (116) afforded norcarane (220).[61] Because there was an accompanying formation of a mixture of the methylcyclohexenes 221 to 223, which appeared to result from indiscriminate insertion into C-H bonds as well as the π-bond, it was concluded that the main product-forming intermediate is an excited state of CH_2I_2 accompanied by free $(CH_2:)$.[61] However, subsequent studies showed that the methylcyclohexenes 221 to 223 are secondary products that can be avoided by stirring the irradiation mixture with a scavenger solution to remove I_2 and HI.[10,62] Under these conditions, norcarane (220) is the exclusive product and can be isolated in excellent yield.[62] The procedure has proved effective for the cyclopropanation of a number of alkenes — especially sterically hindered ones such as 224, which give little or no addition by the traditional Zn(Cu) couple procedures.[62,63] Addition occurs stereoselectively, with retention of stereochemistry. The relative rates of addition to the cyclohexenes 116, 221, and 226 increase with increasing substitution — in contrast to the Zn(Cu) couple method, in which steric effects result in a decreased rate for the tetrasubstituted alkene 226.[64]

Based on the electrophilic nature of the reactive intermediate and on the photobehavior of higher homologs of CH_2I_2 described in the following section, it was proposed that photocyclopropanation involves attack of the intermediate $^+CH_2I$ on the alkene (cf. 227).[62] In support of this, irradiation of CH_2I_2 in the presence of LiBr afforded CH_2IBr and CH_2Br_2. Further support has come from the more recent finding that on irradiation in an Ar matrix, the halomethanes 228a–c undergo equilibration with the contact ion pairs 229a–c.[65] Interestingly, the fluoro analog 228d did not afford even a trace of the photoisomer.

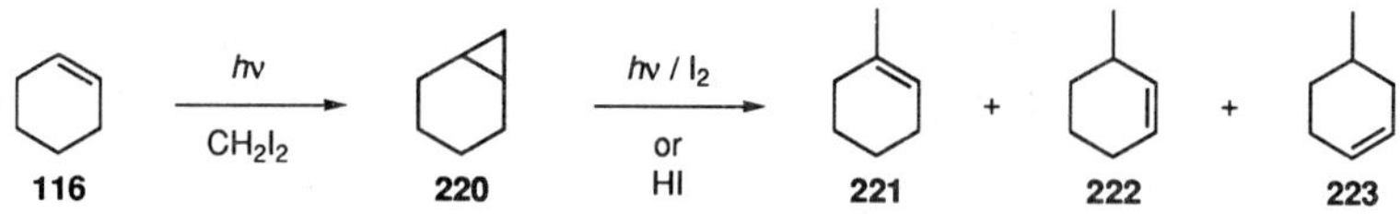
hv
CH2I2
hv / I2
or
HI
+
+
116
220
221
222
223

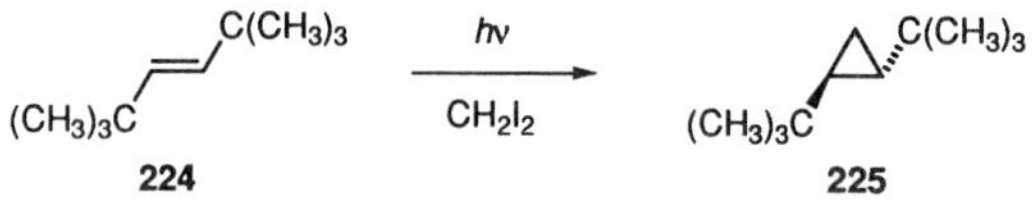
C(CH3)3
(CH3)3C
hv
CH2I2
224
225

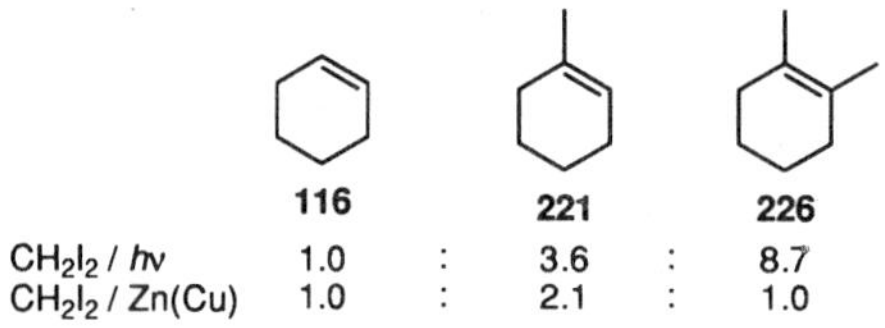
116
221
226
CH2I2 / hv 1.0 : 3.6 : 8.7
CH2I2 / Zn(Cu) 1.0 : 2.1 : 1.0

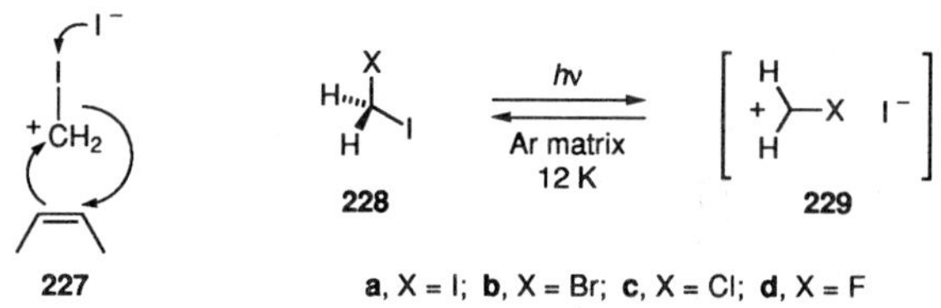
I−
+CH2
227
X
H
I
228
hv
Ar matrix
12 K
H
X
I−
229
a, X = I; b, X = Br; c, X = Cl; d, X = F

Cyclopropanation has also been observed on irradiation of CH_2Br_2 in the presence of alkenes, although in low yield.[66] However, irradiation of diiodo- or dibromohalomethanes in the presence of alkenes affords halocyclopropanes in good yield, as exemplified by formation of the 7-halonorcaranes 230.[66,67] 1,2-Dimethylcyclobutene (231) afforded the vinyl iodides 232, as might be expected from initial addition of the intermediate ^+CHXI.[67]

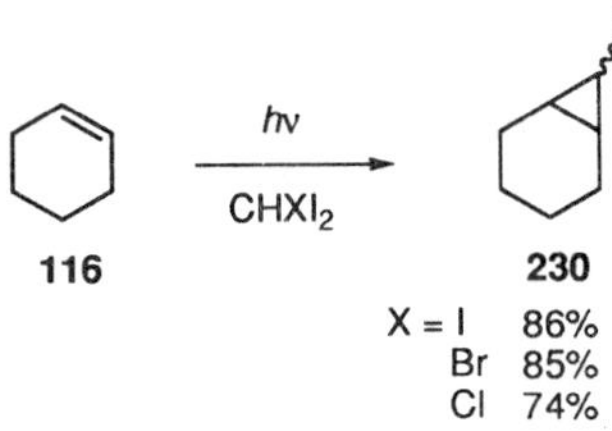
hv
CHXI2
116
230
X = I 86%
Br 85%
Cl 74%

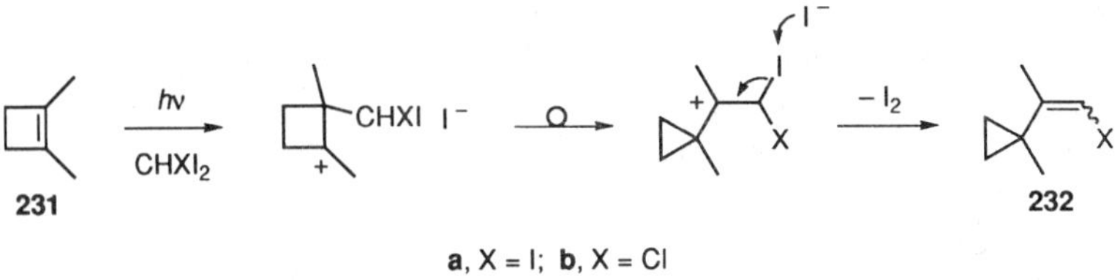
hv
CHXI2
CHXI I−
I−
X
− I2
231
232
a, X = I; b, X = Cl

Higher Homologs. Elimination dominates the photobehavior of higher homologs. Early studies involved diiodide 233, which on irradiation afforded the reduction and elimination products 234 and 158.[10] The dimethyl derivative 235, which is incapable of undergoing elimination, afforded the reduction product 4 almost exclusively.[10] However, it was found more recently that irradiation of diiodide 235 in the more polar solvent $(CH_2Cl)_2$ gives the alkene 5 principally, and irradiation in the polar but nucleophilic solvent CH_3OH affords the acetal 236.[68,69]

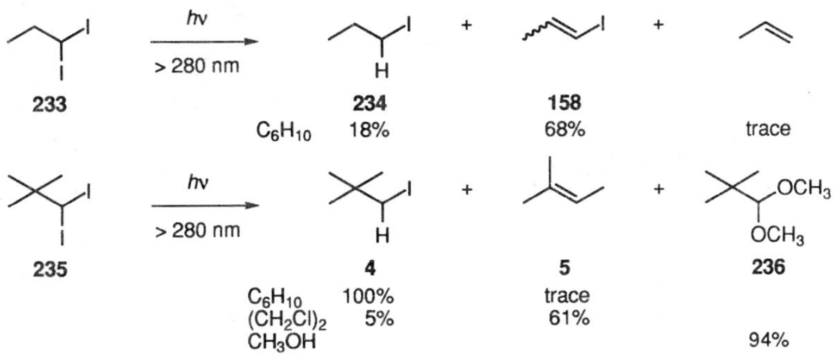

233 → hν, > 280 nm → **234** (H) C6H10 18% + **158** 68% + trace

235 → hν, > 280 nm → **4** (H) + **5** + **236** (OCH3, OCH3)

4: C6H10 100%, (CH2Cl)2 5%, CH3OH
5: trace 61%
236: 94%

More detailed studies were conducted with the diiodide **237**.[68] Irradiation in a variety of solvents afforded predominantly the vinyl iodide **240**, which can arise from either the radical pair **238** or ion pair **239**. It was obtained from the labeled diiodide **237**-d_1 with complete loss of deuterium. In solvents having a dielectric constant greater than 7.5, vinyl iodide **240** was accompanied by methylenecyclohexane (**242**). This is not a secondary product from photoreduction of vinyl iodide **240** since the latter does not absorb under the irradiation conditions. The labeled diiodide **237**-d_1 afforded alkene **242**-d_1 with almost complete retention of deuterium, which was located at the methylene position as expected from hydride rearrangement of the ion pair **239** followed by loss of I_2. Apparently, polar solvents facilitate rearrangement of the ion pair **239**. Although shown here as stepwise, rearrangement and loss of I_2 may occur concertedly, in analogy with the mechanism for the cyclopropanation of alkenes (cf. **227**). In the polar, but nucleophilic, solvent CH_3OH, trapping afforded the α-iodo ether **241a**, which underwent methanolysis subsequently to acetal **241b** as the principal product to the exclusion of methylenecyclohexane (**242**).

The bromo-iodo derivative **243a** afforded mainly the elimination product **245**, whereas the dibromide **243b** gave mainly the reduction product **244** in the solvents $(C_2H_5)_2O$, CH_2Cl_2, and CH_3OH.[68] The cyclopentyl analog **246** displayed behavior analogous to that of diiodide **237** except for giving a mixture of methylenecyclopentane (**195**) and the ring-expanded cyclohexene (**116**), along with the vinyl iodide **193**, in CH_2Cl_2.[68] The cyclobutyl analog **248** afforded only the ring-expanded cyclopentene **250** and the vinyl iodide **249**.[68] Citronellal iodide (**251**) gave a mixture of products **143** and **252** to **254**.[68] Apparently, iodide **253** arises from intramolecular trapping of the α-iodo cation **255**, and carane (**254**) from competing intramolecular cyclopropanation analogous to that described above for CH_2I_2. It is interesting, however, that intermolecular addition of CH_2I_2 to alkenes gave no trapping analogous to the formation of iodide **253**.

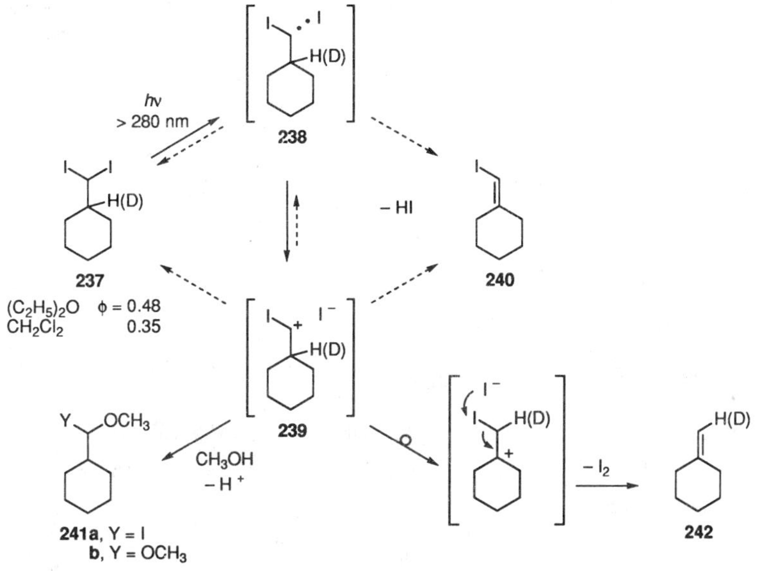

237
$(C_2H_5)_2O$ φ = 0.48
CH_2Cl_2 0.35

hν > 280 nm → **238** → − HI → **240**

239

241a, Y = I
b, Y = OCH3

CH3OH − H⁺

− I2 → **242**

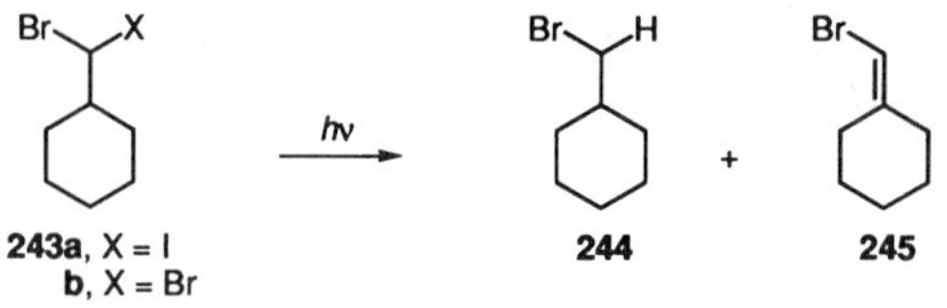

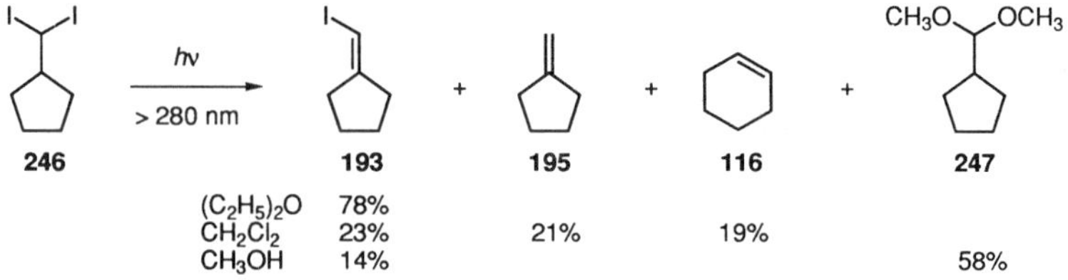

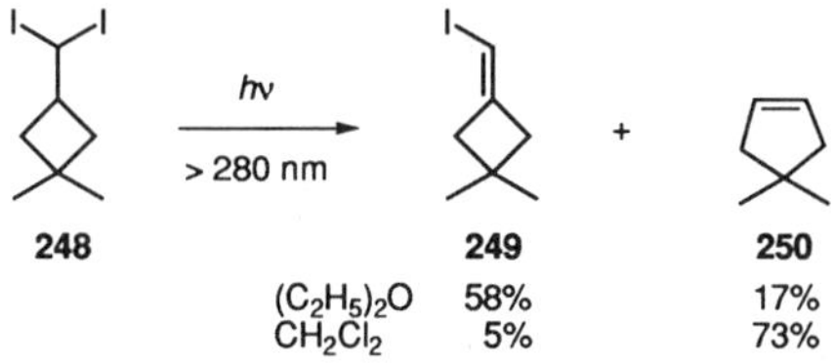

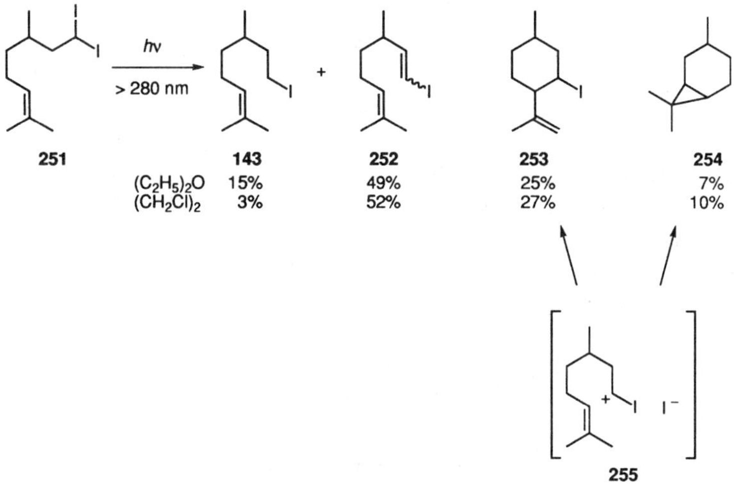

Advantage has been taken of the predominant radical behavior of bromides to effect reduction of the dibromocyclopropanes 256 to the corresponding monobromides 257.[70] Weak absorption by the monobromides 257 under the irradiation conditions permits the selective removal of a single halogen atom.

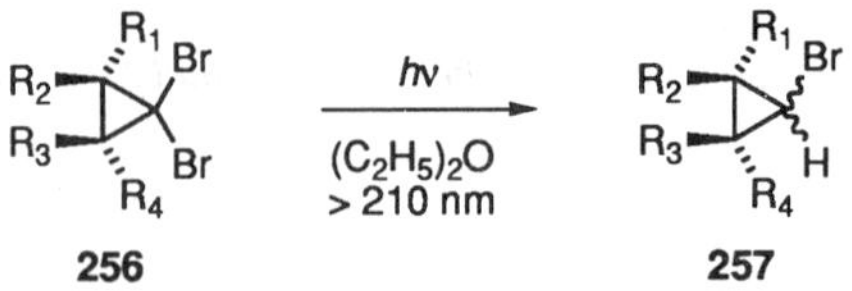

Vinylidene Dihalides

In accord with the trends observed in other systems, the dichloride 258c gave only the reduction products 260 and 242, while the dibromide 258b gave a small amount of the nucleophilic trapping product 259b and the diiodide 258a gave somewhat more of it.[50b] Similar behavior was exhibited by the norbornyl analogs 261, with the dibromide 261b affording only the reduction product 203b and the diiodide **261a** giving a mixture of iodide **203a** and the ether **262a**.[71] At –10°C, diiodide **261a** gave the ether **262a** almost exclusively. On further irradiation, the vinyl halides **203** underwent secondary conversion to camphene [**204** (R = H)], as described above.

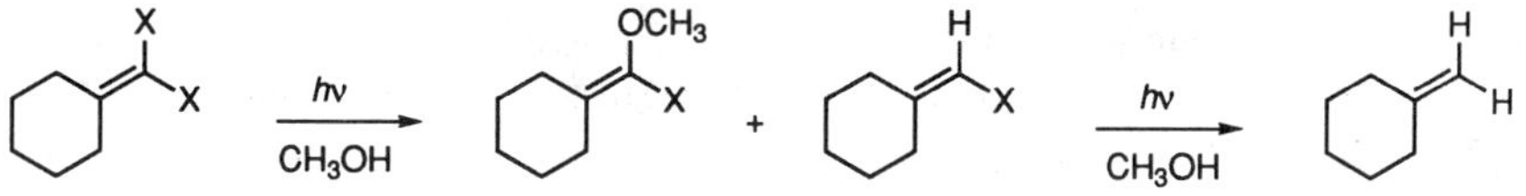

	258a, X = I	**259a**, X = I	**240**, X = I	**242**
	b, X = Br	**b**, X = Br	**245**, X = Br	
	c, X = Cl	**c**, X = Cl	**260**, X = Cl	

	X = I	38%	38%	
	Br	12%	53%	6%
	Cl		59%	8%

| | **261a**, X = I | **262a**, X = I | **203a**, X = I |
| | **b**, X = Br | **b**, X = Br | **b**, X = Br |

X = I (25 °C)	44%	44%
(–10 °C)	90%	
Br (25 °C)		82%

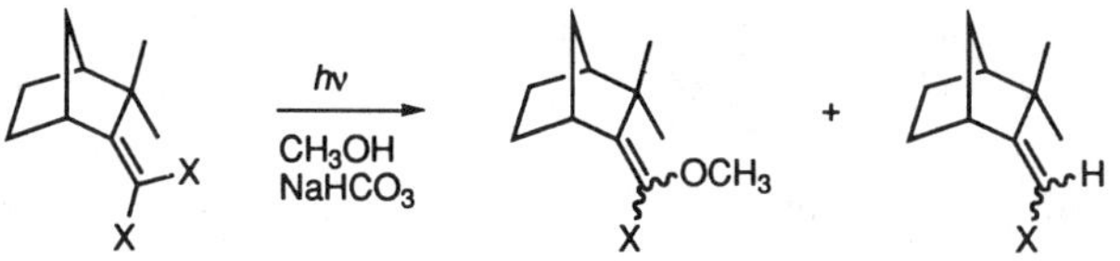

eferences

1. Kropp, P. J., Fryxell, G. E., Tubergen, M. W., Hager, M. W., Harris, G. D., Jr., McDermott, T. P., Jr., and Tornero-Velez, R., Photochemistry of phenyl thioethers and phenyl selenoethers. Radical vs. ionic behavior, *J. Am. Chem. Soc.*, 113, 7300, 1991.

2. Eaborn, C., Safa, K. D., Ritter, A., and Binder, W., Photoinduced ionic and free-radical reactions of some organosilicon iodides, *J. Chem. Soc., Perkin Trans. 2*, 1397, 1982.

3. Kropp, P. J., Photobehavior of alkyl halides in solution: Radical, carbocation, and carbene intermediates, *Acc. Chem. Res.*, 17, 131, 1984.

4. (a) Sammes, P. G., Photochemistry of the C-X group, in *Chemistry of the Carbon-Halogen Bond*, Patai, S., Ed., Wiley, London, 1973, chap. 11; (b) Grimshaw, J. and de Silva, A. P., Photochemistry and photocyclization of aryl halides, *Chem. Soc. Rev.* 1981, 10, 181. (c) Lodder, G., Recent advances in the photochemistry of the carbon-halogen bond, In *The Chemistry of Functional Groups*, Supplement D, Patai, S., Rappoport, Z., Eds., Wiley: London, 1983, Chapter 29. (d) Cristol, S. J., Bindel, T. H., Photosolvolyses and attendant photoreactions involving carbocations, *Org. Photochem.* 1983, 6, 327.

5. (a) Majer, J. R. and Simons, J. P., Photochemical processes in halogenated compounds, *Adv. Photochem.* 1967, 2, 137. (b) Okabe, H., *Photochemistry of Small Molecules*, Wiley-Interscience: New York, 1978.

6. (a) Kimura, K. and Nagakura, S., n → σ* Absorption spectra of saturated organic compounds containing bromine and iodine, *Spectrochim. Acta*, 17, 166, 1961; (b) Balasubramanian, A., Substituent and solvent effects on the n → σ* transition of alkyl iodides, *Ind. J. Chem.*, 1963, 329.

7. Kropp, P. J. and Adkins, R. L., Photochemistry of alkyl halides. 12. Bromides vs. iodides, *J. Am. Chem. Soc.*, 113, 2709, 1991.

8. Shepson, P. B. and Heicklen, J., Photooxidation of ethyl iodide at 22 °C, *J. Phys. Chem.*, 85, 2691, 1981.

9. McCauley, C. E., Hamill, W. H., and Williams, R. R., Jr., Isomerization in the photolysis of alkyl iodides, *J. Am. Chem. Soc.*, 76, 6263, 1954.

10. Neuman, R. C., Jr. and Wolcott, R. G., Photochemistry of 1,1-diiodoalkanes, *Tetrahedron Lett.,* 6267, 1966.

11. Poindexter, G. S. and Kropp, P. J., Photochemistry of alkyl halides. II. Support for an electron transfer process, *J. Am. Chem. Soc.,* 96, 7142, 1974.

12. For a discussion of the concept of "free" cations, see: Keating, J. T. and Skell, P. S., Free carbonium ions, in *Carbonium Ions,* Vol. 2, Olah, G. A. and Schleyer, P. v. R., Eds., Wiley Interscience, New York, 1970, chap. 15.

13. Kropp, P. J., Poindexter, G. S., Pienta, N. J., and Hamilton, D. C., Photochemistry of alkyl halides. 4. 1-Norbornyl, 1-norbornylmethyl, 1- and 2-adamantyl, and 1-octyl bromides and iodides, *J. Am. Chem. Soc.,* 98, 8135, 1976.

14. Leuf, W. and Keese, R., Synthese von 1-hydroxynorbornen aus norcampher, *Chimia,* 36, 81, 1982.

15. Inoue, Y., Fukunaga, T., and Hakushi, T., Direct photolysis of 1-halo-1-hexynes. Lack of ionic behavior, *J. Org. Chem.,* 48, 1732, 1983.

16. Pienta, N. J., Ph.D. Dissertation, University of North Carolina, Chapel Hill, 1978.

17. Kropp, P. J., Worsham, P. R., Davidson, R. I., and Jones, T. H., Photochemistry of alkyl halides. 8. Formation of a bridgehead alkene, *J. Am. Chem. Soc.,* 104, 3972, 1982.

18. Reddy, D. S., Sollott, G. P., and Eaton, P. E., Photolysis of cubyl iodides: Access to the cubyl cation, *J. Org. Chem.,* 54, 722, 1989.

19. See: Hrovat, D. A. and Borden, W. T., *Ab initio* calculations find that formation of cubyl cation requires less energy than formation of 1-norbornyl cation, *J. Am. Chem. Soc.,* 112, 3227, 1990.

20. Perkins, R. R. and Pincock, R. E., Photochemical substitution reactions of adamantyl halides, *Tetrahedron Lett.,* 943, 1975.

21. Sollott, G. P. and Gilbert, E. E., A facile route to 1,3,5,7-tetraaminoadamantane. Synthesis of 1,3,5,7-tetranitroadamantane, *J. Org. Chem.,* 45, 5405, 1980.

22. See also: Kropp, P. J., Gibson, J. R., Snyder, J. J., and Poindexter, G. S., Photochemistry of alkyl halides. 5. 2,4-Dehydroadamantane and protoadamantene from 2-bromo- and 2-iodoadamantane, *Tetrahedron Lett.,* 207, 1978.

23. Davidson, R. I., Tise, F. P., McCraw, G. L., Underwood, G. A., and Kropp, P. J., unpublished results.

24. Kropp, P. J., Jones, T. H., and Poindexter, G. S., Photochemistry of alkyl halides. 1. Ionic vs. radical behavior, *J. Am. Chem. Soc.,* 95, 5420, 1973.

25. Newcomb, M., Sanchez, R. M., and Kaplan, J., Fast halogen abstractions from alkyl halides by alkyl radicals. Quantitation of the processes occurring in and a caveat for studies employing alkyl halide mechanistic probes, *J. Am. Chem. Soc.,* 109, 1195, 1987.

26. Worsham, P. R., Ph.D. Dissertation, University of North Carolina, Chapel Hill, 1977.

27. Roth, R. C. and Binkley, R. W., A mechanistic study of the photoreactions of deoxy iodo sugars, *J. Org. Chem.,* 50, 690, 1985.

28. For a review, see: Binkley, R. W., Photochemical reactions of carbohydrates, *Adv. Carbohydr. Chem. Biochem.,* 38, 105, 1981.

29. Gokhale, P. D., Joshi, A. P., Sahni, R., Naik, V. G., Damodaran, N. P., Nayak, U. R., and Dev, S., Photochemical transformations. I. Reactions of some terpene iodides, *Tetrahedron,* 32, 1391, 1976.

30. Takaishi, N., Miyamoto, N., and Inamoto, Y., Ring enlargement of the photochemically generated *endo*-2,3-trimethylenenorborn-*exo*-2-ylcarbinyl cation. Occurrence of a different reaction pathway from that in sulfuric acid, *Chem. Lett.,* 1251, 1978.

31. Brown, D. M., Underwood, G. A., and Kropp, P. J., unpublished results.

32. Bloom, J. A., Tise, F. P., Suddaby, B. R., and Kropp, P. J., unpublished results.

33. McNeely, S. A., Ph.D. Dissertation, University of North Carolina, Chapel Hill, 1976.

34. Kropp, P. J., Sawyer, J. A., and Snyder, J. J., Photochemistry of alkyl halides. 11. Competing reaction via carbene and carbocationic intermediates, *J. Org. Chem.,* 49, 1583, 1984.

35. McCabe, P. H., de Jenga, C. I., and Stewart, A., Photochemical and assisted cleavages of halo-9-thiabicyclononanes, *Tetrahedron Lett.,* 22, 3681, 1981.

36. (a) Olofson, R. A., Walinsky, S. W., Marino, J. P., and Jernow, J. L., Carbenes from carbonium ions. I. Dithiomethoxymethyl cation and its conversion to tetrathiomethoxyethylene, *J. Am. Chem. Soc.*, 90, 6554, 1968; (b) Curtin, D. Y., Kampmeier, J. A., and O'Connor, B. R., *J. Am. Chem. Soc.*, 87, 863, 1965.

37. (a) Kurz, M. and Rodgers, M., Photochemical aromatic cyclohexylation, *J. Chem. Soc., Chem. Commun.*, 1227, 1985; (b) Kurz, M. E., Noreuil, T., Seebauer, J., Cook, S., Geier, D., Leeds, A., Stronach, C., Barnickel, B., Kerkemeyer, M., Yandrasits, M., Witherspoon, J., and Frank, F. J., Photochemical aromatic alkylation, *J. Org. Chem.*, 53, 172, 1988.

38. Ogata, Y., Tomizawa, K., and Furuta, K., Photochemical reactions of methyl iodide with aromatic compounds, *J. Org. Chem.*, 46, 5276, 1981.

39. Birchall, J. M., Irvin, G. P., and Boyson, R. A., Reactions of trifluoromethyl radicals. I. The photochemical reactions of trifluoroiodomethane with benzene and some halogenobenzenes, *J. Chem. Soc., Perkin Trans. 2*, 435, 1975.

40. Camaggi, C. M., Leardini, R., and Zanirato, P., Photolysis of allyl iodide in aromatic solvents, *J. Org. Chem.*, 42, 1570, 1977.

41. (a) Kimoto, H., Fujii, S., and Cohen, L. A., Photochemical perfluoroalkylation of imidazoles, *J. Org. Chem.*, 47, 2867, 1982; (b) Kimoto, H., Fujii, S., and Cohen, L. A., Photochemical trifluoromethylation of some biologically significant imidazoles, *J. Org. Chem.*, 49, 1060, 1984.

42. (a) Charlton, J. L., Williams, G. J., and Lypka, G. N., The photochemistry of 4-phenyl-1-iodobutane and 4-phenyl-2-iodomethyl-1-butene, *Can. J. Chem.*, 58, 1271, 1980; (b) Bhalerao, V. K., Nanjundiah, B. S., Sonawane, H. R., and Nair, P. M., Photolysis of 2-phenylethyl and 4-phenyl-1-butyl halides in alcoholic solvents, *Tetrahedron*, 42, 1487, 1986; (c) Subbarao, K. V., Damodaran, N. P., and Dev, S., Photochemical transformations. V. Organic iodides (Part 4): Solution photochemistry of 4-phenyl-1-iodobutane and 4-phenyl-1-bromobutane, *Tetrahedron*, 43, 2543, 1987.

43. Saplay, K. M., Sahni, R., Damodaran, N. P., and Dev, S., Photochemical transformations. II. Organic iodides (Part 2): Cetronellyl iodide, 2,3-dihydro-6(Z)-farnesyl, and 2,3-dihydro-6(E)-farnesyl iodides, *Tetrahedron*, 36, 1455, 1980.

44. Saplay, K. M., Damodaran, N. P., and Dev, S., Photochemical transformations. III. Organic iodides (Part 3): Geranyl and neryl iodides and 2(E),6(E)- and 2(Z),6(E)-farnesyl iodides, *Tetrahedron*, 39, 2999, 1983.

45. Neuman, R. C., Jr., Photochemistry of *cis*- and *trans*-1-iodopropene, *J. Org. Chem.*, 31, 1852, 1966.

46. For analogous behavior by iodoethene, see: Roberge, P. C. and Herman, J. A., Photolyse de l'iodure de vinyle en solution dans le tetrachlorure de carbone, *Can. J. Chem.*, 42, 2262, 1964.

47. Neuman, R. C., Jr., and Holmes, G. D., Photolytic formation of isomeric vinyl radicals from *cis*- and *trans*-vinyl iodides, *J. Org. Chem.*, 33, 4317, 1968.

48. Müller, J. P. H., Parlar, H., and Korte, F., Einfache photochemische methode zur deuterierung von olefinen und aromaten, *Synthesis*, 524, 1976.

49. Walia, S., Dureja, P., and Mukerjee, S. K., Photoreductive dehalogenation of 5-bromomethyl-1,2,3,4,7,7-hexachloro-2-norbornene (bromodan) — A cyclodiene insecticide, *Tetrahedron*, 43, 2493, 1987.

50. (a) McNeely, S. A. and Kropp, P. J., Photochemistry of alkyl halides. 3. Generation of vinyl cations, *J. Am. Chem. Soc.*, 98, 4319, 1976; (b) Kropp, P. J., McNeely, S. A., and Davis, R. D., Photochemistry of alkyl halides. 10. Vinyl halides and vinylidene dihalides, *J. Am. Chem. Soc.*, 105, 6907, 1983.

51. Sonawane, H. R., Nanjundiah, B. S., and Rajput, S. I., Photochemistry of vinyl halides. II. Reactions of some conformationally rigid vinyl halides derived from isolongifolene and neoisolongifolene, *Ind. J. Chem.*, 23B, 339, 1984.

52. Sonawane, H. R., Nanjundiah, B. S., and Rajput, S. I., Photochemistry of vinyl halides. I. Radical versus ionic photobehavior of some vinyl halides based on camphene, *Ind. J. Chem.*, 23B, 331, 1984.

53. Sonawane, H. R., Nanjundiah, B. S., Udaykumar, M., and Panse, M. D., Photochemistry of vinyl halides. III. Photolysis of ω-bromo-ω-phenylcamphene in different solvents and its mechanistic implications, *Ind. J. Chem.*, 1985, 24B, 202.

54. (a) For an extensive compilation of references, see: van Ginkel, F. I. M., Cornelisse, J., and Lodder, G., Photoreactivity of some α-arylvinyl bromides in acetic acid. Selectivity toward bromide versus acetate ions as a mechanistic probe, *J. Am. Chem. Soc.,* 113, 4261, 1991. See also: (b) Kitamura, T., Kabashima, T., Kobayashi, S., and Taniguchi, H., Isolation and alcoholysis of an *ipso* adduct, vinylidenecyclohexadiene, from photolysis of 1-(*p*-ethoxyphenyl)vinyl bromide, *Tetrahedron Lett.,* 47, 6141, 1988; (c) Kitamura, T., Nakamura, I., Kabashima, T., Kobayashi, S., and Taniguchi, H., A novel spiro adduct from intramolecular *ipso* substitution in the photolysis of an α-[*p*-(2-hydroxyalkoxy)phenyl]vinyl bromide, *J. Chem. Soc., Chem. Commun.,* 1154, 1989; (d) Kitamura, T., Kobayashi, S., and Taniguchi, H., Photolysis of vinyl halides. Reaction of photogenerated vinyl cations with cyanate and thiocyanate ions, *J. Org. Chem.,* 55, 1801, 1990; (e) Hori, K., Kamada, H., Kitamura, T., Kobayashi, S., and Taniguchi, H., Theoretical and experimental study on the reaction mechanism of photolysis and solvolysis of aryvinyl halides, *J. Chem. Soc., Perkin Trans. 2,* 871, 1992.

55. Suzuki, T., Sonoda, T., Kobayashi, S., and Taniguchi, H., Photochemistry of vinyl bromides: A novel 1,2-aryl group migration, *J. Chem. Soc., Chem. Commun.,* 180, 1976.

56. (a) Sket, B., Zupan, M., and Pollak, A., The Photo-Fritsch-Buttenberg-Wiechell rearrangement, *Tetrahedron Lett.,* 783, 1976; (b) Sket, B. and Zupan, M., Photochemistry of 2-halogeno-1,1-diphenylethylenes. The Photo-Fritsch-Buttenberg-Wiechell rearrangement, *J. Chem. Soc., Perkin Trans. 1,* 752, 1979.

57. See Reference 15 for a compilation of references.

58. Scaiano, J. C., Barra, M., Calabrese, G., and Sinta, R., Photochemistry of 1,2-dibromomethane in solution. A model for the generation of hydrogen bromide, *J. Chem. Soc., Chem. Commun.,* 1418, 1992.

59. Izawa, Y., Takeuchi, M., and Tomioka, H., Photo-dehalogenation of vicinal dihalide to olefin, *Chem. Lett.,* 1297, 1983.

60. Takagi, K. and Ogata, Y., Ultraviolet light induced dechlorination of vicinal polychlorocyclohexanes with triethylamine, *J. Org. Chem.,* 48, 1966, 1983.

61. Blomstrom, D. C., Herbig, K., and Simmons, H. E., Photolysis of methylene iodide in the presence of olefins, *J. Org. Chem.,* 30, 959, 1965.

62. Kropp, P. J., Pienta, N. J., Sawyer, J. A., and Polniaszek, R. P., Photochemistry of alkyl halides. VII. Cyclopropanation of alkenes, *Tetrahedron,* 37, 3229, 1981.

63. Kropp, P. J. and Tise, F. P., Photochemistry of alkenes. 8. Sterically congested alkenes, *J. Am. Chem. Soc.,* 103, 7293, 1981.

64. Rickborn, B. and Chan, J. H.-H., Conformational effects in cyclic olefins. The relative rates of iodomethylzinc addition, *J. Org. Chem.,* 32, 3576, 1967.

65. Maier, G., Reisenauer, H. P., Hu, J., Schaad, L. J., and Hess, B. A., Jr., Photochemical isomerization of dihalomethanes in argon matrices, *J. Am. Chem. Soc.,* 112, 5117, 1990.

66. Marolewski, T. and Yang, N. C., Photochemical addition of polyhalogenomethanes to olefins, *J. Chem. Soc., Chem. Commun.,* 1225, 1967.

67. Yang, N. C. and Marolewski, T. A., The addition of halomethylene to 1,2-dimethylcyclobutene, a methylene-olefin reaction involving a novel rearrangement, *J. Am. Chem. Soc.,* 90, 5644, 1968.

68. Kropp, P. J. and Pienta, N. J., Photochemistry of alkyl halides. 9. Geminal dihalides, *J. Org. Chem.,* 48, 2084, 1983.

69. Moret, E., Jones, C. R., and Grant, B., Photochemistry of organic geminal diiodides, *J. Org. Chem.,* 48, 2090, 1983.

70. Shimizu, N. and Nishida, S., Photochemical reduction of *gem*-dibromocyclopropanes, *Chem. Lett.,* 839, 1977.

71. Sonawane, H. R., Nanjundiah, B. S., and Panse, M. D., Photochemistry of organic halides: Some interesting features of the photobehaviour of vinyl halides and vinylidene dihalides derived from camphene, *Tetrahedron Lett.,* 26, 3507, 1985.

C–X Bond Fission in
Alkene Systems

Sugio Kitamura
Shinshu University

5.1 Definition of the Reaction

This chapter describes the photochemical C-X bond fission in alkene systems that have halogen atoms bonded to the carbon-carbon double bond. The substitution of halogen atoms in alkenes results in a red shift of the π,π^* absorption band because of the interaction between the lone pair on the halogen atom and the π-orbitals.[1,2] However, the UV absorption of aliphatic alkenyl halides, except for the iodides, lies in a region of short wavelength less than 254 nm (Table 85.1). Therefore, the majority of the alkenyl halides studied in these photolyses are the iodides.

Introduction of aromatic substituents to the carbon-carbon double bond leads to considerable π-conjugation, causing the absorption to move to longer wavelength (>254 nm) with a concomitant increase in the extinction coefficient. 1,2-Diaryl- and triarylvinyl halides including the bromides and chlorides have absorptions at wavelengths around 300 nm. Therefore, these systems have been used mostly for synthetic applications because the secondary photochemical reactions can be eliminated by the use of a suitable filter.

The excitation energy obtained by absorption of light is consumed chemically by isomerization, extrusion of some molecules, and bond fission. In this chapter, the bond fission aimed at synthetic application is discussed. The ease of bond fission is proportional to the bond strength of the C-X bond. As shown in Table 85.2, the bond strength of the C-X bond in alkenyl halides is much higher than that of alkyl halides and are intermediate between those of the corresponding ethyl and phenyl halides. The C-X bond increases in strength in the order: I < Br < Cl < F. The bond dissociation energy of vinyl fluoride (479 kJ mol^{-1}) is higher than the energy of the light from low-pressure Hg lamps (253.7 nm = 472 kJ per Einstein) and, accordingly, the direct photochemical cleavage of a carbon-fluorine bond is extremely rare.

5.2 Mechanism of the Reaction

Previously, Sammes[2] and Lodder[18] have reviewed the photochemistry of the carbon-halogen bond. Photolysis of alkenyl halides is well known, but little attention has been paid to the synthetic

Table 85.1   UV Properties of Alkenyl Halides

Compound	λ_{max}, nm (log ε)	Solvent	Ref.
1-Chlorocyclohexene	200 (3.64)	EtOH	3
1-Bromocyclohexene	207.3 (3.65)	95% EtOH	4
1-Iodocycloheptene	263 (2.55)	EtOH	3
E-4-Iodo-3-heptene	249 (~2.60)	—	5
N-[E-(2-Iodoacryloyl)]-2,6-dimethylpyrolidine	249 (4.06)	95% EtOH	6
E-β-Bromostyrene	257 (~4.04)	95% EtOH	7
1-Fluoro-2,2-diphenylethene	244 (4.07)	Cyclohexane	8
1-Chloro-2,2-diphenylethene	255 (4.10)	95% EtOH	9
1-Bromo-2,2-diphenylethene	259 (4.12)	95% EtOH	9
1-Iodo-2,2-diphenylethene	263 (4.15)	95% EtOH	9
1-Bromo-2,2-*bis*(p-methoxyphenyl)ethene	212 (4.54), 252 (4.32), 270 (4.27)	MeOH	10
E-1-Bromo-1,2-diphenylethene	260 (4.0), 300sh (3.7)	MeOH	11
Z-1-Bromo-1,2-diphenylethene	285 (4.2)	MeOH	11
1-Bromo-1,2,2-triphenylethene	231 (4.3), 295 (5.0)	EtOH	12
1-Bromo-1-(p-methoxyphenyl)-2,2-diphenylethene	239 (4.34), 302 (4.06)	Cyclohexane	13
1-Bromo-1,2,2-*tris*(p-methoxyphenyl)ethene	249 (4.36), 310.5 (4.11)	Cyclohexane	13
1-Bromo-1-[p-(2-hydroxyethoxy)phenyl]-2,2-diphenylethene	213 (4.36), 239 (4.27), 301 (4.01)	EtOH	14
1-Bromo-2,2-*bis*(o-methoxyphenyl)-1-phenylethene	284 (9.02)	Cyclohexane	15

Table 85.2   Comparison of Bond Dissociation Energies of the C-X Bond Among Ethyl, Vinyl, and Phenyl Halides

Ethyl halide[16]	D(C-X) (kJ mol^{-1})	Vinyl halide[17]	D(C-X) (kJ mol^{-1})	Phenyl halide[16]	D(C-X) (kJ mol^{-1})
CH_3CH_2-F	452	$CH_2{=}CH$-F	497	Ph-F	527
CH_3CH_2-Cl	335	$CH_2{=}CH$-Cl	372	Ph-Cl	402
CH_3CH_2-Br	285	$CH_2{=}CH$-Br	320	Ph-Br	339
CH_3CH_2-I	222	$CH_2{=}CH$-I	260	Ph-I	272

applications. One advantage for synthetic application of alkenyl halide photochemistry is that the reactions which they undergo allow for its introduction of a C-C double bond into the product molecule.

The well-known principal example in solution is the homolysis of the C-X bond, affording a radical pair composed of a vinyl radical and a halogen atom as represented by the photolysis of alkenyl iodides (Equation 85.1).[2]

$$\begin{array}{c}\diagdown\\\diagup\end{array}C{=}C\begin{array}{c}\diagdown\\\diagup\\\text{I}\end{array}\quad\xrightarrow{h\nu}\quad\begin{array}{c}\diagdown\\\diagup\end{array}C{=}C\begin{array}{c}\diagup\\\diagdown\\\cdot\end{array}\ +\ \text{I}^{\bullet}$$

(85.1)

The main reaction of the vinyl radical is abstraction of a hydrogen or halogen atom, and the dimerization of the vinyl radical is extremely rare because of its high reactivity.

A novel aspect of the photochemical behavior of alkenyl halides is the generation of vinyl cations.[18,19] The detailed mechanism for the formation of the vinyl cations has not yet been fully established. There are two possible paths postulated for cation formation: (1) a homolytic cleavage of the C-X bond and subsequent electron transfer within the radical pair,[20–23] and (2) direct heterolysis of the C-X bond[24] (Equation 85.2). The synthetically interesting and useful characteristics of these processes are (1) the generation of unstable vinyl cations, e.g., α-unsubstituted or cyclic vinyl cations, (2) the mild reaction conditions, and (3) the use of aprotic solvents or less polar solvents.

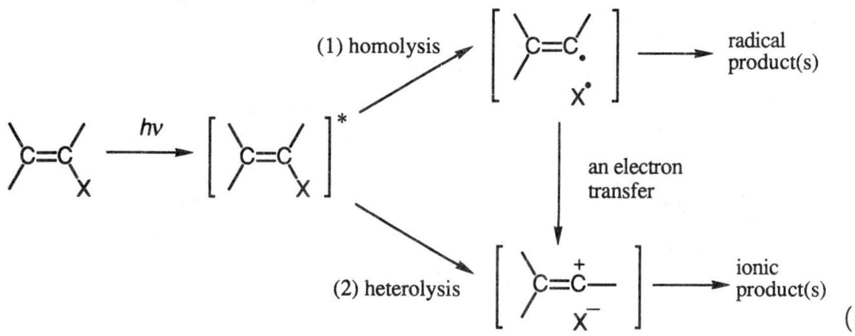

$$(85.2)$$

The reactions that the vinyl cations can undergo are nucleophilic substitution and rearrangement, similar to those of solvolytically generated vinyl cations[25] (Equation 85.3). The photolysis procedures do have the advantage that they can be applied to a wide variety of substrates.

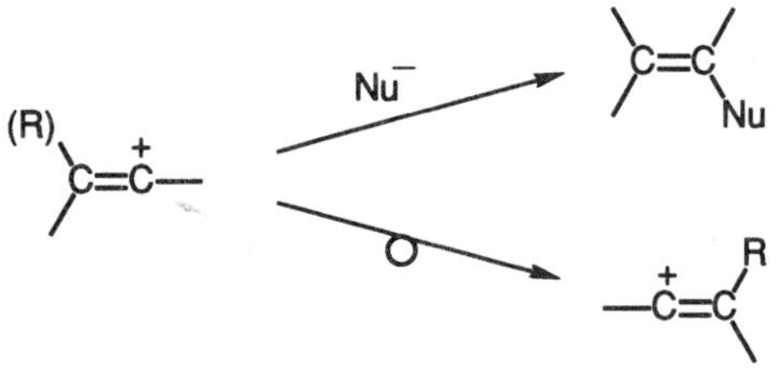

$$(85.3)$$

There were some general rules which can be used to predict the outcome of the irradiation of alkenyl halides. Thus, the photolysis of α-arylvinyl halides leads to stable α-arylvinyl cations and the vinyl cation route is followed predominantly. Such intermediates can be captured by a nucleophilic reagent. Conversely, in the cases that produce unstable vinyl cations, vinyl radicals are formed preferentially or else the formation of both radical and cation intermediates compete. Clearly, the nature of the substituent is important in predicting the reactions of the vinyl cations. Results indicate that the nucleophilic trapping of the vinyl cations follows the order:[26] H ≪ alkyl ≪ Ph < p-MeC$_6$H$_4$ < p-MeOC$_6$H$_4$. The halogen employed is also important with chlorides and bromides being better at undergoing fission than iodides. Fluorides, as mentioned previously, are inactive in photolysis.[22a] In aliphatic alkenyl halides,[3] conversely, iodides afford ionic intermediates more preferentially than do bromides or chlorides. However, the capture by nucleophiles is restricted to the nucleophilic solvent employed.

85.3 Synthetic Applications of the Process

Reduction: Abstraction of a Hydrogen Atom

As described above, synthetic applications using alkenyl halides are divided into two classes: radical and ionic reactions. The most common reaction of the alkenyl radical is abstraction of a hydrogen atom.[5] Although this reaction is useful for reduction of organic halides, its synthetic usefulness is low since most alkenyl halides are prepared from the alkenes and replacement of the halogen by hydrogen simply reforms the starting material. However, deuteration can be carried out successfully by the reaction. Several alkenyl bromides and chlorides are transformed to the deuterated alkenes in good to high yields.[27] An interesting feature of this deuteration is only the site to which the halogen is attached is deuterated. This procedure can also be applied effectively to aryl halides.[27]

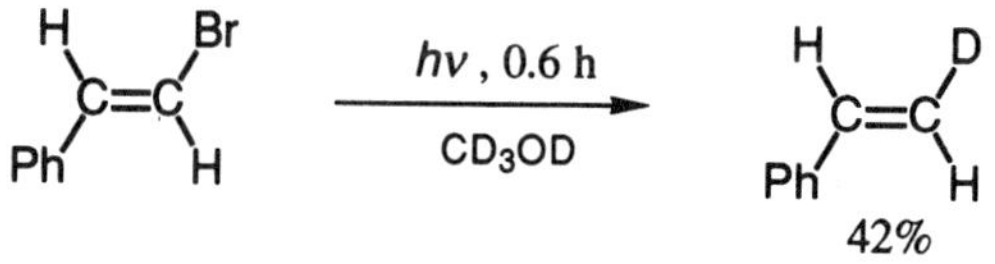

$$(85.4)$$

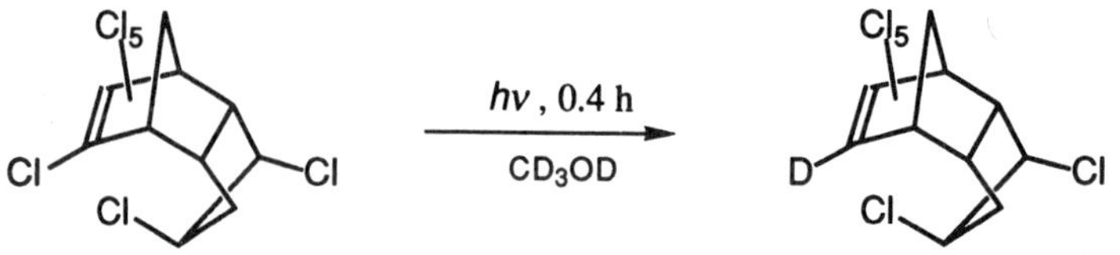

$$(85.5)$$

Other interesting reactions involve the photolysis of β-iodoacrylamides (Equation 85.6), which brings about an intramolecular hydrogen abstraction and subsequent dealkylation, ring opening, of the amine moiety.[6]

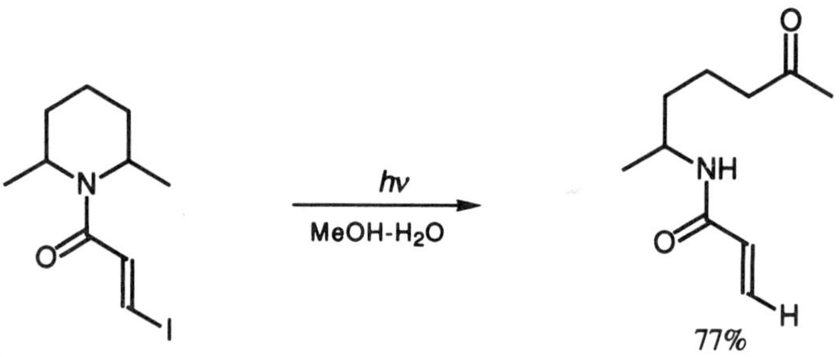

$$(85.6)$$

Intramolecular Cyclization

Intramolecular cyclization by radicals is one of the strategies used to synthesize cyclic compounds. One such example is the intramolecular attack of vinyl radicals at sulfur that is reported to be an efficient synthesis of benzo[*b*]thiophenes[21g,28] (Equation 85.7).

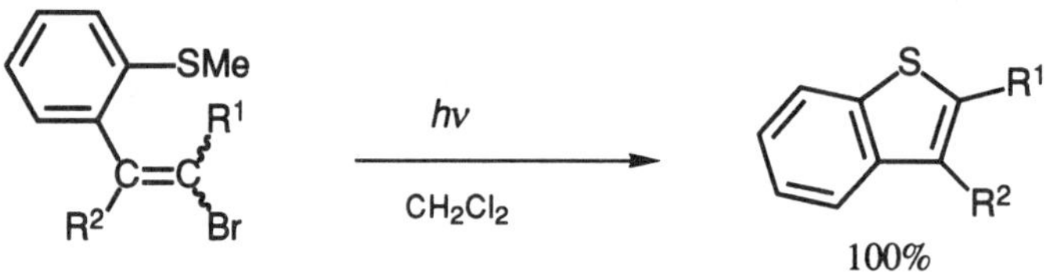

$R^1, R^2 = H, Me, Ph$

$$(85.7)$$

Dimerization

Tadros et al.[29] reported the dimerization of 1-bromo-2,2-*bis*(*p*-methoxyphenyl)ethene in acetic acid under sunlight. However, recent reports concerning this type of compound, 1,1-diaryl-2-haloethenes,[21a,b,g,22a,b] present a more complex reaction system with several processes taking place such as hydrogen abstraction, rearrangement, dimerization, and capture of the solvent.

Substitution by Solvents

Ionic reactions occur preferentially when aliphatic alkenyl iodides and aromatic alkenyl chlorides and bromides are irradiated. The trapping of the vinyl cations generated by irradiation readily takes place in nucleophilic solvents such as alcohols. In these cases, enol ethers or acetals are the main products.[3,20,24,30]

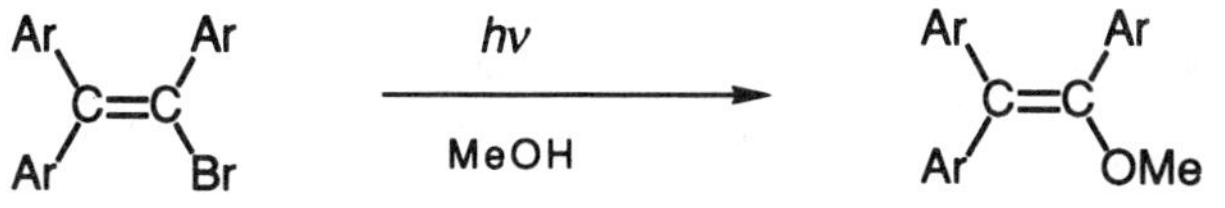

(85.8)

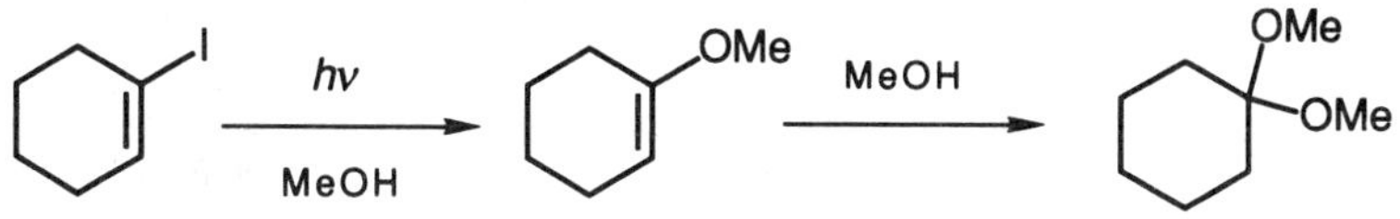

(85.9)

When nitrites, such as acetonitrile, are used on the solvent,[31] a Ritter-type reaction is observed and the cation is trapped to yield an intermediate nitrilium ion. These can cyclize, when there are aryl groups at the β-position of the vinylnitrilium ion, to give isoquinolines in good yield.

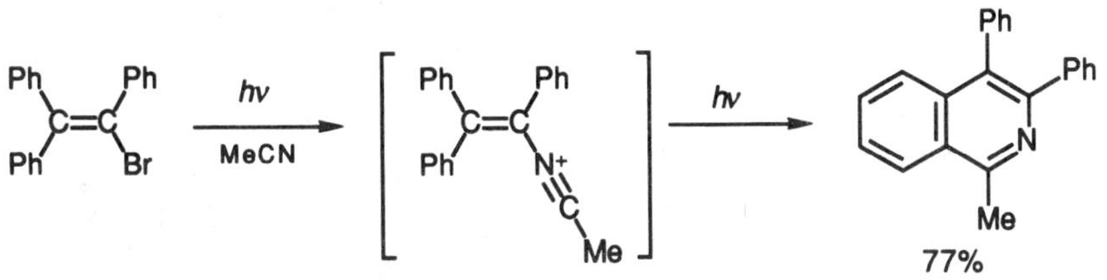

77%   (85.10)

Substitution by Nucleophiles

Other reactions of synthetic utility can be observed when suitable nucleophiles are added to the reaction mixture. Typical examples of these are azide,[26,32] cyanate,[33] thiocyanate,[33] and cyanide anions and thioanisole.[34,39] Reactions using these nucleophiles provide a reaction path to functionalized vinyl derivatives. The transformations are successfully achieved by using triaryl-substituted systems. The reactions with such nucleophiles provide an efficient path to heterocyclic compounds among others. Some examples such as the synthesis of pyrrolines, isoxazolines, isoquinolones, and the thio derivatives are prepared by this method and are shown (Equations 85.11–85.14).

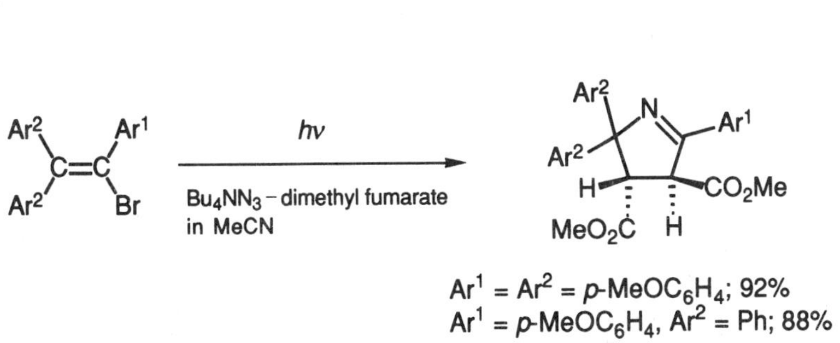

$Ar^1 = Ar^2 = p\text{-}MeOC_6H_4$; 92%   (85.11)
$Ar^1 = p\text{-}MeOC_6H_4$, $Ar^2 = Ph$; 88%

$Ar^1 = Ar^2 = p\text{-}MeOC_6H_4$; 90%
$Ar^1 = p\text{-}MeOC_6H_4$, $Ar^2 = Ph$; 50%   (85.12)

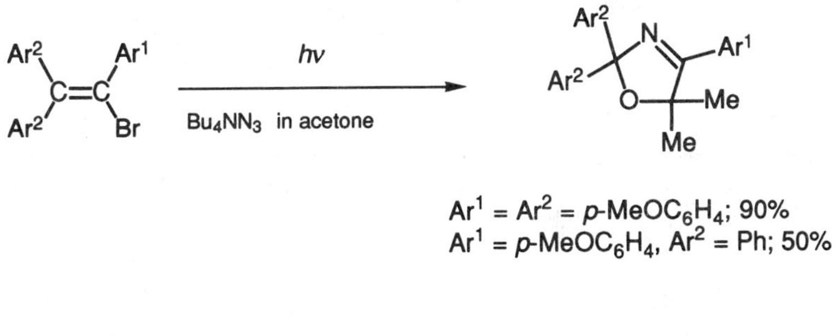

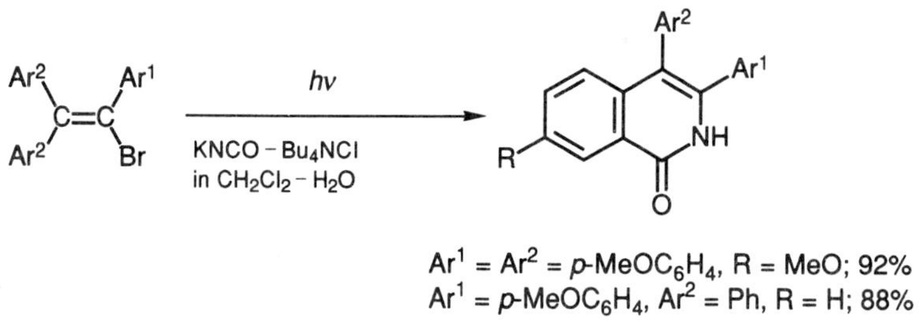

$$\text{(diagram)} \xrightarrow[\substack{\text{KNCO – Bu}_4\text{NCl} \\ \text{in CH}_2\text{Cl}_2 - \text{H}_2\text{O}}]{hv} \text{(diagram)}$$

$$Ar^1 = Ar^2 = p\text{-MeOC}_6H_4, R = \text{MeO}; 92\%$$
$$Ar^1 = p\text{-MeOC}_6H_4, Ar^2 = \text{Ph}, R = \text{H}; 88\%$$

(85.13)

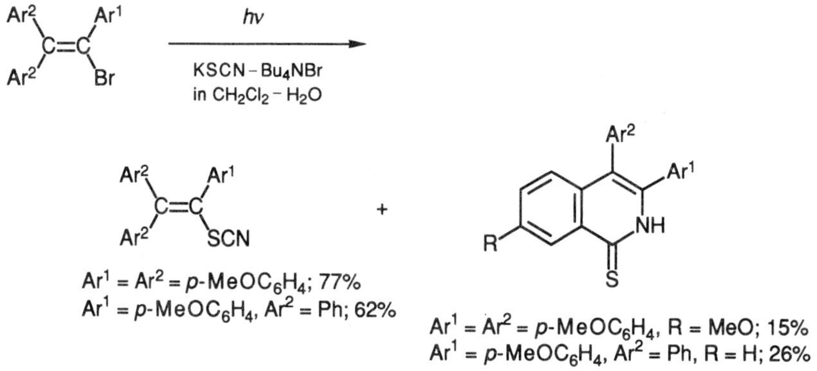

$$\text{(diagram)} \xrightarrow[\substack{\text{KSCN – Bu}_4\text{NBr} \\ \text{in CH}_2\text{Cl}_2 - \text{H}_2\text{O}}]{hv} \text{(diagram)} + \text{(diagram)}$$

$$Ar^1 = Ar^2 = p\text{-MeOC}_6H_4; 77\%$$
$$Ar^1 = p\text{-MeOC}_6H_4, Ar^2 = \text{Ph}; 62\%$$

$$Ar^1 = Ar^2 = p\text{-MeOC}_6H_4, R = \text{MeO}; 15\%$$
$$Ar^1 = p\text{-MeOC}_6H_4, Ar^2 = \text{Ph}, R = \text{H}; 26\%$$

(85.14)

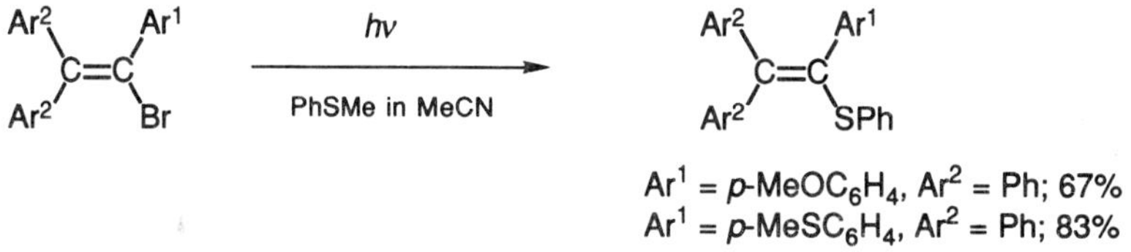

$$\text{(diagram)} \xrightarrow[\text{PhSMe in MeCN}]{hv} \text{(diagram)}$$

$$Ar^1 = p\text{-MeOC}_6H_4, Ar^2 = \text{Ph}; 67\%$$
$$Ar^1 = p\text{-MeSC}_6H_4, Ar^2 = \text{Ph}; 83\%$$

(85.15)

ipso-Substitution

Substitution at the aromatic ring (*ipso*-substitution) is also observed, but only in the cases of *p*-alkoxyphenylvinyl halides.[34,35] The *ipso*-substitution process has been observed in the reaction with cyanide anion as the nucleophile. The irradiation shown in Equation 85.16 brings about double substitution followed by *cis*-stilbene type photocyclization to afford dicyanophenanthrene.[34]

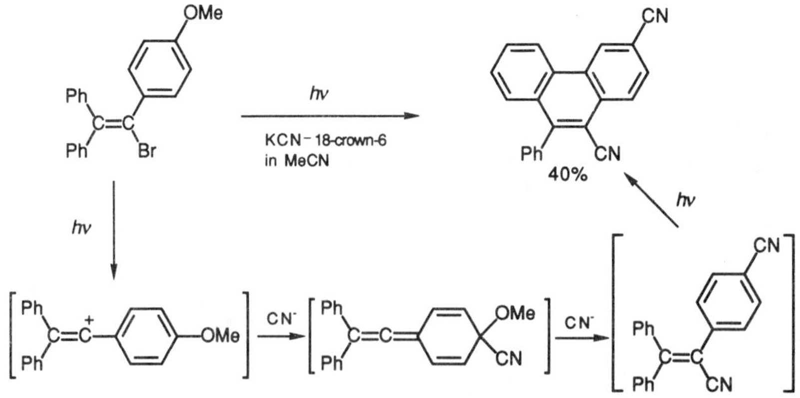

$$\text{(diagram)} \xrightarrow[\substack{\text{KCN – 18-crown-6} \\ \text{in MeCN}}]{hv} \text{(diagram, 40\%)}$$

(85.16)

The intermediate *ipso*-adduct can be isolated in high yield when alkoxide anion is used as the nucleophile and the reaction temperature is controlled between 0 and 5 °C.[35b] The *ipso*-adduct is obtained by intramolecular trapping by a pendant alkoxy chain at the *para*-position. The spiro *ipso*-adducts that are produced are stable at room temperature.[36]

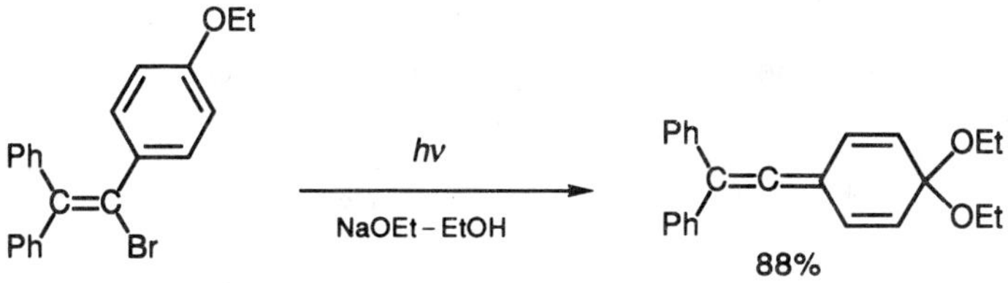

$$(85.17)$$

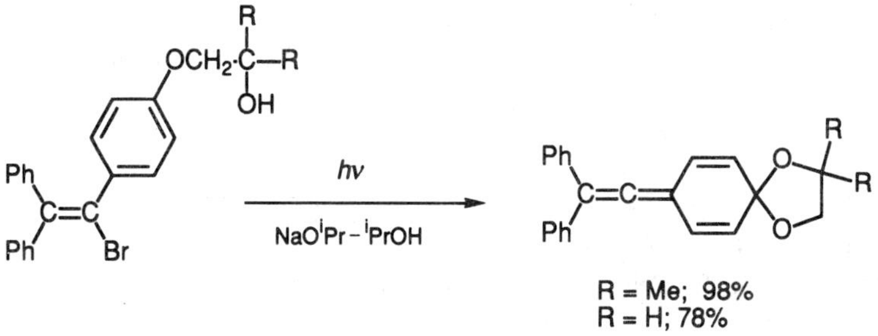

88%

R = Me; 98%
R = H; 78%

$$(85.18)$$

Intramolecular Cyclization

Intramolecular trapping of the vinyl cations provide a useful synthetic approach to several classes of molecule. The presence of an (*ortho*-substituted)aryl group at the β-position facilitates the cyclization. Benzofurans[21c] and dibenz[*b*,*f*]oxepines[37] can be prepared by this method (Equations 85.19 and 85.20).

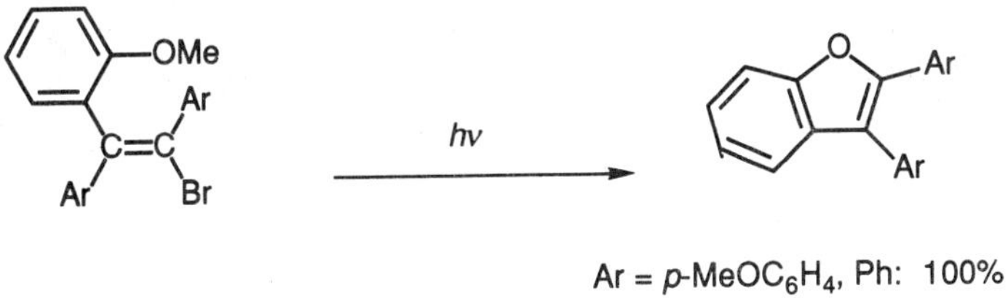

Ar = *p*-MeOC$_6$H$_4$, Ph: 100%

$$(85.19)$$

R = H, Me 100%

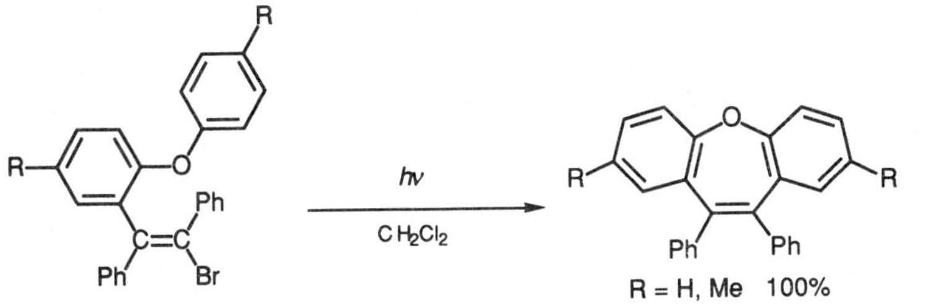

$$(85.20)$$

The β-(*p*-oxidophenyl) group participates to give spiro-cyclopropenes[38] (Equation 85.21).

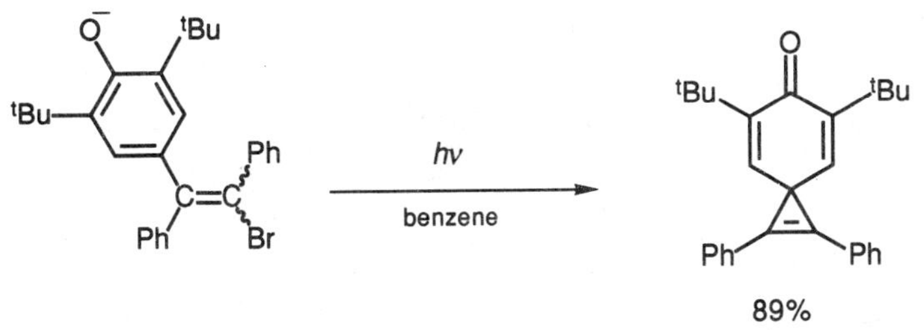

89%

$$(85.21)$$

Acknowledgments

The author is greatly indebted to Profs. H. Taniguchi and S. Kobayashi for discussion on the photolysis of vinyl halides.

References

1. Majer, J. R. and Simons, J. P., Photochemical processes in halogenated compounds, *Advanced Photochemistry*, 2, 137, 1964.

2. Sammes, P. G., Photochemistry of the C-X group, in *Chemistry of The Carbon-Halogen Bond*, Patai, S., Ed., Wiley, New York, 1973, chap. 11.

3. Kropp, P. J., McNeely, S. A., and Davis, R. D., Photochemistry of alkyl halides. 10. Vinyl halides and vinylidene dihalides, *J. Am. Chem. Soc.*, 105, 6907, 1983.

4. Chiurdoglu, G., Ottinger, R., Reisse, J., and Toussaint, A., Etudes spectroscopiques de quelques derives halogenes du cyclohexene, *Spectrochim. Acta*, 18, 215, 1962.

5. Newman, R. C. Jr. and Holmes, G. C., Photolytic formation of isomeric vinyl radicals from *cis*- and *trans*-vinyl iodides, *J. Org. Chem.*, 33, 4317, 1968.

6. Wilson, R. M. and Commons, T. J., Photochemistry of β-iodoacrylamides, *J. Org. Chem.*, 40, 2891, 1975.

7. Grovenstein, E. Jr. and Lee, D. E., The stereochemistry and mechanism of the transformation of cinnamic acid dibromide to β-bromostyrene, *J. Am. Chem. Soc.*, 75, 2639, 1953.

8. Bodenstein, J. and Borden, M. R., Rearrangement accompanying the addition of fluoride to 1,1-diphenylethylene, *Chem. Ind. (London)*, 441, 1958.

9. Pritchard, J. G. and Bothner-By, A. A., Base-induced dehydrohalogenation and rearrangement of 1-halo-2,2-diphenylethylenes in *t*-butyl alcohol. The effect of deuterated solvent, *J. Phys. Chem.*, 64, 1271, 1960.

10. Beltrame, P. and Favini, G., Reaction of halodiphenylethylenes with sodium ethoxide. II. *Gazz. Chim. Ital.*, 93, 757, 1963.

11. Drefahl, G. and Zimmer, C., Untersuchungen über Stilbene. XXXII. Halogenwasserstoff-Addition on Acetylene- und Äthylen-Bindungen, *Chem. Ber.*, 93, 505, 1960.

12. Apelgot, S., Cheutin, A., Mars, S., and Bergers, R., Microsynthesis of radiobromotriphenylethylene, *Bull. Soc. Chim. Fr.*, 533, 1952.

13. Rappoport, Z. and Apeloig, Y., Vinylic cations from solvolysis. II. The stereochemistry of the S_N1 reaction of 1,2-dianisyl-2-phenylvinyl halides, *J. Am. Chem. Soc.*, 91, 6734, 1969.

14. Kitamura, T., Nakamura, I., and Taniguchi, H., unpublished data, 1990.

15. Sonoda, T., Kobayashi, S., and Taniguchi, H., Studies of carbenium ion on unsaturated carbon. II. Exclusive formation of benzofurans via vinyl cations in the solvolytic reactions of 1-aryl-2,2-*bis*(*o*-methoxyphenyl)vinyl halides, *Bull. Chem. Soc. Jpn.*, 49, 2560, 1976.

16. Kerr, J. A., Strengths of chemical bonds, in *CRC Handbook of Chemistry and Physics*, CRC Press, Boca Raton, FL, 1987–1988, F-184.

17. Egger, K. W. and Cocks, A. T., Homopolar and heteropolar bond dissociation energies and heats of formation of radicals and ions in the gas phase. I. Data on organic molecules, *Helv. Chim. Acta*, 56, 1516, 1973.

18. Lodder, G., Recent advances in the photochemistry of the carbon-halogen bond, in *The Chemistry of Functional Groups, Supplement D*, Patai, S. and Rappoport, Z., Eds., John Wiley & Sons, Chichester, 1983, chap. 29.

19. Kropp, P. J., Photobehavior of alkyl halides in solution: Radical, carbocation, and carbene intermediates, *Acc. Chem. Res.*, 17, 131, 1984.

20. McNeely, S. A. and Kropp, P. J., Photochemistry of alkyl halides. 3. Generation of vinyl cations, *J. Am. Chem. Soc.*, 98, 4319, 1976.

21. (a) Suzuki, T., Sonoda, T., Kobayashi, S., Taniguchi, H., Photochemistry of vinyl bromides: a novel 1,2-aryl group migration, *J. Chem. Soc., Chem. Commun.*, 180, 1976; (b) Kitamura, T., Kobayashi, S., and Taniguchi, H., Copper(II) salt-enhanced vinyl cation formation in the photolysis of vinyl bromides, *Chem. Lett.*, 1223, 1978; (c) Suzuki, T., Kitamura, T., Sonoda, T., Kobayashi, S., and Taniguchi, H., Photochemistry of vinyl halides. Formation of benzofurans by photolysis of β-(*o*-methoxyphenyl)vinyl bromides, *J. Org. Chem.*, 46, 5324, 1981; (d) Kitamura, T., Muta, T., Kobayashi, S., and Taniguchi, H., A novel *ortho*-substituent effect on formation of vinyl cations in the photolysis of vinyl bromides, *Chem. Lett.*, 643, 1982; (e) Kitamura, T., Kobayashi, S., and Taniguchi, H., Photochemistry of vinyl halides. Vinyl cation from photolysis of 1,1-diphenyl-2-halopropenes, *J. Org. Chem.*, 47, 2323, 1982; (f) Kitamura, T., Muta, T., Tahara, T., Kobayashi, S., and Taniguchi, H., Photolysis of cyclic arylvinyl halides. Formation of 1,2-benzo-1,3-cycloalkadienyl cations and their rearrangement, *Chem. Lett.*, 759, 1986; (g) Kitamura, T., Kawasato, H., Kobayashi, S., and Taniguchi, H., Exclusive cyclization at sulfur in photolysis of β-[(*o*-arylthio)phenyl]vinyl bromides, *Chem. Lett.*, 839, 1986; (h) Kitamura, T., Kobayashi, S., and Taniguchi, H., Photolysis of vinyl halides. Preferential formation of vinyl cations by copper(II) salts, *J. Am. Chem. Soc.*, 108, 2641, 1986.

22. (a) Sket, B., Zupan, M., and Pollak, A., The photo-Fritsch-Butternberg-Wiechell rearrangement, *Tetrahedron Lett.*, 783, 1976; (b) Sket, B. and Zupan, M., Photochemistry of 2-halogeno-1,1-diphenylethylenes. The photo-Fritsch-Buttenberg-Wiechell rearrangement, *J. Chem. Soc., Perkin Trans. 1*, 752, 1979; (c) Zupanic, N. and Sket, B., The influence of metal(II) salts and solid supports on the nature of the photochemical reaction of 1,1-diphenyl-2-haloethene, *J. Photochem. Photobiol., A.*, 60, 361, 1991.

23. Sonawane, H. R., Nanjundiah, B. S., and Panse, M. D., Photochemistry of organic halides: Some interesting features of the photobehaviour of vinyl halides and vinylidene dihalides derived from camphene, *Tetrahedron Lett.*, 3507, 1985.

24. (a) van Ginkel, F. I. M., Visser, R. J., Varma, C. A. G. O., and Lodder, G., Nanosecond laser flash photolysis of 1-anisyl-2,2-diphenylvinyl bromide in acetonitrile and acetic acid, *J. Photochem.*, 30, 435, 1985; (b) Verbeek, J. M., Cornelisse, J., and Lodder, G., Photolysis of the vinyl bromide 9-(α-bromobenzylidene)fluorene in methanol. Effect of wavelength of irradiation, sodium methoxide and oxygen, *Tetrahedron*, 42, 5679, 1986; (c) van Ginkel, F. I. M., Cornelisse, J., and Lodder, G., Photoreactivity of some α-arylvinyl bromides in acetic acid. Selectivity toward bromide versus acetate ions as a mechanistic probe, *J. Am. Chem. Soc.*, 113, 4261, 1991.

25. (a) Stang, P. J., Rappoport, Z., Hanack, M., and Subramanian, L. R., *Vinyl Cations*, Academic Press, New York, 1979; (b) Rappoport, Z., Vinyl cations, in *Reactive Intermediates*, Vol. 3, Abramovitch, R. A., Ed., Plenum Press, New York, 1983, chap. 7; (c) Hanack, M., *Carbokationen, Carbokation-Radikale*, Georg Thieme Verlag, Stuttgart, 1990.

26. Kitamura, T., Kobayashi, S., and Taniguchi, H., Photochemistry of vinyl halides. Heterocycles from reaction of photogenerated vinyl cations with azide anion, *J. Org. Chem.*, 49, 4755, 1984.

27. Müller, J. P. H., Parlar, H., and Korte, F., Einfache Photochemische Methode zur Deuterierung von Olefinen und Aromaten, *Synthesis*, 524, 1976.

28. Kitamura, T., Kobayashi, S., and Taniguchi, H., Photolysis of β-(*o*-methylthiophenyl)vinyl bromides. A versatile synthesis of benzo[*b*]thiophens, *Chem. Lett.*, 1637, 1988.

29. Tadros, W., Sakla, A. B., and Akhookh, Y., Butadienes and related compounds. III. Further study of the factors bearing the formation of 1,1,4,4-tetraarylbuta-1,3-dienes, *J. Chem. Soc.*, 2701, 1956.

30. Kitamura, T., Kobayashi, S., Taniguchi, H., Fiakpui, C. Y., Lee, C. C., and Rappoport, Z., Degenerate β-aryl rearrangements in photochemically generated triarylvinyl cations, *J. Org. Chem.*, 49, 3167, 1984.

31. Kitamura, T., Kobayashi, S., and Taniguchi, H., Isoquinoline derivatives from the Ritter-type reaction of vinyl cations, *Chem. Lett.*, 1351, 1984.

32. Kitamura, T., Kobayashi, S., and Taniguchi, H., Photolysis of vinyl bromides in the presence of tetrabutylammonium azide: Trapping of a vinyl cation with azide ion, *Tetrahedron Lett.*, 1619, 1979.

33. (a) Kitamura, T., Kobayashi, S., and Taniguchi, H., Reaction of photogenerated vinyl cations with ambident anions, *Chem. Lett.,* 1523, 1984; (b) Kitamura, T., Kobayashi, S., and Taniguchi, H., Photolysis of vinyl halides. Reaction of photogenerated vinyl cations with cyanate and thiocyanate ions, *J. Org. Chem.,* 55, 1801, 1990.

34. Kitamura, T., Murakami, M., Kobayashi, S., and Taniguchi, H., *ipso*-Substitution by cyanide anion in photolysis of 1-(*p*-methoxyphenyl)vinyl bromides, *Tetrahedron Lett.,* 27, 3885, 1986.

35. (a) Kitamura, T., Kabashima, T., Kobayashi, S., and Taniguchi, H., Drastic base dependence of products in photolysis of 1-(*p*-methoxyphenyl)-2,2-diphenylvinyl bromide. *ipso*-Substitution by alkoxide anions, *Chem. Lett.,* 1951, 1988; (b) Kitamura, T., Kabashima, T., Kobayashi, S., and Taniguchi, H., Isolation and alcoholysis of an *ipso* adduct, vinylidenecyclohexadiene, from photolysis of 1-(*p*-ethoxyphenyl)vinyl bromide, *Tetrahedron Lett.,* 29, 6141, 1988; (c) Kitamura, T., Nakamura, I., Kabashima, T., Kobayashi, S., and Taniguchi, H., *ipso*-Substitution by alkoxide ions in photolysis of triarylvinyl halides. Firm evidence for intervention of vinyl cations, *Chem. Lett.,* 9, 1990; (d) Kitamura, T., Kabashima, T., Nakamura, I., Fukuda, T., Kobayashi, S., and Taniguchi, H., *Ipso* substitution of triarylvinyl cations by alkoxide anions, *J. Am. Chem. Soc.,* 113, 7255, 1991.

36. (a) Kitamura, T., Nakamura, I., Kabashima, T., Kobayashi, S., and Taniguchi, H., A novel spiro adduct from intramolecular *ipso* substitution in the photolysis of an α-[*p*-(2-hydroxyalkoxy)-phenyl]vinyl bromide, *J. Chem. Soc., Chem. Commun.,* 1154, 1989; (b) Kitamura, T., Soda, S., Nakamura, I., Fukuda, T., and Taniguchi, H., Importance of β-phenyl group in *ipso* substitution of arylvinyl cations, *Chem. Lett.,* 2195, 1991.

37. (a) Kitamura, T., Kobayashi, S., and Taniguchi, H., Formation of dibenz[*b,f*]oxepins from β-(*o*-aryloxyphenyl)vinyl bromides, *Chem. Lett.,* 547, 1984; (b) Kitamura, T., Kobayashi, S., Taniguchi, H., and Hori, K., Photolytic and solvolytic reactions of β-[*o*-(aryloxy)phenyl]vinyl bromides. Intramolecular arylation of vinyl cations into dibenzoxepins, *J. Am. Chem. Soc.,* 113, 6240, 1991.

38. Ikeda, T., Kobayashi, S., and Taniguchi, H., A new route to spiro[2.5]octa-1,4,7-trien-6-ones, *Synthesis,* 393, 1982.

39. Kitamura, T., Kabashima, T., and Taniguchi, H., Phenylthiolation of arylvinyl bromides by photolysis, *J. Org. Chem.,* 56, 3739, 1991.

86

Photochemical C–X Bond Cleavage in Arenes

Nigel J. Bunce
University of Guelph

86.1 Introduction

The photochemistry of aryl halides (Ar-X) has been studied for over 30 years, and a very large number of citations is available. Previous reviews include Sammes (1973),[1] Grimshaw and de Silva (1981),[2] Lodder,[3] Davidson et al. (1984),[4] and Choudhry et al. (1988).[5] The examples reviewed in this chapter concern solution-phase chemistry, unless noted otherwise.

Photochemical C-X fission in aryl halides may be defined as any photochemical process leading to cleavage of a carbon-halogen bond. The emphasis in this review is on mechanisms of C-X bond fission, of which three mechanistic reaction types may be identified.

1. Homolysis $Ar-X^* \rightarrow Ar^. + X^.$
2. Electron transfer $Ar-X^* \rightarrow (Ar-X)^{.-} \rightarrow AR^. + X^-$
3. Photonucleophilic substitutions $Ar-X^* + Nu:^- \rightarrow Ar-Nu + X^-$

Aryl halides tend to be chemically unreactive and include persistent environmental pollutants such as DDT, PCBs, and chlorinated dibenzo-p-dioxins. Many studies of the photochemistry of halogenated aromatic compounds have been stimulated by environmental concerns, with the goal of understanding whether photolysis is an important sink for these compounds in natural waters[6,7] or in the atmosphere.[8] Numerous practical problems attend these experiments. For example, many pollutants of interest have minimal absorption in the region of the tropospheric solar spectrum (>295 nm), and experiments at environmentally nonrelevant wavelengths such as 254 nm are often more feasible. In addition, most halogenated aromatic compounds have very low solubility in water, and model solvents such as methanol or aqueous acetonitrile are often used. Both the quantum yield of photoreaction (and hence estimates of environmental persistence) and the reaction products may be solvent dependent.

86.2 Reactions Proceeding via Homolysis

Homolytic cleavage of the aryl carbon-halogen bond affords aryl free radicals, which can either arylate a suitable aromatic reaction partner or abstract hydrogen from a hydrogen atom donor (reductive dehalogenation).

$$Ar\text{-}X^* \rightarrow Ar^{\cdot} + X^{\cdot}$$

$$Ar^{\cdot} + R \rightarrow Ar\text{-}H + R^{\cdot}$$

$$or \quad Ar^{\cdot} + Ar'H \rightarrow (ArHAr')^{\cdot} \xrightarrow{-H} Ar\text{-}Ar'$$

However, since arylation and dehalogenation products may form by other mechanisms, product studies alone cannot be used to infer that homolysis has occurred.

Homolysis of photoexcited aryl iodides and bromides has been exploited as a means of producing aryl radicals; in some cases, these reactions have synthetic potential for the formation of biaryls, including the use of intramolecular arylation in order to construct polycyclic systems (reviewed in Reference 2). Reductive dehalogenation has been suggested as a synthetic route to specifically deuterated aromatic compounds, by photolysis of aryl halides in solvents containing abstractable deuterium atoms.[9]

The energetics of homolysis are straightforward. Homolysis requires that the energy of the reactive excited state (usually T_1) be greater than the C-X bond dissociation energy (Table 86.1). Where the aromatic nucleus contains more than one type of halogen, the weakest C-X bond is cleaved preferentially. The rate of dechlorination of chlorobenzenes must be very fast, as shown by 1-bromo-3-*p*-chlorophenoxypropane,[4] which loses Cl exclusively upon photoexcitation; even though the side chain contains the weaker C-Br bond, excitation energy is not transferred to the weaker bond.

The identification of the reactive excited state as T_1 is unequivocal for chlorinated, brominated, and iodinated benzenes because their quantum yields of intersystem crossing $S_1 \rightarrow T_1$ are high, as are their quantum yields of dehalogenation in hydrogen-donating solvents (i.e., $\phi_{isc} + \phi_r > 1$). The data of Table 86.1 show that homolysis is exothermic for triplet iodo- and bromobenzenes, and probably also for chlorobenzenes, although the correct value of the C_6H_5-Cl bond dissociation energy has been the subject of much controversy. Homolysis is thermodynamically unfavorable (and is not seen) for aryl fluorides, in either singlet- or triplet-state reactions. In keeping with the data in Table 86.1, 9,10-dibromoanthracene and 1-chloropyrene (monomer) are unaffected by lengthy photolysis in alkane solvents. The situation is less clear-cut for substances such as 4-chlorobiphenyl and 1-chloronaphthalene, whose photolysis efficiencies are low, but which probably dechlorinate mainly through the intermediacy of excimers.

A number of studies have been made of halobenzenes in the gas phase. Quantum yields of homolysis are high (discussion, Reference 4). In the presence of hydrogen donors such as ethane,[10] reductive dehalogenation occurs, as in solution. The situation is different upon photolysis in air,

Table 86.1   Energetics (kJ mol⁻¹) of C-X Homolysis

Arene System	Triplet Energy	X	$D(C_6H_5\text{-}X)$
Benzene	360	F	520
Naphthalene	255	Cl	350–390
Biphenyl (no *o*-X)	275	Br	335
Biphenyl (one *o*-X)	>285	I	270
Anthracene	180		
Pyrene	200		

when little of the starting material has been accounted for. Low yields of phenols have been identified, presumably the result of trapping aryl radicals by O_2; the remainder of the starting material is assumed to have been mineralized.[8]

Among chlorinated biphenyls, the triplet energy depends upon whether or not the molecule is *ortho*-chlorinated. *ortho*-Substitution raises the energy of the excited state due to partial deconjugation of the biphenyl chromophore, which shifts the absorption spectrum toward the blue; loss of an *ortho*-chlorine relieves steric strain. Among simple chlorobiphenyls (those containing only 1–3 chlorine substituents), those containing *ortho*-chlorines have markedly larger quantum yields of homolysis than those without (see Table 4 of Reference 4). In the environmental context, the efficiency of photolysis of PCBs is determined both by their intrinsic photolability (ϕ_r) and by the overlap of their absorption spectra with the tropospheric solar spectrum, the *o*-chlorobiphenyls having less spectral overlap but higher quantum yields of homolysis.[8]

The regiospecificity of homolysis favors relief of steric strain. For example, photolysis of 1,2,3,5-tetrachlorobenzene in hydrogen-donating media gives predominantly 1,3,5-trichlorobenzene and 1,2,4-trichlorobenzene.[5] Halobiphenyls substituted at an *ortho* and another position cleave the *ortho*-halogen preferentially;[4] however, polychlorodibenzo-*p*-dioxins lose lateral (2,3,7,8) in preference to apical (1,4,5,9) chlorines.[11] Homolytic photodehalogenation has been used commercially to destroy a quantity of 2,3,7,8-tetrachlorodibenzo-*p*-dioxin at an abandoned plant site in Missouri.[12]

86.3 Reactions Proceeding via Excimers and Exciplexes

These reactions are initiated by the full or partial acquisition of an electron by the aryl halide, affording an excimer, an exciplex, or radical ions, depending upon the electron donor and the solvent. In this chapter, the term "excimer" will be used to describe the complexes formed between two arenes, even when they are not identical.

Excimer Reactions

A key mechanistic observation in support of an excimer mechanism is the dependence of the quantum yield of reaction upon the aryl halide concentration: $\phi_r^{-1} \propto [ArX]^{-1}$. Inferential evidence is provided when an excimer can be detected spectroscopically by fluorescence or flash photolysis.

The recent investigation of the photochemistry of 4-bromobiphenyl typifies this mechanism.[13] The reaction affords biphenyl as the sole product, yet the energetics of Table 86.1 make it unlikely that the reaction involves direct homolysis. The participation of the triplet state is clear in that $\phi(S_1 \rightarrow T_1) = 0.98$, while $\phi_r = 0.15$ at high [BpBr]. The involvement of the triplet excimer is shown by the increase of ϕ_r with [BpBr]; $\phi_r^{-1} \propto [BpBr]^{-1}$.

Earlier investigations had shown similar increases of ϕ_r with concentration in the case of 1-chloronaphthalene, but a decrease in the case of chlorobenzene,[14] because in the chlorobenzene series, the excimer cleaves less efficiently than the monomer (for which $\phi_r = 0.54$ in alkanes). Dechlorination via a heteroexcimer is also implicated by studies indicating that benzene "sensitizes" the dechlorination of chlorobenzene.[4] Recent experiments with pentachlorobenzene suggest that three dechlorination pathways are operative: homolysis from both S_1 and T_1, and fragmentation of a triplet excimer.[15]

The reactivity of a heteroexcimer has been exploited as a means of degrading low concentrations of chlorinated pollutants such as polychlorinated biphenyls and polychlorinated benzenes.[16] Copolymers of vinylnaphthalene and styrenesulfonic acid form water-soluble micellar-like copolymers with a hydrophobic core. When the nonpolar chloro-compounds are dissolved in aqueous solutions containing the water-soluble copolymer, they are scavenged into the core where they undergo photodehalogenation. In these reactions, essentially all the incident radiation is absorbed by the naphthalene chromophores, which form heteroexcimers with the polychloro-compound.

The pattern of dechlorination is characteristic of electron transfer rather than direct homolysis (see next section).

Exciplex Reactions Involving Aliphatic Amines

In these reactions, kinetic evidence for the involvement of the amine is shown by the relationship $\phi_r^{-1} \, \alpha \, [\text{amine}]^{-1}$. In cases where the halide has detectable fluorescence, the amine may quench fluorescence according to Stern-Volmer kinetics, and emission from the singlet exciplex is frequently observable in nonpolar solvents.[4]

An important issue is whether the reaction proceeds from the singlet or triplet excited state. Kinetics are often helpful in identifying the excited state. Consider a reaction scheme such as Scheme 86.1, where the possibilities of both singlet and triplet quenching by the amine are included.

$$\text{ArCl} \xrightarrow{\text{hv, } I_{abs}} {}^1\text{ArCl}*$$

$$ {}^1\text{ArCl}* \xrightarrow{\text{k}_1} \text{ArCl}$$

$$ {}^1\text{ArCl}* \xrightarrow{\text{k}_F} \text{ArCl} + \text{hv}'$$

$$ {}^1\text{ArCl}* \xrightarrow{\text{k}_2} {}^3\text{ArCl}$$

$$ {}^1\text{ArX}* + \text{Q} \xrightarrow{\text{k}_3} {}^1\left\{\text{Q}^+ \cdot \text{ArX}^-\right\}$$

$$ {}^1\left\{\text{Q}^+ \cdot \text{ArCl}^-\right\} \xrightarrow{\text{k}_4} \text{ArCl} + \text{Q}$$

$$ {}^1\left\{\text{Q}^. + \cdot \text{ArCl}^-\right\} \xrightarrow{\text{k}_5} \text{products}$$

$$ {}^3\text{ArCl}* \xrightarrow{\text{k}_6} \text{ArCl}$$

$$ {}^3\text{ArCl}* + \text{Q} \xrightarrow{\text{k}_7} {}^3\left\{\text{Q}^+ \cdot \text{ArX}^-\right\}$$

$$ {}^3\left\{\text{Q}^+ \cdot \text{ArCl}^-\right\} \xrightarrow{\text{k}_8} \text{ArCl} + \text{Q}$$

$$ {}^3\left\{\text{Q}^. + \cdot \text{ArCl}^-\right\} \xrightarrow{\text{k}_9} \text{products}$$

SCHEME 86.1

Steady-state analysis gives the following results.

Fluorescence quenching:

$$\left(\phi_F / \phi_F^0\right) = 1 + k_3^1 \tau [\text{Q}] \quad \text{where} \quad {}^1\tau = \left(k_1 + k_2 + k_F\right)^{-1}; K_{SV} = k_3^1 \tau$$

Singlet state reaction, $k_7 \rightarrow 0$:

$$1/\phi_r = \left(1 + k_4 / k_5\right)\left(1 + 1/k_3^1 \tau [\text{Q}]\right)$$

Triplet state reaction, $k_3 \to 0$:

$$1/\phi_r = \left(1/k_2^1\tau\right)\left(1+k_8/k_9\right)\left(1+1/k_7^3\tau[Q]\right) \quad \text{where } {}^3\tau = k_6^{-1}$$

We define K_ϕ as the parameter intercept/slope for a plot of $\phi_r^{-1} \, \alpha \, [\text{amine}]^{-1}$. In the case of a predominantly singlet state reaction, $K_{SV} = K_\phi = k_3^1\tau$, i.e., concordance is seen between K_{SV}, the Stern-Volmer constant for quenching fluorescence, and $K\phi$, the parameter intercept/slope from the plot of $\phi_r^{-1} \, \alpha \, [\text{amine}]^{-1}$. This kinetic behavior has been observed in the diethylamine-assisted photoreduction of 9,10-dihaloanthracenes to 9-haloanthracenes, as shown[17,18] by the results of Table 86.2, where there is excellent agreement between K_{SV} and K_ϕ. The dihaloanthracenes are convenient to study because of the lack of an unassisted reaction.

Unlike diethylamine, monoethylamine failed to quench fluorescence or to induce dehalogenation, while the tertiary alkylamine (Et_3N) quenched fluorescence but induced dehalogenation inefficiently or not at all. The effect on fluorescence is consistent with the order of oxidation potentials of aliphatic amines; the lack of concordance between K_{SV} and K_ϕ in the case of Et_3N suggests a change of mechanism in this case.

These results parallel those obtained upon photolysis of the parent anthracene with amines; namely, neither reaction nor fluorescence quenching with primary alkylamine. Fluorescence quenching and photochemical reaction (reductive addition to give a variety of 9,10-dihydroanthracene derivatives) with dialkylamine, with transfer of the N-H(D) proton; fluorescence quenching and a less efficient chemical reaction with bonding of the α-carbon of Et_3N to the arene.[19]

$$^1\left(ArX^{\delta-} \cdot Et_2NCH_2CH_3^{\delta+}\right) \to \left(ArXH..Et_2NCH^{\cdot}CH_3\right) \to ArH \cdot CH(CH_3)NEt_2$$

According to Scheme 86.1, a predominantly triplet state reaction has $K_{SV} = k_3^1\tau$ but $K_\phi = k_7^3\tau$. Hence, $K_{SV} \neq K_\phi$ is compatible with a triplet-state reaction but not with a singlet-state process.

The photodehalogenation of 4-chlorobiphenyl exemplifies the triplet exciplex mechanism. 4-Chlorobiphenyl is convenient for study because the quantum yield of reduction in the absence of amines is very low ($\phi < 0.001$). In acetonitrile, the addition of 0.1 M amines increased ϕ_r to 0.07 (butylamine), 0.25 (dipropylamine), and 0.49 (triethylamine). Initially,[20] the singlet exciplex mechanism was suggested with Et_3N as the amine ($K_{SV} = 23 \; M^{-1}$; $K_\phi = 20 \; M^{-1}$). However, in later work on the same system, K_ϕ was consistently found to be greater than K_{SV} (Table 86.3), suggesting a predominantly triplet-state reaction,[21] consistent with a high value of $\phi(S_1 \to T_1)$. Furthermore, quenching of the triplet by tertiary amines has been observed experimentally by flash photolysis.[22]

In Table 86.3, note that even though $K_{SV} \neq K_\phi$, the two parameters follow the same trend with changes of amine and solvent, indicating that the relative rates of electron transfer to the excited state are similar for both singlet and triplet excited states.

Table 86.2 Photodehalogenation of 9,10-Dihaloanthracenes with Aliphatic Amines

Halogen	Amine	Solvent	K_{SV}, M^{-1}	K_ϕ, M^{-1}
Cl, Br	$EtNH_2$	Heptane	—	—
Br	Et_2NH	Heptane	6.0	5.5
Br	Et_2NH	Benzene	12	14
Br	Et_2NH	Dioxane	8	9
Br	Et_2NH	Methanol	0.55	0.6
Br	Et_3N	Heptane	10.0	0.9
Br	Et_3N	Benzene	23	—
Br	Et_3N	Dioxane	15	—
Br	Et_3N	Methanol	4.5	—

Table 86.3   Amine-Assisted Photoreduction
of 4-Chlorobiphenyl

Amine	Solvent	K_{SV}, M^{-1}	K_ϕ, M^{-1}
Et$_3$N	Cyclohexane	11	44
Et$_3$N	Methanol	2.7	18
Et$_3$N	Acetonitrile	23	20[a]
Et$_3$N	Acetonitrile	20	83
Et$_3$N	Aq. acetonitrile	11	97
Piperidine	Methanol	0.8	2.2
Piperidine	Acetonitrile	13	61

[a] This value from Ref. 20; all others from Ref. 21.

Photodechlorination of 4-chlorobiphenyl/Et$_3$N systems can be carried out using long-wavelength UV, rather than 254 nm, radiation if anthracene is added as the initial light absorber.[23] Besides biphenyl, photoreduction products of anthracene are formed (Chapter 21). The possibility favored by the authors is endothermic electron transfer between anthracene radical anion and 4-chlorobiphenyl.

$$^1C_{14}H_{10} + Et_3N \rightarrow C_{14}H_{10}^{\cdot -} + Et_3N^{\cdot +}$$

$$C_{14}H_{10}^{\cdot -} + C_{12}H_9Cl \rightarrow C_{14}H_{10} + C_{12}H_9Cl^{\cdot -} \rightarrow C_{12}H_9^{\cdot} + Cl^-$$

However, another possibility is the attack of Et$_3$N on an anthracene-chlorobiphenyl heteroexcimer.

Polychlorinated biphenyls, bromobiphenyls, and 1-bromonaphthalene, all of which undergo efficient intersystem crossing $S_1 \rightarrow T_1$, show enhanced photodehalogenation in the presence of tertiary aliphatic amines, consistent with electron transfer to the triplet excited state. An exception is 2,4,6-trichlorobiphenyl, which showed lower photoefficiency in cyclohexane solution when Et$_3$N was present, a result attributed to its rather efficient photohomolysis ($\phi = 0.21$); the major effect of Et$_3$N is thus to quench this reaction without providing an alternative efficient reaction channel. In other cases, curved plots of ϕ_r vs. [amine] have been interpreted in terms of competing singlet and triplet quenching by the amine.[21]

Two possible sequences of events follow formation of the exciplex or radical ion.
A: Loss of halide followed by hydrogen atom abstraction:

$$ArX^{\cdot -} \rightarrow Ar^{\cdot} + X^- \quad Ar^{\cdot} + RH \rightarrow Ar\text{-}H + R^{\cdot}$$

B: Protonation followed by loss of a halogen atom:

$$ArX^{\cdot -} + H^+ \rightarrow (ArXH)^{\cdot} \rightarrow Ar\text{-}H + X^{\cdot}$$

Deuterated solvents have been used to distinguish these possibilities.[21,24] Mechanism A is predicted to give deuterium incorporation using C$_6$D$_{12}$, CD$_3$OH, etc., while mechanism B should afford deuterium incorporation with CH$_3$OD, CH$_3$CN/D$_2$O, etc.

Mechanism A is implicated by the formation of deuterated product upon photolysis with Et$_3$N/ CD$_3$OH, and by the failure in other cases to observe deuterium incorporation when CH$_3$CN/D$_2$O was used as the solvent. Mechanism B has been suggested for a number of cases (1-chloronaphthalene, 4-chlorobiphenyl, 1-bromonaphthalene, and 9-chloroanthracene), all of which incorporate deuterium into the dehalogenated product when the starting materials are photolyzed with CH$_3$CN/ Et$_3$N/D$_2$O or Et$_3$N/CH$_3$OD. For example, when Et$_2$ND was photolyzed with 9,10-dibromoanthracene in heptane, the product 9-bromoanthracene contained deuterium at the *meso*-position, indicating that hydrogen was transferred from the amine as a proton, rather than as a neutral hydrogen atom.

$$^1\left(ArBr^{\delta-}\cdot HNEt_2^{\delta+}\right)\rightarrow\left(ArBrH..NEt_2\right)\rightarrow ArH+Br^{\cdot}+Et_2N^{\cdot}$$

Finally, a substantial fraction of the hydrogen always originates with the C–H bonds of the amine, indicating in-cage hydrogen transfer, as shown by the failure to effect complete deuteration with, for example, Et_3N/CD_3OD or Et_3N/C_6D_{12}.

The photodechlorination of 1-chloropyrene in acetonitrile is also assisted by Et_3N, but in this case reduction is further assisted by cyanide ion.[25] Flash photolysis suggests the following sequence of intermediates:

$$PyCl\rightarrow PyCl^{\cdot-}\rightarrow PyClH^{\cdot}\rightarrow PyCNH^{\cdot}\rightarrow PyH$$

Singlet 1-chloropyrene gains an electron from Et_3N, followed by in-cage proton transfer. Loss of Cl^- from this intermediate is very inefficient; however, chloride is displaced by the attack of CN^-. $PyCNH^{\cdot}$ does not go on to give pyrenecarbonitrile, but instead loses $CN^{\cdot}$ to complete the reduction.

Polychlorobenzenes undergo intersystem crossing with high efficiency, therefore making T_1 the likely reactive excited state.[24] In these reactions, plots of $\phi^{-1}\,\alpha\,[Et_3N]^{-1}$ are linear at high $[Et_3N]$, but show little Et_3N-dependence at low $[Et_3N]$. This suggests the simultaneous involvement of an amime exciplex mechanism, and another reaction (excimer or triplet homolysis) that does not involve the amine. Most importantly, the distribution of dechlorinated products is different for the unassisted and Et_3N-assisted reactions in acetonitrile,[27] hence the regioselectivity of dechlorination may be used as an indicator of the mechanism (Table 86.4). The intermediacy of the polyhalobenzene anion radical in Et_3N-assisted photolyses has been demonstrated unequivocally by the observation of the "assisted" product spread when the radical anion is generated thermally.[28] Furthermore, the "assisted" product distribution is also obtained in Et_3N-assisted reactions sensitized by acetophenone, thus confirming the triplet as the reactive excited state.

At this point, it is convenient to mention acetone sensitization of the dehalogenation of polychlorobenzenes. For example, upon direct photolysis of 1,2,3,5-tetrachlorobenzene in aqueous CH_3CN, 1,2,4-trichlorobenzene was the major product; acetone sensitization of the same reaction gave predominantly the 1,3,5-isomer. The original hypothesis was that the direct reaction involves S_1 and the acetone-sensitized reaction involves T_1.[5] The suggestion that the direct reaction involves S_1 appears untenable given the high value of $\phi(S_1\rightarrow T_1)$; in addition, the formation of 1,2,4-trichlorobenzene in the "direct" reaction is reminiscent of the product spread under electron-transfer conditions.[27] One possibility is that the acetone-sensitized reaction indeed involves T_1, but that the direct reaction is actually an excimer-mediated reaction.

The regioselectivity of Cl^- ejection following electron transfer to a polychlorobenzene cannot be interpreted simply in terms of the maximal relief of steric strain.[24] Rather, the direction of loss of

Table 86.4 Et_3N-Assisted and Unassisted Reactions of Polychlorobenzenes

Pentachlorobenzene	Products	Unassisted	Et_3N-Assisted
	1,2,3,5-Cl_4	67	25
	1,2,4,5-Cl_4	26	66
	1,2,3,4-Cl_4	7	8
1,2,3,5-Tetrachlorobenzene			
	1,3,5-Cl_3	59	22
	1,2,4-Cl_3	40	73
	1,2,3-Cl_3	<1	5

chlorine is rationalized by considering the stabilization of the radical anion $C_6H_{6-n}Cl_n^{\cdot-}$ by the chlorine substituents. Loss of chloride is fastest from the position where least delocalization of the negative charge is possible.

Triethylamine also causes photoreduction of polyfluorobenzenes in both acetonitrile and alkane solvents.[29] In this case, homolysis cannot be a competing reaction, based on energetic arguments. These reactions appear to be entirely parallel to those with polychlorobenzenes, with electron donation to the polyfluorobenzene, followed by fluoride loss, together with an excimer component to the reaction. The latter is independent of [Et$_3$N], but has $\phi_r^{-1} \propto [ArF]^{-1}$. A different product spread is observed in alkanes compared with acetonitrile, consistent with the relative extents of electron transfer in each solvent. Previous to this work, photodehalogenation of fluoroarenes was almost unknown, although photoaddition and photoreduction to dihydro-derivatives had been observed with fluorobenzene and aliphatic amines.[30]

Exciplex Reactions Involving Aromatic Amines

These reactions are often mechanistically less clear-cut than those with aliphatic amines because both the amine and the aryl halide can potentially absorb light. The reactions of 9,10-dihaloanthracenes with aniline derivatives are unambiguous in this respect, since the use of long-wavelength UV radiation ensures light absorption by the haloarene alone. Photoreduction to the 9-haloanthracene is observed for both X = Cl and X = Br, and the excellent agreement between K_{SV} and K_{ϕ} for PhNH$_2$, PhNHCH$_3$, PhN(CH$_3$)$_2$, and PhNEt$_2$ strongly supports reduction through a singlet exciplex.[31]

Dehalogenation of halobenzenes[32,33] and chlorobiphenyls[34] has been studied under conditions where the radiation was initially absorbed by an *N,N*-dialkylaniline. The aryl halide quenched the fluorescence of the aniline, and in the halobenzene series there was reasonable agreement between K_{SV} and K_{ϕ}. In methanolic solution, the halobenzenes yielded benzene and *o*- (and *p*-)-phenyl-*N,N*-dimethylaniline. Isotopic substitution of the amine (PhN(CD$_3$)$_2$) afforded benzene-d_1 (66%, X = Cl; 42%, X = Br; 12%, X = I). The d_1-product is the result of proton transfer from PhN(CD$_3$)$_2$ within the solvent cage. The photoreaction between *m*-chloronitrobenzene and *N,N*-dimethylaniline is also an electron-transfer process, but in this case it is the nitro group that is reduced, rather than the chlorine.[35]

Miscellaneous Electron Transfer Processes

Like amines, dienes quench many arene singlet and triplet states, a process attributed to the formation of diene/arene exciplexes; these have a much smaller degree of electron transfer than the corresponding arene/amine exciplexes.[36] Ambiguity arises because dienes are used commonly as triplet quenchers, and because simple arenes such as benzenes, naphthalenes, and biphenyls absorb at the same wavelengths as dienes.[34]

9,10-Dichloroanthracene absorbs at longer wavelength than simple dienes. It reacts photochemically with 2,5-dimethyl-2,4-hexadiene by two pathways: in nonpolar solvent, (benzene) a photocycloadduct forms by way of a singlet exciplex,[37] whereas in polar solvents reduction to 9-chloroanthracene is observed and a triplex (C$_{14}$H$_8$Cl$_2$·2DMH) is implicated.[36] Experiments with added D$_2$O show that in polar solvents, reduction involves transfer of a proton to the exciplex rather than loss of chloride ion.

The dehalogenation of aryl halides has been shown to be accelerated by aliphatic sulfides,[38] and an electron-transfer mechanism is implicated by the incorporation of deuterium in the reduction product when the reaction is carried out in the presence of D$_2$O.

The dechlorination of pentachlorobenzene has been studied in micelles of hexadecyltrimethyl-ammonium bromide in water.[39] The predominant reaction at high [C$_6$HCl$_5$] microconcentrations is an excimer-assisted dechlorination, deduced from the relationship ϕ_r^{-1} vs. [C$_6$HCl$_5$]$^{-1}$. At low concentrations, a triplet monomer pathway intervenes. By-products of the reaction include bromotetrachlorobenzenes, which are postulated to arise through trapping the cationic partner of the excimer by Br^-.

Radical Anion Chain Reactions

These reactions are relevant to this review in that they may be photoinduced, but as radical chain processes, many of them can be initiated in other ways. Radical chain dehalogenation of aryl halides has been reviewed recently.[40] The chain propagation sequence for the deiodination of aryl iodides with CH_3O^-/CH_3OH is shown in Scheme 86.2.

$$Ar^{\cdot} + CH_3O^- \rightarrow Ar\text{-}H + CH_2O^{\cdot-}$$

$$CH_2O^{\cdot-} + Ar\text{-}I \rightarrow CH_2\!\uparrow\!O + \left(ArI\right)^{\cdot-}$$

$$\left(ArI\right)^{\cdot-} \rightarrow Ar^{\cdot} + I^-$$

SCHEME 86.2

The species $(ArI)^{\cdot-}$ is very short-lived and deiodinates efficiently. Reactivity diminishes in the order $ArI > ArBr > ArCl$ and is enhanced in cases where there is relief of steric strain. Photochemical dehalogenations of aryl halides with AlH_4^- and BH_4^- have been proposed to follow similar radical chain mechanisms. Evidence for electron transfer from BH_4^- to $ArCl$ has been presented by Freeman and Ramnath.[41]

Several aryl halides have been reported to dehalogenate efficiently with the combination 2-propanol/OH^-. The proposed mechanism[42] is closely similar to Scheme 86.2, with hydroxide ion playing the key role of deprotonating the hydrogen abstracted product of 2-propanol: Scheme 86.2a.

$$ArX^{\cdot-} \rightarrow Ar^{\cdot} + X^-$$

$$Ar^{\cdot} + \left(CH_3\right)_2 CHOH \rightarrow ArH + \left(CH_3\right)_2 C^{\cdot}OH$$

$$\left(CH_3\right)_2 C^{\cdot}OH + OH^- \rightarrow \left(CH_3\right)_2 C\text{-}O^{\cdot-} + H_2O$$

$$\left(CH_3\right)_2 C\text{-}O^{\cdot-} + ArX \rightarrow ArX^{\cdot-} + \left(CH_3\right)_2 C\!\uparrow\!O$$

SCHEME 86.2a

Nucleophilic substitution of halogen by the $S_{RN}1$ mechanism follows a related radical chain pathway (Scheme 86.3), in which the aryl radical $Ar^{\cdot}$ attacks a nucleophile Nu^- rather than abstracting hydrogen. The solvent should not contain easily abstractable hydrogen atoms, in order to depress photodehalogenation.

$$\left(ArX\right)^{\cdot-} \rightarrow Ar^{\cdot} + X^-$$

$$Ar^{\cdot} + Nu^- \rightarrow \left(ArNu\right)^{\cdot-}$$

$$\left(ArNu\right)^{\cdot-} + ArX \rightarrow Ar\text{-}Nu + \left(ArX\right)^{\cdot-}$$

SCHEME 86.3

86.4 Reactions Proceeding via Nucleophilic Substitution

Photochemical nucleophilic displacements have been studied extensively, and much of this work is documented in the exhaustive review by Cornelisse and Havinga.[43] The most detailed studies have been made of methoxynitroarenes, among which it is well established that nucleophilic displacement occurs most readily *meta* to NO_2. Of the various halide nucleofuges, only fluoride appears to behave analogously to methoxy. Thus, *m*-fluoronitrobenzene reacts photochemically with OH^- to give *m*-nitrophenol, but the reaction is not observed for *m*-chloro- or *m*-bromonitrobenzenes. With oxidizable "nucleophiles" such as secondary amines, photoreduction of the nitro group takes place in preference to substitution of halide. However, 1-fluoro-3-nitronaphthalene and other "extended *meta*"-fluoronitronaphthalenes undergo replacement of fluorine upon irradiation with OH^-, OCH_3^-, CN^-, and CH_3NH_2.[44]

Photochemical replacement of a substituent by a nucleophile occurs *o*- and *p*- to electron-donating substituents.[45] These reactions are successful for halides as nucleofuges. An example is the conversion of *p*-chloroanisole to *p*-methoxybenzonitrile upon irradiation with cyanide ion. In the case of the isomeric chlorophenols and chloroanisoles, irradiation with alcohols involves a mixture of substitution (replacement of -Cl by -OR) and dehalogenation.[43] The outcomes of these reactions may be highly solvent dependent, as shown by studies on the replacement of fluoride by CN^- and OH^-, when *o*- and *p*-fluoroanisoles are photolyzed with KCN in water/*t*-butyl alcohol mixtures.[46]

Chlorobenzene is converted to phenol in aqueous solution, in a quantum yield comparable with that of dehalogenation.[8] Evidence has been presented that the corresponding photohydrolyses of the monochlorobiphenyls invlove homolysis, followed immediately by in-cage electron transfer. The aryl cation is trapped by water, but can be diverted back to reactant in the presence of added Cl^-.[47]

$$^3Ar\text{-}Cl \rightarrow (Ar^. .Cl^.) \rightarrow (Ar^+ .Cl^-) \rightarrow \ \rightarrow ArOH$$

Photohydrolysis can therefore be viewed as a variant of photohomolysis. In an organic solvent such as hexane, the radical pair ($Ar^. .Cl^.$) separates and then abstracts hydrogen. In water, there are no abstractable hydrogens, and the high dielectric constant permits electron transfer to occur. An exceptional aspect of the reaction is that both 3- and 4-chlorobiphenyls undergo photoisomerization in parallel with photohydrolysis. This is suggested to proceed by way of valence photoisomerization. Related valence photoisomerizations have been observed occasionally upon photolysis of polychlorobenzenes; for example, photolyses of both 1,2,3- and 1,3,5-trichlorobenzenes in aq. Ch_3CN afford up to 10% of 1,2,4-trichlorobenzene in addition to dechlorination products.[6]

References

1. Sammes, P. G., Photochemistry of the C-X group, in *The Chemistry of the Carbon-Halogen Bond, Part II,* Patai, S., Ed., Wiley, New York, 1973, chap. 11.
2. Grimshaw, J. and de Silva, A. P., Photochemistry and photocyclization of aryl halides, *Chem. Soc. Rev.,* 10, 181, 1981.
3. Lodder, G., Recent advances in the photochemistry of the carbon-halogen bond, in *The Chemistry of Functional Groups, Supplement D,* Patai, S., and Rapoport, Z., Eds., Wiley, New York, chap. 29.
4. Davidson, R. S., Goodin, J. W., and Kemp, G., The photochemistry of aryl halides and related compounds, in *Advances in Physical Organic Chemistry,* Vol. 20, Academic Press, London, 1984, 191.
5. Choudhry, G. G., Webster, G. R. B., and Hutzinger, O., Environmental aquatic photochemistry of chlorinated aromatic pollutants, *Toxicol. Environ. Chem.,* 17, 267, 1988.
6. Dulin, D., Drossman, H., and Mill, T., Products and quantum yields for photolysis of chloroaromatics in water, *Environ. Sci. Technol.,* 20, 72, 1986.

7. Choudhry, G. G., Webster, G. R. B., and Hutzinger, O., Environmentally significant photochemistry of chlorinated benzenes and their derivatives in aquatic systems, *Toxicol. Environ. Chem.*, 13, 27, 1986.

8. Bunce, N. J., Landers, J. P., Langshaw, J.-A., and Nakai, J. S., An assessment of the importance of direct solar degradation of some simple chlorinated benzenes and biphenyls in the vapor phase, *Environ. Sci. Technol.*, 23, 213, 1989.

9. Mansour, M., Parlar, H., and Korte, F., Photoinduzierte deuterierung monosubstituierter dichlorobenzole, *Chemosphere*, 9, 59, 1980.

10. Ichimura, T. and Mori, Y., Photolysis of monochlorobenzene in gas phase, *J. Chem. Phys.*, 58, 288, 1973.

11. Choudhry, G. G. and Hutzinger, O., Photochemical formation and degradation of polychlorinated dibenzofurans and dibenzo-*p*-dioxins, *Residue Rev.*, 84, 113, 1982.

12. Exner, J. H., Johnson, J. D., Ivins, O. D., Wass, M. N., and Miller, R. A., Detoxication of chlorinated dioxins, in *Detoxication of Hazardous Waste*, Exner, J. H., Ed., Ann Arbor Science, Ann Arbor, MI, 1982, chap. 17.

13. Freeman, P. K., Jang, J.-S., and Ramnath, N., The photochemistry of polyhaloarenes. 10. The photochemistry of 4-bromobiphenyl, *J. Org. Chem.*, 56, 6072, 1991.

14. Bunce, N. J., Bergsma, J. P., Bergsma, M. D., De Graaf, W., Kumar, Y., and Ravanal, L., Structure and mechanism in the photoreduction of aryl chlorides in alkane solvents, *J. Org. Chem.*, 45, 3708, 1980.

15. Freeman, P. K., Ramnath, N., and Richardson, A. D., Photochemistry of polyhalobenzenes. 8. The photodechlorination of pentachlorobenzene, *J. Org. Chem.*, 56, 3643, 1991.

16. Nowakowska, M., Sustar, E., and Guillet, J. E., Studies of the antenna effect in polymer molecules. 23. Photosensitized dechlorination of 2,2′,4,4′,6,6′-hexachlorobiphenyl solubilized in an aqueous solution of poly(sodium styrenesulfonic acid-*co*-2-vinylnaphthalene), *J. Am. Chem. Soc.*, 113, 253, 1991.

17. Kulis, Y. Y., Poletaeva, I.-Y., and Kuz'min, M. G., Photochemical reactions of halosubstitution by hydrogen in aryl halides by activity of nucleophilic reagents, *J. Org. Chem. USSR (Engl. trans.)*, 9, 1242, 1973.

18. Soloveichik, O. M. and Ivanov, V. L., Photochemical elimination in dihalo substituted anthracenes in presence of aliphatic amines, *J. Org. Chem. USSR (Engl. trans.)*, 10, 2416, 1974.

19. Yang, N. C. and Libman, J., Chemistry of exciplexes, photochemical addition of secondary amines to anthracene, *J. Am. Chem. Soc.*, 95, 5783, 1973.

20. Ohashi, M., Tsujimoto, K., and Seki, K., Photoreduction of 4-chlorobiphenyl by aliphatic amines, *J. Chem. Soc. Chem. Commun.*, 384, 1973.

21. Bunce, N. J., Photolysis of aryl chlorides with aliphatic amines, *J. Org. Chem.*, 47, 1948, 1982.

22. Beecroft, R. A., Davidson, R. S., and Goodwin, D. C., The amine assisted photo-dehalogenation of halo-aromatic compounds, *Tetrahedron Lett.*, 24, 5673, 1983.

23. Tanaka, Y., Uryu, T., Ohashi, M., and Tsujimoto, K., Dechlorination of 4-chlorobiphenyl mediated by aromatic photocatalysts, *J. Chem. Soc. Chem. Commun.*, 1703, 1987.

24. Davidson, R. S. and Goodin, J. W., Mechanistic aspects of the triethylamine assisted photo-induced dehalogenation of halogeno-aromatic compounds, *Tetrahedron Lett.*, 22, 163, 1981.

25. Lemmetyinen, H., Ovaskainen, R., Nieminen, K., and Sychtchivova, I., Photolysis of pyrene and chloropyrene in the presence of triethylamine in acetonitrile: dehalogenation assisted by potassium cyanide, *J. Chem. Soc. Perkin Trans II*, 113, 1992.

26. Bunce, N. J., Hayes, P. J., and Lemke, M. E., Photolysis of polychlorinated benzenes in cyclohexane solution, *Can. J. Chem.*, 61, 1103, 1983.

27. Freeman, P. K., Srinivasa, R., Campbell, J.-A., and Deinzer, M. L., The photochemistry of polyhaloarenes. 5. Fragmentation pathways in polychlorobenzene radical anions, *J. Am. Chem. Soc.*, 108, 5531, 1986.

28. Freeman, P. K. and Ramnath, N., Photochemistry of polyhaloarenes. 9. Characterization of the radical anion intermediate in the photodehalogenation of polyhalobenzenes, *J. Org. Chem.*, 56, 3646, 1991.

29. Freeman, P. K. and Srinivasa, R., Photochemistry of polyhaloarenes. 6. The fragmentation of polyfluoroarene radical anions, *J. Org. Chem.*, 52, 252, 1987.

30. Gilbert, A. and Krestonosich, S., Excited state substitution and addition reactions of aryl fluorides with aliphatic amines, *J. Chem. Soc., Perkin Trans. 1*, 1393, 1980.

31. Soloveichik, O. M., Ivanov, V. L., and Kuz'min, M. G., Mechanism of photosubstitution of halogen by hydrogen in dihalo-substituted anthracene by activity of amines, *J. Org. Chem. USSR (Engl. trans.)*, 12, 860, 1976.

32. Pac, C., Tosa, T., and Sakurai, H., Photochemical reactions of aromatic compounds. IX. Photochemical reactions of dimethylaniline with halobenzenes, *Bull. Chem. Soc. Jpn.*, 45, 1169, 1972.

33. Grodowski, M. and Latowski, T., A study on a photochemical reaction in the system *N,N*-dimethylaniline-bromobenzene, *Tetrahedron*, 30, 767, 1974.

34. Bunce, N. J. and Gallacher, J. C., Photolysis of aryl chlorides with dienes and with aromatic amines, *J. Org. Chem.*, 47, 1955, 1982.

35. Döpp, D. and Heuber, J., *N*-Demethylation of *N,N*-dimethylaniline by photoexcited 3-nitrochlorobenzene, *Tetrahedron Lett.*, 23, 1553, 1982.

36. Yang, N. C., Yates, R. L., Masnovi, J., Shold, D. M., and Chiang, W., Chemistry of exciplexes. Photocycloadditions of anthracenes to conjugated polyenes, *Pure Appl. Chem.*, 51, 173, 1979.

37. Smothers, W. K., Schanze, K. S., and Saltiel, J., Concerning the diene-induced photodechlorination of chloroaromatics, *J. Am. Chem. Soc.*, 101, 1895, 1979.

38. Beecroft, R. A., Davidson, R. S., Goodwin, D., and Pratt, J. E., Quenching of singlet and triplet excited aromatic hydrocarbons by sulphides: The amine and sulphide enhanced photo-induced degradation of chloro- and cyanoaromatic hydrocarbons, *Tetrahedron*, 40, 4487, 1984.

39. Freeman, P. K. and Lee, Y.-S., Photochemistry of polyhaloarenes. 12. The photochemistry of pentachlorobenzene in micellar media. *J. Org. Chem.*, 57, 2846, 1992.

40. Bunnett, J. F., Radical chain, electron-transfer dehalogenation reactions, *Acc. Chem. Res.*, 25, 2, 1992.

41. Freeman, P. K. and Ramnath, N., Photochemistry of polyhaloarenes. 7. Photodechlorination of pentachlorobenzene in the presence of sodium borohydride, *J. Org. Chem.*, 53, 148, 1988.

42. Nishiwaki, T., Usai, M., Anda, K., and Hida, M., Dechlorination of polychlorinated biphenyls by UV-irradiation. V. Reaction of 2,4,6-trichlorobiphenyl in neutral and alkaline alcoholic solution, *Bull. Chem. Soc. Jpn.*, 52, 821, 1979.

43. Cornelisse, J. and Havinga, E., Photosubstitution reactions of aromatic compounds, *Chem. Rev.*, 75, 353, 1975.

44. Lammers, J. G. and Cornelisse, J., Photoreactions of aromatic compounds. 37. Photosubstitution reactions of nitronaphthalene derivatives, *Isr. J. Chem.*, 16, 299, 1977.

45. Ivanov, V. L. and Eggert, L., Halogen substitution by sulfite ion in 1-hydroxy-4-chloronaphthalene, *J. Org. Chem. USSR (Engl. trans.)*, 19, 2075, 1984.

46. Liu, J. H. and Weiss, R. G., Anomalous effects during aromatic photosubstitutions of 2- and 4-fluoroanisoles in solvent mixtures of water and *tert*-butyl alcohol, *J. Org. Chem.*, 50, 3655, 1985.

47. Orvis, J., Weiss, J., and Pagni, R. M., Further studies on the photoisomerization and hydrolysis of chlorobiphenyls in water. Common ion effect in the photohydrolysis of 4-chlorobiphenyl, *J. Org. Chem.*, 56, 1851, 1991.

87

Photoisomerization Reactions of Chlorinated Hydrocarbon Insecticides

Harun Parlar
University of Kassel

87.1 Introduction

The isomerization reactions of the chlorinated hydrocarbon insecticides, which react from their excited triplet state to isomerization products in high quantum yields, are almost entirely confined to the cyclodiene insecticides.

87.2 Isomerization Reactions of Cyclodiene Insecticides

[2+2]-Photocycloadditions

The [2+2]-photocycloadditions of the cyclodiene insecticides and their derivatives are restricted to those compounds that possess a double bond in an *endo*-position of the ring system in the nonchlorinated part of the molecule, e.g., the chlordenes. Both direct and sensitized irradiation lead to cage isomers in very good yields (Figure 87.1).[1-3]

The C-C distances (C5-C8 = 2.63 Å; C4-C9 = 2.83 Å) taken from molecular models make it extremely likely that the first step of the reaction takes place at the C5-C8 bond. The intermediate biradicals (1′–6′) combine to yield the final products 1a to 6a. In the literature there are reports of determinations of the dependence of quantum yield of the formation of photoproducts on various triplet sensitizers in order to determine the triplet energy of the starting compounds. From these reports, it can be concluded that the triplet energy of the reaction lies between 260 and 278 kJ mol^{-1}. Quenching experiments with *trans*-piperylene (E_T = 247 kJ mol^{-1}) have confirmed these results and yield a value for the rate constant for rearrangement of the triplet (K_r) of 1.40×10^7 s^{-1} on the basis of the Stern-Vollmer plot.

Isodrin (7), that possesses a hexahydro-1,4-*endo,endo*-5,8-dimethanonaphthalene skeleton, is also converted to a photocycloaddition product both on direct irradiation and on sensitization (7a). Sensitization can be carried out with benzophenone and this produces the rearrangement products in almost 95% yield (Figure 87.2). The quantum yield of the reaction is relatively high at 2×10^{-2}.[4]

0-8493-8634-9/95/$0.00+$.50
© 1995 by CRC Press, Inc.

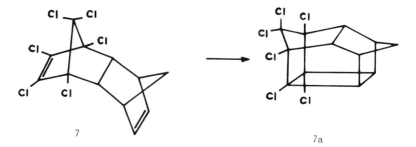

Insecticide	R_1	R_2	R_3	Isomerization product	Triplet sensitizer	Yield* in %	Ref.
Chlordene (1)	H	H	H	Photochlordene (1a)	Acetone	75	1,2
Heptachlor (2)	Cl	H	H	Photoheptachlor (2a)	Acetone	70	1,3,4
1-*exo*-Hydroxychlordene (3)	OH	H	H	Photohydroxychlordene (3a)	Acetone	15	1,5
1-*exo*-Methoxychlordene (4)	CH₃	H	H	Photomethoxychlordene (4a)	Acetone/acetophenone	45	1,6
1-*endo*-Hydroxychlordene (5)	H	OH	H	Photoendohydroxychlordene (5a)	Acetone	30	1
Isoheptachlor (6)	H	H	Cl	Photoisoheptachlor (6a)	Acetone/acetophenone	60	1

* 2×10^{-2} mol solution; $\lambda > 290$ nm, HPK 125 Philips; irradiation time 4 h.

FIGURE 87.1   [2+2]-Photocycloaddition of chlordene derivatives (1–6).

FIGURE 87.2   [2+2]-Photocycloaddition of the cyclodiene insecticide isodrin (7).

[πσ → 2σ] Reactions

This photoinduced reaction, that can be described formally as a (1,3)-sigmatropic proton displace-ment, is generally described as a (πσ → 2σ) reaction, since it is mechanistically different from a synchronous process. Here a π- and a σ-bond interact and form two new σ-bonds, by means of an intramolecular reaction. This has been confirmed by the irradiation of some compounds with acetone-d_6. It was established that the isomerization products did not contain any deuterium.

The (πσ → 2σ) reactions of cyclodiene insecticides[4-11] are limited to those compounds possess-ing a sterically favorable hydrogen atom opposite the chlorinated double bond. Thus, the hexachlorooctahydro-1,4-*endo*-5,8-dimethano-naphthalenes, which include the well-known in-secticides endrin (8) and dihydro-isodrin (9), react on triplet sensitization to yield the rearrange-ment products 8a and 9a (Figure 87.3).[4] As can be seen from Figure 87.4, the triplet energy here is also in the range 260 to 278 kJ mol^{-1}.

In contrast to compounds 8 and 9, which possess a rigid, nonchlorinated norbornane group, the five-membered ring of the polychlorotetrahydromethanoindanes (10 to 13) possesses restricted mobility, particularly at carbon atoms C1-C3. This fact expresses itself in the quantum yield for these compounds, which are considerably less than 10^{-3}. Figure 87.5 lists the (πσ → 2σ) reactions of compounds 10 to 13.

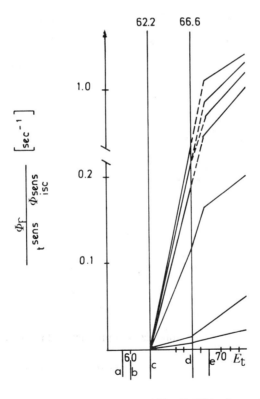

isecticide	R_1	R_2	R_3	R_4	Isomerization product	Triplet sensitizer	Yield* in %	Ref.
ndrin (8)			H	H	Photoendrinketone (8a)	Acetone	80	1
ihydroisodrin (9)	H	H	H	H	Photodihydroisodrin (9a)	Acetophenone	45	1,2

*2×10^{-2} mol solution; $\lambda > 290$ nm, HPK 125 Philips; irradiation time 2 h.

FIGURE 87.3 ($\pi\sigma \to 2\sigma$) Reactions of endrin (8) and dihydroisodrin (9).

FIGURE 87.4 ($\pi\sigma \to 2\sigma$) Reactions of cyclodiene insecticides (9–15) in dioxane in the presence of various triplet sensitizers (Φ = quantum yield for the formation of photoisomerization products 8a and 9a; Φ_{isc}^{sens} = quantum yield of the sensitizer in the intersystem crossing. [a = diacetyl; b = phenanthrene; c = benzophenone; d = acetophenone; e = acetone.]

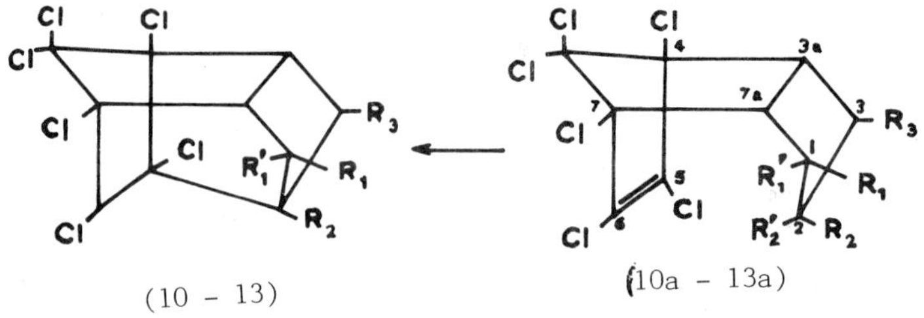

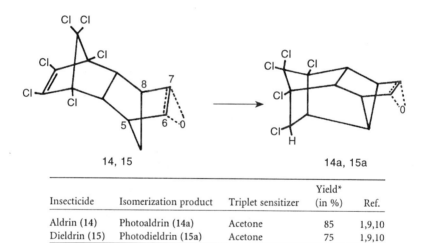

(10 – 13)   (10a – 13a)

Insecticide	R_1	R_1'	R_2	R_2'	R_3	Isomerization product	Triplet Sensitizer	Yield* in %	Ref.
β-Dihydroheptachlor (10)	H	H	Cl	H	H	Photo-β-dihydroheptachlor (10a)	Acetone	42	1,2
α-Chlordane (11)	H	Cl	H	Cl	H	Photo-α-chlordane (11a)	Acetone	30	1,7,8
β-Chlordane (12)	Cl	H	Cl	Cl	H	Photo-β-chlordane (12a)	Acetone	32	7,8
cis-Nonachlor (13)	Cl	Cl	H	H	Cl	Photo-cis-nonachlor (13a)	Acetone	30	7,8

* 2×10^{-2} mol solution; $\lambda > 290$ nm, HPK 125 Philips; irradiation time 6 h.

FIGURE 87.5   ($\pi\sigma \rightarrow 2\sigma$) Reaction of compounds 10 to 13.

14, 15   14a, 15a

Insecticide	Isomerization product	Triplet sensitizer	Yield* (in %)	Ref.
Aldrin (14)	Photoaldrin (14a)	Acetone	85	1,9,10
Dieldrin (15)	Photodieldrin (15a)	Acetone	75	1,9,10

* 2×10^{-2} mol solution; $\lambda > 290$ nm, HPK 125 Philips; irradiation time 5 h.

FIGURE 87.6   ($\pi\sigma \rightarrow 2\sigma$) Reaction of aldrin (14) and dieldrin (15).

The well-known cyclodiene insecticides aldrin (14) and dieldrin (15) (Figure 87.6) also react rapidly in a ($\pi\sigma \rightarrow 2\sigma$) reaction to yield their cage isomers. The introduction of the ethylene bridges between carbon atoms 5 and 8 alters the position of the C-9 carbon atom so that the steric conformation is very favorable for the ($\pi\sigma \rightarrow 2\sigma$) reaction. This in its turn determines the direction of the proton transfers. Here, the protons on carbon atoms C-5 and C-8, that usually compete successfully in the case of compounds 10 to 13, are excluded as potential reaction centers.[4]

Photoreversible and Photoirreversible Double Proton Transfer Reactions of Cyclodiene Insecticides[12,13]

Among group transfer reactions, many examples are known of synchronously occurring suprafacial displacements of two protons from saturated to unsaturated systems. A double group transfer is permitted in the ground state for m + n = 4 q + 2, and in the excited state for m + n = 4 q, where m and n are the numbers of electrons in the participating systems and q is an integer (0, 1, 2, etc.). The same rule also applies to reactions that are antrafacial with respect to one component. In the case of a concerted reaction, on the other hand, there is the possibility that the radical formed in the first stage of the reaction will be stabilized either by recombination or displacement of another proton with the formation of a new double bond (16,17). In principle, it can also be expected that the alkane-alkane pair that is so formed may yield the starting compounds in a reverse reaction.

The irradiation of the ester of 4,5,6,7,8,8-hexachloro-2,3,3a,4,7,7a-hexahydro-4,7-methano-1*H*-inden-1,3-dicarboxylic acids (16) in acetone at −70 °C exclusively yields dimethyl-4,5,6,7,8,8-hexachloro-3a,4,5,6,7,7a-hexahydro-4,7-methano-1*H*-inden-1,3-dicarboxylate (16a), which isomerizes back to the starting compound (16) at temperatures above −30 °C. This reversible reaction can be sensitized by acetone and by acetophenone. In contrast, benzophenone (E_T = 288 kJ mol^{-1}) is unable to accomplish the triplet energy transfer. From this, it can be concluded that the lowest energy triplet state for 16 or 16a must lie between 288 and 319 kJ mol^{-1}. The quantum yields for the formation of 16 and 16a measured in dioxane/acetophenone at 20 °C were $\Phi_1 = 1.98 \times 10^{-2}$ and $\Phi_2 = 0.04 \times 10^{-2}$, respectively, and make it clear that the photoequilibrium is displaced in favor of compound 16a (Figure 87.7).[4,12]

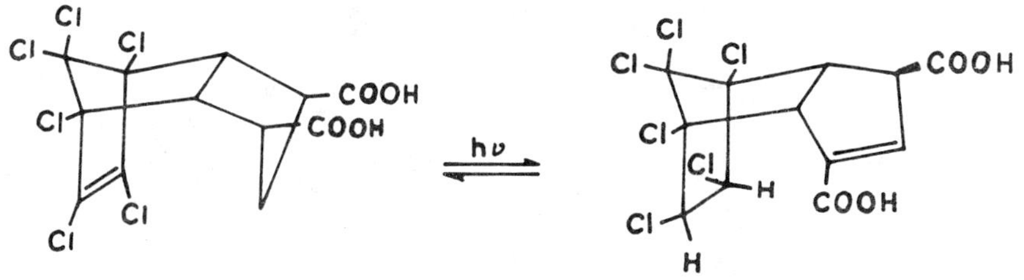

FIGURE 87.7 Photoinduced double proton transfer reactions of aldrindicarboxylic acid (16) in acetone at wavelengths above 290 nm.

A special variant of the double transfer reaction was discovered by irradiation of epoxides 17 and 18 (Figure 87.8). In the first step of the reaction, two protons are transferred leading to an epoxide-ketone rearrangement. The α,β-unsaturated ketones 17a and 18a produced by this are then rearranged to the bridged ketones 17b and 18b in a ($\pi\sigma \rightarrow 2\sigma$) reaction.

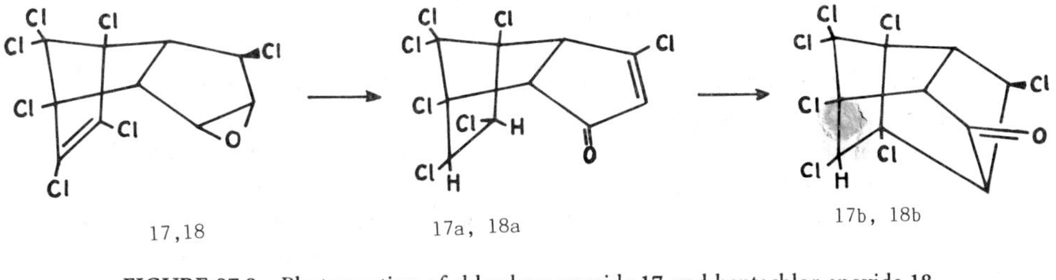

FIGURE 87.8 Photoreaction of chlordene epoxide 17 and heptachlor epoxide 18.

The conversion of epoxides 17 and 18 to the double proton transfer products can be sensitized by acetophenone, whose triplet energy does not, however, suffice for the second reaction. The proton transfer reaction can then be carried out afterwards using acetone as sensitizer. This result indicates the possibility of producing such intermediate products alone by the use of suitable sensitizers. This reaction also makes it possible to synthesize the sterically uniform proton transfer products.

Photo-ene Reactions[4,14]

In contrast to thermally induced ene-reactions, which are some of the best known addition reactions in organic chemistry, there have been only a few reports on photo-ene reactions. Here, as a result of the dominance of [2+2]-cycloadditions, intramolecular photo-ene reactions are less probable than the intermolecular processes. Especially those compounds possessing relatively strain-free and geometrically semiimmobilized π-bonds are suited to the study of such competitive reactions. These include the preparatively readily accessible hexachloro-*endo*-dicyclopentadienes. The literature on the photochemical behavior of these compounds comprises descriptions of [2+2]-cycloadditions leading to cage isomers. However, it has been demonstrated, using the compounds 1 to 4 and 19 to 25 as an example that, after the introduction of suitable substituents at the 1-*exo*-position of hexachloro-*endo*-dicyclopentadiene, a photo-ene reaction can compete successfully with the [2+2]-cycloaddition (Figure 87.9).

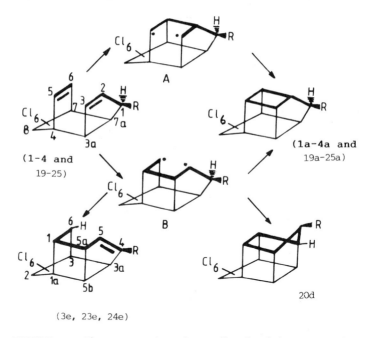

FIGURE 87.9 Photo-ene reactions of some selected cyclodiene insecticides.

The photoisomers 1a to 4a and 19a to 24a, whose production can be attributed to the formation of the biradical intermediates A and B, are not the only possible reaction products. After the sensitized excitation of compounds 3, 23, and 24, it was possible to isolate the ene adducts 3e, 23e, and 24e, which were produced from the biradical intermediate B by abstraction of the sterically favorable *endo*-hydrogen atom followed by recombination of the newly formed radicals. The irradiation experiments carried out with 1-*endo*-*d*-labeled 3 confirmed the strictly intramolecular character of the photo-ene reaction. Thus, after irradiation, deuterium was found to be in the expected 6-*endo*-position of compound 3e. The intramolecular character of the reaction was also confirmed by irradiation of 3e and 23e in acetone-d_6. The investigations that followed revealed clearly that no deuterium had been incorporated in the final product.

The results also indicate that photoreactions of the starting compounds are primarily controlled by C3 → C5 coupling. This relative selectivity is attributed primarily to steric factors. The distances of 2.63 Å for C^3,C^5 and 2.98 Å for C^2,C^6 measured from molecular models emphasize the dependence of the photo process on geometry. The preferred formation of intermediate B is also confirmed by photolysis of **20**. The cage isomer **20a** can be produced via biradical B and then halogen displacement followed by recombination. The absence of the ene-reaction in the other compounds can be by the fact that because the primary biradical B does not possess suitable 1-*exo*-substituents and, hence, is incapable of forming a low-energy system. In contrast, this condition is fulfilled by **2** and **24** so that they react to yield the α,β-unsaturated compounds **23e** and **24e**. The same assumptions apply also to compound **3e**, that is best formulated for the purpose as an enol. The corresponding five-membered ring ketone is isolated as the final product of the photo-ene reaction.

eferences

1. Fischler, H. M. and Korte, F., Sensitized and unsensitized photoisomerization of cyclodiene insecticides, *Tetrahedron Lett.*, 2793, 1969.
2. Vollner, L., Parlar, H., Klein, W., and Korte, F., Ecological chemistry. XXXI. Photoreactions of technical chlordane components, *Tetrahedron*, 27, 501, 1971.
3. Parlar, H. and Korte, F., Reaktionsverhalten von Chlordan in Lösung und in der Gasphase bei UV-Bestrahlung, *Chemosphere*, 1, 125, 1972.
4. Parlar, H., Photochemische Reaktionen der Cyclodieninsectizide. Habilitation, TU Munich, 1980.
5. Knox, J. R., Khalifa, S., Ivie, G. W., and Casida, J. E., Characterization of the photoisomers from *cis*- and *trans*-chlordanes, *trans*-nonachlor, and heptachlor epoxide, *Tetrahedron*, 29, 3869, 1973.
6. Schmitzer, J., Zur Struktur der Photoisomere der Chlordane. Dissertation, TU Munich, 1975.
7. Nitz, S., Parlar, H., and Korte, F., Analytisches Verhalten von endo-Dieldrin und endo-Photodieldrin, *Chemosphere*, 3, 83, 1974.
8. Hustert, K., Parlar, H., and Korte, F., Zur Struktur von Polychlormethano-indenen, *Chemosphere*, 4, 381, 1975.
9. Parsons, A. M. and Moore, D. J., Some reactions of dieldrin and the proton magnetic resonance spectra of the products, *J. Chem. Soc.*, 2026, 1966.
10. Robinson, J., Richardson, A., Bush, B., and Elgar, K. E., Photoisomerization products of dieldrin, *Bull. Environ. Contam. Toxicol.*, 1, 127, 1966.
11. Parlar, H. and Korte, F., Photoreactions of cyclodiene insecticides under simulated environmental conditions, *Chemosphere*, 6, 665, 1977.
12. Parlar, H., Gäb, S., Lahaniatis, E. S., and Korte, F., Photoreversibler Wasserstofftransfer an verbrückten Chlorkohlenwasserstoffen als Konkurrenzschritt zur (πσ-2σ)-Reaktion, *Chem. Ber.*, 108, 3692, 1975.
13. Parlar, H., Mansour, M., and Gäb, S., Photoinduced irreversible hydrogen transfer in chlorinated tetrahydromethanoindanes, *Tetrahedron Lett.*, 1597, 1978.
14. Parlar, H., Intramolekulare Photo-en-Reaktion als Konkurrenzschritt zur (2+2)-Cycloaddition, *Tetrahedron Lett.*, 3885, 1978.

The Photostimulated $S_{RN}1$ Process: Reactions of Haloarenes with Enolates

René Beugelmans
Institut de Chimie des
Substances Naturelles, C.N.R.S.

88.1 The $S_{RN}1$ Process

Background

In 1970, Bunnett[1] proposed a radical chain mechanism to rationalize the unusual leaving group ability and the anomalous (in terms of predicted regioselectivity from a common aryne intermediate) substitution pattern of pairs of isomeric 5- and 6-halo-1,2,4-trimethylbenzene (X = I, Br, Cl) treated by K metal in liquid ammonia. He interpreted these facts by the coupling between an aryl radical Ar· with the amide anion NH_2^-. The term $S_{RN}1$, standing for Substitution Radical Nucleophilic unimolecular reaction, was coined by Bunnett and became widely accepted to designate the mechanism, the principle of which had been discovered earlier, simultaneously and independently by Kornblum[2] and Russell[3] in the field of nitroalkane chemistry. These authors have reviewed $S_{RN}1$ (Al)[4,5] and $S_{RN}1$ (Ar)[6] chemistry, while several reviews on this latter topic have been published by others.[7,8] Variously substituted haloarenes ArX were found to undergo replacement of the leaving group X^- by a wide range of nucleophiles. This chapter, after a brief introduction focused on the $S_{RN}1$ (Ar) mechanism, deals with reactions between haloarenes and enolates, the most important class of nucleophiles (others derived from S, Se, Te, P, As, and Sb[6,7,8] are beyond the scope of this chapter).

Mechanism

In 1973, Rossi and Bunnett[9] reported that the $S_{RN}1$ (Ar) mechanism, like the $S_{RN}1$ (Al), is triggered by light. Indeed, the acetone enolate anion under illumination by Pyrex-filtered UV light was found to behave as an electron donor toward the haloarene ArX and to produce the radical anion ArX·⁻ (Scheme 88.1, Equation 88.1).

0-8493-8634-9/95/$0.00+$.50

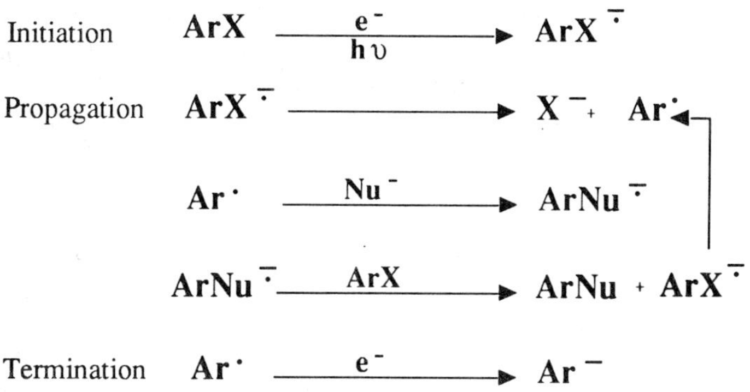

Initiation $\quad$ ArX $\xrightarrow[h\upsilon]{e^-}$ ArX$^{\overline{\cdot}}$

Propagation $\quad$ ArX$^{\overline{\cdot}}$ $\longrightarrow$ X$^-$ + Ar$^{\cdot}$

Ar$^{\cdot}$ $\xrightarrow{\text{Nu}^-}$ ArNu$^{\overline{\cdot}}$

ArNu$^{\overline{\cdot}}$ $\xrightarrow{\text{ArX}}$ ArNu + ArX$^{\overline{\cdot}}$

Termination $\quad$ Ar$^{\cdot}$ $\xrightarrow{e^-}$ Ar$^-$

SCHEME 88.1 The photostimulated $S_{RN}1$ chain process.

The monoelectronic transfer occurs within an excited Charge Transfer Complex (exciplex),[10] (ArX...Nu$^-$)*. Cases are known where an auxiliary nucleophile *Nu'$^-$*, a better electron donor toward ArX than Nu$^-$, is used in a catalytic amount to form the exciplex (ArX...*Nu'$^-$*)* essential for entrainment[11] of a reaction that would not occur solely with Nu$^-$. In the dark, the reaction does not occur, or proceeds very slowly[12] (except in "spontaneous" or "thermally activated" reactions with powerful nucleophiles[13] such as $^-$PPh$_2$), and the catalytic effect of light is therefore commonly used as a mechanistic probe supporting the $S_{RN}1$ process.

The radical anion ArX$^{\overline{\cdot}}$, a species with an odd number of electrons, undergoes fragmentation (Equation 88.2) to give X$^-$ and the aryl radical Ar$^{\cdot}$ whose single electron remains in a σ-orbital and is not delocalized over the orthogonal π-conjugated system. The next step of the process (Equation 88.3) is the *regiospecific* coupling of Ar$^{\cdot}$ with Nu$^-$ (at that stage playing its classical role) to give the radical anion ArNu$^{\overline{\cdot}}$. Catalytic amounts (1 to 10% molar) of electron scavenging substances R$^{\cdot}$, such as di-*t*-but-nitroxide or galvinoxyl which are persistent free radicals, retard the formation of ArNu$^{\overline{\cdot}}$ until R$^{\cdot}$ which combines with Ar$^{\cdot}$ faster than Nu$^-$ (Equation 88.3') is consumed (Scheme 88.2).

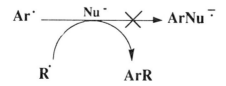

SCHEME 88.2 Radical scavenging.

The $S_{RN}1$ process is also retarded by oxygen, a paramagnetic molecule easily captured by Ar$^{\cdot}$ to give ArOO$^{\cdot}$ and ultimately ArOOH. Retardation by electron-scavenging substances or by a limited quantity of oxygen (usually, the mechanism does not start under air atmosphere) constitutes another classical test indicative of the radical chain process. In some cases, a second electron may be transferred to Ar$^{\cdot}$. This competitive reduction leading to Ar$^-$ (Equation 88.5), frequently encountered near the surface of the cathode[14] but also observed in homogeneous solution typical of photostimulated reactions,[15a] interrupts the radical chain process and decreases the rate and the yield of the reaction.

The radical anion ArNu$^{\overline{\cdot}}$, precursor of the final substitution product ArNu, is deactivated either by reversible cleavage to Ar$^{\cdot}$ and Nu$^-$ or by the bimolecular redox process with ArX (Equation 88.4), which produces simultaneously the neutral ArNu *and* ArX$^{\overline{\cdot}}$. The *concomitant* formation of these species sustains the chain process, clearly established from quantum yield measurements[15b] ($\phi > 1$) and from electrovoltametric studies by Saveant et al.,[16] who observed that a less than stoichiometric amount of electricity is sufficient to promote the reaction at the cathode.

The chain electron transfer is inhibited to various degree by strong electron acceptors like *m*- or *p*-dinitrobenzene (D.N.B.) that are reduced easily by ArX$^-$ or ArNu$^-$ to give the stable radical anion D.N.B.$^-$ (Equation 88.4') and so, the addition of a catalytic amount (1 to 10% molar) of those substances prevents the chain propagation (Scheme 88.3).

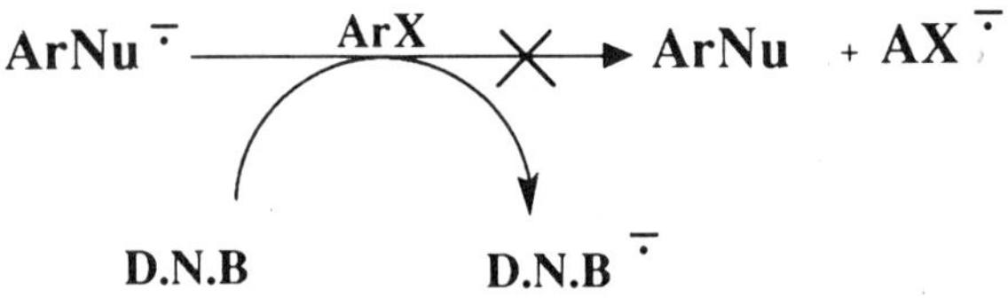

$$\text{ArNu}^{\cdot-} \xrightarrow{\quad \text{ArX} \quad} \!\!\!\times\!\!\!\rightarrow \text{ArNu} + \text{AX}^{\cdot-}$$

D.N.B   **D.N.B**$^{\cdot-}$

SCHEME 88.3   Electron trapping.

The three aforementioned tests — namely light catalysis, radical scavenging, and electron trapping — provide the organic chemist with a set of simple experimental results supporting the $S_{RN}1$ process.

Reactions

Scope

The original features of the mechanism described above bestow on the $S_{RN}1$ (Ar) reactions several advantages over the two classical aromatic substitution reactions. (1) $S_{RN}1$ reactions take place on *unactivated* haloarenes ArX (ArI > ArBr $\gg$ ArCl) and are *compatible* with many functional groups either electron withdrawing (E.W.G.) like CO_2R, COR, $CONH_2$ etc. (with the exception of NO_2) or electron releasing (E.R.G.) like alkyl, OR, NH_2, etc. This flexibility constitutes a saliant advantage over the SNAr mechanism, which does not tolerate E.R.G. and requires activation by strong E.W.G. (usually one to three NO_2 which cannot be removed readily if not desired in further synthetic steps). (2) $S_{RN}1$ reactions are *regiospecific* (no exception has hitherto been reported), while aryne-type reactions, whose synthetic scope is nevertheless broad,[17] often yield mixtures of regioisomers. (3) Some additional and valuable features are worthy of mention: (a) no or weak steric hindrance is observed to the approach of nucleophiles, even at *ortho*-substituted radical sites; (b) good to high yields and short reaction times are often experienced.

Experimental

Solvents. The solvent must (1) dissolve well both the haloarenes that are organic compounds and the nucleophiles that are salts; (2) be inert toward reactants or intermediates involved in the $S_{RN}1$ process and thus nitrobenzene, carbon tetrachloride (which readily accept electrons), or cyclohexane and methanol (which are good hydrogen donors to radicals) have to be discarded; (3) be of low acidity in order to avoid protonation of the nucleophile. Deoxygenation by bubbling an inert gas through the solvent for 10 to 15 min before illumination and an inert atmosphere for performing the reaction are recommended to prevent the aryl radical Ar$^{\cdot}$ being scavenged by O_2. Liquid ammonia at −33 °C or dimethylsulfoxide at room temperature fulfil all the above requirements and are commonly used in $S_{RN}1$ (Ar) chemistry.[7,18]

Photostimulation. Haloarene is mixed with the nucleophile produces a dark-colored C.T.C.; and a Pyrex vessel is thus convenient for illumination by the UV light (in the range of 300–400 nm) which induces the reaction. Notice that $S_{RN}1$ reactions triggered by light are encountered in the literature under various and equivalent headings such as "photocatalyzed", "photoinitiated", or "photostimulated".

38.2 Substitution by Ketone Enolate Anions

Halobenzenes

The photostimulated reaction between halobenzenes and the enolate anion of acetone was the first carbon-carbon bond-forming S$_{RN}$1 reaction reported by Bunnett et al.[9] Iodobenzene is consumed very rapidly to give a good yield of the *mono-β-aryl* ketone **1**, while bromobenzene reacts more slowly to give **1** along with a substantial quantity of the *bis-β* arylated ketone **2**. Diarylation is not a radical chain termination step but may decrease considerably the yield of those mono-β-aryl ketones whose enolate happens to be more nucleophilic toward Ar· than the initial Nu⁻.[15,19]

$$PhCH_2COCH_3 \qquad\qquad Ph_2CHCOCH_3$$

$$1 \qquad\qquad\qquad\qquad 2$$

Unsymmetrical ketones giving two isomeric enolate anions yield a mixture of isomeric products whose proportions reflect the concentration of enolates at equilibrium and the reactivity of each one toward Ar·.[20] Benzene is often formed in high yields from reactions with enolates of cyclic ketone or linearly branched ketones like diisopropyl ketone, which possess β-hydrogen atoms easily abstracted by Ar·.[15,21] Acetophenone enolate, initially reported not to react with bromo- or iodobenzene, even with acetone purposely added for entrainment, was later found by Semmelhack[21] to give a fair yield of substitution product when a different experimental technique (internal UV irradiation) and different conditions (threefold excess of base) were used.

Functionalized Halobenzenes

Shortly after the beginning of their investigations, Bunnett et al.[19] reported that randomly distributed substituents alkyl, CF$_3$, Ph, CO$_2$H, COPh, etc. were compatible with the S$_{RN}$1 mechanism and later, aiming at extending the scope of S$_{RN}$1 (Ar) chemistry, Beugelmans and co-workers undertook a systematic investigation of *o*-functionalized haloarenes as substrates. The topic has been briefly reviewed by this author,[22] and extensively by others.[23,24]

Synthesis of Five-Membered Heterocycles

Indole and Carbazole.   *ortho*-Iodoaniline,[25] predictably a better substrate than *o*-bromoaniline,[26] reacts with ketone enolates to give 2-substituted indoles **4** via the *o*-aminophenylketones **3**, which spontaneously cyclize. Enolate anions from properly protected α-dicarbonyl compounds such as pyruvaldehyde or biacetyl are good nucleophiles that lead to 2-CHO or 2-COCH$_3$ indole derivatives.[27] 1′, 2′, 3′, 4′-Tetrahydrocarbazole **5**[26,27] is obtained by reacting *o*-haloanilines with the cyclohexanone-derived anion, but the yield is only moderate due to competitive reduction of Ar·; and similarly, the carboline ring system **6** is obtained using the *N*-methyl-4-piperidone enolate as the nucleophile.

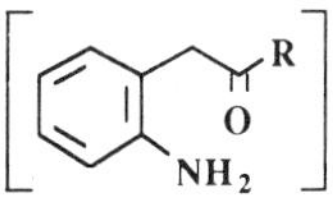

3

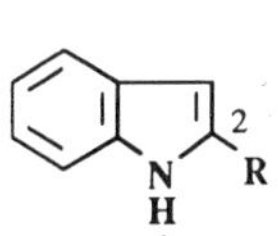

R = CH$_3$; i-C$_3$H$_7$; t-C$_4$H$_9$;

CH(OCH$_3$); C(OCH$_3$)$_2$CH$_3$

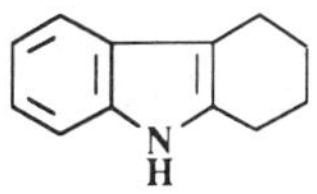

4

5

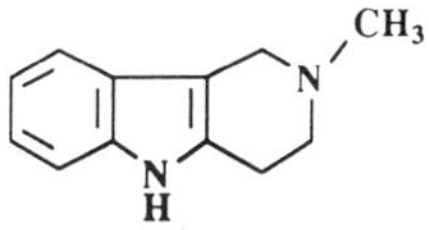

6

Benzofuran.   Oxygen heterocycles are obtained from protected *o*-halophenol reacting as substrate with various ketone enolates to give the *o*-methoxyaryl ketone **7** whose deprotection leads to 2-substituted benzofuran derivatives **8**.[28]

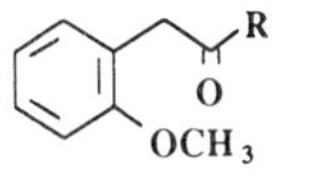

7   R = CH$_3$; i-C$_3$H$_7$; t-C$_4$H$_9$

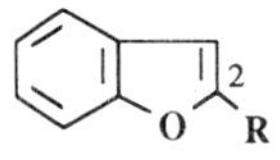

8

Synthesis of Six-Membered Heterocycles

Isoquinoline.   Reactions of *o*-halobenzylamine **9** (R = H) with ketone enolate anions performed by the author's group[29] afford **10** [P1] which leads via [P2] to **11** along with the unexpected 3-alkyl-4-hydroxyquinoline **12** (pathway a). That **12** results from oxidation of [P2] is evidenced by the reaction performed on **9** (R = CO$_2$Et) which gives the stable **10**, isolated and cyclized to **13** in quantitative yield (pathway b). By treating [P1] with palladized charcoal (pathway c), quinoline **11** is quantitatively obtained, while treatment with NaBH$_4$ under an inert atmosphere (pathway d) affords the tetrahydroquinoline **14**.[30] The convergent synthesis of the benzo[*c*]phenanthridine four-ring system is based upon the reaction of *o*-halobenzylamine with the enolate derived from appropriately substituted 1-tetralones[31] (Scheme 88.4). At variance with the model reactions with the acetone enolate, cyclization of **15** followed by clean and spontaneous aromatization of ring C occurs to give **16**, whose dehydrogenation gives the fully aromatic compounds **17**, immediate precursors of nitidine, fagaronine, and avicine alkaloids. The overall yields obtained by this approach (7 to 25%) are comparable to, or better than, those of multistep linear syntheses.

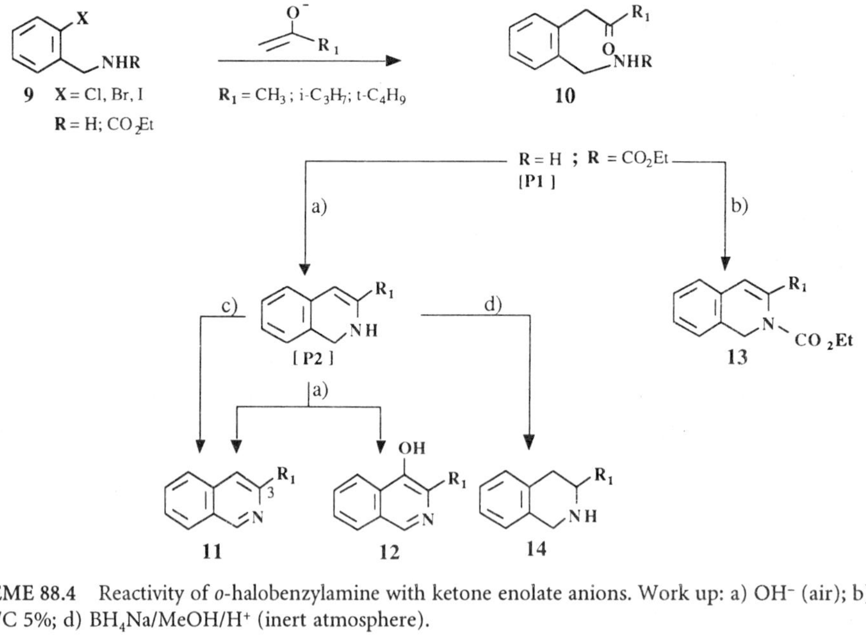

9   X = Cl, Br, I   R$_1$ = CH$_3$; i-C$_3$H$_7$; t-C$_4$H$_9$ 10
     R = H; CO$_2$Et

SCHEME 88.4   Reactivity of *o*-halobenzylamine with ketone enolate anions. Work up: a) OH⁻ (air); b) H⁺; c) Pd/C 5%; d) BH$_4$Na/MeOH/H⁺ (inert atmosphere).

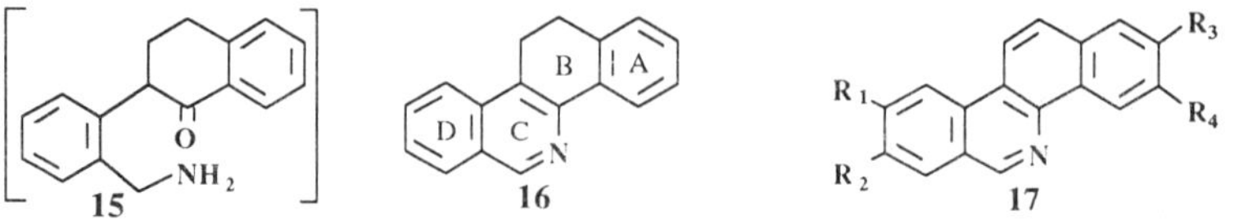

	R$_1$	R$_2$	R$_3$	R$_4$
Nitidine	OCH$_3$	OCH$_3$	-OCH$_2$O-	
Fagaronine	OCH$_3$	OCH$_3$	OH	OCH$_3$
Avicine	-OCH$_2$O-		-OCH$_2$O-	

Isoquinolone. The author and co-workers[32] found that *o*-halobenzamides **18** (R = H) are convenient substrates for $S_{RN}1$ reactions with acetone enolate anion to afford 3-alkylisoquinolones **19** after spontaneous cyclization of the primary product in the medium. A similar reaction carried out on **18** (R = CH_3) provides direct access to isoquinolone alkaloid precursors, in only moderate yield due to competitive reduction of Ar·. As no side reaction of that sort occurred in the model experiment and since the cyclized $S_{RN}1$ product obtained in high yield from primary halobenzamides was quantitatively *N*-alkylated, this two-step route to isoquinolone alkaloids[33] is preferred (Scheme 88.5).

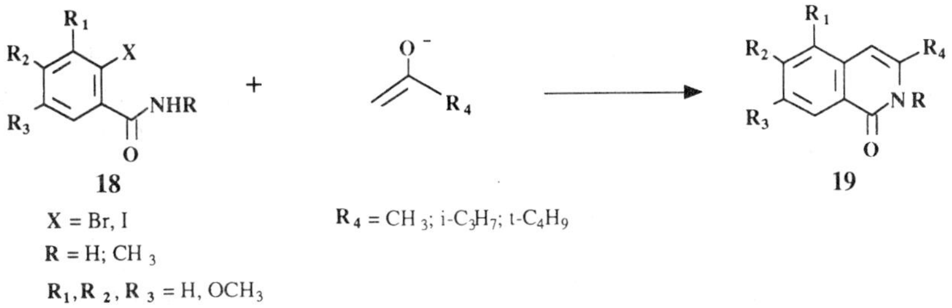
R1
R2
X
NHR
R3
O
18
+
O-
R4
R1
R2
R4
NR
R3
O
19
X = Br, I
R = H; CH 3
R1,R 2, R 3 = H, OCH3
R4 = CH 3; i-C3H7; t-C4H9

SCHEME 88.5 Synthesis of isoquinolones.

An extension of this reaction using the enolate of 2-acetylhomoveratic acid **20** as nucleophile gives access to 3-phenylisoquinolone **21**, whose two carbon side chain can be connected eventually either on the 4-position leading to benzo*[c]*-phenanthridone **22** or to the NH function forming the berbine skeleton **23**. The use of properly substituted substrates allows the synthesis of representatives of both groups of alkaloids.[34]

1- or 2-Naphthyl-3-isoquinolines **24** or **25** are synthesized similarly by reaction of *o*-bromobenzamide with 1- or 2-acetonaphthone-derived enolates.[35]

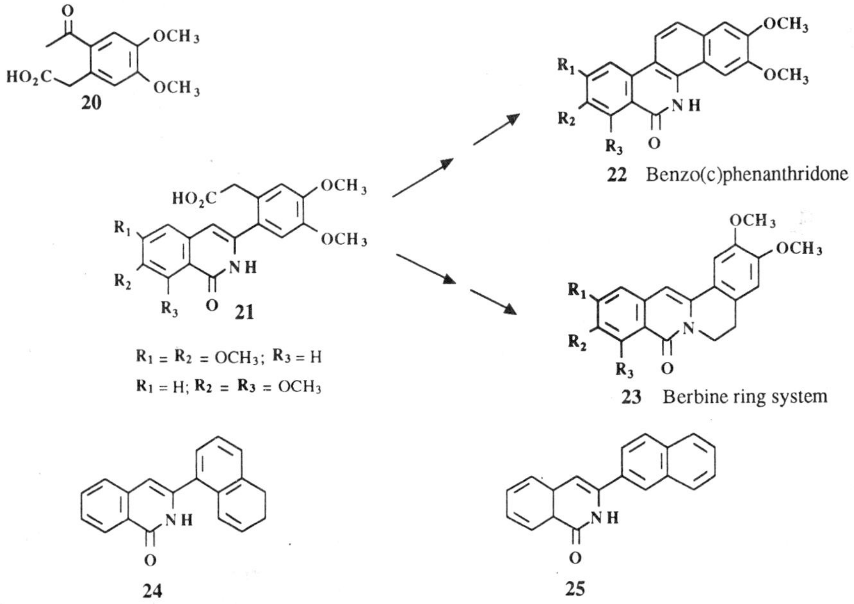
O
HO2C
OCH3
OCH3
20
HO2C
OCH3
OCH3
R1
R2
NH
R3
O
21
R1 = R2 = OCH3; R3 = H
R1 = H; R2 = R3 = OCH3
OCH3
OCH3
R1
R2
NH
R3
O
22 Benzo(c)phenanthridone
OCH3
OCH3
R1
R2
N
R3
O
23 Berbine ring system
NH
O
24
NH
O
25

Isocoumarone. The reaction between *o*-iodotrimethoxybenzoic acid with the acetone-derived enolate[33] leads to **26** and with the 1-tetralone-derived enolate[31] to **27**; **26** and **27** are precursors of isoquinolone or benzo[*c*]phenanthridone-type alkaloids.

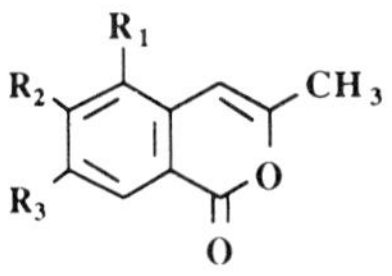

26 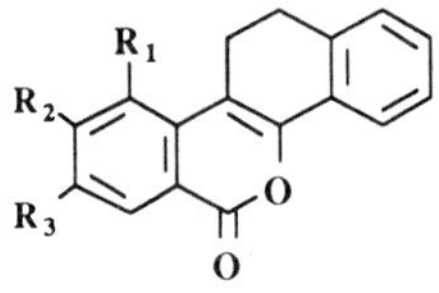   27

R₁, R₂, R₃ = H, OCH₃

Benzopyran. When *o*-, *m*- or *p*-iodo α,α,α-trifluorotoluene is reacted with acetone enolate, only the expected *m*- or *p*-substituted products are obtained, while the *o*-isomer **28** gives the benzo[*c*]pyran **30** together with the expected product **29** that can be converted to **30** by treatment with base in the dark. The latter is the sole product arising from the $S_{RN}1$ reaction performed with a larger excess of nucleophile[36] (Scheme 88.6).

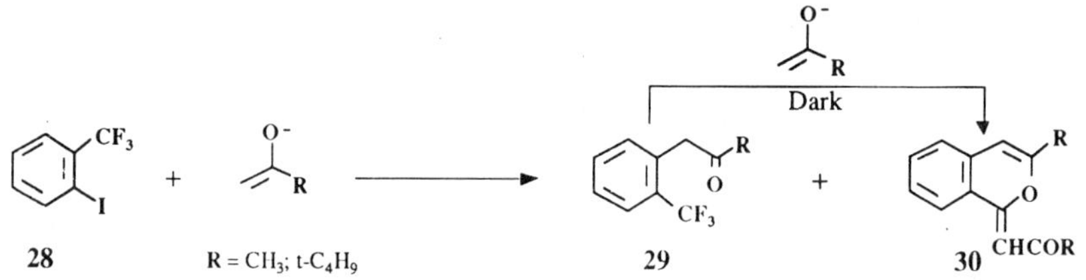

28   R = CH₃; t-C₄H₉   29   30

SCHEME 88.6   Reaction of *o*-iodotrifluorotoluene.

Synthesis of Seven-Membered Heterocycles

The reaction of various ketone enolate anions with *o*-iodohomoveratric acid gives a high yield of the keto acid **31** from which classical manipulations lead to 3-benzoxepines **32** and to 3-benzazepines **33**.[37]

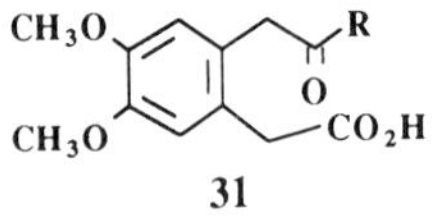

31 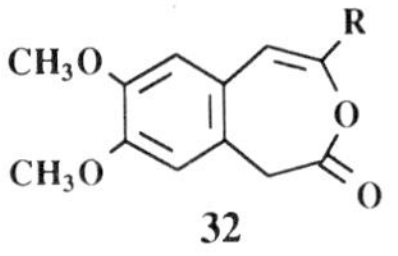   32 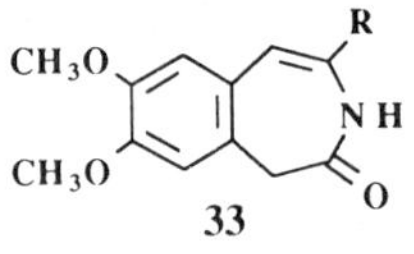   33

Synthesis of Binaphthyl Derivatives

The synthesis of unsymmetrically substituted 2,2′-binaphthyl derivatives **37**, which were needed as unsymmetrical analogues of gossypol **38** (*bis*-diterpene), is based upon the reaction between 2′-haloacetophenone **34** and the 2-acetonaphthone enolate **35**. This route, without isolation of the *o*-acetyldesoxybenzoin intermediate **36**, affords 4-hydroxy-2,2′-binaphthalene **37** in high yield.[38] It is worthy of note that the electronic effects of the *o*-COCH³ group on the substrate **34** are crucial in ensuring for the lability of the nucleofuge. This is seen by comparison with the inertness of bromobenzene or of the carbonyl protected *o*-bromoacetophenone towards such reactions. Unsymmetrical polyhydroxylated 2,2′-binaphthyl derivatives are obtained from $S_{RN}1$ reactions between appropriately substituted 2′-bromoacetophenone and 2-acetonaphthones. More highly substituted derivatives, particularly on positions adjacent to the 2,2′-bond joining the two naphthalene subunits, are obtained by the same strategy,[39] but steric hindrance caused by substituents R₄ and R₅ becomes a limiting factor (Scheme 88.7).

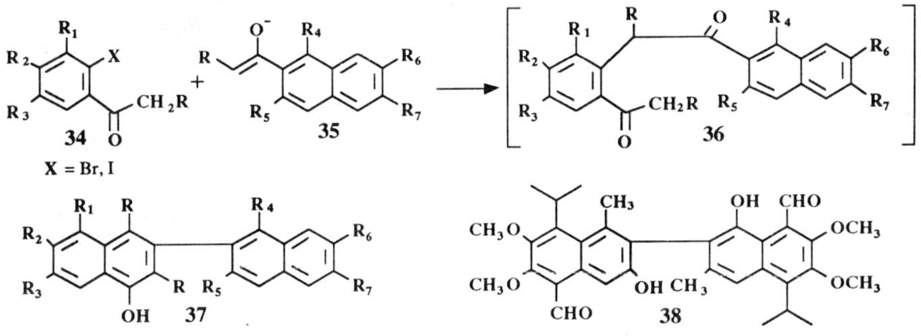

Gossypol

R_1-R_7 = H, OCH_3

R= H; CH_3; OCH_3

SCHEME 88.7 Synthesis of unsymmetrical 2,2′ binaphtyl derivatives.

Synthesis by Intramolecular Reactions

Oxindole. Wolfe et al.[40,41] have reported a general regiospecific synthesis of oxindoles 40 based upon the intramolecular cyclization of carbanions derived from *N*-alkyl- or *N*-acyl-*o*-haloanilines 39. A possible arynic intermediate was discarded by a control reaction on 39 ($R_1 = CH_3$) that cannot give rise to this intermediate and does not react in the dark, but affords high yield of the cyclized product under illumination. Alkylidene oxindoles 42 are obtained from *N*-methyl-α,β-unsaturated anilides 41.

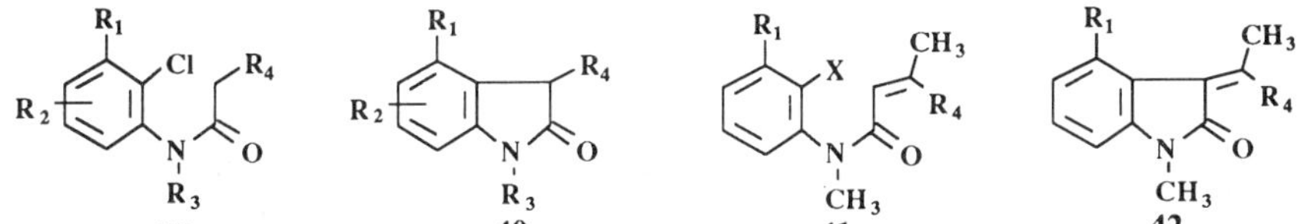

39 40 41 42

R_1, R_2 = H, OCH_3, CH_3

R_3= H; CH_3; CH_2Ph

R_4= H; CH_3; n-C_4H_9; Ph

Isoquinolone. The first isoquinolone synthesis based upon cyclization of the carbanion from 43 (R_3 = CH_3) was performed by Kessar et al.,[42] who synthesized 44a, a precursor to cherylline 44b. Wolfe et al.[41] have investigated further this reaction[41] and have obtained the isoquinolone 44 (R_3 = H) by reacting the *N*-trimethylsilyl-*N*-acyl-*o*-chlorobenzylamine 43 in KNH_2/NH_3.

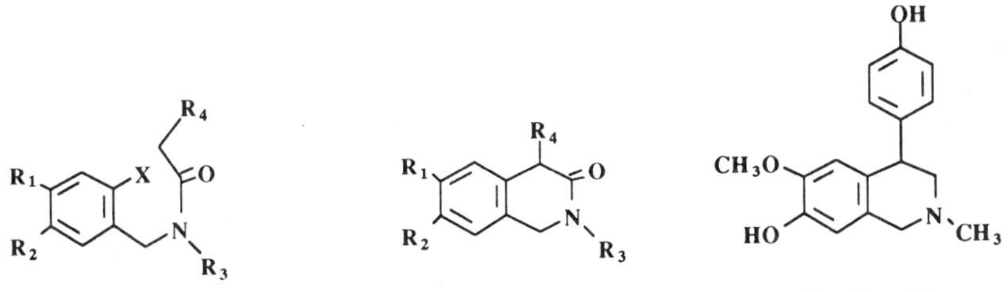

43 44a 44b Cherylline

R_1, R_2 = H, OCH_3; OCH_2Ph

R_3= CH_3, Si(CH $_3$)

R_4= CH_3, C_2H_5, $PhOCH_3$ (p)

Cyclic Ketones. Photostimulated intramolecular cyclization of the ketone enolate anion of iodoarene 45 was first reported by Semmelhack[43,44] as the best method available for the formation of the seven-membered ring in the last step of a total synthesis of cephalotaxinone 46.

Enolate anions derived from *o*-haloketones of various chain length (n = 1, 2, 3) **47** also react to give six-, seven-, eight-, or ten-membered ring systems **48**. The hydrogen atoms β to the carbonyl group (R = H), activated when the enolate in the terminal position is formed, may reduce Ar˙ to Ar⁻. Enolization occurs also in that place and side products **49** are formed, but none of those complications are observed when R = CH₃.

A facile synthesis of [*m-m*]-*meta*cyclophadiones **52** was reported by Fukazawa[45,46] and involves an intermolecular reaction of *m*-bromoketone **50** on itself leading to **51** and finally an intramolecular reaction to yield **52**.

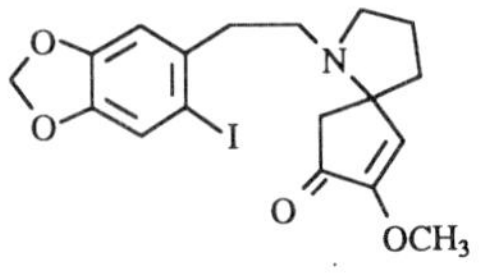

45 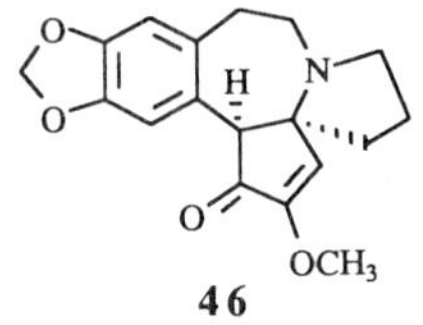   **46**

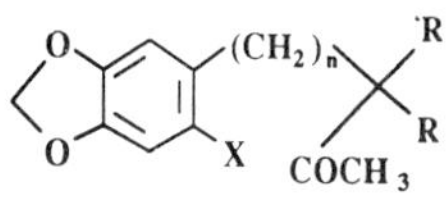

47 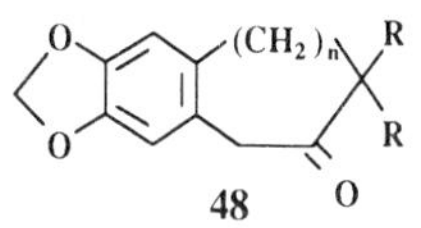   **48** 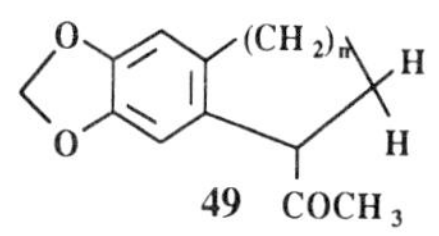   **49**

R = H; CH₃   n= 1, 2, 3

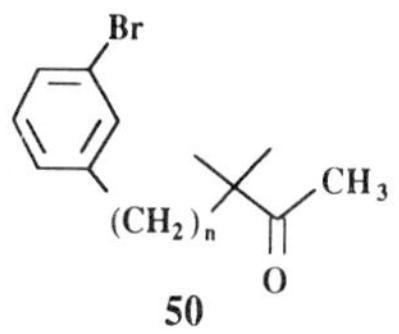

50   **51**   **52**

n= 0- 5

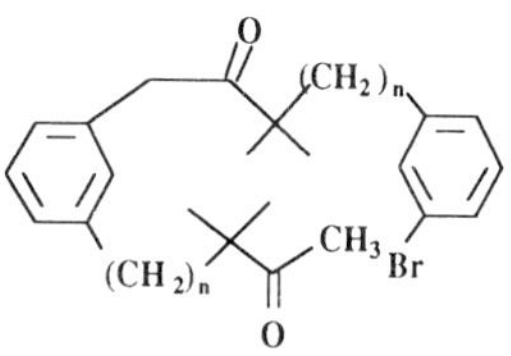

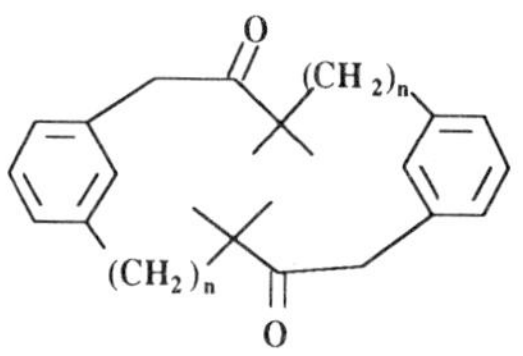

Heteroarenes (Simple and Functionalized)

Five-Membered Heteroarenes

Thiophene.   Bunnett et al.[47] reported that substitution by acetone enolate anion took place on halothiophenes **53** to give **54**. The order of reactivity observed in these reaction was 3-Br > 2-Br > 2-Cl.

Thiazole.   Wolfe et al.[48] have observed that 2-chlorothiazole and analogues **55** treated with pinacolone enolate under photostimulation in liquid NH₃ lead to formation of the mono-2-thiazolyl ketone **56** along with *bis*-2-thiazolyl ketones, while 2-bromothiazole gives only monosubstituted ketones.[48] In contrast, dark reactions or photostimulated reactions do not result in halide displacement, but in carbinol formation (Scheme 88.8).

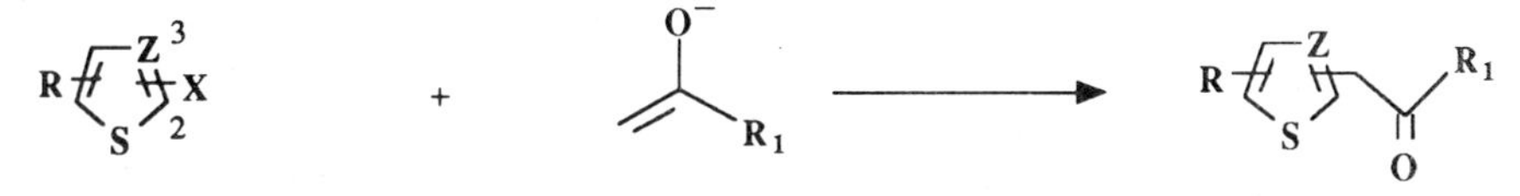

53 Z = CH; R = H	R_1 = CH_3	**54**
X = 2-Cl, 2-Br; 3-Br		
55 Z = N; R = H, CH_3	R_1 = t-C_4H_9	**56**
X = 2-Cl		

SCHEME 88.8   Reactions of halothiophene and halothiazoles.

Six-Membered Heteroarenes

Pyridine. The three isomeric monohalopyridines undergo photostimulated reactions with acetone enolate in the order 2-Br > 3-Br > 4-Br, the leaving group reactivity 2-Br > 2-Cl > 2-F being the same as that found for the benzene derivative. 2-Bromopyridine is more reactive than bromobenzene, as shown by a competitive reaction conducted on an equimolecular mixture of both substrates that yields 2-pyridylacetone and phenylacetone in a ratio 1/0.27.[49] Halopyridines carrying a compatible group *ortho* to the halogen atom allow the synthesis of various heterocyclic compounds:

(1) 2-chloro-3-aminopyridine treated with various ketone enolates gives 4-aza-indoles **57**[26,50] or 1',2',3',4'-tetrahydro-4-aza carbazoles **58**,[50] higher yields of which are obtained by an improved procedure proposed by Fontan.[51] Various *o*-iodoaminopyridines prepared by a versatile method and reacted with ketone enolates by Queguiner et al.[52] afford 5-, 6-, or 7-azaindole derivatives **59**.

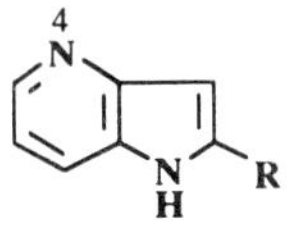

57 4- Azaindole	**58**	**59** 5-; 6-; 7-Azaindoles

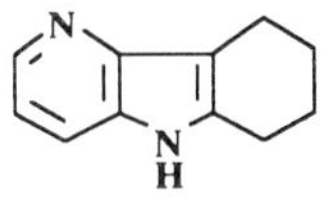

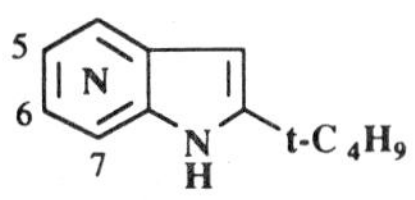

(2) The intramolecular reaction of **60** leading to **61** is the key step of an exceptionally brief synthesis of the azaphenanthrene alkaloid eupolauramine **62** reported by Goehring[53] (Scheme 88.9).

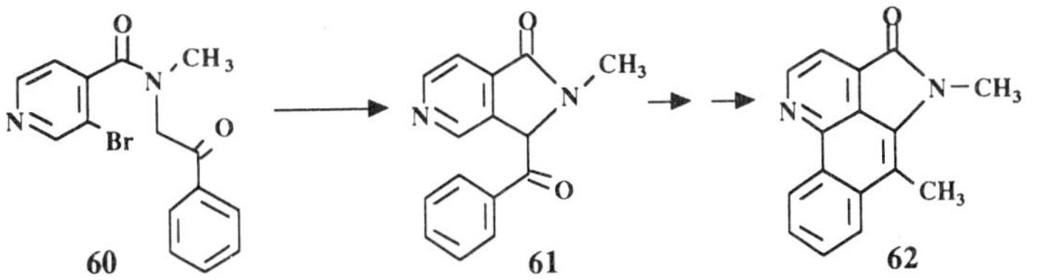

60	**61**	**62**
		Eupolauramine

SCHEME 88.9   Synthesis of eupolauramine.

(3) 2-chloro-3 methoxy- or isopropoxypyridine reacts with ketone enolates[54] to give the $S_{RN}1$ product, whose subsequent deprotection leads to furopyridines **63** via the *o*-hydroxypyridylketone intermediate.

(4) 2-chloro-3-acylpyridine or 3-chloro-4-acylpyridine treated with the 1- or 2-acetonaphthone enolate[35] gives rise to various 1- or 2-naphthyl quinoline **64** or to 1- or 2-naphthyl isoquinoline derivatives **65**.

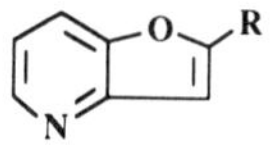

63
R = t-C_4H_9, Ph

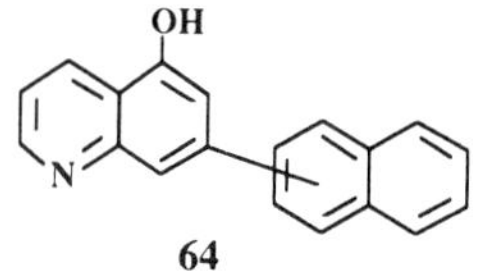

64

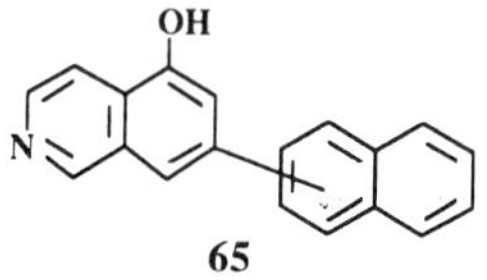

65

Diazines. Variously substituted halopyrimidines **66, 67, 68**, halopyridazine **69**, and halopyrazine **70** react with ketone enolates[55,56] in the order: halopyrazine > halopyridazine > halopyrimidines, the first substrate reacting by thermal activation, the third one requiring UV stimulation, and the second one reacting partly under both kinds of activation. 6-Iodopurine (R = Et) reacts in the dark with various enolates[57] to give **71**, and the reaction, like that of halopyridazine, is catalyzed by UV light. The process also takes place on the more complex 6-iodoadenosine (R = protected ribose).[58]

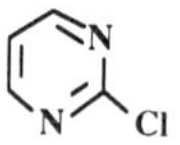

66

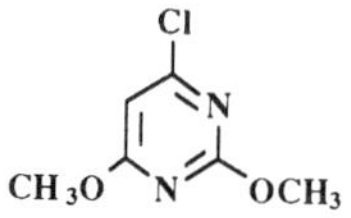

67

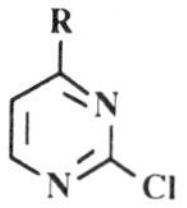

68

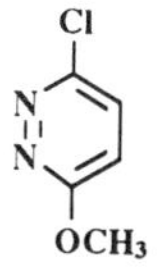

69

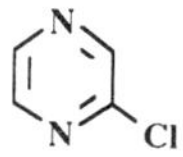

70

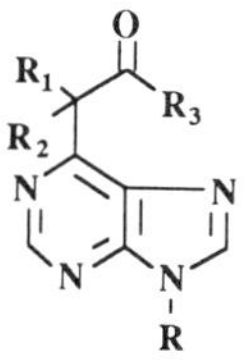

71

R = Et; R_1 = R_2 = H; R_3 = CH_3, Ph, furyl
R_1 - R_3 = $(CH_2)_n$ n= 3, 4; tetralone
R = protected ribose

Condensed Compounds

Haloarenes

Photostimulated reactions of halonaphthalene, haloanthracene, and halophenanthrene with the acetone enolate were reported by Bunnett et al.[19] 1-Chloronaphthalene reacts efficiently to give naphthylacetone in high yield, and no reduction occurs, in contrast with chlorobenzene. Rossi et al.[59] have shown, from comparative studies on 1-chloro- and 1-iodonaphthalene, that initiation by light is selective and superior to that by solvated electrons (K/NH_3), which leads to mixtures of substitution and reduction products.

Haloheteroarenes

2-Chloroquinoline is the sole substrate whose reactivity toward enolates was investigated.[60a,b] From competitive reactions with a mixture of primary and tertiary potassium enolate anions, 2-chloroquinoline appeared to combine preferentially with the tertiary enolate, in contrast with the reaction of halobenzenes. Prats et al.[61a] have reported that 3-iodobenzothiophene, treated with ketone enolates from acetone, acetophenone, and cyclohexanone, undergoes a thermally activated $S_{RN}1$ reaction in competition with an ionic reaction to give the expected β-hetaryl ketones **72**. In contrast, 3-halo-2-amino derivatives (X = Cl, Br, I) react under photostimulated conditions[61b] to give products **73** in low yield since competitive reduction of the aryl radical leading to 2-aminobenzothiophene competes efficiently.

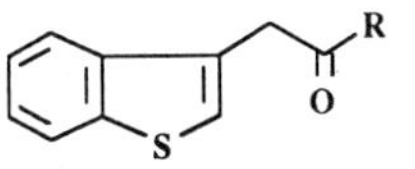

72

R = CH$_3$, Ph

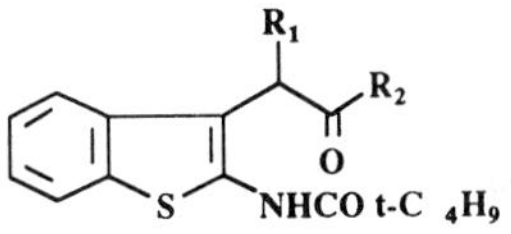

73

R$_1$ = H; R$_2$ = CH$_3$, Ph

R$_1$- R$_2$ = (CH$_2$)$_4$

Dihaloarenes and Dihaloheteroarenes

Dihalobenzenes

o-Dibromobenzene 74 react with pinacolone enolate to give the expected disubstituted product, but with acetone enolate as nucleophile a mixture of acetylmethylindene 76 and 9,10 diacetylanthracene 78 originating from the primary S$_{RN}$1 product 75 is obtained. The product 76 is formed by intramolecular aldol condensation, while 78 results from S$_{RN}$1 arylation of 75 by *o*-dibromobenzene to give 77, which undergoes an intramolecular S$_{RN}$1 ring closure[62] (Scheme 88.10).

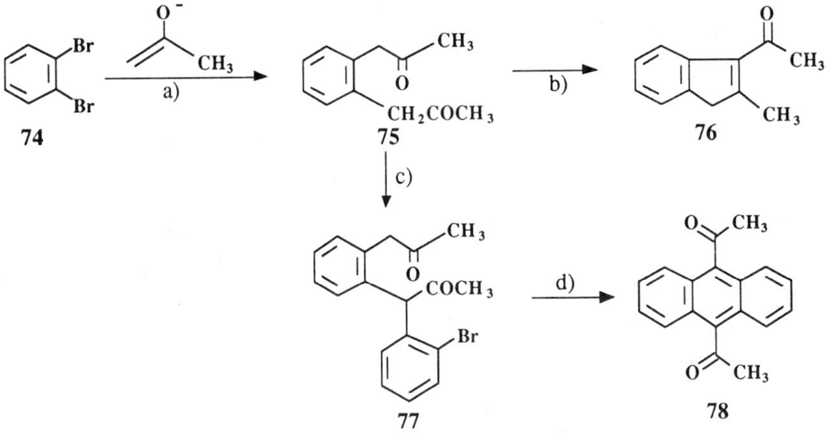

SCHEME 88.10 Sequential S$_{RN}$1 reactions on 1,2-dibromobenzene.

1-Fluoro-3-iodobenzene affords only a monosubstituted product when treated with ketone enolate[21] while 1,4-dichlorobenzene, investigated by Rossi et al.[63], leads to the expected 1,4 disubstituted benzene with the pinacolone derived enolate but to a complex mixture of products when reacted with acetone enolate.

Dihalopyridines

Wolfe et al.[49] have reported that 2,6-dihalopyridine (Br or Cl) gives directly a high yield of 2,6-disubstituted pyridine 79 without the accumulation of the monosubstituted product. This behavior was attributed to extrusion of X$^-$ from the intermediate radical anion X-Pyr-Nu$^-$ being preferred over electron transfer to 2,6-dihalopyridine. 2,3-Dichloro-, 3,5-dichloro-, or dibromopyridine react to give the corresponding disubstituted products 80 when reacted with the pinacolone derived enolate.[64]

Dihaloquinolines

4,7-Dichloroquinoline reacts exclusively at position-4 with the pinacolone enolate to give 81, indicating the higher reactivity of the halogen attached to the heterocycle moiety.[64] The reactivity of quinolines bearing two halogens on the benzenoid ring has been investigated by the Beugelmans' group[54,65] on 5,7-dihalo-8-hydroxyquinoline, whose phenolic function was protected. The 5,7-dibromo derivative, treated with the pinacolone enolate, gives the 5,7-disubstituted product **82**, while the 5,7-dichloro compound affords only the 7-monosubstituted product **84** that is formed from 5-chloro-7-iodo-8-hydroxyquinoline **83** (Vioform) in better yields. Furoquinoline derivatives **85** are obtained from **84** by deprotection of the 8-phenolic function.

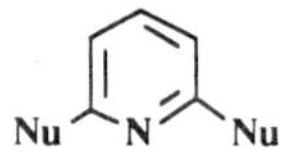

79 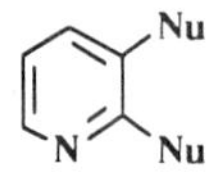   **80** 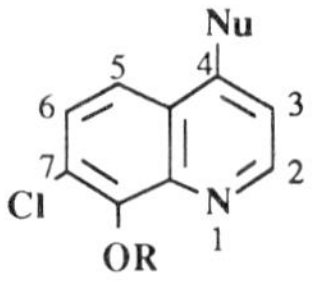   **81** 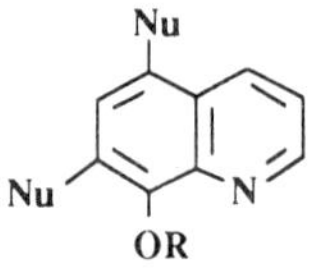   **82**

Nu = CH$_2$CO t-C$_4$H$_9$

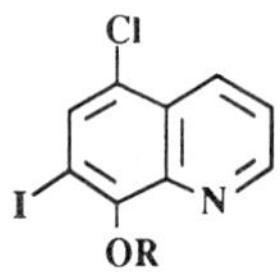

83   **84**   **85**

R = H (Vioform)   R$_1$ = CH$_3$; t-C$_4$H$_9$; furyl; Ph-OCH$_3$;R$_2$= H

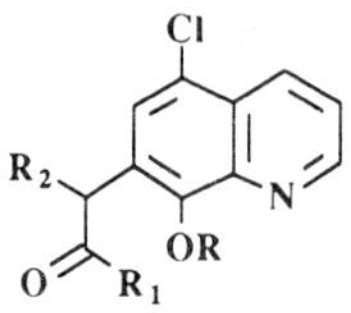

R$_1$R$_2$ = tetralone

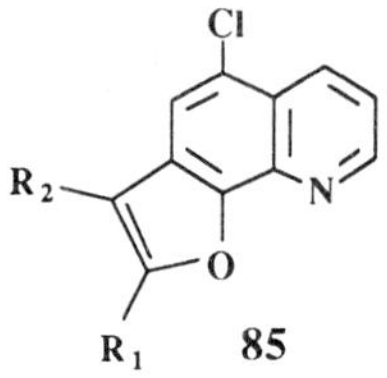

88.3 Other Enolates

In addition to ketone enolates, there are other enolates of interest as nucleophiles for carbon-carbon bonding S$_{RN}$1 (Ar) reactions which will be reviewed briefly here.

Monoactivated Enolates

Aldehydes

ortho-Substituted halobenzenes XC$_6$H$_4$Y (Y = NH$_2$, CH$_2$NH$_2$, OCH$_3$) react with aldehyde-derived enolates to give indole and 3-substituted indoles **86**,[27] isoquinoline,[30] or 2-methoxyphenyl acetaldehyde,[28] together with various amounts of reduction product YC$_6$H$_5$. No reduction takes place with 2-chloro-3-aminopyridine, which yields 4-azaindoles.[51]

Esters

Semmelhack et al.[24] have reported that *t*-butylacetate reacts well with bromobenzene or 4-bromoanisole to give **87**,[66] and that the main limitation arises from further arylation of the primary S$_{RN}$1 product leading to substantial amount of biarylated esters **88**.

Amides

According to Rossi et al.,[67] *N,N*-disubstituted enolate anions RRNCOCH$_2{}^-$ formed in liquid ammonia by acid-base reaction with amide anion react efficiently with halobenzenes (X = Cl, Br, I), 1-chloronaphthalene, and 9-bromophenanthrene to give **89**, but the insolubility of the nucleophile is the limiting factor.

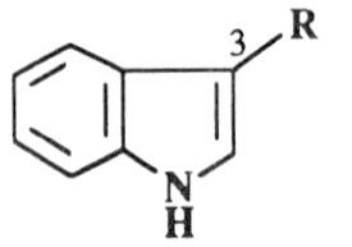

86   R = H, CH$_3$, C$_2$H$_5$

87 ArCH$_2$CO$_2$- t-C$_4$H$_9$

88 Ar$_2$CHCO$_2$- t-C$_4$H$_9$

Ar = Ph, PhOCH$_3$(p)

89 ArCH$_2$CONR$_1$R$_2$

Ar = benzene, naphtalene, anthracene

R$_1$, R$_2$ = Ph, CH$_3$

Nitriles

Bromobenzene was reported by Bunnett et al.[68] to react slowly with $CNCH_2^-$ to give a complex mixture of products, among which $PhCH_2CN$ was found together with $PhCH_2CH_2CN$ and $PhCH_3$. In sharp contrast, Rossi et al.[69,70] discovered that 4-chlorobiphenyl, 4-chlorobenzophenone, condensed haloarenes (naphthalene, phenanthrene), 2-chloropyridine, or halopyrazine give high yields of $ArCH_2CN$. Rossi[71] rationalized all those reactions by considering the reactivity of the intermediate radical anions $Ar CH_2CN^{-\cdot}$. Halopyrazine, halopyrimidine, and haloquinoline were found by Wolfe et al.[72] to react with the phenylacetonitrile anion to afford secondary nitriles 90, which were converted into aryl heteroaryl ketones 91.

HetArCH(CN)Ph   HetArCOPh

90   91

HetAr = Pyridine, Pyrimidine, Pyrazine, Quinoline

Phenols and Naphthols

The anions derived from phenols that, strictly speaking, are not enolates were found to be C-arylated by haloarenes under electrocatalytic conditions.[73] Under photostimulation, a series of variously substituted biaryl[74,75] or heterobiaryl[35] compounds were synthesized in the Beugelmans' group and by Rossi et al.[76] from haloarenes and phenols, naphthols, or other hydroxy aromatic compounds as nucleophiles (Scheme 88.11).

ArX + Ar'OH   ⟶   Ar-Ar'OH (o) or (p)

Ar'X + ArOH   ⟶   Ar'-ArOH (o) or (p)

Ar, Ar'= Benzene, Naphtalene, Heteroarene

SCHEME 88.11   Coupling of aromatic substrates with hydroxyarenes.

Diactivated Enolates

Dianions of β-dicarbonyl compounds were reported to react with 2-chloroquinoline[77] or with 2-bromomesitylene,[52] while it was mentioned that the corresponding monoanions failed to react, even by entrainment with the acetone enolate.[61] Later, the Beugelmans' group[78a] observed that a wide range of β-dicarbonyl monoenolate anions derived from secondary or tertiary malonates, acetoacetate, cyanoacetate, or 1,3-pentadione do react with halobenzene or halopyridine carrying a E.W.G. (CN) and with dihaloquinoline.[65] The outcome of the reaction depends upon the workup since, under basic conditions, retro-Claisen fragmentation causes the loss of $COCH_3$. Thus, 2-cyanophenylacetate 92 is formed by treating 2-bromobenzonitrile 93 with the monoanion of acetylacetate to give 94, circumventing the failure of 93 to react with $-CH_2CO_2Et$. 3-Cyanopyridylacetone 95, obtained in low yield from the reaction of 2-bromo-3-cyanopyridine 96 with acetone enolate, is readily available via 97 by using the 1,3-pentanedione monoanion that allows an efficient synthesis of naphthyridone 98[78b] (Scheme 88.12).

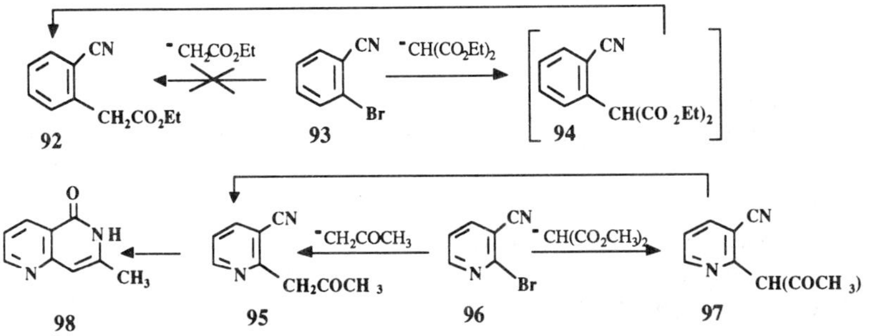

SCHEME 88.12   Monoanions of β-dicarbonyl compounds as nucleophiles.

References

1. Kim, J. K. and Bunnett, J. F., Evidence for a radical mechanism of aromatic nucleophilic substitution, *J. Am. Chem. Soc.*, 92, 7463, 1970.

2. Kornblum, N., Michel, R. E., and Kerber, R. C., Radical anions as intermediates in substitution reactions, *J. Am. Chem. Soc.*, 88, 5660, 1966; Chain reactions in substitution processes which proceed via radical-anion intermediates, *J. Am. Chem. Soc.*, 88, 5662, 1966.

3. Russell, G. A. and Danen, W. C., Coupling reactions of the 2-nitro-2-propyl anion, *J. Am. Chem. Soc.*, 88, 5663, 1966.

4. (a) Kornblum, N., Substitution reactions which proceed via radical anion intermediates, *Angew. Chem. Int. Ed.*, 14, 734, 1975; (b) Kornblum, N., Radical anion reactions of nitro compounds, in *The Chemistry Of The Functional Group. Nitro, Nitroso* Suppl. F., Patai, S., Ed., John Wiley & Sons, New York, 1983, 361.

5. (a) Russel, G. A., Reactions between radical and anions, *J. Chem. Soc.*, Special publications 24, 271, 1970; (b) Free radical chain substitutions involving electron transfer, *A.C.S. Dir. of Petroleum Chem.*, 29, 338, 1984.

6. Bunnett, J. F., Aromatic substitution by the $S_{RN}1$ mechanism, *Acc. Chem. Res.*, 11, 413, 1978.

7. Rossi, R. A. and Rossi, R. H., *Aromatic Substitution by the $S_{RN}1$ Mechanism*, A.C.S. Monograph, 178, Washington, D.C., 1983.

8. Norris, R. K., Nucleophilic coupling with aryl radicals, in *Comprehensive Organic Synthesis*, Vol. 4, Trost, B. M., Ed., Pergamon, Oxford, 1991, 451.

9. Rossi, R. A. and Bunnett, J. F., Photostimulated aromatic $S_{RN}1$ reactions, *J. Org. Chem.*, 38, 1407, 1973.

10. Fox, M. A., Younathan, J., and Frixell, G. E., Photoinitiation of the $S_{RN}1$ reaction by excitation of charge-transfer complexes, *J. Org. Chem.*, 48, 3109, 1983.

11. Scamehorn, R. G. and Bunnett, J. F., Dark reactions of halobenzenes with pinacolone enolate ion. Evidence for a thermally induced aromatic $S_{RN}1$ reaction, *J. Org. Chem.*, 42, 1449, 1977.

12. Scamehorn, R. G., Hardacre, J. M., Lukanich, J. M., and Sharpe, L. R., Thermally initiated $S_{RN}1$ reactions of ketone enolates with iodobenzene in dimethylsulfoxide. Relative reactivities of enolate ions with phenyl radical, *J. Org. Chem.*, 49, 4881, 1984.

13. Swartz, J. E. and Bunnett, J. F., Reactions of halotoluenes with potassium diphenylphosphide. Evidence for a thermally induced aromatic $S_{RN}1$ reaction, *J. Org. Chem.*, 44, 340, 1979.

14. M'Halla, F., Pinson, J., and Saveant, J. M., The solvent as H-atom donor in electrochemical reactions. Reduction of aromatic halides, *J. Am. Chem. Soc.*, 102, 4120, 1980.

15. (a) Bunnett, J. F. and Sundberg, J. E., Photostimulated arylation of ketone enolate ions by $S_{RN}1$ mechanism, *J. Org. Chem.*, 41, 1702, 1976; (b) Hoz, S. and Bunnett, J. F., A quantitative study of the photostimulated reaction of iodobenzene with diethyl phosphite ion, *J. Am. Chem. Soc.*, 99, 4690, 1977.

16. Saveant, J. M., Catalysis of chemical reactions by electrodes, *Acc. Chem. Res.*, 13, 323, 1980.

17. Kessar, S. K., Nucleophilic coupling with arynes, in *Comprehensive Organic Synthesis*, Vol. 4, Trost, B. M., Ed., Pergamon, Oxford, 1991, 483.

18. Bunnett, J. F., Scamehorn, R. G., and Traber, R. P., Solvents for aromatic $S_{RN}1$ reactions, *J. Org. Chem.*, 41, 3677, 1976.

19. Bunnett, J. F. and Sundberg, J. E., On the synthesis of arylacetones by the $S_{RN}1$ arylation of acetone enolate ion, *Chem. Pharm. Bull.*, 23, 2620, 1975.

20. Rossi, R. A. and Bunnett, J. F., Arylation of several carbanions by the $S_{RN}1$ mechanism, *J. Org. Chem.*, 38, 3020, 1973.

21. Semmelhack, M. F. and Bargar, T. M., Cyclization of enolates onto aromatic rings via the photo-$S_{RN}1$ reaction. Preparative and mechanistic aspects, *J. Org. Chem.*, 42, 1481, 1977.

22. Beugelmans, R., L'extension de la $S_{RN}1$: Une nouvelle méthodologie en synthèse hétérocyclique, *Bull. Soc. Chim. Belge*, 93, 547, 1977.

23. Bunnett, J. F., Mitchell, E., and Galli, C., The effect of *ortho* substituents in S$_{RN}$1 reactions. Some synthetic applications, *Tetrahedron*, 41, 4119, 1985.

24. Lablache-Combier, A., Heteroaromatics, in *Photoinduced Electron Transfer*, Part C, Fox, M. A. and Chanon, M., Eds., Elsevier, Amsterdam, 1988.

25. Beugelmans, R. and Roussi, G., New "one-pot" synthesis of indoles under non-acidic conditions (S$_{RN}$1 reaction), *J. Chem. Soc. Chem. Commun.*, 950, 1979.

26. Bard, R. R. and Bunnett, J. F., Indole synthesis via S$_{RN}$1 reactions, *J. Org. Chem.*, 45, 1546, 1980.

27. Beugelmans, R. and Roussi, G., Substitution nucleophile aromatique radicalaire. Nouvelle synthèse du squelette indole, *Tetrahedron*, Suppl. 1, 37, 393, 1981.

28. Beugelmans, R. and Ginsburg, H., New synthesis of benzo[*b*]furans by S$_{RN}$1 reaction of *ortho*-iodoanisole, *J. Chem. Soc., Chem. Commun.*, 508, 1980.

29. Beugelmans, R., Chastanet, J., and Roussi, G., Nouvelle synthèse du squelette dihydro-1,2 isoquinoleine par substitution nucleophile radicalaire en chaîne (S$_{RN}$1), *Tetrahedron Let.*, 23, 2313, 1982.

30. Beugelmans, R., Chastanet, J., and Roussi, G., A new access to the isoquinoline ring system by S$_{RN}$1, *Tetrahedron*, 40, 311, 1984.

31. Beugelmans, R., Chastanet, J., Ginsburg, H., Quintero-Cortes, L., and Roussi, G., Direct synthesis of benzo[*c*]phenanthridines and benzo[*c*]phenanthridones via S$_{RN}$1 reactions, *J. Org. Chem.*, 50, 4933, 1985.

32. Beugelmans, R. and Bois-Choussy, M., One-pot synthesis of 1-oxo-1,2-dihydroisoquinolines (isocarbostyrils) via S$_{RN}$1 (Ar) reactions, *Synthesis*, 729, 1981.

33. Beugelmans, R., Ginsburg, H., and Bois-Choussy, M., S$_{RN}$1 reactions. Synthesis of 3-methyl derivatives of the alkaloids thalactamine, doryanine and 6,7-dimethoxy-N-methyl 1(2H)-isoquinoline, *J. Chem. Soc., Perkin Trans. 1*, 1149, 1982.

34. Beugelmans, R. and Bois-Choussy, M., A common and general access to berberine and benzo[*c*]phenanthridine alkaloids, *Tetrahedron*, 48, 8285, 1992.

35. Beugelmans, R. and Bois-Choussy, M., S$_{RN}$1 based methodology for synthesis of naphthylquinolines and naphthylisoquinolines, *J. Org. Chem.*, 56, 2518, 1991.

36. Bunnett, J. F. and Galli, C., The peculiar behavior of the trifluoromethyl substituent in S$_{RN}$1 processes, *J. Chem. Soc., Perkin Trans. 1*, 2515, 1985.

37. Beugelmans, R. and Ginsburg, H., A novel access to 3-benzazepines and 3-benzoxepines via S$_{RN}$1 reactions, *Heterocycles*, 23, 1197, 1985.

38. Beugelmans, R., Bois-Choussy, M., and Tang, Qian, A general S$_{RN}$1-based method for total synthesis of unsymmetrically hydroxylated 2,2-binaphthalenes, *J. Org. Chem.*, 52, 3880, 1987.

39. Beugelmans, R., Bois-Choussy, M., and Tang, Qian, Synthèses par reaction S$_{RN}$1 de phényl-2-naphtalènes et de binaphtyl-2,2′-dissymetriques. Rôle des facteurs structuraux, *Tetrahedron*, 45, 4203, 1989.

40. Wolfe, J. F., Sleevi, M. C., and Goehring, R. R., Photoinduced cyclization of mono- and dianions of N-acyl-o-chloroanilines. A general oxindole synthesis, *J. Am. Chem. Soc.*, 102, 3646, 1980.

41. Goehring, R. R., Sachdeva, Y. P., Pisipati, J. S., and Wolfe, J. F., Photoinduced cyclizations of mono- and dianions of N-acyl-o-chloroanilines and N-acyl-o-chlorobenzylamines as general methods for the synthesis of oxindoles and 1,4-dihydro-3(2H)-isoquinolines, *J. Am. Chem. Soc.*, 107, 435, 1985.

42. Kessar, S. V., Singh, P., Chawla, R., and Kumar, P., Cyclization of *ortho*-halogenated N-acylbenzylamine: A formal synthesis of (±)-cherylline, *J. Chem. Soc., Chem. Commun.*, 1074, 1981.

43. Semmelhack, M. F., Stauffer, R. D., and Rogerson, T. D., Nucleophilic aromatic substitution via nickel-catalyzed process and via the S$_{RN}$1 reaction. Improved synthesis of cephalotaxinone, *Tetrahedron Lett.*, 13, 4519, 1973.

44. Semmelhack, M. F., Chong, B. P., Staupper, R. D., Rogerson, T. D., Chong, A., and Jones, L. D., Total synthesis of the cephalotaxus alkaloids. A problem in nucleophilic aromatic substitution, *J. Am. Chem. Soc.*, 97, 2507, 1975.

45. Usui, S. and Fukazawa, Y., A facile synthesis of [*m.m*]metacyclophanes, *Tetrahedron Lett.*, 28, 91, 1987.

46. Fukazawa, Y., Takeda, Y., Usui, S., and Kodama, M., Synthesis and conformation of 1,1,10,10-tetramethyl[3.3]metacyclophane, *J. Am. Chem. Soc.*, 110, 7842, 1988.

47. Bunnett, J. F. and Gloor, B. F., Reactions of halothiophenes with acetone enolate and amide ions, *Heterocycles*, 5, special issue 377, 1976.

48. Dilender, S. C., Greenwood, T. D., Hendi, M. S., and Wolfe, J. F., Reactions of 2-halothiazoles with ketone enolates and nitrile carbanions, *J. Org. Chem.*, 51, 1184, 1986.

49. Komin, A. P. and Wolfe, J. F., The $S_{RN}1$ mechanism in heteroaromatic nucleophilic substitution. Photostimulated reactions of halopyridines with ketone enolates, *J. Org. Chem.*, 42, 2481, 1977.

50. Beugelmans, R., Boudet, B., and Quintero, L., Substitution nucleophile aromatique radicalaire. Synthèse directe de 4-azaindoles dans des conditions non acides et douces, *Tetrahedron Lett.*, 21, 1943, 1980.

51. Fontan, R., Galvez, C., and Villadoms, P., $S_{RN}1$ reactions. Synthetic applications to 4-azaindoles, *Heterocycles*, 16, 1473, 1981.

52. Estel, L., Marsais, F., and Queguiner, G., Metallation/$S_{RN}1$ coupling in heterocyclic synthesis. A convenient methodology for ring functionalization, *J. Org. Chem.*, 53, 2740, 1988.

53. Goehring, R. R., An exceptionally brief synthesis of eupolauramine, *Tetrahedron Lett.*, 33, 6045, 1992.

54. Beugelmans, R. and Bois-Choussy, M., A convenient access to furo[3,2-*h*]quinolines and to furo[3,2-*b*]pyridines via $S_{RN}1$ reactions, *Heterocycles*, 26, 1863, 1987.

55. Oosteven, E. A. and Van der Plas, H. C., Reactions of carbon nucleophiles with 4-phenyl and 4-*t*-butyl-5-halogenopyrimidines. On the occurence of an $S_{RN}1$ mechanism, *Rec. Trav. Chim. Pays Bas*, 98, 441, 1979.

56. Carver, D. R., Komin, A. P., Hubbard, J. S., and Wolfe, J. F., $S_{RN}1$ mechanism in heteroaromatic nucleophilic substitution. Reactions involving halogenated pyrimidines, pyridazines and pyrazines, *J. Org. Chem.*, 46, 294, 1981.

57. Nair, V. and Chamberlain, S. D., Novel photoinduced carbon-carbon bond formation in purines, *J. Am. Chem. Soc.*, 107, 2183, 1985.

58. Nair, V. and Chamberlain, S. D., Novel photoinduced functionalized C-alkylations in purine systems, *J. Org. Chem.*, 50, 5069, 1985.

59. Rossi, R. A., de Rossi, R. H., and Lopez, A. F., Reactions of 1-halonaphthalenes with nucleophiles by the $S_{RN}1$ mechanism of aromatic substitution, *J. Am. Chem. Soc.*, 98, 1252, 1976.

60. (a) Hay, J. V., Hudlicky, T., and Wolfe, J. F., The $S_{RN}1$ mechanism in heteroaromatic nucleophilic substitution. Photostimulation and entrainment of the reaction, *J. Am. Chem. Soc.*, 97, 374, 1975; (b) Hay, J. V. and Wolfe, J. F., The $S_{RN}1$ mechanism in heteroaromatic nucleophilic substitution. An investigation of the generality of photostimulated reactions of ketone enolates with 2-chloroquinoline, *J. Am. Chem. Soc.*, 97, 3702, 1975.

61. (a) Prats, M., Galvez, C., and Beltran, L., Study of the reaction of several ketone enolates with 3-iodobenzo[*b*]thiophene under thermally initiated $S_{RN}1$ reaction conditions, *Heterocycles*, 34, 1039, 1992; (b) Beltran, L., Galvez, C., Prats, M., and Salgado, J., Study of the reaction of ketone enolates with 3-halo-2-amino derivatives of benzo[*b*]thiophene under photostimulated $S_{RN}1$ conditions, *J. Heterocyclic Chem.*, 29, 905, 1992.

62. Bunnett, J. F. and Singh, P., $S_{RN}1$ reactions of *o*-dibromobenzene with ketone enolate ions, *J. Org. Chem.*, 46, 5022, 1981.

63. Alonso, R. A. and Rossi, R. A., $S_{RN}1$ mechanism in bifunctional systems, *J. Org. Chem.*, 45, 4760, 1980.

64. Carver, R. R., Greenwood, T. D., Hubbard, J. S., Komin, A. P., Sachdeva, Y. P., and Wolfe, J. F., $S_{RN}1$ mechanism in heteroaromatic nucleophilic substitution. Reactions involving certain dihalogenated π-deficient nitrogen heterocycles, *J. Org. Chem.*, 48, 1180, 1983.

65. Beugelmans, R., Bois-Choussy, M., Gayral, P., and Rigothier, M. C., Synthese par $S_{RN}1$ et évaluation de l'activité amoebicide de nouveaux dérivés quinoléiniques, *Eur. J. Med. Chem.*, 23, 539, 1988.

66. Semmelhack, M. F. and Bargar, T., Photostimulated nucleophilic aromatic substitution for halides with carbon nucleophiles. Preparative and mechanistic aspects, *J. Am. Chem. Soc.*, 102, 7765, 1980.

67. Rossi, R. A. and Alonso, R. A., Photostimulated reactions of *N,N*-disubstituted amide enolate anions with haloarenes by the $S_{RN}1$ mechanism in liquid ammonia, *J. Org. Chem.*, 45, 1239, 1980.

68. Bunnett, J. F. and Gloor, B. F., $S_{RN}1$ phenylation of nitrile carbanions and ensuing reactions. A new route to alkylbenzenes, *J. Org. Chem.*, 38, 4156, 1973.

69. Rossi, R. A., de Rossi, R. H., and Lopez, A. F., Photostimulated arylation of cyanomethyl anion by $S_{RN}1$ mechanism of aromatic substitution, *J. Org. Chem.*, 41, 3371, 1976.

70. Rossi, R. A., de Rossi, R. H., and Pierini, A. B., Reactions of halobenzenes with cyanomethyl anion in liquid ammonia by the $S_{RN}1$ mechanism, *J. Org. Chem.*, 44, 2662, 1979.

71. Rossi, R. A., Phenomenon of radical anion fragmentation in the course of aromatic $S_{RN}1$ reactions, *Acc. Chem. Res.*, 15, 164, 1982.

72. Hermann, C. K. F., Sachdeva, Y. P., and Wolfe, J. F., A new synthesis of aryl hetaryl ketones via $S_{RN}1$ reactions of halogenated heterocycles, *J. Heterocyclic Chem.*, 24, 1061, 1987.

73. Alam, N., Amatore, C., Combellas, C., Thiebault, A., and Verpeaux, J. N., Electrosynthesis of unsymmetrical biaryls using an $S_{RN}1$ type reaction, *Tetrahedron Lett.*, 28, 6171, 1987.

74. Beugelmans, R. and Bois-Choussy, M., Phenoxide and naphthoxide ions as nucleophiles for $S_{RN}1$ reaction. Synthesis of biphenyl and phenylnaphthyl derivatives, *Tetrahedron Lett.*, 29, 1289, 1988.

75. Beugelmans, R., Bois-Choussy, M., and Tang, Qian, Coupling of iodonaphthalene with naphthoxide ions under $S_{RN}1$ conditions. Synthesis of unsymmetrical binaphthyl derivatives, *Tetrahedron Lett.*, 29, 1705, 1988.

76. Pierini, A. B., Baumgartner, M. T., and Rossi, R. A., Photostimulated reactions of aryl iodides with 2-naphthoxide ions by the $S_{RN}1$ mechanism, *Tetrahedron Lett.*, 29, 3429, 1988.

77. Wolfe, J. F., Greene, J. C., and Hudlicky, T., Reaction of dialkali salts of benzoylacetone with 2-chloroquinoline. Evidence for $S_{RN}1$ mechanism in heteroaromatic nucleophilic substitution, *J. Org. Chem.*, 37, 3199, 1972.

78. (a) Beugelmans, R., Bois-Choussy, M., and Boudet, B., New and direct arylation and hetarylation of β-dicarbonyl compounds by $S_{RN}1$, *Tetrahedron*, 38, 3479, 1982; (b) Unpublished results taken from B. Boudet's Thesis, Paris-Sud, Orsay, 1982.

89

Photocyclization of Haloarenes

S. V. Kessar
Panjab University

Anil K. Singh Mankotia
Panjab University

89.1 Introduction

Early work on reactions of radicals generated by irradiation of aryl iodides was reported around 1960 by the schools of Bryce-Smith and Kharasch.[1,2] Since then, photocyclization of haloarenes having a reactive group (G) on an *ortho*-side chain has developed into a widely used synthetic methodology.[3] Such ring closures may be depicted simply by Equation 89.1. However, the actual nature and timing of aryl-halogen bond cleavage can vary widely. It can be a simple homolysis or assistance may come through complexation of ArX with G, as shown in 1 (Scheme 89.1). Photo-induced single electron transfer (SET), from G or an external source to ArX, can also occur to give an ion radical (2), which then loses a halide ion. In the extreme case of electrocylic ring closure (3), full bond formation actually precedes loss of hydrogen halide (Scheme 89.1).

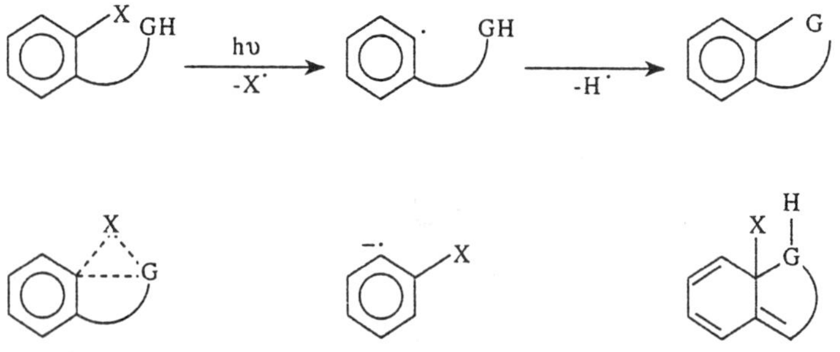

eqn. (1)

(1)   (2)   (3)

SCHEME 89.1

The major factors determining the reaction course are the nature of the excited state, the strength of the C-X bond, and the energy available for its breaking. Thus, iodides react generally through simple homolysis as the C-I bond is weak and dissociative n,σ^* or σ,σ^* states can be populated

1218

0-8493-8634-9/95/$0.00+$.50
© 1995 by CRC Press, Inc.

readily. With aryl bromides, and more so with chlorides, other reaction pathways tend to ingress to a greater extent, especially in solution photochemistry. If prior interaction between Ar and G is necessary, then equilibrium distribution of substrate conformers and the rates of bond rotation on the excited-state surface assume added importance. The following brief survey of synthetic applications of photocyclization of haloarenes is arranged according to the side chain centers undergoing ring closure.[4]

89.2 Cyclization with Aryl and Alkenyl Carbon Centers

Ring closure of a photogenerated aryl radical with an arene ring on the side chain has been used extensively to construct polynuclear systems. For example, aporphines were obtained by photolysis of halogenated 1-benzyltetrahydroisoquinolines (4 → 5).[5,6] These cyclizations are best carried out in acidic solutions or an amide derivative (4b) is employed. Otherwise, electron transfer from nitrogen to the aryl halide can divert the reaction course. However, for amino alcohol 6, hydrocarbon solvents are suitable inasmuch as hydrogen bonding can avert deleterious electron transfer and, at the same time, keep the aryl moieties juxtaposed for cyclization.[7] Similarly, phenolic compounds can be photocyclized in an alkaline medium as the phenoxide moiety competes effectively with nitrogen for intramolecular electron transfer. It also acts as a strong *ortho-para*-director, and the effect has been exploited in the synthesis of proerythrinadienone precursor[9] and some other spiro isoquinoline alkaloids.[8-11] Because of its simplicity and the economy of steps, this approach remains attractive despite modest yield of cyclized product.

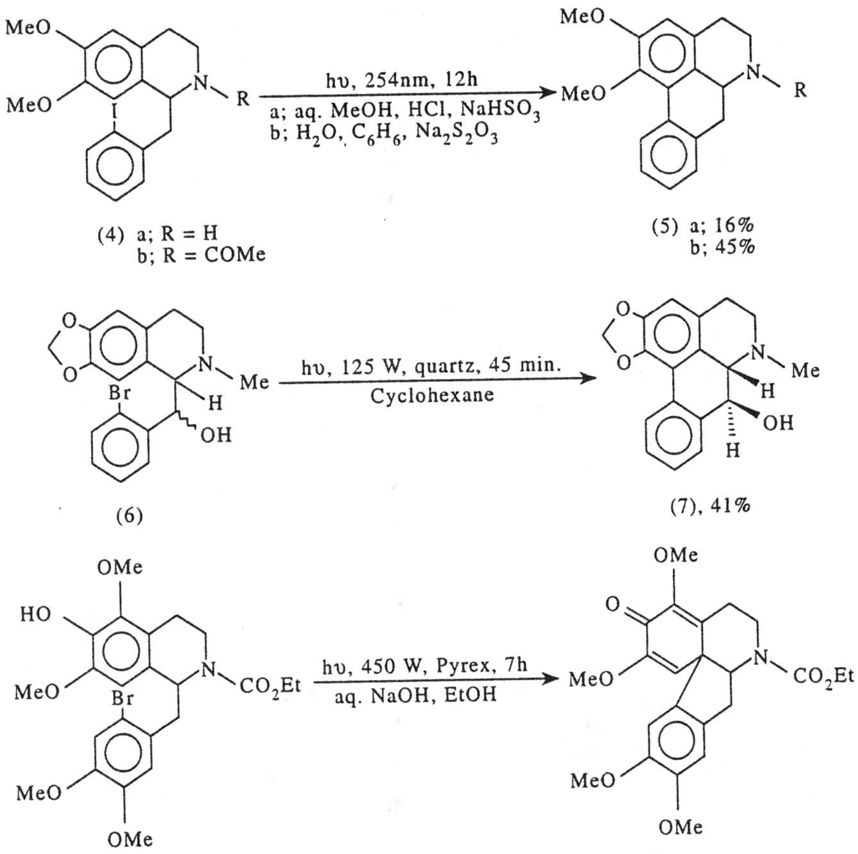

SCHEME 89.2

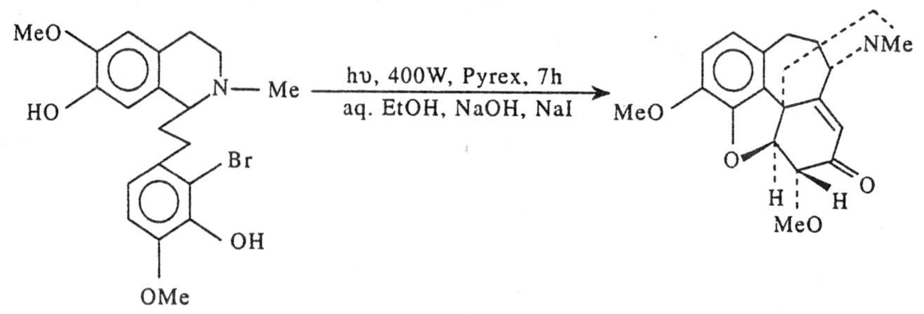

(10)   (11), 2%

SCHEME 89.2 (continued)

Aporphines have also been obtained by photocyclization of stilbenoid compounds (12 → 13),[12-14] although the ability of a halogen to facilitate electrocyclic ring closure at the atom to which it is bonded is not evident in general. In fact, 2-chlorostilbene undergoes oxidative photocyclization to give 1-chlorophenanthrene as the major product. Also, bromine has been used as a blocking group in the synthesis of some helicenes.[15]

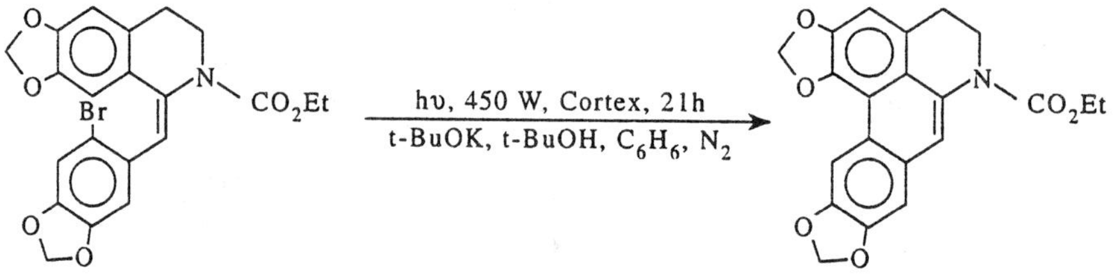

(12)   (13), 72%

SCHEME 89.3

Phenanthridines are accessible through photolysis of *o*-halobenzanilides. Here, chlorides are preferred because, for the cleavage of a C-Cl bond, assisted homolysis (Scheme 89.1) is necessary. It can take place only in the *syn*-conformation (15, R = H, X = Cl) which, in turn, is conducive to cyclization. In bromides and iodides, simple homolysis from the more prevalent *anti*-conformation (14) occurs to give side products.[16] In the corresponding naphthalamides, the excited state is less energetic and even bromides cyclize well,[17] presumably through an assisted homolysis pathway. Substitution of the amide hydrogen by a bulkier group favors the *syn*-conformation and even iodoanilides (15, R = alkyl, X = I) give good yields of ring closed products.[18,19] Locking the aryl groups into a *cis*-configuration, by complex formation (17) or hydrogen bonding (18), has a stronger beneficial effect.[20,21] Photocyclization of enamides (19 → 20) is a similar reaction. However, superior yields are obtained with fluorides than chlorides and this fact implicates an electrocyclic ring closure mechanism.[22] It may be noted that some amide substrates without a halogen substituent (14,19, X = H or OMe) also undergo efficient photocyclization.[23]

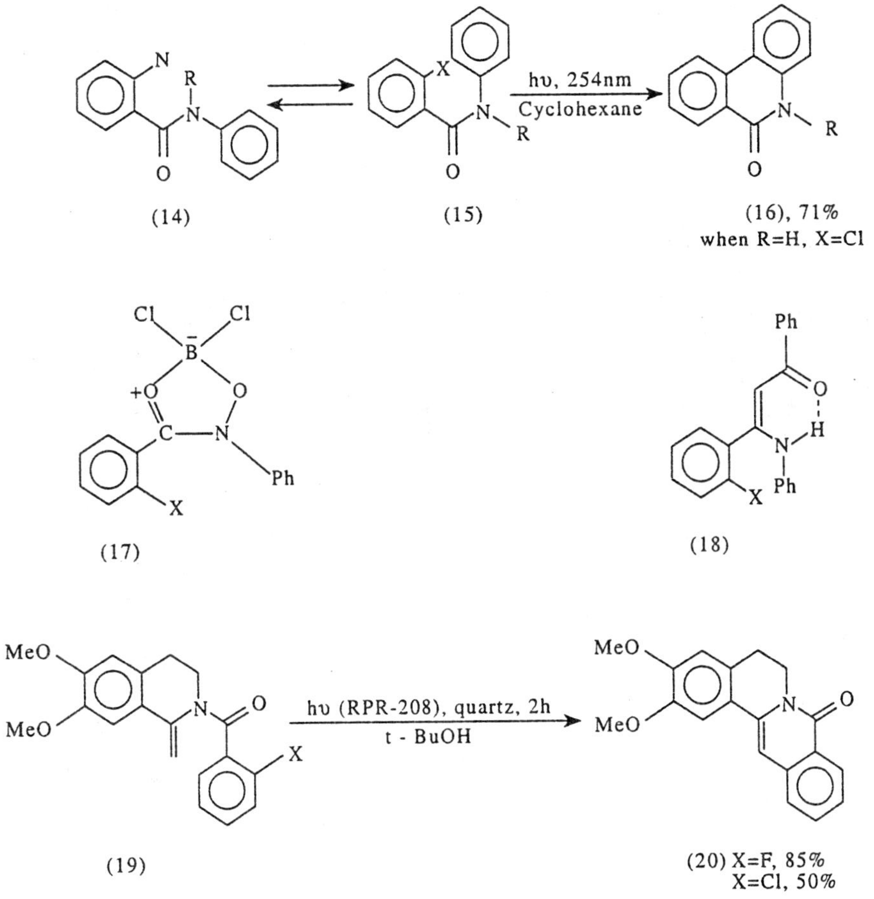
hυ, 254nm
Cyclohexane
(14)
(15)
(16), 71%
when R=H, X=Cl
(17)
(18)
hυ (RPR-208), quartz, 2h
t - BuOH
(19)
(20) X=F, 85%
X=Cl, 50%

SCHEME 89.4

Cleavage of aryl halides through photoinduced electron transfer from amines is widely documented;[24–26] in the conversion of 21 to 22, the added triethylamine perhaps serves such a purpose.[27] An intramolecular version of this process has been proposed for phenanthridine ring closure used in the synthesis of a number of alkaloids (23 → 24).[28,29] Photocyclization of haloarenes has also been employed to obtain a variety of other heterocyclic systems and some examples are given in the scheme below.[30–35]

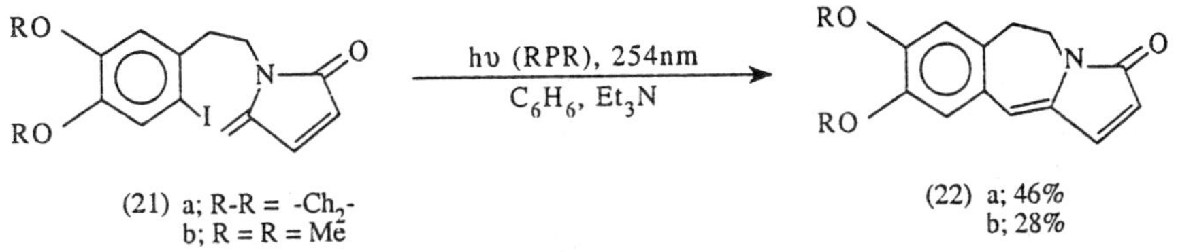
hυ (RPR), 254nm
C6H6, Et3N
(21) a; R-R = -Ch2-
b; R = R = Me
(22) a; 46%
b; 28%

SCHEME 89.5a

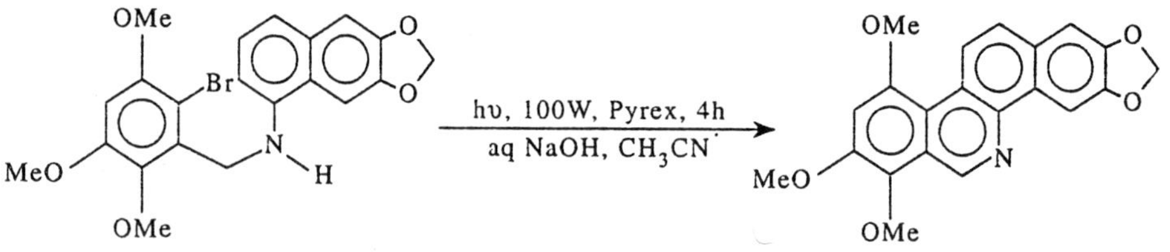

(23)

hυ, 100W, Pyrex, 4h
aq NaOH, CH₃CN

(24), 50%

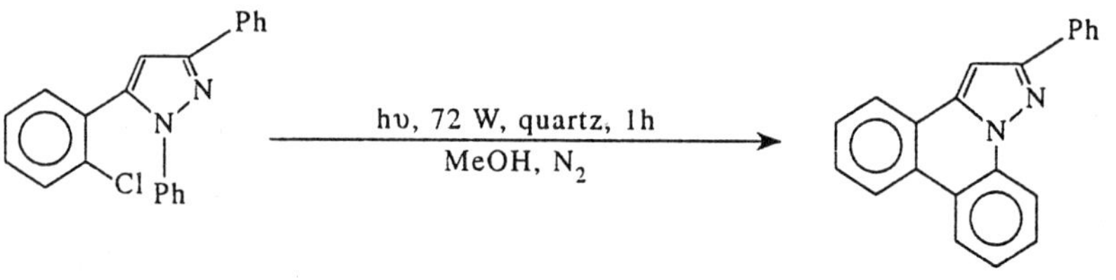

(25)

hυ, 72 W, quartz, 1h
MeOH, N₂

(26), 80%

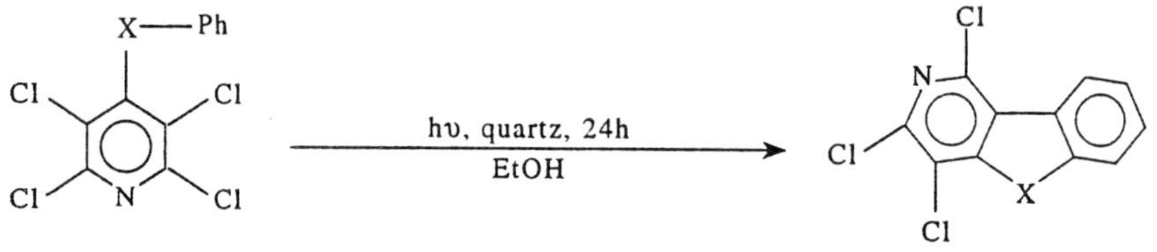

(27)

hυ, quartz, 24h
EtOH

(28) X=S, 68%
X=O, 54%
X=NH, 79%

SCHEME 89.5a (continued)

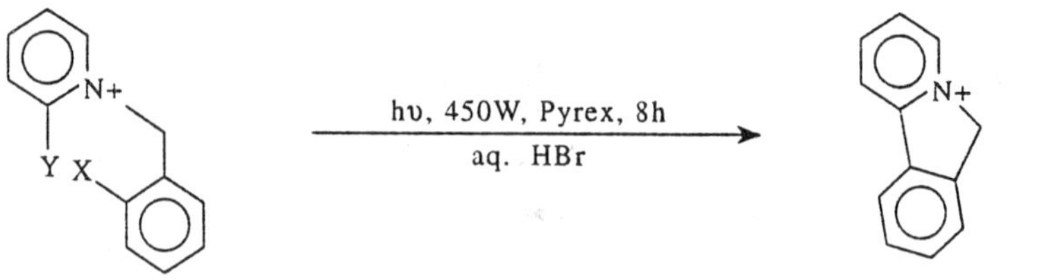

(29) X=Cl or Br, Y=H
X=H, Y=Cl or Br

hυ, 450W, Pyrex, 8h
aq. HBr

(30), 75%
when X=H, Y=Br

SCHEME 89.5b

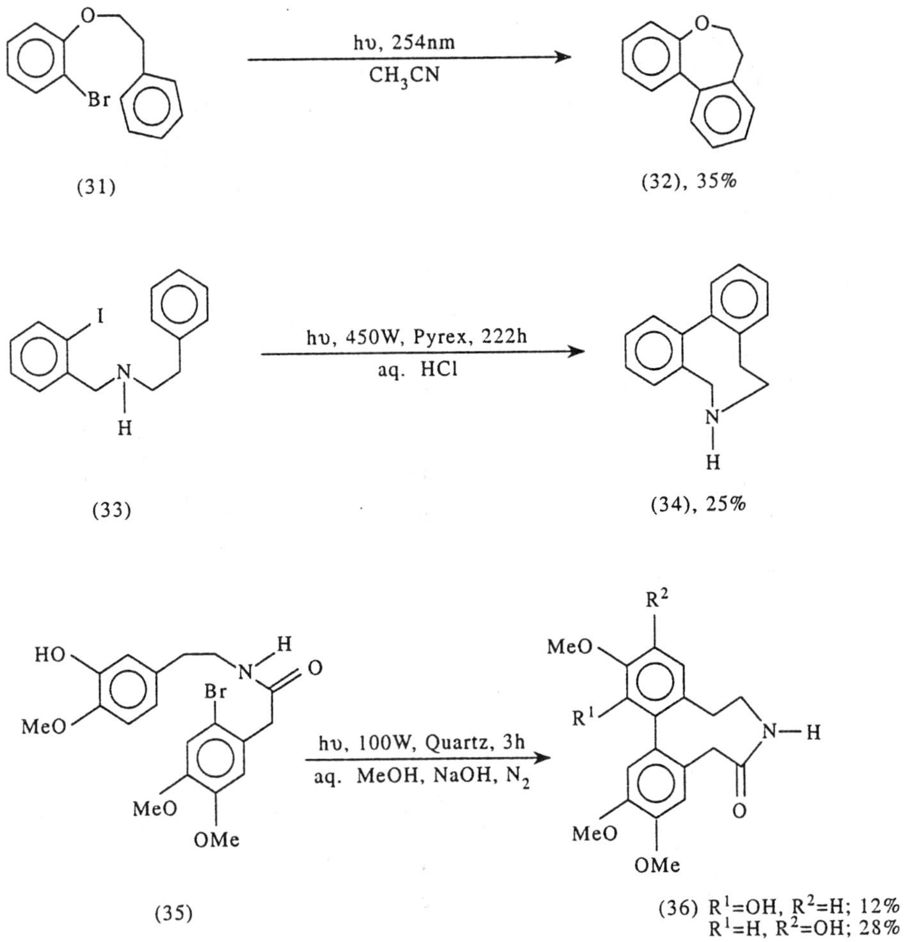

(31)

hυ, 254nm
CH$_3$CN

(32), 35%

(33)

hυ, 450W, Pyrex, 222h
aq. HCl

(34), 25%

(35)

hυ, 100W, Quartz, 3h
aq. MeOH, NaOH, N$_2$

(36) R^1=OH, R^2=H; 12%
R^1=H, R^2=OH; 28%

SCHEME 89.5b (continued)

89.3 Cyclization with Anionic Centers

Displacement of an aryl halogen by a negatively charged center on the side chain does not occur usually in the ground state, but often can be promoted by irradiation. Such photocyclizations may involve (1) electron transfer from the anion to the haloarene, (2) loss of halide ion, and (3) union of the two radical-produced centers. Alternatively, an aryl radical arising by photohomolysis, or through intermolecular SET, may join up with the anionic center to give product radical anion. The latter then transfers, intermolecularly, an electron to the starting aryl halide to sustain a chain S$_{RN}$1 reaction.[36] Often, clear distinction between the two reaction pathways is not possible due to the nonavailability of quantum yield data. In the cyclization of 37 to 38, however, intermolecular electron transfer to the aryl halide from an acetone enolate seems established because the reaction fails in the absence of added acetone.[37]

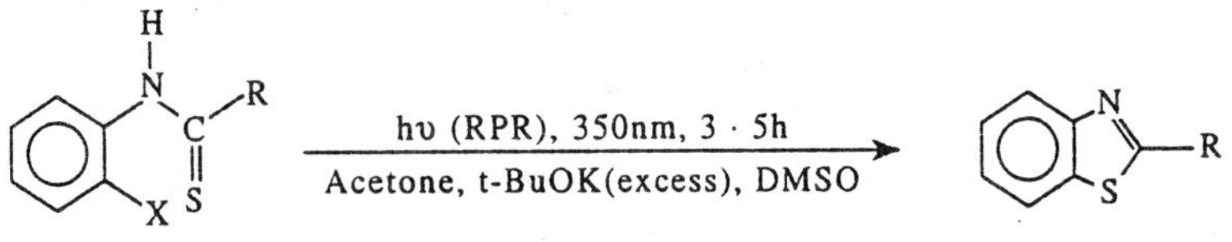

(37)

(38), 100%
when X=1, R=Ph

SCHEME 89.6

Six-, seven-, eight-, and ten-membered ring ketones have also been obtained by light-induced cyclization of enolate anions (39 → 40).[38] This reaction constitutes the key step in an efficient synthesis of cephalotaxine (42).[39] However, a prerequisite for success of such photocyclizations is the absence of hydrogens α to the enolate function; otherwise, these hydrogens **rapidly shift to the aryl radical center.** This problem does not exist in cyclization of amide enolates. For example, chloroanilide 43 furnishes oxindole 44 on irradiation in a basic medium.[40] Suppression of this reaction by small amounts of free-radical inhibitors suggests a chain mechanism. Similarly, 1,4-dihydro-3(2H)-isoquinoline (46) can be obtained from amide 45, and this photoreaction has been used in a novel synthesis of (±) cherylline (46; R^1 = OMe, R^2 = OH, R^3 = Me, R^4 = p-HOC$_6$H$_5$).[40,41]

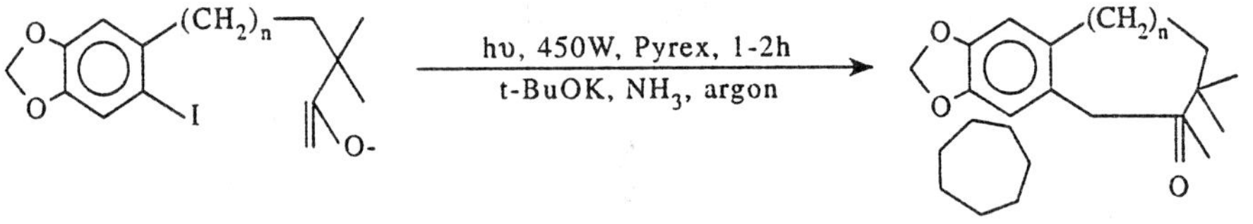

(39)

(40) a; n=0, 99%
    b; n=2, 73%
    c; n=4, 25-35%

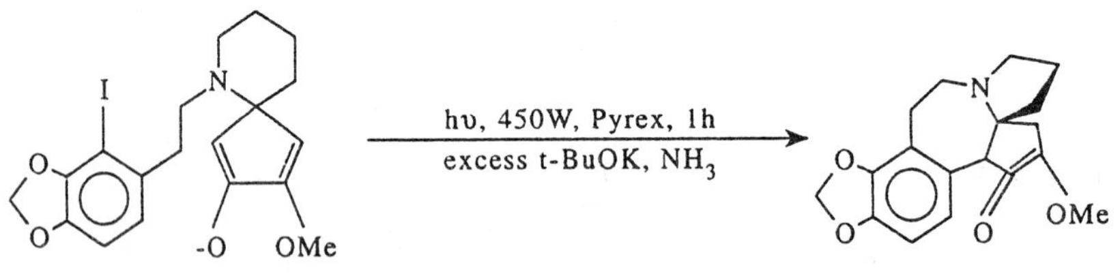

(41)

(42), 94%

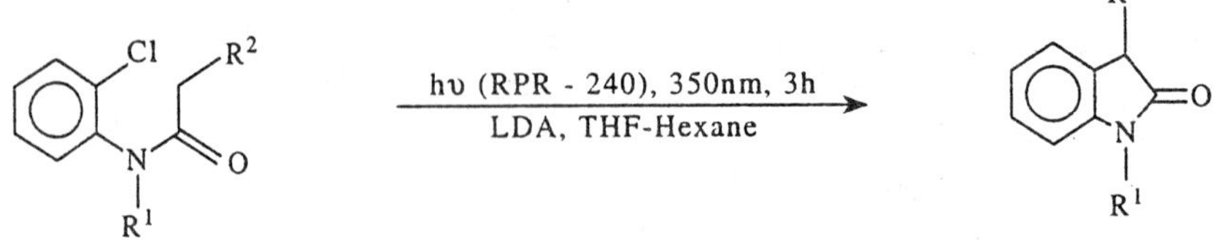

(43)

(44), 60-80%

SCHEME 89.7

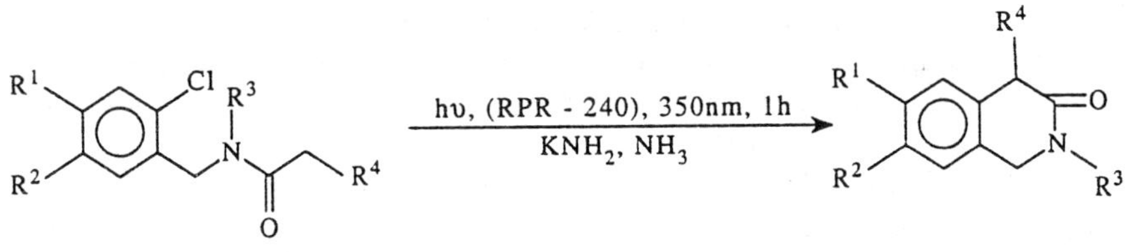

(45) (46), 54-62%

SCHEME 89.7 (continued)

A common and convenient access to berberine and benzo[*c*]phenanthridine alkaloids based on a photoreaction of iodoamide 47 with acetophenone 48 has been reported.[42] The key tricyclic intermediate 49 is considered to arise through arylation of the enolate anion, followed by condensation of the carbonyl group with the amine. An alternate possibility is cyclization of the aryl halide with a carbanion derived by deprotonation of an imine formed initially. In fact, a similar reaction has been realized in the preparation of indole 51 from the Schiff base 50.[43] Finally, readers attention is drawn to the formation of nitrogen heterocycles through the Witkop photocyclization process,[44] although the substrates used, (52), are not haloarenes in a strict sense.

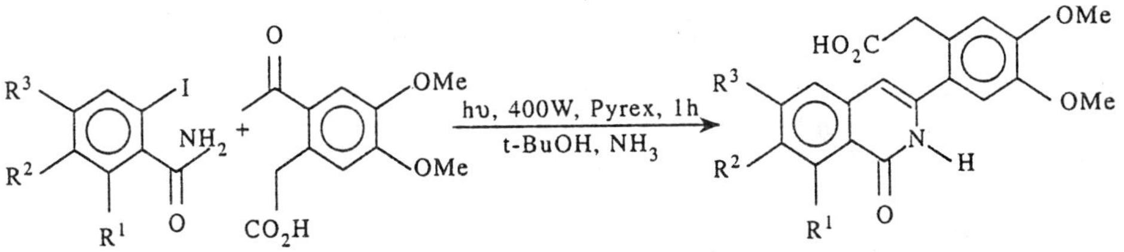

(47) (48) a; R^1=H, R^2=R^3=OMe (49), 70-90%
          b; R^1=R^2=OMe, R^3=H

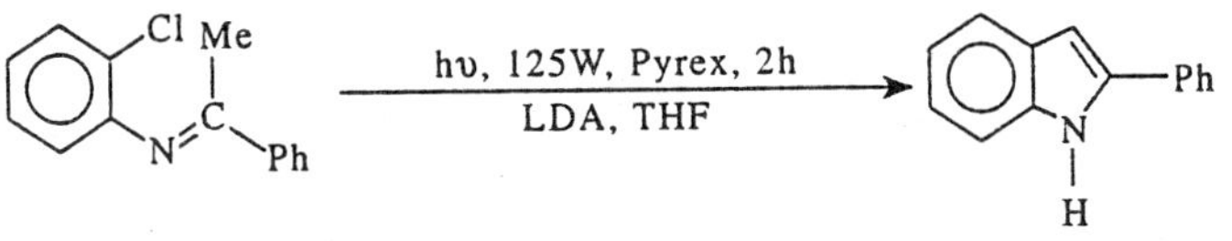

(50) (51), 50%

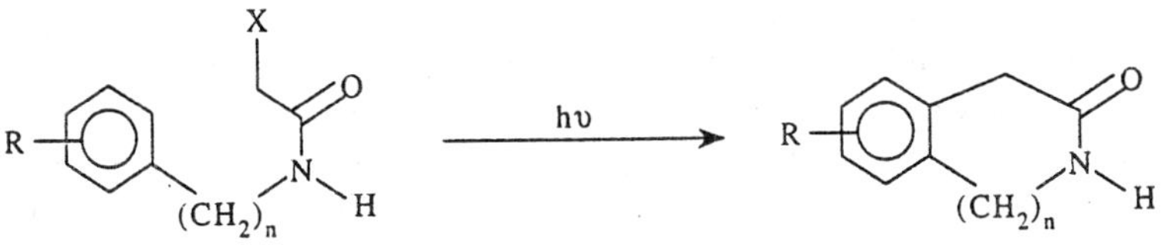

(52) (53)

SCHEME 89.8

References

1. Blair, J. McD., Bryce-Smith, D., and Pengilly, B. W., Liquid-phase photolysis. I. Variation of isomer ratios with radical source in the phenylation of isopropylbenzene. Photolytic generation of phenyl radicals, *J. Chem. Soc.*, 3174, 1959; Blair, J. McD. and Bryce-Smith, D., Liquid-phase photolysis. II. Iodobenzene, *J. Chem. Soc.*, 1788, 1960.

2. Wolf, W. and Kharasch, N., Photolysis of aromatic iodo compounds as a synthetic tool, *J. Org. Chem.*, 26, 283, 1961; Photolysis of iodoaromatic compounds in benzene, *J. Org. Chem.*, 30, 2493, 1965.

3. (a) Grimshaw, J. and de Silva A. P., Photochemistry and photocyclization of aryl halides, *Chem. Soc. Rev.*, 10, 181, 1981; (b) Mattay, J., Photoinduced electron transfer in organic synthesis, *Synthesis*, 233, 1989; (c) Gilbert, A., Photoaddition and photocyclization processes of aromatic compounds, in *Synthetic Organic Photochemistry*, Horspool, W. M., Ed., Plenum, New York, 1984.

4. Whenever possible, information on reaction medium, irradiation time, wattage (mercury arc lamp in an immersion well, unless specified otherwise), and filter, if any, is indicated in the equations.

5. Kupchan, S. M. and Kanojia, R. M., Photochemical synthesis of aporphines, *Tetrahedron Lett.*, 5353, 1966; Neumeyer, J. L., Oh, K. H., Weinhardt, K. K., and Neustadt, R. R., The chemistry of aporphines. IV. Synthesis of aporphines via Reissert alkylation, photochemical cyclization, and the Pschorr cyclization route, *J. Org. Chem.*, 34, 3786, 1969.

6. Kupchan, S. M., Moniot, J. L., Kanojia, R. M., and O'Brien, J. B., Photochemical synthesis of aporphines, *J. Org. Chem.*, 36, 2413, 1971.

7. Kessar, S. V., Gupta, Y. P., and Mohammad, T., Photocyclization of 1-(2'halo-α-hydroxybenzyl)-1,2,3,4-tetrahydroisoquinolines: Synthesis of (±)-oliveridine, *Ind. J. Chem.*, 20B, 984, 1981.

8. Kametani, T., Takahashi, K., Honda, T., Ihara, M., and Fukumoto, K., Studies on the syntheses of heterocyclic compounds. CDLXXXVII. Photolytic synthesis and rearrangement of proerythrina-dienones, *Chem. Pharm. Bull.*, 20, 1793, 1972.

9. Horii, Z., Nakashita, Y., and Iwata, C., Photochemical routes to proaporphines. A new synthesis of pronuciferine, *Tetrahedron Lett.*, 1167, 1971.

10. Horii, Z., Iwata, C., Wakawa, S., and Nakashita, Y., Photolysis of 2-bromo-N-ethyl-4'-hydroxybenzanilide in aqueous alkali, *J. Chem. Soc., Chem. Commun.*, 1039, 1970.

11. Kametani, T., Kohno, T., Charubala, R., and Fukumoto, K., Studies on the synthesis of heterocyclic compounds. CDLXVII. Photolytic synthesis of kreysiginine-type compounds, *Tetrahedron*, 88, 3227, 1972.

12. Cava, M. P. and Libsch, S. S., A total synthesis of Cassamedine, *J. Org. Chem.*, 39, 577, 1974.

13. Cleaver, L., Nimgirawath, S., Ritchie, E., and Taylor, W. C., The alkaloids of Elmerrillia Papuana (*Magnoliaceae*): Structure and synthesis of elmerrillicine, *Aust. J. Chem.*, 29, 2003, 1976.

14. Cava, M. P., Stern, P., and Wakisaka, K., An improved photochemical aporphine synthesis. New synthesis of dicentrine and cassameridine, *Tetrahedron*, 29, 2245, 1973.

15. Mallory, F. B. and Mallory, C. W., Photocyclization of stilbenes and related molecules, in *Organic Reactions*, Dauben, W. G., Ed., Wiley, New York, Vol. 30, 1984, 1.

16. Grimshaw, J. and de Silva, A. P., Photocyclisation of 2-halogenobenzanilides: An extreme example of halogen atom, solvent, and isomer dependence. A practical phenanthridine synthesis, *J. Chem. Soc., Chem. Commun.*, 302, 1980.

17. Kessar, S. V., Singh, G., and Balakrishnan, P., Studies in synthetic photochemistry. I. Synthesis of naphthaphenanthridine alkaloids, *Tetrahedron Lett.*, 2269, 1974; Ninomiya, I., Naito, J., and Ishii, H., *Heterocycles*, 307, 1975.

18. Hey, D. H., Jones, G. H., and Perkins, M. J., Internuclear cyclisation. XXX. The photolysis of 2-iodo-2'-, -3'-, and -4'-methoxy-N-alkylbenzanilides in benzene, *J. Chem. Soc., Perkin Trans. 1*, 1150, 1972.

19. Mondon, A. and Krohn, K., Synthesis of narciprimine and related compounds, *Chem. Ber.*, 105, 3726, 1972; Zee-Cheng, R. K.-Y., Yan, S.-J., and Cheng, C. C., Antileukemic activity of ungeremine and related compounds. Preparation of analogues of ungeremine by a practical photochemical

reaction, *J. Med. Chem.*, 21, 199, 1978; Begley, W. J. and Grimshaw, J., Synthesis of nitidine (8,9-dimethoxy-5-methyl-2,3-methylene-dioxybenzo[*c*]phenanthridinium); A comparison of electro-chemical and photochemical methods, *J. Chem. Soc., Perkin Trans. 1*, 2324, 1977.

20. Grimshaw, J. and de Silva, A. P., The hydrogen bond as a configurational lock in the photocyclisation of dibenzoylmethane *o*-halogenoanils: Wavelength dependence of this reaction, *J. Chem. Soc., Chem. Commun.*, 301, 1980.

21. Prabhakar, S., Lobo, A. M., and Tavares, M. R., Boron Complexes as control synthons in photocyclisations: An improved phenanthridine synthesis, *J. Chem. Soc., Chem. Commun.*, 884, 1978.

22. Lenz, G. R., Enamide photochemistry. Formation of oxyprotoberberines by the elimination of *ortho* substituents in 2-aroyl-1-methylene-1,2,3,4-tetra-hydroisoquinolines, *J. Org. Chem.*, 39, 2839, 1974; Kametani, T., Sugai, T., Shoji, Y., Honda, T., Satoh, F., and Fukumoto, K., Studies on the synthesis of heterocyclic compounds. 698. An alternative protoberberine synthesis; total synthesis of (±)-xylopinine, (±)-schefferine, (±)-nandinine, (±)-corydaline, and (±)-thalictricavine, *J. Chem. Soc., Perkin Trans. 1*, 1151, 1977.

23. Ninomiya, I. and Naita, T., Preparation of heterocyclic compounds, in *Photochemical Synthesis*, Academic Press, London, 1989, chap. 7 and 11; Kanaoka, Y. and Itoh, K., Photocyclization of benzanilides to phenanthridones with elimination of the *ortho*-methoxy-group, *J. Chem. Soc., Chem. Commun.*, 647, 1973.

24. Latowski, T., *Z. Naturforsch. Teil A*, 23, 1127, 1968; Grodowski, M. and Latowski, T., A study on a photochemical reaction in the system *N,N*-dimethylaniline -Bromobenzene, *Tetrahedron*, 30, 767, 1974.

25. Nasieliski, J. and Kirsch-Demesmaeker, A., Photochemistry of aromatic compounds. Mechanism of the amine-accelerated photodebromination of 5-bromo-2-methoxypyrimidine, *Tetrahedron*, 29, 3153, 1973; Ohashi, M., Tsujimoto, K., and Seki, K., Photoreduction of 4-chlorobiphenyl by aliphatic amines, *J. Chem. Soc., Chem. Commun.*, 384, 1973.

26. Beecroft, R. A. Davidson, R. S., and Goodwin, D., The amine assisted photo-dehalogenation of halo-aromatic hydrocarbons, *Tetrahedron Lett.*, 24, 5673, 1983.

27. Tse, I. and Snieckus, V., Photochemical preparation of dihydro-pyrrolo[2,1-*b*][3]benzazepines. A *Cephalotaxus* alkaloid synthon, *J. Chem. Soc., Chem. Commun.*, 505, 1976; Iida, H., Takarai, T., and Kibayashi, C., Facile synthesis of hexahydroapoerysopine via intramolecular photoarylation of β-enamino ketones, *J. Org. Chem.*, 43, 975, 1978.

28. Pac, C., Tosa, T., and Sakurai, H., Photochemical reactions of aromatic compounds. IX. Photo-chemical reactions of dimethylaniline with halobenzenes, *Bull. Chem. Soc. Jpn.*, 45, 1169, 1972.

29. Kessar, S. V., Gupta, Y. P., Dhingra, K., Sharma, G. S., and Narula, S., Studies in synthetic photochemistry-II synthesis of chelilutine and sanguilutine, *Tetrahedron Lett.*, 1459, 1977; Smidrkal, J., *Collect. Czech. Chem. Commun.*, 49, 1412, 1984.

30. Grimshaw, J. and de Silva, A. P., Photocyclization of aryl halides. I. 5-(2-Halogenophenyl)-1,3-diphenylpyrazoles; homolytic fission assisted by radical complexation, *Can. J. Chem.*, 58, 1880, 1980.

31. Bratt, J. and Suschitzky, H., Photochemical cyclisation of substituted polyhalogenopyridines, *J. Chem. Soc., Chem. Commun.*, 949, 1972; Bratt, J., Iddon, B., Mack, A. G., Suschitzky, H., Taylor, J. A., and Wakefield, B. J., Polyhalogenoaromatic compounds. 41. Photochemical dehalogenation and arylation reactions of polyhalogenoaromatic and polyhalogenoheteroaromatic compounds, *J. Chem. Soc., Perkin Trans. 1*, 648, 1980.

32. Fozard, A. and Bradsher, C. K., The synthesis of pyrido-[2,I-*a*]isoindole system by an intramolecu-lar photochemical cyclization, *J. Org. Chem.*, 32, 2966, 1967; Bradsher, C. K. and Voigt, C. F., Electrophilic substitution of the pyrido[2,I-*a*]isoindole system, *J. Org. Chem.*, 36, 1603, 1971.

33. Brightwell, N. E. and Griffin, G. W., Photorearrangement of 9,10-epoxy-9,10-dihydrophenanthrene. Synthesis of 2,3:4,5-dibenzoxepin, *J. Chem. Soc., Chem. Commun.*, 37, 1973.

34. Jeffs, P. W. and Hansen, J. F., Synthesis of a medium ring containing bridge biphenyl by photo-chemically induced intramolecular arylation, *J. Am. Chem. Soc.*, 89, 2798, 1967; Jeffs, P. W.,

Hansen, J. F., and Brine, G. A., Photochemical synthesis of 6,7-dihydro-5H-dibenz[c,e]azepine and 5,6,7,8-tetrahydrodibenz[c,e]-azocine derivatives, *J. Org. Chem.*, 40, 2883, 1975.

35. Ito, K. and Tanaka, H., Studies on the erythrina alkaloids. VIII. Alkaloids of *Erythrina* X bidwillii LINDL. (4). Synthesis of erybidine by photochemical reaction, *Chem. Pharm. Bull.*, 22, 2108, 1974.

36. Moon, M. P., Kamin, A. P., Wolfe, J. F., and Morris, G. F., Photostimulated reactions of 2-bromopyridine and 2-chloroquinoline with nitrile-stabilized carbanions and certain other nucleophiles, *J. Org. Chem.*, 48, 2392, 1983; Bunnett, J. F., Aromatic substitution by the $S_{RN}1$ mechanism, *Acc. Chem. Res.*, 11, 413, 1978; Wolfe, J. F. and Carver, D. R., *Org. Prep. Proced. Int.*, 10, 224, 1978; Chanon, M., Tobe, M. L., et al., A mechanistic concept for inorganic and organic chemistry, *Angew. Chem. Int. Ed.*, 21, 1, 1982; Rossi, R. A. and deRossi, R. H., in *Aromatic Substitution by the $S_{RN}1$ Mechanism*, American Chemical Society, Washington, D.C., 1983; ACS Monogr. No. 1978.

37. Bowman, W. R., Heaney, H., and Smith, P. H. G., Intramolecular aromatic substitution ($S_{RN}1$) reactions: Use of entrainment for the preparation of benzothiazoles, *Tetrahedron Lett.*, 23, 5093, 1982.

38. Semmelhack, M. F. and Bargar, T., Photostimulated nucleophilic aromatic substitution for halides with carbon nucleophiles. Preparative and Mechanistic aspects, *J. Am. Chem. Soc.* 102, 7765, 1980.

39. Semmelhack, M. F., Chong, B. P., Stauffer, R. D., Rogerson, T. D., Chong, A., and Jones, L. D., Total synthesis of the *Cephalotaxus* alkaloids. A problem in nucleophilic aromatic substitution, *J. Am. Chem. Soc.*, 97, 2507, 1975.

40. Goehring, R. R., Sachdeva, Y. P., Pisipati, J. S., Sleevi, M. C., and Wolfe, J. F., Photoinduced cyclization of mono- and dianions of N-acyl-o-chloroanilines and N-acyl-o-chlorobenzylamines as general methods for the synthesis of oxindoles and 1,4-dihydro-3(2H)-isoquinolinones, *J. Am. Chem. Soc.*, 107, 435, 1985.

41. Kessar, S. V., Singh, P., Chawla, R., and Kumar, P., Cyclization of *ortho*-halogenated N-acylbenzylamines: A formal synthesis of (±)-cherylline, *J. Chem. Soc., Chem. Commun.*, 1074, 1981.

42. Beugelmans, R., Ginsburg, H., and Bois-Choussy, M., Studies on $S_{RN}1$ reactions. Synthesis of 3-methyl derivatives of the alkaloids thalactamine, doryanine, and 6,7-dimethoxy-N-methyl-1(2H)-isoquinolone, *J. Chem. Soc., Perkin Trans. 1*, 1149, 1982; Beugelmans, R. and Bois-Choussy, M., One-pot synthesis of 1-oxo-1,2-dihydroisoquinolines (isocarbostyrils) via $S_{RN}1$ (Ar) reactions, *Synthesis*, 729, 1981; Beugelmans, R., Chastanet, J., and Roussi, G., Studies on $S_{RN}1$ reactions. 9. A new access to the isoquinoline ring system, *Tetrahedron*, 40, 311, 1984; Beugelmans, R., Chastanet, J., Ginsburg, H., Quintero-Cortes, L., and Roussi, G., Direct synthesis of benzo[c]-phenanthridines and benzo[c]phenanthridones via $S_{RN}1$ reactions, *J. Org. Chem.*, 50, 4933, 1985.

43. Kessar, S. V., Singh, P., and Dutt, M., Intramolecular reaction of 1-azaallylic anions and arylhalides: A synthesis of isoquinolines, *Ind. J. Chem.*, 28B, 365, 1989.

44. Sundberg, R. J., Chloroacetamide photocyclization and other aromatic alkylations initiated by photo-induced electron transfer, in *Organic Photochemistry*, Padwa, A., Ed., Vol. 6, Marcel Dekker, New York, 1983, 121.

90

Photochemistry of Alkyl Hypohalites

Hiroshi Suginome
Hokkaido University

90.1 A Short Historical Background

Alkyl hypohalites (ROX; X = Cl, Br, I), a class of molecules that can be prepared from the corresponding alcohols by a variety of methods,[1] dissociate to alkoxyl radicals and halogen atoms with high efficiency upon irradiation with light having a wavelength longer than 300 nm (Scheme 90.1).

$$\text{ROX} \xrightarrow{\quad h\nu \quad} \text{RO·} \quad + \quad \text{X·}$$

$$\text{X=Cl, Br, I}$$

SCHEME 90.1

Methyl and ethyl hypochlorites, the first alkyl hypohalites known, were prepared by Sandmeyer in 1885.[2] Chattaway and Backeberg reported subsequently the preparation of alkyl hypochlorites in 1923,[3] such as propyl, isopropyl, and *t*-butyl hypochlorites, and observed that although the primary and secondary hypochlorites are very unstable and decompose rapidly at room temperature, the tertiary hypochlorites (such as *t*-butyl hypochlorite) prepared by the reaction of chlorine, sodium hydroxide, and *t*-butyl alcohol, are moderately stable, pale-yellow liquids that can be distilled without decomposition.[3] Moreover, they found that *t*-butyl hypochlorite decomposed upon exposure to bright sunlight to give methyl chloride and acetone (Scheme 90.2). During the following years, simple alkyl hypochlorites received attention, principally as free-radical chlorinating reagents and oxidants (as summarized in review articles by Anbar and Ginsburg[1a] and Walling[4]). *t*-Butyl hypochlorite converts toluene into benzyl chloride and allylic position of olefins are chlorinated.[5]

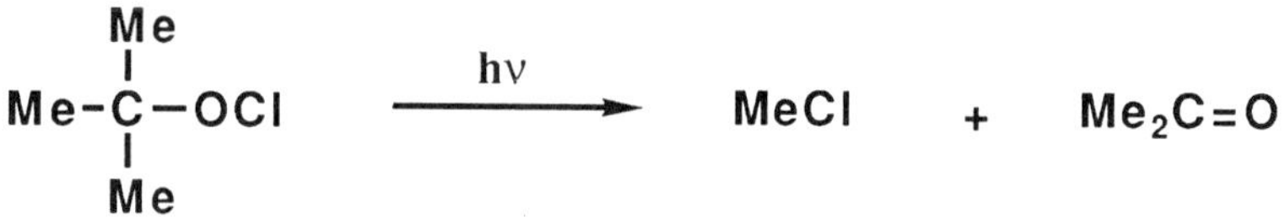

SCHEME 90.2

Since this early work, studies from the late 1950s until the 1960s by Walling,[6-8] Green,[9] Kochi,[10] and their collaborators using *t*-alkyl hypohalites disclosed some fundamental characteristics of the principal radical chain reactions, such as intermolecular hydrogen abstraction and β-scission of intermediate alkoxyl radicals in hypochlorlite photolysis. This work was summarized in a review article by Kochi.[11]

The photochemistry of alkyl hypohalites, however, has attracted considerable attention by organic chemists since the 1960s when Barton and colleagues,[12] a group of Swiss chemists,[13,14] as well as Mills and Petrow,[15] demonstrated that alkyl hypochlorites and hypoiodites can be utilized in a manner parallel to the organic nitrites for the functionalization of an unactivated carbon atom based on a concept delineated by Barton.[16,17] The photolysis of steroidal hypohalites having an appropriate constitution provoked an intramolecular exchange of the halogen of the ROX group with a hydrogen atom attached to a carbon atom in the δ-position (Scheme 90.3). An analogous intramolecular δ-hydrogen abstraction of long-chain hypochlorites was also reported by Walling and Padwa,[18] Green and colleagues,[9] as well as Jenner.[19] Both the mechanistic and synthetic aspects of remote functionalization via the photolysis of hypohalites have been the subjects of comprehensive review articles by Akhtar[20] as well as Heusler and Kalvoda.[1b,15]

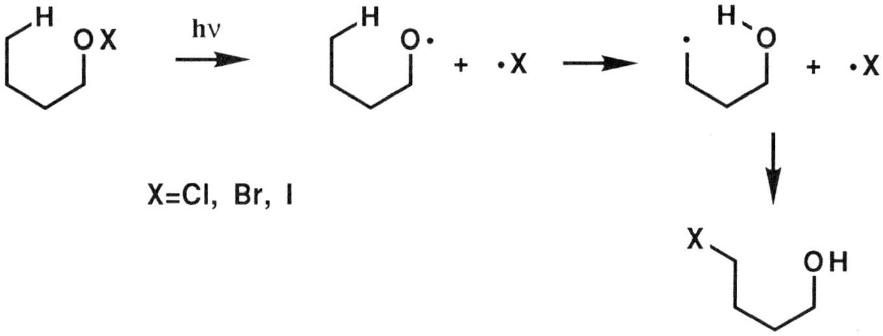

X=Cl, Br, I

SCHEME 90.3

A more recent investigation by Suginome and co-workers has demonstrated that the selective fragmentation of alkoxyl radicals, one of the principal reactions of alkoxyl radicals, can be utilized as the key step in the synthesis of a variety of molecules, including several natural products.[21] Suarez and colleagues have also demonstrated the utility of the (diacetoxyiodo)benzene and iodine procedure in hypoiodite photolysis by their series of studies.[22]

90.2 Photochemistry of *t*-Alkyl Hypochlorites

Alkyl hypochlorites in carbon tetrachloride exhibit two absorption peaks in the UV region: one at about 260 nm and the other at about 310–320 nm.[12,23] 3β-acetoxy-20-methylallopregnan-20-ol hypochlorite in carbon tetrachloride exhibits two absorption maxima at 258 nm (ε, 107) and 318 nm (ε, 9.5). Although the corresponding hypobromites reveal similar absorptions, the peaks are displaced to longer wavelengths; *t*-butyl hypobromite exhibits an absorption maximum at 280 nm (ε, 120) and a long-tail absorption extending into the visible region.[7c] The UV spectra of alkyl hypoiodites, which are unisolated species, have not been reported. Their absorption maxima, however, should appear at even longer wavelengths.

Homolysis of the RO-X bond takes place when alkyl hypohalites in an inert solvent are irradiated with light having a wavelength longer than 300 nm.[24] Homolysis is also initiated by azoisobutyronitrile or thermolysis.[4,6] This characteristic of alkyl hypohalites differs from those of the corresponding nitrites in solution;[17] homolysis of the O-N bond can only be achieved by photolysis.

The competitive nature of the principal reactions of the alkyl hypohalites was investigated using *t*-butyl-hypochlorite[25] and hypobromite.[6c] Walling and collaborators found that *t*-butyl-hypochlorite and hypobromite are efficient free-radical chlorinating agents for saturated aliphatic hydrocarbons.[6] They also investigated the photoinduced allylic chlorination of olefins with *t*-butyl-hypochlorite in great detail.[7] Previously there had been sporadic reports on this from the 1940s through the 1950s.[7] They found that a variety of olefins react with *t*-butyl-hypochlorite via a photoinduced radical chain process to give a good yield of allylic chlorides in preference to an addition to give β-chloroalkyl *t*-butyl-ethers.

Another principal competing reaction of *t*-alkyl-hypohalites in a photoinduced reaction is a β-scission of the alkoxyl radicals to give alkyl halides and ketones. The factors that affect the relative rates of ejection of the alkyl radical in the β-scission of alkoxyl radicals has been extensively investigated by Walling, Green, Kochi, and their collaborators[8–10] by carrying out the photodecomposition of a variety of *t*-alkyl-hypochlorites. The results indicated an increasing β-scission in the order methyl < ethyl < isopropyl < benzyl < *t*-butyl (Scheme 90.5). The high degree of selectivity found in the β-scission of alkoxyl radicals can be attributed to a polar effect in the transition state[10,11] (Scheme 90.6). Studies with some hypochlorites of aliphatic alcohols by Green and collaborators[9] indicated that intermolecular hydrogen abstraction from cyclohexene as the solvent competes poorly with β-scission (Scheme 90.7). The results outlined in Scheme 90.8 also indicate clearly that intramolecular hydrogen abstraction by the alkoxyl radical via a six-membered transition state predominates over fragmentation and intermolecular allylic hydrogen abstraction. For more details concerning this subject, the reader should refer to the original literature and a review article.[11]

$$Me_3C-OX \xrightarrow{h\nu} Me_3C-O\cdot \; + \; X\cdot$$

$$Me_3C-O\cdot \; + \; RH \longrightarrow Me_3C-OH \; + \; R\cdot$$

$$R\cdot \; + \; Me_3C-OX \longrightarrow RX \; + \; Me_3C-O\cdot$$

$$X=Cl, \; Br$$

SCHEME 90.4

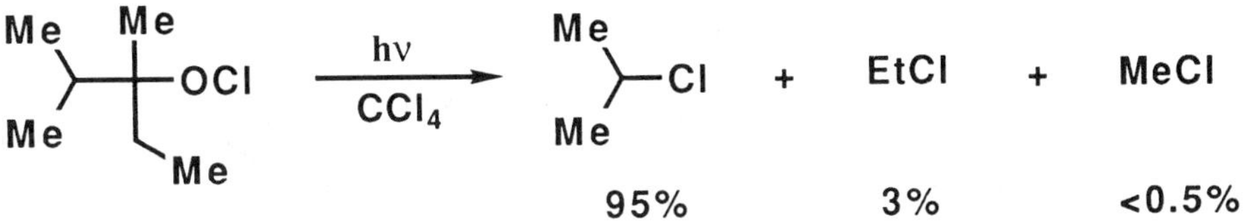

SCHEME 90.5

SCHEME 90.6

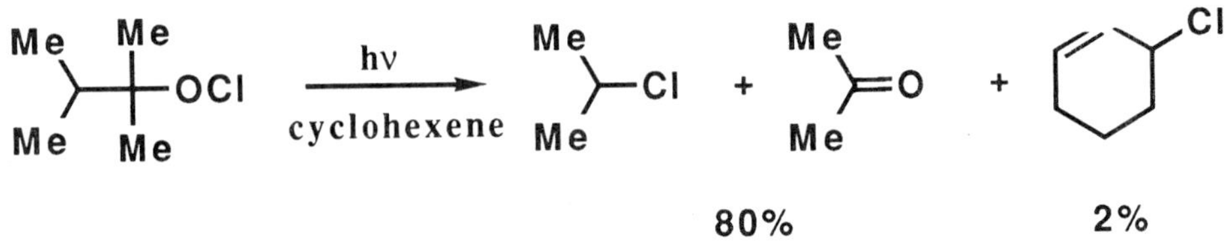

SCHEME 90.7

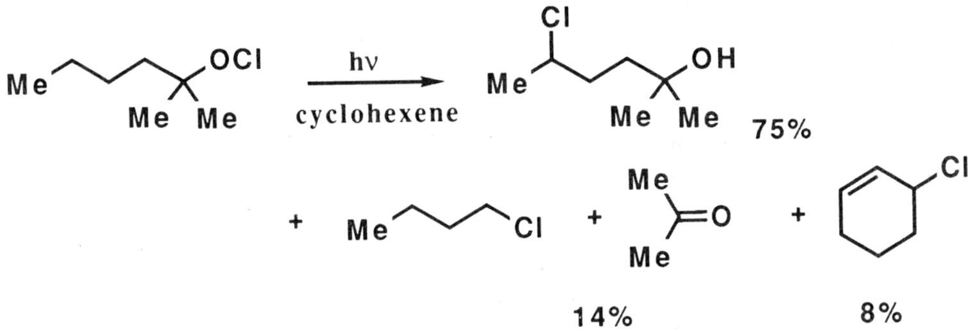

80%   2%

SCHEME 90.8

75%

14%   8%

90.3 Photochemistry of Alkyl Hypoiodites and Synthetic Applications

Among the three competing principal reactions — inter- and intramolecular hydrogen abstractions and β-scission — in the photolysis of alkyl hypohalites, synthetic applications of intramolecular hydrogen abstraction and β-scission of alkoxyl radicals to organic synthesis are described below.

Remote Functionalization by Hypohalites

In 1960, Barton and colleagues suggested that hypohalite photolysis could be used for remote functionalization in a manner analogous to nitrite photolysis.[16,17] Subsequently, Akhtar and Barton[12] demonstrated a successful remote functionalization in the photolysis of steroidal hypochlorite (outlined in Scheme 90.9). The photolysis of 3β-acetoxy-6α-methylcholestan-6β-ol hypochlorite (prepared from the parent 6β-ol and chlorine monoxide) in dry benzene with Pyrex-filtered light gave a chlorohydrin, which in treatment with a base gave 6,19-oxide in 50% yield,[12] through a radical chain process.[6c] Thus, functional groups can be introduced into the unactivated angular 10β- or 13β-methyl groups of steroids by the photolysis of hypohalites, as in the photolysis of nitrites. Scheme 90.10 illustrates the positions of the alkoxyl radicals by which the hydrogen in 10β- and 13β-Me can be abstracted to give the carbon radical. The most favorable O-C distance for the intramolecular abstraction of hydrogen was estimated by a Swiss group to be 2.5 to 2.7 Å on the basis of the results from the remote functionalization of a number of hypoiodites of rigid systems, such as steroids *(vide infra)*, and by an inspection of Dreiding models.[15]

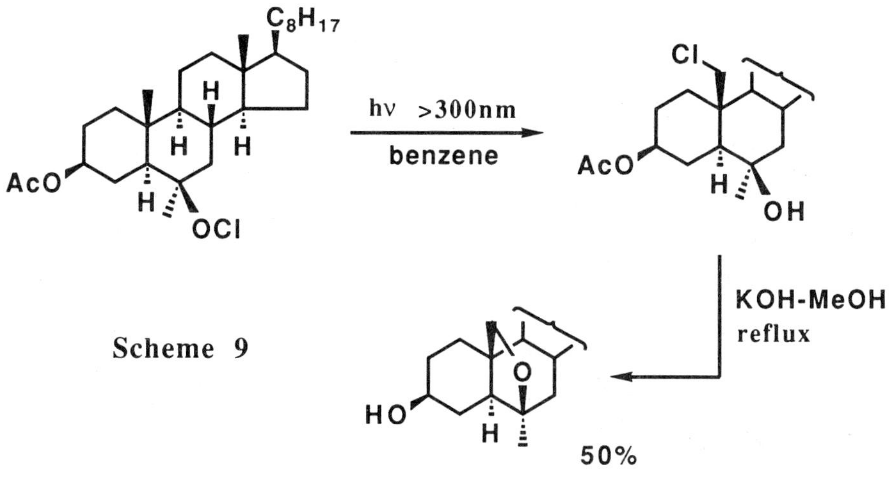

Scheme 9

SCHEME 90.9

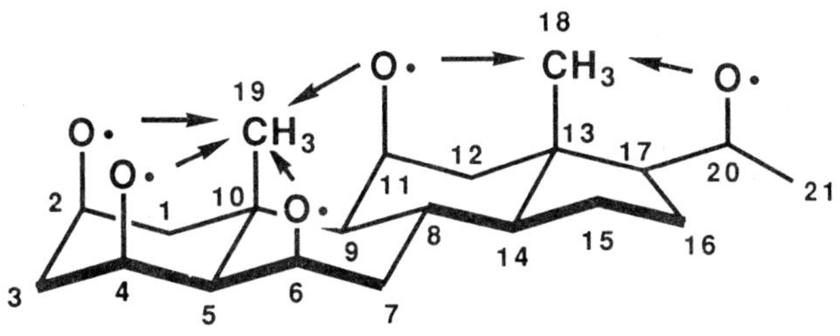

SCHEME 90.10

Although alkyl hypoiodites are unisolated species, they are even more convenient and powerful for remote functionalization;[12,14,15] they can be readily generated by a reaction of alcohols with mercury(II) oxide and iodine,[12] with lead tetraacetate and iodine,[13,14] with iodosobenzene diacetate,[22] or with several other reagents.[12] The stability for the base-catalyzed heterolytic decomposition of secondary hypohalites is of the order hypoiodites > hypobromites > hypochlorites. Moreover, the reactivity of hypoiodites (and iodine) with ketones or carbon-carbon double bonds under neutral conditions decreases in comparison with that of hypochlorites or hypobromites.[15] Remote functionalization can be achieved by irradiating a solution of hypoiodites prepared *in situ* by one of the methods mentioned above. Remote functionalization through the photolysis of alkyl hypoiodite was designated a "hypoiodite reaction" by Heusler and Kalvoda.[15] It is not clear whether the hypoiodite reaction involves a chain process similar to the one postulated for hypochlorite photolysis.

An analogous photoinduced reaction of hypobromites of cedrol and the related bicyclo[3.2.1]octanes generated with mercury(II) oxide and bromine was reported.[26]

In the following, some representative examples among the numerous applications of the hypoiodite reaction to remote functionalization are outlined.

1. *Functionalization of 10β-Me of a steroid from the 6β-ol hypoiodite generated with mercury(II) oxide and iodine[12] (Scheme 90.11).*

A light-induced reaction of steroidal hypoiodites prepared by the reaction of a steroidal alcohol with iodine oxide generated from mercury(II) oxide-iodine generally gives a five-membered oxide that should be formed by an intramolecular S_N2 or S_H2 reaction of an intermediate 19-iodide, as outlined in Scheme 90.11.

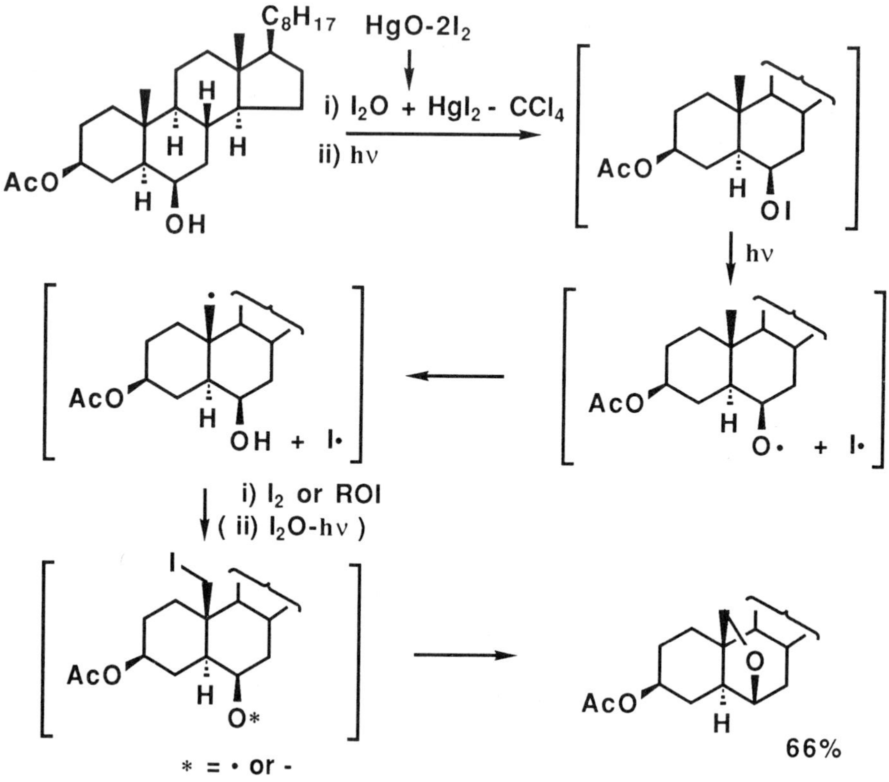

SCHEME 90.11

2. *Functionalization of 10β-Me of a steroid from the 6β-ol hypoiodite generated by lead tetraacetate and iodine (Scheme 90.12).*

The generation of alkyl hypoiodites by the reaction of steroidal alcohols with lead tetraacetate and iodine was found by Meystre and colleagues of the CIBA company in 1961.[14] Among the numerous results reported by them, two typical examples of remote functionalization with this reagent are outlined in Schemes 90.12 and 90.13. The mechanism for the formation of alkyl hypoiodites with lead tetraacetate and iodine could be explained in terms of the initial formation of acetyl hypoiodite, analogous to the formation of acetyl hypochlorite,[27] followed by reactions with steroidal alcohols,[15] as outlined in Scheme 90.14. The products of the remote functionalization of steroids by the hypoiodite reaction are either the five-membered oxide (Scheme 90.12) or the α-lactol acetate (Scheme 90.13), depending on the relative disposition between the relevant alkoxyl radical and the iodine attached to C-18 or C-19 of a conformer of the first-formed intermediate monoiodide. Thus, the five-membered oxide is an exclusive product[14] in the example outlined in Scheme 90.12, since the iodomethyl group can adopt an orientation that is appropriate for an S_H2 or S_N2 displacement of the iodine by the 6β-alkoxyl radical or ion, to give the observed oxide, as shown in Scheme 90.15. On the other hand, the α-lactol acetate is an exclusive product in the example[28] outlined in Scheme

90.13, since the preferred orientation of the iodomethyl group with respect to the 4β-hydroxyl in the first-formed intermediate monoiodide does not allow any displacement of the iodine by the 4β-hydroxyl; however, it is appropriate for the second intramolecular hydrogen abstraction by the 4β-alkoxyl radical to give an iodo-oxide (Scheme 90.16), the iodine of which is displaced by an acetoxy ion to give a lactol acetate. Its oxidation with chromium trioxide then results readily in a transformation into the corresponding lactone. Analogously, the product of the hypoiodite reaction of 20β-hydroxysteroid was the corresponding 18,20β-α-lactol acetate, which was isolated as the corresponding α-lactone by oxidation with chromium trioxide in 72% yield.[29]

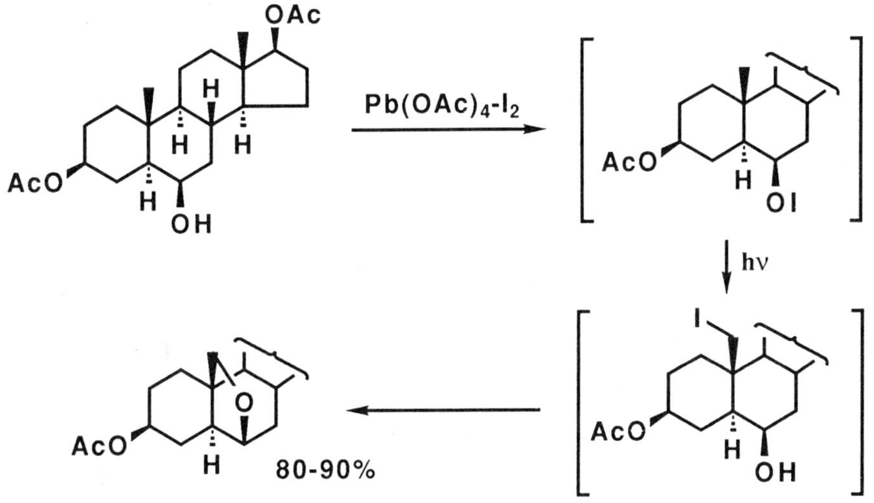

SCHEME 90.12

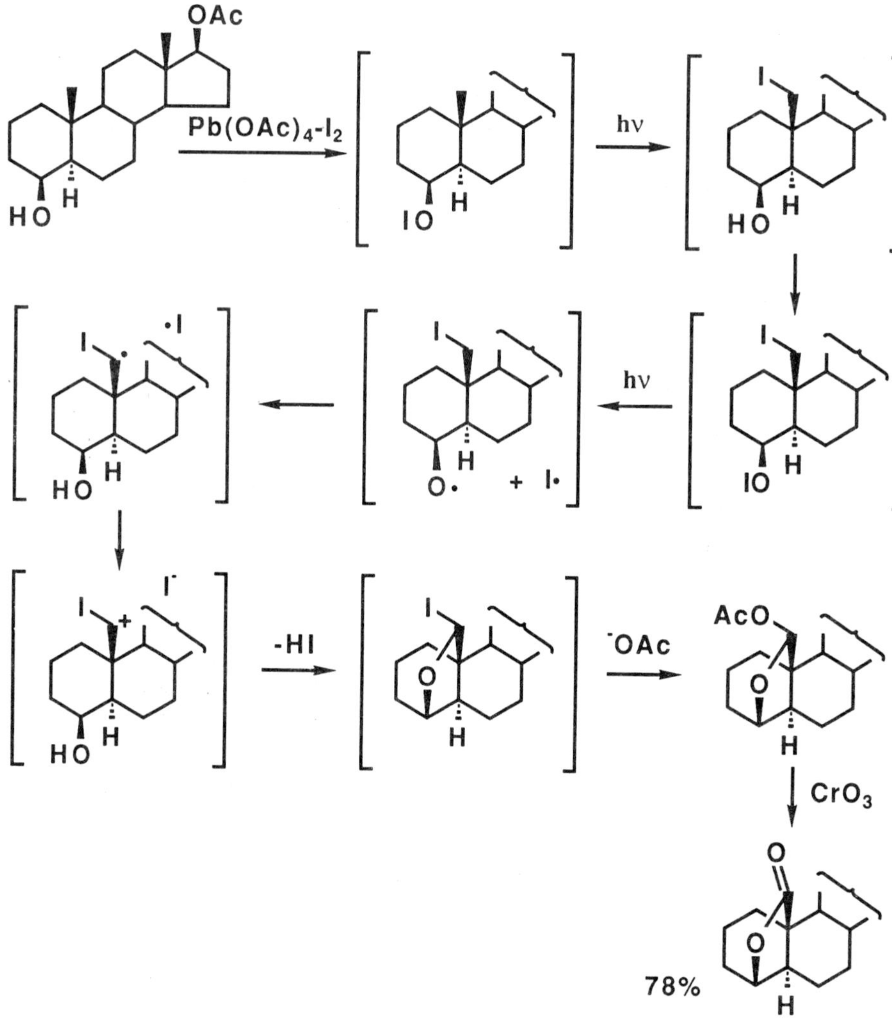

SCHEME 90.13

$$Pb(OAc)_4 + I_2 \longrightarrow Pb(OAc)_2 + 2CH_3COOI$$

$$CH_3COOI \rightleftharpoons CH_3COO^- + I^+$$

$$ROH + I^+ \longrightarrow R{-}\overset{+}{\underset{|\atop H}{O}}{-}I \xrightarrow{AcO^-} ROI + AcOH$$

SCHEME 90.14

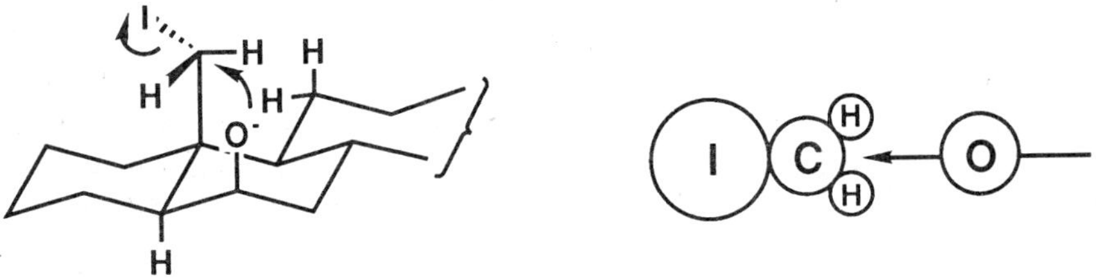

SCHEME 90.15

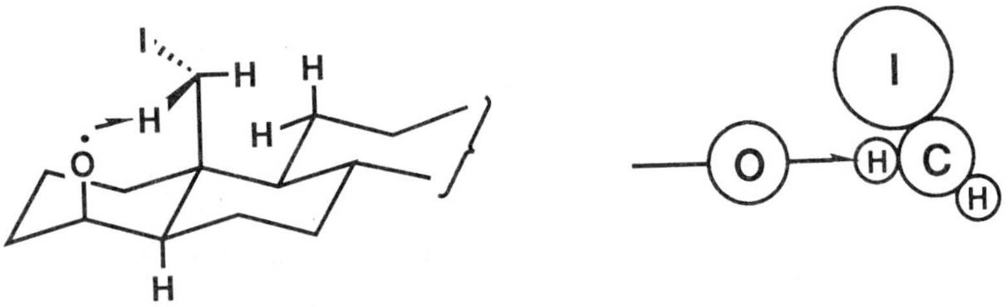

SCHEME 90.16

For a thorough discussion concerning this subject, including more examples and accompanying reactions, the reader should consult the comprehensive review articles cited.[1b,15,20]

3. *"Billiard" reaction[30] in the hypoiodite photolysis.*
The "billiard" reaction — a second hydrogen abstraction from a methyl group located 1,3-diaxially to a carbon radical initially generated by alkoxyl radicals in the hypoiodite reaction — has been observed in certain triterpenoid and diterpenoid skeletons. Scheme 90.17 outlines the reaction path.

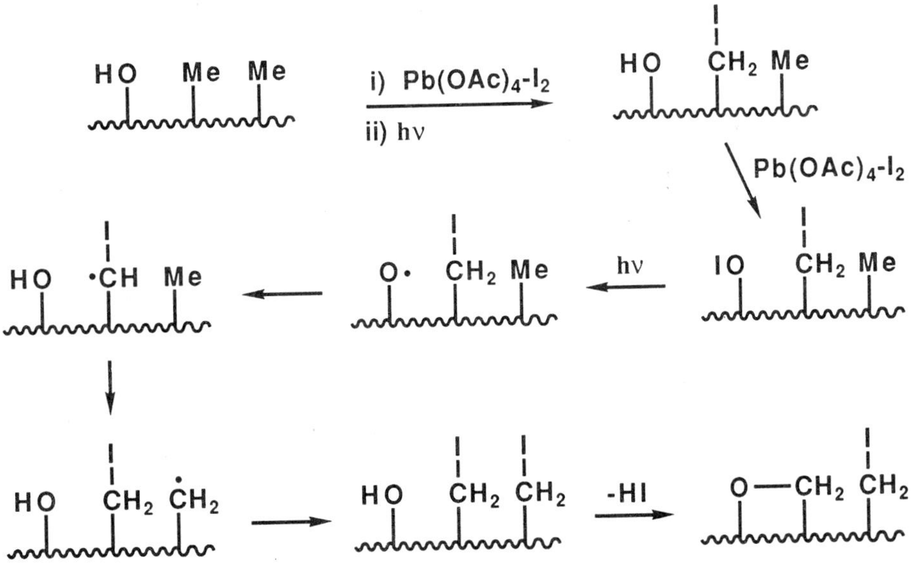

SCHEME 90.17

4. *Functionalization of 10β-Me of steroids by (diacetoxyiodo)benzene (DIB) and iodine.[22]*
Suarez and colleagues[22] have shown that each equivalent of (diacetoxyiodo)benzene (DIB) and iodine in cyclohexane reacts with alcohols to produce the corresponding hypoiodites from which

alkoxyl radicals can be generated by irradiation. The hypoiodites were assumed to be formed by the reaction of the alcohols with acetyl hypoiodite produced *in situ* by the reaction of DIB with iodine (Scheme 90.18). The reaction of the alcohols with acetyl hypoiodite gives the alkyl hypoiodite, as in the case of the lead tetraacetate-iodine procedure. The major products from intramolecular hydrogen abstraction by the reagent were reported to be cyclic ether and monoiodo-alcohol derivatives. However, the products are more complicated in the substrates where double intramolecular hydrogen abstraction can take place.

SCHEME 90.18

Two examples[22] concerning functionalization of the 10β-methyl group of a steroid from the 6β-ol- and 4β-ol-hypoiodites reported by Suarez are outlined in Schemes 90.19 and 90.20 for comparison with the products from the lead tetraacetate-iodine procedure outlined in Schemes 90.12 and 90.13.

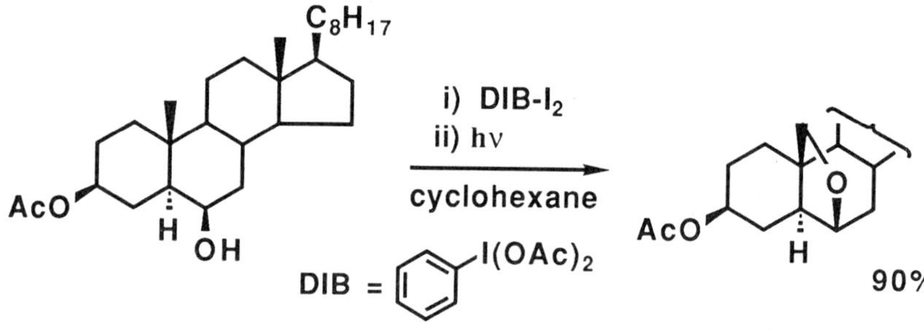

SCHEME 90.19

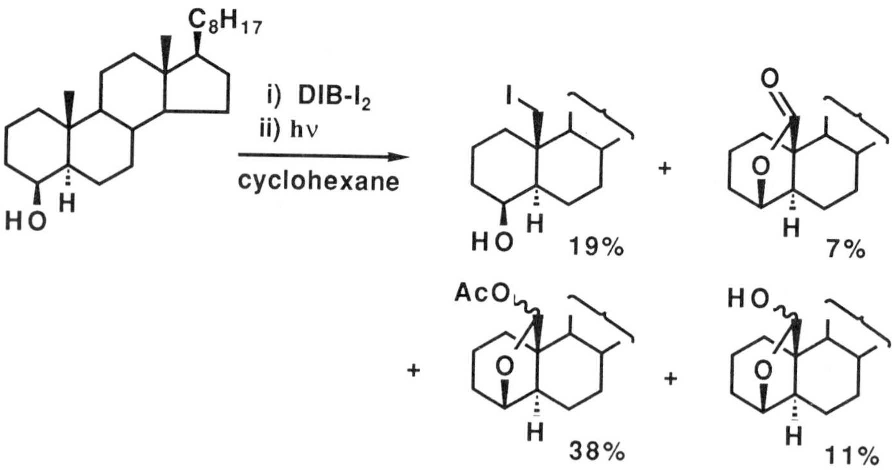

SCHEME 90.20

5. *The oxidative cyanohydrin-cyanoketone rearrangement. (Schemes 90.21 and 90.22).*
An especially interesting example of the hypoiodite reaction is oxidative cyanohydrin-cyanoketone rearrangement.[14,31] which is represented by the general pathway outlined in Scheme 90.21. The hypoiodite reaction of 20-hydroxy-20-cyanosteroids thus gives the 18-cyanoketones arising from a 1,4-shift of the cyano group accompanied by the loss of two atoms of hydrogen (Scheme 90.22).[14]

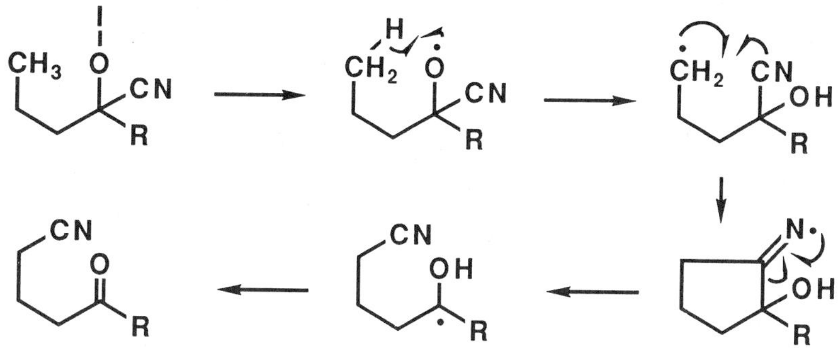

SCHEME 90.21

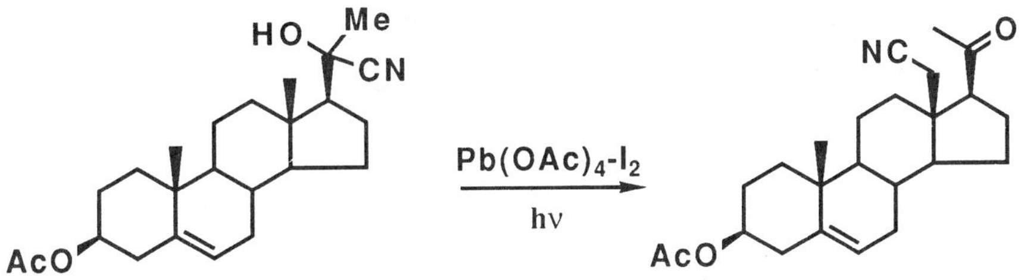

SCHEME 90.22

6. *Intramolecular functionalization via a many-membered cyclic transition state (Schemes 90.23 and 90.24).*

Successful intramolecular hydrogen abstraction via a many-membered cyclic transition state by alkoxyl radicals generated by the photolysis of hypohalites has recently been shown; alkoxyl radicals generated by the irradiation of hypoiodites of 5α-cholestan-7α-yl-(hydroxymethyl)-phenylacetates and -phenylpropionates, respectively, abstracted a hydrogen from the remote C-25 of their cholestane side chain to give novel macrocyclic ether lactones, which gave 5α-cholestane-7α,25-diol by reduction with Na and liquid ammonia, (as outlined in Scheme 90.23[32]). A similar one-step introduction of a carbonyl group to C-15 of the 5α-androstane skeleton, based on a long-range intramolecular hydrogen abstraction by alkoxyl radicals, generated by irradiation of 5α-androstane esters carrying a benzhydryl group in carbon tetrachloride containing mercury(II) oxide and iodine, is outlined in Scheme 90.24.[33] Suginome and co-workers also reported a one-step double introduction of a carbon-carbon double bond and oxygen functions to ring-C of 5α-steroid skeletons, based on a long-range intramolecular hydrogen abstraction by alkoxyl radicals generated from the esters of 5α-cholestan-3α-ol carrying a benzhydryl group.[34]

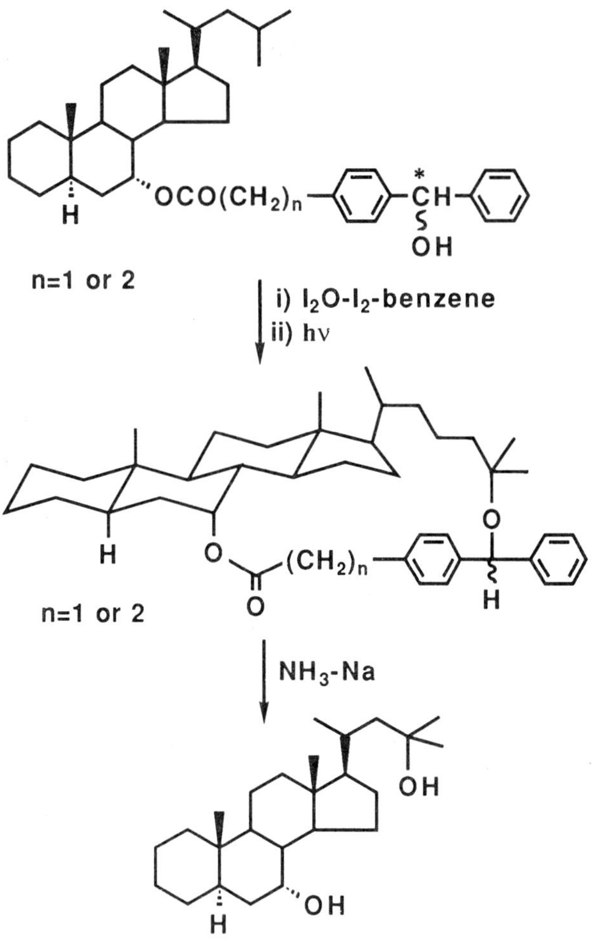

n=1 or 2

i) I$_2$O-I$_2$-benzene

ii) hν

n=1 or 2

NH$_3$-Na

SCHEME 90.23

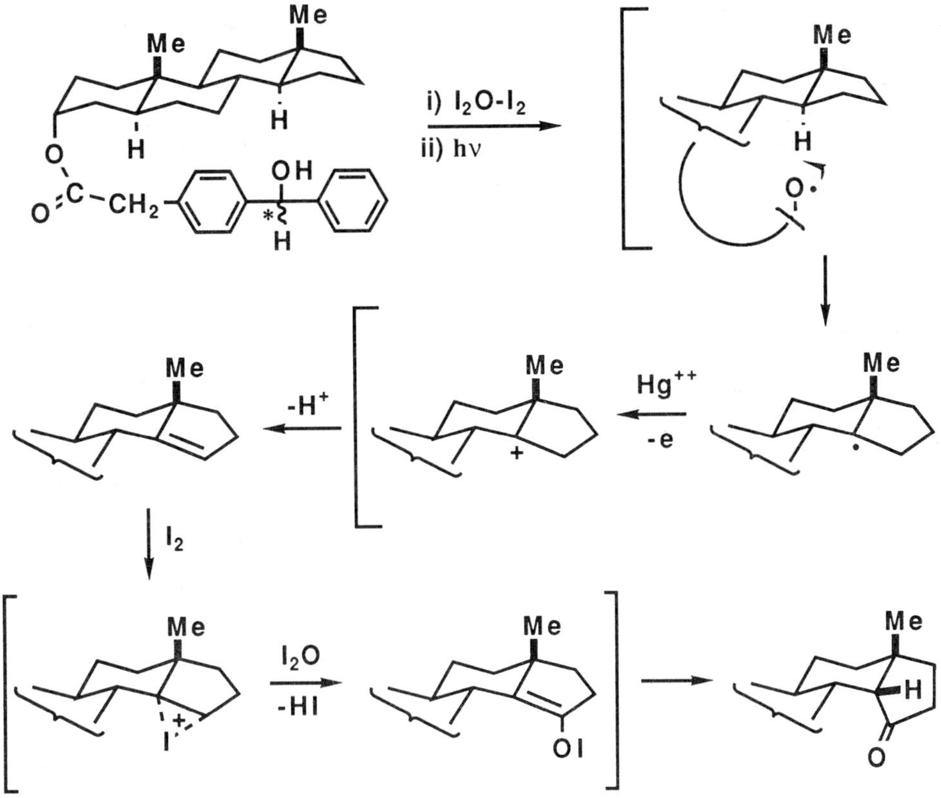

SCHEME 90.24

These results indicate that hypoiodite photolysis is a powerful method that is applicable to even the remote functionalization involving a many-membered transition state, since alkoxyl radicals are repeatedly generated from the regenerated hypoiodites in a solution containing an excess at the reagent.

Synthetic Application of the β-Scission of Alkoxyl Radicals Generated from Hypoiodites

In contrast to the intramolecular hydrogen abstraction of alkoxyl radicals mentioned above, radical fragmentation, another principal reaction, was seldom used in organic synthesis until the 1970s. An analysis of the numerous examples reported in the past and Suginome's results concerning the photochemistry of alkyl hypohalites indicates that the direction of β-scission in unsymmetrical substrates is the outcome of an interplay of multiple factors, such as the relative thermodynamic stability of the resultant radicals, ring strain, and stereoelectronic factors.

This situation is in contrast to the heterolytic cleavage of the bonds (such as the ring-fusion bond) in bicyclic compounds, in which the cleavage requires a bifunctional substrate with strict stereoelectronic constraints.[35] Nevertheless, the β-scission is frequently not only the predominent reaction, but also takes place in a highly selective manner and comprises an exclusive reaction when the alkoxyl radical and the hydrogen to be abstracted are not ideally disposed for intramolecular hydrogen abstraction. The β-scission is especially enhanced in the *hypoiodite* reaction when alkoxyl radicals appear to be repeatedly generated from regenerated parent alcohols by excess reagent. Since the 1980s, Suginome and colleagues have shown that a variety of molecules, including natural products, can be synthesized using β-scission of alkoxyl radicals generated by the photolysis of hypoiodites as the key step.[36] The classes of the molecules synthesized include heterosteroids,[6c,21a]

19- and 18-norsteroids,[21d,e] steroidal lactones,[21t] benzohomotropones,[21g] 18-functionalized steroids,[21h] lignans,[21i] medium sized lactones,[21j] macrolides,[21k,l] phthalides,[21m] naphthalide lignans,[21n] monocyclic lactones,[21o] macrocyclic ketones,[21p] furanoheterocycles,[21r] furanoquinones,[21t] isocoumarins,[21u] sesquiterpene,[21v,w,x] and others.[21y]

In the following, some representative examples, including the formation paths, when they are necessary, are outlined. These selected examples indicate that the β-scission of alkoxyl radicals takes place in a very selective manner and has become an integral part of synthetic strategies. For more detail, the reader should consult the original papers[21] and reviews[36,37] cited.

1. *Transformation of cyclic alcohols into cyclic ethers.[21a]*

Irradiation of the hypoiodites, generated *in situ* by means of the reaction of hydroxysteroids with excess mercury(II) oxide and iodine with Pyrex-filtered light, gives iodo formates that arise from successive reactions triggered by a β-scission of the corresponding alkoxyl radical. These formates can be transformed readily into oxasteroids by treatment with either a complex metal hydride or methyl lithium. The reaction is exemplified by the transformation of androstan-17β-ol into the corresponding oxasteroids, as outlined in Scheme 90.21. The result of ^{18}O-labeling experiments with ^{18}O-labeled mercury(II) oxide as the source of $I_2{}^{18}O$ is also shown in Scheme 90.25.

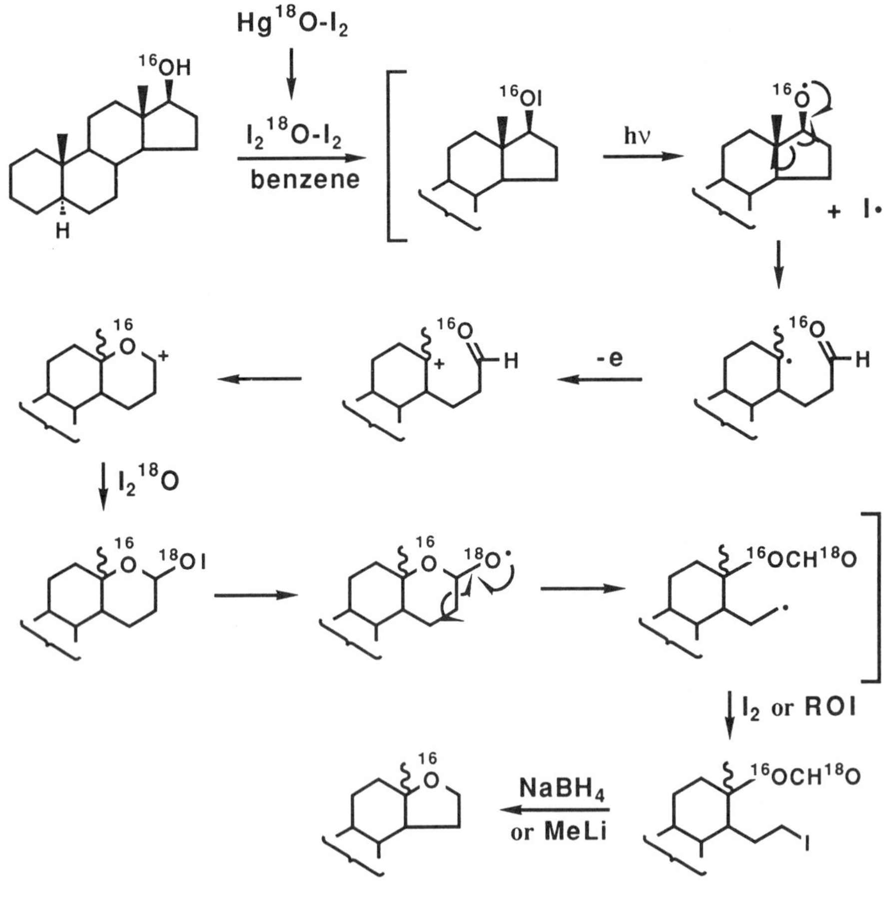

SCHEME 90.25

2. *Four- to Six-step transformations of cyclic ketones into cyclic alkanes with one hetero atom (O, N, S, Te, Se) in their rings.[21b,c]*

Based on the above-mentioned reaction path for the formation of formates from cyclic alcohols, Suginome and co-workers found that the alkoxyl radicals generated by irradiating the hypoiodites generated *in situ* by means of the reaction of the lactol by excess mercury(II) oxide and iodine and pyridine in benzene gives formates that arise from a *regiospecific β-scission of the C-C bond* (as outlined in Scheme 90.26). The formates can then be transformed into cyclic alkanes with one oxygen, sulfur, nitrogen, tellurium, and selenium in their rings in four- to six-step processes. A variety of heterosteroids from the corresponding steroidal ketones were synthesized by these methods. This reaction has recently been applied to the key step in the new synthesis of lignans,[21i] as outlined in Scheme 90.27.

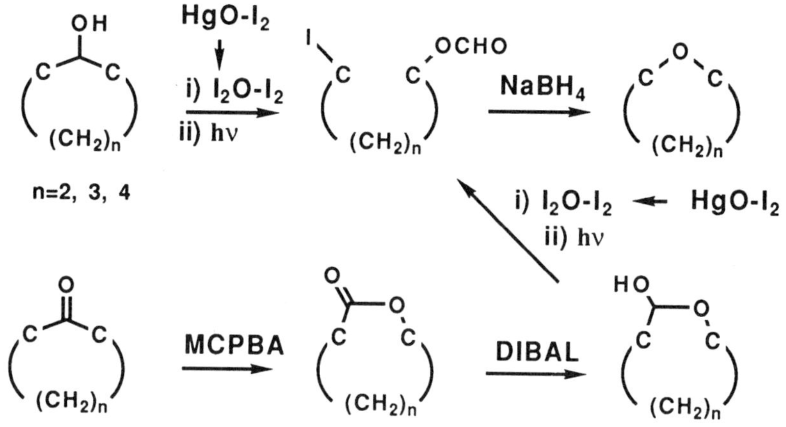

SCHEME 90.26

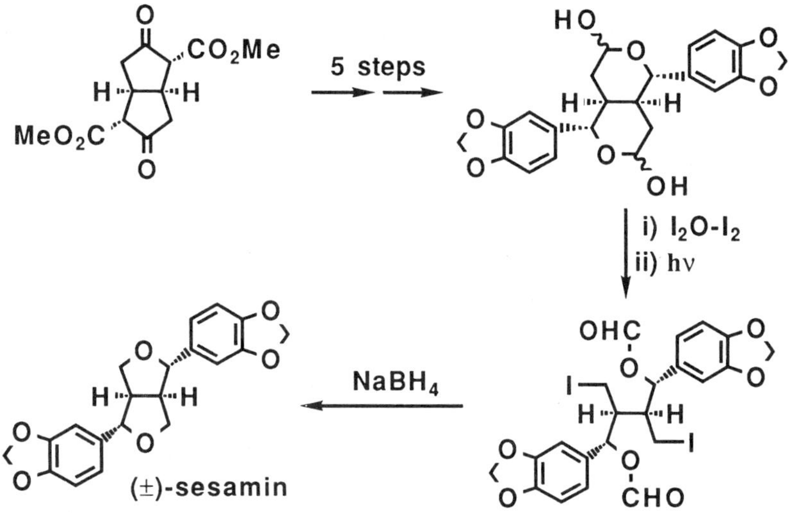

SCHEME 90.27

3. *New general synthesis of medium-ring lactones via a regioselective β-scission of alkoxy radicals generated from the hypoiodites of catacondensed lactols.*[21j]

The above-mentioned regioselective β-scission can be extended further to the synthesis of medium-ring lactones. Irradiation of the hypoiodite of catacondensed lactols gives lactones arising from a regioselective cleavage of the inner bond of catacondensed lactols. The reduction of iodo-lactones with tributyltin hydride gives medium-ring lactones, including a natural 10-membered lactone, (±)-phoracantholide I, in high yields, as outlined in Scheme 90.28. A number of 9- to 11-membered lactones were synthesized from 6/5, 6/6, 7/5, 7/6, and 8/5-fused lactols by this method. In contrast to the established ring expansion involving an ionic cleavage, such as retro-aldolization, the synthesis can be carried out under virtually neutral conditions, and no extra functional group is necessary for the cleavage. Subsequent to our study, similar ring expansion of steroids by (diacetoxyiodo)benzene-iodine procedure has been reported by Suarez and colleagues.[38]

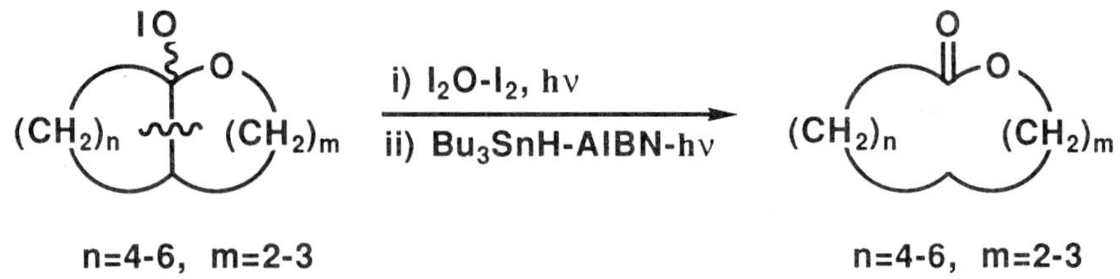

$n=4-6, m=2-3$ $n=4-6, m=2-3$

SCHEME 90.28

4. *New general synthesis of macrolides via a consecutive intramolecular homolytic addition-β-scission of alkoxyl radicals generated from hypoiodites.*[21k,l]

The synthesis of 11- to 17-membered macrolides can be achieved by the photolysis of the hypoiodites of open-hydroxy ketones (as outlined in Scheme 90.29). Here, the iodo-lactones are formed through a consecutive homolytic process, an intramolecular homolytic addition of the alkoxyl radicals to the carbonyl carbon, followed by a regioselective β-scission of the bridged bond of the generated alkoxyl radicals to the macrocyclic radical that traps an iodine atom to give the iodo-lactone.

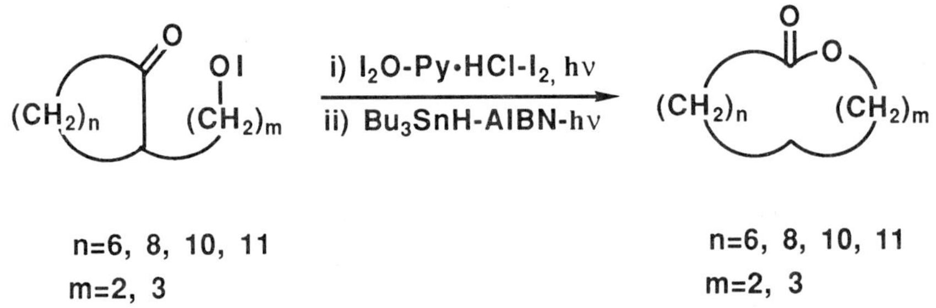

$n=6, 8, 10, 11$
$m=2, 3$ $n=6, 8, 10, 11$
                                        $m=2, 3$

SCHEME 90.29

(±)-Recifeiolide,[211] a natural 12-membered macrolide, was synthesized from cyclooctanone based on this process (as outlined in Scheme 90.30).

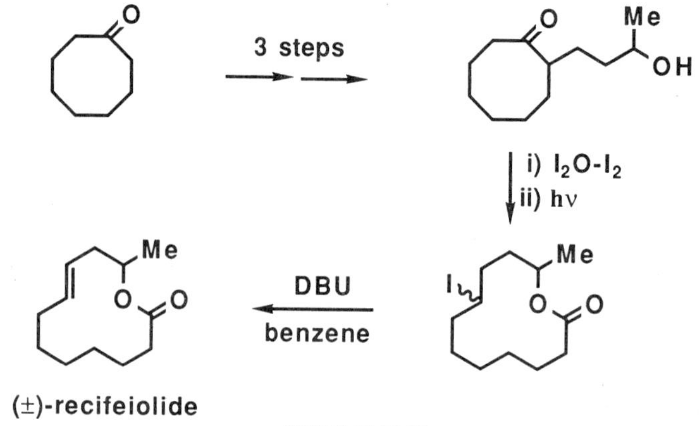

(±)-recifeiolide

SCHEME 90.30

5. *New short-step general synthesis of isobenzofuran-1(3H)-ones (phthalides) based on double β-scission of alkoxyl radicals generated from hypoiodites.*[21m,n]

The photolysis of the hypoiodites of 1-ethylbenzocyclobuten-1-ols, prepared by a treatment of the 1-ols with mercury(II) oxide and iodine, gives the corresponding phthalides in good yields. The pathway is outlined in Scheme 90.31. The principal feature of this process is an intramolecular combination of a carbocation or, less likely, a carbon radical with a carbonyl oxygen generated by a regioselective β-scission of the alkoxyl radicals.

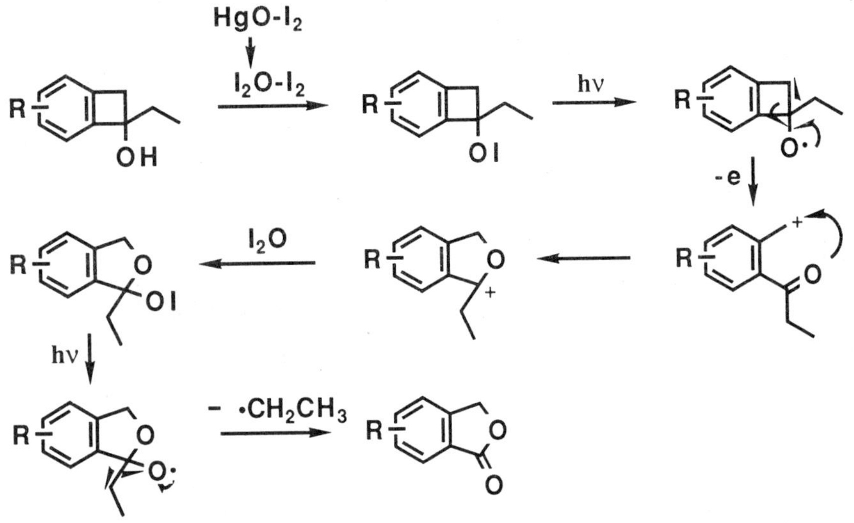

SCHEME 90.31

Among several applications of this process to natural product synthesis, a new synthesis of naphthalide lignans[21n] is outlined in Scheme 90.32.

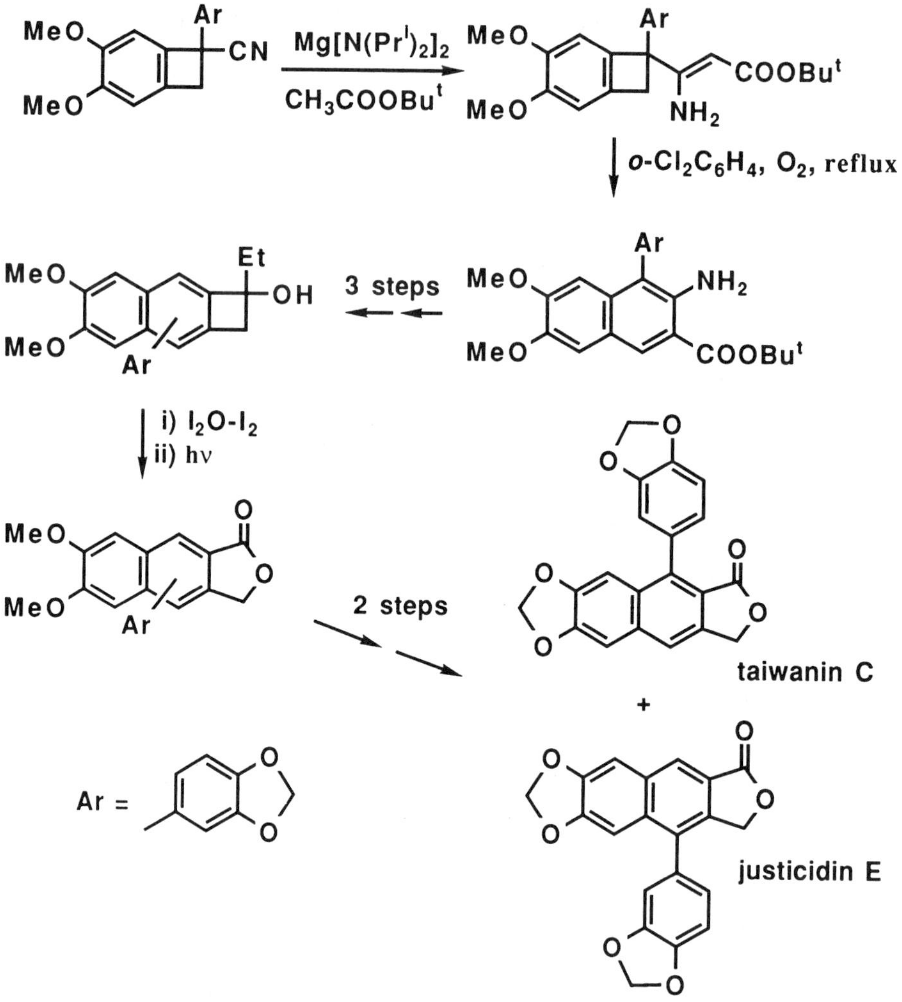

SCHEME 90.32

6. *The general synthesis of macrocyclic ketones based on a ring expansion by the photolysis of the hypoiodites of bicyclic alcohols.[21p,v,39]*

A macrocyclic ketone can be synthesized by the irradiation of hypoiodites prepared from catacondensed bicyclic alcohols analogous to the synthesis of the macrolides mentioned above. The products of the ring expansion are either iodo-ketone and/or olefinic ketones (as outlined in general Scheme 90.33).

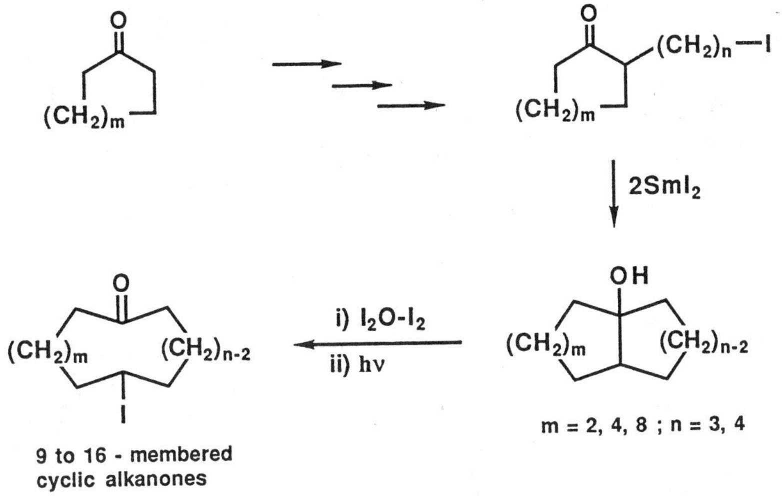

SCHEME 90.33

Schemes 90.34 and 90.35 outline the new synthesis of (±)-muscone[21p] and (±)-caryophyllene[21v] based on this method as the key step.

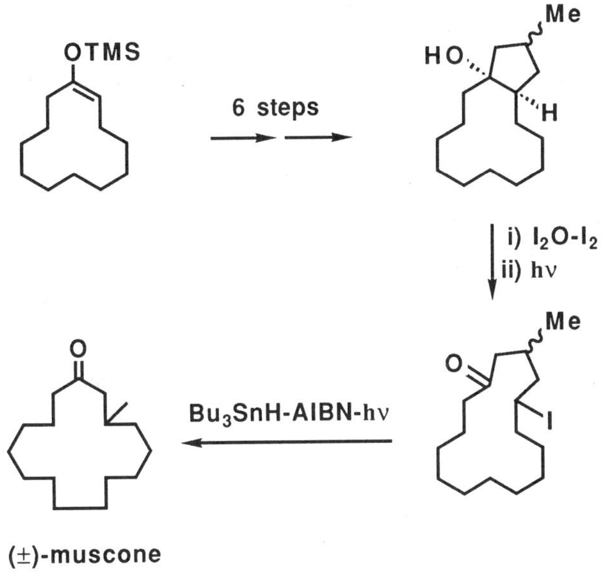

(±)-muscone

SCHEME 90.34

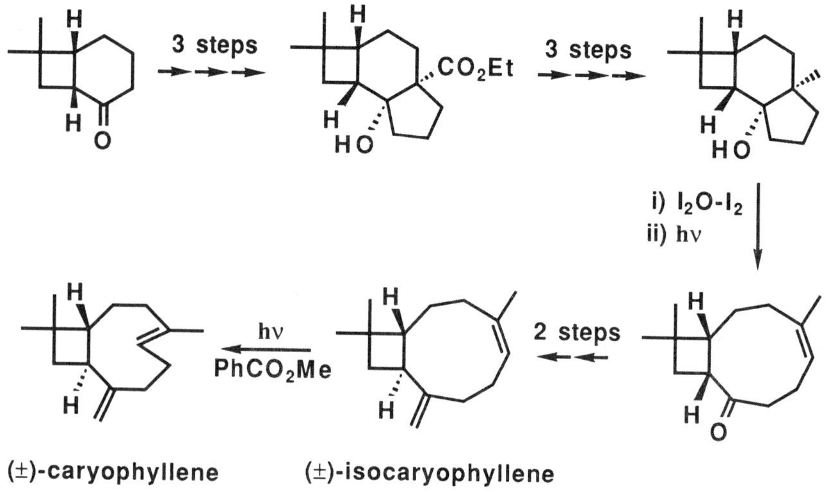

(±)-**caryophyllene**   (±)-**isocaryophyllene**

SCHEME 90.35

7. *The construction of 5/7, 5/8, 6/7, and 6/8 bicyclic systems based on [2+2] Photoaddition,*
   followed by the photolysis of the hypoiodites of the resulting cyclobutanols.[21w,x]

Functionalized bicyclo[x.y.0]alkanes (x = 5 or 6; y = 7 or 8) can be synthesized by the irradiation
of hypoiodites prepared by [2+2] photoaddition of cyclic enones with the trimethylsilyl enol ethers
of cyclic ketones (as outlined in general Scheme 90.36).[21w]

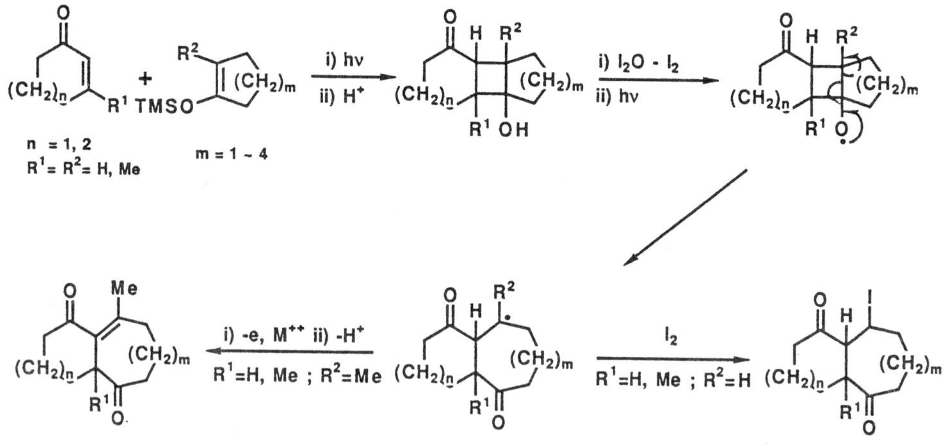

SCHEME 90.36

Scheme 90.37 outlines the new total synthesis of (±)-α-himachalene based on this method as the
key step.[21x]

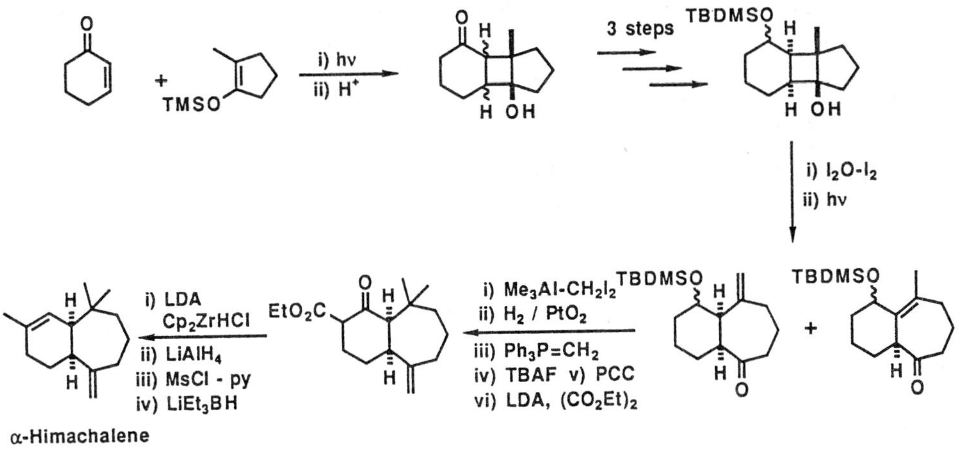

SCHEME 90.37

References

1. For reviews, see (a) Anbar, M. and Ginsburg, D., Organic hypohalites, *Chem. Rev.*, 54, 925, 1954; (b) Heusler, K. and Kalvoda, J., Selective functionalization of the angular methyl group and further transformation to 19-norsteroids, in *Organic Reactions in Steroid Chemistry*, Fried, J. and Edwards, J. A., Eds., van Nostrand Reinhold, New York, 1972, chap. 12.

2. Sandmeyer, T., Ueber den Aethylester der Unterchlorigen Säure, *Ber.*, 18, 1767, 1885; Ueber Aethyl- und Methylhypochlorit, *Ber.*, 19, 857, 1886.

3. Chattaway, F. D. and Backeberg, O. G., Alkyl hypochlorites, *J. Chem. Soc.*, 2999, 1923.

4. For a review, see: Walling, C., *Free Radicals in Solution*, Wiley, New York, 1957, 388.

5. Kenner, J., Oxidation and reduction in chemistry, *Nature*, 156, 370, 1945; Teeter, H. M., Bachman, R. C., Bell. E. W., and Cowan, J. C., Reactions of *tert*-butyl hypochlorite with vegetable oils and derivatives, *Ind. Eng. Chem.*, 41, 849, 1949.

6. (a) Walling, C. and Jacknow, R. B., Positive halogen compounds. I. The radical chain halogenation of hydrocarbons by *t*-butyl hypochlorite, *J. Am. Chem. Soc.*, 82, 6108, 1960; (b) Positive halogen compounds. II. Radical chlorination of substituted hydrocarbons with *t*-butyl hypochlorite, *J. Am. Chem. Soc.*, 82, 6113, 1960.

7. (a) Walling, C. and Thaler, W., Positive halogen compounds. III. Allylic chlorination with *t*-butyl hypochlorite. The stereochemistry of allylic radicals, *J. Am. Chem. Soc.*, 83, 3877, 1961; (b) Walling, C. and Fredricks, P. S., Positive halogen compounds. IV. Radical reactions of chlorine and *t*-butyl hypochlorite with some small ring compounds, *J. Am. Chem. Soc.*, 84, 3326, 1962; (c) Walling, C. and Padwa, A., *J. Org. Chem.*, 27, 2976, 1962.

8. Walling, C. and Padwa, A., Positive halogen compounds. VI. Effects of structure and medium on the β-scission of alkoxyl radicals, *J. Am. Chem. Soc.*, 85, 1593, 1963, and subsequent papers.

9. Greene, F. D., Savitz, M. L., Osterholtz, F. D., Lau, H. H., Smith, W. N., and Zanet, P. M., Decomposition of tertiary alkyl hypochlorites, *J. Org. Chem.*, 28, 55, 1963.

10. (a) Kochi, J. K., Chemistry of alkoxyl radicals: Cleavage reactions, *J. Am. Chem. Soc.*, 84, 1193, 1962; (b) Bacha, J. D. and Kochi, J. K., Polar and solvent effects in the cleavage of *t*-alkoxyl radicals, *J. Org. Chem.*, 30, 3272, 1965.

11. Kochi, J. K., Oxygen radicals, in *Free Radicals*, Vol II, Kochi, J. K., Ed., Wiley, New York, 1973, chap. 23.

12. Akhtar, M. and Barton, D. H. R., Reactions at position 19 in the steroid nucleus. A convenient synthesis of 19-norsteroids, *J. Am. Chem. Soc.*, 86, 1528, 1964.

13. Meystre, Ch., Heusler, K., Kalvoda, J., Wieland, P., Anner, G., and Wettstein, A., Neue Substitutionsreactionen bei Steroiden, *Experientia*, 17, 475, 1961.

14. For reviews, see: (a) Heusler, K. and Kalvoda, J., Intramolecular free-radical reactions, *Angew. Chem. Int. Ed.*, 3, 525, 1964; (b) Kalvoda, J. and Heusler, K., Die Hypoiodit-Reaction, *Synthesis*, 549, 1971.

15. Mills, J. S. and Petrow, V., The rearrangement of steroid hypochlorites: Preparation of 6-methylandrost-5-ene-3β,17β,19-triol, *Chem. Ind. (London)*, 946, 1961.

16. Barton, D. H. R., Geller, J. M., and Pechet, M. M., A new photochemical reaction, *J. Am. Chem. Soc.*, 83, 4076, 1961.

17. See Chapter 10 in this book.

18. Walling, C. and Padwa, A., Intramolecular chlorination with long chain hypochlorites, *J. Am. Chem. Soc.*, 83, 2207, 1961.

19. Jenner, E. L., Intramolecular hydrogen abstraction in primary alkoxyl radical, *J. Org. Chem.*, 27, 1031, 1962.

20. Akhtar, M., Organic nitrites and hypohalites, in *Advances in Photochemistry*, Vol. 2, Noyes, W. A., Hammond, G. S., Pitts, J. N., Jr., Eds., Interscience, New York, 1964.

21. (a) Suginome, H. and Yamada, S., Photoinduced transformations. 73. Transformations of five- (and six-)membered cyclic ethers — A new method of a two-step transformation of hydroxysteroid into oxasteroid, *J. Org. Chem.*, 49, 3753, 1984; (b) Suginome, H. and Yamada, S., Photoinduced transformations. 77. A four-step substitution of a carbonyl group of steroidal ketones by an oxygen atom. A new method for the synthesis of cyclic ethers, *J. Org. Chem.*, 50, 2489, 1985; (c) Suginome, H., Yamada, S., and Wang, J. B., Photoinduced molecular transformations. 108. A versatile substitution of a carbonyl group of steroidal ketones by a heteroatom. The synthesis of aza, oxa-, thia-, selena-, and tellurasteroids, *J. Org. Chem.*, 55, 2170, 1990; (d) Suginome, H., Senboku, H., and Yamada, S., Photoinduced molecular transformations. 112. Transformation of steroids into ring-A-aromatized steroids and 19-norsteroids involving a regioselective β-scission of alkoxyl radicals: Synthesis of two marine natural products, 19-nor-5α-cholestan-3β-ol and 19-norcholest-4-en-3-one, and new synthesis of estrone and 19-nortestosterone, *J. Chem. Soc., Perkin Trans. 1*, 2199, 1990; (e) Suginome, H., Nakayama, Y., and Senboku, H., Photoinduced molecular transformations. 131. Synthesis of 18-norsteroids. Deoxofukujusonorone and the related steroids, based on a selective β-scission of alkoxyl radicals as the key step, *J. Chem. Soc., Perkin Trans. 1*, 1837, 1992; (f) Suginome, H., Wang, J. B., and Satoh, G., Regioselective β-scission of α-oxoalkoxyl radicals: A novel formation of α-hydroxy-ε-lactones by photolysis of steroidal α-oxoalcohol hypoiodites in the presence of mercury(II) oxide and iodine, *J. Chem. Soc., Perkin Trans. 1*, 1553, 1989; (g) Suginome, H., Liu, C. F., Tokuda, M., and Furusaki, A., Photoinduced transformations. 76. Ring expansion through a [2+2] photocycloaddition — β-scission sequence; the photorearrangement of *endo*-4-cyanotricyclo[6.4.0.0²,⁴]dodeca-1(12),6,8,10-tetraen-5-yl hypoiodite to 4-cyanotricyclo[6.4.0.0²,⁴]-dodeca-1(12),6,8,10-tetraen-5-one, *J. Chem. Soc., Perkin Trans. 1*, 327, 1985; (h) Suginome, H. and Nakayama, Y., Photoinduced transformations. 132. A two-step intramolecular transposition of the 7β-acetyl group of pregnan-20-one to C-18 through the formation of cyclobutanols by the reaction of the excited carbonyl, followed by a selective β-scission of alkoxyl radicals generated from them, *J. Chem. Soc., Perkin Trans. 1*, 1843, 1992; (i) Orito, K., Yorita, K., and Suginome, H., *Tetrahedron Lett.*, 32, 7245, 1991; (j) Suginome, H. and Yamada, S., New general synthesis of medium ring-lactones via a regioselective β-scission of alkoxyl radicals generated from catacondensed lactols, *Tetrahedron*, 43, 3371, 1987; (k) Suginome, H. and Yamada, S., Simple new synthesis of macrolides by a four atom ring expansion of cyclic ketones through a consecutive intramolecular homolytic addition-β-scission of alkoxyl radicals. A new entry to the synthesis of 15-pentadecanolides, *Chem. Lett.*, 245, 1988; (l) Suginome, H. and Ihizawa, A., A new synthesis of (±)-recifeiolide based on a consecutive intramolecular homolytic addition and β-scission of alkoxyl radicals; unpublished results; (m) Kobayashi, K., Itoh, M., Sasaki, A., and Suginome, H., New short step general synthesis of isobenzofuran-1(3H)-ones (phthalides) based on a single or double β-scission of alkoxyl radicals

generated from 1-ethylbenzocyclobuten-1-ols: Synthesis of some natural phthalides, *Tetrahedron,* 47, 5437, 1991; (n) Kobayashi, K., Kanno, Y., Seko, S., and Suginome, H., Photoinduced molecular transformations. 135. New synthesis of taiwanin C and justicidin E based on a radical cascade process involving β-scission of alkoxyl radicals generated from 3-aryl-1-ethyl-1,2-dihydrocyclobuta[*b*] naphthalen-1-ols prepared by thermolysis of *(Z)-tert*-butyl 3–3-(bicyclo[4.2.0]octa-1,3,5-trien-7-yl)propenoates, *J. Chem. Soc., Perkin Trans 1,* 3119, 1992; (o) Kobayashi, K., Sasaki, A., Kanno, Y., and Suginome, H., A new route to α-substituted α-lactones and δ-lactones based on the regioselective β-scission of alkoxyl radicals generated from transannular hemiacetals, *Tetrahedron,* 47, 7245, 1991; (p) Suginome, H. and Yamada, S., A simple new synthesis of macrocyclic ketones: A new entry to the synthesis of exaltone and (±)-muscone, *Tetrahedron Lett.,* 28, 3963, 1987; (q) Suginome, H., Liu, C. F., Seko, S., Kobayashi, K., and Furusaki, A., Photoinduced molecular transformations. 100. Formation of furocoumarins and furochromones via a β-scission of cyclobutanoxyl radicals generated from [2+2] photoadducts from 4-hydroxycoumarin and acyclic and cyclic alkenes. X-ray crystal structures of (6aα, 6bα, 9aα, 9bα)-(±)-6b,7,8,9,9a,9b-hexahydro-9b-hydroxybenzo[*b*]cyclopenta[3,4]cyclobuta[1,2-*d*]pyran-6(6a*H*)-one, *cis*-1,1,2,2-tetramethyl-3*H*-benzo[*b*]cyclobuta[*d*]pyran-3-one, *J. Org. Chem.,* 53, 5952, 1988; (r) Suginome, H., Kobayashi, K., Itoh, M., Seko, S., and Furusaki A., Photoinduced molecular transformations. 110. Formation of furoquinolinones via β-scission of cyclobutanoxyl radicals generated from [2+2] photoadducts of 4-hydroxy-2-quinolone and acyclic and cyclic alkenes. X-Ray crystal structure of (6aα, 6aβ, 10aβ, 10bα)-(±)-10b-acetoxy-6a,6b,7,8,9,10,10a,10b-octahydro-5-methylbenzo[3,4]cyclobuta[1,2-*c*]quinolin-6(5*H*)-one, *J. Chem. Soc.,* 55, 4933, 1990; (s) Kobayashi, K., Suzuki, M., and Suginome, H., Photoinduced molecular transformations. 128. Regioselective [2+2] photocycloaddition of 3-acetoxyquinolin-2(1*H*)-one with alkenes and formation of furo[2,3-*c*]quinolin-4(5*H*)-ones, 1-benzazocine-2,3-diones via a β-scission of cyclobutanoxyl radicals generated from the resulting [2+2] photoadducts. X-ray crystal structure of *trans*-5,8,9,10,10a,10b-hexahydro-5-methylcyclopenta[3,4]cyclopropa[1,2-*d*][1]benzazepine-6,7-dione, *J. Org. Chem.,* 57, 599, 1992; (t) Kobayashi, K., Sasaki, A., Takeuchi, H., and Suginome, H., Photoinduced molecular transformations. 127. A new [2+2] photoaddition of 2-amino-1,4-naphthoquinone with vinylarenes and the synthesis of 2,3-dihydronaphtho[1,2-*b*]furan-4,5-diones and 2,3-dihydronaphtho[2,3-*b*]furan-4,9-diones by β-scission of alkoxyl radicals generated from the resulting photoadducts, *J. Chem. Soc., Perkin Trans. 1,* 115, 1992; (u) Kobayashi, K., Konishi, A., Kanno, Y., and Suginome, H., Photoinduced molecular transformations. 137. A new general synthesis of 3-substituted (1*H*)-3,4-dihydrobenzo[2]pyran-1-ones (3,4-dihydroisocoumarins), via radical and photochemical fragmentations as the key step. *J. Chem. Soc., Perkin Trans. 1.,* 111, 1993; (v) Suginome, H., Kondo, T., Gogonea, C. Singh, V., Goto, H., and Ōsawa, E., Photoinduced molecular transformations. 155. General synthesis of macrocyclic ketones based on a ring expansion involving a selective β-scission of alkoxyl radicals, its appication to a new synthesis of (±)-isocaryophyllene and (±)-caryophyllene, and a conformational analysis of the two sesquiterpenes and the radical intermediate in the synthesis by MM3 calculations, *J. Chem. Soc., Perkin Trans. 1,* in press; (w) Suginome, H., Nakayama, Y., Havada, H., Hachiro, H., and Orito, K., A new synthesis of a functionalized bicyclo[5.4.0]undecane skeleton based on a sequence involving [2+2]photoaddition and regioselective β-scission of alkoxyl radicals generated from the resulting cyclobutanols, *J. Chem. Soc. Chem. Commun.,* 451, 1994; (x) Suginome, H. and Nakayama, Y., Photoinduced molecular transformations. 150. A new total synthesis of (±)-α-himachalene based on a sequence involving [2+2] photoaddition and regioselective β-scission of alkoxyl radicals generated from the resulting cyclobutanols, *Tetrahedron,* 26, 7771, 1994; (y) Suginome, H., Takeda, T., Itoh, M., Nakayama, Y., and Kobayashi, K., Photoinduced molecular transformations. 152. Ring expansion based on a sensitized [2+2] photoaddition of enol ethers of cyclic ketones with olefins, followed by a β-scission of alkoxyl radicals generated from the resulting cyclobutanols. Two-carbon ring expansion of β-indanone, β-tetralone and β-suberone, *J. Chem. Soc., Perkin Trans. 1,* in press.

22. (a) Concepcion, J. I., Francisco, C. G., Hernandez, R., Salazar, J. A., and Suarez, E., Intramolecular hydrogen abstraction. Iodosobenzene diacetate, an efficient and convenient reagent for alkoxy radical generation, *Tetrahedron Lett.*, 25, 1953, 1984; (b) de Armas, P., Concepcion, J. I., Francisco, C. G., Hernandez, R., Salazar, J. A., and Suarez, E., Intramolecular hydrogen abstraction. Hypervalent organoiodine compounds, convenient reagents for alkoxyl radical generation, *J. Chem. Soc., Perkin Trans. 1*, 405, 1989, and subsequent papers.

23. Anbar, M. and Dostrovsky, I., Ultraviolet absorption spectra of some organic hypohalites, *J. Chem. Soc.*, 1105, 1954.

24. Gray, P. and Williams, A., The thermochemistry and reactivity of alkoxyl radicals, *Chem. Rev.*, 59, 539, 1959.

25. Teeter, H. M. and Bell, E. W., *tert*-Butyl hypochlorite, *Org. Synth.*, 32, 20, 1952.

26. Bensadoun, N., Brun, P., Casanova, J., and Weagell, B., Stereoelectronic control in the β-fragmentation reaction of alkoxyl radicals. An empirical predictable rule, *J. Chem. Res. (S)*, 236, 1981 and *(M)*, 2601, 1981.

27. Anbar, M. and Dostrovsky, I., A kinetic study of the formation and hydrolysis of *tert*-butyl hypochlorite, *J. Chem. Soc.*, 1094, 1954.

28. Heusler, K., Kalvoda, J., Wieland, P., Anner, G., and Wettstein, A., Reactionen von Steroid-Hypoioditen. IV. Über den Verlauf intramolecularer Substitutions Reactionen, insbezondere bei 2β- and 4β-Hydroxysteroiden, *Helv. Chim. Acta*, 45, 2575, 1962.

29. Meystre, Ch., Heusler, K., Kalvoda, J., Wieland, P., Anner, G., and Wettstein, A., Reactionen von Steroid-Hypoioditen II. Über die Herstellung 18-oxygenierter Pregnenverbindungen, *Helv. Chim. Acta*, 45, 1317, 1962.

30. (a) Wenkert, E. and Mylari, B. L., Intramolecular, long-range oxidation at saturated carbon centers, *J. Am. Chem. Soc.*, 89, 174, 1967; (b) Ceccherelli, P., Curini, M., Marcotullio, M. C., Mylari, B. L., and Wenkert, E., Iodohydrins and tetrahydrofurans from lead tetraacetate-iodine oxidation of terpenic alcohols, *J. Org. Chem.*, 51, 1505, 1986.

31. (a) Kalvoda, J., Meystre, Ch., and Anner, G., Reactionen von Steroid-Hypoioditen. VIII. 1,4-Verschiebung der Nitrilgruppe (18-Cyano-pregnane), *Helv. Chim. Acta*, 49, 424, 1966; (b) Kalvoda, J., Reaction von Steroid-Hypoioditen. IX. Beitrag zum Mechanisms der oxydativen Cyanhydrin-Cyanketon-Umlagerung, *Helv. Chim. Acta*, 51, 267, 1968; (c) Kalvoda, J., A new type of intramolecular group-transfer in steroid photochemistry. A contribution to the mechanism of the oxidative cyanohydrin — cyano-ketone rearrangement, *J. Chem. Soc., Chem. Commun.*, 1003, 1970.

32. Orito, K., Satoh, S., and Suginome, H., A long-range intramolecular functionalization by alkoxyl radicals: A long-range intramolecular hydroxylation of C(25) of cholestane side chain, *J. Chem. Soc., Chem. Commun.*, 1829, 1989; Orito, K., Satoh, S., and Suginome, H., Photoinduced molecular transformations. Part 153. Long-range intramolecular hydroxylation of (C25) of the cholestane side chain, *J. Chem. Soc. Perkin Trans. 1*, in press.

33. (a) Orito, K., Ohto, M., and Suginome, H., A long-range intramolecular functionalization by alkoxyl radicals; A long-range intramolecular oxygenation of C(15) of the androstane skeleton, *J. Chem. Soc., Chem. Commun.*, 1074, 1990; (b) A Wagner-Meerwein rearrangement of the cholestane skeleton induced by a long-range intramolecular hydrogen abstraction by alkoxyl radicals; The first example of long-range intramolecular addition of an alkoxyl radical to a carbon-carbon double bond, *J. Chem. Soc., Chem. Commun.*, 1076, 1990.

34. Orito, K., Ohto, M., Sugawara, N., and Suginome, H., Long-range intramolecular functionalization by alkoxyl radicals: Long-range intramolecular double functionalization of ring C of cholestane and androstane skeletons, *Tetrahedron Lett.*, 31, 5921, 1990.

35. Shibuyama, M., Jaishi, F., and Eschenmoser, A., A fragmentation approach to macrolides: (5-*E*, 9-*E*)-6-methyl-5,9-undecadien-11-olide, *Angew. Chem. Int. Ed.*, 18, 635, 1979.

36. For a review of synthesis via β-scission, see: Remaiah, M., Radical reactions in organic synthesis, *Tetrahedron*, 43, 3541, 1987.

37. Suginome, H., New developments in photoinduced radical reactions in organic synthesis, in *Modern Methodology in Organic Synthesis,* Shono, T., Ed., Kodansha, Tokyo, 1992, 245.
38. Arencibia, M. T., Freire, R., Perales, A., Rodriguez, M. S., and Suarez, E., Hypervalent organoiodine compounds: Radical fragmentation of oxabicyclic hemiacetals. Convenient synthesis of medium-sized and spiro lactones, *J. Chem. Soc., Perkin Trans. 1,* 3349, 1991.
39. Akhtar, M. and Marsh, S., Synthesis of a medium-size ring via alkoxy radical decomposition, *J. Chem. Soc. (C),* 937, 1966.

SECTION II
Photobiology

1

Action Spectroscopy: Methodology

Edward D. Lipson
Syracuse University

1.1 Introduction

Action spectroscopy is a general approach toward identifying the receptor pigment(s) for a particular photobiological response or effect. The early identification of major chromophores, such as rhodopsin and chlorophyll, depended on comparison of the action spectra for vision and photosynthesis, respectively, with the absorption spectra of candidate pigments. It is generally remarkable how few chromophore types with intrinsic photochemistry are employed in photobiological systems.[1,2] A basic list consists of the following: (1) retinal pigments, (2) tetrapyrroles, (3) "cryptochrome" pigments (flavin and pterin chromophores; see below), and (4) hypericin pigments.[2]

The major class that has barely been identified is the group of pigments tentatively called "cryptochrome(s)" that mediate various effects and responses to near UV and blue light, particularly in plants, fungi, and microorganisms.[3–5] The favored chromophore candidates are flavins and pterins;[6,7] pterins have been suggested in recent years to serve as accessory chromophores and may account for the considerable variability found in cryptochrome action spectra in the near-UV region. Pterins have been shown to function in such an accessory role in DNA photolyase, which employs a flavin-type chromophore for the process of DNA photoreactivation.[8] For cryptochrome systems, in the blue region, the traditional alternative to flavins has been β-carotene; however, this has been ruled out in the representative case of phototropism in the zygomycete fungus *Phycomyces blakesleeanus* because caroteneless mutants show normal phototropic sensitivity.[9]

Space limitations prevent inclusion of an appropriate variety of action spectra in this methodological chapter. For examples of action spectra based on fluence-response data, the reader may wish to examine some recent studies of *Phycomyces,* not only of phototropism,[10-12] but also of the light-growth response,[13-14] carotene synthesis,[15] and differentiation,[16] as well as studies of other blue-light systems.[17-19] The more recent studies on *Phycomyces* have used some advanced data analysis methods — including formal error analysis in the curve fitting and parameter estimation — as summarized elsewhere.[20] Collections of representative action spectra for various blue-light effects and responses have been assembled by Presti and Galland[21] and by Galland and Senger.[22] These typically show a peak in the near-UV around 380 nm as well as major peaks in the blue region around 450 and 480 nm. Variations among these action spectra can be attributed to a number of causes, including different methodologies, different organisms, and different photoreceptor systems.

The reader is also referred to the leading photobiology journals, *Photochemistry and Photobiology* and the *Journal of Photochemistry and Photobiology, B: Biology,* both of which regularly carry articles including action spectra. There are also a number of review articles that give various perspectives on action spectroscopy.[20,23-30] For practical advice on light sources and radiometry, the reader should consult standard photobiology texts.[31,32]

1.2 Alternative Types of Action Spectra

The most elementary type of action spectrum consists of a graph of some response or effect as a function of wavelength, under conditions where the photon fluence is maintained constant at all wavelengths. This convenient approach, however, can be confounded by the frequent circumstance that the response may depend nonlinearly on the fluence (or fluence rate). For example, if at some wavelengths the response is in a saturation region, then the action spectrum will be "clipped" (i.e., peaks flattened) at those wavelengths; similarly, other types of nonlinearities will produce other distortions. Accordingly, a preferred way of obtaining an action spectrum — provided the experimenter is willing and able to apply the additional effort needed — is to derive it from fluence-response curves (sometimes called dose-response curves). In this way, one obtains essentially a graph of *sensitivity,* rather than response, as a function of wavelength. Operationally, one determines for each wavelength the photon fluence required to achieve a standard response level, and then plots the reciprocal of that photon fluence as a measure of sensitivity. The graph so obtained is sometimes called an equal-response action spectrum, which is not subject to the nonlinear distortions above and is therefore more likely to represent the absorption spectrum of the responsible chromophore (subject to various considerations specified below).

1.3 Photophysics

After a chromophore absorbs a photon, a number of alternative photophysical events may ensue. These can be represented graphically by a Jablonski diagram depicting transitions among the electronic, vibrational, and rotational states of the pigment molecule.[33] Upon absorption of a photon, the molecule is promoted from the ground state (usually of singlet character, and designated therefore as S_0) to an excited electronic state (also singlet, and designated S_1, S_2, etc.); the molecule generally becomes excited with respect to vibrational and rotational motions as well (the corresponding energy spacings among these states are progressively lower than those between electronic states). The molecule then de-excites rapidly by internal conversion and vibrational relaxation to the lowest excited singlet level S_1 before there is time for photochemistry to take place. Consequently, regardless of the wavelength of the incident photon that excited the molecule, the molecular excitation relaxes rapidly to the bottom vibrational levels based on S_1. Then the relevant photochemistry proceeds either directly from S_1 or else — following intersystem crossing from the singlet to the triplet manifold — from the lowest triplet state T_1, provided that the excitation energy has not been dissipated in the meantime by internal conversion or fluorescence.

1.4 Illumination Units

Fluence-response curves are measured sometimes using broadband rather than monochromatic light, because (1) higher fluence rates are achievable with broadband filters; and (2) unless one is trying to study the wavelength dependence of a particular response or effect, it may be more representative of natural illumination conditions to use broadband light covering the range of sensitivity. For example, in studies of photogravitropism "threshold curves" of *Phycomyces* wild-type and mutant strains,[11,34] it is customary to employ broadband blue illumination at fluence rates extending over a 10-decade range. With broadband illumination, the units pertain to *energy* fluence rather than *photon* fluence. The energy units are joules per square meter ($J\,m^{-2}$) for fluence and joules per square meter per second ($J\,m^{-2}\,s^{-1}$) for fluence rate. In action spectroscopy, however, one must use monochromatic light (occasionally in *combination* with polychromatic background light; see chapter by T. Coohill). It is then preferable to use photon units: photons per square meter (photons m^{-2}) for fluence, and photons per square meter per second (photons $m^{-2}\,s^{-1}$) for fluence rate.

1.5 Conditions for Measurement and Interpretation of Action Spectra

The shape of an action spectrum may be distorted significantly from the absorption spectrum of the responsible pigment. To obviate such distortions, the experimenter should strive to satisfy the following conditions insofar as possible.[23,24]

1. The quantum efficiency (or quantum yield) — defined as the probability that a particular type of event of photobiological or photochemical interest occurs as the result of absorption of a photon — should be independent of wavelength.
2. The absorption spectrum of the receptor pigment should be the same whether measured *in vivo* or *in vitro*.
3. Screening or shading pigments, as well as scattering effects, should not cause significant wavelength-dependent distortion. Scattering is generally stronger at shorter wavelengths, but the trend is a gradual one and normally would not obscure the key features of an action spectrum; instead, scattering tends to superpose a sloping baseline, which can be largely eliminated, if desired, by use of a suitable scattering reference.
5. When an action spectrum is derived from fluence-response curves, the condition of reciprocity should be valid over the range of fluence rates and exposure times used. In other words, the response should depend on the *product* of fluence rate and exposure time, but should not otherwise depend on either factor.
6. Light should not be totally absorbed by the sample for any wavelength under study. More specifically, an ample fraction of the incident light (a practical guideline is 50%) should reach the receptor pigment region under all conditions.

1.6 Reciprocity

For measurements employing continuous light, the stimulus strength is given by the fluence rate itself. With pulse light, however, the fluence can be adjusted by varying the exposure time (pulse width) and/or the fluence rate. Considering the kinetics of photochemical and subsequent dark reactions, it is preferable in most circumstances to maintain the exposure time constant and vary the fluence rate. In general, though, one should establish the range of validity of reciprocity between exposure time and fluence rate (see above).

1.7 Fundamental Derivations

For action spectroscopy with pulse illumination, assume that the response R can be expressed as a function of the product of four variables:

$$R = R(\phi \sigma_1 I_1 \Delta t) \tag{1.1}$$

where ϕ is the quantum efficiency (or quantum yield), σ_1 is the cross section (see below) at wavelength λ_1, I_1 is the fluence rate (note: the traditional symbol I stands for intensity, which is sometimes used informally in place of fluence rate; however, according to strict radiometric terminology, intensity is defined as the power per unit solid angle), and Δt is the exposure time (pulse duration). The fluence is given by $F_1 = I_1 \Delta t$.

To derive an equal-response action spectrum from fluence-response curves, the first step is to specify a "criterion" response level. Then, for each wavelength, one determines the fluence (or fluence rate) needed to produce that standard response. Choices for the criterion response include: (1) some fixed absolute level, (2) a percentage (typically 50%) of the maximum response level, which itself may depend on wavelength, (3) the maximum response (peak) level, for those instances where the response descends at high fluence after reaching a peak, or (4) the absolute threshold fluence rate (often extrapolated downward to the baseline from the linearly rising part of the fluence-response curve; note that an alternative approach that should be avoided would be to try to find the limiting fluence at which there just begins to be a perceptible response; the difficulty is that this measure is highly susceptible to experimental noise and subjective judgment). If fluences F_{λ_1} and F_{λ_2} both elicit the same response level — the criterion response — and if one assumes that Δt is the same in both experiments (or, if reciprocity applies, one can correct for the different values of Δt), then the argument of the function in Equation 1.1 is the same for both experiments at λ_1. If the quantum efficiency ϕ is assumed to be the same at both wavelengths, then

$$\frac{\sigma_{\lambda_1}}{\sigma_{\lambda_2}} = \frac{F_{\lambda_2}}{F_{\lambda_1}} \tag{1.2}$$

Thus, the cross section at any wavelength λ is inversely proportional to the applied fluence that produces the criterion response at that wavelength, or $\sigma_\lambda \propto F_\lambda^{-1}$.

If, instead, the response is measured as a function of fluence rate rather than fluence, then a derivation similar to that above, starting from Equation 1.3

$$R = R(\phi \sigma_1 I_1) \tag{1.3}$$

leads to Equation 1.4 or

$$\frac{\sigma_{\lambda_1}}{\sigma_{\lambda_2}} = \frac{I_{\lambda_2}}{I_{\lambda_1}} \tag{1.4}$$

or $\sigma_\lambda \propto I_\lambda^{-1}$.

1.8 Relationship Between Extinction Coefficient and Cross-section

The extinction coefficient ε is used in conventional spectrophotometry and measured in the traditional units of liters per mole per centimeter (l mol^{-1} cm^{-1}). It can be related by a conversion factor to a quantity from physics, the absorption cross section (introduced in previous section), in traditional units of square centimeters. The ratio between the fluence rate I transmitted through a spectrophotometric sample and the incident intensity I_0 can be derived from the following relations:

$$\frac{I}{I_0} = 10^{-\varepsilon c\ell} = e^{-\sigma n\ell} \tag{1.5}$$

where c is the molar concentration of the pigment, ℓ is the internal path length through the cuvette (usually 1 cm), σ is the absorption cross section, and n is the pigment concentration in units of molecules per cubic centimeter. The expression in the first exponent, $\varepsilon c\ell$, represents the absorbance, A.

The extinction coefficient and the cross section are therefore interrelated by the following conversion formula:

$$\varepsilon = 2.62 \times 10^{20} \sigma \tag{1.6}$$

where ε is in units of liters per mole per centimeter and σ is in centimeters squared. As an example of applying this conversion formula, consider riboflavin (or other flavins), which has an extinction coefficient of 1.25×10^4 l mol^{-1} cm^{-1} (at ~450 nm); then $\sigma = 4.8 \times 10^{-17}$ cm^2. This represents the effective target area the chromophore presents for absorption of light. As another example, rhodopsin, with an extinction coefficient of 4×10^4 l mol^{-1} cm^{-1} (at ~500 nm), has $\sigma = 1.5 \times 10^{-16}$ cm^2.

1.9 Derivation of Action Spectrum from Fluence-Response Curves

Figure 1.1 shows a set of five idealized fluence-response curves. The following discussion will demonstrate how action spectra can be derived from such curves. The hypothetical response is presumed to be measured as a function of the photon fluence; alternatively, fluence rate could have been chosen as the independent variable. The fitting function of the form $ax/(x+b)$, where x stands for the fluence (or fluence rate), is plotted for the five wavelengths on both logarithmic (a) and linear (b) scales for x. In this example, wavelength λ_3 is the most effective one because the least amount of light is required to achieve the criterion response level, chosen here to be half of the maximum response. Conversely, λ_5 is the least effective. So, when this set of "data" is converted into an action spectrum, the ordinate representing the sensitivity, or effectiveness, will be high for λ_3 and low for λ_5.

In the semilogarithmic plot (Figure 1.1a) the curves all have the same sigmoidal shape and differ only by lateral displacement. The symmetrical sigmoidal shape is a property of the hyperbolic saturation function, which is frequently used to fit fluence-response curves and other types of stimulus-response relationships in sensory physiology,[35,36] photochemical kinetics,[37] and in other areas of biophysics and biology, including the well-known Michaelis-Menten enzyme kinetics.

The procedure for deriving the action spectrum (Figure 1.2) from these curves is straightforward. The action spectrum ordinate, usually labeled as relative quantum effectiveness or just effectiveness,

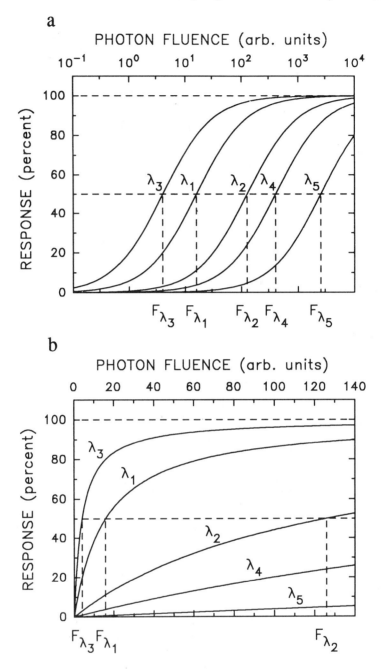

FIGURE 1.1   Generic fluence-response curves for five wavelengths (λ_1 through λ_5) shown on logarithmic (a) and linear (b) scales for photon fluence. The curves represent the hyperbolic saturation function of the form $ax/(x+b)$, where x is the fluence (also denoted in this chapter by the symbol F). On a semilogarithmic scale (a), the curves all have the same sigmoidal shape (see Reference 20). The fluence F needed to produce the standard criterion response (here chosen to be 50%) at each wavelength is shown by the vertical dashed lines. The designation of arbitrary units on the abscissa scales above each graph means that no importance should be given to the absolute numbers used in this example, and, implicitly, that fluence rate could be used instead of fluence if continuous rather than pulse illumination were employed; the actual units of photon fluence would be moles per meter squared. If photon fluence rate were used instead, the units would be moles per meter squared per second.

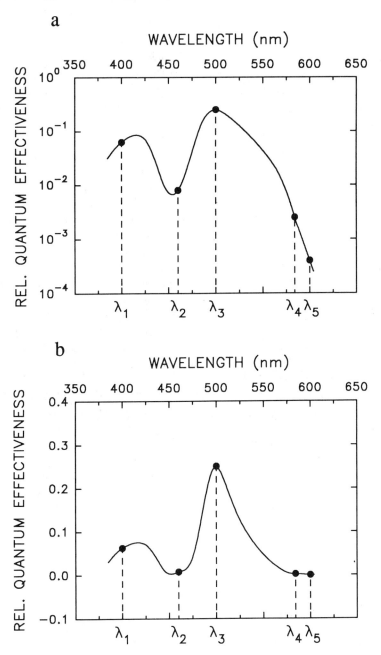

FIGURE 1.2 Five-point action spectrum derived from the fictitious fluence-response curves in Figure 1.1. The relative quantum effectiveness, shown on logarithmic (a) and linear (b) scales, is simply the reciprocal of the photon fluence F required at each wavelength to achieve the criterion response, as indicated on Figure 1.1.

is simply the reciprocal of the fluence required to produce a criterion response (see Equations 1.2 and 1.4). Some authors choose instead to label the ordinate in relative units, for example, by normalizing several action spectra to the wavelength of maximal effectiveness of one of them. However, it is preferable to retain the proper physical units of meters squared per mole. If fluence rate-response curves are used instead as the basis for the action spectra, then the corresponding reciprocal units for the "effectiveness" are meters squared second per mole. In the former case — based on fluence rather than fluence rate — the units of the action spectrum ordinate are reciprocal to the fluence units, which themselves are essentially inverse meters squared (apart from dimensionless entities such as photons, or moles of photons). Consequently, the action spectrum physical units are meters squared, indicating that some kind of area quantity is involved. This dimensionality corresponds specifically to that of a *cross section* (see relationship between extinction coefficient and cross section above).

For a given application, one has to decide between plotting the action spectrum on a logarithmic or linear ordinate. In *Phycomyces,* with a sensitivity range exceeding 10 decades, it is particularly appropriate to plot such action spectra on a logarithmic ordinate.[11,34] This choice is advantageous for comparing action spectra to one another or to absorption spectra of putative receptor pigments. Spectra that differ only by a scale factor will have identical shape on a logarithmic scale (and will be displaced by the logarithm of the scale factor). Now, if such spectra were truly identical, apart from a scale factor, they could be forced to have the same shape on a linear scale too, simply by normalizing them at any wavelength of choice (thereby superposing the spectra completely). The practical problem however is that, in the usual case when spectra are similar but not identical, choosing one wavelength or another for normalization will lead to different relative shapes. Such ambiguity and subjectivity are avoided by doing the comparison on a logarithmic scale. On the other hand, though, using a linear ordinate scale has the advantage that it is the more familiar way of viewing absorption spectra, for example the chart records from commercial spectrophotometers.

The abscissa of an action spectrum is usually presented as a wavelength scale. On the basis of quantum mechanical principles, it would be preferable to use an energy rather than a wavelength scale; the two are related reciprocally according to the relation $E = hc/\lambda$, where E is the energy of a photon of wavelength λ, h is Planck's constant, and c is the speed of light. When an action or absorption spectrum is plotted as a function of energy, the positions of peaks are linearly related to transitions between the energy levels corresponding to molecular states (as on a Jablonski diagram; see above). This is particularly worthwhile in the case of rhodopsin-type pigments in animals — as well as in halophilic bacteria and algae — because such pigments have spectra of standard shape except for lateral shifts in the energy of maximum absorption and vertical shifts in the absolute extinction or effectiveness. Nevertheless, most action and absorption spectra are customarily plotted on a wavelength scale for the abscissa, in keeping with the usual display from spectrophotometers.

1.10 Conclusion

This brief chapter on action spectroscopy has given some practical methodology for obtaining action spectra from fluence-response data. Special mention has been made of applications in the field of blue light effects, in which action spectroscopy has remained particularly active, mainly because the responsible pigments are still being sought.[38,39] The chapter does not delve into action spectroscopy for photochromic receptors, such as phytochrome, a complex topic dealt with in many of the references on action spectroscopy cited in the introduction. Photochromic systems, and others with significant kinetic complexity, call for a mathematical approach termed analytical action spectroscopy by Hartmann.[25] Readers are encouraged to apply formal data analysis techniques, including error analysis, for analyzing experimental results from action spectroscopy and other studies. Details on such methods, including examples of various functions useful in fitting fluence-response data, are available elsewhere.[20]

eferences

1. Delbrück, M., Light and Life III, *Carlsberg Res. Commun.*, 41, 299, 1976.
2. Lipson, E. D. and Horwitz, B. A., Photosensory reception and transduction, *Sensory Recept. Signal Transduct.*, 10, 1, 1991.
3. Senger, H., *The Blue Light Syndrome*, Springer-Verlag, Berlin, 1980.
4. Senger, H., *Blue Light Effects in Biological Systems*, Springer-Verlag, Berlin, 1984.
5. Senger, H., *Blue Light Responses: Phenomena and Occurrence in Plants and Microorganisms*, Vols. I and II, CRC Press, Boca Raton, FL, 1987.
6. Galland, P. and Senger, H., The role of flavins as photoreceptors, *J. Photochem. Photobiol. B: Biol.*, 1, 277, 1988.
7. Galland, P. and Senger, H., The role of pterins in the photoreception and metabolism of plants, *Photochem. Photobiol.*, 48, 811, 1988.
8. Wang, B., Jordan, S., and Schuman Jorns, M., Identification of a pterin derivative in *Escherichia coli* DNA photolyase, *Biochemistry*, 27, 4222, 1988.
9. Presti, D., Hsu, W. J., and Delbrück, M., Phototropism in *Phycomyces* mutants lacking β-carotene, *Photochem. Photobiol.*, 26, 403, 1977.
10. Galland, P. and Lipson, E. D., Action spectra for phototropic balance in *Phycomyces blakesleeanus*: Dependence on reference wavelength and intensity range, *Photochem. Photobiol.*, 41, 323, 1985.
11. Galland, P. and Lipson, E. D., Modified action spectra of photogeotropic equilibrium in *Phycomyces blakesleeanus* mutants with defects in genes *madA*, *madB*, *madC*, and *MadH*, *Photochem. Photobiol.*, 41, 331, 1985.
12. Ensminger, P. A., Chen, X., and Lipson, E. D., Action spectra for photogravitropism of *Phycomyces* wild type and three behavioral mutants (L150, L152, and L154), *Photochem. Photobiol.*, 51, 681, 1990.
13. Ensminger, P. A., Schaefer, H. R., and Lipson, E. D., Action spectra of the light-growth response of *Phycomyces*, *Planta*, 184, 498, 1991.
14. Ensminger, P. A. and Lipson, E. D., Action spectra of the light-growth response in three behavioral mutants of *Phycomyces*, *Planta*, 184, 506, 1991.
15. Bejarano, E. R., Avalos, J., Lipson, E. D., and Cerdá-Olmedo, E., Photoinduced accumulation of carotene in *Phycomyces*, *Planta*, 183, 1, 1990.
16. Corrochano, L. M., Galland, P., Lipson, E. D., and Cerdá-Olmedo, E., Photomorphogenesis in *Phycomyces*: Fluence-response curves and action spectra, *Planta*, 174, 315, 1988.
17. Baskin, T. I. and Iino, M., An action spectrum in the blue and ultraviolet for phototropism in alfalfa, *Photochem. Photobiol.*, 46, 127, 1987.
18. Schmid, R., Idziak, E.-M., and Tuennermann, M., Action spectrum for the blue-light-dependent morphogenesis of hair whorls in *Acetabularia mediterrania*, *Planta*, 171, 96, 1987.
19. De Fabo, E. C., Harding, R. W., and Shropshire, W., Action spectrum between 260 and 800 nanometers for the photoinduction of carotenoid biosynthesis in *Neurospora crassa*, *Plant Physiol.*, 57, 440, 1976.
20. Lipson, E. D., Action Spectroscopy, in *Biophysics of Photoreceptors and Photomovements in Microorganisms*, Lenci, F., Ghetti, F., Colombetti, G., Häder, D.-P., and Song, P.-S., Eds., Plenum, New York, 1991, 293.
21. Presti, D. E. and Galland, P., Photoreceptor biology of *Phycomyces*, in *Phycomyces*, Cerdá-Olmedo, E. and Lipson, E. D., Eds., Cold Spring Harbor Laboratory, Cold Spring Harbor, NY, 1987, 93.
22. Galland, P. and Senger, H., The role of flavins as photoreceptors, *J. Photochem. Photobiol. B: Biol.*, 1, 277, 1988.
23. Jagger, J., *Introduction to Research in Ultraviolet Photobiology*, Prentice-Hall, Englewood Cliffs, NJ, 1967.
24. Shropshire, W., Jr., Action spectroscopy, in *Phytochrome*, Mitrakos, K. and Shropshire, W., Jr., Eds., Academic Press, London, 1972, 162.

25. Hartmann, K. M., Action spectroscopy, in *Biophysics,* 2nd ed., Hoppe, W., Lohmann, W., Markl, H., and Ziegler, H., Eds., Springer-Verlag, Berlin, 1983, 115.

26. Schäfer, E., Fukshansky, L., and Shropshire, W., Jr., Action spectroscopy of photoreversible pigment systems, in *Photomorphogenesis,* Shropshire, W., Jr. and Mohr, H., Eds., Springer-Verlag, Berlin, 1983, 39.

27. Schäfer, E. and Fukshansky, L., Action spectroscopy, in *Techniques in Photomorphogenesis,* Smith, H. and Holmes, M. G., Eds., Academic Press, London, 1984, 109.

28. Galland, P., Action spectroscopy, in *Blue Light Responses: Phenomena and Occurrence in Plants and Microorganisms,* Vol. 2, Senger, H., Ed., CRC Press, Boca Raton, FL, 1987, 37.

29. Coohill, T. P., Action spectra again, *Photochem. Photobiol.,* 54, 859, 1991.

30. Coohill, T. P., Action spectra revisited, *J. Photochem. Photobiol. B: Biol.,* 13, 95, 1992.

31. Smith, K. C., *The Science of Photobiology,* 2nd ed., Plenum Press, New York, 1989.

32. Häder, D.-P. and Tevini, M., *General Photobiology,* Pergamon Press, Oxford, 1987.

33. Grossweiner, L. I., Photophysics, in *The Science of Photobiology,* 2nd ed., Smith, K. C., Ed., Plenum Press, New York, 1989, 1.

34. Galland, P. and Lipson, E. D., Light physiology of *Phycomyces* sporangiophores, in *Phycomyces,* Cerdá-Olmedo, E. and Lipson, E. D., Eds., Cold Spring Harbor Laboratory, Cold Spring Harbor, NY, 1987, 49.

35. Naka, K. I. and Rushton, W. A. H., S-potentials from colour units in the retina of fish (Cyprinidae), *J. Physiol.,* 185, 536, 1966.

36. Williams, T. P. and Gale, J. G., "Compression" of retinal responsivity: V-log I functions and increment thresholds, in *Visual Psychophysics and Physiology,* Armington, J. C., Ed., Academic Press, New York, 1978, 129.

37. Lipson, E. D. and Presti, D., Graphical estimation of cross sections from fluence-response data, *Photochem. Photobiol.,* 32, 383, 1980.

38. Galland, P., Forty years of blue-light research and no anniversary, *Photochem. Photobiol.,* 56, 847, 1992.

39. Lipson, E. D. and Horwitz, B. A., Photosensory reception and transduction, in *Sensory Receptors and Signal Transduction,* Spudich, J. and Satir, B., Eds., Academic Press, New York, 1991, 1.

2

Action Spectroscopy:
Ultraviolet Radiation

homas P. Coohill
ltraviolet Consultants

2.1 Introduction to Ultraviolet Radiation

Ultraviolet (UV) radiation is that portion of the electromagnetic spectrum that extends from the lower wavelength limit of human vision (usually defined as 380 to 400 nm) to wavelengths as short as about 10 nm, where it overlaps the X-ray region. In the natural environment, the shortest wavelength of sunlight that can be routinely measured at the earth's surface is about 290 nm, largely due to the absorption properties of ozone and other atmospheric gases. So, the only environmentally relevant UV region is from 290 to 380 nm. However, artificial UV sources such as certain fluorescent lamps, mercury and xenon arcs, and lasers are readily available and extend the possibility of exposure of biological specimens to UV down to wavelengths of about 190 nm. Below 190 nm, air (oxygen) and water begin to absorb UV heavily, making it difficult to expose biological samples except under extreme conditions (e.g., in a vacuum). Hence, UV photobiology is concerned mainly with the effects on biological processes due to exposure to photons in the wavelength range 190 to 380 nm.[1]

0-8493-8634-9/95/$0.00+$.50
© 1995 by CRC Press, Inc.

2.2 Division of the Ultraviolet Region for Photobiological Studies

Vacuum UV (10 to 190 nm)

Photons in the "vacuum" UV (VUV) are heavily absorbed by water and oxygen (in air), both of which become essentially transparent (more than 50% transmission, for a 1-cm path length) to UV at wavelength above 190 nm. Because of this limited penetration, VUV damage to cells is usually confined to a narrow region (a few micrometers) near the cell surface. The energies of single photons in this region are above 6.5 eV, sufficient to ionize many biomolecules. Since the biological effects due to ionizing photons are different from those due to nonionizing photons, the VUV often causes different types of cellular and molecular damage. Thus, VUV effects can be qualitatively different from those of other UV regions.

UV-C (190 to 290 nm)

The shorter-wavelength end of the UV-C (190 nm) is the wavelength region where air and water become transparent. The longer-wavelength limit (290 nm) is the shortest solar UV wavelength easily measured at the earth's surface. Thus, all of the UV-C is environmentally irrelevant. However, research in the UV-C range was central in elucidating many important features of cell functioning. DNA, the genetic material, has a peak absorption near 260 nm that falls by a factor of six by 290 nm. This fact, combined with the readily available 254-nm "germicidal" UV fluorescent source, allowed for simple experimental molecular manipulation of DNA, and UV-C photobiology was central in establishing the then-new field of molecular biology.

UV-B (290 to 320 nm)

The shorter-wavelength limit (290 nm) of the UV-B region can be defined as the shortest UV wavelength routinely measurable at the earth's surface where it is about 1 million-fold less prevalent than 320-nm radiation. The longer-wavelength limit (320 nm) is where ozone and other atmospheric components begin to (more than 50%) attenuate sunlight appreciably. This absorption prevents much of the significant DNA damage that would result if no ozone layer existed. DNA absorption rapidly decreases toward the longer-wavelength end of this region.[2] Nevertheless, absorption of UV-B photons by DNA contributes to a wide variety of bioeffects. The UV-B is responsible for most of the damage inflicted on organisms by sunlight.

UV-A (320 to 380 nm)

The shorter-wavelength limit of the UV-A (320 nm) can be defined as that wavelength at which ozone becomes transparent. The longer-wavelength limit (380 nm) is where human vision begins and the UV ends. The UV-A region is also biologically effective and causes cellular death, mutation, and DNA damage, although the primary chromophores for these effects may be non-DNA sensitizers that are chemically matched to the photon energy and act as intermediates in relaying the absorbed energy to DNA.[3] UV-A exposures from sunlight are considerable, constituting about 8% of the total sunlight spectrum (compared with less that 0.3% for the UV-B).

Terrestrial Solar UV (290 to 380 nm)

This region is easily defined as the wavelength bracket that is the limit of human vision (380 nm) at the long-wavelength end and the effective limit of the solar UV reaching the earth's surface at the

Table 2.1 Percent Transmission to the Center of Selected Cells and Viruses in the Ultraviolet

Biological sample	Diameter (μm)	Wavelength, nm			
		200	250	300	350
Bacteriophage (T$_2$)	0.1	74	86	100	100
Herpes simplex virus	0.15	66	80	100	100
Bacterial cell	1	33	78	98	100
Yeast cell	5	1.6	69	97	100
Mammalian cell (spherical)	20	10^{-7}	20	91	96
Mammalian tissue (100 μm thick)	—	—	10^{-5}	39	66

Note: All values are approximate; values at λ above 300 nm can vary widely due to the presence of endogenous chromophores. See Reference 7.

short-wavelength (290 nm) end. This is the environmentally relevant UV region and is, of course, just the UV-B plus the UV-A.

2.3 Absorption of UV by Cells

The responses of biological samples to UV can be greatly affected by their absorption properties. Beginning at the cellular level, it can be seen in Table 2.1 that 200-nm UV-C, for example, is absorbed only sightly by viruses (about 30%), more so by bacterial cells (about 70%), and entirely by mammalian cells (more than 99%). This absorption is due mainly to the presence of endogenous molecules that absorb heavily in the UV-C. At longer wavelengths, cell absorption decreases. The absorption of two important biomolecules, the genetic material DNA and protein are compared in Figure 2.1. Both molecules begin to absorb appreciably in the UV-B and substantially in the UV-C. On a weight-by-weight comparison, DNA absorbs about 20-fold as much 260-nm radiation as does protein. Therefore, the absorption of even homogenates of mammalian cells is similar in the wavelength range 220 to 290 nm to that of DNA. The absorbance (a logarithmic function) of such homogenates is seven times higher at 240 nm compared to 300 nm. As the UV wavelength decreases below 220 nm, both proteins and nucleic acids contribute greatly to the total cellular absorption. At wavelengths below 190 nm, water and oxygen absorption prevails. In addition to these macromolecules, some cells contain certain pigments that can alter cellular absorption significantly, even at the longest UV wavelengths. Also, cellular particles and organelles can absorb and scatter UV radiation and, at some wavelengths, shield the center of the cell from a significant portion of the incident beam. Such cytoplasmic screening can alter measured bioeffects to a large degree, especially in the UV-C, and must be considered when cellular exposure is attempted.[4] Such concerns are important because biological cells and tissues will absorb radiation in a wavelength-dependent manner and alter the exposure of the target accordingly. Therefore, the absorption properties of cells are not merely a summation of the individual absorptions of their component molecules.

2.4 Absorption by Tissue

Much of the direct absorption of UV by tissues such as the human skin can be accounted for by the presence of endogenous pigments, especially melanin, hemoglobin, carotenes, and keratin, and exogenous pigments and drugs. Because skin is such a heterogeneous material, UV can be reflected, absorbed, scattered, and rescattered. This means that the direct component of the UV beam is augmented by additions from photons scattered and reflected back into the beam pathway. Hence, at any tissue depth, the sum of the total UV exposure is just the direct plus the diffuse. At wavelengths below 300 nm, absorption by tissue rapidly increases, making all but the first few cell layers of the skin essentially opaque to radiation below about 290 nm.

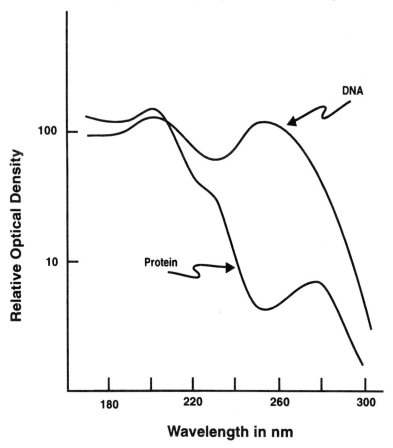

FIGURE 2.1   The absorption spectra of DNA and protein (bovine serum albumin). Both curves are for a 1% solution and 1-cm pathlength.

2.5 Action Spectroscopy

Differences in Effect with Wavelength

An action spectrum (AS) is simply defined as the measurement of a biological effect as a function of wavelength. Crude AS were first used in the 19th century to help identify chlorophyll as the chromophore most responsible for the growth of plants. In this century, more sophisticated methods refined the analysis of AS so that it is now possible, in some instances, to make a reasonable determination of the molecule that is likely to contain the chromophore(s) responsible for the response being studied. It is this latter usage that was among the first methods that pointed to DNA as the genetic material. This was possible because small unicellular organisms, such as bacteria, had relatively high transmission rates for UV and thus meet the rather stringent criteria[5] for a reliable analysis called an analytical action spectrum (AAS). An AAS can only be claimed if the AS corresponds closely to the absorption spectrum of the suspected molecule.

UV-C (190 to 290 nm) Action Spectra

Because some cellular molecules absorb heavily in the UV, certain bioresponses are highly dependent on the wavelengths to which the cell is exposed. In 1930, Gates (Figure 2.2) reported[6] that the AS for bacterial cell death closely followed the absorption of nucleic acid, not protein, which was then widely believed to be the genetic material. In retrospect, this was the first clear evidence that

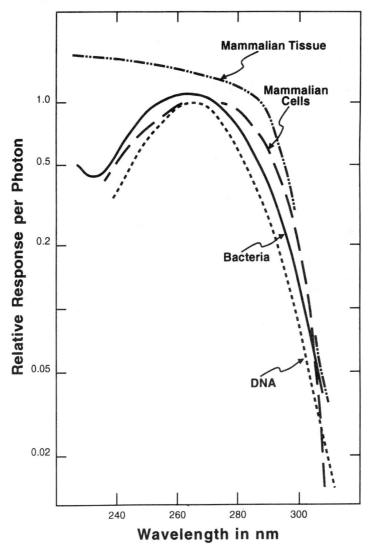

FIGURE 2.2 UV action spectra for the killing of bacteria, mammalian cells, and mammalian tissue. Also represented is the absorption spectrum for DNA. Note that the fine structure inherent in the analytical action spectra for bacteria and mammalian cells is absent in the case of the essentially opaque mammalian tissue.

DNA was the genetic material. It should be remembered that an AS cannot distinguish between DNA and RNA as the target molecules since both have similar absorption spectra. However, an AS for a photoproduct exclusive to DNA, such as the thymine dimer, can. Although AS utilizing small cells is somewhat simple, it is difficult to extend these studies to larger (e.g., mammalian) cells because of the substantial absorption of UV by large cells and tissues (Table 2.1). Two other AS in the UV-C for cell death, one for mammalian tissue and one for individual mammalian cells are shown in Figure 2.2.[5] Experiments utilizing hanging-drop mammalian tissue samples failed to produce an AS with the fine structure of those studies that utilized bacteria because the tissue was essentially opaque to radiation near the peak of DNA absorption (260 nm). Thus, the target molecule (DNA) was shielded to an extent such that the shape of the AS did not parallel the absorption spectrum of the target. The advent of single-cell mammalian culture techniques and the unique flattened geometry that mammalian cells assume when in monolayer culture allowed these studies to begin. Thus, AS for killing of cultured mammalian cells reported data similar to those with bacteria, but the peak was shifted to about 270 nm. This discrepancy can be accounted for by looking at the absorption properties of single mammalian cells and considering, as is the case in bacteria, that the likely target molecule for cell killing is also DNA that resides in the nucleus. This means that the UV beam has to traverse half of the cell on average to strike its target. In bacteria, this distance is small enough to allow one to neglect absorption effects; in mammalian cells, the absorption is substantial. An AS for the production of pyrimidine dimers caused by UV exposure of the DNA in mammalian cells matched the action spectrum for cell death. Therefore, DNA alone was considered responsible for mammalian cell lethality in the 220 to 290-nm region.[7] However, if a detailed knowledge of scattering and absorption events before the beam reaches the target are not known, or if the identification of the target is unsure, then any attempt to "correct" or interpret the data generated is suspect. In some cases, it may be useful to modify the incident beam such that the center of the cell, rather than the cell surface, always receives the same exposure.

Not all UV-C AS follow DNA absorption. For example, the loss of saxitoxin binding to sodium channels in rodent cells, or sodium conduction loss in frog cells or lobster axons, or the UV-induced termination of beating by embryonic chick heart aggregates, all follow an AS similar to the absorption spectra of protein moieties. In addition, UV-C AS for highly pigmented tissues (such as plants) show little correspondence to the absorption of any chromophore. The latter is due to the high degree of absorption of UV-C, which essentially rules out useful AAS in this region for cells and tissues that absorb most of the incident radiation before it can reach the target molecule(s). Thus, UV-C AS are very useful for small or relatively transparent cell studies, but are of very limited use for multicellular or highly pigmented samples. But even if a chromophore cannot be identified, an AS still shows the effect of separate wavelengths on a biological response, and as such, is of some use in predicting the results due to exposure. Studies of human skin exposure to UV-C (e.g., skin cancer or erythema) show on average, an essentially flat response, or in some cases a lessening of effect at shorter wavelengths in the UV-C, that is almost certainly due to limited penetration of the UV beam.

UV-B and UV-A (290 to 380 nm) Action Spectra

It was widely thought that experimental work in the UV-C and UV-B (wavelength range 190 to 320 nm) could be extrapolated to predict photobiological responses in the UV-A (320 to 380 nm). This is not the case, and UV-A AS are much more complex than had been predicted.[3] Here, even events such as cell mutation, which surely involve the genetic material, can be affected at fluences that appear to be below those necessary to affect DNA in these wavelength regions. Details of the divergence of several AS for human cell photoresponses from the DNA absorption spectrum as the curves shift to longer wavelengths in the UV-A are given in Figure 2.3. In the UV-C, cell killing and mutagenesis follow the absorption spectrum of DNA, and carcinogenesis shows a decrease in effect at shorter wavelengths. The UV-B is a transition region from the highly damaging UV-C to the less-damaging UV-A. In the UV-A, all of the biological parameters are affected at levels that are not fully explainable by the absorption properties of DNA. The absolute absorption of moieties in the DNA molecule itself is difficult to measure at wavelengths above 320 nm because of scattering and contamination by extraneous chromophores.[2] It is thought that if the absorption is not intrinsic to DNA, then an intermediate molecule may be involved that absorbs the incident UV-A photon and transfers the effect to DNA.

The motivation for focusing on studies of the effects of UV-A and UV-B — the solar UV wavelength region — is provided by three observations. First, many cells and tissues are more transparent to solar UV, especially in the UV-A; thus, this region should be more amenable to AS analysis. Second, research has suggested that the nature of the primary and secondary chromophore, photoproducts, and mechanisms for cellular response to solar UV appear in some cases to be very different from those elucidated for wavelengths shorter than about 300 nm. For example,

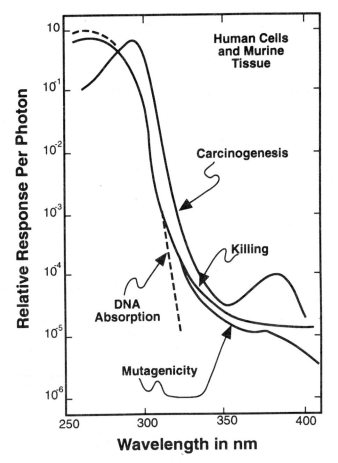

FIGURE 2.3   UV action spectra for human cell killing and mutagenesis and carcinogenic AS for mouse skin. (See text for details.)

in contrast to their marked sensitivity to UV-C, certain photosensitive human cell lines, such as Xeroderma pigmentosum cells (XP), exhibit the same sensitivity to UV-A as normal cells.[8] Third, solar UV impacts a large number of important bioresponses, such as skin cancer, plant growth, cellular survival and mutagenesis, etc. Research in this area should reveal ultimately the nature of biological responses to a portion of the radiation present in a normal environmental setting.[9,10]

2.6 Carcinogenesis

The role of solar UV in human skin cancer has been long established. Squamous and basal cell carcinomas are believed to be the result of chronic UV exposure. The expression of these cancers is highly dependent on skin type and individual predisposition. The carcinogenic data in Figure 2.3 is mainly a composite from the analysis of De Gruijl and van der Leun[11] and incorporates 12 separate studies utilizing mice. This AS shows major effects in the UV-C and UV-B, and falls off rapidly in the UV-A, with an minor peak at about 380 nm. However, due to the large preponderance of UV-A compared to UV-B in solar radiation (about a factor of 35), the contribution of UV-A to carcinogenesis is significant. As shown, however, the major cause of UV-induced cancer is the UV-B component of sunlight.

2.7 Polychromatic Action Spectra (PAS)

Some scientists, perhaps realizing the futility of attempting to construct monochromatic AAS using multicellular organisms, have reported AS that employ polychromatic sources. These studies vary from irradiating the affected system with additional single wavelengths that are added to an ambient background, or generating a set of data using polychromatic sources that employ cut-off filters at successively shorter wavelengths. The polychromatic system is complex and tends to obscure individual chromophores, but it is the closest experimental setup to natural field conditions. A major advantage of using polychromatic radiation in the development of PAS is that interactions of biological responses to different wavelengths (usually unknown) are empirically incorporated into the composite spectrum. In addition, highly pigmented tissues such as plants, have a large variety of chromophores that interact to give a total effect. Here, PAS can give a composite view of the organisms response if a general phenomenon such as plant yield is being studied.

2.8 Effectiveness Spectra (ES)

It is also possible to estimate the damage to a biological system by exposure to UV by combining an AS for a given effect with a known ambient exposure.[12] These are called effectiveness spectra and give at least a first approximation to the effects of UV on the studied system. One such spectrum, ES, is shown in Figure 2.4, which charts the AS for generalized plant damage and the ambient solar

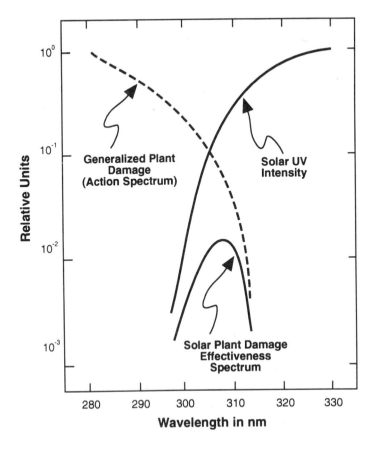

FIGURE 2.4   An AS for generalized plant damage, the terrestrial solar UV intensity spectrum, and a solar plant damage effectiveness spectrum obtained by combining the previous two spectra.

UV. The product of these two curves is the solar plant damage ES. From this chart, one can see that the UV-B region is the most damaging, with a peak at about 308 nm. Although solar intensity increases by a factor of 10 from 310 to 330 nm, the biological response decreases more rapidly over the same wavelength range so that the overall effect is less.

Finally, it should be pointed out that even if all the constraints necessary to produce a detailed AS are adhered to, experimental variables usually limit the reliability of any AS. These variables include, but are not limited to: (1) the spectral purity of the radiation source; (2) the accuracy of the dosimetry measurements; (3) the presence of endogenous or exogenous "nonparticipating" chromophores; (4) the ambient (even microenvironmental) conditions; (5) the time in the life cycle (cellular) or growth cycle (developmental) of the exposed organism; (6) the physical state of the target molecule (e.g., DNA extended or coiled in chromatin); (7) the geometry of the cell or tissue when irradiated; (8) the availability of proper nutrients; (9) water stress in plant studies; and (10) numerous other extraneous conditions.

References

1. Jagger, J., *Introduction to Research in Ultraviolet Photobiology*, Prentice-Hall, Englewood Cliffs, NJ, 1967.
2. Sutherland, J. C. and Griffin, K. P., Absorption spectrum of DNA for wavelengths greater than 300 nm, *Radiat. Res.*, 41, 399, 1981.
3. Coohill, T. P., Peak, M. J., and Peak, J. G., The effects of the ultraviolet wavelengths present in sunlight on human cells *in vitro*, *Photochem. Photobiol.*, 46, 1043, 1987.
4. Coohill, T. P., Knauer, D. J., and Fry, D. G., The wavelength dependence of changes in cell geometry on the sensitivity to ultraviolet radiation of mammalian cellular capacity, *Photochem. Photobiol.*, 30, 565, 1979.
5. Coohill, T. P., Action spectra again?, *Photochem. Photobiol.*, 54, 859, 1990.
6. Gates, F. L., A study of the bactericidal action of ultraviolet light. III. The absorption of ultraviolet light by bacteria, *J. Gen. Physiol.*, 14, 31, 1930.
7. Coohill, T. P., Action spectra for mammalian cells *in vitro*, in *Topics in Photomedicine*, Smith, K. C., Ed., Plenum, New York, 1984, 1.
8. Keyse, S. M., Moss, S. H., and Davies, K. J. G., Action spectra for inactivation of normal and Xeroderma pigmentosum human skin fibroblasts by ultraviolet radiations, *Photochem. Photobiol.*, 37, 307, 1983.
9. Jagger, J., *Solar-UV Actions on Living Cells*, Praeger, New York, 1985.
10. Peak, M. J. and Peak, J. G., Use of action spectra for identifying molecular targets and mechanisms of action of solar ultraviolet light, *Physiol. Plant.*, 58, 367, 1983.
11. de Gruijl, F. R. and van der Leun, J., Action spectra for photocarcinogenesis, in *Biological Responses to UV-A Radiation*, Urbach, F., Ed., Valdenmar, Overland Park, KS, 1992.
12. Coohill, T. P., Ultraviolet action spectra (280 to 380 nm) and solar effectiveness spectra for higher plants, *Photochem. Photobiol.*, 50, 451, 1989.

3

Action Spectroscopy: Photomovement and Photomorphogenesis Spectra

Masakatsu Watanabe
*National Institute for Basic
Biology*

3.1 Introduction

Action spectroscopy (Lipson, this volume) is primarily a crucial methodology for estimating experimentally the absorption spectra, and thus chemical natures, of the (sometimes unidentified) photoreceptor molecules[1] involved in various light-dependent chemical and biological reactions. Technically, recent development and extensive collaborative use of the Okazaki Large Spectrograph (OLS),[2,3] at the National Institute for Basic Biology (NIBB), Okazaki, Japan, and of computerized video image analysis methods[4-7] are especially noteworthy.

For a definition and comprehensive information of photomovement and photomorphogenesis, the reader is referred to the chapters in this volume by Lenci, Poff, and Smith, and also to several excellent textbooks.[8-15]

This chapter aims to provide the reader with a selected set of representative action spectral documentation in a way that enables the reader to grasp easily the overview of the present status of our knowledge. For this aim, after some typical action spectral curves are shown, most of the material is presented in tabular form, closely coupled with figures that show the distribution of spectral sensitivities with respect to phylogenetically arranged taxonomic groups.

0-8493-8634-9/95/$0.00+$.50
© 1995 by CRC Press, Inc.

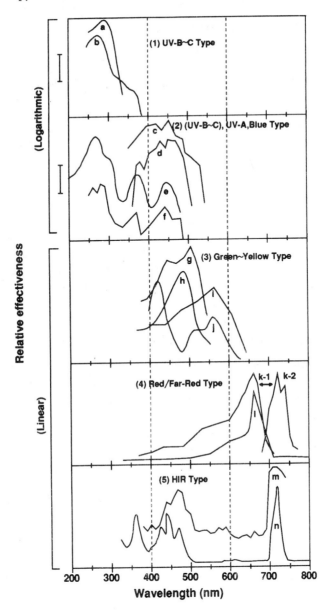

FIGURE 3.1   Major types of photomovement and photomorphogenesis action spectra and their examples. (1) UV-B~C Type: (a) induction of anthocyanin synthesis in *Sorghum* (Angiospermae) (Entry No., EN, 93 in Table 3.1),[106] (b) induction of carotenoid synthesis in *Rhodotorula* (Ascomycetes) (EN 46).[60] (2) (UV-B~C), UV-A, Blue Type: (c) phototaxis in *Ectocarpus* gametes (brown alga) (EN 14),[28] (d) induction of carotenoid synthesis in *Phycomyces* (Zygomycetes) (EN 55),[69] (e) photoavoidance in *Physarum* (slime mold) (EN 9),[23] (f) sexual induction in *Coprinus* (Basidiomycetes) (EN 47).[62] (3) Green Yellow Type: (g) phototaxis in *Chlamydomonas* (green alga) (EN 27),[41] (h) photoavoidance in *Halobacterium* (archaebacterium) (EN 2),[17] (i) phototaxis in *Cryptomonas* (cryptomonad) (EN 16),[30] (j) induction of membrane depolarization in *Paramecium* (protozoa) (EN 19).[33] (4) Red/Far-Red Type: (k-1) induction and (k-2) its cancellation of germination in *Arabidopsis* (Angiospermae) (EN 80),[93] (l) suppression of flowering in *Pharbitis* (Angiospermae) (EN 89).[102] (5) HIR Type: (m) suppression of germination in *Lactuca* (Angiospermae) (EN 86),[99] (n) suppression of hypocotyl growth in *Lactuca* (Angiospermae) (EN 87).[100] These curves were replotted after the sources indicated above. Vertical bars beside the ordinates of (1) and (2) indicate one $\log_{10}$ unit.

3.2 Action Spectra for Photomovement and Photomorphogenetic Responses

Major Types of Photomovement and Photomorphogenetic Action Spectra

Five major types of photomovement and photomorphogenesis action spectra and their examples are shown in Figure 3.1: (1) UV-B~C Type,[1] for which the putative UV-B receptor is not chemically identified yet; (2) (UV-B~C), UV-A, Blue Type,[3,13] for which flavins and pterins are discussed as plausible chromophores of the putative UV-A, Blue receptor(s) (cryptochromes) (Senger, this volume); (3) Green~Yellow Type, for some cases (e.g., g and h in Figure 3.1) of which rhodopsins[14] are established as the photoreceptors;[7,17] (4) Red/Far-Red Type,[1,15] in which "/" means a reversible, mutually canceling effect between red and far-red light, and for which phytochrome, a photoisomerizable chromoprotein, is the established and well-characterized photoreceptor (Furuya; Pratt; Quail, this volume); and (5) HIR Type,[1,15,100] often observed in the case of long-term irradiations of plant materials (HIR standing for high-irradiance reaction), because a strong dependency of the light effect on fluence rate (irradiance) is generally observed.

Distribution of Action Spectral Peak Wavelengths with Respect to Phylogenetically Arranged Taxonomic Groups

The distribution of action spectral peak wavelengths with respect to phylogenetically arranged taxonomic groups is shown in Figure 3.2. Hopefully, together with Table 3.1 below, it will help the reader to think about trends of evolution of photoreceptors, to seek new problems to solve and to find out suitable experimental materials for his or her own research.

List of Action Spectral Data for Photomovement and Photomorphogenetic Responses

A list of about 100 selected sets of action spectral data for photomovement and photomorphogenetic responses in the same phylogenetical order of taxonomic groups as in Figure 3.1 is shown in Table 3.1.

3.3 Prospects

Future evolution of action spectroscopy research would include: (1) the extensive use of photosensitivity mutants, including gene-manipulated organisms[92,109,110] as well as chemical manipulation of contents and characteristics of (sometimes putative) photoreceptors by means of inhibitors and analogs;[7,17] (2) the development and extensive collaborative use of even more efficient, innovative irradiation and observation systems than the ones now available;[3] and (3) the rapid exchange, retrieval, and analysis of action spectral data on the combined bases of a well-organized action spectral database to be constructed and the worldwide digital communication networks that is already available.

Acknowledgments

The author cordially thanks M. Kubota for assistance in preparing Table 3.1 using a relational database program; S. Matsunaga for assistance in preparing Figures 3.1 and 3.2 using computer drawing programs; H. Kawai for phylogenetical arrangement of the taxonomic groups in Figure 3.2 and Table 3.1, and for encouraging the use of a relational database program to handle reasonably

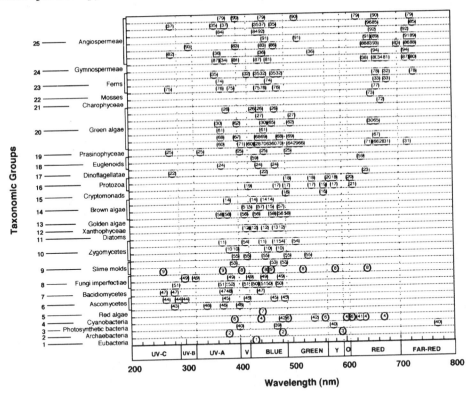

FIGURE 3.2 Distribution of action spectral peak wavelengths with respect to phylogenetically arranged taxonomic groups. The number in each datum point corresponds to the entry number in Table 3.1.

the vast number of data sets from which Table 3.1 was extracted as the essence; H. Kataoka, M. Erata, and M. Sugai for general advice and preparing part of Table 3.1; and K. Yamamiya and K. S. Okamoto for assistance in collecting the references.

A substantial part of the works cited here was supported by the National Institute for Basic Biology (NIBB) Cooperative Research Program for the Use of the Okazaki Large Spectrograph (OLS).

Table 3.1   List of Action Spectral Data for Photomovement and Photomorphogenetic Responses

Entry no.	Organism	Response type	Sign[a]	Wavelength range[b]	Peaks and shoulders (nm)	Ref.
		Photomovement				
	Eubacteria					
1	*Beggiatoa*	Phobic	−	B	430	16
	Archaebacteria					
2	*Halobacterium*	Phobic	−	UV-A,B	380,480	17
3	*Halobacterium*	Phobic	+	G	592	18
	Cyanobacteria					
4	*Anabaena*	Taxis	+	B,O,R	440,600,620–650,670	19
5	*Phormidium*	Taxis	+	Y,R	560,670	20
6	*Phormidium*	Phobic	+	UV-A,B,G,R	390,490,560,610	20
	Red algae					
7	*Porphyridium*	Taxis		V,B	416,443,767–477	21
	Slime molds					
8	*Dictyostelium*	Taxis in amoebae	+	V,B,G,Y,R	405,450,520,580,640	22
9	*Physarum*	Taxis; O₂⁻ generation	−;+	UV-B_C,UV-A,B	260,370,460	23[f]
	Zygomycetes					
10	*Phycomyces*	Tropic balance with reference (450;507 nm)	+	UV-A,B	383,394,455,477;383,455	24
11	*Phycomyces*	Tropism in sporangiophores	+	UV-A,B	370,445,470	25
	Xanthophyceae					
12	*Vaucheria*	Tropism	+	V,B	415,430,450,480	26
13	*Vaucheria*	Chloroplast accumulation	+	B	470	27
	Brown algae					
14	*Ectocarpus*	Taxis in gametes	+	UV-A,B	380,430,450,460	28
15	*Pseudochorda*	Taxis in zoospores		B	420,460	29[f]
	Cryptomonads					
16	*Cryptomonas*	Taxis	+	G	490,560	30
	Protozoa					
17	*Blepharisma*	Phobic	−	B,G,Y	473,491,538,580	31
18	*Blepharisma*	Phobic	−	B,G,Y	493,538,580	32
19	*Paramecium*	Accumulation; membrane depolarization	+;+	B,G	420,560	33
20	*Stentor*	Phobic; membrane depolarization	−;+	G,R	560,610	34
21	*Stentor*	Taxis	−	R	610–620	35
	Dinoflagellates					
22	*Gymnodinium*	Phobic; taxis	−;+	UV-B~C,B	280,450	36

No.	Organism	Process		Spectral region	Wavelengths	Ref.
23	*Peridinium*	Taxis	+	R	640	37
24	Euglenoids / *Euglena*	Phobic	+	UV-A,B	370,440,470	38
25	Prasinophyceae / *Platymonas*	Taxis	+;–	UV-B~C,UV-A,V,B	275,333,405,450,495	39
	Green algae					
26	*Boergesenia*	Tropism	–	UV-A,B	380,430,443,470	40
27	*Chlamydomonas*	Taxis	+	B,G	443,503	41
28	*Cosmarium, Micrasterias*	Phobic; taxis		B,R	440,670	42
29	*Dunaliella*	Phobic; taxis	+,–;+	B,G	510;450–460	43f
30	*Mesotaenium*	Chloroplast rotation		UV-A,B,R(/FR)	366,450,650	44
31	*Mougeotia*	Chloroplast rotation	+/–	R/FR	679,717	45
	Ferns					
32	*Adiantum*	Chloroplast redistribution	+	B,R(/FR)	420,450,480,680	46f
33	*Adiantum*	Tropism; polarotropism	+	R(/FR)	662;680	47f
	Angiospermeae					
34	*Arabidopsis*	Enhancement of phototropism	+	UV-A,R	378,669	48
35	*Avena*	Tropism; growth	+;–	UV-A,B	360,440,470	49
36	*Hordeum*	Streaming in root hair	+	UV-A,B,G	366,433–465;540	50
37	*Medicago*	Tropism	+	UV-B,UV-A,B	280,380,450	51f
38	*Zea*	Induction of geotropism	+	R	640	52

Photomorphogenesis

No.	Organism	Process		Spectral region	Wavelengths	Ref.
	Photosynthetic bacteria					
39	*Rhodobacter, Erythrobacter*	Bchl and carotenoid synthesis	–	B	470	53
40	*Roseobacter*	Bchl and carotenoid synthesis	–	V,Y,FR	400,575,770	54
	Cyanobacteria					
41	*Anabaena*	Germination	+	R	620–630	55
42	*Tolypothrix*	Phycoerythrin synthesis	+	B,G	480,540	56f
	Ascomycetes					
43	*Gelasinospora*	Sexual induction	+	UV-B~C,B	280,350,370,420,440,460,480	57f
44	*Leptosphaerulina*	Sexual induction	+	UV-B~C,UV-A	265,287,300	58
45	*Neurospora*	Conidiation rhythm	–	UV-A,V,B	375,415,465,485	59
46	*Rhodotorula*	Carotenoid synthesis	+	UV-B~C,UV-A,V	280,340,370,400	60f
	Bacidiomycetes					
47	*Coprinus*	Sexual induction	+	UV-B~C,UV-A,B	260,280,370,440	61f
48	*Polyporus*	Styrylpyrone synthesis	+	UV-A,B	380,440	62
	Fungi imperfectiae					
49	*Alternaria*	Conidiation	+;–	UV-B;UV-A,B	300,320;385,420,447,478	63

Table 3.1 (continued)   List of Action Spectral Data for Photomovement and Photomorphogenetic Responses

Entry no.	Organism	Response type	Sign[a]	Wavelength range[b]	Peaks and shoulders (nm)	Ref.
50	Fusarium	Carotenoid synthesis	+	UV-A,B	380,430,455,475	64
51	Mycobacterium	Carotenoid synthesis	+	UV-B~C,UV-A,B	280,370,410,445,470	65
52	Verticillium	Absorbance change and carotenoid synthesis	+	UV-A,B	384	66
	Slime molds					
53	Physarum	Acid from glucose	−	UV-A,B	390,465,485	67
	Zygomycetes					
54	Phycomyces	Macrophore formation	+	UV-A,B,G	383,431,477,514	68
55	Phycomyces	Beta-carotene synthesis	+	UV-A,V,B,G	394,416,450,491,530	69
	Brown algae					
56	Laminaria	Egg production	+	UV-A,V,B	380,412,435,480	70
57	Scytosiphon	Night break of erect thallus formation	−	V,B	414,442,480	71
58	Dictyota	Egg release	+	UV-A,B	366,464,491	72
	Euglenoids					
59	Euglena	Synthesis of Chl; alkaline DNase, etc.	+	B,R	433,631	73
	Green Algae					
60	Acetabularia	Hair whorl formation	+	UV-A,B	370,425,470	74
61	Chlamydomonas	Gametogenesis	+	UV-A,B	370,450	75
62	Chlamydomonas	Division	−	V,G	400,500	76
63	Chlamydomonas	Carbonic anhydrase synthesis	+	B	460	77[f]
64	Chlamydomonas	Rhodopsin synthesis	+	G	500	78
65	Chlamydomonas	Resetting clock in illuminated cells		B,R	450–480,650–670	79[f]
66	Chlamydomonas	Resetting clock in cells in darkness		G,R	520,660	80[f]
67	Chlamydomonas	Flagellation in a mutant	−	V,R	400,660	81[f]
68	Chlorella	Chl and ALA synthesis in glucose bleached	+	UV-A,B	370,440,480	82
69	Scenedesmus	Chl a,b synthesis in autotrophic cells		B,G	450,500	83
70	Scenedesmus	Chl synthesis(LIR[c]) in haterotrophic cells	+	B	450–480	84
71	Scenedesmus	Chl synthesis; cell growth(VLIR[d])	−;+	V,R	408,645	84
	Charophyceae					
72	Nitella	Germination(VLIR[d])	+	R	669	85
	Mosses					
73	Marchantia	Rhizoid formation	+(/−)	R(/FR)	650	86
	Ferns					
74	Adiantum	Division	+	UV-A,B	370,460	87
75	Adiantum	Germination	−	UV-B,UV-A,B	275,390,440	88[f]
76	Adiantum	Apical growth	−	UV-A,B	370,450,470–480	89

No.	Genus	Description	Reaction[a]	Light[b]	Wavelength (nm)	Ref.
77	*Onoclea*	Germination	+/-	R/FR	660/730	90
78	*Dryopteris*	Germination	+/-	R/FR	660/730	91
	Angiospermeae					
79	*Arabidopsis*	Hypocotyl growth in WT; *hy-2*	-	UV-A,B,R,FR;UV-A,B	375,450,625,725;375,450	92[f]
80	*Arabidopsis*	Germination	+/-	R/FR	660/720–740	93
81	*Armoracia*	Adventitious shoots from hairy roots	+	B,R	400,460,680	94[f]
82	*Daucus*	Anthocyanin synthesis in cultured cells	+	UV-B_C	280	95[f]
83	*Egeria*	Membrane hyperpolarization	+	B,R	400,450,650,700	96[f]
84	*Haplopappus*	Anthocyanin synthesis in cultured tissue	+	UV-A,B	372,438	97
85	*Lactuca*	Germination	+/-	R/FR	660/730	98
86	*Lactuca*	Germination (HIR[e])	-	B,FR	470,720	99
87	*Lactuca*	Hypocotyl growth (HIR[e])	-	UV-A,B,FR	363,441,716	100
88	*Lemna*	Flowering	-	R,FR	640,730	101[f]
89	*Pharbitis*	Flowering in etiolated seedlings	-	R	660	102[f]
90	*Phaseolus*	Gene expression (Cab, etc.)	+	V,G,R	400,510,660	103[f]
91	*Raphanus*	Hypocotyl growth	-	B,R,FR	455,515,630–660,705–735	104
92	*Sinapis*	Hypocotyl growth (HIR[e])	-	B,R,FR	448,655,716	105
93	*Sorghum*	Anthocyanin synthesis	+	UV-B,UV-A, B,R	290,385,480,650	106
94	*Triticum*	Flowering in light-grown plants	+	R,FR	660,716	107
95	*Triticum*	Chl a,b synthesis	+	R	650	108

[a] + = positive (phototaxis), step-down or photoattractant (photophobic reaction), induction or stimulation (of photomorphogenetic reactions); - = negative (phototaxis), step-up or photorepellent (photophobic reaction), suppression or inhibition (of photomorphogenetic reactions).

[b] B = blue; G = green; R = red; FR = far-red.

[c] LIR = low-irradiance reaction.

[d] VLIR = very low-irradiance reaction.

[e] HIR = high-irradiance reaction.

[f] Work done at the Okazaki Large Spectrograph (OLS), National Institute for Basic Biology (NIBB), Okazaki, Japan.

References

1. Mohr, H., Criteria for photoreceptor involvement, in *Techniques in Photomorphogenesis,* Smith, H. and Holmes, M. G., Eds., Academic Press, London, 1984, 13.

2. Watanabe, M., Furuya, M., Miyoshi, Y., Inoue, Y., Iwahashi, I., and Matsumoto, K., Design and performance of the Okazaki Large Spectrograph for photobiological research, *Photochem Photobiol.,* 36, 491, 1982.

3. Watanabe, M., High-fluence rate monochromatic light sources, computerized analysis of cell movements, and microbeam irradiation of a moving cell: Current experimental methodology at the Okazaki Large Spectrograph, in *Biophysics of Photoreceptors and Photomovements of Microorganisms,* Colombetti, G., Lenci, F., Haeder, D.-P., and Song, P.-S., Eds., Plenum Press, New York, 1991, 327.

4. Takahashi, T. and Kobatake, Y., Computer-linked automated method for measurement of the reversal frequency in phototaxis of *Halobacterium halobium, Cell Str. Funct.,* 7, 183, 1982.

5. Haeder, D.-P. and Lebert, N., Real time computer-controlled tracking of motile microorganisms, *Photochem Photobiol.,* 42, 509, 1985.

6. Kondo, T., Kubota, M., Aono, Y., and Watanabe, M., A computerized video system to automatically analyze movements of individual cells and its application to the study of circadian rhythms in phototaxis and motility in *Chlamydomonas reinhardtii, Protoplasma,* (Suppl. 1), 185, 1988.

7. Takahashi, T., Yoshihara, K., Watanabe, M., Kubota, M., Johnson, R., Derguini, F., and Nakanishi, K., Photoisomerization of retinal at 13-ene is important for phototaxis of *Chlamydomonas reinhardtii:* Simultaneous measurements of phototactic and photophobic responses, *Biochem. Biophys. Res. Commun.,* 178, 1273, 1991.

8. Colombetti, G., Lenci, F., Haeder, D.-P., and Song, P.-S., Eds., *Biophysics of Photoreceptors and Photomovements of Microorganisms,* Plenum Press, New York, 1991.

9. Haupt, W. and Feinleib, M. E., Eds., *Physiology of Movements, Encyclopedia of Plant Physiology New Series 7,* Springer-Verlag, Heidelberg, 1979.

10. Kendrick, R. E. and Kronenberg, G. H. M., Eds., *Photomorphogenesis in Plants,* Martinus Nijhoff, Dordrecht, 1986.

11. Shropshire, W. Jr. and Mohr, H., Eds., *Photomorphogenesis, Encyclopedia of Plant Physiology New Series 16A,* Springer-Verlag, Berlin, 1983.

12. Shropshire, W. Jr. and Mohr, H., Eds., *Photomorphogenesis, Encyclopedia of Plant Physiology New Series 16B,* Springer-Verlag, Berlin, 1983.

13. Senger, H., Ed., *Blue Light Responses: Phenomena and Occurrence in Plants and Microorganisms,* Vols. I and II, CRC Press, Boca Raton, FL, 1987.

14. Birge, R. R., Photophysics and molecular electronic applications of the rhodopsins., *Annu. Rev. Phys. Chem.,* 41, 683, 1990.

15. Sage, L. C., *Pigment of the Imagination — A History of Phytochrome Research,* Academic Press, San Diego, 1992.

16. Nelson, D. C. and Castenholz, R. W., Light responses of *Beggiatoa, Arch. Microbiol.,* 131, 146, 1982.

17. Takahashi, T., Tomioka, H., Nakamori, Y., Kamo, N., and Kobatake, Y., Phototaxis and the second sensory pigment in *Halobacterium halobium,* in Primary Processes in Photobiology, Kobayashi, T., Ed., Springer-Verlag, Berlin, 1987, 101.

18. Tomioka, H., Takahashi, T., Kamo, N. and Kobatake, Y., Action spectrum of the photoattractant response of *Halobacterium halobium* in early logarithmic growth phase and the role of sensory rhodopsin, *Biochim. Biophys.* Acta, 884, 578, 1986.

19. Nultsch, W., Schuchart, H., and Hoehl, M., Investigations on the photo tactic orientation of *Anabaena variabilis, Arch. Microbiol.,* 122, 85, 1979.

20. Nultsch, W., Photosensing in cyanobacteria, in *Sensory Perception and Transduction In Aneural Organisms,* Colombetti, G., Lenci, F., and Song, P.-S., Eds., Plenum Press, New York, 1985, 147.

21. Nultsch, W. and Schuchart, H., Photo movement of the red alga *Porphyridium cruentum.* 2. Photo taxis, *Arch Microbiol,* 125, 181, 1980.

22. Hong, C. B., Haeder, M. A., Haeder, D.-P., Poff, K. L., Phototaxis in *Dictyostelium discoideum* amoebae, *Photochem. Photobiol.,* 33, 373, 1981.

23. Ueda, T., Mori, Y., Nakagaki, T., and Kobatake, Y., Action spectra for superoxide generation and UV and visible light photoavoidance in plasmodia of *Physarum polycephalum, Photochem. Photobiol.,* 48, 705, 1988.

24. Galland, P. and Lipson, E. D., Action spectra for phototropic balance in *Phycomyces blakesleeanus* dependence on reference wavelength and intensity range, *Photochem. Photobiol.,* 41, 323, 1985.

25. Curry, G. M. and Gruen, H. E., Action spectra for the positive and negative phototropism of phycomyces sporangiophores, *Proc. Natl. Acad. Sci. U.S.A.,* 45, 797, 1959.

26. Kataoka, H., Phototropism in *Vaucheria geminata.* I. The action spectrum, *Plant Cell Physiol.,* 16, 427, 1975.

27. Blatt, M. R., The action spectrum for chloroplast movements and evidence for blue light photo receptor cycling in the alga *Vaucheria, Planta,* 159, 267, 1983.

28. Kawai, H., Mueller, D. G., Foelster, E., and Haeder, D.-P., Phototactic responses in the gametes of the brown alga *Ectocarpus siliculosus, Planta,* 182, 292, 1990.

29. Kawai, H., Kubota, M., Kondo, T., and Watanabe, M., Action spectra for phototaxis in zoospores of the brown alga *Pseudochorda gracilis, Protoplasma,* 161, 17, 1991.

30. Watanabe, M. and Furuya, M., Action spectrum of phototaxis in a cryptomonad alga *Cryptomonas* sp., *Plant Cell Physiol.,* 15, 413, 1974.

31. Scevoli, P., Bisi, F., Colombetti, G., Ghetti, F., Lenci, F., and Passarelli, V., Photomotile responses of *Blepharisma Japonicum.* I. Action spectra determination and time-resolved fluorescence of photoreceptor pigments, *J. Photochem. Photobiol. B: Biol.,* 1, 75, 1987.

32. Matsuoka, T., Matsuoka, S., Yamaoka, Y., Kuriu, T., Watanabe, Y., Takayanagi, M., Kato, Y., and Taneda, K., Action spectra for step-up photophobic response in *Blepharisma, J. Protozool.,* 39, 498, 1992.

33. Matsuoka, K. and Nakaoka, Y., Photoreceptor potential causing phototaxis of *Paramecium bursaria, J. Exp. Biol.,* 137, 477, 1988.

34. Fabczak, S., Fabczak, H., Tao, N., and Song, P.-S., Photosensory transduction in ciliates. 1. An analysis of light-induced electrical and motile responses in *Stentor coeruleus, Photochem. Photobiol.,* 57, 696, 1993.

35. Song, P.-S., Haeder, D.-P., and Poff, K. L., Photo tactic orientation by the ciliate *Stentor coeruleus, Photochem. Photobiol.,* 32, 781, 1980.

36. Forward, R. B., Jr., Phototaxis by the dinoflagellate *Gymnodinium splendens* Lebour, *J. Protozool.,* 21, 312, 1974.

37. Liu, S.-M., Haeder, D.-P., and Ullrich, W., Photoorientation in the freshwater dinoflagellate *Peridinium gatunense* nygaard, *FEMS Microbiol. Ecol.,* 73, 91, 1990.

38. Barghigiani, C., Colombetti, G., Franchini, B., and Lenci, F., Photobehavior of *Euglena gracilis:* Action spectrum for the step-down photophobic response of individual cell, *Photochem. Photobiol.,* 29, 1015, 1979.

39. Halldal, P., Ultraviolet action spectra of positive and negative phototaxis in *Platymonas subcordiformis, Physiol. Plant,* 14, 133, 1961.

40. Ishizawa, K. and Wada, S., Action spectrum of negative photo tropism in *Boergesenia forbesii, Plant Cell Physiol.,* 20, 983, 1979.

41. Nultsch, W., Throm, G., and Rimscha, I. V., Phototaktische Untersuchungen an *Chlamydomonas reinhardii* Dangeard in homokontinuierlicher Kultur, *Arch. Mikrobiol.,* 80, 315, 1971.

42. Wenderoth, K. and Haeder, D.-P., Wavelength dependence of photo movement in desmids, *Planta,* 145, 1, 1979.

43. Wayne, R., Kadota, A., Watanabe, M., and Furuya, M., Photomovement in *Dunaliella salina:* Fluence rate-response curves and action spectra, *Planta,* 184, 515, 1991.

44. Gaertner, R., Die Bewegung des *Mesotaenium*-Chloroplasten im Starklichtbereich. II. Aktionsdichroismus und Wechselwirkungen des Photoreceptors mit Phytochrom, *Z. Pflanzenphysiol.,* 63, 428, 1970.

45. Haupt, W., Die Chloroplastendrehung bei *Mougeotia*. I. Ueber den quantitativen und qualitativen Lichtbedarf der Schwachlichtbewegung, *Planta*, 53, 484, 1959.

46. Yatsuhashi, H., Kadota, A., and Wada, M., Blue-and red-light action in photoorientation of chloroplasts in *Adiantum* protonemata, *Planta*, 165, 43, 1985.

47. Kadota, A., Koyama, M., Wada, M., and Furuya, M., Action spectra for polaro tropism and photo tropism in protonemata of the fern *Adiantum capillus-veneris*, *Physiol. Plant*, 61, 327, 1984.

48. Janoudi, A.-K. and Poff, K. L., Action spectrum for enhancement of phototropism by *Arabidopsis thaliana* seedlings, *Photochem. Photobiol.*, 56, 655, 1992.

49. Elliott, W. M. and Shen-Miller, J., Similarity in dose responses action spectra and red light responses between phototropism and photo-inhibition of growth, *Photochem. Photobiol.*, 23, 195, 1976.

50. Keul, M., Action spectrum of photodinesis within barley root hairs *Hordeum vulgare*, *Z. Pflanzenphysiol.*, 79, 40, 1976.

51. Baskin, T. I. and Iino, M., An action spectrum in the blue and UV for phototropism in alfalfa, *Photochem. Photobiol.*, 46, 127, 1987.

52. Suzuki, T. and Fujii, T., Spectral dependence of the light-induced geotropic response in *Zea* roots, *Planta*, 142, 275, 1978.

53. Takamiya, K.-I., Shioi, Y., Shimada, H., and Arata, H., Inhibition of accumulation of bacteriochlorophyll and carotenoids by blue light in an aerobic photosynthetic bacterium *Roseobacter denitrificans* during anaerobic respiration, *Plant Cell Physiol.*, 33, 1171, 1992.

54. Iba, K. and Takamiya, K.-I., Action spectra for inhibition of light of accumulation of bacteriochlorophyll and carotenoid during aerobic growth of photosynthetic bacteria, *Plant Cell Physiol.*, 30, 471, 1989.

55. Braune, W., C phyco cyanin the main photo receptor in the light dependent germination process of *Anabaena* akinetes, *Arch. Microbiol.*, 122, 289, 1979.

56. Ohki, K., Watanabe, M., and Fujita, Y., Action of near UV and blue light on the photocontrol of phycobiliprotein formation: A complementary chromatic adaptation, *Plant Cell Physiol.*, 23, 651, 1982.

57. Inoue, Y. and Watanabe, M., Perithecial formation in *Gelasinospora reticulispora*. 7. Action spectra in UV region for the photo induction and the photo inhibition of photo inductive effect brought by blue light, *Plant Cell Physiol.*, 25, 107, 1984.

58. Leach, C. M., An action spectrum for light induced sexual reproduction in the ascomycete fungus *Leptosphaerulina trifolii*, *Mycologia*, 64, 475, 1972.

59. Sargent, M. L. and Briggs, W. R., The effect of light on a circadian rhythm of conidiation in *Neurospora*, *Plant Physiol.*, 42, 1504, 1967.

60. Tada, M., Watanabe, M., and Tada, Y., Mechanism of photoregulated carotenogenesis in *Rhodotorula minuta*. VII. Action spectrum for photoinduced carotenogenesis, *Plant Cell Physiol.*, 31, 241, 1990.

61. Durand, R. and Furuya, M., Action spectra for stimulatory and inhibitory effects of UV and blue light on fruit-body formation in *Coprinus congregatus*, *Plant Cell Physiol.*, 26, 1175, 1985.

62. Vance, C. P., Tregunna, E. B., Nambudiri, A. M. D., and Towers, G. H. N., Styryl pyrone biosynthesis in *Polyporus hispidus*. 1. Action spectrum and photo regulation of pigment and enzyme formation, *Biochim. Biophys. Acta*, 343, 138, 1974.

63. Kumagai, T., Action spectra for the blue and near ultraviolet reversible photoaction in the induction of fungal conidiation, *Physiol. Plant*, 57, 468, 1983.

64. Rau, W., Untersuchungen uber die lichtabhangige Carotinoidsynthese. I. Das Wirkungsspektrum von *Fusarium aquaeductum*, *Planta*, 72, 14, 1967.

65. Howes, C. D. and Batra, P. P., Mechanism of photo induced carotenoid synthesis further studies on the action spectrum and other aspects of carotenogenesis, *Arch. Biochem. Biophys.*, 137, 175, 1970.

66. Hsiao, K. C. and Bjorn, L. O., Light induced absorbance changes in the fungus *Verticillium agaricinum*, *Physiol. Plant*, 55, 73, 1982.

67. Schreckenbach, T., Walckhoff, B., and Verfuerth, C., Blue light receptor in a white mutant of *Physarum polycephalum* mediates inhibition of spherulation and regulation of glucose metabolism, *Proc. Natl. Acad. Sci., U.S.A.,* 78, 1009, 1981.

68. Corrochano, L. M., Galland, P., Lipson, E. D., and Cerda-Olmedo, E., Photomorphogenesis in phycomyces fluence-response curves and action spectra, *Planta,* 174, 315, 1988.

69. Bejarano, E. R., Avalos, J., Lipson, E. D., and Cerda-Olmedo, E., Photoinduced accumulation of carotene in phycomyces, *Planta,* 183, 1, 1991.

70. Luening, K. and Dring, M. J., Reproduction growth and photosynthesis of gametophytes of *Laminaria saccharina* grown in blue and red light, *Mar. Biol.,* 29, 195, 1975.

71. Dring, M. J. and Luening, K., A photoperiodic response mediated by blue light in the brown alga *Scytosiphon lomentaria, Planta,* 125, 25, 1975.

72. Kumke, J., Beitraege zur Periodizitaet der Oogon-Entleerung bei *Dictyota dichotoma* (Phaeophyta), *Z. Pflanzenphysiol.,* 70, 191, 1973.

73. Egan, J. M., Jr., Dorsky, D., and Schiff, J. A., Events surrounding the early development of *Euglena gracilis* var. *bacillaris* chloroplasts. 6. Action spectra for the formation of chlorophyll lag elimination in chlorophyll synthesis and appearance of TPN dependent triose phosphate dehydrogenase and alkaline DNase activities, *Plant Physiol.,* 56, 318, 1975.

74. Schmid, R., Idziak, E.-M., and Tunnermann, M., Action spectrum for the blue-light-dependent morphogenesis of hair whorls in *Acetabularia mediterranea, Planta,* 171, 96, 1987.

75. Weissig, H. and Beck, C. F., Action spectrum for the light-dependent step in gametic differentiation of *Chlamydomonas reinhardtii, Plant Physiol.,* 97, 118, 1991.

76. Muenzner, P. and Voigt, J., Blue light regulation of cell division in *Chlamydomonas reinhardtii, Plant Physiol.,* 99, 1370, 1992.

77. Dionisio, M. L., Tsuzuki, M., and Miyachi, S., Blue light induction of carbonic anhydrase activity in *Chlamydomonas reinhardtii, Plant Cell Physiol.,* 30, 215, 1989.

78. Foster, K. W., Saranak, J., and Zarrilli, G., Autoregulation of rhodopsin synthesis in *Chlamydomonas reinhardtii, Proc. Natl. Acad. Sci. U.S.A.,* 85, 6379, 1988.

79. Johnson, C. H., Kondo, T., and Hastings, J. W., Action spectrum for resetting the circadian phototaxis rhythm in the CW 15 strain of *Chlamydomonas.* II. Illuminated cells, *Plant Physiol.,* 97, 1122, 1991.

80. Kondo, T., Johnson, C. H., and Hastings, J. W., Action spectrum for resetting the circadian phototaxis rhythm in the CW15 strain of *Chlamydomonas, Plant Physiol.,* 95, 197, 1991.

81. Nakamura, S., Watanabe, M., Hatase, K., and Kojima, M. K., Light inhibits flagellation in a Chlamydomonas mutant, *Plant Cell Physiol.,* 31, 399–401, 1990.

82. Oh-hama, T. and Senger, H., Spectral effectiveness in chlorophyll and 5 amino levulinic-acid formation during regreening of glucose bleached cells of *Chlorella prothecoides, Plant Cell Physiol.,* 19, 1295, 1978.

83. Thielmann, J., Galland, P., and Senger, H., Action spectra for photosynthetic adaptation in *Scenedesmus obliquus.* I. Chlorophyll biosynthesis under autotrophic conditions, *Planta,* 183, 334, 1991.

84. Thielmann, J. and Galland, P., Action spectra for photosynthetic adaptation in *Scenedesmus obliquus.* II. Chlorophyll biosynthesis and cell growth under heterotrophic conditions, *Planta,* 183, 340, 1991.

85. Sokol, R. C. and Stross, R. G., Phytochrome-mediated germination of very sensitive oospores, *Plant Physiol.,* 100, 1132, 1992.

86. Otto, K.-R. and Halbsguth, W., Stimulation of primary rhizoid formation on gemmae of *Marchantia polymorpha* as caused by light and IAA, *Z. Pflanzenphysiol.,* 80, 197, 1976.

87. Wada, M. and Furuya, M., Action spectrum for the timing of photo induced cell division in *Adiantum capillus-veneris* gametophytes, *Physiol. Plant,* 32, 377, 1974.

88. Sugai, M. and Furuya, M., Action spectrum in UV and blue light region for the inhibition of red-light-induced spore germination in *Adiantum capillus-veneris, Plant Cell Physiol.,* 26, 953, 1985.

89. Kadota, A., Wada, M., and Furuya, M., Apical growth of protonemata in *Adiantum capillus-veneris.* 3. Action spectra for the light effect on dark cessation of apical growth and the intracellular photoreceptive site, *Plant Sci. Lett.,* 15, 193, 1979.

90. Towill, L. R. and Ikuma, H., Photo control of the germination of *Onoclea* spores. 1. Action spectrum, *Plant Physiol.*, 51, 973, 1973.

91. Mohr, H. Die Beeinflussung der Keimung von Farnsporen durch Licht und andere Faktoren, *Planta*, 46, 534, 1956.

92. Goto, N., Yamamoto, K. T., and Watanabe, M., Action spectra for inhibition of hypocotyl growth of wild-type plants and of the hy2 long-hypocotyl mutant of *Arabidopsis thaliana L.*, *Photochem. Photobiol.*, 57, 867, 1993.

93. Shropshire, W., Klein, W. H., and Elstad, V. B., Action spectra of photomorphogenic induction and photoinactivation of germination in *Arabidopsis thaliana*, *Plant Cell Physiol.*, 2, 63, 1961.

94. Saitou, T., Tachikawa, Y., Kamada, H., Watanabe, M., and Harada, H., Action spectrum for light-induced formation of adventitious shoots in hairy roots of horseradish, *Planta*, 189, 590, 1993.

95. Takeda, J. and Abe, S., Light-induced synthesis of anthocyanin in carrot cells in suspension. IV. The action spectrum, *Photochem. Photobiol.*, 56, 69, 1992.

96. Tazawa, M., Shimmen, T., and Mimura, T., Action spectrum of light-induced membrane hyperpolarization in *Egeria densa*, *Plant Cell Physiol.*, 27, 163, 1986.

97. Lackmann, I., Action spectra of anthocyanin synthesis in tissue cultures and seedlings of *Haplopappus gracilis* D., *Planta*, 98, 258, 1971.

98. Borthwick, H. A., Hendreicks, S. B., Toole E. H., and Toole, V. K., Action of light on lettuce-seed germination, *Bot. Gazz.*, 115, 205, 1954.

99. Gwynn, D. and Scheibe, J., An action spectrum in the blue for inhibition of germination of lettuce seed, *Planta*, 106, 247, 1972.

100. Hartmann, K. M., Ein Wirkungsspektrum der Photomorphogenese unter Hochenergiebedingungen und seiner Interpretation auf der Basis des Phytochroms (Hypokotylwachstumshemmung bei *Lactuca sativa L.*), *Z. Naturforsch.*, 22b, 1172, 1967.

101. Lumsden, P. J., Saji, H., and Furuya, M., Action spectra confirm two separate actions of phytochrome in the induction of flowering in *Lemna paucicostata* 441, *Plant Cell Physiol.*, 28, 1237, 1987.

102. Saji, H., Vince-Prue, D., and Furuya, M., Studies on the photoreceptors for the promotion and inhibition of flowering in dark-grown seedlings of *Pharbitis nil* choisy, *Plant Cell Physiol.*, 24, 1183, 1983.

103. Sasaki, Y., Yoshida, K., and Takimoto, A., Action spectra for photogene expression in etiolated pea seedlings, *FEBS Lett.*, 239, 199, 1988.

104. Jose, A. M. and Vince-Prue, D., Action spectra for the inhibition of growth in radish hypocotyls, *Planta*, 136, 131, 1977.

105. Beggs, C. J., Holmes, M. G., Jabben, M., and Schaefer, E., Action spectra for the inhibition of hypocotyl growth by continuous irradiation in light and dark grown *Sinapis alba* seedlings, *Plant Physiol.*, 66, 615, 1980.

106. Yatsuhashi, H., Hashimoto, T., and Shimizu, S., Ultraviolet action spectra for anthocyanin formation in broom sorghum first internodes, *Plant Physiol.*, 70, 735, 1992.

107. Carr-Smith, H. D., Johnson, C. B., and Thomas, B., Action spectrum for the effect of day-extensions on flowering and apex elongation in green light-grown wheat *Triticum aestivum L.*, *Planta*, 179, 428, 1989.

108. Virgin, H. I., Action spectra for chlorophyll formation during greening of wheat *Triticum aestivum* leaves in continuous light, *Physiol. Plant*, 66, 277, 1986.

109. Deng, X.-W., Matsui, M., Wei, N., Wagner, D., Chu, A. M., Feldmann, K., and Quail, P. H., *COP1*, an *Arabidopsis* regulatory gene, encodes a protein with both a zinc-binding motif and a G_{beta} homologous domain, *Cell*, 71, 791, 1992.

110. Chory, J., Out of darkness: Mutants reveal pathways controlling light-regulated development in plants, *Trends Genet.*, 9, 167, 1993.

4

DNA Photochemistry

Bela P. Ruzsicska
University of Sherbrooke

D. G. E. Lemaire
University of Sherbrooke

4.1 Introduction

The formation of photoproducts in DNA is the fundamental cause of the adverse effects of UV irradiation including mutagenesis, carcinogenesis, and cell death. The importance of the UV photolysis of DNA has been recognized for several decades in photobiological research and it serves as the basis for the concern raised by the alarming reductions in the atmospheric ozone layer and the consequent increased exposure to UV-B (290 to 320 nm) radiation at the earth's surface. The object of this review is to elucidate the processes involved in the modification of DNA by direct photolysis beginning with excitation of the nucleic acid bases and ending with the formation of stable photoproducts that are thought to provoke a biological response. Particular emphasis will be placed on UV-B irradiation and how it may affect the photoproduct formation and distribution. Since several reviews have recently appeared with compilation of data, this chapter is intended to be a critical synopsis with the focus on the well-established general features of nucleic acid photo-chemistry and also highlighting the areas of uncertainty. For more detailed information on the photophysics and photochemistry of nucleic acids, the reader is advised to consult several previous reviews.[1–4] Figure 4.1 shows the structures, names, and numbering system of the nucleic acid bases used throughout the chapter.

4.2 Photophysics of Nucleic Acids

Excited States

The spectroscopic difficulties presented by the electronic structure of nucleic acid bases due to the noncorrelation between π-electrons and centers and the presence of nonbonding electrons has been discussed.[2] This review is concerned only with the transitions and states involved in the first UV absorption band (230 to 300 nm) of nucleic acid bases and the conventional photochemical descriptions of π,π^* and n,π^* will be used to correlate the photophysics and the photochemistry.

0-8493-8634-9/95/$0.00+$.50
© 1995 by CRC Press, Inc.

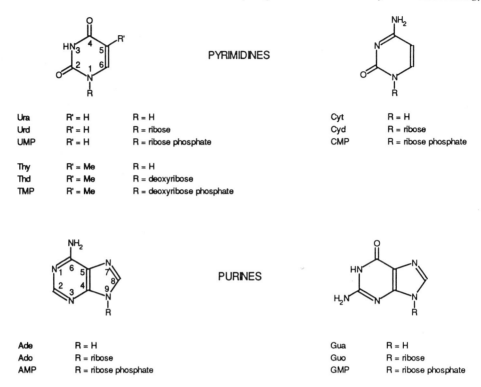

FIGURE 4.1 Abbreviations, structures, and numbering system for bases, nucleosides, and nucleotides. *Bases:* Ura, uracil; Thy, thymine; Cyt, cytosine; Ade, adenine; Gua, guanine. *Nucleosides:* Urd, uridine; Thd, thymidine; Cyd, cytidine; Ado, adenosine; Guo, guanosine. *Nucleotides:* UMP, uridine monophosphate; TMP, thymidine monophosphate; CMP, cytidine monophosphate; AMP, adenosine monophosphate; GMP, guanosine monophosphate.

For thymine and uracil derivatives, several different experiments including polarized absorbance, reflectance and fluorescence studies, as well as linear dichroism indicate that the first absorption band at ~260 nm consists of one π,π* transition.[5–7] Theoretical predictions for n,π* states for uracil derivatives[8,9] have been supported by the crystal absorption spectrum of 1-methyluracil, which is interpreted as indicating a n,π* state roughly isoenergetic with the first π,π* state.[5] Additional evidence that the first singlet π,π* and n,π* states of uracil and thymine have similar energy levels comes from luminescence and flash photolysis experiments that are discussed later.

For the cytosine derivatives, the first absorption band is dominated by a π,π* transition at 270 nm and, because there is a smaller energy gap between the first two π,π* states as compared to thymine and uracil, a second π,π* transition becomes apparent in the absorption band.[10,11] There is no experimental evidence for a low-lying ¹n,π* state from the polarized crystal absorption spectrum; however, UV derivative spectroscopy on cytosine polynucleotides have been interpreted as indicating an n,π* transition at 288 nm.[12]

For the purine bases guanine and adenine, the first absorption band is composed of two π,π* transitions as shown by the polarized absorbance and fluorescence experiments.[13,14] For guanine, the energies of these two transitions are clearly distinct as shown by the shoulder in the absorption band, while for adenine the two π,π* states are nearly degenerate. An n,π* state has been shown to be the lowest singlet excited state in purine[15] and evidence for a low-lying n,π* singlet state in 9-ethylguanine has been presented.[14] A contribution of an n,π* transition to the absorption band in the long-wavelength region (270 to 310 nm) of poly(dA-dT) was suggested by derivative spectroscopy.[16]

In summary, there is some supporting evidence for the near degeneracy of the first singlet π,π^* and n,π^* states for the purine and pyrimidine nucleic acid bases which can be proposed to be the dominant feature of the photophysics and photochemistry of these bases.

Photophysical Parameters of Nucleic Acid Monomers

The excited-state dynamics of nucleic acid bases are dominated by an extremely short singlet excited-state lifetime that is caused by an extremely rapid internal conversion rather than by enhanced intersystem crossing. The resulting very low fluorescence quantum yields (Φ_f) have been quantitated in the last 2 decades with improvements in instrumental sensitivity. The values compiled previously[1,2] show that the nucleosides and nucleotides of guanine, cytosine, and thymine have approximately the same yields, $\Phi_f \sim 1 \times 10^{-4}$, with Ado and AMP having $\Phi_f \sim 0.5 \times 10^{-4}$. When these experimental values are combined with the calculated radiative lifetimes, fluorescent lifetimes of ≤ 1 ps are calculated.[17,18] The observation of rapid ground-state recovery,[19] on the order of several picoseconds, combined with the small intersystem crossing quantum yields (see below), show that an extremely rapid internal conversion, with a rate constant ($k_{ic} \sim 10^{12}$ s^{-1}) close to that of vibrational relaxation, is responsible for the short fluorescence lifetime. One likely explanation for this phenomenon is the near degeneracy of the lowest $^1n,\pi^*$ and $^1\pi,\pi^*$ states that can lead to an enhancement of the nonradiative transition decay rates through vibronic interaction between the states.[2,20] Indirect experimental determinations resulted in fluorescent lifetimes (τ_f) of 3 to 5 ps for the bases, nucleosides, and nucleotides, with some discrepancies obtained for TMP and UMP.[1,21] Direct measurements of τ_f are difficult due to the requirements of short laser pulses (≤ 1 ps) and adequate time resolution in the detecting system. The values reported from a limited number of these measurements are confusing and require comment.

There has been only one measurement on Cyd and Thd that indicated only that the fluorescence lifetime was <100 ps.[22] The very high fluorescence polarizations observed for thymine and cytosine in room-temperature aqueous solution indicate that the fluorescence process is faster than rotational motion and calculations give a limit of ≤ 2 ps for τ_f.[6,23] For adenine and guanine nucleosides and nucleotides, several studies have indicated fluorescence lifetime components of a few nanoseconds.[24–27] A detailed examination on adenosine concluded that the species responsible for the nanosecond lifetime components are likely minor rotational conformers and protonated forms.[26] Protonation, in particular, is an important effect and results in greatly increased fluorescence yields and lifetimes. A good example of this is 7-MeGMP ($pK_a = 7.0$), whose cationic form has been determined to have $\Phi_f = 0.012$[28] and $\tau_f \sim 200$ to 250 ps.[24,25] A similar effect is observed for the purine base tautomers. Fluorescence polarization studies on adenine and guanine indicated that the low polarizations observed were almost certainly due to the fluorescent, long-lived 7-*H* tautomers.[6,29] This suggested that the 9-*H* tautomer has a high fluorescence polarization and consequently a very short τ_f. Room-temperature polarization studies of AMP and GMP revealed that the polarization is high, very near that in low-temperature glass, confirming that the room-temperature lifetimes are a few picoseconds.[30,31] A value of 5 ps was reported for the monomer fluorescence lifetime of 9-methyladenine from picosecond time-resolved spectroscopy,[32] although another study on ATP did not find a short-lived component.[24] Therefore, further lifetime values from direct fluorescence measurements are necessary to confirm the extremely short excited singlet-state lifetimes predicted from steady-state fluorescence measurements.

The analysis of absorbance and fluorescence spectra at room temperature results in the determination of the excited state energies, $^1E_{(0-0')}$.[1] The results show that Cyd, CMP, Thd, and TMP have the same energy (34,000 cm^{-1}) with Guo and GMP being 200 cm^{-1} lower, which is still within the thermal energy range at room temperature. Ado and AMP have significantly higher energies (35,500 cm^{-1}), which results in essentially negligible absorption in the UV-B region. The effect of methylation is an important point to consider for the pyrimidine bases. Methylation of UMP and dCMP, which forms TMP and 5-MedCMP, results in an increase in Φ_f and a decrease in $^1E_{(0-0')}$.[1,33]

The effect of methyl substitution is usually explained in terms of a preferential stabilization of the π,π^* state relative to the n,π^* state.[20,34,35] This would increase the energy separation of the $^1n,\pi^*$ and $^1\pi,\pi^*$ states and decrease the vibronic interaction between them with the result of an increased fluorescent lifetime. 5-MedCMP has been measured to have a $\Phi_f = 8 \times 10^{-4}$ and $^1E_{(0-0')} \sim 32,500$ cm^{-1},[33] which results in 5-MedCMP, a *naturally* occurring nucleotide in DNA (1 mol% in mammalian DNA[36]) having a significantly lower excited singlet state energy than the isoenergetic units TMP, CMP, and GMP. The possible consequences of this for energy transfer in DNA are discussed later.

The second most important decay process for the excited singlet states of nucleic acid bases is intersystem crossing to the triplet state. The base triplet states were first investigated by phosphorescence emission at low temperature with polarization and ESR studies establishing these emissions as being due to $^3\pi,\pi^*$ states.[37] These triplet states have also been characterized by their absorption spectra, which were determined by laser flash photolysis. The pyrimidine bases Thy and Ura, along with their nucleosides and nucleotides, have been the most thoroughly characterized with a determination of the extinction coefficients.[38–41] Recently, the triplet absorption spectra of Cyt, Cyd, and CMP were determined by acetone photosensitization.[42] These results showed that the cytosine derivatives have similar triplet absorption spectra compared to the corresponding uracil derivatives, but the extinction coefficients were an order of magnitude lower.

For guanine and adenine and their nucleosides and nucleotides, the triplet absorption spectra have not been determined by either direct photolysis or photosensitization due to interference from other processes. For example, Ade, Ado, AMP and Gua, Guo, GMP are very easily photoionized in a biphotonic process, and the transient spectra of the resulting radical ions obscure the spectra of the triplets, if any are formed.[43–45] Attempts to observe the triplet absorption spectra by acetone photosensitization were also unsuccessful, apparently due to a hydrogen abstraction reaction being more important than energy transfer.[41] The triplet absorption spectra that have been measured for the purine derivatives, 2-aminopurine, 7-methylguanine, and purine itself show maxima in the 380 to 430-nm region.[41,46,47]

The determination of the triplet absorption spectra permits the direct measurement of the intersystem crossing quantum yields, Φ_{isc}, in aqueous solution at room temperature. The studies on thymine and uracil derivatives have shown that Φ_{isc} is dependent on solvent and excitation wavelength.[39,40,48,49] It was shown that Φ_{isc} increases by a factor of 4 to 5 in room-temperature solution for Thd, TMP, Urd, and UMP in going from a protic to an aprotic solvent.[40] This behavior correlates with the decrease in Φ_f and increase in phosphorescence quantum yield, Φ_p, as a function of the same solvent change in low-temperature frozen solutions.[50] These changes in the photophysical properties as a function of solvent were explained by a change in ordering of $^1\pi,\pi^*$ and $^1n,\pi^*$ states, as shown in Figure 4.2. In aprotic solvents, the lowest excited state is expected to be $^1n,\pi^*$, which would favor intersystem crossing ($^1n,\pi^* \rightarrow {}^3\pi,\pi^*$) due to classical selection rules.[34,51] The change to a protic solvent results in a change in the excited-state ordering since π,π^* states are more stabilized by dipole-dipole and hydrogen-bonding interactions than n,π^* states. The lowest excited state becomes $^1\pi,\pi^*$, which is nearly isoenergetic with $^1n,\pi^*$ and intersystem crossing becomes disfavored relative to the very fast internal conversion. Similar solvent effects on the photophysical properties of 2-aminopurines had been described earlier with similar conclusions,[35,46] but such studies are not available for cytosine, guanine, and adenine derivatives due to the difficulties in obtaining triplet spectra.

The observed effect of Φ_{isc} increasing with decreasing excitation wavelength in aqueous solution[49,52] may be explained by referring to Figure 4.2. Increasing excitation energy will increase the probability of intersystem crossing probably via a $^1\pi,\pi^* \rightarrow {}^3n,\pi^*$ path. The wavelength dependency of Φ_{isc} has been somewhat controversial since this effect necessarily requires that Φ_f will also be wavelength dependent, and it is generally agreed that Φ_f values are wavelength independent.[2] However, the lack of wavelength dependence for Φ_f in room-temperature aqueous solution is not necessarily incompatible with a wavelength-dependent Φ_{isc} given the dominance of internal conversion, the small magnitude of Φ_{isc}, and the even smaller magnitude of Φ_f. Simple calculations

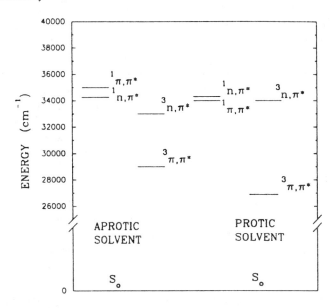

FIGURE 4.2 Possible energy level orderings of excited states of Thd in aprotic and protic solvents assuming $^1n,\pi^*$ to be the lowest excited state in aprotic solvent. $E(^1\pi,\pi^*) = 34,000$ cm^{-1} in H$_2$O taken from Reference 2 and $E(^3\pi,\pi^*) = 26,900$ cm^{-1} taken from Reference 37. All other energy levels are postulated (see Reference 40).

show that for pyrimidines, assuming $\tau_f = 5$ ps, $\Phi_f = 1 \times 10^{-4}$, and Φ_{isc} varying between 0.002 and 0.02, the excited singlet-state lifetime would vary by 2% with the variation of the intersystem crossing rate and, consequently, Φ_f would vary by the same amount. Given the small magnitude of Φ_f, this difference would not be experimentally noticeable. It would be important to verify the wavelength dependency of Φ_{isc} for pyrimidine nucleosides and nucleotides by laser flash photolysis because two studies have indicated that Φ_{isc} decreases dramatically in the UV-B region,[49,52] and this would have important consequences for the involvement of the triplet state in the UV-B photochemistry of DNA.

Values for Φ_{isc} for uracil and thymine derivatives have been compiled[53] and, for Thd and TMP, $\Phi_{isc} = 0.013 \pm 0.003$ with $\lambda_{exc} = 266$ nm. For Cyt, Ade, and Gua derivatives, Φ_{isc} values have only been measured by indirect methods.[49,54] These values, although inconsistent, indicate $\Phi_{isc} < 2 \times 10^{-3}$ for these compounds.

In summary, the examination of Φ_{isc} values indicates that the efficiency of triplet-state formation for the nucleic acid bases by UV photolysis is low. The results suggest that intersystem crossing efficiency may be significantly lower for UV-B photolysis.

Photophysical Properties of Dinucleotides, Polynucleotides, and DNA

The organization of nucleotides into the double-helix structure of DNA with the consequent base stacking results in an electronic interaction between the bases. One well-known effect of this interaction is the hypochromic change of the UV absorbance. This electronic coupling interaction is expected to affect the photophysical processes of the nucleic acid bases due to the possibility of two new processes: exciplex formation and energy transfer.

The fluorescence spectra of di- and polynucleotides have been extensively studied in low-temperature frozen solutions and at room temperature, and emissions from excimers and exciplexes have been observed. The criteria for assignment of an emission as excimeric has been simply an observation of a red-shifted broad emission spectrum clearly different from monomer emissions yet resulting from absorption in the monomer region. It has been recognized that the use of the terms "excimer" and "exciplex" in describing emissions from ordered structures that result from base stacking in di- and polynucleotides does not correspond with the normal definition of an excimer and exciplex, which are defined as an association between an electronically excited molecule and a ground-state molecule that does not exist when both molecules are in their ground state.[34,51,55] For oligonucleotides in room-temperature solution, where there is an equilibrium distribution of stacked and unstacked structures, it is commonly assumed that excitation of an unstacked structure will lead to a similar emission as that seen for the monomeric units. This is because, due to the very short lifetime of the monomeric bases, deactivation occurs before conformational motion can reorient the bases so that an excimer or exciplex can form. Therefore, any changes in the emission spectra between oligonucleotides and mononucleotides is attributed to emission from an excited stacked complex that has been often termed an excimer or exciplex emission in the literature. This usage is described as being valid because the electronic interaction in the stacked complex is at the very weak coupling limit and the absorption spectra of the monomers is essentially unchanged.[55] In other words, the bases are considered to absorb independently. Therefore, changes in the equilibrium distribution between stacked and unstacked species brings about variation in the excited complex emission similar to the concentration dependence seen for true excimer or exciplex emission. With these qualifications in mind, the exciplex emissions observed in di- and polynucleotides can be examined.

First of all, exciplex emission is only seen for certain bases.[56,57] For the dinucleotides pdTpdT and GpG, the absorption and fluorescence spectra are essentially identical to the corresponding mononucleotides in low-temperature frozen solution[56] and the same behavior is observed for dTpdT at room temperature.[58] In contrast, for CpC, poly(rC), ApA, and poly(rA), there are small blue shifts observed in absorption and larger red shifts observed in fluorescence relative to the monomers in low-temperature frozen solutions that were interpreted as showing the presence of stacked species and excimer fluorescence.[56] This interpretation was also suggested by the observed increase in fluorescence quantum efficiency. However, for ApA and poly(rA), in contrast to CpC and poly(rC), significant changes occur in the emission spectrum upon warming to room-temperature solution[22,56,59] with the appearance of a broad emission in the 350 to 500-nm region. Comparison with experimental reference spectra allowed the observed total emission spectra to be resolved into four components, which were assigned originally as monomer fluorescence, excimer fluorescence, monomer phosphorescence, and excimer phosphorescence.[59] A similar analysis was applied to the emission spectra of CpC and poly(rC), resulting in the same conclusions as to the multicomponent nature of the emission.[60] Further studies on the mixed dinucleotides CpA and ApC indicated various amounts of this multicomponent emission.[55] However, the nature of the various components in the emission spectra was reinterpreted. The component assigned previously to the exciplex fluorescence is reinterpreted as exciplex fluorescence from a predominant right-handed stacked conformation. The component assigned previously to the exciplex phosphorescence is reassigned as exciplex fluorescence from a left-handed stacked conformation. This particular interpretation is important with regard to the role of exciplexes in the formation of photoproducts in DNA. It was proposed previously that these exciplexes may be the precursors of some of the photoproducts formed within polynucleotides.[1,22,25] Exciplex formation could play a role in photoproduct formation since it would provide a competing reaction to the internal conversion process discussed before. In addition, cycloaddition reactions of the type that are important in DNA photochemistry often proceed through exciplexes.[61,62] Another possibly important role of exciplex formation is that triplet state formation may be enhanced via an exciplex. This has been detected in a number of exciplexes where it occurs by an electron exchange process.[61] However, in order to establish the importance of fluorescent exciplexes in photoproduct formation, a correlation between the formation of exciplex and photoproduct must be established. No such correlation has so far been observed in di- or polynucleotides. In this regard, it is important to note that exciplex fluorescence is strongly dependent on polynucleotide structure. A good example of this is the comparison between poly(rA) and poly(dA), which exhibit different stacking structures.[63] Low-temperature fluorescence of poly(dA) is quenched about fivefold and red-shifted compared to poly(rA).[21] Since there does not seem to be a room-temperature fluorescence study of poly(dA), it is difficult to compare the fluorescence of the two polynucleotides. However, there has been a fluorescence study of the copolymer poly(dA)·poly(dT) whose structure is known to be a variation of a typical B-DNA helix.[64] This modified B conformation is associated with a set of bifurcating hydrogen bonds along the major groove and maximized base stacking, which provide a particular rigidity for this structure.[65,66] The fluorescent study of this copolymer concluded that the fluorescence spectrum was dominated by monomer-like fluorescence from thymine.[64] In other words, it was concluded that poly(dA), in this copolymer, did not fluoresce, which is in strong contrast to poly(rA).[59] However, it is poly(dA) that is more photoreactive.[67] The results of the study on poly(dA)·poly(dT) are particularly interesting since it has been observed in low-temperature glasses that poly(dA) fluoresces, whereas poly(dT) does not.[57] Of course, the differences in fluorescence behavior may be attributed to different stacking structures in the two cases. Another important point to note is that Φ_f of poly(dA)·poly(dT), which was attributed exclusively to thymine monomer fluorescence, was measured as 4×10^{-4}, four times higher than that observed in TMP.[64] This enhanced fluorescence may be attributed to the described rigidity of the poly(dA)·poly(dT) helix. However, it must be noted that it has been observed several times that the particular stacking of thymine that results in formation of *cis-syn*-cyclobutane dimer is associated with a *quenching* of the thymine fluorescence.[58,68,69] Therefore, the enhanced fluorescence in poly(dA)·poly(dT), if confirmed, should be associated with a decreased photodimerization efficiency.

Another interesting copolymer to examine is poly(dA-dT)·poly(dA-dT). The low-temperature emission spectrum of this copolymer gave two peaks, one at 358 nm assigned to fluorescence, and another at 469 nm assigned to phosphorescence.[57] The room-temperature emission spectrum displayed two peaks, one at 330 nm assigned to thymine monomer fluorescence, and another at 410 nm assigned to exciplex fluorescence.[70] Fluorescence decay measurements indicated several long-lived fluorescence components.[24,25] The complex, poorly understood emission spectral features are probably related to the unusual structural features of this copolymer.[63]

In summary, exciplex fluorescence from polynucleotides is probably related to unusual stacking structures, and there is no evidence that these observed exciplexes are involved in photoproduct formation.

For DNA, the fluorescence spectrum is found to be mostly monomer-like in appearance, although the quantum yield ($\sim 4 \times 10^{-5}$) is approximately one half the value found for an equally absorbing mixture of mononucleotides.[18,71] The DNA fluorescence was also found to have a smaller, broad component ($\lambda_f \sim 450$ nm), which was tentatively assigned to exciplex fluorescence. Time-resolved measurements are in disagreement as to whether there is a long lifetime ($\tau_f \sim 3$ ns) component in the fluorescence of calf thymus DNA.[25,27]

The question of energy transfer in DNA is of interest to determine whether migration of excitation energy is involved in the formation of photoproducts in DNA. Again, one must remember that for energy transfer to occur, it must compete against the other very fast processes that deactivate the excited singlet states, i.e., radiationless internal conversion, intersystem crossing, and exciplex formation. It is assumed that once an exciplex is created, its energy is too low to be transferred to any of the bases in DNA. In other words, the exciplex is an effective singlet excitation trap.[72] Since the quantum yield of DNA fluorescence is quenched compared to the mononucleotides, this indicates that the excited-state lifetime of the bases must be less than a few picoseconds and, consequently, the rate constants for competing processes must be at least 10^{11} to 10^{12} s^{-1}.

The two interactions that are important in energy transfer are (1) coulombic or exciton interaction involving excitation resonances that can be effective over distances of tens of angstroms, and (2) electron exchange interaction involving charge resonance that essentially requires orbital overlap. These mechanisms in electronic energy transfer in nucleic acids have been considered previously.[21,56,72,73] Here, only the basic results will be summarized. From considerations of the distance dependence of both coulombic and electron exchange interactions, it was concluded that transfer between neighboring bases is the only one that needs to be considered.[72] From an estimate of the magnitudes of the two interactions for the separation between stacked bases, it was concluded that the coulombic interaction would be dominant for singlet-singlet energy transfer, whereas the electron exchange interaction dominates triplet-triplet transfer. It has been predicted from the Förster theory of coulombic interaction that considerable transfer is possible for poly(rA) and poly(rC) before vibrational relaxation but very little after relaxation.[72] This latter point is important for UV-B photolysis. Since excitation in the UV-B region will produce vibrationally relaxed excited states, very little energy transfer is predicted. If energy is indeed transferred over several monomer units before being trapped as an exciplex, the polymer fluorescence should be greatly depolarized relative to that of monomers. The experimental results have indicated a variable fluorescence depolarization for poly(rA) and poly(rC) in low-temperature glasses that seems to be dependent on fluorescence wavelength.[56] The high fluorescence polarization seen for excitation near the band origin gives support to the proposal of very little energy transfer for vibrationally relaxed excited singlet states. The fluorescence study of the copolymer poly(dA)·poly(dT), where the fluorescence was attributed exclusively to monomer-like thymine emission, indicated that the fluorescence was partially depolarized compared to monomers.[64]

Sensitized fluorescence experiments with several dye-nucleic acid complexes have shown that singlet excitation energy is transferred to dye molecules when the nucleic acid chromophore is excited by UV light. One study concluded that singlet energy transfer from bases located two to four base pairs from the dye molecule could occur,[74] while another study concluded that fluorescence sensitization due to exciton migration along the DNA chain from other bases is negligible.[75]

It was mentioned previously that if energy transfer occurs in a homopolymer, it could only be trapped at an exciplex. While this is obvious for a homopolymer, it is also true for DNA due to the equivalent singlet state energies of TMP, CMP, and GMP. Of course, there is the possibility that certain exciplexes between the bases will be more stable than others and therefore will be better traps.

Another important point to consider is that modified bases may serve as traps for the singlet-singlet energy transfer. This has been demonstrated for 7-MeGMP, which was shown to be a reasonably efficient irreversible singlet energy trap in DNA via an exciton interaction.[73] This is due to its lower excited singlet-state energy ($\sim 32,300$ cm^{-1}) as compared to TMP, CMP, and GMP. Although 7-MeGMP is not a naturally occurring nucleotide in DNA, 5-MedCMP is present and, as has been shown previously, the available experimental evidence indicates that its singlet state energy is approximately the same as that for 7-MeGMP. Therefore, 5-MedCMP could serve as a singlet energy trap in DNA and this factor, along with a higher UV-B extinction coefficient than that of the other pyrimidine bases, suggests that 5-MedCMP could have a much greater importance in the UV-B photolysis of DNA than its mole percentage would indicate.

As for triplet-triplet energy transfer, it has been well established that energy transfer can occur over many bases and that thymine, due to its low triplet state energy, serves as a trap in DNA.[21,72] Intersystem crossing values have shown that the triplet state is formed in low yield in base monomers; however, it has been proposed from an examination of the fluorescence depolarization of the exciplex fluorescence from stacked dinucleotides that there may be a considerable contribution of charge-transfer resonance to the stability of the exciplex.[55,56,59,60] If this is true, then it would indicate that the electron exchange interaction is important for the singlet exciplex and it may lead to the formation of a triplet exciplex. Although the particular triplet exciplex formed may or may not be photochemically reactive, it would lead to enhanced triplet state formation, energy transfer to thymine, and possible reaction of the thymine base. Evidence for the occurrence of charge transfer resonance and triplet exciplex could be obtained possibly by determining if the triplet-state quantum yield increases for stacked bases such as in Thy-(CH$_2$)$_3$-Thy or poly(dA)·poly(dT).

In summary, the photophysical parameters of nucleic acid bases indicates that they should be fairly unreactive to UV-B photolysis due to the overwhelming predominance of radiationless internal conversion. Singlet-singlet energy transfer does not appear to be an important process for unmodified bases in DNA. The available evidence does not indicate that exciplexes, as observable intermediates, are involved in the formation of photoproducts.

4.3 Photochemistry of Nucleic Acids

Cycloadditions to Form Pyrimidine Cyclobutane Dimers

The most thoroughly characterized photoproduct in DNA is the cyclobutane dimer between adjacent pyrimidine bases, and dimers between thymine-thymine (T<>T), thymine-cytosine (T<>C), cytosine-thymine (C<>T), and cytosine-cytosine (C<>C) have been detected in photolyzed DNA *in vitro* and *in vivo*.[1,76] What is of interest to explore here is how the pyrimidine structure and sequence will influence the quantum efficiency of dimer formation and whether the distribution of dimers from UV-B photolysis is different from conventional UV-C photolysis. The photochemical formation of cyclobutane dimers from thymine and uracil derivatives has been studied extensively.[77] The [2π+2π]-cycloaddition of two double bonds that occurs in a suprafacial-suprafacial manner ($2\pi_s + 2\pi_s$) is, from orbital symmetry arguments, photochemically allowed but thermally forbidden as a concerted process (i.e., from the singlet state).[34] Therefore, cyclobutane dimer formation may proceed from $^1\pi,\pi^*$ or $^3\pi,\pi^*$ states via a concerted or nonconcerted pathway, respectively, and exciplexes may or may not be involved. These are the factors shown to be important for cyclobutane dimer formation in α,β-cyclohexenones,[34,78] models for pyrimidine bases, and therefore important factors to consider in the mechanism of cyclobutane dimer formation in DNA.

In dilute solutions, cyclobutane dimers were shown to arise exclusively from the triplet state of thymine and uracil derivatives.[77] The efficiency of dimerization was shown to be an order of magnitude higher for uracil than for thymine, which was attributed mainly to steric hindrance of dimer formation although the triplet quantum yield of uracil is much higher than that of thymine. In more concentrated solutions, the dimerization quantum yield increased greatly, indicating a greater reaction efficiency for stacked aggregates. This quantum efficiency was in fact higher than the triplet quantum yield that was interpreted as dimerization proceeding from the singlet state of the excited complex.[79–81] When the quantum yields for dimerization of stacked aggregates (Φ_d^s) are compared, the value for 1,3-dimethylthymine ($\Phi_d^s = 0.065$) is found to be higher than that for 1,3-dimethyluracil ($\Phi_d^s = 0.03$). This indicates the well-established feature of cyclodimerization efficiency increasing with electron-donating groups substituent on the double bond.[34] Similar quantum efficiencies have been found for aggregates of Thy ($\Phi_d^s = 0.041$) and Thd ($\Phi_d^s = 0.038$).[82] For cytosine, the quantum efficiency for dimer formation from excited aggregates was calculated to be 0.005, an order of magnitude smaller than for thymine and uracil.[77]

The photolysis of the monomeric base units in solution results in the four isomeric dimers, *cis-syn*, *cis-anti*, *trans-syn*, and *trans-anti* due to the random orientations of the stacked aggregates and collision complexes. However, due to the fixed orientation and overlap of bases in the DNA double helix, only the *cis-syn*-isomer is possible as cyclobutane dimer photoproduct (see Figure 4.3, path A). The *trans-syn*-isomer is formed in single-stranded DNA and oligonucleotides due to the greater conformational flexibility. Actually, due to the glycosidic linkages, two diastereomers are possible for both the *cis-syn*- and *trans-syn*-isomers.[63] The *cis-syn*-isomer has only been found as the *anti-anti*-form, while two *trans-syn*-diastereomers (*syn-anti* and *anti-syn*) have been demonstrated in dinucleotides.[58,83,84] The relative ratio of the two *trans-syn*-diastereomers has been shown to depend on the pyrimidine sequence.

Analysis of the efficiency of photoproduct formation from direct photolysis of small deoxyoligonucleotides, with similar stacking structures as found in DNA, are important to understand DNA photochemistry. There are limited studies of cyclobutane dimer quantum yields in dinucleotides, polynucleotides, and DNA and these are presented in Table 4.1. It must be recognized that the quantum yield values are not corrected for the percentage of stacked complexes. With this correction in mind, the features present in the table can be examined. For the thymine-thymine cyclobutane dimerization, the limiting quantum yield seems to be 0.035 to 0.04, which compares well to the quantum yield of stacked aggregates. In DNA, a value of ~0.017 is measured after corrections for the absorption of TT pairs. This reduced value may be due to the higher rigidity in

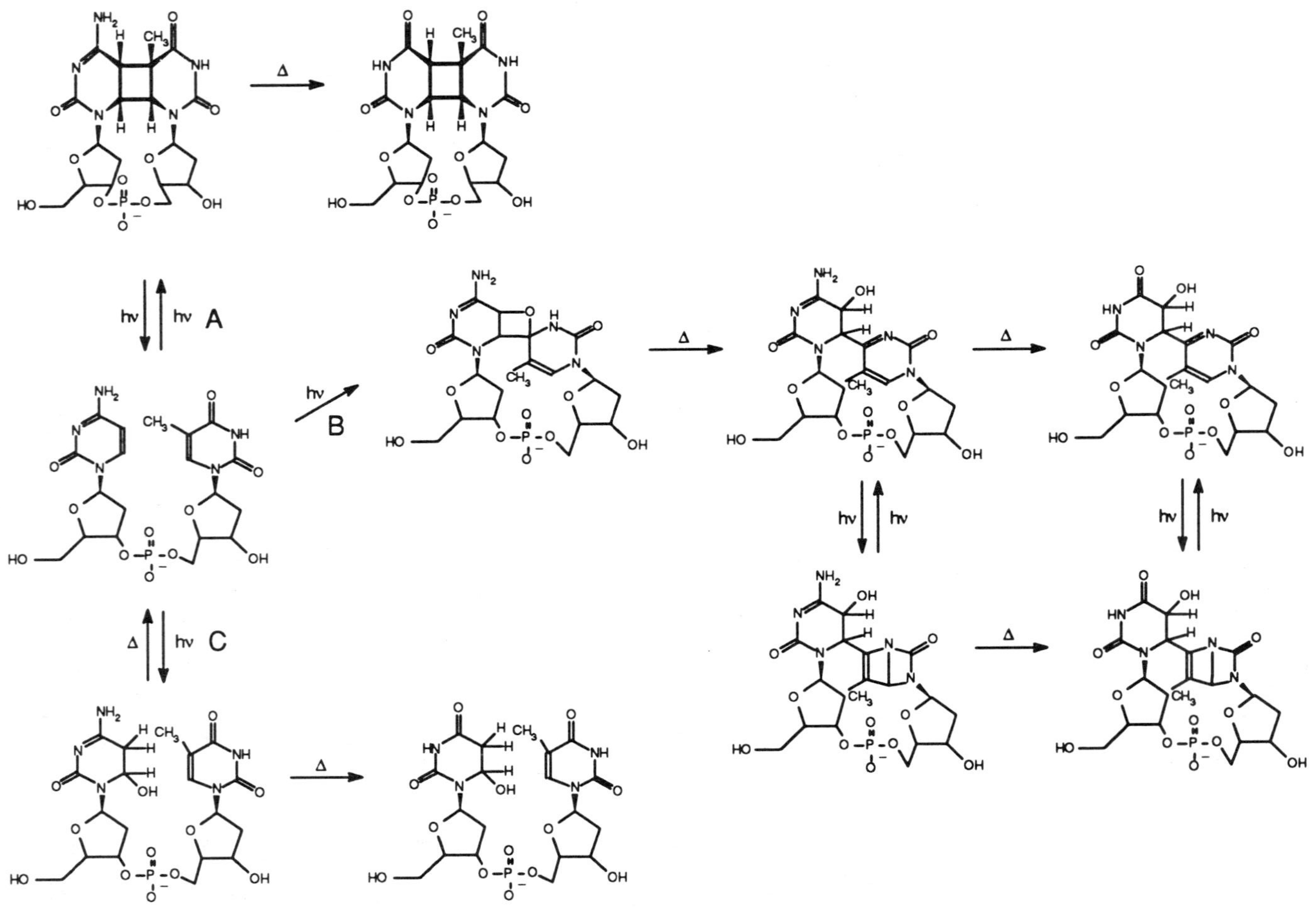

FIGURE 4.3   Reaction scheme demonstrating the three well-characterized photoreactions possible at dipyrimidine sequences (CT is illustrated here), along with the possible secondary thermal and photoreactions. Path A: cyclobutane dimer formation. Path B: [6–4]-photoadduct formation. Path C: cytosine photohydrate formation.

Table 4.1 Pyrimidine Photoproduct Quantum Yields in Di- and Polynucleotides and in DNA

Compound	Φ (c,s-Cyclobutane)	Φ ([6–4]-Photoadduct)	Φ (Photohydrate)
Thy-$(CH_2)_3$-Thy	0.040[69]	—	—
dTpdT	0.016[58]	0.0005[58]	—
	0.010[91]	0.0010[91]	—
dTpdTp	0.011[92,a]	—	—
poly(dT)	0.044[92,a]	—	—
Thy-$(CH_2)_3$-Ura	0.024[69]	—	—
dTpdU	0.0082[58]	0.00045[58]	0.00065[58]
dUpdU	0.007[190]	—	0.018[190]
poly(rU)	0.015[191]	—	0.012[191]
dTpdC	0.0072[58]	0.00074[58]	0.0007[58]
			0.006[192,a]
CpC	0.04[94,b]	—	0.006[94]
poly(rC)	—	—	0.012[193]
DNA			
T<>T	0.015[100]	—	—
	0.019[142]		
T<>C	0.0076[100]	—	—
	0.0026[142]	0.0015[142]	

Note: Where possible, values from long-wavelength excitation, $\lambda > 280$ nm, are presented.
[a] Quantum yield based on absorbance measurements, product not isolated.
[b] Quantum yield strongly wavelength dependent.

the double helix caused by hydrogen bonding. For the thymine-uracil dimerization, the limiting quantum yield seems to be ~0.025. In general, the values show that thymine-uracil and uracil-uracil cyclodimerization are less efficient by a factor of 0.5 to 0.75. This difference in cyclodimerization efficiency could be attributed to the absence of the electron-donating methyl group on the C5-C6 double bond as described previously, assuming the percentage of stacked complexes and the orientation in the stacked complex are similar. This difference in cyclodimerization efficiency could be expected to apply for thymine-cytosine and cytosine-cytosine. However, there are large inconsistencies in the data. For dinucleotides, the Φ_d value measured for CpC, although reported to be strongly wavelength dependent, is much higher than that for dTpdC.[58,94] In contrast, the relative rate of cyclobutane dimer induction in poly(dC) was reported to be 1 to 10% of the rate measured in poly(dT).[85] In DNA, the relative rate of dimer induction from UV-B photolysis, T<>T:(T<>C+C<>T):C<>C, has been measured as 1:1:0.5,[86–90] which is approximately that expected. Therefore, further experiments are warranted on dC oligonucleotides to accurately determine the quantum efficiency of cyclodimerization.

Considering the mechanism of cyclodimerization for thymine-thymine pairs, the evidence indicates that the cyclobutane dimer arises predominantly from the singlet state of the excited complex. The reasons for this are the following: (1) the quantum efficiency of cyclodimerization (Φ_d) for UV-B photolysis is much higher than the estimated triplet quantum yield;[58,91] (2) the cyclodimerization efficiency is not diminished by triplet-state quenchers;[58,68] and (3) the cyclodimerization efficiency is a function of fluorescence quenching and percentage of stacking.[68,92] Of course, there is the possibility that the cyclodimerization proceeds from the triplet excimer after conversion from the singlet excimer, as has been observed in other photocycloadditions.[61,62] However, since there is no triplet quenching, this mechanism requires the observation of the triplet excimer or increased triplet-state yield as a function of stacking for validation. Presently, the evidence points to thymine-thymine dimerization as proceeding from a $^1\pi,\pi^*$ excited complex. If it is assumed that thymine-cytosine and cytosine-cytosine dimerizations are 50 to 75% as efficient as thymine-thymine, then this would suggest a similar mechanism for dimerization involving cytosine. This is because the triplet quantum yield of cytosine, although not well defined, seems to be even lower than that of thymine[49] and therefore it is unlikely that intersystem crossing to the triplet state is involved in cyclodimerization of cytosine. It is interesting to note that Φ_d for CpC seems to be high and excimer formation has been observed for this dinucleotide.[60] Whether this excimer is involved in cyclodimerization may be investigated by correlating excimer fluorescence intensity and dimerization efficiency as a function of stacking. Similar studies would be useful for dCpdC.

Concerning the effect of methylation on the C-5 position of cytosine, it is predicted from the factors previously described that the cyclodimerization efficiency of 5-meCyt should be greater than that of Cyt and should be similar to that of Thy. There is evidence of cyclobutane dimer formation involving 5-meCyt in DNA.[93] It would therefore be important to determine Φ_d for 5-methyl-dC oligonucleotides from UV-B photolysis. For example, 5-MeCyt is found at the HpaII restriction site, C^{m}CGG.[36] The reaction cross section, k_d,* for cyclodimerization at this sequence from UV-B photolysis is predicted to be much greater than the equivalent sequence CTGG due to a larger molar extinction coefficient and singlet energy transfer over 1–2 bases.

Secondary Reactions of Pyrimidine Cyclobutane Dimers

One of the important secondary reactions of pyrimidine cyclobutane dimers is the retrocycloaddition photoreaction to regenerate the pyrimidine base monomers. This photoreversion may proceed by direct photolysis or by photosensitized electron transfer (for a review of the latter process, see the chapter by S. D. Rose). The direct photoreversion is simply the reverse of the $[2\pi + 2\pi]$-cycloaddition and it is a photochemically allowed, concerted process characteristic of cyclobutane dimers.[34,51] This reaction has been usually considered to be important only at wavelengths in the UV-C region due to the absorption spectrum of the T<>T dimer. However, it has also been well established that cytosine-containing cyclobutane dimers have definite absorption in the UV-B region.[58,94–96] Since the quantum yield of pyrimidine cyclobutane photoreversion is quite high, $\Phi_r = 0.7$ to 0.8,[58,96–99] the reaction cross section for this photoreversion will be important in the UV-B photolysis of DNA. This effect is demonstrated by the rapid photoequilibrium reached for cytosine-containing cyclobutane dimers upon UV-B photolysis.[58,86,90,100] However, this photoequilibrium is not always observed.[87] This photoreversion may have a biological effect in UV-B irradiation since the cytosine-containing cyclobutane dimers formed by photolysis mainly in the 295 to 305-nm region should be kept to a low photoequilibrium level by the high fluences present in solar UV-B radiation in the 305 to 315-nm region. In this connection, it is important to note that T<>T, T<>U, and U<>U dimers do not absorb in the UV-B region and, therefore, T<>T dimers will be continually produced. T<>U and U<>U dimers can arise from deamination of the cytosine dimers and thus, this deamination can remove the secondary UV-B photoreversion of cytosine dimers.

The deamination of cytosine cyclobutane dimers has been implicated as an important process in UV-induced mutations associated with SOS repair and photoreactivation.[101–108] Deamination of a 5,6-saturated cytosine, such as in a cyclobutane dimer, is several orders of magnitude faster than that for cytosine;[83,103,109–111] in fact, cytosine deamination is believed to proceed via a 5,6-saturated intermediate.[111] Determination of the *in vivo* deamination rates indicated there were different rates occurring for different dimers,[103] and an analysis of the deamination mechanism in C<>T and T<>C dinucleotides suggested that T<>C, C<>T, and C<>C may deaminate at different rates in DNA due to several factors.[83] In addition, deamination of 5-methylcytosine via cyclobutane dimer formation may be an important process in UV-B mutagenesis. Deamination of a T<>C or C<>C dimer will result in the corresponding uracil dimer and, if the latter dimers are "repaired" by photoreactivation, then a transition of cytosine to uracil will occur. These uracils can be efficiently removed by uracil glycosylase; but if the same process occurs for 5-methylcytosine, then a transition of 5-methylcytosine to thymine will occur. This mutation could become fixed in the absence of a T·G mismatch repair mechanism.[112] Further studies of the deamination efficiency of cytosine- and 5-methylcytosine-containing dimers are therefore warranted.

---

* $k_i = \sigma_i \Phi_i$, where σ_i is the absorption cross section and Φ_i is the quantum yield.

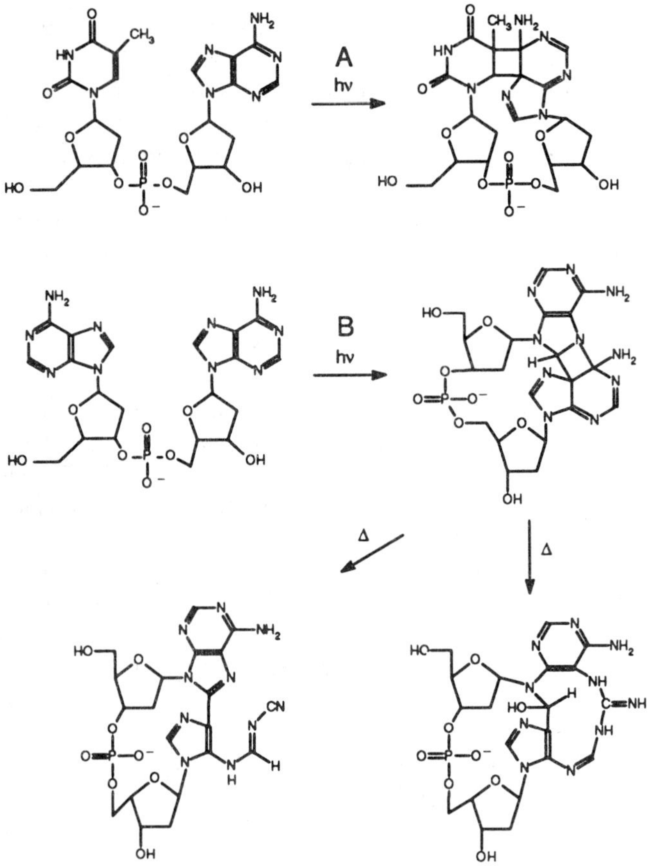

FIGURE 4.4 Reaction scheme demonstrating cycloadditions involving adenine. Path A: cyclobutane dimer formation in a TA sequence. Path B: azetidine dimer formation at a AA sequence. (See References 114 and 121.)

Cycloadditions Involving Adenine

In addition to the photocycloaddition between pyrimidine bases, purine bases (specifically adenine) that were normally considered photochemically unreactive may also undergo photocycloaddition. A photoproduct between thymine and adenine has been found in dinucleotides, polynucleotides, and DNA.[113,114] On the basis of chemical degradation and NMR characterization, the structure of the photodimer was proposed as a cyclobutane linkage between the bonds thymine C6–C5 and adenine C6–C5 (Figure 4.4, path A).[115] The photoproduct was not observed when dTpdA was irradiated at wavelengths >290 nm, either alone or in the presence of acetone as triplet photosensitizer. This strongly indicates that the adenine singlet state is the precursor of the photoproduct, and it is unusual that the thymine $^3\pi,\pi^*$ state cannot form this cyclobutane linkage. This product is also very specific for the sequence and the base type. It is not found in dApdT, dTpdG, dCpdA, or m5dCpdA, but it does seem to be formed in dTpdI.[116] This photoreactivity seems to be quite different from the photochemistry of pyrimidine-purine dinucleotide analogs in which the sugar phosphate group has been replaced by a trimethylene bridge. Previously, it was shown that similar analogs with only pyrimidine bases display a similar photoreactivity as pyrimidine dinucleoside monophosphates. However, for the pyrimidine-purine analogs, studies have indicated that there are two types of [2π+2π]-photocycloadditions that are different from that seen for dTpdA.[117–120] One addition is a cyclobutane formation between pyrimidine C5–C6 and purine C4–C5, and the second is azetidine formation between pyrimidine C5–C6 and purine N7–C8.

Sensitization and quenching experiments suggested that the excited singlet was the reactive state in these photocycloaddition reactions. These products were shown to be unstable in aqueous solution undergoing ring cleavage; in addition, they were shown to undergo photocycloreversions of the cyclobutane or azetidine ring upon irradiation at 254 nm in contrast to the dTpdA photoproduct that is photochemically inert. The reason for this difference in photoreactivity between dinucleoside monophosphates and trimethylene bridged analogs is not clear.

The photodimerization of adjacent adenine bases in oligomers and polymers of deoxyadenylic acid has been reported to occur with a relatively high quantum yield.[67] It has been shown that two distinct photoproducts are formed by adenine photodimerization in dApdA upon 254-nm irradiation in aqueous solution.[121] It was concluded that the primary event in adenine photodimerization involves addition of the N7–C8 double bond of the 5'-adenine across the C5–C6 bond of the 3'-adenine. The azetidine species generated thus acts as a common precursor to both dApdA photoproducts, which are formed by competing modes of azetidine ring scission (Figure 4.4 path B).

Since the photoproducts from dTpdA and dApdA seem to arise from the adenine excited singlet state, these photoproducts are predicted not to be important in the UV-B photolysis of DNA due to the negligible absorption of adenine. However, they may still arise via exciplexes that seem to form preferentially with adenine. The exciplex fluorescence seen with poly(dA-dT) copolymer[70] may be the precursor of the TA photoproduct. However, other evidence is against the involvement of exciplexes. The TA photoproduct is formed with dTpdA, which seems to be unstacked preferentially in solution and not in dApdT, which is more stacked in solution.[122] Also, the fluorescence spectrum of poly(dA)·poly(dT) did not indicate any exciplex that may be involved in the formation of the dApdA photoproduct.[64]

Cycloadditions to form [6–4]-Photoadducts

Another important photoproduct in DNA arises from the $[2\pi+2\pi]$-photocycloaddition of the 3'-base via either the C4-O4 carbonyl of thymine or the C4–N4 imino tautomer of cytosine onto the C5–C6 double bond of the adjacent 5'-base to give oxetane and azetidine intermediates, respectively. These oxetane and azetidine intermediates are unstable and rearrange to give the 4-(5,6-dihydropyrimidin-6-yl)-2-pyrimidone adducts commonly called the [6–4]-photoadducts (see Figure 4.3, path B).

The [6–4]-photoadduct was first characterized partially in the dinucleotide dTpdT.[91,123] Then, a product from acid hydrolysis of UV-irradiated DNA was identified as 6–4'-[pyrimidin-2'-one]-thymine,[124] which arises from deamination of the thymine-cytosine [6–4]-photoadduct. Subsequently, [6–4]-photoadduct formation has been observed for all pyrimidines, their nucleosides and nucleotides, as well as from dinucleotides.[125–130] Analysis of UV-irradiated DNA treated with hot alkali indicated that strand breaks were produced at cytidine positions located 3' to pyrimidine nucleosides.[131] The lesion responsible for this strand breakage was postulated to be the [6–4]-photoadduct.[132] Recent structural studies have confirmed this strand cleavage, although the standard assay conditions may not result in quantitative breakage due to the complex multistep nature of the reaction.[84,133] However, other studies have indicated that [6–4]-photoadducts were not the only photolesions cleaved by hot alkali treatment. It has been shown that the Dewar valence isomers of the [6–4]-photoadducts produced upon UV photolysis of the latter compounds (see Figure 4.3 and below) are much more sensitive to alkali-induced strand breakage.[133,134] Nevertheless, the measurement of these strand breaks gives an indication of the initial formation of [6–4]-photoadducts. Marked differences in the ratio of the four [6–4]-photoadducts have been reported in UV-irradiated DNA.[131,135] Different rates of induction of [6–4]-photoadducts in dinucleotides have been measured,[129,132] and the quantum yields for some of these dinucleotides have been determined (see Table 4.1). What is of interest to examine in terms of DNA photochemistry is the relative ratio of [6–4]-photoadduct formation of each dipyrimidine sequence and also the ratio of cyclobutane dimer to [6–4]-photoadduct for each sequence.

The data indicate that TC and CC are more reactive in forming [6–4]-photoadducts than TT and CT sequences. This indicates that azetidine formation via the C4–N4 imino tautomer of cytosine is more favorable than oxetane formation via the C4–O4 carbonyl group of thymine. This photocycloaddition, when it involves the addition of a carbonyl group onto an alkene, is called the Paterno-Büchi reaction and it normally proceeds through the ^{1,3}n,π^* states of aldehydes and ketones.[34,78] It is of interest to note that while [2π+2π]-photocycloadditions of ketones to alkenes are well characterized, similar additions of imines are relatively rare, which was thought to indicate a poorer reactivity of the excited imino group.[136,137] The formation of [6–4]-photoadducts seems to proceed through excited singlet states since their formation could not be quenched by triplet quenchers and triplet photosensitization did not form these photoadducts.[58,83,109,130,138,139] However, the reaction may proceed via the ^{3}n,π^* state (see Figure 4.2). The involvement of an upper ^{3}n,π^* state was invoked to explain oxetane formation with an excited cyclohexenone reacting with an alkene.[78] For pyrimidine bases, the lifetime of this ^{3}n,π^* state would be extremely short, making it impossible to quench; since it would not be formed with triplet photosensitization, its involvement could be mistaken for the singlet state. However, the quantum yield of the ^{3}n,π^* state should be equal to the $^3\pi,\pi^*$ quantum yield. For cytosine, the only value available for Φ_{isc} is 1.5×10^{-3}, which is equal to the quantum yield of the [6–4]-photoadduct of dTpdC (see Table 4.1, Φ_{6-4} corrected for only cytosine absorption and there is no consideration of stacking percentage or wavelength dependency of Φ_{isc}). The proposal that all the excited ^{3}n,π^* states of the 3′-cytosine react to form the azetidine is highly improbable in as much as deactivation from ^{3}n,π^* to $^3\pi,\pi^*$ is expected to be significant. Therefore, the evidence indicates that the ^{1}n,π^* state of the 3′-pyrimidine is responsible for [6–4]-photoadduct formation. Again, due to the extremely short lifetime of the excited singlet state, this reaction should proceed from the ^{1}n,π^* complex formed from base stacking. Another possibility is the reaction of the π,π^* state of the 5′-pyrimidine onto the carbonyl or imino group of the 3′-pyrimidine.[140] This is considered unlikely because the imino tautomer of cytosine is not considered to be significant. Additional strong evidence that the ^{1}n,π^* state of the 3′-pyrimidine is responsible for cycloaddition comes from methylation. Methylation of the 3′-cytosine in a CC sequence reduces considerably the formation of the [6–4]-photoadduct relative to the cyclobutane dimer.[141] As explained previously, the effect of methylation is to stabilize the π,π^* state relative to the n,π^* state, which would favor cyclobutane formation relative to azetidine or oxetane formation. The same effect is seen in comparing the cyclobutane dimer/[6–4]-photoadduct quantum yield ratio for dTpdT and dTpdU (see Table 4.1). Therefore, the preferential formation of [6–4]-photoadducts for TC and CC over TT and CT may be explained by a greater population of the ^{1}n,π^* state of cytosine relative to thymine. It is important to point out that this reaction is sequence specific. The carbonyl or imino group of the 5′-pyrimidine does not react with the C5–C6 double bond of the 3′-pyrimidine due to steric reasons in the stacked complex found in the right-handed DNA B helix.[132]

The quantum yield ratios of ([6–4]-photoadduct)/(cyclobutane dimer) for some dinucleotides are available in Table 4.1. The ratio for dTpdT is 3% compared to 10% for dTpdC. The ratio for dCpdT should be similar to dTpdT, and the ratio for dCpdC should be similar to dTpdC due to the factors previously discussed. However, it is important to note that these quantum yield ratios may be different from the actual measured product ratios due to the secondary photoreactions of both cyclobutane dimers and [6–4]-photoadducts (see above and below). It is of great importance in the area of UV-B mutagenesis to determine the relative formation of cyclobutane dimers and [6–4]-photoadducts in DNA. The limited studies in this area have shown a wide variation in reported values.[135,142,143] This is probably due to many factors, including the type of assay used, the irradiation state of the DNA, the G:C content of the DNA, and the sequence context. More studies in this area with more sensitive, specific assays are necessary.

Secondary Reactions of [6–4]-Photoadducts

The [6–4]-photoadducts absorb strongly in the near UV with maxima between 312 and 327 nm due to the 2-pyrimidone chromophore. Photolysis in this absorption band was shown to convert the

[6–4]-photoadduct to an unknown photoproduct in the original characterization of the [6–4]-photoadduct of dTpdT.[91] Despite numerous subsequent observations, it was not until 23 years later that the nature of the photoreaction was elucidated. Structural characterization of this photoproduct demonstrated that photolysis leads to an electrocyclization reaction of the pyrimidone that results in a substituted 2-oxo-1,3-diazabicyclo[2.2.0]-hex-5-ene moiety commonly called the Dewar valence isomer (see Figure 4.3).[144–146] It is of interest to examine the photoreactivity of 2-pyrimidones, not only in terms of the photoconversion of the [6–4]-photoadduct but also because the main tautomeric form of cytosine is a 2-pyrimidone structure. Studies on the photochemistry of 1,4,6-trisubstituted-2-pyrimidones showed that the electrocyclization reaction was the sole photoreaction occurring with a high efficiency in nonpolar solvents and a low efficiency in polar solvents.[147] The photocyclization reaction was demonstrated to be photochemically and thermally reversible, and sensitization and quenching studies indicated that the electrocyclization proceeds mainly via the singlet state. These results seem to be in agreement with the disrotatory mode of photocyclization and ring opening for electrocyclic concerted reactions involving $4n\pi$ electrons proceeding from the $^1\pi,\pi^*$ state.[34] However, it has been suggested that the first absorption band consists of an n,π^* transition and the reaction proceeds from the $^3\pi,\pi^*$ state contrary to the sensitization and quenching results.[147,148] Since the molar extinction coefficient is ~5000 M^{-1} cm^{-1}, this indicates that there must also be a π,π^* transition present. The reaction from the $^1\pi,\pi^*$ state should then be favored in polar solvents, again contrary to what is observed.[147] Obviously, the mechanism of this photoreaction is uncertain at the present time.

It was also found that 1,6-dimethyl-4-dimethylamino-2-pyrimidone, which is an analog of cytidine, did not give any photoproducts.[147] However, further studies on the photolysis of cytosine derivatives discovered that cytosine and 5-methylcytosine, as well as their nucleosides and N1-methyl derivatives, undergo photoisomerization reactions to yield the corresponding cis- and trans-3-ureidoacrylonitriles.[149] The mechanism of formation of these products was proposed, to involve initial formation of a N3–C6 Dewar structure by a photoelectrocyclization reaction followed by a ring-opening rearrangement. The quantum yield of this reaction was found to increase in acetonitrile compared to water, again suggesting a $^1n,\pi^* \rightarrow {}^3\pi,\pi^*$ reaction path. It would be of great interest to determine if this photoproduct is formed in DNA.

Returning to the photolysis of [6–4]-photoadducts, examination of the structure (see Figure 4.3) shows that the 3′-moiety can be considered as analogous to a 1,4-dialkyl-2-pyrimidone or 1,4,5-trialkyl-2-pyrimidone corresponding to photoreaction from cytosine and thymine, respectively. Therefore, photocyclization is expected to proceed efficiently and this has been demonstrated.[109,144,146] Due to the fact that [6–4]-photoadducts of dTpdT, dTpdC, dCpdT, and dCpdC have similar absorption spectra, extinction coefficients,[58,132] and quantum yields for the photocyclization reactions ($\Phi_{cyc} \approx 0.02$),[58,109] measurement of Dewar [6–4]-photoadducts in DNA reflects initial formation of [6–4]-photoadducts.[134] Since the 6–4 photoadducts absorb strongly in UV-B and UV-A regions, the photoconversion of 6–4 photoadducts to their Dewar isomers is expected to be important in the UV-B photolysis of DNA. It is therefore important to determine the relative ratios of [6–4]-photoadducts to Dewar isomers in DNA from solar irradiation of cells, especially since the mutagenicity of [6–4]-photoadducts and Dewar isomers seems to be sequence dependent.[108,150,151]

The other biologically relevant secondary reaction of [6–4]-photoadducts and their Dewar isomers is the deamination of those adducts containing 5,6-saturated cytosine that would arise at CC and CT sequences. The deamination reaction would result in the analogous uracil-containing adduct (see Figure 4.3), which is expected to alter greatly the mutagenic potential of these lesions.[108,141,151] Deamination of the 6–4 photoadduct of dCpdT was shown to be slower than the deamination of the cis-syn-cyclobutane dimer.[109]

Photohydrates

The next class of photoproducts of biological importance are the photohydrates (see Figure 4.3, path C). Here, the photoreaction is restricted to the excited pyrimidine base and does not involve addition to a neighboring base. The photohydration reaction has been extensively studied and the important features have been established as follows. First, the excited triplet state is not involved and the reaction is believed to proceed by nucleophilic addition of water exclusively onto the C-6 position of an intermediate derived from the singlet excited state.[152] Second, the overall reaction is a nonstereospecific addition of water across the C5–C6 double bond and the nucleophilic attack can occur on either side of the pyrimidine ring to result in two stereoisomers that are formed in a ratio of 1.3:1 for cytidine and uridine.[153,154] Finally, the relative efficiency for photohydrate formation at neutral pH among the pyrimidine nucleosides is Cyd ≈ Urd ≫ Thd.[152]

The mechanism of photohydration appears to be fundamentally different than the nucleophilic addition reaction of α,β-cyclohexenones, which appears to proceed by stereospecific addition to a strained *trans*-isomer derived from a triplet state.[78] The intermediate derived from the excited singlet state has been postulated to be variously a vibrationally excited ground state or a carbocation resulting from proton transfer to the C-5 position of the excited state.[152,155] The latter mechanism seems the most likely and is essentially the same as the classical acid-catalyzed hydration of alkenes;[156] in fact, acid-catalyzed hydration of uridine has been demonstrated.[157] In addition, nucleophilic attack of bisulfite on Ura and Cyt derivatives to give the 6-sulfite product analogous to the hydrate has been reported.[158,159] The greatly reduced quantum efficiency of thymine photohydration ($\Phi_h \approx 10^{-5}$) compared to cytosine and uracil photohydration ($\Phi_h \approx 10^{-2}$) is due to the methyl group at the C-5 position that results in photohydrate formation being negligible for 5-MeCyt derivatives.[160]

In polynucleotides and DNA, the photohydration reaction will have to compete against all the other possible photoreactions. It is obvious that a stacked structure will favor photoaddition reactions with the neighboring bases, while an unstacked structure will favor photohydration (see Figure 4.3). Selected values for the photohydration quantum yield in di- and polynucleotides are shown in Table 4.1 and it is seen that the photohydrate may be an important photoproduct depending on the oligonucleotide. In DNA, the relative yield of cytosine photohydrates was measured as 1 to 2% of the yield of cyclobutane dimers.[161] The formation of cytosine photohydrates in DNA and polynucleotides has been demonstrated several times since it is recognized and excised by bacterial and mammalian glycosylase activities.[162–165]

As in the case of cyclobutane dimers and [6–4]-photoadducts, secondary reactions of the cytosine photohydrates will be very important in the biological response to this photoproduct. The two reactions, which are considered to be important, are dehydration and deamination. Dehydration reforms the original cytosine, and deamination forms the uracil photohydrate that is much more stable to dehydration than the cytosine photohydrate (see Figure 4.3). The ratio of deamination to dehydration has been measured to be 3 to 14%, depending on solution conditions.[1,152] The reaction mechanism of the deamination of cytosine-containing cyclobutane dimers has been studied in detail[83] and the reaction mechanism for dehydration of uracil photohydrates has also been studied in detail.[166] It is worthwhile to examine the latter in detail for the insights it provides for the corresponding cytosine photohydrate reaction.

The dehydration reaction of photohydrates of uracil derivatives is acid-base catalyzed and, depending upon the N1-substituent, may undergo a spontaneous uncatalyzed reaction.[166] The acid-catalyzed dehydration involves preequilibrium protonation of the 6-OH group and loss of water from the C-6 carbon followed by rate-determining loss of a proton from the C-5 position, a β-elimination reaction of the E1 type.[156] In basic media, there is first a rate-determining proton loss from the C-5 position to form a carbanionic intermediate, followed by elimination of hydroxide ion at the 6-carbon to give the parent uracil derivative, an E1cB elimination reaction. Due to these dehydration mechanisms, there is no deuterium incorporation at C-5 in the product from acid-catalyzed dehydration in D_2O, while there is extensive deuterium incorporation at C-5 from spontaneous and base-catalyzed dehydration.[166] It should be noted that the dehydration of uracil photohydrates proceeds in acidic and neutral media without the formation of intermediates detectable by UV spectroscopy.[152] In contrast, it has been reported that hydrates of uracil glycosides form an intermediate in strongly basic media detected as a UV peak at 290 nm, which has been

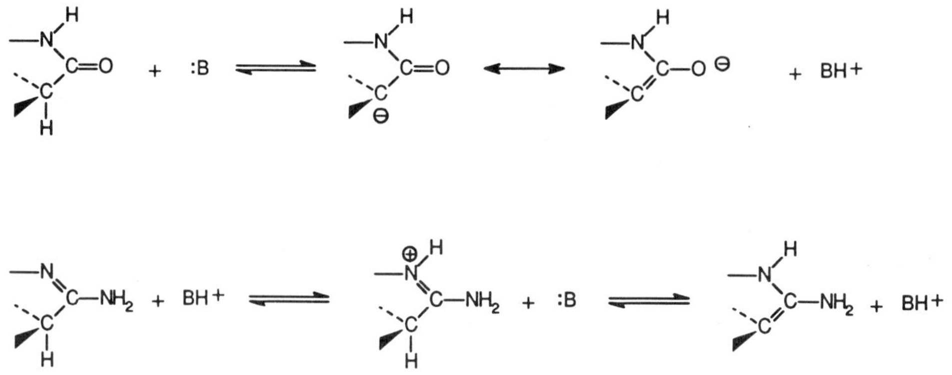

FIGURE 4.5 Mechanism of C(5)-H exchange for 5,6-saturated uracil and cytosine.

proposed to be a ring-opening reaction between the N1-C6 bond.[167] With regard to the cytosine photohydrate, the exchange of protons at the C-5 position is common in many 5,6-dihydrocytosines[154,168–170] and it occurs in acidic and basic media. This difference between the uracil and cytosine photohydrates is due to the presence of the C-4 carbonyl group and amidine group, respectively (see Figure 4.5). The ability of the amidine to accept a proton ($pK_a \sim 4$) renders the α C-H proton much more labile than for the carbonyl group. This accounts for the facile C-5 proton exchange seen for cytosine photohydrates and indicates that the dehydration mechanism is an E1cB elimination in acidic and basic media.[154]

It is of interest to note that the cytosine photohydrate appears to be much more stable to dehydration in the double-stranded copolymer poly(dG-dC) as compared to Cyt, Cyd, and CMP.[171] This stability may be ascribed perhaps to the reduced mobility of H_2O in the hydration shell around the copolymer although, strangely, it was reported also that the uracil photohydrate was less stable in the copolymer than its corresponding monomeric derivatives.

Another secondary thermal reaction that may be important for cytosine photohydrates is depyrimidination. It has been observed that deoxycytidine and some of its derivatives undergo two competing reactions: deamination and depyrimidination.[172,173] Two observations are important here. First, the deamination is postulated to occur via a 5,6-saturated intermediate, i.e., a photohydrate.[111] Second, acid-catalyzed depyrimidination of uridine to uracil occurs via an uridine photohydrate.[157] All of these observations lead inevitably to the conclusion that a third reaction is possible for deoxycytidine photohydrates and that is depyrimidination.

Of additional interest is the observation that broadband UV irradiation of deoxycytidine results in the formation of 2-deoxyribonolactone and an unstable product that is characterized by a 300-nm absorbance band.[174] This latter product resembles the proposed ring-opened intermediate of uridine photohydrate.[167] Although the deoxycytidine photohydrate may lead to depyrimidination, as discussed above, the resulting sugar residue is not expected to be a 2-deoxyribonolactone. This modified sugar residue is usually formed in a radical reaction mechanism involving oxygen, which is initiated by hydrogen abstraction from the C1'-position.[175,176] The involvement of the dC photohydrate and the 300-nm intermediate in the formation of the 2-deoxyribonolactone lesion remain as important questions to be investigated.

Photoadditions of Nucleic Acids with Amino Acids and Related Reactions

There has been a substantial amount of evidence indicating that DNA-protein cross-linking is a major cause of UV-induced damage in biological systems. Several important chromosomal proteins have been shown to be cross-linked to DNA by UV irradiation and the importance of these cross-links in biological responses has been reviewed.[177,178] However, the chemical structures of the cross-links have not been elucidated fully. There have been many investigations of model systems

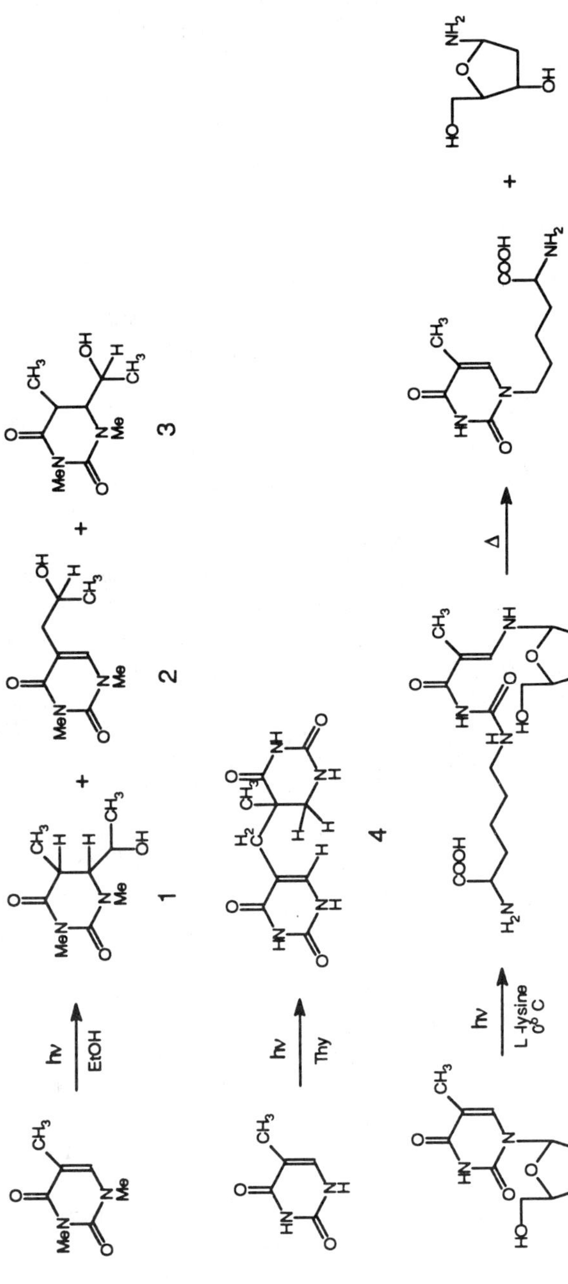

FIGURE 4.6   Examples of radical and nucleophilic addition reactions involving thymine derivatives.

where the structure of photoproducts between nucleic acid bases and amino acids or related substrates have been characterized and a recent review has thoroughly covered this subject.[179] Only a brief summary of these reactions will be presented here concerning the photochemical aspects.

For pyrimidine bases, there are two new types of photoreactions: (1) adduct formation by radicals, and (2) adduct formation via nucleophilic attack on the C-2 carbonyl. These two reactions are illustrated in Figure 4.6. Photolysis of pyrimidine bases in the presence of suitable hydrogen acceptors or hydrogen donors may lead to pyrimidine radicals that then yield stable products by radical combination or proton abstraction. Two types of radicals may be formed: (1) 5,6-dihydropyrimidine radicals (hPyr·) with an odd electron on the pyrimidine ring, and (2) thyminyl radicals derived from thymine by allylic H· abstraction.[130] The 5,6-dihydropyrimidine radicals were proposed to result from an electron-transfer interaction followed by proton transfer, and the thyminyl radical may arise by the H· abstraction from a thymine or its triplet state by radicals.[52] The types of products formed are shown as products 1 to 3 in Figure 4.6. A particularly good example of this type of reaction and the products formed is illustrated by the photoproducts formed between thymine and cysteine, which is believed to be involved in UV-induced cross-linking.[52,180] In addition, there is evidence that these radicals are formed by initial photolysis of tyrosine leading to thymine-tyrosine adducts by radical recombination reactions.[181]

These thymine radicals are also implicated in the formation of a dithymine adduct, 5,6-dihydro-5-(α-thyminyl)-thymine, (product 4 in Figure 4.6), the so-called spore photoproduct formed in DNA.[76,130] However, a recent study of the formation of the spore photoproduct of Thd in frozen aqueous solution suggested a nonradical concerted mechanism.[182]

Radical reactions are also involved in the photochemical reactions of purines with alcohols. Ade, Ado, Gua, and Guo undergo substitution at C-8 when irradiated in the presence of alcohols.[183] The initiation step involves the excitation of the purine base followed by hydrogen abstraction from the alcohol. Whether the singlet or triplet state of the purine base is involved is not certain.

The reaction that seems to be of particular interest in DNA-protein cross-linking is the photo-reaction between pyrimidine bases, particularly thymine, and alkyl amines to give ring-opened adducts by nucleophilic attack of amine at the C-2 carbonyl.[184] This photoreaction does not proceed at acidic pH and the quantum yield increases in the pH region 8 to 12. This result indicates that the excitation of the thymine monoanion, pK_a = 9.8, is responsible for adduct formation. Either a singlet excited state or a vibrationally excited ground state has been suggested to be an intermediate in this photoreaction since triplet photosensitization did not produce the ring-opened products. It was also shown that cytosine and 5-methylcytosine undergo a similar photoreaction, albeit less efficiently, with an alkylamine or an alcohol.[185,186] In fact, for 5-MeCyt and 5-methyl-dCyd, this photoreaction occurs with water resulting in substituted acrylamidines.[160] For deoxynucleosides, the reaction with alkylamines is selective for photoexcited thymidine; however, the reactivity of 5-methyl-dCyd was not compared.[184] This photoreaction has been used to achieve thymine-selective modification of DNA, which leads to efficient release of thymine. Model experiments using dTpdA demonstrated the formation of a ring-opened adduct of the thymine with alkylamine and the subsequent cleavage of the 3'-5'-phosphodiester linkage leading to 5'-dAMP.[184]

The lysine ε-amino group also reacts efficiently with the thymidine monoanion upon photolysis to give a ring-opened adduct at 0°C (product 5 in Figure 4.6) which, upon warming to room temperature, rearranges into a thymine-lysine adduct (product 6 in Figure 4.6).[187] Similar reactions also occur with cytosine and 5-methylcytosine at neutral pH.[188] The photoreaction between the lysine residues in histones and thymine residues in DNA has been demonstrated to occur in UV-irradiated nucleosomes.[187,189]

4.4 Summary

This survey of the photophysics and photochemistry of nucleic acids was concerned with outlining briefly the dominant features that are important in determining the formation of photoproducts in DNA that provoke a biological response with the emphasis placed on UV-B photolysis. It was shown that an outstanding feature is the close proximity of the first $^1n,\pi^*$ and $^1\pi,\pi^*$ states whose interaction is postulated to be responsible for the dominance of radiation internal conversion, thereby rendering the nucleic acids efficient in dissipating the excitation energy. However, the stacking of bases in the DNA double helix allows photochemical reactions via excited complex formation to compete with internal conversion. The effect of base stacking and excitation wavelength on the formation of triplet states remain as important questions to be resolved. Singlet-singlet energy transfer does not seem to be an important process in DNA, but the presence of modified bases such as 5-MeCyt may serve to trap the excitation energy from nearby bases.

The excitation of both $^1n,\pi^*$ and $^1\pi,\pi^*$ states allows for diverse photoreactions to occur with the $^1\pi,\pi^*$ state leading to cyclobutane dimers and the $^1n,\pi^*$ state responsible for [6–4]-photoadduct formation. The formation of a carbocation from a singlet excited state seems to be responsible for photohydrate formation, which is almost exclusively at cytosine. The photoproducts from UV-B photolysis are predicted to be almost entirely at pyrimidine sites since excitation of adenine would not be possible. The ratio of primary photoproducts — cyclobutane dimers, [6–4]-photoadducts, photohydrates and possible ring-fragmentation products of 5-MeCyt — is dependent on the sequence, methylation pattern, and, probably, also dependent on the particular structure of the double helix. Secondary reactions of the primary photoproducts are of special importance to the biological response to UV-B photolysis. Secondary photoreactions in the UV-B region occur for cytosine-containing cyclobutane dimers and [6–4]-photoadducts. The deamination of cytosine is a particularly important reaction since many of the photoproducts contain 5,6-saturated cytosines and this deamination seems to be relevant to UV mutagenesis. Finally, the complexation of proteins with DNA in the organization of nuclear DNA allows other photoreactions to occur.

Acknowledgments

The authors would like to thank Profs. P. R. Callis, M. D. Shetlar, and J.-S. Taylor for reviewing this manuscript and for providing constructive comments and criticisms.

References

1. Cadet, J. and Vigny, P., The photochemistry of nucleic acids, in *Bioorganic Photochemistry, Vol. 1*, Morrison, H., Ed., John Wiley & Sons, New York, 1990, 1.
2. Callis, P. R., Electronic states and luminescence of nucleic acid systems, *Annu. Rev. Phys. Chem.*, 34, 329, 1983.
3. Wang, S. Y., Ed., *Photochemistry and Photobiology of Nucleic Acids, Vols. 1 and 2*, Academic Press, New York, 1976.
4. Nikogosyan, D. N., Two-quantum UV photochemistry of nucleic acids: Comparison with conventional low-intensity UV photochemistry and radiation chemistry, *Int. J. Radiat. Biol.*, 57, 233, 1990.
5. Eaton, W. A. and Lewis, T. P., Polarized single crystal absorption spectrum of 1-methyluracil, *J. Chem. Phys.*, 53, 2164, 1970.
6. Callis, P. R., Polarized fluorescence and estimated lifetimes of the DNA bases at room temperature, *Chem. Phys. Lett.*, 61, 563, 1979.
7. Matsuoka, Y. and Norden, B., Linear dichroism studies of nucleic acid bases in stretched poly(vinyl alcohol) film. Molecular orientation and electronic transition moment directions, *J. Phys. Chem.*, 86, 1378, 1982.
8. Milder, S. J. and Kliger, D. S., Spectroscopy and photochemistry of thiouracils: Implications for the mechanism of photochemistry in tRNA, *J. Am. Chem. Soc.*, 107, 7365, 1985.
9. Hug, W. and Tinoco, I., The electronic spectra of nucleic acid bases. II. Out-of-plane transitions and the structure of the non-bonding orbitals, *J. Am. Chem. Soc.*, 96, 665, 1974.

10. Lewis, T. P. and Eaton, W. A., Polarized single-crystal absorption spectrum of cytosine monohydrate, *J. Am. Chem. Soc.*, 93, 2054, 1971.

11. Callis, P. R. and Simpson, W. T., Polarization of electronic transitions in cytosine, *J. Am. Chem. Soc.*, 92, 3593, 1970.

12. Garriga, P., Garcia-Quintana, D., and Manyosa, J., Study of polynucleotide conformation by resolution-enhanced ultraviolet spectroscopy, *Eur. J. Biochem.*, 210, 205, 1992.

13. Callis, P. R., Fanconi, B., and Simpson, W. T., Polarizations of electronic transitions in 9-ethylguanine, *J. Am. Chem. Soc.*, 93, 6679, 1971.

14. Clark, L. B., Electronic spectra of crystalline 9-ethylguanine and guanine hydrochloride, *J. Am. Chem. Soc.*, 99, 3934, 1977.

15. Chen, H. H. and Clark, L. B., Polarization assignments of the electronic spectrum of purine, *J. Chem. Phys.*, 51, 1862, 1969.

16. Garriga, P., Sági, J., Garcia-Quintana, D., Sabés, M., and Manyosa, J., Conformational isomerizations of the poly(dA-dT) and poly(amino2dA-dT) duplexes involving the unusual X-DNA double helix: A fourth derivative spectrophotometric study, *J. Biomol. Struct. Dynam.*, 7, 1061, 1990.

17. Hauswirth, W. and Daniels, M., Fluorescence of thymine in aqueous solution at 300 K, *Photochem. Photobiol.*, 13, 157, 1971.

18. Vigny, P. and Ballini, J. P., Excited states of nucleic acids at 300 K and electronic energy transfer, in *Excited States in Organic Chemistry and Biochemistry*, Pullman, B. and Goldblum, N., Eds., D. Reidel Publishing, Boston, 1977, 1.

19. Oraevsky, A. A., Sharkov, A. V., and Nikogosyan, D. N., Picosecond study of electronically excited singlet states of nucleic acid components, *Chem. Phys. Lett.*, 83, 276, 1981.

20. Wassam, W. A. and Lim, E. C., "Proximity effects": The effect of methylation and vibrational excitation on bi-exponential decay in pyrimidine, *Chem. Phys.*, 48, 299, 1980.

21. Guéron, M., Eisinger, J., and Lamola, A. A., Excited states of nucleic acids, in *Basic Principles in Nucleic Acid Chemistry*, Ts'o, P. O. P., Ed., Academic Press, New York, 1974, 311.

22. Ballini, J. P., Daniels, M., and Vigny, P., Wavelength-resolved lifetime measurements of emissions from DNA components and poly rA at room temperature excited with synchrotron radiation, *J. Luminesc.*, 27, 389, 1982.

23. Williams, S. A., Renn, C. N., and Callis, P. R., Polarized fluorescence of thymine in neutral aqueous solution at room temperature: Evidence for interference from the anion and for the π,π^* nature of the fluorescence, *J. Phys. Chem.*, 91, 2730, 1987.

24. Rigler, R., Claesens, F., and Kristensen, O., Picosecond fluorescence spectroscopy in the analysis of structure and motion of biopolymers, *Anal. Instrum.*, 14, 525, 1985.

25. Ballini, J. P., Vigny, P., and Daniels, M., Synchrotron excitation of DNA fluorescence. Decay time evidence for excimer emission at room temperature, *Biophys. Chem.*, 18, 61, 1983.

26. Ballini, J. P., Daniels, M., and Vigny, P., Synchrotron-excited time-resolved fluorescence spectroscopy of adenosine, protonated adenosine and *N,N*-dimethyladenosine in aqueous solution, *Eur. Biophys. J.*, 16, 131, 1988.

27. Georghiou, S., Nordlund, T. M., and Saim, A. M., Picosecond fluorescence decay time measurements of nucleic acids at room temperature in aqueous solution, *Photochem. Photobiol.*, 41, 209, 1985.

28. Wilson, R. W. and Callis, P. R., Fluorescent tautomers and the apparent photophysics of adenine and guanine, *Photochem. Photobiol.*, 31, 323, 1980.

29. Ge, G., Zhu, S., Bradrick, T. D., and Georghiou, S., Fluorometric analysis of the long-wavelength absorption band of N-7 methylated GMP into the constituent bands of the two electronic states, *Photochem. Photobiol.*, 51, 557, 1990.

30. Knighton, W. B., M.Sc. thesis, Montana State University, 1980.

31. Williams, S. A., Ph.D. thesis, Montana State University, 1989.

32. Yamashita, M., Kobayashi, S., Torizuka, K., and Sato, T., Observation of the diffusion-free excimer of 9-methyladenine in aqueous solution by picosecond time-resolved spectroscopy, *Chem. Phys. Lett.*, 137, 578, 1987.

33. Gill, J. E., Fluorescence of 5-methylcytosine, *Photochem. Photobiol.*, 11, 259, 1970.

34. Gilbert, A. and Baggot, J., *Essentials of Molecular Photochemistry*, Blackwell Scientific, London, 1991.

35. Smagowicz, J. and Wierzchowski, K. L., Lowest excited states of 2-aminopurine, *J. Luminesc.*, 8, 210, 1974.

36. *DNA Methylation: Biochemistry and Biological Significance*, Springer-Verlag, New York, 1984.

37. Daniels, M., Excited states of the nucleic acids: Bases, mononucleosides, and mononucleotides, in *Photochemistry and Photobiology of Nucleic Acids*, Vol. 1, Wang, S. Y., Ed., Academic Press, New York, 1976, 23.

38. Whillans, D. W. and Johns, H. E., Properties of the triplet states of thymine and uracil in aqueous solution, *J. Am. Chem. Soc.*, 93, 1358, 1971.

39. Salet, C. and Bensasson, R., Studies on thymine and uracil triplet excited state in acetonitrile and water, *Photochem. Photobiol.*, 22, 231, 1975.

40. Salet, C., Bensasson, R., and Becker, R. S., Triplet excited states of pyrimidine nucleosides and nucleotides, *Photochem. Photobiol.*, 30, 325, 1979.

41. Kasama, K., Takematsu, A., and Arai, S., Photochemical reactions of triplet acetone with indole, purine and pyrimidine derivatives, *J. Phys. Chem.*, 86, 2420, 1982.

42. Zuo, Z.-h., Yao, S.-d., Luo, J., Wang, W.-f., Zhong, J.-s., and Lin, N.-y., Laser photolysis of cytosine, cytidine and dCMP in aqueous solution, *J. Photochem. Photobiol. B: Biol.*, 15, 215, 1992.

43. Candeias, L. P. and Steenken, S., Ionization of purine nucleosides and nucleotides and their components by 193-nm laser photolysis in aqueous solution — Model studies for oxidative damage of DNA, *J. Am. Chem. Soc.*, 114, 699, 1992.

44. Arce, R., Characterization of the transient species in the 266-nm laser photolysis of adenine and its derivatives, *Photochem. Photobiol.*, 45, 713, 1987.

45. Arce, R. and Rivera, J., Intermediates produced from the room temperature 266 nm laser photolysis of guanines, *J. Photochem. Photobiol. A: Chem.*, 49, 219, 1989.

46. Wierzchowski, K. L., Berens, K., and Szabo, A. G., Triplet-triplet absorption studies of the intersystem crossing mechanism of 2-aminopurines, *J. Luminesc.*, 10, 331, 1975.

47. Arce, R., Jimenez, L. A., Rivera, V., and Torres, C., Intermediates in the room temperature flash photolysis and low temperature photolysis of purine solutions, *Photochem. Photobiol.*, 32, 91, 1980.

48. Brown, I. H. and Johns, H. E., Photochemistry of uracil. Intersystem crossing and dimerization in aqueous solution, *Photochem. Photobiol.*, 8, 273, 1968.

49. Lamola, A. A. and Eisinger, J., Excited states of nucleotides in water at room temperature, *Biochim. Biophys. Acta*, 240, 313, 1971.

50. Becker, R. S. and Kogan, G., Photophysical properties of nucleic acid components-1. The pyrimidines: Thymine, uracil, *N,N*-dimethyl derivatives and thymidine, *Photochem. Photobiol.*, 31, 5, 1980.

51. Turro, N. J., *Modern Molecular Photochemistry*, Benjamin/Cummings, Menlo Park, CA, 1978.

52. Fisher, G. J., Varghese, A. J., and Johns, H. E., Ultraviolet-induced reactions of thymine and uracil in the presence of cysteine, *Photochem. Photobiol.*, 20, 109, 1974.

53. Görner, H., Transients of uracil and thymine derivatives and the quantum yields of electron ejection and intersystem crossing upon 20 ns photolysis at 248 nm, *Photochem. Photobiol.*, 52, 935, 1990.

54. Nikogosyan, D. N., Angelov, D. A., and Oraevsky, A. A., Determination of parameters of excited states of DNA and RNA bases by laser UV photolysis, *Photochem. Photobiol.*, 35, 627, 1982.

55. Daniels, M., Shaar, C. S., and Morgan, J. P., Sequence-dependent emission from stacked forms of ApC and CpA. Evidence for stacked base (dimer) absorption and left-handed stacked conformation, *Biophys. Chem.*, 32, 229, 1988.

56. Wilson, R. W. and Callis, P. R., Excitons, energy transfer and charge resonance in excited dinucleotides and polynucleotides. A photoselection study, *J. Phys. Chem.*, 80, 2280, 1976.

57. Daniels, M., Excited states of the nucleic acids: polymeric forms, in *Photochemistry and Photobiology of Nucleic Acids*, Vol. 1, Wang, S. Y., Ed., Academic Press, New York, 1976, 109.

58. Lemaire, D. G. E. and Ruzsicska, B. P., Quantum yields and secondary photoreactions of the photoproducts of dTpdT, dTpdC and dTpdU, *Photochem. Photobiol.*, 57, 755, 1993.

59. Morgan, J. P. and Daniels, M., Excited states of DNA and its components at room temperature. III. Spectra, polarization and quantum yields of emissions from ApA and poly rA, *Photochem. Photobiol.*, 31, 101, 1980.

60. Morgan, J. P. and Daniels, M., Excited states of DNA and its components at room temperature. IV. Spectral, polarization and quantum yield studies of emissions from CpC and poly rC, *Photochem. Photobiol.*, 31, 207, 1980.

61. Mattes, S. L. and Farid, S., Photochemical cycloadditions via exciplexes, excited complexes and radical ions, *Acc. Chem. Res.*, 15, 80, 1982.

62. McCullough, J. J., Photoadditions of aromatic compounds, *Chem. Rev.*, 87, 811, 1987.

63. Saenger, W., *Principles of Nucleic acid Structure*, Springer-Verlag, New York, 1984.

64. Ge, G. and Georghiou, S., Room-temperature fluorescence properties of the polynucleotide polydA·polydT, *Photochem. Photobiol.*, 54, 477, 1991.

65. Nelson, H. C. M., Finch, J. T., Luisi, B. F., and Klug, A., The structure of an oligo(dA)·oligo(dT) tract and its biological implications, *Nature*, 330, 221, 1987.

66. Coll, M., Frederick, C. A., Wang, A. H.-J., and Rich, A., A bifurcated hydrogen-bonded conformation in the d(A·T) base pairs of the DNA dodecamer d(CGCAAATTTGCG) and its complex with distamycin, *Proc. Natl. Acad. Sci. U.S.A.*, 84, 8385, 1987.

67. Pörschke, D., A specific photoreaction in polydeoxyadenylic acid, *Proc. Natl. Acad. Sci. U.S.A.*, 70, 2683, 1973.

68. Eisinger, J. and Lamola, A. A., The excited-state precursor of the thymine dimer, *Biochem. Biophys. Res. Commun.*, 28, 558, 1967.

69. Golankiewicz, K. and Strekowski, L., Base stacking interaction and photodimerization in 1,1'-polymethyl*bis*[5-alkyluracils] in solution, *Mol. Photochem.*, 4, 189, 1972.

70. Ge, G. and Georghiou, S., Excited-state properties of the alternating polynucleotide poly(dA-dT)·poly(dA-dT), *Photochem. Photobiol.*, 54, 301, 1991.

71. Aoki, T. I. and Callis, P. R., The fluorescence of native DNA at room temperature, *Chem. Phys. Lett.*, 92, 327, 1982.

72. Eisinger, J. and Lamola, A. A., The excited states of nucleic acids, in *The Excited States of Proteins and Nucleic Acids*, Steiner, R. F. and Weinryb, I., Eds., Plenum Press, New York, 1971, 107.

73. Georghiou, S., Zhu, S., Weidner, R., Huang, C.-R., and Ge, G., Singlet-singlet energy transfer along the helix of a double-stranded nucleic acid at room temperature, *J. Biomol. Struct. Dynam.*, 8, 657, 1990.

74. Sutherland, J. C. and Sutherland, B. M., Ethidium bromide-DNA complex: Wavelength dependence of pyrimidine dimer inhibition and sensitized fluorescence as probes of excited states, *Biopolymers*, 9, 639, 1970.

75. Rayner, D. M., Szabo, A. G., Loutfy, R. O., and Yip, R. W., Singlet energy transfer between nucleic acids bases and dyes in intercalation complexes, *J. Phys. Chem.*, 84, 289, 1980.

76. Patrick, M. H. and Rahn, R. O., Photochemistry of DNA and polynucleotides: Photoproducts, in *Photochemistry and Photobiology of Nucleic Acids*, Vol. 2, Wang, S. Y., Ed., Academic Press, New York, 1976, 1.

77. Fisher, G. J. and Johns, H. E., Pyrimidine photodimers, in *Photochemistry and Photobiology of Nucleic Acids*, Vol. 1, Wang, S. Y., Ed., Academic Press, New York, 1976, 226.

78. Schuster, D. I., The photochemistry of enones, in *The Chemistry of Enones*, Patai, S. and Rappoport, Z., Eds., John Wiley & Sons, New York, 1989, 623.

79. Kleopfer, R. and Morrison, H., Organic photochemistry. XVII. The solution photodimerization of dimethylthymine, *J. Am. Chem. Soc.*, 94, 255, 1972.

80. Otten, J. G., Yeh, C. S., Bryn, S., and Morrison, H., Solution phase photodimerization of tetramethyluracil. Further studies on the photochemistry of ground-state aggregates, *J. Am. Chem. Soc.*, 99, 6353, 1977.

81. Stepien, E., Lisewski, R., and Wierzchowski, K. L., Photochemistry of 2,4-diketopyrimidines. Photodimerization, photohydration and stacking association of 1,3-dimethyluracil in aqueous solution, *Acta Biochim. Pol.*, 20, 313, 1973.

82. Fisher, G. J. and Johns, H. E., Ultraviolet photochemistry of thymine in aqueous solution, *Photochem. Photobiol.*, 11, 429, 1970.

83. Lemaire, D. G. E. and Ruzsicska, B. P., Kinetic analysis of the deamination reactions of cyclobutane dimers of thymidylyl-3′,5′-2′-deoxycytidine and 2′-deoxycytidylyl-3′,5′-thymidine, *Biochemistry*, 32, 2525, 1993.

84. Smith, C. A. and Taylor, J.-S., Preparation and characterization of a set of deoxyoligonucleotide 49-mers containing site-specific *cis*-syn, *trans*-syn-I, (6–4) and Dewar photoproducts of thymidylyl(3′→5′)-thymidine, *J. Biol. Chem.*, 268, 11143, 1993.

85. Mitchell, D. L. and Clarkson, J. M., Induction of photoproducts in synthetic polynucleotides by far and near ultraviolet radiation, *Photochem. Photobiol.*, 40, 735, 1984.

86. Ellison, M. J. and Childs, J. D., Pyrimidine dimers induced in *Escherichia coli* DNA by ultraviolet radiation present in sunlight, *Photochem. Photobiol.*, 34, 465, 1981.

87. Niggli, H. J. and Cerruti, P. A., Cyclobutane-type pyrimidine photodimer formation and excision in human skin fibroblasts after irradiation with 313 nm ultraviolet light, *Biochemistry*, 22, 1390, 1983.

88. Niggli, H. J. and Roethlisberger, R., Sunlight-induced pyrimidine dimers in human skin fibroblasts in comparison with dimerization after artificial UV-irradiation, *Photochem. Photobiol.*, 48, 353, 1988.

89. Mitchell, D. L., Jen, J., and Cleaver, J. E., Sequence specificity of cyclobutane pyrimidine dimers in DNA treated with solar (ultraviolet B) radiation, *Nucl. Acids Res.*, 20, 225, 1992.

90. Carrier, W. L., Lee, W. H., and Regan, J. D., Studies on ultraviolet light induction and excision repair of pyrimidine dimers in human skin fibroblasts, in *Proc. 7th Int. Congr. Radiation Res.*, Broerse, J. J., Barendsen, G. W., Kal, H. B., and van der Kogel, A. J., Eds., Martinus Nijhoff, The Hague, 1983, B2–07.

91. Johns, H. E., Pearson, M. L., LeBlanc, J. C., and Helleiner, C. W., The ultraviolet photochemistry of thymidylyl-(3′-5′)-thymidine, *J. Mol. Biol.*, 9, 503, 1964.

92. Tramer, Z., Wierzchowski, K. L., and Shugar, D., Influence of polynucleotide secondary structure on thymine photodimerization, *Acta Biochim. Pol.*, 16, 83, 1969.

93. Barna, T., Malinowski, J., Holton, P., Ruchirawat, M., Becker, F. F., and Lapeyre, J.-N., UV-induced photoproducts of 5-methylcytosine in a DNA sequence context, *Nucl. Acids Res.*, 16, 3327, 1988.

94. Hariharan, P. V. and Johns, H. E., Photochemical cross sections in cytidylyl-(3′-5′)-cytidine, *Can. J. Biochem.*, 46, 911, 1968.

95. Taguchi, H., Hahn, B.-S., and Wang, S. Y., Photosensitized dimerization of methylcytosine derivatives, *J. Org. Chem.*, 42, 4127, 1977.

96. Varghese, A. J. and Rupert, C. S., Ultraviolet irradiation of cytosine nucleosides in frozen solution produces cyclobutane-type dimeric products, *Photochem. Photobiol.*, 13, 365, 1971.

97. Herbert, M. A., LeBlanc, J. C., Weinblum, D., and Johns, H. E., Properties of thymine dimers, *Photochem. Photobiol.*, 9, 33, 1969.

98. Sztumpf, E. and Shugar, D., Preparation and photoproducts of orotic acid analogs, *Photochem. Photobiol.*, 4, 719, 1965.

99. Pietrzykowski, I. and Shugar, D., Photochemistry of 5-ethyluracil and its glycosides, *Acta Biochim. Pol.*, 17, 361, 1970.

100. Garcés, F. and Davila, C. A., Alterations in DNA irradiated with ultraviolet radiation. I. The formation process of cyclobutylpyrimidine dimers: Cross sections, action spectra and quantum yields, *Photochem. Photobiol.*, 35, 9, 1982.

101. Tessman, I., Liu, S.-K., and Kennedy, M. A., Mechanism of SOS mutagenesis of UV-irradiated DNA: Mostly error-free processing of deaminated cytosine, *Proc. Natl. Acad. Sci. U.S.A.*, 89, 1159, 1992.

102. Tessman, I. and Kennedy, M. A., The two-step model of UV mutagenesis reassessed: Deamination of cytosine in cyclobutane dimers as the likely source of the mutations associated with photoreactivation, *Mol. Gen. Genet.*, 227, 144, 1991.

103. Fix, D. E., Thermal resistance of UV-mutagenesis to photoreactivation in *E. coli* B/r *uvrA ung:* Estimate of activation energy and further analysis, *Mol. Gen. Genet.*, 204, 452, 1986.

104. Fix, D. E. and Bockrath, R., Thermal resistance to photoreactivation of specific mutations potentiated in *E. coli* B/r *ung* by ultraviolet light, *Mol. Gen. Genet.*, 182, 7, 1981.

105. Ruiz-Rubio, M. and Bockrath, R., On the possible role of cytosine deamination in delayed photoreversal mutagenesis targeted at thymine-cytosine dimers in *E. coli, Mutat. Res.*, 210, 93, 1993.

106. Armstrong, J. D. and Kunz, B. A., Photoreactivation implicates cyclobutane dimers as the major promutagenic UVB lesion in yeast, *Mutat. Res.*, 268, 83, 1992.

107. Taylor, J.-S. and O'Day, C. L., *Cis-syn* thymine dimers are not absolute blocks to replication by DNA polymerase I of *Escherichia coli in vitro, Biochemistry*, 29, 1624, 1990.

108. Jiang, N. and Taylor, J.-S., *In vivo* evidence that UV-induced C-T mutations at dipyrimidine sites could result from the replicative bypass of *cis-syn* cyclobutane dimers or their deamination products, *Biochemistry*, 32, 472, 1993.

109. Douki, T. and Cadet, J., Far-UV photochemistry and photosensitization of 2'-deoxycytidylyl-(3'-5')-thymidine: Isolation and characterization of the main photoproducts, *J. Photochem. Photobiol. B: Biol.*, 15, 199, 1992.

110. Frederico, L. A., Kunkel, T. A., and Shaw, B. R., A sensitive genetic assay for the detection of cytosine deamination: Determination of rate constants and the activation energy, *Biochemistry*, 29, 2532, 1990.

111. Garrett, E. R. and Tsau, J., Solvolyses of cytosine and cytidine, *J. Pharm. Sci.*, 61, 1052, 1972.

112. Weibauer, K. and Jiricny, J., Mismatch-specific thymine DNA glycosylase and DNA polymerase beta mediate the correction of G.T mispairs in nuclear extracts from human cells, *Proc. Natl. Acad. Sci. U.S.A.*, 87, 5842, 1990.

113. Bose, S. N., Davies, R. J. H., Sethi, S. K., and McCloskey, J. A., Formation of an adenine-thymine photoadduct in the deoxydinucleoside monophosphate d(TpA) and in DNA, *Science*, 220, 723, 1983.

114. Bose, S. N., Kumar, S., Davies, R. J. H., Sethi, S. K., and McCloskey, J. A., The photochemistry of d(T-A) in aqueous solution and in ice, *Nucl. Acids Res.*, 12, 7929, 1984.

115. Koning, T. M. G., Davies, R. J. H., and Kaptein, R., The solution structure of the intramolecular photoproduct of d(TpA) derived with the use of NMR and a combination of distance geometry and molecular dynamics, *Nucl. Acids Res.*, 18, 277, 1990.

116. Kumar, S. and Davies, R. J. H., The photoreactivity of pyrimidine-purine sequences in some deoxydinucleoside monophosphates and alternating DNA copolymers, *Photochem. Photobiol.*, 45, 571, 1987.

117. Paszyc, S., Skalski, B., and Wenska, G., Photochemical reactions of some pyrimidine-purine dinucleotides analogs, *Tetrahedron Lett.*, 6, 449, 1976.

118. Wenska, G., Paszyc, S., and Skalski, B., Synthesis and photochemical reactions of some pyrimidine-purine dinucleotides analogs, *Bull. Acad. Pol. Sci.*, 24, 517, 1976.

119. Wenska, G., Paszyc, S., and Skalski, B., Photocycloaddition of 6-oxopurines and thymine to products with cyclobutane part structures, *Angew. Chem. Int. Ed.*, 22, 623, 1983.

120. Wenska, G., The photoreactions of thymine with hypoxanthine and imidazole, *Z. Naturforsch.*, 40b, 108, 1984.

121. Kumar, S., Joshi, P. C., Sharma, N. D., Bose, S. N., Davies, R. J. H., Takeda, N., and McCloskey, J. A., Adenine photodimerization in deoxyadenylate sequences: Elucidation of the mechanism through structural studies of a major d(ApA) photoproduct, *Nucl. Acids Res.*, 19, 2841, 1991.

122. Hosur, R. V., Govil, G., Hosur, M. V., and Viswamitra, M. A., Sequence effects in structures of the dinucleotides d-pApT and d-pTpA, *J. Mol. Struct.*, 72, 261, 1981.

123. Pearson, M. L., Ottensmeyer, F. P., and Johns, H. E., Properties of an unusual photoproduct of UV irradiated thymidylyl-thymidine, *Photochem. Photobiol.*, 4, 739, 1965.

124. Wang, S. Y. and Varghese, A. J., Cytosine-thymine addition product from DNA irradiated with ultraviolet light, *Biochem. Biophys. Res. Commun.*, 29, 543, 1967.

125. Varghese, A. J. and Wang, S. Y., Thymine-thymine adduct as a photoproduct of thymine, *Science*, 160, 186, 1968.

126. Rhoades, D. F. and Wang, S. Y., Uracil-thymine adduct from a mixture of uracil and thymine irradiated with ultraviolet light, *Biochemistry*, 9, 4416, 1970.

127. Varghese, A. J., Photochemistry of thymidine on ice, *Biochemistry*, 9, 4781, 1970.

128. Varghese, A. J., Photochemistry of nucleic acids and their constituents, in *Photophysiology*, Vol. 7, Giese, A., Ed., Academic Press, New York, 1973, 207.

129. Hauswirth, W. and Wang, S. Y., Pyrimidine adduct fluorescence in UV irradiated nucleic acids, *Biochem. Biophys. Res. Commun.*, 51, 819, 1973.

130. Wang, S. Y., Pyrimidine bimolecular photoproducts, in *Photochemistry and Photobiology of Nucleic Acids*, Vol. 1, Wang, S. Y., Ed., Academic Press, New York, 1976, 295.

131. Lippke, J. A., Gordon, L. K., Brash, D. E., and Haseltine, W. A., Distribution of UV light-induced damage in a defined sequence of human DNA: Detection of alkaline-sensitive lesions at pyrimidine nucleoside-cytidine sequences, *Proc. Natl. Acad. Sci. U.S.A.*, 78, 3388, 1981.

132. Franklin, W. A., Lo, K. M., and Haseltine, W. A., Alkaline lability of fluorescent photoproducts produced in ultraviolet light-irradiated DNA, *J. Biol. Chem.*, 257, 13535, 1982.

133. Kan, L.-S., Voituriez, L., and Cadet, J., The Dewar valence isomer of the (6–4) photoadduct of thymidylyl-(3′-5′)-thymidine monophosphate: Formation, alkaline lability and conformational properties, *J. Photochem. Photobiol. B: Biol.*, 12, 339, 1992.

134. Mitchell, D. L., Brash, D. E., and Nairn, R. S., Rapid repair kinetics of pyrimidine(6–4)pyrimidone photoproducts in human cells are due to excision rather than conformational change, *Nucl. Acids Res.*, 18, 963, 1990.

135. Brash, D. E. and Haseltine, W. A., UV-induced mutation hotspots occur at DNA damage hot spots, *Nature*, 298, 189, 1982.

136. Nishio, T. and Omote, Y., Photocycloaddition of quinoxalin-2-ones and benzoxazin-2-ones to aryl alkenes, *J. Chem. Soc., Perkin Trans. 1*, 2611, 1987.

137. Padwa, A., Photochemistry of the carbon-nitrogen double bond, *Chem. Rev.*, 77, 37, 1977.

138. Liu, F.-T. and Yang, N. C., Photochemistry of cytosine derivatives. 1. Photochemistry of thymidylyl-(3′-5′)-deoxycytidine, *Biochemistry*, 17, 4865, 1978.

139. Rycyna, R. E. and Alderfer, J. L., UV irradiation of nucleic acids: Formation, purification and solution conformational analysis of the '6–4 lesion' of dTpdT, *Nucl. Acids Res.*, 13, 5959, 1985.

140. Tsuchida, A., Hattori, K., Ohoka, M., and Yamamoto, M., [2+2]-Photocycloaddition of *N*-vinylcarbazole to dimethyl fumarate, *J. Chem. Soc., Perkin Trans. 2*, 1685, 1992.

141. Glickman, B. W., Schaaper, R. M., Haseltine, W. A., Dunn, R. L., and Brash, D. E., The C-C (6–4) UV photoproduct is mutagenic in *Escherichia coli*, *Proc. Natl. Acad. Sci. U.S.A.*, 83, 6945, 1986.

142. Patrick, M. H., Studies on thymine-derived UV photoproducts in DNA-I. Formation and biological role of pyrimidine adducts in DNA, *Photochem. Photobiol.*, 25, 357, 1977.

143. Mitchell, D. L., Allison, J. P., and Nairn, R. S., Immunoprecipitation of pyrimidine(6–4)pyrimidone photoproducts and cyclobutane pyrimidine dimers in UV-irradiated DNA, *Radiat. Res.*, 123, 299, 1990.

144. Taylor, J.-S. and Cohrs, M. P., DNA, light and Dewar pyrimidinones: The structure and biological significance of TpT3, *J. Am. Chem. Soc.*, 109, 2834, 1987.

145. Taylor, J.-S., Garrett, D. S., and Cohrs, M. P., Solution-state structure of the Dewar pyrimidinone photoproduct of thymidylyl-(3′-5′)-thymidine, *Biochemistry*, 27, 7206, 1988.

146. Taylor, J.-S., Lu, H.-L., and Kotyk, J. J., Quantitative conversion of the (6–4) photoproduct of TpdC to its Dewar valence isomer upon exposure to simulated sunlight, *Photochem. Photobiol.*, 51, 161, 1990.

147. Nishio, T., Kato, A., Kashima, C., and Omote, Y., Photochemical electrocyclization of 1,4,6-trisubstituted pyrimidin-2-ones to 2-oxo-1,3-diazabicyclo[2.2.0]hex-5-enes, *J. Chem. Soc., Perkin Trans. 1*, 607, 1980.

148. Lu, Y., Lin, T.-S., and Taylor, J.-S., Electron paramagnetic resonance studies of pyrimidinones, *J. Phys. Chem.,* 94, 4067, 1990.

149. Shaw, A. A. and Shetlar, M. D., 3-Ureidoacrylonitriles: Novel products from the photoisomerization of cytosine, 5-methylcytosine and related compounds, *J. Am. Chem. Soc.,* 112, 7736, 1990.

150. Mitchell, D. L. and Nairn, R. S., The photobiology of the (6–4) photoproduct, *Photochem. Photobiol.,* 49, 805, 1989.

151. LeClerc, J. E., Borden, A., and Lawrence, C. W., The thymine-thymine pyrimidine-pyrimidone(6–4) ultraviolet light photoproduct is highly mutagenic and specifically induces 3′-thymine-to-cytosine transitions in *Escherichia coli, Proc. Natl. Acad. Sci. U.S.A.,* 88, 9685, 1991.

152. Fisher, G. J. and Johns, H. E., Pyrimidine photohydrates, in *Photochemistry and Photobiology of Nucleic Acids,* Vol. 1, Wang, S. Y., Ed., Academic Press, New York, 1976, 169.

153. Wechter, W. J. and Smith, K. C., Nucleic Acids. IX. The structure and chemistry of uridine photohydrate, *Biochemistry,* 7, 4064, 1968.

154. Liu, F.-T. and Yang, N. C., Photochemistry of cytosine derivatives. 2. Photohydration of cytosine derivatives. Proton magnetic resonance study on the chemical structure and property of photohydrates, *Biochemistry,* 17, 4877, 1978.

155. Garner, A. and Scholes, G., Mechanism of the photohydration of pyrimidines: A flash photolysis study, *Photochem. Photobiol.,* 41, 259, 1985.

156. Lowry, T. H. and Richardson, K. S., Addition and elimination reactions, in *Mechanism and Theory in Organic Chemistry,* 3rd ed., Harper & Row, New York, 1987, 567.

157. Prior, J. J. and Santi, D. V., On the mechanism of the acid-catalyzed hydrolysis of uridine to uracil, *J. Biol. Chem.,* 259, 2429, 1984.

158. Hayatsu, H., Wataya, Y., and Kai, K., The addition of sodium bisulfite to uracil and to cytosine, *J. Am. Chem. Soc.,* 92, 724, 1970.

159. Shapiro, R., Servis, R. E., and Welcher, M., Reactions of uracil and cytosine derivatives with sodium bisulfite. A specific deamination method, *J. Am. Chem. Soc.,* 92, 422, 1970.

160. Celewicz, L. and Shetlar, M. B., The photochemistry of 5-methylcytosine and 5-methyl-2′-deoxycytidine in aqueous solution, *Photochem. Photobiol.,* 55, 823, 1992.

161. Mitchell, D. L., Jen, J., and Cleaver, J. E., Relative induction of cyclobutane dimers and cytosine photohydrates in DNA irradiated *in vitro* and *in vivo* with ultraviolet-C and ultraviolet-B light, *Photochem. Photobiol.,* 54, 741, 1991.

162. Vanderhoek, J. Y. and Ceruuti, P. A., The stability of deoxycytidine photohydrates in mononucleotide, oligonucleotides and DNA, *Biochem. Biophys. Res. Commun.,* 52, 1156, 1973.

163. Doetsch, P. W., Helland, D. E., and Haseltine, W. A., Mechanism of action of a mammalian DNA repair endonuclease, *Biochemistry,* 25, 2212, 1986.

164. Weiss, R. B., Gallagher, P. E., Brent, T. P., and Duker, N. J., Cytosine photoproduct-DNA glycosylase in *Escherichia coli* and cultured human cells, *Biochemistry,* 28, 1488, 1989.

165. Boorstein, R. J., Hilbert, T. P., Cadet, J., Cunningham, R. P., and Teebor, G. W., UV-induced pyrimidine hydrates in DNA are repaired by bacterial and mammalian DNA glycosylase activities, *Biochemistry,* 28, 6164, 1989.

166. Prior, J. J., Maley, J., and Santi, D. V., Adducts across the 5,6-double bond of pyrimidines, *J. Biol. Chem.,* 259, 2422, 1984.

167. Fikus, M. and Shugar, D., Alkaline transformations of the photohydrates of some 2,4-diketopyrimidines and their glycosides, *Acta Biochim. Pol.,* 13, 39, 1966.

168. Brown, D. M. and Hewlins, M. J. E., Dihydrocytosine and related compounds, *J. Chem. Soc. (C),* 2050, 1968.

169. DeBoer, G. and Johns, H. E., Hydrogen exchange in photohydrates of cytosine derivatives, *Biochim. Biophys. Acta,* 204, 18, 1970.

170. Hauswirth, W. and Wang, S. Y., Cytidine-C(5)-photoexchange: A kinetic analysis, *Photochem. Photobiol.,* 26, 231, 1977.

171. Boorstein, R. J., Hilbert, T. P., Cunningham, R. P., and Teebor, G. W., Formation and stability of repairable pyrimidine photohydrates in DNA, *Biochemistry,* 29, 10455, 1990.

172. Leutzinger, E. E., Miller, P. S., and Kan, L.-S., Studies on the hydrolysis of 3-methyl-2'-deoxycytidine in aqueous solution. A synthesis of 3-methyl-2'-deoxyuridine, *Biochim. Biophys. Acta*, 697, 243, 1982.

173. Sowers, L. C., Sedwick, W. D., and Shaw, B. R., Hydrolysis of N3-methyl-2'-deoxycytidine: Model compound reactivity of protonated cytosine residues in DNA, *Mutat. Res.*, 215, 131, 1989.

174. Urata, H. and Akagi, M., Photo-induced formation of the 2-deoxyribonolactone-containing nucleotide for d(ApCpA); effects of neighboring bases and modification of deoxycytidine, *Nucl. Acids Res.*, 19, 1773, 1991.

175. von Sonntag, C., DNA model systems: The sugar phosphate moiety, in *The Chemical Basis of Radiation Biology*, Taylor & Francis, London, 1987, 167.

176. Buchko, G. W. and Cadet, J., Identification of 2-deoxy-D-ribono-1,4-lactone at the site of benzophenone photosensitized release of guanine in 2'-deoxyguanosine and thymidylyl-(3',5')-2'-deoxyguanosine, *Can. J. Chem.*, 70, 1827, 1992.

177. Shetlar, M. D., Cross-linking of proteins to nucleic acids by ultraviolet light, *Photochem. Photobiol. Rev.*, 5, 105, 1980.

178. Smith, K. C., The radiation-induced addition of proteins and other molecules to nucleic acids, in *Photochemistry and Photobiology of Nucleic Acids*, Vol. 2, Wang, S. Y., Ed., Academic Press, New York, 1976, 187.

179. Saito, I. and Sugiyama, H., Photoreactions of nucleic acids and their constituents with amino acids and related compounds, in *Bioorganic Photochemistry*, Vol. 1, Morrison, H., Ed., John Wiley & Sons, New York, 1990, 317.

180. Shetlar, M. D. and Hom, K., Mixed products of thymine and cysteine produced by direct and acetone-sensitized photoreactions, *Photochem. Photobiol.*, 45, 703, 1987.

181. Shaw, A. A., Falick, A. M., and Shetlar, M. D., Photoreactions of thymine and thymidine with N-acetyltyrosine, *Biochemistry*, 31, 10976, 1992.

182. Shaw, A. A. and Cadet, J., Radical combination processes under the direct effects of gamma radiation on thymidine, *J. Chem. Soc., Perkin Trans. 2*, 2063, 1990.

183. Steinmaus, H., Rosenthal, I., and Elad, D., Light- and gamma-ray-induced reactions of purines and purine nucleosides with alcohols, *J. Org. Chem.*, 36, 3594, 1971.

184. Saito, I., Sugiyama, H., and Matsuura, T., Photoreaction of thymidine with alkylamines. Application to selective removal of thymine from DNA, *J. Am. Chem. Soc.*, 105, 956, 1983.

185. Shetlar, M. D., Hom, K., Distefano, S., Ekpenyong, K., and Yang, J., Photochemical reactions of cytosine and 5-methylcytosine with methylamine and n-butylamine, *Photochem. Photobiol.*, 47, 799, 1988.

186. Shaw, A. A. and Shetlar, M. D., Ring-opening photoreactions of cytosine and 5-methylcytosine with aliphatic alcohols, *Photochem. Photobiol.*, 49, 267, 1989.

187. Saito, I. and Matsuura, T., Chemical aspects of UV-induced cross-linking of proteins to nucleic acids. Photoreactions with lysine and tryptophan, *Acc. Chem. Res.*, 18, 134, 1985.

188. Dorwin, E., Shaw, A. A., Hom, K., and Shetlar, M. D., Photoexchange products of cytosine and 5-methylcytosine with N^α-acetyl-L-lysine and L-lysine, *J. Photochem. Photobiol. B: Biol.*, 2, 265, 1988.

189. Kurochkina, L. and Kolomijtseva, G., Photo-induced crosslinking of histones H3 and H1 to DNA in deoxyribonucleoprotein: Implication in studying histone-DNA interactions, *Biochem. Biophys. Res. Commun.*, 187, 261, 1992.

190. Helleiner, C. W., Pearson, M. L., and Johns, H. E., The ultraviolet photochemistry of deoxyuridylyl (3' → 5') deoxyuridine, *Proc. Natl. Acad. Sci. U.S.A.*, 50, 761, 1963.

191. Pearson, M. L., Whillans, D. W., LeBlanc, J. C., and Johns, H. E., Dependence on wavelength of photoproduct yields in ultraviolet-irradiated poly U, *J. Mol. Biol.*, 20, 245, 1966.

192. Haug, A., The photochemistry of the dinucleotide TpC, *Photochem. Photobiol.*, 3, 207, 1964.

193. Rhoades, D. F. and Wang, S. Y., Photochemistry of polycytidylic acid, deoxycytidine and cytidine, *Biochemistry*, 10, 4603, 1971.

5

Photosensitized Reactions of DNA

Meyrick J. Peak
Argonne National Laboratory

Jennifer G. Peak
Argonne National Laboratory

5.1 Introduction

Radiation is toxic (kills living cells), mutagenic (changes the genotype of cells, as assayed in the laboratory), and carcinogenic (removes the normal capacity to inhibit growth that most cells in multicellular organisms possess, perhaps by the induction of specific oncogenes). It is commonly accepted that cellular mutagenesis caused by ionizing and nonionizing radiations of different energies results from changes in cellular DNA, and it is generally believed that DNA is also an important target for cell killing resulting from radiation exposure in that the changes induced in the DNA can be lethal if not repaired.[1] This is the case for both ionizing radiation and nonionizing UV radiation in the wavelength region between 200 nm and blue light, in the lower 400-nm region.

In the case of UV radiation, the earliest evidence for the role of DNA in cell killing and mutagenesis came from action spectroscopy. It was observed frequently that action spectra (Figure 5.1) for the biological responses of cells exposed to UV of wavelengths between 220 and 320 nm (<290 nm, UV-C; 290–320 nm, UV-B) superimpose upon the absorption spectrum of DNA.[2] Thus, the probability that a UV-C or UV-B photon will kill a cell or cause a mutation is a direct function of the probability that it will be absorbed by the DNA. The DNA absorbs these photons directly and, as a result, becomes modified, forming pyrimidine photoproducts, in particular cyclobutane dimers and [6–4]-pyrimidine adducts. Thus, DNA is both the chromophore and the target of this radiation.

DNA can also be chemically modified as a result of the absorption of photon energy by a non-DNA chromophore. In this type of reaction, the chromophore itself may not be chemically changed by the absorption of photons, but photon energy is passed to another compound (such as DNA) which becomes changed in turn. This process is known as photosensitization, described further by Harriman,[3] and the chromophore is called a photosensitizer. Photons with wavelengths longer than about 320 nm, which have been shown to cause damage in cellular DNA,[4,5] act via photosensitization reactions. Frequently, molecular oxygen becomes activated in the photosensitization process

0-8493-8634-9/95/$0.00+$.50
© 1995 by CRC Press, Inc.

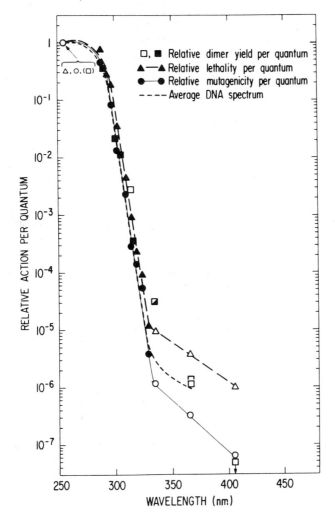

FIGURE 5.1   Action spectra showing the relative efficiencies by which photons of different wavelengths kill and mutate *E. coli* and produce pyrimidine dimers in its DNA, compared with the absorption spectrum of DNA. Biological responses deviate from the DNA spectrum in the UV-A region above 320 nm. Open and closed symbols are different types of photon source. (From Peak et al., *Photochem. Photobiol.,* 40, 613, 1984. Revised with permission.)

to form several reactive oxygen species, and this interaction of light, photosensitizers, and oxygen is also known as photodynamic action. It is these photosensitized reactions with DNA that are the subject of this chapter.

5.2   UV-A Effects on Cells

Although not well characterized because of technical difficulties, the absorption of photons by DNA is known to be negligible at wavelengths longer than about 320 nm.[6] However, it has been known for some time that these photons, extending to blue light in the lower 400-nm range (a wavelength region known as UV-A), can both kill and mutate cells.[7-11] These responses are due to photosensitization reactions whereby naturally occurring cellular sensitizers, with significant absorption in the UV-A region, act as primary chromophores and absorb the photons. The energy transfer from the photon to the sensitizer produces a chemically reactive sensitizer molecule, which possibly may mutate or kill the cell as a result of an interaction with the cellular DNA. The reactive sensitizer may react directly with DNA, but a much more common pathway involves molecular oxygen as an intermediate substrate and the formation of reactive oxygen species that then react with DNA.

5.3 Sensitizers

Any naturally occurring compound that absorbs UV-A photons in a living cell exposed to UV-A radiation has the potential to be a photosensitizer that will absorb photon energy and pass it to DNA, resulting in DNA damage. Several compounds found in mammalian and bacterial cells have been identified clearly as sensitizers. These include porphyrins and their derivatives, which absorb strongly in the Soret region near the lower 400-nm range; reduced nicotinamide coenzymes, such as NADH and NADPH; flavins, such as riboflavin; and a variety of rare RNA bases, such as those found in tRNA of bacteria. It has been shown that all of these sensitizers enhance DNA damage *in vitro* if they are present during exposure to UV-A photons.[12] The photosensitizers that play a role in DNA damage caused by UV-A radiation have not all been identified as yet; for instance, quinones and other participants in the electron transport chain are also possible candidates.

5.4 Mechanism of Photosensitized Reactions

The pathway of energy transfer in a photosensitized reaction is as follows:

$$PS \xrightarrow{\ h\nu\ } PS^* \xrightarrow{\ A\ } A^* + PS \longrightarrow \text{products} \tag{5.1}$$

where PS is the photosensitizer, PS^* is the excited-state photosensitizer, A and A^* are a substrate and an excited substrate, respectively, and $h\nu$ is the photon energy.[13] In the particular case of photosensitized reactions of DNA, the products formed are DNA lesions or photoproducts. Although the substrate (A) that receives the photon energy from the excited sensitizer as shown in Equation 5.1 may be DNA itself, by far the most common substrate in photosensitized reactions of DNA is molecular oxygen, normally abundant in most living cells. Two classes of reaction between excited-state sensitizers and oxygen have been distinguished, both of which produce reactive species of oxygen.

Typically in Type I reactions,[14] the sensitizer reacts first with a substrate that is a hydrogen or electron donor, producing reduced sensitizer radicals that can then react with oxygen. Oxygen is a parastable compound and readily accepts an electron or electrons, thereby becoming reactive itself. When oxygen accepts an electron from a photosensitizer into one of the π-orbitals (one-electron reduction), the superoxide radical anion O_2^- is produced. With the addition of another electron, a peroxide ion is formed that in turn forms hydrogen peroxide, H_2O_2 (Equation 5.2).

$$O_2 \xrightarrow{\ e^-\ } O_2^- \xrightarrow{\ e^-\ } O_2^{2-} \longrightarrow H_2O_2 \tag{5.2}$$

H_2O_2 may split homolytically, producing the very reactive species hydroxyl radical, ·OH.

In the Type II reaction,[14] the photosensitizer transfers its excitation energy directly to ground-state molecular oxygen producing singlet molecular oxygen, $^1O_2$, which has a higher electronic state. This excited species does not possess an extra unpaired electron, but is quite reactive due to removal of the electron spin restriction and can react with many organic molecules, forming peroxides or other oxidized products.

All of these reactive oxygen species can be produced in cells that are exposed to UV-A radiation, and all are capable of reacting with DNA. In particular, ·OH is a potent oxidant that reacts with virtually all organic compounds; evidence supporting its production in biological systems by 365-nm photons was provided by Peak and Peak.[15] It has also been demonstrated recently that $^1O_2$ is sufficiently reactive to produce DNA backbone nicks.[16] The reactivity of O_2^- depends on its concentration and may be due to an equilibrium with its conjugate acid HO_2, the hydroperoxyl radical, a much more reactive species than O_2^- itself.

5.5 DNA Changes Caused by Reactive Oxygen Species

Photosensitized reactions driven by UV-A do not produce significant yields of the pyrimidine photoproducts so typical of the shorter-wavelength photons between 250 and 320 nm, where DNA itself is the chromophore. Instead, as described above, they react with molecular oxygen, giving rise to a spectrum of reactive oxygen species that are similar to those produced by ionizing radiation: a major consequence of ionizing radiation exposure in an aqueous environment is the production of ·OH, formed in this case from the radiolysis of water. Thus, it might be expected that a similar DNA damage spectrum would be produced, and this is indeed the case. Photosensitized reactions produce high yields of DNA single- and double-strand breaks, as well as DNA-to-protein crosslinks, as recently reviewed by Peak et al.[17] A compilation of quantitative data showing the initial rate of accumulation of these DNA changes in the absence of concomitant repair during the exposure periods was presented in the same review. A technique that has proved useful in elucidating these lesions in cellular DNA is that of DNA filter elution.[18–21] However, this and related techniques are only capable of detecting differences in DNA size, and the actual primary products of DNA photooxidations are not well known. (The base guanine may be a preferred target in DNA for photosensitized reactions.[14]) For instance, there is almost no scientific literature that describes the induction by photosensitized reactions of the base changes that are characteristic of ionizing radiation.

5.6 Photosensitized Adduct Formation

Thiolated tRNA bases found in some bacteria often have significant absorption in the UV-A region and can act as sensitizers, both for killing cells and for DNA breakage.[12] The photosensitized reactions of the rare base 4-thiouridine with DNA were investigated further by Ito et al.[22] who irradiated single-stranded M13 phage DNA with UV-A photons in the presence of this base. When the complementary DNA strand was polymerized opposite the exposed strand, synthesis arrests were found to occur opposite all thymine bases, and also the base preceding thymine, regardless of which base it was. This finding, together with other evidence described by these authors, led to the postulate that the synthesis arrests were due to adduct formation between the 4-thiouridine itself and the thymine bases of the DNA. Therefore, Blazek et al.[23] irradiated the free base thymine in the presence of 4-thiouridine and produced such an adduct of thymine, a [5–4]-pyrimidine-pyrimidone photoproduct whose structure is shown in Figure 5.2. A variety of such adducts between DNA bases and rare tRNA bases have been described, most of which are formed in the absence of oxygen. These adduct formations differ from the photodynamic type of sensitized reactions of DNA change in that molecular oxygen is not involved and the sensitizer itself is changed by the reaction. The reactions of the furocoumarins (e.g., the psoralens) with DNA provide another example of photosensitizations that do not require oxygen.

5.7 Cellular Protective Mechanisms

As outlined above, the actual agents that damage DNA in photosensitization reactions are normally the reactive oxygen species that are produced by them. A wide range of strategies exists by which cells and organisms can protect themselves and, most importantly, their genetic material (DNA)

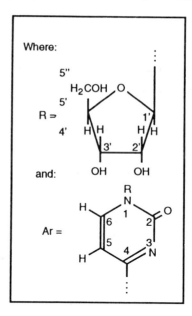

FIGURE 5.2 A new photoproduct formed between 4-thiouridine and thymine exposed to UV-A photons at 334 nm. (From Reference 23. Revised with permission.)

from the resulting oxidative insults. Indeed, such protective mechanisms must have evolved in parallel with the evolution of aerobic metabolism because reactive oxygen species are a major hazard in the oxygen-rich environment inhabited by aerobic organisms. Not only are such reactive species produced by ionizing and photosensitizing radiations, but they are also produced spontaneously or as a result of normal metabolic processes. This topic was reviewed by Fehér et al.[24]

Protective mechanisms fall into three general categories. First, a wide range of constitutive "frontline" protective reagents exists within cells, which compete with DNA for reactive oxygen species, deactivating them before they can react with DNA. This group of protective molecules was fully described by Gutteridge and Halliwell.[25] Fehér et al.[24] divided the constitutive reagents into enzymes and small molecules, often called scavengers. Of prime importance among the approximately 15 protective enzymes so far described are the superoxide dismutases (SODs), catalases, and peroxidases whose actions are shown in Equations 5.3 to 5.5.

$$2O_2^- + 2H^+ \xrightarrow{\text{SOD}} H_2O_2 + O_2 \tag{5.3}$$

$$2H_2O_2 \xrightarrow{\text{catalase}} 2H_2O + O_2 \tag{5.4}$$

$$RH_2^- + H_2O_2 \xrightarrow{\text{peroxidase}} R + 2H_2O \tag{5.5}$$

Many scavengers have been identified, including vitamins C, B, and E as well as readily reducible substances such as cysteine and glutathione. Second, protective mechanisms may be induced in response to oxidative stress; for example, the enzyme heme oxygenase, described below. Third, an extensive array of sophisticated DNA repair systems can restore the integrity of DNA damaged by reactive species that evade the protective net of scavengers.

These protective systems constitute the cellular responses to oxidative insults in general. In a few instances, a specific mechanism has been demonstrated to alleviate the effects of photosensitized reactions per se. These mechanisms are described in the following sections.

Constitutive Agents

In the bacterium *Escherichia coli,* catalase is a constitutive enzyme that protects against the lethal effects of photosensitized reactions and also decreases DNA breakage. This was established by Tuveson and Davenport[26] by the isolation of mutants *(kat)* that are deficient in catalase genes. The sensitivity of these mutants to UV-A radiation is also evidence for a lethal role of the reactive species H_2O_2 in the killing and DNA breakage caused by photosensitized reactions in these bacteria. In mammalian cells by contrast, evidence that catalase is unimportant in protecting against UV-A-induced DNA breakage and cell killing[21,27] supports the opposite conclusion. In mammalian cells, glutathione has been strongly implicated as a protector for cell killing caused by UV-A.[28] Thus far, to our knowledge no experiments have been reported that measured protection by this agent against DNA damage.

Inducible Agents

Tyrrell and co-workers[29–32] have demonstrated recently that the enzyme heme oxygenase is induced strongly in a wide range of mammalian cells that are experimentally exposed to oxidants, including those produced by photosensitizations. This is thought to be an adaptive response whose purpose is to defend against sudden surges in environmental reactive oxygen species. The exact mechanism by which heme oxygenase protects the cell remains unclear; however, a dramatic increase in intracellular levels of heme and related compounds such as biliverdin that might act as scavengers of reactive oxygen species has been reported. Other possible inducible systems in mammalian cells remain to be elucidated. One potential candidate might be the cysteine-rich oligopeptide metallothionein, which has been shown to be induced in response to a variety of environmental insults.[33]

DNA Repair Processes

In normal cells, it is generally accepted that single-strand DNA breaks induced by ionizing radiation are all repaired rapidly (within hours) by ligating enzymes. Several studies have confirmed that this is also the case for the DNA breaks caused by photosensitized reactions; thus, it is believed generally that these lesions are not biologically significant (reviewed in Peak et al.[17]). The importance of this repair process is demonstrated by the finding that cells genetically defective in repair of these DNA lesions are killed much more readily by photosensitized reactions than are normal cells.[10,11] Far less is known about the repair (or chemical nature) of DNA-to-protein crosslinks, the other major type of lesion induced in DNA by photosensitizations. Repair of this lesion does occur, but extremely slowly, extending over hours and even days after the lesions are induced.[34] While this may imply that these DNA events are not important to the well-being of cells, it could also suggest that they are the lesions that kill them, and there is little evidence to support either hypothesis. Further, scant information is available about the repair enzymes involved in this slow repair. Even more enigmatic is the biological impact of DNA double-strand breaks, which are extremely rare events produced by photosensitizations.[20] It seems probable that even one unrepaired event of this type will have a fatal effect, dooming the cell that acquires it.

References

1. Painter, R. B., The role of DNA damage and repair in cell killing induced by ionizing radiation, in *Radiation Biology in Cancer Research,* Meyn, R. E. and Wither, H. R., Eds., Raven Press, New York, 1980, 59.
2. Coohill, T. P., this volume.

3. Harriman, A., this volume.

4. Peak, M. J., Peak, J. G., and Carnes, B. A., Induction of direct and indirect single-strand breaks in human cell DNA by far- and near-ultraviolet radiations: action spectrum and mechanisms, *Photochem. Photobiol.*, 45, 381, 1987.

5. Peak, M. J., Peak, J. G., and Jones, C. A., Different (direct and indirect) mechanisms for the induction of DNA-protein crosslinks in human cells by far- and near-ultraviolet radiations (290 and 405 nm), *Photochem. Photobiol.*, 42, 141, 1985.

6. Sutherland, J. C. and Griffin, K. P., Absorption spectrum of DNA for wavelengths longer than 320 nm, *Radiat. Res.*, 86, 399, 1981.

7. Tyrrell, R. M., Mutation induction by and mutational interaction between monochromatic wavelength radiations in the near-ultraviolet and visible ranges, *Photochem. Photobiol.*, 31, 37, 1980.

8. Tyrrell, R. M., Mutagenic action of monochromatic UV radiation in the solar range on human cells, *Mutat. Res.*, 129, 103, 1984.

9. Jones, C. A., Huberman, E., Cunningham, M. L., and Peak, M. J., Mutagenesis and cytotoxicity in human epithelial cells by far- and near-ultraviolet radiations: action spectra, *Radiat. Res.*, 110, 244, 1987.

10. Churchill, M. E., Peak, J. G., and Peak, M. J., Correlation between cell survival and DNA single-strand breaks repair proficiency in the Chinese hamster ovary cells lines AA8 and EM9 irradiated with 365-nm UVA radiation, *Photochem. Photobiol.*, 53, 229, 1990.

11. Churchill, M. E., Peak, J. G., and Peak, M. J., Repair of near-visible- and blue-light-induced DNA single-strand breaks by the CHO cell lines AA8 and EM9, *Photochem. Photobiol.*, 54, 639, 1991.

12. Peak, J. G., Peak, M. J., and MacCoss, M., DNA Breakage caused by 334-nm ultraviolet light is enhanced by naturally occurring nucleic acid components and nucleotide coenzymes, *Photochem. Photobiol.*, 39, 713, 1984.

13. Kochevar, I. E., Basic concepts in photobiology, in *Photoimmunology*, Parrish, A., Kripke, M. L., and Morison, W. L., Eds., Plenum Press, New York, 1983, 5.

14. Foote, C. S., Photosensitized oxidation and singlet oxygen: Consequences in biological systems, in *Free Radicals in Biology*, Vol. II, Pryor, W. A., Ed., Academic Press, New York, 1976, chap. 3.

15. Peak, M. J. and Peak, J. G., Hydroxyl radical quenching agents afford similar protection against DNA breakage caused by 365-nm UVA or by gamma radiation, *Photochem. Photobiol.*, 51, 649, 1990.

16. Blazek, E. R., Peak, J. G., and Peak, M. J., Singlet oxygen induces frank strand breaks as well as alkali- and piperidine-labile sites in supercoiled plasmid DNA, *Photochem. Photobiol.*, 49, 607, 1989.

17. Peak, M. J., Peak, J. G., and Kochevar, I. K., Solar ultraviolet effects on mammalian cell DNA, in *Oxidative Stress in Dermatology*, Fuchs, W. and Packer, L., Eds., Marcel Dekker, New York, 1993, 169.

18. Kohn, K. W., Ewig, R. A. G., Erickson, L. G., and Zwelling, L. A., Measurement of strand breaks and cross-links by alkaline elution, in *DNA Repair, Vol. 1: A Laboratory Manual of Research Procedures*, Friedberg, E. C. and Hanawalt, P. C., Eds., Marcel Dekker, New York, 1981, 379.

19. Peak, J. G., Peak, M. J., and Blazek, E. R., Improved quantitation of DNA-protein crosslinking caused by 405-nm monochromatic near-UV radiation of human cells, *Photochem. Photobiol.*, 46, 319, 1987.

20. Peak, J. G. and Peak, M. J., Ultraviolet light induces double-strand breaks in DNA of cultured human P3 cells as measured by neutral filter elution, *Photochem. Photobiol.*, 52, 387, 1990.

21. Peak, J. G., Pilas, B., Dudek, E. J., and Peak, M. J., DNA breaks caused by monochromatic 365-nm radiation or hydrogen peroxide and their repair in human epithelioid and xeroderma pigmentosum cells, *Photochem. Photobiol.*, 54, 197, 1991.

22. Ito, A., Robb, F. T., Peak, J. G., and Peak, M. J., Base-specific damage induced by 4-thiouridine photosensitization with 334-nm radiation in M13 phage DNA, *Photochem. Photobiol.*, 47, 231, 1988.

23. Blazek, E. R., Alderfer, J. L., Tabaczynski, W. A., Stamoudis, V. C., Churchill, M. E., Peak, J. G., and Peak, M. J., A 5–4 pyrimidine-pyrimidone photoproduct produced from mixtures of thymine and 4-thiouridine irradiated with 334-nm light, *Photochem. Photobiol.,* 57, 255, 1993.

24. Fehér, J., Csomòs, G., and Vereckei, A., *Free Radical Reactions in Medicine,* Springer-Verlag, Berlin, 1985.

25. Gutteridge, B. and Halliwell, J. M. C., *Free Radicals in Biology and Medicine,* Oxford Science, Oxford, 1985, chap. 3.

26. Tuveson, R. W. and Davenport, R., Control of sensitivity to inactivation by H_2O_2 and broad-spectrum near UV radiation by the *Escherichia coli katF* locus, *J. Bact.,* 168, 13, 1986.

27. Peak, M. J., Jones, C. A., Sedita, B. A., Dudek, E. J., Spitz, D. R., and Peak, J. G., Evidence that hydrogen peroxide generated by 365-nm radiation is not important in mammalian cell killing, *Radiat. Res.,* 123, 220, 1990.

28. Tyrrell, R. M. and Pidoux, M., Endogenous glutathione protects human skin fibroblasts against the cytotoxic action of UVB, UVA, and near-visible radiations, *Photochem. Photobiol.,* 44, 561, 1986.

29. Tyrrell, R. M., Induction of heme oxygenases: A general response to oxidant stress in cultured mammalian cells, *Cancer Res.,* 51, 974, 1991.

30. Keyse, S. M. and Tyrrell, R. M., Induction of the heme oxygenase gene in human skin fibroblasts by hydrogen peroxide and UVA (365 nm) radiation: Evidence for the involvement of the hydroxide radical, *Carcinogenesis,* 11, 787, 1990.

31. Keyse, S. M., Applegate, L. A., Tromvoukis, Y., and Tyrrell, R. M., Oxidant stress leads to transcriptional activation of the human heme oxygenase in cultured skin fibroblasts, *Mol. Cell. Biol.,* 10, 4967, 1990.

32. Applegate, L. A., Luscher, P., and Tyrrell, R. M., Induction of heme oxygenases: A general response to oxidant stress in cultured mammalian cells, *Cancer Res.,* 51, 974, 1991.

33. Hamer, D. H., Metallothionein, *Annu. Rev. Biochem.,* 55, 913, 1986.

34. Peak, M. J. and Peak, J. G., unpublished observations.

6

DNA Damage and Repair

David L. Mitchell
The University of Texas

6.1 Introduction

The amount of solar ultraviolet radiation (UVR) reaching the earth's surface has increased due to chlorofluorocarbon pollution of the upper atmosphere. The consequences of stratospheric deozonation on the human population are difficult to predict, but may include an increased risk of skin cancer and accelerated aging; less direct effects may include deterioration of natural ecosystems and reduction of major food crops.[1]

Due to its maximum absorbance at ≈260 nm, DNA is considered the primary chromophore of UVR. The absorption spectrum of DNA correlates well with photoproduct formation, cell killing, mutation induction, and tumorigenesis. Because UVR induces numerous structural lesions in DNA, sensitive and precise measurements of DNA damage and repair are essential for understanding the cytotoxic and mutagenic effects of specific photoproducts on individual cells and complex organisms.

6.2 Biological Significance

The sun emits energies at wavelengths that range through 11 orders of magnitude, from 10^{-4} nm to 10^{12} nm.[2] The vast majority of this energy is irrelevant biologically; ionizing radiations such as high-energy particles, X-rays, and gamma rays are expended by atomic collisions in the upper atmosphere, and long-wavelength far-IR, microwaves, and radiowaves do not have sufficient energy to influence biochemical reactions. Photodependent biological processes derive their energy from the UV and VIS portions of the solar spectrum. UVR is divided into the UV-C (240 to 290 nm), UV-B (290 to 320 nm), and UV-A (320 to 400 nm) regions. Absorption of UV-C irradiation by stratospheric ozone attenuates these wavelengths greatly resulting in very little light shorter than 300 nm reaching the earth's surface. Hence, although comprising only a negligible portion of sunlight, UV-B is responsible for most of the sun's pathological effects.

UVR is a potent and ubiquitous carcinogen responsible for much of the skin cancer in the human population today.[3] Tumor incidence and mortality correlate with exposure: basal and squamous cell carcinomas are most prevalent on the face and trunk in men, and the face and legs in women; carcinomas increase with decreasing latitude; tumor incidence is increased in individuals working in occupations with high exposure, such as ranchers or fisherman; and the protective action of skin

0-8493-8634-9/95/$0.00+$.50
© 1995 by CRC Press, Inc.

pigmentation results in lower cancer rates in dark-skinned populations compared to lighter-skinned peoples. The importance of UVR damage and its repair in humans is exemplified by genetic diseases that greatly increase the risk of sunlight-induced skin cancer. In one such disease, xeroderma pigmentosum, a failure in the DNA repair process is associated with a major increase in the rate of squamous and basal cell carcinoma and melanoma.

The biological effects of DNA photodamage depend on the type of lesion induced and its genomic location, as well as the nature of the target cell and its developmental state. If the photoproduct is repaired correctly, the DNA is restored to its original state and, after some delay, the cell proceeds with its normal activities. If the damage is not repaired, it may interfere with DNA replication or transcription. Should this occur, cell proliferation will cease or, if the block is situated in a gene required for an essential metabolic function, the cell will die. Alternatively, the damage may not block replication. Bypass of the lesion by a DNA polymerase may produce an incorrect complementary base (i.e., a mutation). This mutation can have several outcomes: (1) it may not alter the genetic code and, hence, not effect normal metabolism; (2) it may produce a truncated or partial RNA transcript and manufacture a dysfunctional protein. If this protein is essential, then the mutation is lethal; or (3) it may result in activation of an oncogene or inactivation of a tumor suppressor gene, thereby initiating the carcinogenic process.

6.3 DNA Photoproducts

Photon absorption (10^{-12} s) converts rapidly a pyrimidine base to an excited state. Various pathways are then available for resolution of this unstable electronic configuration.[4] The major pathway involves rapid dissipation of the energy of the excited base to the ground state (10^{-9} s) by nonradiative transition or by fluorescence, yielding heat or light in the process. Second, the excited base can react with other molecules to form unstable intermediates (i.e., free radicals) or stable photoproducts. Finally, there is a low probability that intersystem crossing, a nonradiative pathway, can transfer a base from the excited singlet state to the excited triplet state. The lifetime of the triplet state is several orders of magnitude longer than the excited singlet state (10^{-3} s), increasing the chance of photoproduct formation. Cyclobutane dimers form through the excited triplet state. Other photoproducts, such as the pyrimidine[6–4]pyrimidone photoproduct (or [6–4]-photoproduct) form by some other mechanism.

Dimerizations between adjacent pyrimidine bases are by far the most prevalent photoreactions resulting from UV-C or UV-B irradiation of DNA (Table 6.1). The relative induction of these photoproducts depends on wavelength, DNA sequence, and protein-DNA interactions. The two major photoproducts are the cyclobutyl pyrimidine dimer and the [6–4]-photoproduct. The [6–4]-photoproduct undergoes a further UV-B-dependent conversion to its valence photoisomer, the Dewar pyrimidinone. In addition to the major photoproducts, rare dimers may also form, such as the adenine-thymine heterodimer.

In contrast to the direct induction of DNA damage by UV-C and UV-B light, UV-A produces damage indirectly through highly reactive chemical intermediates. Similar to ionizing radiation, UV-A radiation generates oxygen and hydroxyl radicals that in turn react with DNA to form monomeric photoproducts, such as cytosine and thymine photohydrates, as well as strand breaks and DNA-protein crosslinks. The relationship between the relative occurrence of these photoproducts and their biological effect depends on the cytotoxic and mutagenic potential of the individual lesion. Hence, even though a photoproduct may occur at a low frequency, its structure and location may elicit a potent biological effect.

6.4 Detection of DNA Damage

A variety of analytical procedures have been developed to measure UVR damage and repair in DNA.[5] Many types of photoproducts have been identified and quantified by chromatographic

Table 6.1   Structures and Relative Induction Frequencies of the Major Photoproducts Induced by UVR

Photoproduct	Structures	Lesions/10^8 Da		% of total photoproducts	
		UV-C	J/m^2 UV-B	UV-C	UV-B
Cyclobutyl pyrimidine dimer	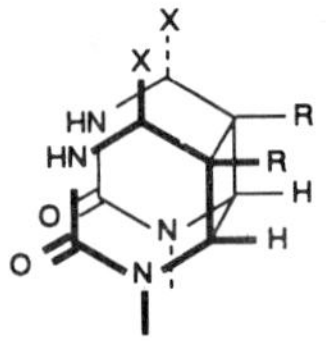	2.3	0.30	77	78
Pyrimidine pyrimidinone [6–4]- photoproduct	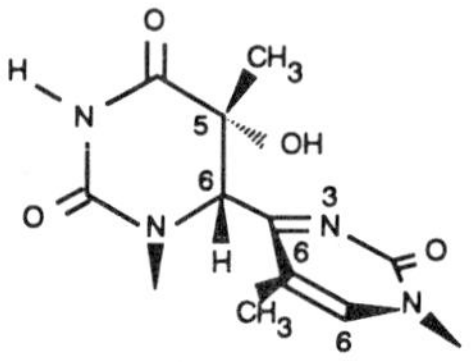	6×10^{-1}	4×10^{-2}	20	10
Dewar pyrimidinone	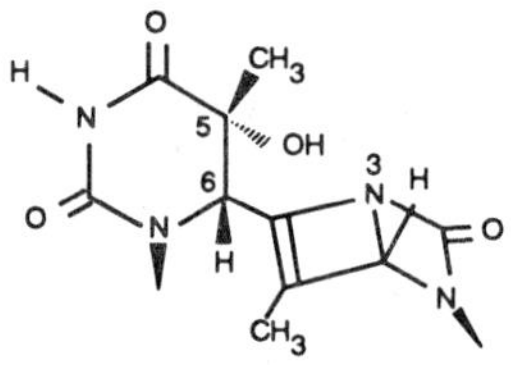	2.3×10^{-2}	$\sim 4 \times 10^{-2}$	0.8	10
Adenine-thymine heterodimer	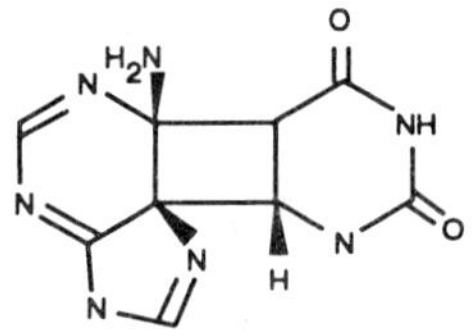	6×10^{-3}	nd	0.2	nd
Cytosine photohydrate	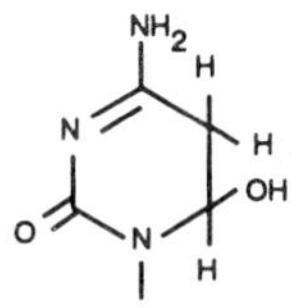	5×10^{-2}	6.6×10^{-3}	~ 2	~ 2
Thymine photohydrates	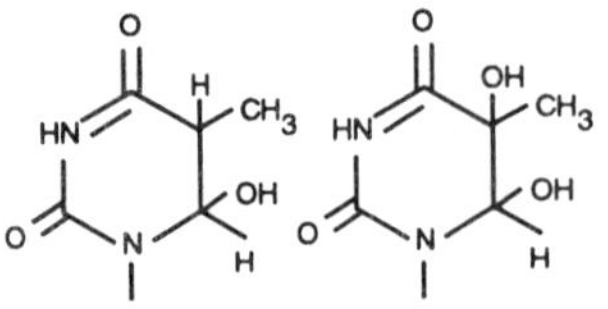				
Single-strand break		$\sim 5 \times 10^{-4}$	$\sim 4 \times 10^{-6}$	<0.1	<0.1
DNA-protein crosslink		$\sim 3 \times 10^{-4}$	$\sim 1 \times 10^{-6}$	<0.1	<0.1

techniques. Since the initial detection of the cyclobutane dimer by two-dimensional paper chromatography, thin-layer chromatography and high-performance liquid chromatography have been adapted to the analysis of this and other types of photodamage. All of these procedures require DNA that has been hydrolyzed to free bases or nucleotides by enzymatic digestion or treatment in strong acid at high temperature.

Many types of photodamage can be converted into single-strand breaks in DNA by enzymatic or biochemical treatment. Frank breaks induced directly in DNA by UVR and breaks produced at sites of photoproducts can be quantified by alkaline gradient centrifugation, alkaline elution, or agarose gel electrophoresis. The sensitivity of these procedures requires maintenance of high molecular weight DNA through the extraction and analytical procedures. *UvrABC* exinuclease, a partial excision repair complex purified from *Escherichia coli*, cleaves DNA on either side of damage produced by exposure to UVR (Figure 6.1, right side). Cleavage of specific photoproducts has been achieved by using purified enzymes that combine a glycolsylase that cuts the base from the sugar leaving an abasic (AP) site, and an apurinic/apyrimidinic endonuclease that cleaves the phosphodiester backbone on either side of the AP site (Figure 6.1, left side). Nonenzymatic cleavage of alkali-labile sites, such as AP sites or Dewar pyrimidinones, also produces quantifiable strand breaks for analysis of induction and repair.

These techniques have recently been adapted to the analysis of UV photoproduct induction and repair at the gene and base sequence levels. Photodamage at the sequence level can be mapped as strand breaks in DNA fragments irradiated with high fluences of UV-C and UV-B light and at the gene level using Southern blot hybridization of DNA fragments separated by agarose gel electrophoresis. Recently, the distribution of photoproducts at the sequence level in a gene promoter irradiated *in vivo* with low UV fluences was analyzed by ligation-mediated polymerase chain reaction.

Polyclonal and monoclonal antibodies recognize and bind a variety of photoproducts, including the cyclobutane dimer, [6–4]-photoproduct, Dewar pyrimidinone, and thymine glycol. Techniques such as immunoprecipitation, enzyme-linked immunoassays, radioimmunoassays, quantitative immunofluorescence, and immuno-electron microscopy are powerful analytical tools; each has its own unique attributes and applications. Unlike chromatography, immunological assays do not require chemical or enzymatic degradation of DNA before analysis; unlike assays that require strand scission, sensitivity does not depend on the molecular weight or purity of sample DNA.

6.5 DNA Repair

The total number of photoproducts remaining in cellular DNA at any time depends on the amount of UVR absorbed and the amount of damage repaired. Damage tolerance strategies are complex and vary greatly among organisms. Nearly all organisms have behaviors or natural features that mitigate exposure of DNA to solar UVR and reduce the amount of photodamage. In addition to habitat selection, avoidance responses, and physical morphologies that attenuate the transmission of UVR to sensitive, internal areas of cells, many organisms have evolved biochemical mechanisms to protect themselves. UVR-absorbing compounds such as melanin and anthocyanin are produced in human skin and in plants; colorless UV-absorbing compounds, such as flavenoids in terrestrial plants, mycosporine amino acids in fungi and mycosporine-like compounds in marine organisms, have also been identified as possible UV-protective chemicals.

Once DNA damage is present, its removal may proceed in whole or in part by at least two well-studied mechanisms, photoenzymatic repair (PER) or excision repair (Figure 6.1).[6] PER involves specific recognition and binding of a small enzyme called a photolyase to a cyclobutane dimer. This enzyme catalyzes the direct reversal of the dimer upon absorption of a photon within the UV-A/VIS range of light. Photoreactivation occurs widely throughout the plant and animal kingdoms. Its presence and operation in placental mammals, particularly in human skin, is a matter of some controversy.

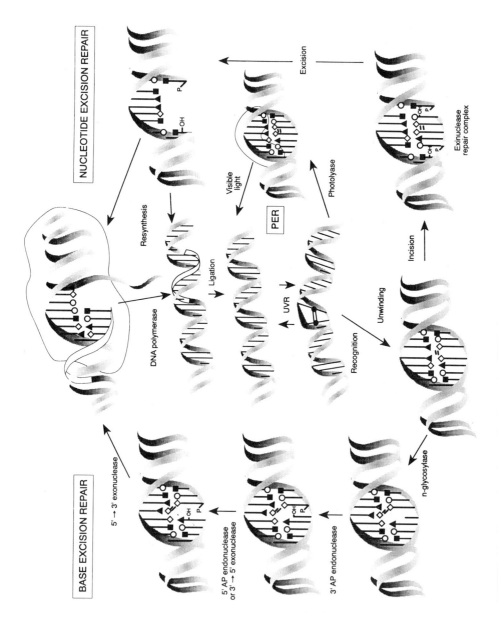

FIGURE 6.1   Enzymatic pathways for the repair of UVR damage in DNA. Shown are photoenzymatic repair (PER), base excision repair, and nucleotide excision repair. $\diamond\!\!-\!\!\diamond$ = cyclobutane dimer.

Organisms that display reduced PER often have a greater capacity for excision repair. The excision repair mechanism utilized by the cell depends on the type of damage encountered and the location of the damage in chromatin. The excision repair pathway begins with the search and recognition of the lesion as a structural distortion in the DNA helix. DNA at the site of damage is unwound or otherwise disassociated from nucleosomal proteins to provide accessibility to repair enzymes. The DNA backbone is then incised at or near the site of the lesion allowing its excision and concomitant resynthesis of the DNA around the damaged site. Finally, the single-strand gap remaining after disengagement of the DNA polymerase complex is ligated.

Base excision repair is initiated by enzymatic recognition of the lesion and scission of the bond between the damaged base and its associated deoxyribose sugar, a process called aglycosylation (Figure 6.1). Examples of *n*-glycosylases include endonuclease III from *E. coli* that repairs photohydrates and endonuclease V from T4 phage that recognizes cyclobutane dimers. In the latter case, cleavage of the *N*-glycosyl bond of the 5'-pyrimidine base leaves an abasic site that is subsequently recognized by an AP endonuclease that cleaves the phosphodiester bond 3' to the abasic site. The remaining abasic site is digested with a 5'-AP endonuclease or 3' → 5'-exonuclease associated with DNA polymerase. The damaged strand is removed and resynthesized by a DNA polymerase complex, and the remaining gap is ligated.

In contrast to base excision repair, nucleotide excision repair recognizes a broad spectrum of DNA damage, including UVR-induced photoproducts. This process has been well studied in *E. coli* and is thought to be an important excision repair pathway in eukaryotes as well. As with base excision repair, the nucleotide excision repair process is initiated by enzymes that recognize and bind the helical distortion created at the damaged site. A repair complex is assembled and cleaves the DNA a few bases to either side of the lesion, leaving a gap. This gap is filled by a DNA polymerase and the strand is ligated to restore the DNA duplex to its original integrity.

6.6 Summary

The clinical effects of defective DNA repair can include increased risk of skin cancer and accelerated aging, as well as neurological and growth disorders. This phenotypic heterogeneity belies the complexity of the excision repair process and presents a formidable task in defining the functions of the various proteins involved in DNA repair. Recent advances in cloning the genes involved in DNA repair in mammals, yeast, and *Drosophila melanogaster* have greatly increased our understanding of the DNA repair process. In addition to defining the components of an ancient and essential molecular process, future studies in DNA repair may help direct the prevention and clinical treatment of sunlight-induced skin cancer and lessen the impact of stratospheric deozonation on the human population.

References

1. Calkins, J., Ed., *The Role of Solar Ultraviolet Radiation in Marine Ecosystems*, NATO Conference Series IV: Marine Sciences, Vol. 7, Plenum Press, New York, 1982.
2. Smith, K. C., Ed., *The Science of Photobiology*, Plenum Press, New York, 1977.
3. Cleaver, J. E. and Kraemer, K. H., Xeroderma pigmentosum, in *The Metabolic Basis of Inherited Disease*, Vol. II, 6th ed., Scriver, C. R., Beaudet, A. L., Sly, W. S., and Valle, D., Eds., McGraw-Hill, 1989, 2949–2971.
4. Wang, S. Y., Ed., *Photochemistry and Photobiology of Nucleic Acids*, Vols. I and II, Academic Press, New York, 1976.
5. Friedberg, E. C. and Hanawalt, P. C., Eds., *DNA Repair: A Laboratory Manual of Research Procedures*, Vols. 1–3, Marcel Dekker, New York.
6. Friedberg, E. C., *DNA Repair*, W. H. Freeman, New York, 1984.

7

DNA Repair: Photochemistry

Seth D. Rose
Arizona State University

7.1 Introduction

Kinds of DNA Damage and the Nature of Photorepair

Photochemical repair of DNA consists principally of the reversion of cyclobutane-type pyrimidine dimers to pyrimidines.[1] The dimers arise by an orbital-symmetry-allowed cycloaddition reaction of two adjacent pyrimidines in a strand of the DNA. The repair reaction is carried out *in vivo* by photolyases, enzymes that bind to the damaged site in a dark reaction and split the dimer in a photochemical step.[2-6] The mechanism of the repair reaction involves photoinduced electron transfer between an enzyme-bound cofactor, reduced flavin adenine dinucleotide, and the enzyme-bound dimer.

Studies of a wide variety of enzymes and model systems have provided mechanistic insights into this fascinating photobiochemical transformation. The results of those studies are summarized in this chapter.

In addition to the relatively large body of work on pyrimidine dimer photorepair, a recent report describes the discovery of a photoreactivating enzyme that repairs pyrimidine[6–4]pyrimidone photoproducts in DNA.[7] There is also evidence for photorepair of [6–4]-photoproducts in UV-irradiated marsupial skin.[8] The structures and biology of [6–4]-photoproducts can be found in References 9 through 12.

General Features of Pyrimidine Dimer Photosplitting

Photoinduced splitting of pyrimidine dimers is known to occur by three processes: (1) direct excitation of the dimer, (2) electron transfer to the dimer, which produces a dimer radical anion, and (3) electron abstraction from the dimer, which produces a dimer radical cation. The splitting

0-8493-8634-9/95/$0.00+$.50
© 1995 by CRC Press, Inc.

of the neutral molecule via the excited state is an orbital-symmetry-allowed pericyclic reaction (i.e., retrocycloaddition). The radical ions, however, undergo the cycloreversion reaction thermally, even though the concerted process is orbital symmetry forbidden. This may be a consequence of a kinetic acceleration of splitting that results from a decrease in the activation barrier of splitting upon electron abstraction from[13] or electron donation to[14] the dimer. This differential effect on the energy of the reactant relative to the transition state may be thought of as follows: electron donation to the dimer raises its energy more than electron donation to the transition state raises its energy. At low temperature, splitting by simple sensitizers is inefficient, which implies that steps subsequent to electron transfer face significant activation (i.e., thermal) barriers.[15,16] Stereochemical and pH effects on splitting efficiency in model systems is pointed out where relevant in the following tables and their footnotes.

The energetics[17] of pyrimidine dimer splitting have been studied, and the conversion of the *cis-syn*-uracil cyclobutane dimer has been found to be -110.0 ± 5.2 kJ mol^{-1}. The exothermicity was attributed to the stabilizing effect of generation of delocalized carbon-carbon double bonds in the pyrimidines and the relief of strain in the polycyclic system.

The tables below summarize the recent data for photosensitized pyrimidine dimer splitting. Data for dimer splitting by sensitizers free in solution, covalently linked to the dimer, or noncovalently complexed to the dimer (or dimer-containing DNA) are tabulated. In many cases, more data can be found in the references indicated and in the literature cited in those references.

Structures of Pyrimidine Dimers

The structures of the dimers derived from uracil and thymine, as well as the abbreviations used in this chapter, are shown in Table 7.1. The *c-s*-T[]T and *t-a*-T[]T isomers are *meso*-compounds, and thus are not chiral. The *t-s*-T[]T and *c-a*-T[]T isomers exhibit enantioisomerism. Cytosine-derived dimers have analogous structures and are prone to hydrolysis to the corresponding uracil dimers; the biochemical consequences of this reaction have been considered.[18,19] Dimers for site-specific incorporation into DNA have been synthesized.[20]

Table 7.1   Structures of the Dimers and Abbreviations Used

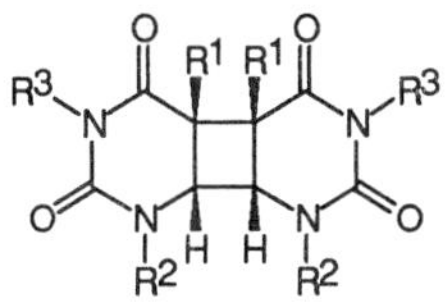

cis-syn 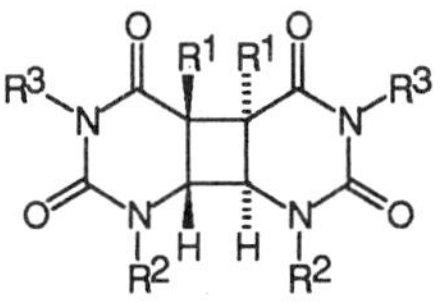   trans-syn

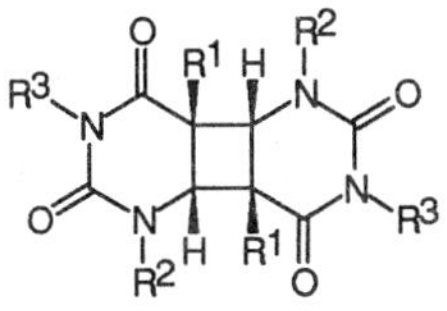

cis-anti 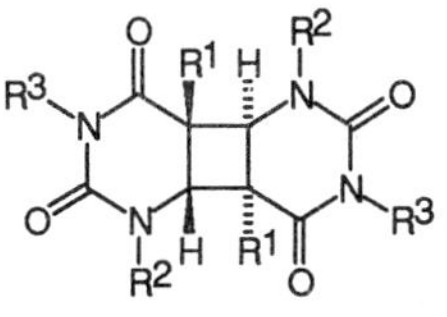   trans-anti

Abbreviation	R^1	R^2	R^3
T[]T	CH$_3$	H	H
DMTD	CH$_3$	CH$_3$	CH$_3$
DMUD	H	CH$_3$	CH$_3$

7.2 Pyrimidine Dimer Splitting by Simple Sensitizers

A wide variety of inorganic and organic species have been found to photosensitize pyrimidine dimer splitting. In this context, the term *simple sensitizers* refers to species that are not part of a larger system with specific dimer-binding or DNA-binding functionalities, such as systems based on molecular recognition, catalytic antibodies, or photolyases. The latter topics are treated in following sections.

Sensitizers Free in Solution

The dimers shown in Table 7.1 have been utilized for studies of dimer splitting by sensitizers free in solution. The sensitizers and their efficiencies of dimer splitting are shown in Table 7.2. Although in many cases the mode of splitting has not been established (i.e., electron abstraction from the dimer to produce radical cations or electron donation to the dimer to produce dimer radical anions), much indirect evidence suggests that electron abstraction is the mode for quinones, oxidized flavins, and aryl cations, whereas electron donation is the likely mode for indoles, aryl ethers and amines, and reduced flavins. The solvated electron, generated by pulse radiolysis, splits *c-s*-DMTD with an efficiency[21] of 0.05. The reduction potential of *c-s*-DMTD has been indirectly estimated at −2.6 V (vs. SCE) by fluorescence quenching measurements in acetonitrile.[22] The irreversible half-peak anodic potential ($E_{p/2}$) of *c-s*-DMTD in acetonitrile is 1.45 V (vs. Ag/AgNO$_3$), and the values for other dimers are reported.[23]

A one-bond-cleaved transient radical cation intermediate, generated by a photoexcited quinone, has been trapped by the rapid loss of an iodine atom (in competition with cleavage of the second cyclobutyl bond), which resulted in a ring-expanded product[24] (Figure 7.1). Isotope effect studies of a pyrimidine dimer split by a quinone were interpreted likewise in terms of stepwise rather than concerted bond cleavage.[25] A uracil-alkene photoadduct also fragmented in a stepwise manner.[26] One-bond-cleaved products were found in a study of quinone-sensitized splitting.[27]

Spectroscopic studies have also been employed in efforts to detect and identify the radical intermediates in splitting. Photochemically induced dynamic nuclear polarization (photo-CIDNP) spectroscopy has allowed detection of radical intermediates in quinone-sensitized splitting.[28,29] Dimer radical cations have also been detected in this way.[30] Splitting by oxidized flavins at high pH[31] requires dideprotonation[32] of thymine dimer; the intermediacy of pyrimidine dimer radicals has been established in that system.[32] Pyrimidine dimer radical anions have also been detected by photo-CIDNP in indole-sensitized splitting[33] in model studies.

Transient absorption spectroscopy following laser flash photolysis allowed detection of radicals in thymine dimer splitting by *N,N*-dimethylaniline[34] as well as in the splitting of *c-s*-DMUD and *t-s*-DMUD by a protonated, oxidized flavin in acetonitrile.[35,36]

Electron paramagnetic resonance studies have detected radicals in tryptophan-sensitized splitting of *c-s*-T[]T.[37,38]

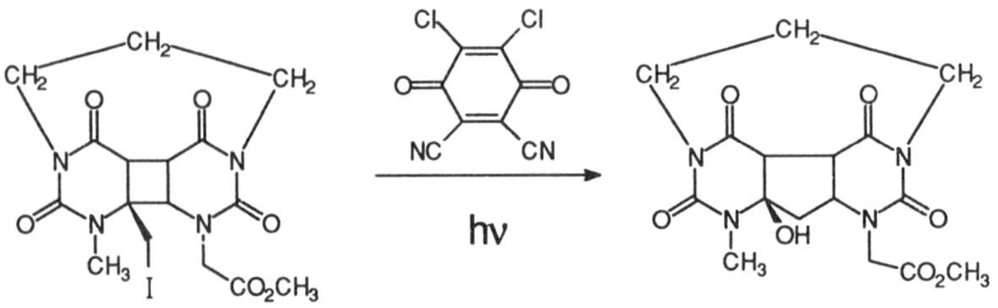

FIGURE 7.1   Loss of iodine and ring expansion provides evidence for a radical cation intermediate in quinone sensitized dimer splitting.[24]

Table 7.2 Dimer Splitting by Sensitizers in Solution

Sensitizer	Dimer	λ_{irr} (nm)	Φ^a	Ref.
	c-s-T[]T	308	0.4 (pH 12)	34
	c-s-T[]T	308	0.1 (pH 7)	34
	c-s-DMTD	308	Not reported	34
	c-s-DMUD	313	0.8^b	42
	c-s-DMUD	313	1.5^c	42
	c-s-DMTD	313	0.1^b	42
	c-s-DMTD	313	200^d (CH$_3$CN, air)	23
	t-s-DMTD	313	12^d (CH$_3$CN, air)	23
	c-s-DMUD	313	100^d (CH$_3$CN, air)	23
	c-s-DMTD	366	0.34	76
	c-s-DMTD	366	0.49 [Mg(ClO$_4$)$_2$]	76
	t-s-DMTD	366	0.39	76
	t-s-DMTD	366	0.59 [Mg(ClO$_4$)$_2$]	76
	c-a-DMTD	366	$\ll 0.01$	76
	c-a-DMTD	366	$\ll 0.01$ [Mg(ClO$_4$)$_2$]	76
				See also 77
	c-s-DMUD	366	1^d	78
	c-s-T[]T	450	10^{-3}–10^{-4} (pH 13.5)^e	31

Table 7.2 (continued)   Dimer Splitting by Sensitizers in Solution

Sensitizer	Dimer	λ_{irr} (nm)	Φ^a	Ref.
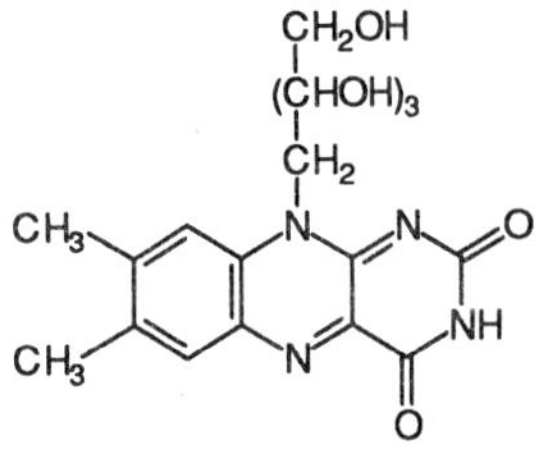 (structure: flavin with CH₂OH–(CHOH)₃–CH₂ chain, CH₃, CH₃)	c-s-T[]T	λ_{max}	Value not reported (>pH 10)^e	31
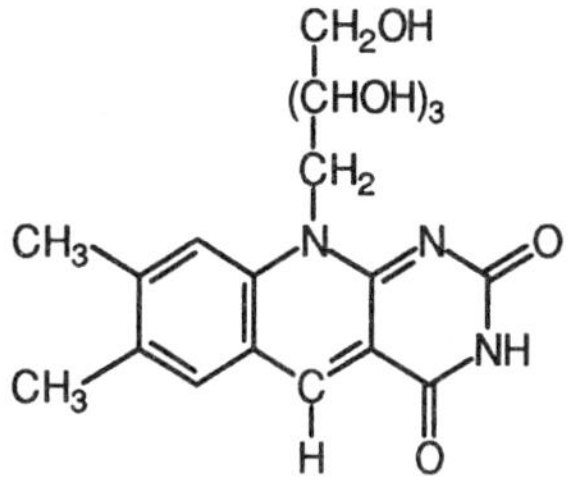 (structure: reduced flavin with CH₂OH–(CHOH)₃–CH₂ chain, CH₃, CH₃, C–H)	c-s-T[]T	400	10^{-3}–10^{-4} (pH 13.5)^e	31
(Also, see later entry.) 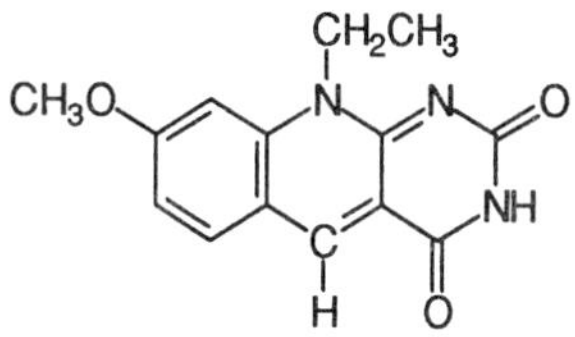 (structure: CH₃O, N–CH₂CH₃, C–H)	c-s-T[]T	383	Value not reported (>pH 10)^e	31
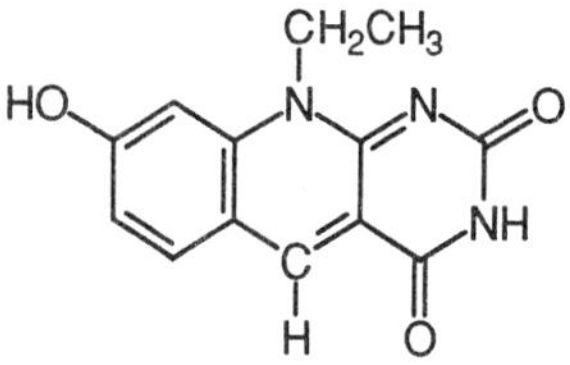 (structure: HO, N–CH₂CH₃, C–H)	c-s-T[]T	λ_{max}	Not detectable (pH 4–12)^e	31^f
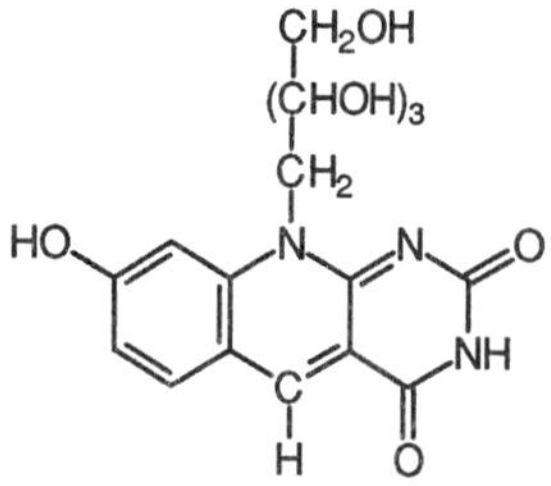 (structure: HO, CH₂OH–(CHOH)₃–CH₂ chain, C–H)	c-s-T[]T	420	Not detectable (pH 4–12)^e	31^f

Table 7.2 (continued)   Dimer Splitting by Sensitizers in Solution

Sensitizer	Dimer	λ_{irr} (nm)	Φ^a	Ref.

CH$_2$OH
(CHOH)$_3$
CH$_2$

R^4 ... N ... N ... O ... NH ... CH$_3$ ... N ... O

| 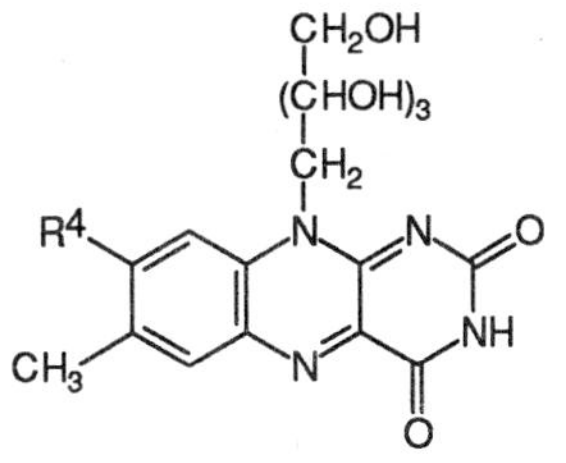 | c-s-T[]T | 470, 490, and 490, resp. | Not detectable^e | 31 |

R^4 = HO-, (CH$_3$)$_2$N-, (CH$_3$)NH-

N(3)-methyllumiflavin				
oxidized	c-s-T[]T	—	31.0% split, pH 11.5^e	79
reduced^h	c-s-T[]T	—	3.0% split, pH 11.5^e	79
1-Deazariboflavin				
oxidized	c-s-T[]T	—	0% split, pH 11.5^e	79
reduced^h	c-s-T[]T	—	15.0% split, pH 11.5^e	79
5-Deazariboflavin (see above)				
oxidized	c-s-T[]T	—	10.6% split, pH 11.5^e	79
reduced^h	c-s-T[]T	—	0% split, pH 11.5^e	79
7,8-Dimethyl-1,10-ethylene-isoalloxazinium perchlorate				
Oxidized	c-s-T[]T	—	2.6% split, pH 10.5^e	79
Reduced^h	c-s-T[]T	—	0% split, pH 10.5^e	79
K$_3$Fe(CN)$_6$	c-s-T[]T	>350^g	30% split/3.5h	80
	DMTD	>350^g	14% split/3.5h	80
	DMUD	>350^g	20% split/3.5h	80
UO$_2$SO$_4$	c-s-T[]T	>350^g	60% split/3.5h	80
	DMTD	>350^g	40% split/3.5h	80
	DMUD	>350^g	80% split/3.5h	80
Hg^{2+}	c-s-T[]T	280	Not reported	81

^{a} (Number of dimers split)/(number of photons absorbed).
^{b} [dimer] = 0.4 mM, pH 10, under Ar.
^{c} [dimer] = 0.7 mM, pH 11, under Ar.
^{d} Φ_∞ (extrapolated to high dimer concentration).
^{e} Anaerobic.
^{f} See also Reference 60.
^{g} Hanovia 450-W with uranium filter.
^{h} Dithionite.

Chain reactions have been observed in dimer splitting by the solvated electron (generated by pulse radiolysis),[39] a reduced flavin,[40] a protonated oxidized flavin,[41] an indole,[42] and polycyclic aromatic hydrocarbon radical cations.[23]

Covalently Linked Sensitizers

In this section, splitting of a dimer is accomplished by irradiation of a covalently linked sensitizer. The quantum yields of splitting are summarized in Tables 7.3 and 7.4, as well as Figure 7.2. The intermediacy of a charge-separated species, consisting of a dimer radical ion and a sensitizer radical ion, is implied by sensitizer fluorescence quenching and transient absorption spectroscopy. (For a linked dimer-flavin in acetonitrile containing a trace of acid, see Reference 36.) In some cases, solvent effects on the quantum yield of splitting are available and are summarized. The increase in splitting efficiency in lower polarity solvents has been interpreted in terms of a possible slowing of

Table 7.3   Dimer Splitting by Covalently Linked Sensitizers

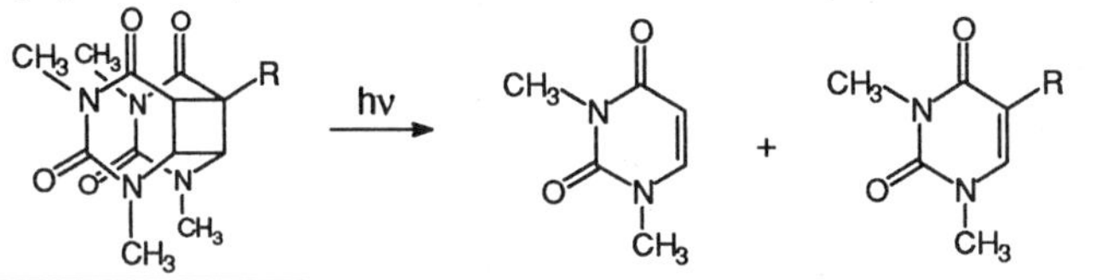

R	λ_{irr} (nm)	Φ^a	Ref.
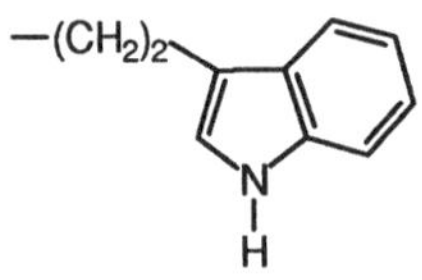 —(CH₂)₂-[indole]	290	0.06 (H₂O)	44[b]
	290	0.15 ((CH₃)₂SO)	
	290	0.19 (CH₃CN)	
	290	0.23 (CH₃OH)	
	290	0.37 ((CH₃CH₂)₂O)	
	290	0.39 (benzene)	
	290	0.41 (1,4-dioxane/isopentane, 5:95)	
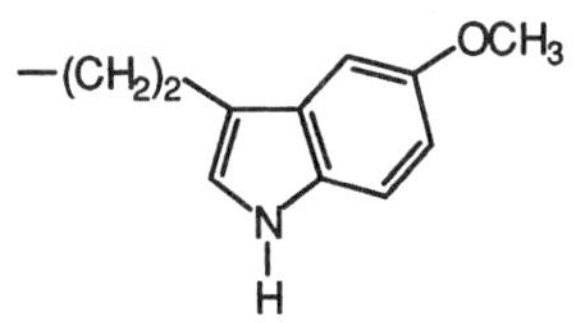 —(CH₂)₂-[indole-OCH₃]	290	0.05 (H₂O)	44[b,c]
	290	0.09 ((CH₃)₂SO)	
	290	0.14 (CH₃CN)	
	290	0.17 (CH₃OH)	
	290	0.29 ((CH₃CH₂)₂O)	
	290	0.30 (benzene)	
	290	0.31 (1,4-dioxane)	
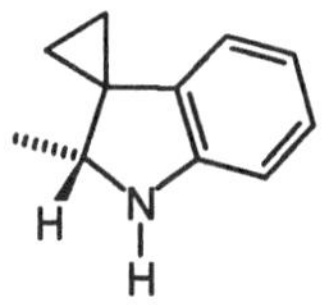 [spiro cyclopropane indoline]	313	0.02 (H₂O)	43[b,d]
	313	0.08 ((CH₃)₂SO)	
	313	0.10 (CH₃CN)	
	313	0.11 (CH₃OH)	
	313	0.24 ((CH₃CH₂)₂O)	
	313	0.19 (1,4-dioxane)	
	313	0.31 (1,4-dioxane/isopentane, 1:99)	
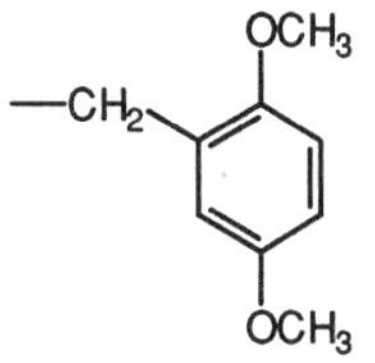 —CH₂-[dimethoxybenzene]	302	0.05 (H₂O)	46[b,e]
	302	0.26 (CH₃CN)	
	302	0.24 (CH₃OH)	
	302	0.20 ((CH₃CH₂)₂O)	
	302	0.24 (1,4-dioxane)	
	302	0.29 (1-butanol)	
	302	0.13 (cyclohexane)	

[a] (Number of dimers split)/(number of photons absorbed).
[b] See the reference for additional solvents.
[c] See Reference 82 for the X-ray structure.
[d] See the reference for *N*-methylated and deuterated derivatives and for another stereoisomer, and see Reference 83 for X-ray structures.
[e] See the reference for a monohydroxy-monomethoxy analogue.

the highly exergonic back electron transfer due to Marcus inverted behavior.[43–46] Transient absorption spectroscopy following laser flash photolysis has been applied to linked dimer-indole systems.[47] Secondary deuterium isotope effects have been employed in a linked dimer-indole system to explore the detailed bond-breaking processes in the dimer radical anion intermediate.[25]

Noncovalently Complexed Sensitizers

Two systems comprise this category. One is based on delivery of the sensitizer to dimer-containing DNA by electrostatic binding to the phosphate groups. The other is based on the principles of molecular recognition.

Table 7.4 Trimethylene-Bridged Dimers Are Split by Covalently Linked Sensitizers

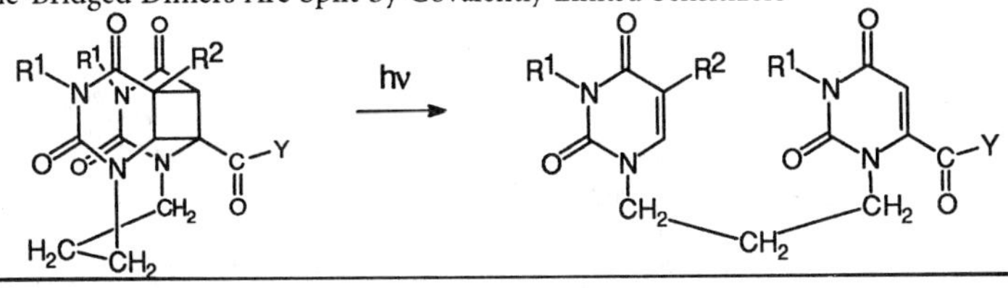

Y	R^1	R^2	λ_{irr} (nm)	Φ^a	Ref.
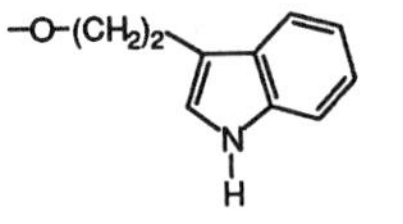 —O-(CH₂)₂-indole	H	CH₃	278	0.05 (N₂O/H₂O)	84
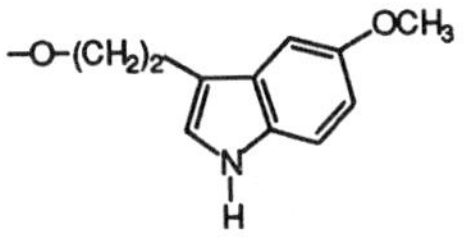 —O-(CH₂)₂-indole-OCH₃	H	CH₃	278	0.05 (N₂O/H₂O)	47
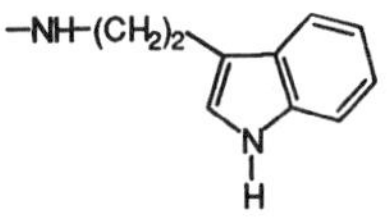 —NH-(CH₂)₂-indole	CH₃	H	290	0.15	85
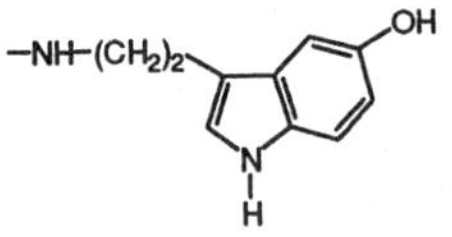 —NH-(CH₂)₂-indole-OH	CH₃	H	313	0.19 (H₂O) 0.20 (CH₃OH) 0.085 (CH₃CN) 0.060 (1,4-dioxane)	85
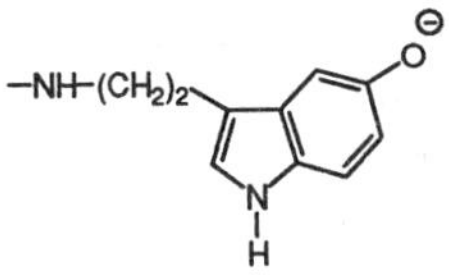 —NH-(CH₂)₂-indole-O⁻	CH₃	H	334	0.34	85

See also Figure 7.2.
^{a} (Number of dimers split)/(number of photons absorbed).

Electrostatic Binding of Simple Peptides to DNA

The tripeptide Lys-Trp-Lys split dimers in UV-irradiated DNA[48] more efficiently than did tryptophan, as a consequence of complexation of the cationic tripeptide with the anionic DNA phosphate groups. (See also Reference 49 for splitting of *c-s*-T[]T by tryptophan and phenylalanine.) Tryptophan-containing proteins split dimers in RNA.[50]

Molecular Recognition of Pyrimidine Dimers by Hydrogen Bonding

Macrocyclic systems with hydrogen bonding capabilities complementary to the hydrogen bonding pattern of a dimer have been prepared. Sensitizers attached to the macrocyclic ring have photosplit the hydrogen-bonded dimer, as shown by the data in Table 7.5.

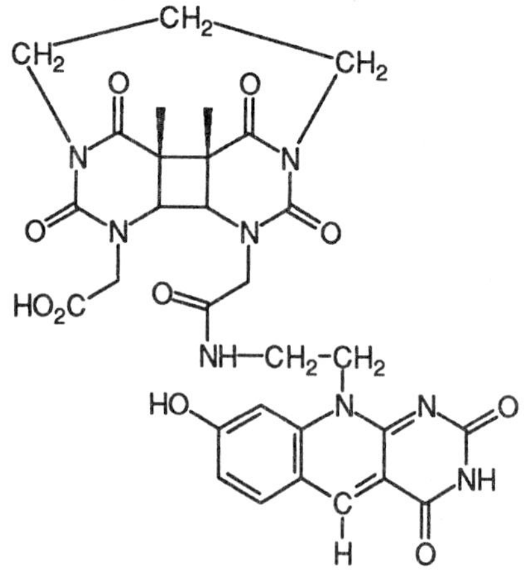

FIGURE 7.2   Intramolecularly photosensitized dimer splitting by a deazaflavin (λ_{irr} = 436 nm, Φ = 6 × 10^{-4}).[89]

7.3 Pyrimidine Dimer Splitting by Proteins

Catalytic Antibodies

Catalytic antibodies have been generated against c-s-T[]T. They bind to dimer-containing DNA and photosplit the dimers, probably by electron donation from a tryptophan residue in the antibody. The quantum yield[51] at 300 nm was 0.08.

Photolyases

Photolyases from a variety of sources have been characterized. For example, see References 52 to 63. The enzyme-bound sensitizer in all dimer-splitting photolyases studied to date is 1,5-dihydroflavin adenine dinucleotide. The excited singlet state[64] of FADH$_2$ apparently donates an electron to the dimer. Isotope effect studies also suggest the intermediacy of a dimer radical anion.[65] Photolyases employ an additional chromophore as a light-gathering antenna and are classified according to whether this second chromophore is a 5-deazaflavin or 5,10-methenyltetrahydrofolate. Energy transfer from excited folate or deazaflavin to FADH$_2$ has been shown to be highly efficient.[66–68] Excitation of the photolyase from *Escherichia coli* at 280 nm results in dimer splitting by direct electron transfer from a tryptophan residue[69] (quantum yield = 0.56).

Transient absorption spectroscopy following laser flash photolysis has been applied to photolyase-catalyzed DNA repair, and radicals have been detected.[70,71] Also, electron paramagnetic resonance spectroscopy has detected radicals in the enzyme-catalyzed DNA repair reaction.[72,73]

Dimer splitting efficiencies of photolyases are summarized in Table 7.6. For recent reviews on the properties of photolyases, see References 2 to 6. For information on substrate and chromophore binding, see References 74 and 75, respectively.

Table 7.5 Dimer Splitting by Noncovalently Bound Chromophores That Employ Molecular Recognition

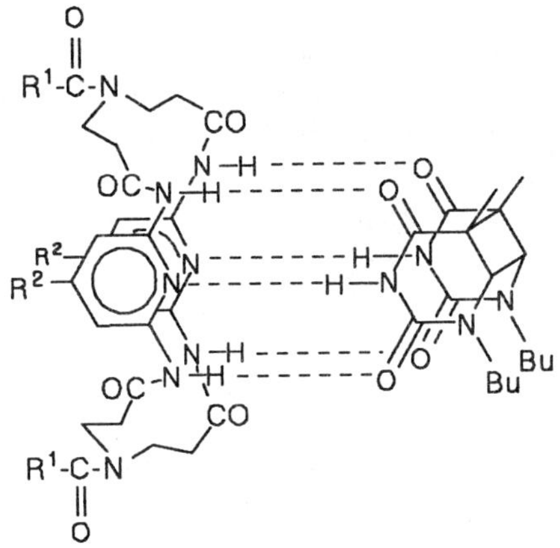

	R¹	R²	λ_{irr} (nm)	Φ^a	Ref.
		H	313	0.01 (Ar-purged CH₃CN)	86
		CH₃CH₂O	313 313	0.11 (CH₃CN) 0.18ᵇ (CH₃OH–CH₃CN, 93:7)	87
		H	436	Not detectable (Ar-purged CH₃CN)	86

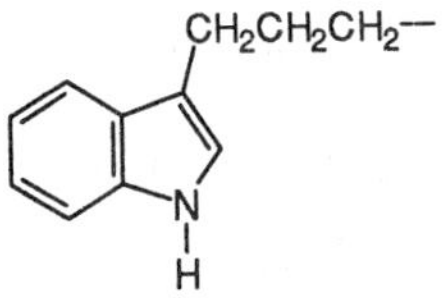

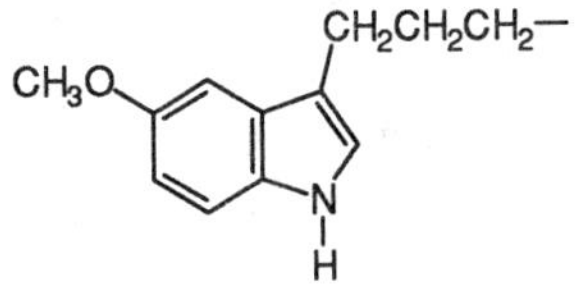

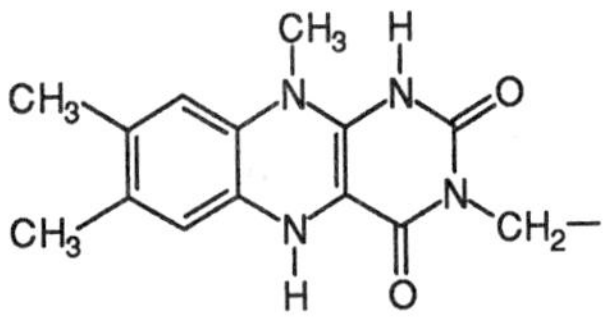

ᵃ (Number of dimers split)/(number of photons absorbed).
ᵇ Extrapolated to saturating dimer concentration.

Table 7.6 Properties of Photolyases

Source	Class	λ_{max} (nm)	Φ	Ref.
Escherichia coli	Folate	384	0.7	66, 67, 88
Saccharomyces cerevisiae	Folate	377	0.5	88
Anacystis nidulans	Deazaflavin	440	0.9	61, 68
Methanobacterium thermoautotrophicum	Deazaflavin	440	0.2–0.4	2, 55

7.4 Conclusion

Many fundamental aspects of the photochemistry of DNA repair have been elucidated, as is clear from the data summarized in this chapter. It is also apparent, however, that much remains to be learned about this fundamental photobiochemical process.

Note added in proof: Since preparation of this chapter, review articles[90-93] on photolyase have appeard, as well as preliminary crystallographic analyses of the enzymes from *E. coli*[94] and *A. nidulans.*[95]

Acknowledgments

I thank Rosemarie F. Hartman, Donna G. Hartzfeld, and M. Scott Goodman for help with the preparation of this manuscript. Financial support from the National Institutes of Health, the American Chemical Society, the Del E. Webb Foundation, and the Burroughs Wellcome Fund are gratefully acknowledged.

References

1. Rupert, C. S., Enzymatic photoreactivation: Overview, in *Molecular Mechanisms for Repair of DNA,* Part A, Hanawalt, P. C. and Setlow, R. B., Eds., Plenum Press, New York, 1975, 73–87.
2. Sancar, A., Photolyase: DNA repair by photoinduced electron transfer, in *Advances in Electron Transfer Chemistry,* Mariano, P. S., Ed., Vol. 2, JAI Press, Greenwich CT, 1992, 215–272.
3. Sancar, A. and Sancar, G. B., DNA repair enzymes, *Annu. Rev. Biochem.,* 57, 29, 1988.
4. Heelis, P. F., Kim, S.-T., Okamura, T., and Sancar, A., The photo repair of pyrimidine dimers by DNA photolyase and model systems, *J. Photochem. Photobiol. B: Biol.,* 17, 219, 1993.
5. Sancar, G. B., DNA photolyases: Physical properties, action mechanism, and roles in dark repair, *Mutat. Res.,* 236, 147, 1990.
6. de Gruijl, F. R. and Roza, L., Photoreactivation in humans, *J. Photochem. Photobiol. B: Biol.,* 10, 367, 1991.
7. Todo, T., Takemori, H., Ryo, H., Ihara, M., Matsunaga, T., Nikaido, O., Sato, K., and Nomura, T., A new photoreactivating enzyme that specifically repairs ultraviolet light-induced (6–4)photoproducts, *Nature,* 361, 371, 1993.
8. Mitchell, D. L., Applegate, L. A., Nairn, R. S., and Ley, R. D., Photoreactivation of cyclobutane dimers and (6–4) photoproducts in the epidermis of the marsupial, *Monodelphis domestica, Photochem. Photobiol.,* 51, 653, 1990.
9. Mitchell, D. L. and Nairn, R. S., The biology of the (6–4) photoproduct, *Photochem. Photobiol.,* 49, 805, 1989.
10. Kan, L.-S., Voituriez, L., and Cadet, J., Nuclear magnetic resonance studies of *cis-syn, trans-syn,* and 6–4 photodimers of thymidylyl-(3′-5′)-thymidine monophosphate and *cis-syn* photodimers of thymidylyl-(3′-5′)-thymidine cyanoethyl phosphotriester, *Biochemistry,* 27, 5796, 1988.
11. Douki, T., Voituriez, L., and Cadet, J., Characterization of the (6–4) photoproduct of 2′-deoxycytidylyl-(3′→5′)-thymidine and of its Dewar valence isomer, *Photochem. Photobiol.,* 53, 293, 1991.
12. Taylor, J.-S., Garrett, D. S., and Cohrs, M. P., Solution-state structure of the Dewar pyrimidinone photoproduct of thymidylyl-(3′→5′)-thymidine, *Biochemistry,* 27, 7206, 1988.
13. Bauld, N. L., Hole and electron transfer catalyzed pericyclic reactions, in *Advances in Electron Transfer Chemistry,* Vol. 2, Mariano, P. S., Ed., JAI Press, Greenwich CT, 1992, 1–66.
14. Hartman, R. F., Van Camp, J. R., and Rose, S. D., Electron delocalization in pyrimidine dimers and the implications for enzyme-catalyzed dimer cycloreversion, *J. Org. Chem.,* 52, 2684, 1987.
15. Kim, S.-T. and Rose, S. D., Activation barriers in photosensitized pyrimidine dimer splitting, *J. Phys. Org. Chem.,* 3, 581, 1990.

16. Kim, S.-T. and Rose, S. D., Thermal requirements of photosensitized pyrimidine dimer splitting, *Photochem. Photobiol.*, 47, 725, 1988.

17. Diogo, H. P., Dias, A. R., Dhalla, A., Minas da Piedade, M. E., and Begley, T. P., Mechanistic studies on DNA photolyase. 4. The enthalpy of cleavage of a model photodimer, *J. Org. Chem.*, 56, 7341, 1991.

18. Lemaire, D. G. and Ruzsicska, B. P., Kinetic analysis of the deamination reactions of cyclobutane dimers of thymidylyl-3′,5′-2′-deoxycytidine and 2′-deoxycytidylyl-3′,5′-thymidine, *Biochemistry*, 32, 2525, 1993.

19. Taylor, J.-S. and Jiang, N., *In vivo* evidence that UV-induced C→T mutations at dipyrimidine sites could result from the replicative bypass of *cis-syn* cyclobutane dimers or their deamination products, *Biochemistry*, 32, 472, 1993.

20. Nadji, S., Wang, C.-I, and Taylor, J. S., Photochemically and photoenzymatically cleavable DNA, *J. Am. Chem. Soc.*, 114, 9266, 1992.

21. Lamola, A. A., Photosensitization in biological systems and the mechanism of photoreactivation, *Mol. Photochem.*, 4, 107, 1972.

22. Yeh, S.-R. and Falvey, D. E., Model studies of DNA photorepair: Energetic requirements for the radical anion mechanism determined by fluorescence quenching, *J. Am. Chem. Soc.*, 114, 7313, 1992.

23. Pac, C., Kubo, J., Majima, T., and Sakurai, H., Structure-reactivity relationships in redox-photosensitized splitting of pyrimidine dimers and unusual enhancing effect of molecular oxygen, *Photochem. Photobiol.*, 36, 273, 1982.

24. Burdi, D. and Begley, T. P., Mechanistic studies on DNA photolyase. 3. The trapping of the one-bond-cleaved intermediate from a photodimer radical cation model system, *J. Am. Chem. Soc.*, 113, 7768, 1991.

25. McMordie, R. A. S. and Begley, T. P., Mechanistic studies on DNA photolyase. 5. Secondary deuterium isotope effects on the cleavage of the uracil photodimer radical cation and anion, *J. Am. Chem. Soc.*, 114, 1886, 1992.

26. Yang, D.-Y. and Begley, T. P., Mechanistic studies on DNA photolyase. VIII. Studies on the fragmentation of the radical anion and cation of a uracil-alkene photoadduct, *Tetrahedron Lett.*, 34, 1709, 1993.

27. Sasson, S. and Elad, D., Photosensitized monomerization of 1,3-dimethyluracil photodimers, *J. Org. Chem.*, 37, 3164, 1972.

28. Roth, H. D. and Lamola, A. A., Cleavage of thymine dimers sensitized by quinones. Chemically induced dynamic nuclear polarization in radical ions, *J. Am. Chem. Soc.*, 94, 1013, 1972.

29. Kemmink, J., Eker, A. P. M., and Kaptein, R., CIDNP detected flash photolysis of *cis,syn* 1,3-dimethylthymine dimer, *Photochem. Photobiol.*, 44, 137, 1986.

30. Young, T., Nieman, R., and Rose, S. D., Photo-CIDNP detection of pyrimidine dimer radical cations in anthraquinonesulfonate-sensitized splitting, *Photochem. Photobiol.*, 52, 661, 1990.

31. Rokita, S. E. and Walsh, C. T., Flavin and 5-deazaflavin photosensitized cleavage of thymine dimer: A model of *in vivo* light-requiring DNA repair, *J. Am. Chem. Soc.*, 106, 4589, 1984.

32. Hartman, R. F., Rose, S. D., Pouwels, P. J. W., and Kaptein, R., Flavin-sensitized photochemically induced dynamic nuclear polarization detection of pyrimidine dimer radicals, *Photochem. Photobiol.*, 56, 305, 1992.

33. Rustandi, R. R. and Fischer, H., CIDNP detection of the 1,3-dimethyluracil dimer radical anion splitting sensitized by 2-methylindole, *J. Am. Chem. Soc.*, 115, 2537, 1993.

34. Yeh, S.-R. and Falvey, D. E., Model studies of DNA photorepair: Radical anion cleavage of thymine dimers probed by nanosecond laser spectroscopy, *J. Am. Chem. Soc.*, 113, 8557, 1991.

35. Pac, C., Miyake, K., Masaki, Y., Yanagida, S., Ohno, T., and Yoshimura, A., Flavin-photosensitized monomerization of dimethylthymine cyclobutane dimer: Remarkable effects of perchloric acid and participation of excited-singlet, triplet, and chain-reaction pathways, *J. Am. Chem. Soc.*, 114, 10756, 1992.

36. Heelis, P. F., Hartman, R. F., and Rose, S. D., Detection of radical ion intermediates in flavin-photosensitized pyrimidine dimer splitting, *Photochem. Photobiol.*, 57, 442, 1993.

37. Balgavý, P. and Šeršeň, F., Tryptophan-photosensitized formation of free radicals in thymine dimer, *Biológia (Bratislava)*, 37, 401, 1982.

38. Balgavý, P., Physical model of photoreactivation, *Acta Universitatis Carolinae — Mathematica et Physica*, 14, 129, 1973.

39. Heelis, P. F., Deeble, D. J., Kim, S.-T., and Sancar, A., Splitting of *cis-syn* cyclobutane thymine-thymine dimers by radiolysis and its relevance to enzymatic photoreactivation, *Int. J. Radiat. Biol.*, 62, 137, 1992.

40. Hartman, R. F. and Rose, S. D., Efficient photosensitized pyrimidine dimer splitting by a reduced flavin requires the deprotonated flavin, *J. Am. Chem. Soc.*, 114, 3559, 1992.

41. Hartman, R. F. and Rose, S. D., A possible chain reaction in photosensitized splitting of pyrimidine dimers by a protonated, oxidized flavin, *J. Org. Chem.*, 57, 2302, 1992.

42. Ceugniet, C., Goodman, M. S., Hartzfeld, D., Hartman, R. F., and Rose, S. D., unpublished results.

43. Kim, S.-T. and Rose, S. D., Pyrimidine dimer splitting in covalently linked dimer-arylamine systems, *J. Photochem. Photobiol. B: Biol.*, 12, 179, 1992.

44. Kim, S.-T., Hartman, R. F., and Rose, S. D., Solvent dependence of pyrimidine dimer splitting in a covalently linked dimer-indole system, *Photochem. Photobiol.*, 52, 789, 1990.

45. Kim, S.-T., Young, T., Goodman, M. S., Forrest, C., Hartman, R. F., and Rose, S. D., Photosensitized pyrimidine dimer splitting in covalently linked dimer-indole systems, *Trends in Photochem. Photobiol.*, 1, 81, 1990.

46. Hartzfeld, D. G. and Rose, S. D., Efficient pyrimidine dimer radical anion splitting in low polarity solvents, *J. Am. Chem. Soc.*, 115, 850, 1993.

47. Young, T., Kim, S.-T., Van Camp, J. R., Hartman, R. F., and Rose, S. D., Transient intermediates in intramolecularly photosensitized pyrimidine dimer splitting by indole derivatives, *Photochem. Photobiol.*, 48, 635, 1988.

48. Charlier, M. and Hélène, C., Photosensitized splitting of pyrimidine dimers in DNA by indole derivatives and tryptophan-containing peptides, *Photochem. Photobiol.*, 21, 31, 1975.

49. Balgavý, P. and Šeršeň, F., Monomerization of thymine dimers photosensitized by aromatic amino acids, *Chemické Zvesti*, 37, 243, 1983.

50. Chen, J., Huang, C. W., Hinman, L., Gordon, M. P., and Deranleau, D. A., Photomonomerization of pyrimidine dimers by indoles and proteins, *J. Theor. Biol.*, 62, 53, 1976.

51. Cochran, A. G., Sugasawara, R., and Schultz, P. G., Photosensitized cleavage of a thymine dimer by an antibody, *J. Am. Chem. Soc.*, 110, 7888, 1988.

52. Kemmink, J., Eker, A. P. M., van der Marel, G. A., van Boom, J. H., and Kaptein, R., Photoreactivation of the thymine dimer containing DNA octamer d(GCGT∧TGCG)· d(CGCAACGC) by the photoreactivating enzyme from *Anacystis nidulans*, *J. Photochem. Photobiol. B: Biol.*, 1, 323, 1988.

53. Eker, A. P. M., Hessels, J. K. C., and Dekker, R. H., Photoreactivating enzyme from *Streptomyces griseus*. VI. Action spectrum and kinetics of photoreactivation, *Photochem. Photobiol.*, 44, 197, 1986.

54. Sabourin, C. L. K. and Ley, R. D., Isolation and characterization of a marsupial DNA photolyase, *Photochem. Photobiol.*, 47, 719, 1988.

55. Kiener, A., Husain, I., Sancar, A., and Walsh, C., Purification and properties of *Methanobacterium thermoautotrophicum* DNA photolyase, *J. Biol. Chem.*, 23, 13880, 1989.

56. Jorns, M. S., Wang, B., Jordan, S. P., and Chanderkar, L. P., Chromophore function and interaction in *Escherichia coli* DNA photolyase: Reconstitution of the apoenzyme with pterin and/or flavin derivatives, *Biochemistry*, 29, 552, 1990.

57. Malhotra, K., Kim, S.-T., Walsh, C., and Sancar, A., Roles of FAD and 8-hydroxy-5-deazaflavin chromophores in photoreactivation by *Anacystis nidulans* DNA photolyase, *J. Biol. Chem.*, 267, 15406, 1992.

58. Eker, A. P. M., Formenoy, L., and de Wit, L. E. A., Photoreactivation in the extreme halophilic archaebacterium *Halobacterium cutirubrum*, *Photochem. Photobiol.*, 53, 643, 1991.

59. Eker, A. P. M., Hessels, J. K. C., and van de Velde, J., Photoreactivating enzyme from the green alga *Scenedesmus acutus*. Evidence for the presence of two different flavin chromophores, *Biochemistry*, 27, 1758, 1988.

60. Eker, A. P. M., Dekker, R. H., and Berends, W., Photoreactivating enzyme from *Streptomyces griseus*. IV. On the nature of the chromophoric cofactor in *Streptomyces griseus* photoreactivating enzyme, *Photochem. Photobiol.*, 33, 65, 1981.

61. Takao, M., Oikawa, A., Eker, A. P. M., and Yasui, A., Expression of an *Anacystis nidulans* photolyase gene in *Escherichia coli*: Functional complementation and modified action spectrum of photoreactivation, *Photochem. Photobiol.*, 50, 633, 1989.

62. Sutherland, B. M., Photoreactivating enzymes, in *The Enzymes*, Part A, Vol. 24, Boyer, P. D., Ed., Academic Press, New York, 1981, 481–515.

63. Sutherland, B. M., Gange, R. W., Freeman, S. E., and Sutherland, J. C., DNA damage and repair in human skin *in situ*, *DNA Damage Repair*, Castellani, A., Ed., Proc. 1st Int. Congress, 1987, Plenum, New York, 1989, 157–168.

64. Jordan, S. P. and Jorns, M. S., Evidence for a singlet intermediate in catalysis by *Escherichia coli* DNA photolyase and evaluation of substrate binding determinants, *Biochemistry*, 27, 8915, 1988.

65. Witmer, M. R., Altmann, E., Young, H., Begley, T. P., and Sancar, A., Mechanistic studies on DNA photolyase. 1. Secondary deuterium isotope effects on the cleavage of 2′-deoxyuridine dinucleotide photodimers, *J. Am. Chem. Soc.*, 111, 9264, 1989.

66. Ramsey, A. J., Alderfer, J. L., and Jorns, M. S., Energy transduction during catalysis by *Escherichia coli* DNA photolyase, *Biochemistry*, 31, 7134, 1992.

67. Ramsey, A. J. and Jorns, M. S., Effect of 5-deazaflavin on energy transduction during catalysis by *Escherichia coli* DNA photolyase, *Biochemistry*, 31, 8437, 1992.

68. Kim, S.-T., Heelis, P. F., and Sancar, A., Energy transfer (Deazaflavin → $FADH_2$) and electron transfer ($FADH_2$ → T<>T) kinetics in *Anacystis nidulans* photolyase, *Biochemistry*, 31, 11244, 1992.

69. Kim, S.-T., Li, Y.-F., and Sancar, A., The third chromophore of DNA photolyase: Trp-277 of *Escherichia coli* DNA photolyase repairs thymine dimers by direct electron transfer, *Proc. Natl. Acad. Sci. U.S.A.*, 89, 900, 1992.

70. Heelis, P. F., Okamura, T., and Sancar, A., Excited state properties of *Escherichia coli* DNA photolyase in the picosecond to millisecond time scale, *Biochemistry*, 29, 5694, 1990.

71. Okamura, T., Sancar, A., Heelis, P. F., Begley, T. P., Hirata, Y., and Mataga, N., Picosecond laser photolysis studies on the photorepair of pyrimidine dimers by DNA photolyase. 1. Laser photolysis of photolyase-2-deoxyuridine dinucleotide photodimer complex, *J. Am. Chem. Soc.*, 113, 3143, 1991.

72. Kim, S.-T., Sancar, A., Essenmacher, C., and Babcock, G. T., Evidence from photoinduced EPR for a radical intermediate during photolysis of cyclobutane thymine dimer by DNA photolyase, *J. Am. Chem. Soc.*, 114, 4442, 1992.

73. Essenmacher, C., Kim, S.-T., Atamian, M., Babcock, G. T., and Sancar, A., Tryptophan radical formation in DNA photolyase: Electron-spin polarization arising from photoexcitation of a doublet ground state, *J. Am. Chem. Soc.*, 115, 1602, 1993.

74. Husain, I., Sancar, G. B., Holbrook, S. R., and Sancar, A., Mechanism of damage recognition by *Escherichia coli* DNA photolyase, *J. Biol. Chem.*, 262, 13188, 1987.

75. Malhotra, K., Baer, M., Li, Y. F., Sancar, G. B., and Sancar, A., Identification of chromophore binding domains of yeast DNA photolyase, *J. Biol. Chem.*, 267, 2909, 1992.

76. Pac, C., Miyamoto, I., Masaki, Y., Furusho, S., Yanagida, S., Ohno, T., and Yoshimura, A., Chloranil-photosensitized monomerization of dimethylthymine cyclobutane dimers and effect of magnesium perchlorate, *Photochem. Photobiol.*, 52, 973, 1990.

77. Wenska, G. and Paszyc, S., Electron-acceptor-sensitized splitting of cyclobutane-type thymine dimers, *J. Photochem. Photobiol. B: Biol.*, 8, 27, 1990.

78. Masaki, Y., Miyake, K., Yanagida, S., and Pac, C., Participation of a chain mechanism in efficient monomerization of dimethylthymine cyclobutane dimer photosensitized by a flavin in the presence of perchloric acid, *Chem. Lett.*, 319, 1992.

79. Jorns, M. S., Photosensitized cleavage of thymine dimer with reduced flavin: A model for enzymic photorepair of DNA, *J. Am. Chem. Soc.*, 109, 3133, 1987.

80. Rosenthal, I., Rao, M. M., and Salomon, J., Transition metal-ion photosensitized monomerization of pyrimidine radicals, *Biochim. Biophys. Acta*, 378, 165, 1975.

81. Perichet, G., Meallier, P., and Pouyet, B., Monomérisation de la thymine dimère par irradiation du complexe thymine dimère-Hg^{2+}, *J. Photochem.*, 29, 375, 1985.

82. Groy, T. L., Kim, S.-T., and Rose, S. D., Structure of 5-[2-(3-indolyl)ethyl]-1,3-dimethyluracil 1,3-dimethyluracil *cis,syn*-cyclobutane photodimer, *Acta Cryst.*, C47, 1898, 1991.

83. Groy, T. L., Kim, S.-T., and Rose, S. D., Structures of two spiro[cyclopropane-1,3′-indoline]-*cis,syn*-1,3-dimethyluracil cyclobutane photodimers, *Acta Cryst.*, C47, 1898, 1991.

84. Van Camp, J. R., Young, T., Hartman, R. F., and Rose, S. D., Photosensitization of pyrimidine dimer splitting by a covalently bound indole, *Photochem. Photobiol.*, 45, 365, 1987.

85. Goodman, M. S., Molecular Recognition and Photosplitting of Pyrimidine Dimers, Ph.D. dissertation, Arizona State University, Tempe, AZ, 1992.

86. Goodman, M. S. and Rose, S. D., Molecular recognition of a pyrimidine dimer and photosensitized dimer splitting by a macrocyclic *bis*(diaminopyridine), *J. Am. Chem. Soc.*, 113, 9380, 1991.

87. Goodman, M. S. and Rose, S. D., Photosensitized pyrimidine dimer splitting by a methoxyindole bound to a dimer-recognizing macrocycle, *J. Org. Chem.*, 57, 3268, 1992.

88. Payne, G. and Sancar, A., Absolute action spectrum of E-FADH$_2$ and E-FADH$_2$-MTHF forms of *Escherichia coli* DNA photolyase, *Biochemistry*, 29, 7715, 1990.

89. Walsh, C., Naturally occurring 5-deazaflavin coenzymes: Biological redox roles, *Acc. Chem. Res.*, 19, 216, 1986.

90. Kim, S.-T. and Sancar, A., Photochemistry, photophysics, and mechanism of pyrimidine dimer repair by DNA photolyase, *Photochem. Photobiol.*, 57, 895, 1993.

91. Sancar, A., Structure and Function of DNA photolyase, *Biochemistry*, 33, 2, 1994.

92. Eker, A. P. M., Yajima, H., and Yasui, A., DNA photolyase from the fungus *Neurospora crassa*. Purification, characterization, and comparison with other photolyases, *Photochem. Photobiol.*, 60, 125, 1994.

93. Begley, T. P., Photoenzymes: a novel class of biological catalysts, *Acc. Chem. Res.*, 27, 394, 1994.

94. Park, H.-W., Sancar, A., and Deisenhofer, J., Crystallization and preliminary crystallographic analysis of *Escherichia coli* DNA photolyase, *J. Mol. Biol.*, 231, 1122, 1993.

95. Miki, K., Tamada, T., Nishida, H., Inaka, K., Yasui, A., deRuiter, P. E., and Eker, A. P. M., Crystallization and preliminary X-ray diffraction studies of photolyase (photoreactivating enzyme) from the cyanobacterium *Anacystis nidulans*, *J. Mol. Biol.*, 233, 167, 1993.

8

Photochemistry of Skin-Sensitizing Psoralens

Sang Chul Shim
Korea Advanced Institute of Science and Technology

8.1 Introduction

Psoralens are heterocyclic aromatic compounds present in numerous plants found throughout the world. Psoralen derivatives are photosensitizers of UV-A, especially from 320 to 380 nm, a range in which cellular nucleic acids and proteins are weakly absorbing if any at all. Because of their photosensitizing properties, psoralens show genotoxic activity[1,2] and this is of special importance because of the use of photoreactive psoralens in photochemotherapy[3–6] for the treatment of psoriasis, vitiligo, micosis fungoides, chronic leukemia, and some infections connected with AIDS[7] and in cosmetology.[8] However, observations on the photomutagenicity and photocarcinogenicity of bifunctional furocoumarins in actual use[1,2,9–11] led to a search for new photoreactive and less genotoxic compounds among the monofunctional derivatives.[12] Psoralens were also used as powerful tools in nucleic acid research concerning structure and function relationships, DNA repair mechanisms, and the genotoxic consequences of defined lesions in DNA.[13,14]

8.2 Photophysical Properties

The intrinsic photoreactivity of psoralens is determined by the electronic structure of the lowest excited states (S_1 and T_1). The UV-A bands of all psoralen derivatives arise from their pyrone moiety. Their extinction coefficients, however, differ appreciably from one compound to another in this wavelength region. Table 8.1 lists the spectroscopic data of various coumarin and psoralen derivatives. The structures of psoralen derivatives in Table 8.1 are given in Figure 8.1.

8.3 Photochemical Reactions

The excited states of psoralens undergo various photochemical reactions with another ground-state psoralen molecule, solvent, and biomolecules via type I and type II mechanisms. The type I mechanism involves the generation of free radicals and the direct photobinding to the substrate.[31] The type II mechanism, on the other hand, involves photooxidations by the formation of singlet oxygen and of superoxide radical anion (photodynamic effects).[32–46]

Table 8.1   Solubility in Water, Absorption (λ_{max}, ε), Fluorescence (Φ_f), and Triplet-State Features (Φ_t, λ_p) of Various Psoralen Derivatives[a]

Psoralens	λ_{max}	ε_{max}	$\varepsilon_{365\,nm}$	λ^f_{max} (nm)	Φ_f	Φ_t	λ_p (nm)	Solubility in water (mg/l)[15,16]	Solvent	Ref.
8-MOP	246	21,000	1150	508	0.0015	0.06		23	Water	
	303	12,014								16–19
	299			460	0.0020	0.04	456.5		Ethanol	
5-MOP	312	14,200	950	460		<0.01		5	Water	16,17
	335			427	0.019	0.1	472		Ethanol	
TMP	250	15,000								
	298	7,950	1680	430	0.056	0.09[b]	446.5	1	Water	16,17,20,
	335									21,22
	337			416	0.044				Ethanol	
4,5'-DMA	298	9,350	125	425				8	Water	16
3-CPs	247									
	318	10,900	6230	460	0.02	0.35	504	13	Water	16,18,23
	365									
DMC	327			450	0.76				Water	16,24,25
	324	15,150		425	0.65	0.072	476		Ethanol	
Ps	330			409	0.019	0.13	456		Water	
	244			444	0.01	0.06		37		16,23
	295								Water/ethanol	
	335									
Angelicin	246		116		0.33			20	Water	26
	300	9,530								
AMT	256									
	303			457	0.06			>10,000	Water	21
	336									
PyPs	310		400	458	0.01	0.04	437[c]		Ethanol	
	327	6,500								27–30
	330	6,700		470	0.01				Water/ethanol	
MePyPs	308	8,000	850	460	0.008	0.02	442[c]		Ethanol	
	330									27–30
	330	8,300		485	0.007				Water-ethanol	
Coumarin	313	3,100		390	0.009		459.8	5,000	Water	17

[a] The structures are given in Figure 8.1.
[b] Measured in methanol.
[c] Measured in *n*-hexane/ethanol.

Photodimerization

Psoralens under UV-A irradiation dimerize under various experimental conditions, such as in the solid state,[47–50] in aqueous or organic solutions,[50–56] or in frozen aqueous solutions.[50,57–60] In general, the dimers are derived from a C4-photocycloaddition engaging the 3,4-double bond of both monomers.

Photoreaction with Biomolecules

The mechanisms of the photobiological activity of psoralens are still not fully understood. However, they have been related to photomodifications and/or photobinding of the cellular macromolecules,[2,61] such as nucleic acids,[14,62] proteins,[63] and membrane lipids.[64] Typically, 8-methoxypsoralen (8-MOP) is covalently bound to proteins (57%), lipids (26%), and DNA (17%) in shaved backs of albino Wistar rats.[65]

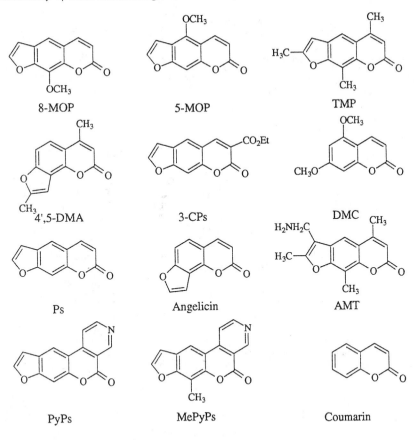

FIGURE 8.1 The structure of various psoralen derivatives.

Photoreaction with Nucleic Acids

The best-known photochemical reactions of psoralens are those with nucleic acids, especially DNA.[66] It appears that the genotoxic effects, as well as the therapeutically important antiproliferative effects, are due mainly to their capacity to induce photoconjugation to DNA. The modification of DNA by psoralens is thought to be a two-step process: the first step being the formation of a complex between psoralens and DNA (intercalated complex),[67,68] followed by the formation of covalent bonds with pyrimidine bases, particularly thymine, upon irradiation with near-UV light. Two types of monoadducts (i.e., 3,4- and 4′,5′-monoadducts) can be formed by the C4-photocycloaddition with the 5,6-double bond of a pyrimidine base of DNA.[66] In particular, in the case of 4′,5′-monoadducts, once these are formed they can absorb a second photon and photoreact further to also engage their 3,4-double bond leading to the formation of interstrand cross-linkages in the DNA.[69] Generally, the stereochemistry of psoralens-thymidine photoadducts in DNA had been confirmed to be *cis-syn*-structures (Table 8.2).

Although excited psoralens react mainly with the pyrimidine bases of DNA and RNA,[70–78] they are able to form adducts with purine nucleosides under certain conditions. In contrast to psoralen-thymidine photoadducts, the photobinding of 4,5′,8-trimethylpsoralen (TMP) proceeds through covalent bond formation between the C(4) of the pyrone ring and ribose 1′-, 5′-, or 4′-carbons in a dry film.[91,92] Similar photoadducts are obtained in the photoreaction of 8-MOP with 2′-deoxyadenosine and of 5,7-dimethoxycoumarin (DMC) with adenosine.[93,94]

Table 8.2   The Photoreaction of Psoralens with DNA or Thymine

Psoralens	DNA or base	Conc. (Ps)	Conc. (DNA or base)	Solvent	C_4-Photoproducts	Ref.
8-MOP	Calf thymus DNA	0.067 mg/ml	0.72 mg/mL	Tris buffer	(f)[a] *cis*,[b]*syn*,[c] (p) *cis,syn*	79,80
		0.019 mmol	0.72 mg/mL	Tris buffer	(f),(p) *cis,syn*-diadducts	81
	Thymidine	— [d]	— [d]	Thin film	(f) *cis,syn*, (f) *trans,syn*, (p) *cis,anti*	82,83
	Thymine	$2 \times 10^{-4}\ M$	$2 \times 10^{-2}\ M$	Frozen water-methanol	(f) *cis,syn*	79
		1 m*M*	50 m*M*	Frozen water-acetonitrile	(p) *cis,syn*, (p) *cis,anti*	84
3-CPs	Salmon sperm DNA		1 mg/ml	Tris buffer	(f) *cis,syn*	85
	Thymidine	— [e]	— [e]	Thin fim	two (f) *cis,syn*	86
TMP	Calf thymus DNA	0.021 mg/ml	0.12 mg/ml	Tris buffer	(f) *cis,syn*, (p) *cis,syn*	80
		0.019 mmol	0.72 mg/ml	Tris buffer	(f),(p) *cis,syn*-diadducts	81
HMT	Calf thymus DNA	0.03 mg/ml	1 mg/ml	Tris-EDTA buffer	Two (f) *cis,syn*, (p)	87
		0.019 mmol	0.72 mg/ml	Tris buffer	(f),(p) *cis,syn*-diadducts	81
Ps	Calf thymus DNA	0.019 mmol	0.72 mg/ml	Tris buffer	(f),(p) *cis,syn*-diadducts	81
4-MA	Salmon testes DNA			Water	(f) *cis,syn*, (p) *cis,anti*	88
	Thymine	0.04 mg/ml	1.2 mg/ml	Water-methanol	Two (p) *cis,anti*	88
4,5'-DMA	Thymine	0.015 mg/ml	0.9 mg/ml	Water	(p) *syn*, (p) *anti*	89
				Frozen water-methanol	(f) *syn*	89
DMC	Thymidine	— [d]	— [d]	Thin film	(p) *trans,syn*, (p) *trans,anti*	90

[a] (f) and (p) designate the thymine binding on the furan and pyrone double bonds, respectively.

[b] *cis* indicates the stereoisomer in which both psoralen and pyrimidine are bonded from the same side of the cyclobutane ring.

[c] *syn* indicates the stereoisomer in which 1-O of psoralen and 1-N of thymine are on the same side of the cyclobutane ring.

[d] Molar ratio of psoralen and thymidine is 1:10.

[e] Ratio of psoralen and thymine is 1:10 by weight.

Photoreaction with Proteins and Membrane Lipids

Photoreactions of psoralens with membranes and proteins may proceed via both type I and type II mechanisms.[32,95] Psoralens show strong affinity for proteins,[65,96] and their excited states are efficiently quenched by amino acids.[71] However, definite proof of photobinding of psoralens to proteins or amino acids has not been reported as yet. Photooxidation of amino acid residues via the type II mechanism is thought to be responsible for inactivation of enzymes,[63,97–104] while the covalent photoconjugation does not seem to play a role in this connection.[97,100] The type II reaction also results in the cross-linking of enzyme subunits of oligomeric proteins such as glutamate dehydrogenase, catalase and alcohol dehydrogenase,[98] erythrocyte ghost proteins,[43] and α-crystalline lens proteins.[105]

The oxidative photoreactions of lipids occur by two mechanisms; that is, by UV-A irradiation in the presence of psoralens[36–38,45,46,106] or by the addition of the previously photooxidized psoralens to aqueous suspension of liposomes.[64] The photoproducts of psoralens and unsaturated fatty acids were isolated and characterized.[64,107–116] [2+2]-Cycloaddition of TMP is observed to occur at the 3,4-bond (pyrone double bond) of TMP and double bond of the oleic acid methyl ester (OAME).[110,116] The methylene group deriving from position-9 (or 10) of OAME and 10-H (9-H) of OAME are on opposite side of the cyclobutane ring, (i.e., *trans*-configuration) while 3-CH$_3$ and the 4-H are on the same face, *cis*-configuration, in all TMP-OAME photoadducts. The *trans*-fatty acid formed by *cis,trans*-isomerization may be an intermediate for *trans,cis*-photoadducts.[116]

References

1. Ben-Hur, E. and Song, P. S., The photochemistry and photobiology of furocoumarins (psoralens), *Adv. Radiat. Biol.*, 11, 131, 1984.
2. Averbeck, D., Yearly review: Recent advances in psoralen phototoxicity mechanism, *Photochem. Photobiol.*, 50, 859, 1989.
3. Parrish, J. A., Stern, P. S., Pathak, M. A., and Fitzpatrick, T. B., Photochemotherapy of skin diseases, in *The Science of Photomedicine*, Regan, J. D. and Parrish, J. A., Eds., Plenum, New York, 1982, 595–624.
4. Parrish, J. A., Fitzpatrick, T. B., Tanenbaum, L., and Pathak, M. A., Photochemotherapy of psoriasis with oral methoxsalen and long-wave ultraviolet light, *New Engl. J. Med.*, 291, 1207, 1974.
5. Musajo, L., Rodighiero, G., Caporale, G., Dall'Acqua, F., Marciani, S., Bordin, F., Baccicahetti, F., and Bevilaqua, R., Photoreactions between skin-photosensitizing furocoumarins and nucleic acids, in *Sunlight and Man*, Pathak, M. A., Ed., Univ. of Tokyo Press, Tokyo, 1974, 335–368.
6. Edelson, R., Berger, C., Gasparro, F., Jegasothy, B., Heald, P., Wintroub, B., Vonderheid, E., Knobler, R., Wolff, K., Plewig, G., Mckiernan, G., Christansen, I., Oster, M., Hönigsmann, H., Wilford, H., Kokoschka, E., Rehle, T., Perez, M., Stingl, G., and Laroche, L., Treatment of cutaneous T-cell lymphoma by extracorporeal photochemotherapy, *New Engl. J. Med.*, 316, 297, 1987.
7. Gorin, J., Lessana-Leibowitch, M., Fortier, P., Leibowitch, J., and Escande, J.-P., Successful treatment of the pruritus of human immunodeficiency virus infection and acquired immunodeficiency syndrome with psoralens plus ultraviolet A therapy, *J. Am. Acad. Dermatol.*, 20, 511, 1989.
8. Suzuki, K., Nakamura, K., and Iwaida, M., Detection and determination of bergapten in bergamot oil and in cosmetics, *J. Soc. Cosmet. Chem.*, 30, 393, 1979.
9. Bridges, B. and Strauss, G., Possible hazards of photochemotherapy for psoriasis, *Nature*, 283, 523, 1980.
10. Song, P. S. and Tapley, K. J., Photochemistry and photobiology of psoralens, *Photochem. Photobiol.*, 29, 1177, 1979.
11. Oginsky, E. L., Green, G. S., Griffith, D. G., and Fowlks, W. L., Lethal photosensitization of bacteria with 8-methoxypsoralen to long wavelength ultraviolet radiation, *J. Bacteriol.*, 78, 821, 1959.
12. Rodighiero, G., Dall'Acqua, F., and Averbeck, D., New psoralen and angelicin derivatives, in *New Psoralen DNA Photobiology*, Vol. 1, Gasparro, F. P., Ed., CRC Press, Boca Raton, FL, 1988, 37–114.
13. Cimino, G. P., Gamper, H. B., Isaacs, S. T., and Hearst, J. E., Psoralens as photoreactive probes of nucleic acid structure and function: Organic chemistry, photochemistry, and biochemistry, *Annu. Rev. Biochem.*, 54, 1151, 1985.
14. Hearst, J. E., Psoralen photochemistry, *Annu. Rev. Biophys. Bioeng.*, 10, 69, 1985.
15. Grossweiner, L. I., Mechanisms of photosensitizations by furocoumarins, *Natl. Cancer Inst. Monogr.*, 66, 47, 1984.
16. Rodighiero, G. and Dall'Acqua F., *In vitro* photoreactions of selected psoralens and methylangelicins with DNA, RNA, and proteins, *Natl. Cancer Inst. Monogr.*, 66, 31, 1984.
17. Mantulin, W. W. and Song, P. S., Excited states of skin-sensitizing coumarins and psoralens. Spectroscopic studies, *J. Am. Chem. Soc.*, 95, 5122, 1973.
18. Averbeck, D. and Moustacchi, E., Genetic effect of 3-carbethoxypsoralen, angelicin, psoralen and 8-methoxypsoralen plus 365-nm irradiation in *Saccharomyces cerevisiae*. Irradiation of reversion, mitotic crossing-over, gene conversion and cytoplasmic "petote" mutations, *Mutat. Res.*, 68, 133, 1979.
19. Gaboriau, F., Vigny, P., Averbeck, D., and Bisagni, E., Spectroscopic study of the dark interaction and of the photoreaction between a new monofunctional psoralen: 3-Carbethoxypsoralen and DNA, *Biochimie*, 83, 899, 1981.
20. Lai, T., Lim, B. T., and Lim, E. C., Photophysical properties of biologically important molecules related to proximity effects: Psoralens, *J. Am. Chem. Soc.*, 104, 7631, 1982.
21. Hearst, J. E., Psoralen photochemistry and nucleic acid structure, *J. Invest. Dermatol.*, 77, 39, 1981.

22. Beaumont, B. C., Parsons, B. J., Navaratnan, S., Phillips, G. O., and Allen, J. C., Laser flash photolysis and fluorescence studies of the interaction of the excited states of furocoumarins with DNA in solution, in *Radiation Biology and Chemistry, Research Developments*, Elsvier, Amsterdam, 1979, 441–451.

23. Vigny, P., Gaboriau, F., Duquesne, M., Bisagni, E., and Averbeck, D., Spectroscopic properties of psoralen derivatives substituted by carbethoxy groups at the 3,4 and/or 4',5' reaction site, *Photochem. Photobiol.*, 30, 557, 1979.

24. Shim, S. C., Kang, H. K., Park, S. K., and Shin, E. J., Photophysical properties of some coumarin derivatives: 5,7-Dimethoxycoumarin, 4',5',-dihydropsoralen, 8-methoxypsoralen (8-MOP) and 8-MOP $\binom{4',5}{5',6}$ THD C_4-cyclomonoadduct (THD=thymidine), *J. Photochem.*, 37, 125, 1987.

25. Land, E. J. and Truscott, T. G., Triplet excited state of coumarin and 4',5'-dihydropsoralen: Reaction with nucleic acid bases and amino acids, *Photochem. Photobiol.*, 29, 861, 1979.

26. Dall'Acqua, F., Vedaldi, D., Guiotto, A., Rodighiero, P., Carlassare, F., Baccichetti, F., and Bordin, F., Methylangelicins: New potential agents for the photochemotherapy of psoriasis. Structure-activity studies on the dark and photochemical interactions with DNA, *J. Med. Chem.*, 24, 806, 1981.

27. Blais, J., Vigny, P., Moron, J., and Bisagni, E., Spectroscopic properties and photoreactivity with DNA of new monofunctional pyridopsoralens, *Photochem., Photobiol.*, 39, 145, 1984.

28. Ronfard-Haret, J. C., Averbeck, D., Bensasson, R. V., Bisagni, E., Land, E. J., and Moron, J., Correlation between the triplet photophysical properties and the photobiological action on yeast of two monofunctional pyridopsoralens, *Photochem. Photobiol.*, 45, 235, 1987.

29. Cosralat, R., Blais, J., Ballini, J.-P., Moysan, A., Cadet, J., Chalvet, O., and Vigny, P., Formation of cyclobutane thymine dimers photosensitized by pyridopsoralens: A triplet-triplet energy transfer mechanism, *Photochem. Photobiol.*, 51, 255, 1990.

30. Papadopoulo, D., Averbeck, D., and Moustacchi, E., Mutagenic effects photoinduced in mammalian cells *in vitro* by two monofunctional pyridopsoralens, *Photochem. Photobiol.*, 44, 31, 1986.

31. Dall'Acqua, F., Caffieri, S., and Rodighiero, G., Photoreactions of furocoumarins (psoralens and angelicins), in *Primary Processes in Biology and Medicine*, Bensasson, E., Jori, G., Land, E. J., and Truscott, T. G., Eds., Plenum, New York, 1985.

32. Midden, W. R., Chemical mechanisms of the bioeffects of furocoumarins: The role of reactions with proteins, lipids and other cellular constituents, in *Psoralen DNA Photobiology*, Vol. II, Gasparro, F. P., Ed., CRC Press, Boca Raton, FL, 1988, 1–49.

33. Dall'Acqua, F., Furocoumarin photochemistry and its main biological implication, *Curr. Probl. Dermatol.*, 15, 137, 1986.

34. Dall'Acqua, F. and Caffieri, S., Recent and selected aspects of furocoumarin photochemistry and photobiology, *Photomed. Photobiol.*, 10, 1, 1988.

35. Musajo, L. and Rodighiero, G., Mode of photosensitizing action of furocoumarins, in *Photophysiology*, Vol. VII, Giese, A. C., Ed., Academic Press, New York, 1972, 115–147.

36. Pathak, M. A. and Joshi, P. C., The nature and molecular basis of cutaneous photosensitivity reactions to psoralen and coal tar, *J. Invest. Dermatol. Suppl.*, 66s, 1983.

37. Joshi, P. C. and Pathak, M. A., Production of active species ($^1O_2$ and O_2^-) by psoralens and ultraviolet radiation (320–400), *Biochim. Biophys. Acta*, 798, 115, 1984.

38. Joshi, P. C. and Pathak, M. A., Production of singlet oxygen and superoxide radicals by psoralens and their biological significance, *Biochim. Biophys. Res. Commun.*, 112, 638, 1983.

39. Carbonare, M. D. and Pathak, M. A., Skin photosensitizing agents and the role of reactive oxygen species in photoaging, *J. Photochem. Photobiol., B: Biol.*, 14, 105, 1992.

40. Dall'Acqua, F., Photosensitizing action of furocoumarins on membrane components and consequent intracellular event, *J. Photochem. Photobiol., B: Biol.*, 8, 235, 1991.

41. Patapenko, A. Ya., Mechanisms of photodynamic effects of furocoumarins, *J. Photochem. Photobiol., B: Biol.*, 9, 1, 1991.

42. Knox, C. N., Land, E. J., and Truscott, T. G., Singlet oxygen generation by furocoumarin triplet states. I. Linear furocoumarins (psoralens), *Photochem. Photobiol.*, 43, 359, 1986.

43. Hornicek, F. J., Malinin, G. I., Glew, W. B., Awret, U., Garcia, J. V., and Nigra, T. P., Photochemical cross-linking of erythrocyte ghost proteins in the presence of 8-methoxy and trimethylpsoralen, *Photobiochem. Photobiophys.*, 9, 263, 1985.

44. Malinin, G. I., Lo, H. K., and Hornicek, F. J., Structural photomodification of erythrocyte ghost by UV-A and psoralen, *Photobiochem. Photobiophys.*, 13, 145, 1986.

45. Blan, Q. A. and Grossweiner, L. I., Singlet oxygen generation by furocoumarins: Effect of DNA and liposomes, *Photochem. Photobiol.*, 45, 177, 1987.

46. Matsuo, I., Yoshino, K., and Ohkido, M., Mechanism of skin surface lipid peroxidation, *Curr. Probl. Dermatol.*, 11, 135, 1983.

47. Rodighiero, G. and Cappellina, V., Recerche sulla fotodimerizzazione di alcune furocumarine, *Gazz. Chim. Ital.*, 91, 103, 1961.

48. Dall'Acqua, F. and Caffieri, S., C_4-cyclomers of psoralen engaging the 4',5'-double bond, *Photochem. Photobiol.*, 45, 13, 1987.

49. Ramasubby, N., Row T. N. G., Venkatesan, K., Ramamurthy, V., and Rao, C. N. R., Photodimerization of coumarins in the solid state, *J. Chem. Soc., Chem. Commun.*, 178, 1982.

50. Marciani, S., Dall'Acqua. F., Rodighiero, P., Gaporale., G., and Rodighiero, G., Ricerche sulla foto-C_4-ciclodimerizzazion di alcumi metilpsoraleni, *Gazz. Chim. Ital.*, 100, 435, 1970.

51. Krauch, C. H. and Farid, S., Photo-cycloaddition mit furocoumarin und furochromonin, *Chem. Ber.*, 100, 1685, 1967.

52. Wessely, F. and Dinjaski, F., Über die lichteinwirkung auf stoffe von typus der furocumarine, *Mh. Chem.*, 64, 131, 1934.

53. Wessely, F. and Kotlan, J., Über die photodimerisierung von furocumarinen und das sphondylin, *Mh. Chem.*, 86, 430, 1955.

54. Caffieri, S. and Dall'Acqua., F. Fotostabilità e fotodimerizzazion della 4,5'-dimetilangelicina, *Chimica Oggi*, 29, 1985.

55. Shim, S. C., Lee, S. S., and Choi, S. J., The C_4-photocyclodimers of 4,5',8-trimethylpsoralen, *Photochem. Photobiol.*, 51, 1, 1990.

56. Muthuramu, K. and Ramamurty, V., Photodimerization of coumarin in aqueous and micellar media, *J. Org. Chem.*, 47, 3976, 1982.

57. Rodighiero, G., Dall'Acqua, F., and Chimenti, G., Fotodimerizzazione della cumaina e di alcune furocumarine in marice di ghiaccio, *Anns. Chim.*, 58, 551, 1968.

58. Hoffman, R., Wells, P., and Morrison, H., Organic photochemistry. XII. Further studies on the mechanism of coumarin photodimerization. Observation of an unusual "heavy atom" effect, *J. Org. Chem.*, 36, 102, 1971.

59. Leenders, L. H., Schouteden, E., and de Schryver, F. C., Photochemistry of nonconjugated bichromophoric systems. Cyclomerization of 7,7'-polymethylenedioxycoumarins and polymethylenedicarboxylic acid (7-coumarinyl) diesters, *J. Org. Chem.*, 38, 957, 1973.

60. Bergman, J., Osaki, K., Schmidt, G. M. J., and Sonntag, F. I., Topochemistry. IV. The crystal chemistry of some *cis*-cinnamic acids, *J. Chem. Soc.*, 2021, 1964.

61. Cadet, J., Vigny, P., and Midden, W. R., Photoreactions of furocoumarins with biomolecules, *J. Photochem. Photobiol., B: Biol.*, 6, 197, 1990.

62. Vigny, P., Gaboriau, F., Voituriez, L., and Cadet, J., Chemical structure of psoralen-nucleic acid photoadducts, J., *Biochimie.*, 67, 317, 1985.

63. Yoshikawa, K., Mori, N., Sakakibara, S., Mizuno, N., and Song, P.-S., Photo-conjugation of 8-methoxypsoralen with proteins, *Photochem. Photobiol.*, 29, 1127, 1979.

64. Caffieri, S., Daga, A., Vedaldi, D., and Dall'Acqua, F., Photoaddition of angelicin to linolenic acid methyl ester, *J. Photochem. Photobiol., B: Biol.*, 2, 515, 1988.

65. Beijersbergen van Henegouwen, G. M. J., Wihn, E. T., Schononderwoerd, S. A., and Dall'Acqua, F., A method for the determination of PUVA-induced *in vivo* irreversible binding of 8-methoxypsoralen

(8-MOP) to epidermal lipids, proteins and DNA/RNA, *J. Photochem. Photobiol., B: Biol.*, 3, 631, 1989.

66. Hearst, J. E., Isaacs, S. T., Kanne, D., Rapoport, H., and Straub, K., The reaction of the psoralen with deoxyribonucleic acid, *Q. Rev. Biophys.*, 17, 1, 1984.

67. Dall'Acqua, F., Terbojevich, M., Marciani, S., Vedaldi, D., and Recher, M., Investigation on the dark interaction between furocoumarins and DNA, *Chem.-Biol. Interact.*, 21, 103, 1978.

68. Rontó, G., Tóth, K., Gáspár, S., and Csik, G., Phage nucleoprotein-psoralen interaction: Quantitative characterization of dark and photoreactions, *J. Photochem. Photobiol., B: Biol.*, 12, 9, 1992.

69. Dall'Acqua, F., Marciani-Mango, S., Zambon, F., and Rodighiero, G., Kinetic analysis of the photoreaction (365 nm) between psoralen and DNA, *Photochem. Photobiol.*, 29, 489, 1979.

70. Song, P.-S., Harter, M. L., Moore, T. A., and Herndon, W. C., Luminescence spectra and photocycloaddition of the excited coumarins to DNA bases, *Photochem. Photobiol.*, 27, 317, 1975.

71. Bensasson, R. V., Land, E. J., and Salet, C., Triplet excited state of furocoumarins: Reaction with nucleic acid bases and amino acids, *Photochem. Photobiol.*, 27, 273, 1978.

72. Bevilacqua, R. and Bordin, F., Photo-C_4-cycloaddition of psoralen and pyrimidine bases: Effect of oxygen and paramagnetic ions, *Photochem. Photobiol.*, 17, 191, 1973.

73. Johnston, B. H., Kung, A. H., Moore, C. B., and Hearst, J. E., Psoralen-DNA photoreaction: Controlled production of mono- and diadducts with nanosecond ultraviolet laser pulses, *Science*, 197, 906, 1977.

74. Salet, C., De Sa E Melo, T. M., Bensasson, R., and Land, E. J., Photophysical properties of aminomethylpsoralen in presence and absence of DNA, *Biochim. Biophys. Acta*, 607, 379, 1980.

75. Beaumont, P. C., Parsons, B. J., Navaratnam, S., Phillips, G. O., and Allen, J. C., The reactivities of furocoumarin excited states with DNA in solution, *Biochim. Biophys. Acta*, 608, 259, 1980.

76. Beaumont, P. C., Parsons, B. J., Navaratnam, S., and Phillips, G. O., A laser flash photolysis and fluorescence study of aminomethyltrimethylpsoralen in the presence and absence of DNA, *Photobiochem. Photobiophys.*, 5, 359, 1983.

77. Shim. S. C. and Kang, H. K., Relative reactivities of the excited states of furocoumarins for [2+2] photocycloaddition reaction with tetramethylethylene, *Photochem. Photobiol.*, 45, 453, 1987.

78. Song, P.-S., Photoreactive states of furocoumarins, *Natl. Cancer Inst. Monogr.*, 66, 15, 1984.

79. Pecker, S., Graves, B., Kanne, D., Rapoport, H., Hearst, J. E., and Kim, S.-H., Structure of a psoralen-thymine monoadduct formed in photoreaction with DNA, *J. Mol. Biol.*, 162, 157, 1982.

80. Kanne, D., Straub, K., Rapoport, H., and Hearst, J. E., Psoralen-deoxyribonucleic acid photoreaction. Characterization of the monoaddition products from 8-methoxypsoralen and 4,5',8-trimethylpsoralen, *Biochemistry*, 21, 861, 1981.

81. Kanne, D., Straub, K., Hearst, J. E., and Rapoport, H., Isolation and characterization of pyrimidine-psoralen-pyrimidine photodiadducts from DNA, *J. Am. Chem. Soc.*, 104, 6754, 1982.

82. Shim, S. C. and Kim, Y. Z. Photoreaction of 8-methoxypsoralen with thymidine, *Photochem. Photobiol.*, 38, 265, 1983.

83. Cadet, J., Voituriez, L., Ulrich, J., Joshi, P. C., and Way, S. Y., Isolation and characterization of the monoheterodimers of 8-methoxypsoralen and thymidine involving pyrone moiety, *Photobiochem. Photobiophys.*, 8, 35, 1984.

84. Joshi, P. C., Way, S. Y., Midden, M. R., Voituriez, L., and Cadet, J., Heterodimers of 8-methoxypsoralen and thymine, *Photobiochem. Photobiophys.*, 8, 51, 1984.

85. Land, E. J., Ruchton, F. A. P., Beddoes, R. L., Bruce, J. C., Cernik, R. J., Dawson, S. C., and Mills, O. S., A 2+2 photo-adduct of 8-methoxypsoralen and thymine: X-ray crystal structure; a model for the reaction of psoralens with DNA in the phototherapy of psoriasis, *J. Chem. Soc., Chem. Commun.*, 1, 22, 1982.

86. Gaboriau, F., Vigny, P., Cadet, J., Voituriez, L., and Bisagni, E., Photoreaction of the monofunctional 3-carbethoxypsoralen with DNA: Identification and conformational study of the predominant *cis-syn* furan-side monoadducts, *Photochem. Photobiol.*, 45, 199, 1987.

87. Straub, K., Kanne, D., Hearst, J. E., and Rapoport, H., Isolation and characterization of pyrimidine-psoralen photoadducts from DNA, *J. Am. Chem. Soc.*, 103, 2347, 1981.

88. Caffieri, S., Lucchini, V., Rodighiero, P., Miolo, G., and Dall'Acqua, F., 3,4 and 4′,5′-photocycloadducts between 4′-methylangelicin and thymine from DNA, *Photochem. Photobiol.*, 48, 573, 1988.

89. Dall'Acqua, F., Caffieri, S., Vedaldi, D., Guiotto, A., and Rodighiero, P., Monofunctional 4′,5′-photocycloadduct between 4′,5′-dimethylangelicin and thymine, *Photochem. Photobiol.*, 37, 373, 1983.

90. Shim, S. C., Koh, H. Y., and Chi, D. Y., Photocycloaddition reaction of 5,7-dimethoxycoumarin to thymidine, *Photochem. Photobiol.*, 34, 177, 1981.

91. Shim, S. C. and Choi, S. J., Photoreaction of 4,5′,8-trimethylpsoralen with adenosine, *Photochem. Photobiol.*, 40, 305, 1990.

92. Yun, M. H., Choi, S. J., and Shim, S. C., A novel photoadduct of 4,5′,8-trimethylpsoralen and adenosine, *Photochem. Photobiol.*, 55, 457, 1992.

93. Cadet, J., Voituriez, L., Nardin, R., Viari, A., and Vigny, P., Isolation and characterization of 8-MOP adducts to the osidic moiety of 2′-deoxyadenosine, *J. Photochem. Photobiol., B: Biol.*, 2, 321, 1988.

94. Cho, T. H., Shim, H. K., and Shim, S. C., Isolation and characterization of the photoadducts of 5,7-dimethoxycoumarin and adenosine, *Photochem. Photobiol.*, 46, 305, 1987.

95. Dall'Acqua, F. and Martelli, P., Photosensitizing action of furocoumarins on membrane components and consequent intracellular events, *J. Photochem. Photobiol., B: Biol.*, 8, 235, 1991.

96. Frederiksen, S., Nielsen, P. E., and Hoyer, P. E., Lysosomes: A possible target for psoralen photodamage, *J. Photochem. Photobiol., B: Biol.*, 3, 437, 1989.

97. Veronese, F. M., Schiavon, O., Beviacqua, R., Bordin, F., and Rodighiero, G., Photoinactivation of enzymes by linear and angular furocoumarins, *Photochem. Photobiol.*, 36, 25, 1982.

98. Schiavon, O. and Veronese, F. M., Extensive crosslinking between subunits of oligomeric products induced by furocoumarins plus UV-A irradiation, *Photochem. Photobiol.*, 43, 243, 1986.

99. Veronese, F. M., Schiavon, O., Bevilacqua, R., Bordin, F., and Rodighiero, G., The effect of psoralens and angelicins on proteins in the presence of UV-A irradiation, *Photochem. Photobiol.*, 34, 351, 1981.

100. Granger, M. and Hélène, C., Photoaddition of 8-methoxypsoralen to *E. coli* DNA polymerase I. Role of psorlalen photoadducts in the photosensitized alterations of Pol. I enzymatic activity, *Photochem. Photobiol.*, 38, 563, 1983.

101. Schiavon, O., Simonic, R., Ronchi, S., Bevilacqua, R., and Veronese, M. The modification of ribonuclease-A by near ultraviolet irradiation in the presence of psoralen, *Photochem. Photobiol.*, 39, 25, 1984.

102. Malinin, G. I., Garcia, J. V., Hornicek, F. J., Glew, W. B., and Nigra, T. B., Selective affinity of 8-methoxypsoralen for erythrocyte ghost proteins during UV-A irradiation, *Photobiochem. Photobiophys.*, 12, 283, 1986.

103. Pathak, M. A. and Krämer, D. M., Photosensitization of skin *in vivo* by furocoumarins (psoralens), *Biochim. Biophys. Acta*, 195, 197, 1969.

104. Granger, M., Tolume, F., and Hélène, C., Photodynamic inhibition of *Escherichia coli* DNA polymerase I, by 8-methoxypsoralen plus near ultraviolet irradiation, *Photochem. Photobiol.*, 36, 175, 1982.

105. Kornhauser, A., Wamer, W. G., Lambert, L. A., and Koch, W. H., Are activated oxygen species involved in PUVA-induced biological effects *in vivo*?, in *Psoralens: Past, Present and Future of Photoprotection and Other Biological Activities*, Fizpatrick, T. G., Forlot, P., Pathak, M. A., and Urbach, F., Eds., John Libbey, Eurotext, Montrouge, France, 1988, 251–260.

106. Salet, C., Moreno, G., and Vinzens, F., Photodynamic effects induced by furocoumarins on a membrane system. Comparison with hematoporphrin, *Photochem. Photobiol.*, 36, 291, 1982.

107. Patapenko, A. Ya. and Sukhorukov, V. L., Photooxidation reactions of psoralens, *Stud. Biophys.*, 101, 89, 1984.

108. Carraro, C. and Pathak, M. A., Studies on the nature of *in vivo* photosensitization reactions by psoralens and porphyrins, *J. Invest. Dermatol.*, 90, 267, 1988.

109. Caffieri, S., Tamborrino, G., and Dall'Acqua, F., Formation of photoadducts between unsaturated fatty acids and furocoumarins, *Med. Biol. Environ.*, 15, 11, 1987.

110. Caffieri, S., Tamborrino, G., and Dall'Acqua, Spectroscopic studies on the C_4-photocycloadducts between psoralen and unsaturated fatty acid methyl esters, *Photomed. Photobiol.*, 10, 111, 1988.

111. Kittler, L. and Löber, G., Furocoumarins: Biophysical investigation on their mode of action, *Stud. Biophys.*, 97, 61, 1983.

112. Kittler, L. and Löber, G., Photoreactions of furocoumarins with membrane constituents. Results with fatty acids and artificial bulayers, *Stud. Biophys.*, 101, 69, 1984.

113. Kittler, L., Midden, R. W., and Wang, S. Y., Interactions of furocoumarins with subunits of cell constituents. Photoreaction of fatty acids and aromatic amino acids with trimethylpsoralen (TMP) and 8-methoxypsoralen (8-MOP), *Stud. Biophys.*, 114, 139, 1986.

114. Caffieri, S., Daga, A., Vedaldi, D., and Dall'Acqua, F., Spectroscopic studies on the C_4-photocycloadducts between psoralen and unsaturated fatty acid methyl ester, *Photomed. Photobiol.*, 10, 111, 1988.

115. Specht, K. G., Kittler, L., and Midden, W. R., A new biological target of furocoumarins: Photochemical formation of covalent adducts with unsaturated fatty acids, *Photochem. Photobiol.*, 47, 537, 1988.

116. Specht, K. G., Midden, W. R., and Chedekel, M. R., Photocycloaddition of 4,5',8-trimethylpsoralen and oleic acid methyl ester: Product structures and reaction mechanisms, *J. Org. Chem.*, 54, 4125, 1989.

9

The Molecular Basis of Psoralen Photochemotherapy

Francesco Dall'Acqua
University of Padova

Daniela Vedaldi
University of Padova

9.1 Psoralen Photochemotherapy

Psoralens, also called furocoumarins, are naturally occurring or synthetic compounds, showing interesting photobiological activities such as skin photosensitization, characterized by the onset of erythema, followed by dark pigmentation.[1] The ancient Hindus, Turks, Egyptians, and Orientals exploited this property in popular medicine for the treatment of depigmented areas of the skin.[2]

Psoralen photochemotherapy is carried out by topical application or oral administration of a psoralen and subsequent irradiation of the patient with UV-A light (320 to 400 nm). A variety of diseases such as psoriasis, mycosis fungoides, lichen planus, urticaria pigmentosa, polymorphous light eruption, alopecia areata, and vitiligo are treated. Some of these diseases are characterized by hyperproliferative conditions, e.g., psoriasis, while vitiligo is manifested by lacking of pigmenting ability of the skin, due to reduced proliferation of melanocytes.[3]

Another type of psoralen photochemotherapy is carried out by photopheresis. This is a process by which peripheral blood is exposed in an extracorporeal flow system to photoactivated 8-methoxypsoralen (8-MOP) for the treatment of disorders caused by aberrant T-lymphocytes. Photopheresis is an effective therapy for cutaneous T-cell lymphoma and for other autoimmune disorders such as pemphigus vulgaris and scleroderma. Clinical trials are in progress for treatment of multiple sclerosis, organ transplant rejection, rheumatoid arthritis, and AIDS.[4]

9.2 Cell Targets and Photochemical Mechanisms

Psoralens are able to photointeract with various biomolecules compartmentalized in the different subcellular structures. In Table 9.1, the main cell targets are reported. Photochemical interactions

0-8493-8634-9/95/$0.00+$.50
© 1995 by CRC Press, Inc.

Table 9.1   Effects of Furocoumarins on Cells

Receptors	Dark and photointeractions with membrane and cytoplasmic receptors
Cell membrane	Lipid peroxidation, formation of cross-links in membrane proteins, C_4-cycloaddition to unsaturated fatty acids and lecithins
Cytoplasm	Photoreactions with proteins, inactivation of enzymes, inactivation of ribosomes, etc.
Nucleus	Photoreactions with DNA and chromatin

with various biomolecules may involve the different mechanisms presented in Scheme 9.1. Other than the classic, oxygen-dependent type I and type II mechanisms, an anoxic mechanism is involved also. This mechanism is responsible for the direct photobinding of psoralens to biomolecules such as nucleic acids and phospholipids.

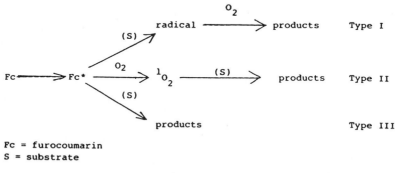

Fc = furocoumarin
S = substrate

SCHEME 9.1

9.3 Interactions with Nucleic Acids

Interactions between psoralens and DNA or nucleic acid helices take place in two steps: (1) formation of a molecular complex in the ground state, in which intercalation of the psoralen inside duplex DNA takes place; and (2) covalent photoconjugation of the complexed psoralen to pyrimidine bases of nucleic acids, forming mono- and diadducts.[5–7]

The initial step is the intercalation of the psoralen into the nucleic acid helix in the ground state, evidenced among other things by linear flow dichroism.[8] In any case, the geometry of intercalation is a factor that strongly affects the successive photobinding to the pyrimidines of the nucleic acids.[9–12] Moreover, the formation of only one particular set of configurational isomers (see later) reflects the stringent restrictions of mode of psoralen intercalation imposed by the double helix conformation.[11] Photocycloaddition to a nucleic acid helix takes place with incident light in the UV-A range, a UV region in which nucleic acids are transparent. The linear structure of psoralen when intercalated inside duplex DNA allows alignment with both its photoreactive double bonds (3,4 and 4′,5′) with two 5,6-double bonds of pyrimidines on opposite strands; when an intercalated psoralen molecule is irradiated with UV-A light source, it gives a C_4-cycloaddition at either the 4′,5′-double bond of the furan ring or the 3,4-double bond of the pyrone ring, with the 5,6-double bond of an adjacent pyrimidine, forming monoadducts.[5,6,11] A second cycloaddition to the complementary nucleic acid strand can take place, yielding an interstrand cross-link.[9,12] This can occur involving a furan-side monoadduct, which can still absorb a photon of wavelength between 300 and 380 nm and a pyrimidine adjacent to the psoralen monoadduct on the opposite strand (see the UV absorption spectra of mono- and diadducts in Figure 9.1). Thus, for cross-link formation, the psoralen intercalation has to occur in either a 5′-purine-pyrimidine-3′ site or in a 5′-pyrimidine-purine-3′ site in the helix.[11] According to Hearst,[13] the formation of monoadduct is associated with a conformational change as a kink in the helices prior to cross-link formation (see Figure 9.2). Quantum yields for the photoreactions at 365 nm between various furocoumarins (7 µg ml^{-1}) and DNA (0.1%) was determined by Rodighiero et al.[14] using a high-intensity grating Baush & Lomb

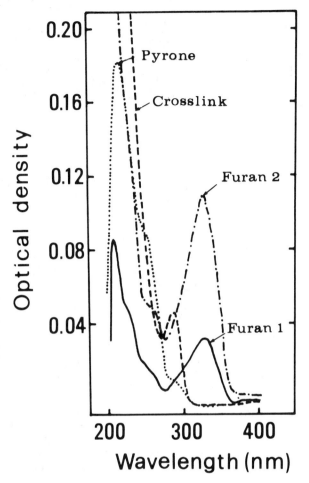

FIGURE 9.1 UV absorption spectra of mono- and diadducts between 8-MOP and thymidine. Pyrone: pyrone-side monoadduct; Furan 1 and 2: diastereomeric furan-side monoadducts. Cross-link: diadduct.

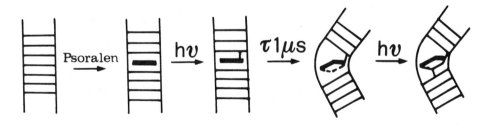

FIGURE 9.2 Conformation duplex DNA changes occurring during cross-link formation.

monochromator; values ranged between 0.032 for 4,8-dimethylpsoralen and 0.002 for angelicin. Tessman et al.[15] determined the quantum yield of initial adduct formation (26 μg ml⁻¹ of 8-MOP in the presence of 730 μg ml⁻¹ calf thymus DNA in water solution containing EDTA [1 mM] and Tris [10 mM]) at 341 nm. While the quantum yield for initial monoadduct formation was 0.008, the quantum yield for cross-link formation was 0.028.

In addition to pyrimidines, purines may also be involved in the photobinding of psoralens to DNA. An early paper[16] reports the covalent photobinding of psoralen to poly A. Cadet et al. isolated and characterized covalent adducts between deoxyadenosine and 8-MOP,[17] while Shim and Yun

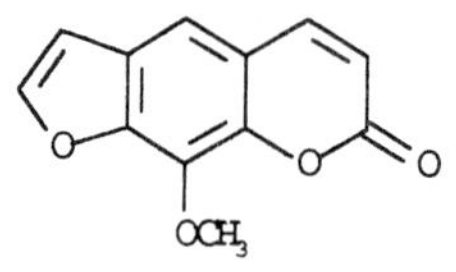

FIGURE 9.3 Molecular structure of 8-MOP (1) and of mono- and diadducts between 8-MOP and thymidine. 2 and 2': diastereomeric furan-side monoadducts; 3: pyron-side monoadduct; and 4: diadduct.

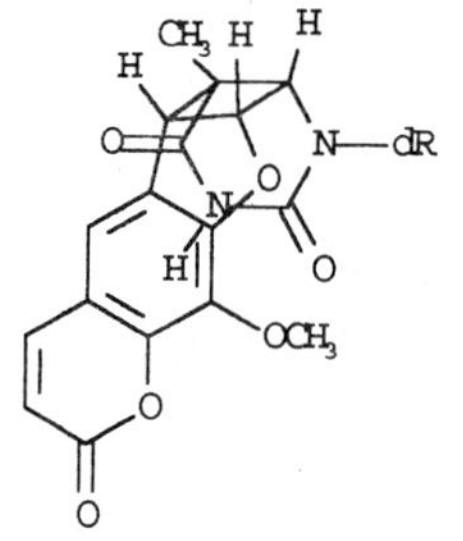

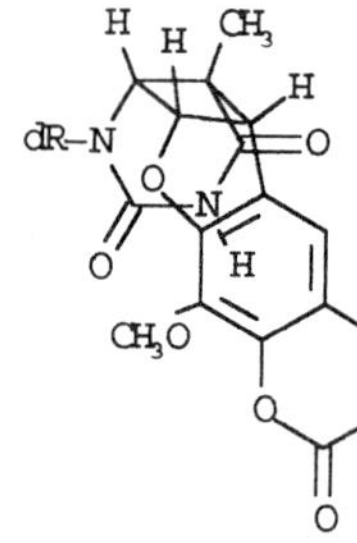

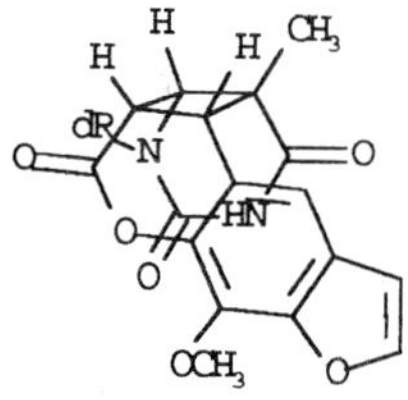

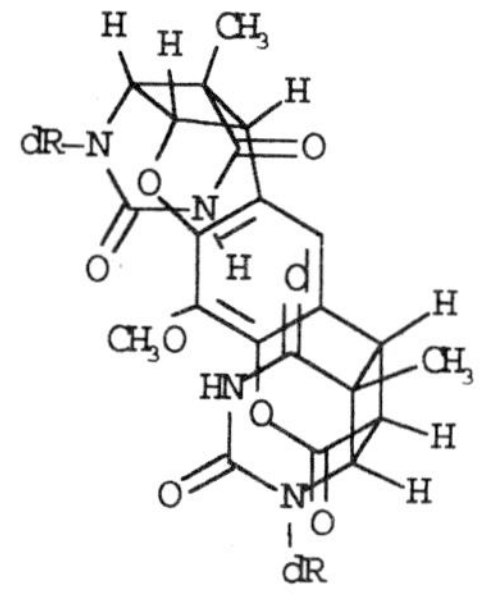

reported photoaddition of furocoumarins to adenosine, in particular, in relation to base sequence selectivity in the furocoumarin-DNA photoaddition reaction.[18]

Psoralen Mono- and Diadduct Formed with Nucleic Acid Helices

The photoadducts between various psoralens and DNA were studied after the isolation of mononucleosides or pyrimidines, by extensive enzymatic or chemical hydrolysis of photoreacted DNA.[11,19] The various products were isolated on reverse-phase HPLC or on TLC, processes that were assisted greatly by the availability of tritium-labeled derivatives.[11] For example, 8-MOP and duplex DNA irradiation, starting with a ratio 8-MOP:base pairs of 1:4 leads to a binding of 25 to 30% of 8-MOP to DNA. The yield of adducts formed is about 25% diadducts, 45% furan-side monoadducts, and 20% pyrone-side monoadducts. A small amount of furan-side deoxyuridine-8-MOP monoadducts (2%), derived from reaction with cytosine, followed by hydrolytic deamination, is also formed.[15] 8-MOP photobinds preferentially to the 5'-TpA cross-linkable site in DNA.[11] In any case, photoadduct formation is wavelength dependent. Irradiation of psoralen-DNA solutions with monochromatic wavelength as long as 400 nm can lead to monoadducts formation.[15,20] Irradiation with wavelength in the range 320 to 370 nm leads to efficient formation of cross-links.[15,20]

The structures of monoadducts and diadducts formed between 8-MOP and DNA are shown in Figure 9.3; the stereochemistry of these adducts is *cis,syn*. 8-MOP can react from either the 3'- or the 5'-face of thymine, depending on the geometry of intercalation. Reaction with either a 5'-TpX

(3'-face) or 3'-TpX (5'-face) sequence leads the formation of a pair of enantiomeric thymine monoadducts. In the case of nucleoside monoadducts, a pair of diastereoisomers are formed, owing to the same chirality of deoxyribose. The furan-side monoadducts, 8-MOP-thymidine, (Figure 9.3, structures 2 and 2') were isolated from DNA as nucleoside adducts, while only one pyrone-side diasteroisomer (structure 3) was characterized. The diadducts (structure 4) are formed as a pair of diastereoisomers with *cis,syn*-stereochemistry. A similar behavior was found with other psoralens; it has been suggested that this is general behavior for all linear furocoumarins.[11]

Sequence Specificity in Photoreactions Between Psoralens and DNA

The sequence specificity in the photoreaction of various psoralens with DNA was investigated using DNA sequencing methodology.[21,22] The 3'-5'-exonuclease activity associated with T4 DNA polymerase serves as a probe to map the psoralen photoaddition (mono- and diadducts) on DNA fragment of defined sequence. The psoralens studied included bifunctional derivatives (8-MOP, 5-MOP, and 4'-(hydroxymethyl)-4,5',8-trimethylpsoralen)[TMP] and monofunctional derivatives (angelicin, 3-carbethoxypsoralen [3-CPs], and three pyridopsoralens). Maps of photochemical binding on two DNA fragments of the lac I gene of *Escherichia coli* were determined for the various compounds. The maps evidence that thymine residues in a GC environment are cold, adjacent thymines are better targets, 5'-TpA sites are strongly preferred vs. 5'-ApT, and alternating $(AT)_n$ sequences are hot spots for photoaddition. The chemical structure of the compounds and their affinity toward DNA may lead to some minor differences in the binding spectrum. These data may contribute to the study of conformational modifications related to the repair and mutagenesis induced by the furocoumarins studied.

Photooxidative Lesions Induced in Purine Derivatives

A photodynamic effect mainly on purine derivatives by some furocoumarins was evidenced by Cadet et al.[23] For example, 3-CPs photooxidizes thymidine and 2'-deoxyguanosine. This photodynamic effect occurs by type I and type II mechanisms as well.[23] Sage et al.[24] studied the oxidative damage photoinduced by 3-CPs and other furocoumarins using DNA sequencing methodology. The psoralens 3-CPs induce photooxidation of guanine residues leading to alkali-labile sites in DNA (revealed by hot piperidine), while 8-MOP, 5-MOP, and angelicin do not. There is a preferential photooxidation of G when located on the 5'-side of CG doublets. Mechanisms operating via both radical (type I) and singlet oxygen are involved in the photooxidation of G residues by 3-CPs. Calf thymus redoxyendonuclease equivalent to endonuclease III of *Escherichia coli*, specific for oxidative DNA damages, recognizes and cleaves DNA at sites of photooxidized G residues. These lesions are apurinic-like lesions. In view of low mutagenicity of 3-CPs, photooxidation products induced by 3-CPs can be effectively repaired by the DNA repair systems. In this connection, Sage suggests that these photooxidative lesions are not markedly genotoxic.[24]

Photodamage induced by psoralen to DNA may explain the antiproliferative activity that characterizes the PUVA therapy of some hyperproliferative skin diseases such as psoriasis, as well as some side effects such as mutagenesis and risk of skin cancer.

9.4 Interactions with Proteins

Psoralens are able to bind in the dark with proteins, and this interaction seems to play a role from a pharmakokinetic point of view.[25] A covalent linkage takes place between psoralens and proteins. A direct covalent binding, involving a type III mechanism, takes place as well as an indirect binding between preirradiated psoralen and proteins;[26,27] in this last case, products of photooxidations of psoralens seem to be involved in this binding, involving both type I and II mechanisms.[26] Oxidation

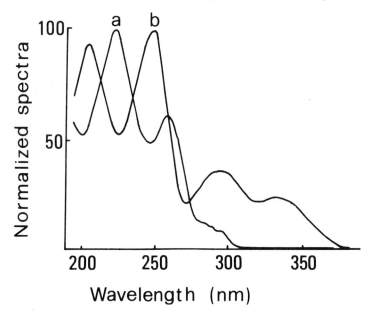

FIGURE 9.4   UV absorption spectra of one cycloadduct between trimethylpsoralen and OAME (a) and of trimethylpsoralen (b).

of aminoacids by psoralens was observed. An HPLC study on tryptic peptides of the photomodified protein indicates that the modification occurs in all oxidizable aminoacids present in the protein.[28] Inactivation of enzymes as well as induction of cross-links in oligomeric proteins were described.[26,29] Different types of photodamage can be induced by psoralen in the proteins or in aminoacids; therefore, different biological consequences can be correlated with these different damages.

9.5 Photolesions Induced to Membrane Constituents

Type I and type II oxygen-dependent, as well as type III oxygen-independent mechanisms are involved in inducing lesions to cell membrane constituents.

Oxygen-Dependent Mechanism

The lesions induced by psoralens to cell membrane constituents involve the formation of cross-links in membrane proteins[30] as well as peroxidation of cholesterol and cholesteryl esters[31] and of unsaturated fatty acids.[32] In this connection, it has been suggested that lipid peroxidation may be responsible for some photobiological effects of furocoumarins such as erythema, inflammation, and skin edema.[32] Another oxygen-dependent process involves the formation of photooxidized furocoumarin derivatives that can react in the dark with several substrates (in particular, membrane constituents), causing damage to the cell. According to Potapenko, these lesions may take into account both therapeutic as well as toxicologic properties of furocoumarins.[33]

Oxygen-Independent Mechanism

An oxygen-independent mechanism, photoinducing lesions to cell membrane constituents, concerns mainly the photoreactions that occur between furocoumarins and unsaturated fatty acids or lecithines. In particular, when an alcohol solution of a furocoumarin is irradiated in the presence of unsaturated fatty acids, a C_4-cycloaddition between the 3,4-double bond of the furocoumarin and the double bond of the fatty acid takes place, forming an adduct with a 1:1 molar ratio[34] (see Figure 9.4). In the photoreaction between TMP and oleic acid methyl ester (OAME), four isomeric

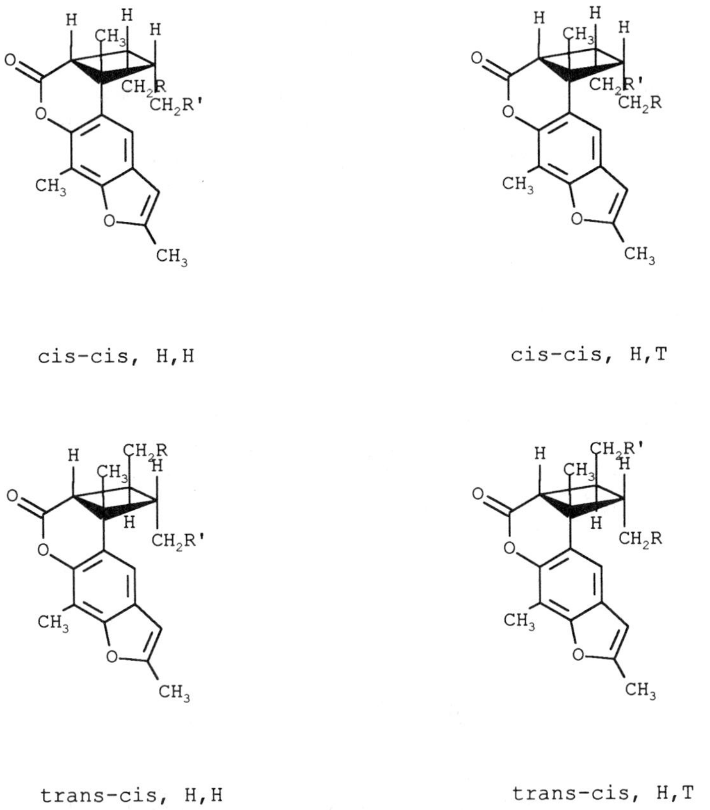

cis-cis, H,H

cis-cis, H,T

trans-cis, H,H

trans-cis, H,T

FIGURE 9.5 Molecular structures of cycloadducts between trimethylpsoralen and OAME (*cis,cis,* H,H and *cis,cis,* H,T) or EAME (*trans,cis,* H,H and *trans,cis,* H,T).

products were isolated by reverse phase HPLC, instead of the eight possible. Two of them, formed in higher yield, derive from the cycloaddition with OAME (having a *cis*-configuration), and show a *cis,cis*-configuration, while the other two derive from eadilic acid methyl ester (EAME; having a *trans*-configuration), and are formed for isomerization of OAME to EAME during the cycloaddition reaction and show a *trans,cis*-configuration.[35] The two pairs of adducts differ one from the other on the basis of the head-to-head or head-to-tail configuration referred to the portion of fatty acid chain with the terminal ester group or with the terminal methyl group (see Figure 9.5). The quantum yield of formation of adducts between TMP and OAME (4 nM TMP and 20 mM OAME in ethanol) was estimated to be 0.003.

Preliminary results show that similar cycloadducts are formed in the photoreaction between lecithin and 8-MOP.[36–39] In fact, by irradiating a mixture of psoralen or 8-MOP and phosphatidylcholine, new photoadducts are formed. They were isolated by HPLC or by TLC and characterized by spectroscopic or NMR measurements and by photosplitting experiments.[36–38] These cycloadducts derive from a cycloaddition between one of the two photoreactive sites of psoralen (3,4 or 4′,5′) and one olefinic bond of the unsaturated fatty acid linked in the 2-position of the glycerol moiety of phosphatidylcholine. The *in vivo* formation of these photoadducts is supported by studies carried out on cultured melanogenic cells and on rat skin treated with psoralens.[39–40] Psoralen-lecithin cycloadducts may be hydrolyzed by phospholipase A$_2$, releasing the unsaturated fatty acid-psoralen cycloadduct.[39] These cycloadducts are able to mimic the secondary messenger role of diacylglycerol (DAG) in terms of activation of protein kinase C (PKC).[37] Similar to DAG, which stimulates melanogenesis in human melanocytes, psoralen-unsaturated fatty acids cycloadducts enhance synthesis of melanin and increase activity of tyrosinase in cultured human melanocytes.[41] This fact may explain the stimulation of melanocyte proliferation by PUVA in human skin.

9.6 Furocoumarin Receptor

Laskin et al.[42] showed the presence of specific, saturable, high-affinity binding sites for 8-MOP on HeLa cells and on other human cell lines and five mouse cell lines. These authors demonstrated that binding of psoralens to their receptors followed by UV-A activation is associated with inhibition of epidermal growth factor (EGF) receptor binding. Inhibition of EGF binding, which requires UV-A light, was rapid and dependent on the dose of UV-A light (0.5 to 2 J cm^{-2}), as well as the concentration of psoralen (10 nM to 1 μM). A number of psoralen derivatives, including 8-MOP, 5-MOP, and 4,5′,8-trimethylpsoralen, when activated by UV-A light, were able to inhibit binding. Inhibition of EGF binding by psoralens was observed in a variety of human and mouse cell lines known to possess psoralen receptors.[43] In the epidermal-derived line PAM 212, at least two populations of receptors with different affinities for EGF were found. Psoralens and UV-A light inhibit selectively binding to higher-affinity EGF receptors. These data demonstrate that the psoralens and UV-A light have different biological effects on cell-surface membranes. Since EGF is a growth regulatory peptide, the ability of psoralens and UV-A light to inhibit EGF binding may underlie the biological effects of these agents in the skins.[44]

References

1. Musajo, L. and Rodighiero, G., The skin-photosensitizing furocoumarins, *Experientia*, 18, 153, 1962.
2. Scott, R. B. and Pathak, M. A., Molecular and genetic basis of furocoumarin reaction, *Mutat. Res.*, 29, 1177, 1979.
3. Parrish, J. A., Stern, R. S., Pathak, M. A., and Fitzpatrick, T. B., Photochemotherapy of skin diseases, in *The Science of Photomedicine*, Regan, J. D. and Parrish, J. A., Eds., Plenum Press, New York, 1982, 595.
4. Edelson, R. L., Photopheresis: A new therapeutic concept, *Yale J. Biol. Med.*, 62, 565, 1989.
5. Musajo, L. and Rodighiero, G., Mode of photosensitizing action of furocoumarins, in *Photophysiology*, Vol. VII, Giese, A. C., Ed., Academic Press, New York, 1972, 115.
6. Dall'Acqua, F., New chemical aspects in the photoreactions between psoralen and DNA, in *Research in Photobiology*, Castellani, A., Ed., Plenum Press, New York, 1977, 245.
7. Musajo, L., Bordin, F., Caporale, G., Marciani, S., and Rigatti, G., Photoreactions at 365 Å between pyrimidine bases and skin-photosensitizing furocoumarins, *Photochem. Photobiol.*, 6, 711, 1967.
8. Dall'Acqua, F., Terbojevic, M., Marciani, S., Vedaldi, D., and Recher, M., Investigation on the dark interactions between furocoumarins and DNA, *Chem.-Biol. Interactions*, 21, 103, 1978.
9. Dall'Acqua, F., Marciani, S., Ciavatta, L., and Rodighiero, G., Formation of inter-strand cross-linkings in the photoreactions between furocoumarins and DNA, *Z. Naturforsch.*, 26, 561, 1971.
10. Cole, R. S., Psoralen monoadducts and inter-strand cross-links in DNA, *Biochim. Biophys. Acta*, 254, 30, 1971.
11. Cimino, G. D., Gamper, H. D., Isaacs, H. T., and Hearst, J. E., Psoralens as photoactive probes for nucleic acids structure and function: Organic chemistry, photochemistry, and biochemistry, *Annu. Rev. Biochem.*, 54, 1151, 1985.
12. Dall'Acqua, F., Marciani, S., and Rodighiero, G., Inter-strand cross-linkages occurring in the photoreaction between psoralen and DNA, *FEBS Lett.*, 9, 121, 1970.
13. Hearst, J. E., Photochemistry of psoralens, *Chem. Res. Toxicol.*, 2, 69, 1989.
14. Rodighiero, G., Musajo, L., Dall'Acqua, F., Marciani, S., Caporale, G., and Ciavatta, L., Mechanism of skin-photosensitization by furocoumarins. Photoreactivity of various furocoumarins with native DNA and with ribosomal RNA, *Biochim. Biophys. Acta*, 217, 40, 1970.
15. Tessman, J. W., Isaacs, S. T., and Hearst, J. E., Photochemistry of the furan-side 8-methoxypsoralen-thymidine monoadduct inside DNA helix. Conversion to diadduct and to pyrone-side monoadduct, *Biochemistry*, 24, 1669, 1985.

16. Krauch, C. H., Kramer, D. M., and Waker, A., Zum wirkungsmechanismus photodynamischer furocumarine photoreaktion von psoralen-(4-^{14}C) mit DNS, RNS, homopolynucleotiden und nucleosiden, *Photochem. Photobiol.*, 6, 341, 1967.

17. Cadet, J., Voituriez, L., Nardin, R., Viari, A., and Vigny, P., A new class of psoralen photoadducts to DNA components: Isolation and characterization of 8-MOP adducts to the osidic moiety of 2'-deoxyadenosine, *J. Photochem. Photobiol., B*, 2, 321, 1988.

18. Shim, S. C. and Yun, M. H., Photochemistry of furocoumarins, *Book of Abstracts of XI Int. on Photobiol.*, Kyoto, 7–12 Sept., Japan, 1992, 168.

19. Musajo, L., Rodighiero, G., Dall'Acqua, F., Bordin, F., Marciani, S., and Bevilacqua, R., Prodotti di fotocicloaddizione a basi pirimidiniche isolati da DNA idrolizzato dopo irradiazione a 3655 Å in presenza di psoralene, *Accad. Naz. Lincei (Rome), Rend. Sc. fis. mat. e nat.*, 42, 457, 1967.

20. Chatterjee, P. K. and Cantor, R. C., Photochemical production of psoralen-DNA monoadducts capable of subsequent photocrosslinking, *Nucl. Acids Res.*, 5, 3619, 1977.

21. Sage, E. and Moustacchi, E., Sequence context effects on 8-MOP photobinding to defined DNA fragments, *Biochemistry*, 26, 3307, 1987.

22. Boyer, V., Moustacchi, E., and Sage, E., Sequence specificity in photoreactions of various psoralen derivatives with DNA: Role in biological activity, *Biochemistry*, 27, 3011, 1988.

23. Cadet, J., Decarroz, C., Voituriez, L., Gaboriau, F., and Vigny, P., Sensitized photoreactions of purine and pyrimidine 2'-deoxyribonucleosides by 8-methoxypsoralen and 3-carbethoxypsoralen, in *Oxygen Radicals in Chemistry and Biology*, Bors, W., Saran, M., and Tait, D. Eds., Waiter de Gruyter & Co., Berlin, 1984, 485.

24. Sage, E., Le Doan, T., Boyer, V., Helland, D. E., Kittler, L., Helene, C., and Moustacchi, E., Oxidative DNA damage photoinduced by 3-CPs and other furocoumarins. Mechanisms of photooxidation and recognition by repair enzymes, *J. Mol. Biol.*, 209, 297, 1989.

25. Veronese, F., M., Bevilacqua, R., Schiavon, O., and Rodighiero, G., Drug-protein interaction: Plasma protein binding by furocoumarins, *Il Farmaco, Ed. Sci.*, 34, 716, 1979.

26. Dall'Acqua, F. and Caffieri, S., Recent and selected aspects of furocoumarin photochemistry and photobiology, *Photomed. Photobiol.*, 10, 1, 1988.

27. Yoshikawa, K., Mori, N., Sakakibara, S., Mizuno, N., and Song, P.-S., Photoconjugation of 8-MOP with proteins, *Photochem. Photobiol.*, 29, 1127, 1979.

28. Schiavon, O., Simonic, R., Ronchi, S., Bevilacqua, R., and Veronese, F. M., The modification of ribonuclease-A by near ultraviolet irradiation in the presence of psoralen, *Photochem. Photobiol.*, 39, 25, 1984.

29. Schiavon, O. and Veronese, F. M., Extensive cross-linking between subunits of oligomeric proteins induced by furocoumarins plus UV-A irradiation, *Photochem. Photobiol.*, 43, 243, 1986.

30. Horniceck, F. J., Malinin, G. I., Glew, W. B., Awnet, U., Garcia, J. V., and Nigra, T. P., Photochemical cross-linking of erythrocyte ghost protein in the presence of 8-methoxypsoralen and trimethylpsoralen, *Photobiochem. Photobiophys.*, 9, 263, 1985.

31. Girotti, A. W., Photosensitized oxidation of cholesterol in biological systems: Reaction pathways, cytotoxic effects and defense mechanism, *J. Photochem. Photobiol. B*, 13, 105, 1992.

32. Midden, R. W., Chemical mechanism of the bioeffects of furocoumarins: The role of reactions with proteins, lipids and other cellular constituents, in *Psoralen DNA Photobiology*, Vol. II, Gasparro, F. P., Ed., CRC Press, Boca Raton, FL, 1988, 2.

33. Potapenko, A. Ya., Mechanism of photosensitizing effects of furocoumarins, *J. Photochem. Photobiol., B*, 9, 1, 1991.

34. Caffieri, S., Vedaldi, D., Daga, A., and Dall'Acqua, F., Photosensitizing furocoumarins: Photocycloaddition to unsaturated fatty acids, in *Psoralens in 1988, Past, Present and Future*, Fitzpatrick, T. B., Forlot, P., Pathak, M. A., and Urbach, F., Eds., John Libbey Eurotext, Montrouge, France, 1988, 137.

35. Specht, K. G., Midden, W. R., and Chedekel, M. R., Photocycloaddition of 4,5',8-trimethylpsoralen and oleic acid methyl ester: Product structures and reaction mechanism, *J. Org. Chem.*, 54, 4125, 1989.

36. Caffieri, S., Zarebska, Z., and Dall'Acqua, F., C₄-cycloaddition reactions between furocoumarins and unsaturated fatty acids or lecithins, in *Frontiers of Photobiology*, Shima, A., Ichahasci, M., Fujiwara, Y., and Takebe, H., Eds., Excerpta Medica, Amsterdam, 1993, 85.

37. Caffieri, S., Ruzzene, M., Guerra, B., Frank, S., Vedaldi, D., and Dall'Acqua, F., Psoralen-fatty acid cycloadducts activate protein kinase C (PKC) in human platelets, *J. Photochem. Photobiol., B*, 22, 253, 1994.

38. Waszkowska, E., Poznanski, J., and Zarebska, Z., Photoaddition of psoralens to phosphatidylcholine in micelles: cell membrane, a target for PUVA therapy, 12th School of Membrane Transport, School Proceedings, Zakopane, Poland, May 4-13, 1994.

39. Antony, F. A., Laboda, M. M., Dowdy, J. C., and Costlow, M. E., 8-Methoxypsoralen forms photoadducts with cellular phospholipids in cultured melanogenic cells, *Photochem. Photobiol.*, 57, 25S, 1993.

40. Caffieri, S., Schoonderwoert, S. A., Daga, A., Dall'Acqua, F., and Beijersbergen van Henegouwen, G. M. J., Photoaddition of 4,6,4'-trimethylangelicin to unsaturated fatty acids: Possible *in vivo* occurrence, *Med. Biol. Environ.*, 17, 797, 1989.

41. Frank, S., Vedaldi, D., Caffieri, S., Schothorst, A., and Dall'Acqua, F., PUVA therapy of vitiligo: psoralen-unsaturated fatty acids cycloadducts stimulate melanogenesis in human melanocytes, *Melanoma Res.*, 4, 34, 1994 (supplement).

42. Laskin, J. D., Lee, E., Yurkov, E. J., Laskin, D. L., and Gallo, M. A., A possible mechanism of psoralen phototoxicity not involving direct interaction with DNA, *Proc. Natl. Acad. Sci., U.S.A.*, 82, 6158, 1985.

43. Laskin, J. D. and Laskin, D. L., Role of psoralen receptors in cell growth regulation, in *Psoralen DNA Photobiology*, Vol. II, Gasparro, F. P., Ed., CRC Press, Boca Raton, FL, 1988, 135.

44. Laskin, J. D. and Lee, E., Psoralen binding and inhibition of epidermal grow factor binding by psoralen/ultraviolet light (PUVA) in human epidermal cells, *Biochem. Pharmacol.*, 41, 125, 1991.

10

8-Methoxypsoralen Molecular Biology

Francis P. Gasparro
Yale University

10.1 Introduction

8-Methoxypsoralen photochemotherapy is used in the treatment of several diseases. The skin is the target of the UV-activated drug when psoriasis and vitiligo are treated.[1] When cutaneous T-cell lymphoma is treated, malignant cells are exposed to UV-A radiation in an extracorporeal field after the separation of white cells from red blood cells by centrifugation.[2] The molecular events involving 8-MOP and UV-A are described in Chapter 9. However, some new aspects of psoralen photochemistry and photobiology (visible activation and photoadduct distribution) are presented in this chapter. When activated with UV-A, 8-MOP forms photoadducts with DNA that inhibit DNA synthesis.[3] In the hyperproliferative keratinocytes characteristic of psoriatic lesions, it is likely that these photoadducts inhibit cell replication and lead to an amelioration of this disease. Although the mechanism of action of photopheresis has not been fully elucidated, photoinduced damage in the target cells has been well-characterized and evidence for immune-augmenting effects has been documented.[4] When the treated cells are reinfused into the patient, there appears to be an auto-vaccination against the pathogenic T-cell clones. This conclusion is based on two kinds of obser-vations. First, several diseases not characterized by hyperproliferation (e.g., rheumatoid arthritis[5] and progressive systemic scleroderma[6]) respond to photopheresis. Second, animal models in which immune responses are assayed directly confirm an immune-modulating effect.[7,8]

Psoralens are tricyclic aromatic compounds (furocoumarins) in which the 2,3-furan bond is fused to the 6,7-bond of the double-ringed coumarin moiety. The extended aromaticity of the resulting structure is responsible for its ability to absorb UV radiation with strong absorption bands near 250 and 300 nm. These latter wavelengths do *not* correspond to the regions that photoactivate the compound most efficiently because these wavelengths are energetic enough to cause photodegradation[9] and photoreversal.[10] Thus, longer UV wavelengths (320 to 400 nm) are typically used for the activation of 8-MOP. Recently, Gasparro et al. showed that the shorter wavelength portion of the visible spectrum could also be used to activate 8-MOP.[11] Although the extinction coefficients in the visible region of the spectrum are very small (see Table 10.1), productive photochemistry can occur. When irradiated with UV-A radiation or VIS light, psoralen is photoactivated and can photobind directly to cellular macromolecules. Other reactions may also occur indirectly as oxygen-dependent events with the production of singlet oxygen.

0-8493-8634-9/95/$0.00+$.50
© 1995 by CRC Press, Inc.

Table 10.1   8-MOP Extinction Coefficients at Selected Wavelengths

Wavelength, nm	300	350	400	420	447
Extinction Coefficient, M^{-1} cm^{-1}	11,800	2016	6.8	0.60	0.22

See Reference 11.

Table 10.2   Quantum Yields for Photoadduct Formation

Reaction	Φ
4′,5′-Monoadduct formation	0.0065
4′,5′-Monoadduct conversion to cross-link	0.028
4′,5′-Monoadduct isomerization	Not calculated
Cross-link photoreversal to 3,4-monoadduct[a]	0.16 (240–266 nm); 0.3 (>280 nm)
3,4-Monoadduct photoreversal[a]	0.02 (<250 nm); 0.007 (295–365 nm)
4′,5′-Monoadduct photoreversal[a]	0.05 (<250 nm); 0.0007 (295–365 nm)

[a] For 4′-hydroxymethyltrimethylpsoralen; other data for 8-MOP.

Source: from Shi, Y.-B. and Hearst, J.A., Wavelength dependence for photoreactions of DNA-psoralen monoadducts, *Biochemistry*, 26, 3786, 1987 and references therein.

The photochemical reactions of psoralens with nucleic acids are the most well-characterized and have been studied for nearly 3 decades. Psoralen molecules form complexes with DNA hydrogen-bonded base pairs in the dark. Following irradiation with UV-A radiation or VIS light, photoadducts can form at the 4′,5′-double bond and perhaps the 3,4-double bond (see below) of the psoralen molecule. Each of these adducts has a cyclobutyl bond involving the 5,6-double bond of a pyrimidine. The extent of photoadduct formation depends on the suitability (base sequence) and accessibility (protein contacts, DNA bending) of intercalation sites between DNA base pairs. The 4′,5′-adducts (furanside) can absorb a second photon and form an interstrand cross-link between two thymines from the adjacent base pairs in a T-A site. These photoreactions have been characterized completely *in vitro* using either calf thymus DNA or various synthetic polynucleotides. Table 10.2 summarizes photochemical quantum yields for the formation of psoralen photoadducts as well as for some photoreversal reactions. Sage et al. used short segments of natural DNA to demonstrate conclusively that repetitive runs of adenine and thymine were most suitable for photoadduct formation under *in vitro* conditions.[12] The determination of these same parameters *in vivo* has been hindered by the lack of techniques with sufficient sensitivity to detect the photoadducts at the parts per million level.

Photoadduct formation is wavelength dependent. 'Monochromatic' 400 nm (bandpass 10 nm) irradiation leads to a preponderance of 4′,5′-monoadducts (>80%) while irradiation at 350 nm produces more crosslinks (~40%) and many fewer 4′,5′-monoadducts. It has been found that 3,4-monoadducts are a very low yield photoadduct. In synthetic polynucleotides, a *split-dose* irradiation protocol (400 nm, followed by 350 nm) also results in the conversion of 4′,5′-monoadducts to 3,4-monoadducts. In recent studies with lymphocytes treated in this manner, the subsequent exposure to 350-nm radiation led to the formation of 3,4-adducts as well as the expected crosslinks.[13] This is consistent with the results of Tessman et al. who demonstrated that 4′,5′-monoadducts can undergo intrahelical isomerization upon absorption of additional photons.[10]

UV-A wavelengths less than 320 nm may cause photoreversal of previously formed adducts and degradation of non-intercalated psoralen molecules. This degradation process is very efficient at 300 nm.[9] The rate of destruction is proportional to the ability of 8-MOP to absorb a particular wavelength. Studies in which different wavelengths of UV and VIS radiation were used to activate 8-MOP have shown that 3,4-monoadducts do not appear to be the result of primary photoreactions. Rather, they appear to form as a result of the absorption of photons by previously formed 4′,5′-monoadducts.[10] A survey of the literature shows no direct evidence for the formation of 3,4-monoadducts as primary photoadducts.

10.2 The Impact of Photoadducts on Cells

Experiments carried out on various cells (human lymphocytes,[14] murine keratinocytes,[15,16] and fibroblasts and bovine aorta smooth muscle cells[11]) as well as synthetic DNA have enabled the quantification and characterization of the photoadducts following the extraction and enzymatic hydrolysis of DNA. The extent of photoadduct formation is reported as the number of photoadducts per million base pairs (mbp). Increasing doses of 8-MOP and UV-A lead to increased yields of photoadducts — with a correlation efficient of 1.0.[14] It has been reported previously that the number of photoadducts formed is directly related to the product of 8-MOP concentration in ng ml^{-1} and the UV-A dose in J cm^{-2} (8-MOP concentrations over the range 10 to 100 ng ml^{-1} and UV-A doses over the range 1–10 J cm^{-2}. At 10 ng ml^{-1} and 1 J cm^{-2}, 0.4 adducts/mbp are formed, and with 100 ng ml^{-1} 8-MOP and 1 J cm^{-2}, ~4 adducts/mbp result. In lymphocytes treated with 100 ng ml^{-1} 8-MOP, 80% of the photoreaction occurs at the 4′,5′-site resulting in the formation of either 4′,5′-MA or cross-links. Using 10 ng ml^{-1} 8-MOP, a twofold increase in the yield of 3,4-MA is seen.[14] This may be a result of more photons being absorbed by 4′,5′-MA (see above) because fewer free 8-MOP molecules are intercalated between the DNA base pairs.

DNA Synthesis and Repair

The inhibition of DNA synthesis is usually attributed to the presence of DNA photoadducts. As discussed above, the degree of inhibition of DNA synthesis in human lymphocytes *in vitro* is related to the product of the 8-MOP dose (ng ml^{-1}) and UV-A (J cm^{-2}) as demonstrated by reduced levels of tritiated thymidine incorporation after mitogen stimulation.[14] Nuclear processes appear to be suppressed completely when the product of the 8-MOP and UV-A doses is greater than 50. An alternate explanation for these results is that these relatively low doses of 8-MOP and UV-A affect the ability of the cells to bind PHA and hence reduce the usual stimulatory effects of PHA on the cells. However, the results of another assay support the previous interpretation. The ability of cells to repair 8-MOP adducts follows a similar titration curve. At combined doses >100 ng ml^{-1}, no repair occurs. Cells treated with lower 8-MOP doses (10 to 20 ng ml^{-1}) are capable of removing photoadducts. The removal rate is 25% in 48 h for cells treated with 10 ng ml^{-1} of 8-MOP and 1 J cm^{-2} UV-A. The removal of these adducts is also associated with recovery of the proliferative activity, as evidenced by the increased levels of tritiated thymidine incorporation after varying repair periods. At doses greater than 300 ng ml^{-1}, the ability of the cells to exclude trypan blue decreases significantly as the damage to cell membranes reaches a critical level. Whether photodamage to cell membranes at lower concentrations could induce more specific, subtle antigenic effects has not been determined.

In metabolically active keratinocytes, inhibition of DNA synthesis or decreased viability was observed in cells containing fewer than 2 adducts per million bases (produced by a combined 8-MOP/UV-A dose of 60). At a combined dose of 30, 0.9 photoadducts were formed and no inhibition of DNA synthesis or decrease viability could be detected.[15] Photoadduct formation was unaltered by level of differentiation or source of keratinocytes (murine or human). At sublethal doses at 8-MOP and UV-A, other cell processes may also be affected. For example, interleukin I production is reduced.[16] In hyperproliferative epidermis characteristic of psoriatic skin, it is assumed that these photoadducts inhibit cell replication. Although it is fashionable to attribute this blockage to DNA cross-links, other psoralens (angular psoralens or angelicins) that cannot form cross-links also block DNA synthesis[17] and lead to therapeutic effects. Thus, 8-MOP monoadducts may also play some role in the therapeutic efficacy of PUVA. Photoadduct removal in keratinocytes has been found to be 54% at 20 h after the treatment of cells with 10 ng ml^{-1} 8-MOP and 1 J cm^{-2} UV-A but reduced to 12% at 20 h when 100 ng ml^{-1} and 1 J cm^{-2} were used. The extent of repair has also been shown to be dependent on calcium concentration, the rate being decreased in low calcium media.[15] Keratinocyte repair is therefore more efficient than lymphocyte repair *in vitro*. This may be due to the increased metabolic activity of keratinocytes.

Extensive repair studies have been carried out in bacterial systems as well as in mammalian cells.[18] Several repair pathways have been described and include excision repair (the most common), post-replication recombination repair, and photoreactivation. 8-MOP photoadducts are known to be repaired, although studies on the repair of discrete adducts has been limited. It has often been assumed that crosslinks would not be repaired. However, an excision-recombination mechanism has been proposed to account for cross-link repair in bacteria.[19] Whether such a mechanism operates in mammalian cells is not known at this time. However, it has been demonstrated clearly that mammalian cells do remove psoralen crosslinks.[20]

Apoptosis

Recently, Marks et al. suggested that the treatment of cells with 8-MOP and UV-A could lead to apoptosis, a process in which the cell actively participates in its own destruction.[21] In this process, enzymes are produced that lead to chromatin condensation, the inhibition of cell-cell interactions, and cytoskeleton disruption. Apoptosis culminates in the phagocytosis of the apoptotic cells by neighboring cells. A hallmark of this process in the production of multiple 180 base-pair units by a nuclear endoculease. Marks et al. showed that 300 ng ml^{-1} 8-MOP and 10 J cm^{-2} UV-A-induced the formation of apoptotic cells (~30%).

Apoptosis has been the focus of new interest among researchers in diverse areas of biological research.[22] The induction of apoptosis by cells treated with 8-MOP and UV-A could lead to programmed cell death of a selected group of cells that in the process of disassembling, could release their proteins, which could lead to the production of new oligopeptide fragments, which may be displayed in surface MHC molecules. These events could result in a higher level of antigenicity.

Mutagenicity

Any agent that modifies DNA has the potential of being mutagenic. UV-A-induced psoralen mutagenesis has been reported in viruses, and in prokaryotic and eukaryotic cells. In V79 Chinese hamster cells,[23] it has been shown that the exposure of "preformed" 8-MOP 4′,5′-monoadducts to additional UV-A radiation (in the absence of free 8-MOP) leads to the formation of cross-links and an increase in the number of mutants. This was interpreted as an indication that cross-link repair occurs and leads to a greater mutation rate as compared to monoadducts. This irradiation regimen for cross-link induction, however, may be flawed. In poly(dA-dT), Tessman et al. have shown that the exposure of "preformed" 4′,5′-monoadducts to additional UV-A radiation causes not only cross-link formation, but also an intrahelical isomerization to 3,4-monoadducts.[10] Therefore, it is also possible that 3,4-monoadducts could also be responsible, at least in part, for the increase in mutagenic activity after the second irradiation step. Olack et al. have shown that this photoisomerization also occurs in lymphocytes treated with the *split-dose* protocol.[13]

For the reasons cited above a study in mutagenesis has been stated in which the extent and distribution of photoadducts are always determined. To limit cross-link formation during the first step of split-dose studies, VIS radiation to activate 8-MOP has been employed.[11] Using longer wavelengths to activate 8-MOP, it is possible to reduce the extent of cross-link formation by nearly an order of magnitude (from ~20% at 400 nm to 2% at 419 nm).

The repair of photoadducts while the DNA is undergoing replication is known as *recombination repair* and may be error-prone and therefore lead to mutations. The base substitutions found are thymine to cytosine and vice versa; thymine or cytosine to adenine or guanine or vice versa. This is of clinical relevance as psoriatic patients using PUVA may require treatment over many years. It has been demonstrated now that there is an association between cumulative exposure to PUVA and increased risk of squamous cell carcinoma.[24]

Do mutagenic events occur during PUVA photochemotherapy? Albertini et al. demonstrated that lymphocytes obtained from patients receiving PUVA for either psoriasis or vitiligo had a significant increase in the number of 6-thioguanine resistant cells[25] which could be attributed to a mutation at the *hprt* locus. These studies were performed on lymphocytes obtained from patients who were receiving PUVA therapy and thus the lymphocytes were only exposed indirectly. Studies are in progress to determine the extent of 6-thioguanine resistance in directly irradiated lymphocytes.

In spite of some fears that the well-known immunosuppressive effects of 8-MOP and UV-A might lead to an enhanced susceptibility to opportunistic infections, this has **not** occurred. Recently, Beer and Zmudska[26] have shown that only *supra-physiological* doses of 8-MOP and UV-A could lead to enhanced HIV promoter activity. They have yet to determine the impact of cumulative effect of repeated incremental doses of 8-MOP and UV-A.

Antigenicity

When lymphocytes are treated with moderate doses of 8-MOP (20 to 200 ng ml^{-1}) and UV-A, the cell membrane remains intact but nuclear processes are inhibited. However, low levels of highly specific cell membrane photomodifications resulting in the formation of new antigenic moieties may have occurred.[27] During photopheresis, unfractionated populations of autologous mononuclear cells, containing an expanded clone(s) of pathogenic T-lymphocytes, can immunize against the activity of the same clone if extracorporeally altered in an appropriate fashion. This has been demonstrated in animal studies using photoactivated effector lymphocytes that were shown to suppress graft rejection temporarily.[28] It is difficult to envision how these modifications alone can lead to an enhanced immunogenic effect in untreated cells. However, they may stimulate the immune system in such a way that cross-recognition of other antigens also occurs.

Tum System

Boon et al. have shown that the potent chemical mutagen, *N*-methyl-*N,N'*-nitronitroso-guanidine or *N*-MNNG, a DNA adduct forming agent, can alter dramatically the antigenicity and subsequently the immunogenicity of a variety of tumor cell lines.[29] P1 cells are normally easily transplantable and highly tumorigenic. Mutagenization of these cells with *N*-MNNG led to tumor growth in immunosuppressed but not in normal syngeneic recipient mice. Significant protection against challenge with the original tumorigenic cell line could be induced by innoculation clones derived from mutagenized cells. Boon et al. have shown that a single point mutation, which caused one amino acid difference in an oligopeptide that is bound to MHC class I molecules, enhanced the antigenicity of the original cell line. Similar molecular events may occur during the treatment of CTCL by photopheresis where the photoactivated psoralen may cause a gene mutation(s) leading to a single amino acid change in the oligopeptides that are displayed by MHC molecules. This in turn would alter antigens on the surface of tumor T-cells and serve as a signal that primes a cytotoxic T-cell response against the tumor cells.

To determine whether 8-MOP and UV-A could enhance immunogenicity by a similar mechanism, P1 cells were treated with 8-MOP (10 ng ml^{-1}) and UV-A (1 J cm^{-2}). More than 99% of the cells were destroyed in each of three successive treatments.[30] Subsequent cloning of the residual cells led to 72 new clones of which 71 were tumorigenic in normal DBA/2 mice. One clone, which was tumorigenic in immunosuppressed mice, failed to demonstrate detectable growth in normal mice during a subsequent 6-week period. Challenge of these resistant mice with the original tumorigenic P1 parent clone resulted in the survival of 40% of the mice at 16 weeks. In a second series of experiments in which the tumor cells were treated with 10 rounds of 8-MOP and UV-A, 102 clones were obtained, 7 of which were nontumorigenic. These clones are currently being tested in cross-protection and adoptive transfer experiments. Preliminary results with 8-MOP and UV-A are consistent with Boon's findings with *N*-MNNG and suggest strongly that 8-MOP/UV-A-induced mutations enhance the recognition of both treated and untreated cells. Adoptive transfer experiments with the original tumor-negative clone support the involvement of cytotoxic cells although we have yet to characterize conclusively any CD8$^+$ cells.

10.3 Summary

Although the field of psoralen photobiology has a 50-year history of modern research, there are many unanswered questions. The biological impact of psoralen photoadducts in DNA first appeared to be straightforward lethal events that inhibited DNA synthesis. Now it appears that these adducts may lead to other more involved effects such as apoptosis and heightened antigenicity.

Acknowledgements

The author thanks Drs. Gerry Olack, Paola Gattolin, Virginia Maxwell, and Peter Glazer for helpful discussions. These studies were supported by gifts from Therakos, Inc. and the New York Cardiac Center, and a grant from the NIH (AR37629).

References

1. Parrish, J. A., Fitzpatrick, T. B., Tannenbaum, L., and Pathank, M. A., Photochemotherapy of psoriasis with oral methoxsalen and long-wave ultraviolet light, *N. Engl. J. Med.*, 291, 1207, 1974.
2. Edelson, R., Berger, C., Gasparro, F., et al., Treatment of cutaneous T-cell lymphoma by extracorporeal photochemotherapy, *N. Engl. J. Med.*, 316, 297, 1987.
3. Hearst, J., Psoralen photochemistry and nucleic acid structure, *J. Invest. Dermatol.*, 77, 39, 1981.
4. Perez, M. I., Berger, C. L., Yamane, Y., John, L., Laroche, L., and Edelson, R. L., Inhibition of antiskin allograft immunity induced by infusions with photoinactivated effector T lymphocytes (PET cells). The congenic model, *Transplantation*, 51, 1283, 1991.
5. Malawista, S. E., Trock, D. H., and Edelson, R. L., Treatment of rheumatoid arthritis by extracorporeal photochemotherapy — A pilot study, *Arthr. Rheum.* 34, 646, 1991.
6. Rook, A. H., Freundlich, B., Jegasothy, B. V., et al., Treatment of systemic sclerosis with extracorporeal photochemotherapy: Results of a multicenter trial, *Arch. Dermatol.*, 128, 337, 1992.
7. Berger, C. L., Experimental murine and primate models for dissection of immunosuppressive potential of photochemotherapy in autoimmune disease and transplantation, *Yale J. Biol. Med.*, 62, 611, 1989.
8. Rose, E. A., Barr, M. L., Xu-H., Pepino, P., Murphy, M. P., McGovern, M. A., Ratner, A. J., Watkins, J. F., Marboe, C. C., and Berger, C. L., Photochemotherapy in human heart transplant recipients at high risk for fatal rejection, *J. Heart Lung Transplant.*, 11, 746, 1992.
9. Taylor, A. and Gasparro, F. P., Extracorporeal photochemotherapy for cutaneous T-cell lymphoma and other diseases, *Sem. Hematol.*, 29, 132, 1992.
10. Tessman, J. W., Isaacs, S. I., and Hearst, J. E., Photochemistry of furan-side 8-methoxy-psoralenthymidine monoadduct inside the DNA helix. Conversion to diadduct and to pyrone-side monoadduct, *Biochemistry*, 24, 1669, 1985.
11. Gasparro, F. P., Gattolin, P., Olack, G., Sumpio, B. E., and Deckelbaum, L. I., Excitation of 8-methoxypsoralen with visible light, *Photochem. Photobiol.*, 57, 000, 1993.
12. Sage, E. and Moustacchi, E., Sequence context effects on 8-methoxypsoralen photobinding to defined DNA fragments, *Biochemistry*, 26, 3307, 1987.
13. Olack, G. A., Gattolin, P., and Gasparro, F. P., Improved high performance liquid chromatographic analysis of 8-methoxypsoralen monoadducts and crosslinks in polynucleotide, DNA and cellular systems: Analysis of split dose protocols. *Photochem. Photobiol.*, 57, 000, 1993.
14. Gasparro, F. P., Bevilacqua, P. M., Goldminz, D., et al., Repair of 8-MOP photoadducts in human lymphocytes, in *DNA Damage and Repair in Human Tissues*, Sutherland, B. M. and Woodhead, A. D., Plenum Press, New York, 1990.
15. Tokura, Y., Edelson, R. L., and Gasparro, F. P., Formation and removal of 8-MOP-DNA photoadducts in keratinocytes: Effects of calcium concentration and retinoids. *J. Invest. Dermatol.*, 96, 942, 1991.

16. Tokura, Y., Yagi, J., Edelson, R. L., et al., Inhibitory effect of 8-MOP plus UVA on IL-1 production by murine keratinocytes. *Photochem. Photobiol.*, 53, 517, 1991.

17. Rodighiero, G., Dall'Acqua, F., and Averbeck, D., New psoralen and angelicin derivatives, in *Psoralen-DNA Photobiology*, Vol. 1, Gasparro, F., Ed., CRC Press, Boca Raton, FL, 1988, 37–133.

18. Sladek, F. M., Munn, M. M., Rupp, W. D., et al., *In vitro* repair of psoralen-DNA crosslinks by RecA, uvrABC, and the 5'-exonuclease of DNA polymerase I, *J. Biol. Chem.*, 264, 6755, 1989.

19. Cortes, F., Morgan, W. F., Varcarel, E. R., Cleaver, J. E., and Wolff, S., Both cross-links and monoadducts induced in DNA by psoralen can lead to sister chromatid exchange formation, *Exptl. Cell. Res.*, 196, 127, 1991.

20. Boesen, J. J. B., Stivenberg, S., Thyssens, C. H. M., Panneman, H., Darroudi, F., Lohman, P. H. M., and Simons, J. W. I. M., Stress response induced by DNA damage leads to specific, delayed and untargeted mutations, *Mol. Gen. Genet.*, 234, 217, 1992.

21. Marks, D. I. and Fox, R. M., Mechanism of photochemotherapy-induced apoptotic cell death lymphoid cells, *Biochem. Cell Biol.*, 69, 754, 1991.

22. Raff, M. C., Social controls on cell survival and cell death, *Nature*, 356, 397, 1992.

23. Babudri, N., Pani, B., Venturini, S., et al., Mutation induction and killing of V79 Chinese hamster cells by 8-methoxypsoralen plus near-ultraviolet light: Relative effects of monoadducts and crosslinks, *Mutat. Res.*, 91, 391, 1981.

24. Stern, R. S., Lange, R., and Members of the Photochemotherapy Follow-up Study, Non-melanoma skin cancer occurring in patients treated with PUVA five to ten years after first treatment, *J. Invest. Dermatol.*, 91, 120, 1988.

25. Strauss, G. H., Albertini, R. J., Krusinskie, P. A., and Baughman, R. D., 6-Thioguanine resistant peripheral blood lymphocytes in humans following long-wave ultraviolet light (PUVA) therapy, *J. Invest. Dermatol.*, 73, 211, 1979.

26. Zmudzka, B. Z., Activation of HIV by UVB radiation and PUVA treatment *in vitro*: An evaluation of the safety of medical procedures and cosmetic applications, *Photochem. Photobiol.*, 55, 89S, 1992.

27. Krylenkov, V. A., Roshshupkin, D. I., Potapenko, A. Y., and Maygin, A. M., Changes in the expression of lymphocyte surface antigens under UV-irradiation effect of antioxidants, *Photobiochem. Photobiophys.*, 8, 19, 1984.

28. Perez, M., Edelson, R. L., Laroche, L., et al., Inhibition of antiskin allograft immunity by infusions with syngeneic photoinactivated effector lymphocytes, *J. Invest. Dermatol.*, 92, 669, 1989.

29. Boon, T., Towards a genetic analysis of tumor rejection antigens, *Adv. Cancer Res.*, 58, 177, 1992.

30. Maxwell, V. M., Malane, M. S., Tigelaar, R. E., and Gasparro, F. P., Treatment with 8-methoxypsoralen and ultraviolet A radiation enhances immune recognition in a murine mastocytoma model, *Ann. N. Y. Acad. Sci.*, in press.

11

Photosensitization in Photodynamic Therapy

Anthony Harriman
University of Texas

11.1 Introduction

Dye-sensitized photoreactions have been used for many purposes in the biomedical sciences,[1] especially in the context of photodynamic therapy (PDT).[2] In photosensitized reactions, light energy is absorbed preferentially by one reagent (the sensitizer) and, by way of multifarious secondary processes, leads to chemical modification of a second reagent (the substrate). Four fundamental mechanisms can be considered to account for the subsequent chemistry; namely, (1) energy, electron, hydrogen atom, or proton transfer reactions, (2) photothermal effects, (3) photoinduced conformational changes, and (4) *in situ* photodegradation of the sensitizer. These chemical reactions, which can have many adverse effects if uncontrolled, may be exploited to destroy selectively or inactive infected cells or macromolecular contaminants in blood supplies. To be effective, however, it is essential that the sensitizer recognizes and concentrates exclusively in the targeted receptor function. This recognition process is the single most important feature of PDT activity, without selective uptake by the target there can be no viable PDT, and the most difficult to engineer. Once the sensitizer is localized at the target site, photochemistry can be initiated provided the sensitizer absorbs in a region where the surrounding biomaterial is transparent. Assuming these daunting requisites can be met, the host biomaterial can be destroyed by a number of processes but it is unlikely that a single reaction will dominate under physiological conditions.

11.2 Mechanisms for Photosensitization

Type I Sensitization

Excitation of an organic photosensitizer results in the initial population of a short-lived excited singlet state that undergoes intersystem crossing to give the corresponding triplet excited state. Except in cases where the sensitizer and substrate are in intimate contact, any subsequent bimolecular photochemical reactions will occur via the triplet state of the dye because of its inherently longer lifetime. The triplet state, by virtue of its excitation energy, is more easily oxidized and reduced than

0-8493-8634-9/95/$0.00+$.50
© 1995 by CRC Press, Inc.

the ground-state molecule and can participate in redox reactions with appropriate biomaterials. Thus, a few common biomaterials are photoreduced readily (e.g., quinones and cytosine) in a process that gives rise to the corresponding π-radical cation of the sensitizer. The complementary reaction in which the sensitizer is reduced to the π-radical anion and the substrate (e.g., hydroquinone, cysteine, and guanine) is oxidized is more common. In either case, the reduced species can transfer an electron to molecular oxygen, forming superoxide ions, while the oxidized species usually undergoes chemical breakdown. The most important of these type I photoreactions is that in which the sensitizer undergoes one-electron reduction since, under physiological conditions, the sensitizer can be recycled via reaction with molecular oxygen. Both the oxidized substrate and resultant superoxide ions (or hydrogen peroxide formed by disproportionation) can cause further damage to the substrate.

One-electron oxidation of biomaterials is often followed by deprotonation of the resultant π-radical cation to form a free radical. If the radical center is localized on a carbon atom, this latter species may react with oxygen to form a peroxyl radical and thereby initiate a chain reaction that could result in widespread oxidative damage. The same species can be formed directly by photo-induced hydrogen atom transfer from substrate to sensitizer. Thus, excitation of Methylene Blue intercalated into DNA results in rapid hydrogen atom abstraction from guanine or adenine.[3] Because the reactants are in very close proximity, reaction occurs by way of the singlet excited state of the sensitizer and hydrogen atom abstraction from guanine occurs on a time scale of ca. 5 ps. The triplet state of intercalated Methylene Blue is unable to abstract a hydrogen atom from the nucleic acid bases because of unfavorable thermodynamics and reacts with molecular oxygen to form singlet molecular oxygen.[4]

Type II Sensitization

Most triplet states transfer excitation energy to molecular oxygen to form singlet molecular oxygen, which reacts rapidly and rather indiscriminately with many kinds of electrophilic biomaterials such as unsaturated lipids, proteins, nucleic acids, etc. Such type II reactions are favored by high local concentrations of dissolved oxygen and, provided the sensitizer is not susceptible to attack by singlet molecular oxygen, very high turnover numbers with respect to the sensitizer can be achieved. To be effective, the sensitizer must possess a relatively high quantum yield for formation of the triplet state, a triplet energy higher than that of $O_2(^1\Delta_g)$ ($E_t = 94$ kJ mol^{-1}), and an inherent triplet lifetime sufficiently long to ensure quantitative reaction with molecular oxygen (i.e., $\tau_t > 100$ µs). A great number of compounds satisfy these requirements easily, especially porphyrins and phthalocyanines; type II reactions are postulated frequently as the dominant mechanism for PDT activity. However, while triplet excited states are detected readily by laser flash photolysis techniques for sensitizers localized inside intact biological cells, there is no convincing experimental support to indicate *in situ* formation of singlet molecular oxygen.

Photothermal Effects

Photothermal effects are ubiquitious in photosensitized processes and arise from nonradiative deactivation of excited states. It is impossible to avoid photothermal effects and there is growing evidence to suggest that this is a very important mechanism for light-induced cytotoxicity. For example, a dye-stained cell illuminated with light will heat up due to nonradiative deactivation of the excited state at a faster rate than an unstained cell. Indeed, it has been estimated[5] that illumination of a cell stained with Merocyanine 540 could increase the internal temperature by about 12°C min^{-1} provided the cell membrane functioned as an adiabatic sink. To be effective, the sensitizer should possess a very short-lived excited singlet state that decays exclusively to the ground state, thereby dissipating all of the excitation energy as heat. It is also important that high-intensity illumination is employed, as might be delivered with a laser, so that the rate of heat release into the system far exceeds the rate of heat migration throughout the entire system.

The possibility that photothermalization can play a major role in PDT has been largely ignored by researchers in the field, probably because it is a "trivial" effect. It is likely, however, that this is the single most important mechanism for photosensitized cell killing.[6] The photothermal effect does not require the involvement of oxygen and extremely high turnover numbers can be achieved for the photosensitizer. The requisites for a good photosensitizer are essentially the opposite of those needed for an effective type I or II photosensitizer, although it remains essential that the dye concentrates preferentially near to the target.

Photoinduced Conformational Changes

The search for sensitizers that absorb in the far red region where biological tissue is transparent has resulted in extensive usage of dyes with linear polymethine chains.[7] Such compounds are prone to isomerize upon promotion to the first excited singlet state. Isomerization can involve large-scale torsional motion in which the terminal subunits are displaced from the ground-state equilibrium position.[8] Consequently, the rate of isomerization depends on frictional forces with adjacent solvent molecules and can be restricted by incorporating the sensitizer in a viscous membrane. Because of the need to achieve some level of solubility in water, however, most sensitizers proposed for PDT are hydrophilic and localize in membranes near the aqueous phase.[9] Photoisomerization of such dyes can cause structural distortions to the lipid membrane and allow reagents to leave or enter the membrane.[10] This, in turn, can induce cytotoxicity. A mechanism of this type does not require the presence of dissolved oxygen or a high triplet yield and it readily facilitates very high turnover numbers with respect to the sensitizer. The only requirement is that the sensitizer possesses an isomerizable bond around which free rotation can occur under illumination.

IN SITU Release of Cytotoxic Reagents

Type I and II photosensitization processes result in formation of highly reactive species in close proximity to the sensitizing molecule. In certain cases, therefore, it is reasonable to suppose that the sensitizer will be attacked by the primary products.[5] This approach can be used to generate highly cytotoxic products by chemical breakdown of the sensitizer in a suicidal fashion. The approach becomes highly practical if the sensitizer exhibits selective uptake by tumors or infected cells compared to the breakdown products.

11.3 Merocyanine 540 As an Illustrative Example

Merocyanine 540 (MC540) has been reported to concentrate within certain types of leukemia cells and to destroy the cells under illumination with visible light.[11] Numerous publications have claimed that light-induced cytotoxicity occurs by way of a type II mechanism in which the triplet excited state of the dye reacts with dissolved oxygen to form singlet molecular oxygen. Other researchers, however, have shown that in homogeneous and microheterogeneous media, the triplet quantum yield for MC540 is extremely low. The dye undergoes efficient photoisomerization from the first excited singlet state[5,8,10] and rapid internal conversion serves to deposit a significant amount of heat into the surrounding lipid. The dye is bleached during illumination; and Gulliya and co-workers[12] have established that some of the resultant photoproducts are highly cytotoxic. It has also been established that MC540 is easily reduced and oxidized and that the ground-state dye is readily attacked by free radicals.[13] In water or when bound to a lipid membrane, MC540 undergoes extensive aggregation and the photochemistry of the various dimers, aggregates, solvates, and adsorbates can differ significantly. The dye binds efficiently to proteins and serum albumins and, again, binding alters the photophysical properties.[14]

Despite such experimental difficulties, it is possible that the important mechanisms for photosensitized cytotoxicity with MC540 will be uncovered. If this feat is achieved, it will be done by careful and systematic investigation rather than by ignoring experimental evidence published by other research groups, ignoring contradictory findings, and claiming again and again that only singlet oxygen production is important. For the rational design of improved photosensitizers, it is crucial that the mechanisms are established; it is meaningless to continue to propose compounds as potential photosensitizers merely because they generate singlet molecular oxygen under illumination in O_2-saturated solution.

11.4 Conclusion

Identifying the overall reaction mechanism for photosensitized cytotoxicity is a very difficult task and it is doubtful if this has been achieved in any given case. The reaction pathways will depend on the properties of the sensitizer, the nature of the biomaterial, and the reaction conditions; there is no need to suppose that one particular mechanism will dominate under any given set of conditions. This is especially true for type I and II sensitized reactions that, more often than not, will operate simultaneously. Sensitizers based on porphyrin or phthalocyanine macrocycles are most likely to react by type I and II mechanisms, but other dyes, like Merocyanine 540, that possess low triplet yields and easily isomerizable structures may well operate by a variety of mechanisms. The most important aspect of sensitized PDT, however, concerns the selectivity with which the sensitizer assimilates into the target region because, without highly selective uptake of dye, the therapeutic process will be unviable. A highly efficient photosensitizer that distributes throughout the system without selectivity has no therapeutic capacity and is a highly dangerous species.

Acknowledgments

This work was supported by the National Institutes of Health (CA53619). The CFKR is supported jointly by the Biotechnology Research Resources Division of the NIH (RR00886) and by The University of Texas at Austin.

References

1. Spikes, J. D., Applications of dye-sensitized photoreactions in neurobiology, *Photochem. Photobiol.*, 54, 1079, 1991.
2. Gomer, C. J., Preclinical examination of first and second generation photosensitizers used in photodynamic therapy, *Photochem. Photobiol.*, 54, 1093, 1991.
3. Atherton, S. J. and Harriman, A., Photochemistry of intercalated methylene blue: Photoinduced hydrogen atom abstraction from guanine and adenine, *J. Am. Chem. Soc.*, in press.
4. Kelly, J. M., van der Putten, W. J. M., and McConnell, D. J., Laser flash spectroscopy of Methylene Blue with nucleic acids, *Photochem. Photobiol.*, 49, 167, 1987.
5. Davila, J., Harriman, A., and Gulliya, K. S., Photochemistry of merocyanine 540: The mechanism of chemotherapeutic activity with cyanine dyes, *Photochem. Photobiol.*, 54, 1, 1991.
6. Jori, G. and Spikes, J. D., Photothermal sensitizers: Possible use in tumor therapy, *J. Photochem. Photobiol.*, B6, 93, 1990.
7. Harriman, A., Luengo, G., and Gulliya, K. S., *In vitro* photodynamic activity of kryptocyanine, *Photochem. Photobiol.*, 52, 735, 1990.
8. Harriman, A., (Photo)isomerization dynamics of merocyanine dyes in solution, *J. Photochem. Photobiol.*, A65, 79, 1992.
9. Dragsten, P. R. and Webb, W. W., Mechanism of the membrane potential sensitivity of the fluorescent membrane probe merocyanine 540, *Biochemistry*, 17, 5228, 1978.
10. Davila, J., Gulliya, K. S., and Harriman, A., Inactivation of tumours and viruses via efficient photoisomerisation, *J. Chem. Soc., Chem. Commun.*, 1215, 1989.
11. Sieber, F., Merocyanine 540, *Photochem. Photobiol.*, 46, 1035, 1987.

12. Gulliya, K. S., Pervaiz, S., Dowben, R. M., and Mathews, J. L., Tumor cell specific dark cytotoxicity of light-exposed merocyanine 540: Implications for systemic therapy without light, *Photochem. Photobiol.*, 52, 831, 1990.
13. Harriman, A., Shoute, L. C. T., and Neta, P., Radiation chemistry of cyanine dyes: Oxidation and reduction of merocyanine 540, *J. Phys. Chem.*, 95, 2415, 1991.
14. Gulliya, K. S., Davila, J., and Harriman, A., The mechanism of LDL-mediated increased uptake of merocyanine 540 by HL-60 cells, *Cancer J.*, 3, 360, 1990.

12

Photodynamic Therapy: Basic and Preclinical Aspects

Giulio Jori
University of Padova

12.1 General Features[1-5]

Photodynamic therapy (PDT) is based on the property of some photosensitizing dyes of being accumulated in significant amounts and of being retained for prolonged periods of time by a variety of solid tumors. At time intervals corresponding to ~24 to 48 h after administration, the concentration of the dye in the tumor is often larger than that found in peritumoral tissues; hence, illumination of the neoplastic lesion with light specifically absorbed by the photosensitizer causes a selective, or at least preferential, damage of the tumor. The main steps involved in PDT of tumors are summarized in Scheme 12.1. The tumor response to the phototreatment is generally massive within 12 to 18 h and results in eschar formation, gradual loss of tumor mass, and eventual reepithelization.

Several factors appear to contribute to the large affinity of photosensitizers for tumor tissues, including a leaky vasculature, lower pH of some tumor compartments, specific interaction of porphyrin carriers (such as lipoproteins) with membrane receptors in malignant cells, and inefficient lymphatic drainage. It is likely that the various factors have different weight in the control of dye uptake and release from tumors with different histological features and biochemical or physiological properties. In any case, present evidence suggests that the maximal tumor-localizing efficiency is typical of dyes having a polycyclic chemical structure and few or no polar functional groups, and hence having a relatively large level of hydrophobicity (e.g., expressed by a *n*-octanol/water partition coefficient larger than 10).

For *in vivo* applications of PDT, preference is usually given to those photosensitizers (porphyrins, chlorins, and their analogues) that exhibit absorption bands in the 600 to 850-nm range, namely in a spectral region where the penetration of light into mammalian tissues is deepest (up to 2 to 3 cm under favorable circumstances). Photoactivation of dyes in tissues is thus obtained by argon-pumped dye lasers or, for irradiations above 680 nm, Ti-sapphire lasers, although adequately filtered noncoherent light sources can be also used for nonendoscopic applications. In the case of

0-8493-8634-9/95/$0.00+$.50
© 1995 by CRC Press, Inc.

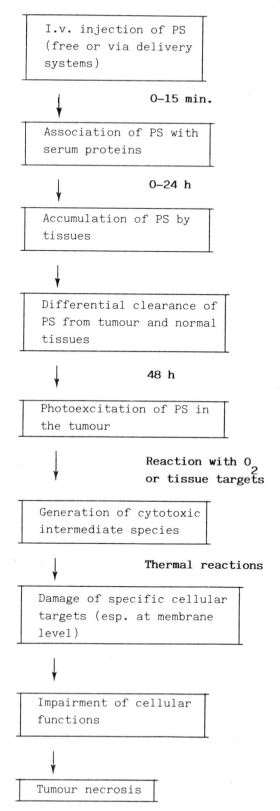

SCHEME 12.1   Main steps involved in the photodynamic therapy of tumors.

Table 12.1 Selected Examples of Tumor-Localizing and -Photosensitizing Agents

Photosensitizer	Absorption max. (nm) in the red	Extinction coefficient $(M^{-1}cm^{-1})$ at max.	Quantum yield for $^1O_2$ generation
Hematoporphyrin	630	3,500	0.51
Photofrin II[a]	630	3,000	0.18
Benzoporphyrin[b] derivative	685	118,000	0.60
meso-Tetra(3-hydroxyphenyl)porphine[b]	642	3,300	0.57
Bacteriochlorin a	785	150,000	0.32
Octaethylpurpurin	690	70,000	0.67
Zn-phthalocyanine[b]	675	150,000	0.48
Zn-phthalocyanine disulfonate	680	120,000	0.43
Si-phthalocyanine	782	430,000	0.19

[a] Photosensitizer presently used in clinical PDT.

[b] Photosensitizer in procinct to be proposed as a second-generation PDT agent.

porphyrins, the action spectrum of PDT is essentially coincident with their lowest energy absorption band peaking at 630 nm.

12.2 The Photosensitizer[6–8]

Until now, almost all clinical applications of PDT have utilized a chemical derivative of hematoporphyrin, HpD, as the photosensitizing agent. HpD is a complex mixture of hematoporphyrin, its mono- and didehydrated products, mono- and di-O-acetylated derivatives, and some covalent oligomers (dimer to hexamer) linked by intermolecular ether, ester, or C–C bonds. Since the oligomeric components show the highest affinity for tumor tissues, a partially purified version of HpD enriched in this material (and known under the commercial name of Photofrin II) was prepared and is now in phase II/III clinical trials in several centers.

The large chemical heterogeneity of Photofrin II and its poor light-absorption properties in the red spectral region (see Table 12.1) stimulated the search for second-generation tumor localizers, which are endowed with a high degree of chemical purity, improved optical properties, and photosensitizing activity at least comparable with that typical of Photofrin II. The latter parameter can be evaluated from the quantum yield of photosensitized generation of the highly cytotoxic oxygen derivative $^1O_2$, which is likely to play a major role in the PDT of tumors. A shift of the dye absorbance toward 700 to 800 nm, where also pigmented tissues display an appreciable transparency to light, and a simultaneous enhancement of the molar absorptivity have been obtained by a variety of approaches, including the addition of phenyl chromophores at the *meso*-carbon atoms of the porphyrin macrocycle, the extension of the 18π-conjugated electron cloud of porphyrin to a 22π- or 26π-electron system by insertion of additional double bonds, or the condensation of the pyrrole moieties with additional aromatic rings. Selected examples are listed in Table 12.1.

12.3 Preclinical Photodynamic Therapy[9–14]

The Delivery System

The selectivity of tumor targeting is often enhanced if the photosensitizer is associated with suitable delivery systems prior to injection. The delivery systems are designed to promote some kind of specific interaction with peculiar features of neoplastic tissues. Promising results have been reported in cell/animal studies by (1) coating polystyrene microspheres with the photosensitizer; (2) covalent attachment of the photosensitizer to monoclonal antibodies directed against antigens present at the surface of tumor cells; and (3) incorporation of the photosensitizer into liposomal vesicles, oil emulsions, or inclusion complexes that favor its transfer to low-density lipoproteins (LDL): these proteins undergo an efficient endocytosis by several types of malignant cells in a receptor-mediated process.

The Transport and Tumor Release of the Photosensitizer

Upon systemic injection, porphyrins and their analogues are readily bound by various serum proteins including albumin, globulins, and lipoproteins. Exceptions are represented by highly water-soluble porphyrins, such as uroporphyrin, which exist largely in the unbound state; however, such porphyrins are not used in PDT due to their low affinity for tumors. The different photosensitizer-protein complexes are characterized by different serum half-life and *in vivo* fate. Thus, albumin-carried dyes disappear from serum with a half-life of ca. 4 h and are mainly released in the vascular stroma, while lipoprotein-bound dyes show a slower rate of serum clearance and largely localize at endocellular loci, such as the plasma, mitochondrial, and lysosomal membranes. No detectable binding of porphyrin photosensitizers to nuclear structures has been observed; this suggests that PDT has a very low mutagenic potential.

Tumor Photosensitization

PDT of tumors in the presence of porphyrins and their analogs requires typically the presence of molecular oxygen. Moreover, several results suggest that the onset of photosensitized tumor damage involves the intermediacy of singlet oxygen generated by energy transfer from triplet dye to ground-state oxygen. Since the latter species is characterized by a very short lifetime in a biological milieu, the subcellular targets involved in the primary stages of the photoprocess are dependent on the distribution of the photosensitizer in the various districts of the tumor tissue. Thus, Photofrin II (as well as other dyes delivered by albumin) induce tumor necrosis mainly via vascular damage: the early injury to endothelial cells and circulating platelets and erythrocytes causes drastic alterations of the vascular wall, leading to hemostasis and thrombosis; the increased permeability of endothelial cells leads to escape of serum proteins and fluid from blood and the ready appearance of edema around the injured vessels. The consequent formation of hypoxic or even anoxic areas can induce an indirect necrosis of neoplastic cells; the tissular damage is enhanced by the release of vasoactive or tissue-lysing agents such as histamine and acid phosphatases from mast cells and neutrophils in the stroma. On the other hand, more hydrophobic dyes, which are transported by lipoproteins, preferentially cause a direct photodamage of malignant cells; early targets of the photoprocess are the mitochondrial and lysosomal membranes, with extensive vesiculation and vacuolization at the cytoplasmic level. Blood capillaries undergo a detectable alteration only at several hours after the end of PDT, thus ensuring a steady supply of oxygen to the irradiated tumor tissue and potentiating the effect of irradiation.

References

1. Neckers, D. C., Ed., *New Directions in Photodynamic Therapy*, SPIE Proc., Washington, D.C., 1988.
2. Gomer, C. J., Photodynamic therapy in the treatment of malignancies, *Semin. Hematol.*, 26, 27, 1989.
3. Dougherty, T. J., Photodynamic therapy: Status and potential, *Oncology*, 3, 67, 1989.
4. Kessel, D., Ed., *Photodynamic Therapy of Neoplastic Disease*, CRC Press, Boca Raton, FL, 1990.
5. Henderson, B. W. and Dougherty, T. J., Eds., *Photodynamic Therapy*, Marcel Dekker, New York, 1992.
6. Bock, G. and Harnett, S., Eds., *Photosensitizing Compounds: Their Chemistry, Biology and Clinical Use*, CIBA Foundation Symp. 146, John Wiley & Sons, Chichester, 1989.

7. Dougherty, T. J., Photosenzitizers: Therapy and detection of malignant tumours, *Photochem. Photobiol.*, 45, 879, 1987.

8. Jori, G., Photosensitized processes *in vivo:* Proposed phototherapeutic applications, *Photochem. Photobiol.*, 52, 439, 1990.

9. Jori, G., Photodynamic therapy of solid tumours, *Radiat. Phys. Chem.*, 30, 375, 1987.

10. Kessel, D., Sites of photosensitization by derivatives of hematoporphyrin, *Photochem. Photobiol.*, 44, 489, 1986.

11. Zhou, C., Mechanisms of tumour necrosis induced by photodynamic therapy, *J. Photochem. Photobiol., B: Biol.*, 3, 299, 1989.

12. Moan, J., Porphyrin-sensitized photodynamic inactivation of cells, *Lasers Med. Sci.*, 1, 5, 1986.

13. Kongshaug, M., Distribution of tetrapyrrole photosensitizers among human plasma proteins, *Int. J. Biochem.*, 2', 1239, 1992.

14. Krammer, B., Hubmer, A., and Hermann, A., Photodynamic effects on the nuclear envelope of human skin fibroblasts, *J. Photochem. Photobiol., B: Biol.*, 17, 109, 1993.

13

Clinical Applications of Photodynamic Therapy

Thomas J. Dougherty
Roswell Park Cancer Institute

Benzoporphyrin Derivative Mono-Acid (BPD) • Chlorin e_6 Aspartic Acid (NPe6) • *m*-Tetrahedroxyphenyl Chlorin (THPC)

13.1 Introduction

Photodynamic therapy (PDT) pertains to light activation of tumor-localized photosensitizers for local destruction of solid cancers. PDT has been reviewed several times previously.[1–3] Its biology and mechanistic aspects are discussed elsewhere in this volume. Briefly, a photosensitizer (generally Photofrin,* a mixture of ether linked nonmetallo porphyrin oligomers)[1] is injected systematically into a patient. Time is allowed for serum clearance and the tumor is then exposed to visible, tissue-penetrating light (630 nm in the case of Photofrin) for *in situ* activation. The putative cytotoxic agent thus generated by energy transfer is singlet oxygen ($^1O_2^*$), a short-lived (microseconds) activated form of molecular oxygen. Interaction of singlet oxygen with cellular components is thought to elicit a host of subsequent reactions involving various free radicals, peroxides, and possibly superoxide. The net result is that profound vascular and cellular effects ensue, causing hemorrhagic necrosis of the tumor tissue. Because Photofrin uptake is not limited to malignant tissue, the proper choice of drug and light dose are important in order to achieve selective destruction of the tumor. The pharmacokinetics of Photofrin have been worked out in detail in mice,[4] and to a lesser extent in man.[5] The photosensitizer is retained for 1 to 2 months in cutaneous tissue, necessitating the avoidance of bright lights (mainly sunlight) for this period. No other systemic effects of the drug have been found. Nonetheless, this avoidable but annoying problem has led to the search for new photosensitizers, some of which have entered Phase I clinical trials (see below). Also, since 630-nm light with Photofrin causes biologic effects only to a depth of 5 to 10 mm, photosensitizers with red absorbances in the 700- to 800-nm range, where theoretically a doubling of depth of biologic destruction may be obtained, are being investigated. It should be noted, however, that interstitial light delivery with one or more fibers also allows for greater volume of tumor destruction, even with Photofrin. However, the main clinical application of PDT is likely to be for treatment of early-stage relatively superficial cancers (see below).

---

* American Cyanamid/Lederle Laboratories, Pearl River, NY

0-8493-8634-9/95/$0.00+$.50
© 1995 by CRC Press, Inc.

13.2 Methodology

In principle, carrying out a PDT treatment is straightforward. The patient is injected intravenously with the photosensitizer and then, after a given time has elapsed, the tumor is illuminated with the appropriate amount and wavelength of light. For treatment of cutaneous lesions, PDT treatment is, in fact, just this easy. For treatment of internal cancers, the light is delivered via fiber optics with special distal ends to provide proper light distribution. In essentially all cases, the light source is a laser, not because laser light is necessary (in fact, the first 35 patients treated at this institute were treated with a lamp with filtration to emit red light), but because of the ability to couple efficiently the small coherent beam into various fiber optics (200 to 400 μM in diameter, generally) for internal delivery. In fact, such fibers make external delivery much more convenient as well. Since the light delivery is critical to PDT and since the dosimetry is different for different organs, each is considered separately below.

Lasers

Gas or solid-state lasers pumping dye lasers are most commonly used. In the past, most investigators used scientific argon-pumped dye lasers adapted to clinical use. These sometimes proved to be unreliable and difficult to use. However, as part of the Phase III clinical trials, a reliable system from Coherent, Lambda Plus (Santa Clara, CA) has been designed and is now available. This laser can be used for essentially all types of PDT treatments, with the exception of some of the more experimental procedures where very large areas are treated requiring higher output, e.g., in the peritoneal or thoracic cavity. A second medical system designed specifically for PDT is available from Laserscope, Santa Clara, CA; it is based on a solid-state frequency doubled Nd-YAG laser (532 nm) pumping a remote dye laser. This system (unlike the Lambda Plus that produces light in a continuous mode) delivers the light in a pulsed mode, but the pulse rate is such that it is quasi-continuous and has been shown to be biologically equivalent to the continuous systems.[6]

Fiber Delivery Systems

Fibers for PDT are available in three basic types from PDT Systems Inc., Santa Barbara, CA, or QLT Phototherapeutics, Vancouver, BC. All are based on 400-μM single quartz fibers with coupling efficiency to the lasers of 80 to 90%. The following are available as Cidex-sterilized fibers.

Micro Lens Fibers (most cutaneous cancers, certain internal cancers)

These fibers provide a homogenous, circular spot the size of which is variable with distance from the tumor. Areas as small as a few millimeters (milliwatts of power) or as large as several centimeters (watts of power) can be treated. When multiple lesions are to be treated, three or four fibers can be used simultaneously by use of beam splitter devices (e.g., from PDT Systems, Inc.). For internal use (e.g., small endobronchial lesions), these fibers are small enough to go through endoscopes.

Sphere Diffuser Fibers (whole bladder treatment, intraoperative procedures, and certain other specialized applications)

These fibers are fitted with small bulbs providing isotropic light distribution and can be fitted through cystoscopes as well as fiber optic endoscopes.

Cylinder Diffuser Fibers (lesions involving a lumen, e.g., bronchus, esophagus, colon, or for interstitial use)

These fibers are fitted with a diffusing tip available in lengths from 0.5 to 3 cm and providing an even output over the length of the diffuser. They have small pointed tips to allow for easy insertion into solid masses when desirable.

13.3 PDT in Clinical Use

The biggest challenge for photodynamic therapy (PDT) is not to determine whether it works or not (that issue appears to have been settled some time ago), or even obtaining health agency approval, but to find its proper place in cancer treatment. Results of both Phase II and Phase III trials indicate that PDT is capable of long-term control of many early-stage cancers and may be used for palliative purposes on advanced cancers. However, it likely will be the former where PDT is most widely accepted, since with PDT certain cancers can be treated relatively noninvasively and/or extensive surgery may be avoided or at least delayed. Examples of such applications follow.

PDT for Treatment of Early-Stage Cancers

CIS of the Bladder

Noninvasive carcinoma *in situ* (CIS) of the urinary bladder can present a challenge to the urologist because of its poor prognosis (often leading to radical cystectomy with its attendant social and physical consequences), particularly when it recurs after the best intravesical chemotherapy. Bacillus Calmette-Guerin (BCG) is an effective agent in control of CIS, with most patients experiencing complete control for several years. However, in a substantial number of cases, this disease will recur (no doubt the causative agent may continue), and BCG is less effective in such situations. In a preliminary analysis of patients with refractory CIS (i.e., having failed intravesical chemotherapy twice, one of which was BCG), Dugan et al.[7] have reported that 50% of patients demonstrated a complete response to PDT after 1 year, thus saving them, at least temporarily, from radical surgery. All patients experienced moderate to severe urinary symptoms (frequency, urgency, spasms, transient or, in some instances, irreversible bladder shrinkage) after PDT. The transient symptoms generally persist for 3 to 5 weeks.

PDT also has been used to prevent recurrence of transitional cell carcinoma of the bladder by prophylactically treating the bladder following tumor resection (Table 13.1).[7] However, since PDT can lead to irreversible bladder shrinkage in rare cases, it remains to be seen if this is an appropriate application, except for patients in whom radical cystectomy has become the only other option.

Early-Stage Lung Cancer

Hayata and colleagues[8] at the Tokyo Medical College reported recently on 96 patients with 109 early-stage lung cancers (83 superficial, 26 nodular) demonstrating excellent long-term control in the vast majority of cases. Superficial lesions up to 0.5 cm in maximum diameter yielded a complete response rate of 89.4%, and nodular lesions of similar size produced 92.3% complete responses. Thus, 19/29 patients in the former group who experienced a complete response are alive, 4 of whom are now 5 years post-PDT. As noted by these investigators, patients with synchronous or metachronous lesions especially may benefit from PDT since resection may not be possible in these patients.

Similarly, Cortese and co-workers at the Mayo Clinic have shown a complete response in 13/14 early-stage lung cancers treated only by PDT with follow-up to several years.[9] This group is currently carrying out a comparative trial in such patients of PDT vs. surgery to determine if preserving lung function (as in PDT) offers an advantage in a group of patients at high risk for additional lung cancers.

Superficial Head and Neck Cancers

Gluckman has evaluated results of PDT both for advanced stage and early stage oral and other head and neck cancers and concluded that PDT is best applied to superficial oral cancers and condemned mucosa.[10] Also, Biel,[11] Monnier,[12] and Feyh[13] have demonstrated excellent response to PDT of CIS and T_1 lesions of the larynx and in papillomatosis, a benign but extremely aggressive and difficult-to-control disease generally affecting children (Table 13.1).

Table 13.1   Photodynamic Therapy for Treatment of Early-Stage Cancers

	CR	PR	NR	Follow-up	Recurr.	Investigators	Ref.
	Response						

	CR	PR	NR	Follow-up	Recurr.	Investigators	Ref.
Bladder cancer							
Stage							
Whole bladder/CIS	8/10	1/10	1/10	to 1 yr	—	Nseyo et al.	29,30
	7/12	2/12	—	to 2 yr	—	Shumaker et al.	31
	27/27	8	8	to 2.5 yr	10	Benson et al.	32
Whole bladder/Ta-T1, <1 cm	22/33	6/33	5/33	—	—	Hisazumi et al.	33
Whole bladder/Ta-T1, >1 cm	4/12	3/12	5/12	—	—		33
Whole bladder/prophylaxis - Ta, T1	12/12	—	—	1 yr	>350 d vs. 93 d for matched controls	Dugan et al.	7
Whole bladder/CIS (refractory)	5/10	5/10	—	1 yr	5/10		7
Lung Cancer							
Site/Stage							
MSB/superficial; 0.5–1.0 cm	34/38	—	—	to 5 yr	3	Hayata et al.	8
MSB/superficial; 1–2 cm	12/14	—	—	to 5 yr	2		8
MSB/superficial; >2 cm	4/14	—	—	to 5 yr	—		8
MSB/nodular; 0.5 cm	12/13	—	—	—	—		8
MSB/nodular; 1–2 cm	3/8	—	—	—	—		8
MSB/nodular; >2 cm	1/4	—	—	—	—		8
MSB/superficial	13/14	1/14	—	to 2 yr	3	Cortese et al.	9
MSB/microinvasive	13/15	—	—	to 3 yr	2	Monnier et al.	12
Head and Neck							
Vocal cord	27/34	—	7	12–48 mo	—	deCorbiere et al.,	34
						Gluckman	10
Oral cavity and oropharynx (early)	11/13	2/13	—	8–40 mo	—		10
Oral/condemned mucosa	7/8	1/8	—	12–48 mo	—		10
Various/early	20/26	5/26	1/26	5–51 mo	—	Wenig et al.	35
Esophagus							
Superficial	12/13	1/13	—	21–32 mo	—	Tian et al.	36
	9/11	2/11	—	8–18 mo	2	Fujinaki et al.	37
	8/21	6/21	7/21	5–22 mo		Calzavara et al.	38
	15/15	—	—	9–60 mo		Monnier et al.	39
Early stage							
Dysplasia	3/3	—	—	6–84 mo	—	Spinelli et al.	14
Type IIa[a]	1/2	1/2	—	6–84 mo	—		14
Type IIb	8/8	—	—	6–84 mo	—		14
Type IIc	2/3	1/3	—	6–84 mo	—		14
Type III	0	1/1	—	6–84 mo	—		14
Skin Cancers							
Head, face, chest, extremities/basal cell ca, primary + recurrent (sites)	133/151	18/151	—	14–29 mo	24/151	Wilson et al.	17
Retreatment of PR + REC	30/30	—	—	3–17 mo.	0/30		17
primary + recurrent (sites) (solitary lesions)	13/13	—	—	>4 yr	—	Keller et al.	18
Breast cancer metastases	2/14	9/14	2/14	3–6 mo	5/14	Schuh	19
	4/20	9/20	7/20	1–18 mo	4/20 (2.5 mo)	Sperduto	21

CIS = carcinoma *in situ.*

MSB = main stem bronchu.

[a] Staged according to Endo, M., Takishita, K., and Yoshino, H., *Surg. Endosc.,* 2, 205–208, 1988.

Esophagus — CIS and Superficial Lesions

Spinelli and co-workers in Milan have treated 16 patients with severe dysplasia and early-stage esophageal cancers, all of whom were considered high surgical risks.[14] Complete response was found in 13/16 patients, with mean follow-up of 35 months (6 to 84 months). Four of the initial responders recurred at 6 to 12 months and were retreated by PDT with complete response.

In a similar study, Monnier and colleagues treated 29 patients, with 26/29 showing complete response at least to 6 months.[12] Hayata's group reports 85 to 90% complete response in 27 cases.[8]

Also, Barrett's esophagus (often occurring in conjunction with CIS) has been successfully treated by PDT with complete response over 5 years.[15,16]

Gastric Cancers

Hayata and colleagues have reported 115 cases of PDT for treatment of early-stage gastric cancers, which indicated clearly that lesions should be smaller than 3 cm in maximum diameter for a high complete response rate, which was 80% or higher for lesions less than 3 cm.[8]

Skin Cancers

Wilson and colleagues have demonstrated a complete response rate of 88 to 90% in 153 lesions in 37 patients (at least 3 years) for basal cell carcinoma lesions involving the face, neck, and chest wall.[17] Morpheaform lesions, especially near or on the nose, require a more aggressive treatment (superficial plus interstitial) for effective deep treatment. Keller has demonstrated similar responses with a similar protocol.[18] In addition to those superficial skin cancers, some very large tumors were found also to be responsive to PDT, although generally requiring multiple sessions over several months. However, in most cases, the patients were spared extensive, potentially disfiguring surgery.

Early-Stage Cutaneous Breast Cancer

In our experience[19] as well as that of McCaughan[20] and Sperduto et al.,[21] PDT can effectively eradicate nodular and/or plaque-type breast carcinoma lesions metastatic to the chest wall provided the lesions are no more than 3 to 5 mm in size and confined to no more than approximately 100 cm². While larger and more widespread tumors can be treated by PDT, and in many cases some eradicated, this generally does not lead to effective palliation for the patient. This is especially true if the area had previously been heavily irradiated, since normal tissue necrosis may result. It is best to use radiation therapy after PDT rather than before, and perhaps chemotherapy as well in this group of patients. Adriamycin, within a few months of PDT, can lead to excessive normal tissue necrosis, especially if the patient has previously received ionizing radiation in the treated area.

13.4 Advanced Cancers

For palliative purposes, PDT can substitute for the Nd-YAG laser for removal of obstructive tumors of the bronchus[20] or esophagus.[22] In a comparative trial for the latter, PDT produced clinical responses with palliation of symptoms with fewer treatments and is considered by some investigators as safer and easier to use than Nd-YAG ablation.[23] Both Nd-YAG laser ablation and PDT are more effective in removing endobronchial tumor than is radiation therapy.[24] Also, a number of investigators including ourselves have noted reduction in pain for some advanced gynecological cancers, although the tumors were not eradicated.

13.5 New Photosensitizers

In an attempt to obviate the cutaneous photosensitivity induced by Photofrin and to maximize tissue penetration, many groups have prepared and tested new photosensitizers in preclinical studies. Essentially, all are based on the basic porphyrin moiety modified to shift the red absorption toward 700 to 800 nm. However, to date, only three such compounds have found their way into Phase I clinical trials. These are described briefly below.

Benzoporphyrin Derivative Mono-Acid (BPD) (Quadra Logics Technologies, Vancouver, BC)

This chlorin-type compound is a mixture of two isomers with essentially equivalent photosensitizing properties and absorbs maximally in the red near 690 nm (ε ~35,000 M^{-1}). For human use, it is formulated in liposomes and is being evaluated in Phase I clinical trials for cutaneous tumors (metastatic breast cancer, basal cell carcinoma). Results to date indicate that 0.5 mg kg^{-1} and 50 J cm^{-2} delivered 3 h postinjection provides tumor control (by biopsy) and some normal tissue damage.[25] Studies continue in optimizing the variables, since it is likely that illuminating some time longer than 3 h postinjection may improve the therapeutic ratio. BPD, like NPe6 discussed below, is a short-acting photosensitizer in that essentially no result is obtained if more than 8 to 10 h elapses between injection and treatment, i.e., when serum levels have dropped. In this regard, it is very unlike Photofrin, which is retained in most tissues for many weeks, even after nearly complete serum clearance. Therefore, cutaneous photosensitivity with BPD is markedly reduced in patients from 4 to 6 weeks for Photofrin to 3 to 5 d for BPD.

Chlorin e$_6$ Aspartic Acid Ester (NPe6) (Nippon Petroleum, Japan)

NPe6 is a chlorin in which one of the three carboxylic acid groups is esterified with aspartic acid, thus introducing one additional carboxylic acid group (total of four). Therefore, this material is more hydrophilic than BPD and can be formulated without the need for liposomes or detergents. However, like BPD it is only effective as an *in vivo* photosensitizer when serum levels are near maximum levels, i.e., a few hours after injection. Cutaneous photosensitivity in man is reported to have been essentially absent after one week. Initial clinical responses have not been reported in the Phase I clinical trials.[26,27]

m-Tetrahydroxyphenyl Chlorin (THPC)

A third new photosensitizer, *m*-tetrahydroxyphenyl chlorin has been used in a few patients with mesothelioma and cutaneous breast metastases.[28] This material appears to be very effective, producing tumor necrosis to depths of 1.0 cm with as little as 0.3 mg kg^{-1} and 10 J cm^{-2} of light at 650 nm. THPC appears[2] to be as persistent as Photofrin in skin, and at least from this standpoint, may not represent an advantage. However, it appears that it may demonstrate greater selectivity, particularly if the light treatment is delayed up to 9 d postinjection.

Two other new photosensitizers, tin etiopurpurin (PDT Systems, Inc.) and zinc phthalocyanine (Ciba-Geigy) are expected to enter Phase I trials in 1993.

References

1. Marcus, S. L., Clinical photodynamic therapy: The continuing evolution, in *Photodynamic Therapy: Principles and Clinical Applications,* Henderson, B. W. and Dougherty, T. J., Eds., Marcel Dekker, New York, 1992, 219.
2. Dougherty, T. J. and Marcus, S. L., Photodynamic therapy, *Eur. J. Cancer,* 28A, 1734, 1992.
3. Henderson, B. W. and Dougherty, T. J., How does photodynamic therapy work?, *Photochem. Photobiol.,* 55, 145, 1992.
4. Bellnier, D. A., Ho, Y. K., Pandey, R. K., Missert, J. R., and Dougherty, T. J., Distribution and elimination of Photofrin II in mice, *Photochem. Photobiol.,* 50, 221, 1989.

5. Brown, S. B., Vernon, D. I., Holroyd, J. A., Marcus, S., Trust, R., Hawkins, W., Shah, A., and Tonelli, A., Pharmacokinetics of photofrin in man, in *Photodynamic Therapy and Biomedical Lasers,* Spinelli, P., Dal Fante, M., and Marchesini, R., Eds., Elsevier Science, Amsterdam, 1992, 475.

6. Ferrario, A., Rucker, N., Ryter, S. W., Doiron, D. R., and Gomer, C. J., Direct comparison of *in vitro* and *in vivo* Photofrin II mediated photosensitization using a pulsed KTP pumped dye laser and a continuous wave argon ion pumped dye laser, *Lasers Surg. Med.,* 11, 404, 1991.

7. Dugan, M., Crawford, E., Nseyo, U., Shumaker, B., Aledia, F., Hoodin, A., Bender, L., Javadpour, N., Prout, G., Reisman, A., and Marcus, S., A randomized trial of observation (OBS) vs. photodynamic therapy (PDT) after transurethral resection (TUR) for superficial papillary bladder carcinoma (SPBC), *Proc. Am. Soc. Clin. Oncol.,* 10, 554, 1991 (Abstract).

8. Hayata, Y., Kato, H., Konaka, C., Okunaka, T., and Furukawa, K., Overview of clinical PDT, in *Photodynamic Therapy and Biomedical Lasers,* Spinelli, P., Dal Fante, M., and Marchesini, R., Eds., Elsevier Science, Amsterdam, 1992, 1.

9. Cortese, D. A., Edell, E. S., Silverstein, M. D., Offord, K., Trastek, V. F., Pairolero, P. C., Allen, M. S., and Deschamps, C., An evaluation of the effectiveness of photodynamic therapy (PDT) compared to surgical resection in early stage roentgenographically occult lung cancer, in *Photodynamic Therapy and Biomedical Lasers,* Spinelli, P., Dal Fante, M., and Marchesini, R., Eds., Elsevier Science, Amsterdam, 1992, 15.

10. Gluckman, J. L., Hematoporphyrin photodynamic therapy: Is there truly a future in head and neck oncology?, Reflections on a 5-year experience, *Laryngoscope,* 101, 36, 1991.

11. Biel, M. A., Photodynamic therapy and the treatment of neoplastic and non-neoplastic diseases of the larynx, in *Photodynamic Therapy and Biomedical Lasers,* Spinelli, P., Dal Fante, M., and Marchesini, R., Eds., Elsevier Science, Amsterdam, 1992, 647.

12. Monnier, P., Fontolliet, C., Wagnieres, G., Braichotte, D., and Van den Bergh, H., Further appraisal of PDI and PDT of early squamous cell carcinomas of the pharynx, oesophagus and bronchi, in *Photodynamic Therapy and Biomedical Lasers,* Spinelli, P., Dal Fante, M., and Marchesini, R., Eds., Elsevier Science, Amsterdam, 1992, 7.

13. Feyh, J., The treatment of larynxpapillomas with the aid of photodynamic therapy, in *Photodynamic Therapy and Biomedical Lasers,* Spinelli, P., Dal Fante, M., and Marchesini, R., Eds., Elsevier Science, Amsterdam, 1992, 653.

14. Spinelli, P., Dal Fante, M., Mancini, A., and Massetti, M., Endoscopic photodynamic therapy of early cancer and severe dysplasia of the esophagus, in *Photodynamic Therapy and Biomedical Lasers,* Spinelli, P., Dal Fante, M., and Marchesini, R., Eds., Elsevier Science, Amsterdam, 1992, 262.

15. Nava, H., Roswell Park Cancer Institute, Buffalo, NY, unpublished data, 1992.

16. Overholt, B., Thompson Cancer Survival Center, Knoxville, TN, unpublished data, 1993.

17. Wilson, B. D., Mang, T. S., Stoll, H., Jones, C., Cooper, M., and Dougherty, T. J., Photodynamic therapy for the treatment of basal cell carcinoma, *Arch. Dermatol.,* 128(12), 1597, 1992.

18. Keller, G. S., Razum, N. J., and Doiron, D. R., Photodynamic therapy for nonmelanoma skin cancer, *Facial. Plast. Surg.,* 6, 180, 1989.

19. Schuh, M., Nseyo, U. O., Potter, W. R., Dao, T. L., and Dougherty, T. J., Photodynamic therapy for palliation of locally recurrent breast carcinoma, *J. Clin. Oncol.,* 5, 1766, 1987.

20. McCaughan, J. S., Jr., Guy, J. T., Hicks, W. M., Laufman, L., Nims, T. A., and Walker, J., Photodynamic therapy for cutaneous and subcutaneous malignant neoplasms, *Arch. Surg.,* 124, 211, 1989.

21. Sperduto, P., Delaney, T. F., Thomas, G., Smith, P., Dachowski, C., Russo, A., Bonner, R., and Glatstein, E., Photodynamic therapy for chest wall recurrences in breast cancer, *Int. J. Radiat. Oncol. Biol. Phys.,* 21, 441, 1991.

22. Dugan, M., McCaughan, J., York, E., Lam, S., Lang, N., Goldberg, M., Mohsenifar, Z., Johnston, M., Fisher, G., Sachdev, O., Grose, M., Chen, S., and Marcus, S., A multicenter Phase III trial of photodynamic therapy (PDT) versus (VS) Nd:YAG Laser in the treatment (Rx) of malignant endobronchial obstruction (EO), *Lasers Med. Sci.,* 7, 215, 1992 (Abstract).

23. McCaughan, J., Hawley, P., and Walker, J., Management of endobronchial tumors. A comparative study, *Semin. Surg. Oncol.,* 5, 38, 1989.

24. Lam, S., Kostashuk, E. C., Coy, E. P., Laukkanen, E., LeRiche, J. C., Mueller, H. A., and Szasz, I. J., A randomized comparative study of the safety and efficacy of photodynamic therapy using Photofrin II combined with palliative radiotherapy versus palliative radiotherapy alone in patients with inoperable obstructive non-small cell bronchogenic carcinoma, *Photochem. Photobiol.,* 46, 893, 1987.

25. Lui, H., Kollias, N., Wimberly, J., and Anderson, R. R., A preliminary report of cutaneous photosensitivity in patients undergoing photodynamic therapy with benzoporphyrin derivative-monoacid ring A, in *Photodynamic Therapy and Biomedical Lasers,* Spinelli, P., Dal Fante, M., and Marchesini, R., Eds., Elsevier Science, Amsterdam, 1992, 459.

26. Allen, R. P., Kessel, D., Tharratt, R. S., and Volz, W., Photodynamic therapy of superficial malignancies with NPe6 in man, in *Photodynamic Therapy and Biomedical Lasers,* Spinelli, P., Dal Fante, M., and Marchesini, R., Eds., Elsevier Science, Amsterdam, 1992, 441.

27. Volz, W. and Allen, R., Cutaneous phototoxicity of NPe6 in man, in *Photodynamic Therapy and Biomedical Lasers,* Spinelli, P., Dal Fante, M., and Marchesini, R., Eds., Elsevier Science, Amsterdam, 1992, 446.

28. Braichotte, D., Wagnieres, G., Philippoz, J.-M., Bays, R., Ris, H.-B., and Van den Bergh, H., Preliminary clinical results on a second generation photosensitizer: *m*THPC, in *Photodynamic Therapy and Biomedical Lasers,* Spinelli, P., Dal Fante, M., and Marchesini, R., Eds., Elsevier Science, Amsterdam, 1992, 461.

29. Nseyo, U. O. and Dougherty, T. J., Photodynamic therapy in the management of resistant bladder cancer, *Lasers Surg. Med.,* 6, 228, 1986 (Abstract).

30. Nseyo, U. O., Dougherty, T. J., and Sullivan, L., Photodynamic therapy in the management of resistant lower urinary tract carcinoma, *Cancer,* 60, 3113, 1987.

31. Shumaker, B. P. and Hetzel, F. W., Clinical laser photodynamic therapy in the treatment of bladder carcinoma, *Photochem. Photobiol.,* 46, 899, 1987.

32. Benson, R. C., Treatment of diffuse transitional cell carcinoma in situ by whole bladder hematoporphyrin derivative photodynamic therapy, *J. Urol.,* 134, 675, 1985.

33. Hisazumi, H., Misaki, T., and Miyoshi, N., Photoradiation therapy of bladder tumors, *J. Urol.,* 130, 685, 1983.

34. deCorbiere, S., Ouayoun, M., Sequert, C., Freche, C., and Chabolle, F., Use of photodynamic therapy in the treatment of vocal cord carcinoma. Retrospective study 1986–1992 on 41 cases, in *Photodynamic Therapy and Biomedical Lasers,* Spinelli, P., Dal Fante, M., and Marchesini, R., Eds., Elsevier Science, Amsterdam, 1992, 656.

35. Wenig, B. L., Kurtzman, D. M., Grossweiner, L. I., Mafee, M. F., Harris, D. M., Lobraico, R. V., Prycz, R. A., and Appelbaum, E. L., Photodynamic therapy in the treatment of squamous cell carcinoma of the head and neck, *Arch. Otolaryngol. Head Neck Surg.,* 116, 1267, 1990.

36. Tian, M. E., Qui, S. L., and Ji, Q., Preliminary results of hematoporphyrin derivative-laser treatment for 13 cases of early esophageal carcinoma, *Adv. Exp. Med. Biol.,* 193, 21, 1985.

37. Fujimaki, M. and Nakayama, K., Endoscopic laser treatment of superficial esophageal cancer, *Semin. Surg. Oncol.,* 2, 248, 1986.

38. Calzavara, F., Tomio, L., Corti, L., Zorat, P. L., Barone, I., Peracchia, A., Norberto, L., D'Arcais, R. F., and Berti, F., Oesophageal cancer treated by photodynamic therapy alone or followed by radiation therapy, *J. Photochem. Photobiol. B. Biol.,* 6, 167, 1990.

39. Monnier, P., Savary, M., Fontolliet, C., Wagnieres, G., Chatelain, A., Cornaz, P., DePeursinge, C., and van den Berg, H., Photodetection and photodynamic therapy of 'early' squamous cell carcinoma of the pharynx, esophagus, and tracheobronchial tree, *Lasers Med. Sci.,* 5, 149, 1990.

14

Photoecology and Environmental Photobiology

Donat-P. Häder
Institut für Botanik und
Pharmazeutische Biologie der
Universität Erlangen-Nürnberg

14.1 Introduction

Photosynthetic plants and microorganisms are faced with an environmental problem in selecting a suitable niche and optimizing their light exposure: on the one hand they have to obtain sufficient solar radiation in order to satisfy their need for energy absorption, and on the other exposure to excessive radiation may inhibit the photosynthetic apparatus or even damage cellular structures. In order to regulate the energy flux and to protect themselves from excessive radiation, plants and microorganisms have developed effective strategies to optimize their conditions for growth and survival.

14.2 Orientation Mechanisms in Plants and Microorganisms

Both higher plants and photosynthetic microorganisms utilize a number of orientation mechanisms to modify their position and exposure to light in order to adapt to the constantly changing environmental conditions.

Orientation Mechanisms in Higher Plants

Even in their embryonic development, higher plants are oriented precisely with their principal root pointing downward and their shoot upward. A recent spaceflight experiment has indicated that this polarity is even a built-in function of the embryo in the seed; the organs continue growing in the previously defined direction.[1] In addition to an orientation in the gravitational field of the earth (gravitropism), the plants start to orient with respect to light as soon as this is available after the penetration of the soil surface. Some of these orientation strategies are controlled by the phytochrome

0-8493-8634-9/95/$0.00+$.50
© 1995 by CRC Press, Inc.

system (see chapters on phytochrome, this volume) others by one or more "blue light receptors" (see chapter by Poff, this volume). In order to optimize the position in space, the main shoot usually orients upward, while lateral branches and leaves are exposed perpendicularly to the incident light. In some species, the leaves even follow the solar path (sun tracking) to increase the absorption of light over the day.[2] In contrast, other plants, exposed to high solar radiation, decrease the amount of incident quanta on their leaves by assuming a vertical position (shadeless forests, e.g., some *Eucalyptus* species). Phototropic orientations are defined by the fact that the direction of movement depends on the stimulus direction (toward, away from, or at a specific angle with respect to the light direction). In contrast, photonastic responses are elicited by sudden changes in the fluence rate; the direction of movement is independent of the light direction and rather a built-in function of a plant organ. As an example, the petals of a flower close toward the center when a cloud covers the sun.

Orientation in Microorganisms

Likewise, motile microorganisms utilize a number of stimuli to optimize their positions in their habitat. In addition to other stimuli including chemical[3] and thermal[4] gradients, the magnetic field of the earth,[5] and even electrical currents,[6] the ecologically most important clues are light[7] and gravity.[8] The responses to light are summarized under the term of "photomovement" (see chapter by Lenci, this volume).[9] In addition to phototaxis, which depends on the light direction (positive or negative or at a specific angle), many cells show photophobic responses that are transient reactions elicited by sudden changes in the fluence rate. The third response to light is photokinesis, which describes the steady-state dependence of the swimming velocity on the fluence rate of the incident light.[10]

Ecological Consequences of Photomovement

Each of these responses can result in the accumulation of organisms in specific microenviron-ments.[11] It is obvious that a directed movement toward the light source (positive phototaxis) causes, for example, flagellates to move upward in the water column, and many organisms show a positive phototaxis at low fluence rates.[12] However, many phytoplankton do not possess the protective capabilities of higher plants to escape the radiation damage by VIS and UV spectral bands of unfiltered solar radiation at the surface of the body of water they occupy; thus, negative phototaxis at higher fluence rates is an ecological necessity. These two antagonistic responses lead to an accumulation of the cells in a band at a specific depth with suitable light conditions that rises and sinks on a diurnal basis and is modulated further by a changing cloud cover.[13] Some dinoflagellates have been found to undergo daily vertical migrations of up to 15 m.[14] In darkness, many cells move upward (guided by negative gravitaxis) to ensure that they are near the surface of the water body at sunrise. This is also of an ecological advantage for photosynthetic organisms when the orienta-tion to light fails, as in turbid waters. In *Euglena*, it could be shown that the gravitational field of the earth is responsible for the upward movement, since in a space experiment under microgravity conditions the cells moved randomly.[15]

Other flagellates utilize different strategies to orient within the water column: several *Peridinium* species (both freshwater and marine) were found to show positive phototaxis as a means of upward movement up to a certain illuminance, which in *P. gatunense* is below 18 klx (Figure 14.1a).[16,17] At higher fluence rates, the cells swim diaphototactically (Figure 14.1b), i.e., perpendicular to the light direction; this behavior is observed at illuminances up to 120 klx, which is higher than tropical solar radiation at the surface at noon. Still other organisms show only diaphototactic orientations.[18] The resulting horizontal movement, however, is modulated by a small upward component (at low fluence rates) or a small downward component (at high fluence rates) so that the cells eventually also move up and down in the water column.

Gliding and swimming prokaryotes utilize photophobic responses to move to and stay in areas of suitable light conditions for growth and survival. The simple reversal at boundaries to dark or

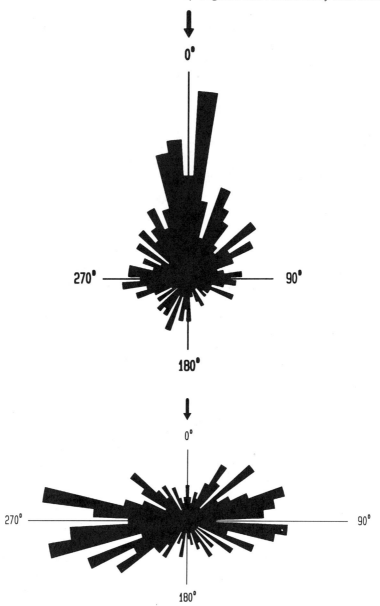

FIGURE 14.1   (a) Positive phototactic orientation (at 555 μE m^{-2} s^{-2}) and (b) diaphototaxis (at 6900 μE m^{-2} s^{-2}) of *Peridium faeroense* (Modified from Eggersdorfer and Häder, Reference 17.)

too bright areas is an effective strategy to adapt to the constantly changing light conditions in the habitat. This primitive, but rather effective mechanism plays an important role in cyanobacteria.[19] As the organisms are very sensitive to subtle changes in the fluence rates, they are capable of fine-tuning their positions in the environment and optimize, e.g., their depths in a microbial mat[20] or outline the shape of a leaf in a puddle. Under experimental conditions this response has even been used to produce a positive of a photographic negative (Figure 14.2).

Photoregulation of Intracellular Movement

Another way of optimizing the amount of light perceived by a plant is to rearrange the position of the chloroplasts within the cell. In many plants, the chloroplasts are constantly moved around by

FIGURE 14.2　Accumulation of *Phormidium* trichomes in bright areas of a photographic negative producing a positive. (From Häder, unpublished results.)

cytoplasmic streaming, which warrants that on average all receive a similar amount of photons. If this movement is induced or accelerated by light, this behavior is called photodinesis.[21] This response is investigated well in the fresh-water plants *Vallisneria* and *Elodea*.[22] Instead of continuous light, a short pulse (seconds or minutes) can be used to induce a photodinetic effect. In either case, the effect is only transient and depends on the irradiance of the continuous light or on the fluence of the pulse. To a first approximation, the action spectrum of photodinesis is consistent with a flavin as the photoreceptor pigment.

The second type of light-controlled intracellular movement is the redistribution or rearrangement of chloroplasts in unidirectional light, which results in well-defined patterns depending on irradiance and direction. Leaves of the moss *Funaria* and fronds of the duckweed *Lemna* are among the most thoroughly investigated plants with respect to chloroplast redistribution,[23] but also filamentous green algae such as *Mougeotia*[24] and macroalgae such as *Dictyota*[25] have been found to utilize this mechanism to optimize the number of absorbed photons. At low fluence rates, the chloroplasts of the moss leaf cells occupy the inner cell walls parallel to the leaf surface and thus receive most of the incident photons (Figure 14.3). At high fluence rates (but also in darkness), they occupy the cell walls perpendicular to the leaf surface, which reduces the absorption cross section. The filamentous *Mougeotia* contains only one chloroplast in each of their barrel-shaped cells. At low fluence rates, the plastid exposes its face to the light source and at higher fluence rate its edge (Figure 14.4). This mechanism also modulates the absorption cross section and thus controls the amount of light perceived. While chloroplast movement in the moss is controlled by a blue-light receptor, it is mediated by the red-absorbing phytochrome in the green alga. The moving force is generated by actin microfilaments, most probably by their interaction with myosin. However, much detail is still unknown, especially concerning the directed movements. In a few marine algae, microtubules appear to be involved in the mechanism of movement; however, this mechanism also is not yet well understood.

14.3　Photoinhibition

In both higher plants and in photosynthetic microorganisms, photosynthesis is controlled by light in a complicated fashion. At high fluence rates, there is the risk that excessive radiation, which cannot be used effectively for photosynthesis, damages cellular structures by a number of mechanisms.

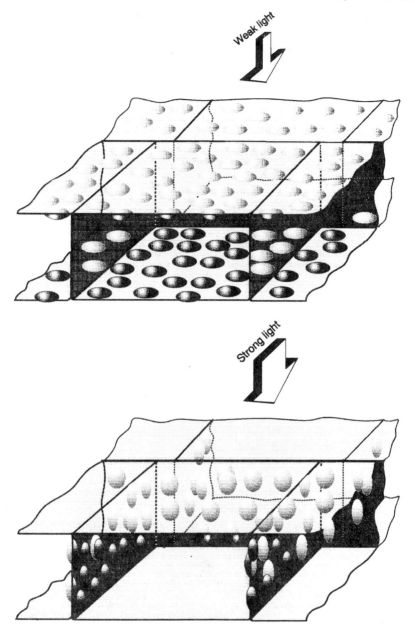

FIGURE 14.3   Orientation of the chloroplasts in leaflet cells of the moss *Funaria* in weak light and in strong light.

Electronically excited pigments can transfer their excitation energy to, e.g., molecular oxygen.[26] Upon energy transfer, this may undergo a transition from the triplet state to the very aggressive singlet state, which has been found to disrupt membranes, oxidize proteins, and affect other cellular structural elements.[27] Higher plants and also some phytoplankton utilize specific carotenoids to dispose of the excess energy by thermal relaxation.[28]

Other mechanisms to avoid photochemical damage by excessive radiation utilize photoinhibition, which regulates the activity of photosynthesis and shuts down oxygen production at extreme fluence rates. This mechanism is not completely understood but the contact site between the two reaction center polypeptides of photosystem II may be involved in photoinhibition.[29] A similar

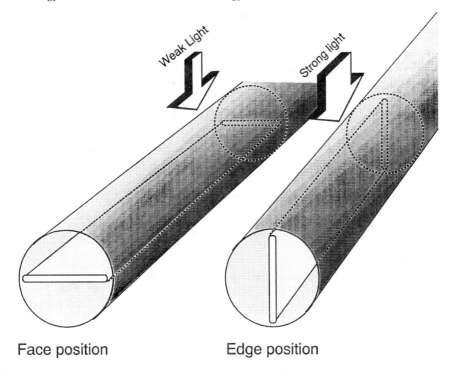

Face position   Edge position

FIGURE 14.4 Orientation of the chloroplast of the filamentous green alga *Mougeotia* in weak light and in strong light.

photoinhibition has been found in phytoplankton[30] and macroalgae.[31] One approach to determine the degree of photoinhibition (both photochemical and nonphotochemical quenching) is the analysis of fluorescence induction curves with an instrument developed by Schreiber.[32]

14.4 Ozone Depletion and Enhanced Solar Ultraviolet Radiation

Ozone is distributed throughout the stratosphere with its highest concentration in a layer between about 15 and 40 km above the surface of the earth. Despite its low concentration (ca. 3 to 4 mm if it could be concentrated under atmospheric pressure), it is an effective filter for short-wavelength UV radiation (UV-B, 280 to 315 nm). Gaseous pollutants of anthropogenic origin such as chlorinated fluorocarbons (CFC) have been found to affect the ozone concentration.[33]

The most drastic and best-known example of stratospheric ozone depletion is the Antarctic ozone hole, which has been documented since 1979 and has increased in size and depth since.[34] In addition to the polar hole, satellite and other data have indicated a general decrease in the stratospheric ozone concentration on a global basis.[35] Future levels of UV-B radiation strongly depend on the scenarios debated currently.[36] Since the mean lifetime of CFCs under stratospheric conditions are on the order of about 100 years and furthermore the substances need about 10 years to be transported into the stratosphere, the pollutants produced and emitted in the past need to be taken into consideration.[37]

In coastal waters with high turbidity and large concentrations of gelbstoff, UV-B penetrates only a few decimeters or meters into the column,[38] while in clear oceanic waters, UV-B has been measured to penetrate to depths of tens of meters.[39] Record penetration was measured in Antarctic waters where 1% of solar UV has been detected at a depth of 35 m.[40]

Solar UV radiation at current and enhanced levels has been found to affect almost all forms of life directly or indirectly. In addition to the effects on human health, adverse UV-B effects have been found in terrestrial plants and in phytoplankton.[41]

UV-B Effects in Higher Plants

The most obvious effect of increased solar UV irradiation is a reduced productivity, both in wild and in cultivated crop plants.[42,43] Solar UV radiation has been found to adversely affect photosynthesis,[44] stomatal movement,[45] growth, and development of a large percentage of plants investigated.[46] However, a number of higher plants are capable of producing shielding substances upon exposure to UV-B to protect themselves from the high levels of solar short-wavelength radiation.[47,48]

It would be deleterious if UV-B were to also affect the orientation mechanisms of higher plants discussed above, as this would reduce the capabilities of the plants to adapt to the constantly changing light conditions in their environment. However, while there are some indications that this is the case, there are no systematic investigations yet.

UV-B Effects in Aquatic Ecosystems

In marine phytoplankton, there is also a strong concern that solar UV-B may have adverse effects.[49] These organisms are not protected by an epidermal layer, they dwell in the top layers of the water column (the euphotic zone), and have been found to be very sensitive to UV radiation.[50–52] Many aquatic ecosystems are under considerable UV-B stress even at current levels.[53,54] Because of the enormous size of these ecosystems, any substantial losses in biomass production will have serious consequences for the intricate food web as well as global climate.[55]

There seem to be several targets of short-wavelength UV radiation in phytoplankton. In addition to DNA, photosynthesis is impaired and pigments are bleached.[56] Also, enzymatic functions are affected, including those for nitrogen incorporation.[57]

Phytoplankton organisms are not equally distributed within the water column, but rather move to specific depths using active motility or bouyancy.[58] UV-B has been found to drastically affect motility and orientation to external stimuli in a number of phytoplankton organisms investigated so far (Figure 14.5).[59] In addition, the orientation mechanisms with respect to light and gravity are equally affected by even short exposure to solar UV radiation.[60,61] This may have potentially deleterious consequences for phytoplankton: cells that are no longer capable of moving and orienting in the water column may be exposed to extreme irradiations at the surface; prolonged exposure to unfiltered solar radiation strongly bleaches the photosynthetic pigments.[62,63] As an alternative, the cells sink in the water column because they are heavier than water so that the radiation is no longer sufficient to sustain photosynthesis.

References

1. Volkmann, D., Behrens, H. M., and Sievers, A., Development and gravity sensing of cress roots under microgravity, *Naturwisschaften*, 73, 438, 1986.
2. Koller, D. and Ritter, S., Diaphototropic responses to polarized light in the solar-tracking leaf of *Lupinus palaestinus* Boiss. (Fabaceae), *J. Plant Physiol.*, 138, 322, 1991.
3. Berg, H. C., Physics of bacterial chemotaxis, in *Sensory Perception and Transduction in Aneural Organisms*, Colombetti, G., Lenci, F., and Song, P.-S., Eds., Plenum Press, New York, 1985, 19.
4. Poff, K. L., Temperature sensing in microorganisms, in *Sensory Perception and Transduction in Aneural Organisms*, Colombetti, G., Lenci, F., and Song, P.-S., Eds., Plenum Press, New York, 1985, 299.
5. Esquivel, D. M. S. and de Barros, H. G. P. L., Motion of magnetotactic microorganisms, *J. Exp. Biol.*, 121, 153, 1986.
6. Mast, S. O., *Light and Behavior of Organisms*, John Wiley & Sons, New York; Chapman & Hall, London, 1911.
7. Nultsch, W. and Häder, D.-P., Photomovement in motile microorganisms II, *Photochem. Photobiol.*, 47, 837, 1988.

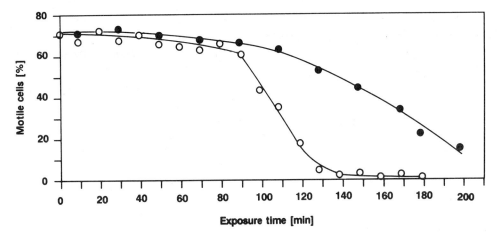

FIGURE 14.5 Inhibition of motility by unfiltered solar radiation (open circles) and by radiation filtered through an ozone cuvette (closed circles) in the marine *Cryptomonas maculata*. (Modified after Häder and Häder, Reference 60.)

8. Häder, D.-P., Polarotaxis, gravitaxis and vertical phototaxis in the green flagellate, *Euglena gracilis*, *Arch. Microbiol.*, 147, 179, 1987.

9. Diehn, B., Feinleib, M., Haupt, W., Hildebrand, E., Lenci, F., and Nultsch, W., Terminology of behavioral responses of motile microorganisms, *Photochem. Photobiol.*, 26, 559, 1977.

10. Häder, D.-P., Photomovement, in *Encyclopedia of Plant Physiology*, New Series, Vol. 7, *Movement*, Haupt, W. and Feinleib, M. E., Eds., Springer, Berlin, 1979, 268.

11. Häder, D.-P., Ecological consequences of photomovement in microorganisms, *J. Photochem. Photobiol. B: Biol.*, 1, 385, 1988.

12. Häder, D.-P., Colombetti, G., Lenci, F., and Quaglia, M., Phototaxis in the flagellates, *Euglena gracilis* and *Ochromonas danica*, *Arch. Microbiol.*, 130, 78, 1981.

13. Häder, D.-P. and Griebenow, K., Orientation of the green flagellate, *Euglena gracilis*, in a vertical column of water, *FEMS Microbiol. Ecol.*, 53, 159, 1988.

14. Burns, N. M. and Rosa, F., *In situ* measurements of the settling velocity of organic carbon particles and ten species of phytoplankton, *Limnol. Oceanogr.*, 2, 855, 1980.

15. Häder, D.-P., Vogel, K., and Schäfer, J., Responses of the photosynthetic flagellate, *Euglena gracilis*, to microgravity, *Microgravity Sci. Technol. III*, 110, 1990.

16. Häder, D.-P., Häder, M., Liu, S.-M., and Ullrich, W., Effects of solar radiation on photoorientation, motility and pigmentation in a freshwater *Peridinium*, *BioSystems*, 23, 335, 1990.

17. Eggersdorfer, B. and Häder, D.-P., Phototaxis, gravitaxis and vertical migrations in the marine dinoflagellate, *Prorocentrum micans*, *Eur. J. Biophys.*, 85, 319, 1991.

18. Rhiel, E., Häder, D.-P., and Wehrmeyer, W., Photoorientation in a freshwater *Cryptomonas* species, *J. Photochem. Photobiol. B: Biol.*, 2, 123, 1988.

19. Nelson, D. C. and Castenholz, R. W., Light responses of *Beggiatoa*, *Arch. Microbiol.*, 131, 146, 1982.

20. Pentecost, A., Effects of sedimentation and light intensity on mat-forming Oscillatoriaceae with particular reference to *Microcoleus lyngbyaceus* Gomont, *J. Gen. Microbiol.*, 130, 983, 1984.

21. Seitz, K., Wirkungsspektren für die Starklichtbewegung der Chloroplasten, die Photodinese und die lichtabhängige Viskositätsänderung bei *Vallisneria spiralis* ssp. *torta*, *Z. Pflanzenphysiol.*, 56, 246, 1967.

22. Haupt, W. and Feinleib, M. E., Eds., *Encyclopedia of Plant Physiology*, New Series 7, *Physiology of Movements*, Springer-Verlag, Berlin, 1979.

23. Haupt, W. and Wagner, G., Chloroplast movement, in *Membranes and Sensory Transduction*, Colombetti, G. and Lenci, F., Eds., Plenum Press, New York, 1984, 331.

24. Haupt, W. and Scheuerlein, R., Chloroplast movement, *Plant Cell Environm.*, 13, 595, 1990.

25. Hanelt, D. and Nultsch, W., The role of chromatophore arrangement in protecting the chromatophores of the brown alga *Dictyota dichotoma* against photodamage, *J. Plant Physiol.*, 138, 470, 1991.

26. Jung, J. and Kim, H.-S., The chromophores as endogenous sensitizers involved in the photogeneration of singlet oxygen in spinach thylakoids, *Photochem. Photobiol.*, 52, 1003, 1990.

27. Valenzeno, D. P., Photomodification of biological membranes with emphasis on singlet oxygen mechanisms, *Photochem. Photobiol.*, 46, 147, 1987.

28. Conn, P. F., Schalch, W., and Truscott, T. G., The singlet oxygen and carotenoid interaction, *J. Photochem. Photobiol. B: Biol.*, 11, 41, 1991.

29. Trebst, A., A contact site between the two reaction center polypeptides of photosystem II is involved in photoinhibition, *Z. Naturforsch.*, 46, 557, 1991.

30. Leverenz, J. W., Falk, S., Pilström, C.-M., and Samuelsson, G., The effects of photoinhibition on the photosynthetic light-response curve of green plant cells *(Chlamydomonas reinhardtii)*, *Planta*, 182, 161, 1990.

31. Nultsch, W., Pfau, J., and Huppertz, K., Photoinhibition of photosynthetic oxygen production and its recovery in the subtidal red alga *Polyneura hilliae*, *Bot. Acta*, 103, 62, 1990.

32. Schreiber, U., Reising, H., and Neubauer, C., Contrasting pH-optima of light-driven O_2- and H_2O_2-reduction in spinach chloroplasts as measured via chlorophyll fluorescence quenching, *Z. Naturforsch.*, 46, 635, 1991.

33. Rowland, F. S., Chlorofluorocarbons and the depletion of stratospheric ozone, *Am. Sci.*, 77, 36, 1989.

34. Schoeberl, M. R. and Hartmann, D. L., The dynamics of the stratospheric polar vortex and its relation to springtime ozone depletions, *Science*, 251, 46, 1991.

35. Brasseur, G. P., A dent outside the hole?, *Nature*, 342, 225, 1989.

36. Solomon, S. and Tuck, A., Evaluating ozone depletion potentials, *Nature*, 348, 203, 1990.

37. Prinn, R. G. and Golombek, A., Global atmospheric chemistry of CFC-123, *Nature*, 344, 47, 1990.

38. Smith, R. C. and Baker, K. S., Penetration of UV-B and biologically effective dose-rates in natural waters, *Photochem. Photobiol.*, 29, 311, 1978.

39. Baker, K. S. and Smith, R. C., Spectral irradiance penetration in natural waters, in *The Role of Solar Ultraviolet Radiation in Marine Ecosystems*, J. Calkins, Ed., Plenum Press, New York, 1982, 233.

40. Gieskes, W. C. and Kraay, G. W., Transmission of ultraviolet light in the Weddell Sea. Report on the first measurements made in Antarctic, *Biomass Newslett.*, 12, 12, 1990.

41. UNEP, Environmental Effects of Ozone Depletion: 1991 Update, Panel Report, 33, 1991.

42. Tevini, M. and Teramura, A. H., UV-B effects on terrestrial plants, *Photochem. Photobiol.*, 50, 479, 1989.

43. Bornman, J. F., UV radiation as an environmental stress in plants, *J. Photochem. Photobiol., B: Biol.*, 8, 337, 1991.

44. Renger, C., Rettig, W., and Gräber, P., The effect of UVB irradiation on the lifetimes of singlet excitons in isolated photosystem II membrane fragments from spinach, *J. Photochem. Photobiol., B: Biol.*, 9, 201, 1991.

45. Negash, L., Wavelength-dependence of stomatal closure by ultraviolet radiation in attached leaves of *Eragrostis tef*: Action spectra under backgrounds of red and blue lights, *Plant Physiol. Biochem.*, 25, 753, 1987.

46. Teramura, A. H., Sullivan, J. H., and Ziska, L. H., Interaction of elevated ultraviolet-B radiation and CO_2 on productivity and photosynthetic characteristics in wheat, rice, and soybean, *Plant Physiol.*, 94, 470, 1990.

47. Teramura, A. H., Tevini, M., Bornman, J. F., Caldwell, M. M., Kulandaivelu, G., and Björn, L. O., Terrestrial Plants, UNEP Environmental Effects Panel Report, 1991, 25.

48. Tevini, M., Braun, J., and Fieser, G., The protective function of the epidermal layer of rye seedlings against ultraviolet-B radiation, *Photochem. Photobiol.*, 53, 329, 1991.

49. Karentz, D., Cleaver, J. E., and Mitchell, D. L., DNA damage in the Antarctic, *Nature*, 350, 28, 1991.

50. El Sayed, S. Z., Productivity of the Southern Ocean: A closer look, *Comp. Biochem. Physiol.*, 90B, 589, 1988.

51. Bidigare, R. R., Potential effects of UV-B radiation on marine organisms of the Southern Ocean: Distributions of phytoplankton and krill during Austral spring, *Photochem. Photobiol.*, 50, 469, 1989.

52. Raven, J. A., Responses of aquatic photosynthetic organisms to increased solar UVB, *J. Photochem. Photobiol., B: Biol.*, 9, 239, 1991.

53. Calkins, J. and Thordardottir, T., The ecological significance of solar UV-B radiations on aquatic organisms, *Nature*, 283, 563, 1980.

54. Smith, R. C., Baker, K. S., Holm-Hansen, O., and Olson, R., Photoinhibition of photosynthesis in natural waters, *Photochem. Photobiol.*, 31, 585, 1980.

55. Smith, R. C., Prezelin, B. B., Baker, K. S., Bidigare, R. R., Boucher, N. P., Coley, T., Karentz, D., MacIntyre, S., Matlick, H. A., Menzies, D., Ondrusek, M., Wan, Z., and Waters, K. J., Ozone depletion: Ultraviolet radiation and phytoplankton biology in Antarctic waters, *Science*, 255, 952, 1992.

56. Häberlein, A. and Häder, D.-P., UV effects on photosynthetic oxygen production and chromoprotein composition in the freshwater flagellate *Cryptomonas* S2, *Acta Protozool.*, 31, 85, 1992.

57. Döhler, G. and Alt, M. R., Assimilation of ^{15}N-ammonia during irradiance with ultraviolet-B and monochromatic light by *Thalassiosira rotula*, *Compt. Rend. Acad. Sci. Paris, Ser. D.*, 308, 513, 1989.

58. Häder, D.-P. and Worrest, R. C., Effects of enhanced solar ultraviolet radiation on aquatic ecosystems, *Photochem. Photobiol.*, 53, 717, 1991.

59. Häder, D.-P. and Häder, M., Inhibition of motility and phototaxis in the green flagellate, *Euglena gracilis*, by UV-B radiation, *Arch. Microbiol.*, 150, 20, 1988.

60. Häder, D.-P. and Häder, M., Effects of solar and artificial UV radiation on motility and pigmentation in the marine *Cryptomonas maculata*, *Env. Exp. Bot.*, 31, 33, 1991.

61. Häder, D.-P. and Liu, S.-L., Effects of artificial and solar UV-B radiation on the gravitactic orientation of the dinoflagellate, *Peridinium gatunense*, *FEMS Microbiol. Ecol.*, 73, 331, 1990.

62. Fischer, M. and Häder, D.-P., UV effects on the pigmentation of the flagellate *Cyanophora paradoxa* — biochemical and spectroscopic analysis, *Eur. J. Protist.*, 28, 163, 1992.

63. Gerber, S. and Häder, D.-P., UV effects on photosynthesis, proteins and pigmentation in the flagellate *Euglena gracilis*: Biochemical and spectroscopic observations, *Biochem. Syst. Ecol.* 20, 485, 1992.

15

Chemistry and Spectroscopy of Chlorophylls

Hugo Scheer
Botanisches Institut der
Universität München

15.1 Introduction

The chlorophylls (Chls) are a group of more than 50 tetrapyrrolic pigments with common structural elements and functions.[1] Chemically, they are defined as Mg-complexes of cyclic tetrapyrroles of the porphyrin (I), chlorin (II,III), or bacteriochlorin (IV) oxidation state bearing a fifth, isocyclic ring. Most Chls are esterified at C-17[4] by phytol (see formula II for numbering), but other alcohols are frequent in bacteriochlorophylls (BChls), and some are free acids (Chl *c*). In the biological definition, only those of the above pigments are chlorophylls, which function in photosynthesis. This excludes precursors or degradation products that conform to the chemical definition, but includes several pheophytins, e.g., Mg-free chlorophylls, which are active in photosynthetic electron transport. Hydroporphyrins functionally not related to chlorophylls occur (as sex-determinant?) in *Bonella viridis,* and are cofactors in many oxidoreductases.

0-8493-8634-9/95/$0.00+$.50
© 1995 by CRC Press, Inc.

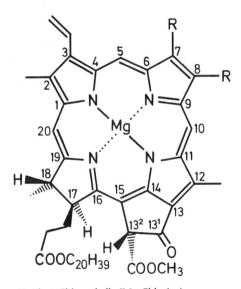

I: Chlorophylls of the *c*-Type (Mg-Porphyrins)

	R_1	R_2
Chl c_1	CH_3	C_2H_5
Chl c_2	CH_3	C_2H_3
Chl c_3	$COOCH_3$	C_2H_3

II: Plant Chlorophylls (Mg-Chlorins)

	R_1	R_2	Remarks
Chl a	CH_3	C_2H_5	
[8-Vinyl]-Chl a	CH_3	C_2H_3	
Chl a'	CH_3	C_2H_5	C-13^2 Epimer
Phe a	CH_3	C_2H_5	Central Mg missing
Chl b	CHO	C_2H_5	
[8-Vinyl]-Chl b	CHO	C_2H_3	
[8-Hydroxyethyl]-Chl a	CH_3	CH_2CH_2OH	
Chl d	CH_3	C_2H_5	CHO at C-3

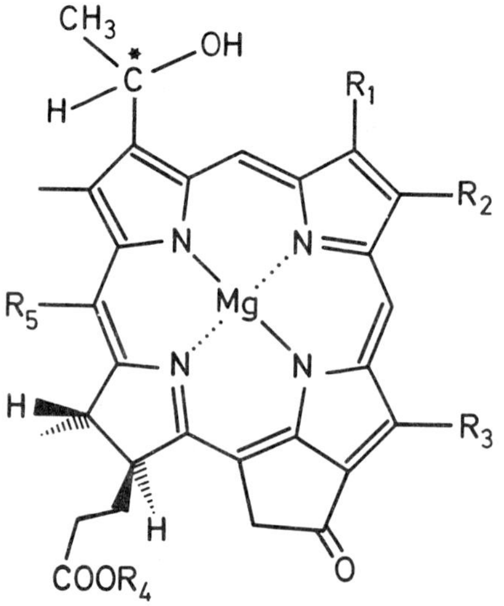

III: Bacteriochlorophylls *c, d, e,* and *f* (Mg-Chlorins)

	R_1	R_2	R_3	R_4	R_5
BChl *c*	CH₃	Variable[a]	Variable[b]	Variable[c]	CH₃
BChl *d*	CH₃	Variable[a]	Variable[b]	Variable[c]	H
BChl *e*	CHO	Variable[a]	Variable[b]	Variable[c]	CH₃
BCHl *f*[d]	CHO	Variable[a]	Variable[b]	Variable[c]	H

[a] Ethyl, propyl, *i*-butyl, or neo-pentyl.
[b] Methyl or ethyl.
[c] C₁₅-isoprenoid, C₂₀-isoprenoid, stearyl, others.
[d] Only one member has been tentatively identified yet in nature.
* Absolute configuration at C-3¹ depends on size of C-8 substituent.

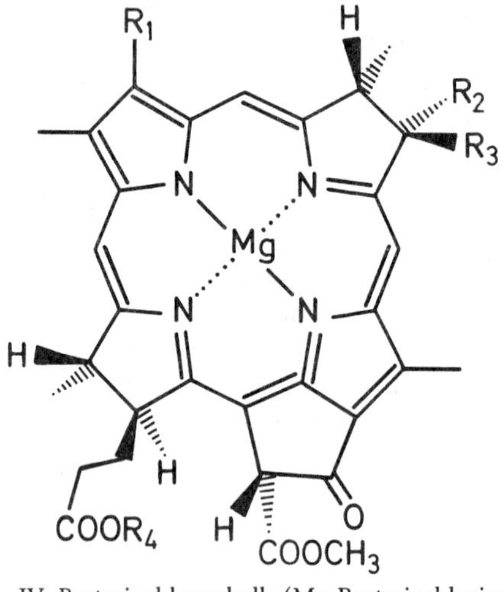

IV: Bacteriochlorophylls (Mg-Bacteriochlorins)

	R_1	R_2	R_3	R_4	Remarks
BChl *a*	COCH₃	H	C₂H₅	Variable[a]	
BChl *a'*	COCH₃	H	C₂H₅	Variable[a]	C-13² epimer
BPhe *a*	COCH₃	H	C₂H₅	Variable[a]	Central Mg missing
BChl *b*	COCH₃	=CHCH₃		Variable[a]	
BChl *b'*	COCH₃	=CHCH₃		Variable[a]	C-13² epimer
BPhe *b*	COCH₃	=CHCH₃		Variable[a]	Central Mg missing
BChl *g*	C₂H₃	=CHCH₃		Farnesyl	
BChl *g'*	C₂H₃	=CHCH₃		Farnesyl	C-13² epimer

[a] Different C₂₀-isoprenoid alcohols, mostly phytol.

15.2 Chlorophyll Structures and Functions

Chl *a* (formula II) is present in all organisms capable of oxygenic photosynthesis, where it occurs in both reaction centers (RC) and in light harvesting complexes (LHC) (see Table 15.1 and Reference 1 for occurrence and functions). It functions as the primary donor in the RC of photosystem (PS) II, and Chl *a* or a closely related pigment is also the primary donor of PS I. Phe *a*, the demetalated Chl *a*, is an early electron acceptor in PS II. A Chl *a*-type pigment is discussed as the first electron acceptor (A_o) in PS I. The 13^2-epimer, Chl *a'*, is slightly less stable than Chl *a* and reversibly interconvertible to it by acid or base catalysis. It is present in a constant ratio in the RC of PS I (2 Chl *a'*/P700), but its precise function (primary donor, acceptor?) is unknown. [8-vinyl]-Chl *a* (often referred to as divinyl-Chl *a*) and the related [8-vinyl]-Chl *b* have been found as major (or sole) chlorophylls in a mutant of *Zea mays* and in marine prochlorophytes. [8-Hydroxyethyl]-Chl *a* is a trace pigment in green bacteria, where it may act as an early electron acceptor. Chl *b* bears a 7-formyl substituent. In the "green" series of oxygenic photosynthetic organisms (prochlorophytes, green [and a few other] algae, green plants), it accompanies Chl *a* as a light-harvesting pigment in about a 1:3 ratio. Chl *d* has been found in extracts from rhodophytes and possibly in *Chlorella*. Its function and location are unknown.

The BChls *c*, *d*, *e*, and (the only once described) *f* (formula III) are a complex group of chlorins that lack the 13^2-carbomethoxy group, and bear an α-hydroxyethyl substituent at position C-3. They are antenna pigments in green, brown, or red sulfur bacteria and in *Chloroflexus*.

Table 15.1   Occurrence and Distribution of Chlorophylls

	Chlorophyll				Bacteriochlorophyll					
	a	*b*	*c*[a]	*d*	*a*	*b*	*c*[a]	*d*[a]	*e*[a]	*g*
Purple bacteria	−	−	−	−	+	+	−	−	−	−
Green/brown bacteria	−	−	−	−	+	−	+	+	+	−
Erythrobacter+other	−	−	−	−	+	−	−	−	−	−
Heliobacteria	−	−	−	−	−	−	−	−	−	+
Prochlorophytes	+	+	−	−	−	−	−	−	−	−
Cyanobacteria	+	−	−	−	−	−	−	−	−	−
Rhodophytes	+	−	−	?	−	−	−	−	−	−
Chromophytes	+	−	+	−	−	−	−	−	−	−
Chlorophytes	+	+	−	?	−	−	−	−	−	−
Micromonadophytes	+	+	+	−	−	−	−	−	−	−
Green plants	+	+	−	−	−	−	−	−	−	−

[a] Mixture of homologues.

Table 15.2   Absorption Maxima of Chlorophylls and Their Pheophytins

Chlorophyll	λ_{max} (nm)	
	Chl	Phe
Chlorophyll *a*	662, 434	667,535,505, 408
Chlorophyll *b*	644, 430	655,525, 412
Chlorophyll *c*	626, 576,444	650,592,579,532,433
Chlorophyll *d*	688, 447	638,586,564,525,417
Bacteriochlorophyll *a*	773, 577,358	749,525, 385,357
Bacteriochlorophyll *b*	794, 580,368	776,258, 398,368
Bacteriochlorophyll *c*	660, 432	664,547,515, 408
Bacteriochlorophyll *d*	646, 458	658,548,505, 406
Bacteriochlorophyll *e*	654, 424	654,534, 439
Bacteriochlorophyll *g*	763,575,470,418,408	753,518, 396,388

BChl *a* (formula IV) occurs in most photosynthetic bacterial RC and is the only BChl in most *Rhodospirillales*, where it acts as antenna pigment, as the primary donor, and probably also as the first acceptor. In several species, it is replaced by BChl *b*. One of the four BChls in RC is also implicated in triplet energy transfer to carotenoids, e.g., in the light protection. There are also several antenna fraction(s) containing BChl *a* in the green bacteria. Two (or sometimes three) molecules of either BPhe *a* or *b* occur in RC of type II from photosynthetic bacteria. One molecule acts as an early electron acceptor; the function of the other is unknown. BChl *g* is present as antenna and reaction center chlorophyll of *Heliobacteria*. It is accompanied by small amounts of BChl *g'*, the C-13^2 epimer, which, like Chl *a'*, may be involved in electron transfer.

The Chls *c* (formula I) are a group of pigments having the fully unsaturated porphyrin macrocycle and are abundant antenna pigments in the chromophyte algae. They generally do not carry a long-chain esterifying alcohol at the C-17 acrylic acid side chain; three Chl *c* structures are currently established.

15.3 Spectroscopy

The absorption spectra of cyclic tetrapyrroles are generally described by the four-orbital model.[2,3] The fully unsaturated porphyrins like Chl *c* (I) have two weakly allowed electronic transitions (Q_y,Q_x) in the visible and two intense ones in the blue or near-UV spectral region (B_y,B_x), which are generally each split into two vibrational bands. The Q_y-band is less forbidden and progressively red-shifted in chlorins (II,III) and bacteriochlorins (IV). These pigments have two regions of intense absorption ($\varepsilon_{max} \approx 10^5$) in the red to near-IR (Q_y) and in the blue to near-UV spectral ranges (Soret or B-bands). The Q_x-band remains weak in the chlorins, but is moderately intense in the bacteriochlorins.[4a] Absorption spectroscopy after extraction is the most common method for (B)Chl quantitation.[4b]

Monomeric (B)Chls are highly fluorescent[5] and phosphorescence has been observed for many pigments.[6,7] Both are quenched *in situ*, and the degree and dynamics of fluorescence quenching are valuable indicators for photosynthetic processes.[8,9]

In (B)Chl proteins, the Q-bands are generally red-shifted and hyperchromic.[10-12] Aggregation has been recognized as one major factor of these changes, based on few X-ray structures,[13] a large number of aggregation studies in various environments,[14-16] and theoretical investigations.[2,20,21] It is generally accompanied by an induction of strong optical activity, with conservative as well as non-conservative excitonic bands. While point-charges (which are important in linear polyenes) seem to play only a minor role, conformational changes can induce considerable red-shifts[17] and optical activity[18] too. Interactions with nearby aromatic amino acids have been recognized as another important factor.[19a,19b]

The intense absorptivities in the visible and NIR spectral regions are an important factor in photosynthetic light-harvesting. However, the bands are quite narrow, and there is only moderate absorptivity in the green spectral region, where the light-flux is high. In antennas, (B)Chls are therefore almost always supplemented by additional light-harvesting pigments.[10,12,19a]

The first excited electronic state of (B)Chls absorbs in the red region close to the ground-state absorption band (^0S-^1S).[22,23] Excited state dynamics have been studied mainly by high-resolution spectroscopy[28] and by nonlinear absorption spectroscopy[22] in solution and in pigment-protein complexes. The absorption spectra of triplets,[24] anion and cation radicals have generally broad and poorly structured absorptions.[2] The cation radical of the primary donor of purple bacterial RC has an intense and narrow near infrared band typical for mixed-valence complexes, and a less intense feature is also seen in oxygenic reaction centers at ≈ 830 nm. Evidence for an electronic absorption band at much longer wavelengths has recently been obtained, this band has been predicted to arise from charge-transfer absorptions in the cation radical.[25] Both radical ions and triplets have been studied widely in reaction centers by optical,[4,24,26-31] magnetic resonance[24,29,32a,32b,33] and vibrational[34] spectroscopy to obtain selective information on the pigments involved in electron transfer.

Many chlorophylls show in solution fairly efficient intersystem crossing from the ^{1}S to the triplet (^{1}T) state.[29] They are accordingly good sensitizers for singlet oxygen. This potentially very damaging effect is neutralized in photosynthetic organisms by efficient triplet and $^1O_2$-quenching by carotenoids.[35] Carotenoidless mutants are, therefore, prone to photodynamic killing at high light intensities in the presence of oxygen. On the other hand, triplet formation and their intense red and NIR absorptions render (B)Chl good candidates for photodynamic therapy (and photodiagnosis) of certain types of cancer.[36a]

15.4 Aggregates

Several modes of aggregation are known in chlorophylls. Since the process has been implicated as an important factor in the stability, structure formation, and spectral properties of (B)Chls in their functional native environments, it has been studied widely.[14–16,36b,c] One type of aggregation can be induced by interactions of the central Mg of one pigment with a peripheral donor group of another, this mechanism is important in nonpolar organic solvents, and has been discussed also as a factor in BChls *c*, *d*, and *e* aggregation in the chlorosome, the major antenna of green and brown bacteria.[37] Dispersive and π,π interactions are probably more important in other aggregation types of (B)Chl, this process has been studied in particular in aqueous-organic solvent mixtures and in micelles.[15] Large micellar aggregates are formed in aqueous systems, and aggregation is also important on surfaces.[14]

15.5 Dark Reactions of the Aromatic Macrocyclic System

A large variety of redox reactions is known for the (B)Chls and related (hydro)porphyrins.[38,39] The reductions involve methine bridges generally followed often by rearrangements of the system. Reductions at the pyrrolenine-peripheral double-bonds are important in the dark biosynthesis of (B)Chls, but little is known about the enzymes or the mechanisms.[40,41] Chemical equivalents of such reactions are only known for simple porphyrins. Oxidation to vic-diols followed by pinacol rearrangement to ketones leads formally to hydroporphyrins, also,[38] and in a more roundabout sequence to the formyl group of Chl *b* and BChls *e* and *f*.[44] Isomerization of endo- into exocyclic double-bonds would yield similar results,[42a] but hithero only the inverse process is experimentally known.[38] The electrochemistry of (B)Chls shows several one-electron redox reactions of the macrocyclic π-system, whereas the pigments derived with other metals show redox reactivities of the metal, the macrocycle, or both.[42b]

Substitution reactions are dependent on the relative electron densities, like in other aromatic systems. Electrophilic substitutions at methine bridges next to reduced rings are studied well.[38] The central Mg is easily lost in the presence of acid. Mg and a variety of other metals are readily (re)introduced into Phes of the chlorin and porphyrin type,[38] whereas BPhes are more difficult to metalate.[43a]

15.6 Reactions of Peripheral Substituents

The peripheral groups of (B)Chls react as they would in similarly substituted simpler aromatic systems.[38] There is a considerable influence of other substituents, whose origin is not fully understood.[43b] A variety of reactions are known at the isocyclic ring, many of them being related to the presence of an enolizable β-ketoester system, and to "peripheral crowding".[44] The β-ketoester system can compete with the central N_4-cavity for chelation of metals. Conjugation reactions with amino-acids, sugars or other biological material is of considerable biochemical interest for affinity labeling[45] and the design of photo-drugs.[36] Useful handles are the 17^4-propionic acid group, C-13^2 via hydroxylation, C-3^1 and in particular C-7^1 (in Chl *b*), and C-8^1 (in BChl *b* and *g*).[46]

15.7 Photochemistry

Light-induced redox and ring-opening reactions have been studied in most detail, because they are important for the biological function, for biosynthesis of Chl *a* and *b*, and possibly also for the breakdown of (B)Chls. Metalloporphyrins and chorins are photoreduced in a two-electron reaction, e.g., by amines at the methine bridges, the former products rearrange to *cis*-chlorins.[39] The light-dependent protochlorophyllide reductase performs a *trans*-reduction at ring D, the structure of intermediate(s) is unknown.[47] A reversible photoreduction of unknown products and mechanism has been characterized for a rare, water-soluble Chl-protein from *Chenopodium*.[48]

One-electron redox reactions are most important in photosynthetic RC.[26] BChls *a,b,g*, and Chl *a* act as electron donors, and both (B)Chls and (B)Phes act as acceptors. (B)Chl-sensitized reductions of redox partners, in particular across membranes, are of considerable interest as model reactions for photosynthesis.[49] These reactions have also been studied in solution in covalently linked systems like caroteno-chlorophyllo-quinones and more elaborate systems.[50,51] One-electron redox processes occur in many (large) (B)Chl aggregates, in particular in the hydrates, and they are probably involved in a variety of oxidative degradation processes subsumed under the term "allomerization".[42]

Most photooxidative ring-opening reactions of Chls of the chlorin-type cleave the macrocycle at the C-20 methine bridge, but cleavage at the C-5 bridge has been reported for a Cd-pheophytin.[52] The hitherto identified products of Chl biodegradation in the dark are also cleaved at this position,[53,54] but it is not clear if the breakdown *in vivo* is a light or dark process. A third type of photoreaction is demetalation (photopheophytinization), of which little is known.[55]

15.8 Biosynthesis

The starting point for the biosynthesis of chlorophylls[40,41,56,57] is δ-amino-levulinate (ALA), which in most photosynthetic organisms is formed from glutamate. Dimerization leads to the monopyrrole, porphobilinogen, which then tetramerizes to protoporphyrin. The specific part of Chl biosynthesis starts with the insertion of Mg, followed by the formation of the isocyclic ring, reduction of the macrocycle (except for Chl *c*), and esterification. In angiosperms, the reduction via the protochlorophyllide reductase, is a photoreaction.[47] Still little is known about the degradation of chlorophylls, both in the parent organisms, or further down the food chain.[58] Several modified cyclic tetrapyrroles have been identified, but the crucial step in removing these phototoxic compounds is probably the ring opening to bile pigments, of which several have recently been identified.[53,54,59]

Acknowledgments

Work of the author was supported by the Deutsche Forschungsgemeinschaft, Bonn.

References

1. Scheer, H., Structure and occurrence of chlorophylls, in *Chlorophylls*, Scheer, H., Ed., CRC Press, Boca Raton, FL, 1992, 3.
2. Plato, M. Möbius, K., and Lubitz, W., Molecular orbital calculations on chlorophyll radical ions, in *Chlorophylls*, Scheer, H., Ed., CRC Press, Boca Raton, FL, 1992, 1015.
3. Hanson, L. K., Molecular orbital theory of monomer pigments, in *Chlorophylls*, Scheer, H., Ed., CRC Press, Boca Raton, FL, 1992, 993.
4. (a) Hoff, A. J. and Amesz, J., Visible absorption spectroscopy of chlorophylls, in *Chlorophylls*, Scheer, H., Ed., CRC Press, Boca Raton, FL, 1992, 723; (b) Porra, R. J., Recent advances and reassessments in chlorophyll extraction and assay procedures for terrestrial, aquatic, and marine

organisms, including recalcitrant algae, in *Chlorophylls*, Scheer, H., Ed., CRC Press, Boca Raton, FL, 1992, 31.

5. Goedheer, J. C., Visible absorption and fluorescence of chlorophyll and its aggregates in solution, in *The Chlorophylls*, Vernon, L. P. and Seely, G. R., Eds., Academic Press, New York, 1966, 147.

6. Takiff, L. and Boxer, S. G., Phosphorescence from the primary electron donor in *Rhodobacter sphaeroides* and *Rhodopseudomonas viridis* reaction centers, *Biochim. Biophys. Acta*, 932, 325, 1988.

7. Takiff, L. and Boxer, S. G., Phosphorescence spectra of bacteriochlorophylls. *J. Am. Chem. Soc.*, 110, 4425, 1988.

8. Karukstis, K. K., Chlorophyll fluorescence as a physiological probe of the photosynthetic apparatus, in *Chlorophylls*, Scheer, H., Ed., CRC Press, Boca Raton, FL, 1992, 669.

9. Schreiber, U., Bilger, W., and Neubauer, C., Chlorophyll fluorescence as a non-intrusive indicator for rapid assessment of *in vivo* photosynthesis, in *Ecological Studies*, Vol. 100, Schulze, E.-D. and Caldwell, M., Eds., Springer Verlag, Berlin, 1993, in press.

10. Thornber, J. P., Morishige, D. T., Anandan, S., and Peter, G., F., Chlorophyll-carotenoid-proteins of higher plant thylakoids, in *Chlorophylls*, Scheer, H., Ed., CRC Press, Boca Raton, FL, 1992, 549.

11. Hawthornthwaite, A. M. and Cogdell, R. J., Bacteriochlorophyll-binding proteins, in *Chlorophylls*, Scheer, H., Ed., CRC Press, Boca Raton, FL, 1992, 493.

12. Hiller, R. G., Anderson, J. M., and Larkum, A. W. D., The chlorophyll-protein complexes of algae, in *Chlorophylls*, Scheer, H., Ed., CRC Press, Boca Raton, FL, 1992, 529.

13. Deisenhofer, J. and Michel, H., Crystallography of chlorophyll proteins, in *Chlorophylls*, Scheer, H., Ed., CRC Press, Boca Raton, FL, 1992, 613.

14. Katz, J. J., Bowman, M. K., Michalski, T. J., and Worcester, D. L., Chlorophyll aggregation: Chlorophyll-water micelles as models for *in vivo* long-wavelength chlorophyll, in *Chlorophylls*, Scheer, H., Ed., CRC Press, Boca Raton, FL, 1992, 211.

15. Scherz, A., Rosenbach-Belkin, V., and Fisher, J. R., E., Chlorophyll aggregates in aqueous solution, in *Chlorophylls*, Scheer, H., Ed., CRC Press, Boca Raton, FL, 1992, 237.

16. Katz, J. J., Shipman, L. L., Cotton, T. M., and Janson, T. R., Chlorophyll aggregation: Coordination interactions in chlorophyll monomers, dimers and oligomers, in *The Porphyrins*, Vol. V, Dolphin, D., Ed., Academic Press, New York, 1978, 401.

17. Medforth, C. J., Senge, M. O., Smith, K. M., Sparks, L. D., and Shelnutt, J. A., Nonplanar distortion modes for highly substituted porphyrins, *J. Am. Chem. Soc.*, 114, 9859, 1992.

18. Wolf, H. and Scheer, H., Stereochemistry and chiroptic properties of Pheophorbides and related compounds, *Ann. N.Y. Acad. Sci.*, 206, 549, 1973.

19. (a) Zuber, H. and Brunisholz, R. A., Structure and function of antenna polypeptides and chlorophyll-protein complexes: Principles and variability, in *Chlorophylls*, Scheer, H., Ed., CRC Press, Boca Raton, FL, 1992, 627; (b) Bylina, E. J. and Youvan, D. C., Protein-chromophore interactions in the reaction center of *Rhodobacter capsulatus*, in *Chlorophylls*, Scheer, H., Ed., CRC Press, Boca Raton, FL, 1992, 705.

20. Pearlstein, R. M., Theoretical interpretation of antenna spectra, in *Chlorophylls*, Scheer, H., Ed., CRC Press, Boca Raton, FL, 1992, 1047.

21. Scherer, P. O. J. and Fischer, S. F., Interpretation of optical reaction center spectra, in *Chlorophylls*, Scheer, H., Ed., CRC Press, Boca Raton, FL, 1992, 1079.

22. Leupold, D., Stiel, H., and Sepiol, J., The S^1 and T^1 spectra of chlorophyll *a* in the visible region and an S^1-bypassing relaxation from two-photon stepwise excited states, *Chem., Phys. Lett.*, 132, 137, 1986.

23. Leupold, D., Ehlert, J., Oberländer, S., and Wiesner, B., S^1 absorption of chlorophyll *a* in the red region, *Chem. Phys. Lett.*, 100, 345, 1983.

24. Clarke, R. H., The chlorophyll triplet state and the structure of chlorophyll aggregates, in *Light Reaction Path of Photosynthesis*, Fong, F. K., Ed., Springer-Verlag, Berlin, 1982, 196.

25. Breton, J., Nabedryk, E., and Parson, W. W., A new infrared electronic transition of the oxidized primary donor in bacterial reaction centers: A way to assess resonance interactions between the bacteriochlorophylls, *Biochemistry*, 31, 7503, 1992.

26. Parson, W. W., Electron transfer in reaction centers, in *Chlorophylls*, Scheer, H., Ed., CRC Press, Boca Raton, FL, 1992, 1153.

27. Boxer, S. G., Goldstein, R. A., Lockhart, D. J., Middendorf, T. R., and Takiff, L., Excited states, electron-transfer reactions and intermediates in bacterial photosynthetic reaction centers, *J. Phys. Chem.*, 93, 8280, 1989.

28. Johnson, S. G., Lee, I., and Small, G. J., Solid state spectral line-narrowing spectroscopies, in *Chlorophylls*, Scheer, H., Ed., CRC Press, Boca Raton, FL, 1992, 739.

29. Angerhofer, A., Chlorophyll triplets and radical pairs, in *Chlorophylls*, Scheer, H., Ed., CRC Press, Boca Raton, FL, 1992, 945.

30. Holzwarth, A. R., Excited-state kinetics in Chlorophyll systems and its relationship to the functional organization of the photosystem, in *Chlorophylls*, Scheer, H., Ed., CRC Press, Boca Raton, FL, 1992, 1125.

31. Sundström, V. and van Grondelle, R., Dynamics of excitation energy transfer in photosynthetic bacteria, in *Chlorophylls*, Scheer, H., Ed., CRC Press, Boca Raton, FL, 1992, 1097.

32. (a) Lubitz, W., EPR and ENDOR studies of Chlorophyll cation and anion radicals, in *Chlorophylls*, Scheer, H., Ed., CRC Press, Boca Raton, FL, 1992, 903; (b) Levanon, H. and Norris, J. R., Triplet state and chlorophylls, in *Light Reaction Path of Photosynthesis*, Fong, F. K., Ed., Springer-Verlag, Berlin, 1982, 152.

33. Hoff, A. J., Optically detected magnetic resonance (ODMR) of triplet states *in vivo*, in *Photosynthesis III — Photosynthetic Membranes and Light Harvesting Systems*, Staehelin, L. A. and Arntzen, C. J., Eds., Springer-Verlag, Berlin, 1986, 400.

34. Lutz, M. and Mäntele, W., Vibrational spectroscopy of chlorophylls, in *Chlorophylls*, Scheer, H., Ed., CRC Press, Boca Raton, FL, 1992, 855.

35. Cogdell, R. J. and Frank, H. A., How carotenoids function in photosynthetic bacteria, *Biochim. Biophys. Acta*, 895, 63, 1987.

36. (a) Spikes, J. D. and Bommer, J. C., Chlorophyll and related pigments as photosensitizers in biology and medicine, in *Chlorophylls*, Scheer, H., Ed., CRC Press, Boca Raton, FL, 1992, 1181; (b) Abraham, R. J. and Rowan, A. E., Nuclear magnetic resonance spectroscopy of chlorophyll, in *Chlorophylls*, Scheer, H., Ed., CRC Press, Boca Raton, FL, 1992, 797; (c) Hunt, J. E. and Michalski, T. J., Desorption-ionization mass spectrometry of chlorophylls, in *Chlorophylls*, Scheer, H., Ed., CRC Press, Boca Raton, FL, 1992, 835.

37. Miller, M., Gillbro, T., and Olson, J., Aqueous aggregates of bacteriochlorophyll *c* as a model for pigment organization in chlorosomes, *Photochem. Photobiol.*, 57, 98, 1993.

38. Hynninen, P. H., Chemistry of chlorophylls: Modifications, in *Chlorophylls*, Scheer, H., Ed., CRC Press, Boca Raton, FL, 1992, 145.

39. Scheer, H., Synthesis and stereochemistry of hydroporphyrins, in *The Porphyrins*, Vol. II, Dolphin, D., Ed., Academic Press, New York, 1978, 1.

40. Leeper, F. J., Intermediate steps in the biosynthesis of chlorophylls, in *Chlorophylls*, Scheer, H., Ed., CRC Press, Boca Raton, FL, 1992, 407.

41. Beale, S. I. and Weinstein, J. D., Biochemistry and regulation of photosynthetic pigment formation in plants and algae, in *Biosynthesis of Tetrapyrroles*, Jordan, P. M., Ed., Elsevier, Amsterdam, 1991, 155.

42. (a) Schaber, P. M., Hunt, J. E., Fries, R., and Katz, J. J., High-performance liquid-chromatography study of the chlorophyll allomerization reaction, *J. Chromatogr.*, 316, 25, 1984; (b) Watanabe, T. and Kobayashi, M., Electrochemistry of chlorophylls, in *Chlorophylls*, Scheer, H., Ed., CRC Press, Boca Raton, FL, 1992, 287.

43. (a) Hartwich, G., Fiedor, L., Ketheder, I., Scherz, A., and Scheer, H., A general method for metalation of bacteriopheophytin by transmetalation, *J. Am. Chem. Soc.*, in press; (b) Struck, A., Cmiel, E., Katheder, I., Schäfer, W., and Scheer, H., Bacteriochlorophylls modified at position C-3: Long-range intramolecular interaction with position C-13^2, *Biochim. Biophys. Acta*, 1101, 321, 1992.

44. Smith, K. M., Chemistry of chlorophylls: Synthesis, in *Chlorophylls*, Scheer, H., Ed., CRC Press, Boca Raton, FL, 1992, 115.

45. Richards, W. R., Walker, C. J., and Griffiths, W. T., The affinity chromatography purification of NADPH:protochlorophyllide oxidoreductase from etiolated wheat, *Photosynthetica*, 21, 462, 1987.

46. Steiner, R., Cmiel, E., and Scheer, H., Chemistry of bacteriochlorophyll *b* — Identification of some (photo)oxidation products, *Z. Naturforsch.* C38, 748, 1983.

47. Griffiths, W. T., Protochlorophyllide Photoreduction, in *Chlorophylls*, Scheer, H., Ed., CRC Press, Boca Raton, FL, 1992, 433.

48. Oku, T. and Tomita, G., The photoconversion of *Chenopodium* chlorophyll protein, *Photochem. Photobiol.*, 25, 199, 1977.

49. Tollin, G., Chlorophyll photochemistry in heterogeneous media, in *Chlorophylls*, Scheer, H., Ed., CRC Press, Boca Raton, FL, 1992, 317.

50. Wasielewski, M. R., Synthetic models for photosynthesis, *Photochem. Photobiol.*, 47, 923, 1988.

51. Wasielewski, M. R., Energy and electron transfer in covalently linked chlorophyll-containing donor-acceptor molecules, in *Chlorophylls*, Scheer, H., Ed., CRC Press, Boca Raton, FL, 1992, 269.

52. Iturraspe, J. and Gossauer, A., Dependence of the regioselectivity of photooxidative ring opening of the chlorophyll macrocycle on the complexed metal ion, *Helv. Chim. Acta*, 74, 1713, 1991.

53. Kräutler, B., Jaun, B., Bortlik, K., Schellenberg, M., and Matile, P., Zum Rätsel des Chlorophyllabbaus: Die Konstitution eines secoporphinoiden Kataboliten, *Angew. Chem.*, 103, 1354, 1991.

54. Engel, N., Jenny, T. A., Mooser, V., and Gossauer, A., Chlorophyll catabolism in Chlorella protothecoides. Isolation and structure elucidation of a red Bilin derivative, *FEBS Lett.*, 293, 131, 1991.

55. Seely, G. R., Photochemistry of chlorophylls *in vitro*, in *The Chlorophylls*, Vernon, L. P. and Seely, G. R., Eds., Academic Press, New York, 1966, 523.

56. Rüdiger, W. and Schoch, S., The last steps of chlorophyll biosynthesis, in *Chlorophylls*, Scheer, H., Ed., CRC Press, Boca Raton, FL, 1992, 451.

57. Smith, K. M., Biosynthesis of bacteriochlorophylls, in *Biosynthesis of Tetrapyrroles*, Jordan, P. M., Ed., Elsevier, Amsterdam, 1991, 237.

58. Brown, S. B., Houghton, J. D., and Hendry, G. F., Chlorophyll breakdown, in *Chlorophylls*, Scheer, H., Ed., CRC Press, Boca Raton, FL, 1992, 465.

59. Gossauer, A., Catabolism of tetrapyrroles, *Chimia*, 48, 352, 1994.

16

Photosynthetic Reaction Centers

Paul Mathis
CEA-SACLAY

16.1 Presentation of Reaction Centers

General Properties, Composition, and Isolation

In plants, algae, and photosynthetic bacteria, the photosynthetic apparatus converts light energy into chemical energy. It has a complex structure that starts with an "antenna" — an ensemble of pigments (chlorins such as the chlorophylls, carotenoids, phycobilins) imbedded in proteins, which absorbs light and funnels excitation energy toward large specific proteins called reaction centers where photoinduced charge separation takes place. Reaction centers (RC) can thus be considered as microscopic photocells that realize the first steps of energy conversion. All elementary processes take place very rapidly (1 ps or less for pigment-pigment energy transfer, less than 3 ps for primary electron transfer), so that the overall quantum yield is higher than 90%, with a small fraction of excitation being lost by fluorescence (less than 2%) or by conversion to the triplet state (less than 4%). For many years, RCs were simply hypothesized as specific sites in the photosynthetic membrane, but progress in membrane biochemistry has led to their effective isolation as complex proteins, made of several polypeptides (from two in the simplest case of nonsulfur green bacteria, to at least ten in PS I or PS II of oxygenic organisms), of several pigments molecules (at least 6 chlorin-type molecules and one or more carotenoids are present in all natural RCs), and several chemical groups implicated in electron transfer (these groups are named redox centers: quinones, hemes, iron-sulfur centers, Mn atoms, etc.).

Several classes of reaction centers are found among the various biological organisms. Their basic mode of functioning is always the same: excitation energy arriving from the antenna is trapped by a primary electron donor P (most often, perhaps always, a dimer of (bacterio)chlorophyll); excitation of P is followed by several steps of electron transfer on the acceptor side, and P+ is then re-reduced by an electron donor. Several categories of RCs will be briefly presented, starting with the best-known RC — purple bacteria.

0-8493-8634-9/95/$0.00+$.50
© 1995 by CRC Press, Inc.

Purple Bacteria

The RC of purple bacteria serves as a general model for all classes of RCs because it was isolated as a pure protein a long time ago, its structural and functional properties have been studied in detail by biochemical and spectroscopic methods, its 3-D structure determined at atomic resolution from crystals, providing a firm basis for site-specific mutations, for theoretical calculations of spectroscopic properties and of electron transfer kinetics, etc.[1,2]

The RC is made of three polypeptides: L, M, H (MW: 32, 34, 28 kDa, respectively). L and M make up the core of the RC; they carry the cofactors: four bacteriochlorophylls, two bacteriopheophytins, two quinones of species-dependent chemical nature, and one nonheme Fe^{2+} atom. In many species, the RC also includes a bound tetraheme c-type cytochrome of about 12 kDa MW (when it is absent, its function of electron donation to P^+ is fulfilled by a soluble c_2 cytochrome). Proteins and cofactors are organized as sketched in Figure 16.1. The RC is endowed with an approximate C_2 symmetry in structure that is not found in the function. The primary donor P is a dimer of bacteriochlorophylls (a so-called special pair in which the two molecules are held in close proximity by the L and M subunits; they are in electronic interaction without establishing any chemical link). Excitation by light brings P to its lowest singlet excited state P^*, permitting a very fast ($t_{1/2} = 2$ ps) electron transfer to the bacteriopheophytin on the L branch (the involvement of a bacteriochlorophyll is still a matter of debate). An electron is then transferred to the quinone Q_A with $t_{1/2} = 200$ ps, and from there to the quinone Q_B ($t_{1/2}$ in the microsecond domain). Quinone Q_A functions as a one-electron carrier, but Q_B functions as a two-electron, two-proton carrier. After its full reduction by two electrons and

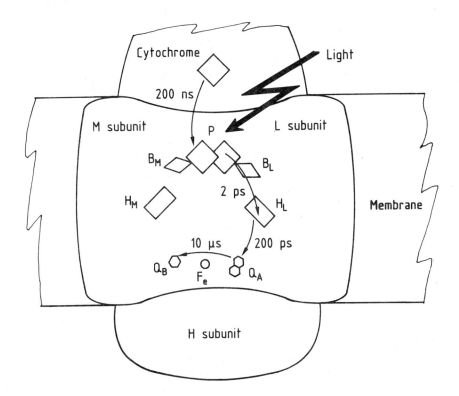

FIGURE 16.1 Schematic representation of the *Rhodopseudomonas viridis* reaction center (inspired by Reference 1), with approximate half-times of the three major steps of electron transfer on the active branch, and of one step from the cytochrome moiety (only partly drawn). Membrane thickness: about 35 Å.

two H^+, hydroquinone leaves the reaction center and diffuses in the membrane to reduce another redox protein, a cytochrome b/c complex, while another oxidized quinone binds at the Q_B site. P^+ is reduced by a c-type cytochrome, which is itself reduced by the b/c complex in a complex mechanism involving proton release. Altogether, the RC functions as a proton pump, pumping protons against their gradient across the membrane, while electron transfer is mostly cyclic.

Photosystem II

In oxygen-evolving organisms (plants, algae, cyanobacteria), the photosynthetic membranes contain two types of RCs, called Photosystem I and Photosystem II. The latter one presents strong analogies with the RC of purple bacteria.[3,4] The inner part of the RC is built from two polypeptides (D1 and D2) of about 32 kDa, which are homologous to L and M. The sequence of electron acceptors is also very similar: pheophytin a → quinone Q_A → quinone Q_B. Both quinones are plastoquinones A, and their functional properties are like their bacterial counterparts. Specific inhibitors block electron transfer by binding at the Q_B site in competition with plastoquinone. Several of them are used as herbicides: dichlorophenyl-dimethylurea and other urea derivatives, triazines, etc. Many of them bind also at the Q_B site in purple bacteria.

The Photosystem II reaction center, however, differs from the RC of purple bacteria in at least three important aspects.

1. It performs water oxidation, with concomitant O_2 evolution and proton release. The equilibrium ($2H_2O \leftrightarrow 4H^+ + 4e^- + O_2$) has an Em of +0.82 V. The reaction is pulled by the oxidized primary donor P^+ (P-680$^+$; its dimeric nature is still under debate). The P-680/P-680$^+$ couple thus must have an Em much more positive than +0.82 V. P-680$^+$ is directly reduced by a tyrosine residue (Tyr 161 of D_1, named Tyr Z). Tyr Z^+ oxidizes a cluster with four manganese atoms, which is the catalytic site for water oxidation. A histidine residue is probably involved in the process, which also requires Ca^{2+} and Cl^- ions. The details of the structure and mechanism of this ensemble are still under intense investigation.
2. This RC is highly complex, since it contains at least ten polypeptides. In addition to D_1 and D_2, there are the two subunits of cytochrome b-559, two chlorophyll-proteins named CP47 and CP43, several hydrophobic polypeptides of unknown function, and three peripheral proteins located at the water oxidizing site (MW: 33, 24, and 16 kDa).
3. The Photosystem II reaction center is susceptible to degradation by light in a process called photoinhibition, which is followed *in vivo* by a recovery requiring protein synthesis. Photoinhibition is usually thought to be an adaptative protective mechanism.

The Photosystem II RC has not yet been crystallized in a manner that could lead to a 3-D structure. The difficulties in the crystallization are presumably due to its complexity and to its inherent instability. However, its partial similarity with purple bacteria is very useful in making hypothetical models for further experimental tests.

Photosystem I

The other reaction center from oxygenic organisms, Photosystem I, is organized on the same basic principles as Photosystem II, but with important differences.[5-7] The core is made of two large polypeptides (PSI-A and PSI-B, of about 83 kDa) that carry a fairly large number of pigment molecules (about 50 chlorophyll a and β-carotene) in addition to the redox active cofactors: the primary donor P-700 (a dimer of chlorophyll a), a putative primary acceptor A_o (a specialized chlorophyll a), the secondary acceptor A_1 (phylloquinone), and a low-potential 4Fe-4S iron-sulfur center (Fx). After excitation, an electron goes from P-700* to A_o, then to A_1, and to Fx, and from there to a small polypeptide (PSI-C) holding two 4Fe-4S centers (F_A, F_B) which serve as a relay

Table 16.1   Sequences of Electron Carriers in Various Classes of Reaction Centers, Ranging from Most Terminal Donors (top) to Most Terminal Acceptors (bottom)

Purple bacteria	Photosystem II	Photosystem I	Green sulfur bacteria (Heliobacteria)
	$(Mn)_4$ +1000	Pc	
Cyt c +350	Tyr Z +1100	or +360	Cyt c +360
		Cyt c	
P(BChl$_2$) +490	P-680 +1200	P-700 +490	P-840 +420
I (BPheo) −600	I (Pheo a) −600	A$_o$ (Chl a ?) −1000	A$_o$ (BChl ?) −1100
		A$_1$(vit K$_1$) −850	A$_1$ (quinone) −900
(MQ) −300			
Q$_A$ or	Q$_A$(PQ) 100	(FeS) X −700	FeS "X" −800
(UQ) 100			
Q$_B$(UQ) +50	Q$_B$(PQ) +50	(FeS)A,B −550	(FeS)"A,B" −600

MQ, UQ, PQ: menaquinone, ubiquinone, plastoquinone, respectively. Pc: plastocyanin. FeS: iron-sulfur center. Other symbols: see text (paragraph 1). The numbers are redox potentials (vs. normal hydrogen electrode) in mV.

toward the physiological acceptor, a soluble 2Fe-2S ferredoxin (replaced by flavodoxin in some cases). Reaction kinetics are still fairly uncertain and even the path mentioned above is not established fully. All electron acceptors have a low Em (see Table 16.1). To complete the electron path, P-700$^+$ is re-reduced by a soluble electron donor (plastocyanin or cytochrome c-553) on the membrane side opposite to that of ferredoxin action.

It is often hypothesized that the Photosystem I RC has a fairly symmetrical structure and that A$_o$ and A$_1$ have a (perhaps inactive) structural counterpart. Crystals of the RC have been obtained and a 3-D structure determined at 6-Å resolution (data from the same laboratory at 4-Å resolution are now being analyzed).[8] The structure shows, in agreement with biochemical data, that Fx is at the interface between PSI-A and PSI-B polypeptides. In addition to the three polypeptides mentioned, the RC contains at least seven additional ones, belonging to three classes: hydrophobic transmembrane polypeptides with unknown function, one polypeptide located at the donor side and serving for the docking of plastocyanin, and at least two polypeptides (PSI-D and E) located at the acceptor side and serving for the docking of PSI-C and of ferredoxin.

Green Sulfur Bacteria and Heliobacteria

Like the purple bacteria, these bacteria have only one type of RC, but it resembles greatly the Photosystem I RC of oxygenic organisms. Their RCs are still poorly known, but the sequence of electron carriers is very similar to that of Photosystem I,[9,10] with small chemical differences in the structure of the chlorin pigments and of the quinones (see Table 16.1). Their polypeptides are also fairly large, like those of Photosystem I. However, recent sequencing led to very surprising results: in green sulfur bacteria[11a] and in heliobacteria,[11b] the RC core is made of two identical polypeptides. This property leads to interesting questions concerning the phylogeny of photosynthetic organisms and the pseudo-symmetry of reaction centers: there might be two identical paths or a unique central path, instead of one active and one inactive path as in purple bacteria. Green thermophilic (nonsulfur) bacteria are more closely related to purple bacteria.[12]

Altogether, it is possible to classify all reaction centers according to the chemical nature and corresponding Em values of their electron acceptors: purple bacteria, green nonsulfur bacteria, and Photosystem II on one side (with two quinones Q$_A$ and Q$_B$, and relatively high Em values); Photosystem I, green sulfur bacteria, and heliobacteria on the other side (with one low-potential quinone and three low-potential iron-sulfur centers).[13]

16.2 Methods for Studying Structural and Functional Properties

Biochemical Methods: Proteins

Reactions centers are complex membrane proteins and their study requires appropriate biochemical methods (only briefly mentioned here):

- Isolation of RCs requires the use of detergents, followed by classical methods of centrifugation, chromatography, etc. (see, e.g., Reference 14). Their polypeptidic composition is analyzed by dissociation of pure RCs with sodium dodecyl sulfate followed by gel electrophoresis under denaturing conditions.

- Sequencing of polypeptides is done either directly for rather small polypeptides or, more generally, by sequencing the genes. Eventual processing of the preprotein can be checked by comparison with N- or C-terminal partial sequences of the polypeptide.

- Crystallization of RCs in view of structure determination by X-ray crystallography requires special methods that have worked successfully for two kinds of purple bacteria (*Rhodopseudomonas viridis* and *Rhodobacter sphaeroides*) and for Photosystem I of cyanobacteria.

- Successful methods for the biochemical manipulation of RCs have been developed recently: site-directed mutagenesis (for three purple bacteria [*R. capsulatus, R. sphaeroides, Rps. viridis*], for several species of cyanobacteria, and for the green alga Chlamydomonas reinhardtii), protein extraction (or denaturation), and reconstitution (which was specially successful for the iron-sulfur centers of Photosystem I and for the peripheral polypeptides of Photosystem II). Selective pigment extraction, modification, and reconstitution are also in progress.

Optical Spectroscopy

Methods of optical spectroscopy are especially important because they provide plentiful information on RCs, where pigments are involved directly in energy or electron transfer, and where redox centers also have characteristic optical properties.[15]

Absorption Spectroscopy

All photosynthetic pigments have intense and distinctive absorption spectra, in relation to the ability of each of them to capture specific parts of the solar light spectrum.[16] Considering chlorophylls and bacteriochlorophylls, these pigments belong to the chemical group of chlorins. Their reduced symmetry compared to porphyrins results in a strong Q_Y band absorbing well into the red or IR, the position of which is quite sensitive to the local environment of the molecule.[17,18] From a single chemical molecule, such as chlorophyll *a* (Figure 16.2) or bacteriochlorophyll *b* (Figure 16.3), complex spectra are thus obtained by pigment-pigment and pigment-protein interactions. This effect is especially clear for the primary donor P in purple bacteria (Figure 16.3). Absorption spectroscopy thus remains a basic technique in separation procedures, to check for the good state and for the purity of samples.

Difference Spectra and Kinetic Properties

Absorption of light induces the formation of excited states and, after charge separation, of reduced or oxidized species. The molecules involved experience a change in their absorption spectrum that can be detected with a high sensitivity by differential spectroscopy.[15] This technique is often coupled to flash excitation and to time-resolved detection (as shown in Figure 16.4), permitting identification of successive partners in energy or electron transfer (from the difference spectra), kinetics of successive steps, and the factors that control the reactions. This methodology is in constant

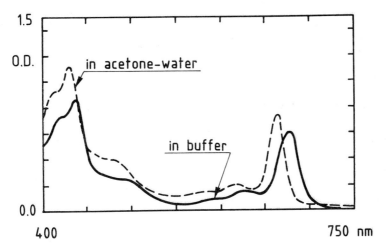

FIGURE 16.2 Absorption spectra of: (a) PS I reaction centers from the cyanobacterium Synechocystis 6803 isolated with the detergent dodecylmaltoside (kindly provided by Dr. H. Bottin); (b) the same material (same pigment concentration) extracted with an acetone-water (80:20) mixture.

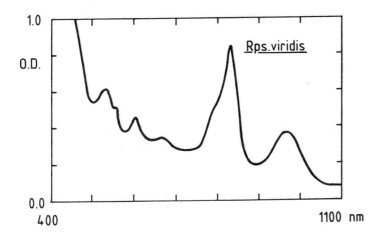

FIGURE 16.3 Absorption spectrum of a suspension of *Rhodopseudomonas viridis* reaction centers (plus 100 μ*M* ascorbate) dispersed by a detergent (LDAO). The absorption band at 970 nm is due to the primary donor P, and the large massif around 820 nm is mainly due to the two bacteriochlorophylls other than P and to the two bacteriopheophytins. (Reaction centers kindly provided by Dr. J. Breton.)

development (see Reference 15 for a detailed presentation), especially in terms of time resolution since the pump-probe methods now provide a resolution of 0.1 ps.[19,20]

Fluorescence

Fluorescence spectroscopy has been used for a long time to investigate energy transfer in photosynthetic membranes.[21] Both spectral and time-resolved properties are considered. Fluorescence properties are still of interest in many cases (e.g., for the screening of photosynthetic mutants in purple bacteria or for studying phycobiliproteins), but their impact is especially important in the study of Photosystem II because the fluorescence yield of chlorophyll *a* in the chloroplast is dependent on several properties of the Photosystem II RC: (1) the redox state of Q_A and, by way of consequence, properties related to electron transfer to Q_B (herbicide binding and resistance, photoinhibition, etc.); (2) properties more related to the water-oxidizing side and to the *in vivo* functioning of photosynthesis, such as the pH inside thylakoids or defects in the manganese cluster.[22,23] Also

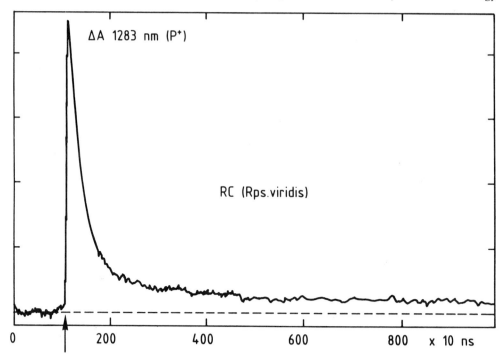

FIGURE 16.4 Absorption change (at 1283 nm) due to P oxidation by a flash, followed by its re-reduction by the high-potential heme of the tetraheme cytochrome in *Rhodopseudomonas viridis* reaction centers at 287 K, at a redox potential of about +200 mV (see Reference 27).

informative on the state of Photosystem II, thermoluminescence is a technique in full development that can be used *in vivo* with cells or leaves, without isolation of RCs or membranes.

Vibrational Spectroscopy

Molecular vibrations are studied by IR absorption and by Raman scattering spectroscopy. Both methods are fairly classical for small molecules. Their utilization for the complex molecules involved in photosynthesis was made possible by recent technological developments in lasers and detectors. Resonance Raman spectroscopy is mostly informative on the mode of binding of cofactors (pigments, redox centers) with polypeptides (see Bocian, this volume, and Reference 24). Fourier transform infrared (FT-IR) spectroscopy is used mainly in a differential mode to investigate the structural changes that are coupled to electron transfer, involving polypeptides, cofactors, water molecules, etc. Most studies have been made on RCs of purple bacteria where differential FT-IR spectra were recorded for the oxidation of P, reduction of bacteriopheophytin, reduction of Q_A or Q_B, and oxidation of cytochrome (see e.g., Reference 25). The difference spectra contain a rich content of information (as exemplified in Figure 16.5) that starts to be unraveled, often by comparison with models *in vitro* or by selective labeling with stable isotopes (^{13}C, ^{15}N, ^{18}O).

Other Optical Techniques

The large number of questions raised by the structure/function relations in RCs and the wealth of their optical properties have led to an intensive use of many methods that cannot be examined in detail here (see Reference 15). A few of them are mentioned here. Linear dichroism with oriented samples or with photoselection gives the relative and absolute orientations of practically all cofactors.[26] Circular dichroism allows the probing of the interaction between pigments or with the protein. A few methods give access to excited-state properties: phosphorescence (for triplet states), hole burning for singlet excited states, the Stark effect (in absorption or fluorescence) for evaluation

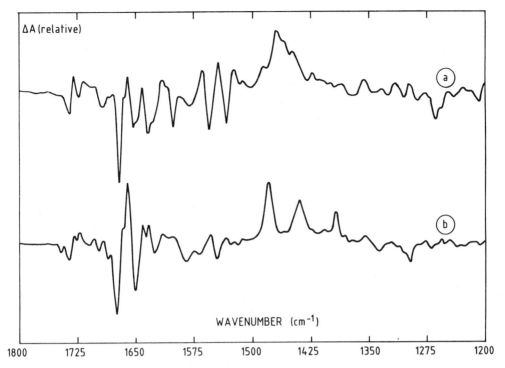

FIGURE 16.5   FT-IR difference spectra (light minus dark) of reaction centers of *Rhodobacter sphaeroides* (a) and *Rhodopseudomonas viridis* (b), at 10°C. The spectra are due to the reduction of Q_A (Q_A^- minus Q_A). Q_A is ubiquinone-10 in (a) and menaquinone in (b). The features between 1380 and 1500 cm^{-1} are attributable to the quinone anion, while the features around 1530 to 1570 cm^{-1} are due to an indirect effect of Q_A reduction on the protein. (Spectra kindly provided by Dr. E. Nabedryk.)

of the charge-transfer character of singlet excited states, etc. Most of these methods are necessary complements to X-ray crystallography, by the kind of information they provide and by the fact that they can be used much more extensively on mutants or on noncrystallized RCs.

Electron Paramagnetic Resonance

EPR is so basic in studies of photosynthetic reaction centers that it should be mentioned here, although very briefly. Detailed presentations have been published for the technique and for its applications to RCs.[28,29] Briefly, it can be said that EPR detects three types of molecular states, all having a non-zero spin: triplet states, radical ions (cations or anions), and centers with transition metals. These states are quite relevant to the functioning of reaction centers.

- The triplet state of P, designated ^{3}P, is populated as a result of charge recombination in the primary radical pair P$^+$I$^-$. Under conditions of normal functioning, this radical pair is very short-lived (well under 1 ns) because of a rapid electron transfer from the primary acceptor I to the next acceptor. When that transfer is made impossible, usually because the acceptor is either already reduced or absent, the radical pair may live approximately 10 to 100 ns and decay primarily by charge recombination with a rather high (10 to 100%) probability of populating ^{3}P. This artifactual reaction is highly informative with regard to interactions between P and the primary electron acceptors, and to structural properties of P itself: orientation, degree of symmetry of the pair, etc. It is thus studied frequently, mainly by EPR spectroscopy.[30]

- Radicals are formed in electron transfer involving organic molecules like chlorins (P, bacteriochlorophylls or bacteriophaeophytins), quinones (Q_A, Q_B, A_1 in PS-I-like RCs), tyrosine

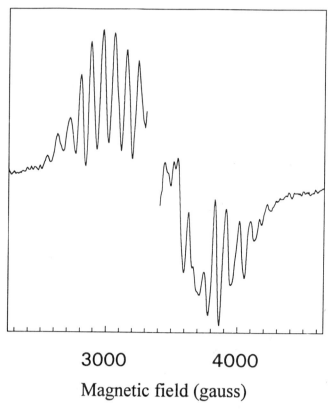

3000   4000
Magnetic field (gauss)

FIGURE 16.6   EPR spectrum measured in Photosystem II membranes at 10 K (X band at 9.45 GHz; modulation amplitude: 25 G). Difference between a sample illuminated at 200 K and a sample kept in darkness. The spectrum is attributed essentially to the S_2 state of the $(Mn)_4$ water-oxidizing enzyme. (Kindly provided by Dr. A. Boussac.)

(Tyr Z or Tyr D in PS II), etc. Classical EPR spectroscopy, now complemented by more elaborate methods such as pulse EPR or ENDOR, is the most efficient way for their study (see examples of applications in References 4, 5, or 9).

- Metal centers. Many of them participate in secondary steps of electron transfer in RCs: iron-sulfur centers (in low-potential RCs), cytochromes (in purple bacteria and PS II), Mn cluster in PS II, etc. Each of them is EPR-detectable at low temperature in one or the other of their redox states: the reduced state for iron-sulfur centers, the oxidized one for cytochromes. The situation is more complex for the Mn cluster, which includes four Mn ions (Figure 16.6). EPR is the best tool for determination of the redox state of the metal center and of some of its structural properties.[4,5,9,28,29]

Electrochemical Properties

Midpoint redox potentials of the redox centers bound to the RC polypeptides are very basic properties related to their functional behavior. They define naturally the possible directions of electron flow, but they are also involved in defining the rates of electron transfer steps and the energetic yield of photoconversion. Redox potentials are measured by very classical methods where a potential is imposed chemically or electrically, and where the redox state of the studied couple is measured, usually by a spectroscopic method such as EPR or absorption spectroscopy.

The approximative redox potentials of redox centers are given in Table 16.1, in a manner that emphasizes similarities between different reaction centers. Some values are not known precisely, mainly because they are either too positive (donor side of PS II) or too negative (first acceptors of PS I and of green sulfur bacteria or heliobacteria). Potentials depending on pH are given for pH = 7.0. The table shows clearly that electron-accepting and -donating sides can be classified into two separate classes that are intermixed. This has been recognized for a long time, but the new findings on green sulfur bacteria and heliobacteria raise interesting questions on the phylogeny of photosynthetic organisms.[13]

16.3 Facts and Questions

The availability of crystallographic 3-D structures makes photosynthetic reaction centers more and more interesting objects. They serve as models for membrane proteins since no other atomic structure has yet been obtained for these. They are also important models for electron-transferring proteins in general because of the possibility to trigger precisely charge separation with short light pulses and to study it by many spectroscopic methods.

It should be kept in mind, however, that the relationships between the structure and the function of reaction centers include a large number of unsolved basic problems that are the motivation of a strong research effort:

- In the best-known purple bacteria, understanding of the very primary electron transfer step(s) is still rudimentary: why is a dimer needed as primary electron donor, what is the function of the monomeric bacteriochlorophyll located on the active branch, what makes possible a high yield of charge separation even at low temperature, why are there two branches in the RC structure and why is only one of them active?

- How can a better knowledge of plant reaction centers be arrived at? To what extent can the knowledge acquired on purple bacteria be used to understand better PS I and PS II? How will it be possible to understand the mechanism of water oxidation?

- Electron transfer from Q_A to Q_B in purple bacteria and in PS II is sensitive to many parameters, including the presence of a metal center (most often Fe^{2+}). How does this step work? How is it coupled to H^+ uptake? Are there privileged paths for H^+ around Q_B?

Quite generally, the description of function is largely limited to the cofactors (pigments and redox centers). Apart from scaffolding, what are the roles of the protein? How are its structural changes coupled to electron or H^+ transfer? In order to receive satisfying answers, all these important questions will require concerted multidisciplinary approaches in the future.

References

1. Deisenhofer, J. and Michel, H., The photosynthetic reaction center from the purple bacterium Rhodopseudomonas viridis, *EMBO J.*, 8, 2149, 1989.
2. Feher, G., Allen, J. P., Okamura, M. Y., and Rees, D. C., Structure and function of bacterial photosynthetic reaction centers, *Nature*, 339, 111, 1989.
3. Mathis, P. and Rutherford, A. W., The primary reactions of photosystems I and II of algae and higher plants, in *Photosynthesis*, Amesz, J., Ed., Elsevier Science, Amsterdam, 1987, chap. 4.
4. Rutherford, A. W., Zimmermann, J. L., and Boussac, A., Oxygen evolution, in *The Photosystems: Structure, Function and Molecular Biology*, Barber, J., Ed., Elsevier, Amsterdam, 1992, chap. 5.
5. Golbeck, J. H. and Bryant, D. A., Photosystem 1, *Current Topics in Bioenergetics*, 16, 83, 1991.
6. Almog, O., Shoham, G., and Nechushtai, R., Photosystem I: Composition, organization and structure, in *The Photosystems: Structure, Function and Molecular Biology*, Barber, J., Ed., Elsevier, Amsterdam, 1992, chap. 11.
7. Sétif, P., Energy transfer and trapping in Photosystem I, in *The Photosystems: Structure, Function and Molecular Biology*, Barber, J., Ed., Elsevier, Amsterdam, 1992, chap. 12.

8. Krauss, N., Hinrichs, W., Witt, I., Fromme, P., Pritzkow, W., Dauter, Z., Betzel, C., Wilson, K. S., Witt, H. T., and Saenger, W., Three-dimensional structure of system 1 of photosynthesis at 6 Å resolution, *Nature*, 361, 326, 1993.

9. Nitschke, W., Sétif, P., Liebl, U., Feiler, U., and Rutherford, A. W., Photosynthetic reaction center of green sulfur bacteria studied by EPR, *Biochemistry*, 29, 3834, 1990.

10. Nitschke, W., Sétif, P., Liebl, U., Feiler, U., and Rutherford, A. W., Reaction center photochemistry of *Heliobacterium chlorum*, *Biochemistry*, 29, 11079, 1990.

11. (a) Büttner, M., Xie, D. L., Nelson, H., Pinther, W., Hauska, G., and Nelson, N., Photosynthetic reaction center genes in green sulfur bacteria and in Photosystem 1 are related, *Proc. Natl. Acad. Sci., U.S.A.*, 89, 8135, 1992; (b) Liebl, U., Mockensturm-Wilson, M., Trost, J. T., Brune, D. C., Blankenship, R. E., and Vermaas, W., Single core polypeptide in the reaction center of the photosynthetic bacterium *Heliobacillus mobilis*: structural implications and relations to other photosystems, *Proc. Natl. Acad. Sci. U.S.A.*, 90, 7124, 1993.

12. Shiozawa, J., Lottspeich, F., Oesterhelt, D., and Feick, R., The primary structure of the *Chloroflexus aurantiacus* reaction center polypeptides, *Eur. J. Biochem.*, 180, 75, 1989.

13. Nitschke, W. and Rutherford, A. W., Photosynthetic reaction centers: Variations on a common structural theme?, *Trends Biochem. Sci.*, 16, 241, 1991.

14. Gingras, G., A comparative review of photochemical reaction center preparations from photosynthetic bacteria, in *The photosynthetic Bacteria*, Clayton, R. K. and Sistrom, W. R., Eds., Plenum Press New York, 1978, chap. 6.

15. Mathis, P., Optical techniques in the study of photosynthesis, in *Methods in Plant Biochemistry*, Vol. 4, Bowyer, J., Ed., Academic Press, New York, 1990, 231–258.

16. Lichtenthaler, U. K., Chlorophylls and carotenoids: Pigments of photosynthetic biomembranes, in *Methods in Enzymology*, 148, 351, 1987.

17. Scheer, H., *Chlorophylls*, CRC Press, Boca Raton, FL, 1991.

18. Hoff, A. J. and Amesz, J., Visible absorption spectroscopy of chlorophylls, in *Chlorophylls*, Scheer, H., Ed., CRC Press, Boca Raton, FL, 1991, 723–738.

19. Martin, J. L., Breton, J., Hoff, A. J., Migus, A., and Antonetti, A., Femtosecond spectroscopy of electron transfer in the reaction center of the photosynthetic bacterium *Rh. sphaeroides* R-26, *Proc. Natl. Acad. Sci., U.S.A.*, 83, 957, 1986.

20. Dressler, K., Umlauf, E., Schmidt, S., Hamm, P., Zinth, W., Buchanan, S., and Michel, H., Detailed studies of the subpicosecond kinetics in the primary electron transfer of reaction centers of *Rhodopseudomonas viridis*, *Chem. Phys. Lett.*, 183, 270, 1991.

21. Govindjee and Fork, D. C., *Light Emission by Plants and Bacteria*, Academic Press, New York, 1985.

22. Horton, P. and Bowyer, J. R., Chlorophyll fluorescence transients, in *Methods in Plant Biochemistry*, Vol. 4, Bowyer, J. R., Ed., Academic Press, New York, 1990, chap. 9.

23. Krause, G. H. and Weiss, E., Chlorophyll fluorescence and photosynthesis: The basics, *Annu. Rev. Plant Physiol. Plant Mol. Biol.*, 42, 313, 1991.

24. Lutz, M. and Mäntele, W., Vibrational spectroscopy of chlorophylls, in *Chlorophylls*, Scheer, H., Ed., CRC Press, Boca Raton, FL, 1991, 855–902.

25. Nabedryk, E., Bagley, K., Thibodeau, D. L., Bauscher, M., Mäntele, W., and Breton, J., A protein conformational change associated with the photoreduction of the primary and secondary quinones in the bacterial reaction center, *FEBS Lett.*, 266, 59, 1990.

26. Breton, J. and Vermeglio, A., Orientation of photosynthetic pigments *in vivo*, in *Photosynthesis: Energy Conversion by Plants and Bacteria*, Govindjee, Ed., Academic Press, New York, 1982, chap. 4.

27. Ortega, J. M. and Mathis, P., Electron transfer from the tetraheme cytochrome to the special pair in isolated reaction centers of *Rhodopseudomonas viridis*, *Biochemistry*, 32, 1141, 1993.

28. Hoff, A. J., Applications of ESR in photosynthesis, *Phys. Rep.*, 54, 75, 1979.

29. Hoff, A. J., *Advanced EPR: Applications in Biology and Biochemistry*, Elsevier, New York, 1989.

30. Budil, D. E. and Thurnauer, M. C., The chlorophyll triplet state as a probe of structure and function in photosynthesis, *Biochim. Biophys. Acta*, 1057, 1, 1991.

17

Photosystem I: Resolution and Reconstitution of Polypeptides and Cofactors

John H. Golbeck
University of Nebraska

17.1   Charge Separation in Photosystem I

Photosynthetic reaction centers (PRCs) such as Photosystem I (PS I) and Photosystem II (PS II) are the specialized pigment-protein complexes in green plants and cyanobacteria that function to convert light to chemical free energy. In all known instances, PRCs consist of dimers of hydrophobic proteins that contain cofactors which absorb light to produce a stable charge separation across a membrane. The similarities in the photosystems lie in the early events in photochemical charge separation. The components of the purple non-sulfur bacterial and PS II RCs are depicted below:

$$P_{870}\, Bph\, Q_A \xrightarrow{h\nu} P_{870}{}^*\, Bph\, Q_A \xrightarrow{2.8\ ps} P_{870}{}^+\, Bph^-Q_A \xrightarrow{230\ ps} P_{870}{}^+\, Bph\, Q_A{}^-$$

$$P_{680}\, Ph\, Q_A \xrightarrow{h\nu} P_{680}{}^*\, Ph\, Q_A \xrightarrow{2.8\ ps} P_{680}{}^+\, Ph^-Q_A \xrightarrow{200\ ps} P_{680}{}^+\, Ph\, Q_A{}^-$$

where P_{860} is the primary electron donor bacteriochlorophyll, Bph is the primary electron acceptor bacteriopheophytin and Q_A is a bound molecule of ubiquinone or menaquinone in the purple bacterial reaction center; P_{680} is the primary electron donor chlorophyll, Ph is the primary electron acceptor pheophytin, and Q_A is a bound molecule of plastoquinone in PS II. The components of the green sulfur bacterial and PS I RCs are depicted below:

$$P_{840}A_0 \xrightarrow{?\ ps} P_{840}\, A_0 \xrightarrow{?\ ps} P_{840}{}^+\, A_0{}^- \xrightarrow{?\ ps} P_{840}{}^+\, A_0\, A_1{}^-\ (?)$$

$$P_{700}A_0A_1 \xrightarrow{<1.5\ ps} P_{700}{}^*\, A_0\, A_1 \xrightarrow{\sim10\ ps} P_{700}{}^+\, A_0{}^-\, A_1 \xrightarrow{35-200\ ps} P_{700}{}^+A_0\, A_1{}^-$$

where P_{840} is the primary electron donor, A_0 is the primary electron acceptor, and A_1 is presumed to be the bound molecule of quinone in the green sulfur bacterial RC; P_{700} is the primary electron donor chlorophyll *a*, A_0 is the primary electron acceptor chlorophyll *a*, and A_1 is bound molecule of phylloquinone in PS I. It is apparent that the same photochemical motif and very similar kinetics are represented in all types of known PRCs.[1,2]

The differences in the two classes of photosystems lie in the redox potentials of the components and in the fate of the electron on the donor and acceptor sides of the photoacts. In PS II, the primary donor P680 has a redox potential (E_m > +0.800 mV) high enough to oxidize water; whereas in PS I, the primary donor P700 has a redox potential (E_m = +430 mV) only high enough to oxidize plastocyanin (and cytochrome c_6). In PS I, the terminal electron acceptors F_A and F_B have redox potentials (E_m = −520 and −580 mV, respectively) low enough to reduce NADP+, whereas in PS II, the terminal electron acceptor, Q_A, is a quinone that has a redox potential only low enough to reduce a second quinone, Q_B, to the level of hydroquinone (through two successive electron transfers). The other difference lies in the mechanism that governs the fate of the electron on the reducing side of the photoacts. In PS II, the electron is retained within the boundaries of the lipid bilayer by reducing a second quinone, Q_B. The resulting plastohydroquinone is released and diffuses within the bilayer to the cytochrome b_6/f complex, where it becomes oxidized.[3] In PS I, the electron is transferred out of the reaction center to a stromal protein, ferredoxin, through the participation of three iron-sulfur clusters, F_X, F_B, and F_A. The reduced ferredoxin reduces NADP+ through a flavin-containing enzyme, ferredoxin:NADP+ oxidoreductase. PS II is therefore considered a "quinone-type" RC in which the bound primary quinone, Q_A, donates its electron to a secondary quinone, Q_B. In contrast, the PS I is considered an "iron-sulfur-type" RC in which the bound primary quinone, A_1, donates its electron to an iron-sulfur cluster, F_X, instead of to the secondary quinone. F_X, in turn, donates its electron to the F_A/F_B iron-sulfur clusters and then on to soluble ferredoxin, FNR, and NADP+ (reviewed in References 4–14).

This chapter focuses on the biochemical resolution of the cyanobacterial PS I complex into a series of subcomplexes in which the electron acceptors are stripped selectively from the PS I RC. The following subsections outline the isolation and properties of the P700-F_A/F_B complex, the P700-F_X core, the P700-A_1 core, the P700-A_0 core, and the reconstitution of the PS I complex from the isolated and purified components.

17.2 Photosystem I Subthylakoid Preparations

P700-F_A/F_B Complex

Figure 17.1 shows a stylized model of the PS I complex within the thylakoid membranes of a cyanobacterium (the high plant RC is similar, except for the addition of PsaG and PsaH, and the absence of PsaM). The PS I complex in green plants also contains LHCI, the four light-harvesting chlorophyll proteins that contain Chl *b* as well as Chl *a*. These added antennae serve to increase the optical cross-section and extend the spectral range to allow more efficient photon capture. The intact P700-F_A/F_B complex is isolated from green plant or cyanobacterial thylakoid membranes with the use of nonionic detergents and purified by ultracentrifugation in a sucrose gradient[15] or by ion-exchange chromatography on Mono Q.[16] The detergent can be either 1% Triton X-100, 30 m*M* dodecyl maltoside (or octyl glucoside), or 1% digitonin. The P700-F_A/F_B complex thus isolated retains the complete complement of bound PS I components, including the terminal electron acceptors F_A and F_B.

The properties of the cyanobacterial P700-F_A/F_B complex are provided in Table 17.1. In terms of structure and function, the detergent-solubilized P700-F_A/F_B complex is nearly equivalent to the thylakoid-bound PRC. The cyanobacterial PS I complex contains ca. 100 Chl/P700, and the polypeptide composition includes PsaA, PsaB, PsaC, PsaD, PsaE, PsaF, PsaI, PsaJ, PsaK, PsaL, and PsaM. The higher plant PS I complex contains ca. 200 Chl/P700 (except for the Triton-isolated

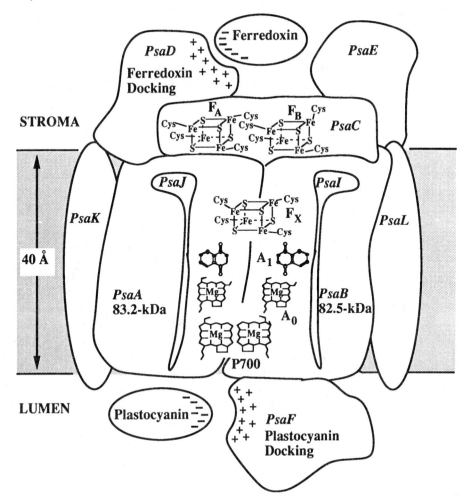

FIGURE 17.1   Proposed architecture for the P700-F_A/F_B complex constructed from biochemical, spectroscopic, and genetic evidence. P700 is located on the lumenal side of the membrane, and the F_X interpolypeptide iron-sulfur cluster is shown close to the stromal side of the membrane. The two molecules of phylloquinone are assumed to be distributed equally between the PsaA and PsaB proteins. The recent crystal structure of the PS I complex at 6-Å resolution shows that the iron-sulfur clusters form an irregular triangle, with the two [4Fe-4S] clusters in PsaC more normal to the membrane than depicted here. Only those polypeptides common to both the eukaryotic and cyanobacterial reaction centers are illustrated. The PsaC, PsaD, and PsaE proteins, located on the stromal side of the membrane, are removed by chaotropic agents, leading to a P700-F_X core. The PsaF protein is located on the lumenal side of the membrane and can be removed with 1% Triton X-100. The PsaI, PsaJ, PsaK, and PsaL proteins are membrane-spanning and have no defined functions.

complex, which strips the LHCI complement from the ca. 100 Chl/P700 PS I RC), and the polypeptide composition includes PsaA, PsaB, PsaC, PsaD, PsaE, PsaF, PsaG, PsaH, PsaI, PsaJ, PsaK, PsaL, and PsaN. The purified P700-F_A/F_B complexes are able to reduce NADP+ (at rates of 600 μmol mg^{-1} Chl/h in cyanobacteria, 800 μmol mg^{-1} Chl/h in green plants) when supplied with the exogenous soluble proteins ferredoxin and ferredoxin:NADP+ oxidoreductase (FNR). In the absence of soluble electron acceptors, the charge recombination between P700+ and [F_A/F_B]$^-$ has a half-life of about 30 ms at room temperature. The low-temperature ESR spectrum shows the photoreduction of F_A at g = 2.05, 1.94, and 1.86 (there is a small population of F_B reduced at g = 2.07, 1.92, and 1.88). When the F_A and F_B reclusters are reduced chemically with sodium hydrosulfite at high pH values (>10.0), or when the complex is illuminated during freezing to

Table 17.1

Preparation	Acceptors	Proteins	Comments
P700-F_A/F_B Complex	A_0,A_1,F_X,F_B,F_A	PsaA,B,C,D,E.,F,I,J,K,L,M	Cyanobacteria only
		PsaA,B,C,D,E.,F,G,H,I,J,K,L,N	Higher plants (contains LHCI if glycosidic detergents used) PsaF is removed with high amounts of Triton X-100
P700-F_X Core	A_0,A_1,F_X	PsaA,B,F,I.J.K,L,M	Cyanobacteria lose PsaC,D,E
		PsaA,B,F,G,(H),I,J,K,L,N	Higher plants lose PsaC,D,E and some PsaH
P700-A_1 Core	A_0,A_1	Same as P_{700}-F_X Core	F_X is oxidatively denatured
P700-A_0 Core	A_0	Same as P_{700}-F_X Core	A_1 is removed or displaced

photoaccumulate both F_A^- and F_B^-, the resulting interaction-type spectrum shows four lines at $g =$ 2.05, 1.94, 1.92, and 1.89 (Figure 17.2). The g values of F_A and F_B differ slightly between cyanobacterial and green plant PS I complexes, and are precise enough be used as a means of identification.[17] Mössbauer[18] and EXAFS[19] spectroscopy support the identification of F_A and F_B as [4Fe-4S] clusters. Redox studies show pH-independent midpoint potentials of –520 mV for F_A and –580 mV for F_B.[20,21] The complex from the thermophilic cyanobacterium *Synechococcus* sp. has been crystallized and the 3-D structure has been determined to 6 Å.[22] The PS I reaction center shows a pseudo C_2 plane of symmetry that runs between P700 and F_X, and the electron density map is interpreted to show the presence of a bridging chlorophyll molecule between P700 and A_0 similar to the bridging bacteriochlorophyll in the purple bacterial reaction center. The three iron-sulfur clusters are positioned in the form of an irregular triangle, implying a serial pathway of electron flow from F_X to ferredoxin. It is not certain which cluster represents F_A or F_B, and it is equally unclear whether one or both clusters participate in the reduction of ferredoxin and/or flavodoxin.

P700-F_X Core

High concentrations of chaotropes affect the loss of the low molecular mass, stromal polypeptides PsaC, PsaD, and PsaE from the P700-F_A/F_B complex without destroying the interpolypeptide F_X cluster. The P700-F_X core is isolated from the P700-F_A/F_B complex with the use of 6.8 M urea or 1.2 M NaI followed by dilution and gel filtration chromatography to remove the low molecular mass polypeptides and chaotropic agents.[23] Table 17.2 shows the types and concentrations of chaotropes that have been used successfully in green plant and cyanobacterial RCs. The initial mechanism of action is most likely the oxidative denaturation of the F_A and F_B iron-sulfur clusters, which leads to a loss of the 3-D structure of PsaC and, hence, its ability to bind to the PS I core. Since PsaD and PsaE require PsaC to bind to PS I,[24] their loss becomes a secondary, rather than primary, effect of the chaotropes.

The properties of the cyanobacterial P700-F_X core are provided in Table 17.1. The primary effect of chaotropes is to oxidatively denature the F_A and F_B iron-sulfur clusters, which results in the selective loss of the PsaC, PsaD, and PsaE polypeptides from the PS I core. The purified P700-F_X core is unable to reduce NADP+,[25] but it can reduce methyl viologen when the concentration of the latter exceeds 100 μM.[26] Light-induced charge separation takes place between P700 and F_X, and the back reaction has a half-life of about 1 ms in the absence of electron acceptors. The ESR spectrum shows the F_X resonances at $g =$ 2.05, 1.86, and 1.76 on illumination at 9K, and less than half the number of spins are measured when the sample is frozen prior to illumination. There is an inefficiency in the A_1^- to F_X electron transfer at low temperature that limits the quantum yield to ~0.3 at 15K.[27] The Mössbauer spectrum[28] and the EXAFS spectrum[29] in the P700-F_X core is compatible with the identification of F_X as a [4Fe-4S] cluster ligated between the PsaA and PsaB polypeptides in PS I.

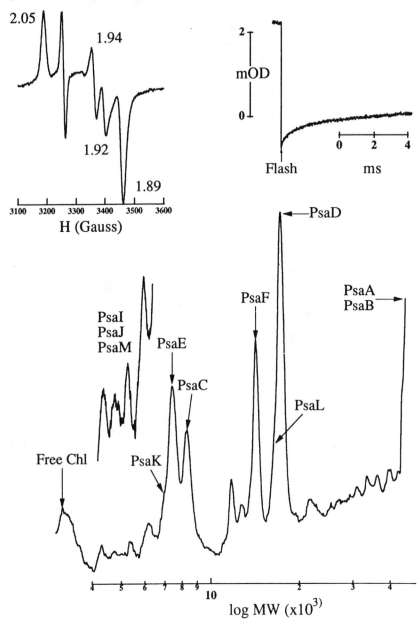

FIGURE 17.2   Spectroscopic properties and polypeptide composition of the P700-F_A/F_B complex. The P700-F_A/F_B complex was isolated from *Synechococcus* sp. PCC 6301 membranes with Triton X-100, followed by ultracentrifugation in a sucrose gradient. Top Left: Low-temperature ESR spectrum of the F_A/F_B iron-sulfur clusters showing principal *g*-values. The sample was suspended in 50 m*M* Tris buffer, pH 8.3, containing 1 m*M* sodium ascorbate and 0.3 m*M* DCPIP at 125 μg Chl per milliliter and illuminated during freezing. Spectrometer conditions: temperature, 19K; microwave power, 20 mW; microwave frequency, 9.128 GHz; receiver gain, 5.0 × 10³; modulation amplitude, 10 G at 100 kHz. Top Right: Room-temperature flash-induced absorption transient of P700. Measurements at 698 nm were performed on 5 μg Chl per milliliter in 50 m*M* Tris buffer, pH 8.3, containing 1.7 m*M* ascorbate and 33 μ*M* DCPIP. Bottom: Laser densitometric tracing of the polypeptides present in the P700-F_A/F_B complex from *Synechococcus* sp. PCC 6301 after separation by SDS-PAGE. The identities of the PsaA, PsaB, PsaC, PsaD, PsaE, PsaF, PsaL, and PsaK proteins were determined by N-terminal amino acid sequencing. The elevated inset shows an enlargement of the very low molecular mass polypeptides in the region from 4.5 to 8 kDa. The PsaI, PsaJ, and PsaM proteins were not sequenced, but are presumed to be present on the basis of analogy with other cyanobacteria. See Reference 24 for details.

Table 17.2   Minimal Chaotrope Concentration Required
for Functional Removal of PsaC, PsaD, and PsaE from the
Photosystem-I Reaction Center

	Synechococcus (mesophilic) (M)	*Synechococcus* (thermophilic) (M)	Spinach (M)
Urea	6.8	6.8	5
NaBr	6.0	6.0	4.7
NaI	2.0	4.0	2.0
NaSCN	2.0	3.5	1.5
NaClO$_4$	2.0	4.0	1.3

P700-A$_1$ Core

The F$_X$ iron-sulfur cluster can be denatured oxidatively to the level of zero-valence sulfur without affecting charge separation between P700 and A$_1$. In this process, the iron is lost and the labile sulfide becomes cross-linked to the cysteines that had formerly held the iron atoms. The denatured cluster is unavailable, therefore, to accept or transfer electrons. The P700-A$_1$ core is isolated by treating the P700-F$_X$ core with 5 mM potassium ferricyanide in the presence of 2 M urea.[30] The urea is necessary to render the F$_X$ cluster accessible to the oxidant, but the concentration of Triton X-100 must be 0.04% or lower, else the phylloquinone will be displaced.

The properties of the cyanobacterial P700-A$_1$ core are provided in Table 17.1. Treatment with K$_3$Fe(CN)$_6$ in the presence of urea results in the total loss of the F$_X$ iron-sulfur cluster, with resulting charge separation occurring between P700 and A$_1$. The purified P700-A$_1$ core is unable to reduce NADP+ or methyl viologen regardless of the concentration. Unlike a P700-F$_A$/F$_B$ complex, where the back reaction between P700$^+$ and A$_1^-$ populates the triplet state of P700 (when the iron-sulfur clusters are reduced[31]), the charge recombination between P700$^+$ and A$_1^-$ in the P700-A$_1$ core repopulates the ground state of P700.[32] The back reaction kinetics in the P700-A$_1$ core are complex, with a 10-μs kinetic phase dominant, but there are multiple slower phases present that may represent a distribution of P700$^+$ A$_1^-$ recombination rates. The reason for the mixed kinetics is not known. The UV/blue difference spectrum for the 10-μs kinetic phase and also for the slower kinetic phases shows a positive absorption band from 340 to 405 nm that resembles that of a semiquinone anion radical [Brettel, K. and Golbeck, J. H., unpublished results]. The electron spin polarized (ESP) ESR spectrum of the P700-A$_1$ core shows a typical |E|A|E|(A) pattern with no evidence of a "late signal" [Van der Est, A., Stehlik, D., and Golbeck, J. H., unpublished results]. The decay of the ESP ESR spectrum of A$_1^-$ in the absence of F$_X$ is due entirely to spin-lattice relaxation rather than electron transfer.

P700-A$_0$ (Triton) Core

The P700-A$_0$ core can be isolated by treating the P700-A$_1$ core with 1% Triton X-100 at 100 μg ml^{-1} Chl at room temperature.[30,32] Nonionic detergents remove or displace the two phylloquinone molecules in the P700-A$_1$ core without destroying the primary charge separation between P700 and A$_0$.

The properties of the cyanobacterial P700-A$_0$ core are provided in Table 17.1. Treatment with Triton X-100 results in the displacement of the A$_1$ acceptor, but no chlorophylls or carotenoids are removed. The purified A$_0$ core is unable to reduce any added electron acceptors except for certain quinone analogs (see below). Light-induced charge separation takes place between P700 and A$_0$, and the back reaction kinetics are dominated by a 25-ns kinetic phase that represents the charge recombination between P700$^+$ and A$_0^-$. A small population of the back reacting electrons (5 to 15%) repopulate the triplet state of P700, which then relaxes to the ground state in 3 to 5 μs. The optical difference spectrum of the 20-ns kinetic phase shows the superposition of $P700^+$ and A_0^-; when the $P700^+$ is subtracted, the spectrum of A_0^- displays a maximum at 760 nm with an extinction coefficient 1.4 to 1.8 times that of $P700^+$, and doublet minima at 412 and 438 nm. The spectrum of the 3-µs kinetic phase shows a broad absorption increase between 730 and 820 nm accompanied by a broad bleaching between 390 and 450 nm, consistent with the relaxation of the radical pair triplet formed in low quantum yield from the back reaction of $P700^+$ with A_0^-. The ESR spectrum of A_0^- is obtained by photoaccumulation in the presence of rapid electron donors to $P700^+$ and is characterized by an isotropic resonance at $g = 2.002$.[33]

17.3 Reconstitution of Photosystem I Proteins and Cofactors

Reconstitution of A_1 in the $P700$-A_0 Core

The protocol for the reconstitution of phylloquinone in hexane-extracted PS I particles and in ether-extracted PS I particles has been described by Biggins[34,35] and Itoh.[36,37] The reconstitution of phylloquinone in the $P700$-A_0 (Triton) core is accomplished using a similar procedure (J. H. Golbeck, unpublished data).

The assay of reconstituted A_1 represents a major challenge because it now appears that while many added quinones and quinone analogs are able to accept an electron from A_0, only some are able to transfer the electron forward to F_X. This implies that the forward transfer of an electron from A_1^- to F_X is a particularly demanding step. A flash kinetic assay of $P700^+$ that measures the suppression of the 25-ns back reaction between $P700^+$ and A_0^- is sometimes used as the assay. However, the measurement with the highest reliability is one that measures forward transfer from the quinone, such as the optical measurement of P430 (F_A/FB) reduction at room temperature or the restoration of the ESP ESR signal at room temperature (for reasons which are Kinetically-based, a "bypass" operates at low temperature that permits direct electron transfer from A_0^- to F_X without the involvement of A_1). Those quinones that can be successfully reconstituted include phyloquinone; 2-methyl-3-decyl-1,4-naphthoquinone; 2-methyl-3-(isoprenyl)2–1,4-naphthoquinone; and 2-methyl-3-(isoprenyl)4–1,4-naphthoquinone. A general rule is that restoration of the ESP signal is possible when the solution reduction potential of the reconstituted acceptor is more electropositive than about −750 mV and structure contains either two aromatic rings (i.e., naphthoquinone) or a benzoquinone (or larger) derivative substituted with an alkyl tail.

Reconstitution of F_X in the $P700$-A_1 Core

The protocol for the reconstitution of iron-sulfur cluster F_X in the $P700$-A_1 core relies on the ability of an artificial iron-sulfur cluster to undergo a ligand exchange reaction with the cysteines on PsaA and PsaB to allow insertion of a [4Fe-4S] cluster. The artificial iron-sulfur clusters are created by adding $FeCl_3$ and Na_2S to an anaerobic solution of 1% β-mercaptoethanol in 25 mM Tris buffer, pH 8.3.[26] The β-mercaptoethanol serves the added function of scising the cys-S^o-S^o-cys bond, which develops as a result of oxidative denaturation of F_X, and makes free cysteine residues available for binding the new iron-sulfur cluster. The reconstitution of F_X is nearly quantitative provided that phylloquinone is not displaced in the $P700$-A_1 core by exposure to excessive detergent (especially Triton X-100). The inorganic agents are removed by ultrafiltration or gel filtration chromatography, leaving a purified, reconstituted $P700$-F_X core. The flash-induced optical transient at 700 nm has a half-life of 1.2 ms, which is typical for the $P700^+$ F_X^- back reaction, and the flash-induced difference spectrum between 400 and 500 nm shows the broad signature of an iron-sulfur protein (when the contribution of $P700^+$ is subtracted). The ESR spectrum of the reconstituted $P700$-F_X core shows a normal F_X spectrum, with a broad set of resonances at $g = 2.04$, 1.86, and 1.76. This protocol can be used to insert Fe^{57} into the F_X cluster for specialized spectroscopic measurements. All attempts to reconstitute the F_X cluster with selenium in place of sulfur have not been successful.

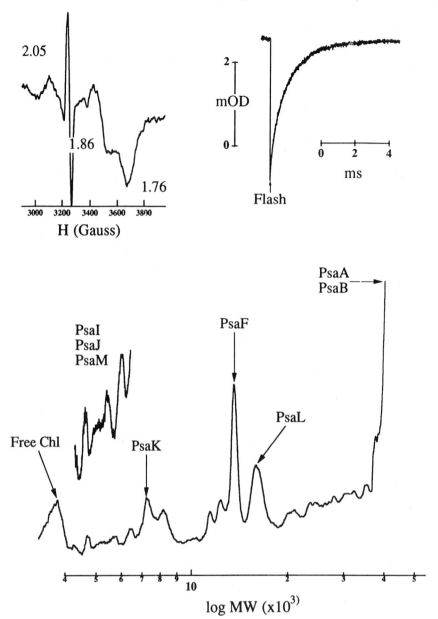

FIGURE 17.3 Spectroscopic properties and polypeptide composition of the P700-F_X core. The P700-F_X core was isolated from the *Synechococcus* sp. PCC 6301 P700-$F_A F_B$ complex with 6.8 *M* urea for 15 min, followed by dialysis and ultracentrifugation in a sucrose gradient. Top Left: Low-temperature ESR spectrum of the F_X iron-sulfur cluster showing principal *g*-values. The sample was suspended in 50 m*M* Tris buffer, pH 8.3 containing 1 m*M* sodium ascorbate and 0.3 m*M* DCPIP at 500 μg Chl per milliliter and illuminated during freezing. The resonances were resolved by subtracting the light-off from the light-on spectrum. Spectrometer conditions: same as Figure 17.2 except temperature, 8K; microwave power, 40 mW; modulation amplitude, 40 G. Top Right: Room-temperature flash-induced absorption transient of the P700-F_X core. Measurements at 698 nm were performed on 5 μg Chl per milliliter in 50 m*M* Tris buffer, pH 8.3, containing 1.7 m*M* ascorbate and 33 μ*M* DCPIP. (C) Laser densitometric tracing of the polypeptides present in the P700-F_X core after separation by SDS-PAGE. The elevated inset shows an enlargement of the very low molecular mass polypeptides in the region from 4.5 to 8 kDa. See Reference 24 for details.

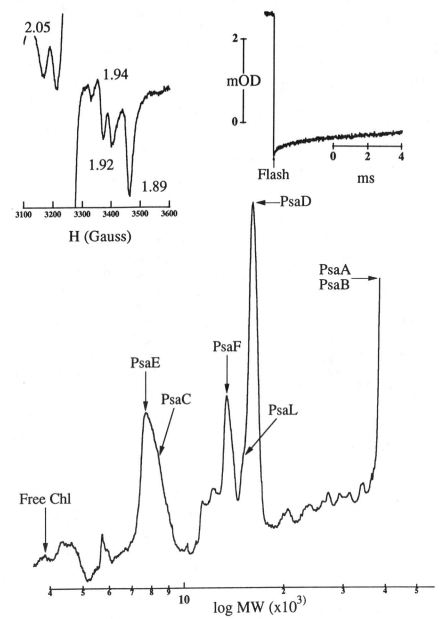

FIGURE 17.4 Spectroscopic properties and polypeptide composition of the reconstituted P700-$F_A F_B$ complex. The reconstitution was performed by incubating the PsaC, PsaD, and PsaE proteins with the P700-F_X core in the presence of $FeCl_3$, $Na_2 S$, and β-ME for 20 h, followed by dialysis, ultrafiltration, and ultracentrifugation in a sucrose gradient. Top Left: ESR spectrum of the F_A / F_B iron-sulfur clusters showing principal *g*-values (see Figure 17.1 legend). Top Right: Room-temperature flash-induced absorption transient of P700 (see Figure 17.1 legend). (C) Laser densitometric tracing of the polypeptides present in the reconstituted P700-F_A / F_B complex after separation by SDS-PAGE. The identities of the proteins were determined by N-terminal amino acid sequencing. The presence of Triton X-100 in the reaction mixture interfered with the separation of the low molecular mass polypeptides in the 4.5- to 8-kDa region. See Reference 24 for details.

Reconstitution of F_A/F_B in the P700-F_X Core

Wild-type or recombinant PsaC apoprotein can be rebound to the P700-F_X core using a protocol similar to the reconstitution of the F_X iron-sulfur cluster. The source can be of either wild-type PsaC purified from higher plant or cyanobacterial thylakoids,[38,39] cloned PsaC that has been expressed in *Escherichia coli*,[40] or recombinant PsaC that has been site-modified.[41] In all cases, PsaD is required to effect stable binding of PsaC on the PS I core. The reconstitution of the P700-F_A/F_B complex is nearly quantitative when wild-type PsaC is used; yields of >90% are typical when PsaD is present in the reconstitution protocol.

The flash-induced optical transient at 700 nm in the reconstituted P700-F_A/F_B complex has a half-life of ca. 20 ms, which is typical for the P700$^+$ [F_X/F_B]$^-$ back reaction, and the flash-induced difference spectrum between 400 and 500 nm shows the broad signature of an iron-sulfur protein (when the contribution of P700$^+$ is subtracted). The ESR properties of the the the reconstituted P700-F_A/F_B complex appear normal, with the photoreduction of F_A predominant when the sample is frozen during darkness and illuminated at 15K. If PsaC is added to the P700-F_X core in the absence of PsaD, F_B, rather than F_A, becomes the acceptor reduced when the sample is frozen during darkness and illuminated at 15K. The PsaC-only reconstituted P700-F_A/F_B complex is unstable to detergents, and PsaC is easily lost in a sucrose gradient containing 1% Triton X-100. PsaE cannot substitute for PsaD in stabilizing PsaC, and it has no added effect on the stability of the complex. This reconstitution protocol can be used with recombinant PsaC and PsaD to produce a mutant PS I complexes with altered properties. For example, the replacement of cysteine-14 with aspartic acid leads to the incorporation of a [3Fe-4S] cluster at the F_B site,[43] which has a midpoint potential of −98 mV [L. Yu, J. Zhao, D. Bryant, and J. H. Golbeck, unpublished results]. In this mutant, F_A is a normal [4Fe-4S] cluster, and it may be possible to study electron transfer in the absence of a fully functional F_B cluster. Conversely, the replacement of cysteine-51 with aspartic acid leads to the incorporation of a [3Fe-4S] cluster at the F_A site, which has a midpoint potential of −90 mV. In this mutant, F_B is a normal [4Fe-4S] cluster, and it may be possible to study electron transfer in the absence of a fully functional F_A cluster.

17.4 Photosystem I Counterparts in Anaerobic Bacteria

The RCs present in green sulfur bacteria of the genus *Chlorbiaceae* (e.g., *Chlorobium limicola* f. *thiosulfatophilum*, *Chlorobium phaeobacteroides*, *Chlorobium vibriforme*) and in the gram-positive bacteria of the genus *Heliobacteriaceae* (e.g., *Heliobacterium chlorum*, *Heliobacterium gestii*, and *Heliobacillus mobilis*) contain bound iron-sulfur cluster similar to F_X, F_B, and F_A.[42,42a-c,43] The relationship to cyanobacterial and higher plant PS I has been confirmed by the recent sequencing of the gene which codes for the reaction center protein. Although the RCs from *Chlorbiaceae*[42] and *Heliobacteriaceae*[43] do not share extensive sequence similarity to PsaA or PsaB of higher plants and cyanobacteria, an important structural motif is represented by the loop between spans VIII and IX that contains two cysteines separated by eight amino acids, five of which are identical to PsaA and PsaB. This sequence occurs in the same relative position as in plant and cyanobacterial PS I and most likely represents the binding site for iron-sulfur center F_X. The RC in Chlorobium and Heliobacterium differs further in that both are considered to be a homodimer instead of a heterodimer, as in plant and cyanobacterial PS I. An open reading frame has been found downstream from the *C. limicola* RC gene that codes for a 24-kDa iron-sulfur protein.[42] The 232 amino acid sequence contains a traditional CxxCxxCxxxCP [4Fe-4S] binding motif and an unusual CxxCxxCxxxxxCP sequence that could represent a second iron-sulfur binding site. Hence, at least in bacteria of the genus *Chlorbiaceae*, the photochemical motif of PS I is preserved from the donor, P840, to the F_A- and F_B-like terminal electron acceptors.

The isolation of the Chlorobium RC is complicated by the fact that the iron-sulfur clusters are unstable to oxygen and, hence, the isolation of the RC must be performed under anaerobic conditions.[46] Nevertheless, detergent-isolated RC complexes have been studied that show charge separation between P840 and the F_A/F_B-like iron-sulfur clusters. In these RC complexes, the room-temperature optical kinetics are complex, and the correlation of the various kinetic phases with bound iron-sulfur clusters and the resident quinone are being made only now. The most trustworthy measurements include measurement of the P840 triplet, and that of ESR studies of RCs oriented on mylar film. The F_A and F_B clusters are easily destroyed by oxidative denaturation in the presence of chaotropic agents, but the iron-sulfur clusters can be partially rebuilt with β-mercaptoethanol, $FeCl_3$, and Na_2S under anaerobic conditions.[47] A P840-F_X core or a P840-A_1 core have not been reported. An electron spin polarized EPR signal has been detected in *Chlorobium*, implying the existence of an A_1 component, and spectroscopic studies on the primary photochemical reactions should be forthcoming in the near future.

References

1. Nitschke W. and Rutherford A. W., Photosynthetic reaction centers: Variations on a common structural theme?, *Trends Biochem., Sci.*, 16, 241, 1991.
2. Golbeck, J. H., Shared thematic elements in photochemical reaction centers, *Proc. Natl. Acad. Sci. U.S.A.*, 90, 1642, 1993.
3. Andersson, B. and Styring, S., Photosystem II, in *Current Topics in Bioenergetics*, Vol. 16, Lee, C. P., Ed., Academic Press, San Diego, CA, 1991, 2–81..
4. Lagoutte, B. and Mathis, P., The Photosystem I reaction center: Structure and photochemistry, *Photochem. Photobiol.*, 49, 833, 1989.
5. Scheller, H. V. and Møller, B. L., Photosystem I polypeptides, *Physiol. Plant.*, 78, 484, 1990.
6. Evans, M. C. W. and Bredenkamp, G., The structure and function of the Photosystem I reaction center, *Physiol. Plant.*, 79, 415, 1990.
7. Golbeck, J. H. and Bryant, D. A., Photosystem I, in *Current Topics in Bioenergetics: Light Driven Reactions in Bioenergetics*, Vol. 16, Lee, C. P., Ed., Academic Press, New York, 1991, 83–177.
8. Bryant, D. A., Photosystem I: polypeptide subunits, genes, and mutants, in *Current Topics in Photosynthesis*, in *The Photosystems: Structure, Function, and Molecular Biology*, Barber, J., Ed., Elsevier, Amsterdam, 1991, 501–519.
9. Golbeck, J. H., Photosystem I, *Ann. Rev. Plant Physiol. Plant Mol. Biol.*, 43, 293, 1992.
10. Ikeuchi, M., Subunit proteins of Photosystem-I, *Plant Cell Physiol.*, 33, 669, 1992.
11. Sétif, P., Energy transfer and trapping in Photosystem I, in *The Photosystems: Structure, Function, and Molecular Biology*, Barber, J., Ed., Elsevier, Amsterdam, 1992, 471–499.
12. Almog, O., Shoham, G., and Nechushtai, R., Photosystem I: Composition, organization, and structure, in *The Photosystems: Structure, Function, and Molecular Biology*, Barber, J., Ed., Elsevier, Amsterdam, 1992, 443–469.
13. Chitnis, P. R. and Nelson, N., Photosystem I, in *Photosynthetic Apparatus: Molecular Biology and Operation*, (Series: Cell Culture and Somatic Cell Genetics of Plants 7), Bogorad, L. and Vasil, I. K., Eds., Academic Press, San Diego, 1992, 177–224.
14. Mathis, P., Photosynthetic reaction centers, *CRC Handbook* (this series), 1993.
15. Golbeck, J. H. and Cornelius, J. M., Photosystem I charge separation in the absence of centers A and B. I. Optical characterization of center A_2 and evidence for its association with a 64-kDa polypeptide, *Biochim. Biophys. Acta*, 849, 16, 1986.
16. Rogner, M., Nixon, P. J., and Diner, B. A., Purification and characterization of Photosystem I and Photosystem II core complexes from wild-type and phycocyanin-deficient strains of the cyanobacterium *Synechocystis* PCC 6803, *J. Biol. Chem.*, 265, 6189, 1990.
17. Mehari, T., Parrett, K. G., Warren, P. V., and Golbeck, J. H., Reconstitution of the iron-sulfur clusters in the isolated F_A/F_B protein: ESR characterization of same-species and cross-species Photosystem I complexes, *Biochim. Biophys. Acta*, 1056, 139, 1991.

18. Evans, E. H., Rush, J. D., Johnson, C. E., and Evans, M. C. W., Mössbauer spectra of Photosystem I reaction centres from the blue-green alga *Chlorogloea fritschii, Biochem. J.,* 182, 861, 1979.

19. McDermott, A. E., Yachandra, V. K., Guiles, R. D., Britt, R. D., Dexheimer, S. L., Sauer, K., and Klein, M. P., Low-potential iron-sulfur centers in Photosystem I: An X-ray absorption spectroscopy study, *Biochemistry,* 27, 4013, 1988.

20. Heathcote, P., Williams-Smith, D. L., Sihra, C. K., and Evans, M. C. W., The role of the membrane-bound iron-sulfur centers A and B in the Photosystem I reaction centre of spinach chloroplasts, *Biochim. Biophys. Acta,* 503, 333, 1978.

21. Ke, B., Hansen, R. E., and Beinert, H., Oxidation-reduction potentials of bound iron-sulfur proteins of Photosystem I, *Proc. Natl. Acad. Sci. U.S.A.,* 70, 2941, 1973.

22. Krauss, N., Hinrichs, W., Witt, I., Fromme, P., Pritzkow, W., Dauter, Z., Betzel, C., Wilson, K. S., Witt, H. T., and Saenger, W., 3-Dimensional structure of System-I of photosynthesis at 6-Angstrom resolution, *Nature,* 361, 326, 1993.

23. Parrett, K. G., Mehari, T., Warren, P., and Golbeck, J. H., Purification and properties of the intact P700 and F_X-containing Photosystem I core, *Biochim. Biophys. Acta,* 973, 324, 1989.

24. Li, N., Warren, P. V., Golbeck, J. H., Frank, G., Zuber, H., and Bryant, D. A., Polypeptide composition of the Photosystem I complex and the Photosystem I core protein from *Synechococcus* sp. PCC 6301, *Biochim. Biophys. Acta,* 1059, 215, 1991.

25. Hanley, J. A., Kear, J., Bredenkamp, G., Li, G., Heathcote, P., and Evans, M. C. W., Biochemical evidence for the role of the bound iron-sulphur centre-A and centre-B in NADP reduction by Photosystem-I, *Biochim. Biophys. Acta,* 1099, 152, (1992).

26. Parrett, K. P., Mehari, T., and Golbeck, J. H., Resolution and reconstitution of the cyanobacterial Photosystem I complex, *Biochim. Biophys. Acta,* 1015, 341, 1990.

27. Sétif, P., Mathis, P., and Vänngård, T., Photosystem I photochemistry at low temperature. Heterogeneity in pathways for electron transfer to the secondary acceptors and for recombination processes, *Biochim. Biophys. Acta,* 767, 404, 1984.

28. Petrouleas, V., Brand, J. J., Parrett, K. P., and Golbeck, J. H., A Mössbauer analysis of the low-potential iron-sulfur center in Photosystem I. Spectroscopic evidence that F_X is a [4Fe-4S] cluster, *Biochemistry,* 28, 8980, 1989.

29. McDermott, A. E., Yachandra, V. K., Guiles, R. D., Sauer, K., Parrett, K. G., and Golbeck, J. H., An EXAFS structural study of F_X, the low potential Fe-S center in Photosystem I, *Biochemistry,* 28, 8056, 1989.

30. Warren, P. V., Parrett, K. G., and Golbeck, J. H., Characterization of a Photosystem I core containing P700 and intermediate electron acceptor A_1, *Biochemistry,* 29, 6545, 1990.

31. Sétif, P. and Bottin, H., Identification of electron-transfer reactions involving the acceptor A_1 of Photosystem I at room temperature, *Biochemistry,* 28, 2689, 1989.

32. Warren, P. V., Golbeck, J. H., and Warden, J. T., Charge recombination between $P700^+$ and A_1 occurs directly to the ground state of P700 in the absence of F_X, F_B and F_A, *Biochemistry,* 32, 849, 1993.

33. McCracken, J. L. and Sauer, K., Electron paramagnetic resonance studies of the primary electron acceptors of Photosystem I, *Biochim. Biophys. Acta,* 724, 83, 1983.

34. Biggins, J., Evaluation of selected benzoquinones, naphthoquinones, and anthraquinones as replacements for phylloquinone in the A_1 acceptor site of the Photosystem I reaction center, *Biochemistry,* 29, 7259, 1990.

35. Rustandi, R. R., Snyder, S. W., Biggins, J., Norris, J. R., and Thurnauer M. C., Reconstitution and exchange of quinones in the A_1 site of Photosystem-I: An electron spin polarization electron paramagnetic resonance study, *Biochim. Biophys. Acta,* 1101, 311, 1992.

36. Itoh, S., Iwaki, M., and Ikegami, I., Extraction of Vitamin K_1 from Photosystem I particles by treatment with diethyl ether and its effects on the A_1-EPR signal and System I photochemistry, *Biochim. Biophys. Acta,* 893, 508, 1987.

37. Iwaki, M. and Itoh, S., Electron transfer in spinach Photosystem I reaction center containing benzo-, naphtho- and anthraquinones in place of phylloquinone, *FEBS Lett.*, 256, 11, 1989.

38. Wynn, R. M. and Malkin, R., Characterization of an isolated chloroplast membrane iron-sulfur protein and its identification as the Photosystem I iron-sulfur A/iron-sulfur B binding protein, *FEBS Lett.*, 229, 293, 1988.

39. Oh-oka, H., Takahashi, Y., Matasubara, H., and Itoh, S., EPR studies of a 9 kDa polypeptide with an iron-sulfur cluster(s) isolated from Photosystem I complex by *n*-butanol extraction, *FEBS Lett.*, 234, 291, 1988.

40. Li, N., Zhao, J. D., Warren, P. V., Warden, J. T., Bryant, D., and Golbeck, J. H., PsaD is required for the stable binding of PsaC to the Photosystem I core protein of *Synechococcus* sp. PCC 6301, *Biochemistry*, 30, 7863, 1991.

41. Zhao, J. D., Li, N., Warren, P. V., Golbeck, J. H., and Bryant, D. A., Site-directed conversion of a cysteine to an aspartate leads to the assembly of a [3Fe-4S] cluster in PsaC of Photosystem I. Photoreduction of F_A is independent of F_B, *Biochemistry*, 31, 5093, 1992.

42. Knaff, D. B. and Malkin, R., Iron-sulfur proteins of the green photosynthetic bacterium *Chlorobium*, *Biochim. Biophys. Acta*, 430, 244, 1976.

42a. Jennings, J. V. and Evans, M. C. W., The irreversible photoreduction of a low potential component at low temperatures in a preparation of the green photosynthetic bacterium *Chlorobium thiosulfatophilum*, *FEBS Lett.*, 75, 33, 1977.

42b. Swarthoff, T., Gast, P., Hoff, A. J., and Amesz, J., An optical and ESR investigation of the reaction center of the green photosynthetic bacterium *Prosthecochloris cestugrii*, *FEBS Lett.*, 130, 93, 1977.

42c. Nitschke, W., Feiler, U., and Rutherford, A. W., Photosynthetic reaction center of the green sulfur bacteria studied by EPR, *Biochemistry*, 29, 3834, 1990.

43. Nitschke, W., Sétif, P., Liebl, U., Feiler, U., and Rutherford, A. W., Reaction center photochemistry of *Heliobacterium chlorum*, *Biochemistry*, 29, 11079, 1990.

44. Büttner, M., Xie, D.-L., Nelson, H., Pinther, W., Hauska, G., and Nelson, N., Photosynthetic reaction center genes in green sulfur bacteria and in Photosystem-1 are related, *Proc. Natl. Acad. Sci. U.S.A.*, 89, 8135, 1992.

45. Trost, J., Brune, D. C., and Blankenship, R. E., Protein sequences and redox titrations indicate that the electron acceptors in reaction centers from heliobacteria are similar to Photosystem-I, *Photosyn. Res.*, 32, 11, 1992.

46. Miller, M., Liu, X., Snyder, S. W., Thurnauer, M., and Biggins, J., Photosynthetic electron-transfer reactions in the green sulfur bacterium *Chlorobium-vibrioforme*: Evidence for the functional involvement of iron-sulfur redox centers on the acceptor side of the reaction center, *Biochemistry*, 31, 4354, 1992.

47. Feiler, U., Nitschke, W., and Michel, H., Characterization of an improved reaction center preparation from the photosynthetic green sulfur bacterium *Chlorobium* containing the FeS centers F(A) and F(B) and a bound cytochrome subunit, *Biochemistry*, 31, 2608, 1992.

18

Thermoluminescence: A Probe of Photosystem II Photochemistry

Yorinao Inoue
The Institute of Physical and
Chemical Research (RIKEN)

18.1 Introduction

Thermoluminescence (TL) is a series of outbursts of light emission at characteristic temperatures that is detected upon warming a preilluminated plant material in darkness. The phenomenon originates from thermally activated recombination of positive and negative charge pairs generated by Photosystem II (PS II) photochemistry and stabilized by low temperature, the reversal of light-driven charge separation: the energy released upon the recombination reexcites the reaction center chlorophyll, resulting in fluorescence emission from itself or from neighbor antenna chlorophyll.[1,2] The phenomenon of TL is common to all photosynthetic organisms capable of oxygen evolution. Recent assignments of positive and negative charges involved in the recombination process have enabled us to use TL as a unique probe of electron transport and redox reactions in PS II.[3]

18.2 Charge Stabilization in PS II Responsible for TL

The individual TL components are distinguished by their emission temperatures on the glow curve, and seven to ten components observed after different protocols of illumination have thus far been identified and named.[3] They are categorized into two groups: those originating from recombination of deeply trapped charge pairs resulting from the reaction center photochemistry of PS II and the others originating from recombination of shallower traps within light-harvesting pigment protein

0-8493-8634-9/95/$0.00+$.50
© 1995 by CRC Press, Inc.

complexes involving no photochemical reaction center. By some unclear reasons, the charge pairs generated by PS I photochemistry are not capable of TL. The TL components of physiological importance are those originating from charge pairs produced by the primary photochemistry in oxygen-evolving PS II and stabilized on respective carriers of positive (hole) and negative (electron) charges around the reaction center.

Upon illuminating a PS II, the reaction center chlorophyll (P_{680}) absorbs one photon and generates a charge-separated state, $P_{680}{}^+$/Pheo$^-$ in 2 to 3 ps. This charge separation is followed by electron transfer from Pheo$^-$ to the first quinone acceptor (Q_A) in 250 to 300 ps to yield Q_A^-. Quinone Q_A is a one-electron acceptor, and its reduction is followed by transfer of the electron to the secondary quinone acceptor, Q_B, in 100 to 200 μs to yield Q_B^-. Due to its rather short lifetime, Q_A^- is not trapped under normal conditions. On the other hand, Q_B is a two-electron acceptor and its reduced form, Q_B^-, is stable for more than tens of seconds at room temperature, so that it constitutes the major negative charge trap detectable by TL. On accepting another electron, Q_B^- is further reduced to Q_B^{2-} in 400 to 500 μs, which is then protonated to form Q_BH_2 to be replaced rapidly by another plastoquinone molecule of the pool, so that Q_B^{2-} cannot serve for TL. For these reasons, the negative charge trap on the acceptor side of PS II is usually either Q_B^- or Q_A^- in the absence or presence, respectively, of herbicides that block the electron transfer between Q_A and Q_B.

Transfer and trapping of positive charges occur on the oxidizing side of PS II. The $P_{680}{}^+$ oxidizes the redox active Tyr161 residue of the D1 protein in nanoseconds to microseconds, and the resulting cation radical Y_Z^+ in turn oxidizes the Mn atom in the water-oxidizing enzyme in microseconds to milliseconds. The water-oxidizing enzyme contains a tetranuclear Mn-cluster that undergoes the so-called S-state transitions involving one unstable and four stable intermediate states designated S_i (i = 0 to 4) with S_1 as the dominant species after dark adaptation. During the S-state transitions, four electrons are successively removed from two molecules of water to yield one molecule of oxygen. Each transition involves both electron abstraction and proton release, but one exceptional transition from S_1 to S_2 does not involve proton release, so that one positive equivalent remains on the S_2 state and S_3 state. The positive equivalents left on these two S states are stable for tens of seconds, and constitute the major positive charge trap detectable by TL.

The PS II reaction centers, having various combinations of positively and negatively charged species, are generated in varying population ratios, depending on the regime of preillumination, illumination temperature, and light-dark history of the sample. This gives rise to remarkably different glow curves from the same PS II preparation, since only the centers having both positive and negative charges on donor and acceptor sides, respectively, are capable of TL, while those lacking either of the positive or negative charges are not. Due to these situations, the intensity of TL exhibits period-four oscillations under illumination with single turnover flashes.[3,4]

18.3 Charge Pairs Assigned to Respective TL Components

The TL components are distinguished by their respective emission bands on the glow curve that exhibit characteristic temperatures for maximum emission. For several of these components, responsible charge pairs have been identified. Following is the current information accumulated about the properties of respective TL bands.

The B Band

The B band that appears at around +30°C is the best characterized TL component. This band is correlated clearly with the water-oxidizing enzyme, in particular with the presence of functionally active Mn, and has been assigned to arise from $S_2Q_B^-$ and $S_3Q_B^-$ charge recombination.[5] When excited by a series of saturating flashes, the intensity of the B band exhibits a period-four oscillation in thylakoids, intact leaves, and in PS II-enriched membrane fragments as well.[3,4] Analysis of the oscillation pattern shows that both the S_2 and S_3 states are involved in generation of this TL component.[4,5] Participation of Q_B^- as the negative charge trap has been shown by modulation of the initial Q_B/Q_B^- ratio by chemical and light pretreatment or by removal and reconstitution of Q_B.[4-6] At or above pH 7.0 to 7.5, both $S_2Q_B^-$ and $S_3Q_B^-$ recombinations give rise to the same TL band with respect to its peak temperature and band shape; but below pH 6.0, they exhibit two distinguishable components denoted B_2 and B_1 bands, respectively, with a few degrees difference in peak temperature. The TL yield of the latter recombination is higher than that of the former by a factor of 1.7 to 2.0 due to an unexplained reason.[7] The emission spectrum of the B band(s) has a maximum at around 690 nm,[20] in accordance with the hypothesis that the $S_{2/3}Q_B^-$ recombination reexcites P_{680} as a reversal of charge separation.[8]

The Q Band

Upon treating PS II with DCMU or other PS II herbicides, the B band is abolished with a concomitant appearance of a new TL band between 0 and +10°C.[9] This band is designated the Q band that arises from $S_2Q_A^-$ charge recombination.[5,9] The conversion of the B band to the Q band is due to inhibition by DCMU of the electron transfer between Q_A and Q_B, which enables stabilization of Q_A^- as a negative charge detectable by TL in place of Q_B^-. The Q band is often used for the analysis of the stabilization of the S_2 state.

The A Band

When a PS II sample, preilluminated with two flashes at room temperature, is cooled and illuminated further with continuous light at 77 K, a clear TL band appears at around −10°C.[7,10] This band has been assigned to arise from $S_3Q_A^-$ charge recombination based on the following considerations: illumination with two flashes at room temperature generates an S_3Q_B state that is incapable of TL due to the absence of negative charge.[11] But the illumination at 77 K oxidizes cyt b_{559} or chlorophyll in place of Mn to generate Q_A^-, which undergoes recombination with the S_3 state previously formed by illumination with two flashes and stabilized by cooling.

The A_T Band

Tris-treated PS II depleted of the functional Mn-cluster emits a TL band also at around −10°C. This band is called the A_T band to distinguish it from the A band having the same emission temperature.[12] The origin of this component is not clear, but it is proposed to arise from charge recombination between Q_A^- and a photooxidized histidine residue of a PS II reaction center protein. The positive equivalent on photooxidized histidine is proposed to be involved in oxidation of Mn^{2+} to Mn^{3+} that takes place during photoligation of Mn in the process of photoactivation of a latent oxygen-evolving system.[12]

The Z_v Band

A minor TL component, denoted as the Z_v band, appears between −80 and −30°C. The emission temperature of this band varies depending on the excitation temperature,[13] usually being higher by 10 to 20°C. This band is abolished by treatment with 1% ethanol and exhibits a period-four oscillation dependent on S-state turnovers,[14] but is reported to be emitted from the D1/D2/cyt b_{559} PS II reaction center preparation depleted of the Mn-cluster.[15] Although it is proposed to arise from the $P_{680}^+Q_A^-$ charge pair, whose trapping depth is modulated by the conformation of the reaction center, its origin is not yet clear. The participation of Q_A^- as the negative counterpart has been confirmed recently by reconstitution of Q_A^- in purified reaction center preparation.[16]

The C Band

A TL component denoted as the C band appears at around +50°C on glow curves from DCMU-treated plant materials.[17] The participation of Q_A^- as the negative counterpart is likely, judging from enhancement of its intensity in the presence of high concentrations of DCMU. It is reported to exhibit a period-four oscillation with maxima in the S_0 and S_1 states, suggesting that this band arises from $S_{0/1}Q_A^-$ charge recombination.[18] Y_D^+ has been proposed as a candidate for the positive equivalent in S_0 and S_1 states.

The Z Band

A strong and broad TL band appears at around −160°C when various plant materials are illuminated with continuous light at 77 K. This component, denoted as the Z band, exhibits an emission maximum at around 730 nm, in contrast to 690 nm of those components arising from PS II photochemistry, and is excited with blue light at a higher yield than with red light.[19] Upon raising the excitation temperature above 77 K, only the higher temperature part of the Z band appears, indicating that the trap for this component has a broad distribution of stabilization free energies. It was shown recently that LHC I and LHC II, having no photochemical reaction center or purified chlorophyll, particularly its aggregated form, emit this band more strongly than PS I or PS II.[20] Thus, it is inferred that the Z band is correlated with neither PS I nor PS II photochemistries.

The TL Bands at Liquid Helium Temperatures

Three new TL components emitting at around 20, 50, and 70 K were recently resolved in the glow curve at liquid helium temperatures, and denoted as Zα, Zβ, and Zγ, respectively.[21] The glow curve at liquid helium temperatures exhibits another component at around 90 K, but this is a different expression of the well-known Z band. The properties of the three components are essentially the same as those of the Z band: being excited preferentially by blue light and emitted more strongly from LHCs and aggregated chlorophyll, suggesting that they originate from charge storage in a chlorophyll molecule interacting with another chlorophyll or solvent molecule.

18.4 Probing the Acceptor Side of PS II by TL

The redox couples of Q_B/Q_B^- and Q_A/Q_A^- have a difference in redox potential of 50 to 70 mV, and this difference manifests itself on TL glow curves as a difference in emission temperature of 25 to 30°C between the B band and Q band, which arise from $S_2Q_B^-$ and $S_2Q_A^-$ charge pairs, respectively. The energetic stability of reduced quinones can be studied easily by monitoring TL.[22]

Removal and Reconstitution of PS II Quinones

Upon extraction of thylakoids with heptane/isobutanol, the B band is converted to the Q band, indicative of extraction of Q_B. Upon addition of dimethylquinone or parabenzoquinone to the extracted thylakoids, a B band with a slightly downshifted emission temperature appears and exhibits an oscillatory behavior, indicative of reconstitution of artificial Q_B.[23] Isolated PS II reaction center (D1/D2/cyt b_{559}) exhibits very low Z_v band. Reconstitution of the reaction center preparation with plastoquinone-9, decylplastoquinone, or duroquinone together with thylakoid lipids enhances the Z_v band intensity, suggesting partial reconstitution of functional Q_A.[16]

Protonation of Quinone Acceptors

At low pH values, the emission temperature of the B_2 band shifts up due to protonation-induced stabilization of Q_B^-. By use of this pH dependence, the pK values for protonation of Q_B in the oxidized and semi reduced forms were determined to be 6.4 and 7.9, respectively.[24] The Q band does not show any pH dependence, indicative of its buried environments in intact PS II. Upon mild trypsination, however, the emission temperature of the Q band exhibits a clear pH-dependent upshift, indicative of the acquisition of its accessibility to protons due to removal of the protein cap around Q_A.[24]

Effects of PS II Herbicides

Treatment of PS II with herbicides converts the B band to Q band due to inhibition of the electron transfer between Q_A and Q_B. Notably, however, the emission temperature of the resulting Q band differs depending on the chemical structure of the herbicide: phenolic-type herbicides induce a Q band emitting between −15 and 0°C, whereas urea/triazine type herbicides induce a Q band between 0 and +10°C.[25] The much lower Q band peak temperature in the presence of phenolic-type herbicides is ascribable either to a lowered redox potential of Q_A/Q_A^- or to the ADRY effect of these chemicals affecting the properties of the S_2 state on donor side of PS II.

Herbicide-resistant mutants exhibit a remarkably downshifted B band. In triazine-resistant mutants of *Eligeron canadensis* L. and *Synechocystis* PCC 6714, the emission temperature of the B band is shifted down by 15°C as compared with that of respective wild-types.[22] This indicates that the redox potential difference between Q_B/Q_B^- and Q_A/Q_A^- is decreased from 70 to 30 mV due to a structural alteration of the Q_B-binding site by the mutation.[22] Recent analysis by use of transformable cyanobacteria reveals that the emission temperature of the B band is strongly sensitive to the replacement of Ser264 of D1 protein by Ala or Gly that induces strong resistance to triazines, but is rather immune to the replacement of Phe255 by Tyr that induces resistance to phenylureas, suggesting a more important role of Ser264 in Q_B-binding.[26]

18.5 Probing the Donor Side of PS II by TL

The generation of the B (B_1 + B_2) and A bands involves the S_2 and S_3 states, so that the emission temperature and oscillatory behavior of those TL bands provide useful information regarding the properties of the S states and the role of various cofactors involved in S-state transitions.

Temperature Dependence of S-State Transitions

Upon lowering the ambient temperature, the S-state transitions become inhibited. From the oscillation of the B band under flash excitation, S_2-to-S_3 and S_3-to-S_4 transitions were shown to be blocked at −35 and −20°C, respectively, while the S_1-to-S_2 transition occurred at −65°C.[27] More recent and precise studies reveal that S_1-to-S_2, S_2-to-S_3, and S_3-to-S_0 transitions are completely blocked at −160, −65, and −40°C, and half-blocked at −95, −45, and −23°C, respectively.[8,28] Upon Ca depletion, the half-blocking temperature of the S_1-to-S_2 transition shifts up remarkably by about 50°C.[40]

Effects of Inhibitors

Substoichiometric concentration of ANT2P gradually abolishes the B band concomitant with a downshift of its emission temperature. In the liquid state, the agent acts as a mobile species: a small amount of the reagent deactivates the S_2 and S_3 states at a large number of centers, while in the frozen state the deactivation is restricted only to the centers to which the agent is bound.[29]

Ammonia interacts with the Mn-cluster in the S_2 state and generates a modified S_2 state. Recombination of the modified S_2 with Q_B^- or Q_A^- exhibits a downward shift of the B_2 or Q band, indicative of NH_3-induced stabilization of the S_2 state due to its lowered redox potential.[30] In contrast to the shift up of the B_2 band, the B_1 band ($S_3Q_B^-$ recombination) shifts down upon NH_3

treatment, and stops oscillation at the S_3 state. It is inferred that destabilization of the S_3 state blocks the S_3-to-S_0 transition, and thereby inhibits oxygen evolution.

Effects of Removal of Extrinsic Proteins

The PS II in higher plants contains a set of three extrinsic proteins of molecular masses of 33, 24, and 18 kDa, which are associated electrostatically on the lumenal surface of PS II and involved in regulating the requirement for inorganic cofactors and stability of the Mn-cluster. By suitable salt-washing followed by mixing with concentrated protein extract, these proteins can be removed reversibly and reconstituted with concomitant inactivation and reactivation of oxygen-evolving activity. Upon removal of the set of proteins, oxygen evolution is almost lost, whereas TL B band can still be induced. However, it exhibits normal oscillation only up to the S_3 state and the final S_3-to-S_0 transition to release molecular oxygen is blocked.[31] The peak position of the B band remains at a normal temperature, suggesting no appreciable change in the stability of the $S_2Q_B^-$ charge pair. In contrast, however, the peak temperature of the Q band is remarkably upshifted by 20 to 25°C, indicative of deeper stabilization of the S_2 state in the absence of the extrinsic proteins.[32] The different stabilization between $S_2Q_B^-$ and $S_2Q_A^-$ charge pairs may be due to a secondary but specific effect on the Q_B-binding site, in addition to the major effect on the S_2 state.

Effects of Removal of Mn

Mn plays a central role in TL as well as in photosynthetic oxygen evolution. Upon removal of two of the four Mn atoms making up the tetranuclear Mn-cluster, the capability of emitting the B and Q bands is lost totally,[31] suggesting that the two labile Mn atoms, rather than the other firmly bound two, are involved in TL as the positive charge carrier, as far as the two TL bands (both related to the S_2 state) are concerned. The TL components thus far detected in complete absence of Mn are as follows: A_T, Z_v, Z, and the three bands emitting at liquid helium temperatures.

Effects of Cl⁻ Depletion

Chloride ion is a cofactor essential to oxygen evolution and its depletion results in interruption of S-state transitions. Removal of Cl⁻ by alkaline shock or by replacement of Cl⁻ by SO_4^{2-} induces 20 and 25°C upward shifts of both the B and Q band,[33] respectively, indicative of increased stabilization energies of the $S_2Q_B^-$ and $S_2Q_A^-$ charge pairs by about 50 and 80 mV, respectively,[32] in the absence of Cl⁻. The similar extent of upward shift found for both the two TL bands indicates that a modification of S_2, but not of Q_B^- or Q_A^-, is the cause for the upward shift, which is well illustrated by the absence of an EPR multiline signal in Cl⁻-depleted PS II, indicative of a structural modulation of the Mn-cluster. Similar but different degrees of upward shifts are induced by replacement of Cl⁻ with F⁻, CH_3COO^-, or NO_3^-, and all these effects are completely reversed by the addition of exogenous Cl⁻ anions.[35] Concomitant with the upward shift of TL peak position, Cl⁻ removal alters the oscillation pattern of the B band. The alteration manifests in two different ways depending on the initial distribution of charge pairs in dark-adapted PS II centers, but commonly suggests an inhibition of S_2-to-S_3 transition.[33,34]

Effects of Ca²⁺ Depletion

Calcium ion (Ca^{2+}) is another inorganic cofactor essential to oxygen evolution. In general, removal of Ca^{2+} results in upshifted TL B and Q bands, indicative of structural modulation of the Mn-cluster.[36] Such modulations are correlated well to the alterations as detected by EPR and X-ray absorption spectroscopy. However, the TL properties of the depleted PS II are variable, depending on the protocol of Ca^{2+} depletion. For example, Ca^{2+} depletion by low pH treatment gives rise to an abnormally modified S_2 state exhibiting elevated peak temperatures of the B or Q band by about

10°C;[36] whereas Ca^{2+} depletion by NaCl wash in the light followed by addition of EDTA generates a similarly modified S_2 state only when the depleted PS II is reconstituted with the 24 kDa extrinsic protein.[37,38] This difference originates from the fact that the PS II depletion by Ca^{2+} by the former method retains (due to reassociation) the 24-kDa protein, while the latter method removes both Ca^{2+} and the 24-kDa protein.[37,38] It is inferred that the extrinsic 24-kDa protein regulates the structure and function of the Mn-cluster in the absence of functional Ca^{2+} through a conformational modulation of the intrinsic protein(s) that bind(s) both Mn and Ca^{2+}. The S_2 state formed in a Ca^{2+}-depleted PS II retaining the 24-kDa protein is called the dark-stable S_2 state, since it has an extremely long half-life of 7 to 8 h in darkness at room temperature when monitored by EPR multiline signal. After such long dark-adaptation, the dark-stable S_2 is incapable of TL due to the absence of negative counterpart. However, when given a 77-K illumination that delivers one electron from cyt b_{559} to Q_A without affecting the Mn-cluster, the dark-stable S_2 recombines with the newly generated Q_A^- to emit TL Q band.[39] Another remarkable property of the dark-stable S_2 is the threshold temperature for its formation, being upward shifted by about 50°C as compared with that of normal S_2 state.[40] Due to this upward shift in threshold temperature, it was considered once that Ca^{2+} depletion inhibits the S_1-to-S_2 transition when monitored by the EPR multiline signal, in which 200- to 210-K illumination is usually employed to induce the S_2 state. TL measurements provided an answer to this argument, but there still remains an argument as to which of the S_2-to-S_3 and S_3-to-S_0 transitions is blocked by Ca^{2+} depletion.[41]

References

1. Inoue, Y. and Shibata, K., Thermoluminescence from photosynthetic apparatus, in *Photosynthesis: Energy Conversion by Plants and Bacteria*, Vol. 1, Govindjee, Ed., Academic Press, New York, 1982, 507.

2. Sane, P. V. and Rutherford, A. W., Thermoluminescence from photosynthetic membranes, in *Light Emission by Plants and Bacteria*, Govindjee, Amesz, J. and Fork, D. C., Eds., Academic Press, New York, 1986, 329.

3. Vass, I. and Inoue, Y., Thermoluminescence in the study of photosystem II, in *The Photosystems: Structure, Function and Molecular Biology. Topics in Photosynthesis*, Vol. 11, Barber, J., Ed., Elsevier, New York, 1992, 259.

4. Inoue, Y., Recent advances in the study of thermoluminescence of photosystem II, in *The Oxygen Evolving System of Photosynthesis*, Inoue, Y., Crofts, A. R., Govindjee, Murata, N., Renger, G., and Satoh, K., Eds., Academic Press, New York, 1983, 439.

5. Rutherford, A. W., Crofts, A. R., and Inoue, Y., Thermoluminescence as a probe of Photosystem II photochemistry: The origin of the flash-induced glow peaks, *Biochim. Biophys. Acta*, 682, 457, 1982.

6. Demeter, S. and Vass, I., Charge accumulation and recombination in Photosystem II studied by thermoluminescence. I. Participation of the primary acceptor Q and secondary acceptor B in the generation of thermoluminescence of chloroplasts, *Biochim. Biophys. Acta*, 764, 24, 1984.

7. Inoue, Y., Charging of the A band of thermoluminescence dependent on the S_3 state in isolated chloroplasts, *Biochim. Biophys. Acta*, 634, 309, 1981.

8. Demeter, S., Rosza, Zs., Vass, I., and Hideg, E., Thermoluminescence study of charge recombination in Photosystem II at low temperatures. II. Oscillatory properties of the Z_v and A thermoluminescence bands in chloroplasts dark-adapted for various time periods, *Biochim. Biophys. Acta*, 809, 379, 1985.

9. Demeter, S., Herczeg, T., Droppa, M., and Horvath, G., Thermoluminescence characterization of granal and agranal chloroplasts of maize, *FEBS Lett.*, 100, 321, 1979.

10. Laufer, A., Inoue, Y., and Shibata, K., Enhanced charging of thermoluminescence A band by a combination of flash and continuous light excitation, in *Proc. Intl. Sym. on Chloroplast Development*. Akoyunoglou, G., Ed., Elsevier, New York, 1978, 379.

11. Koike, H., Siderer, Y., Ono, T., and Inoue, Y., Assignment of thermoluminescence A band to $S_3Q_A^-$ charge recombination: Sequential stabilization of S_3 and Q_A^- by a two-step illumination at different temperatures, *Biochim. Biophys. Acta,* 850, 80, 1986.

12. Ono, T. and Inoue, Y., Biochemical evidence for histidine oxidation in Photosystem II depleted of the Mn-cluster for O_2-evolution, *FEBS Lett.,* 278, 183, 1991.

13. Ichikawa, T., Inoue, Y., and Shibata, K., Characteristics of thermoluminescence bands of intact leaves and isolated chloroplasts in relation to the water-splitting activity in photosysnthesis, *Biochim. Biophys. Acta,* 408, 228, 1975.

14. Vass, I., Rozsa, Zs., and Demeter, S., Flash induced oscillation of the low temperature thermoluminescence band Z_v in chloroplasts, *Photochem. Photobiol.,* 40, 407, 1984.

15. Vass, I., Chapman, D. J., and Barber, J., Thermoluminescence properties of the isolated Photosystem II reaction center, *Photosynth. Res.,* 22, 295. 1989.

16. Chapman, D. J., Vass, I., and Barber, J., Secondary electron transfer reactions of the isolated Photosystem II reaction center after reconstitution with plastoquinone-9 and diacylglycerolipids, *Biochim. Biophys. Acta,* 1057, 391, 1991.

17. Desai, T. S., Sane, P. V., and Tatake, V. G., Thermoluminescence studies on spinach and *Euglena*, *Photochem. Photobiol.,* 21, 345, 1975.

18. Demeter, S., Vass, I., Horvath, G., and Laufer, A., Charge accumulation and recombination in Photosystem II studied by thermoluminescence. II. Oscillation of the C band by flash excitation, *Biochim. Biophys. Acta,* 764, 33, 1984.

19. Arnold, W. and Azzi, J. R., Chlorophyll energy levels and electron flow in photosynthesis, *Proc. Natl. Acad. Sci. U.S.A.,* 43, 105, 1968.

20. Sonoike, K., Koike, H., Enami, I., and Inoue, Y., The emission spectra of thermoluminescence from photosynthetic apparatus, *Biochim. Biophys. Acta,* 1058, 121, 1991.

21. Noguchi, T., Inoue, Y., and Sonoike, K., Thermoluminescence emission at liquid helium temperature from photosynthetic apparatus and purified pigments, *Biochim. Biophys. Acta,* 18, 1992.

22. Demeter, S., Vass, I. Hideg, E., and Sallai, A., Comparative thermoluminescence study of triazine-resistant and -susceptible biotypes of *Erigeron canadensis* L., *Biochim. Biophys. Acta,* 806, 16, 1985.

23. Wydrzynski, T. and Inoue, Y., Modified PSII acceptor side properties upon replacement of the quinone at the Q_B site with 2,5-dimethyl-*p*-benzoquinone and phenyl-*p*-benzoquinone, *Biochim. Biophys. Acta,* 893, 33, 1987.

24. Vass, I. and Inoue, Y., pH dependent stabilization of $S_2Q_A^-$ and $S_2Q_B^-$ charge pairs studied by thermoluminescence, *Photosynth. Res.,* 10, 431, 1986.

25. Vass, I. and Demeter, S., Classification of Photosystem II inhibitors by thermodynamic characterization of the thermoluminescence of inhibitor-treated chloroplasts, *Biochim. Biophys. Acta,* 682, 496, 1982.

26. Gleiter, H., Ohad, N., Hirschberg, J., Fromme, R., Renger, G., Koike, H., and Inoue, Y., An application of thermoluminescence to herbicide studies, *Z. Naturforsch.,* 45C, 353, 1990.

27. Inoue, Y. and Shibata, K., Oscillation of thermoluminescence at medium low temperature, *FEBS Lett.,* 85, 193, 1978.

28. Koike, H. and Inoue, Y., Temperature dependence of the S-state transition in the thermophilic cyanobacterium measured by thermoluminescence, in *Progress in Photosynthesis Research*, Vol. 1, Biggins, J., Ed., Martinus Nijhoff, The Hague, 1978, 645.

29. Renger, G. and Inoue, Y., Studies on the mechanism of ADRY agents on thermoluminescence emission, *Biochim. Biophys. Acta,* 725, 146, 1983.

30. Ono, T. and Inoue, Y., Abnormal S-state turnovers in NH_3-binding Mn centers of photosynthetic O_2 evolving system, *Arch. Biochem. Biophys.,* 275, 440, 1989.

31. Ono, T. and Inoue, Y., S-state turnover in the O_2-evolving system of $CaCl_2$-washed Photosystem II particles depleted of three peripheral proteins as measured by thermoluminescence: Removal of 33 kDa protein inhibits S_3 to S_4 transition, *Biochim. Biophys. Acta,* 806, 331, 1985.

32. Vass, I., Ono, T., and Inoue, Y., Removal of 33 kDa extrinsic protein specifically stabilizes the $S_2Q_A^-$ charge pair in Photosystem II, *FEBS Lett.*, 211, 215, 1987.

33. Homann, P. H., Gleiter, H., Ono, T., and Inoue, Y., Storage of abnormal oxidants 'Σ_1', 'Σ_2', 'Σ_3', in photosynthetic water oxidases inhibited by Cl^--removal, *Biochim. Biophys. Acta*, 850, 10, 1986.

34. Vass, I., Ono, T., Homann, P. H., Gleiter, H., and Inoue, Y., Depletion of Cl^- or 33 kDa extrinsic protein modifies the stability of $S_2Q_A^-$ and $S_2Q_B^-$ charge separation states in Photosystem II, in *Progress in Photosynthesis Research*, Vol. 1, Biggins, J., Ed., Martinus Nijhoff, The Hague, 1987, 649.

35. Ono, T., Nakayama, H., Gleiter, H., Inoue, Y., and Kawamori, A., Modification of the properties of S_2 state in photosynthetic O_2-evolving center by replacement of chloride with other anions, *Arch. Biochem. Biophys.*, 256, 618, 1987.

36. Ono, T. and Inoue, Y., Discrete extraction of the Ca atom functional for O_2 evolution in higher plant Photosystem II by a simple low pH treatment, *FEBS Lett.*, 227, 147, 1985.

37. Ono, T., Izawa, S., and Inoue, Y., Structural and functional modulation of the manganese cluster in Ca^{2+}-depleted Photosystem II induced by binding of the 24 kDa extrinsic protein, *Biochemistry*, 31, 7648, 1992.

38. Homann, P. H. and Madabusi, L., Modification of the thermoluminescence properties of Ca^{2+}-depleted Photosystem II membranes by 23 kDa extrinsic polypeptide and by oligocarboxylic acids, *Photosynth. Res.*, 35, 29, 1993.

39. Ono, T. and Inoue, Y., Abnormal redox reactions in photosynthetic O_2-evolving centers in NaCl/EDTA-washed PSII. A dark stable EPR multiline signal and a new unknown positive charge accumulator, *Biochim. Biophys. Acta*, 1020, 269, 1990.

40. Ono, T. and Inoue, Y., A marked upshift in threshold temperature for the S_1-to-S_2 transition induced by low pH treatment of PSII membranes, *Biochim. Biophys. Acta*, 1015, 373, 1990.

41. Debus, R., The manganese and calcium ions of photosynthetic oxygen evolution, *Biochim. Biophys. Acta*, 1102, 269, 1992.

19

Photoreception and Photomovements in Microorganisms

Francesco Lenci
C.N.R. Istituto Biofisica

19.1 Introductory Remarks

Many freely motile microorganisms like bacteria, unicellular algae, and protozoa are able to detect spatial and temporal variations in the external light field and to react to these environmental stimuli by modifying their movement, usually to achieve the best illumination conditions for their growth and metabolism and/or to avoid harmfully high light intensities.[1] In photomotile responses of microorganisms, light does not play the role of an energy source, but that of an information carrier: the different physical properties of the light stimulus, such as spectral composition, fluence rate, polarization degree, and propagation direction, are perceived by the photoreceptive system and converted into intracellular biophysical/biochemical signals. In all microorganisms, finally, the photosensory transduction chain, connecting the primary light-induced molecular reactions of the photoreceptor pigment and the eventual alteration of the ciliary or flagellar beating pattern, is entirely based on molecular events[1,2] (Figure 19.1).

19.2 Phenomenology of Photomotile Responses of Microorganisms

The main photodependent behaviors of microorganisms are photophobic responses, elicited by sudden increases (step-up) as well as decreases (step-down) in light fluence rate, and phototaxis, resulting from the detection of versus and direction of propagation of the light stimulus,[3] as schematically shown in Table 19.1 and Figure 19.2. Also, in some cases, orientation of cells with respect to the light polarization plane (polarotaxis) has been reported.[4] It is worthwhile remarking that, in true phototaxis, the cell tracks a light source and therefore has to be provided with a photoreceptor system able to sense the vectorial characteristics of the light signal. Schematically, this task can be accomplished either by comparing the light signals detected by two receivers at the same instant (one-instant mechanism) or by a single receiver coupled with a screening device,

0-8493-8634-9/95/$0.00+$.50
© 1995 by CRC Press, Inc.

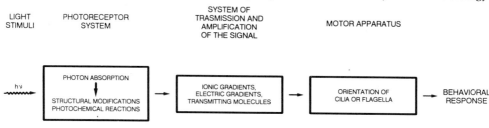

FIGURE 19.1 Block diagram of the photosensory transduction chain for photomotile responses of microorganisms.

Table 19.1

Stimulus	Motile responses
Variations in fluence rate (dI/dt or dI/ds)	Photophobic responses (step-up or step-down)
Light propagation direction $\vec{S} = \vec{E} \wedge \vec{H}$	Phototaxis (positive, negative, transverse)

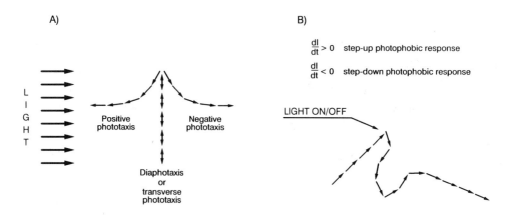

FIGURE 19.2 Schematic representation of the principal photomotile responses of microorganisms. (A): Phototaxis and diaphotaxis; (B): Step-up and step-down photophobic responses.

which, if the cell rotates periodically around an axis, modulates in time the light signal falling on the receptor (two-instant mechanism).[4-6]

Several experimental techniques have been used to study the photobehavior of microorganisms, but most up-to-date studies are accomplished by means of computer-assisted microvideorecording systems. These methods allow tracking of individual cells, analysis of cellular trajectories, in real-time or from videorecording, and measurement of motion parameters such as number of motile cells, mean velocity, distribution of velocities, and distribution of swimming directions for statistically significant numbers of cells.[7]

19.3 Photoreceptor Pigments for Photobehaviors of Microorganisms

To identify photoreceptor pigments, action spectroscopy (even though requiring extreme attention to possible artifacts due, e.g., to the presence of screening pigments, to electronic energy transfer processes, and to incorrect choice of the fluence rate range) plays an unequaled role in linking the spectral sensitivity of the physiological response to the absorption transition probabilities of the pigment. Other experimental approaches that can contribute to answering some questions about the nature and the fate of the photopigment excited state, are, for instance: (1) light-induced absorption change (LIAC), to monitor transformations and/or molecular reactions of the chromophore upon light irradiation; (2) fluorescence polarization and linear dichroism to investigate the spatial arrangement of pigment molecules and, in particular, to check their dichroic orientation; (3) steady-state and time-resolved fluorescence (emission and excitation spectra, quantum yields, lifetimes and rate constants, effects of excited state quenchers) to formulate hypotheses on the reacting electronic state and the primary molecular reactions; (4) flash photolysis to detect and characterize intermediates and thermal steps in the de-excitation pathway of the photoexcited chromophore; (5) *in vivo* absorption/emission microspectroscopy and time-gated fluorescence imaging on intact cells to assess the spectroscopic parameters of photoreceptor pigments in their physiological environment and to search for possible anisotropic distribution of pigment(s) in the cell body.[8]

Apparently, a wide variety of microorganism photopigments exist, absorbing from the near-UV (flavins and pterins) to the blue (flavins, carotenoids), to the green and the yellow-orange (rhodopsins), and to the red (stentorins and blepharismins). Regardless of their chemical nature, however, all photoreceptor pigments have to detect and transduce light signals and be efficient phototransducing devices; these light antennae have to possess a high degree of structural organization, for example, with photopigment molecules embedded in or associated with membranes.[1]

At present, rhodopsins in halobacteria and some flagellated algae and hypericin-type pigments in colored ciliated protozoa seem to have been unambiguously established as photopigments responsible for light-induced motile reactions in these microorganisms. In particular: (1) sensory rhodopsins (SR) I and II trigger step-down (SR I, $\lambda_{max} = 587$ nm) and step-up (SR I, $\lambda_{max} = 373$ nm and SR II, $\lambda_{max} = 487$ nm) photophobic responses in *Halobacterium halobium*;[9,10] (2) in the case of photosynthetic flagellates, a rhodopsin-type retinal protein is the photoreceptor pigment for light-oriented motile behavior of *Chlamydomonas reinhardtii* ($\lambda_{max} = 494$ to 495 nm),[11,12] for photophobic stop-responses of *Haematococcus fluvialis* ($\lambda_{max} = 500$ nm)[13] and for photobehavior of *Euglena gracilis* ($\lambda_{max} = 500$ nm).[14–16] In the case of *Euglena*, however, it has to be mentioned that the old flavin hypothesis[17] has not been abandoned yet, and flavins and pterins are considered to play the role of photoreceptor pigment in this microorganism.[18,19] (3) Stentorins I and II and blepharismin-binding proteins have finally been shown to be responsible for step-up photophobic responses in both the colored ciliated protozoa *Stentor coeruleus* ($\lambda_{max} = 615$ nm)[20,21] and *Blepharisma japonicum* ($\lambda_{max} = 580$ nm for red cells and $\lambda_{max} = 590$ nm for blue cells)[22–24] and for phototaxis in *Stentor*.[25]

In many cases, the nature of the intracellular signal generated by the output of the photoreceptor system (see Figure 19.1) is still to be clarified, but for some microorganisms experimentally well-founded functional models have been set up. In *Halobacterium*, for instance, photoactivated SR I and SR II convey the signal to two methyl-accepting proteins,[9] whereas in *Chlamydomonas* light-activated rhodopsin yields electrical signals that are transmitted to the motor apparatus and modulate the flagellar beating pattern.[26] Similar bioelectric events are thought to be operating in *Haematococcus*.[13] In *Stentor*, finally, and possibly in *Blepharisma* too, the intracellular transduction of the photic stimulus initiates with a change in intracellular pH, in its turn responsible for a cascade of events resulting in Ca^{2+} ions influx.[20,21]

References

1. Lenci, F., Ghetti, F., Colombetti, G., Haeder, D. P., and Song, P. S., Eds., *Biophysics of Photoreceptors and Photomovements in Microorganisms,* Plenum, New York, 1991.
2. Lipson, E. D. and Horwitz, B. A., Photosensory reception and transduction, in *Sensory Receptors and Signal Transduction,* Spudich, J. L. and Satir, B. H., Eds., Wiley, New York, 1991, 1.

3. Diehn, B., Feinleib, M. E., Haupt, W., Hildebrand, E., Lenci, F., and Nultsch, W., Terminology of behavioral responses of motile microorganisms, *Photochem. Photobiol.*, 26, 559, 1977.

4. Colombetti, G. and Marangoni, R., Mechanisms and strategies of photomovements in flagellates, in *Biophysics of Photoreceptors and Photomovements in Microorganisms*, Lenci, F., Ghetti, F., Colombetti, G., Haeder, D. P., and Song, P. S., Eds., Plenum, New York, 1991, 53.

5. Feinleib, M. E., Photomotile responses in flagellates, in *Photoreception and Sensory Transduction in Aneural Organisms*, Lenci, F. and Colombetti G., Eds., Plenum, New York, 1980, 45.

6. Foster, K. W. and Smyth, R. D., Light antennas in phototactic algae, *Microbiol. Rev.*, 44, 572, 1980.

7. Haeder, D.-P. and Vogel, K., Real-time tracking of microorganisms, in *Image Analysis in Biology*, Haeder, D.-P., Ed., CRC Press, Boca Raton, FL, 1992, 289.

8. Lenci, F. and Ghetti, F., Photoreceptor pigments for photomovement of microorganisms: Some spectroscopic and related studies, *J. Photochem. Photobiol. B: Biol.*, 3, 1, 1989.

9. Bogomolni, R. A. and Spudich, J. L., Archebacterial rhodopsins: Sensory and energy transducing membrane proteins, in *Sensory Receptors and Signal Transduction*, Spudich, J. L. and Satir, B. H., Eds., Wiley, New York, 1991, 233.

10. Takahashi, T., Absorption and action spectroscopy of phoborhodopsin (sensory rhodopsin II), in *Biophysics of Photoreceptors and Photomovements in Microorganisms*, Lenci, F., Ghetti, F., Colombetti, G., Haeder, D. P., and Song, P. S., Eds., Plenum, New York, 1991, 249.

11. Foster, K. W., Saranak, J., Patel, N., Zarilli, G., Okabe, M., Kline, T., and Nakanishi, K., A rhodopsin is the functional photoreceptor for phototaxis in the unicellular eucaryote *Chlamydomonas, Nature*, 311, 756, 1984.

12. Hegemann, P., Gaertner, W., and Uhl, R., All-*trans* retinal constitutes the functional chromophore in *Chlamydomonas* rhodopsin, *Biophys. J.*, 60, 1477, 1991.

13. Sineshchekov, O. A., Govorunova, E., G., Der, A., Keszthelyi, L., and Nultsch, W., Photoelectric responses in phototactic flagellated algae measured in cell suspension, *J. Photochem. Photobiol. B: Biol.*, 13, 119, 1992.

14. Gualtieri, P., Barsanti, L., and Passarelli, V., Absorption spectrum of a single isolated paraflagellar swelling of *Euglena gracilis, Biochim. Biophys. Acta*, 993, 293, 1989.

15. Barsanti, L., Passarelli, V., Lenci, P., and Gualtieri, P., Elimination of photoreceptor (paraflagellar swelling) and photoreception in *Euglena gracilis* by means of the carotenoid biosynthesis inhibitor nicotine, *J. Photochem. Photobiol. B: Biol.*, 13, 135, 1992.

16. Gualtieri, P., Pelosi, P., Passarelli, V., and Barsanti, L., Identification of a rhodopsin photoreceptor in *Euglena gracilis, Biochim. Biophys. Acta*, 1117, 55, 1992.

17. Benedetti, P. A. and Lenci, F., *In vivo* microspectrofluorimetry of photoreceptor pigments in *Euglena gracilis, Photochem. Photobiol.*, 26, 315, 1977.

18. Galland, P., Keiner, P., Doernemann, D., Senger, H., Brodhun, B., and Haeder, D.-P., Pterin- and flavin-like fluorescence associated with isolated flagella of *Euglena gracilis, Photochem. Photobiol.*, 51, 675, 1989.

19. Brodhun, B. and Haeder, D.-P., Photoreceptor proteins and pigments in the paraflagellar body of the flagellate *Euglena gracilis, Photochem. Photobiol.*, 52, 876, 1990.

20. Kim, I.-H., Rhee, J. S., Huh, J. W., Florell, S., Faure, B., Lee, K. W., Kahsai, T., Song, P.-S., Tamai, N., Yamazaki, T., and Yamazaki, I., Structure and function of the photoreceptors stentorins in *Stentor coeruleus*. I. Partial characterization of the photoreceptor organelle and stentorins, *Biochim. Biophys. Acta*, 1040, 43, 1990.

21. Song, P.-S., Kim, I.-H., Florell, S., Tamai, N., Yamazaki, T., and Yamazaki, I., Structure and function of the photoreceptor stentorins in *Stentor coeruleus*. II. Primary photoprocess and pico-second time-resolved fluorescence, *Biochim. Biophys. Acta*, 1040, 58, 1990.

22. Scevoli, P., Bisi, F., Colombetti, G., Ghetti, F., Lenci, F., and Passarelli, V., Photomotile responses of *Blepharisma japonicum*. I. Action spectra determination and time-resolved fluorescence of photoreceptor pigments, *J. Photochem. Photobiol., B: Biol.*, 1, 75, 1987.

23. Matsuoka, T., Matsuoka, S., Yamaoka, Y., Kuriu, T., Watanabe, Y., Takayanagi, M., Kato, Y., and Taneda, K., Action spectra for step-up photophobic response in *Blepharisma, J. Protozool.,* 39, 498, 1992.

24. Checcucci, G., Damato, G., Ghetti, F., and Lenci, F., Action spectra of the photophobic response of blue and red forms of *Blepharisma japonicum, Photochem. Photobiol.,* in press.

25. Song, P.-S., Haeder, D.-P., and Poff, K. L., Phototactic orientation by the ciliate *Stentor coeruleus, Photochem. Photobiol.,* 32, 781, 1980.

26. Harz, H., Nonnengaesser, C., and Hegemann, P., The photoreceptor current of the green alga *Chlamydomonas, Phil. Trans. R. Soc. Lond. B,* 338, 39, 1992.

20

Phototropism

Kenneth L. Poff
Michigan State University

Radomir Konjević
University of Belgrade

20.1 Introduction

Phototropism is a growth response directed toward (positive) or away from (negative) light. This response is observed typically in the shoot of a plant or in the supporting "stalk" of the fruiting body of many fungi. Because phototropism is a growth response, curvature toward or away from light is the consequence of unequal growth on the lighted and shaded sides of the responding organ. Phototropism has been studied most intensively in several model systems: the sporangiophores of the fungi *Pilobolus*[1-3] and Phycomyces,[2,3] and shoots of the seedlings of *Avena sativa*,[2,3] *Zea mays*,[2,3] *Helianthus annuus*,[4] and *Arabidopsis thaliana*.[5] In the fungi, phototropism serves to position the reproductive structure to maximize dispersal by wind, rain, and insects. In the green plant, phototropism serves to position the leaves to optimize light reception and photosynthesis. Although photosynthesis may be increased as a consequence of phototropism, light is not used in phototropism in substrate levels as in photosynthesis, but rather for its informational content. The organism obtains information about its environment. It senses the direction from which the light is incident.

To understand the mechanism of phototropism, one would like answers to three questions: What is(are) the identity of the photoreceptor pigment(s)? What is the mechanism for obtaining directional information? What are the steps in the transduction sequence? Only the answer to the second question is known with any certainty.

20.2 Methods

In a typical phototropism experiment, the appropriate material is irradiated unilaterally at a defined fluence rate and for a defined time. Following irradiation, curvature is permitted to develop in darkness and the curvature is measured. Systems have also been devised for monitoring curvature using IR radiation to which the organism is insensitive.[6,7]

0-8493-8634-9/95/$0.00+$.50
© 1995 by CRC Press, Inc.

20.3 Nature of the Photoreceptor Pigment

Action Spectra

Similar action spectra have been measured for *Avena*[8] and *Phycomyces*.[9,10] Based on these action spectra, phototropism appears to be regulated by the blue-light photoreceptor pigment(s) that control many other biological processes.[11] There are typically three peaks in the blue at about 420, 450, and 470 nm, and one in the long UV at about 370 nm.[8–11] In addition, in those instances in which the measurement has been extended into the short UV, there is also a peak at 280 nm.[2] Although the photoreceptor pigment has neither been isolated nor identified, the general agreement has been that the action spectra are the consequence of absorption by a flavoprotein in a hydrophobic environment.[2] The peaks in the blue and that in the long UV are attributed to the flavin chromophore, and the peak at 280 nm is attributed to the protein moiety.[2]

Multiple Photoreceptor Pigments

Phototropism by *Phycomyces* and *Arabidopsis thaliana* has been demonstrated to involve multiple photoreceptor pigments. This has been shown for *Phycomyces* based on differences in action spectra for several mutants and their wild-type parent.[12–14] For *Arabidopsis thaliana*, this conclusion has been based on the wavelength dependence of the shape of the fluence response relationship,[15] and comparison of this dependence in a mutant and its wild-type parent.[16] This mutant of *Arabidopsis thaliana* also shows greatly reduced phosphorylation by light of an approximately 120-kDa, membrane-associated kinase when compared with its wild-type parent.[17] Based on these observations, this kinase is thought to be associated closely with one specific photoreceptor pigment.[17]

Recently, Ahmad and Cashmore[18] cloned the *hy4* gene in *Arabidopsis thaliana* coding for a probable blue light photoreceptor pigment. This gene was cloned from a tDNA-tagged mutant exhibiting impaired blue light suppression of hypocotyl elongation. Based on sequence homology, the protein coded for by this gene is similar to flavoproteins known as DNA photolyases.[18] The protein may contain both flavin- and pterin-binding sites, although the chromophores have not been definitively demonstrated.[18] It is unlikely that this photoreceptor pigment is the same as that of phototropism because phototropism and the blue light suppression of hypocotyl elongation involve genetically distinct pathways.[19] However, photoreceptor pigments for these different processes may be similar. Thus, the cloning of gene for a probable blue light photoreceptor pigment may afford a significant advance toward identification of the photoreceptor pigments for phototropism.

20.4 Detection of Light Direction

There is convincing evidence that both the plant and fungus detect light direction by measuring the difference in the effective fluence on the lighted and shaded sides of the organ. However, the plant and the fungus accomplish this with different mechanisms — the plant by screening and the fungus by refraction.

Screening

A gradient in quantum density is established across the plant shoot by screening of the light through scattering and absorption. Thus, the quantum density is higher on the lighted side than on the shaded side of the shoot. The only pigments that are relevant for establishing the gradient are those absorbing at the wavelengths that are effective in inducing the phototropic response.[20] In etiolated tissues, the major pigments meeting these criteria are the carotenoids.[20–22] The scattering properties of the tissue are determined by the subcellular particles and intercellular air spaces. In spite of the apparently high transverse optical density of a plant shoot, the optical gradient across the shoot is remarkably low.[23] This is a consequence of back-scattering through which the path of the light can be altered many times as it traverses the shoot.[23] The higher quantum density on the lighted side results in a greater photoproduct concentration on this lighted side than on the shaded side of the shoot.[21] In effect, the plant derives information on the light direction from this difference in photoproduct concentration on the lighted and shaded sides of the shoot. Thus far, the method for experimentally manipulating this gradient across the shoot is to decrease the carotenoid concentration with inhibitors or via the use of mutants.[20,22]

Refraction

A gradient in quantum density is established across the phototropic organ of the fungus by focusing the light onto the side of the organ facing away from the light (distal side).[2,21] The light is focused by refraction at the cell-air interface as it passes from the low index of refraction of the air and enters the relatively high index of refraction of the cytoplasm.[24] In the absence of high amounts of scattering and absorption, the quantum density is higher on the distal side of the organ than on the side facing the light (proximal side). The higher quantum density on the distal side results in a greater photoproduct concentration on this side away from the light than on the proximal side facing the light. The fungus derives information on the light direction from this difference in photoproduct concentration on the proximal and distal sides of the responding organ. The effectiveness of the refraction mechanism can be manipulated experimentally in two ways. The focusing advantage can be decreased by submerging the organ in a medium with an index of refraction equal to or greater than the index of refraction of the cytoplasm.[24] The advantage can be increased by decreasing absorption across the responding organ.[21]

20.5 Transduction Sequence

Little is known with certainty concerning the steps in the transduction sequence between photoreception and the final growth response that produces curvature. Since the unequal growth associated with phototropism probably involves perturbations in the normal cellular processes, it has been difficult to obtain definitive evidence that a particular component participates in the tropistic response. For example, Ca^{2+} is thought to be involved in the phototropism transduction sequence,[25] but it is difficult to distinguish between Ca^{2+} fluctuations that cause phototropic curvature, those that happen as a consequence of the unequal growth, and those with no connection to the tropism.

The Cholodny-Went Theory

The Cholodny-Went theory for tropisms in plants is based on the work of Cholodny[26] and Went.[27] The theory explains phototropic curvature as a consequence of the lateral movement of a growth factor (auxin) from the lighted side to the shaded side of the plant. Although considerable controversy continues to be generated by this theory,[28] some direct evidence supports its validity. It has been difficult to establish the lateral transport of a single growth factor and to rule out the involvement of other factors in the response. However, a direct prediction of the Cholodny-Went theory is that growth rate should be increased on the shaded side of the responding organ as a consequence of the increased amount of growth factor on that side. At the same time, growth rate should be decreased on the lighted side as a consequence of the decreased amount of growth factor. Iino and Briggs have demonstrated a simultaneous increase in growth rate on the shaded side and a decrease on the lighted side in corn.[29] Similar data have been reported for Arabidopsis thaliana.[30] Thus, this is in accord with prediction based on the Cholodny-Went theory.

20.6 Adaptation

A major complicating factor in any study of phototropism is the process of adaptation.[31–37] Following an appropriate exposure to light, the organism not only responds phototropically, but also changes its sensitivity and/or responsiveness to a subsequent exposure to light.[33,35,36] Following a single, very short exposure of a dark-adapted plant to light, the response may not be affected by the relatively slow steps of adaptation. However, during longer irradiations (in excess of 4 min for *Arabidopsis thaliana*), the capacity of the plant to process the light and/or to respond phototropically will be altered by adaptation during the irradiation.[38] The processes of adaptation have been shown to cause considerable complications in the fluence-response relationship for phototropism and must be considered in any experiments on phototropism.[5]

20.7 Summary

Multiple photoreceptor pigments control phototropism in fungi and plants. At least one of these pigments appears to be a flavoprotein that mediates the phosphorylation of a membrane-associated kinase. Light direction is measured using the optical geometry of the responding organ. In fungi, light is focused onto the distal side of the organ. In plants, a gradient in photoproduct is produced by screening of light by scattering and absorption in the plant tissue such that more photoproduct is accumulated on the lighted side than on the shaded side of the organ. Although many factors may be involved in the transduction sequence, convincing evidence supports the lateral transport of a growth factor from the lighted side to the shaded side of the organ. The phototropic response is complicated by the processes of adaptation that may occur during the irradiation and alter the capacity of the plant to respond phototropically.

References

1. Kubo, H. and Mihara, H., Phototropic fluence-response curves for *Pilobolus crystallinus* sporangiophore, *Planta*, 174, 174, 1988.
2. Dennison, D., Phototropism, in *Encyclopedia of Plant Physiology, Vol. 7, Physiology of Movements*, Haupt, W. and Feinleib, M., Eds., Springer-Verlag, Berlin, 1979, 506.
3. Pohl, U. and Russo, V. E. A., Phototropism, in *Membranes and Sensory Transduction*, Colombetti, G. and Lenci, F., Eds., Plenum Press, New York, 1984, 231.
4. Bruinsma, J. Karssen, C. M., Benschop, M., and Van Dort, J. B., Hormonal regulation of phototropism in the light-grown sunflower seedling, *Helianthus annuus* L.: Immobility of endogenous indoleacetic acid and inhibition of hypocotyl growth by illuminated cotyledons, *J. Exp. Bot.*, 26, 411, 1975.
5. Janoudi, A.-K. and Poff, K. L., Desensitization and recovery in phototropism of *Arabidopsis thaliana*, *Plant Physiol.*, 101, 1175, 1993.
6. Iino, M. and Carr, D. J., Safelight for photomorphogenetic studies: Infrared and infrared-scope, *Plant Sci. Lett.*, 23, 263, 1981.
7. Orbović, V. and Poff, K. L., Kinetics for phototropic curvature by etiolated seedlings of *Arabidopsis thaliana*, *Plant Physiol.*, 97, 1470, 1991.
8. Thimann, K. V. and Curry, G. M., Phototropism, in *Light and Life*, McElroy, W. D. and Glass, B., Eds., Johns Hopkins University Press, Baltimore, 1961, 646.
9. Curry, G. M. and Gruen, H. E., Action spectra for the positive and negative phototropism of *Phycomyces* sporangiophores, *Proc. Natl. Acad. Sci. U.S.A.*, 45, 797, 1959.
10. Delbruck, M. and Shropshire, W., Jr., Action and transmission spectra of *Phycomyces*, *Plant Physiol.*, 35, 194, 1960.
11. Schmidt, W., Bluelight physiology, *BioScience*, 34, 698, 1984.
12. Galland, P. and Lipson, E. D., Action spectra for phototropic balance in *Phycomyces blakesleeanus* dependence on reference wavelength and intensity range, *Photochem. Photobiol.*, 41, 323, 1985.

13. Galland, P. and Lipson, E. D., Modified action spectra of photogeotropic equilibrium in *Phycomyces blakesleeanus* mutants with defects in genes *madA, madC,* and *madH, Photochem. Photobiol.,* 41, 331, 1985.

14. Galland, P. and Lipson, E. D., Blue-light reception in *Phycomyces* phototropism: Evidence for two photosystems operating in low- and high-intensity ranges, *Proc. Natl. Acad. Sci. U.S.A.,* 84, 104, 1987.

15. Konjević, R., Steinitz, B., and Poff, K. L., Dependence of the phototropic response of *Arabidopsis thaliana* on fluence rate and wavelength, *Proc. Natl. Acad. Sci. U.S.A.,* 86, 9876, 1989.

16. Konjević, R., Khurana, J. P., and Poff, K. L., Analysis of multiple photoreceptor pigments for phototropism in a mutant of *Arabidopsis thaliana, Photochem. Photobiol.,* 55, 789, 1992.

17. Reymond, P., Short, T. W., Briggs, W. R., and Poff, K. L., Light-induced phosphorylation of a membrane protein plays an early role in signal transduction for phototropism in *Arabidopsis thaliana, Proc. Natl. Acad. Sci. U.S.A.,* 89, 4718, 1992.

18. Ahmad, M. and Cashmore, A. R., *HY4* gene of *A. thaliana* encodes a protein with characteristics of a blue-light photoreceptor, *Nature,* 366, 162, 1993.

19. Liscum, E., Young, J. C., Poff, K. L., and Hangarter, R. P., Genetic separation of phototropism and blue light inhibition of stem elongation, *Plant Physiol.,* 100, 267, 1992.

20. Vierstra, R. and Poff, K. L., Role of carotenoids in the phototropic response of corn seedlings, *Plant Physiol.,* 68, 798, 1981.

21. Poff, K. L., Perception of a unilateral light stimulus, *Phil. Trans. R. Soc. Lond.,* B303, 479, 1983.

22. Piening, C. J. and Poff, K. L., Mechanism of detecting light direction in first positive phototropism in *Zea mays* L., *Plant Cell Environ.,* 11, 143, 1988.

23. Vogelmann, T. C. and Haupt, W., The blue light gradient in unilaterally irradiated maize coleoptiles: Measurement with a fiber optic probe, *Photochem. Photobiol.,* 41, 569, 1985.

24. Banbury, G. H., Phototropism of lower plants, in *Encyclopedia of Plant Physiology, Vol. 17,* Ruhland, W., Ed., Springer-Verlag, Berlin, 1959, 530.

25. Gehring, C. A., Williams, D. A., Cody, S. H., and Parish, R. W., Phototropism and geotropism in maize coleoptiles are spatially correlated with increases in cytosolic free calcium, *Nature,* 345, 528, 1990.

26. Cholodny, N., Wuchshormone und Tropismen bei den Pflanzen, *Biol. Zentralbl.,* 47, 604, 1927.

27. Went, F. W., Wuchsstoff und Wachstum, *Recl. Trav. Bot. Nederl.,* 25, 1, 1928.

28. Trewavas, T., Briggs, W. R., Bruinsma, J., Evans, M. L., Firn, R., Hertel, R., Iino, M., Jones, A. M., Leopold, A. C., Pilet, P. E., Poff, K. L., Roux, S. J., Salisbury, F. B., Scott, T. K., Sievers, A., Zieschaug, H. E., and Wayne, R., Forum. What remains of the Cholodny-Went theory?, *Plant Cell Environ.,* 15, 759, 1992.

29. Iino, M. and Briggs, W. R., Growth distribution during first positive phototropic curvature of maize coleoptiles, *Plant Cell Environ.,* 7, 97, 1984.

30. Orbović, V. and Poff, K. L., Growth distribution during phototropism of *Arabidopsis thaliana* seedlings, *Plant Physiol.,* 103, 157, 1993.

31. Blaauw, O. H. and Blaauw-Jansen, G., The phototropic responses of *Avena* coleoptiles, *Acta Bot. Nederl.,* 19, 755, 1970.

32. Shropshire, W., Jr., Stimulus perception, in *Encyclopedia of Plant Physiology, Vol. 7, Physiology of Movements,* Haupt, W. and Feinleib, M., Eds., Springer-Verlag, Berlin, 1979, 10.

33. Iino, M., Kinetic modelling of phototropism in maize coleoptiles, *Planta,* 171, 110, 1987.

34. Iino, M., Desensitization by red and blue light of phototropism in maize coleoptiles, *Planta,* 176, 183, 1988.

35. Galland, P., Photosensory adaptation in aneural organisms, *Photochem. Photobiol.,* 54, 1119, 1991.

36. Janoudi, A.-K. and Poff, K. L., Characterization of adaptation in phototropism of *Arabidopsis thaliana, Plant Physiol.,* 95, 517, 1991.

37. Janoudi, A.-K. and Poff, K. L., Action spectrum for enhancement of phototropism by *Arabidopsis thaliana* seedlings, *Photochem. Photobiol.,* 56, 655, 1992.

38. Janoudi, A.-K., Konjević, R., Apel, P., and Poff, K. L., Time threshold for second positive phototropism is decreased by a preirradiation with red light, *Plant Physiol.,* 99, 1422, 1992.

21

Phytochrome: Molecular Properties

William Parker
University of Nebraska

Pill-Soon Song
University of Nebraska

21.1 Introduction: General

Phytochrome is a ubiquitous red light sensor found in a variety of green plants (recent reviews: see References 1 to 5). Phytochrome adopts two photointerconvertible forms, a red absorbing (Pr) and a far-red absorbing (Pfr) form:

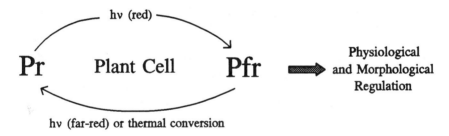

The chromoprotein is found in higher plants (seed producing plants) and in ferns, mosses, and possibly green algae.[6,7] Phytochrome controls a wide variety of developmental and physiological processes in both higher and lower plants. However, most of the information obtained regarding the molecular structure, localization, and signal transduction mechanisms of phytochrome comes from higher plants.

Phytochrome is an approximately 120- to 130-kDa protein (depending on gene and species) that contains the noncyclic tetrapyrrole chromophore phytochromobilin.[8] The tetrapyrrole is bound to the protein with a thioether linkage between a tetrapyrrole ring A side group and a cysteine roughly 320 to 360 residues from the N-terminus (depending on gene and species). Formerly, phytochrome from higher plants has been described as existing in two pools, Type 1 (etiolated or dark-grown phytochrome) and Type 2 (light-grown phytochrome). With the recent progress of molecular biology in the field, phytochromes are described more precisely according to gene family. There are at least three such families, termed *phyA, phyB,* and *phyC.*[9] Enough sequences have been determined

Table 21.1 Published Sequences of Phytochrome

Plant	Phytochrome	No. of residues	Ref.
Corn	phyA	1131	86
Oat	phyA	1128	87,88
Pea	phyA	1124	89
Potato	phyA	1123	90
Zucchini	phyA	1124	91
Rice	phyA	1128	92
Rice	phyB	1171	10
Arabidopsis	phyA	1122	9
Arabidopsis	phyB	1172	9
Arabidopsis	phyC	1111	9
Moss	?	1303	93

Note: The number of amino acid residues is listed for each sequence (no. of res.).

by cDNA analysis to provide a phylogenetic tree of the phytochrome family.[10] Table 21.1 shows a list of the cDNA deduced sequences that have been published.

The phytochrome A *(PhyA)* gene encodes Type 1 phytochrome. This is readily deduced from peptide sequence comparison with cDNA predicted sequences. (For example, see Reference 11.) Phytochrome B is expressed in green plants,[10] is more highly conserved than phytochrome A,[10] and is responsible for at least some Type 2 phytochrome responses.[12] It is now evident that three phytochromes are expressed in oat *in vivo*, only one of which is predominant in etiolated plants.[13] Also, phytochrome C mRNA is produced *in vivo* at about the same levels as is phytochrome B.[9] However, no direct role of phytochrome C in plant photoregulation has been demonstrated unequivocally at this time.

Phytochrome is synthesized in the Pr form and is converted to the Pfr form by red light. Phytochrome A is degraded rapidly upon conversion to Pfr, while Type 2 phytochrome is constitutive.[14] During etiolated growth, phytochrome A (Pr form) is present at a significantly higher concentration than is Type 2 phytochrome. During this time, the plant is maintained in an "infant" state, not producing photosynthetic apparatus and using seed storage proteins as an energy/growth source. Upon irradiation, the phytochrome A pool (Pfr form) is depleted and the plant begins green growth that persists throughout the remaining life cycle of the organism. Both phytochrome A[15,16] and Type 2 phytochrome are capable of light-grown plant photoregulation. However, the roles of different phytochromes in various red light-controlled morphological and physiological processes is still a topic of considerable debate (for reviews, see References 17 and 18). Rapid progress in this area is being facilitated by the use of both mutant and transgenic plants (for review, see Reference 3).

Although the signal transduction pathways of phytochrome are unknown as yet, there is evidence for involvement of protein phosphorylation,[19,20] Ca^{2+}-calmodulin,[21,22] G-proteins,[23,24] different auxins,[25-27] cyclic nucleotides,[28] K^+ channels,[29] and the phosphoinisitol cycle.[28,30] Further, it is likely that the blue light and the red light signal transduction pathways are interactive.[31,32] In short, it is probable that there are numerous transduction pathways affected by phytochromes. The elucidation of these pathways is the subject of intense research in the field.

21.2 Protein Structure and Function

Undegraded phytochrome A has been isolated in high purity from oat,[33-35] rye,[36] rice (some impurities still present),[37] and one dicot, pea.[38] Partial purifications have been performed on zucchini phytochrome A[39] and Type 2 phytochrome from oat[14] and pea.[40] Partially pure phytochrome has also been obtained from wheatgerm,[1] moss,[41] fern,[42] and algal preparations.[43]

Posttranslational modification of phytochrome *in vivo* takes place in the form of chromophore addition, N-terminal modification (methionine removal and acetylation; Reference 11), and phosphorylation.[44-45] Besides chromophore addition, it is not known what (if any) role these modifications play in regulation of the molecule.

Phytochrome A preparations exhibit lyase activity, as they are able to insert phycocyanibilin, phytochromobilin, and phycoerythrobilin autocatalytically.[46] It has also been assumed to have a receptor binding function. This is predicted apparently by sequence homology between a 28-kDa C-terminal region of phytochrome and bacterial sensor proteins.[47,48] These bacterial sensor proteins are kinases, and there is still some question as to whether or not phytochrome A itself is a kinase.[45,49] This issue is complicated further since phytochrome copurifies with a kinase.[50,51]

Pr phytochrome A is presumably a dimer *in vivo* and is sequestered upon conversion to the Pfr form.[52,53] It is not known if this aggregation is important in signal transduction, as it is in other systems.[54] *In vitro*, phytochrome A exists as a dimer at low concentrations,[34,55] but also adopts an oligomeric structure *in vitro*.[56] The protein structure has been studied at low resolution using electron microscopy[57] and small-angle X-ray scattering.[58] The overall appearance of the protein monomer is a dumbbell shape containing two domains. Proteolytic mapping confirms that two domains are present: a 69- to 72-kDa chromophore domain and a 52- to 55-kDa nonchromophore domain.[34]

More detailed domain models have been derived from proteolytic mapping and chromopeptide analysis.[59,60] A compilation of proteolytic mapping studies (mentioned above), Chou-Fasman calculations, and a conservation analysis have yielded a relatively detailed domain model of phytochrome.[61] This domain model includes specific regions responsible for Pr-chromophore interaction, Pfr-chromophore interaction, and subunit (dimer) contact. Recently, Edgerton and Jones[62] used a novel *in vivo* chimeric system to study the subunit contact site between dimers of phytochrome.[62] The contact sites determined by the deletion mutation analysis[62] are not in agreement with those predicted by Romanowski and Song[61] using a sequence symmetry analysis.

Changes in conformation upon Pr to Pfr photoconversion in both chromophore and nonchromophore domains have been found by proteolytic[34,59,60] and kinase studies.[63] However, the only change in secondary structure detectable by circular dichroism occurs in the 6-kDa N-terminal region of the protein.[64-66] This region apparently forms an amphiphilic alpha-helix upon conversion from the Pr to the Pfr form of phytochrome A.[67] Interestingly, mutagenic modification of this region to form a more helix-favoring sequence has produced a physiologically hyperactive phytochrome A.[68] Furthermore, phytochrome A does not have full biological activity *in vivo* without the N-terminal region.[69] These results are in contrast with the C-terminal homology to bacterial sensor proteins, which points toward the C-terminus as the receptor-binding domain.

21.3 Spectral Characteristics

The spectra of both forms of oat phytochrome A is shown in Figure 21.1. The purity of isolated phytochrome A is determined by the absorbance at 666 nm (Pr visible maximum)/the absorbance at 280 nm (tryptophan or protein band). This specific absorbance ratio (SAR) is approximately 1.0 for very pure preparations. The molar extinction coefficient of both oat and rye phytochrome A has been determined by amino acid analysis to be 132 mM^{-1} cm^{-1}.[36]

The Pr species of oat phytochrome A has a red absorbance maximum at 666 nm with a shoulder at 608 nm and a near-UV absorption maximum at 379 nm.[33] These correspond to the Qy, Qx, and Soret transitions, respectively. Oat Pfr has a far-red absorbance maximum at 730 nm (Qy transition) with a shoulder at 674 nm (Qx transition) and a Soret maximum at 400 nm.[33] During isolation, the far N-terminal region (roughly 50 to 70 residues) may be cleaved from the protein, resulting in a 5-nm blue shift of the Pfr Qy band, but little change in the Pr spectrum.[70] In the difference absorbance spectrum ($\Delta A = Pr - Pfr$), the [ΔA 666 nm (Pr)]/[ΔA 730 nm (Pfr)] is 1.06 to 1.07 for intact isolated phytochrome, which is consistent with *in vivo* difference spectra.[33] After

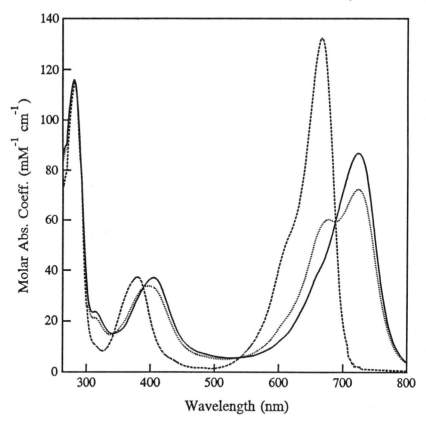

FIGURE 21.1   The spectra of oat phytochrome. The Pr spectra is shown as a dashed line. Both a corrected Pfr spectra (calculated spectra of 100% Pfr; solid line) and a photoequilibrated Pfr spectra (12% residual Pr with 660-nm irradiation; dotted line) are shown.

partial denaturization and/or degradation of the protein, the [ΔA 666 nm/ΔA 730 nm] ratio increases (for example, 1.2 for phytochrome that lacks the N-terminal region). The spectra reported for rice phytochrome have similar absorption characteristics to that of oat.[37]

Pea phytochrome A has Pr and Pfr visible maxima similar to oat phytochrome A (666 to 668 and 728–730 nm, respectively) and a similar [ΔA 666 nm/ΔA 730 nm] (1.08).[38] Rye phytochrome A also has a Pr absorbance maximum at 666 nm and a Soret band at 381 nm.[36] However, the Pfr maximum is red-shifted in rye, with an absorbance maximum at 734 nm.[36] This is in spite of the close relation between rye and oat (both monocots) compared to pea. The determination of absorbance maxima for more phytochrome As will be necessary in order to find an average or range of phytochrome absorbance maxima in the red region. Partial purification of Type 2 phytochrome from oat indicates that the Pr absorbance maximum of this phytochrome is blue-shifted to 660 nm.[14] This suggests that there is likely some diversity in spectral characteristics of the phytochrome family, a point that also awaits further study.

21.4 Photoconversion: Quantum Yield and Intermediates

Upon absorption of a photon by the Pr chromophore, phytochrome is converted to the Pfr form with a quantum efficiency of 15.2% in phosphate buffer.[36] In the same buffer, the reverse reaction takes place with a quantum efficiency of only 6% after absorption of a photon by the Pfr chromophore.[36] These low quantum efficiencies result probably from fast internal conversion from the excited state and reversibility of the initial steps in photoconversion.[71] The quantum yield of Pfr conversion to Pr may be somewhat more dependent on the condition of the phytochrome and the

Table 21.2 Primary Photoprocess and Intermediates
in Pr to Pfr Photoconversion

Method	Name of intermediate	Trapping temp. or lifetime	λ	Ref.
1. Low temperature	lumi-R	<170 K	693	
	meta-Ra	<210 K	663	74
	meta-Rc	<250 K	725	
2. Time-resolved absorbance	lumi-R	nd	nd	
	meta-Ra	nd	nd	75
	meta-Rb'	nd	nd	
	meta-Rc	nd	nd	
	Pfr	nd	nd	
	lumi-R1	7.4 μs	695	73
	lumi-R2	89.5 μs	685	
	meta-Ra1	7.6 ms	660	
	meta-Ra2	42.4 ms	670	
	meta-Rc	266 ms	725	
3. Time-resolved fluorescence				
	Pr*	48 ps		94
	Pr*	35–55 ps		95
	Pr*	48 ps		96
	Pr*	14 ps		97
	prelumi-R*	44 ps		
	Pr*	5–15 ps		98
	prelumi-R*	45–65 ps		

Note: Wavelengths from Zhang et al. (Reference 73) are approximate (nd = not determined).

buffer used, compared to the Pr-to-Pfr process. Quantum yields of photoconversion obtained under different conditions have been summarized previously.[36] The quantum yield of fluorescence for phytochrome has also been determined.[72] The fluorescence yield of the Pfr form is below detection limits, while the fluorescence of the Pr form varies from 1×10^{-3} (near UV or Soret excitation) to 3.5×10^{-3} (red excitation).

The primary photoprocess and intermediates produced during conversion of Pr to Pfr have been analyzed by a number of methods (for review, see Reference 4). It is still not clear whether the Pr-to-Pfr pathway is composed of sequential intermediates or is a more kinetically complex system.[73] The results of low-temperature trapping, time-resolved absorbance spectroscopy, and time-resolved fluorescence spectroscopy are summarized in Table 21.2. Three intermediates are trapped by low temperature.[74] Four[75] and, most recently, five[73] intermediates have been detected by time-resolved absorbance spectroscopy. The primary process involved in photoconversion (involving the excited chromophore, Pr*) has been studied using time-resolved fluorescence (Table 21.2). All studies to date have yielded an electronically excited state lifetime of about 40 ps (Table 21.2). In addition, ultrafast photolysis has indicated the presence of two excited-state species, the first with a 5- to 15-ps lifetime (Table 21.2). Time-resolved fluorescence studies have also detected fluorescing components of about 0.1 and 1.0 ns (references in Table 21.2). These long-lived components comprise generally less than 5% of the total fluorescence and have been attributed to contaminants and/or modified or degraded phytochromes.

The Pfr-to-Pr phototransformation has been studied by low-temperature trapping, revealing two intermediates, lumi-F and meta-F.[74] Unfortunately, the Pfr-to-Pr transformation does not lend itself well to flash photolysis studies because of the poor quantum yield and significant overlap between the Pr and Pfr spectra in the laser excitation range. Also, the excited-state Pfr chromophore does not produce detectable fluorescence. Thus, most of the detailed information concerning photoconversion is limited to the Pr-to-Pfr pathway.

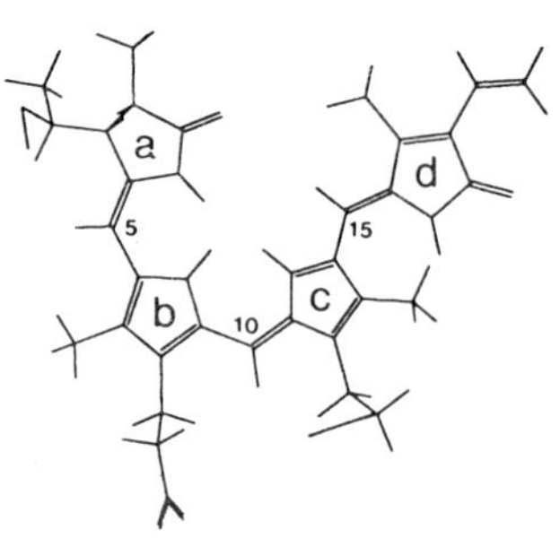

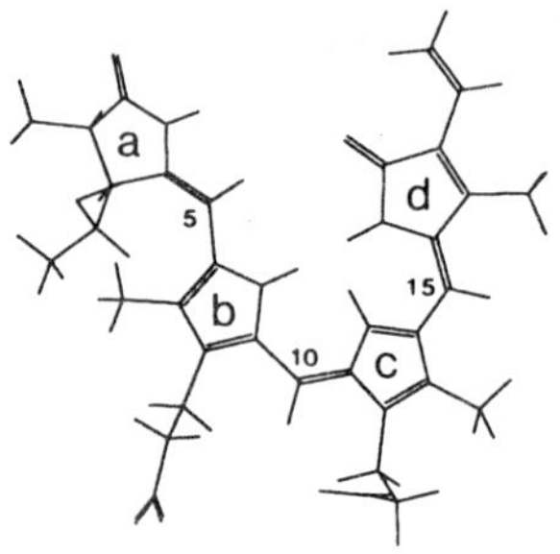

FIGURE 21.2   Two possible conformations/configurations of the Pr phytochrome chromophore. Both chromophores have a Z,Z,Z-configuration. The left-hand chromophore has a C5-*syn*, C10-*syn*, C15-*anti*-conformation, while the chromophore on the right has a C5-*anti*, C10-*syn*, C15-*syn*-conformation.

21.5 Chromophore Conformation

Two possible phytochrome chromophore conformation/configurations of the Pr chromophore are shown in Figure 21.2. ¹H NMR spectroscopy of model chromophores,[76,77] resonance Raman spectroscopy,[78] surface-enhanced resonance Raman spectroscopy, Fourier transform difference infrared spectroscopy,[79] and Fourier transform resonance Raman[80] all point toward a C15-C16 Z (Pr) to E (Pfr) isomerization as the primary photoprocess involved in phytochrome A photoconversion.

Besides the consensus of opinion concerning Z (Pr) and E (Pfr) configurations at the C15-C16 double bond (Figure 21.2), there is little else that is known about the chromophore conformation/configuration in either form (for review, see Reference 81). However, there are other indications as to the state of the chromophore.

1. Fluorescence polarization and calculated oscillator strength ratios of the Pr and Pfr chromophores indicate that the chromophores are neither fully extended (as it is in phycocyanin) nor fully cyclic.[82–84]
2. Based on analysis of chromopeptides, C5-*Z* and C10-*Z* configurations have been assigned for the Pr and Pfr chromophores.[76,77]
3. Although IR spectroscopy indicates that the Pr chromophore is protonated and the Pfr chromophore may be protonated,[79] it is evident from resonance Raman spectroscopy[85] that the Pr chromophore is protonated and the Pfr chromophore is not.
4. Some distortion about the C10 bridge in both the Pr and Pfr chromophores is proposed based on resonance Raman study.[80]
5. A *syn* (Pr) to *anti* (Pfr) conformational change at the C14-C15 bond has been proposed based on comparison of phytochrome A resonance Raman spectra with calculated vibrational frequencies and intensities of model structures.[78] It is proposed that this conformational change occurs simultaneously with the Z to E configurational change.[78] This simultaneous conformational-configurational change is supported by intuitive arguments concerning the chromophore pocket size[4] and the oscillator strength ratios of lumi-R compared to Pr.[74]

21.6 Concluding Remarks

The 100+ phytochrome-related research articles published annually involve plant physiology, signal transduction mechanisms, biophysical/bioanalytical investigation of the chromophore and protein photoconversion dynamics, and protein/chromophore characterization by classical biochemical techniques and molecular biology. Rapid progress is being made in virtually all areas of study, although it is evident that our knowledge of phytochrome structure/function is still in its infancy. The crystallographically determined 3-D structure of the phytochrome has not been reported as of this writing.

Acknowledgments

The authors wish to thank Todd Wells for spectra of highly purified phytochrome and Janelle Jones for help in preparation of the manuscript. The authors' work described in this chapter has been supported by the NIH (RO1-GM36956 to PSS).

References

1. Partis, M. D. and Thomas, B., Phytochromes, *Methods in Plant Biochem.*, 5, 233, 1991.
2. Eilfeld, P. and Haupt, W., Phytochrome, in *Photoreceptor Evolution and Function*, Holmes, M. G., Ed., Academic Press, London, 1991, 203–239.
3. Thomas, B. and Johnson, C., *Phytochrome Properties and Biological Action*, NATO ASI Series, Vol. H 50, Thomas, B. and Johnson, C., Eds., Springer-Verlag, Berlin, 1991, 85–112.
4. Rüdiger, W. and Thümmler, F., Phytochrome, the visual pigment of plants, *Angew. Chem., Int. Ed. Engl.*, 30, 1216, 1991.
5. Quail, P. H., Phytochrome: A light-activated molecular switch that regulates plant gene expression, *Annu. Rev. Genet.*, 25, 389, 1991.
6. Rüdiger, W. and Thümmler, F., Phytochrome in lower plants, in *Phytochrome Properties and Biological Action*, NATO ASI Series, Vol. H 50, Thomas, B. and Johnson, C., Eds., Springer-Verlag, Berlin, 1991, 57–70.
7. Wada, M. and Kadota, A., Photomorphogenesis in lower plants, *Plant Mol. Biol.*, 40, 169, 1989.
8. Lagarias, J. C. and Rapoport, H., Chromopeptides from phytochrome. The structure and linkage of the Pr form of the phytochrome chromophore, *J. Am. Chem. Soc.*, 102, 4821, 1980.
9. Sharrock, R. A. and Quail, P. H., Novel phytochrome sequences in *Arabidopsis thaliana*: Structure, evolution, and differential expression of a plant regulatory photoreceptor family, *Genes Devel.*, 3, 1745, 1989.
10. Dehesh, K., Tepperman, J., Christensen, A. H., and Quail, P. H., PhyB is evolutionarily conserved and constitutively expressed in rice seedling shoots, *Mol. Gen. Genet.*, 225, 305, 1991.
11. Grimm, R., Kellermann, J., Schafer, W., and Rüdiger, W., The amino-terminal structure of oat phytochrome, *FEBS Lett.*, 234, 497, 1988.
12. Nagatani, A., Chory, J., and Furuya, M., Phytochrome B is not detectable in the hy3 mutant of *Arabidopsis*, which is deficient in responding to end-of-day far-red light treatments, *Plant Cell Physiol.*, 32, 1119, 1991.
13. Wang, Y.-C., Stewart, S. J., Cordonnier, M.-M., and Pratt, L. H., *Avena Sativa* L. contains three phytochromes, only one of which is abundant in etiolated tissue, *Planta*, 184, 96, 1991.
14. Pratt, L. H., Stewart, S. J., Shimazaki, Y., and Wang, Y. C., Monoclonal antibodies directed to phytochrome from green leaves of *Avena sativa* L. cross-react weakly or not at all with the phytochrome that is most abundant in etiolated shoots of the same species, *Planta*, 184, 87, 1991.
15. McCormac, A., Whitelam, G., and Smith, H., Light-grown plants of transgenic tobacco expressing an introduced oat phytochrome A gene under the control of a constitutive viral promoter exhibit persistent growth inhibition by far-red light, *Planta*, 188, 173, 1991.
16. Cherry, J. R., Hondred, D. Keller, J. M., Hershey, H. P., and Vierstra, R. D., The use of transgenic plants to study phytochrome domains involved in structure and function, in *Phytochrome Properties and Biological Action*, NATO ASI Series, Vol. H 50, Thomas, B. and Johnson, C., Eds., Springer-Verlag, Berlin, 1991, 113–127.
17. Smith, H., Whitelam, G. C., and McCormac, A. C., Do the members of the phytochrome family have different roles? Physiological evidence from wild-type, mutant and transgenic plants, in

Phytochrome Properties and Biological Action, NATO ASI Series, Vol. H 50, Thomas, B. and Johnson, C., Eds., Springer-Verlag, Berlin, 1991, 217–236.

18. Smith, H. and Whitelam, G. C., Phytochrome, a family of photoreceptors with multiple physiological roles, *Plant Cell Environ.,* 13, 695, 1990.

19. Otto, V. and Schafer, E., Rapid phytochrome-controlled phosphorylation and dephosphorylation in *Avena sativa L., Plant Cell Physiol.,* 29, 1115, 1988.

20. Doshi, A. and Sopory, S. K., Regulation of protein phosphorylation by phytochrome in *Sorghum bicolor, Photochem. Photobiol.,* 55, 465, 1992.

21. Shacklock, P. S., Read, N. D., and Trewavas, A. J., Cytosolic free calcium mediates red light-induced photomorphogenesis, *Nature,* 358, 753, 1992.

22. Lam, E., Benedyk, M., and Chua, N.-H., Characterization of phytochrome regulated gene expression in a photoautotrophic cell suspension: Possible role for calmodulin, *Mol. Cell Biol.,* 9, 4819, 1989.

23. Romero, L. C., Sommer, D., Gotor, C., and Song, P.-S., G-Proteins in etiolated *Avena* seedlings. Possible phytochrome regulation, *FEBS Lett.,* 282, 341, 1991.

24. Bossen, M. E., Kendrick, R. E., and Vrendengerg, W. J., The involvement of a G-protein in phytochrome-regulated, calcium-dependent swelling of etiolated wheat protoplasts, *Physiol. Plant.,* 80, 55, 1990.

25. Martinez-Garcia, J. F. and Garcia-Martinez, J. L., Interaction of gibberellins and phytochrome in the control of cowpea epicotyl elongation, *Physiol. Plant.,* 86, 236, 1992.

26. Behringer, F., Davies, P. J., and Reid, J. B., Phytochrome regulation of stem growth and indole-3-acetic acid levels in the lv and Lv genotypes of *Pisum, Photochem. Photobiol.,* 56, 677, 1992.

27. Kagawa, T. and Sugai, M., Involvement of gibberellic acid in phytochrome-mediated spore germination of the fern *Lygodium japonicum, J. Plant Physiol.,* 138, 299, 1991.

28. Morse, M. J., Crain, R. C., Cote, G. G., and Satter, R. L., Light stimulated phospholipid turnover in *Samanea Samon Pulvini.* Increased levels of diacylglycerol, *Plant Physiol.,* 89, 724, 1989.

29. Lew, R. R., Krasnoshtein, F., Serlin, B. S., and Schauf, C. L., Phytochrome activation of potassium channels and chloroplast rotation in *Mougeotia, Plant Physiol.,* 98, 1511, 1992.

30. Guron, K., Chandok, M. R., and Sopory, S. K., Phytochrome mediated rapid changes in the level of phosphoinositides in etiolated leaves of *Zea Mays, Photochem. Photobiol.,* 56, 691, 1992.

31. Krami, M. and Herrmann, H., Red-blue-interaction in *Mesotaenium* chloroplast movement: Blue seems to stabilize the transient memory of the phytochrome signal, *Photochem. Photobiol.,* 53, 255, 1991.

32. Elmlinger, M. W. and Mohr, H., Coaction of blue/ultraviolet-A light and light absorbed by phytochrome in controlling the appearance of ferredoxin-dependent glutamate synthase in the Scots pine *(Pinus sylvestris L.)* Seedling, *Planta,* 183, 374, 1991.

33. Vierstra, R. D. and Quail, P. H., (1983) Purification and initial characterization of 124 kilodalton phytochrome from *Avena, Biochemistry,* 22, 2498, 1983.

34. Lagarias, J. C. and Mercurio, F. M., Structure-function studies on phytochrome. Identification of light-induced conformational changes in 124-kDa *Avena* phytochrome *in vitro, J. Biol. Chem.,* 260, 2415, 1985.

35. Chai, Y. G., Singh, B. R., Song, P.-S., Lee, J., and Robinson, G. W., Purification and spectroscopic properties of 124 kDa oat phytochrome, *Anal. Biochem.,* 163, 322, 1987.

36. Lagarias, J. C., Kelly, J. M., Cyr, K. L., and Smith, W. O., Jr., Comparative photochemical analysis of highly purified 124 kilodalton oat and rye phytochromes *in vitro, Photochem. Photobiol.,* 46, 5, 1987.

37. Schendel, R., Tong, Z., and Rüdiger, W., Partial proteolysis of rice phytochrome: Comparison with oat phytochrome, *Z. Naturforsch.,* C44, 757, 1989.

38. Nakazawa, M., Yoshida, Y., and Manabe, K., Differences between the surface properties of the Pr and Pfr forms of native pea phytochrome, and their application to a simplified procedure for purification of the phytochrome, *Plant Cell Physiol.,* 32, 1187, 1991.

39. Vierstra, R. D. and Quail, P. H., Spectral characterization and proteolytic mapping of native 120 kilodalton phytochrome from *Cucurbita pepo L., Plant Physiol.,* 77, 990, 1985.

40. Abe, H., Yamamoto, K. T., Nagatani, A., and Furuya, M., Characterization of green tissue specific phytochrome isolated immunochemically from pea seedlings, *Plant Cell Physiol.,* 26, 1387, 1985.

41. Lindemann, P., Braslavsky, S. E., Hartmann, E., and Schaffner, K., Partial purification and initial characterization of phytochrome from the moss *Atrichum undulatum P. Beauv.* grown in light, *Planta,* 178, 436, 1989.

42. Oyama, Y., Yamamoto, K. T., and Wada, M., Phytochrome in the fern, *Adiantum capillus-veneris L.:* Spectrophotometric detection *in vivo* and partial purification, *Plant Cell Physiol.,* 31, 1229, 1990.

43. Kidd, D. G. and Lagarias, J. C., Phytochrome from the green alga *Mesotaenium caldariorum, J. Biol. Chem.,* 265, 7029, 1990.

44. Hunt, R. E. and Pratt, L. H., Partial characterization of undegraded oat phytochrome, *Biochemistry,* 19, 390, 1980.

45. McMichael, R. W., Phytochrome-Mediated Signal Transduction: The Role of Protein Phosphorylation, Ph.D. dissertation, University of California, Davis.

46. Li, L. and Lagarias, J. C., Phytochrome assembly: Defining chromophore structural requirements for covalent attachment and photoreversibility, *J. Biol. Chem.,* 267, 19204, 1992.

47. Schneider-Poetsch, H. A. W. and Braun, B., Proposal on the nature of phytochrome action based on the C-terminal sequences of phytochrome, *J. Plant Physiol.,* 137, 576, 1991.

48. Schneider-Poetsch, H. A. W., Braun, B., Marx, S., and Schaumburg, A., Phytochromes and bacterial sensor proteins are related by structural and functional homologies. Hypothesis on phytochrome-mediated signal-transduction, *FEBS,* 281, 245, 1991.

49. McMichael, R. W. and Lagarias, J. C., Polycation-stimulated phytochrome phosphorylation: Is phytochrome a protein kinase?, *Curr. Top. Plant Biochem. Physiol.,* 9, 259, 1990.

50. Grimm, R., Gast, D., and Rüdiger, W., Characterization of a protein-kinase activity associated with phytochrome from etiolated oat *(Avena sativa L.)* seedlings, *Planta,* 178, 199, 1989.

51. Kim, I.-S., Bai, U., and Song, P.-S., A purified 124-kDa oat phytochrome does not possess a protein kinase activity, *Photochem. Photobiol.,* 49, 319, 1989.

52. Quail, P. H., Marme, D., and Schafer, E., Particle bound phytochrome from maize and pumpkin, *Nature New Biol.,* 245, 189, 1973.

53. Pratt, L. H. and Marme, D., Red-light enhanced phytochrome pellatability. Re-examination and further characterization, *Plant Physiol.,* 58, 686, 1976.

54. Metzger, H., Transmembrane signaling: The joy of aggregation, *J. Immunol.,* 149, 1477, 1992.

55. Jones, A. M. and Quail, P. H., Quaternary structure of 124-kilodalton phytochrome from *Avena sativa L., Biochemistry,* 25, 2987, 1986.

56. Choi, J.-K., Kim, I.-S., Kwon, T.-I., Parker, W., and Song, P.-S., Spectral perturbations and oligomer/monomer formation in 124-kilodalton *Avena* phytochrome, *Biochemistry,* 29, 6883, 1990.

57. Jones, A. M. and Erickson, H. P., Domain structure of phytochrome from *Avena sativa* visualized by electron microscopy, *Photochem. Photobiol.,* 49, 479, 1989.

58. Tokutomi, S. T., Nakasako, M., Sakai, J., Kataoka, M., Yamamoto, K. T., Wada, M., Tokunaga, F., and Furuya, M., A model for the dimeric molecular structure of phytochrome base small-angle X-ray scattering, *FEBS Lett.,* 247, 139, 1989.

59. Jones, A. M., Viestra, R. D., Daniels, S. M., and Quail, P. H., The role of separate molecular domains in the structure of phytochrome from etiolated *Avena sativa L., Planta,* 164, 501, 1985.

60. Grimm, R., Eckerskorn, C., Lottspeich, F., Zenger, C., and Rüdiger, W., Sequence analysis of proteolytic fragments of 124-kilodalton phytochrome from etiolated *Avena sativa L., Planta,* 174, 396, 1988.

61. Romanowski, M. and Song, P.-S., Structural domains of phytochrome deduced from homologies in amino acid sequences, *J. Protein Chem.,* 11, 139, 1992.

62. Edgerton, M. D. and Jones, A. M., Localization of protein-protein interactions between subunits of phytochrome, *Plant Cell,* 4, 161, 1992.

63. Wong, Y. S., Cheng, H. C., Walsh, D. A., and Lagarias, J. C., Phosphorylation of *Avena* phytochrome *in vitro* as a probe of light-induced conformational changes, *J. Biol. Chem.*, 261, 12089, 1986.

64. Chai, Y. G., Song, P.-S., Cordonnier, M.-M., and Pratt, L. H., A photoreversible circular dichroism spectral change in oat phytochrome is suppressed by a monoclonal antibody that binds near its N-terminus and by chromophore modification, *Biochemistry*, 26, 4947, 1987.

65. Vierstra, R. D., Quail, P. H., Hahn, T.-R., and Song, P.-S., Comparison of the protein conformations between different forms (Pr and Pfr) of native (124 kDa) and degraded (118/114 kDa) phytochromes from *Avena sativa*, *Photochem. Photobiol.*, 45, 429, 1987.

66. Sommer, D. and Song, P.-S., Chromophore topography and secondary structure of 124 kilodalton *Avena* phytochrome probed by Zn^{++}-induced chromophore modification, *Biochemistry*, 29, 1943, 1990.

67. Parker, W., Partis, M., and Song, P.-S., N-terminal domain of *Avena* phytochrome: Interactions with sodium dodecyl sulfate micelles and N-terminal chain truncated phytochrome, *Biochemistry*, 31, 9413, 1992.

68. Stockhaus, J., Nagatani, A., Halfter, U., Kay, S., Furuya, M., and Chua, N.-H., Serine to alanine substitutions at the amino-terminal region of phytochrome A result in an increase in biological activity, *Genes Develop.*, 6, 2364, 1992.

69. Cherry, J. R., Hondred, D., Walker, J. M., and Vierstra, R. D., Phytochrome requires the 6-kDa N-terminal domain for full biological activity, *Proc. Natl. Acad. Sci. U.S.A.*, 89, 5039, 1992.

70. Vierstra, R. D. and Quail, P. H., Photochemistry of 124 kilodalton *Avena* phytochrome *in vitro*, *Plant Physiol.*, 72, 264, 1983.

71. Song, P.-S., The molecular topography of phytochrome: Chromophore and apoprotein, *J. Photochem. Photobiol.*, B2, 43, 1988.

72. Columbano, C. G., Braslavsky, S. E., Holzwarth, A. R., and Schaffner, K., Fluorescence quantum yields of 124 kDa phytochrome from oat upon excitation within different absorption bands, *Photochem. Photobiol.*, 52, 19, 1990.

73. Zhang, C.-F., Farrens, D. L., Bjorling, S. C., Song, P.-S., and Kliger, D. S., Time-resolved absorption studies of native etiolated oat phytochrome, *J. Am. Chem. Soc.*, 114, 4569, 1992.

74. Eilfeld, P. and Rüdiger, W., Absorption spectra of phytochrome intermediates, *Z. Naturforsch. C*, 40, 109, 1985.

75. Inoue, Y., Rüdiger, W., Grimm, R., and Furuya, M., The phototransformation pathway of dimeric oat phytochrome from the red-light absorbing form to the far-red-light absorbing form at physiological temperature is composed of four intermediates, *Photochem. Photobiol.*, 52, 1077, 1990.

76. Rüdiger, W., Thümmler, F., Cmiel, E., and Schneider, S. A., Chromophore structure of the physiologically active form (Pfr) of phytochrome, *Proc. Natl. Acad. Sci. U.S.A.*, 80, 6244, 1983.

77. Thümmler, F., Rüdiger, W., Cmiel, E., and Schneider, S., Chromopeptides from phytochrome and phycocyanin. NMR studies of the Pfr and Pr chromophore of phytochrome and *E,Z* isomeric chromophores of phycocyanin, *Z. Naturforsch. C*, 38, 359, 1983.

78. Fodor, S. P. A., Lagarias, J. C., and Mathies, R. A., Resonance Raman analysis of the Pr and Pfr forms of phytochrome, *Biochemistry*, 29, 11141, 1990.

79. Siebert, F., Grimm, R., Rüdiger, W., Schmidt, G., and Scheer, H., Infrared spectroscopy of phytochrome and model pigments, *Eur. J. Biochem.*, 194, 921, 1990.

80. Hildebrandt, P., Hoffmann, A., Lindemann, P., Heibel, G., Braslavsky, S. E., Schaffner, K., and Schrader, B., Fourier Transform Resonance Raman spectroscopy of phytochrome, *Biochemistry*, 31, 7957, 1992.

81. Song, P.-S., Suzuki, S., Kim, I.-D., and Kim, J. H., Properties and evolution of photoreceptors, in *Photoreceptor Evolution and Function*, Holmes, M. G., Ed., Academic Press, San Diego, CA, 1991, 21–63.

82. Song, P.-S., Chae, Q., and Gardner, J. G., Spectroscopic properties and chromophore conformations of the photomorphogenic receptor: Phytochrome, *Biochim. Biophys. Acta*, 576, 479, 1979.

83. Song, P.-S. and Chae, Q., The transformation of phytochrome to its physiologically active form, *Photochem. Photobiol.*, 30, 117, 1979.

84. Parker, W., Goebel, P., Song, P.-S., and Stezowski, J. J., Molecular modeling of phytochrome using constitutive C-phycocyanin from *Fremyella diplosiphon* as a structural template, *Bioconjugate Chem.*, submitted.

85. Mizutani, Y., Tokutomi, S., Aoyagi, K., Horitsu, K., and Kitagawa, T., Resonance Raman study on intact pea phytochrome and its model compounds: Evidence for proton migration during the phototransformation, *Biochemistry*, 30, 10693, 1991.

86. Christensen, A. H. and Quail, P. H., Structure and expression of a maize phytochrome-encoding gene, *Gene*, 85, 381, 1989.

87. Hershey, H. P., Barker, R. F., Idler, K. B., Murray, M. G., and Quail, P. H., Nucleotide sequence and characterization of a gene encoding the phytochrome polypeptide from *Avena*, *Gene*, 61, 339, 1987.

88. Hershey, H. P., Barker, R. F., Idler, K. B., Lissemore, J. L., and Quail, P. H., Analysis of cloned cDNA and genomic sequences for phytochrome: Complete amino acid sequence for two gene products expressed in etiolated *Avena*, *Nucl. Acids Res.*, 13, 8543, 1985.

89. Sato, N., Nucleotide sequence and expression of the phytochrome gene in *Pisum sativum*: Differential regulation by light of multiple transcripts, *Plant Mol. Biol.*, 11, 697, 1988.

90. Heyer, A. and Gatz, C., Isolation and characterization of a cDNA-clone coding for potato type A phytochrome, *Plant Mol. Biol.*, 18, 535, 1992.

91. Lissemore, J. L., Colbert, J. T., and Quail, P. H., Cloning of cDNA for phytochrome from etiolated *Cucurbita* and coordinate photoregulation of the abundance of two distinct phytochrome transcripts, *Plant Mol. Biol.*, 8, 485, 1987.

92. Kay, S. A., Keith, B., Shinozaki, K., and Chua, N.-H., The sequence of the rice phytochrome gene. *Nucl. Acids Res.*, 17, 2865, 1989.

93. Thümmler, F., Dufner, N., Kreisl, P., and Dittrich, P., Phytochrome-Moss *(Ceratodon purpureus)* protein sequence database. Reference number: S20160.

94. Holzwarth, A. R., Wendler, J., Ruzsicska, B. P., Braslavsky, S. E., and Schaffner, K., Picosecond time-resolved and stationary fluorescence of oat phytochrome highly enriched in the native 124 kDa protein, *Biochim. Biophys. Acta*, 791, 265, 1984.

95. Song, P.-S. and Singh, B. R., Primary photoprocess of phytochrome. Picosecond fluorescence kinetics of oat and pea phytochromes, *Biochemistry*, 28, 3265, 1989.

96. Farrens, D. L. and Song, P.-S., Subnanosecond single photon timing measurements using a pulsed diode-laser, *Photochem. Photobiol.*, 54, 313, 1991.

97. Hermann, G., Lippitsch, M. E., Brunner, H., Aussenegg, F. R., and Muller, E., Picosecond dynamics of the excited state relaxations in phytochrome, *Photochem. Photobiol.*, 52, 13, 1990.

98. Holzwarth, A. R., Venuti, E., Braslavsky, S. E., and Schaffner, K., The phototransformation process in phytochrome. I. Ultrafast fluorescence component and kinetic models for the initial Pr to Pfr transformation steps in native phytochrome, *Biochim. Biophys. Acta*, 1140, 59, 1992.

22

Phytochrome Chromophore Biosynthesis

J. Clark Lagarias
University of California

22.1 Introduction

The phytochrome photoreceptor performs a major role as a mediator of photomorphogenetic responses at all stages of the plant life cycle.[1,2] Encoded by a small multigene family, phytochrome consists of two 1100 amino acid subunits to which a linear tetrapyrrole (bilin) chromophore is attached covalently.[3,4] The biosynthesis of holophytochrome therefore requires the convergence of two biosynthetic pathways — one for the apoprotein and another for the free chromophore precursor, phytochromobilin (PΦB). The structure and regulation of the apophytochrome gene family has been reviewed thoroughly.[4] This discussion focuses therefore on the biosynthesis of the phytochrome chromophore, a process that entails the synthesis of PΦB and its assembly with apophytochrome. For an in-depth discussion of phytochrome biosynthesis, the reader should consult the more comprehensive review on this subject.[5]

22.2 *In Vivo* Biosynthetic Studies

Phytochrome is a minor constituent of plant cells. For this reason, the enzymes that are committed to the biosynthesis of the phytochrome chromophore are not expected to be present at high levels. The detection of these enzymes has taken advantage of the discovery of a novel class of compounds that inhibit the synthesis of 5-aminolevulinic acid (ALA), the first committed precursor of all naturally occurring tetrapyrroles. In plants and most bacteria, ALA is synthesized from glutamate via the C5 pathway of tetrapyrrole biosynthesis.[6] The third enzyme of this pathway, glutamate-1-semialdehyde aminotransferase, is strongly inhibited by compounds that were initially developed as suicide inhibitors of the mammalian enzyme, γ-aminobutyric acid α-ketoglutaric acid aminotransferase. For the analysis of PΦB biosynthesis, gabaculine and 4-amino-5-hexynoic acid have

0-8493-8634-9/95/$0.00+$.50
© 1995 by CRC Press, Inc.

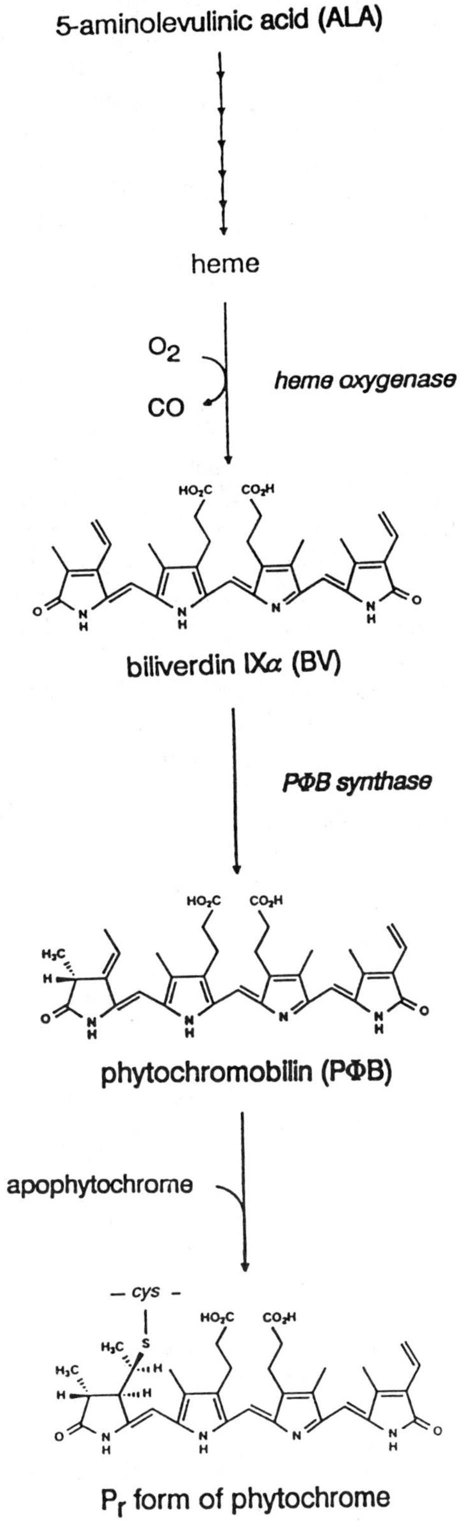

FIGURE 22.1 Proposed pathway of the biosynthesis of the phytochrome chromophore.

been the most useful of this family of inhibitors.[7,8] Etiolated plant seedlings grown in the presence of either compound contain greatly reduced levels of spectrophotometrically active holophytochrome.[7,9,10] Exogenous feeding of ALA, biliverdin IXα (BV) and the PΦB analog phycocyanobilin (PCB) reverses this inhibition in etiolated oat seedlings because they are intermediates in the synthesis of the phytochrome chromophore.[11,12] The incorporation of radiolabeled ALA and BV IXα into holophytochrome has confirmed directly the intermediacy of both compounds in this pathway.[11,12] Based on these *in vivo* experiments and by analogy with the known biosynthetic pathway of PCB in algae,[13] the biochemical pathway for the synthesis of the phytochrome chromophore appears to proceed as shown in Figure 22.1.

22.3 Phytochromobilin Synthase

Recent investigations have addressed the biochemistry of the conversion of BV to PΦB, the proposed precursor of the protein-bound chromophore. Since authentic PΦB was unavailable at the time of these investigations, a coupled assay was developed in order to detect newly synthesized PΦB.[14,15] This assay took advantage of the earlier observation that photoactive holophytochrome is produced during coincubation of apophytochrome and the PΦB analog PCB (see later discussion).[16] With this assay, experimental evidence was obtained for the presence of phytochromobilin synthase, the enzyme that accomplishes the conversion of BV to PΦB, in cucumber plastid preparations.[15] The requirement for reducing power for this chemical interconversion, formally a two-electron reduction, was also demonstrated in this study. In this regard, light-driven conversion of BV to PΦB provided strong evidence that PΦB synthase is plastid-localized.[15] The assignment of the PΦB structure to the product of this reaction was based initially on the holophytochrome difference spectrum that was indistinguishable from that of native oat phytochrome. Recently, the structure of PΦB was confirmed by the serendipitous isolation of authentic 3E-PΦB from phycobiliprotein-containing algae.[17] With this pigment in hand, rapid progress is anticipated now on the fractionation and biochemical characterization of phytochromobilin synthase.

22.4 Enzymology of BV Synthesis in Plants

As shown in Figure 22.1, it is expected that BV is synthesized from heme by the enzyme heme oxygenase. This hypothesis is reasonable, especially in view of the intermediacy of heme in the biosynthesis of phycocyanobilin in the red alga *Cyanidium caldarium*.[13,14] To date, however, direct evidence for heme oxygenase in plants has not been reported. This is likely due to the difficulty in distinguishing between enzymatic and nonenzymatic "coupled" oxidation of heme that requires an HPLC analysis of the isomer composition of the BV product(s). Using the coupled assay protocol described above, it has recently been observed that plastid preparations can convert heme, but not magnesium protoporphyrin, to PΦB.[5] Although this argues in favor of the heme oxygenase-mediated synthesis of BV in plants, this question awaits experimental confirmation. Recent experiments with antibody and cDNA probes to mammalian heme oxygenases suggest that the enzyme that accomplishes the conversion of heme to BV in mammals is quite dissimilar to the plant and algal enzymes (Maines, M. D., personal communication; Beale, S. I., personal communication). In this regard, heme oxygenase utilizes ferredoxin in red algae as opposed to cytochrome P450 reductase for the mammalian enzyme.[18] Owing to the close evolutionary relationship between plants and algae, it is expected that the plant heme oxygenase will be more similar to the algal enzyme. Indeed, all of the enzymes of the PΦB biosynthetic pathway appear to be plastid-localized.[5] For this reason, it is likely that many, if not all, of the earlier steps of PΦB synthesis will share enzymes with the heme and chlorophyll biosynthetic pathways of chloroplasts.[13]

22.5 Holophytochrome Assembly

Phytochrome assembly is autocatalytic, with the C-S lyase activity residing in the phytochrome apoprotein itself. This hypothesis was first suggested by Elich and Lagarias, who demonstrated that immunopurified phytochrome could assemble with PCB *in vitro*.[16] This reaction required no cofactors such as ATP or reduced pyridine nucleotides. Confirmation of this hypothesis was provided by *in vitro* transcription and translation of a full-length cDNA clone that assembled with PCB.[19] In this experiment, the formation of holophytochrome was ascertained by its characteristic light-dependent difference in proteolytic susceptibility. The ability of recombinant apophytochrome expressed in both yeast[20,21] and *E. coli*[21] to assemble with either PCB or PΦB to yield photoactive holophytochromes is also consistent with this hypothesis.

Recombinant apophytochrome has proven especially useful for defining the structural specificity of both bilin and apoprotein components for holophytochrome assembly. Li and Lagarias examined the kinetics of holophytochrome assembly using a zinc-dependent fluorescence blot assay.[22] Five bilins were compared in this study to define the substrate specificity for covalent attachment. Only the ethylidene-containing bilins yielded covalent adducts with recombinant apophytochrome from yeast cells. Kinetic analysis revealed that the bilin binding site on apophytochrome is tailored best to the natural chromophore precursor PΦB. *In vitro* and *in vivo* assembly of recombinant apophytochrome has also proved to be an ideal experimental system for defining the specific amino acids in the apoprotein that are necessary for bilin attachment. Indeed, both deletions and site-specific mutations in apophytochrome can be introduced readily via recombinant DNA protocols. Deletion analyses have revealed already that neither the N-terminus (up to residue 69) nor the entire C-terminal domain of the protein are required for assembly to occur.[5,20,23,24] The catalytic role of specific residues within the bilin lyase domain that is delimited by deletion mutagenesis experiments can now be addressed readily by site-directed mutagenesis. In combination, the analysis of phytochrome assembly with different bilin substrates and mutant apoproteins should lead ultimately to a detailed understanding of the mechanism of the assembly of this important photoreceptor.

22.6 Future Studies

The knowledge obtained from the analysis of phytochrome chromophore biosynthesis will provide insight into the rational design of both chemical and molecular genetic approaches to alter the course of light-mediated plant development. For example, a detailed understanding of the enzymology of the committed steps of the phytochrome chromophore biosynthetic pathway will be invaluable for the design of specific inhibitors of phytochrome accumulation in plants. Chromophore-specific inhibition may prove preferable to gene disruption approaches since it is likely that inactivation of all members of the phytochrome gene family will be effected by this type of inhibitor. It is hoped that chromophore-based inhibitors will also provide biochemical tools for analysis of the molecular mechanism of phytochrome-mediated plant growth and development that will complement molecular genetic approaches. Purification of the committed enzymes of the PΦB biosynthetic pathway (i.e., phytochromobilin synthase and heme oxygenase) should facilitate development of molecular probes for these enzymes. Expression of antisense or modified cDNA clones for these enzymes in transgenic plants may also facilitate the engineering of novel growth and developmental phenotypes into agronomically important crop plants.

Acknowledgments

This work was funded by United States Department of Agriculture Competitive Research Grant No. AMD92–03377.

References

1. Shropshire, W., Jr. and Mohr, H., Eds., Photomorphogenesis, *Encyclopedia of Plant Physiology*, New Series Volume 16A, Springer-Verlag, Berlin, 1983.
2. Kendrick, R. E. and Kronenberg, G. H. M., Eds., *Photomorphogenesis in Plants*, Martinus Nijhoff, Dordrecht, 1986.
3. Furuya, M., Molecular properties and biogenesis of phytochrome I and II, *Adv. Biophys.*, 25, 133, 1989.
4. Quail, P. H., Phytochrome: A light-activated molecular switch that regulates plant gene expression, *Annu. Rev. Genet.*, 25, 389, 1991.
5. Terry, M. J., Wahleithner, J. A., and Lagarias, J. C., The biosynthesis of holophytochrome, submitted to *Arch. Biochem. Biophys.*, 306, 1, 1993.
6. Beale, S. I., Biosynthesis of the tetrapyrrole pigment precursor, δ-aminolevulinic acid, from glutamate, *Plant Physiol.*, 93, 1273, 1990.
7. Gardner, G. and Gorton, H. L., Inhibition of phytochrome synthesis by gabaculine, *Plant Physiol.*, 77, 540, 1985.
8. Elich, T. D. and Lagarias, J. C., 4-Amino-5-hexynoic acid — a potent inhibitor of tetrapyrrole biosynthesis in plants, *Plant Physiol.*, 88, 747, 1988.
9. Jones, A. M., Allen, C. D., Gardner, G. and Quail, P. H., Synthesis of phytochrome apoprotein and chromophore are not coupled obligatorily. *Plant Physiol.*, 81, 1014, 1986.
10. Konomi, K. and Furuya, M., Effects of gabaculine on phytochrome synthesis during imbibition in embryonic axes of *Pisum sativum* L., *Plant Cell Physiol.*, 27, 1507, 1986.
11. Elich, T. D. and Lagarias, J. C., Phytochrome chromophore biosynthesis. Both 5-aminolevulinic acid and biliverdin overcome inhibition by gabaculine in etiolated *Avena sativa* L., seedlings, *Plant Physiol.*, 84, 304, 1987.
12. Elich, T. D., McDonagh, A. F., Palma, L. A., and Lagarias, J. C., Phytochrome chromophore biosynthesis. Treatment of tetrapyrrole-deficient *Avena* explants with natural and non-natural bilatrienes leads to formation of spectrally active holoproteins, *J. Biol. Chem.*, 264, 183, 1989.
13. Beale, S. I. and Weinstein, J. D., Tetrapyrrole metabolism in photosynthetic organisms, in *Biosynthesis of Hemes and Chlorophylls*, Dailey, H. A., Ed., McGraw-Hill, New York, 1990, 287–392.
14. Brown, S. B., Houghton, J. D., and Vernon, D. I., Biosynthesis of phycobilins. Formation of the chromophore of phytochrome, phycocyanin and phycoerythrin, *J. Photochem. Photobiol.*, 5, 3, 1990.
15. Terry, M. J. and Lagarias, J. C., Holophytochrome assembly. Coupled assay for phytochromobilin synthase *in organello*, *J. Biol. Chem.*, 266, 22215, 1991.
16. Elich, T. D. and Lagarias, J. C., Formation of a photoreversible phycocyanobilin-apophytochrome adduct *in vitro*, *J. Biol. Chem.*, 264, 12902, 1989.
17. Cornejo, J., Beale, S. I., Terry, M. J., and Lagarias, J. C., Phytochrome assembly. The structure and biological activity of 2R, 3E-phytochromobilin derived from phycobiliproteins, *J. Biol. Chem.*, 267, 14790, 1992.
18. Rhie, G. and Beale, S. I., Biosynthesis of phycobilins. Ferredoxin-supported NADPH-independent heme oxygenase and phycobilin-forming activities from *Cyanidium caldarium*, *J. Biol. Chem.*, 267, 16088, 1992.
19. Lagarias, J. C. and Lagarias, D. M., Self-assembly of synthetic phytochrome holoprotein *in vitro*, *Proc. Natl. Acad. Sci. U.S.A.*, 86, 5778, 1989.
20. Deforce, L., Tomizawa, K., Ito, N., Farrens, D., and Song, P.-S., *In vitro* assembly of apophytochrome and apophytochrome deletion mutants expressed in yeast with phycocyanobilin, *Proc. Natl. Acad. Sci. U.S.A.*, 88, 10392, 1991.
21. Wahleithner, J. A., Li, L., and Lagarias, J. C., Expression and assembly of spectrally active recombinant holophytochrome. *Proc. Natl. Acad. Sci. U.S.A.*, 88, 10387, 1991.

22. Li, L. and Lagarias, J. C., Phytochrome assembly: Defining chromophore structural requirements for covalent attachment and photoreversibility, *J. Biol. Chem.*, 267, 19204, 1992.

23. Cherry, J. R., Hondred, D., Walker, J. M., and Vierstra, R. D., Phytochrome requires the 6-kDa N-terminal domain for full biological activity, *Proc. Natl. Acad. Sci. U.S.A.*, 89, 5039, 1992.

24. Cherry, J. R., Hondred, D., Walker, J. M., Keller, J. M., Hershey, H. P., and Vierstra, R. D., Carboxy-terminal deletion analysis of oat phytochrome A reveals the presence of separate domains required for structure and biological activity, *Plant Cell*, 5, 565, 1993.

23

Phytochrome Genealogy

Masaki Furuya
Hitachi Ltd.

23.1 Introduction

Our knowledge of the primary structure and biogenesis of phytochromes has improved greatly after that the genes of phytochromes were isolated and sequenced (for reviews, see References 1 and 2). It was very difficult to characterize the structure of large, labile functional proteins like phytochromes by the methods of conventional protein chemistry. The first cloning of phytochrome cDNA was the most essential step in the molecular approaches of phytochromes, and that was succeeded by difference screening of *Avena* cDNA clones using probes either enriched or depleted for phytochrome sequences.[3] Once the DNA sequence of a gene was known, it became significantly easier to clone the gene from other plants. By now, phytochrome genes of wild type *(PHY)* and of mutants[4] *(phy)* are cloned not only from various species of seed plants[5-13] but ferns,[14,15] mosses,[16] and algae,[17] demonstrating that the *PHY*s are able to be classified into a few gene subfamilies. Then, the evolutionary relationship among them was proposed.[10,12,14]

The DNA sequences of 5′-flanking promoter regions and the expression patterns of *PHY*s are quite diverse among the *PHY* gene families. In the past decade, detailed studies on this subject have provided us with an interesting model system of molecular biology.[18,19]

Furthermore, modern techniques of gene engineering have been applied for studies on: (1) *in vitro* assembly of mutated *PHY*s to a chromophore; (2) overexpression of recombinant *PHY*s in transgenic plants; and (3) structure and function analysis of phytochrome molecules.[20]

23.2 Structure of *PHY* Genes

Spectrally active phytochrome (Phy) is well known to be widely distributed throughout eucaryotic green plants.[21] In the 1980s, this was confirmed by cloning of full- or partial-length cDNAs and genomic DNAs of *PHY* with various species from algae to angiosperms.[2,19] The *PHY*s characterized show highly homologous sequences to each other, particularly in the chromophoric domain. In addition, there was evidence for the presence of multiple *PHY* gene families.[22] For example, the *Arabidopsis* genome contains five *PHY* genes, named *PHYA*, *PHYB*, *PHYC*, *PHYD*, and *PHYE*.[10] The sequence of Arabidopsis *PHYA* shows 65 to 80% amino acid sequence similarity with *PHY* genes with rice,[6] maize,[7] pea,[8] and zucchini,[23] whereas the estimated amino acid sequences of apophytochrome A (PHYA), apophytochrome B (PHYB), and apophytochrome C (PHYC) are

0-8493-8634-9/95/$0.00+$.50
© 1995 by CRC Press, Inc.

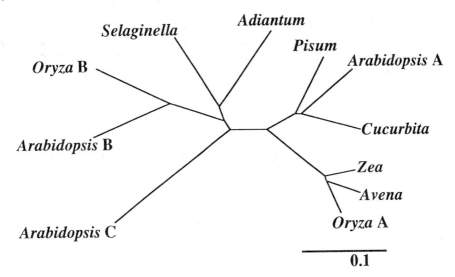

FIGURE 23.1 Unrooted tree of phytochrome genes constructed by the neighbor-joining method[24] (M. Wada and H. Okamoto, unpublished data). Branch lengths show the genetic distances.

equally divergent from each other with ca. 50% homology.[19] The rice genome also contains a single copy of the *PHYB* gene, which shows 73% amino acid sequence similarity with *PHYB* of *Arabidopsis* but is only 50% homologous to rice *PHYA*.[12]

In Gymnosperms, *PHY* cDNAs for *Ginkgo biloba* and *Pinus pulustris* were cloned and partially sequenced by Silverthorne (personal comment), and those genes in Gymnosperms appear not to be highly homologous to any *PHY* subfamilies of Angiosperms. As far as *PHYs* in lower plants are concerned, cDNAs encoding the full length of *PHY* were cloned and sequenced from *Selaginella martensii*[14] and *Adiantum capillus-veneris,*[15] showing that their estimated amino acid sequences has ca. 70% identity with each other, but only 50% identity each with *PHYA*, *PHYB*, and *PHYC* in *Arabidopsis*. In the case of moss *PHYs*, the 1474-bp *PHY* sequence of *Ceratodon* was estimated from five overlapping PCR fragments, and the amino acid sequence deduced had ca. 60% identity with all *PHY* subfamilies in Angiosperms.[16] For algal *PHY* sequences, the internal protein sequence of PHY, purified from dark-adapted *Mesotaenium* cells, was determined;[17] this confirmed that the sequence obtained for the amino acids is approximately 74% similar to *Selaginella*, but is a 50 to 53% homology with PHYA, and 62 to 63% with PHYB of higher plants. A putative *Mougeotia PHY* PCR fragment shows highest homology (>60%) to *PHYB* among reported *PHYs*.[17] There have been no publications yet for the existence of multiple *PHY* gene families in lower plants.

An unrooted phytogenetic tree with full-length PHYs amino acid sequences and their mutation distances (Figure 23.1) was constructed by the neighbor-joining method.[24] It is interesting to note that *PHYs* of the lower plants such as *Adiantum* and *Selaginella* belong to the same branch of PHYB.[15]

23.3 Expression of *PHY* Genes

The expression of *PHYA* is well known to be down-regulated by light and is evident in oats[1,25] and pea,[8] but not always obvious in some other plants.[2] This effect was found for the first time in the early 1980s by the fact[26,27] that the contents of translatable *PHYA*-mRNA measured using an *in vitro* translation system and specific anti-PHYA antibodies were lowered by exposure of oat[28] to white or red light. This phenomenon was confirmed by quantitative hybridization of *PHYA*-mRNA using *PHYA*-cDNA as a probe in oat[29] and rice.[30] Sequence similarities among *cis*-elements in *PHYA* promoters[5,30] indicated the possibility that these elements might be important functionally, and some *trans*-acting factors were proposed to bind to the elements.

The single-copy gene coding for pea PhyA produces three distinct transcripts differing in the length of the 5'-noncoding sequence, as detected by S1 nuclease protection assay.[8] The multiple transcription starting sites of *PHYA* promoters appear to be common among *PHYAs* in dicots,[2] whereas *PHYA* in monocots, like oat and rice, have a single starting site in the promoter region and only one transcript product.[1]

In pea seedlings, the abundance of *PHYA*-mRNA1 and -mRNA2, the shortest and the second shortest transcripts decreased rapidly after brief irradiation with red light, and the red light-induced effect was reversed by subsequent irradiation with far-red light.[31] When the red light-treated seedlings were kept in the dark for a day or longer and then exposed briefly to far-red light, the abundance of *PHYA*-mRNA1 and -mRNA2 increased to similar levels as those observed before the dark period, and the effect was photoreversible repeatedly.[32] This result suggests strongly that stable (type II) Pfr would regulate the abundance of *PHYA*-mRNAs in light-grown pea.

The organ- or tissue-specific expression of *PHYA*-mRNAs in different organs was examined, finding that *PHYA*-mRNA in etiolated rice seedlings is twofold higher in leaves than in roots; whereas, in fully green plants, roots contains higher levels of the mRNA than leaves.[30] The distribution of *PHYA*-mRNA1 and -mRNA2 in etiolated pea seedlings was measured and it was found that the abundance of the *PHYA*-mRNAs is significantly higher around the shoot apex and in the cotyledons than in the other parts.[33] In transgenic *Petunia* of the 5'-upstream region of pea *PHYA*-GUS fusion, the expression of this reporter gene was red/far-red photoreversibly regulated.[34]

In contrast, *PHYB* and *PHYC* are constitutively expressed regardless of the environmental light condition, as shown by Northern blot analysis using transcript-specific probes with *Arabidopsis*[10] and rice.[12] Densitometric analysis shows that *PHYA*-mRNA is fivefold more abundant than *PHYB*-mRNA in etiolated rice seedlings, whereas the two mRNAs are equally abundant in green tissues because of light-induced decline of *PHYA* transcript.[12]

23.4 Application of *PHY* Genes

The cloned cDNAs and genomic DNAs of *PHYs* have been used to solve a wide range of problems in basic and applied biology, among which only a few examples will be introduced in this chapter.

The preparation of Phy apoprotein (PHY) was most difficult since the discovery of Phy, because the chromophore of Phy is attached covalently to a cysteinyl residue of PHY.[20] However, the recent success of expressing recombinant *PHYA*-cDNA in bacteria[35] and yeast,[36–38] and *PHYB*-cDNA in yeast[39] has made it possible to provide sufficient PHYs with or without the chromophore. Using such a method, red/far-red photoreversible adducts of full-length PhyA in pea[37] and oat[38] and PhyB in tobacco[39] were reported. Furthermore, truncation of the N-terminal tail to residue 46 demonstrates that this region is not critical to the chromophore attachment, but that the region to 225 fails to yield holophytochrome *in vitro* under the same conditions.[37] A mutant *phyA* comprised of a deletion of the C-terminus to residue 548 showed bilin incorporation and red/far-red photoreversibility, indicating that the bilin-PHYA assembly still occurred even when the entire C-terminal domain was truncated.[37] Both PHYA and PHYB assemble *in vitro* with the chromophore in a very similar manner, so the kinetic analysis of the *in vitro* assembly with PHYA[40] and PHYB[39] revealed a similar pseudo-first-order rate constant.

Phytochrome structure/biological function analysis has been made extensively using transgenic techniques by expressing a cloned *PHY* gene as well as its mutant *phys* in plants. The biological activity of rice *phyA* in transgenic tobacco seedlings demonstrated clearly that rice PHYA assembles with tobacco phytochromobilin *in vivo* and forms spectrophotometrically active Phy,[41] which regulates hypocotyl elongation in transgenic tobacco seedlings.[42] Since the N-terminus of PHYA in several plant species is always very rich in serine residues, the first ten serine codons of rice *PHYA* were changed to alanine codons. This mutant *phyA* as well as the WT *PHYA* were transferred into the tobacco genome, finding that the serine/alanine mutant responded to dim light with more sensitivity than the WT.[43]

It is clear that (1) the nonchromophoric domain of Phy (C-terminus half of PHYA) is not required for chromophore assembly both *in vitro*[37] and in transgenic plants;[43] (2) recombinant monocot PHY can assemble with phytochromobilin in dicot transgenic plants[41,44,45] and with phococyanobilin prepared from algae;[37] (3) the N-terminus up to 70 amino acid residues does not appear to be crucial for the assembly,[46] but the deletion of N-terminal 80 amino acids or more shows no assembly both *in vivo* and *in vitro*;[43] and (4) biological activity of the transgenes has been demonstrated only when full-length *phy* gene was overexpressed in transgenic plants, while any modification of PHY structure fails to induce such biological functions.[44,46]

23.5 Concluding Remarks

It is evident that molecular biological approaches using *PHY* genes have opened a promising new field to both basic photobiology and applied photosciences. First of all, if the full-length *PHY* genes have not been cloned, it would be impossible to estimate the amino acid sequences of full-length Phys and then subsequently the molecular species of Phys. Second, if the different Phy subfamilies were not known, the inconsistency between spectrophotometrically measured amounts and states of Phy *in vivo* and the physiologically determined responses of Phy[21,22] would never be solved. Third, the amounts of PhyB, PhyC, and others, if any, in plant tissues are so small that the study of their molecular properties by conventional techniques,[47] is not possible but the expression of recombinant *PHY* in bacterial and yeasts and the *in vitro* assembly techniques with *PHY*s to the chromophore have now provided measurable amounts of spectrally active phytochromes for measuring physical and chemical properties. The application of gene engineering to phytochrome studies is still at a preliminary stage, but the techniques will initiate new approaches in this field.

References

1. Quail, P. H., Phytochrome: A light-activated molecular switch that regulates plant gene expression, *Annu. Rev. Genet.*, 25, 389, 1991.
2. Furuya, M., Phytochromes: Their molecular species, gene families, and functions, *Annu. Rev. Plant Physiol. Plant Mol. Biol.*, 44, 617, 1993.
3. Hershey, H. P., Colbert, J. T., Lissemore, J. L., Barker, R. F., and Quail, P. H., Molecular cloning of cDNA for *Avena* phytochrome, *Proc. Natl. Acad. Sci. U.S.A.*, 81, 2332, 1984.
4. Reed, J. W., Nagpal, P., Poole, D. S., Furuya, M., and Chory, J., Mutations in the gene for the red/far-red light receptor phytochrome B alter cell elongation and physiological responses throughout *Arabidopsis* development, *The Plant Cell*, 5, 147, 1993.
5. Hershey, H. P., Barker, R. F., Idler, K. B., Murray, M. G., and Quail P. H., Nucleotide sequence and characterization of a gene encoding the phytochrome polypeptide from *Avena, Gene*, 61, 339, 1987.
6. Kay, S. A., Keith, B., Shinozaki, K., and Chua, N.-H., The sequence of the rice phytochrome gene, *Nucl. Acids Res.*, 17, 2865, 1989.
7. Christensen, A. H., and Quail, P. H., Structure and expression of a maize phytochrome-encoding gene, *Gene*, 85, 381, 1989.
8. Sato, N., Nucleotide sequence and expression of the phytochrome gene in *Pisum sativum:* Differential regulation by light of multiple transcripts, *Plant Mol. Biol.*, 11, 697, 1988.
9. Sharrock, R. A., Lissemore, J. L., and Quail, P. H., Nucleotide and amino acid sequence of a *Cucurbita* phytochrome cDNA clone: Identification of conserved features by comparison with *Avena* phytochrome, *Gene*, 47, 287, 1986.
10. Sharrock, R. A., and Quail, P. H., Novel phytochrome sequences in *Arabidopsis thaliana*: Structure, evolution, and differential expression of a plant regulatory photoreceptor family, *Genes Devel.*, 3, 1745, 1989.

11. Adam, E., Dear, M., Kay, S., Chua, N.-H., and Nagy, F., Sequence of a tobacco *(Nicotiana tabacum)* gene coding for Type A phytochrome, *Plant Phys.*, 101, 1407, 1993.

12. Dehesh, K., Tepperman, J., Christensen, A. H., and Quail, P. H., *phyB* is evolutionarily conserved and constitutively expressed in rice seedling shoots, *Mol. Gen. Genet.*, 225, 305, 1991.

13. Heyer, A., and Gatz, C., Isolation and characterization of a cDNA-clone coding for potato type B phytochrome, *Plant Mol. Biol.*, 20, 589, 1992.

14. Hanelt, S., Braun, B., Marx, S., and Schneider-Poetsch, H. A. W., Phytochrome evolution: A phylogenetic tree with the first complete sequence of phytochrome from a cryptogamic plant, *Photochem. Photobiol.*, 56, 751, 1992.

15. Okamoto, H., Hirano, Y., Abe, H., Tomizawa, K., Furuya, M., and Wada, M., Nucleotide sequence of a cDNA for phytochrome from fern *Adiantum capillus-veneris* L., *Plant Cell Physiol.*, 34, 1329, 1993.

16. Thümmler, F., Beetz, A., and Rüdiger, Phytochrome in lower plants. Detection and partial sequence of a phytochrome gene in the moss *Ceratodon purpureus* using the polymerase chain reaction, *FEBS Lett.*, 275, 125, 1990.

17. Winands, A., Wagner, G., Marx, S., and Schneider-Poetsch, H. A. W., Partial nucleotide sequence of phytochrome from the zygnematophycean green alga *Mougeotia, Photochem. Photobiol.*, 56, 765 1992.

18. Quail, P. H., Gatz, C., Hershey, H. P., Jones, A. M., Lissemore, J. L., Parks, B. M., Sharrock, R. A., Barker, R. F., Idler, K. B., Murray, M. G., Koornneef, M., and Kendrick, R. E., Molecular biology of phytochrome, in *Phytochrome and Photoregulation in Plants*, Furuya, M., Ed., Academic Press, Tokyo, 1987, 23–37.

19. Quail, P. H., Hershey, H. P., Idler, K. B., Sharrock, R. A., Christensen, A. H., Parks, B. M., Somers, D., Bruce, W. B., and Dehesh, K., Phy-gene structure, evolution, and expression, in *Phytochrome Properties and Biological Action*, Thomas, B. and Johnson, C. B., Eds., NATO ASI Series, H50 Springer-Verlag, 1991, 13–38.

20. Furuya, M., and Song, P.-S., Assembly and properties of holophytochrome, in *Photomorphogenesis in Plants*, 2nd ed., Kendrick, R. E. and Kronenberg, G. H. M., Eds., Martinus Nijhoff, Dordrecht, Netherlands, 1994, 105–140.

21. Furuya, M., Biochemistry and physiology of phytochrome, in *Progress in Phytochemistry*, Vol. 1, Reinhold, L. and Liweschitz, Y., Eds., John Wiley & Sons, London, 1968, 347–405.

22. Furuya, M., Molecular properties and biogenesis of phytochrome I and II, *Adv. Biophys.*, 25, 133, 1989.

23. Lissamore, J. L., Colbert, J. T., and Quail, P. H., Cloning of cDNA for phytochrome from etiolated Cucurbita and coordinate photoregulation of the abundance of two distinct phytochrome transcripts, *Plant Mol. Biol.*, 8, 485, 1987.

24. Saito, N. and Nei, M., The neighbour-joining method: A new method for reconstructing phylogenic trees, *Mol. Biol. Evol.*, 4, 406, 1987.

25. Thompson, W. F. and White, M. J. Physiological and molecular studies of light-regulated nuclear genes in higher plants, *Annu. Rev. Plant Physiol. Plant Mol. Biol.*, 42, 423, 1991.

26. Quail, P. H., Phytochrome: A regulatory photoreceptor that controls the expression of its own gene, *Trends Biochem. Sci.*, 9, 450, 1984.

27. Tomizawa, K., Nagatani, A., and Furuya, M., Phytochrome genes: Studies using the tools of molecular biology and photomorphogenetic mutants: Yearly review, *Photochem. Photobiol.*, 52, 265, 1990.

28. Colbert, J. T., Hershey, H. P., and Quail, P. H., Autoregulatory control of translatable phytochrome mRNA levels, *Proc. Natl. Acad. Sci. U.S.A.*, 80, 2248, 1983.

29. Colbert, J. T., Hershey, H. P., and Quail, P. H., Phytochrome regulation of phytochrome mRNA abundance, *Plant Mol. Biol.*, 5, 91, 1985.

30. Kay, S. A., Keith, B., Shinozaki, K., Chye, M.-L., and Chua, N.-H., The rice phytochrome gene: Structure, autoregulated expression, and binding of GT-1 to a conserved site in the 5′ upstream region, *Plant Cell*, 1, 351, 1989.

31. Tomizawa, K., Sato, N., and Furuya, M., Phytochrome control of multiple transcripts of the phytochrome gene in *Pisum sativum., Plant Mol. Biol.*, 12, 295, 1989.

32. Furuya, M., Ito, N., Tomizawa, K., and Schäfer, E., A stable phytochrome pool regulates the expression of the phytochrome I gene in pea seedlings, *Planta*, 183, 218, 1991.

33. Tomizawa, K., Masatsuji, E., Ishii, K., Furuya, M. Distribution of type I phytochrome (*phy*A) RNA1 and RNA2 in etiolated pea seedling, *Photochem. Photobiol., B: Biol.*, 11, 163, 1991.

34. Komeda, Y., Yamashita, H., Sato, N., Tsukaya, H., and Naito, S., Regulated expression of a gene-fusion product derived from the gene for phytochrome I from *Pisum sativum* and the *uidA* gene from *E. coli* in transgenic *Petunia hybrida, Plant Cell Physiol.*, 32, 737, 1991.

35. Tomizawa, K., Ito, N., Komeda, Y., Uyeda, T.Q.P., Takio, K., and Furuya, M., Characterization and intracellular distribution of pea phytochrome I polypeptides expressed in *E. coli, Plant Cell Physiol.*, 32, 95, 1991.

36. Ito, N., Tomizawa, K., and Furuya, M., Production of full-length pea phytochrome A (type 1) apoprotein by yeast expression system, *Plant Cell Physiol.*, 32, 891, 1991.

37. Deforce, L., Tomizawa, K., Ito, N., Farrens, D., Song, P.-S., and Furuya, M., *In vitro* assembly of apophytochrome and apophytochrome deletion mutants expressed in yeast with phycocyanobilin, *Proc. Natl. Acad. Sci. U.S.A.*, 88, 10392, 1991.

38. Wahleithner, J. A., Li, L., and Lagarias, J. C., Expression and assembly of spectrally active recombinant holophytochrome, *Proc. Natl. Acad. Sci. U.S.A.*, 88, 10387, 1991.

39. Kunkel, T., Tomizawa, K., Kern, R., Furuya, M., Chua, N. H., and Schäfer, E., *In vitro* formation of a photoreversible adduct of phycocyanobilin and tobacco apophytochrome B, *Eur. J. Biochem.*, 215, 587, 1993.

40. Li, L. and Lagarias, J. C., Phytochrome assembly: Defining chromophore structural requirements for covalent attachment and photoreversibility. *J. Biol. Chem.*, 267, 19204, 1992.

41. Kay, S. A., Nagatani, A., Keith, B., Deak, M., Furuya, M., and Chua, N.-H., Rice phytochrome is biologically active in transgenic tobacco, *Plant Cell*, 1, 775, 1989.

42. Nagatani, A., Kay, S. A., Deak, M., Chua, N.-H., and Furuya, M., Rice type I phytochrome regulates hypocotyl elongation in transgenic tobacco seedlings, *Proc. Natl. Acad. Sci. U.S.A.*, 88, 5207, 1991.

43. Stockhaus, J., Nagatani, A., Halfter, U., Kay, S., Furuya, M., and Chua, N.-H., Serine-to-alanine substitutions at the amino-terminal region of phytochrome A result in an increase in biological activity, *Genes Devel.*, 6, 2364, 1992.

44. Boylan, M. T. and Quail, P. H., Oat phytochrome is biologically active in transgenic tomatoes, *Plant Cell*, 1, 765, 1989.

45. Keller, J. M., Shanklin, J., Vierstra, R. D., and Hershey, H. P., Expression of a functional monocotyledonous phytochrome in transgenic tobacco, *EMBO J.*, 8, 1005, 1989.

46. Cherry, J. R., Hondred, D., Walker J. M., and Vierstra R. D., Phytochrome requires the 6-kDa N-terminal domain for full biological activity, *Proc. Natl. Acad. Sci. U.S.A.*, 89, 5039, 1992.

47. Furuya, M., Tomizawa, K., Ito, N., Sommer, D., Deforce, L., Konomi, K., Farrens, D., and Song, P.-S., Biogenesis of phytochrome apoprotein in transgenic organisms and its assembly to the chromophore, in *Phytochrome Properties and Biological Action*, Thomas, B. and Johnson, C. B., Eds., NATO ASI Series, Vol. H 50, Springer-Verlag, Berlin, Heidelberg, 1991, 71.

24

Phytochrome: Higher Plant Mutants

Richard E. Kendrick
Wageningen Agricultural University

Maarten Koornneef
Wageningen Agricultural University

24.1 Introduction

Phytochrome is a family of photoreceptors that absorb predominantly in the red (R) and far-red (FR) light region of the spectrum that regulates many aspects of plant growth and development.[1] Phytochrome appears to be involved in the regulation of processes throughout the life cycle of the plant from germination, seedling establishment (de-etiolation after penetration of the soil surface), and shade avoidance to flowering. Phytochromes exist in two forms, one R-absorbing (Pr) and the other FR-absorbing (Pfr), that are repeatedly interconvertible upon exposure to R and FR. Analyses of physiological experiments and studies of phytochrome *in vivo* led to the proposal of more than one pool of phytochrome within a plant.[2] The bulk of the phytochrome that accumulates in etiolated seedlings is labile upon transformation to the Pfr form; this pool has been called labile phytochrome or PI. A second pool of phytochrome is stable in its Pfr form and has been called stable phytochrome or PII. Phytochrome genes have been cloned from a number of plant species and in *Arabidopsis* they have been designated *phyA, phyB, phyC, phyD,* and *phyE*.[3,4] These genes encode the phytochrome apoproteins A, B, C, D, and E and, after insertion of the chromophore, photochemically active holophytochromes A, B, C, D, and E, respectively. There now appears to be general agreement that the bulk labile PI pool in etiolated seedlings is the product of the *phyA* gene, but the stable PII pool has not been critically defined. Whereas phytochrome B has the property of a stable phytochrome and is a component of the PII pool, the contributions of other phytochromes remain unclear.

Phytochromes can work in concert with other photoreceptors that absorb in the UV and blue-light (B) spectral regions.[5] In addition to the different members of the gene family, there appear to be different modes of phytochrome action that have been roughly categorized by the amount of light required to initiate the responses concerned. These are the very low-fluence responses (VLFRs), low-fluence responses (LFRs), and high-irradiance response (HIRs).[6] Such a complexity lends itself to genetic analysis and, in the early 1980s, a genetic approach to photomorphogenesis was initiated with the hope that it could eventually elucidate the underlying molecular and physiological mechanisms.[7-11] Simplistically, mutations resulting in a loss of function would be expected to lead to a phenotype that could be described as exhibiting a simplified photomorphogenesis and careful physiological analysis should enable the consequences of the lesion to be characterized.[9] Clearly, this eventually necessitates the analysis of the consequences of the mutations at the molecular level before firm conclusions can be made. However, the identification of mutations that result in the

0-8493-8634-9/95/$0.00+$.50
© 1995 by CRC Press, Inc.

loss of a series of responses, while retaining others, would give some insight into related subsets of responses within the plant, before anything is known at the molecular level (e.g., several responses under the control of one phytochrome type or regulated by one of the reaction modes might all be reduced or absent). As an example, etiolation affects many different processes in plants, such as changes in elongation, anthocyanin, gene expression, etc., which all appear to be under the "master" control of phytochrome. In this case, a mutant having no phytochrome responsible for this process would be expected to exhibit a complex pleiotropic phenotype.[8,12] What type of mutation can lead to such a phenotype? The phytochrome molecule is composed of chromophore and protein parts and there could be mutations in: (1) the phytochrome gene itself (coding sequence or promoter region) a gene regulating its expression, or stability of the gene product (apophytochrome); (2) the chromophore biosynthesis pathway resulting in chromophore deficiency. The latter type of mutation would presumably not be specific to one member of the phytochrome gene family, since all phytochromes have the same chromophore. It should be possible to rescue (convert to wild-type phenotype) chromophore mutants by feeding a late precursor of the phytochrome chromophore, downstream of the block caused by the mutation (e.g., biliverdin). It would be anticipated that phytochrome mutants would be lethal since de-etiolation is a prerequisite for development of an autotrophic (photosynthetic) green plant. If this is true, then it must be proposed that if such mutations exist they must be leaky, or if they result in the complete removal of one phytochrome gene family member, then another member of the gene family, perhaps present in lower concentration or with lower efficiency, can take over its function.

Another site for potential mutations is at the level of the phytochrome-transduction chain(s). Phenotypically, such a mutant would resemble one that eliminated that member of the gene family (if a specific group of responses is under the control of that family member), yet analysis should reveal the presence of the functional photoreceptor itself (i.e., R/FR reversible holophytochrome). The availability of antibodies has demonstrated the phytochrome apoprotein can be stable in the absence of chromophore[13] and immunological detection alone is not sufficient to indicate a functional photoreceptor. It remains difficult, therefore, to study low levels of phytochrome(s), particularly in light-grown plants. It is hoped that identification of transduction-chain mutations will assist in the elucidation of the mechanism(s) of phytochrome action that has eluded physiologists so far during the 40 years since the photoreceptor was proposed. In the following section, mutations already characterized at the molecular level, as well as putative phytochrome photoreceptor and transduction-chain mutants, are reported.

24.2 Survey of Mutants

Phytochrome mutants have been selected on the basis of: (1) their elongated phenotype in white light (W) or broad-band R or FR at the seedling stage; (2) their exaggerated response to light (hyper-responsive mutants); (3) their development of a "light" phenotype in darkness. The mutants are listed below for each species and are in alphabetical order. Descriptions are given with respect to the corresponding wild-type and, unless otherwise indicated, they are recessive mutations. Mutants associated with flowering and photoperiodism are not included unless there is evidence suggesting that the modification in flowering behavior is a consequence of a change in the phytochrome system.

Arabidopsis thaliana
- *cop1* -Constitutively photomorphogenic;[14] dark-grown seedlings have a short hypocotyl and expanded cotyledons and leaves;[14] chloroplasts develop in roots;[15] epistatic to *hy1*;[14,15] cloned, zinc-binding motif and a G_β homologous domain;[16] chromosome 2.[14]
- *cop9* -Constitutively photomorphogenic, very similar to *cop1*;[17] dark-grown seedlings have a short hypocotyl and expanded cotyledons;[17] light-grown plants severe dwarfs and lethal;[17] photocontrol of seed germination normal.[17]

det1 -De-etiolates in darkness and has a short hypocotyl, anthocyanin and expanded leaves with some chloroplast development;[18] seeds germinate in darkness;[18] epistatic to *hy1*, *hy2*, and *hy6*;[18] chromosome 4.[18]

det2 -De-etiolates in darkness, but less leaf and chloroplast development than *det1*;[19] similar to light regulatory dwarfs *(lrd)*;[20] late flowering;[21] chromosome 2.[19]

fre1 -Elongated hypocotyl in FR;[22] normal in W;[22] normal end-of-day FR (EODFR) response;[22] not rescued by biliverdin;[22] no detectable holophytochrome A;[22] putative *phyA* mutants;[22] might be allelic with *hy8;* chromosome 1.[22]

hy1 -Long hypocotyl in W, R and FR;[7] reduced chlorophyll;[7] early flowering in short days (SD);[21] response to EODFR and supplementary FR during the photoperiod (suppFR);[23] rescued by biliverdin;[24] no spectrophotometrically detectable phytochrome, but apophytochrome present;[13] photoreceptor mutants due to chromophore deficiency;[13,24] chromosome 2.[7]

hy2 -Long hypocotyl in W, R and FR;[7] reduced chlorophyll;[7] early flowering in SD;[21] response to EODFR and suppFR;[23] rescued by biliverdin;[22] no spectrophotometrically detectable phytochrome, but apophytochrome present;[13] photoreceptor mutants due to chromophore deficiency;[13,24] chromosome 3.[7]

hy3 -Long hypocotyl in R;[7] elongated petioles;[23] early flowering in SD;[21] not rescued with biliverdin;[22] no response to EODFR;[23] reduced response to suppFR;[23] cloned;[25] *phyB* photoreceptor mutant, most mutants without apophytochrome B;[25-27] chromosome 2.[7]

hy5 -Long hypocotyl in W, B, R and FR;[7] not rescued with biliverdin;[22] putative transduction chain mutant;[7,8] chromosome 5.[7]

hy6 -Long hypocotyl in W, R and FR;[28] reduced chlorophyll;[28] early flowering in SD; response to EODFR and suppFR; rescued with biliverdin;[29] photoreceptor mutant due to chromophore deficiency.

hy8 -Long hypocotyl in FR, normal in R and W;[30] phytochrome A photoreceptor and/or FR-HIR transduction-chain mutants;[30] *hy8-1* and *hy8-2* no detectable apophytochrome A, *hy8-3* contains holophytochrome A;[30] might be allelic with *fre1.*

Brassica rapa

ein -Elongated internode;[31] no response to EODFR;[31] increased gibberellin (GA) levels;[32] putative phytochrome B photoreceptor mutant.[31]

Cucumber *(Cucumis sativus)*

lh -Long hypocotyl and internodes;[33] loss of R inhibition of elongation growth during de-etiolation;[34] normal FR-HIR;[34] modified phototropism of de-etiolated seedlings;[35] severely reduced EODFR[36,37] and suppFR response;[38,39] putative phytochrome B photoreceptor mutant.[40]

Pea *(Pisum sativum)*

lip -Light-independent photomorphogenesis;[41] etiolated seedlings have short hypocotyls and expanded cotyledons;[41] light regulated gene transcripts present in etiolated seedlings;[41] reduced phytochrome A content;[41] similar to *cop1*, *cop9*, *det1*, and *det2* of *Arabidopsis.*

lv -Slender growing;[42] increased response to GA;[42] lack of EODFR response;[43] reduced flowering response to photoperiod;[43] phytochrome content normal;[44] putative transduction chain mutant.[41]

lw -Dwarf growing;[43] increased responsiveness to continuous W, R, and FR during de-etiolation;[43] exaggerated response to EODFR and suppFR;[43] increased elongation response to photoperiod.[43]

Sorghum *(Sorghum bicolor)*

ma3^R -Tall;[45] reduced tillering;[45] reduced anthocyanin;[45] photoperiod insensitive;[45] increased gibberellin levels;[46] lacks an immunologically detectable protein that is not apophytochrome A;[46,47] putative phytochrome B mutant.[46,47]

Nicotiana plumbaginifolia

hlg1 -Hypocotyl long, green seedling, particularly at low irradiances of W.[48]

eti1 -Etiolated and chlorotic;[48] reduced germination;[48] some rescue with biliverdin;[48] photoreceptor mutant due to chromophore deficiency.[48]

eti2 -Etiolated and chlorotic;[48] reduced germination;[48] some rescue with biliverdin;[48] photoreceptor mutant due to chromophore deficiency.[48]

eti3 -Same as *eti1* and *eti2* except germination normal.[48]

Tomato (Lycopersicon esculentum)

au -*Aurea*, long hypocotyl in W, R, and FR;[49] delayed de-etiolation including chloroplast development;[49] reduced chlorophyll formation;[50] reduced levels of mRNA for plastidic proteins in R;[51,52] yellow leaves;[49] reduced seed germination;[49,53] normal response to EODFR and suppFR;[54,55] translatable *phy* mRNA giving apparently full-length apoprotein *in vitro*;[56] phytochrome A apoprotein reduced, but still detectable;[57] phytochrome A deficient in etiolated seedlings;[49,58] putative, chromophore deficient mutant, particularly in etiolated seedlings; chromosome 1.[49]

yg-2 -Yellow green;[49] similar to *au*, but less extreme;[49] chromosome 12.[49]

hp-1 -High pigment, short hypocotyl with high anthocyanin levels in R;[59,60] higher chlorophyll levels;[60] mature plants somewhat dwarf;[55,61] immature fruit color dark green;[62] higher lycopene content of fruits;[62] putative transduction-chain mutant affecting responsiveness amplification.[63]

hp-2 -High pigment, similar to *hp-1*.[64]

tri -Temporarily R insensitive, long hypocotyl in R, on transfer from darkness temporarily insensitive to R;[65] reduced R-HIR for anthocyanin biosynthesis.[66]

Acknowledgments

We would like to thank Dr. Enrique López-Juez for reading this manuscript critically.

References

1. Kendrick, R. E. and Kronenberg, G. H. M., Eds., *Photomorphogenesis in Plants*, 1986.

2. Furuya, M., Molecular properties and biogenesis of phytochrome I and II, *Adv. Biophys.*, 25, 133, 1989.

3. Sharrock, R. A. and Quail, P. H., Novel phytochrome sequences in *Arabidopsis thaliana*: Structure, evolution, and differential expression of a plant regulatory photoreceptor family, *Genes Devel.*, 3, 1745, 1989.

4. Quail, P. H., Phytochrome: A light-activated molecular switch that regulates plant gene expression, *Annu. Rev. Genet.*, 25, 389, 1991.

5. Mohr, H., Coaction between pigment systems, in *Photomorphogenesis in Plants*, Kendrick, R. E. and Kronenberg, G. H. M., Eds., Martinus Nijhoff, Dordrecht, 1986, 457.

6. Kronenberg, G. H. M. and Kendrick, R. E., The physiology of action, in *Photomorphogenesis in Plants*, Kendrick, R. E. and Kronenberg, G. H. M., Eds., Martinus Nijhoff, Dordrecht, 1986, 99.

7. Koornneef, M., Rolff, E., and Spruit, C. J. P., Genetic control of light-inhibited hypocotyl elongation in *Arabidopsis thaliana* (L.) HEYNH, *Z. Pflanzenphysiol.*, 100, 147, 1980.

8. Koornneef, M. and Kendrick, R. E., A genetic approach to photomorphogenesis, in *Photomorphogenesis in Plants*, Kendrick, R. E. and Kronenberg, G. H. M., Eds., Martinus Nijhoff, Dordrecht, 1986, 546.

9. Koornneef, M., van Tuinen, A., Kerckhoffs, L. H. J., Peters, J. L., and Kendrick, R. E., Photomorphogenetic mutants of higher plants, in *Proc. 14th Int. Conf. Plant Growth Substances*, Karssen, C. M., van Loon, L. C., and Vreugdenhil, D., Eds., Kluwer Academic, Dordrecht, 1992, 54.

10. Kendrick, R. E., and Nagatani, A., Phytochrome mutants, *Plant J.*, 1, 133, 1991.

11. Reed, J. W., Nagpal, P., and Chory, J., Searching for phytochrome mutants, *Photochem. Photobiol.*, 56, 833, 1992.

12. Adamse, P., Kendrick, R. E., and Koornneef, M., Photomorphogenetic mutants of higher plants, *Photochem. Photobiol.*, 48, 833, 1988.

13. Parks, B. M., Shanklin, J., Koornneef, M., Kendrick, R. E., and Quail, P. H., Immunochemically detectable phytochrome is present at normal levels but is photochemically nonfunctional in the *hy1* and *hy2* long hypocotyl mutants of *Arabidopsis, Plant Mol. Biol.*, 12, 425, 1989.

14. Deng, X.-W. and Quail, P. H., Genetic and phenotypic characterization of *cop1* mutants of *Arabidopsis thaliana, The Plant J.*, 2, 83, 1992.

15. Deng, X.-W., Caspar, T., and Quail, P. H., *Cop1*: A regulatory locus involved in light-controlled development and gene expression in *Arabidopsis, Genes Devel.*, 5, 1172, 1991.

16. Deng, X.-W., Matsui, M., Wei, N., Wagner, D., Chu, A. M., Feldmann, K. A., and Quail, P. H., *COP1*, an *Arabidopsis* regulatory gene, encodes a protein with both a zinc-binding motif and a G_β homologous domain, *Cell*, 71, 1, 1992.

17. Wei, N., and Deng, X.-W., *Cop9*: A new genetic locus involved in light-regulated development and gene expression in *Arabidopsis, Plant Cell*, 4, 1507, 1992.

18. Chory, J., Peto, C., Feinbaum, R., Pratt, L. H., and Ausubel, F., *Arabidopsis thaliana* mutant that develops as a light-grown plant in the absence of light, *Cell*, 58, 991, 1989.

19. Chory, J., Nagpal, P., and Peto, C. A., Phenotypic and genetic analysis of *det2*, a new mutant that affects light-regulated seedling development in *Arabidopsis, Plant Cell*, 3, 445, 1991.

20. Feldmann, K., personal communication, 1992.

21. Goto, N., Kumagai, T., and Koornneef, M., Flowering responses to light-breaks in photomorphogenic mutants of *Arabidopsis thaliana*, a long-day plant, *Physiol. Plant.*, 83, 209, 1991.

22. Nagatani, A., Reed, J. W., and Chory, J., Isolation and initial characterization of *Arabidopsis* mutants that are deficient in phytochrome A, *Plant Physiol.*, 102, 269, 1993.

23. Whitelam, G. C. and Smith, H., Retention of phytochrome-mediated shade avoidance response in phytochrome-deficient mutants of *Arabidopsis,* cucumber and tomato, *J. Plant Physiol.*, 139, 119, 1991.

24. Parks, B. M., and Quail, P. H., Phytochrome-deficient *hy1* and *hy2* long hypocotyl mutants of *Arabidopsis* are defective in phytochrome chromophore biosynthesis, *Plant Cell*, 3, 1177, 1991.

25. Reed, J. W., Nagpal, P., Poole, D. S., Furuya, M., and Chory, J., Mutations in the gene for red/far-red light receptor phytochrome B alter cell elongation and physiological responses throughout *Arabidopsis* development, *Plant Cell*, 5, 147, 1993.

26. Nagatani, A., Chory, J., and Furuya, M., Phytochrome B is not detectable in the *hy3* mutant of *Arabidopsis*, which is deficient in responding to end-of-day far-red light treatments, *Plant Cell Physiol.*, 32, 1119, 1991.

27. Somers, D. E., Sharrock, R. A., Tepperman, J. M., and Quail, P. H., The *hy3* long hypocotyl mutant of *Arabidopsis* is deficient in phytochrome B, *Plant Cell*, 3, 1263, 1991.

28. Chory, J., Peto, C. A., Ashbaugh, M., Saganich, R., Pratt, L., and Ausubel, F., Different roles for phytochrome in etiolated and green plants deduced from characterization of *Arabidopsis thaliana* mutants, *Plant Cell*, 1, 867, 1989.

29. Chory, J., A genetic model for light-regulated seedling development in *Arabidopsis, Development*, 115, 337, 1992.

30. Parks, B. M. and Quail, P. H., *hy8*, A new class of *Arabidopsis* long hypocotyl mutants deficient in functional phytochrome A, *Plant Cell*, 5, 39, 1993.

31. Devlin, P. F., Rood, S. B., Somers, D. E., Quail, P. H., and Whitelam, G. C., Photophysiology of the elongated internode *(ein)* mutant of *Brassica rapa, Plant Physiol.*, 100, 1442, 1992.

32. Rood, S. B., Zonewich, K. P., and Bray, D., Growth and development of *Brassica* genotypes differing in endogenous gibberellin content. II. Gibberellin content, growth analysis and cell size, *Physiol. Plant.*, 79, 679, 1990.

33. Adamse, P., Jaspers, P. A. P. M., Kendrick, R. E., and Koornneef, M., Photomorphogenetic responses of a long hypocotyl mutant of *Cucumis sativus* L., *J. Plant. Physiol.*, 127, 481, 1987.

34. Peters, J. L., Kendrick, R. E., and Mohr, H., Phytochrome content and hypocotyl growth of long-hypocotyl mutant and wild-type cucumber seedlings during de-etiolation, *J. Plant Physiol.*, 137, 291, 1991.

35. Ballaré, C. L., Scopel, A. L., Radosevich, S. R., and Kendrick, R. E., Phytochrome-mediated phototropism in de-etiolated seedlings: Occurrence and ecological significance, *Plant Physiol.*, 100, 170, 1992.

36. Adamse, P., Jaspers, P. A. P. M., Bakker, J. A., Wesselius, J. C., Heeringa, G. H., Kendrick, R. E., and Koornneef, M., Photophysiology of a tomato mutant deficient in labile phytochrome, *J. Plant Physiol.*, 87, 264, 1988.

37. López-Juez, E., Nagatani, A., Buurmeijer, W. F., Peters, J. L., Kendrick, R. E., and Wesselius, J. C., Response of light-grown wild-type and *aurea*-mutant tomato plants to end-of-day far-red light, *J. Photochem. Photobiol. B: Biol.*, 4, 391, 1990.

38. Ballaré, C. L., Casal, J. J., and Kendrick, R. E., Responses of light-grown wild-type and long-hypocotyl mutant cucumber seedlings to natural and simulated shade light, *Photochem. Photobiol.*, 54, 819, 1991.

39. Smith, H., Turnbull, M., and Kendrick, R. E., Long-term and rapid elongation growth responses of light-grown plants of the cucumber *lh* mutant to irradiation with supplementary far-red light, *Photochem. Photobiol.*, 56, 607, 1992.

40. López-Juez, E., Nagatani, A., Tomizawa, K.-I., Deak, M., Kern, R., Kendrick, R. E., and Furuya, M., The cucumber long hypocotyl mutant lacks a light-stable PHYB-like phytochrome, *Plant Cell*, 4, 241, 1992.

41. Shannon, F., White, M. J., Edgerton, M. D., Jones, A. M., Elliot, R. C., and Thompson, W. F., Initial characterization of a pea mutant with light-independent photomorphogenesis, *Plant Cell*, 4, 1519, 1992.

42. Reid, J. B. and Ross, J. J., Internode length in *Pisum*. A new gene, *lv*, conferring an enhanced response to gibberellin A$_1$, *Physiol. Plant.*, 72, 595, 1988.

43. Weller, J. L. and Reid, J. B., Photoperiodism and photocontrol of stem elongation in two photomorphogenic mutants of *Pisum sativum* L., *Planta*, 189, 15, 1993.

44. Nagatani, A., Reid, J. B., Ross, J. J., Dunnewijk, A., and Furuya, M., Internode length in *Pisum*. The response to light quality, and phytochrome type I and II levels in *lv* plants, *J. Plant Physiol.*, 135, 667, 1990.

45. Childs, K. L., Pratt, L. H., and Morgan, P. W., Genetic regulation of development in *Sorghum bicolor*. VI. The *ma*$_3$^{R} allele results in abnormal phytochrome physiology, *Plant Physiol.*, 97, 714, 1991.

46. Beall, F. D., Morgan, P. W., Mander, L. N., Miller, F. R., and Babb, K. H., Genetic regulation of development in *Sorghum bicolor*. V. The *ma*$_3$^{R} allele results in giberellin enrichment, *Plant Physiol.*, 95, 116, 1991.

47. Childs, K. L., Cordonnier-Pratt, M.-M., Pratt, L. H., and Morgan, P. W., Genetic regulation of development in *Sorghum bicolor*. VII. The *ma*$_3$^{R} flowering mutant lacks a phytochrome that predominates in green tissue, *Plant Physiol.*, 99, 765, 1992.

48. Kraepiel, Y., Jullien, M., Caboche, M., and Miginiac, E., Etiolated mutants of *Nicotiana plumbaginifolia*, *J. Exp. Bot.*, 43, 23, 1992.

49. Koornneef, M., Cone, J. W., Dekens, R. G., O'Herne-Robers, E. G., Spruit, C. J. P., and Kendrick, R. E., Photomorphogenic responses of long hypocotyl mutants of tomato, *J. Plant Physiol.*, 120, 153, 1985.

50. Ken-Dror, S. and Horwitz, B. A., Altered phytochrome regulation of greening in an *aurea* mutant of tomato, *Plant Physiol.*, 92, 1004, 1990.

51. Oelmüller, R., Kendrick, R. E., and Briggs, W. R., Blue-light mediated accumulation of nuclear-encoded transcripts coding for proteins of the thylakoid membrane is absent in the phytochrome-deficient *aurea*-mutant of tomato, *Plant Mol. Biol.*, 13, 223, 1989.

52. Oelmüller, R. and Kendrick, R. E., Blue light is required for survival of the tomato phytochrome-deficient *aurea* mutant and the expression of four nuclear genes coding for plastidic proteins, *Plant Mol. Biol.*, 16, 293, 1991.

53. Georghiou, K. and Kendrick, R. E., The germination characteristics of phytochrome-deficient *aurea* mutant tomato seeds, *Physiol. Plant.*, 82, 127, 1991.

54. López-Juez, E., Nagatani, A., Buurmeijer, W. F., Peters, J. L., Kendrick, R. E., and Wesselius, J. C., Response of light-grown wild-type and *aurea*-mutant tomato plants to end-of-day far-red light, *J. Photochem. Photobiol. B: Biol.*, 4, 391, 1990.

55. Kerckhoffs, L. H. J., Kendrick, R. E., Whitelam, G. C., and Smith, H., Response of photomorphogenetic tomato mutants to changes in the phytochrome photoequilibrium during the daily photoperiod, *Photochem. Photobiol.*, 56, 611, 1992.

56. Sharrock, R. A., Parks, B. M., Koornneef, M., and Quail, P. H., Molecular analysis of the phytochrome deficiency in an *aurea* mutant of tomato, *Mol. Gen. Genet.*, 213, 9, 1988.

57. Sharma, R., López, E., Nagatani, A., Kendrick, R. E., and Furuya, M., Tomato *aurea* mutant possess spectrally inactive phytochrome A and active phytochrome B, Abstract 273, *Int. Photobiology Congress*, Kyoto, 1992.

58. Parks, B. M., Jones, A. M., Adamse, P., Koornneef, M., Kendrick, R. E., and Quail, P. H., The *aurea* mutant of tomato is deficient in spectrophotometrically and immunochemically detectable phytochrome, *Plant Mol. Biol.*, 9, 97, 1987.

59. Adamse, P., Peters, J. L., Jaspers, P. A. P. M., van Tuinen, A., Koornneef, M., and Kendrick, R. E., Photocontrol of anthocyanin synthesis in tomato seedlings: A genetic approach, *Photochem. Photobiol.*, 50, 107, 1989.

60. Peters, J. L., van Tuinen, A., Adamse, P., Kendrick, R. E., and Koornneef, M., High pigment mutants of tomato exhibit high sensitivity for phytochrome action, *J. Plant Physiol.*, 134, 661, 1989.

61. Peters, J. L., Schreuder, M. E. L., Heeringa, G. H., Wesselius, J. C., Kendrick, R. E., and Koornneef, M., Analysis of the response of photomorphogenetic tomato mutants to end-of-day far-red light, *Acta Hort.*, 305, 67, 1992.

62. Thompson, A. E., Hepler, R. W., and Kerr, E. A., Clarification of the inheritance of high total carotenoid pigments in the tomato, *Am. Soc. Hort. Sci.*, 81, 434, 1962.

63. Peters, J. L., Schreuder, M. E. L., Verduin, S. J. W., and Kendrick, R. E. Physiological characterization of a high pigment mutant of tomato, *Photochem. Photobiol.*, 56, 75, 1992.

64. Soressi, G. P. and Salamini, F., New spontaneous or chemically induced fruitripening tomato mutants, *Tomato Genet. Coop. Rpt.*, 25, 21, 1975.

65. van Tuinen, A., personal communication, 1993.

66. Kerckhoffs, L. H. J., personal communication, 1993.

Note Added in Proof

Several phytochrome mutants in *Arabidopsis* have recently been renamed, see: Quail, P. H., Briggs, W. R., Chory, J., Hangarter, R. P., Harberd, N. P., Kendrick, R. E., Koornneef, M. Parks, B., Sharrock, R. A., Schäfer, E., Thompson, W. F., Whitelam, G. C., Spotlight on phytochrome nomenclature, *Plant Cell*, 6, 468, 1994.

25

Photomorphogenesis: Natural Environment

Harry Smith
University of Leicester

25.1 Introduction

Light serves both as an energy source for photosynthesis and as an external signal regulating the photomorphogenic behavior of plants. The natural light environment is complex and variable, both temporally and spatially, and provides a range of signals to which plant development and metabolism are responsive. Light signals are perceived by informational photoreceptors, the best characterized in plants being the phytochromes, a family of chromoprotein photoreceptors with bilitriene chromophores.[1] Other photoreceptors that transduce environmental light signals include the blue-absorbing flavin or carotenoid photoreceptors and the UV-B absorbing photoreceptors. Evidence on the ecophysiological functions of the phytochrome family is accumulating rapidly, and this chapter concentrates on these phenomena. Phytochrome-mediated signal perception is involved at all stages of plant development, including seed and spore germination, the transition from heterotrophy to photoautotrophy, the induction of the shade avoidance syndrome, and the control of flowering.

25.2 The Natural Light Environment

Global radiation (i.e., the radiation incident upon a horizontal cosine-corrected sensor from the whole hemisphere) is composed of both direct solar radiation and indirect radiation scattered from molecules and particles in the atmosphere (Figure 25.1a).[2] Rayleigh scattering by molecules is inversely proportional to the fourth power of the wavelength, so that light scattered from the sky is enriched in blue. Mie scattering by particles is less wavelength dependent, resulting in neutral "white" light scattering. Modulation of the spectrum is due to both atmospheric absorption and scattering, which are functions of the path length of the solar beam. An example of the daylight

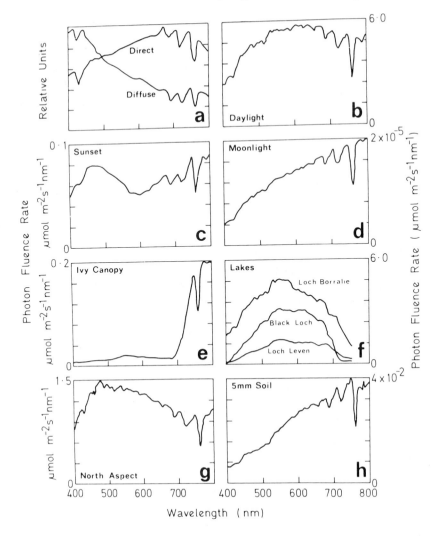

FIGURE 25.1 Typical spectroradiometer scans of the 400- to 800-nm wave band illustrating aspects of the natural light environment that may be ecologically relevant. (From Reference 1.)

spectrum is given in Figure 25.1b; this spectral distribution is modified only marginally by weather or cloud conditions, although the amount of light is obviously subject to large and unpredictable fluctuations. As the solar path length increases either with latitude, or time of day, the relative proportions of direct and scattered radiation varies. At high latitudes, or at dawn and dusk, atmospheric absorption of the direct beam causes a reduction in the longer wavelengths, while the increased proportion of scattered radiation causes a relative increase in the shorter wavelengths (Figure 25.1c). When solar elevation becomes less than ca. $-10°$, scattering and refraction of solar radiation virtually cease to contribute to the global radiation received at the earth's surface, and the only natural sources of radiation are the moon and the stars. Moonlight is reflected solar radiation modified by the optical properties of the moon's surface (the moon has no atmosphere), which cause a monotonic increase in reflectivity with increasing wavelength (Figure 25.1d).

Much larger changes in the incident spectrum are caused by interaction of radiation with absorbing materials at the earth's surface, particularly interaction with vegetation.[3] The photosynthetic pigments, the chlorophylls and carotenoids, absorb light over almost the entire visible spectrum (i.e., *photosynthetically active radiation*, PAR, 400 to 700 nm) but absorb hardly any radiation between 700 and 800 nm (i.e., *far-red*, FR). Thus, virtually all the incoming FR is either transmitted or reflected; i.e., the FR is scattered either through the leaf, or from the surface of the leaf. Figure 25.1e shows a typical daylight spectrum within a vegetation canopy. Other changes in daylight spectrum occur underwater; refraction at the air-water discontinuity leads to the incident light from above being concentrated into a cone of half-angle 48.6°; consequently, a sensor facing upward below but near to the surface inevitably receives a proportion of upwelling radiation reflected back down from the surface. More important phenomena, as far as daylight spectrum is concerned, are scattering and absorption by water itself, and by dissolved molecules or suspended particles. Rayleigh scattering results in the selective attenuation of the blue region of the spectrum of downwelling radiation. Water has strong absorption bands at ca. 730 nm and in the near infrared, and therefore the FR is also attenuated selectively. Thus, in clear water, downwelling radiation is effectively "compressed" with increasing depth into a decreasingly narrow band of wavelengths, usually peaking at or around 500 nm (Figure 25.1f). In turbid waters, major complications are evident. Most natural waters have varying amounts of organically derived material, much of it soluble, that absorbs both blue (B) and red (R) radiation. Also present can be very large amounts of chlorophyllous microorganisms that also absorb strongly in the B and R. In consequence, underwater spectra can be extremely variable. Other situations in which spectral distribution varies from daylight is on a North aspect, where the indirect diffuse radiation predominates (Figure 25.1g), and under the soil, where very small amounts of light penetrate quite surprising depths but in which absorption by soil particles generally attenuates selectively the lower wavelengths (Figure 25.1h). The spectra in Figure 25.1 were chosen to represent environmental signals that may be of ecological significance.

25.3 Perception of Environmental Light Signals by the Phytochromes

The phytochromes are encoded by a small multigene family (the *phy* genes) numbering at least five members (*phyA* to *phyE*) in *Arabidopsis*.[4] Only one phytochrome has been characterized chemically (i.e., phytochrome A, encoded by the *PHYA* gene), but it is assumed that the photochemical characteristics of all the phytochromes are similar. Phytochrome A from oats has an apoprotein of 126,000 M_r, and a bilitriene chromophore. The unique property of the phytochromes is their photochromicity; each phytochrome may exist in either of two photointerconvertible isomeric forms; Pr and Pfr. Both forms absorb over the 400 to 700-nm spectral range, but Pfr has extended absorption up to ca. 800 nm. Upon photon absorption, Pr (λ_{max} 665 nm) is converted to Pfr which, upon photon absorption (λ_{max} 730 nm) is converted to Pr; it is generally assumed that Pr is biologically inactive and that the formation of Pfr leads to the induction of the biological responses. Many photomorphogenic phenomena, particularly in dark-imbibed seeds or dark-grown seedlings but also in light-grown plants, are inducible by a brief pulse of R and reversible by brief FR; this R/FR reversibility is a classic test of mediation by a member of the phytochrome family. In the natural environment, the photochromicity of the phytochromes endows them with the capacity to perceive fluctuations in the relative proportions of R (i.e., ca. 600 to 700 nm) and far-red (FR, i.e., ca. 700 to 800 nm) in the incident radiation. The relationship between phytochrome photoequilibrium (i.e., Pfr/P, or the proportion of total phytochrome present as Pfr at photoequilibrium) and the ratio of R to FR radiation (i.e., R:FR) is a rectangular hyperbola in which the region between R:FR = 0 → 2 represents the most sensitive part of the curve, and at R:FR > 3 the asymptote is reached (Figure 25.2).[5]

Figures 25.1 and 25.2 show that variations in the R:FR ratio are associated with various ecologically or physiologically significant situations. The ratio R:FR in daylight is usually approximately 1.2,[5] and is remarkably constant irrespective of cloud conditions as long as the sun is more than ca. 10° above the horizon. At high latitudes, or near dawn and dusk, R:FR may fall to values as low as 0.7, although cloud on the horizon renders this value very variable. The largest change in R:FR

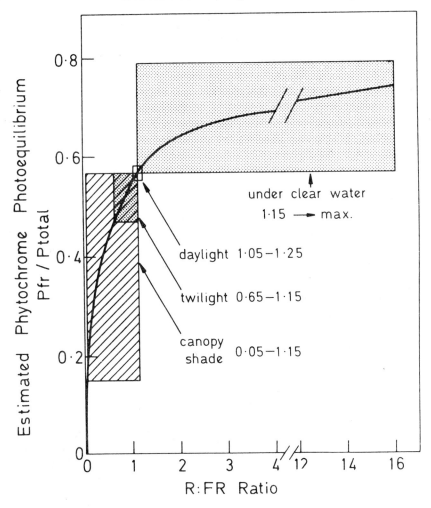

FIGURE 25.2   The relationship between R:FR and estimated phytochrome photoequilibrium (Pfr/P). The blocked area indicate the observed variation in R:FR under certain ecologically relevant environmental conditions. (From Reference 1.)

occurs when daylight interacts with vegetation; within dense vegetation canopies, R:FR can be as low as 0.1. The only situation in which R:FR is increased relative to daylight is underwater, where the absorption of FR by pure water can cause very large increases in R:FR; these are mitigated in turbulent waters, however, by the absorption of R and B by dissolved and suspended chlorophyllous material. The ecological significance of vegetation-induced changes in the daylight spectrum have been extensively investigated, and the role of the phytochromes in perceiving these light signals has been well established.[6]

25.4 Photoregulation of Germination and Seedling Establishment in the Natural Environment

The majority of the physiological literature on photomorphogenesis reports studies of the responses of dark-imbibed seeds or dark-grown seedlings to irradiation with restricted wavelength regions. Thus, in the laboratory, many small seeds are induced to germinate by brief R and reversed by subsequent brief FR, demonstrating that the response is mediated by the phytochromes. Similarly, etiolated seedlings exhibit a major change in developmental pattern when transferred to the light; only some of the phenomena associated with de-etiolation, however, are mediated by the phytochromes. In the natural environment, seed germination and establishment of the resultant seedling as an independent, photoautotrophic individual are crucial phases. Large seeds with plentiful reserves often germinate well in total darkness, but the seeds of many herbs and pioneer species require light for germination. For small seeds with limited reserves, imbibition in darkness normally does not lead to germination, but very small amounts of light will initiate radicle protrusion. Subsequently, in darkness, the seedlings become etiolated, with stem elongation maximized at the expense of leaf and root development. The complete transition to the photoautotrophic mode (de-etiolation) requires exposure of etiolated seedlings to long periods of relatively high irradiance white light, although brief exposures to narrow wavebands (e.g., R) can initiate some of the partial processes involved in de-etiolation.

The scenario is usually as follows. Mature seed is dehydrated and dormant, preventing premature germination and the majority of the population of seed becomes incorporated into the upper layers of the soil together with detritus derived from vegetation; the seeds then become fully hydrated and potentially capable of germination. Germination is from then on prevented by the lack of light penetrating into the soil. Many soils contain large populations of seeds of a wide range of species that can remain dormant for years.[7] Germination is induced by soil disturbance and exposure to light, although other environmental variables such as temperature or nitrate availability can override the light stimulation of germination. The ecological significance is that survival of the species is enhanced by mechanisms that limit resource depletion while at the same time spread the probability of germination over wide time periods. Germination is an all-or-nothing, stochastic process, so conditions that partially favor germination will result in a proportion of the population germinating while the rest remains dormant; the ecological significance of this is obvious also, in that it spreads the chances of survival across the whole population.

The sensitivity of seeds to light is often very high; requiring only a few minutes exposure to R to induce germination. The sensitivity of many weed species may be increased by several orders of magnitude after a period of burial in soil, so that buried seeds are poised to react to the very slightest exposure to light.[8] Since the sensitivity is so high, even very low fluences of FR are sufficient to produce sufficient Pfr to initiate germination, since Pr absorbs FR to a small extent. Thus, although the perception of light is mediated by phytochrome, in this case it appears to be relatively independent of wavelength. On the other hand, exposure of seeds to long-term irradiation of low R:FR inhibits germination. Ecologically, the prevention of germination under low R:FR makes sense, since seeds exposed to light beneath dense vegetation canopies under which normal photoautotrophic growth would be impossible would be prevented from germination. Thus, although buried seed may be stimulated to germinate by very small exposures to light, whether or not the seed will proceed to full germination is determined by the R:FR ratio of the light to which it is subsequently exposed. Therefore, the overall control of germination in nature may involve two separate phytochrome-mediated responses.

Exposure of plants to light in the natural environment is normally long-term, but soil disturbance may represent an exception to this generalization. As soil is disturbed by an animal or by cultivation, seed exposed to light and then covered will be stimulated to germinate. In these circumstances, etiolation becomes of significance in the natural environment. Most small seeds germinate at or close to the soil surface followed by a brief period of etiolation growth, but deeply buried seeds may never make it to the surface. Thus, except for large seeds with ample reserves, etiolation growth is a last-resort strategy. Etiolation appears to be a developmental condition in which the normal (de-etiolation) pattern is repressed, as seems evident from the recent discovery of mutants that show the de-etiolated growth pattern (except for chlorophyll synthesis) in darkness. The *det*[9] and *cop*[10] mutants of *Arabidopsis* behave in darkness as if they were in the light; they are recessive mutations, which implies that light acts to inactivate repressors of photomorphogenesis.

Consistent with this view is that etiolation is only found in the evolutionary more advanced orders of terrestrial plants; bryophytes, ferns and many gymnosperms develop in the dark as they do in the light, even to the extent of making chlorophyll. It seems, therefore, that the angiosperms, in particular, have evolved mechanisms that presumably are of substantial adaptive value that operate to repress the normal growth pattern in the absence of light.

25.5 The Shade Avoidance Syndrome and Proximity Perception

Higher plants exhibit one of two extreme strategies in response to vegetation shade. A relatively small proportion of plants are *shade tolerators*, plants that deploy a conservative utilization of resources, exhibit relatively slow growth rates and are capable of acclimating their photosynthetic machinery to maximize light absorption at low photon flux densities. The other extreme strategy is *shade avoidance*, in which a relatively profligate utilization of resources is coupled to a rapid growth rate and a capacity to respond to shading by a redirection of growth potential from leaf and storage organ development to extension growth of stems and petioles. The net result of shade avoidance is the projection of photosynthetic organs into those parts of the environmental mosaic in which the resource of light is plentiful. Shade avoidance is a syndrome of responses to reduced R:FR that include: enhanced internode and petiole elongation, reduced leaf development, reduced pigment production, enhanced apical dominance and reduced branching, a redirection of assimilates from storage organs into extending stems and petioles, acceleration of flowering, reduction of flower number and of fruit and seed set. The induction of shade avoidance reactions is phytochrome-mediated and is a direct result of the perception of low R:FR in the light environment of plant canopies.[1] Thus, plants utilize information derived from the spectral distribution of radiation to effect responses that favor the interception of high photon flux densities for photosynthesis. Shade avoidance is an effective strategy for life in an herbaceous community, but has limitations for herbs growing on the floor of a dense forest. The two strategies — avoidance and tolerance — are not necessarily mutually exclusive, since some plants display intermediate strategies and appear to be able to adapt to life either in open or shaded habitats, while other plants can exhibit shade avoidance and shade tolerance at different points in their life cycle. In evolutionary terms, shade avoidance appears to be a relatively recent invention, since it is predominantly found in the Angiosperms, although some Gymnosperms show some shade-avoidance characteristics. Ferns, mosses, and liverworts show little if any capacity to react to vegetational shading by characteristic avoidance reactions, and generally cope with shade by tolerance responses.

Studies of shade avoidance responses in controlled environments have demonstrated that they can be induced very quickly in response to a reduction in R:FR. Sensitive linear-displacement transducers have been used to monitor continuously plant extension, and have demonstrated that responses to the addition of FR via fiber-optic probes direct to the growing internodes occur with a 5- to 15-min lag phase and result in upwards of a fivefold increase in extension rate.[11] Similarly, removal of the additional FR, or its negation by additional R fed via bifurcated fiber-optic probes, causes a reduction in extension growth rate with a similar time lag. From such data, and from data of long-term growth rate changes, direct linear relationships between calculated Pfr/P and extension growth rate have been obtained for a number of species ranging from herbaceous seedlings to forest trees.[12] Related studies have shown that plants can react to incipient shading, rather than wait for actual shading to occur.[13] Thus, radiation reflected from neighboring vegetation, which is reduced in R and relatively enriched in FR, is perceived by the phytochromes even when the target plants are exposed simultaneously to high photon flux densities of high R:FR from above. In other words, low flux density radiation propagated more or less horizontally is perceived sensitively even when the plant is exposed to high photon flux densities propagated more or less vertically downward. This phenomenon is most easily explained geometrically, by assuming that the perceptive organs are the more or less vertical internodes that will intercept horizontal radiation effectively and vertical radiation poorly.

25.6 Photoperception of Other Ecologically Relevant Light Signals

Figure 25.1 illustrates certain other situations in which fluctuations in the light environment might be predicted to provide signals of ecological relevance. The possibility that R:FR might provide information on depth of immersion, or more particularly the proximity to the surface, has been considered. The R:FR ratio increases rapidly with depth underwater for relatively clear water, and could provide information theoretically on depth of immersion. However, Figure 25.2 shows that the photochemical properties of the phytochromes are such that there is a very small change in Pfr/P with large increases in depth, indicating that the phytochromes would be insensitive detectors of depth. On the other hand, there will be quite a rapid and large change in Pfr/P as an aquatic plant emerges from a body of water; since many instances of developmental changes at the water/air interface are known (e.g., heterophylly), it is at least possible that phytochrome may be involved in perception of this transition.[1]

A possibly more significant ecological signal is the periodicity of daylight, since both flowering and bud dormancy may be induced by changes in daylength. As the solar elevation decreases below ca. 10°, the R:FR ratio usually decreases, often to levels as low as ca 0.7. This would cause a significant drop in Pfr/P because such a change in R:FR is on the sensitive portion of the hyperbola referred to above. However, end-of-day changes in R:FR are very inconsistent, being susceptible to cloud cover on the horizon. A more telling argument is based on the low photon flux densities that accompany the end-of-day reductions in R:FR. At twilight, the photon flux densities are too low to drive the photoconversions of phytochrome to equilibrium. Consequently, a role for the phytochromes in the perception of environmental end-of-day signals cannot be supported, even though there is ample evidence for phytochrome involvement in photoperiod perception in laboratory experiments.[14] The same argument may be applied to the suggestions that plants are sensitive to moonlight; this is most unlikely inasmuch as the photon flux densities of moonlight are too low to drive the phytochrome photoconversions to equilibrium.[14]

References

1. Smith, H. and Whitelam, G. C., Phytochrome, a family of photoreceptors with multiple physiological roles, *Plant, Cell Environ.*, 13, 695, 1990.
2. Smith, H., Light quality, photoperception and plant strategy, *Annu. Rev. Plant Physiol.*, 33, 481, 1982.
3. Holmes, M. G. and Smith, H., The function of phytochrome in the natural environment. II. The influence of vegetation canopies on the spectral energy distribution of natural daylight, *Photochem. Photobiol.*, 25, 539, 1977.
4. Sharrock, R. A. and Quail, P. H., Novel phytochrome sequences in *Arabidopsis thaliana*: Structure, evolution, and differential expression of a plant regulatory photoreceptor family, *Genes Devel.*, 3, 534, 1989.
5. Smith, H. and Holmes, M. G., The function of phytochrome in the natural environment. III. Measurement and calculation of phytochrome equilibrium, *Photochem. Photobiol.*, 25, 547, 1977.
6. Casal, J. J. and Smith, H.., The function, action and adaptive significance of phytochrome in light-grown plants, *Plant, Cell Environ.*, 12, 855, 1989.
7. Wesson, G. and Wareing, P. F., The induction of light sensitivity in weed seeds by burial, *J. Exp. Bot.*, 20, 414, 1969.
8. Scopel, A. L., Ballaré, C. L., and Sânchez, R. A., Induction of extreme light sensitivity in buried weed seeds and its role in the perception of soil cultivations, *Plant, Cell Environ.*, 14, 501, 1991.
9. Chory, J., Peto, C. A., Feinbaum, R., Pratt, L. H., and Ausubel, F., *Arabidopsis thaliana* mutant that develops as a light-grown plant in the absence of light, *Cell*, 58, 991, 1989.

10. Deng, X. W. and Quail, P. H., Genetic and phenotypic characterization of *cop-1* mutants of *Arabidopsis thaliana*, *Plant.*, 2, 83, 1992.

11. Child, R. and Smith, H., Phytochrome action in light-grown mustard: Kinetics, fluence-rate compensation and ecological significance, *Planta*, 172, 219, 1987.

12. Morgan, D. C. and Smith, H., Linear relationship between phytochrome photoequilibrium and growth in plants under simulated natural radiation, *Nature*, 262, 210, 1976.

13. Ballaré, C. L., Scopel, A. L., and Sânchez, R. A., Far-red radiation reflected from adjacent leaves: An early signal of competition in plant canopies, *Science*, 247, 329, 1990.

14. Smith, H. and Morgan, D. C., The function of phytochrome in nature, in *Encyclopedia of Plant Physiology*, New series, 16B, *Photomorphogenesis*, Shropshire, W., Jr., and Mohr, H., Eds., Springer-Verlag, Berlin, 1983, 401–517.

Tôru Yoshizawa
*The University of
Electro-Communications*

Osamu Kuwata
*The University of
Electro-Communications*

Vision: Photochemistry

26.1 Primary Process

Vision begins with the absorption of photons by visual pigments composed of 11-*cis*-retinylidene chromophore (11-*cis*-retinal) and the protein moiety (opsin). The primary processes of visual pigments have been studied mainly on bovine rhodopsin as a rod pigment, chicken iodopsin (a red light-sensitive pigment) as a cone pigment, and squid or octopus rhodopsin as an invertebrate visual pigment.

The primary process of visual pigment may be defined as shown in Figure 26.1. The photochemical reaction of rhodopsin is characterized by the high quantum yield ($\phi = 0.67$), which is suggestive of rapidity of the reaction.[1] How fast is it? The absorption of a photon by the retinylidene chromophore of rhodopsin gives rise to the transition from the ground state to the excited state (Franck-Condon state) in a few femtoseconds. Since the excited state is a high-energy state, a molecule in the excited state is unstable enough to lose the energy gained by the absorption of a photon by either emitting a fluorescence, causing a photochemical reaction or making a transition to a lower energy state (a triplet state, from which a phosphorescence emits). In order for a rhodopsin molecule in the excited state to cause the photochemical reaction, it must take place prior to the emission of fluorescence or the transition to a lower energy state. In case of rhodopsin, no phosphorescence is observed, but the fluorescence can be detected though it is very weak.[2–4] Accordingly, the photochemical reaction of rhodopsin must occur within a time shorter than the lifetime of the fluorescence (0.1 to 0.05 ps).

What sort of state of rhodopsin is the first photoproduct? Irradiation of bovine rhodopsin at liquid nitrogen temperature (77 K) led to a finding of a photoproduct ($\lambda_{max} = 543$ nm) earlier than lumirhodopsin, which is now called bathorhodopsin followed by several photobleaching products (Figure 26.2 and 26.3).[5–8] Subsequent picosecond laser photolyses confirmed the formation of bathorhodopsin at room temperature (Figures 26.3 and 26.4) with a formation time constant of about 45 ps.[9,10] In early laser experiments, however, the time constant was reported to be within 6 ps, where no multiphoton reactions due to intense excitation laser pulses were taken into consideration.[10–12] Now, the formation of bathorhodopsin from rhodopsin in a rigid solvent frozen at 77 K indicated that an isomerization of the chromophore took place without accompanying conformational changes of the opsin structure.[6] Such an isomerization is unusual because the chromophore in an opsin matrix must be isomerized in a restricted space (a retinal binding site) without moving both ends of it (the β-ionone ring of the chromophore and the α-carbon of lysine-296 residue).[13–15] The mechanism of the isomerization was confirmed experimentally by using a series of fluorinated retinal analogues.[16] A theoretical calculation estimated that the isomerization would occur within 0.2 ps.[17] The suggestion that the *all-trans*-chromophore of bathorhodopsin would be in a twisted

primary process

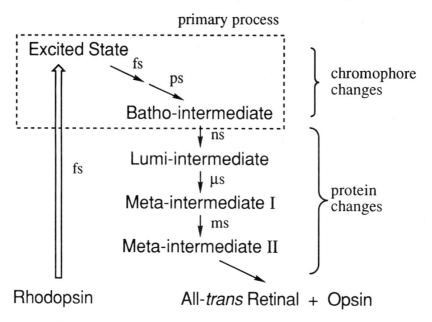

FIGURE 26.1　Photobleaching process of visual pigment. The primary process is shown in the broken line frame. Open and closed arrows denote a photon absorption and a dark process, respectively. All the visual pigments that have been analyzed have their own batho intermediate. In squid and octopus rhodopsin systems, metarhodopsin I and II are called acid and alkaline metarhodopsins, respectively. (Modified from Reference 12.)

form was confirmed by circular dichroism and laser Raman spectroscopies.[6,18–20] The bathochromic shift of rhodopsin to bathorhodopsin would be due to an increase of distance (or charge separation) between the protonated Schiff-base nitrogen and the counter ion.[21,22] A calorimetric measurement displayed that bathorhodopsin is about 35 kcal higher in potential energy than rhodopsin.[23]

In order to examine whether or not any photoproduct earlier than bathorhodopsin is present in the photobleaching process, bovine rhodopsin was irradiated with an orange light (>540-nm light) at liquid helium temperature (about 4 K).[7,8] Under those conditions, a hypsochromic shift was observed owing to the formation of a photoproduct called hypsorhodopsin (λ_{max} = 436 nm) (Figures 26.2 and 26.3).[7,8,24] If one irradiates rhodopsin with blue light (437 nm), a bathochromic shift can be observed owing to formation of bathorhodopsin (Figure 26.3a i). Prolonged irradiation of rhodopsin or isorhodopsin (rhodopsin isomer with 9-cis-chromophore) at 4 K forms a photostationary state composed of rhodopsin, isorhodopsin, hypsorhodopsin, and bathorhodopsin. On warming above 23 K, hypsorhodopsin changes to bathorhodopsin. The similar photochemical behaviors at 4 K were observed in squid (Figure 26.3a ii) and octopus rhodopsins; but in chicken iodopsin (Figure 26.3a iii) no formation of a hypsointermediate was observed.[25–27]

According to laser Raman spectroscopy of octopus rhodopsin and FTIR spectroscopy of bovine rhodopsin at 4°C, hypsorhodopsin is very close to bathorhodopsin in terms of conformation of chromophore and protein moieties including the protonated Schiff base, though the absorption maximum of bovine hypsorhodopsin (436 nm) is quite different from that of the bathorhodopsin (543 nm).[28,29] On the basis of the similarity in absorption maximum to 7-cis-rhodopsin (450 nm), 9-cis-retro-γ-rhodopsin (420 nm) or 9-cis-7,8-dihydrorhodopsin (420 nm), hypsorhodopsin was inferred to have a highly twisted or dissected chromophore at the 7–10 conjugated double bond system.[30–32] In fact, the spectrum of hypsorhodopsin has a small side band around 540 nm, indicating a distortion of the polyene chain of the retinylidene chromophore (Figure 26.2).[24] The formation of hypsorhodopsin at room temperature was not recorded in bovine rhodopsin (Figure 26.3b i) but in squid and octopus rhodopsins, only when an intense pulse was used as the

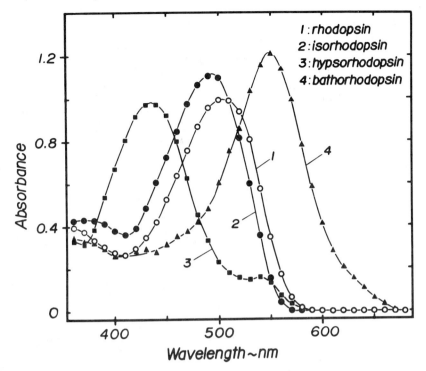

FIGURE 26.2 Absorption spectra of bovine rhodopsin (curve 1), isorhodopsin (curve 2), hypsorhodopsin (curve 3), and bathorhodopsin (curve 4) at liquid helium temperature (4 K).[24] The spectrum of rhodopsin (λ_{max} = 506 nm) is brought arbitrarily to a maximal absorbance of 1.0. Isorhodopsin (λ_{max} = 494 nm), hypsorhodopsin (λ_{max} = 435 nm), and bathorhodopsin (λ_{max} = 548 nm) possess maximal absorbances about 1.12-, 0.98-, and 1.22-fold, respectively, as high as that of rhodopsin. Rhodopsin, isorhodopsin, and bathorhodopsin have their λ_{max} at 498, 485, and 535 nm, respectively, at room temperature, and their λ_{max} at 505, 491, and 543 nm, respectively, at liquid nitrogen temperature (77 K).

excitation pulse which may cause a multiphoton reaction (Figure 26.3b ii).[33] Thus, hypsorhodopsin may be an unphysiological intermediate.

An excitation of bovine, squid, or octopus rhodopsin with a weak laser pulse (wavelength: 532 nm; duration: 25 ps) led to a finding of a precursor of bathorhodopsin that is called *photorhodopsin* (λ_{max} of bovine photorhodopsin = 570 nm) (Figure 26.4).[9] Photorhodopsin changes thermally to bathorhodopsin with a time constant of about 45 ps and converts photochemically to hypsorhodopsin in cases of squid and octopus rhodopsins (Figure 3b ii).[9,10,33] A recent femtosecond laser photolysis (excitation pulse: 35 fs at 500 nm; probe pulse: 10 fs) confirmed the presence of photorhodopsin, which is formed directly from an excited state with a decay time constant of about 200 fs (Figure 26.3c).[34] A similar experiment was carried out in octopus rhodopsin.[35] Picosecond laser analyses on rhodopsin analogues with 11-*cis* double bond of retinylidene chromophore locked by 5-, 7-, or 8-membered ring indicated that the change of rhodopsin to photorhodopsin is a photoisomerization of the 11-*cis*-chromophore to a highly twisted *all-trans* form, which relaxes to bathorhodopsin having another form of twisted *all-trans* chromophore.[4,36]

(a) Low temperature spectrophotometry

i) Bovine rhodopsin[7,8,24] (4 K)

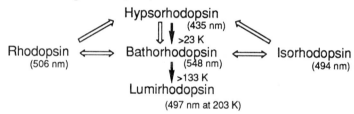

ii) Squid rhodopsin[25] (4 K)

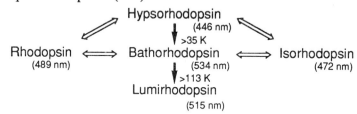

iii) Chicken iodopsin[37] (77 K)

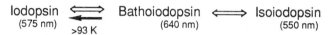

(b) Picosecond laser photolysis (room temperature)

i) Bovine rhodopsin[9,10,12,33]

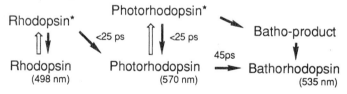

ii) Squid rhodopsin[9,12,33]

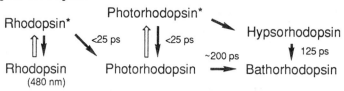

(c) Femtosecond laser photolysis (room temperature)
Bovine rhodopsin[35]

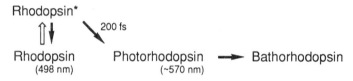

FIGURE 26.3   Reaction diagrams for the primary processes of visual pigments analyzed by low-temperature spectrophotometry (a), and pico- (b) and femtosecond (c) second laser photolyses. Wavelengths in parentheses indicate absorption maxima. Open and closed arrows show photic and dark processes, respectively. The asterisk denotes an excited state. Absolute temperatures and numerals followed by ps or fs express transition temperatures and time constants of dark processes, respectively.

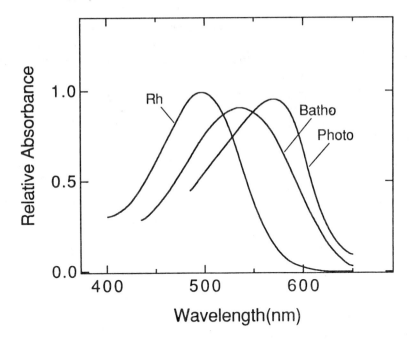

FIGURE 26.4 Absorption spectra of bovine rhodopsin (λ_{max} = 498 nm), bathorhodopsin (λ_{max} = 535 nm), and photorhodopsin (λ_{max} = 570 nm) at room temperature.[38] Maximal absorbance of batho- or photorhodopsin displays about 0.91 or 0.96 times of that of rhodopsin (ε_{max} = 40,600 in digitonin).

References

1. Dartnall, H. J. A., The photosensitivity of visual pigments in the presence of hydroxylamine, *Vision Res.*, 8, 339, 1967.
2. Doukas, A. G., Lu, P. Y., and Alfano, R. R., Fluorescence relaxation kinetics from rhodopsin and isorhodopsin, *Biophys. J.*, 35, 547, 1981.
3. Doukas, A. G., Junnarker, M. R., Alfano, R. R., Callender, R. H., Kakitani, T., and Honig, H., Fluorescence quantum yield of visual pigments: Evidence for subpicosecond isomerization rates, *Proc. Natl. Acad. Sci. U.S.A.*, 81, 4790, 1984.
4. Kandori, H., Matuoka, S., Shichida, Y., Yoshizawa, T., Ito, M., Tsukida, K., Balogh-Nair, V., and Nakanishi, K., Mechanism of isomerization of rhodopsin studied by use of 11-*cis*-locked rhodopsin analogs excited with picosecond laser pulse, *Biochemistry*, 28, 6460, 1989.
5. Yoshizawa, T. and Kitô, Y., Chemistry of the rhodopsin cycle, *Nature*, 182, 1604, 1958.
6. Yoshizawa, T. and Wald, G., Pre-lumirhodopsin and the bleaching of visual pigments, *Nature*, 197, 1279, 1963.
7. Yoshizawa, T., The behaviour of visual pigments at low temperatures, in *Handbook of Sensory Physiology*, Dartnall, H. J. A., Ed., Springer-Verlag, Heidelberg, 1972, 146.
8. Yoshizawa, T. and Horiuchi, S., Studies in intermediates of visual pigments by absorption spectra at liquid helium temperature and circular dichroism at low temperature, in *Biochemistry and Physiology of Visual Pigments*, Langer, H., Ed., Springer-Verlag, Berlin, 1973, 69.
9. Shichida, Y., Matuoka, S., and Yoshizawa, T., Formation of photorhodopsin, a precursor of bathorhodopsin, detected by a picosecond laser photolysis at room temperature, *Photobiochem. Photobiophys.*, 7, 221, 1984.
10. Kandori, H., Matuoka, S., Shichida, Y., and Yoshizawa, T., Dependency of photon density on primary process of cattle rhodopsin, *Photochem. Photobiol.*, 49, 181, 1989.
11. Busch, G. E., Applebury, M. L., Lamola, A. A., and Rentzepis, P. M., Formation and decay of prelumirhodopsin at room temperatures, *Proc. Natl. Acad. Sci. U.S.A.*, 69, 2802, 1972.

12. Yoshizawa, T. and Kandori, H., Primary photochemical events in the rhodopsin molecule, in *Retinal Research*, Osborne, N. N. and Chader, G. J., Eds., Pergamon Press, Oxford, 1991, chap. 2.

13. Matsumoto, H. and Yoshizawa, T., Existence of a β-ionone ring-binding site in the rhodopsin molecule, *Nature*, 258, 525, 1975.

14. Matsumoto, H. and Yoshizawa, T., Recognition of opsin to the longitudinal length of retinal isomers in the formation of rhodopsin, *Vision Res.*, 18, 607, 1978.

15. Ovchinnikov, Yu. A., Rhodopsin and bacteriorhodopsin: Structure-function relationships, *FEBS Lett.*, 179, 1982.

16. Shichida, Y., Ono, T., Yoshizawa, T., Matsumoto, H., Asato, A., Zingoni, J. P., and Liu, R. S. H., Electrostatic interaction between retinylidene chromophore and opsin in rhodopsin studied by fluorinated rhodopsin analogues, *Biochemistry*, 26, 4422, 1987.

17. Warshel, A., Bicycle-pedal model for the first step in the vision process, *Nature*, 260, 679, 1976.

18. Horiuchi, S., Tokunaga, F., and Yoshizawa, T., Circular dichroism of cattle rhodopsin and bathorhodopsin at liquid nitrogen temperatures, *Biochim. Biophys. Acta*, 591, 445, 1980.

19. Fukada, Y., Shichida, Y., Yoshizawa, T., Ito, M., Kodama, A., and Tsukida, K., Studies on structure and function of rhodopsin by use of cyclopentatrienylidene 11-*cis*-locked-rhodopsin, *Biochemistry*, 23, 5826, 1984.

20. Eyring, G., Curry, B., Broek, A., Lugtenburg, J., and Mathies, R., Assignment and interpretation of hydrogen out-of-plane vibrations in the resonance Raman spectra of rhodopsin and bathorhodopsin, *Biochemistry*, 21, 384, 1982.

21. Kawamura, S., Tokunaga, F., and Yoshizawa, T., Absorption spectra of rhodopsin and its intermediates and orientational change of the chromophore, *Vision Res.*, 17, 991, 1977.

22. Honig, B., Callender, R. H., Dimur, U., and Ottolenghi, M., Photoisomerization, energy storage and charge separation: A model for light energy transduction in visual pigments and bacteriorhodopsin, *Proc. Natl. Acad. Sci. U.S.A.*, 76, 2503, 1979.

23. Cooper, A., Energy uptake in the first step of visual excitation, *Nature*, 282, 531, 1979.

24. Yoshizawa, T., Horiuchi, S., Tokunaga, F., and Sasaki, N., Absorption spectrum of cattle hypsorhodopsin, *FEBS Lett.*, 163, 165, 1983.

25. Shichida, Y., Tokunaga, F., and Yoshizawa, T., Squid hypsorhodopsin, *Photochem. Photobiol.*, 29, 343, 1979.

26. Tsuda, M., Tokunaga, F., Ebrey, T. G., Yue, K. T., Marque, J., and Eisenstein, L., Behaviour of octopus rhodopsin and its photoproducts at very low temperatures, *Nature*, 287, 461, 1980.

27. Tsukamoto, Y., Horiuchi, S., and Yoshizawa, T., Photic reactions of chicken iodopsin and rhodopsin at liquid helium temperature, *Vision Res.*, 15, 819, 1975.

28. Pande, A. J., Callender, R. H., Ebrey, T. G., and Tsuda, M., Raman study of the primary photochemistry of visual pigments hypsorhodopsin, *Biophys. J.*, 45, 573, 1984.

29. Sasaki, J., Maeda, A., Shichida, Y., Groesbeek, M., Lugtenburg, J., and Yoshizawa, T., Structure of hypsorhodopsin: Analysis by fourier transform infrared spectroscopy at 10 K, *Photochem. Photobiol.*, 56, 1063, 1992.

30. Kawamura, S., Miyatani, S., Matsumoto, H., Yoshizawa, T., and Liu, R. S. H., Photochemical studies of 7-*cis*-rhodopsin at low temperature. Nature and properties of the bathointermediate, *Biochemistry*, 19, 1549, 1980.

31. Ito, M., Hirata, K., Kodama, A., Tsukida, K., Matsumoto, H., Horiuchi, K., and Yoshizawa, T., Retro-γ-retinals and visual pigment analogues, *Chem. Pharm. Bull.*, 26, 925, 1978.

32. Muto, O., Tokunaga, F., Yoshizawa, T., Kamat, V., Blatchly, H. A., Balogh-Nair, V., and Nakanishi, K., Photochemical reaction of 7,8-dihydrorhodopsin at low temperatures, *Biochim. Biophys. Acta*, 766, 597, 1980.

33. Matuoka, S., Shichida, Y., and Yoshizawa, T., Formation of hypsorhodopsin at room temperature by picosecond green pulse, *Biochem. Biophys. Acta*, 765, 38, 1984.

34. Schoenlein, R. W., Peteanu, L. A., Mathies, R. A., and Shank, C. V., The first step in vision: Femtosecond isomerization of rhodopsin, *Science*, 254, 412, 1991.

35. Taiji, M., Bryl, K., Nakagawa, M., Tsuda, M., and Kobayashi, T., Femtosecond studies of primary photoprocesses in octopus rhodopsin, *Photochem. Photobiol.,* 56, 1003, 1992.
36. Mizukami, T., Kandori, H., Shichida, Y., Cheng, A.-H., Derguini, F., Caldwell, C. G., Bigge, C. F., Nakanishi, K., and Yoshizawa, T., Photoisomerization mechanism of the rhodopsin chromophore: Picosecond photolysis of pigment containing 11-*cis*-locked-8-membered-ring-retinal, *Proc. Natl. Acad. Sci. U.S.A.,* 90, 4072, 1993.
37. Yoshizawa, T. and Wald, G., Photochemistry of iodopsin, *Nature,* 214, 566, 1967.
38. Kandori, H., Shichida, Y., and Yoshizawa, T., Absolute absorption spectra of batho- and photorhodopsins at room temperature. Picosecond laser photolysis of rhodopsin in polyacrylamide gel, *Biophys. J.,* 56, 453, 1989.

27

The Bleaching of Visual Pigments

Thomas G. Ebrey
University of Illinois

Jie Liang
University of Illinois

27.1 Introduction

The focus of this chapter is on the photochemical transformations of visual pigments after the initial photoproduct has been formed. An earlier chapter by Yoshizawa in this volume deals with the initial events in the bleaching of visual pigments. Practically, the chapter will be concerned with processes starting with the decay of bathorhodopsin, the primary photoproduct; batho starts to decay within a nanosecond or so after it is formed by the isomerization of the 11-*cis*-retinal chromophore of the pigment to the *all-trans*-conformation. It was thought originally that batho decayed to the intermediate lumirhodopsin; but more recently, a new product, a blue-shifted intermediate (BSI), has been discovered that seems to lie between batho and lumi. Lumi decays to meta I, which decays to meta II, which decays to meta III, which then decays finally to *all-trans*-retinal plus opsin. Alternative decay pathways, back reactions, and the effects of various photointermediates interacting with other proteins all change this simple picture. Since these transitions between intermediates are all spontaneous at room temperature, each intermediate must have higher free energy than the species to which it decays, with the high free energy of the primary photoproduct deriving ultimately from the energy of the photon.[1]

This chapter will first deal with the characterization of the intermediates. Cone pigments and direct electrical signals associated with the bleaching process will then be examined. In addition to looking at bleaching from the perspective of spectral changes of the chromophore of rhodopsin, it is important to view it as a set of changes involving both the protein and the chromophore. It is believed that the early steps in bleaching, up to the formation of batho and possibly beyond, involve only changes in the chromophore, specifically 11-*cis*- to *all-trans*-isomerization and perhaps twisting or relaxing about other bonds of the chromophore (see Chapter 26 by Yoshizawa). Changes in the pigment subsequent to this are associated implicitly with changes in the protein or with the deprotonation of the chromophore-protein linkage, although other changes in the chromophore cannot be excluded. Current knowledge of these changes, limited though it is, will also be discussed.

0-8493-8634-9/95/$0.00+$.50
© 1995 by CRC Press, Inc.

27.2 The Bleaching Intermediates of Rhodopsin as Defined by Their Spectral Properties

The intermediates of rhodopsin are defined by their absorption (and vibrational) spectra and the temperatures at which they are stable. The transitions between the intermediates are initially characterized by their lifetimes and activation energies. The lifetimes of the latter intermediates depend not only on the usual factors such as temperature and pH, but also on such things as the detergent/lipids present and the species of animal from which the pigment is derived.

The primary photoproduct, bathorhodopsin, absorbs at longer wavelengths than rhodopsin (see Yoshizawa, Chapter 26). In the physiological temperature range, batho decays in about 50 to 100 ns;[2–4] presumably, the product is the blue-shifted intermediate (BSI).[5,6] The BSI has its absorption maxima at ca. 475 nm for bovine rhodopsin;[6] thus far, the vibrational spectrum of the chromophore has been obtained only for the BSI intermediate of 5,6-dihydroretinal artificial pigment.[7] The lifetimes from batho to BSI to lumi are all roughly the same at room temperature, in the range of 50 to 300 ns.[8,9] Batho is stable at about 130K, while BSI is very difficult to detect at low temperatures due to its thermodynamic properties.[5] Bovine lumirhodopsin has its absorption maximum at ca. 495 nm and its extinction is somewhat higher than that of rhodopsin.[10,11]

There is some uncertainty as the rate at which lumi decays to meta I. This may be because these transitions have not been well characterized spectrally, due to the differences in lifetimes for pigments measured in membranes versus in detergent and, to a lesser extent, the way the data are analyzed. The lifetimes reported for this transition are 50 μs at 20°C in frog rods,[2] 100 μs at 27°C in rat rods,[12] and 300 μs at 24°C in bovine rods;[13] in contrast, Lewis et al.[14] and Straume et al.[15] both find lifetimes of 1 to 3 ms at 20°C in bovine rods. Both of these measurements of lumi decay to meta I are suspect, however, due to inadequate instrumental time resolution and/or small signal amplitudes at the monitoring wavelength. More recently, Thorgeirsson et al. (1991)[8,9] have noticed some anomalies in the difference spectra between rhodopsin and its bleaching products in the 1-μs to 1.5-ms time range. They noted that several interpretations of their data are possible but favor the existence of a deprotonated Schiff base precursor of meta I. The notion of a very rapidly forming meta I from lumi, which then quickly decays to meta II could also be consistent with their data.

Meta I absorbs at slightly shorter wavelengths than lumi and has a lower extinction coefficient.[10,14,16] The difference spectra between rhodopsin and meta I have been reported by Baker and Williams[17] and Govindjee et al.[18] The activation energy of the lumi to meta I transition is given as 100 to 130 kJ mol⁻¹ near room temperature.[6] The resonance Raman spectra of meta I and meta II are quite distinctive.[19]

The meta I-to-meta II transition has been perhaps the best studied of the later portion of the bleaching process. This transition involves the deprotonation of the Schiff base of the chromophore. For most species, meta II is formed in ca. 1 to 2 ms at room temperature and probably about 40 to 100 μs at 37°C;[20,21] it is this time constant that could contribute to the kinetics of the receptor potential of rods.[22] The transition from meta I to meta II represents a change from a form of rhodopsin that probably has no special affinity to transducin and cannot activate transducin (by catalyzing the exchange of GDP for GTP), to a form that does bind transducin and can catalyze the exchange of GDP for GTP, activating transducin. This binding greatly alters the equilibrium concentration of the two intermediates, shifting it toward meta II.[23]

The Schiff base must deprotonate in order to form a species that can interact with transducin,[24] but it has been reported that some mutants of rhodopsin can form a meta II-like species without being able to activate transducin.[25] Probably, in this case, a region of the pigment that normally interacts with the transducin has been modified without interfering with the events leading to Schiff base deprotonation. Normally, meta I and meta II are in equilibrium;[16] this equilibrium depends on temperature and pH. Most unusually, even though meta I must lose a proton from its Schiff base to form meta II, there is a net uptake of protons in forming meta II (see, e.g., Parkes and Liebman[26] and the review by Hofmann[23]). Thus meta II formation involves at least two steps — deprotonation of the Schiff base and the uptake of protons. It is not clear if the meta II formed at both these steps can interact with transducin.

Meta II seems to have multiple pathways of decay, either going on to a longer-wavelength intermediate meta III or decaying directly to the free *all-trans*-retinal plus opsin.[27] The decay of meta II *in situ* appears to be quite slow[20,28] and may be the rate-limiting step in the normal regeneration of the visual pigment.

27.3 The Bleaching Intermediates of Cone Pigments

The photochemical transformation of cone pigments are not as well studied as those of the rod pigments. Cone and rod pigments differ in a number of ways. For example, long-wavelength absorbing cone pigments have a Cl⁻ binding site and can be bleached in the dark by hydroxylamine. Currently, only the photochemistry of the chicken red cone pigment iodopsin and the gecko cone-type pigment P521 have been studied.

In chicken cones, low-temperature experiments showed that chicken iodopsin (562 nm) forms bathoiodopsin (640 nm) upon bleaching (see Yoshizawa, Chapter 26). When the temperature is raised to −180°C, bathoiodopsin reverts to iodopsin.[29] However, if iodopsin is in its Cl⁻-deficient state, upon warming to −160°C, bathoiodopsin is converted to a lumiiodopsin absorbing at 518 nm.[30] It has been proposed that the bleaching of iodopsin is accompanied by the release of Cl⁻,[30] but so far there is no direct evidence for this. If so, the reversion of bathoiodopsin to iodopsin at low temperatures may be due to the inhibition of Cl⁻ release at these temperatures.

Hubbard et al. first described the formation of metaiodopsin I upon irradiation of iodopsin at −20°C;[10] its absorption maximum is at ca. 500 nm. Upon warming to −15°C, metaiodopsin I is converted to what is presumably metaiodopsin II. Although no spectrum of metaiodopsin II has been reported, the data are consistent with an absorption maximum in the UV, i.e., 380 nm. Above 0°C, metaiodopsin II is hydrolyzed to *all-trans*-retinal and opsin.[29] Quantitative kinetic studies of iodopsin at physiological temperatures are currently unavailable.

Similar photochemical intermediates have also been found in the gecko cone pigment P521. Low-temperature experiments with digitonin extracts of P521 show that a meta I-type intermediate is formed at −40°C after bleaching.[31a] The reconstructed absorption spectrum of the meta I intermediate of P521 has its λ_{max} at 470 nm. The meta II-type intermediate is observed when the temperature is raised to −3°C; its λ_{max} is at around 380 nm.[31a]

Flash photolysis of the gecko P521 cone pigment in 2% digitonin can resolve the lumi-to-meta I transition; the half decay time is about 1.5 µs at 8.0°C. The kinetics of the meta I-to-meta II transition of gecko P521 has been studied as a function of pH and salt concentration. Although none of the kinetic traces fit with a simple first-order kinetic model, half decay times for a fast and slow component at pH 6.6 and 22°C are estimated to be 4.9 and 340 ms, respectively.[31a] These kinetic time constants are not very different in magnitude to those seen for bovine rhodopsin under similar conditions. (See Reference 31b.)

The results of flash photolysis studies show that low pH and high temperature favor meta II formation, similar to what was found for rod pigments.[16] The transition from meta I to meta II in P521 seems to involve an uptake of protons.[31a] However, the pK of meta II formation in gecko P521 is about 8.7, different from that of ca. 6.4 found in the rod pigment bovine rhodopsin.[16] Recent studies of site-directed mutants of bovine rhodopsin suggest a histidine residue (probably H211) is responsible for setting the meta I-to-meta II equilibrium in bovine rhodopsin.[32] This histidine residue is replaced by a cysteine in all known cone pigments, including gecko P521.[33] It is possible that cysteine can also serve in lieu of the histidine for cone pigments. The pK of 8.7 found in gecko P521 is consistent with this hypothesis.

27.4 Electrical Changes Associated with the Photointermediate Transitions

In bacteriorhodopsin, charge movements can be detected during the photocycle that, in many cases, can be directly associated with transitions between specific spectral intermediates (see review by Trissl, Reference 34). Similar electrical signals were detected from oriented visual pigments (see review by Cone and Pak, Reference 35). Bacteriorhodopsin is contained in the purple membrane that can be oriented in space, for example by a weak electric field, enabling one to record the photoelectric signal with gross external electrodes. For rhodopsin, the orientation of the pigment molecules is accomplished by nature, when it forms the ordered membranes of the photoreceptor cells, which are themselves all ordered in the retina.

Two components of the photovoltage signal have been detected from vertebrate retinas and these have been designated the R1 and R2 components of the early receptor potential (ERP). Component R1 has a very short latency and can be associated plausibly with charges in rhodopsin moving when the pigment goes from its initial state to lumirhodopsin.[12,36–38] It is unclear if the major contributor to the signal is the rhodopsin-to-batho transition or the subsequent changes from batho to lumi. Since it is thought that charges on the chromophore move during photoisomerization, it is not unreasonable to assign R1 to the former transition, as has been done for the earliest phase of the photoelectric signal from bacteriorhodopsin.[34] In contrast, R2 is associated with the meta I-to-meta II transition and probably has a major contribution from the proton uptake events associated with this transition.[35,37,39] The time course of the appearance and disappearance of the photoelectric signals has been used to measure the lifetimes of the intermediates *in situ*. The values obtained are consistent with measurements in detergents but represent a more physiological situation.

In addition, a photoelectric signal can be obtained if a strong flash is absorbed by one of the photobleaching intermediates: meta I, meta II, or meta III. The appearance and the disappearance of these signals can also be used to measure the lifetime of the intermediates.[20,37]

There are two other quite interesting visual pigment systems that have been studied using their ERPs: cones and invertebrate photoreceptors. The former include human, frog, and ground squirrel cones (e.g., References 40 to 42); the most outstanding feature is that the photovoltages closely resemble those seen from rods, suggesting that there are no major differences in the bleaching kinetics of rod and cone pigments, at least as observed using the fast photovoltage method. The fast photovoltages from squid photoreceptors were studied by Hagins and co-workers.[43,44] As noted above, the meta I-to-meta II transition in vertebrates is accompanied by a net uptake of about one proton, while the deprotonation of the Schiff base of octopus rhodopsin during its bleaching at alkaline pH leads to the net release of a proton.[45] So far, the photoelectric signals associated with the proton movement during the bleaching of invertebrate pigment has not been identified unequivocally.

27.5 Changes in the Conformation of Rhodopsin upon Bleaching

Rhodopsin does not activate the G-protein, but its photoproduct meta II does. Thus, there must be a conformational change after light absorption that allows the pigment to interact with the G-protein. Below is a compendium of known light-induced conformational changes in rhodopsin. The driving force for these conformational changes must come from the energy of the photon. There must be similar conformational changes in ligand-activated receptors that also interact with G-proteins (e.g., β-adrenergic and muscarinic ACh receptor) and, in this case, the driving energy must come from ligand binding; very little is known about these binding energy-driven processes (see review by Jackson, Reference 46).

Many of the experimental results regarding these changes are not well specified temporally but probably occur during the slower spectral transitions; few experiments actually distinguish between the conformation of meta I (does not activate the G-protein) and meta II (does activate).

There are a number of light-induced changes in the protein that are not well specified but which nevertheless are clear indications of changes in the protein during the bleaching process. It is also clear that the magnitude and perhaps even the existence of some of these changes depends on the environment of the pigment, especially with respect to pigment in native membranes vs. pigment solubilized in detergent. These changes are

1. Proteolytic enzymes have differential effects[47] on rhodopsin depending on its photolytic state. Thermolysin clips off preferentially the C-terminal 12 amino acids when the pigment is (roughly) in the meta II stage.
2. A second indicator is somewhat trivial — the different absorption maxima for the bleaching intermediates that must signify that the protein has undergone changes near the chromophore during bleaching.
3. Another indicator derives from the chromophore being in the middle of the pigment, 10 to 30 Å away from the cytoplasmic loops of the pigment that interact with the G-protein; thus, the changes in the chromophore and its binding site as a result of isomerization must be propagated to the surface of the pigment by some long-range conformational rearrangement of the pigment upon the formation of meta II.
4. Since rhodopsin is a protonated Schiff base and meta II is an unprotonated one, some change in the protein must allow the pK_a of the Schiff base to change. Indeed, the pK_a of the protonated Schiff base of bovine rhodopsin is estimated to be at least 15,[48] while that of meta I/meta II is 6.6,[16] implying an enormous change in pK_a.
5. Circular dichroism changes[49,50] indicate that the protein changes its conformation at least by the end of the bleaching process.
6. After light absorption, serines and threonines on the C-terminal tail of rhodopsin become available for phosphorylation by a constitutive kinase in the cytoplasm (also true for the β-adrenergic receptor). This requires that the protein changes its conformation so that these groups become available (note that Binder et al.[51] suggest that the enzyme is at least partially activated). Yamamoto and Shichi have reported that this phosphorylation takes place at the meta II stage.[52]
7. A monoclonal antibody has been raised that recognizes a light-dependent structural change in rhodopsin.[53]
8. Conformational changes can be detected by vibrational spectroscopy (FTIR); changes in amide I and II bands as well as carboxyls can be observed.[54-56]
9. Hydroxylamine attacks the Schiff base much more easily after bleaching (it is unclear at which intermediate), suggesting it now is accessible to the Schiff base.
10. Activation entropies for transitions between the bleaching intermediates indicate conformational change.[28]
11. There is a change in the protonation state of groups other than the Schiff base during the bleaching process.[16,26,57]

In addition to these changes in which the amino acid residues involved are not specified, there are a number of experiments in which somewhat or very specific changes have been identified. These changes involve well-defined amino acids residues.

1. There is[58] an increase in the volume of the protein in going from meta I to meta II of 108 ml mol^{-1}. This can be compared with the ionization of a single group of ca. 4 ml mol^{-1}.
2. Note that items 4 and 9 above involve the Schiff base linkage of retinal with a lysine, while 6 involves serines and threonines.
3. In the dark, two cysteines can be alkylated; two additional cysteines can be alkylated only after light absorption. The cysteines most likely to be alkylated are those at positions 140, 222, and 316. It is thought 140 and 316 are accessible in the dark; if so, then probably it is 222 that is the new one alkylated in the light.

4. In linear dichroism experiments, a change at 290 nm is observed upon the formation of meta II, which can be attributed to a single tryptophan rotating with respect to the plane of the membrane.

5. FTIR experiments[55,56,59] all see evidence for protonated carboxyls that change environment and/or protonation state during the meta I/meta II/meta III transformations. The best candidates for these carboxyls (using the bovine rhodopsin numbering system) are Asp-83 and/or Glu-122, both of which are probably usually protonated in the native pigment.[60–62] In addition, Glu-113 may receive a proton from the Schiff base upon the formation of meta II.

In summary, the broad outlines of the bleaching process in visual pigments are fairly well understood. However, almost nothing is known about: what the key changes are that the pigment undergoes in allowing the active form of the pigment, meta II, to be formed; whether the only important thing is the deprotonation of the Schiff base or whether other changes must accompany this change in order to allow the active form of the pigment to be formed; what groups control the pK_a of the Schiff base and what causes them to change, allowing meta II to be formed. Many of these questions would be clarified if a more reliable and detailed structure of rhodopsin were available. However, it should be kept in mind that rhodopsin itself is not the most interesting pigment; that place must be reserved for meta II, the form of the pigment that can interact with the G-protein.

References

1. Honig, B., Ebrey, T. G., Callender, R. H., Dinur, U., and Ottolenghi, M., Photoisomerization, energy storage, and charge separation: A model for light energy transduction in visual pigments and bacteriorhodopsin, *Proc. Natl. Acad. Sci. U.S.A.*, 76, 2503, 1979.

2. Cone, R. A., Rotational diffusion of rhodopsin in the visual receptor membrane, *Nature New Biol.*, 236, 39, 1972.

3. Rosenfeld, T., Alchalal, A., and Ottolenghi, M., Nanosecond laser photolysis of rhodopsin in solution, *Nature*, 240, 482, 1972.

4. Goldschmidt, C. R., Ottolenghi, M., and Rosenfeld, T., Primary processes in photochemistry of rhodopsin at room temperature, *Nature*, 263, 169, 1976.

5. Hug, S. J., Lewis, J. W., Einterz, C. M., Thorgeirsson, T. E., and Kliger, D. S., Nanosecond photolysis of rhodopsin: Evidence for a new, blue-shifted intermediate, *Biochemistry*, 29, 1475, 1990.

6. Lewis, J. W. and Kliger, D. S., Photointermediates of visual pigments, *J. Bioenerg. Biomembr.*, 24, 201, 1992.

7. Ganter, U. M., Kashima, T., Sheves, M., and Siebert, F., FTIR evidence of an altered chromophore-protein interaction in the artificial visual pigment *cis*-5,6-dihydroisorhodopsin and its photoproducts BSI, lumirhodopsin, and metarhodopsin-I, *J. Am. Chem. Soc.*, 113, 4087, 1991.

8. Lewis, J. W., van Kuijk, F. J. G. M., Thorgeirsson, T. E., and Kliger, D. S., Photolysis intermediates of human rhodopsin, *Biochemistry*, 30, 11372, 1991.

9. Thorgeirsson, T. E., Lewis, J. W., Wallace-Williams, S. E., and Kliger, D. S., Photolysis of rhodopsin results in deprotonation of its retinal Schiff's base prior to formation of metarhodopsin II, *Photochem. Photobiol.*, 56, 1135, 1992.

10. Hubbard, R., Brown, P. K., and Kropf, A., Action of light on visual pigments, *Nature*, 183, 442, 1959.

11. Randall, C. E., Lewis, J. W., Hug, S. J., Björling, S. C., Eisner-Shanas, I., Friedman, N., Ottolenghi, M., Sheves, M., and Kliger, D. S., A new photolysis intermediate in artificial and native visual pigments, *J. Am. Chem. Soc.*, 113, 3473, 1991.

12. Cone, R. A., Rhodopsin, receptor potentials, and visual excitation, in *Progress in Photobiology: Proc. VI Int. Congress on Photobiology*, Deutsche Gesellschaft für Lichtforschung e.V., Frankfurt, 1974, 23.

13. Applebury, M. L., Dynamic processes of visual transduction, *Vision Res.*, 24, 1445, 1984.

14. Lewis, J. W., Winterle, J. S., Powers, M. A., Kliger, D. S., and Dratz, E. A., Kinetics of rhodopsin photolysis intermediates in retinal rod disk membranes. I. Temperature dependence of lumirhodopsin and metarhodopsin I kinetics, *Photochem. Photobiol.*, 34, 375, 1981.

15. Straume, M., Mitchell, D. C., Miller, J. L., and Litman, B. J., Interconversion of metarhodopsin I and II: A branched photointermediate decay model, *Biochemistry*, 29, 9135, 1990.

16. Matthews, R. G., Hubbard, R., Brown, P. K., and Wald, G., Tautomeric forms of metarhodopsin, *J. Gen. Physiol.*, 47, 215, 1963.

17. Baker, B. N. and Williams, T. P., Photolysis of metarhodopsin. I. Rate and extent of conversion to rhodopsin, *Vision Res.*, 11, 449, 1971.

18. Govindjee, R., Dancshazy, Z., Ebrey, T. G., Longstaff, C., and Rando, R. R., Photochemistry of methylated rhodopsins, *Photochem. Photobiol.*, 48, 493, 1988.

19. Doukas, A. G., Aton, B., Callender, R. H., and Ebrey, T. G., Resonance Raman studies of bovine metarhodopsin I and metarhodopsin II, *Biochemistry*, 17, 2430, 1978.

20. Ebrey, T. G., The thermal decay of the intermediates of rhodopsin *in situ*, *Vision Res.*, 8, 965, 1968.

21. Baumann, C., The equilibrium between metarhodopsin I and metarhodopsin II in the isolated frog retina, *J. Physiol.*, 279, 71, 1978.

22. Lamb, T. D. and Pugh, E. N., Jr., A quantitative account of the activation steps involved in phototransduction in amphibian photoreceptors, *J. Physiol.*, 449, 719, 1992.

23. Hofmann, K. P., Photoproducts of rhodopsin in the disc membrane, *Photobiochem. Photobiophys.*, 13, 309, 1986.

24. Longstaff, C., Calhoon, R. D., and Rando, R. R., Deprotonation of the Schiff base of rhodopsin is obligate in the activation of the G protein, *Proc. Natl. Acad. Sci. U.S.A.*, 83, 4209, 1986.

25. Franke, R. R., König, B., Sakmar, T. P., Khorana, H. G., and Hofmann, K. P., Rhodopsin mutants that bind but fail to activate transducin, *Science*, 250, 123, 1990.

26. Parkes, J. H. and Liebman, P. A., Temperature and pH dependence of the metarhodopsin I — Metarhodopsin II kinetics and equilibria in bovine rod disk membrane suspensions, *Biochemistry*, 23, 5054, 1984.

27. Blazynski, C. and Ostroy, S. E., Pathways in the hydrolysis of vertebrate rhodopsin, *Vision Res.*, 24, 459, 1984.

28. Ostroy, S. E., Rhodopsin and the visual process, *Biochim. Biophys. Acta*, 463, 91, 1977.

29. Hubbard, R., Bownds, D., and Yoshizawa, T., The chemistry of visual photoreception, *Cold Spring Harbor Symp. Quantitat. Biol.*, 30, 301, 1965.

30. Imamoto, Y., Kandori, H., Okano, T., Fukada, Y., Shichida, Y., and Yoshizawa, T., Effect of chloride ion on the thermal decay process of the batho intermediate of iodopsin at low temperature, *Biochemistry*, 28, 9412, 1989.

31. (a) Liang, J., Govindjee, R., and Ebrey, T. G., Metarhodopsin intermediates of the gecko cone pigment P521, *Biochemistry*, 32, 14187, 1993; (b) Shichida, Y., Okada, T., Kandori, H., Fukada, Y., and Yoshizawa, T., Nanosecond laser photolysis of iodopsin, a chicken red-sensitive cone visual pigment, *Biochemistry*, 32, 10832, 1993.

32. Weitz, C. J. and Nathans, J., Histidine residues regulate the transition of photoexcited rhodopsin to its active confirmation, metarhodopsin II, *Neuron*, 8, 465, 1992.

33. Kojima, D., Okano, T., Fukada, Y., Shichida, Y., Yoshizawa, T., and Ebrey, T. G., Cone visual pigments are present in gecko rod cells., *Proc. Natl. Acad. Sci. U.S.A.*, 89, 6841, 1992.

34. Trissl, H.-W., Photoelectric measurements of purple membranes, *Photochem. Photobiol.*, 51, 793, 1990.

35. Cone, R. A. and Pak, W. L., The early receptor potential, in *Handbook of Sensory Physiology*, Vol. I, Loewenstein, W. R., Ed., Springer-Verlag, New York, 1971, 345.

36. Cone, R. A., Early receptor potential of the vertebrate retina, *Nature*, 204, 736, 1964.

37. Cone, R. A., Early receptor potential: photoreversible charge displacement in rhodopsin, *Science*, 155, 1128, 1967.

38. Trissl, H.-W., On the rise time of the R_1-component of the "early receptor potential": Evidence for a fast light-induced charge separation in rhodopsin, *Biophys. Struct. Mech.*, 8, 213, 1982.

39. Cone, R. A., The early receptor potential, in *Proceedings of the International School of Physics "Enrico Fermi"-Course 43*, Reichardt, W., Ed., Academic Press, New York, 1969, 187.

40. Goldstein, E. B., Early receptor potential of the isolated frog *(Rana pipiens)* retina, *Vision Res.*, 7, 837, 1967.

41. Goldstein, E. R., Visual pigments and the early receptor potential of the isolated frog retina, *Vision Res.*, 8, 953, 1968.

42. Pak, W. L. and Ebrey, T. G., Early receptor potentials of rods and cones in rodents, *J. Gen. Physiol.*, 49, 1199, 1966.

43. Hagins, W. A. and McGaughy, R. E., Molecular and thermal origins of fast photoelectric effects in the squid retina, *Science*, 157, 813, 1967.

44. Hagins, W. A. and McGaughy, R. E., Membrane origin of the fast photovoltage of squid retina, *Science*, 159, 213, 1968.

45. Cooper, A., Dixon, S. F., and Tsuda, M., Photoenergetics of octopus rhodopsin: Convergent evolution of biological photon counters?, *Eur. Biophys. J.*, 13, 195, 1986.

46. Jackson, M. B., Activation of receptors coupled to G-proteins and protein kinases, in *Thermodynamics of Membrane Receptors and Channels*, Jackson, M. B., Ed., CRC Press, Boca Raton, FL, 1993, 295.

47. Kühn, H., Mommertz, O., and Hargrave, P. A., Light-dependent conformational change at rhodopsin cytoplasmic surface detected by increased susceptibility to proteolysis, *Biochim. Biophys. Acta*, 679, 95, 1982.

48. Steinberg, G., Ottolenghi, M., and Sheves, M., pK_a of the protonated Schiff base of bovine rhodopsin. A study with artificial pigments, *Biophys. J.*, 64, 1499, 1993.

49. Shichi, H., Circular dichroism of bovine rhodopsin, *Photochem. Photobiol.*, 13, 499, 1971.

50. Litman, B. J., Rhodopsin: Its molecular substructure and phospholipid interactions, *Photochem. Photobiol.*, 29, 671, 1979.

51. Binder, B. M., Biernbaum, M. S., and Bownds, M. D., Light activation of one rhodopsin molecule causes the phosphorylation of hundreds of others, *J. Biol. Chem.*, 265, 15333, 1990.

52. Yamamoto, K. and Shichi, H., Rhodopsin phosphorylation occurs at metarhodopsin II level, *Biophys. Struct. Mech.*, 9, 259, 1983.

53. Takao, M., Iwasa, T., Yamamoto, H., Takeuchi, T., and Tokunaga, F., Anti-bovine rhodopsin monoclonal antibody recognizes light-dependent structural change, in press, 1994.

54. de Grip, W. J., Gillespie, J., and Rothschild, K. J., Carboxyl group involvement in the meta I and meta II stages in rhodopsin bleaching. A Fourier transform infrared spectroscopic study, *Biochim. Biophys. Acta*, 809, 97, 1985.

55. Rothschild, K. J., Gillespie, J., and de Grip, W. J., Evidence for rhodopsin refolding during the decay of meta II, *Biophys. J.*, 51, 345, 1987.

56. Klinger, A. L. and Braiman, M. S., Structural comparison of metarhodopsin II, metarhodopsin III, and opsin based on kinetic analysis of Fourier transform infrared difference spectra, *Biophys. J.*, 63, 1244, 1992.

57. Nathans, J., Rhodopsin: Structure, function, and genetics, *Biochemistry*, 31, 4923, 1992.

58. Attwood, P. V. and Gutfreund, H., The application of pressure relaxation to the study of the equilibrium between metarhodopsin I and II from bovine retinas, *FEBS Lett.*, 119, 323, 1980.

59. Ganter, U. M., Schmid, E. D., Perez-Sala, D., Rando, R. R., and Siebert, F., Removal of the 9-methyl group of retinal inhibits signal transduction in the visual process. A Fourier transform infrared and biochemical investigation, *Biochemistry*, 28, 5954, 1989.

60. Sakmar, T. P., Franke, R. R., and Khorana, H. G., Glutamic acid-113 serves as the retinylidene Schiff base counterion in bovine rhodopsin, *Proc. Natl. Acad. Sci. U.S.A.*, 86, 8309, 1989.

61. Zhukovsky, E. A. and Oprian, D. D., Effect of carboxylic acid side chains on the absorption maximum of visual pigments, *Science*, 246, 928, 1989.

62. Nathans, J., Determinants of visual pigment absorbance: Role of charged amino acids in the putative transmembrane segments, *Biochemistry*, 29, 937, 1990.

28

Bacterorhodopsin and Rhodopsin

Richard Needleman
Wayne State University School of Medicine

28.1 Bacteriorhodopsin

Introduction

The cytoplasmic membrane of *Halobacterium halobium (salinarium)* contains several retinal proteins unique among prokaryotes but similar to the visual rhodopsin of higher organisms. One of these, bacteriorhodopsin, is an outwardly directed, light-activated proton pump that supports phototrophic growth in the oxygen-deficient conditions common to halobacteria.[1] *H. halobium,* a member of the *Archae,* lives in highly saline environments like the Great Salt Lake of Utah and the brine marshes adjacent to San Francisco Bay where the salinity (3.5 to 5 *M*) greatly exceeds that of sea water and where oxygen and organic nutrients are often in short supply. As might be expected, the synthesis of bacteriorhodopsin is induced by low oxygen and light, and this induction is transcriptionally regulated.[2] Two genes closely linked to the *bop* gene, *brp* and *bat,* are also required for bacteriorhodopsin synthesis, but their function is not known.[3]

Structure and Photoproperties

Bacteriorhodopsin is a small 248-amino acid protein (about 26 kDa) and a member of a larger class of proteins that includes the rhodopsins of algae, invertebrates, and vertebrates, as well as the mammalian G-protein receptors. In common with G-activated membrane receptors, bacteriorhodopsin consists of 7 *trans*-membrane α-helices arranged essentially perpendicular to the membrane surface (all less that 20° from the normal). Bacteriorhodopsin contains retinal, which is linked to lys-216 via a protonated Schiff base; upon photon absorption, its isomerization from *all-trans* to *13-cis* is used to drive proton transport.

Bacteriorhodopsin forms trimers that are arranged as a two-dimensional crystalline lattice in the membrane. This "purple membrane" consists of patches about 0.5-μm diameter, and these membranes (about 75% bacteriorhodopsin by weight) can be obtained easily after cell lysis with low salt and subsequent sucrose gradient centrifugation. The photochemical and structural stability of bacteriorhodopsin in the purple membrane is the basis of its use as a general phototonic material for industrial applications.[4]

0-8493-8634-9/95/$0.00+$.50
© 1995 by CRC Press, Inc.

Henderson and co-workers have been able to use electron cryomicroscopy to determine a high-resolution (3.5 Å parallel and 10 Å perpendicular to the membrane plane) three-dimensional structure of bacteriorhodopsin.[5] This structure has been of critical importance in developing detailed models of proton transport and, because of bacteriorhodopsin's structural similarity to rhodopsin, has also influenced work on vertebrate rhodopsins. The retinal ring is roughly in the middle of the membrane and tilted about 20° to the membrane plane.

Light Adaptation and the Opsin Shift

In the dark, bacteriorhodopsin (i.e., dark-adapted bacteriorhodopsin) is a mixture of 13-*cis*- and *all trans*-retinal, absorbing at 558 nm with only about 35% being in the *all-trans*-form.[6] Light converts bacteriorhodopsin to the "light-adapted" state, absorbing at 568 nm.

Free retinal absorbs at 380 nm, and its protonated Schiff base at 440 nm, yet the absorption maximum of bacteriorhodopsin is at 568 nm, considerably red-shifted from this value. This "opsin shift", the difference between the spectrum of protein-bound retinal and the spectrum of retinal as measured in organic solvents, is caused in part by the electrostatic interaction of charges in the retinal binding pocket of bacteriorhodopsin with retinal.[7] Of particular importance is the complex of charges that interacts to stabilize the Schiff base — the so called "counterion". This counterion is complex, consisting of several residues including asp-85, asp-212, and arg-82.[8,9]

The Photocycle

Absorption of a photon leads to a sequence of reactions in which spectrally distinct forms of retinal appear and decay over a total time of about 50 ms. The discrete states of this "photocycle" are designated by capital letters, i.e., J, K, L, M, N, and O, where subscripts are often used to denote their maximum absorption (e.g., K_{625}, L_{550}, M_{412}, N_{550}, O_{640}, BR_{568}).

While some aspects of the photocycle remain controversial, the following scheme is consistent currently with a broad range of experiments:

$$BR \rightarrow (J) \rightarrow K \leftrightarrow L \leftrightarrow M_1 \rightarrow M_2 \leftrightarrow N \leftrightarrow O \rightarrow BR$$

The $BR \rightarrow J \rightarrow K$ reaction takes place in less than 10 ps, the $K \rightarrow L \leftrightarrow M_1 \rightarrow M_2$ reaction in tens of microseconds, and the $M_2 \leftrightarrow N \leftrightarrow O \rightarrow BR$ reaction in tens of milliseconds. Retinal is *all-trans* in BR_{568} and 13-*cis* in M, but its precise conformation in K and L is not known, being either 13-*cis* or 13,14-*dicis*.[10] It returns to *all-trans* in the $N \leftrightarrow O$ transition.

Thermodynamics of the Photocycle

A more detailed understanding of the events that occur after photon absorption requires an explicit thermodynamic description. To determine thermodynamic state changes, the elementary rate constants are required and not simply the phenomenological time constants. Using a gated optical multichannel analyzer to analyze spectra obtained at various times after flash excitation, Varo and Lanyi were able to determine these constants, and to determine the changes in free energy, entropy, and enthalpy during the photocycle.[11,12] While most of the changes took place near equilibrium, significant changes in the free energy took place during the $M_1 \rightarrow M_2$ transition and during $O \rightarrow BR$.

Lanyi and co-workers proposed initially the existence of two distinct M states, M_1 and M_2 (i.e., substates with identical absorption maxima) in order to resolve a kinetic anomaly in L decay.[13] Subsequently, an early and late M were directly observed in monomeric detergent micelles of bacteriorhodopsin, where these substates are separated by about 4 nm.[14] Additional independent evidence for the existence of M_1 and M_2 has also been obtained.[15,16]

In the photocycle scheme presented above, there are only two irreversible reactions: $M_1 \rightarrow M_2$ and $O \rightarrow BR$. In the $M_1 \rightarrow M_2$ step, there is a decrease in ΔG of at least 13kJ mol^{-1} which is close to the amount required for proton transport. This step is also accompanied by a conversion of ΔH to ΔS.[11] The mechanistic interpretation developed by Lanyi and co-workers considers that this is the critical step for proton transport: it changes the access of the Schiff base from the extracellular side to the intracellular side and increases the pK_a of the Schiff base so that it can accept the proton from asp-96. Thus, in $M_1 \rightarrow M_2$, there is a transfer of energy to the ion gradient and an enthalpy-entropy conversion. It functions therefore as a "molecular switch" that couples the reaction to the proton gradient. A detailed discussion of the switch can be found in recent publications.[11,17,18]

Use of Mutants in Elucidating the Photocycle

The analysis of the mechanism of proton transport has been heavily dependent upon the use of site-directed mutagenesis for the construction of mutant bacteriorhodopsins defective in proton transport or possessing altered photocycles. In a major advance, Khorana and co-workers developed a system for the synthesis of mutant bacteriorhodopsins in *E. coli,* allowing the subsequent purification and reconstitution of the protein into phospholipid vesicles.[19,20] These mutants, and those generated by random mutagenesis *in vivo* by Oesterhelt and co-workers, allowed the identification of the critical roles of asp-85 and asp-96 in the photocycle.[21–25] Recently, using a new system for the transformation of *Halobacteria,*[26] it has been possible to synthesize mutant bacteriorhodopsins in *H. halobium.*[27,28] The major advantage of synthesis of bacteriorhodopsin in *H. halobium,* in addition to its increased yield and simplicity, is that the proteins obtained are much more likely to be folded in their native trimeric form; the properties of mutant proteins obtained in this manner often differ significantly from the same proteins synthesized in *E. coli.*[8,15]

Events of the Photocycle

Sequential shifts in the pK values of aspartate residues triggered by retinal isomerization drive proton transport. The aspartates form a "proton wire" conducting protons from the intracellular to the extracellular side; water also plays a role in this "wire".[29] The Henderson model shows clearly that bacteriorhodopsin has two channels: a narrow cytoplasmic hydrophobic channel above the Schiff base and a wider, hydrophilic channel communicating to the outside below it.[5] In the following, it will be helpful to picture a membrane with the Schiff base in the center, asp-96 directly above it, and asp-85 directly below the Schiff base and between it and the extracellular side. A proton is released from the Schiff base to asp-85, the Schiff base regains this proton from asp-96, and asp-96 is reprotonated from the cytoplasmic side. A proton is transported during this process. Some of the important steps in transport appear as large spectral changes: The $L \rightarrow M$ reaction involves the loss of the Schiff base proton to asp-85, producing the blue-shifted M species. The Schiff base regains the proton from asp-96 in the subsequent $M \rightarrow N$ step and asp-96 is reprotonated from the cytoplasm in $N \rightarrow O$; this last step coincides with the reisomerization of the retinal to *all-trans.* Finally, asp-85 is *deprotonated* to complete the cycle. The release of the proton to the outside is from a so-far unidentified group or groups and occurs at different times in the cycle depending upon pH.[17]

 This, and experimental work of other groups, has led to the following more detailed picture of the photocycle. About 500 fs after the absorption of a photon, bacteriorhodopsin forms the primary photoproduct J in which the retinal has already isomerized from *all-trans* to 13-*cis* or 13,14-*dicis.*[10,30,31] This is followed in a few picoseconds by K, an intermediate that can be trapped at 77K. The quantum yield is high, estimated variously from 0.25 to 0.79.[32,33] A 568-nm photon has an energy of about 210kJ mol^{-1}, but the actual energy level of J is not known. Since a protonmotive force of 180 mV is equivalent to a free energy of -17kJ mol^{-1}, sufficient energy is available for the transport of one or more protons; about 25kJ mol^{-1} has been estimated by Birge and co-workers as the minimum required to pump a proton under ambient conditions.[34] However, only a small amount of this free energy is conserved in K,[18] and the enthalpy of K has been given recently as 48 kJ mol^{-1}.[34] When entropic contributions to K are included, the amount of free energy available is insufficient to pump more than one proton.[33]

A few microseconds after K, the blue-shifted L state is formed in which the strained retinal chain configuration seen in K is relaxed.[31] In L, the energy of the photon has been transferred to the protein and, although asp-96 remains protonated, its O-H bond shows increased hydrogen bonding.[35] The L-to-M transition sees the deprotonation of the Schiff base and the transfer of its proton to asp-85. The pK_a of the Schiff base is 5 to 6 pH units higher than asp-85; this proton transfer therefore requires more than half of the excess free energy retained after photon absorption.[36] In the M↔N transition, the Schiff base is reprotonated from asp-96. Proton transport requires the Schiff base to be accessible to both sides during the photocycle, releasing a proton to the outside and accepting one from the intracellular side. The proton must travel about 12 Å from asp-96 to the Schiff base; this transfer involves the participation of water, which aids the stabilization of the transition state.[29] The pK_a of asp-96 is at least 10, and deprotonation of asp-96 also occurs in a mutant in which the Schiff base does not deprotonate; this suggests that the change in pK_a of asp-96 occurs independently of the protonation state of the Schiff base and therefore doesn't require interaction with the Schiff base as an acceptor.[14,35–37] In the N ↔ O transition, 13-*cis*-retinal becomes *all-trans*. The entropy changes show that the BR_{568} regeneration is now driven by the conformational recovery of the protein.[11]

Bacteriorhodopsin is the simplest and best understood ion pump. The use of retinal as a reporting group, the availability of a large number of applicable spectroscopic techniques (visible, UV, resonance Raman, and FTIR spectroscopy), the ability to produce large quantities of mutant proteins, the existence of a high-resolution structural model, and the possibility of obtaining thermodynamic parameters directly have led to a detailed understanding of this proton pump. It is likely that a complete description of at least the fundamental properties of this transport system will soon be available.

28.2 Rhodopsin

Structure

In dim light, human vision is mediated by rhodopsin, a 348-amino acid (approximately 40kDa) membrane protein and the major protein of rod cells.[38,39] The rod cell contains a stack of thousands of discs, and rhodopsin is present in the plasma membrane of these discs. The protein moiety of rhodopsin is called opsin; 11-*cis*-retinal is covalently bound to opsin in a Schiff base linkage (to lys-296) to form rhodopsin. Rhodopsin is a typical G protein-coupled receptor with the exception that in the initial state, the "ligand", 11-*cis*-retinal is bound to the protein. The detailed three-dimensional structure of the protein is not known, but like bacteriorhodopsin and other G-protein coupled receptors, rhodopsin has seven transmembrane alpha-helices and can be divided into three domains: a membrane domain, a cytoplasmic domain, and an intradiscal domain.

Visual Transduction

The absorption of a photon leads to the isomerization of 11-*cis*- to *all-trans*-retinal, and to a succession of photointermediates, culminating with the formation of meta-rhodopsin II. Amplification of the initial signal (the absorption of a single photon) occurs by the activation of hundreds of transducin molecules by meta-rhodopsin II; activated transducin in turn activates an inactive cGMP-phosophodiesterase that now can convert cGMP to 5′-GMP. As a result of the increased cGMP-phosophodiesterase activity, the cGMP concentration of the cell decreases. The cGMP-dependent cation channel is very sensitive to changes in cGMP concentration, and lowering the concentration hyperpolarizes the cell; this hyperpolarization then appears at the synaptic terminal of the rod cell. Rhodopsin is an exquisitely sensitive light detector, with a high quantum yield for the initial isomerization, and able to detect more than 15% of the incident photons.

The rhodopsin photocycle has the following intermediates:

Rhodopsin (500 nm) $\rightarrow$ Bathorhodopsin (543) nm) $\rightarrow$ Lumirhodopsin (497 nm) $\rightarrow$
Metarhodopsin I (480 nm) $\leftrightarrow$ Metarhodopsin II (380 nm) $\rightarrow$ opsin + *all-trans* retinal

During this photocycle, the Schiff base remains protonated until meta rhodopsin II is formed; as mentioned previously, meta-rhodopsin II is the activated form (R*) of rhodopsin. R* interacts with transducin to initiate the detection cascade.

Transducin comprises three subunits, α, β, and γ (i.e., T$\alpha\beta\gamma$). Metarhodopsin II interacts with T($\alpha\beta\gamma$)·GDP and causes GDP to be exchanged with GTP. The exchange dissociates T into T(α)·GTP + T($\beta\gamma$) and metarhodopsin II continues to dissociate more T($\alpha\beta\gamma$)·GDP molecules. T(α)·GTP activates cGMP-phosophodiesterase by removing the inhibitory γ subunits of PDE forming the activated PDE*. Concentrations of cGMP decrease due to the activation of cGMP-phosophodiesterase; to stop the activation, a timer is built in through the hydrolysis of GTP in T(α)·GTP to form the inactive T(α)·GDP. T(α)·GDP then recombines with T($\beta\gamma$) to form T($\alpha\beta\gamma$)·GDP, ready to interact again with activated rhodopsin. The association of T(α) with T($\beta\gamma$) is also regulated by a phosphoprotein, phosducin.[40]

Turn-off of the receptor occurs by the phosphorylation of rhodopsin by rhodopsin kinase at multiple sites, followed by the binding to these phosphorylated residues by the protein arrestin. Rhodopsin kinase has been cloned recently and its primary sequence shows that it is similar to the β-adrenergic receptor kinase.[41] The interaction of bovine rhodopsin with rhodopsin kinase has shown that phosphorylation is mostly on serine residues and primarily at the most C-terminal amino acids of rhodopsin.[42] Metarhodopsin II is hydrolyzed into opsin and *all-trans*-retinal. After diffusing from the binding site, *all-trans*-retinal is isomerized to 11-*cis*-retinal, which binds to opsin to regenerate rhodopsin. This release step has been been associated recently with the dissociation of arrestin from the complex.[43]

Mutant rhodopsins produced in transiently infected COS cells and in human embryonic kidney cells have been very important in determining structure-function relationships.[38,44] In contrast to bacteriorhodopsin, very little is known about the detailed structural rearrangements that occur in the protein after photon absorption; in particular, the mechanism by which transducin interacts with rhodopsin remains largely unknown. The cytoplasmic domain is, however, the site of the major interactions of rhodopsin with transducin and other enzymes, the intradiscal domain serving a structural role.[45–47] Mutational studies have also shown that two residues in rhodopsin, E134 and R135, are critical for transducin binding and that two cytoplasmic loops, CD and EF, as well as a third region encompassing amino acids 310–321 are also involved.[38,46]

Studies of these *in vitro*-generated mutants have recently been supplemented with studies of mutant rhodopsins in retinitis pigmentosa. Autosomal dominant retinitis pigmentosa was found to be linked to a region of chromosome three that contains the rhodopsin gene.[49] Isolation of the rhodopsin gene by the polymerase chain reaction, with subsequent sequencing, demonstrated that a small proportion of affected individuals had mutations in rhodopsin.[50,51] *In vitro* synthesis of 13 of these mutant rhodopsins in a human cell line found that in 10 the likely defect was in either stability and/or folding.[52]

Studies of bacteriorhodopsin are just beginning to give us detailed knowledge of the conformational changes that ensue after light absorption. Despite the technical advantages of the bacteriorhodopsin system, however, the detailed movements of the protein during the photocycle are still not understood. This knowledge is necessary for an understanding of the switch step in bacteriorhodopsin — the critical step in energy coupling. Similarly, it is important to understand how light absorption and retinal isomerization produces R*. Since rhodopsin is similar in structure (and presumably function) to other G-coupled receptors, it is likely that studies of rhodopsin will contribute to the general question of how protein conformational changes can be coupled to these receptors and to other proteins as well.

References

1. Stoeckenius, W. and Bogomolni, R. A., Bacteriorhodopsin and related pigments of halobacteria, *Annu. Rev. Biochem.,* 52, 587, 1982.
2. Shand, R. F. and Betlach, M., Expression of the *bop* gene cluster of *Halobacterium halobium* is induced by low oxygen tension and by light, *J. Bact.,* 173, 4692, 1991.
3. Betlach, M., Shand, R., and Leong, D., Regulation of the bacterio-opsin gene of a halophilic archaebacterium, *Can. J. Microbiol.,* 35, 134, 1989.
4. Oesterhelt, D., Brauchle, C., and Hampp, N., Bacteriorhodopsin — A biological material for information processing, *Q. Rev. Biophys.,* 24, 425, 1991.
5. Henderson, R., Baldwin, J. M., Ceska, T. A., Zemlin, F., Beckmann, E., and Downing, K. H., Model for the structure of bacteriorhodopsin based on high-resolution electron cryo-microscopy, *J. Mol. Biol.,* 213, 899, 1990.
6. Scherrer, P., Stoeckenius, W., Mathew, M. K., and Sperling, W., Isomerization in dark-adapted bacteriorhodopsin, in *Biophysical Studies of Retinal Proteins,* Ebrey, T. G., Frauenfelder, H., Honig, B., and Nakanishi, K., Eds., University of Illinois Press, Urbana-Champaign, 1987, 206.
7. Nakanishi, K., Balogh-Nair, V., Arnaboldi, M., Tsujimoto, K., and Honig, B., An external point-charge model for bacteriorhodopsin to account for its purple colour, *J. Am. Chem. Soc.,* 102, 7945, 1980.
8. Needleman, R., Chang, M., Ni, B., Varo, G., Fornes, J., White, S. H., and Lanyi, J. K., Properties of asp212 → asn bacteriorhodopsin suggest that asp212 and asp85 both participate in a counterion and proton acceptor complex near the Schiff base, *J. Biol. Chem.,* 266, 11478, 1991.
9. de Groot, H. J. M., Harbison, G. S., Herzfeld, J., and Griffin, R. G., Nuclear magnetic resonance study of the Schiff base in bacteriorhodopsin:counterion effects of the ^{15}N shift anisotropy, *Bio-chemistry,* 28, 3346, 1989.
10. Zhou, F., Windemuth, A., and Schulten, K., Molecular dynamics study of the proton pump cycle of bacteriorhodopsin, *Biochemistry,* 32, 2291, 1993.
11. Varo, G. and Lanyi, J. K., Thermodynamics and energy coupling in the bacteriorhodopsin photocycle, *Biochemistry,* 30, 5016, 1991.
12. Lanyi, J., Proton transfer and energy coupling in the bacteriorhodopsin photocycle, *J. Bioenerg. Biomem.,* 24, 169, 1992.
13. Varo, G. and Lanyi, J. K., Pathways of the rise and decay of the M intermediate of bacteriorhodopsin, *Biochemistry,* 29, 2241, 1990.
14. Varo, G. and Lanyi, J. K., Kinetic and spectroscopic evidence for an irreversible step between deprotonation and reprotonation of the Schiff base in the bacteriorhodopsin photocycle, *Biochem-istry,* 30, 5008, 1991.
15. Varo, G., Zimanyi, L., Chang, M., Ni, B., Needleman, R., and Lanyi, J. K., A residue substitution near the β-ionone ring of the retinal affects the M substates of bacteriorhodopsin, *Biophys. J.,* 61, 820, 1992.
16. Druckmann, S., Friedman, N., Lanyi, J. K., Needleman, R., Ottolenghi, M., and Sheves, M., The back photoreaction of the M intermediate in the photocycle of bacteriorhodopsin Mechanism and evidence for two M species, *Photochem. Photobiol.,* 56, 1041, 1992.
17. Zimanyi, L., Varo, G., Chang, M., Ni, B., Needleman, R., and Lanyi, J. K., Pathways of proton release in the bacteriorhodopsin photocycle, *Biochemistry,* 31, 8535, 1992.
18. Lanyi, J. K., Proton transfer and energy coupling in the bacteriorhodopsin photocycle, *J. Bioenerg. Biomem.,* 24, 169, 1992.
19. Karnik, S. S., Nassal, M., Doi, T., Jay, E., Sgaramella, V., and Khorana, H. G., Structure-function studies on bacteriorhodopsin. II. Improved expression of the bacterio-opsin gene in *Escherichia coli,* *J. Biol. Chem.,* 262, 9255, 1987.
20. Braiman, M. S., Stern, L. J., Chao, B. H., and Khorana, H. G., Structure-function studies on bacteriorhodopsin. IV. Purification and renaturation of bacterio-opsin polypeptide expressed in *Escherichia coli, J. Biol. Chem.,* 262, 9271, 1987.

21. Stern, L. J., Ahl, P. L., Marti, T., Mogi, T., Duanch, M., Berkowitz, S., Rothschild, K. J., and Khorana, H. G., Substitution of membrane-embedded aspartic acids in bacteriorhodopsin causes specific changes in different steps of the photochemical cycle, *Biochemistry*, 28, 10035, 1989.

22. Soppa, J. and Oesterhelt, D., Bacteriorhodopsin mutants of halobacterium spec. grb 1. The 5-bromo-2'-desoxyuridine-selection as a method to isolate point mutants in halobacteria, *J. Biol. Chem.*, 264, 13043, 1989.

23. Gerwert, K., Hess, B., Soppa, J., and Oesterhelt, D., Role of aspartate-96 in proton translocation by bacteriorhodopsin, *Proc. Natl. Acad. Sci. U.S.A.*, 86, 4943, 1989.

24. Butt, H. J., Kendler, K., Bamberg, E., Tittor, J., and Oesterhelt, D., Aspartic acids 96 and 85 play a central role in the function of bacteriorhodopsin as a proton pump, *EMBO J.*, 8, 1657, 1989.

25. Otto, H., Marti, T., Holz, M., Mogi, T., Stern, L. J., Engel, F., Khorana, G. H., and Heyn, M. P., Substitution of amino acids asp-85, asp-212, and arg-82 in bacteriorhodopsin affects the proton release phase of the pump and the pK of the Schiff base, *Proc. Natl. Acad. Sci. U.S.A.*, 87, 1018, 1990.

26. Cline, S. W. and Doolittle, W. F., Efficient transfection of the archaebacterium *Halobacterium haolobium*, *J. Bacteriol.*, 169, 1341, 1987.

27. Ni, B., Chang, M., Duschl, A., Lanyi, J. K., and Needleman, R., An efficient system for the synthesis of bacteriorhodopsin in *Halobacterium halobium*, *Gene*, 90, 167, 1990.

28. Krebs, M. P., Hauss, T., Heyn, M. P., Raj Bhandary, U. L., and Khorana, H. G., Expression of the bacterioopsin gene in *Halobacterium halobium* using a multicopy plasmid, *Proc. Natl. Acad. Sci. U.S.A.*, 88, 859, 1991.

29. Cao, Y., Varo, G., Chang, M., Ni, B., Needleman, R., and Lanyi, J. K., Water is required for proton transfer from aspartate-96 to the bacteriorhodopsin Schiff base, *Biochemistry*, 30, 10972, 1991.

30. Dobler, J., Zinth, W., Kaiser, W., and Oesterhelt, D., Excited-state reaction dynamics of bacteriorhodopsin studied by femtosecond spectroscopy, *Chem. Phys. Lett.*, 144, 215, 1988.

31. Ottolenghi, M. and Sheves, M., Synthetic retinals as probes for the binding site and photoreactions in rhodopsins, *J. Membr. Biol.*, 112, 193, 1989.

32. Mathies, R. A., Lin, S. W., Ames, J. B., and Pollard, W. T., From femtoseconds to biology: Mechanism of bacteriorhodopsin's light-driven proton pump, *Annu. Rev. Biophys. Biophys. Chem.*, 20, 491, 1991.

33. Birge, R. R., Cooper, T. M., Lawrence, A. F., Masthay, M. B., Vasilakis, C., Zhang, C. F., and Zidovetzki, R. A., Spectroscopic, photocalorimetric, and theoretical investigation of the quantum efficiency of the primary event in bacteriorhodopsin, *J. Am. Chem. Soc.*, 111, 4063, 1989.

34. Birge, R. R., Cooper, T. M., Lawrence, A. F., Masthay, M. B., Zhang, C., and Zidovetzki, R., Revised assignment of energy storage in the primary photochemical event in bacteriorhodopsin, *J. Am. Chem. Soc.*, 113, 4327, 1991.

35. Maeda, A., Sasaki, J., Shichida, Y., Yoshizawa, T., Chang, M., Ni, B., Needleman, R., and Lanyi, J. K., Structures of aspartic acid-96 in the L-Intermediate and N-intermediate of bacteriorhodopsin — Analysis by Fourier transform infrared spectroscopy, *Biochemistry*, 31, 4684, 1992.

36. Brown, L. S., Bonet, L., Needleman, R., and Lanyi, J. K., Acid dissociation constants of the Schiff base, asp-85, and arg-82 during the bacteriorhodopsin photocycle, *Biophys. J.*, 65, 124, 1993.

37. Cao, Y., Varo, G., Klinger, A. L., Czajkowsky, D. M., Braiman, M. S., Needleman, R., and Lanyi, J. K., Proton transfer from asp-96 to the bacteriorhodopsin Schiff Base is caused by decrease of the pKa of asp-96 which follows a protein backbone conformation change, *Biochemistry*, 32, 1981, 1993.

38. Khorana, H. G., Rhodopsin, photoreceptor of the rod cell. An emerging pattern for structure and function, *J. Biol. Chem.*, 267, 1, 1992.

39. Stryer, L., Cyclic GMP cascade of vision, *Annu. Rev. Neurosci.*, 9, 87, 1986.

40. Lee, R. H., Ting, T. D., Lieberman, B. S., Tobias, D. E., Lolley, R. N., and Ho, Y. K., Regulation of retinal cGMP cascade by phosducin in bovine rod photoreceptor cells, Interaction of phosducin and transducin, *J. Biol. Chem.*, 267, 25104, 1992.

41. Lorenz, W., Inglese, J., Palczewski, K., Onorator, J. J., Caron, M. G., and Lefkowitz, R. J., The receptor kinase family — Primary structure of rhodopsin kinase reveals similarities to the β-adrenergic-receptor kinase, *Proc. Natl. Acad. Sci. U.S.A.*, 88, 8715, 1991.

42. Brown, N. G., Fowles, C., Sharma, R., and Akhtar, M., Mechanistic studies on rhodopsin kinase — Light-dependent phosphorylation of C-terminal peptides of rhodopsin, *Eur. J. Biochem.*, 208, 659, 1992.

43. Hofmann, K. P., Pulvermuller, A., Buczylko, J., Vanhooser, P., and Palczewski, K., The role of arresin and retinoids in the regeneration pathway of rhodopsin, *J. Biol. Chem.*, 267, 15701, 1992.

44. Oprian, D. D., Molay, R. S., Kaufman, R. J., and Khorana, H. G., Expression of a synthetic bovine rhodopsin gene in monkey kidney cells, *Proc. Natl. Acad. Sci. U.S.A.*, 84, 8874, 1987.

45. Kuhn, H. and Hargrave, P. A., Light-induced binding of guanosinetriphosphatase to bovine rod photoreceptor cells: Effect of limited proteolysis of membranes, *Biochemistry*, 20, 2410, 1981.

46. Konig, K., Arendt, K. A., McDowell, J. M., Kalhlert, M., Hargrave, P. A., and Hoffmann, K. P., Three cytoplasmic loops interact with transducin, *Proc. Natl. Acad. Sci. U.S.A.*, 86, 6878, 1989.

47. Doi, T., Molday, R. S., and Khorana, H. G., Role of intradiscal domain in rhodopsin assembly and function, *Proc. Natl. Acad. Sci. U.S.A.*, 87, 4991, 1990.

48. Franke, R. R., Konig, B., Sakmar, T. P., Khorana, H. G., and Hofmann, K. P., Rhodopsin mutants that bind but fail to activate transducin, *Science*, 250, 123, 1990.

49. McWilliam, P., Farrar, G. J., Kenna, P., and Bradley, D. G., Autosomal dominant retinitis pigmentosa (ADRP): Localization of an ADRP gene to the long arm of chromosome 3, *Genomics*, 5, 619, 1989.

50. Dryja, T. P., McGee, T. L., Reichel, E., Hahn, L. B., Cowley, G. S., Yandell, D. W., Sandberg, M. A., and Benson, E. L., A point mutation of the rhodopsin gene in one form of retinitis pigmentosa, *Nature*, 343, 364, 1990.

51. Dryja, T. P., Hahn, L. B., Cowley, G. S., McGee, T. L., and Benson, E. L., Mutation spectrum of the rhodopsin gene among patients with autosomal dominant retinitis pigmentosa, *Proc. Natl. Acad. Sci. U.S.A.*, 88, 9370, 1991.

52. Sung, C. H., Schneider, B. G., Agarwal, N., Papermaster, D. S., and Nathans, J., Functional heterogeneity of mutant rhodopsins for autosomal dominant retinosa pigmentosa, *Proc. Natl. Acad. Sci. U.S.A.*, 88, 8840, 1991.

29

Rhodopsin and Bacteriorhodopsin: Use of Retinal Analogues As Probes of Structure and Function

Rosalie K. Crouch
Medical University of South Carolina

D. Wesley Corson
Medical University of South Carolina

29.1 Introduction

One of the most fascinating problems in biochemistry today is the question of how a ligand at the core of a receptor triggers the conformational changes that alter the binding of proteins at the periphery of the molecule and thereby initiate a physiological response. This question has become one of more than theoretical interest with the discovery that receptors having a variety of ligands and biological functions are members of a class of receptors that activate GTP-binding proteins.[1] The visual pigment *rhodopsin* is regarded as the prototype for these receptors.[2-7]

29.2 Rhodopsin

The protein primary structure is known for some 20 visual pigments all of which show seven groups of hydrophobic amino acids similar to those found in bovine rhodopsin (Figure 29.1). The cytoplasmic loops are known to function in the activation of the G-protein.[3] The intradiscal loops have been shown to be crucial for rhodopsin assembly and the loops connecting helices II-III and IV-V are highly conserved in rhodopsins, suggesting some specific role,[8] but their function has yet to be determined.[9]

The ligand of rhodopsin, 11-*cis*-retinal, is covalently bound to the protein by a protonated Schiff base linkage to lysine-296 that lies deep within the hydrophobic domain (Figure 29.1). The action of a photon of light is to isomerize the retinal from the 11-*cis*- to the *all-trans*-conformation. Following hydrolysis of the Schiff base linkage and separation of *all-trans*-retinal from opsin, regeneration of the pigment occurs by attachment of a fresh molecule of the chromophore, 11-*cis*-retinal.[10] This cycle provides a convenient means for probing the molecular mechanisms of

0-8493-8634-9/95/$0.00+$.50
© 1995 by CRC Press, Inc.

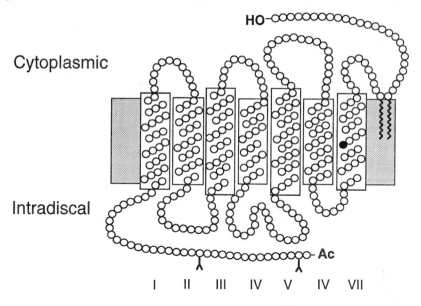

FIGURE 29.1 Primary structure of rhodopsin.

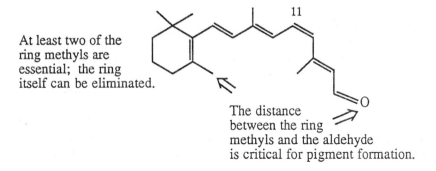

FIGURE 29.2 Sites of retinal interaction with rhodopsin.

transduction with analogue chromophores, a subject of investigation in Crouch and Corson's laboratory for some years.

At the pigment level, it has been established that there are two specific sites of interaction with the protein: (1) two of the ring methyls must be present although the ring itself is not essential;[11] and (2) the aldehyde functionality must be a specific distance from the ring methyls for Schiff base formation (Figure 29.2).[12] Furthermore, fragments of retinal as small as dimethylcyclohexane can fill the binding site and inhibit pigment formation competitively.[13]

Recently, the studies of these analogues have been extended to the living photoreceptor. Retinal has at least two functions in the living receptor. At the time of pigment formation, it deactivates a mechanism by which opsin reduces the sensitivity of receptor below the level expected from pigment depletion.[14,15] In the absence of 11-*cis*-retinal, desensitization by accumulated opsin persists indefinitely.[16–18] This component of desensitization is referred to as *opsin desensitization*. Taken together with the loss of sensitivity resulting from pigment depletion, the total loss of sensitivity is

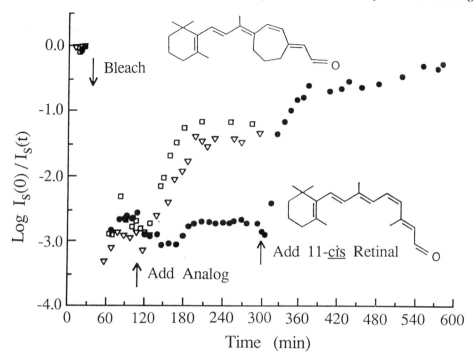

FIGURE 29.3   Sensitization of bleached salamander rods with 11-*cis*-locked retinal (open symbols) and 11-*cis*-retinal (filled symbols). (Adapted from Corson et al., 1990, Reference 18.)

commonly referred to as *bleaching adaptation*. Recent work by Crouch and Corson has demonstrated that this activity of retinal can be separated from that of photoexcitation by means of 11-*cis*-locked analogues of retinal that relieve opsin desensitization but produce an unbleachable and unresponsive pigment (Figure 29.3).[18] Recently, the exciting finding has been made that fragments of retinal as small as β-ionone can change cone opsin back into the dark-adapted conformation and thereby terminate opsin desensitization as if 11-*cis*-retinal were in place.[19]

On photoisomerization, the movement of the chromophore produces changes in the conformation on the cytoplasmic side of the protein that affect binding and activation of the G-protein.[3] The action of the chromophore as a "trigger" for the process does not depend on the formation of a Schiff base linkage to the opsin, as Zhukovsky et al.[20] have shown that retinyl amine can activate the transduction process in a mutant where the critical lysine (296) has been replaced with alanine, which provides space for the amine but no covalent attachment. Recent work has shown that the 9-methyl group of retinal is crucial for the production of a normal quantal response in rods. 9-Desmethylrhodopsin produces a quantal response that is smaller and more prolonged than the normal quantal response.[21] Interestingly, this pigment also exhibits a decreased level of phosphorylation *in vitro* when exposed to light in the presence of rhodopsin kinase.[22]

29.3 Bacteriorhodopsin

The protein bacteriorhodopsin is similar to rhodopsin in that it has the same seven *trans*-membrane helical structure and protonated Schiff base linkage of *all-trans*-retinal to the protein.[23,24] Bacteriorhodopsin is especially attractive for study as it is the only protein contained in the purple membrane and can be efficiently separated by sucrose density gradients.[25] Upon the absorption of a photon of light, a proton is moved from the cytoplasmic side of the protein to the extracellular side in a time scale of about 5 ms. The proton pumping process consists of at least five sequential steps, initiated by light absorbed by the retinal chromophore that isomerizes it from the *all-trans*-to the 13-*cis*-isomer.[26]

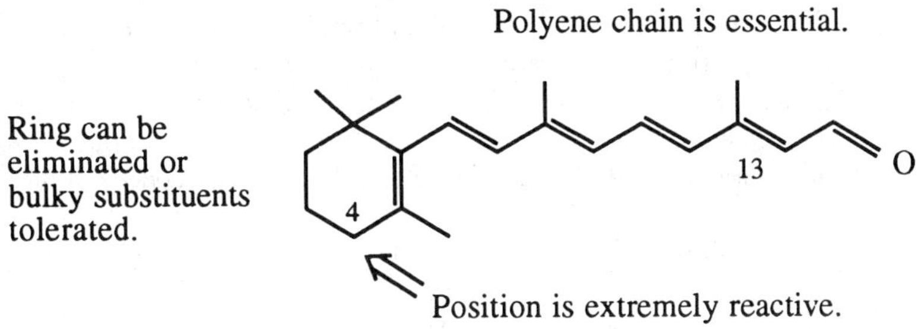

Polyene chain is essential.

Ring can be
eliminated or
bulky substituents
tolerated.

Position is extremely reactive.

FIGURE 29.4   Sites of retinal interaction with bacteriorhodopsin.

Conformational changes of the protein backbone occur during the photocycle of BR. This problem has been studied in many laboratories and has been approached mainly from three directions: (1) study of the native pigment by an array of biophysical techniques and by the addition of various reagents to the protein that serve to create or enhance a measurable property; (2) alteration of the protein by direct derivatization of the protein itself or by site-directed mutagenesis, changing one or more amino acids and characterization of the resulting protein; and (3) alteration of the native chromophore retinal by synthesis of a retinal analogue and then incorporation of that analogue into the apoprotein and subsequent study of these BR derivatives. It is this latter area in which this laboratory has concentrated for the past several years.[27,28]

The analogues that have been studied by Crouch et al. are all aldehydes, which, therefore, can form Schiff base linkages with the lysine-216. From studies of many retinal analogues,[27,28] the following observations are made on the constraints of the chromophore binding site: (1) the binding site of BR is considerably less restrictive than that for the visual pigment rhodopsin; (2) substantial bulk can be tolerated in the ring portion of the retinal; (3) the C-4 position is extremely reactive, the 4-haloretinals being very labile; (4) the cyclohexyl ring is not required for pigment formation or function and can be eliminated altogether; (5) the polyene chain is essential for pigment function although a pigment can be formed with as few as seven of the carbons being present; however, removal of the chain methyl groups or shortening the chain affects the stability and function of the pigments. (Figure 29.4)

29.4 Summary

The two photosensitive retinal binding proteins, rhodopsin and bacteriorhodopsin, contain many features in common including the same chromophore, the general protein structure and an activation by a photon that induces certain conformational changes resulting in their respective physiological responses. Substitution of analogues for the native chromophore has proven useful for probing the structure and function of these proteins.

References

1. Dohlman, H. G., Thorner, J., Caron, M. G., and Lefkowitz, R. J., Model systems for the study of seven-transmembrane-segment receptors, *Annu. Rev. Biochem.*, 60, 653, 1991.
2. Stryer, L., Visual excitation and recovery, *J. Biol. Chem.*, 266, 10711, 1991.
3. Hamm, H. E., Molecular interactions between the photoreceptor G protein and rhodopsin, *Cell Mol. Neurobiol.*, 11, 563, 1991.
4. Khorana, H. G., Rhodopsin, photoreceptor of the rod cell, *J. Biol. Chem.*, 267, 1, 1992.
5. Oprian, D. D., The ligand-binding domain of rhodopsin and other G protein-linked receptors, *J. Bioenerg. Biomembr.*, 24, 211, 1992.
6. Hargrave, P. and McDowell, J. H., Rhodopsin and phototransduction: A model system for G protein-linked receptors, *FASEB J.*, 6, 2323, 1992.

7. Collins, S., Caron, M. G., and Lefkowitz, R. J., From ligand binding to gene expression: New insights into the regulation of G-protein-coupled receptors, *Trends Biochem. Sci.*, 17, 37, 1992.

8. Doi, T., Molday, R. S., and Khorana, H. G., Role of the intradiscal domain in rhodopsin assembly, *Proc. Natl. Acad. Sci., U.S.A.*, 87, 4991, 1990.

9. Applebury, M. L. and Hargrave, P. A., Molecular biology of the visual pigments, *Vision Res.*, 26, 1881, 1986.

10. Rando, R. R., Membrane phospholipids and the dark side of vision, *J. Bioenerg. Biomembr.*, 23, 133, 1991.

11. Crouch, R. K. and Or, Y. S., Opsin pigments formed with acyclic retinal analogues, *FEBS Lett.*, 158, 139, 1983.

12. Matsumoto, H. and Yoshizawa, T., Recognition of opsin to the longitudinal length of retinal isomers in the formation of rhodopsin, *Vision Res.*, 18, 607, 1978.

13. Crouch, R. K., Veronee, C., and Lacy, M., Inhibition of rhodopsin regeneration by cyclohexyl derivatives, *Vision Res.*, 22, 1451, 1982.

14. Rushton, W. A. H., Rhodopsin measurement and dark adaptation in a subject deficient in cone vision, *J. Physiol. (London)*, 156, 193, 1961.

15. Dowling, J. E., Chemistry of visual adaptation in the rat, *Nature*, 188, 114, 1960.

16. Pepperberg, D. R., Brown, P. K., Lurie, M., and Dowling, J. E., Visual pigment and photoreceptor sensitivity in the isolated skate retina, *J. Gen. Physiol.*, 71, 369, 1978.

17. Catt, M., Ernst, W., and Kemp, C. M., The links between rhodopsin bleaching and visual adaptation, *Biochem. Soc. Trans.*, 10, 343, 1982.

18. Corson, D. W., Cornwall, M. C., MacNichol, E. F., Jin, J., Johnson, R., Derguini, F., Crouch, R. K., and Nakanishi, K., Sensitization of bleached rod photoreceptors by 11-*cis*-locked analogues of retinal, *Proc. Natl. Acad. Sci., U.S.A.*, 87, 6823, 1990.

19. Jin, J., Cornwall, M. C., Corson, D. W., Katz, B. M., and Crouch, R. K., Beta-ionone and 9-*cis* C17 aldehyde resensitize cone photoreceptors, *Invest. Ophthal. Vis. Sci.*, 33, 1102, 1992.

20. Zhukovsky, E. A., Robinson, P. R., and Oprian, D. D., Transducin activation by rhodopsin without a covalent bond to the 11-*cis* retinal chromophore, *Science*, 251, 558, 1991.

21. Corson, D. W., Derguini, F., Nakanishi, K., Crouch, R. K., MacNichol, E. F., and Cornwall, M. C., Relief of bleaching adaptation and induction of wavelength dependent response shapes by 9-desmethyl retinal in rods, *Biophys. J.*, 59, 408a, 1991.

22. Morrison, D. F., Ting, T. D., Ho, Y. K., Crouch, R. K., Corson, D. W., and Pepperberg, D. R., Reduced activity of 9-desmethyl rhodopsin in light-dependent phosphorylation, *Biophysical J.*, 64, 211a, 1993.

23. Oesterhelt, D., Meentzen, M., and Schuhmann, L., Reversible dissociation of the purple complex in bacteriorhodopsin and identification of 13-*cis* and all-*trans* retinal as its chromophores, *Eur. J. Biochem.*, 40, 453, 1973.

24. Henderson, R., Baldwin, J. M., Ceska, T. A., Zemlin, F., Beckmann, E., and Downing, K. H., Model for the structure of bacteriorhodopsin based on high-resolution electron cryomicroscopy, *J. Mol. Biol.*, 213, 899, 1990.

25. Oesterhelt, D. and Stoeckenius, W., Isolation of the cell membrane of *Halobacterium halobium* and its fractionation into red and purple membrane, in *Methods in Enzymology*, Vol. XXXI, Fleischer, S. and Packer, L., Eds., Academic Press, 1974, 667.

26. Lanyi, J. K., Proton transfer and energy coupling in the bacteriorhodopsin photocycle, *J. Bioenerg. Biomembr.*, 24, 169, 1992.

27. Crouch, R. K., Analogue pigments of rhodopsin and bacteriorhodopsin; retinal binding site and role of the chromophore in the function of these pigments, in *Chemistry and Biology of Synthetic Retinoids*, Dawson, M. and Okamura, W., Eds., CRC Press, Boca Raton, FL, 1990, 125–146.

28. Beischel, C. J., Mani, V., Govindjee, R., Ebrey, T. G., Knapp, D. R., and Crouch, R. K., Ring oxidized retinals form unusual bacteriorhodopsin analogue pigments, *Photochem. Photobiol.*, 54, 977, 1991.

30

Bacteriorhodopsin: New Biophysical Perspectives

Kenneth J. Rothschild
Boston University

Sanjay Sonar
Boston University

30.1 Introduction

Bacteriorhodopsin (bR), the light-driven proton pump from *Halobacterium salinarium* (formerly *H. halobium*), occupies an important place in current photobiology research. Since the discovery in the early 1970s by Stoeckenius and co-workers[1] of this retinal-containing protein, it has become one of the most intensively studied photoactive proteins and a focus for understanding the molecular mechanisms of active ion transport and energy transduction in biological systems. Much of this interest in bR stems from its unusual properties, including:

1. Vectorial proton translocation coupled to a photocycle consisting of a series of intermediates characterized by different visible absorption.
2. The ease of production of bacteriorhodopsin in large quantity; its thermal[2] and photochemical stability; and the ability to regenerate a structurally intact and functioning protein from its denatured form,[3] even when partially proteolyzed into several fragments.[4,5]
3. The ordering of bacteriorhodopsin into a two-dimensional hexagonal crystalline lattice that can be isolated in the form of purple membrane patches.

These unusual features are also stimulating the development of new biophysical approaches. Examples include resonance Raman spectroscopy[6] and solid-state NMR,[7] which probe the structure

0-8493-8634-9/95/$0.00+$.50

of the retinylidene chromophore, and FTIR difference spectroscopy, which probes structural changes at the level of individual amino acid residues.[8-10] New diffraction methods have also been developed to determine the three-dimensional structure of bacteriorhodopsin.[11,12] The ability to obtain site-directed mutants of bacteriorhodopsin from an *Escherichia coli*-based expression system[13,14] and more recently from the native *H. salinarium*[15-17] has also been essential for current progress, especially when combined with powerful biophysical approaches.

The discovery of a family of bacteriorhodopsin-like proteins including halorhodopsin (hR), a light-driven chloride pump and sensory rhodopsin I (sR-I), a light receptor in phototaxis[18-20] both from the *H. salinarium*, has also broadened interest in bR. Despite the divergent functions of these different retinal proteins, they all undergo a light-driven *all-trans*→13-*cis*-isomerization and conserve several key residues. While rhodopsin, the primary visual receptor, undergoes an 11-*cis* → *all-trans*-retinal isomerization and has only limited primary sequence homology with bR, the intriguing possibility exists that it may share common structural mechanisms with bR as well as hR and sR-I.

The existence of an "opsin shift" (a red shift in the absorption maximum of free retinal chromophore upon binding to the opsin) is another unique feature of interest and is the basis of spectral tuning in rhodopsin. In the case of bacteriorhodopsin, it has been attributed to different factors, including weak H-bonding interactions between the positively charged protonated retinylidene Schiff base and protein counterion(s),[21-25] the 6-S *trans*-conformation of retinal chromophore,[26,27] and the presence of dipolar charge near the β-ionone ring.[23] Recent interest in bacteriorhodopsin also stems from its potential applications in biotechnology. These include optical information processing and real-time interferometry.[28]

In this chapter, the focus is on recent work, with special emphasis on FTIR-based studies aimed at probing the mechanism of proton pumping in bR at the level of individual amino acid residues. There is no attempt, however, to present a comprehensive review of the bacteriorhodopsin literature. Instead, the reader is referred to several recent reviews and articles on bacteriorhodopsin that focus on other aspects of the research, including photochemistry,[29,30] genetic engineering and biochemistry,[13,31] regeneration with retinal analogues and chemical modifications,[32-35] vibrational spectroscopy,[6,9,10,36,37] NMR spectroscopy,[7] and diffraction studies.[11,12,38,39]

30.2 Structure of Bacteriorhodopsin

Primary and Secondary Structure

In 1975, a low-resolution bacteriorhodopsin structure was obtained from electron diffraction[40] on purple membrane patches that revealed the existence of seven closely packed tubes arranged perpendicular to the membrane plane.* The existence of a bundle of transversely oriented α-helices was also detected by IR dichroism.[41] Along with the primary sequence of bacteriorhodopsin[42,43] and neutron diffraction,[44] these data support the existence of seven distinct hydrophobic domains that fold into the membrane as α-helical segments. The transmembrane domain and loop regions have also been defined by immunological[45,46] and proteolytic studies.[47]

Location and Orientation of the Retinal Chromophore

A variety of methods have helped establish the location and orientation of the retinylidene chromophore of bR (for a discussion of the retinal structure, see Section 30.3). After initial disagreement regarding the site of attachment for retinylidene Schiff base linkage,[48] it was conclusively established that it is Lys-216 on the G-helix[49-51] and that the attachment site does not change during the photocycle.[52]

---

* Due to the fact that several of the α-helices are tilted, they do not appear as distinct tubes in the electron density projection.

Because purple membrane can be oriented using a variety of methods including electric, magnetic fields, and isopotential spin drying,[53] the orientation of the retinal can be probed. For example, visible linear dichroism measurements on oriented purple membrane showed that the polyene chain of retinal is tilted at approximately 20° relative to the membrane plane.[53,54] More recently, it was shown by using a combination of photoselection and time-resolved linear dichroism that the chromophore transition dipole tilts out of the membrane plane by 3° in the M intermediate.[55] FTIR dichroism measurements of the retinal hydrogen-out-of-plane modes show that the orientation of the polyene plane is approximately perpendicular to the membrane plane,[56] in agreement with similar studies also using photoselection.[57] By comparing the linear dichroism of bacteriorhodopsin containing a normal and analogue retinal, it was also found that the methyl groups on the polyene chain point toward the cytoplasmic surface of the membrane and that the Schiff base proton points down toward the extracellular surface.[58] Neutron diffraction studies have localized the position of the Schiff base,[59] and recent solid-state two-dimensional NMR has determined that the polyene chain is not straight but has in-plane curvature and possibly an out-of-plane twist.[60] The orientation of the retinal and its direction in the membrane have also been established by elegant photochemical cross-linking experiments using a retinal analog with a diazirine group functionalized to the cyclohexane ring.[61]

A Spectroscopic Model of the bR Active Site

On the basis of results obtained from site-directed mutagenesis,[62,63] FTIR difference spectroscopy on bR mutants, and a variety of other biophysical data,[64–70] a detailed three-dimensional structural model of the bR binding pocket referred to here as the "spectroscopic model" was proposed in 1988.[71] As shown in Figure 30.1A, a key prediction of the model is that the protonated Schiff base interacts with a complex counterion consisting of Asp-85, Asp-212, and Arg-82, thereby maintaining charge neutrality in the active site. The positioning of Asp-85 also requires that Asp-96 be located close to the cytoplasmic surface of bR. Figure 30.1B shows the position of Tyr-185, Trp-182, Trp-189, and Pro-186 which, along with Trp-86 on helix-C,[66,72] are predicted to form part of a "box" that acts to restrict the isomerization of retinal during the photocycle to *all-trans* to *13-cis*.*

An additional factor in the formulation of the spectroscopic model[66] is the homology of residues in the F-helix of bR and the rhodopsin family of proteins. For example, all of these retinal proteins have a Tyr-Pro pair (Tyr-185/Pro-186 in bR) and a tryptophan residue (Trp-182) located one turn higher in the F-helix, except for human rhodopsin blue pigment. Thus, the spectroscopic model predicts that similarities exist between the bR and rhodopsin retinal binding pocket despite the different retinal configuration in rhodopsin.

Electron Diffraction Model

A major step in this field was accomplished in 1990 when Henderson and co-workers succeeded in obtaining a high-resolution electron diffraction structure of bR in 1990.[11] Their work confirms many of the details predicted earlier by the spectroscopic model, including the relative position of the two counterions, Asp-85 and Asp-212, near the Schiff base and Asp-96 near the cytoplasm surface, as well as the relative positioning of the three tryptophan residues surrounding the retinal. While the exact position of the Arg-82 side chain was not specified, its position in the C helix allows it to interact with Asp-212 and Asp-85. Tyr-185 is also located close to the Schiff base and positioned to form a hydrogen bond with Asp-212, consistent with FTIR evidence that Asp-212 and Tyr-185 interact.[73]

---

* In contrast to Figure 30.1A, the C9 and C13-methyl groups of retinal were pointed toward the cytoplasmic side of the membrane and N–H bond toward the extracellular medium, in agreement with more recent evidence.[58]

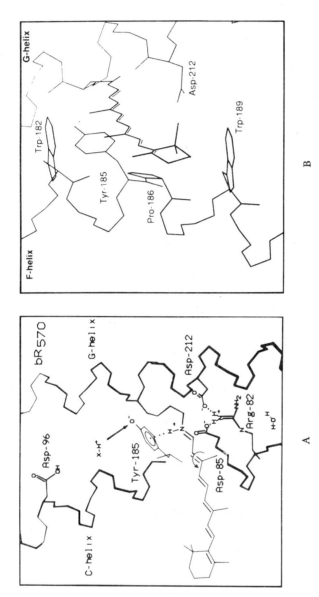

FIGURE 30.1   (A) The structure of the light-adapted state of bR (bR_{570}) proposed on the basis of FTIR difference spectroscopy and site-directed mutagenesis.[71] The Schiff base interacts with a complex counterion consisting of Asp-85, Asp-212, and Arg-82. Tyr-185 was proposed to also be partially ionized[70] and undergo a net protonation on formation of the K intermediate (see text). (B) Three-dimensional structure of the retinal binding pocket containing residues Trp-182, Trp-189, Tyr-185, and Pro-186 that form part of a "box" that acts to constrain the possible conformations of retinal.[66] Trp-86 on helix C (not shown) was also hypothesized to participate in the binding pocket.[72]

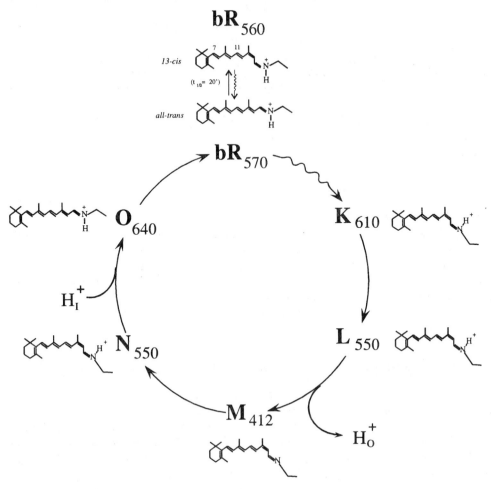

FIGURE 30.2   Different intermediates in the photocycle of bacteriorhodopsin and dark-adapted bR along with the structure of the retinylidene chromophore. The subscripts indicate the approximate λ_{max} of each intermediate. H^+_I and H^+_O denote protons taken up from the cytoplasmic medium and ejected into the extracellular medium, respectively.

The electron diffraction model also goes considerably beyond earlier models by providing detailed information about the position of other residues, including those that might participate in proton translocation pathways. For example, there is a preponderance of hydrophobic residues in the putative cytoplasmic channel that leads from Asp-96 to the Schiff base. Thus, it is unlikely that a bulk water-filled channel exists to conduct a proton during Schiff base reprotonation. On the other hand, a large number of hydrophilic residues appear to exist in the region leading from Asp-85 to the external medium.

30.3 The Bacteriorhodopsin Photocycle

Central to the characterization of proton translocation by bacteriorhodopsin is its unique photocycle. Upon photon absorption, light-adapted bR (bR_{570})rapidly converts into the first metastable intermediate, K_{630}. This is followed by thermal relaxation in the dark of the chromophore back to its initial state through various structural intermediates characterized by different absorption maxima. As shown in Figure 30.2, along with these spectral changes, a proton is released into the external medium during M formation and a second proton is most likely taken up from the inner medium during O formation.

Numerous visible absorption studies at low and room temperature have played a key role in the characterization of the main intermediates in the bR photocycle.[30,74–79] Recently, femtosecond laser spectroscopy has provided unprecedented time resolution of the earliest events.[80,81] While the existence of the intermediates shown in Figure 30.2 are well established, evidence has also been presented for the possibility of additional intermediates and branching.[30,82]

In addition to visible absorption studies, a number of techniques, including resonance Raman spectroscopy, solid-state NMR, FTIR difference spectroscopy, and recently FT-Raman spectroscopy, have contributed to our knowledge of the conformational changes of the retinylidene chromophore during the photocycle as described in the following sections.

Light and Dark Adaptation

Dark-adapted bR consists of two species (bR_{555} and bR_{570}). Illumination produces the functionally active light-adapted bR consisting only of bR_{570}. Extraction of the chromophore shows that the light-adapted state contains an *all-trans*-retinal configuration, whereas dark-adapted bR has a 13-*cis:all-trans* mixture (2:1).[83] Resonance Raman spectroscopy and solid-state NMR, along with the elegant use of isotope labels placed at specific positions on the retinal, have established unambiguously the structure of the chromophore in these two states.[7] bR_{555} contains a 13-*cis*, C=N *syn*-chromophore, while bR_{570} has an *all-trans*, C=N *anti*-chromophore.[26,84–87] These two states also exhibit distinctly different photocycles.[30]

The Primary Photochemical Event: bR→K Transition

The primary photoreaction in the bacteriorhodopsin photocycle is the isomerization of the chromophore from an *all-trans* to 13-*cis*-conformation.[88–92] However, the appearance of intensified bands in both the resonance Raman and FTIR difference spectrum (see Reference 93 and references therein) assigned to the hydrogen-out-of-plane wag mode of hydrogens on the polyene chain indicates that the polyene chain may not be planar due to twisting about C–C bonds. This might occur for example if the protein-chromophore interactions impose constraints that prevent retinal from relaxing fully into a 13-*cis*-configuration. Indeed, several protein bands assigned to Tyr-185, Asp-115, Trp-86, Pro-186, and amide I and II bands appear in the bR→K FTIR difference spectrum, indicating that the protein responds to the initial photoisomerization of the chromophore.[10]

The initial isomerization also weakens the interaction of the Schiff base with its counterion(s) during the bR→K transition. In particular, the frequency of the Schiff base C=N bond can be used as a probe of the environment and protonation state of the Schiff base during the bR photocycle. In the case of the K intermediate, this mode is assigned near 1609 cm^{-1},[22,94] a shift of approximately 30 cm^{-1} below that of the C=N stretch frequency in bR_{570}. In general, several studies show that both increased delocalization of electrons throughout the polyene chain and coupling with the NH in-plane bending mode can have a large effect on the C=N vibrational frequency.[25,95,96]

The L Intermediate

The transition from the K to L intermediate involves a large blue shift in the chromophore from 630 to 550 nm. The resonance Raman spectrum demonstrates that the chromophore is still in a 13-*cis*-configuration with a protonated Schiff base.[97–98] However, because of the absence of intensified HOOP modes, it appears to have relaxed from the twisted form present in the K intermediate.[99] A rotation of the chromophore about the C14–C15 bond has been predicted on the basis of FTIR measurements and on theoretical grounds to explain the drop in the Schiff base pK$_a$.[100,101] However, resonance Raman studies on the L_{550} intermediate combined with isotope labeling indicate that its chromophore has a C14–C15 S-*trans*-structure.[98]

The M Intermediate

Formation of the M intermediate is accompanied by a large shift in the λ_{max} from 550 to 412 nm. Early work by Lewis and co-workers using resonance Raman spectroscopy demonstrated that this color shift is due to deprotonation of the Schiff base.[102] As discussed below, there is strong evidence from FTIR difference spectroscopy[71,103] as well as a number of other biophysical studies (see Section 30.4, Supporting Evidence) that this proton is transferred to the nearby counterion, Asp-85, thereby triggering a release of a proton to the external environment. The chromophore remains in a 13-*cis*-configuration[102,104] and has a 14-S-*trans*-C=N *anti*-configuration about the Schiff base.[87,105] The possibility still exists, however, that more than one form of the M intermediate exists,[82,106] although resonance Raman studies have not yet revealed multiple chromophore structures.[105,107]

The N Intermediate

M decay to the N intermediate is accompanied by a shift of the λ_{max} back to near 550 nm, similar to the λ_{max} of the L intermediate. This intermediate was proposed to exist in early studies of the bR photocycle[108,109] (and references therein), and recently confirmed by several groups.[74,110–112] Resonance Raman spectroscopy shows that it contains a 13-*cis*-chromophore similar to the M intermediates but with a protonated Schiff base.[113] Thus, reprotonation of the bR Schiff base occurs during this step in the photocycle, with a decay time of several milliseconds. Several studies have shown that this proton originates from Asp-96, which deprotonates during this transition.[114–117] A conformational change of the protein also appears to occur between the M and N intermediate.[116–119]

The O Intermediate

The O intermediate has a red-shifted λ_{max} near 640 nm and intensified HOOP modes[112,120] similar to the K intermediate. However, in this case, the chromophore exists in an *all-trans*-configuration. Thus, reisomerization of the chromophore occurs between the N and O intermediates. Recent FTIR studies on the bacteriorhodopsin mutant Y185F, which displays a slow O decay at high pH, indicate that this red-shift may be due to neutralization of the Asp-85 counterion.[121] These observations are further supported by model independent kinetic analyses of FTIR difference spectra.[122]

Evidence for an Equilibrium between bR$_{570}$ and O

Recently, a new method, Fourier transform Raman (FT-Raman) spectroscopy, has been introduced for studying bacteriorhodopsin (see Rath et al.[123] and references therein). This technique uses exciting light in the near-IR region; thus, photoreactions of light-adapted and dark-adapted bR are avoided. In a recent application, the light-dark adaptation of bacteriorhodopsin and the mutant Tyr-185→Phe (Y185F) was investigated at room temperature in solution. Interestingly, in comparison to wild-type bR, both the FT-Raman and resonance Raman spectrum of the light-adapted Y185F displayed new features characteristic of the vibrations of the O chromophore. The presence of the O intermediate along with the normal bR$_{570}$ species in the light-adapted Y185F was also consistent with the results from visible absorption spectroscopy[124,125] and FTIR difference spectroscopy.[126] Further evidence for the existence of an O-like species in Y185F comes from pump-probe Raman difference spectroscopy, where a red pump beam is found to produce a species very similar to the N intermediate in the photocycle.[123] These results demonstrate that an equilibrium exists between bR$_{570}$ and the last intermediate in the bR photocycle, as demanded by simple thermodynamic considerations.[125] In the case of Y185F, this equilibrium is shifted more toward the O intermediate than in normal bR. In support of this picture, recent studies show that high temperature can also produce appreciable levels of the O intermediate in wild-type.[127] A similar phenomena may also occur in the mutant D85N.[128]

30.4 Toward a Proton Pump Mechanism

Detection and Assignment of Individual Amino Acid Residues by FTIR Difference Spectroscopy

It was demonstrated in 1981 that Fourier transform IR difference spectroscopy can detect small conformational changes in both the chromophore and protein components of bR during its photocycle.[8] This technique, along with further developments using polarized light and attenuated total reflection, provides a means to determine changes in the environment and protonation state of key residues in bR as well as in other membrane proteins (for a recent review, see Reference 10). Early studies using this approach focused on the K, KL, L, and M intermediates that are stable at 77, 135, 170, and 250 K, respectively,[91,129,130] and complemented single-wavelength kinetic IR measurements.[131] More recently, the development of time-resolved FTIR techniques[99,116,118,132–134] has permitted measurements of FTIR difference spectra of bR at room temperature and facilitated the study of the N[116,135] and O intermediates.[121] FTIR studies of the N intermediate have also been accomplished at lower temperatures using high pH conditions.[117] The KL has also been examined using time-resolved FTIR[134] and time-resolved IR spectroscopy.[99]

Along with the introduction of FTIR difference spectroscopy to study membrane proteins, progress has been made in assigning bands through the use of amino acid isotope labeling[136–140] and site-directed mutagenesis and site-directed isotope labeling (see Section 30.5, Recent Developments).[67,70,73,103,116,141–143] This has led to an increasingly detailed assignment of bR bands to individual amino acid residues as illustrated in Figure 30.3. Importantly, this type of data, especially in the case of time-resolved FTIR measurements, provides a dynamic map of changes occurring in the protein and chromophore throughout the bR photocycle.

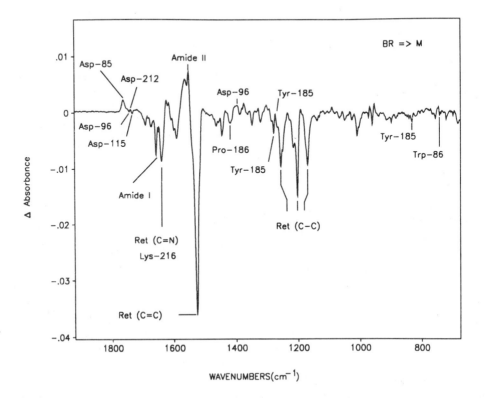

FIGURE 30.3   bR→M FTIR difference spectrum recorded at 250 K, showing assignments of vibrations from the retinylidene chromophore (RET) along with specific amino acids. Assignments are based on isotope labeling and site-directed mutagenesis. (Adapted from Reference 10.)

In the sections below, we describe key elements of a proton transport model proposed on the basis of (1) the available FTIR data, (2) the electron diffraction-derived coordinates of bR,[11] (3) information about the orientation of the retinal and its changes during the photocycle as discussed in the previous section, and (4) a variety of other data, including the results from extensive structure-function studies of bR using site-directed mutagenesis.[13]

Key Molecular Events

The arrangement of several key residues investigated by FTIR difference spectroscopy is shown in Figure 30.4. This model retains most of the key elements of the spectroscopic model of bR[71] (Section 30.2, A Spectroscopic Model of the bR Active Site) and adds several new features based on more recent studies.[144]

1. In the bR_{570} state, Asp-96 is protonated and Asp-212 and Asp-85 are ionized.[71]
2. Upon formation of the K intermediate, there is a change in the environment of Tyr-185 in response to movement of the protonated Schiff base. This involves a *net* increase in the protonation state of Tyr-185,[70,139,145] possibly involving a hydrogen-bonded network extending from Thr-89 to Asp-212.[144]
3. Upon formation of the L intermediate, the environment of the carboxyl group of Asp-96 is disturbed along with its interaction with Thr-46, possibly causing a transient change in its net protonation state.[144]
4. Upon formation of the M intermediate, a proton is transferred from the Schiff base to Asp-85.[71,144,146] As a result of this protonation, the complex counterion involving Asp-85, Arg-82, and Asp-212 is disrupted. The release of Arg-82 from the active site causes a net ejection of a proton into the outer medium.[147] This occurs because of the association of hydroxide ion(s) with the positively charged guanidinium group and not due to a deprotonation of Arg-82. In addition, a transient network of polarized hydrogen bonds is formed that extends from Asp-96 to Asp-212 and includes several water molecules along with Thr-46, Thr-89, and Tyr-185.[144] It is likely that other hydrophilic residues are also involved in this network, which serves in the next step of the photocycle to transfer a proton from Asp-96 to the Schiff base. As a consequence of this network, Asp-212 becomes *partially* protonated and Tyr-185 *partially* deprotonated.
5. Upon formation of the N intermediate, a proton is transferred through the hydrogen-bonded network from Asp-96 to the Schiff base.[114,116,122,146,148,149] Asp-85 remains protonated and Asp-212 partially protonated. A change in the secondary structure of the protein also occurs at this stage of the photocycle.
6. Upon formation of the O intermediate, Asp-96 accepts a proton from the cytoplasmic medium, while the chromophore reisomerizes to an *all-trans*-configuration.[120-122] A reversal also occurs in the protein secondary structure.
7. Upon O decay, Asp-85 deprotonates allowing a re-establishment of the complex counterion environment around the Schiff base.

Supporting Evidence

In support of Asp-85 serving as acceptor of the Schiff base proton, replacement of Asp-85 with the non-proton-accepting residues like Ala and Asn abolishes proton translocation.[62,150] In addition, the photocycles of D85A and D85N exhibit almost no detectable M formation.[63,147] In contrast, D85E exhibits both proton translocation activity[62] and formation of an M intermediate consistent with the ability of Glu to act as a substitute proton acceptor.[63,146,147] Recently, Asp-85 was replaced by a cysteine and modified with either iodoacetic or iodoacetamide to form a non-native residue that also contained a carboxylate at position 85.[151] Again, these modified mutants exhibited proton

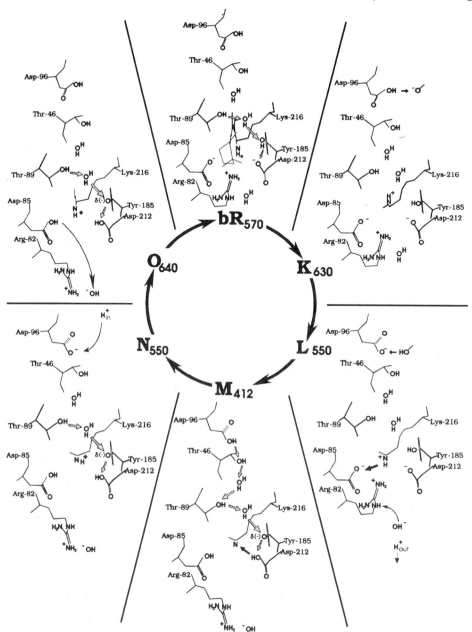

FIGURE 30.4 Model of hydrogen-bonded proton transport network and protonation changes that occur during the bR photocycle. The approximate position of the residues is based on the coordinate map of the electron diffraction-derived bR structure.[11] (See References 10 and 144 for additional details.)

pumping. In contrast, if Asp-85 is replaced by a His residue,[152] no proton pumping is observed over a wide pH range, indicating that a negatively charged residue is required at position 85.

There also exists strong evidence that Asp-85 is located close to the Schiff base and functions as a negative counterion. For example, substitution of Asp-85 with the neutral residues Asn and Ala results in a red-shifted λ_{max} near 600 nm,[62,147,150,153] as predicted on the basis of a point charge model.[154] In addition, both D85A and D85N exhibit Schiff base pK_a values near 7 compared to approximately 11 in wild-type, again an indication that Asp-85 acts as a primary counterion.[147,155] In a recent study based on resonance Raman spectroscopy of both D85N and D85A, it was found that the hydrogen bonding interaction with the Schiff base can be replaced with an equal-strength interaction by Asn-85 but not by Ala-85, consistent with the direct interaction of Asp-85 with the Schiff base.[156]

In support of the role of Arg-82 in proton release, transient pH measurements show that this release is strongly retarded in the mutants R82Q and R82A.[147] Because Arg-82 is replaced by neutral residues in these mutants, proton release might occur via a direct deprotonation of Asp-85 during the late stages of the photocycle. The M intermediate in R82Q and R82A has also been observed to have an accelerated rise kinetics[147,153] and it has been attributed to the Asp-85 counterion having a more negative character and increased affinity for the Schiff base proton.[153]

A Transient Hydrogen-Bonded Proton Wire in bR

On both theoretical and experimental grounds, it has been postulated that proton wires formed from a network of H-bonded residues and water molecules can serve as key functional elements in proteins.[157-160] One characteristic of such a network of polarized hydrogen bonds is the delocalization of protons within the network.[158] For example, even in the case of a hydrogen bond interaction between a high pK_a tyrosine residue and low pK_a aspartate residue, the tyrosine can donate a proton to aspartate and thus become partially deprotonated. For a similar reason, the high pK_a residues in the postulated hydrogen bonded network in bR, which includes Tyr-185, Thr-89, and Thr-46, can donate a proton to Asp-212, becoming *partially* deprotonated in the process.

In support of this model, recent studies have revealed partial protonation of several key residues in bR. For example, a band appearing at 1277 cm^{-1} in both the FTIR light-dark difference spectrum and photocycle difference spectra was assigned to the ionized form of Tyr-185.[70,145,161] However, a recent determination of the intensity of this band in the absolute IR absorption of light-adapted bR reveals that it reflects only a small fraction of ionized tyrosine.[162] This low level, estimated at less than 10% at room temperature (He and Rothschild, unpublished results), may explain why ionized tyrosine has not yet been detected by recent UV resonance Raman[163,164] and solid-state NMR[165] studies.* Thus, while FTIR difference spectroscopy detects *net* changes in the protonation state of Tyr-185, it may bever become fully ionized.

A second example is Asp-212, which may be partially protonated in the M, N, and O intermediates.[121] This is consistent with Asp-212 functioning as part of the relay system for proton transport between Asp-96 and the Schiff base as proposed previously.[70] Partial protonation of Asp-212 might also explain why recent solid-state NMR studies[166,167] detected Asp-85 protonation during formation of the M intermediate but not a similar protonation of Asp-212 (the N and O species were not examined in this study).

Additional evidence for the existence of a transient, H-bonded proton wire in bR comes from FTIR study of the mutant T89D. In this case, an increased protonation of Asp-212 during M formation is indicated by an increased intensity in a band assigned to Asp-212 at 1738 cm^{-1}. Along with this band is the appearance of a new negative band that corresponds to the deprotonation of Asp-89.[144] As shown in the right panel of Figure 30.5, such a transfer of a proton between Asp-89 and Asp-212 is expected if a proton pathway exists between Thr-89 and Asp-212 in wild-type bR.

An attractive feature of an H-bonded proton wire is that light-driven proton transport does not require significant conformational changes to occur in the protein. Instead, the initial movement of the Schiff base due to light-induced isomerization of retinal produces a charge separation that indirectly drives proton transport. Once the K intermediate is formed, its decay back to bR_{570} is prevented because of the high activation barrier for reisomerization of the chromophore to an *all-trans*-configuration. Instead, the system moves to a lower energy state by transferring a proton from the Schiff base to Asp-85 during M formation. Charge flow occurs through two distinct domains, an extracellular pathway that allows the positive charge of the Schiff base that is no longer shielded by its complex counterion to *push* a proton into the external medium and a second intracellular

---

* However, a tyrosinate was detected in a different UV resonance Raman study.[190]

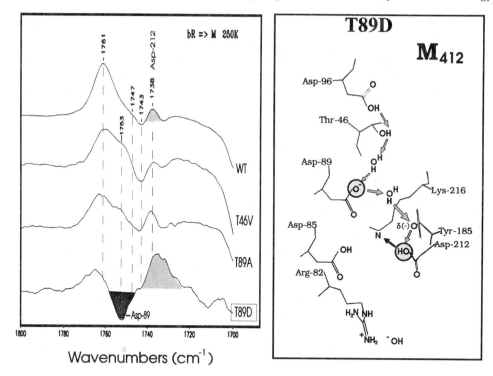

FIGURE 30.5   Left: The bR→M FTIR difference spectra in the 1700- to 1800-cm⁻¹ region for bR mutants with substitutions at position Thr-46 and Thr-89.[144] Right: Model of a hydrogen-bonded proton transport network from Figure 30.4 for the mutant T89D. In this case, the lowered pK_a of Asp-89 relative to Thr-89 causes an increased protonation of Asp-212.

pathway that allows a negative charge, initially residing on Asp-212, to flow toward Asp-96, which then acts to *pull* in a proton from the cytoplasmic medium.

30.5 Recent Developments

Several recent developments have occurred in the application of FTIR difference spectroscopy, which are highlighted below.

Site-Directed Isotope Labeling of bR

FTIR difference spectroscopy has been limited by the absence of a general method for placing isotope labels at specific positions in a protein. A new approach based on site-directed isotope labeling (SDIL) was recently reported.[168-170] SDIL-analogs of bR were produced by expressing bR in an *in vitro* system containing a tyrosine-suppressor tRNA aminoacylated with isotopically labeled tyrosine followed by refolding. SDIL-analogs containing ²H and ¹³C isotopes at both specific tyrosine an backbone peptide carbonyl groups were produced with properties almost identical to native bR. FTIR analysis of the bR→K, L, M, and N steps in the photocycle of these essentially unperturbed SDIL-analogs has led to the identification of structurally active groups that may be involved in proton transport and in the coupling of chromophore isomerization to protein conformational changes.

In the future, it is likely that the SDIL approach will be combined with other biophysical methods besides FTIR to study bR. For example, rotational resonance NMR[171,172] can be used to determine distances (and changes in distances) between specific atoms such as in the Schiff base and Tyr-185. Although the FTIR measurements only require samples in the 10- to 25-μg range, it should be possible to scale-up to larger quantities required for NMR and other biophysical measurements. In general, site-directed isotope labeling should also be applicable to a wide range of other proteins including those involved in enzyme catalysis, ion transport, and signal transduction.

Protein Conformational Changes and Chromophore Isomerization

Several recent studies have focused on the nature of the secondary structural changes during th bR photocycle. Previously, such changes have been detected by circular dichroism,[173] X-ray,[38] and neutron diffraction experiments.[174] More recently, electron diffraction studies of the mutant D96G revealed that the cytoplasmic portion of the F- and G-helices undergo a small tilt away from the membrane normal.[175] Interestingly, this tilt and its reversal might facilitate movement of a proton from Asp-96 to the Schiff base and subsequent proton uptake by Asp-96 from the cytoplasmic medium. Electron diffraction studies also show that significant structural changes occur in wild-type bR upon formation of two different low-temperature substrates of the M intermediate.[176]

Several FTIR studies also show that a significant change occurs in the bR secondary structure and that it is correlated with the M→N transition. In particular, an intense negative band at 1670 cm^{-1} appears in the time-resolved FTIR difference spectrum of bR during formation of the N intermediate[116,118,121] and disappears during N decay.[121,122] The same results are also obtained from low-temperature, steady-state measurements, where the N intermediate is stabilized by either high pH[117,177] or by protein modifications such as Thr-46→Asp,[178] which slow N decay. While these experiments were performed using FTIR transmission methods, which are restricted to thin, hydrated films, similar results were also obtained in the presence of a bulk aqueous medium by using attenuated total reflection (ATR).[169] Significantly, the frequency of the 1670-cm^{-1} band, its insensitivity to H/D exchange, and its dichroism are consistent with its assignment to the amide I mode of membrane-embedded α-helical structure.[178-179] One possibility is that these changes arise from the tilting of the F helix about the Tyr-185/Pro-186 region of bR, which could serve as a hinge.[67,169]

FTIR measurement on mutants of bR also establish that the structural change that is normally correlated with the M→N transition in wild-type bR can also occur under conditions where the Schiff base does not deprotonate or reprotonate.[119,180] For example, the 1670-cm^{-1} band is detected in the photocycle of the mutant Y57D at 250 K, even though decay of the L intermediate is blocked, thereby preventing deprotonation/reprotonation of Schiff base and Asp-96.[181] One implication of these results is that the major structural change in bR is driven by steric interactions produced within the retinal-binding pocket as a direct consequence of chromophore isomerization.

Detection of Water Molecules in the Active Site of bR

Several lines of evidence indicate that one or more water molecules may be located in the active site of bR.[156,181-185] Recently, this was confirmed directly through the assignment of bands in the region above 3500 cm^{-1} to the OH stretch mode of water[186] in the FTIR difference spectra of bR. One or more weakly hydrogen-bonded water molecules were detected that undergo a change in H-bonding as early as formation of the K intermediate.[187] Recently, it was found that the mutation Asp-85→Asn (D85N)[188] and Y57D[187] cause the disappearance of some of these bands, thereby indicating that at least one water molecule is located in the active site. (A second may be located outside of this region.[187]) However, the exact position and number of water molecules still remains to be determined. An additional question is the role of these water(s), which may include participation in a proton wire.[144]

30.6 Future Goals

While great progress has been made in the past few years in elucidating the mechanism of proton transport in bacteriorhodopsin, a complete understanding has not yet been achieved. For example, the path of proton transport through the protein has not yet been established, although as discussed in this chapter the outlines of a hydrogen-bonded proton transport network are emerging. Key questions also remain about the specific mechanism of chromophore reisomerization, protonation changes of Asp groups, and the location and role of secondary structural changes that are detected at different stages of the photocycle.

Progress toward answering these questions is likely to depend on the further development of new biophysical techniques, including time-resolved FTIR,[116,118,132–134] ultrafast time-resolved IR studies,[189] time-resolved X-ray diffraction,[38] and femtosecond visible absorption measurements.[6] The ability of FTIR difference spectroscopy to probe changes in specific amino acid residues in bR during the photocycle should continue to provide details of proton translocation mechanism in bR, especially as methods improve to assign bands in the FTIR difference spectrum.

Acknowledgments

The authors wish to acknowledge the contributions from the many members of the Boston University Molecular Biophysics Group who have helped in the preparation of this chapter, including O. Bousché, C. F. C. Ludlam, M. Coleman, W. Fischer, Y. W. He, X. Liu, J. Olejnik, N. Patel, P. Rath, and T. Russell, as well as many of our collaborators who made possible much of the research described here, including H. G. Khorana, J. Herzfeld, J. Spudich, W. DeGrip, and J. Lugtenburg. This work was supported by grants from the NSF (DMB-9106017), NIH (GMEY47527–01), and ARO (DAAL03-G-0172) to KJR.

References

1. Oesterhelt, D. and Stoeckenius, W., Rhodopsin-like protein from the purple membrane of Halobacterium halobium, *Nature (New Biol.)*, 233, 149, 1971.
2. Shen, Y., Safinya, C. R., Liang, K. S., Ruppert, A. F., and Rothschild, K. J., Stabilization of the membrane protein bacteriorhodopsin to 140°C in two-dimensional films, *Nature*, 366, 48, 1993.
3. Popot, J. L., Gerchman, S. E., and Engelman, D. M., Refolding of bacteriorhodopsin in lipid bilayers. A thermodynamically controlled two-stage process, *J. Mol. Biol.*, 198, 655, 1987.
4. Liao, M. J., Huang, K. S., and Khorana, H. G., Regeneration of native bacteriorhodopsin structure from fragments, *J. Biol. Chem.*, 259, 4200, 1984.
5. Kataoka, M., Kahn, T. W., Tsujiuchi, Y., Engelman, D. M., and Tokunaga, F., Bacteriorhodopsin reconstituted from two individual helices and the complementary five-helix fragment is photoactive, *Photochem. Photobiol.*, 56, 895, 1992.
6. Mathies, R. A., Lin, S. W., Ames, J. B., and Pollard, W. T., From femtoseconds to biology: Mechanism of bacteriorhodopsin's light-driven proton pump, *Annu. Rev. Biophys. Biophys. Chem.*, 20, 491, 1991.
7. Lugtenburg, J., Mathies, R. A., Griffin, R. G., and Herzfeld, J., Structure and function of rhodopsins from solid state NMR and resonance Raman spectroscopy of isotopic retinal derivatives, *Trends Biochem. Sci. (Pers. Ed.)*, 13, 388, 1988.
8. Rothschild, K. J., Zagaeski, M., and Cantore, W. A., Conformational changes of bacteriorhodopsin detected by Fourier transform infrared difference spectroscopy, *Biochem. Biophys. Res. Commun.*, 103, 483, 1981.
9. Gerwert, K., Molecular reaction mechanism of photosynthetic proteins as determined by FTIR-spectroscopy, *Biochim. Biophys. Acta*, 1101, 147, 1992.
10. Rothschild, K. J., FTIR difference spectroscopy of bacteriorhodopsin: Toward a molecular model, *J. Bioenerg. Biomembr.*, 24, 147, 1992.
11. Henderson, R., Baldwin, J. M., Ceska, T. A., Zemlin, F., Beckmann, E., and Downing, K. H., Model for the structure of bacteriorhodopsin based on high-resolution electron cryo-microscopy, *J. Mol. Biol.*, 213, 899, 1990.

12. Hauss, T., Grzesiek, S., Otto, H., Westerhausen, J., and Heyn, M. P., Transmembrane location of retinal in bacteriorhodopsin by neutron diffraction, *Biochemistry*, 29, 4904, 1990.

13. Krebs, M. P. and Khorana, H. G., Mechanism of light-dependent proton translocation by bacteriorhodopsin, *J. Bacteriol.*, 175, 1555, 1993.

14. Miercke, L. J. W., Betlach, M. C., Mitra, A. K., Shand, R. F., Fong, S. K., and Stroud, R. M., Wild-type and mutant bacteriorhodopsins D85N, D96N, and R82Q: Purification to homogeneity, pH dependence of pumping and electron diffraction, *Biochemistry*, 30, 3088, 1991.

15. Soppa, J., Otomo, J., Straub, J., Tittor, J., Meessen, S., and Oesterhelt, D., Bacteriorhodopsin mutants of *Halobacterium* sp. GRB. II. Characterization of mutants, *J. Biol. Chem.*, 264, 13049, 1989.

16. Ni, B., Chang, M., Duschl, A., Lanyi, J., and Needleman, R., An efficient system for the synthesis of bacteriorhodopsin in *Halobacterium halobium*, *Gene*, 90, 169, 1990.

17. Krebs, M. P., Mollaaghababa, R., and Khorana, H. G., Gene replacement in *Halobacterium halobium* and expression of bacteriorhodopsin mutants, *Proc. Natl. Acad. Sci. U.S.A.*, 90, 1987, 1993.

18. Blanck, A. and Oesterhelt, D., The halo-opsin gene. II. Sequence, primary structure of halorhodopsin and comparison with bacteriorhodopsin, *Embo J.*, 6, 265, 1987.

19. Blanck, A., Oesterhelt, D., Ferrando, E., Schegk, E. S., and Lottspeich, F., Primary structure of sensory rhodopsin I, a prokaryotic photoreceptor, *Embo J.*, 8, 3963, 1989.

20. Spudich, J. L. and Bogolmolni, R. A., Sensory rhodopsins of halobacteria, *Annu. Rev. Biophys. Biophys. Chem.*, 17, 193, 1988.

21. Hu, J., Griffin, R. G., and Herzfeld, J., Synergy in the spectral tuning of retinal pigments: Complete accounting of the opsin shift in bacteriorhodopsin, *Proc. Natl. Acad. Sci. U.S.A.*, 91, 8880, 1994.

22. Rothschild, K. J., Roepe, P., Lugtenburg, J., and Pardoen, J. A., Fourier transform infrared evidence for Schiff base alteration in the first step of the bacteriorhodopsin photocycle, *Biochemistry*, 23, 6103, 1984.

23. Spudich, J. L., McCain, D. A., Nakanishi, K., Okabe, M., Shimizu, N., Rodman, H., Honig, B., and Bogomolni, R. A., Chromophore/protein interaction in bacterial sensory rhodopsin and bacteriorhodopsin, *Biophys. J.*, 49, 479, 1986.

24. Lugtenburg, J., Muradin-Szweykowska, M., Heeremans, C., and Pardoen, J. A., Mechanism for the opsin shift of retinal's absorption in bacteriorhodopsin, *J. Am. Chem. Soc.*, 108, 3104, 1986.

25. Baasov, T., Friedman, N., and Sheves, M., Factors affecting the $C=N$ stretching in protonated retinal Schiff base: A model study for bacteriorhodopsin and visual pigments, *Biochemistry*, 26, 3210, 1987.

26. Harbison, G. S., Smith, S. O., Pardoen, J. A., Courtin, J. M. L., Lugtenburg, J., Herzfeld, J., Mathies, R. A., and Griffin, R. G., Solid-state carbon-13 NMR detection of a perturbed 6-s-*trans* chromophore in bacteriorhodopsin, *Biochemistry*, 24, 6955, 1985.

27. McDermott, A. E., Creuzet, F., van der Hoef, K., Levitt, M. H., Herzfeld, J., Lugtenburg, J., and Griffin, R. G., Determination of internuclear distances and the orientation of functional groups by solid-state NMR: Rotational resonance study of the conformation of retinal in bacteriorhodopsin, *Biochemistry*, 33, 6129, 1994.

28. Oesterhelt, D., Bräuchle, C., and Hampp, N., Bacteriorhodopsin: a biological material for information processing, *Q. Rev. Biophys.*, 4, 425, 1991.

29. Birge, R. R., Nature of the primary photochemical events in rhodopsin and bacteriorhodopsin, *Biochim. Biophys. Acta*, 1016, 293, 1990.

30. Hofrichter, J., Henry, E. R., and Lozier, R. H., Photocycles of bacteriorhodopsin in light- and dark-adapted purple membrane studied by time-resolved absorption spectroscopy, *Biophys. J.*, 56, 693, 1989.

31. Lanyi, J. K., Proton translocation mechanisms and energetics in the light-driven pump bacteriorhodopsin, *Biochim. Biophys. Acta*, 1183, 241, 1993.

32. Sonar, S. M. and Singh, A. K., Location of lysine-129 and lysine-40/41 with respect to retinylidene chromophore in bacteriorhodopsin, *Biochim. Biophys. Acta*, 1076, 239, 1991.

33. Singh, A. K. and Sonar, S. M., Modification of carboxyl groups in bacteriorhodopsin: Chemical evidence for the involvement of aspartic acid residues in the structure and function of bacteriorhodopsin, *J Chem. Soc. Perkin Trans. II*, 133, 1993.

34. Crouch, R. K., Studies of rhodopsin and bacteriorhodopsin using modified retinals, *Photochem. Photobiol.*, 44, 803, 1986.

35. Crouch, R. K., Scott, R., Ghent, S., Govindjee, R., Chang, C. H., and Ebrey, T., Properties of synthetic bacteriorhodopsin pigments. Further probes of the chromophore binding site, *Photochem. Photobiol.*, 43, 297, 1986.

36. Rothschild, K. J., Infrared studies of bacteriorhodopsin, *Photochem. Photobiol.*, 47, 883, 1988.

37. Kitagawa, T. and Maeda, A., Vibrational spectra of rhodopsin and bacteriorhodopsin, *Photochem. Photobiol.*, 50, 883, 1989.

38. Koch, M. H. J., Dencher, N. A., Oesterhelt, D., Ploehn, H. J., Rapp, G., and Bueldt, G., Time-resolved X-ray diffraction study of structural changes associated with the photocycle of bacteriorhodopsin, *EMBO J.*, 10, 521, 1991.

39. Glaeser, R. M., Baldwin, J., Ceska, T. A., and Henderson, R., Electron diffraction analysis of the M412 intermediate of bacteriorhodopsin, *Biophys. J.*, 50, 913, 1986.

40. Unwin, P. N. T. and Henderson, R., Molecular structure determination by electron microscopy of unstained crystalline specimens, *J. Mol. Biol.*, 94, 425, 1975.

41. Rothschild, K. J. and Clark, N. A., Polarized infrared spectroscopy of oriented purple membrane, *Biophys. J.*, 25, 473, 1979.

42. Ovchinnikov, Y. A., Abdulaev, N. G., Feigina, M. Y., Kiselev, A. V., Lobanov, N. A., and Nazimov, I. V., Amino acid sequence of bacteriorhodopsin, *Bioorg. Khim.*, 4, 1573, 1978.

43. Khorana, H. G., Gerber, G. E., Herlihy, W. C., Gray, C. P., Anderegg, R. J., Nihei, K., and Biemann, K., Amino acid sequence of bacteriorhodopsin, *Proc. Natl. Acad. Sci. U.S.A.*, 76, 5046, 1979.

44. Engelman, D. M., Henderson, R., McLachlan, A. D., and Wallace, B. A., Path of the polypeptide in bacteriorhodopsin, *Proc. Natl. Acad. Sci. U.S.A.*, 77, 2023, 1980.

45. Kimura, K., Mason, T. L., and Khorana, H. G., Immunological probes for bacteriorhodopsin. Identification of three distinct antigenic sites on the cytoplasmic surface, *J. Biol. Chem.*, 257, 2859, 1982.

46. Ovchinnikov, Y. A., Abdulaev, N. G., Vasilov, R. G., Vturina, I. Y., Kuryatov, A. B., and Kiselev, A. V., The antigenic structure and topography of bacteriorhodopsin in purple membranes as determined by interaction with monoclonal antibodies, *FEBS Lett.*, 179, 343, 1985.

47. Fimmel, S., Choli, T., Dencher, N. A., Bueldt, G., and Wittmann-Liebold, B., Topography of surface-exposed amino acids in the membrane protein bacteriorhodopsin determined by proteolysis and micro-sequencing, *Biochim. Biophys. Acta*, 978, 231, 1989.

48. Bridgen, J. and Walker, I. D., Photoreceptor protein from the purple membrane of *Halobacterium halobium*. Molecular weight and retinal binding site, *Biochemistry*, 15, 792, 1976.

49. Lemke, H. D. and Oesterhelt, D., Lysine 216 is a binding site of the retinyl moiety in bacteriorhodopsin, *FEBS Lett.*, 128, 255, 1981.

50. Mullen, E., Johnson, A. H., and Akhtar, M., The identification of Lys216 as the retinal binding residue in bacteriorhodopsin, *FEBS Lett.*, 130, 187, 1981.

51. Huang, K. S., Liao, M. J., Gupta, C. M., Royal, N., Biemann, K., and Khorana, H. G., The site of attachment of retinal in bacteriorhodopsin. The ε-amino group in Lys-41 is not required for proton translocation, *J. Biol. Chem.*, 257, 8596, 1982.

52. Rothschild, K. J., Argade, P. V., Earnest, T. N., Huang, K. S., London, E., Liao, M. J., Bayley, H., Khorana, H. G., and Herzfeld, J., The site of attachment of retinal in bacteriorhodopsin. A resonance Raman study, *J. Biol. Chem.*, 257, 8592, 1982.

53. Rothschild, K. J., Sanches, R., Hsiao, T. L., and Clark, N. A., A spectroscopic study of rhodopsin alpha-helix orientation, *Biophys. J.*, 31, 53, 1980.

54. Heyn, M. P., Cherry, R. J., and Mueller, U., Transient and linear dichroism studies on bacteriorhodopsin: Determination of the orientation of the 568 nm *all-trans* retinal chromophore, *J. Mol. Biol.*, 117, 607, 1977.

55. Heyn, M. P. and Otto, H., Photoselection and transient linear dichroism with oriented immobilized purple membranes: Evidence for motion of the C(20)-methyl group of the chromophore towards the cytoplasmic side of the membrane, *Photochem. Photobiol.*, 56, 1105, 1992.

56. Earnest, T. N., Roepe, P., Braiman, M. S., Gillespie, J., and Rothschild, K. J., Orientation of the bacteriorhodopsin chromophore probed by polarized Fourier transform infrared difference spectroscopy, *Biochemistry*, 25, 7793, 1986.

57. Fahmy, K., Siebert, F., Grossjean, M. F., and Tavan, P., Photoisomerization in bacteriorhodopsin studied by FTIR, linear dichroism, and photoselection experiments combined with quantum chemical theoretical analysis, *J. Mol. Struct.*, 214, 257, 1989.

58. Lin, S. W. and Mathies, R. A., Orientation of the protonated retinal Schiff base group in bacteriorhodopsin from absorption linear dichroism, *Biophys. J.*, 56, 653, 1989.

59. Heyn, M. P., Westerhausen, J., Wallat, I., and Seiff, F., High-sensitivity neutron diffraction of membranes: location of the Schiff base end of the chromophore of bacteriorhodopsin, *Proc. Natl. Acad. Sci. U.S.A.*, 85, 2146, 1988.

60. Ulrich, A. S., Watts, A., Wallat, I., and Heyn, M. P., Distorted structure of the retinal chromophore in bacteriorhodopsin resolved by ^{1}H-NMR, *Biochemistry*, 33, 5370, 1994.

61. Huang, K. S., Radhakrishnan, R., Bayley, H., and Khorana, H. G., Orientation of retinal in bacteriorhodopsin as studied by crosslinking using a photosensitive analog of retinal, *J. Biol. Chem.*, 257, 13616, 1982.

62. Mogi, T., Stern, L. J., Marti, T., Chao, B. H., and Khorana, H. G., Structure-function studies on bacteriorhodopsin. VII. Aspartic acid substitutions affect proton translocation by bacteriorhodopsin, *Proc. Natl. Acad. Sci. U.S.A.*, 85, 4148, 1988.

63. Stern, L. J., Ahl, P. L., Marti, T., Mogi, T., Dunach, M., Berkowitz, S., Rothschild, K. J., and Khorana, H. G., Substitution of membrane-embedded aspartic acids in bacteriorhodopsin causes specific changes in different steps of the photochemical cycle, *Biochemistry*, 28, 10035, 1989.

64. Mogi, T., Marti, T., and Khorana, H. G., Structure-function studies on bacteriorhodopsin. IX. Substitutions of tryptophan residues affect protein-retinal interactions in bacteriorhodopsin, *J. Biol. Chem.*, 264, 14197, 1989.

65. Mogi, T., Stern, L. J., Chao, B. H., and Khorana, H. G., Structure-function studies on bacteriorhodopsin. VIII. Substitutions of the membrane-embedded prolines 50, 91, and 186: The effects are determined by the substituting amino acids, *J. Biol. Chem.*, 264, 14192, 1989.

66. Rothschild, K. J., Braiman, M. S., Mogi, T., Stern, L. J., and Khorana, H. G., Conserved amino acids in F-helix of bacteriorhodopsin form part of a retinal binding pocket, *FEBS Lett.*, 250, 448, 1989.

67. Rothschild, K. J., He, Y. W., Mogi, T., Marti, T., Stern, L. J., and Khorana, H. G., Vibrational spectroscopy of bacteriorhodopsin mutants: Evidence for the interaction of proline-186 with the retinylidene chromophore, *Biochemistry*, 29, 5954, 1990.

68. Ahl, P. L., Stern, L. J., Mogi, T., Khorana, H. G., and Rothschild, K. J., Substitution of amino acids in helix F of bacteriorhodopsin: Effects on the photochemical cycle, *Biochemistry*, 28, 10028, 1989.

69. Ahl, P. L., Stern, L. J., During, D., Mogi, T., Khorana, H. G., and Rothschild, K. J., Effects of amino acid substitutions in the F helix of bacteriorhodopsin. Low temperature ultraviolet/visible difference spectroscopy, *J. Biol. Chem.*, 263, 13594, 1988.

70. Braiman, M. S., Mogi, T., Stern, L. J., Hackett, N. R., Chao, B. H., Khorana, H. G., and Rothschild, K. J., Vibrational spectroscopy of bacteriorhodopsin mutants. I. Tyrosine-185 protonates and deprotonates during the photocycle, *Proteins: Struct. Funct. Genet.*, 3, 219, 1988.

71. Braiman, M. S., Mogi, T., Marti, T., Stern, L. J., Khorana, H. G., and Rothschild, K. J., Vibrational spectroscopy of bacteriorhodopsin mutants: Light-driven proton transport involves protonation changes of aspartic acid residues 85, 96, and 212, *Biochemistry*, 27, 8516, 1988.

72. Rothschild, K. J., Gray, D., Mogi, T., Marti, T., Braiman, M. S., Stern, L. J., and Khorana, H. G., Vibrational spectroscopy of bacteriorhodopsin mutants: Chromophore isomerization perturbs trytophan-86, *Biochemistry*, 28, 7052, 1989.

73. Rothschild, K. J., Braiman, M. S., He, Y. W., Marti, T., and Khorana, H. G., Vibrational spectroscopy of bacteriorhodopsin mutants. Evidence for the interaction of aspartic acid 212 with tyrosine 185 and possible role in the proton pump mechanism, *J. Biol. Chem.*, 265, 16985, 1990.

74. Dancshazy, Z., Govindjee, R., Nelson, B., and Ebrey, T. G., A new intermediate in the photocycle of bacteriorhodopsin, *FEBS Lett.*, 209, 44, 1986.

75. Chernavskii, D. S., Chizhov, I. V., Lozier, R. H., Murina, T. M., Prokhorov, A. M., and Zubov, B. V., Kinetic model of bacteriorhodopsin photocycle: Pathway from M state to BR, *Photochem. Photobiol.*, 49, 649, 1989.

76. Mueller, K. H., Butt, H. J., Bamberg, E., Fendler, K., Hess, B., Siebert, F., and Engelhard, M., The reaction cycle of bacteriorhodopsin: An analysis using visible absorption, photocurrent and infrared techniques, *Eur. Biophys. J.*, 19, 241, 1991.

77. Varo, G. and Lanyi, J. K., Distortions in the photocycle of bacteriorhodopsin at moderate dehydration, *Biophys. J.*, 59, 313, 1991.

78. Varo, G. and Lanyi, J. K., Thermodynamics and energy coupling in the bacteriorhodopsin photocycle, *Biochemistry*, 30, 5016, 1991.

79. Lozier, R. H., Xie, A., Hofrichter, J., and Clore, G. M., Reversible steps in the bacteriorhodopsin photocycle, *Proc. Natl. Acad. Sci. U.S.A.*, 89, 3610, 1992.

80. Pollard, W. T., Cruz, C. H., Shank, C. V., and Mathies, R. A., Direct observation of the excited-state *cis-trans* photoisomerization of bacteriorhodopsin: Multilevel line shape theory for femtosecond dynamic hole burning and its application, *J. Chem. Phys.*, 90, 199, 1989.

81. Petrich, J. W., Breton, J., Martin, J. L., and Antonetti, A., Femtosecond absorption spectroscopy of light-adapted and dark-adapted bacteriorhodopsin, *Chem. Phys. Lett.*, 137, 369, 1987.

82. Varo, G. and Lanyi, J. K., Pathways of the rise and decay of the M photointermediate(s) of bacteriorhodopsin, *Biochemistry*, 29, 2241, 1990.

83. Scherrer, P., Mathew, M. K., Sperling, W., and Stoeckenius, W., Retinal isomer ratio in dark-adapted purple membrane and bacteriorhodopsin monomers, *Biochemistry*, 28, 829, 1989.

84. Harbison, G. S., Smith, S. O., Pardoen, J. A., Winkel, C., Lugtenburg, J., Herzfeld, J., Mathies, R., and Griffin, R. G., Dark-adapted bacteriorhodopsin contains 13-*cis*,15-*syn* and *all-trans*,15-*anti* retinal Schiff bases, *Proc. Natl. Acad. Sci. U.S.A.*, 81, 1706, 1984.

85. Smith, S. O., Pardoen, J. A., Lugtenburg, J., and Mathies, R. A., Vibrational analysis of the 13-*cis*-retinal chromophore in dark-adapted bacteriorhodopsin, *J. Phys. Chem.*, 91, 804, 1987.

86. Smith, S. O., Braiman, M. S., Myers, A. B., Pardoen, J. A., Courtin, J. M. L., Winkel, C., Lugtenburg, J., and Mathies, R. A., Vibrational analysis of the *all-trans*-retinal chromophore in light-adapted bacteriorhodopsin, *J. Am. Chem. Soc.*, 109, 3108, 1987.

87. Farrar, M. R., Lakshmi, K. V., Smith, S. O., Brown, R., Raap, J., Lugtenburg, J., Griffin, R. G., and Herzfeld, J., Solid-state NMR study of [ε-^{13}C] Lys-bacteriorhodopsin: Schiff base photoisomerization, *Biophys. J.*, 65, 310, 1993.

88. Braiman, M. and Mathies, R., Resonance Raman spectra of bacteriorhodopsin's primary photoproduct: Evidence for a distorted 13-*cis* retinal chromophore, *Proc. Natl. Acad. Sci. U.S.A.*, 79, 403, 1982.

89. Brack, T. L. and Atkinson, G. H., Picosecond time-resolved resonance Raman spectrum of the K-590 intermediate in the room temperature bacteriorhodopsin photocycle, *J. Mol. Struct.*, 214, 289, 1989.

90. Atkinson, G. H., Brack, T. L., Blanchard, D., and Rumbles, G., Picosecond time-resolved resonance Raman spectroscopy of the initial trans to *cis* isomerization in the bacteriorhodopsin photocycle, *Chem. Phys.*, 131, 1, 1989.

91. Rothschild, K. J. and Marrero, H., Infrared evidence that the Schiff base of bacteriorhodopsin is protonated: bR$_{570}$ and K intermediates, *Proc. Natl. Acad. Sci. U.S.A.*, 79, 4045, 1982.

92. Doig, S. J., Reid, P. J., and Mathies, R. A., Picosecond time-resolved resonance Raman spectroscopy of bacteriorhodopsin's J, K, and KL intermediates, *J. Phys. Chem.*, 95, 6372, 1991.

93. Rothschild, K. J., Marrero, H., Braiman, M., and Mathies, R., Primary photochemistry of bacteriorhodopsin: Comparison of Fourier transform infrared difference spectra with resonance Raman spectra, *Photochem. Photobiol.*, 40, 675, 1984.

94. Gerwert, K., Siebert, F., Pardoen, J. A., Winkel, C., and Lugtenburg, J., in *Time-Resolved Vibrational Spectroscopy*, Springer Proc. Phys. 4 ed., Laubereau, A. and Stockburger, M., Eds. Springer-Verlag, Berlin, 1985.

95. Kakitani, H., Kakitani, T., Rodman, H., Honig, B., and Callender, R., Correlation of vibrational frequencies with absorption maxima in polyenes, rhodopsin, bacteriorhodopsin, and retinal analogs, *J. Phys. Chem.*, 87, 3620, 1983.

96. Gilson, H. S. R., Honig, B. H., Croteau, A., Zarrilli, G., and Nakanishi, K., Analysis of the factors that influence the carbon:nitrogen stretching frequency of polyene Schiff bases. Implications for bacteriorhodopsin and rhodopsin, *Biophys. J.*, 53, 261, 1988.

97. Argade, P. V. and Rothschild, K. J., Quantitative analysis of resonance Raman spectra of purple membrane from *Halobacterium halobium*: L_{550} intermediate, *Biochemistry*, 22, 3460, 1983.

98. Fodor, S. P. A., Pollard, W. T., Gebhard, R., van den Berg, E. M. M., Lugtenburg, J., and Mathies, R. A., Bacteriorhodopsin's L_{550} intermediate contains a C14-C15 s-*trans*-retinal chromophore, *Proc. Natl. Acad. Sci., U.S.A.*, 85, 2156, 1988.

99. Sasaki, J., Maeda, A., Kato, C., and Hamaguchi, H., Time-resolved infrared spectral analysis of the KL-to-L conversion in the photocycle of bacteriorhodopsin, *Biochemistry*, 32, 867, 1993.

100. Tavan, P. and Schulten, K., Evidence for a 13,14-*cis* cycle in bacteriorhodopsin, *Biophys. J.*, 50, 81, 1986.

101. Gerwert, K. and Siebert, F., Evidence for light-induced 13-*cis*, 14-s-*cis* isomerization in bacteriorhodopsin obtained by FTIR difference spectroscopy using isotopically labeled retinals, *EMBO J.*, 5, 805, 1986.

102. Lewis, A., Spoonhower, J., Bogomolni, R. A., Lozier, R. H., and Stoeckenius, W., Tunable laser resonance Raman spectroscopy of bacteriorhodopsin, *Proc. Natl. Acad. Sci., U.S.A.*, 71, 4462, 1974.

103. Fahmy, K., Weidlich, O., Engelhard, M., Tittor, J., Oesterhelt, D., and Siebert, F., Identification of the proton acceptor of Schiff base deprotonation in bacteriorhodopsin: A Fourier-transform-infrared study of the mutant Asp85→Glu in its natural lipid environment, *Photochem. Photobiol.*, 56, 1073, 1992.

104. Braiman, M. and Mathies, R., Resonance Raman evidence for an *all-trans* to 13-*cis* isomerization in the proton-pumping cycle of bacteriorhodopsin, *Biochemistry*, 19, 5421, 1980.

105. Ames, J. B., Fodor, S. P. A., Gebhard, R., Raap, J., van den Berg, E. M. M., Lugtenburg, J., and Mathies, R. A., Bacteriorhodopsin's M_{412} intermediate contains a 13-*cis*,14-s-*trans*,15-*anti*-retinal Schiff base chromophore, *Biochemistry*, 28, 3681, 1989.

106. Nakagawa, M., Ogura, T., Maeda, A., and Kitagawa, T., Transient resonance Raman spectra of neutral and alkaline bacteriorhodopsin photointermediates observed with a double-beam flow apparatus: Presence of very fast decaying M_{412}, *Biochemistry*, 28, 1347, 1989.

107. Deng, H., Pande, C., Callender, R. H., and Ebrey, T. G., A detailed resonance Raman study of the M_{412} intermediate in the bacteriorhodopsin photocycle, *Photochem. Photobiol.*, 41, 467, 1985.

108. Lozier, R. H., Bogomolni, R. A., and Stoeckenius, W., Bacteriorhodopsin, a light-driven proton pump in *Halobacterium halobium*, *Biophys. J.*, 15, 955, 1975.

109. Balashov, S. P., Imasheva, E. S., Litvin, F. F., and Lozier, R. H., The N intermediate of bacteriorhodopsin at low temperatures: Stabilization and photoconversion, *FEBS Lett.*, 271, 1, 1990.

110. Kouyama, T., Nasuda-Kouyama, A., Ikegami, A., Mathew, M. K., and Stoeckenius, W., Bacteriorhodopsin photoreaction: Identification of a long-lived intermediate N (P, R350) at high pH and its M-like photoproduct, *Biochemistry*, 27, 5855, 1988.

111. Lakshmi, K. V., Farrar, M. R., Raap, J., Lugtenburg, J., Griffin, R. G., and Herzfeld, J., Solid-state ^{13}C and ^{15}N NMR investigations of the N intermediate of bacteriorhodopsin, *Biochemistry*, 33, 8853, 1994.

112. Ames, J. B. and Mathies, R. A., The role of back-reactions and proton uptake during the N→O transition in bacteriorhodopsin's photocycle: A kinetic resonance Raman study, *Biochemistry*, 29, 7181, 1990.

113. Fodor, S. P. A., Ames, J. B., Gebhard, R., Van den Berg, E. M. M., Stoeckenius, W., Lugtenburg, J., and Mathies, R. A., Chromophore structure in bacteriorhodopsin's N intermediate: Implications for the proton-pumping mechanism, *Biochemistry*, 27, 7097, 1988.

114. Otto, H., Marti, T., Holz, M., Mogi, T., Lindau, M., Khorana, H. G., and Heyn, M. P., Aspartic acid-96 is the internal proton donor in the reprotonation of the Schiff base of bacteriorhodopsin, *Proc. Natl. Acad. Sci. U.S.A.*, 86, 9228, 1989.

115. Drachev, L. A., Kaulen, A. D., Khorana, H. G., Mogi, T., Otto, H., Skulachev, V. P., Heyn, M. P., and Holz, M., Participation of the Asp-96 carboxyl in hydrogen ion transfer along the inward proton-conducting pathway of bacteriorhodopsin, *Biokhimiya (Moscow)*, 54, 1467, 1989.

116. Bousche, O., Braiman, M., He, Y. W., Marti, T., Khorana, H. G., and Rothschild, K. J., Vibrational spectroscopy of bacteriorhodopsin mutants. Evidence that Asp-96 deprotonates during the M→N transition, *J. Biol. Chem.*, 266, 11063, 1991.

117. Pfefferle, J. M., Maeda, A., Sasaki, J., and Yoshizawa, T., Fourier transform infrared study of the N intermediate of bacteriorhodopsin, *Biochemistry*, 30, 6548, 1991.

118. Braiman, M. S., Bousche, O., and Rothschild, K. J., Protein dynamics in the bacteriorhodopsin photocycle: Submillisecond Fourier transform infrared spectra of the L, M, and N photointermediates, *Proc. Natl. Acad. Sci. U.S.A.*, 88, 2388, 1991.

119. Cao, Y., Varo, G., Klinger, A., Czajkowsky, D. M., Braiman, M. S., Needleman, R., and Lanyi, J. K., Proton transfer from Asp-96 to the bacteriorhodopsin Schiff base is caused by a decrease of the pK_a of Asp-96 which follows a protein backbone conformational change, *Biochemistry*, 32, 1981, 1993.

120. Smith, S. O., Pardoen, J. A., Mulder, P. P. J., Curry, B., Lugtenburg, J., and Mathies, R., Chromophore structure in bacteriorhodopsin's O_{640} photointermediate, *Biochemistry*, 22, 6141, 1983.

121. Bousché, O., Sonar, S., Krebs, M. P., Khorana, H. G., and Rothschild, K. J., Time-resolved Fourier transform infrared spectroscopy of the bacteriorhodopsin mutant Tyr-185→Phe: Asp-96 reprotonates during O formation; Asp-85 and Asp-212 deprotonate during O decay, *Photochem. Photobiol.*, 56, 1085, 1992.

122. Heßling, B., Souvignier, G., and Gerwert, K., A model-independent approach to assigning bacteriorhodopsin's intramolecular reactions to photocycle intermediates, *Biophys. J.*, 65, 1929, 1993.

123. Rath, P., Krebs, M. P., He, Y.-W., Khorana, H. G., and Rothschild, K. J., Fourier transform Raman spectroscopy of the bacteriorhodopsin mutant tyr-185→Phe: Formation of a stable O-like species during light adaptation and detection of its transient N-like photoproduct, *Biochemistry*, 32, 2272, 1993.

124. Dunach, M., Marti, T., Khorana, H. G., and Rothschild, K. J., UV-visible spectroscopy of bacteriorhodopsin mutants: Substitution of Arg-82, Asp-85, Tyr-185, and Asp-212 results in abnormal light-dark adaptation, *Proc. Natl. Acad. Sci. U.S.A.*, 87, 9873, 1990.

125. Sonar, S., Krebs, M. P., Khorana, H. G., and Rothschild, K. J., Static and time-resolved absorption spectroscopy of the bacteriorhodopsin mutant Tyr-185→Phe: Evidence for an equilibrium between BR570 and an O-like species, *Biochemistry*, 32, 2263, 1993.

126. He, Y.-W., Krebs, M. P., Fischer, W. B., Khorana, H. G., and Rothschild, K. J., FTIR difference spectroscopy of the bacteriorhodopsin mutant Tyr-185→Phe: Detection of a stable O-like species and characterization of its photocycle at low temperature, *Biochemistry*, 32, 2282, 1993.

127. Fukuda, K. and Kouyama, T., Photoreaction of bacteriorhodopsin at high pH: Origins of the slow decay component of M, *Biochemistry*, 31, 11740, 1992.

128. Turner, G. J., Miercke, L. J. W., Thorgeirsson, T. E., Kliger, D., Betlach, M. C., and Stroud, R. M., Bacteriorhodopsin D85N: Three spectroscopic species in equilibrium, *Biochemistry*, 32, 1332, 1993.

129. Rothschild, K. J., Roepe, P., and Gillespie, J., Fourier transform infrared spectroscopic evidence for the existence of two conformations of the bacteriorhodopsin primary photoproduct at low temperature, *Biochim. Biophys. Acta*, 808, 140, 1985.

130. Siebert, F. and Mantele, W., Investigation of the primary photochemistry of bacteriorhodopsin by low-temperature Fourier-transform infrared spectroscopy, *Eur. J. Biochem.*, 130, 565, 1983.

131. Siebert, F., Maentele, W., and Kreutz, W., Evidence for the protonation of two internal carboxylic groups during the photocycle of bacteriorhodopsin. Investigation by kinetic infrared spectroscopy, *FEBS Lett.*, 141, 82, 1982.

132. Braiman, M. S., Ahl, P. L., and Rothschild, K. J., Millisecond Fourier-transform infrared difference spectra of bacteriorhodopsin's M_{412} photoproduct, *Proc. Natl. Acad. Sci. U.S.A.* 84, 5221, 1987.

133. Gerwert, K., Souvignier, G., and Hess, B., Simultaneous monitoring of light-induced changes in protein side-group protonation, chromophore isomerization, and backbone motion of bacteriorhodopsin by time-resolved Fourier-transform infrared spectroscopy, *Proc. Natl. Acad. Sci. U.S.A.* 87, 9774, 1990.

134. Weidlich, O. and Siebert, F., Time-resolved step-scan FT-IR investigation of the transition from KL to L in the bacteriorhodopsin photocycle: Identification of chromophore twists by assigning hydrogen out-of-plane (HOOP) bending vibrations, *Appl. Spectrosc.*, 47, 1394, 1993.

135. Gerwert, K., Time-resolved FTIR studies on bacteriorhodopsin, *Springer Proc. Phys.*, 68 (*Time-Resolved Vib. Spectrosc. V*), 61, 1992.

136. Engelhard, M., Gerwert, K., Hess, B., and Siebert, F., Light-driven protonation changes of internal aspartic acids of bacteriorhodopsin: An investigation by static and time-resolved infrared difference spectroscopy using [4–13C]aspartic acid labeled purple membrane, *Biochemistry*, 24, 400, 1985.

137. Rothschild, K. J., He, Y. W., Gray, D., Roepe, P. D., Pelletier, S. L., Brown, R. S., and Herzfeld, J., Fourier transform infrared evidence for proline structural changes during the bacteriorhodopsin photocycle, *Proc. Natl. Acad. Sci. U.S.A.*, 86, 9832, 1989.

138. Gerwert, K., Hess, B., and Engelhard, M., Proline residues undergo structural changes during proton pumping in bacteriorhodopsin, *FEBS Lett.*, 261, 449, 1990.

139. Dollinger, G., Eisenstein, L., Lin, S. L., Nakanishi, K., and Termini, J., Fourier transform infrared difference spectroscopy of bacteriorhodopsin and its photoproducts regenerated with deuterated tyrosine, *Biochemistry*, 25, 6524, 1986.

140. Fahmy, K., Weidlich, O., Engelhard, M., Sigrist, H., and Siebert, F., Aspartic acid-212 of bacteriorhodopsin is ionized in the M and N photocycle intermediates: An FTIR study on specifically carbon-13 labeled reconstituted purple membranes, *Biochemistry*, 32, 5862, 1993.

141. Gerwert, K., Hess, B., Soppa, J., and Oesterhelt, D., Role of aspartate-96 in proton translocation by bacteriorhodopsin, *Proc. Natl. Acad. Sci. U.S.A.*, 86, 4943, 1989.

142. Maeda, A., Sasaki, J., Shichida, Y., Yoshizawa, T., Chang, M., Ni, B., Needleman, R., and Lanyi, J. K., Structures of aspartic acid-96 in the L and N intermediates of bacteriorhodopsin: Analysis by Fourier transform infrared spectroscopy, *Biochemistry*, 31, 4684, 1992.

143. Braiman, M. S., Klinger, A. L., and Doebler, R., Fourier transform infrared spectroscopic analysis of altered reaction pathways in site-directed mutants: The D212N mutant of bacteriorhodopsin expressed in *Halobacterium halobium*, *Biophys. J.*, 62, 56, 1992.

144. Rothschild, K. J., He, Y. W., Sonar, S., Marti, T., and Khorana, H. G., Vibrational spectroscopy of bacteriorhodopsin mutants. Evidence that Thr-46 and Thr-89 form part of a transient network of hydrogen bonds, *J. Biol. Chem.*, 267, 1615, 1992.

145. Rothschild, K. J., Roepe, P., Ahl, P. L., Earnest, T. N., Bogomolni, R. A., Das Gupta, S. K., Mulliken, C. M., and Herzfeld, J., Evidence for a tyrosine protonation change during the primary phototransition of bacteriorhodopsin at low temperature, *Proc. Natl. Acad. Sci. U.S.A.*, 83, 347, 1986.

146. Butt, H. J., Fendler, K., Bamberg, E., Tittor, J., and Oesterhelt, D., Aspartic acids 96 and 85 play a central role in the function of bacteriorhodopsin as a proton pump, *EMBO J.*, 8, 1657, 1989.

147. Otto, H., Marti, T., Holz, M., Mogi, T., Stern, L. J., Engel, F., Khorana, H. G., and Heyn, M. P., Substitution of amino acids Asp-85, Asp-212, and Arg-82 in bacteriorhodopsin affects the proton release phase of the pump and the pK of the Schiff base, *Proc. Natl. Acad. Sci. U.S.A.*, 87, 1018, 1990.

148. Holz, M., Drachev, L. A., Mogi, T., Otto, H., Kaulen, A. D., Heyn, M. P., Skulachev, V. P., and Khorana, H. G., Replacement of aspartic acid-96 by asparagine in bacteriorhodopsin slows both the decay of the M intermediate and the associated proton movement, *Proc. Natl. Acad. Sci. U.S.A.*, 86, 2167, 1989.

149. Tittor, J., Soell, C., Oesterhelt, D., Butt, H. J., and Bamberg, E., A defective proton pump, point-mutated bacteriorhodopsin Asp96→Asn is fully reactivated by azide, *EMBO J.*, 8, 3477, 1989.

150. Subramaniam, S., Marti, T., and Khorana, H. G., Protonation state of Asp (Glu)-85 regulates the purple-to-blue transition in bacteriorhodopsin mutants Arg-82→Ala and Asp-85→Glu: The blue form is inactive in proton translocation, *Proc. Natl. Acad. Sci. U.S.A.*, 87, 1013, 1990.

151. Greenhalgh, D. A., Subramaniam, S., Alexiev, U., Otto, H., Heyn, M. P., and Khorana, H. G., Effect of introducing different carboxylate-containing side chains at position 85 on chromophore formation and proton transport in bacteriorhodopsin, *Proc. Natl. Acad. Sci. U.S.A.*, 267, 25734, 1992.

152. Subramaniam, S., Greenhalgh, D., and Khorana, H. G., Aspartic acid 85 in bacteriorhodopsin functions both as proton acceptor and negative counterion to the Schiff base, *Proc. Natl. Acad. Sci. U.S.A.*, 267, 25730, 1992.

153. Thorgeirsson, T. E., Milder, S. J., Miercke, L. J. W., Betlach, M. C., Shand, R. F., Stroud, R. M., and Kliger, D. S., Effects of Asp-96→Asn, Asp-85→Asn, and Arg-82→Gln single-site substitutions on the photocycle of bacteriorhodopsin, *Biochemistry*, 30, 9133, 1991.

154. Nakanishi, K., Balogh-Nair, V., Arnaboldi, M., Tsujimoto, K., and Honig, B., An external point-charge model for bacteriorhodopsin to account for its purple color, *J. Am. Chem. Soc.*, 102, 7945, 1980.

155. Marti, T., Rosselet, S. J., Otto, H., Heyn, M. P., and Khorana, H. G., The retinylidene Schiff base counterion in bacteriorhodopsin, *J. Biol. Chem.*, 266, 18674, 1991.

156. Rath, P., Marti, T., Sonar, S., Khorana, H. G., and Rothschild, K. J., Hydrogen bonding interactions with the Schiff base of bacteriorhodopsin: Resonance Raman spectroscopy of the mutants D85N and D85A, *J. Biol. Chem.*, 268, 17742, 1993.

157. Nagle, J. F. and Mille, M., Molecular models of proton pumps, *J. Chem. Phys.*, 74, 1367, 1981.

158. Merz, H. and Zundel, G., Proton-transfer equilibriums in phenol-carboxylate hydrogen bonds. Implications for the mechanism of light-induced proton activation in bacteriorhodopsin, *Chem. Phys. Lett.*, 95, 529, 1983.

159. Brzezinski, B., Zundel, G., and Kraemer, R., Proton polarizability caused by collective proton motion in intramolecular chains formed by two and three hydrogen bonds. Implications for charge conduction in bacteriorhodopsin, *J. Phys. Chem.*, 91, 3077, 1987.

160. Olejnik, J., Brzezinski, B., and Zundel, G., A proton pathway with large proton polarizability and the proton pumping mechanism in bacteriorhodopsin — Fourier transform difference spectra of photoproducts of bacteriorhodopsin and of its pentademethyl analog, *J. Mol. Struct.*, 271, 157, 1992.

161. Roepe, P. D., Ahl, P. L., Herzfeld, J., Lugtenburg, J., and Rothschild, K. J., Tyrosine protonation changes in bacteriorhodopsin. A Fourier transform infrared study of bR_{548} and its primary photoproduct, *J. Biol. Chem.*, 263, 5110, 1988.

162. He, Y.-W., Krebs, M. P., Herzfeld, J., Khorana, H. G., and Rothschild, K. J., FTIR spectroscopy, site directed mutagenesis and isotope labelling: A new approach for studying membrane proteins, *Proc. SPIE-Int. Soc. Opt. Eng.*, 1575, 109, 1991.

163. Netto, M. M., Fodor, S. P. A., and Mathies, R. A., Ultraviolet resonance Raman spectroscopy of bacteriorhodopsin, *Photochem. Photobiol.*, 52, 605, 1990.

164. Ames, J. B., Ros, M., Raap, J., Lugtenburg, J., and Mathies, R. A., Time-resolved ultraviolet resonance Raman studies of protein structure: Application to bacteriorhodopsin, *Biochemistry*, 31, 5328, 1992.

165. McDermott, A. E., Thompson, L. K., Winkel, C., Farrar, M. R., Pelletier, S., Lugtenburg, J., Herzfeld, J., and Griffin, R. G., Mechanism of proton pumping in bacteriorhodopsin by solid-state NMR: The protonation state of tyrosine in the light-adapted and M states, *Biochemistry*, 30, 8366, 1991.

166. Metz, G., Siebert, F., and Engelhard, M., High-resolution solid state carbon-13 NMR of bacteriorhodopsin: Characterization of [4–13C]Asp resonances, *Biochemistry*, 31, 455, 1992.

167. Metz, G., Siebert, F., and Engelhard, M., Asp85 is the only internal aspartic acid that gets protonated in the M intermediate and the purple-to-blue transition of bacteriorhodopsin. A solid-state carbon-13 CP-MAS NMR investigation, *FEBS Lett.*, 303, 237, 1992.

168. Sonar, S., Lee, C.-P., Coleman, M., Patel, N., Liu, X., Marti, T., Khorana, H. G., RajBhandary, U. L., and Rothschild, K. J., Site-directed isotope labeling and FTIR spectroscopy of bacteriorhodopsin, *Nature Struct. Biol.*, 1, 512, 1994.

169. Ludlam, C. F. C., Sonar, S., Lee, C.-P., Coleman, M., Herzfeld, J., RajBhandary, U. L., and Rothschild, K. J., Site-directed isotope labeling and ATR-FTIR difference spectroscopy of bacteriorhodopsin: The peptide carbonyl group of Tyr 185 is structurally active during the bR→N transition, *Biochemistry*, in press, 1995.

170. Liu, X.-M., Sonar, S., Lee, C.-P., Coleman, M., RajBhandary, U. L., and Rothschild, K. J., Site-directed isotope labeling and FTIR spectroscopy: Assignment of tyrosine bands in the bR→M difference spectrum of bacteriorhodopsin, *Biophys. Chem.*, in press, 1995.

171. Creuzet, F., McDermott, A., Gebhard, R., van der Hoef, K., Spijker-Assink, M. B., Herzfeld, J., Lugtenburg, J., Levitt, M. H., and Griffin, R. G., Determination of membrane protein structure by rotational resonance NMR: Bacteriorhodopsin, *Science*, 251, 783, 1991.

172. Peersen, O. B. Yoshimura, S., Hojo, H., Aimoto, S., and Smith, S. O. Rotational resonance NMR measurements of intranuclear distances in an α-helical peptide, *J. Am. Chem. Soc.*, 114, 4332, 1992.

173. Draheim, J. E. and Cassim, J. Y., Large scale global structural changes of the purple membrane during the photocycle, *Biophys. J.*, 47, 497, 1985.

174. Dencher, N. A., Dresselhaus, D., Zaccai, G., and Bueldt, G., Structural changes in bacteriorhodopsin during proton translocation revealed by neutron diffraction, *Proc. Natl. Acad. Sci. U.S.A.*, 86, 7876, 1989.

175. Subramaniam, S., Gerstein, M., Oesterhelt, D., and Henderson, R., Electron diffraction analysis of structural changes in the photocycle of bacteriorhodopsin, *EMBO J.*, 12, 1, 1993.

176. Han, B.-G., Vonck, J., and Glaeser, R. M., The bacteriorhodopsin photocycle; Direct structural study of two substates of the M-intermediate, *Biophys. J.*, 67, 1179, 1994.

177. Ormos, P., Chu, K., and Mourant, J., Infrared study of the L, M, and N intermediates of bacteriorhodopsin using the photoreaction of M, *Biochemistry*, 31, 6933, 1992.

178. Rothschild, K. J., Marti, T., Sonar, S., He, Y. W., Rath, P., Fischer, W., Bousche, O., and Khorana, H., Asp-96 Deprotonation and transmembrane α-helical structural changes in bacteriorhodopsin, *J. Biol. Chem.*, 268, 27046, 1993.

179. Earnest, T. N., Herzfeld, J., and Rothschild, K. J., Polarized Fourier transform infrared spectroscopy of bacteriorhodopsin. Transmembrane alpha helices are resistant to hydrogen/deuterium exchange, *Biophys. J.*, 58, 1539, 1990.

180. Sasaki, J., Shichida, Y., Lanyi, J. K., and Maeda, A., Protein changes associated with reprotonation of the Schiff base in the photocycle of Asp96→Asn bacteriorhodopsin. The M_N intermediate with unprotonated Schiff base but N-like protein structure, *J. Biol. Chem.*, 267, 20782, 1992.

181. Sonar S., Marti, T., Rath, P., Fischer, W., He, Y.-W., Khorana, H. G., and Rothschild, K. J., A redirected proton pathway in the bacteriorhodopsin mutant Y57D: Evidence for proton translocation without Schiff base deprotonation, *J. Biol. Chem.*, 269, 28851, 1994.

182. Harbison, G. S., Roberts, J. E., Herzfeld, J., and Griffin, R. G., Solid-state NMR detection of proton exchange between the bacteriorhodopsin Schiff base and bulk water, *J. Am. Chem. Soc.*, 110, 7221, 1988.

183. De Groot, H. J. M., Smith, S. O., Courtin, J., van den Berg, E., Winkel, C., Lugtenburg, J., Griffin, R. G., and Herzfeld, J., Solid-state carbon-13 and nitrogen-15 NMR study of the low pH forms of bacteriorhodopsin, *Biochemistry*, 29, 6873, 1990.

184. Hildebrandt, P. and Stockburger, M., Role of water in bacteriorhodopsin's chromophore: resonance Raman study, *Biochemistry*, 23, 5539, 1984.

185. Deng, H., Huang, L., Callender, R., and Ebrey, T., Evidence for a bound water molecule next to the retinal Schiff base in bacteriorhodopsin and rhodopsin: A resonance Raman study of the Schiff base hydrogen/deuterium exchange, *Biophys. J.*, 66, 1129, 1994.

186. Maeda, A., Sasaki, J., Ohkita, Y. J., Simpson, M., and Herzfeld, J., Tryptophan perturbation in the L intermediate of bacteriorhodopsin: Fourier transform infrared analysis with indole-nitrogen-15 shift, *Biochemistry*, 31, 12543, 1992.

187. Fischer, W., Sonar, S., Marti, T., Khorana, H. G., and Rothschild, K. J., Detection of a water molecule in the active site of bacteriorhodopsin: Hydrogen bonding changes during the primary photoreaction, *Biochemistry*, 33, 12757, 1994.

188. Maeda, A., Sasaki, J., Yamazaki, Y., Needleman, R., and Lanyi, J. K., Interaction of Asp-85 with a water molecule and the protonated Schiff base in the L intermediate of bacteriorhodopsin: A Fourier-transform infrared study, *Biochemistry*, 33, 1713, 1994.

189. Diller, R., Iannone, M., Bogomolni, R., and Hochstrasser, R. M., Ultrafast spectroscopy of bacteriorhodopsin, *Biophys. J.*, 60, 286, 1991.

190. Harada, I., Yamagishi, T., Uchida, K., and Takeuchi, H., Ultraviolet resonance Raman spectra of bacteriorhodopsin in the light-adapted and dark-adapted states, *J. Am. Chem. Soc.*, 112, 2443, 1990.

31

The Early Receptor Potential and Its Analogue in Bacteriorhodopsin Membranes

Felix T. Hong
Wayne State University

31.1 Introduction

The early receptor potential (ERP), discovered by Brown and Murakami[1-3] in 1964, is fundamentally different from many commonly encountered bioelectric signals, such as action potentials and sensory receptor generator potentials. Unlike the latter, the ERP is not generated by diffusion of small ions through aqueous membrane channels, but rather by charge displacements in the membrane. It is therefore related closely to the gating current in the squid axon. This difference is highlighted by the comparison between the ERP and the late receptor potential, which is also known as the *a* wave of the electroretinogram (Figure 31.1). The late receptor potential is the electrical manifestation of the photoinduced decrease of the sodium ion current that enters the rod photoreceptor membrane at its outer segment and has a delay of 1.7 ms after flash-light stimulation. The ERP, however, has no detectable delay. The delay was estimated to be less than a microsecond at the time of its discovery, and probably less than a few picoseconds if modern instrumentation is used (cf. Reference 4). The ERP is also distinguished by its resistance to all types of rough treatments imposed on the photoreceptor. The late receptor potential can be eliminated by rendering the photoreceptor anoxic or by treating the retina with concentrated KCl. Yet the ERP persists after these treatments. The ERP depends on the integrity of the photoreceptor membrane and the fixed orientation of rhodopsin in the membrane; heat treatment abolishes the ERP, presumably by disrupting this orientation.

0-8493-8634-9/95/$0.00+$.50
© 1995 by CRC Press, Inc.

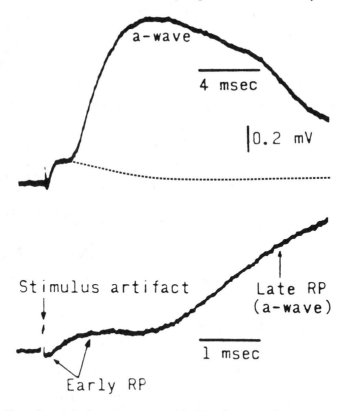

FIGURE 31.1 The early and the late receptor potentials from the retina of a Cynomolgus monkey. Both records were obtained under identical conditions except for the faster time base in the lower record. The stimulus artifact consists of an initial positive-going spike. The small negative-going peak is the R1 component, whereas the delayed positive phase following R1 is the R2 component. The R2 component merges with the late receptor potential after a latency of about 1.7 ms. The dotted line in the upper record shows the decay time course of R2 when the late receptor potential is abolished. (*Source:* Brown, K. T., Watanabe, K., and Murakami, M., *Cold Spring Harbor Symp. Quant. Biol.,* 30, 457, 1965.)

The ERP consists of two distinct molecular components. The slower component, known as the R2 component, can be inhibited reversibly by low temperature (0°C). The faster component, known as the R1 component, persists at −35°C. The two components have opposite polarity: the R1 component is cornea-positive in its polarity, whereas the R2 component is cornea-negative. The physicochemical basis of the ERP has been a subject of intense research since its discovery. The general conclusion is that the ERP is a manifestation of the intramolecular charge separation accompanying the light-induced conformational change. Since the visual pigment maintains a fixed orientation with respect to the membrane, such a process will certainly lead to the formation of a transient array of oriented electric dipoles. This mechanism is commonly known as the dipole mechanism or the oriented dipole mechanism. An alternative mechanism has been considered by Cone[5] and by Ostroy.[6] This mechanism is based on light-induced proton binding by rhodopsin. Cone[3] demonstrated that the appearance of the R2 signal is time-correlated with the conversion of rhodopsin from the metarhodopsin I intermediate to the metarhodopsin II intermediate. The latter reaction involves the uptake of an aqueous proton. This proton binding-based mechanism, which we shall refer to as the interfacial proton transfer mechanism, was abandoned subsequently, presumably because the expected pH dependence of the R2 signal could not be demonstrated experimentally. This apparent lack of pH dependence of the R2 component is a crucial point of discussion in this chapter.

General interest in the ERP gradually subsided in the 1970s because investigators eventually reached a conclusion that the ERP plays no role in the visual transduction process. In fact, the ERP is usually dismissed as an epiphenomenon — a synonym for an evolutionary vestige of little importance. Interest was revived when Trissl and Montal[7] demonstrated an ERP-like photoelectric signal in a reconstituted bacteriorhodopsin membrane. This signal consists of two components:[8] a temperature-insensitive B1 component and a slow but temperature-sensitive B2 component, virtually analogous to the R1 and R2 components, respectively. Bacteriorhodopsin is the sole protein component of the purple membrane of *Halobacterium halobium*.[9] Like rhodopsin, it is a retinal-containing protein with seven transmembrane α-helical loops.[10] However, unlike rhodopsin, bacteriorhodopsin is not a sensor protein but rather a transport protein which converts light energy into a transmembrane protein gradient for photosynthetic purposes. This is similar to the photosynthetic reaction center in *Rhodopseudomonas viridis*. However, unlike the bacterial reaction center, proton transport is carried out by a single polypeptide and without recourse to long-range electron transfers.

The elucidation of the ERP and the ERP-like photoelectric signal requires a combined electrochemical and electrophysiological analysis. The results of our analysis suggest that the ERP might function as a molecular switch for visual transduction. The subject of molecular switches has important technological implications and is discussed in Chapter 32.

31.2 Molecular Mechanism of the ERP and the Equivalent Circuit Analysis

In a 1972 study, Hagins and McGaughy[11] demonstrated that the photocurrent associated with the ERP satisfies the zero time-integral condition. This means that there is no net charge transport across the membrane during the transient of the ERP. This also means that the molecular process responsible for the generation of the ERP photoelectric signal must involve both charge separation *and* charge recombination. The oriented dipole mechanism can be revised readily to accommodate this requirement; the light-induced charge separation generates a transient array of electric dipoles that vanishes upon subsequent charge recombination. This idea of reversible charge displacements is supported by Cone's finding of a photoreversal potential:[5] the membrane, which is first enriched predominantly with metarhodopsin photointermediates by a prior illumination, responds to a second light flash with a signal that resembles the ERP in waveform but has its polarity reversed.

The light-induced oriented dipole mechanism, though consistent with many known observations about rhodopsin, is not the only possible mechanism to generate the ERP. When a proton is bound to rhodopsin during the metarhodopsin I to metarhodopsin II transition mentioned above, the counterions must be left behind in the adjacent diffuse electrical double layer. This event constitutes another kind of charge separation, taking place across the membrane-water interface. The subsequent reverse (back) reaction constitutes the required charge recombination. Experimental evidence in support of this idea was provided by Shevchenko et al.[12,13] These authors showed that illumination causes protons to be bound to the cytoplasmic side of the rod outer segment membrane, but causes no protons to be released from the intradiscal side. Thus, their experiment indicates that rhodopsin is not a proton transporter and that the protons bound to rhodopsin during metarhodopsin II formation must eventually go back to the cytosol.

The electrical equivalent circuits of these two molecular models can be deduced by applying the Gouy-Chapman diffuse double layer theory and conventional chemical kinetics analysis to these models.[14] A snapshot of the charge distribution profiles (shown in Figure 31.2) and the electrical potential profiles (not shown) across the membrane during the transient can be obtained. After linearization, these profiles can then be interpreted as the charging condition of three capacitances: the geometric capacitance due to the membrane dielectric and the two double-layer capacitances. A voltage source E_p' (photoelectromotive force) with an internal resistance of R_p is then added to

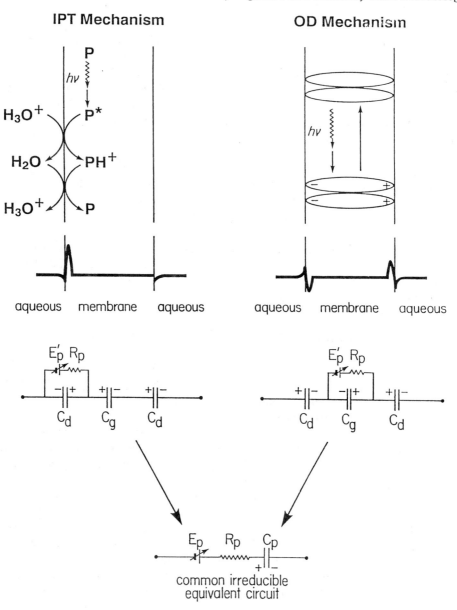

FIGURE 31.2   Two molecular mechanisms for the generation of the ERP and its bacteriorhodopsin analogue. The interfacial proton transfer (IPT) mechanism applies both to cytoplasmic proton binding and extracellular proton release at the membrane surface (only proton binding is shown). The oriented dipole (OD) mechanism applies to charge separation inside the membrane (or rather, inside the pigment molecule). The thick curve across the membrane shows the space charge distribution profile, which, together with the potential profile across the membrane (not shown), allows us to deduce the two microscopic equivalent circuits shown below. The two slightly different circuits are equivalent to the same irreducible equivalent circuit. See text for explanations. (*Source:* Hong, F. T., *Bioelectrochem. Bioenerg.,* 5, 425, 1978.)

each circuit so as to make it possible to generate the desired charge distribution pattern. These two slightly different equivalent circuits are shown under the respective models in Figure 31.2. Despite the detailed difference, these two circuits can be reduced further to the same common irreducible circuits shown at the very bottom of Figure 31.2. This irreducible circuit contains an RC network, consisting of R_p and C_p, that is usually associated with a linear high-pass RC filter. The filter time

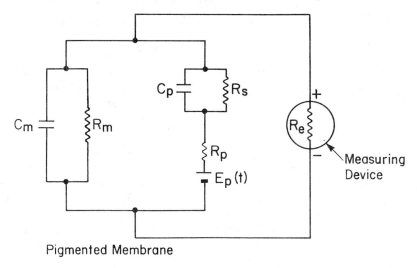

Pigmented Membrane

FIGURE 31.3 Equivalent circuit for a single component of the ERP or its bacteriorhodopsin analogue. The circuit includes the resistive and capacitative parameters of the inert supporting structure and the access impedance of the measuring device. R_e is the access impedance, which includes the input impedance of the measuring device, the electrode impedance, and the electrolyte impedance between the membrane and the electrodes. R_m and C_m are the ionic membrane resistance and the ordinary membrane capacitance of the membrane, with or without the membrane-bound pigment, respectively. C_p is the chemical capacitance. R_p is the internal resistance of the photoelectric voltage source $E_p(t)$, which is a function of the illuminating light power. R_s is the transmembrane resistance encountered by the dc photocurrent. See text for further explanations. (*Source:* Hong, F. T., in *Bioelectrochemistry: Ions, Surfaces, Membranes, Advances in Chemistry Series No. 188,* Blank, M., Ed., American Chemical Society, Washington, D.C., 1980, 211.)

constant is equal to $R_p C_p$. Thus, pulsed light-induced photoelectromotive force (E_p) will generate a photocurrent through this RC filter network. It is not possible for a dc photocurrent to pass through this circuit; only an ac photocurrent can. Upon the stimulation of a brief (<1 µs) light flash, a photocurrent with a spike waveform appears that represents the forward charge separation. Upon cessation of illumination, the photocurrent immediately reverses its polarity because only charge recombination can take place. The reversed phase of the photocurrent gradually declines according to a single exponential decay with the characteristic time constant of $R_p C_p$. The area above the baseline must equal the area below the baseline because these areas represent the total amounts of charges separated and recombined, respectively.

The time course of the photoelectric signal described above occurs under an idealized condition that has seldom been achieved in the laboratory, and the observed photosignal time course seldom follows the time course described above. Instead, the observed relaxation time course is strongly influenced by the particular measurement method being used. An examination of the equivalent circuit explains why.

It has been shown previously[15] that the series capacitance C_p, called the chemical capacitance, shown in Figure 31.2 is physically distinct from the ordinary membrane capacitance. Whereas C_p is connected in series to the photoelectromotive force E_p, the membrane capacitance C_m is connected in parallel. It is necessary to consider the interaction of two RC networks in order to understand the behavior of the photosignal. As shown in Figure 31.3, the RC network due to the lipid membrane, with or without embedded pigment, consists of R_m (membrane resistance associated with ionic conduction) and C_m (ordinary membrane capacitance), whereas the photochemical event is represented by another RC network as derived above and also shown in Figure 31.2. The interaction is absent only when the photosignal is measured under a strictly short circuit condition. The latter condition is much more difficult to achieve in the ERP measurement than in the measurement of action potentials for the following reasons. Because of the presence of a series capacitance, C_p, the source impedance of the photovoltage generator is frequency-dependent and is much reduced in the megahertz range than in the dc and the kilohertz range. A short circuit measurement (data reported as currents) requires that the access impedance, R_e, be much lower than the source impedance. Likewise, an open circuit measurement (data reported as voltages) requires the opposite extreme, i.e., R_e much higher than the source impedance. The frequency dependence of the source impedance implies that a short circuit measurement may be valid under the steady state or in the low frequency range, but may not be valid in the higher frequency range because the access impedance may not be small enough as the source impedance declines in the higher frequency range. The peril of ignoring this important point is exemplified by the use of a commercial picoammeter to measure the short circuit current in a reconstituted rhodopsin membrane. The photocurrent signal so measured was found to be the first derivative of the corresponding open circuit photovoltage.[16] It has been pointed out that such a relationship exists only when the attempted short circuit recording happens to be carried out under the same (open circuit) condition as does the photovoltage.

31.3 Effect of the Access Impedance on the Photosignal Relaxation

The effect of varying the access impedance from zero (i.e., much smaller than the source impedance) to infinity (i.e., much larger than the source impedance) is shown in Figure 31.4 for either brief light flash excitation (i.e., the light pulse is approximately a delta function) or steady step illumination (i.e., illumination with a long rectangular light pulse). The spike waveform that appears upon the onset and the cessation of steady illumination is characteristic of a linear high-pass filter consisting of a resistor and a capacitor. The steady photocurrent after the initial decay of the spike represents the dc photocurrent passing through R_s. In the case of the ERP, R_s is set to infinity because there is no dc photocurrent.

As R_e increases and approaches the value of the source impedance, the photocurrent must split into two fractions: one proceeds to the external measuring device and the other proceeds to charge the membrane capacitance C_m. The relaxation of the photosignal will then be the result of the interaction between the two RC networks mentioned above. The decay of the photocurrent upon flash-light excitation will no longer be a single exponential process, but rather a biexponential process as shown. The intrinsic decay time τ_p, however, can be recovered by deconvolution. The technical detail of deconvolution has been published elsewhere.[17]

As R_e approaches values much larger than that of the source impedance, virtually all photocurrent proceeds to charge the membrane capacitance, C_m. The voltage drop of a diminutive current across a large R_e is thus reported as the open circuit photovoltage. The relaxation of the photosignal will be largely determined by the RC time constant, $R_m C_m$. Under this circumstance, the photosignal relaxation will provide little information about the molecular processes. This is the main reason that the expected pH dependence of the R2 component has never been observed since most reported ERP data were collected under open circuit conditions. Although experimental evidence supporting the latter claim is lacking for the ERP, it is indeed the correct explanation for the ERP-like photosignal in reconstituted bacteriorhodopsin. The B2 signal, which had been shown to be pH insensitive under open circuit conditions, could be reversibly inhibited by low pH under near short circuit conditions.

There are reasons to believe that the ERP and the ERP-like photosignal in reconstituted bacteriorhodopsin membranes are governed by similar molecular processes and can be understood in terms of the same equivalent circuit. For example, the equivalent circuit predicts, and it has been observed[8] that as R_e decreases and approaches the short circuit condition, the B1 component becomes relatively larger than the B2 component and the apparent decay times of both components

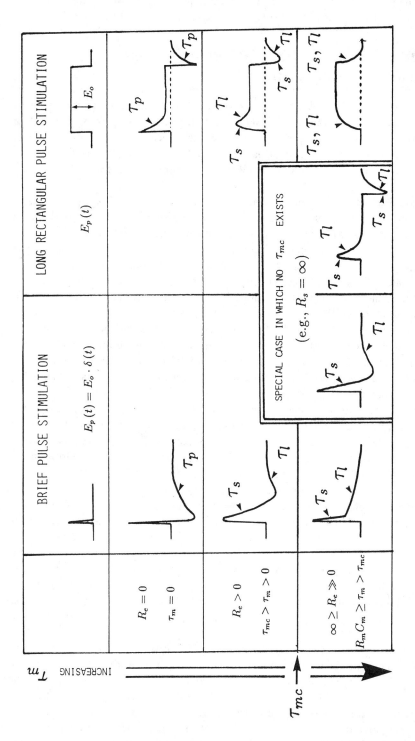

FIGURE 31.4 The dependence of the time course of ac photoelectric responses on the measurement conditions. Two types of light sources are considered: a brief alight pulse and a long rectangular pulse. τ_p is the intrinsic photochemical relaxation time. τ_m is the membrane discharging time constant, which consists primarily of $R_m C_m$ under open circuit conditions, but is replaced by $R_e C_m$ when R_e is reduced below the level of R_m. τ_s and τ_l are the observed apparent relaxation time constants, which vary as the measurement condition changes. (*Source:* Hong, F. T., *Bioelectrochemistry: Ions, Surfaces, Membranes, Advances in Chemistry Series No. 188*, Blank, M., Ed., American Chemical Society, Washington, D.C., 1980, 211.)

become shortened. Similar observations have been reported by Skulachev's group in a reconstituted rhodopsin membrane when the membrane was shunted by a proton ionophore that creates a short circuit path for the photocurrent.[13] Such a change, according to our equivalent circuit analysis, should have the same effect on the photocurrent time course as an external short circuit.

31.4 Component Analysis of the ERP-like Signal in Bacteriorhodopsin Membranes

As mentioned above, the ERP contains at least two components, as does the ERP-like signal in bacteriorhodopsin membranes. Yet the equivalent circuit was derived for a single charge separation process that relaxes in a first-order or a pseudo-first-order decay process. Therefore, the equivalent circuit is not expected to fit an experimental signal that contains more than one relaxation process. In fact, it was impossible to get the circuit to agree with the experimental signal when both B1 and B2 coexist.

Hong et al. have demonstrated previously that the equivalent circuit consistently predicts the time course of a photoelectric signal arising from interfacial electron transfers in a model system that has only a single pseudo-first-order decay process.[18] Thus, the first step toward the component analysis of the ERP or the ERP-like signal is to search for a method that would suppress one component completely but leave the other component intact. The "titration endpoint" of such a search process would be the consistent agreement of the photosignal with the equivalent circuit. Here, this procedure is illustrated with the decomposition of the ERP-like signal in reconstituted bacteriorhodopsin membranes.[17,19]

The original method used by Trissl and Montal[7] to demonstrate the existence of the ERP-like signal uses a thin Teflon film as the substrate on which an oriented layer of purple membrane sheets is deposited. The purple membrane somehow maintains a preferential orientation in such a way that the extracellular side is facing the Teflon substrate. This method allows investigators to observe both the B1 and B2 components.

Okajima[17] took the liberty of modifying the original Trissl-Montal method. Instead, the multiple oriented layers of purple membrane sheets were deposited on the Teflon and allowed to dry over a period of about 4 days or longer. The resulting preparation exhibits a considerably larger amplitude and a total lack of pH dependence. The remaining pH-independent photosignal resembles the original Trissl-Montal signal at low temperature and/or low pH, and is consistently in agreement with the equivalent circuit under a variety of conditions.

That the B1 signal so isolated is indeed a natural entity is supported by the following observations. The time course of the B1 component is completely insensitive to D_2O substitution, to variation of ionic strength, and to chemical modification by fluorescamine. In contrast, the B2 component exhibits a kinetic isotope effect, is sensitive to variation of ionic strength, and can be almost completely eliminated by treatment with fluorescamine, which can only attack the exposed hydrophilic domain of bacteriorhodopsin. Last but not least, the B2 component is completely absent between pH 6 and 11 in a genetic mutant in which the aspartate group at residue 212 is replaced with the asparagine group by site-directed mutagenesis and yet the B1 component remains intact.[20]

In contrast, the conventional approach, used in a number of reports, decompose the photosignal in terms of individual exponential decay terms and associate each of them with a distinct molecular process in the bacteriorhodopsin photocycle. Elsewhere, Hong et al.[21] have documented that such an approach is incompatible with the notion that the ERP is a capacitative signal satisfying the zero time-integral condition. From a practical point of view, such an approach led often to discrepancies between the interpretation of various laboratories. The difference in the decay constants reported by different laboratories may be the apparent consequence of inadvertent variation of the access impedance. Different laboratories often disagree as to which component is sensitive to D_2O exchange and which is not.

31.5 Molecular Interpretation of the ERP-like Signal

Based on the difference of the behavior of the B1 and B2 components, a tentative molecular interpretation can be established. The B2 component is generated most likely by an interfacial proton transfer mechanism; proton binding from the cytosol leads to charge separation, and its back reaction causes charge recombination. By virtue of symmetry, it can be deduced that a similar signal exists, reflecting the proton release event at the extracellular side. The experimental evidence supporting the existence of such a component, which was called B2′, has been documented in the literature.[22]

The B1 component seems rather insensitive to the change of the bathing aqueous solution. It is likely that B1 is generated by a molecular process closer to the chromophore binding pocket that is not in direct contact with the aqueous phase. Some investigators linked the signal with the light-induced deprotonation of the Schiff base that covalently bonds the retinal chromophore to the ε-amino group of lysine 216. The data of Hong et al. do not support this interpretation. First, the B1 component is not affected by overnight D_2O exchange. Second, the B1 component persists in a genetic mutant in which the aspartate at residue 85 is replaced with asparagine.[20] Residue 85 is considered to be the proton acceptor site for the Schiff base proton. Birge[23] has performed a molecular orbital calculation indicating that such a signal may be the manifestation of a dipole moment change when the chromophore undergoes photoisomerization.

31.6 Possible Physiological Role of the ERP

Birge's interpretation of the B1 component is consistent with the idea that photon energy is first stored as charge separation during the ultrafast photoisomerization. This stored energy is then used to drive conformational changes of the protein moiety, which occur on a slower time scale. This may be a recurring theme in retinal proteins. In addition to bacteriorhodopsin, an R1-like component was also found in reconstituted halorhodopsin membrane (called the H1 component). Halorhodopsin is the second most abundant retinal protein in *Halobacterium halobium*. Furthermore, an R2-like component (called H2) is also found in this latter membrane. However, the H2 component is not sensitive to pH change, but instead is sensitive to the chloride ion concentration in the bathing solution.[24]

The significance of the B2 and H2 components is readily understood because binding of charges (protons for bacteriorhodopsin and Cl^- for halorhodopsin) is an obligatory process for transport proteins. The significance of the R2 component of the ERP is more intriguing. As mentioned earlier, Shevchenko et al.[12] have demonstrated that rhodopsin is not a transport protein for protons. Their observation that protons are bound to the cytoplasmic domain rather than the intradiscal domain reinforces our suspicion that the R2 component may play a role in visual transduction after all.

A number of years ago, Hong et al. speculated about the possible existence of a light-induced surface potential during the appearance of the R2 signal as a result of proton binding.[14] It has been pointed out that such a process generates an intense electric field at the vicinity of the membrane surface where an aqueous proton is bound to a photoexcited rhodopsin. This electric field is highly localized since it exists only within the electrical double layer. This electric field also has a very rapid rise time upon illumination because the electrification is governed by ionic cloud relaxation rather than by ionic diffusion. The swift appearance of a highly localized electric field is ideal for a switching process such as visual transduction. The existence of this surface potential was subsequently verified by Cafiso and Hubbell[25] by means of a spin probe.

The observation of Shevchenko et al.[12] is actually consistent with the role of rhodopsin as a photosignal transducer and not a photon energy converter. It is well known that visual transduction incurs an energy amplification of 100,000-fold. Here, light merely acts as a trigger to initiate the transduction process; the energy for transduction has previously been stored. There is no need for rhodopsin to function as an energy converter. The biochemical process of visual transduction has been attributed to a sequence of events known as the cyclic GMP cascade.[26] The cyclic GMP cascade begins with the binding of the precursor of a G-protein (transducin) to the cytoplasmic domain of rhodopsin at the stage of metarhodopsin II. Thus, the R2-related surface potential appears at the right place and at the right time to make it possible for the surface potential to be involved in the switching process.[27] Furthermore, an electrostatic mechanism may also be involved in the termination of the visual transduction process. Hong et al. suspect that the termination is implemented by a delayed light-induced negative surface potential, generated by photophosphorylation of nine amino acid residues (threonine and serine) at the cytoplasmic domain of rhodopsin.[28] This occurs at about the time of the binding of arrestin or deactivation of transducin. It is thus conceivable that the photoreceptor utilizes a positive surface potential to turn on the transduction process while using a subsequent negative surface potential to turn it off. This idea is examined further and expanded in Chapter 32.

31.7 Discussion and Conclusions

Two often-cited reasons for dismissing the ERP's possible physiological role are (1) the ERP is too small and (2) the ERP-like signals are ubiquitous. It is true that ERP-like signals were found in the pigmented epithelium in the eye, the chloroplast membrane, etc. shortly after Brown and Murakami's discovery. Since the introduction of the technique of forming artificial lipid bilayer membranes *in vitro* ca. 1962,[29] many investigators have incorporated dyes or pigments in a lipid bilayer membrane and found ERP-like signals. With the benefit of hindsight, it is readily understood why ERP-like signals are ubiquitous. This is because many dyes are readily photooxidized and the accompanying light-induced rapid charge separation in the form of electron-hole pair production is manifest as an ERP-like signal. It is equally easy to understand why the externally measured ERP is so small. Because the photovoltage (or photocurrent) source resides either inside the membrane (oriented dipole mechanism) or at the membrane surface (interfacial charge transfer mechanism), the photoelectric event is partially screened by the diffuse double layer and therefore appears small when measured with an external device. However, using the right probe, as Cafiso and Hubbell did, the magnitude of the associated surface potential can be quite respectable.

In the above analysis, a light-induced electron transfer as a mechanistic equivalent of a proton transfer is treated tacitly. In fact, the analytical method based on the concept of chemical capacitance was first developed for light-induced electron transfers.[30] It was subsequently extended to light-induced proton transfers because it was found that the ac photoelectric signal arising from interfacial electron transfer in a lipid bilayer possesses all major characteristics of the ERP. Again, with the benefit of hindsight, a symmetric treatment of the mechanistic aspect of electron and proton transfers can be established by describing those processes as being coupled consecutive charge-transfer reactions. A useful general perspective evolves from this point of view.

From a bioenergetic point of view, signals such as B2 and B2' represent processes that are counterproductive; charge recombination has the same effect as internal short circuiting in a battery. Unfortunately, they cannot be completely eliminated but can be minimized. Nature's strategy is a sophisticated scheme of optimization. This fascinating subject is discussed in Chapter 32.

Finally, our analysis, which invokes an unfamiliar concept of chemical capacitance, marks a major departure from the common practice of analyzing relaxation data with a straightforward exponential decomposition. The latter approach has a substantial following in the investigation of the ERP and ERP-like signals. Interested readers should consult a recent review by Trissl[31] (see also references cited in Reference 21). Superficially, the two competing approaches can be viewed as alternate but complementary treatments of the subject, but a detailed comparison reveals that it is not the case. Although the present treatment is more complicated than the alternate one, it is more general than the latter. Here, only a single equivalent circuit model to explain virtually all electric phenomena arising from light-induced charge separation in membranes is used.[32] In contrast, an *ad hoc* model must be concocted for each case in the alternate approach. Furthermore, the latter approach lacks the predictive power of the approach described herein and often leads to conflicting interpretations from different laboratories. In light of increasing interest in the technological implications of the subject, the present approach seems to offer a distinct advantage as a design tool of molecular electronic devices.

Acknowledgments

This research was supported by a contract from the Office of Naval Research (N00014–87-K-0047) and a contract from the Naval Surface Warfare Center (N60921–91-M-G761). The author wishes to acknowledge the contributions of his collaborators, Janos Lanyi, Lowell McCoy, Mauricio Montal, and Richard Needleman. The experimental work that forms the basis of this review was performed by Man Chang, Albert Duschl, Filbert Hong, Sherie Michaile, Baofu Ni, and Ting L. Okajima.

References

1. Brown, K. T. and Murakami, M., A new receptor potential of the monkey retina with no detectable latency, *Nature (London)*, 201, 626, 1964.
2. Brown, K. T., Watanabe, K., and Murakami, M., The early and late receptor potentials of monkey cones and rods, *Cold Spring Harbor Symp. Quant. Biol.*, 30, 457, 1965.
3. Cone, R. A. and Pak, W. L., The early receptor potential, in *Handbook of Sensory Physiology, Vol. 1, Principles of Receptor Physiology*, Loewenstein, W. R., Ed., Springer, Berlin, 1971, 345.
4. Simmeth, R. and Rayfield, G. W., Evidence that the photoelectric response of bacteriorhodopsin occurs in less than 5 picoseconds, *Biophys. J.*, 57, 1099, 1990.
5. Cone, R. A., Early receptor potential: Photoreversible charge displacement in rhodopsin, *Science (Wash.)*, 155, 1128, 1967.
6. Ostroy, S. E., Rhodopsin and the visual process, *Biochim. Biophys. Acta*, 463, 91, 1977.
7. Trissl, H.-W. and Montal, M., Electrical demonstration of rapid light-induced conformational changes in bacteriorhodopsin, *Nature (London)*, 266, 655, 1977.
8. Hong, F. T. and Montal, M., Bacteriorhodopsin in model membranes: A new component of the displacement photocurrent in the microsecond time scale, *Biophys. J.*, 25, 465, 1979.
9. Stoeckenius, W., and Bogomolni, R. A., Bacteriorhodopsin and related pigments of *Halobacteria*, *Annu. Rev. Biochem.*, 51, 587, 1982.
10. Henderson, R., Baldwin, J. M., Ceska, T. A., Zemlin, F., Beckmann, E., and Downing, K. H., Model for the structure of bacteriorhodopsin based on high-resolution electron cryo-microscopy, *J. Mol. Biol.*, 213, 899, 1990
11. Hagins, W. A. and McGaughy, R. E., Molecular and thermal origins of fast photoelectric effects in the squid retina, *Science (Wash.)*, 157, 813, 1967.
12. Shevchenko, T. F., Kalamkarov, G. R., and Ostrovsky, M. A., The lack of H^+ transfer across the photoreceptor membrane during rhodopsin photolysis, *Sensory Systems (U.S.S.R. Acad. Sci.)*, 1, 117, 1987 (in Russian).
13. Ostrovsky, M. A., Animal rhodopsin as a photoelectric generator, in *Molecular Electronics: Biosensors and Biocomputers*, Hong, F. T., Ed., Plenum Press, New York, 1988, 381.
14. Hong, F. T., Mechanisms of generation of the early receptor potential revisited, *Bioelectrochem. Bioenerg.*, 5, 425, 1978.
15. Hong, F. T. and Okajima, T. L., Electrical double layers in pigment-containing biomembranes, in *Electrical Double Layers in Biology*, Blank, M., Ed., Plenum Press, New York, 1986, 129.
16. Trissl, H.-W., Light-induced conformational changes in cattle rhodopsin as probed by measurements of the interface potential, *Photochem. Photobiol.*, 29, 579, 1979.
17. Okajima, T. L. and Hong, F. T., Kinetic analysis of displacement photocurrents elicited in two types of bacteriorhodopsin model membranes, *Biophys. J.* 50, 901, 1986.

18. Hong, F. T., Charge transfer across pigmented bilayer lipid membrane and its interfaces, *Photochem. Photobiol.*, 24, 155, 1976.

19. Michaile, S. and Hong, F. T., Component analysis of the fast photoelectric signal from model bacteriorhodopsin membranes. Part I. Effect of multilayer stacking and prolonged drying, *Bioelectrochem. Bioenerg.*, 33, 135, 1994.

20. Hong, F. T., Hong, F. T., Needleman, R. B., Ni, B., and Chang, M., Modifying the photoelectric behavior of bacteriorhodopsin by site-directed mutagenesis: Electrochemical and genetic engineering approaches to molecular devices, in *Molecular Electronics — Science and Technology*, Aviram, A., Ed., American Institute of Physics, New York, 1992, 204.

21. Hong, F. T., Electrochemical processes in membranes that contain bacteriorhodopsin, in *Membrane Electrochemistry, Advances in Chemistry Series*, No. 235, Vodyanoy, I. and Blank, M., Eds., American Chemical Society, Washington, D.C., 1994, 531.

22. Hong, F. T., Effect of local conditions on heterogeneous reactions in the bacteriorhodopsin membrane: an electrochemical view, *J. Electrochem. Soc.*, 134, 3044, 1987.

23. Birge, R. R., Photophysics and molecular electronic applications of the rhodopsins, *Annu. Rev. Phys. Chem.*, 41, 683, 1990.

24. Michaile, S., Duschl, A., Lanyi, J. K., and Hong, F. T., Chloride ion modulation of the fast photoelectric signal in halorhodopsin thin films, in *Proc. 12th Annu. Int. Conf. IEEE Engineering in Medicine and Biology Society*, Philadelphia, PA, November 1–4, 1990, Oranal, B. and Pedersen, P. C., Eds., Institute of Electrical and Electronic Engineers, Washington, D.C., 1990, 1721.

25. Cafiso, D. S. and Hubbell, W. L., Light-induced interfacial potentials in photoreceptor membrane, *Biophys. J.*, 30, 243, 1980.

26. Stryer, L., Cyclic GMP cascade of vision, *Annu. Rev. Neurosci.*, 9, 87, 1986.

27. Hong, F. T., Relevance of light-induced charge displacements in molecular electronics: design principles at the supramolecular level, *J. Molec. Electron.*, 5, 163, 1989.

28. Wilden, U. and Kühn, H., Light-dependent phosphorylation of rhodopsin: Number of phosphorylation sites, *Biochemistry*, 21, 3014, 1982.

29. Tien, H. T., Membrane photophysics and photochemistry, *Prog. Surface Sci.*, 30, 1, 1989.

30. Hong, F. T. and Mauzerall, D., Interfacial photoreactions and chemical capacitance in lipid bilayers, *Proc. Nat. Acad. Sci., U.S.A.*, 71, 1564, 1974.

31. Trissl, H.-W., Photoelectric measurements of purple membranes, *Photochem. Photobiol.*, 51, 793, 1990.

32. Hong, F. T., Displacement photocurrents in pigment-containing biomembranes: Artificial and natural systems, in *Bioelectrochemistry: Ions, Surfaces, Membranes, Advances in Chemistry Series No. 188*, Blank, M., Ed., American Chemical Society, Washington, D.C., 1980, 211.

32

Molecular Electronic Switches in Photobiology

Felix T. Hong
Wayne State University

32.1 Introduction

Hecht and his co-workers[1] showed that a single visible photon absorbed by a single molecule of rhodopsin is capable of eliciting a neural response of the vertebrate rod photoreceptor. What entails in this process of absorption-excitation coupling is an energy amplification of 100,000-fold: the energy of the absorbed photon is minuscule compared with the energy required to generate a detectable change at the synaptic terminal of a photoreceptor cell. The photon energy serves merely as a trigger to release energy that has been stored prior to the arrival of photons. The photoreceptor response is a highly nonlinear function of the incident photon energy. In this sense, visual transduction, as the amplification process is usually called, is tantamount to the switching of a transistor, or more precisely, a phototransistor, in which variation of the base voltage leads to the massive flow of current between the emitter and the collector. What is remarkable in the photoreceptor function is the sensitivity and the dynamic range of its response.

Photosynthesis, nature's other photobiological function, is the conversion of the incident photon energy into another form of energy that is suitable for sustaining life. Photosynthetic energy conversion starts with light-induced vectorial charge separation in the direction perpendicular to the biological membrane in which the photosynthetic apparatus resides. Just like stored charges on a capacitor, the separated charge pairs have a tendency to recombine unless special provisions are instigated to minimize the recombination. In this regard, a conformational change of the photosynthetic pigment following light absorption may minimize the probability of charge recombination. The conformational change can be regarded as a molecular switch — a conformational switch or, in electronic terminology, a rectification. This is apparently the guiding principle in the design of a class of molecules with twisted intramolecular charge transfer (TICT) states.[2]

Switching functions can also be important in the fine-tuning of photosynthetic function in green plants. It is well known that there are two photosystems in the chloroplast that are linked in a Z-scheme by a mobile charge carrier, plastoquinone (for a concise review, see Reference 3). The two photosystems can be decoupled in response to changing light conditions by a process known as state 1-state 2 transition.[4] The process is mediated by an electrostatic mechanism initiated by photophosphorylation. Phosphorylation is a commonly utilized mechanism in enzyme activation. Phosphorylation of rhodopsin may be important in the deactivation of visual transduction. An electrostatic mechanism may be involved in the initiation and the termination of visual transduction.

In our survey of the role of molecular electronic switching in photobiology, the discussion will be facilitated greatly by also considering the function of bacteriorhodopsin. Bacteriorhodopsin is an integral protein found in the purple membrane of *Halobacterium halobium*.[5] Bacteriorhodopsin is similar in its structure and in its photochemical reaction to the visual pigment rhodopsin. However, bacteriorhodopsin performs a photon-converting function similar to the photosynthetic reaction center of the purple phototrophic bacterium, *Rhodopseudomonas viridis*. It therefore serves as a useful conceptual link between vision and photosynthesis.

In addition to analyzing molecular switches in naturally occurring systems, the use of bacteriorhodopsin and rhodopsin as advanced materials for molecular switch construction will be discussed. However, the present discussion is limited to the use of photoelectric properties of biopigments. Applications based on photochromism will be discussed by Birge et al. in a separate chapter.

32.2 Molecular Switching in Photon Energy Conversion

The primary and essential step in photobiological energy conversion is light-induced charge separation. The absorbed photon energy is utilized to drive photochemical reactions that lead to charge separation in the direction perpendicular to the membrane surface. Because photosynthetic pigments usually maintain a fixed orientation with respect to the membrane, the resulting charge separation is vectorial, i.e., equivalents of positive charges are transferred from the cytoplasmic side of the membrane to the opposite side.

In the case of chlorophyll-based photosynthetic membranes including the chromatophore of *Rhodopseudomonas viridis* and the chloroplast of the green plants, the charge separation is in the form of an electron-hole pair. Inevitably, some separated pairs of charges recombine subsequently and the energy is dissipated as heat; but others eventually span the membrane, and the photon energy is converted into a more stable form of energy in terms of a transmembrane proton gradient. This energy gradient is then utilized in accordance with Mitchell's chemiosmotic theory to synthesize ATP, or utilized directly to drive flagella motion or secondary ion transport across the membrane (symports or antiports).

In the case of the purple membrane of *Halobacterium halobium*, light absorption causes the chromophore retinal to undergo a photoisomerization, which leads to charge separation by virtue of deprotonation of the Schiff base that links the chromophore covalently to the protein moiety of bacteriorhodopsin. Thus, the charge separation is in the form of a proton-"negative hole" pair. As a hole is a vacancy created by losing an electron, a negative hole is a vacancy created by losing a proton. Similar to the chlorophyll-based membranes, the proton energy is converted eventually into a transmembrane proton gradient.

The path of electron transfers in the photosynthetic reaction center is well understood.[6] Electrons are transferred from a bacteriochlorophyll dimer (known as a special pair) through the monomer bacteriochlorophyll* and bacteriopheophytin to a bound quinone (Q_A) in about 200 ps. The electrons are then relayed to a mobile quinone (Q_B) and then to a cytochrome b-c$_1$ complex. While electrons are cycled back to the periplasmic side to refill the hole (oxidized special pair) via action of mobile cytochrome c and four membrane-bound cytochromes, an equivalent number of protons are transferred from the cytoplasmic side to the periplasmic side. The vectorial transfer of electrons can be viewed as coupled consecutive electron transfer reactions via a pathway known as the electron transport chain.

---

* Some investigators question whether the monomer bacteriochlorophyll is actually on the electron transfer path.

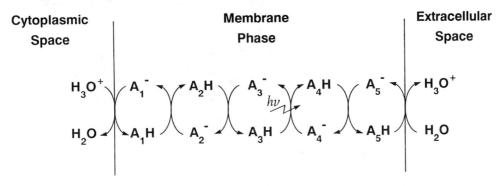

FIGURE 32.1 A model showing coupled consecutive proton transfer reactions across the purple membrane of *Halobacterium halobium*. The actual number of proton-binding sites are not known. For simplicity, only five binding sites (A_1 through A_5) are shown. Site A_3 is the proton binding site at the Schiff base linkage. It is understood that the Schiff base is neutral when unprotonated and positively charged when protonated. Only one of the reactions is light-driven. The reverse reactions are not shown. In spite of the presence of many reverse reactions, a net transport of protons occurs. (*Source:* Hong, F. T., in *Biomedical Engineering in the 21st Century,* Wang, C.-Y., Chen, C.-T., Cheng, C.-K., Huang, Y.-Y., and, Lin, F.-H., Eds., Center for Biomedical Engineering, National Taiwan University College of Medicine, Taipei, 1990, 85.)

The path of proton transfer in bacteriorhodopsin is less well understood, but is most likely to be provided by a series of transmembrane proton binding sites.[7] However, some authors suggested that bound water may also be involved. In analogy to the chromatophore membrane of *R. viridis*, the vectorial transfer of protons can be viewed as coupled consecutive proton transfer reactions via a pathway that can be called the proton transport chain (Figure 32.1).

The general rule that for every chemical reaction there is a reverse reaction takes a toll on the energy conversion efficiency because a reverse reaction for charge-transfer reaction means charge recombination. Charge recombination can be viewed as internal short-circuiting for a photovoltaic cell and will reduce the efficiency of photon energy conversion. However, absence of a reverse reaction is not a necessary condition for a photon energy conversion scheme that is based on coupled consecutive charge transfers in a membrane.

A simple kinetic analysis indicates that a minimum requirement for coupled consecutive charge (electron or proton) transfer reactions along a transmembrane pathway to work as a solar energy converter is that one of the forward charge transfer steps is driven by light. Regardless of the existence of reverse charge transfer (charge recombination) in each and every step, a fraction of absorbed photon energy will be converted into electrochemical energy in the form of a transmembrane proton gradient. This assertion is supported by our model system study performed with an artificial lipid bilayer membrane.[8] This system consists of a single light-driven interfacial electron transfer with an experimentally measurable reverse reaction, but it is capable of carrying out a net electron transfer across the membrane; a dc photocurrent can be detected with steady light illumination.

This minimalist system with a single photon-driven step constitutes the "basal" solar energy conversion system against which the performance of real-life photosynthetic systems will be evaluated. Clearly, the efficiency of solar energy conversion will be enhanced if the reverse reactions in one, if not all, of these steps of charge transfer can be reduced. However, attempts to minimize charge recombination may lead to complications that also reduce the conversion efficiency.

In long-range electron transfers that occur in the photosynthetic apparatus, both the rate of forward and reverse electron transfers are controlled by the distance between the donor-acceptor pair. Therefore, any attempt to minimize the reverse electron transfer rate by increasing this distance also decreases the forward transfer rate, i.e., rate of charge separation. One way to reduce the reverse transfer but to enhance the forward transfer is to increase the forward driving force, i.e., to make the free energy difference, ΔG, more negative or to make the redox potential of the charge-separated state

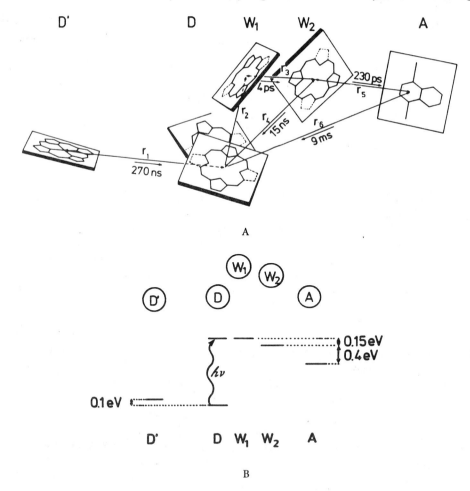

FIGURE 32.2 Spatial arrangement (A) and energy levels (B) of the electron transport chain in the reaction center of *Rhodopseudomonas viridis*. D is the special pair. W_1 is the monomer bacteriochlorophyll. W_2 is bacteriopheophytin. A is the tightly bound quinone Q_A. D' is the secondary electron donor cytochrome. See text for explanation. (*Source:* Kuhn, H., *Phys. Rev. A*, 34, 3409, 1986.)

significantly more positive than the charge-recombined state. However, such a strategy may incur unacceptable energy loss as heat. Besides, excessive increase of the driving force may actually reduce the forward charge transfer rate, as explained by Marcus' theory. It is therefore instructive to see how nature deals with this dilemma.

Kuhn's[9] analysis of the bacterial reaction center of *Rhodopseudomonas viridis* indicated that nature minimizes charge recombination by evolving an intelligent supramolecular structure (Figure 32.2). Kuhn found the spatial positioning of prosthetic groups within the reaction center to be optimal for an efficient forward electron transfer and for a diminished reverse electron transfer. Nature's strategy is to move electrons as far away as possible from the primary donor special pair and also as quickly as possible. This is done efficiently by quantum mechanical tunneling. Since the electron tunneling rate diminishes exponentially with the barrier width, and most single-step electron tunneling can only go as far as 10 to 15 Å, nature utilizes three conjugate molecules to form an extended π-electron system so that the electron can be transferred about 30 Å in an essentially activationless step. The energy levels of the special pair, the monomer bacteriochlorophyll, and bacteriopheophytin are closely matched so as to minimize energy loss by heat dissipation. The transferred electrons are trapped essentially at the fixed quinone, Q_A, for two reasons. First, a

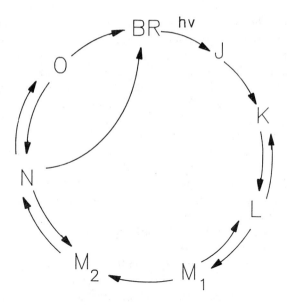

FIGURE 32.3 Bacteriorhodopsin photocycle. This latest version of the bacteriorhodopsin photocycle differs from earlier versions in two important aspects. First, there are two M states in which the forward reaction, M_1 to M_2, is strongly favored. Second, reverse reactions in several transitions are detected. (*Source:* Váró, Gy. and Lanyi, J. K., *Biochemistry (Wash.),* 30, 5016, 1991.)

vibronic relaxation allows the electron to lose about 0.4 eV energy so as to make charge recombination an uphill reaction. Second, the three components of the extended π-electron system are arranged in the shape of a banana. The consequence of such a spatial arrangement is that direct charge recombination by electron tunneling through the shortest path must now go through the σ-bond portion of the protein at the reaction center. The alternative path of charge recombination through the extended π-electron system now requires a thermal activation of 0.4 eV. In addition, the charge recombination is made even less likely by quickly refilling the hole created by photooxidation of the special pair via electron transfer from the secondary electron donor cytochromes. The vibronic relaxation at Q_A can be regarded as a process during which the return electron path is switched from the extended π-electron system to a tunneling barrier which is made of a σ-bonded portion of protein. The switching does not take place until the separated electrons have moved sufficiently far away from the hole so as to make charge recombination by tunneling much less likely.

Bacteriorhodopsin is a single-chain polypeptide without any electron binding prosthetic group. It performs a task that leads to the same end result as in the reaction center: namely, the generation of a transmembrane proton gradient. Apparently, no long-range electron transfers are involved; protons are transferred from the cytoplasmic side to the extracellular side without the help of quinonelike compounds. A superficial comparison reveals little resemblance between bacteriorhodopsin and the reaction center. However, an examination of the underlying principle shows otherwise.

Bacteriorhodopsin undergoes a cyclic photochemical reaction upon light stimulation (Figure 32.3). The first step is a photoisomerization of the chromophore; subsequent conformational changes that are accompanied by shifts of the absorption maximum and a concurrent proton movement from the cytoplasmic side to the extracellular side. These events can be followed photometrically in terms of a photocycle, and electrically in terms of fast photoelectric signals (see Chapter 31). It is generally held that the driving force for the vectorial proton transport in bacteriorhodopsin is the decrease of the pK_a of the proton binding site at the Schiff base (which links the chromophore, *all-trans*-retinal, to the ε-amino group of lysine at residue 216). The decrease of

pK_a is caused in turn by photoisomerization of the chromophore. A proton is then transferred from the Schiff base to the aspartate group of residue 85; the proton vacancy created by this is then refilled with a proton from the aspartate group of residue 96. Consider the general scheme shown in Figure 32.1. This scheme is similar to that for the reaction center except protons replace electrons as charge-carrying species. Both schemes can be described as coupled consecutive charge transfer reactions; the reactants of a reaction are the products of the preceding reaction, and the products, in turn, become the reactants of the subsequent reactions.

Váró and Lanyi's[10] thermodynamic analysis of the photochemical reactions of bacteriorhodopsin revealed that most of these states have closely matched free energy levels (i.e., highly reversible) with the exception of the M1-to-M2 transition, during which there is a significant decrease of the free energy that is mainly due to an entropic change. It is the M1-to-M2 transition that makes the forward proton transfer essentially one-way. This M1-to-M2 transition constitutes a molecular switch that retards greatly the reverse charge transfer and the resultant internal short-circuiting.

Another perspective of the inner working of bacteriorhodopsin as a photoconverter can be achieved by examining the fast photoelectric signals that are a macroscopic manifestation of light-induced charge separation and recombination. A faster B1 component, that is most likely due to light-induced change of the dipole moment as a result of photoisomerization, is pH independent. There are two slower signal components, B2 and B2′, each associated with an interfacial proton transfer reaction at the membrane surface of the cytoplasmic side and the extracellular side, respectively. Both the B2 and the B2′ components are pH dependent. Hong's analysis[11] suggested that the pH dependence of these signals are due mainly to the pK_a shift of their respective proton binding groups at the membrane surfaces under the influence of the pH of the adjacent aqueous phase. Light-induced proton pumping from the intracellular space to the extracellular space eventually causes the intracellular pH to rise and the extracellular pH to fall. The resulting changes in pH (proton concentration) tend to enhance the reverse interfacial proton transfers at the two surfaces by virtue of the law of mass action. Nature managed to alleviate this condition by enhancing the rate of forward proton transfer in response to the pH changes via pH-dependent pK_a of the surface proton binding groups. For example, the rising pH of the intracellular side, as a consequence of proton pumping, will diminish the proton binding rate if the pK_a remains fixed. An increased pK_a will thus enable forward proton pumping to keep its pace. The analysis suggests further that the two interfacial proton transfer reactions are light-assisted, apparently as a result of the global conformational change initiated by the more localized photoisomerization of the chromophore. Such an intelligent adjustment is possible because various parts of the single chain polypeptide constituting the protein moiety of bacteriorhodopsin are coupled together conformationally. Apparently, bacteriorhodopsin is equipped with more than one conformational switch to retard charge recombination. Alternatively, the molecule can be viewed to undergo a single step of conformational switching that leads to many localized switching functions because of conformational coupling. How this task was learned through evolution is an intriguing question but is something to be exploited technologically if investigators plan to breed advanced biomaterials in the laboratory by genetic engineering.

32.3 Molecular Switching in Visual Transduction

Signals similar to the B1, B2, and B2′ components mentioned above were first discovered in the visual photoreceptor membrane of the monkey. It is known as the early receptor potential (ERP) (see Chapter 31). There are two known components, R1 and R2. Like B1, the R1 component is probably related to the photoisomerization of the chromophore of rhodopsin. Likewise, the R2 component, which is analogous to B2, is probably caused by interfacial proton binding to rhodopsin. The R2 component appears at the same time as rhodopsin is converted from the metarhodopsin I intermediate to the metarhodopsin II intermediate. The metarhodopsin I-to-metarhodopsin II transition, like the M1-to-M2 transition in bacteriorhodopsin, consists of a major conformational change in the photobleaching sequence of rhodopsin. The transition thus constitutes a conformational switch that turns on a biochemical amplification process known as the cyclic GMP cascade.

Hong et al. have analyzed the photoelectric events and the underlying molecular processes and have speculated that the switching process during metarhodopsin I-to-metarhodopsin II transition is mediated by an electrostatic mechanism:[12] the light-induced positive surface potential associated with proton binding at the cytoplasmic domain of rhodopsin is the "on" switch, whereas the subsequent development of a negative surface potential as a result of photophosphorylation of the cytoplasmic domain of rhodopsin is the "off" switch.

The appearance of a light-induced surface potential creates an electric field at the vicinity of the membrane and inside the membrane. This electric field is intense because the surface potential declines to the value in the bulk phase within a very short distance of the order of the Debye length. It is also very localized because of charge screening. Thus, this local but intense photoinduced electric field can generate a concentration jump of counterions at the membrane surface within a very short time because the change of local (surface) concentration is governed by ionic cloud relaxation rather than by diffusion.

Based on an analysis of the events associated with binding of the transducin precursor and the appearance of the ERP R2 signal, it is concluded that the appearance of the R2 signal is associated with the appearance of a local but intense electric field at the cytoplasmic domain.[12] This light-induced local electric field may cause metarhodopsin II to bind transducin either by the formation of salt bridges or via a concentration jump of an important regulatory ion yet to be identified.

The validity of such an electrostatic-based trigger mechanism of visual transduction must await experimental verification, but the feasibility of the mechanism may be experimentally assessed in a model system. Drain and co-workers[13] have studied the regulation of ionic current across an artificial lipid bilayer membrane, which is carried by a hydrophobic ion (tetraphenylboride) under the influence of an externally applied transmembrane potential (Figure 32.4). These authors have also built into the membrane a mechanism for generating positive surface potentials at both sides of the membrane by incorporating a magnesium porphyrin. Photoinduced electron transfer from the membrane phase to the two aqueous phases can be elicited if the membrane is bathed in two identical aqueous solutions that contain equal concentrations of an electron donor. The two symmetrical light-induced positive surface potentials do not generate a photovoltaic signal but can lead to a concentration jump of tetraphenylboride ions, resulting in a sudden increase of the ionic current. If, however, the cation tetraphenylborate is replaced by the anion tetraphenylphosphonium, a sudden decrease of ionic current can be detected. The mechanism resembles the switching mechanism found in a conventional field-effect transistor or phototransistor.

For comparison as a basal minimalist system, a simple photosensor can be constructed using either a reconstituted rhodopsin or a bacteriorhodopsin membrane. The ac photoelectric signal elicited in either membrane is a linear function of the stimulating light. A prototype has been constructed by Miyasaka et al. who designed it to be used as a motion detector.[14] The transformation of this basal system into a highly nonlinear one with an enormous dynamic range requires the addition of a switching mechanism possibly mediated by the ERP and an amplification mechanism involving the binding and activation of transducin.

32.4 Molecular Switching as a Means for Resource Allocation in Photosynthesis

The electrostatic mechanism outlined above may be fairly general in the biological world. In fact, phosphorylation is a common mechanism of enzyme activation. Here, photophosphorylation of the light-harvesting chlorophyll-protein complex II (LHC-II) is cited as a concrete example of electrostatic switching (state 1 to state 2 transition). LHC-II also serves as an antenna-pigment

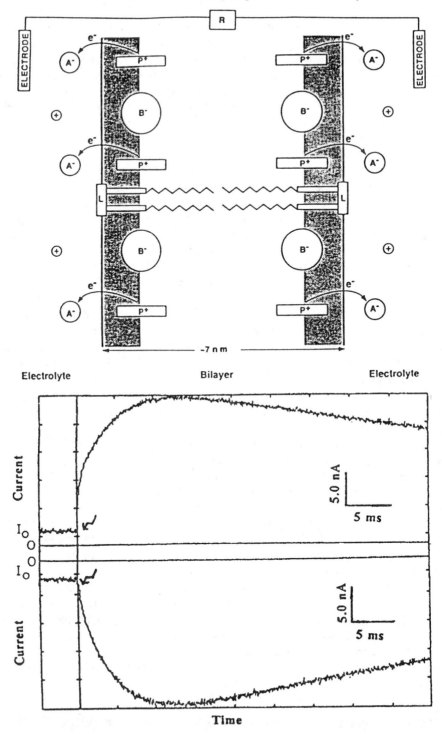

FIGURE 32.4   (A) An experimental prototype showing photoswitching of an ionic current in an artificial bilayer lipid membrane. The lipid bilayer contains magnesium octaethylporphyrin, which is lipid-soluble (3.6 mM). The aqueous phases contain the electron acceptor anthraquinonesulfonate in equal concentrations on both sides (0.1 mM). Tetraphenylboride ions, B$^-$, are partitioned into the membrane at the region of polar head groups in the lipid (the so-called boundary region). Photoactivation of magnesium octaethylporphyrin generates two symmetrical positive surface potentials (or rather, boundary potentials) that increase the surface concentration of B$^-$. (B) The photocurrent induced by excitation from a laser pulse delivered at the time indicated by the arrows. The upper and the lower traces represent ionic currents when +40-mV and −40-mV potentials are applied across the membrane, respectively. (*Source:* Drain, C. M. and Mauzerall, D., *Bioelectrochem. Bioenerg.*, 24, 263, 1990.)

complex to gather photons and to funnel the energy to the reaction center of Photosystem II. Photophosphorylation of LHC-II at its cytosolic side increases the negative charge density.[15] The increasing negative charge density could lead to lateral mobility of LHC-II, and thus affects the coupling of LHC-II to the Photosystem II reaction center via resonant energy transfer. It could also cause both LHC-II and the photosystem II reaction center to move away from Photosystem I. As suggested by Barber,[3] this mechanism may be utilized to regulate the ratio of the cyclic and the noncyclic electron flows in response to changing light conditions. Thus, the state 1-state 2 transition exerts its regulatory action via an electrostatic switching mechanism to allocate dynamically the resources for photosynthesis.

32.5 Discussion and Technological Perspectives

Photosynthesis and vision are intended for diametrically different photobiological functions. In the case of photosynthesis, the purpose is to convert solar energy into a universal form of chemical energy that can be utilized in many cellular functions. The performance of a photosynthetic apparatus is judged by its efficiency of conversion. In contrast, light is used in visual transduction merely as a trigger to regulate energy stored previously as an ionic concentration gradient. Sensitivity and dynamic range instead of efficiency become the primary concern of the performance of a visual photoreceptor. Two different classes of molecules with conjugated molecular structures are selected as chromophores: chlorophylls and bacteriochlorophylls for photosynthetic plants and bacteria, and retinals for visual pigments in the animal kingdom. A direct comparison between photosynthesis and vision was considered previously to be unproductive or irrelevant. With the advent of bacteriorhodopsin, a visual pigment being exploited for photosynthesis, the conceptual gap has been narrowed. In the present chapter, a comparison of the two functions from a mechanistic point of view is attempted. The approach of first establishing a baseline of a minimalist molecular machine is adopted. By so doing, it is easier to appreciate the sophistication of nature's invention. Thus, there is a common theme threading between the two functions. One of them is light-induced rapid charge separation as a means of storing light energy quickly in an intermediate form for driving conformational changes of the proteins. The other turns out to be switching functions. A switching function is used here to enhance the performance of the molecular machine.

The term "switching" strikes a familiar ring for technologically oriented minds. The soaring interest in bacteriorhodopsin, rhodopsin, and the bacterial reaction center is not a mere continuation of scientific efforts of the past generation, but is also motivated by the possibility of using biomaterials as building blocks for electronic devices. As discussed in a separate chapter by Birge et al. in this volume, bacteriorhodopsin-based devices are a few examples that hold promise for commercial viability.

Studies of photobiological membranes are relevant to the development of molecular electronic devices for two reasons. By analyzing and comparing various photobiological systems, it may be possible to reveal nature's secret design principles — an approach to be referred to as reverse engineering.[16] The present chapter represents a modest attempt to do so. The second, but no less important, reason is to develop biomaterials suitable for device construction. In this regard, a new concept of intelligent materials has become a focal point of attention in materials science research. While the definition has been a continuing subject of debate, the position advocated is that intelligence can only be evaluated from the intended function. This is the motivation behind the strategy outlined here to evaluate photobiological functions with reference to a basal minimalist system.

While many biomaterials are of optimal design through billions of years of evolution, they may not be suitable for direct technological applications. After all, nature might never have had in mind our intended applications while establishing the criteria of optimization. Further optimization becomes the task of a new breed of technologists named molecular engineers. Nature apparently favors proteins as building blocks for molecular machines, presumably because of the vast repertoire of conformation states held by a protein of modest complexity. Molecular engineering used to be a tedious task. With the advent of modern recombinant DNA technology, it is now possible to optimize molecular designs by site-directed mutagenesis. For example, Khorana and co-workers[17] have pioneered the technique of producing genetic mutants of bacteriorhodopsin in the laboratory, using *E. coli* as the expression system. Needleman and co-workers[18] extended the method using the native bacterium *Halobacterium halobium,* and improved vastly both the yield and the quality of engineered proteins. Now, the rate-limiting step of future progress is the design of a rapid and efficient procedure for screening the mutants. Techniques as such are not only important for biomaterials research, but also for investigations using a reverse-engineering approach. A potentially rich source of information lies in *H. halobium* itself. This bacterium, that produces bacteriorhodopsin, also produces three additional retinal proteins — halorhodopsin, sensory rhodopsin I, and sensory rhodopsin II (the latter two are also known as sensory rhodopsin and phoborhodopsin, respectively). Halorhodopsin is a light-driven chloride ion pump, while the two sensory rhodopsins are photon sensors used by the bacterium for phototaxis. A detailed comparison of these pigments not only provides clues of past evolution that might not be otherwise available, but also allows us a glimpse of nature's strategy in the design of photobiological machines. Recently, halorhodopsin and sensory rhodopsin I have been expressed in *H. halobium* at high levels (50-fold enhancement in both cases) by Needleman (personal communication) and by Spudich (personal communication), respectively. These pigments, which are minority pigments in native *H. halobium,* can now be obtained in large enough quantities for detailed experimental analysis. These recent accomplishments not only make it feasible for industrial-scale production of these retinal proteins for use as advanced molecular electronic materials, but also raise high hope for investigators adopting the reverse-engineering approach in molecular designs.

Acknowledgments

This research was supported by a contract from the Office of Naval Research (N00014–87–K-0047) and a contract from the Naval Surface Warfare Center (N60921–91-M-G761).

References

1. Hecht, S., Shlaer, S., and Pirenne, M. H., Energy, quanta, and vision, *J. Gen. Physiol.,* 25, 819, 1942.
2. Grabowski, Z. R., Rotkiewicz, K., Siemaiarczuk, A., Cowley, D. J., and Baumann, W., Twisted intramolecular charge transfer states (TICT): A new class of excited states with a full charge separation, *Nouv. J. Chim.,* 3, 443, 1979.
3. Barber, J., Photosynthetic electron transport in relation to thylakoid membrane composition and organization, *Plant Cell Environ.,* 6, 311, 1983.
4. Myers, J., Enhancement studies of photosynthesis, *Annu. Rev. Plant Physiol.,* 22, 289, 1971.
5. Stoeckenius, W. and Bogomolni, R. A., Bacteriorhodopsin and related pigments of *Halobacteria, Annu. Rev. Biochem.,* 51, 587, 1982.
6. Deisenhofer, J. and Michel, H., The photosynthetic reaction center from the purple bacterium *Rhodopseudomonas viridis, Science (Wash.),* 245, 1463, 1989.
7. Henderson, R., Baldwin, J. M., Ceska, T. A., Zemlin, F., Beckmann, E., and Downing, K. H., Model for the structure of bacteriorhodopsin based on high-resolution electron cryo-microscopy, *J. Mol. Biol.,* 213, 899, 1990.
8. Hong, F. T., Charge transfer across pigmented bilayer lipid membrane and its interfaces, *Photochem. Photobiol.,* 24, 155, 1976.
9. Kuhn, H., Electron transfer mechanism in the reaction center of photosynthetic bacteria, *Phys. Rev. A,* 34, 3409, 1986.

10. Váró, Gy. and Lanyi, J. K., Thermodynamics and energy coupling in the bacteriorhodopsin photocycle, *Biochem. (Wash.)*, 30, 5016, 1991.

11. Hong, F. T., Intelligent materials and intelligent microstructures in photobiology, *Nanobiology*, 1, 39, 1992.

12. Hong, F. T., Relevance of light-induced charge displacements in molecular electronics: Design principles at the supramolecular level, *J. Molec. Electron.*, 5, 163, 1989.

13. Drain, C. M. and Mauzerall, D., An example of a working molecular charge sensitive ion conductor, *Bioelectrochem. Bioenerg.*, 24, 263, 1990.

14. Miyasaka, T., Koyama, K., and Itoh, I., Quantum conversion and image detection by a bacteriorhodopsin-based artificial photoreceptor, *Science (Wash.)*, 255, 342, 1992.

15. Kyle, D. J. and Arntzen, C. J., Thylakoid membrane protein phosphorylation selectively alters the local membrane surface charge near the primary acceptor of Photosystem II, *Photobiochem. Photobiophys.*, 5, 11, 1983.

16. Hong, F. T., Retinal proteins in photovoltaic devices, in *Molecular and Biomolecular Electronics, Advances in Chemistry Series No. 240*, Birge, R. R., Ed., American Chemical Society, Washington, D.C., 1994, 527.

17. Gilles-Gonzalez, M. A., Hackett, N. R., Jones, S. J., and Khorana, H. G., Methods for mutagenesis of the bacterio-opsin gene, *Meth. Enzymol.*, 125, 190, 1986.

18. Ni, B., Chang, M., Duschl, A., Lanyi, J., and Needleman, R., An efficient system for the synthesis of bacteriorhodopsin in *Halobacterium halobium*, *Gene*, 90, 169, 1990.

33

Biomolecular Photonics Based on Bacteriorhodopsin

Robert R. Birge
Zhongping Chen
Deshan Govender
Richard B. Gross
Susan B. Hom
K. Can Izgi
Jeffrey A. Stuart
Jack R. Tallent
Bryan W. Vought
Syracuse University

33.1 Introduction

This chapter examines the use of the light-transducing protein bacteriorhodopsin in photonic devices. The principal goal is to provide the reader with a perspective on the use of this protein in selected applications that include holography, spatial light modulators, neural network optical computing, nonlinear optical devices, and optical memories. There are significant advantages inherent in the use of biological molecules, either in their native form, or modified via chemical or mutagenic methods, as active components in photonic devices.[1-3] These advantages derive in large part from the natural selection process, and the fact that nature has solved, through trial and error, problems of a similar nature to those encountered in harnessing organic molecules to carry out logic, switching, or data manipulative functions. Light-transducing proteins such as visual rhodopsin,[3] bacteriorhodopsin,[2] chloroplasts,[4,5] and photosynthetic reaction centers[6] are salient examples of biomolecular systems that have been investigated for photonic applications. Bacteriorhodopsin has received more attention with respect to photonics than any other biomolecule studied to date. The significance of bacteriorhodopsin stems from its biological function as a photosynthetic proton pump in the bacterium *Halobacterium halobium*. A combination of serendipity and natural selection has yielded a native protein with characteristics near optimum for many linear and nonlinear optical applications. Genetic engineering combined with chemical modification and chromophore substitution methods also provides an additional flexibility in the optimization of this material for individual applications.

33.2 Linear Photonic Devices

Applications of bacteriorhodopsin that operate via one-photon absorption are defined for the purposes of this chapter as linear. Most of these devices operate by selecting a pair of the ground

0-8493-8634-9/95/$0.00+$.50
© 1995 by CRC Press, Inc.

states that populate the complex photocycle of the protein. In order to appreciate the characteristics of the photonic devices based on bacteriorhodopsin, it is necessary to understand the nature of the bacteriorhodopsin photocycle. The following section is devoted to a brief overview of the photochemistry of this protein.

Structure and Photochemistry

Bacteriorhodopsin (MW $\cong$ 26,000 Da) is the light-harvesting protein in the purple membrane of a microorganism called *Halobacterium halobium*.[3,7–10] This bacterium thrives in salt marshes where the concentration of salt is roughly six times higher than sea water. The purple membrane, which constitutes a specific functional site in the plasma membrane of the bacterial cell, houses semicrystalline protein trimers in a phospholipid matrix (3:1 protein to lipid). The bacterium synthesizes the purple membrane when the concentration of dissolved oxygen in its surroundings becomes too low to sustain ATP production through aerobic respiration. The light-absorbing chromophore of bacteriorhodopsin is *all-trans*-retinal (Vitamin A aldehyde) (Figure 33.1). It is bound to the protein through a protonated Schiff base linkage to a lysine residue attached to one of the seven α-helices that make up the protein's secondary structure. The absorption of light energy by the chromophore initiates a complex photochemical cycle characterized by a series of spectrally distinct thermal intermediates and a total cycle time of approximately 10 ms (see Figure 33.2). As a result of this process, the protein transports a proton from the intracellular to the extracellular side of the membrane. This light-induced proton pumping generates an electrochemical gradient that the bacterium uses to synthesize ATP in accordance with Mitchell's chemiosmotic model of energy transduction. Thus, bacteriorhodopsin provides the necessary molecular machinery for *H. halobium* to convert light energy into a metabolically useful form of energy vital for the cell's survival under anaerobic conditions. Accordingly, *H. halobium* can switch from aerobic respiration to photosynthesis in response to changing environmental conditions. The fact that the protein must function in the harsh environment of a salt marsh requires a robust protein resistant to both thermal and photochemical damage. The cyclicity of the protein (i.e., the number of times the protein can be photochemically cycled between intermediates before denaturing) exceeds 10^6, a value considerably higher than those observed in known synthetic photochromic materials. This high value is due to the protective features of the integral membrane protein that serves to isolate the chromophore from potentially reactive oxygen, singlet oxygen, and free radicals. Thus, the common misperception that biological materials are too fragile to be used in technological devices does not apply to bacteriorhodopsin. The photonic characteristics of bacteriorhodopsin have been reviewed in detail,[1–3] and this discussion will be selective.

Although the bacteriorhodopsin photocycle is comprised of at least five thermal intermediates (Figure 33.2a), only three of the intermediates (bR, K, and M) have found application in photonic devices as of this review. The absorption spectra of the key intermediates are shown in Figure 33.2b. The unique absorption spectra exhibited by each intermediate is associated with the changing electronic environment of the chromophore binding site during the course of the photocycle. The internally bound chromophore carries a net positive charge and interacts electrostatically with neighboring charged amino acids in the binding site. These interactions in large part determine the protein's photochemical and spectral properties during the photocycle and are therefore targeted areas for genetic engineering and/or chemical modification. When the chromophore absorbs a photon of light energy, an instantaneous ($<10^{-15}$ s) shift of electron density occurs, with negative charge moving along the polyene chain toward the nitrogen atom. This shift in electron density interacts with nearby negatively charged residues and activates a rotation around the C13-C14 double bond, thereby generating a 13-*cis*-chromophore geometry (see Figure 33.1). The result of this photoisomerization process, which occurs in less than 1 ps,[11,12] is the formation of the K spectral

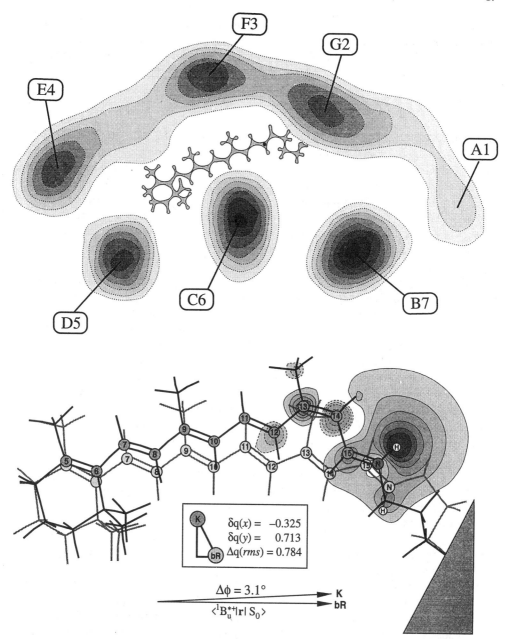

FIGURE 33.1 The chromophore binding site and the primary photochemical event in bacteriorhodopsin. The upper diagram shows electron density profiles (data from Reference 9) of bacteriorhodopsin viewed from the cytoplasmic side showing the seven transmembrane spanning segments and the presumed location of the chromophore in relation to the helices based on the available experimental data.[3] FTIR studies indicate that the polyene chain of the chromophore in bacteriorhodopsin lies roughly perpendicular to the membrane plane.[10] The retinyl chromophore is rotated artificially into the membrane plane to more clearly show the polyene chain (the imine nitrogen is indicated with a solid black circle) and the β-ionylidene ring. The bottom diagram shows a model of the primary photochemical event (bR [gray; underneath] → K [black; above]) and the shift in charge that is associated with the motion of the positively charged chromophore following 13-*trans* → 13-*cis*-photoisomerization. It is believed that the initial photoelectric signal is due primarily to the motion of the chromophore. The conformations of the lysine residue in the bR and K states are tentative.

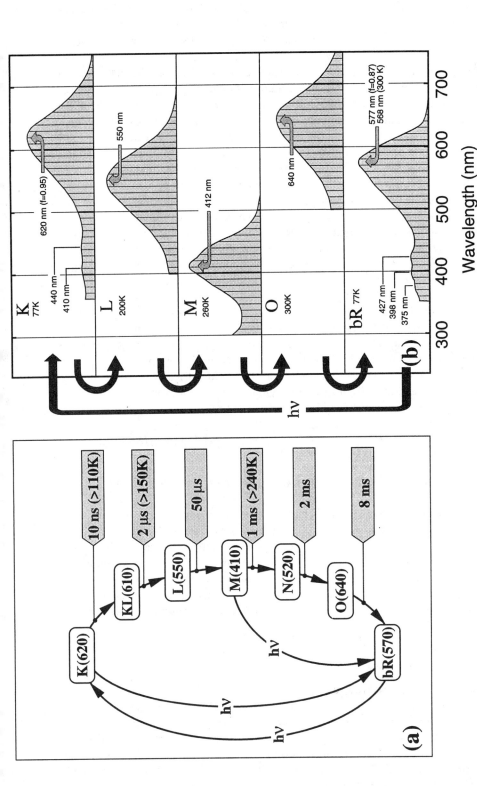

FIGURE 33.2 A simplified model of the light-adapted bacteriorhodopsin photocycle (a) and the electronic (one-photon) absorption spectra of selected intermediates in the photocycle (b). The height of the symbols in (a) is representative of the relative free energy of the intermediates, and the key photochemical transformations relevant to device applications are shown. Note that not all of the intermediates are shown, and that there are in fact two species of M (M_{fast} and M_{slow}), but only one is shown for convenience. (M_{fast} and M_{slow} have virtually identical absorption spectra.) Band maxima are indicated in nanometers. Oscillator strengths (f) determined by log-normal fits of selected λ_{max} bands are indicated in parentheses.

intermediate. The reason for the unusually high isomerization speed is due to a barrierless excited-state potential surface.[3] In this regard, bacteriorhodopsin is the biological analog of high electron mobility transistor (HEMT) devices. The isomerization of the protonated chromophore induces a shift in positive charge perpendicular to the membrane sheet containing the protein and generates a measurable and potentially useful electrical signal. The rise time of this signal is less than 5 ps and correlates with the formation time of K.[13] Another feature of the chromophore *trans,cis*-isomerization process is its photoreversibility. Thus, irradiation of the protein with a wavelength within the absorption band of K results in the reformation of the ground state. Many of the early proposed optical memories based on bacteriorhodopsin utilized the photochemical switching between bR and K. These devices suffered, however, from the requirement that liquid nitrogen temperatures were needed to arrest the photocycle at the K intermediate (e.g. Reference 14). While these devices were potentially efficient and very fast (the bR $\leftrightarrow$ K interconversions take place in a few picoseconds), the use of cryogenic temperatures and the small change in absorption maxima associated with the bR-to-K transition mandated expensive operating hardware and precluded general use.

The most significant photochemical intermediate, both from a physiological and optical application point of view, is the blue light-absorbing M intermediate. The formation of M ensues after a series of protein conformational changes occurring ~50 μs after the absorption of a photon of light by bR. In this stage of the photocycle, the Schiff base proton on the chromophore is transferred to an amino acid of the protein. In doing so, the electrostatic nature of the chromophore and the electric potential of the chromophore binding site is dramatically changed, as reflected in this intermediate's highly blue-shifted absorption spectrum. Under normal conditions, M thermally reverts to the ground state with a time constant of about 10 ms. Most importantly, bR can also be photochemically regenerated from M by the absorption of blue light. This property of a material, where a ground-state photoinitiated reaction results in a relatively long-lived thermal intermediate that can also be photochemically driven back to the ground state, is called *photochromism*. The photochromic properties of bacteriorhodopsin are summarized below:

$$\text{bR } (\lambda_{max} \cong 570 \text{ nm}) \text{ (\textit{State 0})} \quad \underset{\Phi_2 \sim 0.65}{\overset{\Phi_1 \sim 0.65}{\rightleftarrows}} \quad \text{M } (\lambda_{max} \cong 410 \text{ nm}) \text{ (\textit{State 1})}$$

where the quantum yields of the forward reaction (bR to M) and reverse reaction (M to bR) are indicated by Φ_1 and Φ_2, respectively. The quantum yield is a measure of the probability that a reaction will take place after the absorption of a photon of light. A photochromic material possessing a quantum yield of unity and a comparatively long thermal intermediate lifetime is extremely light sensitive. One inherent advantage of bacteriorhodopsin as an optical recording medium is the high quantum efficiency with which it converts light into a state change. Complementing this property is the relative ease with which the thermal decay of M can be prolonged. The M $\rightarrow$ bR thermal transition is highly susceptible to temperature, chemical environment, genetic modification, and chromophore substitution. This property is exploited in many optical devices based on bacteriorhodopsin.

Holographic Optical Recording

Thin films of bacteriorhodopsin fabricated by incorporating the protein into optically transparent polymers and polymer blends have shown good holographic performance and the capability of real-time optical processing.[2,3,15–18] Figure 33.3 shows the results of the theoretically predicted diffraction efficiency of a 6 OD film of chemically enhanced bacteriorhodopsin as a function of varying readout wavelength. An experimental measurement is also shown, indicating that the excellent diffraction efficiency that is predicted can, in fact, be realized experimentally. Holograms can be recorded in pure phase, pure absorption, or mixed modes with recording wavelengths from 400 to 700 nm and with readout wavelengths from 400 to 850 nm. The recording sensitivity at

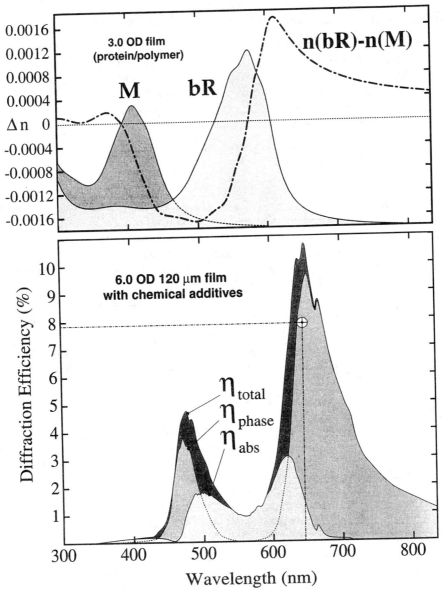

FIGURE 33.3 The change in refractive index associated with the bR → M photoisomerization for a 30-μm film of bacteriorhodopsin with an optical density (OD) of ~3 is shown as a function of wavelength in the upper panel. The refractive index change is expressed as the value for pure bR minus the value for pure M and is calculated by using the Kramers-Kronig transformation. The absorption spectra of bR and M are shown for reference. The diffraction efficiency associated with a 6 OD film for bR (100%) → bR(50%) + M(50%) photoconversion is shown in the lower panel and is calculated based on the observed absorption spectra by using the Kramers-Kronig relationship and Kogelnik approximation.[24] The dot at ~640 nm and ~8% diffraction efficiency represents a recent experimental result from our laboratory using the holographic spatial light modulator described in Figure 33.4.

ambient temperature is in the range 1 to 80 mJ cm⁻². An additional advantage of using this protein as an optical recording medium is its small size (~5 nm diameter) relative to the wavelength of light. This results in diffraction-limited performance (>5000 lines mm⁻¹ for thin films). Diffraction efficiencies can also be improved by using genetically modified proteins,[15] chromophore analogs, and chemical enhancement of the native protein (for reviews, see References 1 to 3).

Fabrication of Protein-Based Optical Recording Devices

There are a number of polymer matrices which can be used to solubilize bacteriorhodopsin, including polyvinyl alcohol, bovine skin gelatin, methylcellulose, and polyacrylamide. Polymeric films containing bacteriorhodopsin are usually sealed from the outside environment to prevent humidity and pH changes that dramatically influence the photochromic properties exhibited by the film. The design shown in Figure 33.4 yields bacteriorhodopsin films with excellent long-term stability and high diffraction efficiencies (see Figure 33.3).

Spatial Light Modulators

Spatial light modulators (SLMs) are integral components in the majority of one-dimensional and two-dimensional optical processing environments. These devices modify the amplitude, intensity, phase, and/or polarization of a spatial light distribution as a function of an external electrical signal and/or the intensity of a secondary light distribution. The observation that a thin film of bacteriorhodopsin can act as a photochromic bistable optical device (either bR $\leftrightarrow$ K or bR $\leftrightarrow$ M photoreactions) or as a voltage-controlled bistable optical device (bR $\leftrightarrow$ M photoreaction) suggests that it has significant potential as the active medium in SLMs.[2,15,16] Soviet scientists were the first to exploit this potential and deserve much of the credit for bringing bacteriorhodopsin to the attention of researchers working in optical engineering.[17–22] The most successful bacteriorhodopsin-based SLM device has been recently demonstrated by German researchers. Their work exploits the bR $\leftrightarrow$ M photoreaction of a mutant protein film in a Fourier optical architectural scheme that implements edge enhancement (spatial frequency filtering) on an input image.[23]

Holographic Associative Memories

Associate memories operate in a fashion quite different from the serial memories that dominate current computer architectures. These memories take an input data block (or image), and independently of the central processor, "scan" the entire memory for the data block that matches the input. In some implementations, the memory will find the closest match if it cannot find a perfect match. Finally, the memory will return the data block in memory that satisfies the matching criteria. Because the human brain operates in a neural, associative mode, many computer scientists believe that the implementation of large-capacity associative memories will be required if we are to achieve artificial intelligence fully. Optical associative memories using Fourier transform holograms have significant potential for applications in optical computer architectures, optically coupled neural network computers, robotic vision hardware, and generic pattern recognition systems. The ability to change rapidly the holographic reference patterns via a single optical input while maintaining both feedback and thresholding increases the utility of the associative memory, and in conjunction with solid-state hardware, opens up new possibilities for high-speed pattern recognition architectures.

One application currently under investigation is the use of bacteriorhodopsin thin films as the holographic storage components in a real-time optical associative memory.[3,16] Our current design is shown in Figure 33.5.[24] The optical design, which employs both feedback and thresholding, is based on the closed-loop autoassociative design of Paek and Psaltis.[25] During the write operation, reference images stored in an electronically addressable spatial light modulator (ESLM) are optically fed into the loop by plane wave illumination (λ_w = 568 nm) from a krypton ion laser. The reference images are stored as Fourier transform holograms on thin polymer films containing bacteriorhodopsin (H1 and H2) (Figure 35.4). For this real-time application, no chemical additives are used to enhance the M state lifetime. Accordingly, the hologram stores the reference image for approximately 10 ms before reverting to the ground state. During the readout operation, the input image (from transparencies or another ESLM) is read into the loop by using the optical imaging system shown in Figure 35.5. Best results are obtained by illuminating the object by using plane

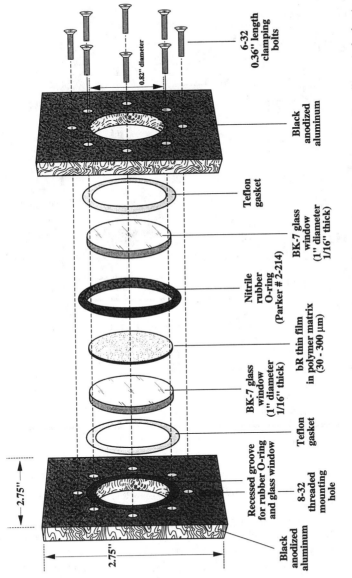

6-32 0.36" length clamping bolts

0.82" diameter

Black anodized aluminum

Teflon gasket

BK-7 glass window (1" diameter 1/16" thick)

Nitrile rubber O-ring (Parker # 2-214)

bR thin film in polymer matrix (30 - 300 µm)

BK-7 glass window (1" diameter 1/16" thick)

Teflon gasket

Recessed groove for rubber O-ring and glass window

8-32 threaded mounting hole

Black anodized aluminum

2.75"

2.75"

FIGURE 33.4 Schematic design of a reversible holographic spatial light modulator based on a thin film of bacteriorhodopsin. A key feature of this design is the use of a compressed nitrile rubber O-ring to seal the protein thin film in order to prevent dehydration of the polymer matrix. The long-term optical and shelf stabilities of the holographic media are excellent.

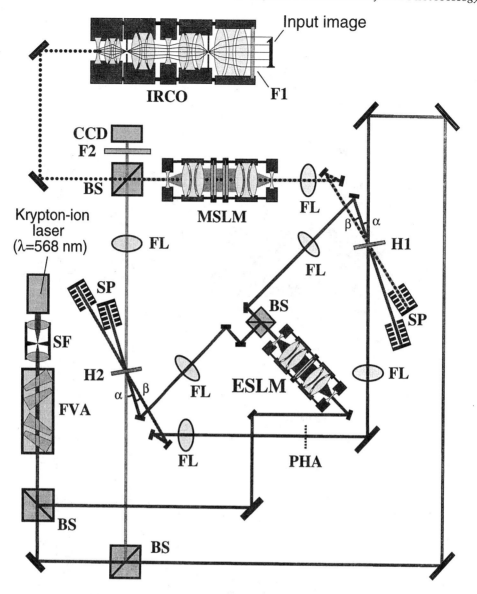

FIGURE 33.5 Schematic diagram of a Fourier transform holographic (FTH) associative memory with read/write FTH reference planes using thin polymer films of bacteriorhodopsin to provide real-time storage of the holograms. The optical design is a modification of one proposed by Izgi.[24] The following symbols are used: BS (beam splitter), CCD (charge coupled device two-dimensional array), CL (condensing lens), ESLM (electronically addressable spatial light modulator), FL (Fourier lens), FVA (Fresnel variable attenuator), F1 (broadband filter for image), F2 (interference filter with transmission maximum at laser wavelength; different from λ_{max} of F1), H1 and H2 (holographic spatial light modulator, Figure 33.4), IRCO (image reduction and condensing optics), MSLM (multichannel plate spatial light modulator), PHA (pin hole array), SF (spatial filter to select TEM_{00}), and SP (beam stop).

wave illumination from a second krypton ion laser operating at a wavelength of 676.5 nm. The input image beam is passed through a microchannel plate spatial light modulator (MSLM) operating in thresholding mode. Thereafter, the Fourier-transformed product of the image reference is formed and retransformed at the plane of a pinhole array (PHA). The resulting correlation patterns are sampled by the pinholes (diameter ~500 μm), which are precisely aligned with the optical axis of the reference images. Light from the pinhole plane is retransformed and superimposed with the reference image stored on the second bacteriorhodopsin hologram (H2). The resulting cross-correlation pattern represents the superposition of all images stored on the multiplexed holograms and is fed back through the microchannel plate spatial light modulator for another iteration. Thus, each image is weighted by the inner product between the pattern recorded on the MSLM from the previous iteration and itself. The output locks onto that image stored in the holograms that produces the largest correlation flux through its aligned pinhole.

The real-time capability of the associative loop is made possible by using bacteriorhodopsin films as the transient holographic medium. The high speed of phototransformation during the write operation (<50 µs), coupled with the quick relaxation time of the M state (~10 ms), allow for framing rates up to 100 frames s^{-1}. Slower or faster framing rates can be attained by simply altering the M lifetime with chemical additives and/or by intermittently erasing the hologram with an external blue-light source. The write and read wavelengths of the krypton ion lasers, as well as the respective angle of incidence, are chosen so as to optimize the diffraction efficiency of the bacteriorhodopsin holograms. During the process of sampling the correlation patterns, it is interesting to note that the inclusion of the pinholes destroys the shift invariance of the optical system. If the input pattern is shifted from its nominal position, the correlation peak shifts as well; the correlation light flux will miss the pinhole. If the pinholes were removed, however, ghost holography would seriously impair image quality. The two apertures within the image reduction and collimation optics (IRCO) serve to provide correct registration, but the input image must still be properly centered in order to generate proper correlation. The problem of shift invariance represents one of the fundamental design issues that will have to be resolved before optical associative memories will reach their full potential. While there are a number of optical "tricks" that can be used to counteract poor registration, the most easily implemented approach is to use the controller of the ESLM to scale and translate the reference images to maximize the correlation light flux as measured by the intensity of the image falling on the CCD output detector.

33.3 Nonlinear Optical Devices

Recent studies have demonstrated the significant potential of organic molecules in applications requiring enhanced second-order[26-38] and third-order[26,28,34,39-45] polarizabilities. Many of these studies have concluded that substituted polyene chromophores are optimal for nonlinear optical applications due to a combination of electronic, conformational, and synthetic advantages. Relatively few studies have been carried out on naturally occurring polyenes, despite the fact that these chromophores have yielded some of the largest second-order[29,31,37] and third-order[39,42,45] polarizabilities. Very few nonlinear applications involving proteins have been reported, but preliminary investigations are encouraging.[29,31,37,45] As demonstrated below, bacteriorhodopsin has interesting nonlinear optical properties that encourage further study.

Two-Photon Properties

The two-photon double-resonance spectrum of light-adapted bacteriorhodopsin in D_2O at room temperature is shown in Figure 33.6.[38] This spectrum is unique relative to other two-photon spectra measured for the visual chromophores and pigments[46-52] in that it exhibits two low-lying band maxima. The lowest energy band maximum at 560 nm ($\delta = 290$ GM; 1 GM = 10^{-50} cm^4 s molec^{-1} photon^{-1}) corresponds within experimental error with the one-photon absorption maximum at 568 nm, and is assigned to the "$^1B_u^{*+}$" $\leftarrow S_0$ transition. The higher energy two-photon band at ~488 nm ($\delta = 120$ GM) does not correspond to a resolved one-photon feature and is assigned to the "$^1A_g^{*-}$" $\leftarrow S_0$ transition. Not only is it surprising to find that both the "$^1A_g^{*-}$" and "$^1B_u^{*+}$" states generate discernible two-photon maxima, but the observation that the "$^1B_u^{*+}$" state two-photon

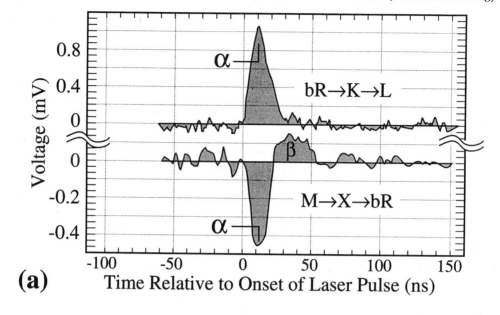

(a)

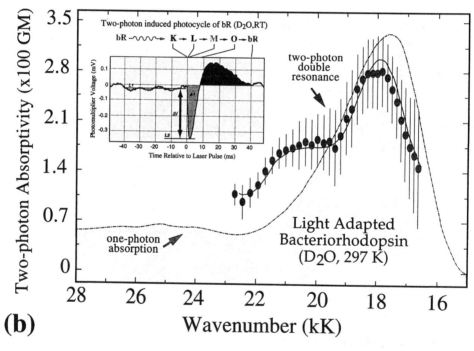

(b)

FIGURE 33.6 Dispersion of the two-photon absorptivity (b) and the two-photon induced photoelectric signals (a) of bR and M states of bacteriorhodopsin. The photoelectric signals are identical under one-photon and two-photon excitation, and consist of two components — a fast component (α) that follows the laser pulse and an oppositely polarized component (β) that has a magnitude and temporal profile that is pH and impedance dependent. In the examples shown, the β component is only observed for the M state signal. Note that the α signals are polarized oppositely for the bR and M states, and can be used to unequivocally assign the state of a bacteriorhodopsin-based memory. Comparison of the one-photon absorption and the two-photon double resonance spectra of light-adapted bacteriorhodopsin is shown in (b). The insert to (b) shows the double resonance signal due to the absorption of the 633-nm laser probe beam associated with the two-photon induced photocycle. Bacteriorhodopsin exhibits an unusually large two-photon absorptivity due to the large change in dipole moment that accompanies electronic excitation (see text). This characteristic can be exploited in the use of the protein in two-photon three-dimensional optical memories (see Figure 33.7).

absorptivity is more than twice as large as that associated with the "$^1A_g^{*-}$" deserves special notice. This result indicates that the two-photon absorptivities are dominated by initial and final state contributions (see discussion in Reference 51), and thus the two-photon absorptivity is proportional to the change in dipole moment upon excitation into the final state. A detailed analysis yields $\Delta\mu$ ("$^1B_u^{*+}$") = 13.5 ± 0.8 D and $\Delta\mu$ ("$^1A_g^{*-}$") = 9.1 ± 4.8 D.[38] Thus, the low-lying, strongly allowed "$^1B_u^{*+}$" state is a $\pi^* \leftarrow \pi$ charge transfer state that has long been recognized as a key contributor to the second-order molecular polarizability of conjugated molecules. Under certain conditions, a low-lying excited state with these properties can enhance the third-order polarizability (see, for example, Reference 40). These issues, as well as the potential of using bacteriorhodopsin in two-photon, three-dimensional optical memories, are discussed below.

Second-Order Polarizability

Values of β for light-adapted bacteriorhodopsin have been determined from second harmonic generation[31] and analysis of the absolute two-photon absorptivity data.[38] The two-photon double resonance determination of β for E_λ = 1.17 eV (λ = 1.06 μ, the Nd:YAG fundamental) yields β = (2250 ± 240) × 10^{30} cm^5 esu^{-1},[38] a value in good agreement with the oriented thin-film measurement of Huang et al.[31] which gave β = 2500 × 10^{30} cm^5 esu^{-1}. The two-photon based value at E_λ = 0.654 eV is β = (329 ± 130) × 10^{30} cm^5 esu^{-1}.

The leading term in the second-order perturbation analysis of the second-order polarizability defines the magnitude of β in terms of the properties of the low-lying, strongly allowed charge transfer excited state (e) relative to those of the ground state (g):

$$\beta \cong \frac{9\, e^2\, h^2}{16\, \pi^2\, m_e} \frac{\Delta\mu_{ge}\, f_{ge}\, \Delta E_{ge}}{\left(\Delta E_{ge}^2 - E_\lambda^2\right)\left(\Delta E_{ge}^2 - 4\, E_\lambda^2\right)} \tag{33.1a}$$

$$\beta \cong 154 \times 10^{-30} \left(cm^5\ esu^{-1}\right)\left(eV^3\ Debye^{-1}\right) \frac{\Delta\mu_{ge}\, f_{ge}\, \Delta E_{ge}}{\left(\Delta E_{ge}^2 - E_\lambda^2\right)\left(\Delta E_{ge}^2 - 4\, E_\lambda^2\right)} \tag{33.1b}$$

where $\Delta\mu_{ge}$ is the change in dipole moment upon excitation, f_{ge} is the oscillator strength, ΔE_{ge} is the transition energy, and Eλ is the photon energy. Equation 33.1b provides a value in (cm^5 esu^{-1}) when $\Delta\mu_{ge}$ is given in Debyes and all energies are given in eV. Bacteriorhodopsin gains its relatively large second-order polarizability from the large change in dipole moment that accompanies excitation into the lowest-lying strongly allowed "$^1B_u^{*+}$" state ($\Delta\mu_{ge}$ = 13.5 ± 0.8 D, see above). If we assume ΔE_{ge} = 2.17 eV and f_{ge} = 0.8, Equation 33.2 predicts β = −1410 × 10^{-30} cm^5 esu^{-1} (ΔE_λ = 1.17 eV) and 280 × 10^{-30} cm^5 esu^{-1} (ΔE_λ = 0.654 eV), in reasonable agreement with the experimental values.[31,38] The simple equation predicts the value of β at ΔE_λ = 1.17 eV is negative because (ΔE_{ge}^2 − 4 E_λ^2) is negative; but this result is artificial due to the neglect of damping. The above results indicate that the large β value at 1.17 eV is due to a near resonance. Although direct measurements of the second-order polarizability of the M state have not been carried out, second harmonic generation studies indicate that it is roughly ten times smaller than that of bR at 1.17 eV (i.e., β[M state] ≈ 230 × 10^{-30} cm^5 esu^{-1} [$\Delta E\lambda$ = 1.17 eV]).[53]

The potential application of the protein as the active element in second harmonic generators is enhanced by the ease with which the protein can be oriented in optically transparent polymer matrices. This capability follows from the fact that the protein has a very large dipole moment (~100 to 500 D depending upon pH) and carries a charge of −3 on the cytoplasmic side. Modest electric fields are known to generate near perfect orientation. The fact that the protein absorbs light at wavelengths throughout the visible region precludes general application. Nevertheless, a number of research groups are investigating the use of bacteriorhodopsin and bacteriorhodopsin analogs for frequency doubling applications in the IR and far-IR.

Third-Order Polarizability

The third-order π-electron polarizability, γ_π, of bacteriorhodopsin in the 0.0- to 1.2-eV optical region has been assigned based on an analysis of the experimental two-photon properties of the low-lying singlet state manifold.[45,54] The following selected values of γ_π (units of 10^{-36} cm^7 esu^{-2}) were observed: $\gamma(0;0,0,0)$ = 2482 ± 327; $\gamma(-3\omega;\omega,\omega,\omega)$ = 2976 ± 385 (ΔE_λ = 0.25 eV), 5867 ± 704 (ΔE_λ = 0.5 eV), 14863 ± 1614 (ΔE_λ = 0.66 eV), 15817 ± 2314 (ΔE_λ = 1.0 eV), 10755 ± 1733 (ΔE_λ = 1.17 eV). In evaluating the third-order properties of bacteriorhodopsin, it is instructive to compare the γ_π values for bacteriorhodopsin with those measured by Hermann and Ducuing for a series of polyenes by using third-harmonic generation.[39] The comparisons suggest that bacteriorhodopsin has a surprisingly large third-order polarizability given the fact that the chromophore has only six double bonds. In fact, it has a γ_π value that is comparable to that measured for dodecapreno β-carotene, a polyene with 11 double bonds.[39] Analysis[54] indicates that bacteriorhodopsin gains a majority of its third-order polarizability due to a Type III enhancement (see Reference 40 for a discussion). When Type III contributions dominate, the zero-frequency third-order polarizability can be approximated by the following equations:

$$\gamma(0;0,0,0) \cong \frac{3\,e^2\,h^2}{2\,\pi^2\,m_e}\frac{\left(\Delta\mu_{ge}\right)^2 f_{ge}}{\left(\Delta E_{ge}\right)^4} \tag{33.2a}$$

$$\gamma(0;0,0,0) \cong \left(256\times10^{-36}\ \text{cm}^7\ \text{esu}^{-2}\right)\left(\text{eV}^4\ \text{Debye}^{-2}\right)\frac{\left(\Delta\mu_{ge}\right)^2 f_{ge}}{\left(\Delta E_{ge}\right)^4} \tag{33.2b}$$

where the parameters are identical to those defined in Equation 33.1. Application of Equation 33.2 predicts $\gamma(0;0,0,0)$ = 1686 × 10^{-36} cm^7 esu^{-2}, a value ~35% lower than the experimental value obtained from an analysis of the two-photon data. Considering the approximations inherent in Equation 33.2 coupled with the experimental error of the two-photon derived value, the agreement is quite good and supports the concept that Type III enhancement dominates the off-resonance third-order polarizability of bacteriorhodopsin. The impact of this term on resonance and near-resonance third-order polarizabilities can be examined by including the following dimensionless energy term:

$$E\left(E_{ge},E_\lambda\right) = \frac{\Delta E_{ge}^6 + \Delta E_{ge}^4\ \Delta E_\lambda^2}{\left(\Gamma^2 + \left(\Delta E_{ge}^2 - 9\,\Delta E_\lambda^2\right)^2\left(\Delta E_{ge}^2 - 4\,\Delta E_\lambda^2\right)^2\left(\Delta E_{ge}^2 - \Delta E_\lambda^2\right)^2\right)^{\frac{1}{2}}} \tag{33.3}$$

where Γ is an energy damping term that prevents the numerous resonances that occur from generating infinities. In principle, the damping factor is equal to the homogeneous linewidth, but larger values are appropriate for inhomogeneously broadened absorption bands. The calculated values based on the product of $\gamma(0;0,0,0) \times E(E_{ge},E_\lambda)$ and assuming Γ^2 = 0.25 eV are 2080 (ΔE_λ = 0.25 eV), 4560 (ΔE_λ = 0.5 eV), 19200 (ΔE_λ = 0.66 eV), 18900 (ΔE_λ = 1.0 eV), 11700 (ΔE_λ = 1.17 eV) with all γ values in cm^7 esu^{-2}. A comparison of the calculated values with the observed values derived from analysis of the two-photon data indicates surprisingly good agreement. It is concluded that Type III enhancement dominates in both resonance and off-resonance energies.

The potential of using bacteriorhodopsin in third-order nonlinear devices has not been explored. Nevertheless, there is one interesting property that might prove useful. It is noted that the γ_π

polarizability of bacteriorhodopsin at zero frequency is predicted to be fairly large $\{\gamma_\pi(0;0,0,0) \cong 2500 \times 10^{-36}\ cm^7\ esu^{-2}\}$. Preliminary studies indicate that the M state has a significantly lower value $\{\gamma_\pi(0;0,0,0) \cong 70 \times 10^{-36}\ cm^7\ esu^{-2}\}$. Thus, the third-order polarizability of the protein can be optically adjusted over a ~35-fold differential. There may be an application for which this characteristic can be exploited.

Nonlinear Optical Volumetric Memories

Although bacteriorhodopsin has competitive second- and third-order polarizabilities, the principal nonlinear property of the protein with respect to device applications is the large two-photon absorptivity. As noted above, the two-photon absorptivity is anomalously large. This observation suggests the potential of using the protein as the photoactive component in volumetric two-photon memories. This application is explored below.

Two-photon three-dimensional optical-addressing architectures offer significant promise for the development of a new generation of ultrahigh-density random access memories.[54–57] These memories read and write information by using two orthogonal laser beams to address an irradiated volume (1 to 50 μm^3) within a much larger volume of a nonlinear photochromic material. Because the probability of a two-photon absorption process scales as the square of the intensity, photochemical activation is limited to a first approximation to regions within the irradiated volume. (Methods to correct for photochemistry outside the irradiated volume are described below.) The three-dimensional addressing capability derives from the ability to adjust the location of the irradiated volume in three dimensions. Two-dimensional optical memories have a storage capacity that is limited to ~$1/\lambda^2$ (where λ is the wavelength), which yields approximately 10^8 bit cm^{-2}. In contrast, three-dimensional memories can approach storage densities of $1/\lambda^3$, which yields storages in the range 10^{11} to 10^{13} bit cm^{-3}. The volumetric memory described below is designed to store 18 Gbytes (1 Gbyte = 10^9 byte) within a data storage cuvette with dimensions of $1.6 \times 1.6 \times 2$ cm. Our current storage capacity is well below the maximum theoretical limit of ~512 Gbyte for the same ~5-cm^3 volume.

Bacteriorhodopsin has four characteristics that combine to yield a comparative advantage as a two-photon volumetric medium.[54] First, as noted above, it has a large two-photon absorptivity due to the highly polar environment of the protein binding site and the large change in dipole moment that accompanies excitation.[38] Second, bacteriorhodopsin exhibits large quantum efficiencies in both the forward and reverse directions. Third, the protein gives off a fast electrical signal when light activated that is diagnostic of its state (see Figure 33.6). Fourth, the protein can be oriented in optically clear polymer matrices, permitting photoelectric state interrogation.[54] The two-photon induced photochromic behavior is summarized in the scheme below.

$$\text{bR }(\textit{State 0})\ (\lambda_{max} \cong 1140\ nm) \underset{(h\omega)^2;\ \Phi_2 \geq 0.65}{\overset{(h\omega)^2;\ \Phi_1 \sim 0.65}{\rightleftarrows}} \text{M }(\textit{State 1})(\lambda_{max} \cong 820\ nm)$$

Arbitrarily, bR is assigned to binary state 0 and M to binary state 1. The chromophore in bR has an unusually large two-photon absorptivity that permits the use of much lower intensity laser excitation to induce the forward photochemistry. The above wavelengths are correct to only ±40 nm, because the two-photon absorption maxima shift as a function of temperature and polymer matrix water content.

The optical design of the two-photon, three-dimensional optical memory[1] is shown in Figure 33.7. The bacteriorhodopsin is contained in a cuvette and is oriented by using electric fields prior to polymerizing the polyacrylamide gel matrix. This orientation is required in order to observe and use the photoelectric signal (Figure 33.6) to monitor the state of the proteins occupying the irradiated volume. A write operation is carried out by firing simultaneously the two 1140-nm lasers

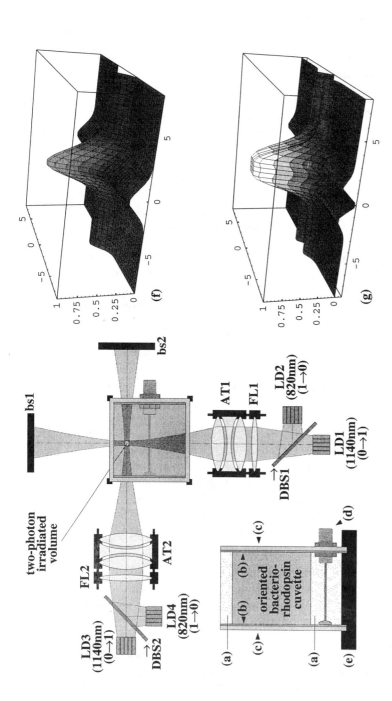

FIGURE 33.7 Schematic diagram of the principal optical components of a two-photon, three-dimensional optical memory-based on bacteriorhodopsin. The write operation involves the simultaneous activation of LD_1 and LD_3 ($0 \rightarrow 1$) or LD_2 and LD_4 ($1 \rightarrow 0$) to induce two-photon absorption within the irradiated volume and partially convert either bR to M ($0 \rightarrow 1$) or M to bR ($1 \rightarrow 0$). The write operation uses a 10-ns pulse'and a pulse simultaneity of 1 ns. The protein is oriented within the cuvette using an electric field prior to polymerization of the polyacrylamide gel. A polymer sealant is then used to maintain the correct polymer humidity. The *SMA* connector is attached to the indium-tin-oxide conducting surfaces on opposing sides of the cuvette and is used to transfer the photoelectric signal to the external amplifiers and box-car integrators. Symbols and letter codes are as follows: (a) sealing polymer, (b) indium-tin-oxide conductive coating; (c) *BK7* optical glass; (d) *SMA* or *OS50* connector; (e) Peltier temperature-controlled base plate (0 to 20°C); *AT* (achromatic focusing triplet); *bs* (beam stop); *DBS* (dichroic beam splitter); *LD* (laser diode); and *FL* (adjustable focusing lens). Computer simulations of the probability of two-photon induced photochemistry (vertical axis) as a function of location relative to the center of the irradiated volume (ΔX_{focus} and ΔY_{focus}) in microns are shown in (f) and (g). The upper right contour plot (f) shows the probability after two 1140-nm laser beams have been simultaneously directed along orthogonal axes crossing at the center of the irradiated volume. The lower right contour plot (g) shows the probability after two 820-nm "cleaning pulses" have been independently directed along the same axes. The maximum conversion probability at x = 0, y = 0 is normalized to unity for both contour plots.

(to write a 1) or the two 820-nm lasers (to write a 0). To eliminate unwanted photochemistry along the laser axes, nonsimultaneous firing of the lasers not used in the original write operation is carried out immediately following the write operation. The position of the cube is controlled in three dimensions using a series of actuators that drive the cube independently in the x, y, or z-directions. For slower speed maximum density applications, electrostrictive micrometers are used. For higher speed, lower density applications, voice-coil actuators are used. Parallel addressing of large data blocks can also be accomplished using holographic lenses or other optical architectures.[56,57]

A key requirement of the two-photon memory is to generate an irradiated volume that is reproducible in terms of *xyz* location over lengths as large as 2 cm. In the present case, the cubes are typically ~1.6 cm in the x- and y-dimensions and ~2 cm in the z-direction (see Figure 33.7). These dimensions are variable up to 2 cm on all sides, and can be as small as 1 cm on a side depending upon the desired storage capacity of the device. By using a set of fixed lasers and lenses, and moving the cube by using orthogonal translation stages, excellent reproducibility can be achieved (± 1 μm for electrostrictive micropositioners, ± 3 μm by using voice-coil actuators). Refractive inhomogeneities that develop within the protein-polymer cube as a function of write cycles adversely affect the ability to position the irradiated volume with reproducibility. This problem is due to the change in refractive index associated with the photochemical transformation (see Figure 33.3). The problem is minimized by operating with a relatively large irradiated volume (30 μm³) and by limiting the photochemical transformation to 60:40 vs. 40:60 in terms of relative bR:M percentages.

Acknowledgments

The research from the authors' laboratory was sponsored in part by grants from the W. M. Keck Foundation, U.S. Air Force Rome Laboratory, the National Institutes of Health, and the Industrial Affiliates Program of the W. M. Keck Center for Molecular Electronics.

References

1. Birge, R. R., Protein based optical computing and optical memories, *IEEE Comput.*, 25, 56, 1992.
2. Oesterhelt, D., Bräuchle, C., and Hampp, N., Bacteriorhodopsin: A biological material for information processing, *Q. Rev. Biophys.*, 24, 425, 1991.
3. Birge, R. R., Photophysics and molecular electronic applications of the rhodopsins, *Annu. Rev. Phys. Chem.*, 41, 683, 1990.
4. Greenbaum, E., Kinetic studies of interfacial photocurrent in platanized chloroplasts, *J. Phys. Chem.*, 96, 514, 1992.
5. Greenbaum, E., Vectorial photocurrents and photoconductivity in metalized chloroplasts, *J. Phys. Chem.*, 94, 6151, 1990.
6. Boxer, S. G., Stocker, J., Franzen, S., and Salafsky, J., Re-engineering photosynthetic reaction centers, in *Molecular Electronics — Science and Technology*, Aviram, A., Ed., American Institute of Physics, New York, 1992, 226.
7. Oesterhelt, D. and Stoeckenius, W., Rhodopsin-like protein from the purple membrane of *Halobacterium halobium*, *Nature (London), New Biol.*, 233, 149, 1971.
8. Birge, R. R., Nature of the primary photochemical events in rhodopsin and bacteriorhodopsin, *Biochim. Biophys. Acta*, 1016, 293, 1990.
9. Hayward, B. S., Grano, D. A., Glaeser, R. M., and Fisher, K. A., Molecular orientation of bacteriorhodopsin within the purple membrane of *Halobacterium halobium*, *Proc. Natl. Acad. Sci. U.S.A.*, 75, 4320, 1978.
10. Earnest, T. N., Roepe, P., Braiman, M. S., Gillespie, J., and Rothschild, K. J., Orientation of the bacteriorhodopsin chromophore probed by polarized Fourier transform infrared difference spectroscopy, *Biochemistry*, 25, 7793, 1986.

11. Mathies, R. A., Lugtenburg, J., and Shank, C. V., From femoseconds to biology: Mechanism of the light-driven proton pump in bacteriorhodopsin, in *Biomolecular Spectroscopy,* Birge, R. R. and Mantsch, H. H., Eds., The International Society for Optical Engineering, Bellingham, WA, 1989, 138.

12. Mathies, R. A., Brito, Cruz, C. H., Pollard, W. T., and Shank, C. V., Direct observation of the femtosecond excited-state *cis-trans* isomerization in bacteriorhodopsin, *Science,* 240, 777, 1988.

13. Simmeth, R. and Rayfield, G. W., Evidence that the photoelectric response of bacteriorhodopsin occurs in less than 5 picoseconds, *Biophys. J.,* 57, 1099, 1990.

14. Birge, R. R., Zhang, C. F., and Lawrence, A. F., Optical random access memory based on bacteriorhodopsin, in *Molecular Electronics,* Hong, F., Ed., Plenum Press, New York, 1989, 369.

15. Hampp, N., Bräuchle, C., and Oesterhelt, D., Bacteriorhodopsin wildtype and variant aspartate-96 to asparagine as reversible holographic media, *Biophys. J.,* 58, 83, 1990.

16. Birge, R. R., Fleitz, P. A., Gross, R. B., Izgi, J. C., Lawrence, A. F., Stuart, J. A., and Tallent, J. R., Spatial light modulators and optical associative memories based on bacteriorhodopsin, *Proc. IEEE EMBS,* 12, 1788, 1990.

17. Bazhenov, V. Y., Soskin, M. S., and Taranenko, V. B., Holographic recording by continuous illumination in a suspension of purple membranes of halobacteria, *Sov. Tech. Phys. Lett.,* 13, 382, 1987.

18. Bazhenov, V. Y., Soskin, M. S., Taranenko, V. B., and Vasnetsov, M. V., Biopolymers for real-time optical processing, in *Optical Processing and Computing,* Arsenault, H. H., Szoplik, T., and Macukow, B., Ed., Academic, New York, 1989, 103.

19. Druzhko, A. B. and Zharmukhamedov, S. K., Biochrome film based on some analogues of bacteriorhodopsin, in *Photosensitive Biological Complexes and Optical Recording of Information,* Ivanitskiy, G. R. and Vsevolodov, N. N., Eds., U.S.S.R. Academy of Sciences, Biological Research Center, Institute of Biological Physics, Pushchino, 1985, 119.

20. Ivanitskiy, G. R. and Vsevolodov, N. N., *Photosensitive Biological Complexes and Optical Recording of Information,* 1–209, U.S.S.R. Academy of Sciences, Biological Research Center, Institute of Biological Physics, Pushchino, 1985.

21. Savranskiy, V. V., Tkachenko, N. V., and Chukharev, V. I., Kinetics of diffraction effectiveness of bacteriorhodopsin, in *Photosensitive Biological Complexes and Optical Recording of Information,* Ivanitskiy, G. R. and Vsevolodov, N. N., Eds., U.S.S.R. Academy of Sciences, Biological Research Center, Institute of Biological Physics, Pushchino, 1985, 97.

22. Vsevolodov, N. N. and Poltoratskii, V. A., Holograms in biochrome, a biological photochromic material, *Sov. Phys. Tech. Phys.,* 30, 1235, 1985.

23. Thoma, R., Hampp, N., Bräuchle, C., and Oesterhelt, D., Bacteriorhodopsin films as spatial light modulators for nonlinear optical filtering, *Opt. Lett.,* 16, 651, 1991.

24. Gross, R. B., Izgi, K. C., and Birge, R. R., Holographic thin films, spatial light modulators and optical associative memories based on bacteriorhodopsin, *Proc. SPIE,* 1662, 186, 1992.

25. Paek, E. G. and Psaltis, D., Optical associative memory using Fourier transform holograms, *Opt. Eng.,* 26, 428, 1987.

26. Chemla, D. S. and Zyss, J., *Nonlinear Optical Properties of Organic Molecules and Crystals,* Academic Press, Orlando, FL, 1987.

27. Dirk, C. W., Twieg, R. J., and Wagniere, G., The contribution of π electrons to second harmonic generation in organic molecules, *J. Am. Chem. Soc.,* 108, 5387, 1986.

28. Hann, R. A. and Bloor, D., *Organic Materials for Non-linear Optics,* Royal Society of Chemistry, London, England, 1989.

29. Huang, J. Y., Lewis, A., and Rasing, T., Second harmonic generation from Langmuir Blodgett films of retinal and retinal Schiff bases: Analysis of monolayer structure and dipolar properties, *J. Phys. Chem.,* 92, 1756, 1988.

30. Huang, J. Y., Lewis, A., and Loew, L., Non-linear optical properties of potential sensitive styryl dyes, *Biophys. J.,* 53, 665, 1988.

31. Huang, J. Y., Chen, Z., and Lewis, A., Second harmonic generation in purple membrane-poly(vinyl alcohol) films: Probing the dipolar characteristics of the bacteriorhodopsin chromophore in bR_{570} and M_{412}, *J. Phys. Chem.*, 93, 3314, 1989.

32. Katz, H. E., Dirk, C. W., Singer, K. D., and Sohn, J. E., Exceptional second-order nonlinear optical susceptibilities in organic compounds, *Mol. Crys. Liq. Cryst. Inc. Nonlin. Opt.*, 157, 525, 1988.

33. Oudar, J. L., Optical nonlinearities of conjugated molecules. Stilbene derivatives and highly polar aromatic compounds, *J. Chem. Phys.*, 67, 446, 1977.

34. Williams, D. J., *Nonlinear Optical Properties of Organic and Polymeric Materials*, American Chemical Society, Washington, D.C., 1985.

35. Spangler, C. W., McCoy, R. K., Birge, R. R., Fleitz, P. A., and Zhang, C. F., The relationship between structure and nonlinear hyperpolarizabilities in donor-acceptor polyenes, in *Molecular Electronics — Science and Technology*, Aviram, A., Ed., Engineering Foundation, New York, 1989, 175.

36. Singer, K. D. and Garito, A. F., Measurements of molecular second order optical susceptibilities using dc induced second harmonic generation, *J. Chem. Phys.*, 75, 3572, 1981.

37. Birge, R. R., Fleitz, P. A., Lawrence, A. F., Masthay, M. A., and Zhang, C. F., Nonlinear optical properties of bacteriorhodopsin. Assignment of second order hyperpolarizabilities of randomly oriented systems by using two-photon spectroscopy, *Mol. Cryst. Liq. Cryst.*, 180, 107, 1990.

38. Birge, R. R. and Zhang, C. F., Two-photon spectroscopy of light adapted bacteriorhodopsin, *J. Chem. Phys.*, 92, 7178, 1990.

39. Hermann, J. P. and Ducuing, J., Third-order polarizabilities of long-chain molecules, *J. Appl. Phys.*, 45, 5100, 1974.

40. Garito, A. F., Helfin, J. R., Wong, K. Y., and Zamani-Khamiri, O., Enhancement of non-linear optical properties of conjugated linear chains through lowered symmetry, in *Organic Materials for Non-linear Optics*, Hann, R. A. and Bloor, D., Eds., Royal Society of Chemistry, London, 1989, 16.

41. de Melo, C. P. and Silbey, R., Non-linear polarizabilities of conjugated chains: Regular polyenes, solitons and polarons, *Chem. Phys. Lett.*, 140, 537, 1987.

42. Pierce, B. M., A theoretical analysis of third-order nonlinear optical properties of linear polyenes and benzene, *J. Chem. Phys.*, 91, 791, 1989.

43. Prasad, P. N., Chopra, P., Carlacci, L., and King, H. F., *Ab initio* calculations of polarizabilities and second hyperpolarizabilities of organic molecules with extended pi-electron conjugation, *J. Phys. Chem.*, 93, 7120, 1989.

44. Ward, J. F. and Elliott, D. S., Measurement of molecular hyperpolarizabilities for ethylene, butadiene, hexatriene and benzene, *J. Chem. Phys.*, 69, 5438, 1978.

45. Birge, R. R., Masthay, M. B., Stuart, J. A., Tallent, J. R., and Zhang, C. F., Nonlinear optical properties of bacteriorhodopsin: Assignment of the third-order polarizability based on two-photon absorption spectroscopy, *Proc. SPIE*, 1432, 129, 1991.

46. Birge, R. R., Bennett, J. A., Pierce, B. M., and Thomas, T. M., Two photon spectroscopy of the visual chromophores. Evidence for a lowest excited ^1A_g-like π,π^* state in *all-trans* retinol (vitamin A), *J. Am. Chem. Soc.*, 100, 1533, 1978.

47. Birge, R. R., Bennett, J. A., Hubbard, L. M., Fang, A. L., Pierce, B. M., Kliger, D. S., and Leroi, G. E., Two-photon spectroscopy of all-trans retinal. Nature of the low-lying states, *J. Am. Chem. Soc.*, 104, 2519, 1982.

48. Birge, R. R., Murray, L. P., Zidovetzki, R., and Knapp, H. M., Two-photon and two-dimensional NMR spectroscopic studies of retinyl Schiff bases, protonated Schiff bases, and Schiff base salts: Evidence for protonation induced π,π^* excited state level ordering reversal, *J. Am. Chem. Soc.*, 109, 2090, 1987.

49. Murray, L. P. and Birge, R. R., Two-photon spectroscopy of the Schiff base of *all-trans* retinal. Nature of the low-lying π,π^* singlet states, *Can. J. Chem.*, 63, 1967, 1985.

50. Birge, R. R., Murray, L. P., Pierce, B. M., Akita, H., Balogh-Nair, V., Findsen, L. A., and Nakanishi, K., Two-photon spectroscopy of locked-11-*cis* rhodopsin: Evidence for a protonated Schiff base in a neutral protein binding site, *Proc. Natl. Acad. Sci. U.S.A.*, 82, 4117, 1985.

51. Birge, R. R., Two-photon spectroscopy of protein bound chromophores, *Acc. Chem. Res.*, 19, 138, 1986.

52. Birge, R. R., Energy levels of the visual chromophores by two-photon spectroscopy, *Meth. Enzymol.*, 88, 522, 1982.

53. Chen, Z., Lewis, A., Takei, H., and Nabenzahl, I., Bacteriorhodopsin oriented in polyvinyl alcohol films as an erasable optical storage medium, *Appl. Opt.*, 30, 5188, 1991.

54. Birge, R. R., Gross, R. B., Masthay, M. B., Stuart, J. A., Tallent, J. R., and Zhang, C. F., Nonlinear optical properties of bacteriorhodopsin and protein based two-photon three-dimensional memories, *Mol. Cryst. Liq. Cryst. Sci. Technol. Sec. B. Nonlinear Opt.*, 3, 133, 1992.

55. Parthenopoulos, D. A. and Rentzepis, P. M., Three-dimensional optical storage memory, *Science*, 245, 843, 1989.

56. Hunter, S., Kiamilev, F., Esener, S., Parthenopoulos, D. A., and Rentzepis, P. M., Potential of two-photon based 3-D optical memories for high performance computing, *Appl. Opt.*, 29, 2058, 1990.

57. Lawrence, A. F. and Birge, R. R., The potential application of optical phased arrays in two-photon three-dimensional optical memories, *Proc. SPIE*, 1773, 401, 1992.

34

Bacterial Bioluminescence: Biochemistry

Shiao-Chun Tu
University of Houston

34.1 Introduction

Most luminous bacteria are of marine origin, existing as free-living or symbiotic organisms. Current knowledge of the biochemistry and molecular biology of bacterial bioluminescence is primarily derived from studies on marine species *Vibrio harveyi* (formerly *Benekea harveyi*), *Vibrio fischeri* (formerly *Photobacterium fischeri*), *Photobacterium phosphoreum*, and *Photobacterium leiognathi*. *V. harveyi* luciferase, in particular, has been the focus for enzymological investigations. More recently, increasing attention has been directed to the nonmarine *Xenorhabdus luminescens* strains Hm and Hb as nematode symbionts and the strain HW isolated from a human wound.

Bacterial luciferases characterized thus far are all α,β-dimers that catalyze Reaction 34.1.

$$FMNH_2 + R\text{–}CHO + O_2 \rightarrow FMN + R\text{–}COOH + H_2O + light \qquad (34.1)$$

Two substrates, reduced riboflavin 5′-phosphate ($FMNH_2$) and a long-chain aliphatic aldehyde (R-CHO), are oxidized by oxygen in a light-emitting process. One oxygen atom is incorporated into the carboxylic acid product and the other oxygen atom is recovered in water. Therefore, bacterial luciferase is classified as a flavin-dependent external monooxygenase. However, luciferase is unique among flavo-monooxygenases in its light-emitting activity.

34.2 Luciferase Purification and Activity Assays

Bacterial luciferase can be obtained in high purity ($\geq$95% homogeneity) and sizable quantities (several hundred milligrams or even grams per preparation). In most cases, luciferase constitutes several percent of the total soluble proteins when wild-type cells reach maximal levels of *in vivo*

0-8493-8634-9/95/$0.00+$.50
© 1995 by CRC Press, Inc.

Table 34.1   Purification of Bacterial Luciferases

Luciferase	Treatment	Purity (%)	Specific activity[a] $[(q\ s^{-1})(A_{280}^{-1})(ml^{-1})(10^{-14})]$	Aldehyde	Light standard[b]	Ref.
V. harveyi	I = DEAE ion exchange + $(NH_4)_2SO_4$	≥90	1.6	Decanal		
			0.2	Dodecanal	Radioisotope	5
	I + aminohexyl-Sepharose	>95	0.18–0.25	Dodecanal	Radioisotope	6
	I + FMN-Agarose	>95[c]	3.5[c]	Decanal	Radioisotope	4
	II = $(NH_4)_2SO_4$ + 2,2-diphenyl-propylamine-Sepharose	>95				1
V. h.(aldehyde⁻)	I + HPLC		1.5	Decanal	Luminol	2
V. fischeri	I		4	Dodecanal	Radioisotope	5
	II	>95				
P. phosphoreum	I		0.4	Dodecanal	Radioisotope	5
P. leiognathi	I + HPLC		12	Tetradecanal	Luminol	2
X. luminescens	I	~80	2.9	Decanal	Radioisotope	7

[a] The activity of *X. luminescens* luciferase was determined by the dithionite assay.[8,9] All other activity values were determined by the standard flavin injection assay.[5] The former assay gives a (~70% or more) higher value.

[b] Light intensity in q·s⁻¹ based on the radioisotope light standard[10] is higher than that based on the luminol light standard.[2]

[c] Results obtained by B. Lei and S.-C. Tu following a modified procedure as described in the text.

bioluminescence. Much higher cellular contents of luciferase can be obtained using cloned luciferase genes and overexpression systems. The purification of luciferases is usually achieved by three main approaches, applied individually or in combination (Table 34.1). The most commonly used approach is ion-exchange chromatography in combination with ammonium sulfate fractionation (Treatment 1 ± aminohexyl column). Another approach relies on binding of luciferase to immobilized hydrophobic ligands such as 2,2-diphenylpropylamine (an aldehyde-competitive inhibitor)[1] and the phenyl group.[2] A third approach is chromatography on FMN matrices. An FMN-based chromatography method was reported some years ago,[3] but was rarely practiced. A new FMN-column material and chromatographic conditions were developed more recently.[4] The latter method, presumably but not necessarily involving FMN as a specific binding ligand, was modified in Tu's laboratory for application of a partially purified sample in and elution by 1.2 *M* phosphate, pH 7.0. When the A_{280} in eluate dropped down to a near-baseline level, luciferase was recovered by elution with 0.1 *M* phosphate, pH 7.0. This modified procedure has been found to be quite effective for obtaining highly purified luciferase (Table 34.1). Molecular sieve chromatography is also useful, but not essential, for improving the luciferase purity. Ion exchange, molecular sieve, and hydrophobic interaction chromatography have all been adapted for HPLC procedures.[2] In general, *P. leiognathi* and *X. luminescens* luciferases are somewhat more difficult to purify than *V. harveyi*, *V. fischeri*, and *P. phosphoreum* luciferases.

The noncovalent binding between α and β subunits appears to be very tight. Isolation of individual subunits from purified luciferase can be achieved by conventional[11] or HPLC[12] ion-exchange chromatography in the presence of a denaturant such as urea.

Luciferase activity assays differ with respect to the method of flavin reduction and the sequence of substrate additions. (1) Standard assay:[1] Enzyme is first mixed with an aldehyde substrate in air-saturated buffer. The activity is initiated by the rapid injection of an $FMNH_2$ solution using a syringe. The flavin can be easily reduced by platinum-catalyzed hydrogenation or photochemically in the presence of EDTA. Luciferases from *V. harveyi*[13] and *X. luminescens*[7] are unusual in their sensitivities to aldehyde inhibition in the standard assay. Bovine serum albumin is usually included in the reaction solution to obtain a broader range of aldehyde concentrations for optimal activity and to minimize the aldehyde inhibition. The luciferases from *V. harveyi* and *X. luminescens* do not exhibit aldehyde inhibition when mixed with $FMNH_2$ either prior to[8] or at the same time with (Lei, B., Cho, K. W. and Tu, S.-C., unpublished results) aldehyde. Therefore, activities expressed under these conditions are roughly twice those observed by the standard assay. (2) Dithionite assay:

Luciferase is first mixed with FMN, and dithionite powder[8] or solution[9] is added to reduce the flavin. To this, an aerobic buffer containing saturated aldehyde is injected rapidly to initiate the bioluminescence. While this method is quite useful for routine luciferase assay, it should be noted that sulfur adduct(s) of flavin can be formed in addition to $FMNH_2$ when dithionite, especially when aged, is used as the reductant.[14,15] (3) Cu(I) assay:[16] To a solution containing FMN and aldehyde, copper(I) plus EDTA is added to reduce the flavin. This solution is then mixed with an aerobic luciferase solution for the activity assay. This new assay is very convenient and excellent for luciferase activity expression. These three major methods are all single-cycle, nonturnover assays. Bioluminescence quickly reaches a maximum and then decays following a near-exponential process. Luciferase activity is usually expressed in quanta per second ($q \cdot s^{-1}$) for the peak intensity or the total quantum output. Using 8- to 14-carbon aliphatic aldehydes as substrates, the peak intensities and decay kinetics vary significantly, but little change occurs in the quantum output. (4) A coupled assay[1] has also been developed for luciferase by coupling with an FMN-NAD(P)H oxidoreductase. This assay gives an approximately steady-state light emission. Thus, it can be adapted for measurement by usual liquid scintillation counters and is particularly useful for bioluminescence spectral measurements.

34.3 Luciferase Structure-Function Relationships

The goal to achieve a detailed elucidation of the structure of bacterial luciferase suffers from a lack of crystallographic or high-resolution NMR data. Therefore, correlations of luciferase structure-function relationships rely on, thus far, low-resolution methods or indirect approaches such as primary sequence comparison, subunit hybridization, chemical modification, affinity and photoaffinity labelings, fluorescence energy transfer, random and site-directed mutageneses, etc. Nonetheless, useful information has indeed been deduced from such studies.

Chemical modifications have been applied fruitfully to structure-function relationship studies of numerous enzymes, including bacterial luciferase. Usually, enzymes are treated with excess reagent for a pseudo-first-order kinetic analysis of the modification (or inactivation) rates. This general practice requires that the reagent is reasonably stable under the experimental conditions in the absence of the target protein. However, a number of commonly used modification reagents, such as diethylpyrocarbonate and methyl *p*-nitrobenzene sulfonate, undergo significant autodecays. The usual semilogarithmic plots of modification (or inactivation) vs. time are curved and would not give the pseudo-first-order rate constants. To solve this problem, a simple method of kinetic analysis has been developed to determine both the pseudo-first-order and the corresponding second-order rate constants for enzyme inactivation by labile reagents.[17]

Structure and Function of Subunits

Primary sequences for the α and β subunits of *V. harveyi*,[18,19] *V. fischeri*,[20–22] *P. phosphoreum*,[23] *P. leiognathi*,[22,24,25] and *X. luminescens*[7,22,26] luciferases, and the α subunit of *Kryptophanaron alfredi* luciferase[27] have been deduced from DNA sequences of the respective *lux*A and *lux*B genes. In these cases, the α subunit contains 354 to 362 amino acid residues, with a molecular weight of 40,108 to 41,389. The β subunit has 324 to 328 residues, corresponding to a molecular weight of 36,349 to 37,684. The α,β dimer molecular weight ranges from 76,457 (*V. harveyi*) to 78,689 (*X. luminescene* Hm). An updated alignment of the protein primary sequences of these luciferase species has been illustrated recently.[23]

It was reported that individual α and β subunits of the *V. harveyi* luciferase can catalyze the bioluminescence reaction with their activities estimated, based on emission peak intensities, to be 10^4 to 10^5 lower than the wild-type enzyme.[28] However, critical information such as the total quantum output, light decay kinetics, subunit concentration dependence, substrate concentration dependence, etc. was not described. It is certain that the α,β dimeric structure is essential for the high quantum yield bioluminescence reaction. Previous hybridization studies using chemically[29] or mutationally[30] modified luciferase subunits indicate that the α subunit is intimately involved in catalysis. Although the specific function of the β subunit remains unclear, the involvement of the β subunit in FMNH$_2$ binding has been indicated experimentally.[31,32] Moreover, (photo)affinity labeling[33,34] and cross-linking[12] results show that the aldehyde binding occurs apparently near the α,β subunit interface.

The domains of luciferase that are important to subunit interaction have never been identified. Recently, various *V. harveyi* luciferase mutants consisting of α'/β, α/β', or α'/β' where α' and β' are each a subunit variant with a selected amino acid residue specifically mutated have been developed.[35] Effects of these single- and double-site mutations on the energetics of thermal and urea inactivations were determined, and the results were analyzed following the principle described by Wells.[36] For all 12 mutants tested, the significant interactions were found to occur between the $\alpha 82$-position and the $\beta 82$-position when the two native histidines were mutated to a Lys/Asp or Asp/Lys pair.[35] Although such results do not suggest necessarily any specific interaction between the native histidine residues at these two positions, they do provide the first indication, at the level of specific residues, that the $\alpha 82$- and $\beta 82$-positions are within a range suitable for significant inter-subunit interactions.

Critical Amino Acid Residues

Most of the enzyme active-site studies on luciferase are aimed at identifying and characterizing critical residues on the α subunit of the *V. harveyi* luciferase. The *V. harveyi* luciferase has 14 cysteine residues, with 8 on the α and 6 on the β subunits. More than 95% inactivation of luciferase occurred as a result of chemical modifications of a particularly reactive cysteine residue,[37] identified later to be the αCys-106.[38,39] An affinity labeling probe, bromodecanal, modifies preferentially a single cysteine on the α subunit, most probably the αCys-106, leading to enzyme inactivation as well as the loss of FMNH$_2$ binding. The cross-linking study showed that the αCys-106 is close to the α,β subunit interface.[12] Paquatte and Tu[39] succeeded in modifying αCys-106 with a relatively small methyl group, and found that the methylated luciferase, although inactivated, retains the ability to bind the FMNH$_2$ and aldehyde substrates. Mutation of the αCys-106 to a serine has no deleterious effect on the activity.[20] Recent site-specific mutagenesis studies by Xi et al.[40] and Baldwin et al.[41] showed that the αCys-106 $\rightarrow$ Ala mutant luciferase (αC106A) retains ~60% of the activity in total quantum output, but the variant with the αCys106 changed to a bulkier valine (αC106V) has only ~2% activity. More specifically, it was found that the deleterious effects of αCys-106 methylation (Cho, K. W. and Tu, S.-C., unpublished results) and mutation to a valine[40] on the bioluminescence activity could be traced to the greatly impeded formation or marked destabilization of a key 4a-hydroperoxy-4a,5-dihydroFMN intermediate (4a-hydroperoxyFMNH or HF-4a-OOH). The substantial activity of the αC106A mutant indicates clearly that the αCys-106 is not involved directly in chemical catalysis. The intriguing question then arises regarding the specific structural/functional roles of αCys-106. A scheme consistent with essentially all experimental findings can be formulated in light of some more-recent results. By equilibrium binding and kinetic experiments (Lei, B., Cho, K. W., and Tu, S.-C., unpublished results), it was found that the *V. harveyi* luciferase has a tight aldehyde substrate site and a weak aldehyde inhibitor site. Moreover, the αCys-106 is at the aldehyde inhibitor site that overlaps with, but is not identical to, the FMNH$_2$ site. The long-recognized aldehyde inhibition of *V. harveyi* luciferase is due to the blockage of FMNH$_2$ binding by the competitive binding of a long-chain aldehyde to the overlapping inhibitor site. The FMNH$_2$ binding is abolished or greatly weakened when the αCys-106 is modified by bulky groups. A slight enlargement of the αCys-106 side chain, e.g., by methylation or mutation to valine, does not abolish the binding of FMNH$_2$ but deleteriously affects the formation, stability, or reactivity of the 4a-hydroperoxyFMNH intermediate. The catalytic activity is even more sensitive to modifications of αCys-106 and, thus far, is retained only when the αCys-106 is converted to a homologous serine or a smaller alanine.

Inactivation of *V. harveyi* luciferase occurs when an unspecified histidyl residue on the α subunit is modified by diethylpyrocarbonate (or ethoxyformic anhydride).[6] In a comparison of α subunit primary structures, five conserved histidyl residues at positions corresponding to 44, 45, 82, 224, and 285 of the *V. harveyi* luciferase were noticed. Selective mutations were carried out for each of these conserved histidines, and a wide range of inactivations was observed with the 10 luciferase variants so generated.[42] The essentialities of the αHis-44 and αHis-45 have been demonstrated by a remarkable 4 to 7 orders of activity reduction resulting from the replacement of either histidine by an alanine (αH44A, αH45A), aspartate (αH44D, αH45D), or lysine (αH45K). On the basis of kinetic analysis, the critical histidine detected previously by chemical modification[6] is probably the αHis-44. Two particularly important findings emerged from this mutagenesis study.[42] First, the bioluminescence activities of the 10 luciferase variants tested were reduced from a few fold to 7 orders of magnitude. Such inactivations, with the apparent exception of αH285A, show a general correlation with the dark decay rates of the corresponding 4a-hydroperoxyFMNH intermediates (designated II) (Figure 34.1). It is known that the dark decay is associated with the C(4a)–O bond scission, whereas the bioluminescence activity is coupled to the fission of the O–O bond of II (Figure 34.1 inset). A recent study on unbound flavin models reveals free energies of ≤16 and 21 kcal mol^{-1} for the O–O and C(4a)–O bond fissions, respectively. Since these free energies do not differ markedly for free flavins, the expression of bioluminescence vs. dark decay appears to rely on the control by binding to luciferase. Hydroperoxide II from the wild-type luciferase has the slowest dark decay and the highest bioluminescence activity. For the histidine-mutated variants, the C(4a)–O bond scission is enhanced to various extents, with concomitant reductions in bioluminescence activity, as a consequence of luciferase structural changes. It should be pointed out, however, that the control of bioluminescence expression may not solely rely on the partition between the O–O and the C(4a)–O bond fission. Second, two species of 4a-hydroperoxy FMNH distinct in bioluminescence activity were detected for the first time. Although αH44A, αH44D, and αH44K have extremely low bioluminescence activities, they form intermediate II in 14 to 45% yields. Moreover, αH44A and αH44D can each form two species of II; the major species is inactive in bioluminescence and decays faster than the minor species which is bioluminescence-active. These two 4a-hydroperoxyFMNH species may be stereoisomers with respect to the chiral C(4a) center.

Chemical modification of a primary amine on α or β results in luciferase inactivation.[43] However, it is unclear whether these two critical amines are associated with N-termini or lysines and, in the latter case, where they are located in the primary sequences. Random mutations have been carried out to generate a series of luciferase mutants.[30] One of such luciferase variants — AK-6 with a reduction of bioluminescence activity by 3 orders of magnitude[44] — has been identified later to be αAsp-113 → Asn.[20] A much lower activity was detected with the αAsp-113 → Lys mutant. The αAsp-113 has been proposed to be near the flavin site.[20]

The α,β subunit interface has been implicated in the binding of flavin and aldehyde substrates. However, very little is known about any critical residues on the β subunit. As mentioned earlier, luciferase can be inactivated by modification of an unspecified primary amine group on either the α or the β subunit.[43] A number of mutants with lesions on the β subunit were generated and found to exhibit significantly decreased thermal stabilities.[30] However, the bioluminescence activities and light decay kinetics of these β mutants exhibit only minor changes at 20°C. In a recent site-specific mutagenesis study, luciferase variants were generated with the βHis-81 and βHis-82 each mutated to an alanine, aspartate, or lysine;[35] 1 to 2 orders of magnitude reductions in quantum yields and 2 to 3 orders of magnitude reductions in $V_{max}/K_{m(flavin)}$ were observed with these β mutants, with the βHis-82 being more critical than the βHis-81. These findings, to our knowledge, provide the first demonstration that specific single-site mutations of the β subunit can markedly affect the luciferase bioluminescence activity.

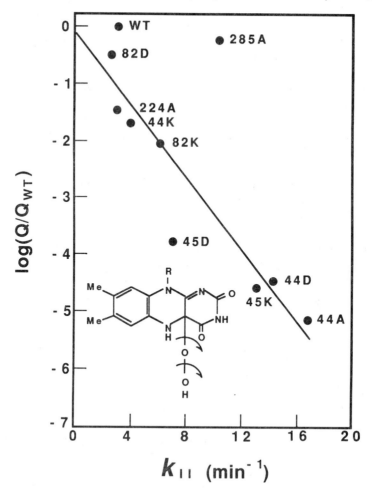

FIGURE 34.1 Correlation of the relative quantum yields with the dark decay rate constants of 4a-hydroperoxyFMNH intermediates formed with wild-type and mutated luciferases. For the designation of variants of *V. harveyi* luciferase, numbers refer to the positions of the original histidyl residues on the α subunit, and the following letters refer to the new residues generated by mutation. Q and Q_{WT} refer to the quantum yields of the mutated and wild-type (WT) luciferases, respectively, and k_{II} is the first-order constant for the dark decay of intermediate II. The inset shows the structure of the intermediate II and the O–O and C(4a)–O bond scissions.

34.4 Reaction Mechanism and Intermediates

Reaction Mechanism

Following an important scheme advanced by Hastings and colleagues,[45,46] the bound $FMNH_2$ reacts with O_2 to generate the 4a-hydroperoxyFMNH (HF-4a-OOH) intermediate II. In the absence of aldehyde, intermediate II decays to form FMN and H_2O_2 with very little light emission. With an aldehyde substrate present, the 4a-peroxyhemiacetalFMNH intermediate III is formed. Intermediate III then proceeds in a Baeyer-Villiger reaction to form a fatty acid and the excited state of 4a-hydroxyFMNH (HF-4a-OH* or IV*). Radiative relaxation of IV* generates ground-state IV and light. Finally, intermediate IV decays to form FMN and water. Many aspects of this scheme have been confirmed experimentally. However, the proposed Baeyer-Villiger reaction for the conversion of III to IV* has no precedent and is not supported by a recent experimental test.[47]

Building on the basis of the scheme by Hastings and colleagues and the concept of chemically induced electron exchange luminescence (CIEEL),[48] an electron (charge) transfer mechanism has been formulated (Scheme 34.1).[49–51] Intermediate II is formed from the bound N(1) deprotonated[52] FMNH⁻ and oxygen. An N(5)-assisted homolysis of the HF-4a-OOH peroxy bond and the subsequent intermolecular hydroxy radical transfer to an aldehyde (process A) result in the formation of a caged pair of alkylhydroxyoxy radical and the novel (HF-4a-O⁻)⁺· radical (IV·). Alternatively, the 4a-peroxyhemiacetalFMNH intermediate III is first formed from II and aldehyde (process B). An intramolecular electron transfer from the N(5) to the peroxy bond also generates the same radical pair. The alkylhydroxyoxy radical is transformed into a carbon-centered alkyldihydroxy radical (process C) which, being a stronger reductant, donates one electron to the (HF-4a-O⁻)⁺· (process D) to form carboxylic acid and the HF-4a-OH* as the primary excited species. The subsequent radiative relaxation of IV* and dehydration of IV remain the same as previously depicted. Scheme 34.1 is presented as an electron transfer mechanism. However, a full electron separation and/or transfer is not essential[53] for steps A (B) through D. Scheme 34.1 can be easily modified to include charge-transfer steps without violating the basic principle.

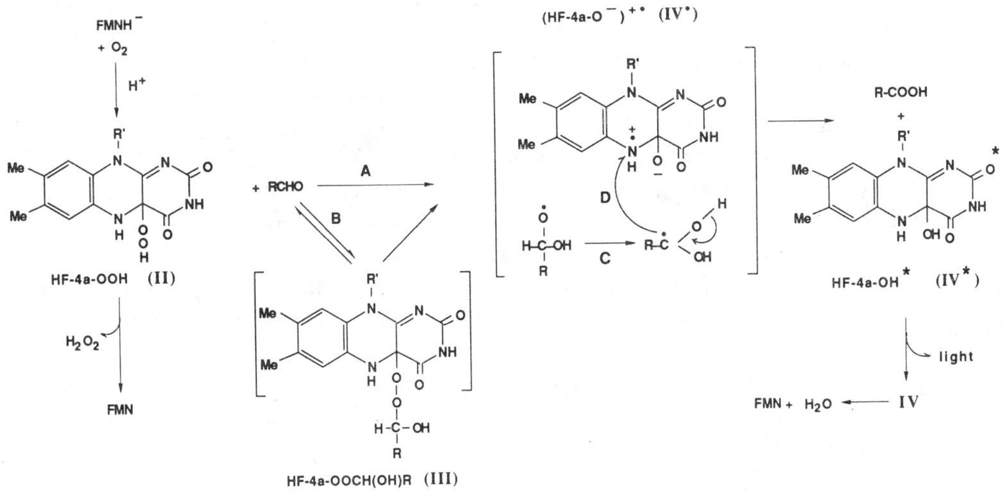

SCHEME 34.1

Supporting Evidence and Characterization of Intermediates

More than 2 decades of work has contributed to the identification and characterization of the 4a-hydroperoxyFMNH intermediate II (Table 34.2). To the author's knowledge, Mager and colleagues[56] were the first to propose in 1966 that O_2 can be covalently added to tetrahydropteridines and dihydroflavins. Initially, these researchers had succeeded in developing 10a-peroxyflavin models.[56,59] The first isolation of the luciferase-bound HF-4a-OOH intermediate was achieved by Hastings et al.[55] in 1973. Their −20°C procedure using an ethylene glycol-aqueous mixed solvent has been modified for a rapid and simpler ≥0 to 4°C operation in aqueous media using non-aldehyde aliphatics as a stabilizing agent.[64,65,68] This latter method has been used widely for stabilizing and isolating the 4a-hydroperoxyflavin intermediate derived from FMNH₂ as well as other reduced flavin derivatives. The proposed 4a-hydroperoxyFMNH identity for luciferase intermediate II is supported well by experimental evidence. The absorption,[55,64,65] fluorescence,[64,67–70] and NMR[71] spectra of the isolated II are all consistent with this structural assignment. The findings that the 5-ethyl-4a-hydroperoxy-4a,5-dihydroflavin model has an absorption spectrum identical to that of II and is active in chemiluminescence upon reacting with aldehyde provide additional strong support.[57,74] The proposed structure of II is supported also by the various ways of the formation of

Table 34.2   Characterization of Luciferase:4a-hydroperoxyflavin
Intermediate and Related Chemical Models

	Reference	
	Luciferase:HF-4a-OOH	Chemical models
Formation		
$FH_2 + O_2$	54, 55	56, 57
$F + H_2O_2$	58	57, 59
$FH\cdot + O_2^-$	60	61, 62
HF-4a-X + HOO$^-$	15	14, 63
Spectroscopy		
Absorption	55, 64, 65	57, 66
Fluorescence	64, 67–70	66
NMR	71	72
Reactivity		
±Ald	65, 73	57, 74
+Ald-O_2	75	

II by reactions of reduced flavin with oxygen,[54,55] oxidized flavin with H_2O_2,[58] flavin semiquinone with superoxide radical,[60] and flavin-4a-adducts with HOO$^-$.[15] The isolated intermediate II decays to oxidized flavin and equimolar H_2O_2 with very little light emission in the absence of aldehyde.[65,73] Alternatively, II reacts with an aldehyde substrate[65,73] in an oxygen-independent process[75] for efficient light emission with much reduced production of H_2O_2.

Two aspects about intermediate II await further delineation. First, when isolated in the dark, II was apparently weakly fluorescent (by itself or due to a slight FMN contamination). However, upon exposure to light within the absorption range (e.g., at 370 nm), II was transformed to a highly fluorescent species with the initial peak at 505 nm shifted to 485 nm (both uncorrected)[67] or from the original 535-nm emission to a 505-nm peak (both corrected).[64,68] Interestingly, both the initial and the light-transformed II exhibit about the same bioluminescence activity upon reacting with aldehyde.[64,67] The difference between these two forms of II at the molecular level is still unclear. It should be noted that years after these initial publications, Lee et al.[70] undertook a study to resolve the so-called "spectral discrepancy" between the blue shift of II fluorescence upon irradiation as observed by Balny and Hastings[67] and a red shift which Lee et al.[70] mistakenly attributed to the work of Tu[64,68] despite the fact the blue shift was reported by Tu also. Second, our recent studies on αHis-44-mutated luciferases show that two forms of II can be formed, with one active and the other inactive in bioluminescence.[42] These two forms are probably the stereoisomers with respect to the C(4a)-center, but a definitive identification requires further studies.

The original proposal of 4a-peroxyhemiacetalFMNH intermediate III makes good chemical sense. However, no direct evidence is available to confirm the formation of III, and the existence of III is not essential to Scheme 34.1. The formation and characterization of flavin 4a-peroxyhemiacetal models have been reported recently.[76] The 4a-hydroxyFMNH (IV) has been detected and characterized by absorption and fluorescence spectroscopy.[77,78] A fluorescent transient reported by Matheson et al.[79] may be the same species as IV.

A key aspect of Scheme 34.1 is the proposed formation of the novel (HF-4a-O$^-$)$^{+\cdot}$ radical. Using 5-ethyl-4a-hydroxy(or methoxy)-3-methyl-4a,5-dihydrolumiflavin (5-EtF-4a-X; X = OH, OMe), the formation of the corresponding flavin radicals was demonstrated first by one-electron electrochemical oxidation of the parent compounds.[80] The formation and accumulation of (5-EtF-4a-X)$^{+\cdot}$ (X = OH, OMe) in hydrophobic[49] and aqueous[81] media by a comproportionation reaction have also been achieved as follows:

$$5\text{-}EtF^+ + HX \rightarrow 5\text{-}EtF\text{-}4a\text{-}X + H^+ \qquad (34.2)$$

$$5\text{-}EtF^+ + 5\text{-}EtF\text{-}4a\text{-}X \rightarrow 5\text{-}EtF\cdot + \left(5\text{-}EtF\text{-}4a\text{-}X\right)^{+\cdot} \qquad (34.3)$$

Table 34.3　Redox Potentials of Flavins and Lumazines in Reference
to Normal Hydrogen Electrode

Redox system	$E_{1/2}$ (V)	Oxidation E_p (V)	Medium	Ref.
(5-EtF-4a-OH)$^+$/5-EtF-4a-OH	1.29		Acetonitrile	80
	1.23		Water	82
(5-EtF-4a-OMe)$^+$/5-EtF-4a-OMe	1.35		Acetonitrile	80
(3-Me-Ac$_4$Rf-4a-OH)$^+$/3-Me-Ac$_4$Rf-4a-OH	1.3^a		Water	82
3-MeF$_{sox}^+$/3-MeF		2.07	Acetonitrile	83
Ac$_4$Rf$_{sox}^+$/Ac$_4$Rf		2.08	Acetonitrile	83
5-EtF$_{sox}^{++}$/5-EtF$^+$		2.44	Acetonitrile	80
Me$_3$Lm$_{sox}^+$/Me$_3$Lm		1.96	Acetonitrile	51
Me$_4$Lm$_{sox}^+$/Me$_4$Lm		1.94	Acetonitrile	51

^{a} Estimated value.

The absorption spectra and redox potentials of (5-EtF-4a-X)$^+$ (X = OH, OMe) were determined first in acetonitrile.[80] Subsequently, the potentials for the one-electron oxidation of 5-EtF-4a-OH and 4a-hydroxy-4a,5-dihydro-3-methyl-tetraacetylrifoflavin (3-Me-Ac$_4$Rf-4a-OH) in water were also determined or estimated.[82] A summary of the redox potentials of these compounds and others to be mentioned later are shown in Table 34.3. Recently, the pK$_a$ values for (5-EtF-4a-OH)$^+$ and (HF-4a-OH)$^+$ hydroxyl groups have been estimated to be 3.7 ± 1.0 and ≤4.0, respectively, whereas a microscopic pK$_a$ of 5.4 has been assigned for the N(5) proton in (HF-4a-OH)$^+$.[82] Accordingly, the zwitterionic species (HF-4a-O$^-$)$^+$ is proposed as the major form of the IV$^.$ intermediate in Scheme 34.1.

Aside from the existence of flavin radicals equivalent to (HF-4a-O$^-$)$^+$, other experimental evidence in support of Scheme 34.1 is also available. Chemical[84] and electrochemical[85] reductions of preformed (5-EtF-4a-X)$^+$ are found to be coupled with chemiluminescence. Further important supporting evidence is obtained through a correlation of the bioluminescence decay kinetics using a series of flavin derivatives with the one-electron oxidation potentials for the N(5)-C(4a) bridged forms of the same series of flavins.[86] The results indicate that negative charge donation by the 4a,5-dihydroflavin is important in regulating the bioluminescence reaction.

Fluorescent Proteins, Emitters, and Primary Excited Species

It is accepted generally that the emitter in the normal luciferase-catalyzed bioluminescence reaction is the singlet-state HF-4a-OH*. However, the identity of the first excited species formed in a chemical step (i.e., the primary excited species) and the mechanism leading to its formation are still subjects of dispute. Lee and colleagues have isolated a lumazine-containing blue fluorescence protein (LP) from several strains of luminous bacteria.[87–89] In the luciferase-catalyzed *in vitro* reaction, the addition of LP results in an enhanced quantum yield, a slight change in the emission kinetics, and a blue shift of the bioluminescence peak from 490 to 476 nm.[87] An FMN-containing yellow fluorescence protein (YFP) has also been isolated.[90,91] Also, the copresence of luciferase and YFP shifts the emission maximum from near 500 to 530 nm, corresponding to the fluorescence of FMN. When present along with luciferase, excited lumazine in LP and excited FMN in YFP appear to function as emitters *in vivo* and *in vitro*. It has been argued that the observed blue shift cannot be attributed to an energy transfer from HF-4a-OH* to the lumazine. Using FMNH$_2$ and 2-thioFMNH$_2$ as substrates, detailed analyses of the emission kinetics and spectrum reveal also that the bioluminescence spectral change caused by YFP cannot be solely attributed to an energy transfer from HF-4a-OH* to YFP.[92] These considerations appear to challenge the identity of HF-4a-OH* as the primary excited species.

Scheme 1 could be modified to account for the LP- and YFP-dependent bioluminescence. HF-4a-OH* can be generated as the primary excited species by a one-electron addition to the lowest antibonding orbital of $(HF\text{-}4a\text{-}O^-)^{+\cdot}$ as depicted in process D of Scheme 34.1. In principle, R-COOH* or an equivalent excited carbonyl product can also be formed as the primary excited species through an electron abstraction from the highest bonding orbital of the alkyldihydroxy radical in the same process D. In this latter case, HF-4a-OH* is formed subsequently as the emitter through energy transfer from the highly energetic R-COOH*. In the presence of LP or YFP, energy transfer from R-COOH* to the bound lumazine or FMN in competition with HF-4a-OH can account for the observed emission spectral changes. In addition to this working hypothesis, the possible formation of an aliphatic carbonyl or other non-flavin primary excited species has also been advocated by several other groups.[93–96]

To elucidate further the central issue regarding the identity of the primary excited species, two fluorescent alkylpolyene aldehyde probes have been developed recently, namely α- and β-parinaraldehyde (α-PAD and β-PAD, respectively), which are active as substrates for the bioluminescence reaction.[97] The polyene moieties of both aldehydes emit fluorescence peaking at about 430 nm. Using *V. harveyi* luciferase and $FMNH_2$, the α-PAD- or β-PAD-initiated luminescence was identical to the normal emission (λ_{max} = 495 nm) showing no additional 430-nm component correlatable with emission from excited α-parinaric acid (α-PAC) or β-parinaric acid (β-PAC). Using 2-thioFMNH$_2$ with *V. harveyi* luciferase or $FMNH_2$ with *V. fischeri* luciferase plus YFP, the replacement of octanal substrate by β-PAD again resulted in no additional 430-nm emission. The lack of emission correlatable with excited α-PAC, β-PAC, or equivalent carbonyl product is not due to the quenching of the polyene moiety by chemical transformation, binding to luciferase, or a 100% energy transfer to HF-4a-OH. These results strongly favor HF-4a-OH* rather than singlet or triplet carbonyl product from aldehyde as the primary excited species in the normal luciferase reaction in the absence of any additional fluorescent protein.

Since HF-4a-OH* is most probably the primary excited species in the normal luciferase reaction, the mechanism of the LP- and YFP-dependent bioluminescence requires further considerations. Two schemes have been proposed in this regard. First, Hastings[98] and Tu et al.[83,84] have proposed that LP and YFP may function as a reactant rather than an energy transfer acceptor in the bioluminescence reaction. Tu's scheme[83,84] can be summarized as follows:

$$HF\text{-}4a\text{-}OOCH(OH)R + Y + H^+ \rightarrow HF\text{-}4a\text{-}OH + R\text{-}CH(OH)O^- + Y^{+\cdot} \quad (34.4)$$

$$R\text{-}CH(OH)O^- \rightarrow R\text{-}C^{\cdot}\text{-}(OH)_2 \quad (34.5)$$

$$R\text{-}C^{\cdot}\text{-}(OH)_2 + Y^{+\cdot} \rightarrow R\text{-}COOH + H^+ + Y^* \quad (34.6)$$

The 4a-hydroperoxyhemiacetalFMNH intermediate III reacts with Y (Y = LP-bound lumazine or YFP-bound FMN) to generate R-CH(OH)O$^-$, Y$^{+\cdot}$, and HF-4a-OH (Equation 34.4). The oxygen-centered R-CH(OH)O$^\cdot$ radical is transformed to the carbon-centered radical R-C$^\cdot$-(OH)$_2$ (Equation 34.5; same as process C in Scheme 34.1). The back electron transfer from R-C$^\cdot$-(OH)$_2$ to Y$^{+\cdot}$ leads to the formation of the excited Y (Equation 34.6). Such a scheme requires the one-electron oxidation of FMN and lumazine to form the corresponding superoxidized species $F_{sox}^{+\cdot}$ and $Lm_{sox}^{+\cdot}$. Using 3-methyllumiflavin (3-MeF), tetraacetylriboflavin (Ac$_4$Rf), 5-ethyl-3-methyllumiflavin (5-EtF$^+$), 6,7,8-trimethyllumazine (Me$_3$Lm), and 3,6,7,8-tetramethyllumazine (Me$_4$Lm), one-electron superoxidations of these compounds have been demonstrated[51,80,83] and their oxidation potentials are shown in Table 34.3. Moreover, chemiluminescence has been observed by reduction of these superoxidized flavins[83] or lumazines (Mager, H. I. X. and Tu, S.-C., unpublished results). A second possible mechanism to account for the LP-dependent bioluminescence was proposed recently.[82] The energy of the caged radical pair shown in Scheme 34.1 is about 90 kcal mol^{-1} above the ground state of HF-4a-OH and R-COOH and about 1 eV in excess of what is needed to populate the lowest singlet state of HF-4a-OH* for the 490-nm λ_{max} emission. This extra energy would be sufficient to populate HF-4a-OH at a higher excited singlet state. In the absence of LP, this higher singlet state can relax to the lower state of HF-4a-OH* as the emitter. In the presence of LP, this higher singlet state is energetically sufficient to sensitize the formation of excited LP by energy transfer, possibly through a collision mechanism. This mechanism requires that the lifetime for the higher singlet state is longer than about 0.1 ns and that a tight complex is formed between luciferase and LP. These two proposed mechanisms await further experimental tests.

Acknowledgments

The support of grants from N.I.H. (GM25953) and The Robert A. Welch Foundation (E-1030) is gratefully acknowledged. The author is indebted to his former and present group members for their invaluable contributions to the bacterial luciferase research project.

References

1. Baldwin, T. O., Holzman, T. F., Holzman, R. B., and Riddle, V. A., Purification of bacterial luciferase by affinity methods, *Meth. Enzymol.,* 133, 98, 1986.
2. O'Kane, D. J., Ahmad, M., Matheson, I. B. C., and Lee, J., Purification of bacterial luciferase by high-performance liquid chromatography, *Meth. Enzymol.,* 133, 109, 1986.
3. Waters, C. A., Murphy, J. R., and Hastings, J. W., Flavin binding by bacterial luciferase: Affinity chromatography of bacterial luciferase, *Biochem. Biophys. Res. Commun.,* 57, 1152, 1974.
4. Gajiwala, K. S., Affinity Chromatography of *Vibrio harveyi* Luciferase, Master thesis, The University of Texas at El Paso, El Paso, 1988.
5. Hastings, J. W., Baldwin, T. O., and Nicoli, M. Z., Bacterial luciferase: Assay, purification, and properties, *Meth. Enzymol.,* 57, 135, 1978.
6. Cousineau, J. and Meighen, E. A., Chemical modification of bacterial luciferase with ethoxyformic anhydride: Evidence for an essential histidyl residue, *Biochemistry,* 15, 4992, 1976.
7. Xi, L., Cho, K.-W., and Tu, S.-C., Cloning and nucleotide sequences of *lux* genes and characterization of luciferase of *Xenorhabdus luminescens* from a human wound, *J. Bacteriol.,* 173, 1399, 1991.
8. Meighen, E. A. and MacKenzie, R. E., Flavine specificity of enzyme-substrate intermediates in the bacterial bioluminescent reaction. Structural requirements of the flavine side chain, *Biochemistry,* 12, 1482, 1973.
9. Tu, S.-C. and Hastings, J. W., Differential effects of 8-anilino-1-naphthalene-sulfonate upon binding of oxidized and reduced flavins by bacterial luciferase, *Biochemistry,* 14, 4310, 1975.
10. Hastings, J. W. and Weber, G., Total quantum flux of isotopic sources, *J. Opt. Soc. Am.,* 53, 1410, 1963.
11. Gunsalus-Miguel, A., Meighen, E. A., Nicoli, M. Z., Nealson, K. H., and Hastings, J. W., Purification and properties of bacterial luciferases, *J. Biol. Chem.,* 247, 398, 1972.
12. Paquatte, O., Fried, A., and Tu, S.-C., Delineation of bacterial luciferase aldehyde site by bifunctional labeling reagents, *Arch. Biochem. Biophys.,* 264, 392, 1988.
13. Hastings, J. W., Weber, K., Friedland, J., Eberhard, A., Mitchell, G. W., and Gunsalus, A., Structurally distinct bacterial luciferases, *Biochemistry,* 8, 4681, 1969.
14. Mager, H. I. X. and Tu, S.-C., Dithionite treatment of flavins: Spectral evidence for covalent adduct formation and effect on *in vitro* bacterial bioluminescence, *Photochem. Photobiol.,* 51, 223, 1990.
15. Tu, S.-C. and Cho, K. W., On the mechanism of dithionite/H_2O_2-induced bacterial bioluminescence, in *Flavins and Flavoproteins 1990,* Curti, B., Ronchi, S., and Zanetti, G., Eds., Walter de Gruyter, Berlin, 1991, 281.
16. Lei, B.-F. and Becvar, J. E., A new reducing agent of flavins and its application to the assay of bacterial luciferase, *Photochem. Photobiol.,* 54, 473, 1991.
17. Paquatte, O. and Tu, S.-C., Kinetic analysis of enzyme inactivation by an autodecaying reagent, *Biochem. Biophys. Acta,* 869, 359, 1986.

18. Cohn, D. H., Mileham, A. J., Simon, M. I., Nealson, K. H., Rausch, S. K., Bonam, D., and Baldwin, T. O., Nucleotide sequence of the *lux*A gene of *Vibrio harveyi* and the complete amino acid sequence of the α subunit of bacterial luciferase, *J. Biol. Chem.*, 260, 6139, 1985.

19. Johnston, T. C., Thompson, R. B., and Baldwin, T. O., Nucleotide sequence of the *lux*B gene of *Vibrio harveyi* and the complete amino acid sequence of the β subunit of bacterial luciferase, *J. Biol. Chem.*, 261, 4805, 1986.

20. Baldwin, T. O., Chen, L. H., Chlumsky, L. J., Devine, J. H., Johnston, T. C., Lin, J.-W., Sugihara, J., Waddle, J. J., and Ziegler, M. M., Structural analysis of bacterial luciferase, in *Flavins and Flavoproteins,* Edmondson, D. E. and McCormick, D. B., Eds., Walter de Gruyter, Berlin, 1987, 621.

21. Foran, D. R. and Brown, W. M., Nucleotide sequence of the *lux*A and *lux*B genes of the bioluminescent marine bacterium *Vibrio fischeri, Nucl. Acids Res.*, 16, 777, 1988.

22. Johnston, T. C., Rucker, E. B., Cochrum, L., Hruska, K. S., and Vandegrift, V., The nucleotide sequence of the *lux*A and *lux*B genes of *Xenorhabdus luminescens* Hm and a comparison of the amino acid sequences of luciferases from four species of bioluminescent bacteria, *Biochem. Biophys. Res. Commun.*, 170, 407, 1990.

23. Meighen, E. A., Molecular biology of bacterial bioluminescence, *Microbiol. Rev.*, 55, 123, 1991.

24. Illarionov, B. A., Protopopova, M. V., Karginov, V. A., Mertvetsov, N. P., and Gitelson, J. I., Nucleotide sequence of part of *Photobacterium leiognathi lux* region, *Nucl. Acids Res.*, 16, 9855, 1988.

25. Lee, C. Y., Szittner, R. B., and Meighen, E. A., The *lux* genes of the luminous bacterial symbiont, *Photobacterium leiognathi,* of the ponyfish. Nucleotide sequence, difference in gene organization, and high expression in mutant *Escherichia coli, Eur. J. Biochem.*, 201, 161, 1991.

26. Szittner, R. and Meighen, E., Nucleotide sequence, expression, and properties of luciferase coded by *lux* genes from a terrestrial bacterium, *J. Biol. Chem.*, 265, 16581, 1990.

27. Haygood, M. G., Relationship of the luminous bacterial symbiont of the Caribbean flashlight fish, *Kryptophanaron alfredi* (family Anomalopidae) to other luminous bacteria based on bacterial luciferase (*lux*A) genes, *Arch. Microbiol.*, 154, 496, 1990.

28. Waddle, J. and Baldwin, T. O., Individual α and β subunits of bacterial luciferase exhibit bioluminescence activity, *Biochem. Biophys. Res. Commun.*, 178, 1188, 1991.

29. Meighen, E. A., Nicoli, M. Z., and Hastings, J. W., Functional differences of the nonidentical subunits of bacterial luciferase. Properties of hybrids of native and chemically modified bacterial luciferase, *Biochemistry*, 10, 4069, 1971.

30. Cline, T. W. and Hastings, J. W., Mutationally altered bacterial luciferase. Implications for subunit functions, *Biochemistry*, 11, 3359, 1972.

31. Watanabe, H., Hastings, J. W., and Tu, S.-C., Activity and subunit functions of immobilized bacterial luciferase, *Arch. Biochem. Biophys.*, 215, 405, 1982.

32. Meighen, E. A. and Bartlet, I., Complementation of subunits from different bacterial luciferases. Evidence for the role of the β subunit in the bioluminescent mechanism, *J. Biol. Chem.*, 255, 11181, 1980.

33. Tu, S.-C. and Henkin, J., Characterization of the aldehyde binding site of bacterial luciferase by photoaffinity labeling, *Biochemistry*, 22, 519, 1983.

34. Fried, A. and Tu, S.-C., Affinity labeling of the aldehyde site of bacterial luciferase, *J. Biol. Chem.*, 259, 10754, 1984.

35. Xin, X., Elucidation of Structural Basis for the Light-Emitting Activity of Bacterial Luciferase by Site-Directed Mutagenesis, Ph.D. dissertation, University of Houston, Houston, 1992.

36. Wells, J., Additivity of mutational effects in proteins, *Biochemistry*, 29, 8509, 1990.

37. Nicoli, M. Z., Meighen, E. A., and Hastings, J. W., Bacterial luciferase. Chemistry of the reactive sulfhydryl, *J. Biol. Chem.*, 249, 2385, 1974.

38. Ziegler, M. M. and Baldwin, T. O., Active center studies on bacterial luciferase: Modification with methyl methanethiolsulfonate, in *Bioluminescence and Chemiluminescence — Basic Chemistry and Analytical Applications,* DeLuca, M. A. and McElroy, W. D., Eds., Academic Press, New York, 1981, 155.

39. Paquatte, O. and Tu, S.-C., Chemical modification and characterization of the alpha subunit cysteine 106 at the *Vibrio harveyi* luciferase active center, *Photochem. Photobiol.*, 50, 817, 1989.

40. Xi, L., Cho, K. W., Herndon, M. E., and Tu, S.-C., Elicitation of an oxidase activity in bacterial luciferase by site-directed mutation of a noncatalytic residue, *J. Biol. Chem.*, 265, 4200, 1990.

41. Baldwin, T. O., Chen, L. H., Chlumsky, L. J., Devine, J. H., and Ziegler, M. M., Site-directed mutagenesis of bacterial luciferase: Analysis of the 'essential' thiol, *J. Biolumin. Chemilumin.*, 4, 40, 1989.

42. Xin, X., Xi, L., and Tu, S.-C., Functional consequences of site-directed mutation of conserved histidyl residues of the bacterial luciferase α subunit, *Biochemistry*, 30, 11255, 1991.

43. Welches, W. R. and Baldwin, T. O., Active center studies on bacterial luciferase: Modification of the enzyme with 2,4-dinitrofluorobenzene, *Biochemistry*, 20, 512, 1981.

44. Anderson, C., Tu, S.-C., and Hastings, J. W., Subunit exchange between and specific activities of mutant bacterial luciferases, *Biochem. Biophys. Res. Commun.*, 95, 1180, 1980.

45. Eberhard, A. and Hastings, J. W., A postulated mechanism for the bioluminescent oxidation of reduced flavin mononucleotide, *Biochem. Biophys. Res. Commun.*, 47, 348, 1972.

46. Hastings, J. W. and Nealson, K. H., Bacterial bioluminescence, *Annu. Rev. Microbiol.*, 31, 549, 1977.

47. Ahrens, M., Macheroux, P., Eberhard, A., Ghisla, S., Branchaud, B. P., and Hastings, J. W., Boronic acids as mechanistic probes for the bacterial luciferase reaction, *Photochem. Photobiol.*, 54, 295, 1991.

48. Schuster, G. B., Chemiluminescence of organic peroxides. Conversion of ground-state reactants to excited-state products by the chemically initiated electron-exchange luminescence mechanism, *Acc. Chem. Res.*, 12, 366, 1979.

49. Mager, H. I. X. and Addink, R., On the role of some flavin adducts as one-electron donors, in *Flavins and Flavoproteins*, Bray, R. C., Engel, P. C., and Mayhew, S. G., Eds., Walter de Gruyter, Berlin, 1984, 37.

50. Mager, H. I. X. and Tu, S.-C., One-electron transfers in flavin systems: Relevance to the postulated CIEEL mechanism in bacterial bioluminescence, in *Flavins and Flavoproteins*, Edmondson, D. E. and McCormick, D. B., Eds., Walter de Gruyter, Berlin, 1987, 583.

51. Tu, S.-C., Mager, H. I. X., Shao, R., Cho, K. W., and Xi, L., Mechanisms of bacterial luciferase and aromatic hydroxylases, in *Flavins and Flavoproteins 1990*, Curti, B., Ronchi, S., and Zanetti, G., Eds., Walter de Gruyter, Berlin, 1991, 253.

52. Vervoort, J., Muller, F., O'Kane, D. J., Lee, J., and Bacher, A., Bacterial luciferase: A carbon-13, nitrogen-15, and phosphorus-31 nuclear magnetic resonance investigation, *Biochemistry*, 25, 8067, 1986.

53. Catalani, L. H. and Wilson, T., Electron transfer and chemiluminescence. Two inefficient systems: 1,4-dimethoxy-9,10-diphenylanthracece peroxide and diphenoyl peroxide, *J. Am. Chem. Soc.*, 111, 2633, 1989.

54. Hastings, J. W. and Gibson, Q. H., Intermediates in the bioluminescent oxidation of reduced flavin mononucleotide, *J. Biol. Chem.*, 238, 2537, 1963.

55. Hastings, J. W., Balny, C., LePeuch, C., and Douzou, P., Spectral properties of an oxygenated luciferase-flavin intermediate isolated by low-temperature chromatography, *Proc. Natl. Acad. Sci. U.S.A.*, 70, 3468, 1973.

56. Berends, W., Posthuma, J., Sussenbach, J. S., and Mager, M. I. X., Mechanism of some flavin-photosensitized reactions, in *Flavins and Flavoproteins*, Slater, E. C., Ed., Elsevier, Amsterdam, 1966, 22.

57. Kemal, C. and Bruice, T. C., Simple synthesis of a 4a-hydroperoxy adduct of a 1,5-dihydroflavine: Preliminary studies of a model for bacterial luciferase, *Proc. Natl. Acad. Sci., U.S.A.*, 73, 995, 1976.

58. Hastings, J. W., Tu, S.-C., Becvar, J. E., and Presswood, R. P., Bioluminesce from the reaction of FMN, hydrogen peroxide, and long chain aldehyde with bacterial luciferase, *Photochem. Photobiol.*, 29, 383, 1979.

59. Mager, H. I. X., Nonenzymatic activation and transfer of oxygen by reduced alloxazines, in *Flavins and Flavoproteins*, Singer, T. P., Ed., Elsevier, Amsterdam, 1976, 23.

60. Kurfürst, M., Ghisla, S., and Hastings, J. W., Bioluminescence emission from the reaction of luciferase-flavin mononucleotide radical with O_2^-, *Biochemistry*, 22, 1521, 1983.

61. Nanni, E. J., Jr., Sawyer, D. T., Ball, S. S., and Bruice, T. C., Redox chemistry of N^5-ethyl-3-methyllumiflavinium cation and N^5-ethyl-4a-hydroperoxy-3-methyllumiflavin in dimethylformamide. Evidence for the formation of the N^5-ethyl-4a-hydroperoxy-3-methyllumiflavin anion via radical-radical coupling with superoxide ion, *J. Am. Chem. Soc.*, 103, 2797, 1981.

62. Anderson, R. F., Flavin-oxygen complex formed on the reaction of superoxide ions with flavosemiquinone radicals, in *Flavins and Flavoproteins*, Massey, V. and Williams, C. H., Jr., Eds., Elsevier, New York, 1982, 278.

63. Murahashi, S.-I., Oda, T., and Masui, Y., Flavin-catalyzed oxidation of amines and sulfur compounds with hydrogen peroxide, *J. Am. Chem. Soc.*, 111, 5002, 1989.

64. Tu, S.-C., Isolation and properties of bacterial luciferase-oxygenated flavin intermediate complexed with long-chain alcohols, *Biochemistry*, 18, 5940, 1979.

65. Tu, S.-C., Isolation and properties of bacterial luciferase intermediates containing different oxygenated flavins, *J. Biol. Chem.*, 257, 3719, 1982.

66. Ghisla, S., Massey, V., Lhoste, J.-M., and Mayhew, S.G., Fluorescence and optical characteristics of reduced flavins and flavoproteins, *Biochemistry*, 13, 589, 1974.

67. Balny, C. and Hastings, J. W., Fluorescence and bioluminescence of bacterial luciferase intermediates, *Biochemistry*, 14, 4719, 1975.

68. Tu, S.-C., Bacterial luciferase 4a-hydroperoxyflavin intermediates: Stabilization, isolation, and properties, *Meth. Enzymol.*, 133, 128, 1986.

69. Hastings, J. W., Ghisla, S., Kurfürst, M., and Hemmerich, P., Fluorescence properties of luciferase peroxyflavins prepared with iso-FMN and 2-thio FMN, in *Bioluminescence and Chemiluminescence: Basic Chemistry and Analytical Applications*, DeLuca, M. and McElroy, W. D., Eds., Academic Press, New York, 1981, 97.

70. Lee, J., O'Kane, D. J., and Gibson, B. G., Dynamic fluorescence properties of bacterial luciferase intermediates, *Biochemistry*, 27, 4862, 1988.

71. Vervoort, J., Muller, F., Lee, J., van den Berg, W. A. M., and Moonen, C. T. W., Identifications of the true carbon-13 nuclear magnetic resonance spectrum of the stable intermediate II in bacterial luciferase, *Biochemistry*, 25, 8062, 1986.

72. Lhoste, J.-M., Favaudon, V., Hastings, J. W., and Ghisla, S., ^{13}C-NMR determinations of the structure of the luciferase flavin oxygen adduct, in *Flavins and Flavoproteins*, Yagi, Y. and Yamano, T., Eds., Japan Scientific Societies Press, Tokyo, 1980, 131.

73. Hastings, J. W. and Balny, C., The oxygenated bacterial luciferase-flavin intermediate. Reaction products via the light and dark pathways, *J. Biol. Chem.*, 250, 7288, 1975.

74. Kemal, C., Chan, T. W., and Bruice, T. C., Chemiluminescent reactions and electrophilic oxygen donating ability of 4a-hydroperoxyflavins: General synthetic method for the preparation of N^5-alkyl-1,5-dihydroflavins, *Proc. Natl. Acad. Sci. U.S.A.*, 74, 405, 1977.

75. Becvar, J. E., Tu, S.-C., and Hastings, J. W., Activity and stability of luciferase-flavin intermediate, *Biochemistry*, 17, 1807, 1978.

76. Merényi, G. and Lind, J., Chemistry of peroxidic tetrahedral intermediates of flavin, *J. Am. Chem. Soc.*, 113, 3146, 1991.

77. Kurfürst, M., Ghisla, S., and Hastings, J. W., Characterization and postulated structure of the primary emitter in the bacterial luciferase reaction, *Proc. Natl. Acad. Sci. U.S.A.*, 81, 2990, 1984.

78. Kurfürst, M., Macheroux, P., Ghisla, S., and Hastings, J. W., Isolation and characterization of the transient, luciferase-bound flavin-4a-hydroxide in the bacterial luciferase reaction, *Biochim. Biophys. Acta*, 924, 104, 1987.

79. Matheson, I. B. C., Lee, J., and Müller, F., Bacterial bioluminescence: Spectral study of the emitters in the *in vitro* reaction, *Proc. Natl. Acad. Sci. U.S.A.*, 78, 948, 1981.

80. Mager, H. I. X., Sazou, D., Liu, Y. H., Tu, S.-C., and Kadish, K. M., Reversible one-electron generation of 4a,5-substituted flavin radical cations: Models for a postulated key intermediate in bacterial bioluminescence, *J. Am. Chem. Soc.*, 110, 3759, 1988.

81. Mager, H. I. X. and Tu, S.-C., Spontaneous formation of flavin radicals in aqueous solution by comproportionation of a flavinium cation and a flavin pseudobase, *Tetrahedron*, 44, 5669, 1988.

82. Merényi, G., Lind, J., Mager, H. I. X., and Tu, S.-C., Properties of 4a-hydroxy-4a,5-dihydroflavin radicals in relation to bacterial bioluminescence, *J. Phys. Chem.*, 96, 10528, 1992.

83. Mager, H. I. X., Tu, S.-C., Liu, Y.-H., Deng, Y., and Kadish, K. M., Electrochemical superoxidation of flavins: Generation of active precursors in luminescent model systems, *Photochem. Photobiol.*, 52, 1049, 1990.

84. Tu, S.-C., Oxygenated flavin intermediates of bacterial luciferase and flavoprotein aromatic hydroxylases: Enzymology and chemical models, in *Advances in Oxygenated Processes*, Vol. 3, Baumstark, A. L., Ed., Jai Press, Greenwich, 1991, 115.

85. Kaaret, T. W. and Bruice, T. C., Electrochemical luminescence with N(5)-ethyl-4a-hydroxy-3-methyl-4a,5-dihydrolumiflavin. The mechanism of bacterial luciferase, *Photochem. Photobiol.*, 51, 629, 1990.

86. Eckstein, J. W. and Ghisla, S., On the mechanism of bacterial luciferase. 4a,5-Dihydroflavins as model compounds for reaction intermediates, in *Flavins and Flavoproteins 1990*, Curti, B., Ronchi, S., and Zanetti, G., Eds., Walter de Gruyter, Berlin, 1991, 269.

87. Gast, R. and Lee, J., Isolation of the *in vivo* emitter in bacterial bioluminescence, *Proc. Natl. Acad. Sci. U.S.A.*, 75, 833, 1978.

88. Koka, P. and Lee, J., Separation and structure of the prosthetic group of the blue fluorescence protein from the bioluminescent bacterium *Photobacterium phosphoreum*, *Proc. Natl. Acad. Sci. U.S.A.*, 76, 3068, 1979.

89. O'Kane, D. J., Karle, V. A., and Lee, J., Purification of lumazine proteins from *Photobacterium leiognathi* and *Photobacterium phosphoreum*: Bioluminescence properties, *Biochemistry*, 24, 1461, 1985.

90. Daubner, S. C., Astorga, A. M., Leisman, G. B., and Baldwin, T. O., Yellow light emission of *Vibrio harveyi* strain Y-1: Purification and characterization of the energy-accepting yellow fluorescent protein, *Proc. Natl. Acad. Sci. U.S.A.*, 84, 8912, 1987.

91. Macheroux, P., Schmidt, K. U., Steinerstauch, P., Ghisla, S., Colepicolo, P., Buntic, R., and Hastings, J. W., Purification of the yellow fluorescent protein from *Vibrio fischeri* and identity of the flavin chromophore, *Biochem. Biophys. Res. Commun.*, 146, 101, 1987.

92. Eckstein, J. W., Cho, K. W., Colepicolo, P., Ghisla, S., Hastings, J. W., and Wilson, T., A time-dependent bacterial bioluminescence emission spectrum in an *in vitro* single turnover system: Energy transfer alone cannot account for the yellow emission of *Vibrio fischeri* Y-1, *Proc. Natl. Acad. Sci. U.S.A.*, 87, 1466, 1990.

93. Hart, R. C. and Cormier, M. J., Recent advances in the mechanisms of bio- and chemiluminescent reactions, *Photochem. Photobiol.*, 29, 209, 1979.

94. Ziegler, M. M. and Baldwin, T. O., Biochemistry of bacterial bioluminescence, *Curr. Top. Bioenerg.*, 12, 65, 1981.

95. Raushel, F. M. and Baldwin, T. O., Proposed mechanism for the bacterial bioluminescence reaction involving a dioxirane intermediate, *Biochem. Biophys. Res. Commun.*, 164, 1137, 1989.

96. Lee, J., Matheson, I. B. C., Muller, F., O'Kane, D. J., Vervoort, J., and Visser, A. J. W. G., The mechanism of bacterial bioluminescence, in *Chemistry and Biochemistry of Flavins and Flavoenzymes*, Vol. 2, Muller, F., Ed., CRC Press, Boca Raton, FL, 1991, 109.

97. Cho, K. W., Tu, S.-C., and Shao, R., Fluorescent polyene aliphatics as spectroscopic and mechanistic probes for bacterial luciferase: Evidence against carbonyl product from aldehyde as the primary excited species, *Photochem. Photobiol.*, 57, 396, 1993.

98. Hastings, J. W., Potrikus, C. J., Gupta, S. C., Kurfürst, M., and Makemson, J. C., Biochemistry and physiology of bioluminescent bacteria, *Adv. Microb. Physiol.*, 26, 235, 1985.

35

Photobiology of Circadian Rhythms

Carl Johnson
Vanderbilt University

35.1 Clocks and Light

Circadian rhythms are the daily expression of endogenous biochemical "clocks" within organisms. In constant conditions (i.e., constant light or constant darkness at a constant temperature), these clocks "freerun" with a period close to, but never exactly, 24 h. Although this freerunning behavior is a salient characteristic, the adaptive value of these clocks is to function as a biochemical wristwatch that informs the organism of the time of day in a natural light/dark cycle. If an organism knows the time, it can anticipate and prepare for the daily transformations of the environment that occur at, e.g., dawn and dusk. To accomplish this purpose, the biochemical clock must be synchronized — or entrained — to the daily cycle of the environment. In its natural setting, the organism's clock entrains to the environmental cycle so that its period becomes exactly 24 h.

What environmental signals entrain circadian oscillators? The daily cycle of light and darkness is the most important time cue, probably because it is the most regular and reliable daily cycle in the environment. Without exception, circadian rhythms can be entrained by light/dark cycles. For many organisms (e.g., nocturnal rodents), entrainment behavior can be modeled very accurately by knowing only two things: (1) the freerunning period, and (2) the phase-dependent responsiveness of the clock to single light (or dark) pulses.[1] This later responsiveness is usually depicted as a "Phase Response Curve," which describes how the clock is reset, or phase-shifted by light (or dark) pulses.[2] The photopigments involved in light-pulse resetting will be discussed in Section 35.3. Light can also affect the freerunning period under constant illumination,[3] as discussed in Section 35.4. Finally, photoreceptive organs that mediate entrainment in animals will be described in Section 35.5.

0-8493-8634-9/95/$0.00+$.50
© 1995 by CRC Press, Inc.

35.2 Action Spectroscopy of Circadian Phase-Resetting: Special Problems

As in other photobiological investigations, action spectroscopy has been a crucial tool for identifying photopigments that are linked to the circadian clockworks. Proper action spectroscopy is based upon fluence response curves, and all the typical caveats and complexities (e.g., univariance, screening) associated with the interpretation of fluence response curves hold for the action spectra of circadian phase-shifting. But, circadian photobiologists can encounter an unusual problem: discontinuous or nonmonotonic fluence response curves. Discontinuous and/or nonmonotonic fluence response curves have been observed in studies of circadian photoresponses of algae[4,5] and fungi.[6] Sometimes the phase-shifting response of circadian clocks to increasing fluence can be positive at low fluence, then level off, and finally become inversely correlated at high fluence.[5]

A discontinuous or nonmonotonic fluence response curve probably means that some process downstream of the photopigment's absorption of light is converting the initially continuous photochemical response into a discontinuous biological response. In the case of clock photoreceptors, the "limit-cycle organization" of the oscillator itself can convert the initially monotonic response into a discontinuous response as the light pulse moves the pacemaker past the singular region. This problem can be circumvented by the appropriate selection of fluence ranges and phases to be tested, as discussed in detail elsewhere.[5]

35.3 Photopigments Involved in Light-Pulse Resetting

Action spectra of clock photoreceptors are listed in Table 35.1. Not all of the studies mentioned in Table 35.1 are of equal quality; in some, the experimental design was not adequate to make firm conclusions about the spectral response of the respective oscillators. For example, several of the action spectra in Table 35.1 use only a single fluence at each wavelength (an equal-intensity action spectrum) instead of the range of fluences required for proper action spectroscopy. This means that neither univariance nor the continuity/monotonicity of the fluence response was tested. This criticism is true for the studies of *Coleus* and *Kalanchoe* (possibly also *Paramecium*). In the case of *Gonyaulax*, only two fluences were checked, so this action spectrum may also have been compromised by this problem. The results of the studies included in Table 35.1 are briefly discussed below.

Blue/Green Action

Clock resetting in *Drosophila*,[7,8] *Pectinophora*,[9] *Periplaneta*,[10] and the bat *Hipposideros*[11] have action spectra that exhibit maximal action in the range 400 to 500 nm. In *Drosophila*, both advance and delay phase-shifts were assayed; they exhibited identical action spectra[7] and were not affected by raising the flies on carotenoid-depleted medium.[12] These experiments and those of other researchers suggest that flavins are candidates for this photoreceptor.[8,13] The action spectrum for clock resetting of *Neurospora*[6] shows action from 375 to 500 nm with a peak at 470 nm. This action spectrum is often interpreted to be due to a flavin photoreceptor; flavin substitution experiments support this notion.[14,15] On the other hand, measurements of action spectra in rodents correspond with that expected of rhodopsin.[16–18] In molluscs, the spectral sensitivity for resetting the circadian clock in the *Bulla* eye[19] is also consistent with the absorption spectrum for rhodopsin.

Red Action

In many plants, phytochrome is involved in circadian entrainment.[20] Assayed by red/far-red reversibility, phytochrome has been implicated in phase-shifting of *Phaseolus*,[13,21] *Hordeum*,[22] *Lemna*,[23] *Albizzia*,[24] and *Samanea*.[24,25] In the latter three cases, blue light can elicit phase-shifts, albeit

Table 35.1   Action Spectra for Light Resetting of Circadian Clocks

Organism	Background illumination[a]	Peaks(s) of action spectrum (nm)	Presumed photopigment	Ref.
Unicells				
Gonyaulax (alga)[b]	DD	475, 650	?	27
Chlamydomonas (alga)	DD	520, 660	?	28
Chlamydomonas (alga)	LL	460–480, 660–680	Chlorophyll	29
Paramecium (protozoan)	DD	350, 430 600–700	?	34, 35
Plants/Fungi				
Neurospora (fungus)	DD	465–470	Flavin (?)	6
Coleus (plant)[b]	Green LL	Blue = delays Red = advances	?	30
Kalanchoe (plant)[b]	DD	366, 600–680	?	32
Several other plants (see text)	Various	Red/far-red	Phytochrome	See text
Invertebrates				
Drosophila (insect)	DD	400–480	Flavin (?)	7
Drosophila (insect)	DD	450–460	Flavin (?)	8
Pectinophora (insect)	DD	400–480	?	9
Periplaneta (insect)[c]	DD	495	Rhodopsin (?)	10
Bulla (mollusc)[c]	DD	500	Rhodopsin	19
Vertebrates				
Mesocricetus (hamster)	DD	500	Rhodopsin	16
Rattus (rat)	DD	530	Rhodopsin (?)	17, 18
Mus (mouse)	DD	500	Rhodopsin	17, 18
Hipposideros (bat)	DD	400–480	Rhodopsin (?)	11

[a] DD = constant darkness; LL = constant light (white unless indicated otherwise).

[b] "Equal-intensity" action spectrum only (or inadequate fluence response curves).

[c] Response was tested at only 4–6 wavelengths, but the data are compatible with a rhodopsin photopigment.

less effectively than red light, and the blue light-induced phase shift can also be reversed by far-red light. *Bryophyllum*[26] is phase-shifted by red light, but far-red only partially reverses this action.

Dual Blue and Red Action

In four photosynthetic organisms — *Gonyaulax*,[27] *Chlamydomonas*,[28,29] *Coleus*,[30] and *Kalanchoe* (= *Bryophyllum*) — both blue and red light phase-shift the rhythm. In *Kalanchoe*, two effects were measured: rhythm initiation by transfers between constant light (LL) and constant darkness (DD),[31] and pulse resetting.[26,32] It is possible that these effects are mediated by two separate photopigments.

In at least one case, this dual blue/red action can be attributed probably to chlorophyll because a photosynthetic inhibitor (DCMU) inhibits the phase-shift by blue or red light in *Chlamydomonas* cells in LL.[29] Thus, components of photosynthesis appear to mediate clock resetting in LL. But for *Chlamydomonas* cells in DD, the action spectrum is different! In DD, green and red light reset the clock and photosynthetic inhibitors are ineffectual.[28] The identity of this latter photoreceptor is unknown, although the green peak may be due to the photopigment identified in a microspectro-photometric study.[33] The sensitivity of the two photosystems is also very different: *Chlamydomonas* cells in DD phase-shift in response to 1/2000 of the fluence needed to reset cells in LL.[28,29]

In the other cases involving blue and red light, the photopigment(s) involved are even less clear. In *Coleus*, red and blue light elicit phase-shifting in different directions: red light advances the rhythm, while blue light delays the rhythm.[30] In the case of *Kalanchoe*, simultaneous irradiation with red and far-red light potentiates the effect of red light alone, which may mean that an HIR-type phytochrome is involved.[26,32] Note, however, that of these studies implicating dual blue and red action, only in the case of *Chlamydomonas* was proper action spectroscopy done, i.e., by testing a range of several fluences. Therefore, the exact shape of some of these "action spectra" is still to be determined.

Dual blue and red action has also been reported for a protozoan that does not contain chlorophyll — a *Chlorella*-less strain of *Paramecium bursaria* has been reported to have action peaks at 350, 430, and 600–700 nm, but the study is difficult to assess because so little information was provided.[34] A more recent investigation of *Chlorella*-less *Paramecium bursaria* has confirmed that both blue and red light will phase-shift, while green light is not effective.[35]

Ultraviolet Action

In both *Gonyaulax*[36] and *Paramecium*,[37] brief pulses of UV light can cause significant phase-shifting. The cells respond differently to UV light than to VIS light: (1) the magnitude of phase-shifting to UV light is not very phase dependent, and (2) the shifts are all advances *(Gonyaulax)* or all delays *(Paramecium)*. Phase resetting by UV light in these unicells has not usually been interpreted to be relevant for entrainment in a natural setting; rather, it has been suggested to result from damage to nucleic acids that may be involved in the biochemical mechanism of circadian oscillators.[36,37] In another case, however, UV-A light appears to entrain the circadian clock of birds in a manner consistent with an entraining role for UV-A in a natural setting.[38]

Reciprocity

Thus far, the identification of some clock photopigments is inconclusive. Nevertheless, it is clear that no single photopigment is used by all circadian pacemakers. Some green plants use a phytochrome, while some animals use a rhodopsin. More action spectroscopy needs to be done on clocks. It may also be worthwhile to test the intensity vs. duration reciprocity of clock photoreceptors. Just as phytochrome, acting in its HIR mode, displays quite different characteristics than that of phytochrome in its classic reversible red/far-red mode, clock photoreceptors may exhibit likewise features not found in other photoreceptive systems. For example, Nelson and Takahashi[39] found efficient reciprocity in the phase-shifting response of nocturnal hamsters for durations between 5 and 60 min. This is an unusually long reciprocity for a visual system based on rhodopsin. Kondo et al[28] found reciprocity between 3 and 300 s for a clock photoreceptor in *Chlamydomonas*. Logically, it makes sense for a clock photopigment to integrate light signals over long durations, so as to minimize their response to environmental noise. Nevertheless, few circadian researchers have quantified the reciprocity of their favorite experimental subject.

Characterization of Phototransduction Pathways

Many clock researchers hope to discover the mechanism of the circadian oscillator by tracing the pathway of phase-shifting information from the photoreceptor to the clockworks.[40] This approach, which might be called following the photon to the clock, has been championed by Eskin.[41] For example, Eskin has shown that the clock of the *Aplysia* eye responds to increases of cyclic GMP in much the same way as it does to light, suggesting that — as in visual transduction within the vertebrate eye — cyclic GMP mediates the effect of light transduction within this clock's photoreceptor.[42] This cGMP-induced resetting may be mediated by new protein synthesis.[43]

Johnson and Nakashima[44] used a similar approach to study light-induced phase resetting in *Neurospora*. Inhibition of protein synthesis was found to prevent light-induced phase-shifts in a dose-dependent manner. As a control, phase-shifting by light was shown to be not inhibited by the drug in mutants whose protein synthetic mechanism was resistant to the drug. Other studies of *Neurospora* entrainment suggest that inositol metabolic pathways[45] and/or ionic flux[46] may also be involved in the phototransduction pathway.

In mammals, a recent breakthrough elucidating the circadian phototransduction pathway in mammals is the discovery that light pulses which cause phase-shifting also induce expression of the c-fos gene within the suprachiasmatic nuclei (SCN) of the brain.[47,48] Much evidence supports the conclusion that the SCN are the loci of the master circadian clock in mammals; c-fos is an immediate-early gene whose protein product is a component of the AP-1 transcriptional factor, which in turn regulates the expression of many other genes. Fluence response and other experiments indicate that photic induction of the c-fos gene and phase-shifting of the circadian clock are positively correlated.[49] These exciting results suggest that photons might be followed to the clock mechanism, even in mammals.

35.4 Influence of Light on Period and Amplitude

The freerunning period in constant white light is different from that in constant darkness.[3] This phenomenon has been modeled as a consequence of the same mechanism that underlies phase-shifting by light pulses,[50] but no one knows if this explanation is correct. The spectral nature of this effect on period has been examined in only a few cases. For example, in *Coleus,* red LL shortens the period, while blue LL lengthens the period;[30] in *Gonyaulax,* the phenomenon is reversed: red LL lengthens the period, while blue LL shortens it.[51] These results have been interpreted to suggest that two photopigments are coupled to the circadian pacemakers in these organisms. Another intriguing observation in *Gonyaulax* is that exogenous creatine and an endogenous substance called gonyauline potentiate the effect of blue light on the freerunning period of the alga, but have no effect in constant red light.[52]

Not only does constant illumination affect the *period* of circadian rhythms, it can also affect the *persistence* of rhythms. For example, in some plants, frequent stimulation of phytochrome appears to be necessary for the maintenance of the rhythm's amplitude. This effect has been observed in *Albizzia,*[53] *Lemna,*[54,55] transgenic tobacco,[56] and perhaps in *Chenopodium.*[57]

35.5 Photoreceptive Organs That Mediate Entrainment in Animals

Animals have recruited a variety of both ocular and *extra*ocular photoreceptors to monitor light/dark conditions so as to entrain their circadian clocks.

Invertebrates

In the molluscs *Aplysia, Bulla, Bursatella,* and *Limax,* ocular photoreceptors are intimately involved in the entrainment pathway (see below, Section 35.5, Eyes As Clocks), while extraocular photoreceptors have an influence also.[58] Among the arthropods, many species are known to entrain after their eyes have been removed. These species include the crayfish *Procambarus,* the scorpion *Androctonus,* the crab *Limulus,* and many insects, including the beetle *Blaps* and the moths *Hyalophora* and *Antheraea.*[58] The eyes may also contribute to entrainment in these arthropods, but extraocular receptors are definitely sufficient. In at least five insects, however, the compound eyes are necessary and extraocular photoreceptors seem to be ignored: these cases are the cockroaches (*Periplaneta, Leucophaea,* and *Blaberus*) and the crickets (*Teleogryllus* and *Gryllus*).[58]

Nonmammalian Vertebrates

Extraretinal photoreceptors are capable of entraining circadian rhythms in fish, amphibians, reptiles, and birds. In fact, no nonmammalian vertebrate that has been examined fails to entrain its clock to a light/dark cycle after blinding.[58] In some cases, the pineal gland is likely to be involved in the extraretinal perception of the light/dark cycle; but in other cases, even the pineal gland is dispensable.[59] The identity of these deep-brain photoreceptors is unknown.[60]

Mammals

The clock photoreceptor in mammals is more simple: it is the eyes. Blinded mammals do not entrain to light/dark cycles. A primary pathway carries the entrainment information from the retina to the suprachiasmatic nuclei (SCN; the site of the master clock). The photopigment that absorbs the light in rodents is rhodopsin (Table 35.1). So far, the story sounds straightforward. New data, however, suggest that a surprise awaits us as to *which* cells in the retina function in entrainment. This statement is based on experiments with a mutant strain of mice in which the retina degenerate so that essentially no rod cells and only a few cone cells remain (and the remaining cone cells do not have outer segments). Nevertheless, entrainment is unaffected in these retinally-degenerate mice in terms of either sensitivity or spectrum.[61] These data suggest that either a very small number of cone cell *bodies* mediate circadian responses to light, or that there may be an as-yet unrecognized cell type within the mammalian retina that is not affected by this mutation, and that normally mediates circadian entrainment.

Eyes As Clocks

One of the fascinating insights of circadian photobiology is that not only are eyes part of entrainment pathways, but sometimes eyes *are* clocks! Spontaneous electrical activity of isolated eyes of the molluscs *Aplysia* and *Bulla* have long been known to oscillate *in vitro*.[62,63] Eyes of the frog *Xenopus* have circadian rhythms of enzyme activity *in vitro* that can be entrained by light/dark cycles.[64] An ocular clock has also been found in the Japanese quail *Coturnix*.[65] Finally, it seems likely that a self-sustained oscillator is also present in the mammalian eye; circadian rhythms of visual sensitivity and disk shedding persist in rat eyes even after the master clock (SCN) has been destroyed by lesioning.[66]

35.6 Conclusion

To have been adaptive, the circadian clockwork must have become linked early in its evolution with a photosensitive process that allowed the clock to entrain to the light/dark cycle of the sun. As described herein, the photoreceptor pigments and organs that the circadian clock has enlisted for entrainment duty are quite diverse among organisms. Does this mean that the biochemical mechanisms of circadian clocks in diverse organisms are similarly divergent? In other words, was the biochemical mechanism invented once, or is circadian behavior the result of convergent evolution? This important question remains unanswered. Nevertheless, the strategy of identifying circadian photoreceptors and tracing their phototransduction pathways is a promising approach toward discovering the biochemical clockworks, a strategy that has already borne fruit in characterizing the anatomical proximity of clock to photoreceptor in some animals.

Moreover, clock photoreceptors sometimes show unusual characteristics, e.g., prolonged reciprocity, that make them interesting in and of themselves. Undoubtedly, other interesting characteristics will be discovered in further investigations of the clock's "eyes". Circadian photobiology has been — and will continue to be — enlightening!

Acknowledgments

The author thanks Dr. Gary Wassmer for his advice and Grace Monty for her expert preparation of the manuscript.

References

1. Pittendrigh, C. S., Circadian systems: Entrainment, in *Handbook of Behavioral Neurobiology, Vol. 4, Biological Rhythms,* Aschoff, J., Ed., Plenum Press, New York, 1981, 95.

2. Johnson, C. H., Phase response curves: What can they tell us about circadian clocks? in *Circadian Clocks from Cell to Human,* Hiroshige, T. and Honma, K., Eds., Hokkaido University Press, Sapporo, 1992, 209.

3. Aschoff, J., Circadian rhythms: Influences of internal and external factors on the period measured in constant conditions, *Z. für Tierpsychologie,* 49, 225, 1979.

4. Johnson, C. H. and Hastings, J. W., Circadian phototransduction: Phase resetting and frequency of the circadian clock of *Gonyaulax* cells in red light, *J. Biol. Rhythms,* 4, 417, 1989.

5. Johnson, C. H. and Kondo, T., Light pulses induce "singular" behavior and shorten the period of the circadian phototaxis rhythm in the CW15 strain of *Chlamydomonas, J. Biol. Rhythms,* 7, 313, 1992.

6. Dharmananda, S., Studies of the Circadian Clock of *Neurospora crassa:* Light-Induced Phase Shifting, Ph. D. thesis, University of California at Santa Cruz, 1980.

7. Frank, K. D. and Zimmerman, W. F., Action spectra for phase shifts of a circadian rhythm in *Drosophila, Science,* 163, 688, 1969.

8. Klemm and Ninnemann, H., Detailed action spectrum for the delay shift in pupae emergence of *Drosophila pseudoobscura, Photochem. Photobiol.,* 24, 369, 1976.

9. Bruce, V. G. and Minis, D. H., Circadian clock action spectrum in a photoperiodic moth, *Science,* 163, 583, 1969.

10. Mote, M. I. and Black, K. R., Action spectrum and threshold sensitivity of entrainment of circadian running activity in the cockroach *Periplaneta americana, Photochem. Photobiol.,* 34, 257, 1981.

11. Joshi, D. and Chandrashekaran, M. K., Spectral sensitivity of the photoreceptors responsible for phase shifting the circadian rhythm of activity in the bat, *Hipposideros speoris, J. Comp. Physiol.,* 156, 189, 1985.

12. Zimmerman, W. F. and Goldsmith, T. H., Photosensitivity of the circadian rhythm and of visual receptors in carotenoid-depleted *Drosophila, Science,* 171, 1167, 1971.

13. Ninneman, H., Photoreceptors for circadian rhythms, *Photochem. Photobiol. Rev.,* 4, 207, 1979.

14. Fritz, B. J., Kasai, S., and Matsui, K., Free cellular riboflavin is involved in phase shifting by light of the circadian clock in *Neurospora crassa, Plant Cell Physiol.,* 30, 557, 1989.

15. Paietta, J. and Sargent, M. L., Modification of blue light photoresponses by riboflavin analogs in *Neurospora crassa, Plant Physiol.,* 72, 764, 1983.

16. Takahashi, J. S., DeCoursey, P. J., Bauman, L., and Menaker, M., Spectral sensitivity of a novel photoreceptive system mediating entrainment of mammalian circadian rhythms, *Nature,* 308, 186, 1984.

17. McGuire, R. A., Rand, W. M., and Wurtman, R. J., Entrainment of the body temperature rhythm in rats: Effect of color and intensity of environmental light, *Science,* 181, 956, 1973.

18. Foster, R. G., Argamaso, S., Coleman, S., Colwell, C. S., and Provencio, I., Photoreceptors regulating circadian behavior: A mouse model, *J. Biol. Rhythms,* 8, S17, 1993.

19. Geusz, M. E. and Page, T. L., An opsin-based photopigment mediates phase shifts of the *Bulla* circadian pacemaker, *J. Comp. Physiol.,* 168, 565, 1991.

20. Lumsden, P. J., Circadian rhythms and phytochrome, *Annu. Rev. Plant Physiol. Plant Mol. Biol.,* 42, 351, 1991.

21. Bunning, E. and Lorcher, L., Regulierung und Auslosung endogentages-periodischer Blattbewegungen durch verschiedene Lichtqualitaten, *Naturwiss,* 44, 472, 1957.

22. Deitzer, G. F. and Frosch, S. H., Multiple action of far-red light in photoperiodic induction and circadian rhythmicity, *Photochem. Photobiol.,* 52, 173, 1990.

23. Hillman, W. S., Entrainment of *Lemna* CO_2 output through phytochrome, *Plant Physiol.,* 48, 770, 1971.

24. Satter, R. L., Guggino, S. E., Lonergan, T. A., and Galston, A. W., The effects of blue and far-red light on rhythmic leaflet movements in *Samanea* and *Albizzia, Plant Physiol.,* 67, 965, 1981.

25. Simon, E., Satter, R. L., and Galston, A. W., Circadian rhythmicity in excised *Samanae pulvini*. II. Resetting the clock by phytochrome conversion, *Plant Physiol.*, 58, 421, 1976.

26. Harris, P. J. C. and Wilkins, M. B., Evidence of phytochrome involvement in the entrainment of the circadian rhythm of carbon dioxide metabolism in *Bryophyllum*, *Planta*, 138, 271, 1978.

27. Hastings, J. W. and Sweeney, B. M., Action spectra for shifting the phase of the rhythm of luminescence in *Gonyaulax polyedra*, *J. Gen. Physiol.*, 43, 697, 1960.

28. Kondo, T., Johnson, C. H., and Hastings, J. W., Action spectrum for resetting the circadian phototaxis rhythm in the CW15 strain of *Chlamydomonas*. I. Cells in darkness, *Plant Physiol.*, 95, 197, 1991.

29. Johnson, C. H., Kondo, T., and Hastings, J. W., Action spectrum for resetting the circadian phototaxis rhythm in the CW15 strain of *Chlamydomonas*. II. Illuminated cells, *Plant Physiol.*, 97, 1122, 1991.

30. Halaban, R., Effects of light quality on the circadian rhythm of leaf movement of a short-day plant, *Plant Physiol.*, 44, 973, 1969.

31. Karve, A., Engelmann, W., and Schoser, G., Initiation of rhythmical petal movements in *Kalanchoe blossfeldiana* by transfer from continuous darkness to continuous light or vice versa, *Planta*, 56, 700, 1961.

32. Schrempf, M., Eigenschaften und Lokalisation des Photorezeptors fur phasenverschiebendes Storlicht bei der Blutenblattbewegung von *Kalanchoe*, Ph. D. dissertation, University of Tubingen, 1975.

33. Crescitelli, F., James, T. W., Erickson, J. M., Loew, E. R., and McFarland, W. N., The eyespot of *Chlamydomonas reinhardtii*: A comparative microspectrophotometric study, *Vision Res.*, 32, 1593, 1992.

34. Ehret, C. F., Action spectra and nucleic acid metabolism in circadian rhythms at the cellular level, in *Cold Spring Harbor Symposia on Quantitative Biology, Vol. 25, Biological Clocks*, Cold Spring Harbor Press, New York, 1960, 149.

35. Miwa, I., personal communication, 1992.

36. Sweeney, B. M., Resetting the biological clock in *Gonyaulax* with UV light, *Plant Physiol.*, 38, 704, 1963.

37. Ehret, C. F., Induction of phase shift in cellular rhythmicity by far ultraviolet and its restoration by visible radiant energy, in *Photoperiodism and Related Phenomena in Plants and Animals*, Withrow, R. B., Ed., AAAS, Washington, 1959, 541.

38. Pohl, H., Ultraviolet radiation: A zeitgeber for the circadian clock in birds, *Naturwisschaften*, 79, 227, 1992.

39. Nelson, D. E. and Takahashi, J. S., Sensitivity and integration in a visual pathway for circadian entrainment in the hamster *(Mesocricetus auratus)*, *J. Physiol.*, 439, 115, 1991.

40. Johnson, C. H. and Hastings, J. W., The elusive mechanism of the circadian clock, *Am. Scient.*, 74, 29, 1986.

41. Eskin, A., Identification and physiology of circadian pacemakers, *Fed. Proc.*, 38, 2570, 1979.

42. Eskin, A., Takahashi, J. S., Zatz, M., and Block, G. D., Cyclic guanosine $3':5'$-monophosphate mimics the effects of light on a circadian pacemaker in the eye of *Aplysia*, *J. Neurosci.*, 4, 2466, 1984.

43. Raju, U., Yeung, S. J., and Eskin, A., Involvement of proteins in light resetting ocular circadian oscillators in *Aplysia*, *Am. J. Physiol.*, 258, R256, 1990.

44. Johnson, C. H. and Nakashima, H., Cycloheximide inhibits light-induced phase shifting of the circadian clock in *Neurospora*, *J. Biol. Rhythms*, 5, 159, 1990.

45. Lakin-Thomas, P. L., Phase resetting of the *Neurospora crassa* circadian oscillator: Effects of inositol depletion on sensitivity to light, *J. Biol. Rhythms*, 7, 227, 1992.

46. Nakashima, H. and Fujimura, Y., Light-induced phase-shifting of the circadian clock in *Neurospora crassa* requires ammonium salts at high pH, *Planta*, 155, 431, 1982.

47. Kornhauser, J. M., Nelson, D. E., Mayo, K. E., and Takahashi, J. S., Regulation of jun-B messenger RNA and AP-1 activity by light and a circadian clock, *Science*, 255, 1581, 1992.

48. Rusak, B., McNaughton, L., Robertson, H. A., and Hunt, S. P., Circadian variation in photic regulation of immediate-early gene mRNAs in rat suprachiasmatic nucleus cells, *Mol. Brain Res.*, 14, 124, 1992.

49. Kornhauser, J. M., Nelson, D. E., Mayo, K. E., and Takahashi, J. S., Photic and circadian regulation of c-fos gene expression in the hamster suprachiasmatic nucleus, *Neuron*, 5, 127, 1990.

50. Daan, S. and Pittendrigh, C. S., A functional analysis of circadian pacemakers in nocturnal rodents. III. Heavy water and constant light: Homeostasis of frequency?, *J. Comp. Physiol.*, 106, 267, 1976.

51. Roenneberg, T. and Hastings, J. W., Two photoreceptors control the circadian clock of a unicellular alga, *Naturwiss*, 75, 206, 1988.

52. Roenneberg, T., Nakamura, H., Cranmer, L. D., III, Ryan, K., Kishi, Y., and Hastings, J. W., Gonyauline: A novel endogenous substance shortening the period of the circadian clock of a unicellular alga, *Experientia*, 47, 103, 1991.

53. Satter, R. L., Applewhite, P. B., Chaudhri, J., and Galston, A. W., PFR phytochrome and sucrose requirement for rhythmic leaf movement in *Albizzia*, *Photochem. Photobiol.*, 23, 107, 1976.

54. Kondo, T., Persistence of the potassium uptake rhythm in the presence of exogenous sucrose in *Lemna gibba* G3, *Plant Cell Physiol.*, 23, 467, 1982.

55. Kondo, T., Phase control of the potassium uptake rhythm by the light signals in *Lemna gibba*, G3, *Pflanzenphysiol. Bd.*, 107, S. 395, 1982.

56. Kay, S. A., Nagatani, A., Keith, B., Deak, M., Furuya, M., and Chua, N.-H., Rice phytochrome is biologically active in transgenic tobacco, *Plant Cell*, 1, 775, 1989.

57. Wagner, E. and Cumming, B. G., Betacyanin accumulation, chlorophyll content, and flower initiation in *Chenopodium rubrum* as related to endogenous rhythmicity and phytochrome action, *Can. J. Bot.*, 48, 1, 1970.

58. Underwood, H. A., Wassmer, G. T., and Page, T. L., Daily and seasonal rhythms, in *Handbook of Comparative Physiology*, in press, 1995.

59. Underwood, H. and Menaker, M., Extraretinal photoreception in lizards, *Photochem. Photobiol.*, 23, 227, 1976.

60. Foster, R. G., Photoreceptors and circadian systems, *Curr. Dir. Psychol. Sci.*, 2, 34, 1993.

61. Foster, R. G., Provencio, I., Hudson, D., Fiske, S., DeGrip, W., and Menaker, M., Circadian photoreception in the retinally degenerate mouse (rd/rd), *J. Comp. Physiol.*, A 169, 39, 1991.

62. Jacklet, J. D., Circadian rhythm of optic nerve impulses recorded in darkness from isolated eye of *Aplysia*, *Science*, 164, 562, 1969.

63. Block, G. D. and Wallace, S., Localization of a circadian pacemaker in the eye of a mollusk, *Bulla*, *Science*, 217, 155, 1982.

64. Cahill, G. M. and Besharse, J. C., Resetting the circadian clock in *Xenopus* eyecups: Regulation of retinal melatonin rhythms by light and D_2 dopamine receptors, *J. Neurosci.*, 11, 2959, 1991.

65. Underwood, H. Barrett, R. K., and Siopes, T., The quail's eye: A biological clock, *J. Biol. Rhythms*, 5, 257, 1990.

66. Remé, C. E., Wirz-Justice, A., and Terman, M., The visual input stage of the mammalian circadian pacemaking system. I. Is there a clock in the mammalian eye?, *J. Biol. Rhythms*, 6, 5, 1991.

Index

Index